22. $\displaystyle\int \frac{dx}{(a^2 + x^2)^2} = \frac{x}{2a^2(a^2 + x^2)} + \frac{1}{2a^3} \tan^{-1} \frac{x}{a} + C$

23. $\displaystyle\int \frac{dx}{(ax^2 + bx + c)^{n+1}} = \frac{2ax + b}{n(4ac - b^2)(ax^2 + bx + c)^n} + \frac{2(2n - 1)a}{n(4ac - b^2)} \int \frac{dx}{(ax^2 + bx + c)^n},$

if $n > 0$ and $b^2 \neq 4ac$

24. $\displaystyle\int \frac{dx}{a^2 - x^2} = \frac{1}{2a} \ln \left| \frac{x + a}{x - a} \right| + C$

25. $\displaystyle\int \frac{dx}{(a^2 - x^2)^2} = \frac{x}{2a^2(a^2 - x^2)} + \frac{1}{2a^2} \int \frac{dx}{a^2 - x^2}$

26. $\displaystyle\int \sqrt{a^2 + x^2}\, dx = \frac{x}{2} \sqrt{a^2 + x^2} + \frac{a^2}{2} \ln (x + \sqrt{x^2 + a^2}) + C$

27. $\displaystyle\int x^2 \sqrt{a^2 + x^2}\, dx = \frac{x(a^2 + 2x^2)\sqrt{a^2 + x^2}}{8} - \frac{a^4}{8} \ln (x + \sqrt{x^2 + a^2}) + C$

28. $\displaystyle\int \frac{\sqrt{a^2 + x^2}}{x}\, dx = \sqrt{a^2 + x^2} - a \ln \left(\frac{a + \sqrt{a^2 + x^2}}{x} \right) + C$

29. $\displaystyle\int \frac{\sqrt{a^2 + x^2}}{x^2}\, dx = -\frac{\sqrt{a^2 + x^2}}{x} + \ln (x + \sqrt{a^2 + x^2}) + C$

30. $\displaystyle\int \frac{dx}{\sqrt{a^2 + x^2}} = \ln (x + \sqrt{a^2 + x^2}) + C$

31. $\displaystyle\int \frac{x^2}{\sqrt{a^2 + x^2}}\, dx = \frac{x}{2} \sqrt{a^2 + x^2} - \frac{a^2}{2} \ln (x + \sqrt{a^2 + x^2}) + C$

32. $\displaystyle\int \frac{dx}{x\sqrt{a^2 + x^2}} = -\frac{1}{a} \ln \left| \frac{a + \sqrt{a^2 + x^2}}{x} \right| + C$

33. $\displaystyle\int \frac{dx}{x^2\sqrt{a^2 + x^2}} = -\frac{\sqrt{a^2 + x^2}}{a^2 x} + C$

34. $\displaystyle\int \sqrt{a^2 - x^2}\, dx = \frac{x}{2} \sqrt{a^2 - x^2} + \frac{a^2}{2} \sin^{-1} \frac{x}{a} + C$

35. $\displaystyle\int x^2 \sqrt{a^2 - x^2}\, dx = \frac{a^4}{8} \sin^{-1} \frac{x}{a} - \frac{1}{8} x \sqrt{a^2 - x^2}(a^2 - 2x^2) + C$

36. $\displaystyle\int \frac{\sqrt{a^2 - x^2}}{x}\, dx = \sqrt{a^2 - x^2} - a \ln \left| \frac{a + \sqrt{a^2 - x^2}}{x} \right| + C$

37. $\displaystyle\int \frac{\sqrt{a^2 - x^2}}{x^2}\, dx = -\sin^{-1} \frac{x}{a} - \frac{\sqrt{a^2 - x^2}}{x} + C$

38. $\displaystyle\int \frac{dx}{\sqrt{a^2 - x^2}} = \sin^{-1} \frac{x}{a} + C$

39. $\displaystyle\int \frac{x^2}{\sqrt{a^2 - x^2}}\, dx = \frac{a^2}{2} \sin^{-1} \frac{x}{a} - \frac{1}{2} x \sqrt{a^2 - x^2} + C$

40. $\displaystyle\int \frac{dx}{x\sqrt{a^2 - x^2}} = -\frac{1}{a} \ln \left| \frac{a + \sqrt{a^2 - x^2}}{x} \right| + C$

41. $\displaystyle\int \frac{dx}{x^2\sqrt{a^2 - x^2}} = -\frac{\sqrt{a^2 - x^2}}{a^2 x} + C$

42. $\displaystyle\int \sqrt{x^2 - a^2}\, dx = \frac{x}{2} \sqrt{x^2 - a^2} - \frac{a^2}{2} \ln |x + \sqrt{x^2 - a^2}| + C$

43. $\displaystyle\int (\sqrt{x^2 - a^2})^n\, dx = \frac{x(\sqrt{x^2 - a^2})^n}{n + 1} - \frac{na^2}{n + 1} \int (\sqrt{x^2 - a^2})^{n-2}\, dx, \qquad n \neq -1$

44. $\int x(\sqrt{x^2 - a^2})^n\,dx = \frac{(\sqrt{x^2 - a^2})^{n+2}}{n+2} + C, \qquad n \neq -2$

45. $\int x^2\sqrt{x^2 - a^2}\,dx = \frac{x}{8}(2x^2 - a^2)\sqrt{x^2 - a^2} - \frac{a^4}{8}\ln|x + \sqrt{x^2 - a^2}| + C$

46. $\int \frac{\sqrt{x^2 - a^2}}{x}\,dx = \sqrt{x^2 - a^2} - a\sec^{-1}\left|\frac{x}{a}\right| + C$

47. $\int \frac{\sqrt{x^2 - a^2}}{x^2}\,dx = -\frac{\sqrt{x^2 - a^2}}{x} + \ln|x + \sqrt{x^2 - a^2}| + C$

48. $\int \frac{dx}{\sqrt{x^2 - a^2}} = \ln|x + \sqrt{x^2 - a^2}| + C$

49. $\int \frac{dx}{(\sqrt{x^2 - a^2})^n} = \frac{x(\sqrt{x^2 - a^2})^{2-n}}{(2-n)a^2} - \frac{n-3}{(n-2)a^2}\int \frac{dx}{(\sqrt{x^2 - a^2})^{n-2}}, \qquad n \neq 2$

50. $\int \frac{x^2}{\sqrt{x^2 - a^2}}\,dx = \frac{x}{2}\sqrt{x^2 - a^2} + \frac{a^2}{2}\ln|x + \sqrt{x^2 - a^2}| + C$

51. $\int \frac{dx}{x\sqrt{x^2 - a^2}} = \frac{1}{a}\sec^{-1}\frac{x}{a} + C$

52. $\int \frac{dx}{x^2\sqrt{x^2 - a^2}} = \frac{\sqrt{x^2 - a^2}}{a^2 x} + C$

53. $\int \sqrt{2ax - x^2}\,dx = \frac{x-a}{2}\sqrt{2ax - x^2} + \frac{a^2}{2}\sin^{-1}\left(\frac{x-a}{a}\right) + C$

54. $\int (\sqrt{2ax - x^2})^n\,dx = \frac{(x-a)(\sqrt{2ax - x^2})^n}{n+1} + \frac{na^2}{n+1}\int (\sqrt{2ax - x^2})^{n-2}\,dx$

55. $\int x\sqrt{2ax - x^2}\,dx = \frac{(x+a)(2x-3a)\sqrt{2ax - x^2}}{6} + \frac{a^3}{2}\sin^{-1}\left(\frac{x-a}{a}\right) + C$

56. $\int \frac{\sqrt{2ax - x^2}}{x}\,dx = \sqrt{2ax - x^2} + a\sin^{-1}\left(\frac{x-a}{a}\right) + C$

57. $\int \frac{\sqrt{2ax - x^2}}{x^2}\,dx = -2\sqrt{\frac{2a-x}{x}} - \sin^{-1}\left(\frac{x-a}{a}\right) + C$

58. $\int \frac{dx}{\sqrt{2ax - x^2}} = \sin^{-1}\left(\frac{x-a}{a}\right) + C$

59. $\int \frac{dx}{(\sqrt{2ax - x^2})^n} = \frac{(x-a)(\sqrt{2ax - x^2})^{2-n}}{(n-2)a^2} + \frac{n-3}{(n-2)a^2}\int \frac{dx}{(\sqrt{2ax - x^2})^{n-2}}$

60. $\int \frac{x\,dx}{\sqrt{2ax - x^2}} = a\sin^{-1}\left(\frac{x-a}{a}\right) - \sqrt{2ax - x^2} + C$

61. $\int \frac{dx}{x\sqrt{2ax - x^2}} = -\frac{1}{a}\sqrt{\frac{2a-x}{x}} + C$

INTEGRALS INVOLVING TRIGONOMETRIC FUNCTIONS

62. $\int \sin ax\,dx = -\frac{1}{a}\cos ax + C$

63. $\int \cos ax\,dx = \frac{1}{a}\sin ax + C$

8 OTHER ELEMENTARY FUNCTIONS 305

9 TECHNIQUES OF INTEGRATION 361

10 INFINITE SERIES OF CONSTANTS 419

11 POWER SERIES 466

12 PLANE CURVES 517

13 POLAR COORDINATES 562

14 SPACE GEOMETRY AND VECTORS 586

15 VECTOR ANALYSIS OF CURVES 648

16 DIFFERENTIAL CALCULUS OF SEVERAL VARIABLES 685

CALCULUS WITH ANALYTIC GEOMETRY

SECOND EDITION

FUNCTIONS AND GRAPHS

1

Much of the material in this first chapter may be familiar to students. The chapter provides a review of the concepts of analytic geometry and function theory that serve as a foundation for the study of calculus. Section 1.5 on graphs of monomial and quadratic functions and on translation of axes may be new material to many students. We recommend that special attention be given to Section 1.5.

1.1 COORDINATES AND DISTANCE

COORDINATES ON THE LINE

We will be concerned only with the calculus involving real numbers. A **real number** is one that can be written as an unending decimal, positive or negative, or zero. For example, $3 = 3.000000\ldots$, $-\frac{2}{3} = -0.666666\ldots$, and $\pi = 3.141592\ldots$ are real numbers. The positive or negative real numbers that can be expressed with only zeros to the right of the decimal points, such as $3 = 3.000000\ldots$, are **integers.** The real numbers that can be expressed as quotients of integers, such as $(-2)/3 = -0.666666\ldots$, with nonzero denominator, are **rational numbers.** Every integer can be expressed in this fashion—for example, $3 = 3/1$ and $-4 = (-8)/2$—so every integer is also a rational number. The real numbers that are not rational are **irrational numbers.** It can be shown that π and $\sqrt{2}$ are irrational numbers. Exercise 48 of Section 10.2 indicates that a "randomly selected" real number is practically sure to be irrational. Precise arithmetic is easiest with rational numbers, so we use them extensively in examples in the text.

It is very useful to identify real numbers with points on the **number line.** Take a line extending infinitely in both directions. Using the real numbers, we can make this line into an infinite ruler (see Fig. 1.1). Label any point on the line with 0 and any point to the right of 0 with 1; this fixes the *scale.* Each positive real number r corresponds to the point a distance r units to the right of 0, while a negative number $-s$ corresponds to the point a distance s units to

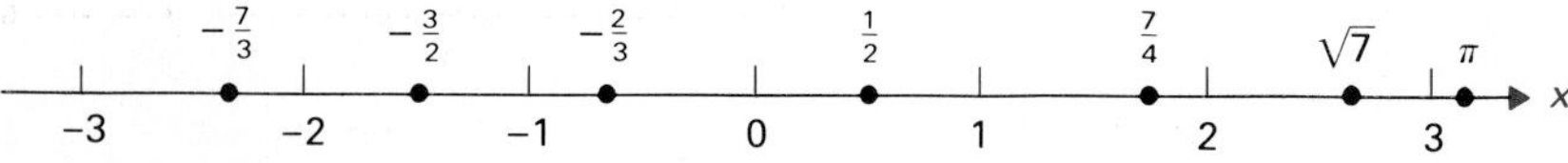

Figure 1.1 The number line.

the left of 0. The arrow on the line indicates the positive direction. The x to the right of the arrow indicates that we think of x as any real number on the line. In this context, x is known as a **real variable,** and the line is called the **x-axis.**

For real numbers r and s, the notation $r < s$ (read "r is less than s") means that r is to the left of s on the number line. The three parts of Fig. 1.2 illustrate that this criterion for $r < s$ is valid whether r and s are positive or negative.

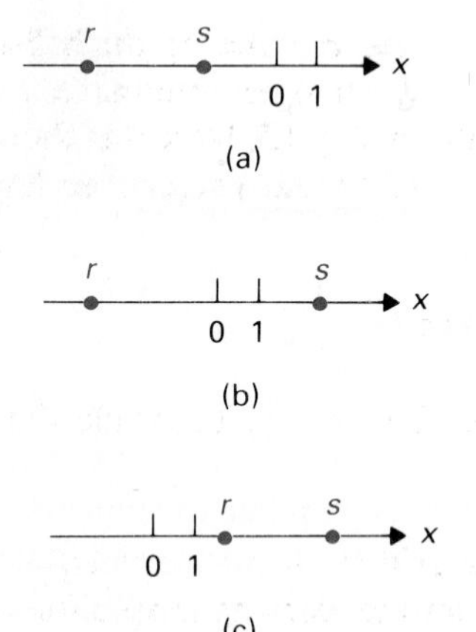

Figure 1.2 $r < s$

EXAMPLE 1 Referring to the line shown in Fig. 1.1, we see that

$$2 < \pi \qquad \text{and} \qquad -2 < 3,$$

because 2 is to the left of π and -2 is to the left of 3. Note also that

$$-3 < -1,$$

because although 3 is larger than 1, the number -3 is to the left of -1 on the line. We also have

$$-\tfrac{2}{3} < \tfrac{1}{2}.$$

Any negative number is less than any positive number. □

This same relation $r < s$ between numbers r and s is sometimes written as $s > r$ (read "s is greater than r"). For example, if we want to consider all numbers greater than 2, we naturally say, "Consider all numbers x greater than 2," mentioning x before 2, and we abbreviate this as "all $x > 2$." The notation $r \leq s$ is read "r is less than or equal to s," while $s \geq r$ is read "s is greater than or equal to r."

EXAMPLE 2 Sketch on the line the points x that satisfy the relation $0 \leq x \leq 2$.

Solution The points are indicated by the colored line and points in Fig. 1.3. Both 0 and 2 satisfy the relation. □

EXAMPLE 3 Sketch on the line the points x that satisfy the relation $-1 < x \leq 1$.

Solution The points are indicated by the colored line and points in Fig. 1.4. This time -1 does not satisfy the relation, while 1 does. □

The collection of points x satisfying a relation of the form $a \leq x \leq b$ is important in calculus. This set of points is the **closed interval** $[a, b]$. The adjective *closed* is used to indicate that both endpoints, a and b, are considered

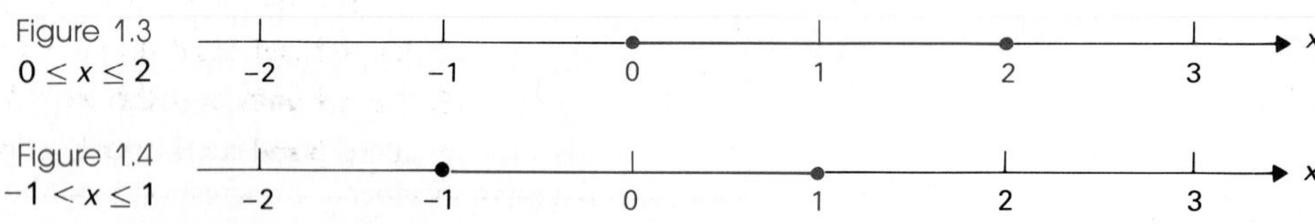

Figure 1.3 $0 \leq x \leq 2$

Figure 1.4 $-1 < x \leq 1$

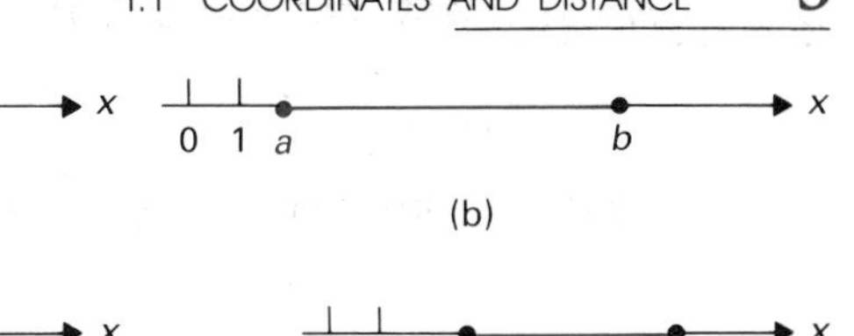

Figure 1.5 (a) Closed $[a, b]$; (b) half-open $[a, b)$; (c) half-open $(a, b]$; (d) open (a, b).

part of the interval. The points x where $a < x \le b$ make up the **half-open interval** $(a, b]$. The half-open interval $[a, b)$ and the open interval (a, b) are defined similarly. These intervals are illustrated in Fig. 1.5. Note that the open interval (a, b) contains no endpoints, in the sense that there is no largest number or smallest number in the interval.

EXAMPLE 4 Sketch the intervals $[0, 2]$ and $(-1, 1]$.

Solution The closed interval $[0, 2]$ is sketched in Fig. 1.3, while Fig. 1.4 shows the half-open interval $(-1, 1]$. □

Appendix 1 summarizes notations and arithmetic laws for inequalities. The box on this page lists four arithmetic laws concerning inequalities that are used in some of the subsequent examples.

EXAMPLE 5 Solutions of an inequality can sometimes be expressed in interval notation. Express the solutions of $-3 \le 4x + 5 \le 7$ in interval notation.

Solution Using the rules for inequalities given below, we obtain

$$-3 \le 4x + 5 \le 7,$$

$$-8 \le 4x \le 2, \qquad \text{Adding } -5$$

$$-2 \le x \le \tfrac{1}{2}, \qquad \text{Multiplying by } \tfrac{1}{4}$$

which is the closed interval $[-2, \frac{1}{2}]$. □

The distance from a point r to the point 0 is known as the **absolute value** of the number r and is denoted by $|r|$. For example,

$$|5| = |-5| = 5,$$

because both 5 and -5 are five units from 0. We see that

$$|r| = \begin{cases} r & \text{if } r \ge 0, \\ -r & \text{if } r < 0. \end{cases}$$

ARITHMETIC LAWS FOR INEQUALITIES

1. If $a \le b$, then $a + c \le b + c$.
2. If $a \le b$ and $c \le d$, then $a + c \le b + d$.
3. If $a \le b$ and $c > 0$, then $ac \le bc$.
4. If $a \le b$ and $c < 0$, then $bc \le ac$.

EXAMPLE 6 Simplify $|6 - |-3|| + |2 - 7|$.

Solution We have $|-3| = 3$, so

$$\begin{aligned}|6 - |-3|| + |2 - 7| &= |6 - 3| + |2 - 7| \\ &= |3| + |-5| = 3 + 5 = 8. \quad \square\end{aligned}$$

EXAMPLE 7 Find all real numbers x such that $|x|/x = 1$.

Solution Now $|x|/x = 1$ is equivalent to $|x| = x$, with $x \neq 0$. For this to be the case, x must be positive, so the solutions consist of all $x > 0$. □

EXAMPLE 8 Solutions of an inequality involving absolute values can sometimes be expressed using interval notation. Express the solutions of $|x - 4| \leq 5$ using interval notation.

Solution Since $|x - 4|$ is the distance from $x - 4$ to the origin, we see that $|x - 4| \leq 5$ means that we must have

$$-5 \leq x - 4 \leq 5.$$

The rules for inequalities given above show that

$$-1 \leq x \leq 9, \qquad \text{Adding 4}$$

which is the closed interval $[-1, 9]$. □

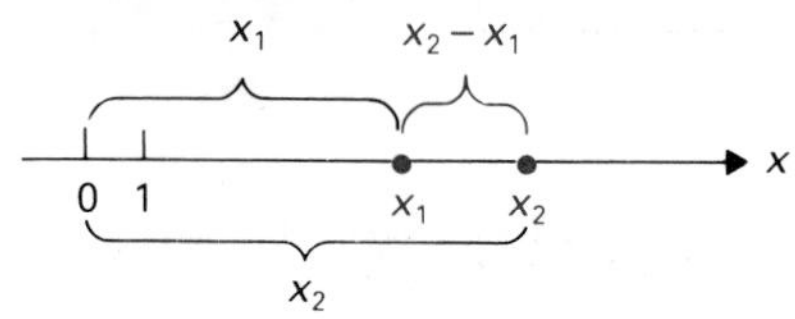

Figure 1.6 Distance between x_1 and x_2 if $x_1 \leq x_2$.

EXAMPLE 9 Express the solutions of $|(3 - 2x)/4| \leq 5$ using interval notation.

Solution Since $|(3 - 2x)/4|$ is the distance from $(3 - 2x)/4$ to the origin, we see that $|(3 - 2x)/4| \leq 5$ means that we must have

$$-5 \leq \frac{3 - 2x}{4} \leq 5.$$

The rules for inequalities given above yield

$$\begin{aligned}
-20 \leq 3 - 2x \leq 20, &\qquad \text{Multiplying by 4} \\
-23 \leq -2x \leq 17, &\qquad \text{Adding } -3 \\
\frac{-23}{-2} \geq x \geq \frac{17}{-2}, &\qquad \text{Multiplying by } -\tfrac{1}{2} \\
-\frac{17}{2} \leq x \leq \frac{23}{2}, &
\end{aligned}$$

which is the closed interval $[-\frac{17}{2}, \frac{23}{2}]$. □

Consider now the distance between any two points on the number line. The distance between the points x_1 and x_2, shown in Fig. 1.6, is surely $x_2 - x_1$. We can easily convince ourselves that for any two points x_1 and x_2, where $x_1 \leq x_2$, the distance between them is $x_2 - x_1$.

Figure 1.7 Distance between -2 and 3.

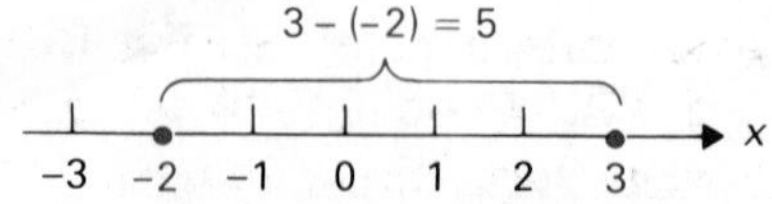

EXAMPLE 10 Find the distance between -2 and 3 on the number line.

Solution Since $-2 < 3$, the distance is $3 - (-2) = 5$, as indicated in Fig. 1.7. □

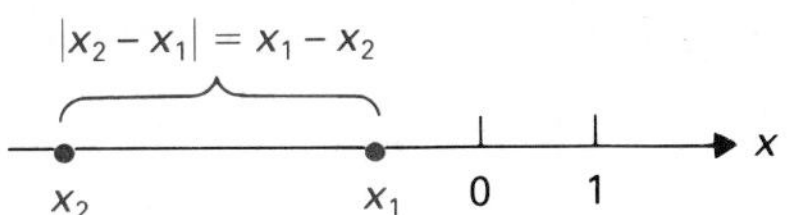

Figure 1.8 Distance between x_1 and x_2 if $x_1 \geq x_2$.

For *any* points x_1 and x_2, the distance between them is either $x_1 - x_2$ or $x_2 - x_1$, whichever is nonnegative. This nonnegative magnitude is, of course, $|x_2 - x_1|$. Thus

> the distance between x_1 and x_2 is $|x_2 - x_1|$.

See Figs. 1.6 and 1.8.

EXAMPLE 11 Use the absolute-value formula to find the distance between -2 and 3.

Solution For the points -2 and 3,

$$|3 - (-2)| = |5| = 5,$$

and also

$$|(-2) - 3| = |-5| = 5. \quad \square$$

Exercise 9 asks you to show that $(a + b)/2$ is the same distance from a as from b, so

> $\dfrac{a + b}{2}$ is the **midpoint** of $[a, b]$.

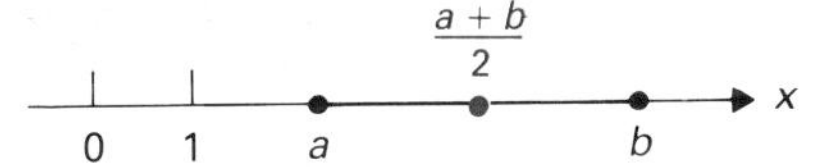

Figure 1.9 Midpoint $(a + b)/2$ of $[a, b]$.

See Fig. 1.9.

EXAMPLE 12 Find the midpoint of the interval $[-\frac{1}{2}, \frac{2}{3}]$.

Solution The midpoint is

$$\frac{-\frac{1}{2} + \frac{2}{3}}{2} = \frac{-\frac{3}{6} + \frac{4}{6}}{2} = \frac{\frac{1}{6}}{2} = \frac{1}{12}. \quad \square$$

Often we need to know not only the distance from x_1 to x_2 but whether x_1 is to the left or right of x_2. The change $x_2 - x_1$ in x-value going from x_1 to x_2 (in that order) is positive if $x_1 < x_2$ and negative if $x_2 < x_1$. Very soon, in calculus we will want to let Δx (read "*delta x*") be such a positive or negative change in x-value. Think, geometrically, of

> the directed distance $\Delta x = x_2 - x_1$

as the *signed length* of the *directed line segment* from x_1 to x_2.

EXAMPLE 13 Find the directed distances from 4 to -2 and from -8 to -5.

Solution The directed distance from $x_1 = 4$ to $x_2 = -2$ is $\Delta x = x_2 - x_1 = -2 - 4 = -6$. To get to -2 from 4, we go six units to the *left*. See Fig. 1.10.

The directed distance from -8 to -5 is $\Delta x = -5 - (-8) = -5 + 8 = 3$. See Fig. 1.11. To get from -8 to -5, we go three units to the *right*. $\square$

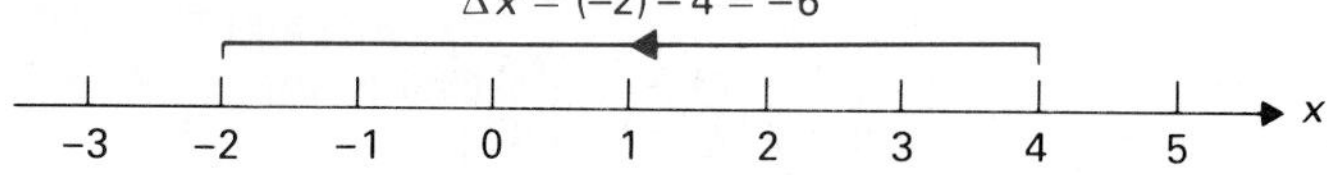

Figure 1.10 Directed distance from 4 to −2.

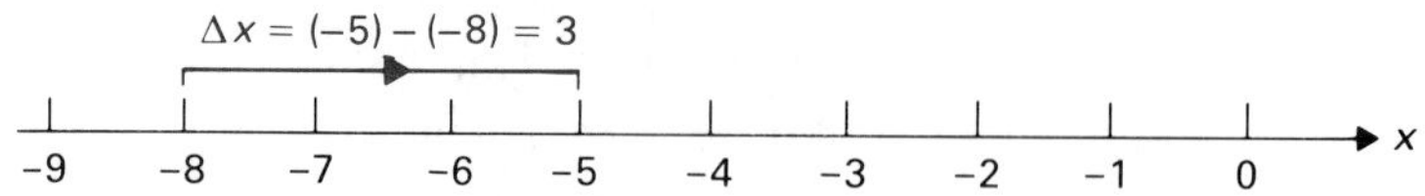

Figure 1.11 Directed distance from −8 to −5.

COORDINATES IN THE PLANE

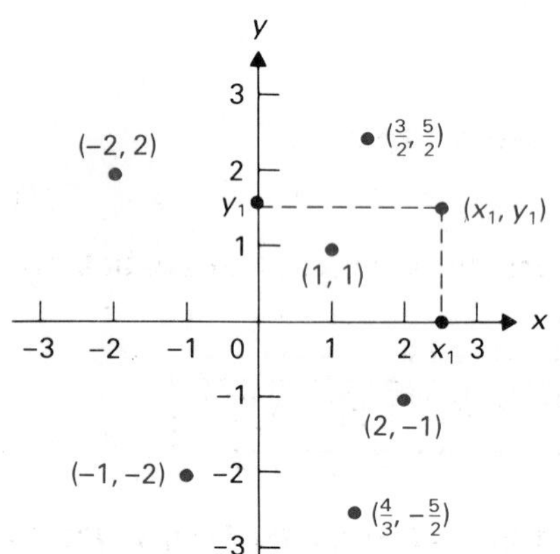

Figure 1.12 Coordinates in the plane.

Take two copies of the number line and place them perpendicular to each other in a plane, so that they intersect at the point 0 on each line (see Fig. 1.12). With each point in the plane, we associate an ordered pair (x_1, y_1) of numbers, as follows: The first number, x_1, gives the left-right position of the point according to the location of x_1 on the horizontal number line. Similarly, the second number, y_1, gives the up-down position of the point according to the location of y_1 on the vertical number line (see Fig. 1.12). Conversely, given any ordered pair of numbers, such as $(2, -1)$, there is a unique point in the plane associated with it.

EXAMPLE 14 Plot the points $(1, 1)$, $(-2, 2)$, $(-1, -2)$ and $(2, -1)$ in the plane.

Solution The positions of the points are shown in Fig. 1.12. □

The solid lines of Fig. 1.12 are the **coordinate axes.** In particular, the horizontal axis is the ***x*-axis** and the vertical axis the ***y*-axis,** according to the labels at the arrows. For the point (x_1, y_1), the number x_1 is the ***x*-coordinate** of the point, and y_1 is the ***y*-coordinate.** The coordinate axes divide the plane into four pieces or **quadrants,** according to the signs of the coordinates of the points. The quadrants are usually numbered as shown in Fig. 1.13. The point $(0, 0)$ is the **origin.** This introduction of coordinates allows us to use numbers and algebra as tools in studying geometry.

Figure 1.13 Quadrants of the plane.

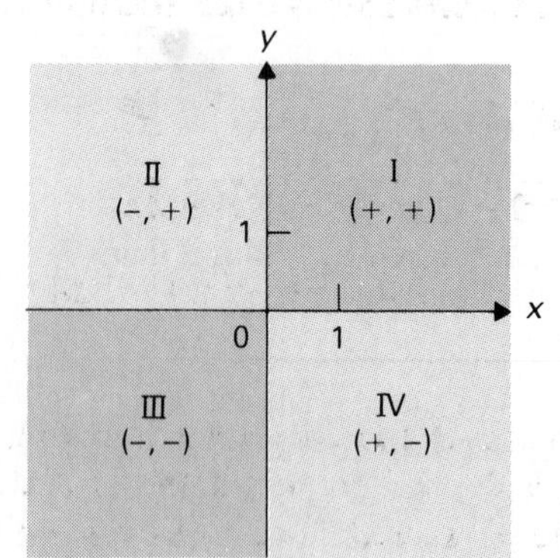

EXAMPLE 15 Sketch the portion of the plane consisting of the points (x, y) satisfying the relation $x \leq 1$.

Solution This portion of the plane is shown colored in Fig. 1.14. □

EXAMPLE 16 Sketch the portion of the plane consisting of the points (x, y) satisfying *both* $-2 \leq x \leq 1$ and $1 \leq y \leq 2$.

Solution This portion of the plane is shown colored in Fig. 1.15. □

We can find the distance between two points (x_1, y_1) and (x_2, y_2) in the plane. Referring to Fig. 1.16, let $\Delta x = x_2 - x_1$ and $\Delta y = y_2 - y_1$, so that $|\Delta x|$ and $|\Delta y|$ are the lengths of the legs of the right triangle shown in the figure. The distance between (x_1, y_1) and (x_2, y_2) is the length d of the hy-

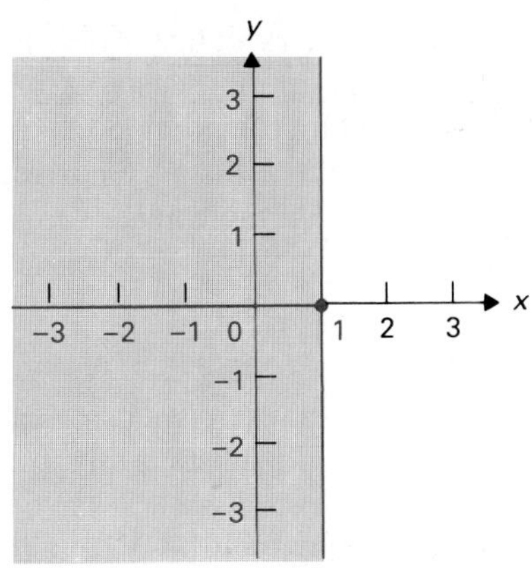

Figure 1.14 $x \leq 1$

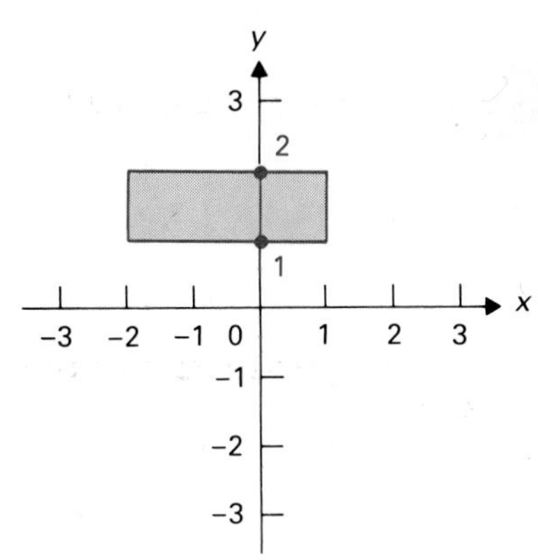

Figure 1.15 $-2 \leq x \leq 1$ and $1 \leq y \leq 2$

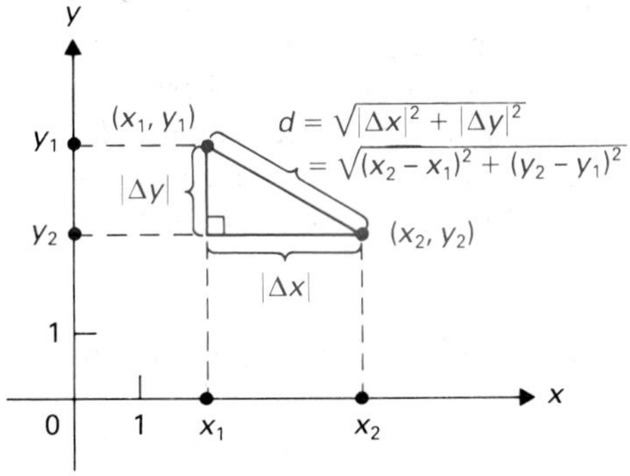

Figure 1.16 Distance between (x_1, y_1) and (x_2, y_2).

potenuse of the triangle. So, by the Pythagorean theorem,

$$d^2 = |\Delta x|^2 + |\Delta y|^2. \tag{1}$$

Since the terms in Eq. (1) are squared, the absolute-value symbols are not needed; so $d^2 = (\Delta x)^2 + (\Delta y)^2$ and

$$d = \sqrt{(\Delta x)^2 + (\Delta y)^2} = \sqrt{(x_2 - x_1)^2 + (y_2 - y_1)^2}.$$

EXAMPLE 17 Find the distance between $(2, -3)$ and $(-1, 1)$ in the plane.

Solution The distance is

$$\sqrt{(-1-2)^2 + [1-(-3)]^2} = \sqrt{(-3)^2 + 4^2}$$
$$= \sqrt{9+16} = \sqrt{25} = 5. \quad \square$$

EXAMPLE 18 A mortar shell has a destructive range of 150 m. The shell explodes at ground level at a point 120 m north and 80 m west of a soldier. Determine whether the soldier is within the destructive range of the explosion.

Figure 1.17 Is $d < 150$?

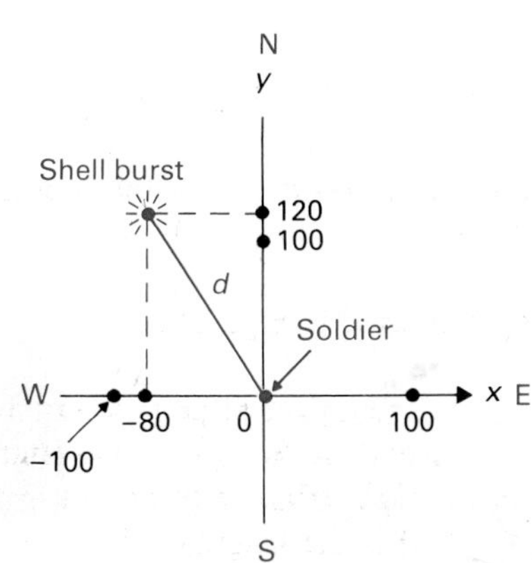

Solution We consider the soldier to be at the origin of a coordinate system with x-axis pointing east and y-axis pointing north, as shown in Fig. 1.17. The shell bursts at the point $(-80, 120)$. The distance from this point to the soldier at $(0, 0)$ is then

$$\sqrt{(80-0)^2 + (120-0)^2} = \sqrt{80^2 + 120^2}$$
$$= \sqrt{6400 + 14{,}400} = \sqrt{20{,}800}\,\text{m}.$$

Since $150 = \sqrt{22{,}500}$, we see that $\sqrt{20{,}800} < 150$, so the soldier is within the destructive range of the explosion. $\square$

So far, our figures with x,y-coordinates used the same scale on both the x-axis and the y-axis. The concepts of calculus will be best illustrated later using axes with equal scales, but of course we may find occasions when it is inconvenient to mark equal scales. For example, if y is the cost of manufacturing x items, and if it costs $2342 to manufacture 3 items, we would be hard pressed to plot the point (3, 2342) using equal scales. In such a case, we might want to let the distance corresponding to 1 item on the x-axis correspond to $1000 on the y-axis.

USING CALCULATORS

While examples and exercises using calculators may be considered optional, we feel that they add substantially to an understanding and appreciation of calculus. We think students will have fun with them, even if they are not an assigned part of classwork. We make no attempt to discuss in detail the pitfalls of using calculators. Students may discover some of them by experimentation.

The calculator exercises that appear in this first chapter are designed to make sure that students know which buttons to push and understand the logic of their own calculator. The interesting use of calculators starts in Chapter 2.

EXAMPLE 19 Find the distance between the points $(\sqrt[3]{17}, \pi^{4/5})$ and $(-\sqrt{5}, \sqrt[4]{63})$.

Solution We must compute

$$\sqrt{(\sqrt[3]{17} + \sqrt{5})^2 + (\pi^{4/5} - \sqrt[4]{63})^2}.$$

The order for pushing the buttons depends on the logic used by the calculator. We compute $\sqrt[3]{17}$ as $17^{1/3}$, compute $\pi^{4/5}$ as $\pi^{0.8}$, and $\sqrt[4]{63}$ as $63^{0.25}$. Of course we always obtain an answer in decimal, or scientific, notation. Our calculator yielded

$$4.81789405$$

for the answer. Another calculator's answer might have fewer or more digits or might differ in the final significant digit. But to two *decimal places*, the answer is surely 4.82, while to five *significant figures*, the answer is 4.8179. □

SUMMARY

1. The closed interval $[a, b]$ consists of all points x such that $a \le x \le b$.
2. The distance from x_1 to x_2 on the line is $|x_2 - x_1|$.
3. The signed length of the directed line segment from x_1 to x_2 is

$$\begin{aligned}\Delta x &= x_2 - x_1\\ &= (\text{Number where you stop}) - (\text{Number where you start}).\end{aligned}$$

4. The midpoint of $[a, b]$ is $(a + b)/2$.
5. The distance between (x_1, y_1) and (x_2, y_2) in the plane is

$$\sqrt{(x_2 - x_1)^2 + (y_2 - y_1)^2}.$$

EXERCISES

In Exercises 1 and 2, sketch as in Figs. 1.3 and 1.4 all points x (if there are any) that satisfy the given relation.

1. a) $2 \le x \le 3$ b) $x^2 = 4$ c) $5 \le x \le -1$

2. a) $x \le 0$ b) $x^2 < 4$ c) $x^2 \le 4$

3. Find the distance between the given points on the number line.
a) 2 and 5 b) -1 and 4 c) -3 and -6

4. Find the distance between the given points on the number line.
a) $-\frac{5}{2}$ and 12 b) $-\frac{8}{3}$ and $-\frac{15}{3}$
c) $\sqrt{2}$ and $-2\sqrt{2}$ d) $\sqrt{2}$ and π

5. Find the number described.
a) $|3 - 5|$ b) $3 - |5|$

6. Find the number described.
a) $|4 - |2 - 7||$ b) $2/|-2|$

7. Find all x such that $|x + 2|/(x + 2) = 1$. (See Example 7.)

8. Find all x such that $|x - 3|/(x - 3) = -1$. (See Example 7.)

9. Show that, for any a and b on the number line, the distance from $(a + b)/2$ to a is the same as the distance from $(a + b)/2$ to b.

10. Find the midpoint of each of the following intervals.
a) $[-1, 1]$ b) $[-1, 4]$ c) $[-\frac{3}{2}, \frac{2}{3}]$

11. Proceed as in Exercise 10.
a) $[-6, -3]$ b) $[-2\sqrt{2}, \sqrt{2}]$ c) $[\sqrt{2}, \pi]$

12. Find the *signed* length Δx of the directed line segment
a) from 2 to 5; b) from -8 to -1.

13. Proceed as in Exercise 12.
a) From 3 to -7 b) From 10 to 2

Each of the relations in Exercises 14 through 25 has as solution all points in a closed interval $[a, b]$. Find the interval in each case. See Appendix 1 if you need to review inequalities.

14. $-2 \le x \le 3$

15. $6 \le 3x \le 12$

16. $4 \le -4x \le 8$

17. $-3 \le x + 2 \le 4$

18. $8 \le x - 3 \le 15$

19. $3 \le 2x - 4 \le 14$

20. $-7 \le 8 - 3x \le 29$

21. $|x| \le 4$

22. $|x - 1| \le 4$

23. $|x + 1| \le 4$

24. $|2x - 3| \le 7$

25. $|8 - 3x| \le 5$

26. Sketch the points (x, y) in the plane satisfying the indicated relations, as in Examples 15 and 16.
a) $x \le y$ b) $x = -y$ c) $y = 2x$ d) $2x \ge y$

27. Proceed as in Exercise 26.
a) $x = 1$ b) $-1 \le x \le 2$
c) $x = -1, -2 \le y \le 3$ d) $x = y, -1 \le x \le 1$

28. Find the coordinates of the indicated point.
a) The point such that the line segment joining it to $(2, -1)$ has the x-axis as perpendicular bisector
b) The point such that the line segment joining it to $(-3, 2)$ has the y-axis as perpendicular bisector

29. Find the coordinates of the indicated point.
a) The point such that the line segment joining it to $(-1, 3)$ has the origin as midpoint
b) The point such that the line segment joining it to $(2, -4)$ has $(2, 1)$ as midpoint

30. Find the distance between the given points.
a) $(-2, 5)$ and $(1, 1)$
b) $(2\sqrt{2}, -3)$ and $(-\sqrt{2}, 2)$

31. Find the distance between the given points.
a) $(2, -3)$ and $(-3, 5)$
b) $(2\sqrt{3}, 5\sqrt{7})$ and $(-4\sqrt{3}, 2\sqrt{7})$

32. It can be shown that a triangle is a right triangle if and only if the sum of the squares of the lengths of two sides equals the square of the length of the third side. The right angle is then opposite the longest side of the triangle. Use this fact and the distance formula to show that $(1, 5)$, $(3, 2)$, and $(7, 9)$ are vertices of a right triangle, and find the area of the triangle.

33. Proceed as in Exercise 32 for the points $(-1, 3)$, $(4, 1)$, and $(1, 8)$.

34. To reach the Edwards' home from the center of town, you drive 2 miles due east on Route 37 and then 5 miles due north on Route 101. Assuming that the surface of the earth near town is approximately flat, find the distance, as the crow flies, from the center of town to the Edwards' home.

35. Refer to Exercise 34. Suppose you drive 6 miles due west on Route 37 and then 4 miles due south on Route 43 to reach the Hammonds' house from the center of town. Find the distance from the Edwards' home to the Hammonds' as the crow flies.

36. Refer to Exercise 34. Charlotte leaves the Edwards' home and jogs south on Route 101 at 7 mph. At the same time, overweight John leaves the center of town and jogs east on Route 37 at 5 mph. Find the distance between them 15 minutes later.

37. Refer to Exercises 34 through 36. Eleanor leaves the Hammonds' house at the same time as Charlotte and jogs north on Route 43 at 6 mph. Find the distance from Eleanor to Charlotte after 15 minutes.

Use a calculator for Exercises 38 through 44.

38. Find the midpoint of $[-2\sqrt{3}, 5\sqrt{7}]$.

39. Find the midpoint of $[\sqrt[5]{23}, -\sqrt[4]{50}]$.

40. Find the signed length of the directed line segment from $22\sqrt{2}$ to π^3.

41. Find the signed length of the directed line segment from $\sqrt[3]{17}$ to $-\sqrt[6]{43}$.

42. Find the distance between $(2, -3)$ and $(4, 1)$.

43. Find the distance between $(-3.7, 4.23)$ and $(8.61, 7.819)$.

44. Find the distance between $(\pi, -\sqrt{3})$ and $(8\sqrt{17}, -\sqrt[3]{\pi})$.

1.2 CIRCLES AND THE SLOPE OF A LINE

CIRCLES

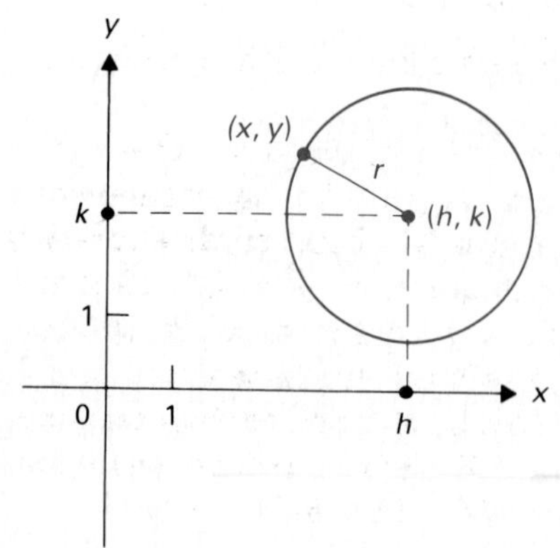

Figure 1.18 The circle $(x-h)^2+(y-k)^2=r^2$.

The **circle** with center (h, k) and radius r consists of all points (x, y) whose distance from (h, k) is r. See Fig. 1.18. Using the formula for the distance from (x, y) to (h, k), we see that this circle consists of all points (x, y) such that

$$\sqrt{(x-h)^2+(y-k)^2}=r. \tag{1}$$

Squaring both sides of Eq. (1), we obtain the equivalent relation

$$\boxed{(x-h)^2+(y-k)^2=r^2.} \tag{2}$$

Equation (2) is known as the *equation of the circle.*

EXAMPLE 1 Find the equation of the circle with center $(-2, 4)$ and radius 5.

Solution From Eq. (2), we see that the circle has equation $[x-(-2)]^2+(y-4)^2=5^2$, or $(x+2)^2+(y-4)^2=25$. The circle is sketched in Fig. 1.19. □

EXAMPLE 2 Find the center and radius of the circle with equation $(x+3)^2+(y+4)^2=18$.

Solution Rewriting the equation as

$$[x-(-3)]^2+[y-(-4)]^2=(\sqrt{18})^2,$$

we see from Eq. (2) that the circle has center $(-3, -4)$ and radius $\sqrt{18}=3\sqrt{2}$. □

Figure 1.19 The circle $(x+2)^2+(y-4)^2=25$.

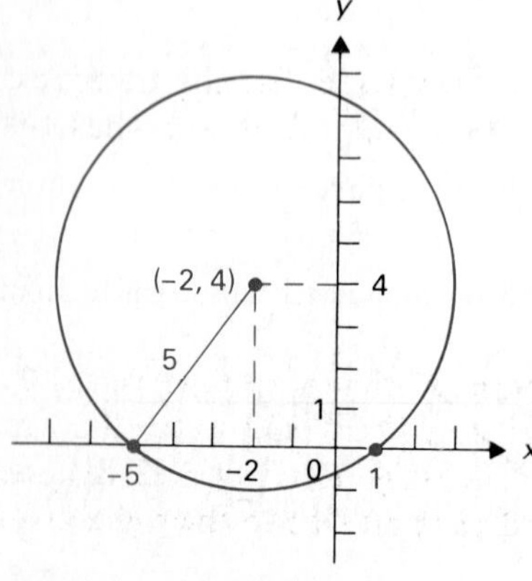

Note that

$$x^2+bx=\left(x^2+bx+\frac{b^2}{4}\right)-\frac{b^2}{4}=\left(x+\frac{b}{2}\right)^2-\frac{b^2}{4}.$$

An expression x^2+bx appearing in an equation can thus be changed to $[x+(b/2)]^2$ by adding $b^2/4$ to both sides of the equation. This algebraic device is known as *completing the square* and is used in the following example.

EXAMPLE 3 Show that $3x^2+3y^2+6x-12y=60$ describes a circle.

Solution We start by dividing by the common coefficient 3 of x^2 and y^2 and obtain

$$x^2+y^2+2x-4y=20.$$

Now we use the algebraic device of completing the square to get our equation

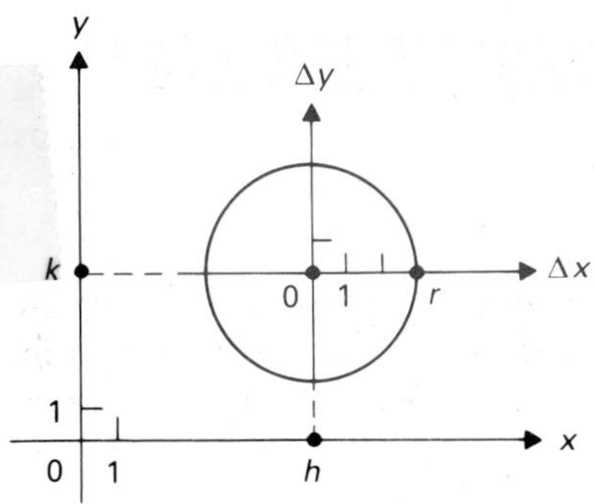

Figure 1.20

$(\Delta x)^2 + (\Delta y)^2 = r^2$

in the form of Eq. (2). The steps are as follows:

$$(x^2 + 2x) + (y^2 - 4y) = 20,$$

$$(x^2 + 2x + 1) + (y^2 - 4y + 4) = 20 + 1 + 4,$$

$$(x + 1)^2 + (y - 2)^2 = 25.$$

Thus our equation describes a circle with center $(-1, 2)$ and radius 5. □

As illustrated by Example 3, every equation of the form $ax^2 + ay^2 + bx + cy = d$ and satisfied by at least one point (x_1, y_1) is the equation of a circle. However, there may be no points of the plane whose coordinates satisfy a particular equation of this type. For example, $x^2 + y^2 = -10$ is not satisfied for any point (x, y) in the plane, for a sum of squares of real numbers cannot be negative. We should try to put any such equation in the form of Eq. (2) to find the center and radius of the circle, as illustrated by Example 3.

There is another way of regarding Eq. (2) that will be very useful to us. If we let $\Delta x = x - h$ and $\Delta y = y - k$, then Eq. (2) becomes

$$(\Delta x)^2 + (\Delta y)^2 = r^2. \tag{3}$$

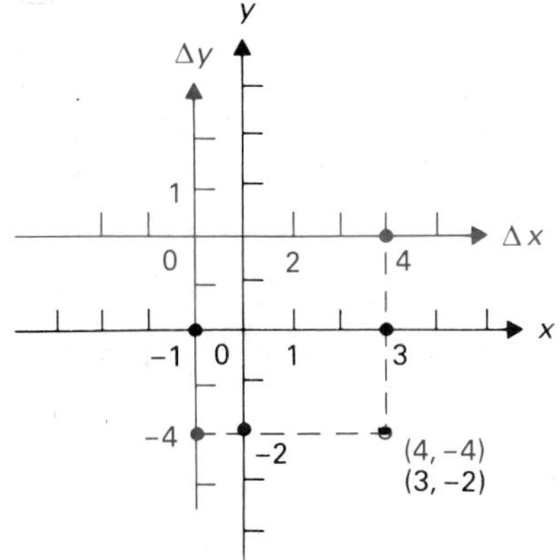

Figure 1.21 $(x, y) = (3, -2)$ has coordinates $(\Delta x, \Delta y) = (4, -4)$.

To interpret Eq. (3) geometrically, we take a new axis called the Δx-axis and similarly a Δy-axis with the point (h, k) as new origin, as shown in Fig. 1.20. Recall that Δx is the directed distance from h to x and Δy is the directed distance from k to y. Thus Eq. (3) is exactly the equation of the circle with respect to our new axes. This device is known as *translation of axes to* (h, k) and will often be useful. We should remember

$$\Delta x = x - h,$$

$$\Delta y = y - k.$$

Translation equations

The equation $x^2 + y^2 = r^2$ describes a circle with center at the origin of the x,y-coordinate system and radius r, while the equation $(\Delta x)^2 + (\Delta y)^2 = r^2$ describes a circle with center at the origin of the $\Delta x,\Delta y$-coordinate system and radius r.

EXAMPLE 4 Let translated $\Delta x,\Delta y$-axes be chosen with origin at the point $(h, k) = (-1, 2)$. Find $\Delta x,\Delta y$-coordinates of the point $(x, y) = (3, -2)$. See Fig. 1.21.

Solution Since $\Delta x = x - h$ and $\Delta y = y - k$, we see that the translated coordinates are $\Delta x = 3 - (-1) = 4$ and $\Delta y = -2 - 2 = -4$. Thus the point is $(\Delta x, \Delta y) = (4, -4)$. □

Figure 1.22 A line of slope 3.

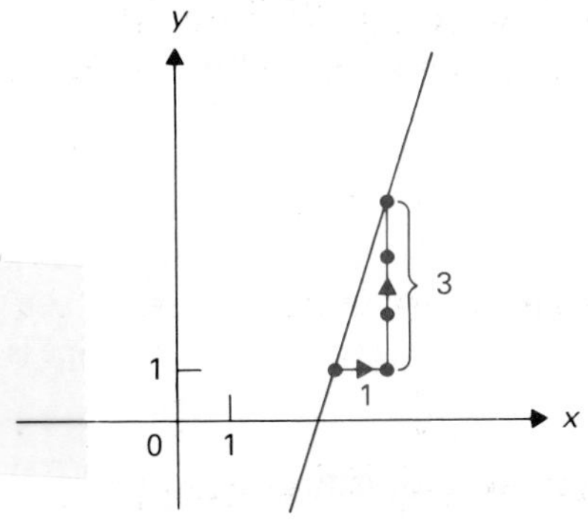

THE SLOPE OF A LINE

The **slope** m of a line is the number of units the line climbs (or falls) vertically for each unit of horizontal change from left to right. To illustrate, if a line climbs upward 3 units for each unit step to the right, as in Fig. 1.22, the line has slope 3. If a line falls 2 units downward per unit step to the right, as in

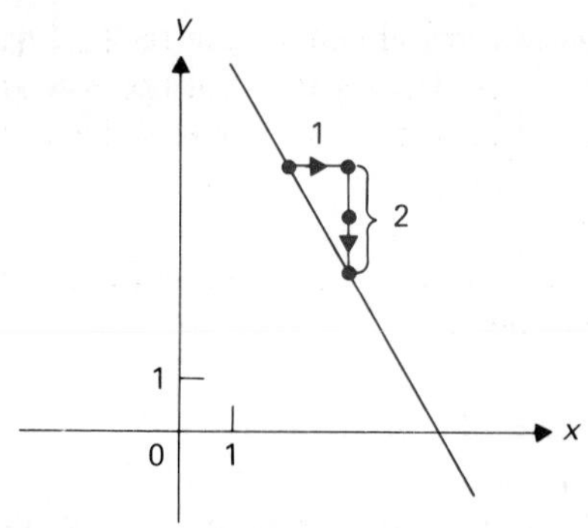

Figure 1.23 A line of slope -2.

Fig. 1.23, the line has slope -2. A horizontal line neither climbs nor falls, so it has slope 0. A vertical line climbs straight up over a single point, so it is impossible to measure how much it climbs per unit horizontal change, as indicated in Fig. 1.24. Consequently,

> the slope of a vertical line is undefined.

EXAMPLE 5 Find the slope of the line through the points (2, 4) and (5, 16).

Solution As we go from (2, 4) to (5, 16), we have $\Delta x = 5 - 2 = 3$ and $\Delta y = 16 - 4 = 12$. Since the line climbed $\Delta y = 12$ units while we went $\Delta x = 3$ units to the right, and since a line climbs at a uniform rate, the amount it climbs per horizontal unit to the right is $\Delta y/\Delta x = \frac{12}{3} = 4$. □

As illustrated in Example 5, we can find the slope m of the line through (x_1, y_1) and (x_2, y_2) if $x_1 < x_2$ by finding Δx and Δy as we go from (x_1, y_1) to (x_2, y_2), and then taking the quotient. So

$$m = \frac{\Delta y}{\Delta x} = \frac{y_2 - y_1}{x_2 - x_1}. \tag{4}$$

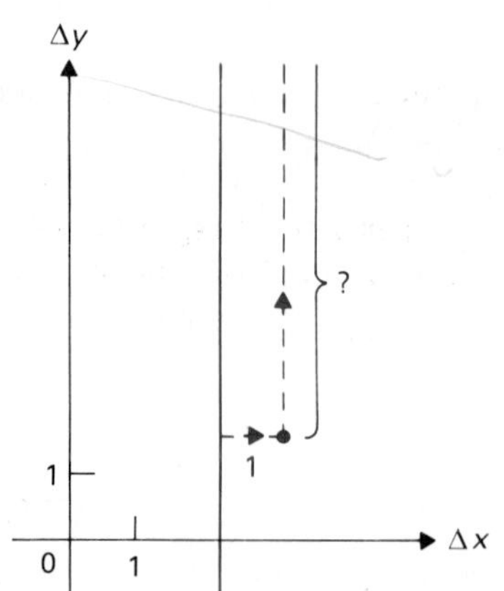

Figure 1.24 Vertical line; slope undefined.

We assume that our line is not vertical, so $x_1 \neq x_2$. If it should happen that $x_2 < x_1$, then to go from left to right we should go from (x_2, y_2) to (x_1, y_1), and we obtain

$$m = \frac{\Delta y}{\Delta x} = \frac{y_1 - y_2}{x_1 - x_2} = \frac{y_2 - y_1}{x_2 - x_1}, \tag{5}$$

which is the same formula as in Eq. (4).

In summary, the slope m of a nonvertical line through two points is given by

$$m = \frac{\Delta y}{\Delta x} = \frac{\text{Difference of } y\text{-coordinates}}{\text{Difference of } x\text{-coordinates in the same order}}. \tag{6}$$

EXAMPLE 6 Find the slope of the line through (7, 5) and $(-2, 8)$.

Solution The slope formula in (6) tells us that the line has slope

$$m = \frac{\Delta y}{\Delta x} = \frac{8 - 5}{-2 - 7} = \frac{3}{-9} = -\frac{1}{3}. \quad \square$$

EXAMPLE 7 Determine the slope of the line through $(-3, 5)$ and $(-3, 11)$.

Solution We have $\Delta x = -3 - (-3) = 0$, so the line is vertical. The slope is *undefined.* □

The next example reinforces our understanding of *slope* as a measure of climb in practical situations.

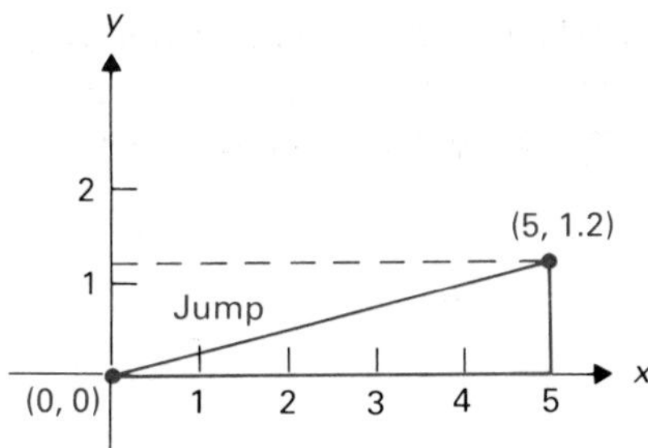

Figure 1.25 Water-ski jump rising 1.2 m over a distance of 5 m.

EXAMPLE 8 A water-ski jump consists of a straight chute that rises 1.2 m above the water over a horizontal distance of 5 m. Taking an x-axis horizontal with origin at the bottom of the jump, find the slope of the jump. Give the physical interpretation of this slope.

Solution As shown in Fig. 1.25, the jump goes from the origin (0, 0) to the point (5, 1.2). By Eq. (6), the slope of the jump is

$$\frac{\Delta y}{\Delta x} = \frac{1.2 - 0}{5 - 0} = 0.24.$$

Thus the jump rises 0.24 m in height for every meter measured horizontally. □

Two lines are parallel precisely when they climb (or fall) at the same rate. Therefore,

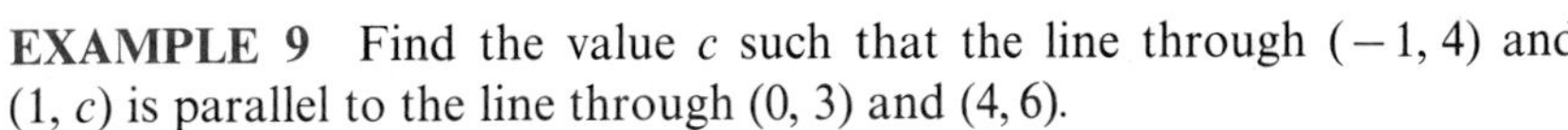
parallel lines have the same slope.

This is illustrated in Fig. 1.26.

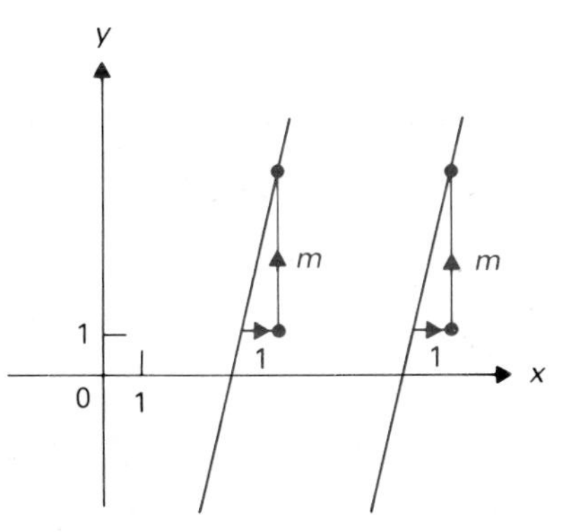

Figure 1.26 Parallel lines have the same slope m.

EXAMPLE 9 Find the value c such that the line through $(-1, 4)$ and $(1, c)$ is parallel to the line through (0, 3) and (4, 6).

Solution The line through $(-1, 4)$ and $(1, c)$ has slope $m_1 = (c - 4)/2$, and the line through (0, 3) and (4, 6) has slope $m_2 = (6 - 3)/(4 - 0) = \frac{3}{4}$. For the lines to be parallel, their slopes m_1 and m_2 must be equal, so

$$\frac{c - 4}{2} = \frac{3}{4},$$

$$c - 4 = \frac{6}{4} = \frac{3}{2},$$

$$c = 4 + \frac{3}{2} = \frac{11}{2}. \quad \square$$

We now derive a condition for two lines to be perpendicular. Let one line have slope m_1 and the other slope m_2. If we choose a new origin at the point of intersection of the lines, as shown in Fig. 1.27, then the points $(\Delta x, \Delta y) = (1, m_1)$ and $(\Delta x, \Delta y) = (1, m_2)$ lie on the lines as shown in the figure. For the angles shown in the figure, we see that

$$\alpha_1 + \beta_1 = 90°, \qquad \text{so} \quad \alpha_1 = 90° - \beta_1.$$

Thus $\alpha_1 + \beta_2 = 90°$ if and only if $(90° - \beta_1) + \beta_2 = 90°$ or if and only if $\beta_2 = \beta_1$. An analogous argument then shows that $\alpha_1 = \alpha_2$, so the triangles are similar. Corresponding sides are then proportional, and from Fig. 1.27, we see that

$$\boxed{\frac{m_1}{1} = \frac{1}{-m_2}, \qquad \text{or} \qquad m_1 m_2 = -1.} \tag{7}$$

The relation (7) is the desired condition for the lines to be perpendicular.

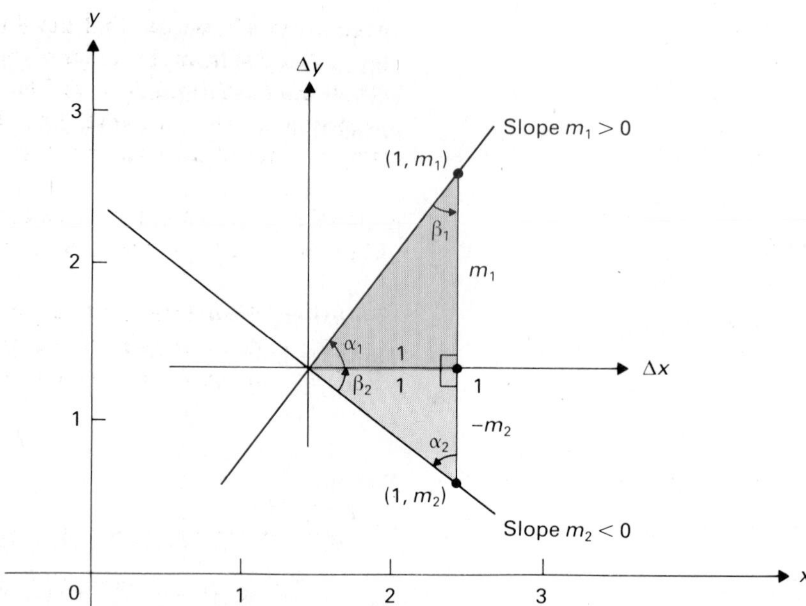

Figure 1.27 The lines are perpendicular when the two triangles are similar, or when $m_1/1 = 1/-m_2$

EXAMPLE 10 Find the slope of a line perpendicular to the line through (6, −5) and (8, 3).

Solution The given line has slope

$$\frac{\Delta y}{\Delta x} = \frac{3 - (-5)}{8 - 6} = \frac{8}{2} = 4.$$

So a perpendicular line has slope $-\frac{1}{4}$. □

In geometric terms, differential calculus is concerned with finding the slope of a line that is "tangent to a curve." Figure 1.28 shows a tangent line to a curve at the point (2, 3) on the curve. We cannot handle this problem in

Figure 1.28 The tangent line to the curve at the point (2, 3).

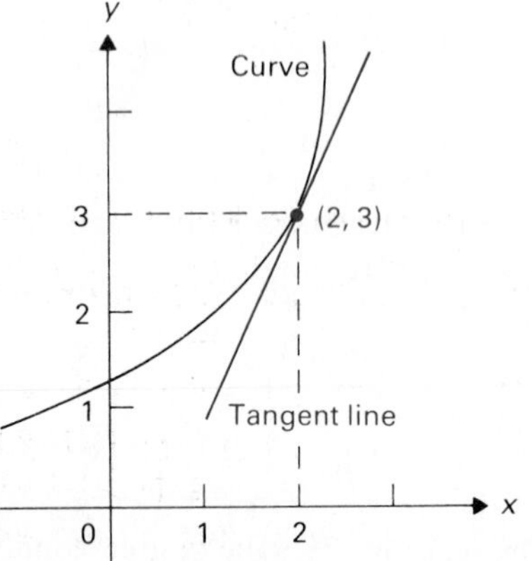

Figure 1.29 Tangent line and perpendicular radial line at (1, 3).

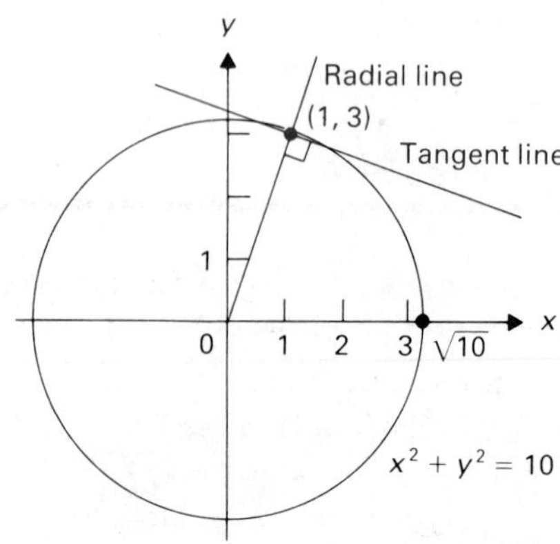

general yet, but we can find the slope of a line tangent to a circle at a point (x, y). The line from the center of the circle through a point (x, y) on the circle is the *radial line* through (x, y). The line *tangent* to the circle at (x, y) is the line perpendicular to the radial line there. Figure 1.29 shows the radial and tangent lines to the circle $x^2 + y^2 = 10$ at the point $(1, 3)$.

EXAMPLE 11 Find the slope of the line tangent to the circle $x^2 + y^2 = 10$ at the point $(1, 3)$ on the circle. See Fig. 1.29.

Solution We use the fact that the tangent line to a circle is perpendicular to the radial line at the point of tangency. The radial line goes through $(0, 0)$ and $(1, 3)$ and so has slope $(3 - 0)/(1 - 0) = 3$. The tangent line thus has slope $-\frac{1}{3}$. □

SUMMARY

1. The circle with center (h, k) and radius r has equation
$$(x - h)^2 + (y - k)^2 = r^2.$$
2. To find the center (h, k) and the radius r of a circle $ax^2 + ay^2 + bx + cy = d$, complete the square on the x-terms and on the y-terms.
3. The relationship between x,y-coordinates and translated $\Delta x,\Delta y$-coordinates with new origin at $(x, y) = (h, k)$ is given by
$$\Delta x = x - h, \qquad \Delta y = y - k.$$
4. A vertical line has undefined slope. If $x_1 \neq x_2$, the line through (x_1, y_1) and (x_2, y_2) has slope
$$m = \frac{\Delta y}{\Delta x} = \frac{y_2 - y_1}{x_2 - x_1}.$$
5. Lines of slopes m_1 and m_2 are

parallel if and only if $m_1 = m_2$,

perpendicular if and only if $m_1 m_2 = -1$, or $m_2 = -\dfrac{1}{m_1}$.

EXERCISES

In Exercises 1 through 4, find the equation of the circle with the given center and radius.

1. center $(0, 0)$, radius 5
2. center $(-1, 2)$, radius 3
3. center $(3, -4)$, radius $\sqrt{30}$
4. center $(-2, -3)$, radius $\sqrt{5}$

In Exercises 5 through 10, find the center and radius of the given circle.

5. $(x - 2)^2 + (y - 3)^2 = 36$
6. $(x + 3)^2 + y^2 = 49$
7. $(x + 1)^2 + (y + 4)^2 = 50$
8. $x^2 + y^2 - 4x + 6y = 3$

9. $x^2 + y^2 + 8x = 9$
10. $4x^2 + 4y^2 - 12x - 24y = -\frac{9}{2}$
11. Find the equation of the circle with center in the second quadrant, tangent to the coordinate axes, and with radius 4.
12. Find the equation of the circle having the line segment with endpoints $(-1, 2)$ and $(5, -6)$ as a diameter.
13. Find the equation of the circle with center $(2, -3)$ and passing through $(5, 4)$.
14. Find the point on the circle $x^2 + y^2 = 25$ that is diametrically opposite the point $(3, 4)$. [*Hint:* Draw a figure.]
15. Find the point on the circle $x^2 + y^2 - 4x + 6y = -12$ that is diametrically opposite the point $(1, -3)$. [*Hint:* Draw a figure.]
16. Find translated $\Delta x, \Delta y$-coordinates of the given points (x, y) with respect to a new origin at $(h, k) = (-3, 2)$.
a) $(4, 6)$ b) $(-1, 3)$ c) $(0, 0)$
17. Proceed as in Exercise 16 for a new origin at $(h, k) = (-4, -1)$.
a) $(-3, -2)$ b) $(5, -3)$ c) $(8, -1)$

In Exercises 18 through 22, find the slope of the line through the indicated points, if the line is not vertical.

18. $(-3, 4)$ and $(2, 1)$
19. $(5, -2)$ and $(-6, -3)$
20. $(3, 5)$ and $(3, 8)$
21. $(0, 0)$ and $(5, 4)$
22. $(-7, 4)$ and $(9, 4)$
23. Find b so that the line through $(2, -3)$ and $(5, b)$ has slope -2.
24. Find a so that the line through $(a, -5)$ and $(3, 6)$ has slope 1.
25. Find the slope of a line perpendicular to the line through $(-3, 2)$ and $(4, 1)$.
26. Find b so that the line through $(8, 4)$ and $(4, -2)$ is parallel to the line through $(-1, 2)$ and $(2, b)$.
27. Find c such that the line through $(3, 1)$ and $(-2, c)$ is perpendicular to the line through $(4, -1)$ and $(3, 2)$.
28. Find the slope of the line through the centers of the circles $x^2 + y^2 + 4x - 2y = 8$ and $x^2 + y^2 - 8x + 6y = 0$.

In Exercises 29 through 32, use slopes to determine whether or not the four points are vertices of (a) a parallelogram, (b) a rectangle.

29. $A(2, 1)$, $B(3, 4)$, $C(-1, 2)$, $D(0, 5)$
30. $A(-1, 4)$, $B(3, 1)$, $C(1, 5)$, $D(1, 0)$
31. $A(3, 1)$, $B(2, -4)$, $C(6, 3)$, $D(5, -2)$
32. $A(4, 0)$, $B(7, 2)$, $C(2, 3)$, $D(5, 4)$
33. Let L be the line through $(-1, 4)$ with slope 5. Find the y-coordinate of the point on L whose x-coordinate is
a) 2, b) -3.
34. Let L be the line through $(2, -3)$ with slope -4. Find the x-coordinate of the point of L with y-coordinate
a) 5, b) 0.

In Exercises 35 through 38, use slopes to determine whether the points A, B, and C are collinear.

35. $A(3, -1)$, $B(5, 3)$, $C(2, -3)$
36. $A(0, 6)$, $B(3, 4)$, $C(6, 1)$
37. $A(1, 4)$, $B(6, 3)$, $C(8, 2)$
38. $A(-1, 3)$, $B(5, -6)$, $C(-3, 6)$
39. A particle moves in the direction of increasing x along the line L through $(0, 0)$ with slope 3. If the particle moves at a constant rate of 2 units/sec, find its position 5 sec after it is at the origin.
40. A particle moves in the direction of decreasing x along the line L through $(-1, 4)$ with slope $-\frac{4}{3}$. If the particle moves at a constant rate of 4 units/sec, find its position 5 sec after it is at $(-1, 4)$.
41. Find the slope of the line tangent to the circle $x^2 + y^2 - 2y = 4$ at the point $(1, -1)$ on the circle. [*Hint:* The tangent line is perpendicular to the radial line.]
42. Find all points on the circle $x^2 + y^2 + 2x - 4y = 15$ where the tangent line has slope 2. [*Hint:* The tangent line is perpendicular to the radial line.]
43. A body travels counterclockwise around the circle $x^2 + y^2 = 25$ at a constant rate of 1 unit/sec. The body flies off the circle at the point $(3, -4)$ and continues at the same speed in the direction tangent to the circle. Find the position of the body 5 sec after it leaves the circle.
44. Show that the line joining the midpoints of two sides of a triangle is parallel to the third side. [*Hint:* Let the vertices of the triangle be $(0, 0)$, $(a, 0)$, and (b, c).]
45. Water freezes at 0° C and 32° F, and it boils at 100° C and 212° F. If points (C, F) are plotted in the plane, where F is the temperature in degrees Fahrenheit corresponding to a temperature of C degrees Celsius, then a straight line is obtained. Find the slope of the line. What does this slope represent in this situation?
46. The Easy Life Prefabricated Homes Company listed its super-deluxe ranch model for \$30,000 in 1960. The company increased the price by the same amount each year and listed the same model for \$90,000 in 1980. Find the slope of the segment drawn through points (Y, C) in the plane, where Y could be any year from 1960 to 1980 and C is the cost of this model ranch house in that year. What does this slope represent in this situation?
47. A house wall is 10 in. thick and is composed of four different types of material, as shown in Fig. 1.30. The graph in the

Figure 1.30 Temperature in a house wall.

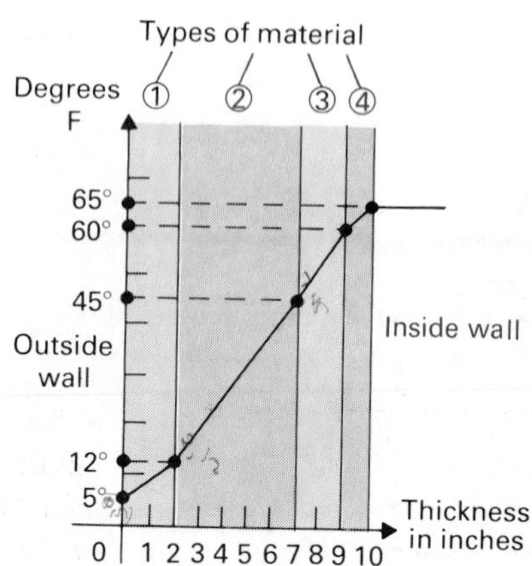

figure shows the temperature in a cross section of the wall when the outside temperature is 5° F and the inside temperature is 65° F. Explain in terms of slopes which of the four types of materials is
a) the most efficient type of insulation,
b) the least efficient type of insulation.

Use a calculator for Exercises 48 through 53. In Exercises 48 through 50, find the center and radius of the given circle.

48. $(x - \pi)^2 + (y - \sqrt{\pi})^2 = 2.736$

49. $x^2 + y^2 + 3.1576x - 1.2354y = 3.33867$

50. $\sqrt{2}x^2 + \sqrt{2}y^2 - \pi^3 x + (\pi^2 + 3.4)y = \sqrt{17}$

In Exercises 51 through 53, find the slope of the line through the indicated points.

51. $(2.367, \pi)$ and $(\sqrt{3.89})$

52. $(\pi^2, \sqrt[3]{19})$ and $(12.378, \sqrt{5.69})$

53. $(\sqrt{2} + \sqrt{3}, \pi - \sqrt{19.3})$ and $(\sqrt{\pi} + 1.45, \sqrt{14} - \sqrt[5]{134})$

1.3 THE EQUATION OF A LINE

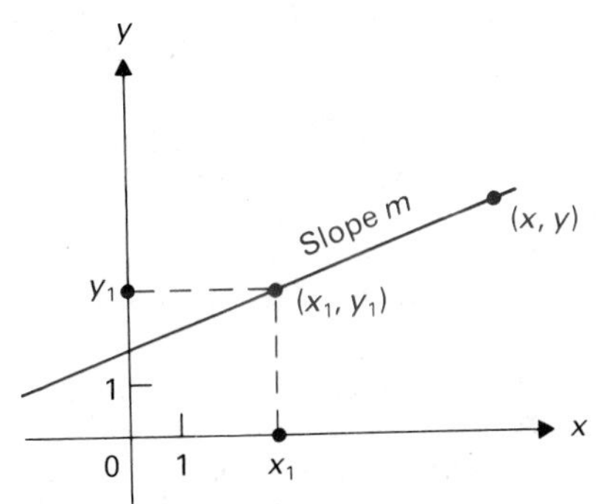

Figure 1.31 (x, y) on the line of slope m through (x_1, y_1).

Let a given line have slope m and pass through the point (x_1, y_1) as shown in Fig. 1.31. Try to find an algebraic condition for a point (x, y) to lie on the line. If the slope of the line that joins (x_1, y_1) and (x, y) is also m, then that line is parallel to the given line, because they have the same slope. But both lines go through (x_1, y_1), so they must coincide. Therefore a condition for (x, y) to lie on the given line is that

$$\frac{y - y_1}{x - x_1} = m, \tag{1}$$

or

$$\boxed{y - y_1 = m(x - x_1).} \tag{2}$$

Equation (2) is the *point–slope form* of the equation of the line.

EXAMPLE 1 Find the equation of the line through $(2, -3)$ with slope 7, shown in Fig. 1.32.

Solution The equation is $y - (-3) = 7(x - 2)$ or $y + 3 = 7(x - 2)$. This equation may be simplified to $y = 7x - 17$. The point $(3, 4)$ lies on this line, since $4 = 7 \cdot 3 - 17$. □

Figure 1.32 Line through $(2, -3)$ with slope 7.

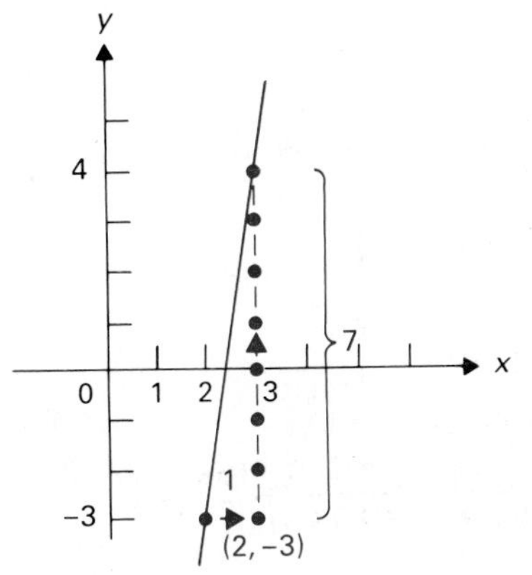

As indicated in Example 1, the point–slope equation (2) can be rewritten in the form

$$\boxed{y = mx + b,} \tag{3}$$

where $b = y_1 - mx_1$. The constant b in Eq. (3) has a nice interpretation. If we set $x = 0$ in Eq. (3), then $y = b$, so the point $(0, b)$ satisfies the equation and thus lies on the line. This point $(0, b)$ is on the y-axis, and b is the ***y*-intercept** of the line. See Fig. 1.33. For this reason, Eq. (3) is the *slope–intercept form* of the equation of the line. If the line crosses the x-axis at $(a, 0)$, then a is the ***x*-intercept** of the line.

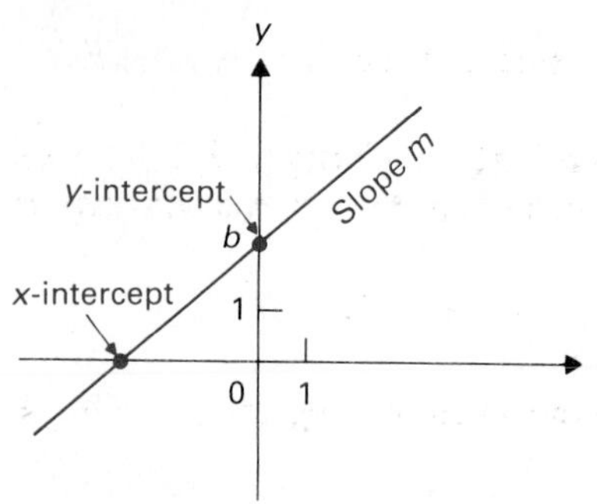

Figure 1.33 The line $y = mx + b$.

EXAMPLE 2 Find the intercepts of the line in Example 1.

Solution The equation is $y = 7x - 17$, so -17 is the y-intercept. To find the x-intercept, we set $y = 0$ and obtain $7x - 17 = 0$, so $x = \frac{17}{7}$. Thus the point $(\frac{17}{7}, 0)$ lies on the line, so $\frac{17}{7}$ is the x-intercept. □

The vertical line through $(a, 0)$ in Fig. 1.34 has undefined slope, so it does not have an equation of the form of Eq. (2) or (3). But surely a condition that (x, y) lie on the line is simply that $x = a$. Of course, $y = b$ is the horizontal line through $(0, b)$ shown in the figure. In any kind of coordinate system, it is important to know what sets of points are obtained by setting the coordinate variables equal to constants. We see that in our rectangular x,y-coordinate system, $x = a$ is a vertical line and $y = b$ is a horizontal line.

EXAMPLE 3 Find the equation of the line through $(-5, 3)$ and $(-5, 7)$.

Solution The slope of this line is undefined since $\Delta x = -5 - (-5) = 0$. Thus the line is vertical, with an equation of the form $x = a$, where a is the x-coordinate of every point on the line. In this case, the equation is $x = -5$. □

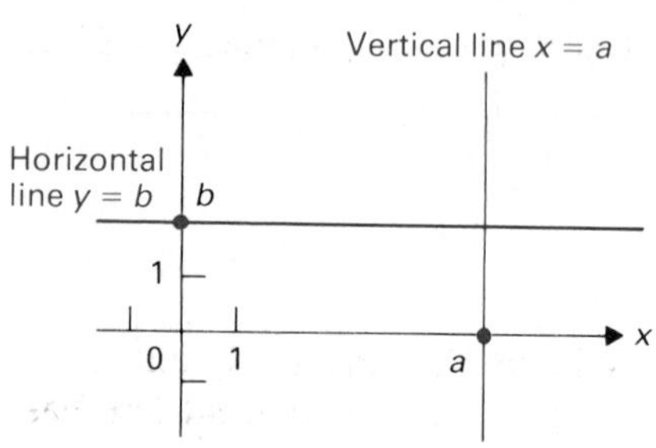

Figure 1.34 Equations of vertical and horizontal lines.

Any time you want to find the equation of a line, say to yourself, "I need to find a *point on the line* and the *slope of the line*." Then use Eq. (2).

EXAMPLE 4 Find the equation of the line through $(-5, -3)$ and $(6, 1)$.

Solution We solve the problem as follows.

Point $(x_1, y_1) = (-5, -3)$

Slope $m = \dfrac{1 - (-3)}{6 - (-5)} = \dfrac{4}{11}$

Equation
$$\begin{array}{ccccc} y - y_1 & = & m(x & - & x_1), \\ \downarrow & & \downarrow & & \downarrow \end{array}$$
$$y - (-3) = \frac{4}{11}[x - (-5)],$$
$$y + 3 = \frac{4}{11}(x + 5)$$

The equation can be simplified to $11y + 33 = 4x + 20$ or $4x - 11y = 13$. □

Figure 1.35 Midpoint of a line segment.

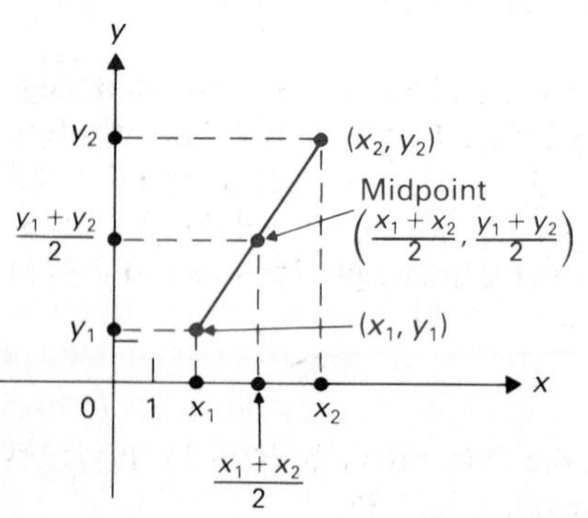

Recall from Section 1.1 that a closed interval $[a, b]$ of the number line has as midpoint $(a + b)/2$. We see from Fig. 1.35 that, in the plane,

> $\left(\dfrac{x_1 + x_2}{2}, \dfrac{y_1 + y_2}{2}\right)$ is the **midpoint** of the line segment joining (x_1, y_1) and (x_2, y_2).

EXAMPLE 5 Find the equation of the perpendicular bisector of the line segment joining $(-1, 4)$ and $(3, 8)$, shown in Fig. 1.36.

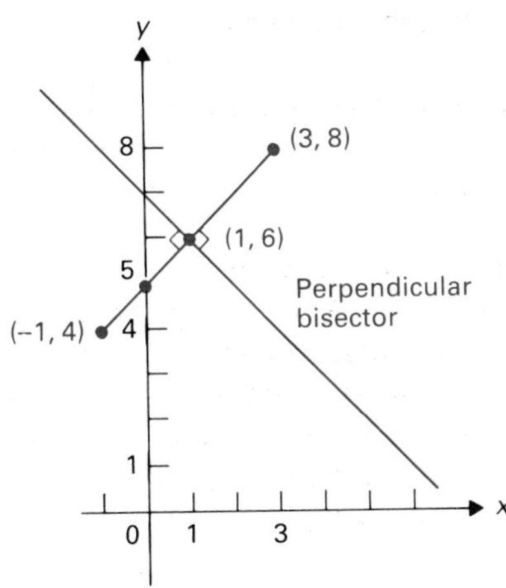

Figure 1.36 Perpendicular bisector of the line segment joining (−1, 4) and (3, 8).

Solution We need to find a point on the line and the slope of the line.

Point The midpoint of the line segment joining $(-1, 4)$ and $(3, 8)$ lies on the bisecting line. From the midpoint formula just given, the midpoint is

$$\left(\frac{-1+3}{2}, \frac{4+8}{2}\right) = \left(\frac{2}{2}, \frac{12}{2}\right) = (1, 6).$$

Slope The line is to be perpendicular to the given line segment, which has slope

$$\frac{\Delta y}{\Delta x} = \frac{8-4}{3-(-1)} = \frac{4}{4} = 1.$$

Thus the desired slope is $m = -1$.

Equation

$$\begin{aligned} y - y_1 &= m(x - x_1), \\ &\downarrow \quad \downarrow \quad \downarrow \\ y - 6 &= -1(x-1), \\ y &= -x + 7 \quad \square \end{aligned}$$

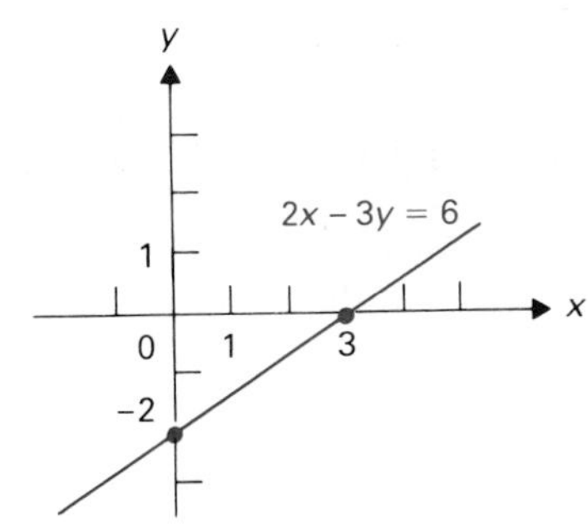

Figure 1.37 Sketch of $2x - 3y = 6$ using intercepts.

EXAMPLE 6 Find the equation of the line tangent to the circle $x^2 + y^2 = 10$ at the point $(1, 3)$.

Solution

Point $(1, 3)$

Slope The line is perpendicular to the radial line through $(0, 0)$ and $(1, 3)$. This radial line has slope $\Delta y/\Delta x = \frac{3}{1} = 3$, so the tangent line has slope $-\frac{1}{3}$.

Equation (We should no longer need to show substitutions as we did in Examples 4 and 5.)

$$y - 3 = -\tfrac{1}{3}(x - 1), \qquad \text{or} \quad 3y + x = 10 \quad \square$$

Finally, observe that every equation $ax + by + c = 0$ where either $a \neq 0$ or $b \neq 0$ is the equation of a line. If $b = 0$, the equation becomes $x = -c/a$, which is a vertical line. If $b \neq 0$, the equation becomes $y = -(a/b)x - c/b$, which is a line with slope $m = -a/b$ and y-intercept $-c/b$.

EXAMPLE 7 Sketch all points (x, y) such that $2x - 3y = 6$.

Solution We have seen students make a table of ten or more points satisfying such an equation, plot the points, and then draw a wobbly line through them. This is unnecessary. We know $2x - 3y = 6$ is the equation of a *line* and is thus determined by just two points. Setting $x = 0$, we obtain $(0, -2)$ as y-intercept. Setting $y = 0$, we obtain $(3, 0)$ as x-intercept. The line is shown in Fig. 1.37. $\square$

Figure 1.38 Line through $(4, -1)$ perpendicular to $2x + 4y = 5$.

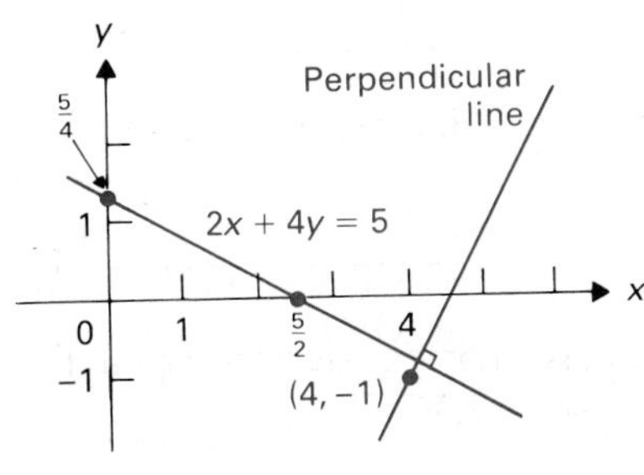

EXAMPLE 8 Find the equation of the line through $(4, -1)$ and perpendicular to the line $2x + 4y = 5$, shown in Fig. 1.38.

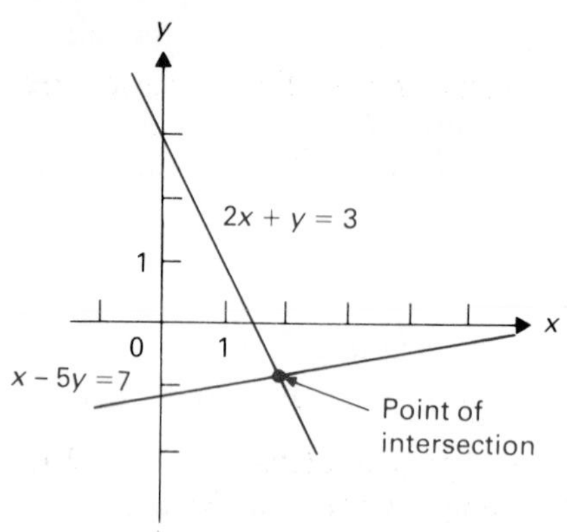

Figure 1.39 Point (2, −1) of intersection of $2x + y = 3$ and $x - 5y = 7$.

Solution

Point $(4, -1)$

Slope The given equation can be rewritten $4y = -2x + 5$, or $y = -\frac{1}{2}x + \frac{5}{4}$ in the form $y = mx + b$. Thus the given line has slope $-\frac{1}{2}$, so the desired perpendicular line has slope 2.

Equation $y + 1 = 2(x - 4)$, or $y = 2x - 9$ □

EXAMPLE 9 Find the point of intersection of the lines $2x + y = 3$ and $x - 5y = 7$, shown in Fig. 1.39.

Solution We must find (x, y) satisfying both equations at once. If we solve the equations simultaneously, we obtain

$$\begin{cases} 2x + y = 3 \\ x - 5y = 7 \end{cases} \qquad \begin{cases} 2x + y = 3 \\ -2x + 10y = -14 \end{cases} \qquad \text{Multiplying by } -2$$

$$11y = -11. \qquad \text{Adding}$$

Thus $y = -1$, so $x = 5y + 7 = -5 + 7 = 2$, and the point of intersection is $(2, -1)$. □

Two quantities P and Q are *linearly related* if each can be computed from the other using a linear equation, such as $Q = mP + b$. Of course, this means geometrically that points (P, Q) having corresponding values of P and Q as coordinates lie on a line. For example, Fahrenheit and Celsius temperatures are linearly related. Exercise 42 asks you to find this relation. We give illustrations in our final two examples.

EXAMPLE 10 For a body immersed in a fluid, the fluid pressure P on the body is linearly related to the depth d of immersion. Suppose for a certain fluid that a depth of 24 cm corresponds to a pressure of 32 g/cm^2. Find the linear relation between P and d.

Solution Since P must be zero when $d = 0$, the point $(d, P) = (0, 0)$ is one point satisfying the desired relation. We are given that another point is $(d, P) = (24, 32)$. Thus we have

Point $(0, 0)$

Slope $m = \dfrac{32 - 0}{24 - 0} = \dfrac{32}{24} = \dfrac{4}{3}$

Linear Relation $P - 0 = \dfrac{4}{3}(d - 0)$, or $3P = 4d$ □

EXAMPLE 11 Suppose a company's revenue remains constant and its costs also remain constant for any two equal periods of time. Let the company's profit over a period of t years after the company started be P. Then P and t are linearly related. If the company has accumulated profits, in thousands of dollars, of 500 after 10 years and of 640 after 12 years, find (a) the linear relations between P and t and (b) the initial cost to start the company.

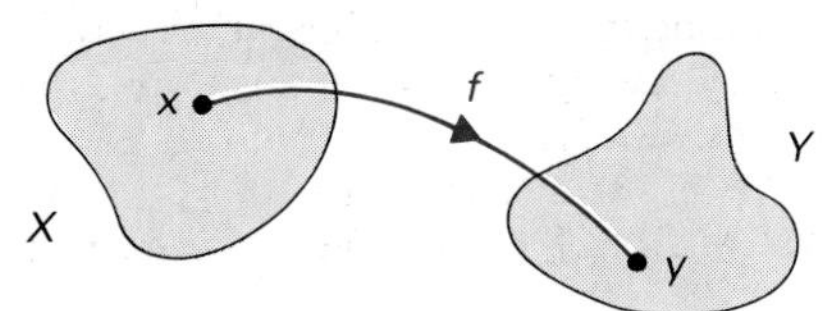

Figure 1.40 Schematic of a function f.

Figure 1.40 attempts to illustrate the definition. Previously in this text, we have used letters only to represent numbers. It is important to understand that a function f is *not* a number but rather should be regarded as a rule of assignment. We have attempted to denote this in Fig. 1.40 by labeling with f the colored arrow that indicates the function assigning to x in X the element y in Y.

Another intuitive picture of a function popular in elementary texts is a "magic box," as in Fig. 1.41. One drops in an element x, and an element y comes out at the end. A scientific calculator is such a magic box. For example, we punch in a value for x and then press the sin x button to "perform the function," and the y-value, where $y = f(x) = \sin x$, is shown as the display.

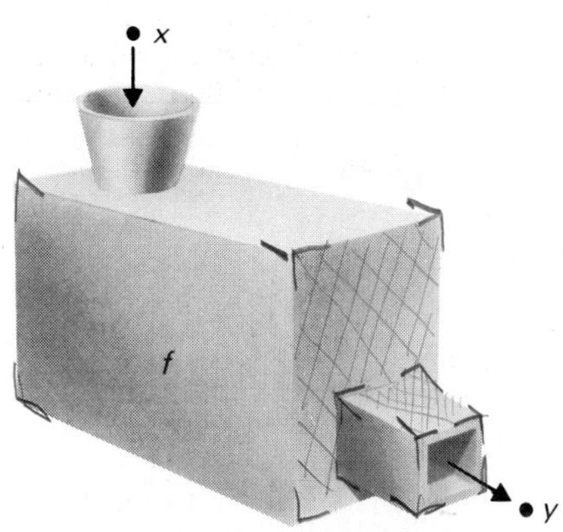

Figure 1.41 Another schematic of a function f.

Let $y = f(x)$ for x in a set X and y in a set Y, as given in Definition 1.1. Since we think of y as depending on x, we call y the **dependent variable** and x the **independent variable.** The set X of all x on which f acts is the **domain** of the function f. The set of all y elements obtained by finding $f(x)$ for all x in X is the **range** of the function f. We will often be sloppy and speak of a function $f(x)$, but technically $f(x)$ is the **value** y of the function f at x, while f is the function.

From now until Chapter 16, we will use function notation and terminology only when both the domain and the range of the function are sets of real numbers. That is, when we write $y = f(x)$, both x and y will be real numbers. Such a function f is a **real-valued function of one real variable.**

When we are talking about some particular function f and writing $y = f(x)$, we have to describe the rule of assignment f that enables us to find the real number y starting with any real number x in the domain of f. Frequently, we will be interested in rules given by formulas. For example, we may discuss

$$y = f(x) = x^2 \qquad \text{for all } x. \qquad \textbf{(1)}$$

Squaring function

The phrase "for all x" indicates that the domain of f consists of the set of *all* real numbers.

EXAMPLE 1 Let $y = f(x) = x^2 + 3$ for all x. Then $f(-2) = 7$, $f(-1) = 4$, $f(0) = 3$, $f(1) = 4$, and $f(2) = 7$. Clearly for this function f, we have $f(-x) = f(x)$ for all x. The range of f consists of all $y \geq 3$. □

EXAMPLE 2 Let $y = f(x) = |x| - 4$ for all x. Then $f(-2) = -2$, $f(-1) = -3$, $f(-\frac{1}{2}) = -\frac{7}{2}$, $f(0) = -4$, $f(\frac{1}{2}) = -\frac{7}{2}$, $f(1) = -3$, and $f(2) = -2$. Note that we can compute $f(\frac{9}{2})$ or $f(\pi)$, for that matter. Sometimes students feel that functions are defined only for integer values of x. This is *not* the case; the description

$$f(x) = |x| - 4 \qquad \text{for all } x$$

means that x can be *any* real number. The range of f consists of all $y \geq -4$. □

According to Definition 1.1, a function f assigns to a number x *exactly one* number y. This *uniqueness* of the number y assigned to x by f is a very important property of the function. We sometimes have occasion to extract square roots, and we introduce

$$y = f(x) = \sqrt{x} \qquad \text{for all } x \geq 0. \tag{2}$$

Square root function

The domain of this function consists of all $x \geq 0$ since square roots of negative numbers are not real numbers. You may be accustomed to thinking of $\sqrt{4}$ as ± 2, since both 2^2 and $(-2)^2$ are equal to 4. However, for $\sqrt{x}$ to describe a *function*, we must consider that $\sqrt{4}$ yields just *one* of these possibilities. *So that* $\sqrt{x}$ *will denote a function, from now on we consider* $\sqrt{x}$ *to be only the* non-negative *square root of* x. Thus we have $\sqrt{4} = 2$, $\sqrt{0} = 0$, $\sqrt{9} = 3$, and so on. If we want the negative square root, we will always use $-\sqrt{x}$.

When a function $y = f(x)$ is defined by a formula and the domain for the independent variable is not specifically given, we always consider the domain to consist of all values of x for which the formula can be evaluated and yields a *real* number. In particular,

> division by zero is not allowed, and square roots (or fourth roots, or any even roots) of negative numbers are not allowed.

EXAMPLE 3 Find the domain of the function f given by the formula $y = f(x) = \sqrt{x - 1}$.

Solution The domain consists of all x such that $x - 1 \geq 0$, or such that $x \geq 1$. The range of f consists of all $y \geq 0$. □

EXAMPLE 4 Find the domain of $y = f(x) = (x^2 - 1)/(x^2 - 9)$.

Solution Note that $f(2) = 3/(-5)$ and $f(5) = \frac{24}{16} = \frac{3}{2}$, but f is not defined at 3 since division by zero is not allowed. The domain of the function consists of all $x \neq \pm 3$. □

Sometimes we will want to discuss more than one function at the same time. In that case, we use different letters to represent different functions. The letters f, g, and h are commonly used to denote functions. We use g in the next two examples to become accustomed to letters other than f.

EXAMPLE 5 Find the domain of $g(x) = x/\sqrt{x - 3}$.

Solution We must be careful that both $x - 3 \geq 0$, so $\sqrt{x - 3}$ is real, and that $\sqrt{x - 3} \neq 0$, so that we are not dividing by zero. Now $x - 3 \geq 0$ means we must have $x \geq 3$, and $\sqrt{x - 3} \neq 0$ means $x \neq 3$. Thus the domain consists of all $x > 3$. □

EXAMPLE 6 At the start of this section we said that the area A of a circle is a function of its radius r. Describe this function.

Solution We will call this function g. Now r is the independent variable and A the dependent variable. We have

$$A = g(r) = \pi r^2 \qquad \text{for } r \geq 0.$$

The domain constraint $r \geq 0$ must be stated because we can't have a circle of negative radius. Of course, we could compute πr^2 for negative values of r. This time, the domain restriction is due to the geometric origin of the function. □

DEFINITION 1.2 Equal functions

Two functions f and g are **equal** if

1. their domains are identical and
2. $f(x) = g(x)$ for each number x in that common domain.

EXAMPLE 7 Show that the functions $f(x) = |x|$ and $g(x) = \sqrt{x^2}$ are equal.

Solution The function f is defined for all x. Since $x^2 \geq 0$ and we can take the square root of any nonnegative number, we see that g has this same domain. For $x \geq 0$, we have $|x| = x = \sqrt{x^2}$, while for $x < 0$, both $|x|$ and $\sqrt{x^2}$ give the positive value $-x$. For example, $\sqrt{(-5)^2} = 5 = |-5|$. Thus f and g are the same function. □

The following example shows that in simplifying a formula describing a function, we have to be very careful not to change the function.

EXAMPLE 8 Determine whether the functions f and g given by

$$f(x) = \frac{x^2 - 1}{x - 1} \qquad \text{and} \qquad g(x) = x + 1$$

are equal.

Solution Since

$$\frac{x^2 - 1}{x - 1} = \frac{(x - 1)(x + 1)}{x - 1},$$

it is tempting to cancel the $(x - 1)$ factors and say that $f(x) = g(x)$, that is, that f and g are the same function. Note, however, that f is not defined at 1, since substituting 1 for x in $(x^2 - 1)/(x - 1)$ leads to division by zero. On the other hand, $g(1) = 1 + 1 = 2$. Thus f and g are not equal functions. Of course, $f(x) = g(x)$ for all $x \neq 1$. If we try to simplify the rule giving $f(x)$, we must write

$$f(x) = x + 1 \qquad \text{for } x \neq 1. \quad \square$$

EXAMPLE 9 Let $f(x) = x^2 - 3x$. Find an expression in terms of Δx for $f(5 + \Delta x)$ and for $[f(5 + \Delta x) - f(5)]/\Delta x$. (We will soon encounter this problem in calculus.)

Solution To compute $f(5 + \Delta x)$, we simply replace x by $5 + \Delta x$ in the formula $x^2 - 3x$ for $f(x)$. We obtain

$$\begin{aligned} f(x) \quad &= \quad x^2 \quad - \quad 3x, \\ \downarrow \qquad & \quad\;\; \downarrow \qquad\qquad \downarrow \\ f(5 + \Delta x) &= (5 + \Delta x)^2 - 3(5 + \Delta x) \\ &= [25 + 10(\Delta x) + (\Delta x)^2] - [15 + 3(\Delta x)] \\ &= 10 + 7(\Delta x) + (\Delta x)^2. \end{aligned}$$

Consequently,

$$\begin{aligned} \frac{f(5 + \Delta x) - f(5)}{\Delta x} &= \frac{[10 + 7(\Delta x) + (\Delta x)^2] - (5^2 - 3 \cdot 5)}{\Delta x} \\ &= \frac{10 + 7(\Delta x) + (\Delta x)^2 - 10}{\Delta x} = \frac{7(\Delta x) + (\Delta x)^2}{\Delta x} \\ &= \frac{\Delta x(7 + \Delta x)}{\Delta x} = 7 + \Delta x \qquad \text{for } \Delta x \neq 0. \quad \square \end{aligned}$$

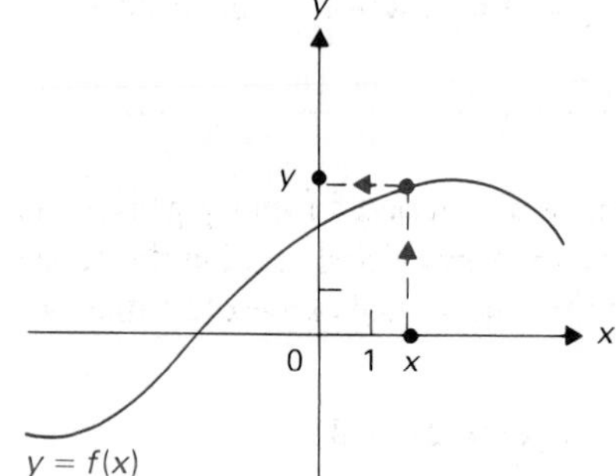

Figure 1.42 The graph of a function f, showing how to obtain the y-value from an x-value.

EXAMPLE 10 A function $y = f(x)$ does not have to be given by one single formula over its entire domain. Different formulas may be used for different parts of the domain. For example,

$$f(x) = \begin{cases} 2x - 4 & \text{for } x \geq 3, \\ |x| & \text{for } -5 < x < 3, \\ 1 + x & \text{for } x \leq -5 \end{cases}$$

describes a function with domain all real numbers x. Here

$$\begin{aligned} f(5) &= 2 \cdot 5 - 4 = 6 && \text{since } 5 \geq 3, \\ f(-2) &= |-2| = 2 && \text{since } -5 < -2 < 3, \\ f(-7) &= 1 + (-7) = -6 && \text{since } -7 \leq -5. \end{aligned}$$

Any device that enables us to compute a single y-value for each x-value in some set X of numbers determines a function having the set X as domain. □

GRAPHS OF FUNCTIONS

Figure 1.43 Not the graph of a function; one x-value must not give more than one y-value.

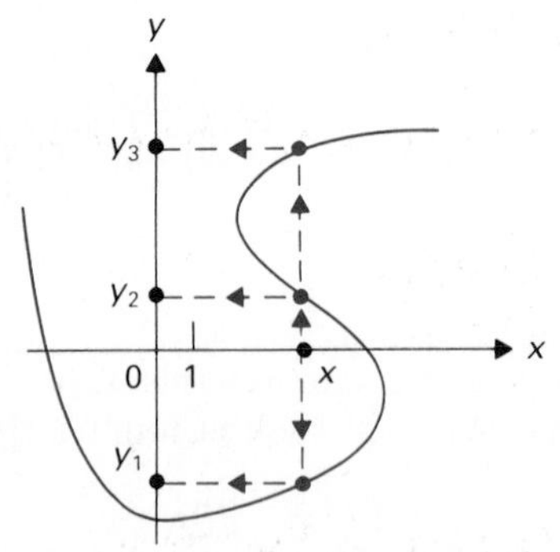

Using the plane analytic geometry in Sections 1.1 through 1.3, we can draw helpful pictures of real-valued functions of one real variable. For such a function f, we may find the points (x, y) in the plane where $y = f(x)$. The points form the **graph** of the function.

The graph of a function f is shown in Fig. 1.42. We can view the graph as giving geometrically the rule for computing the function at each x in the domain. That is, at x, go in a vertical direction until the graph is met, and then go horizontally to find the y such that $y = f(x)$. This geometric technique is indicated by the arrows in Fig. 1.42.

Remember that a function can assign *only one* y-value to each x-value. Thus the curve in Fig. 1.43 is not the graph of a function. For the x-value shown, there are three possible y-values, y_1, y_2, and y_3. A function cannot

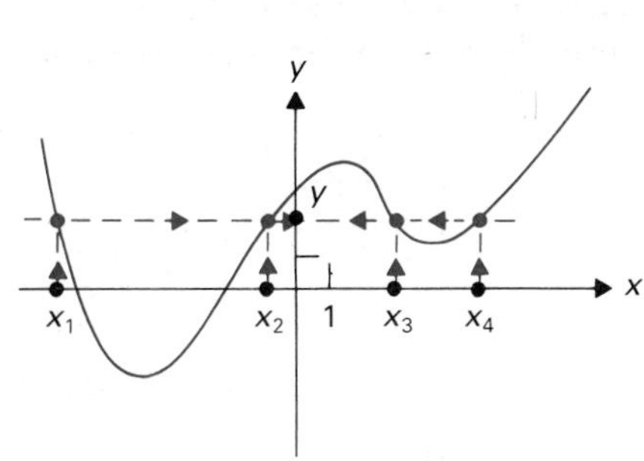

Figure 1.44 The graph of a function; different x-values may give the same y-value.

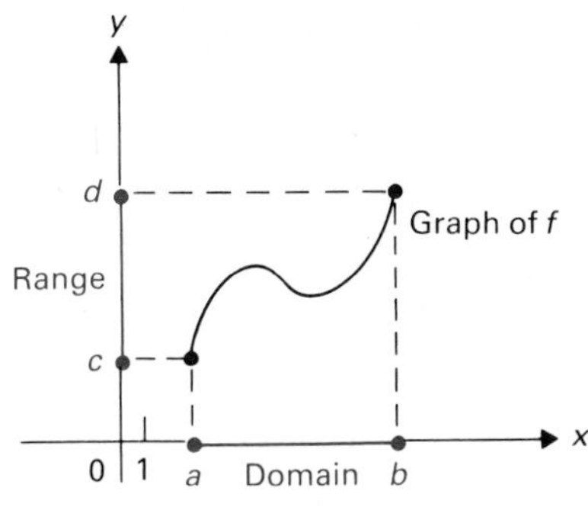

Figure 1.45 Domain of f is $[a, b]$. Range of f is $[c, d]$.

assign three y-values to one x-value. This figure illustrates the following geometric rule:

> A vertical line can intersect the graph of a function in at most one point.

A horizontal line may intersect the graph of a function in many points, as shown in Fig. 1.44. Many different x-values are permitted to give the same y-value. Figure 1.45 illustrates graphically the *domain* and *range* of a function.

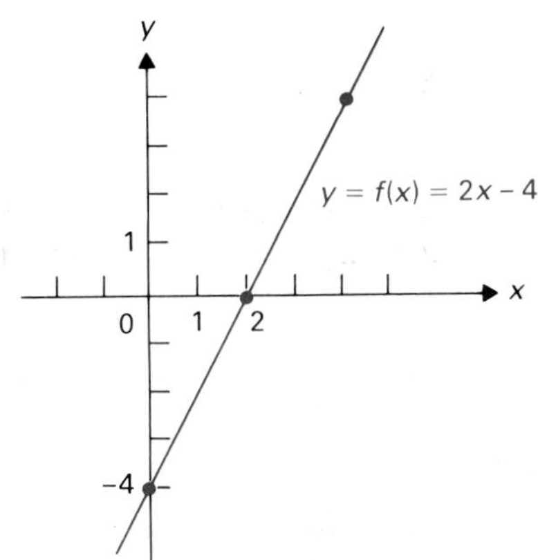

Figure 1.46 The graph of $f(x) = 2x - 4$.

EXAMPLE 11 Sketch the graph of $y = f(x) = 2x - 4$.

Solution The graph of this function is just the graph of the equation $y = 2x - 4$, which is the line with slope 2 and y-intercept -4, shown in Fig. 1.46. □

EXAMPLE 12 Sketch the graph of the squaring function $s = g(t) = t^2$.

Solution The graph is shown in Fig. 1.47. Note the different letters on the axes, representing the letters used in defining the function. This graph will be discussed in more detail in the next section. □

Figure 1.47 The graph of $s = g(t) = t^2$, the *squaring function*.

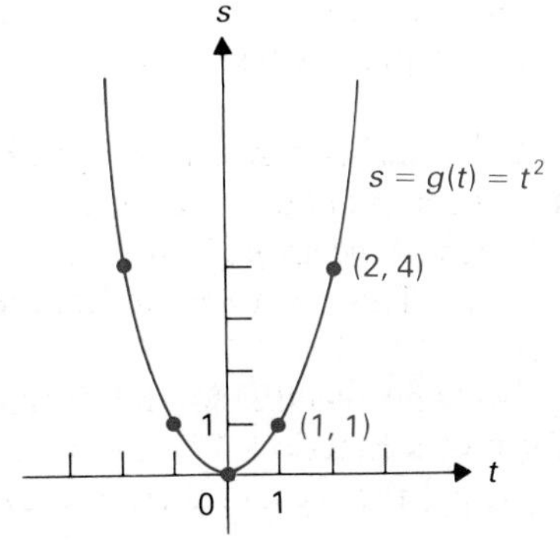

EXAMPLE 13 Sketch the graph of $y = f(x) = |x|$.

Solution The graph of $y = f(x) = |x|$ is shown in Fig. 1.48. Note that $f(x)$ can also be described by

$$f(x) = \begin{cases} x, & \text{if } x \geq 0, \\ -x, & \text{if } x < 0. \end{cases}$$

This graph is one you should remember. □

EXAMPLE 14 Sketch the graph of the function f of Example 10.

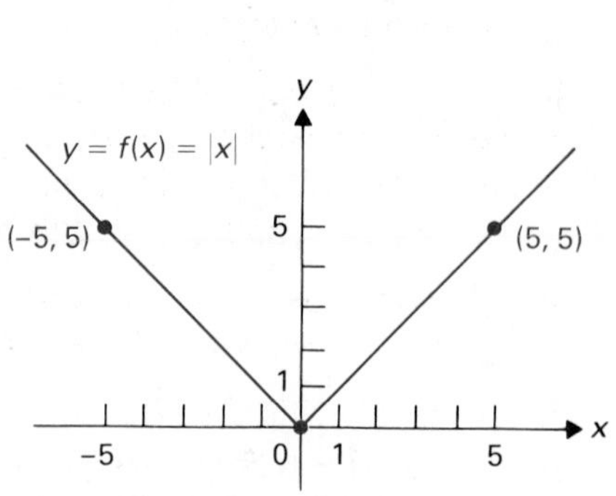

Figure 1.48 The graph of $f(x) = |x|$, the *absolute-value function.*

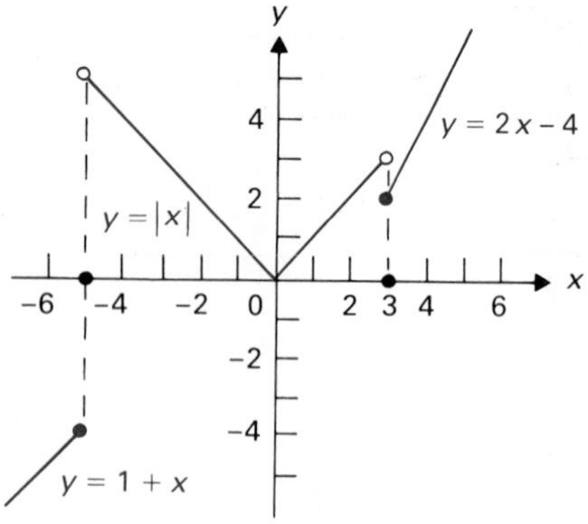

Figure 1.49 The graph of

$$f(x) = \begin{cases} 2x - 4 & \text{for } x \geq 3, \\ |x| & \text{for } -5 < x < 3, \\ 1 + x & \text{for } x \leq -5. \end{cases}$$

Solution The function of Example 10 was defined by

$$f(x) = \begin{cases} 2x - 4 & \text{for } x \geq 3, \\ |x| & \text{for } -5 < x < 3, \\ 1 + x & \text{for } x \leq -5. \end{cases}$$

The graph of f is shown in Fig. 1.49. □

One way to sketch the graph $y = f(x)$ is to make a table of corresponding values of x and y, plot the points, and draw a curve through them. Computing y-values can be tedious, and a calculator is often helpful. The calculator exercises at the end of this section deal with such tables and plots. A computer can easily make a table for many important functions. A video terminal or plotter can provide helpful pictures of graphs of functions.

EXAMPLE 15 Make a table of x- and y-values for $y = f(x) = x^4 - 2x^2 + 3$ with x-values every $\frac{1}{2}$ unit from $x = -2$ to $x = 2$. Then plot the graph.

Figure 1.50 The graph of $f(x) = x^4 - 2x^2 + 3$.

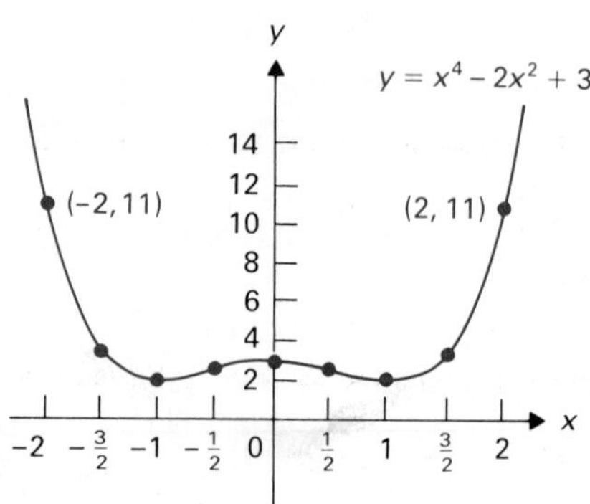

Solution We obtain Table 1.1 and the graph shown in Fig. 1.50. Note that we chose different scales on the axes. □

SUMMARY

1. If $y = f(x)$, then x is the independent variable and y the dependent variable.
2. If $y = f(x)$, the domain of the function f consists of all allowable values of the variable x. The range of f consists of all values obtained for y as x goes through all values in the domain.

Table 1.1

x	$y = x^4 - 2x^2 + 3$
-2	11
$-\frac{3}{2}$	3.5625
-1	2
$-\frac{1}{2}$	2.5625
0	3
$\frac{1}{2}$	2.5625
1	2
$\frac{3}{2}$	3.5625
2	11

3. A function f assumes only *one* value $f(x)$ for each x in its domain. Thus $\pm\sqrt{x}$ is *not* a function.
4. If $y = f(x)$ is described by a formula, the domain of f consists of all x where $f(x)$ can be computed and gives a real number. For us, this usually means just excluding x-values that would lead to division by zero or to taking even roots of negative numbers.
5. Two functions f and g are equal if their domains are identical and $f(x) = g(x)$ for each x in that common domain.
6. The graph of f consists of all points (x, y) such that $y = f(x)$.
7. Graphs can be sketched by making a table of x- and y-values and plotting the points (x, y), although this may be hard work.

EXERCISES

1. Express the volume V of a cube as a function of the length x of an edge of the cube.
2. Express the volume V of a sphere as a function of the radius r of the sphere.
3. Express the area A enclosed by a circle as a function of the perimeter s of the circle.
4. Express the area A of an equilateral triangle as a function of the length x of an edge of the triangle.
5. Express the volume V of a cube as a function of the length d of a diagonal of the cube. (A diagonal of a cube joins a vertex to the *opposite vertex*, which is the vertex farthest away.)
6. Bill starts at a point A at time $t = 0$ and walks in a straight line at a constant rate of 3 mi/hr toward point B. If the distance from A to B is 21 mi, express his distance s from B as a function of the time t, measured in hours.
7. Mary and Sue start from the same point on a level plain at time $t = 0$. Mary walks north at a constant rate of 3 mi/hr, while Sue jogs west at a constant rate of 5 mi/hr. Find the distance s between them as a function of the time t, measured in hours.
8. Smith, who is 6 ft tall, starts at time $t = 0$ directly under a light 30 ft above the ground and walks away in a straight line at a constant rate of 4 ft/sec.
   a) Express the length ℓ of Smith's shadow as a function of the distance x he has walked.
   b) Express the length ℓ of Smith's shadow as a function of the time t, measured in seconds.
   c) Express the distance x walked as a function of the length ℓ of Smith's shadow.
9. Let $f(x) = x^2 - 4x + 1$. Find the following.
   a) $f(0)$ b) $f(-1)$ c) $f(5)$
10. Let $g(t) = t/(1 - t)$. Find the following, if defined.
    a) $g(0)$ b) $g(1)$ c) $g(-1)$
11. Let $h(s) = (s^2 + 1)/(s - 1)$. Find the following.
    a) $h(-2)$ b) $h(-1)$ c) $h(3)$
12. Let $\phi(u) = \sqrt{4 + u^2}$. Find the following.
    a) $\phi(0)$ b) $\phi(2)$ c) $\phi(-\sqrt{12})$
13. Let
$$f(x) = \begin{cases} x^2 - x & \text{for } x < -3, \\ |x| & \text{for } -3 \le x \le 3, \\ 4x - 5 & x > 3. \end{cases}$$
Find the following.
    a) $f(1)$ b) $f(4)$ c) $f(-\frac{1}{2})$ d) $f(-\frac{7}{2})$
14. Let
$$g(t) = \begin{cases} (t + 1)/t & \text{for } t > 0, \\ 4 & \text{for } t = 0, \\ (t + 3)/(t - 1) & \text{for } t < 0. \end{cases}$$
Find the following.
    a) $g(-1)$ b) $g(0)$
    c) $g(-3)$ d) $g(1)$
15. Let $f(x) = x^2$. Find an expression in terms of Δx for each of the following.
    a) $f(2 + \Delta x)$ b) $f(2 + \Delta x) - f(2)$
    c) $\dfrac{f(2 + \Delta x) - f(2)}{\Delta x}$
16. Let $h(s) = s^2 - 3s + 2$. Find an expression in terms of Δs for each of the following.
    a) $h(1 + \Delta s)$ b) $h(1 + \Delta s) - h(1)$
    c) $\dfrac{h(1 + \Delta s) - h(1)}{\Delta s}$
17. Let $g(t) = 1/t$. Find an expression in terms of Δt for each of the following.
    a) $g(-3 + \Delta t)$ b) $g(-3 + \Delta t) - g(-3)$
    c) $\dfrac{g(-3 + \Delta t) - g(-3)}{\Delta t}$

18. Let $f(x) = \sqrt{x-3}$. Find an expression in terms of Δx for each of the following.
a) $f(7 + \Delta x)$ b) $f(7 + \Delta x) - f(7)$
c) $\dfrac{f(7 + \Delta x) - f(7)}{\Delta x}$

In Exercises 19 through 28, find the domain of the given function.

19. $f(x) = \dfrac{1}{x}$
20. $f(x) = \dfrac{1}{x^2 - 1}$
21. $f(x) = \dfrac{x}{x^2 - 3x + 2}$
22. $g(t) = \sqrt{t + 3}$
23. $f(u) = \sqrt{u^2 - 1}$
24. $g(t) = \dfrac{\sqrt{t - 2}}{t^2 - 16}$
25. $h(x) = \sqrt{x - 4}$
26. $k(v) = \dfrac{v^2}{\sqrt{9 - v^2}}$
27. $f(x) = \dfrac{x}{|x| - 1}$
28. $f(t) = \dfrac{t + 1}{\sqrt{|t - 3|}}$

In Exercises 29 through 36, give a simpler description of the function. Be sure you don't change the domain! Describe the domain explicitly where necessary. See Example 8.

29. $f(x) = \dfrac{2x}{x}$
30. $f(x) = \dfrac{x^2 + 3x}{x}$
31. $g(t) = \dfrac{t^2 - 4}{t + 2}$
32. $g(s) = \dfrac{s - 1}{s^2 - 1}$
33. $f(x) = \dfrac{|x|}{x}$
34. $f(x) = \dfrac{x^2 - 3x + 2}{x^2 - 1}$
35. $f(\Delta x) = \dfrac{(2 + \Delta x)^2 - 4}{\Delta x}$
36. $g(\Delta t) = \dfrac{(-3 + \Delta t)^2 + 2(-3 + \Delta t) - 3}{\Delta t}$

37. A portion of the graph of a function f is shown in Fig. 1.51. Estimate each of the following from the graph.
a) $f(0)$ b) $f(1)$ c) $f(-1)$

38. Proceed as in Exercise 37 for the graph in Fig. 1.52.
a) $f(0)$ b) $f(2)$ c) $f(-2)$ d) $f(3)$ e) $f(-3)$

Figure 1.51 Graph for Exercise 37.

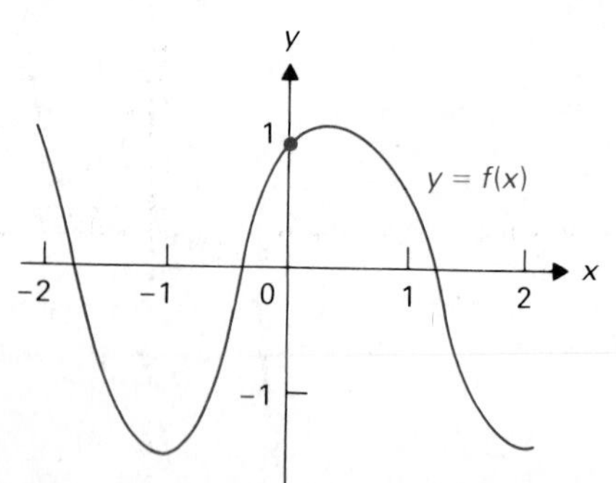

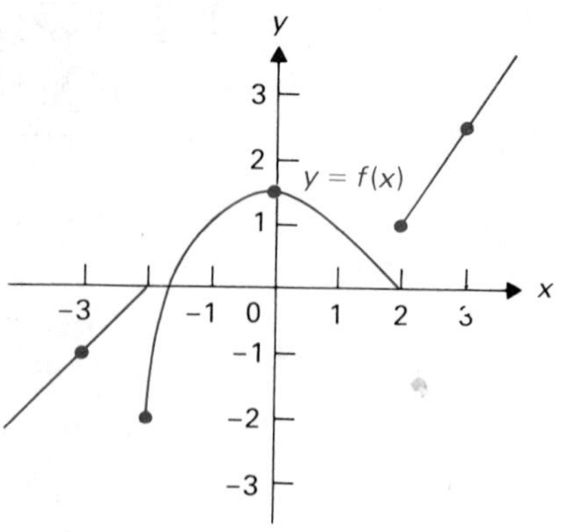

Figure 1.52 Graph for Exercise 38.

In Exercises 39 through 48, sketch the graph of the function over a convenient part of the domain near the origin.

39. $y = f(x) = x - 1$
40. $y = g(x) = (-x)^2$
41. $s = g(t) = t^2 - 4$
42. $y = f(x) = \sqrt{1 - x^2}$
43. $s = f(r) = -\sqrt{1 - r^2}$
44. $y = f(x) = \dfrac{1}{x}$
45. $y = g(x) = \dfrac{1}{x - 2}$
46. $y = h(u) = \dfrac{-1}{u}$
47. $y = f(x) = |x - 3|$
48. $s = g(t) = \dfrac{1}{|t + 1|}$

49. Make a table of x-values and y-values for the function

$$y = f(x) = \frac{x + 1}{x - 1}$$

for $x = -1, -\frac{1}{2}, 0, \frac{1}{2}, \frac{3}{4}, \frac{7}{8}, \frac{9}{8}, \frac{5}{4}, \frac{3}{2}, 2, \frac{5}{2}$, and 3. Plot the points and draw the portion of the graph for all x in the domain such that $-1 \le x \le 3$.

50. Proceed as in Exercise 49 for $y = f(x) = x^3 - 3x^2 + 2$, computing y-values for x every half-unit, starting with $-\frac{3}{2}$ and ending at $\frac{7}{2}$.

Use a calculator for Exercises 51 and 52.

51. Make a table of values for the function

$$f(x) = \frac{x + 1}{\sqrt{x^3 + 1}}$$

using 11 equally spaced x-values starting with $x = 0$ and ending with $x = 10$. Use the data to draw the graph of the function over [0, 10]. [*Note:* Eleven values give ten intervals.]

52. Make a table of values for the function $f(x) = \sin x^2$ using 13 equally spaced x-values starting with $x = 0$ and ending with $x = 3$. Use radian measure. Use the data to draw the graph of the function over [0, 3]. [*Note:* Thirteen values give twelve intervals. While we have not "defined" the function $\sin x^2$ yet, it is defined in the calculator.]

1.5 GRAPHS OF MONOMIAL AND QUADRATIC FUNCTIONS

MONOMIAL FUNCTIONS

The *monomial functions*, also called *power functions*, are those given by the monomials

$$x, \quad x^2, \quad x^3, \quad x^4, \quad x^5, \ldots, \quad x^n, \ldots$$

or constant multiples of them. When a function is given by a formula, we often refer to the formula as the function, to save writing. Thus we may refer to the function $4x^3$ rather than the function f where $f(x) = 4x^3$.

It is important to know the graphs of the monomial functions. We know that the graph of the function x is a straight line of slope 1 through the origin. The graph of the squaring function was shown in Fig. 1.47 on page 27. All the monomial graphs x^n go through the origin (0, 0) and the point (1, 1). If n is even, the graph of x^n goes through $(-1, 1)$, and if n is odd, the graph goes through $(-1, -1)$. The different parts of Fig. 1.53 show the graphs of the first five monomial functions. Note in particular that the larger the value of n,

Figure 1.53 Graphs of the first five monomial functions: (*a*) $y = x$; (*b*) $y = x^2$; (*c*) $y = x^3$; (*d*) $y = x^4$; (*e*) $y = x^5$.

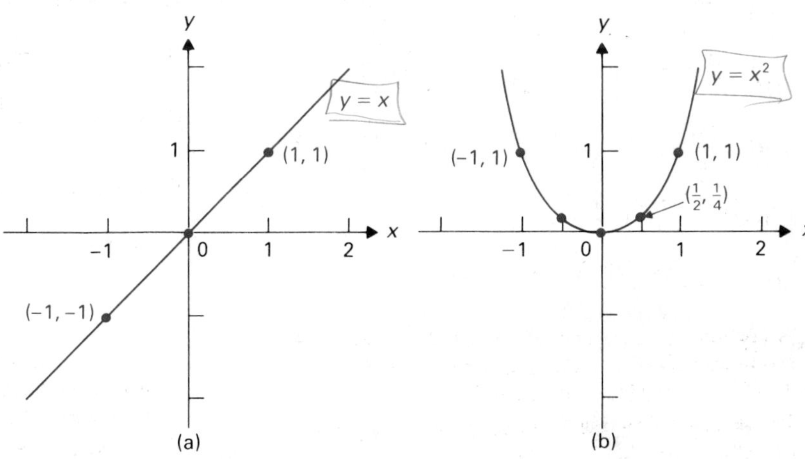

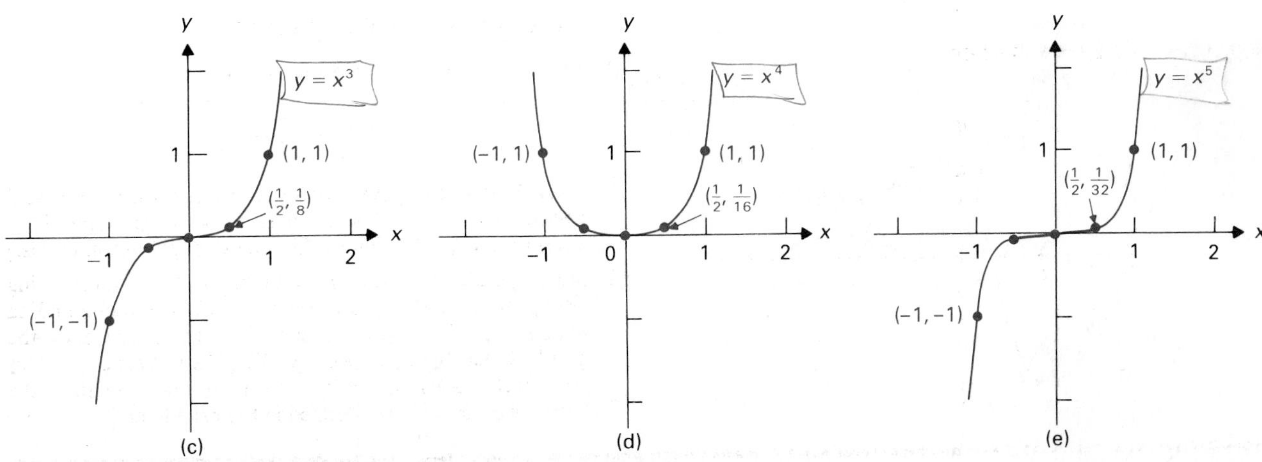

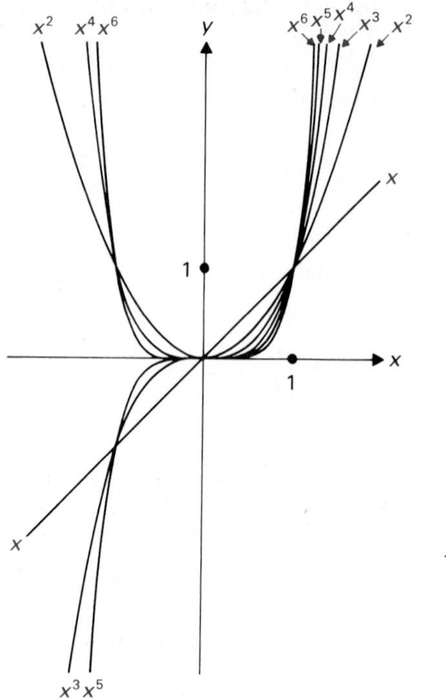

Figure 1.54 Computer-generated graphs of the first six monomial functions.

Figure 1.55 Attempt to draw the graph of x^{100}.

the closer the graph is to the x-axis for $-1 < x < 1$. For example, $(\frac{1}{2})^4 < (\frac{1}{2})^2$, so the graph of x^4 is closer to the x-axis where $x = \frac{1}{2}$ than the graph of x^2 is. If $|x| > 1$, then the larger the value of n, the farther the graph of x^n is from the x-axis. For example, $2^5 > 2^3$, so x^5 is farther from the x-axis than x^3 where $x = 2$. Computer-generated Fig. 1.54 shows the graphs of the functions x, x^2, x^3, x^4, x^5, and x^6 on a single pair of axes. This figure indicates even more clearly that for high powers of n, the graph of x^n clings close to the x-axis for $-1 < x < 1$ and climbs very steeply for $x > 1$.

A calculator gives Table 1.2, which shows that it is hopeless to draw a true graph of x^{100} on a regular-sized piece of paper. An attempt, shown in Fig. 1.55, makes the graph appear as the x-axis for $-1 \leq x \leq 1$ and upward vertical lines at $x = 1$ and $x = -1$.

Table 1.2

x	$y = x^{100}$
0.96	0.017
0.97	0.048
0.98	0.133
0.99	0.366
1.00	1.000
1.01	2.705
1.02	7.245
1.03	19.219
1.04	50.505

Now that we know the graph of x^n, we can easily sketch

$$y - k = (x - h)^n, \quad \textbf{(1)}$$

which is, of course, the graph of the function

$$y = f(x) = k + (x - h)^n.$$

We saw before that if we set $\Delta x = x - h$ and $\Delta y = y - k$, so that Eq. (1) becomes $\Delta y = (\Delta x)^n$, then we can sketch by translating to new $\Delta x, \Delta y$-axes at (h, k). In this graph-sketching context, we will drop the Δ-notation and use

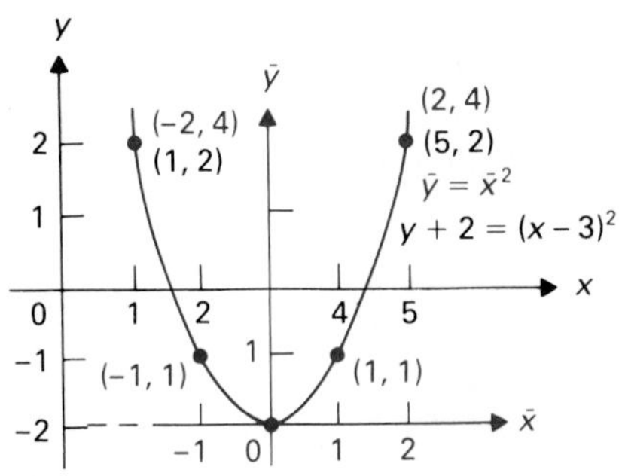

Figure 1.56 Graph of $y + 2 = (x-3)^2$, or $\bar{y} = \bar{x}^2$.

more conventional notation:

$$\bar{x} = x - h, \qquad \bar{y} = y - k.$$

Translation equations

So Eq. (1) becomes $\bar{y} = \bar{x}^n$.

EXAMPLE 1 Sketch the graph of $y + 2 = (x-3)^2$.

Solution If we set $\bar{x} = x - 3$ and $\bar{y} = y + 2$, the equation becomes $\bar{y} = \bar{x}^2$. We translate to $\bar{x},\bar{y}$-axes at $(h, k) = (3, -2)$ and sketch the graph as shown in Fig. 1.56. □

The graph of cx^n is much like the graph of x^n if $c > 0$. The points on cx^n are simply c times as far from the x-axis for each x-value as for x^n.

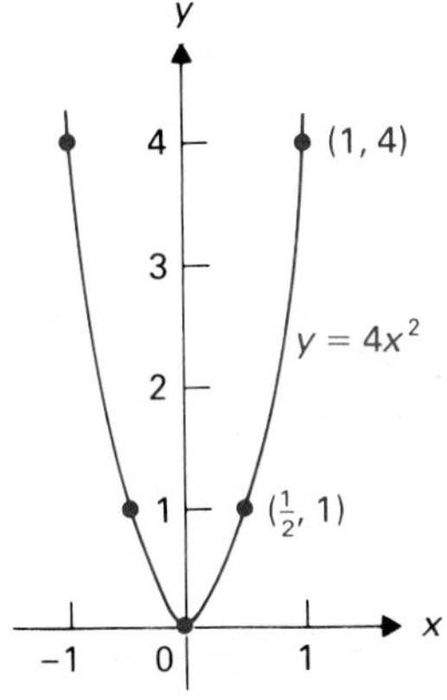

Figure 1.57 Graph of $y = 4x^2$.

EXAMPLE 2 Sketch the graph of $y = 4x^2$.

Solution The graph is shown in Fig. 1.57. When $x = 1$, $y = 4$. □

EXAMPLE 3 Sketch the graph of $y = \frac{1}{4}x^4$.

Solution The graph is shown in Fig. 1.58. When $x = 2$, $y = 4$. □

Of course, if $c < 0$, then the graph of $y = cx^n$ is thrown to the other side of the x-axis.

EXAMPLE 4 Sketch the graph of $y = -\frac{1}{2}x^2$.

Solution The graph is shown in Fig. 1.59. It opens downward rather than upward. □

Figure 1.58 Graph of $y = \frac{1}{4}x^4$.

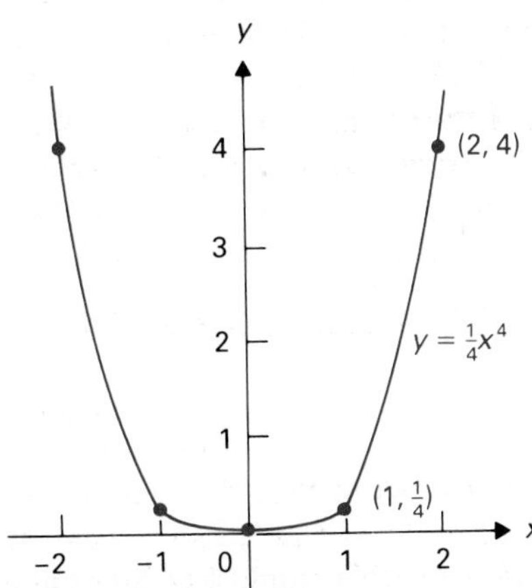

Figure 1.59 Graph of $Y = -\frac{1}{2}x^2$.

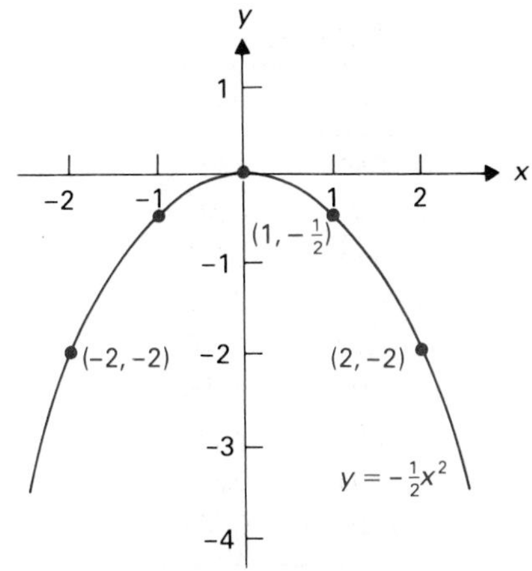

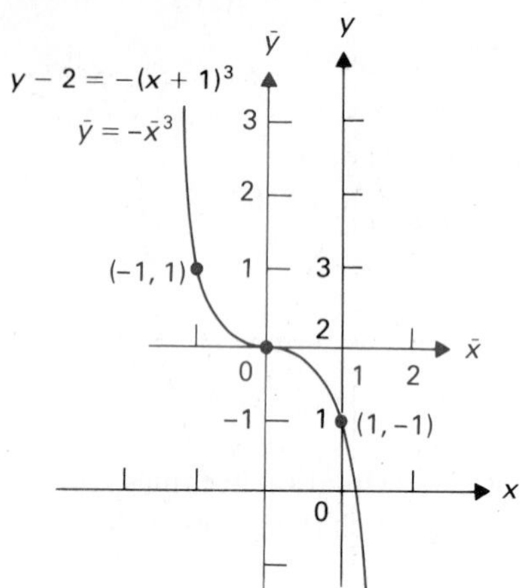

Figure 1.60 Graph of $y - 2 = -(x + 1)^3$, or $\bar{y} = -\bar{x}^3$.

EXAMPLE 5 Sketch the graph of $y = 2 - (x + 1)^3$.

Solution Writing the equation as $y - 2 = -(x + 1)^3$, we let $\bar{x} = x + 1$ and $\bar{y} = y - 2$. The equation then becomes $\bar{y} = -\bar{x}^3$. We translate to $\bar{x},\bar{y}$-axes at $(h, k) = (-1, 2)$ and sketch the graph as shown in Fig. 1.60. □

QUADRATIC FUNCTIONS

A quadratic function f is of the form $f(x) = ax^2 + bx + c$, where $a \neq 0$. Graphs of these functions are called *parabolas.* By completing the square and translating to $\bar{x},\bar{y}$-axes, we can put the equation $y = ax^2 + bx + c$ in the form $\bar{y} = d\bar{x}^2$ for some constant d. That is, the graph of a quadratic function is just a translation of the graph of a quadratic monomial function. The reason for this becomes clear in the following examples.

EXAMPLE 6 Sketch the graph of $y = x^2 - 2x + 3$.

Solution Completing the square on the x-terms, we have

$$y - 3 = x^2 - 2x,$$
$$y - 3 + 1 = x^2 - 2x + 1,$$
$$y - 2 = (x - 1)^2.$$

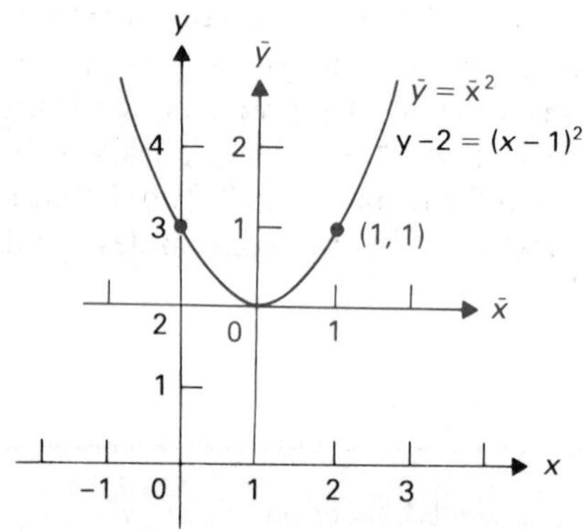

Figure 1.61 Graph of $y - 2 = (x - 1)^2$, or $\bar{y} = \bar{x}^2$.

Setting $\bar{x} = x - 1$ and $\bar{y} = y - 2$, we have the equation $\bar{y} = \bar{x}^2$. We translate to $\bar{x},\bar{y}$-axes at $(h, k) = (1, 2)$ and sketch the graph in Fig. 1.61. □

EXAMPLE 7 Sketch the graph of $y = 3x^2 + 12x + 11$.

Solution As the first step in completing the square on the x-terms, we factor the coefficient 3 of x^2 out of all the x-terms.

$$y - 11 = 3x^2 + 12x,$$
$$y - 11 = 3(x^2 + 4x), \qquad \text{Factoring out 3}$$
$$y - 11 + 12 = 3(x^2 + 4x + 4), \qquad \text{Adding 12 to both sides}$$
$$y + 1 = 3(x + 2)^2.$$

We let $\bar{x} = x + 2$ and $\bar{y} = y + 1$, so the equation becomes $\bar{y} = 3\bar{x}^2$. The parabola is sketched in Fig. 1.62. □

Figure 1.62 Graph of $y + 1 = 3(x + 2)^2$, or $\bar{y} = 3\bar{x}^2$.

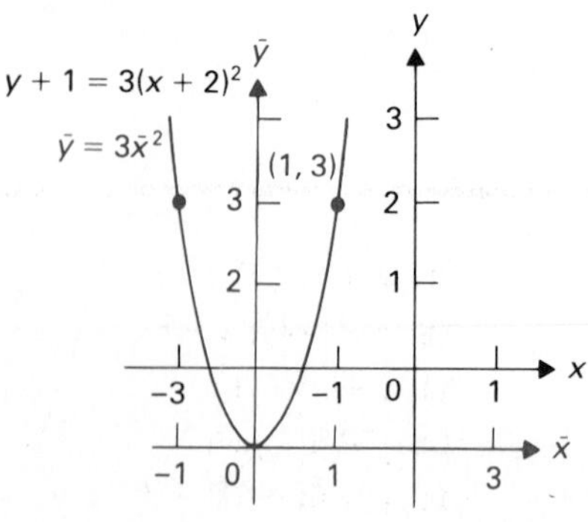

EXAMPLE 8 Sketch the graph of $y = -2x^2 - 6x - 2$.

Solution Factoring the coefficient -2 of x^2 from the x-terms and completing the square, we obtain

$$y + 2 = -2x^2 - 6x,$$
$$y + 2 = -2(x^2 + 3x),$$
$$y + 2 - \tfrac{9}{2} = -2(x^2 + 3x + \tfrac{9}{4}),$$
$$y - \tfrac{5}{2} = -2(x + \tfrac{3}{2})^2.$$

We let $\bar{x} = x + \frac{3}{2}$ and $\bar{y} = y - \frac{5}{2}$ and obtain $\bar{y} = -2\bar{x}^2$. The graph is shown in Fig. 1.63. □

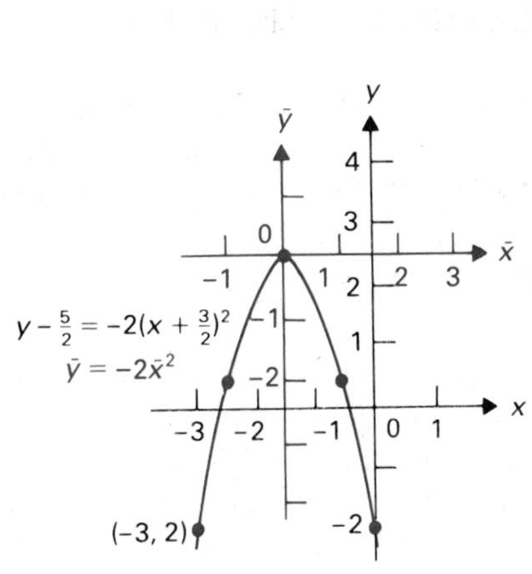

Figure 1.63 Graph of $y - \frac{5}{2} = -2(x + \frac{3}{2})^2$, or $\bar{y} = -2\bar{x}^2$.

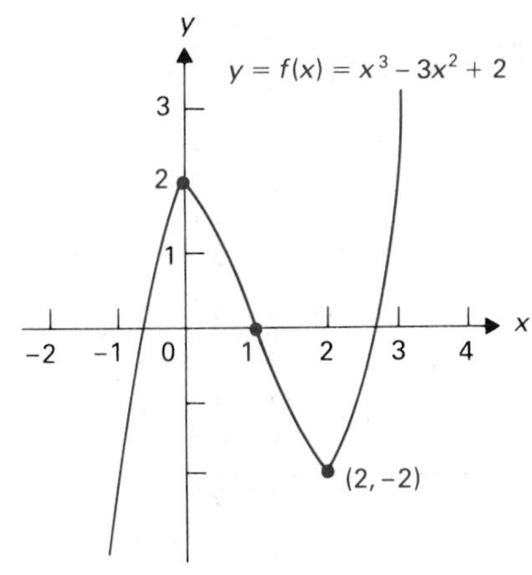

Figure 1.64 Graph of $y = x^3 - 3x^2 + 2$.

We have seen that the graph of any quadratic function $y = f(x) = ax^2 + bx + c$ is a translation of the graph of the monomial function $y = ax^2$. We should mention that this is *not* true for the graphs of polynomial functions of degree greater than 2. Figure 1.64 shows the graph of $y = f(x) = x^3 - 3x^2 + 2$. This graph is *not* a translation of any graph of the form $y = ax^3$. Similarly, the graph of $y = f(x) = x^4 - 2x^3 + 1$, shown in Fig. 1.65, is *not* a translation of any graph of the form $y = ax^4$. The sketches for these polynomial graphs of degree 3 and 4 are derived in Examples 5 and 6 of Section 5.4.

Figure 1.65 Graph of $y = x^4 - 2x^3 + 1$.

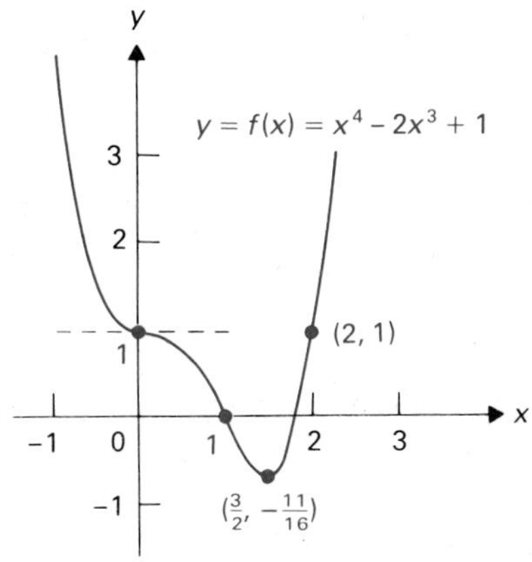

SUMMARY

1. See Fig. 1.53 for the graphs of the monomial functions $x, x^2, x^3, \ldots$.
2. The graph of $y - k = (x - h)^n$ looks like the graph of $y = x^n$, but with the origin translated to (h, k).
3. Every quadratic equation $y = ax^2 + bx + c$, where $a \neq 0$, has a parabola as graph. The parabola can be sketched by completing the square on the x-terms and translating axes to put the equation in the form
$$\bar{y} = d\bar{x}^2.$$

EXERCISES

In Exercises 1 through 28, sketch the graph of $y = f(x)$.

1. $y = 3x$
2. $y = -x^3$
3. $y = -x^4$
4. $y = \dfrac{x^2}{2}$
5. $y = 9x^2$
6. $y = \frac{1}{27}x^3$
7. $y = (x - 1)^2$
8. $y = (x + 2)^3$
9. $y = -(x + 1)^4$
10. $y = -4(x + 3)^2$
11. $y = x^2 + 3$
12. $y = x^3 - 1$
13. $y = 5 - x^4$
14. $y = 6 - \frac{1}{4}x^2$
15. $y = 4 + (x - 2)^2$
16. $y = (x + 1)^3 - 3$

17. $y = -1 - (x + 5)^4$
18. $y = 5 - 2(x + 3)^2$
19. $y = x^2 + 2x + 1$
20. $y = x^2 - 4x + 3$
21. $y = -x^2 - 6x + 5$
22. $y = -x^2 - 10x - 21$
23. $y = 2x^2 - 4x + 6$
24. $y = -3x^2 + 6x - 5$
25. $y = 4x^2 - 12x + 10$
26. $y = 1 - 3x - 3x^2$
27. $y = 8 - 10x + 4x^2$
28. $y = 7x - 2x^2$

29. Sketch the graph of f where $f(x) = |x|$. Can you find *one single* line graph that approximates $f(x)$ well for a short distance on *both* sides at $x = 0$?

EXERCISE SETS FOR CHAPTER 1

Review Exercise Set 1.1

1. a) Find the directed length Δx from -2 to 5.
b) Sketch all points (x, y) in the plane that satisfy $x > y + 1$.

2. a) Find the distance between $(2, -1)$ and $(-4, 7)$.
b) Find the midpoint of the line segment joining $(-1, 3)$ and $(3, 9)$.

3. a) Find the equation of the circle with center $(2, -1)$ and passing through $(4, 6)$.
b) Sketch all points (x, y) in the plane such that

$$(x - 1)^2 + (y + 2)^2 \leq 4.$$

4. a) Find the slope of the line joining $(-1, 4)$ and $(3, 7)$.
b) Find the slope of a line that is perpendicular to the line through $(4, -2)$ and $(-5, -3)$.

5. a) Find the equation of the line through $(-4, 2)$ and $(-4, 5)$.
b) Find the equation of the line through $(-1, 2)$ and parallel to the line $x - 3y = 7$.

6. a) Find the x-intercept and y-intercept of the line $3x + 4y = 12$.
b) Find the point of intersection of the lines $x - 3y = 7$ and $2x - 5y = 4$.

7. Let

$$f(x) = \frac{x^2 - 3x + 2}{x^2 - 5x}.$$

a) Find the domain of f.
b) Find $f(-2)$.

8. Sketch the graph of the function $y = f(x) = 1/x^2$.

9. Sketch the graph of the function $y = f(x) = 3 - (x + 4)^3$.

10. Sketch the graph of the function $y = f(x) = 2x^2 + 8x - 6$.

Review Exercise Set 1.2

1. a) Sketch on the number line all x such that $|x - 1| \leq 2$.
b) Find the midpoint of the interval $[-5.3, 2.1]$.

2. a) Find the distance from $(-6, 3)$ to $(-1, -4)$.
b) Sketch all points (x, y) in the plane such that $x \leq y$ and $x \geq 1$.

3. a) Find the equation of the circle with $(-2, 4)$ and $(4, 6)$ as endpoints of a diameter.
b) Find the center and radius of the circle $x^2 + y^2 - 6x + 8y = 11$.

4. Find c such that the line through $(-1, c)$ and $(4, -6)$ is perpendicular to the line through $(-2, 3)$ and $(4, 7)$.

5. a) Find the equation of the line through $(-1, 4)$ and $(3, 5)$.
b) Find the equation of the vertical line through $(3, -7)$.

6. a) Find the equation of the line through $(-1, 3)$ with y-intercept 5.
b) Find the equation of the line through $(2, 4)$ parallel to the line through $(0, 5)$ and $(2, -3)$.

7. Let $f(x) = \sqrt{25 - x^2}$.
a) Find the domain of f.
b) Find $f(3)$.
c) Sketch the graph of f.

8. Express the distance from the origin to a point (x, y) on the line $2x - 3y = 7$ as a function of x only.

9. Sketch the graph of the function $y = f(x) = 2 + (x - 1)^4$.

10. Sketch the graph of the function $y = f(x) = 4 - 2x^2$.

Review Exercise Set 1.3

1. a) Find $|2 - |7 - 4||$.
b) Express the solution of $|2x - 4| \leq 5$ in interval notation.

2. Find the area of the right triangle with vertices $(-1, 4)$, $(2, 6)$, and $(-3, 7)$.

3. Find the equation of the circle with center $(-1, 4)$ and tangent to the line $x = 5$.

4. Find the slope of the line tangent to the circle $x^2 + y^2 - 8x + 2y = -4$ at the point $(1, -3)$.

5. Find the equation of the line through $(-2, 4)$ perpendicular to the line $3x - 4y = 5$.

6. Find the point of intersection of the lines $2x + 4y = 8$ and $3x + 5y = 3$.

7. Let

$$f(x) = \frac{x^2 - x}{x^2 + x - 2}.$$

a) Find the domain of f.
b) Find a more simple description of $f(x)$.

8. Let $f(x) = x^2 - x + 3$. Express

$$\frac{f(2 + \Delta x) - f(2)}{\Delta x}$$

as a function of Δx in as simple a form as possible.

9. Sketch the graph of the function $y = f(x) = 5 - x^3$.

10. Sketch the graph of the function $y = f(x) = x^2 - 4x + 5$.

Review Exercise Set 1.4

1. a) Sketch all points x on the number line such that $(x - 2)^2 < 9$.
   b) Express the solutions of $|6 - 3x| \leq 12$ in interval notation.
2. Find all values of c such that the point $(c, -3)$ is 8 units from the point $(4, 1)$.
3. Find the equation of the smallest circle with center at $(7, 6)$ that is tangent to the circle $x^2 + y^2 - 2x + 4y = 11$.
4. Use slopes to determine whether the points $(-1, 4)$, $(2, 8)$, and $(-7, -4)$ are collinear.
5. Determine whether the lines $2x - 3y = 7$ and $12x + 8y = 5$ are parallel, perpendicular, or neither.
6. Find the equation of the line through $(-1, 5)$ and the point of intersection of the lines $2x + 3y = 7$ and $5x + y = -2$.
7. Let

$$f(x) = \frac{x^3 + 1}{(1 - |x|)^2}.$$

   Find the following.
   a) $f(0)$ b) $f(-2)$ c) $f(3)$
8. Find the domain of the function g, where $g(t) = (t^2 - 1)/(t\sqrt{t} + 3)$.
9. Sketch the graph of the function $y = f(x) = 5(x + 2)^4$.
10. Sketch the graph of the function $y = f(x) = -6x - 3x^2$.

More Challenging Exercises 1

1. Show that, for all real numbers a and b,
   a) $|a + b| \leq |a| + |b|$,
   b) $|a - b| \geq |a| - |b|$.
2. Prove that, for any numbers a_1, a_2, b_1, and b_2, the following is true.

$$(a_1a_2 + b_1b_2)^2 \leq (a_1{}^2 + b_1{}^2)(a_2{}^2 + b_2{}^2)$$

3. Prove algebraically from Exercise 2 and the formula for distance that the distance from (x_1, y_1) to (x_3, y_3) is less than or equal to the sum of the distance from (x_1, y_1) to (x_2, y_2) and the distance from (x_2, y_2) to (x_3, y_3). This is known as the *triangle inequality* in the plane. [*Hint*: Let

$$a_1 = x_2 - x_1, \quad a_2 = x_3 - x_2,$$
$$b_1 = y_2 - y_1, \quad b_2 = y_3 - y_2,$$

   so that

$$x_3 - x_1 = a_2 + a_1 \quad \text{and} \quad y_3 - y_1 = b_2 + b_1.]$$

4. Show that if two circles

$$x^2 + y^2 + a_1x + b_1y = c_1$$

   and

$$x^2 + y^2 + a_2x + b_2y = c_2$$

   intersect in two points, the line through those points of intersection is

$$(a_2 - a_1)x + (b_2 - b_1)y = c_2 - c_1.$$

5. Use Exercise 4 to find the line through the points of intersection of the following pairs of circles, if they intersect.
   a) $x^2 + y^2 - 4x + 2y = 8$ and $x^2 + y^2 - 2x + 6y = -9$
   b) $x^2 + y^2 - 4x + 2y = 8$ and $x^2 + y^2 + 6x + 4y = 17$
6. Show that the distance from a point (x_1, y_1) to a line $ax + by - c = 0$ is

$$\left|\frac{ax_1 + by_1 - c}{\sqrt{a^2 + b^2}}\right|.$$

7. Use the formula in Exercise 6 to find the distance
   a) from $(-1, 2)$ to $4x - 3y = 10$,
   b) from $(-3, 4)$ to $5x - 12y = 2$.
8. Find the area of the triangle with vertices $(1, -2)$, $(5, 6)$, and $(-4, 3)$. [*Hint*: Find the point at the foot of the altitude from $(-4, 3)$ to the opposite side of the triangle.]
9. Solve the inequality $x^2 + 4x < 1$ for x.
10. Find the equation of the smaller circle tangent to both coordinate axes and passing through the point $(-3, 6)$.
11. Find the distance between the lines

$$x - 2y = 15 \quad \text{and} \quad x - 2y = -3.$$

12. Find the minimum distance between the circles

$$x^2 + y^2 - 2x + 4y = 139$$

   and

$$x^2 + y^2 + 4x - 6y = 3.$$

13. Sketch the graph of the function $y = f(x) = |x^2 - 1|/(x - 1)$.
14. Sketch the graph of the function $y = f(x) = (x^3 - 4x^2)/|x|$.
15. If $f(x) = (2x - 7)/(x + 3)$, find a function g such that $g(f(x)) = x$ for all x in the domain of f.

2 LIMITS

A function $y = f(t)$ described by an algebraic formula is not defined at a point $t = t_1$ if the formula involves a denominator that becomes zero there. For example,

$$y = f(t) = \frac{t^2 - 4}{t - 2}$$

is not defined where $t = 2$. Even though a function f is not defined at $t = t_1$, we often wish to know how function values $f(t)$ behave as t gets *very close to* t_1. Section 2.1 will show that we encounter this problem when we try to find the rate at which something is changing. Calculus is concerned with things that are changing. We will encounter the notion of the *limit* of the function $f(t)$ at a point as we discuss rates of change. With Section 2.1 for motivation, Section 2.2 begins a study of limits, the main topic of this chapter.

The basic notion of differential calculus is the *derivative* of a function. The derivative is introduced in Section 2.1, but detailed study of derivatives and their computation is the topic of Chapter 3. Section 2.1 serves a dual purpose: to provide an early preview of the derivative and to motivate the need for studying limits in the rest of Chapter 2.

2.1 INTUITIVE INTRODUCTION TO LIMITS AND THE DERIVATIVE

RATES OF CHANGE

Suppose a car traveling in a straight line has gone a distance $s = f(t)$ at time t. Let's not worry about the units for s and t; they might be yards and seconds, for example. If the car moves with a *constant* speed of 2, then $s = 2t$ for $t \geq 0$. Consequently, the graph of this motion is the straight line of slope 2 shown in Fig. 2.1. If speed remains constant with value m, then the total distance s

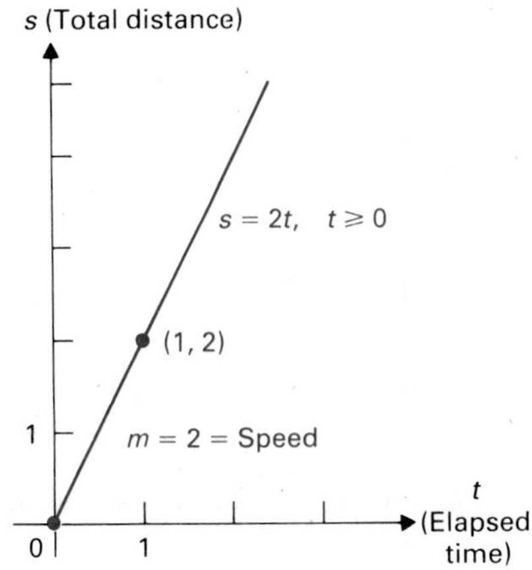

Figure 2.1 The graph of distance s against time t is a line when the speed is constant.

traveled in elasped time t is $s = mt$, which has as graph a line of slope m through the origin.

Now suppose that the graph of the distance function $s = f(t)$ is not a straight line. How could we find the speed at a particular time t_1? We might first estimate the speed at time $t = t_1$ by finding the average speed over a very short time interval starting at time t_1. When driving a car, our speed at any instant is close to our average speed over the next hundredth of a second or our average speed over the last thousandth of a second. We find average speed using the formula

$$\text{Average speed}^* = \frac{\text{Change in distance}}{\text{Corresponding change in time}}. \qquad (1)$$

EXAMPLE 1 The distance traveled by a car at time t is $s = f(t) = t^2$ for $t \geq 0$. Estimate the speed at time $t = 1$ by finding the average speed from time $t = 1$ to time $t = 1.01$.

Solution At time $t = 1$, we have $s = 1^2 = 1$. At time $t = 1.01$, we have $s = (1.01)^2 = 1.0201$. The change in distance is $1.0201 - 1 = 0.0201$. From formula (1), we have

$$\text{Average speed} = \frac{0.0201}{0.01} = 2.01.$$

We expect the speed at time $t = 1$ to be close to 2.01. □

Figure 2.2 shows the graph of $s = f(t) = t^2$ for $t \geq 0$. Since the graph is not a straight line, the speed does not remain constant; the increasing steepness of the graph as t increases indicates that the speed increases as time increases. We are unable to illustrate times $t = 1$ and $t = 1.01$ in Fig. 2.2, for we cannot accurately depict points so close together on the graph. Instead, we show a point where $t = t_1$ and a point where $t = t_1 + \Delta t$. We should think of Δt as being small, in spite of its size in Fig. 2.2. We have from formula (1)

Figure 2.2 Average speed = $\Delta s/\Delta t$, which is the slope m_{sec} of the secant line.

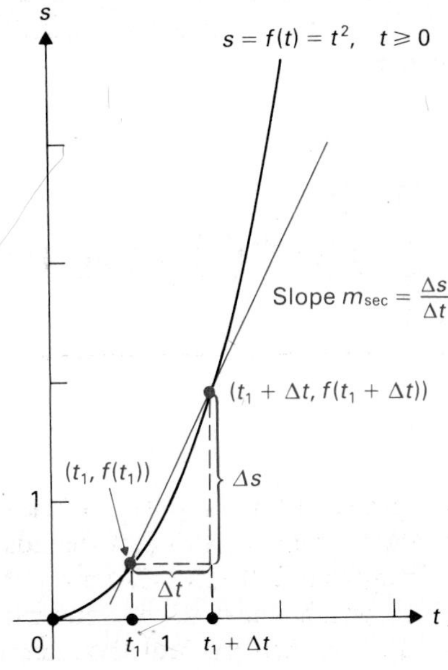

$$\text{Average speed} = \frac{\Delta s}{\Delta t}.$$

But $\Delta s/\Delta t$ is the slope of the secant line through the points $(t_1, f(t_1))$ and $(t_1 + \Delta t, f(t_1 + \Delta t))$ on the graph. We denote this slope by m_{sec}, and we see that from time t_1 to time $t_1 + \Delta t$,

$$\text{Average speed} = m_{\text{sec}} = \frac{f(t_1 + \Delta t) - f(t_1)}{\Delta t}. \qquad (2)$$

We are interested in the exact speed at the instant $t = t_1$. Example 1 estimated it where $t_1 = 1$ by finding the average speed from 1 to $1 + \Delta t$, where $\Delta t = 0.01$. The smaller the size of Δt, the better we expect the approximation of the exact speed to be. Example 2 illustrates that we can think of Δt as being

* We will distinguish between *speed* and *velocity*. Velocity is a signed quantity indicating direction as well as speed. When a body moves vertically near the surface of the earth, for example, it is customary to let positive velocity denote upward motion and negative velocity downward motion. Speed is the absolute value of the velocity, so speed is never negative.

negative as well as positive. Rather than finding the average speed over the next 0.01 seconds, we find the average speed over the last 0.001 seconds.

EXAMPLE 2 Repeat Example 1 for the distance function $s = f(t) = t^2$ when $t = 1$, but find the average speed corresponding to $\Delta t = -0.001$.

Solution Using formula (2), we have

$$\text{Average speed} = \frac{f(1 - 0.001) - f(1)}{-0.001} = \frac{(0.999)^2 - 1^2}{-0.001}$$

$$= \frac{0.998001 - 1}{-0.001} = \frac{-0.001999}{-0.001} = 1.999.$$

We expect the average speed 1.999 to be closer to the exact speed at $t = 1$ than the average speed 2.01 corresponding to $\Delta t = 0.01$, which we found in Example 1. □

If we find average speeds from t_1 to $t_1 + \Delta t$ for smaller and smaller Δt, perhaps running through values 0.01, 0.001, 0.0001, and so on, we should get closer and closer to the exact speed at time t_1. Using formula (2) for average speed, we write

Instantaneous speed at time t_1

$$= \lim_{\Delta t \to 0} \frac{\Delta s}{\Delta t} = \lim_{\Delta t \to 0} \frac{f(t_1 + \Delta t) - f(t_1)}{\Delta t}. \qquad \textbf{(3)}$$

We read $\lim_{\Delta t \to 0}$ as "the limit as Δt approaches zero."

Note in formula (3) that the denominator is zero if $\Delta t = 0$, which is the problem we mentioned in the introduction to this chapter.

EXAMPLE 3 If distance s is given in terms of elapsed time t by $s = f(t) = t^2$, find the instantaneous speed when $t = 1$.

Solution We first find the expression for average speed $\Delta s/\Delta t$ in formula (2), taking $t_1 = 1$. For $\Delta t \neq 0$, we have

$$\frac{\Delta s}{\Delta t} = \frac{f(1 + \Delta t) - f(1)}{\Delta t} = \frac{(1 + \Delta t)^2 - 1^2}{\Delta t}$$

$$= \frac{1 + 2(\Delta t) + (\Delta t)^2 - 1}{\Delta t} = \frac{2(\Delta t) + (\Delta t)^2}{\Delta t}$$

$$= \frac{\Delta t(2 + \Delta t)}{\Delta t} = 2 + \Delta t.$$

From formula (3),

$$\text{Instantaneous speed} = \lim_{\Delta t \to 0} \frac{\Delta s}{\Delta t} = \lim_{\Delta t \to 0} (2 + \Delta t) = 2.$$

That is, as Δt becomes closer and closer to zero, the expression $2 + \Delta t$ becomes closer and closer to 2. Looking back at Examples 1 and 2, we see that the estimate 1.999 in Example 2, with $\Delta t = -0.001$, is indeed a better estimate of the exact speed 2 than the estimate 2.01 in Example 1, with $\Delta t = 0.01$. □

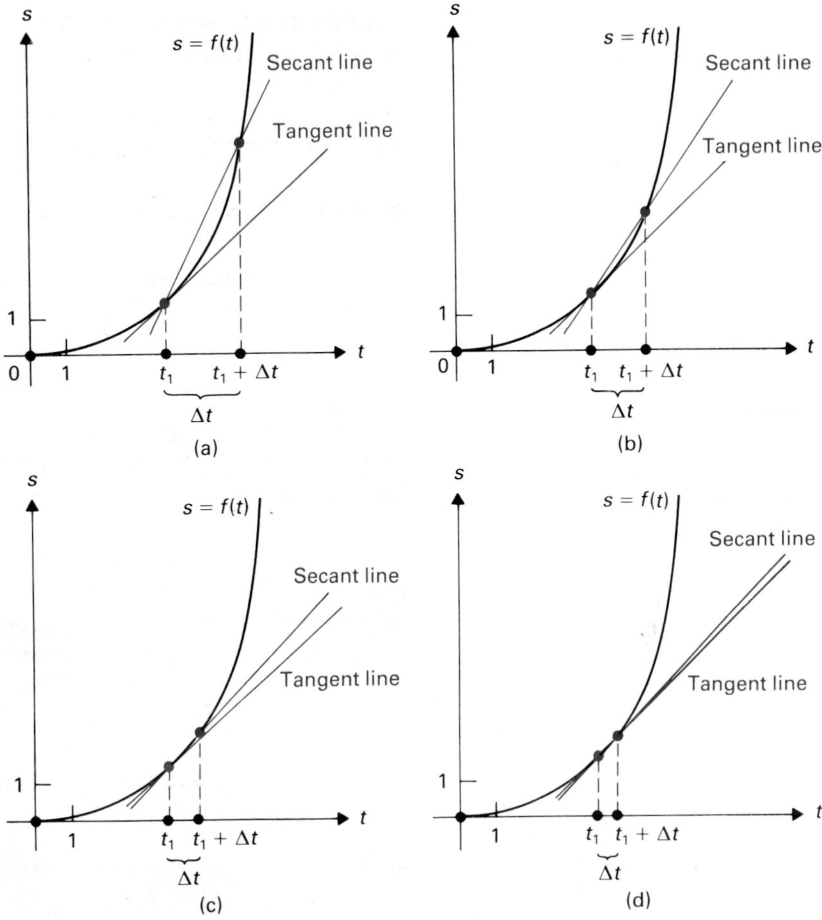

Figure 2.3 As Δt approaches zero, the secant lines approach the tangent line where $t = t_1$.

Figure 2.3 gives geometric pictures of formula (3). The instantaneous speed appears as the limit of average speeds as Δt approaches zero. In each part of Fig. 2.3, the slope m_{sec} of the secant line equals the average speed from time t_1 to time $t_1 + \Delta t$. As Δt approaches zero, the parts of the figure indicate that the secant lines approach the tangent line* to the graph at the point $(t_1, f(t_1))$. The limit of the slopes of the secant lines must then be the slope $m_{\tan}$ of the tangent line.

> Thus the slope of the tangent line gives the instantaneous speed at time t_1.

This tangent line has the same steepness as the graph where $t = t_1$ in Fig. 2.3. We saw that the slope of a *straight-line graph* of distance in terms of time is equal to the constant speed. This should have led us to suspect that the slope of the tangent line would give the speed in the case where the speed did not remain constant.

---

* Our work in geometry has given us an intuitive notion of the tangent line to a curve at a point. Our work in Chapter 3 will show that it can be defined as the limiting position of secant lines. In this introduction, we just build on our intuitive notion.

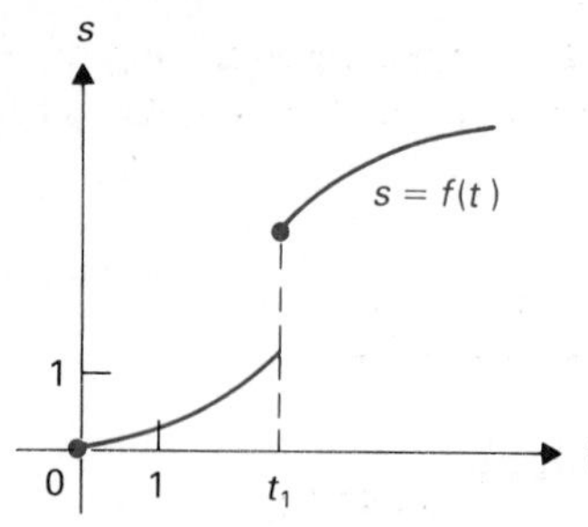

Figure 2.4 The graph has no tangent line at the break where $t = t_1$.

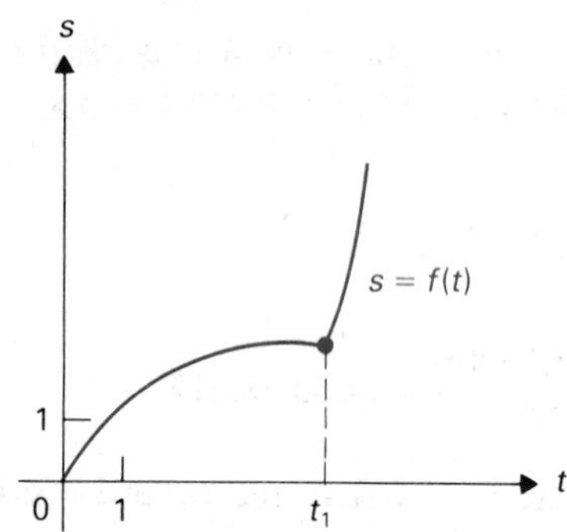

Figure 2.5 The graph has no (unique) tangent line at the sharp point where $t = t_1$.

Some graphs have breaks, as in Fig. 2.4, or sharp points, as in Fig. 2.5. At such points, there is no (unique) tangent line. We will not consider such cases in this section.

AVERAGE AND INSTANTANEOUS RATES OF CHANGE OF $f(x)$

The subject of this chapter is *limits*. The preceding subsection gave one physical example showing how limits appear when computing the rate at which distance changes with respect to time for a body moving in a straight line. We now give a general mathematical formulation that covers the same situation and many others.

As usual, when describing a general mathematical formulation, we change to x,y-notation. We assume that $y = f(x)$ and that the graph of f has a tangent line where $x = x_1$.

The **average rate of change** of y with respect to x from $x = x_1$ to $x = x_1 + \Delta x$, where $\Delta x \neq 0$, is

$$m_{\text{sec}} = \frac{\Delta y}{\Delta x} = \frac{f(x_1 + \Delta x) - f(x_1)}{\Delta x}. \tag{4}$$

The **instantaneous rate of change** of y with respect to x at $x = x_1$ is

$$m_{\text{tan}} = \lim_{\Delta x \to 0} \frac{\Delta y}{\Delta x} = \lim_{\Delta x \to 0} \frac{f(x_1 + \Delta x) - f(x_1)}{\Delta x}. \tag{5}$$

These are very important formulas to remember. In Fig. 2.6, we show secant lines approaching the tangent line at a point where $x = x_1$ for a general function $y = f(x)$, which is not necessarily a graph of motion like the previous figures. The figure is an attempt to collect the idea in the parts of Fig. 2.3 into one single picture.

Figure 2.6 Secant lines approaching the tangent line where $x = x_1$.

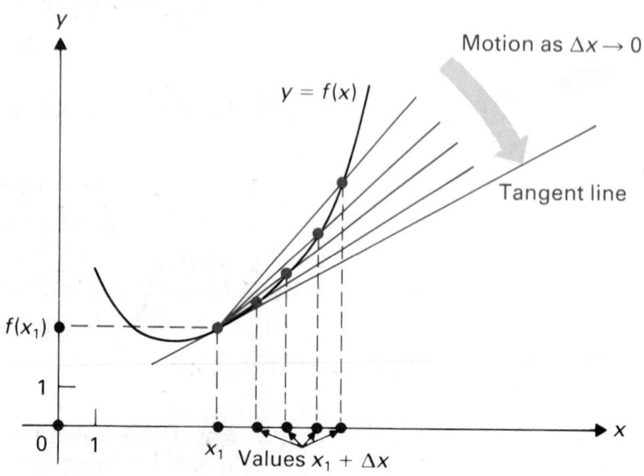

EXAMPLE 4 Let $f(x) = x^3 + 2x$. Find the approximation m_{sec} to the slope of the tangent line to the graph where $x = x_1 = 2$, taking $\Delta x = 0.01$.

Solution Formula (4) for m_{sec} yields

$$m_{\text{sec}} = \frac{[(2.01)^3 + 2(2.01)] - (2^3 + 2 \cdot 2)}{0.01}$$

$$= \frac{12.140601 - 12}{0.01} = \frac{0.140601}{0.01} = 14.0601.$$

We will see in Chapter 3 that the slope m_{tan} is actually 14, so 14.0601 is not a bad approximation. A calculator is handy for computing m_{sec}. □

EXAMPLE 5 Repeat Example 4, but take $\Delta x = 0.001$.

Solution Again we use formula (4) for m_{sec}. We expect a value closer still to 14. The computation was done using a calculator. We give the full display of the calculator:

$$m_{\text{sec}} = \frac{[(2.001)^3 + 2(2.001)] - (2^3 + 2 \cdot 2)}{0.001}$$

$$= \frac{12.014006 - 12}{0.001} = \frac{0.014006001}{0.001} = 14.00600099.$$

This is closer to the true value 14 of m_{tan} at $x_1 = 2$ than the value 14.0601 obtained in Example 4 with the larger $\Delta x = 0.01$. □

EXAMPLE 6 Find the slope m_{tan} of the tangent line to the graph of $f(x) = 4x - 3x^2$ at the point where $x = x_1$.

Solution We use formula (5), but first we compute $m_{\text{sec}} = \Delta y/\Delta x$, simplifying as much as possible. We have, for $\Delta x \neq 0$,

$$m_{\text{sec}} = \frac{f(x_1 + \Delta x) - f(x_1)}{\Delta x}$$

$$= \frac{4(x_1 + \Delta x) - 3(x_1 + \Delta x)^2 - (4x_1 - 3x_1^2)}{\Delta x}$$

$$= \frac{4x_1 + 4(\Delta x) - 3x_1^2 - 6x_1(\Delta x) - 3(\Delta x)^2 - 4x_1 + 3x_1^2}{\Delta x}$$

$$= \frac{4(\Delta x) - 6x_1(\Delta x) - 3(\Delta x)^2}{\Delta x}$$

$$= \frac{\Delta x[4 - 6x_1 - 3(\Delta x)]}{\Delta x} = [4 - 6x_1 - 3(\Delta x)].$$

Then formula (5) yields

$$m_{\text{tan}} = \lim_{\Delta x \to 0} \frac{\Delta y}{\Delta x} = \lim_{\Delta x \to 0} [4 - 6x_1 - 3(\Delta x)] = 4 - 6x_1. \quad \square$$

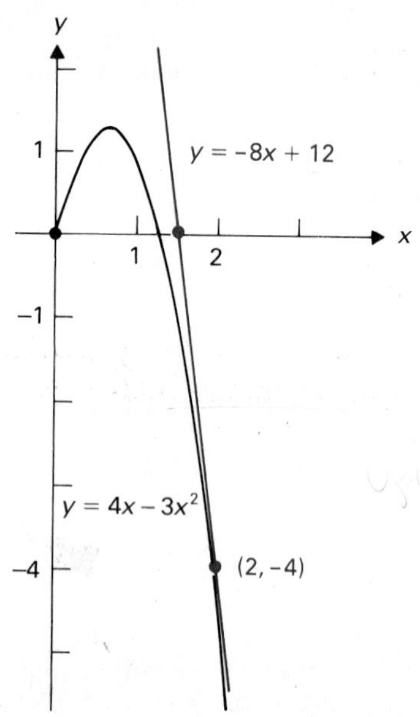

Figure 2.7 Tangent line $y = -8x + 12$ to the parabola $y = 4x - 3x^2$ at the point $(2, -4)$.

EXAMPLE 7 Find the equation of the tangent line to the graph of $y = f(x) = 4x - 3x^2$ at the point where $x = 2$, shown in Fig. 2.7.

Solution We are asked to find the equation of a line, so we need to know a *point* on the line and the *slope* of the line.

Point A point is $(2, f(2)) = (2, -4)$, since $f(2) = 4 \cdot 2 - 3 \cdot 2^2 = -4$.

Slope Since the line is tangent to the graph at $(2, -4)$, the slope is $m_{\tan}$ where $x = x_1 = 2$. Example 6 shows that we have $m_{\tan} = 4 - 6x_1 = 4 - 6 \cdot 2 = -8$.

Equation The equation of the line is

$$y + 4 = -8(x - 2), \qquad \text{or} \qquad y = -8x + 12. \quad \square$$

The trick in computing $m_{\tan}$ using formula (5) is to free a factor Δx in the numerator $f(x_1 + \Delta x) - f(x_1)$ to cancel with the factor Δx in the denominator. For a polynomial function $f(x)$, this is accomplished simply by computing $f(x_1 + \Delta x) - f(x_1)$, as illustrated in Example 6. We illustrate two more cases.

EXAMPLE 8 Find $m_{\tan}$ for $f(x) = 1/(3x)$ at $x = x_1 \neq 0$.

Solution First we find $m_{\sec}$, simplifying as much as possible and trying to cancel the factor Δx from the denominator. For $\Delta x \neq 0$, formula (4) yields

$$m_{\sec} = \frac{\Delta y}{\Delta x} = \frac{f(x_1 + \Delta x) - f(x_1)}{\Delta x}$$

$$= \frac{\dfrac{1}{3(x_1 + \Delta x)} - \dfrac{1}{3x_1}}{\Delta x} = \frac{\dfrac{3x_1 - 3x_1 - 3(\Delta x)}{3(x_1 + \Delta x)(3x_1)}}{\Delta x}$$

$$= \frac{-3(\Delta x)}{\Delta x[3(x_1 + \Delta x)(3x_1)]} = \frac{-3}{3(x_1 + \Delta x)(3x_1)}.$$

In this case, we freed the factor Δx in the numerator by finding a common denominator and computing the difference $f(x_1 + \Delta x) - f(x_1)$. Now we use formula (5) to find $m_{\tan}$, obtaining

$$m_{\tan} = \lim_{\Delta x \to 0} \frac{\Delta y}{\Delta x} = \lim_{\Delta x \to 0} \frac{-3}{3(x_1 + \Delta x)(3x_1)} = \frac{-1}{3x_1^2}. \quad \square$$

EXAMPLE 9 Find $m_{\tan}$ for $f(x) = \sqrt{x + 3}$ at $x = x_1$.

Solution For $\Delta x \neq 0$, formula (4) for $m_{\sec}$ yields

$$m_{\sec} = \frac{\Delta y}{\Delta x} = \frac{f(x_1 + \Delta x) - f(x_1)}{\Delta x}$$

$$= \frac{\sqrt{(x_1 + \Delta x) + 3} - \sqrt{x_1 + 3}}{\Delta x}$$

$$= \frac{\sqrt{x_1 + \Delta x + 3} - \sqrt{x_1 + 3}}{\Delta x} \cdot \frac{\sqrt{x_1 + \Delta x + 3} + \sqrt{x_1 + 3}}{\sqrt{x_1 + \Delta x + 3} + \sqrt{x_1 + 3}}$$

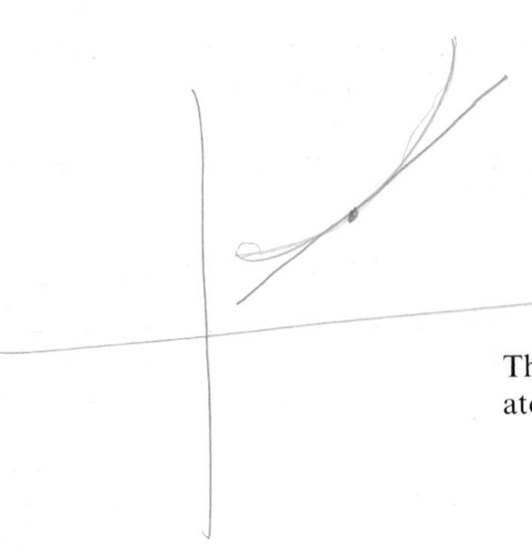

$$= \frac{(x_1 + \Delta x + 3) - (x_1 + 3)}{\Delta x(\sqrt{x_1 + \Delta x + 3} + \sqrt{x_1 + 3})}$$

$$= \frac{\Delta x}{\Delta x(\sqrt{x_1 + \Delta x + 3} + \sqrt{x_1 + 3})}$$

$$= \frac{1}{\sqrt{x_1 + \Delta x + 3} + \sqrt{x_1 + 3}}.$$

This time we freed Δx as a factor in the numerator by rationalizing the numerator to eliminate the radicals there. Now we use formula (5), obtaining

$$m_{\tan} = \lim_{\Delta x \to 0} \frac{\Delta y}{\Delta x}$$

$$= \lim_{\Delta x \to 0} \frac{1}{\sqrt{x_1 + \Delta x + 3} + \sqrt{x_1 + 3}} = \frac{1}{2\sqrt{x_1 + 3}}. \quad \square$$

We have said that calculus deals with changing quantities. The instantaneous rate of change $m_{\tan}$ of $f(x)$ at a point where $x = x_1$, which we have discussed and summarized in formula (5), is called the *derivative of f at* x_1. This derivative is the topic of differential calculus. In Chapter 3 we will study the derivative in greater detail and learn how to compute it easily. There we will introduce the notation $f'(x_1)$ for it. We discussed rates of change here primarily to show why we wish to consider limits, which are the main topic of this chapter.

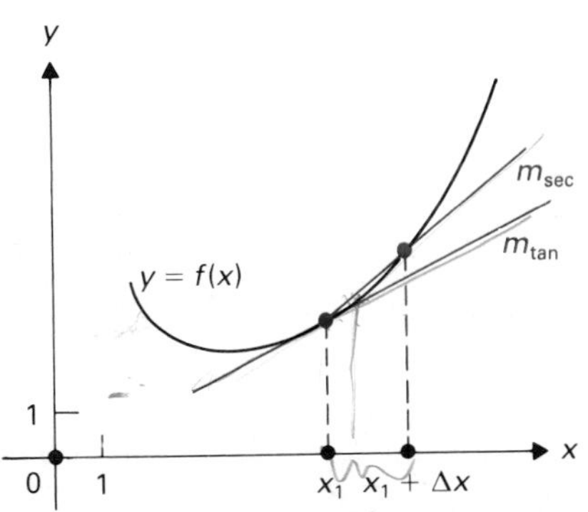

Figure 2.8 Approximation of the slope $m_{\tan}$ by the slope $m_{\sec}$.

CALCULATOR ESTIMATES FOR $m_{\tan}$

Figure 2.8 shows the tangent line to a graph where $x = x_1$ and the secant line to the same graph through the points where $x = x_1$ and $x = x_1 + \Delta x$. Figure 2.9 shows the same graph, the same tangent line where $x = x_1$, and the line through the points on the graph where $x = x_1 - \Delta x$ and $x = x_1 + \Delta x$ for the same Δx as in Fig. 2.8. These points $x_1 - \Delta x$ and $x_1 + \Delta x$ are symmetrically positioned on either side of x_1. It appears from Figs. 2.8 and 2.9 that the symmetrically positioned line in Fig. 2.9 is likely to be more nearly parallel to the tangent line than is the secant line in Fig. 2.8. Let the slope of the symmetrically positioned line be m_{sym}. We expect m_{sym} to be a better estimate for $m_{\tan}$ than $m_{\sec}$ would be, using the same size Δx. Since the symmetric line goes through the points

$$(x_1 - \Delta x, f(x_1 - \Delta x)) \qquad \text{and} \qquad (x_1 + \Delta x, f(x_1 + \Delta x)),$$

Figure 2.9 Better approximation of the slope $m_{\tan}$ by m_{sym}.

its slope is

$$m_{\text{sym}} = \frac{f(x_1 + \Delta x) - f(x_1 - \Delta x)}{2 \cdot \Delta x}. \tag{6}$$

When computing m_{sym}, we use only positive Δx values, because $-\Delta x$ and Δx lead to exactly the same symmetrically positioned line.

EXAMPLE 10 Use Eq. (6) to approximate $m_{\tan}$ for the function $f(x) = x^3 + 2x$, with $x_1 = 2$ and $\Delta x = 0.01$, which we used in Example 4.

Table 2.1

Δx	$m_{sym} = \dfrac{\cos(1+\Delta x) - \cos(1-\Delta x)}{2\cdot\Delta x}$
0.1	−0.8400692342
0.05	−0.8411204157
0.01	−0.8414569603
0.005	−0.8414674787
0.001	−0.8414708454
0.0005	−0.8414709492
0.0001	−0.8414709825

Solution The computation is

$$\frac{f(x_1+\Delta x) - f(x_1-\Delta x)}{2\cdot\Delta x} = \frac{[2.01^3 + 2(2.01)] - [1.99^3 + 2(1.99)]}{2\cdot 0.01}$$

$$= \frac{12.140601 - 11.860599}{0.02}$$

$$= \frac{0.280002}{0.02} = 14.0001.$$

The correct answer is actually 14, so this approximation is significantly better than the approximation 14.0601, which we found in Example 4. And the computation is easy with a calculator. □

Using a calculator and Eq. (6), we can estimate m_{tan} at points of the graphs of functions such as $\sin x$, 2^x, and x^x, which we have not discussed yet but which are defined by our calculators.

EXAMPLE 11 Estimate m_{tan} for $\cos x$ at $x = 1$ radian.

Solution Of course we use m_{sym} for the estimate, as given in Eq. (6). We choose successively smaller Δx-values until the values for m_{sym} seem to stabilize to five or six significant figures. The values we took for Δx and found for m_{sym} are shown in Table 2.1. We expect that m_{tan} is -0.841471 to six significant figures. Later we will see that the actual value of m_{tan} is $-\sin(1) = -0.8414709848$ to ten significant figures. □

Table 2.2

Δx	$m_{sym} = \dfrac{(3+\Delta x)^{(3+\Delta x)} - (3-\Delta x)^{(3-\Delta x)}}{2\cdot\Delta x}$
0.005	56.6637952
0.001	56.66258235
0.0005	56.66254441
0.0001	56.6625325

EXAMPLE 12 Estimate $m_{\tan}$ for $f(x) = x^x$ at $x = 3$.

Solution Again we use m_{sym} in Eq. (6) for our estimate. After working Example 11, it seems reasonable to limit ourselves to the smaller values of Δx and see if we have enough stabilization in the answer. We did this in Table 2.2. It appears that the value of $m_{\tan}$ to six significant figures is 56.6625. The value to ten significant figures can be shown to be 56.66253179. □

You may wonder why we didn't compute m_{sym} in the preceding two examples for much smaller values of Δx, say $\Delta x = 0.0000001$. The reason is that for such Δx, we expect $f(x_1 + \Delta x)$ and $f(x_1 - \Delta x)$ to agree to many significant figures. If our calculator computes to 12 significant figures and $f(x_1 + \Delta x)$ and $f(x_1 - \Delta x)$ agree for 10 significant figures, then $f(x_1 + \Delta x) - f(x_1 - \Delta x)$ has only 2 significant figures in the calculator. The quotient when divided by $2(\Delta x)$ can't be expected to be an accurate estimation of $m_{\tan}$.

SUMMARY

1. When $y = f(x)$, the average rate of change of y per unit increase in x from x_1 to $x_1 + \Delta x$ is

$$\frac{\Delta y}{\Delta x} = \frac{f(x_1 + \Delta x) - f(x_1)}{\Delta x}.$$

   Geometrically, this average rate of change is the slope $m_{\sec}$ of the secant line to the graph through the points where $x = x_1$ and where $x = x_1 + \Delta x$.

2. When $y = f(x)$, the instantaneous rate of change of y with respect to x where $x = x_1$ is

$$\lim_{\Delta x \to 0} \frac{\Delta y}{\Delta x} = \lim_{\Delta x \to 0} \frac{f(x_1 + \Delta x) - f(x_1)}{\Delta x}.$$

   This limit exists when the graph has a tangent line where $x = x_1$, and the limit then equals the slope $m_{\tan}$ of that line.

3. If $s = f(t)$ gives the total distance s traveled in a straight line at elapsed time t, then $m_{\tan}$, where $t = t_1$ is the speed at time t_1.

4. When $y = f(x)$, the instantaneous rate of change $m_{\tan}$ of y with respect to x at x_1 is also called the *derivative of the function f at* x_1 and is denoted by $f'(x_1)$.

5. When $y = f(x)$, the slope of the symmetrically positioned line through the graph where $x = x_1 - \Delta x$ and $x = x_1 + \Delta x$ is

$$m_{\text{sym}} = \frac{f(x_1 + \Delta x) - f(x_1 - \Delta x)}{2 \cdot \Delta x}.$$

In general, m_{sym} is likely to be a better approximation to $m_{\tan}$ than $m_{\sec}$ is for the same value of Δx.

EXERCISES

In Exercises 1 through 10, find the average rate of change m_{sec} given by Eq. (4) for the indicated function, point, and increment.

1. $f(x) = x^2, x_1 = 4, \Delta x = 0.01$
2. $f(x) = 5x, x_1 = 2, \Delta x = 0.005$
3. $f(t) = t^2 - 2t + 4, t_1 = -1, \Delta t = -0.2$
4. $f(x) = 4x - 5x^2, x_1 = -2, \Delta x = 0.01$
5. $f(s) = s^3 - 3s, s_1 = 1, \Delta s = 0.1$
6. $f(x) = 1/x, x_1 = 2, \Delta x = 0.1$
7. $f(u) = u + 1/u, u_1 = -1, \Delta u = -0.001$
8. $f(x) = 4/(x + 1), x_1 = -2, \Delta x = -0.01$
9. $f(x) = \sqrt{x}, x_1 = 4, \Delta x = 0.05$ (Use a calculator.)
10. $f(x) = \sqrt{2x + 5}, x_1 = 2, \Delta x = -0.03$ (Use a calculator.)

In Exercises 11 through 20, find the instantaneous rate of change $m_{\tan}$ at the point x_1 by computing the limit in formula (5) as $\Delta x \to 0$. Compare your answer with the one for the corresponding function in Exercises 1 through 10.

11. $f(x) = x^2, x_1 = 4$
12. $f(x) = 5x, x_1 = 2$
13. $f(x) = x^2 - 2x + 5, x_1 = -1$
14. $f(x) = 4x - 5x^2, x_1 = -2$
15. $f(x) = x^3 - 3x, x_1 = 1$
16. $f(x) = 1/x, x_1 = 2$
17. $f(x) = x + 1/x, x_1 = -1$
18. $f(x) = 4/(x + 1), x_1 = -2$
19. $f(x) = \sqrt{x}, x_1 = 4$
20. $f(x) = \sqrt{2x + 5}, x_1 = 2$

In Exercises 21 through 24, find the equation of the tangent line to the graph at the given point. (Recall that $m_{\tan}$ is the slope of the tangent line.)

21. $y = x^2$ at $(1, 1)$
22. $y = 4x - 3x^2$ at $(2, -4)$
23. $y = 1/x$ at $(2, \frac{1}{2})$
24. $y = \sqrt{x}$ at $(4, 2)$
25. If s is total distance traveled in time t, then $\Delta s/\Delta t$ has the interpretation of the *average speed* over the time interval Δt. Suppose an object travels so that after t hours it has gone $s = f(t) = 3t^2 + 2t$ miles, for $t \geq 0$.
    a) Find the average speed of the object during the two-hour time interval from $t = 3$ to $t = 5$.
    b) Find the average speed of the object during the one-hour time interval from $t = 3$ to $t = 4$.
    c) Find the average speed of the object during the half-hour time interval from $t = 3$ to $t = \frac{7}{2}$.
    d) On the basis of parts (a)–(c), guess the actual speed of the object at time $t = 3$.
26. Referring to Exercise 25, find, from the expression for $m_{\tan}$ as a limit, the exact speed at time $t = 3$.
27. A rock is dropped from 256 ft above the ground. Neglecting air resistance, the distance it has fallen after t sec is $s = 16t^2$ ft, until it hits the ground.
    a) For what time interval is the formula $s = 16t^2$ valid?
    b) Find the *average speed* of the rock during the time interval in part (a).
    c) Find a formula for the instantaneous speed of the rock at any time t in the interval found in part (a).
    d) At what time t is the instantaneous speed of the rock equal to its average speed over the time interval in part (a)?
28. A gas is confined at constant temperature in a container of increasing volume. The pressure of the gas on the walls of the container at time t sec is given by
$$p = \frac{24}{t + 2} \text{ g/cm}^2.$$
    a) Find the average rate of change of pressure from time $t = 0$ to time $t = 2$ sec.
    b) Find the instantaneous rate of change of pressure when $t = 0$.
29. Two positive electrical charges start 5 cm apart when $t = 0$ sec and the distance between them is increased at a constant rate of 1 cm/sec. The force of repulsion between them is given by
$$F = \frac{10}{(t + 5)^2} \text{ dynes.}$$
    a) Find the average rate of change of F from $t = 0$ to $t = 5$ sec.
    b) Find the instantaneous rate of change of F when $t = 5$ sec.

In the remaining exercises, use a calculator and Eq. (6) to approximate $m_{\tan}$ for the given function at the indicated point. You choose Δx. Functions we have not discussed are defined by your calculator. Use radian measure with trigonometric functions.

30. $\sin x$ at $x_1 = 0$
31. $\tan x$ at $x_1 = 1$
32. $\tan x$ at $x_1 = \frac{\pi}{4}$
33. $\sqrt{x^2 + 2x - 3}$ at $x_1 = 2$
34. $\dfrac{\sqrt{x^2 - 4x}}{x + 3}$ at $x_1 = -2$
35. 3^x at $x_1 = 2$
36. $2^{\sqrt{x+4}}$ at $x_1 = 5$
37. x^x at $x_1 = 1.5$

2.2 LIMITS

THE NOTION OF $\lim_{x \to x_1} f(x)$

In the preceding section, we computed limits of functions of Δx as $\Delta x \to 0$. For our general discussion, we will use x, rather than Δx, as the variable and talk about $\lim_{x \to x_1} f(x)$. That is, we consider the behavior of $f(x)$ as x gets closer and closer to x_1 but never becomes equal to x_1. We start with a number of examples to build our intuitive understanding.

Figure 2.10 Graph of $y = (x^2 - 4)/(x - 2) = x + 2$ for $x \neq 2$.

EXAMPLE 1 Try to find

$$\lim_{x \to 2} \frac{x^2 - 4}{x - 2}.$$

That is, can we discover whether $(x^2 - 4)/(x - 2)$ gets very close to some value L as x gets very close to 2?

Solution Note that $(x^2 - 4)/(x - 2)$ is not defined if $x = 2$. For all $x \neq 2$,

$$\frac{x^2 - 4}{x - 2} = \frac{(x - 2)(x + 2)}{x - 2} = x + 2.$$

See Fig. 2.10. Thus

$$\lim_{x \to 2} \frac{x^2 - 4}{x - 2} = \lim_{x \to 2} (x + 2) = 4,$$

since, if x is very close to 2, then $x + 2$ is very close to 4. □

EXAMPLE 2 Find

$$\lim_{x \to -1} \frac{x^2 - 1}{x^2 + 3x + 2}.$$

Solution Our function is not defined at -1. But we have

$$\frac{x^2 - 1}{x^2 + 3x + 2} = \frac{(x + 1)(x - 1)}{(x + 1)(x + 2)} = \frac{x - 1}{x + 2} \qquad \text{for } x \neq -1.$$

See Fig. 2.11. Hence

$$\lim_{x \to -1} \frac{x^2 - 1}{x^2 + 3x + 2} = \lim_{x \to -1} \frac{x - 1}{x + 2} = \frac{-2}{1} = -2.$$

That is, if x is very close to -1, then $x - 1$ is close to -2 and $x + 2$ is close to 1, so their quotient is close to -2. □

Figure 2.11 Graph of $y = (x^2 - 1)/(x^2 + 3x + 2) = (x - 1)/(x + 2)$ for $x \neq -1$.

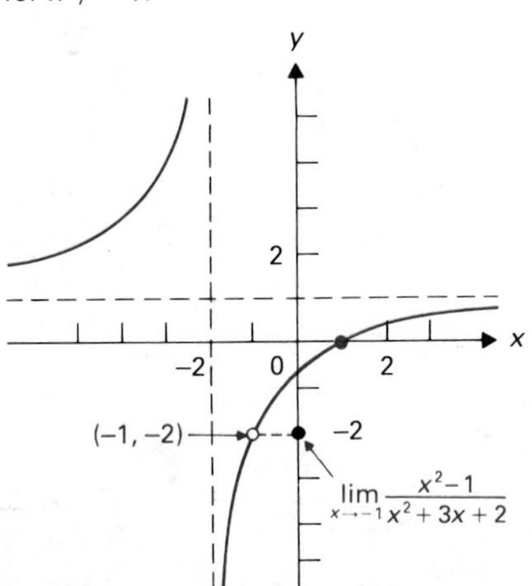

EXAMPLE 3 Find

$$\lim_{x \to 5} \frac{x + 6}{x^2 + 2x}.$$

Solution Here our function is defined at 5. As x becomes very close to 5, the numerator $x + 6$ is close to 11 while the denominator $x^2 + 2x$ is close to 35.

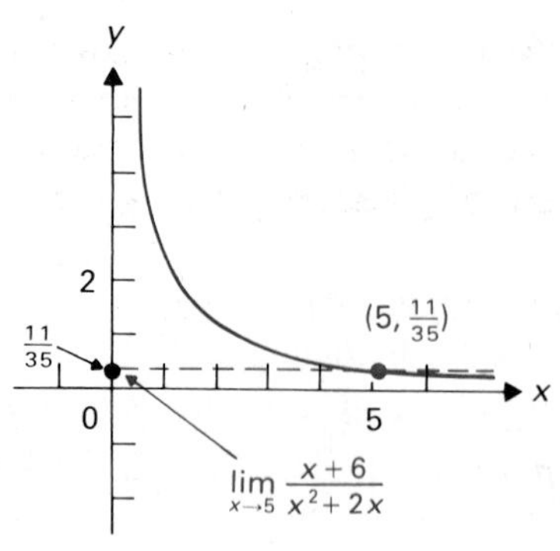

Figure 2.12 Graph of $y = (x+6)/(x^2+2x)$.

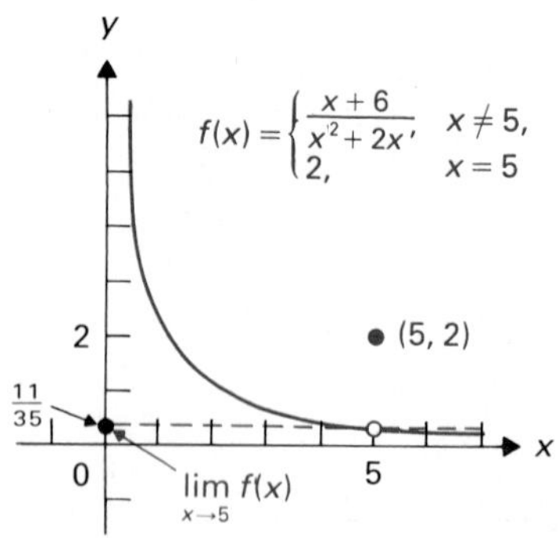

Figure 2.13 The value of $\lim_{x\to 5} f(x)$ does not depend on the value $f(5)$.

Thus the quotient is close to $\frac{11}{35}$, so

$$\lim_{x\to 5} \frac{x+6}{x^2+2x} = \frac{11}{35}.$$

See Fig. 2.12. In this case, the value of the limit was the value of the function at the point, but see the next example! □

EXAMPLE 4 Let

$$f(x) = \begin{cases} \dfrac{x+6}{x^2+2x} & \text{for } x \neq 5, \\ 2 & \text{for } x = 5, \end{cases}$$

and find $\lim_{x\to 5} f(x)$.

Solution See Fig. 2.13; $\lim_{x\to 5} f(x) = \frac{11}{35}$, just as in Example 3. In computing $\lim_{x\to x_1} f(x)$, we think in terms of values $f(x)$ as x gets very close to x_1 *but remains different from* x_1. The value $f(x_1)$ can be changed without having any effect on the limit. □

EXAMPLE 5 Discuss $\lim_{x\to 2} f(x)$ if

$$f(x) = \begin{cases} 3 - x & \text{for } x \geq 2, \\ x + 1 & \text{for } x < 2. \end{cases}$$

Figure 2.14 $\lim_{x\to 2} f(x)$ does not exist since the graph jumps at 2.

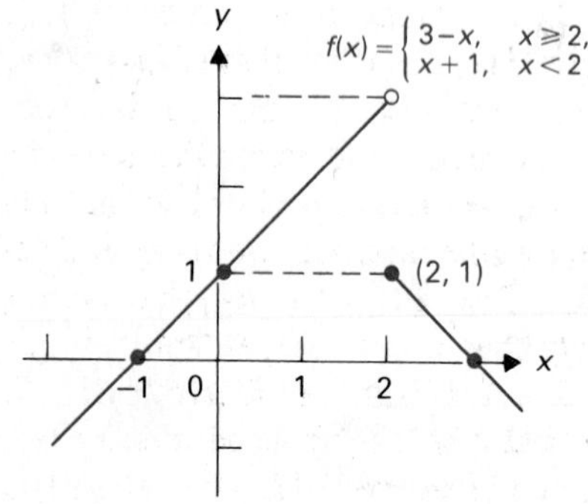

Solution The graph of f is shown in Fig. 2.14; $\lim_{x\to 2} f(x)$ does not exist, because if x is just slightly larger than 2, then $f(x)$ is close to $3 - 2 = 1$, while if x is slightly smaller than 2, $f(x)$ is close to $2 + 1 = 3$. There is no *single* value that $f(x)$ approaches for *all* x sufficiently close to but different from 2. We write "$\lim_{x\to 2} f(x)$ does not exist." □

EXAMPLE 6 Discuss $\lim_{x\to 2} f(x)$ if

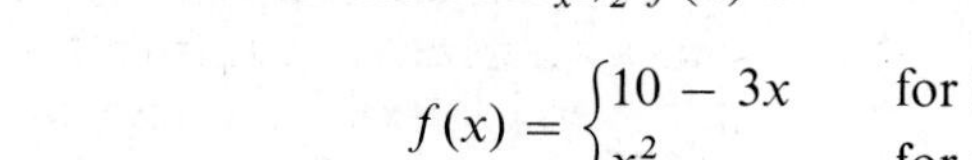

$$f(x) = \begin{cases} 10 - 3x & \text{for } x > 2, \\ x^2 & \text{for } x < 2. \end{cases}$$

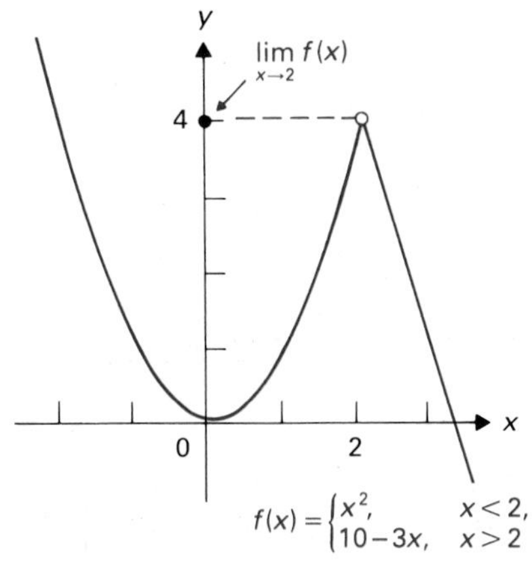

Figure 2.15 As $x \to 2$, $f(x) \to 4$. The graph does not jump at 2.

Solution The graph of f is shown in Fig. 2.15. This time, $\lim_{x\to 2} f(x) = 4$, because $10 - 3x$ is close to 4 for x just larger than 2, and x^2 is also close to 4 for x just smaller than 2. We did not define $f(2)$, but that does not matter. It could be given any value without changing the limit value. □

EXAMPLE 7 Discuss $\lim_{x\to 3} f(x)$ if $f(x) = \sqrt{x-3}\sqrt{3-x}$.

Solution Note that 3 is the *only* point in the domain of f. Thus it is meaningless to consider $\lim_{x\to 3} f(x)$ because we can't evaluate $f(x)$ for any point really close to 3 but not equal to 3. □

DEFINITION OF $\lim_{x\to x_1} f(x)$

This subsection and precise ε,δ-definitions of limits and continuity throughout the rest of this chapter are not used in subsequent chapters. They can be omitted at the discretion of the instructor. We have not marked them with an asterisk because we feel a brief exposure now to these definitions promotes much better understanding of them if students ever encounter them again. The discussion of limits up to this point should have provided an understanding of the real significance of the concept, stated as follows:

> A limit is used to describe the behavior of a function *near* a point but not at the point. The function need not even be defined at the point. If it is defined there, the value of the function at the point does not affect the limit.

Intuitively, $\lim_{x\to x_1} f(x) = L$ means we can make $f(x)$ as *close to* L as we wish by taking any x sufficiently *close to* but different from x_1. This is a vague statement, for what does *close* mean? You may think that $f(x)$ is close to L if $L - 0.1 < f(x) < L + 0.1$, while a friend may say, "No, I want to have $L - 0.00001 < f(x) < L + 0.00001$." It is only proper to say that $\lim_{x\to x_1} f(x) = L$ if *everyone* can be satisfied. So if someone demands to have $L - \varepsilon < f(x) < L + \varepsilon$ for some $\varepsilon > 0$, be it 0.1 or 0.00001, we must be sure that this will be true as long as x is within a certain distance, perhaps 0.05 or 0.003, of x_1 but not equal to x_1. That is, we must be able to find a $\delta > 0$ such that $L - \varepsilon < f(x) < L + \varepsilon$ will be true if $x_1 - \delta < x < x_1 + \delta$, but $x \neq x_1$. Figure 2.16 illustrates the choice of δ from a given ε. We did not bother to define f at x_1 in Fig. 2.16 because the value $f(x_1)$ plays no role in the discussion of $\lim_{x\to x_1} f(x) = L$. We mark $L - \varepsilon$ and $L + \varepsilon$ on the y-axis, because this is where values $f(x)$ are plotted. For the function f in Fig. 2.16 we then travel on horizontal lines from these points until the graph is met and then down to the x-axis, as indicated by the arrows. As long as x is within the colored interval on the x-axis, but not x_1, we see that $f(x)$ will lie between $L - \varepsilon$ and $L + \varepsilon$. We take as δ the *smaller* of the distances from the ends of this colored interval on the x-axis to x_1. Then the set of x-values where $x_1 < \delta < x < x_1 + \delta$, but $x \neq x_1$, surely lies within the interval on the x-axis we found starting from $L - \varepsilon$ and $L + \varepsilon$. Figure 2.17 illustrates that if the given ε is reduced in size, we expect that δ will be reduced in size also. Finally, Fig. 2.18 illustrates that for ε chosen as in Fig. 2.16, a still smaller δ

Figure 2.16 $\lim_{x\to x_1} f(x) = L$. Finding δ from a given ε.

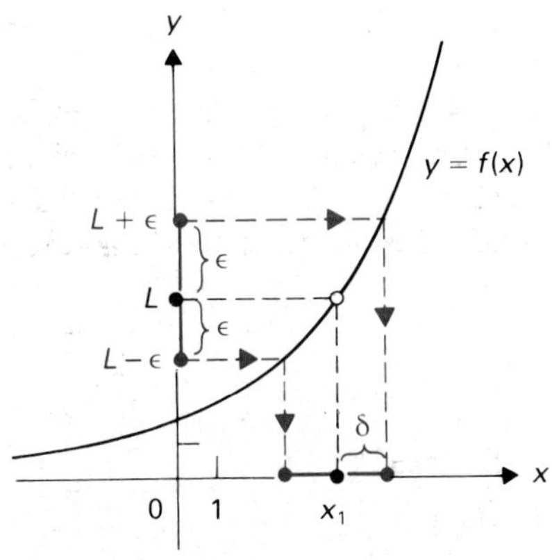

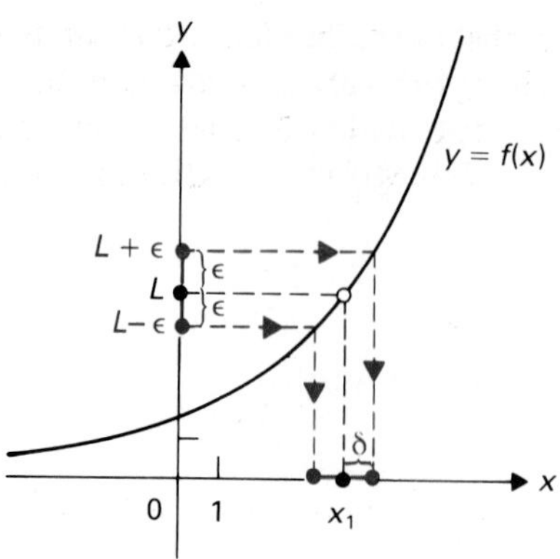

Figure 2.17 A smaller ε may require a smaller δ.

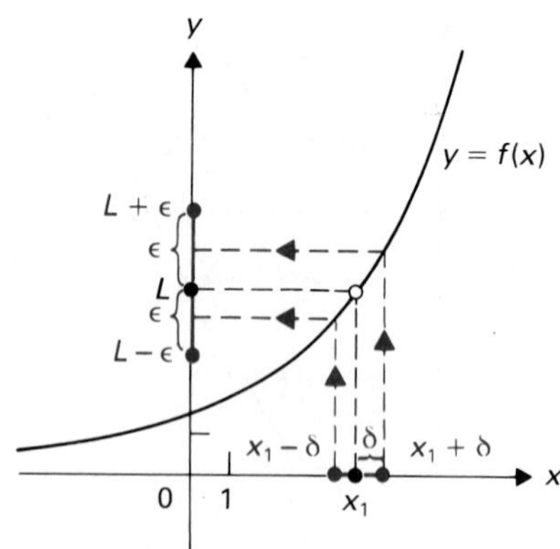

Figure 2.18 If one size for δ is satisfactory (see Fig. 2.16), then any smaller $\delta > 0$ is also satisfactory.

than that in Fig. 2.16 will do. That is, we need to find only *some* $\delta > 0$ that will work, not necessarily the largest possible such δ.

We now formally define $\lim_{x \to x_1} f(x) = L$. [Of course, to talk about $\lim_{x \to x_1} f(x)$ at all, we want to have the domain of f contain points $x \neq x_1$ arbitrarily close to x_1. See Example 7. The first sentence of the definition takes care of this.]

DEFINITION 2.1 Limit of f at x_1

Suppose the domain of f contains points x arbitrarily close to x_1 but different from x_1. Then $\lim_{x \to x_1} f(x) = L$ if for each $\varepsilon > 0$ there exists $\delta > 0$ such that $L - \varepsilon < f(x) < L + \varepsilon$ for any $x \neq x_1$ in the domain of f such that $x_1 - \delta < x < x_1 + \delta$.

The requirement $L - \varepsilon < f(x) < L + \varepsilon$ can also be expressed as $|f(x) - L| < \varepsilon$, while the two conditions $x_1 - \delta < x < x_1 + \delta$, but $x \neq x_1$, can be written as the single condition $0 < |x - x_1| < \delta$. This ε,δ-characterization is very important in theoretical work in mathematics.

EXAMPLE 8 Use the ε,δ-characterization to demonstrate that

$$\lim_{x \to 3} (2x - 2) = 4.$$

This surely should be true; if x is close to 3, then $f(x) = 2x - 2$ is close to $6 - 2 = 4$.

Solution Let $\varepsilon > 0$. We want to be sure that

$$4 - \varepsilon < f(x) < 4 + \varepsilon. \qquad \textbf{(1)}$$

We should mark $4 - \varepsilon$ and $4 + \varepsilon$ on the y-axis (see Fig. 2.19) because this is where $f(x)$ is plotted. Now we have to find $\delta > 0$ and mark $3 - \delta$ and $3 + \delta$ on the x-axis, so that Eq. (1) is true if $3 - \delta < x < 3 + \delta$ but $x \neq 3$. Since the graph of $f(x) = 2x - 2$ is a line with slope 2, each unit of change

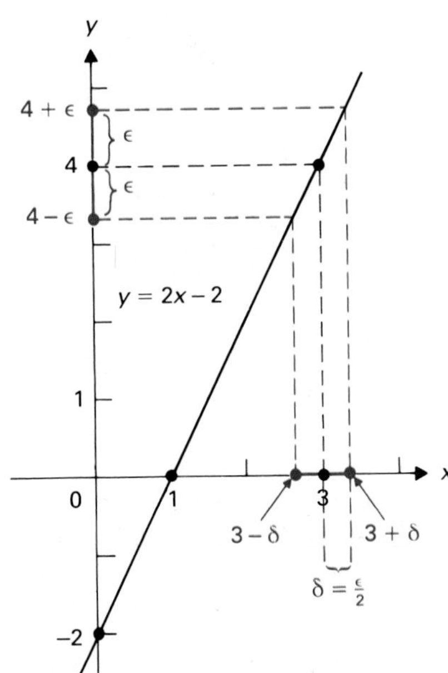

Figure 2.19
$\lim_{x\to 3}(2x-2)=4$.
The graph has slope 2, so δ can be at most $\varepsilon/2$.

on the x-axis produces two units of change of $f(x)$ on the y-axis. Thus a change of ε on the y-axis is produced by a change of only $\varepsilon/2$ on the x-axis, so we may let $\delta = \varepsilon/2$. This was just a geometric argument.

We can arrive at the same result algebraically as follows. We need

$$4 - \varepsilon < 2x - 2 < 4 + \varepsilon,$$

which can be written as

$$6 - \varepsilon < 2x < 6 + \varepsilon, \qquad \text{or} \qquad 3 - \frac{\varepsilon}{2} < x < 3 + \frac{\varepsilon}{2}.$$

Consequently, $\delta = \varepsilon/2$ suffices. Of course, any smaller δ will work also. □

EXAMPLE 9 Give an ε,δ-demonstration that $\lim_{x\to 1}(5-7x) = -2$.

Solution Let $\varepsilon > 0$. We need

$$-2 - \varepsilon < 5 - 7x < -2 + \varepsilon. \qquad \textbf{(2)}$$

Equation (2) is true if and only if

$$-7 - \varepsilon < -7x < -7 + \varepsilon, \qquad \text{or} \qquad 1 + \frac{\varepsilon}{7} > x > 1 - \frac{\varepsilon}{7}.$$

Thus we can take $\delta = \varepsilon/7$, because if x is within $\varepsilon/7$ of 1, then $5 - 7x$ is within ε of -2. □

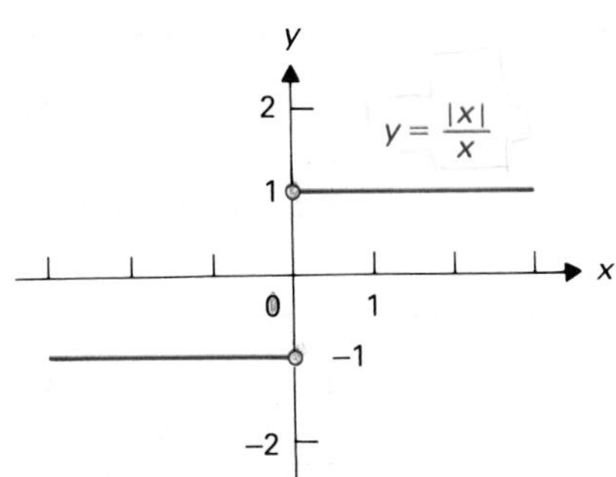

Figure 2.20 $\lim_{x\to 0}|x|/x$ does not exist. The graph jumps 2 units at $x = 0$.

EXAMPLE 10 Give an ε,δ-demonstration that $\lim_{x\to 0}|x|/x$ does not exist.

Solution The graph of $|x|/x$ is shown in Fig. 2.20. For *every* $\delta > 0$,

$$\frac{|x|}{x} = 1, \quad \text{if } 0 < x < \delta, \qquad \text{and} \qquad \frac{|x|}{x} = -1, \quad \text{if } -\delta < x < 0.$$

The points 1 and -1 are two units apart, while for any possible limit L, the numbers $L - \varepsilon$ and $L + \varepsilon$ are 2ε units apart. Consequently, if $\varepsilon < 1$, it is impossible to have

$$L - \varepsilon < \frac{|x|}{x} < L + \varepsilon \qquad \text{for all } -\delta < x < \delta, \quad x \neq 0,$$

for *any* choice of L and $\delta > 0$. Thus given $\varepsilon = \frac{1}{2}$, there exists no $\delta > 0$ such that

$$L - \varepsilon < \frac{|x|}{x} < L + \varepsilon \qquad \text{for } -\delta < x < \delta, \quad x \neq 0,$$

no matter what L might be. This contradicts Definition 2.1, which asserts that for *each* $\varepsilon > 0$, in particular for $\varepsilon = \frac{1}{2}$, such $\delta > 0$ should exist. □

COMPUTING LIMITS

The following is a theorem that we will use frequently, perhaps without realizing it, in computing limits.

THEOREM 2.1 Properties of limits

If $\lim_{x \to x_1} f(x) = L$ and $\lim_{x \to x_1} g(x) = M$, where the domains of f and g contain common points arbitrarily close to x_1 but different from x_1, then

$$\lim_{x \to x_1} (f(x) + g(x)) = L + M, \qquad \text{Sum property} \quad \textbf{(3)}$$

$$\lim_{x \to x_1} (f(x) \cdot g(x)) = L \cdot M, \qquad \text{Product property} \quad \textbf{(4)}$$

$$\lim_{x \to x_1} \left(\frac{f(x)}{g(x)}\right) = \frac{L}{M} \quad \text{if } M \neq 0, \qquad \text{Quotient property} \quad \textbf{(5)}$$

$$\lim_{x \to x_1} \sqrt{f(x)} = \sqrt{L} \quad \text{if } L > 0. \qquad \text{Root property} \quad \textbf{(6)}$$

To illustrate, if $f(x)$ is very near 2 and $g(x)$ is very near 5 when x is close to $x_1 = -1$, then $f(x) + g(x)$ is very near $2 + 5 = 7$, and $f(x) \cdot g(x)$ is very near $2 \cdot 5 = 10$ when x is close to -1. Similarly, $f(x)/g(x)$ is very near $\frac{2}{5}$ and $\sqrt{f(x)}$ is very near $\sqrt{2}$ for such x-values. Theorem 2.1 is certainly intuitively obvious. We used property (5) of the theorem in Examples 1 through 4. You will find an ε,δ-proof of the theorem in any text on advanced calculus.

Here is a sample of what we can do using properties (3), (4), and (5). Surely $\lim_{x \to x_1} x = x_1$. But then, from the product property (4),

$$\lim_{x \to x_1} x \cdot x = x_1 \cdot x_1, \qquad \lim_{x \to x_1} x^3 = \lim_{x \to x_1} x^2 \cdot x = x_1^2 \cdot x_1 = x_1^3, \quad \ldots .$$

From the sum property (3),

$$\lim_{x \to x_1} (x^3 + x^2) = x_1^3 + x_1^2.$$

Also, if $f(x) = 3$ for all x, then surely $\lim_{x \to x_1} f(x) = 3$. Using the product property,

$$\lim_{x \to x_1} 3 \cdot x^2 = 3x_1^2.$$

Similar arguments show that if $f(x)$ is any polynomial function, then $\lim_{x \to x_1} f(x) = f(x_1)$.

By the quotient property (5), if $g(x)$ is also a polynomial function and $g(x_1) \neq 0$, then $\lim_{x \to x_1} f(x)/g(x) = f(x_1)/g(x_1)$. (Such a quotient of polynomial functions is a **rational function**.) Thus the computation of the limit of a rational function as $x \to x_1$ amounts to evaluation of the function at the point x_1, provided that x_1 does not make the denominator of the function zero. We state this as a corollary of Theorem 2.1.

COROLLARY Limits of rational functions

The limit of any polynomial or rational function at a point x_1 in the domain of the function is equal to the value of the function at that point.

The only "bad case" in computing the limit at x_1 of a rational function $f(x)/g(x)$ is the case where $g(x_1) = 0$, so that x_1 is not in the domain of $f(x)/g(x)$. When a denominator becomes zero at a point, one must try some algebraic trick, such as canceling a factor from both numerator and denominator, to try to find the limit at that point. Examples 5 through 7 showed other "bad cases," but those examples were somewhat contrived. We will rarely encounter such functions in practice.

EXAMPLE 11 Find

$$\lim_{x\to 1} \sqrt{\frac{x+3}{x+35}}.$$

Solution By the discussion above, we know that

$$\lim_{x\to 1} \frac{x+3}{x+35} = \frac{1+3}{1+35} = \frac{4}{36} = \frac{1}{9}.$$

By the root property (6), we see that

$$\lim_{x\to 1} \sqrt{\frac{x+3}{x+35}} = \sqrt{\frac{1}{9}} = \frac{1}{3}. \quad \square$$

EXAMPLE 12 Find

$$\lim_{x\to 3} \frac{x^2-9}{x^2-4x+3}.$$

Solution Note that both the numerator and denominator become zero when $x = 3$. We have

$$\lim_{x\to 3} \frac{x^2-9}{x^2-4x+3} = \lim_{x\to 3} \frac{(x-3)(x+3)}{(x-3)(x-1)} = \lim_{x\to 3} \frac{x+3}{x-1} = \frac{6}{2} = 3. \quad \square$$

EXAMPLE 13 Find

$$\lim_{x\to 5} \frac{x^2-25}{x+4}.$$

Solution We have $\lim_{x\to 5} [(x^2-25)/(x+4)] = 0/9 = 0$. It is important to realize that a zero in a *numerator only* causes no problem. It is when the denominator becomes zero that we have to work. A common error is to say that the limit in this example is undefined or that it is $\varnothing$ (the empty set). $\square$

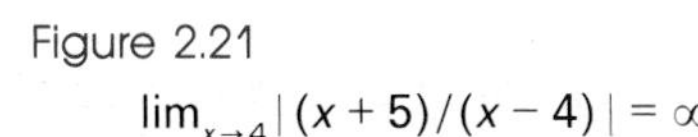

Figure 2.21
$\lim_{x\to 4} |(x+5)/(x-4)| = \infty$

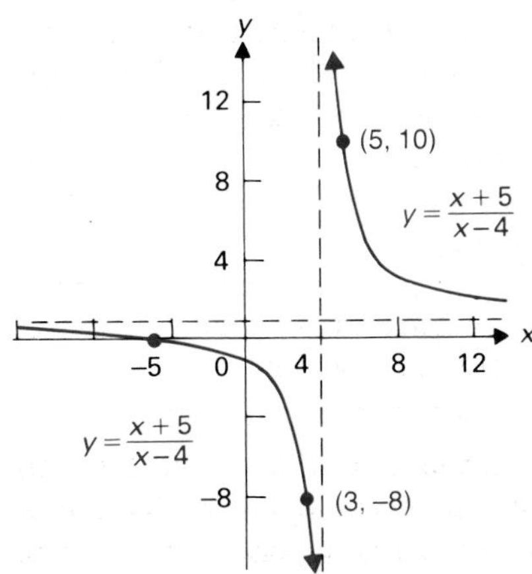

EXAMPLE 14 Discuss the behavior of $f(x) = (x+5)/(x-4)$ as $x \to 4$.

Solution The limit

$$\lim_{x\to 4} \frac{x+5}{x-4}$$

does not exist, because the numerator approaches 9 while the denominator approaches zero. Thus the quotient becomes very large in absolute value as $x \to 4$ (positively large if $x > 4$ and negatively large if $x < 4$). See Fig. 2.21. Symbolically,

$$\lim_{x\to 4} \left| \frac{x+5}{x-4} \right| = \infty. \qquad (7)$$

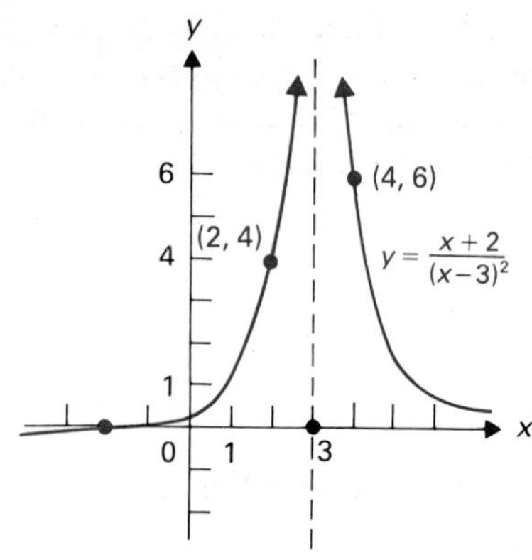

Figure 2.22

$\lim_{x\to 3}[(x+2)/(x-3)^2] = \infty$

This does *not* mean that ∞ (read "infinity") is the limit, but rather that the limit does not exist because the quotient becomes large as $x \to 4$. *The purpose of finding a limit is to describe the behavior of a function near a point, and Eq. (7) does that for us very neatly.* □

EXAMPLE 15 Find

$$\lim_{x\to 3} \frac{x+2}{(x-3)^2}$$

if the limit exists, using the notations ∞ and $-\infty$ if appropriate.

Solution We have $\lim_{x\to 3} [(x + 2)/(x - 3)^2] = \infty$, because the numerator approaches 5 while the denominator approaches zero but is always positive. Thus the quotient becomes large but is always positive. See Fig. 2.22. No absolute-value sign, as used in Example 14, is needed in this case. □

EXAMPLE 16 Proceed as in Example 15 for

$$\lim_{x\to 2} \frac{x-7}{(x-2)^2}.$$

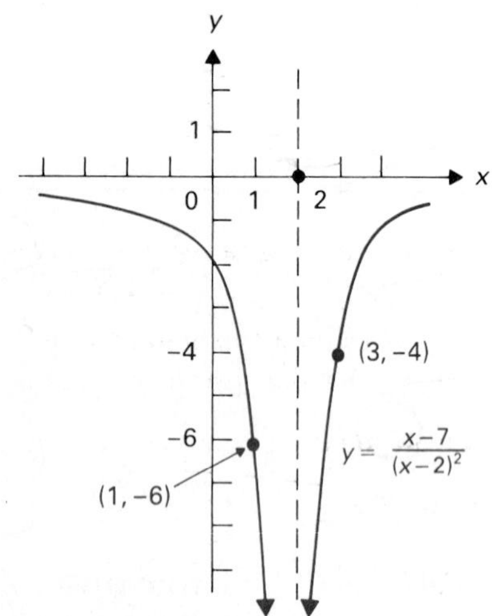

Figure 2.23

$\lim_{x\to 2}[(x-7)/(x-2)^2] = -\infty$

Solution We have $\lim_{x\to 2} [(x - 7)/(x - 2)^2] = -\infty$, because the numerator approaches -5 while the denominator approaches zero but is always positive. Thus the quotient is large but always negative. See Fig. 2.23. □

EXAMPLE 17 Let

$$f(x) = \begin{cases} \dfrac{1}{x-2} & \text{for } x > 2, \\[2ex] \dfrac{1}{x^2 - 2x} & \text{for } 0 < x < 2, \\[2ex] \dfrac{1}{x} & \text{for } x < 0. \end{cases}$$

Find $\lim_{x\to 2} f(x)$ and $\lim_{x\to 0} f(x)$, if they exist, or use the symbols ∞ or $-\infty$ if appropriate.

Solution For x just greater than 2, $f(x) = 1/(x - 2)$ is very large and positive. For x just less than 2, $f(x) = 1/(x^2 - 2x) = 1/[x(x - 2)]$ is very large but negative, since x is positive and $x - 2$ negative. Thus we may write

$$\lim_{x\to 2} |f(x)| = \infty.$$

For x just greater than 0, $f(x) = 1/(x^2 - 2x) = 1/[x(x - 2)]$ is very large but negative. For x just less than 0, $f(x) = 1/x$ is very large and also negative. Thus we may write $\lim_{x\to 0} f(x) = -\infty$. □

LIMITS WITH CALCULATORS

If $f(x_1 + 0.01)$, $f(x_1 - 0.005)$, $f(x_1 + 0.002)$, and $f(x_1 - 0.0001)$ have approximately the same value L, that is a good indication that $\lim_{x\to x_1} f(x) \approx L$. (The symbol $\approx$ means "approximately equals.") The numbers 0.01,

-0.005, 0.002, and -0.0001 can of course be replaced by other values of appropriate, small magnitude, some positive and some negative. The appropriate magnitude of the numbers depends on the function and is best determined by trying a few. The next two examples illustrate this.

Table 2.3

Δx	$f(1 + \Delta x)$
0.01	0.497504146
−0.005	0.5012510442
0.0002	0.49995
−0.00001	0.5000025

EXAMPLE 18 Let

$$f(x) = \frac{\sin(x - 1)}{x^2 - 1},$$

where x is in radian measure. Use a calculator to estimate $\lim_{x \to 1} f(x)$.

Solution In Table 2.3, we show values of $f(1 + \Delta x)$ for some values of Δx. It appears that $\lim_{x \to 1} f(x) \approx 0.5$. □

Table 2.4

Δx	$g(0 + \Delta x)$
0.5	35.52713679
−0.3	41740.27455
0.2	4.4242235×10^{10}
−0.1	1.6358287×10^{43}
0.05	overflow

EXAMPLE 19 Let $g(x) = (1 + x^2)^{1/x^4}$. Use a calculator to estimate $\lim_{x \to 0} g(x)$.

Solution Table 2.4 shows values $g(0 + \Delta x) = g(\Delta x)$. It appears that $\lim_{x \to 0} g(x) = \infty$. □

SUMMARY

1. Limits are used to study the behavior of a function near a point x_1 where the function may not be defined.
2. $\lim_{x \to x_1} f(x) = L$ means that for each $\varepsilon > 0$, there exists some $\delta > 0$ such that $|f(x) - L| < \varepsilon$, provided that $0 < |x - x_1| < \delta$.
3. If $\lim_{x \to x_1} f(x) = L$ and $\lim_{x \to x_1} g(x) = M$, then

$$\lim_{x \to x_1} (f(x) + g(x)) = L + M,$$

$$\lim_{x \to x_1} (f(x) \cdot g(x)) = L \cdot M,$$

$$\lim_{x \to x_1} \left(\frac{f(x)}{g(x)}\right) = \frac{L}{M}, \quad \text{if } M \neq 0,$$

$$\lim_{x \to x_1} \sqrt{f(x)} = \sqrt{L}, \quad \text{if } L > 0.$$

4. Limits as $x \to x_1$ of all polynomial and rational functions, and many other functions defined by a single algebraic formula, can be found by evaluating at x_1 as long as a denominator does not become zero at x_1. If a denominator becomes zero at x_1, try to cancel a factor of the denominator with one in the numerator.
5. The symbols ∞ and $-\infty$ are not numbers but may be used where appropriate with limit expressions to describe the behavior of a function near a point.

EXERCISES

In Exercises 1 through 35, find the indicated limit if it exists. Use the notation ∞ or $-\infty$ to describe the behavior of the function where appropriate.

1. $\lim_{x\to 2} \dfrac{3x-6}{x-2}$

2. $\lim_{t\to 0} \dfrac{4t^2-2t}{t}$

3. $\lim_{u\to 1} \dfrac{|u-1|}{u+1}$

4. $\lim_{u\to 2} \dfrac{|u+2|}{u-4}$

5. $\lim_{x\to 1} \dfrac{x^2-1}{x^2-x}$

6. $\lim_{u\to 1} \dfrac{u^2-1}{u-u^2}$

7. $\lim_{x\to 3} \dfrac{x^2-3x}{x^2-9}$

8. $\lim_{t\to 2} \dfrac{t^2-4}{2t-t^2}$

9. $\lim_{s\to -1} \dfrac{|s+1|}{s+1}$

10. $\lim_{x\to 5} \dfrac{x^2-4x-5}{x^3-5x^2}$

11. $\lim_{u\to -2} \dfrac{u^2-4}{u^2+4}$

12. $\lim_{x\to 0} \dfrac{x^3+x^2+2}{x}$

13. $\lim_{t\to 0} \dfrac{t^3+t^2+2}{t^3+1}$

14. $\lim_{x\to 0} \dfrac{x^4+2x^2}{x^3+x}$

15. $\lim_{s\to 0} \dfrac{s^3-2s^2}{s^4+3s^2}$

16. $\lim_{r\to 0} \dfrac{2r^2-3r}{r^3+4r^2}$

17. $\lim_{x\to 2} \dfrac{x}{x+3}$

18. $\lim_{u\to 1} \dfrac{(u-1)^2}{u-1}$

19. $\lim_{s\to 1} \dfrac{2(s-1)}{(s-1)^2}$

20. $\lim_{x\to -1} \dfrac{x^2+x}{x-1}$

21. $\lim_{t\to -1} \dfrac{t^2+t}{t+1}$

22. $\lim_{x\to 2} \dfrac{x^2-4}{x^2-x-2}$

23. $\lim_{\Delta x\to 0} (2+\Delta x)$

24. $\lim_{\Delta t\to 0} \dfrac{4+\Delta t}{2}$

25. $\lim_{\Delta x\to 0} [(2+\Delta x)^2-4]$

26. $\lim_{\Delta x\to 0} \dfrac{(2+\Delta x)^2-4}{\Delta x}$

27. $\lim_{\Delta t\to 0} \dfrac{[1/(3+\Delta t)]-(1/3)}{\Delta t}$

28. $\lim_{\Delta x\to 0} \dfrac{|\Delta x|}{\Delta x}$

29. $f(x) = \begin{cases} x & \text{for } x<0, \\ 1 & \text{for } x=0, \\ x^2 & \text{for } x>0. \end{cases}$

 a) $\lim_{x\to -2} f(x)$ b) $\lim_{x\to 0} f(x)$ c) $\lim_{x\to 3} f(x)$

30. $g(x) = \begin{cases} x+1 & \text{for } x\ge 1, \\ 2x-4 & \text{for } x<1. \end{cases}$

 a) $\lim_{x\to -2} g(x)$ b) $\lim_{x\to 1} g(x)$ c) $\lim_{x\to 3} g(x)$

31. $g(t) = \begin{cases} \sqrt{t-3} & \text{for } t>3, \\ t^2+1 & \text{for } t<3. \end{cases}$

 a) $\lim_{t\to -1} g(t)$ b) $\lim_{t\to 3} g(t)$ c) $\lim_{t\to 7} g(t)$

32. $g(u) = \begin{cases} |u|/u & \text{for } u>-2,\ u\ne 0, \\ u+1 & \text{for } u<-2, \\ 3 & \text{for } u=-2. \end{cases}$

 a) $\lim_{u\to -3} g(u)$ b) $\lim_{u\to -2} g(u)$
 c) $\lim_{u\to 0} g(u)$ d) $\lim_{u\to 1} g(u)$

33. $f(x) = \begin{cases} -1/x^3 & \text{for } x<0, \\ 10 & \text{for } x=0, \\ 1/x & \text{for } x>0. \end{cases}$

 a) $\lim_{x\to -2} f(x)$ b) $\lim_{x\to 0} f(x)$ c) $\lim_{x\to 3} f(x)$

34. $h(t) = \begin{cases} 1/(t-1) & \text{for } t>1, \\ 1/(t^2-1) & \text{for } -1<t<1, \\ 1/(t+1) & \text{for } t<-1. \end{cases}$

 a) $\lim_{t\to -1} h(t)$ b) $\lim_{t\to 0} h(t)$ c) $\lim_{t\to 1} h(t)$

35. $f(x) = \begin{cases} 1/x & \text{for } x>0, \\ 1/(x^3+2x^2) & \text{for } -2<x<0, \\ 1/(x^2-4) & \text{for } x<-2. \end{cases}$

 a) $\lim_{x\to 2} f(x)$ b) $\lim_{x\to 0} f(x)$ c) $\lim_{x\to -2} f(x)$

In Exercises 36 through 39, explain why it makes no sense to talk about the indicated limit.

36. $\lim_{x\to 2} \sqrt{x^2-9}$

37. $\lim_{x\to -3} \sqrt{-(x+3)^2}$

38. $\lim_{x\to 4} \sqrt{8x-x^2-16}$

39. $\lim_{x\to 0} 1/\sqrt{x^2-9}$

In Exercises 40 through 45, if $\varepsilon>0$ is given, find in terms of ε what size δ must be taken in the ε,δ-characterization of a limit to establish the limit.

40. $\lim_{x\to x_1} x = x_1$

41. $\lim_{x\to x_1} c = c$, where c in $\lim_{x\to x_1} c$ is the constant function f defined by $f(x)=c$ for all x

42. $\lim_{x\to 4} 2x = 8$

43. $\lim_{x\to -2} (-3x) = 6$

44. $\lim_{x\to -3} (14-5x) = 29$

45. $\lim_{x\to -4} (2-\frac{1}{3}x) = \frac{10}{3}$

In Exercises 46 through 53, use a calculator to estimate the limit, if it exists, as illustrated in Examples 18 and 19. Use radian measure for all trigonometric functions.

46. $\lim_{x\to\sqrt{2}} \dfrac{x^2+2\sqrt{2}x-6}{x^2-2}$

47. $\lim_{x\to 3} \dfrac{x-\sqrt{3x}}{27-x^3}$

48. $\lim_{x\to 3} \dfrac{\sin(x-3)}{x^2-9}$

49. $\lim_{x\to 0} (1+x)^{1/x}$

50. $\lim_{x\to 0} \dfrac{\cos x-1}{x^2}$

51. $\lim_{x\to 0} (1+x)^{1/x^2}$

52. $\lim_{x\to\pi/2} (\sin x)^{1/(\pi-2x)}$

53. $\lim_{x\to 0} \dfrac{\sin x^2}{\cos^2 x-1}$

2.3 ONE-SIDED LIMITS AND LIMITS AT INFINITY

$\lim_{x\to x_1+} f(x)$ AND $\lim_{x\to x_1-} f(x)$

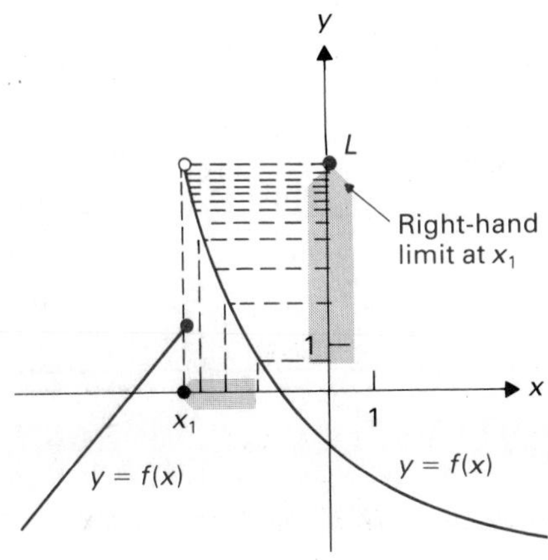

Figure 2.24

$\lim_{x\to x_1+} f(x) = L$

Recall that limits are used to describe the behavior of a function near a point, neglecting the value at the point itself. For the function f in Fig. 2.24, we see that $\lim_{x\to x_1} f(x)$ does not exist. However, as the figure shows, $f(x)$ approaches L if $x \to x_1$ *from the right-hand side only*. This illustration suggests that we introduce the notion of the limit from the right-hand side (or left-hand side) to help describe the behavior of some functions. We denote such a right-hand limit at x_1 by

$$\lim_{x\to x_1+} f(x).$$

Thus for $f(x)$ in Fig. 2.24, we have $\lim_{x\to x_1+} f(x) = L$.

Example 1 involves the function $\sqrt{x}$ as $x \to 0$. Now $\sqrt{x}$ is defined only for $x \geq 0$. While we could work with the regular limit, presented in Section 2.2, we often write it as a right-hand limit in such a case, just to remind us that we can approach zero only from the right-hand side in the domain of the function.

EXAMPLE 1 Find

$$\lim_{x\to 0+} \frac{x - \sqrt{x}}{\sqrt{x}}.$$

Solution Since this function involves $\sqrt{x}$, it is not defined for x negative, but it is defined if $x > 0$. We have

$$\lim_{x\to 0+} \frac{x - \sqrt{x}}{\sqrt{x}} = \lim_{x\to 0+} \frac{\sqrt{x}(\sqrt{x} - 1)}{\sqrt{x}} = \lim_{x\to 0+} (\sqrt{x} - 1) = -1. \quad \square$$

Figure 2.25

$\lim_{x\to 1+} f(x) = 2$; $\lim_{x\to 1-} f(x) = -1$

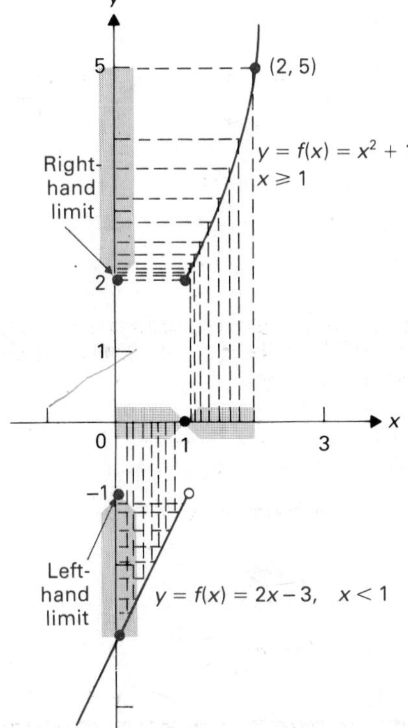

EXAMPLE 2 Let

$$f(x) = \begin{cases} x^2 + 1 & \text{for } x \geq 1, \\ 2x - 3 & \text{for } x < 1. \end{cases}$$

Find $\lim_{x\to 1+} f(x)$.

Solution The graph of $f(x)$ is shown in Fig. 2.25. We are interested in values $f(x)$ for $x > 1$, so we use the formula $x^2 + 1$ to compute the function there. We have

$$\lim_{x\to 1+} f(x) = \lim_{x\to 1+} (x^2 + 1) = 1 + 1 = 2.$$

Note that $\lim_{x\to 1} f(x)$ does not exist. $\square$

With the above two examples as illustration, we give a careful definition of a right-hand limit.

DEFINITION 2.2 Right-hand limit

Suppose the domain of f contains points x arbitrarily close to x_1 but greater than x_1. Then $\lim_{x\to x_1+} f(x) = L$ if, for each $\varepsilon > 0$, there exists $\delta > 0$ such that $|f(x) - L| < \varepsilon$ for any $x \neq x_1$ in the domain of f and between x_1 and $x_1 + \delta$.

Figure 2.26 illustrates Definition 2.2, showing the largest δ that can be chosen corresponding to the given $\varepsilon > 0$.

Of course, we also have the notion of the left-hand limit at x of a function f. The ε,δ-characterization of $\lim_{x\to x_1-} f(x)$ is very similar to that for the right-hand limit in Definition 2.2.

DEFINITION 2.3 Left-hand limit

Suppose the domain of f contains points x arbitrarily close to x_1 but less than x_1. Then $\lim_{x\to x_1-} f(x) = L$ if, for each $\varepsilon > 0$, there exists $\delta > 0$ such that $|f(x) - L| < \varepsilon$ for any $x \neq x_1$ in the domain of f and between $x_1 - \delta$ and x_1.

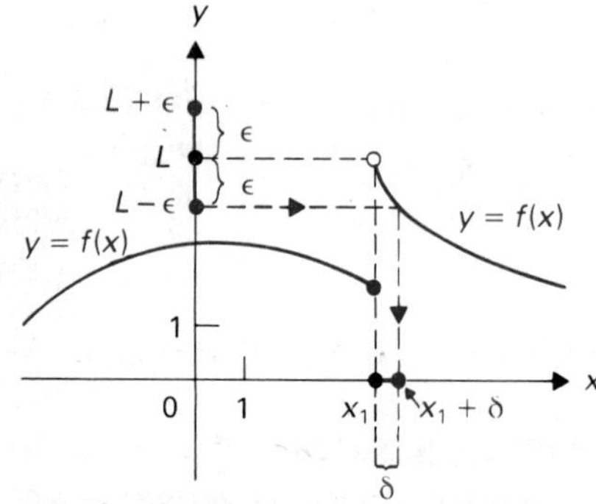

Figure 2.26 For x between x_1 and $x_1 + \delta$, we have $f(x)$ between $L - \varepsilon$ and $L + \varepsilon$.

EXAMPLE 3 Find $\lim_{x\to 1-} f(x)$ for $f(x)$ given in Example 2.

Solution Since we are interested in values $f(x)$ for $x < 1$, we use the formula $2x - 3$ to compute the function there. See Fig. 2.25. We have

$$\lim_{x\to 1-} f(x) = \lim_{x\to 1-} (2x - 3) = -1. \quad \square$$

EXAMPLE 4 Find the left-hand and right-hand limits at $x_1 = 1$ for

$$f(x) = \begin{cases} 2 - x & \text{for } x \geq 1, \\ 2x + 1 & \text{for } x < 1. \end{cases}$$

The graph is shown in Fig. 2.27.

Figure 2.27

$\lim_{x\to 1-} f(x) = 3$; $\lim_{x\to 1+} f(x) = 1$

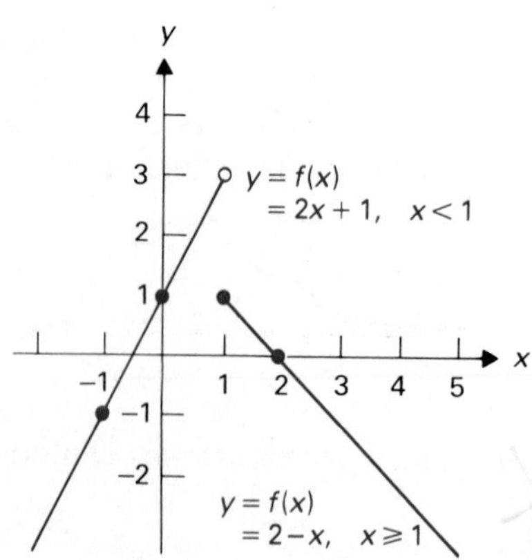

Solution We have

$$\lim_{x\to 1-} f(x) = \lim_{x\to 1-} (2x + 1) = 3,$$

while

$$\lim_{x\to 1+} f(x) = \lim_{x\to 1+} (2 - x) = 1.$$

Of course, the actual value $f(1) = 1$ plays no role in the computation of a limit as $x \to 1$. Note that $\lim_{x\to 1} f(x)$ does not exist, because $f(x)$ does not approach a *single* value as $x \to 1$. $\square$

Let $f(x)$ be defined on both sides of x_1 for values of x arbitrarily close to x_1. As we might guess from Example 4, it is easy to see that

> $\lim_{x\to x_1} f(x)$ exists if and only if $\lim_{x\to x_1+} f(x)$ and $\lim_{x\to x_1-} f(x)$ both exist and are equal.

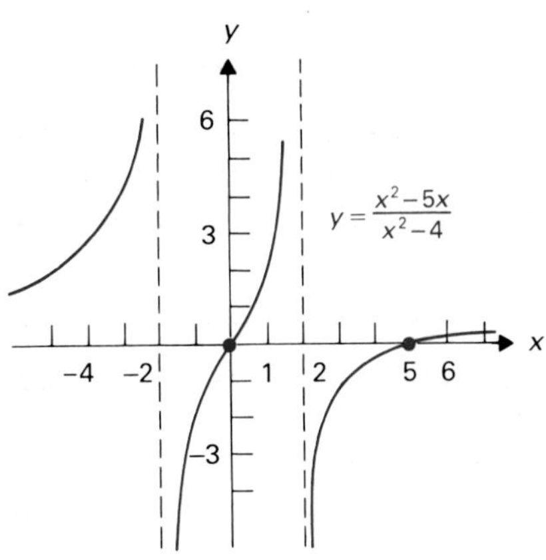

Figure 2.28

$\lim_{x\to 2-}[(x^2-5x)/(x^2-4)] = \infty$

We can use these ideas to show that $m_{\tan}$ does not exist for $f(x) = |x|$ at $x_1 = 0$. Recall from page 42 that

$$m_{\tan} = \lim_{\Delta x\to 0} \frac{f(x_1 + \Delta x) - f(x_1)}{\Delta x}$$

if the limit exists. For $f(x) = |x|$ at $x_1 = 0$, we obtain

$$m_{\tan} = \lim_{\Delta x\to 0} \frac{|0 + \Delta x| - |0|}{\Delta x} = \lim_{\Delta x\to 0} \frac{|\Delta x|}{\Delta x}.$$

Figure 2.20 on page 53 shows that

$$\lim_{\Delta x\to 0+} \frac{|\Delta x|}{\Delta x} = 1, \qquad \text{while} \qquad \lim_{\Delta x\to 0-} \frac{|\Delta x|}{\Delta x} = -1.$$

Thus $\lim_{\Delta x\to 0}(|\Delta x|/\Delta x)$ does not exist, so $m_{\tan}$ does not exist for $|x|$ at $x_1 = 0$.

Just as in Examples 14 through 17 of the preceding section, we can use the symbols ∞ and $-\infty$ to describe the behavior of function values $f(x)$ as x approaches a point x_1 from either the left or the right.

EXAMPLE 5 Find

$$\lim_{x\to 2-} \frac{x^2 - 5x}{x^2 - 4}.$$

The graph is shown in Fig. 2.28.

Solution Here the numerator does not approach zero, but the denominator does. Consequently, the limit will be ∞ or $-\infty$. It is just a question of *sign.* We have $\lim_{x\to 2-}(x^2 - 5x) = -6$, which is negative. On the other hand, for x just a bit less than 2, $x^2 - 4$ is near zero but negative also. Thus the quotient becomes large but is positive, so the limit is ∞. □

EXAMPLE 6 Use right-hand and left-hand limits and the notations ∞ and $-\infty$ to describe the behavior of $f(x) = 1/(x-1)$ as $x \to 1$.

Solution Figure 2.29 shows the graphs of $1/(x-1)$ and $|1/(x-1)|$. We

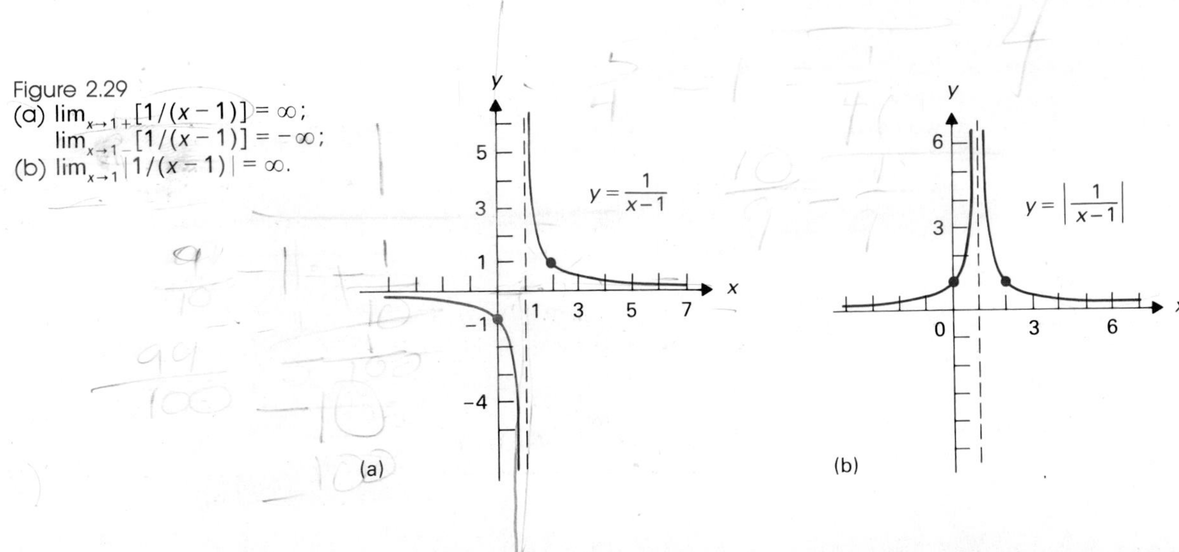

Figure 2.29
(a) $\lim_{x\to 1+}[1/(x-1)] = \infty$;
$\lim_{x\to 1-}[1/(x-1)] = -\infty$;
(b) $\lim_{x\to 1}|1/(x-1)| = \infty$.

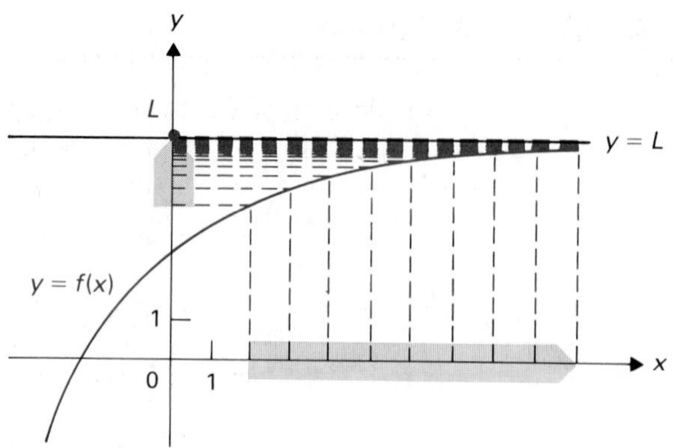

Figure 2.30
$\lim_{x\to\infty} f(x) = L$

easily see that

$$\lim_{x\to 1+} \frac{1}{x-1} = \infty, \qquad \lim_{x\to 1-} \frac{1}{x-1} = -\infty, \qquad \text{and} \qquad \lim_{x\to 1} \left|\frac{1}{x-1}\right| = \infty. \qquad \square$$

LIMITS AT INFINITY

Sometimes we want to know the behavior of $f(x)$ for very large values of x. A natural way to phrase this problem is to discuss the behavior of $f(x)$ "as x approaches ∞." The assertion $\lim_{x\to\infty} f(x) = L$ means that $f(x)$ will be as close to L as we wish if we take *any* sufficiently large x in the domain of f. As shown in Fig. 2.30, the graph of f then must be very close to the horizontal line $y = L$ for large x.

DEFINITION 2.4 Limit at ∞

Let the domain of f contain arbitrarily large values of x. Then $\lim_{x\to\infty} f(x) = L$ if, for each $\varepsilon > 0$, there exists a $K > 0$ such that $|f(x) - L| < \varepsilon$ for any x in the domain of f and greater than K.

Figure 2.31 illustrates Definition 2.4, showing the smallest possible value K corresponding to the given $\varepsilon > 0$ if $\lim_{x\to\infty} f(x) = L$. Of course, K could also be taken to be any larger value than that shown in Fig. 2.31.

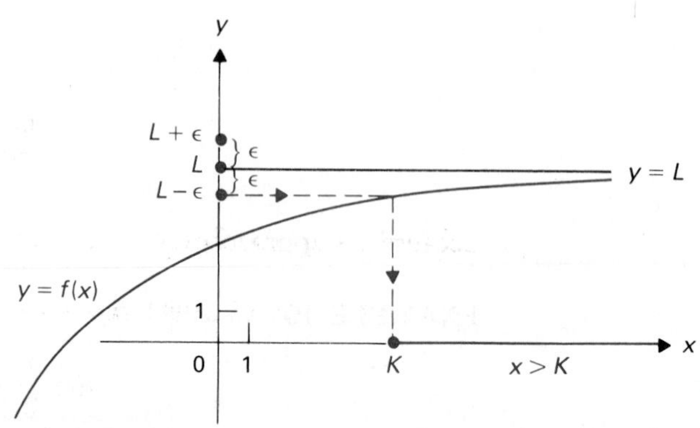

Figure 2.31 For $x > K$, we have $f(x)$ between $L - \varepsilon$ and $L + \varepsilon$.

DEFINITION 2.5 Limit at $-\infty$

Let the domain of f contain values x such that $-x$ is arbitrarily large. Then $\lim_{x\to-\infty} f(x) = L$ if, for each $\varepsilon > 0$, there exists a $K > 0$ such that $|f(x) - L| < \varepsilon$ for any x in the domain of f and less than $-K$.

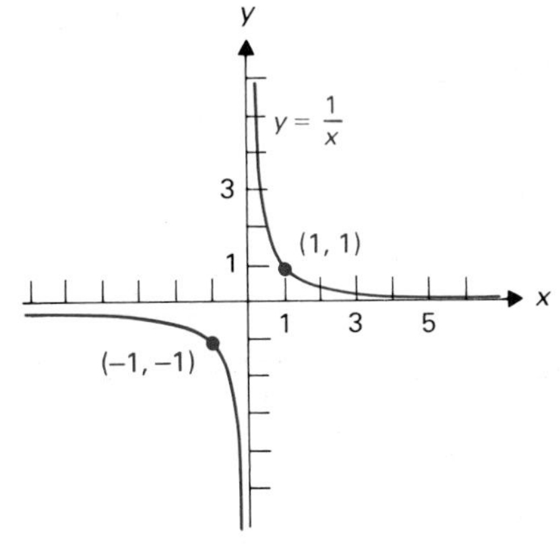

Figure 2.32

$\lim_{x\to\infty}(1/x) = 0$; $\lim_{x\to-\infty}(1/x) = 0$

EXAMPLE 7 Find $\lim_{x\to\infty} (1/x)$ and $\lim_{x\to-\infty} (1/x)$.

Solution The graph of $f(x) = 1/x$ is shown in Fig. 2.32. Clearly we have both

$$\lim_{x\to\infty} \frac{1}{x} = 0 \quad \text{and} \quad \lim_{x\to-\infty} \frac{1}{x} = 0. \quad \square$$

EXAMPLE 8 Find

$$\lim_{x\to\infty} \frac{2x^2 - 3x}{3x^2 + 2}.$$

Solution A nice technique is to factor the monomial terms of highest degree from both the numerator and the denominator. The factorization is not valid when $x = 0$, but we are interested in values of x far from 0. We have

$$\lim_{x\to\infty} \frac{2x^2 - 3x}{3x^2 + 2} = \lim_{x\to\infty} \left(\frac{2x^2}{3x^2} \cdot \frac{1 - \dfrac{3x}{2x^2}}{1 + \dfrac{2}{3x^2}} \right) = \lim_{x\to\infty} \left(\frac{2}{3} \cdot \frac{1 - \dfrac{3}{2x}}{1 + \dfrac{2}{3x^2}} \right)$$

$$= \frac{2}{3} \cdot \frac{1 - 0}{1 + 0} = \frac{2}{3}. \quad \square$$

EXAMPLE 9 Repeat the technique of Example 8 to find

$$\lim_{x\to-\infty} \frac{2x^3 - 3x^2}{2x^2 + 4x - 7}.$$

Solution We have

$$\lim_{x\to-\infty} \frac{2x^3 - 3x^2}{2x^2 + 4x - 7} = \lim_{x\to-\infty} \left(\frac{2x^3}{2x^2} \cdot \frac{1 - \dfrac{3x^2}{2x^3}}{1 + \dfrac{4x}{2x^2} - \dfrac{7}{2x^2}} \right)$$

$$= \lim_{x\to-\infty} \left(\frac{x}{1} \cdot \frac{1 - \dfrac{3}{2x}}{1 + \dfrac{2}{x} - \dfrac{7}{2x^2}} \right) = -\infty,$$

because $x/1$ approaches $-\infty$ and the second factor approaches 1. $\square$

EXAMPLE 10 Use the technique of Example 8 to find

$$\lim_{x\to\infty} \frac{x^2 + 1}{2x - 3x^3}.$$

Solution We have

$$\lim_{x\to\infty}\frac{x^2+1}{2x-3x^3}=\lim_{x\to\infty}\left(\frac{x^2}{-3x^3}\cdot\frac{1+\dfrac{1}{x^2}}{\dfrac{2x}{-3x^3}+1}\right)=\lim_{x\to\infty}\left(\frac{1}{-3x}\cdot\frac{1+\dfrac{1}{x^2}}{\dfrac{-2}{3x^2}+1}\right)=0,$$

because the first factor approaches zero and the second approaches 1. □

The preceding three examples make it clear that the limit of a rational function $f(x)$ as $x \to \infty$ is simply the limit of the quotient of the monomial term of highest degree in the numerator and the one of highest degree in the denominator. We regard these monomial terms of highest degree as *dominating* the other terms of the numerator and denominator as $x \to \infty$ or $x \to -\infty$. We can now simplify the computation of such limits.

EXAMPLE 11 Find

$$\lim_{x\to\infty}\frac{2x^5-3x^2+4x}{8x^3-10x^5}.$$

Solution We have

$$\lim_{x\to\infty}\frac{2x^5-3x^2+4x}{8x^3-10x^5}=\lim_{x\to\infty}\frac{2x^5}{-10x^5}=\frac{2}{-10}=-\frac{1}{5}. \quad □$$

EXAMPLE 12 Find

$$\lim_{x\to\infty}\frac{4x^4+5}{8-3x^3}.$$

Solution We have

$$\lim_{x\to\infty}\frac{4x^4+5}{8-3x^3}=\lim_{x\to\infty}\frac{4x^4}{-3x^3}=\lim_{x\to\infty}\frac{4x}{-3}=-\infty. \quad □$$

APPLICATION TO GRAPHING $f(x) = (ax + b)/(cx + d)$

We have just discussed left-hand and right-hand limits and limits as $x \to \infty$. These limits are useful in graphing rational functions (quotients of polynomial functions). We will discuss graphing in considerable detail in Chapter 5 after we develop some calculus. In this section, we restrict ourselves to graphing a function of the form

$$f(x)=\frac{ax+b}{cx+d}, \qquad \text{where } c \neq 0. \tag{1}$$

If $c = 0$, the graph of the function in Eq. (1) is of course a straight line. Unless the numerator is a multiple of the denominator, the function in Eq. (1) has left-hand and right-hand limits of ∞ or $-\infty$ at $x = -d/c$, where the denominator becomes zero. The vertical line $x = -d/c$ is a **vertical asymptote** of the graph. Also, both

$$\lim_{x\to\infty} \frac{ax+b}{cx+d} = \frac{a}{c} \quad \text{and} \quad \lim_{x\to-\infty} \frac{ax+b}{cx+d} = \frac{a}{c}.$$

The horizontal line $y = a/c$ is a **horizontal asymptote** of the graph. Since $f(x)$ in Eq. (1) is zero if and only if the numerator is zero, we see that $-b/a$ is the **x-intercept** of the graph if $a \neq 0$. If $d \neq 0$, the **y-intercept** is b/d, which is the value obtained when x is zero.

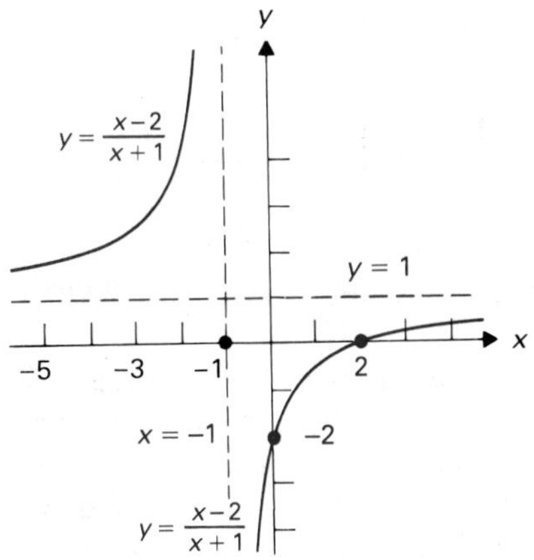

Figure 2.33 Vertical asymptote $x = -1$; horizontal asymptote $y = 1$; x-intercept 2; y-intercept -2; $\lim_{x\to-1+}[(x-2)/(x+1)] = -\infty$; $\lim_{x\to-1-}[(x-2)/(x+1)] = \infty$.

EXAMPLE 13 Sketch $y = (x-2)/(x+1)$, finding the horizontal and vertical asymptotes and the x- and y-intercepts.

Solution The vertical asymptote is the line $x = -1$, since the denominator becomes zero at this value for x. This line is dashed in Fig. 2.33. Since

$$\lim_{x\to\infty} \frac{x-2}{x+1} = \lim_{x\to-\infty} \frac{x-2}{x+1} = 1,$$

the horizontal asymptote is $y = 1$, which is also dashed in Fig. 2.33. The x-intercept is 2, where the numerator is zero. The y-intercept is -2, obtained by setting x equal to zero.

To sketch the graph, we determine the behavior as we approach the vertical asymptote by computing the left-hand and right-hand limits:

$$\lim_{x\to-1-} \frac{x-2}{x+1} = \infty \quad \text{and} \quad \lim_{x\to-1+} \frac{x-2}{x+1} = -\infty.$$

Thus the graph runs up along the asymptote on the left and down along the asymptote at the right. See Fig. 2.33. We can plot a few points to get a reasonably accurate graph. □

EXAMPLE 14 Repeat Example 13 for $y = (-2x+6)/(x+3)$.

Solution Proceeding as in Example 13, we find that our graph has the following properties:

a vertical asymptote $x = -3$ where the denominator is zero,

a horizontal asymptote $y = -2$ since -2 is the limit at ∞ and $-\infty$,

an x-intercept 3 where the numerator is zero,

a y-intercept 2 where x is zero,

$$\lim_{x\to-3+} \frac{-2x+6}{x+3} = \infty, \quad \text{and} \quad \lim_{x\to-3-} \frac{-2x+6}{x+3} = -\infty.$$

The graph is shown in Fig. 2.34. □

Figure 2.34 Vertical asymptote $x = -3$; horizontal asymptote $y = -2$; x-intercept 3; y-intercept 2; $\lim_{x\to-3+}[(-2x+6)/(x+3)] = \infty$; $\lim_{x\to-3-}[(-2x+6)/(x+3)] = -\infty$.

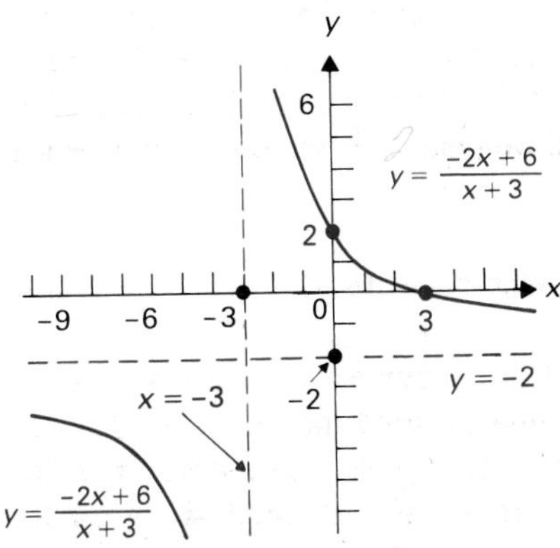

SUMMARY

1. $\lim_{x\to x_1+}$ means the limit as x approaches x_1 from the right-hand side and $\lim_{x\to x_1-}$ means the limit from the left-hand side.
2. The limit as $x\to\infty$ or as $x\to-\infty$ of a rational function is the limit of the quotient of the monomial term of highest degree in the numerator and the monomial term of highest degree in the denominator.

3. For the graph of $y = f(x) = (ax + b)/(cx + d)$, where $c \neq 0$,

a. $x = -d/c$ is the vertical asymptote (unless $ad = bc$),

b. $y = a/c$ is the horizontal asymptote,

c. $-b/a$ is the x-intercept if $a \neq 0$,

d. b/d is the y-intercept if $d \neq 0$,

e. behavior as x approaches the vertical asymptote is determined by the left-hand and right-hand limits there.

EXERCISES

In Exercises 1 through 36, find the indicated limit if it exists. Use the symbols ∞ and $-\infty$ to indicate the behavior near the point where appropriate.

1. $\lim_{x\to 2} \frac{1}{2-x}$ **2.** $\lim_{x\to 2+} \frac{1}{2-x}$

3. $\lim_{x\to 2-} \frac{1}{2-x}$ **4.** $\lim_{t\to 2} \frac{1}{(2-t)^2}$

5. $\lim_{u\to 5+} \frac{u+3}{u^2-25}$ **6.** $\lim_{u\to -5+} \frac{u+3}{u^2-25}$

7. $\lim_{s\to 4+} \frac{s^2+3s+5}{4s-s^2}$ **8.** $\lim_{x\to 4+} \frac{x^2-3x-5}{4x-x^2}$

9. $\lim_{t\to 2-} \frac{t^2-7t+4}{2+t-t^2}$ **10.** $\lim_{x\to 5-} \frac{x^2-25}{x^2-10x+25}$

11. $\lim_{u\to 4+} \frac{4u-u^2}{8u-u^2-16}$ **12.** $\lim_{t\to 2+} \frac{t^2-3t+2}{t^2-4t+4}$

13. $\lim_{x\to 2} \left|\frac{x^2+4}{x-2}\right|$

14. $\lim_{s\to 0+} \left(\frac{3}{s}-\frac{1}{s^2}\right)$

[*Hint:* Write $(3/s)-(1/s^2)$ as a quotient of polynomials.]

15. $\lim_{x\to 0} \left(\frac{1}{x^4}-\frac{1}{x}\right)$ **16.** $\lim_{x\to 0-} \left(\frac{1}{x^3}-\frac{1}{x}\right)$

17. $\lim_{t\to 0-} \left(\frac{t-3}{t^4}+\frac{2}{t^2}\right)$ **18.** $\lim_{x\to 0-} \left(\frac{5+x}{x^2}-\frac{x^2-9}{x^3}\right)$

19. $\lim_{x\to \infty} \frac{x+1}{x}$ **20.** $\lim_{x\to -\infty} \frac{3x^3-2x}{2x^3+3}$

21. $\lim_{t\to -\infty} \frac{|t|}{t}$ **22.** $\lim_{x\to \infty} \frac{x^3+2x}{x^2-3}$

23. $\lim_{x\to \infty} \frac{x^2-2x+1}{x^3+3x-2}$ **24.** $\lim_{x\to -\infty} \frac{x^2-2x+1}{x^3+3x-2}$

25. $\lim_{x\to \infty} \frac{2x^3+4x+2}{8x-5x^2}$ **26.** $\lim_{u\to -\infty} \frac{4u^5-8u^2}{3u^2-8u+2}$

27. $\lim_{t\to \infty} \frac{4t^3-8t^2+3}{7-2t^3}$ **28.** $\lim_{x\to \infty} \frac{4x^{1/2}-2x^{1/3}-2}{3x^{1/3}-5x^{3/4}+5}$

29. $\lim_{x\to -\infty} \frac{8x^{1/3}+4x^{1/5}+3}{5x^{1/9}-7x^{1/5}}$ **30.** $\lim_{x\to \infty} \frac{x^{5/4}-3x^{4/3}+2}{x^{7/6}-3x^{10/9}}$

31. $\lim_{u\to \infty} \frac{u^{2/3}+4u^{5/6}-3}{8u+3u^{5/7}}$ **32.** $\lim_{x\to -\infty} (x^2+3x)$

33. $\lim_{x\to -\infty} (x^3+3x^2)$ **34.** $\lim_{x\to \infty} (x^{1/2}-x^{1/3})$

35. $\lim_{x\to -\infty} (x^{1/5}-x^{1/3})$ **36.** $\lim_{x\to \infty} \left(x-\sqrt{x^2+1}\right)$

37. A rubber ball has the characteristic that, when dropped on a floor from height h, it rebounds to height $h/2$. It can be shown that if the ball is dropped from a height of h ft and allowed to bounce repeatedly, the total distance it has traveled when it hits the floor for the nth time is

$$h + 2h\left[1-\left(\frac{1}{2}\right)^{n-1}\right] \text{ ft.}$$

Find the total distance the ball travels before it stops bouncing if it is dropped from a height of 16 ft.

38. Taking 4000 miles as radius of the earth and 32 ft/sec² as gravitational acceleration at the surface of the earth, and neglecting air resitance, we can show that the velocity v with which a body must be fired upward from the surface of the earth to attain an altitude of s miles is given by the formula

$$v = \frac{8}{\sqrt{5280}}\sqrt{\frac{4000s}{s+(4000)(5280)}} \text{ mi/sec.}$$

Using this formula, find the velocity with which a body must be fired upward to escape the gravitational attraction of the earth.

In Exercises 39 through 48, find the vertical and horizontal asymptotes and the x- and y-intercepts. Sketch the graphs.

39. $y = \frac{1}{x-2}$ **40.** $y = \frac{2}{x+3}$

41. $y = \frac{x-1}{x}$ **42.** $y = \frac{5x+10}{2x}$

43. $y = \frac{2x+4}{x-1}$ **44.** $y = \frac{x-3}{2x+4}$

45. $y = \frac{x-1}{2-x}$ **46.** $y = -\frac{x+4}{2x+2}$

47. $y = \dfrac{2-x}{2x+6}$

48. $y = \dfrac{x}{4-x}$

In Exercises 49 through 54, decide whether the limit exists and estimate its value. Use a calculator to compute the function for at least four values very near the limit point (values of very large magnitude if $x \to \infty$ or $x \to -\infty$). Use radian measure with trigonometric functions.

49. $\lim_{x \to 0+} x^x$

50. $\lim_{x \to 0+} \left(\dfrac{1}{1-x}\right)^{-1/x^2}$

51. $\lim_{x \to 0+} (\cos x)^{1/\tan x}$

52. $\lim_{x \to -\infty} \left(1 + \dfrac{1}{x}\right)^{2x}$

53. $\lim_{x \to \infty} \left(1 + \dfrac{1}{x}\right)^{x}$

54. $\lim_{x \to \infty} \left(1 + \dfrac{1}{x}\right)^{x^2}$

2.4 CONTINUITY

In this section, we will discuss continuous functions. The adjective *continuous* has a meaning in mathematics very close to its meaning in everyday speech. If we say a person talks continuously (or continually), we mean the person talks with no break. Naively stated, a function is *continuous* on an interval if its graph is *unbroken* there and can be sketched without lifting the pencil off the page.

CONTINUOUS FUNCTIONS

In simple terms, a function f is *continuous at a point* x_1 in its domain if $f(x)$ remains close to $f(x_1)$ whenever x is close enough to x_1. That is, moving a small enough distance from x_1 in the domain should result in only a small change in function value. Fig. 2.35 shows the graph of a function continuous at the point x_1 but not continuous at the points x_2, x_3, and x_4 in its domain. For example, in Fig. 2.35 we have $f(x_3) = 6$. If we are at the point x_3 on the x-axis and move any very small nonzero distance to the right, the function value drops at least four units. A graph must not jump or break at x_1 if f is to be continuous at x_1.

An ε,δ-definition of continuity is based very much on the informal discussion just given. The function value $f(x)$ should change only slightly (less than a prescribed ε) if x moves a sufficiently small distance (less than δ) from x_1. Figure 2.36 illustrates the definition.

DEFINITION 2.6 Continuity (ε,δ-version)

A function f is **continuous at a point** x_1 if

1. x_1 is in the domain of f, and
2. for each $\varepsilon > 0$, there exists $\delta > 0$ such that $|f(x) - f(x_1)| < \varepsilon$ if x is in the domain of f and $|x - x_1| < \delta$.

You probably noticed the similarity of Definition 2.6 to Definition 2.1 of limit given on page 52. In Definition 2.6, the value $f(x_1)$ plays the role of the limit L in Definition 2.1.

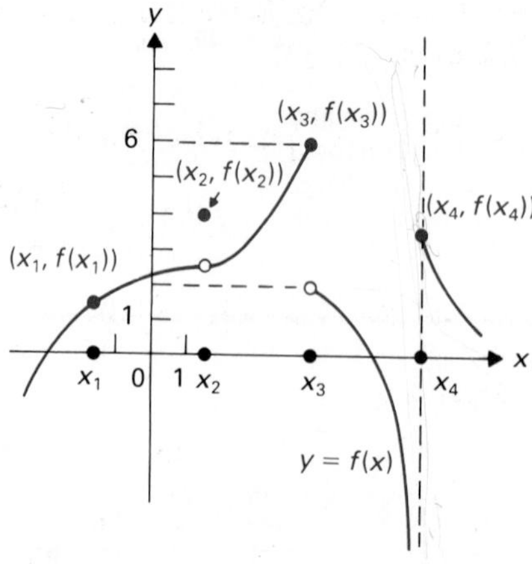

Figure 2.35 f is continuous at x_1 but not continuous at x_2, x_3, or x_4.

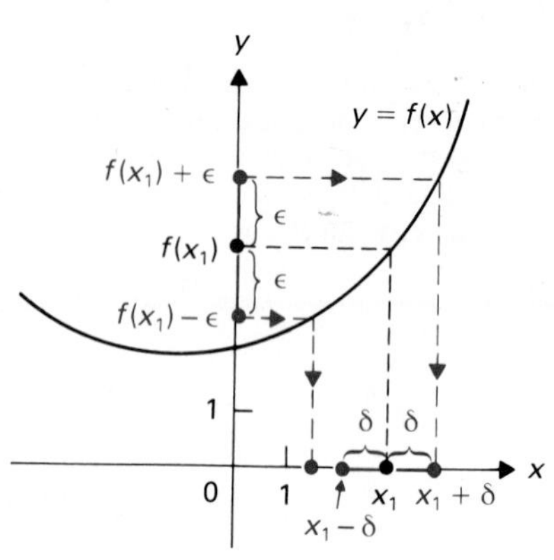

Figure 2.36 If $|x - x_1| < \delta$, then $|f(x) - f(x_1)| < \varepsilon$.

> The value $f(x_1)$ plays an important role in the notion of continuity at x_1, while it plays no role at all in computing the limit at x_1.

Since this chapter is concerned with limits, we will give an alternative definition of continuity in terms of limits and work with this alternative definition. The limit version of the definition is not quite as neat as the ε,δ-version because we can only talk about limits at a point x_1 having other points in the domain of f arbitrarily close to x_1. The footnote to the definition fills this gap.

DEFINITION 2.6 Continuity (limit version)

A function f is **continuous at a point** x_1 if

1. x_1 is in the domain of f,
2. $\lim_{x \to x_1} f(x)$ exists, and
3. $\lim_{x \to x_1} f(x) = f(x_1)$.*

EXAMPLE 1 Determine whether

$$f(x) = \begin{cases} \dfrac{x^2 - 9}{x + 3} & \text{for } x \neq -3, \\ 10 & \text{for } x = -3 \end{cases}$$

is continuous at $x = -3$.

Solution We have

$$\lim_{x \to -3} f(x) = \lim_{x \to -3} \frac{(x - 3)(x + 3)}{x + 3} = -6.$$

* We also define f to be continuous at any point x_1 *in its domain* that *does not* have other points of the domain arbitrarily close to it.

But $f(-3) = 10$. Since $\lim_{x \to -3} f(x) \neq f(-3)$, this function is not continuous at -3. □

EXAMPLE 2 Determine whether

$$g(x) = \begin{cases} x^2 + 2 & \text{for } x > 1, \\ 5x - 1 & \text{for } x \leq 1 \end{cases}$$

is continuous at $x = 1$.

Solution We have

$$\lim_{x \to 1+} g(x) = \lim_{x \to 1+} (x^2 + 2) = 3$$

while

$$\lim_{x \to 1-} g(x) = \lim_{x \to 1-} (5x - 1) = 4.$$

Therefore $\lim_{x \to 1} g(x)$ does not exist, so g is not continuous at $x = 1$. □

EXAMPLE 3 Determine whether

$$h(x) = \begin{cases} \dfrac{x^2 - x - 6}{x - 3} & \text{for } x \neq 3, \\ 5 & \text{for } x = 3 \end{cases}$$

is continuous at $x = 3$.

Solution We have

$$\lim_{x \to 3} h(x) = \lim_{x \to 3} \frac{(x - 3)(x + 2)}{x - 3} = 5 = h(3).$$

Consequently h is continuous at $x = 3$. □

DEFINITION 2.7 Continuous function

A function f is **continuous** if it is continuous at every point in its domain.

Definition 2.6 gave the definition of continuity *at a point*, and Definition 2.7 defines the notion of a function continuous *in its entire domain.*

EXAMPLE 4 Determine whether each of the functions f, g, and h in Examples 1 through 3 is continuous.

Solution The function f in Example 1 is not continuous, since that example shows that f is not continuous at $x = -3$, and -3 is in the domain of f.

The function g in Example 2 is not continuous, since that example shows that g is not continuous at $x = 1$, and 1 is in the domain of g.

We saw in Example 3 that h is continuous at the point 3 in its domain. At all points x other than 3, we have

$$h(x) = \frac{x^2 - x - 6}{x - 3}.$$

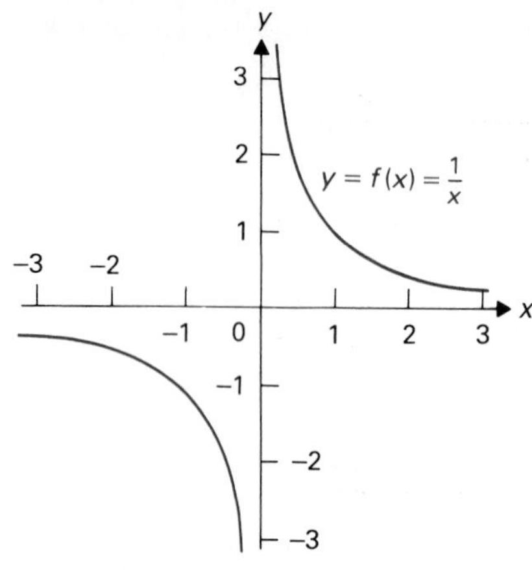

Figure 2.37 The function $f(x) = 1/x$ is continuous at every point in its domain. (Note that 0 is not in the domain.)

The corollary of Theorem 2.1 on page 54 shows that for any point $x_1 \neq 3$, we have

$$\lim_{x \to x_1} h(x) = \lim_{x \to x_1} \frac{x^2 - x - 6}{x - 3} = \frac{x_1^2 - x_1 - 6}{x_1 - 3} = h(x_1),$$

since $(x^2 - x - 6)/(x - 3)$ is a rational function. Thus $h(x)$ is also continuous for all $x \neq 3$. Consequently, it is continuous at all points in its domain and therefore is a continuous function. □

EXAMPLE 5 Show that the function $f(x) = 1/x$ with graph shown in Fig. 2.37 is continuous at every point in its domain.

Solution The domain of f consists of all $x \neq 0$. If $x_1 \neq 0$, then the corollary of Theorem 2.1 on page 54 shows that

$$\lim_{x \to x_1} f(x) = \lim_{x \to x_1} \frac{1}{x} = \frac{1}{x_1} = f(x_1).$$

By Definition 2.6, this means that $f(x) = 1/x$ is continuous at every point x_1 in its domain. □

Looking at Example 5, where $f(x) = 1/x$, we see that it is impossible to define $f(0)$ in a way that would make f continuous at zero. This is the case for any rational function at a point where the denominator is zero, assuming the numerator and denominator have no common factor.

As we might expect, the term *discontinuous* means *not continuous.* Note that we make no classification of a function as continuous or discontinuous at a point not in its domain. The function is not concerned with such points. We now define precisely what we mean when we say that f is *discontinuous at a point* x_1.

DEFINITION 2.8 Discontinuity

A function f is **discontinuous at a point** x_1 if

1. x_1 is in the domain of f, and
2. f is not continuous at x_1.

Note that the function f of Example 1 is discontinuous at $x = -3$, and the function g of Example 2 is discontinuous at $x = 1$. These points are in the domains of the functions, and the functions are not continuous there.

If f and g are both continuous at x_1, then using Theorem 2.1 on page 54, we have

$$\lim_{x \to x_1} (f(x) + g(x)) = \lim_{x \to x_1} f(x) + \lim_{x \to x_1} g(x) = f(x_1) + g(x_1).$$

This shows that $f(x) + g(x)$ is again continuous at x_1. A similar argument can be made for the function $f(x) \cdot g(x)$ and for $f(x)/g(x)$ if $g(x_1) \neq 0$. This gives us the following theorem.

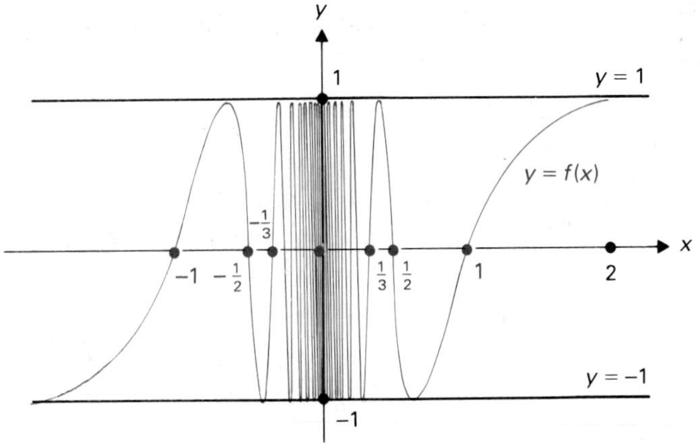

Figure 2.38 $f(0) = 0$. f is not continuous at $x = 0$ since $\lim_{x\to 0} f(x)$ does not exist. Neither the right-hand nor the left-hand limit exists at $x = 0$.

THEOREM 2.2 Properties of continuous functions

Sums, products, and quotients of continuous functions are continuous. (Of course, quotients are not defined wherever denominators are zero.)

The corollary of Theorem 2.1 on page 54 asserts that a limit of a polynomial or rational function at a point in its domain can be computed by simply evaluating the function at the point. This gives us at once a corollary to Theorem 2.2.

COROLLARY Continuity of rational functions

Every polynomial function or rational function is continuous at every point in its domain. (Rational functions are not defined where denominators are zero.)

The function f in Fig. 2.35 is discontinuous at the points x_2, x_3, and x_4 in its domain. The function f in Example 1 is discontinuous at $x = -3$, and the function g in Example 2 is discontinuous at $x = 1$. In each case, the reason for the discontinuity is an abrupt jump, or break, in the graph.

Other types of discontinuities exist. Figure 2.38 shows the graph of a function f such that $f(x) = 0$ at $x = \pm 1, \pm\frac{1}{2}, \pm\frac{1}{3}, \pm\frac{1}{4}, \ldots$. The graph goes alternately up to 1 and down to -1 between the places where it is zero. If we define $f(0) = 0$, then 0 is in the domain of f, but f is not continuous at $x = 0$, since $\lim_{x\to 0} f(x)$ clearly does not exist. Thus f is discontinuous at 0. You may feel that a function such as the one shown in Fig. 2.38 is very contrived and could not be given by a simple formula. However, after we study the function $\sin x$ in Chapter 4, we would have no difficulty in showing that the function

$$f(x) = \begin{cases} \sin(\pi/x) & \text{for } x \neq 0, \\ 0 & \text{for } x = 0 \end{cases}$$

has a graph very similar to that in Fig. 2.38.

Figure 2.39 shows the graph of a function very much like

$$g(x) = \begin{cases} x\sin(\pi/x) & \text{for } x \neq 0, \\ 0 & \text{for } x = 0. \end{cases}$$

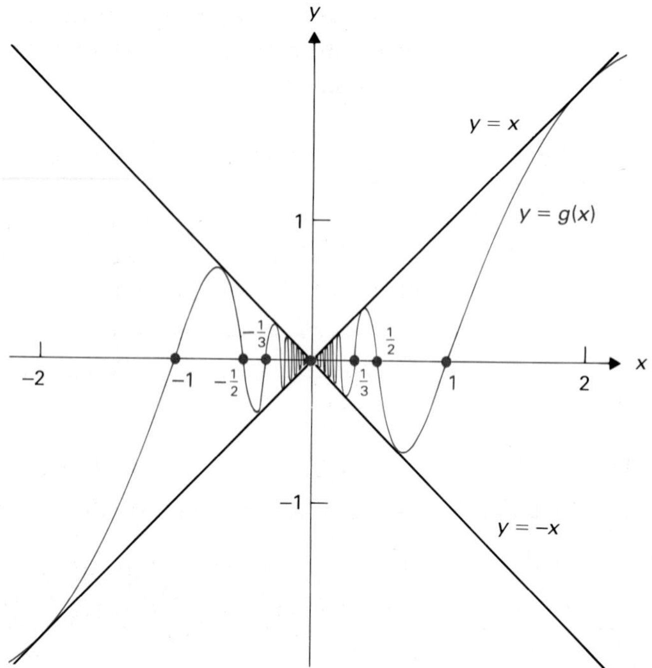

Figure 2.39 $g(0) = 0$. g is continuous at $x = 0$, because $\lim_{x \to 0} g(x) = 0$.

This function g is continuous at $x = 0$ since the oscillations of the graph decrease in height with a limiting height of zero as $x \to 0$.

TWO PROPERTIES OF CONTINUOUS FUNCTIONS

Most functions corresponding to our physical life and activities are continuous. Our height is a continuous function of our age. Our speed, when driving safely, is a continuous function of time. Discontinuities in the physical world are apt to be destructive. For example, suppose we drive into a big tree at a speed of 40 mph. The speed of the leading edge of the front bumper is very nearly a discontinuous function of time at the instant of impact (the tree gives slightly). The speed drops at that instant from 40 mph to 0 mph. The speed of parts of the car farther to the rear decreases more slowly. Air bags attempt to make the driver's speed continuous in such an accident as well as to spread evenly over the upper body the force from the rapid drop in speed.

In Theorems 2.3 and 2.4, which follow, we give two mathematical properties of continuous functions that will be useful in Chapter 5. We will not prove either theorem; careful proofs depend on an important property of the real numbers that is covered in advanced calculus.

Let f be a function continuous at each point of a closed interval $[a, b]$. The graph of such a function f is shown in Fig. 2.40. *Intuitively, the continuity of f on $[a, b]$ means that the graph of f over $[a, b]$ can be sketched with a pencil without taking the pencil off the paper.* This geometric interpretation of continuity of f on $[a, b]$ makes Theorem 2.3 seem very reasonable.

Figure 2.40 f is continuous on $[a, b]$, so its graph can be traced over $[a, b]$ without taking the pencil off the paper.

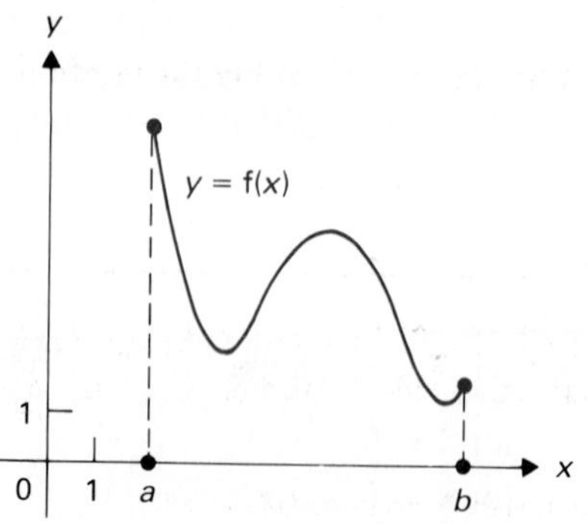

Figure 2.41 shows the graph of a continuous function defined for all x in a closed interval $[a, b]$. The graph is higher at b than at a, that is, $f(a) < f(b)$. Choose any point c between $f(a)$ and $f(b)$ on the y-axis. As the figure indicates, there must be a point x_0 on the x-axis such that $f(x_0) = c$. Think of the horizontal line $y = c$ as a road. The point $(a, f(a))$ is on one side of the

road, and $(b, f(b))$ is on the other side. We can't move continuously from one side of the road to the other without crossing the road somewhere, at some point (x_0, c).

THEOREM 2.3 Weierstrass intermediate-value theorem*

Let f be a continuous function whose domain includes the closed interval $[a, b]$. If c is a number between $f(a)$ and $f(b)$, then there exists some point x_0 in $[a, b]$ such that $f(x_0) = c$.

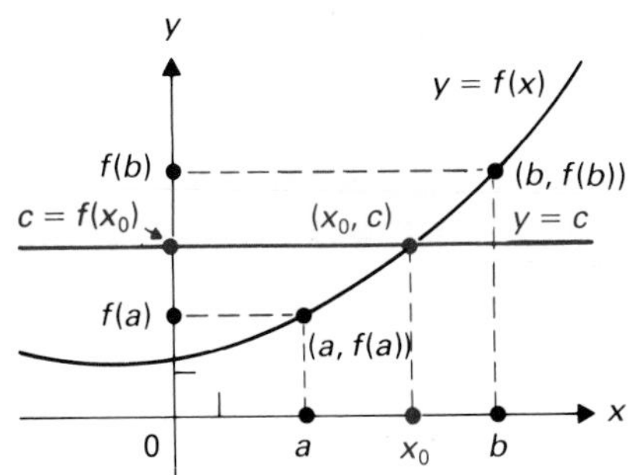

Figure 2.41 The graph of continuous f crosses the line $y = c$ to get from $(a, f(a))$ to $(b, f(b))$.

Figure 2.42 shows that the conclusion of Theorem 2.3 need not hold if the function f is not continuous. The hypothesis of continuity is essential.

EXAMPLE 6 A sapling 12 in. high is planted and grows into a tree 32 ft tall. Show that at some instant, the tree must be exactly 21.357 ft tall.

Solution The height of the tree is a continuous function of time. The tree grows from 1 ft at some time $t = a$ to 32 ft at a later time $t = b$, and 21.357 is between 1 and 32. Theorem 2.3 shows that the tree is 21.357 ft tall at some time t_0. □

Example 6 reinforces our feeling that Theorem 2.3 is intuitively obvious. Example 7 is a bit more sophisticated, and still less obvious applications are given in the exercises.

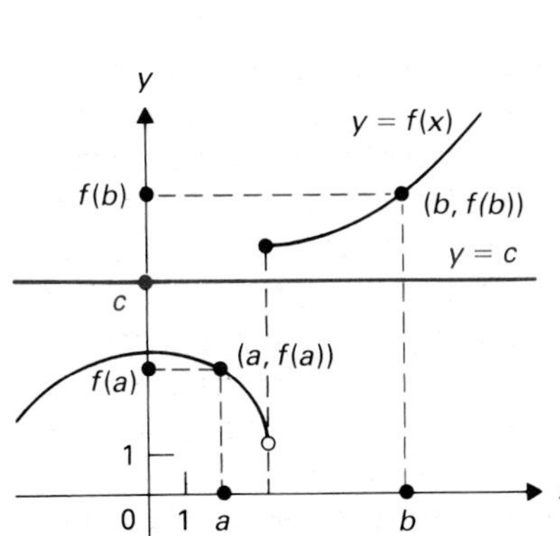

Figure 2.42 The noncontinuous graph need not intersect the line $y = c$ to get from $(a, f(a))$ to $(b, f(b))$.

EXAMPLE 7 Bill Jones has his wife Mary's car wired so that a loud buzzer is activated for 5 seconds each time the speed of her car is exactly 45 mph. Mary has just 35 minutes to drive her car 28 miles to the airport to meet Bill's plane. Show that she can't get there in that time without being buzzed.

Solution To drive 28 mi in 35 min, Mary must average

$$\frac{28}{35/60} = 28 \cdot \frac{12}{7} = 48 \text{ mph}.$$

She can't average 48 mph, starting with speed 0 mph at time $t = a$, without going a bit more than 48 mph at some time $t = b$. Since her speed is a continuous function of time, she will get buzzed at some time $t = c$ between $t = a$ and $t = b$, according to Theorem 2.3. Actually, she will be buzzed at least once more, between time $t = b$ and the time she stops at the airport. □

Our second property of continuous functions concerns the assumption of maximum and minimum values.

DEFINITION 2.9 Extrema

Let S be a subset of the domain of a function f. If x_1 is in S, and $f(x_1) \geq f(x)$ for all x in S, then $M = f(x_1)$ is the **maximum value assumed** by f on S.

* Named for the brilliant German mathematician Karl Weierstrass (1815–1897).

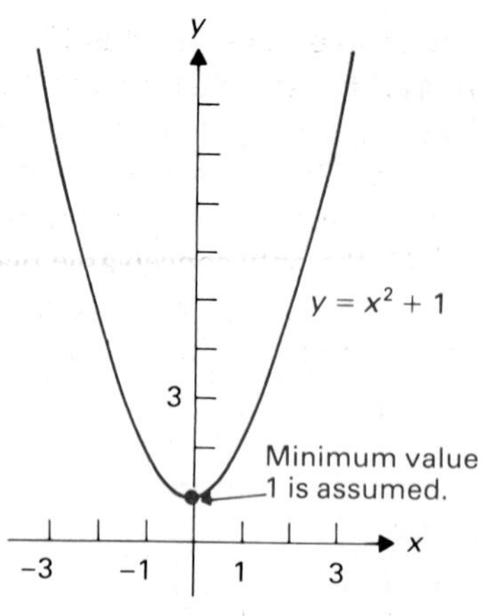

Figure 2.43 $f(x) = x^2 + 1$ assumes the minimum value of 1 where $x = 0$.

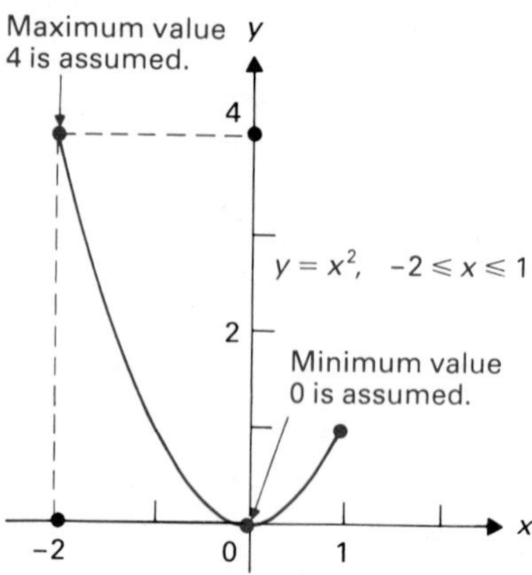

Figure 2.44 $f(x) = x^2$ assumes a minimum value of 0 where $x = 0$ and a maximum value of 4 where $x = -2$ over $[-2, 1]$.

Similarly, if x_2 is in S, and $f(x_2) \leq f(x)$ for all x in S, then $m = f(x_2)$ is the **minimum value assumed** by f on S.

From Fig. 2.43, we see that 1 is the minimum value assumed by the function $x^2 + 1$ on its entire domain. As shown in Fig. 2.44, the minimum value assumed by x^2 on $[-2, 1]$ is 0, and the maximum value assumed is 4.

THEOREM 2.4 Assumption of extreme values

A continuous function assumes both a maximum value M and a minimum value m on any closed interval in its domain.

Figure 2.45 shows that the function f must be continuous or the conclusion of Theorem 2.4 need not be true. The only possible point where a

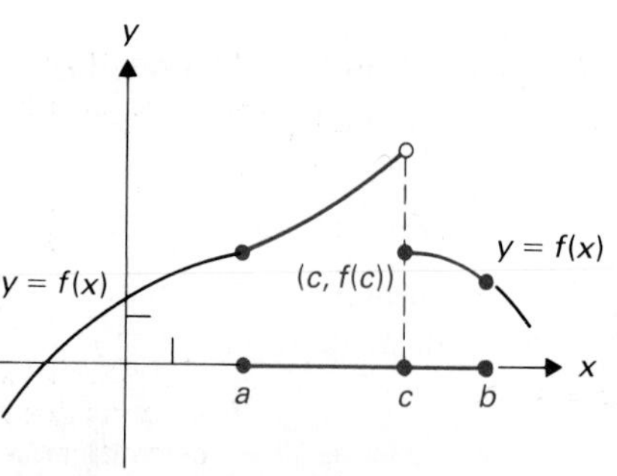

Figure 2.45 $f(x)$ assumes no maximum value over $[a, b]$.

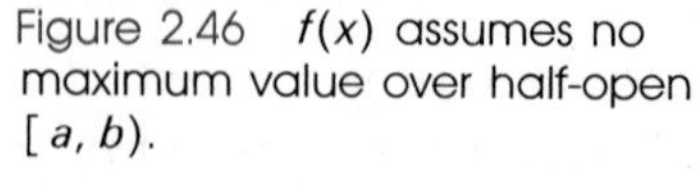

Figure 2.46 $f(x)$ assumes no maximum value over half-open $[a, b)$.

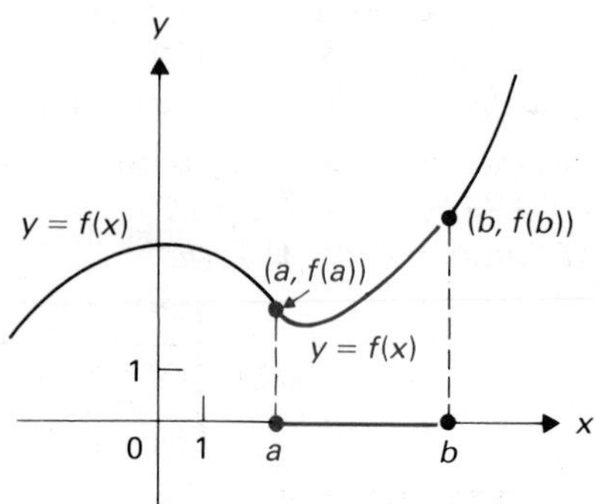

maximum could be assumed in Fig. 2.45 is at the "first point on the x-axis to the left of c." But there is no such point, for if d is to the left of c, then $(c + d)/2$ is halfway between c and d and thus closer to c than d is.

Figure 2.46 shows that the interval must be closed or the conclusion of Theorem 2.4 need not hold. For the function f on the half-open interval $[a, b)$ in Fig. 2.45, the only conceivable place a maximum value could be assumed would be at the "first point on the x-axis to the left of b." As explained in the preceding paragraph, there is no such point.

SUMMARY

1. Let x_1 be a point in the domain of a function f. The function f is continuous at x_1 if $\lim_{x \to x_1} f(x)$ exists and is $f(x_1)$. If f is not continuous at x_1, then it is discontinuous at x_1. A continuous function is one that is continuous at every point in its domain.
2. We make no classification of a function being continuous or discontinuous at a point x_1 unless x_1 is in the domain of f.
3. Sums, products, and quotients of continuous functions are continuous functions. (Quotients are not defined where denominators are zero.) In particular, a rational function is continuous at every point in its domain.
4. Intermediate-value theorem: A continuous function defined on a closed interval $[a, b]$ assumes each value c between $f(a)$ and $f(b)$ at least once in the interval $[a, b]$.
5. A function continuous at each point of a closed interval in its domain assumes a maximum value and also a minimum value on that interval.

EXERCISES

In Exercises 1 through 5, draw the graph of a function f, defined for all x, that satisfies the condition.

1. Continuous at all points but 2, and $\lim_{x \to 2} f(x) = 3$
2. Continuous at all points but 2, and $\lim_{x \to 2} f(x)$ undefined.
3. Not continuous at -1 with $\lim_{x \to -1} f(x) = 1$, but continuous elsewhere with $\lim_{x \to 1} f(x) = 2$
4. Discontinuous at all integer values, but continuous elsewhere
5. Discontinuous at all positive integer values, but continuous elsewhere with $\lim_{x \to 0} f(x) = -2$
6. Is the function f defined by
$$f(x) = \begin{cases} \dfrac{x^2 - 9}{x - 3} & \text{for } x \neq 3, \\ 6 & \text{for } x = 3 \end{cases}$$
continuous? Why?
7. Is the function f defined by
$$f(x) = \begin{cases} \dfrac{4x^2 - 2x^3}{x - 2} & \text{for } x \neq 2, \\ 8 & \text{for } x = 2 \end{cases}$$
continuous? Why?
8. Is the function f defined by
$$f(x) = \begin{cases} x^2 & \text{for } x \geq 2, \\ 8 - 3x & \text{for } -1 \leq x < 2, \\ 12x + 1 & \text{for } x < -1 \end{cases}$$
a) continuous at $x_1 = 2$?
b) continuous at $x_1 = -1$?
c) continuous? Why?

9. Is the function f defined by

$$f(x) = \begin{cases} 3x - x^2 & \text{for } x \geq 3, \\ x - 3 & \text{for } 1 < x < 3, \\ x^2 + 4x - 7 & \text{for } x \leq 1 \end{cases}$$

a) continuous at $x_1 = 3$? b) continuous at $x_1 = 1$?
c) continuous? Why?

10. Alice was 20 in. long when born and grew to a height of 69 in. Use the intermediate-value theorem to argue that at some time in Alice's life, she was exactly 4 ft tall.

11. Give three more everyday applications of the intermediate-value theorem, like the illustration in Exercise 10.

12. Use the intermediate-value theorem to show that on August 4, at some point on the 37° meridian of the earth, there must be exactly 10 hours of daylight. We define daylight to mean that some portion of the sun is above the horizon. A meridian runs from the North Pole to the South Pole and is numbered according to the degrees of longitude, measured from the prime meridian of 0° through Greenwich, England.

13. Prove the following easy corollary of the intermediate-value theorem.

COROLLARY Solution of $f(x) = 0$

Let f be continuous for all x in $[a, b]$ and suppose that $f(a)$ and $f(b)$ have opposite sign. Then the equation $f(x) = 0$ has at least one solution in $[a, b]$.

Exercises 14 through 22 work with the corollary of the intermediate-value theorem stated in Exercise 13.

14. a) Show that $x^3 - 5x^2 + 2x + 6 = 0$ has at least one solution in $[-1, 5]$.
b) Show that the equation in part (a) actually has three solutions in $[-1, 5]$.

15. a) Show that $x^4 + 3x^3 + x + 4 = 0$ has at least one solution in $[-2, 0]$.
b) Does the equation in part (a) have any solution for $x \geq 0$? Explain.

16. Let $f(x) = x^2 - 4x + 3$.
a) Show that $f(0) > 0$ and $f(5) > 0$.
b) Show that $f(x) = 0$ has a solution in $[0, 5]$.
c) Why don't parts (a) and (b) constitute a contradiction of the corollary in Exercise 13?

17. Does $x^2 - 4x + 5 = 0$ have any solution in $[-4, 4]$? Why or why not?

18. Does $-x^2 + 6x - 9 = 0$ have any solution in $[0, 5]$? Why or why not?

19. Let $f(x)$ be continuous for all x. Mark each of the following true or false.

______ a) If $f(a) = f(b)$, then $f(x) = 0$ has no solution in $[a, b]$.
______ b) If $f(a)$ and $f(b)$ have opposite sign, then $f(x) = 0$ must have exactly one solution in $[a, b]$.
______ c) If $f(a)$ and $f(b)$ have opposite sign, then $f(x) = 0$ has at least one solution in $[a, b]$.
______ d) If $f(a)$ and $f(b)$ have the same sign, then $f(x) = 0$ must have either no solution in $[a, b]$ or an even number of solutions in $[a, b]$.
______ e) If $f(a)$ and $f(b)$ have the same sign, then $f(x) = 0$ may have an infinite number of solutions in $[a, b]$.

20. Let f be a polynomial function of odd degree, so that $f(x) = a_n x^n + \cdots + a_1 x + a_0$, where $a_n \neq 0$ and n is an odd integer. Show that $f(x) = 0$ has at least one real solution. [*Hint:* Consider $\lim_{x\to\infty} f(x)$ and $\lim_{x\to-\infty} f(x)$, and then apply the corollary in Exercise 13 to a sufficiently large interval $[-C, C]$.]

21. A car on an oval racetrack passes the flag at the end of the third lap going exactly 96 mph. At the end of the fourth lap, the car is again going exactly 96 mph when passing the flag. Use the corollary in Exercise 13 to show that during the fourth lap, there were two diametrically opposite points of the oval where the car had equal speeds (not necessarily 96 mph, of course). [*Hint:* Let S be the length of the oval track. For $0 \leq x \leq S/2$, let

$$f(x) = \text{Speed at } x - \text{Speed at } \left(x + \frac{S}{2}\right),$$

where x is the distance the car has traveled past the flag during the fourth lap.]

22. A square table with four legs of equal length teeters on diagonally opposite legs when placed on a warped floor. Show, using the corollary in Exercise 13, that by rotating the table less than a quarter turn, you can make all four legs touch the floor, so that the table won't wobble anymore. [*Hint:* Number the legs 1, 2, 3, 4 in a counterclockwise order. Let $f(\theta)$ equal the sum of the distances from legs 1 and 3 to the floor minus the sum of the distances from legs 2 and 4 to the floor, when the table has been rotated counterclockwise through the angle θ degrees for $0 \leq \theta \leq 90$.]

23. Let $f(x) = a_n x^n + a_{n-1}x^{n-1} + \cdots + a_1 x + a_0$, where n is an even positive integer and $a_n > 0$.
a) Find $\lim_{x\to\infty} f(x)$ and $\lim_{x\to-\infty} f(x)$. (Remember, you know how to easily find the limit at infinity of any rational function.)
b) Argue from part (a) that there exists $C > 0$ such that $f(x) > f(0)$ for $|x| > C$.
c) Apply Theorem 2.4 to $[-C, C]$ and show that $f(x)$ assumes a minimum value over the entire x-axis.

24. Give the argument and conclusion analogous to Exercise 23 in the case that $a_n < 0$, assuming the other hypotheses remain the same.

EXERCISE SETS FOR CHAPTER 2

Review Exercise Set 2.1

1. Let $f(x) = 1/x$. Estimate the slope $m_{\tan}$ of the graph where $x = 1$ by finding the slope $m_{\sec}$ through the points where $x = 1$ and where $x = 1 + \frac{1}{2}$.

In Exercises 2 through 7, find the limit, using the notations ∞ and $-\infty$ where appropriate.

2. $\displaystyle\lim_{x\to1} \frac{x^2 - 1}{x - 1}$

3. $\displaystyle\lim_{x\to-2} \frac{x + 1}{(x + 2)^2}$

4. $\lim_{x\to 4} \dfrac{x^2-3x-4}{x^2-16}$

5. $\lim_{x\to -1} \dfrac{x^2+2x+1}{x^2-2x}$

6. $\lim_{x\to \infty} \dfrac{x^4-3x^2}{2x-3x^4}$

7. $\lim_{x\to -\infty} \dfrac{x^4+100x^2}{14-x}$

8. Sketch the graph of $f(x) = 1/(2-x)$.

9. Let

$$f(x) = \begin{cases} \dfrac{x^2-9}{x+3}, & \text{if } x \neq -3 \\ 6, & \text{if } x = -3. \end{cases}$$

Is f a continuous function? If so, why? If not, why not?

10. State the intermediate-value theorem.

Review Exercise Set 2.2

1. Find an expression for $m_{\tan}$ at any point x_1 if $f(x) = 1/(2x+1)$.

In Exercises 2 through 7, find the limit, using the notations ∞ and $-\infty$ where appropriate.

2. $\lim_{x\to 0} \left|\dfrac{x^2-3}{x}\right|$

3. $\lim_{x\to 2} \dfrac{x^2-3x+2}{x^2-4}$

4. $\lim_{x\to 5^-} \dfrac{|x-5|}{x^2-25}$

5. $\lim_{x\to 3} \dfrac{x^3-27}{x^2+9}$

6. $\lim_{x\to \infty} \dfrac{7-5x^2}{x^3+3x}$

7. $\lim_{x\to -\infty} \dfrac{14x^3-7x^2}{8x^3+4x}$

8. Sketch the graph of $f(x) = (2x+6)/x$.

9. Let

$$f(x) = \begin{cases} \dfrac{x^2-4x-5}{x-5}, & \text{if } x > 5, \\ 2x-4, & \text{if } x \leq 5. \end{cases}$$

a) Find $\lim_{x\to 5^-} f(x)$.
b) Find $\lim_{x\to 5^+} f(x)$.
c) Is f continuous at $x = 5$? Why?
d) Is f a continuous function? Why?

10. Show that $x^4 - 5x + 1 = 0$ has a solution in $[0, 1]$.

Review Exercise Set 2.3

1. Find $m_{\tan}$ at any point x_1 if $f(x) = \sqrt{2x-3}$.
2. Compute $m_{\tan}$ where $x_1 = 2$, and then find the equation of the tangent line to the graph of $f(x) = x^2 - 3x$ at the point $(2, -2)$.

In Exercises 3 through 7, find the limit, using the notations ∞ and $-\infty$ where appropriate.

3. a) $\lim_{x\to 2} \dfrac{x^2-4}{|x-2|}$ b) $\lim_{x\to 2} \dfrac{x^2-4}{|x+2|}$

4. $\lim_{x\to -3} \dfrac{x^2+3x}{x^2+2x-3}$

5. $\lim_{x\to -1^+} \dfrac{x^2+x}{x^2+2x+1}$

6. $\lim_{x\to \infty} \dfrac{7-3x^2}{4x^2+3x-2}$

7. $\lim_{x\to -\infty} \dfrac{8x+4x^4}{3x^3-7}$

8. Sketch the graph of $f(x) = (3x-6)/(x+2)$.

9. Is $f(x) = 1/x$ a continuous function? Why or why not?
10. State the theorem on the assumption of maximum and minimum values.

Review Exercise Set 2.4

1. Find $m_{\tan}$ for $f(x) = \sqrt{x} + x$ at any point $x_1 > 0$.

In Exercises 2 through 6, find the limit, using the notations ∞ and $-\infty$ where appropriate.

2. $\lim_{x\to 1} \dfrac{x^2-3x+2}{x+1}$

3. $\lim_{x\to -2} \dfrac{x^3+2x^2}{x^2-x-6}$

4. $\lim_{x\to 3^+} \dfrac{x^2+x-14}{x^2-8x+15}$

5. $\lim_{x\to -2^-} \dfrac{4+x^2}{4-x^2}$

6. $\lim_{x\to \infty} \dfrac{4x^{2/3}-7x}{x^{1/2}+x^{5/4}-4}$

7. Sketch the graph of $f(x) = (x+1)/(2-x)$.
8. Give an ε,δ-argument that if $f(x)$ is continuous at $x_1 = 2$ and $f(-x) = f(x)$ for all x, then $f(x)$ is continuous at -2 also.
9. Let

$$f(x) = \begin{cases} \dfrac{|x|+x}{x+1} & \text{for } x \neq -1, \\ 1 & \text{for } x = -1. \end{cases}$$

Is f continuous at $x_1 = -1$? Why or why not?

10. Does $f(x) = x^2$ assume
a) a maximum value on half-open $(0, 1]$? Why?
b) a minimum value on half-open $(0, 1]$? Why?

More Challenging Exercises 2

We will see later that there is a continuous function $\sin x$ such that

$$\lim_{x\to 0} \frac{\sin x}{x} = 1.$$

Use this information to compute the limits, if they exist, in Exercises 1 through 10.

1. $\lim_{\Delta x\to 0} \dfrac{\sin \Delta x}{\Delta x}$

2. $\lim_{\Delta x\to 0} \dfrac{\sin \Delta x}{|\Delta x|}$

3. $\lim_{x\to 0} \dfrac{\sin 2x}{x}$

4. $\lim_{x\to 0} \dfrac{\sin 2x}{\sin 3x}$

5. $\lim_{x\to \infty} \sin \dfrac{1}{x}$

6. $\lim_{x\to \infty} \left(x \sin \dfrac{1}{x}\right)$

7. $\lim_{x\to \infty} \left(x^2 \sin \dfrac{1}{x}\right)$

8. $\lim_{x\to \infty} \left(x \sin \dfrac{1}{x^2}\right)$

9. $\lim_{x\to \infty} \left(x^2 \sin \dfrac{1}{x^2}\right)$

10. $\lim_{x\to \infty} \left(x^3 \sin \dfrac{1}{x^2}\right)$

11. Students often have trouble understanding the ε,δ-definition of a limit. We are convinced that this difficulty is rooted in logic. The definition uses both the *universal quantifier* ("for each") and the *existential quantifier* ("there exists"),

since the definition states that *for each* $\varepsilon > 0$, *there exists* $\delta > 0 \ldots$ This exercise deals with this logical problem.

a) Negate the statement

For each $\varepsilon > 0$, there exists $\delta > 0$.

That is, write a statement synonymous with "It is not true that for each $\varepsilon > 0$, there exists $\delta > 0$," without just saying, "It is not true that. . . ."

b) Negate the statement

For each apple blossom, there exists an apple.

c) Study your answer to part (a) in light of your answer to part (b), and decide whether your answer to part (a) is correct.

d) Describe what one must do to show that $\lim_{x \to a} f(x) \neq c$.

12. Let f be a function. Classify each of the following "definitions" of the limit of f at a as either correct or incorrect. If it is incorrect, modify it to become correct.

a) The limit of f at a is c if for each $\varepsilon > 0$, there exists $\delta > 0$ such that $|f(x) - c| < \varepsilon$, provided that $|x - a| < \delta$.

b) The limit of f at a is c if for some $\varepsilon > 0$, there exists $\delta > 0$ such that $|f(x) - c| < \varepsilon$, provided that $0 < |x - a| < \delta$.

c) The limit of f at a is c if for each $\varepsilon > 0$, there exists $\delta > 0$ such that $0 < |x - a| < \delta$ implies that $|f(x) - c| < \varepsilon$.

d) The limit of f at a is c if there exist positive numbers ε and δ such that $0 < |x - a| < \delta$ implies that $|f(x) - c| < \varepsilon$.

e) The limit of f at a is c if for $\delta > 0$, $0 < |x - a| < \delta$ implies $|f(x) - c| < \varepsilon$.

f) The limit of f at a is c if $|f(x) - c|$ can be made smaller than any preassigned $\varepsilon > 0$ by restricting x to elements different from a in some small interval $[a - \delta, a + \delta]$.

g) The limit of f at a is c if for each positive integer n, there exists $\delta > 0$ such that

$$|f(x) - c| < \frac{1}{n},$$

provided that $0 < |x - a| < \delta$.

13. Give an ε,δ-proof that if $\lim_{x \to a} f(x) = L$ and $\lim_{x \to a} g(x) = M$, then $\lim_{x \to a} (f(x) + g(x)) = L + M$.

14. Give an example of a function f discontinuous at $x_1 = 1/n$ for every positive integer n but continuous at every other real number. [*Hint:* Be sure f is continuous at $x_1 = 0$.]

15. Classify each of the following "definitions" as correct or incorrect. If it is incorrect, change it to be correct.

a) A function f is continuous at a if for each $\varepsilon > 0$, there exists $\delta > 0$ such that $|x - a| < \delta$ implies that $|f(x) - a| < \varepsilon$.

b) A function f is continuous at a if for each $\varepsilon > 0$, there exists $\delta > 0$ such that $|f(x) - c| < \varepsilon$, provided that $0 < |x - a| < \delta$.

c) A function f is continuous at a if for each $\varepsilon > 0$, there exists $\delta > 0$ such that

$$|f(x) - f(a)| < \varepsilon,$$

provided that $|x - a| < \delta$.

d) Let f be a function. The limit of f at a is $-\infty$ if for some γ, there exists $\delta > 0$ such that $f(x) < \gamma$, provided that $0 < |x - a| < \delta$.

e) Let f be a function. The limit of f as x approaches a from the left is $-\infty$ if for each γ, there exists $\delta > 0$ such that $f(x) < \gamma$, provided that $a - \delta < x \leq a$.

f) A function f is continuous if for each a in its domain and each positive integer n, there exists a positive integer m such that $|f(x) - f(a)| < 1/n$, provided that $|x - a| < 1/m$.

16. Give an example of two functions defined for all x, neither of which is continuous at $x = 2$ but whose sum is continuous at $x = 2$.

17. Repeat Exercise 16 but for the *product* of the functions rather than the sum.

18. Give an example of a function defined for all real numbers x but not continuous at any point.

3 THE DERIVATIVE

This chapter marks the real beginning of our study of calculus and the use of calculus terminology and notation. The importance of calculus lies in its usefulness in studying *dynamic* situations, where quantities are changing, as opposed to *static* situations, where quantities remain constant. Section 2.1 introduced calculus with computations of some instantaneous rates of change.

In Section 3.1, we will review briefly the notions of average and instantaneous rates of change from Section 2.1 and introduce some notation and terminology used in differential calculus. Much of the chapter is devoted to computation of derivatives. We will introduce several formulas that often permit the computation of a derivative without actually working with limits. (However, the formulas themselves will be derived by computing limits.)

3.1 THE DERIVATIVE; DIFFERENTIATION OF POLYNOMIAL FUNCTIONS

THE DERIVATIVE OF A FUNCTION

The derivative is the basic notion of differential calculus. We mentioned the derivative in Section 2.1, as motivation for studying limits. Now we give a more detailed treatment of it.

Let f be defined at x_1 and for at least a short distance on both sides of x_1, say from $x_1 - h$ to $x_1 + h$ for some $h > 0$. The change Δy in $f(x)$ as x changes from x_1 to $x_1 + \Delta x$ is $f(x_1 + \Delta x) - f(x_1)$, as shown in Fig. 3.1. Figure 3.1 also shows the secant line to the graph, having slope

$$m_{\text{sec}} = \frac{\Delta y}{\Delta x}.$$

As shown in Fig. 3.2, the slope $m_{\tan}$ of the tangent line to the graph where $x = x_1$ is given by

$$m_{\tan} = \lim_{\Delta x \to 0} \frac{\Delta y}{\Delta x},$$

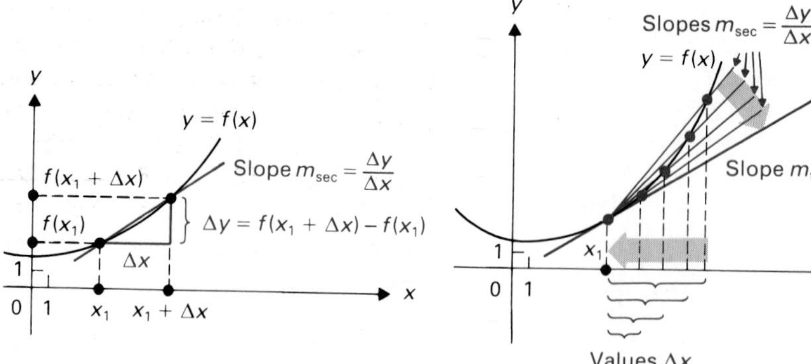

Figure 3.1

$m_{\text{sec}} = \Delta y/\Delta x$

Figure 3.2

$m_{\tan} = \lim_{\Delta x \to 0} \Delta y/\Delta x$

if this limit exists. We will not be using the notations m_{sec} and $m_{\tan}$ from now on. The following two definitions give the usual calculus terminology and notations for these terms.

DEFINITION 3.1 Difference quotient

The **difference quotient** of f from x_1 to $x_1 + \Delta x$ is

$$\frac{\Delta y}{\Delta x} = \frac{f(x_1 + \Delta x) - f(x_1)}{\Delta x}. \qquad \textbf{(1)}$$

It is the **average rate of change** of $f(x)$ with respect to x from x_1 to $x_1 + \Delta x$.

DEFINITION 3.2 Derivative; differentiable function

The **derivative** of f at x_1 is

$$f'(x_1) = \lim_{\Delta x \to 0} \frac{f(x_1 + \Delta x) - f(x_1)}{\Delta x}, \qquad \textbf{(2)}$$

if this limit exists. It is the **instantaneous rate of change** of $f(x)$ with respect to x at x_1. If $f'(x_1)$ exists, then f **is differentiable** at x_1. A **differentiable function** is one that is differentiable at every point x_1 in its domain.

The function f' in the notation $f'(x_1)$ is the *derived function*, and $f'(x)$ is the derivative of f at any point x where the derivative exists.

> The derivative $f'(x)$ is often written as
> $$\frac{dy}{dx}.$$

This notation, which will turn out to be very handy for remembering some formulas, is due to Leibniz. It should be read "the derivative of y with respect to x." At the moment, regard dy/dx as a single symbol, not as a quotient. (A quotient interpretation will appear later in the chapter.) Remember that $f'(x_1)$ is the slope m_{tan} of the line tangent to the graph $y = f(x)$ at $x = x_1$. It also has the interpretation of the instantaneous rate at which y is increasing compared to x at x_1. The notation dy/dx is suggestive of this rate of change of y with respect to x. The most important applications of differential calculus center around this rate-of-change interpretation of dy/dx.

The computation of $f'(x_1)$ from Eq. (2) is exactly the same as the computation of m_{tan} given in Section 2.1. Here are two more illustrations, computing

$$\frac{dy}{dx} = f'(x) = \lim_{\Delta x \to 0} \frac{f(x + \Delta x) - f(x)}{\Delta x}$$

at any point x.

EXAMPLE 1 If $y = f(x) = 1/(x - 1)$, find dy/dx.

Solution

$$\frac{dy}{dx} = f'(x) = \lim_{\Delta x \to 0} \frac{f(x + \Delta x) - f(x)}{\Delta x} = \lim_{\Delta x \to 0} \frac{\dfrac{1}{x + \Delta x - 1} - \dfrac{1}{x - 1}}{\Delta x}$$

$$= \lim_{\Delta x \to 0} \frac{\dfrac{x - 1 - (x + \Delta x - 1)}{(x + \Delta x - 1)(x - 1)}}{\Delta x}$$

$$= \lim_{\Delta x \to 0} \frac{-\Delta x}{\Delta x(x + \Delta x - 1)(x - 1)}$$

$$= \lim_{\Delta x \to 0} \frac{-1}{(x + \Delta x - 1)(x - 1)} = \frac{-1}{(x - 1)^2}. \quad \square$$

EXAMPLE 2 If $g(x) = \sqrt{x + 3}$, find $g'(x)$.

Solution

$$g'(x) = \lim_{\Delta x \to 0} \frac{g(x + \Delta x) - g(x)}{\Delta x} = \lim_{\Delta x \to 0} \frac{\sqrt{x + \Delta x + 3} - \sqrt{x + 3}}{\Delta x}$$

$$= \lim_{\Delta x \to 0} \frac{\sqrt{x + \Delta x + 3} - \sqrt{x + 3}}{\Delta x} \cdot \frac{\sqrt{x + \Delta x + 3} + \sqrt{x + 3}}{\sqrt{x + \Delta x + 3} + \sqrt{x + 3}}$$

$$= \lim_{\Delta x \to 0} \frac{(x + \Delta x + 3) - (x + 3)}{\Delta x(\sqrt{x + \Delta x + 3} + \sqrt{x + 3})}$$

$$= \lim_{\Delta x \to 0} \frac{\Delta x}{\Delta x(\sqrt{x + \Delta x + 3} + \sqrt{x + 3})}$$

$$= \lim_{\Delta x \to 0} \frac{1}{\sqrt{x + \Delta x + 3} + \sqrt{x + 3}} = \frac{1}{2\sqrt{x + 3}}. \quad \square$$

DIFFERENTIATION OF POLYNOMIAL FUNCTIONS

This section presents the first few of many formulas that can be used to compute easily the derivatives of many functions. The process of finding a derivative is *differentiation.* Mastering the technique of differentiation is as important for calculus as mastering the arithmetic operations was for all the mathematics you learned before.

We start by showing that the derivative of a constant function is zero at any point. Let $f(x) = c$, a constant, for all x. Then for any x_1,

$$f'(x_1) = \lim_{\Delta x \to 0} \frac{f(x_1 + \Delta x) - f(x_1)}{\Delta x} = \lim_{\Delta x \to 0} \frac{c - c}{\Delta x} = \lim_{\Delta x \to 0} \frac{0}{\Delta x} = 0.$$

In Leibniz notation,

$$\boxed{\frac{d(c)}{dx} = 0.} \qquad \textbf{(3)}$$

Now suppose $f(x) = x^n$ for a positive integer n. This time the computation of $f'(x_1)$ uses the *binomial theorem* of algebra to expand $(x_1 + \Delta x)^n$. The binomial theorem gives an expanded formula for $(a + b)^n$ in terms of the products,

$$a^n, \quad a^{n-1}b, \quad a^{n-2}b^2, \quad a^{n-3}b^3, \ldots, \quad ab^{n-1}, \quad b^n.$$

Namely,

$$(a + b)^n = a^n + na^{n-1}b + \frac{n(n-1)}{2}a^{n-2}b^2 + \frac{n(n-1)(n-2)}{3 \cdot 2}a^{n-3}b^3 + \cdots + b^n.$$ Binomial theorem

Applying this formula with $a = x_1$ and $b = \Delta x$, we obtain

$$\begin{aligned} f'(x_1) &= \lim_{\Delta x \to 0} \frac{(x_1 + \Delta x)^n - x_1^n}{\Delta x} \\ &= \lim_{\Delta x \to 0} \frac{[x_1^n + nx_1^{n-1}\Delta x + (n(n-1)/2)x_1^{n-2}(\Delta x)^2 + \cdots + (\Delta x)^n] - x_1^n}{\Delta x} \\ &= \lim_{\Delta x \to 0} [nx_1^{n-1} + (n(n-1)/2)x_1^{n-2}\Delta x + \cdots + (\Delta x)^{n-1}] = nx_1^{n-1}. \end{aligned}$$

Since x_1 can be any point, this shows that

$$\boxed{\frac{d(x^n)}{dx} = nx^{n-1}.} \qquad \textbf{(4)}$$

Next, suppose that $u = f(x)$ and $v = g(x)$, and let $h(x) = u + v = f(x) + g(x)$. A change Δx in x produces changes

$$\Delta u = f(x + \Delta x) - f(x) \quad \text{and} \quad \Delta v = g(x + \Delta x) - g(x)$$

in u and v. The change produced in $h(x) = u + v$ is

$$\begin{aligned} h(x + \Delta x) - h(x) &= [(u + \Delta u) + (v + \Delta v)] - (u + v) \\ &= \Delta u + \Delta v. \end{aligned}$$

Suppose now that $f'(x_1)$ and $g'(x_1)$ both exist. Working at x_1 and using Theorem 2.1 on page 54 for the limit of a sum, we find

$$\lim_{\Delta x\to 0}\frac{\text{Change in }(u+v)}{\Delta x}=\lim_{\Delta x\to 0}\frac{\Delta u+\Delta v}{\Delta x}$$

$$=\lim_{\Delta x\to 0}\frac{\Delta u}{\Delta x}+\lim_{\Delta x\to 0}\frac{\Delta v}{\Delta x}=\frac{du}{dx}+\frac{dv}{dx}.$$

That is,

$$\boxed{\frac{d(u+v)}{dx}=\frac{du}{dx}+\frac{dv}{dx}} \tag{5}$$

at any point where the derivatives of both u and v exist.

For the final result before we differentiate any polynomial function, let $u = f(x)$ and consider the function $c \cdot f(x)$ for a constant c. A change of Δx in x produces a change of Δu in u and therefore a change of $c \cdot \Delta u$ in $c \cdot f(x)$. This time we make use of Theorem 2.1 on page 54 for the limit of a product. At any point x where $f'(x)$ exists,

$$\lim_{\Delta x\to 0}\frac{\text{Change in }c\cdot f(x)}{\Delta x}=\lim_{\Delta x\to 0}\frac{c\cdot\Delta u}{\Delta x}$$

$$=\left(\lim_{\Delta x\to 0} c\right)\left(\lim_{\Delta x\to 0}\frac{\Delta u}{\Delta x}\right)=c\cdot\frac{du}{dx},$$

so

$$\boxed{\frac{d(c\cdot u)}{dx}=c\cdot\frac{du}{dx}.} \tag{6}$$

Equations (5) and (6) are very important since they hold for any differentiable functions $u = f(x)$ and $v = g(x)$. We recommend learning them in words, as they are given in the Summary, independent of particular letters such as u and v. They deserve to be stated as a theorem.

THEOREM 3.1 Linearity property

If the functions $u = f(x)$ and $v = g(x)$ are both differentiable at the point x, then so are $u + v = f(x) + g(x)$ and $c \cdot u = c \cdot f(x)$ for any constant c. Furthermore,

$$\frac{d(u+v)}{dx}=\frac{du}{dx}+\frac{dv}{dx}\quad\text{and}\quad\frac{d(c\cdot u)}{dx}=c\cdot\frac{du}{dx}.$$

EXAMPLE 3 Find the derivative of the function $4x^3 - 7x^2$.

Solution Using Eqs. (5), (6), and then (4), we obtain

$$\frac{d(4x^3-7x^2)}{dx}=\frac{d(4x^3)}{dx}+\frac{d(-7x^2)}{dx}=4\frac{d(x^3)}{dx}+(-7)\frac{d(x^2)}{dx}$$

$$=4\cdot 3x^2+(-7)(2x)=12x^2-14x\quad\text{for all } x. \quad \square$$

Example 3 can be generalized to more than two summands to give a very nice formula for the derivative of any polynomial function. Namely,

$$\frac{d(a_n x^n + \cdots + a_2 x^2 + a_1 x + a_0)}{dx} = n a_n x^{n-1} + \cdots + 2a_2 x + a_1. \tag{7}$$

EXAMPLE 4 Find the derivative of $f(x) = 4x^3 - 17x^2 + 3x - 2$.

Solution Using Eq. (7), we find that $f'(x) = 12x^2 - 34x + 3$. □

EXAMPLE 5 Find dy/dx if $y = (2x + 3)^2$.

Solution We have $y = (2x + 3)^2 = 4x^2 + 12x + 9$. Thus $dy/dx = 4(2x) + 12(1) + 0 = 8x + 12$. We give a different method of solving this problem in Section 3.4. □

EXAMPLE 6 Find dy/dx if $y = (x^2 + 4x^3)/5$.

Solution We write $y = \frac{1}{5}(x^2 + 4x^3)$. By Theorem 3.1, we have

$$\frac{dy}{dx} = \frac{1}{5} \cdot \frac{d(x^2 + 4x^3)}{dx} = \frac{1}{5}(2x + 12x^2). \quad \square$$

EXAMPLE 7 Find the average rate of change of $f(x) = x^3 - 2x^2$ over the interval $[-1, 3]$ and the instantaneous rate of change of $f(x)$ where $x = 2$.

Solution We have

$$\text{Average rate of change} = \frac{\Delta y}{\Delta x} = \frac{f(3) - f(-1)}{3 - (-1)}$$
$$= \frac{9 - (-3)}{4} = \frac{12}{4} = 3.$$

The instantaneous rate of change when $x = 2$ is given by $f'(2)$. Now $f'(x) = 3x^2 - 4x$, so $f'(2) = 12 - 8 = 4$. □

EXAMPLE 8 Let $f(x) = |x|$. Show that $f'(0)$ does not exist.

Solution The graph of f is shown in Fig. 3.3. We saw on page 53 that at $x = 0$,

$$\frac{\Delta y}{\Delta x} = \frac{|\Delta x|}{\Delta x} = \begin{cases} 1, & \text{if } \Delta x > 0, \\ -1, & \text{if } \Delta x < 0. \end{cases}$$

Thus $f'(0) = \lim_{\Delta x \to 0} (\Delta y/\Delta x)$ does not exist, because the right-hand limit is 1 while the left-hand limit is -1. This function $|x|$ is perhaps the simplest example of a function that is not differentiable at some point. Note that the graph, shown in Fig. 3.3, has a sharp point where $x = 0$ and no (unique) tangent line there. □

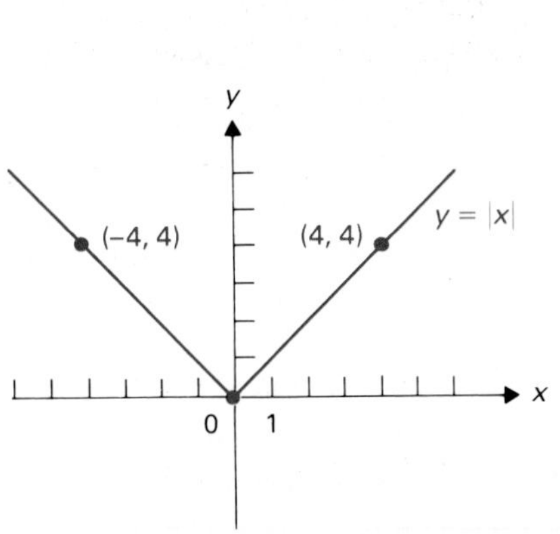

Figure 3.3 $|x|$ does not have a derivative at $x = 0$ where there is a sharp point on the graph.

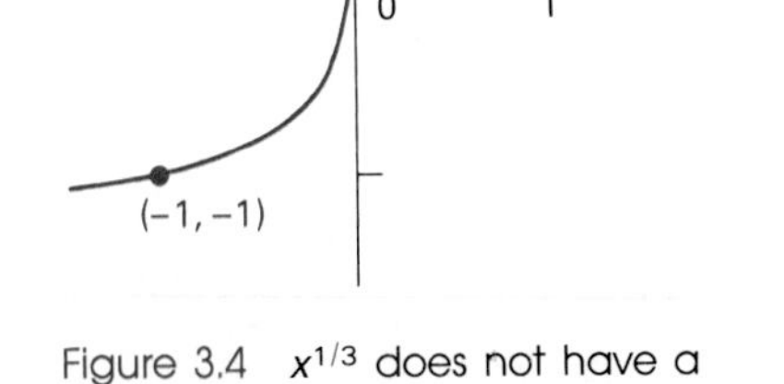

Figure 3.4 $x^{1/3}$ does not have a derivative at $x = 0$ where the graph has a vertical tangent (the y-axis).

EXAMPLE 9 Argue graphically that $f'(0)$ does not exist for the function $f(x) = x^{1/3}$.

Solution The graph of $f(x) = x^{1/3}$ is shown in Fig. 3.4. The tangent line to the graph at the origin is vertical; that is, it is the y-axis. Since a vertical line has no slope, we see that $f'(0)$ does not exist for this function either. We will see later that $f'(x) = 1/(3x^{2/3})$, and of course this expression is undefined when $x = 0$. □

As shown in Example 8, the function $|x|$ is not differentiable at 0, but it is continuous there. Thus continuity does not imply differentiability. However, differentiability does imply continuity, as we proceed to show.

THEOREM 3.2 Differentiable → continuous

If f is differentiable at $x = x_1$, then f is continuous at $x = x_1$.

Theorem 3.2 is easy to show from the definition of the derivative. If $f'(x_1)$ exists, then

$$\lim_{\Delta x \to 0} \frac{f(x_1 + \Delta x) - f(x_1)}{\Delta x} \tag{8}$$

exists. Now make the substitution $x = x_1 + \Delta x$, so $\Delta x = x - x_1$. Then $\Delta x \to 0$ is equivalent to $x \to x_1$, and Eq. (8) becomes

$$\lim_{x \to x_1} \frac{f(x) - f(x_1)}{x - x_1} \tag{9}$$

exists. Since the denominator in Eq. (9) approaches zero as $x \to x_1$, the limit exists only if $\lim_{x \to x_1} (f(x) - f(x_1)) = 0$ also, that is, only if $\lim_{x \to x_1} f(x) = f(x_1)$. Thus f is continuous at $x = x_1$.

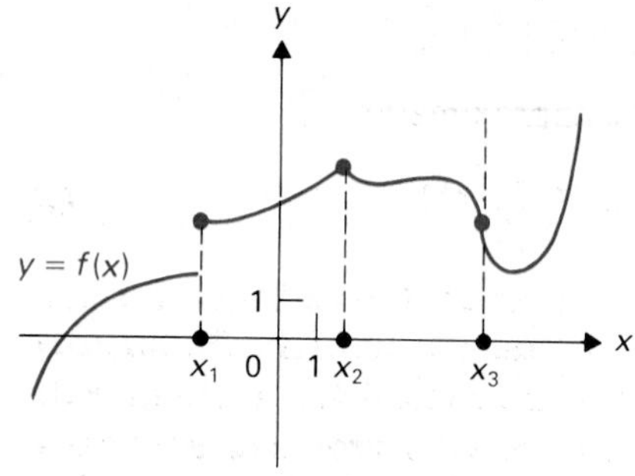

Figure 3.5 $f'(x)$ does not exist at x_1, which is a point of discontinuity; at x_2, where the graph has a sharp point; and at x_3, where there is a vertical tangent.

> In summary, functions will not be differentiable at discontinuities or where their graphs have sharp points or vertical tangents. (Other features can also cause a function to fail to be differentiable.)

Figure 3.5 gives a graphic picture of this summary.

APPLICATIONS

We illustrate applications of the derivative to finding the equations of tangent lines and instantaneous rates of change. Note how easily we can solve such problems after only a few lessons in calculus! The problems would have seemed formidable only one week ago.

EXAMPLE 10 Find the equation of the line tangent to the graph of $y = f(x) = 3x^4 - 2x^2 + 3x - 7$ where $x = 1$.

Solution

Point $(1, f(1)) = (1, -3)$

Slope $f'(1) = (12x^3 - 4x + 3)|_{x=1} = 12 - 4 + 3 = 11$. (The notation $|_{x=1}$ means "evaluated at $x = 1$.")

Equation $y + 3 = 11(x - 1)$, or $y = 11x - 14$ □

EXAMPLE 11 Find all points on the curve $y = 2x^3 - 3x^2 - 12x + 20$ where the tangent line is parallel to the x-axis.

Solution The x-axis is the line $y = 0$ of slope zero. Thus we want to find all points where the tangent line to the given curve has slope zero. This slope is given by the derivative, dy/dx, and we find that

$$\frac{dy}{dx} = 6x^2 - 6x - 12 = 6(x^2 - x - 2) = 6(x - 2)(x + 1).$$

Thus $dy/dx = 0$ where $x = -1$ or $x = 2$. Computing the y-coordinates for those values of x, we find that the desired points are $(-1, 27)$ and $(2, 0)$. □

Let s be the total distance a moving car has traveled in time $t \geq 0$ from its position at time $t = 0$. Note that s need not be the distance of the car from its starting point where $t = 0$, for the car may move on a curved path or back and forth over the same path. Think of s as the reading in the car of an odometer that was set to zero at time $t = 0$. The derivative ds/dt gives the speed of the car, which appears on the speedometer in the car.

EXAMPLE 12 Find the speed at time $t = 3$ if the total distance s traveled by a car from time $t = 0$ is given by $s = t^2 + 2t$ for $t \geq 0$.

Solution The speed of the car when $t = 3$ is

$$\text{Speed} = \left.\frac{ds}{dt}\right|_{t=3} = (2t + 2)|_{t=3} = 6 + 2$$

$$= 8 \text{ (Units distance)/(Unit time).} \quad \square$$

Table 3.1

Δx	$m_{sym} = \dfrac{2^{1+\Delta x} - 2^{1-\Delta x}}{2(\Delta x)}$
0.05	1.386571898
0.01	1.386305462
0.002	1.386294806
0.001	1.386294472

ESTIMATION OF $f'(x_1)$ USING A CALCULATOR

On page 45 we saw that for small Δx, the approximation

$$f'(x_1) \approx m_{sym} = \frac{f(x_1 + \Delta x) - f(x_1 - \Delta x)}{2 \cdot \Delta x} \tag{10}$$

(read $\approx$ as "approximately equals") can be expected to be better than that given by the difference quotient Eq. (1) for the same value of Δx. The approximation Eq. (10) can be used with a computer or calculator to estimate the derivative of a function at any particular point. Of course, unless we know that f is differentiable at x_1, we should compute Eq. (10) with three or four different small values of Δx.

EXAMPLE 13 Use a calculator to attempt to estimate $f'(1)$ if $f(x) = 2^x$.

Solution Using approximation (10) and computing values of m_{sym} for four small values of Δx, we obtain the results shown in Table 3.1. It appears that $f'(1) \approx 1.38629$. □

SUMMARY

Let $f(x)$ be defined for all x between $x_1 - h$ and $x_1 + h$ for some $h > 0$.

1. The derivative of f at x_1 is

$$f'(x_1) = \lim_{\Delta x \to 0} \frac{f(x_1 + \Delta x) - f(x_1)}{\Delta x}.$$

2. If $y = f(x)$, then $f'(x)$ is also written as dy/dx, the derivative of y with respect to x.
3. The derivative of a constant function is the zero function; in symbols,

$$\frac{d(c)}{dx} = 0.$$

4. The derivative of a sum is the sum of the derivatives; in symbols,

$$\frac{d(u + v)}{dx} = \frac{du}{dx} + \frac{dv}{dx}.$$

5. The derivative of a constant times a function is the constant times the derivative of the function; in symbols,

$$\frac{d(c \cdot u)}{dx} = c \cdot \frac{du}{dx}.$$

6. $\dfrac{d(x^n)}{dx} = nx^{n-1}$ for any positive integer n.

7. Speed $= ds/dt$ where s is the total distance traveled at time t.
8. A differentiable function is continuous.
9. (Calculator) If $f'(x_1)$ exists, then

$$f'(x_1) \approx \frac{f(x_1 + \Delta x) - f(x_1 - \Delta x)}{2 \cdot \Delta x}$$

for small Δx.

EXERCISES

In Exercises 1 through 6, find $f'(x)$ using the definition in Eq. (2).

1. $f(x) = x^2 - 3x$
2. $f(x) = 4x^2 + 7$
3. $f(x) = \dfrac{1}{2x + 3}$
4. $f(x) = \dfrac{1}{\sqrt{x}}$
5. $f(x) = \dfrac{x}{x + 1}$
6. $f(x) = \sqrt{2x - 1}$

In Exercises 7 through 19, find the derivative of the given function.

7. $3x - 2$
8. $8x^3 - 7x^2 + 4$
9. $2x^7 + 4x^2 - 3$
10. $15x^3 - 4x^6 + 2x^2 + 5$
11. $\dfrac{x^2 - 3x + 4}{2}$
12. $\dfrac{x^3 - 3x^2 + 2}{4}$
13. $(3x)^4 - (2x)^5$
14. $(x^2 - 2)(x + 1)$
15. $(x^2 + 2x)^2$
16. $x(3x + 2)(3x - 2)$
17. $(2x)^2(3x + 5)$
18. $8x^3 - 3(x + 1)^2 + 2$
19. $\dfrac{x(x - 1)(x + 1)}{3}$

In Exercises 20 through 25, find (a) the average rate of change of the function over the interval and (b) the instantaneous rate of change at the midpoint of the interval.

20. $f(x) = x^2 - 3x$ on $[2, 4]$
21. $g(t) = 2t^3 - 3t^2$ on $[-1, 1]$
22. $f(x) = x^4 - 16x^2 + 2x - 1$ on $[0, 4]$
23. $f(x) = x(3x - 5)^2$ on $[-1, 3]$
24. $f(s) = s^2(s - 1)$ on $[-2, 2]$
25. $f(x) = x(x - 2)(2x - 1)$ on $[0, 1]$

In Exercises 26 through 31, find the equations of (a) the tangent line and (b) the normal line (perpendicular to the tangent line) to the given curve at the indicated point.

26. $y = x^2 - 3x + 2$, where $x = 1$
27. $y = 3x^2 - 2x + 1$, where $x = -2$
28. $y = x^4 - 3x^2 - 3x$, where $x = 2$
29. $y = 2x^3 - 3x$, where $x = 2$
30. $y = (2x + 4)^2$, where $x = -2$
31. $y = x^2(2x + 5)$, where $x = -1$

32. a) Compute $d(1/x)/dx$ assuming that the formula for $d(x^n)/dx$ in Eq. (4) also holds if $n = -1$.
 b) Verify your answer in part (a) by computing the limit of the appropriate difference quotient.
 c) Find
$$\frac{d}{dx}\left(\frac{3}{x} - 2x\right).$$
 d) Find
$$\frac{d}{dx}\left(\frac{1}{4x} - \frac{3}{x} + \frac{2}{5x}\right).$$

33. a) Compute $d(\sqrt{x})/dx$ assuming that the formula for $d(x^n)/dx$ in Eq. (4) also holds if $n = \frac{1}{2}$.
 b) Verify that your answer in part (a) is correct by computing the limit of the appropriate difference quotient.
 c) Find
$$\frac{d}{dx}(3\sqrt{x} - 2x^2).$$
 d) Find
$$\frac{d}{dx}(\sqrt{5x} - \sqrt{7x}).$$

34. An object travels so that after t hours, it has gone $s = f(t) = 4t^3 + 3t^2 + t$ miles for $t \geq 0$. Find the speed of the object as a function of the time t for $t \geq 0$.

35. If the length of an edge of a cube increases at a rate of 1 in./sec, find the (instantaneous) rate of increase of the volume when (a) the edge is 2 in. long, (b) the edge is 5 in. long.

36. Repeat Exercise 35, assuming that the edge of the cube is increasing at a rate of 4 in./sec. (Use Exercise 35 and common sense.)

37. Suppose that when a pebble is dropped into a large tank of fluid, a wave travels outward in a circular ring whose radius increases at a constant rate of 8 in./sec.
 a) Find the area of the circular disk enclosed by the wave 2 sec after the time the pebble hits the fluid.
 b) Find the (instantaneous) rate at which the area of the circular disk enclosed by the wave is increasing 2 sec after the time the pebble hits the fluid. (See Exercises 35 and 36.)

38. Give an example of a continuous function defined for all real numbers but not differentiable where $x = 3$.

39. Repeat Exercise 38, but make the function differentiable at neither -3 nor 3.

40. Show that if $f'(x_1)$ exists, then
$$\lim_{\Delta x \to 0} \frac{f(x_1 + \Delta x) - f(x_1 - \Delta x)}{2 \cdot \Delta x} = f'(x_1).$$
[*Hint:* Use the fact that
$$\frac{f(x_1 + \Delta x) - f(x_1 - \Delta x)}{2 \cdot \Delta x}$$
$$= \frac{1}{2} \cdot \frac{f(x_1 + \Delta x) - f(x_1)}{\Delta x} + \frac{1}{2} \cdot \frac{f(x_1 + (-\Delta x)) - f(x_1)}{-\Delta x}$$
and Theorem 2.1 on page 54.]

41. It is worth noting that
$$\lim_{\Delta x \to 0} \frac{f(x_1 + \Delta x) - f(x_1 - \Delta x)}{2 \cdot \Delta x}$$
may exist while $f'(x_1)$ does not. Give an example of a function $f(x)$ and a point x_1 where this is true. [*Hint:* Consider Example 8.]

Use a calculator and the approximation Eq. (10) in the text to find the derivative of the given function at the indicated point. Use radian measure with all trigonometric functions. You choose Δx.

42. $\sin 2x$ at $x_1 = 0$

43. $[(x + 7)/(x^2 + 5)]^{1/3}$ at $x_1 = 2.374$

44. x^x at $x_1 = 2.36$

45. $\sin(\tan x)$ at $x_1 = -1.3$

46. $(\sin x)^{\cos x}$ at $x_1 = \pi/4$

47. $(x^2 - 3x)^{\sqrt{x}}$ at $x_1 = 4$

3.2 DIFFERENTIATION OF PRODUCTS AND QUOTIENTS

The process of finding $f'(x)$ from $f(x)$ is known as *differentiation*. One nice feature of calculus is that there are formulas that make differentiation easy, at least for the functions used most often. We have already seen how easy it is to find the derivative of a polynomial function. This section gives easy formulas for finding the derivatives of a product $f(x) \cdot g(x)$ and a quotient $f(x)/g(x)$ in terms of the derivatives $f'(x)$ and $g'(x)$. The chain rule in Section 3.4 will complete our list of general differentiation formulas.

Let both f and g be differentiable functions, so that $f'(x)$ and $g'(x)$ exist. If we let $u = f(x)$ and $v = g(x)$, then a change Δx from x to $x + \Delta x$ produces changes Δu in u and Δv in v. Thus

$$f(x + \Delta x) = u + \Delta u \qquad \text{and} \qquad g(x + \Delta x) = v + \Delta v.$$

The change in $f(x) \cdot g(x)$ is therefore

$$f(x + \Delta x) \cdot g(x + \Delta x) - f(x) \cdot g(x) = (u + \Delta u)(v + \Delta v) - uv.$$

Therefore the difference quotient for $f(x) \cdot g(x)$ is

$$\begin{aligned} \frac{\text{Change in } u \cdot v}{\Delta x} &= \frac{(u + \Delta u) \cdot (v + \Delta v) - u \cdot v}{\Delta x} \\ &= \frac{uv + u \cdot \Delta v + v \cdot \Delta u + \Delta u \cdot \Delta v - uv}{\Delta x} \\ &= \frac{u \cdot \Delta v + v \cdot \Delta u + \Delta u \cdot \Delta v}{\Delta x} \\ &= u \cdot \frac{\Delta v}{\Delta x} + v \cdot \frac{\Delta u}{\Delta x} + \frac{\Delta u}{\Delta x} \cdot \Delta v. \end{aligned} \tag{1}$$

We take the limit of Eq. (1) as $\Delta x \to 0$ to find the derivative of uv. Note that $v = g(x)$ is a continuous function at $x = x_1$ since $g'(x_1)$ exists (Theorem 3.2). Thus $\lim_{\Delta x \to 0} \Delta v = 0$. Taking the limit,

$$\begin{aligned} \lim_{\Delta x \to 0} \frac{\text{Change in } uv}{\Delta x} &= \lim_{\Delta x \to 0} \left(u \cdot \frac{\Delta v}{\Delta x} + v \cdot \frac{\Delta u}{\Delta x} + \frac{\Delta u}{\Delta x} \cdot \Delta v \right) \\ &= u \cdot \frac{dv}{dx} + v \cdot \frac{du}{dx} + \frac{du}{dx} \cdot 0 = u \cdot \frac{dv}{dx} + v \cdot \frac{du}{dx}. \end{aligned}$$

This shows that

$$\frac{d(u \cdot v)}{dx} = u \cdot \frac{dv}{dx} + v \cdot \frac{du}{dx} \tag{2}$$

Product rule

at any point where du/dx and dv/dx both exist.

EXAMPLE 1 Find dy/dx in two ways if $y = (2x + 1)(3x - 2)$.

Solution

Method 1 From Eq. (2), we obtain

$$\frac{dy}{dx} = (2x + 1)\frac{d(3x - 2)}{dx} + (3x - 2)\frac{d(2x + 1)}{dx}$$

$$= (2x + 1)3 + (3x - 2)2 = 6x + 3 + 6x - 4 = 12x - 1.$$

Method 2 Note that if we multiply the factors $(2x + 1)(3x - 2)$ first, we obtain $y = 6x^2 - x - 2$, which again yields $dy/dx = 12x - 1$. □

EXAMPLE 2 Find dy/dx if $y = (x^2 + x)(x^3 - 7x^2 + 3x)$.

Solution Working with Eq. (2) and taking $u = x^2 + x$ and $v = x^3 - 7x^2 + 3x$, we find that

$$\frac{dy}{dx} = (x^2 + x)\frac{d(x^3 - 7x^2 + 3x)}{dx} + (x^3 - 7x^2 + 3x)\frac{d(x^2 + x)}{dx}$$

$$= (x^2 + x)(3x^2 - 14x + 3) + (x^3 - 7x^2 + 3x)(2x + 1).$$

This can be simplified algebraically, but such simplification is a waste of time if we just want the derivative at one point. For example,

$$\left.\frac{dy}{dx}\right|_{x=1} = 2(-8) + (-3)3 = -16 - 9 = -25. \quad □$$

EXAMPLE 3 Suppose f is differentiable with $f(3) = -2$ and $f'(3) = 5$. Find dy/dx where $x = 3$ if $y = (x^2 + 4x) \cdot f(x)$.

Solution We have

$$\frac{dy}{dx} = (x^2 + 4x) \cdot f'(x) + (2x + 4) \cdot f(x).$$

Thus

$$\left.\frac{dy}{dx}\right|_{x=3} = 21 \cdot f'(3) + 10 \cdot f(3) = 21 \cdot 5 + 10(-2) = 85. \quad □$$

Now we turn to the differentiation of $f(x)/g(x)$ at $x = x_1$, under the assumptions that $f'(x_1)$ and $g'(x_1)$ exist and $g(x_1) \neq 0$. Again, let $u = f(x)$ and $v = g(x)$ have changes Δu and Δv produced by a change Δx in x. Then

$u/v = f(x)/g(x)$ and

$$\frac{\text{Change in } (u/v)}{\Delta x} = \frac{\dfrac{u + \Delta u}{v + \Delta v} - \dfrac{u}{v}}{\Delta x} = \frac{\dfrac{v(u + \Delta u) - u(v + \Delta v)}{v(v + \Delta v)}}{\Delta x}$$

$$= \frac{\dfrac{v \cdot \Delta u - u \cdot \Delta v}{v(v + \Delta v)}}{\Delta x} = \frac{v \cdot \dfrac{\Delta u}{\Delta x} - u \cdot \dfrac{\Delta v}{\Delta x}}{v(v + \Delta v)}. \qquad (3)$$

We take the limit as $\Delta x \to 0$ to find the derivative of u/v. Note that $\lim_{\Delta x \to 0} (v + \Delta v) = v$, because $v = g(x)$ is continuous at $x = x_1$, since $g'(x_1)$ exists (Theorem 3.2). Taking the limit, we have

$$\lim_{\Delta x \to 0} \frac{\text{Change in } u/v}{\Delta x} = \lim_{\Delta x \to 0} \frac{v \cdot (\Delta u/\Delta x) - u \cdot (\Delta v/\Delta x)}{v(v + \Delta v)}$$

$$= \frac{v \cdot (du/dx) - u \cdot (dv/dx)}{v^2}.$$

This shows that

$$\frac{d(u/v)}{dx} = \frac{v \cdot (du/dx) - u \cdot (dv/dx)}{v^2} \qquad (4)$$

Quotient rule

at any point where du/dx and dv/dx exist and $v \neq 0$.

EXAMPLE 4 Find dy/dx if $y = (x^2 + 1)/(x^3 - 2x)$.

Solution Working with Eq. (4) and taking $u = x^2 + 1$ and $v = x^3 - 2x$, we find that

$$\frac{dy}{dx} = \frac{(x^3 - 2x)[d(x^2 + 1)/dx] - (x^2 + 1)[d(x^3 - 2x)/dx]}{(x^3 - 2x)^2}$$

$$= \frac{(x^3 - 2x)(2x) - (x^2 + 1)(3x^2 - 2)}{(x^3 - 2x)^2} = \frac{-x^4 - 5x^2 + 2}{(x^3 - 2x)^2}. \quad \square$$

EXAMPLE 5 Find dy/dx if $y = (x^2 - 3x)/5$.

Solution If the denominator of a quotient is a constant, we can avoid using the quotient rule when differentiating. In this example, we may write

$$y = \frac{1}{5}(x^2 - 3x)$$

and obtain

$$\frac{dy}{dx} = \frac{1}{5}(2x - 3). \quad \square$$

EXAMPLE 6 Suppose $g(x)$ is differentiable with $g(1) = 4$ and $g'(1) = 2$. Find dy/dx at $x = 1$ if $y = (x^3 - 2x^2)/g(x)$.

Solution We have

$$\frac{dy}{dx} = \frac{g(x)\cdot(3x^2 - 4x) - (x^3 - 2x^2)\cdot g'(x)}{g(x)^2}.$$

Thus

$$\left.\frac{dy}{dx}\right|_{x=1} = \frac{g(1)(-1) - (-1)g'(1)}{g(1)^2} = \frac{4(-1) - (-1)(2)}{4^2}$$
$$= \frac{-2}{16} = -\frac{1}{8}. \quad \square$$

We know that the formula

$$\frac{d(x^n)}{dx} = nx^{n-1} \tag{5}$$

holds if n is a positive integer. It also holds if n is a negative integer, because then $-n$ is positive and, using Eqs. (4) and (5),

$$\frac{d(x^n)}{dx} = \frac{d(1/x^{-n})}{dx} = \frac{x^{-n}\cdot[d(1)/dx] - 1\cdot[d(x^{-n})/dx]}{(x^{-n})^2}$$
$$= \frac{x^{-n}\cdot 0 - (-n)x^{-n-1}}{x^{-2n}}$$
$$= \frac{nx^{-n-1}}{x^{-2n}} = nx^{-n-1+2n} = nx^{n-1}.$$

Thus Eq. (5) holds for any integer n.

EXAMPLE 7 Compute dy/dx if $y = 3/x^5$.

Solution We may compute dy/dx by writing $y = 3\cdot x^{-5}$ and differentiating this expression using Eq. (5). We obtain

$$\frac{dy}{dx} = 3(-5)x^{-6} = \frac{-15}{x^6}.$$

This is easier than using the quotient rule (4). $\square$

In this section, we have seen the following theorem and corollary proved.

THEOREM 3.3 Product and quotient rules

If $u = f(x)$ and $v = g(x)$ are both differentiable functions, then so is $uv = f(x)g(x)$, and

$$\frac{d(uv)}{dx} = u\frac{dv}{dx} + v\frac{du}{dx}.$$

Also, $u/v = f(x)/g(x)$ is differentiable (of course points where $g(x) = 0$ are not allowed), and

$$\frac{d(u/v)}{dx} = \frac{v(du/dx) - u(dv/dx)}{v^2}.$$

COROLLARY Derivative of x^n

We have $d(x^n)/dx = n \cdot x^{n-1}$ for every integer n.

It is best to learn the differentiation formulas in Theorem 3.3 in words rather than with particular letters u and v. Such verbal renditions are given in the Summary. We use "top" for "numerator" and "bottom" for "denominator" in this one instance of the quotient rule because these short words are easier to say and more suggestive. We recommend that you learn these formulas by repeating them over and over in words, off and on, for a week or so.

SUMMARY

1. *Product rule:* The derivative of a product is the first times the derivative of the second, plus the second times the derivative of the first. In symbols,
$$\frac{d(uv)}{dx} = u \cdot \frac{dv}{dx} + v \cdot \frac{du}{dx}.$$
2. *Quotient rule:* The derivative of a quotient is the bottom times the derivative of the top minus the top times the derivative of the bottom, all divided by the bottom squared. In symbols,
$$\frac{d(u/v)}{dx} = \frac{v(du/dx) - u(dv/dx)}{v^2}.$$
3. $d(x^n)/dx = n \cdot x^{n-1}$ for every positive or negative integer n

EXERCISES

In Exercises 1 through 24, find the derivative of the given function. You need not simplify the answers.

1. $3x^2 + 17x - 5$
2. $20x^4 - \frac{3}{2}x^2 + 18$
3. $\dfrac{x^2 - 7}{3}$
4. $\dfrac{x^3 - 2x^2 + 4x}{4}$
5. $\dfrac{3}{x}$
6. $\dfrac{2}{x^3}$
7. $4x^3 - \dfrac{2}{x^2}$
8. $5x + 7 - \dfrac{1}{x^4}$
9. $(x^2 - 1)(x^2 + x + 2)$
10. $(3x^2 - 8x)(x^3 - 7x^2)$
11. $(x^2 + 1)[(x - 1)(x^3 + 3)]$
12. $[(x^2 - 5x)(2x + 3)](8 - 4x^2)$
13. $\left(\dfrac{1}{x^2} - \dfrac{4}{x^3}\right)(2x + 3)$
14. $(x^2 - 4x + 1)\left(\dfrac{1}{x^3} - \dfrac{4}{x}\right)$
15. $\dfrac{4x^2 - 3}{x}$
16. $\dfrac{8x^3 + 2x^2 + x}{x^2}$
17. $\dfrac{x^2 - 2}{x + 3}$
18. $\dfrac{4x^3 - 3x^2}{2x - 3}$
19. $\dfrac{(x^2 + 9)(x - 3)}{x^2 + 2}$
20. $\dfrac{(x^3 + 3x)(8x - 6)}{x^3 - 3x}$
21. $\dfrac{(2x + 3)(x^2 - 4)}{(x - 1)(4x^2 + 5)}$
22. $\dfrac{(8x - 6)(3x^2 - 2x)}{(2x + 1)(x^3 + 7)}$
23. $\left(\dfrac{x + 1}{2x + 3}\right)\left(\dfrac{1}{x} - \dfrac{1}{x^2}\right)$
24. $\left(\dfrac{4}{x^2} - \dfrac{1}{x^3}\right)\left(\dfrac{8x}{x^2 + 2}\right)$

The next chapter will show that there are functions $\sin x$ and $\cos x$ such that
$$\frac{d(\sin x)}{dx} = \cos x \quad \text{and} \quad \frac{d(\cos x)}{dx} = -\sin x.$$
There is also a function $\tan x$ defined by $\tan x = (\sin x)/(\cos x)$. Use these facts and the formulas of this section to differentiate the functions in Exercises 25 through 34.

25. $x(\sin x)$
26. $x^2(\cos x)$
27. $(\sin x)^2$
28. $\sin x \cos x$

29. $\tan x$

30. $\dfrac{\cos x}{\sin x}$

31. $\dfrac{x^3}{\sin x}$

32. $\dfrac{x^4}{\cos x}$

33. $\dfrac{\sin x}{x^2 - 4x}$

34. $\dfrac{x^3 - 3x^2}{\cos x}$

In Exercises 35 through 45, assume that the functions f and g are differentiable for all x.

35. If $f(1) = 3$, $f'(1) = -6$, and $y = x \cdot f(x)$, find $dy/dx|_{x=1}$.
36. If $g(2) = -3$, $g'(2) = 4$, and $y = [g(x)]^2$, find $dy/dx|_{x=2}$.
37. If $f(-1) = 2$, $f'(-1) = 3$, and $y = (x^2 - 3x) \cdot f(x)$, find $dy/dx|_{x=-1}$.
38. If $g(3) = -1$, $g'(3) = 2$, and $y = x/g(x)$, find $dy/dx|_{x=3}$.
39. If $f(5) = -1$, $f'(5) = -4$, and $y = x^2 f(x)/(2x + 3)$, find $dy/dx|_{x=5}$.
40. If $g(1) = 2$, $y = (x^3 - 2x^2)/g(x)$, and $dy/dx|_{x=1} = 4$, find $g'(1)$.
41. If $f(3) = -2$, $y = x \cdot f(x)/(x + 1)$, and $dy/dx|_{x=3} = 5$, find $f'(3)$.
42. Let $u = x \cdot f(x)$ and $v = x^2 f(x)$. If $du/dx|_{x=1} = 5$ and $dv/dx|_{x=1} = 13$, find $f(1)$ and $f'(1)$.
43. Let $u = (x^2 - 3x)f(x)$ and $v = (2x - 3)f(x)$. If $du/dx|_{x=4} = 8$ and $dv/dx|_{x=4} = 3$, find $f(4)$ and $f'(4)$.
44. Let $u = x + f(x)$ and $v = 1/f(x)$. If $du/dx|_{x=2} = 5$ and $dv/dx|_{x=2} = -16$, find all possible values of $f(2)$ and $f'(2)$.
45. Let $u = f(x)g(x)$ and $v = f(x)/g(x)$. If $f(3) = 9$, $g(3) = 1$, $du/dx|_{x=3} = 13$, and $dv/dx|_{x=3} = -23$, find $f'(3)$ and $g'(3)$.

In Exercises 46 through 50, find the equations of the tangent line and the normal line to the graph of the given function at the indicated point.

46. $3x^2 - 2x$ at $(2, 8)$

47. $1/x$ at $(1, 1)$

48. $\left(\dfrac{3}{x} - \dfrac{2}{x^2}\right)$ at $(-1, -5)$

49. $\dfrac{2x + 3}{x - 1}$ at $(0, -3)$

50. $\dfrac{(x^2 - 3)(x + 2)}{(x + 1)}$ at $(0, -6)$

3.3 THE DIFFERENTIAL

Scientists frequently encounter problems involving functions given by very complicated formulas with which it is difficult to work. One frequently used technique is to *linearize* the problem. Geometrically, this amounts to approximating the graph of a function near a point by using the tangent line to the graph there. As indicated in Fig. 3.6, the tangent line to the graph of a function f at $(x_1, f(x_1))$ appears to be the line that clings closest to the graph near that point. The notion of the *differential* of the function f is central to the study of such *local linear approximation*. We will indicate in this section how to use a differential to approximate a solution of an equation $f(x) = 0$. In Section 5.6, we will refine this work to Newton's method, which can often be used to obtain extremely accurate estimates of solutions of complicated equations.

Figure 3.6 Tangent line to $y = f(x)$, where $x = x_1$.

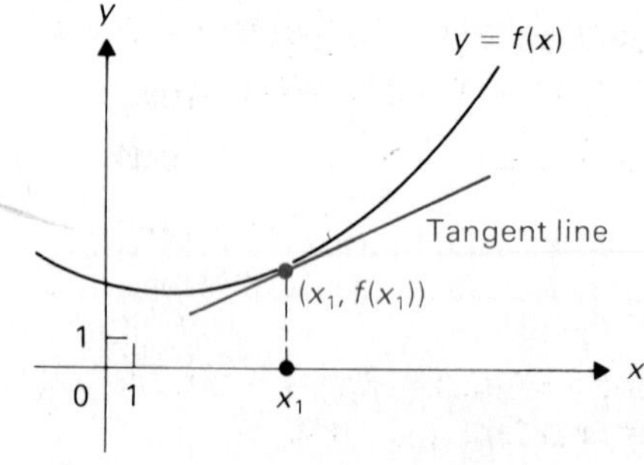

LOCAL LINEAR APPROXIMATION OF A FUNCTION

Figure 3.7 shows the tangent line to the graph of a differentiable function f at a point $(x_1, f(x_1))$. We are interested in how closely the tangent line clings to the graph as x changes from x_1 to $x_1 + \Delta x$. As shown in the figure, this change Δx in x produces a change Δy in $f(x)$ and a change $\Delta y_{\tan}$ in the height of the tangent line. We assume that f is differentiable, so $f'(x_1)$ exists. From Fig. 3.7, we see that the slope of the tangent line at $(x_1, f(x_1))$ is $(\Delta y_{\tan})/(\Delta x)$, so we have

$$f'(x_1) = \frac{\Delta y_{\tan}}{\Delta x}, \qquad \text{or}$$

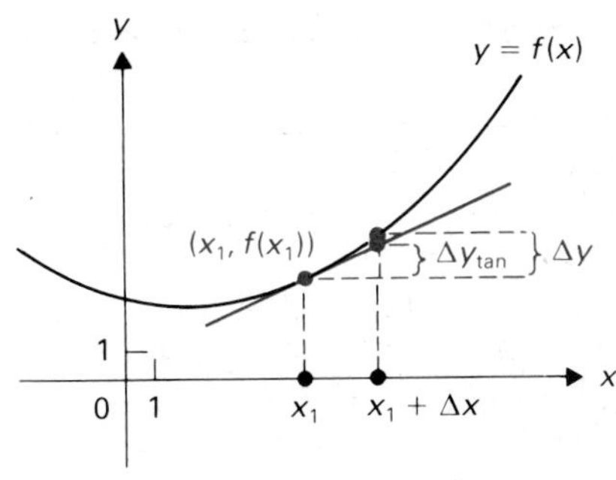

Figure 3.7 Δy = change in height of the graph of f; $\Delta y_{\tan}$ = change in height of the tangent line.

$$\boxed{\Delta y_{\tan} = f'(x_1)\cdot\Delta x.} \tag{1}$$

Figure 3.8 shows the difference $\Delta y - \Delta y_{\tan}$, which can be viewed as the error at $x = x_1 + \Delta x$ in approximating $f(x)$ by the line tangent to the graph where $x = x_1$. This error is a function of Δx, and we denote it by

$$\boxed{E(\Delta x) = \Delta y - \Delta y_{\tan}.} \tag{2}$$

We claim that this error $E(\Delta x)$ is small in comparison with the size of Δx if Δx is itself sufficiently small. Namely, we now show that

$$\boxed{\lim_{\Delta x\to 0}\frac{E(\Delta x)}{\Delta x} = 0.} \tag{3}$$

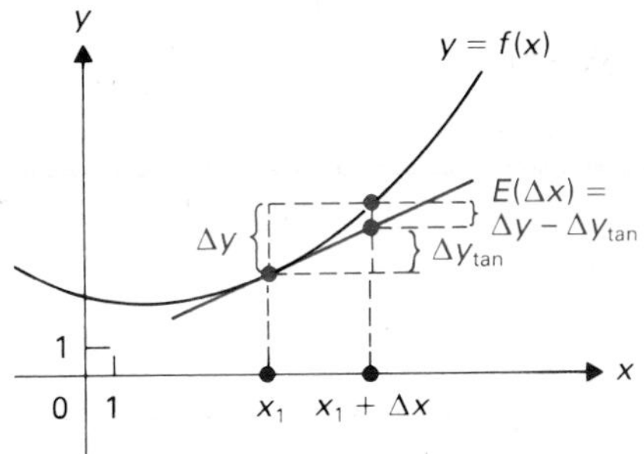

Figure 3.8

$E(\Delta x) = \Delta y - \Delta y_{\tan}$

For the limit of a quotient to approach zero as the denominator approaches zero, the numerator must approach zero significantly faster than the denominator. Thus Eq. (3) tells us that $E(\Delta x)$ is of very small size compared with Δx, provided that Δx is itself sufficiently small. To demonstrate Eq. (3), we use the definition of $E(\Delta x)$ in Eq. (2) and also use Eq. (1):

$$\lim_{\Delta x\to 0}\frac{E(\Delta x)}{\Delta x} = \lim_{\Delta x\to 0}\frac{\Delta y - \Delta y_{\tan}}{\Delta x} = \lim_{\Delta x\to 0}\frac{\Delta y - f'(x_1)\cdot\Delta x}{\Delta x}$$

$$= \lim_{\Delta x\to 0}\left[\frac{\Delta y}{\Delta x} - \frac{f'(x_1)\cdot\Delta x}{\Delta x}\right] = f'(x_1) - f'(x_1) = 0.$$

We now state our result from Eq. (3) as a theorem. To use the notation found in most calculus texts, we first let

$$\varepsilon(\Delta x) = \begin{cases} \dfrac{E(\Delta x)}{\Delta x} & \text{if } \Delta x \neq 0,\\ 0 & \text{if } \Delta x = 0. \end{cases}$$

Then $E(\Delta x) = \varepsilon(\Delta x)\cdot\Delta x$, and Eq. (3) becomes $\lim_{\Delta x\to 0}\varepsilon(\Delta x) = 0$. Since $\Delta y - \Delta y_{\tan} = E(\Delta x)$, we may write $\Delta y = \Delta y_{\tan} + E(\Delta x) = f'(x_1)\cdot\Delta x + \varepsilon(\Delta x)\cdot\Delta x$. Finally, it is conventional to write just ε in place of $\varepsilon(\Delta x)$ in the theorem to make the notation less cumbersome.

THEOREM 3.4 Approximation

Let $y = f(x)$ be differentiable at $x = x_1$, and let $\Delta y = f(x_1 + \Delta x) - f(x_1)$. Then there exists a function ε of Δx, defined for small Δx, such that

$$\Delta y = f'(x_1)\cdot\Delta x + \varepsilon\cdot\Delta x,$$

where $\lim_{\Delta x\to 0}\varepsilon = 0$.

EXAMPLE 1 Illustrate Theorem 3.4 for $y = f(x) = 3x^2 + 2x$ by finding the function ε of Δx and showing directly that $\lim_{\Delta x\to 0}\varepsilon = 0$.

Solution We work at a general point x rather than x_1. From the theorem, we see that

$$\varepsilon = \frac{\Delta y - f'(x)\,\Delta x}{\Delta x}.$$

We have

$$\begin{aligned}\Delta y &= f(x + \Delta x) - f(x)\\ &= [3(x + \Delta x)^2 + 2(x + \Delta x)] - (3x^2 + 2x)\\ &= 3x^2 + 6x(\Delta x) + 3(\Delta x)^2 + 2x + 2(\Delta x) - 3x^2 - 2x\\ &= 6x(\Delta x) + 3(\Delta x)^2 + 2(\Delta x)\end{aligned}$$

and

$$f'(x)\,\Delta x = (6x + 2)\,\Delta x = 6x(\Delta x) + 2(\Delta x).$$

Consequently,

$$\Delta y - f'(x)\,\Delta x = 3(\Delta x)^2.$$

Thus

$$\varepsilon = \frac{\Delta y - f'(x)\,\Delta x}{\Delta x} = \frac{3(\Delta x)^2}{\Delta x} = 3(\Delta x).$$

Of course, $\lim_{\Delta x \to 0} \varepsilon = \lim_{\Delta x \to 0} 3(\Delta x) = 0.$ □

DIFFERENTIAL NOTATION AND APPROXIMATIONS

We have used the notation dy/dx for the derivative,

$$\frac{dy}{dx} = f'(x).$$

If we attempt to view dy/dx as a quotient, then we must think of dx and dy as quantities related in such a way that

$$dy = f'(x)\,dx.$$

DEFINITION 3.3 Differential

Let $y = f(x)$ be differentiable at x_1. The **differential of f at x_1** is the function of the single variable dx given by

$$dy = f'(x_1)\,dx. \tag{4}$$

In Eq. (4), the independent variable is dx, and dy is the dependent variable. If f is differentiable at all points in its domain, then the **differential** dy or df of $y = f(x)$ is

$$dy = f'(x)\,dx, \tag{5}$$

which associates with each point x the differential of f at that point.

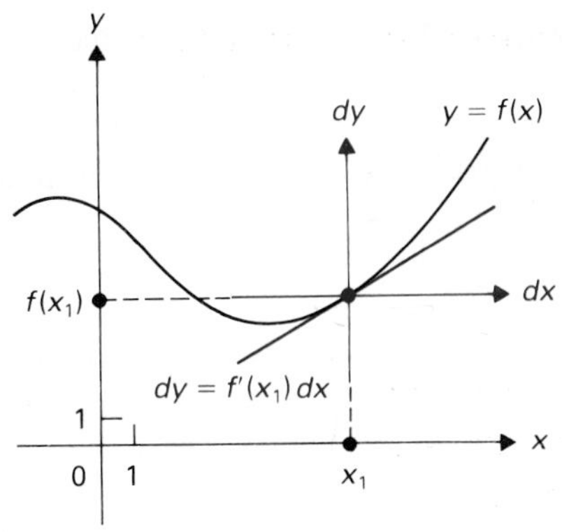

Figure 3.9 The tangent line is $dy = f'(x_1)\,dx$ with respect to local coordinate dx,dy-axes.

EXAMPLE 2 Let $s = g(t) = t^3 - (1/t^2)$. Find the differential ds at $t = 2$.

Solution Using Eq. (4) with g in place of f, t in place of x, and s in place of y, we have

$$ds = g'(2)\,dt.$$

We may write

$$s = g(t) = t^3 - \frac{1}{t^2} = t^3 - t^{-2},$$

so

$$\frac{ds}{dt} = g'(t) = 3t^2 - (-2)t^{-3} = 3t^2 + \frac{2}{t^3}.$$

Therefore

$$g'(2) = 3 \cdot 2^2 + \frac{2}{2^3} = 12 + \frac{1}{4} = \frac{49}{4},$$

so the differential at $t = 2$ is

$$ds = \frac{49}{4}\,dt. \quad \square$$

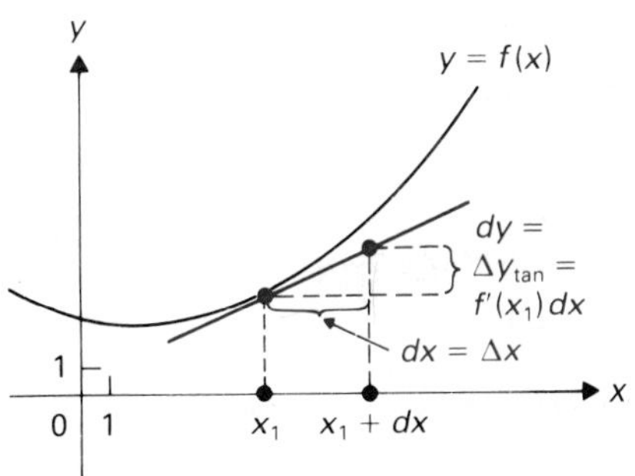

Figure 3.10 If $dx = \Delta x$, then $dy = \Delta y_{\tan} = f'(x_1)\,dx$.

EXAMPLE 3 If $y = f(x) = x^3 - 3x^2$, then the differential is $dy = (3x^2 - 6x)\,dx$. $\square$

The differential of f at x_1 is a *linear function* of the variable dx, because $f'(x_1)$ in Eq. (4) is a *number*, perhaps 3 or -7. In a dx,dy-plane, the equation $dy = 3 \cdot dx$ gives a line through the origin, just as $y = 3x$ is a line through the origin in the x,y-plane. As indicated in Fig. 3.9, if we consider dx and dy to be new variables corresponding to translation of axes to $(x_1, f(x_1))$, then Eq. (4) is precisely the equation of the tangent line to the graph with respect to these new axes.

If we interpret dx as Δx where $x = x_1$, as shown in Fig. 3.10, then Eq. (1) shows that

$$dy = f'(x_1)\,dx = f'(x_1) \cdot \Delta x = \Delta y_{\tan}. \tag{6}$$

We have labeled dy in Fig. 3.10 also. Thus for dx equal to sufficiently small Δx, the differential $dy = f'(x_1)\,dx$ is the approximate change in the height of the graph from x_1 to $x_1 + \Delta x$. Figure 3.11 illustrates the resulting approximation formula:

Figure 3.11 $f(x_1 + dx) \approx f(x_1) + f'(x_1)\,dx$

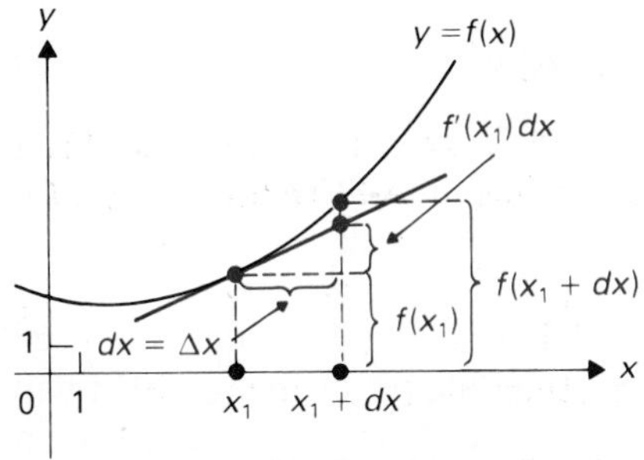

$$f(x_1 + dx) \approx f(x_1) + f'(x_1)\,dx. \tag{7}$$

Differential approximation formula

The difference between the two sides in approximation (7) is

$$[f(x_1 + dx) - f(x_1)] - f'(x_1)\,dx = \Delta y - \Delta y_{\tan} = E(dx).$$

Equation (3) asserts that approximation (7) is a good one for sufficiently small dx, in the sense that the error $E(dx)$ is very small compared with the size of dx.

We illustrate the use of approximation (7) to estimate the value of a function close to a point x_1 where its value is known. Since calculators easily give the value of many functions at *any* point, such illustrations are not of practical importance today, but they do illustrate the approximation formula.

EXAMPLE 4 Estimate $1/0.98^3$, using approximation (7).

Solution Since we are interested in computing the reciprocal of a number cubed, we let $f(x) = 1/x^3 = x^{-3}$. Our estimate will be most accurate if we keep dx as small as possible. Since we can compute $f(1)$ easily, we let $x_1 = 1$ and $dx = -0.02$. Now $f'(x) = -3x^{-4} = -3/x^4$. Then by approximation (7),

$$\begin{aligned} f(x_1 + dx) &= f(0.98) \approx f(1) + f'(1)(-0.02) \\ &= \frac{1}{1^3} - \frac{3}{1^4}(-0.02) = 1.06. \end{aligned}$$

A calculator yields $f(0.98) \approx 1.0625$, so our estimate is quite good. □

EXAMPLE 5 Estimate $2.01^3 - 2.01^2$, using the differential approximation (7).

Solution We let $f(x) = x^3 - x^2$, $x_1 = 2$, and $dx = 0.01$. Then $f'(x) = 3x^2 - 2x$, and approximation (7) becomes

$$\begin{aligned} f(x_1 + dx) &= 2.01^3 - 2.01^2 \\ &\approx (2^3 - 2^2) + (3 \cdot 2^2 - 2 \cdot 2)(0.01) \\ &= 4 + 8(0.01) = 4.08. \end{aligned}$$

A calculator shows that the exact value is 4.080501, so our estimate is quite good. □

EXAMPLE 6 Use approximation (7) to estimate $f(2.05)$ if $f(x) = x^3/(1 + x^2)$.

Solution We let $x_1 = 2$ and $dx = 0.05$. It is easy to compute $f(2)$ and $f'(2)$:

$$f(2) = \frac{8}{5} = 1.6,$$

$$f'(2) = \left.\frac{(1 + x^2)(3x^2) - x^3(2x)}{(1 + x^2)^2}\right|_{x=2} = \frac{5 \cdot 12 - 8 \cdot 4}{5^2} = \frac{28}{25} = 1.12.$$

From approximation (7) we obtain

$$\begin{aligned} f(2 + 0.05) \approx f(2) + f'(2) \cdot (0.05) &= 1.6 + (1.12)(0.05) \\ &= 1.6 + 0.0560 = 1.6560. \end{aligned}$$

A calculator yields $f(2.05) = 8.615125/5.2025 \approx 1.6559587$, so our pencil-and-paper approximation has an error of only 0.0000413. □

Sometimes, rather than estimate the change in function value produced by a change Δx in x, we are interested in estimating the change Δx that will produce a given change Δy in y. We find dx that will produce a change dy in the tangent line equal to the desired Δy in function value and take the resulting

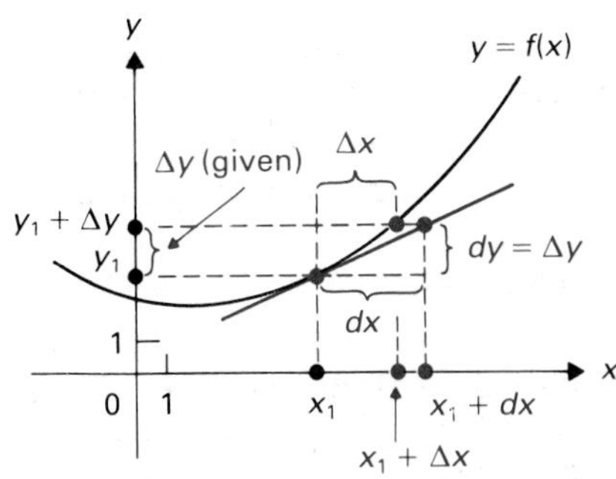

Figure 3.12 Taking $x_1 + dx$ as the approximation for $x_1 + \Delta x$ needed to produce a desired change $\Delta y = dy$ in $f(x)$.

dx as our estimate for Δx. This is illustrated in Fig. 3.12. In these days of easy access to calculators, this application of differentials is much more important than the illustrations of approximation (7) in Examples 4 through 6.

EXAMPLE 7 Estimate the value of x such that $x^3 = 8.05$.

Solution Let $f(x) = x^3$. We have $f(2) = 8$. We set $x_1 = 2$ and find dx that will produce a $dy = f'(x)\,dx$ of 0.05, which is the difference $8.05 - 8$. We have $0.05 = f'(2)\,dx$, so

$$dx = \frac{0.05}{f'(2)} = \frac{0.05}{3 \cdot 2^2} = \frac{0.05}{12} \approx 0.00417.$$

Thus $f(2.00417) \approx 8.05$. A calculator yields $f(2.00417) \approx 8.0501$. □

EXAMPLE 8 A circle has radius 2 ft. By approximately how much should the radius be increased to produce an increase of 0.25 ft^2 in the area of the circle?

Solution We have $A = \pi r^2$, so $dA = (2\pi r)\,dr$. We set $r = 2$ ft and find dr that produces a value 0.25 ft^2 for dA. Thus

$$0.25 = (2\pi \cdot 2)\,dr,$$

so

$$dr = \frac{0.25}{4\pi}\text{ ft.} \quad \square$$

EXAMPLE 9 Estimate a solution of the equation $x^3 - x^2 - 3 = 0$.

Solution Let $f(x) = x^3 - x^2 - 3$. Computing a few values, we find $f(-1) = -5$, $f(0) = -3$, $f(1) = -3$, $f(2) = 1$, and $f(3) = 15$. The intermediate-value theorem on page 73 then shows that the equation has a solution in the interval $[1, 2]$. Since $f(2) = 1$, we find dx that will produce dy of -1 at $x_1 = 2$ for $y = x^3 - x^2 - 3$. (See Fig. 3.13.) Since $dy = (3x^2 - 2x)\,dx$, we obtain

$$-1 = (3 \cdot 2^2 - 2 \cdot 2)\,dx = 8\,dx,$$

so $dx = -\frac{1}{8}$. Thus we expect $2 - \frac{1}{8} = \frac{15}{8}$ to be close to a solution of $x^3 - x^2 - 3 = 0$. A calculator shows that $f(\frac{15}{8}) \approx 0.076$. This technique can be extended to Newton's method for solving an equation, which is developed in Section 5.6. □

Figure 3.13 Finding dx corresponding to $dy = -1$ for the tangent line at (2, 1).

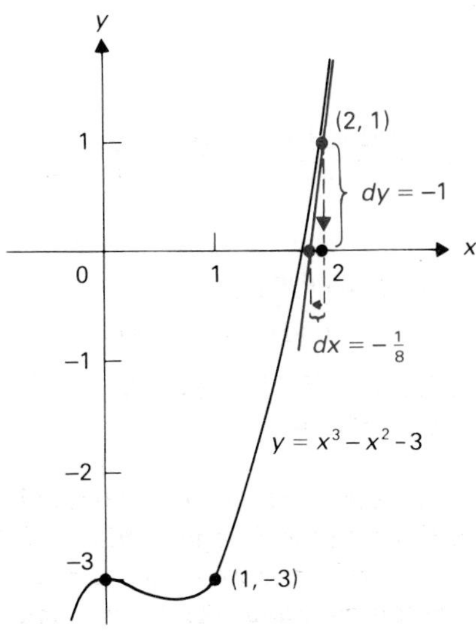

SUMMARY

1. If $f'(x_1)$ exists, then $\Delta y = f'(x_1)\,\Delta x + \varepsilon \cdot \Delta x$ where $\lim_{\Delta x \to 0} \varepsilon = 0$.
2. If $y = f(x)$, then the differential dy is $dy = f'(x)\,dx$.
3. If $f'(x_1)$ exists and is nonzero, the approximation $f(x_1 + dx) \approx f(x_1) + f'(x_1)\,dx$ is a good one for small values of dx.

EXERCISES

In Exercises 1 through 8, find the differential of the given function.

1. $y = f(x) = \dfrac{x}{x+1}$
2. $s = g(t) = t^3 - 2t^2 + 4t$
3. $A = f(r) = \pi r^2$
4. $y = f(x) = (x^2 + 1)(x^2 - x + 2)$
5. $x = h(t) = \dfrac{t^2 + 1}{t^2 - 1}$
6. $y = f(x) = x^3 + \dfrac{4}{x^2}$
7. $V = g(r) = \pi r^3$
8. $y = f(x) = \left(x + \dfrac{1}{x}\right)(x^2 - 3x)$

In Exercises 9 through 14, estimate the indicated quantity using a differential.

9. 0.999^{10}
10. $\dfrac{1}{10.05^5}$
11. $f(1.98)$ if $f(x) = \dfrac{x^3 + 4x}{2x - 1}$
12. $2.03^4 - 2.03^2$
13. $0.95^4 - \dfrac{1}{0.95^4}$
14. $1.03^5 + \dfrac{2}{1.03^2}$
15. Given that the derivative of $\sqrt{x}$ is $1/(2\sqrt{x})$, estimate $\sqrt{101}$. [You may think that $dx = 1$ is too large to give a good estimate, but the graph of $\sqrt{x}$ is turning so slowly at $x = 100$ that the tangent line is close to it for quite a distance.]
16. If $f(x) = x^4 - 3x^2$, then $f(2) = 4$. Use differentials to find the approximate value of x such that $f(x) = 3.98$.
17. If $f(x) = x^3/(x - 2)$, then $f(4) = 32$. Use differentials to find the approximate value of x such that $f(x) = 31.8$.

In Exercises 18 through 21, find the integer x where $|f(x)|$ is smallest and use a differential to estimate a solution of the equation $f(x) = 0$. (See Example 9.)

18. $f(x) = x^3 - 2$
19. $f(x) = x^3 - 7$
20. $f(x) = x^3 - 2x^2 + 18$
21. $f(x) = x^5 + x^3 - 42$
22. Estimate the change in volume of a cylindrical silo 20 ft high if the radius is increased from 3 ft to 3 ft 4 in.
23. Imagine the earth to be a ball of radius 4000 mi, and imagine a string tied around the equator of the earth. The string is cut, and six additional feet of string are inserted.
    a) If the lengthened circle of string were lifted a uniform height above the equator all the way around the earth, estimate how high above the surface of the earth the string would be.
    b) Discuss the accuracy of your estimate in part (a).
24. A ball has radius 4 ft, and a second ball has volume 1 ft^3 greater than the volume of the first ball. Estimate the difference in the surface area of these two balls. (The volume V and surface area A of a ball of radius r are given by $V = \frac{4}{3}\pi r^3$ and $A = 4\pi r^2$.)
25. A rectangle is inscribed in a semicircle of radius 5 ft. Estimate the increase in the area of the rectangle if the length of its base (on the diameter) is increased from 6 ft to 6 ft 2 in. [*Hint:* Let the area be A and the length of the base be x. Express A^2 as a function of x and use $d(A^2) = 2A \cdot dA$.]
26. A sphere has radius 4 ft. What change in radius will result in approximately 2 ft^3 increase in volume? (For a sphere, volume $V = \frac{4}{3}\pi r^3$.)
27. A silo consists of a cylinder of radius 4 ft and height 30 ft surmounted by a hemisphere. If the height of the cylinder remains 30 ft, approximately how much should the radius increase if the total volume is to be increased by 100 ft^3?

Scientists are sometimes interested in the *percent of error* in the measurement of a numerical quantity Q. If the error in computing Q is h, then the **percent of error** is

$$\left|\frac{100h}{Q}\right|.$$

Exercises 28 through 31 deal with estimating percent of error using differentials.

28. For a differentiable function f, let $Q = f(x_1) \neq 0$. Suppose Q is computed by "measuring x_1" and then computing $f(x_1) = Q$.
    a) If a small error of Δx is made in the measurement of x_1, argue that the approximate resulting percent of error in the computed value for Q is $|100f'(x_1)\,\Delta x/f(x_1)|$.
    b) If a small error of k percent of x_1 is made in the measurement of x_1, argue that the approximate resulting percent of error in the computed value for Q is
    $$|kx_1 f'(x_1)/f(x_1)|.$$
29. The radius of a sphere is found by measurement to be 2 ft plus or minus 0.04 ft. Estimate the maximum percent of error in computing the volume of the sphere from this measurement of the radius. (For a sphere, volume $V = \frac{4}{3}\pi r^3$.)
30. One side of an equilateral triangle is found by measurement to be 8 ft with an error of at most 3%. Estimate the maximum percent of error in computing the area of the triangle from this measurement of the length of a side.
31. If it is desired to compute the area of a circle with at most 1% error by measuring its radius, estimate the allowable percent of error that may be made in measuring the radius.
32. A plane flying over the ocean at night is headed straight toward a point A on the coast where a very strong light has been placed at the water line. Visibility is excellent, and the plane is flying at low altitude with the pilot's eyes 264 ft above the ocean. Assuming the radius of the earth is 4000 miles, use differentials to estimate the distance from the light to the plane when the light first becomes visible to the pilot. [*Hint:* If x is the distance from the light to the pilot and y is the distance from the pilot's eyes to the center of the earth, then
$$x^2 = y^2 - 4000^2.$$

(Draw a figure.) You will find that an attempt to estimate $x = \sqrt{y^2 - (4000)^2}$ by a differential for y near 4000 leads to difficulties. A successful technique is to estimate y^2 using a differential, then subtract 4000^2, and finally take the square root, possibly using another differential (see Exercise 15) if the number is not a perfect square.]

33. Let $f(x) = x^2 - 2x$. Find ε as a function of Δx and show directly that

$$\lim_{\Delta x \to 0} \varepsilon = 0.$$

(See Example 1.)

34. Repeat Exercise 33 for the function

$$f(x) = 1/x^2.$$

Use a calculator and a differential to find an approximate solution of the given equation. Use m_{sym} to approximate the necessary derivative.

35. $2^x = 8.3$

36. $3^x = 8.9$

37. $4^{\sqrt{x}} = 17$

38. $x^x = 4.15$

39. $x^{2x} = 15.95$

40. $x^{\cos x} = 6.3$ (Use radian measure. Note that $(2\pi)^{\cos 2\pi} = 2\pi$.)

3.4 THE CHAIN RULE

THE CHAIN RULE FORMULA

Consider the following question.

If Bill is running twice as fast as Mary, and Mary is running three times as fast as Jane, how many times as fast as Jane is Bill running?

Obviously, Bill is running $2 \cdot 3 = 6$ times as fast as Jane. Recall that an interpretation of dy/dx is the *rate of change of y with respect to x*. If we may be permitted to abuse the Leibniz notation, we can express the answer to our problem concerning Bill, Mary, and Jane by

$$\frac{d(\text{Bill})}{d(\text{Jane})} = \frac{d(\text{Bill})}{d(\text{Mary})} \cdot \frac{d(\text{Mary})}{d(\text{Jane})} = 2 \cdot 3 = 6.$$

Now we take a step toward more careful mathematics. Suppose $y = f(x)$ and $x = g(t)$, so that y appears as a **composite function** of t, namely,

$$y = f(g(t)).$$

Composite function

Figure 3.14 gives a schematic picture of the composite function $y = f(g(t))$. The value of the function for a given value of t is computed by "island hopping." We use the bridge $x = g(t)$ in Fig. 3.14 to get from the t-island to the

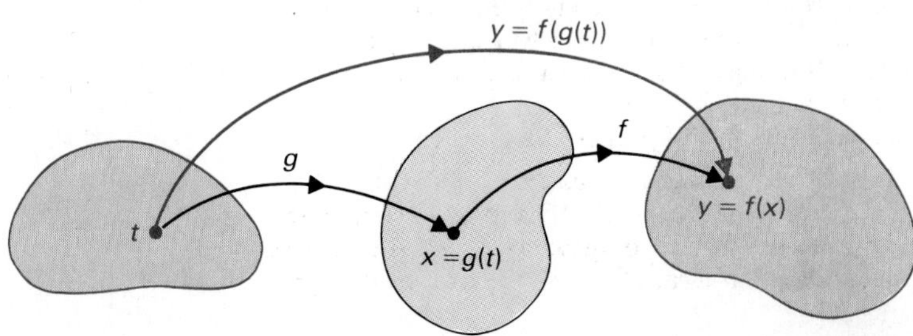

Figure 3.14 The composite function $y = f(g(t))$, or $y = (f \circ g)(t)$.

x-island, and then the bridge $y = f(x)$ to get from the x-island to the y-island. This composite function is defined for all t in the domain of g such that $g(t)$ is in the domain of f. The notation $f \circ g$ is sometimes used for the composite function, so that $(f \circ g)(t) = f(g(t))$. We will use the $y = f(g(t))$ notation.

EXAMPLE 1 Let $y = f(x) = x^2$, and let $x = g(t) = 2t + 1$. Find $f(g(3))$.

Solution We first compute $g(3) = 2 \cdot 3 + 1 = 7$. Then $f(g(3)) = f(7) = 7^2 = 49$. □

EXAMPLE 2 Suppose that $x = h(r) = r^2 + 2r$ and $r = k(s) = s^3 + 5s^2$. Find $x = h(k(s))$, expressing x directly in terms of s.

Solution We have $r = s^3 + 5s^2$, so

$$\begin{aligned} x = h(k(s)) = h(r) = h(s^3 + 2s^2) &= (s^3 + 5s^2)^2 + 2(s^3 + 5s^2) \\ &= s^6 + 10s^5 + 25s^4 + 2s^3 + 10s^2. \quad \square \end{aligned}$$

Let $x = g(t)$ be differentiable at t_1 and let $y = f(x)$ be differentiable at $x_1 = g(t_1)$. The analogy between the following question and the one at the start of this section should be obvious.

If y is increasing twice as fast as x is at x_1, and x is increasing three times as fast as t is at t_1, how many times as fast as t is y increasing when $t = t_1$?

Once again, it is really obvious that y must be increasing $2 \cdot 3 = 6$ times as fast as t is at t_1. In Leibniz notation, this becomes

$$\frac{dy}{dt} = \frac{dy}{dx} \cdot \frac{dx}{dt}. \tag{1}$$

Chain rule

We don't claim to have proved Eq. (1), the chain rule. We will see a more careful argument in a moment. However, good intuitive feeling for a concept is half the battle in an attempt to really understand it.

The chain rule illustrates the advantage of Leibniz notation in remembering formulas. We can remember Eq. (1) by pretending that the "dx's cancel." Again, this must not be regarded as a proof. Before giving a proof of the rule, we illustrate it with a few examples.

EXAMPLE 3 Let $y = x^2 - x$ and $x = t^3$. Find dy/dt when $t = 2$.

Solution Of course, $t = 2$ yields $x = 2^3 = 8$ and $y = 8^2 - 8 = 56$. Using the chain rule, we have

$$\frac{dy}{dt} = \frac{dy}{dx} \cdot \frac{dx}{dt} = (2x - 1)(3t^2).$$

When $t = 2$ and $x = 8$, we obtain

$$\left. \frac{dy}{dt} \right|_{t=2} = 15 \cdot 12 = 180.$$

Alternatively, we could solve the problem by expressing y directly as a function of t only and differentiating, without using the chain rule. From $y = x^2 - x$ and $x = t^3$, we obtain $y = t^6 - t^3$. Then $dy/dt = 6t^5 - 3t^2$, so $dy/dt|_{t=2} = 6 \cdot 32 - 3 \cdot 4 = 192 - 12 = 180$. □

EXAMPLE 4 If the length of an edge of a cube increases at a rate of 4 in./sec, find the rate of increase of the volume per second at the instant when the edge is 2 in. long.

Solution If an edge is of length x, then $V = x^3$. We wish to find the rate of change of V *with respect to time*, that is, dV/dt. Now

$$\frac{dV}{dt} = \frac{dV}{dx} \cdot \frac{dx}{dt} = (3x^2)\frac{dx}{dt}$$

by the chain rule. We are given that $dx/dt = 4$ in./sec. Thus when $x = 2$,

$$\frac{dV}{dt} = 12 \cdot 4 = 48 \text{ in}^3/\text{sec}. \quad \square$$

EXAMPLE 5 Find the derivative of the function $(4x^3 + 7x^2)^{10}$.

Solution We could raise the polynomial $4x^3 + 7x^2$ to the tenth power and obtain one polynomial expression, but that would be hard work. Let $y = (4x^3 + 7x^2)^{10}$ and let $u = 4x^3 + 7x^2$. Then $y = u^{10}$ and, by the chain rule, Eq. (1),

$$\frac{dy}{dx} = \frac{dy}{du} \cdot \frac{du}{dx} = 10u^9(12x^2 + 14x) = 10(4x^3 + 7x^2)^9(12x^2 + 14x).$$

Our problem is solved. □

So far we have seen only an intuitive, rate-of-change explanation of the chain rule, Eq. (1). A more careful argument can be made using Theorem 3.4 on page 95. Let $y = f(x)$ be differentiable at x_1 and $x = g(t)$ be differentiable at t_1 with $g(t_1) = x_1$. Then an increment Δt in t produces a corresponding increment Δx in $x = g(t)$, and Δx in turn produces an increment Δy in $y = f(x)$. From Theorem 3.4 we know that

$$\Delta y = f'(x_1) \cdot \Delta x + \varepsilon \cdot \Delta x, \qquad \text{where } \lim_{\Delta x \to 0} \varepsilon = 0. \tag{2}$$

Dividing Eq. (2) by Δt,

$$\frac{\Delta y}{\Delta t} = f'(x_1) \cdot \frac{\Delta x}{\Delta t} + \varepsilon \cdot \frac{\Delta x}{\Delta t}. \tag{3}$$

Now $x = g(t)$ is continuous at t_1 since g is differentiable there, so as Δt approaches zero, Δx approaches zero also, and, consequently, ε approaches zero. Therefore Eq. (3) yields

$$\left.\frac{dy}{dt}\right|_{t_1} = \lim_{\Delta t \to 0}\left(f'(x_1) \cdot \frac{\Delta x}{\Delta t} + \varepsilon \cdot \frac{\Delta x}{\Delta t}\right) = f'(x_1) \cdot \left.\frac{dx}{dt}\right|_{t_1} + 0 \cdot \left.\frac{dx}{dt}\right|_{t_1}$$

$$= f'(x_1) \cdot \left.\frac{dx}{dt}\right|_{t_1} = \left.\frac{dy}{dx}\right|_{x_1} \cdot \left.\frac{dx}{dt}\right|_{t_1}.$$

We have proved the following theorem.

THEOREM 3.5 Chain rule

Let $y = f(x)$ be differentiable at x_1 and $x = g(t)$ be differentiable at t_1, with $g(t_1) = x_1$. Then the composite function $y = f(g(t))$ is differentiable at t_1, and

$$\left.\frac{dy}{dt}\right|_{t_1} = \left.\frac{dy}{dx}\right|_{x_1} \cdot \left.\frac{dx}{dt}\right|_{t_1}.$$

If $y = f(x)$ and $x = g(t)$ are both differentiable functions, this shows that for $y = f(g(t))$,

$$\frac{dy}{dt} = \frac{dy}{dx} \cdot \frac{dx}{dt}. \tag{4}$$

In the notation $f \circ g$ for the composite function, Eq. (4) becomes

$$\boxed{(f \circ g)'(t_1) = f'(x_1) \cdot g'(t_1).}$$

DERIVATIVE OF A FUNCTION TO A POWER

We know that if n is any integer, then

$$\frac{d(u^n)}{du} = nu^{n-1}. \tag{5}$$

If u is in turn a function of x, then the chain rule, Eq. (4), yields

$$\boxed{\frac{d(u^n)}{dx} = \frac{d(u^n)}{du} \cdot \frac{du}{dx} = nu^{n-1} \cdot \frac{du}{dx}.} \tag{6}$$

Formula (6) is used so often that it is best to memorize it. The Summary gives it in an easily memorized verbal form.

EXAMPLE 6 Find dy/dx if $y = (x^3 - 2x)^5$.

Solution We think of $x^3 - 2x$ as u in formula (6) and obtain

$$\frac{dy}{dx} = 5(x^3 - 2x)^4 \cdot (3x^2 - 2). \quad \square$$

We can extend formula (6) to rational exponents p/q, where p and q are integers and $q \neq 0$. Let u be a differentiable function of x. It can be shown that then $u^{p/q}$ is differentiable, except possibly at points where u becomes zero. Using formula (6) for integer exponents, we have

$$\frac{d((u^{p/q})^q)}{dx} = q(u^{p/q})^{q-1} \cdot \frac{d(u^{p/q})}{dx} \tag{7}$$

and

$$\frac{d(u^p)}{dx} = pu^{p-1} \cdot \frac{du}{dx}. \tag{8}$$

Since $(u^{p/q})^q = u^p$, we obtain from Eqs. (7) and (8)

$$q(u^{p/q})^{q-1}\cdot\frac{d(u^{p/q})}{dx} = pu^{p-1}\cdot\frac{du}{dx},$$

so

$$\frac{d(u^{p/q})}{dx} = \frac{pu^{p-1}}{q(u^{p/q})^{q-1}}\cdot\frac{du}{dx} = \frac{p}{q}\cdot\frac{u^{p-1}}{u^{p-(p/q)}}\cdot\frac{du}{dx}$$

$$= \frac{p}{q}\cdot u^{p-1-p+(p/q)}\cdot\frac{du}{dx}.$$

Finally we obtain

$$\boxed{\frac{d(u^{p/q})}{dx} = \frac{p}{q}u^{(p/q)-1}\cdot\frac{du}{dx}.} \qquad \textbf{(9)}$$

Formula (9) is precisely what we get from formula (6) with $n = p/q$.

EXAMPLE 7 Find the derivative of the function $\sqrt{1 + x^2}$.

Solution From formula (9), we have

$$\frac{d(\sqrt{1 + x^2})}{dx} = \frac{d}{dx}[(1 + x^2)^{1/2}] = \frac{1}{2}(1 + x^2)^{-1/2}\cdot(2x)$$

$$= \frac{x}{\sqrt{1 + x^2}}. \quad \square$$

EXAMPLE 8 Find dy/dx if $y = (x^2 - 2x)^{2/3}\cdot\sqrt{x^3 + 1}$.

Solution We use the product rule for differentiation and formula (9). The answer we obtain may not be elegant, but it is easy to write in one step. We have

$$\frac{dy}{dx} = \left[(x^2 - 2x)^{2/3}\cdot\frac{1}{2}(x^3 + 1)^{-1/2}\cdot 3x^2\right]$$

$$+ \left[\sqrt{x^3 + 1}\cdot\frac{2}{3}(x^2 - 2x)^{-1/3}(2x - 2)\right]. \quad \square$$

The ease with which we can differentiate a fairly complicated function like the one in Example 8 makes us really appreciate our differentiation formulas, especially the chain rule. Imagine what a problem we would have if we had to use the definition of the derivative and compute the limit of the difference quotient. This ease of differentiation is what makes calculus practical.

EXAMPLE 9 Let $y = g(u)$ where g is differentiable for all u. Suppose $u = 1/\sqrt{3x + 4}$ and $dy/dx|_{x=4} = \frac{3}{16}$. If possible, find $g'(4)$ and $g'(\frac{1}{4})$.

Solution By the chain rule,

$$\frac{dy}{dx} = \frac{dy}{du}\cdot\frac{du}{dx}.$$

When $x = 4$, we have $u = 1/\sqrt{3 \cdot 4 + 4} = 1/\sqrt{16} = \frac{1}{4}$. Knowing dy/dx and du/dx when $x = 4$, we can find $dy/du = g'(u)$ when $u = \frac{1}{4}$, but we have no way of finding $g'(4)$. Now $u = (3x + 4)^{-1/2}$, so by formula (9), we have

$$\frac{du}{dx} = -\frac{1}{2}(3x + 4)^{-3/2} \cdot 3 = \frac{-3}{2(3x + 4)^{3/2}}.$$

We then have

$$\left.\frac{dy}{dx}\right|_{x=4} = \left.\frac{dy}{du}\right|_{u=1/4} \cdot \left.\frac{du}{dx}\right|_{x=4},$$

$$\frac{3}{16} = g'\left(\frac{1}{4}\right) \cdot \frac{-3}{2(16)^{3/2}} = g'\left(\frac{1}{4}\right) \cdot \frac{-3}{128}.$$

Thus

$$g'\left(\frac{1}{4}\right) = \frac{3}{16} \cdot \frac{128}{-3} = -8. \quad \square$$

We will see later that the formula

$$\frac{d(u^r)}{dx} = ru^{r-1} \cdot \frac{du}{dx} \tag{10}$$

holds for any real exponent r, and we give a verbal rendition of formula (10) in the Summary.

SUMMARY

1. If $y = f(x)$ and $x = g(t)$ are both differentiable functions, then so is $y = f(g(t))$ and

$$\frac{dy}{dt} = \frac{dy}{dx} \cdot \frac{dx}{dt}.$$

Without Leibniz notation, the formula is $[f(g(t))]' = f'(g(t)) \cdot g'(t)$.

2. The derivative of a function to a constant power is the power times the function to the one less power, times the derivative of the function. In symbols,

$$\frac{d(u^r)}{dx} = ru^{r-1} \cdot \frac{du}{dx}.$$

(We have demonstated this formula only for fractional $r = p/q$.)

EXERCISES

1. If $y = x^2 - 3x$ and $x = t^3 + 1$, find dy/dt when $t = 1$ by
   a) using the chain rule,
   b) expressing y as a function of t only and differentiating.

2. Repeat Exercise 1 for $y = \sqrt{3x + 12}$ and $x = t^2 - 2t$, where $t = 4$.

3. If $u = (v^2 + 3v - 4)^{3/2}$ and $v = w^3 - 3$, find du/dw where $w = 2$ by
   a) using the chain rule,
   b) expressing u directly as a function of w and differentiating.

4. Repeat Exercise 3 for $u = v^3 + 3v^2 - 2v$ and $v = w - 3$ at $w = 2$.

In Exercises 5 through 28, find dy/dx. You need not simplify your answers.

5. $y = (3x + 2)^4$

6. $y = (8x^2 - 17x)^3$

7. $y = (x^2 + 3x)^2(x^3 - 1)^3$

8. $y = (x^3 - 2)^3(2x^2 + 4x)^2$

9. $y = \dfrac{8x^2 - 2}{(4x^2 + 1)^2}$

10. $y = \dfrac{(3x^3 - 2)^2}{4x^3 + 2}$

11. $y = 4x^2 - 2x + 3x^{5/3}$

12. $y = x^3 - 2x^2 - 3\sqrt{x}$

13. $y = \dfrac{1}{\sqrt{x}}$

14. $y = \dfrac{1}{\sqrt[3]{x}}$

15. $y = x^{2/3} + x^{1/5}$

16. $y = 4x^{1/4} + 9x^{7/3}$

17. $y = \sqrt{2x + 1}$

18. $y = (3x - 2)^{4/3}$

19. $y = \dfrac{1}{\sqrt{5x^2 + 10x}}$

20. $y = \dfrac{3}{(4x^3 - 5x)^{3/2}}$

21. $y = \sqrt{x^2 + 1}$

22. $y = (2x^3 - 1)^{2/3}$

23. $y = \dfrac{\sqrt{x}}{x + 1}$

24. $y = \dfrac{x}{\sqrt{x + 1}}$

25. $y = \sqrt{3x + 4}\,(4x + 2)^{2/3}$

26. $y = \dfrac{(x^2 + 4)^{1/3}}{\sqrt{8x - 7}}$

27. $y = \sqrt{2x + 1}\,\dfrac{(4x^2 - 3x)^2}{2x + 5}$

28. $y = \dfrac{\sqrt{x^2 + 4}}{3x - 8}(2x^3 + 1)^2$

Exercises 29 through 34 depend on the fact that $d(\sin x)/dx = \cos x$, which is shown in the next chapter. Use this fact and other differentiation formulas to find dy/dx. You need not simplify your answers.

29. $y = \sin 2x$

30. $y = \sin (x^2)$

31. $y = \sin^3 x$

32. $y = \sin^3(4x + 1)$

33. $y = \sqrt{x + \sin x}$

34. $y = (x^2 + \sin x^3)^4$

In Exercises 35 through 38, assume the functions f and g are differentiable.

35. Let $y = [f(x)]^4$ and suppose $f'(1) = 5$ and $dy/dx|_{x=1} = -160$. Find $f(1)$.

36. Let $y = (f(x) + 3x^2)^3$ and suppose $f(-1) = -5$ and $dy/dx|_{x=-1} = 3$. Find $f'(-1)$.

37. Let $y = x/(x + 1)$ and suppose $x = g(t)$ and $g(5) = 1$. If $dy/dt|_{t=5} = 10$, find $g'(5)$.

38. Suppose $y = (f(x) + 3x)^2$ and $x = t^3 - 2t$. If $f(4) = 6$ and $dy/dt|_{t=2} = 180$, find $f'(4)$.

39. Find the equation of the line tangent to the circle $x^2 + y^2 = 25$ at the point (3, 4).

40. Find the equation of the line normal to the graph of $1/x$ at the point $(2, \frac{1}{2})$.

41. Find the equation of the line tangent to the graph of $\sqrt{2x + 1}$ at the point (4, 3).

42. Use differentials to estimate $\sqrt[3]{26}$.

42. Use differentials to estimate $(4.01)^3 + 3(\sqrt{4.01})$.

44. Use differentials to estimate $[(1.05)^3 + 1]^4$.

45. Use differentials to estimate $\sqrt{63 + 0.95^5}$. Do the estimation in two parts.

3.5 HIGHER-ORDER DERIVATIVES AND MOTION

For an object moving on a straight line, the rate of change of its position with respect to time is called the *velocity*. The rate of change of its velocity with respect to time is called the *acceleration*. The rate of change of the acceleration with respect to time has been given no particular name as far as we know, but we surely can consider it if we wish. Thus if the position of the object on the line (x-axis) is x, then

$$\text{Velocity} = \frac{dx}{dt}, \qquad \text{Acceleration} = \frac{d(dx/dt)}{dt}, \qquad \text{and}$$

$$\text{Rate of change of acceleration} = \frac{d\left(\dfrac{d(dx/dt)}{dt}\right)}{dt}.$$

We open this section by giving a more reasonable notation for such repeated derivatives. We then apply such derivatives to the study of motion on a line or in a plane.

HIGHER-ORDER DERIVATIVES

If $y = f(x)$ is a differentiable function, then $f'(x)$ is again a function of x, the *derived function* f'. We can attempt to find its derivative. The notations are

$$\frac{d(f'(x))}{dx} = \frac{d^2y}{dx^2} = f''(x).$$

The Leibniz notation d^2y/dx^2 is read "the second derivative of y with respect to x." Of course, we can then find the derivative of $f''(x)$, which is the third derivative of $f(x)$. Table 3.2 is a summary of the various notations for derivatives.

EXAMPLE 1 Find the first six derivatives of $y = f(x) = x^4 - 3x^3 + 7x^2 - 11x + 5$.

Solution We have

$$\frac{dy}{dx} = 4x^3 - 9x^2 + 14x - 11, \qquad \frac{d^2y}{dx^2} = 12x^2 - 18x + 14$$

$$\frac{d^3y}{dx^3} = 24x - 18, \qquad \frac{d^4y}{dx^4} = 24, \qquad \frac{d^5y}{dx^5} = 0, \qquad \frac{d^6y}{dx^6} = 0. \quad \square$$

EXAMPLE 2 If $f(x) = (3x + 2)^{5/3}$, find $f'''(2)$.

Solution We obtain

$$f'(x) = \frac{5}{3}(3x + 2)^{2/3} \cdot 3 = 5(3x + 2)^{2/3},$$

$$f''(x) = \frac{10}{3}(3x + 2)^{-1/3} \cdot 3 = 10(3x + 2)^{-1/3},$$

$$f'''(x) = -\frac{10}{3}(3x + 2)^{-4/3} \cdot 3 = -10(3x + 2)^{-4/3}.$$

Thus $f'''(2) = -10(8)^{-4/3} = -10 \cdot 2^{-4} = -\frac{10}{16} = -\frac{5}{8}$. $\square$

Table 3.2 Notations for derivatives of $y = f(x)$

Derivative	f'-notation	y'-notation	Leibniz Notation	D-notation
1st	$f'(x)$	y'	dy/dx	Df
2nd	$f''(x)$	y''	d^2y/dx^2	D^2f
3rd	$f'''(x)$	y'''	d^3y/dx^3	D^3f
4th	$f^{(4)}(x)$	$y^{(4)}$	d^4y/dx^4	D^4f
5th	$f^{(5)}(x)$	$y^{(5)}$	d^5y/dx^5	D^5f
$\vdots$	$\vdots$	$\vdots$	$\vdots$	$\vdots$
nth	$f^{(n)}(x)$	$y^{(n)}$	d^ny/dx^n	D^nf

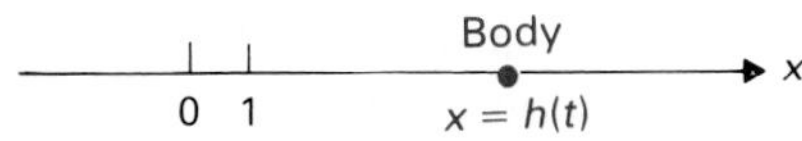

Figure 3.15 The position x of a moving body is a function h of time t.

MOTION ON A LINE

We now consider the motion of a body traveling on a line. We may as well assume that the line is the x-axis. The position of the body on the line at a time t corresponds to a point x on the line. Thus x appears as a function $h(t)$ of t. See Fig. 3.15. If x is a differentiable function of t, then dx/dt is the **velocity** v of the body at time t. Thus

$$\boxed{v = \frac{dx}{dt}.}$$

If $dx/dt > 0$, then a small positive increment Δt in t produces a positive increment Δx in x, so the body is moving in the positive x-direction, while if $dx/dt < 0$, the increment Δx is negative so the body is moving in the negative x-direction. The magnitude of the velocity (without regard to sign) is the **speed**, so that

$$\boxed{\text{Speed} = |v|.}$$

Suppose v is also a differentiable function of t. Since $dx/dt = v$, we have

$$\frac{d^2x}{dt^2} = \frac{dv}{dt} = \text{Rate of change of velocity with respect to time.}$$

This derivative dv/dt of the velocity is the **acceleration** a of the body. Thus

$$\boxed{a = \frac{dv}{dt} = \frac{d^2x}{dt^2}.}$$

If $dv/dt > 0$, then a small positive increment Δt in t produces a positive increment Δv in v, so the velocity is increasing, while if $dv/dt < 0$, the increment Δv is negative, so the velocity is decreasing. For example, if $dx/dt < 0$, then the body is moving in the negative x-direction. If also $d^2x/dt^2 > 0$, then the velocity is increasing, perhaps from -2.1 ft/sec to -1.9 ft/sec as t increases a small amount Δt, so the speed will be decreasing from 2.1 ft/sec to 1.9 ft/sec. The body is moving in the negative x-direction but slowing down since the positive acceleration is opposite to the negative direction of motion.

EXAMPLE 3 Let the position x of a body on a straight line (x-axis) at time $t \geq 0$ be given by

$$x = 4 - \frac{1}{t+1}.$$

Find the velocity and acceleration when $t = 3$.

Solution We have $x = 4 - (t+1)^{-1}$, so

$$v = \frac{dx}{dt} = (t+1)^{-2} \qquad \text{and} \qquad a = \frac{d^2x}{dt^2} = -2(t+1)^{-3}.$$

Thus

$$v|_{t=3} = \frac{1}{16} \qquad \text{and} \qquad a|_{t=3} = \frac{-2}{64} = -\frac{1}{32}.$$

The positive velocity indicates that the body is moving in the positive x-direction when $t = 3$, and the negative acceleration, with sign opposite to that of the velocity, indicates that the body is slowing down at $t = 3$. That is, the *speed* is decreasing. □

EXAMPLE 4 If the position x of a body on a line (x-axis) at time t is given by $x = \sqrt{3t^2 + 4}$, find the velocity, acceleration, and speed when $t = 2$.

Solution We have $x = \sqrt{3t^2 + 4} = (3t^2 + 4)^{1/2}$, so

$$v = \frac{dx}{dt} = \frac{1}{2}(3t^2 + 4)^{-1/2}6t = \frac{3t}{\sqrt{3t^2 + 4}},$$

$$a = \frac{d^2x}{dt^2} = \frac{(\sqrt{3t^2 + 4})3 - 3t(\frac{1}{2})(3t^2 + 4)^{-1/2} \cdot 6t}{3t^2 + 4}$$

$$= \frac{(3t^2 + 4)3 - 9t^2}{(3t^2 + 4)^{3/2}} = \frac{12}{(3t^2 + 4)^{3/2}}.$$

Thus

$$v|_{t=2} = \frac{6}{\sqrt{16}} = \frac{3}{2}, \qquad a|_{t=2} = \frac{12}{16^{3/2}} = \frac{12}{64} = \frac{3}{16}.$$

Now v is positive when $t = 2$, and we have

$$\text{Speed} = |v| = \frac{3}{2}.$$

Thus the body is moving to the right (assuming the x-axis is in the usual horizontal position), and the positive acceleration means that the velocity is increasing. Since the velocity is positive when $t = 2$, the speed is increasing also. □

Let an object move near the surface of the earth on a vertical line (y-axis) perpendicular to the surface of the earth. We indicate such motion in Fig. 3.16. Note that we take the origin of the y-axis at the earth's surface and its positive direction upward. We neglect air resistance and assume that the only force on the object is due to gravity. Exercise 45 of Section 5.9 will show that the position of the object until it hits the earth is given by an equation of the form

$$y = y_0 + v_0 t - \frac{1}{2}gt^2,$$

where y_0, v_0, and g are constants.

We can discover the physical meaning of these constants. When $t = 0$, we see that $y = y_0$, so y_0 is the *initial height* at time $t = 0$. Now

$$v = \frac{dy}{dt} = v_0 - gt.$$

When $t = 0$, we see that $v = v_0$, so v_0 is the *initial velocity* when $t = 0$. Finally,

$$a = \frac{d^2y}{dt^2} = -g,$$

Figure 3.16 An object moving upward near the earth's surface.

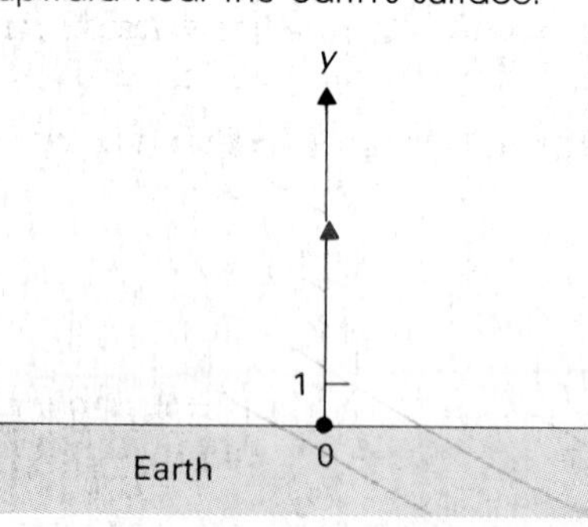

so $-g$ is the acceleration of the body. For acceleration due to gravity near the earth's surface, $g \approx 32$ ft/sec^2 in the British system, and $g \approx 9.8$ m/sec^2 in the metric system.

EXAMPLE 5 A ball is thrown straight upward and is released 2 m above the surface of the earth with velocity 19.6 m/sec at time $t = 0$. Neglecting air resistance, find

a. the height y above the surface at time t before the ball hits the earth,

b. the velocity at time t,

c. the maximum height the ball attains.

Solution

a. The discussion before this example shows that

$$y = 2 + 19.6t - 4.9t^2 \text{ m.}$$

b. We have $v = dy/dt = 19.6 - 9.8t$ m/sec.

c. At the instant the ball has maximum height, it is stopped while its velocity is changing from positive to negative. By the intermediate-value theorem, its velocity must be zero at that instant. This occurs when

$$19.6 - 9.8t = 0, \qquad \text{or} \quad t = \frac{19.6}{9.8} = 2 \text{ sec.}$$

The height then is given by

$$y|_{t=2} = 2 + (19.6)2 - (4.9)4 = 2 + 39.2 - 19.6 = 21.6 \text{ m.} \quad \square$$

MOTION IN A PLANE

Let a body be moving in the x,y-plane, perhaps in the direction of the arrows on the curve shown in Fig. 3.17. The curve need not be the graph of a function. The x-coordinate of the body's position at time t is some function $x = h(t)$, while the y-coordinate of the position is $y = k(t)$. The equations

$$\begin{aligned} x &= h(t), \\ y &= k(t), \end{aligned} \qquad \textbf{(1)}$$

are *parametric equations* of the curve, and t is the time *parameter*. The equation $x = h(t)$ describes the motion on the x-axis of the *x-projection*, which always stays right under (or over) the main body. Similarly, $y = k(t)$ gives the motion on the y-axis of the *y-projection*, which stays opposite the main body. See Fig. 3.17. Then dx/dt and d^2x/dt^2 are the velocity and acceleration of the x-projection, or the *x-components of the velocity and acceleration* of the main body. Similarly, dy/dt and d^2y/dt^2 are the *y-components of the velocity and acceleration.*

Figure 3.17 x- and y-projections of a body moving in the plane on a curve $x = h(t)$, $y = k(t)$.

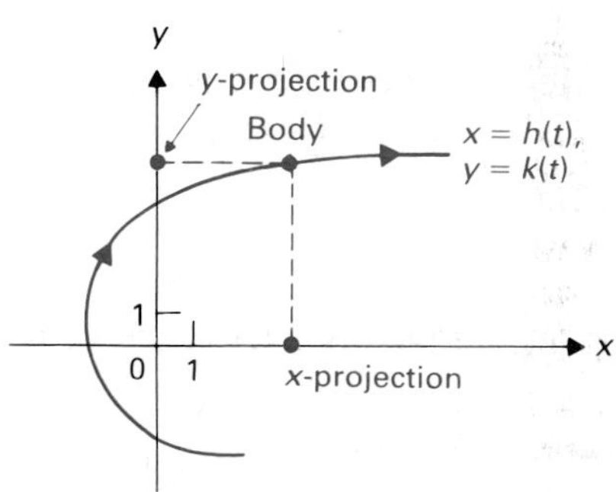

If $x = h(t)$ and $y = k(t)$ are differentiable functions at t and if a piece of the curve from time $t - \Delta t$ to time $t + \Delta t$, for some $\Delta t > 0$, is the graph of a differentiable function of x, then by the chain rule

$$\frac{dy}{dt} = \frac{dy}{dx} \cdot \frac{dx}{dt}.$$

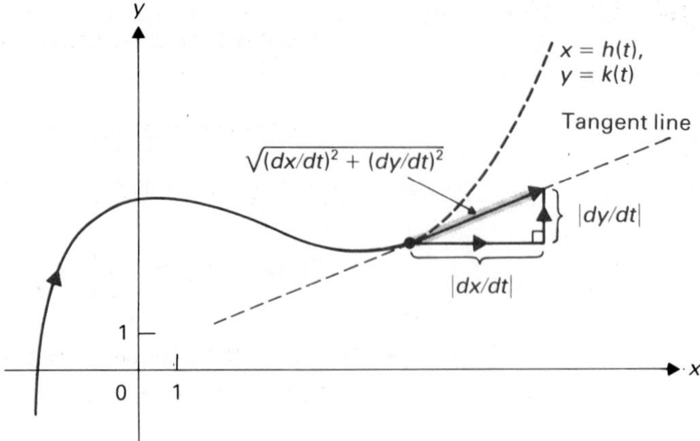

Figure 3.18
Speed = $\sqrt{(dx/dt)^2 + (dy/dt)^2}$

Therefore where $dx/dt \neq 0$, the slope of the curve is given by

$$\boxed{\frac{dy}{dx} = \frac{dy/dt}{dx/dt}.} \tag{2}$$

Again, the handy Leibniz notation makes it easy to remember the formula.

At each instant t, the body is moving in the direction tangent to the curve. That is, if one could command at any instant, "Stop curving and keep going in the same direction you are now and at the same speed," then the body would go off along a tangent line to the curve. If the body were to go off on the tangent line and *keep the same speed* as it had at the instant t, then in one unit of time it would travel a distance $|dx/dt|$ in the x-direction and a distance $|dy/dt|$ in the y-direction. As shown in Fig. 3.18, the distance traveled along the hypotenuse, tangent to the curve, in one unit of time would then be $\sqrt{(dx/dt)^2 + (dy/dt)^2}$. Therefore the speed of the body at time t is

$$\boxed{\text{Speed} = \sqrt{\left(\frac{dx}{dt}\right)^2 + \left(\frac{dy}{dt}\right)^2}.} \tag{3}$$

EXAMPLE 6 Let the position of a body in the plane at time t be given by

$$x = t^3 - 3t,$$
$$y = 2t^2 + 7t.$$

Find the x- and y-components of the velocity and acceleration, the speed, and the slope of the path at any instant t.

Solution From the discussion above, we obtain

$$x\text{-component of velocity} = \frac{dx}{dt} = 3t^2 - 3,$$

$$y\text{-component of velocity} = \frac{dy}{dt} = 4t + 7,$$

$$x\text{-component of acceleration} = \frac{d^2x}{dt^2} = 6t,$$

$$y\text{-component of acceleration} = \frac{d^2y}{dt^2} = 4,$$

$$\text{Speed} = \sqrt{(dx/dt)^2 + (dy/dt)^2} = \sqrt{(3t^2 - 3)^2 + (4t + 7)^2},$$

$$\text{Slope of curve} = \frac{dy}{dx} = \frac{dy/dt}{dx/dt} = \frac{4t + 7}{3t^2 - 3}. \quad \square$$

EXAMPLE 7 If the position of a body in the plane at time t is given by $x = t^2 + 3$, $y = \sqrt{t^2 + 5}$, find the speed when $t = 2$.

Solution We have

$$\frac{dx}{dt} = 2t, \qquad \text{so} \quad \left.\frac{dx}{dt}\right|_{t=2} = 4,$$

$$\frac{dy}{dt} = \frac{1}{2}(t^2 + 5)^{-1/2} \cdot 2t, \qquad \text{so} \quad \left.\frac{dy}{dt}\right|_{t=2} = \frac{2}{3}.$$

Thus

$$\text{Speed}\,|_{t=2} = \sqrt{4^2 + (2/3)^2} = \sqrt{16 + (4/9)} = \frac{\sqrt{148}}{3} = \frac{2\sqrt{37}}{3}. \quad \square$$

EXAMPLE 8 Find the equation of the line normal to the curve given parametrically by

$$x = t^2 + 1,$$
$$y = 2t^3 - 6t,$$

at the point where $t = 2$.

Solution To find the equation of a line, we need to know a point on the line and the slope of the line.

Point When $t = 2$, we see that $x = 5$ and $y = 4$, so the point is $(5, 4)$.

Slope The tangent line has slope

$$\left.\frac{dy}{dx}\right|_{t=2} = \left.\frac{dy/dt}{dx/dt}\right|_{t=2} = \left.\frac{6t^2 - 6}{2t}\right|_{t=2} = \frac{18}{4} = \frac{9}{2}.$$

Therefore the normal line has slope $-\frac{2}{9}$.

Equation $y - 4 = -\frac{2}{9}(x - 5)$, or $9y + 2x = 46$ $\square$

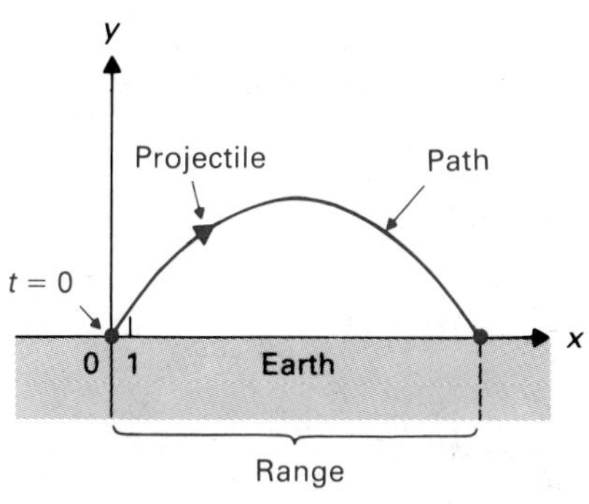

Figure 3.19 The path of a projectile fired from the earth at time $t = 0$.

Figure 3.19 shows the path of a projectile fired at time $t = 0$ from the origin in the plane, taking the x-axis along the earth's surface under the projectile. If air resistance is neglected, the position of the projectile at time t until it hits the earth is given by parametric equations of the form

$$x = v_x t$$
$$y = (v_y)_0 t - \tfrac{1}{2}gt^2,$$

where v_x and $(v_y)_0$ are constants. This is shown in Example 8 of Section 15.1. By arguments like those preceding Example 5 in this section, we easily see that v_x is the x-component of the velocity at all times, while $(v_y)_0$ is the y-component of the velocity when $t = 0$. Again, $-g$ is the acceleration due to gravity. The *muzzle velocity* of the projectile is defined to be the speed when $t = 0$. The *range* is the distance along the x-axis from the origin where the projectile is fired to the point where it hits, as shown in Fig. 3.19.

EXAMPLE 9 Let the position of the projectile just discussed be given in meters by

$$x = 110t, \qquad y = 98t - 4.9t^2$$

until the projectile hits the ground. Find

a. the muzzle velocity of the projectile,

b. the maximum height the projectile attains,

c. the time the projectile hits the ground, and

d. the range of the projectile.

Solution

a. We have

$$\frac{dx}{dt} = 110$$

and

$$\frac{dy}{dt} = 98 - 9.8t.$$

By Eq. (3), we see that

$$\begin{aligned}\text{Muzzle velocity} &= \sqrt{\left(\frac{dx}{dt}\right)^2 + \left(\frac{dy}{dt}\right)^2}\,\Bigg|_{t=0} \\ &= \sqrt{(110)^2 + (98)^2} \\ &= \sqrt{12100 + 9604} \\ &= \sqrt{21704} \approx 147.3 \text{ m/sec.}\end{aligned}$$

b. As in Example 5, we have maximum height when $v_y = 98 - 9.8t = 0$. This occurs when $t = 10$ sec. Thus the maximum height is

$$y|_{t=10} = (98)10 - (4.9)10^2 = 980 - 490 = 490 \text{ m.}$$

c. The projectile hits the earth when $y = 0$, or when

$$t(98 - 4.9t) = 0.$$

This occurs when $t = 0$ or $t = 20$. Since $t = 0$ is the time when the projectile was fired, the projectile hits the ground when $t = 20$ sec.

d. From part (c), we see that the range is given by

$$x|_{t=20} = 110 \cdot 20 = 2200 \text{ m.} \quad \square$$

NUMERICAL APPROXIMATION OF $f''(x_1)$ AND $f'''(x_1)$ USING A CALCULATOR (OPTIONAL)

In Section 3.1, we saw that $f'(x_1)$ can be approximated numerically using

$$f'(x_1) \approx \frac{f(x_1 + \Delta x) - f(x_1 + \Delta x)}{2 \cdot \Delta x} \tag{4}$$

for small Δx. Now we will give formulas for the numerical approximation of $f''(x_1)$ and $f'''(x_1)$. First choose a small Δx and compute:

$$y_1 = f(x_1 - 2 \cdot \Delta x), \qquad y_2 = f(x_1 - \Delta x), \qquad y_3 = f(x_1),$$
$$y_4 = f(x_1 + \Delta x), \qquad y_5 = f(x_1 + 2 \cdot \Delta x).$$

It can be shown that

$$f''(x_1) \approx \frac{-y_1 + 16y_2 - 30y_3 + 16y_4 - y_5}{12(\Delta x)^2} \tag{5}$$

and

$$f'''(x_1) \approx \frac{-y_1 + 2y_2 - 2y_4 + y_5}{2(\Delta x)^3}. \tag{6}$$

These approximations are derived by finding the fourth-degree polynomial through the five points on the graph where x is equal to

$$x_1 - 2 \cdot \Delta x, \quad x_1 - \Delta x, \quad x_1, \quad x_1 + \Delta x, \quad \text{and} \quad x_1 + 2 \cdot \Delta x.$$

Formula (5) gives the second derivative of this polynomial where $x = x_1$, and formula (6) gives the third derivative where $x = x_1$. Note that formula (4) can be written as

$$f'(x_1) \approx \frac{-y_2 + y_4}{2 \cdot \Delta x}$$

in this notation. It can be shown that this is actually the derivative where $x = x_1$ of the second-degree polynomial through the points where $x = x_1 - \Delta x$, x_1, and $x_1 + \Delta x$.

EXAMPLE 10 Use formula (5) with $\Delta x = 1$ to approximate $f''(3)$ if $f(x) = 1/x$. Of course, $\Delta x = 1$ is large, but it makes an easy computation for this noncalculator illustration.

Solution We have $x_1 = 3$ and $\Delta x = 1$, so

$$x_1 - 2 \cdot \Delta x = 1, \qquad x_1 - \Delta x = 2, \qquad x_1 = 3,$$
$$x_1 + \Delta x = 4, \qquad x_1 + 2 \cdot \Delta x = 5,$$
$$y_1 = 1, \qquad y_2 = \tfrac{1}{2}, \qquad y_3 = \tfrac{1}{3}, \qquad y_4 = \tfrac{1}{4}, \qquad y_5 = \tfrac{1}{5}.$$

The approximation (5) becomes

$$f''(3) \approx \frac{-1 + 16 \cdot \frac{1}{2} - 30 \cdot \frac{1}{3} + 16 \cdot \frac{1}{4} - \frac{1}{5}}{12(1)^2}$$

$$= \frac{-1 + 8 - 10 + 4 - 0.2}{12} = \frac{0.8}{12} \approx 0.067.$$

Of course, if $y = x^{-1}$, then $y' = -1 \cdot x^{-2}$ and $y'' = 2x^{-3}$, so $f''(3) = 2/(3)^3 = \frac{2}{27} \approx 0.074$. Our error in approximation is about 0.007. □

SUMMARY

1. If $y = f(x)$, then the second derivative of y with respect to x is

$$\frac{d^2y}{dx^2} = \frac{d(dy/dx)}{dx} = f''(x) = y'' = D^2f,$$

and the nth derivative of y with respect to x is

$$\frac{d^ny}{dx^n} = f^{(n)}(x) = y^{(n)} = D^nf.$$

2. If x is the position of a body on a line (x-axis) at time t, then

$$\text{Velocity} = v = \frac{dx}{dt}, \qquad \text{Acceleration} = a = \frac{dv}{dt} = \frac{d^2x}{dt^2},$$

$$\text{Speed} = |v|.$$

3. Given parametric equations $x = h(t)$ and $y = k(t)$ of motion in a plane,

$$\frac{dx}{dt} = x\text{-component of the velocity},$$

$$\frac{dy}{dt} = y\text{-component of the velocity},$$

$$\sqrt{\left(\frac{dx}{dt}\right)^2 + \left(\frac{dy}{dt}\right)^2} = \text{speed}, \qquad \frac{dy}{dx} = \frac{dy/dt}{dx/dt} = \text{slope of the curve},$$

$$\frac{d^2x}{dt^2} = x\text{-component of the acceleration},$$

$$\frac{d^2y}{dt^2} = y\text{-component of the acceleration}.$$

EXERCISES

In Exercises 1 through 8, find y', y'', and y'''. You need not simplify your answers.

1. $y = x^5 - 3x^4$
2. $y = \sqrt{x}$
3. $y = 1/\sqrt{5x}$
4. $y = x^{2/3}$
5. $y = \sqrt{x^2 + 1}$
6. $y = (3x - 2)^{3/4}$
7. $y = \dfrac{x}{x + 1}$
8. $y = x(x + 1)^4$
9. Find $f''(3)$ if $f(x) = \sqrt{2x + 4}$.
10. Find $g''(-1)$ if $g(t) = 8t^3 - 3t^2 + 14t - 5$.
11. Find $f^{(4)}(2)$ if $f(s) = 1/s$.
12. Find $f^{(4)}(1)$ if $f(x) = x/(x + 1)$.
13. If the position x of a body on an x-axis at time t is given by $x = 3t^3 - 7t$, find the velocity and acceleration of the body when $t = 1$.
14. Repeat Exercise 13 if $x = \sqrt{t^2 + 7}$ and $t = 3$.
15. Let the position x on an x-axis of a body at time t be given by

$$x = 10 - \frac{20}{t^2 + 1} \quad \text{for } t \geq 0.$$

a) Show that, at any time $t \geq 0$, the body has traveled less than 20 units distance since time $t = 0$.
b) Find the velocity of the body as a function of t.
c) Interpret physically the fact that the velocity is positive for all $t > 0$.

16. Let $x = 10/(t + 1)$ be the position of a body on an x-axis for $t \geq 0$.
a) Find the velocity as a function of t.
b) Interpret physically the fact that the velocity is negative for all $t > 0$.
c) Find the acceleration as a function of t.
d) Interpret physically the fact that the acceleration is positive for all $t > 0$.
e) Is the speed increasing or decreasing for $t > 0$?

17. A body is thrown straight up from the surface of the earth at time $t = 0$. Suppose that its height y in feet after t sec is $y = -16t^2 + 48t$.
a) Find the velocity of the body at time t.
b) Find the acceleration of the body at time t.
c) Find the initial velocity v_0 with which the body was thrown upward.
d) At what time does the body reach maximum height? [*Hint:* What is the velocity at maximum height?]
e) Find the maximum height the body attains.
f) From the physics of the problem, during what time interval is the equation $y = -16t^2 + 48t$ valid?

In Exercises 18 through 23, $y = k(t)$ is the position at time t of a body on the y-axis (positive direction upward, as usual). Fill in the blanks with the proper choice: *upward, downward, increasing, decreasing.*

18. If $dy/dt > 0$, the body is moving _____ as t increases.

19. If $d^2y/dt^2 < 0$, the velocity is _____ as t increases.

20. If $d^2y/dt^2 < 0$ and $dy/dt > 0$, the speed of the body is _____ as t increases.

21. If $d^2y/dt^2 < 0$ and $dy/dt < 0$, the speed of the body is _____ and the body is moving _____ as t increases.

22. If $d^2y/dt^2 > 0$ and $dy/dt < 0$, the speed of the body is _____ and the body is moving _____ as t increases.

23. If $d^2y/dt^2 > 0$ and $dy/dt > 0$, the speed of the body is _____ and the body is moving _____ as t increases.

In Exercises 24 through 29, find the speed of the body and the slope of the curve at the indicated time for the given parametric motion.

24. $x = t^3$ $y = t^2$ at time t

25. $x = 1/t^2$, $y = (t + 1)/(t - 1)$ at $t = 2$

26. $x = \sqrt{t^2 + 1}$, $y = (t^2 - 1)^3$ at $t = 1$

27. $x = (2t - 4)^2$, $y = 1 + (1/t)$ at $t = 2$

28. $x = (3t + 7)^{3/2}$, $y = t - 100$ at $t = 3$

29. $x = \sqrt{4t + 5}$, $y = 1/(t^2 - 4t)$ at $t = 5$

30. Find the equation of the tangent line to the curve $x = t^2 - 3t$, $y = \sqrt{t}$, at the point where $t = 4$.

31. Find the equation of the tangent line to the curve $x = 3t + 4$, $y = t^2 - 3t$ at the point where $x = -5$.

32. Find the equation of the normal line to the curve $x = (t^2 - 3)^2$, $y = (t^2 + 3)/(t - 1)$, where $t = 2$.

33. Find the equation of the normal line to the curve $x = \sqrt{2t^2 + 1}$, $y = 4t - 5$, where $y = 3$.

34. Let $x = \sqrt{t}$ and $y = t^3$. Find dy/dx and d^2y/dx^2 in terms of t. [*Hint:*

$$\frac{d^2y}{dx^2} = \frac{d(dy/dx)}{dx} = \frac{[d(dy/dx)]/dt}{dx/dt}.\Big]$$

35. Let $x = t^2$, $y = t^3 - 2t^2 + 5$. Find dy/dx and d^2y/dx^2 when $t = 1$. [See the hint of Exercise 34.]

36. A ball is tossed upward with a velocity of 7 m/sec. The ball is released 1.5 m above the ground at time $t = 0$. Neglecting air resistance, find
a) the velocity when $t = 1$ sec,
b) the speed of the ball when $t = 1$ sec,
c) the maximum height the ball attains.

37. A boy standing on a balcony reaches out and tosses a ball straight upward with velocity 14 m/sec. The ball is released 15 m above the ground at time $t = 0$. Neglecting air resistance, find
a) the velocity when $t = 1$,
b) the speed when $t = 1$,
c) the maximum height the ball attains,
d) the elapsed time until the ball hits the ground.

38. A projectile is fired from the origin in the plane when $t = 0$, as shown in Fig. 3.19. Find the y-component of the velocity when $t = 0$ if the x-component of the velocity then is 40 m/sec and the muzzle velocity is 50 m/sec.

39. A projectile is fired from the origin in the plane when $t = 0$, as shown in Fig. 3.19. Its position in meters at time t sec is given by

$$x = 30t, \qquad y = 49t - 4.9t^2.$$

Find
a) the maximum height the projectile attains,
b) the range of the projectile,
c) the speed of the projectile when it hits the ground.

40. a) Find the speed of the projectile in Exercise 39 at the moment that it attains its maximum height.
b) Make a general statement based on the computation in part (a).

In Exercises 41 through 46, use a calculator to find the second and third derivatives of the given functions at the indicated points, using approximations (5) and (6). You choose Δx.

41. $f(x) = \sqrt{x(x^3 + 2x^2)}$ at $x = 1$

42. $f(x) = \dfrac{\sin x}{x + 2}$ at $x = 3$ (radian measure)

43. $f(x) = 2^x$ at $x = 3$

44. $f(x) = x^x$ at $x = 1.5$

45. $f(x) = (x^2 + 3x)^{\sqrt{x}}$ at $x = 1$

46. $f(x) = x^{\sin x}$ at $x = 2$ (radian measure)

3.6 IMPLICIT DIFFERENTIATION

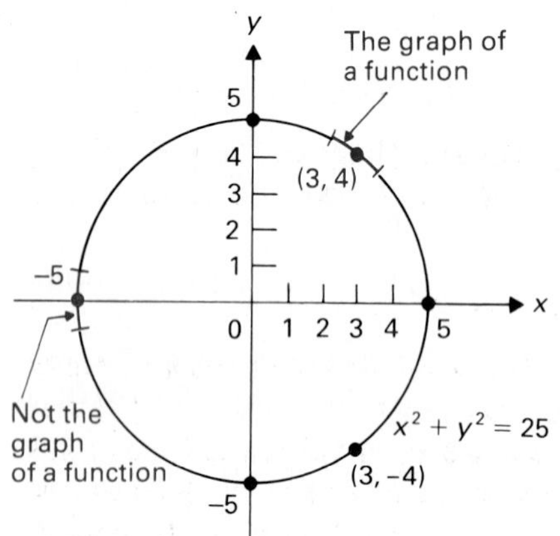

Figure 3.20 Some short arcs of $x^2 + y^2 = 25$ are graphs of functions; others are not.

The graph of a continuous function f of one variable can be viewed as a curve in the plane. The curve has a tangent line at each point where f is differentiable.

Surely it is natural to consider the circle $x^2 + y^2 = 25$ in Fig. 3.20 to be a plane curve. However, this curve is not the graph of a function, since distinct points on the circle may have the same x-coordinate. For example, (3, 4) and (3, −4) are both on the circle. Let us restrict our attention to just a small part of the circle containing (3, 4) and extending a short distance on both sides of (3,4). See the colored curve in Fig. 3.20. This small piece is the graph of a function. In fact, it is part of the graph of the function $y = f(x) = \sqrt{25 - x^2}$, which has as graph the whole upper semicircle. We would like to find the derivative $f'(3)$ of this function, that is, the slope of the tangent line to the circle at the point (3, 4). Of course, we have

$$f'(x) = \frac{1}{2}(25 - x^2)^{-1/2}(-2x) = \frac{-x}{\sqrt{25 - x^2}},$$

so

$$f'(3) = \frac{-3}{\sqrt{16}} = -\frac{3}{4}.$$

We now show another way to find $f'(3)$, the slope of the tangent line to the circle $x^2 + y^2 = 25$ at (3, 4). In this new method, we will not have to solve for y in terms of x. Thinking of y as some function of x near the point with which we are concerned, we differentiate both sides of the equation $x^2 + y^2 = 25$ with respect to x. Since we are differentiating *with respect to* x and since y is a function of x, the chain rule yields

$$\frac{d(y^2)}{dx} = 2y\frac{dy}{dx}.$$

Thus we obtain

$$x^2 + y^2 = 25, \qquad \text{Given}$$

$$2x + 2y\frac{dy}{dx} = 0, \qquad \text{Differentiating}$$

$$\frac{dy}{dx} = -2\frac{2x}{2y} = -\frac{x}{y}, \qquad \text{Solving for } dy/dx$$

$$\left.\frac{dy}{dx}\right|_{(3,4)} = -\frac{3}{4}. \qquad \text{Evaluating at } (x, y) = (3, 4)$$

Figure 3.21 A portion of the curve extending a short distance both ways from (x_1, y_1) is the graph of a function.

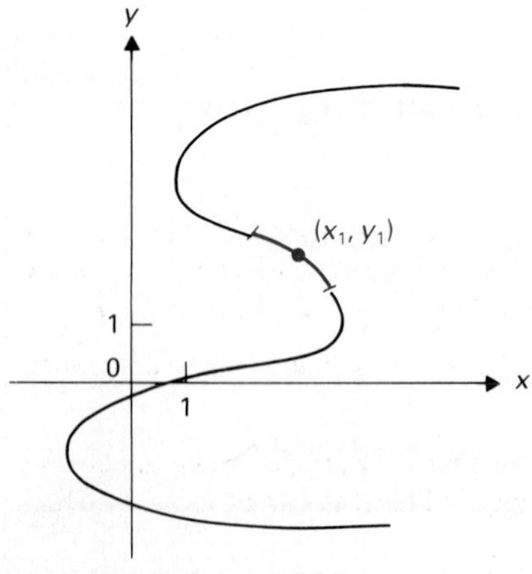

We obtained the same answer as in the preceding paragraph, but no radical appeared in this computation!

The computation we have just completed illustrates our topic in this section. Let a plane curve be given by an x,y-equation. The curve may not be the graph of a function, but a piece of the curve extending out for a way on both sides of a point (x_1, y_1) may be the graph of a function, as shown in Fig. 3.21. In this case, the x,y-equation defines y **implicitly** as a function of x near (x_1, y_1). An equation $y = f(x)$, on the other hand, defines y **explicitly** as a function of x. The terms *implicit* and *explicit* are used to distinguish between these cases. Turning back to our circle,

$x^2 + y^2 = 25$ *defines y implicitly as a function of x near* (3, 4),

while

$y = \sqrt{25 - x^2}$ *defines y explicitly as a function of x near* (3, 4).

For the circle $x^2 + y^2 = 25$, we can solve explicitly for y in terms of x and then find dy/dx. But often it is very difficult to solve an equation for y. For example, it is hard to solve for y if

$$y^5 + 3y^2 - 2x^2 = -4.$$

However, assuming that y is defined implicitly as a differentiable function of x near some point (x, y) on the curve, we can use the chain rule to find dy/dx without solving for y. This technique is *implicit differentiation* and was illustrated above. The labels on the steps of the computation there can be taken as an outline for the general technique. (See the box at the bottom of this page.)

We give several more illustrations of the technique.

EXAMPLE 1 Find dy/dx if $y^5 + 3y^2 - 2x^2 = -4$.

Solution Viewing y as a function of x and using the chain rule, we have

$$\frac{d(y^5)}{dx} = \frac{d(y^5)}{dy} \cdot \frac{dy}{dx} = 5y^4 \frac{dy}{dx} \quad \text{and} \quad \frac{d(3y^2)}{dx} = 6y \frac{dy}{dx}.$$

Differentiating both sides of $y^5 + 3y^2 - 2x^2 = -4$ with respect to x, we obtain

$$5y^4 \frac{dy}{dx} + 6y \frac{dy}{dx} - 4x = 0.$$

We may now solve for dy/dx, obtaining

$$\frac{dy}{dx} = \frac{4x}{5y^4 + 6y}.$$

This formula gives dy/dx at any point (x, y) on the curve where the denominator $5y^4 + 6y$ is nonzero. For example, we can easily see that (2, 1) satisfies

FINDING dy/dx BY IMPLICIT DIFFERENTIATION

Step 1 Write an equation in x and y.

Step 2 Differentiate both sides of the equation with respect to x, thinking of y as some function of x. (When differentiating any y-term, the chain rule must be used, so the derivative of that term will end with the factor dy/dx.)

Step 3 Solve algebraically for dy/dx. (Since dy/dx always appears to the first power only, this is always easy.)

Step 4 If you need the derivative corresponding to some point (x_1, y_1) on the curve, evaluate the expression for dy/dx in Step 3 at that point.

$y^5 + 3y^2 - 2x^2 = -4$ and therefore lies on the curve. Then

$$\left.\frac{dy}{dx}\right|_{(2,1)} = \left.\frac{4x}{5y^4 + 6y}\right|_{(2,1)} = \frac{8}{11}. \quad \square$$

EXAMPLE 2 Find dy/dx if $x^3 + 2x^2y^3 + 3y^4 = 6$.

Solution Implicit differentiation yields

$$3x^2 + 2x^2 \cdot 3y^2 \frac{dy}{dx} + 4xy^3 + 12y^3 \frac{dy}{dx} = 0.$$

Therefore

$$(6x^2y^2 + 12y^3)\frac{dy}{dx} = -(3x^2 + 4xy^3)$$

and

$$\frac{dy}{dx} = -\frac{3x^2 + 4xy^3}{6x^2y^2 + 12y^3}. \quad \square$$

As illustrated by the preceding examples, we obtain a quotient when finding dy/dx by implicit differentiation of an x,y-equation. At certain points (x, y) on the curve, the denominator may become zero. It can be shown that if this calamity does not occur at (x, y) and the curve is a sufficiently "nice" one, then y is implicitly defined as a function of x near (x, y), and implicit differentiation does yield dy/dx.

EXAMPLE 3 Air is blown into a spherical balloon, increasing its volume V. Find the rate of increase of its surface area S with respect to V when V is 288π cm^3.

Solution We wish to compute dS/dV when $V = 288\pi$ cm^3. For a sphere of radius r, we have $V = \frac{4}{3}\pi r^3$ and $S = 4\pi r^2$. To obtain an equation involving just V and S, note that

$$r^3 = \frac{3V}{4\pi} \quad \text{and} \quad r^2 = \frac{S}{4\pi}.$$

Therefore

$$r^6 = (r^3)^2 = \frac{9V^2}{16\pi^2} = (r^2)^3 = \frac{S^3}{64\pi^3}.$$

Then

$$\frac{S^3}{64\pi^3} = \frac{9V^2}{16\pi^2}, \quad \text{so} \quad S^3 = 36\pi V^2.$$

Differentiating this last equation with respect to V, we have

$$3S^2 \frac{dS}{dV} = 72\pi V,$$

$$\frac{dS}{dV} = \frac{72\pi V}{3S^2} = 24\pi \frac{V}{S^2}.$$

When $V = \frac{4}{3}\pi r^3 = 288\pi$, we obtain

$$r^3 = \frac{3}{4}(288) = 3(72) = 216, \qquad \text{so} \quad r = 6.$$

Then $S = 4\pi r^2 = 144\pi$. Our answer is therefore

$$24\pi \left.\frac{V}{S^2}\right|_{\substack{S=144\pi \\ V=288\pi}} = \frac{24\pi \cdot 288\pi}{144^2\pi^2} = \frac{24}{72} = \frac{1}{3}\ \text{cm}^2/\text{cm}^3. \quad \square$$

We now give an application to geometry. First we define the notion of *orthogonal* curves.

DEFINITION 3.4 Orthogonal curves

Two curves are **orthogonal** at a point of intersection if they have perpendicular tangent lines at that point.

EXAMPLE 4 Show that the curve $y - x^2 = 0$ is orthogonal to the curve $x^2 + 2y^2 = 3$ at the point $(1, 1)$ of intersection.

Solution The curves are shown in Fig. 3.22. The slope of the tangent line to $y = x^2$ at $(1, 1)$ is

$$\left.\frac{dy}{dx}\right|_{x=1} = 2x\Big|_{x=1} = 2.$$

Differentiating $x^2 + 2y^2 = 3$ implicitly, we obtain

$$2x + 4y\frac{dy}{dx} = 0,$$

so

$$\frac{dy}{dx} = \frac{-2x}{4y} = \frac{-x}{2y}.$$

Then

$$\left.\frac{dy}{dx}\right|_{(1,1)} = -\frac{1}{2}.$$

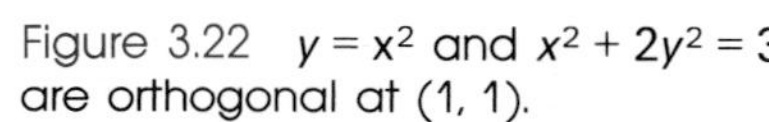
Figure 3.22 $y = x^2$ and $x^2 + 2y^2 = 3$ are orthogonal at (1, 1).

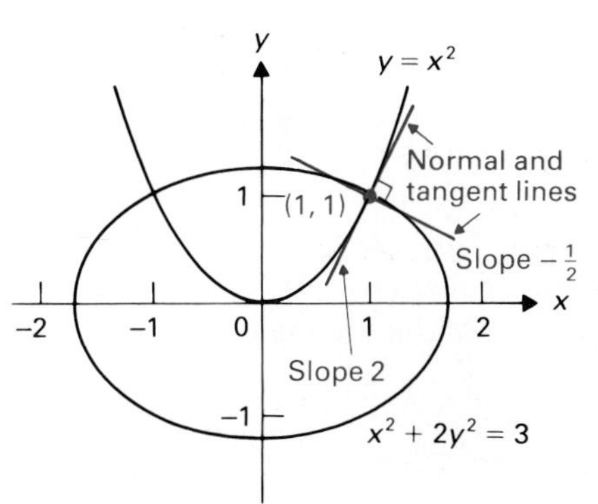

Since the slopes 2 and $-\frac{1}{2}$ are negative reciprocals, the curves are orthogonal at $(1, 1)$. $\square$

EXAMPLE 5 The point $(1, -2)$ lies on the curve $y^3 + 3xy + 4x^3 = -10$. Assuming the curve defines y implicitly as a function $f(x)$ near $(1, -2)$, estimate the point on the curve where $x = 1.05$, using differentials.

Solution Use the approximation formula

$$f(x + dx) \approx f(x) + f'(x)\,dx,$$

where $x = 1$, $f(1) = -2$, and $dx = 0.05$. We must compute $f'(1) = dy/dx|_{x=1}$.

By implicit differentiation,

$$3y^2\frac{dy}{dx} + 3x\frac{dy}{dx} + 3y + 12x^2 = 0,$$

$$\frac{dy}{dx} = \frac{-3y - 12x^2}{3y^2 + 3x},$$

$$\left.\frac{dy}{dx}\right|_{(1,-2)} = \frac{-6}{15} = -\frac{2}{5}.$$

Thus $f(1.05) \approx -2 - \frac{2}{5}(0.05) = -2 - 0.02 = -2.02$. □

Higher-order derivatives of implicit functions can be found by repeated implicit differentiation, as illustrated in the next example,

EXAMPLE 6 Find y'' if $x^2 + y^2 = 2$.

Solution This time we use y' and y'' for the first and second derivatives in the implicit differentiation. For y defined implicitly as a function of x by $x^2 + y^2 = 2$, we obtain

$$2x + 2yy' = 0, \qquad \text{so} \quad y' = \frac{-x}{y}.$$

Differentiating this relation again implicitly with respect to x, we obtain

$$y'' = \frac{y(-1) - (-x)y'}{y^2} = \frac{-y + xy'}{y^2}.$$

Since $y' = -x/y$,

$$y'' = \frac{-y + x(-x/y)}{y^2} = \frac{-y^2 - x^2}{y^3}.$$

Remembering that $x^2 + y^2 = 2$, we have

$$y'' = \frac{-2}{y^3}. \quad \square$$

SUMMARY

Let y be defined implicitly as a function of x. We can find dy/dx by following these steps.

Step 1 Write down an equation in x and y.

Step 2 Differentiate both sides of the equation with respect to x, thinking of y as some function of x and using the chain rule.

Step 3 Solve algebraically for dy/dx in terms of x and y.

EXERCISES

In Exercises 1 through 4, use the following two methods to find dy/dx when $x = x_1$ for y defined implicitly as a function of x near (x_1, y_1). (a) Find y explicitly as a function of x and differentiate, and (b) differentiate implicitly.

1. $x^2 + y^2 = 25$ and $(x_1, y_1) = (-3, 4)$
2. $x - y^2 = 3$ and $(x_1, y_1) = (7, 2)$
3. $y^2 - 2xy + 3x^2 = 1$ and $(x_1, y_1) = (0, -1)$ [*Hint*: Use the quadratic formula to solve $y^2 + (-2x)y + (3x^2 - 1) = 0$ for y.]
4. $2y^2 + 3x^2y - 4x = -2$ and $(x, y) = (1, -2)$ [See the hint to Exercise 3.]

In Exercises 5 through 18, find dy/dx at the given point by implicit differentiation.

5. $x^2 - y^2 = 16; (5, -3)$
6. $x^3 + y^3 = 7; (-1, 2)$
7. $xy = 12; (2, 6)$
8. $xy^2 = 12; (3, -2)$
9. $xy^2 - 3x^2y + 4 = 0; (-1, 1)$
10. $2x^2y^2 - 3xy + 1 = 0; (1, 1)$
11. $(3x + 2y)^{1/2}(x - y) = 8; (4, 2)$
12. $\sqrt{x + y}(x - 2y)^2 = 2; (3, 1)$
13. $3x^2y^3 + 4xy^2 = 6 + \sqrt{y}; (1, 1)$
14. $4xy^4 = 2x^2y^3 - y^{1/3} - 5; (-1, 1)$
15. $\dfrac{x}{x^2 + y^2} = -\dfrac{1}{5}; (-1, 2)$
16. $\dfrac{x^2 - 3y^2}{x + 2y} = 3; (3, -2)$
17. $\dfrac{xy - y^3}{\sqrt{x^2 + 2y}} = 52; (3, -4)$
18. $\dfrac{xy^4 - 3x^2y}{3x^2 + y} = 2; (1, 2)$
19. Find the equations of the tangent and normal lines to the curve $y^2 = x^3 - 2x^2y + 1$ at the point $(2, 1)$.
20. Find the equation of the tangent line to the curve $y^2 - 3x^2 = 1$ at the point $(-1, 2)$.
21. Find the equation of the tangent line to the curve $\sqrt{xy} + 3y = 10$ at the point $(8, 2)$.
22. Find the equation of the normal line to the curve $\sqrt{x + 3y^2} + x^2y = 21$ at the point $(-3, 2)$.
23. Find all points on $y^2 - 3xy + 3x^2 = 4$ where the curve has a horizontal tangent line.
24. Find all points on the curve $3y^2 - 3xy + x^2 = 1$ where the tangent line is vertical.
25. Find all points on the curve $x^2y - xy^2 = 16$ where the curve has a vertical tangent.
26. Find all points on the curve $y^3 + y^2 = x$ where the tangent line has slope 1.
27. If $y = f(x)$ is defined implicitly by $y^3 - 3xy + 2x^2 = 3$ near $(2, 1)$, estimate $f(1.98)$ using differentials.
28. If $y = f(x)$ is defined implicitly by $xy^3 - 2xy + x^4 = 8$ near $(-2, 2)$, estimate $f(-2.1)$ using differentials.
29. If $x^2 - y^2 = 7$, find d^2y/dx^2 at the point $(4, 3)$.
30. Find d^2y/dx^2 at the point $(1, 2)$ if $y^2 - xy = 2$.
31. Find d^2y/dx^2 at the point $(2, -1)$ if $y^3 - xy = 1$.
32. Find d^2y/dx^2 at the point $(-2, 1)$ if $y^4 + x^2y = 5$.
33. Show that the curves $2x^2 + y^2 = 24$ and $y^2 = 8x$ are orthogonal at the point $(2, 4)$ of intersection.
34. Show that for any values of $c \neq 0$ and k, the curves $x^2 + 2y^2 = c$ and $y = kx^2$ are orthogonal at all points of intersection.
35. Show that for all nonzero values of c and k, the curves $y^2 - x^2 = c$ and $xy = k$ are orthogonal at all points of intersection.
36. The volume of a cube is increasing. Use implicit differentiation to find the rate of change of the volume V with respect to the surface area S when $S = 384\ \text{cm}^2$.

EXERCISE SETS FOR CHAPTER 3

Review Exercise Set 3.1

1. a) Define the derivative $f'(x_1)$ of f at the point where $x = x_1$.
   b) Use the definition of the derivative, not differentiation formulas, to find $f'(x)$ if $f(x) = x^2 - 3x$.
2. a) Find the equation of the tangent line to $y = x^3 - 3x^2 + 2$ where $x = 1$.
   b) If the position x of a body on a line (x-axis) at time t is $x = t^2 - (t/3)$, find the velocity when $t = 2$.
3. Find dy/dx if $y = (x^2 - 3x)(4x^3 - 2x + 17)$. You need not simplify your answer.
4. Find dy/dx if $y = (8x^2 - 2x)/(4x^3 + 3)$.
5. Use differentials to estimate 2.05^7.
6. Let $y = 3x^2 - 6x$ and let $x = g(t)$ be a differentiable function such that $g(14) = -2$ and $g'(14) = 8$. Find dy/dt when $t = 14$.
7. Find dy/dx if $y = \sqrt{x^2 - 17x}$.
8. Find dy/dx and d^2y/dx^2 if $y = (4x^3 - 7x)^5$. You need not simplify your answers.
9. Find the equation of the tangent line to the curve with parametric equations $x = 3t^2 - 10t$, $y = \sqrt{t + 1}$, when $t = 3$.
10. If $y^3 + x^2y - 4x^2 = 17$, find dy/dx.

Review Exercise Set 3.2

1. Use the definition of the derivative, not differentiation formulas, to find $f'(x)$ if $f(x) = 1/(2x + 1)$.
2. a) Find the equation of the normal line to $y = 4x^2 - 3x + 2$ at the point $(-1, 9)$.
   b) Let the position x of a body on a line (x-axis) at time t be $x = t^3 + 2t$.
      i) Find the average velocity of the body from time $t = 1$ to time $t = 3$.
      ii) Find the velocity of the body at time $t = 2$.
3. Given that there is a function $\ln x$ such that $d(\ln x)/dx = 1/x$, find dy/dx if $y = (x^2 + 3)(\ln x)$.
4. Find dy/dx if
$$y = (x + 4)/(x^2 - 17).$$
5. If $y = 3x^2 - 6x + 7$, find dy.
6. Find dy/dx if
$$y = \frac{1}{\sqrt[3]{x^3 - 3x + 2}}.$$
7. Let $y = \sqrt{x^2 - 3x}$ and $x = (t + 2)/(t - 1)$. Find dy/dt when $t = 2$.
8. Find d^3y/dx^3 if $y = \sqrt{3x + 4}$.
9. Let the position (x, y) at time t of a body in the plane be given by the parametric equations
$$x = \tfrac{1}{3}t^3 - \tfrac{3}{2}t^2 + 2t - 4,$$
$$y = t^3 - 9t^2.$$
   a) Find all times t when the direction of motion of the body is vertical.
   b) Find all times t when the direction of motion of the body is horizontal.
   c) Find the speed of the body when $t = 4$.
10. Find dy/dx and d^2y/dx^2 at the point $(1, 2)$ if $y^2 - xy = 2$.

Review Exercise Set 3.3

1. Find the equation of the tangent line to the curve $y = x^4 - 3x^2 + 4x$ at the point $(1, 2)$.
2. Let f be differentiable and let $y = x^2 f(x)$. If $dy/dx|_{x=1} = 10$ and $f(1) = 5$, find $f'(1)$.
3. Let g be differentiable and let $y = (x^2 + 1)/g(x)$. If $dy/dx|_{x=3} = -4$ and $g(3) = 2$, find $g'(3)$.
4. The area A of a circle of radius r is given by $A = \pi r^2$. A circle of radius 1 has area $\pi \approx 3.14$. Use differentials to estimate the radius of a circle that has area 3.
5. If $y = x^3 - 2x^2$ and $x = t^2/(t + 3)$, use the chain rule to find dy/dt where $t = 2$.
6. Find d^3y/dx^3 if $y = 6(4x - 2)^{3/2}$.
7. Let the position (x, y) of a body in the plane at time t be given by the parametric equations
$$x = 3t^2 - 2t,$$
$$y = t^3 + 2.$$
   a) Find the x-component of the velocity when $t = 1$.
   b) Find the y-component of the acceleration when $t = 1$.
   c) Find the speed of the body when $t = 1$.
8. Find the equation of the normal line to the curve with parametric equations
$$x = \frac{2t}{t^2 + 1}, \qquad y = 2t - 3$$
when $t = 1$.
9. Find dy/dx at $(4, -1)$ if $\sqrt{xy^2} - 3xy^3 = 14$.
10. Find all points where the curve $y^2 - xy + x^2 = 27$ has a horizontal tangent line.

Review Exercise Set 3.4

1. Find the equation of the normal line to the curve $y = 3x^2 - 4x - 7$ at the point $(2, -3)$.
2. Find the average rate of change of the function $f(x) = x^2 - 4x$ over the interval $[-1, 5]$.
3. Find dy/dx if $y = (x^2 + 3)\sqrt{2x + 4}$.
4. Find dy/dx if $y = (x^2 - 3x)/(2x^3 + 7)$.
5. If we let $dx = \Delta x$, draw the graph of a differentiable function at x_1 and label on the graph the geometric meaning at $x = x_1$ of
   a) Δy b) dy c) $|\Delta y - dy|$
6. Let f be a differentiable function and let $y = (f(x))^3 + 3\sqrt{f(x) + 10}$. If $f(3) = -1$ and $dy/dx|_{x=3} = 18$, find $f'(3)$.
7. Let f be a twice-differentiable function and let $y = g(x) = x^3 f(x)$. If $g(-1) = 4$, $g'(-1) = 2$, and $g''(-1) = -3$, find $f''(-1)$.
8. Find d^2y/dx^2 at $t = 1$, if $x = t^2 + 4$ and $y = 3t + 2$.
9. Find all points (x_1, y_1) on the curve $x^2 + y^2 + 4x - 2y = 20$ where a short arc of the curve with (x_1, y_1) in the center does *not* define y implicitly as a function of x.
10. Find dy/dx if $(x^2 - 3xy)/(x^3 + y^2) = y^3$.

More Challenging Exercises 3

Exercises 1 through 6 indicate that some limits can be recognized as giving derivatives, or at least involving derivatives. Differentiation formulas may then aid in their computation.

1. If $f(x) = x^4 - 3x^2$, find
$$\lim_{\Delta x \to 0} \frac{f(2 + \Delta x) - f(2)}{\Delta x}.$$
2. If $f(x) = x/(x + 1)$, find
$$\lim_{\Delta x \to 0} \frac{f(3 + \Delta x) - f(3)}{2 \cdot \Delta x}.$$
3. If $g(x) = x^{10} - 7x^6 + 5x^4$, find
$$\lim_{\Delta x \to 0} \frac{f(1) - f(1 + 2 \cdot \Delta x)}{\Delta x}.$$
4. If $f(x) = 3x^{10} - 7x^8 + 5x^6 - 21x^3 + 3x^2 - 7$, find
$$\lim_{h \to 0} \frac{f(1 - h) - f(1)}{h^3 + 3h}.$$

5. Let $f(x) = \dfrac{x^3 - 2x^2}{x - 1}$. Find

$$\lim_{h \to 0} \frac{f(2 + 3h) - f(2 + h)}{h}.$$

6. Let $f(x) = \dfrac{x^2}{3x + 4}$. Find

$$\lim_{h \to 0} \frac{f(2h - 1) - f(-3h - 1)}{h}.$$

7. Let f and g be differentiable at $x = a$. Show that, if $f(a) = 0$ and $g(a) = 0$, then $(fg)'(a) = 0$.

8. Find a formula for the derivative of a product

$$f_1(x)f_2(x)f_3(x) \cdots f_n(x)$$

of n functions.

9. Let $p(x)$ be a polynomial function. Show that if $(x - a)^2$ is a factor of the polynomial expression, then both $p(a) = 0$ and $p'(a) = 0$.

10. Generalize the result in Exercise 9.

11. Find the equations of the lines through (4, 10) that are tangent to the graph of $f(x) = (x^2/2) + 4$.

12. Let f be differentiable at $x = a$.
   a) Find A, B, and C such that the polynomial function (parabola) $p(x) = Ax^2 + Bx + C$ passes through the three points on the graph $y = f(x)$ whose x-coordinates are $a - \Delta x$, a, and $a + \Delta x$. (For small Δx, the quadratic function $p(x)$ approximates $f(x)$ near a.)
   b) Compute $p'(a)$ for $p(x)$ from part (a). What familiar expression do you obtain?

13. Let f and g be differentiable functions of x, and let $y = f(x)/g(x)$. Derive the formula for the derivative of the quotient $f(x)/g(x)$ using just implicit differentiation and the rule for differentiating a product.

14. You will see in the next chapter that there exist functions $\sin x$ and $\cos x$ such that

$$\frac{d(\sin x)}{dx} = \cos x \quad \text{and} \quad \frac{d(\cos x)}{dx} = -\sin x.$$

Find the derivatives of the following functions.
   a) $\sin 2x$ b) $\cos 4x^3$
   c) $\sin(\cos x)$ d) $\cos(\sin 3x)$

15. The chain rule states that, under certain conditions,

$$\left.\frac{d[f(g(t))]}{dt}\right|_{t=t_2} = f'(g(t_1)) \cdot g'(t_1).$$

Find a similar formula for

$$\left.\frac{d[f(g(h(t)))]}{dt}\right|_{t=t_1}$$

for suitable functions f, g, and h.

16. Let f and g be differentiable functions satisfying

$$g'(a) \neq 0, \qquad g(a) = b, \qquad \text{and} \qquad f(g(x)) = x.$$

Show that $f'(b) = 1/g'(a)$.

4 THE TRIGONOMETRIC FUNCTIONS

So far, we have learned only how to take derivatives of *algebraic functions*, those that can be defined implicitly by a polynomial equation in x and y. (This includes polynomial and rational functions as well as those generated using radicals.) We have not touched upon the *trigonometric functions*, the *inverse trigonometric functions*, the *logarithmic functions*, or the *exponential functions*. All the functions we mentioned are referred to as *elementary functions*. We develop calculus for the trigonometric functions in this chapter. The remaining elementary functions are treated in Chapter 8.

Students may already be familiar with radian measure, the six elementary trigonometric functions, their graphs, and basic trigonometric identities. In that case, the first two sections of this chapter may be omitted. Calculus for these functions is developed in Section 4.3.

4.1 TRIGONOMETRY REVIEW 1: EVALUATION AND IDENTITIES

This section is devoted to defining and computing the trigonometric functions. We also discuss some properties that the functions satisfy.

THE SIX FUNCTIONS

The circle $u^2 + v^2 = 1$ is shown in Fig. 4.1. It is important to note that the radius of this circle is 1; hence it is called a unit circle. The functions $\sin x$ and $\cos x$ may be evaluated as follows. If $x \geq 0$, we start at the point $(1, 0)$ on the circle and go *counterclockwise* along the circle until we have traveled the distance x on the arc. As shown in the figure, we will stop at some point (u, v) on the circle. Then

$$\sin x = v \qquad \text{and} \qquad \cos x = u. \tag{1}$$

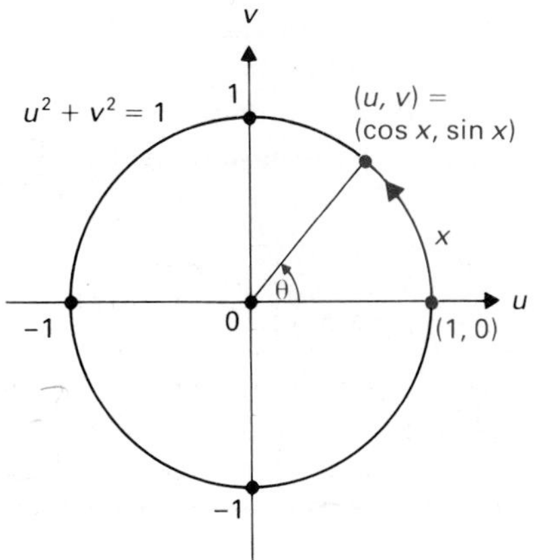

Figure 4.1 Point (u, v) on the unit circle corresponding to directed arc length x measured from (1, 0).

Equations (1) are also valid if $x < 0$, except that then we travel the distance $|x|$ *clockwise* around the circle from (1, 0) to arrive at a point (u, v).

EXAMPLE 1 Find sin 0 and cos 0.

Solution If $x = 0$, we stop at the point (1, 0) on the circle in Fig. 4.1, so $u = 1$ and $v = 0$. Thus $\sin 0 = 0$ and $\cos 0 = 1$. □

EXAMPLE 2 Find $\sin(-\pi/2)$ and $\cos(-\pi/2)$.

Solution With $x = -\pi/2$, we should travel *clockwise* a quarter of the way around the circle in Fig. 4.1, since the circle has circumference 2π. We arrive at the point $(u, v) = (0, -1)$, as shown in Fig. 4.2. Thus $\sin(-\pi/2) = -1$ and $\cos(-\pi/2) = 0$. □

The arc length x shown in Fig. 4.1 is the *radian measure* of the central angle θ shown in the figure. The radian measure of a central angle θ of a circle is

$$\text{Radian measure of } \theta = \frac{\text{Length of intercepted arc}}{\text{Radius}}.$$

Since the radius is 1 in the figure, the radian measure of θ is given by the length x of the arc. For example, a 360° angle has radian measure 2π since the length all the way around the circle is 2π. It is easy to see that

$$\boxed{\text{Radian measure of } \theta = \left(\frac{\pi}{180}\right)(\text{Degree measure of } \theta),}$$

because 180° corresponds to π radians.

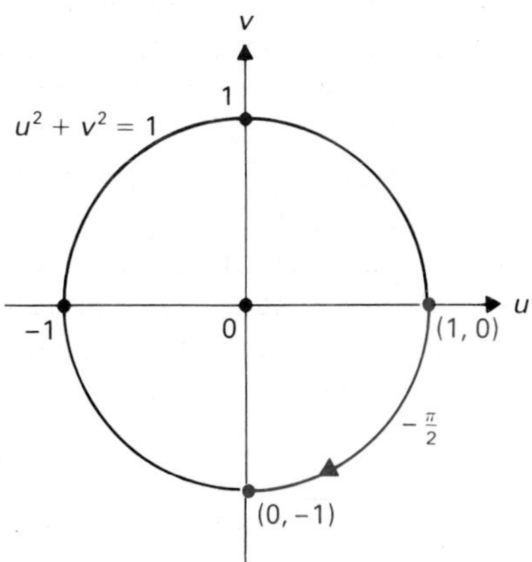

Figure 4.2 Point $(u, v) = (0, -1)$ on the unit circle corresponding to directed arc length $-\pi/2$ measured from (1, 0).

EXAMPLE 3 Find the radian measure of an angle θ having degree measure 240°.

Solution Our conversion equation yields

$$\text{Radian measure of } \theta = \left(\frac{\pi}{180}\right)(240) = \frac{4}{3}\pi. \quad \square$$

From Fig. 4.3, we see that if $0 < \theta < 90°$, then θ is an acute angle of the right triangle shown, and

Figure 4.3 Right triangle of hypotenuse 1 and acute angle θ.

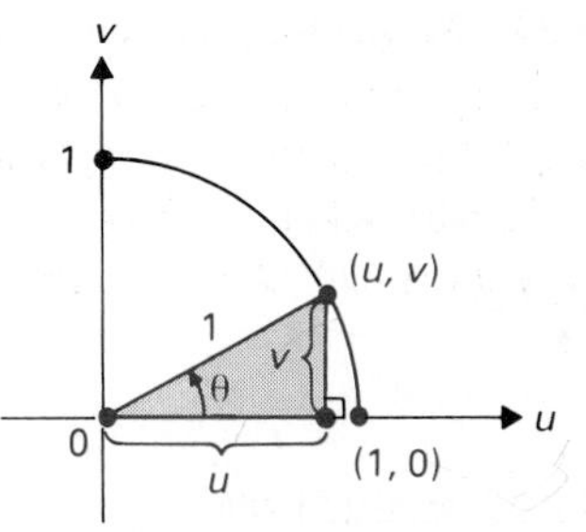

$$\sin\theta = v = \frac{v}{1} = \frac{\text{Opposite side length}}{\text{Hypotenuse length}},$$

while

$$\cos\theta = u = \frac{u}{1} = \frac{\text{Adjacent side length}}{\text{Hypotenuse length}}.$$

You may have learned these definitions before. The advantage of using our definitions given in Eqs. (1) is that we can easily find $\sin x$ if $x > 90°$. For example, $\sin(3\pi/2) = \sin(270°) = -1$ since an arc length of $3\pi/2$ corresponds to the point $(0, -1)$ on the circle.

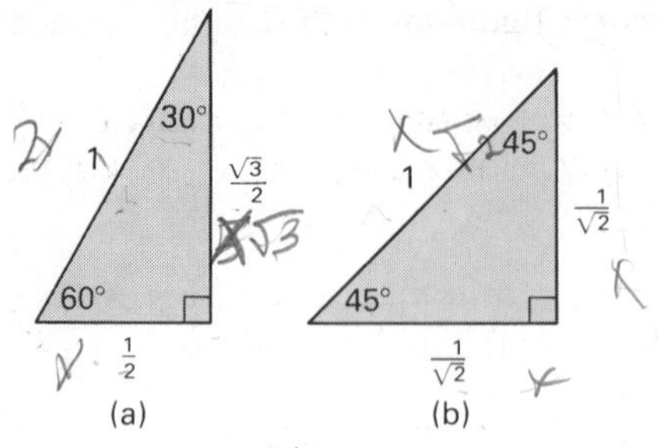

Figure 4.4
(a) $\sin 30° = \sin(\pi/6) = \frac{1}{2} = \cos 60° = \cos(\pi/3)$;
$\sin 60° = \sin(\pi/3) = \sqrt{3}/2 = \cos 30° = \cos(\pi/6)$;
(b) $\sin 45° = \sin(\pi/4) = 1/\sqrt{2} = \cos 45° = \cos(\pi/4)$.

Recall from geometry the lengths of the sides of the right triangles shown in Fig. 4.4. Using these triangles, we easily obtain the values for $\sin x$ and $\cos x$ given in Table 4.1. Note that x is given in *radians* in Table 4.1. We will see in Section 4.3 that radian measure is the easiest to use when differentiating the trigonometric functions.

In the absence of a degree symbol, $\sin x$ and $\cos x$ always signify that x is in radians.

If we wish to use degree measure for some reason, we would write $\sin x°$. Thus we have $\sin 30° = \frac{1}{2}$, but $\sin 30$ means the sine of 30 radians, which our calculator tells us is about -0.988.

Table 4.1

x	$\sin x$	$\cos x$
0	0	1
$\frac{\pi}{6}$	$\frac{1}{2}$	$\frac{\sqrt{3}}{2}$
$\frac{\pi}{4}$	$\frac{1}{\sqrt{2}}$	$\frac{1}{\sqrt{2}}$
$\frac{\pi}{3}$	$\frac{\sqrt{3}}{2}$	$\frac{1}{2}$
$\frac{\pi}{2}$	1	0

EXAMPLE 4 Find $\sin(-2\pi/3)$ and $\cos(-2\pi/3)$.

Solution Since $-2\pi/3$ radians corresponds to $-120° = -90° - 30°$, we see from Fig. 4.5 and triangle (a) in Fig. 4.4 that

$$\sin\left(-\frac{2\pi}{3}\right) = -\frac{\sqrt{3}}{2} \quad \text{and} \quad \cos\left(-\frac{2\pi}{3}\right) = -\frac{1}{2}. \quad \square$$

We will describe the remaining four basic trigonometric functions of x in terms of the point (u, v) in Fig. 4.1.*

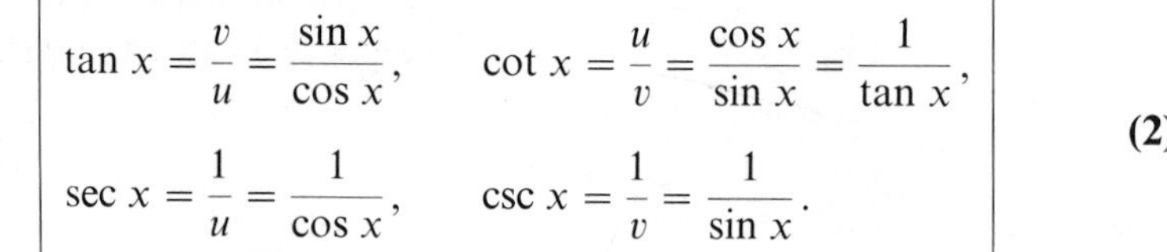

$$\tan x = \frac{v}{u} = \frac{\sin x}{\cos x}, \qquad \cot x = \frac{u}{v} = \frac{\cos x}{\sin x} = \frac{1}{\tan x},$$
$$\sec x = \frac{1}{u} = \frac{1}{\cos x}, \qquad \csc x = \frac{1}{v} = \frac{1}{\sin x}. \qquad \textbf{(2)}$$

The trigonometric functions in Eqs. (2) are defined as quotients of other functions. Their domains thus exclude points where denominators may be zero. For example, $\cot x$ is not defined where $\sin x = 0$, which happens where $x = n\pi$ for any integer n.

EXAMPLE 5 Continuing Example 4, find the other four trigonometric functions of $-2\pi/3$.

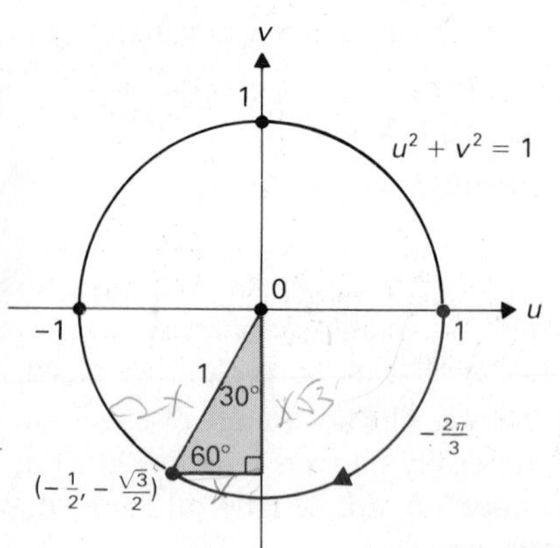

Figure 4.5 Computation of the trigonometric functions at $-2\pi/3$.

Solution From Fig. 4.5, we see that

$$\tan\left(-\frac{2\pi}{3}\right) = \frac{-\sqrt{3}/2}{-1/2} = \sqrt{3},$$

$$\cot\left(-\frac{2\pi}{3}\right) = \frac{1}{\tan(-2\pi/3)} = \frac{1}{\sqrt{3}},$$

$$\sec\left(-\frac{2\pi}{3}\right) = \frac{1}{-1/2} = -2,$$

$$\csc\left(-\frac{2\pi}{3}\right) = \frac{1}{-\sqrt{3}/2} = -\frac{2}{\sqrt{3}}. \quad \square$$

* The trigonometric functions are sometimes called the **circular functions**.

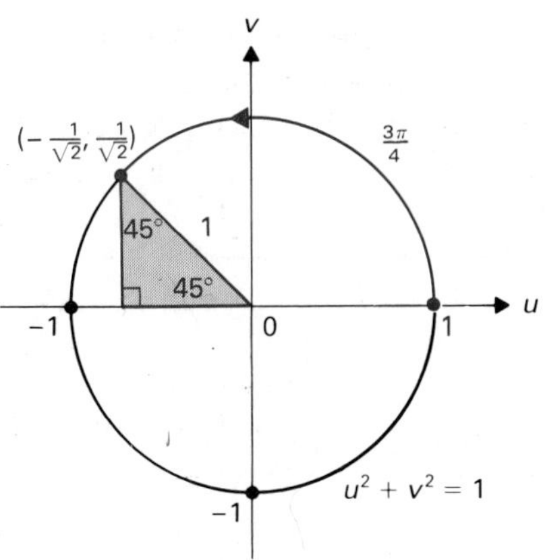

Figure 4.6 Computation of the trigonometric functions at $3\pi/4$.

EXAMPLE 6 Find the six trigonometric functions of $3\pi/4$.

Solution Using triangle (b) of Fig. 4.4, we see that an arc of length $3\pi/4$ starting from (1, 0) brings us to the point $(u, v) = (-1/\sqrt{2}, 1/\sqrt{2})$ on the unit circle. This is shown in Fig. 4.6. Thus

$$\sin\frac{3\pi}{4} = \frac{1}{\sqrt{2}}, \qquad \cos\frac{3\pi}{4} = -\frac{1}{\sqrt{2}},$$

$$\tan\frac{3\pi}{4} = \frac{1/\sqrt{2}}{-1/\sqrt{2}} = -1, \qquad \cot\frac{3\pi}{4} = \frac{-1/\sqrt{2}}{1/\sqrt{2}} = -1,$$

$$\sec\frac{3\pi}{4} = \frac{1}{-1/\sqrt{2}} = -\sqrt{2}, \qquad \csc\frac{3\pi}{4} = \frac{1}{1/\sqrt{2}} = \sqrt{2}. \quad \square$$

IDENTITIES

Referring to Eqs. (1) and Fig. 4.1, we have $\sin x = v$, $\cos x = u$, and $u^2 + v^2 = 1$. Therefore

$$\boxed{\sin^2 x + \cos^2 x = 1.}$$

(Note that $\sin^2 x$ means $(\sin x)^2$.) This is a fundamental trigonometric identity. There are many others, some of which are easily obtained. For example, if we divide both sides of $\sin^2 x + \cos^2 x = 1$ by $\cos^2 x$, then

$$\frac{\sin^2 x}{\cos^2 x} + \frac{\cos^2 x}{\cos^2 x} = \frac{1}{\cos^2 x},$$

or

$$\boxed{\tan^2 x + 1 = \sec^2 x.}$$

EXAMPLE 7 If $-\pi/2 \le \theta \le \pi/2$ and $\sin\theta = -\frac{3}{5}$, find $\tan\theta$.

Solution To have $-\pi/2 \le \theta \le \pi/2$ and $\sin\theta < 0$, we must actually have $-\pi/2 < \theta < 0$. Thus $u = \cos\theta > 0$. From $\sin^2\theta + \cos^2\theta = 1$, we obtain

$$\cos\theta = \pm\sqrt{1 - \sin^2\theta} = \pm\sqrt{1 - \frac{9}{25}} = \pm\sqrt{\frac{16}{25}} = \pm\frac{4}{5}.$$

Since $\cos\theta > 0$, the value $\cos\theta = \frac{4}{5}$ is appropriate. See Fig. 4.7. Then

$$\tan\theta = \frac{\sin\theta}{\cos\theta} = \frac{-3/5}{4/5} = -\frac{3}{4}. \quad \square$$

Figure 4.7 Computation of $\tan\theta$ if $\sin\theta = -\frac{3}{5}$ and $-\pi/2 \le \theta \le \pi/2$.

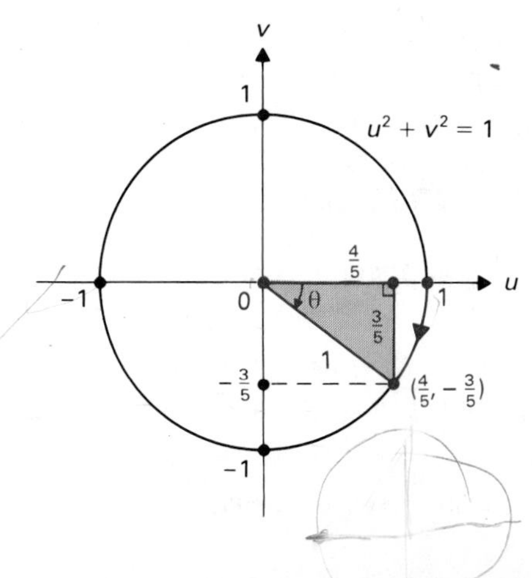

It is not our purpose to give a lot of drill in identities. The Summary lists for reference some of the most important ones. The exercises ask you to prove a few identities that are easy to derive from our definition of $\sin x$ and $\cos x$. We illustrate the technique with an example.

EXAMPLE 8 Show that $\sin(x + \pi) = -\sin x$ and $\cos(x + \pi) = -\cos x$, which are identities (14) and (15) in the Summary.

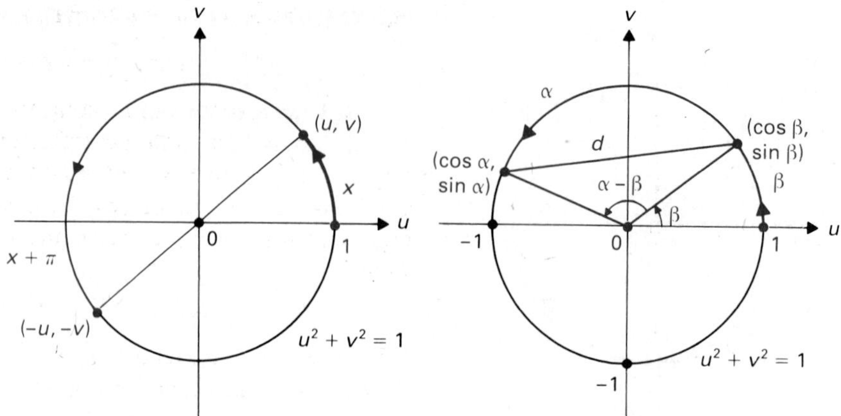

Figure 4.8 Arc x ends at (u, v); Arc $x + \pi$ ends at $(-u, -v)$.

Figure 4.9 Points $(\cos \alpha, \sin \alpha)$ and $(\cos \beta, \sin \beta)$ on the unit circle, corresponding to arcs of lengths α and β.

Solution Refer to Fig. 4.8. Let x correspond to a point (u, v) on the unit circle, so that $\cos x = u$ and $\sin x = v$. Then $x + \pi$ corresponds to the point $(-u, -v)$ on the circle. Consequently,

$$\sin (x + \pi) = -v = -\sin x$$

and

$$\cos (x + \pi) = -u = -\cos x. \quad \square$$

The proofs of identities (6) and (7) for sine and cosine of a sum are not quite as simple as the proof in Example 8 or as those requested in the exercises.

EXAMPLE 9 Show that $\cos (x + y) = \cos x \cos y - \sin x \sin y$, which is identity (7) in the Summary.

Solution Let α and β correspond to points on the upper half of the unit circle, so that $\alpha - \beta$ is the angle shown in Fig. 4.9. Using the formula for the distance d between the two points on the circle, we see from Fig. 4.9 that

$$\begin{aligned} d^2 &= (\cos \alpha - \cos \beta)^2 + (\sin \alpha - \sin \beta)^2 \\ &= \cos^2\alpha - 2 \cos \alpha \cos \beta + \cos^2 \beta \\ &\quad + \sin^2\alpha - 2 \sin \alpha \sin \beta + \sin^2 \beta \\ &= (\sin^2\alpha + \cos^2\alpha) + (\sin^2\beta + \cos^2\beta) \\ &\quad - 2(\cos \alpha \cos \beta + \sin \alpha \sin \beta) \\ &= 2 - 2(\cos \alpha \cos \beta + \sin \alpha \sin \beta). \end{aligned}$$

From Fig. 4.10 and the distance formula, we see that for this same distance d, we have

$$\begin{aligned} d^2 &= [\cos (\alpha - \beta) - 1]^2 + [\sin (\alpha - \beta) - 0]^2 \\ &= \cos^2(\alpha - \beta) - 2 \cos (\alpha - \beta) + 1 + \sin^2(\alpha - \beta) \\ &= [\sin^2(\alpha - \beta) + \cos^2(\alpha - \beta)] + 1 - 2 \cos (\alpha - \beta) \\ &= 2 - 2 \cos (\alpha - \beta). \end{aligned}$$

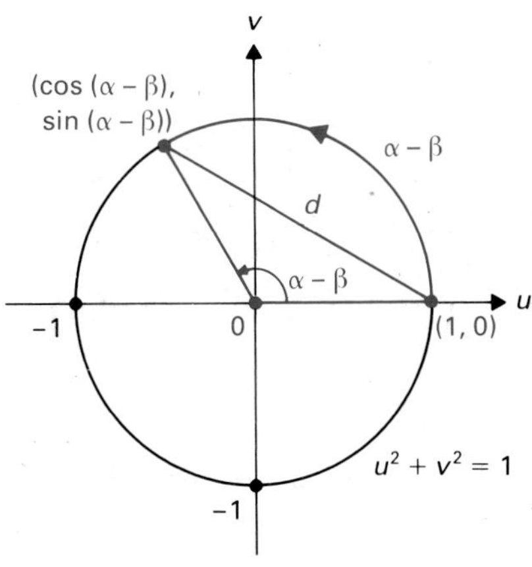

Figure 4.10 Point $(\cos(\alpha-\beta), \sin(\alpha-\beta))$ on the unit circle corresponding to the arc of length $\alpha-\beta$.

Equating the two expressions for d^2, we obtain after cancellation

$$\cos(\alpha-\beta) = \cos\alpha\cos\beta + \sin\alpha\sin\beta.$$

If α corresponds to a point on the lower half of the circle in Fig. 4.9, then $\alpha+\pi$ appears on the upper half of the circle. Then identities (14) and (15), proved in Example 8, can be used with the above formula applied to $\cos[(\alpha+\pi)-\beta]$ to derive the same identity for $\cos(\alpha-\beta)$. A similar argument is made if β corresponds to a point on the lower half of the circle. Thus

$$\cos(\alpha-\beta) = \cos\alpha\cos\beta + \sin\alpha\sin\beta$$

for all α and β. Now let $\alpha = x$ and $\beta = -y$. Using identities (10) and (11) of the Summary (see Exercise 41), we have

$$\begin{aligned}\cos(x+y) &= \cos x\cos(-y) + \sin x\sin(-y)\\ &= \cos x\cos y - \sin x\sin y,\end{aligned}$$

which is identity (7) of the Summary. □

SUMMARY

1. Referring to Fig. 4.11, we have

$$\sin x = v,$$

$$\cos x = u,$$

$$\tan x = \frac{v}{u} = \frac{\sin x}{\cos x},$$

$$\cot x = \frac{u}{v} = \frac{\cos x}{\sin x} = \frac{1}{\tan x},$$

$$\sec x = \frac{1}{u} = \frac{1}{\cos x},$$

$$\csc x = \frac{1}{v} = \frac{1}{\sin x}.$$

Figure 4.11 Point (u, v) on the unit circle corresponding to directed arc length x measured from $(1, 0)$.

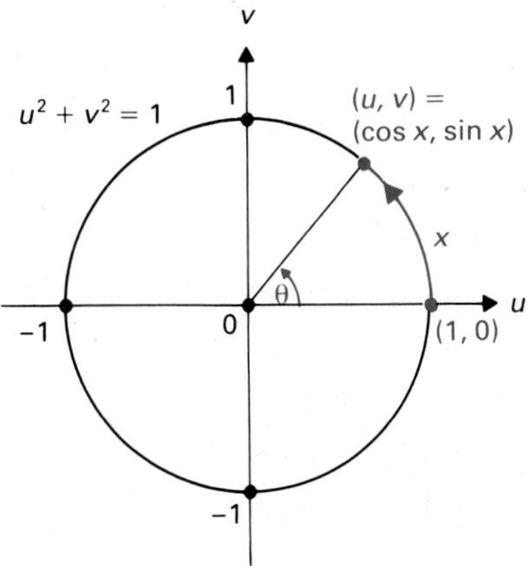

2. The following are some fundamental trigonometric identities in addition to definitions (1) and (2) given in the text.

$$\sin^2 x + \cos^2 x = 1 \tag{3}$$

$$\tan^2 x + 1 = \sec^2 x \tag{4}$$

$$1 + \cot^2 x = \csc^2 x \tag{5}$$

$$\sin(x+y) = \sin x\cos y + \cos x\sin y \tag{6}$$

$$\cos(x+y) = \cos x\cos y - \sin x\sin y \tag{7}$$

$$\sin 2x = 2\sin x\cos x \tag{8}$$

$$\cos 2x = \cos^2 x - \sin^2 x \tag{9}$$

$$\sin(-x) = -\sin x \tag{10}$$

$$\cos(-x) = \cos x \tag{11}$$

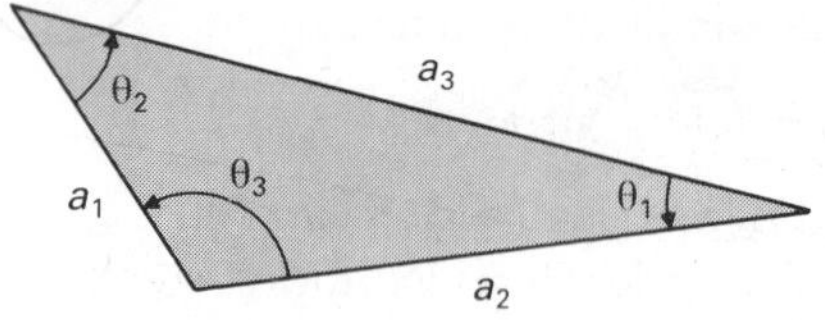

Figure 4.12 Angles and sides of a general triangle.

$$\sin\left(x + \frac{\pi}{2}\right) = \cos x \tag{12}$$

$$\cos\left(x + \frac{\pi}{2}\right) = -\sin x \tag{13}$$

$$\sin(x + \pi) = -\sin x \tag{14}$$

$$\cos(x + \pi) = -\cos x \tag{15}$$

$$\sin(x + 2n\pi) = \sin x \tag{16}$$

$$\cos(x + 2n\pi) = \cos x \tag{17}$$

$$\sin\frac{x}{2} = \pm\sqrt{\frac{1 - \cos x}{2}} \tag{18}$$

$$\cos\frac{x}{2} = \pm\sqrt{\frac{1 + \cos x}{2}} \tag{19}$$

3. Referring to Fig. 4.12, we state two laws.

Law of sines: $\dfrac{\sin\theta_1}{a_1} = \dfrac{\sin\theta_2}{a_2} = \dfrac{\sin\theta_3}{a_3}$

Law of cosines: $a_3^2 = a_1^2 + a_2^2 - 2a_1a_2\cos\theta_3$

EXERCISES

In Exercises 1 through 30, find the indicated value of the function, if the function is defined there.

1. $\sin\frac{\pi}{3}$ **2.** $\cos\frac{3\pi}{2}$

3. $\tan\frac{5\pi}{6}$ **4.** $\sin\frac{4\pi}{3}$

5. $\sec\frac{5\pi}{4}$ **6.** $\tan\frac{3\pi}{4}$

7. $\csc\frac{\pi}{3}$ **8.** $\tan\frac{\pi}{2}$

9. $\cot\pi$ **10.** $\cot\frac{\pi}{2}$

11. $\sec\frac{5\pi}{2}$ **12.** $\csc 2\pi$

13. $\csc\left(-\frac{\pi}{4}\right)$ **14.** $\sin\left(-\frac{2\pi}{3}\right)$

15. $\sec\left(-\frac{3\pi}{4}\right)$ **16.** $\cot\left(-\frac{5\pi}{4}\right)$

17. $\csc\frac{7\pi}{6}$ **18.** $\cot\left(-\frac{2\pi}{3}\right)$

19. $\tan\pi$ **20.** $\cos 3\pi$

21. $\sec 2\pi$ **22.** $\sin 5\pi$

23. $\sin(-3\pi)$ **24.** $\cos(-3\pi)$

25. $\sin\left(-\frac{\pi}{2}\right)$ **26.** $\tan\frac{5\pi}{4}$

27. $\tan\frac{3\pi}{2}$ **28.** $\cot 5\pi$

29. $\sec\frac{9\pi}{4}$ **30.** $\csc\frac{23\pi}{6}$

31. If $-\pi/2 \le \theta < \pi/2$ and $\sin\theta = -\frac{1}{3}$, find $\cos\theta$.

32. If $\pi/2 \le \theta < 3\pi/2$ and $\tan\theta = 4$, find $\sec\theta$.

33. If $0 \le \theta < \pi$ and $\cos\theta = -\frac{1}{5}$, find $\cot\theta$.

34. If $\pi \le \theta < 2\pi$ and $\sec\theta = 3$, find $\tan\theta$.

35. If $\pi/2 \le \theta < 3\pi/2$ and $\sin\theta = \frac{1}{4}$, find $\cot\theta$.

36. If $0 \le \theta < \pi$ and $\cos\theta = \frac{1}{3}$, find $\sin 2\theta$.

37. If $-\pi/2 \le \theta < \pi/2$ and $\sin\theta = -\frac{2}{3}$, find $\sin 2\theta$.

38. If $0 < \theta < \pi/2$ and $\tan\theta = 3$, find $\cos 2\theta$.

39. If $0 < \theta < \pi/2$ and $\sec\theta = 4$, find $\cos 2\theta$.

40. If $0 < \theta < \pi/2$ and $\cos\theta = \frac{1}{3}$, find $\sin 3\theta$.

41. As in Fig. 4.1, let the arc corresponding to x terminate at (u, v).

a) Where does the arc corresponding to $-x$ terminate?
b) Use the result in part (a) to verify identities (10) and (11).

42. As in Fig. 4.1, let the arc corresponding to x terminate at (u, v).
a) Where does the arc corresponding to $x + \pi/2$ terminate?
b) Use the result in part (a) to verify identities (12) and (13).

43. As in Fig. 4.1, let the arc corresponding to x terminate at (u, v).
a) Where does the arc corresponding to $x - \pi/2$ terminate?
b) Use the result in part (a) to derive identities similar to (12) and (13).

In Exercises 44 through 52, use the identities in the Summary to verify each of the following identities.

44. $\tan(-x) = -\tan x$

45. $\sec(-x) = \sec x$

46. $\tan\left(x + \dfrac{\pi}{2}\right) = -\cot x$

47. $\sin\left(x - \dfrac{\pi}{2}\right) = -\cos x$

48. $\cos\left(x - \dfrac{\pi}{2}\right) = \sin x$

49. $\sec\left(x - \dfrac{\pi}{2}\right) = \csc x$

50. $\sin(x - y) = \sin x \cos y - \cos x \sin y$

51. $\cos 2x = 2\cos^2 x - 1 = 1 - 2\sin^2 x$

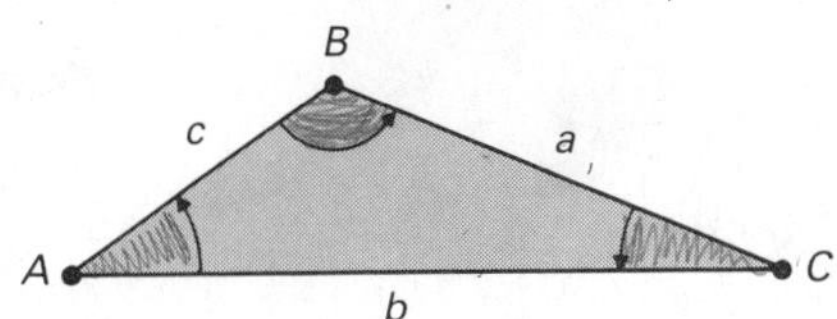

Figure 4.13 Triangle with angles A, B, C, and sides a, b, c.

52. $\tan(x + y) = \dfrac{\tan x + \tan y}{1 - \tan x \tan y}$

Let the angles of a triangle be A, B, and C with corresponding opposite sides of lengths a, b, and c, as shown in Fig. 4.13. In Exercises 53 through 58, find the desired quantity using the law of sines or the law of cosines.

53. For $A = \pi/6$, $a = 5$, $c = 3$, find $\sin C$.

54. For $B = 3\pi/4$, $\sin C = 1/(2\sqrt{2})$, $b = 10$, find c.

55. For $c = 5$, $b = 7$, $A = \pi/4$, find a.

56. For $a = 6$, $c = 4$, $B = 3\pi/4$, find b.

57. For $a = 8$, $c = 4$, $b = 10$, find $\sin C$.

58. For $A = \pi/4$, $a = \sqrt{2}$, $c = \frac{4}{5}$, find $\tan C$.

4.2 TRIGONOMETRY REVIEW 2: GRAPHS OF TRIGONOMETRIC FUNCTIONS

The trigonometric functions are used so frequently that it is best to know their graphs without having to look them up, just as we know the graphs of x, x^2, x^3, x^4, and so on. We have found it feasible to learn the graphs of $\sin x$, $\cos x$, and $\tan x$ and then find the graphs of $\csc x$, $\sec x$, and $\cot x$ by taking reciprocals of the heights. That is, to find the graph of $y = \csc x$, we use the fact that

$$\csc x = \frac{1}{\sin x},$$

so at x where the graph of $\sin x$ has height $\frac{3}{10}$, the height to the graph of $\csc x$ is $\frac{10}{3}$.

GRAPHS OF THE SIX FUNCTIONS

We can easily check that the graph of $y = \sin x$ is that shown in Fig. 4.14. Note that the intercepts on the x-axis are the multiples of π.

Figure 4.14 Graph of $y = \sin x$.

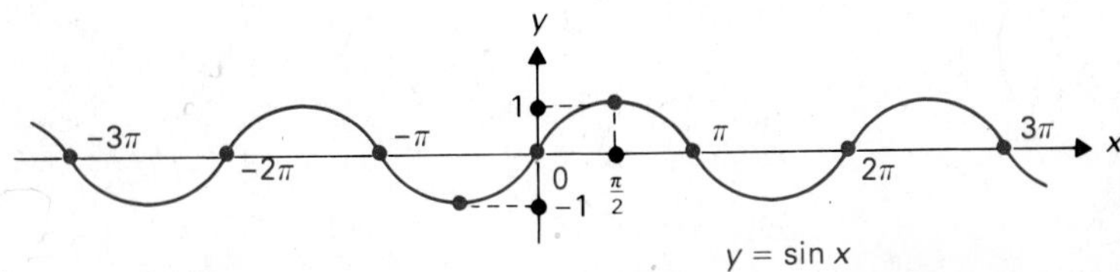

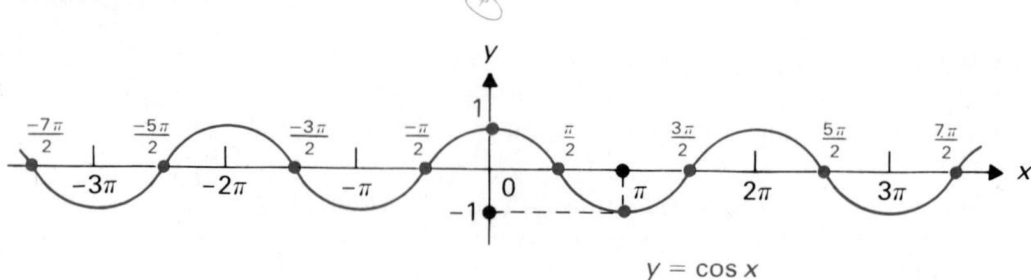

Figure 4.15 Graph of $y = \cos x$.

In view of the identity $\cos x = \sin (x + \pi/2)$, the graph of $\cos x$ can be obtained by translating the graph of $\sin x$ a distance $\pi/2$ to the left. The graph of $\cos x$ is shown in Fig. 4.15.

The graph of $\tan x$ has as height at each point $(\sin x)/(\cos x)$, if $\cos x \neq 0$. From the graphs of $\sin x$ and $\cos x$, we obtain the graph in Fig. 4.16 for $\tan x$. Note that the vertical lines $x = \pm\pi/2$, $x = \pm 3\pi/2$, $x = \pm 5\pi/2, \ldots$ are vertical asymptotes of the graph of $y = \tan x$. For example, we see from Fig. 4.16 that

$$\lim_{x\to\pi/2+} \tan x = -\infty \qquad \text{and} \qquad \lim_{x\to\pi/2-} \tan x = \infty.$$

Taking the reciprocals of the heights of the graph of $\tan x$, we obtain the graph of $\cot x = 1/(\tan x)$ in Fig. 4.17. Figures 4.18 and 4.19 show the graphs of $\sec x = 1/(\cos x)$ and $\csc x = 1/(\sin x)$.

Figure 4.16 Graph of $y = \tan x$.

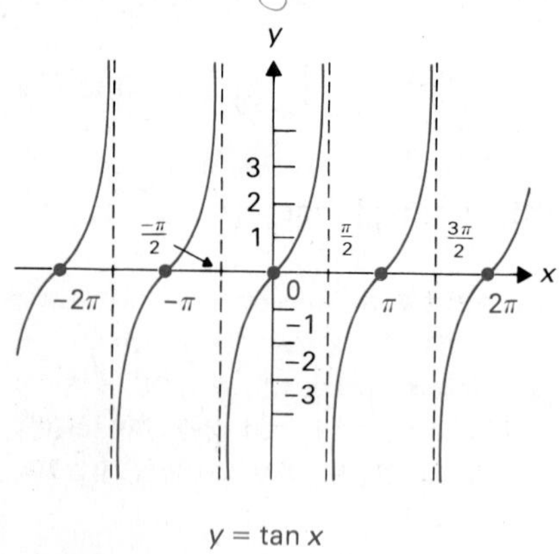

THE GRAPH OF $y = a \cdot \sin [b(x - c)]$

We turn to the graph of $y = a \cdot \sin [b(x - c)]$ and first discuss the effect of the individual constants a, b, and c on the graph.

The graph of $\sin x$ oscillates in height between -1 and 1; the graph has *amplitude* 1. It is obvious that the graph of $a(\sin x)$ oscillates between $-a$ and a, that is, has **amplitude** $|a|$. The constant a controls the amplitude of the graph.

The graph of $\sin x$ repeats itself every 2π units on the x-axis. We say it has *period* 2π. The graph of $\sin bx$ will repeat as soon as bx changes by 2π, which is as soon as x increases by $|2\pi/b|$. Thus $\sin bx$ has **period** $|2\pi/b|$, and b controls the period of the graph.

Figure 4.17 Graph of $y = \cot x$.

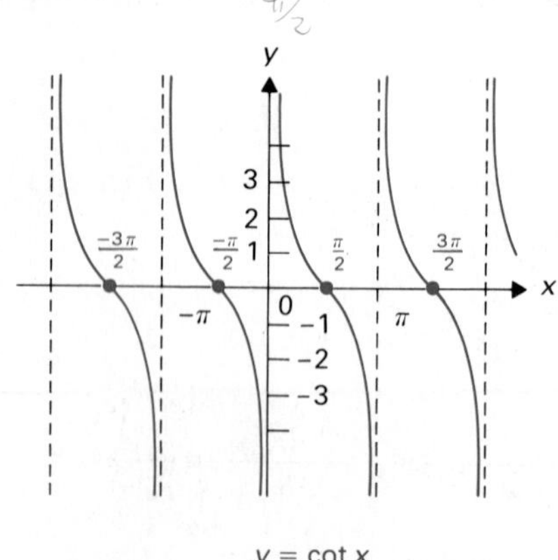

Figure 4.18 Graph of $y = \sec x$.

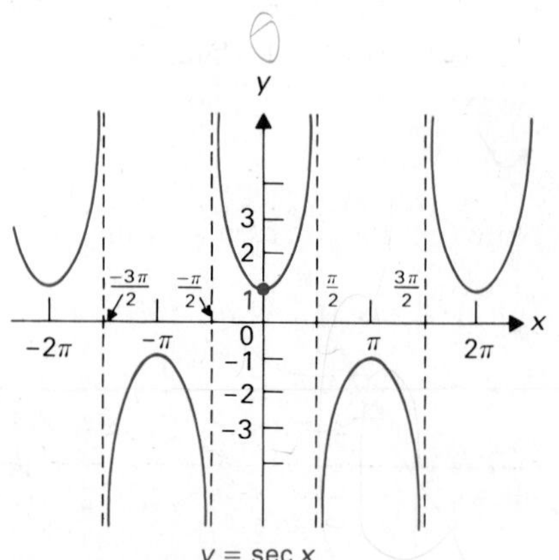

Figure 4.19 Graph of $y = \csc x$.

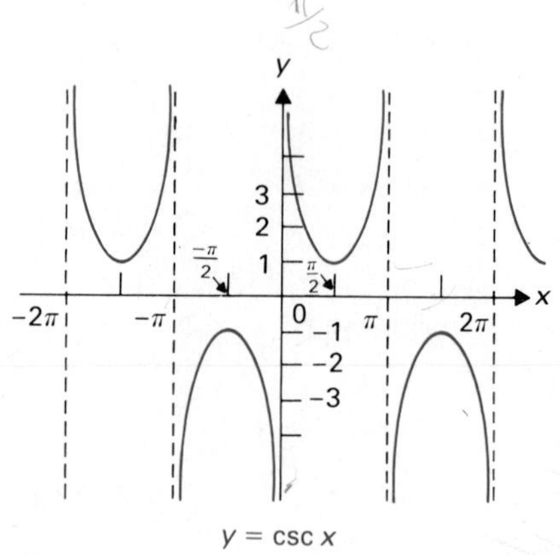

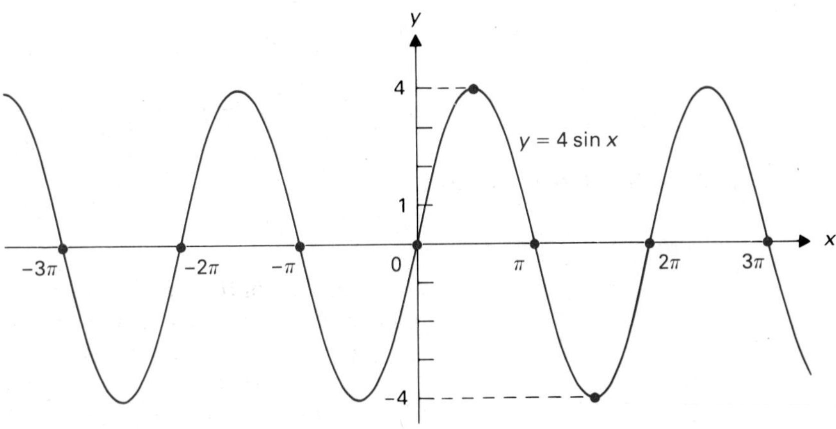

Figure 4.20 The graph of $y = 4 \sin x$ has amplitude 4.

Finally, we turn to the graph of $\sin(x - c)$. We have seen that the substitution $\bar{x} = x - c$, $\bar{y} = y - 0$ amounts to translating axes to the point $(c, 0)$. Thus the graph of $\sin(x - c)$ is the graph of $\sin x$ translated c units to the right. Note that $\sin(x - c)$ is zero when $x = c$, rather than when $x = 0$. The number c is the **phase angle.**

EXAMPLE 1 Sketch the graph of $y = 4 \sin x$.

Solution The graph is shown in Fig. 4.20. The amplitude is 4. □

EXAMPLE 2 Sketch the graph of $y = -\sin \frac{1}{2}x$.

Solution The graph is shown in Fig. 4.21. Here $y = a \sin bx$, with $b = \frac{1}{2}$. The period is $2\pi/(\frac{1}{2}) = 4\pi$. Since $a = -1$, the sign of y is changed from $y = \sin \frac{1}{2}x$, so the graph starts down rather than up to the right of the origin. □

EXAMPLE 3 Sketch the graph of $y = \sin(x - \pi)$.

Solution The graph is shown in Fig. 4.22. The phase angle is π. Setting $\bar{x} = x - \pi$ and $\bar{y} = y - 0$, the graph becomes $\bar{y} = \sin \bar{x}$ with respect to the $\bar{x},\bar{y}$-axes shown in the figure. □

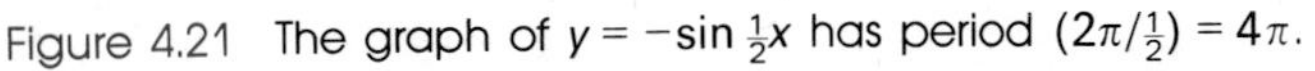

Figure 4.21 The graph of $y = -\sin \frac{1}{2}x$ has period $(2\pi/\frac{1}{2}) = 4\pi$.

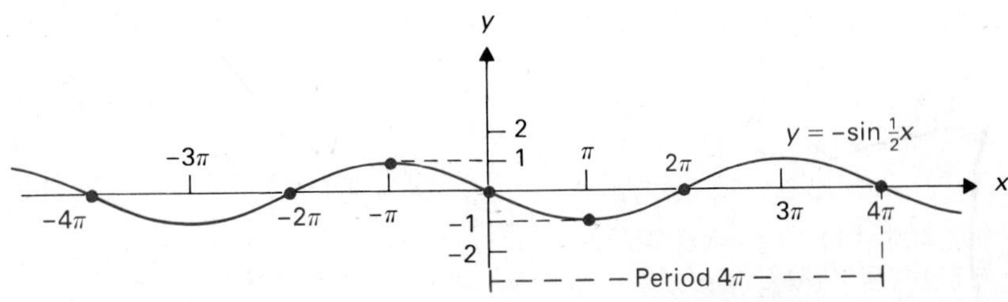

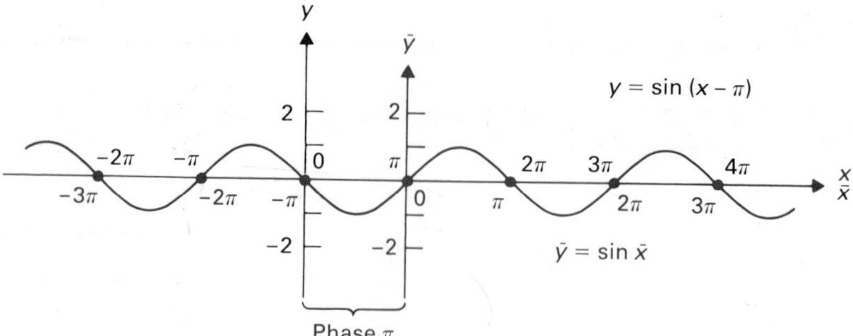

Figure 4.22 The graph of $y = \sin(x - \pi)$ has phase angle π.

EXAMPLE 4 Sketch the graph of $y = 3 \sin(2x + \pi)$.

Solution First we rewrite:

$$y = 3 \sin(2x + \pi) = 3 \sin 2\left[x - \left(-\frac{\pi}{2}\right)\right].$$

The graph has amplitude 3, period $2\pi/2 = \pi$, and phase angle $-\pi/2$. The graph is shown in Fig. 4.23. Think of moving the curve $y = \sin x$ a distance $\pi/2$ to the left and then having it oscillate three times as high (amplitude 3 rather than 1) and twice as fast (period π rather than 2π). □

The effect of a, b, and c on the graph of $y = a \cdot \cos[b(x - c)]$ is precisely the same as for the graph of $a \cdot \sin[b(x - c)]$. Indeed, multiplying any of the six functions by a and replacing x by $b(x - c)$ has an analogous significance.

SUMMARY

1. The graphs of the six trigonometric functions are shown in Figs. 4.14 through 4.19.
2. In the graph of $y = a \cdot \sin[b(x - c)]$, the amplitude $|a|$ controls the height of the oscillation, the period (x-distance for repetition) is $|2\pi/b|$, and c is the phase angle.

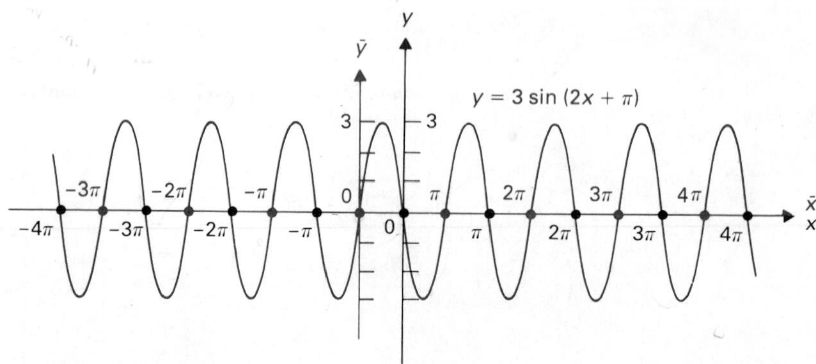

Figure 4.23 The graph of $y = 3 \sin(2x + \pi) = 3 \sin 2[x - (-\pi/2)]$ has amplitude 3, period $2\pi/2 = \pi$, and phase angle $-\pi/2$.

EXERCISES

In Exercises 1 through 16, find the amplitude and period and sketch the graph of the indicated function.

1. $y = 3 \sin x$

2. $y = 4 \cos x$

3. $y = -\frac{1}{2} \cos x$

4. $y = -\frac{1}{3} \sin x$

5. $y = \sin(-x)$

6. $y = \cos(-x)$

7. $y = 3 \sin 3x$

8. $y = 2 \cos\left(\frac{x}{2}\right)$

9. $y = -4 \cos 2x$

10. $y = -6 \sin\left(\frac{x}{3}\right)$

11. $y = -2 \sin\left(x - \frac{\pi}{2}\right)$

12. $y = -3 \cos(x + \pi)$

13. $y = 5 \cos\left(\frac{x}{2} - \frac{\pi}{4}\right)$

14. $y = 3 \sin(4x + \pi)$

15. $y = 5 \sin\left(\frac{x}{4} - \pi\right)$

16. $y = -2 \cos(2x + 5\pi)$

In Exercises 17 through 24, find the period (shortest x-distance for repetition) and sketch the graph of the indicated function.

17. $y = -\tan x$

18. $y = \cot 2x$

19. $y = 3 \sec x$

20. $y = \csc(2x - \pi)$

21. $y = \sin^2 x$

22. $y = 4 \cos^2 x$

23. $y = \tan^2 x$

24. $y = \sec^2 x$

In Exercises 25 through 42, sketch the graph of the function.

25. $y = 1 + \sin x$

26. $y = 3 - 3 \cos x$

27. $y = \cos(2x) - 4$

28. $y = -2 - 3 \sin x$

29. $y = \sin(\pi - x)$

30. $y = -\cos\left(\frac{\pi}{2} - x\right)$

31. $y = 3 \cos(\pi - x)$

32. $y = \sin(\pi - 2x)$

33. $y = x + \sin x$

34. $y = x - \cos x$

35. $y = x + 2 \cos x$

36. $y = 3 \cos x - x$

37. $y = x \sin x$

38. $y = \frac{\cos x}{x}$

39. $y = \sin x + 2 \cos x$ [*Hint:* Sketch $y = \sin x$ and $y = 2 \cos x$ on the same axes and add the heights to the two graphs.]

40. $y = 2 \sin 2x - \cos(x/2)$ [*Hint:* Proceed as in Exercise 39.]

41. $y = \cos x - \sin x$ [*Hint:* Proceed as in Exercise 39.]

42. $y = \sin^2 x + \cos x$ [*Hint:* Proceed as in Exercise 39.]

4.3 DIFFERENTIATION OF TRIGONOMETRIC FUNCTIONS

We are now ready to try to differentiate the trigonometric functions. Since they are not expressed in terms of functions whose derivatives we already know, we will have to go back to the definition of the *derivative as a limit.* When we do this with $\sin x$, we will run into the limits

$$\lim_{\Delta x \to 0} \frac{\sin \Delta x}{\Delta x} \quad \text{and} \quad \lim_{\Delta x \to 0} \frac{\cos(\Delta x) - 1}{\Delta x}.$$

We start by studying these limits, using x rather than Δx as the variable. The limit $\lim_{x \to 0} (\sin x)/x$ is a very important one in mathematics.

After we work out the formulas for derivatives of the trigonometric functions and illustrate them, we will give an application to simple harmonic motion.

$\lim_{x \to 0} (\sin x)/x = 1$

We wish to find

$$\lim_{x \to 0} \frac{\sin x}{x} \quad \text{and} \quad \lim_{x \to 0} \frac{\cos x - 1}{x}.$$

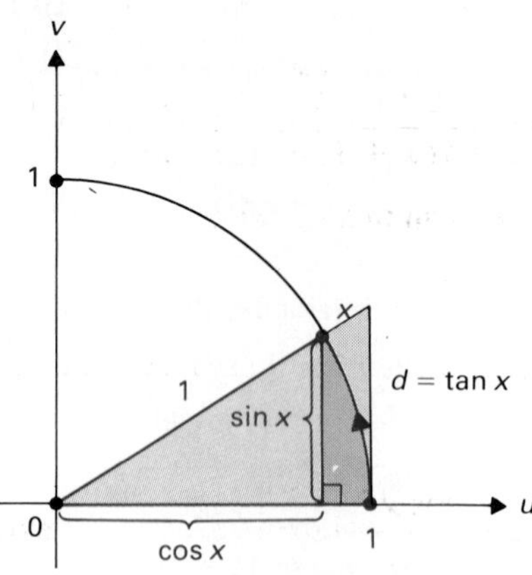

Figure 4.24 In terms of their area, Small triangle < Sector < Large triangle, or

$\frac{1}{2}\sin x \cos x < (x/2\pi)\pi < \frac{1}{2}\tan x.$

Note that the function $(\sin x)/x$ is not defined at 0 and that both numerator and denominator approach zero:

$$\lim_{x\to 0} (\sin x) = \lim_{x\to 0} x = 0.$$

Figure 4.24 shows again part of the circle $u^2 + v^2 = 1$. A positive value of x is indicated by the length of the colored arc. The altitude of the small triangle in the figure is $\sin x$. Since the large and small triangles are similar, if d is the altitude of the large triangle, we have

$$\frac{d}{\sin x} = \frac{1}{\cos x},$$

so $d = \tan x$. Clearly the area of the small triangle in Fig. 4.24 is less than the area of the sector of the circle having arc of length x, which in turn is less than the area of the large triangle. The area of the sector of the circle is the fraction $x/2\pi$ of the area $\pi \cdot 1^2 = \pi$ of the whole circle, so we have

$$\frac{\sin x \cos x}{2} < \frac{x}{2\pi} \cdot \pi < \frac{\tan x}{2}. \tag{1}$$

Multiplying relation (1) by $2/(\sin x)$, we obtain

$$\cos x < \frac{x}{\sin x} < \frac{1}{\cos x}. \tag{2}$$

It is easy to see that relation (2) is valid also for $x < 0$ but near zero; this follows at once from the relations

$$\sin(-x) = -\sin x \quad \text{and} \quad \cos(-x) = \cos x.$$

The definition of cosine and the graph in Fig. 4.15 show that

$$\lim_{x\to 0} (\cos x) = 1,$$

so

$$\lim_{x\to 0} \frac{1}{\cos x} = \frac{1}{1} = 1.$$

But from relation (2), we see that $x/(\sin x)$ is "trapped" between $\cos x$ and $1/(\cos x)$, which both approach 1 as $x \to 0$, so we must have

$$\lim_{x\to 0} \frac{x}{\sin x} = 1.$$

Of course, then

$$\lim_{x\to 0} \frac{\sin x}{x} = \lim_{x\to 0} \frac{1}{[x/(\sin x)]} = \frac{1}{1} = 1.$$

Turning to $\lim_{x\to 0} (\cos x - 1)/x$, for x near zero, we have

$$\frac{\cos x - 1}{x} = \frac{\cos x - 1}{x} \cdot \frac{\cos x + 1}{\cos x + 1}$$

$$= \frac{-\sin^2 x}{x(\cos x + 1)} = -\frac{\sin x}{x} \cdot \frac{\sin x}{\cos x + 1}.$$

Since $\lim_{x\to 0} [(\sin x)/x] = 1$, we have

$$\lim_{x\to 0} \frac{\cos x - 1}{x} = \left[\lim_{x\to 0}\left(-\frac{\sin x}{x}\right)\right]\left(\lim_{x\to 0} \frac{\sin x}{\cos x + 1}\right) = -1 \cdot \frac{0}{2} = 0.$$

These limits are so basic that we summarize them in a theorem.

THEOREM 4.1 Fundamental trigonometric limits

For the functions sin x and cos x,

$$\lim_{x\to 0} \frac{\sin x}{x} = 1 \quad \text{and} \quad \lim_{x\to 0} \frac{\cos x - 1}{x} = 0. \tag{3}$$

EXAMPLE 1 Find $\lim_{x\to 0} [(\sin 5x)/x]$.

Solution Let $u = 5x$, so $x = u/5$. As $x \to 0$, clearly $u \to 0$ also. Then

$$\lim_{x\to 0} \frac{\sin 5x}{x} = \lim_{u\to 0} \frac{\sin u}{u/5} = \lim_{u\to 0}\left(5 \cdot \frac{\sin u}{u}\right) = 5 \cdot 1 = 5.$$

We are usually too lazy to write out the $x = u/5$ substitution. Just remember that

$$\lim_{thing\to 0} \frac{\sin (thing)}{thing} = 1.$$

We then "fix up" our limit to be of this form and write

$$\lim_{x\to 0} \frac{\sin 5x}{x} = \lim_{x\to 0}\left(5 \cdot \frac{\sin 5x}{5x}\right) = 5 \cdot 1 = 5. \quad \square$$

EXAMPLE 2 Find $\lim_{x\to 0} [(1 - \cos 2x)/x]$.

Solution Here we use the second limit in relation (3), thinking of $u = thing = 2x$, as explained in Example 1. We have

$$\lim_{x\to 0} \frac{1 - \cos 2x}{x} = \lim_{x\to 0}\left(-2 \cdot \frac{\cos 2x - 1}{2x}\right) = -2 \cdot 0 = 0. \quad \square$$

EXAMPLE 3 Find $\lim_{x\to 0} [(\sin 4x)/(\sin 3x)]$.

Solution Fixing up the limit to use the first limit in relation (3), as explained in Example 1, we obtain

$$\lim_{x\to 0} \frac{\sin 4x}{\sin 3x} = \lim_{x\to 0}\left(\frac{4}{3} \cdot \frac{\sin 4x}{4x} \cdot \frac{3x}{\sin 3x}\right) = \frac{4}{3} \cdot 1 \cdot 1 = \frac{4}{3}. \quad \square$$

EXAMPLE 4 Find $\lim_{x\to 3} [\sin (x - 3)]/(x^2 - 5x - 6)$.

Solution Here we think of $x - 3$ as the *thing* from Example 1. Then

$$\lim_{x\to 3} \frac{\sin(x-3)}{x^2-5x-6} = \lim_{x\to 3} \frac{\sin(x-3)}{(x-3)(x+2)}$$

$$= \lim_{x\to 3} \left[\frac{\sin(x-3)}{x-3} \cdot \frac{1}{x+2}\right] = 1 \cdot \frac{1}{5} = \frac{1}{5}. \quad \square$$

THE DERIVATIVE OF sin *x*

We must go back to the definition of the derivative

$$f'(x_1) = \lim_{\Delta x\to 0} \frac{f(x_1+\Delta x) - f(x_1)}{\Delta x}$$

to find the derivative of $f(x) = \sin x$. Forming the difference quotient and using identity (6) in the Summary on page 131, we obtain

$$\frac{f(x_1+\Delta x) - f(x_1)}{\Delta x} = \frac{\sin(x_1+\Delta x) - \sin x_1}{\Delta x}$$

$$= \frac{\sin x_1 \cos \Delta x + \cos x_1 \sin \Delta x - \sin x_1}{\Delta x}$$

$$= \frac{\cos x_1 \sin \Delta x + (\sin x_1)(\cos \Delta x - 1)}{\Delta x}$$

$$= \cos x_1 \frac{\sin \Delta x}{\Delta x} + \sin x_1 \frac{\cos \Delta x - 1}{\Delta x}.$$

Therefore, using the limits of relation (3), we have

$$f'(x_1) = \lim_{\Delta x\to 0} \frac{f(x_1+\Delta x) - f(x_1)}{\Delta x}$$

$$= (\cos x_1)\left(\lim_{\Delta x\to 0} \frac{\sin \Delta x}{\Delta x}\right) + (\sin x_1)\left(\lim_{\Delta x\to 0} \frac{\cos \Delta x - 1}{\Delta x}\right)$$

$$= (\cos x_1)(1) + (\sin x_1)(0) = \cos x_1.$$

Thus

$$\frac{d(\sin x)}{dx} = \cos x.$$

If u is a differentiable function of x, then by the chain rule,

$$\frac{d(\sin u)}{dx} = \frac{d(\sin u)}{du} \cdot \frac{du}{dx}.$$

This gives us the differentiation formula

$$\boxed{\frac{d(\sin u)}{du} = (\cos u)\frac{du}{dx}.} \qquad \textbf{(4)}$$

We now give several illustrations of this formula.

EXAMPLE 5 Find the derivative of $y = \sin(x^3)$.

Solution Using formula (4), we have

$$\frac{dy}{dx} = \cos(x^3) \cdot \frac{d(x^3)}{dx} = 3x^2 \cos x^3. \quad \square$$

EXAMPLE 6 Find dy/dx if $y = \sin^5 3x$.

Solution Using the chain rule formula for the derivative of a function to a power, we have

$$\begin{aligned} \frac{d(\sin^5 3x)}{dx} &= 5 \sin^4 3x \cdot \frac{d(\sin 3x)}{dx} \\ &= (5 \sin^4 3x)(\cos 3x)(3) = 15 \sin^4 3x \cos 3x. \quad \square \end{aligned}$$

EXAMPLE 7 Find dy/dx if $y = (\sin 4x)/x$.

Solution The quotient rule for differentiation yields

$$\begin{aligned} \frac{dy}{dx} &= \frac{x[d(\sin 4x)/dx] - (\sin 4x) \cdot 1}{x^2} \\ &= \frac{x(\cos 4x)4 - \sin 4x}{x^2} = \frac{4x \cos 4x - \sin 4x}{x^2}. \quad \square \end{aligned}$$

Now we can finally see why radian measure of an angle is the convenient measure when doing calculus! Let x be radian measure and t be degree measure for an angle. Then

$$x = \frac{\pi}{180} t$$

and

$$\frac{d(\sin x)}{dt} = \frac{d(\sin x)}{dx} \cdot \frac{dx}{dt} = (\cos x)\left(\frac{\pi}{180}\right).$$

In other words, if we used degree measure, we would have the nuisance factor $\pi/180$ in our differentiation formulas for trigonometric functions.

DERIVATIVES OF THE OTHER TRIGONOMETRIC FUNCTIONS

Derivatives of the other five basic trigonometric functions are now easy to find. Since $\cos x = \sin[x + (\pi/2)]$, we have

$$\frac{d(\cos x)}{dx} = \frac{d[\sin(x + \pi/2)]}{dx} = \cos\left(x + \frac{\pi}{2}\right).$$

But $\cos[x + (\pi/2)] = -\sin x$, so

$$\frac{d(\cos x)}{dx} = -\sin x.$$

Using the chain rule,

$$\boxed{\frac{d(\cos u)}{du} = (-\sin u)\frac{du}{dx}.} \tag{5}$$

The formula $d(\cos x)/dx = -\sin x$ can also be obtained by differentiating the identity $\sin^2 x + \cos^2 x = 1$ implicitly:

$$2 \sin x \cos x + 2 \cos x \frac{d(\cos x)}{dx} = 0,$$

$$\frac{d(\cos x)}{dx} = \frac{-2 \sin x \cos x}{2 \cos x} = -\sin x.$$

The other four trigonometric functions are quotients involving only 1 and $\sin x$ and $\cos x$, so their derivatives can be found using the quotient rule for differentiation. For example,

$$\frac{d(\tan x)}{dx} = \frac{d[(\sin x)/(\cos x)]}{dx} = \frac{(\cos x)(\cos x) - (\sin x)(-\sin x)}{\cos^2 x}$$

$$= \frac{\cos^2 x + \sin^2 x}{\cos^2 x} = \frac{1}{\cos^2 x} = \sec^2 x.$$

In a similar fashion, we find that

$$\frac{d(\cot x)}{dx} = -\csc^2 x, \qquad \frac{d(\sec x)}{dx} = \sec x \tan x,$$

and

$$\frac{d(\csc x)}{dx} = -\csc x \cot x.$$

Using the chain rule, we obtain the following formulas analogous to formulas (4) and (5).

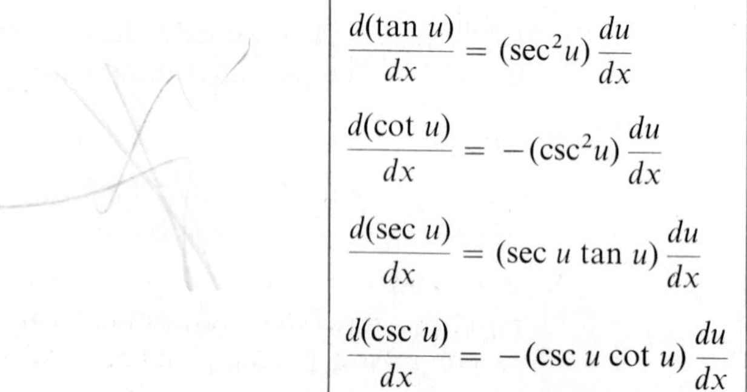

$$\frac{d(\tan u)}{dx} = (\sec^2 u)\frac{du}{dx}$$

$$\frac{d(\cot u)}{dx} = -(\csc^2 u)\frac{du}{dx}$$

$$\frac{d(\sec u)}{dx} = (\sec u \tan u)\frac{du}{dx}$$

$$\frac{d(\csc u)}{dx} = -(\csc u \cot u)\frac{du}{dx}$$

We recommend that these formulas be memorized, together with formulas (4) and (5). Note that if you know the derivatives of the three functions $\sin x$, $\tan x$, and $\sec x$, then the derivatives of each of the corresponding cofunctions $\cos x$, $\cot x$, $\csc x$ are found by changing the derivatives to the cofunctions and changing sign.

EXAMPLE 8 Find dy/dx if $y = x^3 \tan 2x$.

Solution We have

$$\begin{aligned}\frac{dy}{dx} &= x^3 \frac{d(\tan 2x)}{dx} + (\tan 2x)\frac{d(x^3)}{dx} \\ &= x^3(\sec^2 2x)(2) + (\tan 2x)(3x^2) \\ &= 2x^3(\sec^2 2x) + 3x^2(\tan 2x). \quad \square\end{aligned}$$

EXAMPLE 9 Find dy/dx if $y = \sin^2 x/\cos^3 x$.

Solution Using the quotient rule for differentiation, we have

$$\begin{aligned}\frac{dy}{dx} &= \frac{(\cos^3 x)(2 \sin x \cos x) - (\sin^2 x)(3 \cos^2 x)(-\sin x)}{\cos^6 x} \\ &= \frac{(\cos^2 x)(2 \sin x \cos^2 x + 3 \sin^3 x)}{\cos^6 x} \\ &= \frac{2 \sin x \cos^2 x + 3 \sin^3 x}{\cos^4 x}. \quad \square\end{aligned}$$

EXAMPLE 10 Find the equation of the tangent line to the graph of $y = \tan x$ at the point $(\pi/4, 1)$.

Solution

Point $\left(\frac{\pi}{4}, 1\right)$

Slope $\left.\frac{dy}{dx}\right|_{\pi/4} = \sec^2 x\,|_{\pi/4} = (\sqrt{2})^2 = 2$

Equation $y - 1 = 2\left(x - \frac{\pi}{4}\right)$, or $y = 2x + 1 - \frac{\pi}{2}$ $\square$

EXAMPLE 11 Estimate cos 58°, using a differential.

Solution We know $\cos 60° = \frac{1}{2}$. We thus use the formula

$$f(x_0 + dx) \approx f(x_0) + f'(x_0)\, dx,$$

with

$$f(x) = \cos x, \qquad x_0 = \frac{\pi}{3}, \qquad dx = -2 \cdot \frac{\pi}{180} = \frac{-\pi}{90}.$$

Note that we must use *radian* measure in calculus with our trigonometric functions, since our differentiation formulas are given in terms of radian measure. Now $f'(x) = -\sin x$, so we have

$$\begin{aligned}\cos 58° &\approx \cos\left(\frac{\pi}{3}\right) - \sin\left(\frac{\pi}{3}\right) \cdot \frac{-\pi}{90} \\ &= \frac{1}{2} - \frac{\sqrt{3}}{2}\left(\frac{-\pi}{90}\right) = \frac{90 + \sqrt{3}\pi}{180} \approx 0.53023.\end{aligned}$$

A calculator gives $\cos 58° \approx 0.52992$, so our error is about 0.00031. $\square$

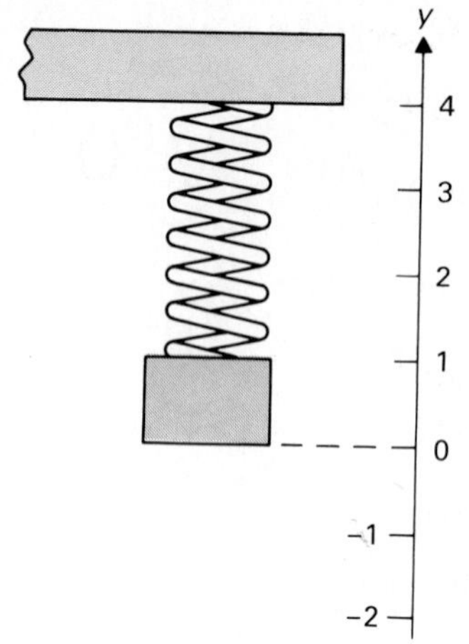

Figure 4.25 A weight hanging by a spring at position of rest $y = 0$.

SIMPLE HARMONIC MOTION

Consider a weight suspended vertically by a spring attached to a beam, as shown in Fig. 4.25. Suppose the weight is raised vertically a distance a, which still keeps the spring stretched a bit, and is then released. The weight bobs up and down. It can be shown that if the weight is released at time $t = 0$, then its position y in Fig. 4.25 at time t is given by

$$y = a \cos \omega t. \tag{6}$$

The constant ω in Eq. (6) depends on the stiffness of the spring. The weight bobs up and down with *amplitude a*, from a units above to a units below its position of rest. (Air resistance and friction eventually stop the motion, but we neglect them.) If a cube of wood, floating in water, is raised slightly and released, a similar motion results.

For the weight on the spring or the cube in the water, a displacement from its position of rest gives rise to a restoring force proportional to the magnitude of the displacement. (It is assumed that the elastic limit of the spring is not exceeded and that the cube is not pushed beneath the water or lifted completely out of it.) Motion governed by this restoring force law is called *simple harmonic motion*. Our work in Section 20.6 will show that the position y in such a situation is always given by

$$y = a \cos \omega t + b \sin \omega t \tag{7}$$

for some constants a and b. Equation (6) is a special case of Eq. (7).

EXAMPLE 12 Let the position (in centimeters) for a body in simple harmonic motion be given by $y = 4 \cos 3t$ after t sec. Find

a. the amplitude of the motion,

b. the velocity when $t = \pi/2$ sec,

c. the speed when $t = \pi/2$ sec,

d. the acceleration when $t = \pi/2$ sec.

Solution Simple harmonic motion is an example of motion on a line, which we treated in Section 3.5.

a. The amplitude of the motion is 4 cm, the coefficient of $\cos 3t$.

b. We have $v = dy/dt = -12 \sin 3t$, so when $t = \pi/2$,

$$v = -12 \sin 3t\,|_{t=\pi/2} = -12 \sin \frac{3\pi}{2} = -12(-1) = 12 \text{ cm/sec}.$$

c. The speed is $|v| = 12$ cm/sec.

d. The acceleration is $d^2y/dt^2 = -36 \cos 3t$, so when $t = \pi/2$, the acceleration is

$$-36 \cos 3t\,|_{t=\pi/2} = -36 \cos \frac{3\pi}{2} = 0 \text{ cm/sec}^2. \quad \square$$

EXAMPLE 13 Let the position in simple harmonic motion be given by Eq. (7) with $\omega = 2$. If $y = 0$ cm and $v = -5$ cm/sec when $t = 0$ sec, find the equation for y.

Solution Putting $y = 0$ and $t = 0$ in Eq. (7), we obtain

$$0 = a \cos 0 + b \sin 0 = a(1) + b(0) = a,$$

so Eq. (7) now becomes

$$y = b \sin 2t.$$

Then

$$v = \frac{dy}{dt} = 2b \cos 2t.$$

Setting $v = -5$ and $t = 0$, we obtain

$$-5 = 2b(\cos 0) = 2b(1) = 2b, \qquad \text{so} \quad b = -\frac{5}{2}.$$

Thus $y = -\frac{5}{2} \sin 2t$ cm at time t sec. □

SUMMARY

1. Limits:

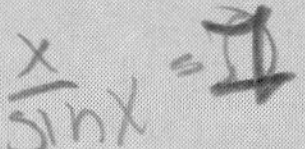

$$\lim_{x \to 0} \frac{\sin x}{x} = 1, \qquad \lim_{x \to 0} \frac{\cos x - 1}{x} = 0$$

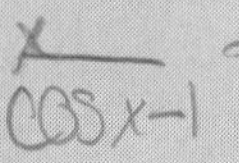

2. Differentiation formulas:

$$\frac{d(\sin u)}{dx} = (\cos u)\frac{du}{dx}, \qquad \frac{d(\cos u)}{dx} = -(\sin u)\frac{du}{dx},$$

$$\frac{d(\tan u)}{dx} = (\sec^2 u)\frac{du}{dx}, \qquad \frac{d(\cot u)}{dx} = (-\csc^2 u)\frac{du}{dx},$$

$$\frac{d(\sec u)}{dx} = (\sec u \tan u)\frac{du}{dx}, \qquad \frac{d(\csc u)}{dx} = (-\csc u \cot u)\frac{du}{dx}$$

3. The position of a body moving on a line (y-axis) in simple harmonic motion is given by an equation of the form

$$y = a \cos \omega t + b \sin \omega t,$$

where a, b, and ω are constants.

EXERCISES

In Exercises 1 through 17, find the indicated limit if it exists, using the symbols ∞ or $-\infty$ where appropriate.

1. $\displaystyle\lim_{x \to 0} \frac{\sin x}{|x|}$

2. $\displaystyle\lim_{x \to 0-} \frac{\sin x}{|x|}$

3. $\displaystyle\lim_{t \to 0} \frac{\sin 2t}{t}$

4. $\displaystyle\lim_{t \to 0} \frac{\sin 2t}{\sin 3t}$

5. $\displaystyle\lim_{x \to 0} \frac{\cos 2x}{\cos 3x}$

6. $\displaystyle\lim_{x \to 0} \frac{\sin 2x}{\cos 3x}$

7. $\displaystyle\lim_{\theta \to 0} (\theta^2 \csc^2 \theta)$

8. $\displaystyle\lim_{x \to 0+} (x \csc^2 x)$

9. $\displaystyle\lim_{v \to 0} \frac{\tan 3v}{v}$

10. $\displaystyle\lim_{u \to 0} \frac{u^2}{\tan^4 u}$

11. $\displaystyle\lim_{x \to 2} \frac{\sin (x - 2)}{x^2 - 4}$

12. $\displaystyle\lim_{t \to -3} \frac{\sin (t + 3)}{t^3 + 3t^2}$

13. $\displaystyle\lim_{t \to -1} \sin \left(\frac{1}{t + 1}\right)$

14. $\displaystyle\lim_{x \to 0} \frac{\cos^2 x - 1}{x^2}$

15. $\lim_{u \to -\pi/2} \frac{\cos u}{u + (\pi/2)}$ [*Hint:* Use $\cos x = \sin(x + \pi/2)$.]

16. $\lim_{x \to \pi} \frac{\sin x}{x - \pi}$

[*Hint:* Use $\sin(-x) = -\sin x$ and $\sin(x + \pi) = -\sin x$.]

17. $\lim_{t \to \pi} \frac{\cos t}{(t - \pi)^2}$

18. Derive the formula for the derivative of cot x.

19. Derive the formula for the derivative of sec x.

20. Derive the formula for the derivative of csc x.

In Exercises 21 through 40, find the derivative of the given function. You need not simplify your answers.

21. $x \cos x$
22. $x^2 \tan x$
23. $(x^2 + 3x) \sec x$
24. $\frac{\csc x}{x}$
25. $\sin^2 x$
26. $\sin 2x$
27. $\sec^2 x$
28. $\sin x \tan x$
29. $\frac{x}{\cot x}$
30. $\frac{x^2 - 2x}{\csc x}$
31. $y = \sin 2x$
32. $y = \sec(3x + 1)$
33. $y = \cos^2(2 - 3x)$
34. $y = \cot^2 x$
35. $y = \sin^2 x \cos^2 x$
36. $y = \tan x \sec 2x$
37. $y = \sqrt{\cot^2 x + \csc^2 x}$
38. $y = \sqrt{8x^2 + \cos^2 x}$
39. $y = \sin(\tan 3x)$
40. $y = \csc(x + \cos x^2)$

In Exercises 41 through 46, find dy/dx at the indicated point using implicit differentiation.

41. $x \cos y + y \sin x = \frac{\pi}{2}$ at $(\pi/2, \pi)$

42. $(\sin x)(\cos y) = \frac{1}{2}$ at $(\pi/4, \pi/4)$

43. $\sin(xy) + 3y = 4$ at $(\pi/2, 1)$

44. $\tan xy = 1$ at $(\pi/4, 1)$

45. $\sec x + \tan y = 1$ at $(0, 0)$

46. $\csc\left(\frac{\pi x y^2}{2}\right) + \sin\left(\frac{\pi y}{2}\right) + y = 3$ at $(1, 1)$

47. Find the equation of the tangent line to $y = \sin x$ when $x = \pi/4$.

48. Find the equation of the normal line to $y = \tan x$ when $x = 3\pi/4$.

49. Find the equation of the normal line to $y = x^2 + 3x - \cos^2(x - \pi/4)$ when $x = 0$.

50. Find the equation of the tangent line to $y = \tan^3 x$ when $x = \pi/4$.

51. Use a differential to estimate sin 31°.

52. Use a differential to estimate tan 128°.

53. Use a differential to estimate $\cos^2 44°$.

54. Use a differential to estimate $\csc^2 59°$.

55. Let the position of a body in simple harmonic motion be given by $y = -5 \sin \pi t$ cm at t sec. Find
a) the amplitude,
b) the velocity when $t = \frac{1}{3}$,
c) the speed when $t = \frac{1}{3}$,
d) the acceleration when $t = \frac{1}{3}$.

56. Let the position of a body in simple harmonic motion be given by $y = 10 \cos 2\pi t$ cm at time t sec. Find
a) the velocity when $t = \frac{1}{4}$,
b) the speed when $t = \frac{1}{3}$,
c) the acceleration when $t = \frac{1}{2}$.

57. Let the position of a body in simple harmonic motion be given by $y = a \cos \pi t + b \sin \pi t$. If $y = 0$ cm and $v = 3$ cm/sec when $t = 0$ sec, find a and b.

58. Let the position of a body in simple harmonic motion be given by $y = a \cos[(\pi/2)t] + b \sin[(\pi/2)t]$. If $y = -2$ cm and $v = -3$ cm/sec when $t = 0$ sec, find the acceleration when $t = \frac{1}{2}$.

EXERCISE SETS FOR CHAPTER 4

Review Exercise Set 4.1

1. Find a) $\tan \frac{5}{6}\pi$, b) $\cos \frac{5}{4}\pi$.

2. If $0 \le \theta < \pi$ and $\sec \theta = -5$, find $\sin \theta$.

3. Use the identity

$$\sin(x - y) = \sin x \cos y - \cos x \sin y$$

to compute sin 15°.

4. Find the amplitude and period, and sketch the graph of $y = 3 \sin(2x - \pi)$.

5. Find the period of $|\sin 3x|$.

6. Find the indicated limit, if it exists.

a) $\lim_{x \to 0} \frac{\sin x}{x^2 + 4x}$ b) $\lim_{x \to 3} \frac{\sin(x^2 - 9)}{x - 3}$

7. Find dy/dx.

a) $y = \sin^3 2x$ b) $y = x^2 \csc x^3$

8. Find dy/dx if $x^2 \cos y + y^3 = -3$.

Review Exercise Set 4.2

1. Find

a) $\sin \frac{11\pi}{6}$, b) $\cot\left(-\frac{4}{3}\pi\right)$.

2. If $\pi/2 \le \theta < 3\pi/2$ and $\tan\theta = 2/3$, find $\sin\theta$.
3. If a triangle has sides of lengths 2, 4, and 5, find $\cos\theta$ if θ is the angle opposite the side of length 5.
4. Find the period and amplitude of $-\frac{1}{2}\cos(x/3)$.
5. Sketch the graph of $y = 3\sec(x/2)$.
6. Find the indicated limit, if it exists.

   a) $\lim_{x\to 0} \dfrac{\sin^2 x}{x^2 + 4x}$

   b) $\lim_{x\to 0} (2x + 4)\cdot \sin\left(\dfrac{1}{x+1}\right)$
7. Find dy/dx.

   a) $y = \tan(x^2 + 1)$ b) $y = \dfrac{\sin^2 x}{x - 4}$
8. Find the equation of the tangent line to $y = 2\cos[x - (\pi/2)]$ at the origin.

Review Exercise Set 4.3

1. Find a) $\sin 23\pi/2$, b) $\csc(-7\pi/4)$.
2. If $\pi/2 \le \theta < 3\pi/2$ and $\tan\theta = \sqrt{3}$, find $\sin\theta$.
3. If a triangle has sides of lengths 2 and 4 with an angle of 30° between them, find the length of the remaining side.
4. Find the amplitude and period and sketch the graph of $y = -2\sin 4x$.
5. Find the period of $y = 5\cos(3x - 2)$.
6. Find the indicated limit, if it exists.

   a) $\lim_{x\to 0} \dfrac{\sin x^2}{x^3 + 4x^2}$ b) $\lim_{x\to 0} \dfrac{1 - \cos^2 x}{x}$
7. Find dy/dx.

   a) $y = \dfrac{x+1}{\cos x}$ b) $\cos(xy) + y^2 = 3$
8. Use a differential to estimate $\sin 28°$.

Review Exercise Set 4.4

1. Find a) $\sec 17\pi$, b) $\cos\dfrac{5\pi}{6}$.
2. If $\cos\theta = \frac{1}{2}$ and $\pi \le \theta < 2\pi$, find θ.
3. A triangle has an angle of 45° opposite a side of length 10. If another angle of the triangle is 30°, find the length of the side opposite this angle.
4. Find all values of c such that the period of $y = \cos[(cx - 3)/5]$ is 4.
5. Find the period and amplitude and sketch the graph of $y = 2\sin[(\pi - x)/2]$.
6. Find the indicated limit, if it exists.

   a) $\lim_{x\to 1} \dfrac{\sin(x-1)}{x - x^2}$ b) $\lim_{t\to 0} (t \csc 4t)$
7. Find dy/dx.

   a) $y = \sin 2x \tan^3 x$ b) $x\sin y - y\sin x = 0$
8. Use a differential to estimate $\cos^2 61°$.

More Challenging Exercises 4

In Exercises 1 through 10, find the indicated limit, if it exists.

1. $\lim_{x\to\infty} x\sin(1/x)$
2. $\lim_{x\to\infty} x^2 \sin(1/x)$
3. $\lim_{x\to 0} \dfrac{\cos 2x - 1}{x}$
4. $\lim_{x\to 0} \dfrac{\cos 3x - 1}{\cos 5x - 1}$
5. $\lim_{x\to 1} \dfrac{\sin(x^2 - 1)}{x - 1}$
6. $\lim_{x\to -3} \dfrac{\sin(x^2 + x - 6)}{x + 3}$
7. $\lim_{x\to 0} \dfrac{\sin(\sin x)}{x}$
8. $\lim_{x\to 0} \dfrac{\sin(\sin^2 x)}{\sin x}$
9. $\lim_{x\to \pi/4} \dfrac{\tan x - 1}{x - \pi/4}$

   [*Hint:* Sometimes a limit can be recognized as the definition of some derivative.]
10. $\lim_{x\to \pi/6} \dfrac{2\sin x - 1}{6x - \pi}$

5 APPLICATIONS OF THE DERIVATIVE

This chapter introduces a few of the many applications of differential calculus to the sciences. It also shows how the derivative can be used in mathematics as an aid to studying functions and their graphs.

5.1 RELATED RATE PROBLEMS

Recall that the derivative dy/dx gives the instantaneous rate of change of y with respect to x. Leibniz notation is very useful in keeping track of just what rate of change we are working with. For example, imagine that a pebble is dropped into a calm pond and a circular wave spreads out from the point where the pebble was dropped. We might be interested in the following rates of change:

$$\frac{dr}{dt} = \text{rate of increase of } \textit{radius} \text{ per unit increase in } \textit{time},$$

$$\frac{dA}{dt} = \text{rate of increase of } \textit{area} \text{ per unit increase in } \textit{time},$$

$$\frac{dA}{dr} = \text{rate of increase of } \textit{area} \text{ per unit increase in } \textit{radius},$$

$$\frac{dC}{dt} = \text{rate of increase of } \textit{circumference} \text{ per unit increase in } \textit{time},$$

and so on. The Leibniz notation helps us remember which rate of change we want.

In many rate-of-change problems, we want to find the time rate of change of a quantity Q if we know the time rate of change of one or more related quantities, say r and s. The box on the next page gives one convenient step-by-step outline to follow in solving *related rate problems*. (The letters may be different from Q, r, and s in a problem, of course.)

The following examples illustrate the use of the outline.

EXAMPLE 1 If the radius r of a circular disk is increasing at the rate of 3 in./sec, find the rate of increase of its area when $r = 4$ in.

Solution Let A be the area and r be the radius of the circular disk.

Step 1 Find dA/dt when $r = 4$ in.

Step 2 Given $dr/dt = 3$ in./sec.

Step 3 $A = \pi r^2$

Step 4 $dA/dt = 2\pi r \cdot dr/dt$

Step 5 When $r = 4$ and $dr/dt = 3$, $dA/dt = 2\pi \cdot 4 \cdot 3 = 24\pi$ in^2/sec. □

EXAMPLE 2 Ship A passes a buoy at 9:00 A.M. and continues on a northward course at a rate of 12 mph. Ship B, traveling at 18 mph, passes the same buoy on its eastward course at 10:00 A.M. the same day. Find the rate at which the distance between the ships is increasing at 11:00 A.M. that day.

Figure 5.1 Positions of ships A and B relative to the buoy.

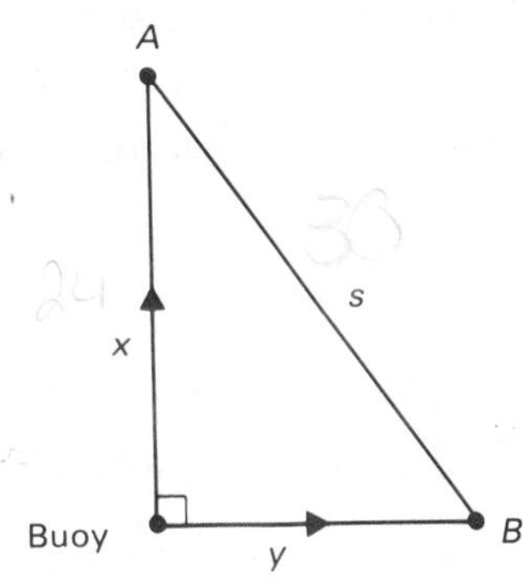

Solution We draw a figure and assign letter variables as shown in Fig. 5.1.

Step 1 Find ds/dt at $t = $ 11:00 A.M.

Step 2 Given $dx/dt = 12$ mph and $dy/dt = 18$ mph.

Step 3 $s^2 = x^2 + y^2$

Step 4 $2s(ds/dt) = 2x(dx/dt) + 2y(dy/dt)$

Step 5 At 11:00 A.M., ship A has traveled for 2 hours at 12 mph since passing the buoy, so it has gone $x = 24$ miles. Similarly, ship B has traveled

---

OUTLINE FOR RELATED RATE PROBLEMS

Read the problem carefully. Draw a figure where appropriate and decide what letter variables to use. Then follow these steps.

Step 1 Decide what rate of change is desired and express it in Leibniz notation:

$$\text{Find } \frac{dQ}{dt} \quad \text{when} \quad t = \_\_\_\_\_.$$

Step 2 Decide what rates of change are given and express these data in Leibniz notation:

$$\text{Given } \frac{dr}{dt} = \_\_\_\_\_ \quad \text{and} \quad \frac{ds}{dt} = \_\_\_\_\_ \quad \text{when} \quad t = \_\_\_\_\_.$$

Step 3 Find an equation relating Q, r, and s. You may have to use a figure or some geometric formula.

Step 4 Differentiate the relation in Step 3 (often implicit differentiation) to obtain a relation between dQ/dt, dr/dt, and ds/dt.

Step 5 Put in values of r, s, and Q and of dr/dt and ds/dt corresponding to the instant when dQ/dt is desired, and solve for dQ/dt.

for 1 hour at 18 mph, so it has gone $y = 18$ miles. Also, $dx/dt = 12$, $dy/dt = 18$, and

$$s = \sqrt{x^2 + y^2} = \sqrt{24^2 + 18^2} = 30.$$

Putting these values into the equation of Step 4, we have

$$2 \cdot 30 \frac{ds}{dt} = 2 \cdot 24 \cdot 12 + 2 \cdot 18 \cdot 18 = 1224,$$

$$\frac{ds}{dt} = \frac{1224}{60} = 20.4 \text{ mph.} \quad \square$$

EXAMPLE 3 Suppose a spherical balloon is being inflated in such a way that the volume of the sphere increases at a constant rate. Show that the rate of increase of the radius is inversely proportional to the surface area of the balloon.

Solution Let the spherical balloon have volume V, radius r, and surface area S at time t.

Step 1 Find dr/dt and show that $dr/dt = k/S$ for some constant k.

Step 2 Given $dV/dt = c$, where c is a constant.

Step 3 The formula for the volume of a sphere is $V = \frac{4}{3}\pi r^3$.

Step 4 $dV/dt = \frac{4}{3} \cdot 3\pi r^2(dr/dt) = 4\pi r^2(dr/dt)$

Step 5 We know that $dV/dt = c$, and the formula for the surface area S of a sphere is $S = 4\pi r^2$. Thus

$$c = S\frac{dr}{dt}, \qquad \text{so} \quad \frac{dr}{dt} = \frac{c}{S},$$

and we are done. $\square$

EXAMPLE 4 Let a parallelogram have sides of lengths 8 and 12 in., and let a vertex angle θ be decreasing at a rate of 2°/min. Find the rate of change of the area of the parallelogram when $\theta = 30°$.

Solution Let A be the area of the parallelogram of altitude h, as shown in Fig. 5.2.

Step 1 Find dA/dt when $\theta = 30° = \pi/6$ radians.

Step 2 Given $d\theta/dt = -2°/\text{min} = -\pi/90$ radians/min. We must use radian measure because the differentiation formulas are in terms of radian measure. The negative sign occurs because θ is decreasing.

Step 3 We have $A = 12h$, and from Fig. 5.2 we see that $h = 8 \sin \theta$. Thus $A = 12(8 \sin \theta) = 96 \sin \theta$.

Step 4 $dA/dt = 96(\cos \theta)(d\theta/dt)$

Step 5 When $\theta = \pi/6$ and $d\theta/dt = -\pi/90$, we have

$$\frac{dA}{dt} = 96\frac{\sqrt{3}}{2} \cdot \frac{-\pi}{90} = -\frac{8\sqrt{3}\pi}{15} \text{ in}^2/\text{min.} \quad \square$$

Figure 5.2 Parallelogram with sides 8 and 12, vertex angle θ, and altitude h.

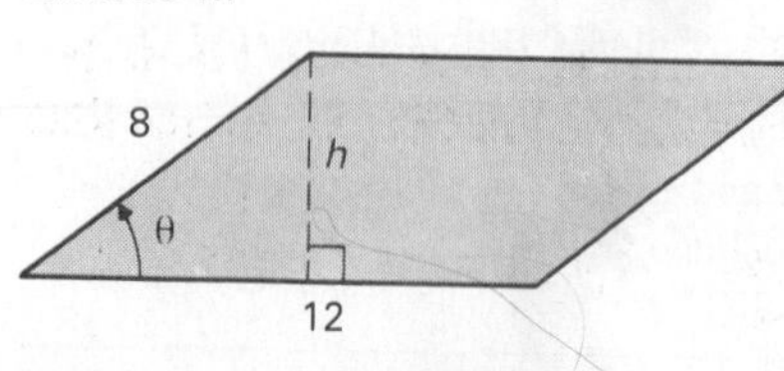

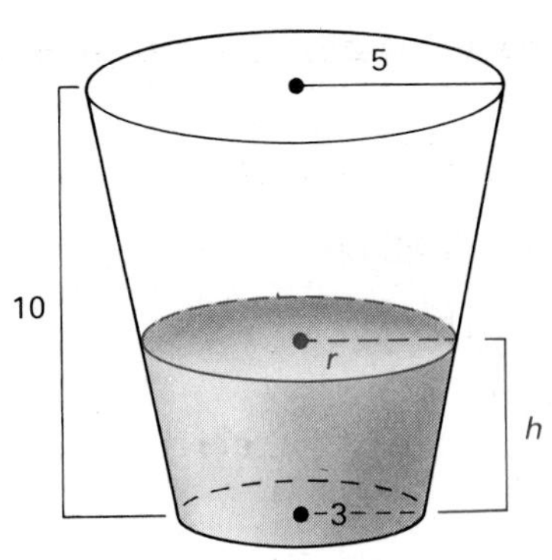

Figure 5.3 Liquid of depth h and top radius r in a glass.

EXAMPLE 5 A gas is confined in a container of variable volume that increases at a constant rate of 10 cm^3/sec. The gas is kept at constant temperature. According to Boyle's law, the pressure of the gas on the walls of the container is inversely proportional to the volume. Suppose the pressure is 80 g/cm^2 when the volume is 400 cm^3. Find the rate of change of the pressure when the volume is 600 cm^3.

Solution Let V be the volume of the container and P be the pressure of the gas.

Step 1 Find dP/dt when $V = 600 \text{ cm}^3$.

Step 2 We are given that $P = 80 \text{ g/cm}^2$ when $V = 400 \text{ cm}^3$ and that $dV/dt = 10 \text{ cm}^3/\text{sec}$ at all times.

Step 3 By Boyle's law,

$$P = \frac{k}{V}$$

for some constant k. The data given in Step 2 allow us to find k. Substituting in $P = k/V$, we have

$$80 = \frac{k}{400}, \qquad \text{so} \quad k = 32{,}000.$$

Thus $P = 32{,}000/V = 32{,}000V^{-1}$.

Step 4 $dP/dt = 32{,}000(-1)V^{-2} \cdot dV/dt = -(32{,}000/V^2)\, dV/dt$

Step 5 Now $dV/dt = 10 \text{ cm}^3/\text{sec}$, so when $V = 600 \text{ cm}^3$, we have

$$\frac{dP}{dt} = -\frac{32{,}000}{600^2} \cdot 10 = -\frac{32}{36} = -\frac{8}{9} \text{ g/cm}^2. \quad \square$$

Figure 5.4 Continuation of the glass in Fig. 5.3 to a cone of altitude $10 + x$.

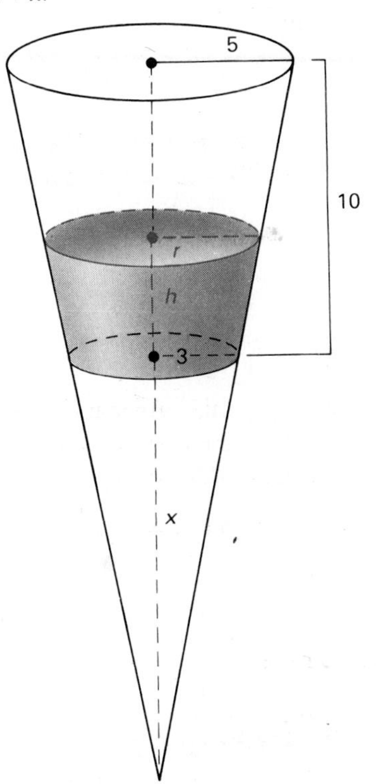

EXAMPLE 6 A drinking glass is in the shape of a truncated cone with base of radius 3 cm, top of radius 5 cm, and altitude 10 cm, as shown in Fig. 5.3. A beverage is poured into the glass at a constant rate of 48 cm^3/sec. Find the rate at which the level of beverage is rising in the glass when it is at a depth of 5 cm.

Solution Let r be the radius at the top of a volume V of beverage in the glass when the depth is h, as shown in Fig. 5.3.

Step 1 Find dh/dt when $h = 5$ cm.

Step 2 Given $dV/dt = 48 \text{ cm}^3/\text{sec}$.

Step 3 We need to find a relation between V and h. First, we express V in terms of r and h. The formula for the volume of a cone is

$$\frac{1}{3}\pi(\text{Radius})^2(\text{Altitude}).$$

If the vertex of the cone before truncation was x cm below the base of the glass, then from similar triangles in Fig. 5.4, we see that

$$\frac{5}{3} = \frac{10 + x}{x}, \qquad 5x = 30 + 3x, \qquad 2x = 30, \qquad x = 15.$$

Then the volume of the beverage is

$$V = \frac{1}{3}\pi r^2(15 + h) - \frac{1}{3}\pi \cdot 3^2 \cdot 15.$$

Again, similar triangles in Fig. 5.4 show that

$$\frac{r}{3} = \frac{x + h}{x}.$$

Since $x = 15$, we have

$$\frac{r}{3} = \frac{15 + h}{15}, \qquad \text{so} \quad r = \frac{15 + h}{5} = 3 + \frac{h}{5}.$$

Thus

$$V = \frac{1}{3}\pi\left(3 + \frac{h}{5}\right)^2(15 + h) - \frac{1}{3}\pi \cdot 3^2 \cdot 15$$

$$= \frac{5}{3}\pi\left(3 + \frac{h}{5}\right)^3 - \frac{1}{3}\pi \cdot 3^2 \cdot 15.$$

Step 4 $dV/dt = 5\pi(3 + h/5)^2 \cdot \frac{1}{5} \cdot dh/dt.$

Step 5 Now $dV/dt = 48\ \text{cm}^3/\text{sec}$ when $h = 5$ cm, so

$$48 = 5\pi(3 + 1)^2 \cdot \frac{1}{5} \cdot \frac{dh}{dt} = 16\pi \frac{dh}{dt}.$$

Thus $dh/dt = 3/\pi$ cm/sec. □

SUMMARY

To solve related rate problems, follow the outline on page 149.

EXERCISES

1. Oil spilled from a tanker anchored in a harbor spreads out in a circular slick. When the radius of the slick is 100 ft, the area of the slick is increasing at 50 ft^2/sec. Find the rate of increase of the radius of the slick at that moment.
2. If the area of a circle is increasing at a constant rate of $200\pi\ \text{in}^2$/sec, find the rate of increase of the circumference when the circumference is 100π inches.
3. Find the rate at which the area of an equilateral triangle is increasing when it is 10 in. on a side, if the length of each side is increasing at the rate of 2 in./min.
4. The vertex angle opposite the base of an isosceles triangle with equal sides of constant length 10 ft is increasing at a rate of $\frac{1}{2}$ radian/min.
   a) Find the rate at which the base of the triangle is increasing when the vertex angle is 60°.
   b) Find the rate at which the area of the triangle is increasing when the vertex angle is 60°.
5. A light tower is located 1 mile directly offshore from a point P on a straight coastline. The beam of light revolves at the rate of $\frac{1}{10}$ radian/sec. Find the rate at which the spot of light on the shore is moving along the coastline when (a) the spot is at P, and (b) the spot is 2 miles from P along the shore.
6. The length of a rectangle is increasing at 6 ft/min while its width is decreasing at a rate of 4 ft/min. Find the rate of change of the area when the length is 100 ft and the width is 30 ft.

7. A rectangle has diagonals of length 50 cm. If the angle θ between the diagonals is increasing at a constant rate of $3°$/min while the diagonals remain the same length, find the rate of change of the area of the rectangle when $\theta = 60°$.

8. Repeat Exercise 7 for a parallelogram with diagonals of constant lengths 50 cm and 30 cm. The other data remain the same.

9. If V volts are measured across a resistance of R ohms in a circuit carrying a current of I amperes, then $V = IR$. At a certain instant, the resistance is 500 ohms and is decreasing at a rate of 2 ohms/min, while the current is 0.5 amp and is increasing at a rate of 0.01 amp/min. Find the rate of change of the voltage at that moment.

10. If a current of I amperes is flowing through a circuit of resistance R ohms, then W watts of power are used, where $W = I^2R$. If current flowing through a constant resistance of 5 ohms is increasing at a constant rate of 0.2 amp/min, find the rate of change of the power used when $I = 4$ amp.

11. An 18-ft ladder is leaning against a vertical wall. If the bottom of the ladder is pulled away from the wall at a constant rate of 3 ft/sec, find the rate at which the top is sliding down the wall when the bottom is 8 ft from the wall.

12. Using the data in Exercise 11, find the rate of change of the area of the triangle formed by the ladder, the ground, and the wall when the bottom of the ladder is 8 ft from the wall.

13. A ship sails out of New York harbor at a rate of 20 ft/sec. The closest it comes to the Statue of Liberty is 1200 ft at a certain time t_0. Find the rate at which the distance from the ship to the statue is increasing 25 sec later. (Assume the ship travels in a straight line during these 25 sec.)

14. An airplane traveling in a straight line at 600 mph at a constant altitude of 4 miles passes directly over your head. Find the rate at which the distance from you to the plane is changing 48 sec later. (Neglect the curvature of the earth.)

15. A particle starts at the origin in the plane and travels on the curve $y = \sqrt{5x + 4} - 2$. If the x-coordinate of the particle increases at a uniform rate of 4 units/sec, find the rate of increase of the y-coordinate 7 sec after the particle starts from the origin.

16. A body travels on the circle $x^2 + y^2 = 25$. When the body is at the point $(3, -4)$, the x-component of the velocity is 2 units/sec. Find the y-component of the velocity at that point.

17. Jim is 6 ft tall and is walking at night straight toward a lighted street lamp at a rate of 5 ft/sec. If the lamp is 20 ft above the ground, find the rate at which his shadow is shortening when he is 30 ft from the lamp post.

18. A triangle has two sides of constant lengths 10 cm and 15 cm. The angle θ between the two sides increases at a constant rate of $9°$/min. Find the rate of increase of the third side of the triangle when $\theta = 60°$.

19. The lengths of two sides of a triangle remain constant at 5 ft and 7 ft while the length of the third side decreases at a constant rate of 6 in./min. Let θ be the angle between the sides of constant lengths. Find the rate at which θ is changing when $\cos\theta = \frac{3}{5}$.

20. Sue is standing on a dock and pulling a boat in to the dock by means of a rope tied to a ring in the bow of the boat. If the ring is 2 ft above water level and her hands are 7 ft above water level, and if she is pulling in the rope at a uniform rate of 2 ft/sec, find the speed with which the boat is approaching the dock when it is 12 ft from the dock.

21. A rocket fired straight up from the ground reaches an altitude of $5t^2$ ft after t sec. An observer whose eyes are 5 ft above the ground watches from 315 ft away. Find the rate at which the angle of elevation from the observer's eyes to the rocket is increasing when the angle is $45°$.

22. A spherical balloon is being inflated so that its volume increases at a constant rate of 8 ft^3/min. Find the rate of increase of the radius when the radius is 3 ft.

23. Assume that when a hard candy ball is dropped in a glass of water, it dissolves at a rate directly proportional to its surface area. Show that the radius of the ball decreases at a constant rate.

24. The trunk of a fir tree is in the shape of a cone. At a certain moment the diameter of the base of the tree is 1 ft and is increasing at $\frac{1}{4}$ in./yr, while the height is 20 ft and is increasing at 6 in./yr. Find the rate of increase of the volume of wood in the trunk of the tree at that moment.

25. Water is being poured into an inverted cone (vertex down) of radius 4 in. and height 10 in. at a rate of 3 in^3/sec. Find the rate at which the area of the base of the cone is increasing when over the vertex is 5 in.

26. Sand is being poured at a rate of 2 ft^3/min to form a conical pile whose height is always three times the radius. Find the rate at which the area of the base of the cone is increasing when the cone is 4 ft high.

27. The vertex angle of a right circular cone of constant slant height l is decreasing at a constant rate of $\frac{1}{5}$ radian/sec. Find the rate at which the volume is changing when the vertex angle is $90°$.

28. In one type of prewalking exerciser, a child in a harness is suspended with feet just touching the floor by a cylinder of elastic rubber with circular cross section. We assume the volume of the cylinder of rubber remains constant while the length and radius change as the child bounces. If the child is bouncing downward at a rate of 30 cm/sec at an instant when the rubber cylinder has length 90 cm and diameter 3 cm, find the rate of change of the diameter of the cylinder at that moment.

29. A bridge goes straight across a river at a height of 60 ft. A car on the bridge traveling at 40 ft/sec passes directly over a boat traveling up the river at 15 ft/sec at time t_0. Find the rate at which the distance between the car and the boat is increasing 3 sec later.

30. In the operation of a tape recorder, a tape of constant thickness is wound from Reel 1 onto Reel 2. Let r_1 be the radius from the center of Reel 1 to the outer edge of the tape on that reel at time t, and let r_2 be similarly defined for Reel 2. Using calculus, show that at any time t,

$$\frac{dr_1/dt}{dr_2/dt} = -\frac{r_2}{r_1}.$$

[*Hint:* Use the fact that the volume of tape remains constant.]

5.2 THE MEAN-VALUE THEOREM

This section presents the mean-value theorem for the derivative and introduces the notions of local maxima and local minima of a function. At the end of the section, we show how the mean-value theorem can be used to give bounds on function values $f(x)$ if bounds are known on $f'(x)$. The theorem will be used in several proofs later in the text.

In Section 2.4, we stated two important properties of continuous functions: the intermediate-value theorem (Theorem 2.3) and the theorem on assumption of maximum and minimum values (Theorem 2.4). We restate Theorem 2.4 here for easy reference.

THEOREM 5.1 Assumption of extreme values

A continuous function assumes both a maximum value M and a minimum value m on any closed interval in its domain.

The function whose graph is shown in Fig. 5.5 assumes its maximum value in $[a, b]$ where $x = b$ and its minimum value where $x = a$. The value $f(x_1)$ is not a maximum for $f(x)$ on $[a, b]$ but is a maximum for $f(x)$ in the immediate vicinity of $x = x_1$. Similarly, $f(x_2)$ in Fig. 5.5 is a minimum value for $f(x)$ in the vicinity of x_2.

DEFINITION 5.1 Local maximum and minimum values

If $f(x_1)$ is a maximum of $f(x)$ for all x in $[x_1 - \Delta x, x_1 + \Delta x]$ for sufficiently small $\Delta x > 0$, then $f(x_1)$ is a **relative maximum** or **local maximum** of $f(x)$. Similarly, $f(x_2)$ in Fig. 5.5 is a **relative minimum** or **local minimum** of $f(x)$.

It appears from Fig. 5.6 that if f is differentiable, then the graph of f has a horizontal tangent line at points corresponding to local maxima or minima. We prove this before we proceed to the main topic of this section.

Figure 5.5 $f(x)$ has a local maximum at x_1 and a local minimum at x_2.

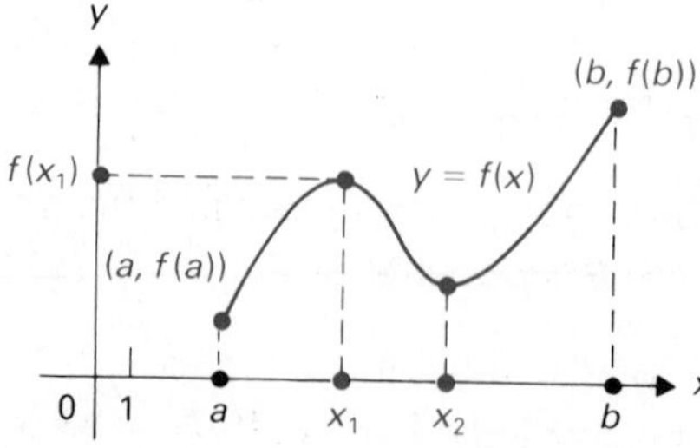

Figure 5.6 Tangent lines are horizontal at local maxima and minima.

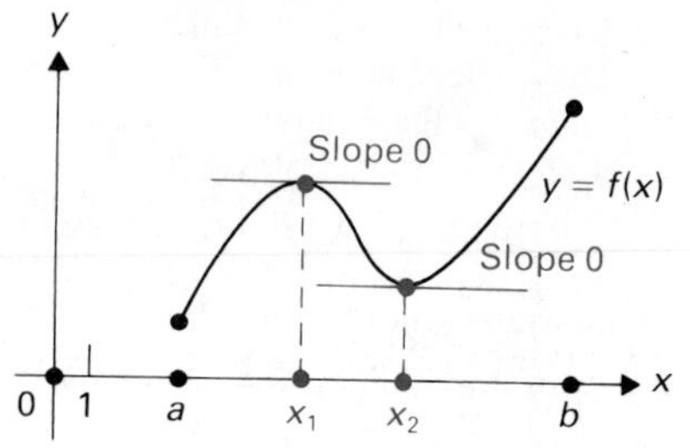

THEOREM 5.2 Derivative at local extrema

If f is differentiable at x_1 and if $f(x_1)$ is a local maximum (or local minimum) of $f(x)$, then $f'(x_1) = 0$.

Proof. Recall that the definition of $f'(x_1)$ is

$$f'(x_1) = \lim_{\Delta x \to 0} \frac{f(x_1 + \Delta x) - f(x_1)}{\Delta x}. \tag{1}$$

If $f(x_1)$ is a maximum, then for small Δx,

$$f(x_1 + \Delta x) \leq f(x_1).$$

Thus the numerator of the difference quotient in Eq. (1) is always negative or zero. If $\Delta x > 0$, the whole quotient is less than or equal to zero, while if $\Delta x < 0$, the whole quotient is greater than or equal to zero. The only way that $f'(x_1)$ can be simultaneously approached by numbers less than or equal to zero and numbers greater than or equal to zero is if $f'(x_1) = 0$. An analogous argument shows that if $f(x_2)$ is minimum and $a < x_2 < b$, then $f'(x_2) = 0$. •

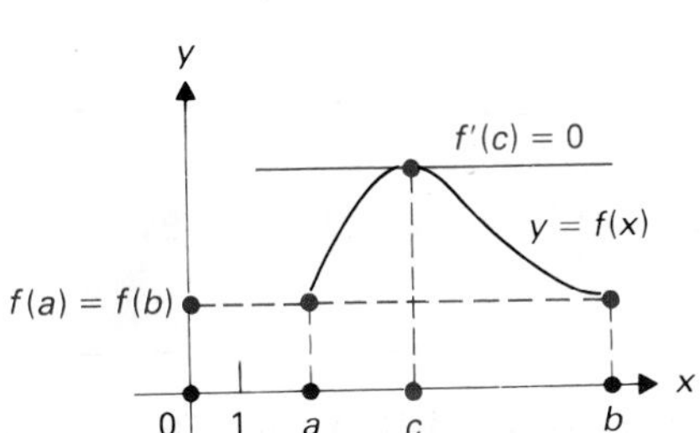

Figure 5.7 Position of c given by Rolle's theorem.

We are now ready to demonstrate Rolle's theorem, a special case of the mean-value theorem. This special case serves as a lemma in the proof of the main theorem.

THEOREM 5.3 Rolle's theorem

If $f(x)$ is continuous in $[a, b]$ and differentiable for $a < x < b$, and if, furthermore, $f(a) = f(b)$, then there exists c where $a < c < b$ such that $f'(c) = 0$.

Proof. Figure 5.7 illustrates Rolle's theorem. By Theorem 5.1, we know that $f(x)$ assumes a maximum value M and a minimum value m in $[a, b]$. If f is a constant function in $[a, b]$ so that $f(x) = f(a) = f(b)$ for all x in $[a, b]$, then of course $f'(c) = 0$ for any c where $a < c < b$. If f is not constant in $[a, b]$, then since $f(a) = f(b)$, either the maximum or the minimum of $f(x)$ must be assumed at a point c where $a < c < b$. From Theorem 5.2, we then know $f'(c) = 0$. This completes the demonstration of Rolle's theorem. •

As indicated by Fig. 5.8, the value c described in Rolle's theorem need not be unique. Both c_1 and c_2 in Fig. 5.8 satisfy the requirement in the theorem.

Figure 5.8 Both c_1 and c_2 satisfy the property described in Rolle's theorem.

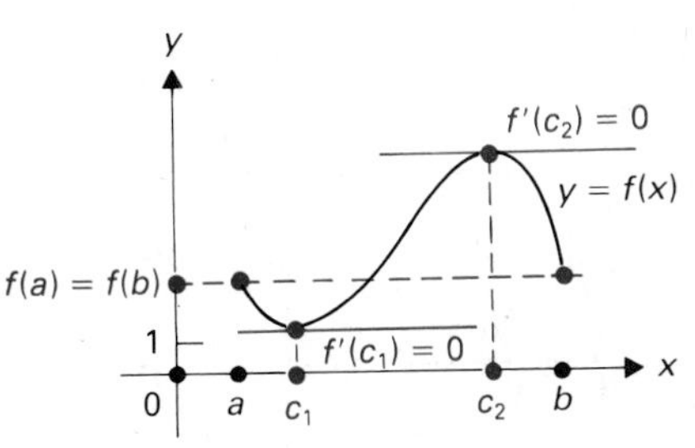

EXAMPLE 1 Show that the hypotheses for Rolle's theorem are satisfied by $f(x) = x^3 - 3x + 4$ on $[-1, 2]$, and find a value c described by the theorem.

Solution As a polynomial function, $f(x)$ is continuous and differentiable everywhere; therefore it is continuous and differentiable on $[-1, 2]$. Also,

$$f(-1) = 6 = f(2).$$

Thus the hypotheses of Rolle's theorem are satisfied.

We have $f'(x) = 3x^2 - 3$, which is zero where

$$3x^2 - 3 = 0, \qquad x^2 = 1, \qquad x = \pm 1.$$

Now -1 does not satisfy the relation $-1 < c < 2$ for c as required by Rolle's theorem. However, 1 does satisfy the relation. Thus $c = 1$, and the value for c is unique in this example. □

EXAMPLE 2 Use the intermediate-value theorem and Rolle's theorem to show that the equation $6x^3 + x^2 + x - 5 = 0$ has exactly one solution in $[0, 1]$.

Solution Let $f(x) = 6x^3 + x^2 + x - 5$. Since $f(0) = -5$ and $f(1) = 3$, the intermediate-value theorem shows that there is at least one value a in $[0, 1]$ such that $f(a) = 0$. Suppose there were also b in $[0, 1]$ such that $f(b) = 0$, and suppose $a < b$. Applying Rolle's theorem to $[a, b]$, we see that $f'(x) = 18x^2 + 2x + 1$ would be zero for some number c where $a < c < b$, and therefore $0 < c < 1$. But clearly $18x^2 + 2x + 1 > 0$ for all $x \geq 0$. Thus no such b can exist, and $x = a$ is the unique solution of $6x^3 + x^2 + x - 5 = 0$ in $[0, 1]$. □

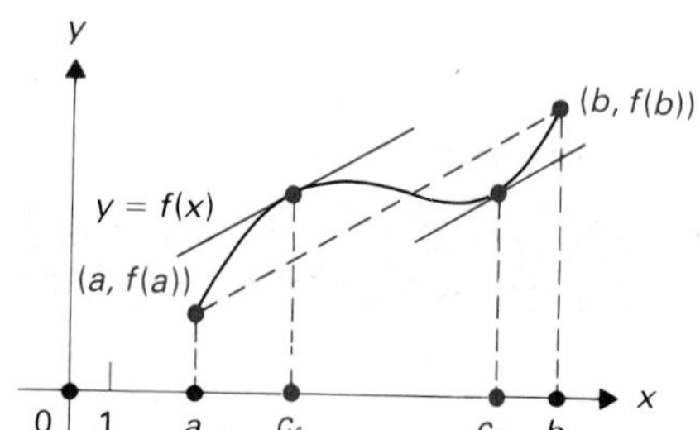

Figure 5.9 Mean-value theorem; tangent lines where $x = c_1$ and $x = c_2$ are parallel to the line through $(a, f(a))$ and $(b, f(b))$.

The mean-value theorem may be regarded as a generalization of Rolle's theorem. It is illustrated in Fig. 5.9. Both theorems assert that for f continuous in $[a, b]$ and differentiable for $a < x < b$, there exists at least one c where $a < c < b$ and where the tangent line to the graph of f is parallel to the line joining the points $(a, f(a))$ and $(b, f(b))$. The slope of this line is

$$\frac{f(b) - f(a)}{b - a},$$

so the conclusion of the mean-value theorem will take the form

$$f'(c) = \frac{f(b) - f(a)}{b - a} \qquad \text{or} \qquad f(b) - f(a) = (b - a)f'(c)$$

for some c where $a < c < b$. There are two such points, c_1 and c_2, in Fig. 5.9.

THEOREM 5.4 Mean-value theorem

Let $f(x)$ be continuous in $[a, b]$ and differentiable for $a < x < b$. Then there exists c where $a < c < b$ such that

$$f(b) - f(a) = (b - a)f'(c). \tag{2}$$

Proof. To obtain this result from Rolle's theorem, we introduce the function g with domain $[a, b]$ whose value at x is indicated in Fig. 5.10. It is easy to see from Fig. 5.10 that we should define g by

$$\begin{aligned} g(x) &= f(x) - \left[f(a) + \frac{f(b) - f(a)}{b - a}(x - a)\right] \\ &= [f(x) - f(a)] - \left[\frac{f(b) - f(a)}{b - a}(x - a)\right]. \end{aligned}$$

Since f is continuous in $[a, b]$ and differentiable for $a < x < b$, we see that

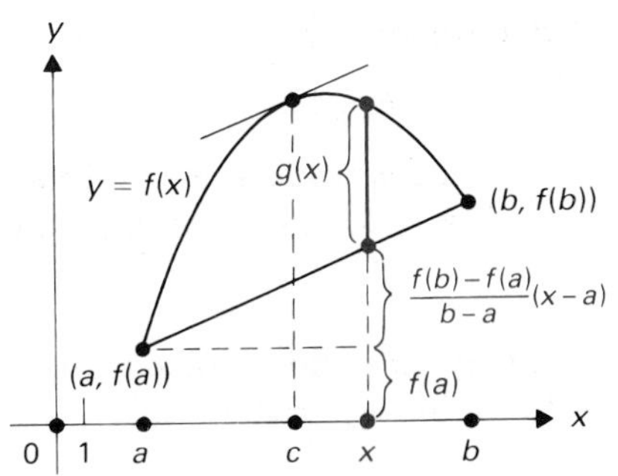

Figure 5.10 Proof of the mean-value theorem:

$$g(x) = f(x) - \left[f(a) + \frac{f(b) - f(a)}{b - a}(x - a) \right]$$

satisfies the hypotheses of Rolle's theorem.

g is also, and $g(a) = g(b) = 0$, so Rolle's theorem applies. Therefore for some c where $a < c < b$, we have $g'(c) = 0$. Now

$$g'(x) = f'(x) - \frac{f(b) - f(a)}{b - a},$$

so

$$0 = g'(c) = f'(c) - \frac{f(b) - f(a)}{b - a},$$

and $f(b) - f(a) = (b - a)f'(c)$. •

EXAMPLE 3 Illustrate the mean-value theorem with $f(x) = x^2$ in the interval $[0, 3]$.

Solution We must find c such that

$$f(3) - f(0) = (3 - 0)f'(c),$$

or such that $9 = 3 \cdot f'(c)$. Now $f'(x) = 2x$, and $9 = 3 \cdot 2c$ when $c = \frac{3}{2}$. Note that, as the mean-value theorem states, we found a value c satisfying $0 < c < 3$. □

The adjective *mean* in the name of Theorem 5.4 is used in its mathematical sense signifying *average* rather than in its everyday sense signifying *ornery*. We know that $(f(b) - f(a))/(b - a)$ is the average (mean) rate of change of $f(x)$ over $[a, b]$. *The mean-value theorem asserts that this average rate of change is assumed at some point.*

EXAMPLE 4 A car accelerates smoothly from 5 mph to 50 mph in 15 sec. Show that at some instant, the acceleration of the car is 3 mi/sec^2.

Solution *Smoothly* means that the velocity is a differentiable function of time. Suppose the velocity is 5 mph at time t_1 and 50 mph at time $t_2 = t_1 + 15$. Then

$$\frac{v(t_2) - v(t_1)}{t_2 - t_1} = \frac{50 - 5}{15} = \frac{45}{15} = 3 \text{ mi/sec}^2$$

is the average acceleration of the car during those 15 sec. By the mean-value theorem, this value 3 mi/sec^2 is the exact acceleration at some instant c during that time interval. □

The mean-value theorem is often used to give bounds on the size of $f(x)$ if we know bounds on the size of $f'(x)$. This application is much more significant than the computation of c illustrated in Example 3.

EXAMPLE 5 Let $f(x)$ be differentiable. Suppose $f(2) = -15$ and $|f'(x)| \leq 3$ for x in $[2, 6]$. Find bounds in terms of x on the size of $f(x)$ in $[2, 6]$.

Solution For any $x \neq 2$ in $[2, 6]$, we have

$$\frac{f(x) - f(2)}{x - 2} = \frac{f(x) + 15}{x - 2} = f'(c)$$

for some c, $2 < c < x$, by the mean-value theorem. But $|f'(c)| \leq 3$ by assumption. Since $x - 2$ is positive, we then have

$$\left|\frac{f(x) + 15}{x - 2}\right| = \frac{|f(x) + 15|}{x - 2} \leq 3.$$

Thus

$$|f(x) + 15| \leq 3(x - 2),$$
$$-3(x - 2) \leq f(x) + 15 \leq 3(x - 2),$$
$$-15 - 3x + 6 \leq f(x) \leq -15 + 3x - 6,$$
$$-3x - 9 \leq f(x) \leq 3x - 21,$$

for x in $[2, 6]$. These are bounds on $f(x)$. For example, we have $-24 \leq f(5) \leq -6$. □

EXAMPLE 6 Let $f(x)$ be differentiable and suppose $f(0) = 2$ and $|f'(x)| \leq 4x$ for x in $[0, 5]$. Find bounds on $f(x)$ in the interval $[0, 5]$.

Solution As in Example 5, we have for x in $[0, 5]$, $x \neq 0$,

$$\frac{|f(x) - 2|}{x} \leq 4x,$$
$$-4x(x) \leq f(x) - 2 \leq 4x(x),$$
$$-4x^2 + 2 \leq f(x) \leq 4x^2 + 2.$$

For example, we have $-34 < f(3) < 38$. □

SUMMARY

1. If f is differentiable, then $f'(x) = 0$ at any point where the function has a local maximum or a local minimum.
2. *Rolle's theorem:* If $f(x)$ is continuous in $[a, b]$ and differentiable for $a < x < b$, and if $f(a) = f(b)$, then there is some number c where $a < c < b$ such that $f'(c) = 0$.
3. *Mean-value theorem:* If $f(x)$ is continuous in $[a, b]$ and differentiable for $a < x < b$, then there is some number c where $a < c < b$ such that

$$\frac{f(b) - f(a)}{b - a} = f'(c).$$

EXERCISES

In Exercises 1 through 6, verify that $f(x)$ satisfies the hypotheses for Rolle's theorem on the given interval, and find all numbers c described by the theorem.

1. $f(x) = x^2 + x + 4$ on $[-3, 2]$
2. $f(x) = -3x^2 + 12x + 4$ on $[0, 4]$
3. $f(x) = \sin x$ on $[0, \pi]$
4. $f(x) = \cos x$ on $[0, 2\pi]$
5. $f(x) = x^3 - 6x^2 + 5x + 3$ on $[1, 5]$
6. $f(x) = (x^2 - 1)/(x + 2)$ on $[-1, 1]$

In Exercises 7 through 10, use the intermediate-value theorem and Rolle's theorem to show that the given equation has exactly one solution in the given interval. (See Example 2.)

7. $x^2 - 3x + 1 = 0$ in $[2, 6]$
8. $10 - 5x - x^2 = 0$ in $[-1, 2]$
9. $\sin x = \frac{1}{4}x$ in $[\pi/2, \pi]$
10. $\cos x = x$ in $[0, \pi/2]$
11. Generalize Rolle's theorem to show that if f is a differentiable function and $f(x)$ is zero at r distinct points in an interval $[a, b]$, then $f'(x)$ is zero for at least $r - 1$ distinct points in $[a, b]$.
12. Let f be a differentiable function and suppose that $f'(x) = 0$ for at most three distinct values of x where $a < x < b$. What is the maximum possible number of distinct solutions of $f(x) = 10$ in $[a, b]$?

In Exercises 13 through 18, illustrate the mean-value theorem for the given function over the given interval by finding all values of c described in the theorem.

13. $f(x) = 3x - 4$ on $[1, 4]$
14. $f(x) = x^3$ on $[-1, 2]$
15. $f(x) = x - 1/x$ on $[1, 3]$
16. $f(x) = x^{1/2}$ on $[9, 16]$
17. $f(x) = \sqrt{1 - x}$ on $[-3, 0]$
18. $f(x) = (x - 1)/(x + 3)$ on $[2, 4]$
19. Let f be a function satisfying the hypotheses of the mean-value theorem on $[a, b]$.
    a) Give the rate-of-change interpretation of
    $$(f(b) - f(a))/(b - a).$$
    b) Give the interpretation of $f'(c)$ as a rate of change.
    c) Restate the mean-value theorem in terms of rates of change.
20. Two towns A and B are connected by a highway ten miles long with a 60-mph speed limit. Mr. Smith is arrested on a speeding charge and admits having driven from A to B in 8 minutes. In a speeding offense, the court is allowed to impose a fine of \$15 plus \$2 for each mph in excess of the limit. Use the mean-value theorem to show that the judge is justified in imposing a fine of \$45 on Mr. Smith. [*Hint:* Use Exercise 19.]
21. For a quadratic function f given by $f(x) = ax^2 + bx + c$, show that the point between x_1 and x_2 where the tangent to the graph of f is parallel to the chord joining $(x_1, f(x_1))$ and $(x_2, f(x_2))$ is halfway between x_1 and x_2. What example in the text illustrates this?
22. Explain the fallacy in the following argument: The Elliot family spent \$99.61 on food during one week, which was an average spending rate of \$14.23 per day. Therefore, by the mean-value theorem, the family must have spent exactly \$14.23 for food on some day that week.
23. Let f be a differentiable function with domain containing $[a, b]$. Suppose that $m \le f'(x) \le M$ in $[a, b]$. Use the mean-value theorem to show that
    $$f(a) + m(x - a) \le f(x) \le f(a) + M(x - a)$$
    for all x in $[a, b]$.

In Exercises 24 through 34, let f be differentiable for all x. Use the mean-value theorem to find bounds on $f(x)$ from the given properties of f and f'. Be sure to state the values of x where the bounds are valid.

24. $f(0) = 3, |f'(x)| \le 3$ for x in $[0, 8]$
25. $f(-2) = 1, |f'(x)| \le 5$ for x in $[-2, 4]$
26. $f(1) = 7, |f'(x)| \le 2\sqrt{x}$ for x in $[1, 3]$
27. $f(0) = 6, 2 \le f'(x) \le 5$ for x in $[0, 6]$
28. $f(10) = -8, -1 \le f'(x) \le 3$ for x in $[10, 100]$
29. $f(2) = 4, f'(x) \le 5$ for x in $[0, 2]$
30. $f(3) = 6, -2 \le f'(x) \le 5$ for x in $[-1, 3]$
31. $f(0) = -4, -1 \le f'(x) \le 4$ for $x \ge 0$
32. $f(0) = 5, 2 \le f'(x) \le 3$ for $x \le 0$
33. $f(0) = 1, 3 \le f'(x) \le 7$ for all x
34. $f(2) = 5, -1 \le f'(x) \le 10$ for all x

5.3 CRITICAL POINTS

Recall once more Theorem 5.1 of the preceding section:

> A continuous function assumes both a maximum value M and a minimum value m on any closed interval in its domain.
>
> ***Assumption of extreme values***

The preceding section used this theorem to prove the mean-value theorem. In this section we tackle the problem of computing M and m. First, we illustrate the theorem.

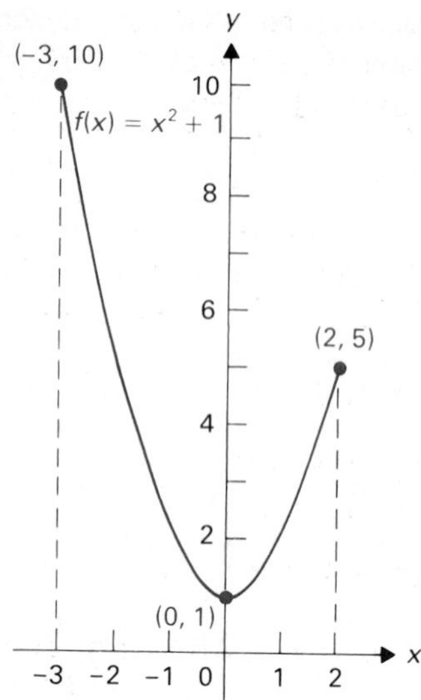

Figure 5.11 $f(x) = x^2 + 1$ has maximum value 10 and minimum value 1 on $[-3, 2]$.

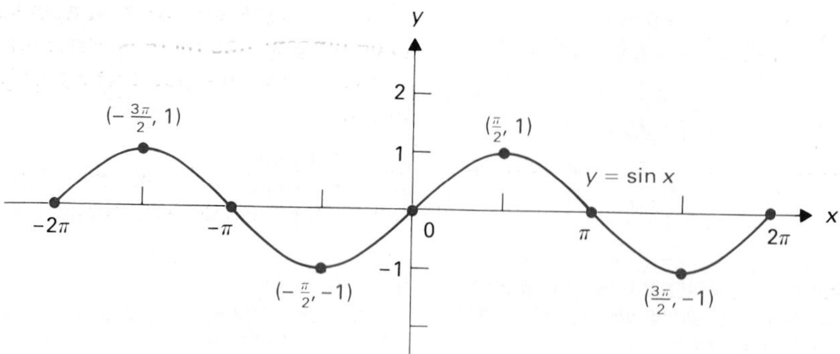

Figure 5.12 $f(x) = \sin x$ has maximum value 1 at $-3\pi/2$ and $\pi/2$ and has minimum value -1 at $-\pi/2$ and $3\pi/2$ on $[-2\pi, 2\pi]$.

EXAMPLE 1 Sketch the graph of $f(x) = x^2 + 1$ and use the graph to find the maximum and minimum values assumed in $[-3, 2]$.

Solution The graph is sketched in Fig. 5.11. We see that the maximum value assumed by $f(x) = x^2 + 1$ in $[-3, 2]$ is 10, which is assumed at $x = -3$. The minimum value is 1, which is assumed at $x = 0$. □

EXAMPLE 2 Repeat Example 1 for the function $f(x) = \sin x$ in $[-2\pi, 2\pi]$.

Solution The graph is sketched in Fig. 5.12. The maximum value assumed by $\sin x$ in $[-2\pi, 2\pi]$ is 1, which is assumed at both $x = -3\pi/2$ and $x = \pi/2$. The minimum value is -1, which is assumed at both $-\pi/2$ and $3\pi/2$. □

Suppose the extreme values M and m of $f(x)$ on $[a, b]$ do not occur at endpoints of the interval. Then M is also a local maximum of the function. If $f(x_1) = M$ and if f is differentiable at x_1, then Theorem 5.2 of the preceding section shows that $f'(x_1) = 0$. A similar analysis holds if f assumes a local minimum at x_2 in the open interval (a, b). Thus if we find all points in the open interval (a, b) where either $f'(x)$ does not exist or where $f'(x) = 0$, we have all candidates for points in this open interval where f may assume a maximum or a minimum value.

DEFINITION 5.2 Critical point

A **critical point** of a function f is a point x_0 in its domain where either $f'(x_0) = 0$ or $f'(x_0)$ does not exist.

EXAMPLE 3 Find all critical points for $f(x) = (x - 1)^{2/3}$.

Solution Note that $f(x)$ is defined for all x. We have

$$f'(x) = \frac{2}{3}(x - 1)^{-1/3} = \frac{2}{3(x - 1)^{1/3}} \qquad \text{for } x \neq 1.$$

Thus $f'(x)$ is never zero, but $f'(1)$ does not exist. Consequently, 1 is a critical point, the only critical point. The graph of $y = f(x)$ is shown in Fig. 5.13.

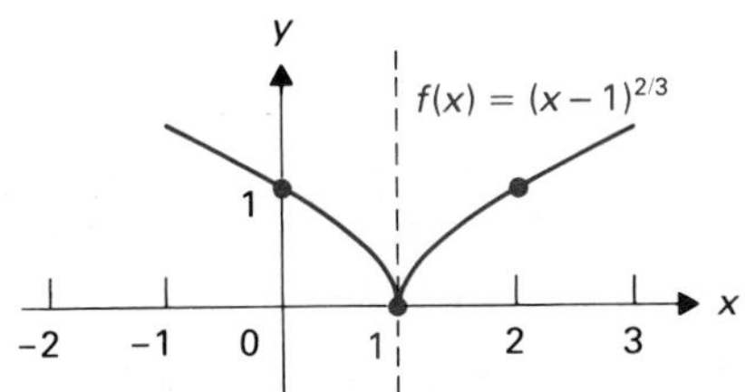

Figure 5.13 $f(x) = (x-1)^{2/3}$ has minimum value 0 at $x = 1$, but $f'(1)$ does not exist since the tangent line there is vertical.

Clearly $f(x)$ assumes a minimum value where $x = 1$, but $f'(1)$ does not exist. (The tangent line there is vertical.) This example illustrates that a derivative need not exist at a point where a function assumes a local maximum or local minimum. □

EXAMPLE 4 Find all critical points of $f(x) = x/(x^2 + 1)$. Then graph the function to decide whether the critical points yield local maxima or local minima.

Solution Again, $f(x)$ is defined for all x. We have

$$f'(x) = \frac{(x^2 + 1)1 - x(2x)}{(x^2 + 1)^2} = \frac{-x^2 + 1}{(x^2 + 1)^2}.$$

We see that $f'(x)$ exists for all x. Now $f'(x) = 0$ when $-x^2 + 1 = 0$, or $x = \pm 1$. Thus the critical points are $x = \pm 1$.

Now $f(x) > 0$ for $x > 0$, and $f(x) < 0$ for $x < 0$. Also, $f(0) = 0$ and

$$\lim_{x \to \infty} \frac{x}{x^2 + 1} = \lim_{x \to -\infty} \frac{x}{x^2 + 1} = 0.$$

Therefore the graph of f must appear as in Fig. 5.14, so $(1, \frac{1}{2})$ corresponds to a local maximum and $(-1, -\frac{1}{2})$ to a local minimum. In this case, $f(1) = \frac{1}{2}$ is actually the maximum value and $f(-1) = -\frac{1}{2}$ the minimum value for the function over its entire domain. □

EXAMPLE 5 Sketch the graph of $f(x) = |\sin x|$ and from the graph find all critical points of f.

Solution The graph is sketched in Fig. 5.15. We see that local maxima equal to 1 occur at

$$x = \pm\frac{\pi}{2}, \pm\frac{3\pi}{2}, \pm\frac{5\pi}{2}, \ldots,$$

where $f'(x) = 0$, so all these points are critical points. It is clear from the

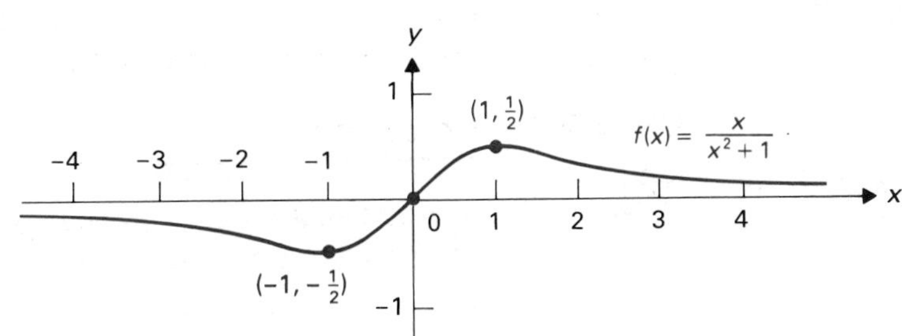

Figure 5.14 $f(x) = x/(x^2 + 1)$ has a maximum of $\frac{1}{2}$ at $x = 1$ and a minimum of $-\frac{1}{2}$ at $x = -1$.

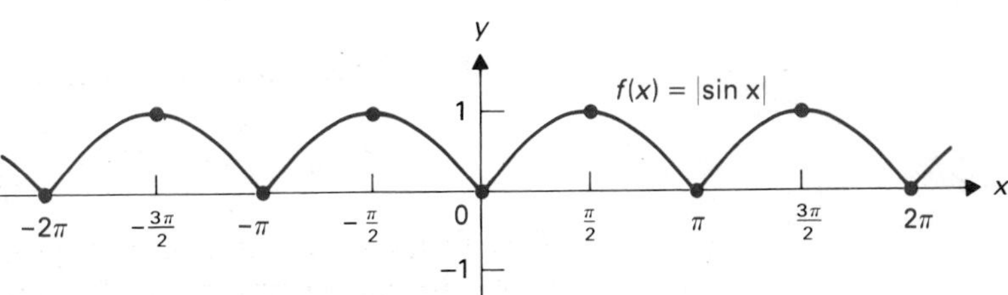

Figure 5.15 $f(x) = |\sin x|$ has maximum values of 1 where $f'(x) = 0$ and has minimum values of 0 where $f'(x)$ does not exist.

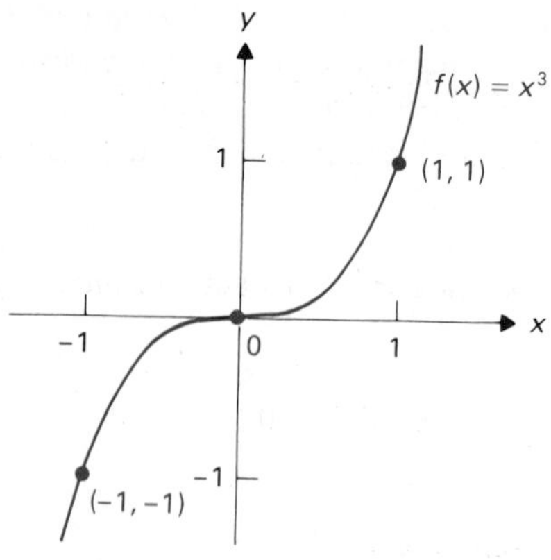

Figure 5.16 $f(x) = x^3$ has neither a local maximum nor a local minimum where $x = 0$, even though $f'(0) = 0$.

graph that $f'(x)$ does not exist where

$$x = 0, \pm\pi, \pm 2\pi, \pm 3\pi, \ldots$$

because the graph has sharp points there. Thus these are also critical points of the function. They correspond to local minima equal to zero. □

EXAMPLE 6 Find all critical points of $f(x) = x^3$, and determine any local maxima or minima by examination of the graph.

Solution We have $f'(x) = 3x^2$ for all x. Of course, $f'(x) = 0$ only if $x = 0$, so $x = 0$ is the only critical point. The familiar graph of x^3 is shown in Fig. 5.16. We see that there is neither a local minimum nor a local maximum where $x = 0$. □

To summarize our work, we have shown that

> a local maximum or local minimum in an open interval must always occur at a critical point.

However, Example 6 shows that

> a function might have some critical points where neither a local maximum nor a local minimum occurs.

We can now describe a method for attempting to find the maximum value M and minimum value m assumed by a continuous function on a closed interval. This is the problem we posed at the start of this section.

> To find the maximum value M and minimum value m assumed by a continuous function f on $[a, b]$:
>
> **Step 1** Find the critical points of f in the open interval (a, b).
>
> **Step 2** Compute $f(x)$ for all x found in Step 1, and also compute $f(a)$ and $f(b)$. The largest value found is the maximum value M and the smallest value is the minimum value m.

This method may break down if f has an infinite number of critical points in $[a, b]$, but for many functions, this problem does not occur.

EXAMPLE 7 Find the maximum and minimum values assumed in $[-\frac{1}{2}, 2]$ by $f(x) = 3x^4 + 4x^3 - 12x^2 + 5$.

Solution

Step 1 Since f is a polynomial function, it is differentiable, so critical points appear only where $f'(x) = 0$. Differentiating, we have

$$\begin{aligned} f'(x) &= 12x^3 + 12x^2 - 24x = 12x(x^2 + x - 2) \\ &= 12x(x + 2)(x - 1). \end{aligned}$$

Thus $f'(x) = 0$ for $x = -2, 0$, and 1. Of these three points, only 0 and 1 are in $[-\frac{1}{2}, 2]$.

Step 2 Computation shows that

$$f(0) = 5, \qquad f(1) = 0, \qquad f(-\tfrac{1}{2}) = \tfrac{27}{16}, \qquad f(2) = 37.$$

Therefore 0 is the minimum value, assumed at $x = 1$, and 37 is the maximum value, assumed at $x = 2$. □

EXAMPLE 8 Find the maximum and minimum values assumed by $f(x) = \sin x - \cos x$ on $[0, \pi]$.

Solution Again, all critical points occur where $f'(x) = 0$.

Step 1 Differentiating, we obtain

$$f'(x) = \cos x + \sin x.$$

Thus $f'(x) = 0$ when $\cos x + \sin x = 0$, that is, when $\sin x = -\cos x$. This condition is equivalent to $\tan x = -1$. In $[0, \pi]$, we have $\tan x = -1$ only at $x = 3\pi/4$.

Step 2 We have

$$f(0) = 0 - 1 = -1, \qquad f\left(\frac{3\pi}{4}\right) = \frac{1}{\sqrt{2}} - \frac{-1}{\sqrt{2}} = \frac{2}{\sqrt{2}} = \sqrt{2},$$

$$f(\pi) = 0 - (-1) = 1.$$

Thus $\sqrt{2}$ is the maximum value, assumed at $x = 3\pi/4$, and -1 is the minimum value, assumed at $x = 0$, over the interval $[0, \pi]$. □

SUMMARY

1. If $f(x)$ is continuous at each point in $[a, b]$, then $f(x)$ assumes a maximum value M and also a minimum value m in $[a, b]$.
2. A critical point of f is a point x_0 in the domain of f where either $f'(x_0) = 0$ or $f'(x_0)$ does not exist.
3. If $f(x)$ is defined for $a < x < b$, then a local maximum or local minimum of f in this open interval can occur only at a critical point.
4. To find the maximum value M and minimum value m assumed by a continuous function f on $[a, b]$, follow the two-step outline given before Example 7.

EXERCISES

1. Mark each of the following true or false.
   ___ a) If x_0 is a critical point of f, then it must be that $f'(x_0) = 0$.
   ___ b) If $f'(x_0) = 0$, then x_0 must be a critical point of f.
   ___ c) If f is differentiable in the open interval (a, b) and has a local maximum at x_0 in (a, b), then we must have $f'(x_0) = 0$.
   ___ d) If the domain of f contains the open interval (a, b) and f has a local minimum at x_0 in (a, b), then x_0 must be a critical point of f.
   ___ e) If the domain of f contains the open interval (a, b) and f has a local minimum at x_0 in (a, b), then we must have $f'(x_0) = 0$.
2. Mark each of the following true or false.
   ___ a) Every function has either a local maximum or a local minimum at each critical point.

_____ b) Either the maximum or the minimum of a continuous function on a closed interval $[a, b]$ must be assumed at one of the endpoints of the interval.

_____ c) It is possible for both the maximum value and the minimum value of a continuous function on a closed interval to be assumed at the same point of the interval.

_____ d) If $f'(x_0) = 0$, then f must have either a local maximum or a local minimum at x_0.

_____ e) A point not in the domain of a function may be a critical point of the function.

3. If $f(x) = 1/x$, then $f'(x) = -1/x^2$. Thus $f'(0)$ does not exist. However, $x_0 = 0$ is not a critical point of f. Why not?

4. If $f(x) = \tan x$, then $f'(x) = \sec^2 x = 1/(\cos^2 x)$. Since $\cos(\pi/2) = 0$, we see that $f'(\pi/2)$ does not exist. Is $x_0 = \pi/2$ a critical point of f? Why or why not?

In Exercises 5 through 20, find all critical points of the given function. Use methods of the preceding chapters to sketch the graph of the function with enough accuracy to decide whether the function has a local maximum, a local minimum, or neither at each critical point.

5. $f(x) = 5 + 4x - 2x^2$

6. $f(x) = x^2 - 4x + 3$

7. $f(x) = 4x^2 - 8x$

8. $f(x) = -x^2 - 6x - 2$

9. $f(x) = 2x^3 - 4$

10. $f(x) = x^3 - 3x^2$

11. $f(x) = x^4 - 1$

12. $f(x) = 3 - x^5$

13. $f(x) = \dfrac{1}{x^2 - 1}$ [*Warning:* See Exercise 3.]

14. $f(x) = \dfrac{1}{x^2 + 1}$

15. $f(x) = \dfrac{-x}{x^2 + 1}$

16. $f(x) = \dfrac{x}{1 - x^2}$

17. $f(x) = \sec 2x$

18. $f(x) = \csc x$

19. $f(x) = |\cos x|$

20. $f(x) = |\tan x|$

In Exercises 21 through 32, find the maximum and minimum values assumed by the function in the given interval.

21. $f(x) = x^4$ in $[-2, 1]$

22. $f(x) = \dfrac{1}{x}$ in $[2, 4]$

23. $f(x) = \dfrac{1}{x^2 + 1}$ in $[-1, 2]$

24. $f(x) = x^3 - 3x^2 + 1$ in
a) $[1, 3]$
b) $[-1, 3]$
c) $[-2, 4]$

25. $f(x) = x^2 + 4x - 3$ in
a) $[-5, -3]$
b) $[-4, -1]$
c) $[-4, 0]$

26. $f(x) = \dfrac{1}{x^2 - 1}$ in $[-3, -2]$

27. $f(x) = \dfrac{x^2 - x + 1}{x^2 + 1}$ in
a) $[-3, -2]$
b) $[-2, 0]$
c) $[0, 2]$
d) $[-2, 2]$

28. $f(x) = \sin x$ in the intervals
a) $\left[0, \dfrac{\pi}{4}\right]$
b) $\left[-\dfrac{\pi}{4}, \dfrac{\pi}{4}\right]$
c) $\left[-\dfrac{3\pi}{4}, \dfrac{3\pi}{4}\right]$

29. $f(x) = \sin x + \cos x$ in the intervals
a) $\left[0, \dfrac{\pi}{2}\right]$
b) $\left[-\dfrac{\pi}{2}, \dfrac{\pi}{2}\right]$
c) $[0, \pi]$
d) $\left[\dfrac{\pi}{2}, 2\pi\right]$

30. $f(x) = \sqrt{3} \sin x - \cos x$ in the intervals
a) $\left[0, \dfrac{\pi}{2}\right]$
b) $[0, \pi]$
c) $\left[0, \dfrac{3\pi}{2}\right]$
d) $[0, 2\pi]$

31. $f(x) = |\cos x|$ in $\left[0, \dfrac{3\pi}{4}\right]$

32. $f(x) = |\tan x|$ in $\left[-\dfrac{\pi}{3}, \dfrac{\pi}{4}\right]$

In Exercises 33 through 36, the height y (in ft) of a body above the surface of the earth at time t sec is given. Find
a) the maximum and minimum height of the body,
b) the maximum and minimum velocity v of the body, and
c) the maximum and minimum speed s of the body
during the given time interval. Recall $s = |v|$.

33. $y = 1600 - 16t^2, 0 \le t \le 5$

34. $y = 100 + 64t - 16t^2, 1 \le t \le 4$

35. $y = 1000 + 160t - 16t^2, 3 \le t \le 6$

36. $y = 100 + 320t + 16t^2, 2 \le t \le 4$

37. Let f be a polynomial function of even degree n such that the coefficient a_n of x^n is positive. Show that $f(x)$ assumes a value that is minimum for all real numbers x. [*Hint:* Consider $\lim_{x\to\infty} f(x)$ and $\lim_{x\to-\infty} f(x)$, and apply Theorem 5.1 to a large interval $[-c, c]$.]

In Exercises 38 through 42, use the result of Exercise 37 and the two-step procedure outlined in the text to find the minimum value assumed by the function for all real numbers x.

38. $3x^2 + 6x - 4$

39. $3x^4 - 8x^3 + 2$

40. $x^4 - 2x^2 + 7$

41. $3x^4 - 4x^3 - 12x^2 + 5$

42. $x^6 - 6x^4 - 8$

5.4 GRAPHING USING THE FIRST DERIVATIVE

In this section, we discuss the significance of the sign of $f'(x)$ when sketching the graph $y = f(x)$. In Section 5.5 we consider the significance of the sign of the second derivative.

INCREASING AND DECREASING FUNCTIONS

It is geometrically obvious that the graph of $y = f(x)$ is rising where $f'(x) > 0$ and is falling where $f'(x) < 0$. This can be proved using the mean-value Theorem. First, we give a definition.

DEFINITION 5.3 Increasing and decreasing functions

Let $f(x)$ and $g(x)$ be defined in an interval. If $f(x_1) < f(x_2)$ whenever $x_1 < x_2$ in the interval, then $f(x)$ is **increasing** in the interval. Similarly, if $g(x_1) > g(x_2)$ whenever $x_1 < x_2$ in the interval, then $g(x)$ is **decreasing** in the interval.

EXAMPLE 1 Determine from the graph in Fig. 5.17 where $f(x)$ is increasing and where it is decreasing.

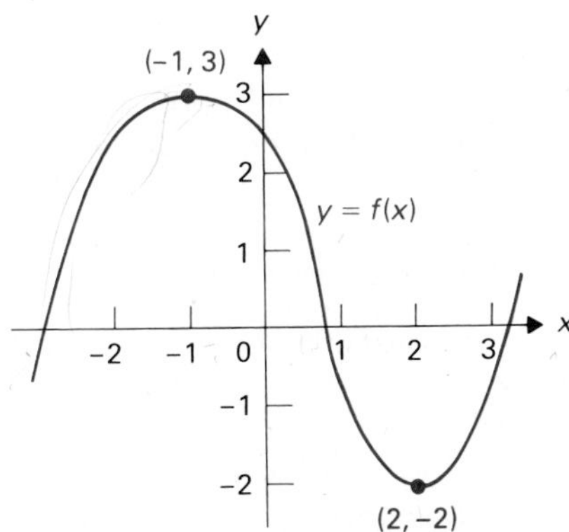

Figure 5.17 $f(x)$ is increasing for $x < -1$, $x > 2$; $f(x)$ is decreasing for $-1 < x < 2$.

Solution From the graph in Fig. 5.17, it appears that $f(x)$ is increasing for $x < -1$, decreasing for $-1 < x < 2$, and increasing for $x > 2$. □

Suppose $f'(x) > 0$ for all x in an interval. Let x_1 and x_2 be in the interval, with $x_1 < x_2$. By the mean-value theorem,

$$\frac{f(x_2) - f(x_1)}{x_2 - x_1} = f'(c) > 0, \tag{1}$$

where $x_1 < c < x_2$. In particular, Eq. (1) shows that $f(x_2) - f(x_1) > 0$, so $f(x_1) < f(x_2)$. Thus $f(x)$ is increasing in the interval. Similarly, if $f'(x) < 0$ for all x in an interval, then $f(x)$ is decreasing in the interval, because if $x_1 < x_2$ in the interval, an analogous argument using Eq. (1) shows that $f(x_1) > f(x_2)$. We summarize in a theorem.

THEOREM 5.5 Test for increasing or decreasing behavior

Let $f(x)$ be differentiable in an interval. If $f'(x) > 0$ throughout the interval, then $f(x)$ is increasing in the interval. If $f'(x) < 0$ throughout the interval, then $f(x)$ is decreasing in the interval.

EXAMPLE 2 Let $f(x) = x^3 - 3x^2 + 2$. Find where $f(x)$ is increasing and where it is decreasing.

Solution We have $f'(x) = 3x^2 - 6x = 3x(x - 2)$, so $f'(x) = 0$ where $x = 0$ or $x = 2$. The points 0 and 2 separate the x-axis into three parts, and

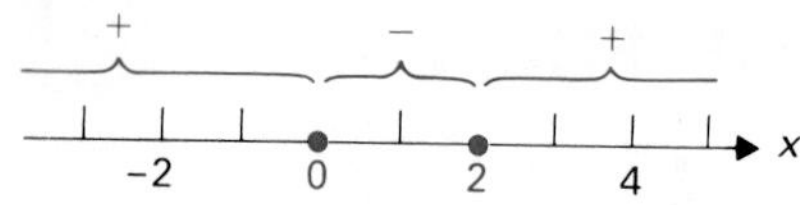

Figure 5.18 The sign of $f'(x) = 3x(x-2)$: $f(x)$ must increase for $0 < x < 2$.

$f'(x)$ must have the same sign throughout each individual part. In Fig. 5.18 we show the sign of $f'(x)$ on each part of the x-axis. For example, if $x < 0$, then $3x < 0$ and $x - 2 < 0$, so $f'(x) = 3x(x-2) > 0$. Thus we see that $f(x)$ is increasing for $x < 0$ or $x > 2$, and $f(x)$ is decreasing for $0 < x < 2$. □

We now discuss whether the sign of $f'(x)$ changes at a point x_0 such that $f'(x_0) = 0$. Suppose $f'(x)$ is a *polynomial* function, and $f'(x_0) = 0$. It can be shown that $(x - x_0)$ is then a factor of the polynomial $f'(x)$. In fact, $(x - x_0)$ might be a multiple factor. For example, if $f'(x) = x^4 - 2x^3 + x^2$, then $f'(0) = 0$ and $f'(1) = 0$, and we find that

$$f'(x) = x^4 - 2x^3 + x^2 = x^2(x-1)^2.$$

In this case, both $(x - 0)$ and $(x - 1)$ are factors of degree 2. The same argument can be made if $f'(x)$ is a *rational* function, since then $f'(x_0) = 0$ only if the numerator polynomial function is zero at $x = x_0$.

Suppose now that $f'(x_0) = 0$ and that we can write

$$f'(x) = (x - x_0)^n g(x),$$

where n is a positive integer, $g(x)$ is continuous, and $g(x_0) \neq 0$. By continuity, $g(x)$ does not change sign as we travel a short distance from one side of x_0 to the other. However, the factor $(x - x_0)$ does change sign as x increases through x_0. If n is even, then $(x - x_0)^n$ does not change sign as x increases through x_0; if n is odd, then $(x - x_0)^n$ does change sign. To summarize:

> If $f'(x) = (x - x_0)^n g(x)$, where $g(x)$ is continuous and $g(x_0) \neq 0$, then $f'(x)$ changes sign as x increases through x_0 if n is odd and does not change sign if n is even.

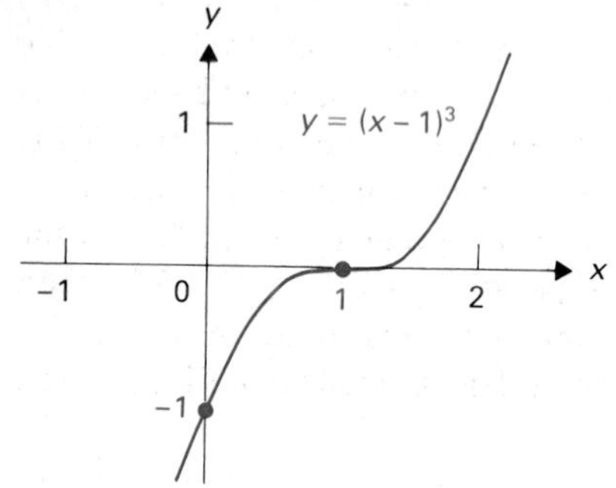

Figure 5.19 $f(x) = (x-1)^3$ is increasing for all x, although $f'(1) = 0$.

EXAMPLE 3 Determine where $f(x) = (x-1)^3$ is increasing and where it is decreasing.

Solution We have $f'(x) = 3(x-1)^2 = 0$ only where $x = 1$. Since the exponent 2 is even, we see that $f'(x)$ does not change sign as x increases through 1. Clearly $f'(x) \geq 0$ for all x, so $f(x)$ is increasing for all x. The graph is shown in Fig. 5.19. □

EXAMPLE 4 If $f'(x) = x^2(x-2)^3(x+1)$, determine where $f(x)$ is increasing and where it is decreasing.

Solution We draw an x-axis and record on the axis the intervals where $f'(x) < 0$ and where $f'(x) > 0$. See Fig. 5.20. Now $f'(x) = 0$ where $x = -1$, 0, or 2. Examination of the expression shows that $f'(x)$ changes sign as x increases through -1 and through 2, but not as x increases through 0. If the factors were multiplied out, the expansion of $f'(x)$ would start with x^6, which must dominate the other terms as $x \to -\infty$. Thus $f'(x)$ is positive for negative x of large magnitude. We work systematically from left to right on the x-axis in Fig. 5.20, first labeling the points -1, 0, and 2. We have

Figure 5.20 The sign of $f'(x) = x^2(x-2)^3(x+1)$: $f(x)$ must increase for $x < -1$ and $x > 2$ and must decrease for $-1 < x < 2$.

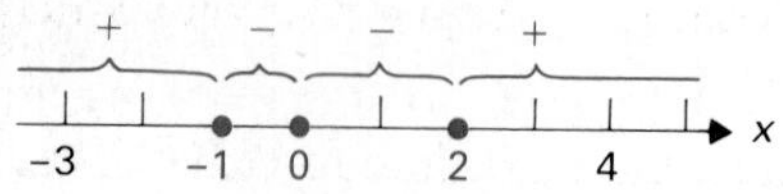

$$f'(x) > 0 \quad \text{for } x < -1,$$

$$f'(x) < 0 \quad \text{for } -1 < x < 0 \text{ since the sign changes at } -1,$$

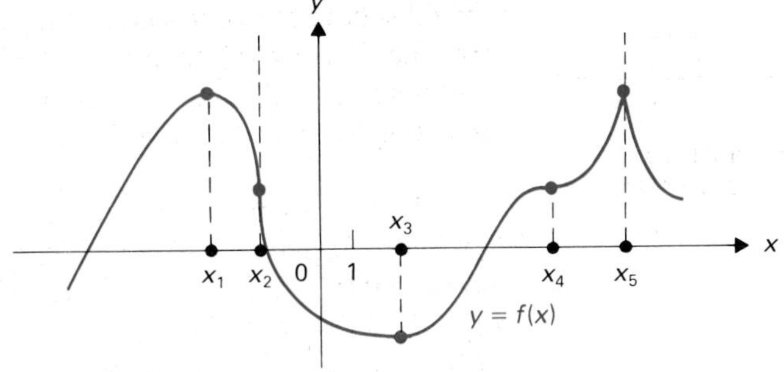

Figure 5.21 Critical points are x_1, x_2, x_3, x_4, x_5. $f'(x)$ changes sign at x_1, x_3, and x_5, where there are local extrema; $f'(x)$ does not change sign at x_2 or x_4.

$$f'(x) < 0 \qquad \text{for } 0 < x < 2 \text{ since the sign does not change at 0,}$$

$$f'(x) > 0 \qquad \text{for } x > 2 \text{ since the sign changes at 2.}$$

Thus $f(x)$ is increasing for $x < -1$ and also for $x > 2$, and $f(x)$ is decreasing for $-1 < x < 2$. □

LOCAL MAXIMA AND LOCAL MINIMA

Our graphical interpretation of the sign of $f'(x)$ helps to determine whether $f(x)$ has a local maximum, a local minimum, or neither at a critical point. Let x_1 be a critical point, so that either $f'(x_1) = 0$ or $f'(x_1)$ does not exist. Suppose $f'(x)$ is defined for a little way on both sides of x_1. Then if $f'(x)$ changes sign as x increases through x_1, we have a local extremum there. If the sign of $f'(x)$ changes from positive to negative as x increases, we have a local maximum as illustrated in Fig. 5.21 for the critical points x_1 and x_5. If the sign of $f'(x)$ changes from negative to positive as x increases through the critical point, we have a local minimum, as at the point x_3 in Fig. 5.21. If the sign of $f'(x)$ does not change, as at critical points x_2 and x_4 in Fig. 5.21, then there is no local extremum there. This analysis leads us to an outline for graphing a function using the sign of the first derivative. The outline is given in the following box.

GRAPHING A FUNCTION $f(x)$ USING THE SIGN OF THE FIRST DERIVATIVE

Step 1 Compute $f'(x)$ and find the critical points of $f(x)$. Mark them on an x-axis and mark the corresponding points on the graph.

Step 2 Determine where $f(x)$ is increasing ($f'(x) > 0$) and where it is decreasing ($f'(x) < 0$).

Step 3 Determine at each critical point whether there is a

local maximum: $f'(x)$ changes sign from positive to negative as x increases;

local minimum: $f'(x)$ changes sign from negative to positive as x increases;

or

neither: $f'(x)$ does not change sign.

Step 4 Sketch the graph using the above information, perhaps plotting a few more points.

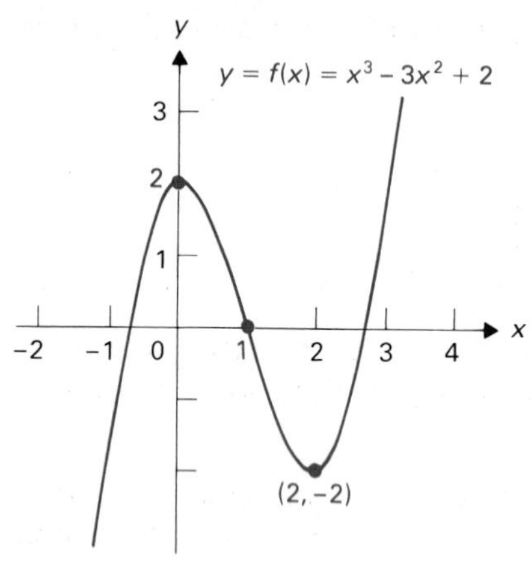

Figure 5.22 $f(x)$ increases for $x < 0$, $x > 2$; $f(x)$ decreases for $0 < x < 2$; local maximum where $x = 0$, local minimum where $x = 2$.

EXAMPLE 5 Sketch the graph of $f(x) = x^3 - 3x^2 + 2$.

Solution

Step 1 We have $f'(x) = 3x^2 - 6x = 3x(x - 2) = 0$ when $x = 0, 2$. These are the critical points. The corresponding points on the graph are $(0, 2)$ and $(2, -2)$.

Step 2 As in Example 2, $f(x)$ is increasing for $x < 0$ and $x > 2$ and $f(x)$ is decreasing for $0 < x < 2$.

Step 3 As x increases through $x = 0$, $f'(x)$ changes sign from positive to negative, so there is a local maximum at $x = 0$. Where $x = 2$, $f'(x)$ changes sign from negative to positive so there is a local minimum there.

Step 4 The graph is shown in Fig. 5.22. □

EXAMPLE 6 Sketch the graph of $f(x) = x^4 - 2x^3 + 1$.

Solution

Step 1 We have $f'(x) = 4x^3 - 6x^2 = 2x^2(2x - 3) = 0$ when $x = 0, \frac{3}{2}$. These are critical points, corresponding to $(0, 1)$ and $(\frac{3}{2}, -\frac{11}{16})$ on the graph of f.

Step 2 We see that $f'(x)$ changes sign as x increases through $\frac{3}{2}$ but not through 0. Since $\lim_{x\to-\infty} f'(x) = -\infty$, we see that $f(x)$ is decreasing for $x < 0$ and $0 < x < \frac{3}{2}$ and that $f(x)$ is increasing for $x > \frac{3}{2}$.

Step 3 Where $x = 0$, we know $f'(x)$ does not change sign, so there is no local extremum there. Where $x = \frac{3}{2}$, $f'(x)$ changes sign from negative to positive as x increases, so there is a local minimum at $\frac{3}{2}$.

Step 4 The graph is shown in Fig. 5.23. Since $f'(0) = 0$, we must make the graph flat, with a horizontal tangent line there. □

We conclude with an example illustrating that the sign of $f'(x)$ may change at a point not in the domain of f.

Figure 5.23 $f(x)$ increases for $x > \frac{3}{2}$; $f(x)$ decreases for $x < 0$, $0 < x < \frac{3}{2}$; local minimum at critical point $x = \frac{3}{2}$, no local extremum at critical point $x = 0$.

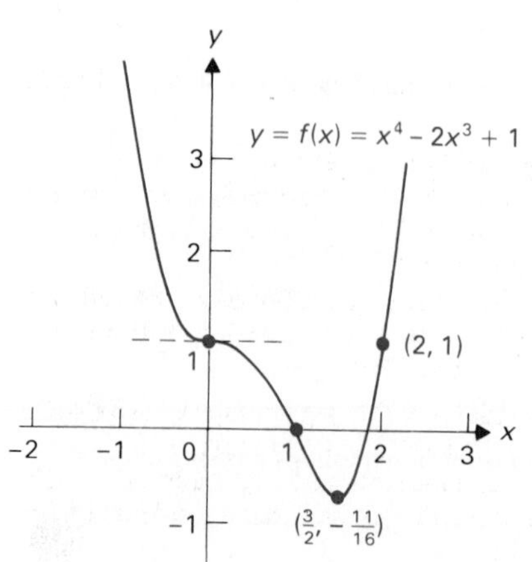

EXAMPLE 7 Sketch the graph of

$$f(x) = \frac{1}{x^2(x - 1)} + 3.$$

Solution

Step 1 Note that $f(x)$ is not defined at $x = 0$ or $x = 1$. We have

$$f'(x) = \frac{-1[x^2 + (x - 1)2x]}{x^4(x - 1)^2} = \frac{-3x^2 + 2x}{x^4(x - 1)^2} = \frac{-3x + 2}{x^3(x - 1)^2}.$$

Thus $f'(x) = 0$ where $3x = 2$, or $x = \frac{2}{3}$. This is the only critical point and corresponds to $(\frac{2}{3}, -\frac{15}{4})$ on the graph.

Step 2 Our formula for $f'(x)$ shows that we have sign changes as x increases through $\frac{2}{3}$ and 0 but not through 1, by the odd/even-exponent

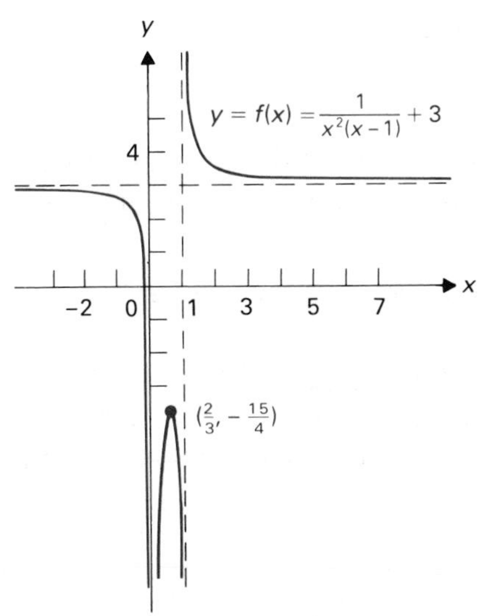

Figure 5.24 $f(x)$ increases for $0 < x < \frac{2}{3}$; $f(x)$ decreases for $x < 0$, $\frac{2}{3} < x < 1$, $x > 1$; local maximum where $x = \frac{2}{3}$, vertical asymptotes $x = 0$ and $x = 1$, horizontal asymptote $y = 3$.

argument. Since $f'(x) < 0$ for negative x of large magnitude, we see that

$f(x)$ is decreasing for $x < 0$,
$f(x)$ is increasing for $0 < x < \frac{2}{3}$,
$f(x)$ is decreasing for $\frac{2}{3} < x < 1$,
$f(x)$ is decreasing for $x > 1$.

Step 3 The critical point where $x = \frac{2}{3}$ corresponds to a local maximum because the sign of $f'(x)$ changes from positive to negative as x increases through $\frac{2}{3}$.

Step 4 The graph is sketched in Fig. 5.24. Note that there are vertical asymptotes $x = 0$ and $x = 1$ and that $y = 3$ is a horizontal asymptote. □

SUMMARY

The first derivative story:

1. If $f'(x) > 0$ throughout an interval, then $f(x)$ is increasing there.
2. If $f'(x) < 0$ throughout an interval, then $f(x)$ is decreasing there.
3. A critical point x_1 where $f'(x_1)$ does not exist or $f'(x_1) = 0$ is a candidate for a point where $f(x)$ has a local maximum or minimum.

   a. If $f'(x)$ changes sign from negative to positive as x increases through x_1, then $f(x_1)$ is a local minimum.

   b. If $f'(x)$ changes sign from positive to negative as x increases through x_1, then $f(x_1)$ is a local maximum.

   c. If $f'(x)$ does not change sign as x increases through x_1, then $f(x_1)$ is neither a local maximum nor a local minimum.

EXERCISES

In Exercises 1 through 14, find the intervals where the function is increasing and the intervals where it is decreasing.

1. $f(x) = x^2 - 6x + 2$

2. $f(x) = 4 - 3x - 2x^2$

3. $f(x) = x^3 + 3x^2 - 9x + 5$

4. $f(x) = x^3 - 6x^2 + 4$

5. $f(x) = 8 + 12x - x^3$

6. $f(x) = 3 - 12x + 6x^2 - x^3$

7. $f(x) = x^3 + 3x^2 + 3x - 5$

8. $f(x) = \dfrac{x}{x^2 + 1}$

9. $f(x) = \dfrac{1}{x + 2}$

10. $f(x) = \dfrac{1}{(x - 1)^2}$

11. $f(x) = \dfrac{x}{x^2 - 1}$

12. $f(x) = \dfrac{x^3}{x^2 - 4}$

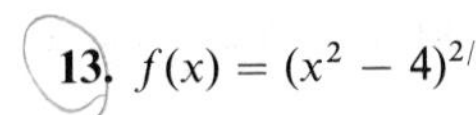

13. $f(x) = (x^2 - 4)^{2/3}$

14. $f(x) = (x^2 - 8)^{1/3}$

15. Find b such that the polynomial function $x^2 + bx - 7$ has a local minimum at 4.

16. Consider the polynomial function $ax^2 + 4x + 13$.
a) Find the value of a such that the function has either a local maximum or a local minimum at 1. Which is it, a maximum or a minimum?
b) Find the value of a such that the function has either a local maximum or a local minimum at -1. Which is it, a maximum or a minimum?

17. Consider the polynomial function f given by $ax^2 + bx + 24$.
a) Find the ratio b/a if f has a local minimum at 2.
b) Find a and b if f has a local minimum at 2 and $f(2) = 12$.
c) Can you determine a and b so that f has a local maximum at 2 and $f(2) = 12$?

18. Sketch the graph of a twice-differentiable function f such that $f(1) = 3$, $f(4) = 1$, f has a local minimum at 1, and f has a local maximum at 4.

19. It is possible for a twice-differentiable function f to satisfy the conditions of Exercise 18 and have no local maximum or minimum for $1 < x < 4$?

In Exercises 20 through 26, find
  a) the intervals where $f(x)$ is increasing,
  b) the intervals where $f(x)$ is decreasing,
  c) x-values where $f(x)$ has a local maximum, and
  d) x-values where $f(x)$ has a local minimum

from the given expression for $f'(x)$.

20. $f'(x) = x(x-1)^2$

21. $f'(x) = (x-3)^2(x+1)^3$

22. $f'(x) = x^2(x+4)^3(x-1)^3$

23. $f'(x) = \dfrac{x^2}{(x-1)(x+2)}$

24. $f'(x) = \dfrac{(x-1)(x+2)}{(x+3)^2}$

25. $f'(x) = \dfrac{x^2(x-4)}{(x-1)(x+2)}$

26. $f'(x) = \dfrac{(x-1)^3(x+1)^2}{x^2(x+4)}$

In Exercises 27 through 30, fill in the blanks.

27. A polynomial function of degree n can have at most ______ local extrema.

28. A polynomial function of even degree n with positive coefficient a_n of x^n can have at most ______ local maxima and at most ______ local minima.

29. A polynomial function of even degree n with negative coefficient a_n of x^n can have at most ______ local maxima and at most ______ local minima.

30. A polynomial function of odd degree n can have at most ______ local maxima and at most ______ local minima.

Exercises 31 through 34 refer to Figs. 5.25 through 5.28, which show graphs of *derivatives* of functions defined and continuous for all x. Mark the parts true or false.

31. If $f'(x)$ has the graph in Fig. 5.25, then
______ a) $f(x)$ has a local minimum where $x = 2$,
______ b) $f(x)$ has a local minimum where $x = 1$,
______ c) $f(x)$ has a local minimum where $x = 3$,
______ d) $f(x)$ has a local maximum where $x = 1$,
______ e) $f(x)$ is increasing for $x > 2$.

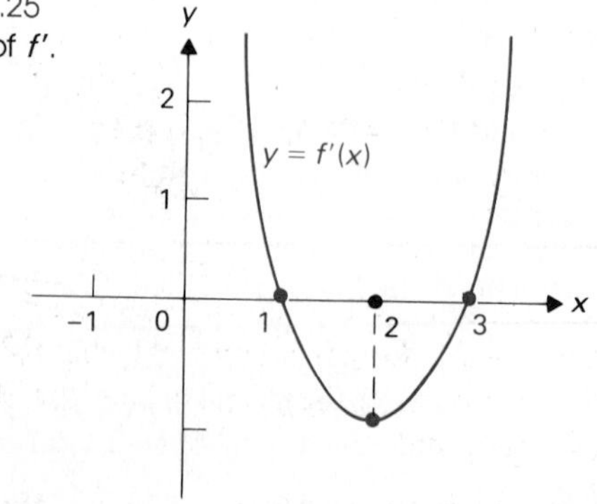

Figure 5.25
Graph of f'.

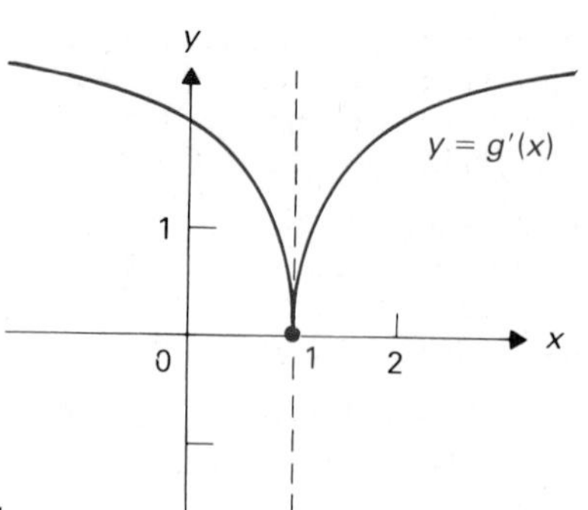

Figure 5.26
Graph of g'.

32. If $g'(x)$ has the graph in Fig. 5.26, then
______ a) $g(x)$ is decreasing for $x < 1$,
______ b) $g(x)$ is increasing for $x > 1$,
______ c) $g(x)$ is increasing for all x,
______ d) $g(x)$ has a local minimum where $x = 1$,
______ e) $g(x)$ has no local extremum where $x = 1$.

33. If $h'(x)$ has the graph in Fig. 5.27, then
______ a) $h(x)$ is decreasing for $0 < x < 4$,
______ b) $h(x)$ has a local maximum where $x = -2$,
______ c) $h(x)$ has a local minimum where $x = 4$,
______ d) $h(x)$ has a local minimum where $x = 0$,
______ e) $h(x)$ is increasing for $x > 2$.

34. If $k'(x)$ has the graph in Fig. 5.28, then
______ a) the graph of $k(x)$ has a vertical tangent line where $x = 2$,

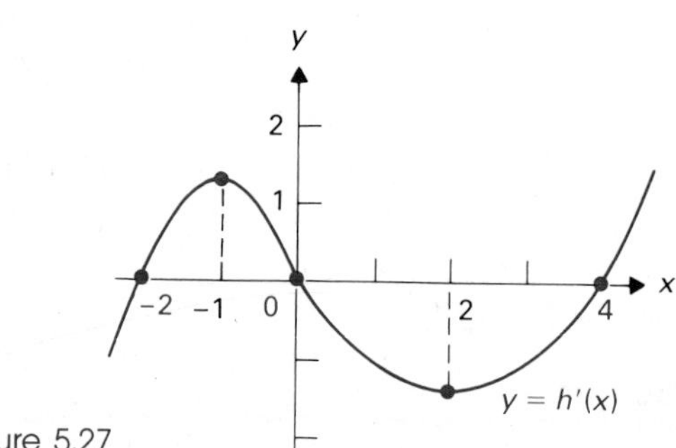

Figure 5.27
Graph of h'.

Figure 5.28
Graph of k'.

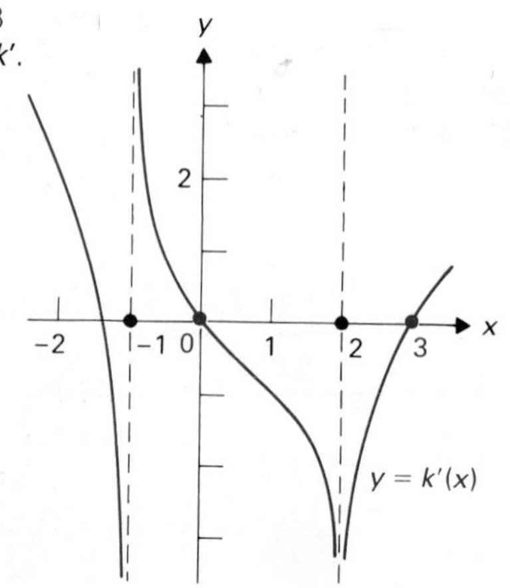

_____ b) $k(x)$ has a local minimum where $x = 2$,
_____ c) $k(x)$ has a local minimum where $x = -1$,
_____ d) $k(x)$ is decreasing for $0 < x < 3$,
_____ e) $k(x)$ has a local maximum where $x = 0$.

In Exercises 35 through 44, use the information given by the first derivative to sketch the graph of the function. The functions are multiples of some of those in Exercises 1 through 14, as noted. You may want to make use of your preceding work.

35. $f(x) = x^2 - 6x + 2$ (Exercise 1)

36. $f(x) = 4 - 3x - 2x^2$ (Exercise 2)

37. $f(x) = \frac{1}{8}(x^3 + 3x^2 - 9x + 5)$ (Exercise 3)

38. $f(x) = \frac{1}{8}(8 + 12x - x^3)$ (Exercise 5)

39. $f(x) = x^3 + 3x^2 + 3x - 5$ (Exercise 7)

40. $f(x) = \dfrac{x}{x^2 + 1}$ (Exercise 8)

41. $f(x) = \dfrac{x}{x^2 - 1}$ (Exercise 11)

42. $f(x) = \dfrac{1}{2} \cdot \dfrac{x^3}{x^2 - 4}$ (Exercise 12)

43. $f(x) = (x^2 - 4)^{2/3}$ (Exercise 13)

44. $f(x) = (x^2 - 8)^{1/3}$ (Exercise 14)

5.5 GRAPHING USING THE SECOND DERIVATIVE

The preceding section applied the first derivative as an aid in sketching the graph of a function. In this section, we discuss the application of the second derivative to sketching the graph of a function. The section concludes with an outline for sketching graphs using the information provided by both the first and the second derivatives.

CONCAVITY

Let $f(x)$ be a twice-differentiable function. Since

$$f''(x) = \frac{d(f'(x))}{dx},$$

we see that $f'(x)$ is increasing throughout any interval where $f''(x) > 0$. Now $f'(x)$ gives the slope of the tangent line at x. For the slope of a line to increase, the line must rotate *counterclockwise*. If the tangent line rotates counterclockwise as x increases, the graph must be bending upward as x increases. This is the case in Fig. 5.29. In this situation, we say the curve is *concave up*. If $f''(x) < 0$, the slope decreases, the tangent line turns *clockwise* as x increases, and the curve is *concave down*. This is the case in Fig. 5.30.

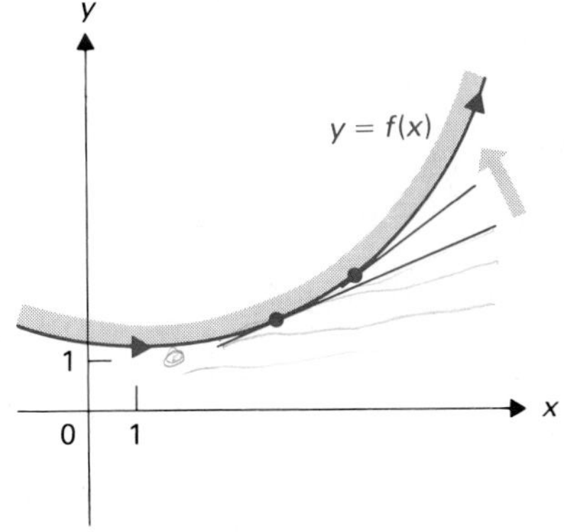

Figure 5.29 $f''(x) > 0$: The tangent line turns counterclockwise as x increases; the graph is concave up.

Points where $f''(x) = 0$ are candidates for places where the concavity changes from concave up to concave down, or vice versa. Such points where the concavity changes are *inflection points*, as illustrated in Fig. 5.31.

We summarize the terminology in a definition.

DEFINITION 5.4 Concavity and inflection point

Let f be a twice-differentiable function. The graph of f is **concave up** in an interval if $f''(x) > 0$ there and is **concave down** in an interval where $f''(x) < 0$. An **inflection point** of a graph is a point on the graph where the concavity changes as x increases through that point.

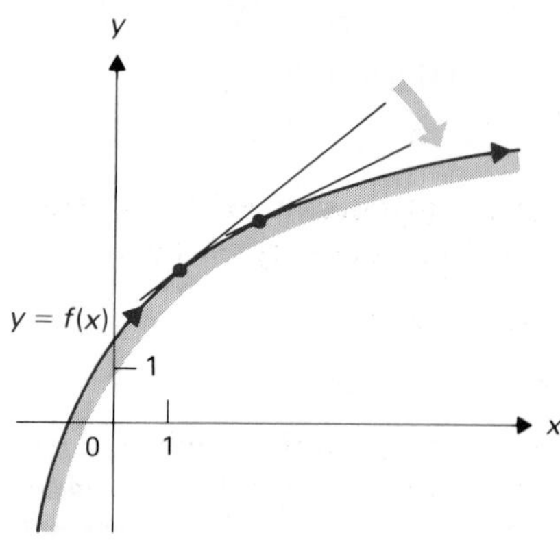

Figure 5.30 $f''(x) < 0$: The tangent line turns clockwise as x increases; the graph is concave down.

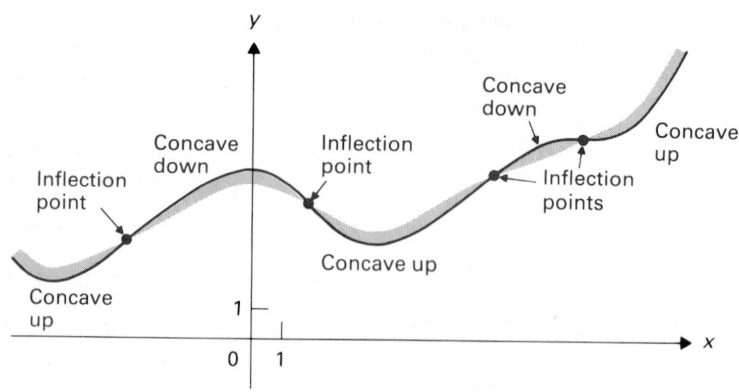

Figure 5.31 Inflection points occur where the concavity changes.

Suppose the graph of $f(x)$ has an inflection point where $x = x_1$, and $f''(x)$ exists and is continuous near x_1. Then we must have $f''(x) > 0$ for x near x_1 on one side and $f''(x) < 0$ for x near x_1 on the other side. By the intermediate-value theorem, we must then have $f''(x_1) = 0$. Thus for functions with continuous second derivatives, we can find all candidates for inflection points by finding where $f''(x) = 0$.

EXAMPLE 1 Discuss concavity and inflection points for $f(x) = x^3 - 3x^2 + 2$.

Solution We have $f'(x) = 3x^2 - 6x$ and $f''(x) = 6x - 6$. Thus $f''(x) = 0$ where $6x - 6 = 0$, or where $x = 1$. Since $6x - 6 < 0$ if $x < 1$, the graph is concave down if $x < 1$. The graph is concave up if $x > 1$ since $f''(x) = 6x - 6 > 0$ there. The concavity does change as x increases through 1, so the point (1, 0) is an inflection point, as indicated in Fig. 5.32. We found the local maximum and local minimum in Example 5 of the preceding section. □

Figure 5.32 Concave down for $x < 1$, concave up for $x > 1$, inflection point (1, 0).

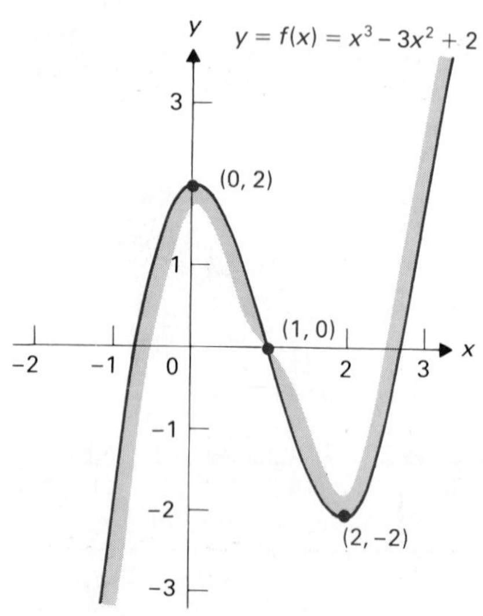

EXAMPLE 2 Discuss concavity and inflection points for $f(x) = x^4$.

Solution If $f(x) = x^4$, then $f'(x) = 4x^3$ and $f''(x) = 12x^2$. Now $f''(0) = 0$, but (0, 0) is not an inflection point since $f''(x) > 0$ on both sides of $x = 0$. Thus the graph in Fig. 5.33 is concave up for $x < 0$ and also for $x > 0$, and there are no inflection points. Actually, (0, 0) is a local minimum since $f'(0) = 0$ and $f'(x) = 4x^3$ is less than zero if $x < 0$ and greater than zero if $x > 0$. □

EXAMPLE 3 Let $f''(x) = x^2(x - 1)^3(x + 2)$. Find the x-coordinates of all inflection points of the graph of f.

Solution We see that $f''(x) = 0$ where $x = -2, 0, 1$. Using the odd/even-exponent argument, we see that $f''(x)$ changes sign as x increases through -2 and 1 but not as x increases through 0. Thus inflection points occur where $x = -2$ and 1. □

EXAMPLE 4 Discuss concavity and inflection points of the graph of $f(x) = x^{1/3} + 1$.

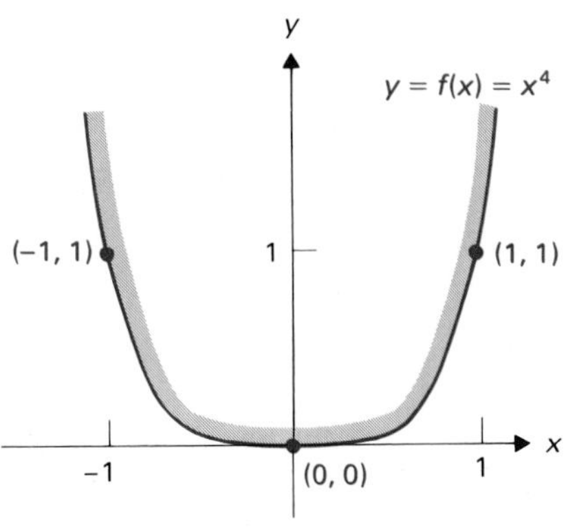

Figure 5.33 Concave up for $x < 0$ and $x > 0$; $f''(0) = 0$, but no inflection point where $x = 0$; $(0, 0)$ is a local minimum.

Solution We have

$$f'(x) = \frac{1}{3}x^{-2/3} \quad \text{and} \quad f''(x) = -\frac{2}{9}x^{-5/3} = \frac{-2}{9x^{5/3}}.$$

Thus $f''(x) \neq 0$ for any nonzero value of x, but $f''(0)$ does not exist. However, $f(x)$ is defined where $x = 0$, and $f(0) = 1$. Testing concavity, we have

$f''(x) > 0$ for $x < 0$, so the graph is concave up there,

$f''(x) < 0$ for $x > 0$, so the graph is concave down there.

Thus concavity does change as we go through $(0, 1)$ on the graph, and we consider $(0, 1)$ to be an inflection point of the graph. The graph is shown in Fig. 5.34. □

THE SECOND DERIVATIVE TEST FOR LOCAL EXTREMA

Let f be a twice-differentiable function. Suppose that $f'(x_1) = 0$ and $f''(x_1) > 0$. Then the graph of f has a horizontal tangent where $x = x_1$, but it is concave up there, so $f(x)$ has a local minimum where $x = x_1$. See Fig. 5.35. On the other hand, if $f'(x_2) = 0$ but $f''(x_2) > 0$, then the graph has a horizontal tangent but is concave down where $x = x_2$, so $f(x)$ has a local maximum there. See Fig. 5.36. Thus if $f'(a) = 0$ but $f''(a) \neq 0$, the sign of $f''(a)$ can be used to determine whether there is a local maximum or local minimum where $x = a$. This method is called the *second derivative test*.

EXAMPLE 5 Use the second derivative test to determine whether $f(x) = x^3 - 3x^2 + 2$ has a local maximum or a local minimum at each point where $f'(x) = 0$.

Solution We have $f'(x) = 3x^2 - 6x = 3x(x - 2) = 0$ if $x = 0$ or $x = 2$. Also, $f''(x) = 6x - 6$. Then $f''(0) = -6 < 0$, so the graph is concave down at $(0, 2)$, which must be a local maximum. Finally, $f''(2) = 12 - 6 = 6 > 0$, so the graph is concave up at $(2, -2)$, which must be a local minimum. The graph was shown in Fig. 5.32. □

Figure 5.34 Concave up for $x < 0$, concave down for $x > 0$, $f''(0)$ does not exist, but $(0, 1)$ is an inflection point.

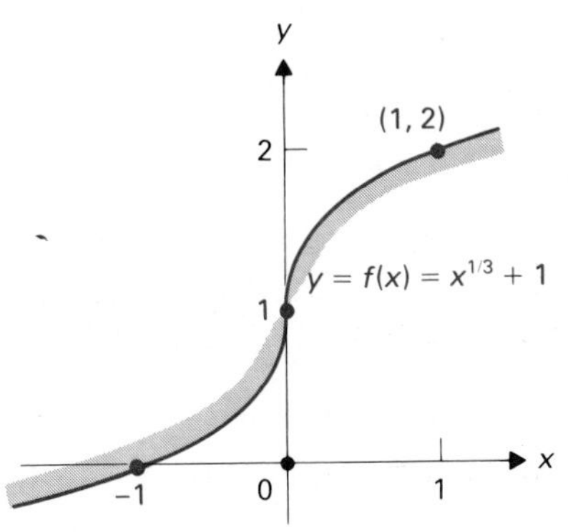

Figure 5.35 $f'(x_1) = 0$, $f''(x_1) > 0$, local minimum where $x = x_1$.

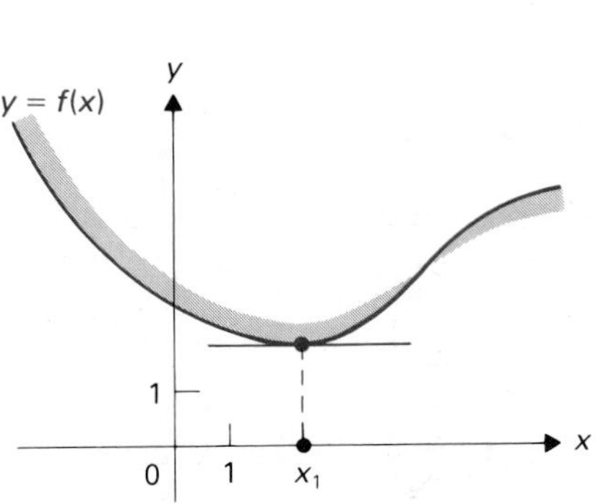

Figure 5.36 $f'(x_2) = 0$, $f''(x_2) < 0$, local maximum where $x = x_2$.

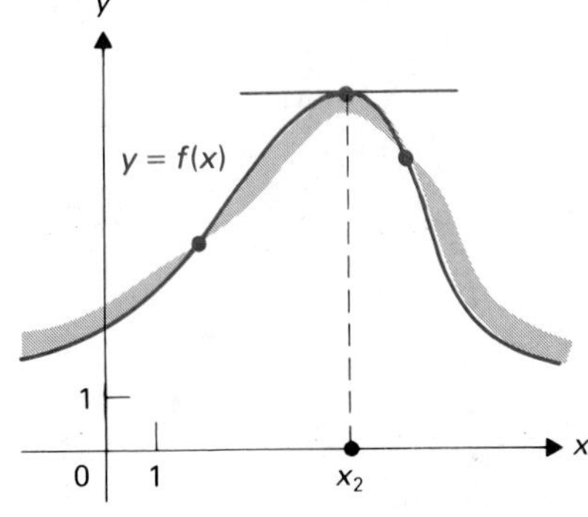

SKETCHING GRAPHS USING DERIVATIVES

We start by summarizing as a theorem the results we have developed in this section.

THEOREM 5.6 Inflection points and the second derivative test

Let f be a function with a continuous second derivative.

1. If $f''(x_1) = 0$ and $f''(x)$ changes sign at x_1 as x increases through x_1, then the graph of f has an inflection point where $x = x_1$.
2. Second derivative test: If $f'(a) = 0$ and $f''(a) \neq 0$, then the graph has a local maximum where $x = a$ if $f''(a) < 0$ and has a local minimum there if $f''(a) > 0$.

The box at the bottom of this page contains an outline for sketching graphs using information given by both the first and the second derivatives.

EXAMPLE 6 Sketch the graph of $f(x) = x + \sin x$, finding all local maxima and minima and all inflection points.

Solution

Step 1 Differentiating, we obtain

$$f'(x) = 1 + \cos x \qquad \text{and} \qquad f''(x) = -\sin x.$$

SKETCHING THE GRAPH OF $f(x)$ USING DERIVATIVES

Step 1 Compute $f'(x)$ and $f''(x)$.

Step 2 Find all critical points, that is, points where $f'(x) = 0$ or does not exist. Determine whether each critical point gives a local maximum, a local minimum, or neither, using the second derivative test or checking for a sign change of the first derivative.

Step 3 (Optional) Find intervals where the graph is increasing ($f'(x) > 0$) or decreasing ($f'(x) < 0$). If you have done an accurate job in Step 2, these intervals will be evident from the graph when you sketch it.

Step 4 Find candidates for inflection points, where $f''(x) = 0$ or where it is not defined or is not continuous. Then check whether $f''(x)$ changes sign as x increases through each candidate point. If a sign change occurs, the point corresponds to an inflection point of the graph.

Step 5 (Optional) Find the intervals where the graph is concave up ($f''(x) > 0$) or concave down ($f''(x) < 0$). If you have done an accurate job in Steps 1, 2, and 4, these intervals will be evident when you sketch the graph.

Step 6 Sketch the graph, using the information in Steps 2 and 5. You may wish to plot a few additional points.

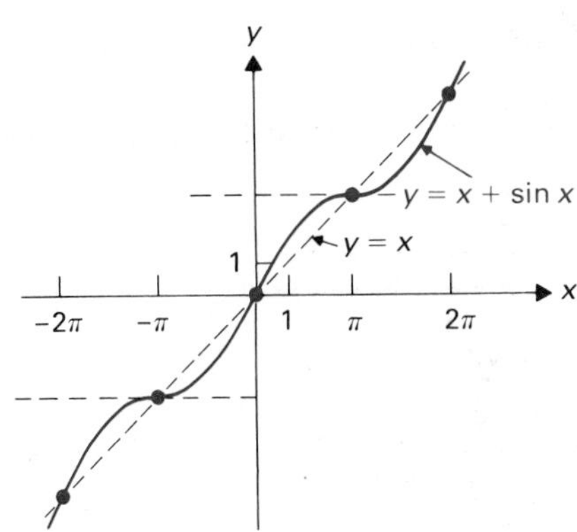

Figure 5.37 No local extrema, inflection points where $x = n\pi$, horizontal tangent lines at inflection points where $x = n\pi$ for n odd.

Step 2 Now $1 + \cos x = 0$ where $\cos x = -1$, which occurs when $x = (2n + 1)\pi$ for any integer n. But $f'(x) = 1 + \cos x$ cannot change sign as x increases through these points because $1 + \cos x \geq 0$ for all x. There are no local maxima or minima.

Step 4 Skipping Step 3 and turning to the second derivative, we find that $f''(x) = -\sin x$ is zero at $x = n\pi$ for all integers n, which include those points where $f'(x)$ is zero. At all these points, $f''(x) = -\sin x$ changes sign as x increases. Thus $x = n\pi$ corresponds to an inflection point for all integers n. If n is odd, then $f'(n\pi) = 0$, and there is a horizontal tangent at the inflection point.

Step 5 The curve is concave up if $-\sin x > 0$, which occurs when $(2n - 1)\pi < x < 2n\pi$ for each integer n. It is concave down for $2n\pi < x < (2n + 1)\pi$.

Step 6 The graph is shown in Fig. 5.37. □

EXAMPLE 7 Sketch the graph of $f(x) = x^4/4 - 4x^3/3 + 2x^2 - 1$, finding all local maxima and minima and all inflection points.

Solution

Step 1 Differentiating, we have

$$f'(x) = x^3 - 4x^2 + 4x \qquad \text{and} \qquad f''(x) = 3x^2 - 8x + 4$$
$$= x(x^2 - 4x + 4) \qquad\qquad = (3x - 2)(x - 2).$$
$$= x(x - 2)^2$$

Step 2 From Step 1, we see that $f'(x) = 0$ when $x = 0$ or $x = 2$. Since the root $x = 0$ comes from the single linear factor x, we see that $f'(x)$ does change sign, from negative to positive since $(x - 2)^2 > 0$, as x increases through 0. Alternatively, $f''(0) = 4 > 0$, so the graph is concave up at $x = 0$. This shows in two ways that $f(x)$ has a local minimum when $x = 0$, corresponding to $(0, -1)$ on the graph.

Although $f'(2) = 0$, the derivative $f'(x)$ does *not* change sign as x increases through 2, since the factor $(x - 2)^2$ has an *even* exponent. Thus there is no local extremum where $x = 2$.

Figure 5.38 Local minimum of -1 where $x = 0$, inflection points $(\frac{2}{3}, -\frac{37}{81})$ and $(2, \frac{1}{3})$, horizontal tangent where $x = 0, 2$.

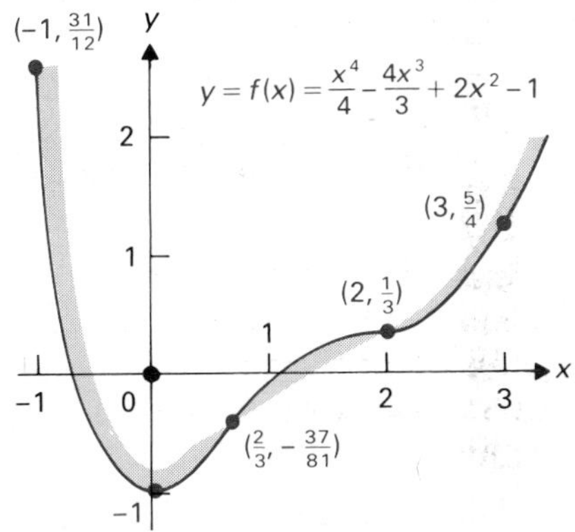

Step 4 From Step 1, we see that candidates for inflection points occur where $x = 2$ and where $x = \frac{2}{3}$. We see that $f''(x)$ does change sign where $x = 2$, because of the odd-powered factor $(x - 2)^1$, so $(2, f(2)) = (2, \frac{1}{3})$ is an inflection point. Another inflection point occurs where $x = \frac{2}{3}$, corresponding to the odd-powered factor $(3x - 2)^1$ in $f''(x)$.

Step 6 The graph is sketched in Fig. 5.38. □

SUMMARY

The second derivative story:

1. If $f''(x) > 0$ throughout an interval, the graph is concave up there.
2. If $f''(x) < 0$ throughout an interval, the graph is concave down there.

3. If $f''(x_2) = 0$, then $(x_2, f(x_2))$ is a candidate for an inflection point of the graph.

   a. If $f''(x)$ changes sign as x increases through x_2, then $(x_2, f(x_2))$ is an inflection point.

   b. If $f''(x)$ does not change sign as x increases through x_2, then $(x_2, f(x_2))$ is not an inflection point.

4. *Second derivative test:* If $f'(x_1) = 0$ and $f''(x_1) \neq 0$, then $f(x_1)$ is a local minimum if $f''(x_1) > 0$ and a local maximum if $f''(x_1) < 0$.

EXERCISES

1. Assume that f is twice differentiable and that f'' is continuous. Mark each of the following true or false.
   ______ a) If f has a local maximum at x_0, then $f'(x_0) = 0$.
   ______ b) If f has a local maximum at x_0, then $f''(x_0) < 0$.
   ______ c) If $f''(x_0) < 0$, then f has a local maximum at x_0.
   ______ d) If $f'(x_0) = 0$ and $f''(x_0) < 0$, then f has a local maximum at x_0.
   ______ e) If $f'(x_0) > 0$, then f is increasing near x_0.
   ______ f) If $f'(x_0) = 0$, then f cannot be increasing near x_0.
   ______ g) If f has an inflection point at x_0, then $f''(x_0) = 0$.
   ______ h) If $f''(x_0) = 0$, then f must have an inflection point at x_0.
   ______ i) If f has a local minimum at x_0, then the tangent line to the graph of f at $(x_0, f(x_0))$ is horizontal.
   ______ j) If f has an inflection point at x_0, then the tangent line to the graph of f at $(x_0, f(x_0))$ may possibly be horizontal.

2. Sketch the graph of a twice-differentiable function f such that $f(0) = 3$, $f'(0) = 0$, $f''(0) < 0$, $f(2) = 2$, $f'(2) = -1$, $f''(2) = 0$, $f(4) = 1$, $f'(4) = 0$, and $f''(4) > 0$.

3. Sketch the graph of a twice-differentiable function f such that $f''(x) < 0$ for $x < 1$, $f(1) = -1$, $f'(1) = 1$, $f''(1) = 0$, $f''(x) > 0$ for $x > 1$, and $f(3) = 4$.

4. Sketch the graph of a twice-differentiable function f such that $f'(x) > 0$ for $x > 2$, $f''(x) > 0$ for $x > 2$, $f''(2) = 0$, $f'(x) < 0$ for $x < 2$, and $f''(x) > 0$ for $x < 2$.

5. Do you think it is possible that, for a twice-differentiable function f, one can have $f(0) = 0$, $f'(0) = 1$, $f''(x) > 0$ for $x > 0$, and $f(1) = 1$? Why?

In Exercises 6 through 26, find
a) the intervals where the graph of f is concave up, and
b) the intervals where the graph of f is concave down
from the given information.

6. $f(x) = 4 - x^2$

7. $f(x) = x^2 - 6x + 4$

8. $f(x) = \dfrac{1}{x+1}$

9. $f(x) = \dfrac{x^3}{3} + x^2 - 3x - 4$

10. $f(x) = \dfrac{x^3}{3} + x^2 + x - 6$

11. $f(x) = \dfrac{x^2}{x-1}$

12. $f(x) = x^4 - 4x + 1$

13. $f(x) = x^5 - 5x + 1$

14. $f(x) = 3x^4 - 4x^3$

15. $f(x) = x^4 - 2x^2 - 2$

16. $f(x) = x + 2 \sin x$

17. $f(x) = x - 2 \cos x$

18. $f''(x) = (x-1)^3(x+2)^2$

19. $f''(x) = x(x+1)^2(x-1)$

20. $f''(x) = x^3(x+2)^2(x-4)$

21. $f''(x) = x(x-1)^3(x+3)^5$

22. $f''(x) = \dfrac{x^2-1}{x^2}$

23. $f''(x) = \dfrac{x^2-x}{x+3}$

24. $f''(x) = \dfrac{x(x-1)^3}{(x+2)^3}$

25. $f''(x) = \dfrac{x^3-2x^2}{(x+1)^2}$

26. $f''(x) = \dfrac{x^3}{(x-1)^2(x+4)}$

Exercises 27 through 30 refer to Figs. 5.39 through 5.42, which show the graphs of *derivatives* of functions defined and continuous for all x. Mark the parts true or false.

27. If $f'(x)$ has the graph in Fig. 5.39, then
   ______ a) the graph of f has an inflection point at $(0, 0)$,
   ______ b) $f(x)$ has a local maximum where $x = 0$,

Figure 5.39
Graph of f'.

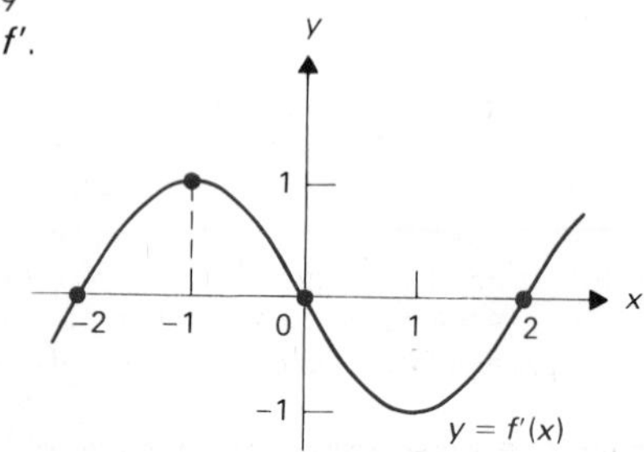

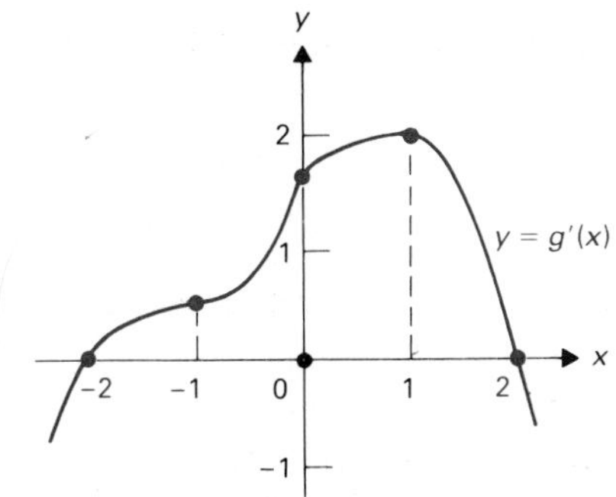

Figure 5.40 Graph of g'.

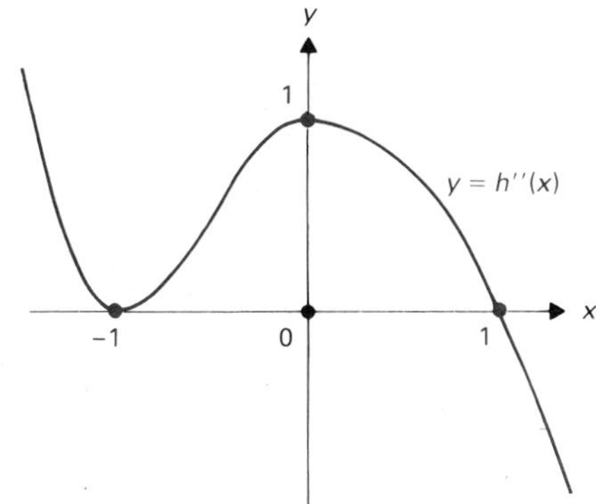

Figure 5.41 Graph of h''.

_____ c) $f(x)$ has a local maximum where $x = -1$,

_____ d) the graph of f has an inflection point where $x = -1$,

_____ e) the graph of f is concave down for $-2 < x < 0$.

28. If $g'(x)$ has the graph in Fig. 5.40, then

_____ a) the graph of g has an inflection point where $x = -1$,

_____ b) the graph of g has an inflection point where $x = 1$,

_____ c) the graph of g is concave down for $x < -1$,

_____ d) the graph of g is concave down for $x > 1$,

_____ e) $g(x)$ has a local minimum where $x = -2$.

29. If $h''(x)$ has the graph in Fig. 5.41, then

_____ a) the graph of h is concave up for $-1 < x < 1$,

_____ b) the graph of h has an inflection point where $x = -1$,

_____ c) the graph of h has an inflection point where $x = 1$,

_____ d) $h'(x)$ has a local maximum where $x = 1$,

_____ e) the graph of $h'(x)$ has an inflection point where $x = -1$.

30. If $k'(x)$ has the graph in Fig. 5.42, then

_____ a) $k(x)$ has a local minimum where $x = -2$,

_____ b) the graph of k has an inflection point where $x = -2$,

_____ c) the graph of k has an inflection point where $x = 1$,

_____ d) $k(x)$ has a local maximum where $x = 1$,

_____ e) the graph of k has an inflection point where $x = 0$.

In Exercises 31 through 42,
a) find all local maxima,
b) find all local minima,
c) find all inflection points, and
d) sketch the graph

of the given function. (Concavity was determined for these functions in Exercises 6 through 17, as noted.)

31. $f(x) = 4 - x^2$ (Exercise 6)

32. $f(x) = x^2 - 6x + 4$ (Exercise 7)

33. $f(x) = \dfrac{1}{x+1}$ (Exercise 8)

34. $f(x) = \dfrac{x^3}{3} + x^2 - 3x - 4$ (Exercise 9)

35. $f(x) = \dfrac{x^3}{3} + x^2 + x - 6$ (Exercise 10)

36. $f(x) = \dfrac{x^2}{x-1}$ (Exercise 11)

37. $f(x) = x^4 - 4x + 1$ (Exercise 12)

38. $f(x) = x^5 - 5x + 1$ (Exercise 13)

39. $f(x) = 3x^4 - 4x^3$ (Exercise 14)

40. $f(x) = x^4 - 2x^2 - 2$ (Exercise 15)

41. $f(x) = x + 2 \sin x$ (Exercise 16)

42. $f(x) = x - 2 \cos x$ (Exercise 17)

In Exercises 43 through 46, sketch the curve with the given equation. [*Hint:* Regard the equation as giving x as a function g of y, and apply the theory in this section with x and y interchanged.]

43. $x = y^2 - 2y + 2$

44. $x = y^3 - 3y^2$

45. $x = \dfrac{1}{y^4 + 1}$

46. $x = \dfrac{y}{y^2 + 1}$

Figure 5.42 Graph of k'.

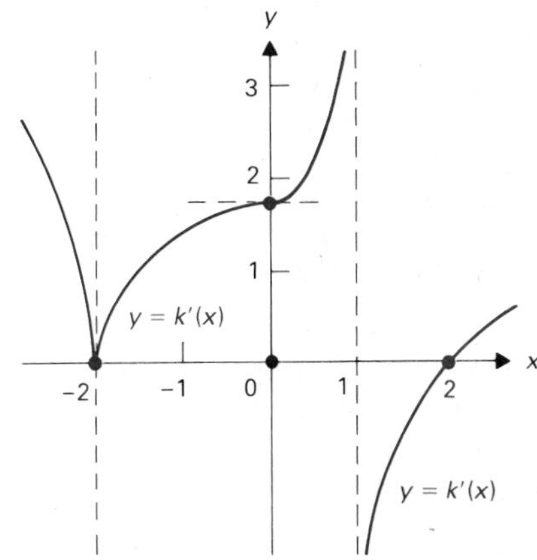

5.6 APPLIED MAXIMUM AND MINIMUM PROBLEMS

There are many situations in which one wishes to maximize or minimize some quantity. For example, manufacturers want to maximize their profit. Builders may want to minimize their costs. Such extremal problems are clearly of great practical importance. We can often solve extremal problems with differential calculus, using the ideas developed in the preceding section.

To find local maxima and minima of a function f, we may proceed as follows. Find all critical points x where either $f'(x) = 0$ or $f'(x)$ does not exist. At critical points where $f'(x) = 0$, check if the derivative changes sign there, or use the second derivative test to determine whether the point corresponds to a local maximum, a local minimum, or neither. Then examine the behavior of $f(x)$ near points where $f'(x)$ does not exist.

The box at the bottom of this page contains a suggested step-by-step procedure to use with applied maximum and minimum problems.

We cannot overemphasize the importance of the second part of Step 1 in the outline—deciding what you wish to maximize or minimize. If you don't know what you are trying to do, you won't have much success doing it. Step 1 sounds like a trivial directive, but many people have floundered aimlessly, writing equation after equation leading nowhere, without knowing where they were heading.

EXAMPLE 1 Find two numbers whose sum is 6 and whose product is as large as possible.

Solution

Step 1 We want to maximize the product P of two numbers.

Step 2 If the two numbers are x and y, then $P = xy$. Since $x + y = 6$, we have $y = 6 - x$, so $P = x(6 - x) = 6x - x^2$.

SOLVING AN APPLIED MAXIMUM/MINIMUM PROBLEM

Step 1 Draw a figure where appropriate and assign letter variables. Decide what you want to maximize or minimize.

Step 2 Express the quantity that you want to maximize or minimize as a function f of *one* other quantity. (You may have to use some algebra to do this.)

Step 3 Find all critical points, where $f'(x) = 0$ or $f'(x)$ does not exist.

Step 4 Decide whether the desired maximum or minimum occurs at one of the points you found in Step 3. Frequently it will be clear that a maximum or minimum exists from the nature of the problem, and if Step 3 gives only one candidate, that is the answer. If there is more than one candidate, or if there are points in the domain of f that are not *inside* an interval where f is differentiable, you may have to make a further examination.

Step 5 Put the answer in the requested form.

Step 3 We have

$$\frac{dP}{dx} = 6 - 2x.$$

Thus $dP/dx = 0$ when $6 - 2x = 0$, or when $x = 3$.

Step 4 Since $d^2P/dx^2 = -2 < 0$, we see that P has a maximum at $x = 3$.

Step 5 We were asked to find the numbers x and y whose sum is 6 and whose product is maximum, not the maximum product. We know that $x = 3$, and consequently $y = 6 - x = 3$ also. Thus our answer is that both numbers must be 3. □

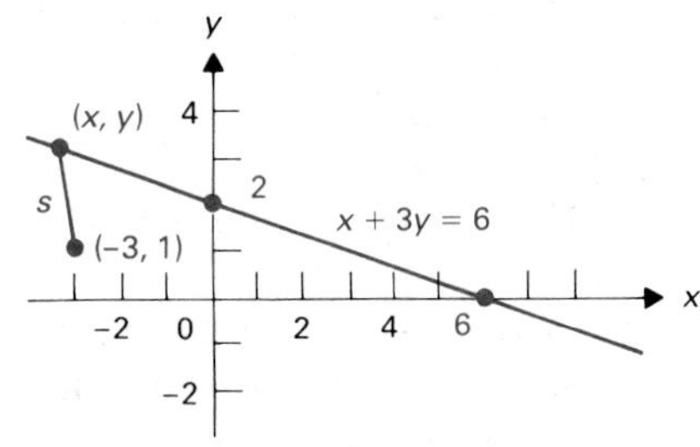

Figure 5.43 The distance s from $(-3, 1)$ to a point (x, y) on $x + 3y = 6$.

EXAMPLE 2 Use the methods of this section to find the point on the line $x + 3y = 6$ that is closest to the point $(-3, 1)$.

Solution

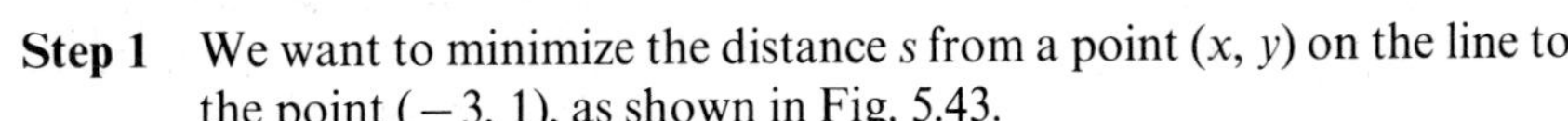

Step 1 We want to minimize the distance s from a point (x, y) on the line to the point $(-3, 1)$, as shown in Fig. 5.43.

Step 2 We have

$$s = \sqrt{(x + 3)^2 + (y - 1)^2}.$$

Since $x = 6 - 3y$, we obtain

$$s = \sqrt{(9 - 3y)^2 + (y - 1)^2}.$$

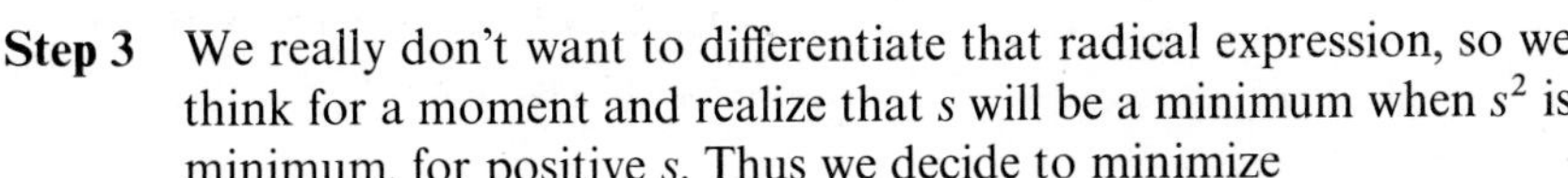

Step 3 We really don't want to differentiate that radical expression, so we think for a moment and realize that s will be a minimum when s^2 is minimum, for positive s. Thus we decide to minimize

$$s^2 = (9 - 3y)^2 + (y - 1)^2.$$

We obtain

$$\begin{aligned}\frac{d(s^2)}{dy} &= 2(9 - 3y)(-3) + 2(y - 1)\\ &= 20y - 56\\ &= 0 \qquad \text{when} \quad y = \frac{56}{20} = \frac{14}{5}.\end{aligned}$$

Step 4 Since our problem obviously has a solution and we found only one candidate, we have the minimum distance when $y = \frac{14}{5}$. (Of course, $d^2(s^2)/dy^2 = 20 > 0$, which also shows that $y = \frac{14}{5}$ yields a minimum.)

Step 5 We were not asked for the minimum distance, but for the point (x, y) where the minimum distance is assumed. Since $y = \frac{14}{5}$, we find that $x = 6 - 3y = 6 - \frac{42}{5} = -\frac{12}{5}$. Thus our answer is the point $(-\frac{12}{5}, \frac{14}{5})$. □

EXAMPLE 3 A manufacturer of dog food wishes to package the product in cylindrical metal cans, each of which is to contain a certain volume V_0 of food. Find the ratio of the height of the can to its radius to minimize the amount of metal, assuming that the ends and side of the can are made from metal of the same thickness.

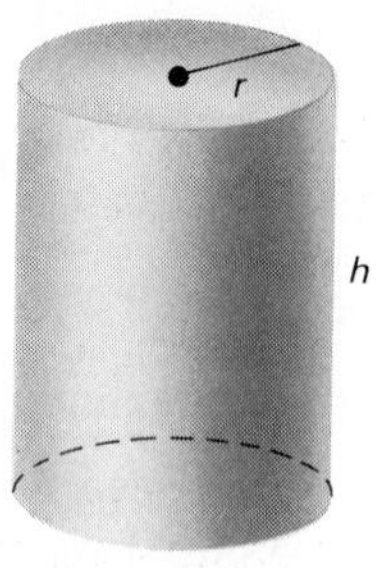

Figure 5.44 Cylindrical can of radius r and height h.

Solution

Step 1 The manufacturer wishes to minimize the surface area S of the can.

Step 2 The surface of the can consists of the two circular disks at the ends and the cylindrical side. If the radius of the can is r and the height is h, the top and bottom disks each have area πr^2 and the cylinder has area $2\pi rh$ (see Fig. 5.44). Thus

$$S = 2\pi r^2 + 2\pi rh.$$

We would like to find h in terms of r to express S as a function of the single quantity r. From $V_0 = \pi r^2 h$, we obtain $h = V_0/\pi r^2$. Thus

$$S = 2\pi r^2 + 2\pi r \frac{V_0}{\pi r^2} = 2\left(\pi r^2 + \frac{V_0}{r}\right).$$

Step 3 We easily find that

$$\frac{dS}{dr} = 2\left(2\pi r - \frac{V_0}{r^2}\right).$$

Thus $dS/dr = 0$ when $2\pi r - (V_0/r^2) = 0$, so $2\pi r^3 = V_0$, and

$$r^3 = \frac{V_0}{2\pi}.$$

Step 4 It is obvious that a minimum for S does exist from the nature of the problem. A can $\frac{1}{32}$ in. high and 6 ft across uses a lot of metal, as does one $\frac{1}{32}$ in. in radius and 1000 ft high. Somewhere between these ridiculous measurements there is a can of reasonable dimensions using the least metal. We found only one candidate in Step 3, so we don't have to look further.

Step 5 We are interested in the ratio h/r. Now $V_0 = \pi r^2 h$, so

$$h = \frac{V_0}{\pi r^2} \quad \text{and} \quad \frac{h}{r} = \frac{V_0}{\pi r^3}.$$

From Step 3, the least metal is used when $r^3 = V_0/2\pi$, so

$$\frac{h}{r} = \frac{V_0}{\pi(V_0/2\pi)} = 2. \quad \square$$

Example 3 illustrates the practical importance of extremal problems. For a cylindrical can of minimal surface area and containing a given volume, we should have $h = 2r$, so the height should equal the diameter. Thus few cylindrical cans on the supermarket shelves represent economical packaging, assuming that the ends and sides are of equally expensive material. The tuna fish cans are usually too short, and the soft drink and beer cans are too high.

EXAMPLE 4 Two straight roads intersect at right angles, one running north and south and the other east and west. Bill jogs north through the intersection at a steady rate of 6 mph. Sue pedals her bicycle west at a steady rate of 12 mph and goes through the intersection 30 minutes after Bill. Find the minimum distance between Bill and Sue.

Solution

Step 1 We denote the roads by x,y-axes with the y-axis pointing north, as is usual on maps. See Fig. 5.45. We wish to minimize the distance s between Bill and Sue.

Step 2 Let time t be measured in hours, with $t = 0$ at the instant when Bill is at the intersection. Since Bill is jogging at a constant rate of 6 mph, he travels $6t$ miles in t hours, so his position at time t is $(0, 6t)$. Sue reaches the origin in the plane at time $t = \frac{1}{2}$ hour. She is riding west, corresponding to going to the *left* on the x-axis, at 12 mph. Thus at time t, she is at the point $-12(t - \frac{1}{2})$ on the x-axis, or at the point $(-12t + 6, 0)$ in the plane. Thus the distance between Sue and Bill is

$$s = \sqrt{(-12t + 6)^2 + 36t^2}.$$

Step 3 As in Example 2, we make our work easier by minimizing

$$s^2 = (-12t + 6)^2 + 36t^2,$$

which is minimum when s is minimum. Then

$$\begin{aligned} d(s^2)/dt &= 2(-12t + 6)(-12) + 72t = 288t - 144 + 72t \\ &= 360t - 144 = 0 \qquad \text{when} \quad t = \tfrac{144}{360} = \tfrac{12}{30} = \tfrac{2}{5}. \end{aligned}$$

Step 4 Our problem obviously has a solution, and we have only one candidate, $t = \frac{2}{5}$, to give that solution. Thus the minimum distance s must occur 24 minutes after Bill goes through the intersection.

Step 5 We were asked to find the minimum distance. Putting $t = \frac{2}{5}$ in the formula for s in Step 2, we obtain

$$\begin{aligned} s &= \sqrt{144(\tfrac{1}{10})^2 + 36(\tfrac{2}{5})^2} = \sqrt{12^2(\tfrac{1}{100} + \tfrac{1}{25})} \\ &= 12\sqrt{\tfrac{5}{100}} = 12/\sqrt{20} = 6/\sqrt{5} \text{ miles} \end{aligned}$$

as our answer. □

EXAMPLE 5 Let a point source of light be at a point L above a flat surface. The illumination that the light provides at a point P on the surface is inversely proportional to the square of the distance s from P to L and directly proportional to $\sin\theta$ where θ is the angle of elevation from P to L. See Fig. 5.46.

Figure 5.45 The distance s between Bill and Sue.

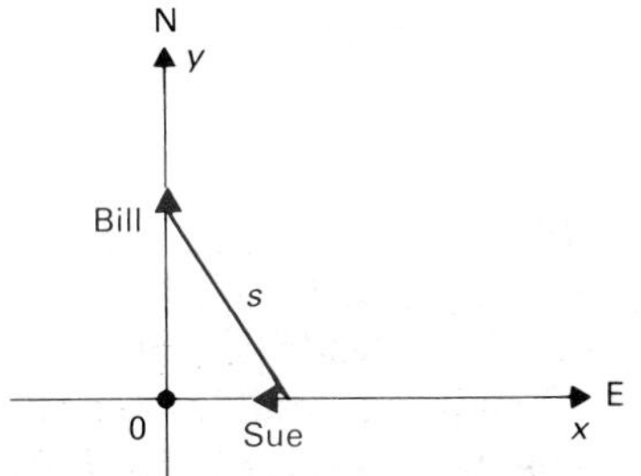

Figure 5.46 The distance s and angle of inclination θ from a point P on a level surface to a light L.

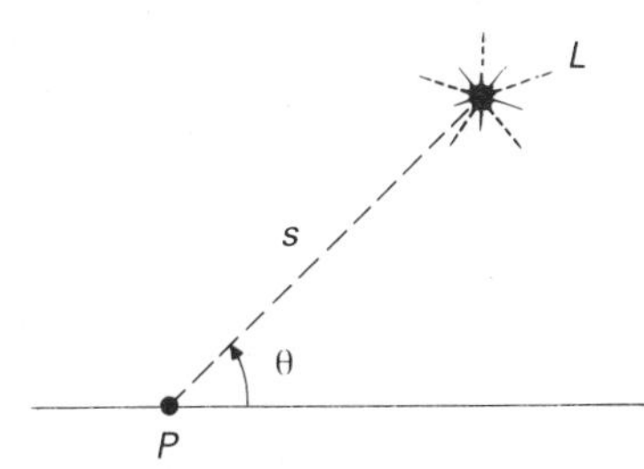

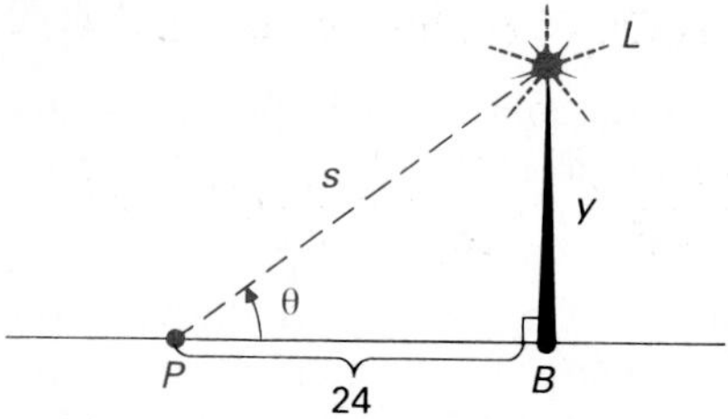

Figure 5.47 Light pole y ft high and 24 ft from P on a level surface.

How high above the sidewalk should a street light bulb be placed to provide maximum illumination on the sidewalk across the street at a point P, which is 24 ft from the base of the light pole?

Solution

Step 1 In Fig. 5.47 the distance from P to the base B of the light pole is 24 ft. We wish to find $y \geq 0$ to maximize the illumination I.

Step 2 Since I is directly proportional to $\sin \theta$ and inversely proportional to s^2, there exists a constant k such that

$$I = k \cdot \frac{1}{s^2} \cdot \sin \theta$$

$$= k \cdot \frac{1}{24^2 + y^2} \cdot \frac{y}{\sqrt{24^2 + y^2}}$$

$$= k \cdot \frac{y}{(576 + y^2)^{3/2}}.$$

Step 3 Differentiating, we obtain

$$\frac{dI}{dy} = k \cdot \frac{(576 + y^2)^{3/2} - y(\frac{3}{2})(576 + y^2)^{1/2}(2y)}{(576 + y^2)^3}$$

$$= k \cdot \frac{\sqrt{576 + y^2}\,(576 + y^2 - 3y^2)}{(576 + y^2)^3}$$

$$= k \cdot \frac{576 - 2y^2}{(576 + y^2)^{5/2}}.$$

Thus $dI/dy = 0$ when

$$576 - 2y^2 = 0,$$

$$y^2 = \tfrac{576}{2} = 288 = 2(144),$$

$$y = \pm 12\sqrt{2}.$$

Step 4 When $y = 0$, $\sin \theta = 0$, so the illumination I is zero. Also, as $y \to \infty$, the illumination $I \to 0$. Thus there is a maximum illumination I for some $y > 0$, and $y = 12\sqrt{2}$ ft is our only candidate.

Step 5 Since we were asked to find the height y and not the maximum illumination, our answer is $y = 12\sqrt{2}$ ft. □

EXAMPLE 6 A piece of wire 20 ft long is cut into two pieces of length x and $20 - x$. The piece of length x is bent into a square, and the piece of length $20 - x$ is bent into a circle, as shown in Fig. 5.48. Find the value of x such that the sum of the areas enclosed by the square and the circle is a maximum.

Figure 5.48 Line segments of lengths x and $20 - x$ bent to form a square of side $x/4$ and a circle of radius $(20 - x)/(2\pi)$.

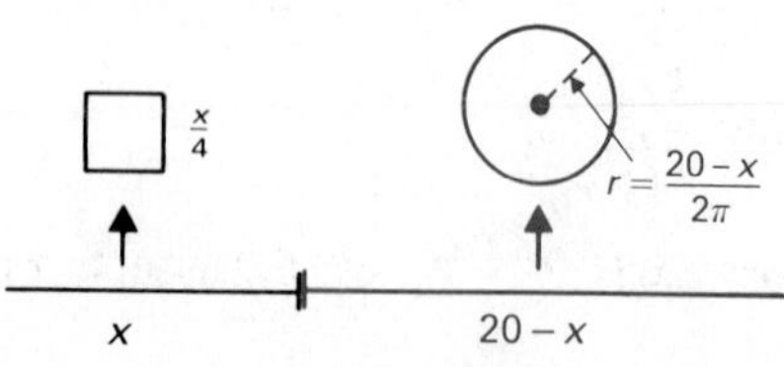

Solution

Step 1 We wish to maximize the sum A of the area A_1 of a square of perimeter x and the area A_2 of a circle of perimeter $20 - x$, subject to $0 \leq x \leq 20$.

Step 2 Since each side of the square has length $x/4$, we have $A_1 = x^2/16$.

Since the radius of the circle is $(20 - x)/(2\pi)$, we see that $A_2 = \pi[(20 - x)/(2\pi)]^2$. Thus

$$A = A_1 + A_2 = \frac{x^2}{16} + \pi\frac{(20 - x)^2}{4\pi^2} = \frac{1}{16}x^2 + \frac{1}{4\pi}(20 - x)^2$$

for $0 \le x \le 20$.

Step 3 We obtain

$$\frac{dA}{dx} = \frac{1}{16}\cdot 2x + \frac{1}{4\pi}\cdot 2(20 - x)(-1) = \frac{1}{16}(2x) + \frac{1}{2\pi}(x - 20)$$

$$= \left(\frac{1}{8} + \frac{1}{2\pi}\right)x - \frac{10}{\pi} = \left(\frac{\pi + 4}{8\pi}\right)x - \frac{10}{\pi}$$

$$= 0 \quad \text{when} \quad x = \frac{10}{\pi}\cdot\frac{8\pi}{\pi + 4} = \frac{80}{\pi + 4} \approx 11.2.$$

Step 4 Since $d^2A/dx^2 = (\pi + 4)/(8\pi) > 0$, we see that the value $x = 80/(\pi + 4)$ corresponds to a local minimum, not a local maximum. Therefore the maximum of A for $0 \le x \le 20$ must be assumed at an endpoint of the interval. When $x = 0$, $A = 20^2/4\pi = 100/\pi$. When $x = 20$, $A = 20^2/16 = 100/4$. Since $100/\pi > 100/4$, we see that the maximum area occurs when $x = 0$.

Step 5 We were asked to find the value of x for maximum total area, so $x = 0$ ft is our answer. The maximum total area is attained when the wire is not cut at all and the whole wire is bent into a circle. Our work shows that the minimum area is attained when $x \approx 11.2$ ft. □

EXAMPLE 7 One vertex of a triangle lies at the center of a circle of radius a, and the other two vertices lie on the circle. Find the size of the vertex angle θ at the center of the circle if the area of the triangle is a maximum.

Solution

Step 1 We wish to find θ to maximize the area A of the triangle shown in in Fig. 5.49.

Step 2 Let the altitude of the triangle be h and the base be b. The area of the triangle is then

$$A = \tfrac{1}{2}bh.$$

From Fig. 5.49, we see that $h = a\cos\theta/2$ and $b = 2a\sin\theta/2$. Thus

$$A = \frac{1}{2}\left(2a\sin\frac{\theta}{2}\right)\left(a\cos\frac{\theta}{2}\right) = a^2\sin\frac{\theta}{2}\cos\frac{\theta}{2}.$$

Figure 5.49 Triangle with vertex angle θ at the center of a circle of radius a, and other vertices on the circle.

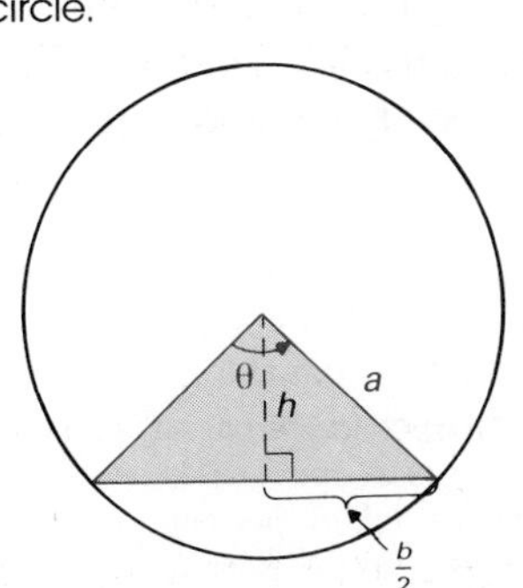

Step 3 We have

$$\frac{dA}{d\theta} = a^2\left[\left(\sin\frac{\theta}{2}\right)\left(-\sin\frac{\theta}{2}\right)\frac{1}{2} + \left(\cos\frac{\theta}{2}\right)\left(\cos\frac{\theta}{2}\right)\frac{1}{2}\right]$$

$$= \frac{a^2}{2}\left(-\sin^2\frac{\theta}{2} + \cos^2\frac{\theta}{2}\right) = 0,$$

where $\sin^2(\theta/2) = \cos^2(\theta/2)$, or $\tan(\theta/2) = 1$, or $\theta/2 = \pi/4$, or $\theta = \pi/2$.

Step 4 The area obviously is a maximum for some θ where $0 < \theta < \pi$, and $\theta = \pi/2$ is the only candidate.

Step 5 We were asked to find the value of θ for maximum area, so $\theta = \pi/2$ is our answer. □

SUMMARY

Suggested outline for solving an applied maximum/minimum problem:

Step 1 Decide what quantity you want to maximize or minimize, drawing a figure where appropriate.

Step 2 Express the quantity as a function f of *one* variable.

Step 3 Find all critical points of f.

Step 4 Decide whether the desired maximum or minimum occurs at one of the points found in Step 3.

Step 5 Put the answer in the requested form.

EXERCISES

1. Find the maximum area a rectangle can have if the perimeter is 20 ft.
2. Generalize Exercise 1 to show that the rectangle of maximum area having a fixed perimeter is a square.
3. Find two positive numbers x and y such that $x + y = 6$ and and xy^2 is as large as possible.
4. Find the positive number x such that the sum of x and its reciprocal is minimum.
5. Find two positive numbers such that their product is 36 and the sum of their cubes is a minimum.
6. Find the maximum distance, measured horizontally, between the graphs of $y = x$ and $y = x^2$ for $0 \leq x \leq 1$.
7. A rectangle has its base on the x-axis and its upper vertices on the parabola $y = 6 - x^2$, as shown in Fig. 5.50. Find the maximum possible area of the rectangle.
8. A trapezoid has endpoints of its lower base at (2, 0) and (−2, 0) on the x-axis, while its upper base is a horizontal chord of the parabola $y = 4 - x^2$. Find the maximum possible area of the trapezoid.
9. Find the maximum possible area of a right triangle with right-angle vertex at (0, 0), legs on the positive x-axis and y-axis, and longest side passing through the point (3, 7).
10. Find the point on the parabola $y = x^2$ that is closest to the point (6, 3).
11. A cardboard box of 108 in^3 volume with a square base and open top is to be constructed. Find the minimum area of cardboard needed. (Neglect waste in construction.)

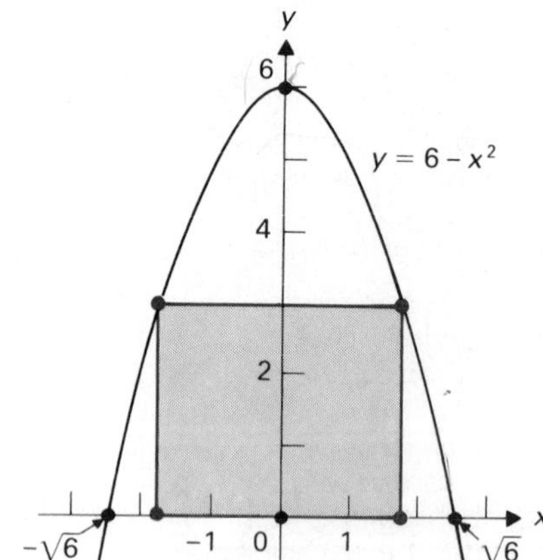

Figure 5.50

12. An open box with a reinforced square bottom and volume 96 ft^3 is to be constructed. If material for the bottom costs three times as much per square foot as material for the sides, find the dimensions of the box of minimum cost. (Neglect waste in construction.)
13. A parcel delivery service will only accept parcels such that the sum of the length and girth is at most 96 in. (Girth is the perimeter of a cross section taken perpendicular to the

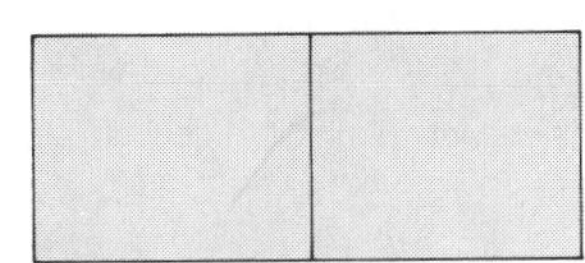
Figure 5.51

longest dimension.) Find the maximum acceptable volume of (a) a rectangular parcel with square ends, (b) a cylindrical parcel.

14. A farmer has 1000 rods of fencing with which to fence three sides of a rectangular pasture; a straight river will form the fourth side of the pasture. Find the dimensions of the pasture of largest area that the farmer can fence.

15. A rancher has 1200 ft of fencing to enclose a double paddock with two rectangular regions of equal areas (Fig. 5.51). Find the maximum area that the rancher can enclose. (Neglect waste in construction and the need for gates.)

16. A couple wish to fence a 6000 ft^2 rectangular plot of their property for their dogs. One side of the rectangle is to border a straight road, and the fence for that side must be ornamental, costing three times as much per foot as the rest of the fence. Find the dimensions of the plot for the most economical fencing.

17. An open gutter of rectangular cross section is to be formed from a long sheet of tin of width 8 in., by bending up the sides of the sheet. Find the dimensions of the cross section of the gutter so formed for maximum carrying capacity.

18. A gardener has some 8-in.-wide planks. He wants to build an irrigation trough with trapezoidal cross section to carry water to his garden. See Fig. 5.52. Find the value of θ in Fig. 5.52 so that the trapezoidal cross section has maximum area (which corresponds to maximum irrigation capacity for the trough).

19. Find the dimensions of the rectangle of maximum area that can be inscribed in a semicircle of radius a. [*Hint:* You may find it easier to maximize the square of the area. Of course, the rectangle having maximum area is the one having the square of its area maximum.]

20. Find the volume of the largest right circular cylinder that can be inscribed in a right circular cone of radius a and altitude b.

21. Find the altitude of the right circular cone of maximum volume that can be inscribed in a sphere of radius a.

22. Find the area of the largest isosceles triangle that can be inscribed in a circle of radius a.

23. Find the maximum possible area of a rectangle inscribed in an equilateral triangle of side a.

Figure 5.52

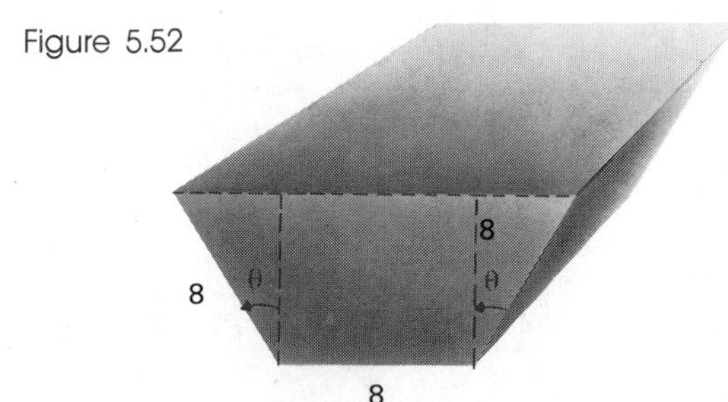

24. Of all sectors of circles where the perimeter of the sector is a fixed constant P_0, find the central angle θ of the sector of maximum possible area, subject to $0 \le \theta \le 2\pi$.

25. A rectangular cardboard poster is to contain 216 in^2 of printed matter with 2-in. margins at the sides and 3-in. margins at the top and bottom. Find the dimensions of the poster using the least cardboard.

26. The wreck of a plane in a desert is 15 mi from the nearest point A on a straight road. A rescue truck starts for the wreck at a point on the road 30 mi distant from A. If the truck can travel at 80 mph on the road and at 40 mph on a straight path in the desert, how far from A should the truck leave the road to reach the wreck in minimum time?

27. A woman in a rowboat is 1 mi from the nearest point A on a straight shore. She wishes to reach a point B, which is 2 mi along the shore from A. If she can row 4 mph and run along the shore at 6 mph, how far from A toward B along the shore should she land to reach B in the minimum time?

28. Suppose in Exercise 27 that the woman can run along the (possibly rocky) shore k times as fast as she can row. Describe, for all values $k > 0$, how far from A toward B she should land to reach B in the minimum time.

29. The illumination provided by a light at a point A is directly proportional to the intensity of the light and inversely proportional to the square of the distance from the point A to the light. Let light L_2 have c times the intensity of light L_1, and let the distance between the lights be a ft. Find the distance from L_1 on the line segment joining L_1 to L_2 where the illumination is minimal.

30. A piece of wire 100 cm long is cut into two pieces. One piece is bent to form a square, and the other is formed into an equilateral triangle. Find how the wire should be cut if the total area enclosed is to be
a) minimum,
b) maximum.

31. Answer Exercise 30 if both pieces of wire are bent into equilateral triangles.

32. A weight suspended from a beam by a spring is bouncing up and down. Its height above the floor at time t sec is $h = 48 + 12 \sin [(\pi/2)t]$ in. Find the maximum and minimum of
a) the heights above the floor that the weight attains,
b) the velocity of the weight,
c) the speed of the weight,
d) the acceleration of the weight.

33. A Norman window is to be built in the shape of a rectangle of clear glass surmounted by a semicircle of colored glass. The total perimeter of the window is to be 36 ft.
a) Find the ratio of the height of the rectangle to the width that maximizes the area of the clear glass in the rectangular portion.
b) If clear glass admits twice as much light per square foot as colored glass, find the ratio of the height of the rectangle to the width to admit the maximum amount of light.
c) If colored glass costs four times as much per square foot as clear glass and the window is to be at least 4 ft wide, find the width of the window of minimum cost.

34. The strength of a beam of rectangular cross section is proportional to the width and the square of the depth. Find the dimensions of the strongest rectangular beam that can be cut from a circular log of radius 9 in.

35. Ship A travels on a due-north course and passes a buoy at 9:00 A.M. Ship B, traveling twice as fast on a due-east course, passes the same buoy at 11:00 A.M. the same day. At what time are the ships closest together?

36. A silo is to have the form of a cylinder capped with a hemisphere. If the material for the hemisphere is twice as expensive per square foot as the material for the cylinder, find the ratio of the height of the cylinder to its radius for the most economical structure of given volume. (Neglect waste in construction.)

37. An open box is to be formed from a rectangular piece of cardboard by cutting out equal squares at the corners and turning up the resulting flaps. Find the size of the corner squares that should be cut out from a sheet of cardboard of length a and width b to obtain a box of maximum volume.

38. The following is a simple economic model for the production of a single perishable item that must be sold on the day produced to avoid a loss from spoilage.

The producer has a basic plant overhead of a dollars per day. Each item produced costs b dollars for ingredients and labor. In addition, if the manufacturer produces x items per day, there is a daily cost of cx^2 dollars, resulting from crowded conditions and inefficient operation as more items are produced. (The value of c is usually quite small, so that cx^2 is of insignificant size until x becomes fairly large). Each day, if a single item were produced, it could be sold that day for A dollars (the initial demand price). However, the price at which every item produced on a given day can be sold drops B dollars for each item produced on that day. (The number B reflects the degree of saturation of the market per item and is usually quite small.)

a) Find an algebraic expression giving the daily profit if the manufacturer produces x items per day.

b) Find, in terms of a, b, c, A, and B, the number x of items the manufacturer should produce each day to maximize the daily profit.

39. Suppose that in the economic model in Exercise 38, the government imposes a tax on the manufacturer of t dollars for each item manufactured.

a) Determine the number x of items the manufacturer should produce each day to maximize the daily profit. [*Hint:* Calculus is not necessary if you worked Exercise 38. Just think how this changes the manufacturer's costs.]

b) Find, in terms of a, b, c, A, and B, the value of t that will maximize the government's return, assuming that the manufacturer maximizes the profit as in part (a).

40. The *probability* p of an event is a number in the interval $[0, 1]$ that measures how likely the event is to occur. For example, a sure event has probability 1, an impossible event has probability zero, and an event as likely to occur as not to occur has probability $\frac{1}{2}$.

Let p be the probability that a particular biased coin produces a head when flipped. We wish to estimate p. The coin is flipped n times, producing a certain sequence containing m heads and $n - m$ tails. It can be shown that if the probability of a head on one flip were x, then the probability of obtaining this particular ordered sequence of heads and tails would be $x^m(1 - x)^{n-m}$. Since this sequence of heads and tails did occur, it is natural to take as estimate for p the value of x in $[0, 1]$ that maximizes $x^m(1 - x)^{n-m}$. Find this estimate for p (the *maximum likelihood estimate*).

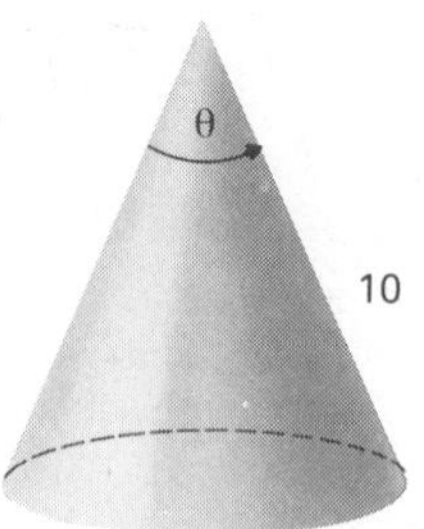

Figure 5.53

41. A fence a ft high is located b ft from the side of a house. Find the length of the shortest ladder that will reach over the fence to the house wall from the ground outside the fence.

42. A right circular cone has slant height 10 in. Find the vertex angle θ (see Fig. 5.53) for the cone of maximum volume.

43. A statue 12 ft high stands on a pedestal 41 ft high. How far from the base of the pedestal on level ground should an observer stand so that the angle θ at his eye subtended by the statue is maximum if the eye of the observer is 5 ft from the ground? [*Hint:* The maximum value of θ occurs when $\tan\theta$ is maximum.]

44. According to Fermat's principle in optics, light travels the path for which the time of travel is minimum. Let light travel with velocity v_1 in medium 1 and velocity v_2 in medium 2, and let the boundary between the media form a plane, as shown in cross section in Fig. 5.54. Show that, according to Fermat's principle, light that travels from A to B in Fig. 5.54 crosses the boundary at a point P such that

$$\frac{\sin\theta_1}{\sin\theta_2} = \frac{v_1}{v_2}.$$

(This is the *law of refraction* or *Snell's law*.)

Figure 5.54 Refraction of light traveling from A to B.

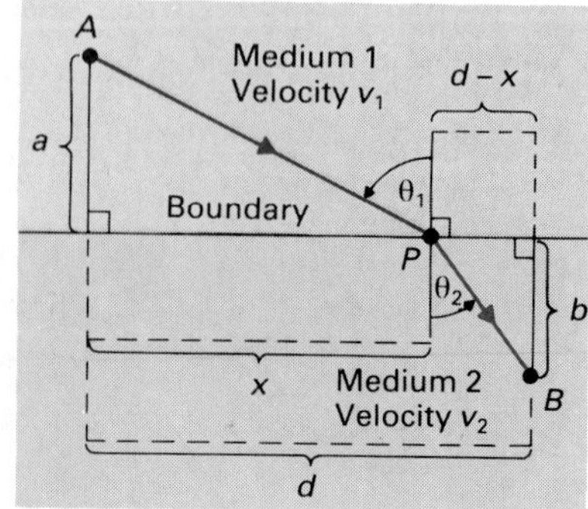

5.7 NEWTON'S METHOD

Mathematics is often used to determine some numerical quantity by finding an equation that the quantity satisfies and then solving the equation. All of us have worked many problems of this type. An equation in an unknown x can always be expressed in the form $f(x) = 0$ by moving everything to the left-hand side. In algebra, we spent a lot of time learning to solve simple polynomial equations, such as $x^2 - x + 6 = 0$. However, we did not learn to solve $x^5 + 7x^3 - 20 = 0$, which we should regard as quite a simple equation. We will use Newton's method to approximate the only real solution of this equation in Example 4 of this section.

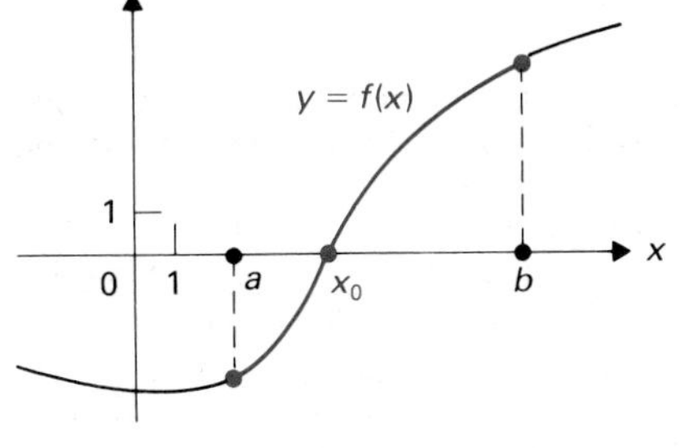

Figure 5.55 If f is continuous in $[a, b]$ and $f(a)$ and $f(b)$ have opposite sign, then $f(x_0) = 0$ for some x_0 between a and b.

Suppose we wish to solve $f(x) = 0$ for a *differentiable* function f. This section describes *Newton's method* for finding successive approximations of a solution.

First, we find a number a_1, which we believe is close to a solution. We might, for example, substitute a few values of x to see where $f(x)$ is close to zero or use a computer to plot the graph. The following corollary of the intermediate-value theorem is often useful. We asked you to prove this corollary to the theorem in Exercise 13 of Section 2.4 (page 76). Figure 5.55 illustrates the corollary.

COROLLARY To the intermediate-value theorem

If f is continuous at every point in $[a, b]$ and if $f(a)$ and $f(b)$ have opposite sign, then $f(x) = 0$ has a solution x_0 where $a < x_0 < b$.

EXAMPLE 1 Use the corollary just stated to locate a solution of $x^3 + x - 1 = 0$ between two consecutive integers.

Solution Let $f(x) = x^3 + x - 1$. Then $f(0) = -1$ and $f(1) = 1$, so by the corollary the equation $x^3 + x - 1 = 0$ has a solution where $0 < x < 1$. □

Figure 5.56 Newton's method for approximating a solution x_0 of $f(x) = 0$, starting with a_1 and finding a_2 and a_3.

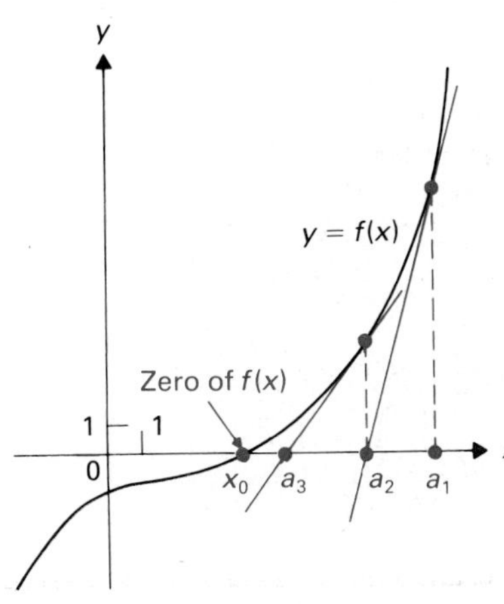

Suppose now we have found our approximate solution a_1 of the equation $f(x) = 0$. Look at the graph in Fig. 5.56. It appears that the tangent line to the graph of f at the point $(a_1, f(a_1))$ intersects the x-axis at a point a_2, which is a better approximation of a solution than a_1. Repeating this construction, starting with a_2, we expect to find a better approximation a_3, and so on.

Referring to Fig. 5.57, we can find a formula for the next approximation a_{i+1} if we know the approximation a_i. The tangent line to $y = f(x)$, where $x = a_i$, goes through the point $(a_i, f(a_i))$ and has slope $f'(a_i)$. Its equation is therefore

$$y - f(a_i) = f'(a_i)(x - a_i). \tag{1}$$

To find the point a_{i+1} where the line crosses the x-axis, set $y = 0$ in Eq. (1) and solve for x:

$$-f(a_i) = f'(a_i)(x - a_i),$$

$$x - a_i = -\frac{f(a_i)}{f'(a_i)}, \qquad x = a_i - \frac{f(a_i)}{f'(a_i)},$$

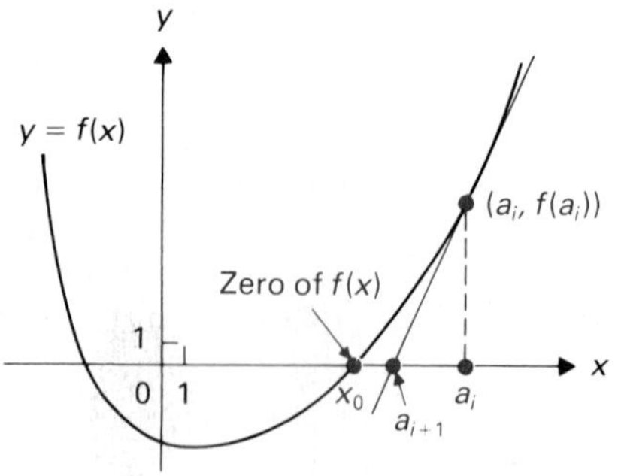

Figure 5.57 Newton's method: The tangent line at $(a_i, f(a_i))$ intersects the x-axis at a_{i+1}.

assuming $f'(a_i) \neq 0$. Thus we have the *Newton's method recursion formula:*

$$a_{i+1} = a_i - \frac{f(a_i)}{f'(a_i)}. \tag{2}$$

In mathematics, a *recursion formula* is one that allows us to compute the next in a sequence of values in terms of the values already found. In the case of Eq. (2), we compute the next approximation using only the last one found.

EXAMPLE 2 Use Newton's method to approximate $\sqrt{2}$ by approximating a solution of $f(x) = x^2 - 2 = 0$.

Solution We have $f'(x) = 2x$. The recursion formula (2) becomes

$$a_{i+1} = a_i - \frac{a_i^2 - 2}{2a_i}.$$

Table 5.1 shows successive approximations starting with $a_1 = 2$. These were easily found using a calculator. Only four iterations gave us at least six-significant-figure accuracy. □

A classical algorithm for computing the square root of a positive number is easily derived using Newton's method. Suppose we wish to compute $\sqrt{c}$ for $c > 0$. Let $f(x) = x^2 - c$. Then $f'(x) = 2x$. If a_i is an approximation for $\sqrt{c}$, then the recursion formula (2) becomes

$$a_{i+1} = a_i - \frac{a_i^2 - c}{2a_i} = a_i - \frac{a_i^2}{2a_i} + \frac{c}{2a_i} = a_i - \frac{a_i}{2} + \frac{c}{2a_i} = \frac{a_i}{2} + \frac{c}{2a_i}.$$

Thus

$$a_{i+1} = \frac{1}{2}\left(a_i + \frac{c}{a_i}\right). \tag{3}$$

It can be shown that the iterates given by Eq. (3) eventually approach $\sqrt{c}$ starting with *any* $a_1 > 0$. Of course, such a method for finding a root is obsolete in view of today's calculators.

EXAMPLE 3 Use a calculator and Eq. (3) to approximate $\sqrt{243}$, starting with $a_1 = 1$.

Table 5.1

i	a_i	a_{i+1}
1	2	$2 - \frac{2}{4} = 1.5$
2	1.5	$1.5 - \frac{0.25}{3} = 1.416666$
3	1.416666	$1.416666 - \frac{0.0069425}{2.833332} = 1.414215$
4	1.414215	$1.414215 - \frac{0.000004}{2.828430} = 1.414214$
5	1.414214	

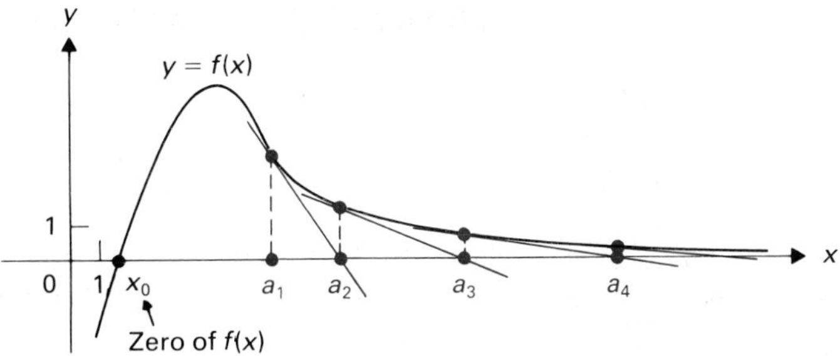

Figure 5.58 The iterates $a_1, a_2, a_3, a_4, \ldots$ approach ∞ rather than x_0.

Solution This choice for a_1 is of course ridiculous. We are just illustrating that Eq. (3) yields $\sqrt{c}$ for any $a_1 > 0$. Results appear in Table 5.2. Our approximations have stabilized to seven significant figures. □

Table 5.2

i	a_i	$a_{i+1} = \frac{1}{2}\left(a_i + \frac{243}{a_i}\right)$
1	1	122
2	122	61.99590
3	61.99590	32.95776
4	32.95776	20.16542
5	20.16542	16.10788
6	16.10788	15.59683
7	15.59683	15.58846
8	15.58846	15.58846
9	15.58846	

Newton's method will not always converge to a solution. For example, if we choose a_1 as shown in Fig. 5.58, the successive a_i approach ∞ rather than the solution of $f(x) = 0$. It is also possible for the a_i to oscillate without converging. Exercises 12 and 13 give examples of such oscillation, which is illustrated in Fig. 5.59.

Newton's method really amounts to repeated approximation by differentials. Recall that

$$dy = f'(a_i)\,dx.$$

If we set $dy = -f(a_i)$, which is the change we desire in y to make $f(x)$ zero, then

$$dx = \frac{-f(a_i)}{f'(a_i)}.$$

Thus we should change x from a_i to

$$a_i + dx = a_i - \frac{f(a_i)}{f'(a_i)},$$

which is precisely formula (2) for a_{i+1} in Newton's method.

It is only reasonable to use a calculator or computer these days to perform the computations involved in Newton's method. Just in case you don't have a calculator, Exercises 1 through 14 are designed to be feasible with pencil and paper. We now give three examples that really require a calculator for solution.

EXAMPLE 4 Using a calculator, find a root of $x^5 + 7x^3 - 20 = 0$. (Our calculator is programmable, which is handy, but by no means essential, for this computation.)

Figure 5.59 The iterates $a_1, a_2, a_3, a_4, \ldots$ oscillate and do not approach x_0.

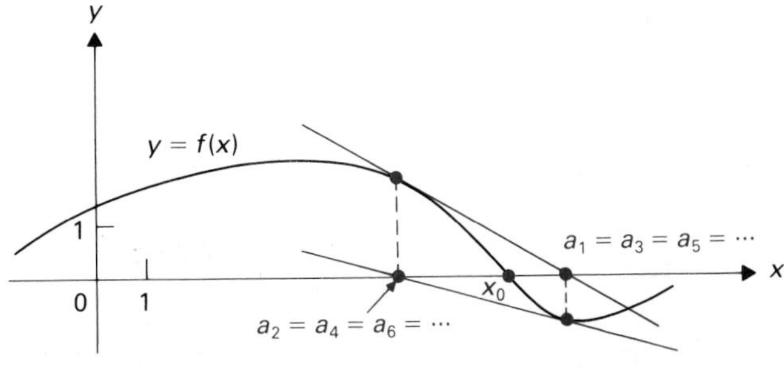

Table 5.3

i	a_i	$f(a_i)$
1	1	−12
2	1.461538462	8.522746187
3	1.335597387	0.9271519711
4	1.318225329	0.0155248157
5	1.317924405	0.0000045805
6	1.317924316	1×10^{-11}
7	1.317924316	1×10^{-11}

Solution Let $f(x) = x^5 + 7x^3 - 20$. Clearly, $f(x) \leq 0$ for $x \leq 1$, while $f(x) \geq 0$ for $x \geq 2$. Since $f(1) = -12$ is closer to zero than $f(2) = 68$, we start with $a_1 = 1$. The recursion formula (2) becomes

$$a_{i+1} = a_i - \frac{a_i^5 + 7a_i^3 - 20}{5a_i^4 + 21a_i^2}.$$

Our calculator yields results given in Table 5.3. We have computed $f(a_i)$ also. Examples can be found where the approximations stabilize to many significant figures without yielding a solution of $f(x) = 0$. (See Exercise 14.) Thus it is a good idea to compute $f(a_i)$ to check that we are indeed converging to a solution. From the data in Table 5.3, we are confident that 1.317924316 is indeed close to a solution. □

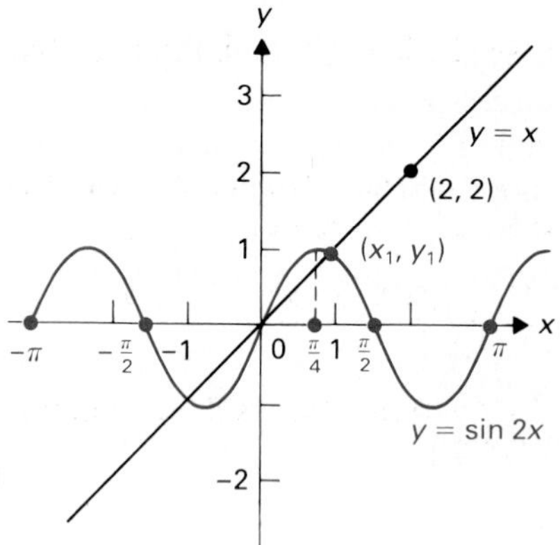

Figure 5.60 $\sin 2x = x$ for some $x_1 > 0$.

EXAMPLE 5 Note that the slope of the graph of $y = x$ is 1 at the origin, while the slope of the graph of $y = \sin 2x$ is 2 at the origin. Figure 5.60 then shows that we must have a solution of $\sin 2x = x$ for $x > 0$. Find this solution using Newton's method.

Solution From the graph in Fig. 5.60, we see that $\sin 2x = x$ for some x where $\pi/4 < x < \pi/2$. We use Newton's method with $f(x) = \sin 2x - x$ and $a_1 = 1$. The recursion relation (2) becomes

$$a_{i+1} = a_i - \frac{\sin(2a_i) - a_i}{2\cos(2a_i) - 1}.$$

Our calculator gave the data in Table 5.4. We are convinced by the table that $\sin 2x = x$ for $x \approx 0.9477471335$. □

EXAMPLE 6 Find a local extremum of $f(x) = x^6 + 3x^5 + 2x^4 + 7x^2 - 10x$ and classify it as a local maximum or local minimum.

Solution We have

$$f'(x) = 6x^5 + 15x^4 + 8x^3 + 14x - 10$$

and

$$f''(x) = 30x^4 + 60x^3 + 24x^2 + 14.$$

We wish to find a solution of $f'(x) = 0$ and check the sign of $f''(x)$ at that solution. Since $f'(0) < 0$ and $f'(1) > 0$, we see that $f'(x) = 0$ for some value of x where $0 < x < 1$. We use Newton's method with $a_1 = 0$. The recursion relation (2) is

$$a_{i+1} = a_i - \frac{6a_i^5 + 15a_i^4 + 8a_i^3 + 14a_i - 10}{30a_i^4 + 60a_i^3 + 24a_i^2 + 14}.$$

Table 5.4

i	a_i	$f(a_i)$
1	1	−0.0907025732
2	0.9504977971	−0.0045200436
3	0.9477558227	−0.0000142334
4	0.9477471336	−0.0000000001
5	0.9477471335	-5×10^{-13}
6	0.9477471335	-2×10^{-13}

Our programmable calculator provided the data in Table 5.5. Our calculator shows that $f''(a_7) \approx 31.90811905$, so we see that $f(x)$ must have a local minimum where $x \approx 0.5285768289$. □

In using Newton's method, we may compute $f'(a_i)$ numerically, using m_{sym}. Of course this can be done on a calculator. Since numerical differentiation can be very inaccurate, we should *always* check $f(a_i)$ to be sure we are indeed approaching a solution of $f(x) = 0$. We give one example that we worked out on our programmable calculator.

Table 5.5

i	a_i	$f'(a_i)$
1	0	−10
2	0.7142857143	7.935681561
3	0.5723744691	1.491926152
4	0.5313126087	0.0876442105
5	0.5285877659	0.0003489823
6	0.5285768291	0.0000000056
7	0.5285768289	0.0

Table 5.6

i	a_i	$f(a_i)$
1	4	156
2	3.744635441	40.3491215
3	3.620734188	5.500400569
4	3.597934159	0.1481299455
5	3.597285529	0.00011535
6	3.597285024	0.0000000004
7	3.597285024	0.0000000004

EXAMPLE 7 Using a calculator and Newton's method, find x such that $x^x = 100$.

Solution Let $f(x) = x^x - 100$. Since $f(3) < 0$ and $f(4) > 0$, there must be a solution x such that $3 < x < 4$. We use Newton's method with

$$f'(a_i) \approx m_{\text{sym}} = \frac{f(a_i + \Delta x) - f(a_i - \Delta x)}{2(\Delta x)},$$

where $\Delta x = 0.00001$. Starting with $a_1 = 4$, we obtained the results in Table 5.6. Thus we see that $x^x = 100$ for $x \approx 3.597285024$. □

SUMMARY

Newton's method to solve $f(x) = 0$: First decide on an approximate solution a_1 of $f(x) = 0$, and then determine successive approximations $a_2, a_3, a_4, \ldots$ of a solution using the recursion formula

$$a_{i+1} = a_i - \frac{f(a_i)}{f'(a_i)}.$$

EXERCISES

It is only reasonable to use a calculator these days in computations involved with Newton's method. However, Exercises 1 through 14 have been devised to be feasible with just pencil and paper.

In Exercises 1 through 4, use Newton's method to estimate the given square root, starting with the given value for a_1 and finding a_3. (See Eq. 3 in the text.)

1. $\sqrt{3}$, $a_1 = 2$
2. $\sqrt{8}$, $a_1 = 3$
3. $\sqrt{6}$, $a_1 = 3$
4. $\sqrt{12}$, $a_1 = 4$
5. Use Newton's method to estimate $\sqrt{7}$ until the difference between successive approximations is less than 0.0005.
6. Repeat Exercise 5, estimating $\sqrt{5}$.
7. Use Newton's method to estimate $\sqrt[3]{6}$, starting with $a_1 = 2$ and finding a_3.
8. Use Newton's method to estimate $\sqrt[3]{9}$, starting with $a_1 = 2$ and finding a_3.
9. Use Newton's method to estimate a solution of $x^3 + x - 1 = 0$, starting with $a_1 = 1$ and finding a_3.
10. Use Newton's method to estimate a solution of $x^3 + 2x - 1 = 0$, starting with $a_1 = 1$ and finding a_3.
11. Use Newton's method to estimate a solution of $x^3 + x^2 - 1 = 0$, starting with $a_1 = 1$ and finding a_3.
12. a) Show that $x^4 + 4x^3 + 4x^2 - x - 1 = 0$ has a solution in $[-1, 0]$.

b) Attempt to find a solution of the equation in part (a) starting with $a_1 = 0$ and using Newton's method. What happens?

13. This exercise shows how oscillatory examples like Exercise 12 can be constructed.

a) Show graphically that if $f(x)$ is differentiable with $f(0) = 1$, $f(1) = -1$, and $f'(0) = f'(1) = -1$, then $f(x) = 0$ has a solution in $[0, 1]$, but Newton's method starting with either $a_1 = 0$ or $a_1 = 1$ will lead to oscillation.

b) Construct a cubic $f(x) = ax^3 + bx^2 + cx + d$ having the properties in part (a). Follow these steps.

i) Find the equation involving a, b, c, d corresponding to $f(0) = 1$.

ii) Repeat part (i) for the requirement $f(1) = -1$.

iii) Repeat part (i) for the requirement $f'(0) = -1$.

iv) Repeat part (i) for the requirement $f'(1) = -1$.

v) Solve the system of four equations in a, b, c, d given by parts (i) through (iv) to find the desired cubic $f(x)$.

14. This exercise shows that it is possible to have the values $a_1, a_2, \ldots, a_n$ generated using Newton's method stabilize to many significant figures without approaching a solution of $f(x) = 0$. Checking values $f(a_i)$ as in Example 4 would guard against reaching a false conclusion in such a case.

a) Let $a_1 > a_2 > \cdots > a_n$ be any n distinct numbers. Let $c_1, c_2, \ldots, c_n$ be any n numbers all greater than 5, and let $d_1, d_2, \ldots, d_n$ be any n numbers. Argue graphically that there exists a differentiable function $f(x)$ such that $f(x) \geq 3$ for all x, while $f(a_i) = c_i$ and $f'(a_i) = d_i$ for $i = 1, 2, \ldots, n$.

b) By part (a), there exists a differentiable function $f(x)$ where $f(x) \geq 3$ for all x and such that $f(1) = f(1.1) = f(1.01) = f(1.001) = \cdots = f(1.00000001) = 10$, while $f'(1.1) = 10/0.09$, $f'(1.01) = 10/0.009$, $f'(0.001) = 10/0.0009, \ldots,$ $f'(1.0000001) = 10/0.00000009$. If Newton's method is used to solve $f(x) = 0$ starting with $a_1 = 1.1$, find $a_1, a_2, \ldots, a_8$. Are significant figures stabilizing? Are you approaching a solution of $f(x) = 0$?

Use a calculator in Exercises 15 through 20. These exercises do not depend on our previous work with calculators in calculus.

15. Use Newton's method to estimate a solution of $f(x) = x^3 + x + 16 = 0$ until the difference between successive approximations is less than 0.000005.

16. Use Newton's method to estimate $\sqrt{17}$, a solution of $x^2 - 17 = 0$, starting with $a_1 = 4$.

17. Use Newton's method to estimate $\sqrt[3]{25}$, a solution of $x^3 - 25 = 0$, starting with $a_1 = 3$.

18. Use Newton's method to estimate a solution of $x - 2 \sin x = 0$, starting with $a_1 = 2$.

19. Argue graphically that $\cos x - x = 0$ has a unique solution. Estimate it using Newton's method.

20. Use Newton's method to estimate the positive solution of $\cos x - x^2 = 0$.

In Exercises 21 through 26, use a calculator or computer and Newton's method to estimate a solution of the given equation. Compute the necessary derivatives using m_{sym}.

21. $2^x = 9$

22. $3^x = 100$

23. $x^x = 5$

24. $x^x = 3000$

25. $2^x + 3^x = 50$

26. $5^x + x^x = 500$

5.8 CALCULUS IN ECONOMICS AND BUSINESS

At many schools, undergraduate majors in economics who intend to do graduate work are encouraged to take many courses from the mathematics department, including three semesters of calculus. Calculus is an important tool in economic theory.

We will talk about a very simplified economic situation. Suppose a company is manufacturing some product in a very ideal situation where there is no competition. This gives the company reasonable control over its own destiny. We further assume that the cost of producing an item, the revenue received from its sale, and the profit made are functions of the number x of units of the product manufactured per unit of time (a month, or a year, and so on). This is a major assumption. It means that advertising decisions, transportation to market, and the like are all functions of this number x of units produced.

In many economic situations, we are interested only in *integer* values of a variable x. After all, a construction company is not going to build 31.347 houses! While some of our functions, such as cost, may be defined for only integer values of x, we will assume that there is some differentiable function $C(x)$ defined throughout a whole interval and giving the cost for all integer values of x in the interval. With this in mind, let

$$C(x) = \text{cost of producing } x \text{ units, per unit time,}$$

$$R(x) = \text{revenue received if } x \text{ units are produced,}$$

$$P(x) = R(x) - C(x) = \text{profit when } x \text{ units are produced.}$$

Economists are interested in the *marginal cost*, *marginal revenue*, and *marginal profit* when x units are produced. In a low-level economics course where calculus is not used, the marginal cost at x is defined to be the cost of producing one additional unit in the time period, so that

$$\text{Marginal cost} = C(x + 1) - C(x) = \frac{C(x + 1) - C(x)}{1}. \qquad \textbf{(1)}$$

From Eq. (1), we see that this marginal cost at x is approximately $C'(x)$, because Eq. (1) gives the approximation

$$C'(x) \approx \frac{C(x + \Delta x) - C(x)}{\Delta x},$$

where $\Delta x = 1$. In a higher-level course, the adjective *marginal* generally signifies a derivative.

DEFINITION 5.5 Marginal quantities

The **marginal cost** is dC/dx, the **marginal revenue** is dR/dx, and the **marginal profit** is dP/dx.

EXAMPLE 1 A company manufactures a popular calculator. Its annual cost function in dollars is

$$C(x) = 90{,}000 + 500x + 0.01x^2,$$

where x is in hundreds of calculators produced per year. The \$90,000 represents the annual capital outlay for the plant, insurance, and such fixed expenses. The coefficient \$500 might represent the cost, exclusive of fixed expenses, of producing 100 calculators if not too many are produced. The term $0.01x^2$ comes into play only as x becomes fairly large and might represent problems caused by crowding, storing excessive inventory, and an increase in the cost of materials if many calculators are produced, so material becomes scarce and its cost increases. Suppose the revenue function is

$$R(x) = 1000x - 0.05x^2.$$

Here $1000x$ appears because the first few calculators produced sell for \$10 each. The term $-0.05x^2$ appears because if x is large there is a glut on the market, so the sales price falls. Find the marginal profit and find how many calculators the company should manufacture for maximum profit.

Solution We have

$$\begin{aligned} P(x) &= R(x) - C(x) \\ &= (1000x - 0.05x^2) - (90{,}000 + 500x + 0.01x^2) \\ &= -90{,}000 + 500x - 0.06x^2. \end{aligned}$$

Then

$$\text{Marginal profit} = \frac{dP}{dx} = 500 - 0.12x.$$

The profit is maximum when $dP/dx = 0$, or when

$$0.12x = 500 \qquad \text{or} \qquad x = \frac{50{,}000}{12} = 4{,}166.66666\ldots.$$

This gives a maximum since $d^2P/dx^2 = -0.12 < 0$. Since x represents hundreds of calculators, we obtain 416,666.66666... calculators. Of course the company isn't going to make $\frac{2}{3}$ of a calculator. A computation on a calculator gives

$$P(4166.66) = P(4166.67) = 951{,}666.6667.$$

From a practical point of view, the company makes the same profit manufacturing 416,666 calculators as manufacturing 416,667 calculators. This profit is \$951,666.67. [Mathematically, $P(4166.67) > P(4166.66)$ by symmetry of the parabola $P(x)$ about its high point where $x = 4166.666666\ldots$.] □

Since $C(x)$ is the cost of producing x units, then

$$\text{Average cost per unit} = \frac{C(x)}{x}. \tag{2}$$

It is interesting to know how many units should be produced to minimize the average cost. Differentiating to minimize $C(x)/x$, we have

$$\frac{d}{dx}\left(\frac{C(x)}{x}\right) = \frac{xC'(x) - C(x)}{x^2}.$$

This is zero when

$$xC'(x) - C(x) = 0 \qquad \text{or} \qquad C'(x) = \frac{C(x)}{x}.$$

Thus *minimum average cost occurs when marginal cost equals average cost.* Of course we should check to be sure we have a minimum and not a local maximum.

EXAMPLE 2 Find how many units should be produced to minimize average cost for the company manufacturing calculators described in Example 1.

Solution The cost function is

$$C(x) = 90{,}000 + 500x + 0.01x^2.$$

Thus $C'(x) = 500 + 0.02x$. Setting $C'(x) = C(x)/x$, we obtain

$$500 + 0.02x = \frac{90{,}000 + 500x + 0.01x^2}{x},$$

or

$$500x + 0.02x^2 = 90{,}000 + 500x + 0.01x^2.$$

Then

$$0.01x^2 = 90{,}000, \qquad \text{so} \quad x^2 = 9{,}000{,}000$$
$$x = \sqrt{9{,}000{,}000} = 3000.$$

Since $\lim_{x\to 0+} C(x)/x = \lim_{x\to\infty} C(x)/x = \infty$, there is a minimum somewhere, and the one candidate we found must be this minimum. This minimum average cost is achieved when 300,000 calculators per year are produced. The minimum average cost is $C(3000)/3000 = \$560$ per hundred calculators, or \$5.60 per calculator. □

We conclude by giving two examples illustrating calculus applied to business problems.

EXAMPLE 3 A large home and auto store sells 9000 auto tires each year. It has *carrying costs* of 50¢ per year for each unsold tire stored (space, insurance, and so on.) The *reorder cost* for each lot ordered is 25¢ per tire plus \$10 for the paperwork of the order. Find how many times per year the store should reorder, and the size of the order, to minimize the sum of these *inventory costs.*

Solution Let x be the lot size of each order, that is, the number of tires ordered at a time. Then

$$\text{Number of orders per year} = \frac{9000}{x}.$$

The average number of unsold tires over the year is then $x/2$. Therefore

$$\text{Carrying costs per year} = 0.5\left(\frac{x}{2}\right) \text{ dollars}$$

and

$$\text{Reorder costs per year} = 10\left(\frac{9000}{x}\right) + 0.25(9000) \text{ dollars.}$$

Consequently, the total inventory costs per year are

$$C(x) = 0.5\left(\frac{x}{2}\right) + 10\left(\frac{9000}{x}\right) + 0.25(9000)$$
$$= \frac{1}{4}x + 90{,}000x^{-1} + 2250 \text{ dollars.}$$

Then

$$C'(x) = \frac{1}{4} - 90{,}000x^{-2} = \frac{1}{4} - \frac{90{,}000}{x^2}.$$

We easily see that $C'(x) = 0$ when $x^2 = 360{,}000$, or when $x = \sqrt{360{,}000} = 600$ tires. Since $C''(x) = 180{,}000x^{-3} > 0$, we do have a minimum. Therefore the tires should be ordered in lots of 600, and there should be $9000/600 = 15$ such orders each year. □

EXAMPLE 4 A fisherman has exclusive rights to a stretch of clam flats. Suppose that conditions are such that on his flats a population of p bushels of clams this year yields a population of $f(p) = 50p - \frac{1}{4}p^2$ bushels next year. (The term $-\frac{1}{4}p^2$ represents a decrease in population due to overcrowding if p is large.) How many bushels should the fisherman dig each year to sustain a maximum harvest year after year?

Solution The fisherman can dig (harvest) $h(p) = f(p) - p$ bushels of clams over the year without depleting the original population of p bushels. He wishes to maximize $h(p)$. We have

$$h(p) = f(p) - p = (50p - \tfrac{1}{4}p^2) - p = 49p - \tfrac{1}{4}p^2.$$

Then

$$h'(p) = 49 - \tfrac{1}{2}p,$$

and $h'(p) = 0$ when $p = 98$. Since $h''(p) = -\frac{1}{2} < 0$, we do have a maximum. Thus the fisherman should first adjust the population of clams to 98 bushels, either by letting them increase without digging or by digging the excess population. Then he can dig

$$h(98) = 49(98) - \tfrac{1}{4}(98)^2 = 2401 \text{ bushels}$$

each year, or about 46 bushels per week, year after year. □

SUMMARY

1. If $C(x)$ is the cost, $R(x)$ the revenue, and $P(x) = R(x) - C(x)$ the profit from producing x units of a product per unit of time, then the marginal cost is dC/dx, the marginal revenue is dR/dx, and the marginal profit is dP/dx.
2. The average cost is $C(x)/x$. Average revenue and average profit are similarly defined.
3. For average cost to be minimum, we must have

$$\text{Marginal cost} = \text{Average cost}.$$

EXERCISES

1. For the company manufacturing calculators described in Example 1, find the following when 2000 units have been manufactured.
   a) the marginal cost
   b) the marginal revenue
   c) the marginal profit
   d) the average profit
2. For the company manufacturing calculators described in Example 1, find how many units should be manufactured for maximum average profit, and find the average profit then.
3. A small company manufactures wood stoves. Its cost $C(x)$ and revenue $R(x)$ when x stoves per year are manufactured are given by

$$C(x) = 10{,}000 + 150x + 0.03x^2,$$
$$R(x) = 250x - 0.02x^2.$$

   a) Find the marginal profit when $x = 100$.
   b) How many stoves should the company make for maximum profit? What is that maximum profit?
4. For the company in Exercise 3, find the number of stoves manufactured for the minimum average cost and for the maximum average profit.

5. A supplier of stove wood finds that with x cords of wood stocked per year, the cost and revenue are

$$C(x) = 500 + 50x + 0.02x^2,$$

$$R(x) = 80x - 0.01x^2.$$

a) Find the marginal revenue when $x = 100$ cords.
b) Find the average revenue when $x = 100$ cords.
c) Find the value of x that maximizes the profit.

6. For the supplier in Exercise 5, find the number of cords of wood for the maximum average profit.

7. Assuming the supplier in Exercise 5 sells only whole cords of wood, find the number of cords sold to produce the maximum average revenue by
a) using calculus blindly,
b) using common sense.
Explain why the calculus method does not work.

8. Show that at any value $x > 0$ where the average profit achieves a maximum (or minimum) greater than zero, the marginal profit must equal the average profit.

9. Give an intuitive argument, using no calculus, that the average cost can be minimum at $x > 0$ only when it equals the marginal cost.

The amount a family saves is a function $S(I)$ of its income I, and the amount it consumes is a function $C(I)$. The *marginal propensity to save* is dS/dI, and the *marginal propensity to consume* is dC/dI. Exercises 10 through 12 deal with this situation.

10. Show that if the entire income is used to either save or consume, then

$$\text{Marginal propensity to consume} = 1 - \text{Marginal propensity to save.}$$

11. The yearly savings of a family in terms of its annual income I is given by

$$S(I) = \frac{2I^2}{5(I + 60{,}000)}.$$

a) Find the marginal propensity to save when $I = \$30{,}000$.
b) If the family consumes all the dollars it does not save, find the marginal propensity to consume when $I = \$30{,}000$.

12. Suppose the annual income of the family in Exercise 11 is extremely large. Find the limiting proportion of income saved as $I \to \infty$ by
a) computing the marginal propensity to save and examining it as $I \to \infty$,
b) examining the behavior of $S(I)$ as $I \to \infty$, as you learned to do in Chapter 2.
c) Would a problem that gave

$$S(I) = \frac{2I^3}{5(I + 60{,}000)}$$

be realistic?

d) Would it be possible to have

$$S(I) = \frac{4I^2}{5(I + 60{,}000)}$$

in the United States?

13. The taxes T paid by a family in a certain country are given as a function of the income I by

$$T(I) = \frac{3I^2}{80{,}000 + 5I}.$$

a) Find the marginal tax rate when $I = 10{,}000$.
b) Describe the minimum and maximum average taxes per unit of income, if they exist.

14. In Exercise 13, find the limiting proportion of the income paid in taxes as $I \to \infty$ by
a) finding the marginal tax as $I \to \infty$,
b) examining the behavior of $T(I)$ as $I \to \infty$, as you learned to do in Chapter 2.

15. An appliance store sells 200 refrigerators each year. It has carrying costs of \$10 per year for each unsold refrigerator stored. The reorder cost for each lot ordered is \$5 per refrigerator plus \$10 for the paperwork of the order. How many times per year should the store reorder, and at what lot size, to minimize these inventory costs?

16. A record store sells 100,000 records each year. It has carrying costs of 25¢ per year for each unsold record, reorder costs averaging 5¢ per record plus \$5 for the paperwork of the order. How many times a year should the store reorder, and in what lot size, to minimize these inventory costs?

17. A college student plans to rent space in a shopping mall during the week before Christmas and sell 2000 Christmas trees. The supplier of trees can either deliver all 2000 trees at once or deliver them in equal lots, but there is a \$40 charge over the cost of the trees for each delivery. The student figures it will cost \$2 per tree on display to rent enough space in the mall for the enterprise. Find the lot size and the number of lots of trees the student should order to minimize these costs.

18. In Example 4 in this section, find the population of clams that will remain constant year after year if left undisturbed.

19. A population of p wild rabbits on a certain farm gives rise to a population of

$$f(p) = 6p - (\tfrac{1}{12})p^2$$

next year if left undisturbed.
a) What population will remain constant year after year if left undisturbed?
b) What is the maximum sustainable number that can be trapped year after year? What is the corresponding population size?

20. A gardener finds that a population of p acceptable-sized Jerusalem artichoke roots in the patch will give rise to a population of $5p - (1/100)p^2$ acceptable-sized roots next year if left undisturbed.
a) What population of acceptable-sized roots will remain constant year after year if left undisturbed?
b) What is the maximum sustainable number of acceptable-sized roots that can be dug year after year? What is the corresponding population size?

5.9 ANTIDERIVATIVES AND DIFFERENTIAL EQUATIONS

ANTIDERIVATIVES

We have spent a lot of time differentiating, that is, finding $f'(x)$ if $f(x)$ is known. Of equal importance is *antidifferentiating*, which is finding $f(x)$ if $f'(x)$ is known. We call $f(x)$ an **antiderivative** of $f'(x)$. Here is one indication of the importance of such a computation. We have seen that if we know the position x of a body on a line at time t, then $dx/dt = v$ is the velocity and $d^2x/dt^2 = a$ is the acceleration at time t. But, in practice, we often know the position and velocity only at the initial time $t = 0$, while we know the acceleration at every time $t \geq 0$. This is because we are often applying a controlled force F to produce the motion, perhaps by controlling the thrust of some motor. By Newton's second law of motion, $F = ma$, where m is the mass of the body. Thus if we know the force F, we know the acceleration $a = F/m$. Antidifferentiation of the acceleration will then find the velocity, and antidifferentiation of the velocity will find the position of the body. We will find the height of a freely falling body in a vacuum, subject only to gravitational acceleration, in Example 10.

We change notation a bit and let $f(x)$ always be the known function and $F(x)$ the desired antiderivative, so that

$$F'(x) = f(x).$$

EXAMPLE 1 Find an antiderivative $F(x)$ of $f(x) = x^2$.

Solution Since differentiation drops the exponent of a monomial by 1, we naturally try x^3 as our antiderivative of x^2. But $d(x^3)/dx = 3x^2$. Thus we see that we should take $F(x) = \frac{1}{3}x^3$ as an antiderivative. Note that $(x^3/3) + 2$ is also an antiderivative of x^2, and indeed, if C is any constant, then

$$F(x) = \frac{x^3}{3} + C$$

is an antiderivative of x^2. □

As illustrated by Example 1, if $F(x)$ is an antiderivative of $f(x)$, then $F(x) + C$ is also an antiderivative for any constant C. In this context, C is called an *arbitrary constant*. The mean-value theorem can be used to show that *all* antiderivatives of $f(x)$ are of the form $F(x) + C$; there are no others. We show this in two steps.

First, we show that if F is differentiable in $[a, b]$ and $F'(x) = 0$ for all x in $[a, b]$, then $F(x) = F(a)$ for all x in $[a, b]$. By the mean-value theorem applied to $[a, x]$ for any x such that $a < x \leq b$,

$$\frac{F(x) - F(a)}{x - a} = F'(c) = 0,$$

where $a < c < x$. Thus $F(x) - F(a) = 0$, so $F(x) = F(a)$, and $F(x)$ is constant in $[a, b]$.

As the second step, we turn to the general case where $F'(x) = f(x)$ for all x in $[a, b]$. Suppose $G'(x) = f(x)$ also for all x in $[a, b]$. Then

$$\frac{d(G(x) - F(x))}{dx} = G'(x) - F'(x) = 0.$$

By the last paragraph, $G(x) - F(x) = G(a) - F(a)$. If we let $C = G(a) - F(a)$, then $G(x) = F(x) + C$. We summarize this work in a theorem.

THEOREM 5.7 Antiderivatives of $f(x)$

If $F'(x) = f(x)$, then all antiderivatives of $f(x)$ are of the form $F(x) + C$ for some constant C.

EXAMPLE 2 Find the general antiderivative of x^2.

Solution The work in Example 1 shows that the general antiderivative of x^2 is $\frac{1}{3}x^3 + C$. □

EXAMPLE 3 Find the general antiderivative of x^n for $n \neq -1$.

Solution As in Example 1, our first attempt to find an antiderivative would be to try x^{n+1}. But $d(x^{n+1}) = (n + 1)x^n$. Thus we should take $x^{n+1}/(n + 1)$ as our trial antiderivative. We see at once that

$$\frac{d\left(\frac{1}{n+1}x^{n+1}\right)}{dx} = x^n.$$

Thus the general antiderivative of x^n is

$$F(x) = \frac{1}{n+1}x^{n+1} + C \qquad \text{for} \quad n \neq -1.$$

The case $n = -1$, which causes the denominator $n + 1$ to become zero, will be treated in Chapter 8. There we will find an antiderivative of $x^{-1} = 1/x$. However, this antiderivative of $1/x$ is not any of the functions we have discussed so far. □

Suppose that f and g are functions with the same domain. If F is an antiderivative of f and G is an antiderivative of g, then it is trivial to check that $F + G$ is an antiderivative of $f + g$ and that cF is an antiderivative of cf. One need only differentiate $F + G$ to obtain

$$\frac{d(F(x) + G(x))}{dx} = \frac{d(F(x))}{dx} + \frac{d(G(x))}{dx} = f(x) + g(x)$$

and differentiate cF to obtain

$$\frac{d(c \cdot F(x))}{dx} = c \cdot \frac{d(F(x))}{dx} = c \cdot f(x).$$

EXAMPLE 4 Use Example 3 and the remarks following the example to find the general antiderivative of a general polynomial function

$$f(x) = a_n x^n + \cdots + a_1 x + a_0.$$

Solution We see at once that the general antiderivative is

$$F(x) = a_n\left(\frac{x^{n+1}}{n+1}\right) + \cdots + a_1\frac{x^2}{2} + a_0x + C.$$

For example, the general antiderivative of $3x^2 + 4x + 7$ is

$$3\left(\frac{x^3}{3}\right) + 4\left(\frac{x^2}{2}\right) + 7x + C = x^3 + 2x^2 + 7x + C. \quad \square$$

EXAMPLE 5 Find the general antiderivative of $f(x) = 2/x^2 + 4/x^3$.

Solution We write $f(x) = 2x^{-2} + 4x^{-3}$. Example 3 and the remarks following it show that the general antiderivative is

$$F(x) = 2\frac{x^{-1}}{-1} + 4\frac{x^{-2}}{-2} + C = \frac{-2}{x} - \frac{2}{x^2} + C. \quad \square$$

EXAMPLE 6 Find the general antiderivatives of $f(x) = \sin ax$ and $g(x) = \cos ax$.

Solution Since the derivative of $\cos x$ is $-\sin x$, we try $-\cos ax$ as an antiderivative of $\sin ax$. However, $d(-\cos ax)/dx = a \sin ax$. We see at once that the general antiderivative of $f(x) = \sin ax$ is

$$F(x) = -\frac{1}{a}\cos ax + C.$$

Similarly, we find the general antiderivative of $g(x) = \cos ax$ is

$$G(x) = \frac{1}{a}\sin ax + C. \quad \square$$

DIFFERENTIAL EQUATIONS

An equation involving a derivative of an "unknown" function that we want to find is a differential equation. Examples of differential equations, written in Leibniz notation, are

$$\frac{dy}{dx} = \sin x, \qquad \frac{d^2y}{dx^2} - 2\frac{dy}{dx} + y = x^2,$$

$$\frac{d^3y}{dx^3} + (\sin x)\left(\frac{dy}{dx}\right)^2 - y^3 = 0.$$

To solve such a differential equation means to determine all functions $y = f(x)$ that will satisfy the equation. That is, y is regarded as an unknown function of x in the equation. Example 6 shows that $y = -\cos x + C$, where C may be any constant, describes all solutions of the first equation given above, because to solve $dy/dx = \sin x$ clearly amounts to finding all antiderivatives of $\sin x$. In Chapter 20, we will see how to solve the second equation given above. We will not learn to solve the third equation above, but clearly one solution is $y = 0$. A differential equation can be expected to have an infinite number of solutions, as our solution $y = -\cos x + C$ of $dy/dx = \sin x$ illustrates.

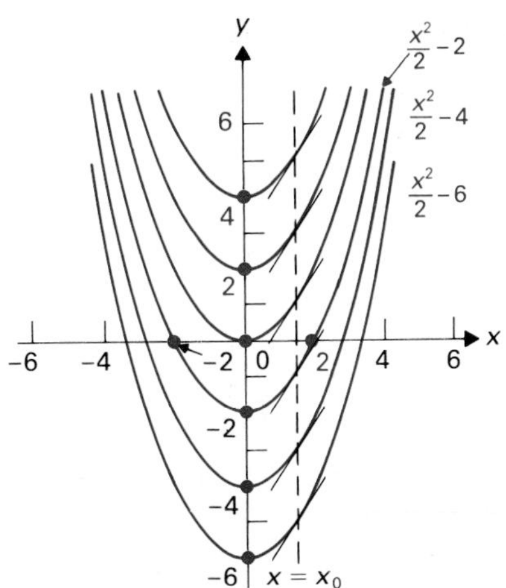

Figure 5.61 Some solutions $x^2/2 + C$ of the differential equation $dy/dx = x$.

We want to find a function $y = F(x)$ that is a solution of the differential equation

$$\frac{dy}{dx} = f(x).$$

If f has an antiderivative, that is, if the differential equation $dy/dx = f(x)$ has a solution, then the general antiderivative of f is the **general solution of this differential equation.**

We can characterize the general solution $F(x) + C$ of the differential equation $dy/dx = f(x)$ geometrically. If $G(x) = F(x) + k$, then the graph of G is simply the graph of F displaced $|k|$ units "upward" if $k > 0$, or $|k|$ units "downward" if $k < 0$. Thus the set $F(x) + C$ of functions may be visualized geometrically as a collection of graphs with the property that any one of them is "congruent" to any other and may be transformed into the other by sliding up or down.

EXAMPLE 7 Sketch some graphs of solutions of the differential equation $dy/dx = x$.

Solution Figure 5.61 shows some of these solutions. All these functions are antiderivatives of $f(x) = x$ and are of the form $(x^2/2) + C$. For antiderivatives F and G of f, the fact that $F'(x_0) = G'(x_0) = f(x_0)$ for each x_0 means that the graphs of F and G have the same slope (that is, parallel tangent lines) over x_0, as illustrated in Fig. 5.61. □

Consider an object moving on a line, which we assume to be the y-axis. As we mentioned at the start of this section, we might know the acceleration of the object, for the acceleration may be caused by some force that we can control. Such motion gives rise to a differential equation of the form

$$\frac{d^2y}{dt^2} = g(t),$$

where $g(t)$ is the acceleration at time t. We want to find the position y of our moving object as a function of t, that is, to find the solution of the differential equation giving the position for this object. A moment's thought shows that this should be physically determined if we know the position where the object started and the velocity with which it started. Suppose that at time $t = 0$, the *initial position* was y_0 and the *initial velocity* was v_0. We refer to these two requirements as *initial conditions* $y(0) = y_0$ and $y'(0) = v_0$ for the desired solution of $d^2y/dt^2 = g(t)$.

The adjective *initial* was undoubtedly first used to describe conditions at time $t = 0$. However, mathematicians currently use the term *initial conditions* with any differential equation to describe a specification of values of the unknown function or its derivatives at any *single value* of the independent variable. Thus we might ask for the solution of

$$\frac{d^2y}{dx^2} - \frac{dy}{dx} - 6y = x$$

that satisfies the *initial conditions* $y(3) = -4$ and $y'(3) = 5$. We will learn how to solve this problem in Chapter 20. Note that both these conditions describe behavior of the solution at the *same point*, $x = 3$. This is characteristic of *initial conditions*. Conditions that describe behavior at *different points* are referred to as *boundary conditions*.

If f is defined at x_0 and a solution F of $dy/dx = f(x)$ exists, then for any y_0 we can find C such that the solution $G(x) = F(x) + C$ satisfies the initial condition expressed by

$$G(x_0) = y_0.$$

Namely, if we want to have

$$y_0 = G(x_0) = F(x_0) + C,$$

we simply take $C = y_0 - F(x_0)$.

EXAMPLE 8 Find the solution $G(x)$ of $dy/dx = x$ such that $G(1) = 3$.

Solution We take $G(x) = (x^2/2) + C$ and require that

$$3 = G(1) = \frac{1^2}{2} + C.$$

We find that $C = 3 - \frac{1}{2} = \frac{5}{2}$. Thus $G(x) = (x^2/2) + \frac{5}{2}$. □

EXAMPLE 9 Find the solution $G(x)$ of $d^2y/dx^2 = \sqrt{x}$ such that $G(1) = 5$ and $G'(1) = -2$.

Solution Let $y = G(x)$. From

$$\frac{d^2y}{dx^2} = \sqrt{x} = x^{1/2},$$

we obtain, using Example 3,

$$\frac{dy}{dx} = G'(x) = \frac{x^{3/2}}{3/2} + C_1 = \frac{2}{3}x^{3/2} + C_1.$$

Since $G'(1) = -2$, we have

$$-2 = \frac{2}{3}(1)^{3/2} + C_1 = \frac{2}{3} + C_1.$$

Thus $C_1 = -2 - \frac{2}{3} = -\frac{8}{3}$, so

$$\frac{dy}{dx} = G'(x) = \frac{2}{3}x^{3/2} - \frac{8}{3}.$$

Example 3 then shows that

$$y = G(x) = \frac{2}{3}\cdot\frac{x^{5/2}}{5/2} - \frac{8}{3}x + C_2 = \frac{4}{15}x^{5/2} - \frac{8}{3}x + C_2.$$

Since $G(1) = 5$, we have

$$5 = \frac{4}{15}(1)^{5/2} - \frac{8}{3} + C_2,$$

so

$$C_2 = 5 - \frac{4}{15} + \frac{8}{3} = 5 + \frac{36}{15} = 5 + \frac{12}{5} = \frac{37}{5}.$$

Thus

$$y = G(x) = \frac{4}{15}x^{5/2} - \frac{8}{3}x + \frac{37}{5}. \quad \square$$

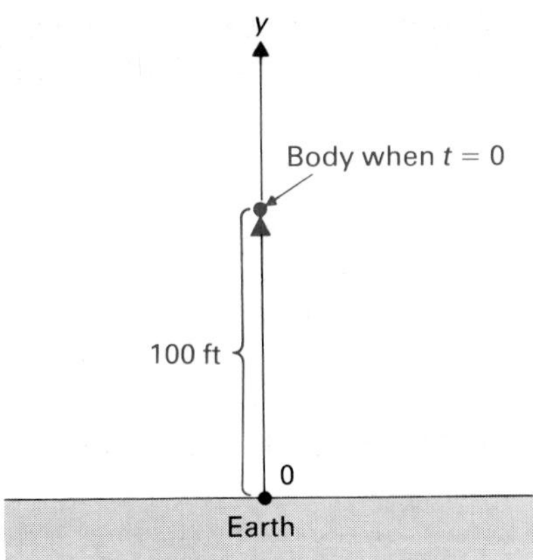

Figure 5.62 Freely falling body: When $t = 0$, then $y = 100$ ft and $v = dy/dt = 30$ ft/sec.

We now give the application to a freely falling body in a vacuum, as promised at the start of this section.

EXAMPLE 10 If air resistance is neglected, the acceleration of a freely falling body near the surface of the earth is known to be about -32 ft/sec^2. (The negative sign is used since the acceleration is downward.) Find the height of a body above the surface of the earth at time t if the body has height 100 ft and (upward) velocity 30 ft/sec at time $t = 0$.

Solution We let y be the height of the body above the surface of the earth, as indicated in Fig. 5.62. We know

$$\frac{d^2y}{dt^2} = -32,$$

so

$$\frac{dy}{dt} = -32t + C_1.$$

When $t = 0$, $dy/dt = 30$, so $C_1 = 30$ and

$$\frac{dy}{dt} = -32t + 30.$$

Then

$$y = -32\frac{t^2}{2} + 30t + C_2 = -16t^2 + 30t + C_2.$$

When $t = 0$, $y = 100$, so $C_2 = 100$ and

$$y = -16t^2 + 30t + 100$$

is the equation giving the height of the body above the surface of the earth. Of course this equation is valid only when $y \geq 0$. We could find the time limits when $y = 0$ by solving the quadratic equation

$$-16t^2 + 30t + 100 = 0.$$

The quadratic formula and a calculator show that the time interval during which the formula is valid is about $-1.733 \leq t \leq 3.608$ sec. □

Exercise 45 asks you to obtain the familiar solution

$$y = y_0 + v_0 t - 16t^2$$

for $d^2y/dt^2 = -32$ with initial conditions $y(0) = y_0$ and $y'(0) = v_0$. This solution gives the position for an object falling freely in a vacuum near the surface of the earth.

SUMMARY

1. If $F'(x) = f(x)$, then $F(x) + C$ is the general antiderivative of $f(x)$, where C is an arbitrary constant.
2. The general antiderivative of x^n is $(x^{n+1})/(n + 1) + C$ if $n \neq -1$.
3. The antiderivative of a sum is the sum of the antiderivatives, and the antiderivative of a constant times a function is the constant times the antiderivative of the function.

EXERCISES

In Exercises 1 through 30, find the general antiderivative of the given function.

1. 2
2. -10
3. $x^2 - 3x + 2$
4. $8x^3 - 2x^2 + 4$
5. $x^{1/2}$
6. $x^{-2/3}$
7. $4x + x^{1/2}$
8. $2x^{3/4} - 4x^{-1/2}$
9. $\frac{1}{x^2} + 3x + 1$
10. $\frac{2}{x^4} - \frac{3}{x^2} + x^2$
11. $\frac{x+1}{x^3}$
12. $\frac{x^2 - 3x + 2}{x^4}$
13. $\frac{3x^2 - 2x + 1}{\sqrt{x}}$
14. $(x^2 + 1)^2$
15. $x(x^2 + 1)$
16. $x^2(x^2 - x + 2)$
17. $\sqrt{x - 1}$
18. $\sqrt{1 - 3x}$
19. $\frac{1}{\sqrt{x - 1}}$
20. $\frac{2}{\sqrt{1 - x}} + \sqrt{2x + 5}$
21. $\sin x$
22. $\cos 3x$
23. $5 \sin 8x$
24. $\sin (2x + 3)$
25. $\cos (2 - 4x)$
26. $\sec^2 3x$
27. $4 \csc^2 2x$
28. $\sec x \tan x$
29. $\csc 4x \cot 4x$
30. $\frac{3 \sin 2x}{\cos^2 2x}$

In Exercises 31 through 40, find the solution $y = F(x)$ of the differential equation that satisfies the given initial conditions.

31. $\frac{dy}{dx} = 8,\ F(2) = -3$
32. $\frac{dy}{dx} = 3x^2 + 2,\ F(0) = 1$
33. $\frac{dy}{dx} = x + \sin x,\ F(0) = 3$
34. $\frac{dy}{dx} = 3 \cos 2x,\ F\left(\frac{\pi}{4}\right) = -1$
35. $\frac{dy}{dx} = x - \sqrt{x},\ F(1) = \pi$
36. $\frac{d^2y}{dx^2} = 0,\ F'(0) = -1,\ F(0) = 3$
37. $\frac{d^2y}{dx^2} = 2,\ F'(0) = -2,\ F(0) = 1$
38. $\frac{d^2y}{dx^2} = 2x,\ F'(1) = -1,\ F(1) = 2$
39. $\frac{d^2y}{dx^2} = -\sin 2x,\ F'(0) = 2,\ F(0) = -1$
40. $\frac{d^2y}{dx^2} = \cos 3x,\ F'(0) = 4,\ F(0) = -2.$

41. Let $d^2y/dx^2 = f(x)$ have $y = F(x)$ as one solution for $a \le x \le b$. Argue that there exists a solution $y = G(x)$ of the differential equation on $[a, b]$ satisfying the *boundary conditions* $G(a) = \alpha$ and $G(b) = \beta$ for any choices of α and β.

42. Find the solution $y = G(x)$ of $d^2y/dx^2 = 0$ satisfying the boundary conditions $G(1) = -3$ and $G(2) = 0$. (See Exercise 41.)

43. Find the solution $y = G(x)$ of $d^2y/dx^2 = \sin x$ satisfying the boundary conditions $G(-\pi/2) = 0$ and $G(\pi/2) = 0$. (See Exercise 41).

44. At time $t = 0$, a man standing on a balcony reaches over and tosses a ball upward at a speed of 32 ft/sec from a height of 48 ft above the ground. Neglecting air resistance,
    a) find the height of the ball as a function of t until the ball hits the ground,
    b) find the time when the ball hits the ground.

45. General equation for free fall in a vacuum: Suppose a body close to the surface of the earth is subject only to the force of gravity; that is, neglect air resistance. It is known that the acceleration of the body is then directed downward with magnitude $g = 32$ ft/sec^2. Thus if y measures the height of the body above the surface at time t, we have $d^2y/dt^2 = -32$. Suppose that at time $t = 0$, the body has *initial velocity* v_0 and *initial position* y_0.
    a) Find the velocity $v = dy/dt$ of the body as a function of t.
    b) Find the position y of the body as a function of t.

46. A body is attached to one end of a spring, whose other end is attached to a beam. The body hangs at rest in suspended position. If the body is pulled down a short distance and released, oscillatory motion results. Neglecting air resistance, the only force on the body is the spring's restoring force, which at each time t is proportional to the body's displacement s from its position at rest. A displacement of 1 unit results in a restoring force of k corresponding units.
    a) Use Newton's law $F = ma$ to find the differential equation governing the motion of such a body of mass m. (Take as origin on a vertical s-axis the position of rest of the body.)
    b) If the body is pulled down a distance of 8 units, held for a moment, and then released at time $t = 0$, give the initial conditions that, together with your answer to part (a), would completely determine the motion of the body. (You are not expected to solve the differential equation.)

47. A body of mass m traveling through a medium (air, water, or the like) encounters a force of resistance from the medium proportional to the speed of the body. Suppose the magnitude $k > 0$ of this constant of proportionality is known, and suppose also that other forces on the body can be neglected. If the body enters the medium with a velocity of 80 ft/sec at time $t = 0$, find the differential equation and initial conditions which determine the distance s the body travels through the medium in time t.

48. A freely falling body of mass m near the surface of the earth encounters a force of air resistance proportional to its velocity in addition to the force produced by gravity. The body is dropped from a position of 5000 ft above the earth at

time $t = 0$. If the air resistance proportionality constant is k (where units are in feet and seconds), find the differential equation and initial conditions that govern the altitude s of the body at time t until the body hits the ground.

49. Under suitable conditions (no predators, disease, and the like), the rate of growth of a population P is proportional to the size of P. If k is the constant of proportionality, give the differential equation that governs this situation.

EXERCISE SETS FOR CHAPTER 5

Review Exercise Set 5.1

1. If the altitude of a right circular cone remains constant at 10 ft while the volume increases at a steady rate of 8 ft^3/min, find the rate of increase of the radius when the radius is 4 ft. (The volume V of a cone of base radius r and altitude h is $V = \frac{1}{3}\pi r^2 h$.)

2. Let f be differentiable for all x. If $f(1) = -2$ and $f'(x) \geq 2$ for x in $[1, 6]$, use the mean-value theorem to show that $f(6) \geq 8$.

3. If $f'(x) = (x - 2)^3(x + 1)^2$, find the intervals where $f(x)$ is
a) increasing, b) decreasing.

4. Find the maximum and minimum values assumed by $x^3 - 3x + 1$ in the interval $[0, 3]$.

5. If $f'(x)$ is as given in Exercise 3, find the intervals where the graph of f is
a) concave up, b) concave down.

6. Sketch the graph of $y = x^3 - 3x^2 + 2$, finding and labeling all local maxima and minima and all inflection points.

7. A right circular cone has constant slant height 12. Find the radius of the base for which the cone has maximum volume. (The volume V of a cone of base radius r and altitude h is $V = \frac{1}{3}\pi r^2 h$.)

8. A solution of $f(x) = x^3 - 3x + 1$ is desired. Since $f(1) = -1$ and $f(2) = 3$, there is a solution between 1 and 2. Starting with 2 as a first approximation of the solution, find the next two approximations given by Newton's method.

9. A company has cost and revenue functions

$$C(x) = 5000 + 1500x + 0.02x^2,$$

$$R(x) = 2000x - 0.5x^{3/2}.$$

a) Find the marginal profit when $x = 10{,}000$.
b) Find the average profit when $x = 10{,}000$.

10. Find the solution of the differential equation $dy/dx = 3x^2 - 4x + 5$ that satisfies the initial condition $y = -2$ when $x = 1$.

Review Exercise Set 5.2

1. If the lengths of two sides of a triangle remain constant at 10 ft and 15 ft while the angle θ between them is increasing at the rate of $\frac{1}{10}$ radian/min, find the rate of increase of the third side when $\theta = \pi/3$.

2. If $f'(x)$ exists for all x and $f(4) = 12$, while $f'(x) \leq -3$ for x in $[-1, 4]$, use the mean-value theorem to find the smallest possible value for $f(-1)$.

3. Let $f(x) = 1/(x^2 - 4)$. Find the intervals where $f(x)$ is
a) increasing, b) decreasing.

4. Find the maximum and minimum values assumed by the function $x^4 - 8x^2 + 4$ in the interval $[-1, 3]$.

5. If $f(x) = x/(x - 2)$, find the intervals where the graph of f is
a) concave up, b) concave down.

6. Sketch the graph of $y = 2x^2 - x^4 + 6$, finding and labeling all relative maxima and minima and all inflection points.

7. A rectangle has base on the x-axis and upper vertices on the parabola $y = 16 - x^2$ for $-4 \leq x \leq 4$. Find the maximum area the rectangle can have.

8. Use Newton's method to estimate $\sqrt[3]{30}$, starting with 3 as first approximation and finding the next approximation.

9. A company manufacturing x units of a certain product per year has a cost function $C(x) = 10{,}000 + 500x + 0.05x^2$ dollars, and a marginal revenue of $MR(x) = 900 - 0.04x$ dollars per year. Find the profit if 100 units are manufactured in a year.

10. A body moving on an s-axis has acceleration $a = 6t - 8$ at time t. If when $t = 1$ the body is at the point $s = 4$ and has velocity $v = -3$, find the position s as a function of time.

Review Exercise Set 5.3

1. If W watts of power are being consumed, then $W = VI$ where the potential is V volts and the current is I amperes. Suppose the power consumption remains constant while, at a certain instant, the current is 15 amp and the potential of 120 volts is dropping at a rate of 0.5 volts/sec. Find the rate of change of the current at that instant.

2. State the mean-value theorem without reference to the text.

3. Suppose $f'(x) = (x + 1)^3(x - 5)^4$. Find all x-values where $f(x)$ has
a) a local maximum, b) a local minimum.

4. Find the maximum and minimum values assumed by $f(x) = x + \sin x$ in the interval $[0, 2\pi]$.

5. If $f'(x) = (x + 1)^3(x - 5)^4$, find all x-values of inflection points.

6. Sketch the portion of the graph of $f(x) = x + \cos x$ for $0 \leq x \leq 2\pi$, finding and labeling all local maxima and minima and all inflection points for x in that interval.

7. Find the area of the largest rectangle that can be inscribed in an equilateral triangle of side a ft.

8. Use a calculator and Newton's method to find a solution of $x^5 + 4x - 10 = 0$ to six significant figures.

9. A ski shop sells 1600 pairs of skis during the year. They have carrying costs of \$5 per year for each pair of skis unsold and order costs averaging 50¢ per pair of skis plus \$10 for the paperwork of each order. Find the lot size in which the skis should be ordered to minimize these inventory costs.

10. Solve the initial value problem $d^2y/dx^2 = x - 3$ where $y = 5$ and $y' = -2$ when $x = 1$.

Review Exercise Set 5.4

1. At a certain instant, an equilateral triangle has sides of length 10 ft that are increasing at the rate of $\frac{1}{2}$ ft/min. Find the rate of increase of the area of the rectangle of maximum area that can be inscribed in the triangle. (See Exercise 7 of Review Exercise Set 5.3.)

2. Suppose $|f'(x)| < x + 2$ for x in $[1, 3]$. If $f(1) = 4$, find bounds in terms of x for $f(x)$ in $[1, 3]$.

3. If

$$f'(x) = \frac{(x + 1)(x - 4)^5}{(x - 2)^3},$$

find all x-values where $f(x)$ has
a) a local maximum, b) a local minimum.

4. Find the maximum and minimum values assumed by $f(x) = x/2 - \sin x$ in the interval $[0, 2\pi]$.

5. If

$$f''(x) = \frac{(x - 2)^3(x + 3)^5}{x^3},$$

find the intervals where the graph of f is
a) concave up, b) concave down.

6. Sketch the portion of the graph of $y = x/2 - \cos x$ for $0 \le x \le 2\pi$, and find and label all local maxima and minima and all inflection points for x in this interval.

7. Find two positive numbers such that their sum is 6 and the product of one by the square of the other is maximum.

8. Use a calculator and Newton's method to find the solution of $\cos x = x/2$ to six significant figures.

9. A family with a trout pond finds that a population of p trout one year produces a population of $51p - 0.1p^2$ five years later if left undisturbed. Find the maximum sustainable harvest of trout over a five-year period.

10. Find the general antiderivative of $f(x) = 3/x^3 - 4x + \cos(x/2)$.

More Challenging Exercises 5

1. A farmer has 1000 ft of fencing. He wishes to use it to fence in one square and one circular enclosure, each having an area of at least 10,000 ft. What size enclosures should he fence to maximize the total area? (Use a calculator.)

2. Show that if f is a twice-differentiable function for $a \le x \le b$ and $f(x)$ assumes the same value at three distinct points in $[a, b]$, then $f''(c) = 0$ for some c where $a < c < b$.

3. State a generalization of the result in Exercise 2.

4. Suppose f is differentiable for $0 \le x \le 10$ and $f(2) = 17$, while $|f'(x)| \le 3$ for $0 \le x \le 10$.
   a) What is the maximum possible value for $f(x)$ for any x in $[0, 10]$?
   b) What is the minimum possible value for $f(x)$ for any x in $[0, 10]$?

5. Maureen is at the edge of a circular lake of radius a mi and wishes to get to the point directly across the lake. She has a boat that she can row at 4 mph, or she can jog along the shore at 8 mph, or she can row to some point and jog the rest of the way. How should she proceed to get across in the shortest time?

6. Town A is located 2 mi back from the bank of a straight river. Town B is located 3 mi from the bank on the same side at a point 15 mi farther downstream. The towns agree to build a pumping station on the bank of the river to supply water to both towns. How far downstream from the point on the bank closest to town A should they build the station to minimize the total length of the water mains to the two towns? (Solve this problem first by using calculus and then by "moving town A across the river" and using geometry. What have you learned?)

7. Show that there exists a polynomial function $f(x)$ of degree 5 such that Newton's method to solve $f(x) = 0$ starting with $x = 0$ yields the repetitive sequence $0, 1, 2, 0, 1, 2, 0, 1, 2, \ldots$.

8. Give an example of a function f and a number a_1 such that an attempt to solve $f(x) = 0$ using Newton's method and starting with a_1 leads to oscillation between positive and negative numbers of ever-increasing magnitude.

THE INTEGRAL 6

The first three sections of this chapter present the main ideas of integral calculus for functions of one variable. These sections culminate in the fundamental theorem of calculus, which appears in Section 6.3. Sections 6.4 through 6.6 are concerned chiefly with computational technique.

6.1 RIEMANN SUMS

In geometric terms, finding a derivative is equivalent to finding the slope of a tangent line to a curve. The main part of this section is devoted to explaining that integral calculus, viewed geometrically, should be concerned with computing areas. As we started to study the derivative, we estimated the slope of a tangent line by computing m_{sec}, the slope of a secant line. Analogously, we will estimate areas by *Riemann sums.*

SUMMATION NOTATION

In this chapter, we will often want to consider a sum of quantities represented by subscripted letters, such as

$$a_1 + a_2 + \cdots + a_n.$$

There is a very useful notation for writing such sums. A sum $a_1 + \cdots + a_n$ is denoted in the *summation notation* by

$$\sum_{i=1}^{n} a_i,$$

read "the sum of the a_i as i runs from 1 to n." The Greek letter Σ (sigma) stands for *sum*; the value of i under the Σ is the **lower limit of the sum** and tells where the sum starts. The value above the Σ is the **upper limit of the sum** and tells where the sum stops. The letter i is the **summation index.** The choice of letter for summation index is of course not significant; the letters i,

j, and k are the most commonly used. Thus

$$a_1 + \cdots + a_n = \sum_{i=1}^{n} a_i = \sum_{j=1}^{n} a_j = \sum_{k=1}^{n} a_k.$$

A typical sum that we will run into in this chapter is

$$f(x_1) + f(x_2) + \cdots + f(x_n) = \sum_{i=1}^{n} f(x_i).$$

You will probably best understand the use of the summation notation by studying some examples. In forming the sum, we simply replace the summation index successively by all integers from the lower limit to the upper limit and add the resulting quantities.

EXAMPLE 1 Write out $\sum_{i=1}^{2} a_i$, $\sum_{i=1}^{3} a_i$, and $\sum_{i=4}^{7} f(a_i)$.

Solution We have

$$\sum_{i=1}^{2} a_i = a_1 + a_2, \qquad \sum_{i=1}^{3} a_i = a_1 + a_2 + a_3,$$

$$\sum_{i=4}^{7} f(a_i) = f(a_4) + f(a_5) + f(a_6) + f(a_7). \quad \square$$

EXAMPLE 2 Compute $\sum_{i=1}^{3} i^2$, $\sum_{j=2}^{4} (j-1)$, and $\sum_{k=0}^{2} (k^2+1)$.

Solution We have

$$\sum_{i=1}^{3} i^2 = 1^2 + 2^2 + 3^2 = 14,$$

$$\sum_{j=2}^{4} (j-1) = (2-1) + (3-1) + (4-1) = 1 + 2 + 3 = 6,$$

$$\sum_{k=0}^{2} (k^2+1) = (0^2+1) + (1^2+1) + (2^2+1) = 1 + 2 + 5 = 8. \quad \square$$

EXAMPLE 3 Show that $\sum_{i=1}^{n} (a_i + b_i) = \sum_{i=1}^{n} a_i + \sum_{i=1}^{n} b_i$.

Solution We have

$$\begin{aligned} \sum_{i=1}^{n} (a_i + b_i) &= (a_1 + b_1) + \cdots + (a_n + b_n) \\ &= (a_1 + \cdots + a_n) + (b_1 + \cdots + b_n) \\ &= \sum_{i=1}^{n} a_i + \sum_{i=1}^{n} b_i. \quad \square \end{aligned}$$

THE INTEGRAL CALCULUS PROBLEM

Suppose a particle moves in the positive direction on the x-axis, starting at the origin at time $t = 0$. If the velocity v remains constant, then $v = x/t$, that is,

$$\text{Velocity} = \frac{\text{Distance}}{\text{Time}}. \tag{1}$$

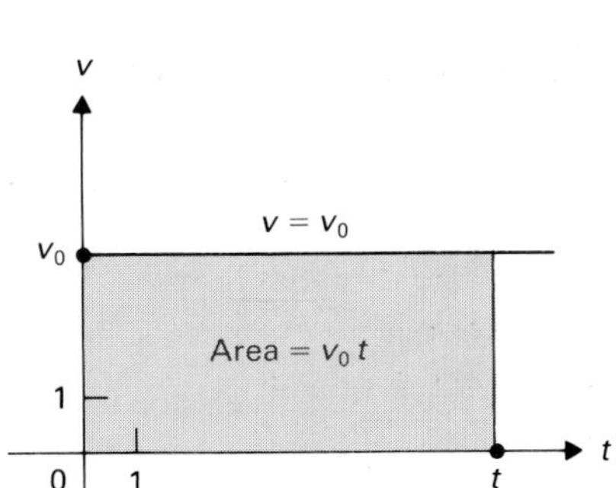

Figure 6.1 For constant velocity v_0, the distance traveled in time t is $v_0 t$, which numerically equals the area of the shaded rectangle.

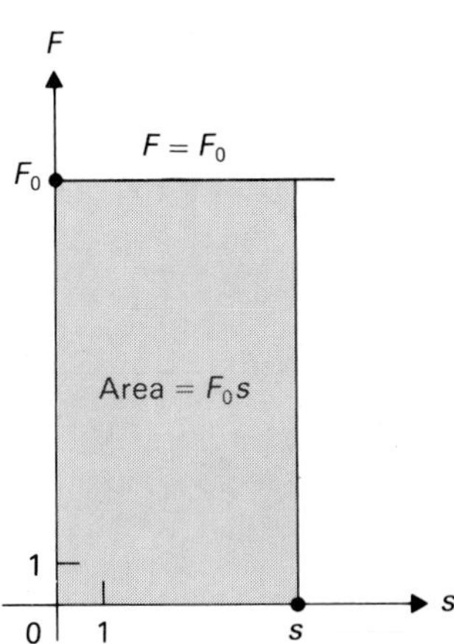

Figure 6.2 For constant force F_0, the work done moving a body a distance s is $W = F_0 s$, which numerically equals the area of the shaded rectangle.

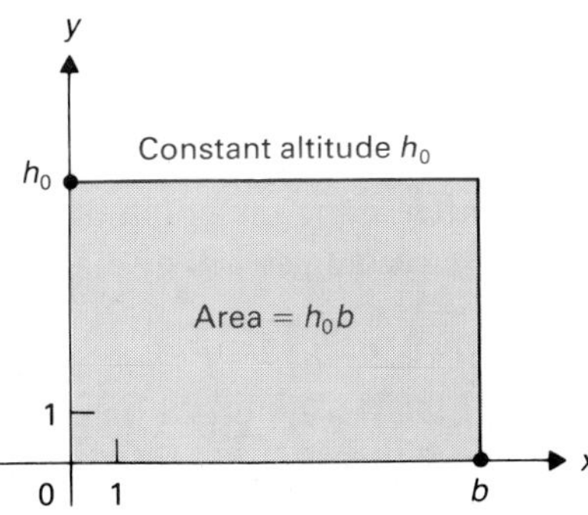

Figure 6.3 The area of a region of base width b and constant altitude h_0 is $A = h_0 b$.

Figure 6.4 Find the area of the region of base width b and nonconstant altitude $h(x)$.

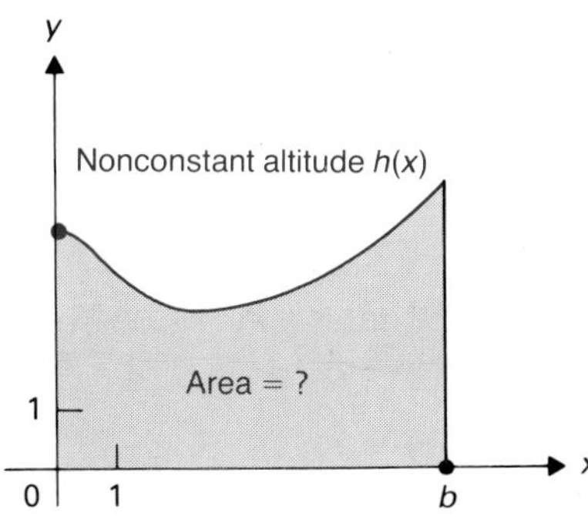

If the velocity does not remain constant, then this *quotient* gives the average velocity, and the velocity at any instant is found by differentiating:

$$v = \frac{dx}{dt}.$$

Calculus is concerned with the study of changing quantities, and differentiation is often appropriate when we would compute a *quotient*, as in formula (1), if quantities remain constant.

Integral calculus is frequently useful with changing quantities in cases where we would compute a *product*, if quantities remain constant. For a particle moving to the right on the x-axis with constant velocity, we see from formula (1) that the distance traveled is given by the *product*

$$\text{Distance} = (\text{Velocity})(\text{Time}).$$

Figure 6.1 shows the graph where a constant velocity v_0 is plotted against time t. We see that the *distance* $x = v_0 t$ traveled in time t is numerically equal to the area of the rectangle shaded in the figure.

For another example, suppose a stalled car is moved s ft by a constant force of F lb in the direction of the car's motion. Then the *work* W done by the force is given by the *product*

$$W = F \cdot s \text{ ft·lb}.$$

That is,

$$\text{Work} = (\text{Force})(\text{Distance}) \tag{2}$$

if the force remains constant. Figure 6.2 shows the graph of a constant force F_0 plotted against distance s. We see that the work $W = F_0 s$ is numerically equal to the area of the rectangle shaded in the figure.

Finally, the area A of a region of base width b and constant altitude h_0 is given by the *product* $A = bh_0$, as shown in Fig. 6.3. Figure 6.4 shows a region of base width b and altitude that varies, depending on where we are in the base. It is not so easy to find the area of the shaded region in Fig. 6.4.

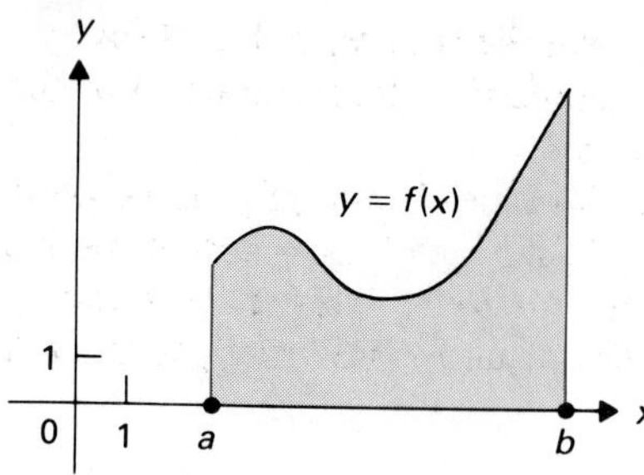

Figure 6.5 The region under $y = f(x)$ over $[a, b]$.

We hope you have been led to guess that the area of the shaded region in Fig. 6.4 might also represent the *work* done by a nonconstant force; just change the labels on the axes in Fig. 6.4 to s and F as in Fig. 6.2. This area can also represent *distance traveled* for motion where the velocity is nonconstant; just change the labels on the axes in Fig. 6.4 to t and v as in Fig. 6.1. We will try to make these interpretations of the area seem even more reasonable later in this section.

ESTIMATING AREAS USING RIEMANN SUMS

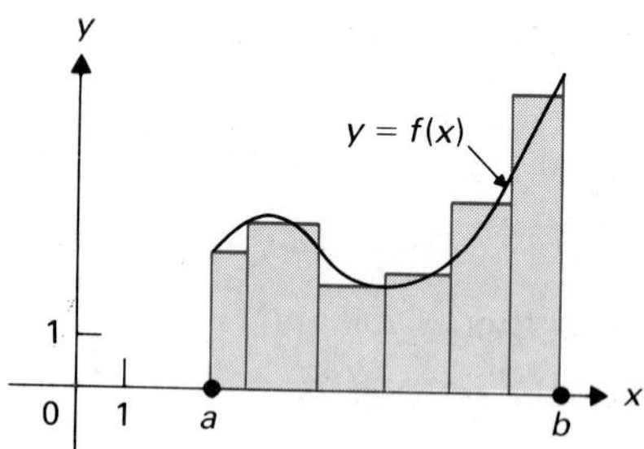

Figure 6.6 The area of the shaded rectangles is an approximation of $A_a^b(f)$.

With the preceding discussion for motivation, we turn to the estimation of the area of a region like the one shaded in Fig. 6.4. Rather than always take the base of the region with its left-hand end at the origin as in Fig. 6.4, we will let the base be any interval $[a, b]$ on the x-axis.

Let f be a continuous function with domain containing $[a, b]$, and suppose that $f(x) \geq 0$ for all x in $[a, b]$. Since the graph of the *continuous* function f over $[a, b]$ is an unbroken curve, it would seem that the notion of the *area of the region under the graph of f over* $[a, b]$, shown shaded in Fig. 6.5 should be meaningful. Let us denote the area of this region by $A_a^b(f)$.

We can find areas of triangles, rectangles, circles, trapezoids, and certain other familiar plane regions, but the region under the graph of f over $[a, b]$ may not be any of these. Since we can find areas of rectangles, we might approximate $A_a^b(f)$ by areas of rectangles with bases on the x-axis and whose tops are horizontal line segments interescting the graph of f, as shown in Fig. 6.6. The sum of the areas of these rectangles will give an approximate value of $A_a^b(f)$.

The bases of the rectangles in Fig. 6.6 divide $[a, b]$ into *subintervals.* To make computations easier, we can divide $[a, b]$ into n subintervals of *equal* lengths. The length of the base of each rectangle is then

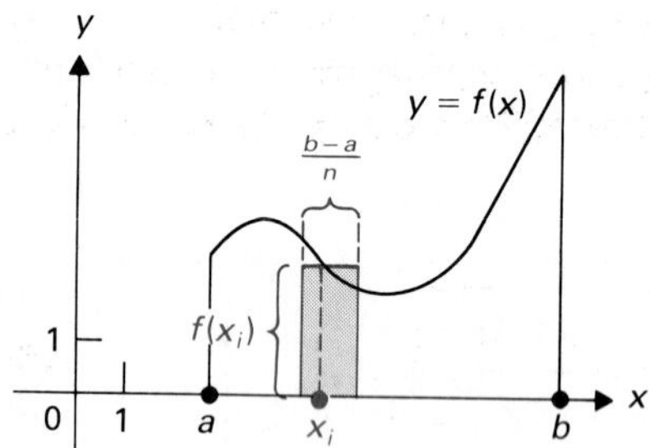

Figure 6.7 The ith rectangle of height $f(x_i)$ and width $(b - a)/n$.

$$\text{Length of base} = \Delta x = \frac{b-a}{n}. \tag{3}$$

The height of the ith rectangle is $f(x_i)$ for some point x_i in the ith subinterval, so

$$\text{Area of the } i\text{th rectangle} = f(x_i) \cdot \frac{b-a}{n}.$$

(See Fig. 6.7.) If we let $\mathscr{S}_n$ be the *sum* of the areas of these rectangles, then

$$\begin{aligned} \mathscr{S}_n &= \frac{b-a}{n} \cdot f(x_1) + \frac{b-a}{n} \cdot f(x_2) + \cdots + \frac{b-a}{n} \cdot f(x_n) \\ &= \frac{b-a}{n}(f(x_1) + f(x_2) + \cdots + f(x_n)) = \frac{b-a}{n}\left(\sum_{i=1}^{n} f(x_i)\right). \end{aligned} \tag{4}$$

DEFINITION Riemann sum

6.1

The sum $\mathscr{S}_n$ shown in Eq. (4) is a **Riemann sum*** of order n for f over $[a, b]$.

* Named for the German mathematician Bernhard Riemann (1826–1866).

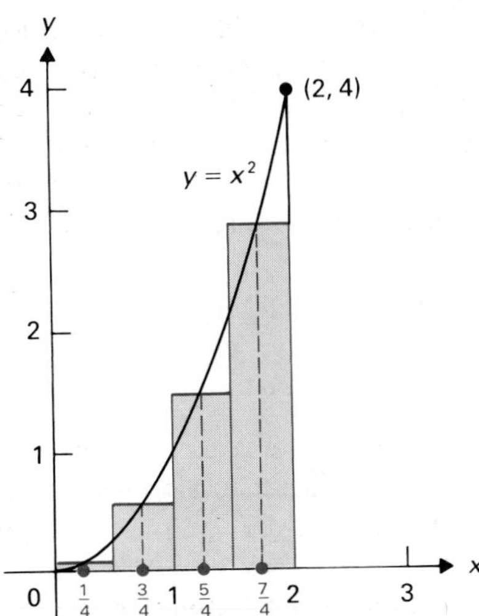

Figure 6.8 $\mathscr{S}_4$ for x^2 on $[0, 2]$ using midpoints is the sum of the areas of the shaded rectangles.

It is important to note that we can form the sum $\mathscr{S}_n$ in Eq. (4) even if $f(x)$ is negative for some x in $[a, b]$; the geometric interpretation of $\mathscr{S}_n$ for such a function will appear in the next section.

Of course, the Riemann sum $\mathscr{S}_n$ in Eq. (4) depends not only on the value of n but also on the function f, the interval $[a, b]$, and the choice of points $x_1, \ldots, x_n$ in the n subintervals into which $[a, b]$ is divided. However, a notation that reflects all these dependencies would be unwieldy, so we will simply use $\mathscr{S}_n$.

EXAMPLE 4 Estimate $A_0^2(x^2)$ by computing $\mathscr{S}_4$, taking for the points x_1, x_2, x_3, and x_4 the midpoints of the first, second, third, and fourth intervals, respectively.

Solution The situation is illustrated in Fig. 6.8. We have

$$x_1 = \tfrac{1}{4}, \qquad x_2 = \tfrac{3}{4}, \qquad x_3 = \tfrac{5}{4}, \qquad x_4 = \tfrac{7}{4},$$

while

$$\frac{b-a}{n} = \frac{2-0}{4} = \frac{1}{2}.$$

The estimate $\mathscr{S}_4$ for $A_0^2(x^2)$ given by Eq. (4) is then

$$\tfrac{1}{2}(\tfrac{1}{16} + \tfrac{9}{16} + \tfrac{25}{16} + \tfrac{49}{16}) = \tfrac{1}{2}(\tfrac{84}{16}) = \tfrac{21}{8} = 2.625.$$

We will see later that the exact value of $A_0^2(x^2)$ is $\tfrac{8}{3}$, so our estimate is not too bad. □

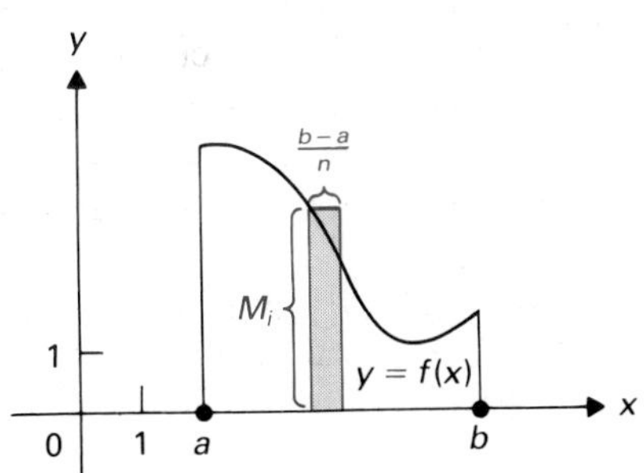

Figure 6.9 The maximum possible contribution to $\mathscr{S}_n$ from the ith subinterval.

There are several natural ways to choose the points x_i in Eq. (4). One way is to select x_i so that $f(x_i)$ is the *maximum value* M_i of $f(x)$ for x in the ith subinterval (see Fig. 6.9). This gives a Riemann sum that we are sure is greater than or equal to $A_a^b(f)$. Another way is to select x_i so that $f(x_i)$ is the *minimum value* m_i of $f(x)$ over the ith subinterval, which gives a Riemann sum that is less than or equal to $A_a^b(f)$ (see Fig. 6.10). We summarize notation, terminology, and relations.

$$S_n = \textbf{Upper Riemann sum} = \frac{b-a}{n}\left(\sum_{i=1}^{n} M_i\right) \tag{5}$$

$$s_n = \textbf{Lower Riemann sum} = \frac{b-a}{n}\left(\sum_{i=1}^{n} m_i\right) \tag{6}$$

$$\mathscr{S}_n = \textbf{General Riemann sum} = \frac{b-a}{n}\left(\sum_{i=1}^{n} f(x_i)\right) \tag{7}$$

$$s_n \le A_a^b(f) \le S_n \tag{8}$$

$$s_n \le \mathscr{S}_n \le S_n \tag{9}$$

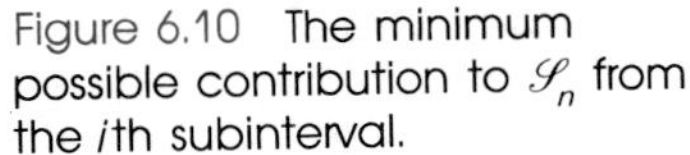
Figure 6.10 The minimum possible contribution to $\mathscr{S}_n$ from the ith subinterval.

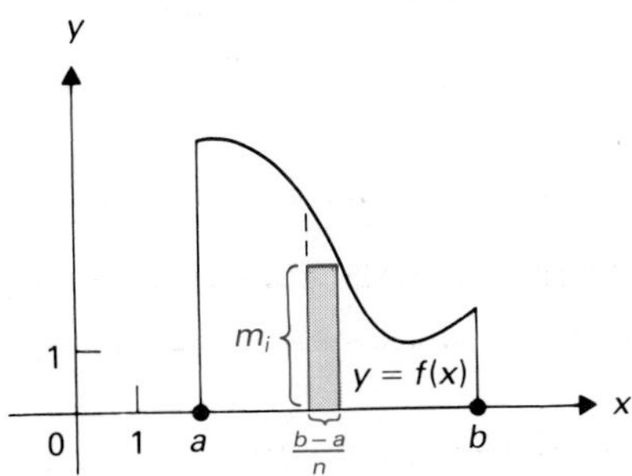

Note that if $f(x)$ is *increasing* on $[a, b]$, then the maximum value M_i is assumed at the right-hand end of the ith subinterval, while the minimum value m_i is assumed at the left-hand end. If $f(x)$ is *decreasing* on $[a, b]$, then the reverse is true.

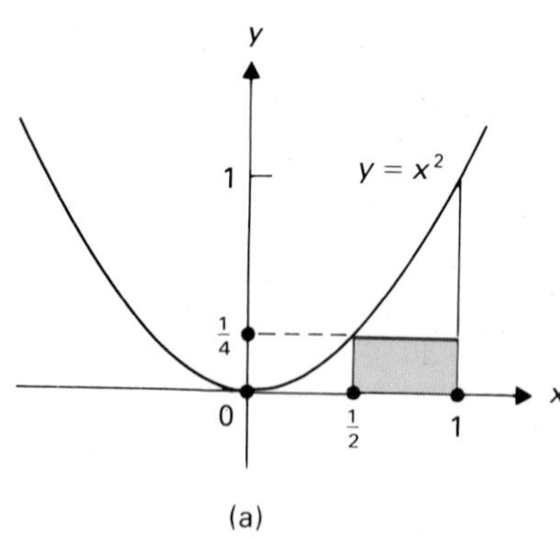

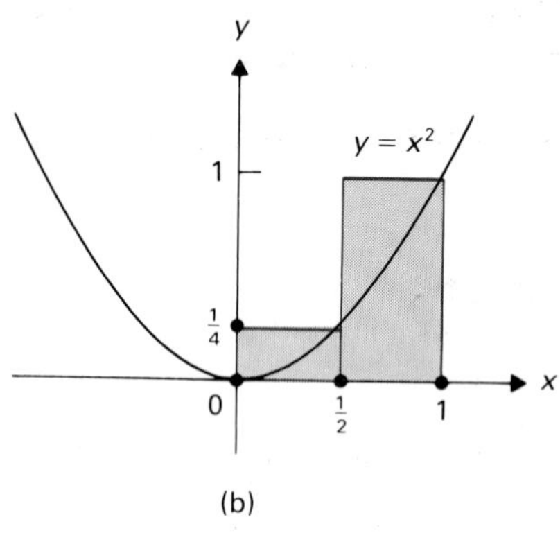

Figure 6.11 (a) The lower sum s_2 for $A_0^1(x^2)$; (b) the upper sum S_2 for $A_0^1(x^2)$.

EXAMPLE 5 Find the upper sum S_2 and lower sum s_2 approximating $A_0^1(x^2)$.

Solution Here $n = 2$, and

$$\frac{b-a}{n} = \frac{1-0}{2} = \frac{1}{2}.$$

It is easily seen from the *increasing* graph of x^2 in Fig. 6.11(a) that

$$m_1 = 0 \quad \text{and} \quad m_2 = \tfrac{1}{4},$$

while

$$M_1 = \tfrac{1}{4} \quad \text{and} \quad M_2 = 1,$$

as shown in Fig. 6.11(b). Thus

$$s_2 = (\tfrac{1}{2})(0 + \tfrac{1}{4}) \leq A_0^1(x^2) \leq (\tfrac{1}{2})(\tfrac{1}{4} + 1) = S_2,$$

so

$$\tfrac{1}{8} \leq A_0^1(x^2) \leq \tfrac{5}{8}.$$

Of course, these bounds on $A_0^1(x^2)$ are very crude. □

EXAMPLE 6 Find the upper sum S_4 and the lower sum s_4 approximating $A_1^3(1/x)$.

Solution Here $n = 4$, and

$$\frac{b-a}{n} = \frac{3-1}{4} = \frac{1}{2}.$$

It is easily seen from the *decreasing* graph of $1/x$ in Fig. 6.12(a) that

$$m_1 = \frac{1}{\frac{3}{2}} = \frac{2}{3}, \qquad m_2 = \frac{1}{2}, \qquad m_3 = \frac{1}{\frac{5}{2}} = \frac{2}{5}, \qquad m_4 = \frac{1}{3},$$

while

$$M_1 = \frac{1}{1} = 1, \qquad M_2 = \frac{1}{\frac{3}{2}} = \frac{2}{3}, \qquad M_3 = \frac{1}{2}, \qquad M_4 = \frac{1}{\frac{5}{2}} = \frac{2}{5},$$

Figure 6.12
(a) The lower sum s_4 for $A_1^3(1/x)$; (b) the upper sum S_4 for $A_1^3(1/x)$.

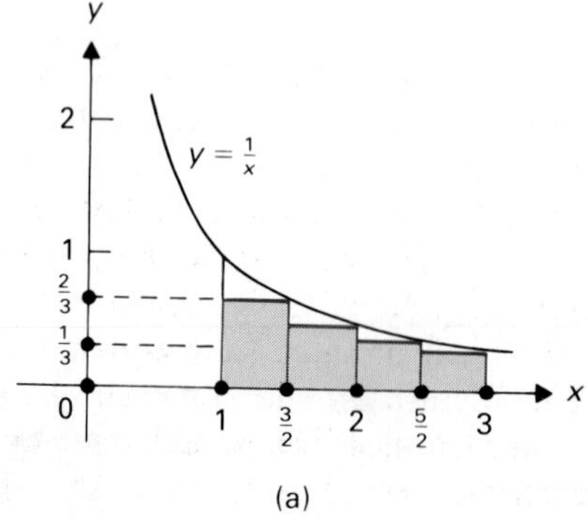

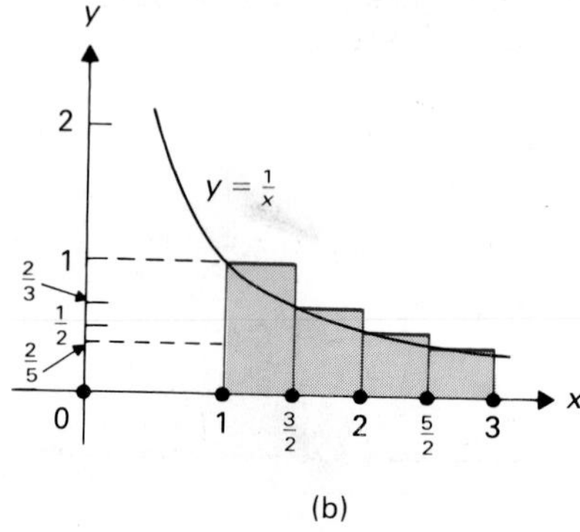

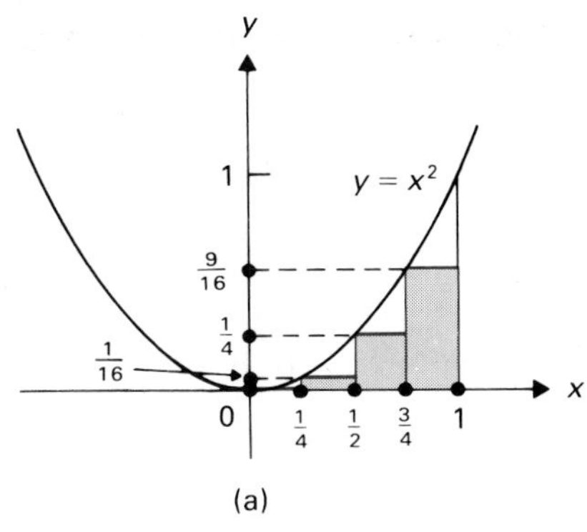

(a)

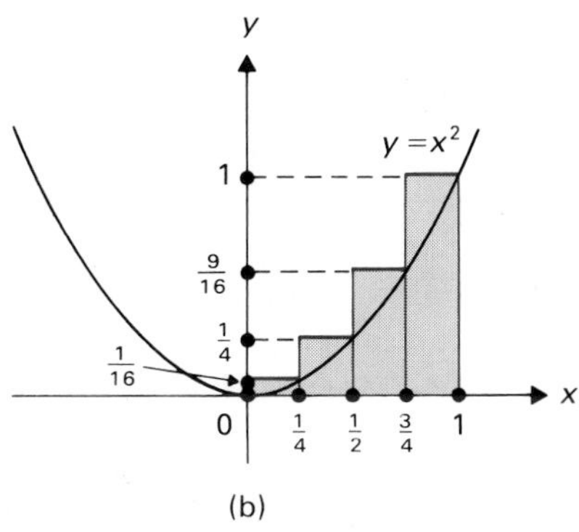

(b)

Figure 6.13 (a) The lower sum s_4 for $A_0^1(x^2)$; (b) the upper sum S_4 for $A_0^1(x^2)$.

as shown in Fig. 6.12(b). Then

$$s_4 = \frac{1}{2}\left(\frac{2}{3} + \frac{1}{2} + \frac{2}{5} + \frac{1}{3}\right) = \frac{1}{2}\left(1 + \frac{1}{2} + \frac{2}{5}\right) = \frac{1}{2}\cdot\frac{19}{10} = \frac{19}{20},$$

and

$$S_4 = \frac{1}{2}\left(1 + \frac{2}{3} + \frac{1}{2} + \frac{2}{5}\right) = \frac{1}{2}\left(\frac{30 + 20 + 15 + 12}{30}\right) = \frac{77}{60}.$$

Therefore

$$\frac{19}{20} \le A_1^3\left(\frac{1}{x}\right) \le \frac{77}{60}.$$

Now $\frac{19}{20} = 0.95$ and $\frac{77}{60} \approx 1.2833$. We will see in Chapter 8 that $A_1^3(1/x) \approx 1.0986123$, which is indeed between the bounds we just found. □

We can improve the bounds for $A_0^1(x^2)$ in Example 5 by taking a larger value of n.

EXAMPLE 7 Estimate $A_0^1(x^2)$ by finding s_4 and S_4.

Solution As indicated in Fig. 6.13(a), we have

$$m_1 = 0, \qquad m_2 = \tfrac{1}{16}, \qquad m_3 = \tfrac{1}{4}, \qquad m_4 = \tfrac{9}{16},$$

while

$$M_1 = \tfrac{1}{16}, \qquad M_2 = \tfrac{1}{4}, \qquad M_3 = \tfrac{9}{16}, \qquad M_4 = 1,$$

as shown in Fig. 6.13(b). Of course,

$$\frac{b - a}{n} = \frac{1 - 0}{4} = \frac{1}{4}.$$

Thus

$$s_4 = (\tfrac{1}{4})(0 + \tfrac{1}{16} + \tfrac{1}{4} + \tfrac{9}{16}) \le A_0^1(x^2) \le (\tfrac{1}{4})(\tfrac{1}{16} + \tfrac{1}{4} + \tfrac{9}{16} + 1) = S_4.$$

The arithmetic works out to give

$$s_4 = \tfrac{7}{32} \le A_0^1(x^2) \le \tfrac{15}{32} = S_4.$$

This is an improvement over Example 5, which gave $\frac{4}{32} \le A_0^1(x^2) \le \frac{20}{32}$.

We might estimate $A_0^1(x^2)$ by averaging $\frac{7}{32}$ and $\frac{15}{32}$, arriving at $\frac{11}{32}$. We will show in Section 6.3 that the exact value of $A_0^1(x^2)$ is $\frac{1}{3}$. Thus $\frac{11}{32}$ is not a bad estimate. □

Figure 6.14 The shaded region has area equal to the work done by the force $F(x)$ over $[a, b]$.

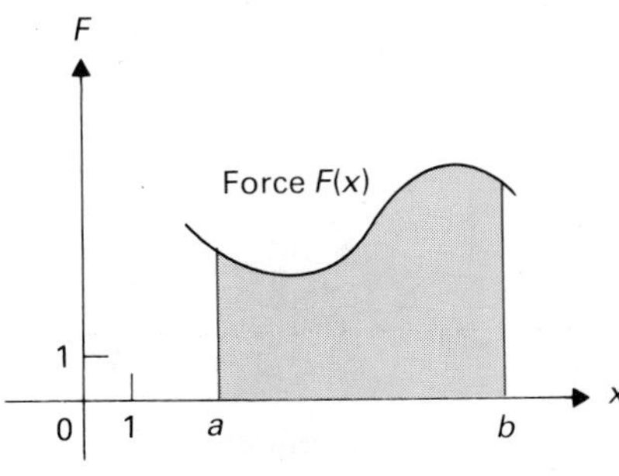

As illustrated by Example 7, we expect a Riemann sum $\mathscr{S}_n$ to be a good estimate for $A_a^b(f)$ if n is large enough. Perhaps you already realize that we should consider $\lim_{n\to\infty} \mathscr{S}_n$ to try to compute $A_a^b(f)$ precisely. We defer passing to this limit until the next section.

Consider the *work* done by a nonconstant force as it moves an object across the interval $[a, b]$. Earlier in this section, we led you to guess that this work should be numerically equal to the area of the region under the graph of the force function, shown shaded in Fig. 6.14. Similarly, we guessed that the *distance* traveled by a particle moving with nonconstant positive velocity is numerically equal to the area under the graph of the velocity function over

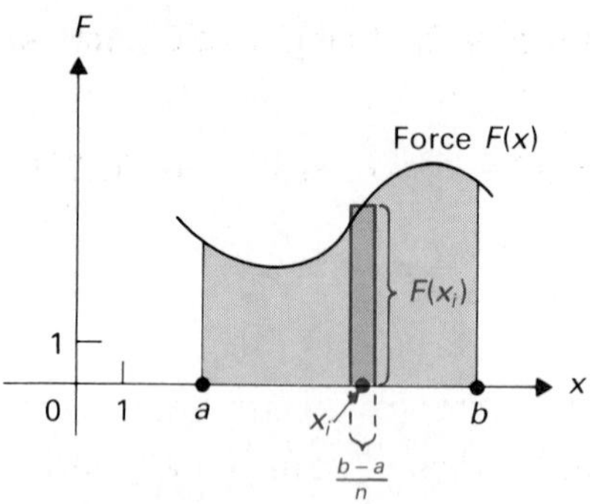

Figure 6.15 $F(x_i)\cdot(b-a)/n$ is the work done by the constant force $F(x_i)$ over the base of the rectangle.

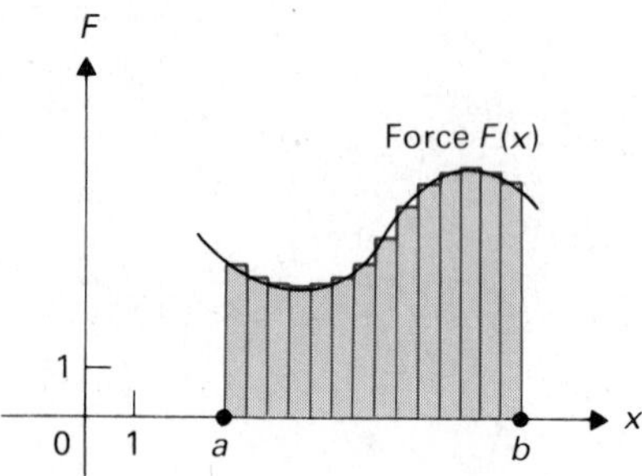

Figure 6.16 In $\mathscr{S}_{14}$, the total work is approximated by adding pieces of work done by constant forces $F(x_i)$ over subintervals.

the elapsed time interval. Let us examine these conjectures in the light of our work with Riemann sums. We treat the work illustration. The argument for the distance example is similar.

Figure 6.15 shows one contribution to a Riemann sum for the area under the graph of the force function in Fig. 6.14. The altitude $F(x_i)$ of the dark-shaded rectangle represents a *constant* force, which approximates the force acting over the distance represented by the base of the rectangle. Thus the work done by the force $F(x)$ as it moves the body along the base of the rectangle is approximately

$$F(x_i)\cdot\frac{b-a}{n},$$

which is precisely the area of the rectangle. We view the Riemann sum given by the areas of the rectangles in Fig. 6.16 as adding up little pieces of work done by these constant forces over the subintervals of $[a, b]$. For large values of n, we expect this Riemann sum to approximate closely the work done by the force over the entire interval $[a, b]$. The same Riemann sum should closely approximate the area under the graph of the force function. Thus we expect the actual work done to be given by the area under the graph.

We conclude with two examples, estimating work done and distance traveled using Riemann sums. We take larger values of n in these examples than in the previous ones and use a calculator to find the Riemann sums. Some of the exercises at the end of the section ask you to use a calculator.

EXAMPLE 8 A stone is moved along a straight line by a force acting in the direction of motion. The force is $F(x) = 4/(1 + x^2)$ lb when the stone has been moved x ft. Use a calculator and the midpoint Riemann sum $\mathscr{S}_{20}$ to estimate the work done in moving the stone the first 5 ft.

Solution We have

$$\frac{b-a}{n} = \frac{5-0}{20} = \frac{1}{4}.$$

Our first midpoint is $x_1 = 0 + \frac{1}{2}\cdot\frac{1}{4} = \frac{1}{8}$, and successive increments of $\frac{1}{4} = 0.25$ give the other midpoints x_i. We wish to compute

$$\mathscr{S}_{20} = \frac{5-0}{20}\sum_{i=1}^{20} F(x_i) = \frac{1}{4}\left(F\left(\frac{1}{8}\right) + F\left(\frac{3}{8}\right) + F\left(\frac{5}{8}\right) + \cdots + F\left(\frac{39}{8}\right)\right).$$

Our calculator gives the approximate value 5.494 ft·lb. □

EXAMPLE 9 Suppose a body traveling in a straight line has a velocity of $v = t^2 + 2t$ cm/sec for $0 \le t \le 10$ sec. Estimate the distance the body travels over these 10 sec by finding the midpoint Riemann sum $\mathscr{S}_{20}$, using a calculator.

Solution We have

$$\frac{b-a}{n} = \frac{10-0}{20} = \frac{1}{2}.$$

Our first midpoint is $t_1 = 0 + \frac{1}{2}/2 = \frac{1}{4}$, and successive increments of $\frac{1}{2} = 0.5$

give the other midpoints. We let $f(t) = t^2 + 2t$. Using a calculator, we compute

$$\mathcal{S}_{20} = \frac{b-a}{n}\sum_{i=1}^{20} f(t_i) = \tfrac{1}{2}(f(0.25) + f(0.75) + f(1.25) + \cdots + f(9.75)).$$

We obtain as answer 433.125 cm. □

SUMMARY

1. To write out or compute a sum that is written symbolically using summation notation starting with

$$\sum_{i=1}^{n},$$

simply replace the summation index i successively by all integers from the lower limit 1 to the upper limit n, and add the resulting quantities. For example,

$$\sum_{i=1}^{5} i^3 = 1^3 + 2^3 + 3^3 + 4^3 + 5^3.$$

2. Using Riemann sums as described below, we can estimate

   a. the work done by a nonconstant force,

   b. the distance traveled by a body with positive nonconstant velocity,

   c. the area $A_a^b(f)$ over $[a, b]$ and under a curve $y = f(x)$ of non-constant height.

Let f be continuous on $[a, b]$, and let $[a, b]$ be divided into n subintervals of equal lengths. Let M_i be the maximum value and m_i the minimum value of $f(x)$ over the ith subinterval, and let x_i be any point of the ith subinterval.

3. $S_n = \dfrac{b-a}{n}(M_1 + M_2 + \cdots + M_n) = \dfrac{b-a}{n}\left(\sum_{i=1}^{n} M_i\right)$

4. $s_n = \dfrac{b-a}{n}(m_1 + m_2 + \cdots + m_n) = \dfrac{b-a}{n}\left(\sum_{i=1}^{n} m_i\right)$

5. $\mathcal{S}_n = \dfrac{b-a}{n}(f(x_1) + f(x_2) + \cdots + f(x_n)) = \dfrac{b-a}{n}\left(\sum_{i=1}^{n} f(x_i)\right)$

6. $s_n \le \mathcal{S}_n \le S_n$

7. For $f(x) > 0$, $s_n \le A_a^b(f) \le S_n$

EXERCISES

In Exercises 1 through 6, write out the sum.

1. $\sum_{i=0}^{3} a_i$

2. $\sum_{j=2}^{6} b_j^2$

3. $\sum_{i=1}^{4} a_{2i}$

4. $\sum_{k=4}^{6} (a_{2k} + b_k^2)$

5. $\sum_{i=1}^{5} c^i$

6. $\sum_{i=2}^{4} 2^{a_i}$

In Exercises 7 through 11, compute the sum.

7. $\sum_{i=0}^{3} (i+1)^2$

8. $\sum_{j=2}^{4} 2^j$

9. $\sum_{i=1}^{3} (2i-1)^2$

10. $\sum_{k=1}^{3} (2^k \cdot 3^{k-1})$

11. $\sum_{j=1}^{4} [(-1)^j \cdot j^3]$

In Exercises 12 through 17, express each sum in summation notation. (Many answers are possible.)

12. $a_1b_1 + a_2b_2 + a_3b_3$
13. $a_1b_2 + a_2b_3 + a_3b_4$
14. $a_1 + a_2^2 + a_3^3 + a_4^4$
15. $a_1^2 + a_2^3 + a_3^4$
16. $a_1b_2^2 + a_2b_4^2 + a_3b_6^2$
17. $a_1^{b_3} + a_2^{b_6} + a_3^{b_9}$

18. Show, as in Example 3, that $\sum_{i=1}^{n} (c \cdot a_i) = c(\sum_{i=1}^{n} a_i)$.

19. Show, as in Example 3, that

$$\sum_{i=1}^{n} (a_i + b_i)^2 = \sum_{i=1}^{n} a_i^2 + 2\left(\sum_{i=1}^{n} a_ib_i\right) + \sum_{i=1}^{n} b_i^2.$$

20. Estimate $A_0^4(x^2)$ using the Riemann sum $\mathscr{S}_2$ where x_i is the midpoint of the ith interval.

21. Estimate $A_1^5(1/x)$ using the Riemann sum $\mathscr{S}_4$ where x_i is the midpoint of the ith interval.

22. Estimate $A_{-1}^1(x^2)$ using the Riemann sum $\mathscr{S}_2$ where x_i is the midpoint of the interval.

23. Find the upper sum S_2 and the lower sum s_2 for x^2 from 0 to 2. (The actual value of $A_0^2(x^2)\,dx$ is $\frac{8}{3}$.)

24. Find the upper sum S_4 and the lower sum s_4 for x^2 from 0 to 2.

25. Estimate $A_1^2(1/x)$ by finding the upper sum S_4 and the lower sum s_4.

26. Give the exact value of $A_{-2}^3(4)$ where 4 is the usual constant function.

27. Estimate $A_1^5(x^3)$ using $\mathscr{S}_4$ where x_i is the midpoint of the ith interval.

28. Estimate $A_0^{\pi}(\sin x)$ by finding the upper sum S_4 and the lower sum s_4.

29. Estimate $A_0^{2\pi}(\sin^2 x)$ using $\mathscr{S}_4$ with midpoints x_i.

30. Find graphically the exact amount of work done in moving a body along the x-axis from 0 ft to $x = 10$ ft if the force in the direction of motion at x ft is $F(x) = 1 + x$ lb.

31. Repeat Exercise 30 if the force at x is $F(x) = 3 + x/5$ lb.

32. Estimate the distance traveled by a body moving in one direction on a straight line with velocity $v(t) = 1 + \sin \pi t$ ft/sec for $0 \le t \le 1$ sec, using $\mathscr{S}_4$ with midpoints. (Use a table for $\sin x$ or a calculator.)

33. Repeat Exercise 32 for $v(t) = 1 + 2t^2$ ft/sec for $0 \le t \le 4$ sec.

Estimation using Riemann sums is of practical importance. The estimates are more accurate for values of n larger than those in the preceding pencil-and-paper exercises. Use a calculator or computer in Exercises 34 through 44 and estimate the indicated area, using $\mathscr{S}_n$ with midpoints x_i for the given value of n.

34. $A_0^2(x^2)$, $n = 10$
35. $A_0^2(\sqrt{x})$, $n = 10$
36. $A_1^4\left(\frac{1}{\sqrt{x}}\right)$, $n = 15$
37. $A_0^{\pi}(\sin x)$, $n = 12$
38. $A_1^2\left(\frac{1}{x}\right)$, $n = 10$
39. $A_0^{2\pi}(\sin^2 x)$, $n = 10$
40. $A_0^1\left(\frac{1}{1 + x^2}\right)$, $n = 10$
41. $A_0^5(\sqrt{25 - x^2})$, $n = 10$
42. $A_0^{2\pi}(\sin^2 x^2)$, $n = 12$
43. $A_0^{4\pi}(\cos^2 \sqrt{x})$, $n = 14$
44. $A_0^{2\pi}\left(\frac{\cos^2 x}{x + 1}\right)$, $n = 20$

6.2 THE DEFINITE INTEGRAL

The preceding section introduced Riemann sums for approximating the area of a region of nonconstant altitude, the work done by a nonconstant force, and the distance traveled at nonconstant velocity. All these problems were geometrically equivalent to finding the area $A_a^b(f)$ of the region under the graph of a positive function f over the interval $[a, b]$. Graphically, it appeared that we obtain the best estimates for $A_a^b(f)$ by computing a Riemann sum $\mathscr{S}_n$ for large values of n. This suggests that we might be able to find exact answers by finding $\lim_{n\to\infty} \mathscr{S}_n$. This section discusses this limit, which is known as a *definite integral.* An optional subsection shows how to find $A_a^b(f)$ by actually computing $\lim_{n\to\infty} \mathscr{S}_n$ if f is a polynomial function of degree ≤ 3. Finally, we present some important properties of definite integrals.

THE DEFINITE INTEGRAL AS $\lim_{n\to\infty} \mathscr{S}_n$

Let f be continuous on $[a, b]$. We recall from the previous section the Riemann sums and the relations they satisfy.

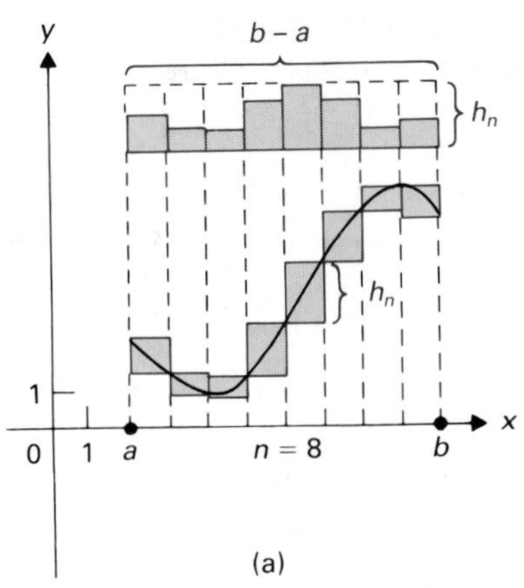

(a)

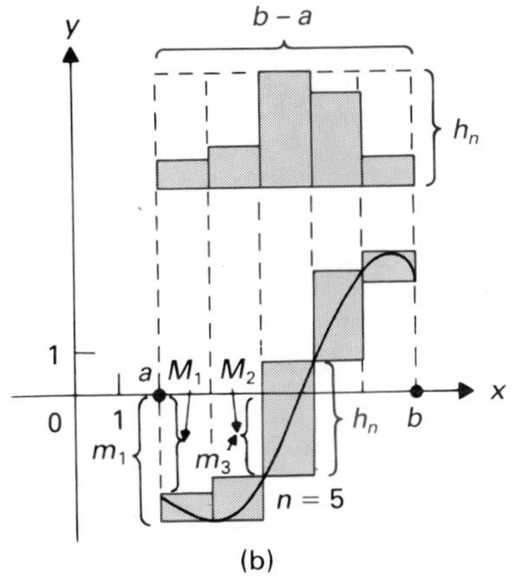

(b)

Figure 6.17 (a) $S_n - s_n$ for $f(x)$ nonnegative in $[a, b]$; (b) $S_n - s_n$ for $f(x)$ sometimes negative in $[a, b]$.

$S_n = \dfrac{b-a}{n}\left(\sum_{i=1}^{n} M_i\right)$	Upper Riemann sum
$s_n = \dfrac{b-a}{n}\left(\sum_{i=1}^{n} m_i\right)$	Lower Riemann sum
$\mathscr{S}_n = \dfrac{b-a}{n}\left(\sum_{i=1}^{n} f(x_i)\right)$	General Riemann sum

$$s_n \le A_a^b(f) \le S_n \tag{1}$$

$$s_n \le \mathscr{S}_n \le S_n. \tag{2}$$

The accuracy of the bounds s_n and S_n for $A_a^b(f)$ can be measured by $S_n - s_n$. Geometrically, $S_n - s_n$ corresponds to the sum of the areas of the small rectangles along the graph shown shaded in Fig. 6.17. (Figure 6.17a illustrates the case $f(x) \ge 0$ for all x in $[a, b]$, and Fig. 6.17b illustrates the case where $f(x)$ is negative for some x in $[a, b]$.) Let h_n be the height of the largest of these small rectangles; that is, h_n is the maximum of $M_i - m_i$. By lining up the shaded rectangles in a row, as shown at the top of Fig. 6.17, we find that the sum of the areas of the small rectangles is less than $h_n(b - a)$, so

$$S_n - s_n \le h_n(b - a). \tag{3}$$

If f is a continuous function, then $f(x)$ is close to $f(x_0)$ for x sufficiently close to x_0, so it is reasonable to expect that, as n becomes large (so that the horizontal dimensions of the small rectangles shaded in Fig. 6.17 become very small), the vertical dimensions become small also. That is, we expect h_n to be close to zero for sufficiently large n, so that

$$\lim_{n\to\infty} h_n = 0.$$

In view of Eq. (3), $\lim_{n\to\infty} h_n = 0$ implies that $S_n - s_n$ is close to zero for sufficiently large n. By Eq. (1), for $f(x) \ge 0$, both S_n and s_n therefore approach $A_a^b(f)$ as n becomes large; that is, the following theorem holds.

THEOREM 6.1 Common limit of Riemann sums

If f is continuous on $[a, b]$, then

$$\lim_{n\to\infty} s_n = \lim_{n\to\infty} \mathscr{S}_n = \lim_{n\to\infty} S_n. \tag{4}$$

DEFINITION 6.2 Definite integral

The common limit of all the Riemann sums in Eq. (4) is the **definite integral of** f **over** $[a, b]$, written $\int_a^b f(x)\,dx$, so

$$\int_a^b f(x)\,dx = \lim_{n\to\infty} s_n = \lim_{n\to\infty} \mathscr{S}_n = \lim_{n\to\infty} S_n. \tag{5}$$

The Leibniz notation $\int_a^b f(x)\,dx$ may be interpreted as follows. We think of the *integral sign* $\int$ as an elongated letter S standing for "sum." If we let

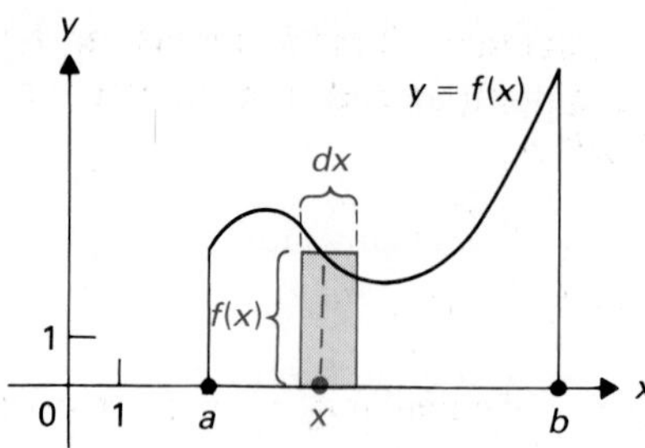

Figure 6.18 Graphic meaning of $f(x)\,dx$ in the notation $\int_a^b f(x)\,dx$.

$dx = \Delta x$, an increment in x, then $f(x)\,dx$ is the area of the rectangle shown in Fig. 6.18. The notation for the integral thus suggests a Riemann sum.

We have drawn the graph of f in most of the figures for the case in which $f(x) \geq 0$ for x in $[a, b]$. If $f(x) < 0$ for x in the ith subinterval of $[a, b]$, then both m_i and M_i are negative, so the contributions $[(b-a)/n]m_i$ to s_n and $[(b-a)/n]M_i$ to S_n are both negative. It is obvious that for a function f where $f(x)$ is sometimes negative and sometimes positive, $\int_a^b f(x)\,dx$ can be interpreted geometrically as the total area given by the portions of the graph above the x-axis, minus the total area given by the portions of the graph below the x-axis (see Fig. 6.19).

EXAMPLE 1 Use geometry to find $\int_1^4 (x-2)\,dx$.

Solution The graph $y = x - 2$ is shown in Fig. 6.20. Since the small shaded triangle is below the x-axis and the large one is above,

$$\int_1^4 (x-2)\,dx = (\text{Area of large triangle}) - (\text{Area of small triangle})$$

$$= \tfrac{1}{2}\cdot 2\cdot 2 - \tfrac{1}{2}\cdot 1\cdot 1 = 2 - \tfrac{1}{2} = \tfrac{3}{2}. \quad \square$$

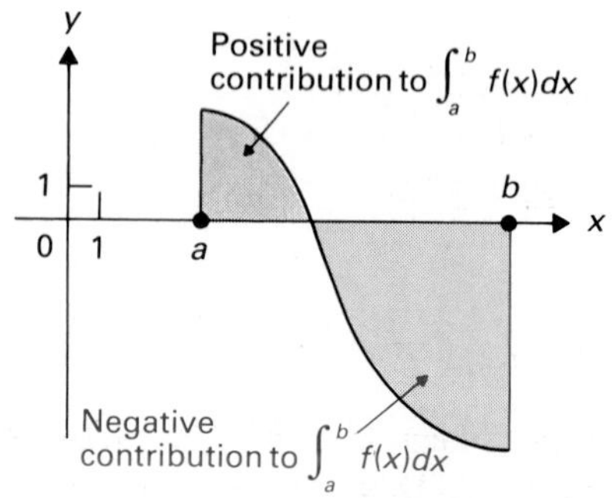
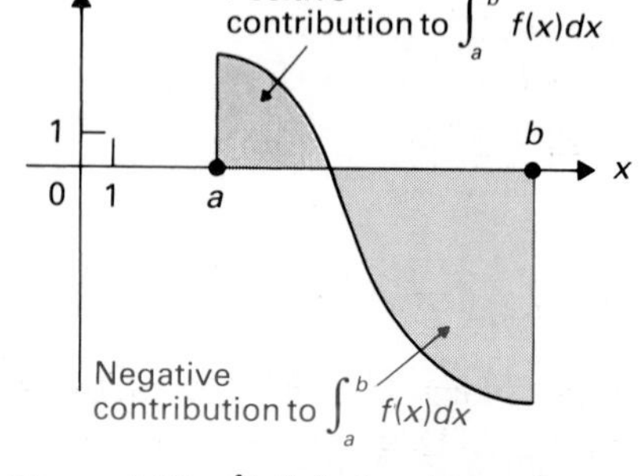

Figure 6.19 $\int_a^b f(x)\,dx$ = Area of black region − Area of red region.

EXAMPLE 2 Find $\int_{-5}^5 \sqrt{25 - x^2}\,dx$ graphically.

Solution Figure 6.21 indicates that the value of the integral is the area of the top half of the disk with center (0, 0) and radius 5. Thus

$$\int_{-5}^5 \sqrt{25 - x^2}\,dx = \frac{1}{2}\pi\cdot 5^2 = \frac{25}{2}\pi. \quad \square$$

We found the exact value of $\int_{-5}^5 \sqrt{25 - x^2}\,dx$ in Example 2. If we estimate the integral using $\mathscr{S}_{20}$ and $\mathscr{S}_{40}$ with midpoints x_i, our calculator gives $\mathscr{S}_{20} \approx 39.4051$ and $\mathscr{S}_{40} \approx 39.3179$. Our calculator also shows that the exact value $25\pi/2$ found in Example 2 is given decimally by $25\pi/2 \approx 39.2699$. This illustrates that $\mathscr{S}_n \to \int_a^b f(x)\,dx$ as $n \to \infty$ and also illustrates that Riemann sums using midpoints may give a fairly good estimate of an integral for quite small values of n.

*COMPUTING INTEGRALS BY LIMITS OF SUMS

Figure 6.20

$\int_1^4 (x-2)\,dx = 2 - \frac{1}{2} = \frac{3}{2}$.

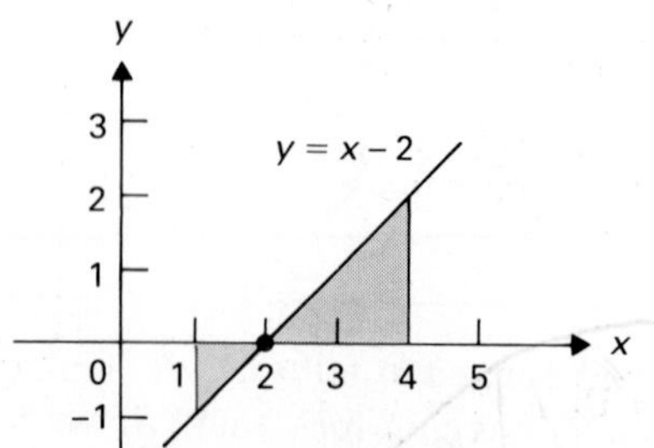

The formulas

$$\sum_{i=1}^n i = \frac{n(n+1)}{2}, \tag{6}$$

$$\sum_{i=1}^n i^2 = \frac{n(n+1)(2n+1)}{6}, \tag{7}$$

and

$$\sum_{i=1}^n i^3 = \left[\frac{n(n+1)}{2}\right]^2 \tag{8}$$

* This subsection can be omitted without loss of continuity.

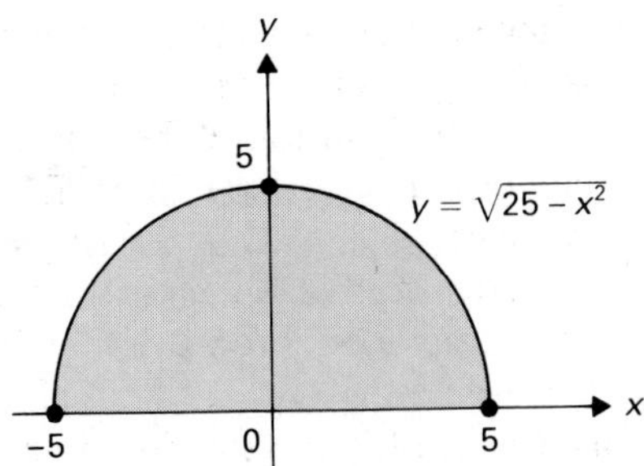

Figure 6.21 $\int_{-5}^{5} \sqrt{25 - x^2}\, dx$ = Area of the shaded half-disk.

can easily be proved using mathematical induction. These formulas can be used to compute definite integrals of linear, quadratic, and cubic polynomial functions, as shown in Examples 3 and 4 below.

EXAMPLE 3 Find the integral $\int_0^4 (5x - 3)\, dx$ using Eq. (6).

Solution Partition [0, 4] into n subintervals of equal length $4/n$. If we take x_i as the right-hand endpoint of the ith subinterval, then

$$x_i = i \cdot \frac{4}{n} = \frac{4i}{n}.$$

Then

$$\mathcal{S}_n = \frac{4}{n} \sum_{i=1}^{n} \left(5 \cdot \frac{4i}{n} - 3\right) = \frac{4}{n}\left[\frac{20}{n}\left(\sum_{i=1}^{n} i\right) - \sum_{i=1}^{n} 3\right]$$
$$= \frac{4}{n}\left[\frac{20}{n} \cdot \frac{n(n+1)}{2} - 3n\right] = \frac{80(n+1)}{2n} - 12 = \frac{40(n+1)}{n} - 12.$$

Therefore

$$\int_0^4 (5x - 3)\, dx = \lim_{n\to\infty} \mathcal{S}_n = \lim_{n\to\infty} \left[\frac{40(n+1)}{n}\right] - 12$$
$$= 40 - 12 = 28. \quad \square$$

EXAMPLE 4 Use Eqs. (6) and (7) to find $\int_2^4 (x^2 + 2x)\, dx$.

Solution Partition [2, 4] into n subintervals of equal length $2/n$, and let x_i be the right-hand endpoint of the ith subinterval. Then

$$x_i = 2 + i \cdot \frac{2}{n} = 2 + \frac{2i}{n},$$

and

$$\mathcal{S}_n = \frac{2}{n} \cdot \sum_{i=1}^{n} \left[\left(2 + \frac{2i}{n}\right)^2 + 2\left(2 + \frac{2i}{n}\right)\right]$$
$$= \frac{2}{n} \cdot \sum_{i=1}^{n} \left(4 + \frac{8i}{n} + \frac{4i^2}{n^2} + 4 + \frac{4i}{n}\right)$$
$$= \frac{2}{n}\left(\sum_{i=1}^{n} 8 + \frac{12}{n} \cdot \sum_{i=1}^{n} i + \frac{4}{n^2} \cdot \sum_{i=1}^{n} i^2\right)$$
$$= \frac{2}{n}\left[8n + \frac{12}{n} \cdot \frac{n(n+1)}{2} + \frac{4}{n^2} \cdot \frac{n(n+1)(2n+1)}{6}\right]$$
$$= 16 + \frac{12n(n+1)}{n^2} + \frac{4n(n+1)(2n+1)}{3n^3}.$$

Taking the limit as $n \to \infty$, we obtain

$$\int_2^4 (x^2 + 2x)\, dx = \lim_{n\to\infty} \mathcal{S}_n = 16 + 12 + \frac{8}{3} = \frac{92}{3}. \quad \square$$

PROPERTIES OF THE DEFINITE INTEGRAL

Let $f(x)$ and $g(x)$ both be continuous in $[a, b]$. We state three properties of the definite integral in Eqs. (9), (10), and (11). Properties (9) and (10) parallel properties of the derivative. When the properties are stated in words, the parallel is striking. We can take our choice of "derivative" or "integral" in the following:

The $\left\{\begin{matrix} \textit{derivative} \\ \textit{integral} \end{matrix}\right\}$ of a sum is the sum of the $\left\{\begin{matrix} \textit{derivatives} \\ \textit{integrals} \end{matrix}\right\}$.

The $\left\{\begin{matrix} \textit{derivative} \\ \textit{integral} \end{matrix}\right\}$ of a constant times a function is the constant times the $\left\{\begin{matrix} \textit{derivative} \\ \textit{integral} \end{matrix}\right\}$ of the function.

THEOREM 6.2 Properties of the integral

Three properties of the definite integral are

$$\int_a^b (f(x) + g(x))\,dx = \int_a^b f(x)\,dx + \int_a^b g(x)\,dx, \tag{9}$$

$$\int_a^b c \cdot f(x)\,dx = c \cdot \int_a^b f(x)\,dx \qquad \text{for any constant } c, \tag{10}$$

$$\int_a^b f(x)\,dx = \int_a^h f(x)\,dx + \int_h^b f(x)\,dx \qquad \text{for any } h \text{ in } [a, b]. \tag{11}$$

All these properties are evident if we think of the interpretation in terms of areas. Figure 6.22 illustrates Eq. (11); surely the area under $f(x)$ from a to b is the area from a to h plus the area from h to b. If you look back at the Riemann sums $\mathcal{S}_n(f)$, $\mathcal{S}_n(g)$, and $\mathcal{S}_n(f + g)$ for $f(x)$, $g(x)$, and their sum, then

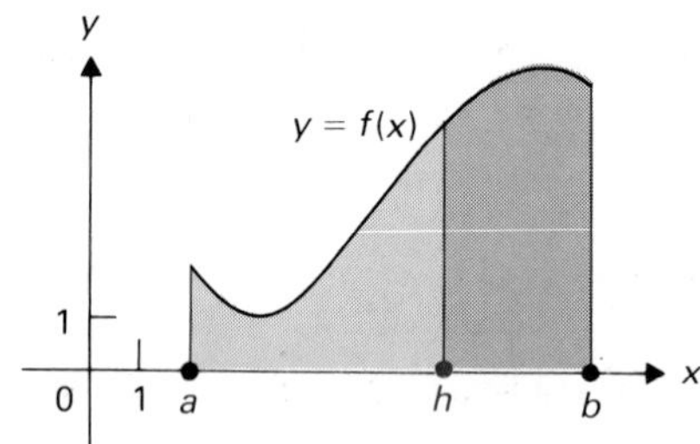

Figure 6.22

$\int_a^b f(x)\,dx = \int_a^h f(x)\,dx + \int_h^b f(x)\,dx$

$$\mathcal{S}_n(f + g) = \frac{b - a}{n}\left(\sum_{i=1}^{n} (f(x_i) + g(x_i))\right)$$

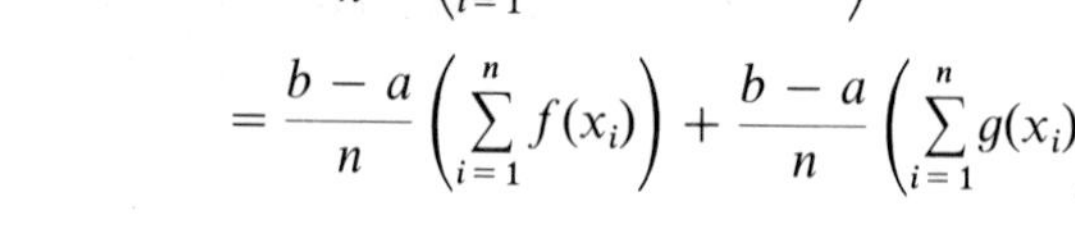

$$= \frac{b - a}{n}\left(\sum_{i=1}^{n} f(x_i)\right) + \frac{b - a}{n}\left(\sum_{i=1}^{n} g(x_i)\right) = \mathcal{S}_n(f) + \mathcal{S}_n(g).$$

Therefore

$$\int_a^b (f(x) + g(x))\,dx = \lim_{n\to\infty} \mathcal{S}_n(f + g) = \lim_{n\to\infty} (\mathcal{S}_n(f) + \mathcal{S}_n(g))$$

$$= \lim_{n\to\infty} \mathcal{S}_n(f) + \lim_{n\to\infty} \mathcal{S}_n(g) = \int_a^b f(x)\,dx + \int_a^b g(x)\,dx.$$

This establishes property (9). Property (10) is established in a similar way.

For another property, note that if $b = a$, then

$$\boxed{\int_a^a f(x)\,dx = 0.} \tag{12}$$

We would like $\int_a^b f(x)\,dx = \int_a^h f(x)\,dx + \int_h^b f(x)\,dx$ to hold whether or not h is between a and b. (Naturally we require that a, b, and h all lie in some interval where $f(x)$ is continuous.) Then if we set $b = a$, we would have

$$0 = \int_a^a f(x)\,dx = \int_a^h f(x)\,dx + \int_h^a f(x)\,dx,$$

which leads to

$$\int_h^a f(x)\,dx = -\int_a^h f(x)\,dx. \qquad \textbf{(13)}$$

We are motivated by Eq. (13) to define $\int_b^a f(x)\,dx$ if $b > a$ by

$$\boxed{\int_b^a f(x)\,dx = -\int_a^b f(x)\,dx.} \qquad \textbf{(14)}$$

We will see again at the end of the next section that Eq. (14) is a convenient definition.

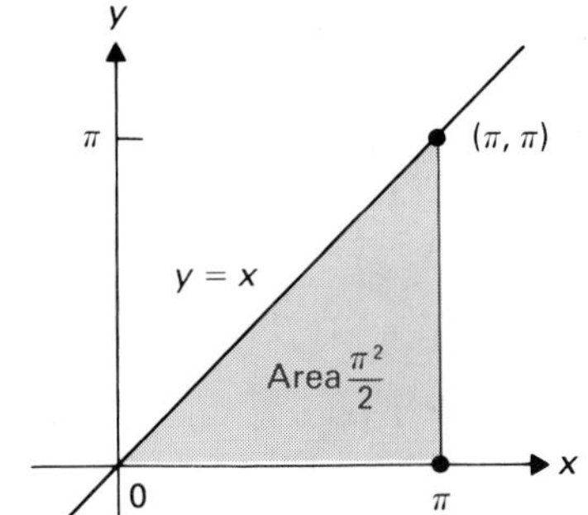

Figure 6.23

$\int_0^\pi x\,dx = (\frac{1}{2}\pi)\pi = \pi^2/2$

EXAMPLE 5 Given that $\int_0^\pi \sin^2 x\,dx = \pi/2$, find $\int_0^\pi (x + \sin^2 x)\,dx$.

Solution From property (9), we have

$$\int_0^\pi (x + \sin^2 x)\,dx = \int_0^\pi x\,dx + \int_0^\pi \sin^2 x\,dx.$$

From Fig. 6.23, we see graphically that $\int_0^\pi x\,dx = \pi^2/2$. Thus

$$\int_0^\pi (x + \sin^2 x)\,dx = \frac{\pi^2}{2} + \frac{\pi}{2} = \frac{\pi}{2}(\pi + 1). \quad \square$$

EXAMPLE 6 If $\int_1^4 f(x)\,dx = -5$ and $\int_1^2 2 \cdot f(x)\,dx = -1$, find $\int_2^4 f(x)\,dx$.

Solution By property (11), we have

$$\int_1^4 f(x)\,dx = \int_1^2 f(x)\,dx + \int_2^4 f(x)\,dx.$$

From property (10), we see that $\int_1^2 2 \cdot f(x)\,dx = 2\int_1^2 f(x)\,dx = -1$, so $\int_1^2 f(x)\,dx = -\frac{1}{2}$. Thus

$$-5 = -\frac{1}{2} + \int_2^4 f(x)\,dx,$$

so $\int_2^4 f(x)\,dx = -\frac{9}{2}$. $\square$

SUMMARY

Let $f(x)$ and $g(x)$ be continuous in $[a, b]$.

1. $\displaystyle\int_a^b f(x)\,dx = \lim_{n\to\infty} S_n = \lim_{n\to\infty} s_n = \lim_{n\to\infty} \mathscr{S}_n$

2. $\displaystyle\int_a^b (f(x) + g(x))\,dx = \int_a^b f(x)\,dx + \int_a^b g(x)\,dx$

3. $\displaystyle\int_a^b c \cdot f(x)\,dx = c \cdot \int_a^b f(x)\,dx$ for any constant c

4. $\displaystyle\int_a^b f(x)\,dx = \int_a^h f(x)\,dx + \int_h^b f(x)\,dx$ for any h in $[a, b]$

5. $\displaystyle\int_b^a f(x)\,dx = -\int_a^b f(x)\,dx$

EXERCISES

1. Let $f(x) = 1 - x$, and consider the interval $[0, 2]$.
   a) Sketch the graph of f over this interval.
   b) Find S_2 for this interval. (Note that $f(x)$ is sometimes negative for x in $[0, 2]$.)
   c) Find s_2 for this interval.
   d) What common value do both S_n and s_n approach as n gets large?
2. a) Find the upper sum S_4 and the lower sum s_4 for x^3 from -1 to 1.
   b) From the graph of x^3, what common number do both S_n and s_n approach as n gets large?

In Exercises 3 through 12, sketch a region whose area is given by the integral. Determine the value of the integral by finding the area of the region.

3. $\displaystyle\int_0^2 x\,dx$
4. $\displaystyle\int_{-1}^2 x\,dx$
5. $\displaystyle\int_1^3 (2x + 3)\,dx$
6. $\displaystyle\int_{-3}^4 (2x - 1)\,dx$
7. $\displaystyle\int_{-1}^1 (x + 1)\,dx$
8. $\displaystyle\int_{-3}^1 (x - 1)\,dx$
9. $\displaystyle\int_{-3}^3 \sqrt{9 - x^2}\,dx$
10. $\displaystyle\int_0^4 \sqrt{16 - x^2}\,dx$
11. $\displaystyle\int_0^4 (3 + \sqrt{16 - x^2})\,dx$
12. $\displaystyle\int_{-3}^3 (5 - \sqrt{9 - x^2})\,dx$
13. Sketch the graph of a function f over the interval $[0, 2]$ for which S_2 is much closer to $\int_0^2 f(x)\,dx$ than the average of S_2 and s_2. (This shows that the averaging technique used in Example 8 of Section 6.1 may not give good results.)
14. Sketch the graph of a function f over the interval $[0, 6]$ for which $S_3 > S_2$. (This shows that while S_n approaches $\int_a^b f(x)\,dx$ for large n, we need not have $S_{n+1} \leq S_n$.)

In Exercises 15 through 30, find the exact value of the integral using geometric arguments, properties of the integral, and the fact (which we will soon show) that $\int_0^\pi \sin x\,dx = 2$.

15. $\displaystyle\int_0^{\pi/2} \sin x\,dx$
16. $\displaystyle\int_{-\pi}^{\pi} \sin x\,dx$
17. $\displaystyle\int_0^{\pi/2} \cos x\,dx$
18. $\displaystyle\int_{-\pi/2}^{\pi/2} \cos x\,dx$
19. $\displaystyle\int_{-2}^2 (x^3 - 3x)\,dx$
20. $\displaystyle\int_{-2}^2 (2x^3 + 4)\,dx$
21. $\displaystyle\int_{-3}^3 (x^5 - 2x^3 - 1)\,dx$
22. $\displaystyle\int_0^\pi \cos x\,dx$
23. $\displaystyle\int_0^{5\pi} \sin x\,dx$
24. $\displaystyle\int_0^{2\pi} |\sin x|\,dx$
25. $\displaystyle\int_{-\pi}^{\pi} |\cos x|\,dx$
26. $\displaystyle\int_{-1}^1 |2x|\,dx$
27. $\displaystyle\int_0^\pi (2 \sin x - 3 \cos x)\,dx$
28. $\displaystyle\int_0^{2\pi} (3|\sin x| + 2|\cos x|)\,dx$
29. $\displaystyle\int_0^{\pi/2} (2x - 3 \cos x)\,dx$
30. $\displaystyle\int_0^\pi (4x + 1 - 2 \sin x)\,dx$

In Exercises 31 through 40, assume that $f(x)$ and $g(x)$ are continuous for all x. Find the indicated integral from the given data.

31. $\displaystyle\int_0^1 (f(x) - g(x))\,dx = 3, \int_0^1 g(x)\,dx = -1, \int_0^1 3 \cdot f(x)\,dx =$ _____
32. $\displaystyle\int_0^1 (f(x) + 2 \cdot g(x))\,dx = 8, \int_0^1 (2 \cdot f(x) - g(x))\,dx = -2, \int_0^1 f(x)\,dx =$ _____
33. $\displaystyle\int_2^5 (f(x) - g(x))\,dx = 1, \int_2^5 2 \cdot g(x)\,dx = 4, \int_2^5 f(x)\,dx =$ _____
34. $\displaystyle\int_1^2 f(x)\,dx = 4, \int_2^4 3 \cdot f(x)\,dx = 5, \int_1^4 2 \cdot f(x)\,dx =$ _____
35. $\displaystyle\int_0^2 g(x)\,dx = 3, \int_1^2 2 \cdot g(x)\,dx = 8, \int_0^1 g(x)\,dx =$ _____
36. $\displaystyle\int_0^4 g(x)\,dx = -3, \int_4^2 3 \cdot g(x)\,dx = -2, \int_0^2 2 \cdot g(x)\,dx =$ _____

37. $\int_1^{10} f(x)\,dx = 7$, $\int_5^{10} 3 \cdot f(x)\,dx = 6$,
$\int_5^1 2 \cdot f(x)\,dx =$ ____

38. $\int_1^4 (f(x) - g(x))\,dx = 10$, $\int_4^1 (f(x) + g(x))\,dx = 3$,
$\int_0^4 g(x)\,dx = 5$, $\int_0^1 g(x)\,dx =$ ____

39. $\int_0^{10} f(x)\,dx = 5$, $\int_0^6 f(x)\,dx = 3$, $\int_{10}^4 2 \cdot f(x)\,dx = 6$,
$\int_6^4 f(x)\,dx =$ ____

40. $\int_1^8 g(x)\,dx = 4$, $\int_5^1 2 \cdot g(x)\,dx = 6$, $\int_2^8 g(x)\,dx = 5$,
$\int_2^5 g(x)\,dx =$ ____

*In Exercises 41 through 48, use Eqs. (6), (7), and (8) and Definition 6.2 to find the value of the definite integral.

*41. $\int_0^3 (x - 2)\,dx$

*42. $\int_0^2 (4x^2 + 5)\,dx$

*43. $\int_3^5 (7 - 2x)\,dx$

*44. $\int_2^5 (x^2 - 4x + 2)\,dx$

*45. $\int_0^1 x^3\,dx$

*46. $\int_0^2 (2x^3 - x^2)\,dx$

*47. $\int_1^2 (x - x^3)\,dx$

*48. $\int_1^3 (4x^3 - 3x^2 + 2x - 10)\,dx$

In Exercises 49 through 52, use a calculator to estimate the given integral using $\mathscr{S}_n$ for the indicated value of n, where x_i is the midpoint of the ith subinterval.

49. $\int_1^2 x^x\,dx$, with $n = 8$

50. $\int_1^3 2^{\sin x}\,dx$, where $n = 10$

51. $\int_0^2 \cos x^2 dx$, with $n = 15$

52. $\int_1^5 x^x\,dx$, with $n = 20$

6.3 THE FUNDAMENTAL THEOREM OF CALCULUS

Our development of the definite integral culminates in this section with the *fundamental theorem of calculus*. The theorem is as important as its name suggests, for it provides us with a technique for finding the exact value of many definite integrals. The theorem also shows that while not every continuous function is differentiable (recall that $|x|$ is not differentiable at $x = 0$), every function continuous on an interval has an antiderivative there.

FINDING THE EXACT VALUE OF $\int_a^b f(x)\,dx$

Let $f(x)$ be continuous in $[a, b]$. In the preceding section, we defined $\int_a^b f(x)\,dx$, and we have had some practice in estimating the integral using Riemann sums. The definite integral has a great variety of important applications, so it is highly desirable to have an easy way to compute $\int_a^b f(x)\,dx$. It turns out that the *exact* value of the integral can be found easily if we can find an *antiderivative* $F(x)$ of $f(x)$. On page 209 of Section 6.1 we did a quotient-versus-product analysis of the derivative versus the integral. Since division and multiplication are opposite, or inverse, operations, it is not surprising that the opposite of differentiation, that is, antidifferentiation, should be involved in finding the integral.

We now state without proof a rule for computing $\int_a^b f(x)\,dx$ exactly for many functions f. Then we illustrate the rule. In the next subsection we state and prove the fundamental theorem of calculus, which includes this rule.

Remember that we always assume that f is continuous in $[a, b]$.

Find $F(x)$ such that $F'(x) = f(x)$. Then

$$\int_a^b f(x)\,dx = F(b) - F(a). \tag{1}$$

Rule for computing $\int_a^b f(x)\,dx$

EXAMPLE 1 Find the exact value of $\int_0^1 x^2\,dx$, which we estimated several times in Section 6.2.

Solution An antiderivative of the function x^2 is $F(x) = x^3/3$. Thus, by rule (1),

$$\int_0^1 x^2\,dx = F(1) - F(0) = \frac{1}{3} - \frac{0}{3} = \frac{1}{3}.$$

After all our work trying to get a good estimate for this definite integral in Section 6.2, we should appreciate the elegance and beauty of this easy computation. □

It is customary to denote $F(b) - F(a)$ by $F(x)]_a^b$. The *upper limit of integration* is b and the *lower limit of integration* is a. For example, we will usually see $\int_0^1 x^2\,dx$ computed as follows:

$$\int_0^1 x^2\,dx = \frac{x^3}{3}\Big]_0^1 = \frac{1}{3} - \frac{0}{3} = \frac{1}{3}.$$

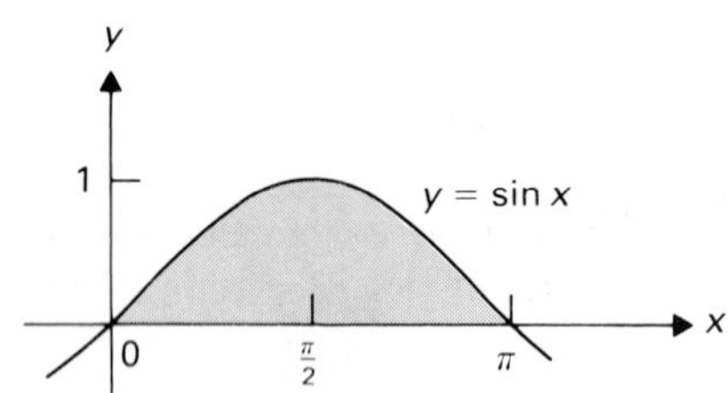

Figure 6.24 Region under one "arch" of $y = \sin x$.

EXAMPLE 2 Find the area of the region under one "arch" of the curve $y = \sin x$, as shown in Fig. 6.24.

Solution Note that $-\cos x$ is an antiderivative of $\sin x$. The area is given by

$$\int_0^\pi \sin x\,dx = -\cos x\Big]_0^\pi = -\cos \pi - (-\cos 0)$$
$$= -(-1) - (-1) = 1 + 1 = 2. \quad □$$

EXAMPLE 3 Sketch the region bounded by the graph of the polynomial function $1 - x^2$ and by the x-axis, and find the area of the region.

Figure 6.25 Region bounded by $y = 1 - x^2$ and the x-axis.

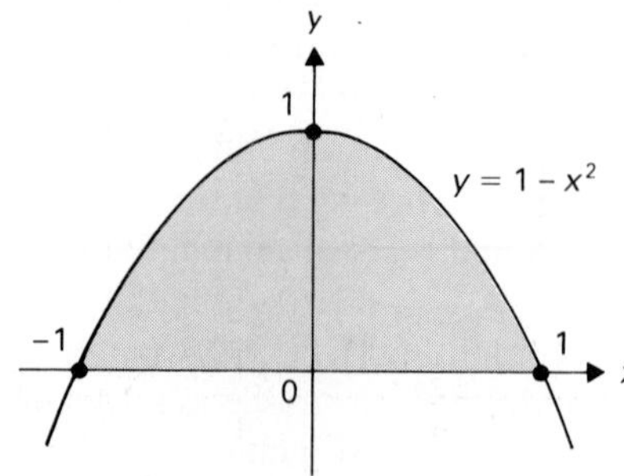

Solution The function $1 - x^2$ obviously has a maximum at $x = 0$ and has decreasing values as x gets farther from zero. The graph crosses the x-axis when $1 - x^2 = 0$, that is, when $x = \pm 1$. The graph is shown in Fig. 6.25, where we have shaded the region whose area we wish to find. Clearly the desired area is $\int_{-1}^1 (1 - x^2)\,dx$ square units, and

$$\int_{-1}^1 (1 - x^2)\,dx = \left(x - \frac{x^3}{3}\right)\Big]_{-1}^1$$
$$= \left(1 - \frac{1}{3}\right) - \left(-1 - \frac{-1}{3}\right) = \frac{2}{3} - \left(-\frac{2}{3}\right) = \frac{4}{3}. \quad □$$

EXAMPLE 4 Find $\int_0^2 \sqrt{36 - 9x^2}\, dx$.

Solution We have not yet learned how to find an antiderivative of $f(x) = \sqrt{36 - 9x^2}$. You might experiment for a bit, but unless you know more calculus than we have covered earlier in this text, you will not find an antiderivative of $\sqrt{36 - 9x^2}$.

Note that $\sqrt{36 - 9x^2} = 3\sqrt{4 - x^2}$, so our integral is equal to

$$3 \int_0^2 \sqrt{4 - x^2}\, dx.$$

This is an integral we can evaluate geometrically, as discussed in the preceding section. As Fig. 6.26 indicates, $\int_0^2 \sqrt{4 - x^2}\, dx$ is the area of a quarter-disk of radius 2. Thus

$$\int_0^2 \sqrt{36 - 9x^2}\, dx = 3 \int_0^2 \sqrt{4 - x^2}\, dx = 3 \cdot \frac{\pi \cdot 2^2}{4} = 3\pi. \quad \square$$

It may seem now that Sections 6.1 and 6.2 on estimating $\int_a^b f(x)\, dx$ were a complete waste of time, in view of our powerful rule (1). However, a clear understanding of Sections 6.1 and 6.2 is needed to recognize applied problems that can be solved using an integral. Also, finding an antiderivative $F(x)$ of $f(x)$ may not be an easy matter, even if $f(x)$ is a fairly simple function. For example, it can be shown that while an antiderivative of $\sin x^2$ exists, it is impossible to express it as an elementary function (a function found by algebraic operations from rational, trigonometric, exponential, or logarithmic functions). For integrals such as $\int_0^1 \sin x^2\, dx$, we have to estimate the integral numerically. The preceding section showed one possible method.

THE FUNDAMENTAL THEOREM OF CALCULUS

Now we will show why antidifferentiation enables us to find the area under the graph of a function $f(x)$. The trick is to consider the integral function

$$F(t) = \int_a^t f(x)\, dx \qquad \text{for } a \le t \le b, \tag{2}$$

so that if $f(x) \ge 0$, $F(t)$ is the area under the graph of f over $[a, t]$, as shown in Fig. 6.27. We will show that $F'(t) = f(t)$, and this will remove any remain-

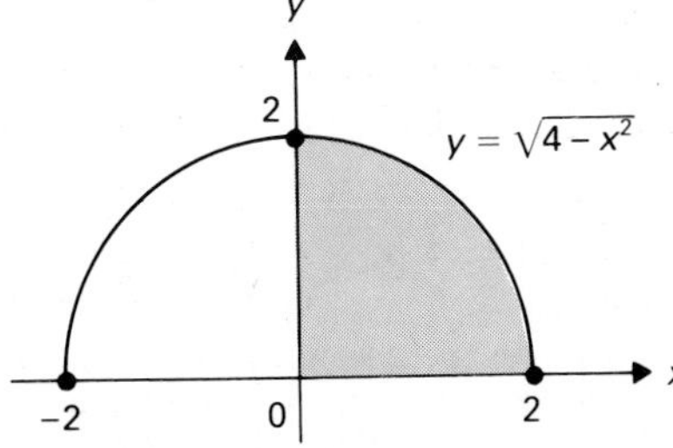

Figure 6.26 Region with area $\int_0^2 \sqrt{4 - x^2}\, dx$.

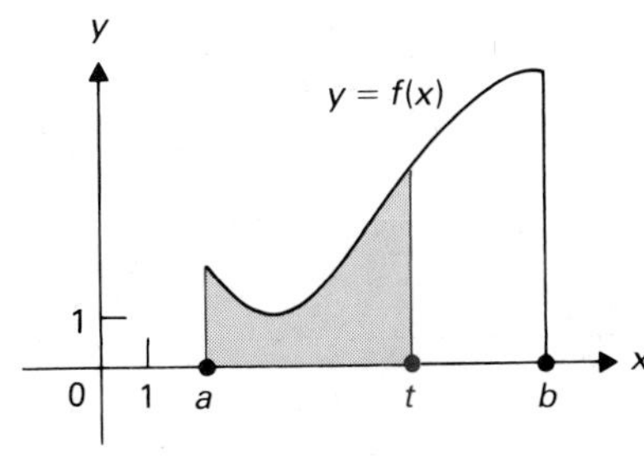

Figure 6.27 Region of area $F(t) = \int_a^t f(x)\, dx$.

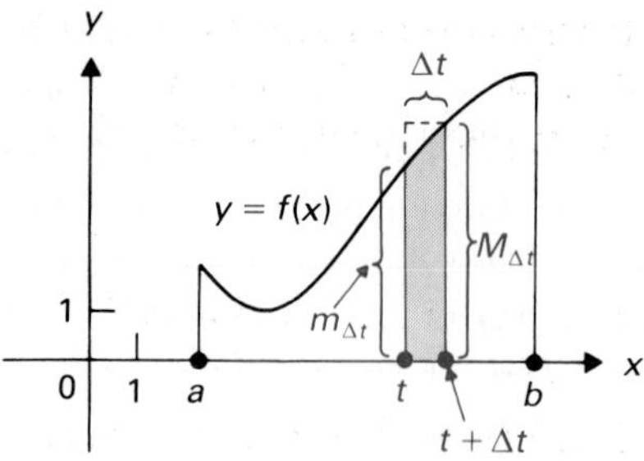

Figure 6.28
$m_{\Delta t} \cdot \Delta t \leq \int_t^{t+\Delta t} f(x)\,dx \leq M_{\Delta t} \cdot \Delta t$

ing mystery.* If the derivative of the integral function F is f, then surely to find F we should find an antiderivative of f.

By a property of the integral,

$$\int_a^{t+\Delta t} f(x)\,dx = \int_a^t f(x)\,dx + \int_t^{t+\Delta t} f(x)\,dx.$$

Using the definition of the derivative, we have

$$F'(t) = \lim_{\Delta t \to 0} \frac{F(t+\Delta t) - F(t)}{\Delta t}$$

$$= \lim_{\Delta t \to 0} \frac{\int_a^{t+\Delta t} f(x)\,dx - \int_a^t f(x)\,dx}{\Delta t} = \lim_{\Delta t \to 0} \frac{\int_t^{t+\Delta t} f(x)\,dx}{\Delta t}. \tag{3}$$

Referring to Fig. 6.28, we see that

$$m_{\Delta t} \cdot \Delta t \leq \int_t^{t+\Delta t} f(x)\,dx \leq M_{\Delta t} \cdot \Delta t, \tag{4}$$

where $m_{\Delta t}$ is the minimum and $M_{\Delta t}$ the maximum value of $f(x)$ in $[t, t+\Delta t]$. Therefore

$$m_{\Delta t} \leq \frac{\int_t^{t+\Delta t} f(x)\,dx}{\Delta t} \leq M_{\Delta t}. \tag{5}$$

Now $\lim_{\Delta t \to 0} m_{\Delta t} = \lim_{\Delta t \to 0} M_{\Delta t} = f(t)$ since f is continuous. We combine Eq. (3) with Eq. (5) to obtain

$$F'(t) = \lim_{\Delta t \to 0} \frac{\int_t^{t+\Delta t} f(x)\,dx}{\Delta t} = f(t). \tag{6}$$

This concludes our demonstration that $F'(t) = f(t)$, which is an important result in its own right.

We can now state the fundamental theorem.

THEOREM 6.3 Fundamental theorem of calculus

Let f be continuous in $[a, b]$.

1. If $F(t) = \int_a^t f(x)\,dx$ for $a \leq t \leq b$, then F is differentiable and $F'(t) = f(t)$, that is,

$$\frac{d}{dt}\left(\int_a^t f(x)\,dx\right) = f(t). \tag{7}$$

2. If $F(x)$ is *any* antiderivative of $f(x)$, then

$$\int_a^b f(x)\,dx = F(b) - F(a). \tag{8}$$

---

* Note that $F(t)$ is defined for $a \leq t \leq b$. Derivatives of $F(t)$ at a and at b would have to be taken from one side only, which we have not discussed. The notion is quite obvious. One forms the usual difference quotient but takes only positive Δt at a and only negative Δt at b. We will gloss over this point in what follows and refer to F as differentiable on $[a, b]$.

Note that Part 1 of the fundamental theorem really asserts the *existence* of an antiderivative for a function f continuous on $[a, b]$. It even gives a formula for one such antiderivative F. We state this existence as a corollary.

COROLLARY

If f is continuous on $[a, b]$, then f has an antiderivative on this interval.

Part 1 of Theorem 6.3 was proved before the theorem was stated. Before we prove Part 2, which is our rule for computing $\int_a^b f(x)\,dx$ given in rule (1), we give specific illustrations of Part 1.

EXAMPLE 5 Find $\frac{d}{dt}\left(\int_0^t \sqrt{x^2+1}\,dx\right)$.

Solution Part 1 of the fundamental theorem tells us immediately that

$$\frac{d}{dt}\left(\int_0^t \sqrt{x^2+1}\,dx\right) = \sqrt{t^2+1}. \quad \square$$

EXAMPLE 6 Find $\frac{d}{dt}\left(\int_t^{-3} \sin^2 x\,dx\right)$.

Solution Part 1 of the fundamental theorem concerns an integral with t as upper limit of integration. Here t is the lower limit. But we may write

$$\int_t^{-3} \sin^2 x\,dx = -\int_{-3}^t \sin^2 x\,dx.$$

Then

$$\frac{d}{dt}\left(\int_t^{-3} \sin^2 x\,dx\right) = -\frac{d}{dt}\left(\int_{-3}^t \sin^2 x\,dx\right) = -\sin^2 t. \quad \square$$

EXAMPLE 7 Find $\frac{d}{dt}\left(\int_0^{t^2} \cos x^2\,dx\right)$.

Solution We let $u = t^2$. Then by the chain rule and Part 1 of the fundamental theorem,

$$\frac{d}{dt}\left(\int_0^{t^2} \cos x^2\,dx\right) = \frac{d}{du}\left(\int_0^{u} \cos x^2\,dx\right)\cdot\frac{du}{dt}$$
$$= (\cos u^2)2t = (\cos t^4)2t. \quad \square$$

EXAMPLE 8 Find

$$\frac{d}{dt}\left(\int_{-2t}^{t} \frac{1}{1+x^2}\,dx\right).$$

Solution Using properties of the integral, the chain rule as in Example 7, and Part 1 of Theorem 6.3, we have

$$\frac{d}{dt}\left(\int_{-2t}^{t} \frac{1}{1+x^2}\,dx\right)$$

$$= \frac{d}{dt}\left(\int_{-2t}^{0} \frac{1}{1+x^2}\,dx + \int_{0}^{t} \frac{1}{1+x^2}\,dx\right) \qquad \text{Property (11), page 220}$$

$$= \frac{d}{dt}\left(-\int_0^{-2t} \frac{1}{1+x^2}\,dx\right) + \frac{1}{1+t^2} \qquad \text{Property (14), page 221}$$

$$= -\frac{1}{1+(-2t)^2}\cdot\frac{d(-2t)}{dt} + \frac{1}{1+t^2} \qquad \text{Chain rule}$$

$$= \frac{-1}{1+4t^2}(-2) + \frac{1}{1+t^2} = \frac{2}{1+4t^2} + \frac{1}{1+t^2}. \quad \square$$

We now prove Part 2 of the fundamental theorem, showing that Eq. (8) can be used to compute $\int_a^b f(x)\,dx$. Let $F(t) = \int_a^t f(x)\,dx$. Then

$$\int_a^b f(x)\,dx = F(b). \tag{9}$$

But $F(a) = \int_a^a f(x)\,dx = 0$, so we can write Eq. (9) as

$$\int_a^b f(x)\,dx = F(b) - F(a). \tag{10}$$

Theorem 6.3 asserted that Eq. (10) is true for *any* antiderivative of $f(x)$, not just our integral function F. Let $G(x)$ be any antiderivative, so $G'(x) = f(x)$. Since $F(x)$ and $G(x)$ have the same derivative $f(x)$, we know from Theorem 5.7 on page 199 that $F(x) = G(x) + C$ for some constant C, and

$$\int_a^b f(x)\,dx = F(b) - F(a) = (G(b) + C) - (G(a) + C) = G(b) - G(a).$$

Hence $\int_a^b f(x)\,dx$ can be computed by finding *any* antiderivative of $f(x)$ and subtracting its value at a from its value at b, as stated in Eq. (8).

Interchanging a and b in Eq. (8) yields

$$\int_b^a f(x)\,dx = F(a) - F(b) = -(F(b) - F(a)) = -\int_a^b f(x)\,dx,$$

which is consistent with our definition of $\int_b^a f(x)\,dx$ in the preceding section.

SUMMARY

Fundamental theorem of calculus: Let f be continuous in $[a, b]$. Then

1. $\frac{d}{dt}\left(\int_a^t f(x)\,dx\right) = f(t)$;

2. if $F'(x) = f(x)$, then $\int_a^b f(x)\,dx = F(b) - F(a)$.

EXERCISES

1. State Part 1 of the fundamental theorem of calculus without referring to the text.

2. State Part 2 of the fundamental theorem of calculus without referring to the text.

In Exercises 3 through 42, use the fundamental theorem, properties of the definite integral, geometry, and, where necessary, the fact that $\int_0^{\pi} \sin^2 x\,dx = \pi/2$ to evaluate the definite integral.

3. $\int_0^1 x^3\,dx$

4. $\int_{-1}^1 x^4\,dx$

5. $\int_0^2 (x^2 + 3x - 1)\,dx$

6. $\int_4^9 \sqrt{x}\,dx$

7. $\int_1^8 x^{1/3}\,dx$

8. $\int_1^2 \frac{dx}{x^2}$

9. $\int_{-2}^{-1} \frac{dx}{x^3}$

10. $\int_1^4 \left(\frac{2}{x^2} + \sqrt{x}\right) dx$

11. $\int_{-3}^{-1} \frac{x-2}{x^3}\,dx$

12. $\int_1^8 \frac{x^{1/3} + x^{1/2}}{x}\,dx$

13. $\int_0^{\pi} \sin x\,dx$

14. $\int_0^{\pi/2} 4\cos x\,dx$

15. $\int_0^{\pi/2} (\sin x + 2\cos x)\,dx$

16. $\int_{\pi/4}^{\pi/2} (\sin x - \cos x)\,dx$

17. $\int_{-\pi/4}^{\pi/4} 3\cos x\,dx$

18. $\int_0^{\pi/4} \sec^2 x\,dx$

19. $\int_{\pi/6}^{\pi/3} \csc^2 x\,dx$

20. $\int_{-\pi/6}^{\pi/3} \sec x \tan x\,dx$

21. $\int_{\pi/6}^{\pi/2} \csc x \cot x\,dx$

22. $\int_1^0 x^2\,dx$

23. $\int_3^{-2} 4\,dx$

24. $\int_3^{-2} (-4)\,dx$

25. $\int_{-1}^{-2} x\,dt$

26. $\int_4^4 \sqrt{x^3+1}\,dx$

27. $\int_1^{-1} \sqrt{1-x^2}\,dx$

28. $\int_0^1 (x + \sqrt{1-x^2})\,dx$

29. $\int_0^{\pi} (\sin x + \sin^2 x)\,dx$

30. $\int_{-\pi}^{\pi} \sin^2 x\,dx$

31. $\int_0^{2\pi} (\cos x + \sin^2 x)\,dx$

32. $\int_0^{\pi} 8\sin^2 x\,dx$

33. $\int_0^{\pi} (2\sin^2 x + 3\sqrt{\pi^2 - x^2})\,dx$

34. $\int_{-\pi/2}^{\pi/2} \cos^2 x\,dx$

35. $\int_0^{\pi} (3x - 4\cos^2 x)/dx$

36. $\int_0^{\pi/2} (2\sin^2 x - \cos^2 x)\,dx$

37. $\int_1^3 \left(\frac{d(x^2)}{dx}\right) dx$

38. $\int_0^1 \left(\frac{d(\sqrt{x^3+1})}{dx}\right) dx$

39. $\int_0^1 \left(\frac{d^2(\sqrt{x^2+1})}{dx^2}\right) dx$

40. $\int_0^2 \left(\frac{d^3(x^2+3x-1)}{dx^3}\right) dx$

41. $\int_0^4 \left(\int_1^x \sqrt{t}\,dt\right) dx$

42. $\int_{-1}^3 \left(\int_x^2 4\,dt\right) dx$

43. Sketch the region bounded by the curve $y = 9 - x^2$ and the x-axis, and find the area of the region.

44. Sketch the region bounded by the curve $y = 2x - x^2$ and the x-axis, and find the area of the region.

45. Sketch the region bounded by the curve $y = x^4 - 16$ and the x-axis, and find the area of the region.

In Exercises 46 through 56, use properties of the definite integral and Theorem 6.3 to find the indicated function of t.

46. $\frac{d}{dt}\left(\int_1^t x^2\,dx\right)$

47. $\frac{d}{dt}\left(\int_{-50}^t \sqrt{x^2+1}\,dx\right)$

48. $\frac{d^2}{dt^2}\left(\int_2^t \sqrt{x^2+1}\,dx\right)$

49. $\frac{d}{dt}\left(\int_t^3 \frac{1}{1+x^2}\,dx\right)$

50. $\frac{d}{dt}\left(\int_2^t \sqrt{x^2+4}\,dx + \int_t^{-1} \sqrt{x^2+4}\,dx\right)$

51. $\frac{d}{dt}\left(\int_{-t}^t \sqrt{3+4x^2}\,dx\right)$

52. $\frac{d^2}{dt^2}\left(\int_{-t}^t \sqrt{3+4x^2}\,dx\right)$

53. $\frac{d}{dt}\left(\int_1^{-t} \sqrt{x^2+1}\,dx\right)$

54. $\frac{d}{dt}\left(\int_1^{3t} \frac{1}{4+x^2}\,dx\right)$

55. $\frac{d}{dt}\left(\int_0^{t^2-3} \sqrt{x^2+1}\,dx\right)$

56. $\frac{d}{dt}\left[\int_{t^2}^{t^3} \frac{1}{4+3x^2}\,dx\right]$

6.4 INTEGRATION AND DIFFERENTIAL EQUATIONS

The fundamental theorem gives a powerful method for computing $\int_a^b f(x)\,dx$. The burden of the computation is placed squarely on finding an antiderivative $F(x)$ of $f(x)$.

THE INDEFINITE INTEGRAL

DEFINITION 6.3 Indefinite integral

The general antiderivative of $f(x)$ is also called the **indefinite integral** of $f(x)$ and is written

$$\int f(x)\,dx$$

without any limits of integration. Computing an antiderivative is called **integration.**

EXAMPLE 1 Find $\int (x^3 + 2x^2 + 1)\,dx$.

Solution We have

$$\int (x^3 + 2x^2 + 1)\,dx = \frac{x^4}{4} + 2\frac{x^3}{3} + x + C,$$

where C is an arbitrary constant. □

According to the preceding section, every function f that has a derivative in $[a, b]$ also has an antiderivative in $[a, b]$. To see this, note that if f is differentiable, it is continuous and by the fundamental theorem, $F(t) = \int_a^t f(x)\,dx$ is an antiderivative of f for t in $[a, b]$. In practice, it is often more difficult to actually compute an antiderivative of a function than to find its derivative. The reason is that there are no simple formulas to handle integration of products, quotients, or composite functions as there are for differentiation. Tables have been prepared that list indefinite integrals of a number of functions frequently encountered. A brief table of this type is found on the endpapers of this book. Such tables are often useful, but no table exists that gives the integral of every continuous function anyone will ever encounter. It can be proved that an antiderivative of an "*elementary function*" (a function formed by algebraic operations from rational, trigonometric, exponential, and logarithmic functions) need not still be an elementary function. This is contrary to the situation for differentiation.

We start your training in integration now. This section presents a small number of integration formulas, which you are expected to remember and to be able to apply faster than you could find the desired formula in a table. For example, you should never have to look up $\int x^2\,dx$ in a table. The integration formulas we now give are those that arise from the differentiation formulas for basic elementary functions. For example, if u is a differentiable function of x, then

$$\frac{d(\tan u)}{dx} = (\sec^2 u) \cdot \frac{du}{dx},$$

so

$$\int (\sec^2 u) \cdot \frac{du}{dx}\,dx = \tan u + C.$$

Since $du = u'(x)\,dx = (du/dx)\,dx$, the preceding integration formula is usually written in the form

$$\int (\sec^2 u)\,du = \tan u + C.$$

The following box contains integration formulas that correspond to our differentiation formulas. Formulas 4 through 10 assume that u is a differentiable function of x.

1. $\int (f(x) + g(x))\,dx = \int f(x)\,dx + \int g(x)\,dx$

2. $\int c \cdot f(x)\,dx = c \cdot \int f(x)\,dx$

3. $\int x^n\,dx = \dfrac{x^{n+1}}{n+1} + C, \qquad n \neq -1$

4. $\int u^n\,du = \dfrac{u^{n+1}}{n+1} + C, \qquad n \neq -1$

5. $\int \sin u\,du = -\cos u + C$

6. $\int \cos u\,du = \sin u + C$

7. $\int \sec^2 u\,du = \tan u + C$

8. $\int \csc^2 u\,du = -\cot u + C$

9. $\int \sec u \tan u\,du = \sec u + C$

10. $\int \csc u \cot u\,du = -\csc u + C$

Formula (3) is a special case of formula (4), which is one of the most often used. The following example illustrates the use of formula (4).

EXAMPLE 2 Find $\int 2x\sqrt{x^2+1}\,dx$.

Solution If we let $u = x^2 + 1$, then $du = 2x\,dx$. From formula (4), we obtain

$$\int 2x\sqrt{x^2+1}\,dx = \int \sqrt{\underbrace{x^2+1}_{u}}\,\underbrace{(2x\,dx)}_{du}$$

$$= \int (u^{1/2})\,du = \frac{u^{3/2}}{\frac{3}{2}} + C$$

$$= \frac{2}{3}(x^2+1)^{3/2} + C. \quad \square$$

EXAMPLE 3 Find $\int x^2 \sin x^3 \, dx$.

Solution If we let $u = x^3$, then $du = 3x^2 \, dx$, so $x^2 \, dx = \frac{1}{3} \, du$. From formula (5), we obtain

$$\int x^2 \sin x^3 \, dx = \int (\sin \underbrace{x^3}_{u})(\underbrace{x^2 \, dx}_{\frac{1}{3} du}) = \int (\sin u)\left(\frac{1}{3} \, du\right) = \frac{1}{3} \int (\sin u) \, du$$

$$= -\frac{1}{3} \cos u + C = -\frac{1}{3} \cos x^3 + C. \quad \square$$

In practice, we usually do not write out the substitution $u = g(x)$ in simple cases like those in Examples 2 and 3, but rather, realizing that the substitution is appropriate, we compute the integral in one step in our head. Suppose, for example, we wish to find $\int \cos 2x \, dx$. If we let $u = 2x$, then $du = 2 \, dx$, and

$$\int \cos 2x \, dx = \int \frac{1}{2} \cdot 2 \cos 2x \, dx = \frac{1}{2} \int (\cos \underbrace{2x}_{u})(\underbrace{2 \, dx}_{du}).$$

We thus find from formula (6) that

$$\int \cos 2x \, dx = \frac{1}{2} \int (\cos \underbrace{2x}_{u})(\underbrace{2 \, dx}_{du}) = \frac{1}{2} \sin 2x + C.$$

We used the colored labels under the integrals to indicate what we were thinking, but we did not actually write u or du in the integral this time.

We illustrate this technique further with more examples.

EXAMPLE 4 Find $\int x^2 \sec x^3 \, dx$.

Solution If $u = x^3$, then $du = 3x^2 \, dx$ and we would like to have an additional factor 3 in the integral so that formula (7) would apply. Of course, the supplied factor 3 for the "fix-up" must be balanced by a factor $\frac{1}{3}$, which we may write outside the integral by formula (2), since it is a constant. Using formula (7), we obtain

$$\int x^2 \sec x^3 \, dx = \frac{1}{3} \int (\sec \underbrace{x^3}_{u})(\underbrace{3x^2 \, dx}_{du}) = \frac{1}{3} \tan x^3 + C. \quad \square$$

EXAMPLE 5 Find $\int \sin 3x \cos^2 3x \, dx$.

Solution If we let $u = \cos 3x$, then $du = -3 \sin 3x \, dx$, and we would like an additional factor (-3) in our integral so that we can apply formula (4). We supply the factor (-3) and the balancing factor $(-\frac{1}{3})$ just as in Example 4. We obtain from formulas (2) and (4)

$$\int \sin 3x \cos^2 3x \, dx = -\frac{1}{3} \int \underbrace{(\cos^2 3x)}_{u^2}\underbrace{(-3 \sin 3x \, dx)}_{du}$$

$$= -\frac{1}{3} \frac{\cos^3 3x}{3} + C = -\frac{1}{9} \cos^3 3x + C. \quad \square$$

As illustrated in the two preceding examples, formula (2) shows that we can "fix up an integral to supply a desired constant factor." *We warn you that a similar technique to supply a variable factor is not valid.* To illustrate,

$$\int x\,dx \neq \frac{1}{x}\int x \cdot x\,dx = \frac{1}{x}\int x^2\,dx,$$

for $\int x\,dx = x^2/2 + C$, while

$$\frac{1}{x}\int x^2\,dx = \frac{1}{x}\left(\frac{x^3}{3} + C\right) = \frac{x^2}{3} + \frac{C}{x} \qquad \text{for } x \neq 0.$$

For example, we can compute $\int x \sin x^2\,dx$, because we can "fix up the desired constant 2" and use formula (5), but at the moment we can't compute $\int \sin x^2\,dx$, because there is no way we can "fix up the variable factor $2x$" needed to apply formula (5).

EXAMPLE 6 Find $\int (x/\cos^2 x^2)\,dx$.

Solution We spot that x is the derivative of x^2, except for a constant factor. Thus we essentially want to integrate $du/\cos^2 u = \sec^2 u\,du$. We have

$$\int \frac{x}{\cos^2 x^2}\,dx = \frac{1}{2}\int (\sec^2 \underbrace{x^2}_{u})(\underbrace{2x\,dx}_{du}) = \frac{1}{2}\tan x^2 + C. \quad \square$$

EXAMPLE 7 Find

$$\int \frac{\cos 3x}{(1 + \sin 3x)^5}\,dx.$$

Solution We immediately spot that $\cos 3x$ is the derivative of $1 + \sin 3x$, except for a constant factor. Thus we essentially want to integrate $du/u^5 = u^{-5}\,du$. We have

$$\begin{aligned}\int \frac{\cos 3x}{(1 + \sin 3x)^5}\,dx &= \frac{1}{3}\int (\underbrace{1 + \sin 3x}_{u})^{-5}(\underbrace{3 \cos 3x\,dx}_{du}) \\ &= \frac{1}{3}\cdot\frac{(1 + \sin 3x)^{-4}}{-4} + C \\ &= \frac{-1}{12(1 + \sin 3x)^4} + C. \quad \square\end{aligned}$$

DIFFERENTIAL EQUATIONS

A differential equation is one containing a derivative or differential. In Section 5.9 we solved some differential equations of the form

$$\frac{dy}{dx} = f(x).$$

Many differential equations cannot be written in this form, where dy/dx equals a function of x alone. (Remember that, when differentiating implicitly, we frequently obtain an expression for dy/dx that involves both x and y.) An example of such a differential equation is

$$\frac{dy}{dx} = \frac{x}{y}.$$

We can solve this differential equation by the following device: *Rewrite the equation with all terms involving y (including dy) on the left and all terms involving x (including dx) on the right:*

$$y \cdot dy = x \cdot dx.$$

Since the differentials $y \cdot dy$ and $x \cdot dx$ are equal, it appears that we must have

$$\int y \cdot dy = \int x \cdot dx, \qquad \text{or} \qquad \frac{y^2}{2} + C_1 = \frac{x^2}{2} + C_2,$$

where C_1 and C_2 are arbitrary constants. This solution may be written as

$$y^2 + 2C_1 = x^2 + 2C_2, \qquad \text{or} \qquad y^2 - x^2 = 2C_1 - 2C_2.$$

Now $2C_1 - 2C_2$ may be any constant, so we can express it as a single arbitrary constant C, obtaining

$$y^2 - x^2 = C$$

as solution of the differential equation. We will consider an equation like this that defines y implicitly as a function of x to be an acceptable form for the solution.

Crucial to this technique is the ability to rewrite the differential equation with all terms involving y on one side and all terms involving x on the other. Such equations are *variables-separable* equations.

EXAMPLE 8 Solve

$$\frac{dy}{dx} = y^2 \sin x.$$

Solution This is a variables-separable equation. The steps of solution are

$$\frac{dy}{y^2} = \sin x\, dx, \qquad \int y^{-2}\, dy = \int \sin x\, dx,$$

$$\frac{y^{-1}}{-1} = -\cos x + C, \qquad \frac{1}{y} = \cos x + C,$$

where we used only one arbitrary constant C and replaced $-C$ by C again in the last step. □

EXAMPLE 9 Solve

$$\frac{dy}{dx} = \frac{\cos x}{\sin y}.$$

Solution Separation of variables leads to

$$\sin y\, dy = \cos x\, dx, \qquad \int \sin y\, dy = \int \cos x\, dx,$$

$$-\cos y = \sin x + C. \quad \square$$

EXAMPLE 10 Find the solution of the differential equation $dy/dx = x^2y^3$ such that $y = -1$ when $x = 1$.

Solution This time, we have an initial-value problem, so we must evaluate the constant in the general solution. First, we separate variables and find the general solution:

$$\frac{dy}{y^3} = x^2\,dx, \qquad \int y^{-3}\,dy = \int x^2\,dx,$$

$$\frac{y^{-2}}{-2} = \frac{x^3}{3} + C, \qquad -\frac{1}{2y^2} = \frac{1}{3}x^3 + C.$$

Setting $x = 1$ and $y = -1$, we obtain $-\frac{1}{2} = \frac{1}{3} + C$, so $C = -\frac{1}{2} - \frac{1}{3} = -\frac{5}{6}$. Thus our desired solution is

$$-\frac{1}{2y^2} = \frac{1}{3}x^3 - \frac{5}{6}. \quad \square$$

The equation

$$\frac{dy}{dx} = x + y$$

is *not* a variables-separable equation. The final chapter of the text discusses solutions of a few more types of differential equations.

SUMMARY

1. The general antiderivative of $f(x)$ is also known as the indefinite integral, $\int f(x)\,dx$.
2. Ten important formulas for integrating are as follows:

$$\int (f(x) + g(x))\,dx = \int f(x)\,dx + \int g(x)\,dx$$

$$\int c \cdot f(x)\,dx = c \cdot \int f(x)\,dx$$

$$\int x^n\,dx = \frac{x^{n+1}}{n+1} + C, \qquad n \neq -1$$

$$\int u^n\,du = \frac{u^{n+1}}{n+1} + C, \qquad n \neq -1$$

$$\int \sin u\,du = -\cos u + C$$

$$\int \cos u\,du = \sin u + C$$

$$\int \sec^2 u\,du = \tan u + C$$

$$\int \csc^2 u\,du = -\cot u + C$$

$$\int \sec u \tan u \, du = \sec u + C$$

$$\int \csc u \cot u \, du = -\csc u + C$$

3. Suppose all the y terms (including dy) can be put on the left side of a differential equation and all the x terms (including dx) on the right, to become

$$g(y)\, dy = f(x)\, dx.$$

Then the solution of the differential equation is

$$\int g(y)\, dy = \int f(x)\, dx + C,$$

where C is an arbitrary constant. No further arbitrary constants need be included when computing $\int g(y)\, dy$ and $\int f(x)\, dx$.

EXERCISES

In Exercises 1 through 40, find the indicated integral without using tables.

1. $\int (x^3 + 4x^2)\, dx$

2. $\int \left(x + \frac{1}{x^2}\right) dx$

3. $\int (x + 1)^5\, dx$

4. $\int (2x + 1)^4\, dx$

5. $\int \frac{dx}{(4x + 1)^2}$

6. $\int \frac{5}{(3 - 7x)^3}\, dx$

7. $\int \frac{x}{(3x^2 + 1)^2}\, dx$

8. $\int \frac{x^2}{(4x^3 - 1)^4}\, dx$

9. $\int \frac{x}{(4 - 3x^2)^3}\, dx$

10. $\int \frac{x}{\sqrt{x^2 + 4}}\, dx$

11. $\int \frac{4x + 2}{\sqrt{x^2 + x}}\, dx$

12. $\int \frac{2x^2}{(x^3 - 3)^{3/2}}\, dx$

13. $\int \frac{dx}{\sqrt{x}(\sqrt{x} + 1)^4}$

14. $\int \frac{dx}{\sqrt{x}(4 - 3\sqrt{x})^2}$

15. $\int x^2(x^3 + 4)^3\, dx$

16. $\int (x^2 + 1)(x^3 + 1)^2\, dx$

17. $\int \cos 3x\, dx$

18. $\int \sin (3x + 1)\, dx$

19. $\int x \cos (x^2 + 1)\, dx$

20. $\int \frac{\sin (\sqrt{x})}{\sqrt{x}}\, dx$

21. $\int \sin x \cos x\, dx$

22. $\int \cos x \sin^2 x\, dx$

23. $\int \sin 4x \cos^3 4x\, dx$

24. $\int \frac{x^2}{\sin^2 x^3}\, dx$

25. $\int x \sec^2 x^2\, dx$

26. $\int \csc 2x \cot 2x\, dx$

27. $\int \tan x \sec^2 x\, dx$

28. $\int \sec^3 2x \tan 2x\, dx$

29. $\int x(\sec^2 x^2)(\tan^3 x^2)\, dx$

30. $\int \frac{x \cot x^2}{\sin x^2}\, dx$

31. $\int \frac{\sin x}{(1 + \cos x)^2}\, dx$

32. $\int \frac{\sec^2 2x}{(4 + \tan 2x)^3}\, dx$

33. $\int \frac{1}{\sec 4x}\, dx$

34. $\int \frac{\sin x}{\cos^2 x}\, dx$

35. $\int \frac{\tan 3x}{\cos^4 3x}\, dx$

36. $\int \frac{\sec^2 x}{\sqrt{1 + \tan x}}\, dx$

37. $\int (\tan 3x)\sqrt{\sec^2 3x + \sec^3 3x}\, dx$

38. $\int \frac{\cot 3x}{\sin 3x}\, dx$

39. $\int \csc^5 2x \cot 2x\, dx$

40. $\int \frac{\cos x}{(1 - \cos^2 x)^2}\, dx$

41. Find $\int \cos^2 x\, dx$. [*Hint:* Consider the "double-angle formula" for cosine.]

42. Find $\int \cos^4 x\, dx$, using the hint in Exercise 41.

In Exercises 43 through 56, solve the differential equation or initial-value problem.

43. $\frac{dy}{dx} = x^2 y^2$

44. $\frac{dy}{dx} = x \cos^2 y$

45. $\dfrac{ds}{dt} = s^2t\sqrt{1+t^2}$

46. $\dfrac{du}{dx} = x^3\sqrt{u}$

47. $\dfrac{dv}{dt} = \sqrt{5v-7}(t\cos t^2)$

48. $\dfrac{dy}{dx} = (\sin^2 y)(\cos^2 x)(\sin x)$

49. $\dfrac{dy}{dx} = \dfrac{\sin^2 y}{\cos^2 x}$

50. $\dfrac{dy}{dx} = \dfrac{(1+x^2)^2}{\sec^2 3y}$

51. $\dfrac{dy}{dx} = xy^3$, $y = 1$ when $x = 0$

52. $\dfrac{dy}{dx} = \cos^2 y$, $y = \dfrac{\pi}{4}$ when $x = 1$

53. $\dfrac{dy}{dx} = \cos x \csc 2y$, $y = \dfrac{\pi}{2}$ when $x = 0$

54. $\dfrac{ds}{dt} = \sqrt{s+1}\sqrt{3t+1}$, $s = 3$ when $t = 5$

55. $\dfrac{dr}{ds} = \dfrac{\sqrt{r^2+1}\sqrt{2s+3}}{r}$, $r = \sqrt{3}$ when $s = 11$

56. $\dfrac{dy}{dx} = \dfrac{\sqrt{1+\cos y}}{\sin y}$, $y = \dfrac{\pi}{2}$ when $x = 3$

6.5 INTEGRATION USING TABLES

Integration using tables is a very important technique, one that should be practiced. You will find a table of integrals on the endpapers of this book. The formulas in the table are given using the variable x, since that is the way they are stated in most tables. For a differentiable function u, it follows, from the chain rule for differentiation and $du = u'(x)\,dx$, that the formulas hold if x is replaced by u and dx by du.

EXAMPLE 1 Compute $\int_0^\pi \sin^2 x\,dx$.

Solution If we worked for a time, we might hit upon the relation

$$\sin^2 x = \tfrac{1}{2}(1 - \cos 2x),$$

which would enable us to find $\int \sin^2 x\,dx$. This indefinite integral is one that at the moment we can probably find more quickly (and more reliably) using the table formula

$$\int \sin^2 ax\,dx = \frac{x}{2} - \frac{\sin 2ax}{4a} + C. \qquad \text{Table formula 64}$$

We take $a = 1$ and obtain

$$\int_0^\pi \sin^2 x\,dx = \left(\frac{x}{2} - \frac{\sin 2x}{4}\right)\Bigg]_0^\pi = \left(\frac{\pi}{2} - \frac{0}{4}\right) - \left(0 - \frac{0}{4}\right) = \frac{\pi}{2}. \quad \square$$

EXAMPLE 2 Find $\int \sin^5 2x\,dx$.

Solution The table integration formula

$$\int \sin^n ax\,dx$$
$$= \frac{-\sin^{n-1} ax \cos ax}{na} + \frac{n-1}{n}\int \sin^{n-2} ax\,dx \qquad \text{Table formula 66}$$

is known as a *reduction formula;* the problem of integrating $\sin^n ax$ is *reduced* to integrating $\sin^{n-2} ax$. One applies this formula repeatedly until the problem is reduced to integrating either sin ax or $\sin^0 ax = 1$. We have

$$\int \sin^5 2x\, dx = \frac{-\sin^4 2x \cos 2x}{5 \cdot 2} + \frac{4}{5}\int \sin^3 2x\, dx$$

$$= \frac{-\sin^4 2x \cos 2x}{10} + \frac{4}{5}\left(\frac{-\sin^2 2x \cos 2x}{3 \cdot 2} + \frac{2}{3}\int \sin 2x\, dx\right)$$

$$= -\frac{\sin^4 2x \cos 2x}{10} - \frac{2 \sin^2 2x \cos 2x}{15}$$

$$- \frac{4}{15}\cos 2x + C. \quad \square$$

We emphasize again that each integration formula in x given in the table can be modified, using the chain rule, to give a more general result. For example, the table formula

$$\int \frac{dx}{x^2\sqrt{a^2 + x^2}} = -\frac{\sqrt{a^2 + x^2}}{a^2 x} + C, \qquad \text{Table formula 33}$$

together with the chain rule for differentiation, yields the formula

$$\int \frac{du}{u^2\sqrt{a^2 + u^2}} = -\frac{\sqrt{a^2 + u^2}}{a^2 u} + C \qquad \textbf{(1)}$$

for any differentiable function u.

EXAMPLE 3 Find

$$\int \frac{\sec^2 x\, dx}{(\tan^2 x)\sqrt{3 + \tan^2 x}}.$$

Solution It is unlikely that we would find this integral with $\tan^2 x$ and $\sec^2 x$ in *any* table. We should notice at once that $\sec^2 x$, which appears as a factor in the numerator, is the derivative of $\tan x$, which appears elsewhere in the integral. This suggests at once that we might be able to simplify things if we let $u = \tan x$. Then $du = \sec^2 x\, dx$, and the integral becomes

$$\int \frac{du}{u^2\sqrt{3 + u^2}}.$$

Taking $a^2 = 3$ in formula (1) and remembering that $u = \tan x$, we obtain

$$\int \frac{\sec^2 x\, dx}{(\tan^2 x)\sqrt{3 + \tan^2 x}} = -\frac{\sqrt{3 + \tan^2 x}}{3 \tan x} + C. \quad \square$$

As in the previous example, *you should train yourself to spot when the integrand contains, as a factor, a function that is the derivative of some other function u in the integrand.* Even if a table of integrals has the formula you need, discovering which formula it is sometimes requires much ingenuity. Facility in integration, as well as facility in differentiation, is developed only by working several hundred problems. The exercises will get you started. First, we give three more examples.

EXAMPLE 4 Find

$$\int \frac{x\,dx}{1 + \sin(3x^2)}.$$

Solution Our table does not contain this integral in its present form. However, we should note at once that x is the derivative of $3x^2$, except for the constant factor 6, which we can "fix up." Thus if we let $u = 3x^2$, we have $du = 6x\,dx$, so $x\,dx = \frac{1}{6}\,du$. Our integral becomes

$$\frac{1}{6}\int \frac{du}{1 + \sin u}.$$

We hunt in the table for an appropriate formula and find

$$\int \frac{dx}{1 + \sin ax} = -\frac{1}{a}\tan\left(\frac{\pi}{4} - \frac{ax}{2}\right) + C. \qquad \text{Table formula 78}$$

With $a = 1$ and u in place of x, this formula becomes

$$\int \frac{du}{1 + \sin u} = -\tan\left(\frac{\pi}{4} - \frac{u}{2}\right) + C.$$

Thus

$$\int \frac{x\,dx}{1 + \sin(3x^2)} = -\frac{1}{6}\tan\left(\frac{\pi}{4} - \frac{3x^2}{2}\right) + C. \quad \square$$

EXAMPLE 5 Find $\int \cos 7x \cos 3x\,dx$.

Solution This time we find the formula

$$\int \cos ax \cos bx\,dx$$

$$= \frac{\sin(a - b)x}{2(a - b)} + \frac{\sin(a + b)x}{2(a + b)} + C, \qquad \text{Table formula 68(c)}$$

which we may use directly with our integral. Taking $a = 7$ and $b = 3$, we obtain

$$\int \cos 7x \cos 3x\,dx = \frac{\sin 4x}{8} + \frac{\sin 10x}{20} + C. \quad \square$$

EXAMPLE 6 Find

$$\int \frac{x^3}{\sqrt{4 + 5x^2}}\,dx.$$

Solution The table does not contain a formula that we can use directly. However, it does contain formulas with $\sqrt{ax + b}$ in the denominator, and we can think of $\sqrt{ax + b}$ as $\sqrt{au + b}$. Thus we try $u = x^2$ to see if our integral can be made to fit such a formula in the table. With $u = x^2$, we have $du = 2x\,dx$, so

$$\int \frac{x^3}{\sqrt{4 + 5x^2}}\,dx = \frac{1}{2}\int \frac{x^2 \cdot 2x}{\sqrt{4 + 5x^2}}\,dx = \frac{1}{2}\int \frac{u}{\sqrt{4 + 5u}}\,du.$$

We hunt for something like this with x in place of u in the table and find

$$\int \frac{x\,dx}{\sqrt{ax+b}} = \frac{2(ax-2b)}{3a^2}\sqrt{ax+b} + C. \qquad \text{Table formula 17}$$

We then use this formula with x replaced by $u = x^2$, $a = 5$, and $b = 4$ and obtain

$$\int \frac{x^3}{\sqrt{4+5x^2}}\,dx = \frac{1}{2}\cdot\frac{2(5x^2-8)}{75}\sqrt{5x^2+4} + C$$

$$= \frac{5x^2-8}{75}\sqrt{5x^2+4} + C. \quad \square$$

SUMMARY

1. All formulas $\int f(x)\,dx$ in the tables should be regarded as formulas $\int f(u)\,du$, where u may be any differentiable function of x.
2. When faced with a complicated integral that you can't find easily in a table, try to see whether some factor of the integrand can be viewed as the derivative of some other portion u. Then try to write the integral in the form $\int f(u)\,du$ and see if you can find $\int f(x)\,dx$ in the table. See Examples 3, 4, and 6.

EXERCISES

In Exercises 1 through 50, find the given integral the fastest way you can, including the use of tables (but excluding looking up the answer in the back of the text).

1. $\int_2^4 x^2\,dx$

2. $\int \frac{3}{x^2\sqrt{4+x^2}}\,dx$

3. $\int \frac{3x}{(\sqrt{4+x^2})^3}\,dx$

4. $\int \frac{1}{x^2\sqrt{x^2-9}}\,dx$

5. $\int \frac{5}{x\sqrt{10x-x^2}}\,dx$

6. $\int 3x\sqrt{2x+5}\,dx$

7. $\int \frac{-2x}{\sqrt{3-7x}}\,dx$

8. $\int 4x(2x+3)^{10}\,dx$

9. $\int x^3\sqrt{4x^2+1}\,dx$ [*Hint:* Use table formula 13.]

10. $\int \frac{x^5}{\sqrt{2x^2+3}}\,dx$ [*Hint:* Use table formula 18.]

11. $\int x^3(4x^2-1)^7\,dx$ [*Hint:* Use table formula 8.]

12. $\int \frac{x^3}{\sqrt{4x^2-1}}\,dx$

13. $\int \frac{dx}{x^3\sqrt{16-x^4}}$

[*Hint:* Multiply numerator and denominator by x.]

14. $\int \frac{dx}{x^3\sqrt{x^4-1}}$

[*Hint:* Multiply numerator and denominator by x.]

15. $\int \frac{dx}{x\sqrt{4x^2-x^4}}$

[*Hint:* Multiply numerator and denominator by x.]

16. $\int \frac{1}{x^3\sqrt{4+x^4}}\,dx$

17. $\int_0^\pi \cos^2 2x\,dx$

18. $\int_0^{\pi/4} \sin^2 x\,dx$

19. $\int \sin^4 x\,dx$

20. $\int \cos^6 2x\,dx$

21. $\int \sin 2x \cos 3x\,dx$

22. $\int_0^\pi \sin 2x \sin 4x\,dx$

23. $\int \sin 4x \cos 4x\,dx$

24. $\int \cos 2x \cos 5x\,dx$

25. $\int \sin^4 x \cos^2 x\,dx$

26. $\displaystyle\int \frac{3}{1 + \cos x}\,dx$ **27.** $\displaystyle\int \frac{x}{1 - \cos x^2}\,dx$

28. $\displaystyle\int x \sin 3x\,dx$ **29.** $\displaystyle\int x \cos 4x\,dx$

30. $\displaystyle\int x^2 \sin 2x\,dx$ **31.** $\displaystyle\int x^3 \cos x\,dx$

32. $\displaystyle\int \sec^2 4x\,dx$ **33.** $\displaystyle\int \tan^2 3x\,dx$

34. $\displaystyle\int x \cot^2 x^2\,dx$ **35.** $\displaystyle\int \tan^4 2x\,dx$

36. $\displaystyle\int x \sec^4 x^2\,dx$ **37.** $\displaystyle\int x^2 \cos^2 x^3\,dx$

38. $\displaystyle\int x(\sin^2 x^2(\cos x^2)\,dx$

39. $\displaystyle\int (\sin x \cos x)(2 \sin x + 4)^8\,dx$

40. $\displaystyle\int \frac{\tan x \sec^2 x}{\sqrt{2 \tan x + 3}}\,dx$ **41.** $\displaystyle\int \frac{\cos x}{\sin^2 x\sqrt{4 - \sin^2 x}}\,dx$

42. $\displaystyle\int \frac{\sin 3x}{\cos^2 3x\sqrt{\cos^2 3x - 25}}\,dx$

43. $\displaystyle\int \frac{\cot 2x}{\sin 2x\sqrt{\sin^2 2x + 9}}\,dx$ **44.** $\displaystyle\int \frac{\tan^2 x \sec^2 x}{\sqrt{3 \tan x + 4}}\,dx$

45. $\displaystyle\int (\sec^2 2x \tan 2x)\sqrt{1 - 4 \sec 2x}\,dx$

46. $\displaystyle\int \frac{\sec^2 x}{\tan^2 x\sqrt{25 - \tan^2 x}}\,dx$ **47.** $\displaystyle\int \frac{\sec^2 x}{\tan^2 x\sqrt{25 + 4\tan^2 x}}\,dx$

48. $\displaystyle\int \frac{\cos x}{\sin^2 x\sqrt{4 \sin^2 x - 1}}\,dx$

49. $\displaystyle\int \frac{\tan 3x\,dx}{\sqrt{4 \sec 3x - \sec^2 3x}}$

[*Hint:* Multiply numerator and denominator by sec $3x$.]

50. $\displaystyle\int \frac{\tan 3x}{\sec 3x\sqrt{25 - 4\sec^2 3x}}\,dx$

[*Hint:* Multiply numerator and denominator by sec $3x$.]

6.6 NUMERICAL METHODS OF INTEGRATION

If we can't find the indefinite integral $\int f(x)\,dx$, even in a table, then the fundamental theorem is of no help in computing $\int_a^b f(x)\,dx$. For example, it can be shown that the integral

$$\int \sqrt{4 - \sin^2 x}\,dx$$

can't be expressed in terms of elementary functions such as polynomials, trigonometric functions, roots, exponentials, and logarithms. Numerical methods for computing a definite integral such as

$$\int_0^1 \sqrt{4 - \sin^2 x}\,dx$$

are therefore very important. With calculators and computers readily available, computation of such integrals is an easy matter if one is willing to settle for a good numerical approximation to the exact value.

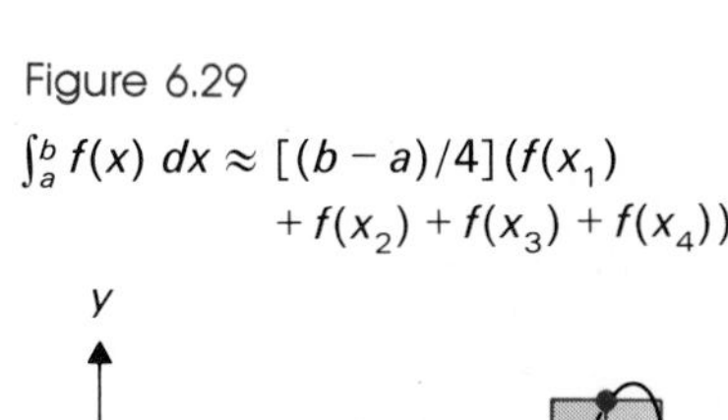

Figure 6.29

$\int_a^b f(x)\,dx \approx [(b-a)/4](f(x_1) + f(x_2) + f(x_3) + f(x_4))$

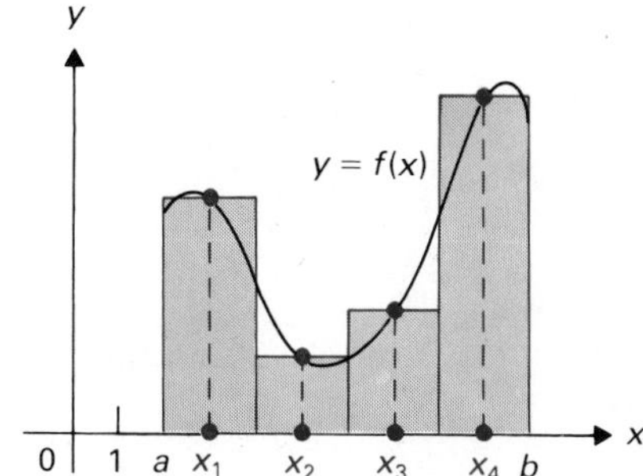

THE RECTANGULAR RULE

One method of approximating $\int_a^b f(x)\,dx$ is to use a midpoint Riemann sum, taking midpoints of the subintervals for the points x_i where we compute $f(x_i)$. The integral is then computed as a sum of areas of rectangles, as indicated in Fig. 6.29. For obvious reasons, approximation by such Riemann sums is also known as the *rectangular rule*.

THEOREM 6.4 Rectangular rule

Let $[a, b]$ be divided into n subintervals of equal length and let x_i be the midpoint of the ith subinterval. Then, for f continuous in $[a, b]$,

$$\int_a^b f(x)\,dx \approx \frac{b-a}{n}[f(x_1) + f(x_2) + \cdots + f(x_n)], \tag{1}$$

and the approximation is very good for sufficiently large n.

We say a word or two about finding the midpoints x_i to use in the rectangular rule. First compute the subinterval length $(b - a)/n$ and then find

$$x_1 = a + \frac{1}{2}\cdot\frac{b-a}{n}.$$

The remaining x_i are then found by incrementing by $(b - a)/n$, that is,

$$x_2 = x_1 + \frac{b-a}{n}, \qquad x_3 = x_2 + \frac{b-a}{n},$$

$$x_4 = x_3 + \frac{b-a}{n}, \qquad \text{and so on.}$$

We know that

$$\int_1^5 x^4\,dx = \frac{x^5}{5}\Big]_1^5 = \frac{5^5}{5} - \frac{1}{5} = 5^4 - \frac{1}{5} = 625 - \frac{1}{5} = 624.8.$$

We will estimate $\int_1^5 x^4\,dx$ using $n = 4$ and the rectangular rule. The purpose is to compare this estimate with later estimates using two other rules.

EXAMPLE 1 Estimate $\int_1^5 x^4\,dx$ using the rectangular rule with $n = 4$.

Solution We have

$$\frac{b-a}{n} = \frac{5-1}{4} = 1.$$

Then

$$x_1 = 1 + \frac{1}{2}\cdot 1 = \frac{3}{2}, \qquad x_2 = \frac{5}{2}, \qquad x_3 = \frac{7}{2}, \qquad x_4 = \frac{9}{2}.$$

Since $f(x) = x^4$, the rectangular rule (1) provides the estimate

$$\int_1^5 x^4\,dx \approx 1\cdot\left[\left(\frac{3}{2}\right)^4 + \left(\frac{5}{2}\right)^4 + \left(\frac{7}{2}\right)^4 + \left(\frac{9}{2}\right)^4\right]$$

$$= \frac{3^4 + 5^4 + 7^4 + 9^4}{16}$$

$$= \frac{9668}{16} = 604.25. \quad \square$$

It should not be necessary to give any more examples using the rectangular rule. We made precisely these estimates in Section 6.1.

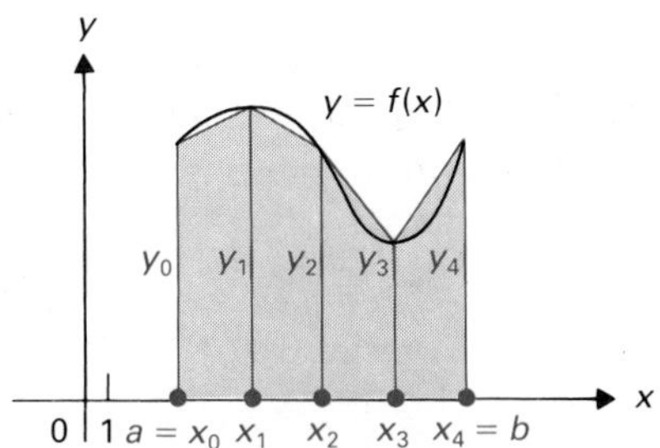

Figure 6.30 Approximation of $\int_a^b f(x)\,dx$ using areas of regions under four chords of the graph.

THE TRAPEZOIDAL RULE

We again divide the interval $[a, b]$ into n subintervals of equal length, but this time we let the *endpoints* of the subintervals be

$$a = x_0, \quad x_1, \quad x_2, \quad \ldots, \quad x_n = b,$$

as illustrated in Fig. 6.30 for $n = 4$. Let the (signed) heights to the graph $y = f(x)$ at these endpoints be

$$y_0 = f(x_0), \quad y_1 = f(x_1), \quad y_2 = f(x_2), \quad \ldots, \quad y_n = f(x_n),$$

as in Fig. 6.30. Consider the chord of the curve $y = f(x)$ joining (x_{i-1}, y_{i-1}) and (x_i, y_i), as shown in Fig. 6.31. The shaded regions in Figs. 6.30 and 6.31 are trapezoids. The area of a trapezoid is the product of its altitude and the average length of the bases, so the area of the shaded trapezoid in Fig. 6.31 is

$$(x_i - x_{i-1})\frac{y_{i-1} + y_i}{2}.$$

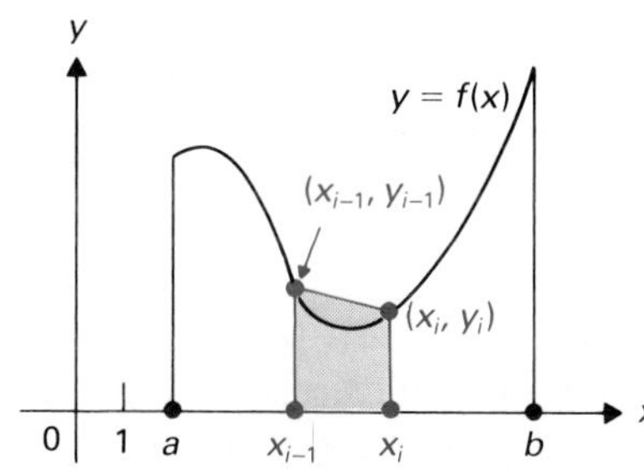

Figure 6.31 Trapezoid of area $(x_i - x_{i-1})[(y_{i-1} + y_i)/2]$.

Since we divide our interval $[a, b]$ into n subintervals of *equal* length,

$$x_i - x_{i-1} = \frac{b-a}{n}.$$

Adding up the area of the trapezoids for $i = 1, \ldots, n$, we obtain

$$\frac{b-a}{n}\left(\frac{y_0 + y_1}{2}\right) + \frac{b-a}{n}\left(\frac{y_1 + y_2}{2}\right) + \frac{b-a}{n}\left(\frac{y_2 + y_3}{2}\right) + \cdots + \frac{b-a}{n}\left(\frac{y_{n-1} + y_n}{2}\right).$$

Factoring out $(b - a)/2n$, we obtain at once the trapezoidal rule.

THEOREM 6.5 Trapezoidal rule

For a continuous function f on $[a, b]$,

$$\int_a^b f(x)\,dx \approx \frac{b-a}{2n}(y_0 + 2y_1 + 2y_2 + \cdots + 2y_{n-1} + y_n) \qquad \textbf{(2)}$$

for sufficiently large values of n.

To find the values y_i in the trapezoidal rule, first find $(b - a)/n$ and then compute x_i where

$$x_0 = a, \quad x_1 = x_0 + \frac{b-a}{n}, \quad x_2 = x_1 + \frac{b-a}{n}, \quad \ldots, \quad x_n = b.$$

Then $y_i = f(x_i)$.

EXAMPLE 2 Use the trapezoidal rule with $n = 4$ to estimate

$$\int_1^3 \frac{1}{x}\,dx.$$

Solution We have

$$\frac{b-a}{n} = \frac{3-1}{4} = \frac{1}{2},$$

so

$$x_0 = 1, \qquad x_1 = \frac{3}{2}, \qquad x_2 = 2, \qquad x_3 = \frac{5}{2}, \qquad x_4 = 3.$$

Thus

$$y_0 = 1, \qquad y_1 = \frac{2}{3}, \qquad y_2 = \frac{1}{2}, \qquad y_3 = \frac{2}{5}, \qquad y_4 = \frac{1}{3}.$$

The trapezoidal rule yields

$$\begin{aligned}\int_1^3 \frac{1}{x}\,dx &\approx \frac{3-1}{2\cdot 4}\left(1 + 2\cdot\frac{2}{3} + 2\cdot\frac{1}{2} + 2\cdot\frac{2}{5} + \frac{1}{3}\right)\\ &= \frac{2}{8}\left(1 + \frac{4}{3} + 1 + \frac{4}{5} + \frac{1}{3}\right)\\ &= \frac{1}{4}\left(\frac{15 + 20 + 15 + 12 + 5}{15}\right)\\ &= \frac{1}{4}\left(\frac{67}{15}\right) = \frac{67}{60} \approx 1.12.\end{aligned}$$

Since the graph of $1/x$ is concave up over $[1, 3]$, it is clear that the approximation by chords is too large. The actual value of the integral to four decimal places is 1.0986. □

EXAMPLE 3 Estimate $\int_1^5 x^4\,dx$ using the trapezoidal rule with $n = 4$, and compare with Example 1.

Solution We have $(b - a)/n = 1$ as in Example 1. This time,

$$x_0 = 1, \qquad x_1 = 2, \qquad x_2 = 3, \qquad x_3 = 4, \qquad x_4 = 5.$$

Thus

$$y_0 = 1, \qquad y_1 = 2^4, \qquad y_2 = 3^4, \qquad y_3 = 4^4, \qquad y_4 = 5^4.$$

The trapezoidal rule yields

$$\int_1^5 x^4\,dx \approx \frac{1}{2}(1 + 2\cdot 2^4 + 2\cdot 3^4 + 2\cdot 4^4 + 5^4) = \frac{1332}{2} = 666.$$

The rectangular estimate 604.25 was closer to the true value 624.8 of the integral than this estimate of 666. □

SIMPSON'S (PARABOLIC) RULE

In the rectangular rule, our approximating rectangles have horizontal lines (constant functions $y = a$) as tops, and the lines usually meet the graph at *one* point. In the trapezoidal rule, the top of a trapezoid may be any line (linear function $y = ax + b$), and the lines usually meet the graph at *two*

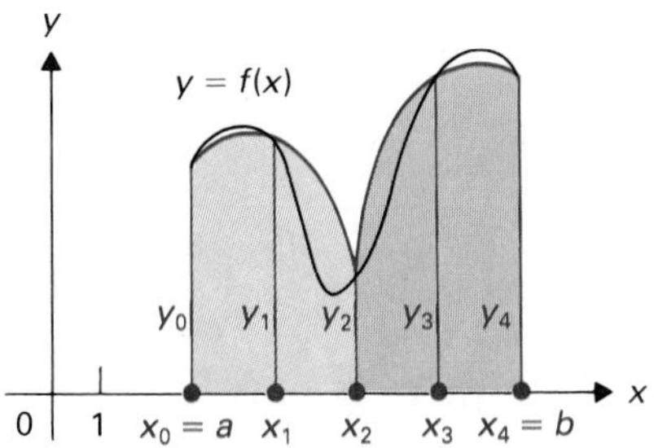

Figure 6.32 Estimating $\int_a^b f(x)\,dx$ by the areas of regions under two parabolas.

points. We can put a quadratic function $y = ax^2 + bx + c$ through *three* points of the graph. The graph of $y = ax^2 + bx + c$ is a *parabola.*

For Simpson's (parabolic) rule, we divide the interval $[a, b]$ into an *even number* n of subintervals of equal length having endpoints x_i for $i = 0, 1, \ldots, n$. We use the notation $y_i = f(x_i)$ as in the trapezoidal rule. We approximate the graph of f over the first *two* subintervals by a parabola through (x_0, y_0), (x_1, y_1), and (x_2, y_2), over the next two subintervals by a parabola through (x_2, y_2), (x_3, y_3), and (x_4, y_4), and so on (see Fig. 6.32). We then estimate the integral $\int_a^b f(x)\,dx$ by taking the areas under the parabolas.

For the case where $x_1 - x_0 = x_2 - x_1$, there is a convenient formula for the area under a parabola through (x_0, y_0), (x_1, y_1), and (x_2, y_2) in terms of the heights y_0, y_1, and y_2. This formula allows us to estimate the integral without explicit computation of the equations of the parabolas approximating the graph of f. (Note that we did not have to find the equations of the linear approximations in the trapezoidal rule either.) It can be shown that if $y = ax^2 + bx + c$ is a parabola, shown in Fig. 6.33, through the points (x_0, y_0), (x_1, y_1), and (x_2, y_2) where $x_1 - x_0 = x_2 - x_1 = h$, then

$$\int_{x_0}^{x_2} (ax^2 + bx + c)\,dx = \frac{h}{3}(y_0 + 4y_1 + y_2). \tag{3}$$

In a moment, we will show why Eq. (3) is true. Of course, the number h becomes $(b - a)/n$ in the approximation of $\int_a^b f(x)\,dx$. Thus Eq. (3) shows that the sum of the areas under the parabolas in approximating $\int_a^b f(x)\,dx$ is

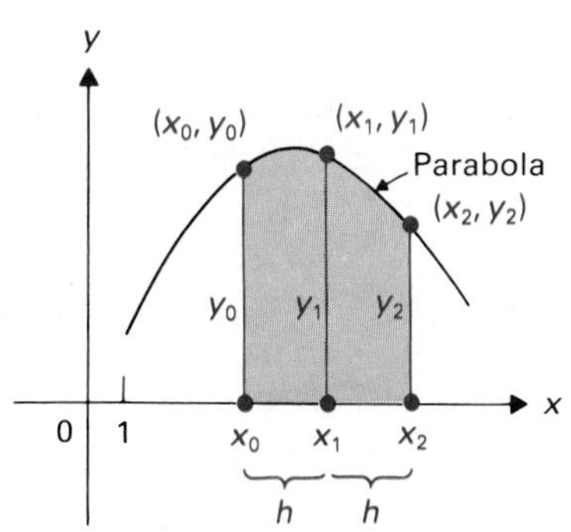

Figure 6.33 The shaded region has area $(h/3)(y_0 + 4y_1 + y_2)$.

$$\frac{b-a}{3n}(y_0 + 4y_1 + y_2) + \frac{b-a}{3n}(y_2 + 4y_3 + y_4) + \cdots$$
$$+ \frac{b-a}{3n}(y_{n-2} + 4y_{n-1} + y_n)$$
$$= \frac{b-a}{3n}(y_0 + 4y_1 + 2y_2 + 4y_3 + 2y_4 + \cdots$$
$$+ 2y_{n-2} + 4y_{n-1} + y_n).$$

This gives us the parabolic rule.

THEOREM 6.6 Simpson's parabolic rule

If f is continuous on $[a, b]$ and if n is even, then

$$\int_a^b f(x)\,dx \approx \frac{b-a}{3n}(y_0 + 4y_1 + 2y_2 + 4y_3 + 2y_4 + \cdots$$
$$+ 2y_{n-2} + 4y_{n-1} + y_n) \tag{4}$$

for n sufficiently large.

The computation of x_i and $y_i = f(x_i)$ for Simpson's rule is identical with that for the trapezoidal rule.

EXAMPLE 4 As in Example 2, take $n = 4$ and estimate $\int_1^3 (1/x)\,dx$.

Solution We have

$$x_0 = 1, \qquad x_1 = \frac{3}{2}, \qquad x_2 = 2, \qquad x_3 = \frac{5}{2}, \qquad x_4 = 3,$$

$$y_0 = 1, \qquad y_1 = \frac{2}{3}, \qquad y_2 = \frac{1}{2}, \qquad y_3 = \frac{2}{5}, \qquad y_4 = \frac{1}{3}.$$

Simpson's rule yields

$$\begin{aligned}\int_1^3 \frac{1}{x}\,dx &\approx \frac{3-1}{3\cdot 4}\left(1 + 4\cdot\frac{2}{3} + 2\cdot\frac{1}{2} + 4\cdot\frac{2}{5} + \frac{1}{3}\right)\\ &= \frac{2}{12}\left(1 + \frac{8}{3} + 1 + \frac{8}{5} + \frac{1}{3}\right)\\ &= \frac{1}{6}\left(\frac{15 + 40 + 15 + 24 + 5}{15}\right)\\ &= \frac{1}{6}\left(\frac{99}{15}\right) = \frac{33}{30} = \frac{11}{10} = 1.1.\end{aligned}$$

The correct answer to four decimal places is 1.0986, and comparison with Example 2 shows that Simpson's rule gives the more accurate estimate for the same value $n = 4$. □

It can be shown that if $f^{(4)}(x)$ exists for $a < x < b$, then the error in approximating $\int_a^b f(x)\,dx$ by Simpson's rule for a value n is given by

$$\left|\frac{(b-a)^5}{180n^4}\,f^{(4)}(c)\right|$$

for some number c between a and b. For example, our error in estimating $\int_1^3 (1/x)\,dx$ by Simpson's rule with $n = 4$ in Example 4 is given by

$$\left|\frac{2^5}{180\cdot n^4}\cdot\frac{24}{c^5}\right| < \frac{2^5}{180\cdot 4^4}\cdot\frac{24}{1} = \frac{1}{60} \approx 0.0167.$$

The difference between our answer 1.1 in Example 4 and the actual value 1.0986 to four decimal places is 0.0014, and we see that our bound for the error is correct.

If $f(x)$ is a polynomial function of degree 3, then $f^{(4)}(x) = 0$ for all x. The above error formula then shows that for any estimate of $\int_a^b f(x)\,dx$ using Simpson's rule, the *exact value* for the integral is obtained, even if we use only $n = 2$. This seems surprising, since Simpson's rule involves approximation by only quadratic polynomials.

EXAMPLE 5 Estimate $\int_1^5 x^4\,dx$ using Simpson's rule with $n = 4$, and compare this estimate with those in Examples 1 and 3.

Solution We have, as in Example 3, $(b - a)/n = 1$ and

$$x_0 = 1, \qquad x_1 = 2, \qquad x_2 = 3, \qquad x_3 = 4, \qquad x_4 = 5,$$

$$y_0 = 1, \qquad y_1 = 2^4, \qquad y_2 = 3^4, \qquad y_3 = 4^4, \qquad y_4 = 5^4.$$

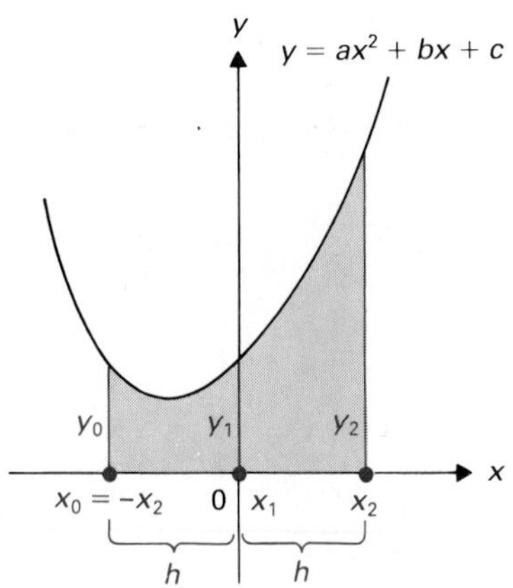

Figure 6.34 Region from x_0 to x_2 under a parabola translated so $x_1 = 0$, $x_0 = -x_2$.

Simpson's rule then yields

$$\int_1^5 x^4\,dx \approx \frac{1}{3}(1 + 4\cdot 2^4 + 2\cdot 3^4 + 4\cdot 4^4 + 5^4)$$

$$= \frac{1876}{3} \approx 625.3333.$$

This is much closer to the true answer 624.8 than the rectangular estimate 604.25 in Example 1 or the trapezoidal estimate 666 in Example 3. □

In general, for a particular value of n, we may expect the rectangular and trapezoidal rules to be of roughly the same magnitude of accuracy. Simpson's rule is apt to give a significantly more accurate estimate for the same value of n, as our examples illustrated. (Of course, we can construct integrals where this is not true; see Exercise 17.) Since Simpson's rule is really no more work than the others, we recommend reaching for it when approximating a definite integral. These days, a calculator or computer is generally available to assist with the computations.

We now give the details of the derivation of Eq. (3), in case you are interested in seeing them. Recall that (x_0, y_0), (x_1, y_1), and (x_2, y_2) are points on the parabola $y = ax^2 + bx + c$. By translating axes if necessary, we can assume that $x_1 = 0$, so $x_0 = -x_2$ (see Fig. 6.34). Then

$$\int_{-x_2}^{x_2} (ax^2 + bx + c)\,dx = \left(\frac{ax^3}{3} + \frac{bx^2}{2} + cx\right)\Big]_{-x_2}^{x_2}$$

$$= \frac{ax_2^3}{3} + \frac{bx_2^2}{2} + cx_2 - \left(-\frac{ax_2^3}{3} + \frac{bx_2^2}{2} - cx_2\right) \tag{5}$$

$$= \frac{2}{3}ax_2^3 + 2cx_2.$$

On the other hand,

$$h = x_1 - x_0 = 0 - (-x_2) = x_2,$$
$$y_0 = f(x_0) = f(-x_2) = ax_2^2 - bx_2 + c,$$
$$y_1 = f(x_1) = f(0) = c,$$
$$y_2 = f(x_2) = ax_2^2 + bx_2 + c.$$

Thus

$$\frac{h}{3}(y_0 + 4y_1 + y_2) = \frac{x_2}{3}(ax_2^2 - bx_2 + c + 4c + ax_2^2 + bx_2 + c)$$

$$= \frac{x_2}{3}(2ax_2^2 + 6c) \tag{6}$$

$$= \frac{2}{3}ax_2^3 + 2cx_2.$$

Comparison of Eqs. (5) and (6) establishes Eq. (3).

SUMMARY

Let f be continuous in $[a, b]$. Let $[a, b]$ be subdivided into n subintervals of equal length $(b - a)/n$.

1. *Rectangular rule:* If x_i is the midpoint of the ith subinterval, then for sufficiently large n,
$$\int_a^b f(x)\,dx \approx \frac{b-a}{n}[f(x_1) + f(x_2) + \cdots + f(x_n)].$$
2. *Trapezoidal rule:* If $x_0 = a, x_1, x_2, \ldots, x_n = b$ are the endpoints of the subintervals and $y_0 = f(x_0), y_1 = f(x_1), y_2 = f(x_2), \ldots, y_n = f(x_n)$, then for sufficiently large n,
$$\int_a^b f(x)\,dx \approx \frac{b-a}{2n}(y_0 + 2y_1 + 2y_2 + \cdots + 2y_{n-1} + y_n).$$
3. *Simpson's rule:* If n is even and $y_0, y_1, y_2, \ldots, y_n$ are as described in the trapezoidal rule, then, for sufficiently large n,
$$\int_a^b f(x)\,dx \approx \frac{b-a}{3n}(y_0 + 4y_1 + 2y_2 + 4y_3 + 2y_4 + \cdots + 2y_{n-2} + 4y_{n-1} + y_n).$$
4. In general, we expect Simpson's rule to be more accurate than either the rectangular rule or the trapezoidal rule for a given value of n.

EXERCISES

It is only realistic to use a calculator or computer these days for approximating an integral. However, Exercises 1 through 15 are designed to be feasible with pencil and paper. Use the indicated rule and value of n to estimate the integral.

1. $\int_1^4 \frac{1}{x}\,dx$, rectangular rule, $n = 3$

2. $\int_1^4 \frac{1}{x}\,dx$, trapezoidal rule, $n = 3$

3. $\int_1^4 \frac{1}{x}\,dx$, Simpson's rule, $n = 2$

4. $\int_1^2 \frac{1}{x}\,dx$

a) trapezoidal rule, $n = 4$
b) Simpson's rule, $n = 4$

5. $\sqrt{3} = \int_0^{\pi/3} 2\cos x\,dx$

a) rectangular rule, $n = 4$
b) Simpson's rule, $n = 4$
(Use a table for cosine.)

6. $\pi = \int_0^1 \frac{4}{1 + x^2}\,dx$, trapezoidal rule, $n = 4$ (Chapter 8 will show that this integral equals π.)

7. $\pi = \int_0^1 \frac{4}{1 + x^2}\,dx$, Simpson's rule, $n = 4$

8. $\int_0^\pi \frac{dx}{2 + \sin^2 x}$, trapezoidal rule, $n = 4$

9. $\int_0^\pi \frac{dx}{2 + \sin^2 x}$, Simpson's rule, $n = 4$

10. $\int_3^7 \frac{dx}{x - 1}$, Simpson's rule, $n = 4$

11. $\int_3^7 \frac{dx}{x - 1}$, trapezoidal rule, $n = 4$

12. $\int_1^2 \frac{x}{x + 1}\,dx$, Simpson's rule, $n = 2$

13. $\int_1^2 \frac{x}{x + 1}\,dx$, trapezoidal rule, $n = 2$

14. $\int_2^4 \frac{x^2}{x + 2}\,dx$, rectangular rule, $n = 4$

15. $\int_2^4 \frac{x^2}{x+2}\,dx$, Simpson's rule, $n = 4$

16. Show that the bound for the error in Simpson's rule given in the text is consistent with the result obtained in Example 5.

17. Give an example of a definite integral and a value of n such that Simpson's rule gives a less accurate estimate than either the rectangular rule or the trapezoidal rule. [*Hint:* Consider $f(x) = |x|$.]

In Exercises 18 through 30, use a computer or calculator to estimate the integral, using the indicated rule and value for n.

18. $\int_0^{\pi} \sqrt{1 + \sin^2 x}\,dx$, Simpson's rule, $n = 10$

19. $\int_1^{21} \frac{1}{x}\,dx$, Simpson's rule, $n = 20$

20. $\pi = \int_0^1 \frac{4}{1+x^2}\,dx$, Simpson's rule, $n = 10$

21. $\pi = \int_0^{1/2} \frac{6}{\sqrt{1-x^2}}\,dx$, trapezoidal rule, $n = 10$

22. $\int_0^{\pi} \sqrt{4 - \sin^2 x}\,dx$, trapezoidal rule, $n = 10$

23. $\int_0^3 (1+x)^x\,dx$, Simpson's rule, $n = 12$

24. $\int_0^4 \frac{dx}{\sqrt{6 - \cos^2 x}}$, Simpson's rule, $n = 8$

25. $\int_0^{\sqrt{\pi}} \sin x^2\,dx$, Simpson's rule, $n = 10$

26. $\int_0^1 \cos(x^2 + 1)\,dx$, Simpson's rule, $n = 10$

27. $\int_0^{\pi/4} \sec x\,dx$, trapezoidal rule, $n = 10$

28. $\int_0^1 2^x\,dx$, Simpson's rule, $n = 20$

29. $\int_0^4 \frac{x}{1 + \sin^2 x}\,dx$, Simpson's rule, $n = 10$

30. $\int_0^2 \frac{x}{4 - \cos^2 x}\,dx$, Simpson's rule, $n = 20$

EXERCISE SETS FOR CHAPTER 6

Review Exercise Set 6.1

1. Find the upper sum S_4 for $f(x) = 1/(x+1)$ over $[2, 6]$.
2. Use geometry to find $\int_0^3 (2 + \sqrt{9 - x^2})\,dx$.
3. Find
$$\frac{d}{dt}\left(\int_1^{\sqrt{t}} \sqrt{2x^2 + x^4}\,dx\right).$$
4. Find the area of the region under the curve $y = x^3 + 2x$ and over the x-axis between $x = 1$ and $x = 2$.
5. If $\int_1^4 f(x)\,dx = 5$ and $\int_4^2 f(x)\,dx = -7$, find $\int_1^2 f(x)\,dx$.
6. Find

a) $\int_0^{\pi/2} \sin 2x\,dx$, b) $\int_0^1 x\sqrt{1-x^2}\,dx$.

7. Find

a) $\int \sin 2x \cos 2x\,dx$, b) $\int \sec^3 x \tan x\,dx$.

8. Find the general solution of the differential equation $dy/dx = x^2\sqrt{y}$.
9. Use the formula
$$\int \frac{x\,dx}{\sqrt{ab + b}} = \frac{2(ax - 2b)}{3a^2}\sqrt{ax + b} + C$$
to find
$$\int_0^{\pi/4} \frac{\sin 4x}{\sqrt{3 + \cos 2x}}\,dx.$$
10. Use Simpson's rule with $n = 4$ to estimate
$$\int_0^2 \frac{1}{x+2}\,dx.$$

Review Exercise Set 6.2

1. Find the Riemann sum $\mathcal{S}_4$ estimating $\int_0^{2\pi} \sin(x/2)\,dx$, using midpoints of the subintervals.
2. Use geometry to find $\int_{-5}^5 (5 - \sqrt{25 - x^2})\,dx$.
3. Find
$$\frac{d}{dt}\left(\int_{-t}^{2t} \sin^2 x\,dx\right).$$
4. Find the area of the region over the x-axis and under the curve $y = 16 - x^2$ between $x = -4$ and $x = 4$.
5. If $\int_{-1}^4 f(x)\,dx = 4$ and $\int_2^4 (3 - f(x))\,dx = 7$, find $\int_2^{-1} f(x)\,dx$.
6. Find

a) $\int_{-1}^1 x(x^2+1)^3\,dx$, b) $\int_0^{\pi} \sin 3x\,dx$.

7. Find

a) $\int \frac{\cos 3x}{\sin^4 3x}\,dx$,

b) $\int \sqrt{\csc x} \cot x\,dx$.

8. Find the solution of the differential equation $dy/dx = y^2 \cos 2x$ such that $y = -3$ when $x = \pi/4$.

9. Use the reduction formula

$$\int \cos^n ax\, dx = \frac{\cos^{n-1} ax \sin ax}{na} + \frac{n-1}{n}\int \cos^{n-2} ax\, dx$$

to find $\int_0^{\pi/8} \cos^4 2x\, dx$.

10. Use the trapezoidal rule with $n = 4$ to estimate

$$\int_{-1}^{1} \frac{1}{x^2+1}\, dx.$$

Review Exercise Set 6.3

1. Compute $\sum_{i=1}^{4} \left(\frac{i}{2}\right)^2$.

2. If $\int_0^5 3 \cdot f(x)\, dx = 2$ and $\int_2^5 f(x)\, dx = -1$, find $\int_2^0 f(x)\, dx$.

3. Find $\int_{-3}^{3} (x^2 - 1 + \sqrt{9 - x^2})\, dx$.

4. Find $(d^2/dt^2)\left(\int_t^3 \cos x^2\, dx\right)$.

5. Find the area of the region under one "arch" of $y = \sin 2x$ and over the x-axis.

6. Find a) $\int_0^{\pi/6} \sec^2 2x\, dx$, b) $\int_1^3 \frac{1}{x^2}\, dx$.

7. Find a) $\int \frac{x^2}{(x^3+1)^4}\, dx$, b) $\int \sqrt{\sec x} \sin x\, dx$.

8. Find the general solution of the differential equation

$$\frac{dy}{dx} = x \cos^2 2y.$$

9. Use the formula

$$\int \frac{dx}{x^2\sqrt{a^2+x^2}} = -\frac{\sqrt{a^2+x^2}}{a^2 x} + C$$

to find

$$\int \frac{dx}{x^{5/2}\sqrt{9+x^3}}.$$

10. Use a calculator and Simpson's rule with $n = 16$ to estimate

$$\int_0^{\pi/4} \tan x\, dx.$$

Review Exercise Set 6.4

1. Find $\mathscr{S}_4$ using midpoints for $f(x) = x + x^2$ over $[0, 1]$.

2. If $\int_0^{10} (3 \cdot f(x) - g(x))\, dx = 5$ and $\int_{10}^{0} (f(x) + g(x))\, dx = 2$, find $\int_0^{10} f(x)\, dx$.

3. Find $\int_0^3 \sqrt{54 - 6x^2}\, dx$.

4. Find $\int \frac{d^3(\sin x^2)}{dx^3}\, dx$.

5. Find the area of the region bounded by $y = x^2 - 4$ and the x-axis.

6. Find a) $\int_{\pi/6}^{\pi/3} \sin 2x\, dx$, b) $\int_1^2 \frac{x^2+1}{x^2}\, dx$.

7. Find a) $\int x^2(4x^3+2)^5\, dx$, b) $\int \sin 3x \sec^2 3x\, dx$.

8. Find the solution of the initial-value problem $dy/dx = x^2y^2$, $y = 2$ when $x = 1$.

9. Use the formula

$$\int \frac{dx}{x^2\sqrt{a^2-x^2}} = -\frac{\sqrt{a^2-x^2}}{a^2 x} + C$$

to find

$$\int \frac{\cos x}{\sin^2 x\sqrt{9 - 4\sin^2 x}}\, dx.$$

10. Use a calculator and Simpson's rule with $n = 12$ to estimate

$$\int_1^3 \frac{x}{1+x^2}\, dx.$$

More Challenging Exercises 6

1. Sketch the graph of a function f over the interval $[0, 6]$ for which $s_3 < s_2$. (This shows that while s_n approaches $\int_a^b f(x)\, dx$ as n increases, we need not have $s_{n+1} \geq s_n$.)
2. Show that, for any continuous function f defined on $[a, b]$, we have $S_{2n} \leq S_n$ for each n. (Compare with Exercise 1.)
3. Show that if f is continuous and increasing on $[0, 1]$ with $f(0) = 0$, then $S_n - s_n = f(1)/n$.
4. Exercise 3 can be generalized easily to give a simple formula for $S_n - s_n$ in the case of a continuous, increasing function on $[a, b]$. Find this formula for $S_n - s_n$.
5. Give a formula for $S_n - s_n$ analogous to that found in Exercise 4 for the case of a continuous, decreasing function on $[a, b]$.
6. Give an example of a function f with domain $[0, 1]$ such that $S_n = 1$ and $s_n = 0$ for *every* positive integer n. (Of course, Theorem 6.1 shows that f could not be continuous, but it is possible to choose f so that it assumes a maximum and a minimum over each subinterval.)

Examples 3 and 4 of Section 6.2 showed how integrals can sometimes be found by evaluating the limit of a sum. Conversely, working backward through the examples, we see how a limit of a sum might be recognized as giving an integral and then evaluated using the fundamental theorem. In Exercises 7 through 12, use an integral to find each of the limits.

7. $\lim_{n\to\infty} \frac{4}{n} \cdot \sum_{i=1}^{n} \frac{4i^2}{n^2}$
8. $\lim_{n\to\infty} \frac{8}{n^4} \cdot \sum_{i=1}^{n} i^3$
9. $\lim_{n\to\infty} \frac{1}{n^{5/2}} \cdot \sum_{i=1}^{n} i\sqrt{i}$
10. $\lim_{n\to\infty} \frac{1}{n^5} \cdot \sum_{k=1}^{n} (1+k)^3$
11. $\lim_{n\to\infty} \frac{1}{n^5} \cdot \sum_{k=1}^{n} (2k+1)^4$
12. $\lim_{n\to\infty} \frac{1}{n^3} \cdot \sum_{k=1}^{n} (k-100)^3$

APPLICATIONS OF THE INTEGRAL

At the start of Chapter 6, we indicated that integral calculus is related to the notion of a *product* in much the same way as differential calculus is related to a *quotient*. Calculus is the mathematics needed to study quantities of changing magnitude. The definite integral is potentially useful in any type of application where, in the case of quantities that do not vary, we would compute a *product*. (Since any quantity can be viewed as the product of itself with 1, this covers a very general situation.) We list here several things that may be given by a product of two quantities, some of which we mentioned before.

Area The area A of a plane region of base width b and constant altitute h (that is, a rectangle) is given by $A = bh$.

Volume For a solid with horizontal cross section of constant area A and with constant height h, the volume V is given by $V = Ah$.

Work If a body is moved a distance s by means of a constant force F acting in the direction of the motion, the work W done in moving the body is given by $W = Fs$.

Distance If a body travels in a straight line for a time t with a constant velocity v, the distance s traveled is given by $s = vt$.

Speed If a body travels in a straight line for a time t with a constant acceleration a, the velocity v attained is given by $v = at$.

Force If the pressure per square unit on a plane region of area A is a constant p throughout the region, then the total force F on the region due to this pressure is given by $F = pA$.

Moment The moment M about an axis of a body of mass m, all points of which are a constant (signed) distance s from the axis, is given by $M = ms$.

Moment of inertia The moment of inertia I about an axis of a body of mass m, all points of which are a constant distance s from the axis, is given by $I = ms^2$.

All these notions can be handled by integral calculus if the factors appearing in the products vary continuously. This chapter is devoted to such applications of the integral.

7.1 AREA AND AVERAGE VALUE

THE AREA OF A PLANE REGION

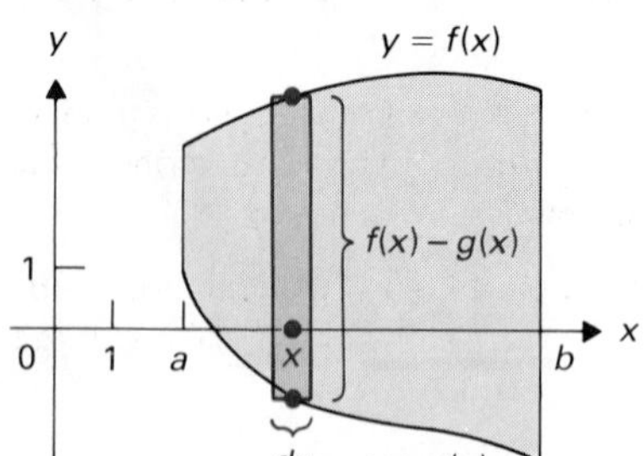

Figure 7.1 Thin vertical rectangle of area $dA = (f(x) - g(x))\,dx$.

This section shows how to find the area of a more general plane region than one under the graph of a continuous function from a to b. Consider, for example, the shaded region in Fig. 7.1 that lies between the graphs of continuous functions f and g from a to b. The area of this region can be estimated by taking thin rectangles like the one shown shaded in color in Fig. 7.1 and adding up the areas of such rectangles over the region. This rectangle is dx units wide and $f(x) - g(x)$ units high, for x as shown in Fig. 7.1. (Note that $g(x)$ is itself negative. The vertical distance between the graphs of f and g over a point x is always $f(x) - g(x)$ if the graph of f is above that of g at x.) Thus the area of this rectangle is

$$dA = (f(x) - g(x))\,dx.$$

We wish to add up the areas of such rectangles and take the limit of the resulting sum as dx grows smaller and the number of rectangles increases. As we know, the limit of such a sum will be $\int_a^b (f(x) - g(x))\,dx$.

In computing the area of a plane region, care must be taken in finding the correct function to integrate, that is, in "setting up the integral." The area of the region bounded by the graphs of continuous functions f and g between a and b is not always $\int_a^b (f(x) - g(x))\,dx$. If $g(x) \le f(x)$ for all x in $[a, b]$, then the correct integral is indeed $\int_a^b (f(x) - g(x))\,dx$. However, if f and g have the graphs shown in Fig. 7.2, then $\int_a^b (f(x) - g(x))\,dx$ is numerically equal to the area of the first region minus the area of the second. The area of the total shaded region in Fig. 7.2 is best obtained by finding the areas of the first and second regions separately and then adding the areas; a correct expression is

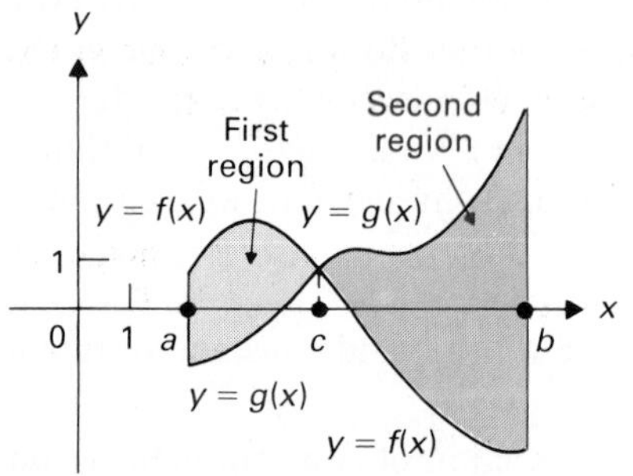

Figure 7.2 Total area is $\int_a^c (f(x) - g(x))\,dx + \int_c^b (g(x) - f(x))\,dx$.

$$\int_a^c (f(x) - g(x))\,dx + \int_c^b (g(x) - f(x))\,dx.$$

The area of the region between the graphs of f and g from a to b can always be expressed as

$$\int_a^b |f(x) - g(x)|\,dx.$$

Indeed, it is appropriate to *define* the area to be this integral. In practice, we do not work with the absolute-value sign; rather we look at a figure and set up one or more integrals to find the desired area. Setting up the integrals is the interesting part of the problem. Computation of the integrals should be regarded as a mechanical chore, much like adding a column of numbers, although most students haven't had as much experience in computing integrals.

Suggested steps for finding the area of a plane region are given in the box on page 253. We illustrate with several examples.

EXAMPLE 1 Find the area of the region bounded by $y = 1$ and $y = x^2$ in two ways, once using a dx-integral and once using a dy-integral.

FINDING THE AREA OF A REGION

Step 1 Sketch the region, finding the points of intersection of bounding curves.

Step 2 On the sketch, draw a typical thin rectangle either vertical of width dx or horizontal of width dy.

Step 3 *Looking at the sketch*, write down the area dA of this rectangle as a product of the width (dx or dy) and the length. Express dA entirely in terms of the variable (x or y) appearing in the differential.

Step 4 Integrate dA between the appropriate (x or y) limits. (By definition of the integral, this amounts to adding up the areas found in Step 3 and taking the limit of the resulting sum.)

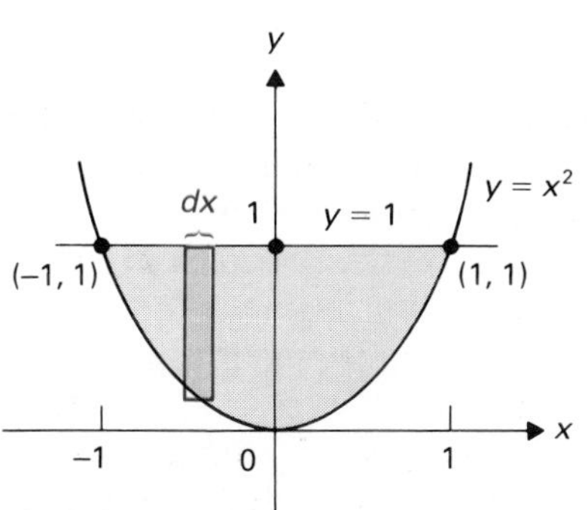

Figure 7.3 Thin vertical rectangle of area $dA = (1 - x^2)\,dx$.

Solution 1 ***dx-integral***

Step 1 Figure 7.3 shows the region bounded by $y = 1$ and $y = x^2$.

Step 2 Figure 7.3 also shows a thin vertical rectangle of width dx for this region.

Step 3 The vertical height of the rectangle in Fig. 7.3 is the y-value at the top minus the y-value at the bottom, which we abbreviate by

$$\text{Vertical height} = y_{\text{top}} - y_{\text{bottom}}.$$

Since we wish to integrate using dx, we must express this height as a function of x. We have $y_{\text{top}} = 1$ and $y_{\text{bottom}} = x^2$, so the height is $1 - x^2$. Thus we have

$$dA = (y_{\text{top}} - y_{\text{bottom}})\,dx = (1 - x^2)\,dx.$$

Step 4 Since we want to add up the areas of our rectangles over the region from $x = -1$ to $x = 1$, we have

$$A = \int_{-1}^{1} (1 - x^2)\,dx = \left(x - \frac{x^3}{3}\right)\Bigg]_{-1}^{1} = \left(1 - \frac{1}{3}\right) - \left(-1 + \frac{1}{3}\right) = \frac{4}{3}.$$

Alternatively, we can make use of the obvious symmetry of our region about the y-axis and write

$$A = 2\int_{0}^{1} (1 - x^2)\,dx = 2\left(x - \frac{x^3}{3}\right)\Bigg]_{0}^{1} = 2\left(1 - \frac{1}{3}\right) - 0 = \frac{4}{3}.$$

We prefer to use such symmetry whenever it enables us to have zero as one of the limits of integration.

Figure 7.4 Thin horizontal rectangle of area $dA = [\sqrt{y} - (-\sqrt{y})]\,dy$.

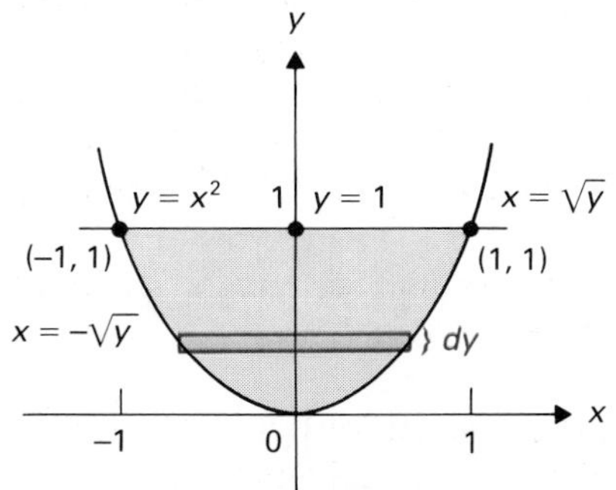

Solution 2 ***dy-integral***

Step 1 The region is sketched in Fig. 7.4.

Step 2 We draw a thin horizontal rectangle of width dy for the region in Fig. 7.4.

Step 3 The horizontal dimension of the rectangle in Fig. 7.4 is the x-value at the right-hand end minus the x-value at the left-hand end, which we abbreviate to

$$\text{Horizontal length} = x_{\text{right}} - x_{\text{left}}.$$

Since we will integrate using dy, we must express this horizontal length as a function of y. Both ends of the rectangle in Fig. 7.4 fall on the curve $y = x^2$, so we have $x_{\text{right}} = \sqrt{y}$ and $x_{\text{left}} = -\sqrt{y}$. Thus

$$dA = (x_{\text{right}} - x_{\text{left}})\,dy = [\sqrt{y} - (-\sqrt{y})]\,dy = 2\sqrt{y}\,dy.$$

Step 4 Since we wish to add up the areas of our rectangles in the y direction from $y = 0$ to $y = 1$, we have

$$A = \int_0^1 2\sqrt{y}\,dy = 2\int_0^1 y^{1/2}\,dy = 2\cdot\frac{2}{3}\,y^{3/2}\Big]_0^1 = \frac{4}{3}\cdot 1^{3/2} - 0 = \frac{4}{3}. \quad \square$$

As indicated in Example 1, it is useful to remember that

$dA = (y_{\text{top}} - y_{\text{bottom}})\,dx,$	For a vertical strip
$dA = (x_{\text{right}} - x_{\text{left}})\,dy.$	For a horizontal strip

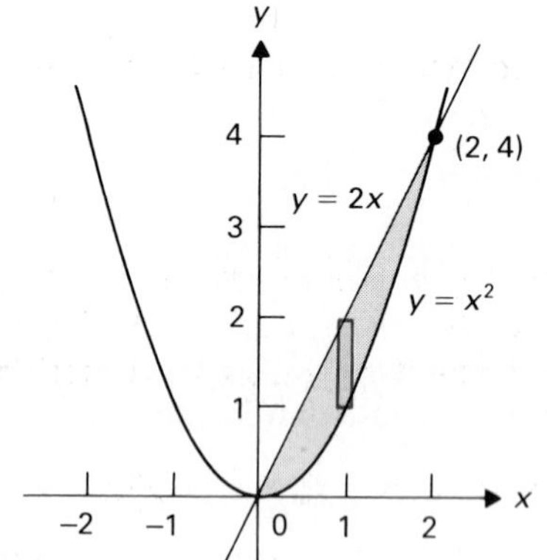

Figure 7.5 Thin vertical rectangle of area $dA = (2x - x^2)\,dx$.

EXAMPLE 2 Find the area of the region bounded by $y = 2x$ and $y = x^2$ in two ways, as in Example 1.

Solution 1 ***dx-integral***

Step 1 The region is sketched in Fig. 7.5.

Step 2 Figure 7.5 shows a thin vertical rectangle of width dx.

Step 3 We need the vertical height of the rectangle in terms of x. Since the line $y = 2x$ is on top of the region and $y = x^2$ is at the bottom, we see that

$$dA = (y_{\text{top}} - y_{\text{bottom}})\,dx = (2x - x^2)\,dx.$$

Step 4 We have

$$A = \int_0^2 (2x - x^2)\,dx = \left(x^2 - \frac{x^3}{3}\right)\Big]_0^2 = \left(4 - \frac{8}{3}\right) - 0 = \frac{4}{3}.$$

Figure 7.6 Thin horizontal rectangle of area $dA = (\sqrt{y} - y/2)\,dy$.

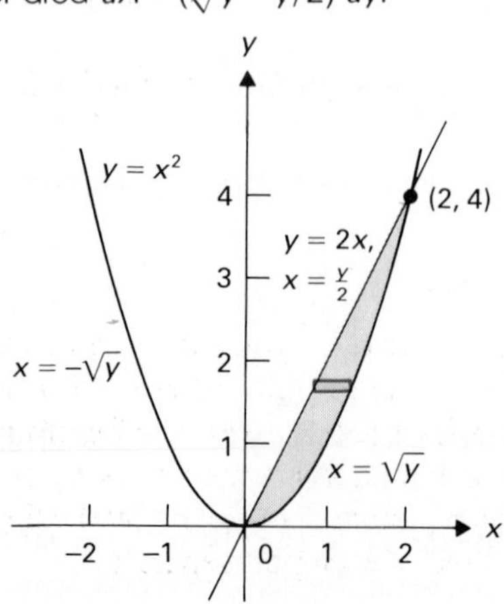

Solution 2 ***dy-integral***

Step 1 The region is sketched in Fig. 7.6.

Step 2 Figure 7.6 shows a thin horizontal rectangle of width dy.

Step 3 We need the horizontal length of the rectangle in terms of y. Since $y = x^2$ is to the right of $y = 2x$ in the region, we have $x_{\text{right}} = \sqrt{y}$ and $x_{\text{left}} = y/2$. Thus

$$dA = (x_{\text{right}} - x_{\text{left}})\,dy = \left(\sqrt{y} - \frac{y}{2}\right)dy.$$

Step 4 We wish to add the areas of the rectangles from $y = 0$ to $y = 4$, so

$$A = \int_0^4 \left(\sqrt{y} - \frac{y}{2}\right) dy = \int_0^4 \left(y^{1/2} - \frac{1}{2}y\right) dy = \left(\frac{2}{3}y^{3/2} - \frac{y^2}{4}\right)\Bigg]_0^4$$

$$= \left(\frac{2}{3}\cdot 8 - \frac{16}{4}\right) = \frac{16}{3} - 4 = \frac{4}{3}. \quad \square$$

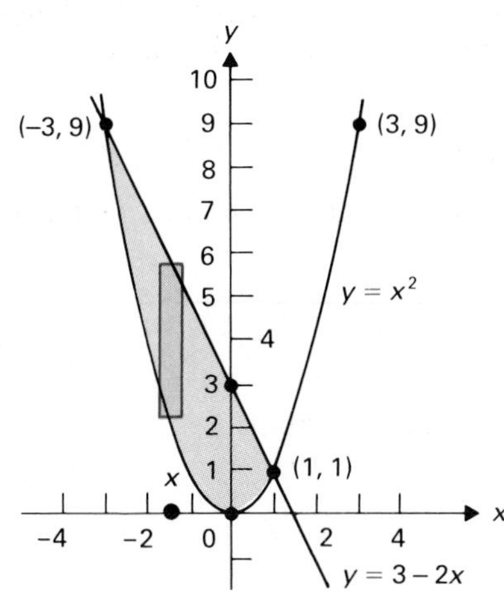

Figure 7.7 Thin vertical rectangle of area $dA = [(3 - 2x) - x^2]\,dx$.

If we want to find the area of a region, we don't bother to compute both the dx integral and the dy integral, as we did in Examples 1 and 2. We set up and compute whichever of the integrals looks easier. In Examples 1 and 2, we would probably have used just the dx integrals, making use of symmetry in Example 1, although the dy integrals were equally simple in those examples. The next two examples show that one integral may be easier to use than the other.

EXAMPLE 3 Find the area of the plane region bounded by the curves with equations $y = x^2$ and $y = 3 - 2x$, using the dx integral.

Solution

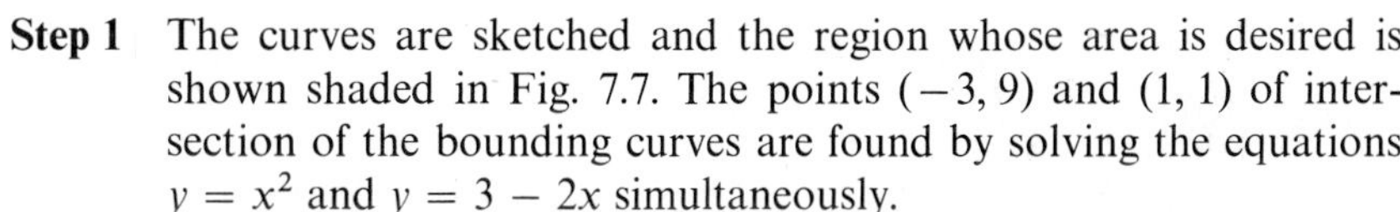

Step 1 The curves are sketched and the region whose area is desired is shown shaded in Fig. 7.7. The points $(-3, 9)$ and $(1, 1)$ of intersection of the bounding curves are found by solving the equations $y = x^2$ and $y = 3 - 2x$ simultaneously.

Step 2 A typical thin vertical rectangle is drawn and shaded in Fig. 7.7.

Step 3 The graph of the function $3 - 2x$ is above the graph of x^2 over this region, so the height of the thin rectangle over a point x is $(3 - 2x) - x^2$. The area of this thin rectangle is therefore $dA = [(3 - 2x) - x^2]\,dx$.

Step 4 Since we wish to add up the areas of the thin rectangles in the x-direction from $x = -3$ to $x = 1$, the appropriate integral is

$$\int_{-3}^{1} [(3 - 2x) - x^2]\,dx = \left(3x - 2\frac{x^2}{2} - \frac{x^3}{3}\right)\Bigg]_{-3}^{1}$$

$$= 3 - 1 - \frac{1}{3} - (-9 - 9 + 9) = \frac{32}{3}. \quad \square$$

Figure 7.8 Upper rectangle has area $dA = [\frac{1}{2}(3 - y) - (-\sqrt{y})]\,dy$; lower rectangle has area $dA = [\sqrt{y} - (-\sqrt{y})]\,dy$.

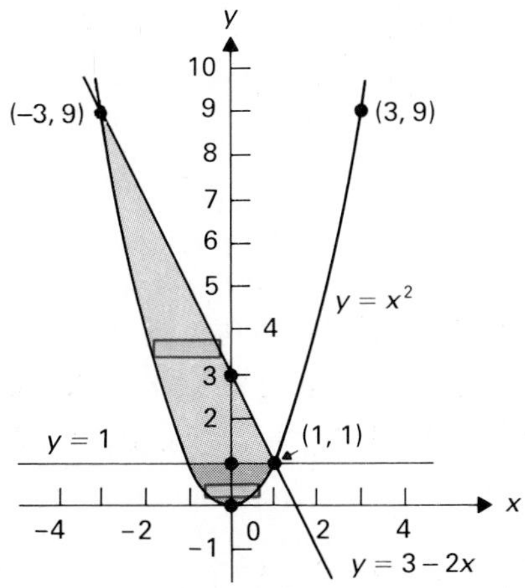

EXAMPLE 4 Solve the same problem as in Example 3 by taking the thin rectangles horizontally.

Solution

Step 1 The appropriate sketch is given in Fig. 7.8.

Step 2 Note that we can't take just one rectangle as typical for the whole region, for if the rectangle lies above the line given by $y = 1$, it is bounded on the right by the line with equation $y = 3 - 2x$, while a rectangle below the line $y = 1$ is bounded on both ends by the curve given by $y = x^2$. We split the region into two parts by the line with equation $y = 1$ and find the area of each part separately.

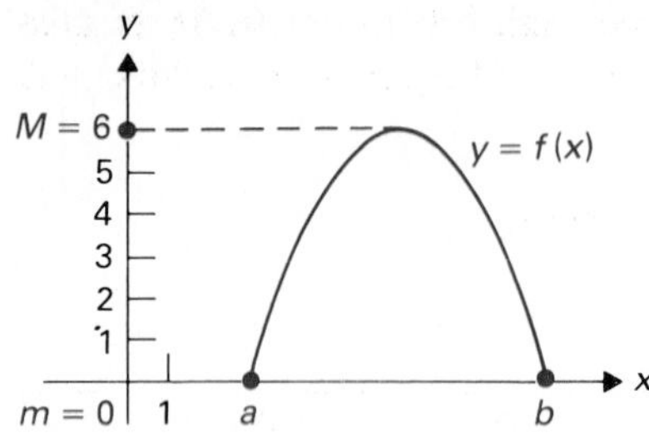

Figure 7.9 $f(a) = f(b) = 0$, but $f(x) > 0$ for $a < x < b$.

Step 3 To find the horizontal dimensions of the rectangles, we must solve for x in terms of y, obtaining $x = (3 - y)/2$ for the line and $x = \pm\sqrt{y}$ for the curve. The upper rectangle has an area of

$$dA = [\tfrac{1}{2}(3 - y) - (-\sqrt{y})]\, dy,$$

while the lower rectangle has an area of

$$dA = [\sqrt{y} - (-\sqrt{y})]\, dy.$$

Step 4 The appropriate integrals are

$$\int_1^9 \left[\frac{1}{2}(3 - y) - (-\sqrt{y})\right] dy = \left(\frac{3}{2}y - \frac{y^2}{4} + \frac{2}{3}y^{3/2}\right)\Bigg]_1^9$$
$$= \left(\frac{27}{2} - \frac{81}{4} + \frac{2}{3}\cdot 27\right) - \left(\frac{3}{2} - \frac{1}{4} + \frac{2}{3}\right)$$
$$= \frac{28}{3},$$

and

$$\int_0^1 [\sqrt{y} - (-\sqrt{y})]\, dy = \int_0^1 2\sqrt{y}\, dy = 2\cdot\frac{2}{3}y^{3/2}\Bigg]_0^1 = \frac{4}{3} - 0 = \frac{4}{3}.$$

Thus the total area is $\frac{28}{3} + \frac{4}{3} = \frac{32}{3}$ square units. Clearly this computation is not as nice as that in Example 3. □

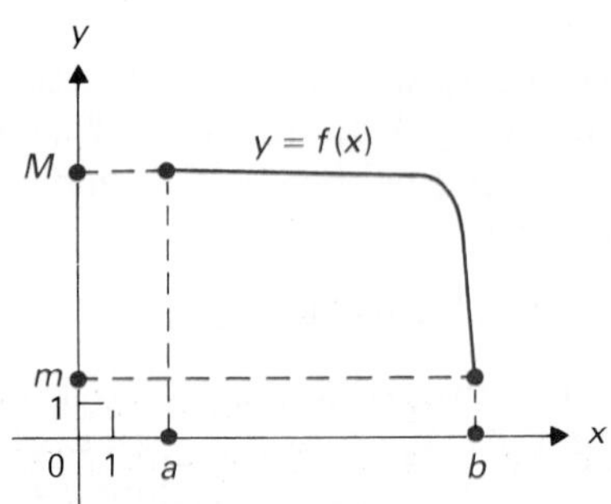

Figure 7.10 $f(x)$ is close to M over most of $[a, b]$.

THE AVERAGE VALUE OF A FUNCTION

Let f be a continuous function with domain containing $[a, b]$, and consider the size of $f(x)$ for x in $[a, b]$. This size can vary tremendously over an interval $[a, b]$, even if f is continuous. But we can try to develop some notion of the *average size of* $f(x)$ *over* $[a, b]$.

Perhaps the first natural attempt to find an average for $f(x)$ over $[a, b]$ is to form the average $(f(a) + f(b))/2$. Upon a little consideration, we see that we should reject such a definition of the average size for $f(x)$ over $[a, b]$, because $(f(a) + f(b))/2$ really reflects the size of $f(x)$ only at a and at b. For f with graph shown in Fig. 7.9, this average $(f(a) + f(b))/2$ is zero, even though $f(x) > 0$ for $a < x < b$.

Probably the next natural attempt to define an average for $f(x)$ over $[a, b]$ is to average the maximum size M and the minimum size m of all the $f(x)$ for x in $[a, b]$, if this maximum M and this minimum m exist. We know that M and m do exist if f is continuous in $[a, b]$. For the function f shown in Fig. 7.9, we have $M = 6$ and $m = 0$, so the average is $(M + m)/2 = \frac{6}{2} = 3$, which may seem like a reasonable value for the average of $f(x)$ over $[a, b]$. But for the function with graph shown in Fig. 7.10, the average of $f(x)$ over $[a, b]$ should surely be closer to M than to m, reflecting the fact that $f(x)$ is close to M over most of the interval $[a, b]$.

Figure 7.11 The area $h(b - a)$ of the rectangle equals the area $\int_a^b f(x)\,dx$ under the graph of f.

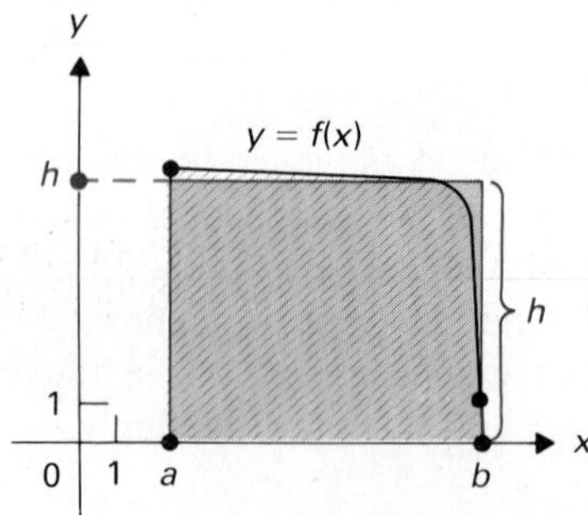

To take care of a situation like that shown in Fig. 7.10, we will regard the *average of* $f(x)$ *over* $[a, b]$ as the height h that a rectangle with base $[a, b]$ must have for the area of the rectangle to be equal to the area of the region under the graph of f over $[a, b]$. This region under the graph of f over $[a, b]$ is shown shaded in Fig. 7.11, where a rectangle of equal area with height h is also indicated. The area of the rectangle is $h(b-a)$, while the area of the shaded region is of course $\int_a^b f(x)\,dx$, so if the areas are to be equal, we must have

$$h = \frac{1}{b-a}\int_a^b f(x)\,dx.$$

DEFINITION 7.1 Average value of a function

Let f be continuous in $[a, b]$. The **average** (or **mean**) **value** of $f(x)$ in $[a, b]$ is

$$\frac{1}{b-a}\int_a^b f(x)\,dx.$$

EXAMPLE 5 Find the average value of x^2 in $[0, 2]$.

Solution We have

$$\int_0^2 x^2\,dx = \frac{x^3}{3}\bigg]_0^2 = \frac{8}{3}.$$

Thus the average value is

$$\frac{1}{2-0}\cdot\frac{8}{3} = \frac{4}{3}. \quad \square$$

EXAMPLE 6 Find the mean value of $\sin x$ in $[0, 3\pi/2]$.

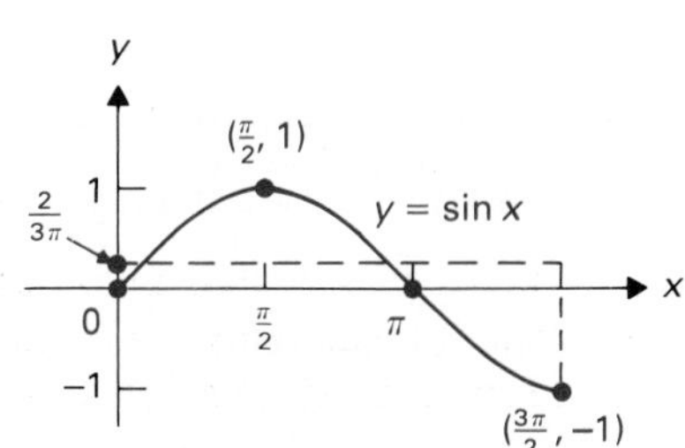

Figure 7.12 $\sin x$ has mean value $2/(3\pi)$ over $[0, 3\pi/2]$.

Solution The function $\sin x$ takes on both positive and negative values in $[0, 3\pi/2]$, as shown in Fig. 7.12, but that does not affect the way we compute its mean value. We have

$$\int_0^{3\pi/2} \sin x\,dx = -\cos x\bigg]_0^{3\pi/2} = 0 - (-1) = 1.$$

Thus the mean value is

$$\frac{1}{\dfrac{3\pi}{2} - 0}\cdot 1 = \frac{2}{3\pi}. \quad \square$$

If m and M are the minimum value and maximum value for $f(x)$ over $[a, b]$, then we know that

$$s_1 = m(b-a) \le \int_a^b f(x)\,dx \le M(b-a) = S_1.$$

Dividing by $(b-a)$, we obtain

$$m \le \frac{1}{b-a}\int_a^b f(x)\,dx \le M,$$

which proves that the mean value of $f(x)$ in $[a, b]$ is between m and M. Since we are assuming that f is continuous on $[a, b]$, it follows by the intermediate-value theorem that

$$\frac{1}{b-a}\int_a^b f(x)\,dx = f(c)$$

for some c where $a < c < b$. The existence of this c is known as the *mean-value theorem for the integral.*

THEOREM 7.1 Mean-value theorem for the integral

Let $f(x)$ be continuous in $[a, b]$. Then there exists a number c, where $a < c < b$, such that

$$\frac{1}{b-a}\int_a^b f(x)\,dx = f(c).$$

EXAMPLE 7 Illustrate Theorem 7.1 for $f(x) = x^2$ in $[0, 2]$.

Solution Example 5 showed that the average value of $f(x)$ in $[0, 2]$ is given by

$$\frac{1}{2-0}\int_0^2 x^2\,dx = \frac{4}{3}.$$

Now $f(x) = x^2 = \frac{4}{3}$ when $x = \pm 2/\sqrt{3}$. The value $2/\sqrt{3}$ does indeed satisfy $0 < 2/\sqrt{3} < 2$, as required by Theorem 7.1, so we have $c = 2/\sqrt{3}$. □

SUMMARY

1. The following is a suggested procedure for finding the area of a region bounded by given curves.

   Step 1 Sketch the region, finding the points of intersection of the bounding curves.

   Step 2 On the sketch, draw a typical thin rectangle either vertical and of width dx or horizontal and of width dy.

   Step 3 Looking at the sketch, write down the area dA of this rectangle as a product of the width (dx or dy) and the length. Express dA entirely in terms of the variable (x or y) appearing in the differential:

   $$dA = (y_{\text{top}} - y_{\text{bottom}})\,dx, \quad \text{or} \quad dA = (x_{\text{right}} - x_{\text{left}})\,dy.$$

   Step 4 Integrate dA between the appropriate (x or y) limits.

2. The mean or average value of $f(x)$ in $[a, b]$ is

   $$\frac{1}{b-a}\int_a^b f(x)\,dx.$$

3. *Mean-value theorem for the integral:* If $f(x)$ is continuous in $[a, b]$, then there exists a number c, where $a < c < b$, such that

   $$\frac{1}{b-a}\int_a^b f(x)\,dx = f(c).$$

EXERCISES

In Exercises 1 through 25, find the total area of the region or regions bounded by the given curves.

1. $y = x^2, y = 4$
2. $y = x^4, y = 1$
3. $y = x, y = x^2$
4. $y = x, y = x^3$
5. $y = x^4, y = x^2$
6. $y = x^2, y = x^3$
7. $y = x^4 - 1, y = 1 - x^2$
8. $y = x^2 - 1, y = x + 1$
9. $x = y^2, y = x - 2$
10. $x = y^2 - 4, y = 2 - x$
11. $y = 1 - x^2, y = x - 1$
12. $y = \sqrt{x}, y = x^2$
13. $y = \sqrt{x}, y = x^4$
14. $x = 2y^2, x = 8$
15. $y = x^2 - 1, y = \sqrt{1 - x^2}$ [*Hint:* Use some known areas.]
16. $y = x^2, y = 10x - 9$
17. $y = x^2/4, y = 2x - 3$
18. $x = y^2 - 1, y = x/4 + 1$
19. $y = \sin x, y = 3 \sin x$ for $0 \le x \le \pi$
20. $y = \sin x, y = \cos x$ for $0 \le x \le \pi/4, x = 0$
21. $y = \sec^2 x$ for $-\pi/2 < x < \pi/2, y = 4$
22. $y = \csc^2 x$ for $0 < x < \pi, y = \frac{4}{3}$
23. $y = \sin 2x, y = \sec^2 x - 1$ for $-\pi/2 < x < \pi/2$
24. $y = \sin x, y = 2x, x = \pi/2$
25. $y = \sin x, y = \sqrt{x}, x = \pi$
26. Express as an integral the area of the smaller region bounded by the curves $x^2 + y^2 = 4$ and $y = -1$.
27. Express as an integral the area of the smaller region bounded by the curves $x = 2y^2$ and $x^2 + y^2 = 68$.
28. Express as one or more integrals the area of the region in the first quadrant bounded by the curves $x^2 + y^2 = 2$, $y = 0$, and $y = x^2$.
29. Express in terms of an integral the area of the larger region bounded by $x^2 + y^2 = 25$ and the line through $(-4, 3)$ and $(3, 4)$.
30. Express in terms of an integral the area of the larger region bounded by $x^2 + y^2 = 25$ and the line $x = -3$.
31. Find the value of c such that the triangle bounded by $y = 2x$, $y = 0$, and $x = 4$ is divided into two regions of equal areas by the line $y = c$.
32. Find the value of c such that the region bounded by $y = x^2$ and $y = 4$ is divided into two regions of equal areas by $y = c$.
33. Find the average value of the function $1 - x^2$ in the interval $[-2, 2]$.
34. Find the average value of $\sin x$ in the interval $[0, \pi]$.
35. Find the value of c such that the average value of the function $x^4 - 1$ in the interval $[-c, c]$ is 0.
36. In one week, a small company makes a profit of \$150 on Monday, \$210 on both Tuesday and Wednesday, \$250 on Thursday, and \$75 on Friday.
  a) Find the company's average daily profit that week.
  b) Try to relate your answer in part (a) to the average value of some function $f(t)$ for $0 \le t \le 5$. Draw a sketch. [*Hint:* Your function will not be continuous, but there should be a natural idea of the area under the graph for $0 \le t \le 5$.]
37. Prove the mean-value theorem for the integral by applying the mean-value theorem for the derivative to $F(t) = \int_a^t f(x)\,dx$ on $[a, b]$.

Exercises 38 through 44 require a calculator or computer. Unless otherwise instructed, use Simpson's rule to estimate an integral giving the indicated area.

38. The area in Exercise 26
39. The area in Exercise 27
40. The area in Exercise 28
41. The area in Exercise 29
42. The area in Exercise 30
43. The area of the region bounded by $y = x^2$ and $y = \sin x$ (Use Newton's method to find one of the limits of integration. You will not need Simpson's rule.)
44. The area of the region bounded by $y = x^2$ and $y = \cos x$ [*Hint:* Proceed as suggested in Exercise 43.]

7.2 VOLUMES USING THE SLAB METHOD

VOLUMES OF REVOLUTION

Let a planar region be revolved in the natural way about a given line (axis) lying in the plane. That is, each point P of the planar region describes a circular orbit, as shown in Fig. 7.13. This orbit bounds a disk having the given axis of revolution as perpendicular axis through its center. The three-dimensional region in Fig. 7.14 consisting of the points on all such orbits is

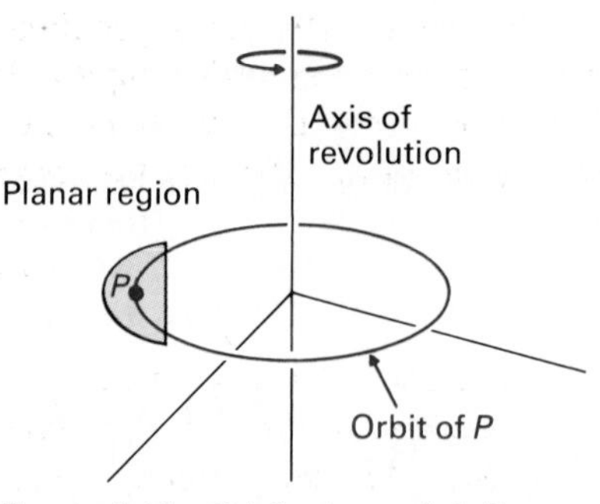

Figure 7.13 Orbit of a point P as a planar region is revolved about an axis.

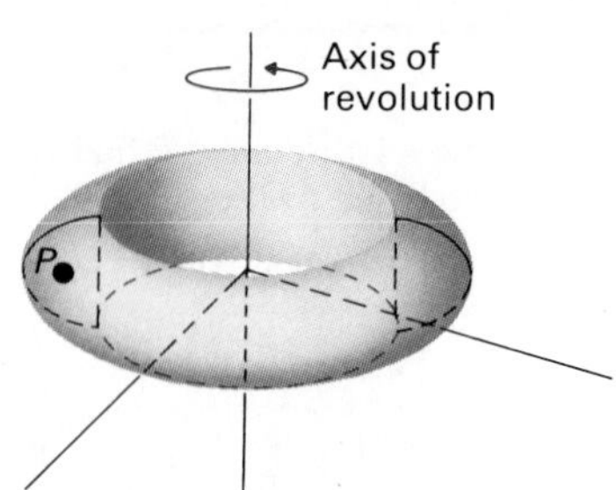

Figure 7.14 Solid of revolution.

the *solid of revolution* generated as the planar region is revolved about the axis. Using integral calculus, we can frequently find the volume of such a solid of revolution.

Consider the case in which a planar region under the graph of a continuous function f from a to b is revolved about the x-axis. Such a region is shown shaded in Fig. 7.15. The treatment for other regions is similar.

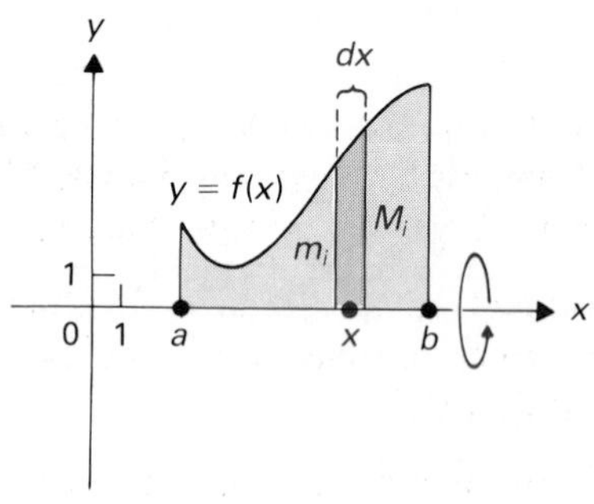

Figure 7.15 Differential strip of width dx with maximum height M_i and minimum height m_i.

Let $[a, b]$ be divided into n subintervals of equal length, as usual. Consider the contribution to the solid of revolution given by revolving the color-shaded strip in Fig. 7.15 about the x-axis. The strip sweeps out a slab, shown in Fig. 7.16, where the minimum radius is m_i and the maximum radius is M_i. The thickness of the slab is

$$\frac{b-a}{n} = dx.$$

The volume of a circular slab of radius r and thickness h is of course $\pi r^2 h$. Since the radius of the slab in Fig. 7.16 varies from m_i to M_i, its volume V_{slab} satisfies

$$\pi m_i^2\left(\frac{b-a}{n}\right) \leq V_{\text{slab}} \leq \pi M_i^2\left(\frac{b-a}{n}\right).$$

Thus for the volume V of our whole solid of revolution, we have the relation

$$\frac{b-a}{n}\left(\sum_{i=1}^{n} \pi m_i^2\right) \leq V \leq \frac{b-a}{n}\left(\sum_{i=1}^{n} \pi M_i^2\right) \tag{1}$$

Figure 7.16 Slab generated by revolving the strip in Fig. 7.15 about the x-axis.

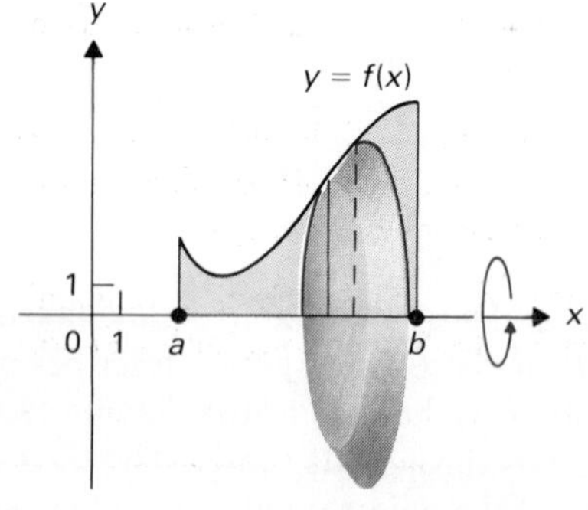

for all n. Now both extremes of the inequality (1) approach $\int_a^b \pi f(x)^2\,dx$ as $n \to \infty$, so our desired volume V of revolution is given by

$$V = \int_a^b \pi f(x)^2\,dx. \tag{2}$$

The notation $\int_a^b \pi(f(x))^2\,dx$ is very helpful and suggestive in this situation. For x as shown in Fig. 7.15, the radius of the slab in Fig. 7.16 is approximately $f(x)$, so the area of one face of the slab is roughly $\pi f(x)^2$. Since the thickness is dx, the approximate volume of the slab is $\pi f(x)^2\,dx$. We then add up all the little contributions to the volume and take the limit as dx approaches zero by applying the integral operator $\int_a^b$. We don't memorize Eq. (2) but rather arrive at the correct integral

$$\int_a^b \pi f(x)^2\,dx$$

by such a geometric argument.

The box on this page contains a suggested outline of steps to follow when using this slab method to find the volume of such a solid of revolution. The steps are similar to those in the outline in the preceding section for finding the area of a planar region.

EXAMPLE 1 Derive the formula $V = \frac{4}{3}\pi a^3$ for the volume of a sphere of radius a from the formula $A = \pi a^2$ for the area of a circular disk of radius a.

Solution

Step 1 A sphere of radius a may be obtained by revolving the region bounded by $y = \sqrt{a^2 - x^2}$ and the x-axis about the x-axis. The region is shaded in Fig. 7.17(a), while Fig. 7.17(b) shows the sphere.

Step 2 A typical thin rectangle perpendicular to the x-axis and of width dx is shaded in color in Fig. 7.17(a). When revolved about the x-axis, the rectangle sweeps out a circular disk slab.

Step 3 The volume of the disk slab is the product of the area of the face of the disk and the thickness of the slab. The area of the face is πy^2 for the point (x, y) shown in Fig. 7.17(a). The thickness is dx, so the differential volume is $dV = \pi y^2\,dx$. We must express this entirely in terms of x. Since $y = \sqrt{a^2 - x^2}$, we have $y^2 = a^2 - x^2$. Thus $dV = \pi(a^2 - x^2)\,dx$.

Figure 7.17 (a) Differential rectangle of width dx and height $y = \sqrt{a^2 - x^2}$; (b) slab of the sphere of revolution generated by the differential rectangle in (a).

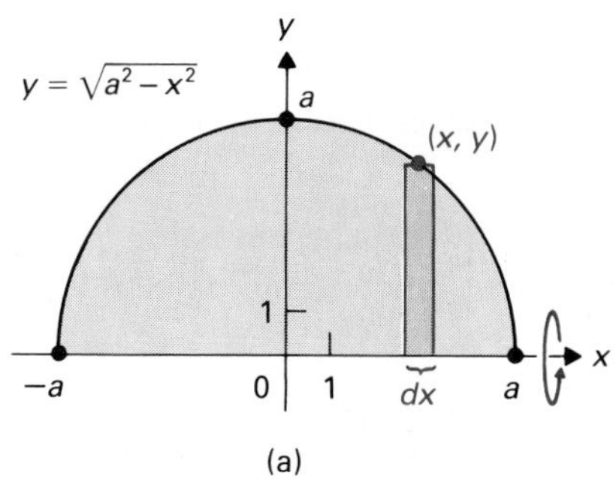

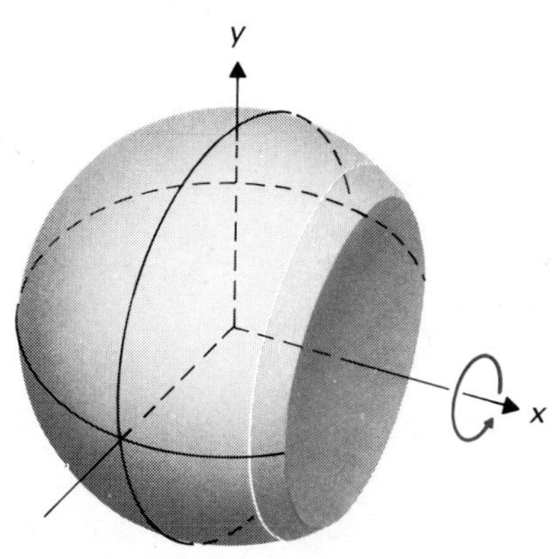

FINDING THE VOLUME OF A SOLID OF REVOLUTION BY THE SLAB METHOD

Step 1 Sketch the planar region that is to be revolved, finding the points of intersection of bounding curves.

Step 2 On the sketch, draw a typical thin rectangle perpendicular to the axis of revolution, that is, either perpendicular to the x-axis and of width dx or perpendicular to the y-axis and of width dy.

Step 3 *Looking at the sketch,* write down the volume dV of the slab swept out as the rectangle is revolved about the given axis. Express dV entirely in terms of the variable (x or y) appearing in the differential (dx or dy).

Step 4 Integrate dV between the appropriate (x or y) limits. (Geometrically, this amounts to adding the volumes found in Step 3 and taking the limit of the resulting sum.)

Step 4 The appropriate integral is

$$V = \int_{-a}^{a} \pi(a^2 - x^2)\,dx = \pi\left(a^2x - \frac{x^3}{3}\right)\Bigg]_{-a}^{a}$$

$$= \pi\left(a^3 - \frac{a^3}{3}\right) - \pi\left(-a^3 + \frac{a^3}{3}\right)$$

$$= \frac{2}{3}\pi a^3 + \frac{2}{3}\pi a^3 = \frac{4}{3}\pi a^3.$$

Alternatively, we could take advantage of the obvious symmetry of the region about the y-axis in Fig. 7.17(a) and write

$$V = 2\int_{0}^{a} \pi(a^2 - x^2)\,dx = 2\pi\left(a^2x - \frac{x^3}{3}\right)\Bigg]_{0}^{a}$$

$$= 2\pi\left(a^3 - \frac{a^3}{3}\right) - 0 = \frac{4}{3}\pi a^3.$$

We prefer to use the symmetry whenever it permits us to have zero as a limit in the integral. □

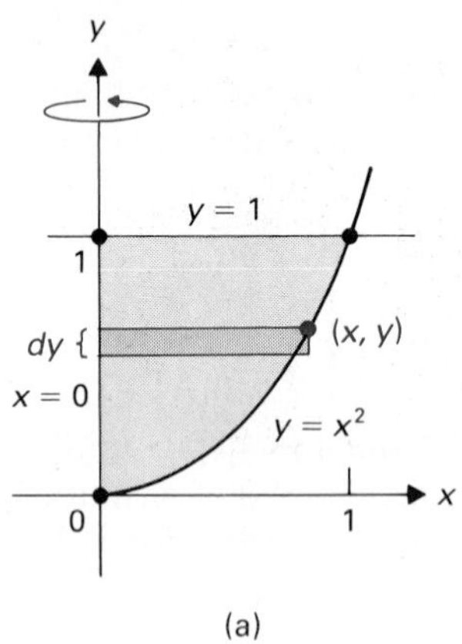

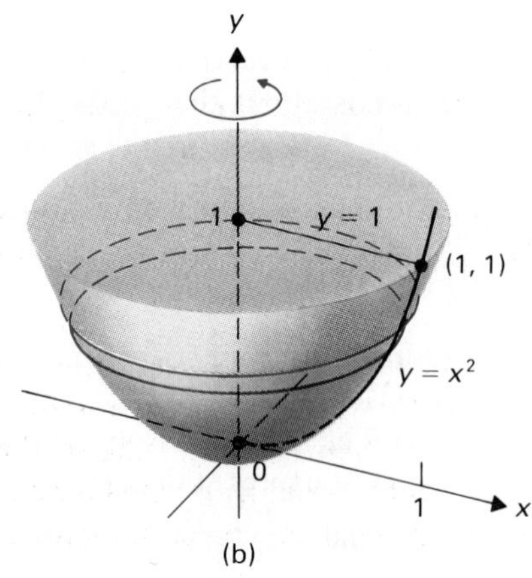

Figure 7.18 (a) Differential rectangle of width dy and length $x = \sqrt{y}$; (b) solid of revolution obtained from (a).

EXAMPLE 2 Let the first-quadrant region bounded by the curves with equations $y = x^2$, $x = 0$, and $y = 1$ be revolved about the y-axis. Find the volume of the resulting solid.

Solution

Step 1 The planar region to be revolved is shown shaded in Fig. 7.18(a). Revolving the region about the y-axis gives a solid bowl, as shown in Fig. 7.18(b).

Step 2 A typical rectangle of width dy is shown shaded in color in Fig. 7.18(a). When revolved about the y-axis, the rectangle sweeps out a thin, circular slab.

Step 3 The volume of such a slab is the product of the area of the circular face and the thickness (or height) of the slab. The area of a face is πx^2, for the point (x, y) shown in Fig. 7.18(a). The thickness of the slab is dy, so the volume of this slab is $dV = \pi x^2\,dy$. We must express the volume $\pi x^2\,dy$ entirely in terms of y. For the point (x, y) in Fig. 7.18(a), we have $x^2 = y$, so the volume of the slab becomes $dV = (\pi y)\,dy$.

Step 4 The appropriate integral is

$$V = \int_{0}^{1} \pi y\,dy = \pi\frac{y^2}{2}\Bigg]_{0}^{1} = \frac{\pi}{2} - 0 = \frac{\pi}{2}. \quad \square$$

The axis of revolution need not fall on a boundary of the region, as illustrated in the next example.

EXAMPLE 3 Find the volume generated when the planar region of Example 2 is revolved about the line $x = -1$.

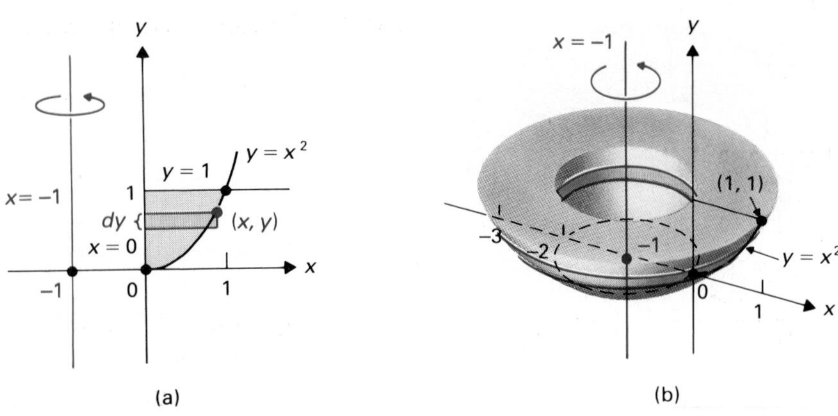

Figure 7.19 (a) Differential rectangle of width dy to be revolved about the line $x = -1$; (b) solid of revolution obtained from (a).

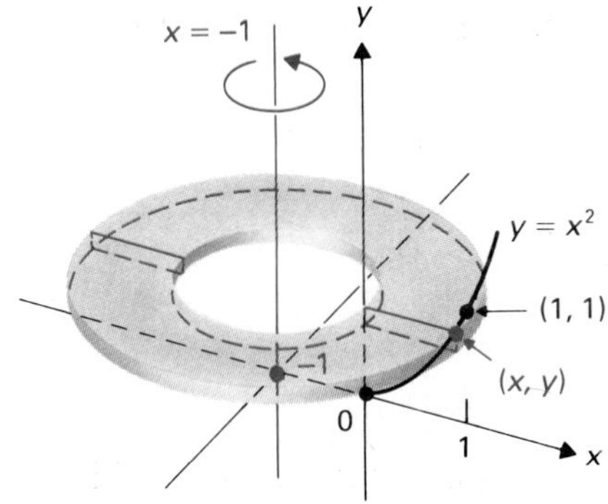

Figure 7.20 Washer slab obtained by revolving the differential rectangle in Fig. 7.19(a) about the line $x = -1$.

Solution

Step 1 The appropriate sketch is given in Fig. 7.19(a). This time, the solid generated is shown in Fig. 7.19(b).

Step 2 A typical rectangle is shown shaded in color in Fig. 7.19(a). When revolved about the line $x = -1$, it sweeps out a thin, annular washer (a disk with a hole in it), as shown in Fig. 7.20.

Step 3 The volume of an annular washer is the volume of the solid slab minus the volume of the hole. The radius of the whole disk slab is $x - (-1) = x + 1$ for the point (x, y) shown in Fig. 7.20, so the solid slab has volume $\pi(x + 1)^2\, dy$. On the other hand, the hole has radius 1 and hence volume $\pi(1)^2\, dy$. Thus the volume of the annular washer is

$$dV = \pi(x + 1)^2\, dy - \pi(1)^2\, dy = \pi(x^2 + 2x)\, dy.$$

We must express the volume $\pi(x^2 + 2x)\, dy$ entirely in terms of y. For the point (x, y) in Fig. 7.20, we have $x^2 = y$, so the volume of the annular washer becomes $dV = \pi(y + 2\sqrt{y})\, dy$.

Step 4 The appropriate integral is

$$V = \int_0^1 \pi(y + 2\sqrt{y})\, dy = \pi\left(\frac{y^2}{2} + \frac{4}{3}y^{3/2}\right)\bigg]_0^1$$
$$= \pi\left(\frac{1}{2} + \frac{4}{3}\right) - \pi(0) = \frac{11}{6}\pi. \quad \square$$

Figure 7.21 (a) Solid disk slab: $dV = \pi r^2(dx \text{ or } dy)$; (b) washer slab: $dV = \pi(R^2 - r^2)(dx \text{ or } dy)$.

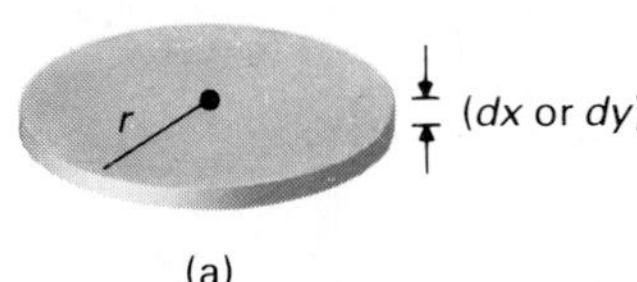

(a)

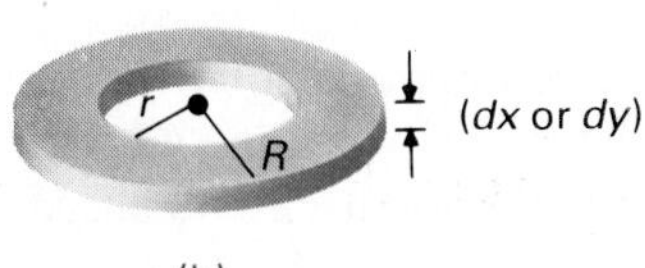

(b)

We have seen that slabs for volumes of revolution are of two types. If the differential rectangle reaches all the way to the axis of revolution, then we have a solid disk slab of radius r, as shown in Fig. 7.21(a). If the differential rectangle does not reach the axis of revolution, we have a washer slab of inner radius r and outer radius R, as shown in Fig. 7.21(b). The following

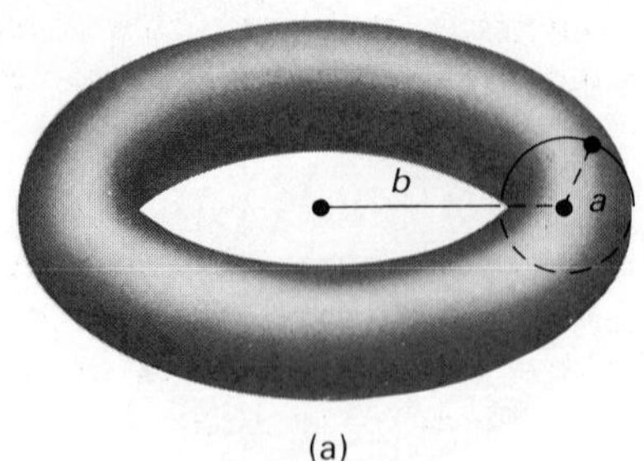

(a)

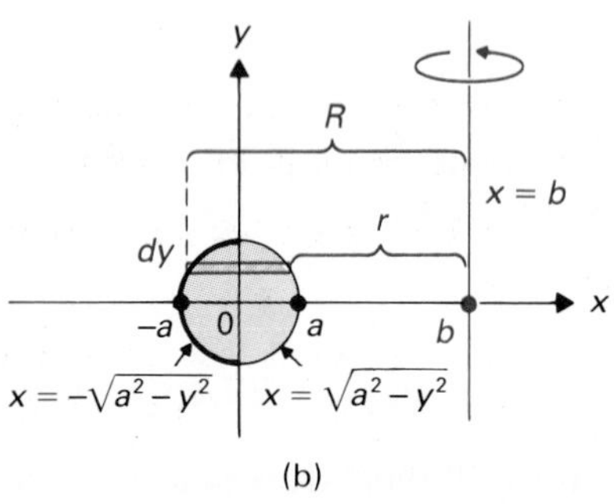

(b)

Figure 7.22 (a) A solid torus with circular cross sections of radius a; (b) differential rectangle of width dy to be revolved about the line $x = b$.

formulas apply:

$dV = \pi r^2\ (dx \text{ or } dy),$	Solid disk slab
$dV = \pi(R^2 - r^2)\ (dx \text{ or } dy).$	Washer slab

Of course, before integrating, we must express r and R in terms of the variable x or y that appears in the differential.

EXAMPLE 4 Find the volume of the solid torus (doughnut) shown in Fig. 7.22(a).

Solution

Step 1 The torus can be generated by revolving the disk $x^2 + y^2 \leq a^2$ about the line $x = b$, as indicated in Fig. 7.22(b).

Step 2 A typical thin rectangle of width dy perpendicular to the line $x = b$ is shaded in color in Fig. 7.22(b).

Step 3 Since our differential rectangle does not reach the axis of revolution $x = b$ in Fig. 7.22(b), we use the formula

$$dV = \pi(R^2 - r^2)\, dy.$$

Since the right-hand semicircle has equation $x = \sqrt{a^2 - y^2}$ and the left-hand one has equation $x = -\sqrt{a^2 - y^2}$ in Fig. 7.22(b), we see that

$$r = b - \sqrt{a^2 - y^2}, \qquad R = b - (-\sqrt{a^2 - y^2}) = b + \sqrt{a^2 - y^2}.$$

Thus

$$\begin{aligned} dV &= \pi[(b + \sqrt{a^2 - y^2})^2 - (b - \sqrt{a^2 - y^2})^2]\, dy \\ &= \pi[(b^2 + 2b\sqrt{a^2 - y^2} + a^2 - y^2) \\ &\quad - (b^2 - 2b\sqrt{a^2 - y^2} + a^2 - y^2)]\, dy \\ &= 4\pi b\sqrt{a^2 - y^2}\, dy. \end{aligned}$$

Step 4 Using the symmetry of the disk in Fig. 7.22(b) about the x-axis, we see that

$$V = 2\int_0^a 4\pi b\sqrt{a^2 - y^2}\, dy = 8\pi b \int_0^a \sqrt{a^2 - y^2}\, dy.$$

We cannot find an antiderivative of $\sqrt{a^2 - y^2}$ at the present time, but we recognize the integral $\int_0^a \sqrt{a^2 - y^2}\, dy$ as giving the area of the quarter-disk shaded in Fig. 7.23. Thus $\int_0^a \sqrt{a^2 - y^2}\, dy = \frac{1}{4}\pi a^2$, so

$$V = 8\pi b(\tfrac{1}{4}\pi a^2) = 2\pi^2 a^2 b. \quad \square$$

Figure 7.23 $\int_0^a \sqrt{a^2 - y^2}\, dy$ is the area of the shaded region.

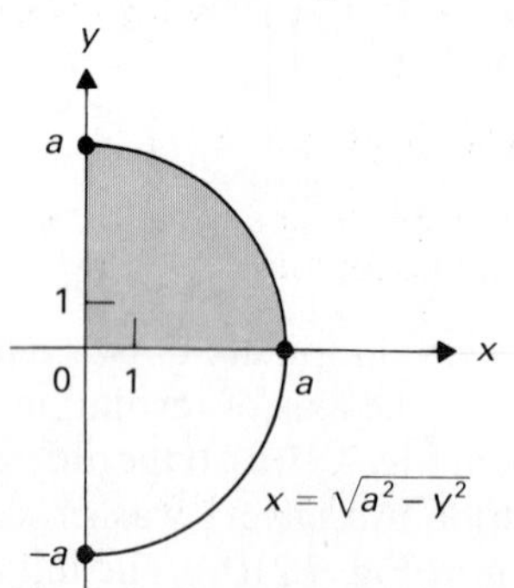

MORE GENERAL VOLUMES BY SLICING

A slab method can often be used to find the volume of a solid that can be sliced up into parallel slabs whose faces have easily computed areas. A solid of revolution is an example of such a solid, for it can be sliced into parallel slabs having circular faces.

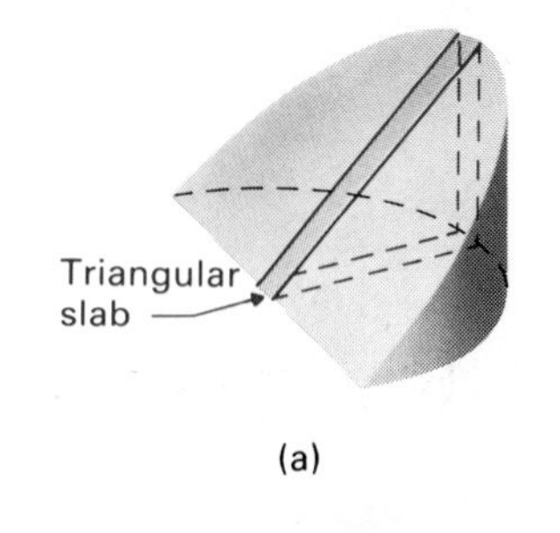

(a)

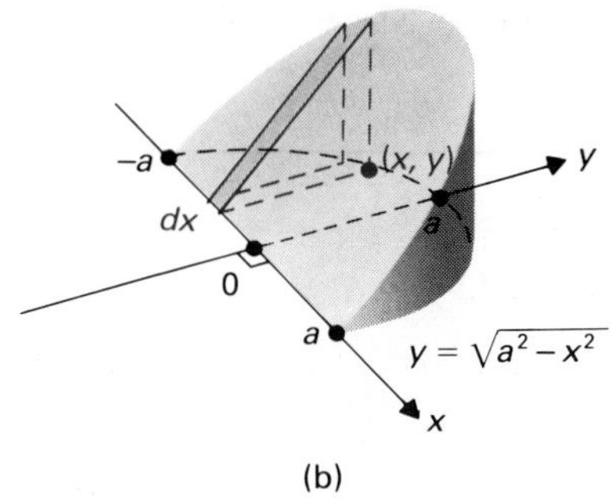

(b)

Figure 7.24 (a) Isosceles right triangular slab on a half-disk base; (b) triangular slab of volume $dV = (y^2/2)\,dx$, where $y = \sqrt{a^2 - x^2}$.

The general technique for finding volumes by slicing is perhaps best illustrated by examples. A step-by-step procedure for finding such volumes is found in the box on this page.

EXAMPLE 5 A wedge of cheese has a semicircular base of diameter $2a$. If the cheese is cut perpendicular to the diameter of the semicircle, the cross section obtained is an isosceles right triangle with the right angle on the semicircle, as shown in Fig. 7.24(a). Find the volume of the cheese.

Solution

Step 1 We may take for the base of the cheese the semicircle bounded by the x-axis and the graph of $\sqrt{a^2 - x^2}$, as shown in Fig. 7.24(b).

Step 2 The slab shown in Fig. 7.24(a) has thickness dx, while its faces are isosceles triangles with legs of length y.

Step 3 The area of such a triangle is $y^2/2$, so the volume of the slab is approximately $dV = (y^2/2)\,dx$. Since $y = \sqrt{a^2 - x^2}$, the slab has volume $[(a^2 - x^2)/2]\,dx$.

Step 4 The volumes of these slabs should be added as x goes from $-a$ to a, so the appropriate integral is

$$\int_{-a}^{a} \frac{1}{2}(a^2 - x^2)\,dx = \frac{1}{2}\left(a^2x - \frac{x^3}{3}\right)\Big]_{-a}^{a}$$

$$= \frac{1}{2}\left[\left(a^3 - \frac{a^3}{3}\right) - \left(-a^3 - \frac{(-a)^3}{3}\right)\right]$$

$$= \frac{1}{2}\left[2a^3 - \frac{2}{3}a^3\right] = \frac{2}{3}a^3. \quad \square$$

EXAMPLE 6 A bookend has as base a half-disk of radius a. Sections of the solid perpendicular to the base and parallel to the diameter are squares. Find the volume of the bookend.

FINDING THE VOLUME OF A SOLID BY THE SLICING METHOD

Step 1 Draw a figure; include an axis perpendicular to the cross section of known area (say, an x-axis).

Step 2 Sketch a cross-sectional slab perpendicular to that x-axis.

Step 3 Express the area $A(x)$ of the face of the cross-sectional slab in terms of its position x on the x-axis. The volume of the slab is then

$$dV = A(x)\,dx,$$

as in Fig. 7.26(c) on page 266.

Step 4 Integrate the expression for dV between appropriate x-limits.

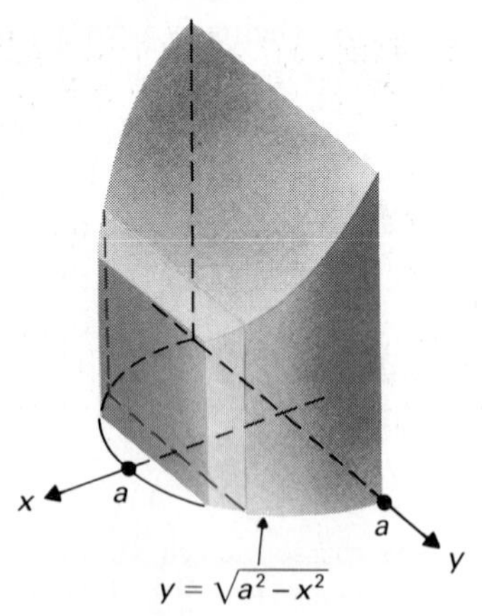

Figure 7.25 Cross section showing a square slab standing on a half-disk base: $dV = (2y)^2\, dx$, where $y = \sqrt{a^2 - x^2}$.

Solution

Step 1 The bookend is shown in Fig. 7.25.

Step 2 A square section is shaded in color.

Step 3 For our choice of axes and the point (x, y) shown in the figure, the square slab has thickness dx and sides of length $2y$. Thus the differential volume of the slab is

$$dV = (2y)^2\, dx = 4y^2\, dx.$$

Step 4 We must express y in terms of x before integrating. We have $y = \sqrt{a^2 - x^2}$, so

$$dV = 4(a^2 - x^2)\, dx.$$

Then

$$V = \int_0^a 4(a^2 - x^2)\, dx = 4\left(a^2 x - \frac{x^3}{3}\right)\Big]_0^a = 4\left(a^3 - \frac{a^3}{3}\right) = \frac{8}{3}a^3. \quad \square$$

SUMMARY

1. A suggested step-by-step outline for finding a volume of revolution by the slab method is as follows:

   Step 1 Sketch the planar region that is to be revolved, finding the points of intersection of bounding curves.

   Step 2 On the sketch, draw a typical thin rectangle perpendicular to the axis of revolution, that is, either perpendicular to the x-axis and of width dx or perpendicular to the y-axis and of width dy.

   Step 3 Looking at the sketch, write down the volume dV of the slab swept out as the rectangle is revolved about the given axis. This slab is either a circular slab of volume

   $$dV = \pi r^2\, (dx \text{ or } dy)$$

   as in Fig. 7.26(a) or a circular washer of volume

   $$dV = \pi(R^2 - r^2)\, (dx \text{ or } dy)$$

   as in Fig. 7.26(b). In either case, express dV entirely in terms of the variable (x or y) appearing in the differential (dx or dy).

   Step 4 Integrate dV between the appropriate (x or y) limits.

2. A suggested step-by-step procedure for finding a volume of a solid with parallel cross sections of known area is as follows:

   Step 1 Draw a figure; include an axis perpendicular to the cross sections of known area (say, an x-axis).

   Step 2 Sketch a cross-sectional slab perpendicular to that x-axis.

Figure 7.26 (a) Solid disk slab; $dV = \pi r^2\, (dx \text{ or } dy)$; (b) Washer slab: $dV = \pi(R^2 - r^2)\, (dx \text{ or } dy)$; (c) Slab with face area $A(x)$ has volume $dV = A(x)\, dx$.

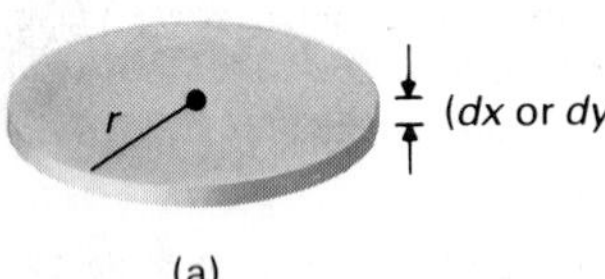

(a)

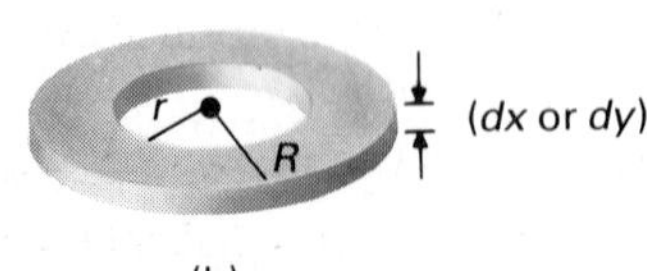

(b)

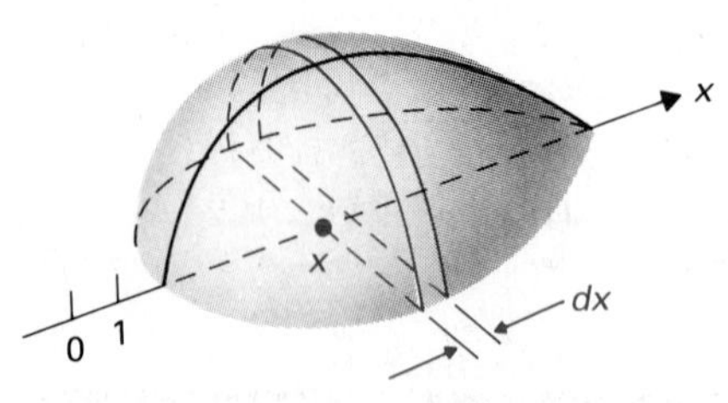

(c)

Step 3 Express the area $A(x)$ of the face of the cross-sectional slab in terms of its position x on the x-axis. The volume of the slab is then

$$dV = A(x)\,dx,$$

as in Fig. 7.26(c).

Step 4 Integrate the expression for dV between appropriate x-limits.

EXERCISES

In Exercises 1 through 18, use the slab method to find the volume of the solid generated by revolving the region with the given boundary about the indicated axis.

1. Bounded by $y = 1 - x^2$, $x = 0$, about the x-axis
2. Bounded by $y = x^2 + 1$, $y = 0$, $x = 1$, $x = -1$ about the x-axis
3. Bounded by $x = \sqrt{3 - y^2}$, $x = 0$ about the y-axis
4. Bounded by $y = \sqrt{x}$, $x = 4$, $y = 0$ about the x-axis
5. Bounded by $y = \sqrt{x}$, $x = 4$, $y = 0$ about the y-axis
6. Bounded by $y = \sqrt{x}$, $y = x^2$ about the x-axis
7. Bounded by $y = x$, $y = x^2$ about the y-axis
8. Bounded by $y = x^2$, $y = 3 - 2x$ about the x-axis
9. Bounded by $y = \sec x$, $y = 0$, $x = \pi/4$, $x = -(\pi/4)$ about the x-axis
10. Bounded by $y = \csc x$, $y = 0$, $x = \pi/6$, $x = 5\pi/6$ about the x-axis
11. Bounded by $y = \sin x$ for $0 \leq x \leq \pi$, $y = 0$ about the x-axis (Use the table of integrals on the endpapers of this book.)
12. Bounded by $y = \sin x$ for $0 \leq x \leq \pi$, $y = 0$ about the line $y = 1$ (Use the table of integrals.)
13. Bounded by $y = \cos x$, $x = \pi/2$, $x = -(\pi/2)$, $y = -1$ about the line $y = -1$ (Use the table of integrals.)
14. Bounded by $y = \sin x$, $y = 2 \sin x$ for $0 \leq x \leq \pi$ about the x-axis (Use the table of integrals.)
15. Bounded by $y = x^2$, $y = x$ about the line $x = 2$
16. Bounded by $y = x^2$, $y = 1$ about the line $y = 2$
17. Bounded by $y = x^2$, $y = 1$ about the line $x = 1$
18. Bounded by $y = 1/x$, $x = 1$, $x = 4$, $y = 0$ about the x-axis
19. Verify the formula $V = \frac{1}{3}\pi r^2 h$ for the volume of a right circular cone of height h with base of radius r. [*Hint*: Revolve the region bounded by the lines with equations $y = (r/h)x$, $y = 0$, and $x = h$, about the x-axis.]
20. Find the volume of the solid generated when the upper half ($y \geq 0$) of the elliptical disk $x^2/a^2 + y^2/b^2 \leq 1$ is revolved about the x-axis.
21. A solid has as base the disk $x^2 + y^2 \leq a^2$. Each plane section of the solid cut by a plane perpendicular to the x-axis is an equilateral triangle. Find the volume of the solid.
22. The base of a certain solid is an isosceles right triangle with hypotenuse of length a. Each plane section of the solid cut by a plane perpendicular to the hypotenuse is a square. Find the volume of the solid.
23. Find the volume of the smaller portion of a solid ball of radius a cut off by a plane b units from the center of the ball, where $0 \leq b \leq a$.
24. A certain solid has as base the plane region bounded by $x = y^2$ and $x = 4$. Each plane section of the solid cut by a plane perpendicular to the x-axis is an isosceles right triangle with the right angle on the graph of $\sqrt{x}$. Find the volume of the solid.
25. Let a solid have a flat base and altitude h. Suppose the area of a cross section parallel to the base and x units above the base is $c(h - x)^2$ for $0 \leq x \leq h$ for some constant c independent of the value of x. (This is true for tetrahedra, pyramids, and cones.) Show that the volume of the solid is $\frac{1}{3}h \cdot$ Area of the base.

In Exercises 26 through 32, use a computer or calculator and numerical techniques (Simpson's rule with $n \geq 10$, Newton's method). Find the volume of the solid generated by revolving the region with the given boundary about the indicated axis.

26. Bounded by $1/(1 + x^2)$, $x = 0$, $x = 1$, $y = 0$ about the x-axis
27. Bounded by $1/(1 + x^2)$, $x = 0$, $x = 1$, $y = 0$ about the line $y = -1$
28. Bounded by $y = 2^x$, $x = 0$, $x = 2$, $y = 0$ about the x-axis
29. Bounded by $6/(1 + \sin x)$, $x = 0$, $x = \pi$, $y = 0$ about the line $y = -1$
30. Bounded by $y = x^2$, $y = \sin x$ about the x-axis
31. Bounded by $y = x^2$, $y = \cos x$ about the line $y = -1$
32. Bounded by $y = x$, $y = \sec x - 1$ for $-\pi/2 < x < \pi/2$ about the x-axis

7.3 VOLUMES OF REVOLUTION: THE SHELL METHOD

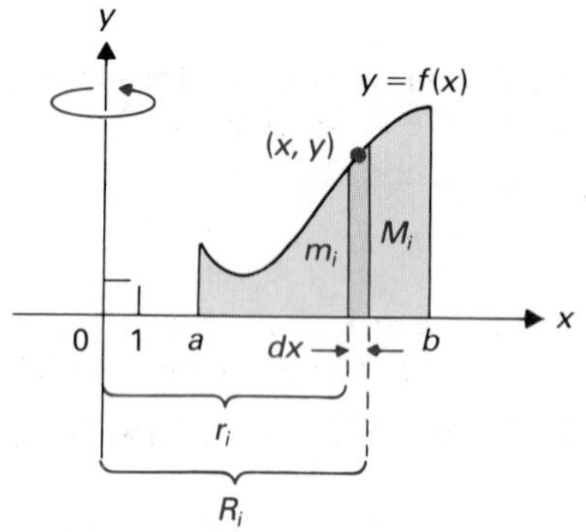

Figure 7.27 Differential strip to be revolved about the y-axis.

We present another method for finding a volume of revolution. Suppose the region shown in Fig. 7.27 is revolved about the y-axis. This time the color-shaded strip in Fig. 7.27 sweeps out a cylindrical shell, as shown in Fig. 7.28. We can find an appropriate integral for the volume using these shells.

The volume of such a cylindrical shell should be approximately the product of the surface area of the cylinder and the thickness of the shell (the wall of the cylinder). The surface area in turn is the product of the perimeter of the circle and the height of the cylinder. For the point (x, y) shown in Fig. 7.27, the perimeter is $2\pi x$ and the height is y. Thus the volume V_{shell} of the cylindrical shell should be about $dV = 2\pi xy\,dx$. Since $y = f(x)$, we would expect, adding up the volumes of these shells and taking the limit of the resulting sum, that the volume of the whole solid of revolution would be

$$\int_a^b 2\pi x f(x)\,dx.$$

The preceding integral was set up using rough estimates, in the manner in which one always sets up an integral for the shell method. A justification of the method can be made as follows. Let m_i and M_i be the minimum and maximum values of $f(x)$ over the ith interval of length $(b - a)/n$ as usual. Referring to Fig. 7.27 and taking first minimum radius r_i and minimum height m_i for the cylindrical shell and then the maximum radius R_i and maximum height M_i, we see that

$$2\pi r_i m_i \frac{b-a}{n} \le V_{\text{shell}} \le 2\pi R_i M_i \frac{b-a}{n}. \tag{1}$$

The expression

$$2\pi R_i M_i \frac{b-a}{n}$$

in Eq. (1) is the value of

$$2\pi x f(x) \frac{b-a}{n}$$

if we evaluate $2\pi x$ at $x_i = R_i$ and $f(x)$ at the (possibly different) point x_i' where $f(x)$ assumes its maximum value M_i in this ith subinterval. A theorem known as Bliss's theorem, which we will not prove, shows that Riemann sums for $g(x)f(x)$ where the two functions f and g are evaluated at possibly *different* points x_i and x_i' in the ith interval still approach the integral. That is,

$$\int_a^b g(x)f(x)\,dx = \lim_{n\to\infty} \frac{b-a}{n}\left(\sum_{i=1}^{n} f(x_i)\,g(x_i')\right).$$

Using this fact and setting $g(x) = 2\pi x$, we see at once from Eq. (1) that

$$V = \int_a^b 2\pi x f(x)\,dx. \tag{2}$$

The outline of steps in Section 7.2 for finding the volume of revolution by the slab method needs only slight modification to serve for computing volumes of solids of revolution by the shell method. See the box on page 269.

Figure 7.28 Cylindrical shell generated by the strip in Fig. 7.27.

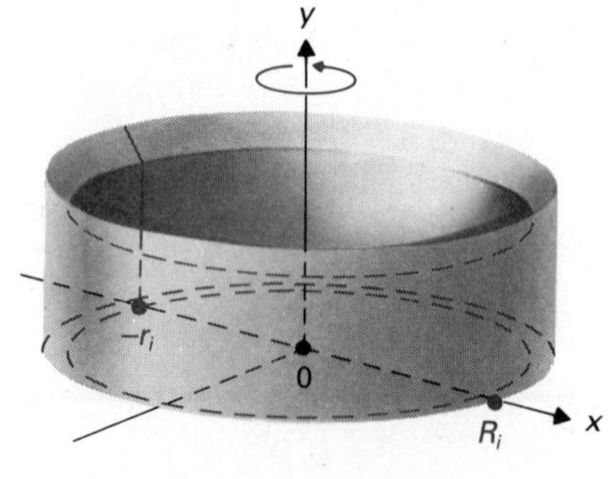

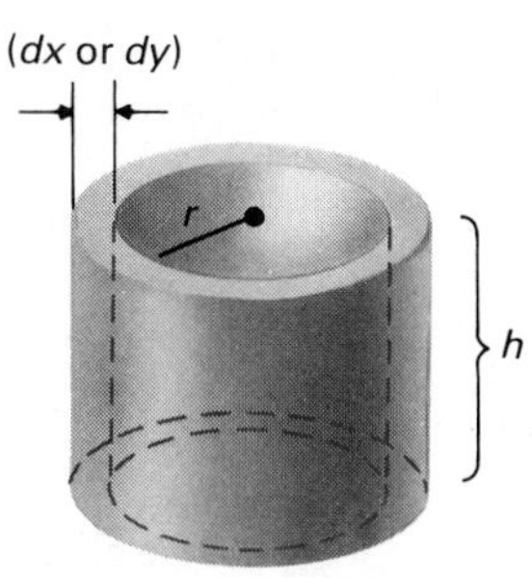

Figure 7.29 Cylindrical shell $dV = 2\pi rh\,(dx \text{ or } dy)$.

FINDING THE VOLUME OF A SOLID OF REVOLUTION BY THE SHELL METHOD

Step 1 Sketch the planar region that is to be revolved, finding the points of intersection of bounding curves.

Step 2 On the sketch, draw a typical thin rectangle parallel to the axis of revolution, that is, either horizontal with width dy or vertical with width dx.

Step 3 *Looking at the sketch,* write down the volume

$$dV = 2\pi rh\,(dx \text{ or } dy)$$

of the shell swept out as the rectangle is revolved about the given axis. (See Fig. 7.29.) Express r and h in terms of the variable (x or y) in the differential.

Step 4 Integrate the differential volume dV in Step 3 between the appropriate (x or y) limits.

EXAMPLE 1 Example 2 of Section 7.2 found the volume of the solid generated by revolving the first-quadrant region bounded by $y = x^2$, $x = 0$, and $y = 1$ about the y-axis. Repeat this computation, using the shell method.

Solution

Step 1 The region is shaded in Fig. 7.30(a).

Step 2 A thin rectangle parallel to the y-axis and of width dx is shaded in color in Fig. 7.30(a). Figure 7.30(b) shows the shell generated by this rectangle.

Figure 7.30 (a) Differential strip of width dx to be revolved about the y-axis; (b) cylindrical shell generated by the strip in (a).

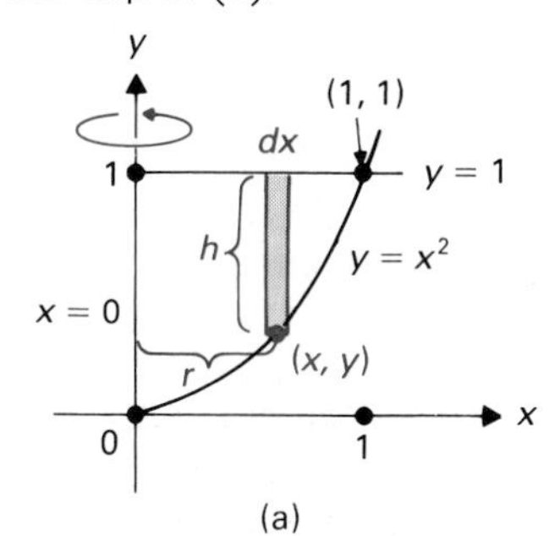

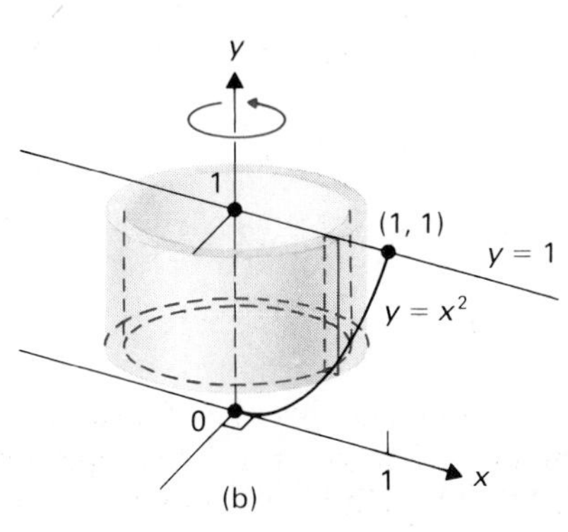

Step 3 We must compute $dV = 2\pi rh\,dx$ in terms of x, for r and h shown in Fig. 7.30(a). Referring to the figure and the point (x, y) shown there, we have $r = x$, while

$$h = y_{\text{top}} - y_{\text{bottom}} = 1 - y = 1 - x^2.$$

Thus

$$dV = 2\pi x(1 - x^2)\,dx = (2\pi x - 2\pi x^3)\,dx.$$

Step 4 The volume is

$$V = \int_0^1 (2\pi x - 2\pi x^3)\,dx = \left(2\pi\frac{x^2}{2} - 2\pi\frac{x^4}{4}\right)\bigg]_0^1$$

$$= \left(\frac{2\pi}{2} - \frac{2\pi}{4}\right) - 0 = \frac{\pi}{2}. \quad \square$$

EXAMPLE 2 Using the shell method, find the volume when the region bounded by $y = 4$ and $y = x^2$ is revolved about the line $y = -1$.

Solution

Step 1 The region is sketched in Fig. 7.31.

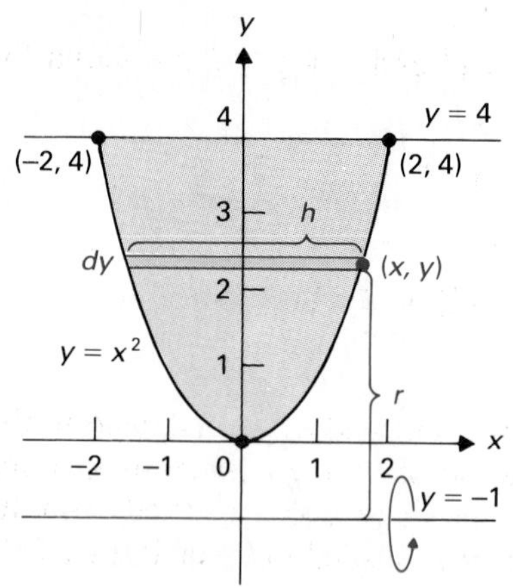

Figure 7.31 Differential strip of width dy to be revolved about the line $y = -1$.

Step 2 A thin rectangle parallel to the line $y = -1$ and of width dy is shaded in color in Fig. 7.31.

Step 3 The volume of the shell is $dV = 2\pi rh\,dy$. We must express r and h in terms of y. In terms of the coordinates of the point (x, y) shown on the curve $y = x^2$, we have $r = y - (-1) = y + 1$, while $h = 2x = 2\sqrt{y}$. Thus

$$dV = 2\pi(y + 1)2\sqrt{y}\,dy = 4\pi(y + 1)\sqrt{y}\,dy.$$

Step 4 The volume is

$$V = \int_0^4 4\pi(y + 1)\sqrt{y}\,dy = 4\pi \int_0^4 (y^{3/2} + y^{1/2})\,dy$$

$$= 4\pi\left(\frac{2}{5}y^{5/2} + \frac{2}{3}y^{3/2}\right)\Bigg]_0^4 = 4\pi\left(\frac{64}{5} + \frac{16}{3}\right) = 64\pi\left(\frac{4}{5} + \frac{1}{3}\right)$$

$$= 64\pi\frac{17}{15} = \frac{1088\pi}{15}. \quad \square$$

EXAMPLE 3 Find the volume of the solid generated when the region bounded by $y = \sin x$ for $0 \le x \le \pi$ and $y = 0$ is revolved about the y-axis.

Solution

Step 1 The region is shaded in Fig. 7.32.

Step 2 A thin rectangle parallel to the y-axis and of width dx is shaded in color in Fig. 7.32.

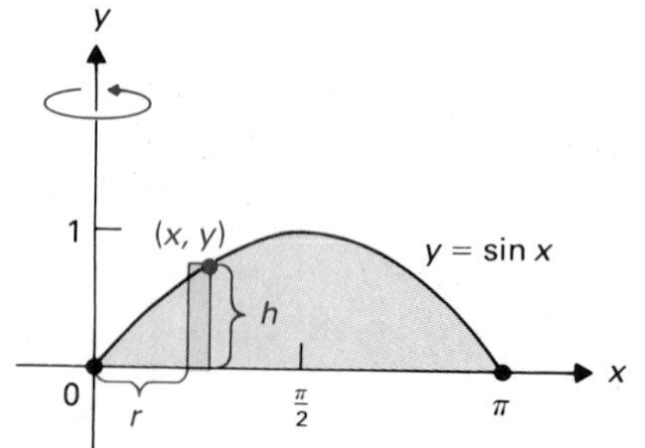

Figure 7.32 Differential strip to generate a shell of volume $dV = 2\pi rh\,dx$.

Step 3 We must compute $dV = 2\pi rh\,dx$ in terms of x. For the point (x, y) shown in Fig. 7.32, we have $r = x$ and $h = y = \sin x$. Thus $dV = 2\pi x(\sin x)\,dx$.

Step 4 The appropriate integral is

$$\int_0^\pi 2\pi x(\sin x)\,dx = 2\pi \int_0^\pi x(\sin x)\,dx.$$

The integral table on the endpapers of this book tells us that

$$\int x(\sin x)\,dx = -x\cos x + \sin x + C.$$

The volume is therefore

$$V = 2\pi \int_0^\pi x(\sin x)\,dx = 2\pi(-x\cos x + \sin x)\Bigg]_0^\pi$$

$$= 2\pi[-\pi(-1) + 0] - 2\pi(0) = 2\pi^2. \quad \square$$

We could not have found the volume in the preceding example using slabs because we do not yet know how to solve $y = \sin x$ for x in terms of y. The shell method was easy. Often one method is easier than the other. The next example also illustrates this.

EXAMPLE 4 Find the volume of the solid generated when the region bounded by $x = y^2$ and $x + y = 2$ is revolved about the line $y = 1$.

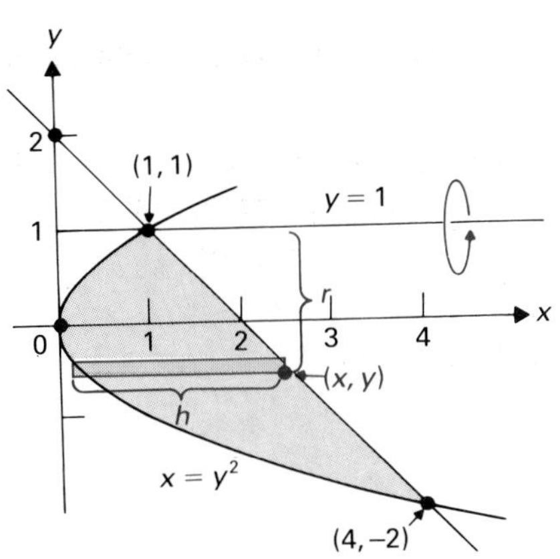

Figure 7.33 Differential strip to generate a shell of volume $dV = 2\pi rh\,dy$.

Solution

Step 1 The points of intersection of $x = y^2$ and $x + y = 2$ are found by solving simultaneously:

$$y^2 + y = 2, \qquad y^2 + y - 2 = 0,$$
$$(y+2)(y-1) = 0, \qquad y = -2 \text{ or } 1,$$
$$(x, y) = (4, -2) \text{ or } (1, 1).$$

The region is shaded in Fig. 7.33.

Step 2 Using the slab method with vertical rectangles perpendicular to the line $y = 1$ would necessitate setting up two dx integrals, one for $0 \le x \le 1$ and one for $1 \le x \le 4$. Using the shell method, with the thin horizontal rectangle of width dy shaded in color in Fig. 7.33, we need only one integral.

Step 3 We must express $dV = 2\pi rh\,dy$ in terms of y. For the point (x, y) in Fig. 7.33, we have

$$r = y_{\text{top}} - y_{\text{bottom}} = 1 - y, \qquad h = x_{\text{right}} - x_{\text{left}} = (2 - y) - y^2.$$

Thus

$$dV = 2\pi(1 - y)(2 - y - y^2)\,dy = 2\pi(2 - 3y + y^3)\,dy.$$

Step 4 The volume is

$$V = \int_{-2}^{1} 2\pi(2 - 3y + y^3)\,dy = 2\pi\left(2y - \frac{3y^2}{2} + \frac{y^4}{4}\right)\Bigg]_{-2}^{1}$$
$$= 2\pi\left(2 - \frac{3}{2} + \frac{1}{4}\right) - 2\pi(-4 - 6 + 4)$$
$$= 2\pi \cdot \frac{3}{4} - 2\pi(-6) = \frac{27\pi}{2} \quad \square$$

SUMMARY

To find a volume of revolution about a horizontal or vertical axis by the shell method, use the following suggested steps.

Step 1 Sketch the planar region that is to be revolved, finding the points of intersection of bounding curves.

Step 2 On the sketch, draw a typical thin rectangle parallel to the axis of revolution, that is, either horizontal with width dy or vertical with width dx.

Step 3 *Looking at the sketch,* write down the volume $dV = 2\pi rh \cdot (dx$ or $dy)$ of the shell swept out as the rectangle is revolved about the given axis. (See Fig. 7.34.) Express r and h in terms of the variable (x or y) in the differential.

Step 4 Integrate the differential volume dV in Step 3 between the appropriate (x or y) limits.

Figure 7.34 Cylindrical shell: $dV = 2\pi rh$ (dx or dy).

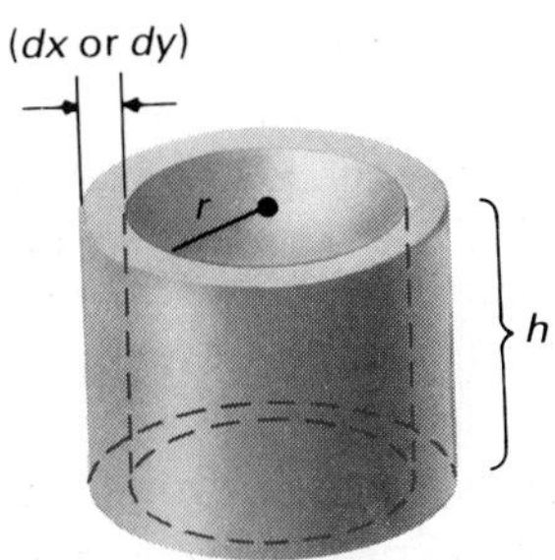

EXERCISES

In Exercises 1 through 14, use the shell method to find the volume of the solid generated by revolving the region with the given boundary about the indicated axis.

1. Bounded by $y = x^2$, $y = 1$ about the line $y = 1$
2. Bounded by $y = x^2$, $y = 1$ about the line $x = 2$
3. Bounded by $y = x$, $y = x^2$ about the y-axis
4. Bounded by $y = x$, $y = x^2$ about the x-axis
5. Bounded by $y = \sqrt{x}$, $x = 4$, $y = 0$ about the y-axis
6. Bounded by $y = \sqrt{x}$, $x = 4$, $y = 0$ about the line $x = 5$
7. Bounded by $y = \sqrt{x}$, $y = x^2$ about the y-axis
8. Bounded by $y = \sqrt{x}$, $y = x^2$ about the x-axis
9. Bounded by $x = y^2$, $y = x - 2$ about the line $y = 2$
10. Bounded by $y = x^2$, $y = 3 - 2x$ about the line $x = 1$
11. Bounded by $y = \cos x$ for $\pi/2 \le x \le 3\pi/2$, $x = 0$ about the y-axis
12. Bounded by $y = \sin x$ for $0 \le x \le \pi$, $x = 0$ about the line $x = -1$
13. Bounded by $y = \sin x$ for $0 \le x \le \pi$, $x = 0$ about the line $x = 2\pi$
14. Bounded by $y = \cos x$ for $-(\pi/2) \le x \le \pi/2$, $x = 0$ about the line $x = 2$
15. Use the method of cylindrical shells to derive the formula $V = \frac{1}{3}\pi r^2 h$ for the volume of a right circular cone of height h with base of radius r. [*Hint:* Revolve the region bounded by the lines $y = (r/h)x$, $y = 0$, and $x = h$ about the x-axis.]
16. Derive the formula $V = \frac{4}{3}\pi a^3$ for the volume of a sphere of radius a, using the method of cylindrical shells.
17. Use the method of cylindrical shells to find the volume of the doughnut (torus) generated by revolving the disk $x^2 + y^2 \le a^2$ about the line $x = b$ for $b > a$. [*Hint:* Part of the integral can be evaluated as the area of a familiar figure.]

In Exercises 18 through 24, use the method of cylindrical shells with a calculator or computer and numerical techniques (Simpson's rule with $n \ge 10$, Newton's method). Find the volume of the solid generated by revolving the region with the given boundary about the indicated axis.

18. Bounded by $y = \tan x$, $y = 0$, $x = \pi/4$ about the y-axis
19. Bounded by $y = \sec x - 1$ for $-(\pi/2) < x < \pi/2$, $y = 1$ about the line $x = 2$
20. Bounded by $y = \sin\sqrt{x}$ for $0 \le x \le \pi^2$, $y = 0$ about the y-axis
21. Bounded by $y = 3^x$, $x = 0$, $x = 2$, $y = 0$ about the line $x = 3$
22. Bounded by $x = 2^y$, $y = 0$, $y = 3$, $x = 0$ about the line $y = -1$
23. Bounded by $y = 10/(x^2 + 1) - 2$, $y = 0$ about the line $x = 3$
24. Bounded by $y = x^2$, $y = \sin x$ about the y-axis

7.4 ARC LENGTH

The preceding three sections presented applications of integral calculus to finding areas and volumes. We now learn how to find the *length of a curve*. Arc length is really an application of both differential and integral calculus. First, we partition the curve into small pieces. Next, these pieces are approximated linearly by short, tangent-line segments using *differential calculus*. Using *integral calculus*, we sum the lengths of the approximating segments and take the limit of the sum as we partition the curve into smaller and smaller pieces.

Figure 7.35 Find the length of the colored arc.

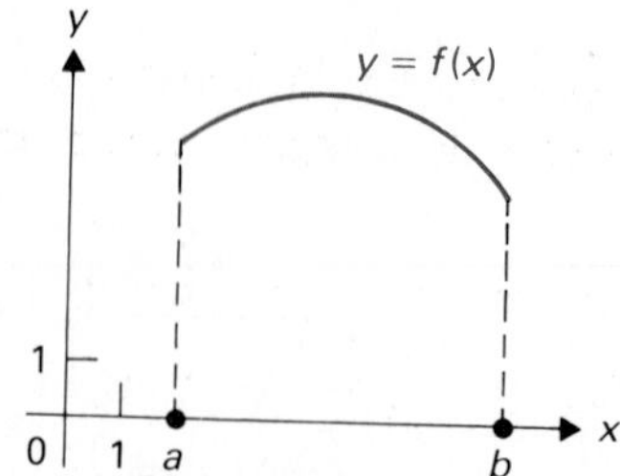

LENGTH OF A GRAPH

A function $y = f(x)$ is **continuously differentiable** if $f'(x)$ exists for all x in the domain of f and if f' is itself a continuous function. We wish to find the length of the portion of the graph of a continuously differentiable function $y = f(x)$ for x in $[a, b]$, as shown in Fig. 7.35. We present two approaches to this problem.

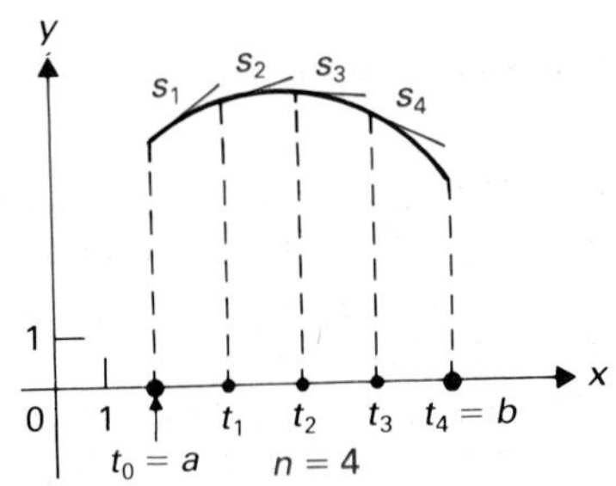

Figure 7.36
Arc length $\approx s_1 + s_2 + s_3 + s_4$.

Approach 1 Partition the interval $[a, b]$ into n subintervals of equal length by points $a = t_0 < t_1 < \cdots < t_n = b$. We take as estimate for the length of the curve the sum of the lengths of line segments tangent to the curve over these subintervals, as illustrated in Fig. 7.36 with $n = 4$. That is, if s_i is the length of the tangent segment over the interval $[t_{i-1}, t_i]$, we take as estimate for the length s of the curve

$$s \approx s_1 + s_2 + \cdots + s_n.$$

Figure 7.37 indicates that if n is large, we would expect this sum to be a good approximation to what we intuitively feel is the length of the curve.

We compute s_i. Since our subintervals are of equal length, we have

$$t_i - t_{i-1} = \frac{b-a}{n}.$$

From Fig. 7.38, we obtain

$$s_i = \sqrt{\left(\frac{b-a}{n}\right)^2 + \left(\frac{b-a}{n} f'(t_{i-1})\right)^2} = \frac{b-a}{n}\sqrt{1 + (f'(t_{i-1}))^2}. \quad (1)$$

We then obtain, as an approximation,

$$s \approx \frac{b-a}{n}\left[\sqrt{1 + f'(t_0)^2} + \sqrt{1 + f'(t_1)^2} + \cdots + \sqrt{1 + f'(t_{n-1})^2}\right]. \quad (2)$$

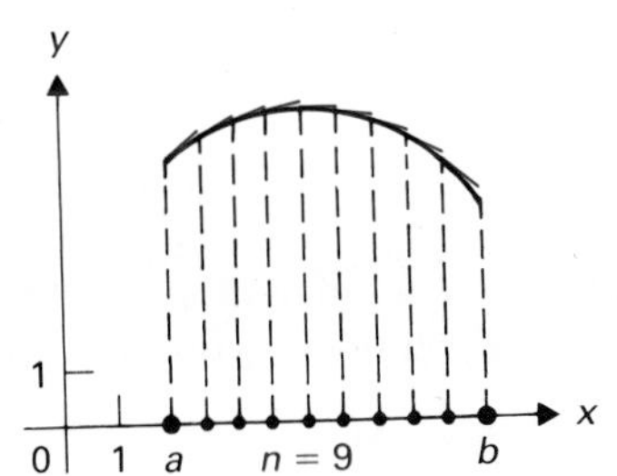

Figure 7.37 The approximation usually improves as n increases.

The expression on the right-hand side of Eq. (2) is a Riemann sum for the function $\sqrt{1 + f'(x)^2}$ from a to b. Since we expect the approximation in Eq. (2) to become accurate as n becomes large, we obtain

$$\boxed{s = \int_a^b \sqrt{1 + f'(x)^2}\, dx} \quad (3)$$

as the length of the curve. This integral exists, because we are assuming that f' is *continuous* in $[a, b]$.

The *arc length function* $s(x)$ giving the length of the arc $y = f(x)$ from a to x is

$$\boxed{s(x) = \int_a^x \sqrt{1 + f'(t)^2}\, dt.} \quad (4)$$

By the fundamental theorem of calculus,

$$\frac{ds}{dx} = \sqrt{1 + f'(x)^2},$$

so

$$ds = \sqrt{1 + f'(x)^2}\, dx. \quad (5)$$

Figure 7.38 Computing s_i by the Pythagorean theorem.

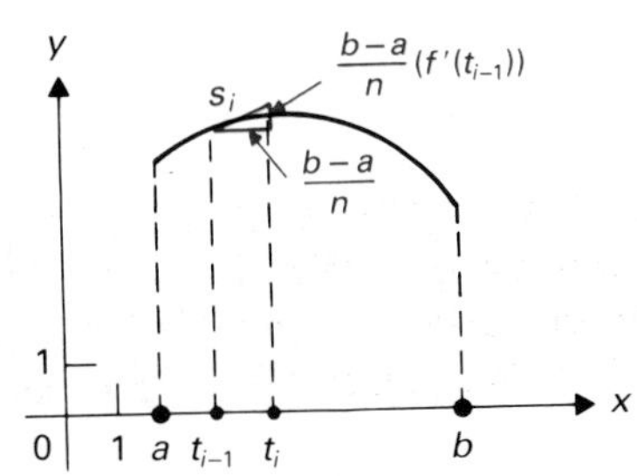

This differential ds is the *differential of arc length*, and Eq. (3) may be written in the form $s = \int_a^b ds$. Writing $f'(x) = dy/dx$, we have these easily remembered forms of ds:

$$\boxed{ds = \sqrt{1 + \left(\frac{dy}{dx}\right)^2}\, dx = \sqrt{(dx)^2 + (dy)^2} = \sqrt{\left(\frac{dx}{dy}\right)^2 + 1}\, dy.} \quad (6)$$

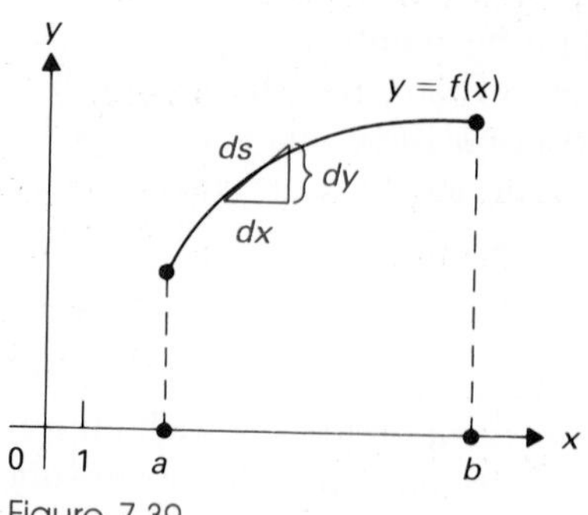

Figure 7.39
Differential right triangle of a graph.

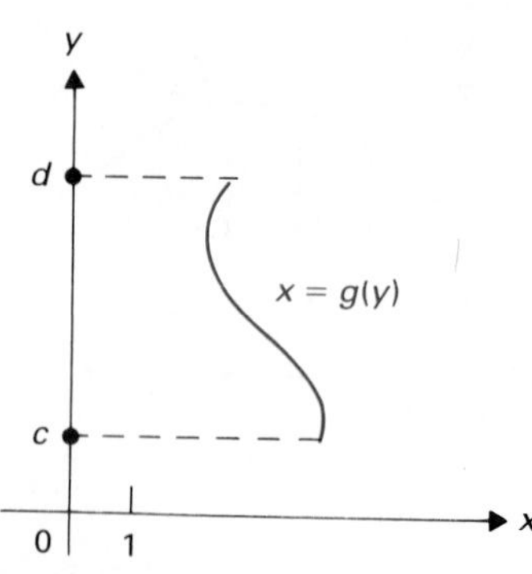

Figure 7.40
Arc length s of $x = g(y)$ for $c \leq y \leq d$ is
$s = \int_c^d \sqrt{1 + g'(y)^2}\, dy$.

The form $ds = \sqrt{(dx)^2 + (dy)^2}$ is easy to remember from Fig. 7.39. We think of ds as the length of a short tangent-line segment to the curve.

The final form for ds in Eq. (6) is used if the arc of a graph is most easily expressed in the form $x = g(y)$ for y in $[c, d]$, as shown in Fig. 7.40. We then have

$$s = \int_c^d \sqrt{1 + g'(y)^2}\, dy. \tag{7}$$

DEFINITION 7.2 Differential of arc length

The right triangle of hypotenuse ds in Fig. 7.39 is a **differential triangle** for $y = f(x)$. The **differential of arc length** is ds, as given in Eq. (6).

This concludes the first approach to arc length.

Approach 2 Once again, we partition $[a, b]$ into n subintervals of equal length. This time, we take as approximation to s the sum of the lengths of the chords shown in Fig. 7.41, where the ith chord joins $(t_{i-1}, f(t_{i-1}))$ and $(t_i, f(t_i))$. The length of this ith chord is

$$\begin{aligned}
&\sqrt{(t_i - t_{i-1})^2 + (f(t_i) - f(t_{i-1}))^2} \\
&= (t_i - t_{i-1})\sqrt{1 + \left(\frac{f(t_i) - f(t_{i-1})}{t_i - t_{i-1}}\right)^2} \\
&= \frac{b - a}{n}\sqrt{1 + \left(\frac{f(t_i) - f(t_{i-1})}{t_i - t_{i-1}}\right)^2}.
\end{aligned} \tag{8}$$

Figure 7.41 Approximating arc length using chords.

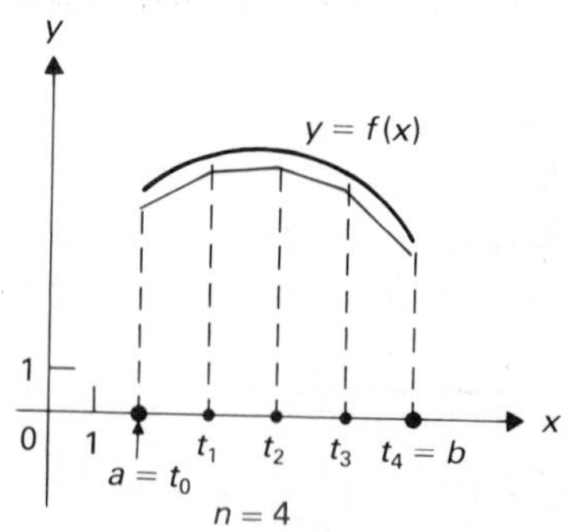

Our hypotheses on f enable us to apply the mean-value theorem to f on $[t_{i-1}, t_i]$, and we have

$$\frac{f(t_i) - f(t_{i-1})}{t_i - t_{i-1}} = f'(x_i)$$

for some point x_i in $[t_{i-1}, t_i]$. From Eq. (8), we obtain, as length of the chord,

$$\frac{b - a}{n}\sqrt{1 + f'(x_i)^2}, \tag{9}$$

which yields the approximation

$$s \approx \frac{b-a}{n}\left(\sqrt{1+f'(x_1)^2} + \sqrt{1+f'(x_2)^2} + \cdots + \sqrt{1+f'(x_n)^2}\right) \qquad \textbf{(10)}$$

for the length of the arc. Again, the right-hand side of Eq. (10) is a Riemann sum of the function $\sqrt{1+f'(x)^2}$ over $[a, b]$, so we again arrive at

$$s = \int_a^b \sqrt{1+f'(x)^2}\, dx$$

for the arc length.

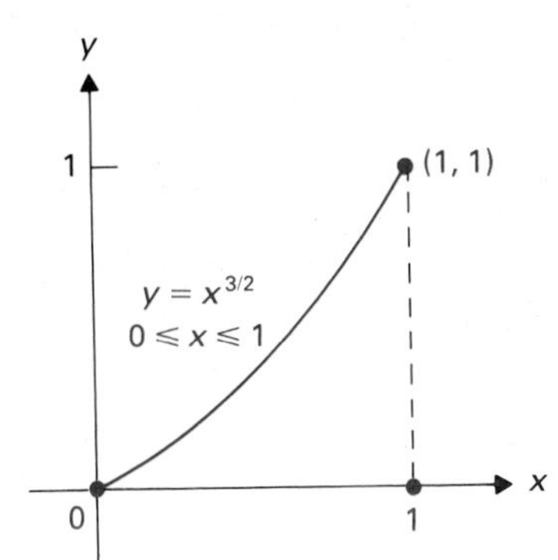

Figure 7.42 The length of the arc is $\int_0^1 \sqrt{1+(9x/4)}\, dx$.

EXAMPLE 1 Find the length of the curve $y = x^{3/2}$ for $0 \le x \le 1$.

Solution The arc whose length is desired is shown in Fig. 7.42. It is not really necessary to sketch the arc in a problem like this. The sketch in Fig. 7.42 will not be used to set up the appropriate integral. When finding areas and volumes, sketches were used for this purpose.

If $y = f(x) = x^{3/2}$, we have

$$\frac{dy}{dx} = f'(x) = \frac{3}{2}x^{1/2},$$

so

$$ds = \sqrt{1+(f'(x))^2}\, dx = \sqrt{1+\tfrac{9}{4}x}\, dx.$$

The length of our curve is therefore

$$\int_0^1 \left(1+\frac{9}{4}x\right)^{1/2} dx = \frac{4}{9}\cdot\frac{(1+\frac{9}{4}x)^{3/2}}{3/2}\Bigg]_0^1 = \frac{4}{9}\cdot\frac{2}{3}\left[\left(\frac{13}{4}\right)^{3/2} - 1\right]$$

$$= \frac{8}{27}\cdot\frac{13^{3/2}}{8} - \frac{8}{27} = \frac{13^{3/2}-8}{27}. \quad \square$$

The radical that appears in Eq. (3) frequently makes evaluation of the integral difficult; examples have to be chosen with care to enable us to compute the integral by taking an antiderivative of $\sqrt{1+f'(x)^2}$. But we know there are numerical methods (such as Simpson's rule) that we can easily use if integration causes difficulty.

EXAMPLE 2 Find the arc length of the portion of the curve $x = 2y^3 + 1/(24y)$ for $1 \le y \le 3$.

Solution We use ds in the form

$$ds = \sqrt{1+\left(\frac{dx}{dy}\right)^2}\, dy.$$

We easily compute

$$\frac{dx}{dy} = 6y^2 - \frac{1}{24y^2}.$$

Thus

$$\begin{aligned} ds &= \sqrt{1 + \left(6y^2 - \frac{1}{24y^2}\right)^2}\, dy = \sqrt{1 + (6y^2)^2 - \frac{1}{2} + \frac{1}{(24y^2)^2}}\, dy \\ &= \sqrt{(6y^2)^2 + \frac{1}{2} + \frac{1}{(24y^2)^2}}\, dy = \sqrt{\left(6y^2 + \frac{1}{24y^2}\right)^2}\, dy \\ &= \left(6y^2 + \frac{1}{24y^2}\right) dy. \end{aligned}$$

Therefore we have

$$\begin{aligned} s &= \int_1^3 \left(6y^2 + \frac{1}{24}y^{-2}\right) dy = \left(2y^3 - \frac{1}{24y}\right)\Bigg]_1^3 \\ &= \left(54 - \frac{1}{72}\right) - \left(2 - \frac{1}{24}\right) = 52 + \frac{2}{72} = 52 + \frac{1}{36} = \frac{1873}{36}. \quad \square \end{aligned}$$

EXAMPLE 3 Find the arc length of the portion of the graph of $y = x^2$ for $0 \le x \le 2$.

Solution We have $dy/dx = 2x$, so

$$ds = \sqrt{1 + \left(\frac{dy}{dx}\right)^2}\, dx = \sqrt{1 + 4x^2}\, dx.$$

Thus

$$s = \int_0^2 \sqrt{1 + 4x^2}\, dx.$$

While formula 26 in the integral table could be used to find an antiderivative of $\sqrt{1 + 4x^2}$, we have not discussed the function ln x, which appears in the antiderivative. Of course, this function is easily evaluated on a computer or calculator. We choose another alternative and compute the integral numerically, using Simpson's rule with $n = 20$. We obtain

$$s = \int_0^2 \sqrt{1 + 4x^2}\, dx \approx 4.6468. \quad \square$$

ARC LENGTH OF A PARAMETRIC CURVE

Consider a curve with parametric equations

$$x = h(t), \qquad y = k(t),$$

where $h'(t)$ and $k'(t)$ are continuous functions of t. A formal manipulation of Leibniz notation from Eq. (6) leads to

$$ds = \sqrt{(dx)^2 + (dy)^2} = \sqrt{\left(\frac{dx}{dt}\right)^2 + \left(\frac{dy}{dt}\right)^2}\, dt. \qquad \textbf{(11)}$$

Thus we expect the total distance s traveled on the curve from time $t = a$ to

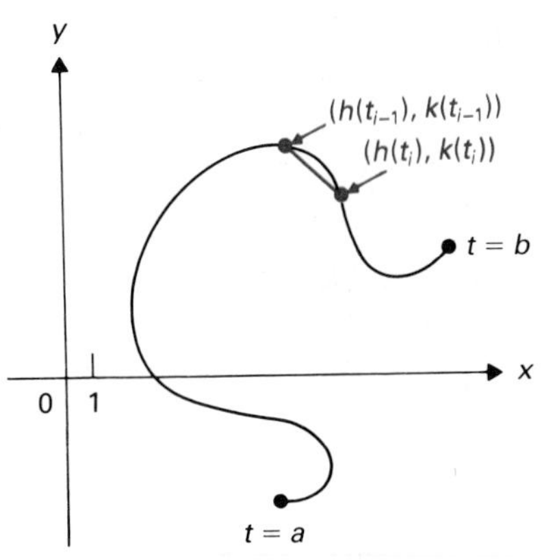

Figure 7.43 Chord of length $\sqrt{(h(t_i)-h(t_{i-1}))^2+(k(t_i)-k(t_{i-1}))^2}$.

time $t = b$ to be

$$s = \int_{t=a}^{t=b} \sqrt{\left(\frac{dx}{dt}\right)^2 + \left(\frac{dy}{dt}\right)^2}\, dt. \tag{12}$$

This is the way we *remember* Eqs. (11) and (12), but it cannot be regarded as a proof.

To prove Eq. (12), partition the interval $[a, b]$ on the t-axis into n subintervals of equal length, so that $a = t_0 < t_1 < \cdots < t_n = b$. We approximate the arc length by adding the lengths of chords of the curve, as in approach 2 above. The ith chord, shown in Fig. 7.43, joins the points

$$(h(t_{i-1}), k(t_{i-1})) \qquad \text{and} \qquad (h(t_i), k(t_i)).$$

The length of this chord is therefore

$$\sqrt{(h(t_i) - h(t_{i-1}))^2 + (k(t_i) - k(t_{i-1}))^2}$$
$$= \sqrt{\left(\frac{h(t_i) - h(t_{i-1})}{t_i - t_{i-1}}\right)^2 + \left(\frac{k(t_i) - k(t_{i-1})}{t_i - t_{i-1}}\right)^2} \cdot \frac{b-a}{n}.$$

By the mean-value theorem, there exist points c_i and c_i' between t_{i-1} and t_i such that this last expression becomes

$$\sqrt{h'(c_i)^2 + k'(c_i')^2} \cdot \frac{b-a}{n}.$$

The total length of the curve from $t = a$ to $t = b$ is therefore approximately

$$\frac{b-a}{n} \cdot \sum_{i=1}^{n} \sqrt{h'(c_i)^2 + k'(c_i')^2}.$$

This sum is *almost* a Riemann sum for the function $\sqrt{h'(t)^2 + k'(t)^2}$ over $[a, b]$. The only problem is the *two* points c_i and c_i' in the ith interval used to evaluate the two summands under the radical. But there is a theorem of Bliss, mentioned in the preceding section, that also says that such a sum approaches the integral of the function as $n \to \infty$. This establishes Eq. (12).

Figure 7.44 The circle $x^2 + y^2 = a^2$ parametrized as $x = a\cos t$, $y = a \sin t$ for $0 \le t \le 2\pi$.

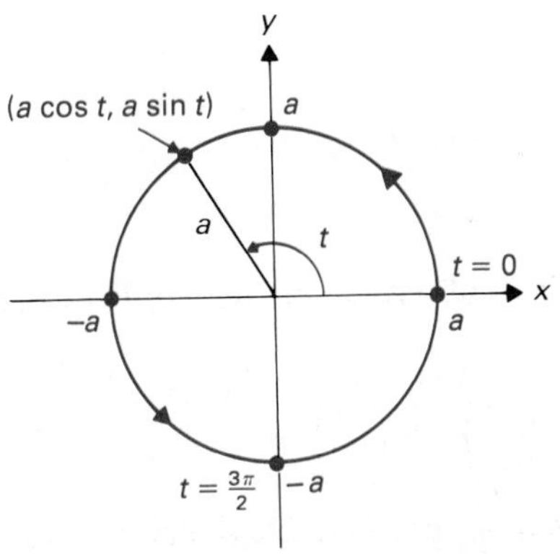

EXAMPLE 4 Verify the formula $C = 2\pi a$ for the circumference of a circle having radius a.

Solution We take as parametric equations for the circle

$$x = a\cos t, \qquad y = a \sin t \qquad \text{for } 0 \le t \le 2\pi.$$

See Fig. 7.44. Since $dx/dt = -a \sin t$ and $dy/dt = a \cos t$, we see from Eq. (12) that the length of the circle is

$$s = \int_0^{2\pi} \sqrt{(-a\sin t)^2 + (a \cos t)^2}\, dt$$
$$= \int_0^{2\pi} \sqrt{a^2}\, dt = a \int_0^{2\pi} (1)\, dt$$
$$= 2\pi a. \quad \square$$

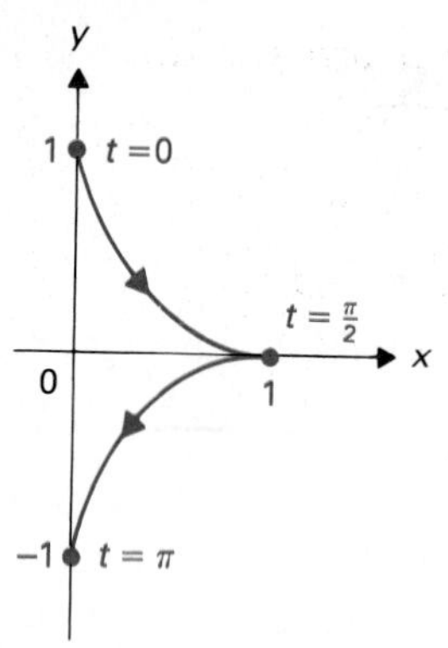

Figure 7.45 The curve $x = \sin^3 t$, $y = \cos^3 t$ for $0 \le t \le \pi$.

EXAMPLE 5 Find the length of the parametric curve $x = \sin^3 t$, $y = \cos^3 t$ for $0 \le t \le \pi$, shown in Fig. 7.45.

Solution This example illustrates a tricky point. We have

$$\frac{dx}{dt} = 3\sin^2 t\cos t, \qquad \frac{dy}{dt} = -3\cos^2 t\sin t.$$

Thus

$$\begin{aligned} ds &= \sqrt{\left(\frac{dx}{dt}\right)^2 + \left(\frac{dy}{dt}\right)^2}\,dt \\ &= \sqrt{9\sin^4 t\cos^2 t + 9\cos^4 t\sin^2 t}\,dt \\ &= \sqrt{(9\sin^2 t\cos^2 t)(\sin^2 t + \cos^2 t)}\,dt \\ &= \sqrt{9\sin^2 t\cos^2 t}\,dt = |3\sin t\cos t|\,dt. \end{aligned}$$

The absolute value is the "tricky point." The formula for ds involves the *positive* square root; the short tangent-line segments have *positive* length. If the expression under the radical turns out to be a perfect square, be sure always to use the *positive* square root. No absolute-value sign was needed in Example 2 because $6y^2 + 1/(24y^2) > 0$.

Continuing our solution, we see that the arc length is given by

$$s = \int_0^\pi |3\sin t\cos t|\,dt.$$

Now $3\sin t\cos t$ is positive for $0 < t < \pi/2$ and negative for $\pi/2 < t < \pi$. Thus

$$\begin{aligned} s &= \int_0^{\pi/2} 3\sin t\cos t\,dt + \int_{\pi/2}^{\pi} -3\sin t\cos t\,dt \\ &= \frac{3\sin^2 t}{2}\bigg]_0^{\pi/2} - \frac{3\sin^2 t}{2}\bigg]_{\pi/2}^{\pi} = \left(\frac{3}{2} - 0\right) - \left(0 - \frac{3}{2}\right) = \frac{6}{2} = 3. \end{aligned}$$

Note that if we had computed $\int_0^\pi 3\sin t\cos t\,dt$, we would have obtained zero, which is obviously incorrect. □

SUMMARY

1. If $f'(x)$ is continuous, then the length of arc of the graph $y = f(x)$ from $x = a$ to $x = b$ is

$$s = \int_a^b \sqrt{1 + f'(x)^2}\,dx.$$

2. If $h'(t)$ and $k'(t)$ are continuous, then the length of the curve $x = h(t)$, $y = k(t)$ traveled from time $t = a$ to time $t = b$ is

$$s = \int_a^b \sqrt{(dx/dt)^2 + (dy/dt)^2}\,dt.$$

3. The differential ds of arc length takes the following forms:

$$ds = \sqrt{(dx)^2 + (dy)^2} = \sqrt{1 + \left(\frac{dy}{dx}\right)^2}\, dx$$

$$= \sqrt{\left(\frac{dx}{dy}\right)^2 + 1}\, dy = \sqrt{\left(\frac{dx}{dt}\right)^2 + \left(\frac{dy}{dt}\right)^2}\, dt.$$

EXERCISES

In Exercises 1 through 17, find the length of the curve with the given x,y-equation or parametric equations.

1. $y^2 = x^3$ from (0, 0) to (4, 8)

2. $y = \frac{1}{3}(x^2 - 2)^{3/2}$ from $x = 2$ to $x = 4$

3. $9x^2 = 16y^3$ from $y = 3$ to $y = 6$, $x > 0$

4. $9(x + 2)^2 = 4(y - 1)^3$ from $y = 1$ to $y = 4$, $x \geq -2$

5. $y = x^{2/3}$ from $x = 1$ to $x = 8$

6. $y = x^3 + 1/(12x)$ from $x = 1$ to $x = 2$

7. $y = x^3/3 + 1/(4x)$ from $x = 1$ to $x = 3$

8. $x = y^4 + 1/(32y^2)$ from $y = -2$ to $y = -1$ [*Hint:* Be sure to take the *positive* square root.]

9. $x = y^5/20 + 1/(3y^3)$ from $y = 1$ to $y = 2$

10. $x^{2/3} + y^{2/3} = a^{2/3}$

11. $x = 4 \sin (t/2)$, $y = 4 \cos (t/2)$ from $t = 0$ to $t = 2\pi$

12. $x = t^2$, $y = \frac{2}{3}(2t + 1)^{3/2}$ from $t = 0$ to $t = 4$

13. $x = a \cos^3 t$, $y = a \sin^3 t$ from $t = 0$ to $t = \pi/2$

14. $x = a \sin^3 2t$, $y = a \cos^3 2t$ from $t = 0$ to $t = \pi$

15. $x = (t - 1)^2$, $y = \frac{8}{3}t^{3/2}$ from $t = 1$ to $t = 2$

16. $x = \frac{2}{5}t^{5/2} + 4$, $y = \frac{1}{2}t^2 - 1$ from $t = 0$ to $t = 3$ (Use the integral table on the endpapers of this book.)

17. $x = \frac{2}{7}t^{7/2} - 2$, $y = (1/\sqrt{3})t^3$ from $t = 0$ to $t = 1$ (Use the integral table.)

Recall the arc length function $s(x) = \int_a^x \sqrt{1 + f'(t)^2}\, dt$ where $y = f(x)$. A similar expression holds if $x = g(y)$ or if $x = h(t)$, $y = k(t)$. In Exercises 18 through 24, approximate the short length of curve indicated using the appropriate arc length function and using approximation by the differential ds.

18. Find the approximate length of the curve $y = \sin x$ from $x = 0$ to $x = 0.03$.

19. Find the approximate arc length of the curve $y = \cos x$ from $x = \pi/6$ to $x = 8\pi/45$.

20. Find the approximate arc length of the curve $y = x^2$ from $x = 3$ to $x = 3.01$.

21. Find the approximate arc length of the curve $x = 2/y^2$ from $y = 1$ to $y = 1.05$.

22. Find the approximate length of the arc of the curve $x = 4t$, $y = \sin t$ from $t = 0$ to $t = 0.05$.

23. Find the approximate length of the arc of the curve $x = 4 \cos t$, $y = 5t + 1$ from $t = 0$ to $t = 0.05$.

24. Find the approximate value of $t_1 > 0$ such that the length of the curve $x = \sin t$, $y = \tan t$ from $t = 0$ to $t = t_1$ is 0.1.

In Exercises 25 through 30, use a calculator to compute the approximate arc length, using Simpson's rule with some $n \geq 10$.

25. The length of $y = x^3$ from $x = 1$ to $x = 5$

26. The length of $y = 1/x$ from $x = \frac{1}{2}$ to $x = 2$

27. The length of $x = \sin y$ from $y = 0$ to $y = \pi$

28. The length of arc of the ellipse $x^2/4 + y^2/9 = 1$.

29. The length of $y = 3 \cos 2x + 4 \sin (x/2)$ from $x = 0$ to $x = \pi$

30. The total length of arc of the curve $x = \sin t$, $y = \cos 2t$ [*Hint:* Make a sketch to find the appropriate limits for the integral.]

7.5 AREA OF A SURFACE OF REVOLUTION

In Section 7.4, we found the length of an arc. In this section, we revolve such arcs about axes and find the *area of the surface* generated.

We assume again that f is a continuously differentiable function. If a portion of the graph of $y = f(x)$ for $a \leq x \leq b$ is revolved about the x-axis, it generates a *surface of revolution* as shown in Fig. 7.46. If we take a small

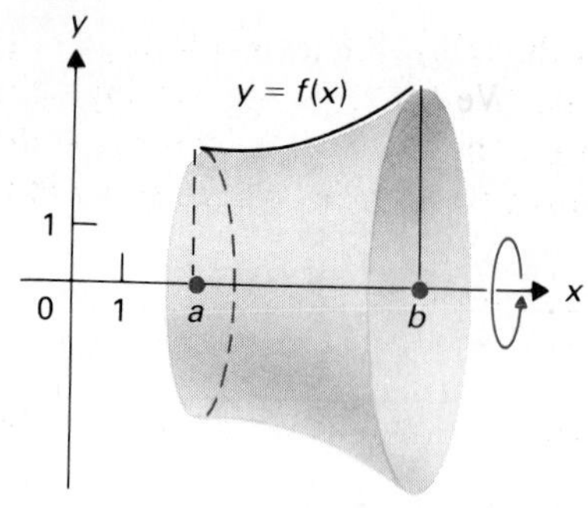

Figure 7.46 Surface generated by revolving $y = f(x)$ about the x-axis.

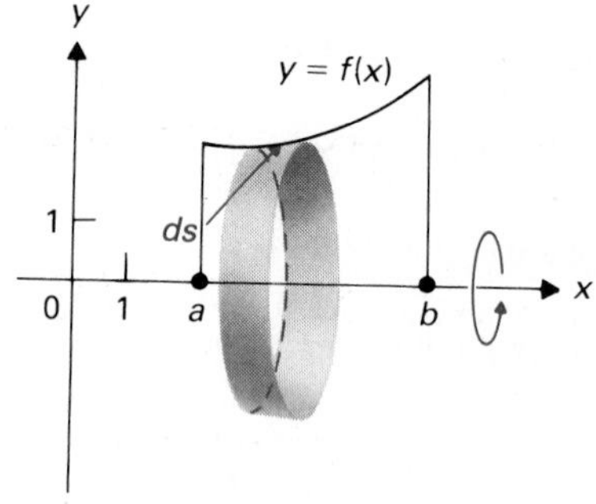

Figure 7.47 Strip of surface generated by the tangent-line segment of length ds.

tangent-line segment of length ds at (x, y) on the curve and revolve this segment about the x-axis as shown in Fig. 7.47, the surface area generated by this segment would seem to be approximately that of a cylinder of radius $|y|$ and height ds, that is, approximately $2\pi|y|\,ds$. Since $y = f(x)$ and $ds = \sqrt{1 + (f'(x))^2}\,dx$, we arrive at the formula

$$S = \int_a^b 2\pi|f(x)|\sqrt{1 + (f'(x))^2}\,dx \tag{1}$$

for the area of the surface of revolution about the x-axis.

The preceding paragraph indicates how we can easily remember formula (1). A justification of the formula can be made along the following lines. Partition the interval $[a, b]$ into n subintervals of equal length. Consider the chords to the curve over the subintervals, as in approach 2 of the previous section. Equation (9) in the last section showed that the length of the ith chord of the curve can be written

$$\frac{b-a}{n}\sqrt{1 + f'(x_i)^2}$$

for some x_i in the ith subinterval. Let c_i and c_i' be chosen in the ith subinterval so that

$|f(c_i)|$ is the maximum value of $|f(x)|$ over the subinterval,

$|f(c_i')|$ is the minimum value of $|f(x)|$ over the subinterval.

Then surely the surface area element generated by revolving this ith chord satisfies

$$2\pi|f(c_i')|\frac{b-a}{n}\sqrt{1 + f'(x_i)^2} \le \text{Area element}$$

$$\le 2\pi|f(c_i)|\frac{b-a}{n}\sqrt{1 + f'(x_i)^2}.$$

By the theorem of Bliss referred to in Sections 7.3 and 7.4, both

$$\frac{b-a}{n}\sum_{i=1}^{n}\left(2\pi|f(c_i')|\sqrt{1 + f'(x_i)^2}\right)$$

and

$$\frac{b-a}{n}\sum_{i=1}^{n}\left(2\pi|f(c_i)|\sqrt{1 + f'(x_i)^2}\right)$$

approach the integral in formula (1) as $n \to \infty$. This justifies formula (1).

If the curve is given in the parametric form $x = h(t)$, $y = k(t)$ for $a \le t \le b$, and if the curve set is traversed just once for t in $[a, b]$, then formula (1) takes the form

$$S = \int_a^b 2\pi|k(t)|\sqrt{(h'(t))^2 + (k'(t))^2}\,dt. \tag{2}$$

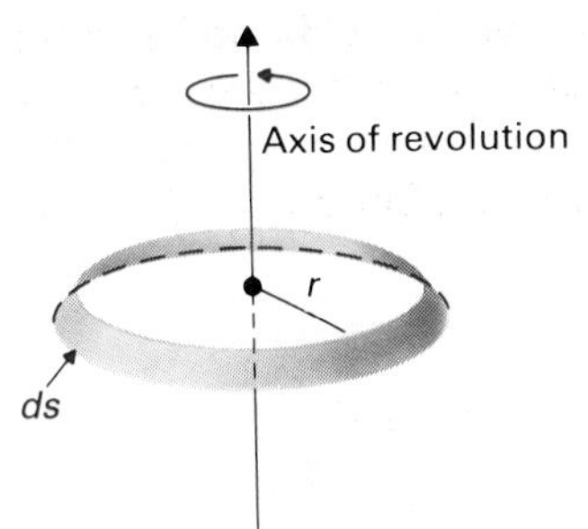

Figure 7.48 $dS = 2\pi r\, ds$

If an arc is revolved about an axis other than the x-axis, formulas (1) and (2) are modified in the obvious way. We "add by integrating" the contributions $2\pi r \cdot ds$ where ds is the length of a tangent-line segment to the curve and r is the radius of the circle through which this tangent segment is revolved. See Fig. 7.48.

The general formula for surface area of revolution is thus

$$\boxed{\text{Surface area} = S = \int_a^b 2\pi \cdot \text{Radius of revolution} \cdot ds,} \tag{3}$$

where a and b are appropriate limits and ds is the differential of arc length.

EXAMPLE 1 Find the area of the surface generated by revolving the arc of $y = x^2$ from (0, 0) to (1, 1) about the y-axis.

Solution We have $dy/dx = 2x$, so

$$ds = \sqrt{1 + 4x^2}\, dx.$$

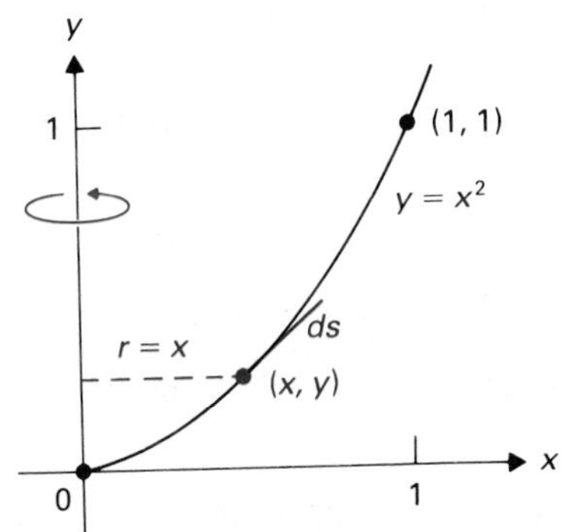

Figure 7.49 Tangent-line segment of length ds is at distance $r = x$ from the x-axis of revolution.

The radius of revolution is x, as indicated in Fig. 7.49. Then formula (3) becomes

$$S = \int_0^1 2\pi x\, ds = \int_0^1 2\pi x\sqrt{1 + 4x^2}\, dx = \frac{2\pi}{8}\int_0^1 8x(1 + 4x^2)^{1/2}\, dx$$

$$= \frac{\pi}{4} \cdot \frac{(1 + 4x^2)^{3/2}}{3/2}\Bigg]_0^1 = \frac{\pi}{6}(1 + 4x^2)^{3/2}\Bigg]_0^1 = \frac{\pi}{6}(5\sqrt{5} - 1). \quad \square$$

EXAMPLE 2 Derive the formula $A = 4\pi a^2$ for the surface area of a sphere having radius a.

Solution We view the sphere as the surface generated by revolving the semicircle $x = a\cos t$, $y = a\sin t$ for $0 \le t \le \pi$ in Fig. 7.50 about the x-axis. Here formula (2) is appropriate, and we obtain

$$S = \int_0^\pi 2\pi a \sin t\sqrt{(-a\sin t)^2 + (a\cos t)^2}\, dt$$

$$= 2\pi a\int_0^\pi (\sin t)\sqrt{a^2}\, dt = 2\pi a^2 \int_0^\pi \sin t\, dt$$

$$= 2\pi a^2(-\cos t)\Big]_0^\pi = 2\pi a^2[1 - (-1)] = 4\pi a^2. \quad \square$$

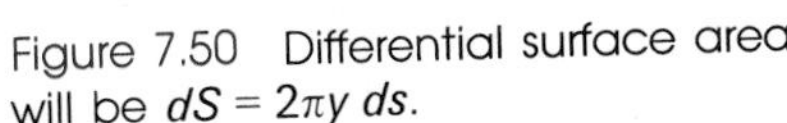

Figure 7.50 Differential surface area will be $dS = 2\pi y\, ds$.

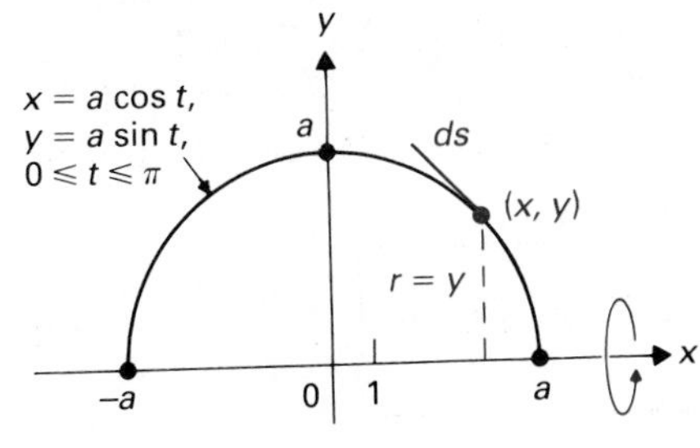

EXAMPLE 3 Find the area of the surface generated by revolving the arc of $y = x^3$ from $(-1, -1)$ to (1, 1) about the x-axis.

Solution We have $dy/dx = 3x^2$, so

$$ds = \sqrt{1 + \left(\frac{dy}{dx}\right)^2}\, dx = \sqrt{1 + 9x^4}\, dx.$$

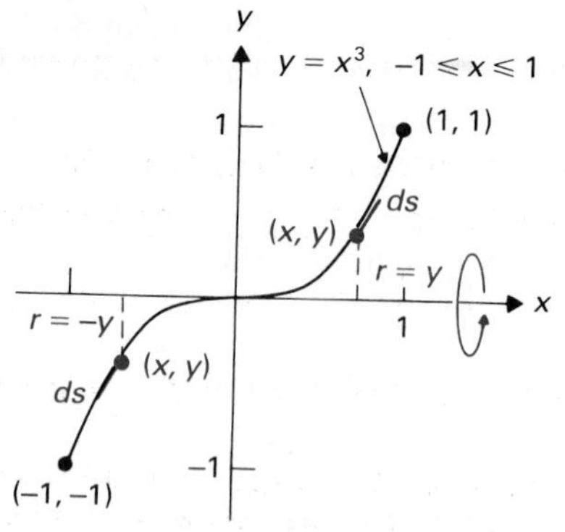

Figure 7.51 $r = y$ for ds above the x-axis; $r = -y$ for ds below the x-axis; thus $r = |y| = |x^3|$.

From Fig. 7.51, we see that $r = |y| = |x^3|$. The absolute-value symbol is crucial here, since $x^3 < 0$ for $-1 \le x < 0$. Thus

$$S = \int_{-1}^{1} 2\pi |x^3| \sqrt{1 + 9x^4}\, dx.$$

Since $x^3 < 0$ for $-1 \le x < 0$, and $x^3 \ge 0$ for $0 \le x \le 1$, we have

$$\begin{aligned} S &= \int_{-1}^{0} 2\pi(-x^3)\sqrt{1 + 9x^4}\, dx + \int_{0}^{1} 2\pi x^3 \sqrt{1 + 9x^4}\, dx \\ &= -\frac{2\pi}{36}\int_{-1}^{0} (1 + 9x^4)^{1/2}(36x^3)\, dx + \frac{2\pi}{36}\int_{0}^{1} (1 + 9x^4)^{1/2}(36x^3)\, dx \\ &= -\frac{\pi}{18}\cdot\frac{2}{3}(1 + 9x^4)^{3/2}\Big]_{-1}^{0} + \frac{\pi}{18}\cdot\frac{2}{3}(1 + 9x^4)^{3/2}\Big]_{0}^{1} \\ &= -\frac{\pi}{27}(1 - 10\sqrt{10}) + \frac{\pi}{27}(10\sqrt{10} - 1) = \frac{2\pi}{27}(10\sqrt{10} - 1). \end{aligned}$$

Of course, we would usually have used symmetry and computed S by

$$S = 2\int_{0}^{1} 2\pi x^3 \sqrt{1 + 9x^4}\, dx,$$

but we wished to call attention to the need to be sure that r is nonnegative. □

EXAMPLE 4 Find the area of the surface generated by revolving the arc of $y = \sin x$ for $0 \le x \le \pi$ about the line $x = 4$.

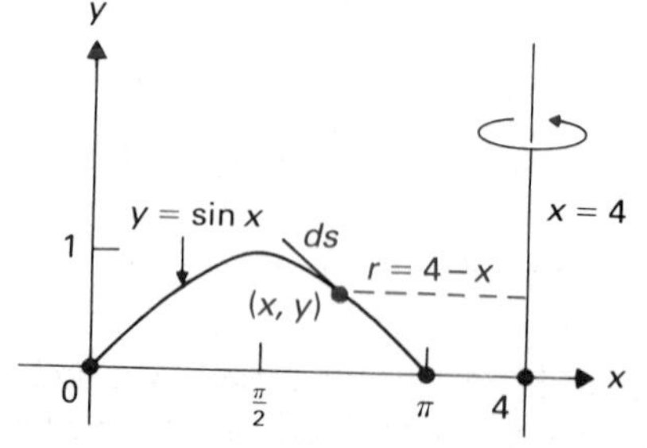

Figure 7.52 $r = x_{\text{right}} - x_{\text{left}} = 4 - x$

Solution We have $dy/dx = \cos x$, so $ds = \sqrt{1 + \cos^2 x}\, dx$. From Fig. 7.52, we see that $r = 4 - x$. Thus

$$S = \int_{0}^{\pi} 2\pi(4 - x)\sqrt{1 + \cos^2 x}\, dx.$$

This integral is one we have to compute numerically. Using Simpson's rule with $n = 20$, we find that

$$S = 2\pi \int_{0}^{\pi} (4 - x)\sqrt{1 + \cos^2 x}\, dx \approx 58.308. \quad \square$$

SUMMARY

If an arc of a smooth curve is revolved about an axis, then the area of the surface generated is

$$S = \int_{a}^{b} 2\pi \cdot \text{Radius of revolution} \cdot ds$$

for appropriate limits a and b.

EXERCISES

In Exercises 1 through 20, find the area of the surface obtained by revolving the given curve about the indicated axis.

1. $y = 4x$ from $(0, 0)$ to $(1, 4)$ about the y-axis

2. $x = 2y$ from $(2, 1)$ to $(6, 3)$ about the x-axis

3. $x = y - 2$ from $y = 2$ to $y = 5$ about the line $y = 3$

4. $y = x - 3$ from $x = 3$ to $x = 5$ about the line $x = -1$

5. $y = x^3$ from $x = -1$ to $x = 2$ about the x-axis

6. $y = \sqrt{x}$ from $x = 1$ to $x = 3$ about the x-axis

7. $y = \frac{2}{3}x^{3/2}$ from $x = 0$ to $x = 2$ about the y-axis (Use the table of integrals on the endpapers of this book.)

8. $y = |x|$ from $x = -1$ to $x = 1$ about the x-axis

9. $y = \frac{1}{3}(x^2 - 2)^{3/2}$ from $x = \sqrt{2}$ to $x = 4$ about the y-axis

10. $y = x^4/4 + 1/(8x^2)$ from $x = 1$ to $x = 2$ about the y-axis

11. $y = x^3 + 1/(12x)$ from $x = 1$ to $x = 2$ about the x-axis

12. $x^{2/3} + y^{2/3} = a^{2/3}$ for $y \geq 0$ about the x-axis

13. $y = \sqrt{a^2 - x^2}$ from $x = -a$ to $x = a$ about the x-axis (Note that this will derive again the formula for the surface area of a sphere of radius a.)

14. $x = a \cos t$, $y = a \sin t$ for $0 \leq t \leq 2\pi$ about the line $x = b$ for $b > a$. (This gives the surface area of a torus.)

15. $x = (t - 1)^2$, $y = \frac{8}{3}t^{3/2}$ from $t = 1$ to $t = 2$ about the line $x = -1$

16. $x = a \cos^3 t$, $y = a \sin^3 t$ from $t = 0$ to $t = \pi/4$ about the x-axis

17. $x = t$, $y = t^2 - 2$ from $t = 0$ to $t = 3$ about the y-axis

18. $x = t^2/2 + 2t$, $y = t + 1$ from $t = 0$ to $t = 2$ about the line $y = -1$

19. $x = 2t + 1$, $y = t^2 - 3t$ from $t = 3$ to $t = 4$ about the line $x = 4$

20. $x = 4 \cos (t/2)$, $y = 4 \sin (t/2)$ from $t = 0$ to $t = 2\pi$ about the line $y = -2$

21. If a cone has base radius r and slant height s, its surface area is πrs. Derive this formula by revolving the segment of the line $y = (r/h)x$ from $x = 0$ to $x = h$ about the x-axis.

22. Derive the formula for the surface area of a frustrum of a cone having base radius R, top radius r, and altitude h, using the methods of this section.

In Exercises 23 through 30, use a calculator or computer to find the area of the surface obtained by revolving the given curve about the indicated axis. Use Simpson's rule with $n \geq 10$.

23. $y = x^2$ from $x = 0$ to $x = 2$ about the x-axis

24. $y = \sin x$ from $x = 0$ to $x = \pi$ about the y-axis

25. $x = y^3 + 3y^2$ from $y = 0$ to $y = 4$ about the y-axis

26. $y = x^2$ from $x = 0$ to $x = 2$ about the line $y = -4$

27. $y = x^2$ from $x = 0$ to $x = 2$ about the line $x = -3$

28. $x = y^2 + 2y$ from $y = 1$ to $y = 5$ about the line $y = 10$

29. $x = \sin y$ from $y = 0$ to $y = \pi$ about the line $x = 2$

30. $y = \tan x$ from $x = -\pi/3$ to $x = \pi/3$ about the line $y = -\pi$

7.6 DISTANCE

We will now apply integral calculus to find the distance traveled by a body moving on a straight line if we know the velocity of the body at any time t. The body may be moving back and forth on the line. In that case, we will have to distinguish clearly whether "distance" means the *resultant distance* from the point where the body starts at the initial time to the point where the body is at a later time or the *total distance* the body travels, as measured, for instance, by an odometer in a car.

If a body travels on a line with constant velocity v, then the distance s traveled after time t is given by the product $|v|t$. The absolute-value sign is used since v is negative if the body is moving in the negative direction on the line. Thus $|v|$ is the *speed* of the body.

EXAMPLE 1 If the velocity at time t sec of a body traveling on a line is $2t + t^2$ ft/sec, find how far the body travels from $t = 2$ sec to $t = 4$ sec, using integral calculus.

Solution Since distance equals the product of speed and time, assuming that speed is constant, we see that at time t, the distance traveled over the next small time interval dt sec is about $(2t + t^2)\,dt$ ft. We wish to add up these distances as t ranges from 2 to 4 and take the limit of the resulting sum as $dt \to 0$. The appropriate integral is

$$\int_2^4 (2t + t^2)\,dt = \left(t^2 + \frac{t^3}{3}\right)\Big]_2^4 = \left(16 + \frac{64}{3}\right) - \left(4 + \frac{8}{3}\right) = \frac{92}{3}.$$

Thus the body travels $\frac{92}{3}$ ft. □

The velocity of a body moving on the x-axis is considered positive if it is moving to the right and negative if it is moving to the left. Thus the integral of the velocity v from t_1 to t_2 will give the total distance traveled toward the right minus the distance traveled toward the left between these times. This *resultant distance* may not give a true picture of how far the body has traveled; it does tell us the *signed* distance of the body at time t_2 from its position at time t_1. To find the actual distance traveled, we must integrate the speed $|v|$. Thus if a body starts at time t_1, stops at time t_2, and has velocity $v(t)$ for $t_1 \le t \le t_2$, then

$$\text{Distance from start to stop} = \int_{t_1}^{t_2} v(t)\,dt,$$

$$\text{Total distance traveled} = \int_{t_1}^{t_2} |v(t)|\,dt.$$

EXAMPLE 2 Suppose the velocity at time t of a body traveling on a line is $\cos(\pi t/2)$ ft/sec. Thus at time $t = 0$, the body is moving at the speed of 1 ft/sec in the positive direction, while at time $t = 2$, the body is moving in the negative direction at the speed of 1 ft/sec. Find the total distance the body travels from time $t = 0$ to time $t = 2$, and also the signed distance from start to stop.

Solution We need to compute

$$\int_0^2 \left|\cos\frac{\pi t}{2}\right| dt.$$

Now $\cos(\pi t/2)$ is positive for $0 \le t < 1$ and negative for $1 < t \le 2$. Thus we have

$$\begin{aligned}\int_0^2 \left|\cos\frac{\pi t}{2}\right| dt &= \int_0^1 \left(\cos\frac{\pi t}{2}\right) dt + \int_1^2 \left(-\cos\frac{\pi t}{2}\right) dt \\ &= \left(\frac{2}{\pi}\sin\frac{\pi t}{2}\right)\Big]_0^1 - \left(\frac{2}{\pi}\sin\frac{\pi t}{2}\right)\Big]_1^2 \\ &= \left(\frac{2}{\pi} - 0\right) - \left(0 - \frac{2}{\pi}\right) = \frac{4}{\pi}.\end{aligned}$$

Therefore the body travels $4/\pi$ ft from time $t = 0$ to time $t = 2$ sec.

The distance from start to stop for this body is

$$\int_0^2 \left(\cos\frac{\pi t}{2}\right) dt = \left(\frac{2}{\pi}\sin\frac{\pi t}{2}\right)\Bigg]_0^2 = \frac{2}{\pi}(0 - 0) = 0.$$

Thus this body returns at $t = 2$ to the point where it started at $t = 0$. □

EXAMPLE 3 Let the acceleration $a = d^2x/dt^2$ of a body traveling on the x-axis be $6t$. If the body starts with initial velocity $v_0 = -3$ at time $t = 0$, find (a) the velocity v as a function of t, (b) the total distances the body travels from time $t = 0$ to $t = 4$.

Solution From $d^2x/dt^2 = 6t$, we obtain

$$v = \frac{dx}{dt} = 3t^2 + C.$$

Since $v = -3$ when $t = 0$, we have $C = -3$, so

$$v = 3t^2 - 3.$$

This answers part (a).

For part (b), we see that v is negative for $0 \le t < 1$ and positive for $1 < t \le 4$. Thus

$$\begin{aligned} s = \int_0^4 |3t^2 - 3|\, dt &= \int_0^1 -(3t^2 - 3)\, dt + \int_1^4 (3t^2 - 3)\, dt \\ &= -(t^3 - 3t)\Big]_0^1 + (t^3 - 3t)\Big]_1^4 \\ &= -(-2 - 0) + [52 - (-2)] \\ &= 2 + 54 = 56. \quad \square \end{aligned}$$

Let the position (x, y) of a body traveling in the plane be given at time t by $x = h(t)$, $y = k(t)$. We assume that $h'(t)$ and $k'(t)$ are continuous. On page 111, we discussed the x-component dx/dt and y-component dy/dt of the velocity and the speed of the body. We showed that the speed is

$$\frac{ds}{dt} = \sqrt{\left(\frac{dx}{dt}\right)^2 + \left(\frac{dy}{dt}\right)^2}. \tag{1}$$

Suppose the speed is not constant. Then the distance the body travels from time t to time $t + dt$ for sufficiently small $dt > 0$ is approximately

$$\sqrt{\left(\frac{dx}{dt}\right)^2 + \left(\frac{dy}{dt}\right)^2}\, dt. \tag{2}$$

Of course we recognize formula (2) as the formula for the differential ds of arc length. This makes sense; ds measures distance tangent to the curve. We now see that from time t_1 to time t_2,

$$\boxed{\text{Total distance traveled} = \int_{t_1}^{t_2} \sqrt{\left(\frac{dx}{dt}\right)^2 + \left(\frac{dy}{dt}\right)^2}\, dt.} \tag{3}$$

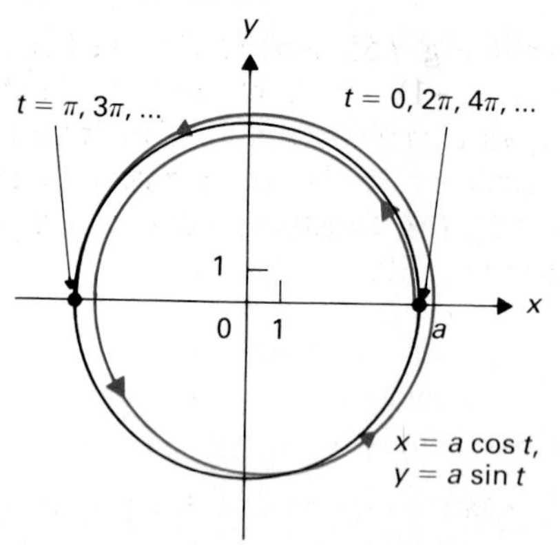

Figure 7.53 Schematic of a body starting at $(a, 0)$ and traveling $\frac{3}{2}$ times counterclockwise around the circle.

The integral (3) need not give the length of arc from the starting point where $t = t_1$ to the stopping point where $t = t_2$. Portions of the curve may be traveled several times. If we think of a car, formula (3) gives us the reading at time t_2 of an odometer set to zero at time t_1. We illustrate with two examples.

EXAMPLE 4 Suppose the position (x, y) of a body at time t is given by $x = a \cos t$, $y = a \sin t$. Find the distance traveled from time $t = 0$ to time $t = 3\pi$.

Solution We have seen several times that $x = a \cos t$, $y = a \sin t$ describes the circle of radius a with center $(0, 0)$. As t increases, the body travels counterclockwise around this circle, starting at $(a, 0)$ when $t = 0$ and going once around every 2π units of time. See Fig. 7.53. Thus from $t = 0$ to $t = 3\pi$, the body travels around the circle $\frac{3}{2}$ times, as indicated schematically by the red arc in Fig. 7.53. Since the circle has perimeter $2\pi a$, we expect to have $s = \frac{3}{2} \cdot 2\pi a = 3\pi a$. The integral (3) bears this out. We have

$$\frac{dx}{dt} = -a \sin t, \qquad \frac{dy}{dt} = a \cos t,$$

$$\sqrt{\left(\frac{dx}{dt}\right)^2 + \left(\frac{dy}{dt}\right)^2} = \sqrt{a^2 \sin^2 t + a^2 \cos^2 t} = \sqrt{a^2 \cdot 1} = a.$$

Thus

$$s = \int_0^{3\pi} a \, dt = at \Big]_0^{3\pi} = 3\pi a. \quad \square$$

EXAMPLE 5 Let the position (x, y) of a body in the plane at time t be given by $x = \cos t$, $y = \cos t$. Find the total distance the body travels from time $t = 0$ to time $t = 10\pi$.

Solution We have

$$\frac{dx}{dt} = \frac{dy}{dt} = -\sin t.$$

Thus

$$\sqrt{\left(\frac{dx}{dt}\right)^2 + \left(\frac{dy}{dt}\right)^2} = \sqrt{\sin^2 t + \sin^2 t} = \sqrt{2 \sin^2 t} = \sqrt{2}\,|\sin t|.$$

The integral (3) becomes

$$s = \int_0^{10\pi} \sqrt{2}\,|\sin t| \, dt = \sqrt{2} \int_0^{10\pi} |\sin t| \, dt.$$

Figure 7.54 Graph of $u = |\sin t|$.

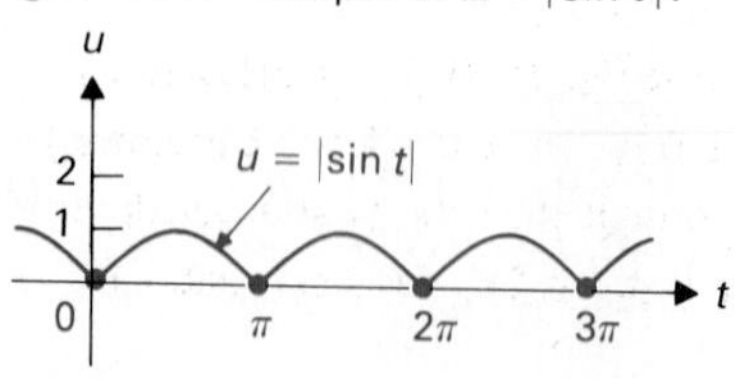

The graph of $u = |\sin t|$ is indicated in Fig. 7.54. Each arc covers a distance of π on the t-axis. Since $\sin t \geq 0$ for $0 \leq t \leq \pi$, we can write

$$s = \sqrt{2} \int_0^{10\pi} |\sin t| \, dt = \sqrt{2} \cdot 10 \int_0^{\pi} \sin t \, dt = 10\sqrt{2}(-\cos t) \Big]_0^{\pi}$$

$$= 10\sqrt{2}[-(-1) - (-1)] = 20\sqrt{2}.$$

We could also verify this computation geometrically. Since $x = \cos t = y$ and $-1 \leq \cos t \leq 1$, the body travels on the segment of the line

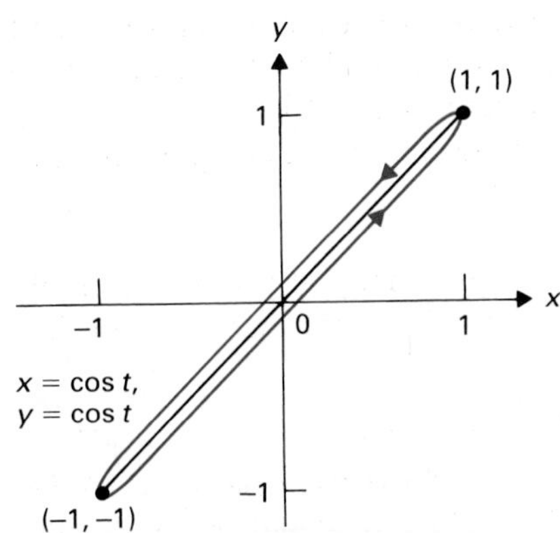

Figure 7.55 A body starting at (1, 1), going to (−1, −1), and then back to (1, 1) travels $4\sqrt{2}$ units on the "round trip."

$y = x$ joining $(-1, -1)$ and $(1, 1)$, shown in Fig. 7.55. When $t = 0$, the body starts at $(1, 1)$. At $t = \pi$, the body is at $(-1, -1)$, and it returns to $(1, 1)$ at $t = 2\pi$. From $t = 0$ to $t = 10\pi$, the body thus makes a total of five "round trips" on this line segment. Since the distance on the line segment from $(1, 1)$ to $(-1, -1)$ is $\sqrt{2^2 + 2^2} = \sqrt{8} = 2\sqrt{2}$, the distance per round trip is $4\sqrt{2}$. Thus the distance for five round trips is $20\sqrt{2}$. □

SUMMARY

1. If a body on a line has velocity $v(t)$ at time t and starts at time t_1 and stops at time t_2, then

$$\text{Distance from start to stop} = \int_{t_1}^{t_2} v(t)\,dt,$$

$$\text{Total distance traveled} = \int_{t_1}^{t_2} |v(t)|\,dt.$$

2. If the position (x, y) of a body in the plane at time t is given by $x = h(t)$, $y = k(t)$, then the total distance the body travels from time $t = t_1$ to $t = t_2$ is

$$s = \int_{t_1}^{t_2} \sqrt{\left(\frac{dx}{dt}\right)^2 + \left(\frac{dy}{dt}\right)^2}\,dt.$$

EXERCISES

In Exercises 1 through 14, the velocity v of a body on a line is given as a function of time t. Find:

a) the distance from beginning point to ending point and
b) the total distance traveled

in the indicated time interval.

1. $v = t - 4, 0 \le t \le 2$
2. $v = t - 3, 1 \le t \le 6$
3. $v = 2t - t^2, 0 \le t \le 4$
4. $v = 3t^2 - 4, 2 \le t \le 3$
5. $v = t^2 - 3t + 2, 0 \le t \le 1$
6. $v = t^2 - 3t + 2, 0 \le t \le 2$
7. $v = t^2 - 3t = 2, 0 \le t \le 3$
8. $v = \sin t, 0 \le t \le 3\pi$
9. $v = \cos t, 0 \le t \le 5\pi$
10. $v = \sin(\pi t/2) + \cos(\pi t/2), 0 \le t \le 1$
11. $v = \sin(\pi t/2) + \cos(\pi t/2), 0 \le t \le 2$
12. $v = \sin(\pi t/2) + \cos(\pi t/2), 0 \le t \le 4$
13. $v = |t - 5|, 0 \le t \le 5$
14. $v = |t - 5|, 0 \le t \le 10$

In Exercises 15 through 20, the acceleration a as a function of time t and the initial velocity v_0 when $t = 0$ are given. Find:

a) the velocity of the body as a function of t and
b) the total distance the body travels

in the indicated time interval.

15. $a = 3, v_0 = 0, 0 \le t \le 2$
16. $a = 2t - 4, v_0 = 3, 0 \le t \le 3$
17. $a = \sin t, v_0 = 0, 0 \le t \le 3\pi/2$
18. $a = -1/\sqrt{t+1}, v_0 = 2, 0 \le t \le 4$
19. $a = 6t - 1/(t+1)^3, v_0 = 2, 0 \le t \le 2$
20. $a = 6/(t+1)^2, v_0 = -3, 0 \le t \le 2$ (Use a calculator.)

In Exercises 21 through 28, the position (x, y) of a body in the plane at time t is given. Find the total distance the body travels during the indicated time interval.

21. $x = 3\cos t, y = -3\sin t, 0 \le t \le \frac{3}{2}\pi$
22. $x = 3\sin t, y = 4\sin t, 0 \le t \le 5\pi$
23. $x = 2 + \sin t, y = \frac{2}{3}(\sin t)^{3/2}, 0 \le t \le \pi$
24. $x = 2\cos^3 t, y = 2\sin^3 t, 0 \le t \le 4\pi$
25. $x = \sin t, y = 3\sin^2 t, 0 \le t \le 21\pi/5$ (Use a calculator.)
26. $x = 3\cos t, y = 4\sin t, 0 \le t \le 10\pi/3$ (Use a calculator.)
27. $x = 2\cos 3t, y = 1 + \sin t, 0 \le t \le 4\pi$ (Use a calculator.)
28. $x = \sin^2 2t, y = \cos^2 3t, 0 \le t \le 2\pi$ (Use a calculator.)

7.7 WORK AND HYDROSTATIC FORCE

WORK

In Section 6.1, we motivated the definite integral by considering work done by a nonconstant force. We review this application now.

If a body is moved a distance s by means of a constant force F in the direction of the motion, then the work W done in moving the body is the product Fs. For example, if a body is being pushed along the x-axis by a constant force of 20 lb directed along the axis, then the work done by the force on the body in pushing from position $x = 0$ to position $x = 10$ is $W = 20 \cdot 10 = 200$ ft · lb.

If the force acting on the body does not remain constant but is a continuous function $F(x)$ of the position x of the body, then the work done over a short interval of length dx is approximately $F(x)\,dx$, for some position x in this interval. Adding all these contributions to the work from starting position $x = a$ to stopping position $x = b$, we obtain

$$W = \int_a^b F(x)\,dx. \tag{1}$$

EXAMPLE 1 By Hooke's law, the force F required to stretch (or compress) a coil spring is proportional to the distance x it is stretched (or compressed) from its natural length. That is, $F = kx$ for some constant k, the *spring constant*. Suppose that a spring is such that the force required to stretch it 1 ft from its natural length is 4 lb. For this spring, $k = 4$. Find the work done in stretching the spring a distance of 4 ft from its natural length.

Solution Work is defined as the product of force and distance, if the force remains constant and is in the direction of the motion. Thus as our spring is stretched an additional small distance dx at a distance x from its natural length, the work done is approximately $F\,dx = 4x\,dx$. We wish to add all these little pieces of work from $x = 0$ to $x = 4$ and take the limit of the resulting sum as $dx \to 0$. The appropriate integral is

$$\int_0^4 4x\,dx = 4\frac{x^2}{2}\bigg]_0^4 = \frac{64}{2} - 0 = 32.$$

Thus 32 ft·lb of work are done. □

EXAMPLE 2 Two electrons a distance s apart repel each other with a force k/s^2, where k is some constant. If an electron remains fixed at the origin on the x-axis, find the work done by the force of repulsion in moving another electron from the point 2 to the point 5 on the axis.

Figure 7.56 Repelling force on the moving electron is k/x^2.

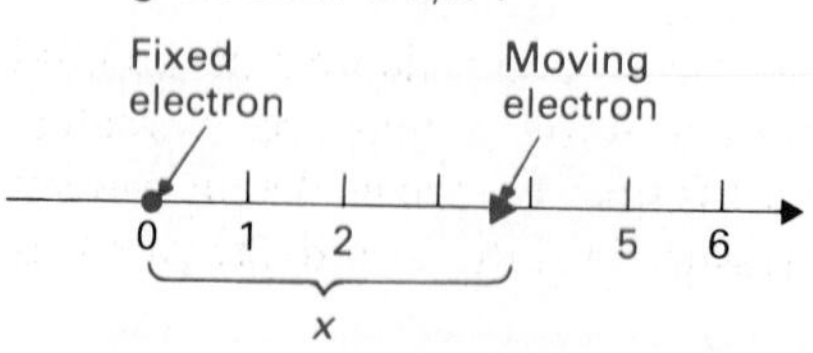

Solution We sketch the situation in Fig. 7.56. When the second electron is at a point x, the force of repulsion is k/x^2. The work done in moving the second electron an additional small distance dx is approximately $(k/x^2)\,dx$. Thus the total work done is

$$W = \int_2^5 \frac{k}{x^2}\,dx = -\frac{k}{x}\bigg]_2^5 = -\frac{k}{5} - \left(-\frac{k}{2}\right) = k\left(\frac{1}{2} - \frac{1}{5}\right) = \frac{3k}{10} \text{ units.} \quad \Box$$

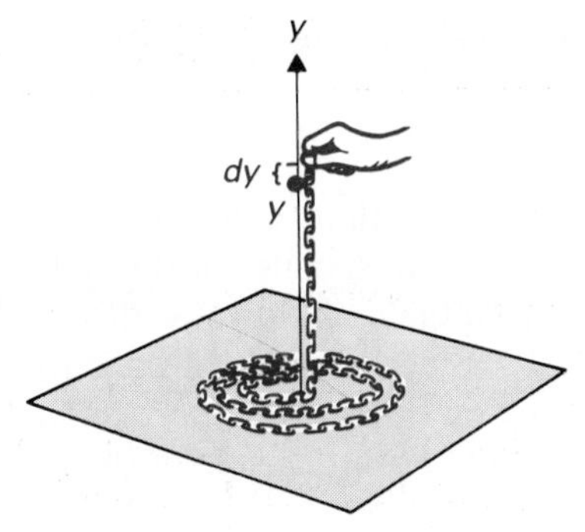

Figure 7.57 Chain weighing 2 lb/ft lifted by a force $2y$ when the end is y ft off the floor; $dV = (2y)\,dy$.

EXAMPLE 3 Suppose a long, heavy, flexible chain lies coiled on the floor. If the chain weighs 2 lb/ft, estimate the work W done in raising one end of the chain vertically 4 ft off the floor.

Solution Since the chain weighs 2 lb/ft, we lift with the force

$$F(y) = 2y \text{ lb}$$

when the end of the chain is y ft off the floor, as shown in Fig. 7.57. The work done as the chain is lifted an additional dy ft off the floor is therefore approximately $(2y)\,dy$. We add these small contributions to the work, taking the limit of the sum as $dy \to 0$, using an integral. Since the end of the chain is lifted 4 ft, the appropriate integral is

$$\int_0^4 (2y)\,dy = y^2\Big]_0^4 = 16 - 0 = 16 \text{ ft} \cdot \text{lb}. \quad \square$$

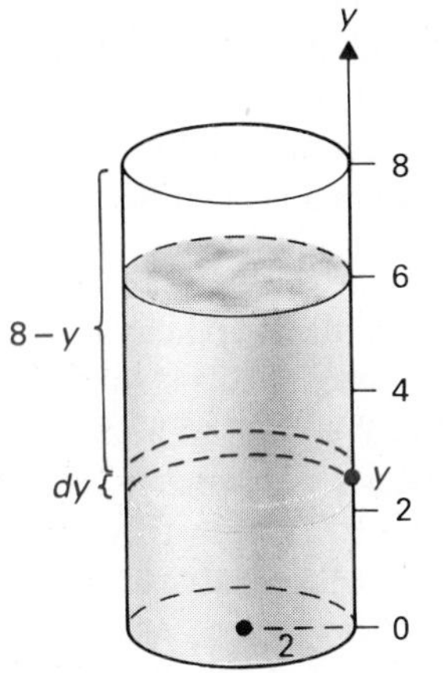

Figure 7.58 The circular slab of water weighs $62.4\,(4\pi)\,dy$ lb, and must be raised $8 - y$ ft to the top of the tank.

EXAMPLE 4 A cylindrical tank of radius 2 ft and height 8 ft is filled with water to a height of 6 ft. The tank is emptied by dropping in a hose, pumping all the water up to the top of the tank, and letting it spill over. If water weighs 62.4 lb/ft^3, find the work done to empty the tank.

Solution We take a y-axis upward from the bottom of the tank, as shown in Fig. 7.58. The circular slab of water at height y and thickness dy shown in the figure has volume

$$dV = \pi 2^2\,dy = 4\pi\,dy \text{ ft}^3.$$

This slab then weights $62.4(4\pi)\,dy$ lb. Therefore a force of $62.4(4\pi)\,dy$ lb is required to raise it, and it must be raised the distance $(8 - y)$ ft to the top of the tank. The work done on such slabs from $y = 0$ to $y = 6$ ft is summed in the usual fashion using an integral. The total work is given by

$$W = \int_0^6 (8 - y)(62.4)(4\pi)\,dy = 249.6\pi \int_0^6 (8 - y)\,dy$$

$$= 249.6\pi\left(8y - \frac{y^2}{2}\right)\Big]_0^6 = 249.6\pi(48 - 18)$$

$$= 249.6\pi(30) = 7488\pi \text{ ft} \cdot \text{lb}. \quad \square$$

Figure 7.59 The column of water s ft high and 1 ft square weighs $62.4s$ lb.

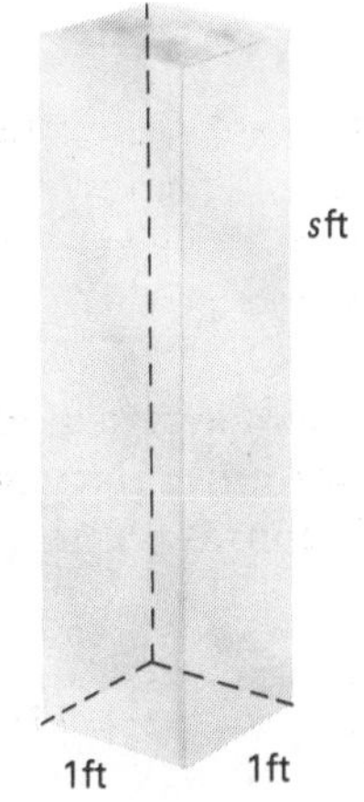

HYDROSTATIC FORCE

If the pressure per square unit on a plane region of area A is a constant p throughout the region, then the total force F on the region due to the pressure is the product pA. Suppose the pressure does not remain constant throughout the region. We then attempt to find the total force by integrating $p\,dA$ where dA is the area of a small piece of the region throughout which the pressure remains roughly constant.

The pressure exerted by the weight of a fluid at a depth s in the fluid acts equally in all directions. A cubic foot of water weighs approximately 62.4 lb, so a column of water s ft high and 1 ft square, such as the one shown in Fig. 7.59, weighs $62.4s$ lb. Thus the pressure in water at a depth of s ft is approximately $62.4s$ lb/ft^2.

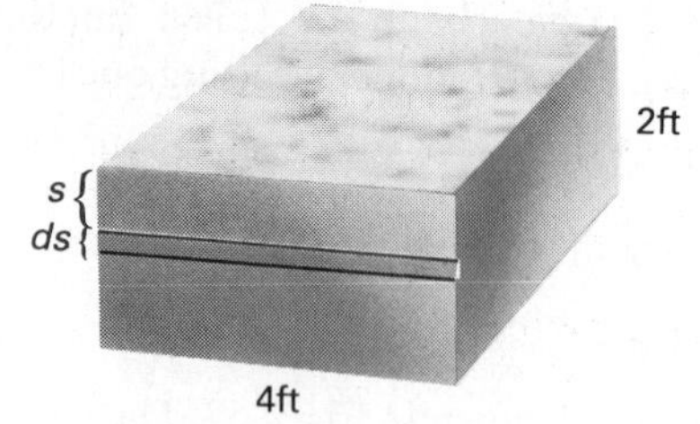

Figure 7.60
Pressure $p = 62.4s$;
Area of strip $= 4\,ds$;
Force on strip $= 62.4s(4)\,ds$

EXAMPLE 5 One side of a rectangular tank measures 4 ft wide by 2 ft high. If the tank is filled with water, find the total force of the water on this side of the tank.

Solution Consider a small horizontal strip of width ds across the side of the tank, as shown in Fig. 7.60. The pressure is almost constant throughout this strip. Let s be the distance from the top of the tank down to this strip. The pressure at that depth is $p = 62.4s\ \text{lb/ft}^2$. Since the area of the strip is $dA = 4\,ds$, the force dF due to water pressure on this strip is approximately $dF = p\,dA = 62.4s(4)\,ds$. To find the total force F on this side of the tank, we integrate, obtaining

$$F = \int_0^2 62.4s(4)\,ds = 4(62.4)\int_0^2 s\,ds = 4(62.4)\left.\frac{s^2}{2}\right]_0^2$$

$$= 4(62.4)(2 - 0) = 8(62.4) = 499.2\ \text{lb.} \quad \square$$

EXAMPLE 6 Suppose a dam 16 ft high has the shape of the region bounded by the curves with equations $y = x^2$ and $y = 16$. Find the total force due to water pressure on the dam when the water is at the top of the dam. (It can be shown that this force depends only on the height of the water at the dam and not on the distance the water is backed up behind the dam.)

Solution The region bounded by the curves $y = x^2$ and $y = 16$ is shaded in Fig. 7.61. For small dy, the pressure at a depth s on the strip heavily shaded in Fig. 7.61 is nearly constant at $62.4s\ \text{lb/ft}^2$. The area of this strip is $2x\,dy$, so the force on it is approximately $62.4s(2x)\,dy$ lb. We want to add up these quantities $62.4s(2x)\,dy$ over the region as y ranges from 0 to 16 and take the limit of the resulting sum as $dy \to 0$. We must express this quantity entirely in terms of y. For the point (x, y) in Fig. 7.61, we have $s = 16 - y$, and $y = x^2$, so $x = \sqrt{y}$. Thus

$$62.4s(2x)\,dy = 62.4(16 - y)2\sqrt{y}\,dy = 124.8(16y^{1/2} - y^{3/2})\,dy.$$

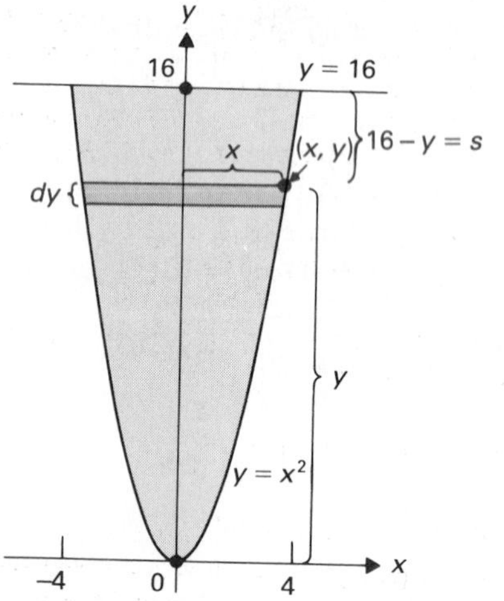

The appropriate integral is

$$\int_0^{16} 124.8(16y^{1/2} - y^{3/2})\,dy = 124.8\left(16\left(\frac{2}{3}\right)y^{3/2} - \frac{2}{5}y^{5/2}\right)\Bigg]_0^{16}$$

$$= 124.8\left(16\left(\frac{2}{3}\right)64 - \left(\frac{2}{5}\right)1024\right) - 0 = 124.8(1024)\left(\frac{2}{3} - \frac{2}{5}\right)$$

$$= 124.8(1024)\frac{4}{15} = 34{,}078.72.$$

Thus the force is 34,078.72 lb. $\square$

Figure 7.61 Force on the strip $= 62.4(16 - y)(2x)\,dy$
$= 62.4(16 - y)(2\sqrt{y})\,dy$.

SUMMARY

1. If the force acting on a body in the direction of its motion along a line is the function $F(x)$ of the position x of the body, then the work done by the force in moving the body from position $x = a$ to position $x = b$ is

$$W = \int_a^b F(x)\,dx.$$

2. The pressure exerted by water at depth s is about $62.4s$ lb/ft^2. Let a submerged vertical plate have its top at a depth a and its bottom at depth b. If the area of a thin horizontal strip of the plate at depth s is dA, then

$$\text{Total force on one side of the plate} = \int_a^b 62.4s\, dA.$$

EXERCISES

1. Find the work done in stretching the spring of Example 1 from 2 ft longer than its natural length to 6 ft longer than its natural length.

2. If a spring has a natural length of 2 ft and a force of 10 lb is required to compress the spring 2 in. (to a length of 22 in.), find the work done in stretching the spring from a length of 26 in. to a length of 30 in.

3. If 36 ft · lb of work are done in stretching a spring 3 ft from its natural length, find the spring constant k.

4. A force of 10 lb stretches a spring 2 ft. The spring is stretched from its natural length until 90 ft · lb of work has been done. How far was the spring stretched?

5. A flexible chain 6 ft long and weighing 2 lb/ft lies coiled on the floor. Find the work done when one end of the chain is raised vertically
a) 3 ft, b) 6 ft.

6. Answer Exercise 5 if the end of the chain is raised vertically 8 ft.

7. A cylindrical tank of radius 2 ft and height 10 ft is full of water and is emptied by dropping in a hose, pumping all the water up to the top of the tank, and letting it spill over. Find the work done to empty the tank.

8. Work Exercise 7 for a conical reservoir of height 10 ft and top radius 2 ft.

9. A rectangular tank of water has a base measuring 4 ft by 8 ft and is 4 ft high. The tank is emptied by pumping the water to the top and letting it spill over the side. If 14,976 ft · lb of work are needed, how deep was the water in the tank?

10. A bucket of cement weighing 100 lb is winched to a height of 30 ft using a cable weighing 1 lb/ft. Find the work done.

11. Work Exercise 10 if the bucket has a hole in it and cement falls out at a uniform rate of $\frac{1}{2}$ lb for each foot it is raised.

12. Two electrons a distance s apart repel each other with a force k/s^2 where k is some constant. If an electron is at the point 2 on the x-axis, find the work done by the force in moving another electron from the point 4 to the point 8.

13. Referring to Exercise 12, suppose one electron is fixed at the origin on the x-axis and another is fixed at the point 1. Find the work done by the forces of repulsion in moving a third electron from the point 2 to the point 6.

14. According to Newton's law of gravitation, the force of attraction of two bodies of masses m_1 and m_2 is $G(m_1m_2/s^2)$, where s is the distance between the bodies and G is the gravitational constant. If the distance between two bodies of masses m_1 and m_2 is a, find the work done in moving the bodies twice as far apart.

15. Let a body of mass m move in the positive direction on the x-axis subject only to a force F acting in the positive direction along the axis. (The magnitude of the force may vary with the time t.) If the body has velocity v, the *kinetic energy* of the body is $\frac{1}{2}mv^2$. Show that the work done by the force from time t_1 to time t_2, for $t_1 < t_2$, is equal to the change in the kinetic energy of the body between these times. [*Hint:* By Newton's second law of motion,

$$F = ma = m\frac{dv}{dt} = m\frac{dv}{dx}\cdot\frac{dx}{dt} = m\frac{dv}{dx}v.$$

Consider both F and v as functions of the position x_t of the body at time t, and express the work

$$W = \int_{x_{t_1}}^{x_{t_2}} F(x)\, dx$$

as an integral involving v, using Newton's law.]

16. A rectangular plate measuring 1 ft by 2 ft is suspended vertically in water with the 2-ft edges horizontal and the upper one at a depth of 5 ft. Find the force of the water on one side of the plate. (Of course, the force on the other side is equal, but in the opposite direction, so the plate does not move.)

17. Work Exercise 16 for a vertically suspended circular plate of radius 2 ft if the top of the plate is at a depth of 4 ft.

18. The top of a rectangular swimming pool measures 20 ft by 40 ft. The pool is 1 ft deep at one end and slopes uniformly to a depth of 11 ft at the other end. Find the force of the water on the bottom of the pool when it is filled.

19. Find the force of the water on one side of the pool in Exercise 18.

20. A vertical dam is in the shape of a semicircle of radius 36 ft, with the diameter of the circle at the top of the dam. Find the force on the dam when the water level is at the top of the dam.

21. A vertical dam is in the shape of an isosceles trapezoid with the long 60-ft base at the top of the dam and the short 30-ft

base at the bottom. The altitude of the trapezoid is 21 ft. Find the force of the water on the dam when the water is at the top of the dam.

22. A cylindrical drum of diameter 2 ft and length 4 ft is filled with water. Find the force on one end of the drum if the drum is lying on its side.

23. For the drum in Exercise 22, find the force on the cylindrical wall of the drum if the drum is standing on end.

24. A sphere of radius 2 ft is submerged in water so that its top is at a depth of 10 ft. Find the force of the water on the surface of the sphere. [*Hint:* Don't try to find some expression to integrate, but use symmetry and a calculus-type argument. For a small piece of surface of area dA on the upper hemisphere at depth $12 - x$, there is a symmetrically located piece of surface of equal area dA at depth $12 + x$ on the lower hemisphere. Find the total force on these two pieces, and then "sum" over the upper hemisphere in the integral sense.]

25. A cylindrical drum 3 ft in diameter and 5 ft long is submerged horizontally in water so that its top is at a depth of 4 ft. Use the method suggested in the hint of Exercise 24 to find the force of water on the cylindrical wall of the drum.

26. For the drum in Exercise 25, find the force of water on one end of the drum, using the method suggested by Exercise 24.

7.8 MASS AND MOMENTS

The *mass* of a body is a numerical measure of the "amount of material" it contains. In the first part of this section, we are interested in bodies that lie in a plane, such as a thin plate or a piece of wire.

The mass density ρ of a flat plate at a point (x, y) is the mass that a piece of the plate of area 1 would have if the plate had everywhere the same composition and thickness as at the point (x, y). A piece of wire has mass density *per unit length* rather than *per unit area.* A body is *homogeneous* if it is of uniform composition throughout, so that its mass density is *constant.* In this case, the mass of a plate or of a wire can be computed by a *product:*

$$\text{Mass of a plate} = \text{Mass density} \cdot \text{Area of the plate},$$

$$\text{Mass of a wire} = \text{Mass density} \cdot \text{Length of the wire}.$$

In general, mass density is a function $\rho(x, y)$ of both x and y, but in this section we will be concerned only with the case where it is either a function $\rho(x)$ of x alone or $\rho(y)$ of y alone. Three-dimensional bodies and more general mass-density functions are treated in Chapter 18. We will use integral calculus to find the mass of a body of *nonconstant* mass density.

MASS

Figure 7.62 Differential strip of the plate has area dA and mass $dm = \rho(x)\, dA$.

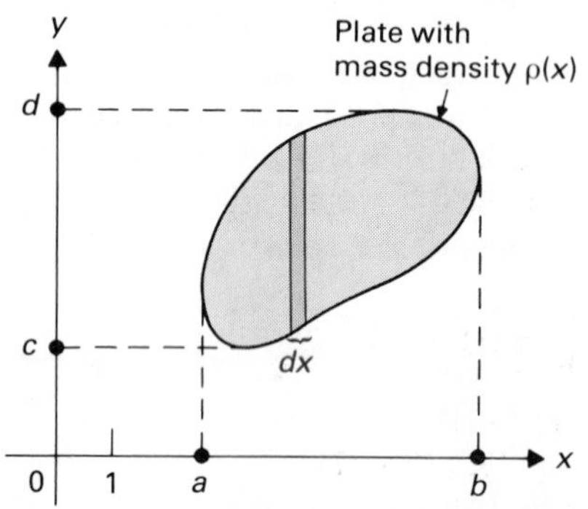

Suppose the mass density of a plate is a continuous function $\rho(x)$ of x only. Then the density varies very little over a sufficiently thin *vertical* strip of the plate of width dx, shown in Fig. 7.62. This is because x-values vary very little over the strip. Thus if the strip has area dA, the mass of the strip is approximately $dm = \rho(x)\, dA$. To find the total mass, we then sum the masses of these thin strips and take the limit of the sums as the strip widths approach zero. For the plate in Fig. 7.62, this leads to the integral formula

$$\boxed{\text{Mass} = m = \int_a^b \rho(x)\, dA.} \qquad \textbf{(1)}$$

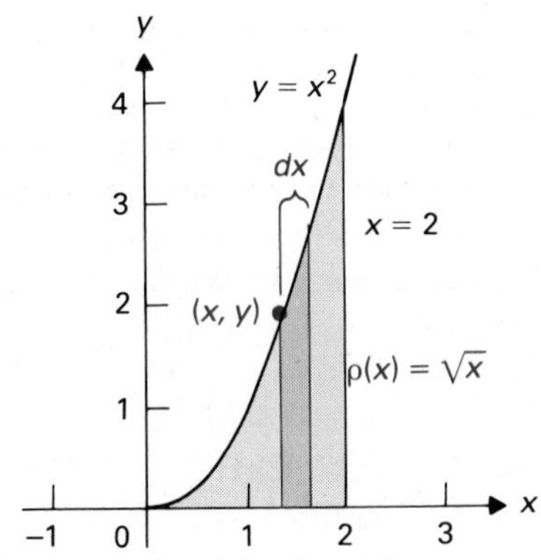

Figure 7.63 Differential strip of the plate has approximate mass $dm = \sqrt{x}(x^2)\,dx$.

Of course, if the mass density of the plate is a continuous function $\rho(y)$ of y only, we take thin *horizontal* strips throughout which y remains roughly constant. Obviously, we then have

$$\text{Mass} = m = \int_c^d \rho(y)\,dA. \tag{2}$$

EXAMPLE 1 Consider a thin sheet of material exactly covering the plane region bounded by the curves $y = x^2$, $x = 2$, and $y = 0$, as shown in Fig. 7.63. Find the mass of the sheet if the mass density at a point (x, y) on the sheet is $\sqrt{x}$.

Solution The mass density is approximately constant with value $\sqrt{x}$ throughout the strip with the color shading in Fig. 7.63. The strip has approximate area $x^2\,dx$, and hence approximate mass $dm = \sqrt{x}(x^2)\,dx$. We wish to add these masses as x goes from 0 to 2 and as $dx \to 0$. The total mass is therefore

$$m = \int_0^2 \sqrt{x}(x^2)\,dx = \int_0^2 x^{5/2}\,dx = \frac{2}{7}x^{7/2}\Big]_0^2$$

$$= \frac{2}{7}\cdot 2^{7/2} - 0 = \frac{16}{7}\sqrt{2}\text{ units.} \quad \square$$

In a similar fashion, if we have a piece of wire lying in the plane, we can find the mass of the wire if its mass density ρ is a function of either x only or of y only. The mass of a very small length of the wire corresponding to a small tangent-line segment of length ds is approximately $\rho\,ds$. To find the mass of the entire wire, we simply integrate $\rho\,ds$. Use the form of ds ending with dx if ρ is a function of x only and ending with dy if ρ is a function of y only.

EXAMPLE 2 Let a wire lie on the curve $x = y^2$ from $y = 0$ to $y = 2$ and have mass density ky for a constant k. Find the mass of the wire.

Solution We have

$$ds = \sqrt{\left(\frac{dx}{dy}\right)^2 + 1}\,dy = \sqrt{4y^2 + 1}\,dy,$$

so

$$m = \int_0^2 ky\sqrt{4y^2 + 1}\,dy = \frac{k}{8}\int_0^2 8y(4y^2 + 1)^{1/2}\,dy$$

$$= \frac{k}{8}\cdot\frac{2}{3}(4y^2 + 1)^{3/2}\Big]_0^2$$

$$= \frac{k}{12}(17\sqrt{17} - 1). \quad \square$$

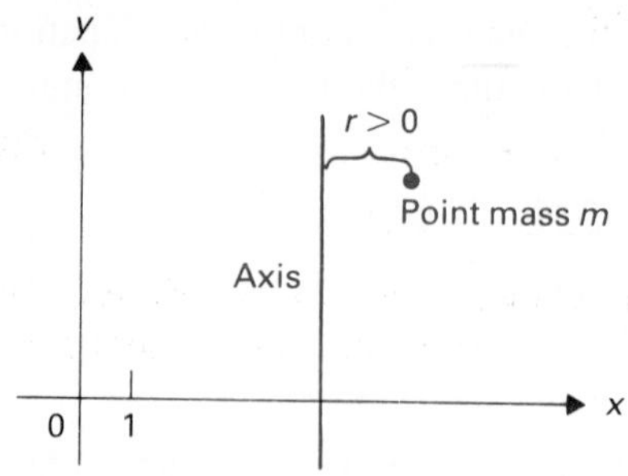

Figure 7.64 A point mass of magnitude m a signed distance r from an axis.

MOMENTS

A line (axis) in the plane divides the plane into two pieces. We will speak of these two pieces as being on one side or the other side of the axis, much as we speak of one side or the other side of a road. Let one side of the axis be designated as the *positive side* and the other as the *negative side*. We use the convention that for a vertical axis, the positive side is on the right, while for a horizontal axis, the positive side is above the axis.

Consider a point mass of magnitude m in the plane, and let r be the *signed* distance from the point mass to some particular axis, as shown in Fig. 7.64. That is, r is positive if the point mass is on the positive side of the axis and negative if it is on the negative side. The **first moment** M, or simply the **moment**, of the point mass about the axis is

$$M = mr.$$

The moment of a body consisting of several points of mass is the sum of the moments of the individual point masses.

Children encounter a problem in moments when they play on a seesaw. Figure 7.65 shows a seesaw consisting of a fiberglass plank of negligible mass supported on a fulcrum. A small child of mass m sits on the left end of the plank, and a larger child of mass M sits on the right end. For the choice of axes and coordinates shown in the figure, the sum of the moments of the children about the y-axis must be zero if the seesaw is to be properly balanced. That is, we must have

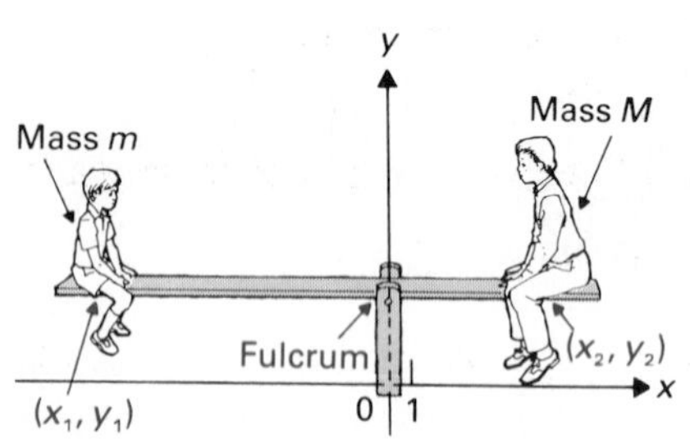

Figure 7.65 For a balanced seesaw, $mx_1 + Mx_2 = 0$, or $m|x_1| = M|x_2|$.

$$mx_1 + Mx_2 = 0.$$

Recall that x_1 is negative and x_2 is positive, so we must have

$$m|x_1| = M|x_2|.$$

Thus if $M = 2m$, we must have $|x_1| = 2|x_2|$, so the plank should be slid along the fulcrum so that the smaller child is twice as far from the fulcrum as the larger child.

EXAMPLE 3 Let a body consist of a point mass of 3 at (1, 4), a point mass of 2 at $(-2, 1)$, and a point mass of 5 at $(3, -4)$. Find (a) the moment M_x of the body about the x-axis and (b) the moment M_y of the body about the y-axis.

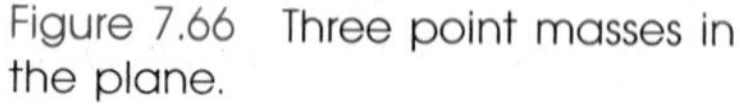

Figure 7.66 Three point masses in the plane.

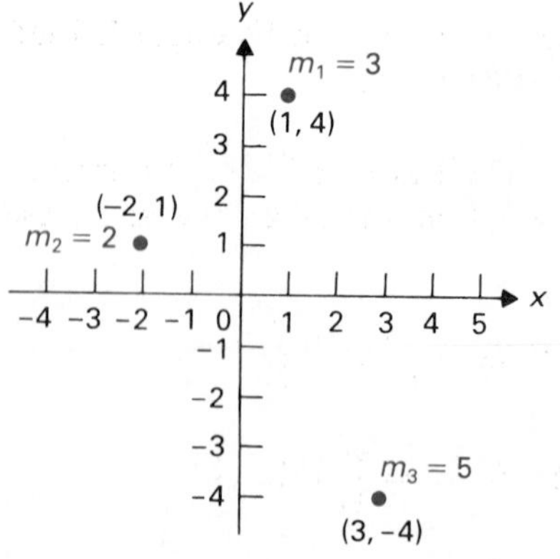

Solution The body is shown in Fig. 7.66. For part (a), the moment M_x about the x-axis is

$$M_x = 3(4) + 2(1) + 5(-4) = -6.$$

For part (b), the moment M_y about the y-axis is

$$M_y = 3(1) + 2(-2) + 5(3) = 14. \quad \square$$

To find the moment about an axis of a thin plate of mass density ρ in the plane, we take a thin strip of the plate *parallel* to the axis and of area dA, as in Fig. 7.67. If ρ is such that it is approximately constant over the strip (for example, a function of x only if the strip is vertical of width dx), then the mass of the strip is approximately $dm = \rho\, dA$. The moment of the strip about the axis is then $r\, dm$, where r is the *signed* distance from the strip to

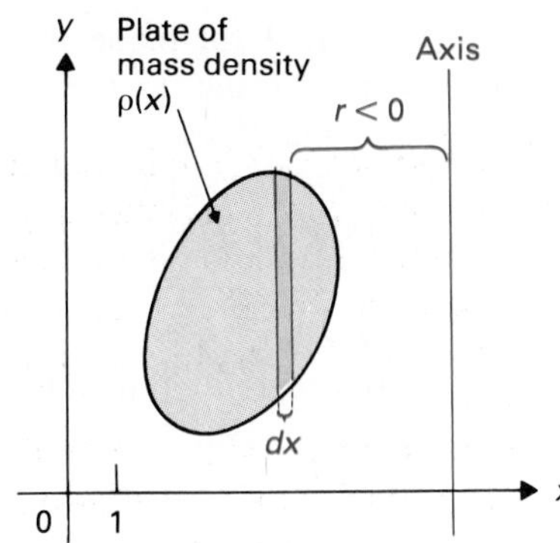

Figure 7.67 The differential strip of plate of area dA has mass $dm = \rho(x)\,dA$ and moment $r\rho(x)\,dA$ about the axis.

the axis. We then have, for the entire plate,

$$\textbf{First moment} = M = \int_a^b r\,dm \tag{3}$$

for appropriate limits a and b. If the body is a wire, then $dm = \rho\,ds$. We will often be interested in the moments M_x about the x-axis and M_y about the y-axis.

EXAMPLE 4 Find the moment M_y about the y-axis of the sheet of material described in Example 1.

Solution All points of the color-shaded strip in Fig. 7.63 are approximately the same distance $r = x$ from the y-axis if dx is small. Since the mass of this strip is approximately $\sqrt{x}(x^2)\,dx = x^{5/2}\,dx$, we see that the moment of this strip about the y-axis is approximately $x(x^{5/2})\,dx$. We wish to add the moments of these strips as x goes from 0 to 2 and as $dx \to 0$. The moment M_y is therefore given by

$$M_y = \int_0^2 x(x^{5/2})\,dx = \int_0^2 x^{7/2}\,dx = \frac{2}{9}x^{9/2}\Big]_0^2 = \frac{2}{9}\cdot 2^{9/2} - 0 = \frac{32}{9}\sqrt{2}. \quad \square$$

The second moment (or moment of inertia) I of one of these plane bodies about an axis is given by

$$\textbf{Second moment} = I = \int_a^b r^2\,dm, \tag{4}$$

and, in general,

$$\textbf{\textit{n}th moment} = \int_a^b r^n\,dm. \tag{5}$$

The second moment I appears in the formula

$$\text{K.E.} = \tfrac{1}{2}I\omega^2 \tag{6}$$

for the kinetic energy of rotation about an axis of a body with moment of inertia I and angular speed of rotation ω. Thus if we wish to double the kinetic energy of a flywheel rotating with a fixed angular speed ω, we might modify the flywheel so that the mass is moved $\sqrt{2}$ times as far from the axis.

EXAMPLE 5 Consider again the sheet of material in Example 1. Find the moment of inertia I_y of this sheet about the y-axis.

Solution We should multiply the mass $\sqrt{x}(x^2)\,dx = x^{5/2}\,dx$ of the strip in Fig. 7.63 by the square of its distance $r = x$ from the y-axis and add these contributions as $dx \to 0$. We arrive at

$$I_y = \int_0^2 x^2(x^{5/2})\,dx = \int_0^2 x^{9/2}\,dx = \frac{2}{11}x^{11/2}\Big]_0^2$$

$$= \frac{2}{11}\cdot 2^{11/2} - 0 = \frac{64}{11}\sqrt{2}. \quad \square$$

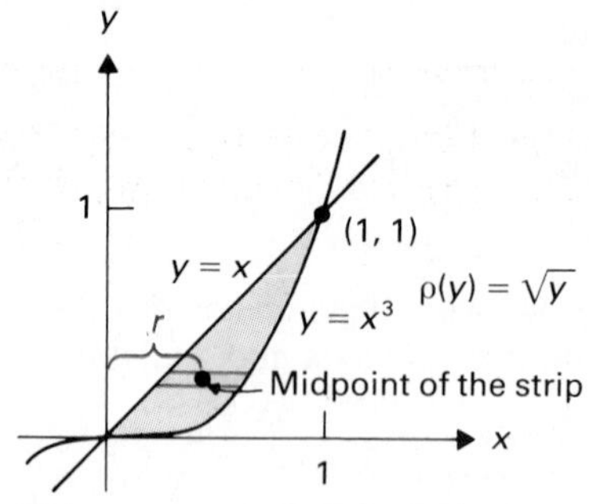

Figure 7.68 This differential strip of mass $dm = \rho\, dA$ has moment $r\rho\, dA$ about the y-axis.

Perhaps we want to find the moment M_y of a plate in the plane about the y-axis, and it is inconvenient to use vertical strips parallel to the axis. Perhaps the mass density $\rho(y)$ is a function of y only and does not remain constant through such a thin vertical strip. We are then forced to use horizontal strips, and the next two examples demonstrate the technique. We ask you to justify the technique in Exercise 14.

EXAMPLE 6 Let a plate cover the first-quadrant region bounded by $y = x^3$ and $y = x$, and let the mass density at a point (x, y) be $\sqrt{y}$. Find the moment M_y about the y-axis.

Solution The mass of the thin horizontal strip shown in Fig. 7.68 is approximately $dm = \rho\, dA = \sqrt{y}(y^{1/3} - y)\, dy$. We consider this mass to be concentrated at the *midpoint* of the strip. We then take as r the x-coordinate of this midpoint, so that

$$r = \frac{x_{\text{right}} + x_{\text{left}}}{2} = \frac{y^{1/3} + y}{2}.$$

The contribution of the strip to the moment M_y is then

$$\frac{y^{1/3} + y}{2} \cdot \sqrt{y}(y^{1/3} - y)\, dy = \frac{\sqrt{y}}{2}(y^{2/3} - y^2)\, dy = \frac{1}{2}(y^{7/6} - y^{5/2})\, dy.$$

Thus the moment is

$$M_y = \int_0^1 \frac{1}{2}(y^{7/6} - y^{5/2})\, dy = \frac{1}{2}\left(\frac{6}{13}y^{13/6} - \frac{2}{7}y^{7/2}\right)\Bigg]_0^1$$

$$= \frac{1}{2}\left(\frac{6}{13} - \frac{2}{7}\right) = \frac{3}{13} - \frac{1}{7} = \frac{8}{91}. \quad \square$$

This technique of considering the mass of a strip to be concentrated at the midpoint of the strip *cannot* be used to compute moments of inertia. *It is valid for first moments only!*

EXAMPLE 7 Find the moment M_x about the x-axis of a plate of constant mass density 3 covering the region bounded by $y = 0$ and $y = \sin x$ for $0 \le x \le \pi$, as shown in Fig. 7.69.

Solution We consider the mass

$$dm = \rho\, dA = 3 \sin x\, dx$$

to be concentrated at the midpoint of the strip, where the distance from the x-axis is $y/2 = (\sin x)/2$. Thus, using the table of integrals, we have

$$M_x = \int_0^\pi \frac{\sin x}{2}(3 \sin x)\, dx = \frac{3}{2}\int_0^\pi \sin^2 x\, dx = \frac{3}{2}\left(\frac{x}{2} - \frac{\sin 2x}{4}\right)\Bigg]_0^\pi$$

$$= \frac{3}{2}\left(\frac{\pi}{2} - 0\right) - \frac{3}{2}(0 - 0) = \frac{3\pi}{4}. \quad \square$$

Figure 7.69 The differential strip contributes

$(y/2)\rho\, dA = [\sin x)/2]3(\sin x)\, dx$

to M_x.

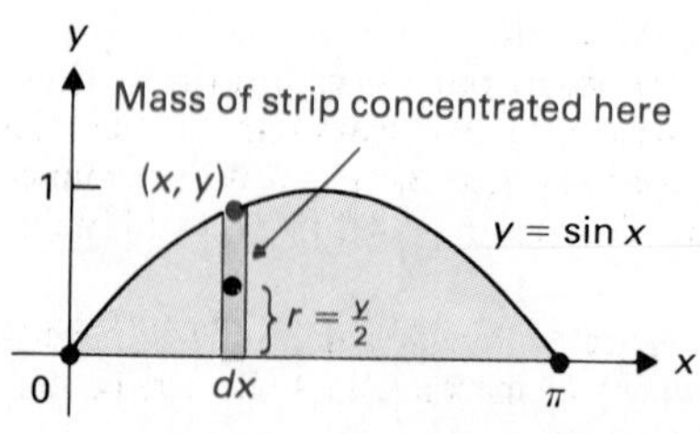

SUMMARY

For mass density ρ and area dA a signed (positive or negative) distance r from an axis, the following properties hold, where a and b are appropriate limits.

1. Total mass $= m = \int_a^b \rho \, dA$
2. First moment $= M = \int_a^b r\rho \, dA$
3. Second moment (of intertia) $= I = \int_a^b r^2\rho \, dA$
4. nth moment $= \int_a^b r^n\rho \, dA$

EXERCISES

In Exercises 1 through 6, find the moment of the body consisting of the given point masses about the indicated axis.

1. Mass 1 at $(-1, 2)$, 4 at $(1, 3)$, 2 at $(-3, -1)$, and 5 at $(4, -2)$ about the y-axis
2. Masses as in Exercise 1 about the x-axis
3. Mass 10 at $(0, 0)$, 4 at $(1, 3)$, 2 at $(-1, 4)$, 5 at $(-2, -3)$, and 1 at $(3, -1)$ about the x-axis
4. Masses as in Exercise 3 about the y-axis
5. Mass 3 at $(-1, 5)$, 4 at $(1, -3)$, 2 at $(3, 2)$, and 5 at $(-4, -1)$ about the line $x = -3$
6. Masses as in Exercise 5 about the line $y = 2$
7. Let a rectangular plate of mass density $\rho(x) = x$ at (x, y) cover the region bounded by $x = 1, x = 3, y = 2$, and $y = 6$. Find the mass of the plate.
8. Find the moment of the plate in Exercise 7 about the y-axis.
9. Find the moment of the plate in Exercise 7 about the line $x = 2$.
10. Find the moment of inertia of the plate in Exercise 7 about the line $x = -1$.
11. Find the moment of the plate in Exercise 7 about the x-axis. [*Hint:* Use the technique of Examples 6 and 7.]
12. Find the moment of the plate in Exercise 7 about the line $y = -1$. [*Hint:* Use the technique of Examples 6 and 7.]
13. Find the moment of the plate in Exercise 7 about the line $y = 4$. [*Hint:* Use the technique of Examples 6 and 7.]
14. This exercise justifies the technique used in Example 6 and 7, where we considered the mass of a differential strip to be concentrated at its midpoint when computing a first moment. Consider a thin rod in the plane placed perpendicular to the y-axis, as shown in Fig. 7.70. Suppose the rod lies above the interval $[a, b]$ on the x-axis and has a constant mass density k per unit length. Figure 7.70 shows two small pieces of rod of equal length dx symmetrically positioned about the midpoint of the rod and at a distance s from the midpoint.
    a) Find
        i) the mass of each of these differential pieces,
        ii) the contribution of each differential piece to the moment about the y-axis,
        iii) the sum of the two contributions in part (ii).
    b) Make an integral-style argument from part (a) that the moment of the entire rod about the y-axis is the product of its mass and the distance $(a + b)/2$ from its midpoint to the y-axis.

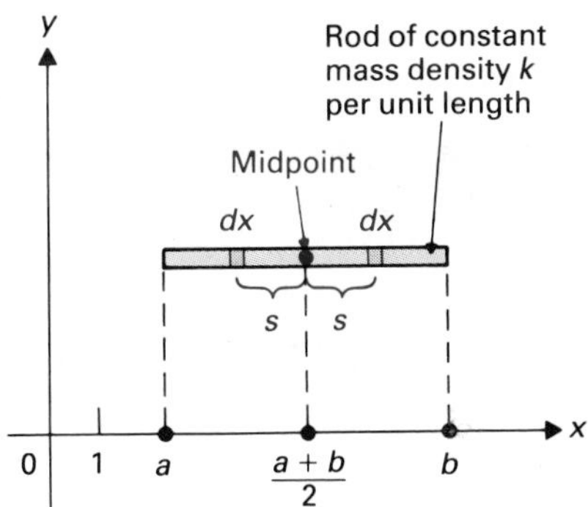

Figure 7.70 Two differential pieces of a rod, located symmetrically about its midpoint.

15. Repeat part (a) of Exercise 14, but find in part (ii) the contributions to the moment of inertia about the y-axis, and

find in part (iii) the sum of these contributions to the moment of inertia. Compare with the results in Exercise 14. This indicates why the mass cannot be considered to be concentrated at the midpoint of a strip in computing second or higher moments about an axis.

16. The triangular region with vertices $(0, 0)$, $(a, 0)$, and $(0, b)$ for $a > 0$ and $b > 0$ is covered by a sheet of material whose mass density at a point (x, y) is ky^2.
   a) Find the mass of the sheet.
   b) Find the first moment of the sheet about the x-axis.

17. Let the mass density at a point (x, y) of a plate covering the semicircular disk bounded by the x-axis and $y = \sqrt{1 - x^2}$ be $2(y + 1)$. Find the mass of the plate.

18. A metal plate is cut from a thin sheet of metal of constant mass density 3. If the plate just covers the first-quadrant region bounded by the curves $y = x^2$, $y = 1$, and $x = 0$, find the first moment of the plate about the y-axis.

19. For the plate in Exercise 18, find its first moment about the axis with equation $x = 2$.

20. For the plate in Exercise 18, find its moment of inertia about the x-axis.

21. Consider a thin rod of length a with constant mass density k (mass per unit length).
   a) Find the first moment of the rod about an axis through an endpoint and perpendicular to the rod.
   b) Find the moment of inertia of the rod about the axis described in part (a).

22. Repeat Exercise 21 in case the mass density is kx^2, where x is the distance along the rod from the axis of rotation.

23. Let a wire lie on $y = x^3$ from $x = 0$ to $x = 2$ and have constant mass density $\rho = 5$. Find the moment M_x of the wire about the x-axis.

24. Let a wire lie on $x = y^{3/2}$ from $y = 1$ to $y = 4$ and have mass density $\rho(y) = 3\sqrt{y}$. Express as an integral the first moment M_y of the wire about the y-axis.

25. Find the moment of inertia of a flat disk of radius a and constant mass density k about an axis perpendicular to the disk through its center. [*Hint:* Add the moments of inertia of concentric circular "washers" by integration.]

26. Express as an integral the moment of inertia of a ball of radius a and constant mass density k (mass per unit volume) about a diameter of the ball. [*Hint:* Add the moments of inertia of cylindrical shells with the diameter as axis.]

27. Show that the first moment of a body in the plane about the line $x = -a$ is $M_y + ma$, where M_y is the moment about the y-axis. (This is known as the *parallel axis theorem.*)

Use a calculator for Exercises 28 through 30.

28. Estimate the moment in Exercise 24.

29. Estimate the moment in Exercise 26 if $a = 4$.

30. A plate of mass density kx at (x, y) covers the region bounded by the x-axis and $y = \sin x$ for $0 \leq x \leq \pi$. Estimate the moment of the plate about the line $y = 3$.

7.9 CENTER OF MASS, CENTROIDS, PAPPUS' THEOREM

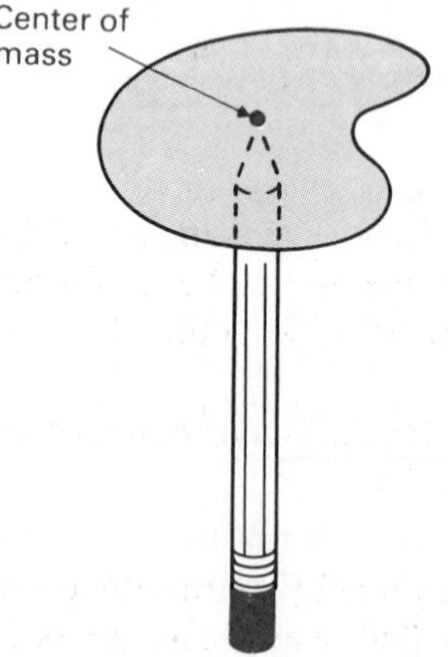

Figure 7.71 The center of mass is the balance point.

It can be proved that for a given body of mass m, there exists a unique point (not necessarily in the body) at which we may consider the entire mass of the body to be concentrated for the computation of *first moments* about *every* axis. This point is the **center of mass** (or **center of gravity**) of the body. If the (signed) distance from the center of mass of the body to an axis is s, then the first moment of the body about the axis is ms. [*We warn you that there is no "center of inertia" for a body where all the mass can be considered to be concentrated for the computation of moments of inertia about every axis.*]

Suppose the body is a thin-plane plate. We can think of the center of mass as the point in the plate where the plate could be balanced in a horizontal position on the point of a pencil, as illustrated in Fig. 7.71. For children on a balanced seesaw, as in Fig. 7.65, the center of mass is over the fulcrum.

The **centroid of a plane region** is defined to be the point that would be the location of the center of mass of a thin sheet of material of *constant density* covering the region. We will see that the location of this centroid is independent of our choice of the constant density.

Consider now a thin sheet of material covering a plane region, or a wire covering a portion of a curve in the plane. The sheet or wire need not

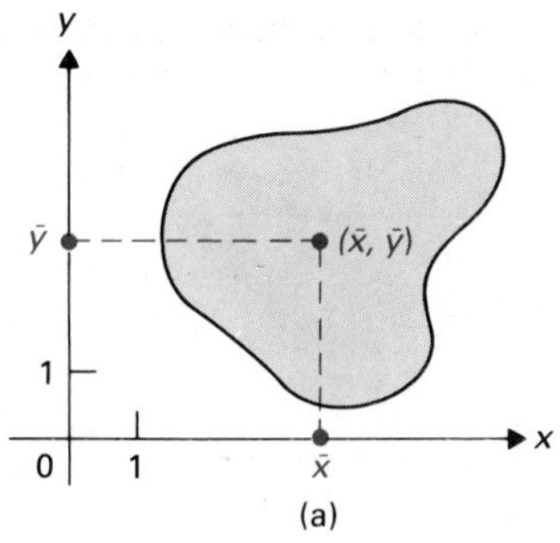

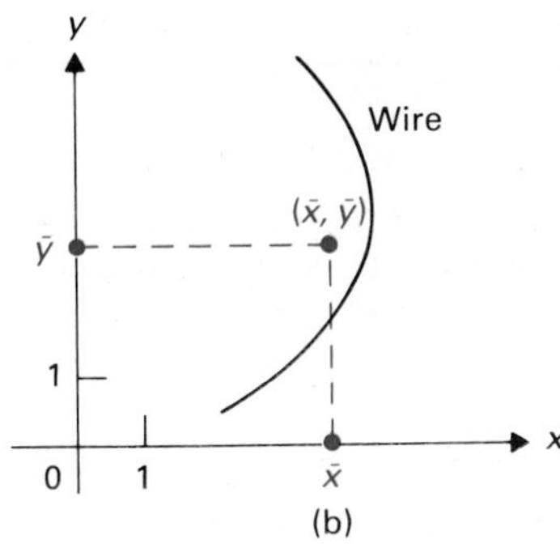

Figure 7.72 (a) Center of mass $(\bar{x}, \bar{y})$ for a thin sheet lying in the plane; (b) the center of mass of a wire in the plane need not lie on the wire.

have constant mass density. Let the center of mass be $(\bar{x}, \bar{y})$, as illustrated for a region in Fig. 7.72(a) and for a wire in Fig. 7.72(b). If we let the mass of the body be m and its moments about the x and y axes be M_x and M_y, respectively, then

$$M_x = m\bar{y},$$

since the (signed) distance from $(\bar{x}, \bar{y})$ to the x-axis is $\bar{y}$. Similarly, $M_y = m\bar{x}$. Thus

$$\boxed{\bar{x} = \frac{M_y}{m} \quad \text{and} \quad \bar{y} = \frac{M_x}{m}.}$$

Since we know how to compute m, M_x, and M_y, we can now compute $(\bar{x}, \bar{y})$.

EXAMPLE 1 Let a thin rectangular sheet of material cover the portion of the plane bounded by $x = 1$, $x = 4$, $y = 3$, and $y = 5$, as shown in Fig. 7.73. If the mass density of the sheet at (x, y) is $\rho(x) = x + 1$, find the center of mass of the sheet.

Solution We simply have to compute three integrals to find m, M_x, and M_y. The vertical differential strip shown in Fig. 7.73 has mass approximately $\rho(x)\,dA = (x + 1)(2)\,dx$, so

$$m = \int_1^4 (2x + 2)\,dx = (x^2 + 2x)\Big]_1^4 = (16 + 8) - (1 + 2) = 21.$$

Also,

$$M_y = \int_1^4 x(2x + 2)\,dx = \int_1^4 (2x^2 + 2x)\,dx = \left(\frac{2x^3}{3} + x^2\right)\Bigg]_1^4$$

$$= \left(\frac{128}{3} + 16\right) - \left(\frac{2}{3} + 1\right) = 57.$$

Considering the mass of the strip to be concentrated at its center (its center of mass!), we see that

$$M_x = \int_1^4 4(2x + 2)\,dx = (4x^2 + 8x)\Big]_1^4 = (64 + 32) - (4 + 8) = 84.$$

Figure 7.73 Differential strip for computing m, M_x, and M_y.

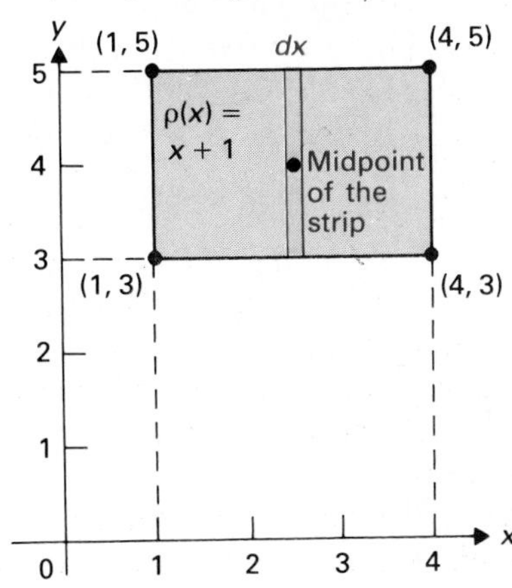

Thus

$$\bar{x} = \frac{M_y}{m} = \frac{57}{21} = \frac{19}{7} \quad \text{and} \quad \bar{y} = \frac{M_x}{m} = \frac{84}{21} = 4.$$

The center of mass is therefore $(\bar{x}, \bar{y}) = (\frac{19}{7}, 4)$.

We could easily have seen that $\bar{y} = 4$ from the symmetry of the body (and of its mass-density function) about the line $y = 4$. We would normally have used this symmetry and not have bothered to compute M_x. □

EXAMPLE 2 Compute the centroid of the half-disk bounded by $y = \sqrt{a^2 - x^2}$ and the x-axis, as shown in Fig. 7.74.

Solution We assume that the half-disk is covered by a thin sheet of constant mass density ρ. By symmetry, $\bar{x} = 0$. To find $\bar{y} = M_x/m$, we need only

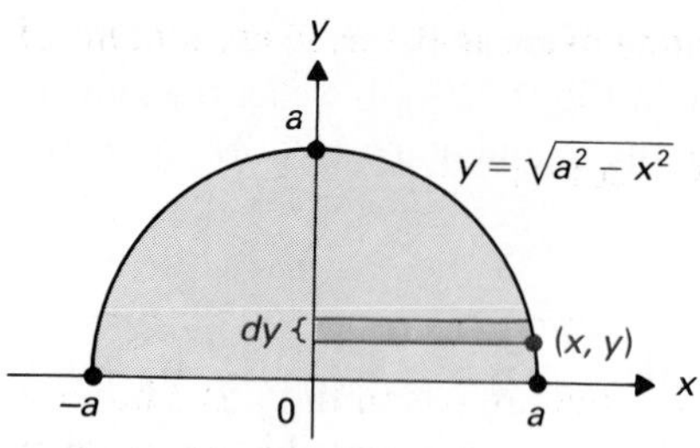

Figure 7.74 Differential strip used in computing M_x.

compute M_x, for we know that the mass m will be the product of the constant density ρ and the area, so that

$$m = \rho \frac{\pi a^2}{2}.$$

By symmetry, to compute M_x we need only compute the moment of the quarter-disk in the first quadrant about the x-axis, and then double it. The moment about the x-axis of the strip shown in Fig. 7.74 is approximately $y\rho x\, dy$. Since $x = \sqrt{a^2 - y^2}$, we arrive at

$$\begin{aligned} M_x &= 2\int_0^a y(\rho\sqrt{a^2 - y^2})\, dy = 2\rho \int_0^a y(a^2 - y^2)^{1/2}\, dy \\ &= \frac{2\rho}{-2}\int_0^a (a^2 - y^2)^{1/2}(-2y)\, dy = -\rho\left(\frac{2}{3}\right)(a^2 - y^2)^{3/2}\Big]_0^a \\ &= -\rho \cdot \frac{2}{3} \cdot 0 - \left(-\rho \cdot \frac{2}{3} \cdot a^3\right) = \frac{2}{3}\rho a^3. \end{aligned}$$

Thus

$$\bar{y} = \frac{M_x}{m} = \frac{\frac{2}{3}\rho a^3}{\pi \rho a^2/2} = \frac{4a}{3\pi}.$$

The centroid of the half-disk is therefore $(0, (4a)/(3\pi))$. □

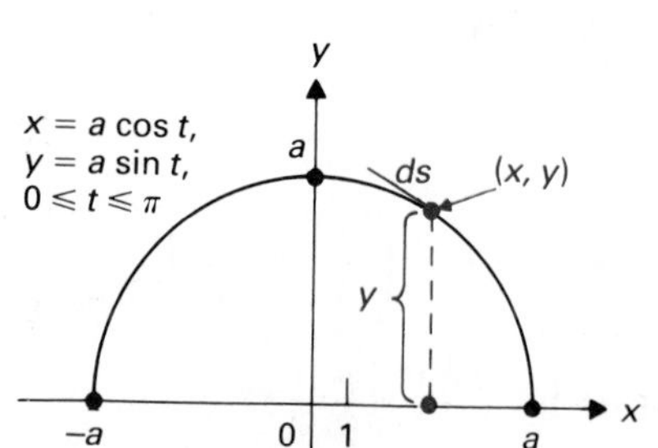

Figure 7.75 Tangent-line segment of length ds to a semicircular wire.

EXAMPLE 3 Find the center of mass of a homogeneous wire bent in the shape of a semicircle of radius a.

Solution We consider the wire to cover the curve

$$x = a\cos t, \qquad y = a \sin t, \qquad 0 \le t \le \pi,$$

shown in Fig. 7.75. Since the wire is homogeneous, we consider its mass density to be a constant k. By symmetry of the wire about the y-axis, we realize that $\bar{x} = 0$.

The mass m of the wire is obviously

$$m = k \cdot \text{Length of the wire} = k(\pi a).$$

We integrate to find M_x, which we need to compute $\bar{y}$. Now

$$\begin{aligned} ds &= \sqrt{\left(\frac{dx}{dt}\right)^2 + \left(\frac{dy}{dt}\right)^2}\, dt = \sqrt{a^2 \sin^2 t + a^2 \cos^2 t}\, dt \\ &= \sqrt{a^2(\sin^2 t + \cos^2 t)}\, dt = \sqrt{a^2}\, dt = a\, dt. \end{aligned}$$

Consequently, $dm = k\, ds = ka\, dt$. Since the distance from the tangent-line segment of length ds in Fig. 7.75 to the x-axis is $y = a \sin t$, we have

$$\begin{aligned} M_x &= \int_0^\pi ka(\sin t)(a)\, dt = ka^2(-\cos t)\Big]_0^\pi \\ &= ka^2[-(-1) - (-1)] = 2ka^2. \end{aligned}$$

Thus

$$\bar{y} = \frac{M_x}{m} = \frac{2ka^2}{\pi a k} = \frac{2a}{\pi}.$$

The center of mass is thus $(\bar{x}, \bar{y}) = (0, 2a/\pi)$. Note that this is *not* a point of the wire. □

The following is known as Pappus' theorem.

THEOREM 7.2 Pappus' theorem

1. If a plane arc of length L is revolved about an axis in the plane but not intersecting the arc, then the area of the surface generated is the product of the length L of the arc and the circumference of the circle described by the centroid of the arc.
2. If a plane region of area A is revolved about an axis in the plane but not intersecting the region, then the volume of the solid generated is the product of the area A and the circumference of the circle described by the centroid of the region.

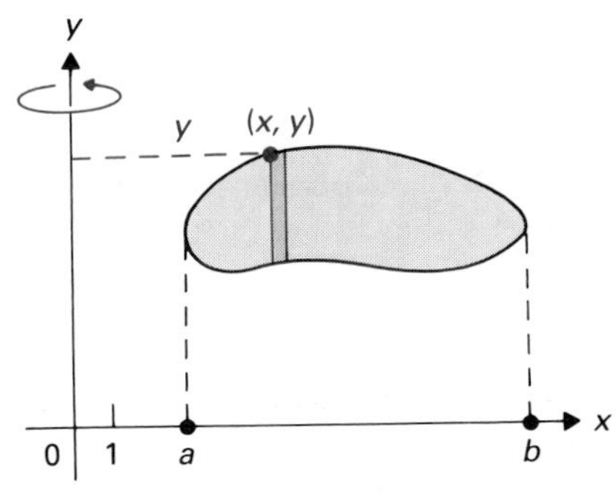

Figure 7.76 Differential strip of area dA.

Obviously these theorems make similar assertions for different dimensional objects. We now show why the second version is true. Consider the region shown in Fig. 7.76, and suppose it is revolved about the y-axis. Using the shell method, we find the volume of the solid tube

$$V = \int_a^b 2\pi x\, dA = 2\pi \int_a^b x\, dA = 2\pi M_y, \tag{1}$$

where dA is the area of the color-shaded strip. But

$$\bar{x} = \frac{M_y}{A}, \tag{2}$$

so $M_y = \bar{x}A$. Thus from Eq. (1) we have

$$V = 2\pi\bar{x} \cdot A, \tag{3}$$

which is the assertion of Pappus' theorem.

EXAMPLE 4 Use Pappus' theorem to find the surface area of a cone.

Solution If the line segment joining $(r, 0)$ and $(0, h)$ is revolved about the y-axis, it generates the surface of a cone (see Fig. 7.77). Let

$$s = \sqrt{r^2 + h^2}$$

be the slant height of the cone. The centroid of the line segment is at the point $(r/2, h/2)$, so, by Pappus' theorem, the surface area S of the cone is

$$S = \left(2\pi \cdot \frac{r}{2}\right)s = \pi rs. \quad \square$$

Figure 7.77 The line segment with centroid $(r/2, h/2)$ generates the surface of the cone.

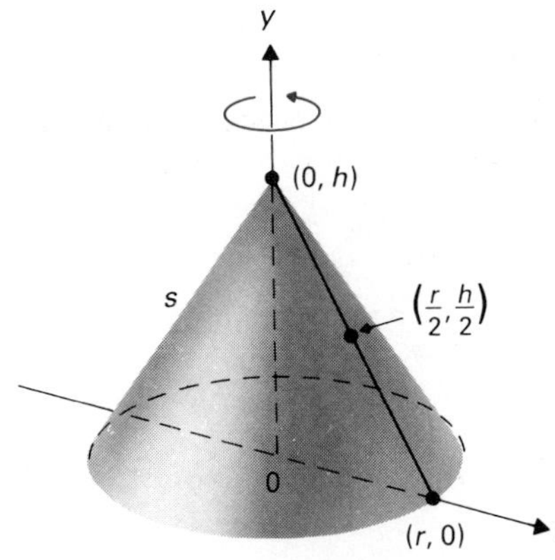

EXAMPLE 5 Use Pappus' theorem to find the centroid of the half-disk in Fig. 7.74, using the formula $V = \frac{4}{3}\pi a^3$ for the volume of the ball generated by revolving the disk about the x-axis.

Solution We know the area of the half-disk is $\pi a^2/2$. If its centroid is $(0, \bar{y})$, then

$$V = \tfrac{4}{3}\pi a^3 = 2\pi\bar{y} \cdot \tfrac{1}{2}\pi a^2 = \pi^2 a^2 \bar{y}.$$

Therefore $\bar{y} = 4a/(3\pi)$, as we found in Example 2. □

SUMMARY

1. The center of mass (centroid) of a plane body (region) is $(\bar{x}, \bar{y})$ where

$$\bar{x} = \frac{M_y}{m} \quad \text{and} \quad \bar{y} = \frac{M_x}{m}.$$

2. *Pappus' theorem:* If a plane region (arc) is revolved about an axis in its plane but not intersecting it, then the volume (area) generated is equal to the product of its area (length) and the circumference of the circle described by its centroid.

EXERCISES

1. Give an example of a plane region whose centroid is not a point in the region.
2. A rectangular sheet of material covers the part of the plane bounded by the coordinate axes and the lines $x = 2$ and $y = 4$. If the mass density of the material at (x, y) is $\sqrt{y}$, find the center of mass.
3. A square sheet of material covers the unit square bounded by the coordinate axes and the lines $x = 1$ and $y = 1$. If the mass density of the material at (x, y) is $x + 1$, find the center of mass.
4. Find the center of mass of a homogeneous sheet of material covering the region bounded by $y = x^2$ and $y = \sqrt{x}$.
5. Find the center of mass of a homogeneous sheet of material covering the region bounded by $y = x^2$ and $y = 1$.
6. Find the centroid of the plane region bounded by $y = x$ and $y = x^3$.
7. Find by integration the centroid of the triangular region whose vertices are $(0, 0)$, $(a, 0)$, $(0, b)$, where $a > 0$ and $b > 0$.
8. Find the centroid of the region bounded by the curves $y = \sqrt{a^2 - x^2}$, $x = a$, and $y = a$, where $a > 0$. [*Hint:* Use symmetry.]
9. Find the center of mass of a sheet of mass density $2y + 3$ that covers the plane region bounded by the curves $y = x^2$ and $y = 4$.
10. Find the centroid of the region bounded by the arc of $y = \sin x$, for $0 \le x \le \pi$, and the x-axis.
11. Find the center of mass of a thin rod on the x-axis covering the interval $[0, 4]$ if the mass density at $(x, 0)$ is $\sqrt{x}$.
12. A wire covers the semicircle $x = a \cos t$, $y = a \sin t$ for $0 \le t \le \pi$. The mass density of the wire at a point (x, y) is y. Find the center of mass of the wire.

Use a calculator for Exercises 13 through 16.

13. Find the center of mass of a sheet of material covering the disk $x^2 + (y - 4)^2 = 16$ if the mass density at (x, y) is $y + 1$.
14. Find the centroid of the region bounded by $y = \tan x$, $y = 0$, and $x = \pi/3$.
15. Find the center of mass of a homogeneous wire covering the curve $y = \sin x$ for $0 \le x \le \pi$.
16. Find the center of mass of a homogeneous wire covering the portion of the parabola $y = x^2$ for $-1 \le x \le 1$.
17. Let the half-disk $0 \le y \le \sqrt{1 - x^2}$ be covered by a sheet of material of constant mass density ρ. Find the first moment of the sheet about the line $x + y = 4$. [*Hint:* You know the center of mass of the sheet from Example 2.]
18. Let the portion of the face of a vertical dam that is covered by water be a plane region of area A ft^2 whose centroid is at a depth s ft below the surface of the water. Show that the force on the dam is $62.4sA$ lb.
19. Let a flat body of constant mass density k cover the unit square with vertices $(0, 0)$, $(1, 0)$, $(0, 1)$, and $(1, 1)$.
    a) Find the moment of inertia of the body about the y-axis.
    b) Find the moment of inertia of the body about the line $x = -a$.
    c) Find a point (x_1, y_1) in the body such that the moment of inertia of the body about either the x-axis or the y-axis is the product of the mass and the square of the distance from (x_1, y_1) to the axis.
    d) Find a point (x_2, y_2) in the body such that the moment of inertia of the body about either the line $x = -a$ or the line $y = -a$ is the product of the mass and the square of the distance from (x_2, y_2) to the line.
    e) Compare the answers to parts (c) and (d), and comment on the result.
20. The *radius of gyration* R of a body about an axis is defined by

$$R = \sqrt{\frac{I}{m}},$$

so that $I = mR^2$.
    a) From the answer $k/3$ to Exercise 19(a), what is the radius of gyration about the y-axis of a homogeneous flat body covering the unit square?
    b) From the answer

$$\frac{k}{3}[(a + 1)^3 - a^3]$$

to Exercise 19(b), what is the radius of gyration about the line $x = -a$ of a homogeneous flat body covering the unit square?

21. Use Pappus' theorem to find the volume generated when the half-disk bounded by $y = \sqrt{1 - x^2}$ and $y = 0$ is revolved about the given line. (The centroid of the disk was found in Example 2.)
a) $y = 1$ b) $y = -1$ c) $y = 2$
d) $x = 2$ e) $x + y = 4$

22. A square of side a is revolved about an axis through a vertex and perpendicular to the diagonal of the square at that vertex. Use Pappus' theorem to find the volume generated.

23. Given that the volume of a right circular cone of altitude h and base of radius r is $(\frac{1}{3})\pi r^2 h$, use Pappus' theorem to find the centroid of the triangular region with vertices $(0, 0)$, $(r, 0)$, and $(0, h)$, for $r > 0$ and $h > 0$.

EXERCISE SETS FOR CHAPTER 7

Review Exercise Set 7.1

1. Find the area of the region bounded by the curves $y = \sqrt{2x}$ and $y = \frac{1}{2}x^2$.

2. Find the volume of the solid generated by revolving the region bounded by $y = 4 - x^2$ and $y = 0$ about the line $y = -1$.

3. The work done in stretching a spring 1 ft from its natural length is 18 ft · lb. Find the work done in stretching it 4 ft from its natural length.

4. Find the arc length of the curve $y = 2(\frac{5}{3} + x)^{3/2}$ from $x = 0$ to $x = \frac{11}{3}$.

5. Find the area of the surface of revolution generated by rotating the arc $y = x^3/\sqrt{3}$ from $x = 0$ to $x = 1$ about the x-axis.

6. Find the total distance traveled on a line from $t = 1$ to $t = 4$ by a body with velocity $v = \sin \pi t$.

7. A thin plate covers the region bounded by $y = 1 - x^2$ and $y = 0$ and has mass density $y + 3$ at the point (x, y). Express as an integral the moment of inertia of the plate about the x-axis.

8. Find the centroid of the region bounded by the curves $y = \frac{1}{2}x^2$ and $y = \sqrt{2x}$.

Review Exercise Set 7.2

1. Find the area of the region bounded by $x = 4 - y^2$ and $x = 3y$.

2. Find the volume of the solid generated by revolving the region bounded by $y = x^2$ and $y = x + 2$ about the line $x = -1$.

3. The face of a dam is in the form of an isosceles triangle with 40-ft base at the top of the dam and equal 30-ft sides at the sides of the dam. Find the total force of the water on the dam when the water is at the top of the dam.

4. Find the value of x such that the length of arc of the circle $x = 5 \cos t$, $y = 5 \sin t$, measured counterclockwise from $(5, 0)$ to (x, y), is 2 units.

5. Find the area of the surface of revolution generated by rotating the arc of the circle $x = 3 \cos t$, $y = 3 \sin t$ from $t = 0$ to $t = \pi/4$ about the y-axis.

6. The velocity of a body traveling on a line is given by $v = 4 - t^2$. If the body starts at time $t = 0$, at what time has the body traveled a total distance of 16 units?

7. Find the first moment of a homogeneous semicircular disk of mass m bounded by $y = \sqrt{25 - x^2}$ and $y = 0$ about the x-axis.

8. Find the center of mass of a thin triangular plate with vertices at $(0, 0)$, $(3, 4)$, and $(0, 8)$ if the mass density of the plate at (x, y) is $x + 1$.

Review Exercise Set 7.3

1. Find the area of the region bounded by $x = 0$, $y = \sin (x/4)$, and $y = \cos (x/4)$ for $0 \le x \le \pi$.

2. Find the volume of the solid generated when the region in Exercise 1 is revolved about the y-axis.

3. The force of attraction between two bodies is k/s^2 lb, where s is the distance between them measured in feet. If the bodies are 10 ft apart, find the work done in separating them an additional 10 ft.

4. Find the arc length of the curve $x = \frac{1}{5}y^{5/2} + (1/\sqrt{y})$ from $y = 1$ to $y = 4$.

5. Find the area of the surface of revolution generated by revolving the arc of $x = \sqrt{y}$ for $1 \le y \le 4$ about the y-axis.

6. The velocity of a body traveling on a line is $v = (2t - 4)$ ft/sec. Find the total distance the body travels from time $t = 0$ to time $t = 10$ sec.

7. A sheet of material covering the region bounded by $y = x^2$ and $y = \sqrt{x}$ has mass density $x + 1$. Find the moment of inertia of the body about the y-axis.

8. Find the center of mass of a thin rod covering the interval $[0, 16]$ on the x-axis if the mass density at $(x, 0)$ is $1 + \sqrt{x}$.

Review Exercise, Set 7.4

1. Find the area of the plane region bounded by $y = x^4$ and $y = 16$.

2. Find the volume of the solid generated when the region in Exercise 1 is revolved about the line $y = 16$.

3. A square of side 2 ft is submerged in water with a diagonal vertical and the top corner of the square at the surface of the water. Find the force of the water on one side of the square.

4. a) Express as an integral the length of the curve $y = \sqrt{x}$ from $x = 1$ to $x = 16$.
b) Use a calculator to evaluate approximately the integral in part (a).

5. a) Express as an integral the surface area generated when the curve $x = 3\cos t$, $y = 2\sin t$ for $0 \le t \le \pi$ is revolved about the line $y = -1$.
   b) Use a calculator to evaluate approximately the integral in part (a).

6. Let the acceleration of a body traveling on a line at time t be $(2t + 2)$ ft/sec^2, and let the initial velocity at time $t = 0$ be -8 ft/sec. Find the total distance the body travels from time $t = 0$ to time $t = 3$ sec.

7. A homogeneous plate of mass density k covers the region bounded by $y = \sin x$ for $0 \le x \le \pi$ and the x-axis. Find the moment of inertia of the plate about the y-axis. (Use a table of integrals.)

8. A plane region of area 10 generates a solid of volume 30 when revolved about the line $x = 2$ and a solid of volume 40 when revolved about the line $y = 3$. If the centroid of the body lies below the line $y = 3$ and to the right of the line $x = 2$, and if the two lines do not intersect the body, find the centroid.

More Challenging Exercises 7

1. Using Pappus' theorem, express *as a single integral* the volume generated by revolving the region bounded by $y = x^2$ and $y = 4$ about the line $y = x - 2$. Evaluate the integral.

2. Find the area of the region in the first quadrant bounded by $y = x^2$, $y = x^2 + 9$, $x = 0$, and $y = 25$.

3. Express as a sum of integrals the volume generated when the region in Exercise 2 is revolved about the line $3x + 4y = -12$.

4. Water flows from a pipe at time t minutes at a rate of $40/(t + 1)^2$ gal/min for $t \ge 0$. Find the amount of water that flows from the pipe during the first hour of flow.

Exercises 5 through 8 illustrate how an integral may be used to estimate a sum.

5. Estimate
$$\sum_{k=1}^{800} \frac{k^{3/2}}{100{,}000}.$$

6. Estimate
$$\sum_{k=1}^{4000} \frac{1000}{(1000 + k)^2}.$$

7. A carpenter has a contract to hang 100 doors in a housing project. It takes him 1 hr to hang the first door. Using the expertise he continually attains, he finds that the time required to hang the nth door after the first is $3\sqrt{n}$ minutes less than 1 hr. Estimate the time required for him to hang all 100 doors.

8. A manufacturer finds that a newly hired employee is able to seal $40 + 2\sqrt[3]{n}$ cartons during the nth hour of work, except that, once a level of 60 cartons per hour is reached, there is no further increase. Estimate the number of cartons the employee can seal during the first 1500 hours of work.

OTHER ELEMENTARY FUNCTIONS

8

This chapter is devoted to enlarging our stockpile of functions for calculus. We introduce the remaining elementary functions, logarithmic, exponential, inverse trigonometric, and hyperbolic. These functions have many important applications, often as solutions of differential equations that arise in the physical sciences.

*8.1 INVERSE FUNCTIONS

This section studies the general notion of inverse functions and their derivatives. It provides a foundation for later introduction of the exponential function as the inverse of the logarithmic function and the study of the inverse trigonometric functions. However, the text is designed so that this section may be omitted; the essential notions are introduced briefly in later sections as they are needed.

THE GENERAL NOTION

Recall that if $y = f(x)$, then the function f is a rule that assigns to each number x in the domain of f a single number y. The graph of a function gives us a geometric way to obtain y from x, as illustrated in Fig. 8.1: We follow the arrows. The curve in Fig. 8.2 is not the graph of a function, because from a single x-value, we can obtain more than one y-value by following the arrows.

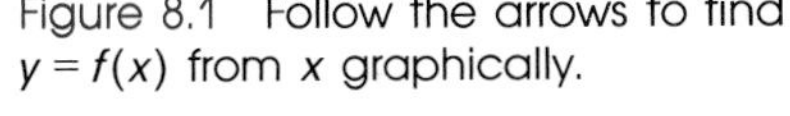

Figure 8.1 Follow the arrows to find $y = f(x)$ from x graphically.

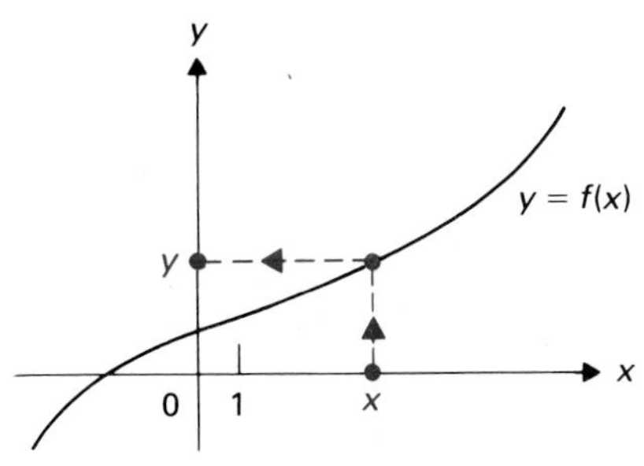

> A vertical line cannot cut the graph of a function in more than one point.

* The remainder of the text is designed so that this section may be omitted at the discretion of the instructor.

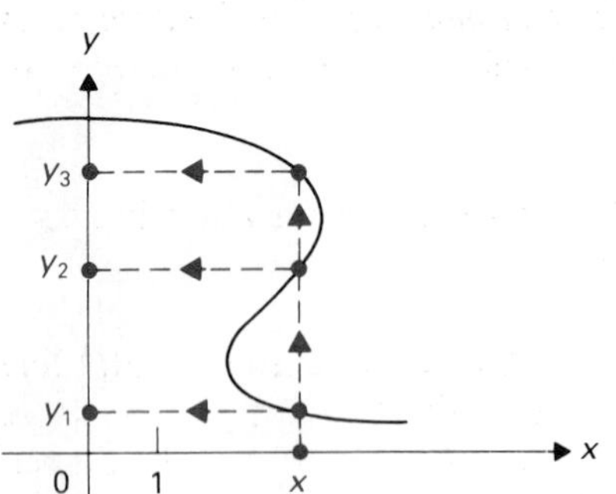

Figure 8.2 Not the graph of a function: Following arrows from x leads to three y-values.

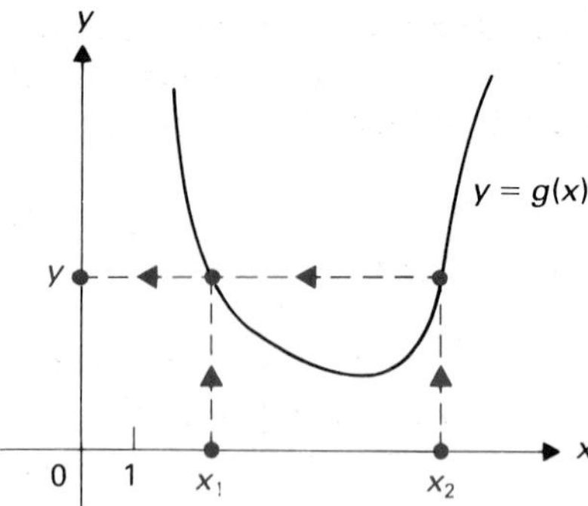

Figure 8.3 The graph of a function: Two x-values may give the same y-value.

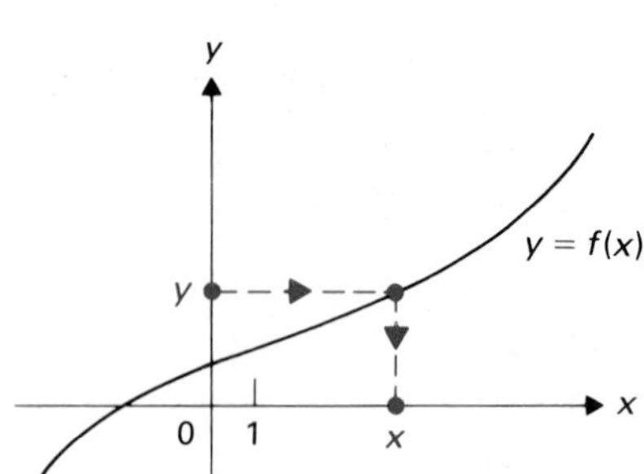

Figure 8.4 x is given as a function of y: Following arrows from a y-value gives just one x-value.

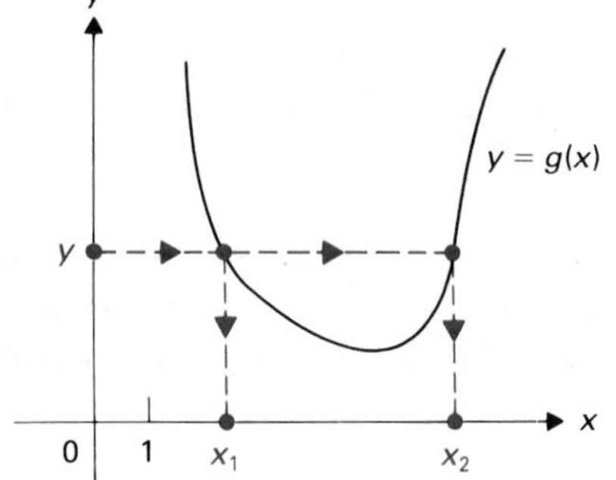

Figure 8.5 x is not given as a function of y: Following arrows from a y-value can give more than one x-value.

However, several different x-values may give rise to the same y-value, as illustrated in Fig. 8.3.

In this section, we are interested in those functions $y = f(x)$ whose graphs can be viewed as also defining x as a function of y. That is, we want to be able to reverse the directions of the arrows shown in Fig. 8.1 and still have a function. Starting with a y-value in the range of f, we want to obtain *just one* x-value. This can be done for the graph of f in Fig. 8.1, as illustrated in Fig. 8.4. However, for the graph in Fig. 8.3, one y-value may lead to two x-values, as illustrated in Fig. 8.5.

> For the graph of $y = f(x)$ also to define x as a function of y, each horizontal line must meet the graph in at most one point.

A function whose graph has this property is called *invertible*. In this case, the function that assigns to each y-value in the range of f the x-value obtained by following the arrows as in Fig. 8.4 is called the *inverse function* of f, denoted by f^{-1}.

> *Caution:* The superscript -1 in f^{-1} must *not* be regarded as an exponent; that is, $f^{-1}(y)$ *is not* $1/f(y)$.

For an invertible function f, we may then write both $y = f(x)$ and $x = f^{-1}(y)$; both functions are defined by the same graph. Let us summarize in a definition.

DEFINITION 8.1 Invertibility

A function f, where $y = f(x)$, is **invertible** if each horizontal line meets the graph of f in at most one point. The function that assigns to each y-value in the range of f the corresponding x-value in the domain of f is the **inverse function** of f and is denoted by f^{-1}.

For an invertible function f, we then have both

$$y = f(x) \quad \text{and} \quad x = f^{-1}(y). \tag{1}$$

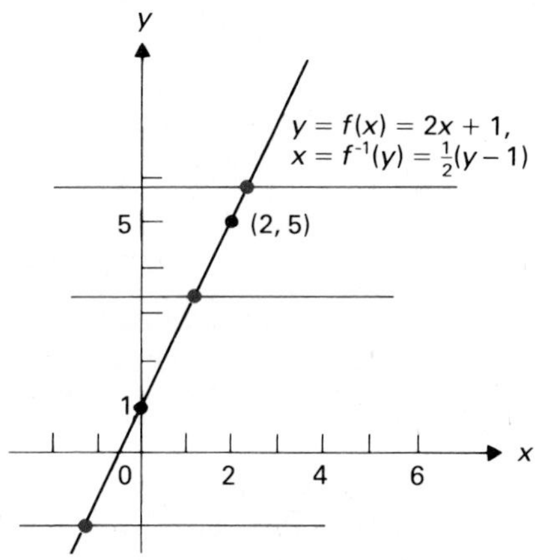

Figure 8.6 $f(x) = 2x + 1$ is invertible: Each horizontal line meets the graph at most once.

The *domain* of f^{-1} is the range of f, and the *range* of f^{-1} is the domain of f. To find f^{-1} from a given f, we attempt to solve $y = f(x)$ for x in terms of y, as illustrated in Example 1.

EXAMPLE 1 Let $y = f(x) = 2x + 1$. Determine graphically whether f is invertible and, if so, express x as $x = f^{-1}(y)$.

Solution The graph of f is shown in Fig. 8.6. Since each horizontal line meets the graph only once, the function f is invertible. To find f^{-1}, we solve $y = 2x + 1$ for x in terms of y. Thus $2x = y - 1$, so $x = \frac{1}{2}(y - 1)$. Consequently,

$$x = f^{-1}(y) = \tfrac{1}{2}(y - 1). \quad \square$$

EXAMPLE 2 Determine whether $y = h(x) = x^2$ is an invertible function and, if so, find $h^{-1}(y)$.

Solution The graph of h is shown in Fig. 8.7. Clearly, it is possible for a horizontal line to meet the graph in more than one point. For example, the line $y = 4$ meets the graph in two points, since $h(2) = 4$ and $h(-2) = 4$. Consequently, h is not invertible. $\square$

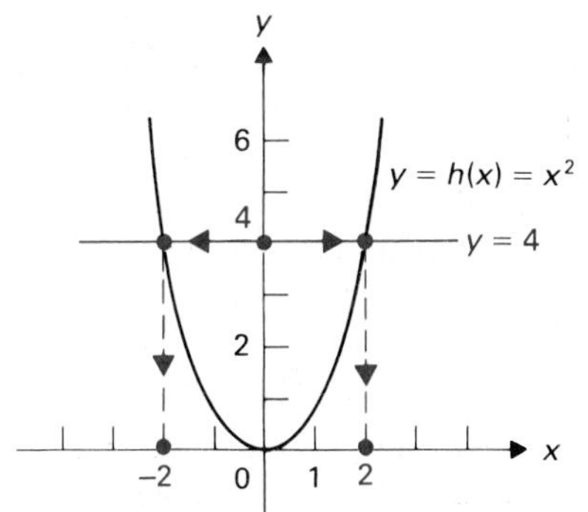

Figure 8.7 $h(x) = x^2$ is not invertible: A horizontal line may meet the graph in more than one point.

EXAMPLE 3 Determine whether $y = g(x) = \sqrt{x}$ is invertible and, if it is, find $g^{-1}(y)$.

Solution The graph of g is shown in Fig. 8.8. Clearly, any horizontal line meets the graph in at most one point. Solving $y = \sqrt{x}$ for x in terms of y, we obtain $x = y^2$. Since the domain of g^{-1} must be the range of g, where $y \geq 0$, we have

$$x = g^{-1}(y) = y^2 \qquad \text{for } y \geq 0. \quad \square$$

EXAMPLE 4 Let f be an invertible function, and suppose that $f(3) = -2$ and $f(4) = 5$. Find (a) $f^{-1}(-2)$, (b) $f(f^{-1}(5))$, and (c) $f^{-1}(f(3))$.

Solution

a) Since $f(3) = -2$, we have $f^{-1}(-2) = 3$.
b) Since $f(4) = 5$, we have $f(f^{-1}(5)) = f(4) = 5$.
c) Since $f(3) = -2$, we have $f^{-1}(f(3)) = f^{-1}(-2) = 3$. $\square$

Figure 8.8 $g(x) = \sqrt{x}$ is invertible: $g^{-1}(y) = y^2$ for $y \geq 0$.

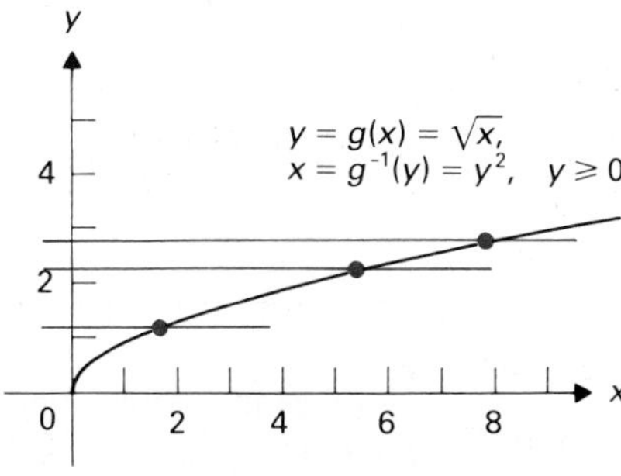

Example 4 illustrates that if f is invertible, then

$$\boxed{f^{-1}(f(x)) = x \qquad \text{and} \qquad f(f^{-1}(y)) = y.} \tag{2}$$

We think of f^{-1} as "undoing" or reversing f, and, similarly, f reverses f^{-1}.

Before we leave this subsection, we introduce one bit of standard terminology. We have seen that invertible functions are those where a single point of the domain gives rise to a single point of the range (definition of a function) and also a single point of the range gives rise to a single point of the domain (definition of invertibility). This single-point-to-single-point concept is abbreviated by the term *one to one.*

DEFINITION 8.2 One-to-one function

A function f is **one to one** if distinct points x_1 and x_2 in the domain are carried by f into distinct points y_1 and y_2 in the range. That is, if $x_1 \neq x_2$, then $f(x_1) \neq f(x_2)$.

Thus the invertible functions are precisely those that are one to one. Indeed, this is usually the way one sees an invertible function defined, rather than with the geometric Definition 8.1 that we gave. However, we want to develop as much geometric understanding as possible.

Let f be an increasing function. Then if $x_1 < x_2$, we have $f(x_1) \neq f(x_2)$, so f is one to one. Similarly, a decreasing function is one to one. This gives the following theorem.

THEOREM 8.1 Invertibility of increasing functions

If f is either an increasing or a decreasing function throughout its domain, then f is invertible.

Since a differentiable function is increasing on an interval throughout which $f'(x)$ is positive and decreasing over an interval where $f'(x)$ is negative, we obtain the following corollary.

COROLLARY The sign of $f'(x)$ and invertibility

Let f be differentiable and let the domain of f be either an interval or the set of all real numbers. If either $f'(x) > 0$ throughout the domain of f or $f'(x) < 0$ throughout the domain, then f is invertible.

GRAPHS OF INVERSE FUNCTIONS

In a moment we will discuss derivatives of inverse functions. We have always discussed derivatives of functions $y = f(x)$, where we think of the *domain* as falling on the horizontal axis and the *range* as part of the vertical axis. The derivative is then the slope of the tangent line to the graph. To make use of this familiar geometric picture as we try to differentiate $x = f^{-1}(y)$, we want to turn the plane so that the y-axis is horizontal and the x-axis is vertical. This can be accomplished by rotating the plane a half-turn about the line $y = x$, as shown in Figs. 8.9(a) and (b). In those figures, we show the graph of $y = f(x) = x^3$ or, equivalently, $x = f^{-1}(y) = y^{1/3}$. In Fig. 8.9(c), we interchange the letters x and y on the axes, so we have the graph of $x = f(y) = y^3$ or, equivalently, $y = f^{-1}(x) = x^{1/3}$. To interpret the derivative of f^{-1} as the usual slope dy/dx of the tangent line, we should work with $y = f^{-1}(x)$.

EXAMPLE 5 Show that $y = f(x) = (x - 1)/(x + 2)$ is invertible, find $f^{-1}(x)$, and sketch the graphs of $y = f(x)$ and $y = f^{-1}(x)$.

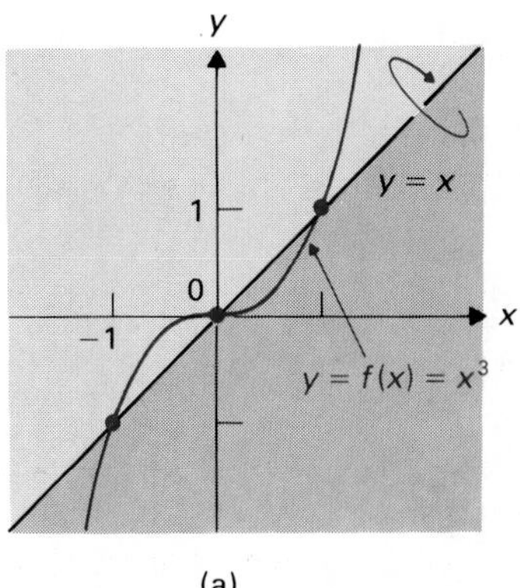

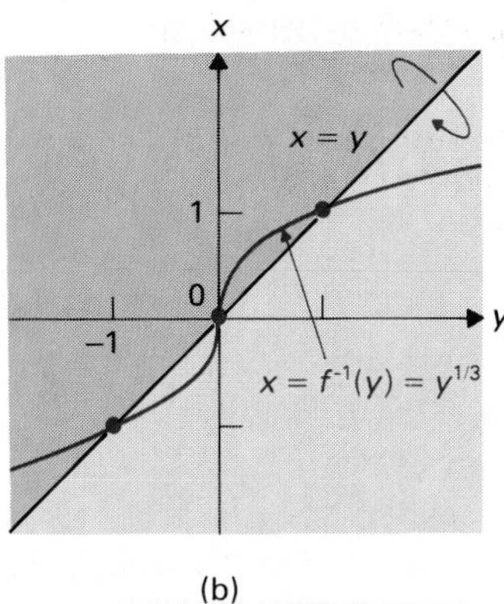

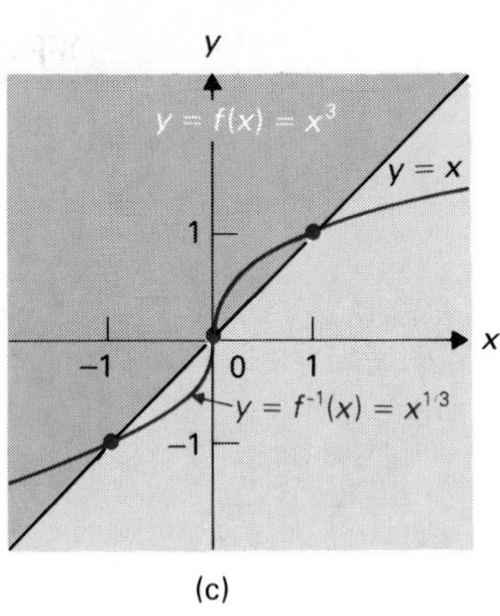

Figure 8.9 Obtaining the graph of $y = f^{-1}(x)$ from the graph of $y = f(x)$, illustrated for $y = f(x) = x^3$. To get from (a) to (b), rotate the plane a half-turn about $y = x$; to get from (b) to (c), interchange the variable names x and y.

Solution From $y = (x - 1)/(x + 2)$, we obtain

$$
\begin{aligned}
y(x + 2) &= x - 1, \\
yx - x &= -2y - 1, \\
x(y - 1) &= -2y - 1 \\
x &= -\frac{2y + 1}{y - 1}.
\end{aligned}
$$

Since each y-value gives at most one x-value, we see that f is indeed invertible. We sketched the graph of $f(x) = (x - 1)/(x + 2)$ in Fig. 8.10, using the methods of Chapter 2. The graph indicates again that f is invertible; each horizontal line meets the graph in at most one point. We see that

$$x = f^{-1}(y) = -\frac{2y + 1}{y - 1},$$

so

$$f^{-1}(x) = -\frac{2x + 1}{x - 1}.$$

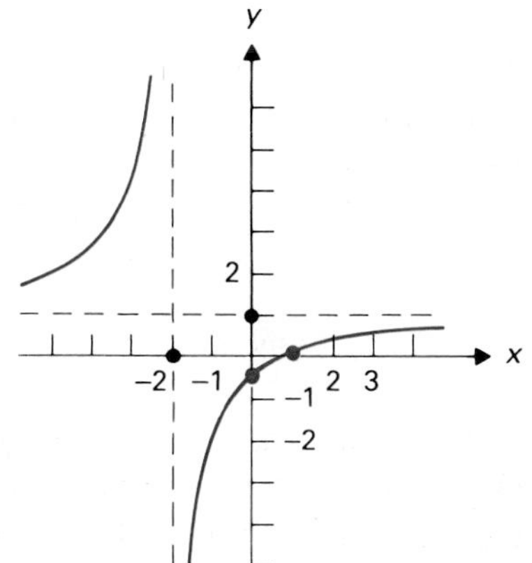

Figure 8.10 Graph of $y = f(x) = (x - 1)/(x + 2)$.

To obtain the graph of $y = f^{-1}(x)$, we just rotate the graph of $y = f(x)$ a half-turn about the line $y = x$, as shown in Fig. 8.11. The white curve in the figure shows where the graph of $y = f(x)$ was before it was rotated to become the graph of $y = f^{-1}(x)$. □

DERIVATIVES OF INVERSE FUNCTIONS

Let f be a differentiable and invertible function, and suppose that $f(a) = b$. Since f is differentiable, $f'(a)$ exists, and the graph of f has a tangent line where $x = a$. From Fig. 8.12, we see that the graph of f^{-1} then has a tangent line where $x = b$. If the tangent line to the graph of f at (a, b) is not horizontal, then the tangent line to the graph of f^{-1} at (b, a) will not be vertical. Under these circumstances, $(f^{-1})'(b)$ should exist. Computing the slopes of the

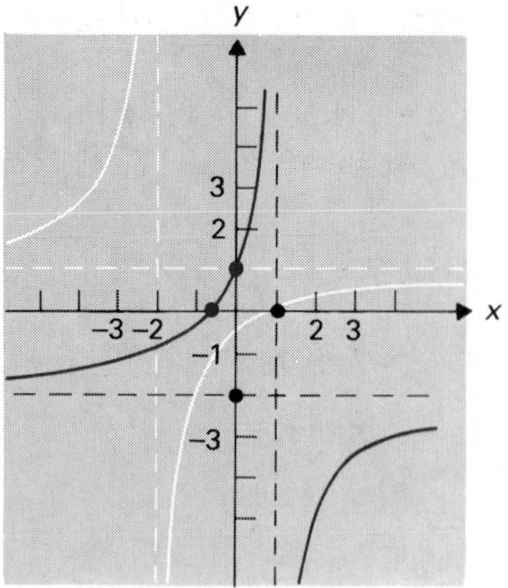

Figure 8.11 White curve: $y = f(x) = (x-1)/(x+2)$; Red curve: $y = f^{-1}(x) = -(2x+1)/(x-1)$.

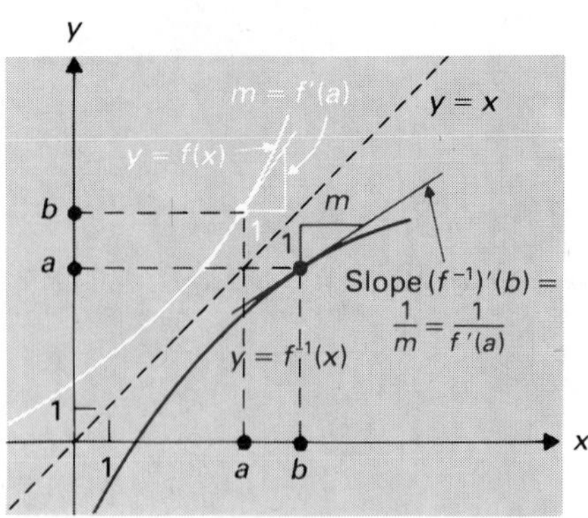

Figure 8.12 If the graph of f has slope $m = f'(a)$ where $x = a$ and $f(a) = b$, then the graph of f^{-1} has slope $1/m = 1/(f'(a))$ where $x = b$.

tangent lines as indicated in Fig. 8.12, we should have

$$(f^{-1})'(b) = \frac{1}{f'(a)}. \tag{3}$$

We can also arrive at Eq. (3) by formal use of implicit differentiation. Let $y = f^{-1}(x)$, so $x = f(y)$. Assuming that dy/dx does exist and differentiating $x = f(y)$ implicitly, we obtain

$$1 = f'(y) \cdot \frac{dy}{dx}, \tag{4}$$

so

$$\frac{dy}{dx} = (f^{-1})'(x) = \frac{1}{f'(y)} \tag{5}$$

if $f'(y) \neq 0$. Since $f'(y) = dx/dy$, we may express Eq. (5) by

$$\frac{dy}{dx} = \frac{1}{dx/dy}. \tag{6}$$

Once more, the correct Leibniz formula is easy to remember using formal rules of algebra.

While we don't claim to have proved the following theorem, we hope we have made it at least plausible.

THEOREM 8.2 The derivative of f^{-1}

Let f be a differentiable and invertible function, and let $f(a) = b$. If $f'(a) \neq 0$, then f^{-1} is differentiable at b and, furthermore,

$$(f^{-1})'(b) = \frac{1}{f'(a)}. \tag{7}$$

EXAMPLE 6 Let $y = f(x) = x^3$. Find $(f^{-1})'(8)$ in two ways.

Solution 1 Now $f(2) = 8$, and $f'(2) = 3x^2|_2 = 12 \neq 0$. Theorem 8.2 shows that f^{-1} is differentiable at 8, and

$$(f^{-1})'(8) = \frac{1}{f'(2)} = \frac{1}{12}.$$

Solution 2 If $y = f(x) = x^3$, then $x = f^{-1}(y) = y^{1/3}$. Then

$$(f^{-1})'(8) = \frac{1}{3}y^{-2/3}\bigg|_8 = \frac{1}{3}\cdot 8^{-2/3} = \frac{1}{3}\cdot\frac{1}{8^{2/3}} = \frac{1}{3}\cdot\frac{1}{4} = \frac{1}{12}. \quad \square$$

EXAMPLE 7 Let $y = f(x) = \sin x$ for $-\pi/2 \le x \le \pi/2$. Show graphically that f is an invertible function, and sketch the graph of $y = f^{-1}(x)$. Find a formula for $(f^{-1})'(x)$ in terms of x, wherever it exists.

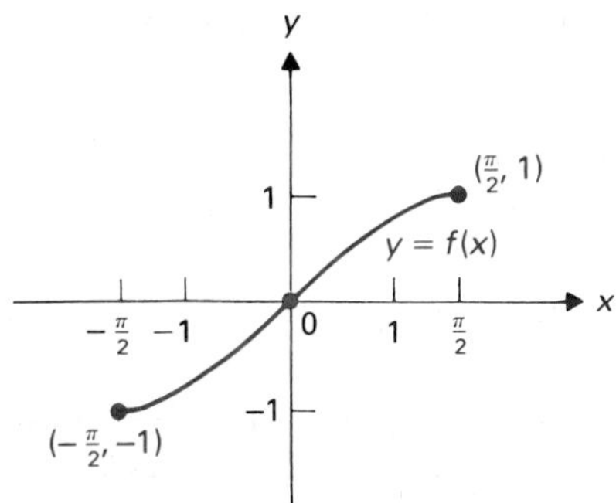

Figure 8.13 Graph of $f(x) = \sin x$ for $-\pi/2 \le x \le \pi/2$.

Solution The graph of f is shown in Fig. 8.13. Note that it is not the graph of the entire function $\sin x$, but just the portion for which $-\pi/2 \le x \le \pi/2$. Since a horizontal line meets this increasing graph in at most one point, we see that f is invertible. The graph of $y = f^{-1}(x)$ is obtained by rotating the graph of f a half-turn about the line $y = x$. The graph is shown in Fig. 8.14.

To find $(f^{-1})'(x)$, we let $y = f^{-1}(x)$. Then $x = f(y) = \sin y$. From Eq. (6), we have

$$\frac{dy}{dx} = (f^{-1})'(x) = \frac{1}{dx/dy} = \frac{1}{\cos y}.$$

We must express $\cos y$ in terms of x to complete the problem. Now $\sin y = x$, and $\cos^2 y + \sin^2 y = 1$. Consequently,

$$\cos y = \pm\sqrt{1 - \sin^2 y} = \pm\sqrt{1 - x^2}.$$

Since $-\pi/2 \le y \le \pi/2$ in Fig. 8.14, we see that $\cos y \ge 0$, so

$$\cos y = \sqrt{1 - x^2}.$$

Thus

$$(f^{-1})'(x) = \frac{1}{\cos y} = \frac{1}{\sqrt{1 - x^2}}.$$

Note that $(f^{-1})'(x)$ is not defined where $x = \pm 1$, although this point is in the domain of f^{-1}. $\square$

Figure 8.14 $f(x) = \sin x$, $-\pi/2 \le x \le \pi/2$; $f^{-1}(x) = \sin^{-1} x$, $-1 \le x \le 1$.

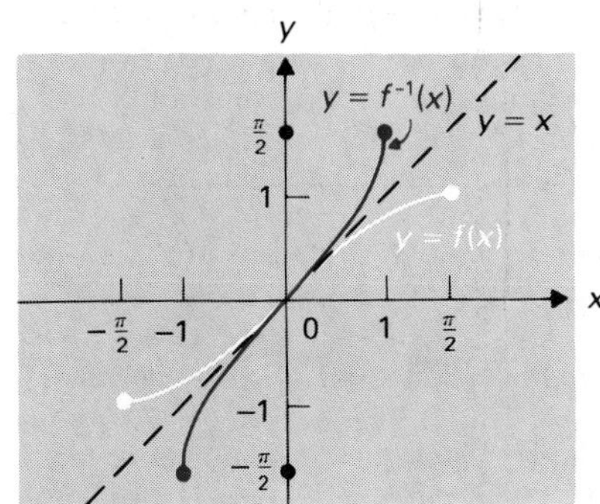

The function f^{-1} of Example 7 is called the *inverse sine function* and is denoted by $\sin^{-1} x$. In the example, we found its derivative. We showed that

$$\boxed{\frac{d(\sin^{-1} x)}{dx} = \frac{1}{\sqrt{1 - x^2}}.} \qquad \textbf{(8)}$$

You may have noticed a fascinating feature of Eq. (8). The derivatives of the six elementary trigonometric functions are again trigonometric expressions. However, the derivative of the function $\sin^{-1} x$ is an *algebraic expression*. Could it be that trigonometry and algebra are really closely related? Chapter 11 will throw more light on this question.

SUMMARY

1. If $y = f(x)$ also defines x as a function of y, then f is invertible and x is the inverse function f^{-1} of y. That is, $x = f^{-1}(y)$.
2. A function $y = f(x)$ is one to one if it always carries two distinct points of the domain into two distinct points of the range, so that if $x_1 \neq x_2$, then $f(x_1) \neq f(x_2)$.
3. A function $y = f(x)$ is invertible if each horizontal line meets the graph in at most one point. Equivalently, f is invertible if it is one to one.
4. If f is invertible, then the domain of f^{-1} is the range of f and the range of f^{-1} is the domain of f. If a is in the domain of f and b in the domain of f^{-1}, then $f^{-1}(f(a)) = a$ and $f(f^{-1}(b)) = b$.
5. The graph of $y = f^{-1}(x)$ may be obtained from the graph of $y = f(x)$ by rotating the plane a half-turn about the line $y = x$.
6. Let f be an invertible function and let $f(a) = b$. If $f'(a)$ exists and is nonzero, then $(f^{-1})'(b)$ exists, and

$$(f^{-1})'(b) = \frac{1}{f'(a)}.$$

In Leibniz notation,

$$\frac{dx}{dy} = \frac{1}{dy/dx}.$$

EXERCISES

In Exercises 1 through 18, determine whether f is invertible. If it is, find $f^{-1}(x)$ and sketch the graph of both $f(x)$ and $f^{-1}(x)$.

1. $f(x) = x - 1$
2. $f(x) = 2x + 4$
3. $f(x) = 3 - x$
4. $f(x) = 8 - 2x$
5. $f(x) = 4$
6. $f(x) = -3$
7. $f(x) = x^2 - 3$
8. $f(x) = 4 - 3x^2$
9. $f(x) = x^3 + 1$
10. $f(x) = 1 - x^3$
11. $f(x) = \sqrt{x}$
12. $f(x) = 3 - 2\sqrt{x}$
13. $f(x) = \sqrt{x^2}$
14. $f(x) = |x| + 1$
15. $f(x) = \dfrac{x - 1}{x + 1}$
16. $f(x) = \dfrac{2x - 1}{x + 2}$
17. $f(x) = \dfrac{x^2}{x + 1}$
18. $f(x) = \dfrac{x - 1}{x^2}$

In Exercises 19 through 26, decide whether f is invertible and, if it is, sketch the graph of both $f(x)$ and $f^{-1}(x)$.

19. $f(x) = \cos x$
20. $f(x) = \tan x$
21. $f(x) = \cos x, 0 \leq x \leq \pi$
22. $f(x) = \tan x, -\pi/2 < x < \pi/2$
23. $f(x) = \sin x, 0 \leq x \leq \pi$
24. $f(x) = \sec x, -\pi/2 < x < \pi/2$
25. $f(x) = \sec x, 0 \leq x \leq \pi, x \neq \pi/2$
26. $f(x) = \cot x, 0 < x < \pi$.
27. Mark each of the following true or false.
    ——— a) Every one-to-one function is invertible.
    ——— b) Every invertible function is one to one.
    ——— c) Every increasing function is invertible.
    ——— d) Every invertible function is either increasing or decreasing over its entire domain.
    ——— e) Every invertible function whose domain is an interval must be either increasing or decreasing over its entire domain.
    ——— f) Every continuous invertible function whose domain is an interval must be either increasing or decreasing over its entire domain.
    ——— g) No polynomial function of even degree is invertible.
    ——— h) Every polynomial function of odd degree is invertible.
    ——— i) Every monomial function of odd degree is invertible.
    ——— j) None of the six elementary trigonometric functions is invertible.

In Exercises 28 through 34, assume all functions are invertible and differentiable. Fill in the blanks.

28. If $f(2) = 3$, then $f^{-1}(3) =$ ______.

29. If $g(3) = -4$, then $g(g^{-1}(-4)) =$ ______.

30. If $h^{-1}(4) = 2$ and $h^{-1}(2) = 5$, then $h(h^{-1}(2)) =$ ______.

31. If $f(3) = -5$ and $f'(3) = 2$, then $(f^{-1})'(-5) =$ ______.

32. If $f(-2) = 4$ and $(f^{-1})'(4) = -3$, then $f'(-2) =$ ______.

33. If $g(4) = 3$ and $g'(1) = 2$, then $(g(g^{-1}))'(3) =$ ______.

34. If $f(4) = 3$ and $f'(3) = -2$, then $(f^{-1}(f))'(4) =$ ______.

35. If $y = f(x) = 2x + 4$, find $(f^{-1})'(3)$ in two ways.

36. If $y = f(x) = x^3 + 1$, find $(f^{-1})'(28)$ in two ways.

37. If $y = g(x) = (x - 1)/(x + 2)$, find $(g^{-1})'(0)$ in two ways.

38. If $y = h(x) = (2x - 3)/(x + 1)$, find $(h^{-1})'(1)$ in two ways.

39. Let $f(x) = \cos x$ for $0 \le x \le \pi$.
a) Show that f is invertible; $f^{-1}(x)$ is denoted by $\cos^{-1}x$.
b) Find the domain of $\cos^{-1}x$.
c) Find $d(\cos^{-1}x)/dx$.
d) Find dy/dx if $y = \cos^{-1}(2x + 1)$.

40. Let $f(x) = \tan x$ for $-\pi/2 < x < \pi/2$.
a) Show that f is invertible; $f^{-1}(x)$ is denoted by $\tan^{-1}x$.
b) Find the domain of $\tan^{-1}x$.
c) Find $d(\tan^{-1}x)/dx$.
d) Find dy/dx if $y = (\tan^{-1}x)^2$.

41. Let $g(x) = \cot x$ for $0 < x < \pi$.
a) Show that g is invertible; $g^{-1}(x)$ is denoted by $\cot^{-1}x$.
b) Find the domain of $\cot^{-1}x$.
c) Find $d(\cot^{-1}x)/dx$.
d) Find dy/dx if $y = \cot^{-1}(x^2)$.

42. Let $h(x) = \sec x$ for $0 \le x < \pi/2$ or $\pi \le x < 3\pi/2$.
a) Show that h is invertible; $h^{-1}(x)$ is denoted by $\sec^{-1}x$.
b) Find the domain of $\sec^{-1}x$.
c) Find $d(\sec^{-1}x)/dx$.
d) Find dy/dx if $y = \sec^{-1}(1/x)$.

43. Let $f(x) = \csc x$ for $0 < x \le \pi/2$ or $\pi < x \le 3\pi/2$.
a) Show that f is invertible; $f^{-1}(x)$ is denoted by $\csc^{-1}x$.
b) Find the domain of $\csc^{-1}x$.
c) Find $d(\csc^{-1}x)/dx$.
d) Find dy/dx if $y = \csc^{-1}(\sqrt{x})$.

8.2 THE FUNCTION ln x

We develop the natural logarithm function $\ln x$ in this section. The following section studies the inverse function of $\ln x$, which we will see can be written in an exponential form e^x for a number $e \approx 2.71828$. These two functions, $\ln x$ and e^x, are among the most important in all of mathematics.

AN INTEGRATION PROBLEM

If n is an integer, we know that

$$\int x^n\,dx = \frac{x^{n+1}}{n+1} + C \qquad \text{for } n \neq -1.$$

As yet, we haven't encountered a function that is an antiderivative of the function $1/x$. We would like to find $\int (1/x)\,dx$.

Of course, $\int_1^2 (1/x)\,dx$ exists, since $1/x$ is continuous in $[1, 2]$. Moreover, the fundamental theorem of calculus (Theorem 6.3, page 226) tells us that $1/x$ does have an antiderivative for $x > 0$, namely F, where

$$F(x) = \int_1^x \frac{1}{t}\,dt \qquad \text{for } x > 0. \tag{1}$$

Figure 8.15 $\ln x = \int_1^x (1/t)\,dt =$ area of the shaded region.

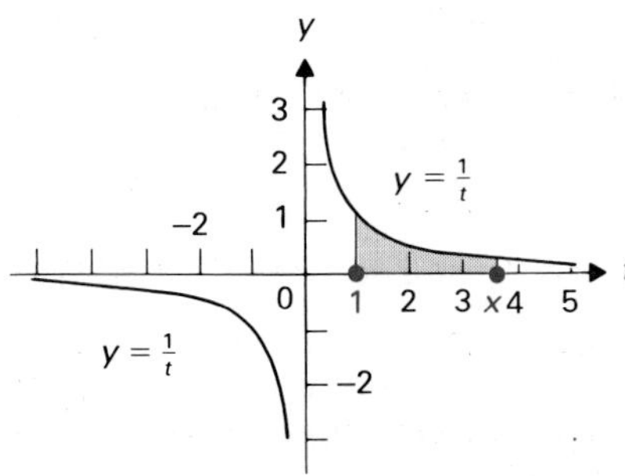

The value $F(x)$ is equal to the area of the shaded region in Fig. 8.15. (If a positive number other than 1 is chosen as lower limit for the integral in Eq. (1), the resulting function differs from F only by a constant.) Thus the fundamental theorem of calculus enables us to "find" an antiderivative F of $1/x$, at least for $x > 0$. This function F is extremely important. You will see in a moment that F has the formal algebraic properties of the logarithm function you studied in previous courses, so the notation "ln" for F in the following definition is appropriate.

DEFINITION 8.3 Logarithm function

The function ln defined by

$$\ln x = \int_1^x \frac{1}{t}\,dt \qquad \text{for } x > 0 \tag{2}$$

is the **(natural) logarithm function**, so $\ln a = \int_1^a (1/t)\,dt$ is the (natural) logarithm of a for any $a > 0$.

From the properties of the integral discussed in Chapter 6, we see at once that $\int_1^1 (1/t)\,dt = 0$, so we have

$$\boxed{\ln 1 = 0.} \tag{3}$$

Properties of the integral also show that

$$\boxed{\ln x \quad \text{is} \quad \begin{cases} >0 \text{ if } x > 1, \\ <0 \text{ if } 0 < x < 1. \end{cases}} \tag{4}$$

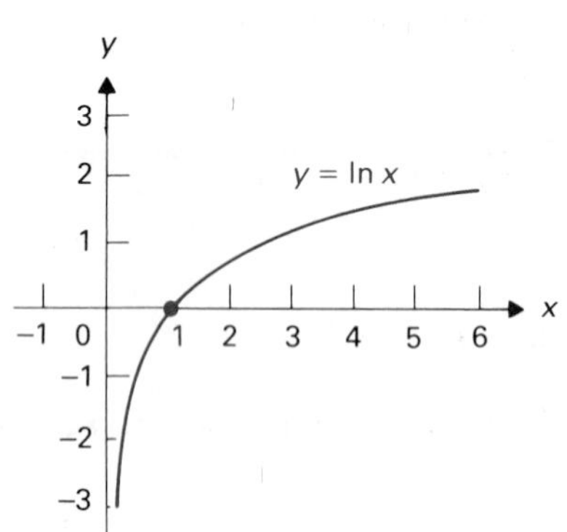

Figure 8.16 The graph of $y = \ln x$.

The graph of $\ln x$ is shown in Fig. 8.16. More justification for this graph appears shortly. A table of some values of $\ln x$ is given in Appendix 2 at the end of the text. Of course, quite accurate values of $\ln x$ can easily be found on a computer or scientific calculator.

THE CALCULUS OF ln x

From the definition of $\ln x$ in Eq. (2) and the fundamental theorem of calculus, we see that

$$\frac{d(\ln x)}{dx} = \frac{1}{x} \qquad \text{for } x > 0. \tag{5}$$

By the chain rule, if $u = g(x)$ where g is a differentiable function, then

$$\boxed{\frac{d(\ln u)}{dx} = \frac{1}{u} \cdot \frac{du}{dx} \qquad \text{for } u > 0.} \tag{6}$$

You may wish to memorize formula (6), which we will use often.

EXAMPLE 1 Find the derivative of $\ln(2x + 1)$.

Solution Using formula (6), we have

$$\frac{d(\ln(2x+1))}{dx} = \frac{1}{2x+1} \cdot \frac{d(2x+1)}{dx} = \frac{2}{2x+1}, \qquad x > -\frac{1}{2}. \quad \square$$

EXAMPLE 2 Find the derivative of ln (sin x).

Solution Using formula (6) again, we obtain

$$\frac{d(\ln(\sin x))}{dx} = \frac{1}{\sin x}\cdot\frac{d(\sin x)}{dx} = \frac{\cos x}{\sin x} = \cot x, \qquad \sin x > 0. \quad \square$$

EXAMPLE 3 Find dy/dx if $y = (\sec x)/(\ln x)^2$.

Solution We use the quotient and power rules for differentiation, as well as formula (6), and find that

$$\frac{dy}{dx} = \frac{(\ln x)^2 \dfrac{d(\sec x)}{dx} - (\sec x)\dfrac{d((\ln x)^2)}{dx}}{(\ln x)^4}$$

$$= \frac{(\ln x)^2(\sec x \tan x) - (\sec x)2(\ln x)(1/x)}{(\ln x)^4}$$

$$= \frac{(\sec x)[(\ln x)(\tan x) + (2/x)]}{(\ln x)^3}. \quad \square$$

EXAMPLE 4 Find dy/dx if $y = (\tan^3 2x)\ln(\sqrt{x}+1)$.

Solution We use the product and power rules for differentiation, as well as formula (6), and find that

$$\frac{dy}{dx} = \tan^3(2x)\frac{d(\ln(\sqrt{x}+1))}{dx} + \ln(\sqrt{x}+1)\frac{d(\tan^3 2x)}{dx}$$

$$= \tan^3(2x)\frac{1}{\sqrt{x}+1}\cdot\frac{1}{2\sqrt{x}} + [\ln(\sqrt{x}+1)]\,3(\tan^2 2x)(\sec^2 2x)2$$

$$= (\tan^2 2x)\left[\frac{\tan 2x}{2x+2\sqrt{x}} + 6(\ln(\sqrt{x}+1))\sec^2 2x\right]. \quad \square$$

From formula (6), we obtain the integration formula

$$\int \frac{du}{u} = \ln u + C \qquad \text{for } u > 0. \tag{7}$$

Now consider $\int (1/x)\,dx$ for $x < 0$. If we let $u = -x$, then $u > 0$ and $du = -1\,dx$. Then

$$\int \frac{1}{x}\,dx = \int \frac{-1\,dx}{-x} = \int \frac{du}{u} = \ln u + C = \ln(-x) + C \qquad \text{for } x < 0.$$

Thus $\int (1/x)\,dx = \ln(-x) + C$ for $x < 0$, so, by the chain rule,

$$\int \frac{du}{u} = \ln(-u) + C \qquad \text{for } u < 0. \tag{8}$$

Formulas (7) and (8) can be collected as one formula, namely

$$\boxed{\int \frac{du}{u} = \ln|u| + C.} \tag{9}$$

Formula (9) is extremely important, and you may wish to memorize it also.

EXAMPLE 5 Find $\int_{-2}^{-1} (1/x)\, dx$.

Solution We have

$$\int_{-2}^{-1} \frac{1}{x}\, dx = \ln|x|\Big]_{-2}^{-1} = \ln|-1| - \ln|-2|$$
$$= \ln 1 - \ln 2 = -\ln 2. \quad \square$$

EXAMPLE 6 Find $\int x/(x^2 + 1)\, dx$.

Solution If we think of $x^2 + 1$ as u, then $du = 2x\, dx$. Multiplying and dividing by 2, we obtain

$$\int \frac{x}{x^2 + 1}\, dx = \frac{1}{2}\int \frac{\overbrace{2x\, dx}^{du}}{\underbrace{x^2 + 1}_{u}} = \frac{1}{2}\ln|x^2 + 1| + C.$$

Since $x^2 + 1$ is always positive, the absolute-value sign is really not needed here. $\square$

We can now integrate $\tan x$ and $\cot x$.

EXAMPLE 7 Find $\int \tan x\, dx$.

Solution We have $\tan x = \sin x/\cos x$, and if we let $u = \cos x$, then $du = -\sin x\, dx$. Thus, fixing up the integral, we have

$$\int \tan x\, dx = \int \frac{\sin x}{\cos x}\, dx$$
$$= -\int \frac{\overbrace{-\sin x\, dx}^{du}}{\underbrace{\cos x}_{u}} = -\ln|\cos x| + C. \quad \square$$

EXAMPLE 8 Find $\int x^2(\cot x^3)\, dx$.

Solution We have

$$x^2(\cot x^3) = \frac{x^2(\cos x^3)}{\sin x^3},$$

and we should spot that the derivative of the denominator in this quotient is the numerator, except for a constant factor 3 that we can "fix up." That is, if $u = \sin x^3$, then $du = (\cos x^3)3x^2\,dx$. Multiplying and dividing by 3 to fix up the integral to fit formula (9), we obtain

$$\int x^2(\cot x^3)\,dx = \frac{1}{3}\int \frac{\overbrace{(\cos x^3)3x^2\,dx}^{du}}{\underbrace{\sin x^3}_{u}} = \frac{1}{3}\ln|\sin x^3| + C. \quad \square$$

Finally, we would like to be able to integrate $\ln x$. We easily find that

$$\frac{d(x(\ln x) - x)}{dx} = x\frac{1}{x} + \ln x - 1 = \ln x.$$

Thus

$$\int \ln x\,dx = x(\ln x) - x + C. \tag{10}$$

Therefore, by the chain rule,

$$\boxed{\int (\ln u)\,du = u(\ln u) - u + C} \tag{11}$$

for a differentiable function u. You may well ask how we knew we should try $x(\ln x) - x$ as a candidate for an antiderivative of $\ln x$. You would probably arrive at $x(\ln x) - x$ after a bit of experimentation. There is also a technique of integration called "integration by parts," which can be used to obtain $x(\ln x) - x$ without guessing. Until you learn how to integrate by parts, you should consult a table for formula (11) if you need it.

EXAMPLE 9 Find $\int x \ln(x^2 + 1)\,dx$.

Solution By formula (11) with $u = x^2 + 1$ and $du = 2x\,dx$, we have

$$\int x \ln(x^2+1)\,dx = \frac{1}{2}\int \ln \underbrace{(x^2+1)}_{u}\underbrace{(2x\,dx)}_{du}$$

$$= \frac{1}{2}[(x^2+1)\ln(x^2+1) - (x^2+1)] + C. \quad \square$$

Many integration formulas found in a table involve the logarithm function, and we can now handle these formulas.

SOME PROPERTIES OF ln x

We will show that the function $\ln x$ has properties that relate multiplication with addition. These relationships make $\ln x$ a very important function indeed.

THEOREM Properties of ln x

8.3 For any $a > 0$ and $b > 0$ and for any rational number (fraction) r, we have

$$\ln(ab) = \ln a + \ln b, \tag{12}$$

$$\ln(a/b) = \ln a - \ln b, \tag{13}$$

$$\ln(a^r) = r(\ln a). \tag{14}$$

Theorem 8.3 makes our use of the name *logarithm* for this function seem more reasonable. You are already familiar with a logarithm function that satisfies formulas (12), (13), and (14). You may wonder how we could ever derive these properties from the definition of $\ln x$ in Eq. (2) as an *integral*. The way it is done is fascinating. To show formula (12), we show that the functions

$$f(x) = \ln(ax) \qquad \text{and} \qquad g(x) = \ln a + \ln x$$

are identical. We can then obtain formula (12) by setting $x = b$.

> This two-step method is often used to show that two differentiable functions $f(x)$ and $g(x)$ having an interval as common domain are the same:
>
> **Step 1** Show that $f' = g'$.
>
> **Step 2** Show that $f(c) = g(c)$ at *one* point c in the domain of the functions.

From Step 1, we can conclude that

$$(f - g)' = f' - g' = 0,$$

so we must have

$$f(x) - g(x) = k$$

for some constant k and all x in the domain of the functions. Since $f(c) = g(c)$ by Step 2, we obtain

$$0 = f(c) - g(c) = k,$$

so $k = 0$ and $f(x) = g(x)$ for all x in the domain of the functions.

We can apply this technique to demonstrate formulas (12), (13), and (14). Let $f(x) = \ln ax$ and let $g(x) = \ln a + \ln x$ for $x > 0$. Then

$$f'(x) = \frac{1}{ax} \cdot a = \frac{1}{x}, \qquad g'(x) = 0 + \frac{1}{x} = \frac{1}{x},$$

so $f'(x) = g'(x)$, and Step 1 is satisfied. For Step 2, we find that

$$f(1) = \ln a, \qquad g(1) = \ln a + \ln 1 = \ln a,$$

since Eq. (3) tells us that $\ln 1 = 0$. Therefore $f(x) = g(x)$ for all $x > 0$. In particular, $f(b) = g(b)$; that is,

$$\ln(ab) = \ln a + \ln b,$$

which establishes formula (12).

It is convenient to establish formula (14) next. Let $h(x) = \ln(x^r)$ and $k(x) = r(\ln x)$ for $x > 0$. Then

$$h'(x) = \frac{1}{x^r}(rx^{r-1}) = \frac{r}{x}, \qquad k'(x) = r\frac{1}{x} = \frac{r}{x},$$

so Step 1 is satisfied. For Step 2, we have

$$h(1) = \ln(1^r) = \ln 1 = 0, \qquad k(1) = r(\ln 1) = r \cdot 0 = 0.$$

Therefore $h(x) = k(x)$ for all $x > 0$; in particular, taking $x = a$, we have

$$\ln(a^r) = r(\ln a),$$

which is formula (14).

From formula (14), we see that

$$\ln\left(\frac{1}{b}\right) = \ln(b^{-1}) = -1(\ln b) = -\ln b.$$

Using formula (12), we then obtain

$$\ln\left(\frac{a}{b}\right) = \ln\left(a\frac{1}{b}\right) = \ln a + \ln\left(\frac{1}{b}\right) = \ln a - \ln b,$$

which establishes formula (13).

The properties of ln in Theorem 8.3 can sometimes be used to simplify the differentiation of a function defined in terms of ln.

EXAMPLE 10 Differentiate $\ln\sqrt{(x^2+1)(2x+3)}$.

Solution Using formulas (12) and (14) we obtain

$$\frac{d(\ln\sqrt{(x^2+1)(2x+3)})}{dx} = \frac{d(\frac{1}{2}[\ln(x^2+1) + \ln(2x+3)])}{dx}$$

$$= \frac{1}{2}\left(\frac{1}{x^2+1}\cdot 2x + \frac{1}{2x+3}\cdot 2\right)$$

$$= \frac{x}{x^2+1} + \frac{1}{2x+3}, \qquad x > -\frac{3}{2}. \quad \square$$

EXAMPLE 11 Find dy/dx if $y = \ln\left(\frac{\tan x}{x^2+1}\right)$.

Solution Using formula (13), we have

$$y = \ln\left(\frac{\tan x}{x^2+1}\right) = \ln(\tan x) - \ln(x^2+1).$$

Then

$$\frac{dy}{dx} = \frac{1}{\tan x}\cdot\sec^2 x - \frac{1}{x^2+1}\cdot 2x = \frac{\sec^2 x}{\tan x} - \frac{2x}{x^2+1}, \qquad \tan x > 0. \quad \square$$

We now use the calculus of ln x to discuss its graph. First, the second derivative of ln x is $-1/x^2 < 0$, so the graph is always concave down. Since $d(\ln x)/dx = 1/x$ and $(1/x) > 0$ for $x > 0$, we see that ln x is an increasing function. However, since $\lim_{x\to\infty}(1/x) = 0$, we also see that the graph of $\ln x$ becomes closer and closer to horizontal as x becomes large. We would like to discover whether $\ln x$ becomes large as x becomes large, or whether $\ln x$ remains less than some constant c for all $x > 1$. Similarly, we would like to know the behavior of $\ln x$ near 0. These limits give the answer:

$$\lim_{x\to\infty} \ln x = \infty, \qquad \lim_{x\to 0+} \ln x = -\infty.$$

We obtain these limits easily from formula (14) as follows. Surely $\ln 2 > 0$, since $\ln 2 = \int_1^2 (1/x)\,dx$. In fact, estimating $\int_1^2 (1/x)\,dx$ by the lower sum s_1, we find that

$$\ln 2 > \tfrac{1}{2}. \tag{15}$$

Therefore by formula (14),

$$\ln(2^n) = n(\ln 2) > n\left(\frac{1}{2}\right) = \frac{n}{2}.$$

As $n \to \infty$, so does $n/2$. Thus for large values 2^n of x, we see that $\ln x$ is also large, so $\lim_{x\to\infty} \ln x = \infty$.

From formulas (13) and (15), we see that

$$\ln\left(\tfrac{1}{2}\right) = \ln 1 - \ln 2 = 0 - \ln 2 < -\tfrac{1}{2}. \tag{16}$$

Using formula (14), we have

$$\ln\left(\frac{1}{2}\right)^n = n\left(\ln\frac{1}{2}\right) < n\left(-\frac{1}{2}\right) = -\frac{n}{2}.$$

As n gets large, $(\frac{1}{2})^n \to 0$ and $-n/2 \to -\infty$. Thus $\lim_{x\to 0+} \ln x = -\infty$. These limits justify our sketch of the graph of $\ln x$ in Fig. 8.16.

We can use Newton's method or Simpson's rule with a calculator or computer to find solutions of equations or definite integrals involving $\ln x$.

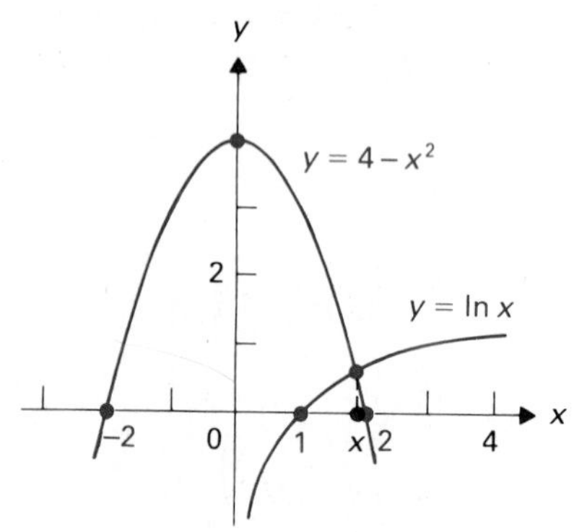

Figure 8.17 $\ln x = 4 - x^2$ has a solution x in [1, 2].

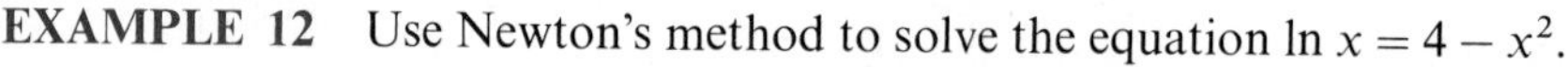

EXAMPLE 12 Use Newton's method to solve the equation $\ln x = 4 - x^2$.

Solution The graph in Fig. 8.17 shows that there will be a solution between $x = 1$ and $x = 2$. We apply Newton's method to $f(x) = \ln x - 4 + x^2$ starting with $a_1 = 2$. The recursion relation is

$$a_{i+1} = a_i - \frac{f(a_i)}{f'(a_i)} = a_i - \frac{\ln a_i - 4 + a_i^2}{\dfrac{1}{a_i} + 2a_i}.$$

Table 8.1 gives the results. We see that $x = 1.841097058$ is very close to a solution of the equation. □

Table 8.1

i	a_i	$f(a_i)$
1	2	0.6931471806
2	1.845967293	0.0205986658
3	1.841101837	0.000020193
4	1.841097058	$1.4 \cdot 10^{-11}$
5	1.841097058	$6 \cdot 10^{-12}$

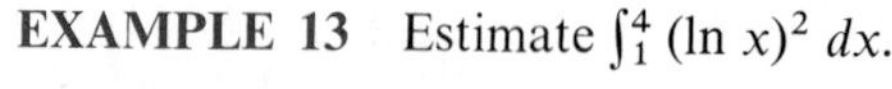

EXAMPLE 13 Estimate $\int_1^4 (\ln x)^2\,dx$.

Solution We use Simpson's rule with $n = 20$ and a calculator and obtain

$$\int_1^4 (\ln x)^2\,dx \approx 2.5969. \quad □$$

SUMMARY

1. $\ln x = \int_1^x \frac{1}{t}\,dt$ for $x > 0$
2. The graph of $\ln x$ is shown in Fig. 8.16.
3. $\frac{d(\ln u)}{dx} = \frac{1}{u} \cdot \frac{du}{dx}$ for $u > 0$
4. $\int \frac{du}{u} = \ln |u| + C$
5. $\ln (ab) = \ln a + \ln b$ for $a, b > 0$
6. $\ln (a/b) = \ln a - \ln b$ for $a, b > 0$
7. $\ln (a^r) = r(\ln a)$ for $a > 0$

EXERCISES

1. Is the function ln continuous? How do you know?
2. Estimate ln 2 by finding the upper sum S_4 for $1/x$ on $[1, 2]$.
3. Find the equation of the tangent line to the graph of $\ln x$ at the point $(1, 0)$.
4. Sketch the graph of $y = \ln |x|$.
5. Sketch the graph of $x = \ln y$.
6. Sketch the graph of $y = \ln (x - 2)$.
7. Sketch the graph of $y = \ln (3 - x)$.
8. Use a differential to estimate ln 1.01.

In Exercises 9 through 16, use formulas (12), (13), and (14) to estimate the quantity, given that $\ln 2 \approx 0.7$ and $\ln 3 \approx 1.1$.

9. $\ln \frac{1}{3}$
10. $\ln 6$
11. $\ln 8$
12. $\ln 12$
13. $\ln \frac{3}{4}$
14. $\ln 27$
15. $\ln \sqrt{3}$
16. $\ln \sqrt[3]{4}$

In Exercises 17 through 34, find the derivative of the given function. Use properties of $\ln x$, where applicable, to simplify the differentiation.

17. $\ln (3x + 2)$
18. $\ln \sqrt{x}$
19. $\ln x^3$
20. $\ln (\tan x)$
21. $\ln (\cos^2 x)$
22. $\ln \left(\frac{2x + 3}{x^2 + 4}\right)$
23. $(\ln x)^2$
24. $(\ln x)(\sin x)$
25. $\ln \sqrt{3x^3 - 4x}$
26. $\frac{\ln x}{x^2}$
27. $\ln (\sec x \tan x)$
28. $\ln ((x^2 + 4x)^2(3x - 2)^3)$
29. $\ln (\cos^2 x \sin^3 2x)$
30. $\tan (\ln x)$
31. $\ln (\ln x)$
32. $\ln (x\sqrt{2x + 3})$
33. $\cos (\ln x)$
34. $\ln \left(\frac{\cos^2 x}{\sin^3 2x}\right)$

In Exercises 35 through 48, find the given integral without using tables.

35. $\int \frac{1}{x + 1}\,dx$
36. $\int \frac{1}{2x + 3}\,dx$
37. $\int \tan 2x\,dx$
38. $\int \cot 3x\,dx$
39. $\int \frac{x}{x^2 + 1}\,dx$
40. $\int \frac{1}{(x + 1)^2}\,dx$
41. $\int \frac{\ln x}{x}\,dx$
42. $\int \frac{1}{x(\ln x)}\,dx$
43. $\int \frac{1}{x \ln (x^2)}\,dx$
44. $\int \frac{\sec^2 x}{\tan x}\,dx$
45. $\int \frac{x}{1 - 4x^2}\,dx$
46. $\int \frac{1}{(3x + 1)^2}\,dx$
47. $\int \frac{\sin x}{1 + \cos x}\,dx$
48. $\int \frac{\sin x}{\cos^2 x}\,dx$

In Exercises 49 through 56, find the given integral, using tables if necessary.

49. $\int_0^1 \frac{x}{x + 1}\,dx$
50. $\int_0^1 \frac{x}{(2x + 1)^2}\,dx$
51. $\int_1^2 \frac{x + 1}{x}\,dx$
52. $\int_0^1 \frac{1}{4 - x}\,dx$
53. $\int_1^4 \frac{1}{x\sqrt{9 + x^2}}\,dx$
54. $\int_1^3 \frac{\sqrt{9 - x^2}}{x}\,dx$
55. $\int_0^{\pi/8} \sec 2x\,dx$
56. $\int_{\pi/3}^{\pi/2} \csc \frac{x}{2}\,dx$

57. a) Show that for $t > 1$, we have $\ln t < 2(\sqrt{t} - 1)$. [*Hint:* Compare $\int_1^t (1/x)\,dx$ with $\int_1^t (1/\sqrt{x})\,dx$ for $t > 1$.]
b) Using part (a), find $\lim_{x\to\infty} [(\ln x)/x]$.
c) Using part (b), find $\lim_{x\to 0+} [x(\ln x)]$. [*Hint:* Let $x = 1/u$ and find the limit as $u \to \infty$.]

Use a calculator or computer in Exercises 58 through 64.

58. Estimate $\ln 5 = \int_1^5 (1/x)\,dx$ using Simpson's rule with $n = 20$. What is the error?

59. Solve $\ln x = x - 2$.

60. Solve $(\ln x)^2 - 4 \ln x - 8 = 0$.

61. Solve $\ln x = \sin x$.

62. Find the solution of $\ln(\sin x) = -2$ where $0 < x < \pi/2$.

63. Estimate $\int_0^1 \ln(\cos x)\,dx$.

64. Estimate $\int_0^1 \ln(x^2 + 1)\,dx$.

8.3 THE FUNCTION e^x

THE EXPONENTIAL FUNCTION

The graph of $y = \ln x$ is shown again in Fig. 8.18. In the preceding section, we showed that

$$\lim_{x\to\infty} \ln x = \infty \quad \text{and} \quad \lim_{x\to 0+} \ln x = -\infty.$$

Since $d(\ln x)/dx = 1/x > 0$ for $x > 0$, we also see that $\ln x$ is an increasing function. Since $\ln x$ is differentiable, Theorem 3.2 on page 85 shows that $\ln x$ is continuous. The intermediate-value theorem then shows that, given any y-value, there is a *unique* x-value such that $y = \ln x$. (See the arrows in Fig. 8.19.) That is, we can consider x to be defined as a *function* of y for all y by the relation $y = \ln x$. For the moment, we will denote this new function by $x = \exp(y)$, the *exponential function* of y.

Whenever $y = f(x)$ also defines x as a function g of y, the functions f and g are called *inverses* of each other. Inverse functions were discussed in detail in Section 8.1.

Figure 8.18 The graph of $y = \ln x$.

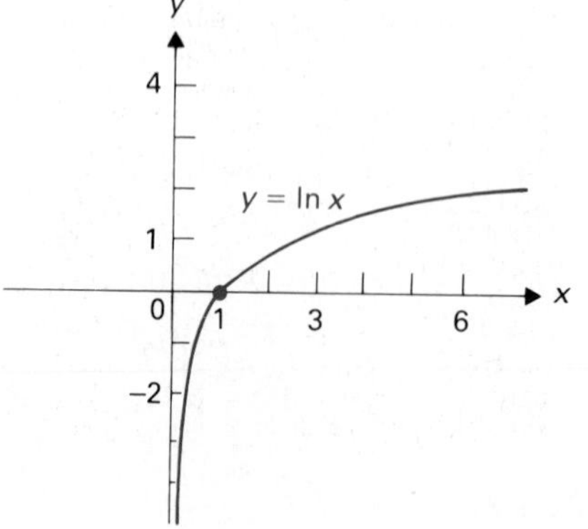

Figure 8.19 Given a y-value, there is a unique x-value such that $y = \ln x$.

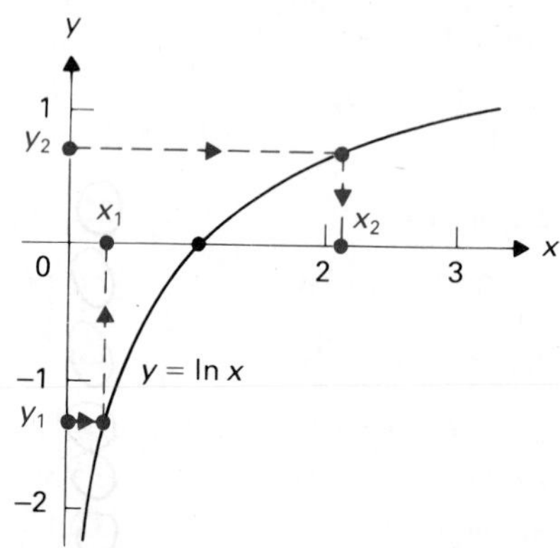

DEFINITION 8.4 Exponential function

The **(natural) exponential function** $x = \exp(y)$ is the inverse of the natural logarithm function $y = \ln x$.

Now exp (1) is the unique number whose natural logarithm is 1. This number is denoted by e, so we have $e = \exp(1)$, or

$$\boxed{\ln e = 1.}$$

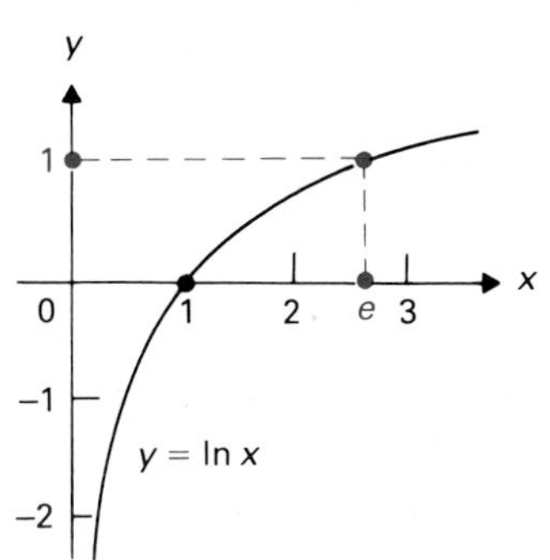

Figure 8.20 ln $e = 1$.

Figure 8.20 shows the number e such that $\ln e = 1$. This number e is one of the most important numbers in mathematics. We can use a calculator and Newton's method with $f(x) = (\ln x) - 1$ to estimate the solution e of $f(x) = 0$. The recursion relation is

$$a_{i+1} = a_i - \frac{f(a_i)}{f'(a_i)} = a_i - \frac{(\ln a_i) - 1}{1/a_i},$$

and a calculator shows that

$$e \approx 2.718281828.$$

The number e, like the number π, is one of those numbers we encounter naturally in mathematics. Both π and e are *irrational numbers;* that is, neither can be expressed as a quotient of integers.

When working with functions, we are accustomed to using x as the independent variable. Thus we would like to express the exponential function in the form $y = \exp(x)$. Figure 8.21 shows how to obtain the graph of $y = \exp(x)$ from the graph of $y = \ln x$. First, the letters on the axes are interchanged, as in Fig. 8.21(a), so that the usual logarithmic curve becomes $x = \ln y$ rather than $y = \ln x$. This curve is now the graph of $y = \exp(x)$, but with the x-axis vertical and the y-axis horizontal. The plane is then rotated a half-turn about the line $y = x$ to bring the axes into their usual position. This gives the graph of $y = \exp(x)$ shown in Fig. 8.21(b).

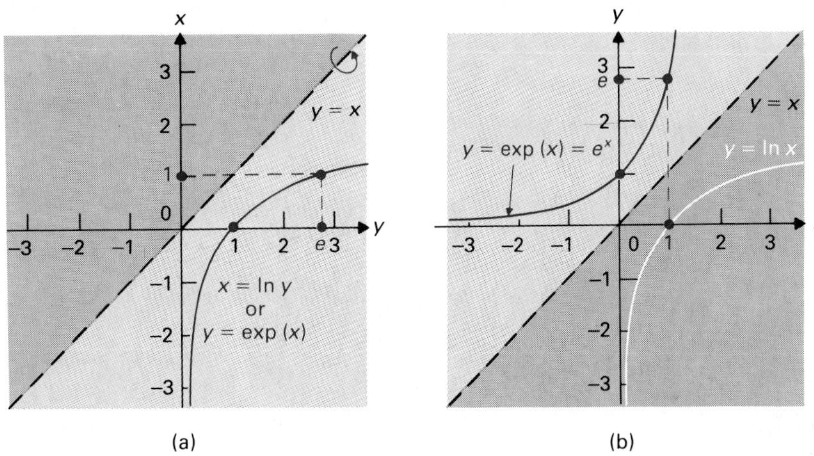

Figure 8.21 Obtaining the graph of $y = e^x$ from the graph of $y = \ln x$. To get from (a) to (b), rotate a half-turn about the line $y = x$.

Using one of the properties of $\ln x$, we see that

$$\ln (e^r) = r(\ln e) = r(1) = r \qquad \textbf{(1)}$$

for any rational number r. Since exp is the inverse of the function ln, we see at once from Eq. (1) that

$$\exp (r) = \exp (\ln e^r) = e^r \qquad \textbf{(2)}$$

for any rational number r. Equation (2) explains the use of the name *exponential*, for e^r is the number e to the *exponent* r; that is, it is e raised to the rth power. We have not defined a^x except for x a rational number. Equation (2) suggests that we *define* e^x to be $\exp x$ *for any real number* x; note the labeling of the graph in Fig. 8.21(b). From now on, we will use the more intuitive notation e^x to denote the exponential function.

Values of the function e^x can be found using a calculator or computer. On some calculators, e^x is computed as the inverse logarithm of x, using an inverse key and a $\ln x$ key. On others, it may be computed directly, and $\ln x$ may have to be computed using an inverse key. On a computer, $\exp(x)$ often stands for e^x in the program language.

PROPERTIES OF e^x

As indicated in Fig. 8.21(b), the exponential function e^x is an increasing function and

$$\boxed{e^x > 0 \qquad \text{for all } x.} \qquad \textbf{(3)}$$

This is a useful fact to remember. Since the inverse relations $y = \ln x$ and $x = e^y$ hold simultaneously, we see that $x = e^y = e^{\ln x}$, so

$$\boxed{e^{\ln x} = x \qquad \text{for } x > 0.} \qquad \textbf{(4)}$$

Similarly,

$$\boxed{\ln (e^x) = x \qquad \text{for all } x} \qquad \textbf{(5)}$$

follows from the simultaneous relationships $y = e^x$ and $x = \ln y$. We see at once from Eq. (4) that

$$\boxed{\ln x \text{ is the power to which } e \text{ must be raised to yield } x.}$$

This is sometimes a useful way to think of $\ln x$. Perhaps you have studied the function $\operatorname{Log} x$, defined, for $x > 0$, as the power to which 10 must be raised to equal x.

EXAMPLE 1 Simplify $\ln (1/e^5)$.

Solution By a property of $\ln x$ and Eq. (5), we have

$$\ln \left(\frac{1}{e^5}\right) = \ln (1) - \ln (e^5) = 0 - 5 = -5. \quad \square$$

Equation (4) enables us to give a definition of a^b for any $a > 0$ and *any* real number b. Since

$$a = e^{\ln a},$$

we *define* a^b by

$$\boxed{a^b = e^{b(\ln a)} \qquad \text{for } a > 0.} \tag{6}$$

The usual laws of exponents hold for the function e^x, namely,

$$e^a \cdot e^b = e^{a+b}, \tag{7}$$

$$\frac{e^a}{e^b} = e^{a-b}, \tag{8}$$

$$(e^a)^b = e^{ab}, \tag{9}$$

To verify Eqs. (7) and (8), we simply take the natural logarithm of each side of the equation; $\ln x$ is an increasing function, so $\ln x_1 = \ln x_2$ must imply $x_1 = x_2$. Now

$$\ln (e^a \cdot e^b) = \ln (e^a) + \ln (e^b) = a + b,$$

while

$$\ln (e^{a+b}) = a + b$$

also. Therefore $e^a \cdot e^b = e^{a+b}$. Similarly,

$$\ln \left(\frac{e^a}{e^b}\right) = \ln (e^a) - \ln (e^b) = a - b,$$

and

$$\ln (e^{a-b}) = a - b,$$

which establishes Eq. (8). Finally, by our definition in Eq. (6),

$$(e^a)^b = e^{b(\ln e^a)} = e^{ba} = e^{ab}.$$

EXAMPLE 2 Simplify $(e^2)^{\ln 3}$.

Solution Equations (9) and (4) yield

$$(e^2)^{\ln 3} = e^{2(\ln 3)} = (e^{\ln 3})^2 = 3^2 = 9. \quad \square$$

THE DERIVATIVE OF e^x

Since $\ln x$ is differentiable, the graph of $y = \ln x$ has a tangent line at each point. This tangent line is never horizontal since $d(\ln x)/dx = 1/x \neq 0$ for any $x > 0$. Since the graph of e^x is obtained from $y = \ln x$ by rotating the plane a half-turn about the line $y = x$, we see that the graph of $y = e^x$, shown in Fig. 8.21(b), has a nonvertical tangent line at each point. Thus e^x must also be differentiable.

If $y = e^x$, then $x = \ln y$. Implicit differentiation of $x = \ln y$ with respect to x yields

$$1 = \frac{1}{y} \cdot \frac{dy}{dx},$$

so

$$\frac{dy}{dx} = y = e^x.$$

We have shown that

$$\frac{d(e^x)}{dx} = e^x.$$

Thus *the exponential function e^x is unchanged by differentiation.* This is one of the reasons the exponential function is so extremely important. If u is a differentiable function of x, then, by the chain rule,

$$\boxed{\frac{d(e^u)}{dx} = e^u \cdot \frac{du}{dx}.} \tag{10}$$

You should memorize formula (10). *We warn you not to confuse formula* (10), *which gives the "derivative of a constant to a function power," with the formula* $d(u^n)/dx = nu^{n-1} \cdot (du/dx)$ *for the "derivative of a function to a constant power."*

EXAMPLE 3 Find the derivative of $e^{\sin x}$.

Solution We have

$$\frac{d(e^{\sin x})}{dx} = e^{\sin x} \cdot \frac{d(\sin x)}{dx} = e^{\sin x} \cos x. \quad \square$$

EXAMPLE 4 Find dy/dx if $y = e^{\sqrt{x}}$.

Solution We have

$$\frac{dy}{dx} = e^{\sqrt{x}} \cdot \frac{d(\sqrt{x})}{dx} = e^{\sqrt{x}} \cdot \frac{1}{2\sqrt{x}} = \frac{e^{\sqrt{x}}}{2\sqrt{x}}. \quad \square$$

Let u be a differentiable function of x such that $u > 0$. We can now show that our old formula

$$\frac{d(u^r)}{dx} = ru^{r-1} \cdot \frac{du}{dx}$$

holds for *every* real number r. (It was only shown for *rational* r before.) We have

$$\frac{d(u^r)}{dx} = \frac{d(e^{r(\ln u)})}{dx} = e^{r(\ln u)} \cdot \frac{d(r(\ln u))}{dx} = e^{r(\ln u)} \cdot r \cdot \frac{1}{u} \cdot \frac{du}{dx}$$

$$= ru^r \cdot \frac{1}{u} \cdot \frac{du}{dx} = ru^{r-1} \cdot \frac{du}{dx}.$$

THE INTEGRAL OF e^x

From $d(e^x)/dx = e^x$, we obtain

$$\int e^x \, dx = e^x + C, \tag{11}$$

and from formula (10),

$$\boxed{\int e^u \, du = e^u + C} \tag{12}$$

for a differentiable function u. You should learn formula (12) also.

EXAMPLE 5 Find $\int xe^{x^2} \, dx$.

Solution Taking $u = x^2$ so that $du = 2x \, dx$ and fixing up the integral, we find that

$$\int xe^{x^2} \, dx = \frac{1}{2} \int e^{\overbrace{x^2}^{u}} \cdot \underbrace{2x \, dx}_{du} = \frac{1}{2} e^{x^2} + C. \quad \square$$

EXAMPLE 6 Find

$$\int \frac{e^{2x}}{1 - e^{2x}} \, dx.$$

Solution We should spot at once that the derivative of the denominator of the integrand is the numerator, except for a constant factor -2. Thus if $u = 1 - e^{2x}$, we have

$$du = -e^{2x} \cdot 2 \, dx,$$

so

$$\int \frac{e^{2x}}{1 - e^{2x}} \, dx = -\frac{1}{2} \int \frac{\overbrace{(-e^{2x} \cdot 2) \, dx}^{du}}{\underbrace{1 - e^{2x}}_{u}} = -\frac{1}{2} \int \frac{du}{u}$$

$$= -\frac{1}{2} \ln |u| + C = -\frac{1}{2} \ln |1 - e^{2x}| + C. \quad \square$$

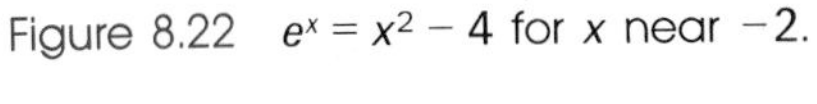

Figure 8.22 $e^x = x^2 - 4$ for x near -2.

EXAMPLE 7 Use Newton's method to approximate the solution of $e^x = x^2 - 4$.

Solution The sketch in Fig. 8.22 indicates that there is a solution close to $x = -2$. We let $f(x) = e^x - x^2 + 4$ and use the recursion relation

$$a_{i+1} = a_i - \frac{e^{a_i} - a_i^2 + 4}{e^{a_i} - 2a_i},$$

with $a_1 = -2$. The data we found in Table 8.2 using a calculator indicate that the desired solution is approximately $x = -2.032488404$.

Table 8.2

i	a_i	$f(a_i)$
1	-2	0.1353352832
2	-2.032726556	-0.0009993378
3	-2.032488416	-0.000000053
4	-2.032488404	$-4 \cdot 10^{-12}$
5	-2.032488404	$-4 \cdot 10^{-12}$

SUMMARY

1. e is the unique number such that $\ln e = 1$.
2. $y = e^x$ is the inverse relation to $x = \ln y$.
3. $e^{\ln x} = x$ for $x > 0$
4. $\ln e^x = x$ for all x
5. $e^a e^b = e^{a+b}$
6. $e^a/e^b = e^{a-b}$
7. $(e^a)^b = e^{ab}$
8. $\dfrac{d(e^u)}{dx} = e^u \cdot \dfrac{du}{dx}$
9. $\int e^u\,du = e^u + C$

EXERCISES

In Exercises 1 through 12, use properties of logarithm and exponential functions to simplify the given expression.

1. $\ln (e^2)$
2. $e^{\ln 3}$
3. $\ln (e^{\ln 1})$
4. $\ln (e^{1/2})$
5. $\ln \left(\dfrac{1}{e^2}\right)$
6. $e^{(1+\ln 2)}$
7. $e^{(\ln 2+\ln 3)}$
8. $e^{\ln 3-\ln 4}$
9. $e^{4(\ln 2)}$
10. $\ln \left(\dfrac{\sqrt{e}}{e^2}\right)$
11. $(e^2)^{\ln 3}$
12. $\ln (\ln e)$

In Exercises 13 through 28, find the derivative of the given function.

13. e^{2x}
14. xe^x
15. $e^{2x} \sin x$
16. $e^x + e^{-x}$
17. $e^x(\ln 2x)$
18. $\tan (e^x)$
19. $e^{\sec x}$
20. $e^{(x^2+x)}$
21. $e^{1/x}$
22. $x^2 e^{-x^2}$
23. $\dfrac{e^x + 1}{e^x}$
24. $\dfrac{e^{2x}}{1 + e^x}$
25. $\dfrac{e^{\sin x}}{\cos x}$
26. $e^{x^2}(\ln x)$
27. $\dfrac{e^{-x^2}}{x + 1}$
28. $\dfrac{e^x - e^{-x}}{e^x + e^{-x}}$

In Exercises 29 through 40, find the given integral without the use of tables.

29. $\int_0^{\ln 3} e^{2x}\,dx$
30. $\int x^2 e^{x^3}\,dx$
31. $\int (\sin x) e^{\cos x}\,dx$
32. $\int_0^{\ln 2} (e^x - e^{-x})\,dx$
33. $\int \dfrac{e^x}{1 + e^x}\,dx$
34. $\int \dfrac{e^{\sqrt{x}}}{\sqrt{x}}\,dx$
35. $\int \dfrac{e^x + e^{-x}}{e^x - e^{-x}}\,dx$
36. $\int e^{\ln \sqrt{x}}\,dx$
37. $\int (e^{3x} + e^{2x})^{1/2}\,dx$
38. $\int \dfrac{e^{2x}}{\sqrt{e^{2x} - 4}}\,dx$
39. $\int e^{1+\ln x}\,dx$
40. $\int e^{3+\ln (\sin x)}\,dx$
41. Sketch the graph of e^{-x}.
42. Sketch the graphs of e^x and e^{x^2} on the same axes.
43. Sketch the graphs of e^{-x} and e^{-x^2} on the same axes.
44. Find all values of m such that $y = e^{mx}$ is a solution of the differential equation

$$\frac{d^2y}{dx^2} - 5\frac{dy}{dx} + 6y = 0.$$

45. Repeat Exercise 44 for the differential equation

$$\frac{d^2y}{dx^2} - 4y = 0.$$

46. Repeat Exercise 44 for the differential equation

$$\frac{d^3y}{dx^3} - 9\frac{dy}{dx} = 0.$$

Use a calculator for Exercises 47 through 54. Find all solutions of the given equation. In Exercises 51 through 54, we suggest substituting $w = e^x$, solving for w, and then finding x.

47. $e^x = 10 - x$

48. $e^{3x} = 1000 - 2x$

49. $e^{-x} = \ln x$

50. $e^{-x} = 12 - x^2$

51. $e^{2x} + 7e^x - 3 = 0$

52. $e^{3x} - 9e^{2x} + 10 = 0$

53. $e^{(e^x)} + 3e^x - 10 = 0$

54. $e^{(e^x)} - e^{2x} - 2 = 0$

8.4 OTHER BASES AND LOGARITHMIC DIFFERENTIATION

We have just studied the function e^x where $e \approx 2.718281828$. You may wonder why we did not start by studying calculus for more familiar exponential functions, like 2^x or 10^x. This section fills that gap and also shows that for calculus, e is really the most convenient base to use for an exponential function. The situation is much like that for the trigonometric functions, where *radian measure* is the most convenient measure to use with calculus.

EXPONENTIAL AND LOGARITHM FUNCTIONS WITH OTHER BASES

The function e^x is sometimes called the *exponential function with base e*. Using the functions e^x and $\ln x$, we can easily define an exponential function a^x with any base $a > 0$. Since $a = e^{\ln a}$ and $(e^{\ln a})^x = e^{(\ln a)x}$, it is natural to *define*

$$\boxed{a^x = e^{(\ln a)x}}$$

for any real number x. We made this definition in the preceding section.

Now $y = e^x$ is an increasing function with range all $y > 0$. It follows that if $a > 1$ so that $\ln a > 0$, then $y = e^{(\ln a)x} = a^x$ is an increasing function. If $0 < a < 1$ so that $\ln a < 0$, then $y = a^x$ is a decreasing function, again with range $y > 0$. Thus if $a \neq 1$ and $y > 0$, there is a unique x such that $y = e^{(\ln a)x} = a^x$. The function that assigns to each such $y > 0$ the corresponding x-value is the inverse of the function $y = a^x$. This inverse function is the *logarithm function with base a*, written $\log_a$. We summarize these observations in a definition.

DEFINITION 8.5 Functions with other bases

The **exponential function with base** $a > 0$ is defined by

$$a^x = e^{(\ln a)x}. \tag{1}$$

For positive $a \neq 1$, the inverse of a^x is the **logarithm function with base** a and is denoted by $\log_a x$.

Thus $\ln = \log_e$ is the logarithm function with base e. For the functions a^x and $\log_a x$, we have the relations

$$a^{\log_a x} = x, \tag{2}$$

$$\log_a (a^x) = x. \tag{3}$$

Before studying calculus, you may have worked with the function $\text{Log}\, x = \log_{10} x$. With $a = 10$, Eq. (2) becomes

$$10^{\log_{10} x} = x,$$

so that $\log_{10} x$ is the power to which 10 must be raised to yield x.

For $a > 0$ and any numbers b and c, the three properties

$$a^b \cdot a^c = a^{b+c}, \tag{4}$$

$$\frac{a^b}{a^c} = a^{b-c}, \tag{5}$$

$$(a^b)^c = a^{bc} \tag{6}$$

of the function a^x follow at once from the definition of a^x in Eq. (1) and the properties of e^x shown in the preceding section. From these properties of a^x, we easily obtain, for $b > 0$ and $c > 0$,

$$\log_a (bc) = \log_a b + \log_a c, \tag{7}$$

$$\log_a \left(\frac{b}{c}\right) = \log_a b - \log_a c, \tag{8}$$

$$\log_a (b^c) = c(\log_a b). \tag{9}$$

If $f(x)$ and $g(x)$ are functions, it is natural to *define* the function $f(x)^{g(x)}$ for all x such that $f(x) > 0$ by

$$f(x)^{g(x)} = e^{(\ln f(x))\cdot g(x)}.$$

CALCULUS USING OTHER BASES

The derivatives of the functions introduced in this section are easily found since they are defined in terms of the exponential function. Indeed, we suggest that you simply express all such functions in terms of the base e when differentiating or integrating, rather than memorize the formulas below. That is, repeat the derivation of the formula each time. To illustrate:

$$\frac{d(a^x)}{dx} = \frac{d(e^{(\ln a)x})}{dx} = e^{(\ln a)x} \cdot \frac{d((\ln a)x)}{dx}$$

$$= (a^x)(\ln a) = (\ln a)(a^x).$$

This gives the formula

$$\frac{d(a^x)}{dx} = (\ln a)(a^x). \tag{10}$$

If u is a differentiable function of x, the chain rule then yields

$$\frac{d(a^u)}{dx} = (\ln a)a^u \cdot \frac{du}{dx}. \tag{11}$$

EXAMPLE 1 Find the derivative of 2^x.

Solution We have

$$\frac{d(2^x)}{dx} = \frac{de^{(\ln 2)x}}{dx} = (\ln 2)e^{(\ln 2)x} = (\ln 2)2^x. \quad \square$$

EXAMPLE 2 Find dy/dx if $y = 5^{\sin x}$.

Solution Formula (11) yields

$$\frac{dy}{dx} = (\ln 5)5^{\sin x} \cdot \cos x. \quad \square$$

EXAMPLE 3 Find the derivative of x^x.

Solution We have

$$\begin{aligned}\frac{d(x^x)}{dx} = \frac{de^{(\ln x)x}}{dx} &= e^{(\ln x)x}\left(\ln x + x \cdot \frac{1}{x}\right)\\ &= x^x(\ln x + 1). \quad \square\end{aligned}$$

Turning to integration, we now find $\int a^x\,dx$. We change to base e and fix up the integral:

$$\begin{aligned}\int a^x\,dx = \int e^{(\ln a)x}\,dx &= \frac{1}{\ln a}\int (e^{(\ln a)x} \cdot \ln a)\,dx\\ &= \frac{1}{\ln a}e^{(\ln a)x} + C = \frac{1}{\ln a}a^x + C.\end{aligned}$$

Using the chain rule, we obtain the formula

$$\int a^u\,du = \frac{1}{\ln a}a^u + C. \tag{12}$$

EXAMPLE 4 Find $\int x10^{x^2}\,dx$.

Solution Setting $u = x^2$, we have $du = 2x\,dx$. Then we fix up the integral to fit formula (12) and obtain

$$\int x10^{x^2}\,dx = \frac{1}{2}\int 10^{x^2}(2x\,dx) = \frac{1}{2}\frac{1}{\ln 10} \cdot 10^{x^2} + C. \quad \square$$

The derivative of the function $\log_a$ can be found by implicit differentiation, since formula (10) gives the derivative of the inverse function a^x. We leave as an exercise (Exercise 49) the demonstration that

$$\frac{d(\log_a x)}{dx} = \frac{1}{\ln a} \cdot \frac{1}{x} \tag{13}$$

The chain rule version of formula (13) is given in the Summary to this section.

Probably you will have little occasion to use calculus with exponential or logarithm functions to bases other than e. With base $a = e$, the annoying constant $\ln a$ becomes $\ln e = 1$, and the calculus is easy to handle. This is one of the reasons e is such an important number in calculus, just as π is important for calculus. If u is in radians and x is in degrees, then $u = (\pi/180)x$. If we use *degree measure* x for the trigonometric functions, then by the chain rule,

$$\frac{d(\sin x)}{dx} = (\cos x)\left(\frac{\pi}{180}\right), \qquad x \text{ in degrees.}$$

Radian measure is the only one that eliminates a nuisance factor like $\pi/180$ for calculus of the trigonometric functions. Similarly, $a = e$ is the only base for an exponential function that eliminates the necessity for the factor $\ln a$ in the differentiation of a^x, since e is the *unique* number x satisfying $\ln x = 1$.

LOGARITHMIC DIFFERENTIATION

Rather than use the technique of the preceding section to differentiate a^x, x^x, and so on, you might prefer *logarithmic differentiation.* To differentiate $y = x^x$ by this method, we take the logarithm of both sides of the equation and then differentiate implicitly. Illustrating with the function $y = x^x$, we obtain

$$\begin{aligned} y &= x^x, \\ \ln y &= \ln (x^x), \\ \ln y &= x(\ln x), \\ \frac{1}{y} \cdot \frac{dy}{dx} &= x \cdot \frac{1}{x} + \ln x = 1 + \ln x, \\ \frac{dy}{dx} &= y(1 + \ln x) = x^x(1 + \ln x). \end{aligned}$$

We have solved the problem very easily. Logarithmic differentiation is handy for such exponentials and also for products and quotients. Here is another illustration.

EXAMPLE 5 Find dy/dx if $y = 2^x \cdot (\sin x)^{\cos x}$.

Solution The technique of logarithmic differentiation yields

$$\begin{aligned} \ln y &= \ln (2^x) + \ln ((\sin x)^{\cos x}), \\ \ln y &= x(\ln 2) + (\cos x) \cdot \ln (\sin x), \\ \frac{1}{y} \cdot \frac{dy}{dx} &= (\ln 2) + (\cos x) \cdot \frac{1}{\sin x} \cdot (\cos x) + (\ln (\sin x))(-\sin x), \end{aligned}$$

$$\frac{dy}{dx} = y\left[(\ln 2) + \frac{\cos^2 x}{\sin x} - (\sin x)(\ln(\sin x))\right],$$

$$\frac{dy}{dx} = 2^x(\sin x)^{\cos x}[(\ln 2) + \cos x \cot x - (\sin x)(\ln(\sin x))].$$

The answer isn't very pretty, but it was not hard to find. □

EXAMPLE 6 Find dy/dx if $y = x^x/(x+1)^2$.

Solution Using logarithmic differentiation, we have

$$\ln y = \ln(x^x) - \ln(x+1)^2,$$

$$\ln y = x(\ln x) - 2\ln(x+1),$$

$$\frac{1}{y}\cdot\frac{dy}{dx} = x\cdot\frac{1}{x} + (\ln x) - \frac{2}{x+1},$$

$$\frac{dy}{dx} = y\left(1 + (\ln x) - \frac{2}{x+1}\right),$$

$$\frac{dy}{dx} = \frac{x^x}{(x+1)^2}\left(1 + \ln(x) - \frac{2}{x+1}\right). \quad □$$

You must not think that logarithmic differentiation can be used only with the exponential functions introduced in this chapter. We can use the technique to advantage with problems that we could have solved in Chapter 4.

EXAMPLE 7 Find dy/dx if

$$y = \frac{x^3\sin^2 x}{(x+1)(x-2)^2}.$$

Solution Using logarithmic differentiation, we have

$$\ln y = 3(\ln x) + 2\ln(\sin x) - \ln(x+1) - 2\ln(x-2),$$

$$\frac{1}{y}\cdot\frac{dy}{dx} = \frac{3}{x} + 2\frac{-\cos x}{\sin x} - \frac{1}{x-1} - \frac{2}{x-2},$$

$$\frac{dy}{dx} = \frac{x^3\sin^2 x}{(x+1)(x-1)^2}\left(\frac{3}{x} - 2\cot x - \frac{1}{x-1} - \frac{2}{x-2}\right).$$

This solution looks quite different from the one we would have found in Chapter 4 using the product and quotient rules, but it is no less valid. □

The Summary gives some formulas we proved, and some we didn't.

SUMMARY

1. $a^x = e^{(\ln a)x}$
2. $\log_a x$ is the inverse of the exponential function with base a.
3. $a^{\log_a x} = x$

4. $\log_a(a^x) = x$

5. $\log_a x = \frac{1}{\ln a}(\ln x)$

6. $\frac{d(a^u)}{dx} = (\ln a)a^u \cdot \frac{du}{dx}$

7. $\int a^u\, du = \frac{1}{\ln a} a^u + C$

8. $\frac{d(\log_a u)}{dx} = \frac{1}{\ln a} \cdot \frac{1}{u} \cdot \frac{du}{dx}$ for $u > 0$

9. Logarithmic differentiation simplifies the differentiation of products, quotients, and exponentials.

EXERCISES

In Exercises 1 through 14, use properties of logarithm and exponential functions to simplify the given expression.

1. $2^{\log_2 5}$
2. $3^{-\log_3 7}$
3. $3^{\log_5 1}$
4. $\log_4 64$
5. $\log_{25} 5$
6. $\log_3(3^4)$
7. $4^{\log_2 3}$
8. $2^{\log_4 2}$
9. $\log_2(4^3)$
10. $\log_9(3^8)$
11. $2^{3(\log_2 4)}$
12. $\log_{10}(2^{\log_2 10})$
13. $\log_3(2^{5(\log_2 3)})$
14. $2^{(\ln 3)(\log_2 e)}$

In Exercises 15 through 32, find the derivative of the given function.

15. 2^{3x}
16. 10^{x^2+2x}
17. $x^2 3^{2x-1}$
18. $\frac{4^{1/x}}{x}$
19. $2^x 3^{2x}$
20. $x(3^{\sin x})$
21. $\log_{10} 2x$
22. $\log_2(x^2 + 1)$
23. $\log_2(\sin x)$
24. $\log_{10}\left(\frac{1}{x}\right)$
25. $\log_5 \frac{2x+1}{x}$
26. $\log_3(x \sin x)$
27. $\log_4(x^x)$
28. x^{2x}
29. $(\sin x)^x$
30. $x^{\tan x}$
31. $\cos(x^x)$
32. $(x - 1)^{x^2}$

In Exercises 33 through 40, use logarithmic differentiation to find the derivative of the given function.

33. $2^x 3^{2x}$
34. $7^x \cdot 8^{-x^2} \cdot 100^x$
35. $5^{x^2}/7^x$
36. $\frac{5^{\sin x}}{2^{1/x}}$
37. $x^2 e^x \cos x$
38. $\frac{x^2 3^x}{\tan x}$
39. $(x^2 + 1)\sqrt{2x + 3}(x^3 - 2x)$
40. $x^{\sin x}/(\cos x)^x$

In Exercises 41 through 48, find the given integral without the use of tables.

41. $\int \frac{e^{\sqrt{x}}}{\sqrt{x}}\, dx$
42. $\int 3^{-x}\, dx$
43. $\int x(5^{3x^2})\, dx$
44. $\int 2^{\sin x} \cos x\, dx$
45. $\int \frac{2^x}{1 + 2^x}\, dx$
46. $\int \frac{3^x}{(1 - 3^x)^3}\, dx$
47. $\int 2^{(2^x)} \cdot 2^x\, dx$
48. $\int 3^{(3^x+1)} \cdot 3^x\, dx$

49. Show that $d(\log_a x)/dx = 1/[(\ln a)x]$.

50. Explain why the integration formula

$$\int \frac{1}{(\ln a)u}\, du = \log_a |u| + C$$

is not of much importance.

51. Sketch the graphs of $y = 2^x$ and $y = 3^x$ on the same set of axes.

52. Sketch the graphs of $y = 3^{-x}$ and $y = 5^{-x}$ on the same set of axes.

53. Discuss the effect of the size of a on the graph of a^x. Distinguish the cases $0 < a < 1$, $a = 1$, and $a > 1$.

54. Let $f(x) = x^x$.
a) Find all local maxima and local minima of f.
b) Show that the graph of f is concave up.
c) Given that $\lim_{x\to 0} x^x = 1$, find $\lim_{x\to 0} f'(x)$.
d) Using the preceding parts of this exercise, sketch the graph of f.

Use a computer or calculator for Exercises 55 through 60.

55. Find all solutions of $x^{2x} - 5(x^x) - 6 = 0$.

56. Find all solutions of $x^x = \cos x$.

57. Find all solutions of $x^x = 2^x + 4$.

58. Estimate $\int_1^3 x^x\,dx$.

59. Estimate $\int_1^{10} \log_{10} \sqrt{x}\,dx$.

60. Estimate $\int_\pi^{2\pi} x^{\sin x}\,dx$.

8.5 APPLICATIONS TO GROWTH AND DECAY

GROWTH AND THE EQUATION $dy/dt = ky$

There are many physical situations where the rate at which a quantity changes is proportional to its magnitude. If the magnitude y is given by $y = f(t)$ at time t, then we must have

$$\frac{dy}{dt} = ky, \qquad \text{or} \qquad f' = k \cdot f. \tag{1}$$

Here are descriptions of four situations where differential equations like Eq. (1) occur.

Application 1 It is known that the rate of decay of any particular radioactive element is proportional to the amount of the element present. This means that if $Q = f(t)$ gives the quantity of the element present at time t, then

$$\frac{dQ}{dt} = -cQ, \tag{2}$$

where c is a positive constant of proportionality. (The minus sign occurs because the quantity Q present is *decreasing* as time increases.) Since Eq. (2) can be written in the form $f' = -c \cdot f$, we see that Eq. (2) is an equation of the form of Eq. (1), where $k = -c$ is a negative constant.

Application 2 Suppose a body traveling through a medium (like air or water) is subject only to a force of retardation proportional to the velocity of the body through the medium. Let $v = f(t)$ be the velocity at time t. By Newton's second law of motion, the force F is ma, where m is the mass of the body and $a = dv/dt$ is the acceleration. Thus $m(dv/dt) = -cv$, where c is some positive constant of proportionality, so

$$\frac{dv}{dt} = -\frac{c}{m} v. \tag{3}$$

Since $v = f(t)$, we may write Eq. (3) in the form $f' = (-c/m) \cdot f$, which is again an equation of the form of Eq. (1), with $k = -c/m$.

Application 3 Under "ideal" conditions with no overcrowding, predators, or disease, the rate of growth of a population P (whether it be people or bacteria) is proportional to the size of the population. This means that if $P = f(t)$ is the size of the population at time t, then

$$\frac{dP}{dt} = cP, \tag{4}$$

for some constant c. Equation (4) can be written as $f' = c \cdot f$, which is again of the form of Eq. (1).

Application 4 Banks often offer savings accounts where the interest is "compounded continuously." If the amount in such a savings account at time t years is $S = f(t)$, and if the interest rate is c percent per year, then

$$\frac{dS}{dt} = \frac{c}{100} S.$$

This is again a differential equation of the form of Eq. (1), where $k = 0.01c$.

SOLUTIONS OF THE EQUATION $dy/dt = ky$

We can easily check that, for any constant A, a solution of the equation $dy/dt = ky$ is given by

$$y = f(t) = Ae^{kt}. \tag{5}$$

Namely, from Eq. (5)

$$\frac{dy}{dt} = Ae^{kt} \cdot k = k(Ae^{kt}) = ky.$$

We now show that the functions Ae^{kt} are the only solutions of the differential equation (1). Starting with $dy/dt = ky$, we separate the variables and solve the equation thus:

$$\frac{dy}{dt} = ky, \qquad \frac{dy}{y} = k\,dt, \qquad \int \frac{dy}{y} = \int k\,dt,$$

$$\ln|y| = kt + C, \qquad |y| = e^{kt+C} = e^{kt} \cdot e^C,$$

$$y = \pm e^C \cdot e^{kt}.$$

Now $\pm e^C$ may be any constant except 0. A check shows that $y = 0$ is a solution of $dy/dt = ky$. We lost this solution when we divided by y to form the second step of the solution. Therefore if we let A be any constant, our solution can be written as

$$y = Ae^{kt},$$

which is what we wished to show.

We know now that each of the functions $f(t)$ for the physical situations described in Applications 1 through 4 must be one of the functions Ae^{kt} for some constants A and k. Note that for $f(t) = Ae^{kt}$, we have $f(0) = A$. Thus A has the physical interpretation of the *initial quantity* y_0 at time $t = 0$. The solution thus takes the form

$$\boxed{y = y_0 e^{kt}.} \tag{6}$$

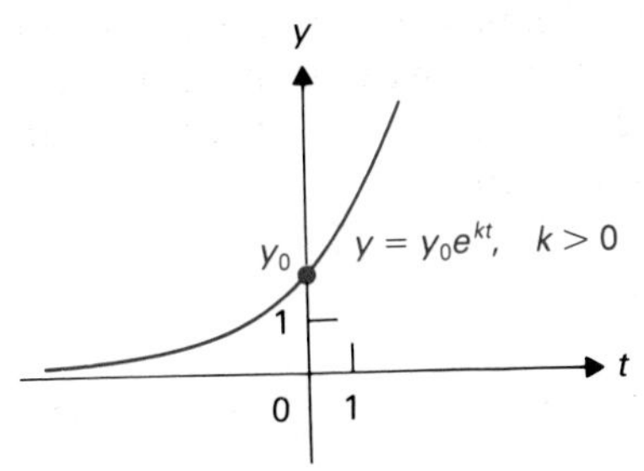

Figure 8.23 Exponential growth occurs if $k > 0$.

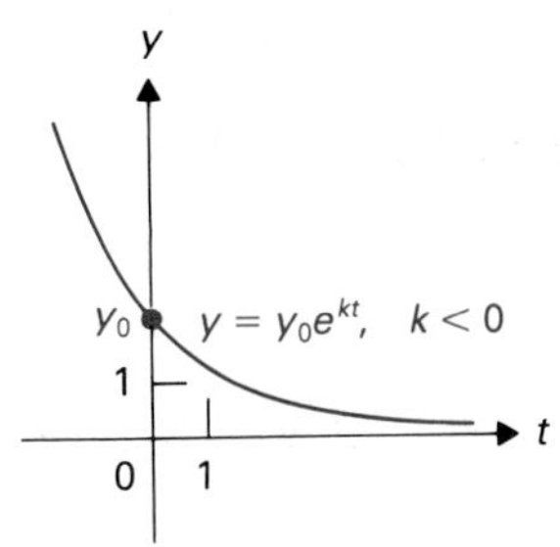

Figure 8.24 Exponential decay occurs if $k < 0$.

The two constants y_0 and k can be determined if the value of $f(t)$ is known at two different times t_1 and t_2. The sign of k determines whether $f(t)$ is *increasing* (if $k > 0$) or *decreasing* (if $k < 0$) as the time t increases; this is indicated by Figs. 8.23 and 8.24.

APPLICATIONS

We illustrate the preceding discussion with some specific examples.

EXAMPLE 1 Show that the time t_h required for half of an initial amount of a radioactive element to decay is independent of the initial amount of the element present. (This time t_h is the *half-life* of the element.)

Solution If an amount Q of the element is present at time t and if Q_0 is the initial amount present at time $t = 0$, then we know by Application 1 that $dQ/dt = -cQ$ for some positive constant c. Our work above then shows that

$$Q = Q_0e^{-ct}.$$

Since $Q = \frac{1}{2}Q_0$ when $t = t_h$, we have

$$\tfrac{1}{2}Q_0 = Q_0e^{-ct_h}.$$

so

$$e^{ct_h} = 2.$$

Taking the (natural) logarithm of each side of this equation, we obtain

$$ct_h = \ln 2,$$

so

$$t_h = \frac{1}{c}(\ln 2).$$

Thus t_n depends only on the constant c and is independent of the initial amount Q_0 of the element present. □

EXAMPLE 2 A certain culture of bacteria grows at a rate proportional to its size. If the size doubles in 3 days, find the time required for the culture to increase to 100 times its original size.

Solution If we let $P = f(t)$ be the size of the culture after t days, then $dP/dt = kP$ for some constant k. The form of the solution (6) shows that

$$P = P_0e^{kt},$$

where P_0 is the initial size of the culture. When $t = 3$ days, $P = 2P_0$, so

$$2P_0 = P_0e^{3k},$$

$$2 = e^{3k},$$

$$\ln 2 = 3k,$$

$$k = \frac{1}{3}(\ln 2).$$

Thus we have

$$P = P_0 e^{(1/3)(\ln 2)t}.$$

We wish to find t when $P = 100P_0$. We have

$$100P_0 = P_0 e^{(1/3)(\ln 2)t},$$

$$100 = e^{(1/3)(\ln 2)t},$$

$$\ln 100 = \frac{1}{3}(\ln 2)t,$$

$$t = \frac{3(\ln 100)}{\ln 2} \approx 19.93 \text{ days.} \quad \square$$

EXAMPLE 3 Suppose a body traveling through a medium is subject only to a force of retardation proportional to the velocity. If the body is traveling at 100 ft/sec at time $t = 0$ and at 20 ft/sec at time $t = 3$ sec, find the velocity of the body at time 9 sec, and also find the total distance the body travels through the medium.

Solution From Application 2, we know that $dv/dt = -kv$ for some positive constant k, so by the preceding section,

$$v = v_0 e^{-kt}.$$

From the data, the initial velocity v_0 is 100 ft/sec, so

$$v = 100e^{-kt}.$$

Also, when $t = 3$, we have $v = 20$, so

$$20 = 100e^{-3k}. \tag{7}$$

We can use Eq. (7) to find k. We find that $e^{3k} = 5$, so taking logarithms, we have

$$3k = \ln 5 \qquad \text{and} \qquad k = \frac{1}{3}(\ln 5).$$

Thus

$$v = 100e^{-(\ln 5)t/3} = 100 \cdot 5^{-t/3}.$$

When $t = 9$, we obtain

$$v\big|_{t=9} = 100 \cdot 5^{-9/3} = \frac{100}{125} = \frac{4}{5} \text{ ft/sec.}$$

If s is the distance traveled by the body from time $t = 0$, then

$$\frac{ds}{dt} = v = 100e^{-(\ln 5)t/3}.$$

Integrating, we find that

$$s = -\frac{300}{\ln 5} e^{-(\ln 5)t/3} + C$$

for some constant C. Now $s = 0$ when $t = 0$, so we have

$$0 = -\frac{300}{\ln 5} \cdot 1 + C$$

$$C = \frac{300}{\ln 5}.$$

Hence, the distance s as a function of t is given by

$$s = -\frac{300}{\ln 5} \cdot 5^{-t/3} + \frac{300}{\ln 5}.$$

As t approaches ∞, $5^{-t/3}$ approaches 0 and s approaches 300/ln 5. Thus, the body travels a total distance of 300/ln 5 ft through the medium. □

EXAMPLE 4 Savings with interest compounded continuously increase at a rate proportional to the amount in the savings account. If $S(t)$ is the amount in the account after t years, and if interest is compounded continuously at $c\%$, then

$$\frac{dS}{dt} = \frac{c}{100} S.$$

Find how long it takes savings to triple at 5%, compounded continuously.

Solution The differential equation becomes

$$\frac{dS}{dt} = \frac{5}{100} S = \frac{1}{20} S.$$

Therefore

$$S = S_0 e^{t/20}.$$

When $S = 3S_0$ so the original amount S_0 has been tripled, then

$$3S_0 = S_0 e^{t/20},$$

$$3 = e^{t/20},$$

$$\ln 3 = \frac{t}{20},$$

$$t = 20(\ln 3) \approx 21.97.$$

Thus money triples in just under 22 years if compounded continuously at 5%. □

Figure 8.25 Attempt to draw the graph of t^{100}.

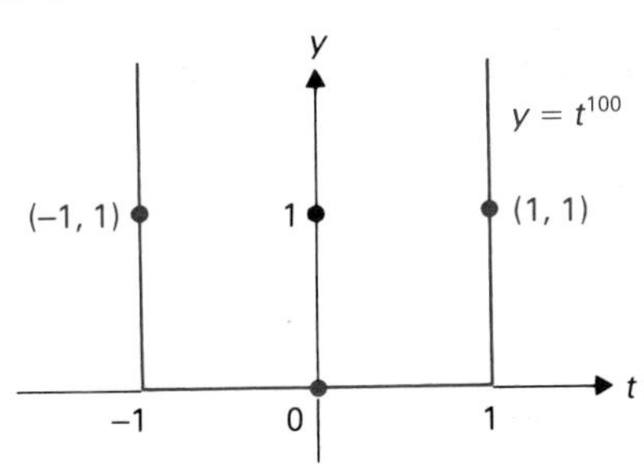

UNRESTRICTED EXPONENTIAL GROWTH

We hope you understand how rapidly the exponential function e^t increases as $t \to \infty$. In Chapter 1, we attempted to draw the graph of $y = x^{100}$ to demonstrate how rapidly it increases for $x \geq 1$. Figure 8.25 shows our best attempt to draw the graph of $f(t) = t^{100}$. It looks as though it increases faster than e^t, but the graph only shows the behavior near the origin. If t is large enough, then e^t is much larger than t^{100} (or than t^{1000}, for that matter). Suppose $t = 10{,}000$. Then $e^{10{,}000}$ is the product of 10,000 factors e, while

$10{,}000^{100}$ is the product of only 100 factors, each of size 10,000. Our calculator shows that

$$e^{10,000} = (e^{100})^{100} \approx (2.6881171 \cdot 10^{43})^{100} \approx 8.8068 \cdot 10^{42} \cdot 10^{4300}.$$

Thus

$$e^{10,000} \approx 8.8068 \cdot 10^{4,342} \qquad \text{and} \qquad 10{,}000^{100} = 10^{400}.$$

Any possibility of unrestricted exponential growth e^t, corresponding to large values of t, is a cause of great concern.

EXAMPLE 5 According to the 1971 *World Book Encyclopedia*, the world population in 1971 was about 3,692,000,000 and was increasing at an annual rate of about 1.9%. Suppose the rate of increase of population is proportional to the population, which continues to increase at the continuous rate of 1.9% annually. If the surface of the earth is a sphere of radius 4000 mi, estimate the year when there will be one person for every square yard of surface area of the earth, including oceans as well as continents.

Solution If the population t years after 1971 is $P = f(t)$, then

$$\frac{dP}{dt} = \frac{1.9}{100}P = 0.019P.$$

Solution (6) shows that then

$$P = 3{,}692{,}000{,}000e^{0.019t}.$$

We want to find t when P is the number of square yards of surface area of the earth. Since the area of a sphere is $A = 4\pi r^2$, we would have at that time

$$P = 4\pi\left(4000 \cdot \frac{5280}{3}\right)^2 \text{ yd}^2.$$

Using a calculator to solve

$$4\pi\left(4000 \cdot \frac{5280}{3}\right)^2 = 3{,}692{,}000{,}000e^{0.019t},$$

we have

$$e^{0.019t} = \frac{4\pi(4000 \cdot 5280)^2}{9 \cdot 3{,}692{,}000{,}000},$$

$$0.019t = \ln\left[\frac{4\pi(4000 \cdot 5280)^2}{9 \cdot 3{,}692{,}000{,}000}\right]$$

$$\approx \ln 168{,}691 \approx 12.0358,$$

$$t \approx 633.4646 \text{ years.}$$

This shows that there would be one person for every square yard of the earth's surface in the year 1971 + 633 = 2604. Taken in the context of the history of the human race, the year 2604 might be like "next month" as we view our own lives. □

The preceding example explains the great concern many people have about the "population explosion." Some denounce the concern, saying that we can build cities in space to avoid overcrowding on earth. Exercise 27 asks you to compute the year when there would be one person for every 3 cubic yards of space out as far as the moon, under the growth assumptions in Example 5. It takes a bit longer; the year would be 4029. In the context of the history of our race, that might be like "next season" to us.

We turn to an economic discussion. Government officials strive to achieve an "ideal," long-term, stable economic growth of about 3% annually, after correction for inflation. Such growth would be exponential and would lead to absurdities, as our next example illustrates.

EXAMPLE 6 In 1982, roughly 7,000,000 cars were manufactured in the United States. Suppose the number manufactured were to increase continuously at an annual rate of 3%. If the average weight of a car is 0.7 ton, find the year when the entire mass of the earth, which weighs about $6.6 \cdot 10^{20}$ tons, would have to be used to manufacture that year's cars in the United States.

Solution If $N = f(t)$ is the number of cars manufactured t years after 1982 in the United States, then we have

$$\frac{dN}{dt} = 0.03N,$$

so

$$N = 7{,}000{,}000e^{0.03t}.$$

We wish to find t when $N = 6.6 \cdot 10^{20}/0.7$. Solving

$$\frac{6.6 \cdot 10^{20}}{0.7} = (7 \cdot 10^6)e^{0.03t}$$

using a calculator, we have

$$e^{0.03t} = \frac{6.6 \cdot 10^{14}}{4.9},$$

$$0.03t = \ln\left(\frac{6.6 \cdot 10^{14}}{4.9}\right) \approx 32.534,$$

$$t \approx 1084.5 \text{ years.}$$

Thus all the earth would be used to make just that year's cars in the year $1084 + 1982 = 3066$, leaving no earth on which to drive the cars! □

The preceding example illustrates that "boom" periods of 3% growth must lead inevitably to the painful "bust" periods. We who have studied calculus can see for ourselves the inevitable consequences of creating an economic condition governed by $dQ/dt = kQ$ for $k > 0$. However, zero growth, or even more realistic linear growth, like having the gross national product increase by the equivalent of a flat $500,000,000 each year, is not a popular idea. Until the population as a whole has studied $dQ/dt = kQ$, we will continue to suffer the "boom and bust."

SUMMARY

1. The general solution of the differential equation $dy/dt = ky$ is $y = Ae^{kt}$. The arbitrary constant A is the initial value y_0 of y, when $t = 0$.
2. There are several important situations in which the rate of growth (or decay) of some quantity is proportional to the amount $Q(t)$ present at

time t. Examples are decay of a radioactive substance, growth of a population, and savings with interest compounded continuously. In all these cases, the differential equation $dQ/dt = kQ$ governs the growth (or decay).

3. Long-term exponential growth can lead to catastrophe.

EXERCISES

1. If the half-life of a radioactive element is 1600 yr, how long does it take for $\frac{1}{3}$ of the original amount to decay?
2. A certain culture of bacteria grows at a rate proportional to the size of the culture. If the culture triples in size in the first 2 days, find the factor by which the culture increases in size in the first 10 days.
3. A culture of bacteria increases at a rate proportional to the size of the culture. If 5000 bacteria are initially present and the culture increases to 12,000 bacteria after 1 day, find how many days before 10,000,000 bacteria are present.
4. Suppose half of a radioactive substance decomposes in 50 years. Find how long it takes for $\frac{31}{32}$ of it to decompose by
   a) reasoning that half decomposes in the first 50 years, and then half of the remaining half, or $\frac{1}{4}$ more, decomposes during the next 50 years, and so on, and
   b) working with the exponential equation describing the amount remaining at time t.
5. A population grows at a rate proportional to its size and doubles in 10 years. If the initial population is 50, find the number of years required for it to increase to 12,800 by
   a) reasoning that after the first 10 years the population would be 100, after the next 10 it doubles again to 200, and so on, and
   b) working with the exponential equation describing the population at time t.
6. A population grows at a rate proportional to the number present. If the population doubles in 20 days, find the factor by which the population increases after 10 days
   a) without using the exponential equation giving the population at time t, and
   b) using the exponential equation.
7. Suppose the population in Exercise 6 triples in 20 days. Find the factor by which it increases in 5 days
   a) without using the exponential equation giving the population at time t, and
   b) using the exponential equation.
8. If the half-life of a radioactive substance is 50 years, how long does it take for 100 lb of the substance to completely decompose?
9. Find how long it takes savings to double at 5.5% interest, compounded continuously.
10. Mr. and Mrs. Brice put $10,000 into a savings account in 1970. They plan to leave it there until 1990, when they will withdraw it and all the accrued interest to make a down payment on a new house. During that 20-year period, their savings will earn 5% interest, compounded continuously. How much will the bank pay the Brices in 1990?
11. When you worked Exercise 10, you discovered that the Brices will receive $27,182.82 in 1990. During the time the bank has the Brices' money, the bank loans it continually for home mortgages, and receives what amounts to 9% interest, compounded continuously. If it costs the bank an average of $200 per year to handle its investments arising from the Brices' $10,000 deposit, how much profit will the bank have made after paying the Brices in 1990?
12. Answer Exercise 11 if the bank instead continually uses the money generated by the Brices' $10,000 for car loans at 12% interest, compounded continuously. Assume the same expenses as in Exercise 11.
13. What percent annual interest rate, compounded continuously, must a bank offer on a savings account if an undisturbed and untaxed balance doubles every 8 years?
14. A sum of money deposited in a savings account grows undisturbed at a continuously compounded annual rate of c%. If the account has a balance of $10,000 after 5 years and a balance of $12,000 after 7 years, find
   a) the value of c and
   b) the initial amount S_0 deposited in the account.
15. Suppose the average worker's annual salary in 1985 was $20,000. Suppose salaries increase continuously at a rate proportional to their size. If the rate of increase is kept constant at 5% annually, find the year in which the average worker's salary will be
   a) $1,000,000,
   b) $1,000,000,000.
   c) What might be done to keep workers "happy" with raises but also prevent working with such cumbersome large numbers?
16. If you buy a $2000 IRA (individual retirement account) on your 20th birthday, and if it earns 9% annual interest, compounded continuously, find the cash value of the IRA on your 65th birthday by
   a) assuming the 9% per year means per calendar year and
   b) assuming the 9% per year means per banker's year, which is 360 days. Assume there are 11 leap years between your 20th and 65th birthdays, and neglect any calendar adjustment in the year 2000.
17. With reference to Exercise 16, suppose inflation runs at 5% annually (compounded continuously) while you are between age 20 and 65, and you would not have had to pay taxes on

the original \$2000 if you had not invested it in the IRA when you were 20.

a) How much money would you need at age 65 to equal the purchasing power of the \$2000 when you were 20?

b) Assuming that the cash value of your IRA at age 65 is \$122,000 and you then have to pay \$48,000 deferred taxes on that amount, by what factor has the purchasing power of your original investment increased during the 45 years it was invested?

18. A tank initially contains a solution of 100 gal of brine with a salt concentration of 2 lb/gal. Fresh water is added at a rate of 4 gal/min, and the brine is drawn off at the bottom of the tank at the same rate. Assume that the concentration of salt in the solution is kept homogeneous by stirring. If $f(t)$ is the number of pounds of salt in solution in the tank at time t, show that f satisfies a differential equation of the form of Eq. (1).

19. With reference to Exercise 18, find the amount of salt in solution in the tank after 25 min.

20. With reference to Exercise 18, find the time when there is just 1 lb of salt left in the tank.

21. With reference to Exercise 18, find the amount of salt in the tank after 1 hr if fresh water flows in and brine is drawn off at the rate of 2 gal/min, while the other data remain the same.

22. A body traveling in a medium is retarded by a force proportional to its velocity. If the velocity of the body after 4 sec is 80 ft/sec and the velocity after 6 sec is 60 ft/sec, find the initial velocity of the body.

23. In Exercise 22, find the total distance the body travels through the medium.

24. Newton's law of cooling states that a body placed in a colder medium will cool at a rate proportional to the difference between its temperature and the temperature of the surrounding medium. Assume the temperature of the surrounding medium remains constant and let $f(t)$ be the difference between the temperature of the body and the temperature of the surrounding medium at time t. Show that f satisfies a differential equation of the form of Eq. (1).

25. With reference to Exercise 24, a body having a temperature of 90° is placed in a medium whose temperature is kept constant at 40°. If the temperature of the body after 15 min is 70°, find its temperature after 40 min.

26. Describe all solutions of the differential equation $d^2y/dt^2 = k(dy/dt)$, where k is a constant. [*Hint:* Let $u = dy/dt$.]

27. Referring to Example 5 in the text, suppose it were possible to have people live in space near the earth, say as far out as the moon. Use a calculator to estimate the year in which there would be one person for every 3 cubic yards of such space, assuming the moon is 250,000 mi from the earth.

8.6 THE INVERSE TRIGONOMETRIC FUNCTIONS

THE INVERSE SINE FUNCTION $\sin^{-1} x$

The graph of $y = \sin x$ is shown in Fig. 8.26. The function $\sin x$ does not have an inverse because, given a y-value c where $-1 \le c \le 1$, there are many values of x such that $\sin x = c$, as indicated in Fig. 8.26. In Fig. 8.27, we picked out by color one unbroken portion of the graph of $\sin x$ that covers the range $-1 \le y \le 1$ just once. We naturally picked out the unbroken arc closest to the origin. We use this colored arc in Fig. 8.27 to define the inverse sine function. That is, given a y-value where $-1 \le y \le 1$, there is a unique x-value where $-\pi/2 \le x \le \pi/2$ such that $\sin x = y$. We denote this inverse sine function by $x = \sin^{-1} y$ or $x = \arcsin y$. We will use the first notation,

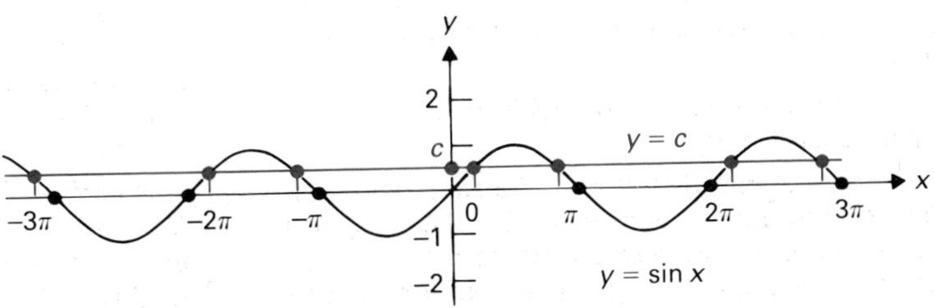

Figure 8.26 The function $\sin x$ has no inverse because one y-value c may correspond to many x-values.

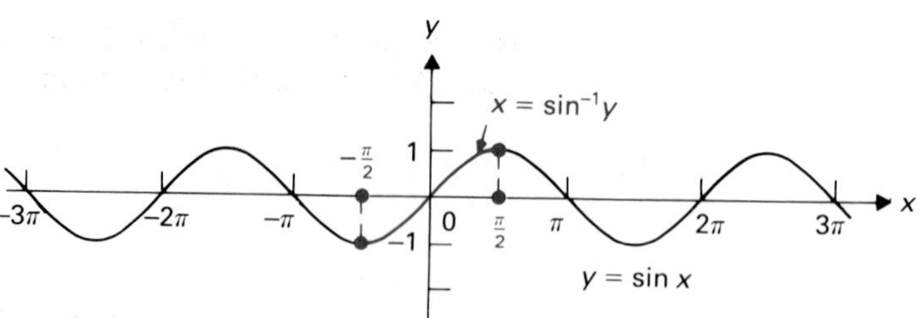

Figure 8.27 The arc of $y = \sin x$ where $-\pi/2 \le x \le \pi/2$ is used to define the inverse sine function.

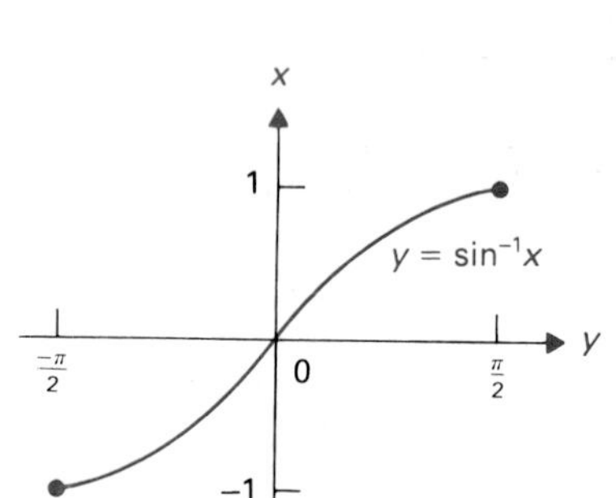

Figure 8.28 Changing axis labels from Fig. 8.27 gives the graph $y = \sin^{-1}x$.

$x = \sin^{-1}y$, and we emphasize that the -1 *must not be considered as an exponent.*

The colored arc in Fig. 8.27 is the graph of $x = \sin^{-1}y$. To obtain the graph of $y = \sin^{-1}x$, we first interchange the x and y labels on the axes, as shown in Fig. 8.28. Then we rotate the plane a half-turn about the line $y = x$ to bring the axes into their usual position, with the x-axis horizontal, as shown in Fig. 8.29. From Fig. 8.29, we see that the domain of $\sin^{-1}x$ is $-1 \le x \le 1$, while the range is $-\pi/2 \le y \le \pi/2$. We will see in a moment that it is very important to remember this range.

EXAMPLE 1 Find $\sin^{-1}\frac{1}{2}$.

Solution Since $\sin(\pi/6) = \frac{1}{2}$ and $\pi/6$ falls in the range $-\pi/2 \le y \le \pi/2$ of $\sin^{-1}x$, we see that $\sin^{-1}\frac{1}{2} = \pi/6$. □

EXAMPLE 2 Find $\sin^{-1}(-\sqrt{3}/2)$.

Solution We see that $\sin(2\pi/3) = -\sqrt{3}/2$, but $2\pi/3$ does not fall in the range $-\pi/2 \le y \le \pi/2$ of $\sin^{-1}x$. The angle falling within this range whose sine is $-\sqrt{3}/2$ is $-\pi/3$. Thus $\sin^{-1}(-\sqrt{3}/2) = -\pi/3$. □

The easiest way to find values of $\sin^{-1}x$ is to use a calculator. When radian mode is used, the calculator provides values in the correct range from $-\pi/2$ to $\pi/2$.

We develop calculus for $y = \sin^{-1}x$, and in the process we see why we are so fussy about the range of $\sin^{-1}x$. If $y = \sin^{-1}x$, then $x = \sin y$, and we find dy/dx by differentiating $x = \sin y$ implicitly:

$$x = \sin y,$$

$$1 = (\cos y)\frac{dy}{dx},$$

$$\frac{dy}{dx} = \frac{1}{\cos y}.$$

Figure 8.29 Rotating the plane in Fig. 8.28 a half-turn about the line $y = x$ brings the axes into their usual position.

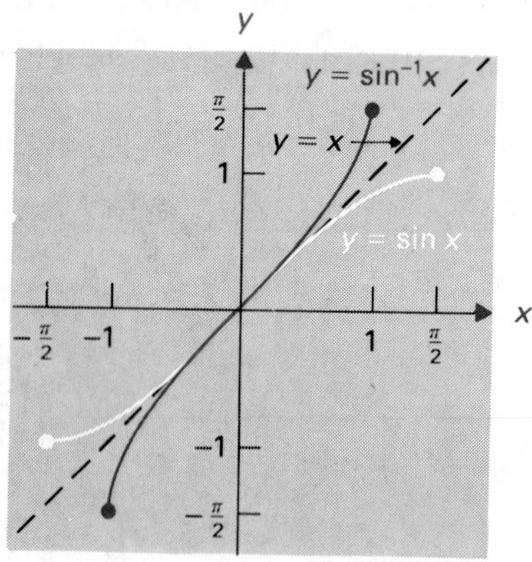

We would like to express dy/dx in terms of x rather than y. Since $\sin y = x$, we have

$$\cos y = \pm\sqrt{1 - \sin^2 y} = \pm\sqrt{1 - x^2}.$$

Since $-\pi/2 \le y \le \pi/2$, *we see that* $\cos y \ge 0$, *so the choice*

$$\cos y = \sqrt{1 - x^2}$$

is appropriate. Our differentiation formula thus depends on our choice of range for $\sin^{-1}x$. We then obtain

$$\frac{dy}{dx} = \frac{1}{\sqrt{1 - x^2}},$$

so that

$$\frac{d(\sin^{-1}x)}{dx} = \frac{1}{\sqrt{1 - x^2}}.$$

Of course, if u is a differentiable function of x, then by the chain rule,

$$\boxed{\frac{d(\sin^{-1}u)}{dx} = \frac{1}{\sqrt{1 - u^2}} \cdot \frac{du}{dx}.} \tag{1}$$

EXAMPLE 3 Find dy/dx if $y = \sin^{-1}2x$.

Solution From Eq. (1), we have

$$\frac{d(\sin^{-1}2x)}{dx} = \frac{1}{\sqrt{1 - (2x)^2}} \cdot \frac{d(2x)}{dx} = \frac{1}{\sqrt{1 - 4x^2}} \cdot 2 = \frac{2}{\sqrt{1 - 4x^2}}. \qquad \square$$

From Eq. (1), we obtain the important integration formula

$$\boxed{\int \frac{du}{\sqrt{1 - u^2}} = \sin^{-1}u + C.}$$

We now know one reason to study the inverse sine and the other inverse trigonometric functions. We obtain new and important integration formulas.

Frequently, we encounter an integral of the form

$$\int \frac{dx}{\sqrt{a^2 - x^2}},$$

where $a^2 \neq 1$. Rewriting this integral to take the form above, we have

$$\int \frac{dx}{\sqrt{a^2 - x^2}} = \frac{1}{a} \int \frac{dx}{\sqrt{(a^2/a^2) - (x^2/a^2)}} = \int \frac{(1/a)\,dx}{\sqrt{1 - (x/a)^2}}.$$

Taking $u = x/a$, so $du = (1/a)\,dx$, and using the integration formula above, we obtain

$$\int \frac{dx}{\sqrt{a^2 - x^2}} = \sin^{-1}\left(\frac{x}{a}\right) + C.$$

If u is a differentiable function of x, the chain rule then gives us the useful formula

$$\boxed{\int \frac{du}{\sqrt{a^2 - u^2}} = \sin^{-1}\left(\frac{u}{a}\right) + C.} \tag{2}$$

EXAMPLE 4 Find

$$\int \frac{dx}{\sqrt{1-9x^2}}.$$

Solution We "fix up" the integral to fit formula (2), with $a = 1$ and $u = 3x$, obtaining

$$\int \frac{dx}{\sqrt{1-9x^2}} = \int \frac{dx}{\sqrt{1-(3x)^2}} = \frac{1}{3}\int \frac{\overbrace{3\,dx}^{du}}{\sqrt{1-\underbrace{(3x)}_{u}^2}}$$

$$= \frac{1}{3}\sin^{-1}(3x) + C. \quad \square$$

EXAMPLE 5 Find

$$\int_{-1.6}^{1.2} \frac{dx}{\sqrt{4-x^2}}.$$

Solution Formula (2) shows that

$$\int_{-1.6}^{1.2} \frac{dx}{\sqrt{4-x^2}} = \sin^{-1}\left(\frac{x}{2}\right)\Bigg]_{-1.6}^{1.2}$$

$$= \sin^{-1}(0.6) - \sin^{-1}(-0.8)$$

$$\approx 0.6435 - (-0.9273) = 1.5708. \quad \square$$

THE OTHER INVERSE TRIGONOMETRIC FUNCTIONS

We have developed the function $\sin^{-1}x$ in detail. We will now do the other inverse trigonometric functions all at once and leave the derivation of the appropriate calculus formulas as exercises.

Figure 8.30 shows the graphs of all the elementary trigonometric functions. None of them has an inverse, in the sense that a horizontal line $y = c$ may meet any of the graphs in more than one point. The colored arcs in Fig. 8.30 show the portions of the graphs that are selected to define the inverse functions. Note that for the functions $\sec x$ and $\csc x$, it is impossible to pick one *unbroken* arc that covers both the values $y \geq 1$ and the values $y \leq -1$.

Figure 8.31 shows the graphs of the inverse functions of the independent variable x. Table 8.3 summarizes the basic information concerning them. We have discussed the information for $\sin^{-1}x$ in detail. The domains and ranges for the other inverse functions are apparent from Fig. 8.31. We ask you to establish the differentiation formulas in the exercises.

In Table 8.3, note that the derivatives of $\cos^{-1}x$, $\cot^{-1}x$, and $\csc^{-1}x$ are just the negatives of the derivatives of $\sin^{-1}x$, $\tan^{-1}x$, and $\sec^{-1}x$, respectively. This means that the "inverse cofunctions" give us no additional integration formulas. That is, we would normally write

$$\int \frac{-1}{\sqrt{1-x^2}} = -\int \frac{dx}{\sqrt{1-x^2}} = -\sin^{-1}x + C$$

rather than

$$\int \frac{-1}{\sqrt{1-x^2}} = \cos^{-1}x + C.$$

The values of the inverse trigonometric functions falling in the y-ranges shown in Table 8.3 are called the **principal values** of those functions. There is no uniform agreement as to the principal values for $\sec^{-1}x$ and $\csc^{-1}x$. Some people prefer the second-quadrant range $\pi/2 < y \le \pi$ for principal values of $\sec^{-1}x$ where $x \le -1$. With that choice, we have $\sec^{-1}x = \cos^{-1}(1/x)$. Since calculators normally have keys to compute only $\sin^{-1}x$, $\cos^{-1}x$, and $\tan^{-1}x$, that choice might seem to be the proper one for today's technology. However, with that choice, we would have $d(\sec^{-1}x) = \pm 1/(x\sqrt{x^2-1})$, where the minus sign would occur if $x \le -1$. We ask you to show this in Exercise 63. Also, in Section 9.6 we will want to write $\sqrt{\sec^2 t - 1} = \sqrt{\tan^2 t} = \tan t$ for all t in the principal value range of $\sec^{-1}x$. This is true for our choice of principal values, where t is a first- or third-quadrant angle. However, it would not be true for t a second-quadrant angle, because $\tan t$ is negative in the second quadrant. We chose what we feel is the lesser of two evils: more difficult evaluation of $\sec^{-1}x$ using a calculator.

Figure 8.30 Portions of the graphs of trigonometric functions used to define their inverse functions: (a) $y = \sin x$; (b) $y = \cos x$; (c) $y = \tan x$; (d) $y = \cot x$; (e) $y = \sec x$; (f) $y = \csc x$.

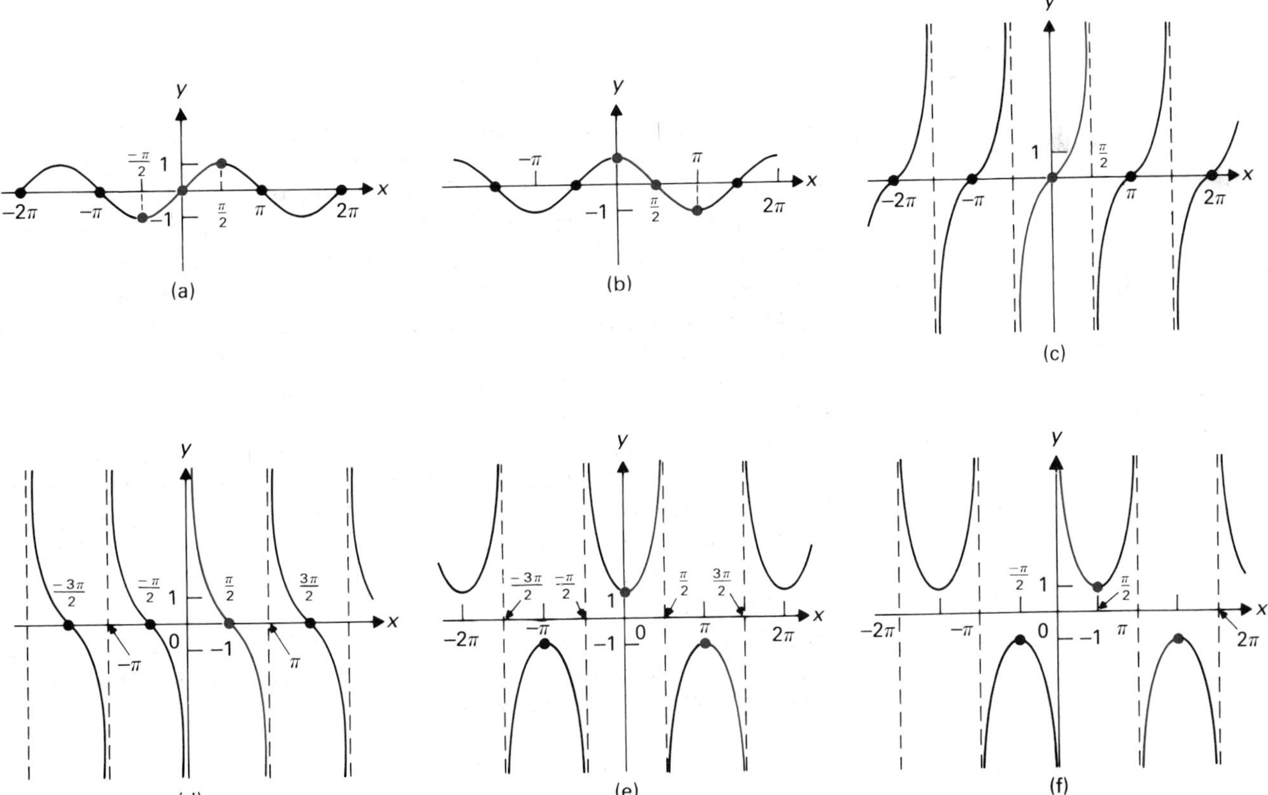

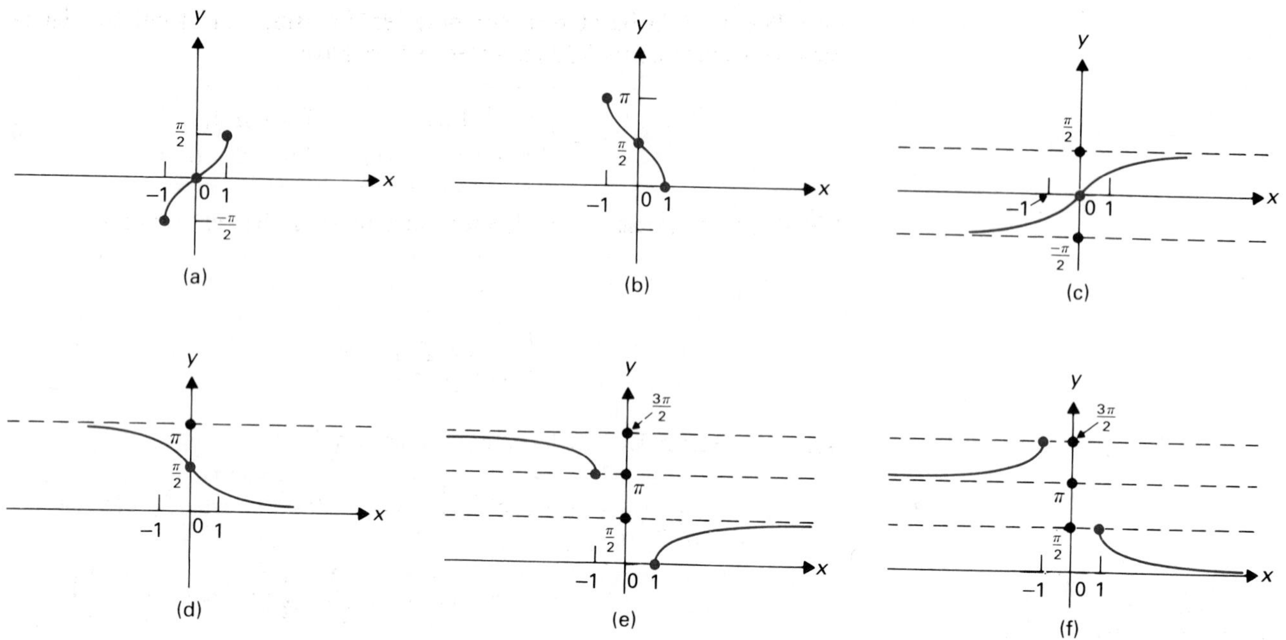

Figure 8.31 Graphs of the inverse trigonometric functions: (a) $y = \sin^{-1}x$; (b) $y = \cos^{-1}x$; (c) $y = \tan^{-1}x$; (d) $y = \cot^{-1}x$; (e) $y = \sec^{-1}x$; (f) $y = \csc^{-1}x$.

Table 8.3 Inverse trigonometric functions

Function	Domain	Range	Derivative
$\sin^{-1}x$	$[-1, 1]$	$\left[-\frac{\pi}{2}, \frac{\pi}{2}\right]$	$\frac{1}{\sqrt{1-x^2}}$
$\cos^{-1}x$	$[-1, 1]$	$[0, \pi]$	$\frac{-1}{\sqrt{1-x^2}}$
$\tan^{-1}x$	All x	$-\frac{\pi}{2} < y < \frac{\pi}{2}$	$\frac{1}{1+x^2}$
$\cot^{-1}x$	All x	$0 < y < \pi$	$\frac{-1}{1+x^2}$
$\sec^{-1}x$	$\begin{cases} x \geq 1 \\ x \leq -1 \end{cases}$	$0 \leq y < \frac{\pi}{2}$ $\pi \leq y < \frac{3\pi}{2}$	$\frac{1}{x\sqrt{x^2-1}}$
$\csc^{-1}x$	$\begin{cases} x \geq 1 \\ x \leq -1 \end{cases}$	$0 < y \leq \frac{\pi}{2}$ $\pi < y \leq \frac{3\pi}{2}$	$\frac{-1}{x\sqrt{x^2-1}}$

We want to be able to compute $\sec^{-1}x$ using our calculator. Since $\sec\theta = 1/(\cos\theta)$, Fig. 8.32 indicates that we have

$$\sec^{-1}x = \begin{cases} \cos^{-1}(1/x) & \text{for } x \geq 1, \\ 2\pi - \cos^{-1}(1/x) & \text{for } x \leq -1. \end{cases} \quad (3)$$

We illustrate the use of Eq. (3) with a calculator in the next example.

EXAMPLE 6 Find

$$\int_{-5}^{-2} \frac{1}{x\sqrt{x^2-1}}\,dx.$$

Solution From Table 8.3 and Eq. (3), we have

$$\int_{-5}^{-2} \frac{1}{x\sqrt{x^2-1}}\,dx = \sec^{-1}x\bigg]_{-5}^{-2} = \sec^{-1}(-2) - \sec^{-1}(-5)$$

$$= \left(2\pi - \cos^{-1}\left(-\frac{1}{2}\right) - \left(2\pi - \cos^{-1}\left(-\frac{1}{5}\right)\right)\right)$$

$$= \cos^{-1}\left(-\frac{1}{5}\right) - \cos^{-1}\left(-\frac{1}{2}\right)$$

$$\approx 1.7722 - 2.0944 = -0.3222. \quad \square$$

Figure 8.32
(a) $\theta = \sec^{-1}x = \cos^{-1}(1/x)$ for $x \geq 1$;
(b) for $x \leq -1$, $\theta = \sec^{-1}x$,
$\phi = \cos^{-1}(1/x)$, $\theta = 2\pi - \phi$.

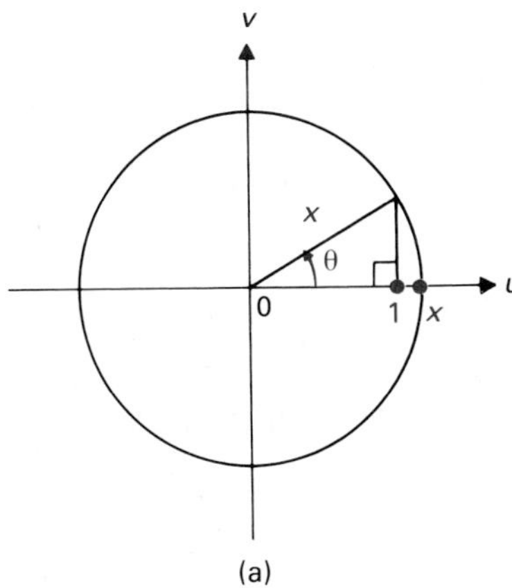

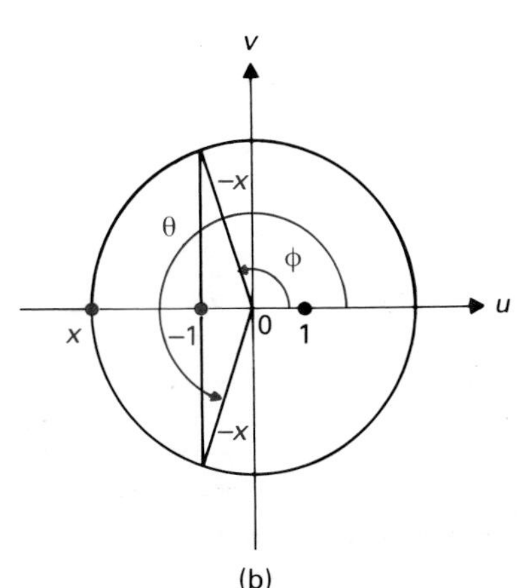

The following box lists the formulas arising from Table 8.3 and the chain rule. The integration formulas provide a motivation for studying the calculus of the inverse trigonometric functions. Formulas (11) and (12) are obtained from formulas (6) and (8), just as we obtained formula (10) from formula (4) earlier in this section.

DIFFERENTIATION FORMULAS

4. $\dfrac{d(\sin^{-1}u)}{dx} = \dfrac{1}{\sqrt{1-u^2}}\cdot\dfrac{du}{dx}$

5. $\dfrac{d(\cos^{-1}u)}{dx} = \dfrac{-1}{\sqrt{1-u^2}}\cdot\dfrac{du}{dx}$

6. $\dfrac{d(\tan^{-1}u)}{dx} = \dfrac{1}{1+u^2}\cdot\dfrac{du}{dx}$

7. $\dfrac{d(\cot^{-1}u)}{dx} = \dfrac{-1}{1+u^2}\cdot\dfrac{du}{dx}$

8. $\dfrac{d(\sec^{-1}u)}{dx} = \dfrac{1}{u\sqrt{u^2-1}}\cdot\dfrac{du}{dx}$

9. $\dfrac{d(\csc^{-1}u)}{dx} = \dfrac{-1}{u\sqrt{u^2-1}}\cdot\dfrac{du}{dx}$

INTEGRATION FORMULAS

10. $\displaystyle\int \frac{du}{\sqrt{a^2-u^2}} = \sin^{-1}\left(\frac{u}{a}\right) + C$

11. $\displaystyle\int \frac{du}{a^2+u^2} = \frac{1}{a}\tan^{-1}\left(\frac{u}{a}\right) + C$

12. $\displaystyle\int \frac{du}{u\sqrt{u^2-a^2}} = \frac{1}{a}\sec^{-1}\left(\frac{u}{a}\right) + C$

EXAMPLE 7 Find

$$\int_0^{1/2} \frac{dx}{\sqrt{1-x^2}}.$$

Solution We have

$$\int_0^{1/2} \frac{1}{\sqrt{1-x^2}}\,dx = \sin^{-1}x\bigg]_0^{1/2} = \sin^{-1}\frac{1}{2} - \sin^{-1}0$$

$$= \frac{\pi}{6} - 0 = \frac{\pi}{6}. \quad \square$$

EXAMPLE 8 Find

$$\int_0^3 \frac{1}{1+4x^2}\,dx.$$

Solution Using formula (11) with $a = 1$ and $u = 2x$, so $du = 2\,dx$, we find that

$$\int_0^3 \frac{1}{1+4x^2}\,dx = \frac{1}{2}\int_0^3 \frac{2}{1+(2x)^2}\,dx = \frac{1}{2}\tan^{-1}2x\bigg]_0^3$$

$$= \frac{1}{2}\tan^{-1}6 - \frac{1}{2}\tan^{-1}0 = \frac{1}{2}\tan^{-1}6.$$

If we desire a decimal approximation for $\tan^{-1}6$, we use a calculator and find that $\tan^{-1}6 \approx 1.406$. $\square$

EXAMPLE 9 Find $\int_0^1 \sqrt{4-x^2}\,dx$.

Solution Using formula 34 from the integral table on the endpapers of this book, we find that

$$\int_0^1 \sqrt{4-x^2}\,dx = \left(\frac{x}{2}\sqrt{4-x^2} + \frac{4}{2}\sin^{-1}\frac{x}{2}\right)\bigg]_0^1$$

$$= \left(\frac{1}{2}\sqrt{3} + 2\sin^{-1}\frac{1}{2}\right) - 0$$

$$= \frac{\sqrt{3}}{2} + 2\frac{\pi}{6} = \frac{\sqrt{3}}{2} + \frac{\pi}{3}. \quad \square$$

SUMMARY

1. A summary of the six inverse trigonometric functions is given in Table 8.3.
2. A summary of differentiation and integration formulas is given in the box on page 349.

EXERCISES

In Exercises 1 through 18, find the given quantity.

1. $\sin^{-1}1$

2. $\cos^{-1}\left(\frac{-1}{\sqrt{2}}\right)$

3. $\tan^{-1}(-\sqrt{3})$

4. $\csc^{-1}2$

5. $\sec^{-1}\left(\frac{-2}{\sqrt{3}}\right)$

6. $\sin^{-1}\left(\frac{\sqrt{3}}{-2}\right)$

7. $\cot^{-1}1$

8. $\sec^{-1}1$

9. $\csc^{-1}(-1)$

10. $\cos^{-1}(-1)$

11. $\sec^{-1}(-1)$

12. $\csc^{-1}\left(\frac{-2}{\sqrt{3}}\right)$

13. $\tan^{-1}0$

14. $\sin^{-1}(-1/\sqrt{2})$

15. $\sec^{-1}(-\sqrt{2})$

16. $\tan^{-1}1$

17. $\tan^{-1}(-1)$

18. $\sin^{-1}(-1/2)$

19. Find $\lim_{x\to\infty} \tan^{-1}x$.

20. Find $\lim_{x\to-\infty} \tan^{-1}x$.

21. Find $\lim_{x\to\infty} \sec^{-1}x$.

22. Find $\lim_{x\to-\infty} \sec^{-1}x$.

23. Find $\lim_{x\to\infty} \cot^{-1}x$.

24. Find $\lim_{x\to-\infty} \cot^{-1}x$.

In Exercises 25 through 35, find the derivative of the given function.

25. $\sin^{-1}(2x)$

26. $\cos^{-1}(x^2)$

27. $\tan^{-1}(\sqrt{x})$

28. $x \sec^{-1}x$

29. $\csc^{-1}\left(\frac{1}{x}\right)$

30. $(\sin^{-1}x)(\cos^{-1}x)$

31. $(\tan^{-1}2x)^3$

32. $\sec^{-1}(x^2+1)$

33. $\frac{1}{\tan^{-1}x}$

34. $\sqrt{\csc^{-1}x}$

35. $(x+\sin^{-1}3x)^2$

36. A careless mathematics professor asked his calculus class on their final examination to find dy/dx if $y = \sin^{-1}(1+x^2)$. What is wrong with this problem?

In Exercises 37 through 41, find the integral without the use of tables.

37. $\int_{-1}^{1} \frac{1}{1+x^2}\,dx$

38. $\int \frac{1}{\sqrt{1-9x^2}}\,dx$

39. $\int_{-1}^{\sqrt{3}} \frac{1}{\sqrt{1-(x^2/4)}}\,dx$

40. $\int_{2/\sqrt{3}}^{2} \frac{1}{x\sqrt{x^2-1}}\,dx$

41. $\int \frac{1}{3x\sqrt{4x^2-1}}\,dx$

In Exercises 42 through 49, find the integral without using a table, but use your calculator or computer in the final step to obtain a decimal value for the integral, as in Examples 5 and 6 in the text.

42. $\int_0^{0.75} \frac{dx}{\sqrt{1-x^2}}$

43. $\int_{-0.2}^{0.3} \frac{dx}{1+x^2}$

44. $\int_2^{10} \frac{dx}{x\sqrt{x^2-1}}$

45. $\int_{-3}^{-8} \frac{dx}{x\sqrt{x^2-1}}$

46. $\int_0^3 \frac{dx}{\sqrt{16-x^2}}$

47. $\int_{-2}^{4} \frac{dx}{4+x^2}$

48. $\int_6^{10} \frac{dx}{x\sqrt{x^2-4}}$

49. $\int_0^2 \frac{\cos x}{\sqrt{4-\sin^2 x}}\,dx$

In Exercises 50 through 57, use tables, if necessary, to find the definite integral.

50. $\int_0^1 \frac{1}{\sqrt{4-x^2}}\,dx$

51. $\int_{-1}^{1} \sqrt{1-x^2}\,dx$

52. $\int_0^2 x^2\sqrt{16-x^2}\,dx$

53. $\int_1^{\sqrt{3}} \frac{\sqrt{4-x^2}}{x^2}\,dx$

54. $\int_{-4}^{-4/\sqrt{3}} \frac{\sqrt{x^2-4}}{x}\,dx$

55. $\int_2^3 \frac{1}{\sqrt{4x-x^2}}\,dx$

56. $\int_0^1 \sqrt{2x-x^2}\,dx$

57. $\int_2^3 x\sqrt{4x-x^2}\,dx$

58. Derive the formula for $d(\cos^{-1}x)/dx$.

59. Derive the formula for $d(\tan^{-1}x)/dx$.

60. Derive the formula for $d(\cot^{-1}x)/dx$.

61. Derive the formula for $d(\sec^{-1}x)/dx$.

62. Derive the formula for $d(\csc^{-1}x)/dx$.

63. In defining $\sec^{-1}x$, some mathematicians prefer to choose branches of the graph of secant so that the range (set of principal values) of $\sec^{-1}x$ is $0 \le y < \pi/2$ or $\pi/2 < y \le \pi$. With this definition, we see that the relation $\sec^{-1}x = \cos^{-1}(1/x)$ holds.

a) Show by an example that the relation $\sec^{-1}x = \cos^{-1}(1/x)$ does not hold for the definition of $\sec^{-1}x$ in Table 8.3.

b) Show that if $\sec^{-1}x$ is defined to have range described in this exercise, then the formula for the derivative of the inverse secant changes from that in Table 8.3 to

$$\frac{d(\sec^{-1}x)}{dx} = \frac{1}{|x|\sqrt{x^2-1}}.$$

8.7 THE HYPERBOLIC FUNCTIONS

The hyperbolic functions are defined in terms of exponential functions, and their inverses can be expressed in terms of logarithm functions. (The inverse of the exponential function is the logarithm function, so this is not surprising!) This section gives just a brief summary of hyperbolic functions. Our goal is to give enough familiarity with them so that they won't appear mysterious or difficult if you encounter them in your work.

THE FUNCTIONS

The hyperbolic functions have many points of similarity with the trigonometric functions. The reason for this similarity is shown in a course in *complex variables.* There are six basic hyperbolic functions, just as there are six basic trigonometric functions; the hyperbolic functions are the *hyperbolic sine*, denoted by $\sinh x$, the *hyperbolic cosine*, or $\cosh x$, the *hyperbolic tangent*, or $\tanh x$, and so on. We usually pronounce "$\sinh x$" as though it were spelled "cinch x."

We will define the hyperbolic functions in terms of exponential functions; this is very different from the way we defined the trigonometric functions. However, if you study *complex analysis* later, you will discover that the trigonometric functions can also be defined in terms of exponential functions of a "complex variable." Table 8.4 gives the definitions of the hyperbolic functions. We have to remember the definitions for $\sinh x$ and $\cosh x$, and then we proceed exactly as for the trigonometric functions.

It is easy to see that

$$\boxed{\cosh^2 x - \sinh^2 x = 1} \tag{1}$$

for all x, namely,

$$\left(\frac{e^x + e^{-x}}{2}\right)^2 - \left(\frac{e^x - e^{-x}}{2}\right)^2 = \frac{e^{2x} + 2 + e^{-2x}}{4} - \frac{e^{2x} - 2 + e^{-2x}}{4}$$

$$= \frac{2}{4} + \frac{2}{4} = 1.$$

Equation (1) is the basic relation for hyperbolic functions and plays a role

Table 8.4 The six hyperbolic functions

Function	Definition	Function	Definition
$\sinh x$	$\frac{e^x - e^{-x}}{2}$	$\cosh x$	$\frac{e^x + e^{-x}}{2}$
$\tanh x$	$\frac{\sinh x}{\cosh x}$	$\coth x$	$\frac{\cosh x}{\sinh x}$
$\operatorname{sech} x$	$\frac{1}{\cosh x}$	$\operatorname{csch} x$	$\frac{1}{\sinh x}$

similar to the relation $\sin^2 x + \cos^2 x = 1$ for the trigonometric functions. For the trigonometric functions, the point $(\cos t, \sin t)$ lies on the circle with equation $x^2 + y^2 = 1$ for all t, while for the hyperbolic functions, the point $(\cosh t, \sinh t)$ lies on the curve with equation $x^2 - y^2 = 1$, which is called a *hyperbola*. This explains the name "hyperbolic functions."

The graphs of the six hyperbolic functions are easily sketched. To sketch the graph of $\sinh x$, we simply take half the difference in height between the graphs of e^x and e^{-x}, while for $\cosh x$ we average the heights of these two graphs. Since $\lim_{x\to\infty} e^{-x} = 0$, we see that the graphs of both $\sinh x$ and $\cosh x$ are close to the graph of $e^x/2$ for large x. The graphs of the six hyperbolic functions are sketched in Fig. 8.33.

The hyperbolic functions occur naturally in the physical sciences. For example, a flexible cable suspended from two supports of equal height hangs in a "catenary curve." It can be shown that with suitable choice of axes and scale, the curve has the equation $y = a \cosh (x/a)$ for some constant a.

Note from the graphs that $\sinh x$, $\tanh x$, $\coth x$, and $\operatorname{csch} x$ all yield a unique x for each y, so the inverse hyperbolic functions $\sinh^{-1} y$, $\tanh^{-1} y$, $\coth^{-1} y$, and $\operatorname{csch}^{-1} y$ exist. While $\cosh x$ and $\operatorname{sech} x$ do not satisfy this condition, we abuse terminology just as we did for trigonometric functions and pick the branches of their graphs indicated by the darker curves in Fig. 8.33(b) and (e) to define the functions $\cosh^{-1} y$ and $\operatorname{sech}^{-1} y$.

Figure 8.33 Graphs of the six hyperbolic functions: (a) $y = \sinh x$; (b) $y = \cosh x$; (c) $y = \tanh x$; (d) $y = \coth x$; (e) $y = \operatorname{sech} x$; (f) $y = \operatorname{csch} x$.

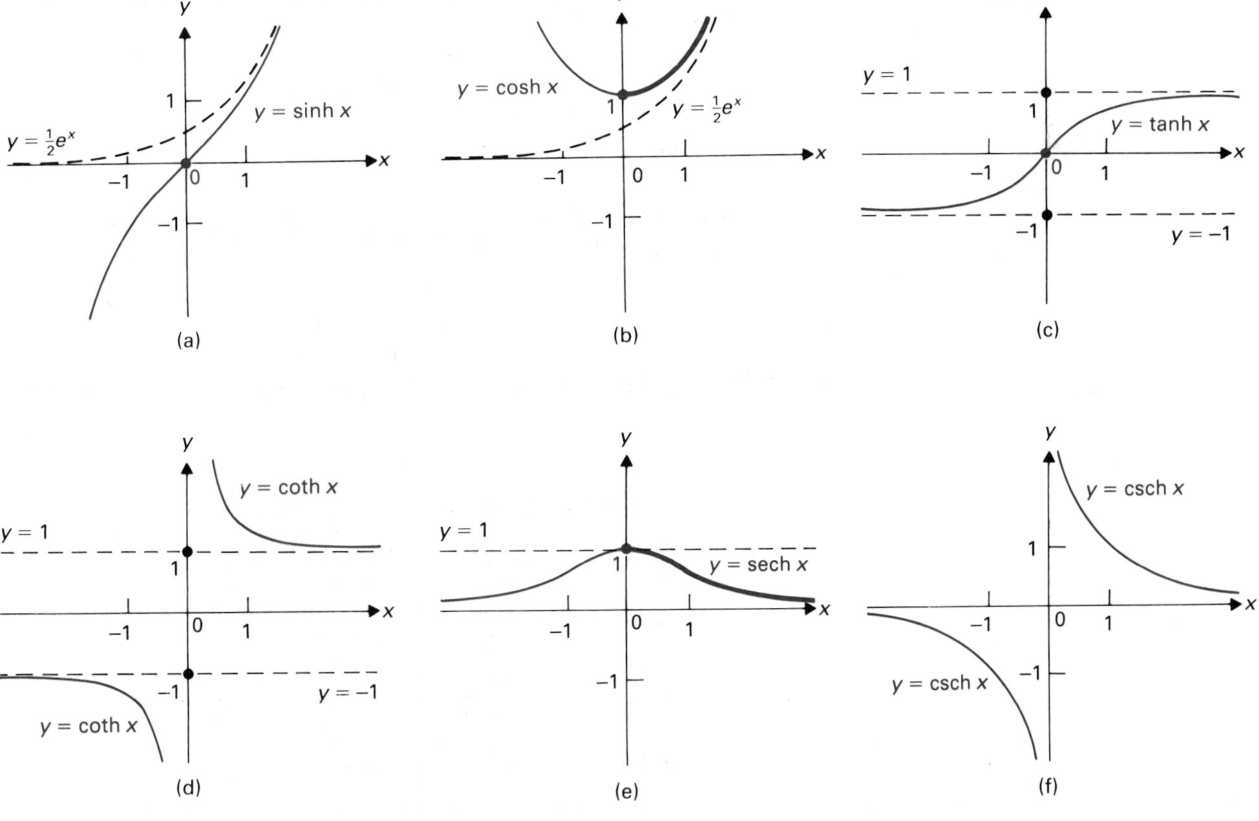

CALCULUS OF THE HYPERBOLIC FUNCTIONS

Calculus of the hyperbolic functions is easily developed, since the functions are defined in terms of the exponential function. Turning to the derivative of sinh x, we find that

$$\begin{aligned}\frac{d(\sinh x)}{dx} &= \frac{d}{dx}\left(\frac{e^x - e^{-x}}{2}\right)\\ &= \frac{1}{2}\cdot\frac{d}{dx}(e^x - e^{-x}) = \frac{1}{2}(e^x + e^{-x}) = \cosh x.\end{aligned}$$

Similarly, we find that

$$\frac{d(\cosh x)}{dx} = \sinh x.$$

The derivatives of the remaining four hyperbolic functions can then be found using Eq. (1) just as for the corresponding trigonometric functions. For example,

$$\begin{aligned}\frac{d(\tanh x)}{dx} &= \frac{d}{dx}\left(\frac{\sinh x}{\cosh x}\right)\\ &= \frac{\cosh x \cosh x - \sinh x \sinh x}{\cosh^2 x}\\ &= \frac{1}{\cosh^2 x} = \operatorname{sech}^2 x.\end{aligned}$$

The left-hand portion of Table 8.5 gives the derivatives of the hyperbolic functions. These differentiation formulas parallel those for the derivatives of the trigonometric functions, except for occasional differences in sign.

EXAMPLE 1 Find dy/dx if $y = \sinh(x^2 + 1)$.

Solution Using Table 8.5 and the chain rule, we have

$$\frac{dy}{dx} = \cosh(x^2 + 1)\cdot\frac{d(x^2 + 1)}{dx} = 2x\cosh(x^2 + 1). \quad \square$$

EXAMPLE 2 Find dy/dx if $y = \sinh^2 3x \tanh^3 4x$.

Solution Using Table 8.5, differentiation rules, and the chain rule, we have

$$\begin{aligned}\frac{dy}{dx} &= (\sinh^2 3x)(3\tanh^2 4x \operatorname{sech}^2 4x)4\\ &\quad + (\tanh^3 4x)(2\sinh 3x \cosh 3x)3\\ &= 12\sinh^2 3x \tanh^2 4x \operatorname{sech}^2 4x\\ &\quad + 6\sinh 3x \cosh 3x \tanh^4 4x. \quad \square\end{aligned}$$

In the preceding section, we studied the inverse trigonometric functions because they led to important integration formulas, such as

$$\int \frac{du}{\sqrt{1 - u^2}} = \sin^{-1} u + C. \tag{2}$$

Table 8.5 Derivatives of the hyperbolic functions

Function	Derivative	Function	Derivative
$\sinh x$	$\cosh x$	$\sinh^{-1}x$	$\dfrac{1}{\sqrt{1+x^2}}$
$\cosh x$	$\sinh x$	$\cosh^{-1}x$	$\dfrac{1}{\sqrt{x^2-1}}, \quad x>1$
$\tanh x$	$\operatorname{sech}^2 x$	$\tanh^{-1}x$	$\dfrac{1}{1-x^2}, \quad \|x\|<1$
$\coth x$	$-\operatorname{csch}^2 x$	$\coth^{-1}x$	$\dfrac{1}{1-x^2}, \quad \|x\|>1$
$\operatorname{sech} x$	$-\operatorname{sech} x \tanh x$	$\operatorname{sech}^{-1}x$	$\dfrac{-1}{x\sqrt{1-x^2}}, \quad 0<x<1$
$\operatorname{csch} x$	$-\operatorname{csch} x \coth x$	$\operatorname{csch}^{-1}x$	$\dfrac{-1}{\|x\|\sqrt{1+x^2}}$

The inverse hyperbolic functions also lead to integration formulas.

If $y = f(x) = \sinh x$, then $x = \sinh^{-1}y$ and dx/dy can be obtained by differentiating $y = \sinh x$ implicitly with respect to y:

$$y = \sinh x,$$

$$1 = (\cosh x)\frac{dx}{dy},$$

$$\frac{dx}{dy} = \frac{1}{\cosh x} = \frac{1}{\sqrt{1+\sinh^2 x}} = \frac{1}{\sqrt{1+y^2}}.$$

Interchanging x and y so that x is the dependent variable, we obtain

$$\frac{d(\sinh^{-1}x)}{dx} = \frac{1}{\sqrt{1+x^2}}.$$

The derivatives of the other inverse hyperbolic functions are found similarly and are given in the right-hand portion of Table 8.5. These differentiation formulas can, of course, be combined with the chain rule in the usual way.

EXAMPLE 3 Find the derivative of $\sinh^{-1}(\tan x)$.

Solution We have

$$\frac{d(\sinh^{-1}(\tan x))}{dx} = \frac{1}{\sqrt{1+\tan^2 x}} \cdot \sec^2 x$$

$$= \frac{\sec^2 x}{|\sec x|} = |\sec x|. \quad \square$$

The formula $d(\sinh^{-1}x)/dx = 1/\sqrt{1+x^2}$ leads to the integration formula

$$\int \frac{du}{\sqrt{1+u^2}} = \sinh^{-1}u + C. \tag{3}$$

This formula should surely be as important as formula (2). However, $\sinh^{-1}x$ can be expressed in terms of the logarithm function, and with a calculator or computer, the function $\ln x$ is very easy to evaluate. Most calculators do not have a $\sinh^{-1}x$ key. Thus one generally consults an integral table and uses in place of formula (3)

$$\int \frac{du}{\sqrt{a^2 + u^2}} = \ln (u + \sqrt{a^2 + u^2}) + C. \tag{4}$$

We now show that we indeed have

$$\sinh^{-1}x = \ln (x + \sqrt{1 + x^2}),$$

as you would guess by comparing formulas (3) and (4). The exercises ask you to imitate the procedure below to find the logarithmic expressions for the other inverse hyperbolic functions.

EXAMPLE 4 Show that $\sinh^{-1}x = \ln (x + \sqrt{1 + x^2})$.

Solution Let $y = \sinh^{-1}x$. Then

$$x = \sinh y = \frac{e^y - e^{-y}}{2},$$

so

$$e^y - e^{-y} = 2x.$$

Multiplying this equation by e^y, we obtain

$$e^{2y} - 1 = 2xe^y,$$

$$(e^y)^2 - 2x(e^y) - 1 = 0. \tag{5}$$

We regard Eq. (5) as a quadratic equation in the unknown e^y and solve it using the quadratic formula:

$$e^y = \frac{2x \pm \sqrt{4x^2 + 4}}{2} = x \pm \sqrt{x^2 + 1}.$$

Since $e^y > 0$ for all y, we see that the plus sign is appropriate, so

$$e^y = x + \sqrt{x^2 + 1}.$$

Then

$$\sinh^{-1}x = y = \ln (x + \sqrt{x^2 + 1}). \quad \square$$

EXAMPLE 5 Use formula (4) and a calculator to find the decimal value of

$$\int_0^5 \frac{x}{\sqrt{1 + x^4}}\,dx.$$

Solution We can "fix up" our integral to fit formula (4) with $a = 1, u = x^2$, and $du = 2x$, as follows:

$$\int_0^5 \frac{x\,dx}{\sqrt{1+x^4}} = \frac{1}{2}\int_0^5 \frac{\overbrace{2x\,dx}^{du}}{\sqrt{1+(\underbrace{x^2}_{u})^2}}$$

$$= \frac{1}{2}\ln\left(x^2+\sqrt{1+x^4}\right)\Big]_0^5$$

$$= \frac{1}{2}\ln\left(25+\sqrt{626}\right) - \frac{1}{2}\ln 1$$

$$\approx 1.9562 - 0 = 1.9562. \quad \square$$

From the formulas for the derivatives of the inverse hyperbolic functions found in Table 8.5, we obtain the integration formulas listed in the Summary. The Summary also gives the logarithmic form of the integration formulas. The reason for the split in the formula for $\int [1/(a^2 - u^2)]\,du$ is that $\tanh^{-1}x$ and $\coth^{-1}x$ have algebraically identical derivatives, but the domain of $\tanh^{-1}x$ is $|x| < 1$ while the domain of $\coth^{-1}x$ is $|x| > 1$. Since

$$\frac{1}{1-u^2} = \frac{\frac{1}{2}}{1-u} + \frac{\frac{1}{2}}{1+u},$$

we have, as the logarithmic alternative,

$$\int \frac{du}{1-u^2} = -\frac{1}{2}\ln|1-u| + \frac{1}{2}\ln|1+u| + C$$

$$= \frac{1}{2}\ln\left|\frac{1+u}{1-u}\right| + C.$$

EXAMPLE 6 Find

$$\int \frac{1}{x\sqrt{4-x^2}}\,dx.$$

Solution Using the next to last formula in the Summary, we have

$$\int \frac{1}{x\sqrt{4-x^2}}\,dx = -\frac{1}{2}\operatorname{sech}^{-1}\left|\frac{x}{2}\right| + C = -\frac{1}{2}\ln\left|\frac{2+\sqrt{4-x^2}}{x}\right| + C. \quad \square$$

SUMMARY

1. The hyperbolic functions are defined in Table 8.4.
2. The graphs of the hyperbolic functions are shown in Fig. 8.33.
3. Differentiation formulas are in Table 8.5.
4. Integration formulas given by the inverse hyperbolic functions (and

also by logarithm functions) are as follows.

$$\int \frac{du}{\sqrt{a^2 + u^2}} = \sinh^{-1}\left(\frac{u}{a}\right) + C = \ln\left(u + \sqrt{a^2 + u^2}\right) + C$$

$$\int \frac{du}{\sqrt{u^2 - a^2}} = \cosh^{-1}\left(\frac{u}{a}\right) + C = \ln\left(u + \sqrt{u^2 - a^2}\right) + C$$

$$\int \frac{du}{a^2 - u^2} = \begin{cases} \dfrac{1}{a}\tanh^{-1}\left(\dfrac{u}{a}\right) + C, & |u| < a \\ \dfrac{1}{a}\coth^{-1}\left(\dfrac{u}{a}\right) + C, & |u| > a \end{cases}$$

$$= \frac{1}{2a}\ln\left|\frac{a + u}{a - u}\right| + C$$

$$\int \frac{du}{u\sqrt{a^2 - u^2}} = -\frac{1}{a}\operatorname{sech}^{-1}\left|\frac{u}{a}\right| + C = -\frac{1}{a}\ln\left|\frac{a + \sqrt{a^2 - u^2}}{u}\right| + C$$

$$\int \frac{du}{u\sqrt{a^2 + u^2}} = -\frac{1}{a}\operatorname{csch}^{-1}\left|\frac{u}{a}\right| + C = -\frac{1}{a}\ln\left|\frac{a + \sqrt{a^2 + u^2}}{u}\right| + C$$

EXERCISES

In Exercises 1 through 6, prove the given relation for hyperbolic functions.

1. $1 - \tanh^2 x = \operatorname{sech}^2 x$

2. $\coth^2 x - 1 = \operatorname{csch}^2 x$

3. $\sinh(-x) = -\sinh x$

4. $\cosh(-x) = \cosh x$

5. $\sinh(x + y) = \sinh x \cosh y + \cosh x \sinh y$

6. $\cosh(x + y) = \cosh x \cosh y + \sinh x \sinh y$

7. Use Exercises 5 and 6 to find "double-angle" formulas for $\sinh 2x$ and $\cosh 2x$.

8. If $\sinh a = -\frac{3}{4}$, find $\cosh a$, $\tanh a$, and $\operatorname{csch} a$.

In Exercises 9 through 17, derive the formula for the derivative of the given function. In each exercise, you may use the answer to any previous exercise.

9. $\cosh x$ **10.** $\coth x$

11. $\operatorname{sech} x$ **12.** $\operatorname{csch} x$

13. $\cosh^{-1} x$ **14.** $\tanh^{-1} x$

15. $\coth^{-1} x$ **16.** $\operatorname{sech}^{-1} x$

17. $\operatorname{csch}^{-1} x$

In Exercises 18 through 35, find the derivative of the given function.

18. $\sinh(2x - 3)$ **19.** $\cosh(x^2)$

20. $\tanh^2 3x$ **21.** $\operatorname{sech}(\sqrt{x})$

22. $\coth(e^{2x})$ **23.** $\operatorname{csch}(\ln x)$

24. $\operatorname{sech}^3 3x$ **25.** $\sinh^2 x \cosh^2 x$

26. $(e^{-x} + \tanh x)^2$ **27.** $\sinh^{-1}(2x)$

28. $\cosh^{-1}(\sec x)$ **29.** $\tanh^{-1}(\sin 2x)$

30. $\coth^{-1}(e^{2x+1})$ **31.** $\operatorname{sech}^{-1}(x^2)$

32. $\operatorname{csch}^{-1}(\tan x)$ **33.** $e^x \sinh^{-1}(e^x)$

34. $\cosh^{-1}(x^2 + 1)$ **35.** $\tanh^{-1}(\cot 3x)$

In Exercises 36 through 44, compute the given integral without using tables.

36. $\int \tanh x\, dx$ **37.** $\int \coth x\, dx$

38. $\int \sinh^2 x \cosh x\, dx$ **39.** $\int \sinh(3x + 2)\, dx$

40. $\int x \cosh(x^2)\, dx$ **41.** $\int \operatorname{sech}^2 3x\, dx$

42. $\int \frac{1}{(e^x + e^{-x})^2}\,dx$

43. $\int \frac{e^x + e^{-x}}{e^x - e^{-x}}\,dx$

44. $\int (\cosh 2x)(e^{\sinh 2x})\,dx$

In Exercises 45 through 59, use tables, if necessary, to compute the given integral.

45. $\int_0^{1/2} \frac{x}{1 - x^2}\,dx$

46. $\int_0^{1/4} \frac{1}{1 - 4x^2}\,dx$

47. $\int \frac{x}{\sqrt{1 + 4x^2}}\,dx$

48. $\int \frac{1}{\sqrt{4 + x^2}}\,dx$

49. $\int \frac{1}{\sqrt{1 - e^{2x}}}\,dx$

50. $\int \frac{\sin 2x}{(\cos 2x)\sqrt{1 + \cos^2 2x}}\,dx$

51. $\int_0^1 \sqrt{4 + x^2}\,dx$

52. $\int_0^4 x^2\sqrt{9 + x^2}\,dx$

53. $\int \frac{\sqrt{16 + x^2}}{x}\,dx$

54. $\int \frac{\sqrt{2 + x^2}}{3x^2}\,dx$

55. $\int \frac{\sin^2 x \cos x}{\sqrt{9 + \sin^2 x}}\,dx$

56. $\int_3^5 x^2\sqrt{x^2 - 9}\,dx$

57. $\int \frac{e^{3x}}{\sqrt{e^{2x} - 16}}\,dx$

58. $\int x \sinh 2x\,dx$

59. $\int \cosh^3 4x\,dx$

In Exercises 60 through 64, use the technique of Example 4 in the text to derive the logarithmic expression for the given inverse hyperbolic function.

60. $\cosh^{-1}x$

61. $\tanh^{-1}x$

62. $\coth^{-1}x$

63. $\text{sech}^{-1}x$

64. $\text{csch}^{-1}x$

EXERCISE SETS FOR CHAPTER 8

Review Exercise Set 8.1

1. a) Give the definition of $\ln x$ as an integral.
b) Sketch the graph of $\ln(x/2)$.

2. a) Find dy/dx if $y = \ln(x^2 + 1)$.
b) Find $\int [(\sin x)/(1 + \cos x)]\,dx$.

3. a) Find dy/dx if $y = e^{\tan x}$.
b) Evaluate $\int_0^{\ln 2} e^{3x}\,dx$, simplifying your answer as much as possible.

4. a) Differentiate 2^{3x+4}.
b) Differentiate x^{5x}, $x > 0$.

5. a) Solve for x if $(2^x)(3^{x+1}) = 4^{-x}$.
b) Simplify $2^{\log_4 9}$.

6. If savings are compounded continuously at 5% interest, how long does it take for the savings to double?

7. a) Evaluate $\sin^{-1}(-1/2)$.
b) Evaluate $\tan^{-1}(-\sqrt{3})$.

8. a) Find dy/dx if $y = \sin^{-1}(\sqrt{x})$.
b) Evaluate $\int_{-1}^{\sqrt{3}} [1/(1 + x^2)]\,dx$.

9. a) Give the definition of $\sinh x$ in terms of the exponential function.
b) Differentiate $\text{sech}^3\, 2x$.

***10.** If $f(x) = (x - 1)/(x + 2)$, find $f^{-1}(x)$.

Review Exercise Set 8.2

1. a) How is the number e defined?
b) Sketch the graph of e^{-x^2}.

2. a) Find dy/dx if $y = \ln(\tan x)$.
b) Find $\int [x^2/(4 + x^3)]\,dx$.

3. a) Find dy/dx if $y = e^{\sin^{-1}x}$.
b) Find $\int (\sec^2 3x)e^{\tan 3x}\,dx$.

4. a) Differentiate 10^{x^2-x}.
b) Differentiate $(\cos x)^{\sin x}$.

5. a) Solve for x if $\ln x - 2(\ln x^3) = \ln 10$.
b) Simplify $25^{\log_5 7}$.

6. The only force on a body traveling on a line is a force of resistance proportional to its velocity. If the velocity is 100 ft/sec at time $t = 0$ and 50 ft/sec at time $t = 10$, find
a) the velocity of the body as a function of t,
b) the distance traveled as a function of t.

7. Evaluate:
a) $\cos^{-1}(\frac{1}{2})$
b) $\sec^{-1}(-2)$

8. a) Find dy/dx if $y = \tan^{-1}4x$.
b) Evaluate $\int_{-1/4}^{1/4} (1/\sqrt{1 - 4x^2})\,dx$.

9. a) Sketch the graph of $y = \cosh(x + 1)$.
b) Differentiate $\coth^3(2x + 1)$.

***10.** If $f(x)$ is invertible and $f(2) = 4$, while $f'(2) = 3$, find $(f^{-1})'(4)$.

Review Exercise Set 8.3

1. a) Find x such that $\ln x = 500$.
b) Find x such that $\ln x = -4$.

2. a) Find dy/dx if $y = \ln[x^3(x + 1)^2(\sin x)]$.
b) Find $\int \cot 2x\,dx$.

3. a) Find dy/dx if $y = e^{2x}\cos x$.
b) Find $\int \frac{e^x - 1}{e^x}\,dx$.

4. a) Simplify $3^{\log_9 4}$.
b) Simplify $\log_4 16^3$.

5. a) Find dy/dx if $y = 3^{\sin x}$.

b) Find dy/dx if $y = \dfrac{3^x \cdot 4^{x^3}}{5^x}$.

6. Suppose $f'(t) = 3 \cdot f(t)$. Find $f(5)/f(2)$.

7. Evaluate
a) $\tan^{-1}(-1)$, b) $\sin^{-1}(-1)$.

8. a) Find dy/dx if $y = \sec^{-1}(x^2)$.
b) Find $\displaystyle\int_{-\sqrt{2}}^{\sqrt{2}} \frac{dx}{4 + x^2}$.

9. a) Give the definition of $\cosh x$ in terms of the exponential function.
b) Find $\int_0^1 \cosh x\, dx$ in terms of e.

*10. Show that $f(x) = x + \sin x$ is an invertible function.

Review Exercise Set 8.4

1. a) Simplify $(1/\sqrt{e})^{\ln 5}$.
b) Solve for x if $3^x/5^{x-1} = 20$.

2. a) Find dy/dx if $y = \ln(\ln 4x)$.
b) Find $\int_{e^2}^{e^6} [dx/x(\ln x)]$.

3. a) Find dy/dx if $y = \tan(e^{\sqrt{x}})$.
b) Find $\int_{\ln 2}^{\ln 6} e^{2x+1}\, dx$, simplifying your answer as much as possible.

4. a) Find $\int (\sin x)5^{\cos x}\, dx$.
b) Find dy/dx if $y = x^{\sqrt{x}}$, $x > 0$.

5. a) How would you find $\int_1^5 x^x\, dx$?
b) Compute the integral in part (a).

6. A radioactive substance has a half-life of 30 years. If 40 lb are left after 100 years, find the initial amount present.

7. a) Find $\cos^{-1}(-\sqrt{3}/2)$.
b) Find $\lim_{x\to\infty} \tan^{-1}[(1 - x^2)/2]$.

8. a) Find dy/dx if $y = (\tan^{-1} 2x)^3$.
b) Find $\displaystyle\int_{\sqrt[4]{2}}^{\sqrt{2}} \frac{x\, dx}{x^2\sqrt{x^4 - 1}}$.

9. Find the expression for $\sinh^{-1} 2x$ in terms of the logarithm function.

*10. a) Is $\sin x$ an invertible function? Why?
b) Is $\sin^{-1} x$ an invertible function? Why?

More Challenging Exercises 8

1. Solve $e^{2x} - 3e^x + 2 = 0$.

2. Outline how it could be proved from our definition of $\ln x$ and of e that $2.7 < e < 2.8$.

3. Show from the definition of $\ln x$ as an integral that, for each integer $n > 1$,

$$\frac{1}{2} + \frac{1}{3} + \cdots + \frac{1}{n} < \ln n < 1 + \frac{1}{2} + \frac{1}{3} + \cdots + \frac{1}{n-1}.$$

4. Show that

$$\lim_{\Delta x \to 0} \frac{a^{\Delta x} - 1}{\Delta x} = \ln a \qquad \text{for } a > 0.$$

[*Hint:* You know $d(a^x)/dx$. Use the definition of the derivative for $f(x) = a^x$.]

5. Using Exercise 4, find $\lim_{h\to 0} [(a^{2h} - 1)/h]$.

6. Show that, for any particular integer n_0, e^x is larger than x^{n_0} if x is sufficiently large.

7. How does the size of e^x compare with the size of $f(x)$ for any particular polynomial function f if x is large? [*Hint:* Use Exercise 6.]

TECHNIQUES OF INTEGRATION

9

We had some practice integrating in Sections 6.4 and 6.5. Perhaps we should review the integration formulas given in Chapters 6 and 8. To save turning pages, they are collected in the box on page 362. We omit those involving the hyperbolic functions and their inverses. Formulas not mentioned before for $\int \sec x\, dx$ and $\int \csc x\, dx$ are included. These formulas are easily verified by differentiation.

One important method of integration is the use of a table of integrals. Proper use of a table of integrals is so important that we included a section on this topic when the integral was introduced in Chapter 6. Very large computer programs have been developed to find formal derivatives and indefinite as well as definite integrals and to perform many other tasks of calculus. It may well be that such programs will be generally accessible in the near future and will largely replace the use of tables as an efficient and reliable way to find many indefinite integrals.

You should learn how to *transform* certain integrals that do not appear in a table into integrals that you know or that a table contains. The first three sections of this chapter give three such methods: integration by parts, partial-fraction decomposition, and substitution techniques. Sections 9.4, 9.5, and 9.6 describe how to integrate, without the use of tables, many of the most frequently encountered functions. These sections continue to develop your facility in the general techniques presented in Sections 9.1, 9.2, and 9.3, especially the technique of substitution.

Remember that if you are unable to integrate a function to find a *definite* integral, it is easy to use Simpson's rule in these days of calculators and computers. The importance of such a nice numerical method of solution cannot be overemphasized.

9.1 INTEGRATION BY PARTS

For differentiation, there are certain rules and formulas that enable us to compute derivatives of constant multiples, sums, products, and quotients of functions whose derivatives are known. While it is easy to compute antiderivatives of constant multiples and sums of functions with known antiderivatives, *there are no simple rules or formulas for integrating products or quotients of functions with known antiderivatives.* In this section, we present our only formula for integrating a product of functions. The technique is known as *integration by parts.* The formula is used not only as a tool for integration but also in some theoretical considerations.

INTEGRATION FORMULAS

1. $\int (u + v)\, dx = \int u\, dx + \int v\, dx + C$
2. $\int cu\, dx = c \int u\, dx + C$
3. $\int u^n\, du = \dfrac{u^{n+1}}{n+1} + C, \qquad n \neq -1$
4. $\int \dfrac{1}{u}\, du = \ln |u| + C$
5. $\int \dfrac{du}{\sqrt{a^2 - u^2}} = \sin^{-1}\left(\dfrac{u}{a}\right) + C$
6. $\int \dfrac{du}{a^2 + u^2} = \dfrac{1}{a} \tan^{-1}\left(\dfrac{u}{a}\right) + C$
7. $\int \dfrac{du}{u\sqrt{u^2 - a^2}} = \dfrac{1}{a} \sec^{-1}\left(\dfrac{u}{a}\right) + C$
8. $\int \sin u\, du = -\cos u + C$
9. $\int \cos u\, du = \sin u + C$
10. $\int \sec^2 u\, du = \tan u + C$
11. $\int \csc^2 u\, du = -\cot u + C$
12. $\int \sec u \tan u\, du = \sec u + C$
13. $\int \csc u \cot u\, du = -\csc u + C$
14. $\int \sec u\, du = \ln |\sec u + \tan u| + C$
15. $\int \csc u\, du = -\ln |\csc u + \cot u| + C$
16. $\int e^u\, du = e^u + C$
17. $\int a^u\, du = \dfrac{1}{\ln a} a^u + C$

THE FORMULA

The formula for integration by parts is easily obtained from the formula for the derivative of a product. Recall that, if $u = f(x)$ and $v = g(x)$ are differentiable functions, then

$$\frac{d(uv)}{dx} = u\frac{dv}{dx} + v\frac{du}{dx}, \tag{1}$$

or, in differential notation,

$$d(uv) = u\,dv + v\,du. \tag{2}$$

From formula (2), we obtain

$$u\,dv = d(uv) - v\,du, \tag{3}$$

so

$$\int u\,dv = \int [d(uv) - v\,du] = \int d(uv) - \int v\,du.$$

Since $\int d(uv) = uv$, we obtain

$$\boxed{\int u\,dv = uv - \int v\,du,} \tag{4}$$

which is the formula for *integration by parts.*

THE TECHNIQUE

Suppose we are faced with the problem of integrating a product of two functions, that is, of finding

$$\int f(x)\cdot g(x)\,dx = \int \underbrace{g(x)}_{u}\cdot\underbrace{(f(x)\,dx)}_{dv} = \int \underbrace{f(x)}_{u}\cdot\underbrace{(g(x)\,dx)}_{dv}. \tag{5}$$

Suppose that F and G are antiderivatives of f and g, respectively, so that

$$F'(x) = f(x) \qquad \text{and} \qquad G'(x) = g(x).$$

Then formula (4) allows us to transform the integral (5) as follows:

> The problem of finding $\int f(x)\cdot g(x)\,dx$ may be reduced to finding either
>
> $$\int F(x)\cdot g'(x)\,dx \qquad \text{or} \qquad \int f'(x)\cdot G(x)\,dx. \tag{6}$$
>
> That is, we can change the problem of integrating a product to integrating the derivative of one factor times an antiderivative of the other.
>
> ***Integration by parts***

EXAMPLE 1 Find $\int x \sin x \, dx$.

Solution The two factors are x and $\sin x$, and, according to integrals (6), we can reduce the problem to finding either

$$\int \frac{x^2}{2}(\cos x)\, dx \qquad \text{or} \qquad \int 1(-\cos x)\, dx.$$

The first of these integrals is worse than the original problem, but the second one is easy. Accordingly, we wish to differentiate x and integrate $\sin x$. When using formula (4), we will write out the substitution as

$$\begin{array}{ll} u = x & dv = \sin x \, dx \\ du = dx & v = -\cos x. \end{array}$$

Formula (4) tells us:

> The integral of the product $u\,dv$ in the top row of the array equals the product uv of the functions at the ends of the dashed diagonal minus the integral of the product $v\,du$ in the bottom row.

Thus we obtain

$$\begin{aligned} \int x \sin x \, dx &= x(-\cos x) - \int 1(-\cos x)\, dx \\ &= -x \cos x + \int \cos x \, dx \\ &= -x \cos x + \sin x + C. \quad \square \end{aligned}$$

EXAMPLE 2 Find $\int xe^{3x}\, dx$.

Solution We wish the first factor x were not present. It would be changed to a factor 1 by differentiation. Thus we make the substitution

$$\begin{array}{ll} u = x & dv = e^{3x}\, dx \\ du = dx & v = \frac{1}{3}e^{3x}. \end{array}$$

Using formula (4), we find that

$$\int xe^{3x}\, dx = x \cdot \frac{1}{3}e^{3x} - \int \frac{1}{3}e^{3x}\, dx = \frac{1}{3}xe^{3x} - \frac{1}{9}e^{3x} + C. \quad \square$$

EXAMPLE 3 Find $\int x^3\sqrt{1 + x^2}\, dx$.

Solution We don't know how to integrate $\sqrt{1 + x^2}$, and if we try differentiating $\sqrt{1 + x^2}$ and integrating x^3 using formula (4), we only make things worse. We borrow one x-factor of x^3 to put with $\sqrt{1 + x^2}$, for we can integrate $x\sqrt{1 + x^2}$. This leads to the substitution

$$\begin{array}{ll} u = x^2 & dv = x\sqrt{1 + x^2}\, dx \\ du = 2x\, dx & v = \frac{1}{3}(1 + x^2)^{3/2}. \end{array}$$

Formula (4) shows that

$$\int x^3\sqrt{1+x^2}\,dx = x^2\cdot\frac{1}{3}(1+x^2)^{3/2} - \frac{1}{3}\int 2x(1+x^2)^{3/2}\,dx$$

$$= \frac{x^2}{3}(1+x^2)^{3/2} - \frac{1}{3}\cdot\frac{2}{5}(1+x^2)^{5/2} + C$$

$$= \frac{x^2}{3}(1+x^2)^{3/2} - \frac{2}{15}(1+x^2)^{5/2} + C. \quad \square$$

EXAMPLE 4 Compute $\int x^2e^x\,dx$.

Solution Integration by parts of the product of x^2 and e^x leads to the computation of either $\int 2xe^x\,dx$ or $\int (x^3/3)e^x\,dx$. The former integral $\int 2xe^x\,dx$ is simpler than our original integral, and this new integral can obviously be further reduced to $\int 2e^x\,dx$ by an additional application of the formula for integration by parts. Thus we make the substitution

$$u = x^2 \qquad dv = e^x\,dx$$

$$du = 2x\,dx \qquad v = e^x$$

and obtain

$$\int x^2e^x\,dx = x^2e^x - \int 2xe^x\,dx. \tag{7}$$

To compute the remaining integral $\int 2xe^x\,dx$, we integrate by parts again. (Do not be confused by the use of u again for the function $2x$ rather than x^2; it is customary.) Our substitution this time is

$$u = 2x \qquad dv = e^x\,dx$$

$$du = 2\,dx \qquad v = e^x$$

and Eq. (7) yields

$$\int x^2e^x\,dx = x^2e^x - \int 2xe^x\,dx = x^2e^x - \left(2xe^x - \int 2e^x\,dx\right)$$

$$= x^2e^x - 2xe^x + 2e^x + C. \quad \square$$

Any function can be regarded as the product of itself and 1. Thus for $\int f(x)\,dx$ we can always try the substitution

$$u = f(x) \qquad dv = 1\,dx$$

$$du = f'(x)\,dx \qquad v = x.$$

EXAMPLE 5 Find $\int \ln x\,dx$ using integration by parts.

Solution The substitution

$$u = \ln x \qquad dv = 1\,dx$$

$$du = \frac{1}{x}\,dx \qquad v = x$$

yields

$$\int \ln x\,dx = x \ln x - \int x \cdot \frac{1}{x}\,dx$$
$$= x \ln x - \int 1\,dx = x \ln x - x + C. \quad \square$$

EXAMPLE 6 The inverse trigonometric functions can be integrated by parts. Find $\int \sin^{-1}x\,dx$.

Solution We set

$$u = \sin^{-1}x \qquad dv = 1\,dx$$
$$du = \frac{1}{\sqrt{1 - x^2}}\,dx \qquad v = x.$$

Then

$$\int \sin^{-1}x\,dx = x \sin^{-1}x - \int \frac{x}{\sqrt{1 - x^2}}\,dx$$
$$= x \sin^{-1}x + \frac{1}{2}\int -2x(1 - x^2)^{-1/2}\,dx$$
$$= x \sin^{-1}x + \frac{1}{2}\cdot\frac{(1 - x^2)^{1/2}}{1/2} + C$$
$$= x \sin^{-1}x + \sqrt{1 - x^2} + C. \quad \square$$

EXAMPLE 7 Compute $\int e^x \sin x\,dx$.

Solution This time, both substitutions for integration by parts lead to $\pm\int e^x \cos x\,dx$, which seems to be just as hard as our original problem. The following is an effective trick. We make the substitution

$$u = e^x \qquad dv = \sin x\,dx$$
$$du = e^x\,dx \qquad v = -\cos x$$

and obtain

$$\int e^x \sin x\,dx = -e^x \cos x + \int e^x \cos x\,dx. \tag{8}$$

We then integrate by parts again, *continuing to let u be the exponential term*, so that

$$u = e^x \qquad dv = \cos x\,dx$$
$$du = e^x\,dx \qquad v = \sin x.$$

We obtain from Eq. (8)

$$\int e^x \sin x\,dx = -e^x \cos x + \left(e^x \sin x - \int e^x \sin x\,dx\right)$$
$$= (e^x \sin x - e^x \cos x) - \int e^x \sin x\,dx. \tag{9}$$

We may now solve Eq. (9) for $\int e^x \sin x\,dx$, and we find that

$$\int e^x \sin x\,dx = \frac{1}{2}(e^x \sin x - e^x \cos x) + C. \quad \square$$

The technique illustrated in Example 6 involves finding an equation in which the integral is regarded as an unknown and then solving for the integral. We give another illustration of this technique.

EXAMPLE 8 Find $\int \sec^3 ax\, dx$.

Solution Since we can integrate $\sec^2 ax$, we try the substitution

$$u = \sec ax \qquad dv = \sec^2 ax\, dx$$

$$du = a \sec ax \tan ax \qquad v = \frac{1}{a} \tan ax.$$

We obtain

$$\int \sec^3 ax\, dx = \frac{1}{a} \sec ax \tan ax - \int \sec ax \tan^2 ax\, dx.$$

Using the identity $\tan^2 ax = \sec^2 ax - 1$ and the integration formula (14) in the box on page 362, we obtain

$$\begin{aligned}\int \sec^3 ax\, dx &= \frac{1}{a} \sec ax \tan ax - \int (\sec ax)(\sec^2 ax - 1)\, dx \\ &= \frac{1}{a} \sec ax \tan ax - \int \sec^3 ax\, dx + \int \sec ax\, dx \\ &= \frac{1}{a} \sec ax \tan ax - \int \sec^3 ax\, dx \\ &\quad + \frac{1}{a} \ln|\sec ax + \tan ax|.\end{aligned}$$

Solving for $\int \sec^3 ax\, dx$, we have

$$2 \int \sec^3 ax\, dx = \frac{1}{a} \sec ax \tan ax + \frac{1}{a} \ln|\sec ax + \tan ax| + C,$$

so

$$\int \sec^3 ax\, dx = \frac{1}{2a} (\sec ax \tan ax + \ln|\sec ax + \tan ax|) + C. \quad \square$$

EXAMPLE 9 Integration by parts is a basic technique for finding reduction formulas. Derive the reduction formula

$$\int \sin^n x\, dx = -\frac{1}{n} \sin^{n-1} x \cos x + \frac{n-1}{n} \int \sin^{n-2} x\, dx \quad \text{for } n \geq 2.$$

Solution We form the substitution

$$u = \sin^{n-1} x \qquad dv = \sin x\, dx$$

$$du = (n-1) \sin^{n-2} x \cos x\, dx \qquad v = -\cos x,$$

which yields

$$\int \sin^n x\, dx = -\sin^{n-1} x \cos x - \int (n-1)(\sin^{n-2} x)(-\cos^2 x)\, dx. \quad \textbf{(10)}$$

From Eq. (10) and the identity $-\cos^2 x = \sin^2 x - 1$, we have

$$\int \sin^n x\, dx = -\sin^{n-1} x \cos x - \int (n-1)(\sin^{n-2} x)(\sin^2 x - 1)\, dx$$

$$= -\sin^{n-1} x \cos x - (n-1) \int \sin^n x\, dx$$

$$+ (n-1) \int \sin^{n-2} x\, dx. \qquad \textbf{(11)}$$

From Eq. (11), we obtain

$$n \int \sin^n x\, dx = -\sin^{n-1} x \cos x + (n-1) \int \sin^{n-2} x\, dx,$$

from which we obtain the desired reduction formula for $\int \sin^n x\, dx$ on division by n. □

When integrating by parts, we have to decide what part of the integrand to call u and what part to call dv. Sometimes, we wish to lower a power of x or to change an integrand involving a logarithmic or inverse trigonometric function to an algebraic expression. The preceding examples indicate that we should then let u be the power of x or the logarithmic or inverse trigonometric function.

SUMMARY

1. Integration by parts is used for finding $\int f(x) \cdot g(x)\, dx$ when the product of the derivative of one of the functions times an antiderivative of the other gives an easier integration problem.
2. If we make the substitution

$$\begin{array}{ll} u = f(x) & dv = g(x)\, dx \\ du = f'(x)\, dx & v = \int g(x)\, dx, \end{array}$$

then $\int f(x) \cdot g(x)\, dx$ equals the product uv of the functions at the ends of the dashed diagonal minus the integral of the product $v\, du$ of the terms in the bottom row. In symbols,

$$\int u\, dv = uv - \int v\, du.$$

EXERCISES

In Exercises 1–46, find the indicated integral without the use of tables.

1. $\int x \cos x\, dx$
2. $\int x \sin 5x\, dx$
3. $\int x e^{3x}\, dx$
4. $\int x^3 \cos x^2\, dx$
5. $\int x^3 \sin 2x^2\, dx$
6. $\int x^2 e^{-x^3}\, dx$

7. $\int x^2 \sin x\,dx$

8. $\int x^3 e^x\,dx$

9. $\int x^3 e^{x^2}\,dx$

10. $\int x^5 \sin x^3\,dx$

11. $\int x \tan^2 x\,dx$

12. $\int x \csc^2 3x\,dx$

13. $\int x \sec^2 2x\,dx$

14. $\int x^3\sqrt{4 - x^2}\,dx$

15. $\int \frac{x^3}{\sqrt{1 + 2x^2}}\,dx$

16. $\int \frac{x^3}{\sqrt[3]{9 - x^2}}\,dx$

17. $\int x^5\sqrt{1 + x^2}\,dx$

18. $\int \frac{x^2}{(1 + x^2)^2}\,dx$

19. $\int \frac{x^3}{(1 + x^2)^2}\,dx$

20. $\int \ln x^2\,dx$

21. $\int x(\ln x)\,dx$

22. $\int_1^e x^3(\ln x)\,dx$

23. $\int \ln(1 + x)\,dx$

24. $\int (\ln x)^2\,dx$

25. $\int \ln(1 + x^2)\,dx$

26. $\int_0^{1/2} \sin^{-1}(2x)\,dx$

27. $\int \cos^{-1} x\,dx$

28. $\int \cot^{-1} 5x\,dx$

29. $\int \tan^{-1} 3x\,dx$

30. $\int x \sec^{-1} x\,dx$

31. $\int x \tan^{-1} x\,dx$

32. $\int x \cot^{-1} 3x\,dx$

33. $\int x^2 \sin^{-1} x\,dx$

34. $\int \sin(\ln x)\,dx$

35. $\int \cos(\ln x)\,dx$

36. $\int e^{ax}\cos bx\,dx$

37. $\int e^{ax}\sin bx\,dx$

38. A reduction formula for $\int (\ln ax)^n\,dx$

39. A formula for $\int \sqrt{ax + b}/x^2\,dx$ in terms of

$$\int [1/(x\sqrt{ax + b})]\,dx$$

40. A reduction formula for $\int x^n e^{ax}\,dx$

41. A reduction formula for $\int x^n \sin ax\,dx$

42. A reduction formula for $\int x^n \cos ax\,dx$

43. A reduction formula for $\int \cos^n ax\,dx$

44. A reduction formula for $\int \sin^n ax \cos^m ax\,dx$, which reduces the power of $\sin ax$ by 2

45. A reduction formula for $\int \tan^n ax\,dx$

46. A reduction formula for $\int \sec^n ax\,dx$

9.2 INTEGRATION OF RATIONAL FUNCTIONS BY PARTIAL FRACTIONS

We would not find a formula for

$$\int \frac{x^6 - 2}{x^4 + x^2}\,dx$$

in most tables. This section shows how this integral, and any other integral of a rational function (quotient of polynomial functions), can be transformed into a sum of integrals that we know how to find or that are found in many tables. The complete technique involves four steps, which we state in abbreviated form as follows:

Step 1	Perform long division if necessary.
Step 2	Factor the denominator into irreducible polynomials.
Step 3	Find the partial fraction decomposition.
Step 4	Integrate the result of Step 3.

Each step is explained in detail in a separate subsection.

STEP 1: PERFORM LONG DIVISION IF NECESSARY

We wish to find

$$\int \frac{f(x)}{g(x)}\,dx,$$

where $f(x)$ and $g(x)$ are polynomial functions. If the polynomial $f(x)$ in the numerator has degree greater than or equal to the degree of the denominator polynomial $g(x)$, perform polynomial long division and write

$$\frac{f(x)}{g(x)} = q(x) + \frac{r(x)}{g(x)} \tag{1}$$

for polynomials $q(x)$ and $r(x)$, where the degree of $r(x)$ is less than the degree of $g(x)$. Since the polynomial function $q(x)$ is easy to integrate, we have essentially reduced the problem of integrating $f(x)/g(x)$ to integrating $r(x)/g(x)$.

EXAMPLE 1 Do Step 1 for

$$\int \frac{x^6 - 2}{x^4 + x^2}\,dx.$$

Solution Polynomial long division yields

$$\begin{array}{r|l} & x^2 - 1 \\ \hline x^4 + x^2 & x^6 \qquad\qquad -2 \\ & x^6 + x^4 \\ \hline & -x^4 \\ & -x^4 - x^2 \\ \hline & x^2 - 2. \end{array}$$

Thus

$$\int \frac{x^6 - 2}{x^4 + x^2}\,dx = \int \left(x^2 - 1 + \frac{x^2 - 2}{x^4 + x^2}\right) dx. \quad \square$$

EXAMPLE 2 Do Step 1 for

$$\int \frac{x^2 + x - 3}{(x - 2)(x^2 - 1)}\,dx.$$

Solution Since the degree 2 of the numerator polynomial is already less than the degree 3 of the denominator polynomial, no long division is necessary. $\square$

STEP 2: FACTOR THE DENOMINATOR INTO IRREDUCIBLE FACTORS

For Step 2, we factor the denominator polynomial $g(x)$ in Eq. (1) into a product of (possibly repeated) linear and irreducible quadratic factors, so that the quadratic factors cannot be factored further into a product of real linear factors. *It is a theorem of algebra that such a factorization exists.*

EXAMPLE 3 Continuing with the integral in Example 1, do Step 2 for

$$\int \left(x^2 - 1 + \frac{x^2 - 2}{x^4 + x^2} \right) dx.$$

Solution The factorization of the denominator $x^4 + x^2$ into irreducible linear and quadratic factors is given by

$$x^4 + x^2 = x \cdot x(x^2 + 1).$$

Example 5 describes a test that establishes that $x^2 + 1$ cannot be factored further, but you probably know this already. □

EXAMPLE 4 Continuing with the integral in Example 2, do Step 2 for

$$\int \frac{x^2 + x - 3}{(x - 1)(x^2 - 1)}\, dx.$$

Solution The factorization of $(x - 2)(x^2 - 1)$ into irreducible factors is given by

$$(x - 2)(x^2 - 1) = (x - 2)(x - 1)(x + 1). \quad \square$$

EXAMPLE 5 Factor $x^3 + 1$ into irreducible factors.

Solution As you learned in algebra, if $x = a$ is a root of a polynomial equation $g(x) = 0$, then $x - a$ is a factor of $g(x)$. Since $x = -1$ is a root of $x^3 + 1 = 0$, we see that $x - (-1) = x + 1$ is a factor of $x^3 + 1$. The remaining factor can be found by long division:

$$\begin{array}{r|l}
 & x^2 - x + 1 \\ \hline
x + 1 & x^3 \qquad\qquad + 1 \\
 & \underline{x^3 + x^2} \\
 & \quad -x^2 \\
 & \quad \underline{-x^2 - x} \\
 & \qquad\qquad x + 1 \\
 & \qquad\qquad \underline{x + 1} \\
 & \qquad\qquad\quad 0.
\end{array}$$

Thus

$$x^3 + 1 = (x + 1)(x^2 - x + 1). \tag{2}$$

We now have to decide whether $x^2 - x + 1$ is irreducible. A quadratic polynomial $ax^2 + bx + c$ is irreducible precisely when $ax^2 + bx + c = 0$ has no real roots. The quadratic formula shows that this is the case precisely when $b^2 - 4ac < 0$. For $x^2 - x + 1$, we have

$$b^2 - 4ac = (-1)^2 - 4 \cdot 1 \cdot 1 = -3 < 0,$$

so $x^2 - x + 1$ is irreducible. Thus Eq. (2) is the factorization of $x^3 + 1$ into irreducible factors. □

STEP 3: FIND THE PARTIAL FRACTION DECOMPOSITION

It is a theorem of algebra (whose proof we will not attempt) that $r(x)/g(x)$ in Eq. (1) can be written as a sum of *partial fractions*, which arise from the factors of $g(x)$ as follows.

A linear factor $ax + b$ of $g(x)$ of multiplicity 1 gives rise to a single term

$$\frac{A}{ax + b} \tag{3}$$

for some constant A, while a linear factor with multiplicity n, say $(ax + b)^n$, gives rise to a sum of such terms, one for each power of the factor up to the nth power. For example, a multiple factor $(ax + b)^4$ gives rise to the sum

$$\frac{A}{(ax + b)^4} + \frac{B}{(ax + b)^3} + \frac{C}{(ax + b)^2} + \frac{D}{(ax + b)} \tag{4}$$

for some constants A, B, C, and D.

An irreducible quadratic factor $ax^2 + bx + c$ of $g(x)$ of multiplicity 1 gives rise to a single term

$$\frac{Ax + B}{ax^2 + bx + c} \tag{5}$$

for constants A and B, while such a factor of multiplicity n gives rise to a sum of such terms, one for each power of the factor up to the nth power. For example, a multiple factor $(ax^2 + bx + c)^3$ gives rise to the sum

$$\frac{Ax + B}{(ax^2 + bx + c)^3} + \frac{Cx + D}{(ax^2 + bx + c)^2} + \frac{Ex + F}{ax^2 + bx + c} \tag{6}$$

for some constants A, B, C, D, E, and F.

EXAMPLE 6 Continuing the integral in Examples 1 and 3, do Step 3 for

$$\int \left[x^2 - 1 + \frac{x^2 - 2}{x^2(x^2 + 1)} \right] dx.$$

Solution As Eqs. (4) and (5) indicate, there should be constants A, B, C, and D such that

$$\frac{x^2 - 2}{x^2(x^2 + 1)} = \frac{A}{x^2} + \frac{B}{x} + \frac{Cx + D}{x^2 + 1}. \tag{7}$$

Adding the quotients on the right-hand side of Eq. (7), we see that

$$\frac{x^2 - 2}{x^2(x^2 + 1)} = \frac{A(x^2 + 1) + Bx(x^2 + 1) + (Cx + D)x^2}{x^2(x^2 + 1)}. \tag{8}$$

The two numerators in Eq. (8) must be equal, so

$$A(x^2 + 1) + Bx(x^2 + 1) + (Cx + D)x^2 = x^2 - 2. \qquad (9)$$

Numerator equation

We must find constants A, B, C, and D so that the numerator equation (9) holds identically for all x. If a linear factor appears in our factorization of the denominator, then we can find one unknown constant easily by setting x in Eq. (9) equal to the value that makes that factor zero. Thus setting $x = 0$, which makes the linear factor x of the denominator in Eq. (8) equal to zero, we obtain from Eq. (9)

$$A = -2. \qquad \text{Setting } x = 0$$

Since we have no more *linear* factors of the denominator, we must find the values for B, C, and D by another method. These remaining constants are found by equating coefficients of like powers of x on the two sides of the numerator equation (9). We proceed:

$$B = 0, \qquad \text{Equating } x\text{-coefficients}$$

$$A + D = 1, \quad D = 1 - A = 1 + 2 = 3, \qquad \text{Equating } x^2\text{-coefficients}$$

$$B + C = 0, \quad C = -B = 0. \qquad \text{Equating } x^3\text{-coefficients}$$

Thus using Eq. (7), we have

$$\int \left[x^2 - 1 + \frac{x^2 - 2}{x^2(x^2 + 1)}\right] dx = \int \left(x^2 - 1 + \frac{-2}{x^2} + \frac{0}{x} + \frac{0x + 3}{x^2 + 1}\right) dx$$

$$= \int \left(x^2 - 1 - \frac{2}{x^2} + \frac{3}{x^2 + 1}\right) dx. \quad \square$$

EXAMPLE 7 Continuing with the integral in Examples 2 and 4, do Step 3 for

$$\int \frac{x^2 + x - 3}{(x - 2)(x + 1)(x - 1)}\, dx.$$

Solution From formula (3), we have

$$\frac{x^2 + x - 3}{(x - 2)(x + 1)(x - 1)} = \frac{A}{x - 2} + \frac{B}{x + 1} + \frac{C}{x - 1}$$

$$= \frac{A(x + 1)(x - 1) + B(x - 2)(x - 1) + C(x - 2)(x + 1)}{(x - 2)(x + 1)(x - 1)}$$

for some constants A, B, and C. Equating numerators yields

$$A(x + 1)(x - 1) + B(x - 2)(x - 1) + C(x - 2)(x + 1) = x^2 + x - 3. \qquad (10)$$

Numerator equation

We set x equal to each value that makes a linear factor zero in Eq. (10), obtaining

$$6B = -3, \quad B = -\tfrac{1}{2}, \qquad \text{Setting } x = -1$$

$$-2C = -1, \quad C = \tfrac{1}{2}, \qquad \text{Setting } x = 1$$

$$3A = 3, \quad A = 1. \qquad \text{Setting } x = 2$$

Thus

$$\int \frac{x^2 + x - 3}{(x-2)(x+1)(x-1)}\,dx = \int \left(\frac{1}{x-2} - \frac{\frac{1}{2}}{x+1} + \frac{\frac{1}{2}}{x-1}\right) dx. \quad \square$$

STEP 4: PERFORM THE INTEGRATION

EXAMPLE 8 To conclude the work in Examples 1, 3, and 6, find

$$\int \left(x^2 - 1 - \frac{2}{x^2} + \frac{3}{x^2+1}\right) dx.$$

Solution The desired integral is

$$\frac{x^3}{3} - x + \frac{2}{x} + 3\tan^{-1}x + C.$$

Thus, referring back to Example 1, we see that

$$\int \frac{x^6 - 2}{x^4 + x^2}\,dx = \frac{x^3}{3} - x + \frac{2}{x} + 3\tan^{-1}x + C. \quad \square$$

EXAMPLE 9 To conclude the work in Examples 2, 4, and 7, find

$$\int \left(\frac{1}{x-2} - \frac{\frac{1}{2}}{x+1} + \frac{\frac{1}{2}}{x-1}\right) dx.$$

Solution The desired integral is

$$\ln|x-2| - \frac{1}{2}\ln|x+1| + \frac{1}{2}\ln|x-1| + C$$

$$= \ln|x-2| + \frac{1}{2}\ln\left|\frac{x-1}{x+1}\right| + C.$$

Referring back to Example 2, we see that

$$\int \frac{x^2 + x - 3}{(x-2)(x^2-1)}\,dx = \ln|x-2| + \frac{1}{2}\ln\left|\frac{x-1}{x+1}\right| + C. \quad \square$$

In this final step of the process, we may have to find an integral of the form

$$\int \frac{A}{(ax+b)^n}\,dx \qquad \text{or} \qquad \int \frac{Bx + C}{(ax^2 + bx + c)^n}\,dx$$

for some integer value of $n \geq 1$. The box on page 375 gives a complete list of the integration formulas we may need for Step 4.

BASIC FORMULAS FOR INTEGRATION OF RATIONAL FUNCTIONS

11. $\displaystyle\int \frac{1}{ax+b}\,dx = \frac{1}{a}\ln|ax+b| + C$

12. $\displaystyle\int \frac{1}{(ax+b)^n}\,dx = \frac{-1}{a(n-1)}\cdot\frac{1}{(ax+b)^{n-1}} + C, \qquad n \neq 1$

13. $\displaystyle\int \frac{1}{ax^2+bx+c}\,dx = \frac{2}{\sqrt{4ac-b^2}}\tan^{-1}\frac{2ax+b}{\sqrt{4ac-b^2}} + C, \qquad b^2 < 4ac$ New

14. $\displaystyle\int \frac{2ax+b}{ax^2+bx+c}\,dx = \ln|ax^2+bx+c| + C$

15. $\displaystyle\int \frac{1}{(ax^2+bx+c)^{n+1}}\,dx = \frac{2ax+b}{n(4ac-b^2)(ax^2+bx+c)^n} + \frac{2(2n-1)a}{n(4ac-b^2)}\int \frac{1}{(ax^2+bx+c)^n}\,dx, \qquad 4ac \neq b^2$ New

16. $\displaystyle\int \frac{2ax+b}{(ax^2+bx+c)^n}\,dx = \frac{-1}{(n-1)(ax^2+bx+c)^{n-1}} + C, \qquad n \neq 1$

Formulas (11) and (14) are easily obtained from the familiar formula

$$\int \frac{du}{u} = \ln|u| + C, \tag{17}$$

and you should not need to look up formula (11) or (14). For example, using formula (17), we have

$$\int \frac{2}{3x+5}\,dx = \frac{2}{3}\int \frac{3}{3x+5}\,dx = \frac{2}{3}\ln|3x+5| + C.$$

The formulas (12) and (16) are instances of the familiar formula

$$\int u^n\,du = \frac{u^{n+1}}{n+1} + C \qquad \text{for } n \neq -1, \tag{18}$$

and you should not need to refer to tables for formula (12) or (16) either. This leaves just formula (13) and the reduction formula (15), which you may wish to look up when you need them. We illustrate the use of formulas (13) and (15).

EXAMPLE 10 Find

$$\int \frac{x-2}{(x^2-5x+7)^2}\,dx.$$

Solution We first "fix up" the integral so that part of the numerator is the derivative $2x - 5$ of the factor $x^2 - 5x + 7$ in the denominator, and we can use formula (16). We have

$$\begin{aligned}\int \frac{x-2}{(x^2-5x+7)^2}\,dx &= \frac{1}{2}\int \frac{2x-4}{(x^2-5x+7)^2}\,dx \\ &= \frac{1}{2}\int \frac{(2x-5)+1}{(x^2-5x+7)^2}\,dx\end{aligned}$$

$$= \frac{1}{2}\int \frac{2x-5}{(x^2-5x+7)^2}\,dx + \frac{1}{2}\int \frac{1}{(x^2-5x+7)^2}\,dx$$

$$= \frac{1}{2}\cdot\frac{-1}{(x^2-5x+7)} + \frac{1}{2}\int \frac{1}{(x^2-5x+7)^2}\,dx.$$

Using formulas (15) and (13), we have

$$\frac{1}{2}\int \frac{1}{(x^2-5x+7)^2}\,dx = \frac{1}{2}\cdot\frac{2x-5}{1\cdot 3(x^2-5x+7)} + \frac{1}{2}\cdot\frac{2\cdot 1\cdot 1}{1\cdot 3}\int \frac{1}{x^2-5x+7}\,dx$$

$$= \frac{2x-5}{6x^2-30x+42} + \frac{1}{3}\cdot\frac{2}{\sqrt{3}}\tan^{-1}\left(\frac{2x-5}{\sqrt{3}}\right) + C.$$

Putting everything together, we have

$$\int \frac{x-2}{(x^2-5x+7)^2}\,dx = \frac{-1}{2x^2-10x+14} + \frac{2x-5}{6x^2-30x+42} + \frac{2}{3\sqrt{3}}\tan^{-1}\left(\frac{2x-5}{\sqrt{3}}\right) + C. \quad \square$$

COMMENTS AND ADDITIONAL EXAMPLES

We should mention that the actual *execution* of the technique described in this section may break down at Step 2, factoring the denominator. While our algebraic theory assures us that the denominator polynomial $g(x)$ in Eq. (1) does have a factorization into linear and irreducible quadratic factors, *finding* such a factorization may be a very tough job. Roots of $g(x) = 0$, and hence *linear* factors of $g(x)$, could be found by Newton's method. But factorization is the only problem; all the other steps are mechanical chores.

EXAMPLE 11 Find

$$\int \frac{13-7x}{(x+2)(x-1)^3}\,dx.$$

Solution Since the degree of the numerator is less than that of the denominator, long division as in Step 1 is unnecessary, and the denominator is already factored for Step 2. For the partial fraction decomposition (Step 3), let

$$\frac{13-7x}{(x+2)(x-1)^3} = \frac{A}{x+2} + \frac{B}{(x-1)^3} + \frac{C}{(x-1)^2} + \frac{D}{x-1}.$$

The numerator equation is

$$\begin{aligned} A(x-1)^3 + B(x+2) + C(x+2)(x-1)& \\ + D(x+2)(x-1)^2 = 13 - 7x.& \end{aligned} \tag{19}$$

We find as many of the unknown constants as possible using the zeros, -2, and 1 of the linear factors in the denominator. We have, from Eq. (19),

$$-27A = 27, \qquad A = -1, \qquad \text{Setting } x = -2$$

$$3B = 6, \qquad B = 2. \qquad \text{Setting } x = 1$$

To find C and D, we need two equations containing them found by equating coefficients. The easiest equations to find are often those corresponding to the terms of highest degree and to the constant terms. Computing the constant terms of course amounts to setting $x = 0$. We obtain, from Eq. (19),

$$A + D = 0, \quad D = -A = 1, \qquad \text{Equating } x^3\text{-coefficients}$$

$$-A + 2B - 2C + 2D = 13, \qquad \text{Setting } x = 0$$

$$\begin{aligned} 2C &= -A + 2B + 2D - 13 \\ &= 1 + 4 + 2 - 13 \\ &= -6, \\ C &= -3. \end{aligned}$$

Hence

$$\begin{aligned} &\int \frac{13 - 7x}{(x+2)(x-1)^3}\,dx \\ &\qquad = \int \left[\frac{-1}{x+2} + \frac{2}{(x-1)^3} + \frac{-3}{(x-1)^2} + \frac{1}{x-1}\right] dx \\ &\qquad = -\ln|x+2| - \frac{1}{(x-1)^2} + \frac{3}{x-1} + \ln|x-1| + C. \quad \square \end{aligned}$$

EXAMPLE 12 Find

$$\int \frac{3x^4 + 2x^3 + 8x^2 + x + 2}{x^5 + 2x^3 + x}\,dx.$$

Solution Again, long division is unnecessary, and factoring the denominator yields

$$x^5 + 2x^3 + x = x(x^2+1)^2.$$

For the partial fraction decomposition, we let

$$\frac{3x^4 + 2x^3 + 8x^2 + x + 2}{x^5 + 2x^3 + x} = \frac{A}{x} + \frac{Bx + C}{(x^2+1)^2} + \frac{Dx + E}{x^2 + 1}.$$

The numerator equation is

$$\begin{aligned} A(x^2+1)^2 + (Bx + C)x + (Dx + E)(x^3 + x)& \\ = 3x^4 + 2x^3 + 8x^2 + x + 2.& \end{aligned} \tag{20}$$

Working to find the constants, we have, from Eq. (20),

$$A = 2, \qquad \text{Setting } x = 0$$

$$A + D = 3, \quad D = 3 - A = 3 - 2 = 1, \qquad \text{Equating } x^4\text{-coefficients}$$

$$E = 2 \qquad \text{Equating } x^3\text{-coefficients}$$

$$2A + B + D = 8, \quad B = 8 - 2A - D = 8 - 4 - 1 = 3, \qquad \text{Equating } x^2\text{-coefficients}$$

$$C + E = 1, \quad C = 1 - E = 1 - 2 = -1. \qquad \text{Equating } x\text{-coefficients}$$

Thus we have

$$\int \frac{3x^4 + 2x^3 + 8x^2 + x + 2}{x(x^2 + 1)^2}\,dx = \int \left[\frac{2}{x} + \frac{3x - 1}{(x^2 + 1)^2} + \frac{x + 2}{x^2 + 1}\right] dx. \tag{21}$$

Using formulas (16), (15), and (13), we obtain

$$\begin{aligned}\int \frac{3x - 1}{(x^2 + 1)^2}\,dx &= \frac{3}{2}\int \frac{2x}{(x^2 + 1)^2}\,dx - \int \frac{1}{(x^2 + 1)^2}\,dx \\ &= \frac{3}{2}\cdot\frac{-1}{x^2 + 1} - \left[\frac{2x}{1(4)(x^2 + 1)} + \frac{2 \cdot 1}{1 \cdot 4}\int \frac{1}{x^2 + 1}\,dx\right] \\ &= \frac{-3}{2(x^2 + 1)} - \frac{x}{2(x^2 + 1)} - \frac{1}{2}\tan^{-1}x + C. \end{aligned} \tag{22}$$

We also find that

$$\begin{aligned}\int \frac{x + 2}{x^2 + 1}\,dx &= \frac{1}{2}\int \frac{2x}{x^2 + 1}\,dx + 2\int \frac{1}{x^2 + 1}\,dx \\ &= \frac{1}{2}\ln|x^2 + 1| + 2\tan^{-1}x + C. \end{aligned} \tag{23}$$

Then Eqs. (21), (22), and (23) yield

$$\begin{aligned}\int \frac{3x^4 + 2x^3 + 8x^2 + x + 2}{x(x^2 + 1)^2}\,dx &= 2\ln|x| - \frac{3}{2(x^2 + 1)} - \frac{x}{2(x^2 + 1)} - \frac{1}{2}\tan^{-1}x \\ &\quad + \frac{1}{2}\ln|x^2 + 1| + 2\tan^{-1}x + C \\ &= \frac{1}{2}\ln|x^6 + x^4| - \frac{x + 3}{2(x^2 + 1)} \\ &\quad + \frac{3}{2}\tan^{-1}x + C. \quad \square \end{aligned}$$

SUMMARY

To integrate a quotient of polynomials, the following steps are suggested.

Step 1 Perform polynomial long division, if necessary, to reduce the problem to integrating a rational function whose numerator has degree less than the degree of the denominator.

Step 2 Factor the denominator into linear and irreducible quadratic factors.

Step 3 Obtain the partial fraction decomposition of the rational function.

Step 4 Integrate the summands in the resulting decomposition of the original rational function, using formulas (11) through (16) in this section.

EXERCISES

Use the methods described in this section to find the integral.

1. $\displaystyle\int \frac{x-1}{x^2}\,dx$

2. $\displaystyle\int \frac{x^5-4x^3+2x^2-7}{x^3}\,dx$

3. $\displaystyle\int \frac{x^3-2x}{x+1}\,dx$

4. $\displaystyle\int \frac{x^4+3x^2}{2x-3}\,dx$

5. $\displaystyle\int \frac{x^3+2x+1}{x^2+1}\,dx$

6. $\displaystyle\int \frac{x^4-x}{x^2+4}\,dx$

7. $\displaystyle\int \frac{1}{x^2-1}\,dx$

8. $\displaystyle\int \frac{1}{x^2-x}\,dx$

9. $\displaystyle\int \frac{x^3+4}{x^2-4}\,dx$

10. $\displaystyle\int \frac{x^3+x^2-x+8}{x^2+2x}\,dx$

11. $\displaystyle\int \frac{x^4-2x^2+6}{x^2-3x+2}\,dx$

12. $\displaystyle\int \frac{x^4-x^2-4x-2}{x^3-x}\,dx$

13. $\displaystyle\int \frac{3x+4}{x^3-2x^2-3x}\,dx$

14. $\displaystyle\int \frac{2x^4+6x^3-10x^2-3x+2}{x^3+3x^2-6x-8}\,dx$

15. $\displaystyle\int \frac{x^3+4x^2-5x-4}{x^3+3x^2-x-3}\,dx$

16. $\displaystyle\int \frac{2x+1}{x^3-x^2}\,dx$

17. $\displaystyle\int \frac{x-1}{x^3+2x^2+x}\,dx$

18. $\displaystyle\int \frac{x^4+1}{x^4+2x^3}\,dx$

19. $\displaystyle\int \frac{x^2-3}{(x-2)(x+1)^3}\,dx$

20. $\displaystyle\int \frac{2x^2+2x-2}{x^3+2x}\,dx$

21. $\displaystyle\int \frac{5x^3-x^2+4x-12}{x^4-4x^2}\,dx$

22. $\displaystyle\int \frac{2x^2-x+3}{x^3-x^2+x-1}\,dx$

23. $\displaystyle\int \frac{6x^2+3x+1}{x^3-x^2+x-1}\,dx$

24. $\displaystyle\int \frac{3x^3+3x^2-5x+7}{x^4-1}\,dx$

25. $\displaystyle\int \frac{2x^3+5x^2-2x+16}{x^4-2x^2-8}\,dx$

26. $\displaystyle\int \frac{x+3}{x^2-5x+7}\,dx$

27. $\displaystyle\int \frac{5x-4}{3x^2-4x+2}\,dx$

28. $\displaystyle\int \frac{x+4}{(x^2-x+1)^2}\,dx$

29. $\displaystyle\int \frac{3x-1}{(x^2-2x+2)^2}\,dx$

30. $\displaystyle\int \frac{x^3+2x^2+3x+1}{x^4+2x^2+1}\,dx$

31. $\displaystyle\int \frac{x^3-x^2+2x+1}{(x^2-x+1)^2}\,dx$

32. $\displaystyle\int \frac{-x^3+10x^2-19x+22}{(x-1)^2(x^2+3)}\,dx$

33. $\displaystyle\int \frac{3x^6+9x^4-4x^3+7x^2+3}{x^2(x^2+1)^3}\,dx$

9.3 SUBSTITUTION

Various substitution techniques for integration are presented in this section and in Sections 9.4 and 9.6. This section deals primarily with *algebraic substitutions*, often made to change an integral to a more familiar form or to attempt to eliminate a radical from an integral. Trigonometric substitutions are studied in Sections 9.4 and 9.6.

Chapter 6 introduced the type of substitution technique illustrated in the following example.

EXAMPLE 1 Find $\int x(x^2 + 3)^4\, dx$.

Solution We note that $x\, dx$ is the differential of $x^2 + 3$, except for a constant factor. We let $u = x^2 + 3$, so $du = 2x\, dx$ and $x\, dx = \frac{1}{2}\, du$. Then

$$\int x(x^2 + 3)^4\, dx = \int u^4 \frac{1}{2}\, du = \frac{1}{2} \cdot \frac{u^5}{5} + C = \frac{1}{10}(x^2 + 3)^5 + C. \quad \square$$

Success in this technique depends on spotting a factor of the integrand to serve as u' for some expression u that already appears in the integral; then $u'\, dx = du$. If $f(x)$ can be expressed as $g(u)u'$ and if $\int g(u)\, du = G(u) + C$, then

$$\frac{d(G(u))}{dx} = \frac{d(G(u))}{du} \cdot \frac{du}{dx}$$
$$= g(u)u' = f(x),$$

so $G(u)$ is indeed an antiderivative of $f(x)$.

Here is another substitution example.

EXAMPLE 2 Find $\int x\sqrt{x + 1}\, dx$ by substitution.

Solution The radical is causing us trouble, so we let $u = \sqrt{x + 1}$ to eliminate it. Then $u^2 = x + 1$, so

$$x = u^2 - 1 \qquad \text{and} \qquad dx = 2u\, du.$$

(*This substitution for dx must be computed. A common error is just to replace dx by du.*) Therefore

$$\int x\sqrt{x + 1}\, dx = \int (u^2 - 1)u \cdot 2u\, du = 2 \int (u^4 - u^2)\, du$$
$$= 2\left(\frac{u^5}{5} - \frac{u^3}{3}\right) + C$$
$$= 2\left[\frac{(\sqrt{x + 1})^5}{5} - \frac{(\sqrt{x + 1})^3}{3}\right] + C$$
$$= 2(\sqrt{x + 1})^3\left(\frac{x + 1}{5} - \frac{1}{3}\right) + C$$
$$= 2(\sqrt{x + 1})^3\left(\frac{x}{5} - \frac{2}{15}\right) + C. \quad \square$$

In the type of substitution illustrated in Example 2, we let $x = h(u)$, so $dx = h'(u)\,du$. We then write

$$\int f(x)\,dx = \int f(h(u)) \cdot h'(u)\,du = G(u) + C. \tag{1}$$

Again,

$$\frac{d(G(u))}{dx} = \frac{d(G(u))}{du} \cdot \frac{du}{dx} = f(h(u)) \cdot h'(u) \cdot \frac{du}{dx} = f(h(u)) \cdot \frac{dx}{du} \cdot \frac{du}{dx}$$
$$= f(h(u)) = f(x),$$

so $G(u) = G(h^{-1}(x))$ is indeed an antiderivative of $f(x)$. After obtaining the indefinite integral $G(u) + C$ in Eq. (1), we want to express $G(u)$ in terms of the original variable x. To do this, we solve the substitution $x = h(u)$ for u in terms of x and substitute the resulting expression for u in $G(u)$.*

As mentioned in Example 2, a common error when substituting $x = h(u)$ is to replace dx by du instead of replacing dx by $h'(u)\,du$. *Don't forget to compute dx.*

An *algebraic substitution* $x = h(u)$ is one where $h(u)$ is a function involving only arithmetic operations and roots (no trigonometric functions, for example).

EXAMPLE 3 Use an algebraic substitution to find the integral

$$\int x^3\sqrt{4 - x^2}\,dx.$$

Solution To eliminate the radical, we let $u = \sqrt{4 - x^2}$ so that $u^2 = 4 - x^2$. Implicit differentiation yields $2u\,du = -2x\,dx$, so

$$x\,dx = -u\,du.$$

We thus have

$$\int x^3\sqrt{4 - x^2}\,dx = \int x^2\sqrt{4 - x^2}\,x\,dx = \int (4 - u^2)u(-u\,du)$$
$$= \int (-4u^2 + u^4)\,du = -\frac{4u^3}{3} + \frac{u^5}{5} + C.$$

Since $u = \sqrt{4 - x^2}$, we obtain

$$\int x^3\sqrt{4 - x^2}\,dx = -\frac{4(4 - x^2)^{3/2}}{3} + \frac{(4 - x^2)^{5/2}}{5} + C. \quad \square$$

EXAMPLE 4 Find

$$\int \frac{x^3}{(x^2 - 1)^{3/2}}\,dx.$$

---

* To solve $x = h(u)$ for u in terms of x, the function h must have an inverse. Thus we should restrict our substitutions to invertible functions with continuous derivatives. Strictly speaking, we really should think of our substitution as $u = h^{-1}(x)$, defining what u is to be in terms of x, since x is given. We did this in Examples 1 and 2. We then go to $x = h(u)$ to compute dx. However, we will soft-pedal these theoretical considerations and feel free to substitute $x = h(u)$ in those cases where it seems most natural.

Solution We let $u = \sqrt{x^2 - 1}$, so $u^2 = x^2 - 1$. Then $2u\,du = 2x\,dx$, so $u\,du = x\,dx$. We then obtain

$$\int \frac{x^3}{(x^2-1)^{3/2}}\,dx = \int \frac{x^2(x\,dx)}{(x^2-1)^{3/2}} = \int \frac{(u^2+1)(u\,du)}{u^3}$$

$$= \int \frac{u^3+u}{u^3}\,du = \int \left(1 + \frac{1}{u^2}\right) du$$

$$= u - \frac{1}{u} + C = \sqrt{x^2-1} - \frac{1}{\sqrt{x^2-1}} + C. \quad \square$$

We should not think that all integrals with radicals will yield to algebraic substitution. Look at the following example.

EXAMPLE 5 Find $\int x^4\sqrt{9-x^2}\,dx$.

Attempted Solution It is natural to try the substitution $u = \sqrt{9-x^2}$. Then $u^2 = 9 - x^2$, so $2u\,du = -2x\,dx$ or $x\,dx = -u\,du$. We obtain

$$\int x^4\sqrt{9-x^2}\,dx = \int x^3\sqrt{9-x^2}\,x\,dx = \int x^3u(-u\,du).$$

But since $u^2 = 9 - x^2$, we have $x^2 = 9 - u^2$, so

$$x^3 = x^2 \cdot x = (9-u^2)\sqrt{9-u^2}$$

and our integral becomes

$$\int x^3u(-u\,du) = \int (9-u^2)\sqrt{9-u^2}\,u(-u\,du)$$

$$= -\int (9u^2 - u^4)\sqrt{9-u^2}\,du.$$

We have not eliminated the radical and clearly have just as tough an integral. We will see how to find this integral in Section 9.6. $\square$

Sometimes a substitution can be used to simultaneously eliminate different fractional powers from an integrand.

EXAMPLE 6 Find

$$\int \frac{x^{1/2}}{1+x^{1/3}}\,dx.$$

Solution To eliminate the fractional powers $x^{1/3}$ and $x^{1/2}$ at the same time, we let $u = x^{1/6}$. Then $x = u^6$, so that $dx = 6u^5\,du$ and

$$\int \frac{x^{1/2}}{1+x^{1/3}}\,dx = \int \frac{u^3}{1+u^2}\,6u^5\,du.$$

Application of the method of partial fractions yields

$$6\int \frac{u^8}{u^2+1}\,du = 6\int\left(u^6 - u^4 + u^2 - 1 + \frac{1}{u^2+1}\right) du$$

$$= 6\left(\frac{u^7}{7} - \frac{u^5}{5} + \frac{u^3}{3} - u + \tan^{-1}u\right) + C.$$

Since $u = x^{1/6}$, we have

$$\int \frac{x^{1/2}}{1 + x^{1/3}}\,dx = 6\left(\frac{x^{7/6}}{7} - \frac{x^{5/6}}{5} + \frac{x^{1/2}}{3} - x^{1/6} + \tan^{-1}x^{1/6}\right) + C. \quad \square$$

Suppose now that a substitution $x = h(u)$ is involved in the computation of a *definite* integral $\int_a^b f(x)\,dx$. Then we may choose to change the x-limits, a and b, to u-limits and evaluate the integral as soon as it is found in terms of u, using these u-limits. The u-interval for integration is generally different from the x-interval $[a, b]$. Alternatively, we can find an indefinite integral in terms of x as in the preceding examples and then compute the definite integral using the x-limits. We illustrate both approaches in the next example.

EXAMPLE 7 Find

$$\int_0^3 \frac{x}{\sqrt{x+1}}\,dx.$$

Solution 1 We let $u = \sqrt{x+1}$, so $x + 1 = u^2$. Then $dx = 2u\,du$. When $x = 0$, we see that $u = \sqrt{0+1} = 1$. Similarly, when $x = 3$, we have $u = \sqrt{3+1} = 2$. Then

$$\int_0^3 \frac{x}{\sqrt{x+1}}\,dx = \int_{u=1}^2 \frac{(u^2-1)(2u\,du)}{u} = \int_{u=1}^2 (2u^2 - 2)\,du$$

$$= \left(2\frac{u^3}{3} - 2u\right)\Big]_1^2 = \left(\frac{16}{3} - 4\right) - \left(\frac{2}{3} - 2\right)$$

$$= \frac{14}{3} - 2 = \frac{8}{3}.$$

Solution 2 Using the same substitution and computation of an indefinite integral as in Solution 1, we have

$$\int_0^3 \frac{x}{\sqrt{x+1}}\,dx = \int_{x=0}^3 (2u^2 - 2)\,du = \left(2\frac{u^3}{3} - 2u\right)\Big]_{x=0}^3$$

$$= \left(2\frac{(x+1)^{3/2}}{3} - 2\sqrt{x+1}\right)\Big]_0^3$$

$$= \left(2\cdot\frac{8}{3} - 2\cdot 2\right) - \left(\frac{2}{3} - 2\right)$$

$$= \frac{14}{3} - 2 = \frac{8}{3}. \quad \square$$

While we have used only algebraic substitutions so far in this section, the substitution technique is very general. We deal with other substitutions in the following sections. Finally, we present one example involving a non-algebraic substitution.

EXAMPLE 8 Find

$$\int e^{\sin^{-1}x}\,dx.$$

Solution This integral looks pretty hopeless, but we can try to simplify it by the substitution

$$u = \sin^{-1}x.$$

Then $x = \sin u$, and $dx = \cos u \, du$. The integral becomes

$$\int e^{\sin^{-1}x}dx = \int e^{u} \cos u \, du.$$

A table (or integration by parts) yields

$$\int e^{u} \cos u \, du = \frac{e^{u}}{2}(\cos u + \sin u) + C.$$

Since $u = \sin^{-1}x$, we know that $-\pi/2 \leq u \leq \pi/2$, so $\cos u \geq 0$. We then have

$$\sin u = x \qquad \text{and} \qquad \cos u = \sqrt{1 - \sin^2 u} = \sqrt{1 - x^2}.$$

Thus we have

$$\int e^{\sin^{-1}x}\, dx = \frac{e^{\sin^{-1}x}}{2}(x + \sqrt{1 - x^2}) + C. \quad \square$$

SUMMARY

1. If $x = h(u)$, then $\int f(x)\, dx$ becomes $\int f(h(u)) \cdot h'(u)\, du$. If the latter integral can be computed as $G(u) + C$, then solve the original substitution for u in terms of x and substitute in $G(u) + C$ to obtain $\int f(x)\, dx$ in terms of x.
2. Integration of expressions involving fractional powers may yield to algebraic substitution designed to eliminate all the fractional powers.
3. Suppose a substitution $x = h(u)$ is used in finding a definite integral $\int_a^b f(x)\, dx$. We may wish to find the u-limits corresponding to the x-limits, a and b, and evaluate the definite integral using those u-limits as soon as we find an indefinite integral in terms of u.

EXERCISES

1. Integrals that yield easily to one technique often yield easily to another. Show that $\int x\sqrt{x + 1}\, dx$ of Example 2 can be found using integration by parts.

2. Repeat Exercise 1 for $\int x^3\sqrt{4 - x^2}\, dx$ of Example 3.

In Exercises 3 through 37, find the integral. For the definite integrals, change to limits for the substitution variable, as illustrated in Example 7, Solution 1.

3. $\displaystyle\int \frac{x}{\sqrt{1 + x}}\, dx$

4. $\displaystyle\int \frac{\sqrt{x - 1}}{x}\, dx$

5. $\displaystyle\int x^2\sqrt{4 + x}\, dx$

6. $\displaystyle\int x\sqrt{3 - 2x}\, dx$

7. $\displaystyle\int \frac{\sqrt{x}}{\sqrt{x} + 1}\, dx$

8. $\displaystyle\int \frac{\sqrt{x} - 1}{\sqrt{x}}\, dx$

9. $\displaystyle\int \frac{\sqrt{x + 2} - 1}{\sqrt{x + 2} + 1}\, dx$

10. $\displaystyle\int \frac{x^2}{\sqrt{x - 1}}\, dx$

11. $\displaystyle\int \frac{3x + 2}{\sqrt{x + 1}}\, dx$

12. $\displaystyle\int_2^5 \frac{x}{\sqrt{x - 1}}\, dx$

13. $\displaystyle\int_{-3}^{0} x\sqrt{1 - x}\, dx$

14. $\displaystyle\int_1^4 \frac{\sqrt{x}}{1 + \sqrt{x}}\, dx$

15. $\int \frac{x}{\sqrt{1+x^2}}\,dx$

16. $\int \frac{x-3}{\sqrt{4-x^2}}\,dx$

27. $\int_0^3 \frac{x^3}{\sqrt{x^2+16}}\,dx$

28. $\int \frac{\sqrt{x}}{1+x^{1/4}}\,dx$

17. $\int \frac{x^3}{\sqrt{1+x^2}}\,dx$

18. $\int \frac{\cos x}{\sqrt{4-\sin^2 x}}\,dx$

29. $\int \frac{x+\sqrt{x}}{x^{1/4}-1}\,dx$

30. $\int \frac{x^{1/2}}{4+x^{1/3}}\,dx$

19. $\int x^3(x^2+1)^{3/2}\,dx$

20. $\int \frac{x^3+2x}{\sqrt{x^2-2}}\,dx$

31. $\int \frac{x^{2/3}}{1+x^{1/2}}\,dx$

32. $\int \frac{dx}{x^{1/2}+x^{1/3}}$

21. $\int \frac{x^3-x}{\sqrt{x^2+8}}\,dx$

22. $\int \frac{x^5}{\sqrt{1-x^3}}\,dx$

33. $\int e^{\cos^{-1}x}dx$

34. $\int xe^{\cos^{-1}x}dx$

23. $\int \frac{x^3}{(x^2-1)^{5/2}}\,dx$

24. $\int x^5\sqrt{9+x^3}\,dx$

35. $\int xe^{\sin^{-1}x}dx$

36. $\int_0^1 e^{\sin^{-1}x}dx$

25. $\int_3^5 x\sqrt{25-x^2}\,dx$

26. $\int_0^4 x^3\sqrt{9+x^2}\,dx$

37. $\int_0^1 e^{\cos^{-1}x}dx$

*9.4 INTEGRATION OF RATIONAL FUNCTIONS OF sin x AND cos x

Section 9.2 described how the integration of any rational function can (theoretically) be accomplished. We show in this section how the integration of any rational expression (quotient of polynomials) in sin x and cos x can be reduced to the integration of a rational function by substitution. Since tan x, cot x, sec x, and csc x can all in turn be written as rational expressions in sin x and cos x, the technique will actually provide us with a method of integrating any rational expression in the trigonometric functions.

We make the substitution

$$\boxed{t = \tan\frac{x}{2},} \tag{1}$$

so that

$$\boxed{x = 2\tan^{-1}t,} \tag{2}$$

which was discovered by some ingenious person. Using identities for trigonometric functions, we then obtain

$$\sin x = 2\sin\frac{x}{2}\cos\frac{x}{2} = 2\tan\frac{x}{2}\cos^2\frac{x}{2} = 2\frac{\tan(x/2)}{\sec^2(x/2)}$$

$$= 2\frac{\tan(x/2)}{1+\tan^2(x/2)} = \frac{2t}{1-t^2}.$$

Similarly,

$$\cos x = 2\cos^2\frac{x}{2} - 1 = \frac{2}{\sec^2(x/2)} - 1 = \frac{2}{1+\tan^2(x/2)} - 1$$

$$= \frac{2}{1+t^2} - 1 = \frac{1-t^2}{1+t^2}.$$

* This section is not used elsewhere in the text and may be omitted.

Finally,

$$dx = 2\frac{1}{1+t^2}\,dt.$$

Collecting these formulas in one place, we have

$$\boxed{\begin{aligned} \sin x &= \frac{2t}{1+t^2}, \\ \cos x &= \frac{1-t^2}{1+t^2}, \\ dx &= \frac{2}{1+t^2}\,dt. \end{aligned}} \tag{3}$$

Clearly formulas (3) may be used to convert the integration of a rational expression in $\sin x$ and $\cos x$ to the integration of a rational function of t.

We should not blindly make substitution (3) with every integral of a rational expression in $\sin x$ and $\cos x$. First, we should look for a shorter method. Also, we can try to find the integral in a table. We illustrate with examples.

EXAMPLE 1 Find $\int (1 - \cos x)/(1 + \cos x)\,dx$.

Solution This integral can be found using substitution (3), but it is easier to use the "trick"

$$\begin{aligned} \int \frac{1-\cos x}{1+\cos x}\,dx &= \int \frac{(1-\cos x)^2}{(1+\cos x)(1-\cos x)}\,dx \\ &= \int \frac{(1-\cos x)^2}{1-\cos^2 x}\,dx \\ &= \int \frac{1 - 2\cos x + \cos^2 x}{\sin^2 x}\,dx \\ &= \int (\csc^2 x - 2\csc x\cot x + \cot^2 x)\,dx \\ &= \int (\csc^2 x - 2\csc x\cot x + \csc^2 x - 1)\,dx \\ &= \int (2\csc^2 x - 2\csc x\cot x - 1)\,dx \\ &= -2\cot x + 2\csc x - x + C. \quad \square \end{aligned}$$

EXAMPLE 2 Find $\int 1/(2 + \cos x)\,dx$.

Solution Using formula (80) from the table of integrals on the endpapers of this book, we easily find the integral to be

$$\int \frac{1}{2+\cos x}\,dx = \frac{2}{\sqrt{3}}\tan^{-1}\left(\frac{1}{\sqrt{3}}\tan\frac{x}{2}\right) + C. \quad \square$$

EXAMPLE 3 Find $\int (\sin x)/(\cos^2 x)\,dx$, using substitution (3) and partial fractions, to illustrate the technique.

Solution From substitution (3), we have

$$\int \frac{\sin x}{\cos^2 x}\,dx = \int \frac{\dfrac{2t}{1+t^2}}{\left(\dfrac{1-t^2}{1+t^2}\right)^2} \cdot \frac{2}{1+t^2}\,dt = \int \frac{4t}{(1-t^2)^2}\,dt$$

$$= \int \frac{4t}{(t-1)^2(t+1)^2}\,dt.$$

Forming a partial fraction decomposition, we have

$$\frac{4t}{(t-1)^2(t+1)^2} = \frac{A}{(t-1)^2} + \frac{B}{t-1} + \frac{C}{(t+1)^2} + \frac{D}{t+1}.$$

The numerator equation is

$$\begin{aligned} A(t+1)^2 + B(t-1)(t+1)^2 + C(t-1)^2 \\ + D(t+1)(t-1)^2 = 4t. \end{aligned} \tag{4}$$

Finding the constants, we obtain from Eq. (4)

$$\begin{aligned} 4C &= -4, \quad C = -1, && \text{Setting } t = -1 \\ 4A &= 4, \quad A = 1, && \text{Setting } t = 1 \\ B + D &= 0, \quad B = -D, && \text{Equating } t^3\text{-coefficients} \\ A - B + C + D &= 0, \quad -B + D = 0, && t = 0 \\ & 2D = 0, \quad D = 0, \quad B = 0. \end{aligned}$$

Thus

$$\int \frac{4t}{(t-1)^2(t+1)^2}\,dt = \int \left[\frac{1}{(t-1)^2} - \frac{1}{(t+1)^2}\right] dt$$

$$= -\frac{1}{t-1} + \frac{1}{t+1} + C.$$

Since $t = \tan(x/2)$ from Eq. (1), we obtain

$$\int \frac{\sin x}{\cos^2 x}\,dx = -\frac{1}{\tan(x/2) - 1} + \frac{1}{\tan(x/2) + 1} + C.$$

(Of course, the integral is found more easily as $\int (\sin x)/(\cos^2 x)\,dx = \int \tan x \sec x\,dx = \sec x + C$.) □

EXAMPLE 4 Find the integral

$$\int \frac{1}{3\sin x + 4\cos x}\,dx.$$

Solution This integral is not found in the table on the endpapers of this book. Using the variable transformation in substitution (3), we have

$$\begin{aligned}\int \frac{1}{3\sin x + 4\cos x}\,dx \\ &= \int \frac{1}{3[2t/(1+t^2)] + 4[(1-t^2)/(1+t^2)]}\cdot\frac{2}{1+t^2}\,dt \\ &= \int \frac{2}{6t + 4 - 4t^2}\,dt = -\int \frac{dt}{2t^2 - 3t - 2} \\ &= -\int \frac{dt}{(2t+1)(t-2)} = -\int\left(\frac{-2/5}{2t+1} + \frac{1/5}{t-2}\right)dt \\ &= \frac{1}{5}\ln|2t+1| - \frac{1}{5}\ln|t-2| + C.\end{aligned}$$

Since $t = \tan(x/2)$ by Eq. (1), we obtain finally

$$\int \frac{dx}{3\sin x + 4\cos x} = \frac{1}{5}\ln\left|2\left(\tan\frac{x}{2}\right) + 1\right| - \frac{1}{5}\ln\left|\left(\tan\frac{x}{2}\right) - 2\right| + C. \quad \square$$

Examples 3 and 4 indicate that substitution (3) often leads to awkward computations. In particular, we prefer to compute a *definite integral* of this type using Simpson's rule rather than using substitution (3) to find an antiderivative.

EXAMPLE 5 Find $\int_0^{\pi/2} 1/(3\sin x + 4\cos x)\,dx$.

Solution Our calculator has a module of packaged programs, including one for Simpson's rule. We did this problem on our calculator, using that program with $n = 20$, and we timed ourselves:

$$\int_0^{\pi/2} \frac{1}{3\sin x + 4\cos x}\,dx \approx 0.3583536427.$$

It took us 87 seconds from the moment we turned the calculator on. Our calculator is slow compared with a computer. Using the answer to Example 4, we obtain 0.3583518938 as a more accurate value for the integral. Our Simpson's rule approximation was accurate enough for most purposes. $\square$

SUMMARY

1. A rational function of $\sin x$ and $\cos x$ can be integrated using the substitution

$$\sin x = \frac{2t}{1+t^2}, \qquad \cos x = \frac{1-t^2}{1+t^2}, \qquad dx = \frac{2}{1+t^2}\,dt.$$

Use the method of partial fractions to integrate the resulting rational function of t and, after integration, substitute $t = \tan(x/2)$.

2. Always look for an easier method before making the preceding substitution.
3. Simpson's rule is often a fast way to find a definite integral.

EXERCISES

1. Find $\int \sec dx = \int (1/\cos x)\, dx$ using the method described in this section.
2. Find $\int \csc x\, dx = \int (1/\sin x)\, dx$ using the method described in this section.

In Exercises 3 through 15, find the indicated indefinite integral by any method, including the use of the tables in the text and on the endpapers.

3. $\displaystyle\int \frac{\sin x}{1 + \cos x}\, dx$
4. $\displaystyle\int \frac{dx}{1 + \sin x}$
5. $\displaystyle\int \frac{dx}{2 - \sin x}$
6. $\displaystyle\int \frac{dx}{1 + \sin x + \cos x}$
7. $\displaystyle\int \frac{dx}{\sin x + \tan x}$
8. $\displaystyle\int \frac{dx}{\sin x + \cos x}$
9. $\displaystyle\int \frac{1 - \sin x}{1 + \cos x}\, dx$
10. $\displaystyle\int \sqrt{1 - \cos 2x}\, dx$
11. $\displaystyle\int \frac{\sin x}{\sin x + \cos x}\, dx$
12. $\displaystyle\int \frac{\cos x - \sin x}{\sin x}\, dx$
13. $\displaystyle\int \frac{dx}{1 + \tan x}$
14. $\displaystyle\int \frac{dx}{4 \sin x + 3 \cos x}$
15. $\displaystyle\int \frac{dx}{4 \sin x - 3 \cos x}$

9.5 INTEGRATION OF POWERS OF TRIGONOMETRIC FUNCTIONS

INTEGRATION OF ODD POWERS OF sin x AND cos x

An integral of the form

$$\int \sin^m x \cos^n x\, dx,$$

where either m or n is an odd positive integer, can be computed by the following device. If m is odd, "save" one factor $\sin x$ and change all other factors of $\sin^2 x$ to $1 - \cos^2 x$. The result is an integral of the form

$$\int f(\cos x) \sin x\, dx,$$

where $f(\cos x)$ is a polynomial in $\cos x$. This integral can easily be found. If n is odd, a similar procedure can be followed, "saving" one factor $\cos x$ and changing all other factors $\cos^2 x$ to $1 - \sin^2 x$.

EXAMPLE 1 Find $\int \sin^3x \cos^2x\, dx$.

Solution We have

$$\begin{aligned}\int \sin^3x \cos^2x\, dx &= \int \sin x(1 - \cos^2x) \cos^2x\, dx \\ &= \int (\cos^2x - \cos^4x) \sin x\, dx \\ &= -\frac{1}{3}\cos^3x + \frac{1}{5}\cos^5x + C. \quad \square\end{aligned}$$

EXAMPLE 2 Find $\int \cos^5x\, dx$.

Solution We have

$$\begin{aligned}\int \cos^5x\, dx &= \int (1 - \sin^2x)^2\cos x\, dx \\ &= \int (1 - 2\sin^2x + \sin^4x) \cos x\, dx \\ &= \sin x - \frac{2}{3}\sin^3x + \frac{1}{5}\sin^5x + C. \quad \square\end{aligned}$$

An integral $\int \sin^mx \cos^nx\, dx$ can be found using this technique if either m or n is not an integer, as long as the other is an odd integer.

EXAMPLE 3 Find

$$\int \frac{\sin^3 2x}{(\cos 2x)^{3/2}}\, dx.$$

Solution We have

$$\begin{aligned}\int \frac{\sin^3 2x}{(\cos 2x)^{3/2}}\, dx &= \int \frac{(1 - \cos^2 2x)}{(\cos 2x)^{3/2}} (\sin 2x\, dx) \\ &= -\frac{1}{2}\int [(\cos 2x)^{-3/2} - (\cos 2x)^{1/2}](-2\sin 2x\, dx) \\ &= -\frac{1}{2}[-2(\cos 2x)^{-1/2} - \frac{2}{3}(\cos 2x)^{3/2}] + C \\ &= \frac{1}{\sqrt{\cos 2x}} + \frac{1}{3}(\cos 2x)^{3/2} + C. \quad \square\end{aligned}$$

INTEGRATION OF EVEN POWERS OF $\sin x$ AND $\cos x$

An integral of the form

$$\int \sin^mx \cos^nx\, dx,$$

where both m and n are nonnegative *even* integers, is more difficult to find than the case where either m or n is odd. This time, we use the trigonometric identities

$$\sin^2 ax = \frac{1 - \cos 2ax}{2}, \qquad \cos^2 bx = \frac{1 + \cos 2bx}{2}, \tag{1}$$

which are easily derived from the more familiar relation $\cos 2x = \cos^2 x - \sin^2 x$. (See Exercise 1.) Identities (1) are used repeatedly until we obtain just first powers of cosine functions. The technique is best illustrated by examples.

EXAMPLE 4 Find $\int \sin^4 x\, dx$.

Solution Using identities (1) repeatedly, we have

$$\begin{aligned}\int \sin^4 x\, dx &= \int (\sin^2 x)^2\, dx = \int \left(\frac{1 - \cos 2x}{2}\right)^2 dx \\ &= \frac{1}{4}\int (1 - 2\cos 2x + \cos^2 2x)\, dx \\ &= \frac{1}{4}\int \left(1 - 2\cos 2x + \frac{1 + \cos 4x}{2}\right) dx \\ &= \frac{1}{8}\int (3 - 4\cos 2x + \cos 4x)\, dx \\ &= \frac{3}{8}x - \frac{1}{4}\sin 2x + \frac{1}{32}\sin 4x + C. \quad \square\end{aligned}$$

EXAMPLE 5 Find

$$\int \sin^2 x \cos^2 x\, dx.$$

Solution We use the relation $\sin x \cos x = \frac{1}{2}\sin 2x$ followed by identities (1) and obtain

$$\begin{aligned}\int \sin^2 x \cos^2 x\, dx &= \frac{1}{4}\int \sin^2 2x\, dx = \frac{1}{4}\int \frac{1 - \cos 4x}{2}\, dx \\ &= \frac{1}{8}x - \frac{1}{32}\sin 4x + C. \quad \square\end{aligned}$$

INTEGRATION OF EXPRESSIONS SUCH AS $\sin mx \cos nx$

The relations

$$\sin mx \sin nx = \tfrac{1}{2}[\cos (m - n)x - \cos (m + n)x], \tag{2}$$

$$\sin mx \cos nx = \tfrac{1}{2}[\sin (m - n)x + \sin (m + n)x], \tag{3}$$

$$\cos mx \cos nx = \tfrac{1}{2}[\cos (m - n)x + \cos (m + n)x] \tag{4}$$

enable us to integrate $\sin mx \sin nx$, $\sin mx \cos nx$, and $\cos mx \cos nx$. Formulas (2), (3), and (4) are easily derived from the more familiar formulas

$$\sin(mx + nx) = \sin mx \cos nx + \cos mx \sin nx, \tag{5}$$

$$\cos(mx + nx) = \cos mx \cos nx - \sin mx \sin nx. \tag{6}$$

See Exercise 2.

EXAMPLE 6 Find $\int \sin 2x \cos 3x \, dx$.

Solution We have, using formula (3),

$$\begin{aligned}\int \sin 2x \cos 3x \, dx &= \frac{1}{2}\int [\sin(-x) + \sin 5x] \, dx\\ &= \frac{1}{2}\int (-\sin x + \sin 5x) \, dx\\ &= \frac{1}{2}\cos x - \frac{1}{10}\cos 5x + C. \quad \square\end{aligned}$$

INTEGRATION OF POWERS OF tan x AND cot x AND EVEN POWERS OF sec x AND csc x

We make use of the identities

$$\boxed{1 + \tan^2 x = \sec^2 x \quad \text{or} \quad \tan^2 x = \sec^2 x - 1} \tag{7}$$

and

$$\boxed{1 + \cot^2 x = \csc^2 x \quad \text{or} \quad \cot^2 x = \csc^2 x - 1.} \tag{8}$$

To integrate a power of $\tan x$ or an even power of $\sec x$, we use identity (7) and the fact that $d(\tan x) = \sec^2 x \, dx$.

To integrate an even power of $\sec x$, we "save" one factor $\sec^2 x$ to serve in du and change all the other powers of $\sec^2 x$ to powers of $u = \tan x$.

EXAMPLE 7 Find $\int \sec^6 x \, dx$.

Solution We obtain

$$\begin{aligned}\int \sec^6 x \, dx &= \int (1 + \tan^2 x)^2 \sec^2 x \, dx\\ &= \int (1 + 2\tan^2 x + \tan^4 x) \sec^2 x \, dx\\ &= \int \sec^2 x \, dx + 2\int \tan^2 x \sec^2 x \, dx + \int \tan^4 x \sec^2 x \, dx\\ &= \tan x + 2\frac{\tan^3 x}{3} + \frac{\tan^5 x}{5} + C. \quad \square\end{aligned}$$

We now illustrate how to integrate a power of $\tan x$.

EXAMPLE 8 Find $\int \tan^5 x\, dx$.

Solution We obtain

$$\begin{aligned}
\int \tan^5 x\, dx &= \int \tan^3 x\,(\sec^2 x - 1)\, dx \\
&= \int \tan^3 x \sec^2 x\, dx - \int \tan^3 x\, dx \\
&= \frac{\tan^4 x}{4} - \int \tan x\,(\sec^2 x - 1)\, dx \\
&= \frac{\tan^4 x}{4} - \int \tan x \sec^2 x\, dx + \int \tan x\, dx \\
&= \frac{\tan^4 x}{4} - \frac{\tan^2 x}{2} + \int \frac{\sin x}{\cos x}\, dx \\
&= \frac{\tan^4 x}{4} - \frac{\tan^2 x}{2} - \ln|\cos x| + C. \quad \square
\end{aligned}$$

Obviously, powers of $\cot x$ and even powers of $\csc x$ can be integrated in a similar fashion using identity (8) and $d(\cot x) = -\csc^2 x\, dx$.

REDUCTION FORMULAS FOR POWERS OF tan x, cot x, sec x, AND csc x

Odd powers of $\sec x$ and $\csc x$ are tough to integrate. They can be done using integration by parts. Example 8 on page 367 showed how to find $\int \sec^3 ax\, dx$. We ask you to experiment with $\int \csc^3 x\, dx$ in Exercise 33. Of course, an alternative way to integrate any power of $\tan x$, $\cot x$, $\sec x$, or $\csc x$ is to use a table and make repeated use of the appropriate reduction formula:

$$\int \tan^n ax\, dx = \frac{\tan^{n-1} ax}{a(n-1)} - \int \tan^{n-2} ax\, dx, \qquad n \neq 1, \tag{9}$$

$$\int \cot^n ax\, dx = -\frac{\cot^{n-1} ax}{a(n-1)} - \int \cot^{n-2} ax\, dx, \qquad n \neq 1, \tag{10}$$

$$\int \sec^n ax\, dx = \frac{\sec^{n-2} ax \tan ax}{a(n-1)} + \frac{n-2}{n-1}\int \sec^{n-2} ax\, dx, \qquad n \neq 1, \tag{11}$$

$$\int \csc^n ax\, dx = -\frac{\csc^{n-2} ax \cot ax}{a(n-1)} + \frac{n-2}{n-1}\int \csc^{n-2} ax\, dx, \qquad n \neq 1. \tag{12}$$

This will reduce the problem to the integration of the function to the first power (or to integration of a constant). Then we can use

$$\int \tan x \, dx = -\ln|\cos x| + C, \tag{13}$$

$$\int \cot x \, dx = \ln|\sin x| + C, \tag{14}$$

$$\int \sec x \, dx = \ln|\sec x + \tan x| + C, \tag{15}$$

$$\int \csc x \, dx = -\ln|\csc x + \cot x| + C, \tag{16}$$

which we have already seen. You are asked to derive the reduction formulas (9) through (12) in the exercises, using identities (7) and (8) and integration by parts.

EXAMPLE 9 Find $\int \sec^5 2x \, dx$.

Solution We have, using formula (11),

$$\begin{aligned}\int \sec^5 2x \, dx &= \frac{\sec^3 2x \tan 2x}{2 \cdot 4} + \frac{3}{4}\int \sec^3 2x \, dx \\ &= \frac{\sec^3 2x \tan 2x}{8} + \frac{3}{4}\left(\frac{\sec 2x \tan 2x}{2 \cdot 2} + \frac{1}{2}\int \sec 2x \, dx\right) \\ &= \frac{\sec^3 2x \tan 2x}{8} + \frac{3}{16}\sec 2x \tan 2x \\ &\quad + \frac{3}{16}\ln|\sec 2x + \tan 2x| + C. \quad \square\end{aligned}$$

Finally, remember always to see whether there is an easy way to find an integral before blindly reaching for a "rule." Consider, for example, three methods for finding $\int \tan x \sec^2 x \, dx$.

Method 1

$$\int \tan x \sec^2 x \, dx = \int \frac{\sin x}{\cos x} \cdot \frac{1}{\cos^2 x} \, dx = \int \frac{\sin x}{\cos^3 x} \, dx$$

could be done by the $t = \tan(x/2)$ substitution presented in the preceding section. However, that would be a lot of work.

Method 2

$$\int \tan x \sec^2 x \, dx = \int \tan x(1 + \tan^2 x) \, dx = \int (\tan x + \tan^3 x) \, dx$$

could be done using formulas (9) and (13). This would be an improvement over Method 1.

Method 3 Since $d(\tan x) = \sec^2 x \, dx$, then of course

$$\int \tan x \sec^2 x \, dx = \frac{\tan^2 x}{2} + C.$$

This is the easiest method of all.

SIMPSON'S RULE AGAIN

We emphasize that if a calculator or computer program is at hand, Simpson's rule is often a faster way to find a definite integral, even if the necessary integration formulas are contained in a convenient table.

EXAMPLE 10 Find $\int_0^\pi \sec^7(x/4)\,dx$.

Solution We could use a reduction formula and find $\int \sec^7(x/4)\,dx$ and then evaluate it from zero to π to find the definite integral. However, our calculator contains a built-in program for Simpson's rule. Timing ourselves from the moment we turned the calculator on, we found that

$$\int_0^\pi \sec^7\left(\frac{x}{4}\right) dx \approx 8.998128946$$

in 89 seconds flat. We couldn't do it that fast using a table. We used $n = 20$ in Simpson's rule. Taking a larger value of n for comparison, we find that the $n = 20$ approximation was accurate to four significant figures. □

SUMMARY

1. All the indefinite integrals discussed in this section can easily be found using an integral table. Definite integrals that would require much work with tables are often computed faster using Simpson's rule.
2. Odd powers of $\sin x$ can be integrated by "saving" one factor $\sin x$ to serve in du and changing all other factors $\sin^2 x$ to powers of $u = \cos x$ using $\sin^2 x = 1 - \cos^2 x$. Odd powers of $\cos x$ are integrated similarly.
3. Even powers of $\sin x$ or $\cos x$ can be integrated using

$$\sin^2 ax = \frac{1 - \cos 2ax}{2}$$

and

$$\cos^2 bx = \frac{1 + \cos 2bx}{2}$$

repeatedly.
4. Formulas (2), (3), and (4) of this section can be used to integrate expressions involving $\sin mx \sin nx$, $\cos mx \cos nx$, and $\sin mx \cos nx$.
5. Powers of $\tan x$ and $\cot x$ and even powers of $\sec x$ and $\csc x$ can be integrated using

$$1 + \tan^2 x = \sec^2 x$$

and

$$1 + \cot^2 x = \csc^2 x.$$

6. All powers of $\tan x$, $\cot x$, $\sec x$, and $\csc x$ can be integrated using the reduction formulas (9) through (12) and formulas (13) through (16) of this section.

EXERCISES

1. Derive identities (1) from the relation $\cos 2x = \cos^2 x - \sin^2 x$ and other trigonometric relations.
2. Derive formulas (2), (3), and (4) from formulas (5) and (6) and other well-known trigonometric relations.

In Exercises 3 through 35, compute the indicated integrals without the use of tables.

3. $\int \sin^2 x \cos^3 x \, dx$
4. $\int \sin^3 x \cos^4 x \, dx$
5. $\int \sin^4 x \cos^5 x \, dx$
6. $\int \frac{\sin^3 x}{\cos^2 x} \, dx$
7. $\int \sin^3 2x \sqrt{\cos 2x} \, dx$
8. $\int \frac{\cos^3 x}{\sqrt{\sin x}} \, dx$
9. $\int \frac{\cos^3 x}{(\sin x)^{3/2}} \, dx$
10. $\int \sin^2 3x \, dx$
11. $\int \cos^2 5x \, dx$
12. $\int \sin^2 3x \cos^2 3x \, dx$
13. $\int \cos^4 2x \, dx$
14. $\int \sin^4 x \cos^2 x \, dx$
15. $\int \sin^6 x \, dx$
16. $\int \cos^6 x \, dx$
17. $\int \frac{1}{\sin^2 x} \, dx$
18. $\int \frac{1}{\cos^2 x} \, dx$
19. $\int \sec^4 x \, dx$
20. $\int \csc^4 x \, dx$
21. $\int \sec^2 x \tan^2 x \, dx$
22. $\int \sec^4 x \tan^2 x \, dx$
23. $\int \sec^2 x \tan^4 x \, dx$
24. $\int \frac{\cos^2 x}{\sin^4 x} \, dx$
25. $\int \frac{\sin 3x}{\cos^3 3x} \, dx$
26. $\int \tan^3 x \, dx$
27. $\int \tan^4 x \, dx$
28. $\int \frac{dx}{\cos^6 x}$
29. $\int \tan^7 2x \, dx$
30. $\int \tan^3 x \sec^2 x \, dx$
31. $\int \csc^4 x \cot x \, dx$
32. $\int \csc^4 5x \cot^3 5x \, dx$
33. $\int \csc^3 x \, dx$ [*Hint:* Use integration by parts.]
34. $\int \csc^5 x \, dx$ [*Hint:* Use integration by parts.]
35. $\int \sec^5 x \, dx$ [*Hint:* Use integration by parts.]

In Exercise 36 through 41, derive the indicated reduction formula.

36. Formula (9) of this section
37. Formula (10) of this section
38. Formula (11) of this section
39. Formula (12) of this section
40. $\int \sin^n ax \, dx = \frac{-\sin^{n-1} ax \cos ax}{na} + \frac{n-1}{n} \int \sin^{n-2} ax \, dx$

    [*Note:* This is the best way to integrate powers of sin x.]
41. $\int \cos^n ax \, dx = \frac{\cos^{n-1} ax \sin ax}{na} + \frac{n-1}{n} \int \cos^{n-2} ax \, dx$

    [*Note:* This is the best way to integrate powers of cos x.]

9.6 TRIGONOMETRIC SUBSTITUTION

INTEGRALS INVOLVING $\sqrt{a^2 \pm x^2}$ AND $\sqrt{x^2 - a^2}$

The familiar trigonometric identities

$$a^2 - a^2 \sin^2 t = a^2 \cos^2 t \tag{1}$$

$$a^2 + a^2 \tan^2 t = a^2 \sec^2 t, \tag{2}$$

$$a^2 \sec^2 t - a^2 = a^2 \tan^2 t \tag{3}$$

are sometimes useful in eliminating radicals from integrals involving $\sqrt{a^2 - x^2}$, $\sqrt{a^2 + x^2}$, or $\sqrt{x^2 - a^2}$. Suppose, for illustration, we wish to eliminate a radical expression $\sqrt{x^2 - a^2}$ from an integral. A substitution where $x = a \sec t$ would result in

$$\sqrt{x^2 - a^2} = \sqrt{a^2 \sec^2 t - a^2} = \sqrt{a^2 \tan^2 t}, \tag{4}$$

according to identity (3). We would like to replace $\sqrt{a^2 \tan^2 t}$ by $a(\tan t)$, but we should be a little careful. The radical $\sqrt{a^2 \tan^2 t}$ is positive, while $a(\tan t)$ might be positive or negative. We assume $a > 0$ and focus our attention on the sign of $\tan t$. We should think of the substitution that yields $x = a \sec t$ as really being defined by

$$t = \sec^{-1}\left(\frac{x}{a}\right), \tag{5}$$

since x was the given variable and t is to be defined. By our definition of the principal values of the inverse secant function, we have

$$0 \le t < \frac{\pi}{2} \quad \text{or} \quad \pi \le t < \frac{3\pi}{2}. \tag{6}$$

For these values of t in Eq. (6), the function $\tan t$ is nonnegative. Thus for the substitution $t = \sec^{-1}(x/a)$, we indeed have

$$x = a \sec t, \qquad dx = a \sec t \tan t,$$
$$\sqrt{x^2 - a^2} = \sqrt{a^2 \tan^2 t} = a(\tan t),$$

and we have eliminated the radical.

Table 9.1 shows the trigonometric substitutions suggested by identities (1), (2), and (3). Although the second column of the table gives the actual definition of the substitution, we always think in terms of the x-replacement column when deciding which substitution to choose to eliminate the radical. We checked that the last row of table entries is correct in the preceding paragraph. In Exercises 1 and 2, we ask you to check that the first two rows in the radical-replacement column are correct.

Substitutions using Table 9.1 result in integrals of trigonometric functions. These integrals are sometimes easy, can often be found using methods of the preceding section, or (as a last resort) may be integrated using the $t = \tan(x/2)$ substitution of Section 9.4.

EXAMPLE 1 Find

$$\int x^3 \sqrt{4 - x^2}\, dx,$$

which is not found in the brief integral table on the endpapers of this book.

Table 9.1

Radical	Substitution	x-replacement	Radical replacement	dx-replacement
$\sqrt{a^2 - x^2}$	$t = \sin^{-1}(x/a)$	$x = a \sin t$	$a \cos t$	$dx = a \cos t\, dt$
$\sqrt{a^2 + x^2}$	$t = \tan^{-1}(x/a)$	$x = a \tan t$	$a \sec t$	$dx = a \sec^2 t\, dt$
$\sqrt{x^2 - a^2}$	$t = \sec^{-1}(x/a)$	$x = a \sec t$	$a \tan t$	$dx = a \sec t \tan t\, dt$

Solution If we use the substitution $x = 2 \sin t$, then $dx = 2 \cos t\, dt$, and we obtain

$$\begin{aligned}\int x^3\sqrt{4 - x^2}\, dx &= \int 8 \sin^3 t\,(\sqrt{4 - 4\sin^2 t})(2\cos t)\, dt\\ &= \int 8\sin^3 t\,(4\cos^2 t)\, dt\\ &= 32\int \sin t\,(1 - \cos^2 t)(\cos^2 t)\, dt\\ &= 32\int (\cos^2 t - \cos^4 t)\sin t\, dt\\ &= -\frac{32\cos^3 t}{3} + \frac{32\cos^5 t}{5} + C.\end{aligned}$$

From $x = 2 \sin t$, we obtain

$$\begin{aligned}\cos t &= \sqrt{1 - \sin^2 t}\\ &= \sqrt{1 - \left(\frac{x}{2}\right)^2} = \frac{1}{2}\sqrt{4 - x^2},\end{aligned}$$

so we obtain for our integral

$$\int x^3\sqrt{4 - x^2}\, dx = -\frac{4(4 - x^2)^{3/2}}{3} + \frac{(4 - x^2)^{5/2}}{5} + C. \quad \square$$

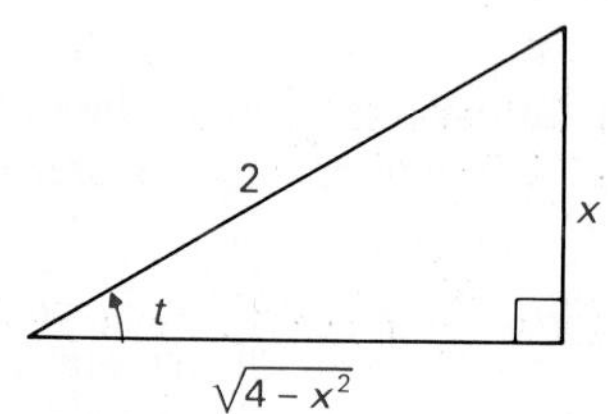

Figure 9.1 Triangle for $x = 2 \sin t$.

When we make a substitution such as $x = 2 \sin t$ in Example 1 and then integrate, the answer obtained is in terms of trigonometric functions of t rather than a function of x. The triangle shown in Fig. 9.1 is helpful in figuring out the trigonometric functions of t in terms of x. If $x = 2 \sin t$, then $\sin t = x/2$, so we label the side opposite the angle t with x and the hypotenuse with 2. The remaining side is then found using the Pythagorean theorem.

EXAMPLE 2 Compute

$$\int x^2\sqrt{1 + x^2}\, dx.$$

Solution This time, we use the substitution $x = \tan t$, so that $dx = \sec^2 t\, dt$. The triangle in Fig. 9.2 shows the relation between t and x. We obtain

$$\begin{aligned}\int x^2\sqrt{1 + x^2}\, dx &= \int \tan^2 t\sqrt{1 + \tan^2 t}\,\sec^2 t\, dt\\ &= \int \tan^2 t\,\sec^3 t\, dt = \int (\sec^2 t - 1)\sec^3 t\, dt\\ &= \int (\sec^5 t - \sec^3 t)\, dt.\end{aligned}$$

Figure 9.2 Triangle for $x = \tan t$.

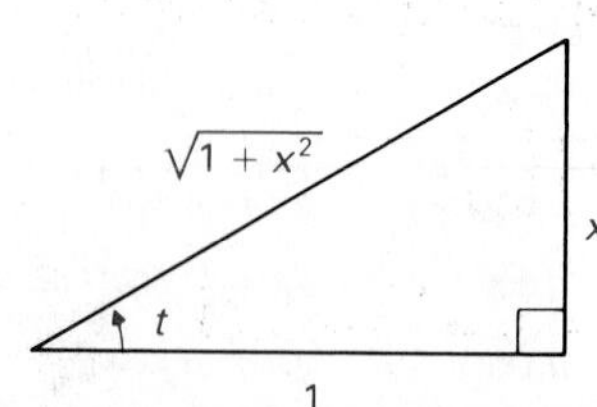

We use reduction formula 98 for powers of secant from the table on the endpapers and obtain

$$\int \sec^5 t\, dt = \frac{\sec^3 t\tan t}{4} + \frac{3}{4}\int \sec^3 t\, dt.$$

Our integral thus takes the form

$$\int(\sec^5 t - \sec^3 t)\,dt = \frac{\sec^3 t\tan t}{4} - \frac{1}{4}\int \sec^3 t\,dt$$

$$= \frac{\sec^3 t\tan t}{4} - \frac{1}{4}\left(\frac{\sec t\tan t}{2} + \frac{1}{2}\int \sec t\,dt\right)$$

$$= \frac{\sec^3 t\tan t}{4} - \frac{1}{8}\sec t\tan t$$

$$- \frac{1}{8}\ln|\sec t + \tan t| + C.$$

From Fig. 9.2, we obtain $\sec t = \sqrt{1+x^2}$. Thus we have

$$\int x^2\sqrt{1+x^2}\,dx = \frac{x(1+x^2)^{3/2}}{4} - \frac{1}{8}x\sqrt{1+x^2}$$

$$- \frac{1}{8}\ln|\sqrt{1+x^2} + x| + C. \quad \square$$

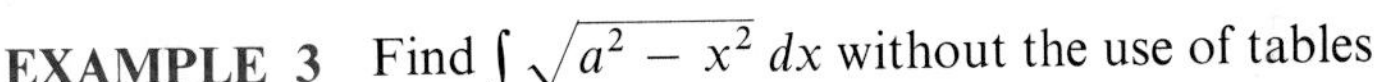

EXAMPLE 3 Find $\int\sqrt{a^2-x^2}\,dx$ without the use of tables.

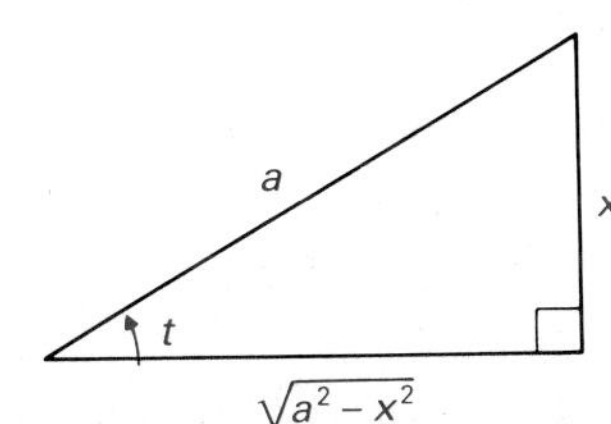

Figure 9.3 Triangle for $x = a\sin t$.

Solution We let $x = a\sin t$, so that $dx = a\cos t\,dt$. Figure 9.3 shows the triangle. Then

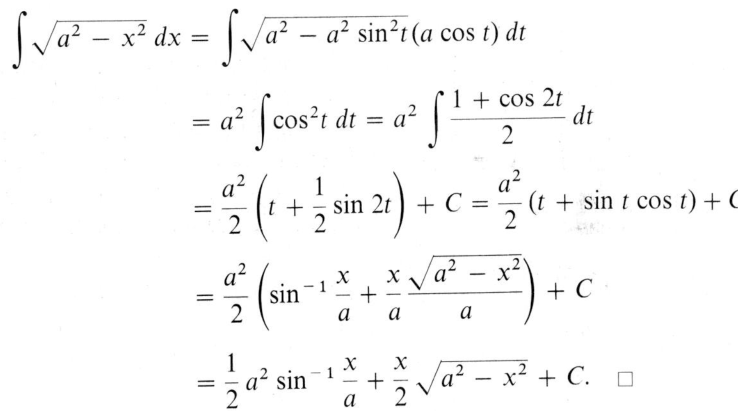

$$\int\sqrt{a^2-x^2}\,dx = \int\sqrt{a^2 - a^2\sin^2 t}\,(a\cos t)\,dt$$

$$= a^2\int\cos^2 t\,dt = a^2\int\frac{1+\cos 2t}{2}\,dt$$

$$= \frac{a^2}{2}\left(t + \frac{1}{2}\sin 2t\right) + C = \frac{a^2}{2}(t + \sin t\cos t) + C$$

$$= \frac{a^2}{2}\left(\sin^{-1}\frac{x}{a} + \frac{x}{a}\frac{\sqrt{a^2-x^2}}{a}\right) + C$$

$$= \frac{1}{2}a^2\sin^{-1}\frac{x}{a} + \frac{x}{2}\sqrt{a^2-x^2} + C. \quad \square$$

EXAMPLE 4 Find $\int_0^2\sqrt{16-x^2}\,dx$.

Solution The indefinite integral $\int\sqrt{16-x^2}\,dx$ is the special case of Example 3 where $a = 4$. Thus we use $x = 4\sin t$, corresponding to the substitution $t = \sin^{-1}(x/4)$. The x-limit zero then corresponds to the t-limit $\sin^{-1}0 = 0$, and the x-limit 2 corresponds to the t-limit $\sin^{-1}(1/2) = \pi/6$. Using the third line of the computation in Example 3, we see that

$$\int_0^2\sqrt{16-x^2}\,dx = \frac{16}{2}\left(t + \frac{1}{2}\sin 2t\right)\Big]_{t=0}^{\pi/6}$$

$$= 8\left(\frac{\pi}{6} + \frac{1}{2}\cdot\frac{\sqrt{3}}{2}\right) - 0 = \frac{4\pi}{3} + 2\sqrt{3}. \quad \square$$

INTEGRALS INVOLVING $\sqrt{ax^2 + bx + c}$, WHERE $a \neq 0$

By completing the square, we may be able to handle the integral of an expression involving $\sqrt{ax^2 + bx + c}$ by the methods of this section. We illustrate with two examples.

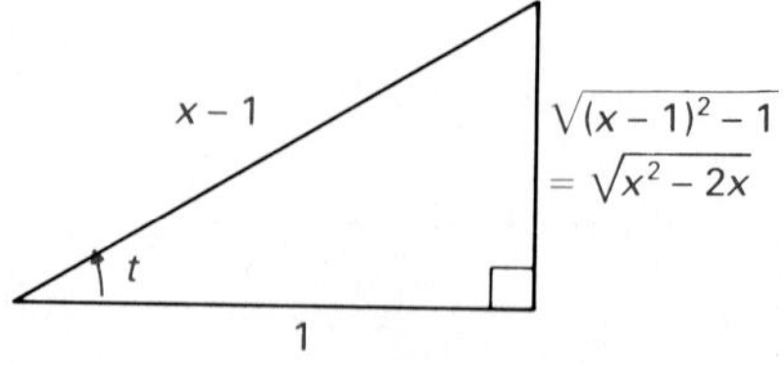

Figure 9.4 Triangle for $x - 1 = \sec t$.

EXAMPLE 5 Find $\int \sqrt{x^2 - 2x}\, dx$ without using tables.

Solution We have

$$\int \sqrt{x^2 - 2x}\, dx = \int \sqrt{(x-1)^2 - 1}\, dx.$$

We let $x - 1 = \sec t$, so $dx = \sec t \tan t\, dt$. Figure 9.4 shows the triangle. Then

$$\begin{aligned}\int \sqrt{(x-1)^2 - 1}\, dx &= \int \sqrt{\sec^2 t - 1}\, \sec t \tan t\, dt \\ &= \int \tan^2 t \sec t\, dt = \int (\sec^2 t - 1) \sec t\, dt \\ &= \int (\sec^3 t - \sec t)\, dt.\end{aligned}$$

We should know $\int \sec t\, dt$. Example 8 on page 367 shows that

$$\int \sec^3 t\, dt = \frac{1}{2} \sec t \tan t + \frac{1}{2} \ln|\sec t + \tan t| + C.$$

We now obtain for our original integral

$$\begin{aligned}\int \sqrt{x^2 - 2x}\, dx &= \int \sec^3 t\, dt - \int \sec t\, dt \\ &= \frac{1}{2} \sec t \tan t - \frac{1}{2} \ln|\sec t + \tan t| + C \\ &= \frac{1}{2}(x-1)\sqrt{(x-1)^2 - 1} \\ &\quad - \frac{1}{2} \ln|(x-1) + \sqrt{(x-1)^2 - 1}| + C \\ &= \frac{1}{2}[(x-1)\sqrt{x^2 - 2x} \\ &\quad - \ln|x - 1 + \sqrt{x^2 - 2x}|] + C. \quad \square\end{aligned}$$

EXAMPLE 6 Find $\int (1/\sqrt{1 + 4x - x^2})\, dx$.

Solution Completing the square, we have

$$\int \frac{dx}{\sqrt{1 + 4x - x^2}} = \int \frac{dx}{\sqrt{5 - (x-2)^2}}.$$

We use $x - 2 = \sqrt{5}\sin t$, so that $dx = \sqrt{5}\cos t\, dt$. Then

$$\int \frac{dx}{\sqrt{5-(x-2)^2}} = \int \frac{\sqrt{5}\cos t}{\sqrt{5 - 5\sin^2 t}}\, dt = \int \frac{\sqrt{5}\cos t}{\sqrt{5}\cos t}\, dt$$

$$= \int 1 \cdot dt = t + C = \sin^{-1}\frac{x-2}{\sqrt{5}} + C. \quad \square$$

SUMMARY

1. If $x = a\sin t$, then $\sqrt{a^2 - x^2} = a\cos t$.
2. If $x = a\tan t$, then $\sqrt{a^2 + x^2} = a\sec t$.
3. If $x = a\sec t$, then $\sqrt{x^2 - a^2} = a\tan t$.
4. To eliminate the radical from $\sqrt{ax^2 + bx + c}$ for $a \neq 0$, complete the square and then use the appropriate substitution above.

EXERCISES

1. Verify that the radical replacement given in Table 9.1 for the substitution $t = \sin^{-1}(x/a)$ is correct. (Worry about the sign of the radical, as we did following Eq. 4.)
2. Repeat Exercise 1 for the substitution $t = \tan^{-1}(x/a)$ in Table 9.1.

It is our experience that students often have as much trouble choosing the appropriate trigonometric substitution and then transforming the integral as they have in finally performing the integration. In Exercises 3 through 24, give the appropriate substitution to eliminate the radical and then find the transformed integral. Do not perform the integration.

3. $\int \sqrt{9 - x^2}\, dx$
4. $\int \frac{x^2}{\sqrt{4 + x^2}}\, dx$
5. $\int \frac{dx}{\sqrt{x^2 - 16}}$
6. $\int (25 - x^2)^{3/2}$
7. $\int \frac{x+2}{\sqrt{x^2 - 1}}\, dx$
8. $\int \frac{dx}{x\sqrt{4 - x^2}}$
9. $\int \sqrt{1 + 4x^2}\, dx$
10. $\int x^2\sqrt{9x^2 - 1}\, dx$
11. $\int \frac{x^4}{\sqrt{1 - 16x^2}}\, dx$
12. $\int \sqrt{16 - 4x^2}\, dx$
13. $\int \frac{dx}{(9 + 5x^2)^{3/2}}$
14. $\int (8x^2 - 4)^{5/2}\, dx$
15. $\int \sqrt{2 + 3x^2}\, dx$
16. $\int \sqrt{x^2 - 2x + 5}\, dx$
17. $\int \sqrt{x^2 + 4x + 3}\, dx$
18. $\int \frac{x\, dx}{\sqrt{3 - 2x - x^2}}$
19. $\int \frac{x^2\, dx}{\sqrt{-2x - x^2}}$
20. $\int \sqrt{4x^2 + 6x + 5}\, dx$
21. $\int \frac{dx}{\sqrt{8 - 4x - 3x^2}}$
22. $\int e^x\sqrt{e^{2x} + 4e^x + 5}\, dx$
23. $\int \frac{e^x\, dx}{\sqrt{15 + 2e^x - e^{2x}}}$
24. $\int e^{2x}\sqrt{e^{2x} + 6e^x + 5}\, dx$

In Exercises 25 through 46, find the integral without using tables.

25. $\int \frac{x}{\sqrt{1 + x^2}}\, dx$
26. $\int \frac{x - 3}{\sqrt{4 - x^2}}\, dx$
27. $\int_0^{\sqrt{3}} \frac{x^3}{\sqrt{1 + x^2}}\, dx$
28. $\int (16 - x^2)^{3/2}\, dx$
29. $\int_1^{\sqrt{5}} x^3\sqrt{x^2 - 1}\, dx$
30. $\int \sqrt{4 - x^2}\, dx$
31. $\int \frac{x}{\sqrt{5 + x^2}}\, dx$
32. $\int_0^1 \frac{dx}{\sqrt{x^2 + 1}}$
33. $\int \frac{x - 1}{\sqrt{x^2 - 16}}\, dx$
34. $\int \frac{dx}{x\sqrt{x^2 - 4}}$

35. $\int_0^{\sqrt{6}} \sqrt{1+4x^2}\,dx$

36. $\int \frac{dx}{(4-x^2)^{3/2}}$

37. $\int \frac{x^3\,dx}{(9+x^2)^{5/2}}$

38. $\int_4^5 x^3(x^2-16)^{3/2}\,dx$

39. $\int e^{4x}\sqrt{1-e^{2x}}\,dx$

40. $\int \frac{e^{4x}}{\sqrt{1+e^{2x}}}\,dx$

41. $\int_{-1}^0 \sqrt{x^2+2x+2}\,dx$

42. $\int \sqrt{x^2-6x+8}\,dx$

43. $\int_{1/2}^1 \frac{dx}{\sqrt{2x-x^2}}$

44. $\int \frac{dx}{4x^2+8x+12}$

45. $\int \frac{x}{\sqrt{x^2-4x+5}}\,dx$

46. $\int_{-1}^0 \frac{3x-2}{x^2+2x+2}\,dx$

9.7 IMPROPER INTEGRALS

In this section and the next, we depart from our study of integration techniques and introduce improper integrals. Geometrically, improper integrals are concerned with areas of regions that are infinitely wide or infinitely high in altitude. Some of the outcomes of our study may seem very surprising. Exercise 31 will give an example of a solid of finite volume but infinite surface area! Just imagine, such a solid could be cast using a finite amount of plaster of paris, but there is not enough red paint in the world to paint it red. To make it red, we would have to mix red coloring in the plaster before we cast the solid. Of course, there is one catch in this: Since the surface area is infinite, there is not enough material in the entire solar system to build a mold in which to cast the solid! We find such ideas intriguing.

BASIC TYPES OF IMPROPER INTEGRALS

Thus far, our discussion of a definite integral $\int_a^b f(x)\,dx$ has been confined to integrals of continuous functions over a closed interval. We now turn to such integrals as

$$\int_a^\infty f(x)\,dx \qquad \text{and} \qquad \int_a^b f(x)\,dx,$$

where $\lim_{x\to b^-} |f(x)| = \infty$. These are *improper integrals*. If $f(x) \geq 0$, the first integral is an attempt to find the area under the graph "from a to ∞," as shown in Fig. 9.5. The second integral is an attempt to find the area of the shaded region in Fig. 9.6.

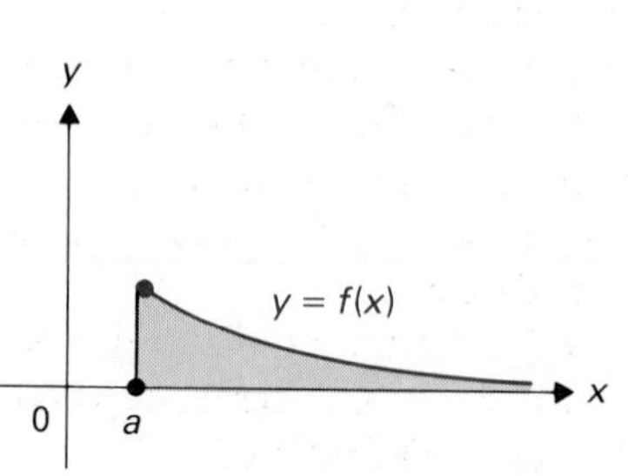

Figure 9.5 $\int_a^\infty f(x)\,dx$ is the area of the shaded region, if it is finite.

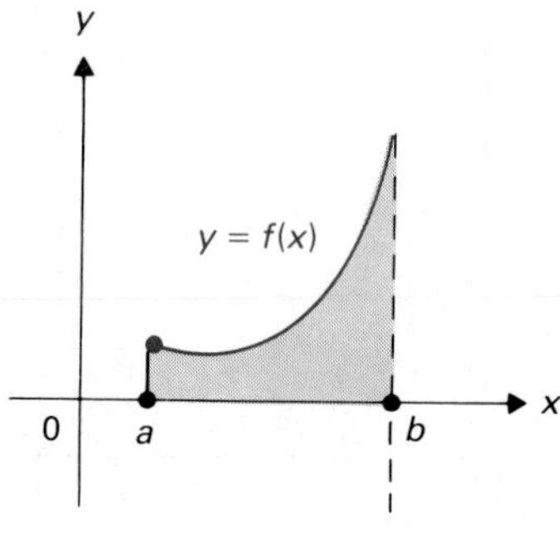

Figure 9.6 $\int_a^b f(x)\,dx$ is the area of the shaded region, if it is finite.

If f is continuous for all $x > a$, as shown in Fig. 9.7, then $\int_a^h f(x)\,dx$ for $h > a$ makes perfectly good sense. The definition that follows is surely the natural attempt to make sense out of $\int_a^\infty f(x)\,dx$.

DEFINITION 9.1 Improper integrals $\int_a^\infty f(x)\,dx$ and $\int_{-\infty}^a f(x)\,dx$

Let f be continuous for $x > a$. The **improper integral** $\int_a^\infty f(x)\,dx$ is defined by

$$\int_a^\infty f(x)\,dx = \lim_{h\to\infty} \int_a^h f(x)\,dx. \tag{1}$$

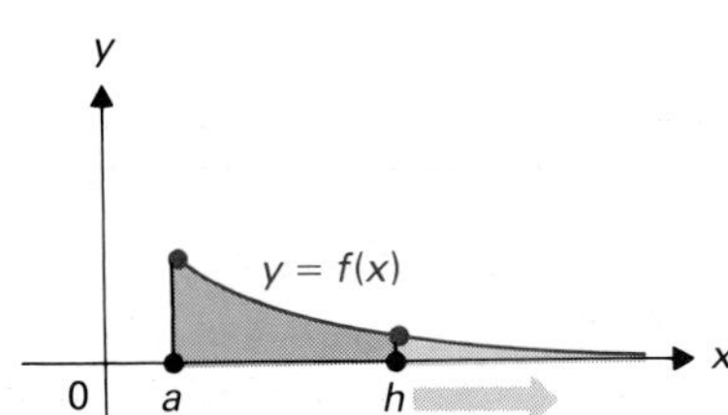

Figure 9.7
$\int_a^\infty f(x)\,dx = \lim_{h\to\infty} \int_a^h f(x)\,dx$

If this limit exists, the integral is said to **converge**, whereas the integral **diverges** if the limit is not a finite number.

The **improper integral** $\int_{-\infty}^a f(x)\,dx$ is defined in a similar fashion by

$$\int_{-\infty}^a f(x)\,dx = \lim_{h\to -\infty} \int_h^a f(x)\,dx \tag{2}$$

and **converges** or **diverges** according to whether the limit exists.

EXAMPLE 1 Find the improper integral $\int_1^\infty (1/x^2)\,dx$, if it converges.

Solution We have

$$\int_1^\infty \frac{1}{x^2}\,dx = \lim_{h\to\infty} \int_1^h \frac{1}{x^2}\,dx = \lim_{h\to\infty} -\frac{1}{x}\bigg]_1^h$$

$$= \lim_{h\to\infty} \left[-\frac{1}{h} - (-1)\right] = 0 + 1 = 1.$$

Thus the integral converges to 1. □

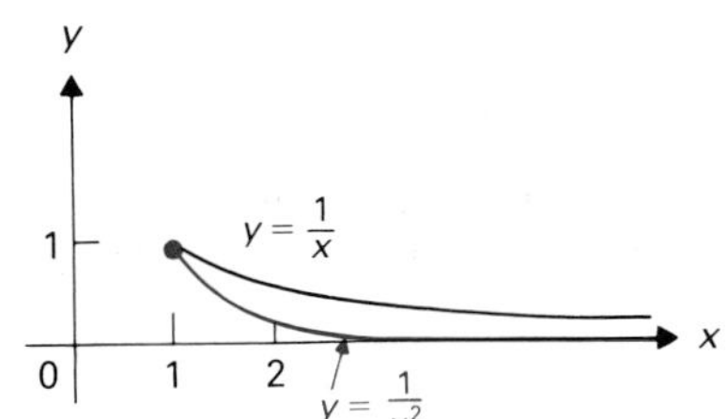

Figure 9.8 $1/x^2 < 1/x$ for $x > 1$

EXAMPLE 2 Find the improper integral $\int_1^\infty (1/x)\,dx$, if it converges.

Solution We have

$$\int_1^\infty \frac{1}{x}\,dx = \lim_{h\to\infty} \int_1^h \frac{1}{x}\,dx = \lim_{h\to\infty} \ln|x|\bigg]_1^h = \lim_{h\to\infty} \ln|h| - \ln 1 = \infty.$$

Thus the integral diverges. □

The graphs of $1/x^2$ and $1/x$ are shown in Fig. 9.8. Note that $1/x^2 < 1/x$ for $x > 1$. Intuitively, we see that for a decreasing function f with graph of the type shown in Fig. 9.5, the integral $\int_a^\infty f(x)\,dx$ converges only if the graph approaches the x-axis fast enough as $x \to \infty$. Equivalently, $f(x)$ must approach zero fast enough as $x \to \infty$. Example 1 showed that $1/x^2$ does approach zero fast enough as $x \to \infty$ to make $\int_1^\infty (1/x^2)\,dx$ converge. Example 2 showed that $1/x$ does not approach zero fast enough as $x \to \infty$ to make $\int_1^\infty (1/x)\,dx$ converge.

If f is continuous for $a \le x < b$, as shown in Fig. 9.9, then $\int_a^h f(x)\,dx$ for $a \le h < b$ makes perfectly good sense. We give the following definition analogous to Definition 9.1.

Figure 9.9
$\int_a^b f(x)\,dx = \lim_{h\to b^-} \int_a^h f(x)\,dx$

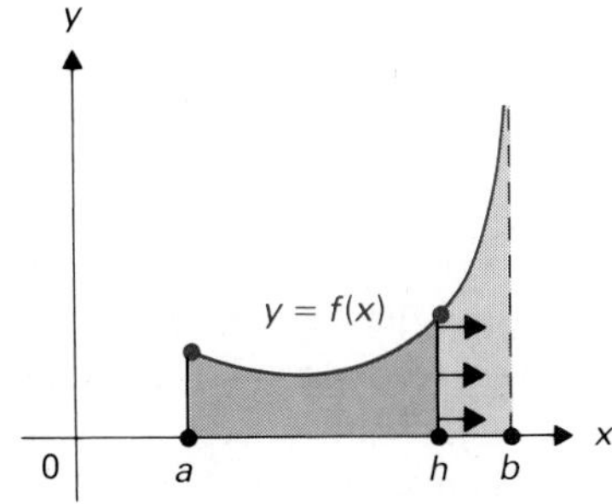

DEFINITION 9.2 Improper integral $\int_a^b f(x)\,dx$

Let f be continuous for $a \le x < b$, while f is either undefined or not continuous at $x = b$. The **improper integral** $\int_a^b f(x)\,dx$ is defined by

$$\int_a^b f(x)\,dx = \lim_{h\to b-} \int_a^h f(x)\,dx. \tag{3}$$

If the limit exists, the integral is said to **converge,** while the integral **diverges** if the limit is not a finite number.

EXAMPLE 3 Find the improper integral $\int_0^1 1/(1 - x)\,dx$, if it converges.

Solution We have

$$\int_0^1 \frac{1}{1-x}\,dx = \lim_{h\to 1-} \int_0^h \frac{1}{1-x}\,dx = \lim_{h\to 1-} -\ln|1-x|\Big]_0^h$$
$$= \lim_{h\to 1-} -\ln|1-h| + \ln 1 = \infty.$$

Thus the integral diverges. □

EXAMPLE 4 Find the improper integral $\int_0^1 (1/\sqrt{1-x})\,dx$, if it converges.

Solution We have

$$\int_0^1 \frac{1}{\sqrt{1-x}}\,dx = \lim_{h\to 1-} \int_0^h \frac{1}{\sqrt{1-x}}\,dx = \lim_{h\to 1-} -2\sqrt{1-x}\Big]_0^h$$
$$= \lim_{h\to 1-} -2\sqrt{1-h} + 2 = 2.$$

Thus the integral converges to 2. □

Suppose that f is continuous for $a < x \le b$, while f is either undefined or not continuous at $x = a$, as shown in Fig. 9.10. By analogy with Definition 9.2, we define

$$\boxed{\int_a^b f(x)\,dx = \lim_{h\to a+} \int_h^b f(x)\,dx,} \tag{4}$$

if the limit exists.

Figures 9.6, 9.9, and 9.10 all illustrate improper integrals where $|f(x)| \to \infty$ as we approach a limit of integration. There are other possibilities.

Figure 9.10

$\int_a^b f(x)\,dx = \lim_{h\to a+} \int_h^b f(x)\,dx$

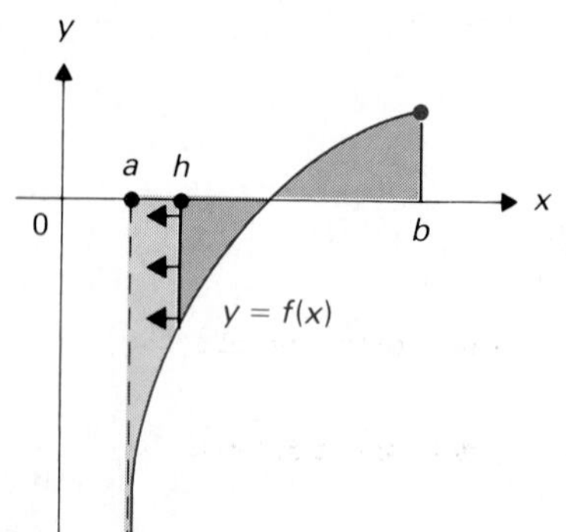

EXAMPLE 5 Discuss $\int_0^1 \sin(1/x)\,dx$.

Solution The integral is improper since $\sin(1/x)$ is not defined where $x = 0$. An attempt to sketch the graph of $\sin(1/x)$ for $0 < x \le 1$ is shown in Fig. 9.11. All the waves of the $\sin x$ graph for $x > 1$ are thrown between 0 and 1 by the reciprocal transformation $1/x$. We do not yet have the technique to decide whether $\int_0^1 \sin(1/x)\,dx$ converges. Techniques introduced in Section 10.5 can be used to show that the integral does converge. □

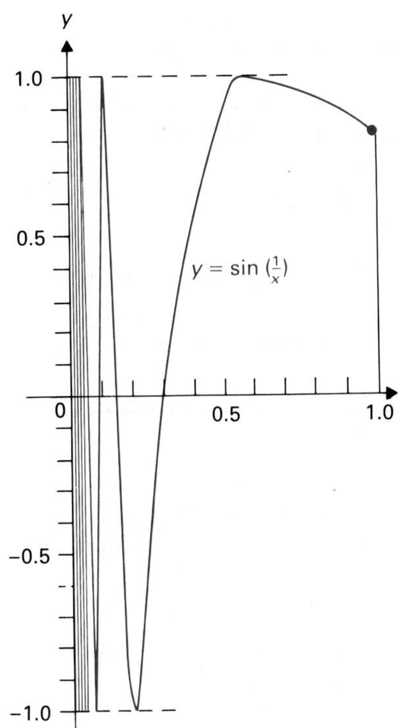

Figure 9.11 Attempt to graph $y = \sin(1/x)$ for $0 < x \le 1$. The graph oscillates an infinite number of times as $x \to 0$.

EXAMPLE 6 Discuss $\int_0^1 (\sin x)/x\, dx$.

Solution The integral is improper since $(\sin x)/x$ is not defined where $x = 0$. On page 139 in Section 4.3, we saw that

$$\lim_{x\to 0} \frac{\sin x}{x} = 1.$$

Thus if we let

$$g(x) = \begin{cases} \dfrac{\sin x}{x} & \text{for } 0 < x \le 1, \\ 1 & \text{for } x = 0, \end{cases}$$

then $g(x) = (\sin x)/x$ for $0 < x \le 1$, and g is continuous for $0 \le x \le 1$. Since

$$\int_h^1 \frac{\sin x}{x}\, dx = \int_h^1 g(x)\, dx \qquad \text{for } 0 < h \le 1,$$

we see that

$$\lim_{h\to 0+} \int_h^1 \frac{\sin x}{x}\, dx = \lim_{h\to 0+} \int_h^1 g(x)\, dx = \int_0^1 g(x)\, dx,$$

so $\int_0^1 (\sin x)/x\, dx$ does converge. Using Simpson's rule with $n = 40$ and a calculator to estimate $\int_{0.0001}^1 [(\sin x)/x]\, dx$, we find

$$\int_0^1 \frac{\sin x}{x}\, dx \approx 0.946. \quad \square$$

OTHER TYPES OF IMPROPER INTEGRALS

Letting f be continuous except where indicated, we list the basic types of improper integrals, presented in the previous subsection, in the box at the bottom of this page.

We consider ∞, $-\infty$, b, and a in integrals (5), (6), (7), and (8), respectively, to be "bad points" causing the integral to be improper. Some improper integrals involve more than one "bad point." For an improper integral involving only a finite number of "bad points," we can express the integral as a sum of integrals of the basic types (5) through (8). Such an integral

BASIC TYPES OF IMPROPER INTEGRALS

5. $\displaystyle\int_a^\infty f(x)\, dx = \lim_{h\to\infty} \int_a^h f(x)\, dx$

6. $\displaystyle\int_{-\infty}^b f(x)\, dx = \lim_{h\to -\infty} \int_h^b f(x)\, dx$

7. $\displaystyle\int_a^b f(x)\, dx = \lim_{h\to b-} \int_a^h f(x)\, dx,$ $\quad f$ not continuous at b

8. $\displaystyle\int_a^b f(x)\, dx = \lim_{h\to a+} \int_h^b f(x)\, dx,$ $\quad f$ not continuous at a

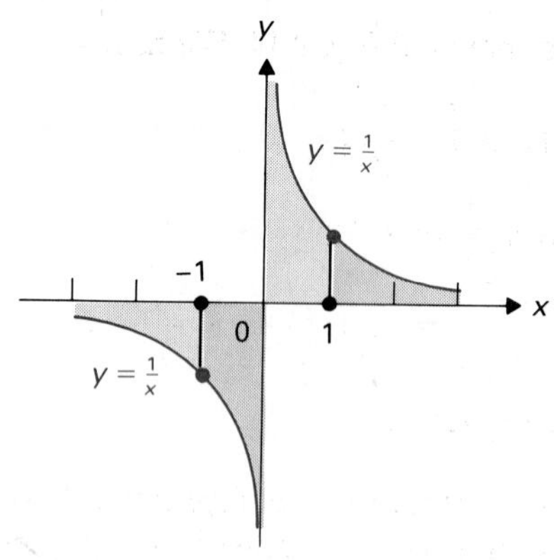

Figure 9.12 Breaking $\int_{-\infty}^{\infty} (1/x)\,dx$ into a sum of four basic types of improper integrals.

converges only if *each* of the basic types of which it is composed converges. If even one of the basic types involved diverges, then the whole integral is said to **diverge.** This is easier to illustrate by examples than to express in words.

EXAMPLE 7 Express $\int_{-\infty}^{\infty} (1/x)\,dx$ as a sum of basic types and determine whether the integral converges or diverges.

Solution The graph of $1/x$ is shown in Fig. 9.12. "Bad points" for $\int_{-\infty}^{\infty} (1/x)\,dx$ are $-\infty$, zero, and ∞. The integral can be expressed as a sum of basic types by

$$\int_{-\infty}^{\infty} \frac{1}{x}\,dx = \int_{-\infty}^{-1} \frac{1}{x}\,dx + \int_{-1}^{0} \frac{1}{x}\,dx + \int_{0}^{1} \frac{1}{x}\,dx + \int_{1}^{\infty} \frac{1}{x}\,dx.$$

The points -1 and 1 simply serve as convenient "good points" to separate the "bad points." We could replace -1 by -10 and 1 by 57, for example, if we wished. Example 2 showed that $\int_1^{\infty} (1/x)\,dx$ diverges. Consequently, $\int_{-\infty}^{\infty} (1/x)\,dx$ diverges. □

EXAMPLE 8 Express $\int_{-\infty}^{\infty} 1/(1 + x^2)\,dx$ as a sum of basic types and determine whether the integral converges or diverges.

Solution The graph of $1/(1 + x^2)$ is shown in Fig. 9.13. We have

$$\begin{aligned}\int_{-\infty}^{\infty} \frac{dx}{1+x^2} &= \int_{-\infty}^{0} \frac{dx}{1+x^2} + \int_{0}^{\infty} \frac{dx}{1+x^2}\\ &= \lim_{h\to-\infty} \int_{h}^{0} \frac{dx}{1+x^2} + \lim_{h\to\infty} \int_{0}^{h} \frac{dx}{1+x^2}\\ &= \lim_{h\to-\infty} \tan^{-1}x\Big]_h^0 + \lim_{h\to\infty} \tan^{-1}x\Big]_0^h\\ &= \lim_{h\to-\infty} (0 - \tan^{-1}h) + \lim_{h\to\infty} (\tan^{-1}h - 0)\\ &= 0 - \left(-\frac{\pi}{2}\right) + \left(\frac{\pi}{2} - 0\right) = \pi.\end{aligned}$$

Thus the integral converges to π. □

EXAMPLE 9 Express $\int_0^{\infty} (1/x^2)\,dx$ as a sum of basic types, and determine whether the integral converges.

Figure 9.13
$\int_{-\infty}^{\infty} [1/(1 + x^2)]\,dx$
$= \int_{-\infty}^{0} [1/(1 + x^2)]\,dx$
$+ \int_0^{\infty} [1/(1 + x^2)]\,dx$

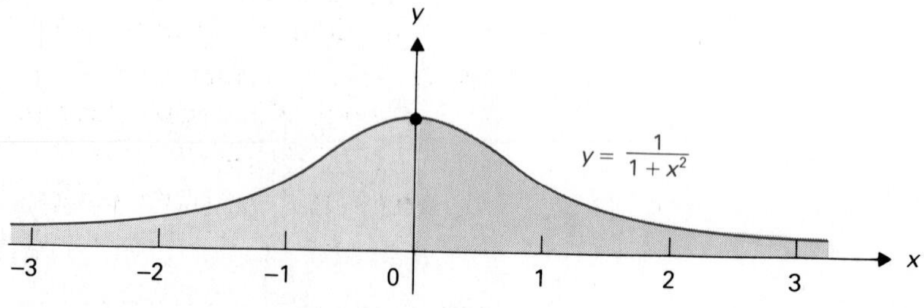

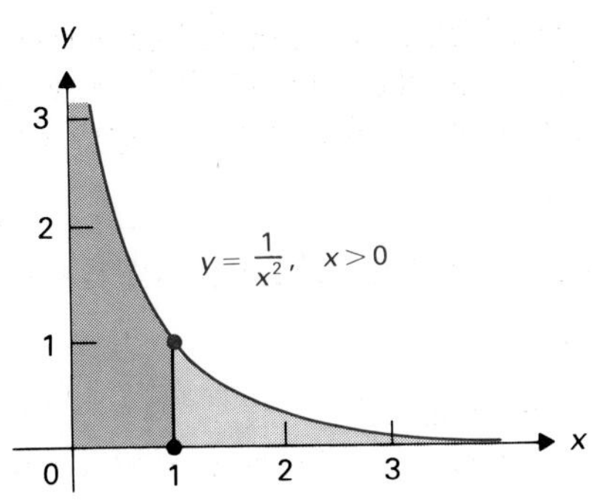

Figure 9.14
$\int_0^\infty (1/x^2)\,dx$
$= \int_0^1 (1/x^2)\,dx + \int_1^\infty (1/x^2)\,dx$

Solution The graph of $1/x^2$ for $x > 0$ is shown in Fig. 9.14. We have

$$\int_0^\infty \frac{1}{x^2}\,dx = \int_0^1 \frac{1}{x^2}\,dx + \int_1^\infty \frac{1}{x^2}\,dx.$$

Example 1 showed that $\int_1^\infty (1/x^2)\,dx$ converges to 1. Now

$$\int_0^1 \frac{1}{x^2}\,dx = \lim_{h\to 0+} \int_h^1 x^{-2}\,dx = \lim_{h\to 0+} \left.\frac{-1}{x}\right]_h^1 = \lim_{h\to 0+}\left(-1+\frac{1}{h}\right) = \infty.$$

Thus $\int_0^1 (1/x^2)\,dx$ diverges, so $\int_0^\infty (1/x^2)\,dx$ diverges. □

Our mathematical life has abruptly become more complicated. Now we have to examine every definite integral and decide whether it is improper before we try to evaluate it. The following example shows the danger if we do not do this but just blindly integrate and evaluate.

EXAMPLE 10 Find $\int_{-1}^{1} (1/x)\,dx$.

Wrong Solution We have

$$\int_{-1}^1 \frac{1}{x}\,dx = \ln|x|\Big]_{-1}^1 = (\ln 1) - (\ln 1) = 0 - 0 = 0.$$

Correct Solution The integral is improper because $1/x$ is not defined where $x = 0$. Expressing the integral as a sum of basic types, we have

$$\int_{-1}^1 \frac{1}{x}\,dx = \int_{-1}^0 \frac{1}{x}\,dx + \int_0^1 \frac{1}{x}\,dx.$$

Now

$$\int_{-1}^0 \frac{1}{x}\,dx = \lim_{h\to 0-} \ln|x|\Big]_{-1}^h = \lim_{h\to 0-} (\ln|h| - \ln 1)$$
$$= \lim_{h\to 0-} \ln|h| = -\infty.$$

Thus $\int_{-1}^1 (1/x)\,dx$ diverges. □

SUMMARY

1. $\int_a^\infty f(x)\,dx = \lim_{h\to\infty} \int_a^h f(x)\,dx$
2. $\int_{-\infty}^b f(x)\,dx = \lim_{h\to -\infty} \int_h^b f(x)\,dx$
3. If f is not continuous at b, then $\int_a^b f(x)\,dx = \lim_{h\to b-} \int_a^h f(x)\,dx$.
4. If f is not continuous at a, then $\int_a^b f(x)\,dx = \lim_{h\to a+} \int_h^b f(x)\,dx$.
5. The improper integrals just listed are said to converge if the limits are finite. Otherwise, they diverge.
6. An improper integral involving a finite number of "bad points" can be expressed as a sum of the basic types (1 through 4 above). The entire integral converges only if each individual basic type involved converges.
7. You should now examine every definite integral to see whether or not it is improper before you attempt to evaluate it.

EXERCISES

In Exercises 1 through 12, express the improper integrals as a sum of basic types described in Eqs. (5) through (8) in the box on page 405, using limits. You need not establish the convergence or divergence of the integrals.

1. $\int_0^1 \frac{dx}{x^2 - 1}$
2. $\int_0^{\pi} \tan x \, dx$
3. $\int_1^{\infty} \frac{dx}{x - 1}$
4. $\int_0^1 \frac{dx}{x^2 + x}$
5. $\int_{-\infty}^{\infty} e^{-x^2} \, dx$
6. $\int_{-1}^{1} \frac{dx}{x^2 - 4}$
7. $\int_0^2 \frac{dx}{x^3 + 1}$
8. $\int_0^4 \frac{dx}{x^2 - 2x - 3}$
9. $\int_0^{\infty} \frac{dx}{\sqrt{x} + x^2}$
10. $\int_{-\infty}^{\infty} \frac{dx}{\sqrt[3]{x} + x^3}$
11. $\int_0^{2\pi} \sec x \, dx$
12. $\int_0^{\pi} \frac{dx}{\sin x - \cos x}$

In Exercises 13 through 28, determine whether or not the improper integral converges, and find its value if it converges.

13. $\int_1^{\infty} \frac{1}{x^3} \, dx$
14. $\int_1^{\infty} \frac{1}{\sqrt{x}} \, dx$
15. $\int_0^{\infty} \frac{1}{x^2 + 1} \, dx$
16. $\int_{-\infty}^{0} \frac{x^2}{x^3 + 1} \, dx$
17. $\int_{-\infty}^{0} e^x \, dx$
18. $\int_0^1 \frac{1}{x^{2/3}} \, dx$
19. $\int_0^2 \frac{1}{\sqrt{2 - x}} \, dx$
20. $\int_1^2 \frac{1}{(x - 1)^2} \, dx$
21. $\int_{-1}^{1} \frac{1}{x^{2/5}} \, dx$
22. $\int_0^{\infty} \frac{1}{\sqrt{x}} \, dx$
23. $\int_{-\infty}^{\infty} |x| e^{-x^2} \, dx$
24. $\int_0^1 \frac{1}{\sqrt{2x - x^2}} \, dx$
25. $\int_{-\infty}^{\infty} e^{-x} \cos x \, dx$
26. $\int_0^{\infty} e^{-x} \sin x \, dx$
27. $\int_0^{\pi/2} \tan x \, dx$
28. $\int_0^{\pi/2} \sqrt{\cos x \cot x} \, dx$

29. Show that $\int_1^{\infty} (1/x^p) \, dx$ converges if $p > 1$ and diverges if $p \leq 1$.

30. Show that

$$\int_a^b \left[\frac{1}{(x - a)^p}\right] dx \quad \text{and} \quad \int_a^b \left[\frac{1}{(b - x)^p}\right] dx$$

converge if $p < 1$ and diverge if $p \geq 1$.

31. Consider the region under the graph of $1/x$ over the half-line $x \geq 1$.
    a) Does the region have finite area?
    b) Show that the unbounded solid obtained by revolving the region about the x-axis has finite volume, and compute this volume.
    c) Does the solid of finite volume described in part (b) have finite surface area?

32. Consider the region under the graph of $1/\sqrt{x}$ over the half-open interval $0 < x \leq 1$.
    a) Does the region have finite area?
    b) Show that the unbounded solid obtained by revolving the region about the y-axis has finite volume, and compute this volume.
    c) Does the solid of finite volume described in part (b) have finite surface area?

33. The position (x, y) of a body in the plane at time t is given by $x = e^{-t} \cos t$, $y = e^{-t} \sin t$. Find the total distance the body travels throughout time $t \geq 0$.

34. Two electrons a distance s apart repel each other with a force k/s^2, where k is some constant. If an electron is fixed at the point 2 on the x-axis, find the work done by the force in moving another electron from the point 4 "all the way out to infinity."

35. With reference to Exercise 34, suppose one electron is fixed at the point 0 and another is fixed at the point 1. Find the work done by the forces in moving a third electron from the point 2 "all the way out to infinity."

36. A body outside the earth is attracted to the earth by a gravitational force inversely proportional to the square of the distance from the body to the center of the earth. Assume the earth is a ball of radius 4000 mi. Find the work done in moving a body weighing 10 lb at the surface of the earth from the surface of the earth to beyond the earth's gravitational field (that is, "to infinity"). Neglect the influence of other celestial bodies.

37. Suppose one could build a spaceship that could achieve a velocity v that increases exponentially as a function of the distance s traveled, so that $v = Ae^{ks}$ for some positive constants k and A.
    a) Show that such a spaceship could travel an infinite distance in a finite time. [*Hint:* Since the time required for a journey is Distance/Velocity if the velocity is constant, we see that the time required to travel a distance c is $\int_0^c ds/v(s)$ if the velocity is not constant.]
    b) Show that for such a spaceship, its acceleration dv/dt is directly proportional to the square of its velocity.
    c) Show that one could still achieve an infinite distance in a finite time with a velocity increasing more slowly than an exponential function of the total distance traveled.

9.8 TESTS FOR CONVERGENCE OF IMPROPER INTEGRALS

This section closely parallels work that we will be doing with infinite series in Chapter 10. The ideas are very important and different from anything we have done so far. Developing these notions here for improper integrals as well as in Chapter 10 for series gives us more opportunity to understand them and work with them.

THE p-INTEGRALS

We will develop methods for trying to establish the convergence or divergence of an improper integral without actual integration. Basically, we will compare the integral in question with another improper integral whose behavior (convergence or divergence) we already know. Of course, we can do this only if we have at least a small stockpile of improper integrals whose behavior we know. Theorem 9.1 provides such a stockpile. The improper integrals appearing in the theorem are known as the "p-integrals." You were asked to prove the theorem in Exercises 29 and 30 of the preceding section.

THEOREM 9.1 p-integrals

1. For $a > 0$ and $b < 0$, the improper integrals $\int_a^\infty (1/x^p)\,dx$ and $\int_{-\infty}^b (1/x^p)\,dx$ converge if $p > 1$ and diverge if $p \leq 1$.
2. The improper integrals $\int_a^b [1/(b - x)^p]\,dx$ and $\int_a^b [1/(x - a)^p]\,dx$ converge if $p < 1$ and diverge if $p \geq 1$.

Intuitively, we think of $p = 1$ in Theorem 9.1 as being the *breaking point* between convergent and divergent integrals. Figure 9.15 shows the graph of $1/x$ and graphs of $1/x^p$ for $p > 1$ and for $p < 1$. The theorem asserts that the area of the dark-shaded region, corresponding to $p > 1$, remains finite if taken all the way out to infinity, while the area of the total shaded region becomes infinite. Figure 9.16 similarly illustrates Theorem 9.1 for $\int_a^b [1/(x - a)^p]\,dx$, but this time the region has finite area if $p < 1$.

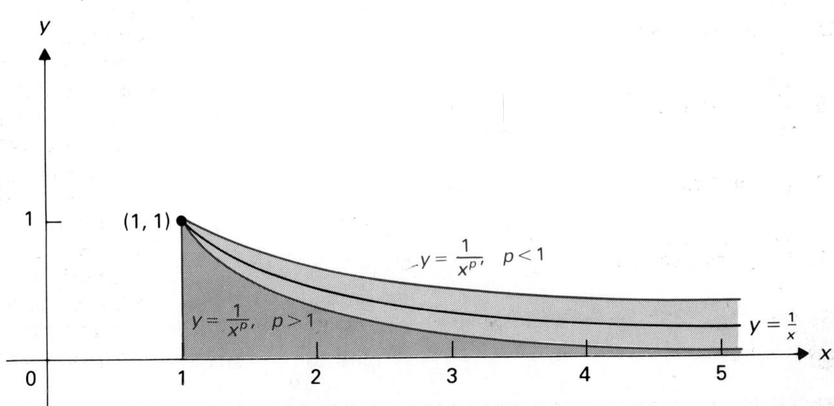

Figure 9.15 The dark-shaded region has finite area out to infinity; the total shaded region has infinite area out to infinity.

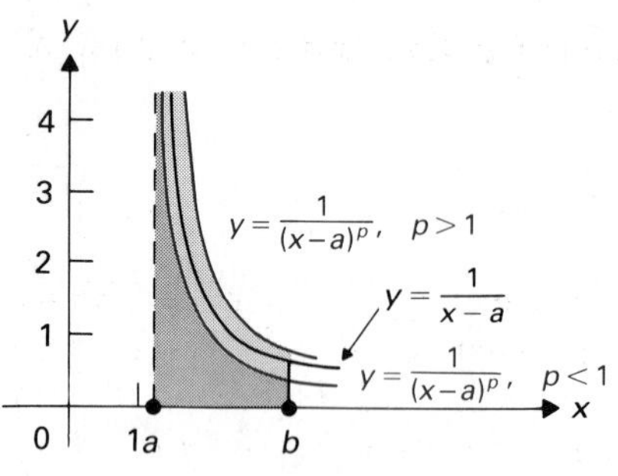

Figure 9.16 The dark-shaded region has finite area up to infinity; the total shaded region has infinite area up to infinity.

EXAMPLE 1 Establish the convergence or divergence of the integral $\int_1^\infty [1/(x-1)^2]\,dx$.

Solution Splitting the integral into basic types, we have

$$\int_1^\infty \frac{dx}{(x-1)^2} = \int_1^2 \frac{dx}{(x-1)^2} + \int_2^\infty \frac{dx}{(x-1)^2}.$$

By part 2 of Theorem 9.1, the integral $\int_1^2 [1/(x-1)^2]\,dx$ diverges as a p-integral with $p = 2$. Thus $\int_1^\infty [1/(x-1)^2]\,dx$ diverges. □

EXAMPLE 2 Establish the convergence or divergence of

$$\int_{-3}^\infty \frac{dx}{\sqrt{x+3}}.$$

Solution Splitting the integral into basic types, we have

$$\int_{-3}^\infty \frac{dx}{\sqrt{x+3}} = \int_{-3}^0 \frac{dx}{\sqrt{x+3}} + \int_0^\infty \frac{dx}{\sqrt{x+3}}.$$

By part 2 of Theorem 9.1, we see that

$$\int_{-3}^0 \frac{dx}{\sqrt{x+3}} = \int_{-3}^0 \frac{dx}{(x+3)^{1/2}}$$

converges as a p-integral at $a = -3$ with $p = 1/2$.

Now $\int_0^\infty (1/\sqrt{x+3})\,dx$ is not precisely in the form of part 1 of Theorem 9.1. However, we can put it in that form by the substitution $u = x + 3$, $du = dx$. Then

$$\int_0^\infty \frac{dx}{\sqrt{x+3}} = \int_3^\infty \frac{du}{\sqrt{u}},$$

which diverges as a p-integral at ∞ with $p = 1/2$. Thus $\int_{-3}^\infty (1/\sqrt{x+3})\,dx$ also diverges. □

Figure 9.17 (a) $\lim_{x\to\infty} F(x) = c$; (b) $\lim_{x\to\infty} F(x) = \infty$.

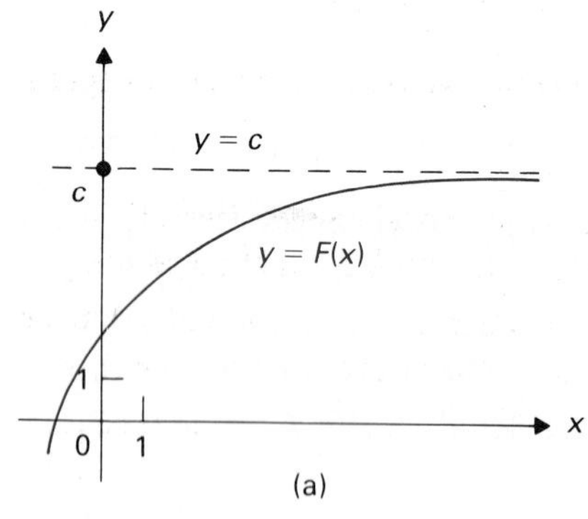

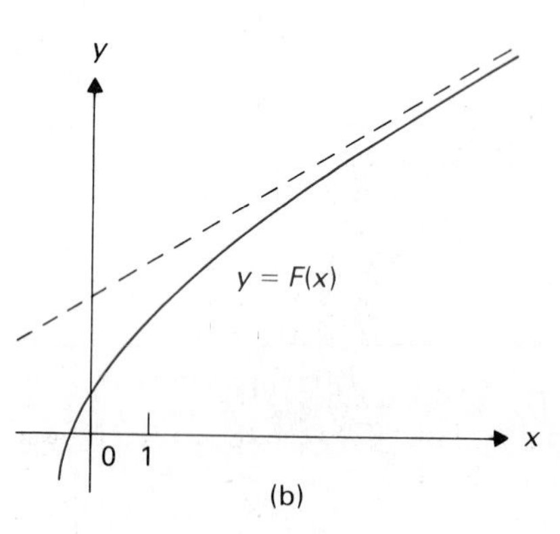

COMPARISON TESTS

Our work here really depends on the following fundamental property of the real numbers. A proof of the property is given in more advanced courses, where the real numbers are "constructed."

> Let $F(x)$ be defined for all sufficiently large x, and suppose $F(x_1) \le F(x_2)$ if $x_1 < x_2$. (Such a function is monotone increasing.) Then either $\lim_{x\to\infty} F(x) = c$ for some finite number c or $\lim_{x\to\infty} F(x) = \infty$.
>
> ***Fundamental property of the real numbers***

Figure 9.17 illustrates these two possible types of behavior for $F(x)$. The fundamental property certainly seems reasonable. It says that something that is increasing either remains finite, in which case it approaches a limiting magnitude c from below, or else becomes arbitrarily large, that is, approaches ∞.

Suppose now that f is a continuous function and $f(x) \geq 0$ for all $x \geq a$. Then

$$F(h) = \int_a^h f(x)\,dx$$

is a monotone-increasing function of h. Consequently, either $\lim_{h\to\infty} F(h) = c$ for a finite number c or $\lim_{h\to\infty} F(h) = \infty$. That is, either $\int_a^\infty f(x)\,dx$ converges to a finite number c or diverges to ∞. The same result holds true for the other basic types of improper integrals, described in Section 9.7.

THEOREM 9.2 Improper integrals of nonnegative functions

Let f be a continuous function such that $f(x) \geq 0$ for all x in the domain of f. Then each of the basic types of improper integrals of $f(x)$, as described in Section 9.7, either converges to a finite number c or diverges to ∞.

We are now ready for the main goal of this section, comparison tests. The following comparison test is illustrated in Fig. 9.18.

THEOREM 9.3 Standard comparison test (SCT)

1. Let f and g be continuous, and suppose $0 \leq f(x) \leq g(x)$ for all $x \geq a$.

   a. If $\int_a^\infty g(x)\,dx$ converges, then $\int_a^\infty f(x)\,dx$ converges.
   b. If $\int_a^\infty f(x)\,dx$ diverges, then $\int_a^\infty g(x)\,dx$ diverges.

2. Analogous results hold for the other basic types of improper integrals of nonnegative functions. That is:

   a. If the improper integral of the larger function converges, then the corresponding improper integral of the smaller function converges.
   b. If the improper integral of the smaller function diverges, then the corresponding improper integral of the larger function diverges.

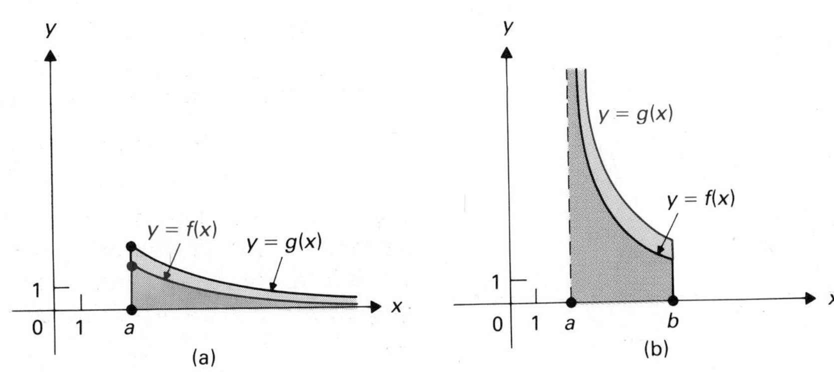

Figure 9.18 (a) If $\int_a^\infty g(x)\,dx$ converges, then $\int_a^\infty f(x)\,dx$ converges; (b) if $\int_a^b f(x)\,dx$ diverges, then $\int_a^b g(x)\,dx$ diverges.

Proof. We demonstrate part 1 of the theorem. The proof of part 2 is analogous. Suppose $0 \le f(x) \le g(x)$ for $x \ge a$. We have

$$\int_a^h f(x)\,dx \le \int_a^h g(x)\,dx. \tag{1}$$

Now suppose $\int_a^\infty g(x)\,dx$ converges, so that

$$\lim_{h\to\infty} \int_a^h g(x)\,dx = c$$

for some finite number c. Then by Eq. (1), $\lim_{h\to\infty} \int_a^h f(x)\,dx = \infty$ is impossible, so by Theorem 9.2, the integral $\int_a^\infty f(x)\,dx$ converges also. If, on the other hand, $\int_a^\infty f(x)\,dx$ diverges to ∞, then Eq. (1) shows immediately that $\int_a^\infty g(x)\,dx$ must diverge also. •

EXAMPLE 3 Show that

$$\int_4^\infty \frac{dx}{x^2 + 10x + 23}$$

converges.

Solution We have

$$\frac{1}{x^2 + 10x + 23} < \frac{1}{x^2} \qquad \text{for } x \ge 4.$$

Now $\int_4^\infty (1/x^2)\,dx$ converges as a p-integral at ∞ with $p = 2$. Thus $\int_4^\infty [1/(x^2 + 10x + 23)]\,dx$ converges by the SCT. □

EXAMPLE 4 Show that

$$\int_0^2 \frac{(x + 7)}{(x - 2)}\,dx$$

diverges.

Solution We have

$$\frac{1}{x - 2} < \frac{x + 7}{x - 2} \qquad \text{for } 0 \le x < 2.$$

Now $\int_0^2 [1/(x - 2)]\,dx$ diverges as a p-integral at $b = 2$ with $p = 1$. Thus $\int_0^2 [(x + 7)/(x - 2)]\,dx$ diverges by the SCT. □

Now suppose we wish to establish the convergence or divergence of $\int_5^\infty [1/(x^2 - 3)]\,dx$. We have

$$\frac{1}{x^2} < \frac{1}{x^2 - 3} \qquad \text{for } x \ge 5. \tag{2}$$

Now $\int_5^\infty (1/x^2)\,dx$ converges as a p-integral at ∞ with $p = 2$. However, the inequality (2) goes the wrong way for the SCT.

> An integral larger than a convergent integral may either converge or diverge, depending on how much larger it is. Similarly, an integral smaller than a divergent integral may either converge or diverge, depending on how much smaller it is.

We are suspicious that $1/(x^2 - 3)$ is not enough larger than $1/x^2$ to cause divergence of $\int_5^\infty [1/(x^2 - 3)]\, dx$. The limit comparison test (LCT), which we now present, shows that this is the case. While the new test seems to be much more powerful and useful than the SCT of Theorem 9.3, we will see that it is really an easy corollary of the theorem.

THEOREM 9.4 Limit comparison test (LCT)

1. Let f and g be continuous and let $f(x) > 0$ and $g(x) > 0$ for $x \geq a$. If

$$\lim_{x \to \infty} \frac{f(x)}{g(x)} = c > 0$$

for some positive constant c, then $\int_a^\infty f(x)\, dx$ and $\int_a^\infty g(x)\, dx$ behave the same way; that is, they both converge or both diverge.

2. Analogous results hold for the other basic types of improper integrals, corresponding to other possible "bad points" $-\infty$, a, or b. Let $f(x)$ and $g(x)$ be positive and continuous. If

$$\lim_{x \to \text{"bad point"}} \frac{f(x)}{g(x)} = c > 0,$$

then the improper integrals of $f(x)$ and of $g(x)$ with the same limits of integration either both converge or both diverge.

Proof. Again, we just prove part 1. We have

$$\lim_{x \to \infty} \frac{f(x)}{g(x)} = c > 0.$$

Thus $f(x)/g(x)$ is very close to c for x large enough. That is, there is some number η such that

$$\frac{c}{2} < \frac{f(x)}{g(x)} < 2c \qquad \text{for } x \geq \eta,$$

so

$$\frac{c}{2} \cdot g(x) < f(x) < 2c \cdot g(x) \qquad \text{for } x \geq \eta. \tag{3}$$

Now

$$\int_a^\infty f(x)\, dx = \int_a^\eta f(x)\, dx + \int_\eta^\infty f(x)\, dx,$$

and we see that $\int_a^\infty f(x)\, dx$ converges or diverges precisely when $\int_\eta^\infty f(x)\, dx$ converges or diverges, since $\int_a^\eta f(x)\, dx$ is finite. The same holds true for $\int_a^\infty g(x)\, dx$. Now suppose $\int_\eta^\infty f(x)\, dx$ converges. Then by Eq. (3) and the SCT, we see that $\int_\eta^\infty (c/2)g(x)\, dx$ converges, and consequently $\int_\eta^\infty g(x)\, dx$ converges, since $c \neq 0$. On the other hand, if $\int_\eta^\infty f(x)\, dx$ diverges, then by Eq. (3) and the SCT, the integral $\int_\eta^\infty 2c \cdot g(x)\, dx$ diverges, and consequently $\int_\eta^\infty g(x)\, dx$ diverges, since $c \neq 0$. This concludes the proof of part 1. •

We now give examples illustrating the great power of the SCT and the LCT.

EXAMPLE 5 Establish the convergence or divergence of the integral $\int_5^\infty [1/(x^2 - 3)]\, dx$, discussed prior to Theorem 9.4.

Solution We let $f(x) = 1/(x^2 - 3)$ and $g(x) = 1/x^2$. We have

$$\lim_{x\to\infty} \frac{f(x)}{g(x)} = \lim_{x\to\infty} \frac{\dfrac{1}{x^2 - 3}}{\dfrac{1}{x^2}} = \lim_{x\to\infty} \frac{x^2}{x^2 - 3} = \lim_{x\to\infty} \frac{x^2}{x^2} = 1 > 0.$$

(Recall from Section 2.3 that the limit as $x \to \infty$ of a quotient of polynomials in x equals the limit of the quotient of the monomial terms of highest degree.) Since $\int_5^\infty (1/x^2)\, dx$ converges as a p-integral at ∞ with $p = 2$, we see that $\int_5^\infty [1/(x^2 - 3)]\, dx$ converges by the LCT. □

Example 5 indicates that for a basic type of improper integral at ∞ or $-\infty$, we can replace any polynomial expression by the monomial term of greatest degree, according to the LCT. The LCT is very powerful.

EXAMPLE 6 Establish the convergence or divergence of

$$\int_1^\infty \frac{x^3 - 2x^2 + x - 1}{2x^4 + 3x + 2}\, dx.$$

Solution As $x \to \infty$, the LCT shows that this integral behaves like

$$\int_1^\infty \frac{x^3}{2x^4}\, dx = \frac{1}{2}\int_1^\infty \frac{dx}{x}.$$

(See the remarks preceding this example.) Now $\int_1^\infty (1/x)\, dx$ diverges as a p-integral at ∞ with $p = 1$. Thus the given integral diverges also, by the LCT. □

EXAMPLE 7 Establish the convergence or divergence of

$$\int_0^\infty \frac{dx}{\sqrt{x + x^2}}.$$

Solution Splitting the integral into its basic types, we have

$$\int_0^\infty \frac{dx}{\sqrt{x + x^2}} = \int_0^1 \frac{dx}{\sqrt{x + x^2}} + \int_1^\infty \frac{dx}{\sqrt{x + x^2}}.$$

By the remarks preceding Example 6 and by the LCT, we see that

$$\int_1^\infty \frac{dx}{\sqrt{x + x^2}}$$

behaves like $\int_1^\infty (1/x^2)\, dx$, which converges. Now near zero, a sum of powers of x is dominated by its term of *lowest* degree. Thus we expect $\sqrt{x + x^2}$ to behave like $\sqrt{x}$ near zero. We check this out carefully, using the LCT, this one time:

$$\lim_{x\to 0} \frac{1/\sqrt{x}}{\dfrac{1}{\sqrt{x} + x^2}} = \lim_{x\to 0} \frac{\sqrt{x} + x^2}{\sqrt{x}} = \lim_{x\to 0} \frac{\sqrt{x}(1 + x^{3/2})}{\sqrt{x}}$$

$$= \lim_{x\to 0} (1 + x^{3/2}) = 1 > 0.$$

Thus

$$\int_0^1 \frac{dx}{\sqrt{x} + x^2}$$

behaves like $\int_0^1 (1/\sqrt{x})\, dx$, which converges as a p-integral at $a = 0$ with $p = 1/2$.

Since both of the basic types of integrals converge, we see that our original integral $\int_0^\infty [1/(\sqrt{x} + x^2)]\, dx$ converges. □

EXAMPLE 8 Establish the convergence or divergence of

$$\int_0^\infty \frac{|\sin x|}{x^{4/3}}\, dx.$$

Solution Splitting the integral into its basic types, we have

$$\int_0^\infty \frac{|\sin x|}{x^{4/3}}\, dx = \int_0^1 \frac{|\sin x|}{x^{4/3}}\, dx + \int_1^\infty \frac{|\sin x|}{x^{4/3}}\, dx.$$

Now $|\sin x| \le 1$, so

$$\frac{|\sin x|}{x^{4/3}} \le \frac{1}{x^{4/3}} \qquad \text{for } x \ge 1. \tag{4}$$

Now $\int_1^\infty (1/x^{4/3})\, dx$ converges as a p-integral at ∞ with $p = \frac{4}{3}$. By Eq. (4) and the SCT, we see that $\int_1^\infty [|\sin x|/x^{4/3}]\, dx$ converges.

If $0 \le x \le 1$, we have $\sin x \ge 0$, so the absolute value is not needed as we work near zero. We have

$$\frac{\sin x}{x^{4/3}} = \frac{\sin x}{x} \cdot \frac{1}{x^{1/3}}.$$

Since $\lim_{x\to 0} [(\sin x)/x] = 1$, we see by the LCT that $\int_0^1 [(\sin x)/x^{4/3}]\, dx$ behaves like $\int_0^1 (1/x^{1/3})\, dx$, which converges as a p-integral at $a = 0$ with $p = \frac{1}{3}$. Thus $\int_0^1 [(\sin x)/x^{4/3}]\, dx$ converges.

Since both of the basic types of integrals of which it is composed converge, we see that our original integral $\int_0^\infty [|\sin x|/x^{4/3}]\, dx$ converges. □

EXAMPLE 9 Establish the convergence or divergence of

$$\int_3^\infty \frac{dx}{\sqrt{x^3 - 2x^2 - 3x}}.$$

Solution Since $x^3 - 2x^2 - 3x = x(x + 1)(x - 3)$, we see that the improper integral has the splitting

$$\int_3^\infty \frac{dx}{\sqrt{x^3 - 2x^2 - 3x}} = \int_3^4 \frac{dx}{\sqrt{x^3 - 2x^2 - 3x}} + \int_4^\infty \frac{dx}{\sqrt{x^3 - 2x^2 - 3x}}$$

into basic types. Now

$$\frac{1}{\sqrt{x^3 - 2x^2 - 3x}} = \frac{1}{\sqrt{x^2 + x}\sqrt{x - 3}}.$$

Since $\lim_{x \to 3} (1/\sqrt{x^2 + x}) = 1/\sqrt{12} > 0$, the LCT shows that

$$\int_3^4 (1/\sqrt{x^3 - 2x^2 - 3x})\, dx$$

behaves like $\int_3^4 (1/\sqrt{x - 3})\, dx$, which converges as a p-integral at $a = 3$ with $p = \frac{1}{2}$.

The LCT shows that $\int_4^\infty (1/\sqrt{x^3 - 2x^2 - 3x})\, dx$ behaves like $\int_4^\infty (1/x^{3/2})\, dx$, which converges as a p-integral at ∞ with $p = \frac{3}{2}$. Thus $\int_3^\infty (1/\sqrt{x^3 - 2x^2 - 3x})\, dx$ converges. □

SUMMARY

1. For $a > 0$ and $b < 0$, the improper integrals $\int_a^\infty (1/x^p)\, dx$ and $\int_{-\infty}^b (1/x^p)\, dx$ converge if $p > 1$ and diverge if $p \leq 1$.
2. The improper integrals $\int_a^b [1/(x - a)^p]\, dx$ and $\int_a^b [1/(b - x)^p]\, dx$ converge if $p < 1$ and diverge if $p \geq 1$.
3. *Standard comparison test (SCT)*: Let $0 \leq f(x) \leq g(x)$. If an improper integral of $g(x)$ converges, then the improper integral of $f(x)$, having the same limits, converges. If an improper integral of $f(x)$ diverges, then the same improper integral of $g(x)$ diverges.
4. *Limit comparison test (LCT)*: Let $f(x) > 0$ and $g(x) > 0$. Suppose basic types of improper integrals of $f(x)$ and $g(x)$ have the same limits of integration, and the two integrals have a common "bad point" of $-\infty$, a, b, or ∞. If

$$\lim_{x \to \text{"bad point"}} \frac{f(x)}{g(x)} = c > 0,$$

then the improper integrals of $f(x)$ and $g(x)$ either both converge or both diverge.

EXERCISES

In Exercises 1 through 40, decide whether the integral converges or diverges. Be prepared to give reasons for your answer. You need not compute the value of the integral if it converges.

1. $\int_{-\infty}^{2} \frac{dx}{\sqrt{2 - x}}$
2. $\int_{-3}^{10} \frac{dx}{\sqrt{x + 3}}$
3. $\int_{-1}^{1} \frac{dx}{x^{2/3}}$
4. $\int_{-\infty}^{\infty} \frac{dx}{x^{2/3}}$
5. $\int_{-\infty}^{\infty} \frac{dx}{x^{4/3}}$
6. $\int_{2}^{\infty} \frac{x^3 + 3x + 2}{x^5 - 8x}\, dx$
7. $\int_{1}^{\infty} \frac{x^4 - 3x^2 + 8}{x^5 + 7x}\, dx$
8. $\int_{1}^{\infty} \frac{x^2 + 2}{\sqrt{x^5 + 3x^2}}\, dx$
9. $\int_{1}^{\infty} \frac{x^2 + 3x + 1}{\sqrt{x^7 + 2x}}\, dx$
10. $\int_{3}^{\infty} \frac{dx}{x^2 - 3x - 4}$
11. $\int_{-\infty}^{\infty} \frac{dx}{x^2 + 4x + 4}$
12. $\int_{-\infty}^{\infty} \frac{dx}{x^2 + 4x + 5}$
13. $\int_{2}^{\infty} \frac{dx}{x^2 - x}$
14. $\int_{2}^{\infty} \frac{dx}{x^2 - 4}$

15. $\int_2^{\infty} \frac{dx}{\sqrt{x^2-4}}$

16. $\int_2^{\infty} \frac{dx}{(x^2-4)^{2/3}}$

17. $\int_0^{\infty} \frac{dx}{x+x^2}$

18. $\int_0^{\infty} \frac{dx}{x^{2/3}+x^{3/2}}$

19. $\int_{-\infty}^{\infty} \frac{dx}{|x^2-4|}$

20. $\int_{-\infty}^{\infty} \frac{dx}{(x^2-4)^{2/3}}$

21. $\int_0^2 \frac{dx}{\sqrt{2x-x^2}}$

22. $\int_0^2 \frac{dx}{\sqrt{2x^2-x^3}}$

23. $\int_0^1 \frac{dx}{\sqrt{x}-1}$

24. $\int_0^{\infty} \frac{dx}{x^2-\sqrt{x}}$

25. $\int_2^{\infty} \frac{dx}{x(\ln x)}$

26. $\int_1^2 \frac{dx}{x(\ln x)}$

27. $\int_1^2 \frac{dx}{x\sqrt{\ln x}}$

28. $\int_2^{\infty} \frac{dx}{x(\ln x)^2}$

29. $\int_1^{\infty} \frac{dx}{x[\sqrt{\ln x}+(\ln x)^2]}$

30. $\int_1^{\infty} \frac{|\cos x|}{x^2}\,dx$

31. $\int_0^{\infty} \frac{|\cos x|}{x^2}\,dx$

32. $\int_0^{\pi} \frac{|\cos x|}{\sqrt{x}}\,dx$

33. $\int_0^{\pi} \frac{\sin x}{x^{3/2}}\,dx$

34. $\int_0^{\infty} \frac{|\sin x|}{x^{3/2}}\,dx$

35. $\int_0^{\pi} \frac{\sin x}{x^{5/2}}\,dx$

36. $\int_0^{\pi} \frac{\sqrt{\sin x}}{x^{7/4}}\,dx$

37. $\int_0^{\pi} \frac{\sin x}{x^{7/4}}\,dx$

38. $\int_0^{\pi/4} \frac{\tan x}{x^{3/2}}\,dx$

39. $\int_0^{\pi/4} \frac{\tan x}{x^2}\,dx$

40. $\int_0^{\pi/2} \frac{\tan x}{x^{3/2}}\,dx$

The following theorem is sometimes useful for establishing convergence of an improper integral of a function which assumes both positive and negative values.

THEOREM Absolute convergence

9.5

Let f be continuous. If an improper integral of $|f(x)|$ converges, then the improper integral with the same limits of $f(x)$ also converges. In this case, we say the integral of $f(x)$ **converges absolutely.**

In Exercises 41 through 50, decide whether or not the given integral converges absolutely.

41. $\int_1^{\infty} \frac{\sin x}{x^2}\,dx$

42. $\int_0^{\infty} \frac{\cos x}{x^2}\,dx$

43. $\int_0^{\infty} \frac{\sin x}{x^2}\,dx$

44. $\int_0^{\infty} \frac{\sin x}{x^{3/2}}\,dx$

45. $\int_0^{\infty} \frac{\cos x^2}{x^2+1}\,dx$

46. $\int_0^{\infty} \frac{\sin x^2}{x^2}\,dx$

47. $\int_0^{\infty} \frac{\sin x^2}{x^3}\,dx$

48. $\int_0^{\infty} \frac{\sin x^2}{x^{5/2}}\,dx$

49. $\int_0^1 \frac{\sin(1/x)}{\sqrt{x}}\,dx$

50. $\int_0^{\infty} \frac{\sin(1/x)}{\sqrt{x}+x}\,dx$

EXERCISE SETS FOR CHAPTER 9

Review Exercise Set 9.1

In Exercises 1 through 11, find the given integral without using tables.

1. $\int x^2 \cos 3x\,dx$

2. $\int \sin^{-1} 3x\,dx$

3. $\int \frac{5x+40}{2x^2-7x-15}\,dx$

4. $\int \frac{x}{\sqrt{x+3}}\,dx$

5. $\int \frac{x^{1/3}}{x^{1/6}+1}\,dx$

6. $\int \frac{1}{\sin x+\cos x}\,dx$

7. $\int \sin^5 2x\,dx$

8. $\int \cos^2 x \sin^2 x\,dx$

9. $\int \sec^4 3x \tan^2 3x\,dx$

10. $\int \cot^3 2x\,dx$

11. $\int \sqrt{16-9x^2}\,dx$

12. Find $\int_1^{\infty} [1/(1+x^2)]\,dx$, if the integral converges.

13. Determine the convergence or divergence of the following improper integrals.

a) $\int_{-\infty}^{\infty} \frac{1}{x^2}\,dx$ b) $\int_2^{\infty} \frac{x+7}{x^2-1}\,dx$ c) $\int_1^2 \frac{\sqrt{2-x}}{x^2-4}\,dx$

Review Exercise Set 9.2

In Exercises 1 through 11, find the given integral without using tables.

1. $\int x \ln x\,dx$

2. $\int e^x \sin 2x\,dx$

3. $\int \frac{8x^2+3x-7}{4x^3-12x^2+x-3}\,dx$

4. $\int \frac{(x-1)^{3/2}}{x}\,dx$

5. $\int \frac{(x-1)^{1/2}-7}{3-(x-1)^{1/4}}\,dx$

6. $\int \frac{\sin x}{\sin x-\cos x}\,dx$

7. $\int \cos^3 x \sin^2 x\,dx$

8. $\int \cos^4 2x\,dx$

9. $\int \cot^2 x \csc^4 x\,dx$

10. $\int \cot^4 x\,dx$

11. $\int \frac{x^2}{\sqrt{4-x^2}}\,dx$

12. Find $\int_0^2 \frac{x}{\sqrt{2-x}}\,dx$, if the integral converges.

13. Determine the convergence or divergence of the following improper integrals.

a) $\int_0^{\infty} \frac{1}{x^3}\,dx$ b) $\int_{-1}^{4} \frac{x}{(x-1)^{2/3}}\,dx$

c) $\int_{-\infty}^{\infty} \frac{1}{1+x^3}\,dx$

Review Exercise Set 9.3

In Exercises 1 through 10, find the given integral without using tables.

1. $\int x \sin^2 x\,dx$ **2.** $\int x^5 e^{x^2}\,dx$

3. $\int \frac{x^3-3x+2}{x^2-1}\,dx$ **4.** $\int \frac{x^3}{\sqrt{x^2+4}}\,dx$

5. $\int \frac{dx}{1+2\sin x}$ **6.** $\int \frac{\sin^3 x}{\cos^2 x}\,dx$

7. $\int \sin^2 4x\,dx$ **8.** $\int \tan^3 x \sec^2 x\,dx$

9. $\int \sec^3 2x\,dx$ **10.** $\int \frac{dx}{\sqrt{9-4x^2}}$

11. Find $\int_2^{\infty} \frac{dx}{x(\ln x)^2}$, if the integral converges.

12. Find $\int_{-\infty}^{\infty} \frac{e^x}{1+e^{2x}}\,dx$, if the integral converges.

13. Determine the convergence or divergence of the following improper integrals.

a) $\int_0^{\infty} \frac{dx}{\sqrt{x(x+1)}}$ b) $\int_0^{1} \frac{x}{\sin^2 x}\,dx$

c) $\int_{-\infty}^{\infty} \frac{\cos^2 x}{1+x^2}\,dx$

Review Exercise Set 9.4

In Exercises 1 through 11, find the given integral without using tables.

1. $\int x \cos^3 x\,dx$ **2.** $\int \tan^{-1} 2x\,dx$

3. $\int \frac{x^4+1}{x^3+x}\,dx$ **4.** $\int \frac{\sqrt{x-1}}{1+\sqrt{x-1}}\,dx$

5. $\int \frac{dx}{x\sqrt{9-x^2}}$ **6.** $\int \frac{dx}{2\cos x - 1}$

7. $\int \frac{\cos^3 x}{\sin^5 x}$ **8.** $\int x \cos^2 3x\,dx$

9. $\int \sec^4 x \tan x\,dx$ **10.** $\int \sqrt{4+x^2}\,dx$

11. $\int \frac{dx}{\sqrt{8-2x-x^2}}$

12. Find $\int_0^{\infty} \frac{dx}{\sqrt{x+x^{3/2}}}$, if the integral converges.

13. Determine the convergence or divergence of the following improper integrals.

a) $\int_1^{2} \frac{\sin x}{\sqrt{x-1}}\,dx$ b) $\int_0^{\infty} \frac{\sin^2(1/x)}{\sqrt{x}}\,dx$

c) $\int_{-\infty}^{\infty} \frac{dx}{1+x^5}$

More Challenging Exercises 9

1. Let $f''(x)$ be continuous for all x. Show that

$$f(x) = f(0) + f'(0)x + \int_0^x f''(t)(x-t)\,dt.$$

[*Hint:* Show that $\int_0^x f'(t)\,dt = f(x) - f(0)$. Then integrate $\int_0^x f'(t)\,dt$ by parts, taking $u = f'(t)$, $dv = dt$, and letting $v = t - x$.]

2. Continue Exercise 1 to show that if $f'''(x)$ is continuous for all x, then

$$f(x) = f(0) + f'(0)x + \frac{f''(0)}{2}x^2 + \frac{1}{2}\int_0^x f'''(t)(x-t)\,dt.$$

3. Let $f(x)$ be a continuous function for $x \geq 0$. Show that if $\lim_{x\to\infty} f(x)$ exists and $\int_0^{\infty} f(x)\,dx$ converges, then $\lim_{x\to\infty} f(x) = 0$.

4. Let $g(x)$ be continuous for $x \geq a$, and let $c \neq 0$. Show that $\int_a^{\infty} g(x)\,dx$ converges if and only if $\int_a^{\infty} cg(x)\,dx$ converges.

5. Let $f(x)$ be continuous for all x. The **Cauchy principal value** of $\int_{-\infty}^{\infty} f(x)\,dx$ is defined to be $\lim_{h\to\infty} \int_{-h}^{h} f(x)\,dx$, if the limit exists.

a) Give an example of a continuous function f such that the Cauchy principal value of $\int_{-\infty}^{\infty} f(x)\,dx$ exists, but $\int_{-\infty}^{\infty} f(x)\,dx$ diverges.

b) Show that if $\int_{-\infty}^{\infty} f(x)\,dx$ converges, then the Cauchy principal value of $\int_{-\infty}^{\infty} f(x)\,dx$ exists and is $\int_{-\infty}^{\infty} f(x)\,dx$.

6. Give an example of a continuous function f such that $f(x) \geq 0$ for $x \geq 0$, and $\int_0^{\infty} f(x)$ converges, but $\lim_{x\to\infty} f(x)$ does not exist. [*Hint:* Let the graph of f have narrower and narrower upward "spikes" of unit height as $x \to \infty$, and let $f(x) = 0$ between the spikes.]

7. Give an example of a continuous unbounded (assuming arbitrarily large values) function f satisfying the other conditions of the preceding exercise.

INFINITE SERIES OF CONSTANTS 10

10.1 SEQUENCES

MOTIVATION FOR SERIES AND SEQUENCES

It is too bad that not all functions are polynomials. Calculus of the polynomial functions, even integration, is very easy to handle. However, many functions we encounter are not polynomial functions. For example, $\sin x$ is not a polynomial function, because $\sin x$ is always between -1 and 1, while every nonconstant polynomial function is unbounded.

We will see in the next chapter that many elementary functions such as $\sin x$, while not polynomial functions, can be viewed as "infinite polynomial functions," that is, as polynomials with an infinite number of terms. We will see that

$$\sin x = x - \frac{x^3}{3!} + \frac{x^5}{5!} - \frac{x^7}{7!} + \frac{x^9}{9!} - \frac{x^{11}}{11!} + \cdots. \tag{1}$$

(Recall that

$$1! = 1, \qquad 2! = 2 \cdot 1 = 2, \qquad 3! = 3 \cdot 2 \cdot 1 = 6,$$

and, in general, $n!$ [read "n factorial"] is defined by

$$n! = n(n-1)(n-2) \cdots 3 \cdot 2 \cdot 1.)$$

Then, replacing x by x^2, we obtain

$$\sin x^2 = x^2 - \frac{x^6}{3!} + \frac{x^{10}}{5!} - \frac{x^{14}}{7!} + \frac{x^{18}}{9!} - \frac{x^{22}}{11!} + \cdots. \tag{2}$$

It can be shown that we can't find $\int \sin x^2\, dx$ as a combination of elementary functions. However, we can use Eq. (2) to integrate $\sin x^2$ as an "infinite polynomial."

If we use Eq. (1) to compute $\sin 1$ (the sine of one radian), replacing x by 1, then we have to add up the infinite sum

$$1 - \frac{1}{3!} + \frac{1}{5!} - \frac{1}{7!} + \frac{1}{9!} - \frac{1}{11!} + \cdots \tag{3}$$

to find the answer. This leads us to study infinite sums such as (3), or *infinite series*, as they are called. Of course, we really have this idea embodied in our notation for numbers; for example,

$$\frac{1}{3} = 0.33333\ldots = \frac{3}{10} + \frac{3}{100} + \frac{3}{1000} + \frac{3}{10000} + \frac{3}{100000} + \cdots$$

and

$$\begin{aligned} e &= 2.71828\ldots \\ &= 2 + \frac{7}{10} + \frac{1}{100} + \frac{8}{1000} + \frac{2}{10000} + \frac{8}{100000} + \cdots. \end{aligned}$$

EXAMPLE 1 Use Eq. (1) and a calculator to estimate sin 1.

Solution Our calculator shows that

$$\begin{aligned} 1 &= 1.0000000000 \\ 1 - \frac{1}{3!} &= 0.8333333333 \\ 1 - \frac{1}{3!} + \frac{1}{5!} &= 0.8416666667 \\ 1 - \frac{1}{3!} + \frac{1}{5!} - \frac{1}{7!} &= 0.841468254 \\ 1 - \frac{1}{3!} + \frac{1}{5!} - \frac{1}{7!} + \frac{1}{9!} &= 0.8414710097 \\ 1 - \frac{1}{3!} + \frac{1}{5!} - \frac{1}{7!} + \frac{1}{9!} - \frac{1}{11!} &= 0.8414709846 \\ 1 - \frac{1}{3!} + \frac{1}{5!} - \frac{1}{7!} + \frac{1}{9!} - \frac{1}{11!} + \frac{1}{13!} &= 0.8414709848 \\ &\vdots \end{aligned}$$

From Eq. (1), we expect the *sequence of numbers* in the column to the right of the equal signs to get closer and closer to sin 1 as we go down the column. Our calculator gives sin 1 = 0.8414709848, so our last computation from Eq. (1) gives sin 1 as accurately as our calculator can display it. □

Calculators and computers do not have huge tables for sin x stored in them. When we use them to evaluate sin x, they simply evaluate some suitable algebraic formula that approximates sin x correct to the number of digits the calculator works with. Example 1 indicates how this might be done.

EXAMPLE 2 Try to find the "sum" of the infinite series of constants

$$1 + \frac{1}{2} + \frac{1}{4} + \frac{1}{8} + \frac{1}{16} + \cdots. \tag{4}$$

Solution We try to add up all the numbers in Eq. (4), and we obtain successively

$$1 = 1$$

$$1 + \frac{1}{2} = \frac{3}{2}$$

$$1 + \frac{1}{2} + \frac{1}{4} = \frac{7}{4}$$

$$1 + \frac{1}{2} + \frac{1}{4} + \frac{1}{8} = \frac{15}{8}$$

$$\frac{15}{8} + \frac{1}{16} = \frac{31}{16}$$

$$\vdots$$

This leads us to consider the *sequence of numbers*

$$1, \quad \frac{3}{2}, \quad \frac{7}{4}, \quad \frac{15}{8}, \quad \frac{31}{16}, \quad \ldots. \tag{5}$$

Figure 10.1 indicates that each number in the sequence is halfway between the preceding number and 2. Thus it is clear that the numbers in the sequence (5) get very close to 2 as we continue through the sequence. This suggests that the "sum" of the infinite series (4) should be 2. □

From Examples 1 and 2, it is natural to try to define the sum of an infinite *series* of numbers by examining the *sequence* of numbers obtained by adding more and more terms of the series. In the following subsection we discuss sequences and their limits. These ideas will then be used to discuss series in Section 10.2.

SEQUENCES

An example of a sequence was given in (5) above. We naively think of a sequence as an endless row of numbers separated by commas, symbolically,

$$a_1, \quad a_2, \quad a_3, \quad \ldots, \quad a_n, \quad \ldots, \tag{6}$$

where each a_i is a real number. Sequence (6) is also written $\{a_n\}$ for brevity. Since a sequence has one term for each positive integer, we can also consider a sequence to be a real-valued function ϕ with domain the set of positive integers. For sequence (6), we then have

$$\phi(1) = a_1, \qquad \phi(2) = a_2, \qquad \phi(3) = a_3, \qquad \ldots.$$

We may regard sequence (5) as a function ϕ where

$$\phi(1) = 1, \qquad \phi(2) = \frac{3}{2}, \qquad \phi(3) = \frac{7}{4}, \qquad \ldots.$$

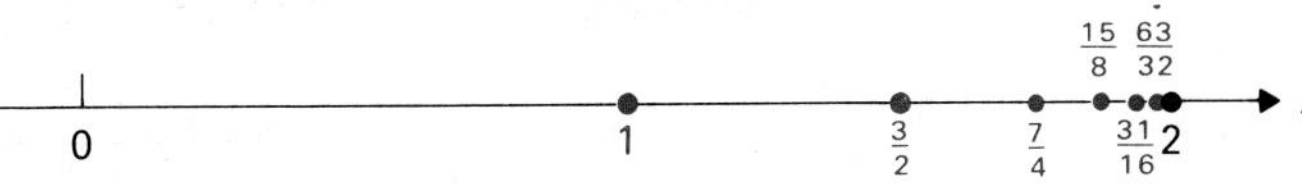

Figure 10.1 The sequence 1, $\frac{3}{2}$, $\frac{7}{4}$, $\frac{15}{8}$, $\frac{31}{16}$, $\frac{63}{32}$, ... converges to 2.

DEFINITION 10.1 Sequence

A **sequence** is a real-valued function ϕ with domain the set of positive integers. If $\phi(n) = a_n$, the sequence is denoted by $\{a_n\}$ or by $a_1, a_2, \ldots, a_n, \ldots$.

We want to find the limit of a sequence if it exists. Naively, the limit of $\{a_n\}$ is c if the terms a_n get just as close to c as we wish, provided that n is large enough. This is a vague statement similar to the one we used to introduce $\lim_{x\to a} f(x) = c$. The more precise definition that follows is very much like the one for the limit of a function $f(x)$ as $x \to \infty$.

DEFINITION 10.2 Limit of a sequence

The **limit of a sequence** $\{a_n\}$ is c if for each $\varepsilon > 0$, there exists an integer N such that $|a_n - c| < \varepsilon$ if $n > N$. We write "$\lim_{n\to\infty} \{a_n\} = c$" or "$\lim_{n\to\infty} a_n = c$," and we say that the sequence $\{a_n\}$ **converges to** c. A sequence that has no limit **diverges.**

We leave to the exercises (see Exercise 38) the easy proof that a sequence can't converge to two different values.

Figure 10.2 illustrates Definition 10.2. We mark c on the x-axis and then mark $c - \varepsilon$ and $c + \varepsilon$. The definition asserts that we must have all terms beyond some point in the sequence falling within the bracketed interval between $c - \varepsilon$ and $c + \varepsilon$. The N in the definition is simply a position indicator, giving the position in the sequence beyond which all terms, that is, a_{N+1}, $a_{N+2}, a_{N+3}, \ldots$, must be in the bracketed interval. As illustrated in the next example, the smaller ε is, the larger we expect N to have to be.

EXAMPLE 3 Figure 10.3 shows some terms of a sequence $\{a_n\}$ that has limit 4. Assume that each term of the sequence is closer to 4 than the preceding term is. Find the smallest possible N in Definition 10.2 corresponding to $\varepsilon = 3$, $\varepsilon = 1$, and $\varepsilon = \frac{1}{2}$.

Solution

$\varepsilon = 3$ We need to have terms fall between $4 - 3 = 1$ and $4 + 3 = 7$. Now a_3 is not between 1 and 7, but a_4 is. Since we are told that terms of the sequence become consecutively closer to 4, we see that we can take $N = 3$, since for $n > 3$ we have a_n between 1 and 7.

$\varepsilon = 1$ We need to have terms between $4 - 1 = 3$ and $4 + 1 = 5$. Figure 10.3 shows that a_{19} is not between 3 and 5, but a_{20} is. Thus we have $N = 19$.

$\varepsilon = \frac{1}{2}$ We need to have terms between $4 - \frac{1}{2} = \frac{7}{2}$ and $4 + \frac{1}{2} = \frac{9}{2}$. Figure 10.3 shows that a_{21} is not between $\frac{7}{2}$ and $\frac{9}{2}$, but a_{22} is. Thus we have $N = 21$. □

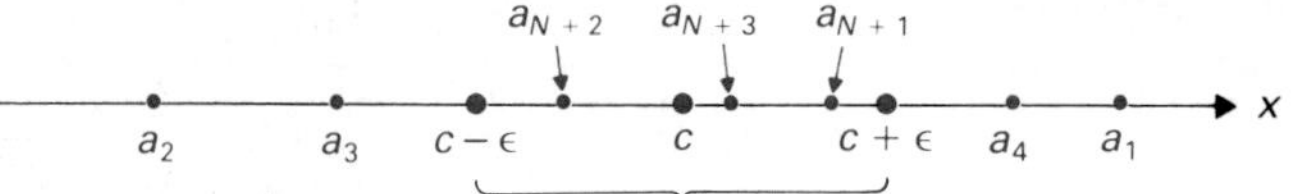

Figure 10.2 Terms $a_{N+1}, a_{N+2}, a_{N+3}, \ldots$, falling between $c - \varepsilon$ and $c + \varepsilon$ for Definition 10.2.

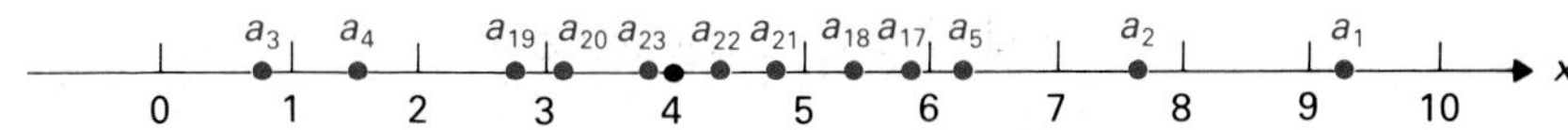

Figure 10.3 Some terms of a sequence $\{a_n\}$ that converges to 4.

EXAMPLE 4 For the sequence

$$1, \quad \frac{3}{2}, \quad \frac{7}{4}, \quad \frac{15}{8}, \quad \frac{31}{16}, \quad \ldots$$

in sequence (5), describe how to find an N in Definition 10.2 for a given $\varepsilon > 0$.

Solution For this sequence, which has limit 2, $a_1 = 1$ is 1 less than 2, $a_2 = \frac{3}{2}$ is $\frac{1}{2}$ less than 2, $a_3 = \frac{7}{4}$ is $\frac{1}{4}$ less than 2, and so on, as illustrated in Fig. 10.1. We then see that

$$a_n = 2 - \frac{1}{2^{n-1}}.$$

Thus

$$|a_n - 2| = \frac{1}{2^{n-1}},$$

so, given $\varepsilon > 0$, if we choose N so that $(\frac{1}{2})^{N-1} < \varepsilon$, then

$$|a_n - 2| = \frac{1}{2^{n-1}} < \varepsilon$$

for $n > N$. Solving $(\frac{1}{2})^{N-1} < \varepsilon$ for N, we obtain

$$\left(\frac{1}{2}\right)^{N-1} < \varepsilon,$$

$$(N-1)\left(\ln \frac{1}{2}\right) < \ln \varepsilon, \qquad \text{Taking ln of both sides}$$

$$(N-1)(-\ln 2) < \ln \varepsilon, \qquad \ln\left(\tfrac{1}{2}\right) = \ln 1 - \ln 2$$

$$N - 1 > -\frac{\ln \varepsilon}{\ln 2},$$

$$N > 1 - \frac{\ln \varepsilon}{\ln 2}.$$

For example, with $\varepsilon = 0.001$, our calculator gives $N > 1 - (-9.966) = 10.966$. Thus we can take $N = 11$ for $\varepsilon = 0.001$. □

Frequently, one describes a sequence $\{a_n\}$ by giving a formula in terms of n for the "nth term" a_n of the sequence. For example, the sequence discussed in Example 4 is the sequence where $a_n = 2 - (\frac{1}{2})^{n-1}$, or, more briefly, the sequence

$$\left\{2 - \frac{1}{2^{n-1}}\right\}$$

Example 4 shows that

$$\lim_{n\to\infty}\left(2 - \frac{1}{2^{n-1}}\right) = 2 - 0 = 2.$$

Finding the limit of a sequence $\{a_n\}$, where a_n is given by a formula involving n, is much like finding $\lim_{x\to\infty} f(x)$. We do not expect you to exhibit N corresponding to ε to prove convergence to the limit in your routine work.

EXAMPLE 5 Find the limit of the sequence

$$\left\{\frac{1}{n}\right\} = 1, \quad \frac{1}{2}, \quad \frac{1}{3}, \quad \frac{1}{4}, \quad \ldots, \quad \frac{1}{n}, \quad \ldots.$$

Solution We must find $\lim_{n\to\infty} (1/n)$. Clearly

$$\lim_{n\to\infty} \frac{1}{n} = 0.$$

Thus the sequence converges to 0. □

EXAMPLE 6 Establish the convergence or divergence of the sequence

$$\{(-1)^{n+1}\} = 1, \quad -1, \quad 1, \quad -1, \quad 1, \quad -1, \quad \ldots.$$

Solution The sequence diverges, since the terms of the sequence do not approach and stay close to any number c as we continue along the sequence. We leave an ε,N-proof of this fact to the exercises (see Exercise 42). □

EXAMPLE 7 Find the limit of the sequence $\{(3n^2 - 3)/(5n - 2n^2)\}$, if it converges.

Solution We must find

$$\lim_{n\to\infty} \frac{3n^2 - 3}{5n - 2n^2}.$$

We use the technique we would use to find $\lim_{x\to\infty} f(x)$ for $f(x) = (3x^2 - 3)/(5x - 2x^2)$. Our work in Section 2.3 showed that for a quotient of polynomials, we can compute the limit at ∞ and $-\infty$ by keeping only the monomial terms of highest degree. Thus

$$\lim_{n\to\infty} \frac{3n^2 - 3}{5n - 2n^2} = \lim_{n\to\infty} \frac{3n^2}{-2n^2} = -\frac{3}{2},$$

so the sequence converges to $-\frac{3}{2}$. □

EXAMPLE 8 Find the limit of the sequence $\{n \sin (2/n)\}$, if it converges.

Solution We know that

$$\lim_{x\to 0} \frac{\sin x}{x} = 1.$$

Thus

$$\lim_{n\to\infty} \frac{\sin (2/n)}{2/n} = 1.$$

We have

$$\lim_{n\to\infty}\left[n \sin\left(\frac{2}{n}\right)\right] = \lim_{n\to\infty} 2\left[\frac{\sin (2/n)}{2/n}\right] = 2\cdot 1 = 2.$$

Thus the sequence converges to 2. □

Just as for a function, we have the notions of $\lim_{n\to\infty} a_n = \infty$ and $\lim_{n\to\infty} a_n = -\infty$. We ask you to state these definitions for sequences in Exercises 1 and 2.

> Let us emphasize that a sequence is said to *converge* if and only if it has a *finite* limit. A sequence that does not converge to a finite limit *diverges*.

EXAMPLE 9 Establish the convergence or divergence of $\{n^2\}$ and of $\{(-n^2 + 2)/(n + 1)\}$.

Solution We have

$$\lim_{n\to\infty} n^2 = \infty$$

while

$$\lim_{n\to\infty}\frac{-n^2 + 2}{n + 1} = \lim_{n\to\infty}\frac{-n^2}{n} = \lim_{n\to\infty}(-n) = -\infty.$$

Thus the sequence $\{n^2\}$ diverges to ∞ and the sequence $\{(-n^2 + 2)/(n + 1)\}$ diverges to $-\infty$. □

EXAMPLE 10 Discuss the convergence or divergence of the sequence $\{(-1)^n \cdot n\}$.

Solution The sequence

$$\{(-1)^n \cdot n\} = -1,\quad 2,\quad -3,\quad 4,\quad -5,\quad 6,\quad -7,\quad \ldots$$

diverges. However, it does not diverge to either ∞ or $-\infty$, since for n even, the terms approach ∞ and for n odd, the terms approach $-\infty$. □

EXAMPLE 11 Find the limit of the sequence

$$\left\{\frac{\cos n}{\ln n}\right\}, \qquad n > 1,$$

if the sequence converges.

Solution Of course the restriction $n > 1$ is made since $\ln 1 = 0$ and we do not wish zero in a denominator. Now $|\cos n| \leq 1$ and we know that $\ln n$ becomes arbitrarily large as $n \to \infty$, although it increases slowly. Thus the terms of the sequence eventually become arbitrarily close to zero, with some terms negative and others positive. The sequence thus converges to zero. □

A calculator or computer can be used to try to estimate the limit of a sequence, just as we tried to estimate $\lim_{x\to\infty} f(x)$ in Section 2.3.

EXAMPLE 12 Use a calculator to estimate the limit of the sequence

$$\left\{\left(1+\frac{1}{n}\right)^n\right\}.$$

Solution Our calculator yields the following data:

n	$(1 + 1/n)^n$
1,000	2.716923932
10,000	2.718145926
1,000,000	2.718280469
100,000,000	2.718281828
1,000,000,000	2.718281828

We feel that $\lim_{n\to\infty} (1 + 1/n)^n \approx 2.718281828$. Since this number equals e to all the decimal places shown, we are suspicious that $\{(1 + 1/n)^n\}$ converges to e. This is indeed the case, and we will show it in the next chapter. This is a famous limit. □

SUMMARY

1. A sequence is an endless row of numbers

$$a_1, \quad a_2, \quad a_3, \quad \ldots, \quad a_n, \quad \ldots$$

separated by commas. (Repetition of numbers is allowed.) Alternatively, it is a real-valued function with domain the set of positive integers.

2. $\text{Lim}_{n\to\infty} a_n = c$ if for each $\varepsilon > 0$, there exists a positive integer N depending on ε such that $|a_n - c| < \varepsilon$ if $n > N$.
3. If $\lim_{n\to\infty} a_n$ exists (is a *finite* number c), then the sequence converges; otherwise it diverges.

EXERCISES

1. Define what is meant by $\lim_{n\to\infty} a_n = \infty$.

2. Define what is meant by $\lim_{n\to\infty} b_n = -\infty$.

In Exercises 3 through 36, find the limit of the sequence if the sequence converges or has limit ∞ or $-\infty$.

3. $\left\{\dfrac{1}{2n}\right\}$

4. $1, 0, \frac{1}{2}, 0, \frac{1}{3}, 0, \frac{1}{4}, 0, \ldots$

5. $\frac{1}{2}, \frac{2}{3}, \frac{3}{4}, \frac{4}{5}, \ldots$

6. $\left\{\dfrac{n^2+1}{3n}\right\}$

7. $\left\{\dfrac{2n-\sqrt{n}}{n}\right\}$

8. $\left\{\dfrac{3n^2-2n}{2n^3}\right\}$

9. $\left\{\dfrac{3n}{\sqrt{n}+100}\right\}$

10. $\left\{\dfrac{n-n^3}{n^2}\right\}$

11. $\left\{\left(-\dfrac{1}{2}\right)^n\right\}$

12. $\{\sin^n 1\}$

13. $1, 2, 1, -3, 1, 4, 1, -5, 1, 6, \ldots$

14. $\{e^n\}$

15. $\{e^{-n/(\sqrt{n}+1)}\}$

16. $\{e^{(n^2+1)/(n-2n^2)}\}$

17. $\{e^{1/\sqrt{n}}\}$

18. $\{e^{\sin n}\}$

19. $\{\ln n\}$

20. $\left\{\dfrac{1}{\ln(n+1)}\right\}$

21. $\left\{\ln\left(\dfrac{2n^2+1}{3n+2n^2}\right)\right\}$

22. $\left\{\ln\left(\frac{3n+n^3}{n^4+2}\right)\right\}$

23. $\left\{\frac{\ln n}{n^2}\right\}$

24. $\left\{\frac{\sin n}{\ln n}\right\}, n > 1$

25. $\{\sin n\}$

26. $\left\{\sin\frac{1}{n}\right\}$

27. $\left\{\frac{\cos n}{n}\right\}$

28. $\left\{\frac{\sin^2 n}{n}\right\}$

29. $\left\{\frac{\sin(1/n)}{n}\right\}$

30. $\left\{n \sin\left(\frac{1}{n}\right)\right\}$

31. $\left\{n^2 \sin\left(\frac{1}{n}\right)\right\}$

32. $\left\{n \sin\left(\frac{3}{n}\right)\right\}$

33. $\left\{2n \sin\left(\frac{3}{5n}\right)\right\}$

34. $\left\{(n^2 - 3n)\sin\left(\frac{4}{n^2}\right)\right\}$

35. $\left\{(n^3 - 2n^2)\sin\left(\frac{4n}{(2+n^4)}\right)\right\}$

36. $\left\{n \sin^2\left(\frac{1}{n}\right)\right\}$

37. A ball dropped from a height of 40 ft bounces repeatedly, each time to half its height on the preceding bounce. In the next section, we will be able to show that the total distance the ball has traveled at the instant it bounces for the nth time is $40(3 - 1/2^{n-2})$ ft. Find the total distance the ball bounces if it is allowed to bounce forever.

38. Show that a sequence $\{a_n\}$ can't converge to two different limits.

39. Give an ε,N-proof that the sequence $1, 1, 1, \ldots, 1, \ldots$ converges.

40. Give an ε,N-proof that the sequence $1, -\frac{1}{2}, \frac{1}{3}, -\frac{1}{4}, \frac{1}{5}, -\frac{1}{6}, \ldots$ converges.

41. Give an ε,N-proof that the sequence

$$\frac{1}{\sqrt{2}}, \quad \frac{1}{\sqrt{3}}, \quad \frac{1}{\sqrt{4}}, \quad \ldots, \quad \frac{1}{\sqrt{n+1}}, \quad \ldots$$

converges.

42. Give an ε,N-proof that the sequence

$$1, \quad -1, \quad 1, \quad -1, \quad 1, \quad \ldots, \quad (-1)^{n+1}, \quad \ldots$$

diverges.

43. Give an ε,N-proof that the sequence $1, 2, 3, 4, \ldots, n, \ldots$ diverges.

44. We will often use the fact that if $|a| < 1$, then $\{a^n\}$ converges to zero. Convince yourself that $\lim_{n\to\infty} a^n = 0$ without giving a formal ε,N-proof. [*Suggestion:* Find $\lim_{n\to\infty} \ln(|a|^n)$.]

45. Give an ε,N-proof of the limit in Exercise 44. Give a formula for N in terms of ε.

46. It can be shown that $\lim_{n\to\infty}(1 + 1/n)^n = e$. Use a calculator to find the smallest integer N such that

$$|(1 + 1/N)^N - e| < \varepsilon$$

for the following values of ε.

a) 0.1　　b) 0.01
c) 0.001　　d) 0.0001

In Exercises 47 through 51, attempt to find the limit of the sequence using a calculator or computer. Exercises 50 and 51 indicate one type of problem that may arise.

47. $\left\{\left(1 - \frac{1}{n}\right)^n\right\}$

48. $\left\{\left(1 + \frac{1}{2n}\right)^n\right\}$

49. $\{(1 + n)^{1/n}\}$

50. $\{n(\sqrt[n]{2} - 1)\}$

51. $\left\{n^2\left[1 - \cos\left(\frac{1}{n}\right)\right]\right\}$

10.2 SERIES

THE SUM OF A SERIES

It is mathematically imprecise to define a sequence of constants to be an endless row of numbers

$$a_1, \quad a_2, \quad \ldots, \quad a_n, \quad \ldots \tag{1}$$

separated by commas, although this is the way we sometimes think of a sequence in our work. In Section 10.1, we defined a sequence to be a *function*. It would be similarly imprecise to define an infinite series of constants to be an endless row of numbers

$$a_1 + a_2 + \cdots + a_n + \cdots \tag{2}$$

with plus signs between them, although this is often the way we think of a series in our work. We saw in the previous section that in attempting to find the "sum" of the series (2), we are led to consider the limit of the sequence

$$s_1, \quad s_2, \quad \ldots, \quad s_n, \quad \ldots, \tag{3}$$

where $s_1 = a_1$, $s_2 = a_1 + a_2$, and, in general, $s_n = a_1 + \cdots + a_n$. From a mathematical standpoint, the precise thing to do is to define the series (2) to *be* the sequence in (3); we have already based the notion of a sequence on the notion of a function.

DEFINITION 10.3 Infinite series

Let $a_1, a_2, \ldots, a_n, \ldots$ be a sequence and let $s_n = a_1 + \cdots + a_n$ for all positive integers n. The **infinite series**

$$\sum_{n=1}^{\infty} a_n = a_1 + a_2 + \cdots + a_n + \cdots$$

is the sequence $\{s_n\}$. The number a_n is the ***n*th term of the series** $\sum_{n=1}^{\infty} a_n$, and s_n is the ***n*th partial sum of the series.**

DEFINITION 10.4 Sum of a series

If $\sum_{n=1}^{\infty} a_n$ is a series, then the **sum of the series** is the limit of the sequence $\{s_n\}$ of partial sums, if this limit exists. If $\lim_{n\to\infty} s_n$ is a finite number c, then the series $\sum_{n=1}^{\infty} a_n$ **converges to** c. If $\lim_{n\to\infty} s_n$ is ∞, $-\infty$, or is undefined, then the series **diverges.**

EXAMPLE 1 Determine the convergence or divergence of the series

$$0 + 0 + 0 + 0 + \cdots.$$

Solution Computing partial sums, we have $0 = 0$, $0 + 0 = 0$, $0 + 0 + 0 = 0, \ldots$. Thus the sequence of partial sums is

$$0, \quad 0, \quad 0, \quad 0, \quad 0, \quad \ldots,$$

which converges to zero. Thus the series converges to zero. □

EXAMPLE 2 Determine the convergence or divergence of the series

$$1 - 1 + 1 - 1 + 1 - 1 + \cdots.$$

Solution We find that the sequence of partial sums is

$$1, \quad 0, \quad 1, \quad 0, \quad 1, \quad 0, \quad \ldots.$$

Since this sequence diverges, the series diverges. □

HARMONIC AND GEOMETRIC SERIES

In the next section, we will develop comparison tests for convergence or divergence of series, similar to the comparison tests for improper integrals in Section 9.8. Such comparison tests are useful only if we have a stockpile of series whose behavior (convergence or divergence) is known. The series discussed now contribute to such a stockpile.

The series

$$\sum_{n=1}^{\infty} \frac{1}{n} = 1 + \frac{1}{2} + \frac{1}{3} + \frac{1}{4} + \cdots + \frac{1}{n} + \cdots \tag{4}$$

is the **harmonic series.** This series diverges, as the following argument shows. We group together certain terms of the series, using parentheses, so that the series appears as

$$(1) + \left(\frac{1}{2}\right) + \left(\frac{1}{3} + \frac{1}{4}\right) + \left(\frac{1}{5} + \cdots + \frac{1}{8}\right) + \left(\frac{1}{9} + \cdots + \frac{1}{16}\right) + \cdots. \tag{5}$$

The sum of the terms in each parenthesis in series (5) is $\geq \frac{1}{2}$; for example,

$$\frac{1}{5} + \frac{1}{6} + \frac{1}{7} + \frac{1}{8} > \frac{1}{8} + \frac{1}{8} + \frac{1}{8} + \frac{1}{8} = \frac{1}{2}.$$

This shows that for the harmonic series, we have

$$s_{2^n} \geq 1 + \frac{1}{2} n,$$

and it is then clear that the series diverges to ∞.

We turn now to the series

$$\sum_{n=0}^{\infty} ar^n = a + ar + ar^2 + \cdots + ar^n + \cdots. \tag{6}$$

The series (6) is the **geometric series** with **initial term** a and **ratio** r; it is a very important series. We have

$$s_n = a + ar + \cdots + ar^{n-1},$$

and, consequently,

$$rs_n = ar + \cdots + ar^{n-1} + ar^n.$$

Subtracting, we obtain

$$s_n - rs_n = a - ar^n,$$

so if $r \neq 1$,

$$s_n = \frac{a - ar^n}{1 - r} = \frac{a}{1 - r} - \frac{ar^n}{1 - r}.$$

Thus

$$\lim_{n\to\infty} s_n = \lim_{n\to\infty} \left(\frac{a}{1 - r} - \frac{ar^n}{1 - r}\right) = \frac{a}{1 - r} - a\left(\lim_{n\to\infty} \frac{r^n}{1 - r}\right). \tag{7}$$

The value of the limit in Eq. (7) depends on the size of r. We have

$$\lim_{n\to\infty} \frac{r^n}{1 - r} = \begin{cases} 0, & \text{if } -1 < r < 1, \\ -\infty, & \text{if } r > 1, \\ \text{undefined}, & \text{if } r \leq -1. \end{cases}$$

If $r = 1$, then the series (6) reduces to

$$a + a + \cdots + a + \cdots, \tag{8}$$

which obviously does not converge if $a \neq 0$. Thus the geometric series (6), where $a \neq 0$, converges if $|r| < 1$ and diverges if $|r| \geq 1$. For $|r| < 1$, the sum of the series is found from Eq. (7) to be

$$\lim_{n\to\infty} s_n = \frac{a}{1-r} - a\left(\lim_{n\to\infty} \frac{r^n}{1-r}\right) = \frac{a}{1-r} - 0 = \frac{a}{1-r}.$$

We summarize the results of this section in a theorem for easy reference.

THEOREM 10.1 Harmonic and geometric series

1. The *harmonic series* $\sum_{n=1}^{\infty} (1/n)$ diverges to ∞.
2. The *geometric series*

$$\sum_{n=0}^{\infty} (ar^n) \begin{cases} \text{converges to } \dfrac{a}{1-r} & \text{if } |r| < 1, \\ \text{diverges} & \text{if } |r| \geq 1 \text{ and } a \neq 0. \end{cases}$$

EXAMPLE 3 Find the sum of the series

$$\sum_{n=0}^{\infty} \frac{1}{2^n} = 1 + \frac{1}{2} + \frac{1}{4} + \cdots + \frac{1}{2^n} + \cdots$$

discussed on pages 420–421.

Solution This is a geometric series with $a = 1$ and $r = \frac{1}{2}$. By Theorem 10.1, the series converges to

$$\frac{a}{1-r} = \frac{1}{1-\frac{1}{2}} = 2,$$

which we guessed to be the case in Section 10.1. □

EXAMPLE 4 Find the sum of the series

$$\sum_{n=2}^{\infty} \frac{(-1)^n}{3^n},$$

if it converges.

Solution This series

$$\frac{1}{9} - \frac{1}{27} + \frac{1}{81} - \frac{1}{243} + \cdots$$

is a geometric series with initial term $a = \frac{1}{9}$ and ratio $r = -\frac{1}{3}$. Since $|r| = \frac{1}{3} < 1$, Theorem 10.1 shows that the series converges to

$$\frac{a}{1-r} = \frac{\frac{1}{9}}{1-(-\frac{1}{3})} = \frac{\frac{1}{9}}{\frac{4}{3}} = \frac{1}{9}\cdot\frac{3}{4} = \frac{1}{12}. \quad \square$$

EXAMPLE 5 Find the sum of the series

$$\sum_{n=0}^{\infty} \frac{2^{3n}}{7^n},$$

if the series converges.

Solution Since $2^{3n} = (2^3)^n = 8^n$, we have

$$\sum_{n=0}^{\infty} \frac{2^{3n}}{7^n} = \sum_{n=0}^{\infty} \frac{8^n}{7^n} = \sum_{n=0}^{\infty} \left(\frac{8}{7}\right)^n.$$

This is a geometric series with ratio $r = \frac{8}{7} > 1$, so by Theorem 10.1, the series diverges. □

EXAMPLE 6 Express the repeating decimal 4.277777... as a rational number (quotient n/m of integers where $m \neq 0$).

Solution We have

$$4.277777\ldots = \frac{42}{10} + \left(\frac{7}{100} + \frac{7}{1000} + \frac{7}{10000} + \cdots\right).$$

The sum in brackets is a geometric series with initial term $a = \frac{7}{100}$ and ratio $r = \frac{1}{10}$. Thus

$$4.277777\ldots = \frac{42}{10} + \frac{\frac{7}{100}}{1 - \frac{1}{10}} = \frac{42}{10} + \frac{7}{100} \cdot \frac{10}{9} = \frac{42}{10} + \frac{7}{90}$$

$$= \frac{378 + 7}{90} = \frac{385}{90} = \frac{77}{18}. \quad □$$

Exercises 42 through 48 continue the idea of Example 6 and indicate that the rational numbers are precisely the real numbers whose unending decimal expansion has a repeating pattern.

ALGEBRA OF SEQUENCES AND SERIES

There are natural ways to "add" two sequences or two series and to "multiply by a constant" a sequence or a series.

DEFINITION 10.5 Sequence algebra

Let $\{s_n\}$ and $\{t_n\}$ be sequences and let c be any number. The sequence $\{s_n + t_n\}$ with nth term $s_n + t_n$ is the **sum of** $\{s_n\}$ **and** $\{t_n\}$, while the sequence $\{cs_n\}$ with nth term cs_n is the **product of the constant** c **and** $\{s_n\}$.

DEFINITION 10.6 Series algebra

Let $\sum_{n=1}^{\infty} a_n$ and $\sum_{n=1}^{\infty} b_n$ be series of constants, and let c be any number. The series $\sum_{n=1}^{\infty} (a_n + b_n)$ with nth term $a_n + b_n$ results from **adding the series** $\sum_{n=1}^{\infty} a_n$ **and** $\sum_{n=1}^{\infty} b_n$, while the series $\sum_{n=1}^{\infty} (ca_n)$ with nth term ca_n results from **multiplying the series** $\sum_{n=1}^{\infty} a_n$ **by the constant** c.

It is important to note that if s_n is the nth partial sum of $\sum_{n=1}^{\infty} a_n$ and t_n is the nth partial sum of $\sum_{n=1}^{\infty} b_n$, then the nth partial sum of $\sum_{n=1}^{\infty} (a_n + b_n)$ is $s_n + t_n$, while the nth partial sum of $\sum_{n=1}^{\infty} (ca_n)$ is cs_n.

Throughout this chapter, we are interested primarily in whether sequences (and series) converge or diverge. After the preceding definitions, we at once ask ourselves whether the sequence obtained by adding two convergent sequences still converges and whether the sequence obtained by multiplying a convergent sequence by a constant still converges. We then ask the analogous questions for series. The answers to these questions are contained in the following theorem and its corollaries; the answers are really intuitively obvious.

THEOREM 10.2 Convergence of sums and multiples

If the sequence $\{s_n\}$ converges to s and the sequence $\{t_n\}$ converges to t, then the sequence $\{s_n + t_n\}$ converges to $s + t$ and the sequence $\{cs_n\}$ converges to cs for all c.

Proof. Let $\varepsilon > 0$ be given. Find an integer N_1 such that $|s_n - s| < \varepsilon/2$ for $n > N_1$, and find an integer N_2 such that $|t_n - t| < \varepsilon/2$ for $n > N_2$. Let N be the maximum of N_1 and N_2. Then for $n > N$, we have simultaneously

$$-\frac{\varepsilon}{2} < s_n - s < \frac{\varepsilon}{2}, \qquad -\frac{\varepsilon}{2} < t_n - t < \frac{\varepsilon}{2}.$$

Adding, we obtain

$$-\varepsilon < (s_n + t_n) - (s + t) < \varepsilon, \qquad \text{or} \qquad |(s_n + t_n) - (s + t)| < \varepsilon$$

for $n > N$. Thus the sequence $\{s_n + t_n\}$ converges to $s + t$.

If $c = 0$, then $\{cs_n\}$ is the sequence $0, 0, 0, \ldots$, which clearly converges to $0 = 0 \cdot s$. If $c \neq 0$, we can find N_3 such that

$$|s_n - s| < \frac{\varepsilon}{|c|}$$

for $n > N_3$. Then for $n > N_3$, we have

$$-\frac{\varepsilon}{|c|} < s_n - s < \frac{\varepsilon}{|c|}.$$

Multiplying by c, we obtain for $c < 0$ as well as $c > 0$,

$$-\varepsilon < cs_n - cs < \varepsilon,$$

so $|cs_n - cs| < \varepsilon$ for $n > N_3$. Thus the sequence $\{cs_n\}$ converges to cs. •

COROLLARY 1 Convergence and series algebra

If $\sum_{n=1}^{\infty} a_n$ converges to a and $\sum_{n=1}^{\infty} b_n$ converges to b, then $\sum_{n=1}^{\infty} (a_n + b_n)$ converges to $a + b$ and $\sum_{n=1}^{\infty} (ca_n)$ converges to ca for all c.

Proof. The proof is immediate from the preceding theorem and the observation that if $\{s_n\}$ is the sequence of partial sums of $\sum_{n=1}^{\infty} a_n$ and $\{t_n\}$ is the sequence of partial sums of $\sum_{n=1}^{\infty} b_n$, then $\{s_n + t_n\}$ is the sequence of partial sums of $\sum_{n=1}^{\infty}(a_n + b_n)$ and $\{cs_n\}$ is the sequence of partial sums of $\sum_{n=1}^{\infty}(ca_n)$. •

COROLLARY 2 Divergence and series algebra

If $\sum_{n=1}^{\infty} a_n$ diverges, then for any $c \neq 0$, the series $\sum_{n=1}^{\infty}(ca_n)$ diverges also.

Proof. If $\sum_{n=1}^{\infty}(ca_n)$ converges, then by Corollary 1, the series

$$\sum_{n=1}^{\infty} \frac{1}{c}(ca_n) = \sum_{n=1}^{\infty} a_n$$

would also converge, contrary to hypothesis. •

EXAMPLE 7 Establish the convergence or divergence of the series $\sum_{n=1}^{\infty} 1/(2n)$.

Solution The series diverges by Corollary 2, for it is the divergent harmonic series multiplied by $\frac{1}{2}$. □

EXAMPLE 8 Find the sum of the series

$$\sum_{n=0}^{\infty}\left(\frac{1}{2^n} - \frac{2}{3^n}\right),$$

if it converges.

Solution The two geometric series $\sum_{n=0}^{\infty}(\frac{1}{2})^n$ and $\sum_{n=0}^{\infty}(\frac{1}{3})^n$ with ratios $\frac{1}{2}$ and $\frac{1}{3}$ both converge. Corollary 1 then shows that

$$\sum_{n=0}^{\infty}\left(\frac{1}{2^n} - \frac{2}{3^n}\right) = \sum_{n=0}^{\infty}\left(\frac{1}{2}\right)^n - 2\sum_{n=0}^{\infty}\left(\frac{1}{3}\right)^n$$

converges. Now

$$\sum_{n=0}^{\infty}\frac{1}{2^n} \quad \text{converges to } \frac{1}{1-\frac{1}{2}} = 2$$

and

$$\sum_{n=0}^{\infty}\frac{1}{3^n} \quad \text{converges to } \frac{1}{1-\frac{1}{3}} = \frac{3}{2},$$

so

$$\sum_{n=0}^{\infty}\left(\frac{1}{2^n} - \frac{2}{3^n}\right)$$

converges to $2 + (-2)\frac{3}{2} = 2 - 3 = -1$. □

EXAMPLE 9 Find the sum of the series

$$\sum_{n=1}^{\infty}\frac{5^n + 3 \cdot 2^{3n}}{9^n},$$

if it converges.

Solution Now

$$\frac{5^n + 3 \cdot 2^{3n}}{9^n} = \left(\frac{5}{9}\right)^n + 3\left(\frac{8}{9}\right)^n.$$

Since $\sum_{n=1}^{\infty} (\frac{5}{9})^n$ and $\sum_{n=1}^{\infty} (\frac{8}{9})^n$ both converge, we see by Corollary 1 that

$$\begin{aligned}\sum_{n=1}^{\infty} \frac{5^n + 3 \cdot 2^{3n}}{9^n} &= \sum_{n=1}^{\infty} \left(\frac{5}{9}\right)^n + 3 \sum_{n=1}^{\infty} \left(\frac{8}{9}\right)^n \\ &= \frac{\frac{5}{9}}{1 - (\frac{5}{9})} + 3 \cdot \frac{\frac{8}{9}}{1 - (\frac{8}{9})} \\ &= \frac{5}{9} \cdot \frac{9}{4} + 3 \cdot \frac{8}{9} \cdot \frac{9}{1} = \frac{5}{4} + 24 = \frac{101}{4}. \quad \square\end{aligned}$$

You may be expecting an example on determining the convergence or divergence of a series using a calculator or computer. In general, this is not possible to do. If we *know* a series converges, a calculator might be useful to estimate the sum of the series, assuming we know that it converges "fast enough." To illustrate the difficulty of establishing the convergence or divergence of a series using a calculator, consider the harmonic series $\sum_{n=1}^{\infty} (1/n)$. In Section 10.4, we will see how to show that

$$9.21 \leq \sum_{n=1}^{10000} \frac{1}{n} \leq 10.21.$$

This indicates it would be impossible to establish the divergence to ∞ of $\sum_{n=1}^{\infty} (1/n)$ using a calculator. The convergence or divergence of $\sum_{n=1}^{\infty} a_n$ can't be determined by the values of any *finite* number of partial sums, and this is all a computer can find.

SUMMARY

1. Associated with a series

$$\sum_{n=1}^{\infty} a_n = a_1 + a_2 + a_3 + \cdots + a_n + \cdots$$

is the sequence of partial sums $\{s_n\}$, where

$$s_1 = a_1, \qquad s_2 = a_1 + a_2, \qquad s_3 = a_1 + a_2 + a_3, \quad \ldots$$

The series converges to the sum c if $\lim_{n \to \infty} s_n = c$, and the series diverges if $\{s_n\}$ is a divergent sequence.

2. Two sequences (series) may be added term by term, and each may be multiplied by a constant. If each of two sequences (series) converges, then their sum converges to the sum of the limits. A constant multiple of a convergent sequence (series) converges to that multiple of the limit of the original sequence (series).
3. The harmonic series $\sum_{n=1}^{\infty} 1/n$ diverges.
4. The geometric series $\sum_{n=0}^{\infty} (ar^n)$ converges to $a/(1 - r)$ if $|r| < 1$ and diverges if $|r| \geq 1$ and $a \neq 0$.

EXERCISES

1. Consider the series $1 + 0 - 1 + 0 + 1 + 0 - 1 + 0 + \cdots$. Find the indicated partial sums.
a) s_1 b) s_2 c) s_3 d) s_4
e) s_8 f) s_{15} g) s_{122}

2. Find the first four partial sums of the harmonic series $\sum_{n=1}^{\infty} (1/n)$.

3. Find the first five terms $a_1, \ldots, a_5$ of the series having as sequence of partial sums $\{s_n\} = \frac{1}{2}, \frac{1}{3}, \frac{1}{4}, \frac{1}{5}, \frac{1}{6}, \ldots$.

4. Find the first six terms of the series having as sequence of partial sums $\{s_n\} = \frac{1}{2}, \frac{2}{3}, \frac{3}{4}, \frac{4}{5}, \frac{5}{6}, \frac{6}{7}, \ldots$.

In Exercises 5 through 27, determine whether the series converges or diverges and find the sum of the series if the series converges.

5. $\displaystyle\sum_{n=1}^{\infty} \frac{1}{n}$

6. $\displaystyle\sum_{n=0}^{\infty} \frac{1}{3^n}$

7. $\displaystyle\sum_{n=1}^{\infty} \frac{1}{2^n}$

8. $\displaystyle\sum_{n=2}^{\infty} \frac{1}{2^n}$

9. $\displaystyle\sum_{n=1}^{\infty} \frac{-1}{2n}$

10. $\displaystyle\sum_{n=1}^{\infty} (-1)^n$

11. $\displaystyle\sum_{n=1}^{\infty} \frac{n+1}{n}$

12. $\displaystyle\sum_{n=1}^{\infty} \frac{3}{10^n}$

13. $\displaystyle\sum_{n=1}^{\infty} \frac{4}{(-2)^n}$

14. $\displaystyle\sum_{n=1}^{\infty} \frac{1}{n+1}$

15. $\displaystyle\sum_{n=0}^{\infty} e^{-2n}$

16. $\displaystyle\sum_{n=0}^{\infty} \frac{1}{(-5)^{n+3}}$

17. $\displaystyle\sum_{n=0}^{\infty} \frac{3^{2n+1}}{8^n}$

18. $\displaystyle\sum_{n=0}^{\infty} \frac{3^{2n+1}}{10^n}$

19. $\displaystyle\sum_{n=1}^{\infty} \frac{2^n}{3^{2n+1}}$

20. $\displaystyle\sum_{n=0}^{\infty} 5(\ln 2)^n$

21. $\displaystyle\sum_{n=0}^{\infty} 4\left(\ln \frac{1}{2}\right)^n$

22. $\displaystyle\sum_{n=0}^{\infty} 2\left(\ln \frac{1}{3}\right)^n$

23. $\displaystyle\sum_{n=0}^{\infty} \left(\frac{1}{2^n} + \frac{1}{3^n}\right)$

24. $\displaystyle\sum_{n=0}^{\infty} \left(7 \cdot \frac{1}{3^n} - 4 \cdot \frac{1}{2^n}\right)$

25. $\displaystyle\sum_{n=0}^{\infty} \frac{2^n + 3^n}{4^n}$

26. $\displaystyle\sum_{n=0}^{\infty} \frac{3^{n+1} - 7 \cdot 5^n}{10^n}$

27. $\displaystyle\sum_{n=1}^{\infty} \frac{8^n + 9^n}{10^n}$

28. Variant of Achilles and the tortoise: A driver is speeding down a straight road at 70 mph. A state trooper 1 mi behind is chasing the speeder at a speed of 80 mph.
a) Assuming the speeds remain the same, how long will it take the trooper to catch the speeder? (You should not need to use series to solve this part.)
b) The speeder, being an optimistic philosopher, reasons as follows:

When the trooper reaches my present position, I will no longer be here but at a new position, so he will not have caught me. When he reaches my new position, I will again have moved on to another new position, so he still will not have caught me. Since I can continue that argument an infinite number of times, the trooper will never catch me.

Find the flaw in the speeder's argument.

29. A ball is dropped from a height of 30 ft. Each time it hits the ground, it rebounds to $\frac{1}{3}$ of the height it attained on the preceding bounce. Find the total distance the ball travels if it is allowed to bounce forever.

30. An unscrupulous mathematician tells a farmer, "I will work for you in your fields today all day for just one penny, as long as you agree to hire me for a total of 25 days this month and to double my wages each day." The farmer agrees.
a) Express the amount the farmer must pay the mathematician for the 25 days of work as the first 25 terms of an infinite series.
b) Find the sum of the 25 terms of part (a). [*Hint:* Use the formula for the nth partial sum of a geometric series in Eq. (7) of the text.]

31. Show that if one could build a spaceship with the property that the time required to travel the next mile is k times the time required to cover the last mile for some fixed positive constant $k < 1$, then one could travel an infinite distance in a finite time.

32. Consider the series

$$\sum_{n=1}^{\infty} \left(\frac{1}{n} - \frac{1}{n+1}\right).$$

a) Compute the first four partial sums of the series.
b) Find a formula for the nth partial sum s_n of the series. (The series is known as a "telescoping series." Can you guess why?)
c) Show that the series converges, and find the sum of the series.

33. Consider the series

$$\sum_{n=1}^{\infty} \left(\ln \frac{n+1}{n}\right).$$

a) Show that the series can be viewed as a telescoping series (see Exercise 32, and compute the nth partial sum s_n. [*Hint:* Use a property of the function ln.]
b) Show that the series diverges.

34. Following the idea of Exercise 32, find the sum of the series

$$\sum_{n=1}^{\infty} \left(\frac{1}{n} - \frac{1}{n+2}\right).$$

35. Following the idea of Exercise 32, find the sum of the series

$$\sum_{n=1}^{\infty} \left(\frac{1}{n} - \frac{1}{n+3}\right).$$

36. a) Find the partial fraction decomposition of $1/(n^2 + 3n + 2)$.
b) Following the idea of Exercise 32, find the sum of the series

$$\sum_{n=1}^{\infty} \frac{1}{n^2 + 3n + 2}.$$

37. a) Find the partial fraction decomposition of $1/(n^2 - 1)$.
b) Following the idea of Exercise 32, find the sum of the series

$$\sum_{n=4}^{\infty} \frac{1}{n^2 - 1}.$$

38. Find the fallacy in the following argument. If $\{s_n\}$ converges to c and $\{t_n\}$ converges to d, then $\{s_n t_n\}$ converges to cd. If s_n is the nth partial sum of $\sum_{n=1}^{\infty} a_n$, and t_n is the nth partial sum of $\sum_{n=1}^{\infty} b_n$, then $s_n t_n$ is the nth partial sum of $\sum_{n=1}^{\infty} a_n b_n$. Therefore if $\sum_{n=1}^{\infty} a_n$ converges to c, and $\sum_{n=1}^{\infty} b_n$ converges to d, then $\sum_{n=1}^{\infty} a_n b_n$ converges to cd.

39. Give an example of two divergent sequences whose sum converges.

40. Give an example of two divergent series whose sum converges.

41. If $\sum_{n=1}^{\infty} a_n$ converges and $\sum_{n=1}^{\infty} b_n$ diverges, what can be said concerning the convergence or divergence of $\sum_{n=1}^{\infty} (a_n + b_n)$?

In Exercises 42 through 47, find a rational number equal to the given repeating decimal. (See Example 6.)

42. 0.222222 . . .

43. 8.2333333 . . .

44. 11.31888888 . . .

45. 0.1212121212 . . .

46. 7.1465465465 . . .

47. 273.14653653653 . . .

48. Exercises 42 through 47 indicate that every repeating decimal represents a rational number. Show, conversely, that the decimal expansion of any rational number tails off in a repeating pattern. [*Hint:* Think of using long division to find the decimal form of a rational number. Argue that some "remainder" obtained in the division must eventually repeat and that the "quotients" obtained must therefore eventually occur in a repeating pattern.]

49. Consider the real numbers 4.13000000 . . . and 4.12999999 . . . .
a) Which number do you think is larger?
b) Show that the numbers are equal by arguing that their difference must be zero. [*Hint:* If $(4.13000000\ldots) - (4.12999999\ldots) > 0$, then its decimal expansion must have a nonzero digit at some finite number of places beyond the decimal point. Show that this is impossible.]
c) Show that the numbers are equal by summing a geometric series to show that $4.12999999\ldots = \frac{413}{100}$.

10.3 COMPARISON TESTS

The comparison tests in this section parallel very closely the comparison tests for convergence or divergence of an improper integral of a positive function, discussed in Section 9.8. In Section 10.4, we show that the convergence or divergence of many series depends on the convergence or divergence of an improper integral, so it is not surprising that the tests for improper integrals can be carried over to series. The comparison tests in this section will provide our most general tests for convergence or divergence of a series. All the tests developed in the rest of this chapter will be based on these comparison tests.

We open the section by showing that the first few hundred, or million, terms of a series do not affect its convergence or divergence. We indicated this in Section 10.2, when we explained that computation of a finite number of partial sums can never be used as a criterion for convergence or divergence of a series.

INSERTION OR DELETION OF TERMS IN A SERIES

The insertion or deletion of a finite number of terms in a series cannot affect whether the series converges or diverges, although if the series converges, the sum of the series may be affected. In particular, suppose the first few terms of a series do not conform to a pattern present in the rest of the series. Then we can neglect those first few terms when studying the convergence or divergence of the series. After an example, we state this property of series as a theorem and leave the easy proof as an exercise.

EXAMPLE 1 Find the sum of the series

$$\pi - 3 + 17 + 1 + \frac{1}{2} + \frac{1}{4} + \frac{1}{8} + \frac{1}{16} + \cdots.$$

Solution The geometric series

$$1 + \frac{1}{2} + \frac{1}{4} + \frac{1}{8} + \frac{1}{16} + \cdots \tag{1}$$

converges to 2. The series

$$\pi - 3 + 17 + 1 + \frac{1}{2} + \frac{1}{4} + \frac{1}{8} + \frac{1}{16} + \cdots,$$

obtained from series (1) by inserting three additional terms at the beginning, also converges and clearly must converge to

$$(\pi - 3 + 17) + 2 = \pi + 16. \quad \square$$

THEOREM 10.3 Changing a finite number of terms

Suppose that each of two series $\sum_{n=1}^{\infty} a_n$ and $\sum_{n=1}^{\infty} b_n$ contains all but a finite number of terms of the other in the same order. That is, suppose that there exist N and k such that $b_n = a_{n+k}$ for all $n > N$. Then either both series converge or both series diverge.

The condition in the preceding theorem that the common terms in the two series be "in the same order" is very important. Later work will show that the series

$$1 - \frac{1}{2} + \frac{1}{3} - \frac{1}{4} + \frac{1}{5} - \frac{1}{6} + \frac{1}{7} - \frac{1}{8} + \frac{1}{9} - \frac{1}{10} + \frac{1}{11} - \cdots$$

converges, and it can be proved that a *divergent* series can be found that contains *exactly the same terms* but in a different order.

IF $\lim_{n\to\infty} a_n \neq 0$, THEN $\sum_{n=1}^{\infty} a_n$ DIVERGES

The following theorem is very important and is frequently misused. It shows that some series diverge, but it can *never* be used to show convergence of a series.

THEOREM 10.4 Divergence of $\sum_{n=1}^{\infty} a_n$ if $\lim_{n\to\infty} a_n \neq 0$

If the series $\sum_{n=1}^{\infty} a_n$ converges, then $\lim_{n\to\infty} a_n = 0$. Equivalently, if $\lim_{n\to\infty} a_n \neq 0$, then $\sum_{n=1}^{\infty} a_n$ diverges.

Proof. Let $\sum_{n=1}^{\infty} a_n$ converge to c, and let $\varepsilon > 0$ be given. Then there exists a positive integer N such that for all $n > N$, the nth partial sum s_n satisfies

$$|s_n - c| < \frac{\varepsilon}{2}.$$

In particular, if $n > N + 1$, we have

$$|s_{n-1} - c| < \frac{\varepsilon}{2} \qquad \text{and} \qquad |s_n - c| < \frac{\varepsilon}{2}.$$

Since s_{n-1} and s_n are both within the distance $\varepsilon/2$ of c, they must be a distance at most ε from each other. That is,

$$|s_n - s_{n-1}| < \varepsilon$$

for $n > N + 1$. But

$$s_n - s_{n-1} = (a_1 + \cdots + a_{n-1} + a_n) - (a_1 + \cdots + a_{n-1}) = a_n,$$

so $|a_n| < \varepsilon$ for $n > N + 1$. Hence $\lim_{n\to\infty} a_n = 0$. •

EXAMPLE 2 Establish the convergence or divergence of the series

$$\sum_{n=1}^{\infty} \frac{n^2}{5n^2 + 100n}.$$

Solution Theorem 10.4 tells us that the series

$$\sum_{n=1}^{\infty} \frac{n^2}{5n^2 + 100n}$$

diverges, for

$$\lim_{n\to\infty} \frac{n^2}{5n^2 + 100n} = \frac{1}{5} \neq 0. \quad \square$$

EXAMPLE 3 Establish the convergence or divergence of the series

$$\sum_{n=1}^{\infty} \frac{4 - n^{3/2}}{n + 5}.$$

Solution We have

$$\lim_{n\to\infty} \frac{4 - n^{3/2}}{n + 5} = \lim_{n\to\infty} \frac{-n^{3/2}}{n} = \lim_{n\to\infty} -\sqrt{n} = -\infty.$$

Thus the series diverges, by Theorem 10.4. □

EXAMPLE 4 Establish the convergence or divergence of the series $\sum_{n=0}^{\infty} \sin n$.

Solution Since $3\pi/4 - \pi/4 = \pi/2 > 1$, we see that the interval $[\pi/4, 3\pi/4]$ must contain at least one integer n. (Of course, it contains both 1 and 2.) A similar argument shows that, for each positive integer m, the interval $[2m\pi + (\pi/4), 2m\pi + (3\pi/4)]$ contains at least one integer n. As shown in Fig. 10.4, for any integer n in one of these intervals, we have $\sin n \geq 1/\sqrt{2}$. Thus $\lim_{n\to\infty} \sin n \neq 0$, so $\sum_{n=0}^{\infty} \sin n$ diverges, by Theorem 10.4. □

EXAMPLE 5 Find the flaw in this argument: Since $\lim_{n\to\infty} 1/n = 0$, the series $\sum_{n=1}^{\infty} 1/n$ converges, by Theorem 10.4.

Solution Theorem 10.4 makes no assertion whatever about the convergence or divergence of a series $\sum_{n=0}^{\infty} a_n$ if $\lim_{n\to\infty} a_n = 0$. The theorem only says

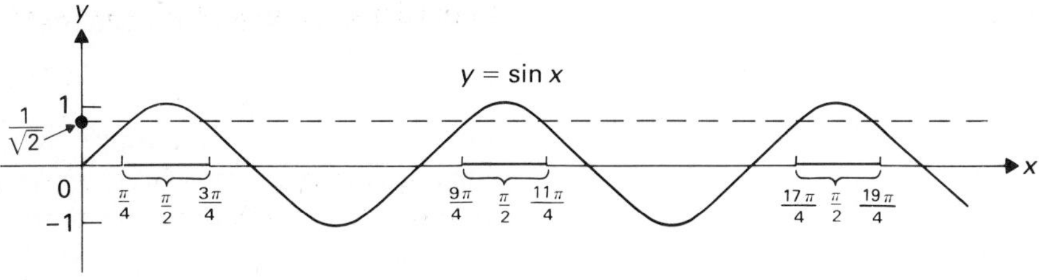

Figure 10.4 Since $\pi/2 > 1$, each colored interval contains an integer n such that $\sin n > 1/\sqrt{2}$.

that the series $\sum_{n=0}^{\infty} a_n$ diverges if $\lim_{n\to\infty} a_n \neq 0$. That is, Theorem 10.4 gives only a *necessary condition* (and not a *sufficient condition*) for a series $\sum_{n=0}^{\infty} a_n$ to converge. Thus the theorem can *never* be used to demonstrate convergence of a series. It can sometimes be used to show divergence, as in Examples 2 through 4. Of course, the series $\sum_{n=1}^{\infty} 1/n$ is the harmonic series, and we know it diverges. □

THE STANDARD COMPARISON TEST (SCT)

During the course of this chapter, we will be introducing certain "tests" for convergence or divergence of a series of constants. Most of these tests depend on the *comparison* of the series with another series that is either known to converge or known to diverge. (This is one reason why it is important to build a "stockpile" of series whose convergence or divergence is known.) Our work in the rest of this section parallels our treatment of comparison tests for improper integrals in Section 9.8.

Comparison tests for series with nonnegative terms follow easily from the following fundamental property of the real numbers. A proof of this property is given in more advanced courses where the real numbers are "constructed."

> Let $\{s_n\}$ be a sequence of numbers such that $s_{n+1} \geq s_n$ for $n = 1, 2, 3, \ldots$. (Such a sequence is **monotone increasing**.) Then either $\{s_n\}$ converges to some c or $\lim_{n\to\infty} s_n = \infty$.
>
> ***Fundamental property of the real numbers***

Our fundamental property asserts that the only way a monotone-increasing sequence $\{s_n\}$ can *fail* to converge is to diverge to ∞. Note that if $\sum_{n=1}^{\infty} a_n$ is a series of *nonnegative* terms, then the sequence $\{s_n\}$ of partial sums is monotone increasing.

THEOREM 10.5 Standard comparison test (SCT)

Let $\sum_{n=1}^{\infty} a_n$ and $\sum_{n=1}^{\infty} b_n$ be series of nonnegative terms such that $a_n \leq b_n$ for $n = 1, 2, 3, \ldots$. If $\sum_{n=1}^{\infty} b_n$ converges, then $\sum_{n=1}^{\infty} a_n$ converges also, while if $\sum_{n=1}^{\infty} a_n$ diverges, then $\sum_{n=1}^{\infty} b_n$ diverges also.

Proof. Let s_n be the nth partial sum of $\sum_{n=1}^{\infty} a_n$ and let t_n be the nth partial sum of $\sum_{n=1}^{\infty} b_n$. Suppose $\sum_{n=1}^{\infty} b_n$ converges to c, so that $\lim_{n\to\infty} t_n = c$.

From $a_n \leq b_n$, we see at once that

$$s_n \leq t_n$$

for $n = 1, 2, 3, \ldots$. Since $\lim_{n\to\infty} t_n = c$ and $s_n \leq t_n$, we see that $\lim_{n\to\infty} s_n = \infty$ is impossible, so by the fundamental property, the sequence $\{s_n\}$ must converge also. This means that $\sum_{n=1}^{\infty} a_n$ converges.

Suppose that $\sum_{n=1}^{\infty} a_n$ diverges. Since $\{s_n\}$ is monotone increasing and diverges, we must have $\lim_{n\to\infty} s_n = \infty$ by the fundamental property. Since $t_n \geq s_n$, obviously

$$\lim_{n\to\infty} t_n = \infty,$$

so $\sum_{n=1}^{\infty} b_n$ diverges also. •

Theorem 10.5 is sometimes summarized by saying that, for series of nonnegative terms, a series "smaller" than a known convergent series also converges, while a series "larger" than a known divergent series must diverge also. If a series is "smaller" than a known divergent series, it may either converge or diverge, depending on how much "smaller" it is. Similarly, a series that is "larger" than a known convergent series may converge or diverge, depending on how much "larger" it is. It is important for us to remember that the comparison test works only in the stated direction.

> We can never establish convergence using a divergent test series or establish divergence using a convergent test series.

EXAMPLE 6 Establish the convergence or divergence of the series

$$\sum_{n=1}^{\infty} \frac{1}{2^{n-1} + 1}.$$

Solution We know that the series

$$1 + \frac{1}{2} + \frac{1}{4} + \frac{1}{8} + \cdots + \frac{1}{2^{n-1}} + \cdots$$

converges. Therefore the "smaller" series

$$\frac{1}{2} + \frac{1}{3} + \frac{1}{5} + \frac{1}{9} + \cdots + \frac{1}{2^{n-1} + 1} + \cdots$$

converges, by the SCT. □

EXAMPLE 7 Establish the convergence or divergence of the series $\sum_{n=1}^{\infty} 1/\sqrt{n}$.

Solution We know that the harmonic series

$$1 + \frac{1}{2} + \frac{1}{3} + \frac{1}{4} + \cdots + \frac{1}{n} + \cdots$$

diverges. Since $\sqrt{n} \leq n$, we have $(1/n) \leq (1/\sqrt{n})$. Therefore the "larger" series

$$1 + \frac{1}{\sqrt{2}} + \frac{1}{\sqrt{3}} + \frac{1}{\sqrt{4}} + \cdots + \frac{1}{\sqrt{n}} + \cdots$$

diverges, by the SCT. □

EXAMPLE 8 Establish the convergence or divergence of the series

$$\sum_{n=0}^{\infty} \frac{\sin^2 n}{2^n}.$$

Solution Since $\sin^2 n \leq 1$, we see that

$$\frac{\sin^2 n}{2^n} \leq \frac{1}{2^n}.$$

Now $\sum_{n=0}^{\infty} 1/2^n$ converges as a geometric series with ratio $\frac{1}{2}$. Thus $\sum_{n=0}^{\infty} (\sin^2 n)/2^n$ converges, by the SCT. □

Since a finite number of terms can be inserted in a series or deleted from it without affecting its convergence or divergence (Theorem 10.3), we can weaken the hypotheses of the SCT and require only that $a_n \leq b_n$ for all but a finite number of positive integers n.

THE LIMIT COMPARISON TEST (LCT)

The series

$$\sum_{n=1}^{\infty} \frac{n^2 + 3n}{2n^3 - n^2}$$

diverges, because

$$\frac{n^2 + 3n}{2n^3 - n^2} > \frac{n^2}{2n^3} = \frac{1}{2n},$$

and $\sum_{n=1}^{\infty} 1/(2n)$ diverges since it is "half" the harmonic series. Rather than fuss about such a precise inequality for the SCT, the mathematician usually says that the given series diverges since $(n^2 + 3n)/(2n^3 - n^2)$ is of order of magnitude $n^2/(2n^3) = 1/(2n)$ for large n, and $\sum_{n=1}^{\infty} 1/(2n)$ diverges. The justification for this argument is given in the following theorem.

THEOREM 10.6 Limit comparison test (LCT)

Let $\sum_{n=1}^{\infty} a_n$ and $\sum_{n=1}^{\infty} b_n$ be series of nonnegative terms, with $a_n \neq 0$ for all sufficiently large n, and suppose that $\lim_{n\to\infty} (b_n/a_n) = c > 0$. Then the two series either both converge or both diverge.

Proof. Since $\lim_{n\to\infty} (b_n/a_n) = c > 0$, there exists N such that for $n > N$, we have

$$\frac{c}{2} < \frac{b_n}{a_n} < \frac{3c}{2},$$

or

$$\frac{c}{2}a_n < b_n < \frac{3c}{2}a_n. \qquad \textbf{(2)}$$

If $\sum_{n=1}^{\infty} a_n$ converges, then $\sum_{n=1}^{\infty} (3c/2)a_n$ converges by Corollary 1 of Theorem 10.2 (page 432). Since inequality (2) provides a "comparison test" for all but a finite number of terms b_n, we see from the SCT and the remarks following Example 8 that $\sum_{n=1}^{\infty} b_n$ converges also. Similarly, if $\sum_{n=1}^{\infty} a_n$ diverges, then $\sum_{n=1}^{\infty} (c/2)a_n$ diverges, so by inequality (2) and the SCT, we see that $\sum_{n=1}^{\infty} b_n$ diverges. •

We saw in Section 2.3 that the limit at ∞ of a quotient of polynomial functions in x is equal to the limit of the quotient of the dominating monomial terms of highest degree in the numerator and denominator. Of course, the same is true for the limit at ∞ of a quotient of polynomials in n.

EXAMPLE 9 Determine convergence or divergence of the series

$$\sum_{n=1}^{\infty} \frac{2n^3 - 3n^2}{7n^4 + 100n^3 + 7}.$$

Solution For large n, the nth term of the series is of order of magnitude $2n^3/(7n^4)$, or $2/(7n)$. More precisely,

$$\lim_{n\to\infty} \frac{(2n^3 - 3n^2)/(7n^4 + 100n^3 + 7)}{2/(7n)} = \lim_{n\to\infty} \frac{14n^4 - 21n^3}{14n^4 + 200n^3 + 14} = 1.$$

Thus by the LCT, our given series diverges, since

$$\sum_{n=1}^{\infty} \frac{2}{7n}$$

is $\frac{2}{7}$ times the harmonic series and therefore diverges. □

We give two more examples of the way in which the LCT is used to determine convergence or divergence.

EXAMPLE 10 Establish the convergence or divergence of the series

$$\sum_{n=1}^{\infty} \frac{2n^2 + 4}{(n^2 + n)2^n}.$$

Solution Since

$$\lim_{n\to\infty} \frac{2n^2 + 4}{n^2 + n} = \lim_{n\to\infty} \frac{2n^2}{n^2} = 2 > 0,$$

the LCT shows that our series behaves like $\sum_{n=1}^{\infty} 1/2^n$, which converges since it is a geometric series with ratio $\frac{1}{2}$. Thus both series converge. □

EXAMPLE 11 Establish the convergence or divergence of

$$\sum_{n=1}^{\infty} \frac{(4n^3 + 5)\sin(1/n)}{n^2\, 3^n}.$$

Solution We know that

$$\lim_{n\to\infty}\left[n \sin\left(\frac{1}{n}\right)\right] = \lim_{n\to\infty}\frac{\sin(1/n)}{1/n} = 1.$$

Thus

$$\lim_{n\to\infty}\frac{(4n^3+5)\sin(1/n)}{n^2} = \lim_{n\to\infty}\left[\frac{4n^3+5}{n^3}\, n \sin\left(\frac{1}{n}\right)\right]$$
$$= 4\cdot 1 = 4 > 0.$$

By the LCT, this shows that the given series behaves like $\sum_{n=1}^{\infty} 1/3^n$, which converges as a geometric series with ratio $\frac{1}{3}$. □

SUMMARY

1. The convergence or divergence of a series (or sequence) is not changed by inserting, deleting, or altering any *finite* number of terms.
2. If $\lim_{n\to\infty} a_n \neq 0$, then $\sum_{n=1}^{\infty} a_n$ diverges.
3. *Standard comparison test (SCT):* If $0 \le a_n \le b_n$ for all sufficiently large n, and $\sum_{n=1}^{\infty} b_n$ converges, then $\sum_{n=1}^{\infty} a_n$ converges. On the other hand, if $\sum_{n=1}^{\infty} a_n$ diverges, then $\sum_{n=1}^{\infty} b_n$ diverges.
4. *Limit comparison test (LCT):* If $a_n > 0$ for sufficiently large n and $b_n \ge 0$ for all n, and $\lim_{n\to\infty}(b_n/a_n) = c > 0$, then the series $\sum_{n=1}^{\infty} a_n$ and $\sum_{n=1}^{\infty} b_n$ either both converge or both diverge.

EXERCISES

In Exercises 1 through 4, determine whether the given series converges or diverges; if it is convergent, find its sum.

1. $1 - 2 + \frac{1}{2} + \sqrt{5} + \frac{1}{4} + \frac{1}{8} + \frac{1}{16} + \cdots + \frac{1}{2^{n-3}} + \cdots,$

where $n \ge 5$

2. $1 + \frac{1}{2} + \frac{1}{4} + \frac{1}{8} + \frac{1}{16} + \frac{1}{32} + \frac{2}{32} + \frac{3}{32} + \cdots + \frac{n-5}{32} + \cdots,$

where $n \ge 6$

3. $\sqrt{2} + 1 - \frac{1}{3} - \sqrt{3} + \frac{1}{9} - \frac{1}{27} + \frac{1}{81} - \cdots$

$+ (-1)^{n+1}\frac{1}{3^{n-3}} + \cdots,$ where $n \ge 5$

4. $2 - 5 + 7 - \frac{1}{2} + 1 - \frac{1}{4} + \frac{1}{3} - \frac{1}{8} + \frac{1}{9} - \frac{1}{16}$

$+ \frac{1}{27} - \frac{1}{32} + \frac{1}{81} - \cdots$

$(1/3^{(n-5)/2}$ for odd $n \ge 5$; $-1/2^{(n/2-1)}$ for even $n \ge 4)$

5. Mark each of the following true or false.

_____ a) The SCT and LCT as stated in the text hold only for series of nonnegative terms.

_____ b) The test "if $\lim_{n\to\infty} a_n \neq 0$, then $\sum_{n=1}^{\infty} a_n$ diverges" holds only for series of nonnegative terms.

_____ c) If $\lim_{n\to\infty} a_n = 0$, then $\sum_{n=1}^{\infty} a_n$ converges.

_____ d) If $\sum_{n=1}^{\infty} a_n$ converges, then $\lim_{n\to\infty} a_n = 0$.

_____ e) If $a_n \le b_n$, and $\sum_{n=1}^{\infty} b_n$ converges, then $\sum_{n=1}^{\infty} a_n$ converges.

_____ f) If $0 \le a_n \le b_n$, and $\sum_{n=1}^{\infty} b_n$ converges, then $\sum_{n=1}^{\infty} a_n$ converges.

_____ g) If $0 \le a_n \le b_n$, and $\sum_{n=1}^{\infty} a_n$ diverges, then $\sum_{n=1}^{\infty} b_n$ diverges.

_____ h) If $a_n \le b_n \le 0$, and $\sum_{n=1}^{\infty} b_n$ converges, then $\sum_{n=1}^{\infty} a_n$ converges.

_____ i) If $a_n \le b_n \le 0$, and $\sum_{n=1}^{\infty} b_n$ diverges, then $\sum_{n=1}^{\infty} a_n$ diverges.

_____ j) If $a_n \le b_n \le 0$, and $\sum_{n=1}^{\infty} a_n$ converges, then $\sum_{n=1}^{\infty} b_n$ converges.

In Exercises 6 through 42, classify the series as convergent or divergent, and indicate a reason for your answer. (You should do all these exercises and try to develop facility in ascertaining the convergence or divergence of the series at a glance.)

6. $\sum_{n=1}^{\infty} \frac{1}{10n}$

7. $\sum_{n=1}^{\infty} \frac{\sqrt{n}}{n \cdot 2^n}$

8. $\sum_{n=1}^{\infty} (\ln n)$

9. $\sum_{n=2}^{\infty} \frac{\cos^2 n}{2^n}$

10. $\sum_{n=3}^{\infty} \frac{1}{n-1}$

11. $\sum_{n=1}^{\infty} \frac{1}{n+2^n}$

12. $\sum_{n=1}^{\infty} \frac{1}{\sqrt[3]{n}}$

13. $\sum_{n=1}^{\infty} \frac{n^3}{3n^3+n^2}$

14. $\sum_{n=1}^{\infty} \left(\frac{1}{n} - \frac{1}{2^n}\right)$

15. $\sum_{n=1}^{\infty} \frac{|\sin n|}{5^n}$

16. $\sum_{n=1}^{\infty} \frac{\sqrt{n}}{n+17}$

17. $\sum_{n=1}^{\infty} \frac{8^n+9^n}{10^n}$

18. $\sum_{n=1}^{\infty} \frac{n^2-3n}{4n^3+n^2}$

19. $\sum_{n=1}^{\infty} \frac{n+2}{n^2+3}$

20. $\sum_{n=1}^{\infty} \frac{1}{\sqrt[n]{100}}$

21. $\sum_{n=1}^{\infty} \frac{n^2 e^{-2n}}{3n^2-2n}$

22. $\sum_{n=1}^{\infty} \frac{e^n}{n^3+4n}$

23. $\sum_{n=1}^{\infty} \frac{1}{\sin^2 n}$

24. $\sum_{n=1}^{\infty} \frac{2^n}{3^{2n+1}}$

25. $\sum_{n=1}^{\infty} \sqrt[n]{n}$

26. $\sum_{n=1}^{\infty} \frac{1}{\sqrt[n]{n}}$

27. $\sum_{n=1}^{\infty} \frac{n^2-1}{10n^2(\ln 2)^n}$

28. $\sum_{n=1}^{\infty} \frac{\sqrt{n+1000}}{\sqrt{n}(\ln 3)^n}$

29. $\sum_{n=1}^{\infty} \frac{(n^2+1)\cos^2 n}{4n^2+3n+2}$

30. $\sum_{n=2}^{\infty} \left(\frac{\ln n}{\ln n^2}\right)^n$

31. $\sum_{n=1}^{\infty} n \sin\left(\frac{1}{n^2}\right)$

32. $\sum_{n=1}^{\infty} \frac{n^{3/2}-1}{1000n^2+300n+50}$

33. $\sum_{n=1}^{\infty} \left(\frac{n-1}{2n+3}\right)^n$

34. $\sum_{n=1}^{\infty} \frac{100^n}{n^n}$

35. $\sum_{n=1}^{\infty} \left(\frac{n+30}{2n-1}\right)^n$

36. $\sum_{n=1}^{\infty} \left[n^2 \sin\left(\frac{2n}{n^3+n^2}\right)\right]^n$

37. $\sum_{n=1}^{\infty} \frac{n(\sin^2 n)}{50n+100}$

38. $\sum_{n=1}^{\infty} \frac{\sqrt{n^5+5n-1}}{4n^3+7}$

39. $\sum_{n=1}^{\infty} (n^2+3) \sin\left(\frac{1}{n^2}\right)$

40. $\sum_{n=1}^{\infty} \left(\frac{2n^3+n^2+1}{2n^3+100n}\right)^n$

41. $\sum_{n=1}^{\infty} \frac{100^{2n+3}}{n^{n/2}}$

42. $\sum_{n=1}^{\infty} \frac{n^2 \cdot 2^{2n-10}}{(n^2+100n)3^n}$

10.4 THE INTEGRAL AND RATIO TESTS

The tests given in this section depend on the SCT, which can be viewed as the most general test we consider.

THE INTEGRAL TEST

We discussed convergence of improper integrals of the form $\int_1^\infty f(x)\,dx$ in Sections 9.7 and 9.8. If the nth term a_n of a series $\sum_{n=1}^\infty a_n$ is given by a formula in terms of n, then we may replace n by x in that formula and obtain a function $f(x)$. For example, the series $\sum_{n=1}^\infty (1/n^2)$ gives rise to the function $f(x) = 1/x^2$ in this way. For many important series, there is a close relationship between the behavior of $\sum_{n=1}^\infty a_n$ and the behavior of $\int_1^\infty f(x)\,dx$. This relationship is explained in the following theorem.

THEOREM 10.7 The integral test

Let $\sum_{n=1}^{\infty} a_n$ be a series of nonnegative terms. Suppose also that f is a continuous function for $x \geq 1$ such that

1. $f(n) = a_n$ for $n = 1, 2, 3, \ldots$
2. f is monotone decreasing for $x \geq 1$; that is, $f(x_1) \geq f(x_2)$ if $x_1 \leq x_2$ for $x_1 \geq 1$ and $x_2 \geq 1$.

Then $\sum_{n=1}^{\infty} a_n$ converges if and only if $\int_1^{\infty} f(x)\,dx$ converges.

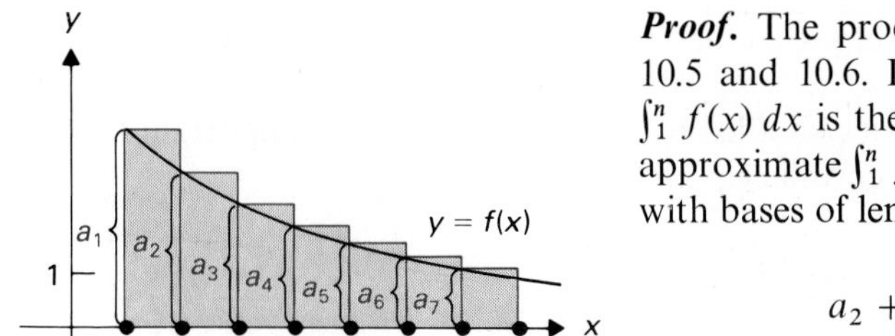

Figure 10.5
Upper sum approximation:
$\int_1^8 f(x)\,dx \leq a_1 + a_2 + \cdots + a_7$.

Proof. The proof follows easily from our previous work and from Figs. 10.5 and 10.6. Let s_n be the nth partial sum of the series $\sum_{n=1}^{\infty} a_n$. Now $\int_1^n f(x)\,dx$ is the area of the region under the graph of f over $[1, n]$. If we approximate $\int_1^n f(x)\,dx$ by the upper and lower sums given by the rectangles with bases of length 1 in Figs. 10.5 and 10.6, we obtain

$$a_2 + \cdots + a_n \leq \int_1^n f(x)\,dx \leq a_1 + a_2 + \cdots + a_{n-1} \tag{1}$$

Note that the relation (1) depends on the fact that f is monotone decreasing for $x \geq 1$. As illustrated in Fig. 10.7, the relation (1) need not hold if $f(x)$ is not monotone decreasing. From the relation (1), we obtain

$$s_n - a_1 \leq \int_1^n f(x)\,dx \leq s_{n-1}. \tag{2}$$

Since $f(x) \geq 0$ for $x \geq 1$, the sequence $\{\int_1^n f(x)\,dx\}$ is clearly monotone increasing. Therefore, if $\int_1^{\infty} f(x)\,dx$ converges, then by the fundamental property on page 439, the sequence $\{\int_1^n f(x)\,dx\}$ must converge. From the relation (2), using a comparison test, we find that the sequence $\{s_n - a_1\}$ converges. But then the sequence $\{s_n\}$ converges, by Theorem 10.2 (page 432), so the series $\sum_{n=1}^{\infty} a_n$ converges.

On the other hand, if $\{s_n\}$ converges, then the sequence $\{\int_1^n f(x)\,dx\}$ must converge by the relation (2), using the SCT. Since $F(t) = \int_1^t f(x)\,dx$ is a monotone-increasing function of t, it is clear that $\int_1^{\infty} f(x)\,dx$ must converge also. •

Figure 10.6
Lower sum approximation:
$a_2 + a_3 + \cdots + a_8 \leq \int_1^8 f(x)\,dx$.

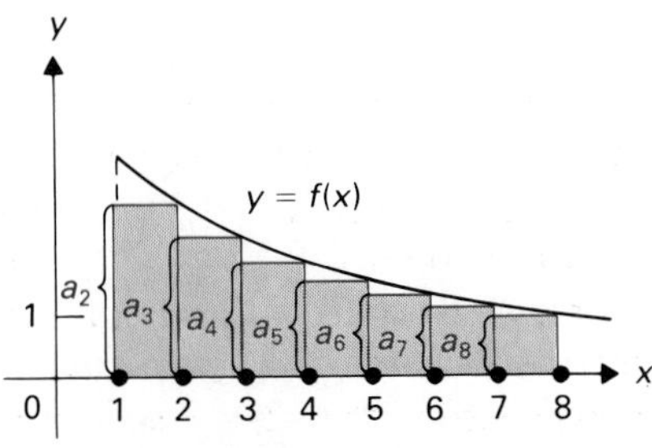

Figure 10.7
$\int_1^n f(x)\,dx \leq a_1 + a_2 + \cdots + a_{n-1}$
need not hold if $f(x)$ is not monotone decreasing.

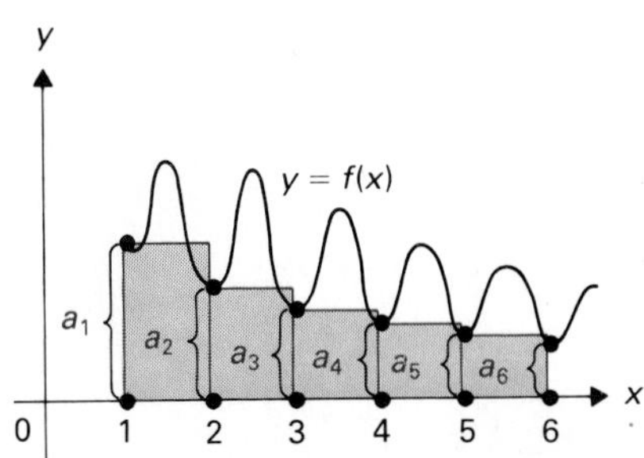

We regard the integral test as asserting that for $\sum_{n=1}^{\infty} a_n$ and f as described, the series "behaves like" $\int_1^{\infty} f(x)\,dx$, as far as convergence or divergence is concerned.

EXAMPLE 1 Show the divergence of the harmonic series $\sum_{n=1}^{\infty} (1/n)$ using the integral test.

Solution If $f(x) = 1/x$, then the conditions of the integral test are satisfied. Now

$$\lim_{h\to\infty} \int_1^h \frac{1}{x}\,dx = \lim_{h\to\infty} (\ln x)\Big]_1^h = \lim_{h\to\infty} (\ln h - \ln 1) = \lim_{h\to\infty} (\ln h) = \infty.$$

Thus the integral diverges, so the series diverges also. □

One type of series handled by the integral test is so important that we treat it formally in a corollary.

COROLLARY *p*-series test

Let p be any real number. The series $\sum_{n=1}^{\infty} (1/n^p)$ converges if $p > 1$ and diverges for $p \le 1$.

Proof. If $p > 1$, then define f by $f(x) = 1/x^p$. Then f is a continuous function that satisfies the conditions of the integral test, and the behavior of the series $\sum_{n=1}^{\infty} (1/n^p)$ is the same as the behavior of $\int_1^{\infty} (1/x^p)\,dx$. By Theorem 9.1 (page 409), we know that $\int_1^{\infty} (1/x^p)\,dx$ converges for $p > 1$.

If $p \le 1$, then $n^p \le n$, so $1/(n^p) \ge (1/n)$, and the series diverges by comparison with the harmonic series. •

The preceding result is especially useful when combined with the LCT (Section 10.3), as illustrated now.

EXAMPLE 2 Establish the convergence or divergence of

$$\sum_{n=1}^{\infty} \frac{n + 20}{n^3 + 20n + 1}.$$

Solution By the LCT, the series behaves like $\sum_{n=1}^{\infty} 1/n^2$, which is a "$p$-series" with $p = 2$, and hence converges. □

As Example 2 indicates, the LCT and the p-series test enable us to determine at a glance whether or not $\sum_{n=1}^{\infty} a_n$ converges if a_n is a quotient of polynomial functions in n. The same technique is valid when radical expressions are involved.

EXAMPLE 3 Determine the convergence or divergence of

$$\sum_{n=1}^{\infty} \frac{(2n^5 + 4n^2 + 2)^{1/3}}{5n^3 + 2n}.$$

Solution Keeping only the monomial terms of highest degree and neglecting their coefficients, which is permissible by the LCT, we see that the series behaves like

$$\sum_{n=1}^{\infty} \frac{(n^5)^{1/3}}{n^3} = \sum_{n=1}^{\infty} \frac{n^{5/3}}{n^3} = \sum_{n=1}^{\infty} \frac{1}{n^{4/3}}.$$

This series converges by the p-series test, with $p = 4/3 > 1$. □

EXAMPLE 4 Establish the convergence or divergence of

$$\sum_{n=1}^{\infty} \sin\left(\frac{2}{n^2 + 4n}\right).$$

Solution Motivated by the fact that $\lim_{x\to 0} [(\sin x)/x] = 1$, we write

$$\sum_{n=1}^{\infty} \sin\left(\frac{2}{n^2 + 4n}\right) = \sum_{n=1}^{\infty} \frac{2 \sin\left(\dfrac{2}{n^2 + 4n}\right)}{(n^2 + 4n)\dfrac{2}{n^2 + 4n}}.$$

The LCT then shows that the series behaves the same as

$$\sum_{n=1}^{\infty} \frac{2}{n^2 + 4n},$$

or, using the LCT again, as

$$\sum_{n=1}^{\infty} \frac{1}{n^2}.$$

This series converges as a p-series with $p = 2 > 1$. □

ESTIMATING THE SUM OF A SERIES OF MONOTONE-DECREASING TERMS BY AN INTEGRAL

The proof of the integral test really gives more information than indicated by the statement of the test because the relation (2) gives a very useful estimate for the partial sum s_n of a series, provided that a function f can be found satisfying the hypotheses of the test. Namely, using the relation (2) and the fact that $s_{n-1} \le s_n$, we obtain

$$\int_1^n f(x)\,dx \le s_n \le \left(\int_1^n f(x)\,dx\right) + a_1. \tag{3}$$

If the series and integral converge, then we obtain from the relation (3) upon letting $n \to \infty$,

$$\int_1^{\infty} f(x)\,dx \le \sum_{n=1}^{\infty} a_n \le a_1 + \int_1^{\infty} f(x)\,dx. \tag{4}$$

We illustrate the use of the relations (3) and (4) by examples.

EXAMPLE 5 Use the relation (3) to estimate the partial sums s_n of the harmonic series $\sum_{n=1}^{\infty} (1/n)$.

Solution From the relation (3), with $f(x) = 1/x$, we see that for the partial sum s_n of the harmonic series $1 + \frac{1}{2} + \frac{1}{3} + \cdots$, we have

$$\int_1^n \frac{1}{x}\,dx \le s_n \le \left(\int_1^n \frac{1}{x}(x)\,dx\right) + 1.$$

Since $\int_1^n (1/x)\,dx = \ln n - \ln 1 = \ln n$, we have

$$\ln n \le s_n \le (\ln n) + 1.$$

In particular, since $\ln 10{,}000 \approx 9.21$, we obtain the estimate

$$9.21 \le \sum_{n=1}^{10{,}000} \frac{1}{n} \le 10.21$$

stated on page 434. □

EXAMPLE 6 Use the relation (4) to estimate $\sum_{n=1}^{\infty} (1/n^2)$.

Solution We have

$$\int_1^\infty \frac{1}{x^2}\,dx = \lim_{h\to\infty} -\frac{1}{x}\bigg]_1^h = \lim_{h\to\infty}\left(-\frac{1}{h} + 1\right) = 1.$$

We then obtain from the relation (4),

$$1 \le \sum_{n=1}^{\infty} \frac{1}{n^2} \le 1 + 1 = 2. \quad \square$$

We may replace our lower summation index and integral limit 1 in relation (4) by any positive integer r and obtain

$$\boxed{\int_r^\infty f(x)\,dx \le \sum_{n=r}^{\infty} a_n \le a_r + \int_r^\infty f(x)\,dx.} \qquad \textbf{(5)}$$

Suppose $a_r \le \varepsilon$. Since $\sum_{n=1}^{\infty} a_n = \sum_{n=1}^{r-1} a_n + \sum_{n=r}^{\infty} a_n$, the relation (5) shows that

$$\boxed{\sum_{n=1}^{\infty} a_n \approx \sum_{n=1}^{r-1} a_n + \int_r^\infty f(x)\,dx,} \qquad \textbf{(6)}$$

and the sum of the series exceeds this approximation by at most ε. We could use a calculator or computer to find $\sum_{n=1}^{r-1} a_n$ if r is not too large.

EXAMPLE 7 Continuing Example 6, find an estimate for $\sum_{n=1}^{\infty} (1/n^2)$ with error ≤ 0.001.

Solution To have $a_r = 1/r^2 \le 0.001$, we must have $r^2 \ge 1000$, or $r \ge \sqrt{1000} \approx 31.6$. Thus we may take $r = 32$. Our calculator shows that $\sum_{n=1}^{31} (1/n^2) \approx 1.61319$. Now

$$\int_{32}^\infty \frac{1}{x^2}\,dx = \lim_{h\to\infty} -\frac{1}{x}\bigg]_{32}^h = \lim_{h\to\infty}\left(-\frac{1}{h} + \frac{1}{32}\right) = \frac{1}{32}.$$

Then the relation (6) shows that

$$\sum_{n=1}^{\infty} \frac{1}{n^2} \approx 1.61319 + \frac{1}{32} \approx 1.64444,$$

and the error is at most 0.001. □

THE RATIO TEST

Recall that if there is any hope that $\sum_{n=1}^{\infty} a_n$ converges, we must have $\lim_{n\to\infty} a_n = 0$. The ratio test studies the rate at which terms a_n decrease as n increases and is based on a comparison of the series $\sum_{n=1}^{\infty} a_n$ with a suitable geometric series. The (geometric) decrease from the term a_n to the next term a_{n+1} in $\sum_{n=1}^{\infty} a_n$ can be measured by the *ratio* a_{n+1}/a_n.

THEOREM 10.8 Ratio test

Let $\sum_{n=1}^{\infty} a_n$ be a series of positive terms and suppose that $\lim_{n\to\infty} (a_{n+1}/a_n)$ exists and is r. Then $\sum_{n=1}^{\infty} a_n$ converges if $r < 1$ and diverges if $r > 1$. (If $r = 1$, further information is necessary to determine the convergence or divergence of the series.)

Proof. Suppose $\lim_{n\to\infty} (a_{n+1}/a_n) = r < 1$. Find $\delta > 0$ such that

$$r < r + \delta < 1.$$

For example, we could let $\delta = (1 - r)/2$. Since $\lim_{n\to\infty} (a_{n+1}/a_n) = r$, there exists an integer N such that

$$r - \delta < \frac{a_{n+1}}{a_n} < r + \delta$$

for $n > N$. Multiplying by a_n, we find that

$$a_{n+1} < (r + \delta)a_n \tag{7}$$

for $n > N$. Using relation (7) repeatedly, starting with $n = N + 1$, we obtain

$$a_{N+2} < (r + \delta)a_{N+1},$$

$$a_{N+3} < (r + \delta)a_{N+2} < (r + \delta)^2 a_{N+1}$$

$$a_{N+4} < (r + \delta)a_{N+3} < (r + \delta)^3 a_{N+1}, \ldots .$$

Thus each term of the series

$$a_{N+1} + a_{N+2} + a_{N+3} + \cdots + a_{N+k} + \cdots \tag{8}$$

is less than or equal to the corresponding term of the series

$$a_{N+1} + a_{N+1}(r+\delta) + a_{N+1}(r+\delta)^2 + \cdots + a_{N+1}(r+\delta)^{k-1} + \cdots. \tag{9}$$

But series (9) is a geometric series with ratio $r + \delta < 1$ and converges. By the SCT, we then see that the series (8) converges. Since the series (8) is $\sum_{n=1}^{\infty} a_n$ with a finite number of terms deleted, Theorem 10.3 (page 437) shows that $\sum_{n=1}^{\infty} a_n$ converges also.

Now suppose that $r > 1$. Then for all sufficiently large n,

$$\frac{a_{n+1}}{a_n} > 1;$$

that is,

$$a_{n+1} > a_n$$

for n sufficiently large. This means that after a while, the terms of $\sum_{n=1}^{\infty} a_n$ *increase*. Thus $\lim_{n\to\infty} a_n = 0$ is impossible, so by Theorem 10.4 (page 437), the series $\sum_{n=1}^{\infty} a_n$ must diverge. •

We should comment on the observation in parentheses in the statement of the ratio test. For the series $\sum_{n=1}^{\infty} (1/n^2)$, we have

$$r = \lim_{n\to\infty} \frac{a_{n+1}}{a_n} = \lim_{n\to\infty} \frac{1/(n+1)^2}{1/n^2} = \lim_{n\to\infty} \left(\frac{n}{n+1}\right)^2 = 1,$$

and for the series $\sum_{n=1}^{\infty} (1/n)$, we also have

$$r = \lim_{n\to\infty} \frac{a_{n+1}}{a_n} = \lim_{n\to\infty} \frac{1/(n+1)}{1/n} = \lim_{n\to\infty} \frac{n}{n+1} = 1.$$

In both cases, the *limiting ratio* r exists and is 1, but by the p-series test, the series $\sum_{n=1}^{\infty} (1/n^2)$ converges, while the harmonic series $\sum_{n=1}^{\infty} (1/n)$ diverges. This illustrates that for a series having a limiting ratio $r = 1$, more information must be obtained before the convergence or divergence of the series can be determined.

The ratio test is especially useful in handling series whose nth term a_n is given by a formula involving a constant to the nth power (such as 3^n), or involving a factor $n!$ where

$$n! = (n)(n-1)\cdots(3)(2)(1).$$

It is worth mentioning that for any constant $c > 1$, the exponential c^n increases much faster as $n \to \infty$ than a polynomial in n of any degree s. This is true in the strong sense that

$$\lim_{n\to\infty} \frac{c^n}{n^s} = \infty \qquad \text{and} \qquad \lim_{n\to\infty} \frac{n^s}{c^n} = 0$$

for any s. We can easily convince ourselves that this is true by taking a logarithm and showing that $\lim_{n\to\infty} \ln(c^n/n^s) = \infty$. It is also worth mentioning that $n!$ increases much faster than c^n, once again in the strong sense that

$$\lim_{n\to\infty} \frac{n!}{c^n} = \infty \qquad \text{and} \qquad \lim_{n\to\infty} \frac{c^n}{n!} = 0.$$

Write

$$\frac{n!}{c^n} = \frac{n}{c}\cdot\frac{n-1}{c}\cdot\frac{n-2}{c}\cdots\frac{r}{c}\cdots\frac{3}{c}\cdot\frac{2}{c}\cdot\frac{1}{c},$$

where r is the largest integer such that $r < 2c$. Then

$$\frac{n!}{c^n} \geq 2\cdot 2\cdot 2\cdots 2\cdot\frac{r}{c}\cdot\frac{r-1}{c}\cdots\frac{3}{c}\cdot\frac{2}{c}\cdot\frac{1}{c} = 2^{n-r}\left(\frac{r}{c}\cdots\frac{3}{c}\cdot\frac{2}{c}\cdot\frac{1}{c}\right).$$

Consequently, $\lim_{n\to\infty}(n!/c^n) = \infty$.

Summarizing, for large n, we expect $n!$ to "dominate" c^n for any c, and c^n for $c > 1$ in turn "dominates" any polynomial in n for large n.

We now give some applications of the ratio test that illustrate these ideas.

EXAMPLE 8 Find the limiting ratio r of the series

$$\sum_{n=1}^{\infty} \frac{2^n}{n!}.$$

Solution We have

$$r = \lim_{n\to\infty} \frac{a_{n+1}}{a_n} = \lim_{n\to\infty} \frac{[2^{n+1}/(n+1)!]}{2^n/n!} = \lim_{n\to\infty} \left(\frac{2^{n+1}}{(n+1)!} \cdot \frac{n!}{2^n}\right)$$

$$= \lim_{n\to\infty} \frac{2 \cdot n!}{(n+1)n!} = \lim_{n\to\infty} \frac{2}{n+1} = 0.$$

Since r exists and is $0 < 1$, we see that the series converges.

Note our use of the relation $(n+1)! = (n+1)n!$. This relation is worth remembering.

Example 8 illustrates that $n!$ increases with n so much faster than 2^n that $\sum_{n=1}^{\infty}(2^n/n!)$ converges. You should try to develop an intuitive feeling for such behavior of a series. Note that the numerator 2^n of the nth term of the series contributed the 2 in the numerator of the ratio. The $n!$ in the denominator contributed the $n + 1$ in the denominator of the ratio, before the limit was taken. □

EXAMPLE 9 Determine convergence or divergence of the series

$$\sum_{n=1}^{\infty} \frac{n^{25}}{3^n}.$$

Solution We know that 3^n increases more rapidly than n^{25} for large n, and we wonder whether 3^n increases rapidly enough to completely dominate n^{25} and make our series behave like a geometric series with ratio $\frac{1}{3}$.

We find that

$$r = \lim_{n\to\infty} \frac{a_{n+1}}{a_n} = \lim_{n\to\infty} \frac{(n+1)^{25}/3^{n+1}}{n^{25}/3^n} = \lim_{n\to\infty} \frac{(n+1)^{25}}{n^{25}} \cdot \frac{3^n}{3^{n+1}}$$

$$= \lim_{n\to\infty} \left(\frac{n+1}{n}\right)^{25} \frac{1}{3} = 1^{25} \cdot \frac{1}{3} = \frac{1}{3}.$$

Thus r exists and is $\frac{1}{3} < 1$, so our series does indeed converge. □

Example 9 shows that if n is large enough, 3^n dominates n^{25} to such an extent that $\sum_{n=1}^{\infty}(n^{25}/3^n)$ converges and, indeed, "behaves" like a geometric series with ratio $\frac{1}{3}$. The 3^n in the denominator of the series contributed the 3 in the denominator of the ratio.

Example 9 also illustrates that a "polynomial part" of a formula giving the nth term a_n of a series just contributes a factor 1 to the limiting ratio and hence is never significant in a ratio test. (Note the contribution $1^{25} = 1$ from n^{25} in our computation.) To see this in more detail, let $p(n)$ be a polynomial function of n. Then $p(n + 1)$ and $p(n)$ are polynomials of the same degree and even have the same coefficients of the terms of highest degree. As shown in Section 2.3, we then must have

$$\lim_{n\to\infty} \frac{p(n + 1)}{p(n)} = 1.$$

EXAMPLE 10 Establish the convergence or divergence of the series

$$\sum_{n=1}^{\infty} \frac{n!}{n^{100} \cdot 5^n}.$$

Solution Example 8 indicates that the term $n!$ will contribute to the ratio a factor $n + 1$ after cancellation, in the same (numerator or denominator) position. The same example indicates that 5^n will contribute a factor 5 in the same (numerator or denominator) position in the ratio. We just showed that the monomial term n^{100} will contribute a factor 1 after taking the limit in the ratio test. Thus the limiting ratio for this series will be

$$\lim_{n\to\infty} \frac{n + 1}{5} = \infty.$$

Therefore the series diverges. The $n!$ dominates both n^{100} and 5^n. □

EXAMPLE 11 Determine the convergence or divergence of the series

$$\sum_{n=2}^{\infty} \frac{100^n(n^2 + 3)}{n!}.$$

Solution An analysis like that in Example 10 shows that the limiting ratio will be

$$\lim_{n\to\infty} \frac{100}{n + 1} = 0 < 1.$$

Thus the series converges. The $n!$ dominates both 100^n and $n^2 + 3$. □

EXAMPLE 12 Establish the convergence or divergence of the series

$$\sum_{n=1}^{\infty} \frac{n^5 \cdot 2^{n+1}}{3^n}.$$

Solution We may rewrite our series as

$$2 \sum_{n=1}^{\infty} \frac{n^5 \cdot 2^n}{3^n}.$$

Our arguments above show that the limiting ratio will be $\frac{2}{3} < 1$, because the n^5 contributes a factor 1, the 2^n a factor 2 in the numerator, and the 3^n a factor 3 in the denominator. Thus the series converges. The geometric terms 2^n and 3^n dominate n^5. □

EXAMPLE 13 Establish the convergence or divergence of the series

$$\sum_{n=1}^{\infty} \frac{2^{3n-1}}{3^{2n+4}}.$$

Solution We may write the series as

$$\left(\frac{1}{2}\cdot\frac{1}{3^4}\right)\sum_{n=1}^{\infty}\frac{2^{3n}}{3^{2n}} = \frac{1}{2\cdot 3^4}\sum_{n=1}^{\infty}\frac{8^n}{9^n}.$$

We see at once that the series converges as a geometric series with ratio $\frac{8}{9}$. □

EXAMPLE 14 Establish the convergence or divergence of the series

$$\sum_{n=1}^{\infty} \frac{(2n)!}{100^n \cdot n!}.$$

Solution Since we have not worked out a ratio test involving a factor $(2n)!$, we will have to write out that portion of the ratio to see what this factor contributes. Since 100^n contributes 100 in the denominator of the ratio and $n!$ contributes $n + 1$ in the denominator, we have

$$\lim_{n\to\infty}\frac{a_{n+1}}{a_n} = \lim_{n\to\infty}\left[\frac{1}{(100)(n+1)}\cdot\frac{(2n+2)!}{(2n)!}\right].$$

Now $(2n + 2)! = (2n + 2)(2n + 1)(2n)!$, so we obtain upon cancellation

$$\lim_{n\to\infty}\frac{a_{n+1}}{a_n} = \lim_{n\to\infty}\frac{(2n+2)(2n+1)}{100(n+1)} = \infty.$$

Thus the series diverges. □

SUMMARY

1. *Integral test:* Let $\sum_{n=1}^{\infty} a_n$ be a series of nonnegative terms. Suppose also that $f(x)$ is a continuous function for $x \geq 1$ such that

   a) $f(n) = a_n$ for $n = 1, 2, 3, \ldots$

   b) f is monotone decreasing for $x \geq 1$; that is, $f(x_1) \geq f(x_2)$ if $x_1 \leq x_2$ for $x_1 \geq 1, x_2 \geq 1$.

   Then $\sum_{n=1}^{\infty} a_n$ converges if and only if $\int_1^{\infty} f(x)\,dx$ converges.

2. *p-series test:* Let p be a real number. The series $\sum_{n=1}^{\infty} (1/n^p)$ converges if $p > 1$ and diverges for $p \leq 1$.

3. If $\sum_{n=1}^{\infty} a_n$ and $f(x)$ satisfy the conditions of the integral test and the series converges, then its sum may be estimated by

$$\sum_{n=1}^{\infty} a_n \approx \sum_{n=1}^{r-1} a_n + \int_r^{\infty} f(x)\,dx.$$

   The actual value of $\sum_{n=1}^{\infty} a_n$ exceeds this estimate by at most a_r.

4. *Ratio test:* Let $\sum_{n=1}^{\infty} a_n$ be a series of positive terms and suppose that $\lim_{n\to\infty} (a_{n+1}/a_n)$ exists and is r. Then $\sum_{n=1}^{\infty} a_n$ converges if $r < 1$ and

diverges if $r > 1$. (If $r = 1$, further information is necessary to determine the convergence or divergence of the series.)

5. Suppose a_n in $\sum_{n=1}^{\infty} a_n$ is given by a formula involving n, containing factors in either the numerator or denominator of the form b^n, $n!$, or $p(n)$, where $p(n)$ is a polynomial function of n. The individual contributions of these factors in the ratio test are as follows.

b^n: Contributes a factor b in the same (numerator or denominator) position.

$n!$: Contributes a factor $n + 1$ in the same position, before $\lim_{n\to\infty}$ is computed.

$p(n)$: Contributes a factor 1 after $\lim_{n\to\infty}$ is computed and thus may be neglected if there are other factors.

Using these ideas, we see at a glance that

$$\sum_{n=1}^{\infty} \frac{2^n \cdot n^4}{n!} \quad \text{leads to} \quad r = \lim_{n\to\infty} \frac{2}{n+1} = 0$$

in the ratio test, while

$$\sum_{n=1}^{\infty} \frac{n!(n^7 + n^3)}{(n+2)! \cdot 5^n} \quad \text{leads to} \quad r = \lim_{n\to\infty} \frac{n+1}{(n+3)\cdot 5} = \frac{1}{5}.$$

EXERCISES

While you should be able to ascertain the convergence or divergence of each series in Exercises 1 through 6 at a glance, use the integral test, for practice, to discover whether or not the series converges.

1. $\sum_{n=1}^{\infty} \frac{1}{3n}$

2. $\sum_{n=1}^{\infty} \frac{1}{n^2 + 1}$

3. $\sum_{n=1}^{\infty} \frac{1}{4n - 1}$

4. $\sum_{n=1}^{\infty} \frac{n+1}{n^2 + 2n - 2}$

5. $\sum_{n=1}^{\infty} \frac{n}{(n+1)^3}$

6. $\sum_{n=1}^{\infty} \frac{1}{e^n}$

7. Note that $1/n^2 \le 1/[n(\ln n)] \le 1/n$ for $n \ge 3$, and $\sum_{n=2}^{\infty} (1/n^2)$ converges while $\sum_{n=2}^{\infty} (1/n)$ diverges. Use the integral test to show that $\sum_{n=2}^{\infty} 1/[n(\ln n)]$ diverges, and try to file this result in your memory with other series that you know diverge.

8. Use the integral test to show that the series $\sum_{n=2}^{\infty} 1/[n(\ln n)^2]$ converges.

While you should be able to ascertain the convergence or divergence of each series in Exercises 9 through 15 at a glance, write out the ratio test, as in Examples 8 and 9, to determine whether or not the series converges.

9. $\sum_{n=1}^{\infty} \frac{n^2 \cdot 2^n}{n!}$

10. $\sum_{n=1}^{\infty} \frac{n^3 + 3n}{2^n}$

11. $\sum_{n=1}^{\infty} \frac{5^{n+1}}{n^3 \cdot 4^{n+2}}$

12. $\sum_{n=1}^{\infty} \frac{(n+1)!}{100^{n+10}}$

13. $\sum_{n=1}^{\infty} \frac{n^2 \cdot 5^{n+1}}{3^{2n-1}}$

14. $\sum_{n=1}^{\infty} \frac{(n+3)(n+7)}{n!}$

15. $\sum_{n=1}^{\infty} \frac{(n+5)!}{n^2 \cdot n! \cdot 2^n}$

16. Obviously $n^n > n!$ for large n. Let's discover whether n^n dominates $n!$ enough to make $\sum_{n=1}^{\infty} n!/n^n$ converge.
   a) Show that if n is even, then
   $$\frac{n!}{n^n} < \left(\frac{1}{2}\right)^{n/2} = \left(\frac{1}{\sqrt{2}}\right)^n.$$
   [*Hint:* After you are "halfway through the $n!$," the factors in $n!$ are less than $n/2$.]
   b) Show that if n is odd, then $n!/n^n \le (\frac{1}{2})^{(n-1)/2} = (1/\sqrt{2})^{n-1}$. [*Hint:* After you are "down to $(n-1)/2$" in the $n!$, the ratio of each remaining factor of $n!$ to n is less than $(n-1)/(2n)$.]
   c) From parts (a) and (b), conclude that n^n does indeed dominate $n!$ to such an extent that $\sum_{n=1}^{\infty} n!/n^n$ converges.

In Exercises 17 through 48, try to determine by "inspection" (without writing out any computations, as illustrated in Examples 10 through 13) whether or not the series converges. Make use of the results in Exercises 7, 8, and 16 where appropriate. Write out a formal test only if you get stuck.

17. $\sum_{n=1}^{\infty} \frac{3n^2 + 3n}{n^4 + 2}$

18. $\sum_{n=1}^{\infty} \frac{n}{\sqrt{n^3 + 3n}}$

19. $\sum_{n=1}^{\infty} \frac{1}{\sqrt{n + 3}}$

20. $\sum_{n=1}^{\infty} \frac{n^{10}}{2^n}$

21. $\sum_{n=1}^{\infty} \frac{\ln n}{n}$

22. $\sum_{n=2}^{\infty} \frac{n - 1}{n^2(\ln n)}$

23. $\sum_{n=1}^{\infty} \frac{1}{\ln (n^2 + 4n)}$

24. $\sum_{n=1}^{\infty} \frac{n^n}{n^3 \cdot n!}$

25. $\sum_{n=1}^{\infty} \frac{\sqrt{3n^2 + 6n - 1}}{4n^3 - 3n}$

26. $\sum_{n=1}^{\infty} \frac{n^3 \cdot 4^n}{n!}$

27. $\sum_{n=1}^{\infty} \frac{2^{3n+1}}{n!}$

28. $\sum_{n=1}^{\infty} \frac{3^{2n+4}}{10^n}$

29. $\sum_{n=1}^{\infty} \frac{2^{3n-1}}{8^{n+6}}$

30. $\sum_{n=1}^{\infty} \frac{n!}{(2n)!}$

31. $\sum_{n=1}^{\infty} \frac{n^2 \cdot n!}{(n + 3)!}$

32. $\sum_{n=1}^{\infty} \frac{n! + 3^n}{(n + 1)!}$

33. $\sum_{n=1}^{\infty} \frac{n! + 2^n}{(n + 2)!}$

34. $\sum_{n=1}^{\infty} \frac{(n + 3)! - n!}{2^n}$

35. $\sum_{n=1}^{\infty} \frac{(n + 3)! - (n + 1)!}{2^n \cdot (n + 2)!}$

36. $\sum_{n=1}^{\infty} \frac{(n + 3)! - (n - 1)!}{(n + 4)!}$

37. $\sum_{n=1}^{\infty} \frac{(n + 3)! - (n + 1)!}{(n + 5)!}$

38. $\sum_{n=1}^{\infty} \frac{n!}{3^{2n}}$

39. $\sum_{n=1}^{\infty} \frac{n!}{3^{n^2}}$

40. $\sum_{n=1}^{\infty} \frac{1}{2^{\ln n}}$

41. $\sum_{n=1}^{\infty} \frac{1}{3^{\ln n}}$

42. $\sum_{n=1}^{\infty} \frac{1}{2^{\ln n^2}}$

43. $\sum_{n=1}^{\infty} \frac{1}{(\sqrt{2})^{\ln n^2}}$

44. $\sum_{n=2}^{\infty} \frac{1}{n(\ln n)^{3/2}}$

45. $\sum_{n=1}^{\infty} \sin \left(\frac{n}{n^2 + 1}\right)$

46. $\sum_{n=1}^{\infty} \sin \left(\frac{\sqrt{n}}{n^2 + 1}\right)$

47. $\sum_{n=1}^{\infty} \frac{(2n)!}{3^{n^2}}$

48. $\sum_{n=1}^{\infty} \frac{n^n}{3^{n^2}}$

49. Find n such that the partial sum s_n of the harmonic series ≥ 1000. (See Example 5.)

In Exercises 50 through 55, use a calculator or computer to estimate the sum of the series to the accuracy indicated.

50. $\sum_{n=1}^{\infty} \frac{1}{n^3}$, error ≤ 0.0001

51. $\sum_{n=0}^{\infty} \frac{1}{n^2 + 1}$, error ≤ 0.001

52. $\sum_{n=0}^{\infty} \frac{8n}{n^4 + 1}$, error ≤ 0.001

53. $\sum_{n=1}^{\infty} \frac{1}{n^{3/2}}$, error ≤ 0.01

54. $\sum_{n=1}^{\infty} \frac{10}{n^4}$, error ≤ 0.00001

55. $\sum_{n=2}^{\infty} \frac{5}{n(\ln n)^2}$, error ≤ 0.01

10.5 ALTERNATING SERIES TEST; ABSOLUTE CONVERGENCE

ALTERNATING SERIES

We have established several tests for the convergence of a series of nonnegative terms. Analogous tests may be used for series all of whose terms are less than or equal to zero, since $\sum_{n=1}^{\infty} a_n$ converges to s if and only if $\sum_{n=1}^{\infty} (-a_n)$ converges to $-s$. A finite number of negative (or positive) terms can be neglected in establishing the convergence or divergence of a series. However, if a series contains an infinite number of positive and an infinite number of negative terms, the situation becomes more complicated. One type of series often encountered is an *alternating series*, in which the terms are alternately positive and negative. For example, the series

$$1 - 2 + 3 - 4 + 5 - 6 + \cdots$$

is an alternating series. This series diverges since the nth term does not approach 0 as $n \to \infty$.

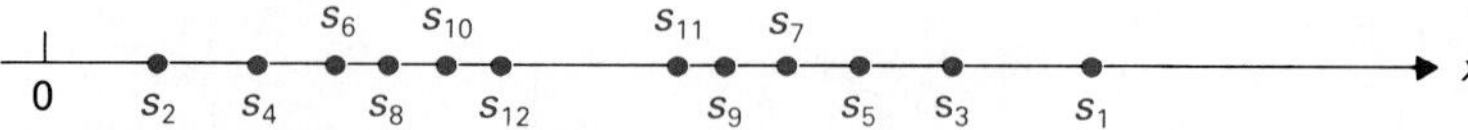

Figure 10.8 Partial sum sequence terms of an alternating series $\sum_{n=1}^{\infty} a_n$, where $|a_{n+1}| < |a_n|$.

The following test establishes the convergence of certain alternating series. While the class of series covered by the test may seem very restrictive, such series occur quite often in practice.

THEOREM 10.9 Alternating series test

Let $\sum_{n=1}^{\infty} a_n$ be a series such that

1. the series is alternating,
2. $|a_{n+1}| \leq |a_n|$ for all n, and
3. $\lim_{n\to\infty} a_n = 0$.

Then the series converges.

An ε,N-proof of the alternating series test will be given in a moment. First, we give an intuitive explanation of why a series that satisfies this test converges.

Suppose $\sum_{n=1}^{\infty} a_n$ satisfies the alternating series test, and suppose $a_1 > 0$. To make our argument simpler, we suppose that condition 2 of the test is satisfied in the strong form $|a_{n+1}| < |a_n|$. Now $s_1 = a_1$, and

$$s_2 = s_1 + a_2 < s_1,$$

since $a_2 < 0$. However, since $|a_2| < |a_1|$, we see that s_2 is still positive, as shown in Fig. 10.8. Next

$$s_3 = s_2 + a_3 > s_2,$$

since $a_3 > 0$. Since $|a_3| < |a_2|$, we see that $s_3 < s_1$. In a similar fashion we find that the partial sums satisfy

$$s_2 < s_4 < s_6 < s_8 < \cdots < \cdots < s_7 < s_5 < s_3 < s_1,$$

as shown in Fig. 10.8. The sequence of partial sums thus oscillates back and forth. The size of the oscillation from s_{n-1} to s_n is $|a_n|$, which decreases by our assumption that $|a_{n+1}| < |a_n|$. If the magnitude of the oscillations does not approach zero, then we could have the situation shown in Fig. 10.9,

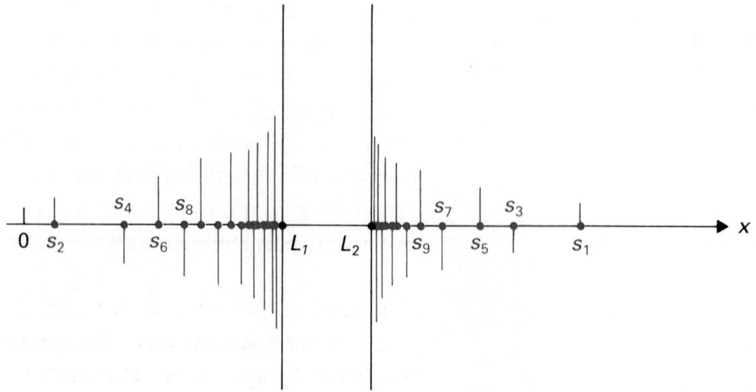

Figure 10.9 Possible behavior for alternating $\sum_{n=1}^{\infty} a_n$ if $|a_{n+1}| < |a_n|$, but oscillations remain large.

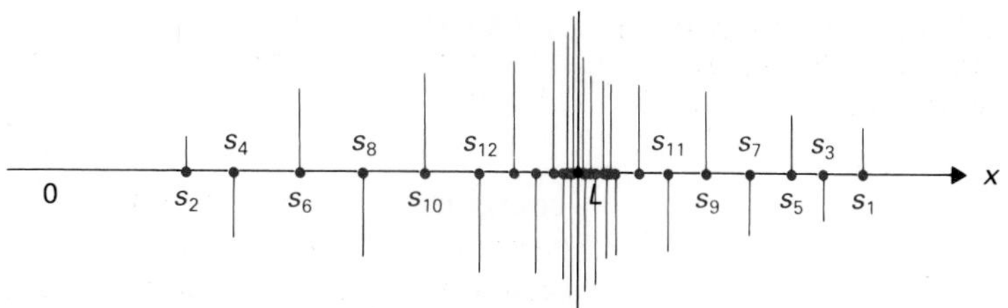

Figure 10.10 Behavior of an alternating series $\sum_{n=1}^{\infty} a_n$ if $|a_{n+1}| < |a_n|$ and $\lim_{n\to\infty} a_n = 0$.

where the oscillations always bridge a gap between two numbers L_1 and L_2, and the sequence $\{s_n\}$ would not converge. However, condition 3 of the test tells us that this cannot happen because $\lim_{n\to\infty} a_n = 0$. Thus the situation must be as shown in Fig. 10.10, where $\{s_n\}$, and therefore $\sum_{n=1}^{\infty} a_n$, approach a limit L.

As shown in Fig. 10.10, the partial sums s_n jump back and forth across L as n increases. We obtain at once the following error estimate.

> If $\sum_{n=1}^{\infty} a_n$ satisfies the alternating series test and converges to L, then
>
> $$|s_n - L| \le |a_{n+1}|.$$
>
> That is, the error in approximating L by the sum of the first n terms of the series is at most the magnitude of the next term.
>
> ***Error estimate***

EXAMPLE 1 Show that the alternating harmonic series

$$\sum_{n=1}^{\infty} (-1)^n \cdot \frac{1}{n}$$

converges.

Solution The series is alternating, $1/(n + 1) < 1/n$, and $\lim_{n\to\infty} (1/n) = 0$. Thus the conditions of the alternating series test are satisfied, and the series converges. Exercise 37 in Section 11.3 will show that the sum of the series is ln 2. □

EXAMPLE 2 Show that the series

$$\sum_{n=1}^{\infty} (-1)^n \cdot \frac{1}{n - (7/2)}$$

converges.

Solution The first few terms of the series are

$$\frac{2}{5} - \frac{2}{3} + 2 + 2 - \frac{2}{3} + \frac{2}{5} - \frac{2}{7} + \frac{2}{9} - \frac{2}{11} + \cdots.$$

The series satisfies the alternating series test starting with the fourth term. Since the convergence or divergence of a series is not affected by a finite number of terms, the series converges. □

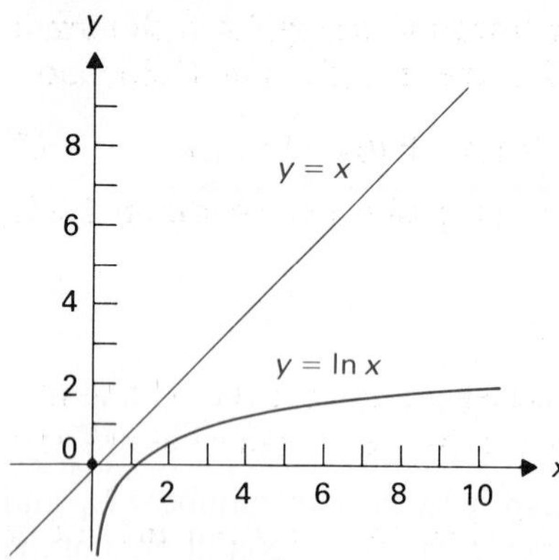

Figure 10.11 x increases so much faster than $\ln x$ that $\lim_{x\to\infty}(\ln x)/x = 0$.

EXAMPLE 3 Argue as best you can that the series

$$\sum_{n=1}^{\infty}(-1)^n\frac{\ln n}{n}$$

converges.

Solution The series is alternating. Since

$$\frac{d}{dx}\left(\frac{\ln x}{x}\right) = \frac{x\cdot(1/x) - \ln x}{x^2} = \frac{1-\ln x}{x^2} < 0 \qquad \text{for } x > e,$$

we have $|a_{n+1}| < |a_n|$ if $n \geq 3$. Section 11.4 will enable us to show carefully that $\lim_{n\to\infty}[(\ln n)/n] = 0$. Surely this seems reasonable. The graph of $\ln x$, with slope approaching 0 as $x \to \infty$, increases extremely slowly compared with the graph of x, as shown in Fig. 10.11. For example, we have $\ln(10^{10})/10^{10} = 10\ln(10)/10^{10} \approx 2.3\cdot 10^{-9}$. The series satisfies the alternating series test for $n \geq 3$ and consequently converges. □

EXAMPLE 4 Show that $\sum_{n=1}^{\infty}(-1)^n/n^3$ converges, and find the sum of the series with error ≤ 0.001.

Solution The series satisfies the alternating series test, so it converges. Using the error estimate given above and our calculator, we find that

$$\sum_{n=1}^{\infty}(-1)^n\frac{1}{n^3} \approx -1 + \frac{1}{2^3} - \frac{1}{3^3} + \frac{1}{4^3} - \frac{1}{5^3} + \frac{1}{6^3} - \frac{1}{7^3} + \frac{1}{8^3} - \frac{1}{9^3}$$
$$\approx -0.9021,$$

with error at most $1/10^3 = 0.001$. □

We emphasize that *all three* conditions of the alternating series test must hold before we can conclude that the series converges. We leave as exercises the construction of examples to show that if any one of these three conditions is dropped, a series can be found that satisfies the remaining two conditions but diverges. (See Exercises 1, 2, and 3.)

PROOF OF THE ALTERNATING SERIES TEST

Proof. Let s_n be the nth partial sum of the series $\sum_{n=1}^{\infty} a_n$. We may suppose that $a_1 > 0$, so that

$$a_1 > 0, \qquad a_2 < 0, \qquad a_3 > 0, \quad \ldots.$$

(A similar proof will hold if $a_1 < 0$.) We have

$$\begin{aligned} s_2 &= (a_1 + a_2),\\ s_4 &= (a_1 + a_2) + (a_3 + a_4),\\ s_6 &= s_4 + (a_5 + a_6),\\ s_8 &= s_6 + (a_7 + a_8),\\ &\vdots \end{aligned}$$

and the sequence $s_2, s_4, s_6, s_8, \ldots$ is monotone increasing, for each term in parentheses is nonnegative by condition 2 of the test. (See Fig. 10.8.) Since

$$s_{2n} = a_1 + (a_2 + a_3) + \cdots + (a_{2n-2} + a_{2n-1}) + a_{2n} \tag{1}$$

and each term in parentheses in Eq. (1) is nonpositive by condition 2, and since $a_{2n} < 0$, we see that

$$s_{2n} \leq a_1.$$

Thus $s_2, s_4, s_6, s_8, \ldots$ is a monotone-increasing sequence that is bounded above and therefore converges to a number L by the fundamental property (page 439).

We claim that $\sum_{n=1}^{\infty} a_n$ converges to L. Let $\varepsilon > 0$ be given and find N such that $|a_n| < \varepsilon/2$ for $n > N$ and also $|s_{2m} - L| < \varepsilon/2$ for $2m > N$. This is possible by condition 3 and the preceding paragraph. If n is even and $n > N$, then $|s_n - L| < \varepsilon/2 < \varepsilon$ by choice of N. If n is odd and $n > N$, then

$$s_n = s_{n+1} - a_{n+1},$$

so

$$|s_n - L| = |(s_{n+1} - L) - a_{n+1}| \leq |s_{n+1} - L| + |a_{n+1}| \leq \frac{\varepsilon}{2} + \frac{\varepsilon}{2} = \varepsilon$$

by choice of N. Thus $\{s_n\}$ converges to L, so $\sum_{n=1}^{\infty} a_n$ converges and has sum L. •

ABSOLUTE CONVERGENCE

Let $\sum_{n=1}^{\infty} a_n$ be a series containing both positive and negative terms. The series $\sum_{n=1}^{\infty} |a_n|$ contains only nonnegative terms; we could apply some of the tests developed in the preceding sections to $\sum_{n=1}^{\infty} |a_n|$, and we might be able to establish its convergence or divergence. The next theorem shows that if $\sum_{n=1}^{\infty} |a_n|$ converges, then $\sum_{n=1}^{\infty} a_n$ converges also. It is important to note that if $\sum_{n=1}^{\infty} |a_n|$ diverges, then $\sum_{n=1}^{\infty} a_n$ may diverge or may converge. We illustrate this following the theorem.

THEOREM 10.10 Absolute convergence test

Let $\sum_{n=1}^{\infty} a_n$ be any infinite series. If $\sum_{n=1}^{\infty} |a_n|$ converges, then $\sum_{n=1}^{\infty} a_n$ converges.

Proof. We define a new series $\sum_{n=1}^{\infty} u_n$ by replacing the negative terms of $\sum_{n=1}^{\infty} a_n$ by zeros, and a new series $\sum_{n=1}^{\infty} v_n$ by replacing the positive terms of $\sum_{n=1}^{\infty} a_n$ by zeros. That is, we let

$$u_n = \begin{cases} a_n, & \text{if } a_n \geq 0, \\ 0, & \text{if } a_n < 0, \end{cases} \quad \text{and} \quad v_n = \begin{cases} a_n, & \text{if } a_n \leq 0, \\ 0, & \text{if } a_n > 0. \end{cases}$$

Note that $\sum_{n=1}^{\infty} a_n = \sum_{n=1}^{\infty} (u_n + v_n)$.

Suppose that $\sum_{n=1}^{\infty} |a_n|$ converges. Now $\sum_{n=1}^{\infty} u_n$ is a series of nonnegative terms and $u_n \leq |a_n|$, so $\sum_{n=1}^{\infty} u_n$ converges by a comparison test. Similarly, $\sum_{n=1}^{\infty} (-v_n)$ is a series of nonnegative terms and $-v_n \leq |a_n|$, so

$\sum_{n=1}^{\infty} (-v_n)$ converges by a comparison test also. We then find that the series $(-1)\sum_{n=1}^{\infty} (-v_n) = \sum_{n=1}^{\infty} v_n$ converges, and therefore the series

$$\sum_{n=1}^{\infty} (u_n + v_n) = \sum_{n=1}^{\infty} a_n$$

converges. This is what we wished to prove. •

EXAMPLE 5 Determine the convergence or divergence of the series

$$1 + \frac{1}{2^2} - \frac{1}{3^2} + \frac{1}{4^2} + \frac{1}{5^2} - \frac{1}{6^2} + \frac{1}{7^2} + \frac{1}{8^2} - \frac{1}{9^2} + \cdots,$$

with two positive terms followed by a negative term.

Solution The series does not satisfy the alternating series test. However, the series does converge, for the corresponding series of absolute values is the series $\sum_{n=1}^{\infty} (1/n^2)$, which is a convergent p-series with $p = 2 > 1$. □

We emphasize again that the absolute convergence test does not say anything about the behavior of $\sum_{n=1}^{\infty} a_n$ if $\sum_{n=1}^{\infty} |a_n|$ diverges. To illustrate, both the series $1 - 2 + 3 - 4 + 5 - 6 + \cdots$ and the corresponding series $1 + 2 + 3 + 4 + 5 + 6 + \cdots$ of absolute values diverge since the nth terms do not approach zero. *However, the alternating harmonic series*

$$1 - \frac{1}{2} + \frac{1}{3} - \frac{1}{4} + \frac{1}{5} - \frac{1}{6} + \cdots$$

converges (by the alternating series test), while the corresponding series of absolute values is the harmonic series

$$1 + \frac{1}{2} + \frac{1}{3} + \frac{1}{4} + \frac{1}{5} + \frac{1}{6} + \cdots,$$

which diverges. We will describe terminology used in this connection.

DEFINITION 10.7 Absolute and conditional convergence

A series $\sum_{n=1}^{\infty} a_n$ **converges absolutely** (or is **absolutely convergent**) if the series $\sum_{n=1}^{\infty} |a_n|$ converges. If $\sum_{n=1}^{\infty} a_n$ converges and $\sum_{n=1}^{\infty} |a_n|$ diverges, then $\sum_{n=1}^{\infty} a_n$ **converges conditionally** (or is **conditionally convergent**).

Note that every convergent series of nonnegative terms is absolutely convergent, since it is identical with the corresponding series of absolute values.

EXAMPLE 6 Classify the series $\sum_{n=1}^{\infty} (-1)^n/n!$ as absolutely convergent, conditionally convergent, or divergent.

Solution The series $\sum_{n=1}^{\infty} (1/n!)$ converges by the ratio test. Thus $\sum_{n=1}^{\infty} (-1)^n/n!$ is absolutely convergent. □

We have often heard students say that a series converges conditionally "because it satisfies the alternating series test."

> *Satisfying the alternating series test is not enough to guarantee that a series converges conditionally. The series of absolute values must also diverge.*

For example, the series $\sum_{n=1}^{\infty} (-1)^n/n!$ of Example 6 satisfies the alternating series test. However, the series $\sum_{n=1}^{\infty} 1/n!$ of absolute values converges, so the series is absolutely convergent.

EXAMPLE 7 Classify the alternating harmonic series $\sum_{n=1}^{\infty} (-1)^n/n$ as absolutely convergent, conditionally convergent, or divergent.

Solution The harmonic series $\sum_{n=1}^{\infty} (1/n)$ of absolute values diverges. The alternating harmonic series $\sum_{n=1}^{\infty} (-1)^n/n$ converges by the alternating series test. Thus the series is conditionally convergent. □

When classifying a series as absolutely convergent, conditionally convergent, or divergent, we should *first* check whether or not the series converges absolutely. If it does converge absolutely, we are done. If not, we try the alternating series test. It is a waste of time to check the alternating series test first.

EXAMPLE 8 Classify the series

$$\sum_{n=1}^{\infty} (-1)^n \frac{n^2 + n}{n^4 + 3n^2 + 1}$$

as absolutely convergent, conditionally convergent, or divergent.

Solution The series of absolute values behaves like $\sum_{n=1}^{\infty} (1/n^2)$ by the LCT, and this series converges as a p-series with $p = 2 > 1$. Thus the original series converges absolutely. □

EXAMPLE 9 Classify the series

$$\sum_{n=1}^{\infty} (-1)^n \frac{2n^2 + 1}{n^2 + 100n}$$

as absolutely convergent, conditionally convergent, or divergent.

Solution The series diverges absolutely since

$$\lim_{n \to \infty} \frac{2n^2 + 1}{n^2 + 100n} = 2 \neq 0.$$

Thus $\lim_{n \to \infty} a_n \neq 0$, so the original series $\sum_{n=1}^{\infty} a_n$ also diverges. That the series is alternating is irrelevant. □

EXAMPLE 10 Classify the series $\sum_{n=1}^{\infty} (-1)^n \sin(1/n)$ as absolutely convergent, conditionally convergent, or divergent.

Solution We have

$$\sum_{n=1}^{\infty} \sin\left(\frac{1}{n}\right) = \sum_{n=1}^{\infty} \frac{1}{n} \cdot \frac{\sin(1/n)}{1/n}.$$

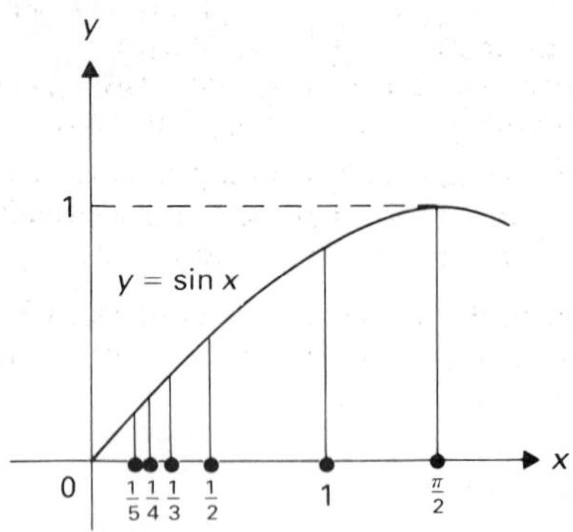

Figure 10.12
$\sin (1/(n+1)) < \sin (1/n)$,
$\lim_{n \to \infty} \sin (1/n) = 0$

By the LCT, this series behaves like $\sum_{n=1}^{\infty} (1/n)$, which diverges. Thus the series does not converge absolutely. A glance at the graph of $\sin x$ in Fig. 10.12 shows that the series is alternating, $\sin [1/(n + 1)] < \sin (1/n)$, and $\lim_{n \to \infty} \sin (1/n) = 0$. Thus the series satisfies the alternating series test and converges. Since it did not converge absolutely, the series is conditionally convergent. □

SUMMARY

1. An alternating series is one containing alternately positive and negative terms.
2. *Alternating series test:* Let $\sum_{n=1}^{\infty} a_n$ be a series such that

   a. the series is alternating,

   b. $|a_{n+1}| \leq |a_n|$ for all n, and

   c. $\lim_{n \to \infty} a_n = 0$.

   Then the series converges.
3. For a series $\sum_{n=1}^{\infty} a_n$ that satisfies the alternating series test, the estimate of $\sum_{n=1}^{\infty} a_n$ by the nth partial sum s_n has error of magnitude at most $|a_{n+1}|$.
4. *Absolute convergence test:* Let $\sum_{n=1}^{\infty} a_n$ be any infinite series. If $\sum_{n=1}^{\infty} |a_n|$ converges, then $\sum_{n=1}^{\infty} a_n$ converges; the series is said to be absolutely convergent.
5. A conditionally convergent series is one that converges but is not absolutely convergent.

EXERCISES

1. Give an example of a divergent series that satisfies conditions 1 and 2 of the alternating series test.
2. Give an example of a divergent series that satisfies conditions 1 and 3 of the alternating series test.
3. Give an example of a divergent series that satisfies conditions 2 and 3 of the alternating series test.
4. Show that every convergent series of nonpositive terms converges absolutely.
5. Mark each of the following true or false.
   a) Every convergent series is absolutely convergent.
   b) Every absolutely convergent series is convergent.
   c) If $\sum_{n=1}^{\infty} a_n$ is conditionally convergent, then $\sum_{n=1}^{\infty} |a_n|$ diverges.
   d) If $\sum_{n=1}^{\infty} |a_n|$ diverges, then $\sum_{n=1}^{\infty} a_n$ is conditionally convergent.
   e) Every alternating series converges.
   f) Every convergent alternating series is conditionally convergent.

In Exercises 6 through 39, classify the series as either absolutely convergent, conditionally convergent, or divergent.

6. $\sum_{n=1}^{\infty} (-1)^n \frac{1}{4n}$
7. $\sum_{n=1}^{\infty} (-1)^n \frac{1}{4n^2}$
8. $\sum_{n=1}^{\infty} (-1)^{n+1} \frac{\cos^2 n}{2^n}$
9. $\sum_{n=1}^{\infty} \frac{-1}{\sqrt[3]{n}}$
10. $\sum_{n=4}^{\infty} \frac{2 - n^2}{(n-3)^2}$
11. $\sum_{n=1}^{\infty} (-1)^n \frac{n}{n^2 - 10{,}001}$
12. $\sum_{n=1}^{\infty} (-1)^{n-1} \frac{n^2}{n^4 - 10{,}001}$
13. $\sum_{n=1}^{\infty} \frac{(-1)^{2n+1}}{\sqrt{n}}$
14. $\sum_{n=1}^{\infty} (-1)^{2n} \frac{n^2 + 3n}{n^3 + 4}$

15. $\sum_{n=1}^{\infty}(-1)^n\frac{\sin^2 n}{n^2}$

16. $\sum_{n=1}^{\infty}(-1)^{n+1}\frac{\sin^2 n}{n^{3/2}}$

17. $\sum_{n=1}^{\infty}\frac{\cos n}{n^{3/2}}$

18. $\sum_{n=1}^{\infty}\frac{\sin n}{n^2}$

19. $\sum_{n=1}^{\infty}(-1)^n\frac{1}{\sin^2 n}$

20. $\sum_{n=2}^{\infty}(-1)^n\frac{1}{\ln n}$

21. $\sum_{n=2}^{\infty}(-1)^n\frac{1}{n(\ln n)}$

22. $\sum_{n=2}^{\infty}(-1)^n\frac{1}{n(\ln n)^2}$

23. $\sum_{n=1}^{\infty}(-1)^n\frac{\ln n}{n}$

24. $\sum_{n=1}^{\infty}\frac{\ln(1/n)}{n}$

25. $\sum_{n=2}^{\infty}(-1)^n\frac{\ln(n^n)}{n^2}$

26. $\sum_{n=1}^{\infty}(-1)^n\frac{n^3\cdot 2^n}{n!}$

27. $\sum_{n=1}^{\infty}(-1)^n\frac{n!}{100^n}$

28. $\sum_{n=1}^{\infty}(-1)^n\frac{n!}{n^n}$

29. $\sum_{n=1}^{\infty}(-1)^{n+1}\frac{n^4\cdot 3^{2n}}{10^n}$

30. $\sum_{n=1}^{\infty}(-1)^{n-1}\frac{(n+1)!-n!}{(n+3)!}$

31. $\sum_{n=1}^{\infty}(-1)^{n+2}\frac{(n+2)!-n!}{(n+3)!}$

32. $\sum_{n=1}^{\infty}(-1)^{n+1}n\sin\left(\frac{1}{n}\right)$

33. $\sum_{n=1}^{\infty}(-1)^{n+3}\frac{\sin(1/n)}{n}$

34. $\sum_{n=1}^{\infty}(-1)^n\csc\left(\frac{1}{n}\right)$

35. $\sum_{n=1}^{\infty}(-1)^n\frac{\csc(1/n)}{n}$

36. $\sum_{n=1}^{\infty}(-1)^n\frac{\csc(1/n)}{n^2}$

37. $\sum_{n=1}^{\infty}\frac{\sin(n\pi/2)}{n}$

38. $\sum_{n=1}^{\infty}\frac{1}{\sqrt{n}}\tan\left[\frac{(2n+1)\pi}{4}\right]$

39. $\sum_{n=1}^{\infty}\frac{1}{n}\sin\left[\frac{(2n+1)\pi}{4}\right]$

40. Show that $\sum_{n=1}^{\infty}(1/n)\sin(n\pi/4)$ converges. [*Hint:* Express the series as a sum of convergent series.]

41. Let $\sum_{n=1}^{\infty}a_n$ be a series and let $\sum_{n=1}^{\infty}u_n$ be the series of positive terms (and zeros) and $\sum_{n=1}^{\infty}v_n$ the series of negative terms (and zeros) defined in the proof of Theorem 10.10.
   a) Show that if $\sum_{n=1}^{\infty}u_n$ and $\sum_{n=1}^{\infty}v_n$ both converge, then $\sum_{n=1}^{\infty}a_n$ converges.
   b) Show that if one of $\sum_{n=1}^{\infty}u_n$ and $\sum_{n=1}^{\infty}v_n$ converges while the other diverges, then $\sum_{n=1}^{\infty}a_n$ diverges.
   c) Show by an example that it is possible that $\sum_{n=1}^{\infty}a_n$ converges while both $\sum_{n=1}^{\infty}u_n$ and $\sum_{n=1}^{\infty}v_n$ diverge.

42. Give an example of a conditionally convergent series that is not an alternating series and does not become an alternating series upon deletion of a finite number of terms.

In Exercises 43 through 48, find the smallest value of n for which you know that the series can be estimated by s_n with error indicated, using number 3 of the Summary.

43. $\sum_{n=1}^{\infty}(-1)^n\frac{1}{n}$, error ≤ 0.01

44. $\sum_{n=1}^{\infty}(-1)^n\frac{1}{n^2}$, error ≤ 0.01

45. $\sum_{n=1}^{\infty}(-1)^n\frac{1}{n^3}$, error ≤ 0.008

46. $\sum_{n=1}^{\infty}(-1)^n\frac{1}{n^2+200}$, error ≤ 0.0001

47. $\sum_{n=1}^{\infty}(-1)^n\frac{1}{n!}$, error ≤ 0.001

48. $\sum_{n=1}^{\infty}(-1)^n\frac{1}{(2n+1)!}$, error ≤ 0.0001

In Exercises 49 through 54, use a calculator or computer to find the sum of the series with error of magnitude indicated.

49. $\sum_{n=1}^{\infty}(-1)^n\frac{1}{n^2}$, error ≤ 0.001

50. $\sum_{n=1}^{\infty}(-1)^n\frac{1}{n^4}$, error ≤ 0.0001

51. $\sum_{n=1}^{\infty}(-1)^n\frac{10}{n^2+1}$, error ≤ 0.01

52. $\sum_{n=2}^{\infty}(-1)^n\frac{1}{(\ln n)^3}$, error ≤ 0.05

53. $\sum_{n=1}^{\infty}(-1)^n\frac{1}{n!}$, error ≤ 0.0001

54. $\sum_{n=1}^{\infty}(-1)^n\frac{1}{(2n+1)!}$, error ≤ 0.0001

EXERCISE SETS FOR CHAPTER 10

Review Exercise Set 10.1

1. Define what is meant by $\lim_{n\to\infty}a_n = c$.

2. Find the limit of the given sequence if it converges or has limit ∞ or $-\infty$.

   a) $\left\{\frac{4n^2-2n}{3-5n^2}\right\}$ b) $\left\{\frac{n^3}{e^n}\right\}$

3. Find the sum of the series

$$\sum_{n=0}^{\infty}\frac{3^{n+1}}{4^{n-1}}$$

if the series converges.

4. Express the repeating decimal 4.731313131... as a fraction.

In Problems 5 and 6, classify the series as convergent or divergent, and give a reason for your answer.

5. a) $\sum_{n=1}^{\infty} \frac{n^2 + 3n}{400n + n^2}$ b) $\sum_{n=2}^{\infty} \frac{n^2 - 1}{1 + n^3}$

6. a) $\sum_{n=1}^{\infty} \frac{1}{\sqrt{n + 17}}$ b) $\sum_{n=1}^{\infty} \frac{\sqrt{n + 3}}{n^2 + 4n}$

7. Use the integral test to establish the convergence or divergence of

$$\sum_{n=2}^{\infty} \frac{1}{n^2 - n}.$$

8. Write out the ratio test to establish the convergence or divergence of

$$\sum_{n=1}^{\infty} \frac{n^2 \cdot 2^n}{e^{n+1}}.$$

In Problems 9 and 10, classify the series as absolutely convergent, conditionally convergent, or divergent, and indicate the reason for your answer.

9. a) $\sum_{n=1}^{\infty} \frac{(-1)^n}{\sqrt{n + 3}}$ b) $\sum_{n=2}^{\infty} \frac{(-1)^n}{n(\ln n)^2}$

10. a) $\sum_{n=1}^{\infty} (-1)^n \frac{\cos(1/n)}{n^{3/2}}$ b) $\sum_{n=1}^{\infty} (-1)^n \frac{3^n}{n^{100}}$

Review Exercise Set 10.2

1. Give an ε,N-proof that the sequence $\{(n - 1)/n\}$ converges.

2. Find the limit of the given sequence if it converges or has limit ∞ or $-\infty$.

a) $\left\{\frac{\sqrt{n} - n^2}{3n - 7}\right\}$ b) $\left\{\frac{1}{n} \sin\left(\frac{1}{n}\right)\right\}$

3. A ball has the property that when it is dropped, it rebounds to $\frac{1}{3}$ of its height on the previous bounce. Find the height from which the ball must be dropped if the total distance it is to travel is 60 ft.

4. If possible, find r such that the sum of the series

$$3 - 3r + 3r^2 - 3r^3 + \cdots + (-1)^n 3r^n + \cdots$$

is 7.

In Problems 5 and 6, classify the series as convergent or divergent, and give a reason for your answer.

5. a) $\sum_{n=1}^{\infty} \frac{n^2 + 3}{n^2 \cdot 3^{n-1}}$ b) $\sum_{n=1}^{\infty} \frac{3^n + 4^n}{5^n}$

6. a) $\sum_{n=1}^{\infty} \frac{n^2 + \sin n}{n^4 + 3n}$ b) $\sum_{n=1}^{\infty} \frac{1}{\cos^2 n}$

7. Use the integral test to establish the convergence or divergence of

$$\sum_{n=2}^{\infty} \frac{n + 1}{n^2 + 1}.$$

8. Write out the ratio test to establish the convergence or divergence of

$$\sum_{n=1}^{\infty} \frac{n!}{100^n}.$$

In Problems 9 and 10, classify the series as absolutely convergent, conditionally convergent, or divergent, and indicate the reason for your answer.

9. $\sum_{n=2}^{\infty} \frac{(-1)^n}{n(\ln n)}$ b) $\sum_{n=1}^{\infty} (-1)^n \frac{\cos^2 n}{n^{3/2}}$

10. a) $\sum_{n=1}^{\infty} (-1)^n \frac{n^n}{n!}$ b) $\sum_{n=1}^{\infty} (-1)^n \frac{\ln n}{n}$

Review Exercise Set 10.3

1. Give a careful definition of $\lim_{n\to\infty} a_n = \infty$.

2. Find the limit of the given sequence if it converges or has limit ∞ or $-\infty$.

a) $\left\{\frac{n^2 + 1}{n} - \frac{n^3 + n^2}{n^2 + 1}\right\}$ b) $\left\{\frac{3^n - 5^n}{4^n}\right\}$

3. Find the sum of the series

$$\sum_{n=0}^{\infty} \frac{3^n - 2^{n+2}}{5^n},$$

if it converges.

4. Express the repeating decimal 5.2107107107107... as a fraction.

In Problems 5 and 6, classify the series as convergent or divergent, and give a reason for your answer.

5. a) $\sum_{n=1}^{\infty} \frac{\sqrt{n^5 + 4n}}{(n^2 + n + 3)^2}$

b) $\sum_{n=1}^{\infty} \sin\left(\frac{1}{n^2 + 1}\right)$

6. a) $\sum_{n=1}^{\infty} \frac{3^{2n+1}}{(n + 4)9^n}$

b) $\sum_{n=1}^{\infty} \frac{(n + 3)! - (n + 1)!}{(n + 4)!}$

7. State the integral test, giving all hypotheses.

8. State the ratio test, giving all hypotheses.

In Problems 9 and 10, classify the series as absolutely convergent, conditionally convergent, or divergent, and indicate the reason for your answer.

9. a) $\sum_{n=1}^{\infty} (-1)^n \frac{\sqrt{n}}{n^2 + 1}$

b) $\sum_{n=1}^{\infty} \frac{\sin(3n\pi/2)}{n}$

10. a) $\sum_{n=1}^{\infty} \frac{(-3)^{n+1}}{(\sqrt{3})^{3n}}$

b) $\sum_{n=1}^{\infty} (-1)^n \frac{\ln(n^2)}{n}$

Review Exercise Set 10.4

1. Give an example of sequences $\{s_n\}$ and $\{t_n\}$, both of which diverge, such that $\{s_n + t_n\}$ converges to 3.

2. Find the limit of the given sequence if it converges or has limit ∞ or $-\infty$.

   a) $\left\{\dfrac{\sin n}{\sqrt{n}}\right\}$

   b) $\left\{\dfrac{\csc^2(1/n)}{n^{3/2}}\right\}$

3. A ball has the property that when it is dropped, it rebounds to k times its height on the preceding bounce. If the ball travels a total distance of 400 ft when dropped from a height of 100 ft, find the value of k.

4. Is 3.0100100010000100001 . . . a rational number? Why or why not?

In Problems 5 and 6, classify the series as convergent or divergent, and give a reason for your answer.

5. a) $\displaystyle\sum_{n=1}^{\infty} \frac{(3^n - 2^n)^2}{7^n}$

   b) $\displaystyle\sum_{n=1}^{\infty} \frac{1}{(2.7)^{\ln n}}$

6. a) $\displaystyle\sum_{n=1}^{\infty} \frac{(n+4)! + (n+3)!}{(n+1)! \cdot 2^{n/2}}$

   b) $\displaystyle\sum_{n=2}^{\infty} \frac{\sqrt{\ln(n-1)}}{n(\ln n)^2}$

7. Use the integral test to establish the convergence or divergence of

$$\sum_{n=1}^{\infty} \frac{e^n}{1 + e^{2n}}.$$

8. Write out the ratio test to establish the convergence or divergence of

$$\sum_{n=1}^{\infty} \frac{4^{3n}}{n^2 7^{2n+4}}.$$

In Problems 9 and 10, classify the series as absolutely convergent, conditionally convergent, or divergent, and indicate a reason for your answer.

9. a) $\displaystyle\sum_{n=1}^{\infty} (-1)^n \frac{1}{n^4 \cdot 3^{-n}}$

   b) $\displaystyle\sum_{n=1}^{\infty} \frac{\cos n\pi}{\sqrt{n}}$

10. a) $\displaystyle\sum_{n=1}^{\infty} (-1)^n \ln\left(\frac{1}{n^2}\right)$

    b) $\displaystyle\sum_{n=1}^{\infty} (-1)^n \frac{(n!)^2 \cdot 5^n}{(3n)!}$

More Challenging Exercises 10

1. A sequence $\{t_n\}$ is monotone decreasing if $t_{n+1} \le t_n$ for $n = 1, 2, 3, \ldots$. Assuming the fundamental property of the real numbers stated in the text, show that a monotone-decreasing sequence $\{t_n\}$ either converges to a number d or $\lim_{n\to\infty} t_n = -\infty$.

2. Prove Theorem 10.3. [*Hint:* Let s_n be the nth partial sum of $\sum_{n=1}^{\infty} a_n$ and t_n be the nth partial sum of $\sum_{n=1}^{\infty} b_n$. Deduce that there exist an integer N and a number c such that $t_n = s_{n+k} + c$ for $n > N$, and show that if $\{s_n\}$ converges to a, then $\{t_n\}$ converges to $a + c$.]

3. Generalize Theorem 10.6 as follows: Let $\sum_{n=1}^{\infty} a_n$ and $\sum_{n=1}^{\infty} b_n$ be series of nonnegative terms with $a_n \ne 0$ for sufficiently large n. Show that if there exist constants $m > 0$ and $M > 0$ such that $m < b_n/a_n < M$ for all sufficiently large n, then the two series either both converge or both diverge.

4. A sequence $\{s_n\}$ is a *Cauchy sequence* if for each $\varepsilon > 0$, there exists an integer N such that

$$|s_n - s_m| < \varepsilon,$$

provided that both $n > N$ and $m > N$. Prove that every convergent sequence $\{s_n\}$ is a Cauchy sequence. (The converse is also true but is harder to prove.)

Each of two series is said to be a **rearrangement** of the other if the two series contain exactly the same terms, but the terms do not necessarily appear in the same order. Exercises 5 through 9 deal with this concept. It can be shown that rearranging an absolutely convergent series does not change its convergence or divergence or its sum if it converges. However, let $\sum_{n=1}^{\infty} a_n$ be a series such that $\lim_{n\to\infty} a_n = 0$ and such that the series $\sum_{n=1}^{\infty} u_n$ of nonnegative terms (and zeros) and the series $\sum_{n=1}^{\infty} v_n$ of nonpositive terms (and zeros), defined in the proof of Theorem 10.10, both diverge. Describe how to construct a rearrangement of $\sum_{n=1}^{\infty} a_n$ that has the given behavior.

5. Converges to 17
6. Converges to -50
7. Diverges to ∞
8. Diverges to $-\infty$
9. Diverges and has partial sums alternately increasing to ≥ 15 and decreasing to ≤ -6

11 POWER SERIES

11.1 POWER SERIES

THE FUNCTION REPRESENTED BY A POWER SERIES

Among the most important series are *power series*, and most of our work will be with these series. Power series are precisely the "infinite polynomials" that we mentioned at the start of Chapter 10. We will see that

$$\sin x = x - \frac{x^3}{3!} + \frac{x^5}{5!} - \frac{x^7}{7!} + \frac{x^9}{9!} - \frac{x^{11}}{11!} + \cdots$$

for any value of x. We will also see that

$$\frac{1}{x} = 1 - (x-1) + (x-1)^2 - (x-1)^3 + (x-1)^4 - (x-1)^5 + \cdots$$

for any x such that $0 < x < 2$. The series for $\sin x$ is in powers of $x = (x - 0)$, and the series for $1/x$ is in powers of $(x - 1)$.

DEFINITION 11.1 Power series centered at x_0

A **power series centered at** x_0 is a series of the form

$$\sum_{n=0}^{\infty} a_n(x - x_0)^n = a_0 + a_1(x - x_0) + \cdots + a_n(x - x_0)^n + \cdots. \tag{1}$$

The constants a_i are the **coefficients of the series;** in particular, a_0 is the **constant term of the series.**

It will be convenient to change terminology slightly and to consider the constant term a_0 of Eq. (1) to be the *zeroth term of the series* and the term $a_n(x - x_0)^n$ to be the *n*th *term.* With this convention, the *n*th term of a power series becomes the term with exponent n.

At each value of x, the power series (1) becomes a series of constants that may or may not converge. The sum of such a convergent series of constants generally varies with the value of x and is a function of x. This function is the **sum function** of the series. The set of all values of x for which the power series converges is the **domain of the sum function** defined by the series; the value of the function at each such point is the sum of the series for that value of x.

The power series (1) should be regarded as an attempt to describe a function *locally*, near x_0. In translated coordinates with $\Delta x = x - x_0$, the series becomes $\sum_{n=0}^{\infty} a_n(\Delta x)^n$. The series (1) converges for $x = x_0$ because at x_0, the series becomes

$$a_0 + a_1 \cdot 0 + a_2 \cdot 0 + \cdots + a_n \cdot 0 + \cdots,$$

which converges to a_0. It is possible that this is the only point at which the series converges, as Example 1 shows, but such series are of little importance for us.

EXAMPLE 1 Show that $\sum_{n=1}^{\infty} (n!)x^n$ converges only for $x = 0$.

Solution If $x = a$, we obtain the series $\sum_{n=1}^{\infty} (n!)a^n$. As shown in Section 10.5, the ratio test for absolute convergence yields the limiting ratio

$$\lim_{n\to\infty} |(n + 1)a| = \infty, \qquad \text{if } a \neq 0.$$

Thus if $a \neq 0$, the terms of the series eventually increase in magnitude, so the nth term does not approach zero. Therefore the series converges only if $x = a = 0$. □

THE RADIUS OF CONVERGENCE OF A POWER SERIES

Theorem 11.1 shows that the domain of convergence of a power series centered at x_0 has x_0 as center point. The theorem is stated and proved in the case where $x_0 = 0$, that is, for a series $\sum_{n=0}^{\infty} a_n x^n$. See the remark after the proof of the theorem.

THEOREM 11.1 Absolute convergence

If a power series $\sum_{n=0}^{\infty} a_n x^n$ converges for $x = c \neq 0$, then the series converges absolutely for all x such that $|x| < |c|$. If the series diverges at $x = d$, then the series diverges for all x such that $|x| > |d|$.

Proof. Suppose that $\sum_{n=0}^{\infty} a_n c^n$ converges, and let $|b| < |c|$. Since $\sum_{n=0}^{\infty} a_n c^n$ converges, we must have $\lim_{n\to\infty} a_n c^n = 0$; in particular, $|a_n c^n| < 1$, or

$$|a_n| < \frac{1}{|c^n|}$$

for n sufficiently large. We then obtain

$$|a_n b^n| = |a_n| \cdot |b^n| < \left|\frac{b^n}{c^n}\right| = \left|\frac{b}{c}\right|^n$$

for n sufficiently large. Recall that we are assuming $|b/c| < 1$. Thus the series $\sum_{n=0}^{\infty} |a_n b^n|$ is, term for term, less than the convergent geometric series

$\sum_{n=0}^{\infty} |b/c|^n$ for n sufficiently large, so $\sum_{n=0}^{\infty} |a_n b^n|$ converges by the comparison test. This shows that $\sum_{n=0}^{\infty} a_n x^n$ converges absolutely for all x such that $|x| < |c|$.

The assertion regarding divergence is really the contrapositive of the assertion we just proved for convergence. Suppose $\sum_{n=0}^{\infty} a_n d^n$ diverges, and let $|h| > |d|$. Convergence of $\sum_{n=0}^{\infty} a_n h^n$ would imply convergence of $\sum_{n=0}^{\infty} a_n d^n$ by the first part of our proof. But this would contradict our hypothesis, so $\sum_{n=0}^{\infty} a_n h^n$ diverges also. •

We stated and proved the theorem for a power series centered at $x_0 = 0$. It is immediate from the theorem that if $\sum_{n=0}^{\infty} a_n(\Delta x)^n$ converges for $\Delta x = c$, then the series converges for $|\Delta x| < |c|$. Putting $\Delta x = x - x_0$ as usual, we see that if a series $\sum_{n=0}^{\infty} a_n(x - x_0)^n$ converges at $x = x_0 + c$, that is, for $\Delta x = c$, then it converges for all x such that $|x - x_0| < |c|$.

The important corollary that follows seems very plausible from our theorem. A careful proof depends on the fundamental property of the real numbers (page 439) and is not given here.

COROLLARY Radius of convergence

For a power series $\sum_{n=0}^{\infty} a_n(x - x_0)^n$, exactly one of the three following alternatives holds.

- **a.** The series converges at x_0 only.
- **b.** The series converges for all x.
- **c.** There exists R such that the series converges for $|x - x_0| < R$, that is, for $x_0 - R < x < x_0 + R$, and diverges for $|x - x_0| > R$.

The number R that appears in case (c) of the corollary is the **radius of convergence of the series.** In case (a), it is natural to say that the radius of convergence is zero and in case (b) we say that the radius of convergence is ∞.

In case (c), the behavior of the series at the endpoints of the interval $x_0 - R < x < x_0 + R$ depends on the individual series; certain series converge at both endpoints, others diverge at both endpoints, and some converge at one endpoint and diverge at the other.

EXAMPLE 2 Find the radius of convergence of the geometric series $\sum_{n=0}^{\infty} x^n$.

Solution Our work in Section 10.2 shows that this series converges to $1/(1 - x)$ for $|x| < 1$. The radius of convergence of the series is 1. The series diverges for $x = 1$ and $x = -1$, since the nth term does not approach 0 as $n \to \infty$ at these points. □

For many power series, the radius of convergence can be computed by using the ratio test. The technique is best illustrated by examples. By Theorem 11.1, if a power series converges at $x - x_0 = c$, it converges absolutely for $|x - x_0| < |c|$, so we try to compute the limit of the *absolute value* of the ratio.

EXAMPLE 3 Find the radius of convergence of the series

$$\sum_{n=1}^{\infty} \frac{x^n}{n \cdot 2^n}.$$

Solution The absolute value of the ratio of the $(n + 1)$st term to the nth is

$$\left|\frac{x^{n+1}/[(n + 1)2^{n+1}]}{x^n/(n \cdot 2^n)}\right| = \left|\frac{x^{n+1}}{(n + 1)2^{n+1}} \cdot \frac{n \cdot 2^n}{x^n}\right| = \left|\frac{nx}{(n + 1)2}\right|.$$

The limit as $n \to \infty$ is

$$\lim_{n\to\infty} \left|\frac{nx}{(n + 1)2}\right| = \left|\frac{x}{2}\right|.$$

Thus the series converges for $|x/2| < 1$, or for $|x| < 2$. The radius of convergence is therefore 2, and the series converges at least for $-2 < x < 2$.

To see what happens at the endpoint 2 of the interval $-2 < x < 2$, examine the series

$$\sum_{n=1}^{\infty} \frac{2^n}{n \cdot 2^n} = \sum_{n=1}^{\infty} \frac{1}{n}.$$

This is the harmonic series, and it diverges. At the endpoint -2, we obtain as series

$$\sum_{n=1}^{\infty} \frac{(-2)^n}{n \cdot 2^n} = \sum_{n=1}^{\infty} (-1)^n \cdot \frac{1}{n},$$

which is the convergent alternating harmonic series. Thus the series converges for $-2 \le x < 2$. This interval is the **interval of convergence of the series.** □

Example 3 illustrates the usual procedure for finding the *radius* and *interval* of convergence for a power series where the limit of the ratio exists. The two series of constants corresponding to the endpoints of the interval must be examined separately.

> The ratio test never determines the behavior at the endpoints of the interval of convergence, for the limit of the ratio at these endpoints will be 1.

We illustrate with an example for a power series centered at a point $x_0 \neq 0$.

EXAMPLE 4 Determine the interval of convergence of the series

$$\sum_{n=1}^{\infty} \frac{(x - 3)^{2n}}{n^2 \cdot 5^n},$$

which is a power series centered at $x_0 = 3$.

Solution For the ratio, we obtain

$$\left|\frac{(x - 3)^{2(n+1)}}{(n + 1)^2 \cdot 5^{n+1}} \cdot \frac{n^2 \cdot 5^n}{(x - 3)^{2n}}\right| = \left|\frac{n^2(x - 3)^2}{(n + 1)^2 \cdot 5}\right|.$$

We have

$$\lim_{n\to\infty}\left|\frac{n^2(x-3)^2}{(n+1)^2\cdot 5}\right| = \left|\frac{(x-3)^2}{5}\right|.$$

Thus the series converges if $|(x-3)^2/5| < 1$, or if $|x-3|^2 < 5$. This is equivalent to $|x-3| < \sqrt{5}$. The radius of convergence about $x_0 = 3$ is thus $\sqrt{5}$, and the series converges at least for $3-\sqrt{5} < x < 3+\sqrt{5}$.

Turning to the endpoints, at $3-\sqrt{5}$ our series becomes

$$\sum_{n=1}^{\infty}\frac{(-\sqrt{5})^{2n}}{5^n\cdot n^2} = \sum_{n=1}^{\infty}\frac{5^n}{5^n\cdot n^2}$$
$$= \sum_{n=1}^{\infty}\frac{1}{n^2}.$$

This series converges as a p-series with $p = 2 > 1$. The same series is obtained at $3+\sqrt{5}$, so the series converges at both endpoints, and the interval of convergence is

$$[3-\sqrt{5}, 3+\sqrt{5}]. \quad \square$$

The preceding examples indicate that the term $(x-x_0)^n$ in a power series $\sum_{n=0}^{\infty} a_n(x-x_0)^n$ contributes $|x-x_0|$ to the limiting absolute-value ratio, the term $(x-x_0)^{2n}$ contributes $|x-x_0|^2$, and so on. Of course, we expect this from our work in Section 10.4. We should be able to use the technique described in number 5 of the Summary on page 454 to find the radius of convergence for many power series without actually writing out the ratio test in detail.

EXAMPLE 5 Find the interval of convergence of the series $\sum_{n=1}^{\infty}(x^n/n!)$.

Solution Since x^n contributes x to the ratio and $n!$ contributes $n+1$, the limit of the absolute-value ratio is

$$\lim_{n\to\infty}\left|\frac{x}{n+1}\right| = 0 < 1$$

for all x. Thus the series converges for all x; its interval of convergence is $-\infty < x < \infty$. $\square$

EXAMPLE 6 Find the interval of convergence of the series

$$\sum_{n=1}^{\infty}\frac{(-2)^n(x+3)^n}{n}.$$

Solution Now $(-2)^n$ contributes -2 and $(x+3)^n$ contributes $x+3$ to the ratio. Also, n contributes 1 after taking the limit as $n\to\infty$. We see that the absolute-value limiting ratio is

$$|(-2)(x+3)| = 2|x+3|.$$

Thus the series converges for

$$2|x+3| < 1 \quad \text{or} \quad |x+3| < \tfrac{1}{2}.$$

The radius of convergence is $\frac{1}{2}$, and the series converges for

$$-3 - \tfrac{1}{2} < x < -3 + \tfrac{1}{2}, \qquad \text{or} \qquad -\tfrac{7}{2} < x < -\tfrac{5}{2}.$$

For $x = -\frac{7}{2}$, we obtain the series

$$\sum_{n=1}^{\infty} \frac{(-2)^n(-\frac{1}{2})^n}{n} = \sum_{n=1}^{\infty} \frac{1}{n},$$

which diverges. For $x = -\frac{5}{2}$, we obtain the series

$$\sum_{n=1}^{\infty} \frac{(-2)^n(\frac{1}{2})^n}{n} = \sum_{n=1}^{\infty} \frac{(-1)^n}{n},$$

which converges. Thus the interval of convergence is $-\frac{7}{2} < x \leq -\frac{5}{2}$. □

GEOMETRIC POWER SERIES

The series

$$a + ar + ar^2 + ar^3 + \cdots + ar^n + \cdots$$

converges to $a/(1 - r)$ if $|r| < 1$. We can sometimes "work backwards" using this sum formula, starting with a function and finding a geometric power series that equals the function in its interval of convergence.

EXAMPLE 7 Find a power series centered at $x_0 = 0$ that equals $2/(1 + x)$ in its interval of convergence.

Solution We put $2/(1 + x)$ in the form $a/(1 - r)$ by writing

$$\frac{2}{1 + x} = \frac{2}{1 - (-x)}.$$

We recognize this as the sum of a geometric series with initial term 2 and ratio $-x$. The series is

$$2 - 2x + 2x^2 - 2x^3 + 2x^4 - \cdots + (-1)^n 2x^n + \cdots.$$

The series converges to $2/(1 + x)$ for $|x| < 1$. □

EXAMPLE 8 Find a power series centered at $x_0 = 0$ that equals $5/(3 - x)$ in its interval of convergence.

Solution We put $5/(3 - x)$ into the form $a/(1 - r)$:

$$\frac{5}{3 - x} = \frac{1}{3} \cdot \frac{5}{1 - \dfrac{x}{3}} = \frac{5/3}{1 - \dfrac{x}{3}}.$$

This is the sum of a geometric series with initial term $\frac{5}{3}$ and ratio $x/3$. The series is

$$\frac{5}{3} + \frac{5}{3^2}x + \frac{5}{3^3}x^2 + \frac{5}{3^4}x^3 + \cdots + \frac{5}{3^{n+1}}x^n + \cdots = \sum_{n=0}^{\infty} \frac{5}{3}\left(\frac{x}{3}\right)^n.$$

The series converges for $|x/3| < 1$, or for $|x| < 3$. □

EXAMPLE 9 Find a power series centered at $x_0 = 1$ that equals $5/(3 - x)$ in its interval of convergence.

Solution As in Example 8, we put $5/(3 - x)$ in the form $a/(1 - r)$, but this time we want the ratio r in terms of $x - 1$ rather than x. We have

$$\frac{5}{3-x} = \frac{5}{3 - [(x-1)+1]} = \frac{5}{2-(x-1)}$$

$$= \frac{1}{2} \cdot \frac{5}{1 - \dfrac{x-1}{2}} = \frac{5/2}{1 - \dfrac{x-1}{2}}.$$

This is the sum of a geometric series with initial term $\frac{5}{2}$ and ratio $(x - 1)/2$. The series is

$$\frac{5}{2} + \frac{5}{2^2}(x-1) + \frac{5}{2^3}(x-1)^2 + \cdots + \frac{5}{2^{n+1}}(x-1)^n + \cdots$$

$$= \sum_{n=0}^{\infty} \frac{5}{2}\left(\frac{x-1}{2}\right)^n,$$

which converges for $|(x - 1)/2| < 1$, or for $|x - 1| < 2$. □

SUMMARY

1. A power series centered at x_0 is a series of the form

$$a_0 + a_1(x - x_0) + a_2(x - x_0)^2 + \cdots + a_n(x - x_0)^n + \cdots.$$

   The function having as domain all values of x for which the series converges, and having as value at each such x the sum of the series, is called the sum function defined by the series.
2. The series in number (1) may converge for $x = x_0$ only, or it may converge for all x, or it may converge if $|x - x_0| < R$ and diverge if $|x - x_0| > R$ for some $R > 0$. Such a number R is the radius of convergence of the series.
3. The radius R of convergence of a power series can often be found by:

   a. forming the absolute value of the ratio of the $(n + 1)$st term divided by the nth term,

   b. computing the limit of this ratio as $n \to \infty$,

   c. setting the resulting limit < 1,

   d. solving the resulting inequality for $|x - x_0|$, obtaining an expression of the form $|x - x_0| < R$. The radius of convergence is then R.

4. To determine the interval of convergence of a power series centered at x_0 after the radius R of convergence has been found, substitute $x = x_0 - R$ and $x = x_0 + R$ to obtain two series of constants. Test these series for convergence to determine which of the endpoints, $x_0 - R$ or $x_0 + R$, should be included with $|x - x_0| < R$ to obtain the interval

of convergence. The ratio test should never be tried at these endpoint series, for the limiting ratio will always be 1.

5. If a function $f(x)$ can be written in the form $a/(1 - r)$ where both a and r are monomial functions of $x - x_0$, then we can find a geometric power series centered at x_0 whose sum equals $f(x)$ in its interval of convergence.

EXERCISES

In Exercises 1 through 8, use the ratio test to find the radius of convergence of the series. Then find the interval of convergence (including endpoints).

1. $\sum_{n=1}^{\infty} \frac{x^n}{n}$
2. $\sum_{n=1}^{\infty} \frac{x^{2n+1}}{n}$
3. $\sum_{n=1}^{\infty} (-1)^n \frac{x^n}{n}$
4. $\sum_{n=0}^{\infty} (-1)^n \frac{x^{2n}}{n+3}$
5. $\sum_{n=0}^{\infty} \frac{(x-1)^n}{n^2+1}$
6. $\sum_{n=0}^{\infty} \frac{(x+2)^n}{3^n}$
7. $\sum_{n=1}^{\infty} \frac{(x+1)^n}{n \cdot 5^{n+2}}$
8. $\sum_{n=2}^{\infty} \frac{(-1)^n(x-3)^{n+1}}{(\ln n) \cdot 2^{n-1}}$

In Exercises 9 through 27, proceed as above, but this time try to find the radius of convergence without explicitly writing out the ratio, as illustrated after Example 4 in the text; write out the ratio test only if you have to.

9. $\sum_{n=0}^{\infty} \frac{x^n}{3^n}$
10. $\sum_{n=1}^{\infty} \frac{2^{n+1}x^n}{n \cdot 3^n}$
11. $\sum_{n=0}^{\infty} \frac{(n+1)x^n}{(n^3+4)2^{2n+1}}$
12. $\sum_{n=2}^{\infty} \frac{(-2)^n x^n}{n(\ln n)}$
13. $\sum_{n=0}^{\infty} \frac{\sqrt{n}(x-1)^n}{n+1}$
14. $\sum_{n=1}^{\infty} \frac{(x-2)^{2n}}{n \cdot 9^n}$
15. $\sum_{n=1}^{\infty} \frac{n^2(x+4)^n}{n!}$
16. $\sum_{n=0}^{\infty} n^2(x-3)^n$
17. $\sum_{n=1}^{\infty} n!(x+5)^{2n+1}$
18. $\sum_{n=1}^{\infty} \frac{(x+4)^{n+1}}{n^2 \cdot 3^n}$
19. $\sum_{n=1}^{\infty} \frac{(-2)^n(x+3)^{n+1}}{\sqrt{n}}$
20. $\sum_{n=1}^{\infty} \frac{3^n(x-2)^{2n+1}}{n!}$
21. $\sum_{n=1}^{\infty} \frac{(2x-4)^n}{n^{3/2} \cdot 3^n}$
22. $\sum_{n=1}^{\infty} \frac{(3x-12)^{2n}}{\sqrt{n}}$
23. $\sum_{n=0}^{\infty} \frac{(2x+6)^n}{(n+1)3^n}$
24. $\sum_{n=1}^{\infty} \frac{(2x-1)^n}{n^2 \cdot 2^{2n+1}}$
25. $\sum_{n=1}^{\infty} \frac{n!(3x+6)^n}{(n+2)!}$
26. $\sum_{n=0}^{\infty} \frac{(5x-10)^{2n}}{(n+2)4^n}$
27. $\sum_{n=0}^{\infty} (-1)^n \frac{\sqrt{n}(2x+4)^{2n+1}}{n+5}$

In Exercises 28 through 31, give a power series that has the given interval as interval of convergence, including or excluding endpoints as indicated. (Many answers are possible.)

28. $[0, 6]$
29. $1 < x < 4$
30. $-2 < x \leq 4$
31. $-5 \leq x < -1$

In Exercises 32 through 46, find a geometric power series centered at the indicated point x_0 that equals the given function in its interval of convergence.

32. $\frac{1}{1+x}$ at $x_0 = 0$
33. $\frac{1}{1-x^2}$ at $x_0 = 0$
34. $\frac{3}{1-2x}$ at $x_0 = 0$
35. $\frac{5}{3x-1}$ at $x_0 = 0$
36. $\frac{2x}{x^3-1}$ at $x_0 = 0$
37. $\frac{2}{4-x}$ at $x_0 = 0$
38. $\frac{3x}{9+x}$ at $x_0 = 0$
39. $\frac{4x^2}{2+3x}$ at $x_0 = 0$
40. $\frac{2x^3}{3-2x}$ at $x_0 = 0$
41. $\frac{1}{1+x}$ at $x_0 = 1$
42. $\frac{3}{2+x}$ at $x_0 = -1$
43. $\frac{2}{6+x}$ at $x_0 = 2$
44. $\frac{2}{3+2x}$ at $x_0 = -1$
45. $\frac{(x-3)^2}{4-2x}$ at $x_0 = 3$
46. $\frac{2(x+2)^3}{3x-5}$ at $x_0 = -2$

Exercises 47 through 56 deal with series of the form

$$\sum_{n=0}^{\infty} \frac{a_n}{(x-x_0)^n}.$$

47. Show that if $\sum_{n=0}^{\infty} (a_n/x^n)$ converges at $x = c$, then the series converges absolutely for all x such that $|x| > |c|$. [*Hint:* Let $u = 1/x$ and apply Theorem 11.1.]
48. Show that if $\sum_{n=0}^{\infty} (a_n/x^n)$ diverges as $x = d$, then it diverges for all x such that $|x| < |d|$. [*Hint:* Use Exercise 47.]

In Exercises 49 through 56, apply the absolute-value ratio test and find all values of x for which the series converges, including "endpoints."

49. $\sum_{n=0}^{\infty} \frac{2^n}{x^n}$

50. $\sum_{n=0}^{\infty} \frac{3^n}{(n+1)x^n}$

53. $\sum_{n=1}^{\infty} \frac{4^{n+3}}{n^2(x-2)^n}$

54. $\sum_{n=1}^{\infty} \frac{(-1)^n 5^{4n-1}}{n(x+3)^{2n}}$

51. $\sum_{n=1}^{\infty} \frac{n!}{x^n}$

52. $\sum_{n=1}^{\infty} \frac{5^n}{n!x^n}$

55. $\sum_{n=1}^{\infty} \frac{16^n}{(3x-12)^{2n}}$

56. $\sum_{n=1}^{\infty} \frac{n!+4n}{(n+2)!(2x+6)^n}$

11.2 TAYLOR'S FORMULA

Often we want to study the behavior of function values $f(x)$ for x near a point x_0. For example, we might be studying motion at times t near a time t_0 or studying the current in an electric circuit for values R of the resistance near a value R_0. In this section, we will approximate functions near a point x_0 by polynomial functions.

THE TAYLOR POLYNOMIAL OF DEGREE n

We have seen that a power series $\sum_{n=0}^{\infty} a_n(x-x_0)^n$ converges in an interval having x_0 as center point. The sum function of the series is easiest to evaluate at $x = x_0$, where we obtain the value a_0, the constant term of the series.

If $f(x) = 4 - x + 2x^2 + x^3$, then $f(0) = 4$ is the easiest value of $f(x)$ to compute. Also, $f'(0) = -1$ is the easiest value to compute of $f'(x) = -1 + 4x + 3x^2$, and $f''(0) = 4$ is the easiest value to compute of $f''(x) = 4 + 6x$. We think of $f(x)$ as being expressed by a polynomial formula **centered at** zero.

On the other hand, if $g(x) = 2 + 3(x+1) - 2(x+1)^2 - 4(x+1)^3$, then $g(-1) = 2$ is the easiest value of $g(x)$ to find. Also, $g'(-1) = 3$ is the easiest value to find for $g'(x) = 3 - 4(x+1) - 12(x+1)^2$, and $g''(-1) = -4$ is the easiest value to find for $g''(x) = -4 - 24(x+1)$. Here, $g(x)$ is expressed by a polynomial formula centered at -1.

The next example indicates that any polynomial function can be expressed by a polynomial formula centered at x_0 for any real number x_0.

EXAMPLE 1 Express $f(x) = 4 - x + 2x^2 + x^3$ by a polynomial formula centered at $x_0 = 2$.

Solution We simply replace x in our given formula by $(x-2)+2$ and expand using the binomial theorem. We have

$$\begin{aligned} f(x) &= 4 - x + 2x^2 + x^3 \\ &= 4 - [(x-2)+2] + 2[(x-2)+2]^2 + [(x-2)+2]^3 \\ &= 4 - [(x-2)+2] + 2[(x-2)^2 + 4(x-2) + 4] \\ &\quad + [(x-2)^3 + 6(x-2)^2 + 12(x-2) + 8] \\ &= (4 - 2 + 8 + 8) + (-1 + 8 + 12)(x-2) \\ &\quad + (2+6)(x-2)^2 + (x-2)^3 \\ &= 18 + 19(x-2) + 8(x-2)^2 + (x-2)^3. \quad \square \end{aligned}$$

Our goal in this section and the next is to try to find a power series centered at x_0 that is equal to a given function throughout some interval having center x_0. In this section, we approach the problem by approximating the function near x_0 by a polynomial expression of degree n centered at x_0. In Section 11.3 we move to series by taking the limit of our approximations as $n \to \infty$.

Let $f(x)$ be a function defined in a neighborhood of x_0, that is, for $x_0 - h < x < x_0 + h$ for some $h > 0$, and let $f^{(n)}(x_0)$ exist. We want to find the polynomial

$$g(x) = a_0 + a_1(x - x_0) + a_2(x - x_0)^2 + \cdots + a_n(x - x_0)^n$$

of degree n centered at x_0, which is the best approximation to $f(x)$ near x_0. Surely we want to require that $g(x_0) = f(x_0)$ and that $g'(x_0) = f'(x_0)$, so that the functions have the same slope at x_0. If we also require that $g''(x_0) = f''(x_0)$, then the rates of change of slope will be the same; the graphs will bend or curve at the same rate at x_0. It seems reasonable to require that $g(x)$ have as many as possible of the same derivative values at x_0 as $f(x)$ has. We determine the coefficients $a_0, a_1, a_2, \ldots, a_n$ to make this true. Computing, we easily find that

$$\begin{aligned}
g(x) &= a_0 + a_1(x - x_0) + a_2(x - x_0)^2 + \cdots + a_n(x - x_0)^n, \\
g'(x) &= a_1 + 2a_2(x - x_0) + 3a_3(x - x_0)^2 + \cdots + na_n(x - x_0)^{n-1}, \\
g''(x) &= 2a_2 + 3 \cdot 2a_3(x - x_0) + \cdots + n(n-1)a_n(x - x_0)^{n-2}, \\
&\vdots \\
g^{(n)}(x) &= n(n-1)(n-2)\cdots 3 \cdot 2 \cdot 1 \cdot a_n = n!a_n.
\end{aligned}$$

Then

$$\begin{aligned}
g(x_0) &= a_0, \\
g'(x_0) &= a_1, \\
g''(x_0) &= 2a_2, \\
g'''(x_0) &= 3 \cdot 2a_3 = 3!a_3, \\
&\vdots \\
g^{(n)}(x_0) &= n!a_n.
\end{aligned}$$

Setting these equal to the corresponding derivatives of $f(x)$ at x_0, we have

$$\begin{aligned}
f(x_0) &= a_0, & & a_0 = f(x_0), \\
f'(x_0) &= a_1, & & a_1 = f'(x_0), \\
f''(x_0) &= 2a_2, & & a_2 = \frac{1}{2!}f''(x_0), \\
f'''(x_0) &= 3!a_3, & \text{which lead to} \quad & a_3 = \frac{1}{3!}f'''(x_0), \\
&\vdots & & \vdots \\
f^{(n)}(x_0) &= n!a_n, & & a_n = \frac{1}{n!}f^{(n)}(x_0).
\end{aligned}$$

DEFINITION Taylor polynomials

11.2

The polynomial function

$$g(x) = T_n(x) = f(x_0) + f'(x_0)(x - x_0) + \frac{f''(x_0)}{2!}(x - x_0)^2$$

$$+ \frac{f'''(x_0)}{3!}(x - x_0)^3 + \cdots + \frac{f^{(n)}(x_0)}{n!}(x - x_0)^n \qquad \textbf{(1)}$$

$$= \sum_{i=0}^{n} \frac{f^{(i)}(x_0)}{i!}(x - x_0)^i$$

is the nth **Taylor polynomial for** $f(x)$ **at** x_0. (We define $f^{(0)}(x) = f(x)$ and $0! = 1$ so that we may include the constant term $f(x_0)$ in our formal sum.)

In Definition 11.2, we said "the nth Taylor polynomial" rather than "the Taylor polynomial of degree n" since it is possible that $f^{(n)}(x_0) = 0$ so that the degree of $T_n(x)$ is less than n. These polynomials are named in honor of the English mathematician Brook Taylor (1685–1731).

EXAMPLE 2 Find the Taylor polynomial $T_3(x)$ at $x_0 = 2$ for the polynomial function $f(x) = 4 - x + 2x^2 + x^3$ of Example 1.

Solution Computing derivatives, we obtain

$$f(2) = (4 - x + 2x^2 + x^3)|_2 = 4 - 2 + 8 + 8 = 18,$$

$$f'(2) = (-1 + 4x + 3x^2)|_2 = -1 + 8 + 12 = 19,$$

$$f''(2) = (4 + 6x)|_2 = 4 + 12 = 16,$$

$$f'''(2) = 6|_2 = 6.$$

Formula (1) then yields

$$T_3(x) = 18 + 19(x - 2) + \frac{16}{2!}(x - 2)^2 + \frac{6}{3!}(x - 2)^3$$

$$= 18 + 19(x - 2) + 8(x - 2)^2 + (x - 2)^3.$$

Note that this is the same polynomial expression centered at $x_0 = 2$ that we obtained in Example 1 for $f(x)$. The polynomial formula centered at x_0 for a polynomial function must of course yield the same derivatives at every point as the expression centered at zero. The argument before Definition 11.2 then shows that if $f(x)$ is a polynomial function of degree n, then the polynomial expression for $f(x)$ centered at x_0 must be $T_n(x)$ at x_0, as given by formula (1). □

EXAMPLE 3 Find the Taylor polynomial $T_7(x)$ for the function sin x at $x_0 = 0$.

Solution We have to compute sin 0 and the derivatives of orders less than or equal to 7 of sin x at $x_0 = 0$. For $f(x) = \sin x$, we obtain $f(0) = \sin 0 = 0$, while

$$f'(0) = \cos 0 = 1, \qquad f''(0) = -\sin 0 = 0,$$

$$f'''(0) = -\cos 0 = -1, \qquad f^{(4)}(0) = \sin 0 = 0.$$

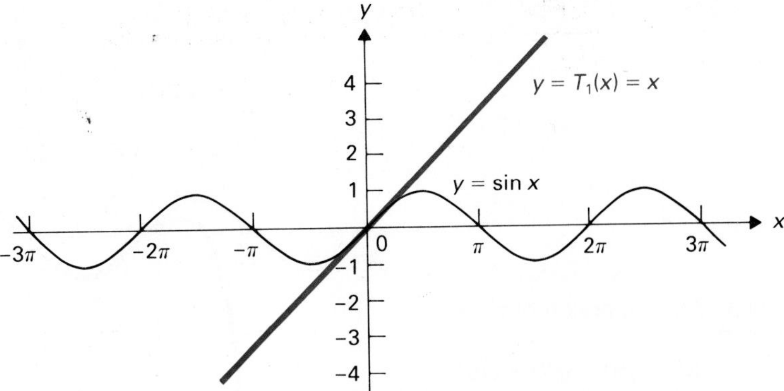

Figure 11.1 Approximation of $\sin x$ near zero by $T_1(x) = x$.

At this point, we note that since $f^{(4)}(x) = f(x) = \sin x$, the derivatives will start repeating. Thus the derivatives of $\sin x$ at 0, starting with the first derivative, are

$$1,\ 0,\ -1,\ 0,\ 1,\ 0,\ -1,\ 0,\ 1,\ 0,\ -1,\ 0,\ \ldots$$

for as far as we wish to take them. Since $x_0 = 0$, we have $x - x_0 = x$, and the seventh Taylor polynomial is

$$T_7(x) = 0 + 1 \cdot x + \frac{0}{2!}x^2 + \frac{-1}{3!}x^3 + \frac{0}{4!}x^4 + \frac{1}{5!}x^5 + \frac{0}{6!}x^6 + \frac{-1}{7!}x^7$$

$$= x - \frac{x^3}{3!} + \frac{x^5}{5!} - \frac{x^7}{7!}. \quad \square$$

Figures 11.1–11.4 show the approximation to $\sin x$ by $T_n(x)$ at $x_0 = 0$ for $n = 1, 3, 9$, and 19. The graphs of the approximations seem to cling to the graph of $\sin x$ farther and farther out as n increases. Think what might happen in the next section when we let n approach infinity!

EXAMPLE 4 Find the Taylor polynomial $T_8(x)$ for the function $\cos x$ at $x_0 = \pi$.

Solution As in Example 3, we easily find that $\cos \pi = -1$ and that the derivatives of $\cos x$ at π, starting with the first derivative, are

$$0,\ 1,\ 0,\ -1,\ 0,\ 1,\ 0,\ -1,\ 0,\ 1,\ 0,\ -1,\ \ldots$$

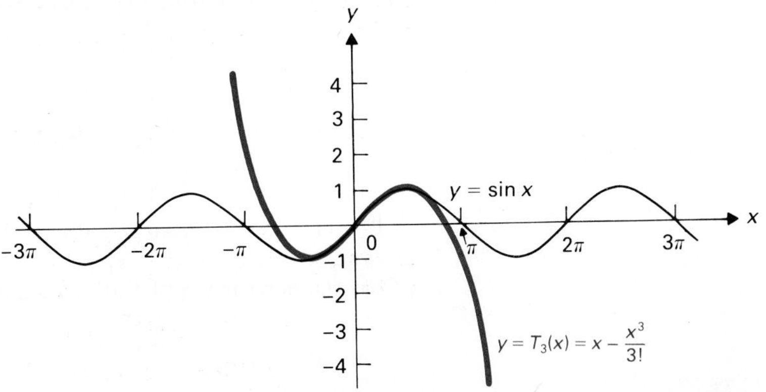

Figure 11.2 Approximation of $\sin x$ near zero by $T_3(x) = x - x^3/3!$.

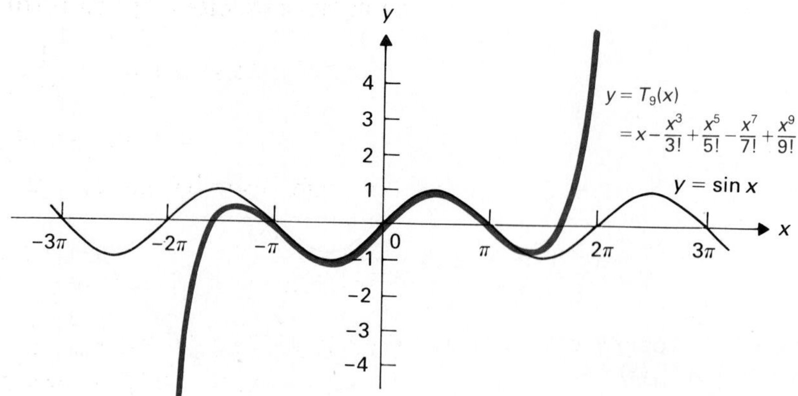

Figure 11.3 Approximation of $\sin x$ near zero by $T_9(x) = x - x/3! + x/5! - x/7! + x/9!$

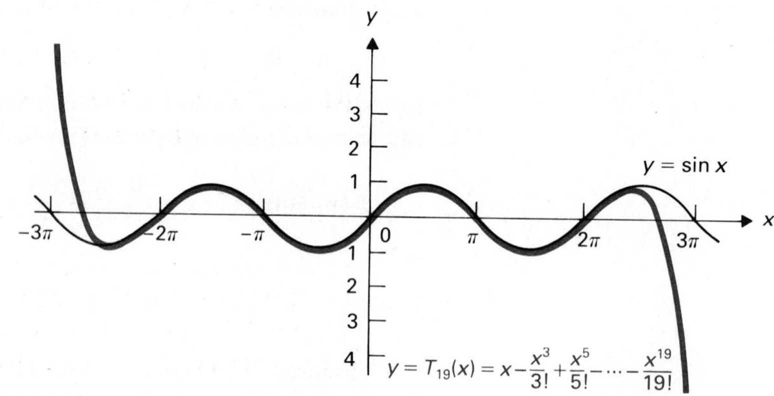

Figure 11.4 Approximation of $\sin x$ near zero by $T_{19}(x) = x - x^3/3! + x^5/5! - x^7/7! + \cdots - x^{19}/19!$

for as far as we wish to take them. Dropping the terms with coefficients zero, we obtain the polynomial

$$T_8(x) = -1 + \frac{(x-\pi)^2}{2!} - \frac{(x-\pi)^4}{4!} + \frac{(x-\pi)^6}{6!} - \frac{(x-\pi)^8}{8!}. \quad \square$$

EXAMPLE 5 Find the fifth Taylor polynomial $T_5(x)$ at $x_0 = 0$ for f, where $f(x) = 1/(1-x) = (1-x)^{-1}$.

Solution Computing derivatives, we obtain

$$\begin{aligned} f'(x) &= (1-x)^{-2}, \\ f''(x) &= 2(1-x)^{-3}, \\ f'''(x) &= 3 \cdot 2(1-x)^{-4} = 3!(1-x)^{-4}, \\ f^{(4)}(x) &= 4!(1-x)^{-5}, \\ f^{(5)}(x) &= 5!(1-x)^{-6}, \ldots . \end{aligned}$$

Thus the derivatives of f at zero, starting with the first derivative, are

$$1!, \quad 2!, \quad 3!, \quad 4!, \quad 5!, \quad 6!, \quad 7!, \quad 8!, \quad \ldots$$

for as far as we wish to go. Since $f(0) = 1$, we obtain

$$T_5(x) = 1 + x + \frac{2!}{2!}x^2 + \frac{3!}{3!}x^3 + \frac{4!}{4!}x^4 + \frac{5!}{5!}x^5$$
$$= 1 + x + x^2 + x^3 + x^4 + x^5$$

as the fifth Taylor polynomial. □

TAYLOR'S THEOREM

Let $f(x)$ and x_0 be as described in the last section, so that we can consider the Taylor polynomial $T_n(x)$. We expect to have

$$f(x) \approx T_n(x)$$

for x close to x_0. Naturally, we are interested in the accuracy of this approximation. We would hope that the *error* $E_n(x)$ given by

$$E_n(x) = f(x) - T_n(x) \tag{2}$$

is small if x is close to x_0. The following theorem gives some information on the size of $E_n(x)$, and for this reason the theorem is extremely important.

THEOREM 11.2 Taylor's theorem

Let $f(x)$ be defined for $x_0 - h < x < x_0 + h$, let its derivatives of orders less than or equal to $n + 1$ exist throughout that interval, and let $E_n(x) = f(x) - T_n(x)$. Then for each x in that interval, there exists a number c depending on x and strictly between x and x_0 (for $x \neq x_0$) such that

$$E_n(x) = \frac{f^{(n+1)}(c)}{(n+1)!}(x - x_0)^{n+1}. \tag{3}$$

Exercise 1 asks you to show that for $n = 0$, Taylor's theorem actually reduces to the mean-value theorem, presented on page 156 of Chapter 5. We will prove Taylor's theorem using Rolle's theorem, which you recall is a special case of the mean-value theorem, but we defer the proof to the end of this section.

Taylor's theorem is really very easy to remember. For f, x, and x_0 as described in the theorem, we have

$$\boxed{f(x) = T_n(x) + E_n(x) = T_n(x) + \frac{f^{(n+1)}(c)}{(n+1)!}(x - x_0)^{n+1}.}$$

Note that $E_n(x)$ is precisely what we would have for the "next" term of degree $n + 1$ in the Taylor polynomial $T_{n+1}(x)$, *except that the derivative* $f^{(n+1)}$ *must be computed at some* c *between* x_0 *and* x *instead of at* x_0.

We illustrate a few types of applications of Taylor's theorem. We are usually concerned with obtaining a *bound* B on the size of $E_n(x)$, that is, with finding a number $B > 0$ such that

$$|E_n(x)| \leq B.$$

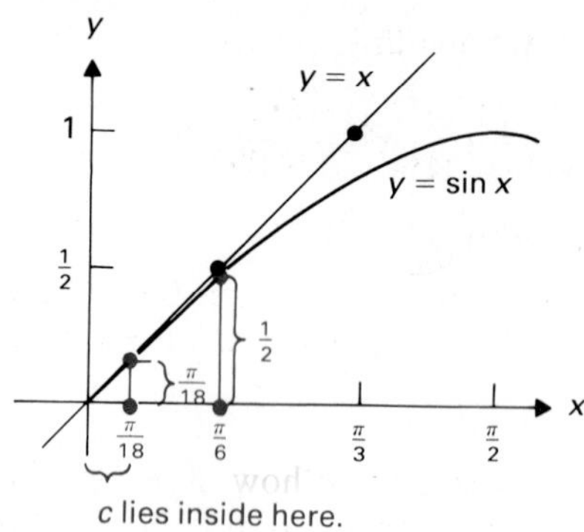

Figure 11.5 For $0 < c < \pi/18$, $\sin c < \sin(\pi/6) = \frac{1}{2}$; also,

$\sin c < \sin(\pi/18) < \pi/18$.

EXAMPLE 6 Estimate $\sin 10° = \sin(\pi/18)$, using $T_3(x)$ at $x_0 = 0$. Then find a bound for the absolute error, $|E_3(\pi/18)|$.

Solution (Remember that we developed calculus for trigonometric functions using *radian measure.*) The computations in Example 3 show that at $x_0 = 0$,

$$T_3(x) = x - \frac{x^3}{3!}.$$

Then

$$T_3\left(\frac{\pi}{18}\right) = \frac{\pi}{18} - \frac{1}{6}\left(\frac{\pi}{18}\right)^3 = 0.1736468.$$

Now if $f(x) = \sin x$, then $f^{(4)}(x) = \sin x$ also. Then from formula (3),

$$\left|E_3\left(\frac{\pi}{18}\right)\right| = \left|\frac{\sin c}{4!}\left(\frac{\pi}{18}\right)^4\right| \tag{4}$$

for some c where $0 < c < \pi/18$. Our task is to find a reasonable bound on $|\sin c|$ for $0 < c < \pi/18$. Of course $|\sin c| < 1$, but that is a rather crude bound for c in this interval. From Fig. 11.5, we see that $|\sin c| < \sin(\pi/6) = \frac{1}{2}$. We could also use the fact that $\sin x \le x$ for $x \ge 0$, as shown in the figure. Thus $|\sin c| < \sin(\pi/18) < \pi/18$. Thus we obtain, from Eq. (4),

$$\left|E_3\left(\frac{\pi}{18}\right)\right| < \frac{1}{4!}\left(\frac{\pi}{18}\right)^5 \approx 0.0000068.$$

Our calculator shows that $\sin(\pi/18) \approx 0.1736482$, so the actual error in our approximation is about 0.0000014, which is indeed less than the bound 0.0000068 that we obtained.

We could improve our bound on the error by noting that $T_4(x) = T_3(x)$ for $\sin x$ at $x_0 = 0$, as shown by our computations in Example 3. Thus $T_4(\pi/18) = T_3(\pi/18)$, but formula (3) for $E_4(\pi/18)$ and the fact that $|\cos x| \le 1$ show that

$$\left|E_4\left(\frac{\pi}{18}\right)\right| = \left|\frac{\cos c}{5!}\left(\frac{\pi}{18}\right)^5\right| < \frac{1}{5!}\left(\frac{\pi}{18}\right)^5 \approx 0.0000014.$$

Note the extra factor 5 in the denominator from the 5!, versus the 4! in the estimate in the preceding paragraph. This bound for the error is closer to the actual error than the one obtained using $E_3(\pi/18)$. □

EXAMPLE 7 Consider approximating sin 46° by using a Taylor polynomial $T_n(x)$ for $f(x) = \sin x$ at $x_0 = \pi/4$. Find a value of n such that the error in the approximation would be less than 0.00001.

Solution We need to have $|E_n(46\pi/180)| < 0.00001$. Since the derivatives of $\sin x$ are all either $\pm\sin x$ or $\pm\cos x$, we have $|f^{(n+1)}(c)| < 1$ for any n and c. Since $x - x_0 = \pi/180$, we have, from Eq. (2),

$$\left|E_n\left(\frac{46\pi}{180}\right)\right| < \frac{1}{(n+1)!}\left(\frac{\pi}{180}\right)^{n+1}.$$

Since $(\pi/180) < 1/50$, we surely have

$$\left|E_n\left(\frac{46\pi}{180}\right)\right| < \frac{1}{(n+1)!}\left(\frac{1}{50}\right)^{n+1} = \frac{1}{(n+1)!5^{n+1}\cdot 10^{n+1}}.$$

We try a few values of n and discover that for $n = 2$ we obtain

$$\left|E_2\left(\frac{46\pi}{180}\right)\right| < \frac{1}{3!5^3 \cdot 10^3} = \frac{1}{6 \cdot 125 \cdot 1000} < \frac{1}{100000}$$
$$= 0.00001.$$

Thus we may use $n = 2$ with safety to achieve the desired accuracy. (Our calculator shows that $\sin 46° \approx 0.7193398$, while $T_2(46\pi/180) \approx 0.71934042$.) □

The preceding examples give some indication of how tables for the trigonometric functions were constructed. Also, rather than use space in the memory of an electronic computer to store tables of trigonometric functions, the manufacturer builds into the computer a program that the machine uses to estimate values of these functions as it needs them, as we estimated values in Examples 6 and 7.

Estimating $f(x)$ near x_0 using $T_1(x)$ at x_0 amounts to estimating $f(x)$ using the tangent line to its graph at x_0. This must then amount to estimating using differentials, which we discussed in Chapter 3. Indeed, if we set $x - x_0 = dx$, so $x = x_0 + dx$, we obtain

$$T_1(x) = f(x_0) + f'(x_0)(x - x_0) = f(x_0) + f'(x_0)\,dx.$$

We then have the approximation formula

$$f(x) = f(x_0 + dx) \approx T_1(x_0 + dx) = f(x_0) + f'(x_0)\,dx, \tag{5}$$

which is precisely the formula for approximation using differentials. Theorem 11.2 now tells us that the error in approximating by differentials is

$$\boxed{E_1(x_0 + dx) = \frac{f''(c)}{2!}(dx)^2} \tag{6}$$

for some c between x_0 and $x_0 + dx$.

EXAMPLE 8 Use a differential to estimate $\sqrt{101}$, and find a bound for the error.

Solution We let $f(x) = x^{1/2}$, $x_0 = 100$, and $dx = 1$. Our estimate using a differential is simply

$$T_1(x_0 + dx) = f(x_0) + f'(x_0)\,dx.$$

Since $f'(x) = \frac{1}{2}x^{-1/2}$, we obtain

$$T_1(101) = \sqrt{100} + \frac{1}{2} \cdot \frac{1}{\sqrt{100}} \cdot 1 = 10 + \frac{1}{20} = 10.05$$

as our estimate for $\sqrt{101}$.

Now $f''(x) = (-\frac{1}{4})x^{-3/2}$, so by the error (6), we have

$$|E_1(101)| = \frac{1}{2!} \cdot \left|-\frac{1}{4}c^{-3/2}\right| \cdot 1 = \frac{1}{8} \cdot \frac{1}{c^{3/2}}$$

for some c where $100 < c < 101$. But for such c, the largest values of $1/c^{3/2}$

must occur where $c^{3/2}$ is smallest, that is, for c close to 100, so

$$\frac{1}{c^{3/2}} < \frac{1}{100^{3/2}} = \frac{1}{1000}.$$

Thus

$$|E_1(101)| < \frac{1}{8} \cdot \frac{1}{1000} = \frac{1}{8000} = 0.000125.$$

Therefore our estimate 10.05 for $\sqrt{101}$ is accurate to at least three decimal places. □

The preceding example illustrates again the *typical procedure* for finding a bound on $|E_n(x)|$. We don't know the *precise value of* c, so we attempt to find the *maximum possible value* that $|f^{(n+1)}(c)|$ can have for any c *over the entire interval* from x_0 to x. If the exact maximum of $|f^{(n+1)}(c)|$ is hard to find, we find the best easily computed bound we can for $|f^{(n+1)}(c)|$.

EXAMPLE 9 Consider $T_n(x)$ for e^x at $x_0 = 0$. Show that for *every* real number a, we have

$$\lim_{n\to\infty} |e^a - T_n(a)| = 0.$$

This will show that for any a, we can approximate e^a accurately by $T_n(a)$ if n is large enough.

Solution Since all derivatives of $f(x) = e^x$ are again e^x, we have

$$E_n(a) = \frac{e^c}{(n+1)!} a^{n+1}$$

for some c between 0 and a. For $a > 0$ we have $e^c < e^a$, and for $a \leq 0$ we have $e^c \leq e^0 = 1$. Thus

$$|e^a - T_n(a)| = |E_n(a)| \leq \begin{cases} \left|\dfrac{e^a \cdot a^{n+1}}{(n+1)!}\right| & \text{for } a > 0, \\[2ex] \left|\dfrac{a^{n+1}}{(n+1)!}\right| & \text{for } a \leq 0. \end{cases}$$

Now $(n + 1)!$ dominates $|a^n|$ as $n \to \infty$, so we have

$$\lim_{n\to\infty} \frac{|a|^{n+1}}{(n+1)!} = 0,$$

which shows that $\lim_{n\to\infty} |E_n(a)| = 0$. □

EXAMPLE 10 Consider $T_n(x)$ for $f(x) = \ln x$ at $x_0 = 1$. Show that $\lim_{n\to\infty} |(\ln 2) - T_n(2)| = 0$, so that $\ln 2$ can be approximated by $T_n(2)$ when n is large enough.

Solution Computing $f^{(n+1)}(x)$, we find that

$$f(x) = \ln x,$$

$$f'(x) = \frac{1}{x},$$

$$f''(x) = -\frac{1}{x^2},$$

$$f'''(x) = \frac{2}{x^3},$$

$$f^{(4)}(x) = -\frac{3!}{x^4},$$

$$f^{(5)}(x) = \frac{4!}{x^5},$$

$$\vdots$$

$$f^{(n+1)}(x) = \frac{(-1)^n n!}{x^{n+1}}.$$

Then for some c where $1 < c < 2$, we have

$$|(\ln 2) - T_n(2)| = |E_n(2)| = \left|\frac{(-1)^n n!(2-1)^{n+1}}{c^{n+1}(n+1)!}\right|$$

$$< \frac{n!}{1^{n+1}(n+1)!} = \frac{1}{n+1}.$$

Since $\lim_{n\to\infty} [1/(n+1)] = 0$, we must have $\lim_{n\to\infty} |(\ln 2) - T_n(2)| = 0$. □

PROOF OF TAYLOR'S THEOREM

Proof. For $x = x_0$, the theorem is obvious. Choose $x \neq x_0$ where $x_0 - h < x < x_0 + h$; we think of x as remaining fixed throughout this proof; in particular, we consider x to be a constant in any differentiation. For such a fixed x, there exists a unique number A such that

$$f(x) = f(x_0) + f'(x_0)(x - x_0) + \frac{f''(x_0)}{2!}(x - x_0)^2 + \cdots$$

$$+ \frac{f^{(n)}(x_0)}{n!}(x - x_0)^n + \frac{A}{(n+1)!}(x - x_0)^{n+1}. \tag{7}$$

Namely, we could solve Eq. (7) for A. We now define a function $F(t)$ of the variable t by

$$F(t) = f(x) - f(t) - f'(t)(x - t) - \frac{f''(t)}{2!}(x - t)^2$$

$$- \cdots - \frac{f^{(n)}(t)}{n!}(x - t)^n - \frac{A}{(n+1)!}(x - t)^{n+1}.$$

We see from Eq. (7) that A was chosen in such a way that $F(x_0) = 0$. Note that $F(x) = 0$ also. We now apply Rolle's theorem to $F(t)$ and deduce that $F'(c) = 0$ for some c between x_0 and x. Now

$$F'(t) = -f'(t) - [f'(t)(-1) + (x - t)f''(t)]$$

$$- \left[f''(t)(x - t)(-1) + \frac{f'''(t)}{2!}(x - t)^2\right]$$

$$- \cdots - \left[\frac{f^{(n)}(t)}{(n-1)!}(x - t)^{n-1}(-1) + \frac{f^{(n+1)}(t)}{n!}(x - t)^n\right]$$

$$- \frac{A}{n!}(x - t)^n(-1).$$

Many terms cancel in this expression for $F'(t)$; in fact, the first term in each square bracket cancels with the last term of the preceding square bracket. After this cancellation, we are left with only

$$F'(t) = -\frac{f^{(n+1)}(t)}{n!}(x-t)^n + \frac{A}{n!}(x-t)^n;$$

thus

$$F'(c) = -\frac{f^{(n+1)}(c)}{n!}(x-c)^n + \frac{A}{n!}(x-c)^n = 0,$$

and, dividing through by the common factor $(x-c)^n/n!$, we obtain

$$-f^{(n+1)}(c) + A = 0,$$

so

$$A = f^{(n+1)}(c).$$

Substituting this expression for A in Eq. (7), we obtain Taylor's theorem. •

SUMMARY

1. If $f(x)$ has derivatives of orders less than or equal to n at x_0, then the nth Taylor polynomial for $f(x)$ at x_0 is

$$T_n(x) = f(x_0) + f'(x_0)(x-x_0) + \frac{f''(x_0)}{2!}(x-x_0)^2 + \frac{f'''(x_0)}{3!}(x-x_0)^3 + \cdots + \frac{f^{(n)}(x_0)}{n!}(x-x_0)^n = \sum_{i=0}^{n} \frac{f^{(i)}(x_0)}{i!}(x-x_0)^i.$$

The polynomial $T_n(x)$ and the function $f(x)$ have the same derivatives of orders less than or equal to n at x_0.

2. A polynomial function $f(x)$ of degree n expressed in powers of x can also be expressed as a polynomial centered at x_0, that is, in powers of $x - x_0$. Such an expression can be found either by replacing x by $(x - x_0) + x_0$, or by computing $T_n(x)$ at x_0.

3. Taylor's theorem states that, under suitable conditions,

$$f(x) = T_n(x) + E_n(x),$$

where the error expression (remainder term) $E_n(x)$ is given by

$$E_n(x) = \frac{f^{(n+1)}(c)}{(n+1)!}(x-x_0)^{n+1}$$

for some c (depending on x) between x_0 and x.

EXERCISES

1. State the special case of Taylor's theorem for $n = 0$. What familiar theorem do you obtain?

In Exercises 2 through 13, find the polynomial formula centered at the indicated point x_0 for the given polynomial function. Use the method in either Example 1 or Example 2 where indicated.

2. $x - 2$ at $x_0 = 3$, Example 1
3. $2x + 4$ at $x_0 = -1$, Example 1
4. $\frac{1}{2}x + 5$ at $x_0 = 3$, Example 2
5. $-5x + 2$ at $x_0 = -2$, Example 2
6. $3x^2 + 2x - 4$ at $x_0 = 1$, Example 1
7. $\frac{1}{2}x^2 - 3x + 1$ at $x_0 = -2$, Example 1
8. $3x^2 - x - 1$ at $x_0 = 5$, Example 2
9. $-4x^2 + 3x - 2$ at $x_0 = -3$, Example 2
10. $2x^3 - x^2 + 4x - 2$ at $x_0 = 2$
11. $x^4 - 3x$ at $x_0 = -1$
12. $3x^4 - 4x^3 + 2x^2 - x + 1$ at $x_0 = 1$
13. $-5x^4 + 2x^3 - 3x^2 + 2x - 1$ at $x_0 = -2$
14. Show that if f is a polynomial function of degree r, then $T_n(x) = T_r(x)$ at any point x_0 for $n \geq r$.

In Exercises 15 through 26, find the indicated Taylor polynomial for the function at the point.

15. $T_{10}(x)$ for $\cos x$ at $x_0 = 0$
16. $T_5(x)$ for $1/(1 + x)$ at $x_0 = 0$
17. $T_6(x)$ for $\sin x$ at $x_0 = \pi/2$
18. $T_2(x)$ for $\tan x$ at $x_0 = 0$
19. $T_6(x)$ for $1/x$ at $x_0 = 1$
20. $T_3(x)$ for $\sqrt{x}$ at $x_0 = 4$
21. $T_4(x)$ for $x^2 + x - \cos x$ at $x_0 = 0$
22. $T_5(x)$ for $3x + 2\cos x$ at $x_0 = \pi/2$
23. $T_8(x)$ for $x^3 + e^x$ at $x_0 = 0$
24. $T_5(x)$ for $3x - e^x$ at $x_0 = -1$
25. $T_4(x)$ for $e^x + \ln(1 + x)$ at $x_0 = 0$
26. $T_{10}(x)$ for $x^5 + e^{-x}$ at $x_0 = 0$
27. a) Find the Taylor polynomial $T_7(x)$ for $\sin 2x$ at $x_0 = 0$.
    b) Compare your answer in part (a) with Example 3. What do you notice?
28. a) Find the Taylor polynomial $T_3(x)$ for $\sin x^2$ at $x_0 = 0$.
    b) Compare your answer in part (a) with Example 3, as in Exercise 7(b).
    c) Guess the Taylor polynomial $T_{10}(x)$ for $\sin x^2$ at $x_0 = 0$.
29. a) Find the Taylor polynomial $T_3(x)$ at $x_0 = 0$ for $1/(1 - x^2)$.
    b) Compare your answer in part (a) with Example 5. What do you notice?
    c) Guess the Taylor polynomial $T_8(x)$ at $x_0 = 0$ for $1/(1 - x^2)$.
30. a) Find $T_5(x)$ for $3/(1 + 2x)$ at $x_0 = 0$.
    b) Find a geometric power series centered at zero that equals $3/(1 + 2x)$ in its interval of convergence.
    c) Compare your answers in parts (a) and (b).
31. a) Find $T_4(x)$ for $2/(1 + x)$ at $x_0 = 1$.
    b) Find a geometric power series centered at 1 that equals $2/(1 + x)$ in its interval of convergence.
    c) Compare your answers in parts (a) and (b).
32. a) Find $T_3(x)$ for $5x/(2 + 4x)$ at $x_0 = 0$.
    b) Find a geometric power series centered at zero that equals $5x/(2 + 4x)$ in its interval of convergence.
    c) Compare your answers in parts (a) and (b).
33. Let f and g be functions that have derivatives of orders less than or equal to n at zero, and let $g(x) = f(cx)$. Show that if $T_n(x)$ is the nth Taylor polynomial for f at zero, then $T_n(cx)$ is the nth Taylor polynomial for g at zero.
34. Let f and g be functions that have derivatives of orders less than or equal to n at x_0. Show that the nth Taylor polynomial for $f + g$ at x is the sum of the Taylor polynomials for f and g at x_0.
35. a) Estimate 2.98^3 using a differential.
    b) Find a bound for the error in your estimate in part (a).
36. a) Estimate $\sqrt[3]{28}$ using a differential.
    b) Find a bound for the error in your estimate in part (a).
37. a) Using a differential, estimate the change in volume of a cylindrical silo 20 ft high if the radius is increased from 3 ft to 3 ft 1 in.
    b) Find a bound for the error in your estimate in part (a).
38. a) Using a differential, estimate the change in the volume of a spherical ball of radius 2 ft if the radius is increased by 1 in.
    b) Find a bound for the error in your estimate in part (a).
39. Repeat Exercise 38(a) and (b), but estimate the increase in volume using the Taylor polynomial $T_2(r)$ at $r_0 = 2$, rather than estimating using a differential.
    c) What would be the error if you estimated using $T_3(r)$?
40. Find a bound for the error if $T_8(x)$ at $x_0 = 0$ is used to estimate $\cos 3°$ by
    a) finding a bound for $E_8(\pi/60)$;
    b) finding a bound for $E_9(\pi/60)$. Why can one use $E_9(\pi/60)$ as a bound for the error in $T_8(\pi/60)$?
41. a) Estimate $\tan 2°$ using the Taylor polynomial $T_2(x)$ at $x_0 = 0$.
    b) Find a bound for the error in your estimate in part (a).
42. a) Estimate $\sqrt{1.05} + 0.96^3$ using two Taylor polynomials of degree 2.
    b) Find a bound for the error in your estimate in part (a).
43. a) Using a calculator, estimate e by using $T_8(x)$ for e^x at $x_0 = 0$.
    b) Find a bound for the error in part (a).
44. a) Using a calculator, estimate $\sin 15°$ using $T_6(x)$ for $\sin x$ at $x_0 = 0$.
    b) Find a bound for the error in part (a).
45. a) Using a calculator, estimate $\cosh 1$ using $T_7(x)$ for $\cosh x$ at $x_0 = 0$.
    b) Find a bound for the error in part (a).
46. a) Using a calculator, estimate $\sinh 1$ using $T_8(x)$ for $\sinh x$ at $x_0 = 0$.
    b) Find a bound for the error in part (a).

47. Use a calculator to find the smallest value of n such that the terms of degree less than or equal to n of the series for $\sin x$ about $x_0 = 0$ can be used to find $\sin 1$ with error less than 10^{-6}. How does this value of n compare with that given by the error term in Taylor's formula if all derivatives of $\sin x$ at $x = c$ are replaced by 1 in bounding the error?

48. Repeat Exercise 48, but use the series for e^x about $x_0 = 0$ to estimate $\sqrt{e}$, and answer the second part if all derivatives e^c of e^x are replaced by 2 in bounding the error.

In Exercises 49 through 57, show that $\lim_{n\to\infty} |f(x) - T_n(x)| = 0$ for the indicated function $f(x)$, point x_0 where $T_n(x)$ is centered, and indicated values of x.

49. $f(x) = e^{-x}$, $x_0 = 0$, all x

50. $f(x) = \sin x$, $x_0 = 0$, all x

51. $f(x) = \cos x$, $x_0 = 0$, all x

52. $f(x) = \sinh x$, $x_0 = 0$, all x

53. $f(x) = \cosh x$, $x_0 = 0$, all x

54. $f(x) = e^x - e^{3x}$, $x_0 = 0$, all x

55. $f(x) = \ln x$, $x_0 = 1$, $1 \le x \le 2$

56. $f(x) = x^{3/2}$, $x_0 = 1$, $1 \le x \le 2$

57. $f(x) = \sqrt{x + 2}$, $x_0 = 0$, $0 \le x \le 2$

11.3 TAYLOR SERIES; REPRESENTATION OF A FUNCTION

TAYLOR SERIES

Taylor's theorem suggests that, for a function having derivatives of *all* orders in a neighborhood $x_0 - h < x < x_0 + h$ of x_0, we consider the power series

$$\sum_{n=0}^{\infty} \frac{f^{(n)}(x_0)}{n!}(x - x_0)^n$$

centered at x_0. This series is the **Taylor series of $f(x)$ at x_0**. (If $x_0 = 0$, the series is often called the **Maclaurin series of $f(x)$**.)

When we speak of "the Taylor series of f at x_0," we will understand that f has derivatives of all orders in some neighborhood of x_0. The Taylor series **represents $f(x)$ at x_1** if the series converges to $f(x_1)$ where $x = x_1$. The Taylor series of a function f at x_0 certainly represents f at x_0; because it converges at x_0 to its constant term, $f(x_0)$.

Unfortunately, it is not always true that the Taylor series represents f throughout the neighborhood of x_0 where the derivatives exist. In extreme cases, the series may represent f only at the point x_0 itself. One can show that the function f defined by

$$f(x) = \begin{cases} e^{-1/x^2} & \text{for } x \neq 0, \\ 0 & \text{for } x = 0, \end{cases}$$

has derivatives of all orders everywhere; in particular, one can show that $f^{(n)}(0) = 0$ for all n. The Taylor series of f at $x_0 = 0$ is therefore

$$0 + 0x + 0x^2 + 0x^3 + \cdots + 0x^n + \cdots,$$

which represents f only at 0.

A necessary and sufficient condition for the Taylor series of f at x_0 to represent f at a point $x_1 \neq x_0$ is easily obtained from Taylor's theorem.

THEOREM Representation

11.3

Let f have derivatives of all orders in a neighborhood $x_0 - h < x < x_0 + h$ of x_0. The Taylor series of f at x_0 represents f at x_1 in this neighborhood if and only if $\lim_{n\to\infty} E_n(x_1) = 0$, where

$$E_n(x_1) = \frac{f^{(n+1)}(c)}{(n+1)!}(x_1 - x_0)^{n+1}$$

is the error term in Taylor's theorem (page 479).

Proof. For the Taylor series, the nth partial sum s_n at $x = x_1$ is given by

$$s_n(x_1) = T_n(x_1) = \sum_{i=0}^{n} \frac{f^{(i)}(x_0)}{i!}(x_1 - x_0)^i.$$

By Taylor's theorem, we have

$$f(x_1) - s_n(x_1) = f(x_1) - T_n(x_1) = E_n(x_1).$$

Thus for any $\varepsilon > 0$, we have $|f(x_1) - s_n(x_1)| < \varepsilon$ if and only if $|E_n(x_1)| < \varepsilon$, so the conditions $\lim_{n\to\infty} s_n(x_1) = f(x_1)$ and $\lim_{n\to\infty} E_n(x_1) = 0$ are equivalent. •

The following corollary of Theorem 11.3 will usually suffice for our purposes.

COROLLARY Simultaneously bounded derivatives

Let f have derivatives of all orders in $x_0 - h < x < x_0 + h$. If there is a number $B > 0$ such that $|f^{(n)}(x)| \le B$ for all positive integers n and for all x such that $x_0 - h < x < x_0 + h$, then the Taylor series of f at x_0 represents f throughout the neighborhood $x_0 - h < x < x_0 + h$.

Proof. By hypothesis, we have, for $x_0 - h < x_1 < x_0 + h$,

$$\begin{aligned}\lim_{n\to\infty} |E_n(x_1)| &= \lim_{n\to\infty} \left| f^{(n+1)}(c)\frac{(x_1 - x_0)^{n+1}}{(n+1)!}\right| \\ &\le B\left(\lim_{n\to\infty} \frac{|x_1 - x_0|^{n+1}}{(n+1)!}\right) \\ &= B \cdot 0 = 0,\end{aligned}$$

so, by Theorem 11.3, the Taylor series of f represents f at x_1. •

EXAMPLE 1 Find a power series centered at $x_0 = 0$ representing e^x.

Solution All derivatives of the function e^x are again e^x, and the Taylor series of e^x at $x_0 = 0$ is easily found to be $\sum_{n=0}^{\infty} (x^n/n!)$. Since e^x is bounded by e^b in every interval $-b < x < b$, the corollary shows that the series converges to e^x for all x; that is,

$$\boxed{e^x = 1 + x + \frac{x^2}{2!} + \frac{x^3}{3!} + \cdots + \frac{x^n}{n!} + \cdots} \qquad \textbf{(1)}$$

for all x. Remember this series for e^x; it occurs frequently. □

EXAMPLE 2 Find power series centered at $x_0 = 0$ representing $\sin x$ and $\cos x$.

Solution Derivatives of sine and cosine are again just sine and cosine functions, and these are bounded by 1 everywhere. The Taylor series for these functions at any x_0 therefore represent them everywhere. Computing the Taylor series at $x_0 = 0$, we easily find that

$$\boxed{\sin x = x - \frac{x^3}{3!} + \frac{x^5}{5!} - \frac{x^7}{7!} + \cdots + (-1)^n \frac{x^{2n+1}}{(2n+1)!} + \cdots} \tag{2}$$

and

$$\boxed{\cos x = 1 - \frac{x^2}{2!} + \frac{x^4}{4!} - \frac{x^6}{6!} + \cdots + (-1)^n \frac{x^{2n}}{(2n)!} + \cdots} \tag{3}$$

for all x. Remember these series also. □

DIFFERENTIATION AND INTEGRATION OF POWER SERIES

We now turn to the study of functions that are represented throughout some neighborhood of x_0 by a power series at x_0. Such functions are very important and are called analytic at x_0. A function is **analytic** if it is analytic at each point in its domain.

EXAMPLE 3 Show that e^x, $\sin x$, and $\cos x$ are analytic.

Solution Examples 1 and 2 show that these functions are analytic at $x_0 = 0$. The arguments in those examples concerning the bounds on their derivatives can be made in a neighborhood of any point. Thus e^x, $\sin x$, and $\cos x$ are analytic everywhere. □

It can be shown that if f is analytic at x_0 and is represented by a power series throughout $x_0 - h < x < x_0 + h$, then f is analytic at every point in $x_0 - h < x < x_0 + h$. Thus, since the Taylor series for e^x, $\sin x$, and $\cos x$ at $x_0 = 0$ represent these functions for all x, we see again that these three functions are analytic at each point. Of course, the coefficients are different in two Taylor series representing one function like e^x if the series are centered at different points.

The calculus of analytic functions ("infinite polynomial functions") is much like the calculus of the polynomial functions. If f is analytic at x_0, then f has derivatives of all orders at x_0, and these derivatives can be computed by differentiating the series at x_0 representing f, just as we would differentiate a polynomial. Antiderivatives can be found similarly. The next theorem states this formally. A proof is left to a more advanced course.

THEOREM 11.4 Term-by-term differentiation and integration

Let f be analytic at x_0 and let $\sum_{n=0}^{\infty} a_n(x - x_0)^n$ represent f in $x_0 - h < x < x_0 + h$, so that

$$f(x) = \sum_{n=0}^{\infty} a_n(x - x_0)^n$$

in that neighborhood. Then f has derivatives of all orders throughout this neighborhood, and the derivatives at any x in the neighborhood may be computed by differentiating the series term by term. Thus

$$f'(x) = \sum_{n=1}^{\infty} n \cdot a_n(x - x_0)^{n-1}$$

for $x_0 - h < x < x_0 + h$. The indefinite integral of f in this same neighborhood can be obtained by integrating the series term by term:

$$\int f(x)\,dx = C + \sum_{n=0}^{\infty} \frac{a_n}{n+1}(x - x_0)^{n+1},$$

where C is an arbitrary constant.

EXAMPLE 4 Illustrate term-by-term differentiation for the series for e^x at $x_0 = 0$.

Solution By Eq. (1), we have

$$e^x = 1 + x + \frac{x^2}{2!} + \frac{x^3}{3!} + \cdots + \frac{x^n}{n!} + \cdots.$$

The derivative of this series is

$$0 + 1 + \frac{2x}{2!} + \frac{3x^2}{3!} + \cdots + \frac{nx^{n-1}}{n!} + \cdots = 1 + x + \frac{x^2}{2!} + \cdots$$
$$+ \frac{x^{n-1}}{(n-1)!} + \cdots,$$

which is the same series and again represents e^x. We expect this since the derivative of e^x is again e^x. □

EXAMPLE 5 Illustrate Theorem 11.4 by differentiating and integrating the series representing $1/(1 - x)$ near $x_0 = 0$.

Solution The function $1/(1 - x)$ is analytic at zero, for we know that the geometric series

$$\sum_{n=0}^{\infty} x^n = 1 + x + x^2 + x^3 + \cdots + x^n + \cdots$$

converges to $1/(1 - x)$ in $-1 < x < 1$. Thus

$$\frac{d}{dx}\left(\frac{1}{1-x}\right) = \frac{1}{(1-x)^2}$$

is analytic at zero, and we obtain by differentiation

$$\frac{1}{(1-x)^2} = \sum_{n=1}^{\infty} nx^{n-1}$$
$$= 1 + 2x + 3x^2 + 4x^3 + \cdots + nx^{n-1} + \cdots$$

for $-1 < x < 1$.

Since $\int 1/(1-x)\,dx = -\ln|1-x| + C$, we find by integrating the geometric series that

$$-\ln(1-x) = k + \sum_{n=0}^{\infty} \frac{x^{n+1}}{n+1}$$

$$= k + x + \frac{x^2}{2} + \frac{x^3}{3} + \cdots + \frac{x^n}{n} + \cdots$$

for some constant k and $-1 < x < 1$. Putting $x = 0$, we find that $k = -\ln 1 = 0$, so

$$\ln(1-x) = -x - \frac{x^2}{2} - \frac{x^3}{3} - \cdots - \frac{x^n}{n} - \cdots \qquad \textbf{(4)}$$

for $-1 < x < 1$. □

UNIQUENESS OF POWER SERIES REPRESENTATION

It is an important fact that if f is analytic at x_0, then there is *only one* power series at x_0 that represents f throughout a neighborhood of x_0. This must be the Taylor series, for the coefficients of a power series representing the function are determined by the derivatives of the function to be precisely the coefficients in the Taylor series, as indicated on pages 475 and 476.

Using this uniqueness, we can find the Taylor series for many functions f without differentiating f to compute coefficients. We illustrate with examples.

EXAMPLE 6 Find the power series at $x_0 = 0$ representing $\sin x^2$.

Solution We know that

$$\sin x = x - \frac{x^3}{3!} + \cdots + (-1)^n \frac{x^{2n+1}}{(2n+1)!} + \cdots$$

for all x. Replacing x by x^2, we have

$$\sin x^2 = x^2 - \frac{x^6}{3!} + \cdots + (-1)^n \frac{x^{4n+2}}{(2n+1)!} + \cdots \qquad \textbf{(5)}$$

for all x. Therefore the series (5) must be the Taylor series for $\sin x^2$ at $x_0 = 0$. If we try to find the Taylor series by differentiating $\sin x^2$ repeatedly, we quickly appreciate the easy way we obtained series (5). □

EXAMPLE 7 If $f(x) = \sin x^2$, find $f^{(6)}(0)$.

Solution The power series for $\sin x^2$ at $x_0 = 0$ is given in series (5). We know this series must be the Taylor series, so the coefficient of x^6 must be the Taylor coefficient $f^{(6)}(0)/6!$. From series (5), we then obtain

$$\frac{f^{(6)}(0)}{6!} = -\frac{1}{3!},$$

so

$$f^{(6)}(0) = -\frac{1}{3!}\cdot 6! = -6\cdot 5\cdot 4 = -120. \quad □$$

EXAMPLE 8 Find the Taylor series for $\ln(1-x)$ at $x_0 = 0$.

Solution In Eq. (4), we obtained a power series centered at $x_0 = 0$ that represented $\ln(1-x)$ for $-1 < x < 1$. By uniqueness, this must be the Taylor series. □

EXAMPLE 9 Find the Taylor series for $\tan^{-1}x$ at $x_0 = 0$.

Solution Now $d(\tan^{-1}x)/dx = 1/(1+x^2)$. We can represent $1/(1+x^2)$ as the sum of a geometric series with initial term 1 and ratio $-x^2$:

$$\frac{1}{1+x^2} = 1 - x^2 + x^4 - x^6 + \cdots + (-1)^n x^{2n} + \cdots \tag{6}$$

for $-1 < x < 1$. Integrating Eq. (6), we find that

$$\tan^{-1}x = k + x - \frac{x^3}{3} + \frac{x^5}{5} - \frac{x^7}{7} + \cdots + (-1)^n \frac{x^{2n+1}}{2n+1} + \cdots$$

for some constant k. Putting $x = 0$, we see that $k = 0$, so

$$\tan^{-1}x = x - \frac{x^3}{3} + \frac{x^5}{5} - \frac{x^7}{7} + \cdots + (-1)^n \frac{x^{2n+1}}{2n+1} + \cdots \tag{7}$$

for $-1 < x < 1$. The series (7) must be the Taylor series of $\tan^{-1}x$ at $x_0 = 0$. If we try to compute the series (7) by differentiating $\tan^{-1}x$ repeatedly, we quickly appreciate the easy way we obtained the series (7). □

EXAMPLE 10 Let $f(x) = \tan^{-1}(x^2)$. Find $f^{(10)}(0)$.

Solution From the series (7), we see that the Taylor series for $\tan^{-1}(x^2)$ must start

$$\tan^{-1}(x^2) = x^2 - \frac{x^6}{3} + \frac{x^{10}}{5} - \frac{x^{14}}{7} + \cdots.$$

Using the formula for the Taylor coefficient of x^{10}, we obtain

$$\frac{f^{(10)}(0)}{10!} = \frac{1}{5},$$

so

$$f^{(10)}(0) = \frac{10!}{5}. \quad \square$$

We take a moment to give a preview of coming attractions, if you ever study *functions of complex variables*. The function $1/(1-x)$ can be expressed as the sum of a geometric power series near $x_0 = 0$ by

$$\frac{1}{1-x} = 1 + x + x^2 + x^3 + \cdots + x^n + \cdots. \tag{8}$$

This series converges only for $-1 < x < 1$, but we are not surprised because it represents the function $1/(1-x)$, which "blows up" at $x = 1$. The function $1/(1+x^2)$ has derivatives of all orders everywhere and can be shown to be analytic at every point x_0. However, its Taylor series (6) at $x_0 = 0$ still only converges to $1/(1+x^2)$ for $-1 < x < 1$. To fully appreciate why this happens, one must study complex analysis. The function $1/(1 + x^2)$ "blows up" at $x = i$ and $x = -i$, and in complex analysis one sees that the numbers i and $-i$ have distance 1 from $x_0 = 0$. It is for this reason that the radius of convergence of the Taylor series for $1/(1 + x^2)$ at $x_0 = 0$ is only 1.

MULTIPLICATION AND DIVISION OF POWER SERIES

Two power series at x_0 representing functions f and g in $x_0 - h < x < x_0 + h$ can be multiplied and divided as "infinite" polynomials, to yield power series representing the functions fg and f/g in neighborhoods of x_0, with the obvious restriction that $g(x_0) \neq 0$. We state this as a theorem without proof. The functions f and g can be approximated at each point of $x_0 - h < x < x_0 + h$ as closely as we like by polynomial partial sums of the series, so the theorem seems reasonable.

THEOREM 11.5 Series multiplication and division

Let series $\sum_{n=0}^{\infty} a_n(x - x_0)^n$ and $\sum_{n=0}^{\infty} b_n(x - x_0)^n$ converge to functions f and g, respectively, in $x_0 - h < x < x_0 + h$. Then the product series (called the **Cauchy product**)

$$a_0 b_0 + (a_0 b_1 + a_1 b_0)(x - x_0) + (a_0 b_2 + a_1 b_1 + a_2 b_0)(x - x_0)^2 + \cdots,$$

whose nth coefficient is $\sum_{i=0}^{n} (a_i b_{n-i})$, represents fg throughout $x_0 - h < x < x_0 + h$. Also, if $b_0 = g(x_0) \neq 0$, the series

$$\frac{a_0}{b_0} + \frac{a_1 b_0 - a_0 b_1}{b_0^2}(x - x_0) + \cdots,$$

obtained by long division, represents f/g in some neighborhood $x_0 - \delta < x < x_0 + \delta$.

We are accustomed to polynomial long division, where we write the polynomials with the terms of highest degree first and divide only until we obtain a remainder of lower degree than the divisor. *In series long division, we write the terms of lowest degree first and the division may never terminate;* symbolically,

$$\begin{array}{r|l}
 & \dfrac{a_0}{b_0} + \dfrac{a_1 b_0 - a_0 b_1}{b_0^2}(x - x_0) + \cdots \\
\hline
b_0 + b_1(x - x_0) + \cdots & a_0 \quad + \quad a_1(x - x_0) + \cdots \\
 & a_0 \quad + \dfrac{a_0 b_1}{b_0}(x - x_0) + \cdots \\
\cline{2-2}
 & 0 + \dfrac{a_1 b_0 - a_0 b_1}{b_0}(x - x_0) + \cdots
\end{array}$$

etc.

An illustration of such division appears in Example 12 below.

EXAMPLE 11 Use series multiplication to find the first few terms of the Taylor series for $e^x \sin x$ at $x_0 = 0$.

Solution Since

$$e^x = 1 + x + \frac{x^2}{2} + \frac{x^3}{6} + \frac{x^4}{24} + \frac{x^5}{120} + \cdots$$

and

$$\sin x = x - \frac{x^3}{6} + \frac{x^5}{120} - \cdots,$$

we obtain

$$\begin{aligned} e^x \sin x &= x + x^2 + \left(\frac{x^3}{2} - \frac{x^3}{6}\right) + \left(\frac{x^4}{6} - \frac{x^4}{6}\right) \\ &\quad + \left(\frac{x^5}{24} - \frac{x^5}{12} + \frac{x^5}{120}\right) + \cdots \\ &= x + x^2 + \frac{x^3}{3} - \frac{x^5}{30} + \cdots \end{aligned}$$

for all x. □

EXAMPLE 12 Find the Taylor series for $(1 + x^2)/(1 - x)$ at $x_0 = 0$ in two ways.

Solution The computation using series division is

$$\begin{array}{r|l} & 1 + x + 2x^2 + 2x^3 + \cdots \\ \hline 1 - x & 1 \qquad + x^2 \\ & 1 - x \\ \hline & \quad x + x^2 \\ & \quad x - x^2 \\ \hline & \qquad 2x^2 \\ & \qquad 2x^2 - 2x^3 \\ \hline & \qquad\quad 2x^3 \\ & \qquad\quad 2x^3 - 2x^4 \\ \hline & \qquad\qquad 2x^4 \end{array}$$

etc.

We thus obtain

$$\frac{1 + x^2}{1 - x} = 1 + x + 2x^2 + 2x^3 + 2x^4 + \cdots + 2x^n + \cdots$$

for $-1 < x < 1$. Alternatively, we could multiply the geometric series $1 + x + x^2 + \cdots + x^n + \cdots$ for $1/(1 - x)$ by $1 + x^2$. The computation is

$$\begin{array}{r} 1 + x + x^2 + x^3 + \cdots + x^n + x^{n+1} + \cdots \\ \times\, 1 + x^2 \\ \hline 1 + x + x^2 + x^3 + \cdots + x^n + x^{n+1} + \cdots \\ x^2 + x^3 + \cdots + x^n + x^{n+1} + \cdots \\ \hline 1 + x + 2x^2 + 2x^3 + \cdots + 2x^n + 2x^{n+1} + \cdots \end{array}$$

and we obtain the same series, as we must by uniqueness. If we try to compute the Taylor series for $(1 + x^2)/(1 - x)$ at $x_0 = 0$ by repeated differentiation, we quickly appreciate the easy ways we found it here. □

Sometimes we can recognize an elementary function that a given series represents.

EXAMPLE 13 Find the elementary function represented by the series

$$10x + 2x^2 - \frac{x^3}{3!} + \frac{x^5}{5!} - \frac{x^7}{7!} + \cdots + (-1)^n \frac{x^{2n+1}}{(2n+1)!} + \cdots$$

for $n \geq 1$.

Solution If the first two terms, $10x + 2x^2$, of the series were replaced by x, we would have the series for $\sin x$. Since

$$10x + 2x^2 = x + (9x + 2x^2),$$

we see that the given series represents $9x + 2x^2 + \sin x$. □

EXAMPLE 14 Find the elementary function represented by the series

$$3 + x - \frac{x^3}{3} + \frac{x^5}{5} - \frac{x^7}{7} + \cdots + (-1)^n \frac{x^{2n+1}}{2n+1} + \cdots$$

for $n \geq 1$ and $|x| < 1$.

Solution We can get rid of the troublesome denominators by differentiating. Let the given series represent $f(x)$. Then

$$f'(x) = 1 - x^2 + x^4 - x^6 + \cdots + (-1)^n x^{2n} + \cdots.$$

We recognize this series as geometric with initial term 1 and ratio $-x^2$. Thus

$$f'(x) = \frac{1}{1 - (-x^2)} = \frac{1}{1 + x^2}.$$

Integrating, we find that $f(x) = k + \tan^{-1}x$ for some k and $|x| < 1$. From our original series for $f(x)$, we see that $f(0) = 3$, so $3 = k + \tan^{-1}(0) = k$. Thus the series represents $3 + \tan^{-1}x$ for $|x| < 1$. □

SUMMARY

1. If $f(x)$ has derivatives of all orders at x_0, then
$$\sum_{n=0}^{\infty} \frac{f^{(n)}(x_0)}{n!}(x - x_0)^n$$
is the Taylor series of $f(x)$ at x_0. (Here $0! = 1$ and $f^{(0)}(x) = f(x)$.)
2. The Taylor series of $f(x)$ at x_0 represents $f(x)$ at x_1 if and only if $\lim_{n\to\infty} E_n(x_1) = 0$. This condition will always hold if all derivatives between x_0 and x_1 are bounded by the same constant B.
3. A function is analytic at x_0 if it is represented by some power series in some neighborhood of x_0. It is analytic if it is analytic at each point in its domain.

4. If $f(x)$ is represented by a power series in an open interval, then $f'(x)$ is represented by the term-by-term derivative of that power series, and $\int f(x)\,dx$ by the term-by-term antiderivative of that power series, plus an arbitrary constant.
5. The only power series at x_0 that can represent $f(x)$ in a neighborhood of x_0 is the Taylor series of $f(x)$.
6. Series at x_0 representing $f(x)$ and $g(x)$ in a common neighborhood of x_0 may be multiplied (as infinite polynomials) to obtain the series representing $f(x)g(x)$ in that neighborhood and divided if $g(x_0) \neq 0$ to represent $f(x)/g(x)$ in *some* neighborhood of x_0.
7. It is convenient to remember the following series.

$$e^x = 1 + x + \frac{x^2}{2!} + \frac{x^3}{3!} + \cdots + \frac{x^n}{n!} + \cdots \qquad \text{for all } x$$

$$\sin x = x - \frac{x^3}{3!} + \frac{x^5}{5!} - \cdots + (-1)^n \frac{x^{2n+1}}{(2n+1)!} + \cdots \qquad \text{for all } x$$

$$\cos x = 1 - \frac{x^2}{2!} + \frac{x^4}{4!} - \cdots + (-1)^n \frac{x^{2n}}{(2n)!} + \cdots \qquad \text{for all } x$$

$$\frac{1}{1-x} = 1 + x + x^2 + x^3 + \cdots + x^n + \cdots \qquad \text{for } -1 < x < 1$$

$$\ln(1-x) = -x - \frac{x^2}{2} - \frac{x^3}{3} - \cdots - \frac{x^n}{n} - \cdots \qquad \text{for } -1 < x < 1$$

EXERCISES

1. Mark each of the following true or false.
   a) If f has derivatives of all orders throughout some neighborhood of x_0, then f is analytic at x_0.
   b) If f is analytic at x_0, then f has derivatives of all orders throughout some neighborhood of x_0.
   c) If f and g are analytic at x_0, then $f + g$ is analytic at x_0.
   d) Every power series represents a function that is analytic at every point (except possibly endpoints) of the interval of convergence of the series.
   e) There is at most one power series at x_0 that represents a given function f at x_0.
   f) There is at most one power series at x_0 that represents a given function f throughout some neighborhood of x_0.
2. Is the function $\sqrt{x}$ analytic at $x_0 = 0$? Why?

In Exercises 3 through 24, find as many terms of the Taylor series of the function at the given point as you conveniently can in the easiest way you can.

3. $x^2 + e^x$ at $x_0 = 0$
4. $1 + x^3 - \sin x$ at $x_0 = 0$
5. $x \sin x$ at $x_0 = 0$
6. $\cos x$ at $x_0 = \pi$
7. $\dfrac{x}{1-x}$ at $x_0 = 0$
8. e^{-x^2} at $x_0 = 0$
9. $\cos x^3$ at $x_0 = 0$
10. $\dfrac{2x + 3x^2}{1 + 4x}$ at $x_0 = 0$
11. $e^x \cos x$ at $x_0 = 0$
12. e^{3x} at $x_0 = 0$
13. $\dfrac{1}{(1+x)^2}$ at $x_0 = 0$
14. $\ln(\cos x)$ at $x_0 = 0$
15. $\sec x$ at $x_0 = 0$
16. $\ln x$ at $x_0 = 2$
17. $\dfrac{e^x}{1-x}$ at $x_0 = 0$
18. $\dfrac{e^x + e^{-x}}{2}$ at $x_0 = 0$
19. $\dfrac{e^x - e^{-x}}{2}$ at $x_0 = 0$
20. $\sqrt{x}$ at $x_0 = 1$
21. $\dfrac{1}{e^x}$ at $x_0 = 0$
22. $\dfrac{x-1}{x^3}$ at $x_0 = 1$
23. $\sec x \tan x$ at $x_0 = 0$
24. $\dfrac{1}{2-x}$ at $x_0 = 0$

In Exercises 25 through 32, use series methods to find the indicated derivative of the given function.

25. $f(x) = \dfrac{x^2}{1 + x^2}$, find $f^{(4)}(0)$

26. $f(x) = \sin(3x^2)$, find $f^{(6)}(0)$

27. $f(x) = e^{-x^2/4}$, find $f^{(4)}(0)$

28. $f(x) = \tan^{-1}\left(\dfrac{x^3}{2}\right)$, find $f^{(3)}(0)$

29. $f(x) = \tan^{-1}\left(\dfrac{x^2}{4}\right)$, find $f^{(10)}(0)$

30. $f(x) = \ln(1 - 2x^2)$, find $f^{(6)}(0)$

31. $f(x) = x^4 \sin 2x$, find $f^{(5)}(0)$

32. $f(x) = \dfrac{\tan^{-1} x^2}{1 - x^2}$, find $f^{(4)}(0)$

33. a) Find the terms for $n \leq 5$ of the Taylor series of $\sin x \cos x$ at $x_0 = 0$ by series multiplication.
b) Find the Taylor series of $\sin x \cos x$ at $x_0 = 0$ by the use of the identity $\sin x \cos x = (\sin 2x)/2$.

34. Find the Taylor series of $1/x$ at $x_0 = 2$ by
a) differentiating $1/x$ repeatedly to compute the coefficients,
b) using the identity
$$\frac{1}{x} = \frac{1}{2} \cdot \frac{1}{1 - [-(x - 2)/2]}$$
and expanding in a geometric series.

35. Use the technique suggested by Exercise 34(b) to find the Taylor series of $1/x$ at $x = -1$; you have to find the appropriate identity.

36. Find the terms for $n \leq 3$ of the Taylor series of $\tan x$ at $x_0 = 0$ by dividing the series for $\sin x$ by the series for $\cos x$.

37. a) Obtain the series expansion
$$\ln(1 + x) = x - \frac{x^2}{2} + \frac{x^3}{3} - \frac{x^4}{4} + \cdots + (-1)^{n+1}\frac{x^n}{n} + \cdots$$
for $-1 < x < 1$. [*Hint:* You may use Eq. (4) or integrate the geometric series for $1/(1 + x) = 1/(1 - (-x))$.]
b) Show that the series in part (a) converges for $x = 1$.
c) Show that the alternating harmonic series
$$1 - \tfrac{1}{2} + \tfrac{1}{3} - \tfrac{1}{4} + \cdots$$
converges to $\ln 2$. [*Hint:* Since 1 is an endpoint of the interval of convergence of the series in part (a), Theorem 11.4 cannot be used. You must check $\lim_{n\to\infty} E_n(1)$ for the function $\ln(1 + x)$.]

38. Find the series at $x_0 = 0$ representing the function f defined by
$$f(x) = \int_0^x \ln(1 - t)\, dt$$
for $-1 < x < 1$.

39. Find the series expansion at $x_0 = 0$ for the indefinite integral of e^{x^2}.

40. Find the series at $x_0 = 0$ representing the function f defined by
$$f(x) = \int_0^x [1/(1 - t^3)]\, dt$$
for $-1 < x < 1$.

41. Find the series at $x_0 = 0$ representing the function f defined by $f(x) = \pi + \int_0^x \cos t^2\, dt$ for all x.

42. Find the series at $x_0 = 0$ representing the function f defined by
$$f(x) = \int_0^x \frac{(3 + t^2)}{(1 + 2t)}\, dt$$
for $-\frac{1}{2} < x < \frac{1}{2}$.

43. a) Proceeding purely formally, find the Taylor series at $x_0 = 0$ of e^{ix} and of e^{-ix}, where $i^2 = -1$.
b) From part (a), "derive" Euler's formula $e^{ix} = \cos x + i(\sin x)$.
c) From part (a), "derive" the formula $e^{-ix} = \cos x - i(\sin x)$.
d) From parts (b) and (c), find formulas for $\cos x$ and $\sin x$ in terms of the complex exponential function.
e) Compare the formulas for $\sin x$ and $\cos x$ found in part (d) with the formulas for $\sinh x$ and $\cosh x$ in terms of the exponential function.

Exercises 44 through 55 give you practice in series recognition. The given series represents a familiar elementary function in its interval of convergence. Find the function.

44. $1 - x + x^2 - x^3 + \cdots + (-1)^n x^n + \cdots$

45. $1 + x + x^2 - x^3 + x^4 - x^5 + \cdots + (-1)^n x^n + \cdots$ for $n \geq 2$

46. $1 - x + \dfrac{x^2}{2!} - \dfrac{x^3}{3!} + \dfrac{x^4}{4!} - \cdots + (-1)^n \dfrac{x^n}{n!} + \cdots$

47. $1 + 3x - \dfrac{x^2}{2!} + \dfrac{x^4}{4!} - \cdots + (-1)^n \dfrac{x^{2n}}{(2n)!} + \cdots$ for $n \geq 1$

48. $1 - \dfrac{x}{2} + \dfrac{x^2}{4} - \dfrac{x^3}{8} + \cdots + \dfrac{(-1)^n}{2^n} x^n + \cdots$

49. $1 + x - \dfrac{x^2}{2!} - \dfrac{x^3}{3!} + \dfrac{x^4}{4!} + \dfrac{x^5}{5!} - \dfrac{x^6}{6!} - \dfrac{x^7}{7!} + \cdots$

50. $1 - x^3 + x^6 - x^9 + x^{12} - \cdots + (-1)^n x^{3n} + \cdots$

51. $1 + 2x + \dfrac{2! + 1}{2!} x^2 + \dfrac{3! + 1}{3!} x^3 + \cdots + \dfrac{n! + 1}{n!} x^n + \cdots$ for $n \geq 2$

52. $-1 + 2x - 3x^2 + 4x^3 - \cdots + (-1)^{n+1}(n + 1)x^n + \cdots$
[*Hint:* Integrate the series.]

53. $2 + 3 \cdot 2x + 4 \cdot 3x^2 + 5 \cdot 4x^3 + \cdots + (n + 2)(n + 1)x^n + \cdots$

54. $x^3 - \dfrac{x^7}{5!} + \dfrac{x^9}{7!} - \cdots + (-1)^n \dfrac{x^{2n+1}}{(2n - 1)!} + \cdots$ for $n \geq 3$

55. $4x^4 - 8x^6 + 16x^8 - 32x^{10} + \cdots + (-1)^{n+1} 2^{n+1} x^{2n+2} + \cdots$ for $n \geq 2$

11.4 INDETERMINATE FORMS

We pause in our study of series to give an application to the evaluation of indeterminate forms. The first indeterminate form we met was in the definition of the derivative,

$$f'(x_0) = \lim_{\Delta x \to 0} \frac{f(x_0 + \Delta x) - f(x_0)}{\Delta x}.$$

If we assume that f is continuous at x_0, the quotient whose limit is to be found has numerator and denominator both approaching zero. As we know from our many computations of derivatives, $f'(x_0)$ might be 2 or 3 or -50. In other words, knowing that the numerator and denominator of the quotient both approach zero does *not determine* the value of the limit, which is thus called of *indeterminate* type.

We will present two methods to compute such limits, a series method and l'Hôpital's rule, which we will prove by series methods. L'Hôpital's rule could have been proved in Chapter 5, after we studied the mean-value theorem. However, using series to analyze the behavior of a quotient is a much more transparent technique than applying l'Hôpital's rule. That is, it will be really easy to see *why* the series technique works. Demonstrating l'Hôpital's rule using series will make it clear why l'Hôpital's rule works, even in cases where the rule must be applied several times to find the limit.

TYPES OF INDETERMINATE FORMS

Note that

$$\lim_{x \to 1} \frac{(x-1)^2}{x-1} = 0, \qquad \lim_{x \to 1} \frac{2(x-1)}{x-1} = 2, \qquad \text{and} \qquad \lim_{x \to 1} \frac{(x-1)^2}{(x-1)^4} = \infty.$$

In each of these three cases, the function whose limit is being evaluated is undefined at $x = 1$, and a formal substitution of 1 in the numerator and denominator of the quotient leads to the expression "$0/0$" in each case. These examples show that one could define $0/0$ to be 0, 2, or ∞ with equal justification. It is for this reason that one does not attempt to define $0/0$; the expression "$0/0$" is an example of an *indeterminate form*. Similarly, the limits

$$\lim_{x \to 2+} \frac{1/(x-2)}{2/(x-2)} = \frac{1}{2} \qquad \text{and} \qquad \lim_{x \to 2+} \frac{1/(x-2)}{3/(x-2)^2} = 0,$$

where both the numerator and denominator in each limit approach ∞ as $x \to 2+$, show that we should consider ∞/∞ to be an indeterminate form.

Note that $0/\infty$ is not an indeterminate form, for if $\lim_{x \to a} f(x) = 0$ and $\lim_{x \to a} g(x) = \infty$, then $\lim_{x \to a} f(x)/g(x) = 0$. Also, $2/0$ is not an indeterminate form, for if $\lim_{x \to a} f(x) = 2$ and $\lim_{x \to a} g(x) = 0$, then $\lim_{x \to a} f(x)/g(x)$ is always undefined, and the quotient $f(x)/g(x)$ becomes large in absolute value as x approaches a.

> The *indeterminate quotient forms* are
>
> $$\frac{0}{0} \qquad \text{and} \qquad \pm\frac{\infty}{\infty}.$$

Turning to products, we find that the limits

$$\lim_{x\to 1+} (x-1)\frac{3}{x-1} = 3 \qquad \text{and} \qquad \lim_{x\to 1+} (x-1)^2\left(\frac{2}{x-1}\right) = 0$$

show that we should consider $0 \cdot \infty$ to be an indeterminate form.

> The *indeterminate product forms* are
>
> $0 \cdot \infty,\quad 0(-\infty),\quad \infty \cdot 0,\quad \text{and}\quad (-\infty)0.$

From

$$\lim_{x\to a+} \left(\frac{1}{x-a} - \frac{1}{x-a}\right) = 0$$

and

$$\lim_{x\to a+} \left(\frac{1}{x-a} - \frac{1+2a-2x}{x-a}\right) = \lim_{x\to a+} \frac{2(x-a)}{x-a} = 2, \tag{1}$$

we see that $\infty - \infty$ should be considered an indeterminate form.

> The *indeterminate sum and difference forms* are
>
> $(-\infty) + \infty \quad \text{and} \quad \infty - \infty.$

Finally, there are indeterminate exponential forms arising from expressions of the form $\lim_{x\to a} f(x)^{g(x)}$. An example of such an indeterminate form is furnished by

$$\lim_{x\to\infty} \left(1 + \frac{1}{x}\right)^x. \tag{2}$$

Note that as $x \to \infty$, the function $1 + 1/x > 1$ but is approaching 1. We know that 1 raised to any power again yields 1 but that a number just greater than 1 raised to a sufficiently large power can yield a very large number. Thus it is impossible to know the value of the limit in the expression (2) without some more work. The limit is of indeterminate form 1^∞. We will show in Example 5 that

$$\lim_{x\to\infty} \left(1 + \frac{1}{x}\right)^x = e.$$

This is a very famous limit.

Let us try to find all types of indeterminate exponential forms arising from $\lim_{x\to a} f(x)^{g(x)}$. Recall that we have defined the exponential r^s for all s only if $r > 0$; hence we assume that $f(x) > 0$ for $x \neq a$. Since the logarithm function is continuous and is the inverse of the exponential function, we see that if $\lim_{x\to a} \ln(f(x)^{g(x)}) = b$, then $\lim_{x\to a} f(x)^{g(x)} = e^b$. Now

$$\ln[f(x)^{g(x)}] = g(x)\ln(f(x)),$$

so for $\lim_{x\to a} f(x)^{g(x)}$ to give rise to an indeterminate exponential form, the product $g(x)\ln(f(x))$ must give rise to one of the indeterminate product forms $0 \cdot \infty$, $\infty \cdot 0$, $0(-\infty)$, or $(-\infty)0$ at $x = a$. The product $g(x)\ln(f(x))$ gives rise to $0 \cdot \infty$ at $x = a$ if $\lim_{x\to a} g(x) = 0$ and $\lim_{x\to a} \ln(f(x)) = \infty$, in which case $\lim_{x\to a} f(x) = \infty$ also. Thus $0 \cdot \infty$ gives rise to the indeterminate exponential form ∞^0. Similarly, the product form $\infty \cdot 0$ gives rise to the exponential form 1^∞, while $0(-\infty)$ gives rise to the form 0^0 and $(-\infty)0$ gives rise to $1^{-\infty}$.

> The *indeterminate exponential forms* are
>
> $0^0, \quad 1^\infty, \quad 1^{-\infty}, \quad \text{and} \quad \infty^0.$

FINDING LIMITS HAVING INDETERMINATE FORM BY SERIES METHODS

A limit corresponding to an indeterminate form is usually computed by trying to convert the problem to a limit corresponding to the indeterminate quotient form 0/0. For example, if $\lim_{x\to a} f(x) = 0$ and $\lim_{x\to a} g(x) = \infty$, then

$$\lim_{x\to a} f(x) \cdot g(x) = \lim_{x\to a} \frac{f(x)}{1/g(x)},$$

and the second limit corresponds to the indeterminate form 0/0. A mathematically terrifying but mnemonically helpful way to remember how to convert a $0 \cdot \infty$-type problem to a 0/0-type problem is to write

$$0 \cdot \infty = \frac{0}{1/\infty} = \frac{0}{0}.$$

Similarly, the mnemonic device

$$\frac{\infty}{\infty} = \frac{1/\infty}{1/\infty} = \frac{0}{0}$$

enables us to convert an ∞/∞-type problem to a 0/0-type problem. One usually tries to compute a limit corresponding to an indeterminate sum or difference form such as $\infty - \infty$ by the technique in Eq. (1), where the problem was again converted to a 0/0-type limit problem. We just saw that limits having indeterminate exponential form can be reduced to limits having indeterminate product form by taking logarithms, and we have just shown how a $0 \cdot \infty$-type limit problem can be converted to a 0/0-type problem. Thus we concentrate on the 0/0-type problem.

Let f and g be analytic at a, and suppose

$$\lim_{x\to a} f(x) = \lim_{x\to a} g(x) = 0.$$

We wish to find $\lim_{x\to a} f(x)/g(x)$. We assume that neither $f(x)$ nor $g(x)$ is identically zero throughout an entire neighborhood of a. Since f and g are analytic at a, we have, for x in some sufficiently small neighborhood of a,

$$f(x) = a_r(x-a)^r + a_{r+1}(x-a)^{r+1} + \cdots$$

and

$$g(x) = b_s(x-a)^s + b_{s+1}(x-a)^{s+1} + \cdots,$$

where a_r and b_s are the first nonzero coefficients in the series for f and g, respectively, at a. Now for a polynomial function or infinite series centered at a, *the nonzero monomial term of lowest degree dominates near a.* Thus we expect the behavior of $f(x)$ near a to be described by $a_r(x - a)^r$ and the behavior of $g(x)$ to be described by $b_s(x - a)^s$. Consequently, we expect that

$$\lim_{x \to a} \frac{f(x)}{g(x)} = \lim_{x \to a} \frac{a_r(x - a)^r}{b_s(x - a)^s}.$$

Our guess is easily demonstrated to be correct:

$$\lim_{x \to a} \frac{f(x)}{g(x)} = \lim_{x \to a} \frac{a_r(x - a)^r + a_{r+1}(x - a)^{r+1} + \cdots}{b_s(x - a)^s + b_{s+1}(x - a)^{s+1} + \cdots}$$

$$= \lim_{x \to a} \left[\frac{a_r(x - a)^r}{b_s(x - a)^s} \cdot \frac{1 + \dfrac{a_{r+1}}{a_r}(x - a) + \cdots}{1 + \dfrac{b_{s+1}}{b_s}(x - a) + \cdots} \right]$$

$$= \lim_{x \to a} \frac{a_r(x - a)^r}{b_s(x - a)^s}.$$

Of course, a limit of the form

$$\lim_{x \to a} \frac{a_r(x - a)^r}{b_s(x - a)^s}$$

is very easy to find. We have three cases.

Case 1 If $r > s$, the limit is zero.

Case 2 If $r = s$, the limit is a_r/b_s.

Case 3 If $r < s$, then

$$\lim_{x \to a+} \frac{a_r(x - a)^r}{b_s(x - a)^s} \quad \text{and} \quad \lim_{x \to a-} \frac{a_r(x - a)^r}{b_s(x - a)^s}$$

are ∞ or $-\infty$. The sign is determined by the signs of a_r and b_s and by whether $s - r$ is even or odd.

THEOREM 11.6 Dominating terms

Let f and g be analytic at a, and let a_r and b_s be the first nonzero coefficients in the series for f and g, respectively, at a. Then

$$\lim_{x \to a} \frac{f(x)}{g(x)} = \lim_{x \to a} \frac{a_r(x - a)^r}{b_s(x - a)^s}.$$

The same is true for one-sided limits at a.

EXAMPLE 1 Give a series derivation of the fundamental limit

$$\lim_{x \to 0} \frac{\sin x}{x} = 1.$$

Solution Now $\sin x$ is analytic at zero; in fact,

$$\sin x = x - \frac{x^3}{3!} + \frac{x^5}{5!} - \frac{x^7}{7!} + \cdots$$

for all x. By Theorem 11.6,

$$\lim_{x\to 0} \frac{\sin x}{x} = \lim_{x\to 0} \frac{x}{x} = 1. \quad \square$$

EXAMPLE 2 Compute $\lim_{x\to 0} (\cos x - 1)/x$ by series methods.

Solution We have

$$\cos x = 1 - \frac{x^2}{2!} + \frac{x^4}{4!} - \frac{x^6}{6!} + \cdots$$

for all x, so

$$\cos x - 1 = -\frac{x^2}{2!} + \frac{x^4}{4!} - \frac{x^6}{6!} + \cdots$$

for all x. By Theorem 11.6,

$$\lim_{x\to 0} \frac{\cos x - 1}{x} = \lim_{x\to 0} \frac{-x^2/2}{x} = \lim_{x\to 0} \frac{-x}{2} = 0. \quad \square$$

EXAMPLE 3 Compute

$$\lim_{x\to 0+} (\cot x)(\ln (1 - x)),$$

which corresponds to the indeterminate form $\infty \cdot 0$.

Solution We convert to a 0/0-type problem:

$$\lim_{x\to 0+} (\cot x)(\ln (1 - x)) = \lim_{x\to 0+} \frac{\ln (1 - x)}{1/(\cot x)} = \lim_{x\to 0+} \frac{\ln (1 - x)}{\tan x}.$$

We saw, in Example 5 on page 489, that

$$\ln (1 - x) = -x - \frac{x^2}{2} - \frac{x^3}{3} - \cdots$$

for $-1 < x < 1$, and series division shows that

$$\tan x = \frac{\sin x}{\cos x} = x + \frac{x^3}{3} + \cdots$$

for x sufficiently near 0. Our theorem then yields

$$\lim_{x\to 0+} \frac{\ln (1 - x)}{\tan x} = \lim_{x\to 0+} \frac{-x}{x} = -1. \quad \square$$

EXAMPLE 4 Compute

$$\lim_{x\to 0+} \left(\frac{2}{x} - \frac{x + 1}{x - x^2}\right),$$

which is an $(\infty - \infty)$-type problem.

Solution We have

$$\lim_{x\to 0+}\left(\frac{2}{x}-\frac{x+1}{x-x^2}\right)=\lim_{x\to 0+}\frac{2(x-x^2)-x(x+1)}{x^2-x^3}$$

$$=\lim_{x\to 0+}\frac{x-3x^2}{x^2-x^3}=\lim_{x\to 0+}\frac{x}{x^2}=\infty. \quad \square$$

Using the fact that $\lim_{x\to\infty} f(x)=\lim_{t\to 0+} f(1/t)$, we can often compute a limit at ∞ by series methods.

EXAMPLE 5 Compute the important limit

$$\lim_{x\to\infty}\left(1+\frac{1}{x}\right)^x,$$

which corresponds to the indeterminate form 1^∞.

Solution We make use of the relation $\lim_{x\to\infty} f(x)=\lim_{t\to 0+} f(1/t)$. Taking logarithms to handle the exponential indeterminate form, we reduce the computation to

$$\lim_{x\to\infty}\left[x\cdot\ln\left(1+\frac{1}{x}\right)\right]=\lim_{t\to 0+}\frac{1}{t}\ln(1+t)=\lim_{t\to 0+}\frac{\ln(1+t)}{t}.$$

Example 5 on page 489 shows that

$$\ln(1+t)=t-\frac{t^2}{2}+\frac{t^3}{3}-\frac{t^4}{4}+\cdots$$

for $-1<t<1$, so

$$\lim_{t\to 0+}\frac{\ln(1+t)}{t}=\lim_{t\to 0+}\frac{t}{t}=1.$$

Thus the limit of the logarithm approaches 1, so

$$\lim_{x\to\infty}\left(1+\frac{1}{x}\right)^x=e^1=e.$$

This is a famous limit. $\square$

L'HÔPITAL'S RULE

We present an alternative method, l'Hôpital's rule, for finding the limit of a 0/0-type form as $x\to a$. Again, we give a series explanation, which is valid only for functions analytic at a. The method is valid for some functions that are not analytic at a. A sequence of easy exercises leads you through a demonstration of l'Hôpital's rule, using the mean-value theorem. (See Exercises 51 through 53.) However, we feel the series approach is more transparent.

Let $f(x)$ and $g(x)$ be analytic at a, and suppose that $f(a)$ and $g(a)$ are both zero. Let $a_r(x-a)^r$ be the first nonzero term in the series for $f(x)$ at a, and $b_s(x-a)^s$ the first nonzero term of the series for $g(x)$ at a, as in the preceding subsection. *Since we are assuming that $f(a)=g(a)=0$, we know that $r>0$ and $s>0$.* Now a series representing an analytic function can be differentiated term by term to obtain the series representing the derivative of the function. Thus we see from Theorem 11.6 that

$$\lim_{x \to a} \frac{f'(x)}{g'(x)} = \lim_{x \to a} \frac{ra_r(x-a)^{r-1}}{sb_s(x-a)^{s-1}}. \tag{3}$$

We claim that

$$\lim_{x \to a} \frac{ra_r(x-a)^{r-1}}{sb_s(x-a)^{s-1}} = \lim_{x \to a} \frac{a_r(x-a)^r}{b_s(x-a)^s} \tag{4}$$

in all cases, including one-sided limits. *Remember that* $r > 0$ *and* $s > 0$.

Case 1 If $r > s$, then both limits in Eq. (4) are zero.

Case 2 If $r = s$, then both limits in Eq. (4) are a_r/b_s.

Case 3 If $r < s$ and the limits in Eq. (4) are one-sided, then both limits are either ∞ or $-\infty$. The sign is determined in each case by the signs of a_r and b_s and whether $(s-1)-(r-1)$ and $s-r$, respectively, are even or odd. But $(s-1)-(r-1) = s-r$, so the same limit of ∞ or $-\infty$ is obtained in each case.

Theorem 11.6 together with Eqs. (3) and (4) gives us at once the following theorem.

THEOREM 11.7 L'Hôpital's rule

Let $f(x)$ and $g(x)$ be analytic at a and let $f(a) = 0$ and $g(a) = 0$. Then

$$\lim_{x \to a} \frac{f(x)}{g(x)} = \lim_{x \to a} \frac{f'(x)}{g'(x)}. \tag{5}$$

This result holds for one-sided limits also.

EXAMPLE 6 Compute $\lim_{x \to 0} [(\sin x)/x]$ using l'Hôpital's rule.

Solution Since $(\sin x)|_0 = 0$ and $x|_0 = 0$, l'Hôpital's rule applies and we have

$$\lim_{x \to 0} \frac{\sin x}{x} = \lim_{x \to 0} \frac{\cos x}{1} = \frac{1}{1} = 1. \quad \square$$

Sometimes $\lim_{x \to a} f'(x)/g'(x)$ is again a 0/0-type problem. If $f(x)$ and $g(x)$ are analytic at a, so are $f'(x)$ and $g'(x)$, so we can apply l'Hôpital's rule again, obtaining

$$\lim_{x \to a} \frac{f(x)}{g(x)} = \lim_{x \to a} \frac{f'(x)}{g'(x)} = \lim_{x \to a} \frac{f''(x)}{g''(x)}.$$

One usually applies l'Hôpital's rule repeatedly *until a limit is obtained that is no longer an indeterminate form.* If $r = s > 0$ in Theorem 11.6, then by the formula for the Taylor coefficients, we see that

$$\lim_{x \to a} \frac{f(x)}{g(x)} = \frac{a_r}{b_s} = \frac{a_r}{b_r} = \frac{f^{(r)}(a)/r!}{g^{(r)}(a)/r!} = \frac{f^{(r)}(a)}{g^{(r)}(a)}.$$

L'Hôpital's rule repeated r times finds this limit by computing this quotient of derivatives.

EXAMPLE 7 Use l'Hôpital's rule to find $\lim_{x\to 0} (x - \sin x)/x^3$.

Solution We have

$$\lim_{x\to 0} \frac{x - \sin x}{x^3} = \lim_{x\to 0} \frac{1 - \cos x}{3x^2} = \lim_{x\to 0} \frac{\sin x}{6x} = \lim_{x\to 0} \frac{\cos x}{6} = \frac{1}{6}. \quad \square$$

> We emphasize that l'Hôpital's rule must not be applied to compute $\lim_{x\to a} f(x)/g(x)$ unless the quotient $f(a)/g(a)$ is an indeterminate form. To illustrate,
>
> $$\lim_{x\to 0} \frac{x^2}{\cos x} = \frac{0}{1} = 0,$$
>
> and the following "l'Hôpital's-rule computation" is
>
> $$\text{WRONG: } \lim_{x\to 0} \frac{x^2}{\cos x} = \lim_{x\to 0} \frac{2x}{-\sin x} = \lim_{x\to 0} \frac{2}{-\cos x} = \frac{2}{-1} = -2.$$

There are many useful variants of l'Hôpital's rule. In particular, if $\lim_{x\to a} f(x) = \infty$ and $\lim_{x\to a} g(x) = \infty$, then it can be shown that

$$\lim_{x\to a} \frac{f(x)}{g(x)} = \lim_{x\to a} \frac{f'(x)}{g'(x)}$$

under suitable conditions on the functions f and g. We can also compute one-sided limits of $f(x)/g(x)$ as well as limits at ∞ or $-\infty$ by the l'Hôpital's rule procedure with suitable conditions. Finally, if $\lim_{x\to a} f'(x)/g'(x) = \infty$, we can show that $\lim_{x\to a} f(x)/g(x) = \infty$, also under suitable conditions on f and g. We will use these variants freely and will call them all "l'Hôpital's rule."

EXAMPLE 8 Compute $\lim_{x\to\infty} x^{1/x}$.

Solution Taking logarithms, we try to compute

$$\lim_{x\to\infty} \frac{1}{x}(\ln x) = \lim_{x\to\infty} \frac{\ln x}{x}.$$

This is a limit at ∞ of type ∞/∞, and we find, by l'Hôpital's rule,

$$\lim_{x\to\infty} \frac{\ln x}{x} = \lim_{x\to\infty} \frac{1/x}{1} = \frac{0}{1} = 0.$$

Since $\ln x^{1/x} \to 0$ at ∞, we must have

$$\lim_{x\to\infty} x^{1/x} = e^0 = 1. \quad \square$$

EXAMPLE 9 Find $\lim_{x\to 0+} x^2 e^{1/x}$.

Solution This is a $0 \cdot \infty$-type form, and we write

$$\lim_{x\to 0+} x^2 e^{1/x} = \lim_{x\to 0+} \frac{x^2}{e^{-1/x}}$$

to convert it to a 0/0-type form. Then

$$\lim_{x\to 0+} \frac{x^2}{e^{-1/x}} = \lim_{x\to 0+} \frac{2x}{e^{-1/x}\cdot 1/x^2} = \lim_{x\to 0+} \frac{2x^3}{e^{-1/x}}.$$

This limit is worse; obviously we are going "the wrong way." Starting again and converting to an ∞/∞-type form, we have

$$\begin{aligned}\lim_{x\to 0+} x^2 e^{1/x} &= \lim_{x\to 0+} \frac{e^{1/x}}{1/x^2} \\ &= \lim_{x\to 0+} \frac{e^{1/x}(-1/x^2)}{-2/x^3} = \lim_{x\to 0+} \frac{e^{1/x}}{2/x} \\ &= \lim_{x\to 0+} \frac{e^{1/x}(-1/x^2)}{-2/x^2} = \lim_{x\to 0+} \frac{e^{1/x}}{2} = \infty. \quad \square\end{aligned}$$

EXAMPLE 10 We repeat the previous example but use the substitution $x = 1/t$.

Solution We have

$$\lim_{x\to 0+} x^2 e^{1/x} = \lim_{t\to\infty} \frac{1}{t^2} e^t = \lim_{t\to\infty} \frac{e^t}{t^2} = \lim_{t\to\infty} \frac{e^t}{2t} = \lim_{t\to\infty} \frac{e^t}{2} = \infty.$$

Clearly, this is easier than the computation in Example 9. □

EXAMPLE 11 Find $\lim_{x\to\infty} x^{100}/e^x$.

Solution This is an ∞/∞-type indeterminate form. Using l'Hôpital's rule repeatedly, we obtain

$$\lim_{x\to\infty} \frac{x^{100}}{e^x} = \lim_{x\to\infty} \frac{100x^{99}}{e^x} = \lim_{x\to\infty} \frac{100\cdot 99x^{98}}{e^x} = \cdots = \lim_{x\to\infty} \frac{100!}{e^x} = 0. \quad \square$$

We have mentioned before that e^x increases faster than any monomial function x^n as $x \to \infty$. Example 11 showed how this can be demonstrated by l'Hôpital's rule.

SUMMARY

1. Limits of sums, products, quotients, and exponentials involving two functions are called indeterminate forms if they lead formally to an expression of one of the following types:

$$\frac{0}{0}, \quad \frac{\infty}{\infty}, \quad 0\cdot\infty, \quad \infty\cdot 0, \quad (-\infty)+\infty,$$

$$\infty-\infty, \quad 0^0, \quad 1^\infty, \quad 1^{-\infty}, \quad \infty^0.$$

2. Product indeterminate forms are reduced to the quotient type by the algebraic device symbolized by

$$0\cdot\infty = \frac{0}{1/\infty} = \frac{0}{0} \quad \text{or} \quad 0\cdot\infty = \frac{\infty}{1/0} = \frac{\infty}{\infty}.$$

Exponential types are reduced to product types by taking a logarithm; as a symbolic illustration, $\ln(0^0) = 0(\ln 0) = 0(-\infty)$.

3. Let f and g be analytic at a and let a_r and b_s be the first nonzero coefficients in the series for f and g, respectively, at a. Then

$$\lim_{x\to a}\frac{f(x)}{g(x)} = \lim_{x\to a}\frac{a_r(x-a)^r}{b_s(x-a)^s}.$$

4. *L'Hôpital's rule:* Let $f(x)$ and $g(x)$ be analytic at a and let $f(a) = 0$ and $g(a) = 0$. Then

$$\lim_{x\to a}\frac{f(x)}{g(x)} = \lim_{x\to a}\frac{f'(x)}{g'(x)}.$$

This result holds for one-sided limits also.

5. The l'Hôpital's rule procedure can be used to find limits of ∞/∞-type forms as well as 0/0-type forms.

6. Check carefully that a form is indeterminate before applying l'Hôpital's rule. Blind use of the method on limits of other quotients can lead to incorrect results.

EXERCISES

In Exercises 1 through 10, find the limit by series methods.

1. $\lim_{x\to 0}\frac{e^x-1}{x}$
2. $\lim_{x\to 0}\frac{\sin x - x}{e^{x^2}-1}$
3. $\lim_{x\to 0}\frac{\sin x^2}{e^x-1-x}$
4. $\lim_{x\to 0}\frac{x[x^2-\ln(1-x^2)]}{x-\sin x}$
5. $\lim_{x\to 0+}\frac{\ln(1-x)}{x^2}$
6. $\lim_{x\to 0}\left(\frac{\sin x}{x\cos x - x} - \frac{1}{\cos x - 1}\right)$
7. $\lim_{x\to 0}\left(\frac{1}{x} - \frac{1}{\sin x}\right)$
8. $\lim_{x\to 0}(\cot 2x^2)[\ln(1-x^2)]$
9. $\lim_{x\to\infty}\left(1-\frac{1}{x^2}\right)^{\cot(1/x)}$
10. $\lim_{x\to 0+}\left(\frac{1}{1-x}\right)^{-1/x^2}$

In Exercises 11 through 20, find the limit by l'Hopital's rule.

11. $\lim_{x\to\infty}\frac{x}{e^x}$
12. $\lim_{x\to\infty}\frac{x^{20}}{e^x}$
13. $\lim_{x\to\infty}\frac{\ln x}{x}$
14. $\lim_{x\to\infty}\frac{(\ln x)^2}{x}$
15. $\lim_{x\to\infty}\frac{(\ln x)^{100}}{x}$
16. $\lim_{x\to 0+}x^x$
17. $\lim_{x\to 0+}x(\ln x)$
18. $\lim_{x\to 1}\frac{\ln x}{1-x}$
19. $\lim_{x\to\pi/2}\frac{\cos x}{x-(\pi/2)}$
20. $\lim_{x\to 3}\frac{x-\sqrt{3x}}{27-x^3}$

In Exercises 21 through 50, find the limit by any valid method.

21. $\lim_{x\to 0}\frac{\cos x}{x^2}$
22. $\lim_{x\to 0+}\frac{\ln(\sin x)}{\csc x}$
23. $\lim_{x\to 0}\frac{\sin x}{e^x-e^{-x}}$
24. $\lim_{x\to 0}\frac{\ln(1-x)}{e^x-e^{-x}}$
25. $\lim_{x\to 0}\frac{\sin x^2}{\cos^2 x - 1}$
26. $\lim_{x\to 0}\frac{\ln(\cos x)}{\sin^2 x}$
27. $\lim_{x\to\infty}\frac{e^{2x}}{e^x}$
28. $\lim_{x\to 1+}\frac{\ln(x^2-1)}{\ln(3x^2+3x-6)}$
29. $\lim_{x\to 1+}(1-x)[\ln(\ln x)]$
30. $\lim_{x\to 0+}\frac{\ln(e^x-1)}{\ln x}$
31. $\lim_{x\to\pi/2}\frac{e^x-e^{\pi/2}(x-\pi/2)}{\cos^2 x}$
32. $\lim_{x\to\pi/2+}\frac{\sec x}{x-\pi/2}$
33. $\lim_{x\to\pi}\frac{\sin^2 x}{(x-\pi)^2}$
34. $\lim_{x\to-\pi/2}\frac{1+\sin x}{\cos^2 x}$
35. $\lim_{x\to 1}\frac{(\ln x)^2}{x^2-2x+1}$
36. $\lim_{x\to\infty}\left(1-\frac{1}{x}\right)^x$
37. $\lim_{x\to\infty}\left(1+\frac{2}{x}\right)^x$
38. $\lim_{x\to\infty}\left(\frac{x}{x+1}\right)^x$
39. $\lim_{x\to\infty}\left(1+\frac{1}{x^2}\right)^x$
40. $\lim_{x\to\infty}\left(1-\frac{1}{x}\right)^{x^2}$
41. $\lim_{x\to(\pi/2)+}(\cos x)\ln\left(x-\frac{\pi}{2}\right)$
42. $\lim_{x\to\infty}\left(1-\frac{1}{x}\right)^{2x}$

43. $\lim_{x\to\infty}\left(1+\frac{1}{x}\right)^{x^2}$

44. $\lim_{x\to\pi/2}(\sin x)^{1/(\pi-2x)}$

45. $\lim_{x\to\infty}\left(\frac{3x+2}{2x}\right)^{x}$

46. $\lim_{x\to\infty}(1+x^2)^{1/x}$

47. $\lim_{x\to\infty}(x+e^x)^{3/x}$

48. $\lim_{x\to0+}(\sec x)^{\cot x}$

49. $\lim_{x\to0+}x^3e^{1/x}$

50. $\lim_{x\to1+}(x-1)^{\ln x}$

51. *Cauchy mean-value theorem:* Let f and g be continuous on $[a, b]$ and differentiable for $a < x < b$. Let $g'(x) \neq 0$ for $a < x < b$. Let

$$h(x) = (f(b) - f(a))g(x) - (g(b) - g(a))f(x).$$

a) Show that $h(x)$ satisfies the hypotheses for Rolle's theorem (page 155) on $[a, b]$. (Don't forget the continuity and differentiability hypotheses.)
b) Deduce from Rolle's theorem that there exists c, where $a < c < b$, such that

$$(f(b) - f(a))g'(c) = (g(b) - g(a))f'(c).$$

c) Using the hypotheses of this problem and Rolle's theorem, show that $g(b) - g(a) \neq 0$.
d) *Cauchy mean-value theorem:* Show that part (b) can be rewritten

$$\frac{f(b)-f(a)}{g(b)-g(a)} = \frac{f'(c)}{g'(c)}$$

for some c, where $a < c < b$.

52. *L'Hôpital's rule:* Let f and g be continuous and differentiable in a neighborhood $x_0 - h < x_0 < x_0 + h$ of x_0. Suppose $f(x_0) = g(x_0) = 0$. Let $g'(x) \neq 0$ in this neighborhood, except possibly at x_0. Finally, suppose $\lim_{x\to x_0} f'(x)/g'(x)$ exists or is ∞ or $-\infty$.

a) By applying the Cauchy mean-value theorem (Exercise 51) to the interval $[x_0, x]$, where $x_0 < x < x_0 + h$, show that

$$\lim_{x\to x_0+}\frac{f(x)}{g(x)} = \lim_{x\to x_0}\frac{f'(x)}{g'(x)}.$$

b) By applying the Cauchy mean-value theorem to the interval $[x, x_0]$ for $x_0 - h < x < x_0$, show that

$$\lim_{x\to x_0-}\frac{f(x)}{g(x)} = \lim_{x\to x_0-}\frac{f'(x)}{g'(x)}.$$

c) *L'Hôpital's rule:* Deduce that $\lim_{x\to x_0}(f(x))/(g(x)) = \lim_{x\to x_0}(f'(x))/(g'(x))$.

53. *L'Hôpital's rule at* ∞: Suppose $\lim_{x\to\infty}(f(x))/(g(x))$ corresponds to a 0/0-type indeterminate form. Let $x = 1/u$, $F(u) = f(1/u)$, and $G(u) = g(1/u)$.
a) Show that $\lim_{u\to0+}(F(u))/(G(u))$ corresponds to a 0/0-type indeterminate form.
b) *L'Hôpital's rule at* ∞: Assuming that the hypotheses in Exercise 52 for l'Hopital's rule hold for F and G at 0, show that

$$\lim_{x\to\infty}\frac{f(x)}{g(x)} = \lim_{x\to\infty}\frac{f'(x)}{g'(x)}.$$

In Exercises 54 through 59, try to find the limit in the indicated exercise using a calculator or computer rather than the methods of this section.

54. Exercise 1

55. Exercise 2

56. Exercise 9

57. Exercise 15 (This should be illuminating!)

58. Exercise 23

59. Exercise 47

11.5 THE BINOMIAL SERIES; COMPUTATIONS

In this final section on series, we present another important class of series, the binomial series. We then show how series can be used to compute definite integrals like $\int_0^1 \sin x^2\,dx$, where the corresponding indefinite integral cannot be expressed in terms of elementary functions.

THE BINOMIAL SERIES

The binomial theorem states that for any numbers a and b and any positive integer n, we have

$$(a+b)^n = a^n + na^{n-1}b + \frac{n(n-1)}{2!}a^{n-2}b^2 + \cdots + \frac{n(n-1)\cdots(n-k+1)}{k!}a^{n-k}b^k + \cdots + b^n. \quad (1)$$

EXAMPLE 1 Expand $(a + b)^4$ by the binomial theorem.

Solution We have

$$(a + b)^4 = a^4 + 4a^3b + \frac{4 \cdot 3}{2 \cdot 1} a^2b^2 + \frac{4 \cdot 3 \cdot 2}{3 \cdot 2 \cdot 1} ab^3 + \frac{4 \cdot 3 \cdot 2 \cdot 1}{4 \cdot 3 \cdot 2 \cdot 1} b^4$$

$$= a^4 + 4a^3b + 6a^2b^2 + 4ab^3 + b^4. \quad \square$$

We can write Eq. (1) in the form

$$\boxed{(a + b)^n = \sum_{k=0}^{n} \binom{n}{k} a^{n-k}b^k,} \tag{2}$$

where we define the *binomial coefficients* $\binom{n}{k}$ by

$$\binom{n}{k} = \begin{cases} \dfrac{n(n-1)\cdots(n-k+1)}{k!} & \text{for } k > 0, \\ 1 & \text{for } k = 0. \end{cases} \tag{3}$$

If we let $a = 1$ and $b = x$ in Eq. (2), we obtain

$$\boxed{(1 + x)^n = \sum_{k=0}^{n} \binom{n}{k} x^k.} \tag{4}$$

EXAMPLE 2 Expand $(1 + x)^5$.

Solution We have

$$(1 + x)^5 = 1 + 5x + \frac{5 \cdot 4}{2 \cdot 1} x^2 + \frac{5 \cdot 4 \cdot 3}{3 \cdot 2 \cdot 1} x^3 + \frac{5 \cdot 4 \cdot 3 \cdot 2}{4 \cdot 3 \cdot 2 \cdot 1} x^4 + \frac{5 \cdot 4 \cdot 3 \cdot 2 \cdot 1}{5 \cdot 4 \cdot 3 \cdot 2 \cdot 1} x^5$$

$$= 1 + 5x + 10x^2 + 10x^3 + 5x^4 + x^5. \quad \square$$

The sum in Eq. (4) must (by uniqueness) be the Taylor series for $(1 + x)^n$ at $x_0 = 0$. We could also verify this directly by differentiation; we easily find that

$$D^k(1 + x)^n = n(n - 1) \cdots (n - k + 1)(1 + x)^{n-k},$$

so

$$D^k(1 + x)^n|_{x=0} = n(n - 1) \cdots (n - k + 1).$$

Since n is a positive *integer*, $(1 + x)^n$ is a polynomial of degree n, so $D^k(1 + x)^n = 0$ for $k > n$. For n a positive integer, we thus expect $\binom{n}{k} = 0$ for $k > n$, and this is easily seen from Eq. (3); namely,

$$\binom{n}{k} = \frac{n(n - 1) \cdots (n - n) \cdots (n - k + 1)}{k!} = \frac{0}{k!} = 0$$

for $k > n$.

We generalize Eq. (3), and define the **binomial coefficient** $\binom{p}{k}$ for any real number p and integer $k \geq 0$ to be given by

$$\binom{p}{k} = \begin{cases} \dfrac{p(p-1)\cdots(p-k+1)}{k!} & \text{for } k > 0, \\ 1 & \text{for } k = 0. \end{cases}$$

EXAMPLE 3 Find the binomial coefficient

$$\binom{3/2}{5}.$$

Solution We have

$$\binom{3/2}{5} = \frac{\frac{3}{2}\cdot\frac{1}{2}\cdot -\frac{1}{2}\cdot -\frac{3}{2}\cdot -\frac{5}{2}}{5\cdot 4\cdot 3\cdot 2\cdot 1} = \frac{\frac{-3}{2^5}}{4\cdot 2\cdot 1} = \frac{-3}{8\cdot 32} = \frac{-3}{256}. \quad \square$$

Note that if p is not a nonnegative integer, then no factor in the numerator of $\binom{p}{k}$ is zero, even if $k > p$. We are led to consider the series analogue of the sum (4) for any real number p.

DEFINITION 11.3 Binomial series

For any p, the **binomial series for** $(1 + x)^p$ is

$$\sum_{k=0}^{\infty} \binom{p}{k} x^k = 1 + px + \frac{p(p-1)}{2!}x^2 + \cdots + \frac{p(p-1)\cdots(p-k+1)}{k!}x^k + \cdots. \tag{5}$$

(We use k rather than n for summation index to avoid confusion with the use of n in Eq. 1.)

It is easily checked that the series (5) is the Taylor series for $(1 + x)^p$ at $x_0 = 0$; we easily find that

$$\begin{aligned} D^k(1 + x)^p|_{x=0} &= p(p-1)\cdots(p-k+1)(1 + x)^{p-k}|_{x=0} \\ &= p(p-1)\cdots(p-k+1). \end{aligned}$$

We would like to find the radius of convergence of the binomial series (5) and to determine whether the series does represent $(1 + x)^p$ in a neighborhood of zero. Of course, if p is a nonnegative integer, then the series contains only a finite number of terms with nonzero coefficients, has radius of convergence ∞, and represents $(1 + x)^p$ by sum (4). We now suppose that p is not a nonnegative integer and compute the ratio

$$\frac{\binom{p}{k+1}x^{k+1}}{\binom{p}{k}x^k} = \frac{[(p(p-1)\cdots(p-k))/(k+1)!]x^{k+1}}{[(p(p-1)\cdots(p-k+1))/k!]x^k} = \frac{p-k}{k+1}x.$$

We obtain

$$\lim_{k\to\infty}\left|\frac{\binom{p}{k+1}x^{k+1}}{\binom{p}{k}x^k}\right| = \lim_{k\to\infty}\left|\frac{p-k}{k+1}x\right| = |-x| = |x|.$$

The radius of convergence of the binomial series for p not a nonnegative integer is therefore 1; the series converges if $|x| < 1$ or if $-1 < x < 1$.

To show that the series (5) represents $(1 + x)^p$ for $-1 < x < 1$, we could try to show that $\lim_{k\to\infty} E_k(x_1) = 0$ for $-1 < x_1 < 1$, but the following argument is less tedious. We set

$$f(x) = \sum_{k=0}^{\infty}\binom{p}{k}x^k \tag{6}$$

for $-1 < x < 1$. Then we may differentiate term by term to obtain

$$f'(x) = \sum_{k=0}^{\infty}\binom{p}{k}kx^{k-1} = \sum_{k=1}^{\infty}\frac{p(p-1)\cdots(p-k+1)}{(k-1)!}x^{k-1}. \tag{7}$$

From the series (6) and (7), we can easily verify that

$$pf(x) = (1 + x)f'(x) \tag{8}$$

(see Exercise 49). From Eq. (8), we obtain

$$\int\frac{f'(x)}{f(x)}\,dx = \int\frac{p}{1+x}\,dx, \tag{9}$$

or, taking the antiderivatives,

$$\ln|f(x)| = p\ln|1 + x| + C = \ln|1 + x|^p + C \tag{10}$$

for $-1 < x < 1$. From the series (6), we see that $f(0) = 1$, so putting $x = 0$ in Eq. (10), we obtain

$$0 = \ln 1 = \ln(1 + 0)^p + C = \ln 1 + C = C.$$

Thus $C = 0$ and $\ln|f(x)| = \ln|1 + x|^p$. Therefore $|f(x)| = |1 + x|^p$, so $f(x) = \pm(1 + x)^p$. However, $f(0) = 1$, so the positive sign is appropriate, and

$$f(x) = (1 + x)^p$$

for $-1 < x < 1$, which is what we wished to show. We summarize these results in a theorem.

THEOREM 11.8 Binomial representation

If p is not a nonnegative integer, then the binomial series $\sum_{k=0}^{\infty}\binom{p}{k}x^k$ has radius of convergence 1 and represents $(1 + x)^p$ for $-1 < x < 1$. For a nonnegative integer n, the (finite) binomial series $\sum_{k=0}^{n}\binom{n}{k}x^k$ converges for all x to $(1 + x)^n$.

EXAMPLE 4 Find the binomial series for $(1 + x)^{1/2}$ and use it to estimate $\sqrt{\frac{3}{2}}$.

Solution We have

$$(1+x)^{1/2} = 1 + \frac{1}{2}x + \frac{\frac{1}{2}\cdot -\frac{1}{2}}{2!}x^2 + \frac{\frac{1}{2}\cdot -\frac{1}{2}\cdot -\frac{3}{2}}{3!}x^3 + \cdots$$

$$= 1 + \frac{1}{2}x - \frac{1}{2^2\cdot 2!}x^2 + \frac{3}{2^3\cdot 3!}x^3 - \cdots$$

$$+(-1)^{k-1}\frac{3\cdot 5\cdot 7\cdots(2k-3)}{2^k\cdot k!}x^k + \cdots,$$

for $k \geq 1$ and for $-1 < x < 1$. Putting $x = \frac{1}{2}$ and using the terms of exponent less than or equal to 3, we obtain the estimate

$$\sqrt{\frac{3}{2}} \approx 1 + \frac{1}{4} - \frac{1}{32} + \frac{1}{128} \approx 1.2266.$$

When $x = \frac{1}{2}$, our series is alternating with terms of decreasing size, so the error in our estimate is less than the next term $5/2^{11} \approx 0.0024$. The actual value of $\sqrt{\frac{3}{2}}$ to six decimal places is 1.224745. □

EXAMPLE 5 Find the terms of degree less than or equal to 9 of the series for $\sin^{-1}x$ at $x_0 = 0$.

Solution Differentiating and using the binomial series, we have

$$\frac{d(\sin^{-1}x)}{dx} = \frac{1}{\sqrt{1-x^2}} = (1-x^2)^{-1/2}$$

$$= 1 - \frac{1}{2}(-x^2) + \frac{-\frac{1}{2}\cdot -\frac{3}{2}}{2\cdot 1}(-x^2)^2$$

$$+ \frac{-\frac{1}{2}\cdot -\frac{3}{2}\cdot -\frac{5}{2}}{3\cdot 2\cdot 1}(-x^2)^3$$

$$+ \frac{-\frac{1}{2}\cdot -\frac{3}{2}\cdot -\frac{5}{2}\cdot -\frac{7}{2}}{4\cdot 3\cdot 2\cdot 1}(-x^2)^4 + \cdots$$

$$= 1 + \frac{1}{2}x^2 + \frac{3}{8}x^4 + \frac{5}{16}x^6 + \frac{35}{128}x^8 + \cdots$$

for $-1 < x < 1$. Integrating, we find that

$$\sin^{-1}x = k + x + \frac{1}{6}x^3 + \frac{3}{40}x^5 + \frac{5}{112}x^7 + \frac{35}{1152}x^9 + \cdots$$

for some constant k. Since $\sin^{-1}0 = 0$, we see that $k = 0$, so

$$\sin^{-1}x = x + \frac{1}{6}x^3 + \frac{3}{40}x^5 + \frac{5}{112}x^7 + \frac{35}{1152}x^9 + \cdots$$

for $-1 < x < 1$. □

ESTIMATING INTEGRALS

Suppose we wish to compute $\int_a^b f(x)\,dx$. Perhaps it is hard to find an elementary function that is an antiderivative of f, even if f is a known elementary function. For example, it can be shown that no antiderivative of e^{-x^2} is an elementary function. To estimate $\int_a^b f(x)\,dx$, we might take a power series $\sum_{n=0}^{\infty} a_n(x - x_0)^n$ that converges to f in $[a, b]$ and compute

$$\int_a^b \left[\sum_{n=0}^{\infty} a_n(x - x_0)^n\right] dx$$

instead. The theory of power series shows that this integral can be computed by term-by-term integration. As another alternative, we might just use a partial sum s_n of the series to estimate f in $[a, b]$, and estimate $\int_a^b f(x)\,dx$ by $\int_a^b s_n(x)\,dx$. In this case, we would like a bound for our error. If f and g are any continuous functions defined on $[a, b]$, it follows easily from the definition of the definite integral that

$$\text{if } |f(x) - g(x)| < \varepsilon \quad \text{for all } x \text{ in } [a, b],$$
$$\text{then } \left|\int_a^b f(x)\,dx - \int_a^b g(x)\,dx\right| < \varepsilon(b - a)$$

(see Exercise 50). In the examples that follow, we illustrate both estimation by integrating a series and estimation by integrating a partial sum.

EXAMPLE 6 Estimate $\int_0^1 e^{-x^2}\,dx$ by integrating a series.

Solution Replacing x by $-x^2$ in the well-known series for e^x, we find that

$$e^{-x^2} = 1 - x^2 + \frac{x^4}{2!} - \frac{x^6}{3!} + \cdots + (-1)^n \frac{x^{2n}}{n!} + \cdots$$

for all x. Thus

$$\begin{aligned}
\int_0^1 e^{-x^2}\,dx &= \int_0^1 \left(1 - x^2 + \frac{x^4}{2!} - \frac{x^6}{3!} + \frac{x^8}{4!} - \cdots + (-1)^n \frac{x^{2n}}{n!} + \cdots\right) dx \\
&= \left(x - \frac{x^3}{3} + \frac{x^5}{5 \cdot 2!} - \frac{x^7}{7 \cdot 3!} + \frac{x^9}{9 \cdot 4!} - \cdots \right. \\
&\quad \left.\left. + \frac{(-1)^n x^{2n+1}}{(2n + 1)n!} + \cdots\right)\right]_0^1 \\
&= 1 - \frac{1}{3} + \frac{1}{5 \cdot 2!} - \frac{1}{7 \cdot 3!} + \frac{1}{9 \cdot 4!} - \cdots + \frac{(-1)^n}{(2n + 1)n!} + \cdots.
\end{aligned}$$

Our answer is the sum of an infinite series of constants and may in turn be approximated by a partial sum of the series. Since the series is alternating with terms of decreasing size, the error in approximating this series by a partial sum is less than the size of the next term. Thus

$$\int_0^1 e^{-x^2}\,dx \approx 1 - \frac{1}{3} + \frac{1}{5 \cdot 2!} - \frac{1}{7 \cdot 3!} + \frac{1}{9 \cdot 4!} \approx 0.7475,$$

with error less than

$$\frac{1}{11 \cdot 5!} \approx 0.0008. \quad \square$$

EXAMPLE 7 We know that

$$\int_0^{1/2} (1 - x^2)^{-1/2}\, dx = \sin^{-1} x\Big]_0^{1/2} = \sin^{-1}\frac{1}{2} - \sin^{-1} 0 = \frac{\pi}{6} - 0 = \frac{\pi}{6}.$$

Integrate a partial sum of the binomial series representing $(1 - x^2)^{-1/2}$ to estimate $\pi/6$ and then $\pi = 6(\pi/6)$.

Solution We have

$$\begin{aligned}(1 - x^2)^{-1/2} &= 1 + -\frac{1}{2}\cdot -x^2 + \frac{-\frac{1}{2}\cdot -\frac{3}{2}}{2!}(-x^2)^2 \\ &\quad + \frac{-\frac{1}{2}\cdot -\frac{3}{2}\cdot -\frac{5}{2}}{3!}(-x^2)^3 + \cdots \\ &= 1 + \frac{1}{2}x^2 + \frac{3}{2^2\cdot 2!}x^4 + \frac{3\cdot 5}{2^3\cdot 3!}x^6 \\ &\quad + \left(\frac{3\cdot 5\cdot 7}{2^4\cdot 4!}x^8 + \frac{3\cdot 5\cdot 7\cdot 9}{2^5\cdot 5!}x^{10} + \cdots\right)\end{aligned}$$

for $-1 < x < 1$. The portion of the series that we have placed in parentheses has all coefficients less than or equal to 1 and is therefore, term for term, less than the series

$$x^8 + x^{10} + \cdots = x^8(1 + x^2 + x^4 + \cdots).$$

Since we will be integrating for $0 \le x \le \frac{1}{2}$, we have

$$\begin{aligned}x^8(1 + x^2 + x^4 + \cdots) &\le \frac{1}{2^8}\left(1 + \frac{1}{4} + \frac{1}{4^2} + \cdots\right) \\ &= \frac{1}{2^8}\cdot\frac{1}{1 - \frac{1}{4}} = \frac{1}{2^8}\cdot\frac{4}{3} = \frac{1}{192}.\end{aligned}$$

Thus we have

$$\begin{aligned}\frac{\pi}{6} = \int_0^{1/2}(1 - x^2)^{-1/2}\, dx &\approx \int_0^{1/2}\left(1 + \frac{1}{2}x^2 + \frac{3}{8}x^4 + \frac{15}{48}x^6\right) dx \\ &= \left(x + \frac{1}{6}x^3 + \frac{3}{40}x^5 + \frac{15}{336}x^7\right)\Big]_0^{1/2} \\ &= \frac{1}{2} + \frac{1}{48} + \frac{3}{1280} + \frac{15}{43008} \approx 0.52353,\end{aligned}$$

with error at most

$$\frac{1}{192}\left(\frac{1}{2} - 0\right) = \frac{1}{384} \approx 0.00260.$$

Therefore,

$$\pi = 6\left(\frac{\pi}{6}\right) \approx 6(0.52353) = 3.14118$$

with error at most $6(0.00260) = 0.01560$. Since the value of π to five decimal places is 3.14159, our actual error in estimating π was only about 0.00041. The reason we obtained such a crude bound for the error is that we estimated the series $x^8 + x^{10} + \cdots$ throughout $[0, \frac{1}{2}]$ by its greatest value at $x = \frac{1}{2}$. □

SUMMARY

1. If p is not a nonnegative integer, then the binomial series $\sum_{k=0}^{\infty} \binom{p}{k} x^k$ has radius of convergence 1 and represents $(1 + x)^p$ for $-1 < x < 1$. For a nonnegative integer n, the (finite-length) binomial series $\sum_{k=0}^{n} \binom{n}{k} x^k$ converges for all x to $(1 + x)^n$.
2. Integrals $\int_a^b f(x)\,dx$ such as $\int_a^b e^{-x^2}\,dx$, which cannot be found in the usual way, can sometimes be estimated by taking a series expression for $f(x)$ and integrating the series.

EXERCISES

In Exercises 1 through 12, compute the binomial coefficient.

1. $\binom{4}{2}$ **2.** $\binom{7}{3}$ **3.** $\binom{8}{7}$ **4.** $\binom{10}{0}$

5. $\binom{5}{5}$ **6.** $\binom{5}{7}$ **7.** $\binom{3.5}{3}$ **8.** $\binom{-1.5}{2}$

9. $\binom{7.5}{0}$ **10.** $\binom{0.5}{4}$ **11.** $\binom{1/3}{3}$ **12.** $\binom{-4}{4}$

13. What is the Taylor series for $(1 + x)^0$ at $x_0 = 0$? Check that the coefficient of x^k in this series is indeed the binomial coefficient $\binom{0}{k}$ as defined on page 509.

In Exercises 14 through 23, find the first five terms of the binomial series that represents the given function in a neighborhood of zero. Give the radius of convergence in each case.

14. $(1 + x)^{3/2}$ **15.** $(1 + x)^{-2}$

16. $(1 - x)^{1/2}$ **17.** $(1 - x^2)^{1/3}$

18. $\left(1 + \frac{x}{2}\right)^{2/3}$ **19.** $(1 + 3x)^{5/3}$

20. $(4 + x)^{1/2}$ [*Hint:* $(4 + x)^{1/2} = 2[1 + x/4]^{1/2}$.]

21. $(25 - x^2)^{-1/2}$

22. $(8 - x)^{-1/3}$

23. $(8 + x^3)^{4/3}$

24. Estimate $\sqrt{2}$ by finding a partial sum with $x = \frac{1}{2}$ of a binomial series that represents $(1 - x)^{-1/2}$ in a neighborhood of the origin.

In Exercises 25 through 28, indicate how you could estimate the given quantity by using a suitable binomial series. (You need not compute an estimate.) To illustrate, Exercise 24 shows how $\sqrt{2}$ could be estimated (rather inefficiently) using a binomial series.

25. $\sqrt{2/3}$ **26.** $\sqrt{5}$

27. $\sqrt[3]{2}$ **28.** $\sqrt[4]{17}$

In Exercises 29 through 38, find as many terms of the series for the given function at $x_0 = 0$ as you conveniently can.

29. $\sinh^{-1} x$ [*Hint:* $D(\sinh^{-1} x) = 1/\sqrt{1 + x^2}$.]

30. $\sin^{-1} x^2$ **31.** $\dfrac{x}{\sqrt{1 - x^2}}$

32. $(1 + x^2)\sqrt{1 - x}$ **33.** $\dfrac{\sqrt{1 + x}}{1 - x}$

34. $\dfrac{\sqrt{1 - x}}{1 + x}$ **35.** $\sqrt{1 + x}\sqrt{1 + 2x}$

36. $x^2\sqrt{16 - x^2}$ **37.** $\dfrac{x^3}{\sqrt{4 - x^2}}$

38. $\dfrac{\sqrt{1 - x^2}}{\sqrt{4 + x^2}}$

In Exercises 39 through 48, estimate the integral by series methods and find a bound for your error. (You need not simplify your answers or put them in decimal form.)

39. $\int_0^1 \sin x^2\,dx$ **40.** $\int_0^{1/2} e^{x^2}\,dx$

41. $\int_0^1 x^3 \cos x^2\,dx$ **42.** $\int_0^{1/2} (1 + x^2)^{1/3}\,dx$

43. $\int_0^{0.1} e^{-x^2}\,dx$

44. $\int_0^1 \sqrt{16 - x^4}\,dx$ $\left[\textit{Hint: } \sqrt{16 - x^4} = 4\sqrt{1 - \frac{x^4}{16}}.\right]$

45. $\int_0^{1/2} \cos x^3\,dx$

46. $\int_0^1 e^{-x^3}\,dx$

47. $\int_{1/2}^{3/2} \cos(x-1)^2\,dx$

48. $\int_0^1 x^2\sqrt{25-x^2}\,dx$

49. Verify Eq. (8) on page 510.

50. Use the definition of $\int_a^b f(x)\,dx$ to show that if f and g are continuous in $[a, b]$ and $|f(x)-g(x)| < \varepsilon$ for all x in $[a, b]$, then

$$\left|\int_a^b f(x)\,dx - \int_a^b g(x)\,dx\right| < \varepsilon(b-a).$$

Use a calculator or computer for Exercises 51 and 52.

51. Find the absolute numerical difference between the estimate for $\int_0^1 e^{x^2}\,dx$ using Simpson's rule with $n = 20$ and integrating the first eight nonzero terms of the series for e^{x^2}.

52. Repeat Exercise 51 for $\int_0^1 \cos x^2\,dx$, but use the first five nonzero terms of the series for $\cos x^2$.

EXERCISE SETS FOR CHAPTER 11

Review Exercise Set 11.1

1. Find the interval of convergence, including endpoints, of the series

$$\sum_{n=1}^{\infty} \frac{n}{(n^2+1)3^n}(x+5)^n.$$

2. Find an expression for the nth term $a_n(x-x_0)^n$ of a power series having as interval of convergence $-1 \le x < 3$. (Many answers are possible.)

3. Find the fourth Taylor polynomial $T_4(x)$ for $f(x) = \sin x$ at $x_0 = \pi/3$.

4. Approximate $\ln 1.04$ using the Taylor polynomial $T_3(x)$ for $\ln x$ at $x_0 = 1$, and find a bound for the error.

5. a) Define the notion of an analytic function of a real variable.
   b) Find the Taylor series for $x/(1-x^2)$ at $x_0 = 0$ in the easiest way you can.

6. Use series long division to find the first three terms of the series for $(\sin x)/e^x$ at $x_0 = 0$.

7. Use series methods to find

$$\lim_{x\to 0} \frac{e^{x^3}-1}{x-\sin x}.$$

8. Find $\lim_{x\to 0^+} [1 + (2/\sin x)]^x$.

9. Use the binomial series expansion to give the first five nonzero terms of the series for $\sqrt{1+x^2}$ at $x_0 = 0$.

10. Use series techniques to find $\int_0^1 \sin x^2\,dx$, with error at most 0.001.

Review Exercise Set 11.2

1. Find the interval of convergence, including endpoints, of the series

$$\sum_{n=1}^{\infty} \frac{(2x+1)^{2n}}{n^2\cdot 3^n}.$$

2. Find the radius of convergence of the series

$$\sum_{n=1}^{\infty} \frac{(x-4)^{2n+1}}{3^{4n-3}}.$$

3. Find the fourth Taylor polynomial $T_4(x)$ for $f(x) = \cos 2x$ at $x_0 = \pi/4$.

4. Approximate $\sinh 0.1$ using the Taylor polynomial $T_4(x)$ at $x_0 = 0$, and find a bound for the error.

5. Find the Taylor series for $x^3 \tan^{-1} x^2$ at $x_0 = 0$ in the easiest way you can.

6. Use series multiplication to find the first four nonzero terms of the Taylor series for $e^x \sin 2x$ at $x_0 = 0$.

7. Use series methods to find

$$\lim_{x\to 0} \frac{\cos x^2 - 1}{x \sin x^3}.$$

8. Find $\lim_{x\to\infty} [(2x+3)/2x]^x$.

9. Give the first four terms of the binomial series representing $(1-x/2)^{-1/2}$ in a neighborhood of $x_0 = 0$.

10. Use series methods to estimate

$$\int_0^1 \frac{1-\cos x^2}{x}\,dx,$$

with error at most 0.001.

Review Exercise Set 11.3

1. Find the interval of convergence, including endpoints, of the series

$$\sum_{n=0}^{\infty} \frac{100^n(x-1)^n}{n!}.$$

2. Find the radius of convergence of the series

$$\sum_{n=0}^{\infty} \frac{(4x-2)^{2n}}{3^n}.$$

3. Find the fourth Taylor polynomial $T_4(x)$ for $f(x) = e^{2x}$ centered at $x_0 = 1$.

4. Approximate $\sqrt{e}$ using the Taylor polynomial $T_4(x)$ for e^x centered at $x_0 = 0$, and find a bound for the absolute error.

5. If $f(x) = x^4/(1-2x^2)$, find $f^{(10)}(0)$.

6. Find

$$\sum_{n=1}^{\infty} \frac{3^n}{n!}.$$

7. Find

$$\lim_{x\to 0} \frac{x(e^{x^5} - 1)}{x^2 - \sin x^2}.$$

8. Find

$$\lim_{x\to \pi/2} \left(x - \frac{\pi}{2}\right)^2 \tan x.$$

9. Give the first four terms of the Taylor series representing $(8 + x)^{2/3}$ near $x_0 = 0$.

10. Use series techniques to estimate $\int_0^{1/2} \cos x^3 \, dx$, with error at most 0.00001.

Review Exercise Set 11.4

1. Find the interval of convergence, including endpoints, of the series

$$\sum_{n=0}^{\infty} \frac{n!(x-3)^{5n}}{1000^n}.$$

2. Find a power series with interval of convergence $0 \le x \le 100$. (Many answers are possible.)

3. Express $x^4 - 3x^3 + 2x^2 + x - 6$ as a polynomial centered at -1.

4. Use a differential to estimate $9.05^{3/2}$, and find a bound for the absolute error.

5. Find as many terms as you conveniently can of the Taylor series for $(1 - x^2)/(1 + x^2)$ at $x_0 = 1$.

6. Find the function represented near 0 by the series

$$x + \frac{x^3}{3} + \frac{x^5}{5} + \frac{x^7}{7} + \frac{x^9}{9} + \cdots.$$

7. Find

$$\lim_{x\to 0} \frac{(\sin x^4) - x^4(\cos x^4)}{x^{12}}.$$

8. Find

$$\lim_{x\to \infty} \left(\frac{2x}{3x+1}\right)^{\sqrt{x}}.$$

9. Compute

$$\binom{5/3}{4}.$$

10. Find the first four terms of the Taylor series for $x^4/\sqrt{1-x}$ at $x_0 = 0$.

More Challenging Exercises 11

In Exercises 1 through 10, find the radius of convergence of the series.

1. $\displaystyle\sum_{n=1}^{\infty} \frac{n^n}{n!} x^n$

2. $\displaystyle\sum_{n=1}^{\infty} \frac{n^n}{e^{2n} \cdot n!} x^n$

3. $\displaystyle\sum_{n=1}^{\infty} \frac{n^n}{(n!)^2} x^n$

4. $\displaystyle\sum_{n=1}^{\infty} [1 + (-1)^n] x^n$

5. $\displaystyle\sum_{n=1}^{\infty} [1 + (-1)^n]^n x^n$

6. $\displaystyle\sum_{n=1}^{\infty} [2 + (-1)^n 7]^n x^{2n}$

7. $\displaystyle\sum_{n=1}^{\infty} \frac{(2n)!}{n^n} x^n$

8. $\displaystyle\sum_{n=1}^{\infty} \frac{(2n)!}{(n!)^2} x^n$

9. $\displaystyle\sum_{n=1}^{\infty} \frac{2 \cdot 5 \cdot 8 \cdots (3n-1)}{3 \cdot 7 \cdot 11 \cdots (4n-1)} x^n$

10. $\displaystyle\sum_{n=1}^{\infty} \frac{4 \cdot 7 \cdot 10 \cdots (3n+1)}{n^n} x^n$

PLANE CURVES 12

Sections 12.1 through 12.4 give a treatment of ellipses, hyperbolas, and parabolas, complete with foci, directrices, eccentricities, vertices, major axes, and rotation of axes. We have tried to write these sections to provide as much flexibility as possible, allowing anywhere from zero to four lessons on this material.

1 Lesson Do only Section 12.1 on sketching with translation of axes. This is useful for sketching quadric surfaces later.

2 Lessons Do Section 12.1 and either Section 12.2 on synthetic definitions or Section 12.3 on rotation of axes.

3 Lessons Do Sections 12.1, 12.2, and 12.3.

4 Lessons Do all four sections, 12.1–12.4.

Section 12.4 on applications is especially suitable for outside reading, and we recommend that it be left as such. Sections 12.5 and 12.6 deal with parametric curves and curvature, and should not be omitted.

12.1 SKETCHING CONIC SECTIONS

Let two congruent right circular cones in space be placed vertex to vertex to form a double cone, as shown in Fig. 12.1. A plane can intersect this cone in three types of curves. Figure 12.1(a) shows the closed-curve type of intersection, which is an *ellipse*. Figure 12.1(b) shows the intersection giving a two-piece, open-ended curve, which is a *hyperbola*. The one-piece, open-ended curve in Fig. 12.1(c) is a *parabola*.

In this section, we sketch these curves, starting with their equations with respect to carefully chosen x- and y-axes in the plane of intersection.

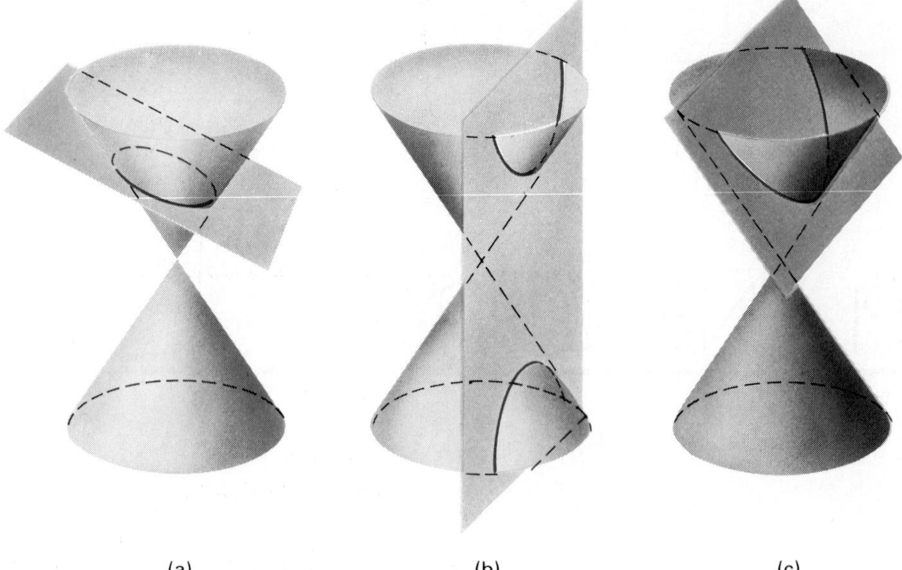

Figure 12.1 Conic sections: (a) elliptic section; (b) hyperbolic section; (c) parabolic section.

THE ELLIPSE

The equation

$$\boxed{\frac{x^2}{a^2} + \frac{y^2}{b^2} = 1} \tag{1}$$

describes an ellipse, shown in Fig. 12.2. Setting $x = 0$, we see that the curve meets the y-axis at b and $-b$. Setting $y = 0$, we find that the x-intercepts are a and $-a$. If $a = b$, the ellipse becomes a circle.

Recall from page 33 that if we choose translated $\bar{x},\bar{y}$-axes at a new origin (h, k), then

$$\bar{x} = x - h \qquad \text{and} \qquad \bar{y} = y - k.$$

It follows that an equation of the form

$$\frac{(x - h)^2}{a^2} + \frac{(y - k)^2}{b^2} = 1 \tag{2}$$

can be rewritten as

$$\frac{\bar{x}^2}{a^2} + \frac{\bar{y}^2}{b^2} = 1$$

with respect to the translated $\bar{x},\bar{y}$-axes. This shows that Eq. (2) describes an ellipse with center at (h, k), as sketched in Fig. 12.3.

Figure 12.2 An ellipse $x^2/a^2 + y^2/b^2 = 1$.

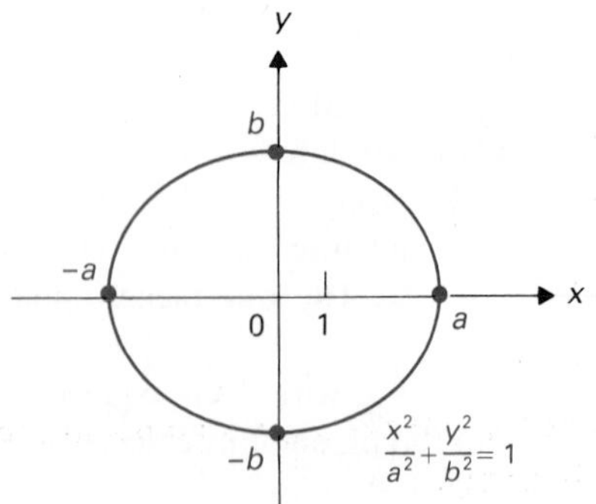

EXAMPLE 1 Sketch the curve $9x^2 + 4y^2 = 36$.

Solution Dividing through by 36, we obtain

$$\frac{x^2}{4} + \frac{y^2}{9} = 1,$$

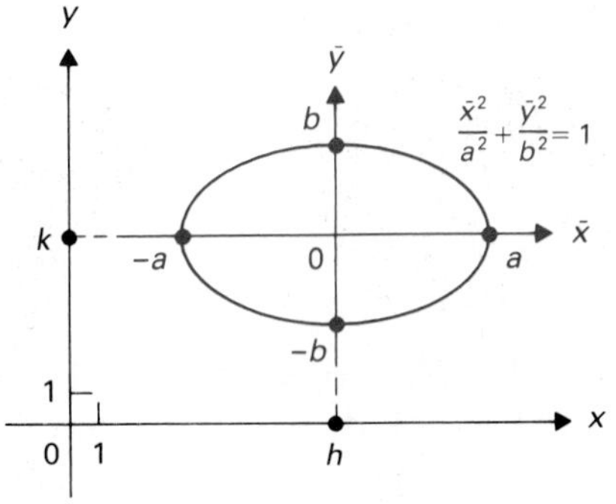

Figure 12.3 The ellipse
$(x-h)^2/a^2 + (y-k)^2/b^2 = 1$
becomes
$\bar{x}^2/a^2 + \bar{y}^2/b^2 = 1$
using translated axes at (h, k).

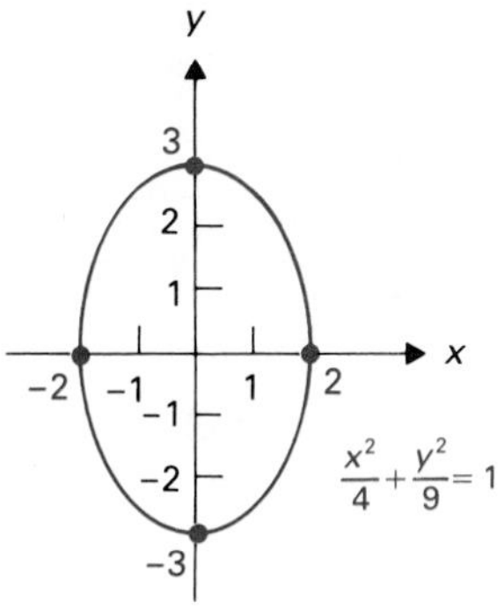

Figure 12.4 The ellipse
$9x^2 + 4y^2 = 36$.

which is the ellipse shown in Fig. 12.4. The x-intercepts are $\pm\sqrt{4} = \pm 2$, and the y-intercepts are $\pm\sqrt{9} = \pm 3$. □

EXAMPLE 2 Sketch the curve with equation

$$x^2 + 3y^2 - 4x + 6y = -1.$$

Solution Completing the square, we obtain

$$(x^2 - 4x) + 3(y^2 + 2y) = -1,$$
$$(x-2)^2 + 3(y+1)^2 = 4 + 3 - 1 = 6,$$
$$\frac{(x-2)^2}{6} + \frac{(y+1)^2}{2} = 1,$$

which is in the form of Eq. (2). Setting $\bar{x} = x - 2$ and $\bar{y} = y + 1$, we obtain the equation

$$\frac{\bar{x}^2}{6} + \frac{\bar{y}^2}{2} = 1.$$

This is the equation of an ellipse with center $(h, k) = (2, -1)$, as shown in Fig. 12.5. □

Figure 12.5 The ellipse
$x^2 + 3y^2 - 4x + 6y = -1$.

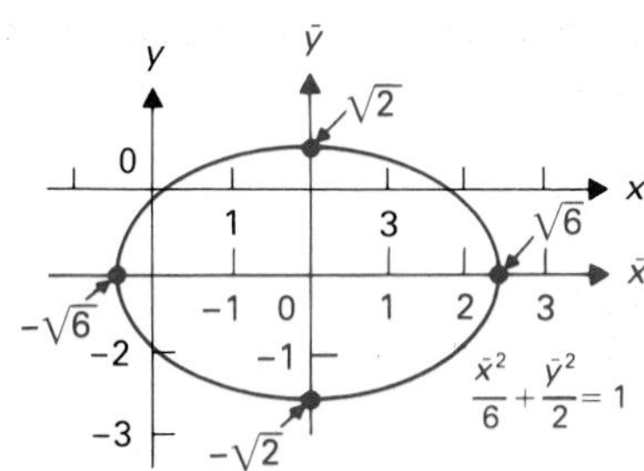

EXAMPLE 3 Describe the curve with equation

$$2x^2 + 5y^2 - 4x + 10y = -8.$$

Solution Completing the square, we obtain

$$2(x^2 - 2x) + 5(y^2 + 2y) = -8,$$
$$2(x-1)^2 + 5(y+1)^2 = 2 + 5 - 8 = -1.$$

Setting $\bar{x} = x - 1$ and $\bar{y} = y + 1$, we have

$$2\bar{x}^2 + 5\bar{y}^2 = -1.$$

Now for any point $(\bar{x}, \bar{y})$, we must have $2\bar{x}^2 + 5\bar{y}^2 \geq 0$, so no points in the plane satisfy the equation. Our "curve" is empty. □

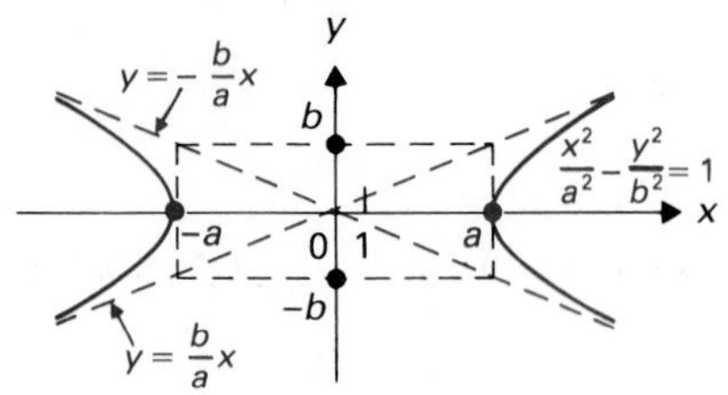

Figure 12.6 A hyperbola $x^2/a^2 - y^2/b^2 = 1$.

THE HYPERBOLA

The equation

$$\boxed{\frac{x^2}{a^2} - \frac{y^2}{b^2} = 1} \tag{3}$$

describes a hyperbola, shown in Fig. 12.6. To sketch this hyperbola, proceed as follows. Mark $\pm a$ on the x-axis and $\pm b$ on the y-axis. Then draw the rectangle crossing the axes at those points, and draw and extend the diagonals of that rectangle. The hyperbola has these diagonals

$$y = \pm \frac{b}{a} x$$

as *asymptotes*. The hyperbola crosses the x-axis at $\pm a$ but does not meet the y-axis, because if $x = 0$, then Eq. (3) reduces to $y^2 = -b^2$, which has no real solutions. If we solve Eq. (3) for y, we obtain

$$y = \pm b\sqrt{(x^2/a^2) - 1}$$

and

$$\lim_{x\to\infty} \left[b\sqrt{(x^2/a^2) - 1} - \frac{b}{a}x \right] = 0,$$

which shows that the lines $y = \pm(b/a)x$ are indeed asymptotes of the curve.

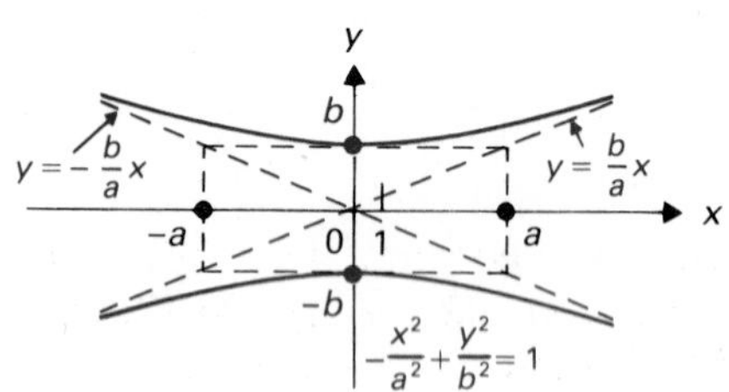

Figure 12.7 A hyperbola $-(x^2/a^2) + y^2/b^2 = 1$.

If the negative sign in Eq. (3) appears with the other term x^2/a^2 instead, as in

$$-\frac{x^2}{a^2} + \frac{y^2}{b^2} = 1, \tag{4}$$

then the asymptotes are still $y = \pm(b/a)x$, but now the hyperbola crosses the y-axis at $\pm b$ and does not meet the x-axis, as shown in Fig. 12.7.

EXAMPLE 4 Sketch the hyperbola

$$\frac{y^2}{4} - \frac{x^2}{9} = 1.$$

Solution The sketch is shown in Fig. 12.8. □

Figure 12.8 The hyperbola $y^2/4 - x^2/9 = 1$ with asymptotes $y = \pm\frac{2}{3}x$.

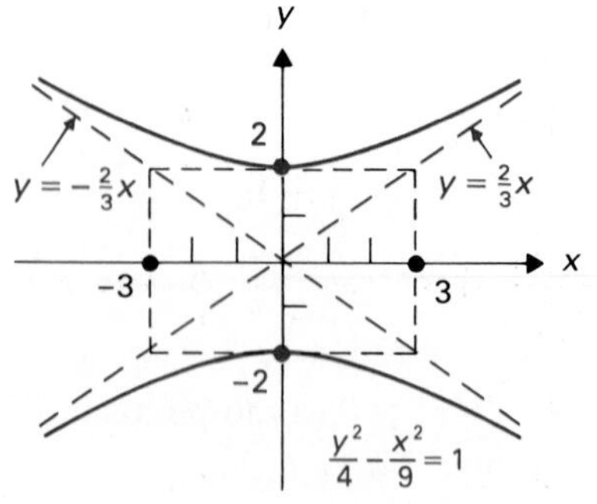

The device of completing the square can be used again to sketch a hyperbola whose center is not (0, 0).

EXAMPLE 5 Sketch $2x^2 - 3y^2 - 4x - 6y = 13$.

Solution Completing the square, we obtain

$$2(x^2 - 2x) - 3(y^2 + 2y) = 13,$$

$$2(x - 1)^2 - 3(y + 1)^2 = 2 - 3 + 13 = 12,$$

$$\frac{(x - 1)^2}{6} - \frac{(y + 1)^2}{4} = 1.$$

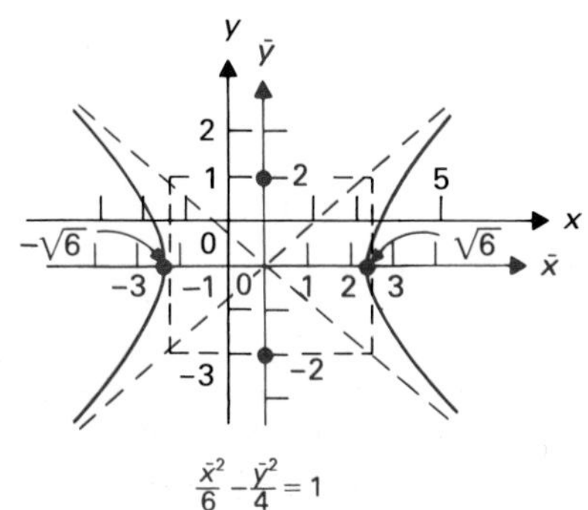

$\frac{\bar{x}^2}{6} - \frac{\bar{y}^2}{4} = 1$

Figure 12.9 The hyperbola $2x^2 - 3y^2 - 4x - 6y = 13$ centered at $(h, k) = (1, -1)$.

Setting $\bar{x} = x - 1$ and $\bar{y} = y + 1$, we have

$$\frac{\bar{x}^2}{6} - \frac{\bar{y}^2}{4} = 1.$$

We see that we have a hyperbola with center $(h, k) = (1, -1)$ and crossing the $\bar{x}$-axis at $\pm\sqrt{6}$, as shown in Fig. 12.9. □

EXAMPLE 6 Sketch the curve $3x^2 - 4y^2 - 6x - 8y = 1$.

Solution Completing the square, we have

$$3(x^2 - 2x) - 4(y^2 + 2y) = 1,$$

$$3(x - 1)^2 - 4(y + 1)^2 = 3 - 4 + 1 = 0.$$

Setting $\bar{x} = x - 1$ and $\bar{y} = y + 1$, we obtain the equation

$$3\bar{x}^2 - 4\bar{y}^2 = 0,$$

$$\bar{y}^2 = \frac{3}{4}\bar{x}^2,$$

$$\bar{y} = \pm\frac{\sqrt{3}}{2}\bar{x}.$$

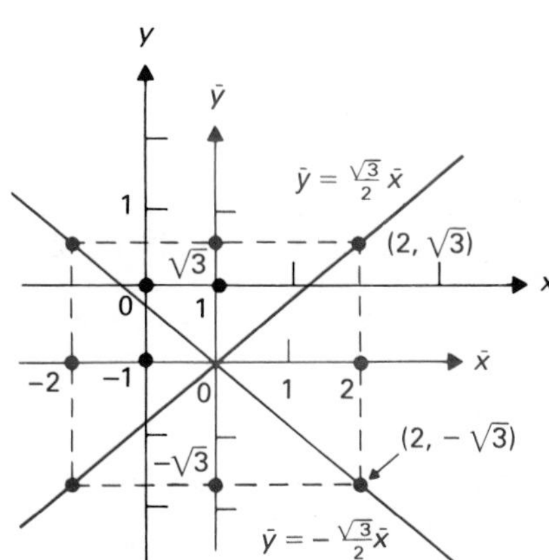

Figure 12.10 The degenerate hyperbola $3x^2 - 4y^2 - 6x - 8y = 1$, $\bar{x}^2/4 - \bar{y}^2/3 = 0$, $\bar{y} = \pm(\sqrt{3}/2)\bar{x}$.

We see that our "curve" consists of the two intersecting lines shown in Fig. 12.10. We view these intersecting lines as a *degenerate hyperbola.* Such a pair of intersecting lines is obtained if we intersect the cone in Fig. 12.1 with a vertical plane through the origin. □

THE PARABOLA

The curves

$$y = 4x^2, \qquad y = -x^2, \qquad x = \frac{1}{4}y^2, \qquad x = -4y^2$$

are parabolas with vertices at the origin, as shown in Fig. 12.11. The size of the coefficient of the quadratic term controls how fast the parabola "opens." If the vertex of $y = 4x^2$ were moved to (h, k), the equation of the curve would become

$$y - k = 4(x - h)^2.$$

As indicated in Section 1.5, any polynomial equation in x and y that is quadratic in one of the variables and linear in the other describes a parabola.

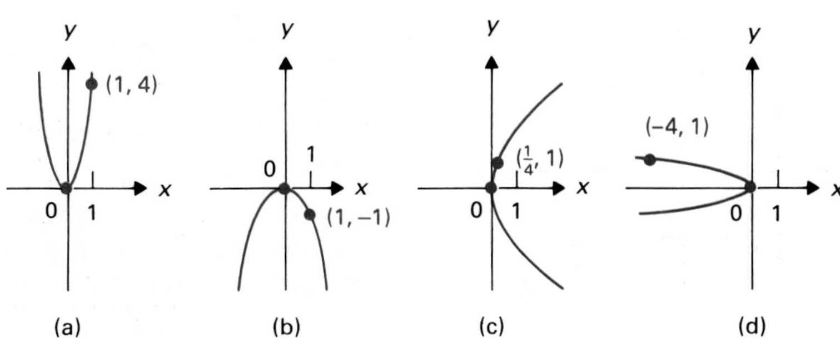

Figure 12.11 Four types of parabolas with vertices at the origin: (a) $y = 4x^2$; (b) $y = -x^2$; (c) $x = \frac{1}{4}y^2$; (d) $x = -4y^2$.

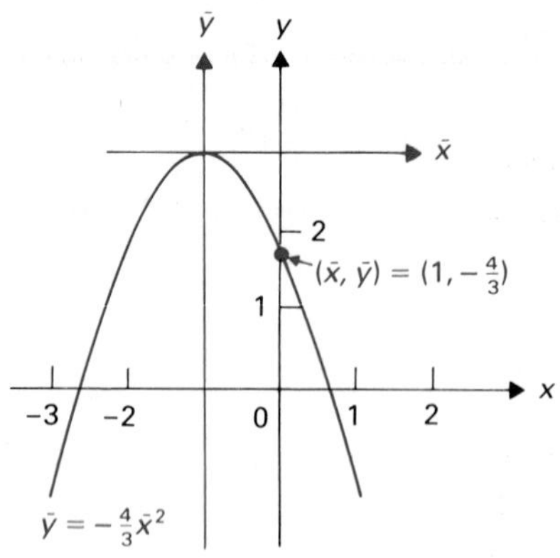

Figure 12.12 The parabola $3y + 4x^2 + 8x = 5$.

EXAMPLE 7 Sketch $3y + 4x^2 + 8x = 5$.

Solution Completing the square yields

$$3y + 4(x^2 + 2x) = 5,$$

$$3y + 4(x + 1)^2 = 4 + 5 = 9,$$

$$3y - 9 = -4(x + 1)^2,$$

$$3(y - 3) = -4(x + 1)^2,$$

$$y - 3 = -\frac{4}{3}(x + 1)^2.$$

Setting $\bar{x} = x + 1$ and $\bar{y} = y - 3$, we obtain

$$\bar{y} = -\frac{4}{3}\bar{x}^2.$$

This is a parabola with vertex at $(h, k) = (-1, 3)$ and opening downward, as shown in Fig. 12.12. □

SUMMARY

1. An equation of the form

$$\frac{(x - h)^2}{a^2} + \frac{(y - k)^2}{b^2} = 1$$

describes an ellipse with center at (h, k), meeting the $\bar{x}$-axis at $\pm a$ and the $\bar{y}$-axis at $\pm b$, where $\bar{x} = x - h$ and $\bar{y} = y - k$.

2. Equations of the form

$$\frac{(x - h)^2}{a^2} - \frac{(y - k)^2}{b^2} = 1$$

or

$$-\frac{(x - h)^2}{a^2} + \frac{(y - k)^2}{b^2} = 1$$

describe hyperbolas with center at (h, k). To sketch them, draw the rectangle with center (h, k) crossing at right angles the $\bar{x}$-axis at $\pm a$ and the $\bar{y}$-*axis* at $\pm b$. Draw and extend the diagonals of the rectangle. The first hyperbola crosses the $\bar{x}$-axis at $\pm a$, while the second hyperbola crosses the $\bar{y}$-axis at $\pm b$. Both hyperbolas have the diagonals of the rectangle as asymptotes.

3. Equations of the form

$$y - k = c(x - h)^2 \qquad \text{and} \qquad x - h = c(y - k)^2$$

describe parabolas with vertices at (h, k). The magnitude $|c|$ controls how fast the parabola "opens," and the sign of c determines the direction in which it opens.

4. Any equation of the form $ax^2 + cy^2 + dx + ey + f = 0$ represents a (possibly degenerate) conic section. The curve may be sketched by completing the square to obtain one of the forms described above.

EXERCISES

In Exercises 1 through 30, sketch the given curve.

1. $x^2 + 4y^2 = 16$
2. $x^2 - 4y^2 = 16$
3. $4y^2 - x^2 = 16$
4. $x^2 = -4y$
5. $x = 4y^2$
6. $9x^2 = -y$
7. $9x = y^2$
8. $4x^2 - 9y^2 = -36$
9. $4x^2 - 9y^2 = 36$
10. $x^2 + 4y^2 = -4$
11. $5x^2 + 2y^2 = 50$
12. $2x^2 - 3y^2 + 4x + 12y = 0$
13. $x^2 - y^2 + 6x + 2y = -3$
14. $4x^2 - 8x + 2y = 5$
15. $x^2 - 4x - 4y = 0$
16. $x^2 + 9y^2 - 4x + 18y = -4$
17. $x^2 + 4y^2 + 2x - 8y = -1$
18. $x^2 + y^2 - 4y = 9$
19. $4x^2 + 4y^2 - 8x + 8y = 1$
20. $x^2 + 2y^2 - 4x + 12y = -24$
21. $3x^2 + y^2 + 6x = -5$
22. $4y^2 + x + 8y = 0$
23. $9x^2 - 18x + y = -4$
24. $3x^2 - y^2 + 18x + 4y = -23$
25. $x^2 - 4y^2 + 2x + 16y = 15$
26. $x^2 - y^2 + 4x - 2y = 1$
27. $4x^2 + 2y^2 + 8x - 4y = -6$
28. $4x^2 + y^2 + 6y = -5$
29. $3x^2 - 4y^2 - 6x - 8y = 0$
30. $-x^2 + 2y^2 + 2x + 8y = -1$
31. For every real number t, the point $(\cos t, \sin t)$ lies on the circle $x^2 + y^2 = 1$. The functions sine and cosine are *circular functions*. Can you think of a reason why the hyperbolic sine and the hyperbolic cosine are called "hyperbolic functions"?

12.2 GEOMETRIC DEFINITIONS OF CONIC SECTIONS

In this section, we characterize the conic sections in geometric terms rather than defining them by their equations, as we did in Section 12.1. Definitions of these curves without use of coordinates are sometimes called *synthetic characterizations* of the conic sections. Section 12.4 will show that these synthetic characterizations provide some applications of conic sections that are not immediately apparent from their definitions in terms of equations.

THE PARABOLA

DEFINITION 12.1 Parabola

A **parabola** in the plane consists of all points equidistant from a fixed line ℓ (the **directrix**) and a fixed point F (the **focus**) not on the line. The point V on the parabola closest to the directrix is the **vertex** of the parabola.

Figure 12.13 shows a parabola with focus F, directrix ℓ, and vertex V. To describe the curve analytically, we choose axes through the vertex, as indicated in Fig. 12.14, so that the focus F has coordinates $(p, 0)$ and the directrix is the line $x = -p$. If (x, y) is on the parabola, then the definition states that

$$\sqrt{(x - p)^2 + y^2} = x + p.$$

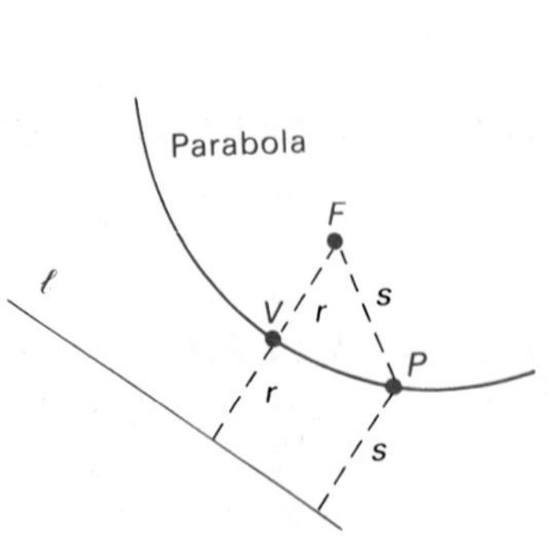

Figure 12.13 Each point P on a parabola is the same distance from its focus F as from its directrix ℓ.

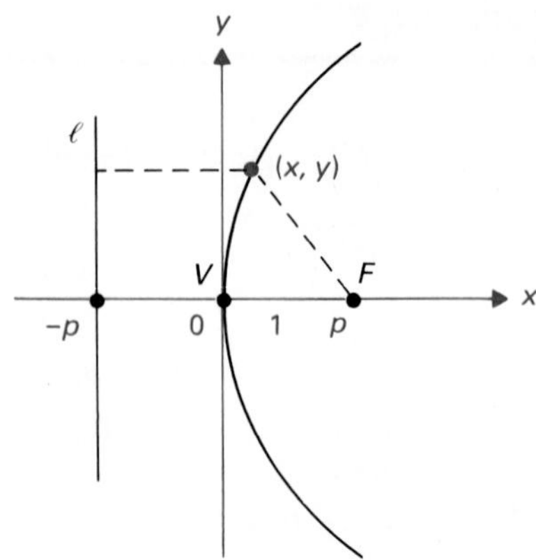

Figure 12.14 Axes chosen so that F is $(p, 0)$, V is $(0, 0)$, and ℓ is $x = -p$.

Simplifying, we obtain

$$(x - p)^2 + y^2 = (x + p)^2,$$

$$x^2 - 2xp + p^2 + y^2 = x^2 + 2xp + p^2,$$

$$\boxed{y^2 = 4px.} \tag{1}$$

Equation (1) is the *standard form* for the equation of the parabola.

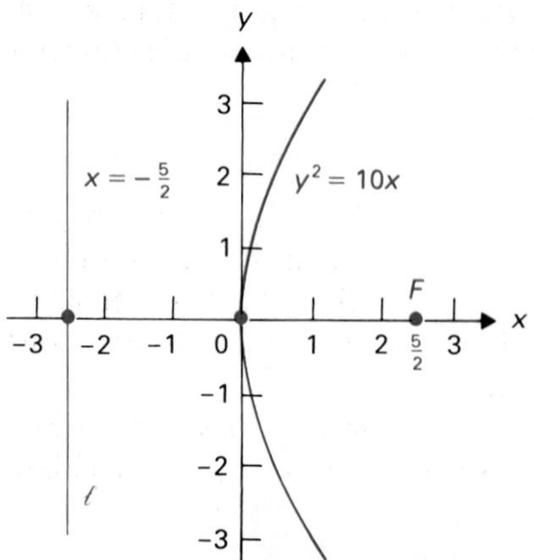

Figure 12.15 Focus and directrix for the parabola $y^2 = 10x$.

EXAMPLE 1 Find the focus and directrix for $y^2 = 10x$.

Solution Here, $4p = 10$, so $p = \frac{5}{2}$. The focus of the parabola is $(\frac{5}{2}, 0)$ and the directrix is the line $x = -\frac{5}{2}$. The parabola, its focus, and its directrix are shown in Fig. 12.15. □

Of course, $x^2 = 4py$ is the parabola "opening upward" with focus $(0, p)$ and directrix $y = -p$. Similarly, $x^2 = -4py$ is the parabola "opening downward" with focus $(0, -p)$ and directrix $y = p$.

Our work in Section 12.1 shows that an equation in x and y that is quadratic in one variable and linear in the other describes a parabola. It may be necessary to complete a square and translate axes to bring the equation into standard form.

Figure 12.16 The parabola
$$x^2 + 4x + 2y = 2,$$
or $(x + 2)^2 = -3(y - 2)$.

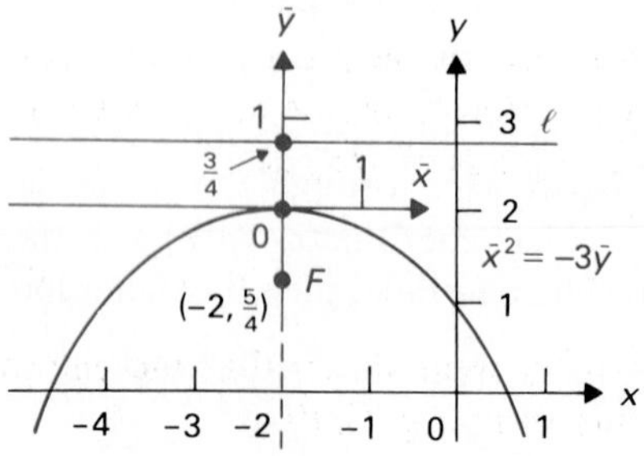

EXAMPLE 2 Sketch $x^2 + 4x + 3y = 2$, showing its vertex, focus, and directrix.

Solution Completing the square, we obtain

$$(x + 2)^2 + 3y = 2 + 4 = 6,$$

$$(x + 2)^2 = -3y + 6,$$

$$(x + 2)^2 = -3(y - 2).$$

This parabola has vertex at $(-2, 2)$. We set $\bar{x} = x + 2$ and $\bar{y} = y - 2$, corresponding to translated axes through the point $(h, k) = (-2, 2)$. Our equation then becomes

$$\bar{x}^2 = -3\bar{y}.$$

Introducing the factor 4 to find p, we rewrite this equation as

$$\bar{x}^2 = -4\left(\frac{3}{4}\right)\bar{y}.$$

This time, $p = \frac{3}{4}$. With $\bar{x}, \bar{y}$-axes at $(-2, 2)$, the equation became $\bar{x}^2 = -4p\bar{y}$. The focus is at $(x, y) = (-2, \frac{5}{4})$, and the directrix is $y = \frac{11}{4}$. The parabola is sketched in Fig. 12.16. □

THE ELLIPSE

DEFINITION 12.2 Ellipse

An **ellipse** in the plane consists of all points the sum of whose distances from two fixed points F_1 and F_2 (the **foci**) remains a constant value $2a$ (greater than the distance between the foci).

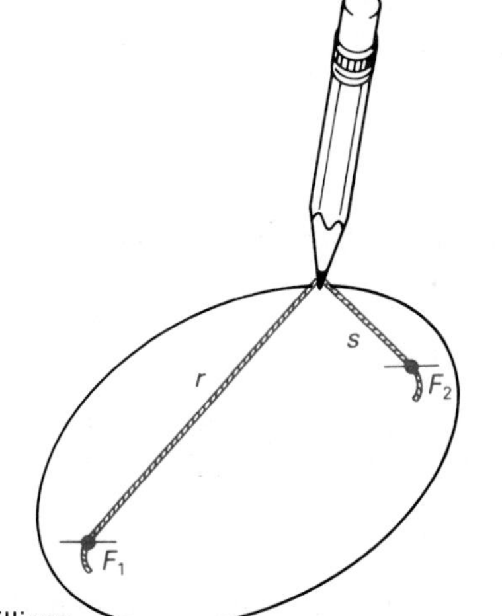

Figure 12.17 An ellipse is traced by the pencil if the string is kept taut.

Figure 12.17 shows an ellipse with foci F_1 and F_2. This ellipse may be drawn as follows: Take a piece of string a bit more than $2a$ units long and fasten near the ends at F_1 and F_2 with staples, so that the length of string between the staples is exactly $2a$. A pencil placed to keep the string taut as in Fig. 12.17 will then trace out the ellipse.

To describe an ellipse analytically, we choose axes as shown in Fig. 12.18, so that the foci are $(-c, 0)$ and $(c, 0)$. Then, for (x, y) on the ellipse and r and s in the figure, we obtain

$$\begin{aligned}
r + s &= 2a, \\
\sqrt{(x-c)^2 + y^2} + \sqrt{(x+c)^2 + y^2} &= 2a, \qquad \textbf{(2)} \\
\sqrt{(x-c)^2 + y^2} &= 2a - \sqrt{(x+c)^2 + y^2}, \\
x^2 - 2cx + c^2 + y^2 &= 4a^2 - 4a\sqrt{(x+c)^2 + y^2} \\
&\quad + x^2 + 2cx + c^2 + y^2, \\
a\sqrt{(x+c)^2 + y^2} &= a^2 + cx, \\
a^2(x^2 + 2cx + c^2 + y^2) &= a^4 + 2a^2cx + c^2x^2, \\
(a^2 - c^2)x^2 + a^2y^2 &= a^4 - a^2c^2 = a^2(a^2 - c^2). \qquad \textbf{(3)}
\end{aligned}$$

Figure 12.18 Axes chosen so that the origin is at the center and the foci are the points $(\pm c, 0)$ on the x-axis.

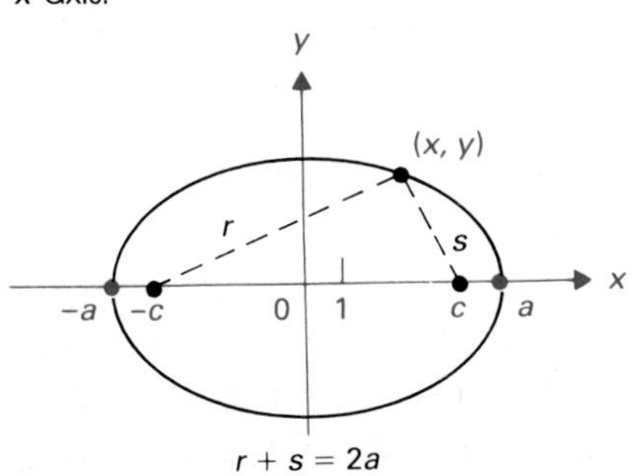

Since the point where the ellipse intersects the positive y-axis is equidistant from the foci, its distance from each focus must be a, as shown in Fig. 12.19. We see from the figure that $a > c$. Thus we may set

$$b^2 = a^2 - c^2. \qquad \textbf{(4)}$$

Then $a^2 = b^2 + c^2$, and the Pythagorean theorem shows that the ellipse intersects the y-axis at the point $\pm b$, as shown in Fig. 12.19.

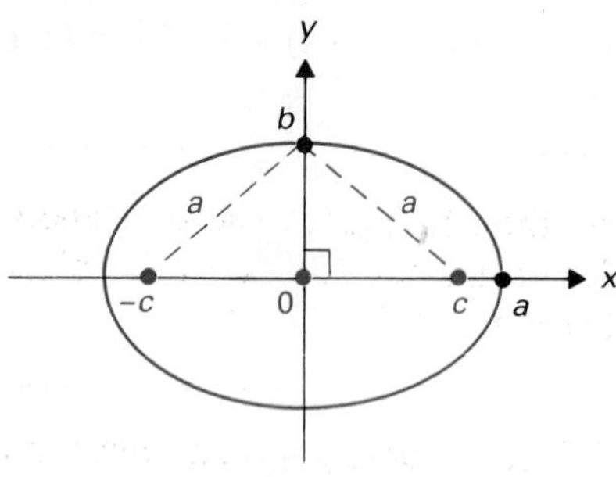

Figure 12.19 $a^2 = b^2 + c^2$.

We now substitute b^2 for $a^2 - c^2$ in Eq. (3), which then simplifies to become

$$b^2x^2 + a^2y^2 = a^2b^2,$$

$$\boxed{\frac{x^2}{a^2} + \frac{y^2}{b^2} = 1.} \tag{5}$$

Equation (5) is the *standard form* of the equation of the ellipse. The line segment from $(-a, 0)$ to $(a, 0)$ in Fig. 12.19 is the **major axis** of the ellipse, and the segment from $(0, -b)$ to $(0, b)$ is the **minor axis.**

EXAMPLE 3 Find the foci and the lengths of major and minor axes for the ellipse

$$\frac{x^2}{25} + \frac{y^2}{9} = 1.$$

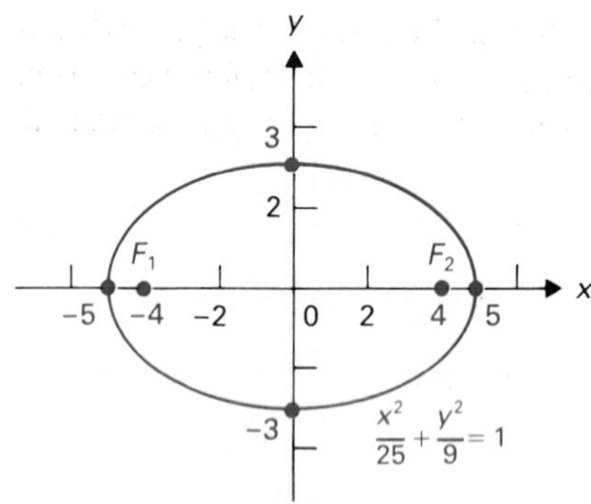

Figure 12.20 The ellipse $x^2/25 + y^2/9 = 1$ has foci $(\pm 4, 0)$.

Solution Here $a^2 = 25$ and $b^2 = 9$. From Eq. (4), we obtain

$$c^2 = a^2 - b^2. \tag{6}$$

Thus $c^2 = 25 - 9 = 16$, so $c = 4$. The foci are $(-4, 0)$ and $(0, 4)$. The length of the major axis is $2a = 10$, while the length of the minor axis is $2b = 6$. The ellipse and its foci are shown in Fig. 12.20. □

EXAMPLE 4 Sketch the ellipse $9x^2 + 4y^2 - 54x + 16y = -61$.

Solution Completing the square, we obtain

$$9(x^2 - 6x) + 4(y^2 + 4y) = -61,$$
$$9(x - 3)^2 + 4(y + 2)^2 = -61 + 81 + 16 = 36,$$
$$\frac{(x - 3)^2}{4} + \frac{(y + 2)^2}{9} = 1,$$
$$\frac{\bar{x}^2}{4} + \frac{\bar{y}^2}{9} = 1,$$

where we chose translated $\bar{x}, \bar{y}$-axes at $(x, y) = (3, -2)$. *This time the major axis of the ellipse is vertical.*

Figure 12.21 The ellipse $9x^2 + 4y^2 - 54x + 16y = -61$ has center $(3, -2)$ and foci $(3, -2 \pm \sqrt{5})$.

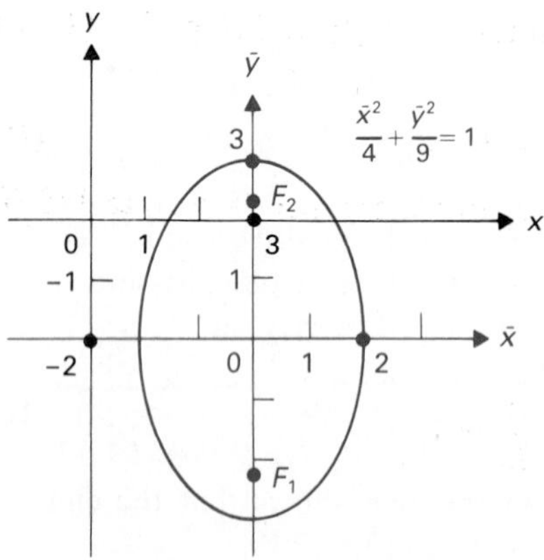

> We should always take as a^2 the *larger* of the two numbers in the denominator when the equation is brought into standard form.

(We did not bother to do this in Section 12.1.) Thus $a^2 = 9$, $b^2 = 4$, and $c^2 = a^2 - b^2 = 5$, so $c = \sqrt{5}$. The foci are then $(3, -2 - \sqrt{5})$ and $(3, -2 + \sqrt{5})$. The major axis is of length $2a = 6$, and the minor axis is of length $2b = 4$. The ellipse and its foci are shown in Fig. 12.21. □

THE HYPERBOLA

DEFINITION 12.3 Hyperbola

A **hyperbola** in the plane consists of all points the difference of whose distances from two fixed points F_1 and F_2 (the **foci**) remains at a constant value of $2a$ (which must be less than the distance between the foci).

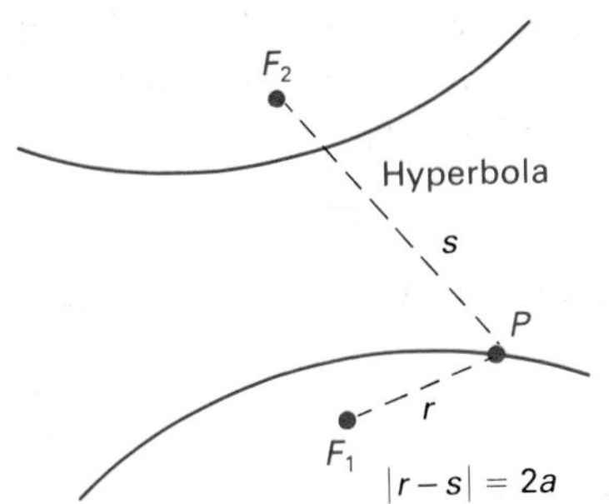

Figure 12.22 A hyperbola with foci F_1 and F_2. The difference of the distances from P to F_1 and to F_2 is $2a$.

Figure 12.22 shows a hyperbola with foci F_1 and F_2. To determine the curve analytically, we choose axes as indicated in Fig. 12.23, so that the foci are at $(-c, 0)$ and $(c, 0)$. This time, $a < c$, which we ask you to prove from Definition 12.3 in Exercise 1. For a point (x, y) on the hyperbola, and r and s in Fig. 12.23, we then have $|r - s| = 2a$, or

$$\left|\sqrt{(x-c)^2 + y^2} - \sqrt{(x+c)^2 + y^2}\right| = 2a. \tag{7}$$

The simplification of Eq. (7) is similar to that of Eq. (2); there is just a difference in sign, which disappears the second time the two sides of the equation are squared. In Exercise 2, we ask you to show that Eq. (3) is again obtained. This time $c > a$, so we set

$$b^2 = c^2 - a^2 \tag{8}$$

and obtain from Eq. (3)

$$-b^2x^2 + a^2y^2 = a^2(-b^2)$$

or

$$\boxed{\frac{x^2}{a^2} - \frac{y^2}{b^2} = 1.} \tag{9}$$

Equation (9) is the *standard form* of the equation of a hyperbola. We discussed the sketching of the hyperbola, and in particular its *asymptotes*, in Section 12.1.

EXAMPLE 5 Sketch the hyperbola

$$\frac{x^2}{16} - \frac{y^2}{9} = 1,$$

showing the asymptotes and foci.

Solution The hyperbola has asymptotes $y = \pm(\frac{3}{4})x$. From Eq. (8), we obtain

$$c^2 = a^2 + b^2, \tag{10}$$

so $c = \sqrt{16 + 9} = 5$. The foci of the hyperbola are then $(-5, 0)$ and $(5, 0)$. The sketch is given in Fig. 12.24. □

Figure 12.23 Axes chosen so that the foci are $(\pm c, 0)$ on the x-axis.

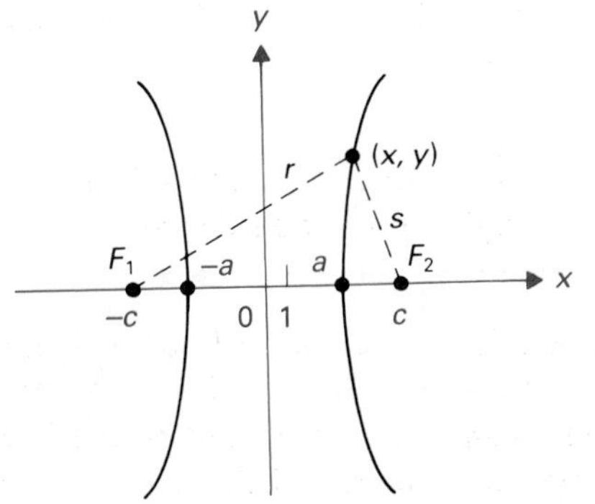

EXAMPLE 6 Sketch the hyperbola

$$-\frac{x^2}{144} + \frac{y^2}{25} = 1,$$

showing the asymptotes and foci.

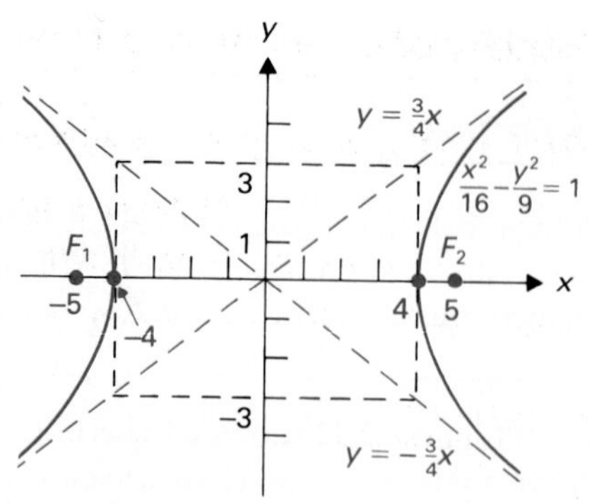

Figure 12.24 The foci of $x^2/16 - y^2/9 = 1$ are at $(\pm 5, 0)$.

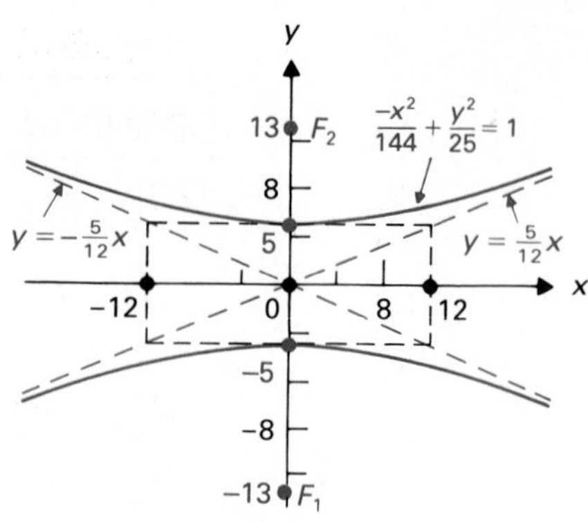

Figure 12.25 The foci of $-x^2/144 + y^2/25 = 1$ are at $(0, \pm 13)$.

Solution This hyperbola has foci on the y-axis.

> For a hyperbola, we should think of a^2 as the number under the quadratic variable appearing with the *positive* sign in the standard form.

(We did not bother to do this in Section 12.1.) So this time $a^2 = 25$ and $b^2 = 144$. From Eq. (8),

$$c = \sqrt{a^2 + b^2} = \sqrt{25 + 144} = \sqrt{169} = 13.$$

Thus the foci are $(0, -13)$ and $(0, 13)$. The asymptotes are $y = \pm(5/12)x$. The sketch is shown in Fig. 12.25. □

ECCENTRICITY AND DIRECTRICES

For the ellipse and hyperbola, each focus has an associated line as directrix, just as for the parabola. Figure 12.26 shows the foci F_1 and F_2 with their associated directrices ℓ_1 and ℓ_2, respectively. These **directrices** are the lines $x = \pm a^2/c$.

We leave to Exercise 3 the demonstration of the following theorem.

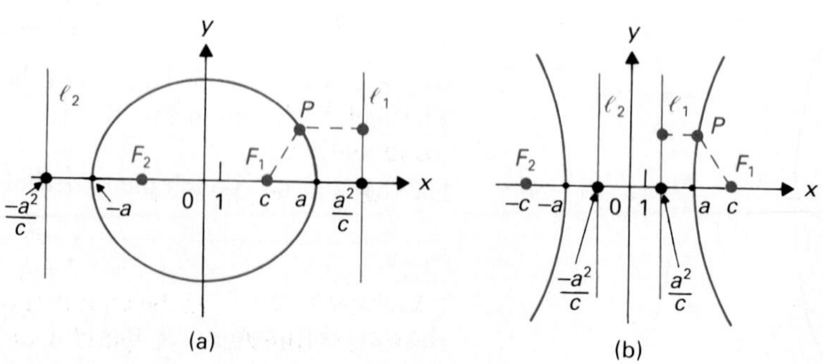

Figure 12.26 Directrices for an ellipse or hyperbola in standard position cross the axes at $\pm a^2/c$.

THEOREM 12.1 Property of an ellipse or a hyperbola

If P is a point on an ellipse or a hyperbola, then

$$\frac{\text{Distance from } P \text{ to a focus}}{\text{Distance from } P \text{ to the associated directrix}} = \frac{c}{a}. \quad \textbf{(11)}$$

DEFINITION 12.4 Eccentricity

The number

$$e = \frac{c}{a} \quad \textbf{(12)}$$

is called the **eccentricity** of the ellipse or hyperbola.

EXAMPLE 7 Find the eccentricity and directrices for the hyperbola

$$\frac{x^2}{16} - \frac{y^2}{9} = 1.$$

Solution We have $a = 4$, and $c = 5$ as in Example 5. Thus $e = \frac{5}{4}$, and the directrices are the lines $x = \pm a^2/c = \pm \frac{16}{5}$. □

EXAMPLE 8 Find the eccentricity and directrices for the ellipse

$$\frac{x^2}{25} + \frac{y^2}{9} = 1.$$

Solution We have $a = 5$ and $c = 4$, so $e = c/a = \frac{4}{5}$. The directrices are the lines $x = \pm a^2/c = \pm \frac{25}{4}$. □

In view of Eqs. (11) and (12) and Definition 12.1 on page 523, we see that all three conic sections can be defined concisely, as follows.

DEFINITION 12.5 Synthetic descriptions

Let F be a point in the plane and ℓ a line not containing F. Let e be any positive number. The curve in the plane consisting of all points P such that

$$\frac{\text{Distance from } P \text{ to } F}{\text{Distance from } P \text{ to } \ell} = e$$

is

1. a **parabola** if $e = 1$,
2. an **ellipse** if $e < 1$,
3. a **hyperbola** if $e > 1$.

EXAMPLE 9 Find the equation of the hyperbola with a focus at $(-1, 2)$, associated directrix $y = 1$, and with eccentricity $\frac{3}{2}$.

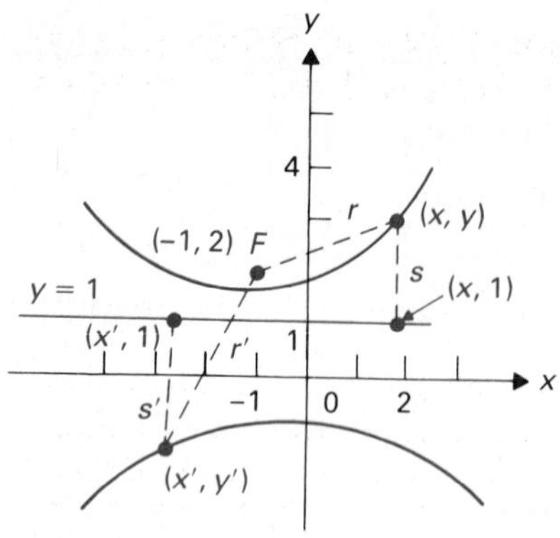

Figure 12.27 $r/s = r'/s' = e = \frac{3}{2}$.

Solution We use Definition 12.5, and we see from Fig. 12.27 that a point (x, y) lies on the hyperbola if and only if

$$\frac{\sqrt{(x+1)^2 + (y-2)^2}}{|y-1|} = \frac{3}{2}.$$

Squaring, we obtain

$$(x+1)^2 + (y-2)^2 = \frac{9}{4}(y-1)^2$$

or

$$4x^2 - 5y^2 + 8x + 2y + 11 = 0. \quad \square$$

EXAMPLE 10 Find the equation of the ellipse with center at the origin, major axis of length 10, and a directrix $y = \frac{25}{3}$.

Solution Now $2a = 10$, so $a = 5$. Referring to the chart in the Summary, we see that the directrices are $y = \pm a^2/c$, so we have $a^2/c = \frac{25}{3}$. Thus

$$c = \frac{3}{25}a^2 = 3.$$

Then

$$b^2 = a^2 - c^2 = 25 - 9 = 16,$$

so $b = 4$. The ellipse has equation $x^2/b^2 + y^2/a^2 = 1$, or

$$\frac{x^2}{16} + \frac{y^2}{25} = 1,$$

and is shown in Fig. 12.28. $\square$

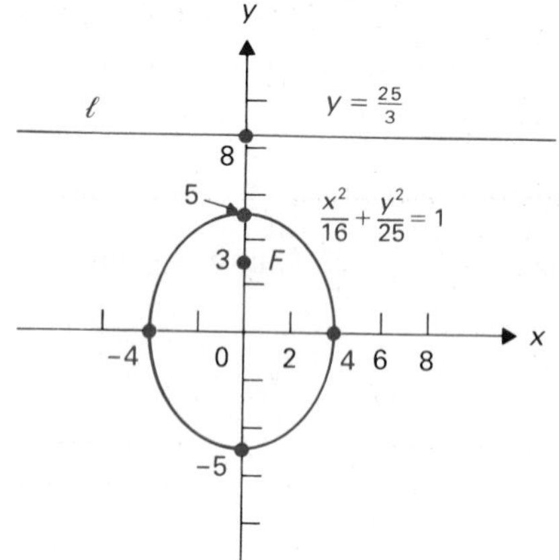

Figure 12.28 The ellipse with center (0, 0), major axis of length 10, and a directrix $y = \frac{25}{3}$.

EXAMPLE 11 Find the equation of the hyperbola with center at the origin, foci at $(\pm\sqrt{5}, 0)$, and directrices $x = \pm 1/\sqrt{5}$.

Solution Referring to the chart in the Summary, we see that $c = \sqrt{5}$ and $a^2/c = 1/\sqrt{5}$. Consequently, $a^2 = 1$. Then from $c = \sqrt{a^2 + b^2}$, we have $b^2 = c^2 - a^2 = 5 - 1 = 4$. Since the hyperbola crosses the x-axis, its equation is $x^2/a^2 - y^2/b^2 = 1$, or

$$\frac{x^2}{1} - \frac{y^2}{4} = 1.$$

The hyperbola is shown in Fig. 12.29. $\square$

Figure 12.29 The hyperbola with center at (0, 0), foci $(\pm\sqrt{5}, 0)$, and directrices $x = \pm 1/\sqrt{5}$.

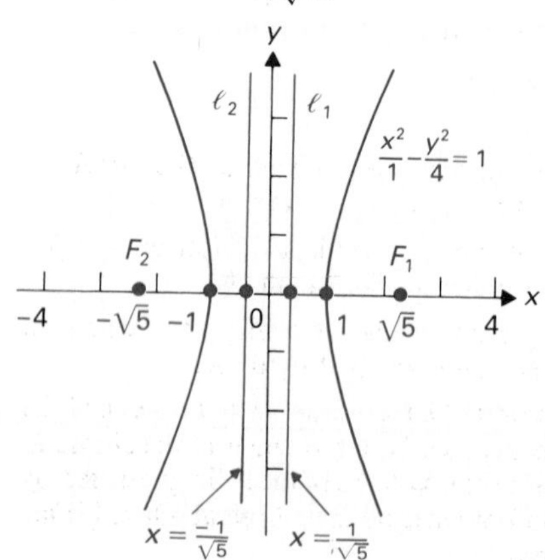

SUMMARY

1. An ellipse in the plane consists of all points such that the sum of their distances from two fixed points (foci) remains constant.
2. A hyperbola in the plane consists of all points such that the difference of their distances from two fixed points (foci) remains constant.
3. Let a fixed point F (focus) and a line ℓ (directrix) not through F be given in the plane. For a point P in the plane, let $|PF|$ be the distance

from P to F and $|PD|$ be the distance from P to ℓ. Let $e > 0$ (eccentricity) be given. The curve consisting of all points P such that $|PF|/|PD| = e$ is

a. a parabola if $e = 1$,

b. an ellipse if $e < 1$,

c. a hyperbola if $e > 1$.

4. The following chart gives other information about the conic sections.

Type of curve	Standard forms	Focus	Directrix	Eccentricity
Parabola	$y^2 = 4px$	$(p, 0)$	$x = -p$	$e = 1$
	$y^2 = -4px$	$(-p, 0)$	$x = p$	
	$x^2 = 4py$	$(0, p)$	$y = -p$	
	$x^2 = -4py$	$(0, -p)$	$y = p$	
Ellipse a^2 is the *larger* number in the denominator. $c = \sqrt{a^2 - b^2}$	$\frac{x^2}{a^2} + \frac{y^2}{b^2} = 1$	$(c, 0)$ $(-c, 0)$	$x = a^2/c$ $x = -a^2/c$	$e = \frac{c}{a} < 1$
	$\frac{x^2}{b^2} + \frac{y^2}{a^2} = 1$	$(0, c)$ $(0, -c)$	$y = a^2/c$ $y = -a^2/c$	
Hyperbola a^2 is the number in the denominator of the *plus* term. $c = \sqrt{a^2 + b^2}$	$\frac{x^2}{a^2} - \frac{y^2}{b^2} = 1$	$(c, 0)$ $(-c, 0)$	$x = a^2/c$ $x = -a^2/c$	$e = \frac{c}{a} > 1$
	$-\frac{x^2}{b^2} + \frac{y^2}{a^2} = 1$	$(0, c)$ $(0, -c)$	$y = a^2/c$ $y = -a^2/c$	

EXERCISES

1. Show from Definition 12.3 that $a < c$ for a hyperbola. [*Hint:* Use the fact that the sum of the lengths of two sides of a triangle is greater than the length of the third side.]

2. Show that Eq. (7) of the text again leads to Eq. (3), as asserted in the text.

3. Starting from the equation

$$\frac{x^2}{a^2} + \frac{y^2}{b^2} = 1$$

for an ellipse, prove that Eq. (11) of the text holds. That is, prove Theorem 12.1. [*Hint:* Using the equation of the ellipse and $b^2 = a^2 - c^2$, express everything on the left in Eq. (11) in terms of x, a, and c. Square, and show that it simplifies to c^2/a^2.] Observe that the same computation is valid for the hyperbola, since two sign changes cancel each other.

In Exercises 4 through 15, find the foci and directrices of the conic section with the given equation, and sketch the curve, showing the foci and directrices.

4. $\frac{x^2}{9} + \frac{y^2}{25} = 1$

5. $\frac{x^2}{25} + \frac{y^2}{4} = 1$

6. $\frac{x^2}{25} - \frac{y^2}{16} = 1$

7. $\frac{x^2}{4} - \frac{y^2}{9} = -1$

8. $x^2 = 12y$

9. $y^2 = -8x$

10. $x^2 + 2x + 4y^2 - 16y + 13 = 0$

11. $4x^2 + y^2 - 16x + 6y = 11$

12. $3y^2 + 12y - 4x^2 + 3 = 0$

13. $x^2 - 9y^2 - 2x - 18y = 44$

14. $y^2 - 4x + 14y + 57 = 0$

15. $x^2 + 6x + 3y = 0$

16. Find the x-intercept of the line tangent to the parabola $y^2 = 4px$ at a point (x_0, y_0) on the parabola.

17. Find the area of the region bounded by an ellipse with major axis of length $2a$ and minor axis of length $2b$.

18. Find the volume generated by revolving the region in Exercise 17 about the major axis of the ellipse.

19. For the ellipse in Exercise 17, let ℓ be a line in the plane of the ellipse such that the distance from the center of the ellipse to ℓ is s, where $s > a$. Find the volume of the solid generated by revolving the region bounded by the ellipse about ℓ. [*Hint:* Use Pappus' theorem.]

In Exercises 20 through 28, find the equation of the second-degree curve with the given focus, directrix, and eccentricity e.

20. Focus $(0, 0)$, directrix $x = -2$, $e = 3$

21. Focus $(-1, 2)$, directrix $x = 1$, $e = 2$

22. Focus $(0, 0)$, directrix $y = 4$, $e = \frac{1}{2}$

23. Focus $(-1, 2)$, directrix $x = 1$, $e = \frac{2}{3}$

24. Focus $(0, 0)$, directrix $x = 3$, $e = 1$

25. Focus $(-2, 2)$, directrix $y = 3$, $e = 1$

26. Focus $(-2, 3)$, directrix $x - y = 6$, $e = \frac{3}{2}$. [*Hint:* We will show later that the distance from (x_0, y_0) to the line $ax + by + c = 0$ is $|ax_0 + by_0 + c|/\sqrt{a^2 + b^2}$.]

27. Focus $(-1, 1)$, directrix $3x + 4y = 1$, $e = 1$

28. Focus $(1, 2)$, directrix $4x - 3y = 2$, $e = 5$

29. Find the equation of the ellipse with center at the origin, foci at $(\pm 3, 0)$, and eccentricity $\frac{1}{2}$.

30. Find the equation of the ellipse with center at the origin, foci at $(0, \pm 1)$, and directrices $y = \pm 4$.

31. Find the equation of the hyperbola with center at the origin, foci at $(\pm 4, 0)$, and directrices $x = \pm 2$.

32. Find the equation of the hyperbola with center at the origin, foci at $(0, \pm 6)$, and eccentricity 2.

33. Find the equation of the parabola with vertex at the origin and focus at $(3, 0)$.

34. Find the equation of the parabola with vertex at the origin and directrix $y = 5$.

35. Find the equation of the ellipse with center at $(-1, 2)$, a focus at $(-1, 4)$, and eccentricity $\frac{1}{2}$.

36. Find the equation of the hyperbola with center at $(1, -3)$, a focus at $(1, 0)$, and a directrix at $(1, -2)$.

37. Find the equation of the parabola with vertex at $(-5, 2)$ and and focus at $(-5, 0)$.

38. Find the equation of the parabola with vertex at the origin and focus at $(1, 1)$.

39. The chord of a parabola that passes through the focus and is parallel to the directrix is the *latus rectum* of the parabola. Show that the latus rectum of the parabola $y^2 = 4px$ has length $4p$.

40. Consider ellipses of the form $(x^2/a^2) + (y^2/b^2) = 1$, where a is held constant and $b \to a$.

a) What does the limiting locus of such ellipses approach as $b \to a$?

b) What is the limiting position of the foci as $b \to a$?

c) What is the limiting position of the directrices as $b \to a$?

d) What is the limiting value of the eccentricity?

41. Answer Exercise 40 if $a \to \infty$ while b is held constant, rather than $b \to a$.

42. Answer Exercise 40 for the hyperbolas $(x^2/a^2) - (y^2/b^2) = 1$ if a is held constant and $b \to \infty$.

43. Answer Exercise 40 for the hyperbolas $(x^2/a^2) - (y^2/b^2) = 1$ if b is held constant and $a \to \infty$.

44. Answer Exercise 40 for the hyperbolas $(x^2/a^2) - (y^2/b^2) = 1$ if a is held constant and $b \to 0$.

45. Answer Exercise 40 for the hyperbolas $(x^2/a^2) - (y^2/b^2) = 1$ if b is held constant and $a \to 0$.

12.3 CLASSIFICATION OF SECOND-DEGREE CURVES

In this section, we will classify and sketch the various types of plane curves that are described by a second-degree equation of the form

$$Ax^2 + Bxy + Cy^2 + Dx + Ey + F = 0, \tag{1}$$

where at least one of A, B, and C is nonzero. We will see that each such curve can be characterized as either an *ellipse*, a *hyperbola*, or a *parabola*.

ROTATION OF AXES

If the coefficient B of xy in Eq. (1) is 0, then the curve can be sketched quite easily by completing the squares on x and on y and choosing translated coordinates. We did this in Section 12.1, and we will review it in the course of this section. We now show that by *rotating axes*, we can get rid of the term Bxy.

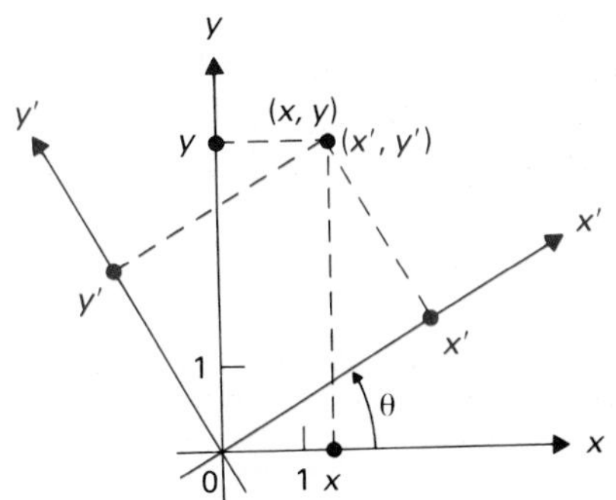

Figure 12.30 Rotation of axes through θ to new x',y'-axes.

Choose a new set of perpendicular axes, an x'-axis and a y'-axis, through the origin 0 of the Euclidean plane. Let the x'-axis make an angle θ with the x-axis, as shown in Fig. 12.30. We take the same scale on our new axes as on our old axes. Each point of our Euclidean plane then has two types of rectangular coordinates, (x, y) and (x', y'), as shown in Fig. 12.30. We would like to obtain formulas for the x,y-coordinates of a point in terms of its x',y'-coordinates, and vice versa. The following equations are easily obtained from Fig. 12.31:

$$\boxed{\begin{aligned} x &= x' \cos\theta - y' \sin\theta, \\ y &= x' \sin\theta + y' \cos\theta. \end{aligned}} \qquad \textbf{(2)}$$

For a curve with Eq. (1), we can use Eqs. (2) to obtain the equation of the same curve in terms of the x',y'-coordinate system. Substitution yields

$$\begin{aligned} &A(x'\cos\theta - y'\sin\theta)^2 + B(x'\cos\theta - y'\sin\theta)(x'\sin\theta + y'\cos\theta) \\ &\quad + C(x'\sin\theta + y'\cos\theta)^2 + D(x'\cos\theta - y'\sin\theta) \\ &\quad + E(x'\sin\theta + y'\cos\theta) + F = 0. \end{aligned} \qquad \textbf{(3)}$$

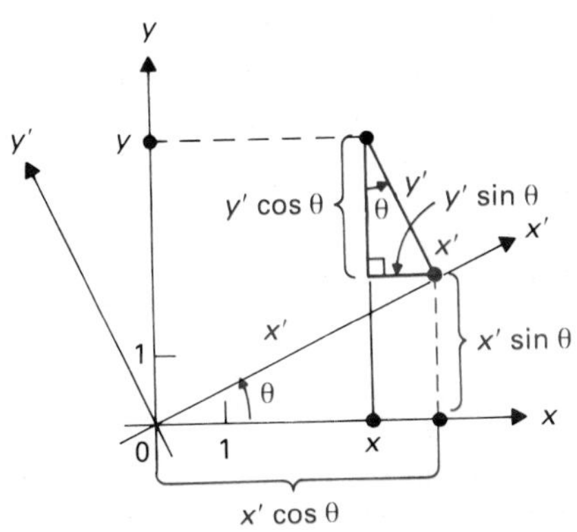

Figure 12.31 $x = x'\cos\theta - y'\sin\theta$; $y = x'\sin\theta + y'\cos\theta$.

If the terms of Eq. (3) are multiplied out and collected, we obtain an equation of the form

$$A'x'^2 + B'x'y' + C'y'^2 + D'x' + E'y' + F' = 0, \qquad \textbf{(4)}$$

where, in particular,

$$\boxed{\begin{aligned} A' &= A\cos^2\theta + B\sin\theta\cos\theta + C\sin^2\theta, \\ B' &= A(-2\sin\theta\cos\theta) + B(\cos^2\theta - \sin^2\theta) \\ &\quad + C(2\sin\theta\cos\theta) \\ &= (C - A)\sin 2\theta + B\cos 2\theta, \\ C' &= A\sin^2\theta + B(-\sin\theta\cos\theta) + C\cos^2\theta. \end{aligned}} \qquad \textbf{(5)}$$

Several interesting results can be obtained from the relations (5). In particular, we see that $B' = 0$ if and only if

$$(C - A)\sin 2\theta + B\cos 2\theta = 0.$$

Thus if $B \neq 0$, we see that

$$\boxed{B' = 0 \qquad \text{if } \cot 2\theta = \frac{A - C}{B}.} \qquad \textbf{(6)}$$

Equation (6) shows that to rotate axes so that the x',y'-form of Eq. (1) has zero coefficient of $x'y'$, we should rotate through an angle θ such that

$$2\theta = \cot^{-1}\frac{A - C}{B}.$$

If $B \neq 0$, there will be exactly one such angle θ where $0 < \theta < \pi/2$.

From the relations (5), it is easy to compute directly (see Exercises 2 and 3) that

$$A' + C' = A + C \qquad \text{and} \qquad B'^2 - 4A'C' = B^2 - 4AC.$$

That is, these quantities are *invariants* under rotation of axes.

We summarize in a theorem.

THEOREM 12.2 Eliminating the x,y-term

If the axes of the Euclidean plane are rotated through an angle θ as shown in Fig. 12.30, then the x', y'-form of Eq. (1) has zero as coefficient B' of $x'y'$ if

$$\cot 2\theta = \frac{A - C}{B}.$$

Furthermore, $A + C$ and $B^2 - 4AC$ remain invariant. That is, $A + C = A' + C'$ and $B^2 - 4AC = B'^2 - 4A'C'$.

To employ rotation of axes, we need to find Eqs. (2). First we find $\cot 2\theta = (A - C)/B$. Sometimes 2θ and θ are angles such that $\sin \theta$ and $\cos \theta$ are familiar. If this is not the case, find $\cos 2\theta$ from $\cot 2\theta$ and then use the relations

$$\sin \theta = \sqrt{\frac{1 - \cos 2\theta}{2}} \qquad \text{and} \qquad \cos \theta = \sqrt{\frac{1 + \cos 2\theta}{2}}. \tag{7}$$

EXAMPLE 1 Eliminate the "cross term" xy in the equation $x^2 - xy + y^2 - 4x + 5 = 0$ by rotation of axes.

Solution Here $A = 1$, $B = -1$, and $C = 1$, so

$$\cot 2\theta = \frac{1 - 1}{-1} = 0.$$

Therefore $2\theta = \pi/2$ and $\theta = \pi/4$. The transformation equations (2) then become

$$x = \frac{1}{\sqrt{2}}x' - \frac{1}{\sqrt{2}}y', \qquad y = \frac{1}{\sqrt{2}}x' + \frac{1}{\sqrt{2}}y'.$$

Substituting, we obtain

$$\frac{1}{2}(x' - y')^2 - \frac{1}{2}(x' - y')(x' + y') + \frac{1}{2}(x' + y')^2$$
$$- 2\sqrt{2}(x' - y') + 5 = 0.$$

We now obtain

$$\frac{1}{2}x'^2 + \frac{3}{2}y'^2 - 2\sqrt{2}x' + 2\sqrt{2}y' + 5 = 0,$$

or

$$x'^2 + 3y'^2 - 4\sqrt{2}x' + 4\sqrt{2}y' + 10 = 0$$

as an x', y'-equation for our curve with zero for coefficient of $x'y'$. □

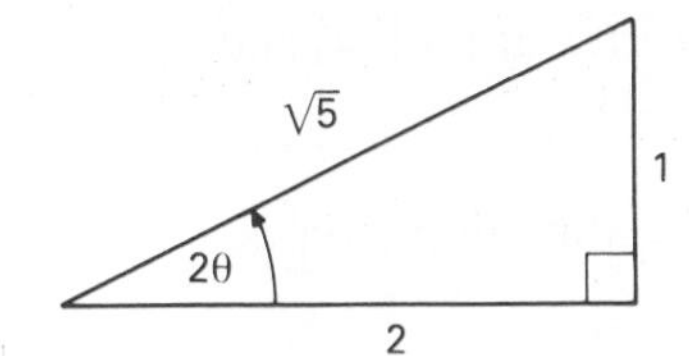

Figure 12.32 If $\cot 2\theta = 2$, then $\cos 2\theta = 2/\sqrt{5}$.

EXAMPLE 2 Find the substitution that would eliminate the "cross term" $2xy$ in the equation $3x^2 + 2xy - y^2 - 4x + 5y = 20$.

Solution From Eq. (6) we see that we should choose θ such that

$$\cot 2\theta = \frac{A - C}{B} = \frac{3 - (-1)}{2} = 2.$$

From Fig. 12.32, we then obtain $\cos 2\theta = 2/\sqrt{5}$. The relations (7) then yield

$$\sin \theta = \sqrt{\frac{1 - (2/\sqrt{5})}{2}} = \sqrt{\frac{\sqrt{5} - 2}{2\sqrt{5}}},$$

$$\cos \theta = \sqrt{\frac{1 + (2/\sqrt{5})}{2}} = \sqrt{\frac{\sqrt{5} + 2}{2\sqrt{5}}}.$$

Equations (2) become

$$x = \sqrt{\frac{\sqrt{5} + 2}{2\sqrt{5}}}\,x' - \sqrt{\frac{\sqrt{5} - 2}{2\sqrt{5}}}\,y',$$

$$y = \sqrt{\frac{\sqrt{5} - 2}{2\sqrt{5}}}\,x' + \sqrt{\frac{\sqrt{5} + 2}{2\sqrt{5}}}\,y'.$$

Using a calculator, we find that

$$x \approx 0.97x' - 0.23y', \qquad y \approx 0.23x' + 0.97y'. \quad \square$$

THE CLASSIFICATION

The discussion above has shown that in studying the curve given by Eq. (1), we can assume (by rotation of axes if necessary) that the equation is of the form

$$Ax^2 + Cy^2 + Dx + Ey + F = 0, \tag{8}$$

where not both A and C are zero.

Case 1 If $A = 0$ and $D \neq 0$, then Eq. (8) becomes

$$Cy^2 + Dx + Ey + F = 0. \tag{9}$$

Our work in Section 12.1 showed that this is the equation of a parabola, and we can easily sketch it by completing the square and translating axes. A similar analysis can be made if $C = 0$ and $E \neq 0$.

EXAMPLE 3 Sketch the parabola $x^2 + 2x + y - 3 = 0$.

Solution Completing the square, we obtain

$$(x^2 + 2x) + y = 3,$$

$$(x + 1)^2 + y = 1 + 3 = 4,$$

$$(x + 1)^2 = -(y - 4).$$

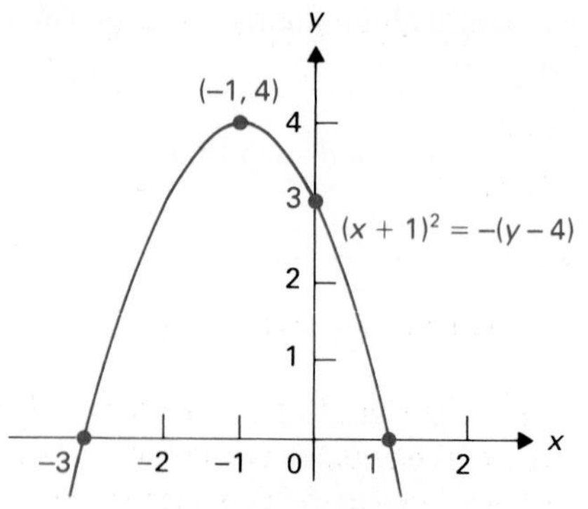

Figure 12.33 The parabola $x^2 + 2x + y - 3 = 0$.

This is the equation of a parabola having vertex at $(-1, 4)$ and opening in the negative y-direction, as shown in Fig. 12.33. □

If A and D are both zero in Eq. (8), the equation becomes

$$Cy^2 + Ey + F = 0.$$

This equation either has empty solution set or describes a straight line parallel to the x-axis or two such straight lines. If C and E are both zero in Eq. (8), similar cases may occur. It is natural to consider straight lines that occur in this fashion to be *degenerate parabolas*.

EXAMPLE 4 Graph the solution set in the plane of $y^2 - 2y - 8 = 0$.

Solution We have $y^2 - 2y - 8 = (y - 4)(y + 2)$, so

$$y^2 - 2y - 8 = 0 \qquad \text{if } y = 4 \text{ or } y = -2.$$

Thus we obtain as solution set a degenerate parabola consisting of the two parallel lines shown in Fig. 12.34. □

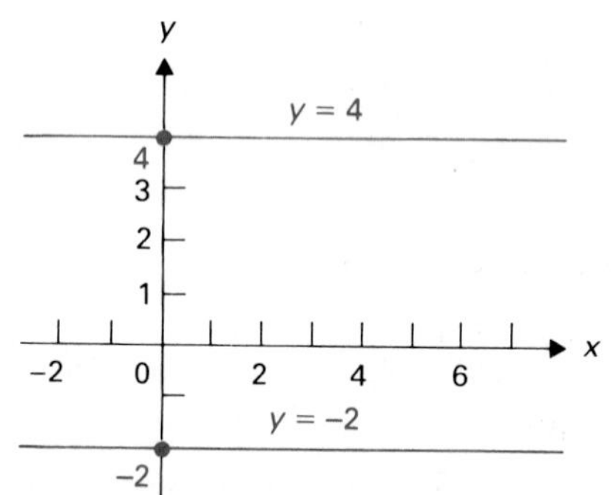

Figure 12.34 The degenerate parabola $y^2 - 2y - 8 = 0$.

EXAMPLE 5 Describe the solution set in the plane of the equation $x^2 + 2x + 2 = 0$.

Solution Since $x^2 + 2x + 2 = (x + 1)^2 + 1$, we see that $x^2 + 2x + 2 = 0$ when $(x + 1)^2 = -1$. Since a square cannot be negative, we have the empty solution set in this case. □

Case 2 Suppose now that in Eq. (8), A and C are either both positive or both negative. Our work in Section 12.1 shows that the equation then either describes an ellipse, a circle, or a single point or has empty solution set.

EXAMPLE 6 Sketch the curve $x^2 + 4y^2 + 2x - 24y + 33 = 0$.

Solution Completing the square, we obtain

$$(x^2 + 2x) + 4(y^2 - 6y) = -33,$$

$$(x + 1)^2 + 4(y - 3)^2 = 1 + 36 - 33 = 4,$$

$$\frac{(x + 1)^2}{2^2} + \frac{(y - 3)^2}{1^2} = 1.$$

This is the equation of an ellipse with center at $(-1, 3)$, as shown in Fig. 12.35. □

Figure 12.35 The ellipse $x^2 + 4y^2 + 2x - 24y + 33 = 0$.

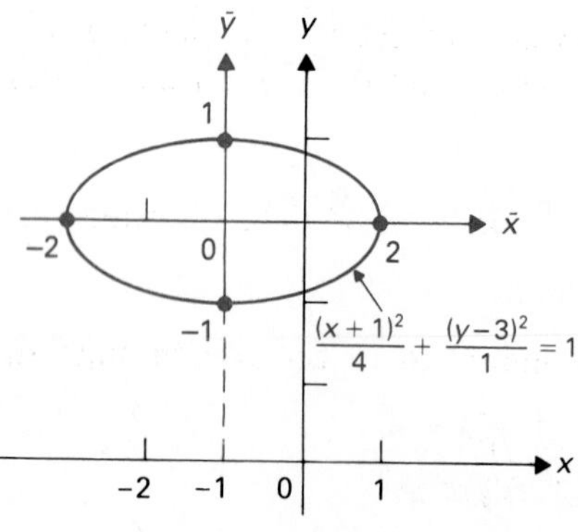

EXAMPLE 7 Find the solution set of $2x^2 + y^2 - 4x + 4y + 6 = 0$.

Solution Completing the square, we obtain

$$2(x^2 - 2x) + (y^2 + 4y) = -6,$$

$$2(x - 1)^2 + (y + 2)^2 = 2 + 4 - 6 = 0.$$

Thus the only solution is the single point $(1, -2)$. □

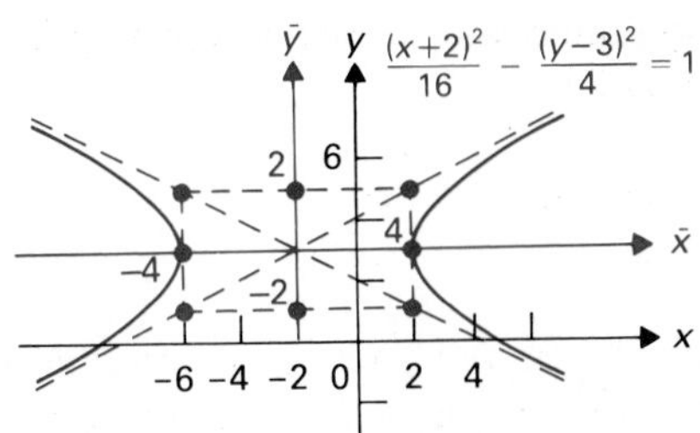

Figure 12.36 The hyperbola $x^2 - 4y^2 + 4x + 24y - 48 = 0$.

EXAMPLE 8 Find the solution set of $x^2 + y^2 - 2x + 6y + 11 = 0$.

Solution Completing the square, we obtain

$$(x^2 - 2x) + (y^2 + 6y) = -11,$$
$$(x - 1)^2 + (y + 3)^2 = 1 + 9 - 11 = -1.$$

Since a sum of squares cannot be negative, the solution set is empty. □

Case 3 Now suppose that A and C have opposite signs in Eq. (8). Our work in Section 12.1 shows that in this case, the equation describes either a hyperbola or two intersecting lines, which we consider to be a degenerate hyperbola.

EXAMPLE 9 Sketch the curve $x^2 - 4y^2 + 4x + 24y - 48 = 0$.

Solution Completing the square, we obtain

$$(x^2 + 4x) - 4(y^2 - 6y) = 48,$$
$$(x + 2)^2 - 4(y - 3)^2 = 4 - 36 + 48 = 16,$$
$$\frac{(x + 2)^2}{4^2} - \frac{(y - 3)^2}{2^2} = 1.$$

This curve is the hyperbola sketched in Fig. 12.36. □

EXAMPLE 10 Sketch the set in the plane described by

$$y^2 - 2x^2 - 4x + 6y + 7 = 0.$$

Solution Completing the square, we obtain

$$-2(x^2 + 2x) + (y^2 + 6y) = -7,$$
$$-2(x + 1)^2 + (y + 3)^2 = -2 + 9 - 7 = 0,$$
$$(y + 3)^2 = 2(x + 1)^2,$$
$$y + 3 = \pm\sqrt{2}(x + 1).$$

Translating axes to $(-1, -3)$ using $\bar{x} = x + 1$ and $\bar{y} = y + 3$, we obtain the equation

$$\bar{y} = \pm\sqrt{2}\bar{x}.$$

These equations describe two intersecting lines, shown in Fig. 12.37. □

Figure 12.37 The degenerate hyperbola $y^2 - 2x^2 - 4x + 6y + 7 = 0$.

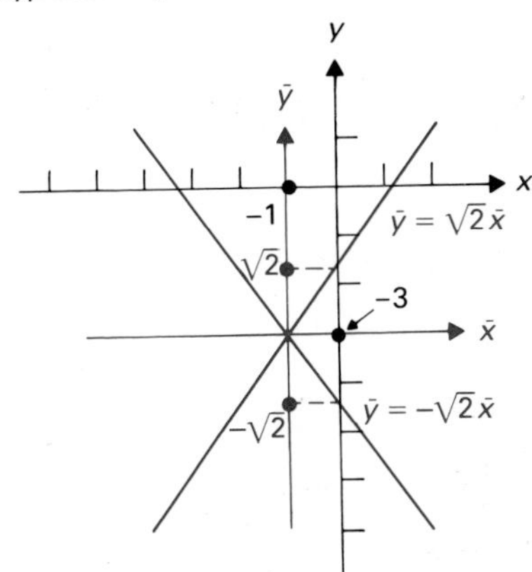

We now give one example involving both rotation of axes and translation of axes. While the theory is clear, there is apt to be a fair amount of arithmetic involved in its application.

EXAMPLE 11 Sketch the curve $xy - 3x + y - 4 = 0$, showing the x,y-axes, the x',y'-axes and the $\bar{x},\bar{y}$-axes.

Solution From Eq. (6) we see that we should rotate the axes through an angle θ such that

$$\cot 2\theta = \frac{A - C}{B} = \frac{0}{1} = 0.$$

Thus we may take $2\theta = \pi/2$ and $\theta = \pi/4$. Equations (2) for rotation of axes then become

$$x = \frac{1}{\sqrt{2}}x' - \frac{1}{\sqrt{2}}y' = \frac{1}{\sqrt{2}}(x' - y'),$$

$$y = \frac{1}{\sqrt{2}}x' + \frac{1}{\sqrt{2}}y' = \frac{1}{\sqrt{2}}(x' + y').$$

We then obtain

$$\begin{aligned} xy - 3x + y - 4 &= 0, \\ \frac{1}{\sqrt{2}}(x' - y')\frac{1}{\sqrt{2}}(x' + y') - \frac{3}{\sqrt{2}}(x' - y') + \frac{1}{\sqrt{2}}(x' + y') - 4 &= 0, \\ \frac{1}{2}(x')^2 - \frac{1}{2}(y')^2 - \sqrt{2}x' + 2\sqrt{2}y' - 4 &= 0, \\ x'^2 - y'^2 - 2\sqrt{2}x' + 4\sqrt{2}y' - 8 &= 0, \\ (x'^2 - 2\sqrt{2}x') - (y'^2 - 4\sqrt{2}y') &= 8, \\ (x' - \sqrt{2})^2 - (y' - 2\sqrt{2})^2 &= 2 - 8 + 8 = 2, \\ \frac{(x' - \sqrt{2})^2}{2} - \frac{(y' - 2\sqrt{2})^2}{2} &= 1. \end{aligned}$$

We now let $\bar{x} = x' - \sqrt{2}$ and $\bar{y} = y' - 2\sqrt{2}$ and obtain the equation

$$\frac{\bar{x}^2}{2} - \frac{\bar{y}^2}{2} = 1,$$

which is a hyperbola.

To sketch the curve, we *first* rotate to x', y'-axes as shown in Fig. 12.38. Then we translate to $\bar{x}, \bar{y}$-axes at the point $(x', y') = (\sqrt{2}, 2\sqrt{2})$, as shown in Fig. 12.39. Finally, we sketch the hyperbola, also shown in Fig. 12.39. □

Figure 12.38 Rotation of axes through $\theta = \pi/4$ to x', y'-axes.

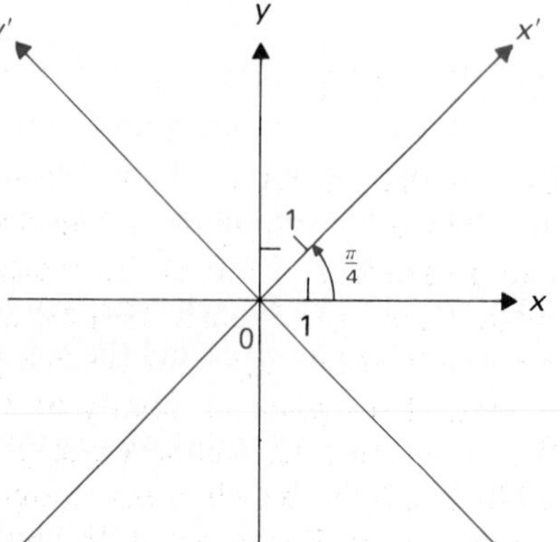

Figure 12.39 The hyperbola $xy - 3x + y - 4 = 0$.

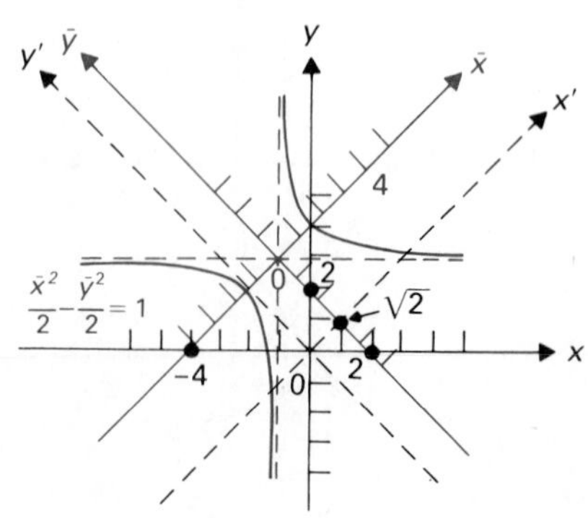

CLASSIFICATION WITHOUT ROTATION OF AXES

We have seen that if either A or C is zero in Eq. (8), we have a parabola, while we have an ellipse if A and C have the same sign and a hyperbola if A and C have the opposite sign. These cases are easily separated by considering $B^2 - 4AC$. Recall that $B^2 - 4AC$ is invariant under rotation of axes (Theorem 12.2). If $B = 0$ as in Eq. (8) so that $B^2 - 4AC = -4AC$, we see that

$$\begin{array}{lll} \text{if} \quad -4AC = 0, & \text{then} & A \text{ or } C = 0; \\ \text{if} \quad -4AC < 0, & \text{then} & A \text{ and } C \text{ have the same sign;} \\ \text{if} \quad -4AC > 0, & \text{then} & A \text{ and } C \text{ have opposite sign.} \end{array}$$

We obtain at once:

THEOREM 12.3 Classification

The second-degree equation

$$Ax^2 + Bxy + Cy^2 + Dx + Ey + F = 0$$

represents a (possibly empty or degenerate)

1. parabola if $B^2 - 4AC = 0$,
2. ellipse if $B^2 - 4AC < 0$,
3. hyperbola if $B^2 - 4AC > 0$.

EXAMPLE 12 Classify the curve with equation

$$3x^2 + 4xy + y^2 - 8x + 7y - 7 = 0.$$

Solution Since $B^2 - 4AC = 4^2 - 4 \cdot 3 \cdot 1 = 16 - 12 = 4 > 0$, we see that the equation describes a hyperbola. □

EXAMPLE 13 Classify $x^2 - 6xy + 9y^2 - 2x + 7y = 50$.

Solution Since $B^2 - 4AC = (-6)^2 - 4 \cdot 1 \cdot 9 = 0$, we see that the equation describes a parabola. □

A PROJECTIVE VIEW OF THE CONIC SECTIONS

The Euclidean plane can be enlarged to form a *projective plane*, which provides an intriguing view of the distinction between an ellipse, a hyperbola, and a parabola. To form the projective plane, we adjoin to our Euclidean plane a *line ℓ_∞ at infinity*. It is postulated that each line ℓ in our old Euclidean plane intersects ℓ_∞ at exactly one point, the *point at infinity on the line ℓ*. (Think of the point at infinity on a line as being a common point at both "ends" of the line. That is, we may visualize a line in the projective plane as a "huge circle," which looks straight to us nearby, where we can see it, but which is in reality closed at "infinity.") It is further postulated that any two

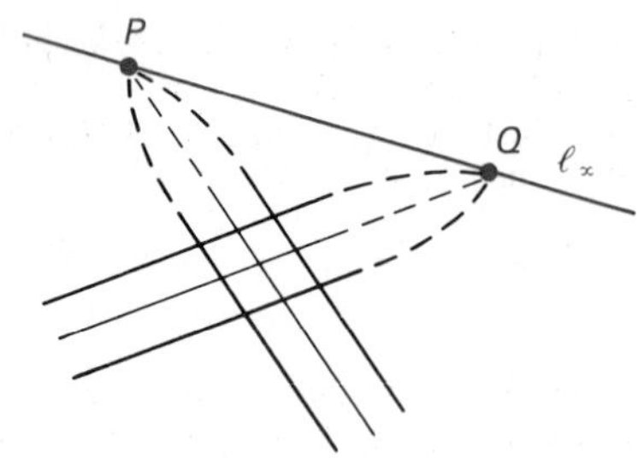

Figure 12.40 Families of parallel Euclidean lines going through points P and Q on the line ℓ_∞.

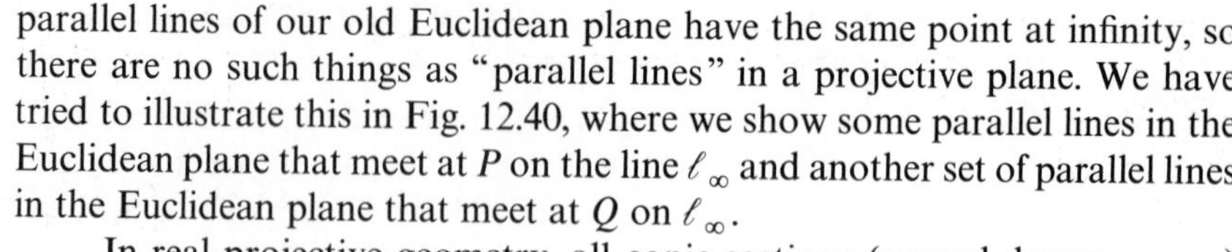

parallel lines of our old Euclidean plane have the same point at infinity, so there are no such things as "parallel lines" in a projective plane. We have tried to illustrate this in Fig. 12.40, where we show some parallel lines in the Euclidean plane that meet at P on the line ℓ_∞ and another set of parallel lines in the Euclidean plane that meet at Q on ℓ_∞.

In real projective geometry, all conic sections (second-degree curves) look alike, and the different appearance in the old Euclidean part of the plane is just due to the way the curves meet the line ℓ_∞, as shown in Fig. 12.41. The ellipse does not meet the line at infinity; it lies totally within the old Euclidean plane and can be seen in its entirety by a Euclidean bug. The hyperbola is cut into two pieces by the line at infinity, and a Euclidean bug can see only these two pieces. The parabola is tangent to the line at infinity, and a Euclidean bug sees only one piece whose extremities are approaching parallel.

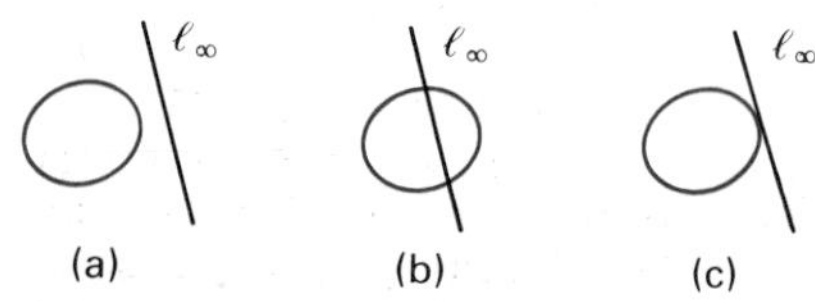

Figure 12.41 The conic sections are classified according to the way they meet ℓ_∞: (a) an ellipse; (b) a hyperbola; (c) a parabola.

SUMMARY

1. If new x', y'-axes are obtained by rotating the old x, y-axes counterclockwise through the angle θ, then
$$x = x' \cos \theta - y' \sin \theta, \qquad y = x' \sin \theta + y' \cos \theta.$$
2. Starting with $Ax^2 + Bxy + Cy^2 + Dx + Ey + F = 0$, rotation of axes through θ such that
$$\cot 2\theta = \frac{A - C}{B}$$
gives an x', y'-equation in which the coefficient of $x'y'$ is zero.
3. For the equation in number 2, $A + C$ and $B^2 - 4AC$ are invariants under rotation of axes.
4. The equation in number 2 describes
    - **a.** an ellipse if $B^2 - 4AC < 0$,
    - **b.** a hyperbola if $B^2 - 4AC > 0$,
    - **c.** a parabola if $B^2 - 4AC = 0$.

EXERCISES

1. Express x' and y' in terms of x and y for the case of rotation of axes shown in Fig. 12.30. [*Hint:* You can solve Eqs. (2) for x' and y', but it is easier to think of a rotation through the angle $-\theta$.]
2. Show from relations (5) that $A + C$ is invariant under rotation, as stated in Theorem 12.2.
3. Show from relations (5) that $B^2 - 4AC$ is invariant under rotation, as stated in Theorem 12.2.

In Exercises 4 through 9, find (a) the angle θ through which axes should be rotated so that the coefficient of $x'y'$ is zero, and (b) the equations for x and y in terms of x' and y'. Use relations (7) where necessary.

4. $2x^2 - 3xy + 2y^2 - 4x + y = 10$
5. $\sqrt{3}x^2 + 3xy - 5x + 7y = 18$
6. $3xy + \sqrt{3}y^2 + 10x - 3y = 36$
7. $2x^2 + 2\sqrt{2}xy + y^2 - 5x + 4y = 8$
8. $5x^2 + 24xy - 2y^2 - 8y = 4$
9. $2x^2 - 4\sqrt{2}xy - 5y^2 + 14x = 25$

In Exercises 10 through 15, obtain an equation of the given curve in x',y'-coordinates formed by rotating axes so that the coefficient of $x'y'$ is zero.

10. $xy = -3$

11. $x^2 - 3xy + y^2 = 5$

12. $2x^2 + \sqrt{3}xy + y^2 - 2x = 20$

13. $x^2 + \sqrt{3}xy + 2y^2 - 4y = 18$

14. $x^2 + xy = 10.$

15. $5x^2 + 3xy + y^2 + x = 12.$

In Exercises 16 through 25, use Theorem 12.3 to classify the curve with the given equation as a (possibly empty or degenerate) ellipse, hyperbola, or parabola.

16. $2x^2 + 8xy + 8y^2 - 3x + 2y = 13$

17. $y^2 + 4xy - 5x^2 - 3x = 12$

18. $-x^2 + 5xy - 7y^2 - 4y + 11 = 0$

19. $xy + 4x - 3y = 8$

20. $2x^2 - 3xy + y^2 - 8x + 5y = 30$

21. $x^2 + 6xy + 9y^2 - 2x + 14y = 10$

22. $4x^2 - 2xy - 3y^2 + 8x - 5y = 17$

23. $8x^2 + 6xy + 2y^2 - 5x = 25$

24. $x^2 - 2xy + 4x - 5y = 6$

25. $2x^2 - 3xy + 2y^2 - 8y = 15$

26. Can you think of a "use" for the invariant $A + C$ given in Theorem 12.2?

27. The equation $x^2 + 4xy + y^2 - 2x + 3y = 11$ is transformed into one of the form $A'x'^2 + C'y'^2 + D'x' + E'y' + F' = 0$ by rotation of axes. Using just the invariants $B^2 - 4AC$ and $A + C$, find all possible values for A' and C'.

28. Answer Exercise 27 for the equation $3x^2 - 2\sqrt{2}xy + 4y^2 + 3x = 15.$

In Exercises 29 through 32, rotate axes to simplify the equation, complete squares, and sketch the curve, showing the x,y-axes, the x',y'-axes, and the $\bar{x},\bar{y}$-axes. That is, do the complete job.

29. $x^2 + 2xy + y^2 - 2\sqrt{2}x + 6\sqrt{2}y = 6$

30. $xy - x - 2y = -6$

31. $2x^2 + 4xy - y^2 - 12\sqrt{5}x = -6$

32. $3x^2 + 4xy + 3y^2 + 2\sqrt{2}x - 12\sqrt{2}y = -29$

33. Supplement relations (5) in the text by showing that $D' = D\cos\theta + E\sin\theta$, $E' = E\cos\theta - D\sin\theta$, and $F' = F$.

In Exercises 34 through 37, use relations (5), Exercise 33, and a calculator to find the equation without an $x'y'$-term obtained from the given equation by rotation of axes. [*Hint:* Don't bother with relations (7). Find $\cot 2\theta$, then $\tan 2\theta$, then 2θ, then θ, then $\sin\theta$, then $\cos\theta$, and then compute A', C', D', E', and F'.]

34. $4x^2 + 5xy + 3y^2 - 7x + 2y - 14 = 0$

35. $3x^2 - 4xy - 7y^2 + 8x + y - 10 = 0$

36. $\sqrt{17}x^2 - 3xy + \sqrt{5}y^2 - \sqrt{5}x + \sqrt{3}y - 3 = 0$

37. $\frac{1}{3}x^2 + 2.3xy - \sqrt{10}y^2 - x + 3y - 5 = 0$

12.4 WHY STUDY CONIC SECTIONS?

It is quite proper to ask why we should bother to classify all the second-degree plane curves. After all, it is clearly pretty hopeless to attempt to classify successively all plane curves of degree three, then all of degree four, and so on. The plane curves of degree one (the lines) are certainly important; they are the graphs of linear functions, and calculus deals with approximation using these functions. The classification of the second-degree curves is classical and quite elegant, but that is not in itself sufficient reason to include it in this text. However, the second-degree plane curves do have important physical applications.

ORBITS IN CENTRAL FORCE FIELDS

Consider a moving body that is subject only to a force of attraction toward a single fixed point. The fixed point is considered to be the *center* of the system, and an attraction of this type constitutes a *central force field.*

Suppose the force F of attraction is inversely proportional to the square of the distance d of the body from the fixed point, so that $F = k/d^2$. This is the case, for example, for the force of gravitational attraction. It can be

Table 12.1

Planet	Eccentricity	Period	Mean distance from sun (millions of miles)
Mercury	0.2056234	87.967 days	36.0
Venus	0.0067992	224.701 days	67.3
Earth	0.0167322	365.256 days	93.0
Mars	0.0933543	1.881 years	141.7
Jupiter	0.0484108	11.862 years	483.9
Saturn	0.0557337	29.458 years	887.1
Uranus	0.0471703	84.015 years	1785.0
Neptune	0.0085646	164.788 years	2797.0
Pluto	0.2485200	247.697 years	3670.0

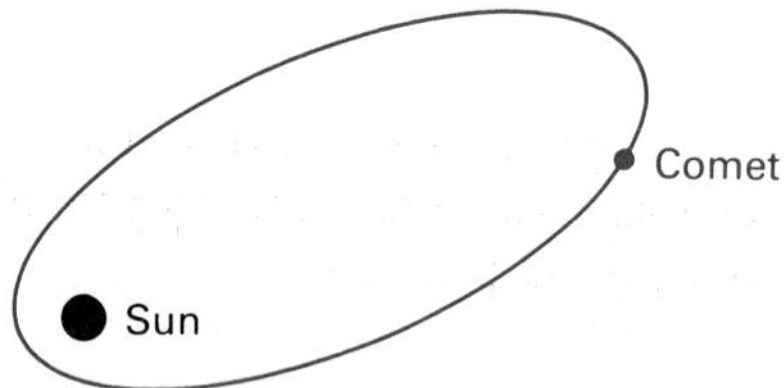

Figure 12.42 A comet in elliptical orbit about the sun.

shown from the laws of Newtonian mechanics that the orbit (path) on which the body travels due to such a force field is a second-degree plane curve, that is, either an ellipse, a hyperbola, or a parabola, having the center of the force field system as focus. For example, if we neglect the force upon the earth due to celestial bodies other than the sun, then we are in a central force field situation with center at the sun. The earth has an elliptical orbit about the sun, with the sun at one focus of the ellipse. Orbits of all the planets about the sun are ellipses with a focus at the sun. Also, a satellite that we send up about our earth travels in an elliptical orbit. The orbits of communication satellites, which "hang" over one spot on the equator, have to be very nearly circular. The smaller the eccentricity of an ellipse, the more nearly circular the ellipse, because $b^2 = a^2 - c^2$ yields $(b/a)^2 = 1 - (c/a)^2 = 1 - e^2$. Table 12.1 gives the eccentricities of the orbits of the planets.

The orbits of comets about our sun are of two types. Some comets have "narrow" elliptical orbits with the sun at a focus that is comparatively close to one "end" of the ellipse, as illustrated in Fig. 12.42 (see Exercise 1). One such comet is Halley's comet, which takes about 75 or 76 years to travel once around its orbit and which was seen, when it came to this "end" of its orbit near the sun, in 1910. Another type of comet enters our solar system from "outer space," is attracted by our sun in accordance with the central force field principle, and has such velocity that when it has passed the sun, it continues on and escapes into outer space again. Such a comet is traveling on a hyperbolic or parabolic path while in our solar system (see Fig. 12.43). (The path will almost certainly be a hyperbola, for in order for it to be a parabola, the comet would have to enter our solar system in precisely the right way to get a solar orbit of eccentricity *exactly* 1, barely escaping capture.)

Figure 12.43 A comet in hyperbolic orbit about the sun.

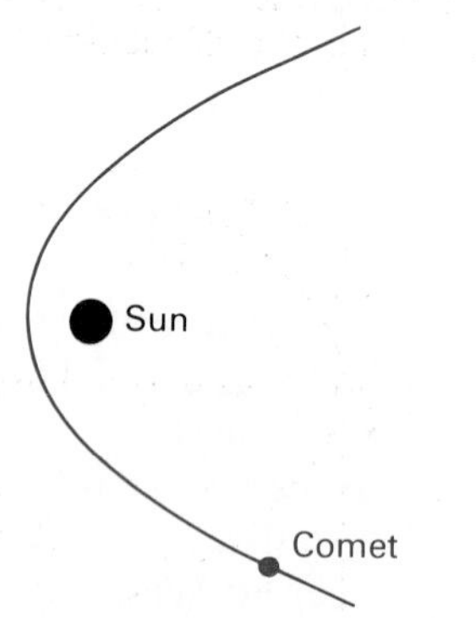

APPLICATIONS TO OPTICS AND SOUND

The second-degree plane curves have important applications to reflected energy waves (such as light and sound). It has been found that when such a wave meets a reflecting barrier, the directions of the wave as it meets the barrier and as it is reflected make equal angles with the normal to the barrier (see Fig. 12.44). If the barrier is curved, then the normal is considered to be perpendicular to the tangent line (or plane) of the barrier (see Fig. 12.45).

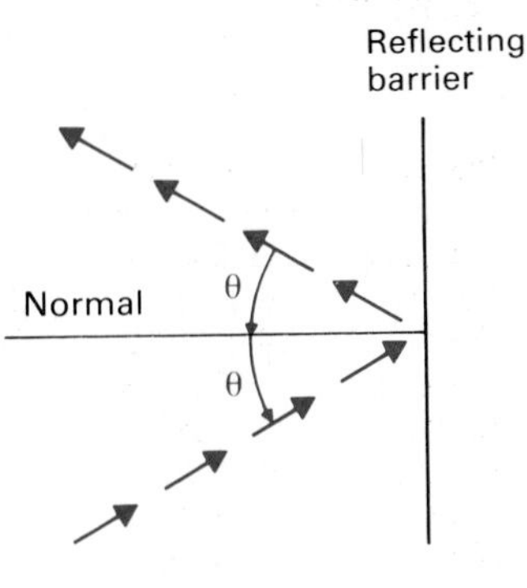

Figure 12.44 Energy waves reflected by a straight barrier.

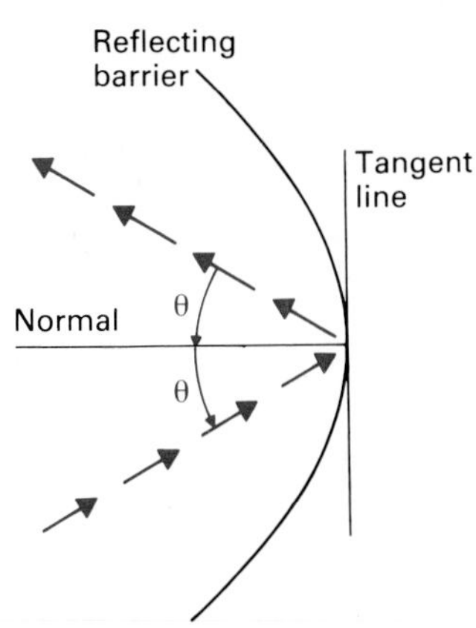

Figure 12.45 Energy waves reflected by a curved barrier.

The conic sections have special geometric properties that yield important applications for these physical situations. We state these properties without proof. Proofs are not difficult and are good exercises.

THEOREM 12.4 Reflecting properties of conic sections

1. *Ellipse and hyperbola:* The lines joining a point (x, y) on an ellipse or hyperbola to the foci make equal angles with the normal to the curve at (x, y). (See Figs. 12.46 and 12.47.)
2. *Parabola:* The line joining a point (x, y) on a parabola to the focus and the line through (x, y) parallel to the axis of the parabola make equal angles with the normal to the parabola. (See Fig. 12.48.)

Consider the *paraboloid* formed by revolving the parabola $y^2 = 4px$ about the x-axis. If the inside of this paraboloid is silvered to reflect light, then the beams from a light source at the focus of the paraboloid will be reflected in rays parallel to the axis of the paraboloid (see Fig. 12.49). This principle is used in a searchlight. Conversely, light emanating from a source

Figure 12.46 Energy at one focus of an ellipse is reflected to the other focus.

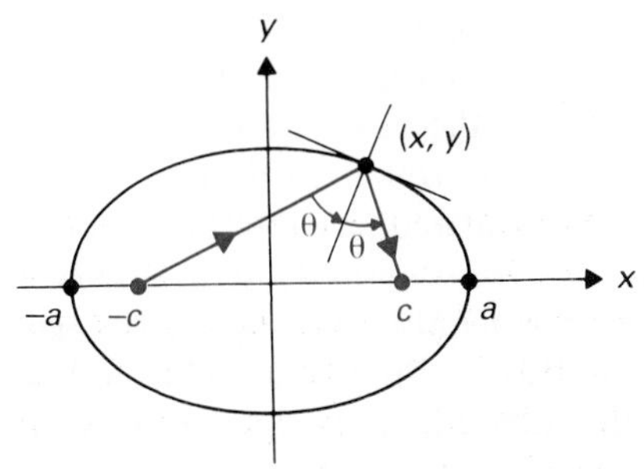

Figure 12.47 A normal to a hyperbola reflects energy from one focus to the other.

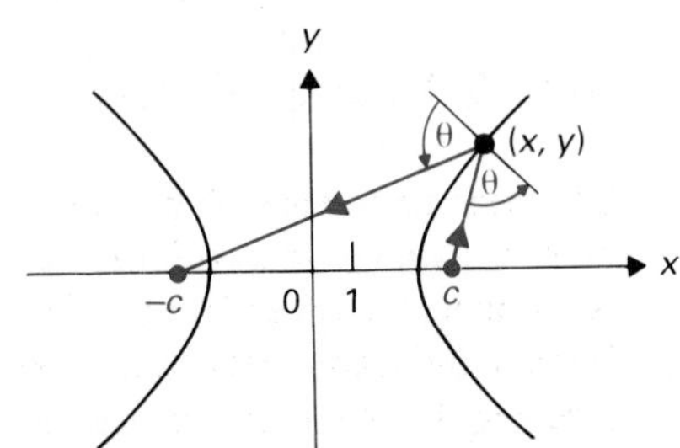

Figure 12.48 Energy at a focus of a parabola is reflected parallel to the axis of the parabola.

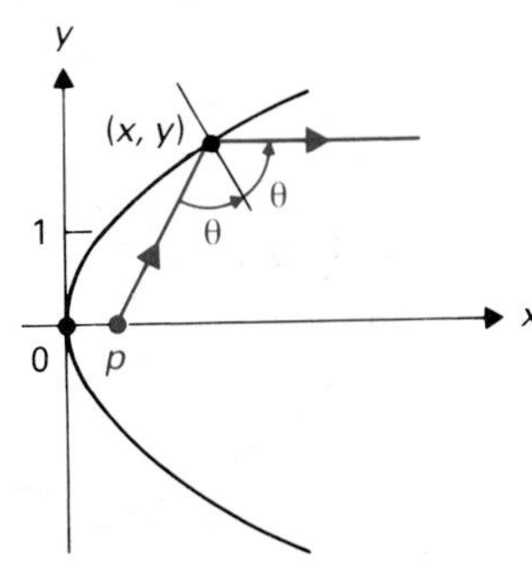

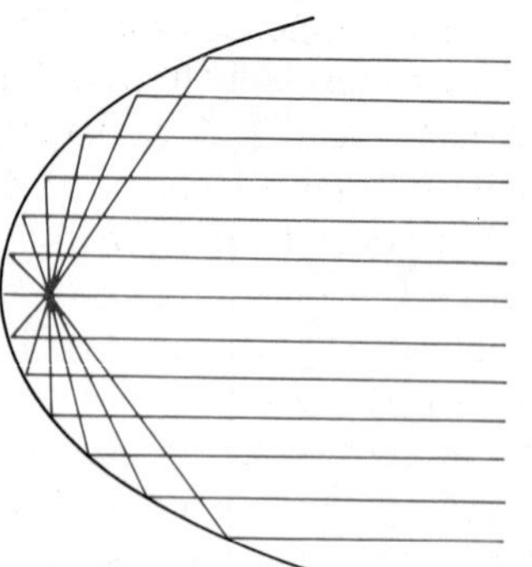

Figure 12.49 A cross section of a paraboloid: Light at the focus is reflected in parallel rays.

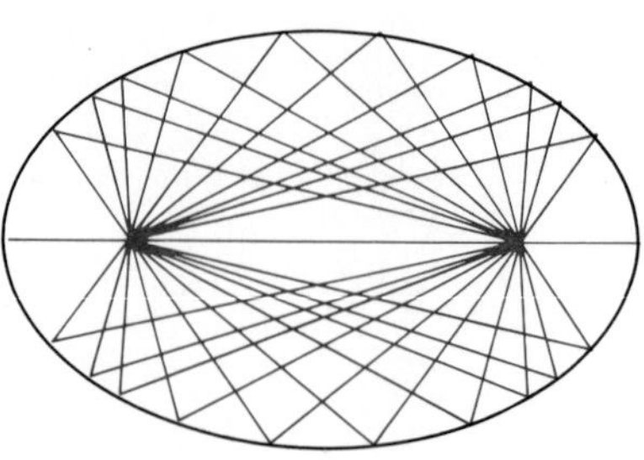

Figure 12.50 A cross section of an ellipsoid: Light at one focus is reflected to the other focus.

at a great distance out from the axis enters the silvered paraboloid in essentially parallel rays and is all reflected to the focus. This principle is used in telescopes, such as the one at Mount Palomar in California, which has a parabolic mirror of diameter 200 in. The large antennas used in radio astronomy are also constructed in the shape of a paraboloid, and small parabolic reflectors are used in sending and receiving radio and television signals.

Let an *ellipsoid* be formed by revolving an ellipse about its major axis. If the inside of the ellipsoid is silvered, then light emanating in all directions from a source at one focus will all be reflected to the other focus (see Fig. 12.50). Definition 12.2 on page 525 shows that the distance traveled from one focus to the other via reflection is always the length $2a$ of the major axis, independent of the point of reflection. Thus, sound that emanates in all directions from one focus of an ellipsoid will not only all be reflected to the other focus but will all reach the other focus *at the same time*. This is the principle behind a "whisper gallery," with walls and ceiling forming a portion of an ellipsoid; a whisper at one focus can be heard distinctly at a great distance at the other focus.

APPLICATION TO NAVIGATION

In World War I, the following scheme was sometimes used to determine the location of an enemy cannon. Three observers would synchronize their watches and then move to observation points A, B, and C with known coordinate locations. These observers at A, B, and C would note the precise time the cannon fired. Knowing the speed of sound, they could then calculate the difference of the distances from A and B to the cannon. For example, if the observer at A heard the sound 3 sec before the observer at B, then the cannon must have been about 3300 ft farther from B than from A, for sound travels at about 1100 ft/sec. If A and B were considered foci, the cannon must lie on a hyperbola with foci A and B and $2a = 3300$. Similarly, if the observer at C heard the same firing 2 sec before the observer at B, the cannon was also on a hyperbola having B and C as foci, and where $2a = 2200$. The cannon then must have been at the intersection of these two hyperbolas, as illustrated in Fig. 12.51.

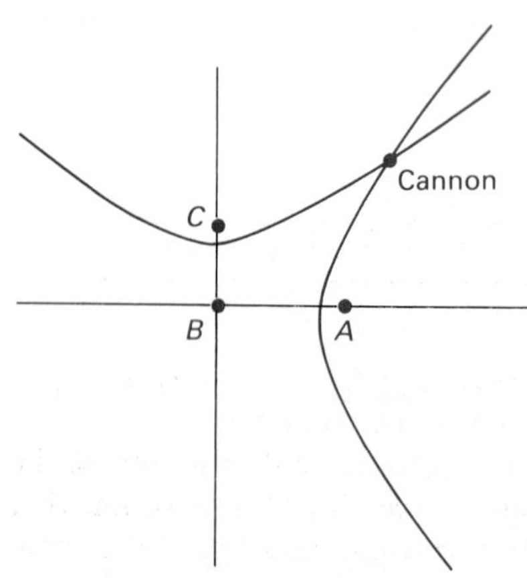

Figure 12.51 A cannon located at the intersection of two hyperbolas by observers at A, B, and C.

The same principle as used in locating the cannon is currently used in LORAN (for long-range navigation). This time, radio waves, which travel at 186,000 mi/sec, are used rather than sound waves traveling only 1100 ft/sec. Consequently, time differences must be measured very accurately, using sensitive equipment, in microseconds (millionths of seconds). A master station at A transmits a signal pattern, and two remote stations at B and C transmit the same pattern, but with a known time delay measured in microseconds. Equipment on board a ship measures the number of microseconds delay in the signal patterns received from A and B. This locates the ship on a hyperbola with foci at A and B. The delay in the signals received from A and C provide another hyperbola, and the ship must lie on the intersection of these two hyperbolas. If all the equipment is in good order, a fisherman can determine his position using LORAN within about 50 yards.

SUMMARY

1. Bodies traveling subject only to gravitational attraction of a single, much larger body move in conic-section orbits.
2. The reflective properties of conic sections (given in Theorem 12.4) make them useful in optics and acoustics.
3. The hyperbola has important applications to navigation.

EXERCISES

1. Show that if the major axis of an ellipse is very large compared with the minor axis, then the foci of the ellipse are comparatively near the "ends."
2. Show that the closest point on an ellipse to a focus is the point nearest the focus on the end of the major axis.
3. The *apogee* of an earth satellite is its maximum altitude above the surface of the earth during orbit, and its *perigee* is its minimum altitude during orbit. Using Exercise 2, argue that for a satellite in elliptical earth orbit with major axis of length $2a$, we have $2a =$ Diameter of earth + Apogee + Perigee.
4. Prove the reflecting property of the hyperbola.
5. Prove the reflecting property of the parabola.
6. An ellipse and a hyperbola are *confocal* if they have the same foci. Using the reflecting properties, give a synthetic proof (not using coordinates) that an ellipse is perpendicular to a confocal hyperbola at each point of intersection. [*Hint:* Show that the normals to the curves are perpendicular at a point of intersection.]

12.5 PARAMETRIC CURVES REVIEWED

On page 111, we first introduced parametric equations $x = h(t)$, $y = k(t)$, which can be viewed as giving the location of a body on a curve at time t. In this section, we discuss sketching parametric curves, and then we review the formulas for the slope and arc length of such curves. We always assume that $h(t)$ and $k(t)$ are continuous functions. In discussing the slope of the curve

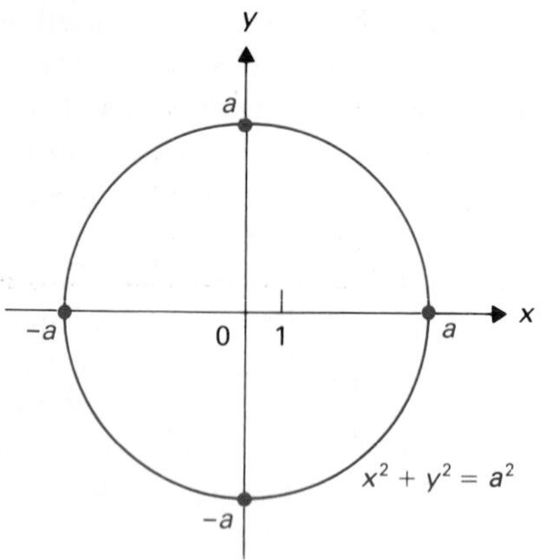

Figure 12.52 The circle $x^2 + y^2 = a^2$.

and arc length, we assume further that $h'(t)$ and $k'(t)$ exist and are continuous. If $h'(t)$ and $k'(t)$ are also never simultaneously zero, the curve has no sharp points. Such a curve is called *smooth*.

SKETCHING AND PARAMETRIZING

Curves given parametrically can sometimes be sketched by eliminating the parameter t to obtain a (we hope familiar) x, y-equation.

EXAMPLE 1 Sketch the plane curve given parametrically by

$$x = a \cos t, \qquad y = a \sin t.$$

Solution Squaring and adding the equations, we find that

$$x^2 + y^2 = a^2 \cos^2 t + a^2 \sin^2 t = a^2(\cos^2 t + \sin^2 t) = a^2 \cdot 1 = a^2.$$

Thus each point (x, y) of the curve lies on the circle $x^2 + y^2 = a^2$, shown in Fig. 12.52. We may view these parametric equations as picking up the t-axis and wrapping it around and around this circle, with $t = 0$ placed at $(a, 0)$. □

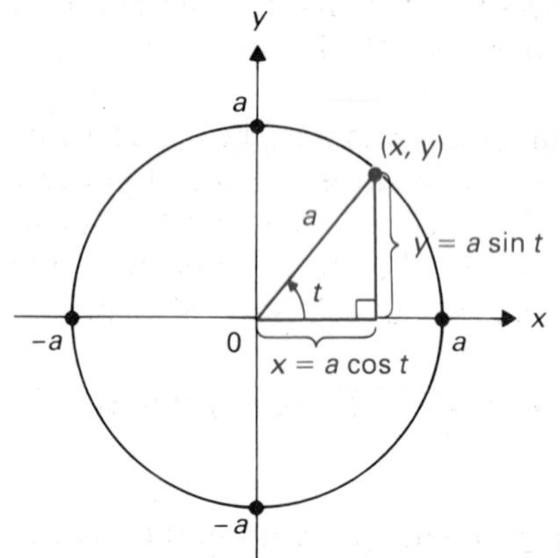

Figure 12.53 Parametrizing a circle by the central angle t.

One interpretation of the parameter t in Example 1 is the angle shown in Fig. 12.53. We easily see from the figure that

$$\boxed{x = a \cos t, \qquad y = a \sin t.} \tag{1}$$

We will frequently want to parametrize a circle, and we should remember the parametrization (1). We should also remember that $t = 0$ corresponds to the point $(a, 0)$, and increasing t corresponds to going *counterclockwise* around the circle.

When sketching a parametric curve $x = h(t)$, $y = k(t)$ using an x, y-equation obtained by eliminating the parameter, we must be careful. The nature of the functions $x = h(t)$ and $y = k(t)$ may have placed restrictions on x and y that are not apparent from the x, y-equation.

EXAMPLE 2 Sketch the parametric curve

$$x = \sin t, \qquad y = \sin t.$$

Solution Elimination of t from the parametric equations yields $y = x$. However, since $-1 \le \sin t \le 1$ for all values of t, we see that the points of the curve given by these parametric equations all lie on the segment of the line $y = x$ joining $(-1, -1)$ and $(1, 1)$, as shown in Fig. 12.54. An object whose position in the plane at time t is given by these equations travels back and forth along this line segment. □

Figure 12.54 The parametric curve $x = \sin t$, $y = \sin t$.

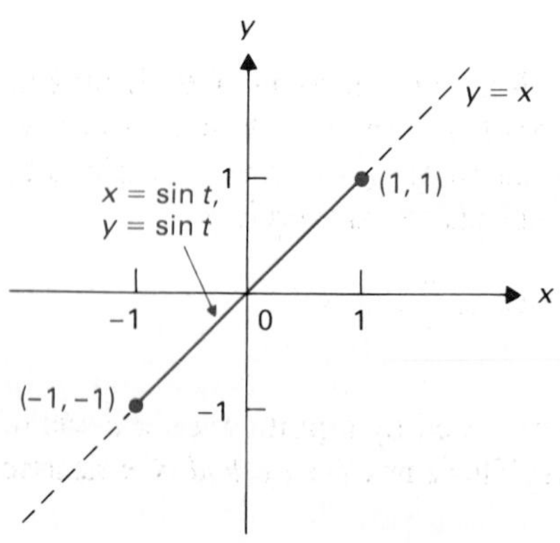

EXAMPLE 3 Sketch the parametric curve

$$x = t^2, \qquad y = t^4 + 1.$$

Solution Since $t^2 = x$, we see that

$$y = t^4 + 1 = (t^2)^2 + 1 = x^2 + 1.$$

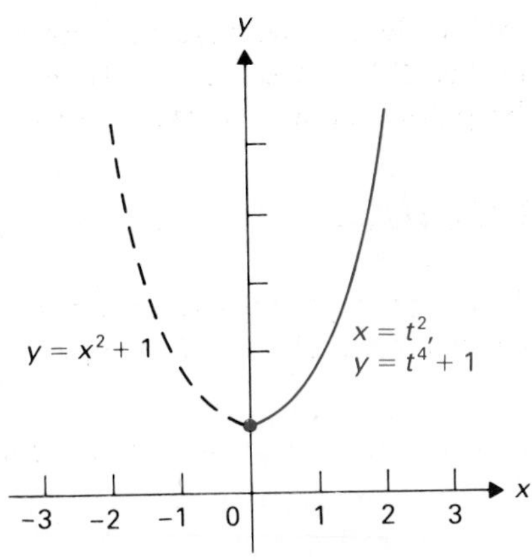

Figure 12.55 The parametric curve $x = t^2$, $y = t^4 + 1$.

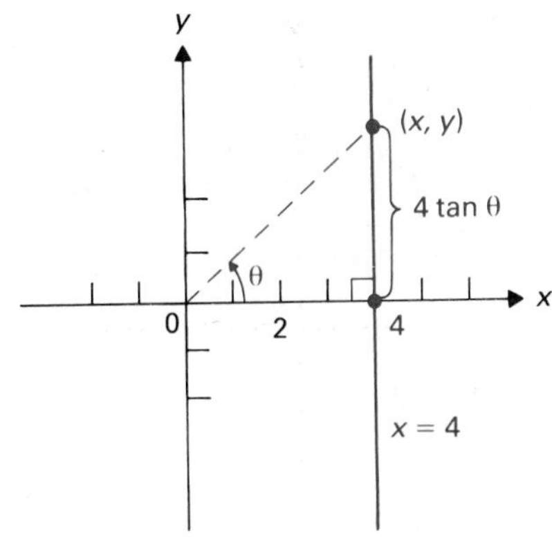

Figure 12.56 Parametrizing the line $x = 4$ using the angle of inclination θ.

However, since $t^2 \geq 0$, we see that $x \geq 0$, so we obtain only the right-hand half of the parabola $y = x^2 + 1$, as shown in Fig. 12.55. □

Sometimes a curve given by an equation in x and y can be parametrized in terms of some quantity (parameter) arising naturally from the curve or from some physical consideration. For such a parametrization to be accomplished, there must be a *unique* point on the curve corresponding to each value of the parameter. The next three examples illustrate this idea.

EXAMPLE 4 Parametrize the parabola $y = x^2$, taking as parameter the slope m of the parabola at each point.

Solution Our problem is to express both x and y in terms of the slope m. At a point (x, y) on the parabola, the slope is given by $m = y' = 2x$, so $x = m/2$. Since $y = x^2$, we obtain the parametrization

$$x = \tfrac{1}{2}m, \qquad y = \tfrac{1}{4}m^2$$

of the parabola. □

EXAMPLE 5 Parametrize the line $x = 4$, taking as parameter the angle of inclination θ at the origin from the x-axis to the point (x, y) on the line, as shown in Fig. 12.56.

Solution Our problem is to express both x and y in terms of θ. From Fig. 12.56, we see that $x = 4$, while $y/4 = \tan\theta$, so that $y = 4\tan\theta$. Since we want the parameter θ to be the angle of inclination, we should restrict θ to the interval $-\pi/2 < \theta < \pi/2$. Our parametrization is therefore

$$x = 4, \qquad y = 4\tan\theta \qquad \text{for } -\frac{\pi}{2} < \theta < \frac{\pi}{2}. \quad \square$$

EXAMPLE 6 Consider the plane curve traced by a point P on a circle of radius a as the circle rolls along the x-axis. This curve is a *cycloid*. We assume

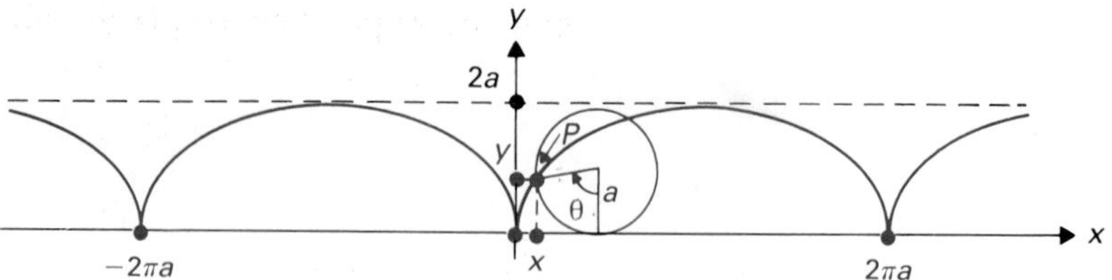

Figure 12.57 A cycloid to be parametrized using the angle θ through which the circle has rolled from the origin.

that the point P on the circle touches the x-axis at the points zero and $\pm 2n\pi a$ as the circle rolls along, as shown in Fig. 12.57.

Parametrize this cycloid in terms of the angle θ through which the circle has rolled, starting with $\theta = 0$ when P is at the origin, as shown in Fig. 12.57.

Solution We assume that θ is taken to be positive as the circle rolls to the right. From the detail of the rolling circle shown in Fig. 12.58, we easily obtain

$$x = a\theta - a\sin\theta, \qquad y = a - a\cos\theta$$

as the desired parametric equations. □

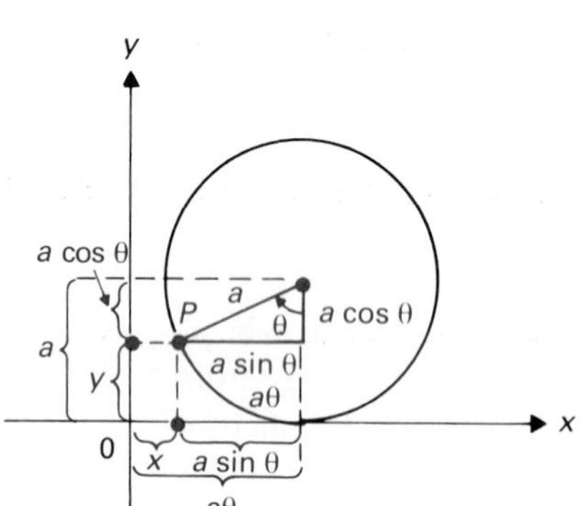

Figure 12.58
$x = a\theta - a\sin\theta,$
$y = a - a\cos\theta.$

The cycloid discussed in Example 6 has some very fascinating physical properties. Let a point Q be given in the fourth quadrant, and imagine a drop of water placed at $(0, 0)$ that slides (without friction) along a curve from the origin $(0, 0)$ to Q, subject only to a force mg downward due to gravity, as shown in Fig. 12.59. What shape should the curve have so that the drop slides from $(0, 0)$ to Q in the least possible time? (This is the *brachistochrone problem*; the name comes from two Greek words meaning "shortest time.") Initially, we might think that the drop should slide along the straight-line segment joining $(0, 0)$ and Q, but after a little thought it might seem reasonable that the curve should drop more steeply at first and allow the drop to gain speed more quickly. It can be shown that the "smooth" curve (with no sharp corners) corresponding to the shortest time is a portion of a single arc of an inverted cycloid

$$x = a\theta - a\sin\theta, \qquad y = a\cos\theta - a$$

generated by a point P on a circle rolling on the *under* side of the x-axis, with P starting at the origin. The cycloid is thus the solution to the brachistochrone problem.

It can be shown that the inverted cycloid is also the solution of the *tautochrone problem* (meaning "same time"), for if a drop of water is placed at a point other than the low point on an arc of an inverted cycloid, the time required for it to slide to the low point of the arc is independent of the initial point where the drop is placed.

Figure 12.59 A drop of water at P sliding from (0, 0) to Q subject only to the force mg of gravity.

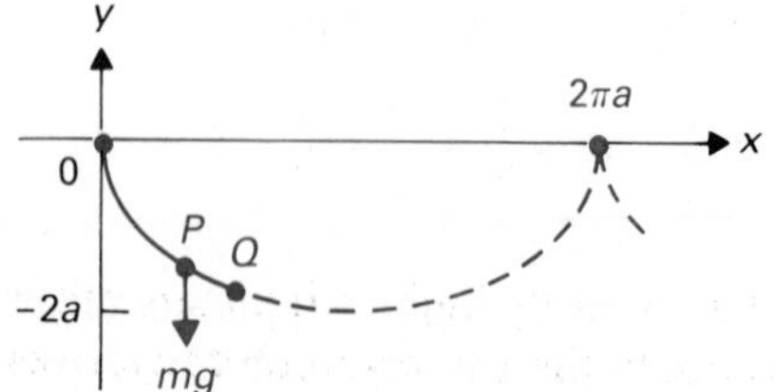

REVIEW OF SLOPE

Now assume that $x = h(t)$ and $y = k(t)$ both have continuous derivatives. Recall from our previous work that the slope of the parametric curve at a point corresponding to t_1 is given by

$$\text{Slope} = \left.\frac{dy}{dx}\right|_{t_1} = \left.\frac{dy/dt}{dx/dt}\right|_{t_1}, \tag{2}$$

if $dx/dt \neq 0$ at t_1.

EXAMPLE 7 Find the tangent and normal lines to the curve $x = \cos t$, $y = \sin t$ at the point $(\sqrt{3}/2, \frac{1}{2})$ corresponding to $t_1 = \pi/6$.

Solution We have

$$\left.\frac{dy}{dx}\right|_{t=\pi/6} = \left.\frac{dy/dt}{dx/dt}\right|_{t=\pi/6} = \frac{\cos(\pi/6)}{-\sin(\pi/6)} = \frac{\sqrt{3}/2}{-\frac{1}{2}} = -\sqrt{3}.$$

The tangent line thus has x,y-equation $y = -\sqrt{3}x + 2$ and the normal line has equation $y = (1/\sqrt{3})x$. □

If dx/dt is zero and dy/dt nonzero at $t = t_1$, then the curve has a vertical tangent at the corresponding point (x_1, y_1).

EXAMPLE 8 Find all vertical tangent lines to the parametric curve

$$x = t^2 - 4t + 3, \qquad y = t - 1.$$

Solution We have

$$\frac{dx}{dt} = 2t - 4 \qquad \text{and} \qquad \frac{dy}{dt} = 1.$$

Thus $dx/dt = 0$ when $t = 2$, and $dy/dt \neq 0$ there. Consequently, we have a vertical tangent line when $t = 2$, or at the point $(x, y) = (-1, 1)$. The equation of this vertical line is of course $x = -1$. □

It may happen that both dx/dt and dy/dt are zero at $t = t_1$. In that case, the slope is given by the limit of the quotient (2) as $t \to t_1$ if the limit exists.

EXAMPLE 9 Find the slope when $t = 0$ of the parametric curve

$$x = t^2, \qquad y = \cos t.$$

Solution If $x = t^2$ and $y = \cos t$, then

$$\frac{dy}{dx} = \frac{dy/dt}{dx/dt} = \frac{-\sin t}{2t}.$$

The slope of the curve when $t = 0$ is given by

$$\lim_{t\to 0} \frac{-\sin t}{2t} = \lim_{t\to 0} -\frac{1}{2}\cdot\frac{\sin t}{t} = -\frac{1}{2}\cdot 1 = -\frac{1}{2}. \quad \square$$

If $h(t)$ and $k(t)$ have derivatives of sufficiently high order, then it is a simple (but often tedious) matter to compute the higher-order derivatives

d^2y/dx^2, d^3y/dx^3, and so on. They can be computed successively using the following formulas:

$$\frac{d^2y}{dx^2} = \frac{\dfrac{d(dy/dx)}{dt}}{dx/dt}, \qquad \frac{d^3y}{dx^3} = \frac{\dfrac{d(d^2y/dx^2)}{dt}}{dx/dt},$$
$$\frac{d^4y}{dx^4} = \frac{\dfrac{d(d^3y/dx^3)}{dt}}{dx/dt},$$

and so on.

EXAMPLE 10 Find d^2y/dx^2 for the curve $x = \cos t$, $y = \sin t$.

Solution We have

$$\frac{dy}{dx} = \frac{dy/dt}{dx/dt} = \frac{\cos t}{-\sin t} = -\cot t.$$

Then

$$\frac{d^2y}{dx^2} = \frac{\dfrac{d(dy/dx)}{dt}}{dx/dt} = \frac{\csc^2 t}{-\sin t} = -\csc^3 t. \quad \square$$

EXAMPLE 11 Find d^3y/dx^3 if $x = t^2 - 3t + 4$ and $y = t - 1$.

Solution We have

$$\frac{dy}{dx} = \frac{dy/dt}{dx/dt} = \frac{1}{2t-3} = (2t-3)^{-1}.$$

Then

$$\frac{d^2y}{dx^2} = \frac{\dfrac{d(dy/dx)}{dt}}{dx/dt} = \frac{-1(2t-3)^{-2}\cdot 2}{2t-3} = -2(2t-3)^{-3}.$$

Finally,

$$\frac{d^3y}{dx^3} = \frac{\dfrac{d(d^2y/dx^2)}{dt}}{dx/dt} = \frac{6(2t-3)^{-4}\cdot 2}{2t-3} = \frac{12}{(2t-3)^5}. \quad \square$$

REVIEW OF ARC LENGTH

Recall from page 277 that if $x = h(t)$ and $y = k(t)$ have continuous derivatives, then the arc length of the curve from $t = t_1$ to $t = t_2$ is given by

$$\text{Arc length} = \int_{t_1}^{t_2} \sqrt{\left(\frac{dx}{dt}\right)^2 + \left(\frac{dy}{dt}\right)^2}\, dt, \tag{3}$$

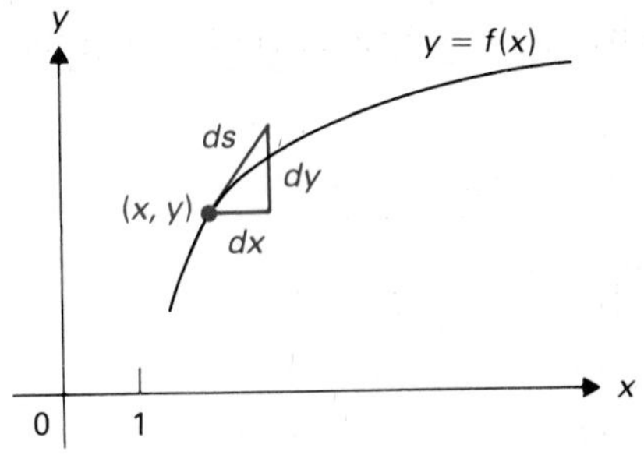

Figure 12.60 The differential triangle of a graph $y = f(x)$ at (x, y).

and $ds = \sqrt{(dx/dt)^2 + (dy/dt)^2}\,dt$ is the *differential of arc length*. We may consider ds to be the length of a short piece of tangent-line segment to the curve, corresponding to a change dx in x, as shown in Fig. 12.60.

EXAMPLE 12 Verify the formula $C = 2\pi a$ for the circumference of a circle whose radius is a.

Solution We take as parametric equations for the circle

$$x = a\cos t, \qquad y = a\sin t, \qquad \text{for } 0 \leq t \leq 2\pi.$$

Since $dx/dt = -a\sin t$ and $dy/dt = a\cos t$, we see from Eq. (3) that the length of the circle is

$$C = \int_0^{2\pi} \sqrt{(-a\sin t)^2 + (a\cos t)^2}\,dt$$

$$= \int_0^{2\pi} \sqrt{a^2}\,dt = a\int_0^{2\pi} (1)\,dt = 2\pi a. \quad \square$$

EXAMPLE 13 Use Simpson's rule with $n = 20$ and a calculator to estimate the length of the parametric curve $x = t^2$, $y = \sin t$ for $0 \leq t \leq 2\pi$.

Solution We have

$$ds = \sqrt{\left(\frac{dx}{dt}\right)^2 + \left(\frac{dy}{dt}\right)^2}\,dt = \sqrt{4t^2 + \cos^2 t}\,dt.$$

Our calculator shows that

$$\text{Arc length} = \int_0^{2\pi} \sqrt{4t^2 + \cos^2 t}\,dt \approx 40.051. \quad \square$$

SUMMARY

1. A parametric curve can sometimes be sketched by eliminating the parameter to obtain an x, y-equation, but care must be taken to note any restrictions on x and y imposed by the parametric equations.
2. If $x = h(t)$ and $y = k(t)$, then

$$\frac{dy}{dx} = \frac{dy/dt}{dx/dt}, \qquad \frac{d^2y}{dx^2} = \frac{\dfrac{d(dy/dx)}{dt}}{dx/dt}, \qquad \frac{d^3y}{dx^3} = \frac{\dfrac{d(d^2y/dx^2)}{dt}}{dx/dt},$$

and so on.

3. The differential of arc length is $ds = \sqrt{(dx/dt)^2 + (dy/dt)^2}\,dt$ and

$$\text{Arc length} = \int_{t_1}^{t_2} ds = \int_{t_1}^{t_2} \sqrt{\left(\frac{dx}{dt}\right)^2 + \left(\frac{dy}{dt}\right)^2}\,dt.$$

EXERCISES

In Exercises 1 through 12, sketch the curve having the given parametric representation.

1. $x = \sin t$, $y = \cos t$ for $0 \le t \le \pi$

2. $x = 2\cos t$, $y = 3\sin t$ for all t

3. $x = \tan t$, $y = \sec t$ for $-\dfrac{\pi}{2} < t < \dfrac{\pi}{2}$

4. $x = t^2$, $y = t + 1$ for all t

5. $x = t^2$, $y = t^3$ for all t

6. $x = \sin t$, $y = \cos 2t$ for $0 \le t \le 2\pi$

7. $x = t - 3$, $y = t^2 + 1$ for all t

8. $x = 3\cosh t$, $y = 2\sinh t$ for all t

9. $x = t - 1$, $y = \ln t$ for $0 < t < \infty$

10. $x = \sin t$, $y = 1 + \sin^2 t$ for all t

11. $x = t^2$, $y = t^2 + 1$ for all t

12. $x = t^2$, $y = t^4$ for all t

13. Parametrize the curve $y = e^x$ taking as parameter the slope m of the curve at (x, y).

14. Parametrize the curve $y = \sqrt{x}$ taking as parameter the slope s of the normal to the curve at (x, y).

15. Parametrize the cubic $y = x^3$ in terms of the second derivative s at the point (x, y) on the curve.

16. Parametrize the curve $y = \sin x$ in terms of degree measure ϕ of the angle with radian measure x.

17. Parametrize the half-line $y = 3$, $x \ge 0$, taking as parameter the angle of inclination θ at the origin from the x-axis to the point (x, y) on the half-line.

18. Parametrize the half-line in Exercise 17, taking as parameter the distance s from (x, y) to the origin.

19. Can you parametrize the entire graph of $y = x^4$ taking as parameter
a) the first derivative m at a point (x, y)?
b) the second derivative s at a point (x, y)?
c) the third derivative t at a point (x, y)?
d) the fourth derivative f at a point (x, y)?
Give reasons for your answers.

20. Can you parametrize the curve $y = x^2$ by taking as parameter the distance from (x, y) to $(0, 0)$? Why?

21. Reparametrize the arc $x = a\cos\theta$, $y = a\sin\theta$ for $\pi/4 \le \theta \le 3\pi/4$, taking as parameter the slope m of the arc.

22. Parametrize the curve $y = \sqrt{x}$ taking as parameter the distance d from (x, y) to $(0, 0)$.

23. A circular disk of radius a rolls along the x-axis. Find parametric equations of the curve traced by a point P on the disk a distance b from the center of the disk, where $0 \le b \le a$. Assume that the point P falls on the interval $[0, a]$ on the y-axis when the disk touches the origin, and take as parameter the angle θ through which the disk rolls from its position at the origin. (This curve is a *trochoid*. For $b = a$, we obtain the cycloid, while $b = 0$ yields the line $y = a$.)

In Exercises 24 through 28, find the slope of the curve, with the given parametric representation, at the point on the curve corresponding to the indicated value of the parameter.

24. The cycloid $x = a\theta - a\sin\theta$, $y = a - a\cos\theta$, $\theta = \pi/4$

25. $x = t^2 - 3t$, $y = \sin 2t$, where $t = 0$

26. $x = \sinh t$, $y = \cosh 2t$, where $t = 0$

27. $x = e^t$, $y = \ln(t + 1)$, where $t = 0$

28. $x = t^3 - 3t^2 + 3t - 5$, $y = 4(t - 1)^3$, where $t = 1$

29. Find all points where the parametric curve $x = 3t - 1$, $y = t^3 - 3t^2 - 9t + 1$ has a horizontal tangent.

30. Find all points where the parametric curve $x = t^2 + 2t + 3$, $y = e^t - t$ has a horizontal tangent.

31. Find all points where the parametric curve $x = t^2 - 2t$, $y = e^t - t$ has a vertical tangent.

32. Find all points where the curve $x = \cos 2t$, $y = \sin t$ has a vertical tangent.

33. Find the minimum distance from a point on the parametric curve $x = t - 3$, $y = t + 1$ to the origin.

34. Find the maximum distance from a point on the parametric curve $x = 1/(1 + t^2)$, $y = 2t/(1 + t^2)$ to the origin.

In Exercises 35 through 37, the parametric equations represent a curve that is the graph of a function in a neighborhood of the point corresponding to the indicated value of the parameter. Find d^2y/dx^2 at this point.

35. $x = t^2$, $y = t^3 - 2t^2 + 5$, where $t = 1$

36. $x = \sin 3t$, $y = e^t$, where $t = 0$

37. $x = \ln(t + 3)$, $y = \cos 2t$, where $t = 0$

38. If $x = 3t^2 - 1$ and $y = 4t + 1$, find d^3y/dx^3 in terms of t.

39. If $x = 3t + 2$ and $y = \sin 2t$, find d^4y/dx^4 in terms of t.

40. If $x = 2t - 5$ and $y = e^{3t}$, find d^5y/dx^5 in terms of t.

In Exercises 41 through 43, find the length of the curve with the given parametric representation.

41. $x = t^2$, $y = \frac{2}{3}(2t + 1)^{3/2}$ from $t = 0$ to $t = 4$

42. $x = t$, $y = \ln(\cos t)$ from $t = 0$ to $t = \pi/4$

43. $x = a\cos^3 t$, $y = a\sin^3 t$ from $t = 0$ to $t = \pi/2$

44. Find the length of one arch of the cycloid $x = a\theta - a\sin\theta$, $y = a - a\cos\theta$. [*Hint:* Integrate $\int \sqrt{1 - \cos\theta}\, d\theta$ by multiplying numerator and denominator by $\sqrt{1 + \cos\theta}$.]

In Exercises 45 through 48, use Simpson's rule to estimate the length of the indicated parametric arc.

45. $x = t^3 - 3t$, $y = 1/(t + 2)$ for $-1 \le t \le 1$

46. $x = \sinh t$, $y = \cosh t$ for $0 \le t \le 2$

47. $x = \sin t$, $y = \cos 3t$ for $0 \le t \le 2\pi$

48. $x = \ln t$, $y = e^t$ for $1 \le t \le 4$

12.6 CURVATURE

WHAT IS CURVATURE?

We are interested in the rate at which a plane curve bends (or "curves") as we travel along the curve. We will attempt to give a numerical measure of such a rate of turning at a point on the curve; this number will be the *curvature of the curve at the point*. Thus the curvature of a curve at a point is to be a number; the more the curve bends at the point, the larger the number will be. It would be natural for us to expect curvature to satisfy the following three conditions.

Condition 1 Since a straight line does not bend at all, we would like its curvature at each point to be zero.

Condition 2 A circle "curves" at a uniform rate; we would like the curvature of a circle at one point to be the same as the curvature of the circle at every other point on the circle and the same as the curvature at each point of any other circle of the same radius. This would allow us to speak of the *curvature of a circle of radius a*.

Condition 3 The smaller the radius of a circle, the more the circle "curves" at each point. We would like the curvature of a small circle to be greater than the curvature of a circle of larger radius.

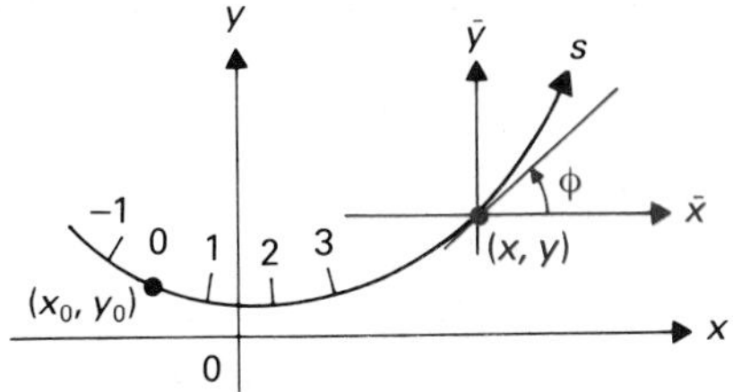

Figure 12.61 $\kappa = |d\phi/ds|$ measures the rate the curve bends at (x, y).

Consider a curve defined by one or more differentiable functions and having a tangent line at each point. We choose a point (x_0, y_0) on our curve from which we measure arc length s along the curve, so that (x_0, y_0) corresponds to $s = 0$, as shown in Fig. 12.61. Let ϕ be the angle shown in the figure, measured from the positive $\bar{x}$-axis at (x, y) to the direction corresponding to increasing s along the tangent line. The rate at which the curve bends at (x, y) can be described in terms of the rate at which the tangent line is turning as we travel along the curve at (x, y) and this in turn can be measured by the rate at which the angle ϕ is increasing at (x, y). We want curvature to be intrinsic to the point set or *trace* of the curve and not dependent on the rate at which we may be traveling along the curve. Thus it would not be appropriate to let the curvature be the rate of change of ϕ per unit change in *time* as we travel along the curve. It is intuitively apparent that the notion of arc length is intrinsic to the trace, and we let the curvature of the curve at a point be the rate of change of ϕ per unit change in arc length along the curve at that point.

DEFINITION 12.6 Curvature

The **curvature** κ of a curve at (x, y) is $|d\phi/ds|$, where ϕ is the angle in Fig. 12.61 and s is arc length measured along the curve.*

---

* Some texts define the curvature to be the signed quantity $d\phi/ds$, but we take a definition that specializes the notion of curvature for a space curve, whuch will be introduced in Chapter 15. The interpretation of the sign of $d\phi/ds$ is explained later in this section.

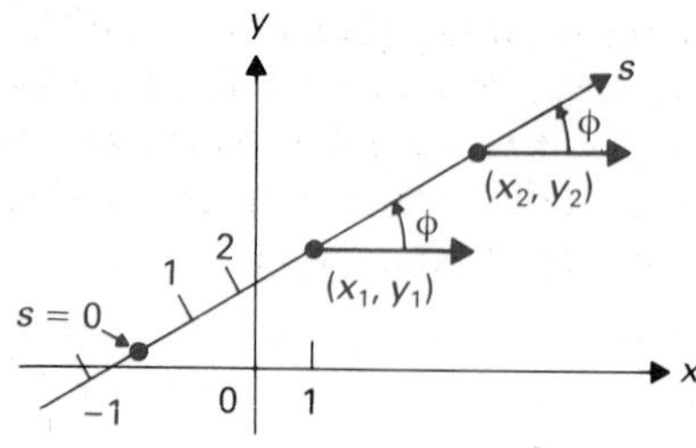

Figure 12.62 ϕ remains constant for a straight line, so $\kappa = |d\phi/ds| = 0$.

EXAMPLE 1 Find the curvature of a straight line.

Solution Since the angle ϕ along a straight line is a constant function of the arc length s (see Fig. 12.62), we have $d\phi/ds = 0$, so the curvature of a straight line at each point is zero. Thus Condition 1 above is satisfied by our definition of curvature. □

EXAMPLE 2 Find the curvature of a circle of radius a at a point on the circle.

Solution Measure arc length on the circle $x^2 + y^2 = a^2$ in the counterclockwise direction, starting with $s = 0$ at $(a, 0)$, as shown in Fig. 12.63. From this figure, we see that at a point (x, y) on the circle corresponding to a central angle θ and hence to an arc length $s = a\theta$, we have $\phi = \theta + n(\pi/2)$. (The value of n depends on the quadrant of θ; for θ in the first quadrant as in Fig. 12.63, we have $n = 1$.) Therefore

$$\phi = \theta + n\frac{\pi}{2} = \frac{1}{a}s + n\frac{\pi}{2},$$

so

$$\kappa = \frac{d\phi}{ds} = \frac{1}{a}.$$

Thus the curvature at any point of a circle of radius a is the reciprocal $1/a$ of the radius. We see that Conditions 2 and 3 above are satisfied by our definition of curvature. □

The preceding example suggests the following definition.

DEFINITION 12.7 Radius of curvature

If a curve has curvature $\kappa \neq 0$ at a point, then the **radius of curvature** of the curve at this point is $\rho = 1/\kappa$.

THE FORMULA FOR THE CURVATURE OF $y = f(x)$

Figure 12.63 s at (x, y) is $a\theta$; $\phi = \theta + \pi/2 = s/a + \pi/2$; $\kappa = |d\phi/ds| = 1/a$.

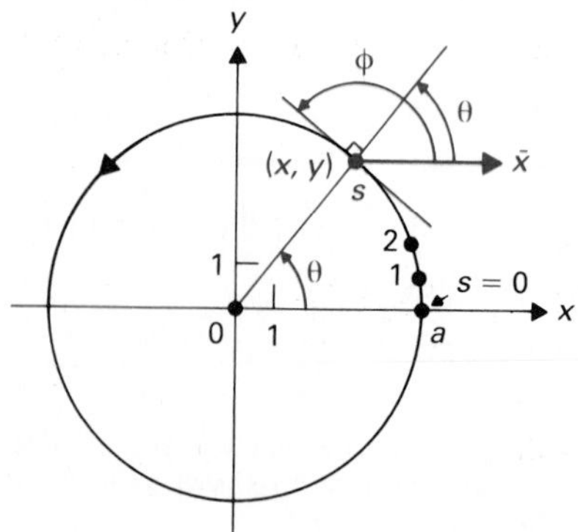

Let a curve be the graph $y = f(x)$ for a *twice*-differentiable function f, and let the direction of increasing arc length measured from (x_0, y_0) be the direction of increasing x, so that

$$s = \int_{x_0}^{x} \sqrt{1 + f'(t)^2}\, dt. \tag{1}$$

For the angle ϕ shown in Fig. 12.61, we have

$$\tan\phi = f'(x)$$

at any point (x, y) on the graph. Thus in a neighborhood of (x, y),

$$\phi = b + \tan^{-1} f'(x), \tag{2}$$

where the constant b occurs since ϕ may not fall in the principal value range of the inverse tangent function. (In Fig. 12.61, we see that $b = 0$.) Since f

is a twice-differentiable function of x, we see from Eq. (2) that ϕ is a differentiable function of x. From Eq. (1) we see that s is a differentiable function of x; in a neighborhood of a point where $ds/dx \neq 0$, it can then be shown that x appears as a differentiable function of s. Under these conditions, ϕ then appears as a differentiable function of s, and, by the chain rule,

$$\frac{d\phi}{ds} = \frac{d\phi}{dx} \cdot \frac{dx}{ds}. \tag{3}$$

From Eq. (2),

$$\frac{d\phi}{dx} = \frac{1}{1 + (f'(x))^2} \cdot f''(x), \tag{4}$$

and from Eq. (1),

$$\frac{ds}{dx} = \sqrt{1 + (f'(x))^2}.$$

Thus

$$\frac{dx}{ds} = \frac{1}{\sqrt{1 + (f'(x))^2}}, \tag{5}$$

and, from Eqs. (3), (4), and (5),

$$\kappa = \left|\frac{d\phi}{ds}\right| = \left|\frac{f''(x)}{[1 + (f'(x))^2]^{3/2}}\right|. \tag{6}$$

It is easy to check that $d\phi/ds > 0$ where, when traveling along the curve in the direction of increasing s, the curve bends to the left (ϕ is increasing), and $d\phi/ds < 0$ where the curve bends to the right (ϕ is decreasing). See Exercise 17.

THEOREM 12.5 Curvature of a graph

The curvature of the graph of a twice-differentiable function f at a point (x, y) on the graph is given by

$$\kappa = \left|\frac{d\phi}{ds}\right| = \left|\frac{f''(x)}{[1 + (f'(x))^2]^{3/2}}\right|. \tag{7}$$

(Geometrically, $d\phi/ds$ is positive at a point if, when traveling along the curve in the direction of increasing s, the curve bends to the left, and $d\phi/ds$ is negative if the curve bends to the right.)

EXAMPLE 3 Find the curvature and radius of curvature of the parabola $y = x^2$ at the origin.

Solution For $f(x) = x^2$, we have

$$\kappa = \left|\frac{f''(x)}{[1 + (f'(x))^2]^{3/2}}\right| = \frac{2}{(1 + 4x^2)^{3/2}},$$

so

$$\kappa|_{(0,0)} = 2.$$

The radius of curvature is $\rho = 1/\kappa = \frac{1}{2}$. □

At a point P on a curve where the radius of curvature is ρ, go along the normal line on the concave side of the curve the distance ρ, arriving at a point C on the normal. The circle with center C and radius ρ is clearly the circle that best fits the curve at P, as shown in Fig. 12.64.

DEFINITION 12.8 Osculating circle and center of a curvature of a curve

Let a curve have radius of curvature ρ at a point (x, y). The **osculating circle of the curve at** (x, y) is the circle of radius ρ through (x, y) having the same tangent as the curve at (x, y) and having center on the concave side of the curve. The center of this circle is the **center of curvature of the curve at** (x, y).

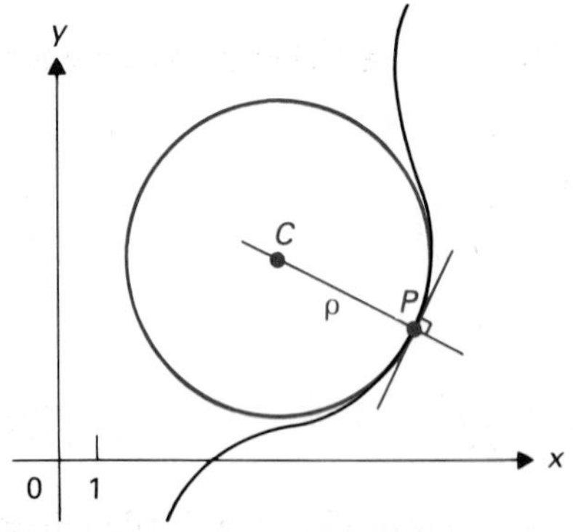

Figure 12.64 The osculating circle is the one best fitting the curve at P.

EXAMPLE 4 Find the center of curvature and the equation of the osculating circle to the graph of $y = x^2$ at the origin.

Solution From Example 3 and Fig. 12.65, we see that the center of curvature of the parabola $y = x^2$ at $(0, 0)$ is $(0, 1/2)$, and the osculating circle has equation

$$x^2 + \left(y - \frac{1}{2}\right)^2 = \frac{1}{4}. \quad \square$$

EXAMPLE 5 Find the curvature and radius of curvature of the ellipse $x^2/25 + y^2/9 = 1$ at the point $(0, 3)$.

Solution Near $(0, 3)$, the ellipse defines y as a function f of x. We find $f'(x) = dy/dx$ and $f''(x) = d^2y/dx^2$ for Eq. (7) by implicit differentiation:

$$\frac{x^2}{25} + \frac{y^2}{9} = 1, \qquad \frac{2x}{25} + \frac{2y}{9} \cdot \frac{dy}{dx} = 0,$$

$$\frac{dy}{dx} = -\frac{9x}{25y}, \qquad \frac{d^2y}{dx^2} = -\frac{25y \cdot 9 - 9x[25(dy/dx)]}{(25y)^2}.$$

We obtain

$$\left.\frac{dy}{dx}\right|_{(0,3)} = 0 \quad \text{and} \quad \left.\frac{d^2y}{dx^2}\right|_{(0,3)} -\frac{25 \cdot 3 \cdot 9 - 0}{(25 \cdot 3)^2} = \frac{3}{25}.$$

Thus

$$\kappa = \frac{d^2y/dx^2}{[1 + (dy/dx)^2]^{3/2}} = \frac{\frac{3}{25}}{(1 + 0^2)^{3/2}} = \frac{3}{25}.$$

The radius of curvature is $\rho = 1/\kappa = \frac{25}{3}$. $\square$

Figure 12.65 The osculating circle to $y = x^2$ at $(0, 0)$.

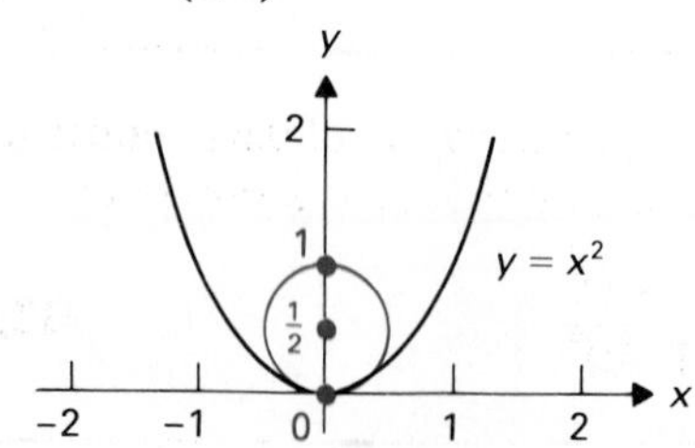

EXAMPLE 6 Find the center of curvature and the equation of the osculating circle at the point $(0, 3)$ of the ellipse in Example 5.

Solution From Fig. 12.66, we see that the center of curvature is the point $(0, 3 - \frac{25}{3}) = (0, -\frac{16}{3})$. The osculating circle then has equation

$$(x - 0)^2 + \left(y + \frac{16}{3}\right)^2 = \left(\frac{25}{3}\right)^2, \quad \text{or} \quad 3x^2 + 3y^2 + 32y^2 = 123. \quad \square$$

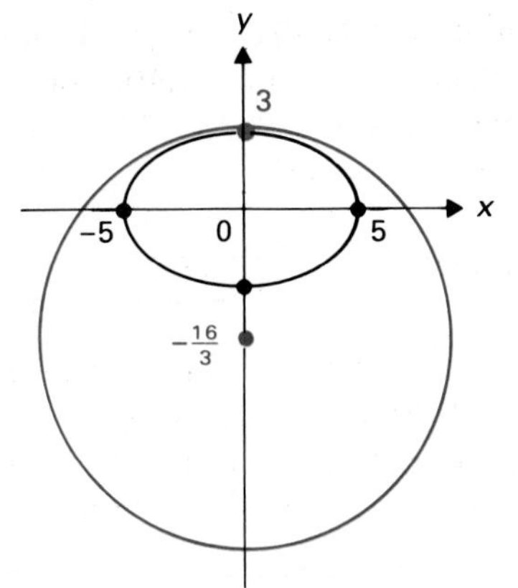

Figure 12.66 The osculating circle to the ellipse $x^2/25 + y^2/9 = 1$ at (0, 3).

If the normal line to a curve at a point is not horizontal or vertical, we may not be able to find the center of curvature from ρ by inspection, as we did in Examples 4 and 6. Formulas given just before Exercise 30 are convenient for finding the center of curvature in such a case.

THE CURVATURE FORMULA IN PARAMETRIC FORM

Let a curve be given parametrically by $x = h(t)$, $y = k(t)$, where h and k are twice-differentiable functions of t. We have previously defined the arc length function

$$s(t) = \int_{t_0}^{t} \sqrt{h'(u)^2 + k'(u)^2}\, du. \tag{8}$$

If $ds/dt \neq 0$, then it can be shown that t appears as a differentiable function of s, with

$$\frac{dt}{ds} = \frac{1}{\sqrt{(h'(t))^2 + (k'(t))^2}}. \tag{9}$$

For the angle ϕ shown in Fig. 12.61, we obtain

$$\phi = b + \tan^{-1}\frac{k'(t)}{h'(t)} \tag{10}$$

as analogue of Eq. (2), so ϕ is a differentiable function of t. Under these conditions,

$$\begin{aligned}\frac{d\phi}{ds} &= \frac{d\phi}{dt}\cdot\frac{dt}{ds}\\ &= \frac{1}{1 + (k'(t)/h'(t))^2}\cdot\frac{h'(t)k''(t) - k'(t)h''(t)}{(h'(t))^2}\cdot\frac{1}{\sqrt{(h'(t))^2 + (k'(t))^2}} \qquad (11)\\ &= \frac{h'(t)k''(t) - k'(t)h''(t)}{[(h'(t))^2 + (k'(t))^2]^{3/2}}.\end{aligned}$$

Formula (11) looks less cumbersome if we let

$$\dot{x} = \frac{dx}{dt} = h'(t) \qquad \text{and} \qquad \dot{y} = \frac{dy}{dt} = k'(t)$$

and denote second derivatives with respect to t by $\ddot{x}$ and $\ddot{y}$. This *Newtonian notation* appears often in physics texts. We then obtain formula (12), given in the following theorem.

THEOREM 12.6 Curvature of a parametric curve

Let h and k be twice-differentiable functions. The curvature of the parametric curve $x = h(t)$, $y = k(t)$ is given at a point corresponding to a value t by

$$\kappa = \left|\frac{d\phi}{ds}\right| = \left|\frac{\dot{x}\ddot{y} - \dot{y}\ddot{x}}{(\dot{x}^2 + \dot{y}^2)^{3/2}}\right|. \tag{12}$$

EXAMPLE 7 Find the curvature of the ellipse $x = 3\cos t$, $y = 4\sin t$ at the point (3, 0) corresponding to $t = 0$, taking the direction of increasing s counterclockwise to coincide with the direction of increasing t.

Solution From formula (12), we obtain

$$\kappa = \left|\frac{\dot{x}\ddot{y} - \dot{y}\ddot{x}}{(\dot{x}^2 + \dot{y}^2)^{3/2}}\right| = \left|\frac{12\sin^2 t - (-12\cos^2 t)}{(9\sin^2 t + 16\cos^2 t)^{3/2}}\right|.$$

For $t = 0$ corresponding to the point (3, 0), we obtain

$$\kappa|_{t=0} = \frac{12}{16^{3/2}} = \frac{12}{64} = \frac{3}{16}. \quad \square$$

EXAMPLE 8 Find the curvature and radius of curvature of the curve $x = t^2 - 3t$, $y = t^3$ at the point where $t = 1$.

Solution We have

$$\dot{x}|_1 = (2t - 3)|_1 = -1, \qquad \dot{y}|_1 = 3t^2|_1 = 3,$$
$$\ddot{x}|_1 = 2, \qquad \ddot{y}|_1 = 6t|_1 = 6.$$

Thus from formula (12),

$$\kappa = \left|\frac{-1\cdot 6 - 3\cdot 2}{(1 + 9)^{3/2}}\right| = \left|\frac{-12}{10\sqrt{10}}\right| = \frac{6}{5\sqrt{10}}.$$

The radius of curvature is $\rho = 1/\kappa = 5\sqrt{10}/6$. $\square$

Sometimes we may want to find the curvature of a graph given in the form $x = g(y)$ rather than $y = f(x)$. In such a case,

$$\kappa = \left|\frac{g''(y)}{[1 + (g'(y))^2]^{3/2}}\right|, \tag{13}$$

which is a formula parallel to Eq. (6). Of course, we would expect this to be the case; κ is intrinsic to the trace of the curve and independent of the choice of axes. For another way of justifying Eq. (13), we put $x = g(y)$ in parametric form as

$$x = g(t), \qquad y = t.$$

Then

$$\dot{x} = g'(t), \qquad \ddot{x} = g''(t), \qquad \dot{y} = 1, \qquad \ddot{y} = 0.$$

Formula (12) yields

$$\kappa = \left|\frac{-1\cdot g''(t)}{[(g'(t))^2 + 1]^{3/2}}\right| = \left|\frac{g''(t)}{[1 + (g'(t))^2]^{3/2}}\right|.$$

Since $t = y$, we at once obtain Eq. (13).

EXAMPLE 9 Find the curvature and radius of curvature of the curve $x = y^3 - 3y^2 + 2$ at the point $(-2, 2)$.

Solution We have

$$\left.\frac{dx}{dy}\right|_2 = (3y^2 - 6y)|_2 = 12 - 12 = 0,$$

$$\left.\frac{d^2x}{dy^2}\right|_2 = (6y - 6)|_2 = 12 - 6 = 6.$$

Thus from Eq. (13),

$$\kappa = \left|\frac{6}{(1 + 0^2)^{3/2}}\right| = 6 \quad \text{and} \quad \rho = \frac{1}{\kappa} = \frac{1}{6}. \quad \square$$

SUMMARY

1. Curvature $\kappa = |d\phi/ds|$, where ϕ is the angle (measured counterclockwise) from the horizontal to the tangent to the curve, and s is arc length.
2. If $y = f(x)$, then
$$\kappa = \left|\frac{d^2y/dx^2}{[1 + (dy/dx)^2]^{3/2}}\right|.$$
3. If $x = g(y)$, then
$$\kappa = \left|\frac{d^2x/dy^2}{[1 + (dx/dy)^2]^{3/2}}\right|.$$
4. If $x = h(t)$ and $y = k(t)$, then
$$\kappa = \left|\frac{\dot{x}\ddot{y} - \dot{y}\ddot{x}}{(\dot{x}^2 + \dot{y}^2)^{3/2}}\right|,$$
where $\dot{x} = dx/dt$, $\ddot{x} = d^2x/dt^2$, and $\dot{y}$ and $\ddot{y}$ are similarly defined.
5. The radius of curvature ρ at a point is equal to $1/\kappa$.
6. The osculating circle to a curve at a point is the circle with center on the concave side of the curve, radius $\rho = 1/\kappa$, and tangent to the curve.

EXERCISES

In Exercises 1 through 16, find the curvature κ at the indicated point of the curve having the given x,y-equation or parametric equations.

1. $y = \sin x$ at $(\pi/2, 1)$
2. $xy = 1$ at $(1, 1)$
3. $y = \ln x$ at $(1, 0)$
4. $y = \dfrac{1}{1 + x^2}$ at $(0, 1)$
5. $y = x^4 - 3x^3 + 4x^2 - 2x + 2$ at $(1, 2)$
6. $x = e^y - 2y$ at $(1, 0)$
7. $x = \sin 2y - 3\cos 2y$ at $\left(3, \dfrac{\pi}{2}\right)$
8. $x^2 - y^2 = 16$ at (5. 3)
9. $x = \sqrt{25 - y^2}$ at $(3, -4)$
10. $x^2y + 2xy^2 = 3$ at $(1, 1)$
11. $\dfrac{x^2}{16} - \dfrac{y^2}{25} = 1$ at $(4, 0)$ [*Hint:* The hyperbola defines x as a function of y near $(4, 0)$.]
12. $x = t^2 + 3t$, $y = t^4 - 3t^2$, where $t = -1$
13. $x = 4\sin t$, $y = 5\cos t$, where $t = 0$
14. The cycloid $x = a(\theta - \sin\theta)$, $y = a(1 - \cos\theta)$, where $\theta = \pi$
15. $x = e^t$, $y = t^2$, where $t = 0$
16. $x = \cosh t$, $y = \sinh t$, where $t = 0$

17. Give rate-of-change arguments that $d\phi/ds$ at a point of a curve is *positive* if the curve bends to the left as you travel in the direction of increasing s at the point and is *negative* if the curve bends to the right.
18. Discuss the curvature at the origin of the graphs of the monomial functions ax^n for integers $n \geq 1$.
19. What is the curvature of the graph of a twice-differentiable function at an inflection point on the graph?
20. The osculating circle to a given curve at a point (x, y) has the same tangent line and the same radius of curvature as the given curve at (x, y).
    a) Show that the given curve and the circle have the same curvature κ at (x, y).
    b) Argue from part (a) that if both the given curve and the circle are graphs of functions in a neighborhood of (x, y), then these two functions have the same first derivatives and the same second derivatives for this value x.
21. Prove that the only twice-differentiable functions whose graphs have curvature zero at each point are those functions having straight lines as graphs.
22. Find the equation of the osculating circle to the curve $y = \ln x$ at $(1, 0)$.
23. Find the equation of the osculating circle to the curve $x^2 + y^2 - 4x + 2y = 0$ at the point $(3, 1)$.
24. Find the equation of the osculating circle to the hyperbola $(x^2/4) - (y^2/9) = 1$ at the point $(-2, 0)$.
25. Let κ_n be the curvature of $y = x^n$ at $(1, 1)$. Find $\lim_{n\to\infty} \kappa_n$.
26. Find the point(s) where the curve $y = x^2$ has maximum curvature.
27. Find the point(s) where the curve $y = x^3$ has (a) maximum curvature, (b) minimum curvature.
28. For $n > 2$, the curve $y = x^n$ has two points of maximum curvature, one point A_n in the first quadrant and one point B_n in either the second or third quadrant.
    a) Find the point in the plane that A_n approaches as $n \to \infty$; that is, find $\lim_{n\to\infty} A_n$.
    b) Find $\lim_{n\to\infty} B_n$.
    [*Hint:* Don't try to solve this problem analytically. Just think in terms of the graphs.]
29. Find the point of maximum curvature of $x = \cosh t$, $y = \sinh t$.

Let the graph $y = f(x)$ of a twice-differentiable function f have nonzero curvature at a point (x, y). It can be shown that the coordinates (α, β) of the center of curvature of the graph at (x, y) are given by

$$\alpha = x - y'\frac{1 + (y')^2}{y''}, \qquad \beta = y + \frac{1 + (y')^2}{y''}.$$

In Exercises 30 through 33, use the formulas above to find the equation of the osculating circle to the given curve at the indicated point.

30. $y = x^2$ at $(1, 1)$
31. $xy = 1$ at $(1, 1)$
32. $y = \tan x$ at $(\pi/4, 1)$
33. $x = t^2$, $y = t^3$, where $t = -1$

The set of all centers of curvature of a given curve is the **evolute of the curve.** The given curve is an **involute** of this set of centers of curvature. (While a curve has only one evolute, a single set may have many involutes, that is, it may be the evolute of many curves, as illustrated by the next exercise.)

34. a) What is the evolute of a circle?
    b) What are the involutes of a point?
35. Find parametric equations for the evolute of the parabola $y = x^2$. Use the formulas given before Exercise 30.
36. Find parametric equations for the evolute of the ellipse $x = 2 \sin t$, $y = \cos t$, using the formulas given before Exercise 30.

EXERCISE SETS FOR CHAPTER 12

Review Exercise Set 12.1

If you did Section 12.1 but not Sections 12.2, 12.3, and 12.4, replace Exercises 1, 2, and 3 by 1.1, 1.2, and 1.3.

1.1. Sketch the curve $x^2 - 4y^2 = 16$.

1.2. Sketch the curve $x + 4y^2 - 8y = -12$.

1.3. Sketch the curve $3x^2 + y^2 - 12x + 4y = -4$.

1. Find the equation of the conic section with focus $(3, 2)$, directrix $x = -1$, and eccentricity 2.
2. Find the foci and directrices of the conic section with equation $2x^2 + 4x + y^2 - 2y + 1 = 0$, and sketch the curve.
3. Classify the curve as a (possibly empty or degenerate) ellipse, hyperbola, or parabola.
    a) $x^2 - 3xy + 2y^2 - 3x + 4y = 7$
    b) $x^2 + 2xy + y^2 - 2x + 4y = 11$
    c) $x^2 + 3xy + 3y^2 - 8x + 6y = 0$
4. Sketch the curve described parametrically by $x = t^2$, $y = t^2 - 1$.
5. Find dy/dx and d^2y/dx^2 at $t = 1$ if $x = t^2$, $y = t^3 - 2t$.
6. Parametrize the curve $y = x^3$, taking as parameter the y-coordinate h at each point on the curve.
7. Express as an integral the length of the curve $x = \sin t$, $y = t^2 - 2$ from $t = 0$ to $t = 3$.
8. Find the curvature of the parabola $y = x^2$ at the point $(2, 4)$.
9. Find the curvature of the curve $x = \sin 2t$, $y = \cos 3t$ at the point $t = \pi/6$.
10. Find the equation of the osculating circle to the curve $y = \cos x$ at the point $(0, 1)$.

Review Exercise Set 12.2

If you did Section 12.1 but not Sections 12.2, 12.3, and 12.4, replace Exercises 1, 2, and 3 by 1.1, 1.2, and 1.3.

1.1. Sketch the curve $x^2 + 10x + 4y = 3$.

1.2. Sketch the curve $2x^2 - y^2 - 4x + 6y = 1$.

1.3. Sketch the curve $4x^2 + y^2 - x + 3y = \frac{111}{64}$.

1. Find the equation of the conic section with focus $(-1, 1)$, directrix $y = 2$, and eccentricity $\frac{1}{3}$.

2. Find the foci, directrices, and asymptotes of the hyperbola $4x^2 - 16x - y^2 - 2y = -11$, and sketch the curve.

3. Rotate axes, complete squares, and then sketch the curve $x^2 - 2xy + y^2 + \sqrt{2}x - 3\sqrt{2}y = 4$.

4. Sketch the curve described parametrically by $x = 1 + 3 \sin t$, $y = 2 - \cos t$.

5. Find dy/dx and d^2y/dx^2 at $t = \pi/3$ if $x = \sin t$, $y = \cos 2t$.

6. Parametrize the curve $y = e^{-x/2}$, taking as parameter the slope m of the curve at each point.

7. Find the point (x_0, y_0) on the curve $x = t$, $y = \frac{2}{3}t^{3/2}$ such that the distance from the origin to (x_0, y_0) measured along the curve is 42.

8. Find the curvature of the curve $y = \sin x$ at the point where $x = \pi/4$.

9. Find the curvature of the curve $x = t^2$, $y = t^3 + 2t$ at the point where $t = 1$.

10. Find the equation of the osculating circle to the curve $x = 2 \cos t$, $y = 3 \sin t$ at the point $t = \pi/2$.

Review Exercise Set 12.3

If you did Section 12.1 but not Sections 12.2, 12.3, and 12.4, replace Exercises 1, 2, and 3 by 1.1, 1.2, and 1.3.

1.1. Sketch the curve $4x^2 + 9y^2 = 36$.

1.2. Sketch the curve $x^2 + 3y^2 - 2x + 12y + 13 = 0$.

1.3. Sketch the curve $y^2 - 4x^2 + 8x - 6y = 11$.

1. Give two geometric definitions of an ellipse.

2. Find the equation of the ellipse with major axis the interval $[0, 10]$ on the y-axis and one focus at $(0, 8)$.

3. Write the equations for rotation of axes that eliminate the $x'y'$-term when substituted in $x^2 - 2xy + 3y^2 - 4x + 5y = 15$.

4. Sketch the parametric curve $x = 4 \sin^2 t$, $y = 4 \cos^2 t$.

5. Find d^4y/dx^4 in terms of t if $x = t^2$ and $y = t^4 + 3t^3$.

6. Reparametrize the curve $x = t^2 - 1$, $y = t^4 + 2t^2 - 3$, taking as parameter the slope m of the curve at each point.

7. Use Simpson's rule and a calculator to estimate the length of the curve $x = t^2$, $y = t + \ln t$ for $1 \le t \le 3$.

8. Find the curvature of the curve $x = (\ln y)^2$ at the point $(0, 1)$.

9. Find the point where the graph of $y = e^x$ has maximum curvature.

10. Find the radius of curvature of the parametric curve $x = 1/t$, $y = t^2 + 3t$, where $t = 2$.

Review Exercise Set 12.4

If you did Section 12.1 but not Sections 12.2, 12.3, and 12.4, replace Exercises 1, 2, and 3 by 1.1, 1.2, and 1.3.

1.1. Sketch the curve $4x^2 - 8x + 2y = 0$.

1.2. Sketch the curve $x^2 - 4y^2 + 2x - 8y - 3 = 0$.

1.3. Sketch the curve $2x^2 + 3y^2 - 4x - 27y + 30 = 0$.

1. Find the equation of the conic section with focus $(1, 1)$, directrix $y = -3$, and eccentricity 1.

2. Find the foci and the directrices of the ellipse $x^2 + 4y^2 - 4x + 8y = 8$.

3. Classify the curve as a (possibly empty or degenerate) ellipse, hyperbola, or parabola.

a) $2x^2 - 8xy + 8y^2 - 3x + 5y = -1$

b) $3x^2 + 4xy + y^2 - 2x + y = 5$

c) $5x^2 + 6xy + 2y^2 + 4x - 3y = 10$

4. Sketch the curve defined parametrically by $x = e^t + 1$, $y = e^{3t}$.

5. Find the point(s) on the parametric curve $x = t^2$, $y = t^3 + 4t^2$ where the slope is 1.

6. Parametrize the curve $y = e^x$, taking as parameter t the square of the y-coordinate.

7. Find the length of the parametric curve $x = t^2 + 5$, $y = t^3 - 7$ from $t = 0$ to $t = 3$.

8. Find all points on $y = x^2$ where the curvature is $1/\sqrt{2}$.

9. Find the equation of the osculating circle to the parabola $x^2 + 4x + 2y - 6 = 0$ at its vertex.

10. Find the curvature of the parametric curve $x = \sinh 2t$, $y = \cosh 3t$ at the point where $t = 0$.

More Challenging Exercises 12

1. Let f be a twice-differentiable function. Show that $|\kappa(x)| \le |f''(x)|$, where $\kappa(x)$ is the curvature of the graph at $(x, f(x))$.

2. Let $f(x) = 1 + x + (x/2)^2 + \cdots + (x/n)^n + \cdots$. Find the curvature of the graph where $x = 0$.

3. Find the length of arc of the curve $x = a \cos^3 t$, $y = a \sin^3 t$ for $0 \le t \le 2\pi/3$.

13 POLAR COORDINATES

13.1 THE POLAR COORDINATE SYSTEM

This section introduces polar coordinates and describes how to change back and forth from rectangular to polar coordinates. An optional part at the end of the section describes the equations of conic sections in polar coordinates. Optional Section 15.3 will use this information about the conic sections in studying motion due to a central force field.

POLAR COORDINATES

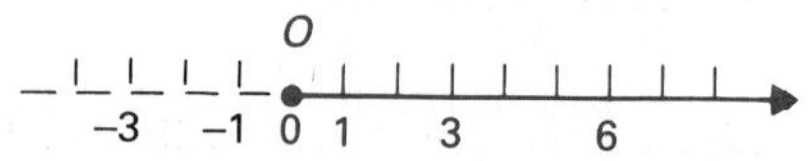

Figure 13.1 The polar axis.

We now describe the polar coordinate system and its relation to x,y-coordinates. Choose any point O in the Euclidean plane as the *pole* for our coordinate system and any half-line emanating from O as the *polar axis*. It is conventional in sketching to let the polar axis be a horizontal half-line, extending to the right, as shown in Fig. 13.1. We choose a scale on the polar axis, as indicated in the figure.

Let P be any point in the plane, and rotate the polar axis through an angle θ so that the rotated axis passes through the point P, as indicated in Fig. 13.2. We will always let positive values of θ correspond to *counterclockwise* rotation. The point P falls at a number r on the scale of the rotated axis, and the r and θ are **polar coordinates** for the point P. We denote P by (r, θ).

In a cartesian coordinate system, each point in the plane corresponds to a *unique* ordered pair (x, y) of numbers. A feature of a polar coordinate system is that each point has an *infinite* number of polar coordinate pairs. As indicated in Fig. 13.3, if a point P has polar coordinates (r, θ) it also has polar coordinates $(r, \theta + 2n\pi)$ for each integer n. Figure 13.4 shows that the same point P also has polar coordinates $(-r, \theta + \pi + 2n\pi)$ for all integers n. All polar coordinates of a point (r, θ) different from O are of the form

$$(r, \theta + 2n\pi) \qquad \text{or} \qquad (-r, \theta + \pi + 2n\pi).$$

Figure 13.2 The polar axis rotated to go through P.

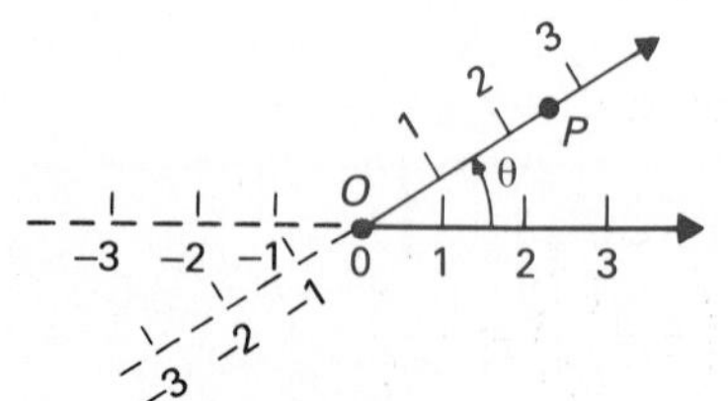

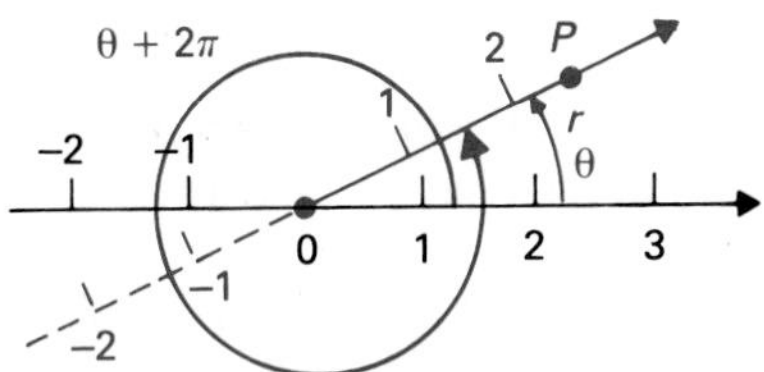

Figure 13.3 P has polar coordinates $(r, \theta + 2n\pi)$ for all integers n.

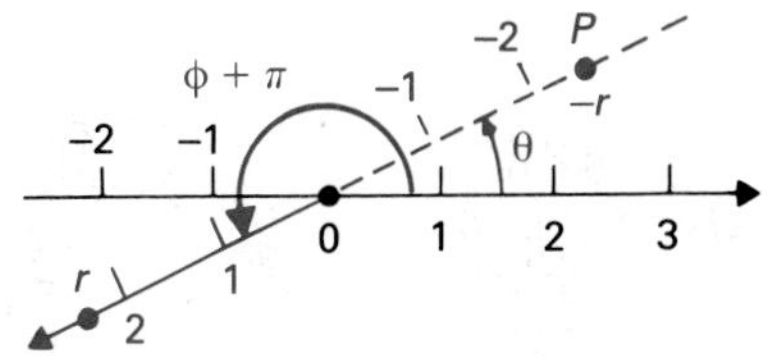

Figure 13.4 P also has coordinates $(-r, \theta + \pi + 2n\pi)$ for all integers n.

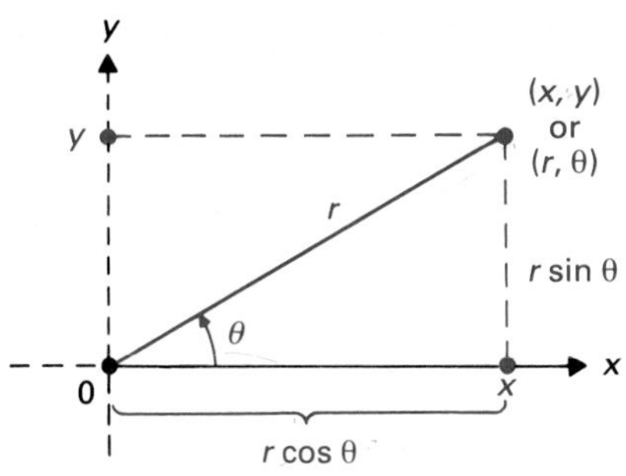

Figure 13.5
$x = r\cos\theta$; $y = r\sin\theta$;
$r^2 = x^2 + y^2$; $\theta = \tan^{-1}(y/x)$.

Note that the pole O has coordinates $(0, \theta)$ for every real number θ.

We will simultaneously consider a cartesian and a polar coordinate system for the plane. It is conventional to let the pole be at the Euclidean origin $(0, 0)$ with the polar axis falling on the positive x-axis. We will always follow this convention. From Fig. 13.5, we see that the cartesian x,y-coordinates can be expressed in terms of polar r,θ-coordinates by the equations

$$\boxed{\begin{aligned} x &= r\cos\theta, \\ y &= r\sin\theta. \end{aligned}} \tag{1}$$

It also follows at once from the figure that

$$\boxed{\begin{aligned} r^2 &= x^2 + y^2, \\ \theta &= \tan^{-1}\frac{y}{x}, \end{aligned}} \tag{2}$$

provided that we select θ such that $-\pi/2 < \theta < \pi/2$. Equations (1) and (2) are useful in changing from one coordinate system to the other.

EXAMPLE 1 Find cartesian x,y-coordinates for the point with polar coordinates $(r, \theta) = (2, \pi/3)$, shown in Fig. 13.6.

Solution From Eqs. (1), we find that

$$(x, y) = \left(2\cos\frac{\pi}{3}, 2\sin\frac{\pi}{3}\right) = (1, \sqrt{3}). \quad \square$$

Figure 13.6
$x = 2\cos(\pi/3) = 1$,
$y = 2\sin(\pi/3) = \sqrt{3}$.

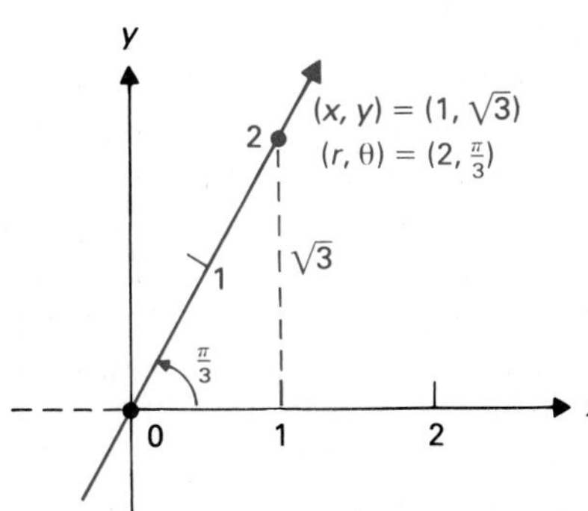

EXAMPLE 2 Find all polar r,θ-coordinates for the point with cartesian coordinates $(x, y) = (-\sqrt{3}, 1)$.

Solution From Eqs. (2), the point with cartesian coordinates $(x, y) = (-\sqrt{3}, 1)$ has polar coordinates (r, θ) such that

$$r^2 = x^2 + y^2 = 4$$

and

$$\theta = \tan^{-1}\frac{y}{x} = \tan^{-1}\left(-\frac{1}{\sqrt{3}}\right) = -\frac{\pi}{6}.$$

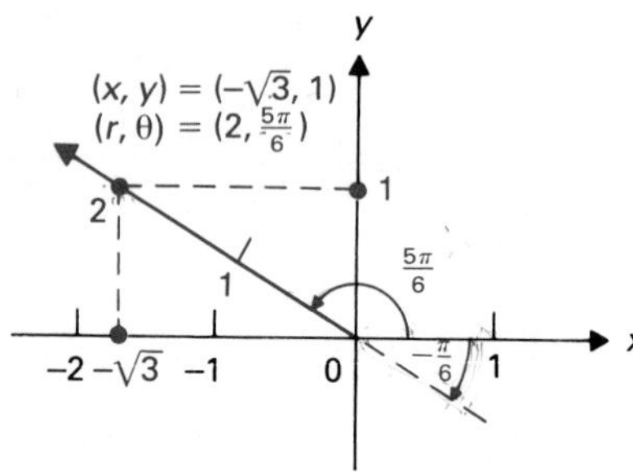

Figure 13.7 Cartesian $(-\sqrt{3}, 1)$ is polar $(2, 5\pi/6)$ or $(-2, -\pi/6)$.

Since $(-\sqrt{3}, 1)$ lies in the second cartesian quadrant, we easily see that $r = -2$, and the point has polar coordinates $(r, \theta) = (-2, -\pi/6)$. Of course, the point also has polar coordinates

$$\left(-2, -\frac{\pi}{6} + 2n\pi\right) \quad \text{and} \quad \left(2, \frac{5\pi}{6} + 2n\pi\right)$$

for all integers n. This also may be seen graphically, in Fig. 13.7. □

EXAMPLE 3 The force of attraction between two bodies has magnitude directly proportional to the product of their masses and inversely proportional to the square of the distance between them. A satellite of mass m is in planar orbit about the earth. Imagine the earth to be at the pole for polar coordinates in this plane. If the earth has mass M, express the magnitude of the force the earth exerts on the satellite in terms of the polar coordinate position (r, θ) of the satellite.

Solution With the earth at the pole, the distance from the earth to the satellite is given by the polar coordinate r. Thus the force of attraction has magnitude

$$\frac{GmM}{r^2}$$

for some constant G (the constant of gravitational attraction). (In Section 15.3, we will make a more detailed study of this force, using polar coordinates.) □

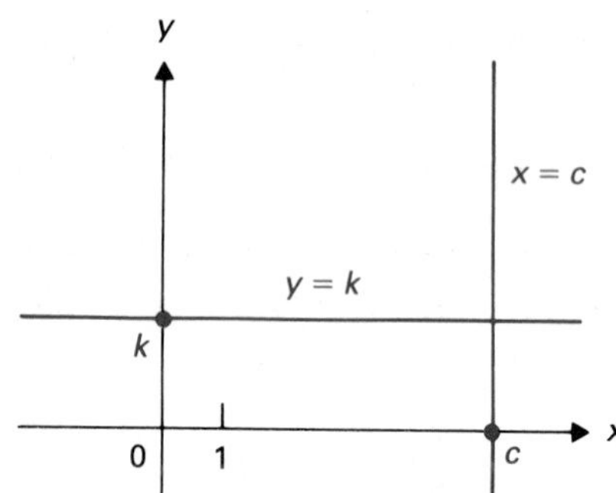

Figure 13.8 Level coordinate "curves" in cartesian coordinates.

In a coordinate system for the plane, among the first things we examine are the *level coordinate curves* of the system, obtained by putting the coordinate variables equal to constants. In a cartesian system, a level curve is either a vertical line $x = c$ or a horizontal line $y = k$. See Fig. 13.8. In a polar coordinate system, a level curve is either $r = a$, which is a circle about the pole of radius a, or $\theta = \beta$, which is a line through the pole, as indicated in Fig. 13.9. This suggests that polar coordinates might be useful in handling circles and lines through the origin.

Equations (1) and (2) can be used to express a cartesian characterization of a curve in polar form, and vice versa.

EXAMPLE 4 Find a polar equation for the vertical line $x = c$, shown in Fig. 13.10.

Solution Since $x = r \cos \theta$, we obtain from $x = c$

$$r \cos \theta = c \qquad \text{or} \qquad r = c \sec \theta.$$

The line is traced once for $-\pi/2 < \theta < \pi/2$. □

Figure 13.9 Level coordinate "curves" in polar coordinates.

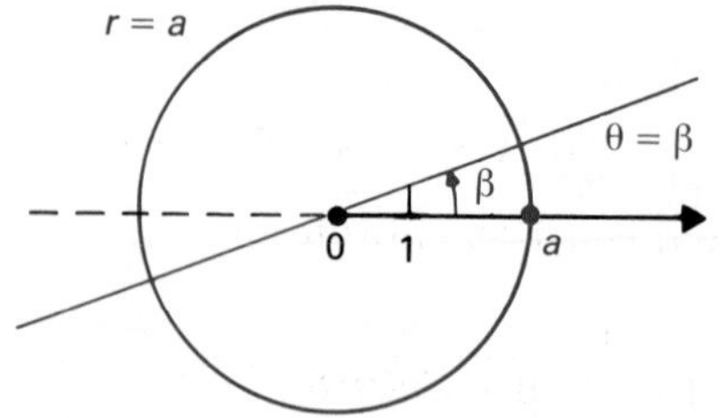

EXAMPLE 5 Sketch the polar curve $r = a \sin \theta$ by changing it to cartesian form.

Solution If $r \neq 0$, our equation is equivalent to $r^2 = ar \sin \theta$. Since $r = 0$ if $\theta = 0$ in our original equation, we see that our new equation actually gives

Figure 13.10 Cartesian: $x = c$; polar: $r = c \sec \theta$.

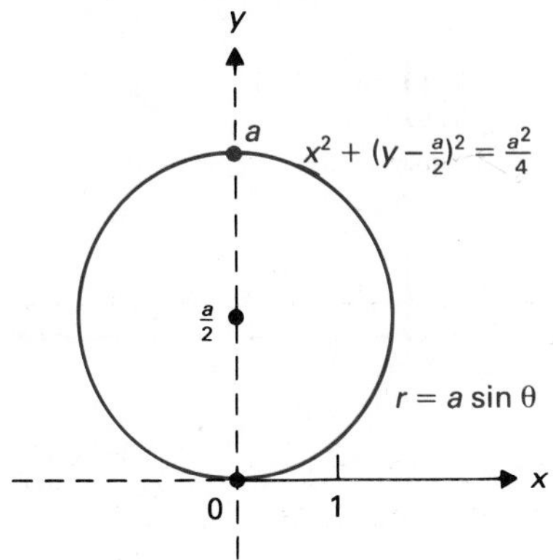

Figure 13.11 Polar: $r = a \sin \theta$; cartesian: $x^2 + y^2 = ay$.

exactly the original curve. By Eqs. (1) and (2), the equation $r^2 = ar \sin \theta$ has cartesian form

$$x^2 + y^2 = ay, \qquad \text{or} \qquad x^2 + \left(y - \frac{a}{2}\right)^2 = \frac{a^2}{4}.$$

We see that the curve is the circle with center at the point $(x, y) = (0, a/2)$ and radius $a/2$, as shown in Fig. 13.11.

We could have found the x, y-equation without multiplying $r = a \sin \theta$ through by r, although it is a bit more work. Starting with $r = a \sin \theta$, Eqs. (2) yield

$$\sqrt{x^2 + y^2} = a \sin\left(\tan^{-1} \frac{y}{x}\right).$$

From the triangle in Fig. 13.12, we see that $\sin(\tan^{-1} y/x) = y/\sqrt{x^2 + y^2}$. Thus our equation becomes

$$\sqrt{x^2 + y^2} = a \frac{y}{\sqrt{x^2 + y^2}},$$

or

$$x^2 + y^2 = ay,$$

which is the form we obtained before. □

Figure 13.12 $\theta = \tan^{-1}(y/x)$; $\sin \theta = y/\sqrt{x^2 + y^2}$.

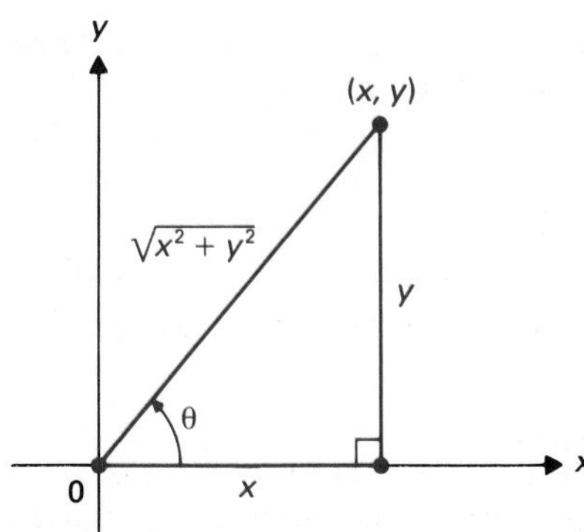

EXAMPLE 6 Find the polar coordinate equation of the ellipse with x, y-equation $x^2 + 3y^2 = 10$.

Solution We use Eqs. (1) and obtain

$$r^2 \cos^2\theta + 3(r^2 \sin^2\theta) = 10, \qquad \text{or} \qquad r^2(\cos^2\theta + 3 \sin^2\theta) = 10.$$

Since $\cos^2\theta = 1 - \sin^2\theta$, we may write our equation in the form

$$r^2(1 + 2 \sin^2\theta) = 10, \qquad \text{or} \qquad r^2 = \frac{10}{1 + 2 \sin^2\theta}. \quad \square$$

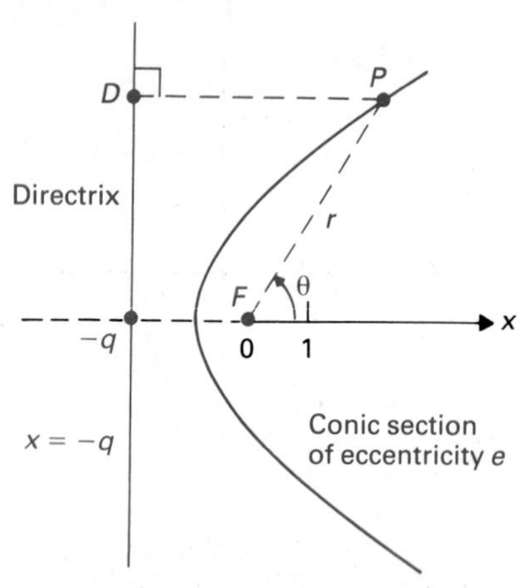

Figure 13.13 $e = \overline{FP}/\overline{DP}$ by the eccentricity definition of a conic section.

*CONIC SECTIONS IN POLAR COORDINATES

We describe equations of the conic sections in polar coordinates. Let a conic section of eccentricity e have focus at the pole O and have the line $x = -q$ as directrix, as shown in Fig. 13.13. Using the notation of that figure, we have

$$\overline{FP} = e(\overline{DP}).$$

Now

$$\overline{FP} = r \qquad \text{and} \qquad \overline{DP} = q + (\overline{FP}) \cos \theta = q + r \cos \theta.$$

Thus the polar equation of our conic section is

$$r = e(q + r \cos \theta)$$

or

$$r = \frac{qe}{1 - e \cos \theta}. \tag{3}$$

From Eq. (3), we see that every equation of the form

$$r = \frac{k}{1 - e \cos \theta}, \tag{4}$$

where $k > 0$ and $e > 0$, describes a conic section of eccentricity e and focus at the pole, because if we let $q = k/e$, we obtain Eq. (3). (Note that the solution of Example 6 was not of this form because the *center* of the ellipse rather than the focus was at the poles.)

EXAMPLE 7 Classify and sketch the conic section with polar equation

$$r = \frac{4}{1 - \cos \theta}.$$

Solution From Eq. (3), we see that $e = 1$, so we have a parabola. Then $4 = qe = q$, so the directrix $x = -q$ becomes $x = -4$. The parabola is shown in Fig. 13.14. We can sketch it roughly by remembering that the parabola consists of all points equidistant from the focus and directrix. □

EXAMPLE 8 Classify and sketch the conic section with polar equation

$$r = \frac{4}{1 - 2 \cos \theta}.$$

Figure 13.14 The parabola $r = 4/(1 - \cos \theta)$.

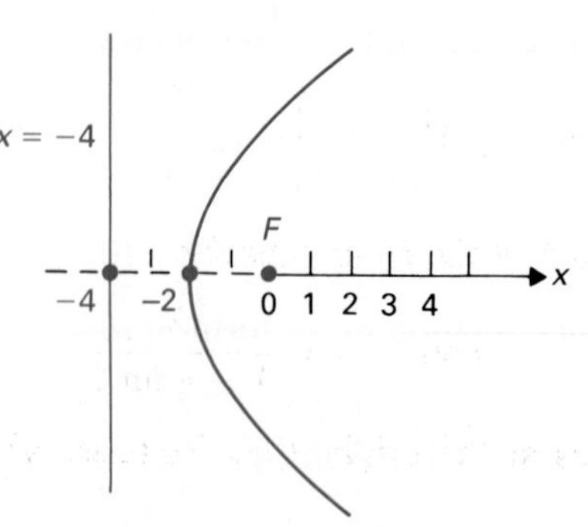

Solution From Eq. (3), we see that $e = 2$, so we have a hyperbola. Now $4 = qe = q \cdot 2$, so $q = 2$. Thus the hyperbola has as directrix $x = -2$. The hyperbola is shown in Fig. 13.15. We can sketch it roughly by remembering that it consists of all points twice as far from the focus as from the directrix. □

EXAMPLE 9 Find the x,y-equations of the asymptotes of the hyperbola with polar equation

$$r = \frac{8}{1 - 2 \cos \theta}.$$

* This subsection may be omitted if Section 15.3 is also to be omitted.

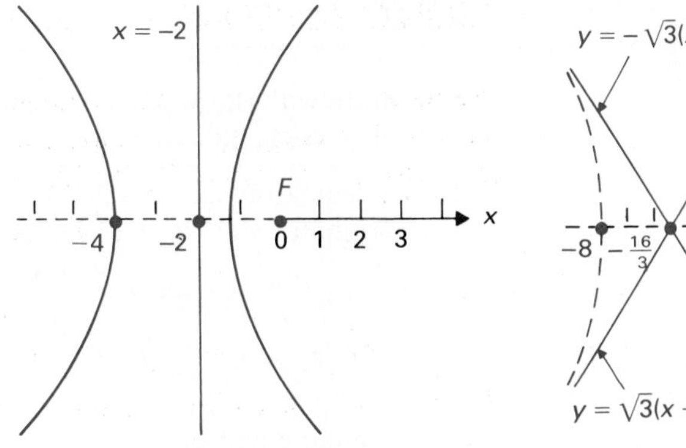

Figure 13.15 The hyperbola $r = 4/(1 - 2\cos\theta)$.

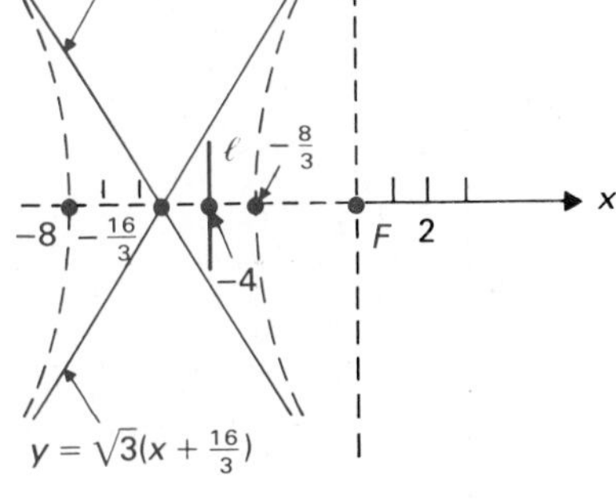

Figure 13.16 Asymptotes of the hyperbola $r = 8/(1 - 2\cos\theta)$.

Solution We see from Eq. (3) that $e = 2$. Then $8 = qe = q \cdot 2$, so $q = 4$. A directrix is $x = -4$, as shown in Fig. 13.16.

We can find where the hyperbola intersects the x-axis by setting $\theta = 0$ and π. We obtain

$$\theta = 0, \quad r = \frac{8}{1 - 2} = -8; \qquad \theta = \pi, \quad r = \frac{8}{1 + 2} = \frac{8}{3}.$$

Thus the hyperbola intersects the x-axis at the (x, y) points $(-8, 0)$ and $(-\frac{8}{3}, 0)$. The center of the hyperbola has x-coordinate

$$\frac{-8 + (-\frac{8}{3})}{2} = \frac{-\frac{32}{3}}{2} = -\frac{16}{3}.$$

Thus

$$a = -\frac{8}{3} - \left(-\frac{16}{3}\right) = \frac{8}{3} \qquad \text{and} \qquad c = \frac{16}{3}.$$

For this hyperbola,

$$b^2 = c^2 - a^2 = \left(\frac{16}{3}\right)^2 - \left(\frac{8}{3}\right)^2 = \left(\frac{8}{3}\right)^2(4 - 1) = \frac{64}{3}.$$

Thus $b = 8/\sqrt{3}$. Since the center of the hyperbola is at $(-\frac{16}{3}, 0)$, the x,y-equations of the asymptotes are

$$y - 0 = \pm\frac{b}{a}\left(x + \frac{16}{3}\right)$$

$$y = \pm\frac{8/\sqrt{3}}{8/3}\left(x + \frac{16}{3}\right) = \pm\sqrt{3}\left(x + \frac{16}{3}\right). \quad \square$$

In the exercises, we ask you to show that the polar equations

$$r = \frac{k}{1 + e\cos\theta}, \qquad r = \frac{k}{1 - e\sin\theta}, \qquad \text{and} \qquad r = \frac{k}{1 + e\sin\theta}$$

also describe conic sections with a focus at the origin. (See Exercises 49 through 51.)

SUMMARY

1. A point with the polar coordinates (r, θ) also has as polar coordinates $(r, \theta + 2n\pi)$ and $(-r, \theta + \pi + 2n\pi)$ for every integer n.
2. The equations for changing from x,y-coordinates to polar r,θ-coordinates and vice versa are

$$x = r\cos\theta, \qquad r^2 = x^2 + y^2,$$

$$y = r\sin\theta, \qquad \theta = \tan^{-1}\frac{y}{x}, \quad -\frac{\pi}{2} < \theta < \frac{\pi}{2}.$$

*3. The polar equation

$$r = \frac{k}{1 - e\cos\theta} \qquad \text{for } k > 0, \quad e > 0,$$

has as locus a conic section of eccentricity e and focus at the pole. The directrix is the line $x = -q$, where $q = k/e$.

EXERCISES

In Exercises 1 through 10, find the cartesian coordinates of the points with the given polar coordinates.

1. $(4, \pi/4)$ **2.** $(0, 5\pi/7)$
3. $(6, -\pi/2)$ **4.** $(3, 5\pi/6)$
5. $(-2, \pi/4)$ **6.** $(-1, 2\pi/3)$
7. $(-4, 3\pi)$ **8.** $(-2, 11\pi/4)$
9. $(0, \pi/12)$ **10.** $(3, -\pi/4)$

In Exercises 11 through 20, find all polar coordinates of the points with the given cartesian coordinates.

11. $(2, 2)$ **12.** $(-1, \sqrt{3})$
13. $(0, 0)$ **14.** $(-3, -3)$
15. $(3, -4)$ **16.** $(-3, 0)$
17. $(-2, -3)$ **18.** $(0, -5)$
19. $(2\sqrt{3}, -2)$ **20.** $(-1, 2)$

In Exercises 21 through 30, find the polar-coordinate equation of the curve with the given x,y-equation.

21. $2x + 3y = 5$ **22.** $y^2 = 8x$
23. $x^2 - y^2 = 1$ **24.** $x^2 + y^2 = 3$
25. $2x^2 + y^2 = 4$ **26.** $x^3 + xy^2 = 3$
27. $x^2y + y^3 = -4$ **28.** $x^2 + y^2 - 8x = 0$
29. $x^2 + y^2 + 4y = 0$ **30.** $y = -x$

In Exercises 31 through 40, find the x,y-equation of the curve with the given polar coordinate equation.

31. $r = 8$ **32.** $r = 2a\cos\theta$
33. $r = -3a\sin\theta$ **34.** $r = 4\csc\theta$
35. $r = -3\sec\theta$ **36.** $r^2 = 2a\sin\theta$
37. $r^2 = -2a\cos\theta$ **38.** $r^3 = \sin\theta$
39. $r^3 = 4\csc\theta$ **40.** $r^4 = 3\sec^2\theta$

41. Show that the distance d in the plane between points having polar coordinates (r_1, θ_1) and (r_2, θ_2) is

$$\sqrt{r_1^2 - 2r_1r_2\cos(\theta_1 - \theta_2) + r_2^2}.$$

[*Hint:* Use the distance formula in cartesian form and Eqs. (1).]

***42.** Find the polar coordinate equation of the parabola with focus at the pole and directrix $x = -3$.

***43.** Find the polar coordinate equation of the hyperbola of eccentricity 3 with focus at the pole and directrix $x = -6$.

***44.** Find the eccentricity and directrix of the conic section with polar coordinate equation $r = 8/(1 - 4\cos\theta)$.

***45.** Repeat Exercise 44 for $r = 5/(2 - 3\cos\theta)$.

***46.** Repeat Exercise 44 for $r = 2/(5 - 2\cos\theta)$.

***47.** Find the length of the major axis of the ellipse with polar equation

$$r = \frac{8}{1 - \frac{1}{2}\cos\theta}.$$

***48.** Find the length of the minor axis of the ellipse in Exercise 47.

***49.** Show that the polar coordinate equation of the conic section of eccentricity e with focus at the pole and directrix $y = -q$ is $r = qe/(1 + e\sin\theta)$.

***50.** Show that the polar coordinate equation of the conic section of eccentricity e with focus at the pole and directrix $x = q$ is $r = qe/(1 + e\cos\theta)$.

***51.** Find the polar coordinate equation of the conic section of eccentricity e, focus at the pole, and directrix $y = q$.

13.2 SKETCHING CURVES IN POLAR COORDINATES

SKETCHING CURVES

We turn now to the problem of sketching a curve given in polar form without changing back to cartesian form. We concern ourselves primarily with equations of the form $r = f(\theta)$, where $f(\theta)$ involves just a sine or cosine function to the first power. We would like to plot points corresponding to local maximum or local minimum values of r, to find where the curve is nearest to or farthest from the origin. The technique is best illustrated by examples.

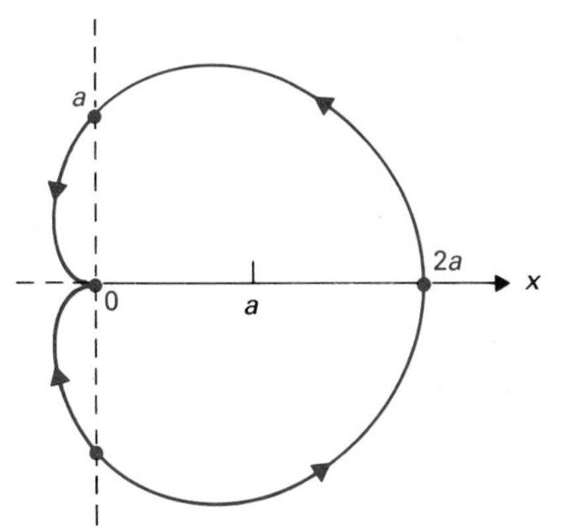

Figure 13.17 The cardioid $r = a(1 + \cos\theta)$.

EXAMPLE 1 Sketch the *cardioid* $r = a(1 + \cos\theta)$.

Solution Note that $r \geq 0$ for all θ since $\cos\theta \geq -1$. Thus we may always measure r along our *positive* rotated polar axis.

Since θ appears in the equation only in "$\cos\theta$," we plot points (r, θ) "every 90°," starting with $\theta = 0$. The "90° lines" are marked with dashes in Fig. 13.17. Of course, our curve $r = a(1 + \cos\theta)$ repeats itself after θ runs through 2π radians.

We find that for $\theta = 0$, we have $r = a(1 + \cos 0) = 2a$; we mark this point on the polar axis. As θ increases to $\pi/2$, $\cos\theta$ decreases from 1 to zero and r decreases from $2a$ to a. This enables us to sketch the first-quadrant portion of our curve, as shown in Fig. 13.17. To establish the actual shape of the curve in the first quadrant, we could plot (r, θ) for a few more values of θ, say $\theta = \pi/6$, $\pi/4$, and $\pi/3$. We will not bother to plot these points but will content ourselves with rough sketches. We will see in Section 13.4 that the curve makes an angle of 45° with the y-axis where it crosses it.

As θ increases from $\pi/2$ to π, we see that $\cos\theta$ decreases from 0 to -1, so r decreases from a to zero; this enables us to sketch the second-quadrant portion of the curve. The curve actually comes into the origin tangent to the negative x-axis. A common error is to draw the curve near the origin as shown in Fig. 13.18. If this were the curve, r would have value zero for $\alpha \leq \theta \leq \beta$, and, of course, $r = 0$ only if $\theta = \pi$.

Figure 13.18 An *incorrect* sketch of $r = a(1 + \cos\theta)$ at the origin: The curve should come in tangent to the negative x-axis there.

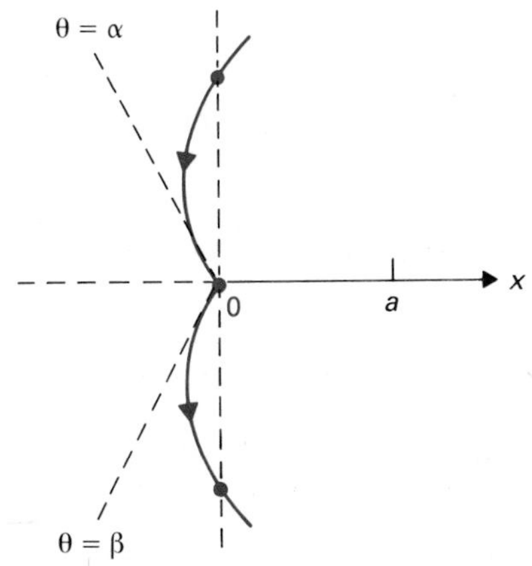

In a similar fashion, we see that r increases from 0 to a through the third quadrant, and from a to $2a$ through the fourth quadrant. The origin of the name *cardioid* for this curve is clear from the shape of the curve. The arrows in Fig. 13.17 indicate the direction of increasing θ along the curve. □

EXAMPLE 2 Sketch the *four-leaved rose* $r = a \sin 2\theta$.

Solution Since $a \sin 2\theta$ assumes both positive and negative values, we see that r can be both positive and negative.

> To help you follow the figures, we draw portions of the curve corresponding to $r \geq 0$ in color and portions corresponding to $r < 0$ in black in this section.

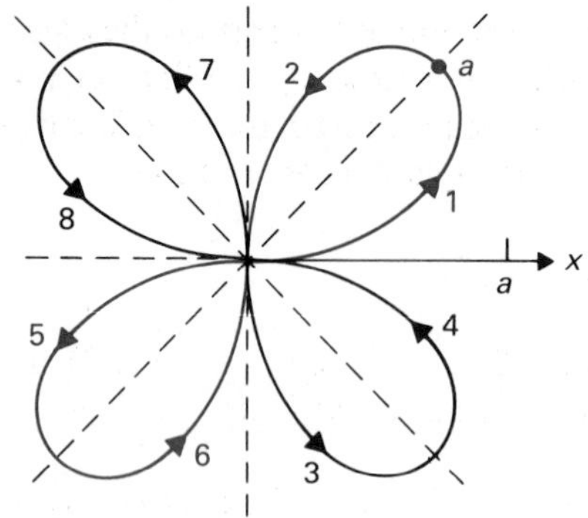

Figure 13.19 The four-leaved rose $r = a \sin 2\theta$.

We are interested in plotting points where r assumes maximum and minimum values, or becomes zero; this occurs every 90° for 2θ, or every 45° for θ. Thus we plot the curve "every 45°" and start by marking the dashed "45° lines" in Fig. 13.19.

As θ increases from 0 to $\pi/4$, we see that r increases from zero to a; this gives the arc of the curve we have numbered 1 in Fig. 13.19. As θ increases from $\pi/4$ to $\pi/2$, r decreases from a to zero, giving the arc numbered 2.

Now as θ increases from $\pi/2$ to $3\pi/4$, we see that r runs through the *negative* values from zero to $-a$, and our curve therefore lies in the "diagonally opposite wedge" and forms the black arc numbered 3 in the figure. We can check that as θ continues to increase up to 2π, we obtain in succession the arcs numbered 4 through 8 for each 45° increment in θ. The arrows on the arcs indicate the direction of increasing θ. □

EXAMPLE 3 Sketch the *spiral of Archimedes*, $r = \theta$.

Solution This time, the curve does not repeat after θ runs through 2π radians, for r increases without bound as θ increases. For $\theta \geq 0$, we have $r \geq 0$, and we obtain the colored spiral in Fig. 13.20. For $\theta < 0$, we have $r < 0$, and we obtain the black spiral in the figure. □

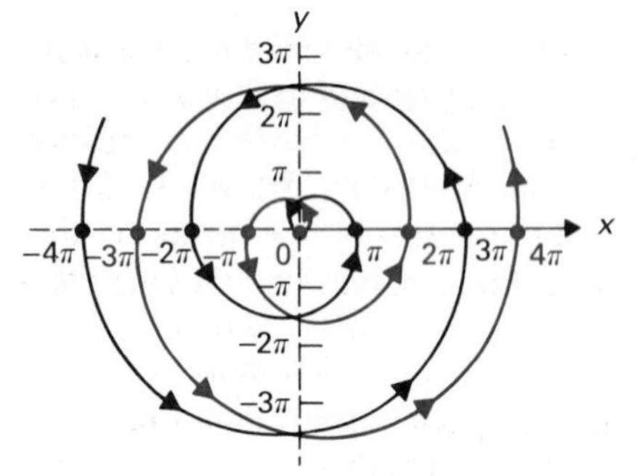

Figure 13.20 The spiral of Archimedes $r = \theta$.

EXAMPLE 4 Sketch the *limaçon* $r = a(1 + 2\cos\theta)$.

Solution We first note that r will be negative when $\cos\theta < -\frac{1}{2}$. A moment's thought shows that $\cos\theta < -\frac{1}{2}$ when $2\pi/3 < \theta < 4\pi/3$, so we sketch the rays $\theta = 2\pi/3$ and $\theta = 4\pi/3$ in Fig. 13.21. We also mark the 90° rays in the figure, since we want to let θ jump 90° at a time.

As θ increases from 0 to $2\pi/3$, we see that r decreases from $3a$ to zero, with value a when $\theta = \pi/2$. The curve approaches the origin tangent to the ray $\theta = 2\pi/3$.

As θ increases from $2\pi/3$ to $4\pi/3$, we see that r runs through negative values, with value $-a$ when $\theta = \pi$. This gives us the black portion of the curve in Fig. 13.21.

Finally, as θ increases from $4\pi/3$ to 2π, we see that r increases from zero to $3a$, with value a when $\theta = 3\pi/2$. This completes the sketch of the curve in Fig. 13.21. □

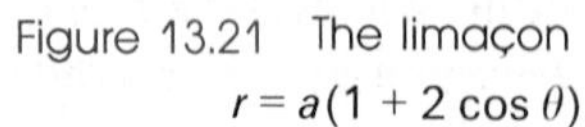

Figure 13.21 The limaçon $r = a(1 + 2\cos\theta)$.

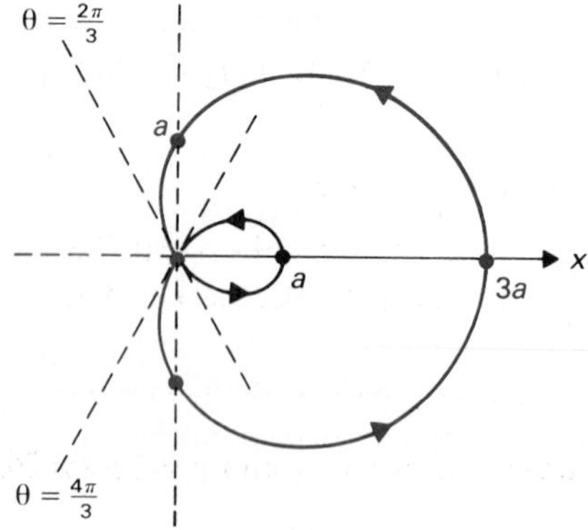

EXAMPLE 5 Sketch the *lemniscate* $r^2 = a^2 \sin 2\theta$.

Solution Since r^2 can't be negative, we note that there will be no curve when $\sin 2\theta < 0$, which occurs for $\pi/2 < \theta < \pi$ and $3\pi/2 < \theta < 2\pi$. Note also that when $\sin 2\theta > 0$, we may have r either negative or positive, for $r = \pm a\sqrt{\sin 2\theta}$. We should sketch this curve "every 45°," since as θ jumps 45°, 2θ jumps 90°. We dash the 45° rays in Fig. 13.22.

We first sketch the part of the curve corresponding to $r \geq 0$, where $r = a\sqrt{\sin 2\theta}$. As θ increases from 0 to $\pi/4$, r increases from zero to a. Then r decreases to zero again at $\theta = \pi/2$. There is no curve for $\pi/2 < \theta < \pi$. As θ increases from π to $5\pi/4$, r increases from zero to a. Then r decreases from a to zero as θ increases from $5\pi/4$ to $3\pi/2$. Finally, there is no curve for $3\pi/2 < \theta < 2\pi$.

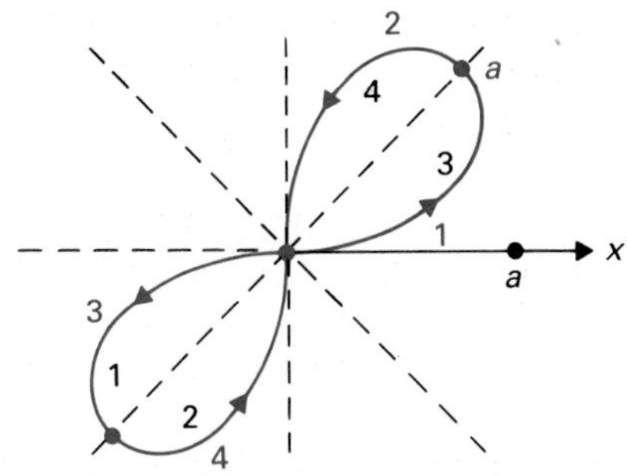

Figure 13.22 The lemniscate $r^2 = a^2 \sin 2\theta$.

For $r = a\sqrt{\sin 2\theta}$, we obtained in succession as θ increased the color-numbered arcs 1, 2, 3, and 4 in Fig. 13.22. We can easily see that for $r = -a\sqrt{\sin 2\theta}$, we obtain in succession the black-numbered arcs 1, 2, 3, and 4 as θ increases. Thus negative values of r do not give any additional curve. □

EXAMPLE 6 Sketch the curve $r = a \sin(\theta/2)$.

Solution We must sketch this curve for $0 \le \theta \le 4\pi$, since $\sin(\theta/2)$ has period 4π. We plot the curve every 180°, because $\theta/2$ jumps 90° when θ jumps 180°.

As θ increases from zero to π, we see that r increases from 0 to a, giving the arc numbered 1 in Fig. 13.23. Then r decreases from a to zero as θ increases from π to 2π giving the arc numbered 2.

For $2\pi < \theta < 4\pi$, we see that r is negative. The arc numbered 3 in Fig. 13.23 is traced as θ increases from 2π to 3π, and the arc numbered 4 is traced as θ increases from 3π to 4π. □

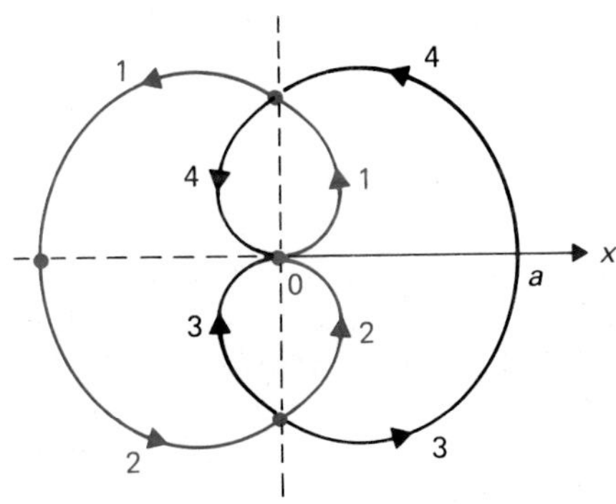

Figure 13.23 The curve $r = a \sin(\theta/2)$.

INTERSECTIONS OF CURVES IN POLAR COORDINATES

The problem of finding intersections of curves in polar coordinates is more complicated than in cartesian coordinates, since a point may have many polar coordinates. For example, the origin lies on both the cardioid $r = a(1 + \cos\theta)$ in Fig. 13.17 and the limaçon $r = a(1 + 2\cos\theta)$ in Fig. 13.21. However, there is no *one* set of coordinates (r, θ) for the origin that satisfies *both* equations, since $(0, \pi)$ is on the cardioid and $(0, 2\pi/3)$ and $(0, 4\pi/3)$ are on the limaçon.

> The origin is a point of intersection of two polar curves if, for each curve separately, r can be zero for some value of θ.

Figure 13.24 shows the points of intersection of the circle $r = a$ and the rose $r = 2a \sin 2\theta$. There are eight points of intersection in all. The points where a colored arc of the rose meets the circle have coordinates (r, θ) for $r > 0$ satisfying both $r = a$ and $r = 2a \sin 2\theta$ simultaneously. However, the points where a black arc of the rose meets the circle have r-coordinate a on the circle and r-coordinate $-a$ on the rose. Thus simultaneous algebraic solution of $r = a$ and $r = 2a \sin 2\theta$ would miss the four points of intersection in quadrants II and IV in Fig. 13.24. Taking these problems into account, we come up with the following rule.

Figure 13.24 The eight points of intersection of the circle $r = a$ and the rose $r = 2a \sin 2\theta$.

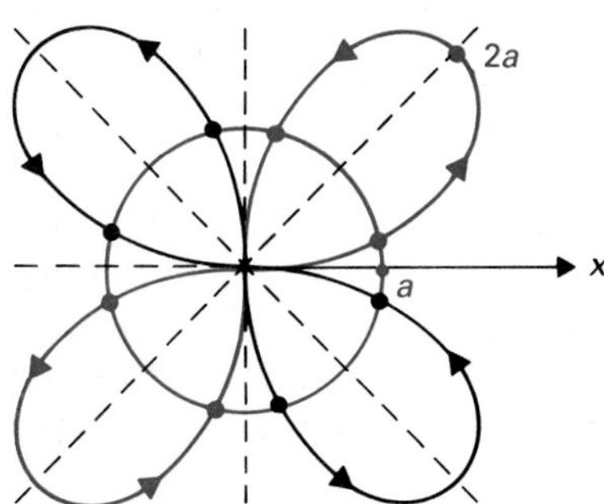

> *To find intersections of polar curves:* Find (r, θ) satisfying the first equation for which some points
>
> $$(r, \theta + 2n\pi) \qquad \text{or} \qquad (-r, \theta + \pi + 2n\pi)$$
>
> satisfy the second equation. Check separately to see if the origin lies on both curves, that is, if r can be zero for both curves. Use a sketch as aid when possible.

EXAMPLE 7 Find the points of intersection of $r = a$ and $r = 2a \sin 2\theta$, shown in Fig. 13.24.

Solution First we solve $a = 2a \sin 2(\theta + 2n\pi)$. Since $\sin(2\theta + 4n\pi) = \sin 2\theta$, we may simplify to

$$a = 2a \sin 2\theta, \qquad \text{or} \quad \sin 2\theta = \tfrac{1}{2}.$$

We want all values of θ from zero to 2π, so we should find all 2θ from zero to 4π such that $\sin 2\theta = \frac{1}{2}$. We have

$$2\theta = \frac{\pi}{6}, \frac{5\pi}{6}, \frac{13\pi}{6}, \frac{17\pi}{6}, \qquad \text{so} \quad \theta = \frac{\pi}{12}, \frac{5\pi}{12}, \frac{13\pi}{12}, \frac{17\pi}{12}.$$

This gives us the points

$$\left(a, \frac{\pi}{12}\right), \quad \left(a, \frac{5\pi}{12}\right), \quad \left(a, \frac{13\pi}{12}\right), \quad \left(a, \frac{17\pi}{12}\right)$$

as the points of intersection in quadrants I and III of Fig. 13.24.

Now we form the equation

$$a = -2a \sin 2(\theta + \pi + 2n\pi).$$

Since $\sin 2(\theta + \pi + 2n\pi) = \sin 2\theta$ again, we obtain

$$a = -2a \sin 2\theta, \qquad \text{or} \quad \sin 2\theta = -\tfrac{1}{2}.$$

This time we have

$$2\theta = \frac{7\pi}{6}, \frac{11\pi}{6}, \frac{19\pi}{6}, \frac{23\pi}{6}, \qquad \text{so} \quad \theta = \frac{7\pi}{12}, \frac{11\pi}{12}, \frac{19\pi}{12}, \frac{23\pi}{12}.$$

These yield the points of intersection

$$\left(a, \frac{7\pi}{12}\right), \quad \left(a, \frac{11\pi}{12}\right), \quad \left(a, \frac{19\pi}{12}\right), \quad \left(a, \frac{23\pi}{12}\right)$$

in quadrants II and IV in Fig. 13.24.

Finally, we check whether the origin is a point of intersection. Since $r \neq 0$ on the circle $r = a$, we see that the origin is not a point of intersection. □

EXAMPLE 8 Find all points of intersection of

$$r = 3 + 2 \sin \theta \qquad \text{and} \qquad r = \cos 2\theta.$$

Solution We first try to find coordinates (r, θ) that satisfy the first equation, while coordinates $(r, \theta + 2n\pi)$ satisfy the second equation. That is, we find solutions of

$$3 + 2 \sin \theta = \cos 2(\theta + 2n\pi).$$

We obtain

$$\begin{aligned} 3 + 2 \sin \theta &= \cos(2\theta + 4n\pi) \\ &= \cos 2\theta = 1 - 2 \sin^2\theta. \end{aligned}$$

This yields

$$\sin^2\theta + \sin \theta + 1 = 0,$$

or, solving by the quadratic formula,

$$\sin \theta = \frac{-1 \pm \sqrt{1 - 4}}{2},$$

which gives no real solutions.

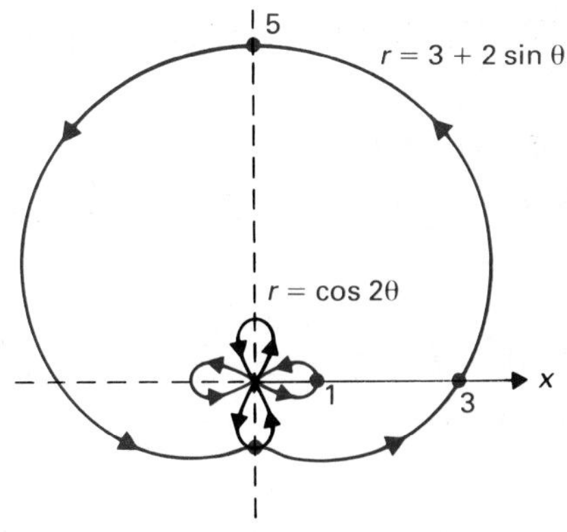

Figure 13.25 The point of intersection of $r = \cos 2\theta$ and $r = 3 + 2 \sin \theta$.

Now we try to find coordinates (r, θ) satisfying the first equation, while coordinates $(-r, \theta + \pi + 2n\pi)$ satisfy the second. We obtain the equation

$$\begin{aligned} 3 + 2 \sin \theta &= -\cos(2\theta + 2\pi + 4n\pi) = -\cos 2\theta \\ &= 2 \sin^2\theta - 1, \end{aligned}$$

which yields

$$\sin^2\theta - \sin \theta - 2 = 0$$

$$(\sin \theta - 2)(\sin \theta + 1) = 0.$$

Since $\sin \theta = 2$ is impossible, we are left with $\sin \theta = -1$, or $\theta = 3\pi/2$. Substituting in $r = 3 + 2 \sin \theta$, we obtain the point $(1, 3\pi/2)$ of intersection.

Finally we check whether the origin lies on both curves. Now $3 + 2 \sin \theta$ is never zero, since $-1 \leq \sin \theta \leq 1$, so the origin does not lie on even the first curve. Thus the point $(1, 3\pi/2)$ is the only point of intersection of the curves, which are sketched in Fig. 13.25. □

SUMMARY

1. When plotting a polar curve $r = f(\theta)$, plot those points where r assumes relative maximum or minimum values or becomes zero. For example, $r = 4 \cos 3\theta$ should be plotted in increments of 30° in θ, starting with $\theta = 0$.
2. To find points of intersection of two polar curves, find (r, θ) satisfying the first equation for which some points $(r, \theta + 2n\pi)$ or $(-r, \theta + \pi + 2n\pi)$ satisfy the second equation. Check separately to see if the origin lies on both curves, that is, if r can be zero. Sketch the curves.

EXERCISES

In Exercises 1 through 30, sketch the curve having the given polar equation.

1. $r = a\theta$ (spiral of Archimedes)

2. $r\theta = a$ (hyperbolic spiral)

3. $r = ae^{\theta}$ (logarithmic spiral)

4. $r = a(1 + \sin \theta)$

5. $r = a(1 - \cos \theta)$

6. $r = 1 + 2 \sin \theta$

7. $r = 3 + 2 \cos \theta$

8. $r = 2 + 3 \cos \theta$

9. $r = a(4 - 2 \sin \theta)$

10. $r = 2a \cos \theta$

11. $r = 2a \sin \theta$

12. $r = 3 \csc \theta$

13. $r = -2 \sec \theta$

14. $r^2 = 9 \sec^2\theta$

15. $r = a \cos 2\theta$

16. $r = a \cos 3\theta$

17. $r = a \sin 3\theta$

18. $r = a \sin 4\theta$

19. $r = a \sin(\theta/2)$

20. $r = a \cos(\theta/2)$

21. $r = a \cos(3\theta/2)$

22. $r = 4$

23. $r^2 = a^2 \cos 2\theta$

24. $r^2 = a^2 \sin 3\theta$

25. $r^2 = a^2 \cos 3\theta$

26. $r = a(1 + \cos(\theta/2))$

27. $r = a(1 + \cos 2\theta)$

28. $r = a(1 - \sin 3\theta)$

29. $r = a(2 + \cos 3\theta)$

30. $r = a(2 + |\sin 3\theta|)$

31. Let $r = 2 - \csc \theta$.
a) Show that $y = 2/(2 - r) - 1$. [*Hint*: Write $\csc \theta = 1/\sin \theta$, and remember that $r \sin \theta = y$.]
b) Show that the line $y = -1$ is a horizontal asymptote of the curve by computing $\lim_{r \to \infty} y$ from part (a).
c) Sketch the curve.

32. Following the idea of Exercise 31, sketch the polar curve $r = \sqrt{2} + \sec \theta$.

33. Let $r = a(1 + \cos \theta)$.
a) Recall that $x = r \cos \theta$ and $y = r \sin \theta$. Express x and y in terms of θ only.
b) Regarding your answer to part (a) as parametric equations of a curve, find the slope of the tangent line to the cardioid where $\theta = \pi/6$.

34. As in Exercise 33, find the slope of the polar curve $r = a \cos \theta$ where $\theta = \pi/3$.

35. Following the idea of Exercise 33, find all points where the cardioid $r = a(1 + \cos\theta)$ has a vertical tangent.

36. Following the idea of Exercise 33, find all points where the cardioid $r = a(1 + \cos\theta)$ has a horizontal tangent.

In Exercises 37 through 46, find all points of intersection of the given curves.

37. $r = \theta/\pi$ and $r = \frac{1}{4}$

38. $\theta = \pi/6$ and $r = 2$

39. $r = a\sin\theta$ and $r = a\cos\theta$

40. $r = 2a\sin\theta$ and $r = a$

41. $r = a$ and $r^2 = 2a^2\sin 2\theta$

42. $r = a\sin 2\theta$ and $r = a\cos\theta$

43. $r = a$ and $r = 2a\cos 2\theta$

44. $r = 1 - \cos\theta$ and $r = 1 + 2\cos\theta$

45. $r = \cos(\theta/2)$ and $r = \sqrt{3}/2$

46. $r = a\cos 2\theta$ and $r = a(1 + \cos\theta)$

13.3 AREA IN POLAR COORDINATES

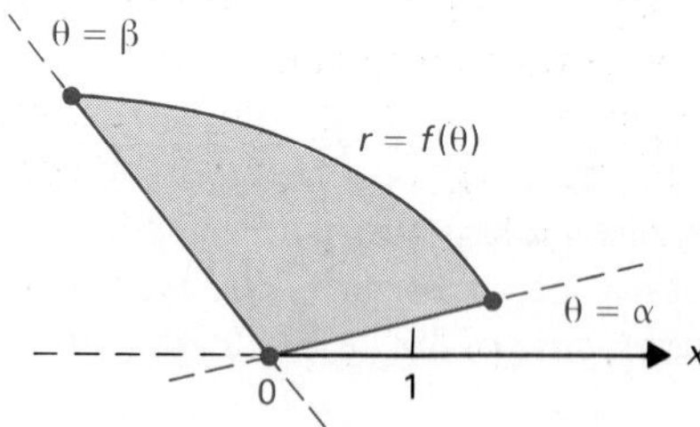

Figure 13.26 The region bounded by $r = f(\theta)$ and the rays $\theta = \alpha$, $\theta = \beta$.

Let $r = f(\theta)$ be a continuous function of θ. We want to find the area A of a region bounded by the polar curve $r = f(\theta)$ and by two rays $\theta = \alpha$ and $\theta = \beta$, as shown in Fig. 13.26. Our treatment parallels the development of the integral in Chapter 6.

Assume $r \geq 0$ for θ in $[\alpha, \beta]$. We divide the interval $[\alpha, \beta]$ into n equal subintervals of size $d\theta = (\beta - \alpha)/n$, as shown in Fig. 13.27 for $n = 8$. Let M_i be the maximum value of r for θ in the ith subinterval, as indicated in Fig. 13.28. In that figure, we have shaded in color a wedge of constant radius M_i and central angle $d\theta$. This wedge has area equal to the sector of the circle shown in Fig. 13.29. The area of this sector is the fraction $d\theta/(2\pi)$ of the area of the entire circle, so it has area

$$\frac{d\theta}{2\pi} \cdot \pi M_i^2 = \frac{1}{2} M_i^2 \, d\theta.$$

For the area in Fig. 13.26, we see immediately that

$$\text{Area} \leq \sum_{i=1}^{n} \frac{1}{2} M_i^2 \, d\theta = \frac{\beta - \alpha}{n} \cdot \sum_{i=1}^{n} \frac{1}{2} M_i^2. \tag{1}$$

In exactly the same fashion, we let m_i be the minimum value of r for θ in the ith subinterval, as indicated in Fig. 13.30. We then find that

$$\frac{\beta - \alpha}{n} \cdot \sum_{i=1}^{n} \frac{1}{2} m_i^2 \leq \text{Area}. \tag{2}$$

Figure 13.27 Subdivision of $[\alpha, \beta]$ into $n = 8$ equal subintervals of length $d\theta$.

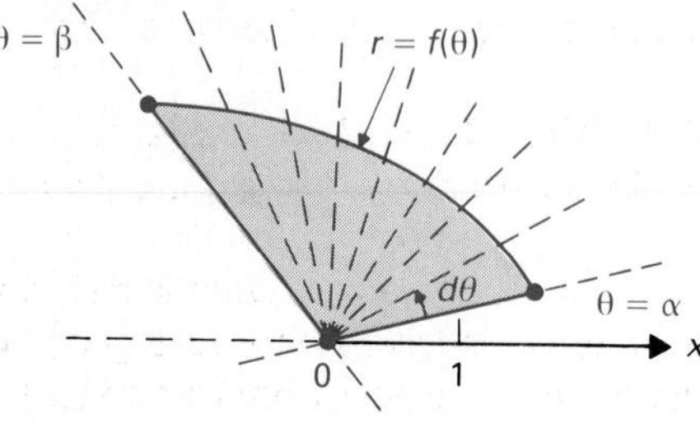

Figure 13.28 Wedge of constant radius M_i and central angle $d\theta$.

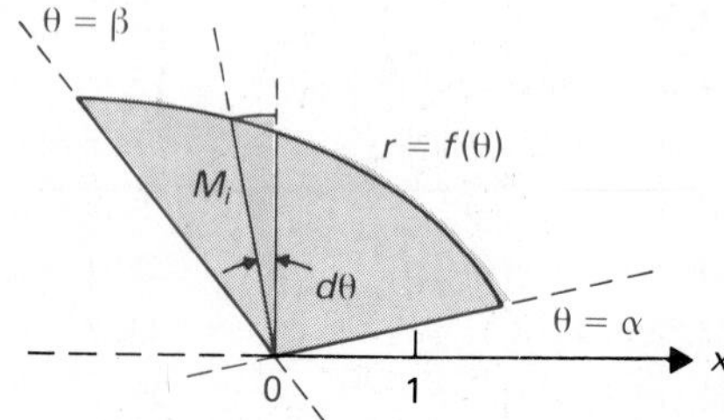

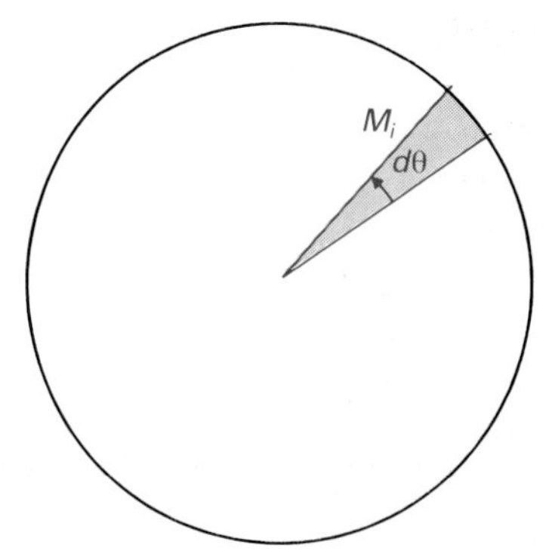

Figure 13.29 The area of the sector is $(d\theta/2\pi)\cdot\pi M_i^2 = \frac{1}{2}M_i^2\,d\theta$.

We recognize the sums in relations (1) and (2) as the upper and lower sums for the function $\frac{1}{2}(f(\theta))^2$. As $n \to \infty$, both of these sums approach $\int_\alpha^\beta \frac{1}{2}(f(\theta))^2\,d\theta$. Thus relations (1) and (2) show that

$$\text{Area} = \int_\alpha^\beta \frac{1}{2}r^2\,d\theta = \int_\alpha^\beta \frac{1}{2}(f(\theta))^2\,d\theta, \tag{3}$$

at least for the case where $r \geq 0$. In case the curve in Fig. 13.26 actually corresponds to negative values of r, we would let M_i and m_i be the maximum and minimum values, respectively, of $|r| = |f(\theta)|$. Since $|r|^2 = r^2$, we again obtain the formula in formula (3).

EXAMPLE 1 Find the area of the region bounded by the cardioid $r = a(1 + \cos\theta)$, shown in Fig. 13.31.

Solution We think of the differential wedge shaded in color in Fig. 13.31 as having area $\frac{1}{2}r^2\,d\theta$. This is the way we remember formula (3).

Since the curve is traced completely as θ varies from zero to 2π, we want to add the areas of these wedges with an integral as θ goes from zero to 2π. Expressing r in terms of θ, we obtain from formula (3)

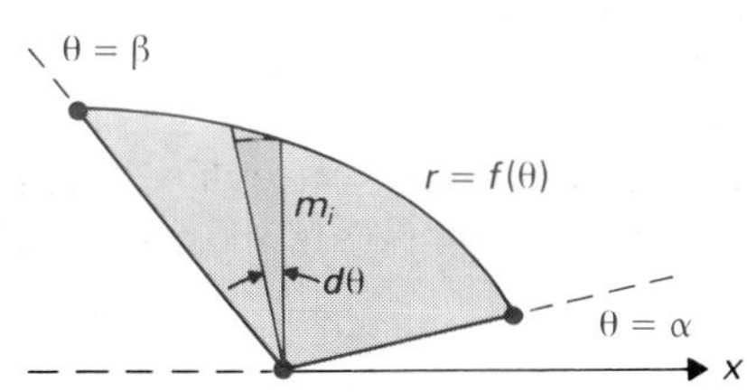

Figure 13.30 Wedge of constant radius m_i and central angle $d\theta$.

$$\begin{aligned}
\text{Area} &= \int_0^{2\pi} \frac{1}{2}r^2\,d\theta = \int_0^{2\pi}\frac{1}{2}[a(1+\cos\theta)]^2\,d\theta \\
&= \frac{1}{2}a^2\int_0^{2\pi}(1 + 2\cos\theta + \cos^2\theta)\,d\theta \\
&= \frac{1}{2}a^2\int_0^{2\pi}\left[1 + 2\cos\theta + \frac{1}{2}(1+\cos 2\theta)\right]d\theta \\
&= \frac{1}{2}a^2\int_0^{2\pi}\left(\frac{3}{2} + 2\cos\theta + \frac{1}{2}\cos 2\theta\right)d\theta \\
&= \frac{1}{2}a^2\left(\frac{3}{2}\theta + 2\sin\theta + \frac{1}{4}\sin 2\theta\right)\Bigg]_0^{2\pi} \\
&= \frac{1}{2}a^2\left(\frac{3}{2}\cdot 2\pi\right) - \frac{1}{2}a^2\cdot 0 = \frac{3}{2}\pi a^2. \quad \square
\end{aligned}$$

A common mistake is to always integrate from 0 to 2π when finding the area of a region using polar coordinates.

Figure 13.31 Differential wedge of area $\frac{1}{2}r^2\,d\theta$.

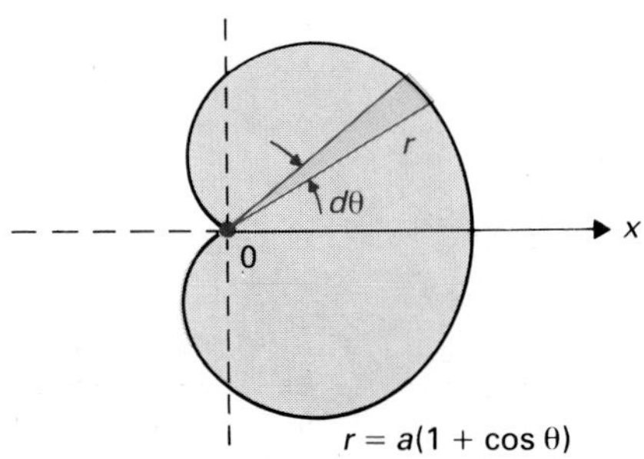

> Think carefully just what interval for θ is needed for the differential wedges to sweep out the region just once.

The following example illustrates the problem.

EXAMPLE 2 Find the area of the region bounded by the circle $r = 2a\cos\theta$, shown in Fig. 13.32, by integration in polar coordinates.

Solution We know the area is πa^2, since the circle has radius a. Now the entire circle is traced once as θ goes from zero to π. (The circle would be traced *twice* as θ goes from zero to 2π.) Thus to get the area of the circle, we

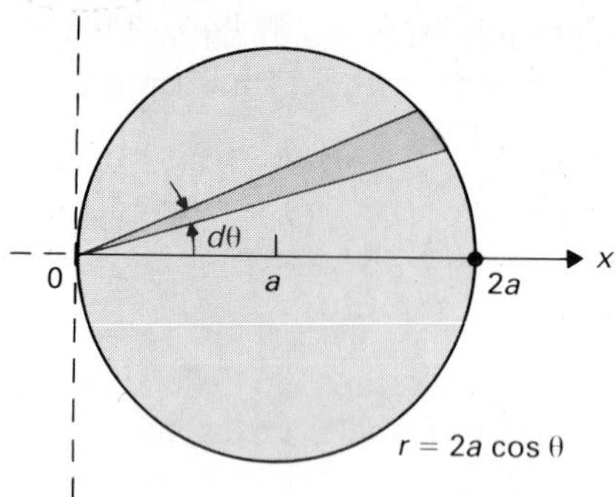

Figure 13.32 Differential wedge of area $\frac{1}{2}r^2\,d\theta$.

may integrate as θ goes from zero to π. We have

$$\begin{aligned}\text{Area} &= \int_0^{\pi} \frac{1}{2}(2a\cos\theta)^2\,d\theta = 2a^2\int_0^{\pi}\cos^2\theta\,d\theta \\ &= 2a^2\int_0^{\pi}\frac{1}{2}(1+\cos 2\theta)\,d\theta = a^2\left(\theta + \frac{1}{2}\sin 2\theta\right)\Big]_0^{\pi} \\ &= a^2\pi - 0 = \pi a^2.\end{aligned}$$

Clearly, if we had integrated blindly from zero to 2π, we would have obtained the incorrect result $2\pi a^2$. □

Symmetry is often useful, as illustrated in the next example.

EXAMPLE 3 Find the total area of the regions bounded by the lemniscate $r^2 = a^2\cos 2\theta$ shown in Fig. 13.33.

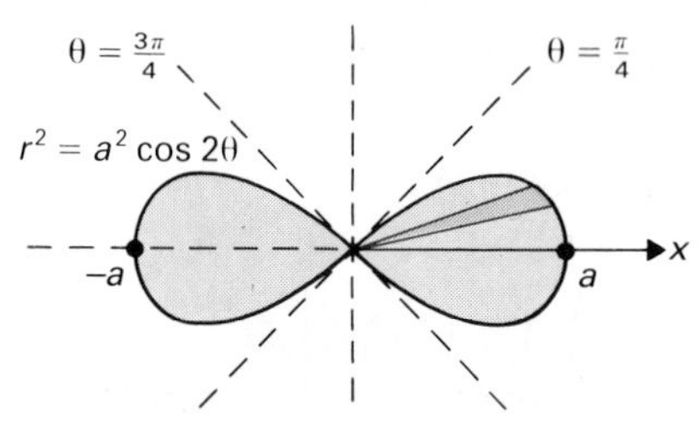

Figure 13.33 Differential wedge of area $\frac{1}{2}r^2\,d\theta$.

Solution By symmetry, we may find the area of the first-quadrant portion of the region, and multiply by 4. We thus obtain from relation (1)

$$\begin{aligned}4\int_0^{\pi/4}\frac{1}{2}r^2\,d\theta &= 4\int_0^{\pi/4}\frac{1}{2}a^2\cos 2\theta\,d\theta = 2a^2\int_0^{\pi/4}\cos 2\theta\,d\theta \\ &= 2a^2\,\frac{1}{2}\sin 2\theta\Big]_0^{\pi/4} = a^2\left(\sin\frac{\pi}{2}\right) - a^2(\sin 0) = a^2\end{aligned}$$

as our desired area. Note that r is not defined for $\pi/4 < \theta < 3\pi/4$ and $5\pi/4 < \theta < 7\pi/4$. Blindly integrating from 0 to 2π would give the incorrect answer zero. □

EXAMPLE 4 Find the area of the region that is inside the cardioid $r = a(1+\cos\theta)$ but outside the circle $r = a$, shown shaded in Fig. 13.34.

Solution We may double the area of the first-quadrant portion of the region, which is traced out as θ varies from zero to $\pi/2$. The area of the dark partial wedge of central angle $d\theta$ shown in Fig. 13.34 is approximately

$$\frac{1}{2}(r_{\text{cardioid}})^2\,d\theta - \frac{1}{2}(r_{\text{circle}})^2\,d\theta,$$

and we obtain

Figure 13.34 Differential region of area $\frac{1}{2}(r_{\text{cardioid}})^2\,d\theta - \frac{1}{2}(r_{\text{circle}})^2\,d\theta$.

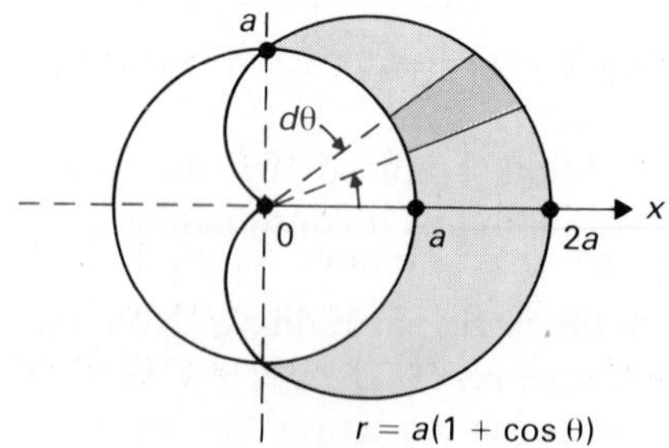

$$\begin{aligned}2\cdot\frac{1}{2}\int_0^{\pi/2}&[a^2(1+\cos\theta)^2 - a^2]\,d\theta \\ &= \int_0^{\pi/2}(a^2 + 2a^2\cos\theta + a^2\cos^2\theta - a^2)\,d\theta \\ &= a^2\int_0^{\pi/2}(2\cos\theta + \cos^2\theta)\,d\theta \\ &= a^2\left(2\sin\theta + \frac{\theta}{2} + \frac{\sin 2\theta}{4}\right)\Big]_0^{\pi/2} \\ &= a^2\left(2 + \frac{\pi}{4}\right) - a^2(0) = 2a^2 + \frac{\pi a^2}{4}\end{aligned}$$

as the desired area. □

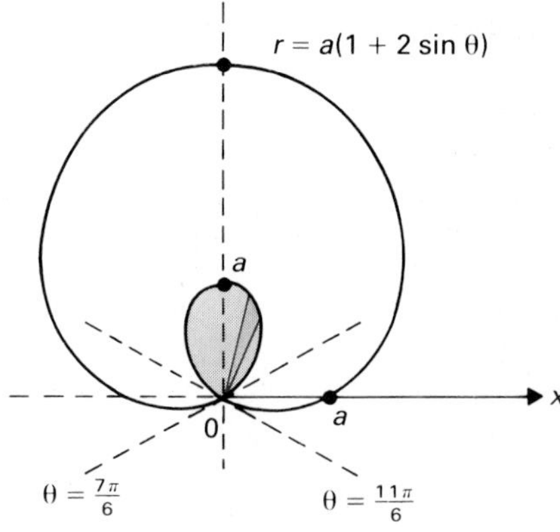

Figure 13.35 Differential wedge of area $\frac{1}{2}r^2\,d\theta$.

EXAMPLE 5 Find the area of the region bounded by the inner loop of the limaçon $r = a(1 + 2\sin\theta)$, shown in Fig. 13.35.

Solution We may double the area inside the first-quadrant part of the loop, swept out as θ goes from $7\pi/6$ to $3\pi/2$, or we can simply integrate from $7\pi/6$ to $11\pi/6$. Since we can evaluate trigonometric functions more easily at $3\pi/2$ than at $11\pi/6$, we select the first alternative. We have

$$\begin{aligned}
\text{Area} &= 2\int_{7\pi/6}^{3\pi/2} \frac{1}{2}[a(1 + 2\sin\theta)]^2\,d\theta \\
&= a^2\int_{7\pi/6}^{3\pi/2} (1 + 4\sin\theta + 4\sin^2\theta)\,d\theta \\
&= a^2\int_{7\pi/6}^{3\pi/2} [1 + 4\sin\theta + 2(1 - \cos 2\theta)]\,d\theta \\
&= a^2\int_{7\pi/6}^{3\pi/2} (3 + 4\sin\theta - 2\cos 2\theta)\,d\theta \\
&= a^2(3\theta - 4\cos\theta - \sin 2\theta)\Big]_{7\pi/6}^{3\pi/2} \\
&= a^2\left(\frac{9\pi}{2} - 0 - 0\right) - a^2\left[\frac{7\pi}{2} - 4\left(-\frac{\sqrt{3}}{2}\right) - \frac{\sqrt{3}}{2}\right] \\
&= a^2\left(\pi - \frac{3\sqrt{3}}{2}\right). \quad \square
\end{aligned}$$

SUMMARY

To find the area of a region bounded by polar curves:

Step 1 Draw a figure.

Step 2 Draw polar rays corresponding to a small increment $d\theta$ in θ.

Step 3 Write down the area of the resulting wedge. A wedge with vertex at the origin, small central angle $d\theta$, and going out to $r = f(\theta)$ has approximate area $dA = \frac{1}{2}r^2\,d\theta = \frac{1}{2}f(\theta)^2\,d\theta$.

Step 4 Integrate between the appropriate limits.

EXERCISES

In Exercises 1 through 21, find the area of the indicated region, using integration in polar coordinates.

1. The region bounded by $r = a$ (Does this problem constitute a "proof" that the area of a disk of radius a is πa^2? Why?)
2. The region bounded by the spiral $r = a\theta$ for $\pi/2 \le \theta \le 3\pi/2$ and the rays $\theta = \pi/2$ and $\theta = 3\pi/2$
3. The region bounded by the spiral $r = ae^\theta$ for $0 \le \theta \le \pi$ and the x-axis
4. The region bounded by $r\theta = 1$ for $\pi/2 \le \theta \le \pi$ and the rays $\theta = \pi/2$ and $\theta = \pi$
5. The region inside the cardioid $r = a(1 + \sin\theta)$
6. The region bounded by $r = 3 + 2\cos\theta$
7. A region bounded by one leaf of the rose $r = a\cos 2\theta$
8. A region bounded by one leaf of the rose $r = a\sin 4\theta$
9. A region inside one whole loop, from (0, 0) back to (0, 0), of $r = a(\cos 3\theta/2)$

10. The total region inside $r^2 = a^2 \sin 2\theta$
11. The total region inside $r^2 = a^2 \sin 3\theta$
12. The region inside both the circles $r = 2a \cos \theta$ and $r = 2a \sin \theta$
13. The total region inside the rose $r = 2a \cos 2\theta$ but outside the circle $r = a$
14. The region inside both the cardioid $r = a(1 + \cos \theta)$ and $r = a$
15. The region between the loops of the limaçon

$$r = a(1 - 2 \cos \theta)$$

16. The region in the upper half-plane having a portion of $r = \sin(\theta/2)$ as its total boundary
17. The region in the right half-plane having a portion of $r = a \sin (\theta/2)$ as its total boundary
18. The smaller region bounded by $r = 4$ and $r = 2 \sec \theta$
19. The region bounded by $r = a(4 - \sin^2\theta)$
20. The total region bounded by $r = a|\cos \theta|$
21. The region inside $r = a(3 + |\sin 3\theta|)$

In Exercises 22 through 26, use a calculator and Simpson's rule to estimate the area of the region.

22. The region inside the ellipse $r = 3/(2 - \cos \theta)$
23. The region inside the ellipse $r = 6/(5 + \sin \theta)$
24. The region inside $r = 2 + \cos (\theta^2/2\pi)$, $0 \leq \theta \leq 2\pi$
25. The region bounded by $r = 10/(2 + \cos^2\theta)$
26. The region bounded by $r = (4 + \sin^2\theta)/(2 + \cos^4\theta)$

13.4 THE TANGENT LINE AND ARC LENGTH

THE SLOPE OF THE TANGENT LINE

Let $r = f(\theta)$, where f is differentiable. We would like to find the slope dy/dx of the tangent line to the graph at a point where y appears as a differentiable function of x.

Recall that

$$x = r \cos \theta, \qquad y = r \sin \theta.$$

Substituting $f(\theta)$ for r, we have

$$\boxed{\begin{aligned} x &= f(\theta) \cdot \cos \theta, \\ y &= f(\theta) \cdot \sin \theta, \end{aligned}} \tag{1}$$

Figure 13.36 Some tangent lines to the cardioid $r = a(1 + \cos \theta)$.

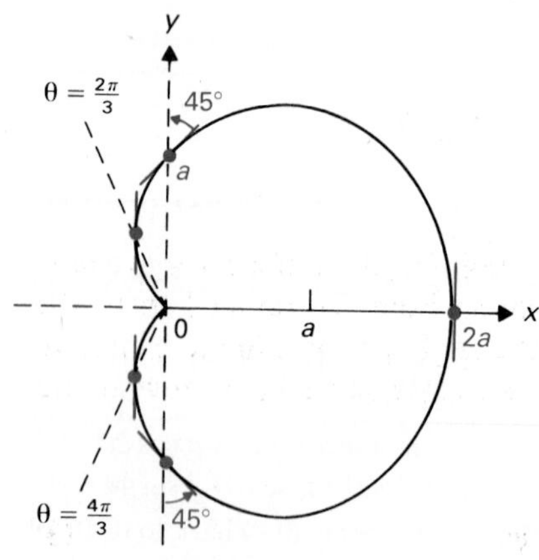

Equations (1) can be regarded as parametric equations for the curve $r = f(\theta)$, where θ is the parameter. From Section 12.5, we see that

$$\boxed{\frac{dy}{dx} = \frac{dy/d\theta}{dx/d\theta},} \tag{2}$$

where $dx/d\theta \neq 0$. We actually anticipated Equations (1) and (2) in Exercises 33 through 36 of Section 13.2.

EXAMPLE 1 Show that the cardioid $r = a(1 + \cos \theta)$ in Fig. 13.36 makes an angle of 45° with the y-axis where it crosses it.

Solution We want to show that the tangent lines to the cardioid at the points $(r, \theta) = (a, \pi/2)$ and $(a, 3\pi/2)$ have slope ± 1. Equations (1) become

$$x = a(1 + \cos\theta)(\cos\theta) = a(\cos\theta + \cos^2\theta),$$
$$y = a(1 + \cos\theta)(\sin\theta) = a(\sin\theta + \sin\theta\cos\theta).$$

From Eq. (2), we have

$$\frac{dy}{dx} = \frac{dy/d\theta}{dx/d\theta} = \frac{a(\cos\theta - \sin^2\theta + \cos^2\theta)}{a(-\sin\theta - 2\sin\theta\cos\theta)}.$$

Thus

$$\left.\frac{dy}{dx}\right|_{\theta=\pi/2} = \frac{a(0 - 1 + 0)}{a(-1 - 0)} = 1$$

and

$$\left.\frac{dy}{dx}\right|_{\theta=3\pi/2} = \frac{a(0 - 1 + 0)}{a(1 - 0)} = -1. \quad \square$$

EXAMPLE 2 Find all points where the tangent line to the cardioid $r = a(1 + \cos\theta)$ is vertical.

Solution From Eq. (2), we see that the tangent line is vertical at any point where $dx/d\theta = 0$ and $dy/d\theta \neq 0$. From Example 1, we see that

$$\frac{dx}{d\theta} = a(-\sin\theta - 2\sin\theta\cos\theta) = a(\sin\theta)(-1 - 2\cos\theta) = 0$$

when $\theta = 0, \pi, 2\pi/3$, and $4\pi/3$. We easily see that $dy/d\theta \neq 0$ at $\theta = 0, 2\pi/3$, and $4\pi/3$. However, $dy/d\theta = 0$ when $\theta = \pi$, so this point requires further examination. From Example 1 and l'Hôpital's rule, we obtain

$$\left.\frac{dy}{dx}\right|_{\theta=\pi} = \lim_{\theta\to\pi} \frac{\cos\theta - \sin^2\theta + \cos^2\theta}{-\sin\theta - 2\sin\theta\cos\theta}$$

$$= \lim_{\theta\to\pi} \frac{-\sin\theta - 2\sin\theta\cos\theta - 2\cos\theta\sin\theta}{-\cos\theta + 2\sin^2\theta - 2\cos^2\theta} = \frac{0}{-1} = 0.$$

Thus the tangent line at $(r, \theta) = (0, \pi)$ is horizontal and not vertical. The tangent lines at $(r, \theta) = (2a, 0)$, $(a/2, 2\pi/3)$, and $(a/2, 4\pi/3)$ are vertical, as shown in Fig. 13.36. $\square$

THE ANGLE ψ FROM THE RADIAL LINE TO THE TANGENT LINE

Figure 13.37 $\phi = \theta + \psi$.

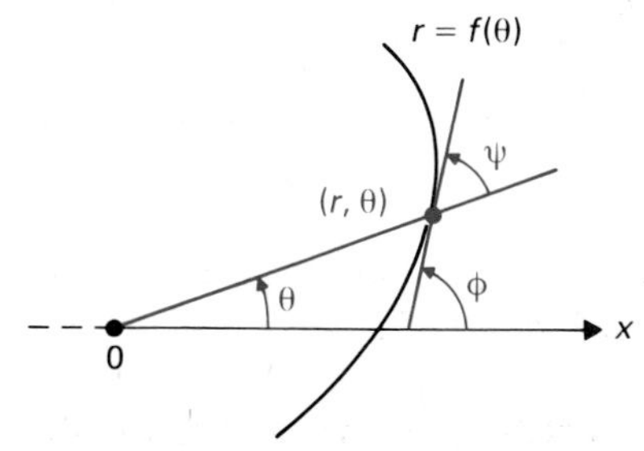

Let $r = f(\theta)$ where f is differentiable. At a point (r, θ) on the curve, we now find the angle ψ measured counterclockwise from the radial line to the tangent line, as shown in Fig. 13.37. For the angle ϕ shown in the figure, we have $\tan\phi = dy/dx$. We just saw how to find dy/dx, so we know how to find $\tan\phi$ and therefore ϕ.

By plane geometry, we know that

$$\phi = \theta + \psi. \tag{3}$$

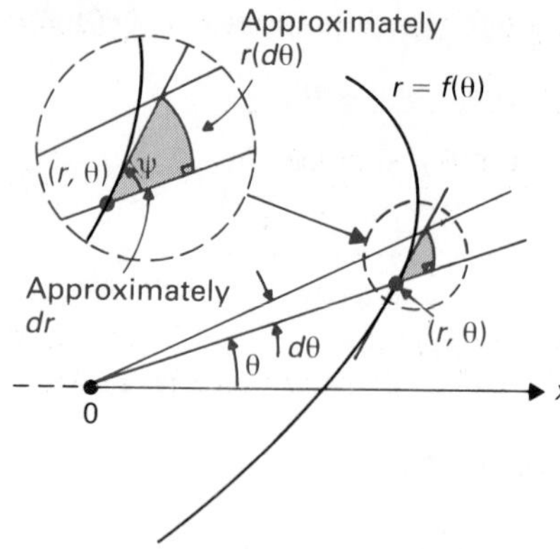

Figure 13.38 The differential triangle of polar coordinates.

We will show that

$$\boxed{\tan \psi = \frac{r}{dr/d\theta} = \frac{f(\theta)}{f'(\theta)},} \tag{4}$$

if $f'(\theta) \neq 0$. Figure 13.38 indicates how formula (4) can be remembered. The figure shows the *differential right triangle* for polar coordinates. This triangle has as hypotenuse the segment of the tangent line at (r, θ) between the radial lines corresponding to θ and $\theta + d\theta$. The side of the triangle opposite the angle ψ is really a short arc of a circle, but we pretend it is a straight line of approximate length $r\, d\theta$. For small $d\theta$, the legs of the "right triangle" are approximately of lengths dr and $r\, d\theta$, which at once suggests that

$$\tan \psi = \frac{r\, d\theta}{dr} = \frac{r}{dr/d\theta}.$$

For a careful derivation of formula (4), note from Eq. (3) that

$$\tan \psi = \tan (\phi - \theta) = \frac{\tan \phi - \tan \theta}{1 + \tan \phi \tan \theta}. \tag{5}$$

Now $\tan \phi = dy/dx$, and from the parametric equations

$$x = r \cos \theta = f(\theta) \cos \theta, \qquad y = r \sin \theta = f(\theta) \sin \theta,$$

we obtain

$$\tan \phi = \frac{dy}{dx} = \frac{dy/d\theta}{dx/d\theta} = \frac{r \cos \theta + (dr/d\theta) \sin \theta}{-r \sin \theta + (dr/d\theta) \cos \theta}. \tag{6}$$

Substituting in Eq. (5) the value for $\tan \phi$ found in Eq. (6), and putting $\tan \theta = (\sin \theta)/(\cos \theta)$, we obtain a compound quotient that we can easily show yields

$$\begin{aligned} \tan \psi &= \frac{r \cos^2\theta + (dr/d\theta) \sin \theta \cos \theta + r \sin^2\theta - (dr/d\theta) \sin \theta \cos \theta}{-r \sin \theta \cos \theta + (dr/d\theta) \cos^2\theta + r \sin \theta \cos \theta + (dr/d\theta) \sin^2\theta} \\ &= \frac{r}{dr/d\theta}. \end{aligned}$$

Figure 13.39 $\beta = \pi - \psi$.

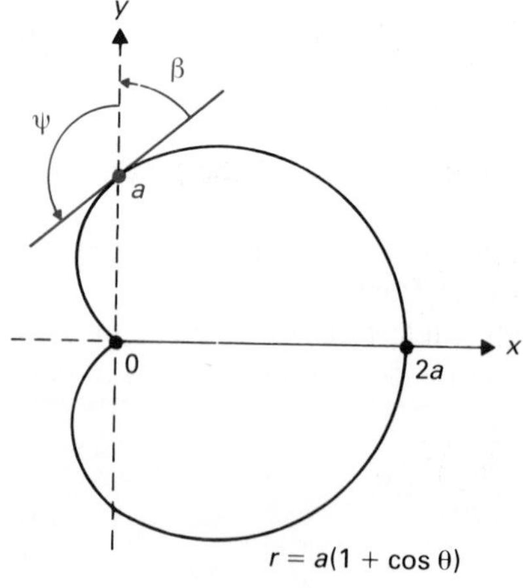

EXAMPLE 3 Show again that the cardioid $r = a(1 + \cos \theta)$ meets the y-axis at 45° by finding the acute angle β shown in Fig. 13.39.

Solution From Fig. 13.39, we see that we can reduce our problem to finding the angle ψ, for the angle β shown in the figure is then given by $\beta = \pi - \psi$.

We have

$$\tan \psi = \left.\frac{r}{dr/d\theta}\right|_{\theta=\pi/2} = \left.\frac{a(1 + \cos \theta)}{-a \sin \theta}\right|_{\pi/2} = \frac{a}{-a} = -1.$$

Thus $\psi = 3\pi/4$, and $\beta = \pi - 3\pi/4 = \pi/4$. □

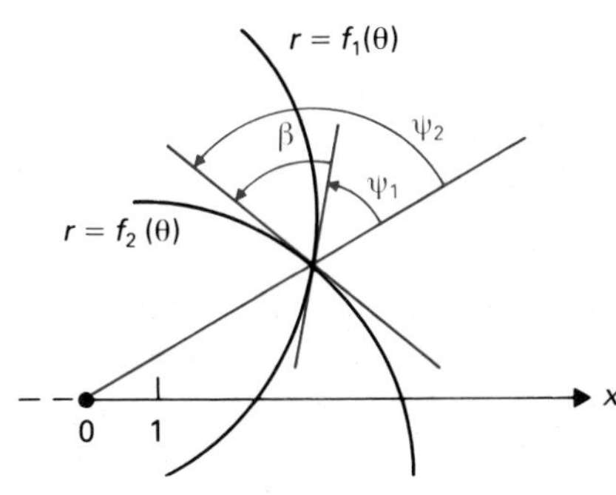

Figure 13.40
$\beta = \psi_2 - \psi_1$.

From Fig. 13.40, we see that the angle β between curves $r = f_1(\theta)$ and $r = f_2(\theta)$ can be computed by finding

$$\tan \beta = \tan(\psi_2 - \psi_1) = \frac{\tan \psi_2 - \tan \psi_1}{1 + \tan \psi_2 \tan \psi_1}. \tag{7}$$

EXAMPLE 4 Find the angle β between the cardioid $r = 2 - 2\cos\theta$ and the curve $r = 2 + \cos\theta$ at the point $(r, \theta) = (2, \pi/2)$, shown in Fig. 13.41.

Solution From Fig. 13.41 and formula (4), we have

$$\tan \psi_1 = \left.\frac{r}{dr/d\theta}\right|_{\pi/2} = \left.\frac{2 - 2\cos\theta}{2\sin\theta}\right|_{\pi/2} = \frac{2}{2} = 1$$

and

$$\tan \psi_2 = \left.\frac{r}{dr/d\theta}\right|_{\pi/2} = \left.\frac{2 + \cos\theta}{-\sin\theta}\right|_{\pi/2} = \frac{2}{-1} = -2.$$

From Eq. (7), we have

$$\tan \beta = \frac{-2 - 1}{1 + (-2)(1)} = \frac{-3}{-1} = 3,$$

so $\beta = \tan^{-1} 3 \approx 71.565°$. □

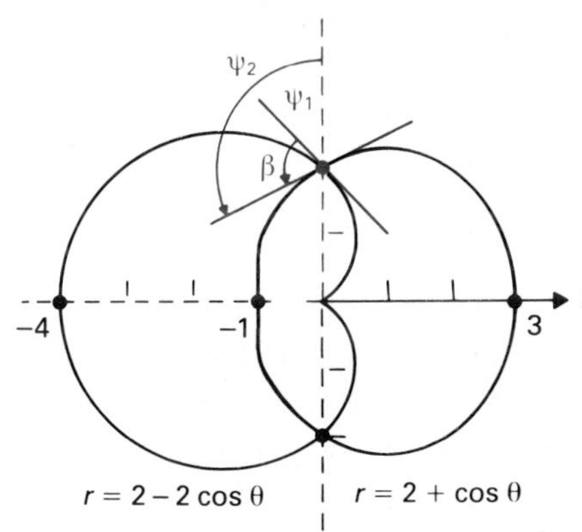

Figure 13.41 The angle $\beta = \psi_2 - \psi_1$ of intersection of the curves.

ARC LENGTH IN POLAR COORDINATES

In Fig. 13.42, we shade again the differential right triangle shown in Fig. 13.38. From this triangle, we obtain the estimate

$$ds = \sqrt{(dr)^2 + (r\,d\theta)^2} = \sqrt{\left(\frac{dr}{d\theta}\right)^2 + r^2}\,d\theta \tag{8}$$

for the length of the tangent-line segment to the curve. To estimate arc length, we may add lengths of tangent-line segments, and from Eq. (8), we expect the arc length of a continuously differentiable polar curve $r = f(\theta)$ from (r_1, θ_1) to (r_2, θ_2) to be given by

$$\text{Arc length} = \int_{\theta_1}^{\theta_2} \sqrt{\left(\frac{dr}{d\theta}\right)^2 + r^2}\quad d\theta. \tag{9}$$

Figure 13.42 From the differential triangle, $ds = \sqrt{(dr)^2 + (r\,d\theta)^2}$.

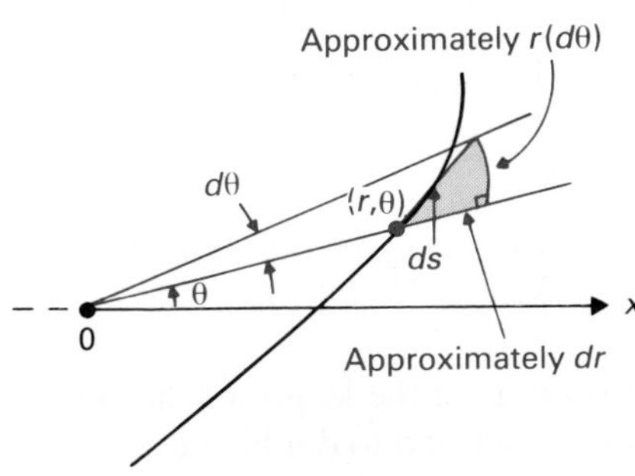

For a careful derivation of Eq. (8), simply note that the polar curve $r = f(\theta)$ is defined parametrically by

$$x = r\cos\theta = f(\theta)\cos\theta,$$

$$y = r\sin\theta = f(\theta)\sin\theta,$$

and use the parametric formula

$$ds = \sqrt{\left(\frac{dx}{d\theta}\right)^2 + \left(\frac{dy}{d\theta}\right)^2}\,d\theta.$$

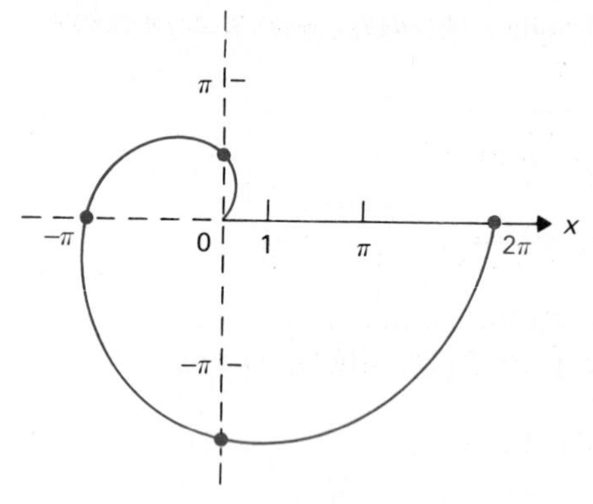

Figure 13.43 The spiral $r = \theta$ for $0 \le \theta \le 2\pi$.

We have

$$\frac{dx}{d\theta} = -f(\theta)\sin\theta + f'(\theta)\cos\theta,$$

$$\frac{dy}{d\theta} = f(\theta)\cos\theta + f'(\theta)\sin\theta,$$

and we obtain

$$\begin{aligned} ds &= \sqrt{\left(\frac{dx}{d\theta}\right)^2 + \left(\frac{dy}{d\theta}\right)^2}\, d\theta \\ &= \sqrt{(f(\theta))^2(\sin^2\theta + \cos^2\theta) + (f'(\theta))^2(\sin^2\theta + \cos^2\theta)}\, d\theta \\ &= \sqrt{(f(\theta))^2 + (f'(\theta))^2}\, d\theta \\ &= \sqrt{r^2 + \left(\frac{dr}{d\theta}\right)^2}\, d\theta. \end{aligned}$$

EXAMPLE 5 Find the length of the spiral $r = \theta$ shown in Fig. 13.43 from $\theta = 0$ to $\theta = 2\pi$.

Solution The arc length is given by the integral

$$\begin{aligned} \int_0^{2\pi} \sqrt{r^2 + \left(\frac{dr}{d\theta}\right)^2}\, d\theta &= \int_0^{2\pi} \sqrt{\theta^2 + 1}\, d\theta \\ &= \left(\frac{\theta}{2}\sqrt{1 + \theta^2} + \frac{1}{2}\ln\left(\theta + \sqrt{1 + \theta^2}\right)\right)\Bigg]_0^{2\pi} \\ &= \pi\sqrt{1 + 4\pi^2} + \frac{1}{2}\ln\left(2\pi + \sqrt{1 + 4\pi^2}\right). \quad \square \end{aligned}$$

EXAMPLE 6 Find the area of the surface generated when the cardioid $r = a(1 + \cos\theta)$ shown in Fig. 13.36 is revolved about the x-axis.

Solution The surface area is given by

$$\begin{aligned} \int_0^{\pi} 2\pi y\, ds &= \int_0^{\pi} 2\pi(r\sin\theta)\sqrt{r^2 + \left(\frac{dr}{d\theta}\right)^2}\, d\theta \\ &= \int_0^{\pi} 2\pi a(1 + \cos\theta)(\sin\theta)\sqrt{a^2(1 + \cos\theta)^2 + a^2\sin^2\theta}\, d\theta \\ &= 2\pi a^2 \int_0^{\pi} (1 + \cos\theta)(\sin\theta)\sqrt{2 + 2\cos\theta}\, d\theta \\ &= 2\sqrt{2}\pi a^2 \int_0^{\pi} (1 + \cos\theta)^{3/2}(\sin\theta)\, d\theta \\ &= -2\sqrt{2}\pi a^2 \frac{2}{5}(1 + \cos\theta)^{5/2}\Bigg]_0^{\pi} = \frac{32}{5}\pi a^2. \quad \square \end{aligned}$$

Figure 13.44 The curve $r = a(2 - \cos\theta)$.

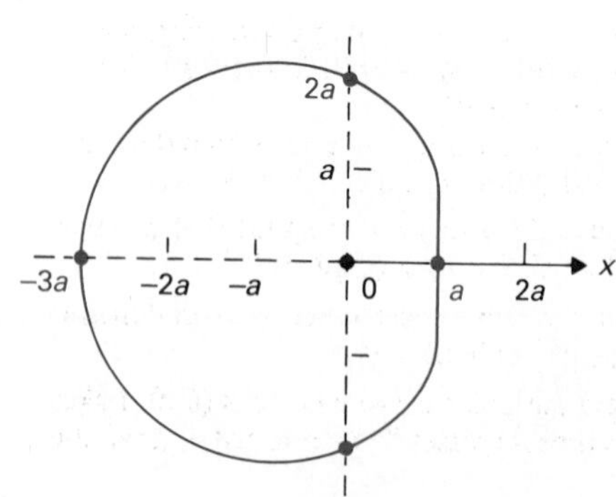

EXAMPLE 7 Use a calculator and Simpson's rule to estimate the arc length of the curve $r = a(2 - \cos\theta)$ in Fig. 13.44.

Solution It is probably more accurate to estimate the length of the curve for $0 \le \theta \le \pi$ using a value for n in Simpson's rule and to double the answer than to estimate the integral from zero to 2π using the same n. Now $dr/d\theta = a \sin \theta$, so the length is given by

$$2 \int_0^{\pi} \sqrt{a^2 \sin^2\theta + a^2(2 - \cos \theta)^2}\, d\theta$$

$$= 2a \int_0^{\pi} \sqrt{\sin^2\theta + 4 - 4 \cos \theta + \cos^2\theta}\, d\theta$$

$$= 2a \int_0^{\pi} \sqrt{5 - 4 \cos \theta}\, d\theta \approx (13.36489)a. \quad \square$$

SUMMARY

1. If $r = f(\theta)$, then $x = f(\theta) \cos \theta$ and $y = f(\theta) \sin \theta$. Also,

$$\frac{dy}{dx} = \frac{dy/d\theta}{dx/d\theta}.$$

2. If $r = f(\theta)$ is differentiable and ψ is the angle from the radius vector to the tangent of the polar curve $r = f(\theta)$, then

$$\tan \psi = \frac{r}{dr/d\theta} = \frac{f(\theta)}{f'(\theta)}.$$

3. The angle β between polar curves $r = f_1(\theta)$ and $r = f_2(\theta)$ at a point of intersection is given by

$$\tan \beta = \frac{\tan \psi_2 - \tan \psi_2}{1 + \tan \psi_1 \tan \psi_2}.$$

4. If $r = f(\theta)$ is continuously differentiable, then the arc length of the polar curve from θ_1 to θ_2 is

$$s = \int_{\theta_1}^{\theta_2} \sqrt{\left(\frac{dr}{d\theta}\right)^2 + r^2}\, d\theta.$$

EXERCISES

1. Find the slope of the circle $r = a \sin \theta$, where $\theta = \pi/3$.
2. Find the slope of the limaçon $r = 1 + 2 \cos \theta$, where $\theta = 5\pi/6$.
3. Find the slope of the spiral $r = \theta$, where $\theta = \pi/2$.
4. Find the slope of the curve $r = 1 + \cos(\theta/2)$, where $\theta = \pi/2$.
5. Find all points where the circle $r = a \cos \theta$ has slope 1.
6. Find all points where the limaçon $r = 1 + 2 \sin \theta$ has a horizontal tangent.
7. Find all points where the limaçon $r = 1 + 2 \sin \theta$ has a vertical tangent. [*Hint:* Express $dx/d\theta$ in terms of $\sin \theta$ and solve $dx/d\theta = 0$ by the quadratic formula.]
8. The cardioid $r = a + a \cos \theta$ has a sharp point when $\theta = \pi$.
   a) Show that the curve $r = a + b \cos \theta$, with $a > b > 0$, has no sharp point where $\theta = \pi$.
   b) Find the values of b, where $0 < b < a$, for which the curve $r = a + b \cos \theta$ still has a "dimple" where $\theta = \pi$.
9. Find the acute angle that the hyperbolic spiral $r\theta = a$ makes with the y-axis at the point $(r, \theta) = (2a/\pi, \pi/2)$.
10. Find the angle that the polar curve $r = 2 + 3 \sin \theta$ makes with the x-axis at the point $(r, \theta) = (2, 0)$.
11. Find the acute angle the polar curve $r = a \cos(\theta/2)$ makes with the y-axis each time it crosses it at a point other than the origin.

12. Let ψ be the angle at (r, θ) on the spiral $r = a\theta$, as described in the text. Find $\lim_{\theta \to \infty} \psi$.
13. Show that the spiral $r = ae^{\theta}$ makes a constant angle with the radial line at each point on the spiral, and find the angle.
14. Find the angle between the circles $r = 2a \cos \theta$ and $r = 2a \sin \theta$ at the point $(r, \theta) = (a\sqrt{2}, \pi/4)$ of intersection.
15. Find the angle between the circle $r = a$ and the four-leaved rose $r = 2a \cos 2\theta$ at the point of intersection $(r, \theta) = (a, \pi/6)$.
16. Find the acute angle between the cardioid $r = a(1 + \cos \theta)$ and the limaçon $r = a(1 + 2 \cos \theta)$ at the point $(a, \pi/2)$.
17. Find the acute angle at which the polar curve $r = a \sin (\theta/2)$ meets itself corresponding to $\theta = \pi/2$ and $\theta = 7\pi/2$.
18. Find the length of the parabolic spiral $r = a\theta^2$ from $\theta = 0$ to $\theta = 2\pi$.
19. Find the total length of the cardioid $r = a(1 + \sin \theta)$. [*Hint:* Evaluate the integral by multiplying the integrand by $\sqrt{2 - 2 \sin \theta}/\sqrt{2 - 2 \sin \theta}$.]
20. Express as an integral the length of the polar curve $r = a \cos (\theta/2)$ from $\theta = 0$ to $\theta = \pi$.
21. Express as an integral the total length of the three-leaved rose $r = a \sin 3\theta$.
22. Find the area of the surface generated when the circle $r = 2a \sin \theta$ is revolved about the x-axis.
23. Find the area of the surface generated when the cardioid $r = 2 + 2 \sin \theta$ is revolved about the y-axis.
24. Express as an integral the area of the surface generated when the arc of the spiral $r = \theta$ from $\theta = 0$ to $\theta = \pi$ is revolved about the y-axis.
25. Let f be a twice-differentiable function. Show that the curvature κ of the curve $r = f(\theta)$ at a point (r, θ) is given by the formula

$$\kappa = \left| \frac{(f(\theta))^2 + 2(f'(\theta))^2 - f(\theta)f''(\theta)}{[(f(\theta))^2 + (f'(\theta))^2]^{3/2}} \right|.$$

26. Use the formula in Exercise 25 to find the curvature of the cardioid $r = a(1 + \cos \theta)$, where $\theta = 0$.
27. Use the formula in Exercise 25 to find the radius of curvature of the limaçon $r = a(1 + 2 \sin \theta)$, where $\theta = 3\pi/2$.

Use a calculator and Simpson's rule in Exercises 28 through 32.

28. Find the arc length in Exercise 20.
29. Find the arc length in Exercise 21.
30. Find the surface area in Exercise 24.
31. Find the area of the surface generated when the polar curve $r = 3 + 2 \sin \theta$ is revolved about the line $y = -2$.
32. Find the area of the surface generated when the limaçon $r = 1 + 2 \cos \theta$ is revolved about the line $x = -1$.

EXERCISE SETS FOR CHAPTER 13

Review Exercise Set 13.1

1. Find *all* polar coordinates of the point $(-\sqrt{3}, 1)$.
2. Find x,y-coordinates of the point $(r, \theta) = (-5, \pi/3)$.
3. Find the polar coordinate equation of the ellipse $4x^2 + 9y^2 = 1$.
4. Find the x,y-equation of the polar curve $r = \sin \theta + \cos \theta$.
5. Sketch the curve with polar coordinate equation $r = a(1 + 2 \sin \theta)$.
6. Find all points of intersection of $r^2 = a^2 \sin \theta$ and $r = a/\sqrt{2}$.
7. Find the area inside $r = a(1 + \frac{1}{2} \sin \theta)$ and outside the circle $r = a$.
8. Find the slope of the circle $r = 2a \sin \theta$, where $\theta = 2\pi/3$.
9. Find the angle between $r^2 = a^2 \sin \theta$ and $r = a/\sqrt{2}$ at their first-quadrant point of intersection.
10. Find the length of arc of the spiral $r = e^{2\theta}$ from $\theta = 0$ to $\theta = 2\pi$.

Review Exercise Set 13.2

1. Find *all* polar coordinates of the point $(1, -1)$.
2. Find the polar coordinate equation of the hyperbola $x^2 - y^2 + 4x = 9$.
3. Find the x,y-equation of the polar curve $r^2 = 2 + \sin 2\theta$.

*4. Find the eccentricity and directrix of the conic section $r = 3/(2 - 5 \cos \theta)$.

5. Sketch the curve with polar coordinate equation $r = a \sin 2\theta$.
6. Find all points of intersection of $r^2 = a^2 \cos 2\theta$ and $r = a/\sqrt{2}$.
7. Find the total area enclosed by the polar curve $r^2 = 1 + \sin^2\theta$.
8. Find the points where the curve $r = 1 - \cos \theta$ has a horizontal tangent line.
9. Find the angle between the cardioids $r = a(1 + \cos \theta)$ and $r = -a(1 + \cos \theta)$ at their point of intersection in the upper half-plane.
10. Find the length of arc of the curve $r = a \cos^2(\theta/2)$ from $\theta = 0$ to $\theta = \pi/2$.

Review Exercise Set 13.3

1. Find *all* polar coordinates of the point $(x, y) = (-4, -4\sqrt{3})$.
2. Find x,y-coordinates of the point $(r, \theta) = (3, 3\pi/4)$.
3. Find the x,y-equation of the polar curve $r^2 = \cos 2\theta$.
4. Sketch the curve with polar coordinate equation $r = -3 \sec \theta$.
5. Sketch the curve with polar coordinate equation $r = 1 + \cos (\theta/2)$.

6. Find all points of intersection of $r = 2a \sin \theta$ and $r = a$.

7. Find the total area of the region inside the curve $r^2 = 1 + \cos 2\theta$.

8. Find the slope of the curve $r = a \cos (\pi/2)$, where $\theta = \pi/3$.

9. Find the acute angle that the lemniscate $r^2 = 4 \sin 2\theta$ makes with the ray $\theta = \pi/6$.

10. Use a calculator to estimate the total arc length of the curve $r = 5 \sin 4\theta$.

Review Exercise Set 13.4

1. Find *all* polar coordinates for the point $(x, y) = (0, 0)$.

2. Find x,y-coordinates of the point $(r, \theta) = (0, 3\pi/17)$.

3. Find the polar coordinate equation of the line $x - 3y = 5$.

*4. Find the length of the minor axis of the ellipse

$$r = 8/(3 - \cos \theta).$$

5. Sketch the polar curve $r = 4 - 3|\cos \theta|$.

6. Sketch the polar curve $r = 1 + \cos 3\theta$.

7. Find the total area of the regions enclosed by the curve $r = a|\sin \theta|$.

8. Find the slope of the tangent line to the tip of the leaf of the rose $r = a \sin 3\theta$ lying in the first quadrant.

9. Use a calculator to estimate the total arc length of the curve $r = 2 + \sin 2\theta$.

10. Use a calculator to estimate the area of the surface generated when the cardioid $r = 1 + \cos \theta$ is revolved about the line $y = 2$.

More Challenging Exercise 13

1. Fly A is located at $(x, y) = (1, 1)$, while fly B is at $(-1, 1)$, fly C is at $(-1, -1)$, and fly D is at $(1, -1)$. The flies all crawl at the same rate of 1 unit distance per unit time. The flies all start crawling at the same instant, with fly A always crawling directly toward B, B directly toward C, C directly toward D, and D directly toward A.
   a) Find the point at which the flies meet.
   b) How long do the flies crawl before they meet?
   c) Find the polar coordinate equation of the path traveled by fly A. Sketch the path. [*Hint:* Find the angle ψ at a point on this path, and solve the differential equation $r = (\tan \psi)\, dr/d\theta$.]
   d) Find the length of the path traveled by fly A.
   e) What physiological problem will fly A encounter as it travels the path found in part (c) in the time found in part (b)?

14 SPACE GEOMETRY AND VECTORS

This chapter begins our introduction to analytic geometry in space. Sections 14.3 through 14.5 present the notion of a *vector* and develop some *vector algebra*. Chapter 14 plays a similar role in the study of multivariable calculus to the role Chapter 1 played in studying single-variable calculus. The main difference is our introduction here of vector methods. It is not essential to use vector methods for multivariable calculus, but they are very neat and convenient. Vector methods started appearing in the undergraduate calculus sequence in the 1940s, but they were known many years before that. They could have been used perfectly well in Chapter 1, in which case Chapter 16 would seem a much more natural extension of the single-variable calculus. However, it is traditional to introduce them at this point rather than ask the beginning calculus student to be concerned with them. They have a long history of use in physics and engineering.

14.1 COORDINATES IN SPACE

RECTANGULAR COORDINATES

We know how to describe the location of a point in the plane using an ordered pair (x, y) of real numbers. The location of a point in space can be described using an ordered triple (x, y, z) of numbers. We set up a *rectangular* (or *cartesian*) system of coordinates as follows. Select some point of space as *origin* and imagine three coordinate axes, any two of which are perpendicular, through this point. Figure 14.1 shows only half of each of these x-, y-, and z-axes for clarity. The plane containing the x- and y-coordinate axes is the *x, y-coordinate plane*, and the x, z-coordinate plane and y, z-coordinate plane are similarly defined.

The three coordinate planes naturally divide space into eight parts or *octants* according to whether the coordinates are positive or negative.

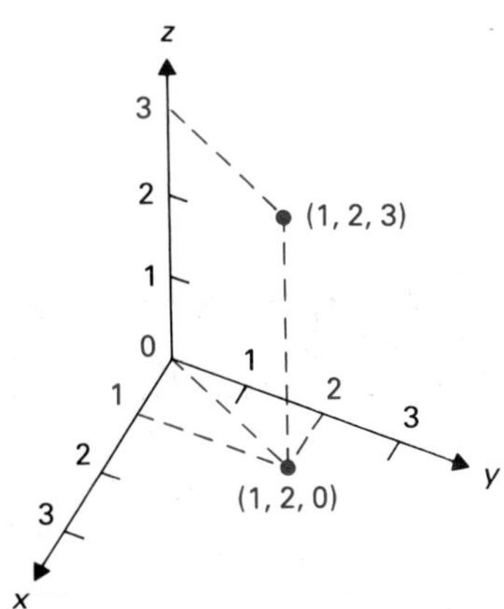

Figure 14.1 The cartesian point (1, 2, 3) in space.

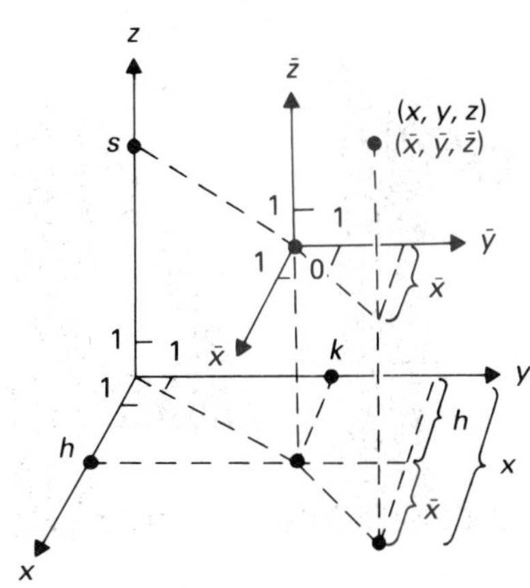

Figure 14.2 Axes translated to (h, k, s); $\bar{x} = x - h$, $\bar{y} = y - k$, $\bar{z} = z - s$.

Symbolically, the possible combinations of coordinate signs are

$$(+,+,+),\quad (+,+,-),\quad (+,-,+),\quad (+,-,-),$$
$$(-,+,+),\quad (-,+,-),\quad (-,-,+),\quad (-,-,-).$$

The portion where all coordinates are positive, that is, the $(+, +, +)$ part, is called the *first octant*. We have never seen any attempt to number the other octants.

We can use translated $\bar{x},\bar{y},\bar{z}$-coordinates in space, just as we used $\bar{x},\bar{y}$-coordinates in the plane. Figure 14.2 shows translated axes at a new origin (h, k, s). The equations

$$\boxed{\bar{x} = x - h, \qquad \bar{y} = y - k, \qquad \bar{z} = z - s} \qquad (1)$$

for transforming coordinates are natural extensions of the formulas we are familiar with for translation in the plane. We simply have one more equation to handle the third coordinate.

Figure 14.3 The distance from (x, y, z) to $(0, 0, 0)$ is $\sqrt{x^2 + y^2 + z^2}$.

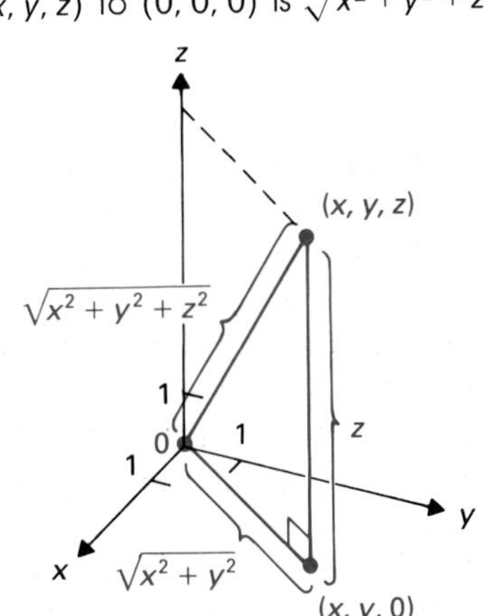

From Fig. 14.3, it is clear that the distance from the origin to the point (x, y, z) is $\sqrt{x^2 + y^2 + z^2}$. We can compute the distance from (x_1, y_1, z_1) to a point (x_2, y_2, z_2). At (x_1, y_1, z_1), we take translated $\bar{x},\bar{y},\bar{z}$-coordinates as shown in Fig. 14.4. From Eqs. (1), the point $(x, y, z) = (x_2, y_2, z_2)$ has translated coordinates $(\bar{x}, \bar{y}, \bar{z}) = (x_2 - x_1, y_2 - y_1, z_2 - z_1)$. The distance from this point to the new origin is then $\sqrt{\bar{x}^2 + \bar{y}^2 + \bar{z}^2}$, so we see that the distance d between (x_1, y_1, z_1) and (x_2, y_2, z_2) is given by

$$\boxed{d = \sqrt{(x_2 - x_1)^2 + (y_2 - y_1)^2 + (z_2 - z_1)^2}.} \qquad (2)$$

This is an easily remembered extension of the formula for the distance between points (x_1, y_1) and (x_2, y_2) in the plane.

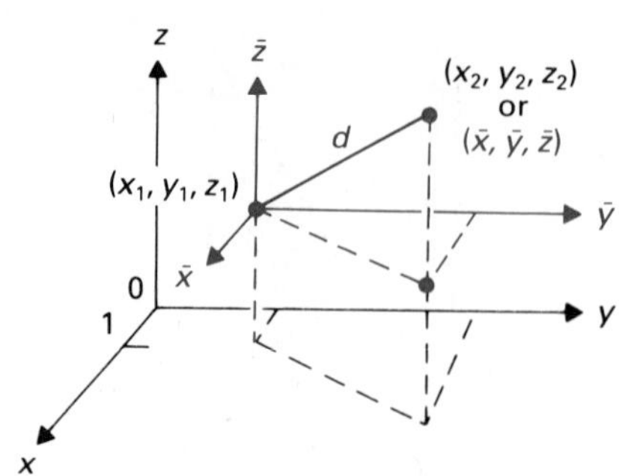

Figure 14.4
$d = \sqrt{\bar{x}^2 + \bar{y} + \bar{z}^2}$
$= \sqrt{(x_2 - x_1)^2 + (y_2 - y_1)^2 + (z_2 - z_1)^2}$

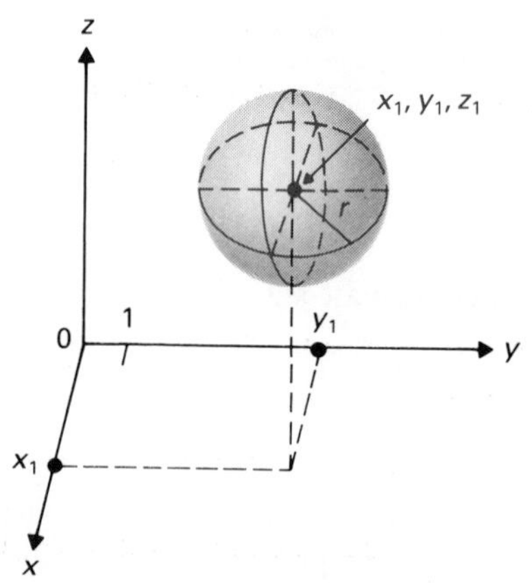

Figure 14.5 Sphere with center (x_1, y_1, z_1) and radius r.

EXAMPLE 1 Find the distance between $(-1, 3, 2)$ and $(1, 1, 3)$.

Solution By Eq. (2), the distance is

$$\sqrt{[1-(-1)]^2 + (1-3)^2 + (3-2)^2} = \sqrt{2^2 + (-2)^2 + 1^2}$$
$$= \sqrt{9} = 3. \quad \square$$

The set of all points (x, y, z) a fixed distance r from a point (x_1, y_1, z_1) is a *sphere of radius r with center at* (x_1, y_1, z_1), as shown in Fig. 14.5. From the distance formula, we see that (x, y, z) lies on this sphere if and only if

$$\sqrt{(x - x_1)^2 + (y - y_1)^2 + (z - z_1)^2} = r,$$

or

$$\boxed{(x - x_1)^2 + (y - y_1)^2 + (z - z_1)^2 = r^2.} \qquad \textbf{(3)}$$

Equation (2) is thus the equation of a sphere. By completing the square, we easily see that any equation

$$x^2 + y^2 + z^2 + ax + by + cz = d$$

describes a sphere, if the equation has a nonempty solution set in space.

EXAMPLE 2 Find the center and radius of the sphere

$$x^2 + y^2 + z^2 - 6x + 4y = -9$$

and then sketch the sphere.

Figure 14.6 The sphere $(x-3)^2 + (y+2)^2 + z^2 = 4$.

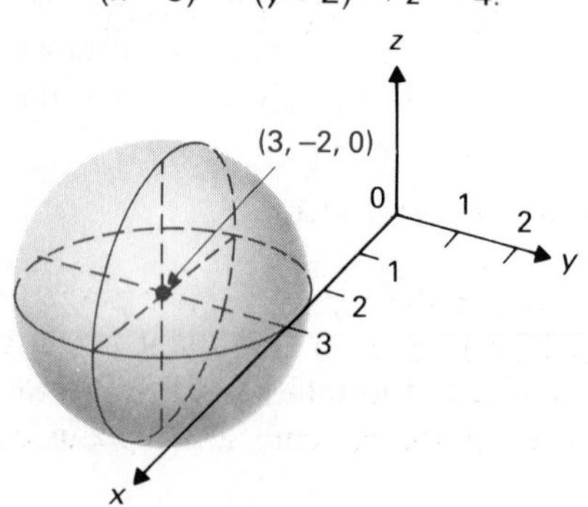

Solution The steps for completing the square are

$$(x^2 - 6x) + (y^2 + 4y) + z^2 = -9,$$
$$(x-3)^2 + (y+2)^2 + (z-0)^2 = -9 + 9 + 4 = 4.$$

Thus the sphere has center $(3, -2, 0)$ and radius 2. Its sketch is shown in Fig. 14.6. $\square$

EXAMPLE 3 Find the solution set of the equation

$$x^2 + y^2 + z^2 + 2x - 4z = -6.$$

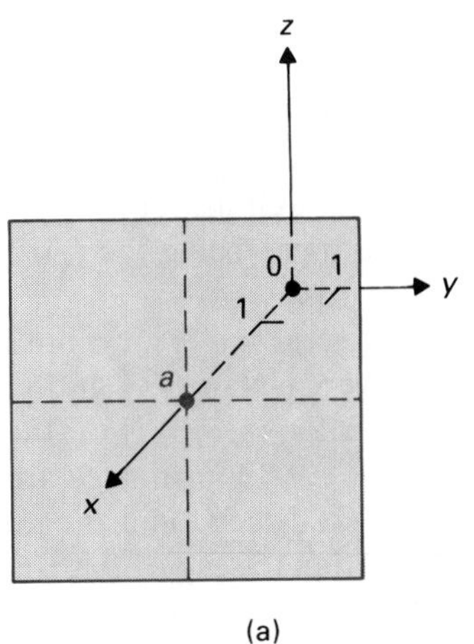

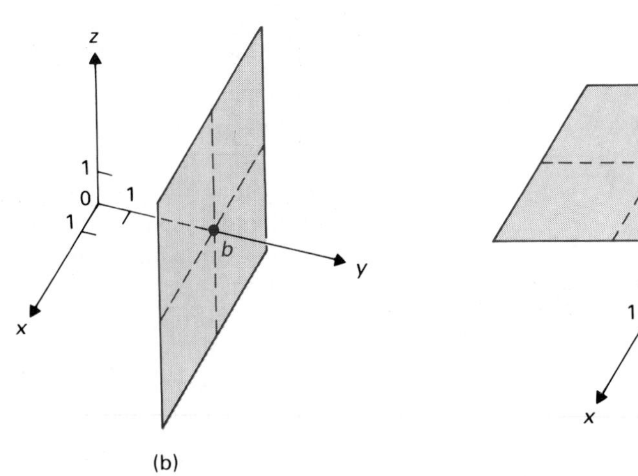

Figure 14.7 Level coordinate sets for rectangular coordinates: (a) the plane $x = a$; (b) the plane $y = b$; (c) the plane $z = c$.

Solution Completing the square, we obtain

$$(x^2 + 2x) + y^2 + (z^2 - 4z) = -6,$$
$$(x + 1)^2 + y^2 + (z - 2)^2 = -6 + 1 + 4 = -1.$$

Since a sum of squares cannot be negative, we see that the solution set is empty. □

In any coordinate system, we want to know the *level coordinate sets.* Such a set is described by setting one of the coordinate variables equal to a constant; it consists of the points where the coordinate achieves that constant "level." Level coordinate sets are the geometric configurations that the coordinate system can handle most easily, since they have the simplest possible equations. In x,y-coordinates for the plane, the equations $x = a$ and $y = b$ describe a vertical line and a horizontal line, respectively. From Fig. 14.7(a), we see that in space, $x = a$ describes a vertical plane, parallel to the y,z-coordinate plane. Similarly, Figs. 14.7(b) and (c) show that $y = b$ gives a vertical plane parallel to the x,z-coordinate plane, and $z = c$ describes a horizontal plane, parallel to the x,y-coordinate plane.

In the cartesian plane, a single linear equation gives a *line.* We will show later that in space, a single linear equation describes a *plane.* The level coordinate planes $x = a$, $y = b$, $z = c$ are our first illustrations of this. We give another illustration.

Figure 14.8 The plane $x + y = 1$ in space.

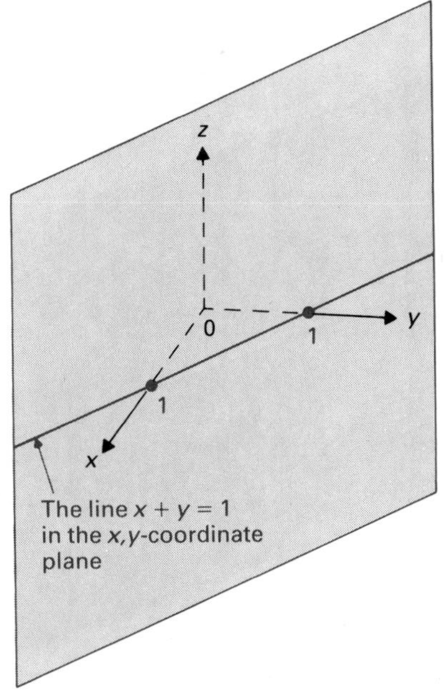

EXAMPLE 4 Sketch the solution set of $x + y = 1$ in space.

Solution First we draw the line $x + y = 1$ in the x,y-coordinate plane in Fig. 14.8. The equation places no restriction on z, so any point directly above or below this line also satisfies $x + y = 1$. The solution set is thus the vertical plane containing the line $x + y = 1$ in the x,y-coordinate plane, as shown in Fig. 14.8. □

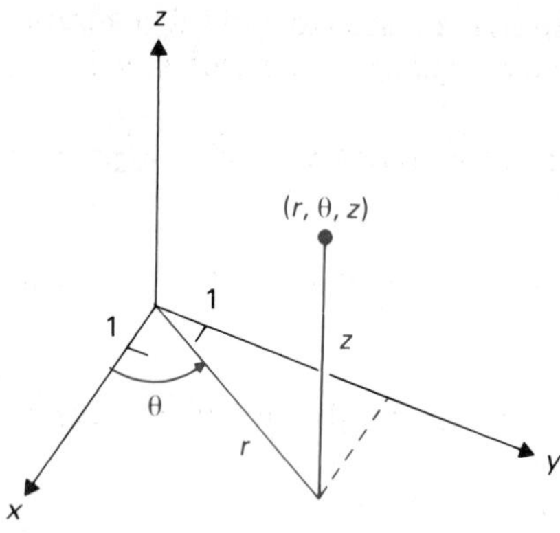

Figure 14.9 r, θ, and z in cylindrical coordinates.

CYLINDRICAL COORDINATES

We can also locate a point in space by specifying the x,y-coordinates of its position using polar r,θ-coordinates and specifying its height by the z-coordinate. The point then has coordinates (r, θ, z) as well as coordinates (x, y, z). Of course, the coordinates r and θ are not unique, being the usual polar coordinates. Such coordinates are shown in Fig. 14.9.

Figure 14.10(a) shows that the level coordinate set $r = a$ is a cylinder of radius a about the z-axis, for $r = a$ has a circle in the x,y-plane as solution set and there is no restriction placed on z. Consequently, (a, θ, z) is on the solution set for all θ and all z. For this reason, these r,θ,z-coordinates are called *cylindrical coordinates*. Figure 14.10(b) shows that the level coordinate set $\theta = \alpha$ is a vertical plane. Of course, the level coordinate set $z = c$ is the plane shown in Fig. 14.7(c).

Since we know how to change from polar r,θ-coordinates to x,y-coordinates in the plane, we know how to change back and forth from cylindrical to rectangular coordinates in space. Namely,

$$x = r\cos\theta, \qquad y = r\sin\theta, \qquad z = z \tag{4}$$

and

$$r^2 = x^2 + y^2, \qquad \theta = \tan^{-1}\left(\frac{y}{x}\right)[+n\pi], \qquad z = z. \tag{5}$$

The $[+n\pi]$ in Eqs. (5) appears in case the value for the inverse tangent functions does not lie in the appropriate quadrant. Suppose, for example, we always wish to have θ in the range $0 \le \theta < 2\pi$. Then for a point (x, y) in the

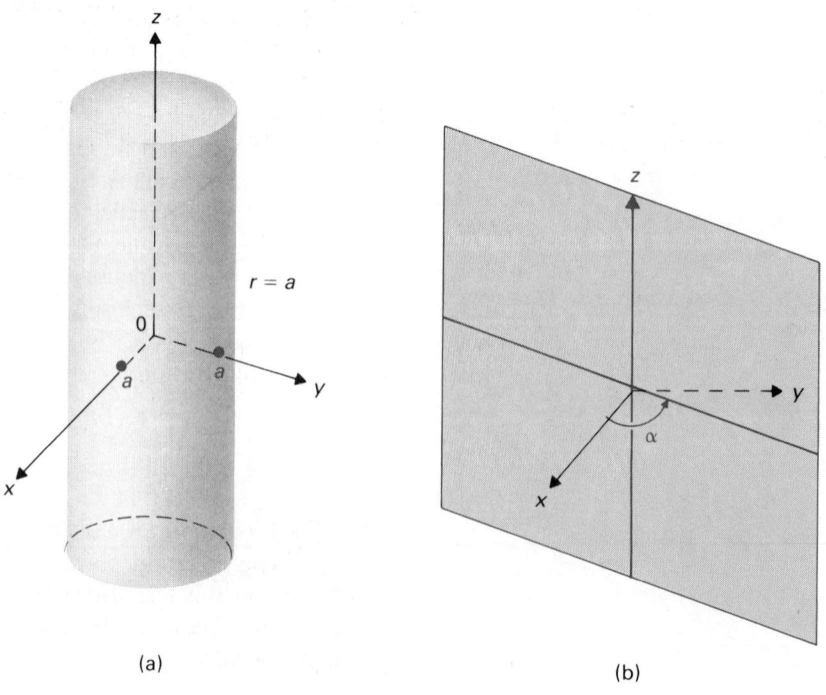

Figure 14.10 Level coordinate sets for cylindrical coordinates: (a) the cylinder $r = a$; (b) the plane $\theta = \alpha$.

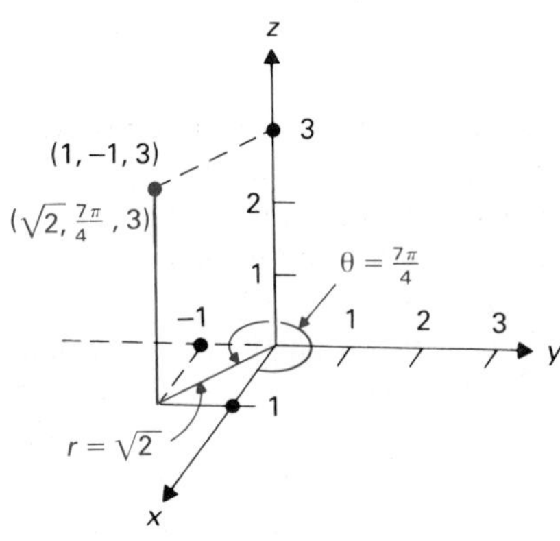

Figure 14.11 $(x, y, z) = (1, -1, 3)$, $(r, \theta, z) = (\sqrt{2}, 7\pi/4, 3)$

first quadrant, we may take $n = 0$. For points in the second or third quadrant, we need $n = 1$, while for a point in the fourth quadrant, we need $n = 2$.

EXAMPLE 5 Transform the equation of the sphere $x^2 + y^2 + z^2 = 4$ to a cylindrical coordinate form.

Solution Since $x^2 + y^2 = r^2$, the sphere is described by $r^2 + z^2 = 4$ in cylindrical coordinates. □

EXAMPLE 6 Find cylindrical coordinates of the point $(x, y, z) = (1, -1, 3)$.

Solution From Eqs. (5) and Fig. 14.11, we see that $r^2 = 1^2 + (-1)^2 = 2$, so $r = \sqrt{2}$. Then $\theta = \tan^{-1}(y/x) = \tan^{-1}(-1) = -\pi/4$. Of course, $z = 3$. The point is thus $(\sqrt{2}, -\pi/4, 3)$ or $(\sqrt{2}, 7\pi/4, 3)$ in cylindrical coordinates. □

SPHERICAL COORDINATES

Another useful coordinate system in space is the *spherical coordinate* system, where a point has coordinates (ρ, ϕ, θ), as indicated in Fig. 14.12. The coordinate ρ is the length of the line segment joining the point and the origin. We will restrict ourselves to values of $\rho \geq 0$, considering ρ to be a nondirected distance. This is in contrast to our use of r in polar coordinates. We will be using spherical coordinates chiefly with integrals, and $\rho \geq 0$ will suffice for that purpose.

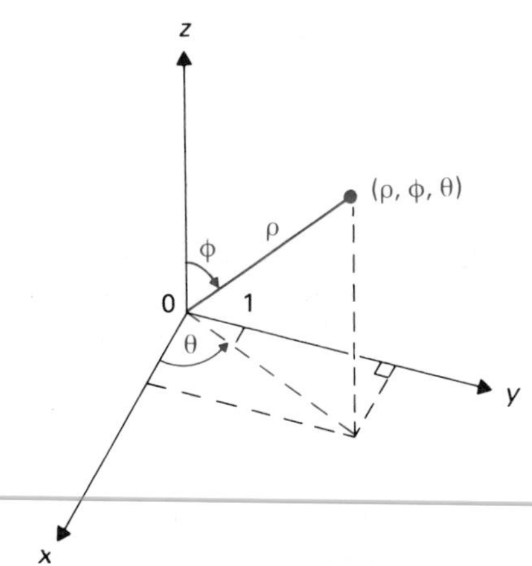

Figure 14.12 ρ, ϕ, and θ, in spherical coordinates.

The angle ϕ is the angle from the z-axis to the line segment from the origin to the point, shown in Fig. 14.12. We will also consider ϕ to be a nondirected measurement, and we see from Fig. 14.12 that we may always take ϕ in the range $0 \leq \phi \leq \pi$. Finally, θ is the same angle as in cylindrical coordinates. Later, when integrating using spherical coordinates, we generally let θ have values from 0 to 2π. However, we may on occasion use negative values of θ when using cylindrical coordinates.

The level coordinate surface $\rho = a$ is the sphere shown in Fig. 14.13(a); hence the name "spherical coordinates." The level coordinate surface $\phi = \beta$ is the cone shown in Fig. 14.13(b). The level coordinate set $\theta = \alpha$ is the plane

Figure 14.13 Level coordinate sets for spherical coordinates: (a) the sphere $\rho = a$; (b) the cone $\phi = \beta$.

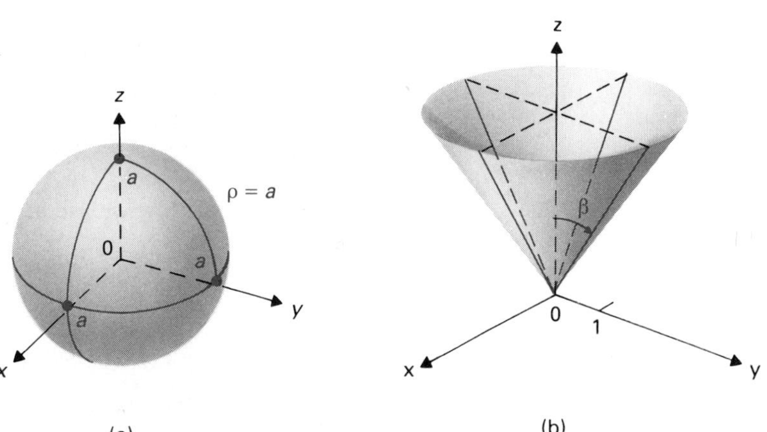

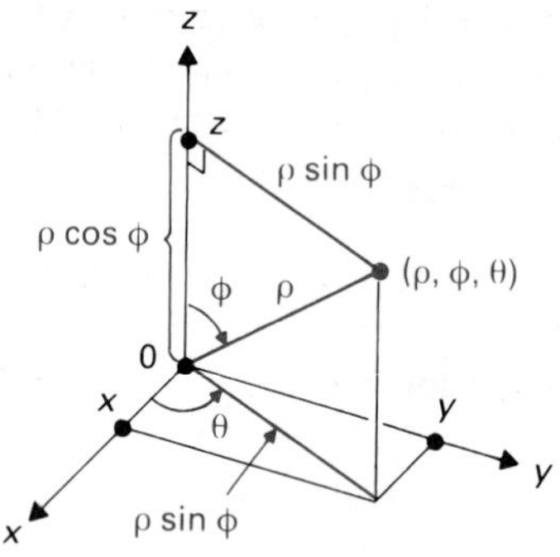

Figure 14.14 Relations between rectangular and spherical coordinates.

shown in Fig. 14.10(b). Spherical coordinates are especially useful when dealing with spheres or cones.

We need to express x, y, and z in terms of ρ, ϕ, and θ, and conversely. From Fig. 14.14, we easily see that

$$\boxed{x = \rho \sin\phi \cos\theta, \qquad y = \rho \sin\phi \sin\theta, \qquad z = \rho\cos\phi.} \tag{6}$$

Since ρ is the distance from the point (x, y, z) to the origin, it is obvious that $\rho^2 = x^2 + y^2 + z^2$. From Fig. 14.14, we can see that

$$\boxed{\rho = \sqrt{x^2 + y^2 + z^2}, \quad \phi = \cos^{-1}\left(\frac{z}{\sqrt{x^2+y^2+z^2}}\right), \quad \theta = \tan^{-1}\left(\frac{y}{x}\right)[+\pi].} \tag{7}$$

The $[+\pi]$ in Eqs. (7) is included in case the value of the inverse function does not lie in the appropriate quadrant.

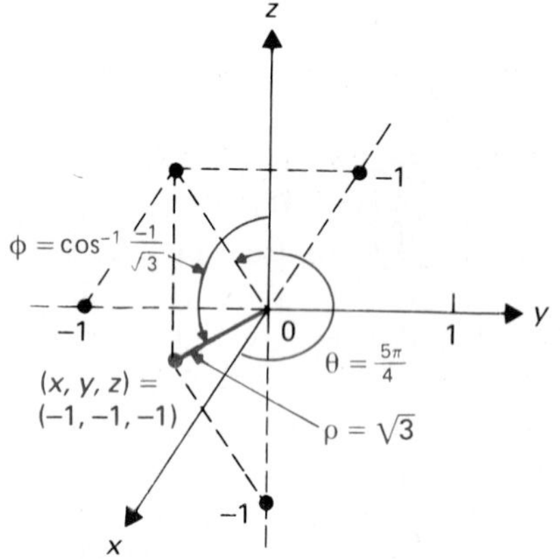

Figure 14.15 Spherical coordinates $\left(\sqrt{3}, \cos^{-1}\frac{-1}{\sqrt{3}}, \frac{5\pi}{4}\right)$ for $(x, y, z) = (-1, -1, -1)$.

EXAMPLE 7 Find spherical coordinates of the point $(x, y, z) = (-1, -1, -1)$.

Solution We obtain from Eqs. (7)

$$\rho = \sqrt{1^2 + 1^2 + (-1)^2} = \sqrt{3},$$

$$\phi = \cos^{-1}\left(\frac{-1}{\sqrt{1+1+1}}\right) = \cos^{-1}\left(\frac{-1}{\sqrt{3}}\right),$$

$$\theta = \tan^{-1}\left(\frac{-1}{-1}\right) + \pi = \tan^{-1}1 + \pi = \frac{5\pi}{4}.$$

Our choice of $\tan^{-1}1 + \pi$ rather than $\tan^{-1}1$ for θ comes from the fact that we require $\rho > 0$ for spherical coordinates. As shown in Fig. 14.15, the point $(-1, -1, -1)$ lies under the *third* quadrant of the x, y-coordinate plane. So we *must* have a third-quadrant angle for θ. Since $\tan^{-1}1$ is a *first*-quadrant angle, we must add π. Thus the spherical coordinates are $(\sqrt{3}, \cos^{-1}(-1/\sqrt{3}), 5\pi/4)$, as shown in Fig. 14.15. □

Figure 14.16 Spherical $(\sqrt{2}, \pi, \pi/2)$ is rectangular $(0, 0, -\sqrt{2})$.

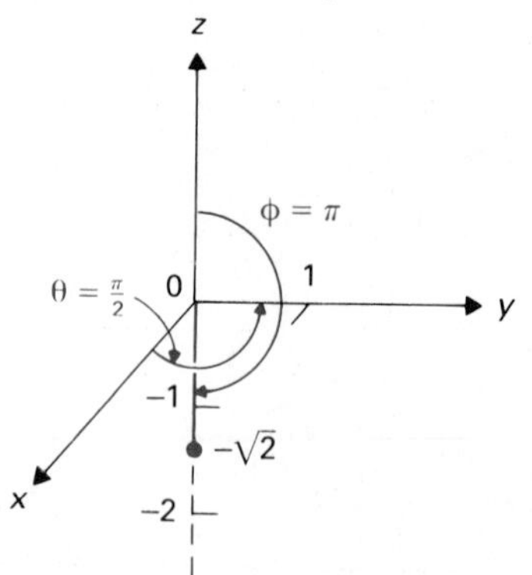

EXAMPLE 8 Find x,y,z-coordinates of the point

$$(\rho, \phi, \theta) = (\sqrt{2}, \pi, \pi/2).$$

Solution This time we use Eqs. (6) and obtain

$$x = \rho\sin\phi\cos\theta = \sqrt{2}\cdot 0 \cdot 0 = 0$$

$$y = \rho\sin\phi\sin\theta = \sqrt{2}\cdot 0\cdot 1 = 0$$

$$z = \rho\cos\phi = \sqrt{2}\cdot(-1) = -\sqrt{2}$$

Thus the point is $(x, y, z) = (0, 0, -\sqrt{2})$, as shown in Fig. 14.16. □

Both Examples 7 and 8 could have been solved just by referring to the figures, without using Eqs. (6) and (7). It is a good idea to do some simple problems from the figures occasionally to be sure you understand the geometric meaning of ρ, ϕ, and θ. Some exercises ask you to do this.

EXAMPLE 9 Find the spherical-coordinate equation of the plane $z = 2$.

Solution From Eqs. (6), we see that the equation becomes $\rho \cos \phi = 2$, or $\rho = 2 \sec \phi$. □

SUMMARY

1. In rectangular coordinates for space, a point has coordinates (x, y, z).
2. The distance from (x_1, y_1, z_1) to (x_2, y_2, z_2) in rectangular coordinates is
$$d = \sqrt{(x_2 - x_1)^2 + (y_2 - y_1)^2 + (z_2 - z_1)^2}.$$
3. The rectangular equation of a sphere in space with center (x_1, y_1, z_1) and radius r is
$$(x - x_1)^2 + (y - y_1)^2 + (z - z_1)^2 = r^2.$$
4. In cylindrical coordinates for space, a point has coordinates (r, θ, z), where r and θ are the usual polar coordinates in the x,y-plane.
5. Transformation from rectangular to cylindrical coordinates and vice versa is accomplished using the formulas
$$x = r\cos\theta, \qquad r = \sqrt{x^2 + y^2},$$
$$y = r\sin\theta, \qquad \text{and} \qquad \theta = \tan^{-1}\left(\frac{y}{x}\right)\ [+\pi],$$
$$z = z, \qquad z = z.$$
6. In spherical ρ,ϕ,θ-coordinates for a point in space,

   ρ is the distance to the origin,

   ϕ is the angle from the positive z-axis to the ray from the origin, where $0 \le \phi \le \pi$,

   θ is the same angle as for cylindrical coordinates.
7. Transformation from rectangular to spherical coordinates and vice versa is accomplished using the formulas
$$x = \rho\sin\phi\cos\theta, \qquad \rho = \sqrt{x^2 + y^2 + z^2},$$
$$y = \rho\sin\phi\sin\theta, \qquad \text{and} \qquad \phi = \cos^{-1}\left(\frac{z}{\sqrt{x^2 + y^2 + z^2}}\right),$$
$$z = \rho\cos\phi, \qquad \theta = \tan^{-1}\left(\frac{y}{x}\right)\ [+\pi].$$

EXERCISES

In Exercises 1 through 8, sketch all points in space satisfying the given equation.

1. $x = 2$ 2. $z = 3$ 3. $x = y$ 4. $y^2 = 1$
5. $x = y = z$ 6. $x = z$ 7. $y + z = 1$ 8. $x - y = 1$

While we have defined neither a line nor a plane in space, use your geometric intuition in Exercises 9 through 14 to find the desired point.

9. The point such that the line segment joining it to $(-2, 1, -4)$ is bisected by and perpendicular to the plane $x = 0$.
10. The point such that the line segment joining it to $(-1, \pi, \sqrt{2})$ has the origin as midpoint.
11. The point such that the line segment joining it to $(-1, 4, -3)$ has $(-1, 2, -3)$ as midpoint.
12. The point in the plane $y = 2$ that is closest to the point $(-1, -5, 2)$.
13. The point on the z-axis closest to the point $(3, -1, 2)$.
14. The midpoint of the line segment joining $(-1, 4, 3)$ and $(3, -2, 5)$.
15. Let $(-1, 2, 1)$ be chosen as origin for $\bar{x},\bar{y},\bar{z}$-coordinates. Express each point in terms of these new translated coordinates.
 a) $(1, -2, 1)$ b) $(-3, 4, 0)$ c) $(5, -1, 2)$

In Exercises 16 through 20, find the distance between the given points.

16. $(-1, 0, 4)$ and $(1, 1, 6)$ 17. $(2, -1, 3)$ and $(0, 1, 7)$
18. $(3, -1, 0)$ and $(-1, 2, 6)$ 19. $(-2, 1, 4)$ and $(-3, 2, 8)$
20. $(0, 1, 8)$ and $(-3, 1, 12)$
21. Find the equation of the sphere with center $(-1, 2, 4)$ and passing through the point $(2, -1, 5)$.
22. Find the equation of the sphere having $(-1, 2, 6)$ and $(1, 6, 0)$ as endpoints of a diameter.
23. Find the center and radius of the given sphere.
 a) $x^2 + y^2 + z^2 - 2x + 2y = 0$
 b) $x^2 + y^2 + z^2 - 6x - 4y + 8z = -4$
24. Find *all* cylindrical coordinates of the point $(1, 1, 1)$.
25. Find all cylindrical coordinates of the cartesian point $(1, -\sqrt{3}, 5)$.
26. Sketch in space the locus of $\theta = \pi/4$ in cylindrical coordinates.
27. Sketch in space the locus of $r = 2$ in cylindrical coordinates.
28. a) Transform the x,y,z-equation $x^2 + y^2 = 9$ into cylindrical coordinates.
 b) Using your answer to part (a), sketch all points in space satisfying the equation.

In Exercises 29 through 42, use a figure rather than transformation equations to find the indicated coordinates. (For cylindrical coordinates, give r-values ≥ 0 and θ-values from 0 to 2π.)

29. Rectangular coordinates for cylindrical $(\sqrt{2}, \pi/4, 3)$
30. Rectangular coordinates for cylindrical $(4, 2\pi/3, -2)$
31. Rectangular coordinates for cylindrical $(0, \pi, -1)$
32. Cylindrical coordinates for rectangular $(0, 1, 1)$
33. Cylindrical coordinates for rectangular $(-1, 1, -5)$
34. Cylindrical coordinates for rectangular $(\sqrt{3}, -1, 2)$
35. Spherical coordinates for rectangular $(1, 0, 0)$
36. Spherical coordinates for rectangular $(0, 0, -4)$
37. Spherical coordinates for rectangular $(1, 1, 1)$
38. Spherical coordinates for rectangular $(-3, -4, 5)$
39. Rectangular coordinates for spherical $(2, \pi/4, -\pi)$
40. Rectangular coordinates for spherical $(0, 3\pi/4, \pi/6)$
41. Rectangular coordinates for spherical $(4, \pi/2, \pi/3)$
42. Rectangular coordinates for spherical $(6, 2\pi/3, \pi)$
43. Express the cylindrical coordinates r, θ, and z in terms of the spherical coordinates ρ, ϕ, and θ.
44. Express the spherical coordinates ρ, ϕ, and θ in terms of the cylindrical coordinates r, θ, and z.

In Exercises 45 through 52, use the transformation equations to find the indicated coordinates. (For cylindrical coordinates, give r-values ≥ 0 and θ-values from 0 to 2π.)

45. Spherical coordinates for rectangular $(1, 2, -2)$
46. Spherical coordinates for rectangular $(3, 4, 0)$
47. Spherical coordinates for rectangular $(0, 1, -\sqrt{3})$
48. Cylindrical coordinates for rectangular $(1, -1, 3)$
49. Cylindrical coordinates for rectangular $(-\sqrt{3}, 1, 4)$
50. Rectangular coordinates for spherical $(2, \pi/4, 2\pi/3)$
51. Rectangular coordinates for spherical $(1, 5\pi/6, 4\pi/3)$
52. Rectangular coordinates for cylindrical $(3, 7\pi/6, -5)$
53. Find the spherical coordinate equation for $x^2 + y^2 + z^2 = 25$ by using the transformation equations.
54. Sketch in space the locus of $\phi = \pi/4$ in spherical coordinates.
55. Find the volume of the region described in spherical coordinates by $2 \leq \rho \leq 5$ and $0 \leq \phi \leq \pi/2$.
56. Describe in terms of x,y,z-coordinates all points in space where the cylindrical r-coordinate is equal to the spherical ρ-coordinate.
57. Describe in terms of spherical coordinates all points in space where $x + y = 1$.
58. Describe in terms of spherical coordinates all points in space where $z^2 = 4$.
59. A boat traveling down the middle of a straight river approaches a bridge for a highway going straight across the river 20 ft above the river. When the boat is 60 ft from the point directly under the middle of the bridge, a car, driven in the middle of the road, is approaching the bridge at a point 30 ft from the center of the bridge. Find the distance from the boat to the car at that instant.
60. Suppose the boat in Exercise 59 is traveling 10 ft/sec and the car 40 ft/sec. Find the rate of change of the distance between them at the instant given in Exercise 59,

14.2 CYLINDERS AND QUADRIC SURFACES

CYLINDERS

The points in space satisfying a single equation in x, y, and z usually form a surface. If one of the variables x, y, and z is missing from the equation, then the surface is a cylinder. A **cylinder** in space is a surface that can be generated by a line that moves along a plane curve, keeping a fixed direction. The parallel lines (rulings) on the cylinder corresponding to positions of the generating line are the *elements of the cylinder*. To illustrate, suppose the variable x is missing in an equation $F(x, y, z) = 0$. If (a, b, c) lies on the surface, then so will (x, b, c) for all x, that is, the line through (a, b, c) parallel to the x-axis. Consequently, the surface is a cylinder with elements parallel to the x-axis and meeting the y,z-plane in the curve $F(0, y, z) = 0$. Similar results hold if y or z is missing.

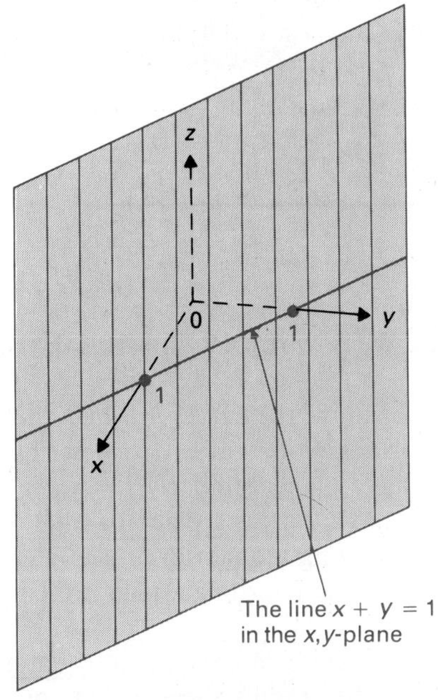

Figure 14.17 The plane $x + y = 1$ viewed as a cylinder with rulings parallel to the z-axis.

The surface in space described by an equation in two of the variables x, y, and z is a cylinder. The cylinder is generated by a line parallel to the axis of the missing variable as it moves along the curve the equation describes in the coordinate plane of present variables.

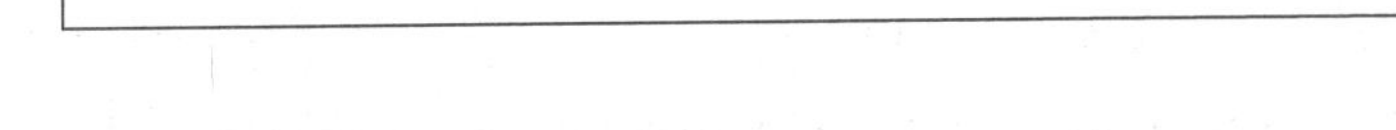

EXAMPLE 1 Sketch the *cylinder* $x + y = 1$ in space.

Solution This was Example 4 on page 589 of the preceding section. We obtained the vertical plane shown with rulings in Fig. 14.17. The plane may be regarded as a cylinder generated by a line parallel to the z-axis moving along the line $x + y = 1$ in the x,y-plane. The rulings show some of the positions of the generating line. □

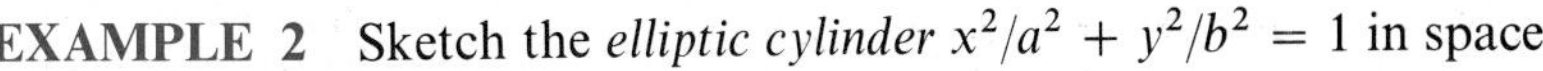

EXAMPLE 2 Sketch the *elliptic cylinder* $x^2/a^2 + y^2/b^2 = 1$ in space.

Solution This cylinder is sketched in Fig. 14.18. It is generated by a line parallel to the z-axis moving along the ellipse $(x^2/a^2) + (y^2/b^2) = 1$ in the x,y-plane. □

Figure 14.18 The elliptic cylinder $x^2/a^2 + y^2/b^2 = 1$.

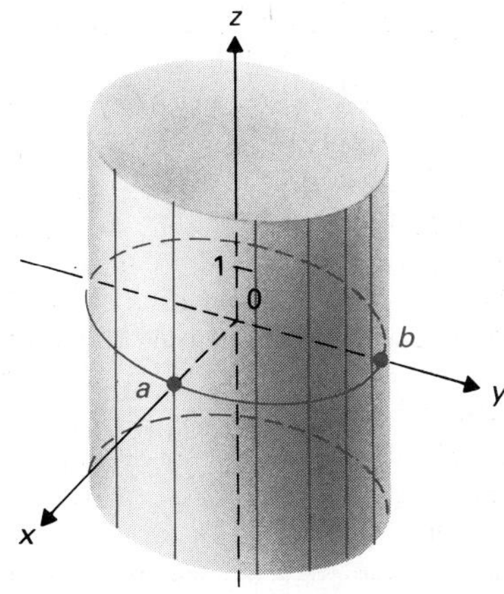

EXAMPLE 3 Sketch the *parabolic cylinder* $z^2 = 4py$ in space.

Solution The cylinder, shown in Fig. 14.19, is generated by a line parallel to the x-axis moving along the parabola $z^2 = 4py$ in the y,z-plane. □

QUADRIC SURFACES

A **quadric surface** in space is the solution set of a polynomial equation in x, y, and z of degree two. We have seen that if one of the three variables is missing, the surface is a cylinder, which is fairly easy to sketch. The cylinders

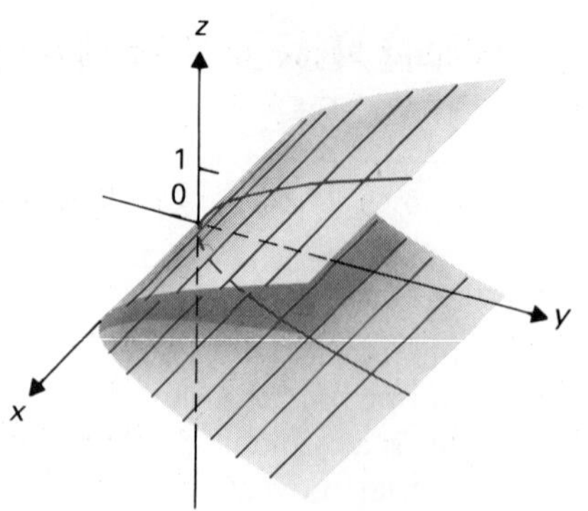

Figure 14.19 The parabolic cylinder $z^2 = 4py$.

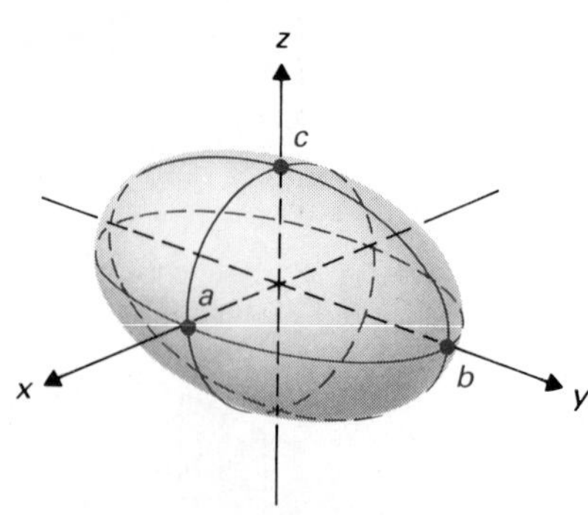

Figure 14.20 The ellipsoid $x^2/a^2 + y^2/b^2 + z^2/c^2 = 1$.

in Examples 2 and 3 are quadric surfaces. We now restrict ourselves to the case where all three variables are present.

As an aid in sketching such surfaces, examine the curves in which they intersect planes parallel to the coordinate planes. Note that $x = x_0$ describes in space a plane parallel to the y,z-coordinate plane. Similarly $y = y_0$ is a plane parallel to the x,z-coordinate plane, and $z = z_0$ is a plane parallel to the x,y-coordinate plane. Each such plane meeting a quadric surface intersects it in an ellipse, hyperbola, or parabola. The coordinate planes themselves, $x = 0$, $y = 0$, and $z = 0$, are especially useful.

As for second-degree plane curves, the device of completing squares and choosing a new origin can often be used to simplify the sketching of a quadric surface. We assume that this now causes no difficulty, and the examples that follow start with equations where the completion of squares is unnecessary. These examples exhibit some types of quadric surfaces.

EXAMPLE 4 Sketch the *ellipsoid* $x^2/a^2 + y^2/b^2 + z^2/c^2 = 1$.

Solution The surface is sketched in Fig. 14.20. We set $x = 0$, which corresponds to intersecting the surface with the y,z-plane, and obtain the equation

Figure 14.21 The elliptic paraboloid $z = x^2/a^2 + y^2/b^2$.

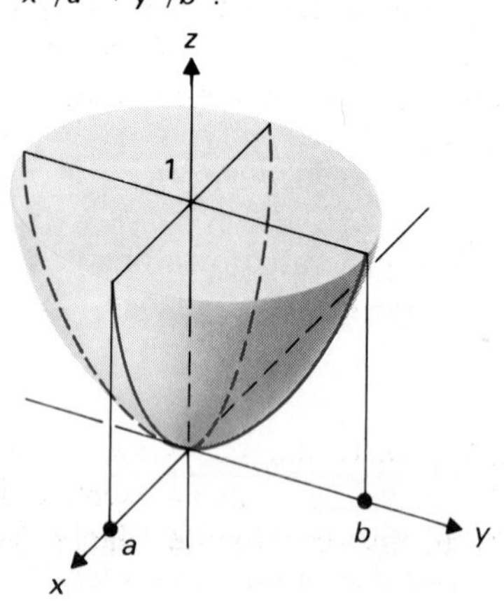

$$\frac{y^2}{b^2} + \frac{z^2}{c^2} = 1,$$

which is an ellipse. This ellipse is the *trace* of the surface in the y,z-plane.

Now set $x = x_0$, which corresponds to intersecting the surface with the plane $x = x_0$, parallel to the y,z-plane. The equation becomes

$$\frac{y^2}{b^2} + \frac{z^2}{c^2} = 1 - \frac{x_0^2}{a^2}.$$

If $|x_0| > a$, the right-hand side of the equation is negative, and the intersection of the surface with the plane $x = x_0$ is empty. If $-a < x_0 < a$, we obtain the equation of an ellipse in the plane $x = x_0$. The largest ellipse occurs when $x_0 = 0$, and the ellipses become smaller as $|x_0| \to a$. When $x_0 = \pm a$, we have the points $(\pm a, 0, 0)$ as the only solution.

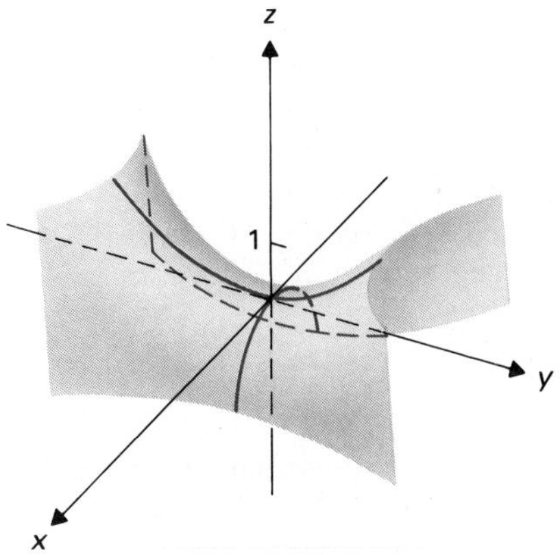

Figure 14.22 The hyperbolic paraboloid $z = y^2/b^2 - x^2/a^2$.

Similar results hold for planes $y = y_0$ if $-b < y_0 < b$ and for planes $z = z_0$ if $-c < z_0 < c$. □

EXAMPLE 5 Sketch the *elliptic paraboloid* $z = x^2/a^2 + y^2/b^2$.

Solution The sketch is shown in Fig. 14.21. The plane $z = z_0$ does not meet the surface if $z_0 < 0$; it meets the surface in a point if $z_0 = 0$; and it meets the surface in an ellipse if $z_0 > 0$. The surface thus has, as horizontal cross sections, ellipses that expand as we go higher.

If we set $x = x_0$, we obtain the equation

$$z = \frac{x_0^2}{a^2} + \frac{y^2}{b^2},$$

which is linear in z and quadratic in y. This must be the equation of a parabola. In particular, $x_0 = 0$ yields the trace parabola $b^2z = y^2$ in the y,z-plane. Setting $y = 0$, we obtain the trace parabola $a^2z = x^2$ in the x,z-plane. □

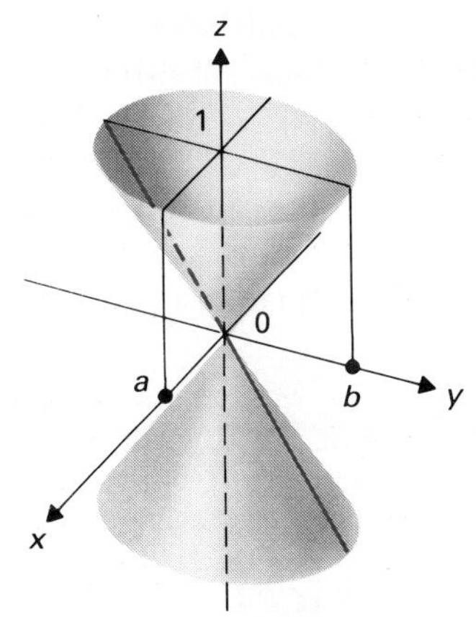

Figure 14.23 The elliptic cone $z^2 = x^2/a^2 + y^2/b^2$.

EXAMPLE 6 Sketch the *hyperbolic paraboloid* $z = y^2/b^2 - x^2/a^2$.

Solution The sketch is given in Fig. 14.22. This is not an easy surface for a person without artistic ability to sketch. We find that a quadric surface is fairly easy to sketch if the equation can be put in a form where one side consists of a *sum* of two quadratic terms. As the other side assumes constant values, we obtain elliptical sections, and elliptical sections are easy to sketch and visualize. But this cannot be done with the equation in this example.

The plane $z = z_0$ meets the surface in a hyperbola that "opens" in the y-direction if $z_0 > 0$ and in the x-direction if $z_0 < 0$, while the plane $z = 0$ meets the surface in the degenerate hyperbola consisting of two intersecting lines. A plane $x = x_0$ meets the surface in a parabola "opening upward" while a plane $y = y_0$ meets the surface in a parabola "opening downward." □

EXAMPLE 7 Sketch the *elliptic cone* $z^2 = x^2/a^2 + y^2/b^2$.

Solution The surface is shown in Fig. 14.23. Note that we gave the equation in a form where one side consists of a sum of two quadratic terms, as discussed in the preceding example. Setting $z = z_0$ thus yields elliptical sections; the larger $|z_0|$, the larger the ellipse. Setting $y = 0$, we obtain the lines $z = \pm x/a$ in the x,z-plane, and setting $x = 0$, we have the lines $z = \pm y/b$ in the y,z-plane. Setting x or y equal to nonzero constants gives hyperbolic sections. (Recall that a hyperbola is a *conic section*.) □

Figure 14.24 The hyperboloid of two sheets $z^2/c^2 - 1 = x^2/a^2 + y^2/b^2$.

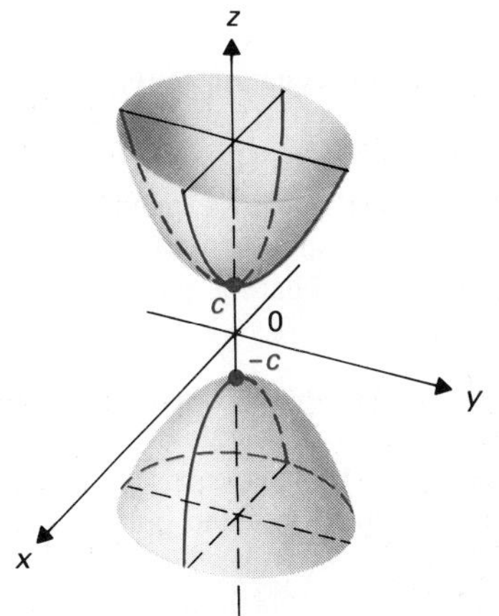

EXAMPLE 8 Sketch the *hyperboloid of two sheets*

$$\frac{z^2}{c^2} - 1 = \frac{x^2}{a^2} + \frac{y^2}{b^2}.$$

Solution The sketch is shown in Fig. 14.24. Clearly, the planes $z = z_0$ have empty intersection with the surface if $|z_0| < c$. We leave as an exercise the discussion of sections by planes parallel to the coordinate planes (see Exercise 1). □

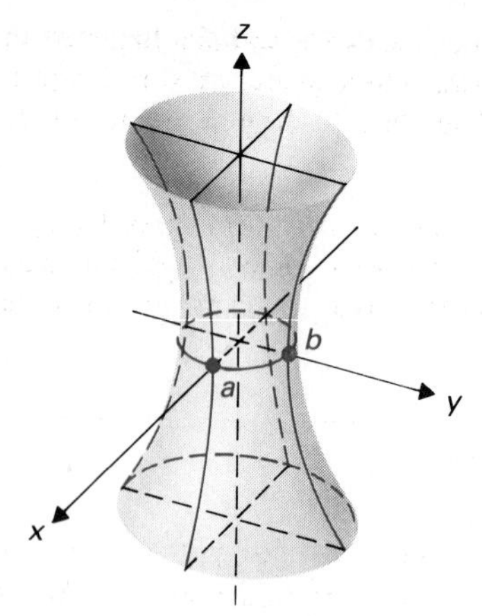

Figure 14.25 The hyperboloid of one sheet $z^2/c^2 + 1 = x^2/a^2 + y^2/b^2$.

EXAMPLE 9 Sketch the *hyperboloid of one sheet*

$$\frac{z^2}{c^2} + 1 = \frac{x^2}{a^2} + \frac{y^2}{b^2}.$$

Solution The surface is sketched in Fig. 14.25. Again, we leave the discussion of planar sections as an exercise (see Exercise 2). □

SKETCHING QUADRIC SURFACES

The box at the bottom of this page gives some steps that are helpful in sketching quadric surfaces.

EXAMPLE 10 Sketch the surface $4x = y^2$ in space.

Solution Since z is missing, this is a cylinder; indeed, a parabolic cylinder since the equation describes a parabola in the x,y-plane. We draw the parabola in the x,y-plane and copies of it above and below in parallel planes (see Fig. 14.26). Then we draw some of the rulings parallel to the axis of the missing variable, that is, the z-axis. □

EXAMPLE 11 Sketch the surface $x^2 + y + z^2 = 0$.

Solution This is not a cylinder. We rearrange the equation in the form

$$-y = x^2 + z^2.$$

We see at once that for $y > 0$, we have no surface. For $y < 0$, we have circular

Figure 14.26 The parabolic cylinder $4x = y^2$.

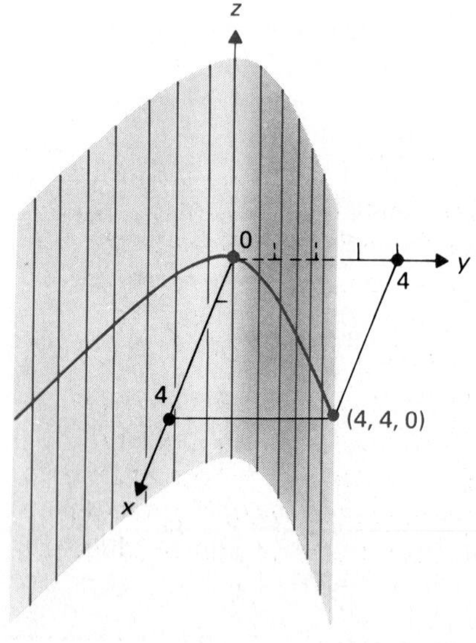

SKETCHING QUADRIC SURFACES

Step 1 Complete squares and translate axes if any variable is present in both linear and quadratic terms.

Step 2 Decide whether the surface is a cylinder (one variable missing). If it is, sketch the curve the equation describes in the coordinate plane of the present variables. Then draw a copy or two of the curve in parallel planes. Finally, draw some of the element lines (rulings) parallel to the axis of the missing variable. See Example 10.

Step 3 If the surface is not a cylinder, try to arrange the equation so that one side consists of a *sum* of two quadratic terms. If this can be done, then setting the variable on the other side equal to constants yields elliptical or circular (or empty) sections. Visualize how these elliptical sections expand or shrink for different values of the constants. Then set the variables in the sum of the quadratic terms equal to zero individually, to see what the curves look like through the "ends" of the ellipses. See Example 11.

Step 4 If neither step 2 nor 3 is the case and there are two quadratic terms, then the surface is a hyperbolic paraboloid. With Fig. 14.22 as a guide, sketch the parabola opening "upward," in the positive direction of the axis of the linear variable. (Of course the saddle surface is oriented to stand on end if this linear variable is x or y, so "upward" might be "sideways.") Similarly, sketch the "downward"-opening parabola, which meets the first one at right angles at their common vertex. Then draw the two hyperbolas that meet the ends of the two parabolas you have drawn; a total of four curves is involved. Finally, join the ends of the "upper" hyperbola to the ends of the "lower" ones by straight-line segments.

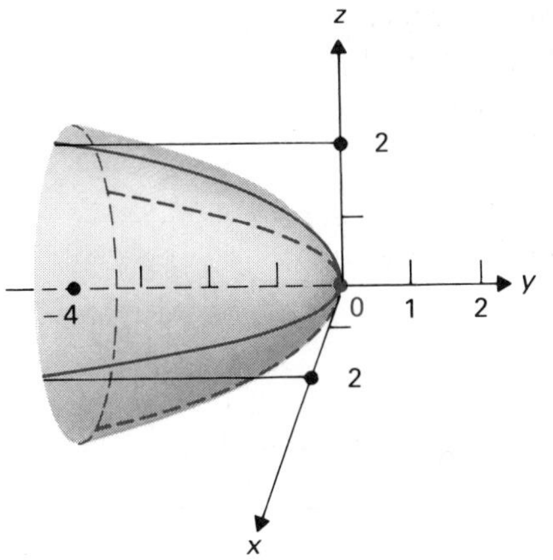

Step 5 As a general rule, try to visualize the entire surface before sketching it. Dash all axes and curves until you are sure what you can "see" and make solid, and determine what is behind another part of the surface and thus must be left dashed.

Figure 14.27 The circular paraboloid $-y = x^2 + z^2$.

sections. These circular sections expand as $y \to -\infty$. The planes $x = 0$ and $z = 0$ intersect the surface in the parabolas $y = -z^2$ and $y = -x^2$, respectively. We have the circular paraboloid shown in Fig. 14.27. □

We conclude with an example in which it is necessary to complete the square.

EXAMPLE 12 Sketch the surface $-16x^2 + 4y^2 - z^2 - 8y + 4z = 0$.

Solution We complete squares and arrange our equation as follows:

$$-16x^2 + 4(y^2 - 2y) - (z^2 - 4z) = 0,$$

$$-16x^2 + 4(y-1)^2 - (z-2)^2 = 4 - 4 = 0,$$

$$4(y-1)^2 = 16x^2 + (z-2)^2,$$

$$4\bar{y}^2 = 16\bar{x}^2 + \bar{z}^2,$$

where $\bar{x} = x$, $\bar{y} = y - 1$, $\bar{z} = z - 2$. In Fig. 14.28, we take $\bar{x},\bar{y},\bar{z}$-axes at (0, 1, 2). Planes $\bar{y} = c$ meet the surface in elliptical sections, which increase in size as $|c|$ increases. Setting $\bar{x} = 0$, we see that the $\bar{y},\bar{z}$-plane meets the surface in the two lines $\bar{z} = \pm 2\bar{y}$. Setting $\bar{y} = 0$, we obtain only the point $(\bar{x}, \bar{y}, \bar{z}) = (0, 0, 0)$. Setting $\bar{z} = 0$, we obtain the two lines $\bar{y} = \pm 2\bar{x}$ in the $\bar{x},\bar{y}$-plane. The surface is an elliptic cone. □

Figure 14.28 The elliptic cone $4(y-1)^2 = 16x^2 + (z-2)^2$.

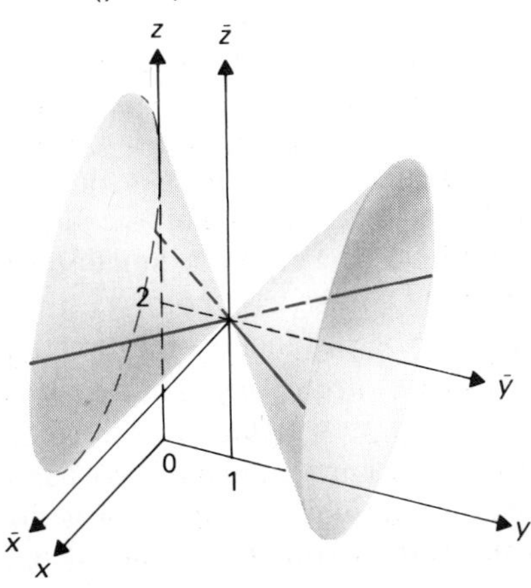

SUMMARY

1. An equation containing only two of the three space variables x, y, z has as graph a cylinder with the axis of the cylinder parallel to the axis of the missing variable. To illustrate, suppose x and y are present in the equation. The cylinder intersects the x,y-coordinate plane in the curve given by the equation and is parallel to the z-axis.
2. To sketch quadric surfaces, first complete squares where possible. If possible, put the equation in a form where one side is a sum of two quadratic terms, yielding elliptical cross sections as the other side assumes constant values. Sketch the trace curves in the coordinate planes $x = 0$, $y = 0$, and $z = 0$. See Figs. 14.18 through 14.25 for the possible types of surfaces.

EXERCISES

1. Describe the curves of intersection of the hyperboloid of two sheets in Example 8 with planes parallel to the coordinate planes.
2. Describe the curves of intersection of the hyperboloid of one sheet in Example 9 with planes parallel to the coordinate planes.

In Exercises 3 through 12, sketch the given cylinder. Draw some rulings on the cylinder.

3. $y^2 + z^2 - 4 = 0$
4. $x^2 + 2x + y^2 = 0$
5. $y^2 - z = 0$
6. $xz - 1 = 0$
7. $4x - y^2 + 2y + 3 = 0$
8. $x + 2z = 4$
9. $x + z = -2$
10. $y = e^{-x}$
11. $x = \sin y$
12. $x = -\ln z$

In Exercises 13 through 30, sketch the surface in space having the given equation and give the name of the surface, as in Figs. 14.20 through 14.25.

13. $y^2 - x^2 - z^2 = 0$
14. $x^2 + 4y^2 - z^2 - 8x + 16 = 0$
15. $x^2 + y^2 - z = 0$
16. $4x^2 + 9y^2 - 6z = 0$
17. $9x^2 + 36y^2 + 4z^2 - 36 = 0$
18. $x^2 + y^2 + z^2 - 2x + 4y - 6z - 2 = 0$
19. $x^2 - y^2 - 2y - z = 0$
20. $x^2 + y^2 - z^2 + 2x - 8 = 0$
21. $16x^2 + 16y^2 - z^2 - 64y + 73 = 0$
22. $36x - 9y^2 - 16z^2 = 0$
23. $\dfrac{x^2}{4} - \dfrac{y^2}{9} + z^2 + 1 = 0$
24. $\dfrac{x^2}{4} - \dfrac{y^2}{25} - \dfrac{z^2}{9} + 4 = 0$
25. $\dfrac{x^2}{4} + \dfrac{y^2}{9} + z - 3 = 0$
26. $\dfrac{x^2}{4} - \dfrac{y^2}{9} + z - 1 = 0$
27. $2x^2 + 3y^2 + 4z^2 - 24 = 0$
28. $x^2 - 4y^2 + 16z^2 = 0$
29. $\dfrac{x^2}{4} - \dfrac{y^2}{25} - \dfrac{z^2}{9} - 4 = 0$
30. $x^2 - 4y + z^2 - 8 = 0$

In Exercises 31 through 45, give the descriptive name of the quadric surface without sketching it.

31. $x - 4y^2 - 2z^2 + 3 = 0$
32. $x^2 - 3y^2 + 4z^2 - 2x + 6y - 8z + 2 = 0$
33. $x^2 - 3y^2 + 4z^2 - 2x + 6y - 8z + 3 = 0$
34. $x^2 - 3y^2 + 4z^2 - 2x + 6y - 8z + 1 = 0$
35. $x + 4y^2 - 3z^2 - 2y + z - 5 = 0$
36. $x^2 + 2y^2 + 4z^2 - 2x + 2 = 0$
37. $x^2 + 2y^2 + 4z^2 - 2x + 1 = 0$
38. $x^2 + 2y^2 + 4z^2 - 2x = 0$
39. $x^2 - z^2 + 2x - 4z + 3 = 0$
40. $2x^2 - y^2 - z^2 = 0$
41. $x^2 + 2z^2 - 2x = 0$
42. $2x^2 - 3y - 5z^2 + 6z - 4 = 0$
43. $3x - 2y^2 + 5y - 3 = 0$
44. $3x^2 - 4y + 3z^2 - 6x + 8 = 0$
45. $2x^2 + 3y^2 - z^2 + 2z = 0$

14.3 VECTORS AND THEIR ALGEBRA

Represented geometrically, a vector is nothing more than an arrow that has a *length* and points in a *direction*. We will focus for a moment on the idea of direction. If $y = f(x)$, the derivative $f'(x_0)$ gives the rate the function increases at x_0 *in the direction of increasing* x. Of course, at a point x_0, there are only two directions one can go in the domain of the function, the direction of *increasing* x (to the right) and the direction of *decreasing* x (to the left). Obviously if $f'(x_0) = 3$, the function is *increasing* at a rate of 3 as we go to the right and is *decreasing* at a rate of 3 (or increasing at a rate of -3) as we go to the left. With only these two directions to worry about, it was not necessary to introduce vectors to talk about direction in calculus of one variable.

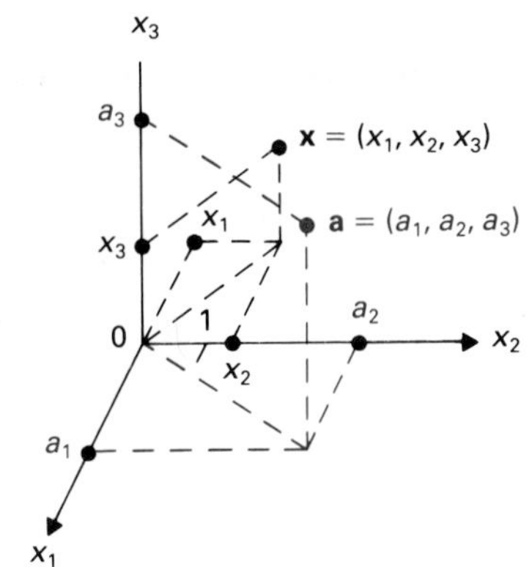

Figure 14.29 The points **a** and **x** in space.

On the other hand, if $z = f(x, y)$ is a function of two variables, and (x_0, y_0) is a point in the domain, then there are infinitely many directions we can travel from (x_0, y_0) in the x,y-plane. In Chapter 16, we will see how to find the rate at which $f(x, y)$ changes as we travel in *any* of these directions. Vectors will be very handy when finding all those derivatives.

For another illustration, we have described the *direction* of a line in the plane by its *slope m*. There is only one catch: A vertical line has no slope, so the slope does not really do the entire job for us. But vertical lines pose no problem for vectors. We can consider an arrow pointing straight up (or down) as easily as in any other direction.

Vectors have long been used in physics and engineering to describe *direction* and *magnitude* (represented by the *length* of the vector). The magnitude and direction of a force acting on a body is often represented by an arrow, as we will explain later in this section. Also, the magnitude and direction of the electric current induced in a wire as it moves through a magnetic field may be represented by a vector.

VECTOR NOTATION AND TERMINOLOGY

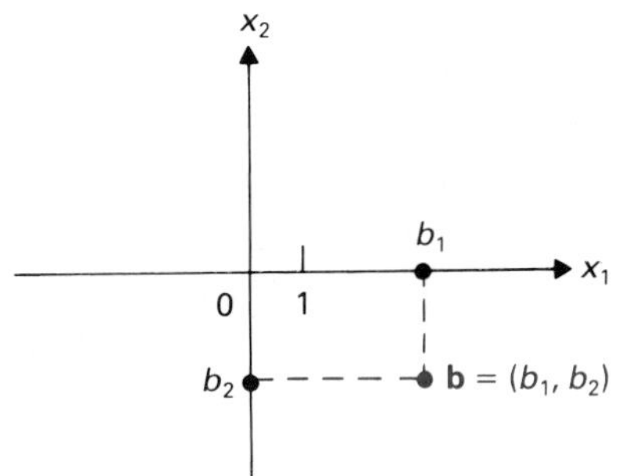

Figure 14.30 The point **b** in the plane.

We start out with notation considerations. In this chapter, we will want to think in terms of the *first*, *second*, or *third* coordinate of a point. This suggests changing notation for coordinates to index the coordinate position. We often write

$$(a_1, a_2) \text{ in place of } (a, b),$$

$$(x_1, x_2) \text{ in place of } (x, y),$$

$$(a_1, a_2, a_3) \text{ in place of } (a, b, c),$$

$$(x_1, x_2, x_3) \text{ in place of } (x, y, z).$$

As we work with more coordinates, lengthy notations such as (a_1, a_2, a_3) are time-consuming to write and may cause printing problems if a number of such notations appear in a single formula. We will often use a single boldface letter **a** to denote such a point (a_1, a_2, a_3). The number of coordinates will always be either explicitly stated or clear from the context. For example, the point **a** in space is (a_1, a_2, a_3) and the point **x** is (x_1, x_2, x_3), as shown in Fig. 14.29. The point **b** in the plane is (b_1, b_2), as shown in Fig. 14.30, and so on. We suggest that in written work, you use $\vec{a}$ with an arrow over the letter in place of the boldface letter. This is *vector notation*, and points correspond to vectors, as we will explain. We use a boldface zero for the origin; for example, in space $\mathbf{0} = (0, 0, 0)$.

Figure 14.31 The vector **a** in space.

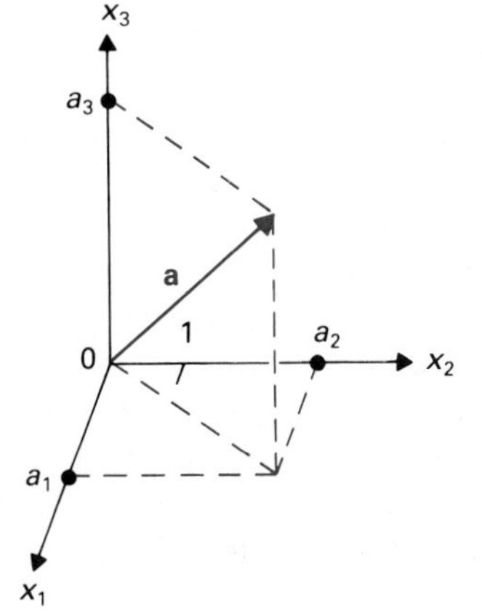

Each point **a** in space (or the plane) naturally describes a numerical *magnitude*, the distance $\sqrt{a_1^2 + a_2^2 + a_3^2}$ from **0** to **a**, and a *direction*, the direction from **0** to **a**. In the terminology of classical mechanics, any quantity that has associated with it a magnitude and a direction is called a *vector*. We will use this classical terminology and consider **a** to be a **vector** in space as well as a point in space.

From the vector viewpoint, we focus our attention on the *magnitude* and *direction* described by **a**. It is natural to associate with a vector **a** an arrow from the origin to the point **a**, as in Fig. 14.31. The length of the arrow is the *magnitude* of the vector **a**, and the arrow points in the *direction* of **a**.

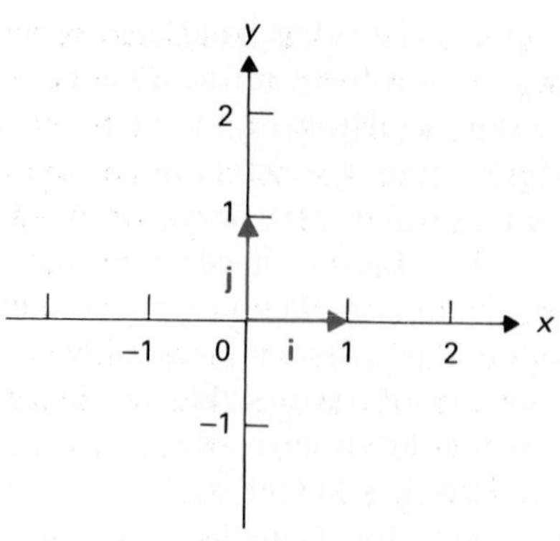

Figure 14.32 Unit coordinate vectors **i** and **j** in the plane.

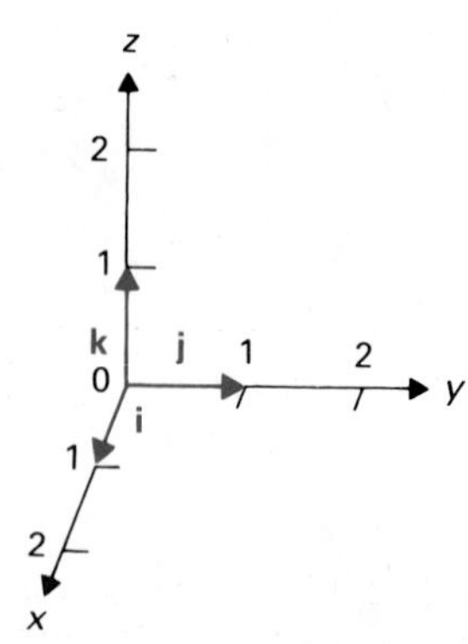

Figure 14.33 Unit coordinate vectors **i**, **j**, and **k** in space.

We call **0** the **zero vector;** it has magnitude zero, but it is the only vector that has no unique direction associated with it. The vectors

$$\mathbf{i} = (1, 0) \qquad \text{and} \qquad \mathbf{j} = (0, 1)$$

are the **unit coordinate vectors** in the plane, shown in Fig. 14.32, while

$$\mathbf{i} = (1, 0, 0), \qquad \mathbf{j} = (0, 1, 0), \qquad \text{and} \qquad \mathbf{k} = (0, 0, 1)$$

are the **unit coordinate vectors** in space, shown in Fig. 14.33.

We emphasize that the mathematical definition of a *vector* is identical to the definition of a *point;* each is an ordered collection of real numbers. The names "vector" and "point" indicate different geometric interpretations for such a collection. If you ask mathematicians to show pictorially the *vector* (1, 2) in the plane, they will draw the arrow indicating length and direction shown in Fig. 14.34. On the other hand, if you ask them to show pictorially the *point* (3, −2), they will make the large dot shown in the figure, just indicating a position.

The vector $\mathbf{b} = (1, 3)$ in the plane is shown in Fig. 14.35. If we choose translated $\bar{x}, \bar{y}$-coordinates at the point $(x, y) = (-4, 2)$, then the point

Figure 14.34 Draw an arrow for a vector and a dot for a point.

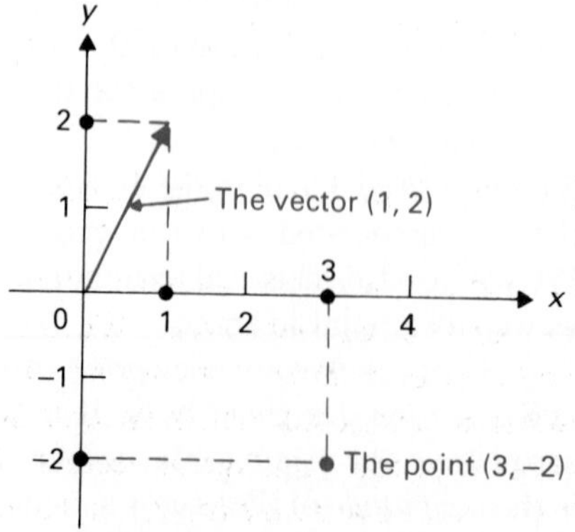

Figure 14.35 The two vectors **b** = (1, 3) have the same magnitude and direction and are considered equal.

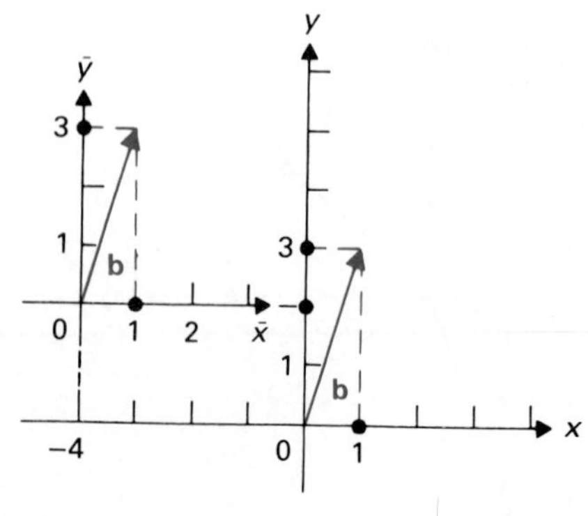

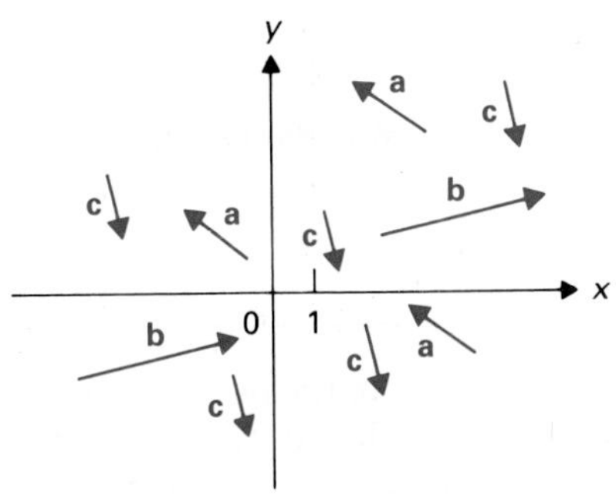

Figure 14.36 Vectors labeled **a** are equal, as are vectors labeled **b** and vectors labeled **c**.

$(x, y) = (-3, 5)$ has translated coordinates $(\bar{x}, \bar{y}) = (1, 3)$. The vector from the translated origin to $(\bar{x}, \bar{y}) = (1, 3)$ has the same magnitude and direction as **b**; the arrows have the same length and are parallel. Since we consider magnitude and direction to be the distinguishing characteristics of a vector, it is natural to call this vector, having $(x, y) = (-4, 2)$ as point of origin, **b** also. That is, we consider two vectors to be **equal** if they have the same magnitude and direction, even though they emanate from different points. Figure 14.36 gives an illustration of this. Sometimes the term "free vectors" is used to indicate this independence from the point of origin.

The following summarizes what we have done and gives a bit more terminology and notation.

> Each triple $\mathbf{a} = (a_1, a_2, a_3)$ is both a point in space and a **vector** in space. The **length** of vector **a**, denoted by $|\mathbf{a}|$, is $\sqrt{a_1^2 + a_2^2 + a_3^2}$. The number a_i is the ***i*th component** of the vector. Any vector of length 1 is a **unit vector;** in particular, $\mathbf{i} = (1, 0, 0)$, $\mathbf{j} = (0, 1, 0)$, $\mathbf{k} = (0, 0, 1)$ are the unit coordinate vectors in space. The vector $\mathbf{0} = (0, 0, 0)$ is the **zero vector.** Similar terminology is used for a vector $\mathbf{a} = (a_1, a_2)$ in the plane. Two vectors are **equal** if they have the same magnitude and direction; their points of origin may be different.

EXAMPLE 1 Show that $\mathbf{a} = (0, 1)$ and $\mathbf{b} = (\frac{1}{2}, \sqrt{3}/2)$ are both unit vectors.

Solution A unit vector is one of length 1. We indeed have

$$|\mathbf{a}| = \sqrt{0^2 + 1^2} = 1$$

and

$$|\mathbf{b}| = \sqrt{\left(\frac{1}{2}\right)^2 + \left(\frac{\sqrt{3}}{2}\right)^2} = \sqrt{\frac{4}{4}} = 1. \quad \square$$

EXAMPLE 2 Find all values of c such that $\mathbf{a} = (\frac{1}{3}, -\frac{2}{3}, c)$ is a unit vector.

Solution We must have

$$1 = |\mathbf{a}| = \sqrt{\left(\frac{1}{3}\right)^2 + \left(-\frac{2}{3}\right)^2 + c^2} = \sqrt{\frac{5}{9} + c^2}.$$

Thus $\frac{5}{9} + c^2 = 1$, so $c^2 = 1 - \frac{5}{9} = \frac{4}{9}$, and $c = \pm\frac{2}{3}$. $\square$

THE ALGEBRA OF VECTORS

Addition of numbers is a very important operation. It was the first topic in mathematics we ever studied. We think of the real numbers as filling up a line. Addition of numbers on the line has a very important generalization to addition in the plane and in space. Addition in space or in the plane is usually phrased in the language of vectors. We will follow this convention. Recall that in space, **a** means (a_1, a_2, a_3) and **b** means (b_1, b_2, b_3).

DEFINITION 14.1 Vector addition

Let **a** and **b** be vectors in space. The **sum** of **a** and **b** is the vector in space defined by

$$\mathbf{a} + \mathbf{b} = (a_1 + b_1, a_2 + b_2, a_3 + b_3).$$

The analogous notion of sum holds for vectors in the plane. Note that vector addition is defined only for two vectors with the *same number of components*.

EXAMPLE 3 Let $\mathbf{a} = (0, -1, 3)$ and $\mathbf{b} = (4, 2, -6)$. Find $\mathbf{a} + \mathbf{b}$.

Solution We have $\mathbf{a} + \mathbf{b} = (0, -1, 3) + (4, 2, -6) = (4, 1, -3)$. □

We can interpret vector addition geometrically, in terms of arrows. We make our sketches in the plane, but similar arguments hold for vectors in space.

There are two ways that we can arrive at the arrow representing $\mathbf{a} + \mathbf{b}$ from arrows representing **a** and representing **b**. One way is to find the diagonal of the parallelogram that has (0, 0) as a vertex and the arrows represented by **a** and **b** as sides emanating from (0, 0). As indicated in Fig. 14.37, the diagonal arrow of the parallelogram that starts at (0, 0) represents the vector $\mathbf{a} + \mathbf{b}$.

For an alternative method to represent $\mathbf{a} + \mathbf{b}$, choose the point (a_1, a_2) as new origin and draw the arrow representing the *translated vector* **b**. This arrow starts at the tip of the original vector **a** and goes to the point with translated coordinates (b_1, b_2). The tip of this translated vector **b** then falls at the same point as the tip of the desired vector $\mathbf{a} + \mathbf{b}$ emanating from the original origin, (0, 0), as shown in Fig. 14.38. As we remarked before, we consider the vector **b** from the original origin and the vector **b** emanating from (a_1, a_2) in Fig. 14.38 to be equal. They have the same magnitude and direction.

We turn now to another operation in the vector algebra. When dealing with vectors in space (or the plane), we often refer to a real number as a **scalar** to distinguish it from the vectors. As our second algebraic operation involving vectors, we define the product of a scalar r and a vector $\mathbf{a} = (a_1, a_2, a_3)$ in space.

Figure 14.37 **a + b** as the diagonal of the parallelogram.

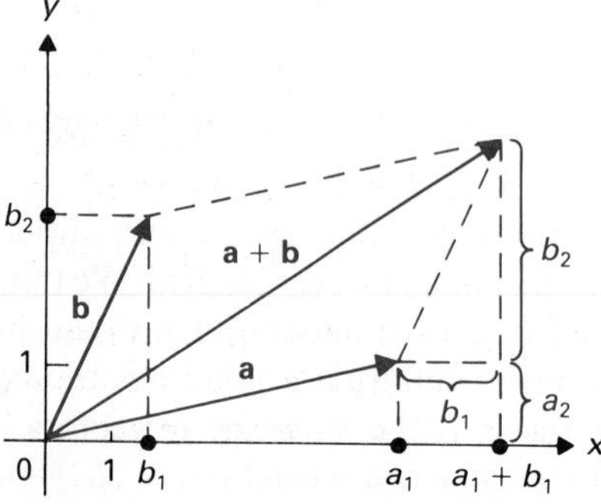

Figure 14.38 Finding **a + b** by starting **b** at the tip of **a**.

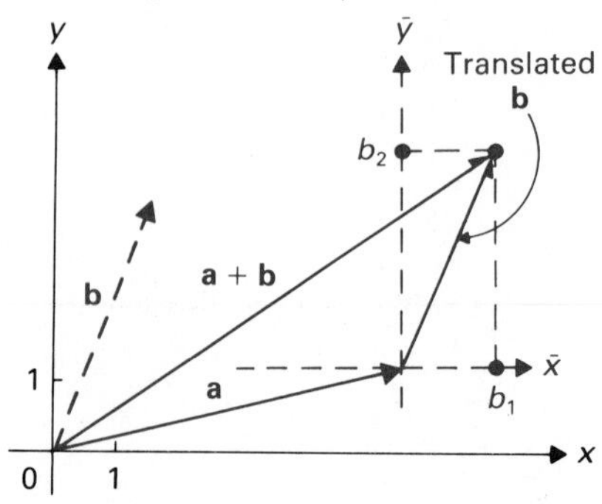

DEFINITION 14.2 Scalar multiplication

The **product** $r\mathbf{a}$ of the scalar r and the vector $\mathbf{a}$ is the vector

$$r\mathbf{a} = (ra_1, ra_2, ra_3).$$

Again, the analogous notion holds for the product of a vector in the plane by the scalar r.

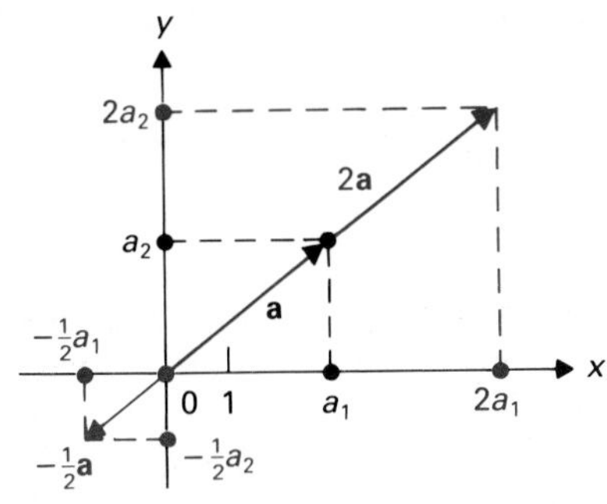

Figure 14.39 Multiplying the vector **a** by 2 and by $-\frac{1}{2}$.

EXAMPLE 4 Let $\mathbf{a} = (3, -1, 4)$. Find $2\mathbf{a}$.

Solution We have $2\mathbf{a} = 2(3, -1, 4) = (6, -2, 8)$. □

From Definition 14.2, we see at once that for any real number r and vector $\mathbf{a} = (a_1, a_2, a_3)$, we have

$$|r\mathbf{a}| = |(ra_1, ra_2, ra_3)| = \sqrt{(ra_1)^2 + (ra_2)^2 + (ra_3)^2}$$
$$= \sqrt{r^2} \cdot \sqrt{a_1^2 + a_2^2 + a_3^2}.$$

But $\sqrt{r^2} = |r|$, and $\sqrt{a_1^2 + a_2^2 + a_3^2} = |\mathbf{a}|$. Consequently,

$$\boxed{|r\mathbf{a}| = |r| \cdot |\mathbf{a}|.} \qquad \textbf{(1)}$$

Thus if we wish to describe $r\mathbf{a}$ in terms of *length* and *direction*, Eq. (1) shows that the length of the product $r\mathbf{a}$ is $|r|$ times the length of $\mathbf{a}$. Figure 14.39 indicates that we should consider $r\mathbf{a}$ to have the same direction as $\mathbf{a}$ if $r > 0$ and the opposite direction if $r < 0$.

EXAMPLE 5 Let $|\mathbf{a}| = 5$. Describe the magnitude of $3\mathbf{a}$ and $-7\mathbf{a}$, and compare their direction with that of $\mathbf{a}$.

Solution From Eq. (1), we have

$$|3\mathbf{a}| = |3| \cdot |\mathbf{a}| = 3 \cdot 5 = 15$$

and

$$|-7\mathbf{a}| = |-7| \cdot |\mathbf{a}| = 7 \cdot 5 = 35.$$

Now $3 > 0$ and $-7 < 0$, so $3\mathbf{a}$ has the same direction as $\mathbf{a}$, and $-7\mathbf{a}$ has the opposite direction. □

Figure 14.40 $\mathbf{a} = a_1\mathbf{i} + a_2\mathbf{j} + a_3\mathbf{k}$

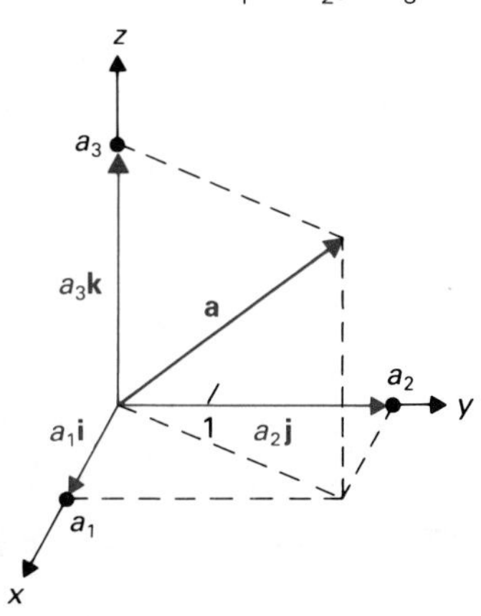

Note that for every vector $\mathbf{a} = (a_1, a_2, a_3)$ in space, we have

$$\begin{aligned} \mathbf{a} &= (a_1, a_2, a_3) \\ &= a_1(1, 0, 0) + a_2(0, 1, 0) + a_3(0, 0, 1) \\ &= a_1\mathbf{i} + a_2\mathbf{j} + a_3\mathbf{k}, \end{aligned}$$

as illustrated in Fig. 14.40. Similarly, in the plane,

$$\mathbf{b} = (b_1, b_2) = b_1(1, 0) + b_2(0, 1) = b_1\mathbf{i} + b_2\mathbf{j}.$$

Such **i**, **j**, **k** expressions for vectors are used extensively. We will use this notation, wherever it is not too cumbersome, to indicate that we are thinking in terms of the "vector interpretation" of the ordered triple or pair of numbers.

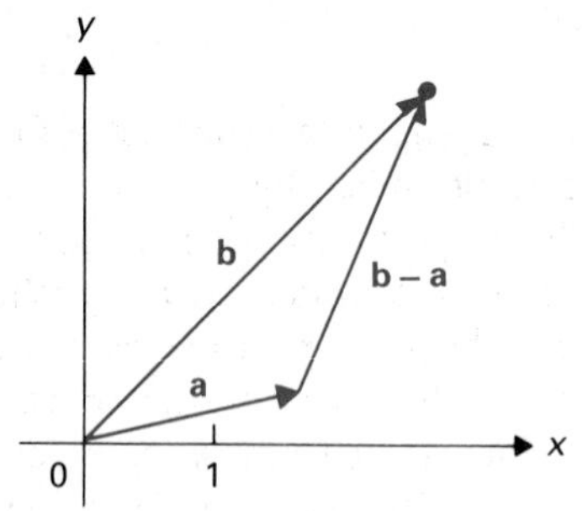

Figure 14.41 **b** – **a** goes from the tip of **a** to the tip of **b**, for

$$\mathbf{a} + (\mathbf{b} - \mathbf{a}) = \mathbf{b}.$$

Note that nonzero vectors **a** and **b** are *parallel* if and only if $\mathbf{b} = r\mathbf{a}$ for some real number r. Thus $r\mathbf{a}$ is a vector parallel to **a** of length $|r| \cdot |\mathbf{a}|$ and having the *same* direction as **a** if $r > 0$ and the *opposite* direction if $r < 0$.

EXAMPLE 6 Show that the vectors $\mathbf{a} = \mathbf{i} - 3\mathbf{j}$ and $\mathbf{b} = 2\mathbf{i} - 6\mathbf{j}$ in the plane are parallel and have the same direction.

Solution If they are parallel, then $\mathbf{b} = r\mathbf{a}$ for some r. Looking at the coefficients of **i** in the two vectors, we see we would have to have $r = 2$. We find that $2\mathbf{a} = 2(\mathbf{i} - 3\mathbf{j}) = 2\mathbf{i} - 6\mathbf{j} = \mathbf{b}$, so the vectors are parallel. Since $2 > 0$, they have the same direction. □

We may now define the *difference* $\mathbf{b} - \mathbf{a}$ of vectors **a** and **b** by

$$\mathbf{b} - \mathbf{a} = \mathbf{b} + (-1)\mathbf{a}.$$

We have already defined vector addition and the product $(-1)\mathbf{a}$. Since $\mathbf{a} + (\mathbf{b} - \mathbf{a}) = \mathbf{b}$, we see that

> $\mathbf{b} - \mathbf{a}$ is the vector that when added to **a** yields **b**.

A translated coordinate representation of $\mathbf{b} - \mathbf{a}$, starting from the tip of **a** as translated origin, therefore ends at (b_1, b_2), as shown in Fig. 14.41.

EXAMPLE 7 Find the vector that, as an arrow, starts from the point $(1, -4)$ and ends at the point $(-3, 2)$ in the plane.

Solution We let $\mathbf{a} = \mathbf{i} - 4\mathbf{j}$ and $\mathbf{b} = -3\mathbf{i} + 2\mathbf{j}$. We have just seen that $\mathbf{b} - \mathbf{a}$ is the vector starting from the point $\mathbf{a} = (1, -4)$ that reaches to the point $\mathbf{b} = (-3, 2)$. Thus the desired vector is

$$\mathbf{b} - \mathbf{a} = (-3\mathbf{i} + 2\mathbf{j}) - (\mathbf{i} - 4\mathbf{j}) = -4\mathbf{i} + 6\mathbf{j}. \quad \square$$

In summary, vector addition, subtraction, and multiplication by a scalar are easy to do; we simply carry out the corresponding numerical operations in each component.

We list some algebraic laws that hold for vector algebra. These laws are easy to prove. Their demonstration is left to the exercises (see Exercises 48 through 52).

THEOREM 14.1 Properties of vector operations

For all vectors **a**, **b**, and **c** in space or the plane, and all scalars r and s, the following laws hold.

a. $(\mathbf{a} + \mathbf{b}) + \mathbf{c} = \mathbf{a} + (\mathbf{b} + \mathbf{c})$	Associativity of addition
b. $\mathbf{a} + \mathbf{b} = \mathbf{b} + \mathbf{a}$	Commutativity of addition
c. $r(s\mathbf{a}) = (rs)\mathbf{a}$	Associativity of multiplication by scalars
d. $(r + s)\mathbf{a} = r\mathbf{a} + s\mathbf{a}$	A right distributive law
e. $r(\mathbf{a} + \mathbf{b}) = r\mathbf{a} + r\mathbf{b}$	A left distributive law

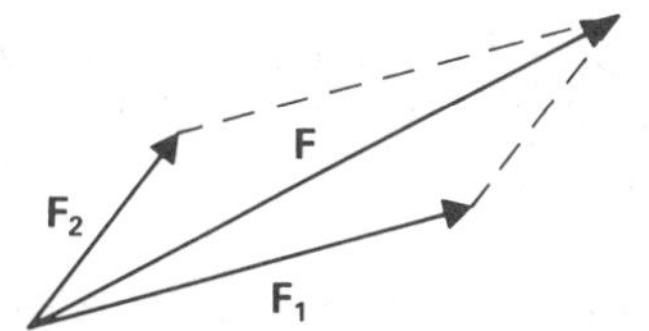

Figure 14.42 Forces $\mathbf{F}_1$ and $\mathbf{F}_2$ applied to an object are equivalent to the single force $\mathbf{F} = \mathbf{F}_1 + \mathbf{F}_2$.

A PHYSICAL MODEL FOR VECTORS

When studying motion, a physicist may use a vector to represent a *force*. Suppose, for example, that we are moving some object by pushing it. The motion of the object is influenced by the *direction* in which we are pushing and how *hard* we are pushing. Thus the force with which we are pushing can be conveniently represented by a vector, which we regard as having the *direction* in which we are pushing and a *length* representing how hard we are pushing. If we double the force with which we push, the force vector doubles in length; this corresponds to the multiplication of the force vector by the scalar 2.

Suppose that two people are pushing on an object with forces that correspond to vectors $\mathbf{F}_1$ and $\mathbf{F}_2$, as shown in Fig. 14.42. It can be shown that the motion of the object due to these combined forces is the same as the motion that would result if only one person were pushing with a force expressed by a vector that is the diagonal of the parallelogram with arrow vectors $\mathbf{F}_1$ and $\mathbf{F}_2$ as adjacent sides (see Fig. 14.42). Thus this *resultant force vector* is precisely the vector $\mathbf{F}_1 + \mathbf{F}_2$.

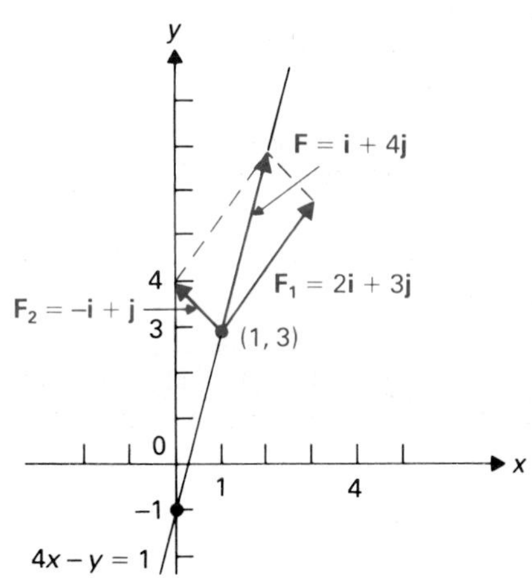

Figure 14.43 $\mathbf{F} = \mathbf{F}_1 + \mathbf{F}_2$ gives the direction of motion.

EXAMPLE 8 A body in the plane is traveling subject only to the two constant forces $\mathbf{F}_1 = 2\mathbf{i} + 3\mathbf{j}$ and $\mathbf{F}_2 = -\mathbf{i} + \mathbf{j}$. The body passes through the point (1, 3). Find the equation of the line on which the body travels.

Solution The resultant force vector is

$$\mathbf{F} = \mathbf{F}_1 + \mathbf{F}_2 = (2\mathbf{i} + 3\mathbf{j}) + (-\mathbf{i} + \mathbf{j}) = \mathbf{i} + 4\mathbf{j},$$

as shown in Fig. 14.43. The body moves in the direction given by this resultant vector, along the line containing the arrow of this vector viewed as starting at the point (1, 3). This line has slope

$$\frac{\mathbf{j}\text{-coefficient}}{\mathbf{i}\text{-coefficient}} = \frac{4}{1} = 4.$$

The equation of the line is then

$$y - 3 = 4(x - 1), \qquad \text{or} \qquad 4x - y = 1. \quad \square$$

PERPENDICULAR VECTORS

It is very important for us to know when two directions are perpendicular. There is a very easy criterion for this in terms of vectors. Any three points in space not all on the same line determine a plane. Imagine that Fig. 14.44 gives a picture of such a plane determined by three points, (0, 0, 0), (a_1, a_2, a_3), and (b_1, b_2, b_3). The vectors $\mathbf{a}$ and $\mathbf{b}$ of Fig. 14.44 are perpendicular if and only if the Pythagorean relation holds, that is, if and only if

$$|\mathbf{a}|^2 + |\mathbf{b}|^2 = |\mathbf{a} - \mathbf{b}|^2. \tag{2}$$

Figure 14.44 For a 90° angle at (0, 0, 0), we must have $|\mathbf{a} - \mathbf{b}|^2 = |\mathbf{a}|^2 + |\mathbf{b}|^2$.

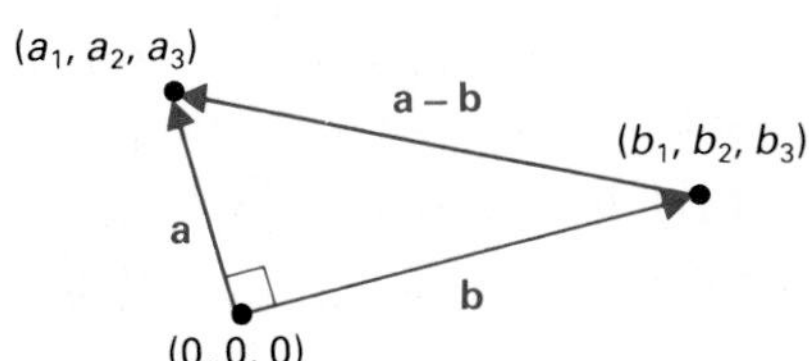

By definition of the length of a vector, we see that this is true when

$$a_1^2 + a_2^2 + a_3^2 + b_1^2 + b_2^2 + b_3^2 = (a_1 - b_1)^2 + (a_2 - b_2)^2 + (a_3 - b_3)^2. \tag{3}$$

Squaring the terms on the right-hand side of Eq. (3) and canceling, we obtain the condition

$$0 = -2a_1b_1 - 2a_2b_2 - 2a_3b_3, \qquad \text{or} \qquad a_1b_1 + a_2b_2 + a_3b_3 = 0.$$

Thus the vectors **a** and **b** in space are *perpendicular* if and only if

$$a_1b_1 + a_2b_2 + a_3b_3 = 0. \tag{4}$$

Of course, the corresponding result with just two components holds for vectors in the plane.

EXAMPLE 9 Determine whether or not the vectors $\mathbf{a} = -\mathbf{i} + 3\mathbf{j} + 2\mathbf{k}$ and $\mathbf{b} = 5\mathbf{i} - \mathbf{j} + 4\mathbf{k}$ in space are perpendicular.

Solution We have

$$-1 \cdot 5 + 3 \cdot -1 + 2 \cdot 4 = -5 - 3 + 8 = 0,$$

so **a** and **b** are indeed perpendicular. □

According to condition (4), the zero vector **0** is perpendicular to *every* vector in space. For this reason, it is convenient to think of **0** as having *all directions* rather than *no direction.*

SUMMARY

Formulas are given for vectors in space, but are equally valid for vectors in the plane.

1. Each point $\mathbf{a} = (a_1, a_2, a_3)$ can also be considered as a vector having direction from the origin to the point (a_1, a_2, a_3) and length $|\mathbf{a}| = \sqrt{a_1^2 + a_2^2 + a_3^2}$. A unit vector has length 1. The zero vector is $\mathbf{0} = (0, 0, 0)$.
2. Addition of vectors is given by adding corresponding coordinates, that is, $\mathbf{a} + \mathbf{b} = (a_1, a_2, a_3) + (b_1, b_2, b_3) = (a_1 + b_1, a_2 + b_2, a_3 + b_3)$.
3. In space, the unit coordinate vectors are often written

$$\mathbf{i} = (1, 0, 0), \qquad \mathbf{j} = (0, 1, 0), \qquad \mathbf{k} = (0, 0, 1),$$

so $(a_1, a_2, a_3) = a_1\mathbf{i} + a_2\mathbf{j} + a_3\mathbf{k}$. In the plane, one uses just $\mathbf{i} = (1, 0)$ and $\mathbf{j} = (0, 1)$.

4. Multiplication of a vector **a** by a scalar (real number) r is given by

$$r\mathbf{a} = r(a_1\mathbf{i} + a_2\mathbf{j} + a_3\mathbf{k}) = (ra_1)\mathbf{i} + (ra_2)\mathbf{j} + (ra_3)\mathbf{k}.$$

5. $\mathbf{b} - \mathbf{a} = (b_1 - a_1)\mathbf{i} + (b_2 - a_2)\mathbf{j} + (b_3 - a_3)\mathbf{k}$
6. $\mathbf{a} + \mathbf{b}$ is the vector along the diagonal of the parallelogram having **a** and **b** as coterminous edges.
7. $\mathbf{b} - \mathbf{a}$ is the vector starting at the tip of **a** and reaching to the tip of **b**.
8. Two nonzero vectors **a** and **b** are parallel if $\mathbf{b} = r\mathbf{a}$ for some scalar r.
9. Two vectors **a** and **b** are perpendicular if $a_1b_1 + a_2b_2 + a_3b_3 = 0$.

EXERCISES

1. Let $\mathbf{a} = 2\mathbf{i} - \mathbf{j}$ and $\mathbf{b} = -3\mathbf{i} - 2\mathbf{j}$ be vectors in the plane. Sketch, using arrows, the vectors $\mathbf{a}$, $\mathbf{b}$, $\mathbf{a} + \mathbf{b}$, $\mathbf{a} - \mathbf{b}$, and $-(4/3)\mathbf{a}$.
2. Repeat Exercise 1 for the vectors $\mathbf{a} = 3\mathbf{i} + \mathbf{j}$ and $\mathbf{b} = -\mathbf{i} + 2\mathbf{j}$.

In Exercises 3 through 10, let $\mathbf{a} = -\mathbf{i} + 3\mathbf{j} - 2\mathbf{k}$, $\mathbf{b} = 4\mathbf{i} - \mathbf{k}$, and $\mathbf{c} = -3\mathbf{i} - \mathbf{j} + 2\mathbf{k}$. Compute the indicated vector.

3. $3\mathbf{a}$
4. $-2\mathbf{c}$
5. $\mathbf{a} + \mathbf{b}$
6. $3\mathbf{b} - 2\mathbf{c}$
7. $\mathbf{a} + 2(\mathbf{b} - 3\mathbf{c})$
8. $3(\mathbf{a} - 2\mathbf{b})$
9. $4(3\mathbf{a} + 5\mathbf{b})$
10. $2\mathbf{a} - (3\mathbf{b} - \mathbf{c})$

In Exercises 11 through 16, let $\mathbf{a} = 3\mathbf{i} - 2\mathbf{j} + 2\mathbf{k}$ and $\mathbf{b} = -\mathbf{i} + 4\mathbf{j} + \mathbf{k}$. Compute the indicated quantity.

11. $|\mathbf{a}|$
12. $|\mathbf{a} + \mathbf{b}|$
13. $|\mathbf{a}| + |\mathbf{b}|$
14. $|-2\mathbf{a}|$
15. $|\mathbf{b}| - |\mathbf{a}|$
16. $|\mathbf{b} - 3\mathbf{a}|$

In Exercises 17 through 20, find the vector represented by an arrow reaching from the first point to the second.

17. From $(-1, 3)$ to $(4, 2)$
18. From $(-5, -8)$ to $(3, -2)$
19. From $(-3, 2, 5)$ to $(4, -2, -6)$
20. From $(3, -6, -2)$ to $(1, 4, -8)$

In Exercises 21 through 26, determine whether the pair of vectors is parallel, perpendicular, or neither. If two vectors are parallel, state whether they have the same or opposite directions.

21. $3\mathbf{i} - \mathbf{j}$ and $4\mathbf{i} + 12\mathbf{j}$
22. $-2\mathbf{i} + 6\mathbf{j}$ and $4\mathbf{i} - 12\mathbf{j}$
23. $3\mathbf{i} - \mathbf{j}$ and $4\mathbf{i} + 3\mathbf{j} + 2\mathbf{k}$
24. $2\mathbf{i} - 3\mathbf{j} + \mathbf{k}$ and $8\mathbf{i} + 2\mathbf{j} - 10\mathbf{k}$
25. $\sqrt{2}\mathbf{i} + \sqrt{18}\mathbf{j} - \sqrt{8}\mathbf{k}$ and $2\mathbf{i} + 6\mathbf{j} - 4\mathbf{k}$
26. $6\mathbf{i} - 9\mathbf{j} + 3\mathbf{k}$ and $-4\mathbf{i} + 6\mathbf{j} - 2\mathbf{k}$

In Exercises 27 through 36, find all values of c, if there are any, such that the given statement is true.

27. $2\mathbf{i} + c\mathbf{j}$ is parallel to $4\mathbf{i} + 6\mathbf{j}$
28. $2\mathbf{j} + c\mathbf{j}$ is parallel to $-5\mathbf{i} + 3\mathbf{j}$
29. $2\mathbf{i} + c\mathbf{j}$ is parallel to $3\mathbf{i}$
30. $2\mathbf{i} + c\mathbf{j}$ is parallel to $3\mathbf{j}$
31. $2\mathbf{i} + c\mathbf{j}$ is perpendicular to $3\mathbf{i} + 4\mathbf{j}$
32. $2\mathbf{i} + c\mathbf{j}$ is perpendicular to $8\mathbf{i} - c\mathbf{j}$
33. $2\mathbf{i} + c\mathbf{j}$ is perpendicular to $8\mathbf{i} + c\mathbf{j}$
34. $c\mathbf{i} + 2\mathbf{j} - \mathbf{k}$ is perpendicular to $\mathbf{j} - 4\mathbf{k}$
35. $c\mathbf{i} + 2\mathbf{j} - \mathbf{k}$ is perpendicular to $\mathbf{i} - 3\mathbf{k}$
36. $c\mathbf{i} + 2\mathbf{j} - \mathbf{k}$ is perpendicular to $-5\mathbf{i} + \mathbf{j} + 2\mathbf{k}$
37. Find a unit vector in space parallel to $\mathbf{i} - \mathbf{j} + 3\mathbf{k}$ and having the same direction.
38. Find two unit vectors in the plane perpendicular to $3\mathbf{i} - 4\mathbf{j}$.
39. Find two unit vectors in space that are not parallel and each of which is perpendicular to $-2\mathbf{i} + \mathbf{j} + 2\mathbf{k}$.
40. Find a unit vector parallel to the line tangent to $y = x^2$ at $(1, 1)$ and pointing in the direction of increasing x.
41. Find a unit vector normal to the tangent line to $y = x^4$ at $(-1, 1)$ and pointing in the direction of increasing y.
42. A body in the plane is moving subject only to the constant force vectors $\mathbf{F}_1 = \mathbf{i} + 3\mathbf{j}$ and $\mathbf{F}_2 = 2\mathbf{i} - \mathbf{j}$. If the body travels through $(1, 2)$, find the equation of the line on which it travels.
43. Repeat Exercise 42 for the force vectors $\mathbf{F}_1 = -3\mathbf{i} + 2\mathbf{j}$ and $\mathbf{F}_2 = 5\mathbf{i} - 4\mathbf{j}$ and the point $(-1, -3)$.
44. Repeat Exercise 42 for the force vectors $\mathbf{F}_1 = 3\mathbf{i} + 4\mathbf{j}$, $\mathbf{F}_2 = -\mathbf{i} + 2\mathbf{j}$, and $\mathbf{F}_3 = -3\mathbf{i} - 2\mathbf{j}$ and the point $(2, -1)$.
45. Repeat Exercise 42 for the force vectors $\mathbf{F}_1 = \mathbf{i} - 3\mathbf{j}$, $\mathbf{F}_2 = 4\mathbf{i} - 2\mathbf{j}$, and $\mathbf{F}_3 = -5\mathbf{i} + 5\mathbf{j}$ and the point $(-1, 4)$.
46. Show that $(1, -5)$, $(9, -11)$, and $(4, -1)$ are vertices of a right triangle in the plane.
47. Show that $(1, -1, 4)$, $(3, -2, 4)$, $(-4, 2, 6)$, and $(-2, 1, 6)$ are vertices of a parallelogram in space.

In Exercises 48 through 52, let $\mathbf{a} = a_1\mathbf{i} + a_2\mathbf{j} + a_3\mathbf{k}$, $\mathbf{b} = b_1\mathbf{i} + b_2\mathbf{j} + b_3\mathbf{k}$, and $\mathbf{c} = c_1\mathbf{i} + c_2\mathbf{j} + c_3\mathbf{k}$ be any vectors, and let r and s be any scalars. Prove the indicated properties.

48. $(\mathbf{a} + \mathbf{b}) + \mathbf{b} = \mathbf{a} + (\mathbf{b} + \mathbf{c})$
49. $\mathbf{a} + \mathbf{b} = \mathbf{b} + \mathbf{a}$
50. $r(s\mathbf{a}) = (rs)\mathbf{a}$
51. $(r + s)\mathbf{a} = r\mathbf{a} + s\mathbf{a}$
52. $r(\mathbf{a} + \mathbf{b}) = r\mathbf{a} + r\mathbf{b}$

14.4 THE DOT PRODUCT OF VECTORS

It is often important to know the *angle* between two vectors, viewing the vectors as arrows starting from the same point. The direction a sailboat is sailing and the direction from which the wind is blowing can both be represented by vectors. The angle between these vectors measures how close the boat is sailing into the wind.

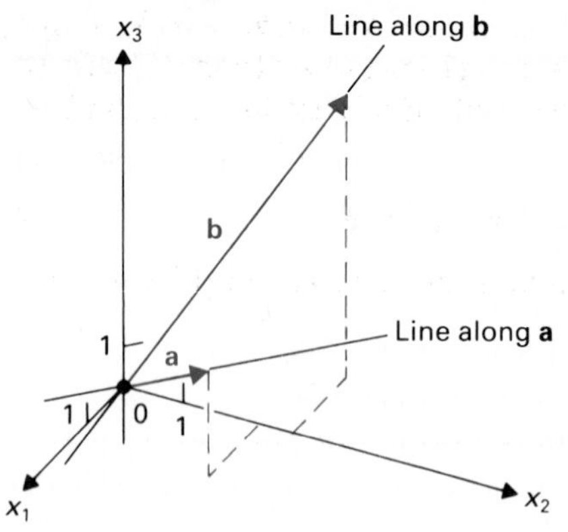

Figure 14.45 Nonparallel vectors **a** and **b** in space.

We know how to determine if two vectors are parallel (at angles of 0 or π) and whether two vectors are perpendicular (at an angle of $\pi/2$). In this section, we will derive an easy formula for the angle between any two nonzero vectors. This formula will involve the dot product of the vectors, which is a scalar quantity easily computed from the components of the vectors. The dot product will also enable us to find what part of a force vector can be considered to be acting in a particular direction. For example, if a number of people are pushing a stalled car, some may not be able to get directly behind the car to push but may be pushing at an angle. How much of their effort actually goes toward moving the car forward? The dot product will enable us to answer this question.

In Section 14.5, we take up another way of multiplying two vectors, their cross product. The cross product is a *vector* quantity, while the dot product is a *scalar*.

THE DOT PRODUCT

Let $\mathbf{a} = a_1\mathbf{i} + a_2\mathbf{j} + a_3\mathbf{k}$ and $\mathbf{b} = b_1\mathbf{i} + b_2\mathbf{j} + b_3\mathbf{k}$ be nonparallel and nonzero vectors in space. We can view **a** and **b** geometrically as arrows starting from the origin, as shown in Fig. 14.45. The vector **a** gives the direction for a line through the origin, labeled "the line along **a**" in Fig. 14.45. Similarly, the vector **b** gives the direction for the line along **b**. These two intersecting lines in space determine a plane, as indicated in Fig. 14.46. The points in this plane are precisely those points $\mathbf{x} = (x_1, x_2, x_3)$ that can be expressed in the form $\mathbf{x} = t\mathbf{a} + s\mathbf{b}$ for some scalars t and s.

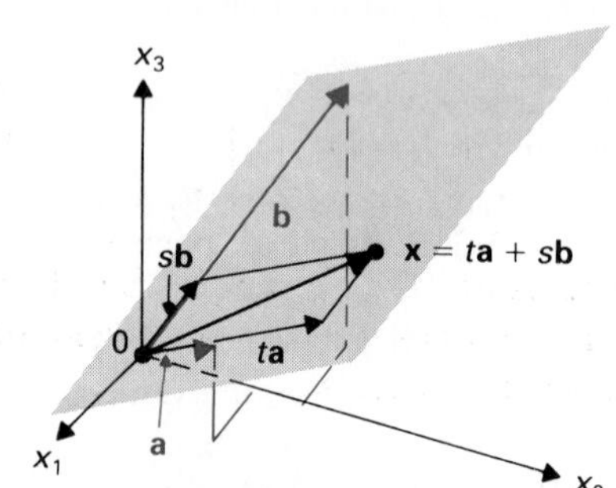

Figure 14.46 The plane in space determined by **a** and **b** at the origin; every point **x** in the plane can be written as **x** = *t***a** + *s***b**.

In Fig. 14.47, we take the plane in Fig. 14.46 as the plane of our page. We are interested in finding the angle θ between **a** and **b** as shown in Fig. 14.47. To find θ, we apply the law of cosines to the triangle shown in that figure and obtain

$$d^2 = |\mathbf{a}|^2 + |\mathbf{b}|^2 - 2|\mathbf{a}||\mathbf{b}|\cos\theta. \tag{1}$$

From the preceding section, we know that $d^2 = |\mathbf{b} - \mathbf{a}|^2$. We can easily compute $|\mathbf{b} - \mathbf{a}|^2$, $|\mathbf{a}|$, and $|\mathbf{b}|$ in terms of the components a_i and b_i of the vectors **a** and **b**, so we can use Eq. (1) to find $\cos\theta$ in terms of these components. We have

$$d^2 = |\mathbf{b} - \mathbf{a}|^2 = (b_1 - a_1)^2 + (b_2 - a_2)^2 + (b_3 - a_3)^2,$$

while

$$|\mathbf{a}|^2 = a_1^2 + a_2^2 + a_3^2 \qquad \text{and} \qquad |\mathbf{b}|^2 = b_1^2 + b_2^2 + b_3^2.$$

Substituting these quantities in Eq. (1), squaring out the terms $(b_i - a_i)^2$ in d^2, and canceling the terms a_i^2 and b_i^2 from both sides of the resulting equation, we obtain

$$-2a_1b_1 - 2a_2b_2 - 2a_3b_3 = -2|\mathbf{a}||\mathbf{b}|\cos\theta. \tag{2}$$

Thus

$$\cos\theta = \frac{a_1b_1 + a_2b_2 + a_3b_3}{|\mathbf{a}||\mathbf{b}|}. \tag{3}$$

Figure 14.47 The angle θ between **a** and **b**.

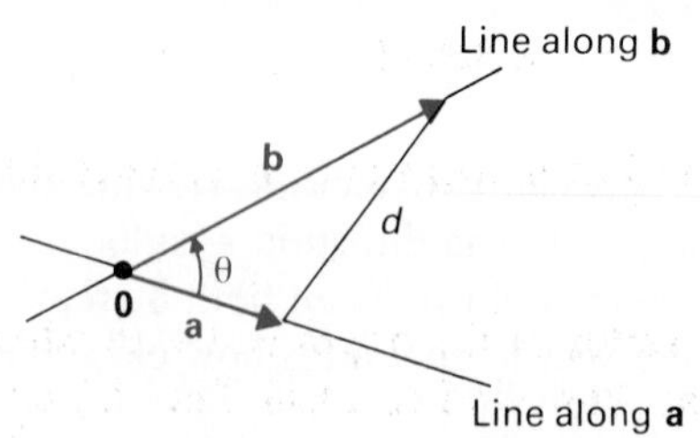

The numerator in Eq. (3) is familiar; we saw on page 608 that vectors **a** and **b** are perpendicular if and only if $a_1b_1 + a_2b_2 + a_3b_3 = 0$. Note that this is consistent with Eq. (3); the vectors should be perpendicular if and only if $\theta = \pi/2$ or $\theta = 3\pi/2$, so that $\cos\theta = 0$. The number $a_1b_1 + a_2b_2 + a_3b_3$ appearing in Eq. (3) is so important that it is given a special name, the *dot* (or *scalar*, or *inner*) *product of* **a** *and* **b**. (The result of this product of **a** and **b** is a *scalar*.) From Eq. (3), we obtain

$$a_1b_1 + a_2b_2 + a_3b_3 = |\mathbf{a}||\mathbf{b}|\cos\theta. \quad \textbf{(4)}$$

Note that if either **a** or **b** is **0**, so that θ is undefined, then $|\mathbf{a}|$ or $|\mathbf{b}|$ is zero, and Eq. (4) still holds formally.

DEFINITION 14.3 Dot product

The **dot product a · b** of **a** and **b** is

$$\mathbf{a} \cdot \mathbf{b} = a_1b_1 + a_2b_2 + a_3b_3. \quad \textbf{(5)}$$

Equation (4) shows that

$$\boxed{\mathbf{a} \cdot \mathbf{b} = |\mathbf{a}||\mathbf{b}|\cos\theta,} \quad \textbf{(6)}$$

where θ is the angle between **a** and **b**. This gives us a geometric interpretation of the dot product. If we are interested in finding the angle θ, we rewrite Eq. (6) as

$$\boxed{\cos\theta = \frac{\mathbf{a} \cdot \mathbf{b}}{|\mathbf{a}||\mathbf{b}|}, \quad \text{or} \quad \theta = \cos^{-1}\frac{\mathbf{a} \cdot \mathbf{b}}{|\mathbf{a}||\mathbf{b}|}.} \quad \textbf{(7)}$$

Of course, this same work could be done with vectors in the plane having only two components. The notion of the dot product is defined for any two vectors having the same number of components.

EXAMPLE 1 Compute **a · b** for the vectors

$$\mathbf{a} = \mathbf{i} - 4\mathbf{j} + 3\mathbf{k} \quad \text{and} \quad \mathbf{b} = 6\mathbf{i} - 2\mathbf{j} - \mathbf{k}.$$

Solution We have

$$\mathbf{a} \cdot \mathbf{b} = (1)(6) + (-4)(-2) + (3)(-1) = 6 + 8 - 3 = 11. \quad \square$$

EXAMPLE 2 Find the angle between the vectors $\mathbf{a} = \mathbf{i} - 4\mathbf{j}$ and $\mathbf{b} = 3\mathbf{i} + 2\mathbf{j}$ in the plane.

Solution We have

$$\theta = \cos^{-1}\frac{\mathbf{a} \cdot \mathbf{b}}{|\mathbf{a}||\mathbf{b}|} = \cos^{-1}\frac{3 - 8}{\sqrt{17}\sqrt{13}} = \cos^{-1}\frac{-5}{\sqrt{17}\sqrt{13}}$$

$$\approx 109.65°. \quad \square$$

Figure 14.48 $\mathbf{d} = \mathbf{i} + \mathbf{j} + \mathbf{k}$ is directed along the diagonal of the cube.

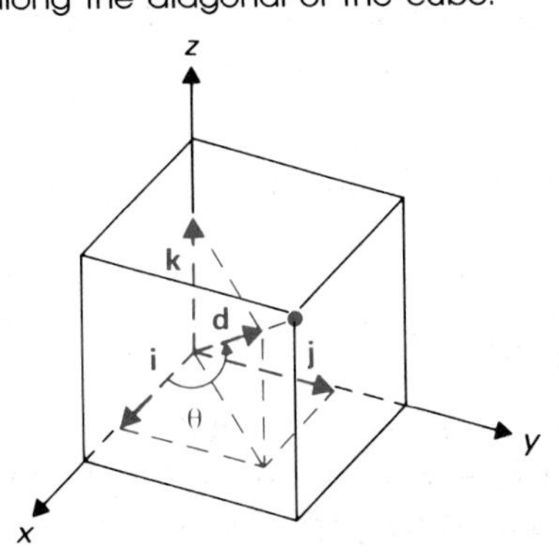

EXAMPLE 3 Find the angle θ that the diagonal of a cube in space makes with an edge of the cube.

Solution We may take a cube with a vertex at the origin and with edges falling on the positive coordinate axes, as shown in Fig. 14.48. Then **i**, **j**, and

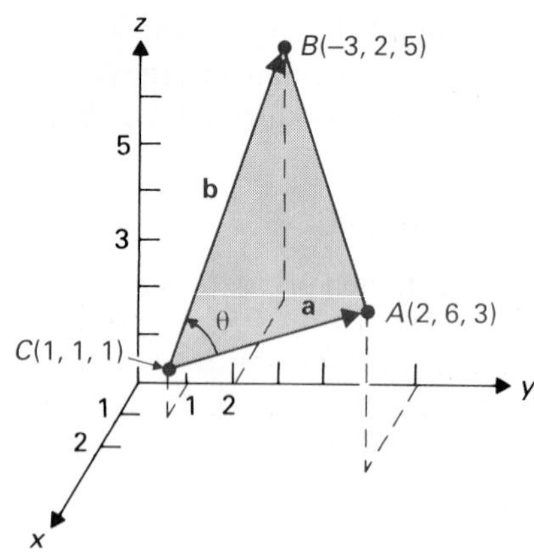

Figure 14.49 Angle $ACB = \theta$ of the triangle with the given vertices.

k are vectors along edges of the cube, while a vector along a diagonal is

$$\mathbf{d} = \mathbf{i} + \mathbf{j} + \mathbf{k}.$$

We have

$$\theta = \cos^{-1} \frac{\mathbf{i} \cdot \mathbf{d}}{|\mathbf{i}|\,|\mathbf{d}|} = \cos^{-1} \frac{1}{1 \cdot \sqrt{3}}$$

$$= \cos^{-1} \frac{1}{\sqrt{3}} \approx 54.74°. \quad \square$$

EXAMPLE 4 Find the angle BCA of the triangle in space with vertices $A(2, 6, 3)$, $B(-3, 2, 5)$, and $C(1, 1, 1)$.

Solution We find the vector **a** reaching from C to A and the vector **b** reaching from C to B, as shown in Fig. 14.49. From the preceding section, we know that

$$\mathbf{a} = \text{Vector from } C \text{ to } A = \mathbf{i} + 5\mathbf{j} + 2\mathbf{k},$$

$$\mathbf{b} = \text{Vector from } C \text{ to } B = -4\mathbf{i} + \mathbf{j} + 4\mathbf{k}.$$

From Eq. (7), we have

$$\cos\theta = \frac{\mathbf{a} \cdot \mathbf{b}}{|\mathbf{a}|\,|\mathbf{b}|} = \frac{-4 + 5 + 8}{\sqrt{30}\sqrt{33}} = \frac{9}{3\sqrt{110}} = \frac{3}{\sqrt{110}}.$$

Then

$$\theta = \cos^{-1} \frac{3}{\sqrt{110}} \approx 73.38°. \quad \square$$

EXAMPLE 5 Find the acute angle at which the line $3x + y = 10$ meets the parabola $y = x^2$ at the point $(2, 4)$.

Solution We try to find a vector **a** with direction tangent to the parabola at (2, 4) and a vector **b** along the line at (2, 4), as shown in Fig. 14.50. The angle between the line and the parabola is then the angle θ between **a** and **b** in the figure.

The slope of the parabola at (2, 4) is given by

$$\left.\frac{dy}{dx}\right|_2 = 2x|_2 = 4,$$

so we see that the vector

$$\mathbf{a} = \mathbf{i} + 4\mathbf{j}$$

has direction tangent to the parabola. The line $y = -3x + 10$ has slope -3, so the vector $\mathbf{i} - 3\mathbf{j}$ is directed along the line. It appears from Fig. 14.50 that we want a vector with negative **i**-component for **b**, for we want the acute angle θ. Thus we take

$$\mathbf{b} = -\mathbf{i} + 3\mathbf{j}.$$

Equation (7) then yields

$$\cos\theta = \frac{\mathbf{a} \cdot \mathbf{b}}{|\mathbf{a}|\,|\mathbf{b}|} = \frac{-1 + 12}{\sqrt{17}\sqrt{10}} = \frac{11}{\sqrt{170}}.$$

Figure 14.50 The angle θ between $y = x^2$ and $3x + y = 10$ at (2, 4).

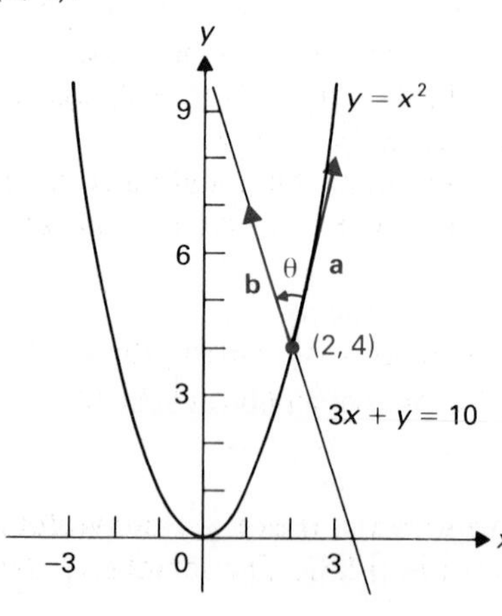

Thus

$$\theta = \cos^{-1} \frac{11}{\sqrt{170}} \approx 32.47°. \quad \square$$

ALGEBRAIC PROPERTIES OF THE DOT PRODUCT

Theorem 14.2 lists some of the algebraic properties of the dot product. We observe the usual convention that an algebraic operation written in multiplicative notation is performed before one written in additive notation, in the absence of parentheses. For example,

$$\mathbf{a} \cdot \mathbf{b} + \mathbf{a} \cdot \mathbf{c} = (\mathbf{a} \cdot \mathbf{b}) + (\mathbf{a} \cdot \mathbf{c}).$$

THEOREM 14.2 Properties of the dot product

Let **a**, **b**, and **c** be vectors with the same number of components and let r be a scalar. Then

a.	$\mathbf{a} \cdot \mathbf{a} \geq 0$ and $\mathbf{a} \cdot \mathbf{a} = 0$ if and only if $\mathbf{a} = \mathbf{0}$	Nonnegative property
b.	$\mathbf{a} \cdot \mathbf{b} = \mathbf{b} \cdot \mathbf{a}$	Commutative property
c.	$\mathbf{a} \cdot (\mathbf{b} + \mathbf{c}) = \mathbf{a} \cdot \mathbf{b} + \mathbf{a} \cdot \mathbf{c}$	Distributive property
d.	$(r\mathbf{a}) \cdot \mathbf{b} = \mathbf{a} \cdot (r\mathbf{b}) = r(\mathbf{a} \cdot \mathbf{b})$	Homogeneous property
e.	$\mathbf{a} \cdot \mathbf{a} = \lvert\mathbf{a}\rvert^2$	Length property
f.	$\mathbf{a} \cdot \mathbf{b} = 0$ if and only if **a** and **b** are perpendicular	Perpendicular property

Properties (a), (b), (c), and (d) are proved easily from formula (5) for $\mathbf{a} \cdot \mathbf{b}$ in terms of the components of **a** and of **b**. Illustrating with the proof of property (a), we have, for vectors in space,

$$\mathbf{a} \cdot \mathbf{a} = a_1a_1 + a_2a_2 + a_3a_3 = a_1^2 + a_2^2 + a_3^2 \geq 0,$$

and this sum of squares is zero if and only if each $a_i = 0$, that is, if and only if $\mathbf{a} = \mathbf{0}$. We leave the proofs of properties (b), (c), and (d) to the exercises (see Exercises 37, 38, and 39).

Properties (e) and (f) are really restatements of previous definitions in the notation of the dot product. We defined the length of a vector **a** in space to be $\sqrt{a_1^2 + a_2^2 + a_3^2} = \sqrt{\mathbf{a} \cdot \mathbf{a}}$, and we also defined **a** and **b** to be perpendicular vectors if and only if $\mathbf{a} \cdot \mathbf{b} = a_1b_1 + a_2b_2 + a_3b_3 = 0$. Recall that we defined the vector **0** to be perpendicular to every vector.

The properties of the dot product are very important, and all sorts of consequences can be easily derived from them. We will give a geometric illustration.

Figure 14.51 The parallelogram has $\mathbf{a} + \mathbf{b}$ and $\mathbf{a} - \mathbf{b}$ as vector diagonals.

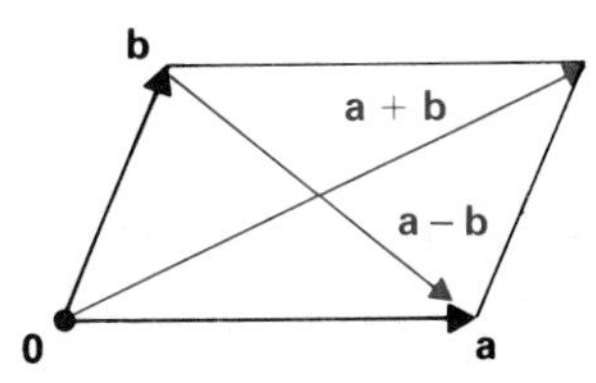

EXAMPLE 6 Show that the sum of the squares of the lengths of the diagonals of a parallelogram is equal to the sum of the squares of the lengths of the sides. (This is the *parallelogram relation.*)

Solution We take our parallelogram with a vertex at the origin and vectors **a** and **b** as coterminous sides, as shown in Fig. 14.51. The lengths of the diagonals are then $|\mathbf{a}+\mathbf{b}|$ and $|\mathbf{a}-\mathbf{b}|$. Using Theorem 14.2, we have

$$\begin{aligned}|\mathbf{a}+\mathbf{b}|^2+|\mathbf{a}-\mathbf{b}|^2 &= (\mathbf{a}+\mathbf{b})\cdot(\mathbf{a}+\mathbf{b})+(\mathbf{a}-\mathbf{b})\cdot(\mathbf{a}-\mathbf{b})\\ &= \mathbf{a}\cdot\mathbf{a}+2\mathbf{a}\cdot\mathbf{b}+\mathbf{b}\cdot\mathbf{b}+\mathbf{a}\cdot\mathbf{a}-2\mathbf{a}\cdot\mathbf{b}+\mathbf{b}\cdot\mathbf{b}\\ &= 2(\mathbf{a}\cdot\mathbf{a})+2(\mathbf{b}\cdot\mathbf{b})\\ &= 2|\mathbf{a}|^2+2|\mathbf{b}|^2,\end{aligned}$$

which is what we wished to prove. You may think that we have used only property (e) of Theorem 14.2, but we also used properties (b), (c), and (d), as we ask you to show in Exercise 40. □

VECTOR PROJECTION

Let **a** and **b** be nonzero vectors in space. Then

$$\frac{1}{|\mathbf{a}|}\mathbf{a} \quad \text{and} \quad \frac{1}{|\mathbf{b}|}\mathbf{b}$$

are unit vectors in the directions of **a** and **b**, respectively. We write such vectors as

$$\frac{\mathbf{a}}{|\mathbf{a}|} \quad \text{and} \quad \frac{\mathbf{b}}{|\mathbf{b}|}$$

in what follows; for a nonzero scalar r and vector **a**, we define

$$\boxed{\frac{\mathbf{a}}{r}=\frac{1}{r}\mathbf{a}.}$$

In Fig. 14.52, we again imagine that the plane containing both **a** and **b** is the plane of the page and show the angle θ between **a** and **b**. In the figure, we have also labeled the *vector projection of* **a** *onto* **b**, or *on the line along* **b**. This vector has the direction of **b** if θ is an acute angle and the direction of $-\mathbf{b}$ if θ is an obtuse angle. The length of this vector projection is the distance from the origin **0** to the foot of the perpendicular dropped to the line along **b** from the tip of the vector **a**.* From Fig. 14.52, this length is $|\mathbf{a}|\cos\theta$, so the vector projection is

$$\begin{aligned}(|\mathbf{a}|\cos\theta)\frac{\mathbf{b}}{|\mathbf{b}|} &= |\mathbf{a}||\mathbf{b}|(\cos\theta)\frac{\mathbf{b}}{|\mathbf{b}|^2}\\ &= \frac{\mathbf{a}\cdot\mathbf{b}}{|\mathbf{b}|^2}\mathbf{b}=\left(\frac{\mathbf{a}\cdot\mathbf{b}}{\mathbf{b}\cdot\mathbf{b}}\right)\mathbf{b},\end{aligned}$$

where we have used the fact that $\mathbf{b}\cdot\mathbf{b}=|\mathbf{b}|^2$ from property (e) of Theorem 14.2. The number

$$|\mathbf{a}|(\cos\theta)=\frac{|\mathbf{a}||\mathbf{b}|\cos\theta}{|\mathbf{b}|}=\frac{\mathbf{a}\cdot\mathbf{b}}{|\mathbf{b}|}$$

Figure 14.52 Vector projection of **a** onto **b** is $|\mathbf{a}|(\cos\theta)(\mathbf{b}/|\mathbf{b}|)$; scalar component of **a** along **b** is $|\mathbf{a}|(\cos\theta)$.

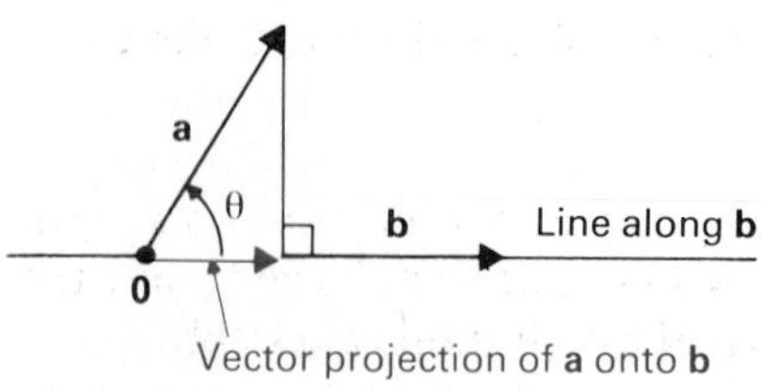

* For this reason, this vector projection is also called the orthogonal (perpendicular) projection.

is the (signed) length of the vector projection. In summary, if $\mathbf{b} \neq \mathbf{0}$, then

$$\frac{\mathbf{a} \cdot \mathbf{b}}{\mathbf{b} \cdot \mathbf{b}}\mathbf{b} = \textbf{Vector projection of } \mathbf{a} \text{ onto } \mathbf{b}, \tag{8}$$

$$\frac{\mathbf{a} \cdot \mathbf{b}}{|\mathbf{b}|} = \textbf{Scalar component of } \mathbf{a} \text{ along } \mathbf{b}. \tag{9}$$

EXAMPLE 7 Find the vector projection of $\mathbf{a} = 2\mathbf{i} - \mathbf{j} + 3\mathbf{k}$ onto $\mathbf{b} = \mathbf{i} - 3\mathbf{j} - \mathbf{k}$ and the scalar component of $\mathbf{a}$ along $\mathbf{b}$.

Solution The vector projection of $\mathbf{a}$ onto $\mathbf{b}$ is

$$\frac{\mathbf{a} \cdot \mathbf{b}}{|\mathbf{b}|^2}\mathbf{b} = \frac{2 + 3 - 3}{1 + 9 + 1}(\mathbf{i} - 3\mathbf{j} - \mathbf{k}) = \frac{2}{11}(\mathbf{i} - 3\mathbf{j} - \mathbf{k})$$
$$= \frac{2}{11}\mathbf{i} - \frac{6}{11}\mathbf{j} - \frac{2}{11}\mathbf{k}.$$

The scalar component of $\mathbf{a}$ along $\mathbf{b}$ is

$$\frac{\mathbf{a} \cdot \mathbf{b}}{|\mathbf{b}|} = \frac{2}{\sqrt{11}}. \quad \square$$

If $\mathbf{b}$ is a unit vector so that $|\mathbf{b}| = 1$, then the vector projection of $\mathbf{a}$ onto $\mathbf{b}$ takes the simpler form

$$(\mathbf{a} \cdot \mathbf{b})\mathbf{b}.$$

EXAMPLE 8 Show that the vector projection of $\mathbf{a} = 3\mathbf{i} - 4\mathbf{j}$ along $\mathbf{i}$ is $3\mathbf{i}$ and along $\mathbf{j}$ is $-4\mathbf{j}$.

Solution By Eq. (8), the vector projection of $\mathbf{a}$ along $\mathbf{i}$ is

$$(\mathbf{a} \cdot \mathbf{i})\mathbf{i} = (3 + 0)\mathbf{i} = 3\mathbf{i}$$

and along $\mathbf{j}$ is

$$(\mathbf{a} \cdot \mathbf{j})\mathbf{j} = (0 - 4)\mathbf{j} = -4\mathbf{j}.$$

Of course the scalar components of $\mathbf{a}$ along $\mathbf{i}$ and $\mathbf{j}$ are just 3 and -4 respectively. $\square$

Figure 14.53 Force vector **F** broken into perpendicular projections.

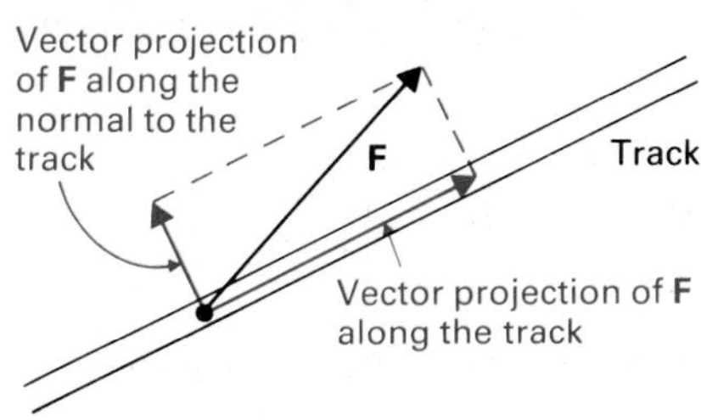

Suppose a body is constrained to move along a track, as shown in Fig. 14.53. It can be shown that if a force $\mathbf{F}$ is applied to the body, as shown in the figure, then the body moves along the track as though $\mathbf{F}$ were replaced by its vector projection along the track. This projection along the track is the portion of the force vector that actually moves the body along the track. The vector projection of $\mathbf{F}$ along the perpendicular to the track acts to try to tip the body off the track. Note that $\mathbf{F}$ in Fig. 14.53 is the sum of these two vector projections.

EXAMPLE 9 Let a body in the plane be constrained to move on a track along the line $y = 2x$. If a force vector $\mathbf{F} = \mathbf{i} + 5\mathbf{j}$ acts on the body, find the magnitude of the force that actually moves the body along the track.

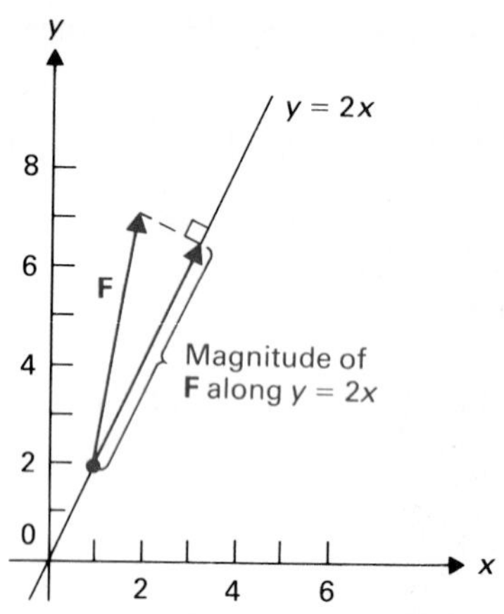

Figure 14.54 The scalar component (or magnitude) of **F** along $y = 2x$.

Solution Since we are asked for the magnitude of the force along the track, we want the scalar component of **F** along the line $y = 2x$. A vector along the line is $\mathbf{a} = \mathbf{i} + 2\mathbf{j}$, as shown in Fig. 14.54. By Eq. (9), the scalar component of **F** along the line is

$$\frac{\mathbf{F} \cdot \mathbf{a}}{|\mathbf{a}|} = \frac{1 + 10}{\sqrt{5}} = \frac{11}{\sqrt{5}}. \quad \square$$

SUMMARY

1. Let **a** and **b** be vectors in space or in the plane. The dot (or scalar) product of **a** and **b** is the number
$$\mathbf{a} \cdot \mathbf{b} = a_1b_1 + a_2b_2 + a_3b_3.$$
2. The dot product of **a** and **b** is described geometrically by
$$\mathbf{a} \cdot \mathbf{b} = |\mathbf{a}||\mathbf{b}| \cos \theta,$$
where θ is the angle between **a** and **b**.
3. Algebraic properties of the dot product are listed in Theorem 14.2 on page 613.
4. The vector projection of **a** onto **b** if $\mathbf{b} \neq \mathbf{0}$ is the vector
$$\frac{\mathbf{a} \cdot \mathbf{b}}{\mathbf{b} \cdot \mathbf{b}}\mathbf{b},$$
and the number $(\mathbf{a} \cdot \mathbf{b})/|\mathbf{b}|$ is the component of **a** along **b**.

EXERCISES

In Exercises 1 through 10, find the angle between the vectors.

1. $\mathbf{i} + 4\mathbf{j}$ and $-8\mathbf{i} + 2\mathbf{j}$
2. $3\mathbf{i} + 2\mathbf{j} - 2\mathbf{k}$ and $4\mathbf{j} + \mathbf{k}$
3. $\mathbf{k}$ and $\mathbf{i} - \mathbf{k}$
4. $3\mathbf{i} + 4\mathbf{j}$ and $-\mathbf{i}$
5. $\left(\cos \frac{\pi}{3}\right)\mathbf{i} - \left(\sin \frac{\pi}{3}\right)\mathbf{j}$ and $\left(\cos \frac{\pi}{4}\right)\mathbf{i} + \left(\sin \frac{\pi}{4}\right)\mathbf{j}$.
[*Hint:* Make use of a trigonometric identity.]
6. $\left(\sin \frac{\pi}{6}\right)\mathbf{i} + \left(\cos \frac{\pi}{6}\right)\mathbf{j}$ and $\left(\cos \frac{\pi}{4}\right)\mathbf{i} + \left(\sin \frac{\pi}{4}\right)\mathbf{j}$.
[*Hint:* As in Exercise 5.]
7. $\left(\sin \frac{\pi}{3}\right)\mathbf{i} + \left(\cos \frac{\pi}{3}\right)\mathbf{j}$ and $\left(\cos \frac{\pi}{5}\right)\mathbf{i} - \left(\sin \frac{\pi}{5}\right)\mathbf{j}$.
[*Hint:* As in Exercise 5.]
8. $\mathbf{i} - 3\mathbf{j} + 4\mathbf{k}$ and $10\mathbf{i} + 6\mathbf{j} + 2\mathbf{k}$
9. $3\mathbf{i} - \mathbf{j} + 2\mathbf{k}$ and $-9\mathbf{i} + 3\mathbf{j} - 6\mathbf{k}$
10. $\mathbf{i} - 2\mathbf{j} + 2\mathbf{k}$ and $-3\mathbf{i} - 6\mathbf{j} - 2\mathbf{k}$
11. Find the angle ACB of the triangle with vertices $A(0, 1, 6)$, $B(2, 3, 0)$, and $C(-1, 3, 4)$.
12. Find the angle between the line through $(-1, 2, 4)$ and $(3, 4, 0)$ and the line also through $(-1, 2, 4)$ that passes through $(5, 7, 2)$.
13. Find the acute angle between the lines $x - 2y = -4$ and $2x + 3y = 13$ at their point of intersection.
14. Find two lines in the plane through $(1, -4)$ that intersect the line $y = -2x + 7$ at an angle of 45°.
15. Find the acute angle between the curves $y = x^2$ and $y = 2 - x^2$ at a point of intersection.
16. Find the acute angle between the curves $y = x^2$ and $y = x^3$ at the point $(1, 1)$ of intersection.

In Exercises 17 through 24, find the vector projection of the first vector onto the second vector and the scalar component of the first vector along the second.

17. $6\mathbf{i} - 3\mathbf{j} + 12\mathbf{k}$ on $\frac{1}{3}\mathbf{i} - \frac{2}{3}\mathbf{j} + \frac{2}{3}\mathbf{k}$
18. $\mathbf{i} + 3\mathbf{j} + 4\mathbf{k}$ on $\mathbf{j}$
19. $\mathbf{j}$ on $\mathbf{i} + 3\mathbf{j} + 4\mathbf{k}$
20. $2\mathbf{i} - \mathbf{j}$ on $-2\mathbf{i} + 3\mathbf{j}$
21. $3\mathbf{i} + \mathbf{j} - 2\mathbf{k}$ on $4\mathbf{i} + 2\mathbf{j} + 7\mathbf{k}$
22. $\mathbf{a} = -\mathbf{i} + \mathbf{j} + 3\mathbf{k}$ on $\mathbf{b} = 3\mathbf{i} - 2\mathbf{j} + \mathbf{k}$

23. $\mathbf{i} + \mathbf{j} + \mathbf{k}$ on $\mathbf{i} + \mathbf{j}$

24. $\mathbf{i} + \mathbf{j}$ on $\mathbf{i} + \mathbf{j} + \mathbf{k}$

25. Let $\mathbf{a}$ and $\mathbf{b}$ be vectors with $\mathbf{b} \neq \mathbf{0}$ and let $\mathbf{c}$ be the vector projection of $\mathbf{a}$ onto $\mathbf{b}$. Show that $\mathbf{a} - \mathbf{c}$ is perpendicular to $\mathbf{b}$.

26. A person is pushing a box across a floor with a force vector of magnitude 20 lb directed at an angle of 30° below the horizontal. Find the magnitude of the component of the force vector actually moving the box along the floor.

27. A body in the plane is constrained to move on the curve $y = 3x^2$. The body is moved in the direction of increasing x by a constant force vector $\mathbf{F} = 50\mathbf{j}$. Find the magnitude of the component of $\mathbf{F}$ actually moving the body along the curve at the instant when the body is at (1, 3).

28. a) Answer Exercise 27 if the body is at the point (x, y) on $y = 3x^2$.
b) What is the magnitude of the component of $\mathbf{F}$ actually moving the body when it is at the origin?
c) Find the limit of the magnitude of the component of $\mathbf{F}$ moving the body as $x \to \infty$.

29. A 100-lb weight is suspended by a rope passed through an eyelet on top of the weight and making angles of 30° with the vertical, as shown in Fig. 14.55. Find the tension (magnitude of the force vector) along the rope. [*Hint:* The sum of the force vectors along the two halves of the rope at the eyelet must be an upward vertical vector of magnitude 100.]

30. a) Answer Exercise 29 if each half of the rope makes an angle of α with the vertical at the eyelet.
b) Find the tension in the rope if both sides are vertical ($\alpha = 0$).
c) What happens if an attempt is made to stretch the rope out straight (horizontal) while the 100-lb weight hangs on it?

31. Suppose a weight of 100 lb is suspended by *two* ropes *tied* at an eyelet on top of the weight, as shown in Fig. 14.56. Let the angles the ropes make with the vertical be α and β, as shown in the figure. Let the tension in the ropes be T_1 for the right-hand rope and T_2 for the left-hand rope.
a) Argue that the force vector $\mathbf{F}_1$ shown in Fig. 14.56 is $T_1(\sin \alpha)\mathbf{i} + T_1(\cos \alpha)\mathbf{j}$.
b) Find the corresponding expression for $\mathbf{F}_2$ in terms of T_2 and β.
c) If the system is in equilibrium, $\mathbf{F}_1 + \mathbf{F}_2 = 100\mathbf{j}$, so $\mathbf{F}_1 + \mathbf{F}_2$ must have $\mathbf{i}$-component 0 and $\mathbf{j}$-component 100. Write two equations reflecting this fact, using your answers to parts (a) and (b).
d) Find T_1 and T_2 if $\alpha = 45°$ and $\beta = 30°$.

Figure 14.55 Both halves of the rope make an angle of 30° with the vertical.

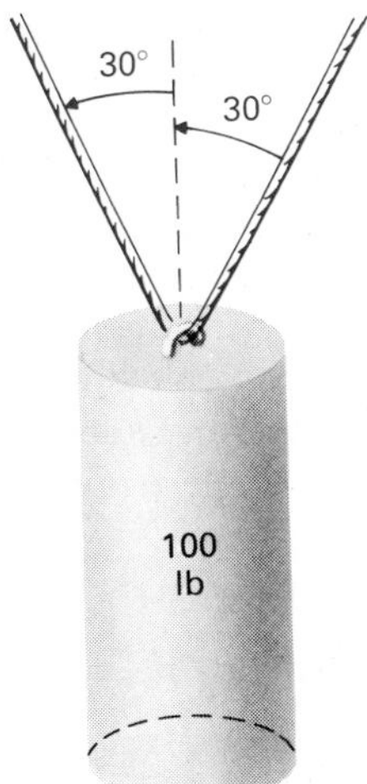

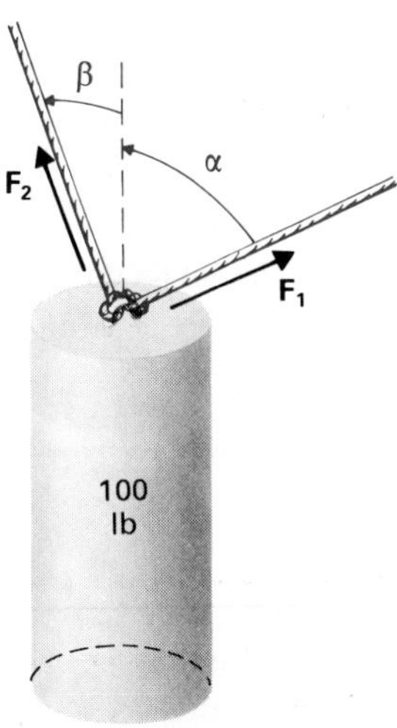

Figure 14.56 Two ropes tied at the eyelet and making angles of α and β with the vertical.

32. Use vector methods to show that the diagonals of a rhombus (parallelogram with equal sides) are perpendicular. [*Hint:* Use a figure like Fig. 14.51 and show that $(\mathbf{a}+\mathbf{b})\cdot(\mathbf{a}-\mathbf{b})=0$.]

33. Use vector methods to show that the midpoint of the hypotenuse of a right triangle is equidistant from the three vertices. [*Hint:* See Fig. 14.57. Show that

$$\left|\frac{\mathbf{a}+\mathbf{b}}{2}\right| = \left|\frac{\mathbf{a}-\mathbf{b}}{2}\right|.]$$

34. Show that the vectors $|\mathbf{a}|\mathbf{b} + |\mathbf{b}|\mathbf{a}$ and $|\mathbf{a}|\mathbf{b} - |\mathbf{b}|\mathbf{a}$ are perpendicular.

Figure 14.57 Vector $(\mathbf{a} + \mathbf{b})/2$ to the midpoint of the hypotenuse.

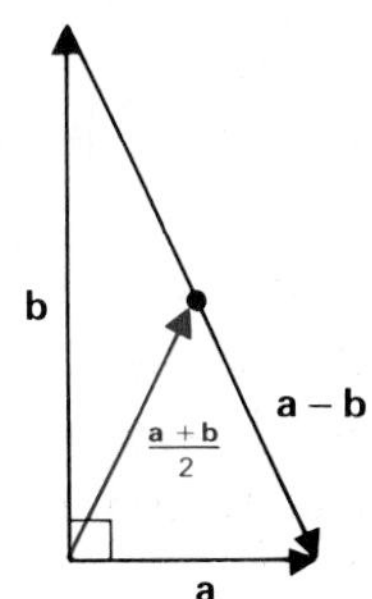

35. Show that the vector $|\mathbf{a}|\mathbf{b} + |\mathbf{b}|\mathbf{a}$ bisects the angle between **a** and **b**.

36. Show that for **a**, **b**, and **c**, the equation

$$\mathbf{a} \cdot \mathbf{b} = \mathbf{a} \cdot \mathbf{c},$$

where $\mathbf{a} \neq \mathbf{0}$, need not imply $\mathbf{b} = \mathbf{c}$.

37. Using the formula for the dot product, show that

$$\mathbf{a} \cdot \mathbf{b} = \mathbf{b} \cdot \mathbf{a}$$

for all **a** and **b** in space.

38. Using the formula for the dot product, show that

$$\mathbf{a} \cdot (\mathbf{b} + \mathbf{c}) = \mathbf{a} \cdot \mathbf{b} + \mathbf{a} \cdot \mathbf{c}$$

for all **a**, **b**, and **c** in space.

39. Using the formula for the dot product, show that $(r\mathbf{a}) \cdot \mathbf{b} = \mathbf{a} \cdot (r\mathbf{b}) = r(\mathbf{a} \cdot \mathbf{b})$ for all **a** and **b** in space and all scalars r.

40. Find where properties (b), (c), and (d) given in Theorem 14.2 were used in the proof of the parallelogram relation in Example 4.

14.5 THE CROSS PRODUCT AND TRIPLE PRODUCTS

In the preceding section, we introduced the *dot product* of two vectors. The dot product is a *scalar* quantity. Using the dot product, we can easily find the angle between vectors, and vector and scalar projections of one vector along another.

In this section, we study the *cross product* of two vectors in space. The cross product is a *vector* quantity. In this chapter, we use the cross product to find a vector perpendicular to each of two given vectors. The cross product of two vectors is easily computed using a symbolic *determinant*. We open this section with the information about matrices and determinants that we need to understand the cross product. The section closes with an application of the dot and cross products to finding the area of a triangle in space and the volume of a skew box in space.

REVIEW OF 2×2 AND 3×3 DETERMINANTS

A *square matrix* is a square array of numbers. For example,

$$\begin{pmatrix} -3 & 4 \\ 2 & 6 \end{pmatrix}$$

is a 2 × 2 (read "two-by-two") matrix, and

$$\begin{pmatrix} -1 & 0 & 4 \\ 2 & 1 & 0 \\ 3 & -4 & 5 \end{pmatrix}$$

is a 3 × 3 matrix. Each square matrix has associated with it a number, called the *determinant* of the matrix. The determinant is denoted by vertical lines rather than large parentheses on the sides of the array. The determinant of a 2 × 2 matrix is defined to be

$$\begin{vmatrix} a_1 & a_2 \\ b_1 & b_2 \end{vmatrix} = a_1b_2 - a_2b_1. \qquad (1)$$

The determinant of a 3 × 3 matrix is defined in terms of the determinants of 2 × 2 matrices, as follows:

$$\begin{vmatrix} a_1 & a_2 & a_3 \\ b_1 & b_2 & b_3 \\ c_1 & c_2 & c_3 \end{vmatrix} = a_1\begin{vmatrix} b_2 & b_3 \\ c_2 & c_3 \end{vmatrix} - a_2\begin{vmatrix} b_1 & b_3 \\ c_1 & c_3 \end{vmatrix} + a_3\begin{vmatrix} b_1 & b_2 \\ c_1 & c_2 \end{vmatrix}. \quad \textbf{(2)}$$

Formula (2) is easily remembered. The coefficients of the three determinants on the right-hand side of formula (2) are the entries of the first row of the original 3 × 3 matrix, with alternate plus and minus signs. The first determinant on the right-hand side in formula (2) is the determinant of the 2 × 2 matrix obtained by crossing out the row and column in which the coefficient a_1 appears in the 3 × 3 matrix. The second determinant is obtained by crossing out the row and column in which a_2 appears, and so on.

EXAMPLE 1 Find

$$\begin{vmatrix} 7 & -2 \\ 4 & 3 \end{vmatrix}.$$

Solution We have

$$\begin{vmatrix} 7 & -2 \\ 4 & 3 \end{vmatrix} = 7 \cdot 3 - (-2)4 = 21 + 8 = 29. \quad \square$$

EXAMPLE 2 Find

$$\begin{vmatrix} 2 & 3 & 5 \\ -4 & 2 & 6 \\ 1 & 0 & 3 \end{vmatrix}.$$

Solution We have

$$\begin{aligned} \begin{vmatrix} 2 & 3 & 5 \\ -4 & 2 & 6 \\ 1 & 0 & 3 \end{vmatrix} &= 2\begin{vmatrix} 2 & 6 \\ 0 & 3 \end{vmatrix} - 3\begin{vmatrix} -4 & 6 \\ 1 & 3 \end{vmatrix} + 5\begin{vmatrix} -4 & 2 \\ 1 & 0 \end{vmatrix} \\ &= 2(6 - 0) - 3(-12 - 6) + 5(0 - 2) \\ &= 12 + 54 - 10 = 56. \quad \square \end{aligned}$$

The only facts we need to know about determinants for our work with this text are given in the following theorem.

THEOREM 14.3 Properties of determinants

a. If any two rows of a square matrix are the same, then the determinant of the matrix is zero.

b. If any two rows of a square matrix are interchanged, the determinant of the new matrix differs from the determinant of the original matrix only in sign.

We will actually need special cases of Theorem 14.3 only for 3×3 matrices, where for part (a) one of the two identical rows is the first row and for part (b) the second and third rows are interchanged. We ask you to prove these special cases in Exercises 9 and 10 by computing that

$$\begin{vmatrix} a_1 & a_2 & a_3 \\ a_1 & a_2 & a_3 \\ c_1 & c_2 & c_3 \end{vmatrix} = 0, \qquad \begin{vmatrix} a_1 & a_2 & a_3 \\ b_1 & b_2 & b_3 \\ a_1 & a_2 & a_3 \end{vmatrix} = 0, \qquad \text{and}$$

$$\begin{vmatrix} a_1 & a_2 & a_3 \\ b_1 & b_2 & b_3 \\ c_1 & c_2 & c_3 \end{vmatrix} = -\begin{vmatrix} a_1 & a_2 & a_3 \\ c_1 & c_2 & c_3 \\ b_1 & b_2 & b_3 \end{vmatrix}.$$

THE CROSS PRODUCT OF VECTORS

Let $\mathbf{a} = a_1\mathbf{i} + a_2\mathbf{j} + a_3\mathbf{k}$ and $\mathbf{b} = b_1\mathbf{i} + b_2\mathbf{j} + b_3\mathbf{k}$ be vectors in space.

DEFINITION 14.4 Cross product of vectors

The **cross product a × b** of **a** and **b** is the vector found by computing a symbolic determinant as follows:

$$\mathbf{a} \times \mathbf{b} = \begin{vmatrix} \mathbf{i} & \mathbf{j} & \mathbf{k} \\ a_1 & a_2 & a_3 \\ b_1 & b_2 & b_3 \end{vmatrix} = \begin{vmatrix} a_2 & a_3 \\ b_2 & b_3 \end{vmatrix}\mathbf{i} - \begin{vmatrix} a_1 & a_3 \\ b_1 & b_3 \end{vmatrix}\mathbf{j} + \begin{vmatrix} a_1 & a_2 \\ b_1 & b_2 \end{vmatrix}\mathbf{k}. \qquad (3)$$

This cross product is also known as the **vector product,** for $\mathbf{a} \times \mathbf{b}$ is a vector quantity.

EXAMPLE 3 Find $\mathbf{a} \times \mathbf{b}$ if $\mathbf{a} = 3\mathbf{i} - 2\mathbf{j} + \mathbf{k}$ and $\mathbf{b} = -2\mathbf{i} + 3\mathbf{j} + 4\mathbf{k}$.

Solution We have

$$\mathbf{a} \times \mathbf{b} = \begin{vmatrix} \mathbf{i} & \mathbf{j} & \mathbf{k} \\ 3 & -2 & 1 \\ -2 & 3 & 4 \end{vmatrix} = \begin{vmatrix} -2 & 1 \\ 3 & 4 \end{vmatrix}\mathbf{i} - \begin{vmatrix} 3 & 1 \\ -2 & 4 \end{vmatrix}\mathbf{j} + \begin{vmatrix} 3 & -2 \\ -2 & 3 \end{vmatrix}\mathbf{k}$$

$$= -11\mathbf{i} - 14\mathbf{j} + 5\mathbf{k}. \quad \square$$

EXAMPLE 4 Show that

$$\mathbf{i} \times \mathbf{j} = \mathbf{k}, \qquad \mathbf{j} \times \mathbf{k} = \mathbf{i}, \qquad \text{and} \qquad \mathbf{k} \times \mathbf{i} = \mathbf{j}.$$

Solution We show that $\mathbf{i} \times \mathbf{j} = \mathbf{k}$. The other two computations are similar. We have

$$\mathbf{i} \times \mathbf{j} = \begin{vmatrix} \mathbf{i} & \mathbf{j} & \mathbf{k} \\ 1 & 0 & 0 \\ 0 & 1 & 0 \end{vmatrix} = \begin{vmatrix} 0 & 0 \\ 1 & 0 \end{vmatrix}\mathbf{i} - \begin{vmatrix} 1 & 0 \\ 0 & 0 \end{vmatrix}\mathbf{j} + \begin{vmatrix} 1 & 0 \\ 0 & 1 \end{vmatrix}\mathbf{k}$$

$$= 0\mathbf{i} + 0\mathbf{j} + \mathbf{k} = \mathbf{k}. \quad \square$$

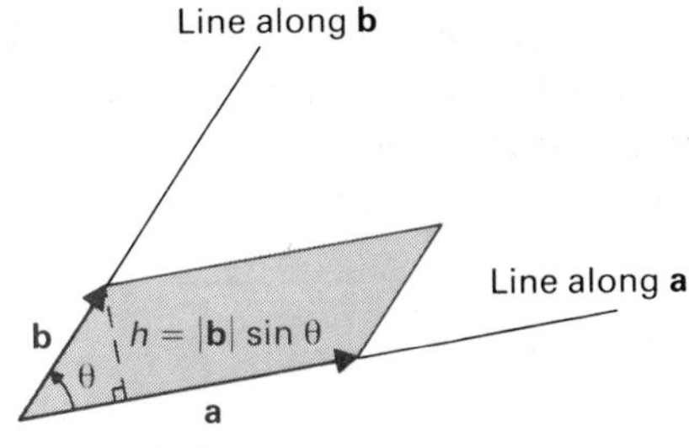

Figure 14.58
Area = $|\mathbf{a}|h = |\mathbf{a}||\mathbf{b}|$ (sin θ).

The important geometric properties of any vector are its *length* and its *direction*. In Fig. 14.58, we take the plane determined by the lines along **a** and **b** as the plane of the page of the text and shade the parallelogram having **a** and **b** as two edges. We claim that the length $|\mathbf{a} \times \mathbf{b}|$ is equal to the area of this shaded parallelogram, and we compute the area to verify this. Referring to Fig. 14.58, we have

$$\begin{aligned}\text{Area} &= (\text{Length of base})(\text{Altitude}) \\ &= |\mathbf{a}| \cdot h = |\mathbf{a}| \cdot |\mathbf{b}| \sin \theta.\end{aligned}$$

Therefore

$$\begin{aligned}\text{Area}^2 &= |\mathbf{a}|^2 |\mathbf{b}|^2 \sin^2\theta \\ &= |\mathbf{a}|^2 |\mathbf{b}|^2 (1 - \cos^2\theta) \\ &= |\mathbf{a}|^2 |\mathbf{b}|^2 - (|\mathbf{a}| \cdot |\mathbf{b}| \cos \theta)^2 \\ &= |\mathbf{a}|^2 |\mathbf{b}|^2 - (\mathbf{a} \cdot \mathbf{b})^2 \\ &= (a_1^2 + a_2^2 + a_3^2)(b_1^2 + b_2^2 + b_3^2) - (a_1b_1 + a_2b_2 + a_3b_3)^2.\end{aligned}$$

A bit of straightforward pencil-pushing shows that this all boils down to

$$\begin{aligned}\text{Area}^2 &= (a_2b_3 - a_3b_2)^2 + (a_1b_3 - a_3b_1)^2 + (a_1b_2 - a_2b_1)^2 \\ &= \begin{vmatrix} a_2 & a_3 \\ b_2 & b_3 \end{vmatrix}^2 + \begin{vmatrix} a_1 & a_3 \\ b_1 & b_3 \end{vmatrix}^2 + \begin{vmatrix} a_1 & a_2 \\ b_1 & b_2 \end{vmatrix}^2.\end{aligned}$$

But this is the square of the length of the cross product $\mathbf{a} \times \mathbf{b}$, so $\text{Area}^2 = |\mathbf{a} \times \mathbf{b}|^2$ and area $= |\mathbf{a} \times \mathbf{b}|$. This shows that

$$\boxed{|\mathbf{a} \times \mathbf{b}| = \text{Area of parallelogram} = |\mathbf{a}| \cdot |\mathbf{b}| \sin \theta.} \qquad \textbf{(4)}$$

Finally, we want to know the direction of $\mathbf{a} \times \mathbf{b}$. First, we see that $\mathbf{a} \times \mathbf{b}$ is perpendicular to both **a** and **b** and therefore perpendicular to the plane containing the parallelogram shaded in Fig. 14.58. We need only show that $\mathbf{a} \cdot (\mathbf{a} \times \mathbf{b}) = 0$ and $\mathbf{b} \cdot (\mathbf{a} \times \mathbf{b}) = 0$. Referring to Eq. (3), where $\mathbf{a} \times \mathbf{b}$ is defined, we see that

$$\mathbf{a} \cdot (\mathbf{a} \times \mathbf{b}) = a_1 \begin{vmatrix} a_2 & a_3 \\ b_2 & b_3 \end{vmatrix} - a_2 \begin{vmatrix} a_1 & a_3 \\ b_1 & b_3 \end{vmatrix} + a_3 \begin{vmatrix} a_1 & a_2 \\ b_1 & b_2 \end{vmatrix}.$$

But this is equal to the determinant

$$\begin{vmatrix} a_1 & a_2 & a_3 \\ a_1 & a_2 & a_3 \\ b_1 & b_2 & b_3 \end{vmatrix} = 0,$$

which is zero since the first and second rows are the same (Theorem 14.3). Thus $\mathbf{a} \cdot (\mathbf{a} \times \mathbf{b}) = 0$ and **a** is perpendicular to $\mathbf{a} \times \mathbf{b}$. A similar argument shows that $\mathbf{b} \cdot (\mathbf{a} \times \mathbf{b}) = 0$; this time the determinant has its first and third rows the same.

We now know that $\mathbf{a} \times \mathbf{b}$ has length $|\mathbf{a}| \cdot |\mathbf{b}| \sin \theta$ and direction perpendicular to the plane determined by **a** and **b**. There are two vectors of this length perpendicular to the plane; one is the negative of the other. One of them is $\mathbf{a} \times \mathbf{b}$, and the other is $-(\mathbf{a} \times \mathbf{b})$, which we claim equals $\mathbf{b} \times \mathbf{a}$. We can easily see why $\mathbf{a} \times \mathbf{b} = -\mathbf{b} \times \mathbf{a}$. Referring to Eq. (3), we see that the determinants of the 2×2 matrices used to find $\mathbf{b} \times \mathbf{a}$ are the ones used to find $\mathbf{a} \times \mathbf{b}$ with the rows interchanged. Theorem 14.3 thus shows at once that

$$\boxed{\mathbf{b} \times \mathbf{a} = -(\mathbf{a} \times \mathbf{b}).} \tag{5}$$

We summarize what we have done and a bit more in a theorem.

THEOREM 14.4 Magnitude and direction of $\mathbf{a} \times \mathbf{b}$

The vector $\mathbf{a} \times \mathbf{b}$ has length given by

$$|\mathbf{a} \times \mathbf{b}| = |\mathbf{a}| \cdot |\mathbf{b}| \sin \theta,$$

where θ satisfying $0 \leq \theta \leq \pi$ is the angle between **a** and **b**. The direction of $\mathbf{a} \times \mathbf{b}$ is perpendicular to both **a** and **b** in the direction in which the thumb of the right hand points as the fingers curl through θ from **a** to **b**. This manner of describing the direction of $\mathbf{a} \times \mathbf{b}$ is known as the "right-hand rule." It is illustrated in Fig. 14.59.

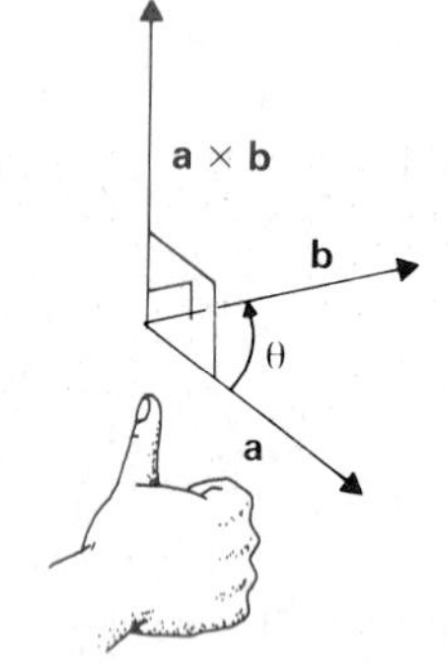

Figure 14.59 Right-hand rule for $\mathbf{a} \times \mathbf{b}$, $0 < \theta < \pi$.

The only part of the theorem that we have not proved is the "right-hand rule" part. We will not prove this; we can easily illustrate it using

$$\mathbf{i} \times \mathbf{j} = \mathbf{k}, \qquad \mathbf{j} \times \mathbf{k} = \mathbf{i}, \qquad \text{and} \qquad \mathbf{k} \times \mathbf{i} = \mathbf{j},$$

which were shown in Example 4.

EXAMPLE 5 Find all possible cross products of two of the vectors **i**, **j**, and **k**.

Solution From Example 4, we have

$$\mathbf{i} \times \mathbf{j} = \mathbf{k}, \qquad \mathbf{j} \times \mathbf{k} = \mathbf{i}, \qquad \text{and} \qquad \mathbf{k} \times \mathbf{i} = \mathbf{j}. \tag{6}$$

Therefore, from Eq. (5),

$$\mathbf{j} \times \mathbf{i} = -\mathbf{k}, \qquad \mathbf{k} \times \mathbf{j} = -\mathbf{i}, \qquad \text{and} \qquad \mathbf{i} \times \mathbf{k} = -\mathbf{j}. \quad \square \tag{7}$$

> We can remember Eqs. (6) and (7) by writing the sequence
>
> $$\mathbf{i},\ \mathbf{j},\ \mathbf{k},\ \mathbf{i},\ \mathbf{j},\ \mathbf{k}.$$
>
> The cross product of two consecutive vectors in left-to-right order is the next one to the right, while the cross product in right-to-left order is the negative of the next vector to the left.

Equation (4) gives us an easy way to find the area of a parallelogram or a triangle.

Figure 14.60 Parallelogram with adjacent edges $\mathbf{i} + 3\mathbf{j}$ and $4\mathbf{i} + 2\mathbf{j}$.

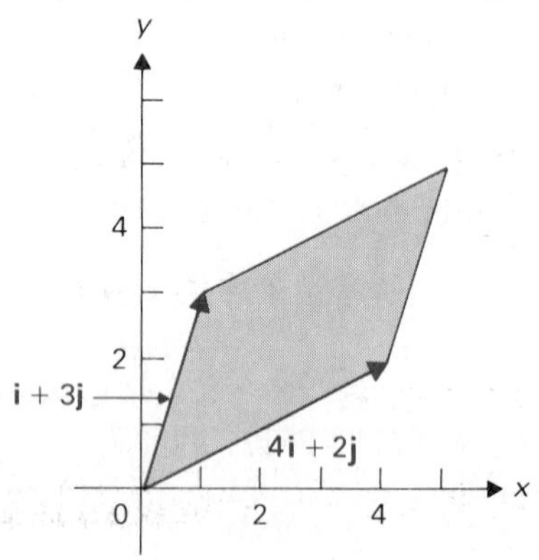

EXAMPLE 6 Find the area of the parallelogram in the plane having $\mathbf{i} + 3\mathbf{j}$ and $4\mathbf{i} + 2\mathbf{j}$ as coterminous edges.

Solution In Fig. 14.60, we have shown a parallelogram with these vectors emanating from the origin as edges. Equation (4) involves the cross product, which is defined for vectors in space. However, we can view the x,y-plane as

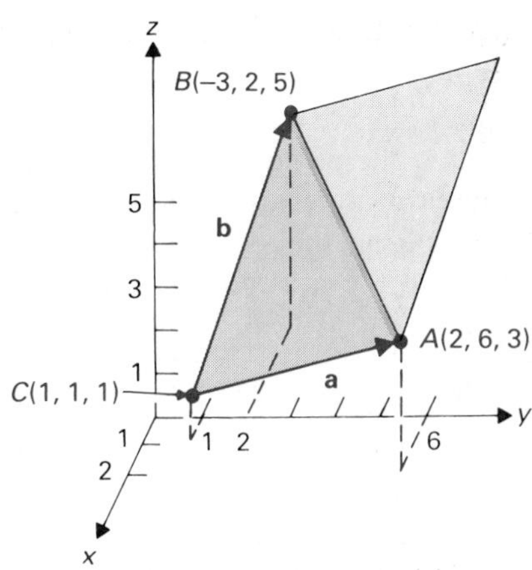

Figure 14.61 Triangle with vertices (2, 6, 3), (−3, 2, 5), and (1, 1, 1).

part of space and the given vectors as

$$\mathbf{a} = \mathbf{i} + 3\mathbf{j} + 0\mathbf{k} \qquad \text{and} \qquad \mathbf{b} = 4\mathbf{i} + 2\mathbf{j} + 0\mathbf{k}.$$

We then have

$$\mathbf{a} \times \mathbf{b} = \begin{vmatrix} \mathbf{i} & \mathbf{j} & \mathbf{k} \\ 1 & 3 & 0 \\ 4 & 2 & 0 \end{vmatrix} = 0\mathbf{i} + 0\mathbf{j} - 10\mathbf{k}.$$

From Eq. (4), we have

$$\text{Area of parallelogram} = |\mathbf{a} \times \mathbf{b}| = \sqrt{0^2 + 0^2 + (-10)^2} = 10. \quad \square$$

EXAMPLE 7 Find the area of the triangle in space with vertices $A(2, 6, 3)$, $B(-3, 2, 5)$, and $C(1, 1, 1)$.

Solution The triangle is shown in Fig. 14.61. We let **a** be the vector from C to A and **b** the vector from C to B, as shown in the figure. Then

$$\mathbf{a} = \mathbf{i} + 5\mathbf{j} + 2\mathbf{k} \qquad \text{and} \qquad \mathbf{b} = -4\mathbf{i} + \mathbf{j} + 4\mathbf{k}.$$

The area of the triangle is half the area of the parallelogram having **a** and **b** as coterminous edges. We find that

$$\mathbf{a} \times \mathbf{b} = \begin{vmatrix} \mathbf{i} & \mathbf{j} & \mathbf{k} \\ 1 & 5 & 2 \\ -4 & 1 & 4 \end{vmatrix} = 18\mathbf{i} - 12\mathbf{j} + 21\mathbf{k}.$$

Then

$$\text{Area of triangle} = \frac{1}{2}|\mathbf{a} \times \mathbf{b}| = \frac{1}{2}\sqrt{18^2 + 12^2 + 21^2}$$

$$= \frac{1}{2}\sqrt{909} \approx 15.075. \quad \square$$

EXAMPLE 8 Find a unit vector perpendicular to both $\mathbf{a} = \mathbf{i} + \mathbf{j} + 2\mathbf{k}$ and $\mathbf{b} = -\mathbf{i} + \mathbf{j} - 3\mathbf{k}$.

Solution By Theorem 14.4, the vector $\mathbf{a} \times \mathbf{b}$ is perpendicular to both **a** and **b**. Now

$$\mathbf{a} \times \mathbf{b} = \begin{vmatrix} \mathbf{i} & \mathbf{j} & \mathbf{k} \\ 1 & 1 & 2 \\ -1 & 1 & -3 \end{vmatrix} = -5\mathbf{i} + \mathbf{j} + 2\mathbf{k}.$$

To form a unit vector, we multiply $\mathbf{a} \times \mathbf{b}$ by the reciprocal of its length

$$|\mathbf{a} \times \mathbf{b}| = |-5\mathbf{i} + \mathbf{j} + 2\mathbf{k}| = \sqrt{25 + 1 + 4} = \sqrt{30}.$$

We obtain as answer

$$-\frac{5}{\sqrt{30}}\mathbf{i} + \frac{1}{\sqrt{30}}\mathbf{j} + \frac{2}{\sqrt{30}}\mathbf{k}.$$

Another possible answer is $(5/\sqrt{30})\mathbf{i} - (1/\sqrt{30})\mathbf{j} - (2/\sqrt{30})\mathbf{k}$, which corresponds to computing $\mathbf{b} \times \mathbf{a}$ rather than $\mathbf{a} \times \mathbf{b}$. $\square$

EXAMPLE 9 Given that $2x - 3y + z = 4$ is the equation of a plane in space, find a vector perpendicular to the plane.

Solution We find three points A, B, and C in the plane but not all on the same line. Surely the points in the plane over (or under) the points (0, 0, 0), (1, 0, 0), and (0, 1, 0) in the x,y-plane can't be on the same line. To find these points in the given plane, we substitute the x,y-coordinates in the equation and solve for z. We obtain the points

$$A(0, 0, 4), \qquad B(1, 0, 2), \qquad \text{and} \qquad C(0, 1, 7).$$

The vectors **a** from C to A and **b** from C to B can be viewed as lying in the plane, and $\mathbf{a} \times \mathbf{b}$ is thus perpendicular to the plane. We have

$$\mathbf{a} = -\mathbf{j} - 3\mathbf{k} \qquad \text{and} \qquad \mathbf{b} = \mathbf{k} - \mathbf{j} - 5\mathbf{k}.$$

Thus

$$\mathbf{a} \times \mathbf{b} = \begin{vmatrix} \mathbf{i} & \mathbf{j} & \mathbf{k} \\ 0 & -1 & -3 \\ 1 & -1 & -5 \end{vmatrix} = 2\mathbf{i} - 3\mathbf{j} + \mathbf{k}$$

is one possible vector perpendicular to the plane $2x - 3y + z = 4$. All other possible answers are of the form $r(2\mathbf{i} - 3\mathbf{j} + \mathbf{k})$ for some r.

Note that the coefficients of **i**, **j**, and **k** in $2\mathbf{i} - 3\mathbf{j} + \mathbf{k}$ are the coefficients of x, y, and z in $2x - 3y + z = 4$. We will see in Section 14.7 that the coefficients of x, y, and z are always also coefficients of **i**, **j**, and **k** for a vector perpendicular to a plane. □

EXAMPLE 10 Show that if **a** and **b** are parallel, then $\mathbf{a} \times \mathbf{b} = \mathbf{0}$.

Solution If **a** and **b** are parallel, then the angle θ between them is either 0 or π. By Theorem 14.4, the magnitude of $\mathbf{a} \times \mathbf{b}$ is then

$$|\mathbf{a} \times \mathbf{b}| = |\mathbf{a}|\,|\mathbf{b}| \sin \theta = |\mathbf{a}|\,|\mathbf{b}| \cdot 0 = 0.$$

Thus $\mathbf{a} \times \mathbf{b}$ must be the zero vector. □

Equation (5) shows that taking cross products is not a commutative operation. However, it is true that

$$\boxed{\mathbf{a} \times (\mathbf{b} + \mathbf{c}) = \mathbf{a} \times \mathbf{b} + \mathbf{a} \times \mathbf{c}} \tag{8}$$

and

$$\boxed{(k\mathbf{a}) \times \mathbf{b} = \mathbf{a} \times (k\mathbf{b}) = k(\mathbf{a} \times \mathbf{b})} \tag{9}$$

for any vectors **a**, **b**, **c** in space and any scalar k. You can easily prove Eqs. (8) and (9) as exercises.

TRIPLE PRODUCTS

We now know two ways of taking a product of vectors **a** and **b** in space. We can find $\mathbf{a} \cdot \mathbf{b}$, which is a scalar, or $\mathbf{a} \times \mathbf{b}$, which is a vector. It is natural to try to multiply three vectors, **a**, **b**, and **c**, in space. The product $\mathbf{a} \cdot (\mathbf{b} \cdot \mathbf{c})$

makes no sense, for $\mathbf{a}$ is a vector and $\mathbf{b} \cdot \mathbf{c}$ is a scalar. However, $\mathbf{a} \times (\mathbf{b} \times \mathbf{c})$ makes sense, for both $\mathbf{a}$ and $\mathbf{b} \times \mathbf{c}$ are vectors. This product $\mathbf{a} \times (\mathbf{b} \times \mathbf{c})$ is the *triple vector product*. It is most easily computed using the formula

$$\boxed{\mathbf{a} \times (\mathbf{b} \times \mathbf{c}) = (\mathbf{a} \cdot \mathbf{c})\mathbf{b} - (\mathbf{a} \cdot \mathbf{b})\mathbf{c},} \qquad \textbf{(10)}$$

which we ask you to establish in Exercise 42. In Exercise 47, you are asked to convince yourself that

$$\mathbf{a} \times (\mathbf{b} \times \mathbf{c}) \neq (\mathbf{a} \times \mathbf{b}) \times \mathbf{c}$$

unless **a**, **b**, and **c** are carefully chosen. That is, the triple cross product is not associative.

EXAMPLE 11 If $\mathbf{a} = 2\mathbf{i} - 3\mathbf{j} + 4\mathbf{k}$ while $\mathbf{b} = 3\mathbf{i} - \mathbf{j} - 2\mathbf{k}$ and $\mathbf{c} = -3\mathbf{i} - 5\mathbf{j} - \mathbf{k}$, find $\mathbf{a} \times (\mathbf{b} \times \mathbf{c})$.

Solution From Eq. (10), we have

$$\mathbf{a} \times (\mathbf{b} \times \mathbf{c}) = (\mathbf{a} \cdot \mathbf{c})\mathbf{b} - (\mathbf{a} \cdot \mathbf{b})\mathbf{c} = 5\mathbf{b} - \mathbf{c} = 18\mathbf{i} - 9\mathbf{k}. \quad \square$$

The product $\mathbf{a} \cdot (\mathbf{b} \times \mathbf{c})$ also makes sense, but the answer this time is a scalar. Consequently $\mathbf{a} \cdot (\mathbf{b} \times \mathbf{c})$ is called the *triple scalar product*. Let **a**, **b**, and **c** be as shown in Fig. 14.62. The vectors are coterminous edges of a box, shown shaded in the figure. For such a box, we have

$$\text{Volume} = (\text{Area of base})\,(\text{Altitude}).$$

Using the vectors and angles shown in Fig. 14.62, we see that

$$\text{Volume} = (\text{Area of base})\,(|\mathbf{a}| \cos \phi).$$

Now the area of the base is $|\mathbf{b} \times \mathbf{c}|$, by Eq. (4). Therefore

$$\text{Volume} = |\mathbf{b} \times \mathbf{c}| \cdot |\mathbf{a}| \cos \phi.$$

But $\mathbf{b} \times \mathbf{c}$ is perpendicular to the base of the box, and ϕ is the angle between $\mathbf{b} \times \mathbf{c}$ and **c**. Therefore

$$|\mathbf{b} \times \mathbf{c}| \cdot |\mathbf{a}| \cos \phi = \mathbf{a} \cdot (\mathbf{b} \times \mathbf{c}).$$

If the order of **b** and **c** is reversed, we obtain $\mathbf{a} \cdot (\mathbf{c} \times \mathbf{b}) = -\mathbf{a} \cdot (\mathbf{b} \times \mathbf{c})$, but of course the box given by **a**, **c**, **b** is the same one as that given by **a**, **b**, **c**. Therefore we have the formula

$$\boxed{\text{Volume} = |\mathbf{a} \cdot (\mathbf{b} \times \mathbf{c})|.} \qquad \textbf{(11)}$$

Figure 14.62 Box with **a**, **b**, **c** as edges and volume $\mathbf{a} \cdot (\mathbf{b} \times \mathbf{c})$.

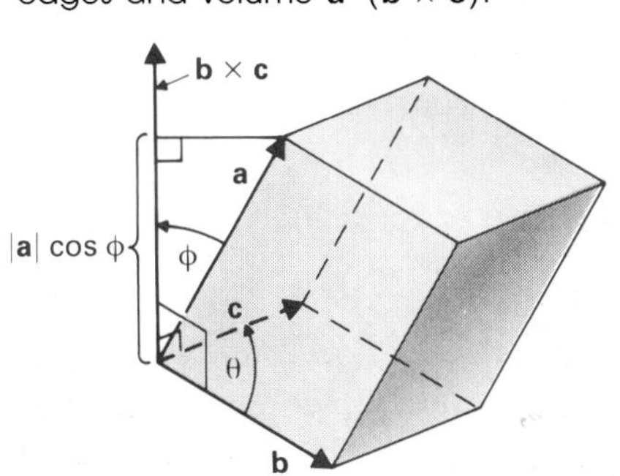

There is a very easy way to compute $\mathbf{a} \cdot (\mathbf{b} \times \mathbf{c})$. Form the matrix having the vectors **a**, **b**, and **c** as the first, second, and third rows, respectively. Then $\mathbf{a} \cdot (\mathbf{b} \times \mathbf{c})$ is the determinant of this matrix. To see why this is true, note that, if $\mathbf{a} = a_1\mathbf{i} + a_2\mathbf{j} + a_3\mathbf{k}$ and $\mathbf{d} = d_1\mathbf{i} + d_2\mathbf{j} + d_3\mathbf{k}$, then $\mathbf{a} \cdot \mathbf{d} = d_1a_1 + d_2a_2 + d_3a_3$ can be found by formally replacing the **i**, **j**, and **k** in the expression for **d** by a_1, a_2, and a_3, respectively. If we do this for

$$\mathbf{d} = \mathbf{b} \times \mathbf{c} = \begin{vmatrix} \mathbf{i} & \mathbf{j} & \mathbf{k} \\ b_1 & b_2 & b_3 \\ c_1 & c_2 & c_3 \end{vmatrix}$$

to form $\mathbf{a} \cdot \mathbf{d} = \mathbf{a} \cdot (\mathbf{b} \times \mathbf{c})$, we obtain

$$\mathbf{a} \cdot (\mathbf{b} \times \mathbf{c}) = \begin{vmatrix} a_1 & a_2 & a_3 \\ b_1 & b_2 & b_3 \\ c_1 & c_2 & c_3 \end{vmatrix}. \tag{12}$$

EXAMPLE 12 Find the volume of the box in space having as coterminous edges the vectors $\mathbf{a} = \mathbf{i} - 2\mathbf{j} + \mathbf{k}$, $\mathbf{b} = 2\mathbf{i} + 3\mathbf{j} - 2\mathbf{k}$, and $\mathbf{c} = -\mathbf{i} + 3\mathbf{j} - 2\mathbf{k}$.

Solution We have

$$\mathbf{a} \cdot (\mathbf{b} \times \mathbf{c}) = \begin{vmatrix} 1 & -2 & 1 \\ 2 & 3 & -2 \\ -1 & 3 & -2 \end{vmatrix} = 1(0) - (-2)(-6) + 1(9) = -3.$$

Therefore,

$$\text{Volume} = |\mathbf{a} \cdot (\mathbf{b} \times \mathbf{c})| = |-3| = 3. \quad \square$$

SUMMARY

1. $$\begin{vmatrix} a_1 & a_2 \\ b_1 & b_2 \end{vmatrix} = a_1 b_2 - a_2 b_1,$$

$$\begin{vmatrix} a_1 & a_2 & a_3 \\ b_1 & b_2 & b_3 \\ c_1 & c_2 & c_3 \end{vmatrix} = a_1 \begin{vmatrix} b_2 & b_3 \\ c_2 & c_3 \end{vmatrix} - a_2 \begin{vmatrix} b_1 & b_3 \\ c_1 & c_3 \end{vmatrix} + a_3 \begin{vmatrix} b_1 & b_2 \\ c_1 & c_2 \end{vmatrix}$$

2. $$\mathbf{a} \times \mathbf{b} = \begin{vmatrix} \mathbf{i} & \mathbf{j} & \mathbf{k} \\ a_1 & a_2 & a_3 \\ b_1 & b_2 & b_3 \end{vmatrix}$$

3. The cross product $\mathbf{a} \times \mathbf{b}$ has length $|\mathbf{a}| \cdot |\mathbf{b}| \sin \theta$, where θ is the angle between **a** and **b**. This length equals the area of the parallelogram having **a** and **b** as adjacent sides.

4. The direction of $\mathbf{a} \times \mathbf{b}$ is perpendicular to the plane of **a** and **b** and in the direction given by the right-hand rule, illustrated in Fig. 14.59.

5. For any vectors **a**, **b**, **c** in space and any scalar k,

$$\mathbf{a} \times \mathbf{b} = -\mathbf{b} \times \mathbf{a},$$

$$\mathbf{a} \times (\mathbf{b} + \mathbf{c}) = \mathbf{a} \times \mathbf{b} + \mathbf{a} \times \mathbf{c},$$

$$(k\mathbf{a}) \times \mathbf{b} = \mathbf{a} \times (k\mathbf{b}) = k(\mathbf{a} \times \mathbf{b}).$$

6. $$\mathbf{i} \times \mathbf{j} = \mathbf{k}, \quad \mathbf{j} \times \mathbf{k} = \mathbf{i}, \quad \mathbf{k} \times \mathbf{i} = \mathbf{j},$$

while

$$\mathbf{j} \times \mathbf{i} = -\mathbf{k}, \quad \mathbf{k} \times \mathbf{j} = -\mathbf{i}, \quad \mathbf{i} \times \mathbf{k} = -\mathbf{j}$$

7. The triple scalar product $\mathbf{a}\cdot(\mathbf{b}\times\mathbf{c})$ and the triple vector product $\mathbf{a}\times(\mathbf{b}\times\mathbf{c})$ are most easily computed using

$$\mathbf{a}\cdot(\mathbf{b}\times\mathbf{c}) = \begin{vmatrix} a_1 & a_2 & a_3 \\ b_1 & b_2 & b_3 \\ c_1 & c_2 & c_3 \end{vmatrix}$$

and

$$\mathbf{a}\times(\mathbf{b}\times\mathbf{c}) = (\mathbf{a}\cdot\mathbf{c})\mathbf{b} - (\mathbf{a}\cdot\mathbf{b})\mathbf{c}.$$

8. $|\mathbf{a}\cdot(\mathbf{b}\times\mathbf{c})|$ is the volume of the box having **a**, **b**, and **c** as adjacent edges.

EXERCISES

In Exercises 1 through 8, find the indicated determinant.

1. $\begin{vmatrix} -1 & 3 \\ 5 & 0 \end{vmatrix}$ **2.** $\begin{vmatrix} -1 & 0 \\ 0 & 7 \end{vmatrix}$

3. $\begin{vmatrix} 0 & -3 \\ 5 & 0 \end{vmatrix}$ **4.** $\begin{vmatrix} 21 & -4 \\ 10 & 7 \end{vmatrix}$

5. $\begin{vmatrix} 1 & 4 & -2 \\ 3 & 13 & 0 \\ 2 & -1 & 3 \end{vmatrix}$ **6.** $\begin{vmatrix} 2 & -5 & 3 \\ 1 & 3 & 4 \\ -2 & 3 & 7 \end{vmatrix}$

7. $\begin{vmatrix} 1 & -2 & 7 \\ 0 & 1 & 4 \\ 1 & 0 & 3 \end{vmatrix}$ **8.** $\begin{vmatrix} 2 & -1 & 1 \\ -1 & 0 & 3 \\ 2 & 1 & -4 \end{vmatrix}$

9. Show by direct computation that

a) $\begin{vmatrix} a_1 & a_2 & a_3 \\ a_1 & a_2 & a_3 \\ c_1 & c_2 & c_3 \end{vmatrix} = 0,$ b) $\begin{vmatrix} a_1 & a_2 & a_3 \\ b_1 & b_2 & b_3 \\ a_1 & a_2 & a_3 \end{vmatrix} = 0.$

10. Show by direct computation that

$$\begin{vmatrix} a_1 & a_2 & a_3 \\ b_1 & b_2 & b_3 \\ c_1 & c_2 & c_3 \end{vmatrix} = -\begin{vmatrix} a_1 & a_2 & a_3 \\ c_1 & c_2 & c_3 \\ b_1 & b_2 & b_3 \end{vmatrix}.$$

In Exercises 11 through 16, find $\mathbf{a}\times\mathbf{b}$.

11. $\mathbf{a} = 2\mathbf{i} - \mathbf{j} + 3\mathbf{k}$, $\mathbf{b} = \mathbf{i} + 2\mathbf{j}$

12. $\mathbf{a} = -5\mathbf{i} + \mathbf{j} + 4\mathbf{k}$, $\mathbf{b} = 2\mathbf{i} + \mathbf{j} - 3\mathbf{k}$

13. $\mathbf{a} = -\mathbf{i} + 2\mathbf{j} + 4\mathbf{k}$, $\mathbf{b} = 2\mathbf{i} - 4\mathbf{j} - 8\mathbf{k}$

14. $\mathbf{a} = \mathbf{i} - \mathbf{j} + \mathbf{k}$, $\mathbf{b} = 3\mathbf{i} - 2\mathbf{j} + 7\mathbf{k}$

15. $\mathbf{a} = 2\mathbf{i} - 3\mathbf{j} + 5\mathbf{k}$, $\mathbf{b} = 4\mathbf{i} - 5\mathbf{j} + \mathbf{k}$

16. $\mathbf{a} = -2\mathbf{i} + 3\mathbf{j} - \mathbf{k}$, $\mathbf{b} = 4\mathbf{i} - 6\mathbf{j} + \mathbf{k}$

In Exercises 17 through 21, find the area of the parallelogram having the given vectors as edges. If the vectors are in the plane, regard them as being in space with the coefficient of **k** being zero.

17. $-\mathbf{i} + 4\mathbf{j}$ and $2\mathbf{i} + 3\mathbf{j}$ **18.** $-5\mathbf{i} + 3\mathbf{j}$ and $\mathbf{i} + 7\mathbf{j}$

19. $\mathbf{i} + 3\mathbf{j} - 5\mathbf{k}$ and $2\mathbf{i} + 4\mathbf{j} - \mathbf{k}$

20. $2\mathbf{i} - \mathbf{j} + \mathbf{k}$ and $\mathbf{i} + 3\mathbf{j} - \mathbf{k}$

21. $\mathbf{a} = \mathbf{i} - 4\mathbf{j} + \mathbf{k}$ and $\mathbf{k} = 2\mathbf{i} + 3\mathbf{j} - 2\mathbf{k}$

In Exercises 22 through 29, find the area of the given geometric configuration.

22. The triangle with vertices $(-1, 2)$, $(3, -1)$, and $(4, 3)$

23. The triangle with vertices $(3, -4)$, $(1, 1)$, and $(5, 7)$

24. The triangle with vertices $(2, 1, -3)$, $(3, 0, 4)$, and $(1, 0, 5)$

25. The triangle with vertices $(3, 1, -2)$, $(1, 4, 5)$, and $(2, 1, -4)$

26. The triangle in the plane bounded by the lines $y = x$, $y = -3x + 8$, and $3y + 5x = 0$

27. The parallelogram with vertices $(1, 3)$, $(-2, 6)$, $(1, 11)$, and $(4, 8)$

28. The parallelogram with vertices $(1, 0, 1)$, $(3, 1, 4)$, $(0, 2, 9)$, and $(-2, 1, 6)$

29. The parallelogram in the plane bounded by the lines $x - 2y = 3$, $x - 2y = 8$, $2x + 3y = -1$, and $2x + 3y = -5$

In Exercises 30 through 33, find $\mathbf{a}\cdot(\mathbf{b}\times\mathbf{c})$ and $\mathbf{a}\times(\mathbf{b}\times\mathbf{c})$.

30. $\mathbf{a} = \mathbf{i} + 2\mathbf{j} - 3\mathbf{k}$, $\mathbf{b} = 4\mathbf{i} - \mathbf{j} + 2\mathbf{k}$, and $\mathbf{c} = 3\mathbf{i} + \mathbf{k}$

31. $\mathbf{a} = -\mathbf{i} + \mathbf{j} + 2\mathbf{k}$, $\mathbf{b} = \mathbf{i} + \mathbf{k}$, and $\mathbf{c} = 3\mathbf{i} - 2\mathbf{j} + 5\mathbf{k}$

32. $\mathbf{a} = \mathbf{i} - 3\mathbf{k}$, $\mathbf{b} = -\mathbf{i} + 4\mathbf{j}$, and $\mathbf{c} = \mathbf{i} + \mathbf{j} + \mathbf{k}$

33. $\mathbf{a} = 4\mathbf{i} - \mathbf{j} + 2\mathbf{k}$, $\mathbf{b} = 3\mathbf{i} + 5\mathbf{j} - 2\mathbf{k}$, and $\mathbf{c} = \mathbf{i} - 3\mathbf{j} + \mathbf{k}$

In Exercises 34 through 37, find the volume of the box having the given vectors as adjacent edges.

34. $-\mathbf{i} + 4\mathbf{j} + 7\mathbf{k}$, $3\mathbf{i} - 2\mathbf{j} - \mathbf{k}$, and $4\mathbf{i} + 2\mathbf{k}$

35. $2\mathbf{i} + \mathbf{j} - 4\mathbf{k}$, $3\mathbf{i} - \mathbf{j} + 2\mathbf{k}$, and $\mathbf{i} + 3\mathbf{j} - 10\mathbf{k}$

36. $-2\mathbf{i} + \mathbf{j}$, $3\mathbf{i} - 4\mathbf{j} + \mathbf{k}$, and $\mathbf{i} - 2\mathbf{k}$

37. $3\mathbf{i} - \mathbf{j} + 4\mathbf{k}$, $\mathbf{i} - 2\mathbf{j} + 7\mathbf{k}$, and $5\mathbf{i} - 3\mathbf{j} + 10\mathbf{k}$

In Exercises 38 through 41, find the volume of the tetrahedron having the given vertices. (Think how the volume of a tetrahedron having three vectors from one point as edges is related to the volume of the box having the same three vectors as adjacent edges.)

38. $(-3, 0, 1)$, $(4, 2, 1)$, $(0, 1, 7)$, and $(1, 1, 1)$

39. $(0, 1, 1)$, $(8, 2, -7)$, $(3, 1, 6)$, and $(-4, -2, 0)$

40. $(-1, 1, 2)$, $(3, 1, 4)$, $(-1, 6, 0)$, and $(2, -1, 5)$

41. $(-1, 2, 4)$, $(2, -3, 0)$, $(-4, 2, -1)$, and $(0, 3, -2)$

42. Let $\mathbf{a} = a_1\mathbf{i} + a_2\mathbf{j} + a_3\mathbf{k}$, $\mathbf{b} = b_1\mathbf{i} + b_2\mathbf{j} + b_3\mathbf{k}$, and $\mathbf{c} = c_1\mathbf{i} + c_2\mathbf{j} + c_3\mathbf{k}$. Verify that $\mathbf{a} \times (\mathbf{b} \times \mathbf{c}) = (\mathbf{a} \cdot \mathbf{c})\mathbf{b} - (\mathbf{a} \cdot \mathbf{b})\mathbf{c}$. (This dull problem involves merely a lot of tedious algebra.)

43. Use the properties of the cross product to find a formula, similar to that in the preceding exercise, for the computation of $(\mathbf{a} \times \mathbf{b}) \times \mathbf{c}$.

44. Use the results of Exercises 42 and 43 to express $(\mathbf{a} \times \mathbf{b}) \times (\mathbf{c} \times \mathbf{d})$ in each of the forms $h\mathbf{a} + k\mathbf{b}$ and $r\mathbf{c} + s\mathbf{d}$ for scalars h, k, r, and s.

45. Show that, for any vectors **a**, **b** in space, $\mathbf{a} \cdot (\mathbf{a} \times \mathbf{b}) = 0$.

46. Show that $\mathbf{i} \times (\mathbf{i} \times \mathbf{k}) = -\mathbf{k}$, while $(\mathbf{i} \times \mathbf{i}) \times \mathbf{k} = \mathbf{0}$, illustrating that the cross product is not associative.

47. Consider the triple vector products $\mathbf{a} \times (\mathbf{b} \times \mathbf{c})$ and $(\mathbf{a} \times \mathbf{b}) \times \mathbf{c}$ for vectors **a**, **b**, and **c** in space.
a) Argue geometrically that $\mathbf{a} \times (\mathbf{b} \times \mathbf{c})$ is a vector in the plane containing **b** and **c**.
b) Argue geometrically that $(\mathbf{a} \times \mathbf{b}) \times \mathbf{c}$ is a vector in the plane containing **a** and **b**.
c) Use parts (a) and (b) to argue that the cross product is not associative.

14.6 LINES

This section develops parametric equations for lines in space. If we had covered vectors in Chapter 1, we could have presented lines in the plane parametrically also. The present section then would seem really trivial; we would simply put in one new equation for the z-component. Using parametric equations, it is easy to describe a line (a one-dimensional "flat" piece) in space of any finite dimension.

Figure 14.63 The line L through (a_1, a_2, a_3) with parallel vector **d**.

PARAMETRIC EQUATIONS FOR A LINE

We are accustomed to thinking of a line as being determined by two points. While this is perfectly correct, it will be useful for us to think of a line as being determined by one *point* on the line and the *direction* of the line. Of course, the direction of a line can be specified in terms of a nonzero vector parallel to the line.

Let (a_1, a_2, a_3) be a point on a line L in space. If the origin $(0, 0, 0)$ is not on L, there is a unique plane in space containing the origin and L. If the origin lies on L, there are many such planes.

In Fig. 14.63, we take the plane of the page to be a plane containing L and the origin. The figure shows the point (a_1, a_2, a_3) on the line L and a vector $\mathbf{d} = d_1\mathbf{i} + d_2\mathbf{j} + d_3\mathbf{k}$ parallel to the line. We consider any such vector **d** to be a *direction vector* for the line. The line L is the unique line through (a_1, a_2, a_3) parallel to **d**.

Figure 14.64 $\mathbf{x} = \mathbf{a} + t\mathbf{d}$

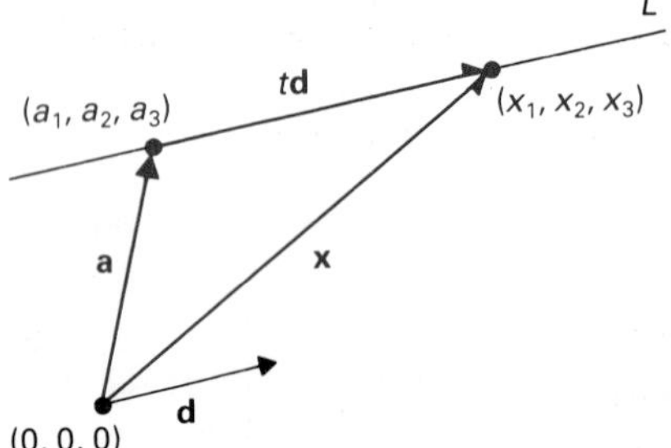

We want to find all points (x_1, x_2, x_3) on the line L. (Subscripted coordinates make things clearer.) In Fig. 14.64, we show the vectors **a** from $(0, 0, 0)$ to (a_1, a_2, a_3) and **x** from $(0, 0, 0)$ to a point (x_1, x_2, x_3) on the line. The vector along the line from (a_1, a_2, a_3) to (x_1, x_2, x_3) is parallel to **d**. Consequently, it must be of the form $t\mathbf{d}$ for some scalar t, as shown in Fig. 14.64. From the figure, we have

$$\boxed{\mathbf{x} = \mathbf{a} + t\mathbf{d}.} \qquad (1)$$

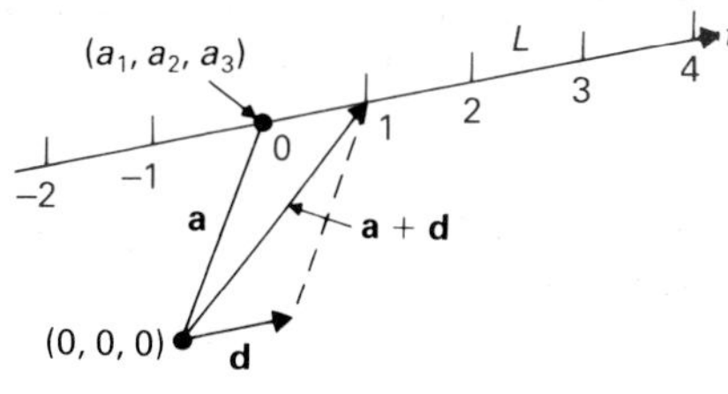

Figure 14.65 The line L becomes a t-axis.

Equation (1) is a **vector equation of the line.** As t runs through all values, we obtain all vectors from the origin to points on the line. Equation (1) makes the line L into a t-axis, as shown in Fig. 14.65. When $t = 0$, Eq. (1) gives the vector **a** to the point (a_1, a_2, a_3) on the line, so this point corresponds to the t-origin, as shown in Fig. 14.65. The point corresponding to $t = 1$ is the tip of the vector $\mathbf{a} + 1 \cdot \mathbf{d}$, and so on. Of course, the t-scale on L will be different from the scale on the coordinate axes unless **d** is a unit vector.

If we write Eq. (1) in coordinate form, we obtain

$$(x_1, x_2, x_3) = (a_1, a_2, a_3) + t(d_1, d_2, d_3). \tag{2}$$

Equating first, second, and third coordinates, we obtain

$$\boxed{\begin{aligned} x_1 &= a_1 + d_1 t, \\ x_2 &= a_2 + d_2 t, \\ x_3 &= a_3 + d_3 t \end{aligned}} \tag{3}$$

as scalar parametric equations of the line.

We summarize our work in a definition. We go back to x, y, z in place of x_1, x_2, x_3 for ease in writing as we work problems.

DEFINITION 14.5 Line in space

Let (a_1, a_2, a_3) be a point and let $\mathbf{d} = d_1\mathbf{i} + d_2\mathbf{j} + d_3\mathbf{k}$ be a vector. The **line** through (a_1, a_2, a_3) with **direction vector d** is the set of all points (x, y, z) such that, for some scalar t,

$$\begin{aligned} x &= a_1 + d_1 t, \\ y &= a_2 + d_2 t, \\ z &= a_3 + d_3 t. \end{aligned} \tag{4}$$

Equations (4) are **parametric equations** for the line, and t is the **parameter.**

In Chapter 1, we found the single linear equation for a nonvertical line in the plane by following the step sequence

Point

Slope (5)

Equation

Students are sometimes drilled that a single linear equation gives a line. This is true in the plane, but not in space. We will see in the next section that a single linear equation in space gives a plane. Lines in space of any dimension are easily described parametrically. Let us use $\|$ as a symbol for "parallel."

> Our step sequence for finding parametric equations of a line is
>
> ***Point***
>
> ***$\|$ vector*** (6)
>
> ***Equations***

EXAMPLE 1 Find parametric equations of the line through $(-1, 0, 2)$ having $\mathbf{d} = 2\mathbf{i} - 3\mathbf{j} + \mathbf{k}$ as direction vector.

Solution Following the step sequence (6), we have from Eqs. (4)

Point $(-1, 0, 2)$

$\|$ ***vector*** $2\mathbf{i} - 3\mathbf{j} + \mathbf{k}$

Equations $x = -1 + 2t,$

$$y = 0 - 3t,$$

$$z = 2 + t \quad \square$$

EXAMPLE 2 Find the point on the line in Example 1 with z-coordinate 6.

Solution This example is to emphasize that as t runs through all values, we obtain all points on the line. To have z-coordinate 6, we must have

$$6 = 2 + t, \qquad \text{or} \quad t = 4.$$

When $t = 4$, we obtain the point $(7, -12, 6)$ on the line. $\square$

The next example shows how to find the line through two points.

EXAMPLE 3 Find parametric equations of the line through $(1, -2, 3)$ and $(0, 1, 4)$.

Solution We have two points from which to choose. For direction vector, we may take the vector from $(1, -2, 3)$ to $(0, 1, 4)$.

Point $(1, -2, 3)$

$\|$ ***vector*** $-\mathbf{i} + 3\mathbf{j} + \mathbf{k}$

Equations $x = 1 - t,$

$$y = -2 + 3t,$$

$$z = 3 + t$$

Parametric equations for a line are not unique. We can use *any* point on the line, and *any* vector $\mathbf{d}$ parallel to the line. To illustrate, with $t = 2$, we see that $(-1, 4, 5)$ is on the line in this example. Multiplying the direction vector $-\mathbf{i} + 3\mathbf{j} + \mathbf{k}$ by -3, we obtain

$$x = -1 + 3t, \qquad y = 4 - 9t,$$

$$z = 5 - 3t$$

as another system of parametric equations for the same line. $\square$

Different systems of parametric equations for a line correspond geometrically to different choices for origin and scale on the line as a t-axis in Fig. 14.65.

EXAMPLE 4 Find parametric equations for the line $x - 3y = 7$ in the plane.

Solution We find any point on the line; $(4, -1)$ will do. The line has slope $\frac{1}{3}$, so $\mathbf{i} + (\frac{1}{3})\mathbf{j}$ is parallel to the line. To eliminate fractions, we multiply by 3 and take the parallel vector $3\mathbf{i} + \mathbf{j}$.

Point $(4, -1)$

$\|$ ***vector*** $3\mathbf{i} + \mathbf{j}$

Equations $x = 4 + 3t,$

$$y = -1 + t \quad \square$$

Two lines in the plane intersect unless they are parallel. Two lines in space usually don't intersect. The next two examples show how to find points of intersection.

EXAMPLE 5 Find the point of intersection of the lines

$$\begin{aligned} x &= 1 + 2t, \\ y &= 2 - 3t, \end{aligned} \quad \text{and} \quad \begin{aligned} x &= 5 + 3s, \\ y &= 3 - 4s \end{aligned}$$

in the plane.

Solution We used s as parameter for the second line, because we can't expect the point of intersection to correspond to the same values of the parameters on the two lines. (We will find in a moment that the lines intersect at the point $(-37, 59)$, which corresponds to $t = -19$ in the first line and $s = -14$ in the second. Trying to use the same parameter t for both lines would lead to hopeless algebraic confusion.)

Equating x- and y-coordinates, we have

$$\begin{aligned} 1 + 2t &= 5 + 3s, \\ 2 - 3t &= 3 - 4s, \end{aligned} \quad \text{or} \quad \begin{aligned} 2t - 3s &= 4, \\ -3t + 4s &= 1. \end{aligned}$$

Multiplying the first equation by 3 and the second by 2 and adding, we have

$$-s = 14 \quad \text{or} \quad s = -14.$$

If we trust our work, we can simply put $s = -14$ in the second given system of equations, obtaining

$$\begin{aligned} x &= 5 + 3(-14) = -37, \\ y &= 3 - 4(-14) = 59, \end{aligned}$$

giving the point $(-37, 59)$ of intersection. As a check, we find that when $s = -14$, we have $t = -19$, which also yields $(-37, 59)$ when substituted in the equations for the first line. $\square$

EXAMPLE 6 Determine whether the line through $(-1, 2, 0)$ and $(2, 1, 4)$ intersects the line through $(1, 2, 1)$ and $(-14, 5, -14)$.

Solution We obtain parametric equations for the two lines as in Example 3.

First line	***Second line***
Point $(-1, 2, 0)$	***Point*** $(1, 2, 1)$
$\|$ ***vector*** $3\mathbf{i} - \mathbf{j} + 4\mathbf{k}$	$\|$ ***vector*** $-15\mathbf{i} + 3\mathbf{j} - 15\mathbf{k}$ or $-5\mathbf{i} + \mathbf{j} - 5\mathbf{k}$

Equations $x = -1 + 3t,$ $y = 2 - t,$ $z = 4t$ ***Equations*** $x = 1 - 5s,$ $y = 2 + s,$ $z = 1 - 5s$

Equating coordinates, we have

$$\begin{aligned} -1 + 3t &= 1 - 5s, \\ 2 - t &= 2 + s, \\ 4t &= 1 - 5s, \end{aligned} \qquad \text{or} \qquad \begin{aligned} 3t + 5s &= 2, \\ -t - s &= 0, \\ 4t + 5s &= 1. \end{aligned}$$

We should be able to solve for t and s using just two of these three equations. Subtracting the third equation from the first, we have

$$-t = 1, \quad \text{or} \quad t = -1.$$

Then from the first equation, we obtain $s = 1$. We would not expect these values for t and s to satisfy the second equation, $-t - s = 0$, but of course this example was fixed so they do. Thus these two lines do intersect, and, setting $t = -1$ in the equations for the first line, we obtain $(-4, 3, -4)$ as the point of intersection. □

EXAMPLE 7 Find the point on the line through (2, 3, 1) and (−1, 4, 2) that is closest to the origin.

Solution Parametric equations for the line are found by

Point (2, 3, 1)

∥ ***vector*** $-3\mathbf{i} + \mathbf{j} + \mathbf{k}$

Equations $x = 2 - 3t,$ $y = 3 + t,$ $z = 1 + t$

We want to minimize $s = \sqrt{x^2 + y^2 + z^2}$. Equivalently, we minimize $s^2 = x^2 + y^2 + z^2$ to avoid working with the radical. We have

$$s^2 = x^2 + y^2 + z^2 = (2 - 3t)^2 + (3 + t)^2 + (1 + t)^2.$$

Taking a derivative, we have

$$\frac{d(s^2)}{dt} = 2(2 - 3t)(-3) + 2(3 + t) + 2(1 + t) = 22t - 4 = 0,$$

when $t = \frac{2}{11}$. Since the second derivative 22 is positive, this corresponds to the minimum distance. The point corresponding to $t = \frac{2}{11}$ is $(\frac{16}{11}, \frac{35}{11}, \frac{13}{11})$. □

THE ANGLE BETWEEN INTERSECTING LINES

One nice thing about *parametric* equations for a line is that the direction of the line is easily found from the equations. Naturally two lines are *parallel* if they have parallel direction vectors and are *perpendicular* if they meet and have perpendicular direction vectors. Of course, two lines in the plane either are parallel or intersect, but this need not be true in space.

DEFINITION Angle

14.6

The **angle between intersecting lines** is the angle between their direction vectors.

EXAMPLE 8 Find the angle between the lines

$$x = 3 + 4t, \quad y = -2 + t, \qquad \text{and} \qquad x = -1 - 3t, \quad y = 5 + 12t$$

in the plane.

Solution Direction vectors for these lines are $\mathbf{d} = 4\mathbf{i} + \mathbf{j}$ and $\mathbf{d}' = -3\mathbf{i} + 12\mathbf{j}$. Now $\mathbf{d} \cdot \mathbf{d}' = -12 + 12 = 0$, so the vectors, and therefore the lines, are perpendicular. □

EXAMPLE 9 Find the acute angle between the lines

$$x = -2 + 2t, \quad y = 3 - 4t, \quad z = -4 + t, \qquad \text{and} \qquad x = -2 - t, \quad y = 3 + 2t, \quad z = -4 + 3t,$$

which meet at the point $(-2, 3, -4)$ in space.

Solution Direction vectors for these lines are $\mathbf{d} = 2\mathbf{i} - 4\mathbf{j} + \mathbf{k}$ and $\mathbf{d}' = -\mathbf{i} + 2\mathbf{j} + 3\mathbf{k}$. Now $\mathbf{d} \cdot \mathbf{d}' = -2 - 8 + 3 = -7$. We change the direction vector of the first line to $-\mathbf{d} = -2\mathbf{i} + 4\mathbf{j} - \mathbf{k}$ to obtain a positive dot product, corresponding to an acute angle θ. Then

$$\cos\theta = \frac{(-\mathbf{d}) \cdot \mathbf{d}'}{|-\mathbf{d}| \cdot |\mathbf{d}'|} = \frac{7}{\sqrt{21}\sqrt{14}} = \frac{1}{\sqrt{6}}.$$

Therefore,

$$\theta = \cos^{-1}\left(\frac{1}{\sqrt{6}}\right) \approx 65.91°. \quad \square$$

LINE SEGMENTS

We describe how to find the points on the line segment from (a_1, a_2, a_3) to (b_1, b_2, b_3) in space. (Of course, we could drop one coordinate and work in the plane.) In Fig. 14.66, we consider the plane of the page to be the plane containing the origin and the line through the two points. We take as direction vector for this line

$$\mathbf{d} = (b_1 - a_1)\mathbf{i} + (b_2 - a_2)\mathbf{j} + (b_3 - a_3)\mathbf{k}.$$

Figure 14.66 The line segment from (a_1, a_2, a_3) to (b_1, b_2, b_3).

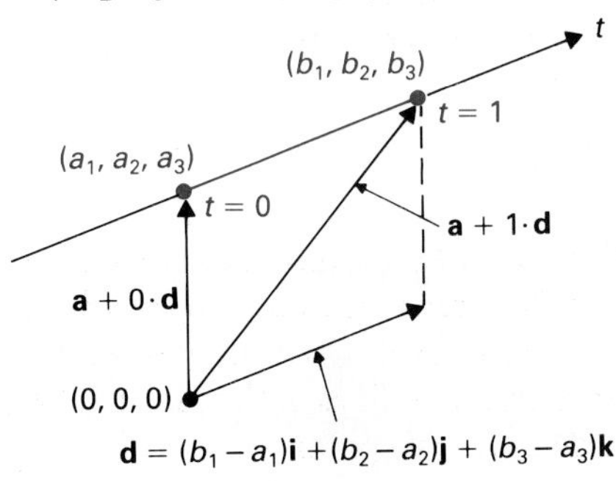

Parametric equations for the line are

$$x = a_1 + (b_1 - a_1)t,$$
$$y = a_2 + (b_2 - a_2)t,$$
$$z = a_3 + (b_3 - a_3)t.$$

Our line becomes a t-axis, as shown. From the figure (or equations), we see that $t = 0$ corresponds to (a_1, a_2, a_3) and $t = 1$ to (b_1, b_2, b_3) on the line. Thus we obtain all points on the line segment between these points by taking all values of t where $0 \le t \le 1$. Thinking of the line as a t-axis, we see that $t = \frac{1}{2}$ gives the midpoint of the line segment, $t = \frac{1}{3}$ gives the point one-third of the way from (a_1, a_2, a_3) to (b_1, b_2, b_3), and so on. We summarize in a definition.

DEFINITION 14.7 Segments and midpoints

Let (a_1, a_2, a_3) and (b_1, b_2, b_3) be two points in space (or the plane). The **line segment** joining them consists of all points (x, y, z) such that, for some value of t, where $0 \le t \le 1$,

$$\begin{aligned} x &= a_1 + (b_1 - a_1)t, \\ y &= a_2 + (b_2 - a_2)t, \\ z &= a_3 + (b_3 - a_3)t. \end{aligned} \qquad (7)$$

The **midpoint** of the line segment is the point

$$\left(\frac{a_1 + b_1}{2}, \frac{a_2 + b_2}{2}, \frac{a_3 + b_3}{2}\right).$$

EXAMPLE 10 Find the midpoint of the line segment joining $(-1, 3, 2)$ and $(3, 1, -1)$.

Solution The midpoint is

$$\left(\frac{-1 + 3}{2}, \frac{3 + 1}{2}, \frac{2 - 1}{2}\right) = \left(1, 2, \frac{1}{2}\right). \quad \square$$

EXAMPLE 11 Find the point one-third of the way from $(-1, 3, 2)$ to $(3, 1, -1)$ on the line segment joining them.

Solution Equations (7) become

$$\begin{aligned} x &= -1 + 4t, \\ y &= 3 - 2t, \\ z &= 2 - 3t. \end{aligned}$$

Setting $t = \frac{1}{3}$, we obtain the point $(\frac{1}{3}, \frac{7}{3}, 1)$. $\square$

SUMMARY

1. A line in the plane or space is determined by a point on the line and a vector $\mathbf{d}$ parallel to the line.
2. If a line goes through a point (a_1, a_2, a_3) and has direction given by the nonzero vector $\mathbf{d} = d_1\mathbf{i} + d_2\mathbf{j} + d_3\mathbf{k}$, then parametric equations for

the line are

$$x = a_1 + d_1 t,$$
$$y = a_2 + d_2 t,$$
$$z = a_3 + d_3 t.$$

All values of the parameter t give all points on the line.

3. Parametric equations of the line through (a_1, a_2, a_3) and (b_1, b_2, b_3) are

$$x = a_1 + (b_1 - a_1)t,$$
$$y = a_2 + (b_2 - a_2)t,$$
$$z = a_3 + (b_3 - a_3)t.$$

The first two equations are used for a line in the plane. The points obtained by restricting t to the interval $0 \leq t \leq 1$ make up the line segment joining (a_1, a_2, a_3) and (b_1, b_2, b_3). The midpoint is obtained when $t = \frac{1}{2}$.

EXERCISES

In Exercises 1 through 24, find parametric equations for the indicated line.

1. Through $(3, -2)$ with direction vector $\mathbf{d} = -8\mathbf{i} + 4\mathbf{j}$

2. Through $(-4, -1)$ with direction vector $\mathbf{d} = 2\mathbf{i} + \mathbf{j}$

3. Through $(4, -1, 0)$ with direction vector $\mathbf{d} = \mathbf{i} - 3\mathbf{j} + \mathbf{k}$

4. Through $(0, 0, 0)$ with direction vector $\mathbf{d} = 2\mathbf{i} - 3\mathbf{j} + \mathbf{k}$

5. Through $(1, -1, 1)$ with direction vector $\mathbf{d} = \mathbf{i} + 4\mathbf{k}$

6. Through $(-1, 4, 0)$ with direction vector $\mathbf{d} = \mathbf{j}$

7. Through $(1, -4)$ parallel to the line $x = 3 - 2t$, $y = 4 + 5t$

8. Through $(2, -3)$ perpendicular to the line $x = 4 - t$, $y = 3 + 2t$

9. Through $(3, -1, 4)$ parallel to the line $x = 3 - 2t$, $y = 4 + 3t$, $z = 6 - 8t$

10. Through $(4, -1, 2)$ parallel to the line $x = -2t$, $y = 5$, $z = 3 + 4t$

11. The line $x - 2y = 5$ in the plane

12. The line $7x - 3y = 2$ in the plane

13. Through $(-1, 5)$ and perpendicular to $3x + y = 7$ in the plane

14. Through $(-2, -3)$ and parallel to $4x - 5y = 8$ in the plane

15. Through $(-1, 4)$ and $(2, 6)$

16. Through $(3, -5)$ and $(6, -2)$

17. Through $(-1, 4, 0)$ and $(8, 1, -4)$

18. Through $(8, -1, 2)$ and $(0, 0, 0)$

19. Through $(4, -1, 3)$ and $(-1, 4, 6)$

20. Through $(2, -1, 4)$ and $(-1, 1, -1)$

21. Through $(2, 1, 4)$ and perpendicular to the x,z-coordinate plane

22. Through $(-1, 4, 6)$ and perpendicular to the y,z-coordinate plane

23. Through $(-1, 2, 3)$ perpendicular to both the lines $x = -1 + 3t$, $y = 2$, $z = 3 - t$ and $x = -1 - t$, $y = 2 + 3t$, $z = 3 + t$. [*Hint:* Use a cross product.]

24. Through $(2, -1, 4)$ perpendicular to both the lines $x = 2 + 3t$, $y = -1 - 2t$, $z = 4 + t$ and $x = 2 - t$, $y = -1 + t$, $z = 4 - 5t$. [*Hint:* As in Exercise 23.]

25. Find the point on the line through $(-1, 5)$ and $(2, 6)$ having x-coordinate -7.

26. Find the point on the line through $(1, -2)$ and $(7, -5)$ having y-coordinate 7.

27. Find the point on the line through $(-1, 4, 6)$ and $(2, 1, 3)$ having y-coordinate -2.

28. Find the point on the line through $(2, -1, 2)$ and $(0, 1, 5)$ having z-coordinate 7.

In Exercises 29 through 34, find all points of intersection of the given pair of lines, if there are any.

29. $x = 5 + 3t$, $y = -6 - 5t$ and $x = -5 + 4s$, $y = 5 - s$ in the plane

30. $x = 6 - 2t$, $y = -4 + 4t$ and $x = -3 + 3s$, $y = 7 - 6s$ in the plane

31. $x = -3 + 3t$, $y = -11 + 9t$ and $x = 12 - 6s$, $y = -6 + 2s$ in the plane

32. $x = 4 + t$, $y = 2 - 3t$, $z = -3 + 5t$ and $x = 11 + 3s$, $y = -9 - 4s$, $z = -4 - 3s$ in space

33. $x = 11 + 3t$, $y = -3 - t$, $z = 4 + 3t$ and $x = 6 - 2s$, $y = -2 + s$, $z = -15 + 7s$ in space

34. $x = -2 + t$, $y = 3 - 2t$, $z = 1 + 5t$ and $x = 7 - 3s$, $y = 1 + 2s$, $z = 4 - s$ in space

In Exercises 35 through 38, find the acute angle between the pairs of intersecting lines.

35. $x = 3 - 4t$, $y = 2 + 3t$ and $x = 5 - t$, $y = 7 + 2t$ in the plane

36. $x = 2 - 3t$, $y = 4 + t$ and $x = 3 - t$, $y = 2 + 6t$ in the plane

37. $x = 2 - 4t$, $y = -1 - t$, $z = 3 + 5t$ and $x = 2 - t$, $y = -1 + 2t$, $z = 3$ in space

38. $x = -2t$, $y = 3 - 2t$, $z = 1$ and $x = -3t$, $y = 3 + 4t$, $z = 1 - 3t$ in space

39. Find the acute angle that the line $x = t$, $y = t$, $z = t$ in space makes with the coordinate axes.

40. Let a line in space pass through the origin and make angles of α, β, and γ with the three positive coordinate axes. Show that $\cos^2\alpha + \cos^2\beta + \cos^2\gamma = 1$.

41. Find the cosines of the angles that the direction vector of the line

$$x = a_1 + d_1t,$$
$$y = a_2 + d_2t,$$
$$z = a_3 + d_3t$$

makes with the vectors **i**, **j**, and **k** along the positive coordinate axes. These cosines are called the *direction cosines* of the line. They are usually denoted by $\cos\alpha$, $\cos\beta$, and $\cos\gamma$, respectively. (See Exercise 40.)

42. Find the point on the line $x = 3 + t$, $y = 4 - 2t$ in the plane closest to the origin.

43. Find the point on the line $x = 5 - t$, $y = 3 - 4t$ in the plane closest to the point $(-1, 4)$.

44. Find the point on the line $x = t$, $y = 4 - 3t$, $z = 5 + 2t$ in space closest to the origin.

45. Find the point on the line $x = 5 + t$, $y = 2 - 3t$, $z = 1 + t$ in space closest to the point $(-1, 4, 1)$.

46. Find the midpoint of the line segment joining $(3, -1, 6)$ and $(0, -3, -1)$.

47. Find the midpoint of the line segment joining $(2, -1, 4)$ and $(3, 7, -2)$.

48. Find the point in space one-fourth of the way from $(-2, 1, 3)$ to $(0, -5, 2)$ on the line segment joining them.

49. Find the point in space two-thirds of the distance from $(3, -1, 2)$ to $(2, 5, 8)$ on the line segment joining them.

50. Find the point in the plane on the line segment joining $(-1, 3)$ and $(2, 5)$ that is twice as close to $(-1, 3)$ as to $(2, 5)$.

51. Let (a_1, a_2, a_3) and (b_1, b_2, b_3) be any two points in space. Show that the line segment joining the points consists of all points (x_1, x_2, x_3), where $x_i = (1 - t)a_i + tb_i$ for $0 \le t \le 1$, $i = 1, 2, 3$.

52. Find the point in space on the line through $(0, 2, 1)$ and $(2, 1, 3)$ that is 5 units beyond $(2, 1, 3)$ in the direction of increasing x. [*Hint*: Think of making the line into a t-axis with $(2, 1, 3)$ the origin, as in Fig. 14.65, but use a *unit* direction vector **d**.]

53. Find the point on the line through $(-1, 3, 2)$ and $(4, -1, 0)$ that is $\sqrt{5}$ units from $(-1, 3, 2)$ in the direction of decreasing y. [*Hint*: As in Exercise 52.]

54. Find the point on the line $x = -1 + 2t$, $y = 4 + 5t$, $z = -3 - t$ that is 5 units from the point $(-5, -6, -1)$ in the direction of increasing z. [*Hint:* As in Exercise 52.]

55. Find the point on the line $x = 2 + 3t$, $y = 4 - t$, $z = 6$ that is 4 units from $(-1, 5, 6)$ in the direction of decreasing x. [*Hint:* As in Exercise 52.]

14.7 PLANES

THE EQUATION OF A PLANE

Let (a_1, a_2, a_3) be a point in space and, for the moment, use subscripted variable notation. The line with parametric equations

$$x_1 = a_1 + d_1t, \qquad x_2 = a_2 + d_2t, \qquad x_3 = a_3 + d_3t$$

can be characterized as the set of all points (x_1, x_2, x_3) such that the vector

$$\mathbf{x} - \mathbf{a} = (x_1 - a_1)\mathbf{i} + (x_2 - a_2)\mathbf{j} + (x_3 - a_3)\mathbf{k}$$

is *parallel* to the direction vector $\mathbf{d} = d_1\mathbf{i} + d_2\mathbf{j} + d_3\mathbf{k}$ of the line. See Fig. 14.67. Consider now the set of all (x_1, x_2, x_3) such that the vector

$$\mathbf{x} - \mathbf{a} = (x_1 - a_1)\mathbf{i} + (x_2 - a_2)\mathbf{j} + (x_3 - a_3)\mathbf{k}$$

Figure 14.67 Points (x_1, x_2, x_3), where $(\mathbf{x} - \mathbf{a}) \| \mathbf{d}$ lie on a line.

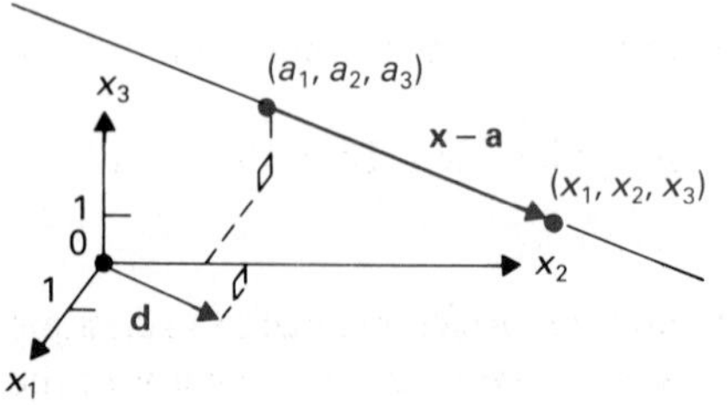

is *perpendicular* to a vector $\mathbf{d} = d_1\mathbf{i} + d_2\mathbf{j} + d_3\mathbf{k}$, where not all d_i are zero. These two vectors are perpendicular if and only if

$$d_1(x_1 - a_1) + d_2(x_2 - a_2) + d_3(x_3 - a_3) = 0. \tag{1}$$

We will try to see whether this concept gives some familiar geometric configuration in space and also look at the similarly defined set with equation $d_1(x_1 - a_1) + d_2(x_2 - a_2) = 0$ in the plane. From Fig. 14.68, we see that all such points (x_1, x_2) in the plane lie on a line through (a_1, a_2) that is perpendicular to $\mathbf{d}$. From Fig. 14.69, we see that all such points (x_1, x_2, x_3) in space lie in a plane through (a_1, a_2, a_3). This suggests the following definition.

DEFINITION 14.8 Plane in space

Let (a_1, a_2, a_3) be a point in space and $\mathbf{d} = d_1\mathbf{i} + d_2\mathbf{j} + d_3\mathbf{k}$ a nonzero vector. The **plane** through (a_1, a_2, a_3) with normal vector $\mathbf{d}$ is the set of all points (x, y, z) satisfying

$$d_1(x - a_1) + d_2(y - a_2) + d_3(z - a_3) = 0. \tag{2}$$

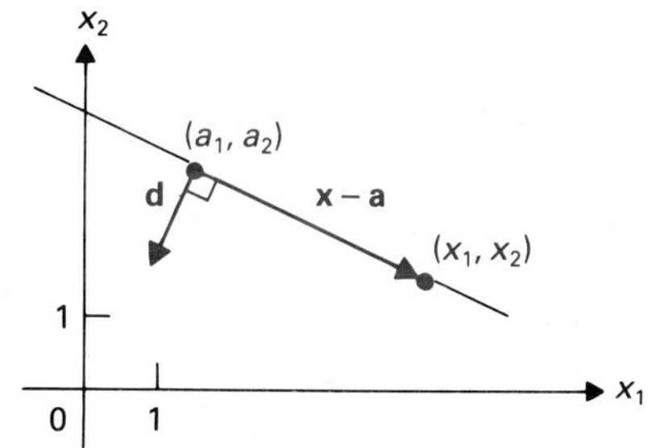

Figure 14.68 Points (x_1, x_2) in the plane where $(\mathbf{x} - \mathbf{a}) \perp \mathbf{d}$ lie on a line.

We emphasize that the vector giving "direction" for the plane in Fig. 14.69 is *normal* to the plane. We used the symbol $\|$ for "parallel," and we will use the symbol $\perp$ for "perpendicular." The terms "normal" and "orthogonal" are also used for perpendicular. You should get accustomed to all three terms.

From Definition 14.8, we see that a step sequence for finding an equation of a plane is

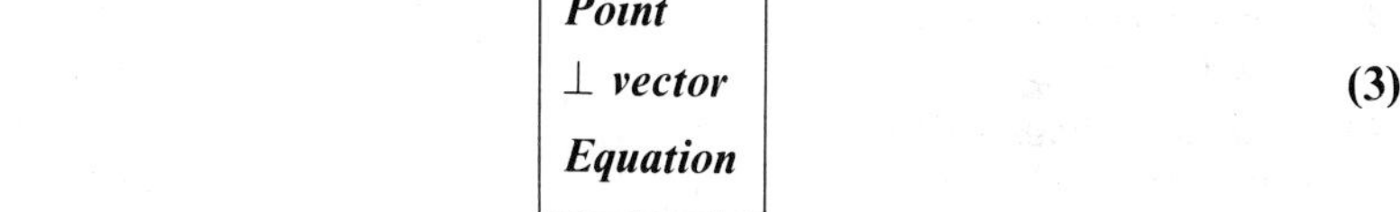

Point
⊥ vector
Equation (3)

EXAMPLE 1 Find an equation of the plane in space that passes through $(-1, 2, 1)$ and has the normal vector $\mathbf{d} = \mathbf{i} - 3\mathbf{j} + 2\mathbf{k}$.

Solution Following the step sequence (3), we have

Point $(-1, 2, 1)$

⊥ vector $\mathbf{i} - 3\mathbf{j} + 2\mathbf{k}$

Equation $1[x - (-1)] + (-3)(y - 2) + 2(z - 1) = 0$

This equation may be written in the form

$$x - 3y + 2z = -5. \quad \square$$

Figure 14.69 Points (x_1, x_2, x_3) in space where $(\mathbf{x} - \mathbf{a}) \perp \mathbf{d}$ lie on a plane.

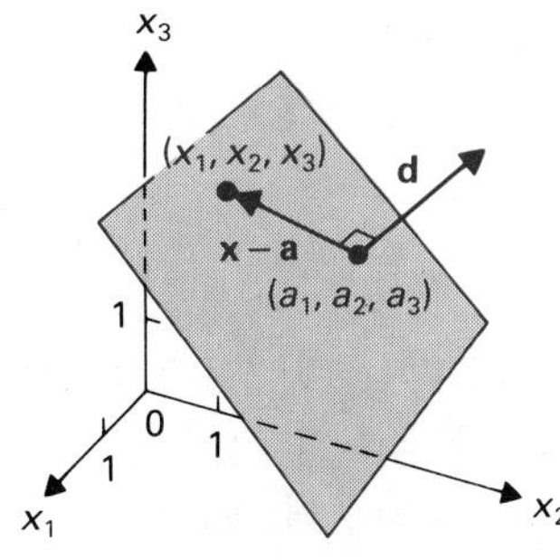

Of course, if $\mathbf{d}$ is a normal vector to a plane, then $r\mathbf{d}$ is also a normal vector to the same plane for all nonzero r.

Just as in Example 1, Eq. (2) can always be rewritten in the form

$$d_1x + d_2y + d_3z = c,$$

where $c = d_1a_1 + d_2a_2 + d_3a_3$ is a real number. Referring back to Example 1, we would normally write our final answer $x - 3y + 2z = -5$ without the intermediate step. Since the coefficients of x, y, and z are the coefficients in the normal vector, we would at once write

$$x - 3y + 2z = \_\_\_\_. \tag{4}$$

Then we substitute the coordinates of the point $(-1, 2, 1)$ in Eq. (4) to find the constant for the right-hand side. Substitution yields $(-1) - 3(2) + 2(1) = -5$, so we obtain

$$x - 3y + 2z = -5. \tag{5}$$

We can write this answer (5) down at once, doing all the steps in our head.

Now we will show that every linear equation of the form $d_1x + d_2y + d_3z = c$, where not all d_i are zero, indeed has as solution set some plane in space. We may suppose that $d_1 \neq 0$. We choose any numbers a_2 and a_3 and let

$$a_1 = \frac{c - d_2a_2 - d_3a_3}{d_1}.$$

Then $d_1a_1 + d_2a_2 + d_3a_3 = c$. Now for any point (x, y, z) such that $d_1x + d_2y + d_3z = c$, we have the two equations

$$d_1x + d_2y + d_3z = c,$$

$$d_1a_1 + d_2a_2 + d_3a_3 = c.$$

Subtracting, we obtain

$$d_1(x - a_1) + d_2(y - a_2) + d_3(z - a_3) = 0.$$

Thus (x, y, z) lies on the plane through (a_1, a_2, a_3) with normal vector $\mathbf{d} = d_1\mathbf{i} + d_2\mathbf{j} + d_3\mathbf{k}$. Dropping one coordinate, we see we can view the solution set in the x,y-plane of a linear equation in x and y similarly.

EXAMPLE 2 Find vectors normal and parallel to the line $x - 3y = 4$ in the plane.

Solution A normal vector is $\mathbf{d} = \mathbf{i} - 3\mathbf{j}$, having as coefficients of $\mathbf{i}$ and $\mathbf{j}$ the coefficients of x and y, respectively. A vector parallel to the line is a vector orthogonal to $\mathbf{d}$. We can easily find a vector orthogonal to one in the plane by interchanging the coefficients of $\mathbf{i}$ and $\mathbf{j}$ and then changing the sign of one of them. (Check that the dot product of the vectors is then zero.) Thus $3\mathbf{i} + \mathbf{j}$ is orthogonal to $\mathbf{d} = \mathbf{i} - 3\mathbf{j}$, so $3\mathbf{i} + \mathbf{j}$ is parallel to the line. □

EXAMPLE 3 Describe the plane $z = 0$ in light of the preceding discussion.

Solution The plane $z = 0$ contains the origin. It has $\mathbf{k}$ as a normal vector. Thus it must be the x, y-coordinate plane. □

We can easily check that the following definitions coincide with our intuitive ideas of parallel and perpendicular.

DEFINITION 14.9 Parallel and perpendicular planes

Two planes in space are **parallel** if normal vectors for the two planes are parallel to each other. The planes are **perpendicular** if their normal vectors are orthogonal to each other. The **angle between** two planes is the angle between their normal vectors. A line and a plane are **parallel** if a parallel vector for the line is orthogonal to a normal vector for the plane. A line and a plane are **perpendicular** if a parallel vector for the line is also a normal vector for the plane.

COMPUTATIONS

The following examples illustrate a few of the many types of problems we can now solve.

> *Remember*:
>
> **a.** To find parametric equations for a line, we need to know a point on the line and a parallel vector to the line.
>
> **b.** To find an equation of a plane, we need to know a point in the plane and a normal vector to the plane.

EXAMPLE 4 Find parametric equations for the line in space passing through the point $(1, 2, -1)$ and perpendicular to the plane with equation

$$3x + 5y - z = 6.$$

Solution We know a point $(1, 2, -1)$ on the line, and we need a direction vector for the line. Since the line is to be perpendicular to the plane given by $3x + 5y - z = 6$, we see that $\mathbf{d} = 3\mathbf{i} + 5\mathbf{j} - \mathbf{k}$ is a direction vector for the line. Following the step sequence for a line, we have

Point $(1, 2, -1)$

$\|$ ***vector*** $3\mathbf{i} + 5\mathbf{j} - \mathbf{k}$

Equations $x = 1 + 3t,$

$y = 2 + 5t,$

$z = -1 - t$ □

EXAMPLE 5 Find the equation of the plane through $(-1, 5, 2)$ and parallel to the plane $x - 3y + 5z = 8$.

Solution By Definition 14.9, parallel planes have their normal vectors parallel to each other. Thus we can take the coefficients of x, y, and z in the given equation as coefficients of **i**, **j**, and **k** for the desired normal vector. We have

Point $(-1, 5, 2)$

$\|$ ***vector*** $\mathbf{i} - 3\mathbf{j} + 5\mathbf{k}$

Equation $x - 3y + 5z = -6$ □

Example 5 makes it clear that as c varies, the equation $x - 3y + 5z = c$ gives parallel planes. We can think of the planes as "sliding along" perpendicular to the line along the vector $\mathbf{d} = \mathbf{i} - 3\mathbf{j} + 5\mathbf{k}$ as c varies.

EXAMPLE 6 Find all points of intersection in space of the plane with equation

$$3x + 5y - z = -2$$

and the line with parametric equations

$$x = -3 + 2t, \qquad y = 4 + t, \qquad z = -1 - 3t.$$

Solution If (x, y, z) lies on both the line and the plane, then, substituting, we must have

$$3(-3 + 2t) + 5(4 + t) - (-1 - 3t) = -2,$$

so

$$14t = -14 \quad \text{and} \quad t = -1.$$

Thus the only point of intersection is $(-5, 3, 2)$. □

EXAMPLE 7 Find the intersection of the plane

$$x - y + 2z = 10$$

with the line

$$x = 3 + 2t, \qquad y = 4 + 8t, \qquad \text{and} \qquad z = -1 + 3t.$$

Solution Proceeding as in Example 6, we obtain

$$(3 + 2t) - (4 + 8t) + 2(-1 + 3t) = 10,$$
$$0 \cdot t = 13.$$

This equation has no solutions, so the line is parallel to the plane but not in the plane.

If the plane instead has equation $x - y + 2z = -3$, we would obtain the equation

$$0 \cdot t = 0,$$

which is satisfied for all values of t. In this case, the line lies in the plane $x - y + 2z = -3$. □

EXAMPLE 8 Find the acute angle θ between the planes $x - 3y + z = 4$ and $3x + 2y + 4z = 6$.

Solution Normal vectors for the planes are

$$\mathbf{d} = \mathbf{i} - 3\mathbf{j} + \mathbf{k} \qquad \text{and} \qquad \mathbf{d}' = 3\mathbf{i} + 2\mathbf{j} + 4\mathbf{k}.$$

Now $\mathbf{d} \cdot \mathbf{d}' = 3 - 6 + 4 = 1$, so

$$\cos\theta = \frac{\mathbf{d} \cdot \mathbf{d}'}{|\mathbf{d}| \cdot |\mathbf{d}'|} = \frac{1}{\sqrt{11}\sqrt{29}}.$$

Therefore

$$\theta = \cos^{-1}\left(\frac{1}{\sqrt{11}\sqrt{29}}\right) \approx 86.79°. \quad □$$

EXAMPLE 9 Find the equation of the plane in space that contains the points $(1, -1, 1)$, $(2, 3, -4)$, and $(0, 1, -2)$.

Solution To find the equation of a plane, we want to know a vector normal to the plane. Taking $(1, -1, 1)$ as new origin, we see that the vectors

from $(1, -1, 1)$ to $(2, 3, -4)$

and

from $(1, -1, 1)$ to $(0, 1, -2)$

lie in the plane. Thus

$$\mathbf{i} + 4\mathbf{j} - 5\mathbf{k} \qquad \text{and} \qquad -\mathbf{i} + 2\mathbf{j} - 3\mathbf{k}$$

are vectors in the desired plane. A normal vector to the plane must be perpendicular to both of these vectors; their cross product will be such a vector. The symbolic determinant

$$\begin{vmatrix} \mathbf{i} & \mathbf{j} & \mathbf{k} \\ 1 & 4 & -5 \\ -1 & 2 & -3 \end{vmatrix} = -2\mathbf{i} + 8\mathbf{j} + 6\mathbf{k}$$

shows that $\mathbf{i} - 4\mathbf{j} - 3\mathbf{k}$ is a normal vector to the plane. Following the step sequence (3), we have

Point $(1, -1, 1)$

$\perp$ ***vector*** $\mathbf{i} - 4\mathbf{j} - 3\mathbf{k}$

Equation $x - 4y - 3z = 2$ □

EXAMPLE 10 Find parametric equations for the line of intersection of the planes $x - 2y + z = 4$ and $2x + 3y - z = 7$.

Solution We show two intersecting planes (not the given ones) in Fig. 14.70. The figure indicates that a parallel vector to the line of intersection is orthogonal to both the normal vectors to the two planes. Such a vector can be found by taking the cross product of the normal vectors to the planes. We obtain

$$\begin{vmatrix} \mathbf{i} & \mathbf{j} & \mathbf{k} \\ 1 & -2 & 1 \\ 2 & 3 & -1 \end{vmatrix} = -\mathbf{i} + 3\mathbf{j} + 7\mathbf{k}.$$

Figure 14.70 The vector **d** along the line of intersection is perpendicular to the normals **n** and **N** to the planes.

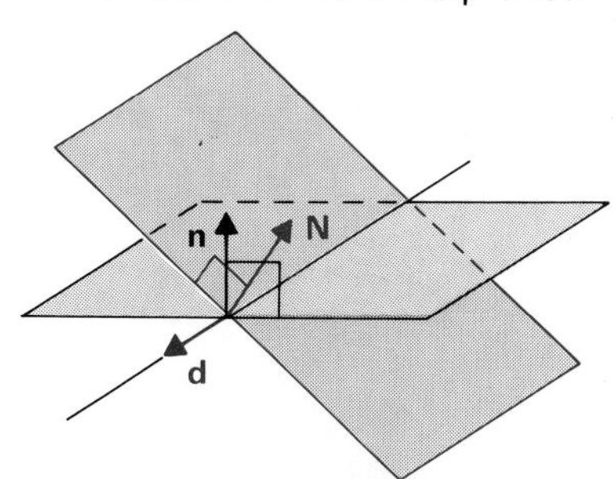

We still have to find a point on the line. We need a point in both planes. Adding the equations for the planes, we have

$$\begin{aligned} x - 2y + z &= 4 \\ 2x + 3y - z &= 7 \\ \hline 3x + y \quad &= 11. \end{aligned}$$

We can take $x = 3$ and $y = 2$ to satisfy this equation. Substituting back into either of the original equations, we find that $z = 5$. Therefore we obtain as answer

Point (3, 2, 5)
$\parallel$ ***vector*** $-\mathbf{i} + 3\mathbf{j} + 7\mathbf{k}$
Equations $x = 3 - t,$
$y = 2 + 3t,$
$z = 5 + 7t$ □

THE DISTANCE FROM A POINT TO A PLANE

Let $d_1x + d_2y + d_3z = c$ be the equation of a plane in space, and let (a_1, a_2, a_3) be a point in space. We would like to find the distance from (a_1, a_2, a_3) to the plane. This distance is measured along the line through (a_1, a_2, a_3) and perpendicular to the plane, since this gives the shortest distance.

Let (b_1, b_2, b_3) be any point on the plane, so that

$$d_1b_1 + d_2b_2 + d_3b_3 = c. \tag{6}$$

From Fig. 14.71, we see that the distance s we are after is the magnitude of the scalar component of the vector

$$\mathbf{v} = (a_1 - b_1)\mathbf{i} + (a_2 - b_2)\mathbf{j} + (a_3 - b_3)\mathbf{k}$$

from (b_1, b_2, b_3) to (a_1, a_2, a_3) along $\mathbf{d} = d_1\mathbf{i} + d_2\mathbf{j} + d_3\mathbf{k}$. The scalar component of $\mathbf{v}$ along $\mathbf{d}$ is

$$\begin{aligned}\frac{\mathbf{d}\cdot\mathbf{v}}{|\mathbf{d}|} &= \frac{d_1(a_1 - b_1) + d_2(a_2 - b_2) + d_3(a_3 - b_3)}{\sqrt{d_1^2 + d_2^2 + d_3^2}} \\ &= \frac{d_1a_1 + d_2a_2 + d_3a_3 - (d_1b_1 + d_2b_2 + d_3b_3)}{\sqrt{d_1^2 + d_2^2 + d_3^2}} \\ &= \frac{d_1a_1 + d_2a_2 + d_3a_3 - c}{\sqrt{d_1^2 + d_2^2 + d_3^2}},\end{aligned}$$

Figure 14.71 The distance s from (a_1, a_2, a_3) to the plane is the scalar projection of $\mathbf{v}$ along $\mathbf{d}$.

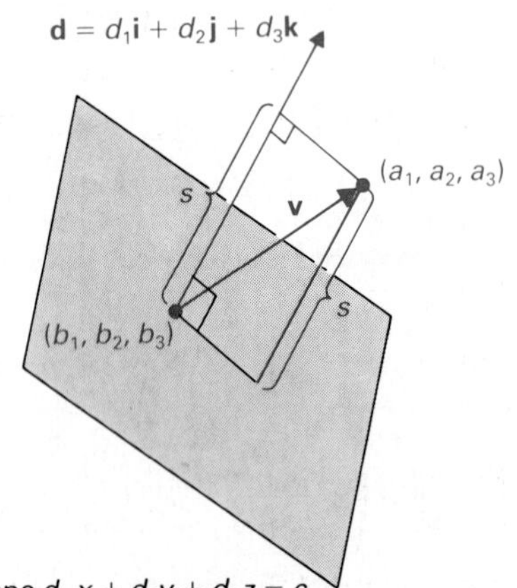

where we used Eq. (6) to obtain the last expression. Thus the distance s from (a_1, a_2, a_3) to $d_1x + d_2y + d_3z = c$ is given by

$$\boxed{s = \frac{|d_1a_1 + d_2a_2 + d_3a_3 - c|}{\sqrt{d_1^2 + d_2^2 + d_3^2}}.} \tag{7}$$

We remember Eq. (7) by the step sequence:

Step 1 Write the equation as $d_1x + d_2y + d_3z - c = 0$. Drop the "= 0."

Step 2 Substitute (a_1, a_2, a_3) for (x, y, z).

Step 3 Take the absolute value.

Step 4 Divide by $\sqrt{d_1^2 + d_2^2 + d_3^2}$.

Of course, the same derivation using two components gives the formula

$$\frac{|d_1a_1 + d_2a_2 - c|}{\sqrt{d_1^2 + d_2^2}}$$

as the distance in the x,y-plane from (a_1, a_2) to the line $d_1x + d_2y = c$.

EXAMPLE 11 Find the distance from the point $(-9, 6, 2)$ to the plane with equation $5x - 2y + z = 2$.

Solution The preceding discussion shows that this distance is

$$\frac{|-45 - 12 + 2 - 2|}{\sqrt{25 + 4 + 1}} = \frac{|-57|}{\sqrt{30}} = \frac{57}{\sqrt{30}}. \quad \square$$

EXAMPLE 12 Find the distance from $(-1, 4)$ to $x - 3y = 8$ in the plane.

Solution We obtain from Eq. (7) the distance

$$\frac{|-1 - 12 - 8|}{\sqrt{1 + 9}} = \frac{|-21|}{\sqrt{10}} = \frac{21}{\sqrt{10}}. \quad \square$$

SUMMARY

1. A plane is determined by a point in the plane and a normal (perpendicular) vector.
2. The plane through (a_1, a_2, a_3) and having normal vector $d_1\mathbf{i} + d_2\mathbf{j} + d_3\mathbf{k}$ has an equation

$$d_1(x - a_1) + d_2(y - a_2) + d_3(z - a_3) = 0.$$

3. The equation $d_1x + d_2y + d_3z = c$, where not all d_i are zero, has as solution set in space a plane with $d_1\mathbf{i} + d_2\mathbf{j} + d_3\mathbf{k}$ as normal vector.
4. The line $d_1x + d_2y = c$ in the plane has $d_1\mathbf{i} + d_2\mathbf{j}$ as normal vector.
5. Planes are parallel if their normal vectors are parallel to each other, and perpendicular if these vectors are perpendicular. The angle between two planes is the angle between their normal vectors.
6. The distance from a point (a_1, a_2, a_3) to the plane $d_1x + d_2y + d_3z = c$ is

$$\frac{|d_1a_1 + d_2a_2 + d_3a_3 - c|}{\sqrt{d_1^2 + d_2^2 + d_3^2}}$$

Similarly, the distance from (a_1, a_2) to the line $d_1x + d_2y = c$ in the plane is

$$\frac{|d_1a_1 + d_2a_2 - c|}{\sqrt{d_1^2 + d_2^2}}.$$

EXERCISES

In Exercises 1 through 29, find the equation of the indicated plane in space.

1. Through $(-1, 4, 2)$ with normal vector $\mathbf{i} - 2\mathbf{j} + \mathbf{k}$
2. Through $(1, -3, 0)$ with normal vector $\mathbf{j} + 4\mathbf{k}$
3. Through $(2, -1, 3)$ with normal vector $\mathbf{i}$
4. Through $(1, 2, -5)$ with normal vector $\mathbf{i} + 2\mathbf{j} - 4\mathbf{k}$
5. Through $(3, -4, 2)$ with normal vector $4\mathbf{i} - 2\mathbf{j} + \mathbf{k}$
6. Through $(3, -1, 4)$ and parallel to $2x - 4y + 5z = 7$
7. Through $(1, 1, -1)$ and parallel to $4x - 3y + 4z = 10$
8. Through $(5, -2, 1)$ and parallel to $-3x + 4y - z = 13$
9. Through $(-1, 4, -3)$ and orthogonal to the line $x = 3 - 7t$, $y = 4 + t$, $z = 2t$
10. Through $(2, -2, 4)$ and orthogonal to the line $x = t$, $y = 4 - t$, $z = 5$
11. Through $(-2, 1, 5)$ and orthogonal to the line $x = 3 - 2t$, $y = 3t$, $z = 4 - t$
12. Through the origin and orthogonal to the line through $(-1, 3, 0)$ and $(2, -4, 3)$
13. Through $(1, -1, 1)$ and orthogonal to the line through $(4, -1, 2)$ and $(3, 1, -5)$
14. Through $(0, 3, -1)$ and orthogonal to the line through $(2, -1, 7)$ and $(4, 2, -6)$
15. Through the points $(1, 0, 0)$, $(0, 1, 0)$, and $(0, 0, 1)$
16. Through the points $(1, 0, 1)$, $(-1, 2, 0)$, and $(0, 1, 3)$
17. Through the points $(2, 1, 5)$, $(-1, -2, 1)$, and $(3, -1, 4)$
18. Through the points $(-1, 4, 2)$, $(1, 2, -1)$, and $(3, -2, 0)$
19. Through the points $(2, -3, 1)$, $(5, -2, 7)$, and $(-1, 1, 0)$
20. Through the points $(3, -2, 3)$, $(1, 1, 2)$, and $(-1, -5, -2)$
21. Containing the lines $x = -1 + 3t$, $y = 4 + t$, $z = 2 - 2t$ and $x = -1 + s$, $y = 4$, $z = 2 + 7s$
22. Containing the lines $x = 4 - t$, $y = 3 + 7t$, $z = 1 + 2t$ and $x = 4 + 2s$, $y = 3 + s$, $z = 1 - 3s$
23. Containing the lines $x = -1 + t$, $y = 3 - 2t$, $z = 2$ and $x = -1 - 3s$, $y = 3 + 5s$, $z = 2 + 4s$
24. Containing the line $x = 3 + 2t$, $y = 1 - t$, $z = 4 - 2t$ and the point $(3, -1, 2)$
25. Containing the line $x = 2 - t$, $y = 4 + 5t$, $z = 3 - 7t$ and the point $(4, -1, 3)$
26. Containing the line $x = 3t$, $y = 5$, $z = 4 + t$ and the point $(-2, 1, 4)$
27. Containing $(-1, 4, 3)$ and the line of intersection of the planes $2x - 3y + 4z = 7$ and $x - 2z = 5$
28. Containing $(2, -1, 5)$ and the line of intersection of the planes $3x - 4y + z = 5$ and $2x + y - 2z = 4$
29. Containing $(-5, 3, 8)$ and the line of intersection of the planes $x + z = 4$ and $y = 2$

In Exercises 30 through 36, find the acute angle between the given lines or planes.

30. $x - 3y = 7$ and $2x + 4y = 1$ in the plane
31. $3x + 2y = -7$ and $6x + 4y = 2$ in the plane
32. $3x + 4y - z = 1$ and $x - 2y = 3$ in space
33. $4x - 7y + z = 3$ and $3x + 2y + 2z = 17$ in space
34. $x = 4$ and $x + z = 2$ in space
35. $x - 2y + 2z = 3$ and $x = 2 - 6t$, $y = 4 + 3t$, $z = 1 - 2t$ in space
36. $3x - z = 4$ and $x = -t$, $y = 2 + 2t$, $z = 3 - t$ in space

In Exercises 37 through 40, find all points of intersection of the given line and plane in space.

37. $x = 5 + t$, $y = -3t$, $z = -2 + 4t$ and $x - 3y + 2z = -35$
38. $x = 2$, $y = 4 - 3t$, $z = 3t$ and $4x - 2y + 3z = 30$
39. $x = 3 - t$, $y = 4 - 5t$, $z = 6 - 3t$ and $2x - 4y + 6z = 5$
40. $x = 3t$, $y = 4 - t$, $z = 5 - 5t$ and $6x - 2y + 4z = 12$

In Exercises 41 through 46, find parametric equations of the indicated line.

41. Through $(-2, 1, 0)$ and orthogonal to the plane $x - 2y + 4z = 3$
42. Through $(1, 4, -3)$ and orthogonal to the plane $2x + y - 7z = 1$
43. Through $(1, -1, 3)$ and orthogonal to the plane $4x - 3y + z = 11$
44. The line of intersection of $x - 3y + z = 7$ and $3x + 2y + z = -1$
45. The line of intersection of $4x - 2y + z = 3$ and $-3x + 2y + z = 4$
46. The line of intersection of $x + y - 2z = 7$ and $2x + y - z = -2$

In Exercises 47 through 52, find the distance from the given point to the given line or plane.

47. $(-1, 3)$ to $3x - 4y = 5$
48. $(2, 4)$ to $x - 3y = 4$
49. $(2, -1)$ to the line with parametric equations $x = 3 - t$, $y = 2 + 4t$. [*Hint:* Obtain the x,y-equation for the line by eliminating t from the parametric equations.]
50. $(1, 3, -1)$ to $2x + y + z = 4$
51. $(-1, 4, 3)$ to $x - 2y + 2z = 3$
52. $(4, 1, -2)$ to $3x + 6y - 2z = 4$
53. A homogeneous plate in the plane of mass density 3 occupies the disk $(x - 1)^2 + (y + 4)^2 \leq 36$. Find the (unsigned) first moment of the plate about the line $2x - y = -10$. [*Hint:* Don't integrate! Use Pappus' theorem.]
54. Show that the equation of the line in the plane through two distinct points (a_1, a_2) and (b_1, b_2) is given by

$$\begin{vmatrix} x & y & 1 \\ a_1 & a_2 & 1 \\ b_1 & b_2 & 1 \end{vmatrix} = 0.$$

[*Hint:* Do not expand the determinant, but argue that a linear equation in x and y is obtained and that the equation is satisfied if $(x, y) = (a_1, a_2)$ or $(x, y) = (b_1, b_2)$.]

55. Use the method suggested by Exercise 54 to find the equation of the line in the plane through (1, −4) and (2, 3).

56. Let $\mathbf{d} = d_1\mathbf{i} + d_2\mathbf{j} + d_3\mathbf{k}$ be a vector with all coefficients not equal to zero.
   a) Show that the line through (a_1, a_2, a_3) with parallel direction vector $\mathbf{d}$ consists of all (x, y, z) such that

$$\frac{x - a_1}{d_1} = \frac{y - a_2}{d_2} = \frac{z - a_3}{d_3}.$$

   [*Hint:* Write parametric equations for the line and eliminate the parameter.]
   b) From part (a), find three planes containing the line. [The equations in part (a) are called *symmetric equations* of the line. The line appears as an intersection of planes. Some texts emphasize these symmetric equations. We consider the parametric equations to be much better, since one or two zero coefficients in $\mathbf{d}$ cause no trouble. Whenever we have tried using symmetric equations, some of our students write the line through (1, −1, 4) with direction vector $4\mathbf{i} - \mathbf{k}$ as

$$\frac{x - 1}{4} = \frac{y + 1}{0} = \frac{z - 4}{-1},$$

which is of course nonsense. A correct symmetric description is

$$\frac{x - 1}{4} = \frac{z - 4}{-1}, \qquad y = -1.]$$

EXERCISE SETS FOR CHAPTER 14

Review Exercise Set 14.1

1. a) Find the distance between (−1, 2, 4) and (3, 8, −2).
   b) Find the equation of the sphere having (1, −3, 5) and (−3, 5, 7) as endpoints of a diameter.
2. a) Sketch the solution set of the cylindrical coordinate equation $r = 4$.
   b) Find spherical coordinates of the point $(x, y, z) = (0, 1, -1)$.
3. Sketch the surface

$$\frac{x^2}{4} + \frac{y^2}{9} = 1 + z^2.$$

4. a) Find the length of the vector $2\mathbf{i} - 3\mathbf{j} + \mathbf{k}$.
   b) Find c_2 so that the vectors $\mathbf{i} - c_2\mathbf{j} + 3\mathbf{k}$ and $4\mathbf{i} + 2\mathbf{j} - 7\mathbf{k}$ are perpendicular.
5. Find the angle between the vectors $2\mathbf{i} - 3\mathbf{j} + \mathbf{k}$ and $-\mathbf{i} + 2\mathbf{j} + 3\mathbf{k}$.
6. Find the vector projection of $2\mathbf{i} - 3\mathbf{j} + \mathbf{k}$ on $3\mathbf{i} - 4\mathbf{k}$.
7. Find $\mathbf{a} \times \mathbf{b}$ if $\mathbf{a} = \mathbf{i} + 2\mathbf{j} - 3\mathbf{k}$ and $\mathbf{b} = 4\mathbf{i} - \mathbf{j} + \mathbf{k}$.
8. Find the volume of the box in space having the vectors $\mathbf{i} - 3\mathbf{j} + \mathbf{k}$, $2\mathbf{i} - 3\mathbf{j} + 2\mathbf{k}$, $-3\mathbf{i} + 5\mathbf{j} - 2\mathbf{k}$ as coterminous edges.
9. If $\mathbf{a} = \mathbf{i} - 2\mathbf{j} + \mathbf{k}$, $\mathbf{b} = -2\mathbf{i} + 3\mathbf{j} - \mathbf{k}$, and $\mathbf{c} = 4\mathbf{i} - 2\mathbf{j} + 2\mathbf{k}$, compute $(\mathbf{a} \times \mathbf{b}) \times \mathbf{c}$.
10. Find parametric equations of the line through (−5, 1, 2) and (2, −1, 7).
11. Find parametric equations of the line perpendicular to the plane $x - 2y + 3z = 8$ and passing through the origin.
12. Find the equation of the plane passing through (−1, 1, 4) and perpendicular to the line $x = 2 + t$, $y = 3 - 4t$, $z = -7 + 2t$.
13. Find the distance from (−1, 1, 3) to the plane

$$2x + y - 2z = 4.$$

14. Classify the given planes as parallel, perpendicular, or neither parallel nor perpendicular.
   a) $x - 3y + 4z = 7$, $-3x + y + 2z = 11$
   b) $2x + y + 3z = 8$, $4x + 7y - 5z = -3$
   c) $-4x + 6y - 12z = 7$, $2x - 3y + 6z = 5$
15. Find the point of intersection of the line $x = 2 - t$, $y = 4 + t$, $z = 1 - 3t$, with the plane $x - 3y + 4z = -1$.

Review Exercise Set 14.2

1. a) Find c so that the distance from $(c, -2, 3)$ to (1, 4, −5) is 15.
   b) Find the center and radius of the sphere $x^2 + y^2 + z^2 - 2x + 4z = 4$.
2. a) Find cylindrical coordinates of $(x, y, z) = (1, -\sqrt{3}, -2)$.
   b) Find rectangular coordinates of $(\rho, \phi, \theta) = (3, 5\pi/6, 3\pi/4)$.
3. a) Sketch the surface $z = y^2 + 1$.
   b) Sketch the surface

$$\frac{y}{4} = \frac{x^2}{4} + \frac{z^2}{9}.$$

4. Let $\mathbf{a} = \mathbf{i} - 2\mathbf{j} + 3\mathbf{k}$, $\mathbf{b} = 4\mathbf{i} - \mathbf{j} + 2\mathbf{k}$, and $\mathbf{c} = 2\mathbf{i} - 3\mathbf{k}$. Find s such that $\mathbf{a} + s\mathbf{b}$ is perpendicular to $\mathbf{c}$.
5. Find the angle between the vectors $\mathbf{i} - 3\mathbf{j} + \mathbf{k}$ and $2\mathbf{i} - \mathbf{j} + 2\mathbf{k}$.
6. Find the vector projection of $3\mathbf{i} - 4\mathbf{j} + 2\mathbf{k}$ onto $\mathbf{i} - 2\mathbf{j} + \mathbf{k}$.
7. Let $\mathbf{a}$ and $\mathbf{b}$ be perpendicular unit vectors in space. Simplify each of the following as much as possible from this information.
   a) $(\mathbf{a} \times \mathbf{b}) \cdot \mathbf{a}$ b) $(\mathbf{a} \times \mathbf{b}) \times \mathbf{a}$ c) $\mathbf{a} \times (\mathbf{b} \times \mathbf{b})$
8. Find the area of the parallelogram in space having as adjacent edges $\mathbf{i} + 2\mathbf{j} - \mathbf{k}$ and $2\mathbf{i} - 3\mathbf{j} + 4\mathbf{k}$.
9. Find the volume of the box in space with one corner at (−1, 3, 4) and adjacent corners at (0, −1, 2), (3, −1, 4), and (−1, 2, −1).
10. Find all points on the line $x = 1 - 2t$, $y = 3 + 4t$, $z = 2 - t$, whose distance from (1, 6, 3) is $\sqrt{18}$.

11. Find parametric equations of the line through $(-1, 5, 2)$ and $(3, -1, 4)$.
12. Find the equation of the plane through $(-2, 5, 4)$ and parallel to the plane $3x - 4y + 7z = 0$.
13. Find the distance from $(-2, 1)$ to the line $x = 4 + t$, $y = 2 - 3t$ in the plane.
14. Find the point of intersection of the line $x = 4 + t, y = 3 - t$, $z = 2 + 3t$ with the plane $2x + y + 3z = -3$.
15. Find the equation of the plane containing the two lines

$$x = 2 + t, \quad y = -1 + 2t, \quad z = 3 - 4t$$

and

$$x = 2 - 3t, \quad y = -1 + 4t, \quad z = 3 - t$$

that intersect at $(2, -1, 3)$.

Review Exercise Set 14.3

1. Find the equation of the sphere having the line segment from $(-1, 4, 2)$ to $(-3, -2, 6)$ as a diameter.
2. Find cylindrical coordinates of the point with spherical coordinates $(4, 5\pi/6, 3\pi/4)$.
3. Sketch the surface $x^2 + 4y^2 - z^2 - 2x + 16y + 17 = 0$.
4. Find a unit vector having direction from $(-1, 3, 2)$ toward $(1, 4, 0)$.
5. Find all values of c such that the angle between $\mathbf{i} - \mathbf{j} + \mathbf{k}$ and $\mathbf{i} + \mathbf{j} + c\mathbf{k}$ is $\pi/3$.
6. Find the scalar component of $\mathbf{i} - 3\mathbf{j} + 2\mathbf{k}$ along $2\mathbf{i} + \mathbf{j} - 3\mathbf{k}$.
7. If $\mathbf{a} = \mathbf{i} - 3\mathbf{j} + \mathbf{k}$, $\mathbf{b} = 2\mathbf{i} - \mathbf{k}$, and $\mathbf{c} = 4\mathbf{i} + \mathbf{j} - 3\mathbf{k}$, find $\mathbf{b} \times (\mathbf{a} \times \mathbf{c})$.
8. Find the area of the triangle in the plane with vertices $(-1, 4)$, $(2, -3)$, and $(4, -2)$.
9. Find the area of the triangle in space with vertices $(-1, 3, 0)$, $(1, 0, 5)$, and $(-5, 3, -2)$.
10. Find the point on the line $x = 2 - 3t, y = 4 + t, z = 2 - 5t$ having z-coordinate 17.
11. Find the point of intersection (if any) of the lines $x = -4 + 2t$, $y = 3 - 7t$, $z = 1 - 3t$ and $x = 6 + 2s$, $y = -2 + 3s$, $z = 7 + 4s$.
12. Find the equation of the plane through the points $(-1, 3, 2)$, $(4, 0, 1)$, and $(0, -3, 5)$.
13. Find the equation of the plane containing both $(-1, 2, 0)$ and the line $x = 3 + 4t, y = 2 - t, z = t$.
14. Find the point where the line of intersection of the planes $x - y + 2z = 4$ and $2x + y - z = -4$ meets the plane $3x - 4y + 5z = 8$.
15. Find all values of c so that the point $(-3, 2, 1)$ is 4 units from the plane $2x - y + 2z = c$.

Review Exercise Set 14.4

1. Find all values c such that $(-1, 3, 0)$, $(2, 1, -2)$, and $(1, 1, c)$ are vertices of a right triangle with right angle at the third point.
2. Find spherical coordinates of the point with cylindrical coordinates $(2, 2\pi/3, -2)$.
3. Sketch the surface $x = -\sqrt{y}$ in space.
4. Find all unit vectors in the plane orthogonal to $3\mathbf{i} - 4\mathbf{j}$.
5. Find all values of c such that the length of $\mathbf{i} - 2\mathbf{j} + c\mathbf{k}$ is 5.
6. A body in the plane is constrained to move on the curve $y = x^3 - x^2 + 2x$, subject to the force vector $\mathbf{F} = \mathbf{i} + \mathbf{j}$. Find the scalar component of $\mathbf{F}$ actually moving the body on the curve at the instant when $x = 2$.
7. Let $\mathbf{a}$ and $\mathbf{b}$ be parallel unit vectors with opposite direction in space. Simplify each of the following as much as possible from this information.
   a) $\mathbf{a} \cdot \mathbf{b}$ b) $\mathbf{a} \times \mathbf{b}$
   c) $(\mathbf{a} \cdot \mathbf{a})\mathbf{b}$ d) $|\mathbf{b} - \mathbf{a}|$
8. Find the area of the tetrahedron in space with vertices $(-1, 4, 1)$, $(0, 0, 3)$, $(5, -1, 2)$, and $(1, 4, -3)$.
9. Find all values of c such that the box having the vectors $\mathbf{i} - 2\mathbf{j} + \mathbf{k}$, $3\mathbf{i} + 4\mathbf{j} + 2\mathbf{k}$, and $c\mathbf{i} - 3\mathbf{j} + 5\mathbf{k}$ as coterminous edges has area 20.
10. Find parametric equations of the line through $(-1, 2, 7)$ and orthogonal to the plane through $(1, -4, 2)$, $(2, -3, 4)$, and $(0, 1, 0)$.
11. Find parametric equations of the line in the plane through $(-1, 3)$ and parallel to the line $x - 2y = 4$.
12. Find the equation of the plane through $(-1, 4, 3)$ and parallel to the plane $2x - 3y + z = 11$.
13. Find the equation of the plane containing the point $(1, -2, 4)$ and the line of intersection of the planes $2x - 3y + z = 4$ and $x - 3y + 2z = 5$.
14. Find the equation of the plane through the origin and orthogonal to the line through $(-1, 4, 3)$ and $(2, 0, -1)$.
15. A homogeneous plate of mass density 4 covers the disk $x^2 + y^2 - 2x + 4y \leq 11$ in the plane. Find the unsigned moment of the body about the line with parametric equations $x = -5 + t$, $y = 4t$.

More Challenging Exercises 14

1. Find the distance between the planes $x - 2y + 3z = 10$ and $4x - 8y + 12z = -7$.
2. Find the (shortest) distance between the lines $x = 1 - t$, $y = 2 + 3t$, $z = -1 + 2t$, and $x = 4 + 2t$, $y = -3 + t$, $z = -2 + 5t$.
3. Find the equation of the plane containing the intersection of the sphere with center $(-1, 2, 4)$ and radius 5 with the sphere with center $(1, -1, 3)$ and radius 3.
4. Give a geometric discussion of the accuracy of the statement "Three simultaneous linear equations in three unknowns have a unique solution."
5. Find the shortest distance from the line $x = 7 - 3t$, $y = -5 + 4t$, $z = 6 + 2t$ to the sphere $x^2 + y^2 + z^2 = 9$.
6. Consider the parallelogram in the plane having $\mathbf{a} = a_1\mathbf{i} + a_2\mathbf{j}$ and $\mathbf{b} = b_1\mathbf{i} + b_2\mathbf{j}$ as contiguous edges emanating from the origin, as in Fig. 14.58. Show that the area of this parallelogram is the absolute value of the determinant

$$\begin{vmatrix} a_1 & a_2 \\ b_1 & b_2 \end{vmatrix}.$$

7. We motivated consideration of $\mathbf{a} \cdot \mathbf{b}$ by finding the cosine of the angle between two vectors $\mathbf{a}$ and $\mathbf{b}$. For an alternative

approach, suppose we had simply *defined*, with no motivation,

$$\mathbf{a} \cdot \mathbf{b} = (a_1\mathbf{i} + a_2\mathbf{j} + a_3\mathbf{k}) \cdot (b_1\mathbf{i} + b_2\mathbf{j} + b_3\mathbf{k}) = a_1b_1 + a_2b_2 + a_3b_3. \quad \textbf{(1)}$$

The properties (a) through (d) of Theorem 14.2 (Section 14.4) are easy to show from the definition (1). We then attempt to *define* the angle θ between the vectors **a** and **b** by

$$\theta = \cos^{-1} \frac{\mathbf{a} \cdot \mathbf{b}}{\sqrt{\mathbf{a} \cdot \mathbf{a}}\sqrt{\mathbf{b} \cdot \mathbf{b}}}.$$

But we must first know that

$$\left| \frac{\mathbf{a} \cdot \mathbf{b}}{\sqrt{\mathbf{a} \cdot \mathbf{a}}\sqrt{\mathbf{b} \cdot \mathbf{b}}} \right| \leq 1 \quad \textbf{(2)}$$

for this definition to be meaningful. This inequality is known as the *Schwarz inequality* and is usually written as

$$|\mathbf{a} \cdot \mathbf{b}| \leq \|\mathbf{a}\| \, \|\mathbf{b}\|, \quad \textbf{(3)}$$

where $\|\mathbf{a}\|$ is simply another notation for $|\mathbf{a}| = \sqrt{\mathbf{a} \cdot \mathbf{a}}$, used to avoid confusing the length of a vector with the absolute value of a scalar. Prove the Schwarz inequality (3), using just the definition (1) and properties (a) through (d) of Theorem 14.2. [*Hint:* Use the fact that $(\mathbf{a} - x\mathbf{b}) \cdot (\mathbf{a} - x\mathbf{b}) \geq 0$ for all values x by property (a) of Theorem 14.2, and use the quadratic formula.]

8. Continuing Exercise 7, prove from the Schwarz inequality the *triangle inequality*

$$\|\mathbf{a} + \mathbf{b}\| \leq \|\mathbf{a}\| + \|\mathbf{b}\|$$

for all vectors **a** and **b** in space. [If you draw a figure, you will see that this inequality can be viewed as asserting that the length of one side of a triangle is less than or equal to the sum of the lengths of the other two sides; hence the name.]

15 VECTOR ANALYSIS OF CURVES

This chapter delays for a moment the development of multivariable calculus to give a few introductory lessons on the differential geometry of curves in the plane or in space. This topic provides an immediate application of the notion of a *vector* introduced in Section 14.4. In Sections 15.1 and 15.2, we develop calculus for vectors and use it to study motion in the plane and in space. Section 15.3 gives an important application to physics and astronomy. In Section 15.4, we make a more detailed study of motion in space and space curves.

15.1 VELOCITY AND ACCELERATION VECTORS

THE POSITION AND VELOCITY VECTORS

Let $x = h(t)$ and $y = k(t)$ be a parametric representation of a smooth curve, so that h' and k' exist, are continuous, and are never simultaneously zero. We may view these parametric equations as giving the position of a body in the plane at time t. We let $\mathbf{r}$ be the vector from the origin to a point (x, y) on the curve, shown in Fig. 15.1. This vector

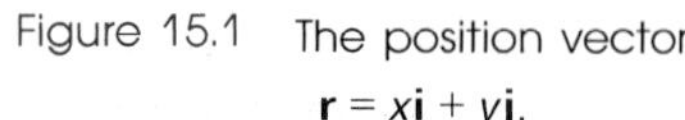

Figure 15.1 The position vector $\mathbf{r} = x\mathbf{i} + y\mathbf{j}$.

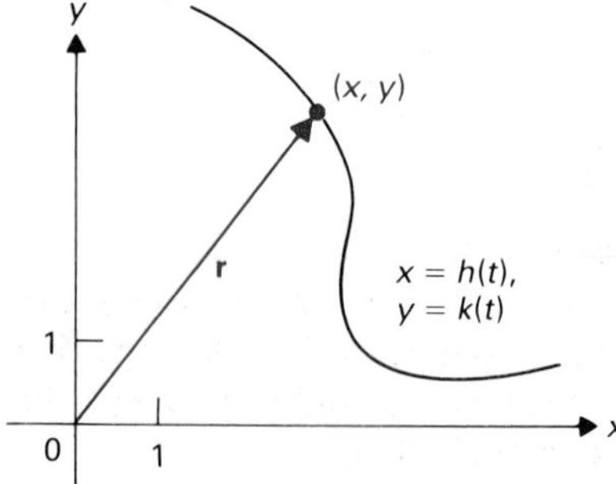

$$\boxed{\mathbf{r} = x\mathbf{i} + y\mathbf{j} = h(t)\mathbf{i} + k(t)\mathbf{j}} \qquad \textbf{(1)}$$

is the **position vector** of the body at time t.

We can view Eq. (1) as describing a function f that carries a portion of a t-axis into the x,y-plane, so that

$$(x, y) = f(t) = (h(t), k(t)).$$

Such a function, having as values points (which may also be regarded as vectors) in a space of dimension greater than one, is often called a **vector-valued function.**

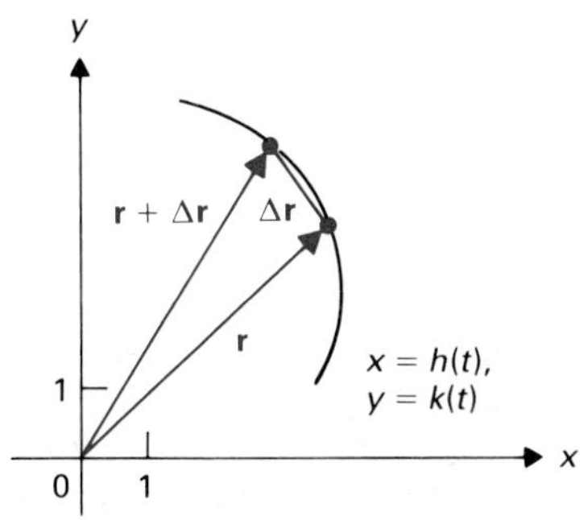

Figure 15.2 The increment vector $\Delta\mathbf{r}$.

A change Δt in t produces a change $\Delta\mathbf{r}$ in $\mathbf{r}$, where $\Delta\mathbf{r}$ is as shown in Fig. 15.2. The vector $\Delta\mathbf{r}$ is directed along a chord of the curve. Let the notation $\Delta\mathbf{r}/\Delta t$ be understood to mean the scalar $1/\Delta t$ times $\Delta\mathbf{r}$. We may write $\Delta\mathbf{r} = (\Delta x)\mathbf{i} + (\Delta y)\mathbf{j}$. Then

$$\frac{\Delta\mathbf{r}}{\Delta t} = \frac{\Delta x}{\Delta t}\mathbf{i} + \frac{\Delta y}{\Delta t}\mathbf{j}.$$

Taking the limit as $\Delta t \to 0$, we define the **vector derivative** by

$$\boxed{\frac{d\mathbf{r}}{dt} = \lim_{t\to 0}\frac{\Delta\mathbf{r}}{\Delta t} = \frac{dx}{dt}\mathbf{i} + \frac{dy}{dt}\mathbf{j}.} \tag{2}$$

Note that dx/dt and dy/dt exist, since we are assuming that $x = h(t)$ and $y = k(t)$ are differentiable functions.

We now study the vector $d\mathbf{r}/dt$. The first things we want to know about a vector are its length and direction. From Eq. (2), we have

$$\boxed{\left|\frac{d\mathbf{r}}{dt}\right| = \sqrt{\left(\frac{dx}{dt}\right)^2 + \left(\frac{dy}{dt}\right)^2}.} \tag{3}$$

From the formula

$$s(t) = \int_{t_0}^{t}\sqrt{\left(\frac{dx}{dt}\right)^2 + \left(\frac{dy}{dt}\right)^2}\,dt$$

for the arc length function, we obtain

$$\frac{ds}{dt} = \sqrt{\left(\frac{dx}{dt}\right)^2 + \left(\frac{dy}{dt}\right)^2}, \tag{4}$$

so Eq. (3) yields

$$\boxed{\left|\frac{d\mathbf{r}}{dt}\right| = \frac{ds}{dt}.} \tag{5}$$

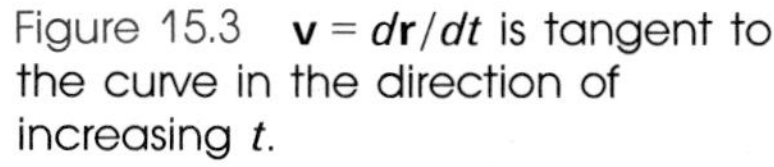
Figure 15.3 $\mathbf{v} = d\mathbf{r}/dt$ is tangent to the curve in the direction of increasing t.

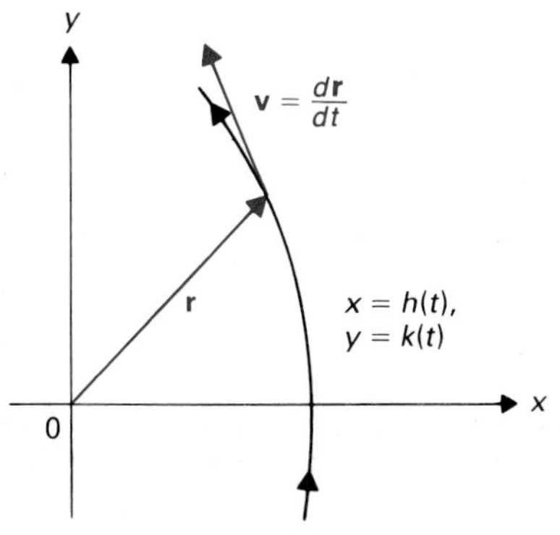

Now ds/dt has the physical interpretation of the instantaneous rate of change of arc length per unit change in time as the body travels along the curve, so ds/dt is the **speed** of the body along the curve.

Turning to the direction of $d\mathbf{r}/dt$, the limiting direction of $\Delta\mathbf{r}$ as $\Delta t \to 0$ is tangent to the curve. Consequently $d\mathbf{r}/dt$ has direction tangent to the curve in the direction given by increasing t. See Fig. 15.3. We can also see this from Eq. (2), since the "slope" of the vector arrow $d\mathbf{r}/dt$ is

$$\frac{dy/dt}{dx/dt} = \frac{dy}{dx},$$

which is the slope of the tangent to the curve.

Summarizing, at any instant the vector $d\mathbf{r}/dt$ points in the *direction* in which the body is moving and has *length* equal to the speed of the body. We think of *velocity* as a directed quantity, reflecting not only *speed* but the *direction* of motion. We see that $d\mathbf{r}/dt$ does precisely that.

DEFINITION 15.1 Velocity vector

If $\mathbf{r} = h(t)\mathbf{i} + k(t)\mathbf{j}$ is the position vector of a body traveling on a smooth curve, then

$$\mathbf{v} = \frac{d\mathbf{r}}{dt} = h'(t)\mathbf{i} + k'(t)\mathbf{j} \tag{6}$$

is the **velocity vector** (or simply the **velocity**) of the body at time t.

EXAMPLE 1 Let the position of a body in the plane at time t be given by $x = \cos t$, $y = \sin t$. Find the position vector, velocity vector, and speed at time t.

Solution Since $x^2 + y^2 = \cos^2 t + \sin^2 t = 1$, we know the body travels on the circle of radius 1 with center at the origin, shown in Fig. 15.4. The position vector is

$$\mathbf{r} = (\cos t)\mathbf{i} + (\sin t)\mathbf{j}.$$

The velocity vector is

$$\mathbf{v} = \frac{d\mathbf{r}}{dt} = (-\sin t)\mathbf{i} + (\cos t)\mathbf{j}.$$

Since $\mathbf{r}$ is along a radial line of the circle and $\mathbf{v}$ is tangent to the circle, $\mathbf{r}$ and $\mathbf{v}$ should be orthogonal. We easily check this by computing

$$\mathbf{r} \cdot \mathbf{v} = (\cos t)(-\sin t) + (\sin t)(\cos t) = 0.$$

Finally, the speed is given by

$$\text{Speed} = |\mathbf{v}| = \sqrt{\sin^2 t + \cos^2 t} = 1.$$

Thus the speed is constant for this motion on the circle. In Fig. 15.4, we were careful to make $|\mathbf{v}| = |\mathbf{r}| = 1$. □

Since $\mathbf{v}$ is tangent to the curve $x = h(t)$, $y = k(t)$ in the direction corresponding to increasing t, we see that, for $|\mathbf{v}| \neq 0$,

$$\boxed{\mathbf{t} = \frac{1}{|\mathbf{v}|}\mathbf{v}} \tag{7}$$

is a *unit* vector tangent to the curve in the direction of increasing t. Since $|\mathbf{v}| = ds/dt$, we obtain, from Eq. (7),

$$\boxed{\mathbf{v} = \frac{ds}{dt}\mathbf{t}.} \tag{8}$$

Equation (8) is a compact description of the velocity vector $\mathbf{v}$ as tangent to the curve and of magnitude ds/dt.

Exactly the same analysis can be made for a space curve

$$x = h(t), \qquad y = k(t), \qquad z = g(t),$$

Figure 15.4 $\mathbf{r} = (\cos t)\mathbf{i} + (\sin t)\mathbf{j}$; $\mathbf{v} = (-\sin t)\mathbf{i} + (\cos t)\mathbf{j}$; $\mathbf{v} \perp \mathbf{r}$ and $|\mathbf{v}| = 1$

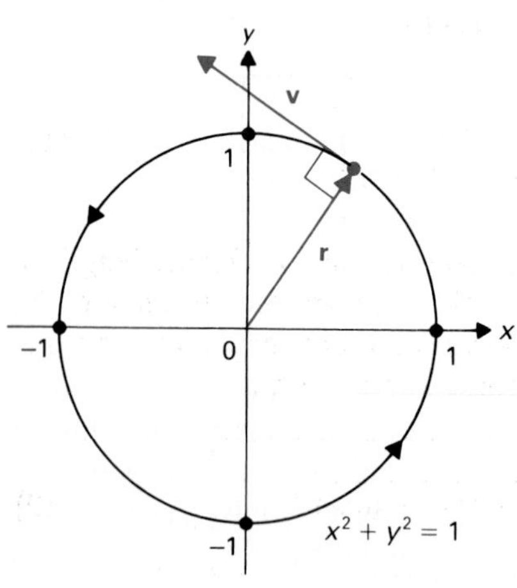

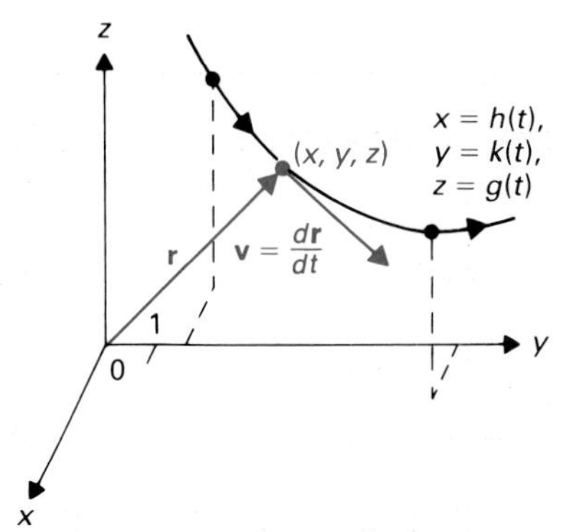

Figure 15.5 Position vector **r** and velocity vector **v** to a curve in space.

where $h(t)$, $k(t)$, and $g(t)$ all have continuous derivatives. See Fig. 15.5. The position vector is

$$\mathbf{r} = h(t)\mathbf{i} + k(t)\mathbf{j} + g(t)\mathbf{k}, \tag{9}$$

and the velocity vector is

$$\mathbf{v} = \frac{d\mathbf{r}}{dt} = \frac{dx}{dt}\mathbf{i} + \frac{dy}{dt}\mathbf{j} + \frac{dz}{dt}\mathbf{k}. \tag{10}$$

Since the direction of $\mathbf{v}$ is determined as the limiting direction of a chord of the curve corresponding to a vector $\Delta\mathbf{r}$ as $\Delta t \to 0$, we again see that $\mathbf{v}$ is tangent to the curve and points in the direction given by increasing t.

EXAMPLE 2 Find the equation of the line tangent to the space curve with parametric equations

$$x = t^2, \qquad y = \frac{1}{t^2 + 1}, \qquad \text{and} \qquad z = \ln t$$

at the point where $t = 1$.

Solution To find parametric equations of a line, we need to know a point on the line and a direction vector parallel to the line. Setting $t = 1$, we find the point $(1, \frac{1}{2}, 0)$. Taking the position vector

$$\mathbf{r} = t^2\mathbf{i} + \frac{1}{t^2 + 1}\mathbf{j} + (\ln t)\mathbf{k},$$

we find that

$$\mathbf{v} = \frac{d\mathbf{r}}{dt} = (2t)\mathbf{i} + \frac{-2t}{(t^2 + 1)^2}\mathbf{j} + \frac{1}{t}\mathbf{k}$$

is tangent to the curve. We can find the desired direction vector by finding $\mathbf{v}|_{t=1}$. Thus we have

Point $(1, \frac{1}{2}, 0)$
∥ vector $2\mathbf{i} - \frac{2}{4}\mathbf{j} + \mathbf{k}$
Equations $x = 1 + 2t,\ y = \frac{1}{2} - \frac{1}{2}t,\ z = t$ □

We now turn to the arc length of a space curve with position vector (9). A small change dt in t produces approximate changes dx in x, dy in y, and dz in z, so the approximate change in arc length is:

$$\begin{aligned} ds &= \sqrt{(dx)^2 + (dy)^2 + (dz)^2} \\ &= \sqrt{(dx/dt)^2 + (dy/dt)^2 + (dz/dt)^2}\, dt. \end{aligned} \tag{11}$$

(A rigorous demonstration of Eq. 11 can be given just as for arc length of parametrically described plane curves, which was treated in Chapter 7.) Thus Eq. (11) gives the *differential of arc length* for a space curve; the length of arc as a function of t, measured from a point where $t = t_0$, is

$$s(t) = \int_{t_0}^{t} \sqrt{(dx/dt)^2 + (dy/dt)^2 + (dz/dt)^2}\, dt. \tag{12}$$

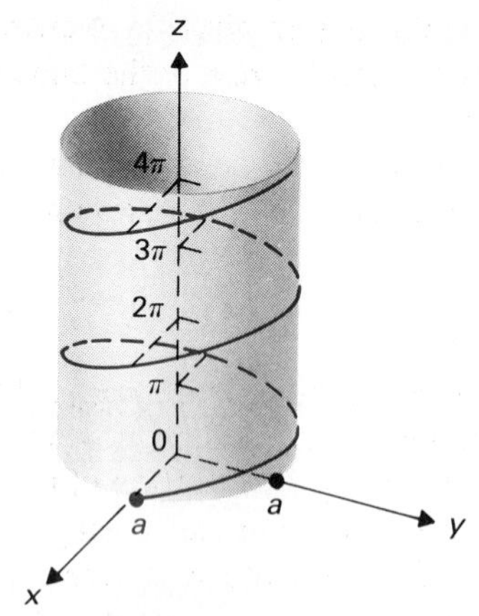

Figure 15.6 The helix $x = a \cos t$, $y = a \sin t$, $z = t$.

Once again, the length of the velocity vector is the speed, for

$$|\mathbf{v}| = \sqrt{(dx/dt)^2 + (dy/dt)^2 + (dz/dt)^2} = \frac{ds}{dt}. \tag{13}$$

The unit tangent vector $\mathbf{t}$ is again given by Eq. (7), and Eq. (8) is still valid.

EXAMPLE 3 The space curve

$$x = a \cos t, \quad y = a \sin t, \quad z = t$$

lies on the cylinder $x^2 + y^2 = a^2$. This curve, which is a *helix*, is sketched in Fig. 15.6. Find the vectors $\mathbf{r}$ and $\mathbf{v}$, the speed, and the length of one "turn" of the helix about the cylinder.

Solution We have

$$\mathbf{r} = (a \cos t)\mathbf{i} + (a \sin t)\mathbf{j} + t\mathbf{k},$$

$$\mathbf{v} = -(a \sin t)\mathbf{i} + (a \cos t)\mathbf{j} + \mathbf{k},$$

$$\frac{ds}{dt} = |\mathbf{v}| = \sqrt{a^2 \cos^2 t + a^2 \sin^2 t + 1} = \sqrt{a^2 + 1}.$$

The length of one "turn" of the helix is therefore

$$\int_0^{2\pi} \sqrt{a^2 + 1}\, dt = 2\pi\sqrt{a^2 + 1}. \quad \square$$

THE ACCELERATION VECTOR

When we studied motion on a line in Section 3.4, we let $x = h(t)$ describe the position of a body on a line at time t. We defined the velocity to be the signed quantity $h'(t)$. Of course, we would now use the vector notation $\mathbf{v} = h'(t)\mathbf{i}$. We then defined the acceleration to be $h''(t)$. Again, we would now use $\mathbf{a} = h''(t)\mathbf{i}$. The following definition is an extension to the plane and space. The definition is given for the plane case but is equally valid, with one more component, for space.

DEFINITION 15.2 Acceleration vector

If the position vector of a body at time t is $\mathbf{r} = h(t)\mathbf{i} + k(t)\mathbf{j}$, where h and k are twice-differentiable functions of t, then

$$\mathbf{a} = \frac{d\mathbf{v}}{dt} = \frac{d^2\mathbf{r}}{dt^2} = \frac{d^2x}{dt^2}\mathbf{i} + \frac{d^2y}{dt^2}\mathbf{j} \tag{14}$$

is the **acceleration vector** (or simply the **acceleration**) of the body at time t.

As in the case of the velocity vector, we would like to know the *magnitude* and *direction* of the acceleration vector. From Eq. (14), we obtain

$$|\mathbf{a}| = \sqrt{(h''(t))^2 + (k''(t))^2}. \tag{15}$$

We leave a detailed discussion of the direction of the acceleration vector $\mathbf{a}$ to

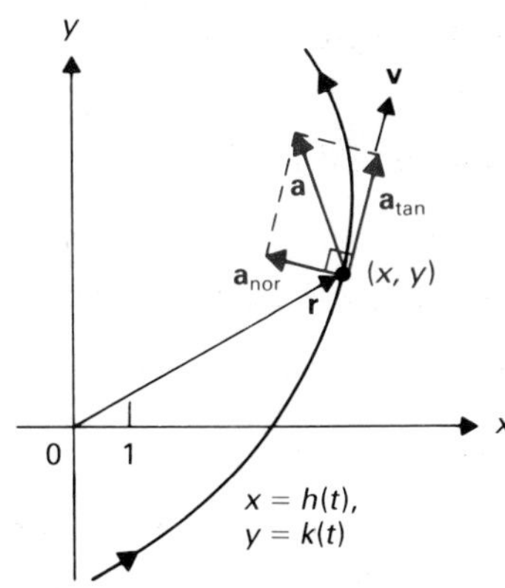

Figure 15.7 $\mathbf{a} = \mathbf{a}_{tan} + \mathbf{a}_{nor}$; speed is increasing.

the next section. We mention now only the information given by Newton's law, $\mathbf{F} = m\mathbf{a}$. Here $\mathbf{F}$ is the *force vector*, $\mathbf{a}$ is the *acceleration* of the body on which $\mathbf{F}$ acts, and m is the *mass* of the body.

Look at the plane motion in Fig. 15.7. At a point where $\mathbf{v} \neq \mathbf{0}$, we can break the acceleration vector $\mathbf{a}$ into vector components $\mathbf{a}_{tan}$ along $\mathbf{v}$ and $\mathbf{a}_{nor}$ orthogonal to $\mathbf{v}$, such that

$$\mathbf{a} = \mathbf{a}_{tan} + \mathbf{a}_{nor}. \tag{16}$$

To cause the body to travel a *curved* path, the vector component $m\mathbf{a}_{nor}$ of the force vector $\mathbf{F}$ must be nonzero and must point toward the concave side of the curve. A nonzero vector component $m\mathbf{a}_{tan}$ of $\mathbf{F}$ having the direction of $\mathbf{v}$, as in Fig. 15.7, causes the speed to *increase*. A nonzero component $m\mathbf{a}_{tan}$ of $\mathbf{F}$ having direction opposite to $\mathbf{v}$, as shown in Fig. 15.8, causes a *decrease* in speed. If the speed remains constant, then $\mathbf{a}_{tan}$ should be $\mathbf{0}$, and $\mathbf{a} = \mathbf{a}_{nor}$ is normal to the curve at each point, and it points toward the concave side. All this seems clear from the physical meaning of $\mathbf{F} = m\mathbf{a}$. The same analysis can be made in space.

EXAMPLE 4 Discuss the direction and magnitude of the acceleration vector $\mathbf{a}$ for the motion on the circle in Example 1 with position vector $\mathbf{r} = (\cos t)\mathbf{i} + (\sin t)\mathbf{j}$.

Solution We saw in Example 1 that

$$\mathbf{v} = \frac{d\mathbf{r}}{dt} = (-\sin t)\mathbf{i} + (\cos t)\mathbf{j}$$

and $|\mathbf{v}| = 1$. We have

$$\mathbf{a} = (-\cos t)\mathbf{i} - (\sin t)\mathbf{j},$$

so $|\mathbf{a}| = \sqrt{\cos^2 t + \sin^2 t} = 1$. We easily see that $\mathbf{a} \cdot \mathbf{v} = 0$, so $\mathbf{a} = \mathbf{a}_{nor}$ is orthogonal to $\mathbf{v}$ in this case where the speed is constant. Note that $\mathbf{a} = -\mathbf{r}$, so that $\mathbf{a}$ is directed toward the center of the circle in Fig. 15.4. □

EXAMPLE 5 Let the position vector of a body in the plane at time t be given by

$$\mathbf{r} = (t^2 + t)\mathbf{i} - t^3\mathbf{j}.$$

Find the vector components $\mathbf{a}_{tan}$ and $\mathbf{a}_{nor}$ of the acceleration vector that are tangent and normal to the curve at time $t = 1$. Determine whether the speed is increasing or decreasing there.

Figure 15.8 $\mathbf{a} = \mathbf{a}_{tan} + \mathbf{a}_{nor}$; speed is decreasing.

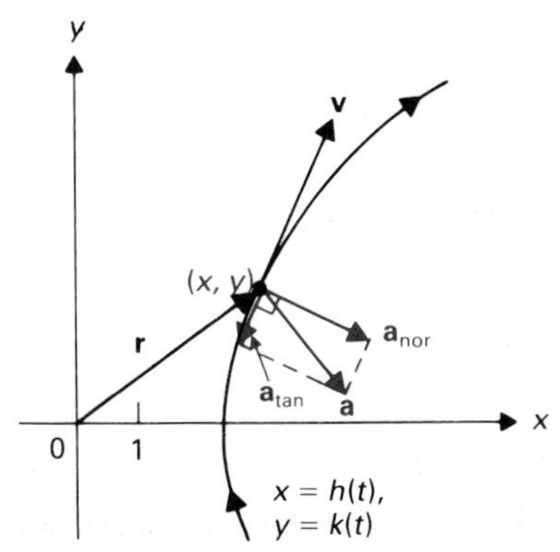

Solution We obtain

$$\mathbf{r}|_{t=1} = [(t^2 + t)\mathbf{i} - t^3\mathbf{j}]|_1 = 2\mathbf{i} - \mathbf{j},$$

$$\mathbf{v}|_{t=1} = [(2t + 1)\mathbf{i} - 3t^2\mathbf{j}]|_1 = 3\mathbf{i} - 3\mathbf{j},$$

$$\mathbf{a}|_{t=1} = [2\mathbf{i} - (6t)\mathbf{j}]|_1 = 2\mathbf{i} - 6\mathbf{j}.$$

Now $\mathbf{a}_{tan}$ is the vector projection of $\mathbf{a}$ along $\mathbf{v}$, since $\mathbf{v}$ is tangent to the curve. Thus

$$\mathbf{a}_{tan} = \left(\frac{\mathbf{a} \cdot \mathbf{v}}{\mathbf{v} \cdot \mathbf{v}}\right)\mathbf{v}.$$

When $t = 1$,

$$\mathbf{a}_{\tan}|_{t=1} = \frac{6 + 18}{9 + 9}(3\mathbf{i} - 3\mathbf{j}) = \frac{4}{3}(3\mathbf{i} - 3\mathbf{j}) = 4\mathbf{i} - 4\mathbf{j}.$$

Since $\mathbf{a}_{\text{nor}} = \mathbf{a} - \mathbf{a}_{\tan}$, we see that

$$\mathbf{a}_{\text{nor}}|_{t=1} = (2\mathbf{i} - 6\mathbf{j}) - (4\mathbf{i} - 4\mathbf{j}) = -2\mathbf{i} - 2\mathbf{j}.$$

Since $\mathbf{a} \cdot \mathbf{v} = 24 > 0$, we see that $\mathbf{a}_{\tan}$ and $\mathbf{v}$ have the same direction, so the speed is increasing. □

EXAMPLE 6 Find the acceleration vector $\mathbf{a}$ and its magnitude $|\mathbf{a}|$ for the motion in Example 3 on a helix, where the position vector is

$$\mathbf{r} = (a \cos t)(\mathbf{i} + (a \sin t)\mathbf{j} + t\mathbf{k}.$$

Solution Taking first and second derivatives of $\mathbf{r}$, we have

$$\mathbf{v} = (-a \sin t)\mathbf{i} + (a \cos t)\mathbf{j} + \mathbf{k},$$

$$\mathbf{a} = (-a \cos t)\mathbf{i} - (a \sin t)\mathbf{j}.$$

Then $|\mathbf{a}| = \sqrt{a^2 \cos^2 t + a^2 \sin^2 t} = a$. Note that for this motion also, $\mathbf{a} \cdot \mathbf{v} = 0$, so $\mathbf{a}$ is orthogonal to $\mathbf{v}$. We therefore expect the speed to be constant, and Example 3 showed this to be the case. □

We have seen how to find the velocity and acceleration vectors if the position vector is known. Equally important is finding the velocity and position vectors if the acceleration vector is known. Many times, we know what force vector is being applied, and we want to find the position vector of the body. Of course, this can be done by integrating.

EXAMPLE 7 Let the force vector

$$\mathbf{F} = (12t)\mathbf{i} + (4 \sin t)\mathbf{j} + (2 \cos 2t)\mathbf{k}$$

act on a body of mass 2. If the body has initial position and velocity vectors

$$\mathbf{r}|_{t=0} = \mathbf{i} - 2\mathbf{j} + \mathbf{k} \qquad \text{and} \qquad \mathbf{v}|_{t=0} = -3\mathbf{i} - 2\mathbf{k}$$

when $t = 0$, find the velocity and position vectors at time t.

Solution Since $\mathbf{F} = m\mathbf{a}$ and $m = 2$, we see that

$$\mathbf{a} = (6t)\mathbf{i} + (2 \sin t)\mathbf{j} + (\cos 2t)\mathbf{k}.$$

Integrating, we obtain

$$\mathbf{v} = (3t^2 + C_1)\mathbf{i} - (2 \cos t + C_2)\mathbf{j} + (\tfrac{1}{2} \sin 2t + C_3)\mathbf{k}.$$

When $t = 0$, we must have

$$\mathbf{v}|_{t=0} = C_1\mathbf{i} - (2 + C_2)\mathbf{j} + C_3\mathbf{k} = -3\mathbf{i} - 2\mathbf{k}.$$

Thus $C_1 = -3$, $C_2 = -2$, and $C_3 = -2$. We now have

$$\mathbf{v} = (3t^2 - 3)\mathbf{i} - (2 \cos t - 2)\mathbf{j} + (\tfrac{1}{2} \sin 2t - 2)\mathbf{k}.$$

Thus

$$\mathbf{r} = (t^3 - 3t + C_4)\mathbf{i} - (2 \sin t - 2t + C_5)\mathbf{j} - (\tfrac{1}{4} \cos 2t + 2t + C_6)\mathbf{k}.$$

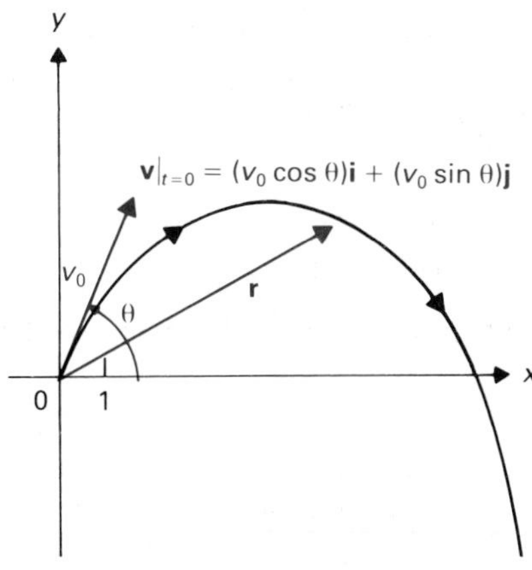

Figure 15.9 Path of a projectile fired from the origin with muzzle speed v_0 at the angle of inclination θ.

When $t = 0$, we must have

$$\mathbf{r}|_{t=0} = C_4\mathbf{i} - C_5\mathbf{j} - (\tfrac{1}{4} + C_6)\mathbf{k} = \mathbf{i} - 2\mathbf{j} + \mathbf{k},$$

so $C_4 = 1$, $C_5 = 2$, and $C_6 = -\frac{5}{4}$. Therefore

$$\mathbf{r} = (t^3 - 3t + 1)\mathbf{i} - (2\sin t - 2t + 2)\mathbf{j} - (\tfrac{1}{4}\cos 2t + 2t - \tfrac{5}{4})\mathbf{k}. \quad \square$$

EXAMPLE 8 A projectile is fired from a cannon with muzzle speed v_0 ft/sec at an angle of elevation θ with the horizontal. Neglecting air resistance, the acceleration after firing remains constant at $-32\mathbf{j}$, where units are in ft/sec^2. (See Fig. 15.9.) Find the path of the projectile.

Solution We choose our origin at the muzzle of the cannon, as shown in Fig. 15.9. Finding the position vector at time t is a good way to describe the path of the projectile. When $t = 0$, we have

$$\mathbf{r}|_{t=0} = 0\mathbf{i} + 0\mathbf{j} \quad \text{and} \quad \mathbf{v}|_{t=0} = (v_0 \cos\theta)\mathbf{i} + (v_0 \sin\theta)\mathbf{j}.$$

From $\mathbf{a} = -32\mathbf{j}$, we obtain

$$\mathbf{v} = C_1\mathbf{i} + (C_2 - 32t)\mathbf{j}.$$

When $t = 0$, we must have

$$\mathbf{v}|_{t=0} = C_1\mathbf{i} + C_2\mathbf{j} = (v_0 \cos\theta)\mathbf{i} + (v_0 \sin\theta)\mathbf{j},$$

so $C_1 = v_0 \cos\theta$ and $C_2 = v_0 \sin\theta$. Thus

$$\mathbf{v} = (v_0 \cos\theta)\mathbf{i} + (v_0 \sin\theta - 32t)\mathbf{j}.$$

Consequently,

$$\mathbf{r} = [(v_0 \cos\theta)t + C_3]\mathbf{i} + [(v_0 \sin\theta)t - 16t^2 + C_4]\mathbf{j}.$$

When $t = 0$, we must have

$$\mathbf{r}|_{t=0} = C_3\mathbf{i} + C_4\mathbf{j} = 0\mathbf{i} + 0\mathbf{j},$$

so $C_3 = C_4 = 0$. Thus

$$\mathbf{r} = [(v_0 \cos\theta)t]\mathbf{i} + [(v_0 \sin\theta)t - 16t^2]\mathbf{j}. \quad \textbf{(17)}$$

Of course this position vector is only valid until the projectile hits the earth. See the next example. □

EXAMPLE 9 Suppose the projectile in Example 8 is fired from a cannon at the edge of a cliff, with the muzzle of the cannon 96 ft above a flat plain below. If the muzzle speed is 160 ft/sec and the angle of fire is 30° above the horizontal, find the distance from the base of the cliff where the projectile hits. (This distance is called the *range*.)

Solution The projectile hits the ground when the **j**-coefficient of the position vector is -96. From Eq. (17) in Example 8, taking $v_0 = 160$ ft/sec and $\sin\theta = \sin 30° = \frac{1}{2}$, we see that

$$160 \cdot \tfrac{1}{2}t - 16t^2 = -96,$$

$$16t^2 - 80t - 96 = 0,$$

$$t^2 - 5t - 6 = 0,$$

$$(t - 6)(t + 1) = 0.$$

Thus the projectile hits the ground when $t = 6$ sec. When $t = 6$, the **i**-coefficient of **r** is

$$160 \cdot \frac{\sqrt{3}}{2} \cdot 6 = 480\sqrt{3} \text{ ft.}$$

This is the range of the projectile. □

DIFFERENTIATION OF PRODUCTS OF VECTORS

The previous discussion has shown that differentiation of a vector can be accomplished by just differentiating each component of the vector. Let **a** and **b** be differentiable vector functions of t in space, and let $f(t)$ be a differentiable scalar function. We can form the products $f(t)\mathbf{a}$, $\mathbf{a} \cdot \mathbf{b}$, and $\mathbf{a} \times \mathbf{b}$. As we might guess, the usual product formula holds for differentiation of all these products. Namely,

$$\frac{d(f(t)\mathbf{a})}{dt} = f(t)\frac{d\mathbf{a}}{dt} + f'(t)\mathbf{a}, \tag{18}$$

$$\frac{d(\mathbf{a} \cdot \mathbf{b})}{dt} = \mathbf{a} \cdot \frac{d\mathbf{b}}{dt} + \frac{d\mathbf{a}}{dt} \cdot \mathbf{b}, \tag{19}$$

$$\frac{d(\mathbf{a} \times \mathbf{b})}{dt} = \mathbf{a} \times \frac{d\mathbf{b}}{dt} + \frac{d\mathbf{a}}{dt} \times \mathbf{b}. \tag{20}$$

*Since the cross product is not commutative, we must be careful to always have the "**a**-factor" first, before the "**b**-factor" in* Eq. (20).

We give the proof of Eq. (19). Proofs of the others are similar. Using properties of the dot product $\mathbf{a} \cdot \mathbf{b}$, we see that a change Δt in t produces a change

$$\begin{aligned}(\mathbf{a} + \Delta\mathbf{a}) \cdot (\mathbf{b} + \Delta\mathbf{b}) - \mathbf{a} \cdot \mathbf{b} &= \mathbf{a} \cdot \mathbf{b} + \mathbf{a} \cdot \Delta\mathbf{b} + \Delta\mathbf{a} \cdot \mathbf{b} + \Delta\mathbf{a} \cdot \Delta\mathbf{b} - \mathbf{a} \cdot \mathbf{b} \\ &= \mathbf{a} \cdot \Delta\mathbf{b} + \Delta\mathbf{a} \cdot \mathbf{b} + \Delta\mathbf{a} \cdot \Delta\mathbf{b}\end{aligned}$$

in $\mathbf{a} \cdot \mathbf{b}$. Thus the derivative $d(\mathbf{a} \cdot \mathbf{b})/dt$ is

$$\begin{aligned}\frac{d(\mathbf{a} \cdot \mathbf{b})}{dt} &= \lim_{\Delta t \to 0} \left(\mathbf{a} \cdot \frac{\Delta\mathbf{b}}{\Delta t} + \frac{\Delta\mathbf{a}}{\Delta t} \cdot \mathbf{b} + \frac{\Delta\mathbf{a}}{\Delta t} \cdot \Delta\mathbf{b}\right) \\ &= \mathbf{a} \cdot \frac{d\mathbf{b}}{dt} + \frac{d\mathbf{a}}{dt} \cdot \mathbf{b} + \frac{d\mathbf{a}}{dt} \cdot \mathbf{0} \\ &= \mathbf{a} \cdot \frac{d\mathbf{b}}{dt} + \frac{d\mathbf{a}}{dt} \cdot \mathbf{b},\end{aligned}$$

which is Eq. (19).

We ask you to illustrate Eqs. (18), (19), and (20) in the exercises.

SUMMARY

Assume that you have parametric equations $x = h(t)$, $y = k(t)$ of a plane curve, or $x = h(t)$, $y = k(t)$, $z = g(t)$ for a space curve, where all the functions of t are twice differentiable.

1. The position vector to the curve is

$$\mathbf{r} = h(t)\mathbf{i} + k(t)\mathbf{j} \quad \text{for a plane curve,}$$

$$\mathbf{r} = h(t)\mathbf{i} + k(t)\mathbf{j} + g(t)\mathbf{k} \quad \text{for a space curve.}$$

2. The velocity vector is

$$\mathbf{v} = \frac{d\mathbf{r}}{dt} = \frac{dx}{dt}\mathbf{i} + \frac{dy}{dt}\mathbf{j} \quad \text{for a plane curve,}$$

$$\mathbf{v} = \frac{d\mathbf{r}}{dt} = \frac{dx}{dt}\mathbf{i} + \frac{dy}{dt}\mathbf{j} + \frac{dz}{dt}\mathbf{k} \quad \text{for a space curve.}$$

3. The length of the velocity vector is

$$|\mathbf{v}| = \frac{ds}{dt} = \text{Speed along the curve.}$$

4. The distance traveled along a space curve from t_0 to t is

$$s(t) = \int_{t_0}^{t} \sqrt{\left(\frac{dx}{dt}\right)^2 + \left(\frac{dy}{dt}\right)^2 + \left(\frac{dz}{dt}\right)^2}\, dt.$$

5. The direction of $\mathbf{v}$ is tangent to the curve and in the direction corresponding to increasing t.

6. The acceleration vector is

$$\mathbf{a} = \frac{d^2\mathbf{r}}{dt^2} = \frac{d\mathbf{v}}{dt} = \frac{d^2x}{dt^2}\mathbf{i} + \frac{d^2y}{dt^2}\mathbf{j} \quad \text{for a plane curve,}$$

$$\mathbf{a} = \frac{d^2\mathbf{r}}{dt^2} = \frac{d\mathbf{v}}{dt} = \frac{d^2x}{dt^2}\mathbf{i} + \frac{d^2y}{dt^2}\mathbf{j} + \frac{d^2z}{dt^2}\mathbf{k} \quad \text{for a space curve.}$$

7. We may write $\mathbf{a} = \mathbf{a}_{\tan} + \mathbf{a}_{\text{nor}}$ where $\mathbf{a}_{\tan}$ and $\mathbf{a}_{\text{nor}}$ are the vector components of $\mathbf{a}$ tangent and normal to the curve. We have

$$\mathbf{a}_{\tan} = [(\mathbf{a}\cdot\mathbf{v})/(\mathbf{v}\cdot\mathbf{v})]\mathbf{v} \quad \text{and} \quad \mathbf{a}_{\text{nor}} = \mathbf{a} - \mathbf{a}_{\tan}.$$

8. For differentiable vector functions $\mathbf{a}$ and $\mathbf{b}$ of t, and a differentiable scalar function $f(t)$, we have

$$\frac{d(f(t)\mathbf{a})}{dt} = f(t)\frac{d\mathbf{a}}{dt} + f'(t)\mathbf{a},$$

$$\frac{d(\mathbf{a}\cdot\mathbf{b})}{dt} = \mathbf{a}\frac{d\mathbf{b}}{dt} + \frac{d\mathbf{a}}{dt}\mathbf{b},$$

and

$$\frac{d(\mathbf{a}\times\mathbf{b})}{dt} = \mathbf{a}\times\frac{d\mathbf{b}}{dt} + \frac{d\mathbf{a}}{dt}\times\mathbf{b}.$$

EXERCISES

1. Let $\mathbf{r}(t)$ be the position vector from the origin to a point $(x, y) = (h(t), k(t))$ on a smooth curve. Show that if s is the arc length along the curve defined as usual, then $d\mathbf{r}/ds = \mathbf{t}$, where $\mathbf{t}$ is the *unit* tangent vector to the curve in the direction of increasing t.

2. Let $\mathbf{b}$ be a differentiable vector function of t.
   a) Show that if $|\mathbf{b}|$ is a constant function of t, then $\mathbf{b}$ is orthogonal to $d\mathbf{b}/dt$. [*Hint:* Differentiate $\mathbf{b} \cdot \mathbf{b}$.]
   b) State the interpretation of part (a) if $\mathbf{b} = \mathbf{r}$, the position vector of a body in motion.
   c) State the interpretation of part (a) if $\mathbf{b} = \mathbf{v}$, the velocity vector of a body in motion.

In Exercises 3 through 13, the position vector $\mathbf{r}$ at time t of a body moving on a curve is given. Find the following at the indicated time t_0: (a) the velocity vector of the body, (b) the speed of the body, (c) the acceleration vector of the body.

3. $\mathbf{r} = 2t\mathbf{i} + (3t - 1)\mathbf{j}$ at $t_0 = 0$

4. $\mathbf{r} = (3t + 1)\mathbf{i} + t^2\mathbf{j}$ at $t_0 = 1$

5. $\mathbf{r} = (\sin t)\mathbf{i} + (\cos 2t)\mathbf{j}$ at $t_0 = \pi$

6. $\mathbf{r} = e^t\mathbf{i} + t^2\mathbf{j}$ at $t_0 = 0$

7. $\mathbf{r} = (\ln t)\mathbf{i} + (\cosh(t - 1))\mathbf{j}$ at $t_0 = 1$

8. $\mathbf{r} = (e^t \sin t)\mathbf{i} + (e^t \cos t)\mathbf{j}$ at $t_0 = 0$

9. $\mathbf{r} = (1/t)\mathbf{i} + (1/t^2)\mathbf{j}$ at $t_0 = 1$

10. $\mathbf{r} = (\ln(\sin t))\mathbf{i} + (\ln(\cos t))\mathbf{j}$ at $t_0 = \pi/4$

11. $\mathbf{r} = t^2\mathbf{i} + t^3\mathbf{j} - (t + 1)\mathbf{k}$ at $t_0 = 2$

12. $\mathbf{r} = e^t\mathbf{i} - (\sin t)\mathbf{j} + (\cos t)\mathbf{k}$ at $t_0 = 0$

13. $\mathbf{r} = (1 + 1/t)\mathbf{i} + (1 - 1/t)\mathbf{j} + t^2\mathbf{k}$ at $t_0 = 1$

14. Prove that if we let $\mathbf{a}_{\text{tan}} = [(\mathbf{a} \cdot \mathbf{v})/(\mathbf{v} \cdot \mathbf{v})]\mathbf{v}$ and define $\mathbf{a}_{\text{nor}} = \mathbf{a} - \mathbf{a}_{\text{tan}}$, then $\mathbf{a}_{\text{nor}}$ is indeed orthogonal to $\mathbf{v}$. [*Hint:* Compute a dot product.]

In Exercises 15 through 20, find tangential and normal vector components $\mathbf{a}_{\text{tan}}$ and $\mathbf{a}_{\text{nor}}$ of the acceleration vector $\mathbf{a}$ at the given point, so that $\mathbf{a} = \mathbf{a}_{\text{tan}} + \mathbf{a}_{\text{nor}}$. State whether the speed is increasing or decreasing at that point. See Example 5.

15. $\mathbf{r} = (t^2 - 3t)\mathbf{i} + t^3\mathbf{j}$, where $t = -1$

16. $\mathbf{r} = te^t\mathbf{i} - (\cos t)\mathbf{j}$, where $t = 0$

17. $\mathbf{r} = (1/t)\mathbf{i} + (\ln t)\mathbf{j}$, where $t = 1$

18. $\mathbf{r} = t^2\mathbf{i} + t^3\mathbf{j} - t^4\mathbf{k}$, where $t = -1$

19. $\mathbf{r} = (\sin t)\mathbf{i} + (\cos 2t)\mathbf{j} - t^2\mathbf{k}$, where $t = \pi$

20. $\mathbf{r} = (t^3 - 4t^2)\mathbf{i} + (8/t)\mathbf{j} - (t^2 - 4t)\mathbf{k}$, where $t = 2$

In Exercises 21 through 24, find parametric equations of the line tangent to the given space curve at the given point.

21. $x = t + \cosh t$, $y = \sinh t$, $z = t^3$, where $t = 0$

22. $x = \sin t$, $y = \cos t$, $z = t$, where $t = \pi$

23. $x = 1/(t^2 + 1)$, $y = 8/t$, $z = t^3 - 2$, where $t = 1$

24. $x = \sinh 2t$, $y = \ln(t^2 + t + 1)$, $z = 1/(t - 1)$, where $t = 0$

In Exercises 25 through 28, find the equation of the plane normal to the given space curve at the given point.

25. The curve and point in Exercise 21

26. The curve and point in Exercise 22

27. The curve and point in Exercise 23

28. The curve and point in Exercise 24

In Exercises 29 through 33, find the length of the indicated portion of the space curve. (Estimate using a calculator and Simpson's rule in Exercises 32 and 33.)

29. $x = a \cos 2t$, $y = t$, $z = a \sin 2t$ for $0 \le t \le 2\pi$

30. $x = 2t$, $y = t^2$, $z = -t^2$ for $0 \le t \le 2$

31. $x = 3t^2$, $y = 2t^3$, $z = 3t$ for $0 \le t \le 4$

32. $x = t^2$, $y = 2t + 1$, $z = \sin \pi t$ for $0 \le t \le 1$

33. $x = \sqrt{t^2 + 4}$, $y = t^3$, $z = 1/(1 + t^2)$ for $0 \le t \le 2$

34. Let $\mathbf{a} = (t \sin t)\mathbf{i} - t^2\mathbf{j} + te^t\mathbf{k}$ and let $f(t) = 1/t$. Illustrate Eq. (18) of the text by computing $d(f(t)\mathbf{a})/dt$ in two ways.

35. Let $\mathbf{a} = t^2\mathbf{i} - (3t + 1)\mathbf{j}$ and $\mathbf{b} = (2t)\mathbf{i} - t^3\mathbf{j}$ in the plane. Illustrate Eq. (19) of the text by computing $d(\mathbf{a} \cdot \mathbf{b})/dt$ in two ways.

36. Let $\mathbf{a} = 3t\mathbf{i} - (4t + 1)\mathbf{j} + t^2\mathbf{k}$ and $\mathbf{b} = (t^2 - 2)\mathbf{i} + 5t\mathbf{j} - 6t\mathbf{k}$ in space. Illustrate Eq. (20) of the text by computing $d(\mathbf{a} \times \mathbf{b})/dt$ in two ways.

Exercises 37 through 43 refer to the path of a projectile having muzzle speed v_0 ft/sec and fired with angle of inclination θ to the horizontal. See Examples 8 and 9. Air resistance is neglected in all these problems.

37. Show that the path of the projectile is a parabola.

38. Show that the projectile returns to the same altitude at which it was fired at time $(v_0 \sin \theta)/16$ sec.

39. If a projectile is fired at ground level over a level plain, show that its range is $(v_0^2 \sin 2\theta)/32$ ft.

40. Use the result of Exercise 39 to find the angle θ of inclination for maximum ground-level range with a fixed muzzle speed v_0.

41. A baseball player bats a ball at an angle of inclination 15°, and the ball is caught at the same height 256 ft away. Find the speed with which the ball left the bat.

42. Show that the maximum altitude attained by a projectile fired at ground level is $(v_0^2 \sin^2\theta)/64$ ft.

43. A projectile fired at ground level 800 ft from the bottom of a 200-ft cliff is to reach the top of the cliff. Find the minimum muzzle speed v_0 required and the corresponding angle θ of inclination at which the projectile should be fired.

44. The vector $\mathbf{r} = a(\cos t)\mathbf{i} + a(\sin t)\mathbf{j}$ is the position vector for motion on the circle $x^2 + y^2 = a^2$ where the body makes one revolution around the circle in 2π units of time. Modify $\mathbf{r}$ so that the body makes β revolutions per unit time.

45. A body travels at constant speed making β revolutions per second around a circle of radius a. Show that the magnitude of the force required to hold the body on the circle is directly proportional to the mass m, radius a, and β^2.

46. A child whirls a $\frac{1}{2}$-lb stone tied to a string around and around his head in a circle of radius 4 ft. If the string breaks at any pull exceeding 25 lb, find the maximum number of revolutions per second that the stone can make without breaking the string. (A mass of 1 slug weighs 32 lb.)

15.2 THE MAGNITUDES $|\mathbf{a}_{tan}|$ AND $|\mathbf{a}_{nor}|$

Suppose we are passengers in a car and we are instructing the driver which way to go at an intersection. We rarely say, "turn north," or "turn south," or "continue west," using a global coordinate system for our earth. We are much more likely to say, "turn right," or "turn left," or "continue straight ahead." We have in mind two perpendicular axes that travel right along with us: the "ahead/behind" axis and the "left/right" axis.

In the context just discussed, we think of an "ahead" unit vector and a "left" unit vector as forming an *orthogonal frame of unit vectors* at each instant. (We could equally well have chosen "ahead" and "right," or "behind" and "left," and so on.) In this section, we discuss such an orthogonal frame of two unit vectors $\mathbf{t}$ and $\mathbf{n}$, at each point of a smooth curve described by a position vector function $\mathbf{r}(t)$. As you might guess, $\mathbf{t}$ will point in a direction *tangent* to the curve, and $\mathbf{n}$ will point in a *normal* direction to the curve. To a bug crawling on the curve, $\mathbf{t}$ seems to point *ahead* at any instant, while $\mathbf{n}$ points either *right* or *left*. We will also describe $\mathbf{t}$ and $\mathbf{n}$ for space curves. Of course, a bug traveling on a curve in space, with no gravity to provide a global notion of "down," needs another vector orthogonal to both $\mathbf{t}$ and $\mathbf{n}$ to complete a space frame. In Section 15.4, we will define this other vector to be $\mathbf{b} = \mathbf{t} \times \mathbf{n}$ and make a more detailed study of the space situation.

After discussing the orthogonal frame $\mathbf{t}$ and $\mathbf{n}$ for curves in the plane or in space, we will express the velocity vector $\mathbf{v}$ and the acceleration vector $\mathbf{a}$ associated with a moving body in terms of this moving orthogonal frame.

THE ORTHOGONAL VECTOR FRAME t AND n

Let $h''(t)$ and $k''(t)$ be continuous, and let $\mathbf{r} = h(t)\mathbf{i} + k(t)\mathbf{j}$ be the position vector for motion in the plane. We wish to work on an arc of the curve where $\mathbf{v} \neq \mathbf{0}$ at any point of the arc. This means that $h'(t)$ and $k'(t)$ must never be simultaneously zero on the arc, so that the arc is *smooth*. If $\mathbf{v} = \mathbf{0}$ at some point, then the body in motion comes to a stop at that point and could proceed in a new direction, making a sharp point on the arc. This explains the use of the term *smooth*. For example, $\mathbf{r} = t^3\mathbf{i} + t^2\mathbf{j}$ is not smooth where $t = 0$. The curve has x,y-equation $y^3 = x^2$ or $y = x^{2/3}$ and has a sharp point at the origin, as shown in Fig. 15.10. In space, where $\mathbf{r} = h(t)\mathbf{i} + k(t)\mathbf{j} + g(t)\mathbf{k}$, we require that $h'(t)$, $k'(t)$, and $g'(t)$ never be simultaneously zero for an arc to be smooth.

Figure 15.10 $\mathbf{v} = 3t^2\mathbf{i} + 2t\mathbf{j}$ is $\mathbf{0}$ where $t = 0$; there is a sharp point at the origin.

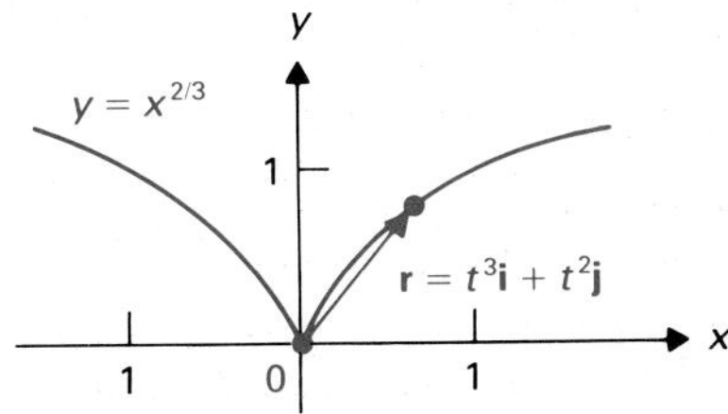

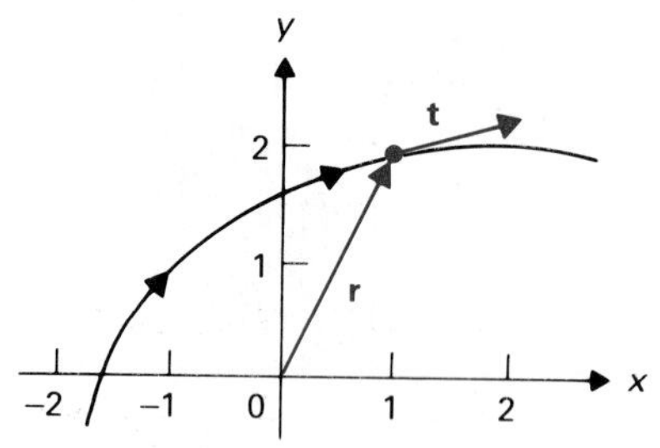

Figure 15.11 Unit tangent vector $\mathbf{t} = \mathbf{v}/|\mathbf{v}|$.

On a smooth arc of a curve in the plane or in space, we let

$$\mathbf{t} = \frac{1}{|\mathbf{v}|}\mathbf{v}. \tag{1}$$

Then $\mathbf{t}$ is a unit vector tangent to the curve in the direction corresponding to increasing t, shown in Fig. 15.11. To a bug crawling along the curve, $\mathbf{t}$ looks to be "straight ahead" at any instant. Since $\mathbf{t}$ is a unit vector, we have

$$|\mathbf{t}| = 1.$$

Now $\mathbf{t}$ can be viewed as a vector-valued function of the parameter t, the *arc length* s along the curve, or other possible parameters. Exercise 2 of the preceding section showed that the derivative of a vector-valued function of constant magnitude is always orthogonal to the vector. This is so important that we state it again as a theorem and give the easy proof. While t is used as parameter in the theorem, the argument is the same for any parameter.

THEOREM 15.1 Derivatives of vectors of constant magnitude

Let $\mathbf{b}(t)$ be a differentiable vector-valued function of constant magnitude. Then $d\mathbf{b}/dt$ is orthogonal to $\mathbf{b}$.

Proof. From the preceding section, we know that

$$\frac{d(\mathbf{b}\cdot\mathbf{b})}{dt} = \mathbf{b}\cdot\frac{d\mathbf{b}}{dt} + \frac{d\mathbf{b}}{dt}\cdot\mathbf{b} = 2\left(\mathbf{b}\cdot\frac{d\mathbf{b}}{dt}\right).$$

But $\mathbf{b}\cdot\mathbf{b} = |\mathbf{b}|^2$ is constant by hypothesis, so its derivative is zero. Consequently, $\mathbf{b}\cdot(d\mathbf{b}/dt) = 0$, so $d\mathbf{b}/dt$ is orthogonal to $\mathbf{b}$. •

We are now ready to define the second unit vector $\mathbf{n}$ of the orthogonal vector frame $\mathbf{t}$, $\mathbf{n}$ that we carry along with us as we travel a curve in the plane or in space. Since $\mathbf{t}$ has constant length 1, Theorem 15.1 shows that we could define $\mathbf{n}$ using the derivative of $\mathbf{t}$ with respect to any parameter that determines $\mathbf{t}$. It will be convenient for us to use the arc length parameter s, for we are going to run into the *curvature* of the curve. Recall from page 553 that the curvature κ of a plane curve is defined to be

$$\kappa = \left|\frac{d\phi}{ds}\right|,$$

where ϕ at any point on the curve is the angle from the horizontal counterclockwise to the tangent line to the curve at that point.

Let $\mathbf{t}$ be the unit vector tangent to a curve in the plane or space as defined by Eq. (1). By Theorem 15.1, $d\mathbf{t}/ds$ is orthogonal to $\mathbf{t}$. If $d\mathbf{t}/ds \neq \mathbf{0}$, we *define* a unit vector $\mathbf{n}$ orthogonal to $\mathbf{t}$ by

$$\mathbf{n} = \frac{d\mathbf{t}/ds}{|d\mathbf{t}/ds|}. \tag{2}$$

Figure 15.12 $\Delta\mathbf{t} = \mathbf{t}(s + \Delta s) - \mathbf{t}(s)$ points toward the concave side of the curve.

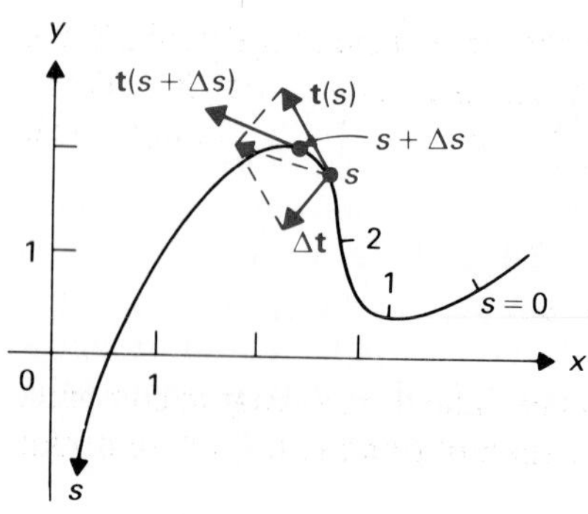

Figure 15.12 shows $\Delta\mathbf{t}$ for a plane curve. As indicated in the figure,

$$\Delta\mathbf{t} = \mathbf{t}(s + \Delta s) - \mathbf{t}(s)$$

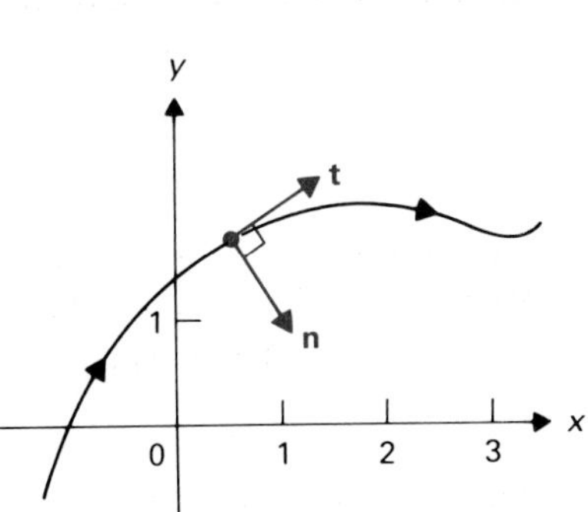

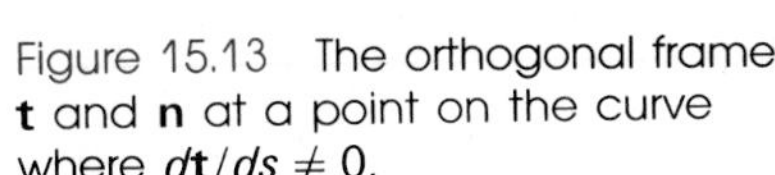
Figure 15.13 The orthogonal frame **t** and **n** at a point on the curve where $d\mathbf{t}/ds \neq 0$.

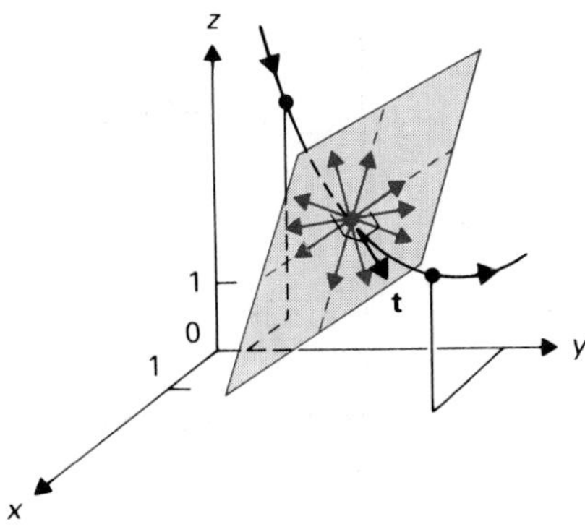

Figure 15.14 In the plane normal to the curve, every vector is orthogonal to **t**.

points toward the *concave* side of the curve. Thus $d\mathbf{t}/ds = \lim_{\Delta s \to 0} \Delta\mathbf{t}/\Delta s$ is a vector orthogonal to **t** on the *concave* side of the curve, and the unit vector **n** in that direction therefore also points to the concave side, as shown in Fig. 15.13.

We could give an example, computing **n** at a point on a curve, using definition (2) for **n**. However, we will find an easier method for computing **n** in a moment, so we delay this illustration.

It is a bit more complicated to describe just which way **n** points for a space curve, because any unit vector in the entire plane normal to **t**, shown in Fig. 15.14, is a vector orthogonal to **t**. The vector **n** is called the **principal normal vector.** The plane containing **t** and **n** at a point is called the **osculating plane** at that point. In Fig. 15.15, we show a triangle having as two sides $\mathbf{t}(s)$ and $\mathbf{t}(s + \Delta s)$ starting at the point s and, as third side, $\Delta\mathbf{t} = \mathbf{t}(s + \Delta s) - \mathbf{t}(s)$. This triangle determines a plane. As $\Delta s \to 0$, it seems reasonable that the limiting position of these planes is the plane most nearly containing the curve near the point s. Since $d\mathbf{t}/ds = \lim_{\Delta s \to 0} \Delta\mathbf{t}/\Delta s$ and **n** and $d\mathbf{t}/ds$ are parallel, we see that this limiting plane is our osculating plane, containing **t** and **n**.

We would like to compute the equation of an osculating plane. To find the equation of any plane, we want to know a point in the plane and a normal vector to the plane. How can we find such a normal vector? One way is to take a cross product of two nonparallel vectors in the plane, say **t** and **n**. Since **v** and **t** are parallel, we can use **v** and **n** just as well. Recall that

$$\frac{d(f(s)\mathbf{t})}{ds} = f(s)\frac{d\mathbf{t}}{ds} + f'(s)\mathbf{t}.$$

This equation shows that $d(f(s)\mathbf{t})/ds$ lies in the same plane as $d\mathbf{t}/ds$ and **t**. Since **v** and **t** differ by a scalar function factor, this argument shows that $d\mathbf{v}/ds$ lies in the same plane as **t** and $d\mathbf{t}/ds$, that is, $d\mathbf{v}/ds$ is also in the osculating plane. Finally, since

$$\frac{d\mathbf{v}}{ds} = \frac{dt}{ds}\frac{d\mathbf{v}}{dt} = \frac{dt}{ds}\mathbf{a},$$

we see that **a** lies in the osculating plane. Since **v** and **a** are so easy to compute, we take the cross product of them, rather than of **t** and **n**, to find a normal vector to the osculating plane.

Figure 15.15 The triangle with sides $\mathbf{t}(s)$, $\mathbf{t}(s + \Delta s)$, and $\Delta\mathbf{t}$.

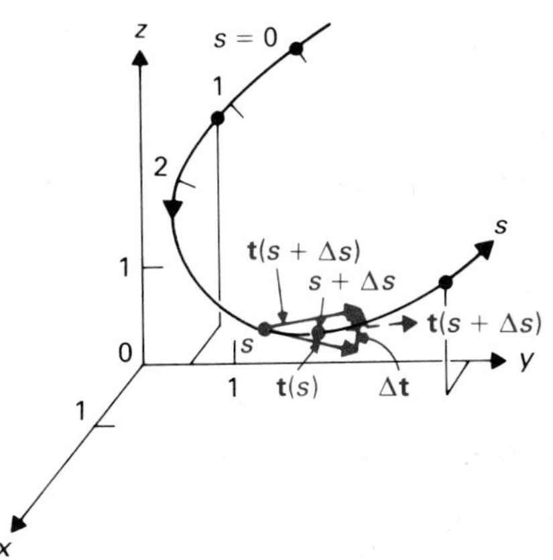

EXAMPLE 1 Find the equation of the osculating plane to the twisted cubic $x = 3t^2$, $y = 2t^3$, $z = 3t$, where $t = 1$.

Solution We have

$$\mathbf{r}|_1 = 3t^2\mathbf{i} + 2t^3\mathbf{j} + 3t\mathbf{k}|_1 = 3\mathbf{i} + 2\mathbf{j} + 3\mathbf{k},$$
$$\mathbf{v}|_1 = 6t\mathbf{i} + 6t^2\mathbf{j} + 3\mathbf{k}|_1 = 6\mathbf{i} + 6\mathbf{j} + 3\mathbf{k},$$
$$\mathbf{a}|_1 = 6\mathbf{i} + 12t\mathbf{j}|_1 = 6\mathbf{i} + 12\mathbf{j}.$$

Then $\frac{1}{3}\mathbf{v} = 2\mathbf{i} + 2\mathbf{j} + \mathbf{k}$ and $\frac{1}{6}\mathbf{a} = \mathbf{i} + 2\mathbf{j}$ lie in the osculating plane. Consequently,

$$\begin{vmatrix} \mathbf{i} & \mathbf{j} & \mathbf{k} \\ 2 & 2 & 1 \\ 1 & 2 & 0 \end{vmatrix} = -2\mathbf{i} + \mathbf{j} + 2\mathbf{k}$$

is normal to the plane. We have

Point $(3, 2, 3)$

⊥ *vector* $-2\mathbf{i} + \mathbf{j} + 2\mathbf{k}$

Equation $-2x + y + 2z = 2$ □

$|\mathbf{a}_{\tan}|$ AND $|\mathbf{a}_{\text{nor}}|$

In Section 15.1, we learned to break the acceleration vector **a** into tangential and normal vector components $\mathbf{a}_{\tan}$ and $\mathbf{a}_{\text{nor}}$, where

$$\mathbf{a}_{\tan} = \frac{\mathbf{a}\cdot\mathbf{v}}{\mathbf{v}\cdot\mathbf{v}}\mathbf{v} \quad \text{and} \quad \mathbf{a}_{\text{nor}} = \mathbf{a} - \mathbf{a}_{\tan}. \tag{3}$$

These vector components are easy to compute. In terms of our orthogonal frame of vectors **t** and **n**, we can write

$$\mathbf{a} = |\mathbf{a}_{\tan}|\mathbf{t} + |\mathbf{a}_{\text{nor}}|\mathbf{n}. \tag{4}$$

We now study $|\mathbf{a}_{\tan}|$ and $|\mathbf{a}_{\text{nor}}|$ a bit more.

For a smooth arc of a curve in the plane, let ϕ be the angle that **t** makes with the horizontal, shown in Fig. 15.16. Recall that the curvature κ is given by

$$\kappa = \left|\frac{d\phi}{ds}\right|. \tag{5}$$

Figure 15.16 $\mathbf{t} = (\cos\phi)\mathbf{i} + (\sin\phi)\mathbf{j}$; $d\mathbf{t}/d\phi = (-\sin\phi)\mathbf{i} + (\cos\phi)\mathbf{j}$.

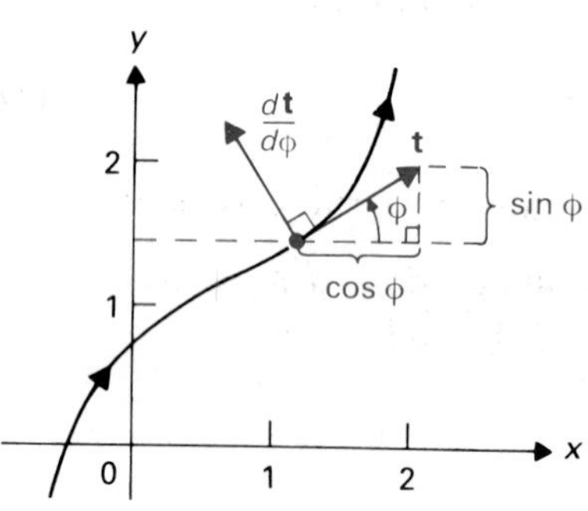

We may write **t** as

$$\mathbf{t} = (\cos\phi)\mathbf{i} + (\sin\phi)\mathbf{j}. \tag{6}$$

Then

$$\frac{d\mathbf{t}}{d\phi} = (-\sin\phi)\mathbf{i} + (\cos\phi)\mathbf{j}.$$

By Theorem 15.1, $d\mathbf{t}/d\phi$ is orthogonal to **t**, and it is a unit vector since

$\sin^2\phi + \cos^2\phi = 1$. Recall the trigonometric identities

$$\sin\left(\phi + \frac{\pi}{2}\right) = \cos\phi \quad \text{and} \quad \cos\left(\phi + \frac{\pi}{2}\right) = -\sin\phi.$$

Thus $d\mathbf{t}/d\phi$ may be obtained by replacing ϕ by $\phi + \pi/2$ in Eq. (6). This shows that $d\mathbf{t}/d\phi$ results if we rotate $\mathbf{t}$ *counterclockwise* through 90°. Of course $d\mathbf{t}/d\phi$ is either $\mathbf{n}$ or $-\mathbf{n}$.

By the chain rule, we see that

$$\frac{d\mathbf{t}}{ds} = \frac{d\phi}{ds}\frac{d\mathbf{t}}{d\phi}. \tag{7}$$

If the arc bends to the left as s increases, then ϕ is increasing, so $d\phi/ds = \kappa$, and $d\mathbf{t}/d\phi = \mathbf{n}$, since the concave side of the arc is on the left. If the arc bends to the right, then $d\phi/ds = -\kappa$ and $d\mathbf{t}/d\phi = -\mathbf{n}$. Thus, in either case, Eq. (7) shows that

$$\frac{d\mathbf{t}}{ds} = \frac{d\phi}{ds}\frac{d\mathbf{t}}{d\phi} = \kappa\mathbf{n}. \tag{8}$$

We have not defined curvature for a space curve. Recall that we defined the principal normal vector $\mathbf{n}$ for a space curve to be

$$\mathbf{n} = \frac{d\mathbf{t}/ds}{|d\mathbf{t}/ds|}, \tag{9}$$

if $d\mathbf{t}/ds \neq \mathbf{0}$. Equation (8) shows that in the plane, we have

$$\boxed{\kappa = \left|\frac{d\mathbf{t}}{ds}\right|.} \tag{10}$$

We take Eq. (10) as the *definition* of curvature for a space curve.

DEFINITION 15.3 Curvature of a space curve

The **curvature** κ at a point on a smooth arc of a space curve is given by $\kappa = |d\mathbf{t}/ds|$.

Thus for both plane and space curves we have

$$\boxed{\frac{d\mathbf{t}}{ds} = \kappa\mathbf{n}.} \tag{11}$$

Recall that for both a plane curve and a space curve,

$$\mathbf{v} = \frac{ds}{dt}\mathbf{t}. \tag{12}$$

Differentiating Eq. (12), using the product and chain rules, we have

$$\mathbf{a} = \frac{d\mathbf{v}}{dt} = \frac{ds}{dt}\frac{d\mathbf{t}}{dt} + \frac{d^2s}{dt^2}\mathbf{t} = \frac{ds}{dt}\left(\frac{ds}{dt}\frac{d\mathbf{t}}{ds}\right) + \frac{d^2s}{dt^2}\mathbf{t}$$

$$= \frac{d^2s}{dt^2}\mathbf{t} + \left(\frac{ds}{dt}\right)^2\frac{d\mathbf{t}}{ds}.$$

Using Eq. (11), we then obtain

$$\mathbf{a} = \frac{d^2s}{dt^2}\mathbf{t} + \kappa\left(\frac{ds}{dt}\right)^2\mathbf{n}. \tag{13}$$

Comparing Eqs. (13) and (4), we see that

$$|\mathbf{a}_{\tan}| = \frac{d^2s}{dt^2} \quad \text{and} \quad |\mathbf{a}_{\text{nor}}| = \kappa\left(\frac{ds}{dt}\right)^2. \tag{14}$$

The relations in Eq. (14) have a nice physical interpretation. By Newton's second law of motion, the force vector $\mathbf{F}$ governing a body of mass m moving on a plane curve with acceleration $\mathbf{a}$ is given by $\mathbf{F} = m\mathbf{a}$. From Eq. (14), we see that the component of the force tangential to the path of the body controls the rate of change of the speed of the body, for

$$m|\mathbf{a}_{\tan}| = m\frac{d^2s}{dt^2} = m\frac{d(ds/dt)}{dt}.$$

On the other hand, the component of the force normal to the direction of a body controls the curvature of the path on which the body travels. This seems intuitively reasonable. The formula for $|\mathbf{a}_{\text{nor}}|$ in Eq. (14) shows that, when we are driving a car around an unbanked curve, the force normal to the curve exerted by the road on the wheels must be proportional to the product of the curvature of the curve and the *square* of the speed of the car. The larger the curvature, the more force is required. If the speed of the car is doubled, the force required normal to the curve to "hold the car on the road" is *quadrupled*. If we take a sharp unbanked curve too fast, then the available force due to friction normal to the wheels is not sufficient for the car to make the curve; we skid and go off the road.

COMPUTATIONS

We summarize in Table 15.1 easy ways to compute all the things defined in this section and the preceding. one. Everything is easily found from $\mathbf{v}$ and $\mathbf{a}$, which are themselves usually easy to compute. The computations should be made in the order in the table.

The formulas in Table 15.1 can easily be verified from the work we have done. To illustrate, Eq. (14) tells us that $|\mathbf{a}_{\text{nor}}| = \kappa(ds/dt)^2$. Since $(ds/dt)^2 = |\mathbf{v}|^2 = \mathbf{v} \cdot \mathbf{v}$, we obtain the formula for κ in Eq. (24).

EXAMPLE 2 Let the position vector of a body in the plane at time t be

$$\mathbf{r} = (1 + \cos t - \sin t)\mathbf{i} + (\sin t + \cos t)\mathbf{j}.$$

Find all the things listed in Table 15.1, except the equation of the osculating plane, at any time t.

Solution From Eqs. (16) and (17),

$$\mathbf{v} = (-\sin t - \cos t)\mathbf{i} + (\cos t - \sin t)\mathbf{j},$$

$$\text{Speed} = |\mathbf{v}| = \sqrt{2\sin^2 t + 2\cos^2 t} = \sqrt{2}.$$

Table 15.1 Formulas for computations

Position vector	$\mathbf{r}$ (given)	**(15)**
Velocity vector	$\mathbf{v} = \dfrac{d\mathbf{r}}{dt}$	**(16)**
Speed	$\lvert\mathbf{v}\rvert = \dfrac{ds}{dt}$	**(17)**
Acceleration vector	$\mathbf{a} = \dfrac{d\mathbf{v}}{dt} = \dfrac{d^2\mathbf{r}}{dt^2}$	**(18)**
Tangential vector component of **a**	$\mathbf{a}_{\text{tan}} = \left[\dfrac{(\mathbf{a} \cdot \mathbf{v})}{(\mathbf{v} \cdot \mathbf{v})}\right]\mathbf{v}$	**(19)**
Normal vector component of **a**	$\mathbf{a}_{\text{nor}} = \mathbf{a} - \mathbf{a}_{\text{tan}}$	**(20)**
Unit tangent vector if $\mathbf{v} \neq \mathbf{0}$	$\mathbf{t} = \dfrac{\mathbf{v}}{\lvert\mathbf{v}\rvert}$	**(21)**
Principal normal vector if $\mathbf{a}_{\text{nor}} \neq \mathbf{0}$	$\mathbf{n} = \dfrac{\mathbf{a}_{\text{nor}}}{\lvert\mathbf{a}_{\text{nor}}\rvert}$	**(22)**
Rate of change of speed	$\dfrac{d^2s}{dt^2} = \pm\lvert\mathbf{a}_{\text{tan}}\rvert$, negative if $\mathbf{a} \cdot \mathbf{v} < 0$	**(23)**
Curvature if $\mathbf{v} \neq \mathbf{0}$	$\kappa = \dfrac{\lvert\mathbf{a}_{\text{nor}}\rvert}{(\mathbf{v} \cdot \mathbf{v})} = \dfrac{\lvert\mathbf{a}_{\text{nor}}\rvert}{\lvert\mathbf{v}\rvert^2}$	**(24)**
Osculating plane if **v** and **a** are not parallel	Point (given), normal vector $\mathbf{v} \times \mathbf{a}$	**(25)**

By Eq. (18),

$$\mathbf{a} = (-\cos t + \sin t)\mathbf{i} - (\sin t + \cos t)\mathbf{j}.$$

We easily see that $\mathbf{a} \cdot \mathbf{v} = 0$. Thus Eq. (19) yields

$$\mathbf{a}_{\text{tan}} = \mathbf{0}.$$

From Eq. (20), we have

$$\mathbf{a}_{\text{nor}} = \mathbf{a} - \mathbf{0} = (-\cos t + \sin t)\mathbf{i} - (\sin t + \cos t)\mathbf{j}.$$

Equation (21) shows that

$$\mathbf{t} = \frac{1}{\sqrt{2}}\mathbf{v} = \frac{1}{\sqrt{2}}[(-\sin t - \cos t)\mathbf{i} + (\cos t - \sin t)\mathbf{j}].$$

We easily see that

$$|\mathbf{a}_{\text{nor}}| = \sqrt{2\sin^2 t + 2\cos^2 t} = \sqrt{2}.$$

Equation (22) becomes

$$\mathbf{n} = \frac{1}{\sqrt{2}}[(-\cos t + \sin t)\mathbf{i} - (\sin t + \cos t)\mathbf{j}].$$

Since $ds/dt = \sqrt{2}$, we see that

$$\frac{d^2s}{dt^2} = 0.$$

Finally, Eq. (24) tells us that

$$\kappa = \frac{\sqrt{2}}{(\sqrt{2})^2} = \frac{1}{\sqrt{2}}.$$

Our curve has constant curvature $1/\sqrt{2}$, so it must be a circle of radius $\sqrt{2}$. □

Generally, the computations at a single point are less tedious than the computations for any time t that we did in the preceding example.

EXAMPLE 3 Find all the things listed in Table 15.1 for the twisted cubic $x = 3t^2$, $y = 2t^3$, $z = 3t$ of Example 1, where $t = 1$.

Solution From Example 1, we know that when $t = 1$,

$$\mathbf{r} = 3\mathbf{i} + 2\mathbf{j} + 3\mathbf{k},$$

$$\mathbf{v} = 6\mathbf{i} + 6\mathbf{j} + 3\mathbf{k},$$

$$\mathbf{a} = 6\mathbf{i} + 12\mathbf{j}.$$

Then

$$\text{Speed} = |\mathbf{v}| = \sqrt{36 + 36 + 9} = \sqrt{81} = 9.$$

We have

$$\mathbf{a}_{\tan} = \frac{\mathbf{a} \cdot \mathbf{v}}{\mathbf{v} \cdot \mathbf{v}}\mathbf{v} = \frac{108}{81}\mathbf{v} = \frac{4}{3}\mathbf{v} = 8\mathbf{i} + 8\mathbf{j} + 4\mathbf{k},$$

and

$$\mathbf{a}_{\text{nor}} = \mathbf{a} - \mathbf{a}_{\tan} = -2\mathbf{i} + 4\mathbf{j} - 4\mathbf{k}.$$

Now

$$\mathbf{t} = \frac{1}{|\mathbf{v}|}\mathbf{v} = \frac{1}{9}\mathbf{v} = \frac{2}{3}\mathbf{i} + \frac{2}{3}\mathbf{j} + \frac{1}{3}\mathbf{k},$$

and

$$\mathbf{n} = \frac{1}{|\mathbf{a}_{\text{nor}}|}\mathbf{a}_{\text{nor}} = \frac{1}{6}\mathbf{a}_{\text{nor}} = -\frac{1}{3}\mathbf{i} + \frac{2}{3}\mathbf{j} - \frac{2}{3}\mathbf{k}.$$

We have

$$\frac{d^2s}{dt^2} = |\mathbf{a}_{\tan}| = 12,$$

since $\mathbf{a} \cdot \mathbf{v} > 0$, and

$$\kappa = \frac{|\mathbf{a}_{\text{nor}}|}{|\mathbf{v}|^2} = \frac{6}{81} = \frac{2}{27}.$$

We found the equation of the osculating plane in Example 1 to be

$$-2x + y + 2z = 2. \quad \square$$

SUMMARY

Let **t** and **n** be tangent and normal unit vectors to a twice-differentiable smooth plane curve at a point, with **t** pointing in the direction corresponding to increasing t and **n** pointing toward the concave side of the curve.

1. $d\mathbf{t}/ds = \kappa\mathbf{n}$, where κ is the curvature of the curve

2. $\mathbf{v} = (ds/dt)\mathbf{t}$

3. $\mathbf{a} = d^2\mathbf{r}/dt^2 = (d^2s/dt^2)\mathbf{t} + \kappa(ds/dt)^2\mathbf{n}$,

so the tangential and normal scalar components of acceleration are

$$|\mathbf{a}_{tan}| = \frac{d^2s}{dt^2} \quad \text{and} \quad |\mathbf{a}_{nor}| = \kappa\left(\frac{ds}{dt}\right)^2.$$

Now consider a twice-differentiable smooth space curve.

4. $\mathbf{t} = \mathbf{v}/|\mathbf{v}|$

5. The principal normal vector **n** is the unit vector in the direction of $d\mathbf{t}/ds$ if $d\mathbf{t}/ds \neq \mathbf{0}$.

6. The osculating plane is the plane containing **t** and **n** at a point if $d\mathbf{t}/ds \neq \mathbf{0}$. It is the plane that most nearly contains the curve close to the point.

7. The curvature $\kappa = |d\mathbf{t}/ds|$

8. Number 3 also holds for space curves.

9. See Table 15.1 for easy ways to compute these things.

EXERCISES

In Exercises 1 through 5, find the indicated quantity when $t = 0$ for the motion in the plane having position vector

$$\mathbf{r} = a(\sin 2t)\mathbf{i} + b(\cos 3t)\mathbf{j}.$$

1. The vector **t**

2. The vector **n**

3. The rate of change of speed

4. $|\mathbf{a}_{nor}|$

5. The curvature

In Exercises 6 through 10, find the indicated quantity when $t = 1$ for the motion in the plane having position vector

$$\mathbf{r} = (t^2 + 2t)\mathbf{i} + \left(4t + \frac{1}{t}\right)\mathbf{j}.$$

6. The vector **t**

7. The vector **n**

8. The rate of change of speed

9. $|\mathbf{a}_{nor}|$

10. The curvature

In Exercises 11 through 15, find the indicated quantity when $t = 0$ for the motion in the plane having position vector

$$\mathbf{r} = e^{2t}\mathbf{i} + \ln(1 + t)\mathbf{j}.$$

11. The vector **t**

12. The vector **n**

13. The rate of change of speed

14. $|\mathbf{a}_{nor}|$

15. The curvature

In Exercises 16 through 22, find the indicated quantity when $t = 1$ for the motion in space with position vector

$$\mathbf{r} = \left(\frac{1}{t}\right)\mathbf{i} + \left(\frac{1}{t^2}\right)\mathbf{j} + \left(\frac{2}{t}\right)\mathbf{k}.$$

16. The vector **t**

17. The speed

18. The vector **n**

19. The equation of the osculating plane

20. The rate of change of speed

21. $|\mathbf{a}_{\text{nor}}|$

22. The curvature

In Exercises 23 through 29, find the indicated quantity when $t = 0$ for the motion in space with position vector

$$\mathbf{r} = (e^t \sin t)\mathbf{i} + (e^t \cos t)\mathbf{j} + e^{2t}\mathbf{k}.$$

23. The vector $\mathbf{t}$

24. The speed

25. The vector $\mathbf{n}$

26. The equation of the osculating plane

27. The rate of change of speed

28. $|\mathbf{a}_{\text{nor}}|$

29. The curvature

In Exercises 30 through 36, find the indicated quantity when $t = \pi/4$ for the motion in space with position vector

$$\mathbf{r} = [\ln (\sin t)]\mathbf{i} + [\ln (\cos t)]\mathbf{j} + (\sin 2t)\mathbf{k}.$$

30. The vector $\mathbf{t}$

31. The speed

32. The vector $\mathbf{n}$

33. The equation of the osculating plane

34. The rate of change of speed

35. $|\mathbf{a}_{\text{nor}}|$

36. The curvature

37. Show that if the curvature of a space curve is zero at a point, then the osculating plane to the curve at the point is not defined.

38. Consider the motion in space with position vector $\mathbf{r} = (3 + 7t)\mathbf{i} + (4 - t)\mathbf{j} + (1 + t)\mathbf{k}$. Without using any calculus, explain why there is no unique osculating plane at any point of the path.

39. A body in space of mass 3 is moving subject to the force vector

$$\mathbf{F} = 6(\cos t)\mathbf{i} + 9(\sin t)\mathbf{j}.$$

If $\mathbf{v}|_{t=0} = \mathbf{i} - 3\mathbf{j} + 4\mathbf{k}$, find
a) the rate of change of speed when $t = \pi$;
b) the curvature of the path when $t = \pi$.

40. A body of mass 2 in space moves subject to the force vector

$$\mathbf{F} = \frac{24}{t^4}\mathbf{j} - \frac{12}{t^3}\mathbf{k}$$

for $t \geq 1$. If $\mathbf{v}|_{t=2} = 2\mathbf{i} + \mathbf{j}$, find
a) the speed when $t = 2$;
b) the rate of change of speed when $t = 2$;
c) the curvature of the path when $t = 2$.

*15.3 POLAR VECTOR ANALYSIS AND KEPLER'S LAWS

*VELOCITY AND ACCELERATION IN POLAR COORDINATES

At each point in the polar coordinate plane except the pole (origin), we let $\mathbf{u}_r$ be the unit vector pointing directly away from the origin. We then let $\mathbf{u}_\theta$ be the unit vector normal to $\mathbf{u}_r$ in the direction of increasing θ, as illustrated in Fig. 15.17. As shown in Fig. 15.18, we have

$$\mathbf{u}_r = (\cos \theta)\mathbf{i} + (\sin \theta)\mathbf{j}, \tag{1}$$

while

$$\mathbf{u}_\theta = (-\sin \theta)\mathbf{i} + (\cos \theta)\mathbf{j}. \tag{2}$$

Figure 15.17 Orthogonal unit vectors $\mathbf{u}_r$ and $\mathbf{u}_\theta$ at points other than the origin.

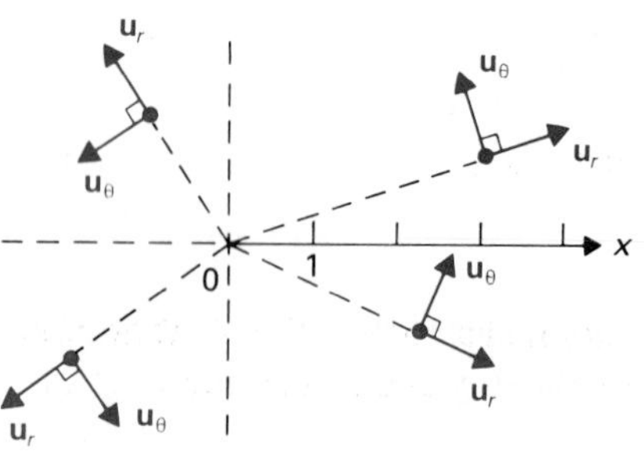

EXAMPLE 1 Express $\mathbf{u}_r$ and $\mathbf{u}_\theta$ at the point $(r, \theta) = (5, \pi/3)$ in terms of $\mathbf{i}$ and $\mathbf{j}$.

Solution From Eqs. (1) and (2), we have

$$\mathbf{u}_r = \left(\cos \frac{\pi}{3}\right)\mathbf{i} + \left(\sin \frac{\pi}{3}\right)\mathbf{j} = \frac{1}{2}\mathbf{i} + \frac{\sqrt{3}}{2}\mathbf{j}$$

* Sections marked with an asterisk may be omitted without loss of continuity. Some may want to cover just the one subsection on vectors in polar coordinates.

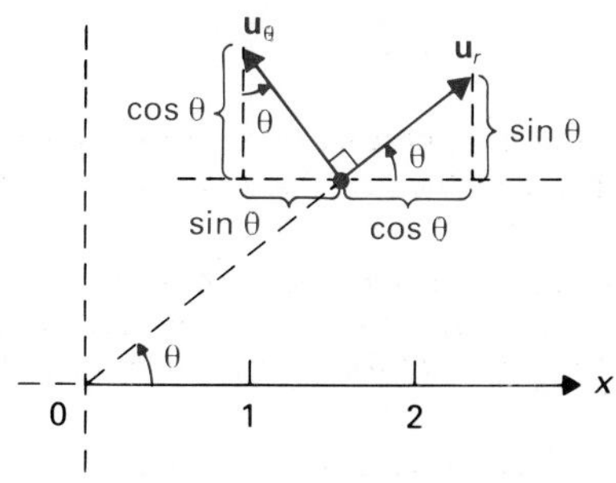

Figure 15.18 $\mathbf{u}_r = (\cos\theta)\mathbf{i} + (\sin\theta)\mathbf{j}$; $\mathbf{u}_\theta = (-\sin\theta)\mathbf{i} + (\cos\theta)\mathbf{j}$.

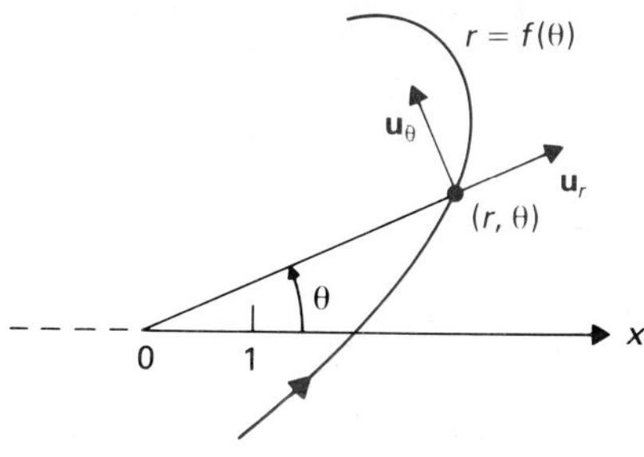

Figure 15.19 The vectors $\mathbf{u}_r$ and $\mathbf{u}_\theta$ on the path of a body in motion.

and

$$\mathbf{u}_\theta = \left(-\sin\frac{\pi}{3}\right)\mathbf{i} + \left(\cos\frac{\pi}{3}\right)\mathbf{j} = -\frac{\sqrt{3}}{2}\mathbf{i} + \frac{1}{2}\mathbf{j}. \quad \square$$

Let $r = f(\theta)$, where f is a twice-differentiable function, and consider a body moving along the curve $r = f(\theta)$. We wish to express the velocity and acceleration vectors of the body at time t in terms of the unit vector $\mathbf{u}_r$ directed away from the origin and the perpendicular unit vector $\mathbf{u}_\theta$ in the direction of increasing θ, as shown in Fig. 15.19. From Eqs. (1) and (2), we at once obtain

$$\frac{d\mathbf{u}_r}{d\theta} = (-\sin\theta)\mathbf{i} + (\cos\theta)\mathbf{j} = \mathbf{u}_\theta \tag{3}$$

and

$$\frac{d\mathbf{u}_\theta}{d\theta} = (-\cos\theta)\mathbf{i} + (-\sin\theta)\mathbf{j} = -\mathbf{u}_r. \tag{4}$$

We can now obtain the desired formulas for the velocity vector $\mathbf{v}$ and acceleration vector $\mathbf{a}$ by differentiating the position vector

$$\boxed{\mathbf{r} = r\mathbf{u}_r} \tag{5}$$

with respect to t. Using a product rule, the chain rule, and Eq. (3), we obtain from Eq. (5)

$$\mathbf{v} = \frac{d\mathbf{r}}{dt} = r\frac{d\mathbf{u}_r}{dt} + \frac{dr}{dt}\mathbf{u}_r = r\frac{d\theta}{dt}\frac{d\mathbf{u}_r}{d\theta} + \frac{dr}{dt}\mathbf{u}_r = r\frac{d\theta}{dt}\mathbf{u}_\theta + \frac{dr}{dt}\mathbf{u}_r,$$

so

$$\boxed{\mathbf{v} = \dot{r}\mathbf{u}_r + r\dot{\theta}\mathbf{u}_\theta,} \tag{6}$$

where a dot over a variable denotes a derivative with respect to time. Differentiating Eq. (6), using a product rule, chain rules, and Eqs. (3) and (4), we have

$$\begin{aligned}\mathbf{a} = \frac{d\mathbf{v}}{dt} &= \dot{r}\frac{d\mathbf{u}_r}{dt} + \ddot{r}\mathbf{u}_r + r\dot{\theta}\frac{d\mathbf{u}_\theta}{dt} + (r\ddot{\theta} + \dot{r}\dot{\theta})\mathbf{u}_\theta \\ &= \dot{r}\frac{d\theta}{dt}\frac{d\mathbf{u}_r}{d\theta} + \ddot{r}\mathbf{u}_r + r\dot{\theta}\frac{d\theta}{dt}\frac{d\mathbf{u}_\theta}{d\theta} + (r\ddot{\theta} + \dot{r}\dot{\theta})\mathbf{u}_\theta \\ &= \dot{r}\dot{\theta}\mathbf{u}_\theta + \ddot{r}\mathbf{u}_r + r\dot{\theta}^2(-\mathbf{u}_r) + (r\ddot{\theta} + \dot{r}\dot{\theta})\mathbf{u}_\theta,\end{aligned}$$

which yields

$$\boxed{\mathbf{a} = (\ddot{r} - r\dot{\theta}^2)\mathbf{u}_r + (r\ddot{\theta} + 2\dot{r}\dot{\theta})\mathbf{u}_\theta.} \tag{7}$$

EXAMPLE 2 The polar coordinate position of a body in the plane at time $t \geq 0$ is given by $(r, \theta) = (t^2, \sqrt{t})$. Find the velocity and acceleration vectors at time $t > 0$.

Solution We have

$$r = t^2, \qquad \theta = t^{1/2},$$

$$\dot{r} = 2t, \qquad \dot{\theta} = \frac{1}{2}t^{-1/2},$$

$$\ddot{r} = 2, \qquad \ddot{\theta} = -\frac{1}{4}t^{-3/2}.$$

Using Eqs. (6) and (7), we obtain

$$\mathbf{v} = 2t\mathbf{u}_r + \frac{1}{2}t^{3/2}\mathbf{u}_\theta$$

and

$$\mathbf{a} = \left(2 - \frac{1}{4}t\right)\mathbf{u}_r + \left(-\frac{1}{4}\sqrt{t} + 2\sqrt{t}\right)\mathbf{u}_\theta = \left(2 - \frac{1}{4}t\right)\mathbf{u}_r + \frac{7}{4}\sqrt{t}\,\mathbf{u}_\theta. \quad \square$$

EXAMPLE 3 For the motion in Example 2, find the speed when $t = 4$.

Solution We know that

$$\text{Speed} = |\mathbf{v}|.$$

From Example 2,

$$\mathbf{v}|_{t=4} = 8\mathbf{u}_r + 4\mathbf{u}_\theta.$$

Since $\mathbf{u}_r$ and $\mathbf{u}_\theta$ are *orthogonal unit* vectors, the speed when $t = 4$ is

$$\text{Speed} = |8\mathbf{u}_r + 4\mathbf{u}_\theta| = \sqrt{64 + 16} = \sqrt{80} = 4\sqrt{5}. \quad \square$$

EXAMPLE 4 Find the rate of change of speed for the body in Example 2 when $t = 4$.

Solution We have from Example 2,

$$\text{Speed} = \frac{ds}{dt} = |\mathbf{v}| = \sqrt{4t^2 + (t^3/4)}.$$

The rate of change of speed is then

$$\frac{d^2s}{dt^2} = \frac{1}{2}\frac{8t + (3t^2/4)}{\sqrt{4t^2 + (t^3/4)}}.$$

When $t = 4$, we obtain

$$\left.\frac{d^2s}{dt^2}\right|_{t=4} = \frac{1}{2}\frac{32 + 12}{\sqrt{64 + 16}} = \frac{22}{4\sqrt{5}} = \frac{11}{2\sqrt{5}}. \quad \square$$

EXAMPLE 5 For the body in Example 2, find the curvature of the path when $t = 4$.

Solution From Example 2, we have

$$\mathbf{a}|_{t=4} = \mathbf{u}_r + \frac{7}{2}\mathbf{u}_\theta.$$

Then

$$|\mathbf{a}|^2|_{t=4} = 1 + \frac{49}{4} = \frac{53}{4}.$$

From our work in the previous section, we know that

$$|\mathbf{a}|^2 = |\mathbf{a}_{\tan}|^2 + |\mathbf{a}_{\text{nor}}|^2 = \left(\frac{d^2s}{dt^2}\right)^2 + \left(\kappa\left(\frac{ds}{dt}\right)^2\right)^2.$$

From Examples 3 and 4, when $t = 4$, this equation becomes

$$\frac{53}{4} = \frac{121}{20} + (\kappa \cdot 80)^2,$$

so

$$(\kappa \cdot 80)^2 = \frac{53}{4} - \frac{121}{20} = \frac{265 - 121}{20} = \frac{144}{20}.$$

Thus

$$\kappa = \frac{1}{80}\frac{12}{\sqrt{20}} = \frac{3}{40\sqrt{5}}. \quad \square$$

*CENTRAL FORCE FIELDS AND KEPLER'S LAWS

In this section, we work in polar coordinates and show how we can derive the planetary laws of Johannes Kepler (1571–1630) from Newton's law of gravitation. These laws are as follows:

The force F of attraction between two bodies of masses m and M is given by

$$F = \frac{GmM}{d^2},$$

where G is a "universal" constant of gravitational attraction and d is the distance between the bodies.

Newton's law of universal gravitation

1. The orbit of each planet is an ellipse with the sun at a focus.
2. A line segment joining the sun and a particular planet sweeps out regions of equal areas during equal time intervals. (See Fig. 15.20.)
3. Let the elliptical orbit of a planet have a major axis of length $2a$ and let T be the **period** of the planet (that is, the time required for one complete revolution about the sun). Then the ratio a^3/T^2 is the same for all planets of our sun.

Kepler's laws

Figure 15.20 Region swept out from time t_1 to t_2, referred to in Kepler's second law.

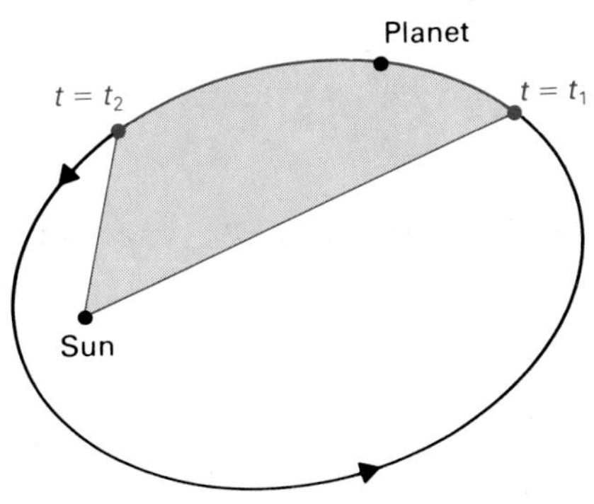

We derive Kepler's laws from Newton's law. This reverses the historical order. Newton actually derived his inverse square law from Kepler's laws.

Kepler, in turn, formulated his laws on the basis of his analysis of astronomical observations by Tycho Brahe (1546–1601). After some exercises illustrating the material in the text, we give a sequence of exercises that constitute a detailed outline for the derivation of Newton's inverse square law from Kepler's three laws.

We will consider a body of mass M to be fixed at the pole O of a polar coordinate system and will examine the motion of a body of mass m, which is subject only to the force of gravitational attraction toward the body at O, as specified by Newton's law. Newton's second law of motion in vector form is

$$\boxed{\mathbf{F} = m\mathbf{a}.}$$

Since the force is directed entirely in the direction $-\mathbf{u}_r$ toward O, we see from Eq. (7) that

$$\boxed{r\ddot{\theta} + 2\dot{r}\dot{\theta} = 0.} \quad \textbf{(8)}$$

We obtain, from Eq. (7) and Newton's universal law of gravitation,

$$m(\ddot{r} - r\dot{\theta}^2) = -\frac{mMG}{r^2},$$

or

$$\boxed{\ddot{r} - r\dot{\theta}^2 = -\frac{MG}{r^2} = -\frac{k}{r^2},} \quad \textbf{(9)}$$

where $k = MG > 0$ is a constant independent of the mass m of the moving body.

We derive Kepler's laws from Eqs. (8) and (9). Kepler's second law is the easiest one to derive.

Derivation of Kepler's second law From Eq. (8), we see that

$$\frac{d}{dt}\left(\frac{1}{2}r^2\dot{\theta}\right) = \frac{1}{2}r^2\ddot{\theta} + r\dot{r}\dot{\theta} = \frac{1}{2}r(r\ddot{\theta} + 2\dot{r}\dot{\theta}) = 0.$$

Thus

$$r^2\dot{\theta} = K \quad \textbf{(10)}$$

for some constant K. Now the area A of the region swept out by the line segment joining O and the moving body from time $t = t_0$ where $\theta = \theta_0$ to time t is given by

$$A = \int_{\theta_0}^{\theta(t)} \frac{1}{2}r^2\,d\theta = \int_{\theta_0}^{\theta(t)} \frac{1}{2}(f(\theta))^2\,d\theta.$$

Thus

$$\frac{dA}{dt} = \frac{dA}{d\theta}\frac{d\theta}{dt} = \frac{1}{2}r^2\frac{d\theta}{dt} = \frac{1}{2}r^2\dot{\theta} = \frac{1}{2}K. \quad \textbf{(11)}$$

From Eq. (11), we have

$$A = \frac{1}{2}Kt + C,$$

from which it follows at once that equal time intervals correspond to equal increments in A.

Derivation of Kepler's first law From Eq. (10), we have

$$r\dot{\theta}^2 = \frac{(r^2\dot{\theta})^2}{r^3} = \frac{K^2}{r^3}. \tag{12}$$

Using Eq. (9), we then have

$$\ddot{r} - \frac{K^2}{r^3} = -\frac{k}{r^2}. \tag{13}$$

Multiplication of Eq. (13) by $\dot{r}$ yields

$$\dot{r}\ddot{r} - K^2\frac{\dot{r}}{r^3} = -k\frac{\dot{r}}{r^2}. \tag{14}$$

Integrating Eq. (13) with respect to t and multiplying by 2, we obtain

$$\dot{r}^2 + K^2\frac{1}{r^2} = 2k\frac{1}{r} + C. \tag{15}$$

If we now let $p = 1/r$, then $r = 1/p$ and $\dot{r} = -\dot{p}/p^2$. We may then write Eq. (15) as

$$\frac{\dot{p}^2}{p^4} + K^2p^2 = 2kp + C. \tag{16}$$

From $r^2\dot{\theta} = K$ in Eq. (10), we obtain $\dot{\theta}/p^2 = K$, so

$$K^2\left(\frac{dp}{d\theta}\right)^2 = \frac{1}{p^4}\left(\frac{d\theta}{dt}\right)^2\left(\frac{dp}{d\theta}\right)^2 = \frac{1}{p^4}\left(\frac{dp}{dt}\right)^2 = \frac{\dot{p}^2}{p^4}. \tag{17}$$

From Eqs. (16) and (17),

$$K^2\left[\left(\frac{dp}{d\theta}\right)^2 + p^2\right] = 2kp + C. \tag{18}$$

Differentiation of Eq. (18) with respect to θ yields

$$K^2\left[2\left(\frac{dp}{d\theta}\right)\frac{d^2p}{d\theta^2} + 2p\frac{dp}{d\theta}\right] = 2k\frac{dp}{d\theta},$$

or

$$2K^2\frac{dp}{d\theta}\left(\frac{d^2p}{d\theta^2} + p - \frac{k}{K^2}\right) = 0. \tag{19}$$

If $dp/d\theta = 0$, then $p = 1/r = a$ for some constant a, so $r = 1/a$ and the orbit of the body is a circle with center O, which is in accordance with Kepler's first law. Suppose

$$\frac{d^2p}{d\theta^2} + p - \frac{k}{K^2} = 0. \tag{20}$$

In Section 20.6 we will see that the solution of the differential equation

$$\frac{d^2p}{d\theta^2} + p = \frac{k}{K^2} \tag{21}$$

describing the motion of the body must be given by

$$p = A\cos\theta + B\sin\theta + \frac{k}{K^2}$$

$$= \sqrt{A^2+B^2}\left(\frac{A}{\sqrt{A^2+B^2}}\cos\theta + \frac{B}{\sqrt{A^2+B^2}}\sin\theta\right) + \frac{k}{K^2} \qquad \textbf{(22)}$$

for some constants A and B. We can find an angle γ such that

$$\frac{A}{\sqrt{A^2+B^2}} = -\cos\gamma \qquad \text{and} \qquad \frac{B}{\sqrt{A^2+B^2}} = -\sin\gamma. \qquad \textbf{(23)}$$

Let $E = \sqrt{A^2+B^2}$. From Eqs. (22) and (23), we obtain

$$p = -E\cos(\theta-\gamma) + \frac{k}{K^2}. \qquad \textbf{(24)}$$

By choosing a new polar axis, if necessary, we can assume that $\gamma = 0$, so that Eq. (24) becomes

$$\frac{1}{r} = -E\cos\theta + \frac{k}{K^2}, \qquad \textbf{(25)}$$

which yields

$$r = \frac{1}{(k/K^2) - E\cos\theta}$$

$$= \frac{K^2/k}{1-(K^2E/k)\cos\theta}. \qquad \textbf{(26)}$$

Note that $K^2E/k > 0$. Setting $e = K^2E/k$, we obtain finally

$$r = \frac{e(1/E)}{1-e\cos\theta}, \qquad \textbf{(27)}$$

which is the equation of a conic section with focus at the pole O. In particular, if the orbit is a closed path (as for the planets orbiting our sun) the orbit must be an ellipse.

Derivation of Kepler's third law By integration, we easily find that the area of the region enclosed by the ellipse $(x^2/a^2) + (y^2/b^2) = 1$ of major axis $2a$ and minor axis $2b$ is πab. If T is the period of the orbit of our planet, then Eq. (11) shows that the area swept out by the line segment joining O and the planet during one revolution is given by

$$\int_0^T \frac{dA}{dt}\,dt = \int_0^T \frac{1}{2}K\,dt = \frac{1}{2}Kt\Big]_0^T = \frac{1}{2}KT.$$

Figure 15.21 Ellipse with focus at the pole; $2a = f(0) + f(\pi)$.

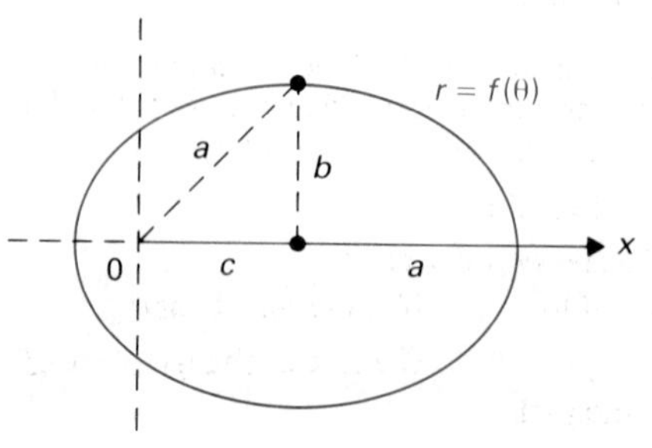

Thus we have

$$\frac{1}{2}KT = \pi ab. \qquad \textbf{(28)}$$

Now refer to Fig. 15.21. For an ellipse $r = f(\theta)$ with focus at the pole and major axis of length $2a$ along the x-axis, as shown, we see that $2a = r|_{\theta=0} + r|_{\theta=\pi}$. From Eq. (26),

$$2a = \frac{K^2}{k-EK^2} + \frac{K^2}{k+EK^2} = \frac{2kK^2}{k^2-E^2K^4}. \qquad \textbf{(29)}$$

Letting b and c be as indicated in Fig. 15.21, we have

$$c = a - r|_{\theta=\pi} = \frac{kK^2}{k^2 - E^2K^4} - \frac{K^2}{k + EK^2} = \frac{kK^2}{k^2 - E^2K^4} - \frac{K^2(k - EK^2)}{k^2 - E^2K^4}$$

$$= \frac{EK^4}{k^2 - E^2K^4}. \tag{30}$$

A computation then yields

$$\frac{b^2}{K^2} = \frac{a^2 - c^2}{K^2} = \frac{k^2K^4 - E^2K^8}{K^2(k^2 - E^2K^4)^2} = \frac{K^2}{k^2 - E^2K^4} = \frac{a}{k}. \tag{31}$$

Equations (28) and (31) now yield

$$T^2 = 4\pi^2 a^2 \frac{b^2}{K^2} = 4\pi^2 \frac{a^3}{k},$$

so

$$\boxed{\frac{a^3}{T^2} = \frac{k}{4\pi^2} = \frac{GM}{4\pi^2},} \tag{32}$$

which yields Kepler's third law, because $GM/(4\pi^2)$ is independent of the mass m of the planet orbiting the sun of mass M.

SUMMARY

1. For unit vectors $\mathbf{u}_r$ and $\mathbf{u}_\theta$ in polar coordinates,

$$\text{Position vector} = \mathbf{r} = r\mathbf{u}_r,$$

$$\text{Velocity vector} = \mathbf{v} = \dot{r}\mathbf{u}_r + r\dot{\theta}\mathbf{u}_\theta,$$

$$\text{Acceleration vector} = \mathbf{a} = (\ddot{r} - r\dot{\theta}^2)\mathbf{u}_r + (r\ddot{\theta} + 2\dot{r}\dot{\theta})\mathbf{u}_\theta.$$

2. Newton's and Kepler's laws are given on page 671.

EXERCISES

In sequential Exercises 1 through 5, find the indicated quantity when $t = 1$ for the motion in the plane where the polar coordinate position at time t is $(r, \theta) = (t^3 - 2t, t - 1)$.

*1. $\mathbf{u}_r$ and $\mathbf{u}_\theta$ in terms of **i** and **j**

*2. The velocity and acceleration vectors
   a) in terms of $\mathbf{u}_r$ and $\mathbf{u}_\theta$ b) in terms of **i** and **j**

*3. The speed *4. The rate of change of speed

*5. The curvature of the path

In sequential Exercises 6 through 10, find the indicated quantity when $t = \pi/3$ for the motion in the plane where the polar coordinate position at time t is $(r, \theta) = (2a \cos t, t)$.

*6. $\mathbf{u}_r$ and $\mathbf{u}_\theta$ in terms of **i** and **j**

*7. The velocity and acceleration vectors
   a) in terms of $\mathbf{u}_r$ and $\mathbf{u}_\theta$ b) in terms of **i** and **j**

*8. The speed *9. The rate of change of speed

*10. The curvature of the path

In sequential Exercises 11 through 15, find the indicated quantity when $t = 5\pi/6$ for the motion in the plane where the polar coordinate position at time t is $(r, \theta) = (a + a \sin t, t)$.

*11. $\mathbf{u}_r$ and $\mathbf{u}_\theta$ in terms of **i** and **j**

*12. The velocity and acceleration vectors
   a) in terms of $\mathbf{u}_r$ and $\mathbf{u}_\theta$ b) in terms of **i** and **j**

*13. The speed *14. The rate of change of speed

*15. The curvature of the path

*16. The value of the universal gravitational constant G was calculated by Cavendish in 1798. Explain how one might use Eq. (32) to calculate the mass of the sun.

*17. Argue from Eq. (32) that the period of a body in an elliptical orbit in a given central force field can be found if the length of the major axis of the ellipse is known.

*18. The *apogee* of an earth satellite is its maximum distance from the surface of the earth, and the *perigee* is the minimum distance from the surface. Use Eq. (32) to argue that the period of an earth satellite is completely determined by the apogee and perigee of the satellite. [*Hint:* Let the radius of the earth be R and the length of the major axis of the elliptical orbit be $2a$. Find a relation between R, a, the apogee, and the perigee.]

*19. Would you expect the ratio a^3/T^2 of Eq. (32) to be the same for the earth in revolution about the sun as for a moon in revolution about the planet Jupiter? Explain.

Exercises 20 through 33 deal with the derivation of the inverse square property and the formulation of Newton's law of universal gravitation from Kepler's laws. Suppose that planets describe planar orbits about the sun in accordance with Kepler's laws, and let the position of a certain planet with respect to a polar coordinate system with pole at the sun be given by $r = r(t)$, $\theta = \theta(t)$.

*20. Deduce from Kepler's second law that if $A = A(t)$ is the area of the region swept out by the line segment joining the sun and the planet from time t_0 to time t, then $dA/dt = \beta$ for some constant β.

*21. Deduce from Exercise 20 and the formula for area in polar coordinates that $r^2\dot{\theta} = 2\beta$.

*22. Deduce from Exercise 21 that $r\ddot{\theta} + 2\dot{r}\dot{\theta} = 0$.

*23. Deduce from Exercise 22 that the acceleration of the planet is toward the sun at all times. (This shows we are indeed in a "central force field" situation.)

*24. Let the orbit of the planet be an ellipse

$$r = \frac{B}{1 - e\cos\theta}$$

for $B > 0$. (This is Kepler's first law.) Show that

$$\dot{r}(1 - e\cos\theta) + re\dot{\theta}\sin\theta = 0.$$

*25. Multiply the result of Exercise 24 by r and use Exercise 21 to show that

$$B\dot{r} + 2\beta e\sin\theta = 0.$$

*26. Deduce by differentiating the result in Exercise 25 and using Exercise 21 that

$$\ddot{r} = -\frac{4\beta^2}{r^2}\frac{e\cos\theta}{B} = \frac{4\beta^2}{r^2}\left(\frac{1}{r} - \frac{1}{B}\right).$$

*27. Deduce from Exercise 21 that $r\dot{\theta}^2 = 4\beta^2/r^3$, and conclude from Exercise 26 that

$$\ddot{r} - r\dot{\theta}^2 = -\frac{4\beta^2}{B}\cdot\frac{1}{r^2}.$$

*28. Use the result of Exercise 23 to argue that the force on a planet of mass m is of magnitude

$$\frac{4m\beta^2/B}{r^2}$$

and is directed toward the sun. (This gives the inverse square property.)

*29. Let $2a$ be the length of the major axis and $2b$ the length of the minor axis of the ellipse in Exercise 24 and show that

$$B = a(1 - e^2) \quad \text{and} \quad b = a\sqrt{1 - e^2}.$$

*30. Show from Exercise 20 that if the planet has elliptical orbit of period T, then the area of the ellipse is βT.

*31. Deduce from Exercise 30 that $\beta = \pi ab/T$.

*32. Deduce from Exercises 29 and 31 that the force in Exercise 28 may be written as

$$\frac{4\pi^2 a^3 m}{T^2 r^2}.$$

*33. Conclude from Exercise 32 and Kepler's third law that the force of attraction of the sun per unit mass of planet is independent of the planet ("universal"), depending only on the distance from the planet to the sun.

*34. The ratio a^3/T^2 can be measured astronomically for various situations (a planet orbiting the sun or a moon orbiting a planet) in our solar system. Show that astronomical indications that a^3/T^2 is proportional to the mass at the center of the central force field, together with Exercises 32 and 33, would lead to prediction of Newton's universal law of gravitation.

*15.4 THE BINORMAL VECTOR AND TORSION

Section 15.2 introduced the orthogonal unit vector frame, **t** and **n**, for both curves in the plane and space curves. In space, we need another unit vector, orthogonal to both **t** and **n**, to complete a three-dimensional space frame. In this section, we introduce such a vector, the *binormal vector* **b**.

Recall that at a point on a space curve, the vectors **t** and **n** lie in the osculating plane to the curve at that point, which is the plane that most nearly contains the curve near the point. The vector **b** is then normal to the osculating plane. The rate that **b** "turns" as we travel along the curve is a measure of how quickly the curve twists away from the osculating plane. We are led to introduce the notion of the *torsion* (twist) of a space curve. The section concludes with the *Frenet formulas*, which summarize the relationships of the vectors **t**, **n**, and **b** to each other.

In this section, we always assume that we are working on a smooth arc of a space curve with parametrizing functions having continuous derivatives of all orders that we desire to take.

*THE BINORMAL VECTOR

We list for quick reference the vector geometry concepts we developed for space curves in Section 15.2.

$\mathbf{r} = x\mathbf{i} + y\mathbf{j} + z\mathbf{k}$	Position vector
$\mathbf{v} = \dfrac{d\mathbf{r}}{dt}$	Velocity
$\dfrac{ds}{dt} = \lvert\mathbf{v}\rvert$	Speed
$\mathbf{a} = \dfrac{d\mathbf{v}}{dt} = \dfrac{d^2\mathbf{r}}{dt^2}$	Acceleration
$\mathbf{t} = \left(\dfrac{1}{\lvert\mathbf{v}\rvert}\right)\mathbf{v} \quad \text{if} \quad \mathbf{v} \neq \mathbf{0}$	Unit tangent vector
$\mathbf{n} = \dfrac{d\mathbf{t}/ds}{\lvert d\mathbf{t}/ds\rvert} \quad \text{if} \quad \dfrac{d\mathbf{t}}{ds} \neq \mathbf{0}$	Principal normal vector
$\kappa = \left\lvert\dfrac{d\mathbf{t}}{ds}\right\rvert$	Curvature
$\mathbf{a}_{\text{tan}} = \left(\dfrac{d^2s}{dt^2}\right)\mathbf{t} = \dfrac{\mathbf{v}\cdot\mathbf{a}}{\mathbf{v}\cdot\mathbf{v}}\mathbf{v}$	Tangential vector component of **a**
$\mathbf{a}_{\text{nor}} = \mathbf{a} - \mathbf{a}_{\text{tan}} = \kappa\left(\dfrac{ds}{dt}\right)^2\mathbf{n}$	Normal vector component of **a**
Plane at the point with normal vector $\mathbf{t} \times \mathbf{n}$ or $\mathbf{v} \times \mathbf{a}$.	Osculating plane

We illustrate the computation of **t** and **n** again for a helix.

EXAMPLE 1 Find **t** and **n** for the helix with position vector

$$\mathbf{r} = (a \cos t)\mathbf{i} + (a \sin t)\mathbf{j} + t\mathbf{k}.$$

Solution We have

$$\mathbf{v} = (-a \sin t)\mathbf{i} + (a \cos t)\mathbf{j} + \mathbf{k},$$
$$\mathbf{a} = (-a \cos t)\mathbf{i} - (a \sin t)\mathbf{j}.$$

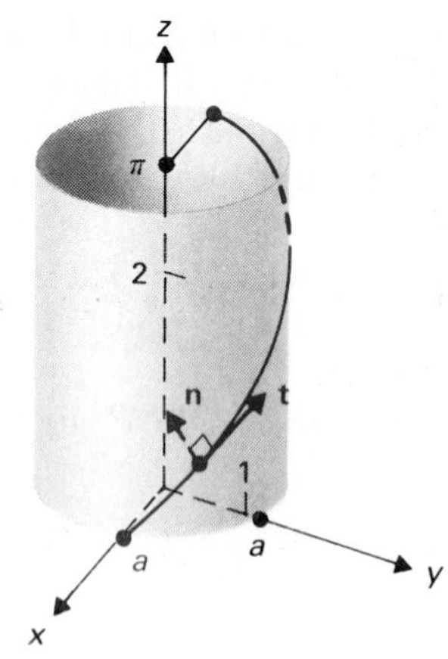

Figure 15.22 **t** and **n** on the helix $x = a \cos t$, $y = a \sin t$, $z = t$.

Since $|\mathbf{v}| = \sqrt{a^2 \sin^2 t + a^2 \cos^2 t + 1} = \sqrt{a^2 + 1}$, we have

$$\mathbf{t} = \frac{1}{|\mathbf{v}|}\mathbf{v} = \frac{1}{\sqrt{a^2 + 1}}[(-a \sin t)\mathbf{i} + (a \cos t)\mathbf{j} + \mathbf{k}].$$

Now $\mathbf{a} \cdot \mathbf{v} = a^2 \sin t \cos t - a^2 \sin t \cos t = 0$, so $\mathbf{a}$ is normal to $\mathbf{v}$. Since $|\mathbf{a}| = a$, we see in this case that

$$\mathbf{n} = \frac{1}{|\mathbf{a}|}\mathbf{a} = \frac{1}{a}[-(a \cos t)\mathbf{i} - (a \sin t)\mathbf{j}]$$

$$= -(\cos t)\mathbf{i} - (\sin t)\mathbf{j}.$$

Note that $\mathbf{n}$ is directed toward the z-axis, as shown in Fig. 15.22. □

Now consider a point on the space curve where $d\mathbf{t}/ds \neq 0$, which is equivalent to saying that $\kappa \neq 0$. At such a point, **t** and **n** are both defined. We then define the **binormal vector b** at the point to be

$$\boxed{\mathbf{b} = \mathbf{t} \times \mathbf{n}.} \qquad \textbf{(1)}$$

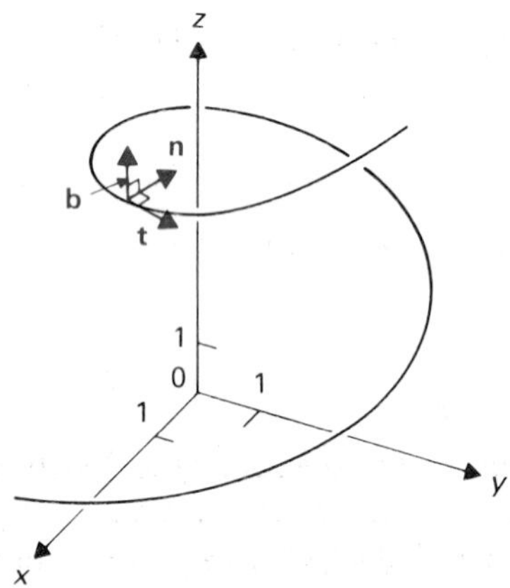

Figure 15.23 Orthogonal local unit vector frame **t**, **n**, **b** on a curve.

Therefore

$$|\mathbf{b}| = |\mathbf{t}||\mathbf{n}| \sin 90° = 1,$$

so **b** is a unit vector perpendicular to both **t** and **n**. The sequence

$$\mathbf{t}, \quad \mathbf{n}, \quad \mathbf{b}$$

of vectors is a right-hand triple of orthogonal unit vectors at each point on the space curve and may be regarded as a local coordinate 3-frame at each point on the curve (see Fig. 15.23). To a bug crawling in the curve in the direction of increasing t so that **t** points "straight ahead" and **n** points "left," it appears that **b** points "up." Of course,

$$\boxed{\mathbf{b} = \mathbf{t} \times \mathbf{n}, \qquad \mathbf{t} = \mathbf{n} \times \mathbf{b}, \qquad \text{and} \qquad \mathbf{n} = \mathbf{b} \times \mathbf{t}.} \qquad \textbf{(2)}$$

EXAMPLE 2 Find **b** for the helix in Example 1.

Solution We see from the example that

$$\mathbf{b} = \mathbf{t} \times \mathbf{n} = \frac{1}{\sqrt{a^2 + 1}}\begin{vmatrix} \mathbf{i} & \mathbf{j} & \mathbf{k} \\ -a \sin t & a \cos t & 1 \\ -\cos t & -\sin t & 0 \end{vmatrix}$$

$$= \frac{1}{\sqrt{a^2 + 1}}[(\sin t)\mathbf{i} - (\cos t)\mathbf{j} + a\mathbf{k}]. \quad □$$

Note that at a point on a space curve, **b** is a normal vector to the osculating plane to the curve. Recall that $\mathbf{v} \times \mathbf{a}$ is also such a normal vector. We argue that $\mathbf{v} \times \mathbf{a}$ has the same direction as **b**. Now the osculating plane doesn't really contain the curve near the point, but the curve clings very close to it for a short distance. In Fig. 15.24, we suppose the osculating plane is the plane of the page, and we draw in this plane a short piece of the curve that is closest to the space curve. Since **a** and **n** both point toward the concave side of this curve, we see that $\mathbf{b} = \mathbf{t} \times \mathbf{n}$ and $\mathbf{v} \times \mathbf{a}$ have the same direction. Thus we can com-

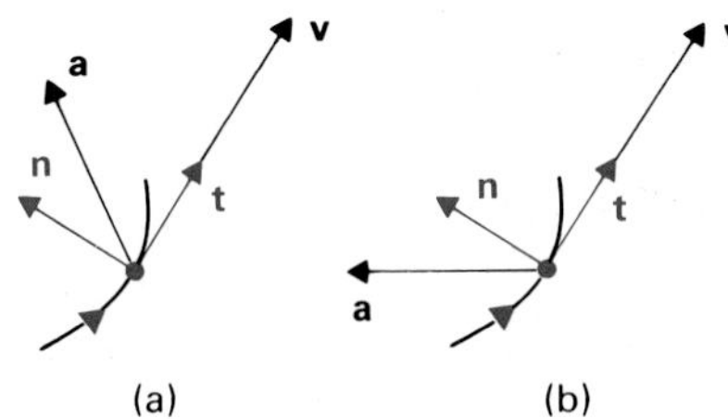

Figure 15.24 **t** × **n** and **v** × **a** have the same direction: (a) increasing speed case; (b) decreasing speed case.

pute **b** easily without going after **t** and **n**, using

$$\boxed{\mathbf{b} = \frac{1}{|\mathbf{v} \times \mathbf{a}|}(\mathbf{v} \times \mathbf{a}).} \tag{3}$$

EXAMPLE 3 Find **b** when $t = 1$ for the space curve with position vector

$$\mathbf{r} = (t^2 - 3t)\mathbf{i} + t^3\mathbf{j} - (t^2 + 2t)\mathbf{k}.$$

Solution We have

$$\mathbf{v} = (2t - 3)\mathbf{i} + 3t^2\mathbf{j} - (2t + 2)\mathbf{k} \qquad \text{and} \qquad \mathbf{a} = 2\mathbf{i} + 6t\mathbf{j} - 2\mathbf{k}.$$

When $t = 1$, we obtain

$$\mathbf{v} = -\mathbf{i} + 3\mathbf{j} - 4\mathbf{k} \qquad \text{and} \qquad \mathbf{a} = 2\mathbf{i} + 6\mathbf{j} - 2\mathbf{k}.$$

Consequently, when $t = 1$,

$$\mathbf{v} \times \mathbf{a} = \begin{vmatrix} \mathbf{i} & \mathbf{j} & \mathbf{k} \\ -1 & 3 & -4 \\ 2 & 6 & -2 \end{vmatrix} = 18\mathbf{i} - 10\mathbf{j} - 12\mathbf{k} = 2(9\mathbf{i} - 5\mathbf{j} - 6\mathbf{k}).$$

Finally, when $t = 1$,

$$\mathbf{b} = \frac{1}{|\mathbf{v} \times \mathbf{a}|}(\mathbf{v} \times \mathbf{a}) = \frac{1}{\sqrt{142}}(9\mathbf{i} - 5\mathbf{j} - 6\mathbf{k}). \qquad \square$$

*TORSION

Now **b** is normal to the osculating plane and really determines the *direction* of the plane. Since **b** is a unit vector, its length does not change as we go along the curve, but its direction changes. The rate of change $d\mathbf{b}/ds$ of **b** per unit length of curve should be a measure of the twist of the osculating plane as we travel along the curve.

Recall that $d\mathbf{t}/ds = \kappa\mathbf{n}$. Viewing **t**, **n**, and **b** as functions of arc length s and differentiating, we obtain

$$\frac{d\mathbf{b}}{ds} = \frac{d(\mathbf{t} \times \mathbf{n})}{ds} = \mathbf{t} \times \frac{d\mathbf{n}}{ds} + \frac{d\mathbf{t}}{ds} \times \mathbf{n}$$

$$= \mathbf{t} \times \frac{d\mathbf{n}}{ds} + (\kappa\mathbf{n}) \times \mathbf{n} = \mathbf{t} \times \frac{d\mathbf{n}}{ds}. \tag{4}$$

Since **n** is a vector of constant magnitude 1, Theorem 15.1 on page 660 shows that **n** is orthogonal to $d\mathbf{n}/ds$. Equation (4) shows that $d\mathbf{b}/ds$ has direction orthogonal to both **t** and $d\mathbf{n}/ds$. Since **n** is orthogonal to both **t** and $d\mathbf{n}/ds$, we see that $d\mathbf{b}/ds$ is parallel to **n**. Hence

$$\boxed{\frac{d\mathbf{b}}{ds} = -\tau\mathbf{n}} \tag{5}$$

for some (positive or negative) constant τ. Since $d\mathbf{b}/ds$ measures the rate at which the osculating plane *twists* per unit change in arc length, the number τ is called the **torsion** of the curve at the point. The minus sign is introduced in

Eq. (5) so that *positive* torsion corresponds to the vector **b** turning in the direction of $-\mathbf{n}$ (like a right-threaded screw) as we travel the curve in the direction given by the tangent vector **t**.

EXAMPLE 4 Find the torsion τ for the helix in Examples 1 and 2.

Solution From Example 1, we see that

$$\frac{ds}{dt} = |\mathbf{v}| = \sqrt{a^2 + 1}, \qquad \text{so} \quad \frac{dt}{ds} = \frac{1}{\sqrt{a^2 + 1}}.$$

From Example 2, we know that

$$\mathbf{b} = \frac{1}{\sqrt{a^2 + 1}}[(\sin t)\mathbf{i} - (\cos t)\mathbf{j} + a\mathbf{k}].$$

By the chain rule and Example 1,

$$\begin{aligned}\frac{d\mathbf{b}}{ds} = \frac{dt}{ds}\frac{d\mathbf{b}}{dt} &= \frac{1}{\sqrt{a^2 + 1}}\frac{1}{\sqrt{a^2 + 1}}[(\cos t)\mathbf{i} + (\sin t)\mathbf{j}] \\ &= \frac{1}{a^2 + 1}[(\cos t)\mathbf{i} + (\sin t)\mathbf{j}] = -\frac{1}{a^2 + 1}\mathbf{n}.\end{aligned}$$

Thus the torsion of the helix is constant, independent of the point on the curve, and is given by

$$\tau = \frac{1}{a^2 + 1}. \quad \square$$

*FORMULAS FOR COMPUTING κ AND τ

We have had some practice computing κ using the formula $\kappa = |\mathbf{a}_{\text{nor}}|/|\mathbf{v}|^2$ in Section 15.2. For the helix in Example 4, we found τ easily from its definition in Eq. (5). We give alternative formulas for κ and τ that are often easier to use. The sequence of Exercises 27 through 30 leads to the formulas

$$\boxed{\kappa = \frac{|\dot{\mathbf{r}} \times \ddot{\mathbf{r}}|}{|\dot{\mathbf{r}}|^3} \quad \text{and} \quad \tau = \frac{(\dot{\mathbf{r}} \times \ddot{\mathbf{r}}) \cdot \dddot{\mathbf{r}}}{|\dot{\mathbf{r}} \times \ddot{\mathbf{r}}|^2}.} \tag{6}$$

EXAMPLE 5 Find the curvature κ and torsion τ for the curve with position vector

$$\mathbf{r} = t^2\mathbf{i} - (3t + 1)\mathbf{j} + t^3\mathbf{k}$$

at the point where $t = 1$.

Solution Now

$$\begin{aligned}\dot{\mathbf{r}} &= 2t\mathbf{i} - 3\mathbf{j} + 3t^2\mathbf{k}, \\ \ddot{\mathbf{r}} &= 2\mathbf{i} + 6t\mathbf{k}, \\ \dddot{\mathbf{r}} &= 6\mathbf{k}.\end{aligned}$$

Thus at $t = 1$,

$$\dot{\mathbf{r}} = 2\mathbf{i} - 3\mathbf{j} + 3\mathbf{k},$$
$$\ddot{\mathbf{r}} = 2\mathbf{i} + 6\mathbf{k},$$
$$\dddot{\mathbf{r}} = 6\mathbf{k};$$

$$\dot{\mathbf{r}} \times \ddot{\mathbf{r}} = \begin{vmatrix} \mathbf{i} & \mathbf{j} & \mathbf{k} \\ 2 & -3 & 3 \\ 2 & 0 & 6 \end{vmatrix} = -18\mathbf{i} - 6\mathbf{j} + 6\mathbf{k},$$

$$|\dot{\mathbf{r}} \times \ddot{\mathbf{r}}| = 6\sqrt{9 + 1 + 1} = 6\sqrt{11},$$
$$|\dot{\mathbf{r}}| = \sqrt{4 + 9 + 9} = \sqrt{22}.$$

Hence

$$\kappa = \frac{|\dot{\mathbf{r}} \times \ddot{\mathbf{r}}|}{|\dot{\mathbf{r}}|^3} = \frac{6\sqrt{11}}{22\sqrt{22}} = \frac{3}{11\sqrt{2}}$$

and

$$\tau = \frac{(\dot{\mathbf{r}} \times \ddot{\mathbf{r}}) \cdot \dddot{\mathbf{r}}}{|\dot{\mathbf{r}} \times \ddot{\mathbf{r}}|^2} = \frac{(-18)(0) + (-6)(0) + 6 \cdot 6}{36 \cdot 11}$$
$$= \frac{36}{36 \cdot 11} = \frac{1}{11}. \quad \square$$

*THE FRENET FORMULAS

Recall that

$$\frac{d\mathbf{t}}{ds} = \kappa\mathbf{n} \quad \text{and} \quad \frac{d\mathbf{b}}{ds} = -\tau\mathbf{n}. \tag{7}$$

We would like to find $d\mathbf{n}/ds$, so that we know the rate of change of all three of our local coordinate unit vectors with respect to arc length as we move along the curve. We differentiate

$$\mathbf{n} = \mathbf{b} \times \mathbf{t}$$

with respect to s and obtain

$$\frac{d\mathbf{n}}{ds} = \mathbf{b} \times \frac{d\mathbf{t}}{ds} + \frac{d\mathbf{b}}{ds} \times \mathbf{t}. \tag{8}$$

Making use of Eqs. (2) and (7), we obtain

$$\frac{d\mathbf{n}}{ds} = \mathbf{b} \times (\kappa\mathbf{n}) - (\tau\mathbf{n}) \times \mathbf{t} = \kappa(\mathbf{b} \times \mathbf{n}) - \tau(\mathbf{n} \times \mathbf{t}) = -\kappa\mathbf{t} + \tau\mathbf{b}. \tag{9}$$

Equations (7) and (9) are known as the *Frenet formulas*. We collect them in one place.

$$\frac{d\mathbf{t}}{ds} = \kappa\mathbf{n}, \qquad \frac{d\mathbf{n}}{ds} = -\kappa\mathbf{t} + \tau\mathbf{b}, \qquad \frac{d\mathbf{b}}{ds} = -\tau\mathbf{n}$$

Frenet formulas

In Exercise 35, you are asked to use the Frenet formulas to show that the shape of a space curve is completely determined by the curvature and the torsion as functions of the arc length traveled.

SUMMARY

1. For a space curve with unit tangent vector **t** and principal normal vector **n**, the unit binormal vector is defined by $\mathbf{b} = \mathbf{t} \times \mathbf{n}$.
2. **b** may be computed using

$$\mathbf{b} = \frac{1}{|\mathbf{v} \times \mathbf{a}|}(\mathbf{v} \times \mathbf{a}).$$

3. The torsion τ of the curve is defined by

$$\frac{d\mathbf{b}}{ds} = -\tau\mathbf{n}.$$

4. Curvature κ and torsion τ may also be found from the position vector **r** using

$$\kappa = \frac{|\dot{\mathbf{r}} \times \ddot{\mathbf{r}}|}{|\dot{\mathbf{r}}|^3} \quad \text{and} \quad \tau = \frac{(\dot{\mathbf{r}} \times \ddot{\mathbf{r}}) \cdot \dddot{\mathbf{r}}}{|\dot{\mathbf{r}} \times \ddot{\mathbf{r}}|^2}.$$

5. The Frenet formulas are

$$\frac{d\mathbf{t}}{ds} = \kappa\mathbf{n}, \qquad \frac{d\mathbf{n}}{ds} = -\kappa\mathbf{t} + \tau\mathbf{b}, \qquad \frac{d\mathbf{b}}{ds} = -\tau\mathbf{n}.$$

EXERCISES

***1.** The torsion of a straight line is not defined by $d\mathbf{b}/ds = -\tau\mathbf{n}$ since **n** and **b** are not defined for a line in space. What do you think the torsion of a straight line should be defined to be?

Exercises 2 through 6 are concerned with the space curve

$$x = 2t, \quad y = t^2, \quad z = -t^2.$$

***2.** Find the vectors **t**, **n**, and **b** at a point on the curve in terms of the parameter t.

***3.** Find the curvature κ of the curve in terms of the parameter t.

***4.** Find the equation of the osculating plane to the curve for a value t_0 of the parameter.

***5.** Find the torsion τ of the curve in terms of the parameter t.

***6.** From the answers to the two preceding exercises, it appears that the curve lies in a plane. Deduce this directly from the equations of the curve.

Exercises 7 through 10 are concerned with the twisted cubic

$$x = 3t, \quad y = 3t^2, \quad z = 2t^3.$$

Find the indicated quantity when $t = 1$.

***7.** The binormal vector **b**

***8.** The equation of the osculating plane

***9.** The curvature κ

***10.** The torsion τ

Exercises 11 through 15 are concerned with the space curve

$$x = \frac{1}{t}, \quad y = \frac{1}{t^2}, \quad z = \frac{2}{t},$$

where $t > 0$. Find the indicated quantity when $t = 1$.

***11.** The binormal vector **b**

***12.** The equation of the osculating plane

***13.** The curvature κ

***14.** The torsion τ

Exercises 15 through 18 are concerned with the space curve

$$x = e^t \sin t, \quad y = e^t \cos t, \quad z = e^{2t}.$$

Find the indicated quantity when $t = 0$.

***15.** The binormal vector **b**

***16.** The equation of the osculating plane

***17.** The curvature κ

***18.** The torsion τ

Exercises 19 through 22 are concerned with the space curve

$$x = \ln(\sin t), \quad y = \ln(\cos t), \quad z = \sin 2t$$

for $0 < t < \pi/2$. Find the indicated quantity when $t = \pi/4$.

***19.** The binormal vector **b**

***20.** The equation of the osculating plane

***21.** The curvature κ

***22.** The torsion τ

Exercises 23 through 26 are concerned with the space curve

$$x = \cos t, \quad y = e^{2t}, \quad z = (t+1)^3.$$

Find the indicated quantity when $t = 0$.

***23.** The binormal vector $\mathbf{b}$

***24.** The equation of the osculating plane

***25.** The curvature κ

***26.** The torsion τ

***27.** Show that $\ddot{\mathbf{r}} = \ddot{s}\mathbf{t} + \dot{s}^2\kappa\mathbf{n}$. [*Hint:* Differentiate $\dot{\mathbf{r}} = \dot{s}\mathbf{t}$.]

***28.** Show that $\dddot{\mathbf{r}} = (\dddot{s} - \dot{s}^3\kappa^2)\mathbf{t} + (3\dot{s}\ddot{s}\kappa + \dot{s}^2\dot{\kappa})\mathbf{n} + \dot{s}^3\kappa\tau\mathbf{b}$. [*Hint:* Differentiate the result in Exercise 27.]

***29.** Deduce from the preceding two exercises that
a) $\dot{\mathbf{r}} \times \ddot{\mathbf{r}} = \dot{s}^3\kappa\mathbf{b}$, b) $(\dot{\mathbf{r}} \times \ddot{\mathbf{r}}) \cdot \dddot{\mathbf{r}} = \dot{s}^6\kappa^2\tau$.

[*Hint:* Since $\mathbf{t}$, $\mathbf{n}$, $\mathbf{b}$ form a right-hand perpendicular unit 3-frame at each point, they may be used in the role of $\mathbf{i}$, $\mathbf{j}$, and $\mathbf{k}$ in computations of the cross product.]

***30.** Deduce from the preceding exercise that

$$\text{a) } \kappa = \frac{|\dot{\mathbf{r}} \times \ddot{\mathbf{r}}|}{|\dot{\mathbf{r}}|^3}, \qquad \text{b) } \tau = \frac{(\dot{\mathbf{r}} \times \ddot{\mathbf{r}}) \cdot \dddot{\mathbf{r}}}{|\dot{\mathbf{r}} \times \ddot{\mathbf{r}}|^2}.$$

***31.** Show that if we let $\boldsymbol{\delta} = \tau\mathbf{t} + \kappa\mathbf{b}$ (the *Darboux vector*), then the Frenet formulas take the symmetric form

$$\frac{d\mathbf{t}}{ds} = \boldsymbol{\delta} \times \mathbf{t}, \quad \frac{d\mathbf{n}}{ds} = \boldsymbol{\delta} \times \mathbf{n}, \quad \frac{d\mathbf{b}}{ds} = \boldsymbol{\delta} \times \mathbf{b}.$$

***32.** Show that

$$\left|\frac{d\mathbf{n}}{ds}\right|^2 = \left|\frac{d\mathbf{t}}{ds}\right|^2 + \left|\frac{d\mathbf{b}}{ds}\right|^2.$$

***33.** Show that if the curvature of a smooth space curve is zero at each point, then the curve is a straight line. [*Hint:* Use the chain rule to show that $\mathbf{t} = d\mathbf{r}/ds$. Argue that $\kappa = 0$ then implies that x, y, and z are linear functions of the parameter s of arc length.]

***34.** Show that a space curve with torsion zero at each point lies in a plane. [*Hint:* Deduce that $\mathbf{b}$ is a constant vector. Show that $d(\mathbf{b} \cdot \mathbf{r})/ds = 0$, and conclude that $\mathbf{b} \cdot (\mathbf{r}(t) - \mathbf{r}(t_0)) = 0$, so that the curve lies in the plane through the point where $t = t_0$ having orthogonal vector $\mathbf{b}$.]

***35.** *Fundamental theorem:* Use the Frenet formulas to show that two space curves having the same curvature and torsion for each value of the arc length parameter s are congruent. [*Hint:* You may suppose that the position vectors of the curves are given by $\mathbf{r}(s)$ and $\bar{\mathbf{r}}(s)$, and that

$$\mathbf{r}(0) = \bar{\mathbf{r}}(0), \quad \mathbf{t}(0) = \bar{\mathbf{t}}(0), \quad \mathbf{n}(0) = \bar{\mathbf{n}}(0),$$

and

$$\mathbf{b}(0) = \bar{\mathbf{b}}(0).$$

You must then show that $\mathbf{r}(s) = \bar{\mathbf{r}}(s)$. Set $w = \mathbf{t} \cdot \bar{\mathbf{t}} + \mathbf{n} \cdot \bar{\mathbf{n}} + \mathbf{b} \cdot \bar{\mathbf{b}}$, and show that $dw/ds = 0$. Show that $\mathbf{w}(0) = 3$, and conclude that

$$\mathbf{t} = \bar{\mathbf{t}}, \quad \mathbf{n} = \bar{\mathbf{n}}, \quad \text{and} \quad \mathbf{b} = \bar{\mathbf{b}}$$

for all values of s. From $\mathbf{t} = \bar{\mathbf{t}}$, deduce that $\mathbf{r} = \bar{\mathbf{r}} + \mathbf{c}$, and, taking $s = 0$, that $\mathbf{c} = \mathbf{0}$, so that $\mathbf{r}(s) = \bar{\mathbf{r}}(s)$.]

EXERCISE SETS FOR CHAPTER 15

Review Exercise Set 15.1

Exercises 1 through 8 concern a body moving in the plane with position at time t described parametrically by $x = \sin 2t$, $y = \cos t$. Find the indicated quantity when $t = \pi/4$.

1. The position vector $\mathbf{r}$

2. The velocity vector $\mathbf{v}$

3. The speed

4. The acceleration vector $\mathbf{a}$

5. The tangential vector component of $\mathbf{a}$

6. The rate of change of speed

7. The normal vector component of $\mathbf{a}$

8. The curvature κ

***9.** If the polar coordinate position of a body in the plane at time t is given by $(r, \theta) = (\sqrt{t}, \sqrt{3t+4})$, find the velocity and acceleration vectors in terms of the unit vectors $\mathbf{u}_r$ and $\mathbf{u}_\theta$ at time $t = 4$.

***10.** State Kepler's laws.

Review Exercise Set 15.2

Exercises 1 through 8 concern a body moving in space with position at time t described parametrically by $x = t$, $y = 3 \sin t$, $z = -3 \cos t$. Find the indicated quantity when $t = \pi/4$.

1. The velocity vector $\mathbf{v}$

2. The speed

3. The equation of the osculating plane

4. The rate of change of speed

5. The unit normal vector $\mathbf{n}$

6. The curvature κ

***7.** The unit binormal vector $\mathbf{b}$

***8.** The torsion τ

***9.** What type of curve is traveled by a body moving subject to a central force field?

10. Explain how the curvature of a space curve is defined.

Review Exercise Set 15.3

Exercises 1 through 6 concern the motion of a body in the plane with position vector

$$\mathbf{r} = [\ln(\sin 2t)]\mathbf{i} + [\ln(\cos 2t)]\mathbf{j}.$$

Find the indicated quantity when $t = \pi/8$.

1. The speed
2. The unit tangent vector $\mathbf{t}$
3. The unit normal vector $\mathbf{n}$
4. The rate of change of speed
5. The magnitude $|\mathbf{a}_{\text{nor}}|$ of the normal vector component of $\mathbf{a}$
6. The curvature κ
7. Let a body be moving in the plane subject to some force. Show that if

$$\frac{d^2s/dt^2}{\kappa(ds/dt)^2} = 1$$

at all times t, then the force must be directed at an angle of 45° to the direction of motion of the body at every time t.

8. For a body traveling a circular track at 20 ft/sec, a force of 500 lb perpendicular to the track is required to keep the body from leaving the track. If the maximum force the track can exert against the body in this perpendicular direction is 4500 lb, how fast can the body travel without leaving the track?

*9. Give the polar formula for $\mathbf{u}_r$ and $\mathbf{u}_\theta$ components of acceleration for a body traveling a curve in the plane.

*10. State Newton's law of universal gravitation.

Review Exercise Set 15.4

Exercises 1 through 8 concern the motion of a body in space with position vector

$$\mathbf{r} = t^3\mathbf{i} + 2t\mathbf{j} + t^2\mathbf{k}.$$

Find the indicated quantity when $t = -1$.

1. The speed
2. The unit tangent vector $\mathbf{t}$
3. The tangential vector component $\mathbf{a}_{\tan}$ of $\mathbf{a}$
4. The rate of change of speed
5. The equation of the osculating plane

*6. The unit binormal vector $\mathbf{b}$

7. The curvature κ

*8. The torsion τ

*9. Two satellites A and B are traveling elliptic orbits about the same body. If the major axis of A's orbit is four times the length of the major axis of B's orbit, and if A's period is 48 hours, find B's period.

*10. Explain how the torsion of a space curve is defined.

DIFFERENTIAL CALCULUS OF SEVERAL VARIABLES 16

16.1 PARTIAL DERIVATIVES

FUNCTIONS OF SEVERAL VARIABLES

Until now, we have managed to work primarily with functions of a single variable, usually denoted by $y = f(x)$. It is amazing how much we were able to accomplish with this restriction, for so many things depend on more than one parameter. To take a practical illustration, a tailor making a new suit for a person needs to know many parameters, including measurements of waist, hips, inseam, sleeve, chest, shoulders, and so on.

Pictures helped us understand calculus of functions of one variable. We often sketched a graph of a function f, consisting of points (x, y) in the plane where $y = f(x)$. The next-easiest case is a function of two variables, which we will generally write as $z = f(x, y)$. The helpful graph for such a function consists of all points (x, y, z) in space, where $z = f(x, y)$. We will also deal with functions of three or more variables. A function of three variables is often written as $w = f(x, y, z)$. This time, the graph consists of points (x, y, z, w) in four-dimensional space such that $w = f(x, y, z)$. This is almost hopeless to sketch in a plane, and we will make no attempt to do so. However, the concept is no more mysterious than functions of one or two variables.

When discussing functions of many variables, we often turn to subscripted notation and write $y = f(x_1, x_2, \ldots, x_n)$ for a function of n variables. We will see that it is sometimes convenient to view y as a function of a single *vector variable* $\mathbf{x} = (x_1, x_2, \ldots, x_n)$ and to use the notation $y = f(\mathbf{x})$.

We could make a long list of functions of more than one variable that appear naturally. For example, the volume of a cylinder of radius r and altitude h is $V = f(r, h) = \pi r^2 h$. We will refer to one particular illustrative situation often as we develop calculus of several variables. Suppose that at a particular moment the temperature at a point in space is a function of just the position in space. If rectangular coordinates are introduced, we may denote this temperature function by $T(x, y, z)$. If we wish to illustrate a two-variable situation, we use $T(x, y)$ for the temperature at a point in the x, y-plane.

In one simple type of situation, temperature in space is entirely controlled by a single source of heat at the origin. The temperature is then inversely proportional to the square of the distance from the origin, that is

$$T(x, y, z) = \frac{k}{x^2 + y^2 + z^2}$$

for some constant k. This is due to the fact that a unit of heat emanating from the origin is spread evenly over the surface of a sphere having area $4\pi r^2$ at the distance r from the origin. If we view the x, y-plane as part of space, then the temperature at a point in the plane is given by

$$T(x, y) = \frac{k}{x^2 + y^2}.$$

EXAMPLE 1 Suppose the temperature in space depends only on a single source of heat at the origin, and suppose the temperature at $(1, -1, 3)$ is 23°C. Find the temperature at the point $(2, -3, 1)$.

Solution We have

$$T(x, y, z) = \frac{k}{x^2 + y^2 + z^2}.$$

Now

$$23 = T(1, -1, 3) = \frac{k}{1 + 1 + 9} = \frac{k}{11},$$

so $k = 23 \cdot 11 = 253$. Thus

$$T(2, -3, 1) = \frac{253}{4 + 9 + 1} = \frac{253}{14} \approx 18.07°\text{C}. \quad \square$$

CONTINUITY AND GRAPHS OF FUNCTIONS OF TWO VARIABLES

Let $z = f(x, y)$ be a function of two independent variables x and y. For example, perhaps $z = f(x, y) = x^2 - 3xy$. We informally mentioned the *graph* of such a function above.

DEFINITION 16.1 Graph

The **graph** of $z = f(x, y)$ consists of all points (x, y, z) in space such that $z = f(x, y)$.

The function is *continuous at a point* if sufficiently small changes Δx in x and Δy in y produce only very small changes Δz in z. We give a formal definition of continuity.

DEFINITION 16.2 Continuity

The function $f(x, y)$ is **continuous at a point** (x_0, y_0) in its domain if, given any $\varepsilon > 0$, there is some $\delta > 0$ such that $|f(x, y) - f(x_0, y_0)| < \varepsilon$ for all (x, y) satisfying $\sqrt{(x - x_0)^2 + (y - y_0)^2} < \delta$. If f is continuous at each point in its domain, then it is a **continuous function.**

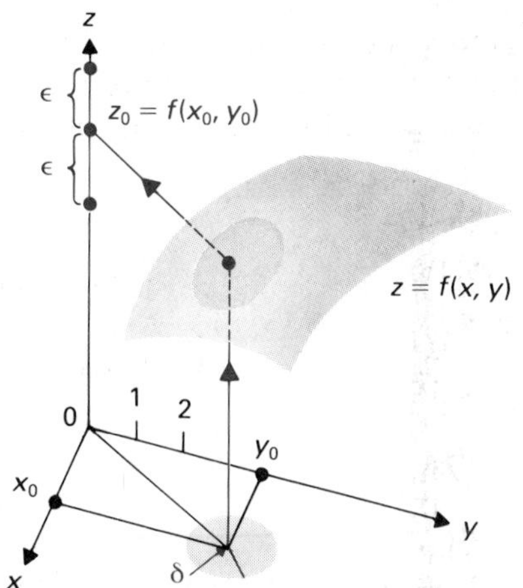

Figure 16.1 All points in the red disk in the x,y-plane must be carried by f into the red interval on the z-axis.

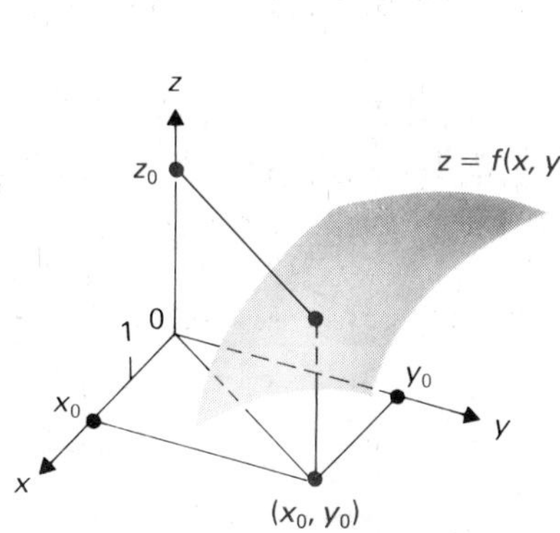

Figure 16.2 The graph of a continuous function $f(x, y)$.

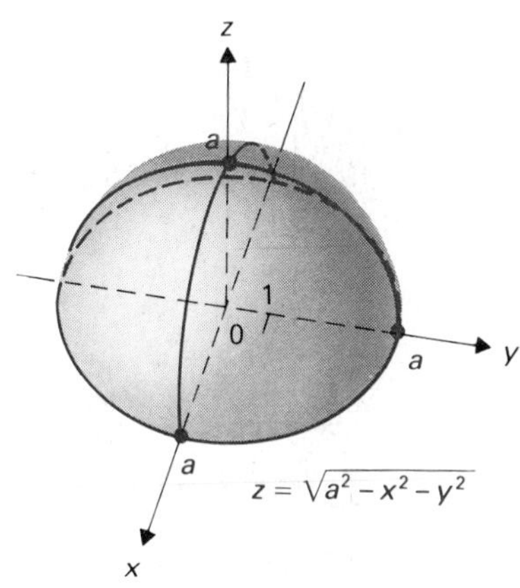

Figure 16.3 The graph of $f(x, y) = \sqrt{a^2 - x^2 - y^2}$.

Figure 16.1 shows how we should interpret the measurement ε on the z-axis and the measurement δ in the x, y-plane. The definition asserts that the function f carries points in the x, y-plane that are within the distance δ of (x_0, y_0) to points on the z-axis within the distance ε of $z = f(x_0, y_0)$. Once again, it can be shown that all the elementary functions (rational functions, trigonometric functions, exponential and logarithmic functions) are continuous. For example, $x^3 - 2xy$, $\cos(xy)$, and $e^{x^2+y^2}$ are continuous functions of x and y.

We will not talk anymore about continuity, but simply state that, if a function $f(x, y)$ is continuous, then its graph can be regarded as a surface lying over (or sometimes under, depending on the sign of z) its domain in the x, y-plane. A picture of such a surface is shown in Fig. 16.2.

EXAMPLE 2 Sketch the graph of $z = f(x, y) = \sqrt{a^2 - x^2 - y^2}$.

Solution Since $x^2 + y^2 + z^2 = a^2$ is the equation of a sphere with center at the origin and radius a we see that $z = f(x, y) = \sqrt{a^2 - x^2 - y^2}$ has the upper hemisphere as its graph. The graph is shown in Fig. 16.3. □

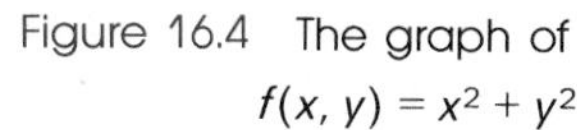
Figure 16.4 The graph of $f(x, y) = x^2 + y^2$.

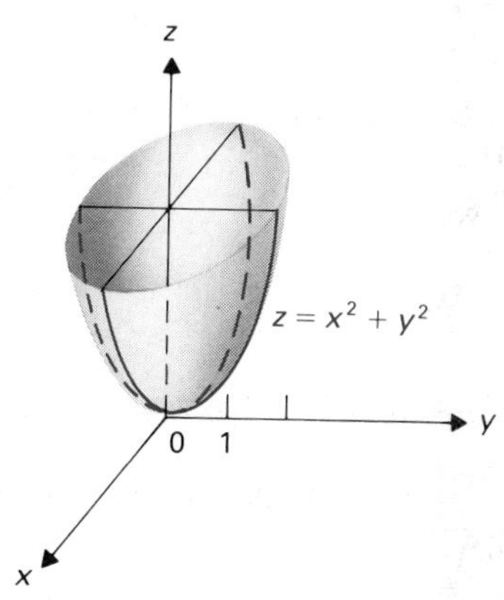

EXAMPLE 3 Sketch the graph of $z = f(x, y) = x^2 + y^2$.

Solution The graph is the circular paraboloid shown in Fig. 16.4. One of the reasons we spent time sketching quadric surfaces was to be able to sketch the graphs of a few nonlinear functions of two variables. Of course, the graph of every linear function $z = g(x, y) = ax + by + c$ is a plane in space. □

PARTIAL DERIVATIVES

If $y = f(x)$, the derivative dy/dx gives the rate of change of y with respect to x. For a function of one variable, there are only two directions we can travel from a point in the domain, in the direction of increasing x or in the

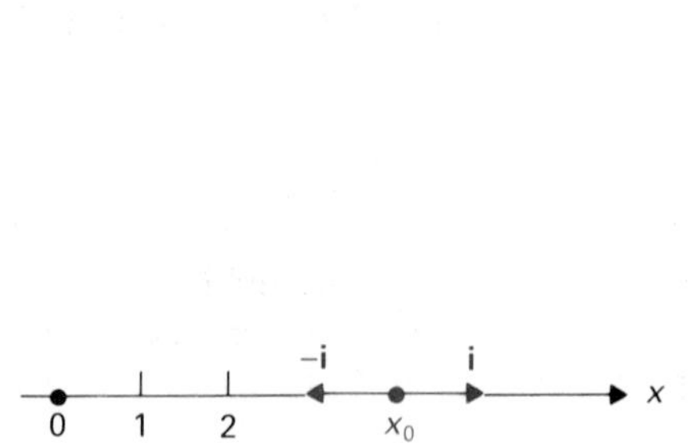

Figure 16.5 **i** and $-$**i** give the only directions on the x-axis from x_0.

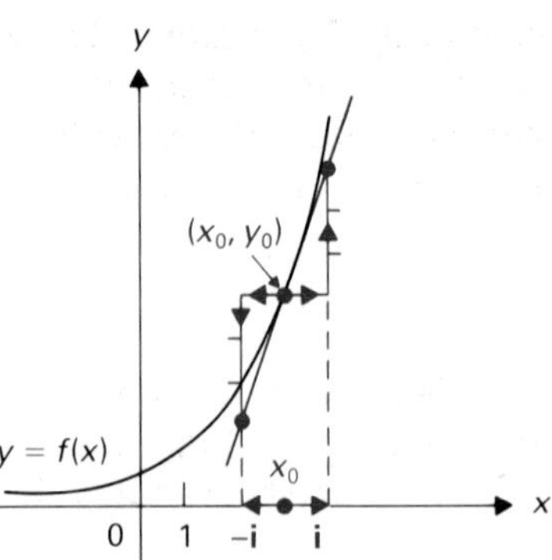

Figure 16.6 Rate of change of $f(x)$ at x_0 is 3 in the direction **i**, -3 in the direction $-$**i**.

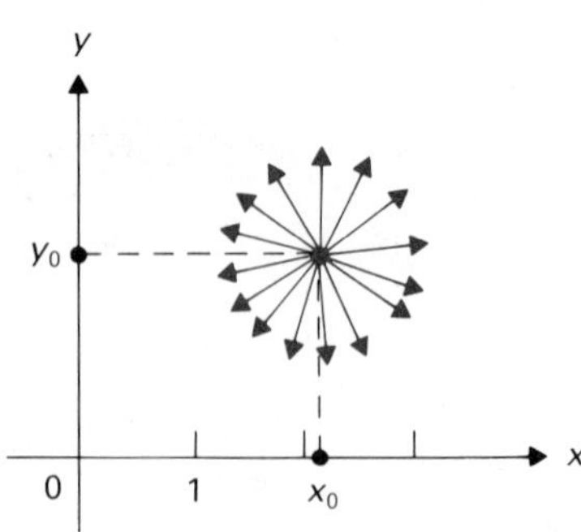

Figure 16.7 There are many directions from (x_0, y_0) in the plane.

direction of decreasing x. We can describe these two directions by the vectors **i** and $-$**i**, respectively, as shown in Fig. 16.5. Suppose the rate of change of $y = f(x)$ at x_0 is 3 in the direction of **i**, so that $f'(x_0) = 3$ as in Fig. 16.6. Then the rate of change in the direction of $-$**i** is -3; that is, the function is *decreasing* at an instantaneous rate of 3 units per unit *decrease* in x there.

If $z = f(x, y)$, we can also attempt to find some rates of change. This time there are many directions we can go from a point (x_0, y_0) in the domain. See Fig. 16.7. We will eventually show how to find the rate at which z changes in each of these directions.

First we will find the rate of change in the direction parallel to the x-axis given by increasing x as y is held constant. That is, we will find the rate of change at (x_0, y_0) in the direction given by the vector **i** at that point. Suppose $z = f(x, y)$ has as graph the surface shown in Fig. 16.8. The portion of the surface over the line $y = y_0$ in the x,y-plane is a curve on the surface, as shown in the figure. On the curve, the height z appears as a function $g(x)$ of x only, as indicated in Fig. 16.9. This function $g(x)$ is, of course, found by setting $y = y_0$ in $z = f(x, y)$, that is,

$$z = g(x) = f(x, y_0).$$

Since we have a function of one variable $z = g(x)$, we can easily find the rate of change $g'(x)$ of z with respect to x. This time we denote the rate of

Figure 16.8 The portion of the graph where $y = y_0$.

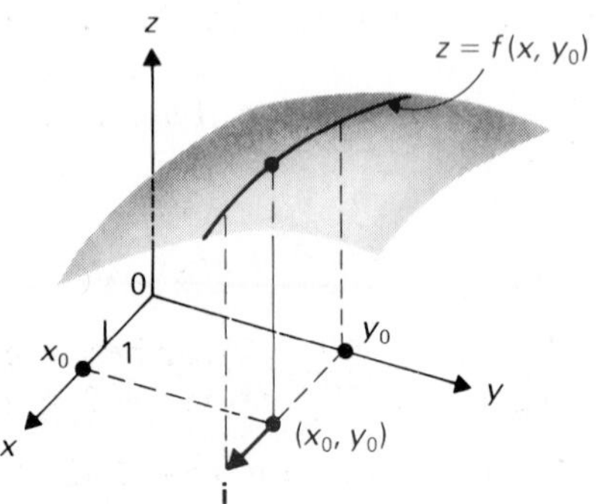

Figure 16.9 The graph of $z = g(x) = f(x, y_0)$.

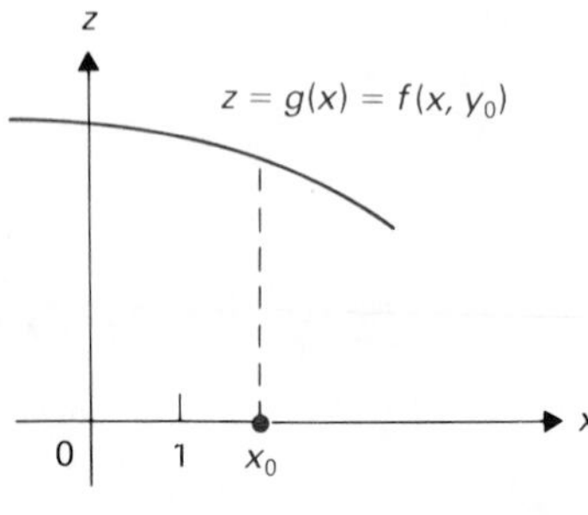

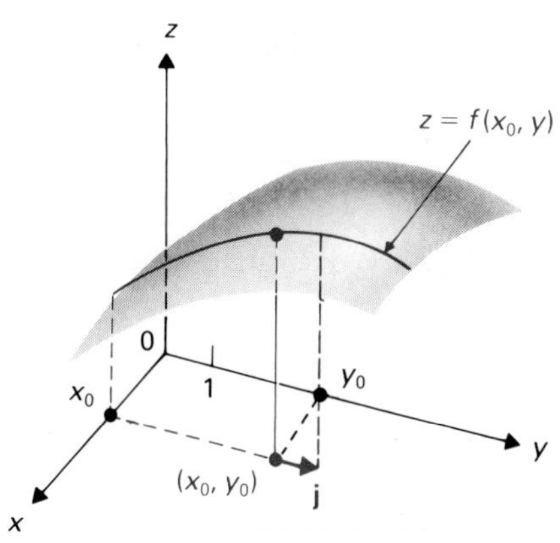

Figure 16.10 The portion of the graph where $x = x_0$.

change by $\partial z/\partial x$ rather than dz/dx. The "round d's" remind us that z is actually a function of more than one variable. We call $\partial z/\partial x$ the **partial derivative** of z with respect to x. The value of this derivative at $x = x_0$ (recall that $y = y_0$) is written

$$\left.\frac{\partial z}{\partial x}\right|_{(x_0, y_0)} \quad \text{or} \quad \left.\frac{\partial f}{\partial x}\right|_{(x_0, y_0)} \quad \text{or} \quad f_x(x_0, y_0).$$

We use the function $z = h(y) = f(x_0, y)$ obtained by holding x constant at x_0 to find the rate of change $h'(y)$ of $z = f(x, y)$ at (x_0, y_0) in the direction given by the vector **j** at that point. See Fig. 16.10. The notations are

$$\left.\frac{\partial z}{\partial y}\right|_{(x_0, y_0)} \quad \text{or} \quad \left.\frac{\partial f}{\partial y}\right|_{(x_0, y_0)} \quad \text{or} \quad f_y(x_0, y_0).$$

The derivative of $g(x) = f(x, y_0)$ at x_0 is defined as the *limit*

$$\lim_{\Delta x \to 0} \frac{g(x_0 + \Delta x) - g(x_0)}{\Delta x}.$$

Note that $g(x_0 + \Delta x) - g(x_0) = f(x_0 + \Delta x, y_0) - f(x_0, y_0)$. We are led to the following definition of the partial derivatives as limits.

DEFINITION 16.3 Partial derivative

Let $f(x, y)$ be defined everywhere inside some sufficiently small circle with (x_0, y_0) as center. Then the **partial derivative of f with respect to x** is

$$\left.\frac{\partial f}{\partial x}\right|_{(x_0, y_0)} = f_x(x_0, y_0) = \lim_{\Delta x \to 0} \frac{f(x_0 + \Delta x, y_0) - f(x_0, y_0)}{\Delta x}$$

and the **partial derivative of f with respect to y** is

$$\left.\frac{\partial f}{\partial y}\right|_{(x_0, y_0)} = f_y(x_0, y_0) = \lim_{\Delta y \to 0} \frac{f(x_0, y_0 + \Delta y) - f(x_0, y_0)}{\Delta y},$$

if these limits exist.

COMPUTATION OF PARTIAL DERIVATIVES

We can regard $f_x(x_0, y_0)$ as the derivative at x_0 of the function g given by $g(x) = f(x, y_0)$, that is, the function of one real variable x obtained from f by keeping $y = y_0$ and allowing only x to vary. This means that partial derivatives can be computed using the techniques for finding derivatives of functions of one variable. This is illustrated in the following examples.

EXAMPLE 4 Let f be the polynomial function given by $f(x, y) = x^2 + 3xy + 2y^3$. Find

$$f_x(1, 2) = \left.\frac{\partial f}{\partial x}\right|_{(1, 2)} \quad \text{and} \quad f_y(1, 2) = \left.\frac{\partial f}{\partial y}\right|_{(1, 2)}.$$

Solution Now $f_x(1, 2)$ can be viewed as the derivative at $x = 1$ of the function g obtained by setting $y = 2$ in the polynomial $x^2 + 3xy + 2y^3$. That is,

$$g(x) = x^2 + 3x(2) + 2(2^3) = x^2 + 6x + 16.$$

The derivative of $x^2 + 6x + 16$ is $2x + 6$, and the value of $2x + 6$ when $x = 1$ is 8. Hence

$$f_x(1, 2) = \left.\frac{\partial f}{\partial x}\right|_{(1,2)} = 8.$$

Similarly, putting $x = 1$, we find that $f_y(1, 2)$ is the derivative, when $y = 2$, of $1 + 3y + 2y^3$. Since

$$\frac{d(1 + 3y + 2y^3)}{dy} = 3 + 6y^2,$$

we obtain

$$f_y(1, 2) = \frac{\partial f}{\partial y}(1, 2) = 3 + 6 \cdot 2^2 = 27. \quad \square$$

We could simplify the computation of the partial derivatives in Example 4 by noting that we can compute $f_x(x, y)$ for any point (x, y) by differentiating $x^2 + 3xy + 2y^3$ "with respect to x only," treating y as a constant. The notation "$\partial f/\partial x$" is a practical way to indicate this derivative "with respect to x."

EXAMPLE 5 Repeat Example 4, using the technique just described.

Solution Differentiating $f(x, y) = x^2 + 3xy + 2y^3$ with respect to x, regarding y as a constant, we obtain

$$\frac{\partial f}{\partial x} = \frac{\partial(x^2 + 3xy + 2y^3)}{\partial x} = 2x + 3y,$$

for $\partial(2y^3)/\partial x = 0$, since we think of y as a constant. Similarly,

$$\frac{\partial f}{\partial y} = \frac{\partial(x^2 + 3xy + 2y^3)}{\partial y} = 3x + 6y^2.$$

Therefore

$$\frac{\partial f}{\partial x}(1, 2) = 2 \cdot 1 + 3 \cdot 2 = 8 \quad \text{and} \quad \frac{\partial f}{\partial y}(1, 2) = 3 \cdot 1 + 6 \cdot 2^2 = 27. \quad \square$$

In practice, we always find partial derivatives by the technique illustrated in Example 5, and the ∂-notation is obviously suggestive here; we differentiate "with respect to x" or "with respect to y." For instance, to find $\partial f/dx$, just differentiate f with respect to x only, pretending that y is constant.

EXAMPLE 6 Find $\partial z/\partial x$ and $\partial z/\partial y$ if $z = xe^{xy^2}$.

Solution Regarding y as a constant, we obtain

$$\frac{\partial z}{\partial x} = x \cdot \frac{\partial(e^{xy^2})}{\partial x} + e^{xy^2} \cdot \frac{\partial(x)}{\partial x} = x \cdot e^{xy^2} \cdot y^2 + e^{xy^2} \cdot 1$$

$$= e^{xy^2}(xy^2 + 1).$$

Similarly, regarding x as a constant, we have

$$\frac{\partial z}{\partial y} = x \cdot e^{xy^2} \cdot 2xy = 2x^2ye^{xy^2}. \quad \square$$

As derivatives, $\partial f/\partial x$ and $\partial f/\partial y$ give the rates of change of $f(x, y)$ as we go in the directions of the vectors **i** and **j**, respectively. We give two-variable approaches to some problems that we could solve also by one-variable techniques.

EXAMPLE 7 Find the rate of change of the volume with respect to the radius r of a right circular cylinder when the radius is 4 ft if the height remains constant at 20 ft.

Solution The volume of the cylinder is given by the formula

$$V = f(r, h) = \pi r^2 h.$$

The rate of change of V per unit increase in r as h is held constant is precisely $\partial V/\partial r$. When $r = 4$ and $h = 20$, we have

$$\begin{aligned} \left.\frac{\partial V}{\partial r}\right|_{(4,20)} &= 2\pi rh|_{(4,20)} = 2\pi \cdot 4 \cdot 20 \\ &= 160\pi \text{ ft}^3/(\text{Ft increase in } r). \quad \square \end{aligned}$$

EXAMPLE 8 The temperature at a point in the x, y-plane is given by

$$T(x, y) = \frac{100}{x^2 + y^2} \text{ degrees centigrade.}$$

Find the rate of change in temperature with respect to y only, that is, in the direction of **j**, at the point (3, 4).

Solution Since x remains constant at 3 in the direction of **j**, we see that our answer is given by

$$\left.\frac{\partial T}{\partial y}\right|_{(3,4)} = \left.\frac{0 - 100(2y)}{(x^2 + y^2)^2}\right|_{(3,4)} = \frac{-800}{25^2} = -\left(\frac{32}{25}\right)°\text{C}. \quad \square$$

FUNCTIONS OF MORE VARIABLES

If (x_0, y_0, z_0) is a point in the domain of a function $w = f(x, y, z)$, then we can try to find the partial derivatives of w in the directions of **i**, **j**, and **k** at (x_0, y_0, z_0). To find the derivative in the direction given by **i**, we should set $y = y_0$ and $z = z_0$, allowing only x to vary. That is, we hold y and z constant and differentiate just with respect to x to find

$$\left.\frac{\partial w}{\partial x}\right|_{(x_0, y_0, z_0)} \quad \text{or} \quad f_x(x_0, y_0, z_0).$$

Similarly, we can find $\partial w/\partial y$ and $\partial w/\partial z$.

EXAMPLE 9 Find the three partial derivatives for $w = x \sin yz$.

Solution We easily see that

$$\frac{\partial w}{\partial x} = \sin yz, \qquad \frac{\partial w}{\partial y} = xz \cos yz, \qquad \frac{\partial w}{\partial z} = xy \cos yz. \quad \square$$

For functions of two or more variables, it is sometimes handy to use the subscripted notation presented in Chapter 14, where a point in the plane is (x_1, x_2) and a point in space is (x_1, x_2, x_3). The vector notation $\mathbf{x}$ may be used to represent either of these points; the dimension should be clear from the context. Thus a function of three variables may be written

$$y = f(\mathbf{x}) = f(x_1, x_2, x_3).$$

Of course, then we can compute all the partial derivatives

$$\frac{\partial y}{\partial x_1}, \quad \frac{\partial y}{\partial x_2}, \quad \frac{\partial y}{\partial x_3},$$

and $\partial y/\partial x_i$ is also written $\partial f/\partial x_i$ or $f_{x_i}(x_1, x_2, x_3)$.

EXAMPLE 10 Find f_{x_1} and f_{x_3} if $f(x) = x_1 x_2^2/x_3$.

Solution Regarding x_2 and x_3 as constants, we have

$$f_{x_1} = \frac{\partial f}{\partial x_1} = \frac{\partial[(x_2^2/x_3)x_1]}{\partial x_1} = \frac{x_2^2}{x_3}.$$

Regarding x_1 and x_2 as constants, we have

$$f_{x_3} = \frac{\partial f}{\partial x_3} = \frac{\partial[x_1 x_2^2(1/x_3)]}{\partial x_3}$$

$$= x_1 x_2^2 \cdot \frac{-1}{x_3^2} = -\frac{x_1 x_2^2}{x_3^2}. \quad \square$$

HIGHER-ORDER DERIVATIVES

Let $z = f(x, y)$. Then $\partial z/\partial x = f_x(x, y)$ is again a function of two variables and we can attempt to compute its partial derivatives with respect to either x or y. These are *second-order derivatives* of our original function $f(x, y)$. The notations are

$$\frac{\partial}{\partial x}\left(\frac{\partial z}{\partial x}\right) = \frac{\partial^2 z}{\partial x^2} = f_{xx}(x, y) \quad \text{and} \quad \frac{\partial}{\partial y}\left(\frac{\partial z}{\partial x}\right) = \frac{\partial^2 z}{\partial y\, \partial x} = f_{xy}(x, y).$$

The derivative f_{xy} is sometimes called a *mixed* or *cross* second partial derivative, since derivatives with respect to more than one variable are involved.

Of course, we could equally well find first $\partial z/\partial y$ and then the second-order derivatives

$$\frac{\partial}{\partial y}\left(\frac{\partial z}{\partial y}\right) = \frac{\partial^2 z}{\partial y^2} = f_{yy}(x, y) \quad \text{and} \quad \frac{\partial}{\partial x}\left(\frac{\partial z}{\partial y}\right) = \frac{\partial^2 z}{\partial x\, \partial y} = f_{yx}(x, y).$$

Note that f_{xy} means $(f_x)_y$ while f_{yx} means $(f_y)_x$.

It is a theorem that, if $z = f(x, y)$ has continuous second partial derivatives, then the "mixed" partial derivatives are equal, that is,

$$\boxed{\frac{\partial^2 z}{\partial y\, \partial x} = \frac{\partial^2 z}{\partial x\, \partial y}.}$$

This hypothesis of continuity is true for all elementary functions with which we will work.

EXAMPLE 11 Illustrate that $f_{xy} = f_{yx}$ for $f(x, y) = \sin(x^2y)$.

Solution Now

$$f_x(x, y) = 2xy \cos(x^2y),$$

so

$$f_{xy}(x, y) = -2x^3y \sin(x^2y) + 2x \cos(x^2y).$$

Differentiating in the other order, we have

$$f_y(x, y) = x^2 \cos(x^2y),$$

so

$$f_{yx}(x, y) = -2x^3y \sin(x^2y) + 2x \cos(x^2y).$$

Thus $f_{xy} = f_{yx}$. □

Of course, we can continue taking derivatives. Thus $\partial^3 f/\partial y^2\, \partial x$ means the third partial derivative of $f(x, y)$, first with respect to x and then twice more with respect to y. We have

$$\boxed{\frac{\partial^3 f}{\partial y^2\, \partial x} = \frac{\partial^3 f}{\partial y\, \partial x\, \partial y} = \frac{\partial^3 f}{\partial x\, \partial y^2}}$$

for functions with continuous partial derivatives of order 3. Similar notations are used for functions of more than two variables. We state conditions for equality of mixed partial derivatives in the general case as a formal theorem. The proof is not given here.

THEOREM 16.1 Equality of mixed partials

Two mixed partial derivatives of the same order are equal if all partial derivatives of that order are continuous and the total number of differentiations with respect to each variable is the same in one mixed partial as in the other.

Occasionally, Theorem 16.1 can be used to simplify the computation of a higher-order derivative.

EXAMPLE 12 Let $w = f(x, y, z) = xy/(y^2 + z^4)$. Find f_{yzxx}.

Solution The notation f_{yzxx} means to differentiate first with respect to y, then with respect to z, and finally twice with respect to x. By Theorem 16.1, we can change the order and compute instead f_{xxyz}. Since

$$f_x = \frac{y}{y^2 + z^4} \qquad \text{and} \qquad f_{xx} = 0,$$

we see that $f_{xxyz} = 0$, so $f_{yzxx} = 0$ also. □

SUMMARY

1. If $z = f(x, y)$, then $\partial z/\partial x = f_x(x, y)$ is the partial derivative of $f(x, y)$ in the direction **i** corresponding to increasing x as y is held constant. It is computed by the usual differentiation methods, regarding y as a constant. The partial derivative $\partial z/\partial y = f_y(x, y)$ is similarly defined and computed.
2. If $y = f(\mathbf{x}) = f(x_1, x_2, x_3)$, then $\partial y/\partial x_i$ is computed by differentiating with respect to x_i only, treating all other variables as though they were constants in the differentiation.
3. If $z = f(x, y)$, then second-order partial derivatives are

$$\frac{\partial}{\partial x}\left(\frac{\partial z}{\partial x}\right) = \frac{\partial^2 z}{\partial x^2} = f_{xx}(x, y),$$

$$\frac{\partial}{\partial y}\left(\frac{\partial z}{\partial y}\right) = \frac{\partial^2 z}{\partial y^2} = f_{yy}(x, y),$$

together with the mixed partial derivatives

$$\frac{\partial}{\partial y}\left(\frac{\partial z}{\partial x}\right) = \frac{\partial^2 z}{\partial y\, \partial x} = f_{xy}(x, y)$$

and

$$\frac{\partial}{\partial x}\left(\frac{\partial z}{\partial y}\right) = \frac{\partial^2 z}{\partial x\, \partial y} = f_{yx}(x, y).$$

For all the functions that we will encounter, we have $\partial^2 z/\partial x\, \partial y = \partial^2 z/\partial y\, \partial x$. Similar notations are used for more variables and higher-order derivatives. For all the functions we will use, mixed partials of the same order are equal if differentiation with respect to each variable occurs the same total number of times. For example, if $w = f(x, y, z)$, we have $\partial^4 w/\partial x\, \partial y\, \partial x\, \partial z = \partial^4 w/\partial z\, \partial y\, \partial x^2$.

EXERCISES

In Exercises 1 through 21, find $f_x(x, y)$ and $f_y(x, y)$ for the given function $f(x, y)$. You need not simplify the answers.

1. $3x + 4y$ **2.** xy

3. $x^2 + y^2$ **4.** $xy^3 + x^2y^2$

5. $e^{x/y}$ **6.** $\dfrac{x^2 + 3x + 1}{y}$

7. $xy^2 + \dfrac{3x^2}{y^3}$ **8.** $(3xy^2 - 2x^2y)^3$

9. $(x^2 + 2xy)(y^3 + x^2)$ **10.** $(xy)^3(x^2 - y^3)^2$

11. $\dfrac{x^2 + y^2}{x^2 + y}$ **12.** $\sin xy$

13. $\tan (x^2 + y^2)$ **14.** $e^{xy} \cos x^2$

15. $e^{xy^2} \sec (x^2y)$ **16.** $\ln (2x + 3y)$

17. $\ln (2x + y) \cdot \cot y^2$ **18.** $(\sin x^2 + \cos y^2)^5$

19. $y \sec^3 x + xy^2$ **20.** $\ln [\sin (xy)]$

21. $\tan^{-1}(xy^2)$

In Exercises 22 through 30, find the rate of change of the function in the indicated direction at the given point.

22. $f(x, y) = \sqrt{x^2 - y^2}$ at $(5, -3)$ in the direction of **j**

23. $f(x, y) = \dfrac{x + y}{y^2}$ at $(1, -1)$ in the direction of **i**

24. $f(x, y, z) = \dfrac{x}{\sqrt{x^2 + y^2 + z^2}}$ at $(-2, 1, -2)$ in the direction of **j**

25. $f(x, y, z) = x \sin y^2 z$ at $(2, 3, 0)$ in the direction of **k**

26. $f(x, y, z) = x^2e^{xz}$ at $(3, -2, 4)$ in the direction of **j**

Solution We have

$$\frac{\partial z}{\partial x} = 2x + 3y \quad \text{and} \quad \frac{\partial z}{\partial y} = 3x + 6y^2,$$

so at (1, 2, 23), we have

$$\left.\frac{\partial z}{\partial x}\right|_{(1,2)} = 2 + 6 = 8$$

and

$$\left.\frac{\partial z}{\partial y}\right|_{(1,2)} = 3 + 24 = 27.$$

Therefore a vector normal to the plane is $8\mathbf{i} + 27\mathbf{j} - \mathbf{k}$. The equation of the plane is found by

Point (1, 2, 23)
$\perp$ ***vector*** $8\mathbf{i} + 27\mathbf{j} - \mathbf{k}$
Equation $8x + 27y - z = 39$ □

Of course, we can also find parametric equations of the normal line to a surface if we know a point and a vector parallel to the line, that is, normal to the surface.

EXAMPLE 2 Find parametric equations of the normal line to the surface $z = x^2 + 3xy + 2y^3$ of Example 1 at the point (1, 2, 23).

Solution By Example 1, we have

Point (1, 2, 23)
$\|$ ***vector*** $8\mathbf{i} + 27\mathbf{j} - \mathbf{k}$
Equations $x = 1 + 8t,$
$y = 2 + 27t,$
$z = 23 - t$ □

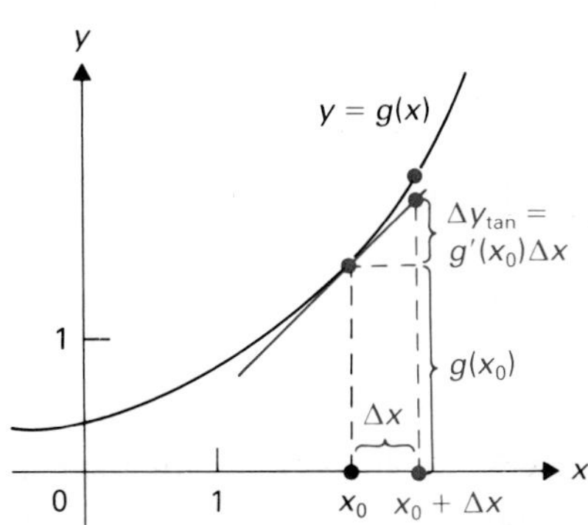

Figure 16.15 Recall the approximation $g(x_0 + \Delta x) \approx g(x_0) + g'(x_0)\,\Delta x$.

APPROXIMATIONS

In Chapter 3, we saw how to approximate $g(x_0 + \Delta x)$, where $g(x_0)$ is easily computed and $\Delta x = dx$ is small. Geometrically, we replaced the curve by its tangent line at $(x_0, g(x_0))$. We then found the value y on this tangent line corresponding to $x = x_0$, as shown in Fig. 16.15.

We are now working with a plane tangent to the surface graph of $z = f(x, y)$ at $(x_0, y_0, f(x_0, y_0))$. To strengthen our geometric intuition in preparation for the differential in the next section, we treat geometrically the analogous approximation of $f(x_0 + \Delta x, y_0 + \Delta y)$, where $f(x_0, y_0)$ is easily computed and Δx and Δy are small. This approximation will be treated again in the language of differentials in the next section, where we will describe conditions on f that guarantee that the approximation is a good one for sufficiently small values Δx and Δy.

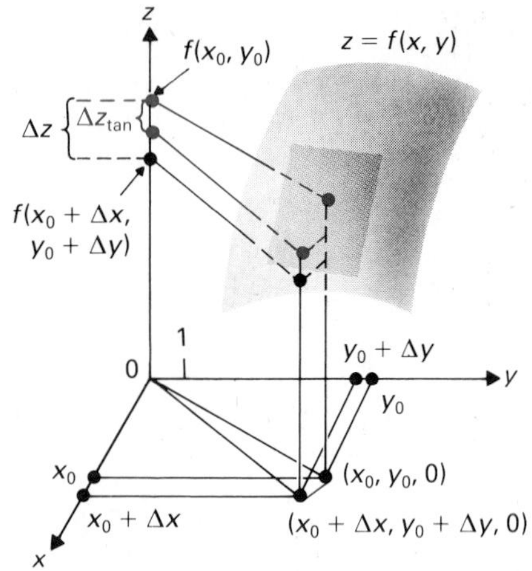

Figure 16.16 $\Delta z = f(x_0 + \Delta x, y_0 + \Delta y) - f(x_0, y_0)$; Δz_{tan} = (Change in height of the tangent plane).

Let $z = f(x, y)$. The equation of the tangent plane at a point (x_0, y_0, z_0) on the graph is found as follows:

Point (x_0, y_0, z_0)

⊥ vector $f_x(x_0, y_0)\mathbf{i} + f_y(x_0, y_0)\mathbf{j} - \mathbf{k}$

Equation $f_x(x_0, y_0)(x - x_0) + f_y(x_0, y_0)(y - y_0) - (z - z_0) = 0$

We write this equation in the form

$$z - z_0 = f_x(x_0, y_0)(x - x_0) + f_y(x_0, y_0)(y - y_0). \quad \textbf{(1)}$$

Let us set $\Delta x = x - x_0$, $\Delta y = y - y_0$, and $\Delta z_{tan} = z - z_0$ in Eq. (1). We use Δz_{tan} rather than Δz since it represents the change in z for the tangent plane, and we regard Δz as representing the change in z for $z = f(x, y)$. See Fig. 16.16. Then Eq. (1) becomes

$$\boxed{\Delta z_{tan} = f_x(x_0, y_0)\Delta x + f_y(x_0, y_0)\Delta y.} \quad \textbf{(2)}$$

We can regard Eq. (2) as the equation of the tangent plane with respect to local Δx, Δy, Δz-axes, as shown in Fig. 16.17.

Recall that for a function $y = g(x)$ of one variable, we have $\Delta y_{tan} = g'(x_0)\,\Delta x$, which leads to the approximation

$$g(x_0 + \Delta x) \approx g(x_0) + \Delta y_{tan} = g(x_0) + g'(x_0)\,\Delta x$$

shown in Fig. 16.15. Correspondingly, we now have

$$f(x_0 + \Delta x, y_0 + \Delta y) \approx f(x_0, y_0) + \Delta z_{tan}. \quad \textbf{(3)}$$

From Eqs. (2) and (3), we obtain the approximation formula

$$\boxed{\begin{aligned} f(x_0 + \Delta x, y_0 + \Delta y) \approx f(x_0, y_0) &+ f_x(x_0, y_0)\,\Delta x \\ &+ f_y(x_0, y_0)\,\Delta y, \end{aligned}} \quad \textbf{(4)}$$

where the approximation is best for small values of Δx and Δy.

EXAMPLE 3 Estimate $1.05^2 \cdot 2.99^3$.

Solution Let $f(x, y) = x^2y^3$. We know $f(1, 3) = 1^2 \cdot 3^3 = 27$. Let $\Delta x = 0.05$ and $\Delta y = -0.01$. Now

$$f_x = 2xy^3 \qquad \text{and} \qquad f_y = 3x^2y^2.$$

Then by the approximation (4), we have

$$\begin{aligned} f(1.05, 2.99) &= f(1 + \Delta x, 3 + \Delta y) \\ &\approx 27 + f_x(1, 3)(0.05) + f_y(1, 3)(-0.01) \\ &= 27 + 54(0.05) + 27(-0.01) = 27 + 2.70 - 0.27 \\ &= 29.43. \end{aligned}$$

Actually, $1.05^2 \cdot 2.99^3 = 29.4708161475$. □

Figure 16.17 The tangent plane has equation $\Delta z = f_x(x_0, y_0)\,\Delta x + f_y(x_0, y_0)\,\Delta y$.

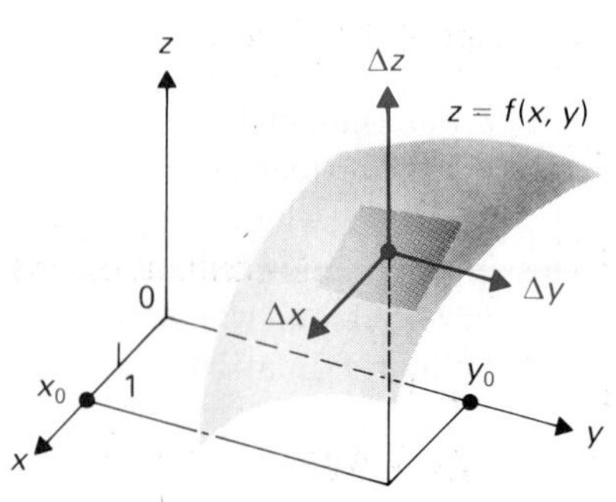

EXAMPLE 4 Find the approximate change in volume of a right circular cone of base radius 4 and altitude 10 if the radius of the base is decreased by 0.05 and the altitude is increased by 0.2.

Solution The volume V of a cone is given by

$$V = f(r, h) = \frac{1}{3}\pi r^2 h.$$

The approximate change in V is given by

$$\Delta V_{\tan} = \frac{\partial V}{\partial r}\Delta r + \frac{\partial V}{\partial h}\Delta h = \frac{2}{3}\pi r h\,\Delta r + \frac{1}{3}\pi r^2\,\Delta h.$$

Taking

$$r = 4, \qquad \Delta r = -0.05, \qquad h = 10, \qquad \text{and} \qquad \Delta h = 0.2,$$

we obtain as answer

$$\begin{aligned}\Delta V_{\tan} &= \left(\frac{2}{3}\pi \cdot 40\right)(-0.05) + \left(\frac{1}{3}\pi \cdot 16\right)(0.2) \\ &= \frac{2}{3}\pi(-2 + 1.6) \\ &= \frac{2}{3}\pi(-0.4) = -\frac{8\pi}{30}. \quad \square\end{aligned}$$

Similar results hold for functions of more variables. For example, if $w = f(x, y, z)$, then the analogue of the approximation (4) is

$$\boxed{\begin{aligned}f(x_0 + \Delta x, y_0 + \Delta y, z_0 + \Delta z) \approx f(x_0, y_0, z_0) &+ f_x(x_0, y_0, z_0)\,\Delta x \\ &+ f_y(x_0, y_0, z_0)\,\Delta y \\ &+ f_z(x_0, y_0, z_0)\,\Delta z.\end{aligned}} \qquad (5)$$

Such formulas can be written more compactly using vector notation with subscripted variables. For example, the approximation near $\mathbf{a} = (a_1, a_2, a_3)$ of

$$y = f(\mathbf{x}) = f(x_1, x_2, x_3)$$

corresponding to a change $\mathbf{\Delta x} = (\Delta x_1, \Delta x_2, \Delta x_3)$ is given by

$$\boxed{f(\mathbf{a} + \mathbf{\Delta x}) \approx f(\mathbf{a}) + f_{x_1}(\mathbf{a})\,\Delta x_1 + f_{x_2}(\mathbf{a})\,\Delta x_2 + f_{x_3}(\mathbf{a})\,\Delta x_3,} \qquad (6)$$

for a vector $\mathbf{\Delta x}$ of sufficiently small magnitude. The use of vector notation and scripted variables often makes the structure of formulas easier to follow. Formulas just like (6) hold for approximation of functions of one or two variables also; it just depends on the subscript of the "last x."

EXAMPLE 5 Estimate $\sqrt{2.01^2 + 1.98^2 + 1.05^2}$.

Solution We let $f(x, y, z) = \sqrt{x^2 + y^2 + z^2}$ and use approximation (5) with

$$x_0 = 2, \qquad y_0 = 2, \qquad z_0 = 1$$

$$\Delta x = 0.01, \qquad \Delta y = -0.02, \qquad \Delta z = 0.05.$$

We easily find that

$$f_x(2, 2, 1) = \left.\frac{x}{\sqrt{x^2 + y^2 + z^2}}\right|_{(2,2,1)} = \frac{2}{3},$$

and similarly,

$$f_y(2, 2, 1) = \frac{2}{3}$$

and

$$f_z(2, 2, 1) = \frac{1}{3}.$$

Then approximation (5) becomes

$$\sqrt{2.01^2 + 1.98^2 + 1.05^2} \approx 3 + \frac{2}{3}(0.01) + \frac{2}{3}(-0.02) + \frac{1}{3}(0.05)$$

$$= 3 + \frac{0.03}{3} = 3.01. \quad \square$$

Problems like those in Examples 3, 4, and 5 are common in calculus texts but are totally impractical today with our calculators and computers. It is a triviality to compute that

$$\sqrt{2.01^2 + 1.98^2 + 1.05^2} \approx 3.010481689$$

using a calculator. It takes much less time than working through Example 5. Such examples serve only to illustrate the approximations (4) and (5). It is much more practical to use the approximation (3) to estimate changes Δx and Δy producing a desired change Δz.

EXAMPLE 6 A conical pile of sand has base radius 6 ft and is 15 ft high. An additional 10 ft^3 of sand is dropped onto the top of the pile to form a new conical pile. If the altitude of the pile increases by 1 in., estimate the change in the radius of the pile.

Solution The volume of a cone is given by $V = \frac{1}{3}\pi r^2 h$. We want to estimate Δr to provide a change $\Delta V = 10$ and $\Delta h = \frac{1}{12}$ when $r = 6$ and $h = 15$. As in Example 4, we have

$$\Delta V_{\tan} = \frac{2}{3}\pi rh\,\Delta r + \frac{1}{3}\pi r^2\,\Delta h.$$

Putting in the appropriate values, we find that

$$10 = \frac{2}{3}\pi \cdot 90\,\Delta r + \frac{1}{3}\pi \cdot 36\,\frac{1}{12} = 60\pi\,\Delta r + \pi.$$

Thus $60\pi\,\Delta r = 10 - \pi$, so

$$\Delta r = \frac{10 - \pi}{60\pi} \approx 0.036 \text{ ft},$$

or about 0.437 in. $\square$

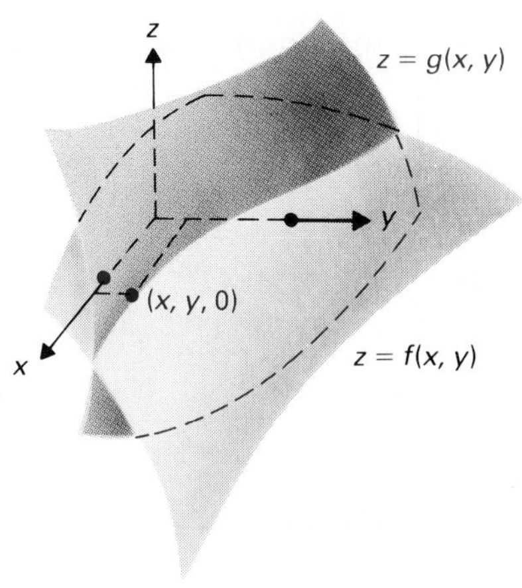

Figure 16.18 The curve of intersection of the surfaces $z = f(x, y)$ and $z = g(x, y)$.

*NEWTON'S METHOD FOR FUNCTIONS OF TWO VARIABLES

Let $z = f(x, y)$. We will write $\Delta f_{\tan}$ for $\Delta z_{\tan}$ here. Then Eq. (2) becomes

$$\Delta f_{\tan} = f_x(x_0, y_0)\,\Delta x + f_y(x_0, y_0)\,\Delta y. \tag{7}$$

If we use Eq. (7) to produce a desired change $\Delta f_{\tan}$ at a point (x_0, y_0), there are generally many choices for Δx and Δy. If the partial derivatives are nonzero, we can set Δx equal to any number and solve for Δy. Indeed, we expect to be able to solve *two* equations like Eq. (7) *simultaneously* for Δx and Δy.

Let $z = g(x, y)$ be another function. The surface graphs of $f(x, y)$ and $g(x, y)$ generally intersect in a curve, as shown in Fig. 16.18. This curve may go through the x,y-plane at one point, many points, or perhaps no point at all. To find such points amounts to solving the system of simultaneous equations

$$\begin{cases} f(x, y) = 0, \\ g(x, y) = 0. \end{cases} \tag{8}$$

Simultaneous equations like Eq. (8) are often very tough to solve if they are not linear. But

$$\begin{cases} \Delta f_{\tan} = f_x(x_0, y_0)\,\Delta x + f_y(x_0, y_0)\,\Delta y, \\ \Delta g_{\tan} = g_x(x_0, y_0)\,\Delta x + g_y(x_0, y_0)\,\Delta y \end{cases} \tag{9}$$

is a *linear* system in Δx and Δy and is *easy* to solve!

Newton's method for two variables makes use of the system (9). The steps of the method are as follows:

Step 1 Try to find a reasonably decent first approximate solution point (a_1, b_1) for the system (8).

Step 2 Set $\Delta f_{\tan} = -f(a_1, b_1)$ and $\Delta g_{\tan} = -g(a_1, b_1)$ in system (9), obtaining the system

$$\begin{cases} f_x(a_1, b_1)\,\Delta x + f_y(a_1, b_1)\,\Delta y = -f(a_1, b_1), \\ g_x(a_1, b_1)\,\Delta x + g_y(a_1, b_1)\,\Delta y = -g(a_1, b_1). \end{cases} \tag{10}$$

Step 3 Solve the linear system (10) for Δx and Δy.

Step 4 Let $a_2 = a_1 + \Delta x$ and $b_2 = b_1 + \Delta y$, to obtain the next approximate solution point (a_2, b_2) for the system (8).

Step 5 Start Step 2 again with (a_2, b_2) in place of (a_1, b_1), and so on.

In Exercise 39, we ask you to show that recursion formulas for a_{i+1} and b_{i+1} in terms of a_i and b_i are

$$\begin{aligned} a_{i+1} &= a_i - \left(\frac{f\cdot g_y - g\cdot f_y}{f_x g_y - g_x f_y}\right)\bigg|_{(a_i, b_i)}, \\ b_{i+1} &= b_i - \left(\frac{g\cdot f_x - f\cdot g_x}{f_x g_y - g_x f_y}\right)\bigg|_{(a_i, b_i)}. \end{aligned} \tag{11}$$

* This topic formerly was omitted from calculus texts since it involves quite a bit of computation. With our calculators and computers, it is now entirely feasible. We urge students at least to read it to appreciate the real importance of the approximations (4) and (5).

EXAMPLE 7 Use one iteration of Newton's method to estimate a solution point of the system

$$\begin{cases} f(x, y) = x^2 + 3xy + y^2 - 9 = 0, \\ g(x, y) = x^3 - y^3 + 10 = 0. \end{cases}$$

Solution A few moments of experimentation shows that $f(1, 2) = 2$ and $g(1, 2) = 3$. Since these function values are not too big, we take $(a_1, b_1) = (1, 2)$ as our first-approximation solution point.

We must also compute f_x, f_y, g_x and g_y at (1, 2). We have

$$f_x(1, 2) = (2x + 3y)|_{(1,2)} = 8,$$

$$f_y(1, 2) = (3x + 2y)|_{(1,2)} = 7,$$

$$g_x(1, 2) = 3x^2|_{(1,2)} = 3,$$

$$g_y(1, 2) = -3y^2|_{(1,2)} = -12.$$

We could form the system (10) and solve it, but we prefer to simply substitute in the recursion formulas (11). We obtain

$$a_2 = 1 - \frac{(2)(-12) - (3)(7)}{(8)(-12) - (3)(7)} = 1 - \frac{-45}{-117} = \frac{72}{117} \approx 0.6154$$

and

$$b_2 = 2 - \frac{(3)(8) - (2)(3)}{-117} = 2 - \frac{18}{-117} = \frac{252}{117} \approx 2.1538.$$

Thus we obtain as answer $(a_2, b_2) = (72/117, 252/117)$. Our calculator shows that $f(a_2, b_2) \approx -0.006$ and $g(a_2, b_2) \approx 0.241$. □

EXAMPLE 8 Use a calculator and Newton's method to estimate a solution point of the system of equations in Example 7, starting with $(a_1, b_1) = (1, 2)$.

Solution We have a programmable calculator, which is very convenient for this problem. We use some of its memory registers as follows.

Register no.:	1	2	3	4	5	6	7	8
Quantity:	a_i	b_i	f	g	f_x	f_y	g_x	g_y

That is, we program the calculator to compute $f(a_i, b_i)$ using the values in registers 1 and 2 and then store the result in register 3. Then the program computes $g(a_i, b_i)$ from the values in registers 1 and 2 and stores the result in register 4, and so on. After register 8 is computed, the program computes a_{i+1} and b_{i+1} using formulas (11), storing them in registers 1 and 2, respectively. We then arrange for the run to stop momentarily so we can copy down a_{i+1}, b_{i+1}, $f(a_i, b_i)$, and $g(a_i, b_i)$, which are available from registers 1, 2, 3, and 4, respectively. The program then resets the calculator to the start of the program to find the next iterate. Having programmed our calculator to perform this sequence of steps, we enter 1 in register 1 and 2 in register 2 and then run the program. We obtain the data in Table 16.1. Thus we see that

$$(0.6031892778, 2.170081386)$$

is very close to a solution point of the system. □

Table 16.1

i	a_i	b_i	$f(a_i, b_i)$	$g(a_i, b_i)$
1	1.	2.	2.	3.
2	0.6153846154	2.153846154	−0.0059171598	0.241238052
3	0.6030896945	2.170176343	−0.000184495	−0.0014502708
4	0.6031892745	2.170081389	−0.0000000094	−0.0000000408
5	0.6031892778	2.170081386	-7×10^{-12}	1×10^{-11}
6	0.6031892778	2.170081386		

The iterates in formulas (11) may not converge to a solution, so it is best to always check $f(a_i, b_i)$ and $g(a_i, b_i)$ as in Table 16.1 to be sure they are approaching zero.

Of course, Newton's method can be used to solve a system of three equations

$$\begin{cases} f(x, y, z) = 0, \\ g(x, y, z) = 0, \\ h(x, y, z) = 0 \end{cases}$$

in three unknowns. The problem is reduced at each iteration to solving a system of three *linear* equations in Δx, Δy, and Δz. In general, we can attempt the method with any system that has the same number of equations as unknowns.

SUMMARY

1. A vector normal to the surface given by $z = f(x, y)$ at any point (x, y, z) on the surface is

$$\frac{\partial z}{\partial x}\mathbf{i} + \frac{\partial z}{\partial y}\mathbf{j} - \mathbf{k}.$$

2. In view of the vector just described, it is easy to find the tangent plane and normal line to the graph of the function, for they are determined by the point on the graph and a vector perpendicular to the graph.

3. The approximate value of $f(x_0 + \Delta x, y_0 + \Delta y)$ given by the height of the tangent plane over $(x_0 + \Delta x, y_0 + \Delta y)$ is

$$f(x_0 + \Delta x, y_0 + \Delta y) \approx f(x_0, y_0) + f_x(x_0, y_0)\,\Delta x + f_y(x_0, y_0)\,\Delta y$$

for sufficiently small Δx and Δy.

4. If $y = f(\mathbf{x}) = f(x_1, x_2, x_3)$ while $\mathbf{a} = (a_1, a_2, a_3)$ and

$$\mathbf{\Delta x} = (\Delta x_1, \Delta x_2, \Delta x_3),$$

then

$$f(\mathbf{a} + \mathbf{\Delta x}) \approx f(\mathbf{a}) + f_{x_1}(\mathbf{a})\,\Delta x_1 + f_{x_2}(\mathbf{a})\,\Delta x_2 + f_{x_3}(\mathbf{a})\,\Delta x_3$$

for $\mathbf{\Delta x}$ of sufficiently small magnitude.

***5.** In Newton's method for finding approximate solutions (a_i, b_i) of simultaneous equations $f(x, y) = 0$ and $g(x, y) = 0$, the recursion relations are

$$a_{i+1} = a_i - \frac{f \cdot g_y - g \cdot f_y}{f_x g_y - g_x f_y} \quad \text{and} \quad b_{i+1} = b_i - \frac{g \cdot f_x - f \cdot g_x}{f_x g_y - g_x f_y},$$

where the functions and their derivatives are evaluated at (a_i, b_i).

EXERCISES

In Exercises 1 through 6, find the equation of the plane tangent to the graph of the given function at the indicated point.

1. $f(x, y) = xy + 3y^2$ at $(-2, 3, 21)$

2. $f(x, y) = \sin xy$ at $(1, \pi/2, 1)$

3. $f(x, y) = x^2/(x + y)$ at $(2, 2, 1)$

4. $f(x, y) = \ln (x^2 + y)$ at $(1, 0, 0)$

5. $f(x_1, x_2) = x_1^2 x_2 - 4x_1 x_2^3$ at $(-2, 1, 12)$

6. $f(x_1, x_2) = x_1^2 e^{x_1 x_2}$ at $(3, 0, 9)$

In Exercises 7 through 12, find parametric equations of the line normal to the surface at the indicated point.

7. $z = x^3 y + xy^2$ at $(1, -1, 0)$

8. $z = \ln (x^2 + y^2)$ at $(3, 4, \ln 25)$

9. $z = xy/(x + y)$ at $(2, -3, 6)$

10. $z = y \tan^{-1}(xy)$ at $(1, -1, \pi/4)$

11. $x_3 = \sqrt{x_1^2 + 4x_2^2}$ at $(-3, 2, 5)$

12. $x_3 = x_2 \sinh (x_1 + x_2)$ at $(-2, 2, 0)$

13. Find the point on the surface $z = x^2 + y^2$ where the tangent plane is parallel to the plane $6x - 4y + 2z = 5$.

14. Find the point on the graph of $f(x, y) = x^2 - y^2$ where the tangent plane is parallel to the plane $3x + 4y + z = 6$.

15. Find the point on the surface $z = xy$ where the normal line is parallel to the line $x = 3 - 2t$, $y = 4 + 5t$, $z = 3 + 3t$.

16. Find the point on the graph of $f(x, y) = xy + x^2$ where the normal line is parallel to the line $x = 4 - 3t$, $y = 5 + 8t$, $z = 1 - 2t$.

17. Find all points on the surface $z = x^2 y$ where the tangent plane is orthogonal to the line $x = 2 - 6t$, $y = 3 - 12t$, $z = 2 + 3t$.

18. Find all points on the surface $z = x/y^2$ where the normal line is orthogonal to the plane $12x + 6y - 3z = 1$.

In Exercises 19 through 22, use calculus to estimate the indicated quantity.

19. $\sqrt{(4.04)(0.95)}$

20. $(2.01)(1.98)^3 + (2.01)^2(1.98)$

21. $(\cos 1°)(\tan 44°)$

22. $\sqrt{1.97^2 + 2.02^2 + 1.05^2}$

23. A rectangular box has inside measurements of 14-in. width, 20-in. length, and 8-in. height. Estimate the volume of material used in construction of the box if the sides and bottom are $\frac{1}{8}$ in. thick and the box has no top.

24. A cylindrical silo with a hemispherical cap has volume

$$V = \pi r^2 h + \tfrac{2}{3}\pi r^3,$$

where h is the height and r the radius of the cylinder. Estimate the change in volume of a silo of 6-ft radius and 30-ft height if the radius is increased by 4 in. and the height is decreased by 6 in.

25. The magnitude of the centripetal acceleration of a body moving on a circle of radius r with constant speed v is v^2/r. A body is moving at a speed of 10 ft/sec on a circle of radius 2 ft. Find the approximate change in the magnitude of the acceleration if the speed is increased 0.5 ft/sec and the radius is decreased 1 in.

26. For the body in Exercise 25, suppose the magnitude of the centripetal force is kept constant. Find the approximate change in the velocity if the radius r is increased by 1 in.

27. The pressure P in lb/ft^2 of a certain gas at temperature T degrees in a container of variable volume V ft^3 is given by $P = 8T/V$. Suppose the temperature is 20° and the volume is 10 ft^3. If the temperature is increased by 0.5°, estimate the change in volume that produces a decrease in pressure of 0.25 lb/ft^2.

28. The voltage drop V, measured in volts, across a certain conductor of variable resistance R ohms is IR, where I, measured in amperes, is the current flowing through the conductor. At a certain instant, a current of 1 amp is flowing through a resistance of 200 ohms. If the resistance is decreased by 2 ohms and the voltage is decreased by 4 volts, estimate the change in the current.

If $z = f(x, y)$, then small changes Δx in x and Δy in y produce approximately a $100(\Delta z_{\tan})/z$ percent change in z. Exercises 29 through 32 deal with this phrasing in terms of percent change.

29. Find the approximate percent change in $z = x^2 + xy$ when $x = 2$ and $y = 4$ if x is increased by 5% and y is decreased by 6%.

30. If each dimension of a rectangular box is increased by a%, where a is small, find the approximate percent change in the volume of the box.

31. If both the radius of the base and the altitude of a right circular cylinder are increased by a%, where a is small, find the approximate percent change in the volume of the cylinder.

32. Find the approximate percent change in the volume of a right circular cone if the altitude is increased by 2% and the radius decreased by 0.5%.

In Exercises 33 through 38, use Newton's method to find the next approximate solution (a_2, b_2) to the given system of two equations, starting with the given approximate solution (a_1, b_1).

***33.** $(a_1, b_1) = (6, -1)$, $\begin{cases} x^2 + xy - 30 = 0 \\ xy^2 - 2y^3 - 20 = 0 \end{cases}$

***34.** $(a_1, b_1) = (10, 0)$, $\begin{cases} x + y - 10 = 0 \\ x^2y + 5 = 0 \end{cases}$

***35.** $(a_1, b_1) = (1, 1)$, $\begin{cases} x^3 - 3x^2y - 3y^3 + 4 = 0 \\ x^2y - y^3 - 1 = 0 \end{cases}$

***36.** $(a_1, b_1) = (0, -2)$, $\begin{cases} e^{xy} - y - 4 = 0 \\ x^2 - 2y^2 + 5 = 0 \end{cases}$

***37.** $(a_1, b_1) = (0, 0)$, $\begin{cases} e^x - e^y + 1 = 0 \\ x + 3y - 2 = 0 \end{cases}$

***38.** $(a_1, b_1) = (0, 2)$, $\begin{cases} 8 \sin(xy) - e^{xy} + 2 = 0 \\ x^3 - y^3 + 6 = 0 \end{cases}$

***39.** Derive the recursion formulas (11) for the two-variable case of Newton's method.

In Exercises 40 through 45, use a programmable calculator or a computer and Newton's method. Determine accuracy to as many places as you can for a solution point for the indicated system, starting from the given first approximation.

***40.** The system and point in Exercise 33

***41.** The system and point in Exercise 34

***42.** The system and point in Exercise 35

***43.** The system and point in Exercise 36

***44.** The system and point in Exercise 37

***45.** The system and point in Exercise 38

16.3 THE DIFFERENTIAL AND THE GRADIENT VECTOR

In the preceding section, we learned how to find the equation of a tangent plane to the graph of a function $f(x, y)$, and we regarded the plane as giving a good linear approximation to the function f near the point of tangency. We then practiced using this approximation. In the present section, we will justify our use of the tangent-plane approximation and give some terminology and notation.

THE LINEAR APPROXIMATION THEOREM

It is instructive to recall our work in Section 3.3 for a function $y = f(x)$ of one variable. The hypotheses that f is defined in an interval centered at x_0 and that $f'(x_0)$ exists were all we needed to show that $f(x)$ has a good linear approximation near x_0. Recall that

$$\Delta y = f(x_0 + \Delta x) - f(x_0) \tag{1}$$

and

$$\Delta y_{\tan} = f'(x_0)\,\Delta x. \tag{2}$$

We let

$$E(\Delta x) = \Delta y - \Delta y_{\tan}, \tag{3}$$

so that $E(\Delta x)$ is the error in the approximation of Δy by $\Delta y_{\tan}$. We showed that

$$\lim_{\Delta x \to 0} \frac{E(\Delta x)}{\Delta x} = 0. \tag{4}$$

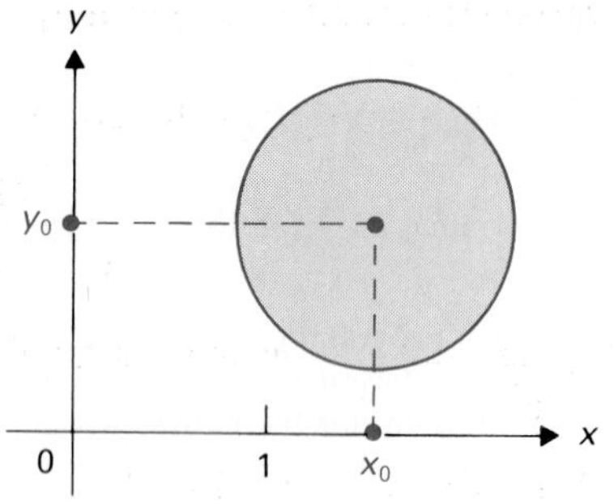

Figure 16.19 Disk centered at (x_0, y_0).

Equation (4) means that the error in approximating Δy by $\Delta y_{\tan}$ is not merely small as $\Delta x \to 0$ but is even small compared with the size of Δx. We defined the function $\varepsilon_1(\Delta x)$ by

$$\varepsilon_1(\Delta x) = \begin{cases} \dfrac{E(\Delta x)}{\Delta x} & \text{for } \Delta x \neq 0, \\ 0 & \text{for } \Delta x = 0. \end{cases} \tag{5}$$

From Eq. (4), we have

$$\lim_{\Delta x \to 0} \varepsilon_1(\Delta x) = 0. \tag{6}$$

Then

$$\begin{aligned} f(x_0 + \Delta x) = f(x_0) + \Delta y &= f(x_0) + \Delta y_{\tan} + E(\Delta x) \\ &= f(x_0) + f'(x_0)\,\Delta x + E(\Delta x), \end{aligned}$$

so

$$\boxed{\begin{aligned} &f(x_0 + \Delta x) = f(x_0) + f'(x_0)\,\Delta x + \varepsilon_1(\Delta x)\cdot \Delta x, \\ &\text{where } \lim_{\Delta x \to 0} \varepsilon_1(\Delta x) = 0. \end{aligned}} \tag{7}$$

The relation (7) was the crux of Theorem 3.4 on page 95 and was crucial to our proof of the chain rule in Section 3.4.

We wish to parallel this development for a function $z = f(x, y)$ of two variables. Let $f(x, y)$ be defined in a disk centered at (x_0, y_0), as shown in Fig. 16.19. With a function of one variable, we needed *only the existence* of the derivative at the center point x_0 to derive the relation (7). The big difference for two or more variables is the need to strengthen this hypothesis.

> We now assume that f_x and f_y exist and are continuous throughout a disk centered at (x_0, y_0).

We proceed to develop the analogues of Eqs. (1) through (7). We let

$$\Delta z = f(x_0 + \Delta x, y_0 + \Delta y) - f(x_0, y_0) \tag{8}$$

and

$$\Delta z_{\tan} = f_x(x_0, y_0)\cdot \Delta x + f_y(x_0, y_0)\cdot \Delta y. \tag{9}$$

We also let

$$E(\Delta x, \Delta y) = \Delta z - \Delta z_{\tan}, \tag{10}$$

so that $E(\Delta x, \Delta y)$ is the error in the approximation of Δz by $\Delta z_{\tan}$. Our goal is to show that $E(\Delta x, \Delta y)$ becomes small compared with the size of Δx and Δy as both $\Delta x \to 0$ and $\Delta y \to 0$, analogous to Eq. (4).

Now

$$\begin{aligned} \Delta z &= f(x_0 + \Delta x, y_0 + \Delta y) - f(x_0, y_0) \\ &= [f(x_0 + \Delta x, y_0 + \Delta y) - f(x_0, y_0 + \Delta y)] \\ &\quad + [f(x_0, y_0 + \Delta y) - f(x_0, y_0)]. \end{aligned} \tag{11}$$

Figure 16.20 The change in $f(x, y)$ going from (x_0, y_0) to $(x_0 + \Delta x, y_0 + \Delta y)$ is the sum of the changes involving $(x_0, y_0 + \Delta y)$ as an intermediate point.

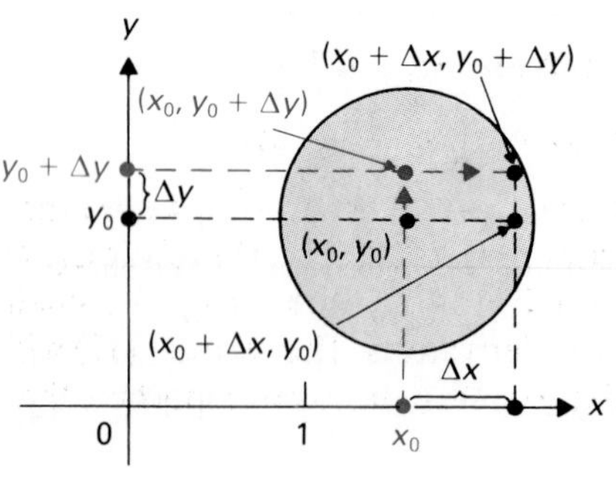

Relation (11) is a "subtract-and-add trick" using the point $(x_0, y_0 + \Delta y)$ shown in Fig. 16.20. The point $(x_0 + \Delta x, y_0)$ could have been used equally

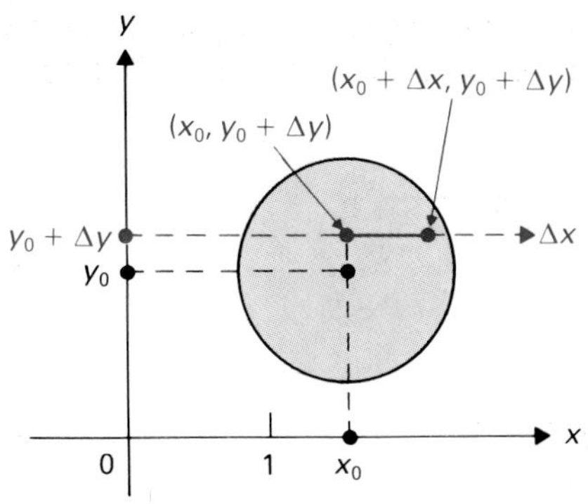

Figure 16.21 $f(x, y)$ is a function of the variable Δx only on the line segment from $(x_0, y_0 + \Delta y)$ to $(x_0 + \Delta x, y_0 + \Delta y)$.

well. Let the function $\varepsilon_1(\Delta x, \Delta y)$ of the variables Δx and Δy be given by

$$\varepsilon_1 = \begin{cases} \dfrac{f(x_0 + \Delta x, y_0 + \Delta y) - f(x_0, y_0 + \Delta y)}{\Delta x} - f_x(x_0, y_0) & \text{for } \Delta x \neq 0, \\ 0 & \text{for } \Delta x = 0. \end{cases} \tag{12}$$

(We suppress the Δx and Δy in $\varepsilon_1(\Delta x, \Delta y)$ here and in some of the sequel for reasons of space.) From Eq. (12), we see that

$$f(x_0 + \Delta x, y_0 + \Delta y) - f(x_0, y_0 + \Delta y) = f_x(x_0, y_0) \cdot \Delta x + \varepsilon_1 \cdot \Delta x. \tag{13}$$

The numerator of the quotient appearing in Eq. (12) is a function of the single variable Δx on the line segment shown in Fig. 16.21, where Δy is constant. We can apply the mean-value theorem to this function over this line segment. We obtain

$$\frac{f(x_0 + \Delta x, y_0 + \Delta y) - f(x_0, y_0 + \Delta y)}{\Delta x} = f_x(c, y_0 + \Delta y) \tag{14}$$

for some point $(c, y_0 + \Delta y)$ on the line segment, shown in Fig. 16.21. Using Eq. (12), we therefore have

$$\varepsilon_1 = f_x(c, y_0 + \Delta y) - f_x(x_0, y_0), \quad \text{where } c \text{ is between } x_0 \text{ and } x_0 + \Delta x.$$

Since f_x is assumed to be continuous, we see that

$$\lim_{\Delta x, \Delta y \to 0} \varepsilon_1 = f_x(x_0, y_0) - f_x(x_0, y_0) = 0.$$

Now let

$$\varepsilon_2 = \begin{cases} \dfrac{f(x_0, y_0 + \Delta y) - f(x_0, y_0)}{\Delta y} - f_y(x_0, y_0) & \text{for } \Delta y \neq 0, \\ 0 & \text{for } \Delta y = 0. \end{cases} \tag{15}$$

Then

$$f(x_0, y_0 + \Delta y) - f(x_0, y_0) = f_y(x_0, y_0) \cdot \Delta y + \varepsilon_2 \cdot \Delta y. \tag{16}$$

An argument using the mean-value theorem like the argument above, but applied to the function of Δy on the line segment joining (x_0, y_0) to $(x_0, y_0 + \Delta y)$ in Fig. 16.22, shows that $\lim_{\Delta x, \Delta y \to 0} \varepsilon_2 = 0$. Substituting the expressions in Eqs. (13) and (16) for the bracketed expressions in relation (11), we see that

$$\boxed{\begin{gathered} \Delta z = f_x(x_0, y_0) \cdot \Delta x + f_y(x_0, y_0) \cdot \Delta y + \varepsilon_1 \cdot \Delta x + \varepsilon_2 \cdot \Delta y, \\ \text{where } \lim_{\Delta x, \Delta y \to 0} \varepsilon_1 = \lim_{\Delta x, \Delta y \to 0} \varepsilon_2 = 0. \end{gathered}} \tag{17}$$

Figure 16.22 $f(x, y)$ is a function of the variable Δy only on the line segment from (x_0, y_0) to $(x_0, y_0 + \Delta y)$.

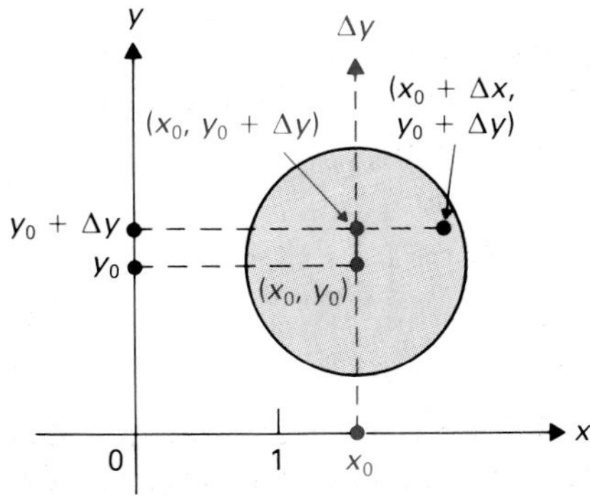

Our error for the approximation of Δz by $\Delta z_{\tan}$ is thus

$$E(\Delta x, \Delta y) = \varepsilon_1(\Delta x, \Delta y) \cdot \Delta x + \varepsilon_2(\Delta x, \Delta y) \cdot \Delta y,$$

and since ε_1 and ε_2 both approach zero as $\Delta x \to 0$ and $\Delta y \to 0$, we see that this error is small compared with the size of Δx and Δy. Thus the relation (17) achieves our goal and justifies our approximation of Δz by $\Delta z_{\tan}$ for small Δx and Δy if f has continuous first partial derivatives. The relation (17) will enable us to prove the chain rule in Section 16.4 and is so important that we state it as a theorem.

THEOREM 16.2 Linear approximation

Let $f(x, y)$ have continuous partial derivatives f_x and f_y inside a circle of radius $r > 0$ with center (x_0, y_0). Then there exist functions $\varepsilon_1(\Delta x, \Delta y)$ and $\varepsilon_2(\Delta x, \Delta y)$, defined for $(\Delta x)^2 + (\Delta y)^2 < r^2$, such that

$$\lim_{\Delta x, \Delta y \to 0} \varepsilon_1 = \lim_{\Delta x, \Delta y \to 0} \varepsilon_2 = 0$$

and such that

$$f(x_0 + \Delta x, y_0 + \Delta y) = f(x_0, y_0) + f_x(x_0, y_0)\,\Delta x + f_y(x_0, y_0)\,\Delta y + \varepsilon_1(\Delta x, \Delta y)\cdot\Delta x + \varepsilon_2(\Delta x, \Delta y)\cdot\Delta y.$$

EXAMPLE 1 Find $\varepsilon_1(\Delta x, \Delta y)$ and $\varepsilon_2(\Delta x, \Delta y)$ in Theorem 16.2 for $f(x, y) = x^2 + xy$ at the point (1, 2). Show directly that

$$\lim_{\Delta x, \Delta y \to 0} \varepsilon_1 = \lim_{\Delta x, \Delta y \to 0} \varepsilon_2 = 0.$$

Solution Now $f_x = 2x + y$, so $f_x(1, 2) = 4$. By Eq. (12), we have

$$\begin{aligned}\varepsilon_1(\Delta x, \Delta y) &= \frac{f(1 + \Delta x, 2 + \Delta y) - f(1, 2 + \Delta y)}{\Delta x} - 4 \\ &= \frac{[(1 + \Delta x)^2 + (1 + \Delta x)(2 + \Delta y)] - [1^2 + 2 + \Delta y]}{\Delta x} - 4 \\ &= \frac{1 + 2(\Delta x) + (\Delta x)^2 + 2 + 2(\Delta x) + \Delta y + (\Delta x)(\Delta y) - 3 - \Delta y - 4(\Delta x)}{\Delta x} \\ &= \frac{(\Delta x)^2 + (\Delta x)(\Delta y)}{\Delta x} = \Delta x + \Delta y.\end{aligned}$$

Since $f_y(1, 2) = 1$, we see from Eq. (15) that

$$\begin{aligned}\varepsilon_2(\Delta x, \Delta y) &= \frac{f(1, 2 + \Delta y) - f(1, 2)}{\Delta y} - 1 \\ &= \frac{1 + 2 + \Delta y - 3}{\Delta y} - 1 = \frac{\Delta y - \Delta y}{\Delta y} = 0.\end{aligned}$$

Then $\lim_{\Delta x, \Delta y \to 0} \varepsilon_1 = \lim_{\Delta x, \Delta y \to 0} (\Delta x + \Delta y) = 0$, and of course

$$\lim_{\Delta x, \Delta y \to 0} \varepsilon_2 = 0. \quad \square$$

If $w = f(x, y, z)$ has continuous first partial derivatives in a ball having (x_0, y_0, z_0) as center, the analogue of Theorem 16.2 holds with functions ε_1, ε_2, and ε_3 of Δx, Δy, and Δz. In fact, the hypothesis of continuous first partial derivatives is sufficient to guarantee that a function $f(x_1, x_2, \ldots, x_n)$ of n variables has good local linear approximations.

THE DIFFERENTIAL AND THE GRADIENT VECTOR

If $y = f(x)$, the *differential* dy of f at x_0 is the linear function

$$dy = f'(x_0)\,dx \tag{18}$$

of the independent variable dx. Note that Eq. (18) is the equation of the tangent line to the graph of $y = f(x)$, where $x = x_0$, with respect to local dx,dy-axes at the point (x_0, y_0), as shown in Fig. 16.23. The analogue of Fig. 16.23 for a function of two variables is Fig. 16.24, and we see that

$$dz = f_x(x_0, y_0)\,dx + f_y(x_0, y_0)\,dy \tag{19}$$

is the equation with respect to local dx,dy,dz-axes of the plane tangent to $z = f(x, y)$ at (x_0, y_0, z_0), if the tangent plane exists. This suggests the following definition.

DEFINITION 16.4 Differential

Let $z = f(x, y)$ have partial derivatives f_x and f_y at (x_0, y_0). Let dx and dy be independent variables. The **differential** dz (or df) **of** f **at** (x_0, y_0) is

$$dz = f_x(x_0, y_0)\,dx + f_y(x_0, y_0)\,dy.$$

If f_x and f_y exist at all points in the domain of f, then the **differential** dz (or df) is

$$dz = f_x(x, y)\,dx + f_y(x, y)\,dy.$$

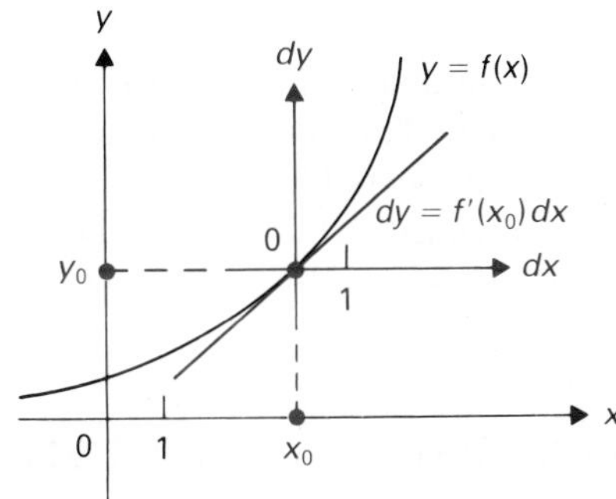

Figure 16.23 $dy = f'(x_0)\,dx$ is the equation of the tangent line at (x_0, y_0).

The differential in Definition 16.4 is sometimes called the *total differential.*

EXAMPLE 2 Find the differential dz if $z = x^3y + xy^2$.

Solution We have

$$\frac{\partial z}{\partial x} = 3x^2y + y^2 \qquad \text{and} \qquad \frac{\partial z}{\partial y} = x^3 + 2xy,$$

so

$$dz = (3x^2y + y^2)\,dx + (x^3 + 2xy)\,dy. \quad \square$$

The natural extension of Definition 16.4 is made for functions of more than two variables. If $y = f(\mathbf{x}) = f(x_1, x_2, \ldots, x_n)$, then

$$dy = f_{x_1}\,dx_1 + f_{x_2}\,dx_2 + \cdots + f_{x_n}\,dx_n. \tag{20}$$

Figure 16.24 $dz = f_x(x_0, y_0)\,dx + f_y(x_0, y_0)\,dy$ is the equation of the tangent plane at (x_0, y_0, z_0).

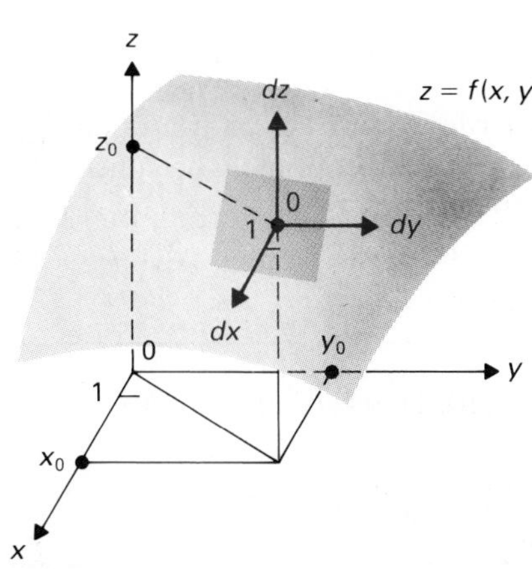

EXAMPLE 3 Find dy if $y = f(x_1, x_2, x_3) = x_1^2 + x_2e^{x_3}$.

Solution We have $f_{x_1} = 2x_1$, $f_{x_2} = e^{x_3}$, and $f_{x_3} = x_2e^{x_3}$, so

$$dy = 2x_1\,dx_1 + e^{x_3}\,dx_2 + x_2e^{x_3}\,dx_3. \quad \square$$

EXAMPLE 4 Find dw if $w = z\tan^{-1}(xy)$.

Solution We have

$$\frac{\partial w}{\partial x} = \frac{yz}{1 + x^2y^2}, \quad \frac{\partial w}{\partial y} = \frac{xz}{1 + x^2y^2}, \quad \text{and} \quad \frac{\partial w}{\partial z} = \tan^{-1}(xy).$$

Consequently,

$$dw = \frac{yz}{1 + x^2y^2}\,dx + \frac{xz}{1 + x^2y^2}\,dy + \tan^{-1}(xy)\,dz. \quad \square$$

Recall that a function of one variable is *differentiable at a point* if the derivative exists there. If functions f and g of one variable are differentiable, then the chain rule asserts that the composite function $f(g(x))$ is also differentiable. We would like this to be true for functions of two or more variables. We will need the relation (17) to prove the chain rule, and since it does not follow from the mere existence of first partial derivatives, we take the relation (17) as the defining criterion for a function of two variables to be differentiable at a point.

DEFINITION Differentiable function

16.5

A function $f(x, y)$ is **differentiable at** (x_0, y_0) if it is defined and its first partial derivatives exist in a disk having (x_0, y_0) as center and if functions $\varepsilon_1(\Delta x, \Delta y)$ and $\varepsilon_2(\Delta x, \Delta y)$ exist so that relation (17) holds. The function is **differentiable** if it is differentiable at every point in its domain.

The definition asserts that a function is differentiable at a point if it has a good local linear approximation there. In view of Theorem 16.2, functions having continuous first partial derivatives are differentiable.

The approximation problems of the previous section can all be rephrased in terms of "approximation by differentials," and the approximation formula becomes

$$\boxed{f(x_0 + dx, y_0 + dy) \approx f(x_0, y_0) + f_x(x_0, y_0)\,dx + f_y(x_0, y_0)\,dy.} \qquad (21)$$

EXAMPLE 5 Approximate $2.98^4/1.03^2$ using a differential.

Solution We use Eq. (21) with $f(x, y) = x^4/y^2$, $(x_0, y_0) = (3, 1)$, $dx = -0.02$, and $dy = 0.03$. Then

$$f_x(3, 1) = \left.\frac{4x^3}{y^2}\right|_{(3,1)} = 108 \quad \text{and} \quad f_y(3, 1) = \left.\frac{-2x^4}{y^3}\right|_{(3,1)} = -162.$$

We obtain from Eq. (21),

$$\begin{aligned}\frac{2.98^4}{1.03^2} &\approx 81 + 108(-0.02) - 162(0.03)\\ &= 81 - 2.16 - 4.86 = 73.98. \quad \square\end{aligned}$$

For $z = f(x, y)$, we have discussed *partial* derivatives at a point. We proceed to define a *vector* that we will see can appropriately be considered to be the "*total* derivative" of a function at a point.

DEFINITION Gradient vector

16.6

Let $z = f(x, y)$ and let f_x and f_y exist at (x_0, y_0). The **gradient vector** $\nabla f(x_0, y_0)$ is the vector

$$\nabla f(x_0, y_0) = f_x(x_0, y_0)\mathbf{i} + f_y(x_0, y_0)\mathbf{j}.$$

The symbol ∇ in the preceding definition is read "del." It is often convenient to think of ∇ as a symbolic operator,

$$\nabla = \frac{\partial}{\partial x}\mathbf{i} + \frac{\partial}{\partial y}\mathbf{j}.$$

To find ∇ of a function, we take its partial with respect to x times $\mathbf{i}$ plus its partial with respect to y times $\mathbf{j}$.

EXAMPLE 6 If $z = f(x, y) = x^2y + xy^3$, find ∇f at $(1, -2)$.

Solution We have

$$f_x(1, -2) = (2xy + y^3)|_{(1,-2)} = -12$$

and

$$f_y(1, -2) = (x^2 + 3xy^2)|_{(1,-2)} = 13.$$

Thus $\nabla f(1, -2) = -12\mathbf{i} + 13\mathbf{j}$. □

The notion of the gradient vector extends naturally to a function of more than two variables; we simply have more components in the vector.

EXAMPLE 7 Find ∇g for $w = g(x, y, z) = (x + y)/z$.

Solution We have

$$g_x = \frac{1}{z}, \quad g_y = \frac{1}{z}, \quad \text{and} \quad g_z = -\frac{x+y}{z^2}.$$

Thus

$$\nabla g = \frac{1}{z}\mathbf{i} + \frac{1}{z}\mathbf{j} - \frac{x+y}{z^2}\mathbf{k}. \quad \square$$

We now indicate why ∇f plays the role of a total derivative of f. We know that f_x and f_y give the rates of change of $f(x, y)$ in only two of the many possible directions at a point. Namely, f_x gives the rate of change in the direction of $\mathbf{i}$, and f_y in the direction of $\mathbf{j}$. We will see in Section 16.5 that if we know ∇f at a point for a differentiable function f, then we can easily find the rate of change of f in *every* possible direction at that point. Thus we think of ∇f as the *derivative* (or *total derivative*) at the point.

For another insight into the role of ∇f, recall that if $y = f(x)$, then

$$dy = f'(x)\,dx. \tag{22}$$

We will use subscript notation for a function $y = f(\mathbf{x}) = f(x_1, x_2)$ of two variables. At a point $\mathbf{x} = (x_1, x_2)$, we have

$$\nabla f(\mathbf{x}) = f_{x_1}(\mathbf{x})\mathbf{i} + f_{x_1}(\mathbf{x})\mathbf{j}.$$

If we let the differential vector $\mathbf{dx}$ be

$$\mathbf{dx} = dx_1\mathbf{i} + dx_2\mathbf{j},$$

then the differential $dy = f_{x_1}(\mathbf{x})\,dx_1 + f_{x_1}(\mathbf{x})\,dx_1$ appears as the dot product

$$dy = \nabla f(\mathbf{x}) \cdot \mathbf{dx}. \tag{23}$$

Comparing Eqs. (22) and (23), we see that ∇f seems to play the role of $f'(x)$. It would be wonderful if mathematicians used either the notation ∇f in place of f' for a function of one variable or the notation $\mathbf{f}'$ in place of ∇f for the gradient vector. However, that is not the way the notation developed.

The gradient vector ∇f is the first vector we have introduced without promptly discussing its magnitude and direction. We will show in Section 16.5 that $\nabla f(x_0, y_0)$ points in the direction of maximum rate of change of the function f at (x_0, y_0) and has magnitude equal to that maximum rate of change.

SUMMARY

1. If f_x and f_y are continuous, then

$$f(x_0 + \Delta x, y_0 + \Delta y) = f(x_0, y_0) + f_x(x_0, y_0)\,\Delta x + f_y(x_0, y_0)\,\Delta y + \varepsilon_1 \cdot \Delta x + \varepsilon_2 \cdot \Delta y,$$

where ε_1 and ε_2 are functions of Δx and Δy that both approach zero as $\Delta x \to 0$ and $\Delta y \to 0$.

2. The differential dz or df of $z = f(x, y)$ at a point (x_0, y_0) is

$$dz = f_x(x_0, y_0)\,dx + f_y(x_0, y_0)\,dy.$$

3. The differential of $y = f(\mathbf{x}) = f(x_1, x_2, \ldots, x_n)$ is

$$dy = f_{x_1}(\mathbf{x})\,dx_1 + f_{x_2}(\mathbf{x})\,dx_2 + \cdots + f_{x_n}(\mathbf{x})\,dx_n.$$

4. If a function has continuous first partial derivatives, its differential at a point is a good local linear approximation to the function near that point.
5. The gradient vector ∇f of $f(x, y)$ at a point (x_0, y_0) is

$$\nabla f(x_0, y_0) = f_x(x_0, y_0)\mathbf{i} + f_y(x_0, y_0)\mathbf{j}.$$

More components are added for functions of more variables.

EXERCISES

Sequential Exercises 1 through 5 deal with the function $z = f(x, y) = \sqrt{2|x||y|}$.

1. What is the value of the function on the x-axis and on the y-axis?
2. Find $f_x(0, 0)$ and $f_y(0, 0)$.
3. What plane contains the point $(0, 0, f(0, 0))$ and has as normal vector $f_x(0, 0)\mathbf{i} + f_y(0, 0)\mathbf{j} - \mathbf{k}$?
4. Show that the portions of the graph lying over the lines $y = x$ and $y = -x$ in the x,y-plane consist of four straight rays leaving the origin at an angle of inclination of 45° with the x,y-plane.
5. Write relation (17) at $(x_0, y_0) = (0, 0)$ for $\Delta y = \Delta x$, and deduce that $f(x, y)$ is not differentiable at $(0, 0)$.

In Exercises 6 through 10, find the functions $\varepsilon_1(\Delta x, \Delta y)$ and $\varepsilon_2(\Delta x, \Delta y)$ defined in Eqs. (12) and (15), and show directly that they approach zero as $\Delta x \to 0$ and $\Delta y \to 0$.

6. $f(x, y) = x^2 + y^2$ at $(0, 0)$
7. $f(x, y) = x^2 + y^2$ at $(1, -1)$
8. $f(x, y) = x/y$ at $(1, 1)$
9. $f(x, y) = 1/(xy)$ at $(-1, 1)$
10. $f(x, y) = x^2 - 2xy + 3x - y^2$ at a general point (x, y)

In Exercises 11 through 16, find the differential of the indicated function at the given point.

11. $f(x, y) = x^3y^2 + 2x^2y$ at $(2, 3)$
12. $f(x, y) = x^2 - 2y^2 + \tan(xy) + (1/x)$ at $(-2, 0)$

13. $f(x, y) = 2x + x \cos y + x \sin y$ at $(3, \pi)$
14. $f(x, y, z) = x^2 + 2yz$ at $(4, -1, 2)$
15. $f(x, y, z) = \ln (xy) + e^{yz} + \sin (xz)$ at $(2, 4, \pi)$
16. $f(x) = 3x^2 + \sec x + \ln x$ at π
17. Let $f(x, y, z) = xy + \sin z$. Use a differential to estimate $f(0.98, 2.03, 0.05)$.
18. Let $f(x, y) = e^{xy} + \sin (xy) + 4y$. Use a differential to estimate $f(0.02, 4.97)$.
19. Let $z = f(x, y) = x/y^2$. When $x = 1$ and $y = 2$, use a differential to estimate the change in x required to keep z constant if y is decreased by 0.05.
20. Let $z = f(x, y) = x/(x + y)$. When $x = -1$ and $y = 3$, use a differential to estimate the change in x that produces an increase of 0.2 in z if y is increased by 0.05.

In Exercises 21 through 26, find the gradient vector of the given function at the indicated point.

21. $f(x, y) = x^3 - 3xy^2$ at $(1, 1)$
22. $f(x, y) = xe^y$ at $(2, 0)$
23. $f(x, y, z) = (x - y)/z$ at $(1, -1, 3)$
24. $f(x, y, z) = xy^3 + ye^{xz}$ at $(-1, 2, 0)$
25. $f(x) = \cos (x/2) + 2 \sin x$ at $x = \pi$
26. $f(x, y, z) = xy^2 + \tan^{-1}(xz)$ at $(-1, 4, 1)$

In Exercises 27 through 30, describe all points in the plane where the gradient vector of $f(x, y) = x^2 + y^2 + 2x$ has the given property.

27. A unit vector
28. Parallel to the vector $\mathbf{i} - 2\mathbf{j}$
29. Orthogonal to the vector $3\mathbf{i} - 4\mathbf{j}$
30. Orthogonal to the position vector from the origin to the point

In Exercises 31 through 34, describe all points in space where the gradient vector of $f(x, y, z) = xy + z^2$ has the given property.

31. Magnitude 2
32. Parallel to the vector $2\mathbf{i} - 3\mathbf{j} + \mathbf{k}$
33. Orthogonal to the vector $\mathbf{i} - 3\mathbf{j} + 2\mathbf{k}$
34. Directed toward the origin

16.4 CHAIN RULES

Let $y = f(x)$ and $x = g(t)$ both be differentiable functions. In Section 3.4, we showed that the derivative of the *composite* function $y = f(g(t))$ can be found by the *chain rule*

$$\frac{dy}{dt} = \frac{dy}{dx}\frac{dx}{dt}.$$

In this section, we study chain rules for derivatives of composite functions composed of functions of more than one variable. The rules are a bit more complicated than for the single-variable case. However, they are easily mastered. At the end of this section, we indicate how vector notation could be used to make the chain rules all look just like the one above for the single-variable case.

Suppose $z = f(x, y)$ and $x = g_1(t)$, $y = g_2(t)$ so t determines z. Suppose also that the derivatives $\partial z/\partial x$, $\partial z/\partial y$, dx/dt, and dy/dt all exist and are continuous. We want to find dz/dt, since z appears as a composite function of the one variable t. The rate of change of z with respect to t can be expressed in terms of the rates at which z changes with respect to x and to y and the rates at which x and y change with respect to t. This is surely not too surprising. The total rate of change of z is the sum of the rates of change due to the changing quantities x and y. These rates of change due to x and y individually are, by the chain rule in Chapter 3,

$$\frac{\partial z}{\partial x}\cdot\frac{dx}{dt} \quad \text{and} \quad \frac{\partial z}{\partial y}\cdot\frac{dy}{dt}.$$

Thus, although we have not proved it yet, the valid formula

$$\boxed{\frac{dz}{dt} = \frac{\partial z}{\partial z} \cdot \frac{dx}{dt} + \frac{\partial z}{\partial y} \cdot \frac{dy}{dt}} \tag{1}$$

seems reasonable.

We can prove the formula (1) from our work in the last section. Theorem 16.2 shows that

$$\Delta z = f_x(x, y)\,\Delta x + f_y(x, y)\,\Delta y + \varepsilon_1 \cdot \Delta x + \varepsilon_2 \cdot \Delta y, \tag{2}$$

where $\varepsilon_1 \to 0$ and $\varepsilon_2 \to 0$ as $\Delta x \to 0$ and $\Delta y \to 0$. Thus

$$\begin{aligned}
\frac{dz}{dt} = \lim_{\Delta t \to 0} \frac{\Delta z}{\Delta t} &= \lim_{\Delta t \to 0} \left(f_x(x, y) \frac{\Delta x}{\Delta t} + f_y(x, y) \frac{\Delta y}{\Delta t} + \varepsilon_1 \frac{\Delta x}{\Delta t} + \varepsilon_2 \frac{\Delta y}{\Delta t} \right) \\
&= f_x(x, y) \frac{dx}{dt} + f_y(x, y) \frac{dy}{dt} + 0 \cdot \frac{dx}{dt} + 0 \cdot \frac{dy}{dt} \\
&= \frac{\partial z}{\partial x} \frac{dx}{dt} + \frac{\partial z}{\partial y} \frac{dy}{dt},
\end{aligned}$$

which substantiates formula (1).

A similar argument shows that, if $w = f(x, y, z)$ and $x = g_1(t)$, $y = g_2(t)$, and $z = g_3(t)$, then

$$\boxed{\frac{dw}{dt} = \frac{\partial w}{\partial x} \frac{dx}{dt} + \frac{\partial w}{\partial y} \frac{dy}{dt} + \frac{\partial w}{\partial z} \frac{dz}{dt}.}$$

In subscripted notation, if $y = f(\mathbf{x}) = f(x_1, x_2, x_3)$ and $x_1 = g_1(t)$, $x_2 = g_2(t)$, $x_3 = g_3(t)$, then

$$\boxed{\frac{dy}{dt} = \frac{\partial y}{\partial x_1} \frac{dx_1}{dt} + \frac{\partial y}{\partial x_2} \frac{dx_2}{dt} + \frac{\partial y}{\partial x_3} \frac{dx_3}{dt}.}$$

EXAMPLE 1 Let $z = x^2 + (x/y)$, where $x = t^2 - 3t$ and $y = 3t - 5$. Find dz/dt, when $t = 2$, in two ways:

a. by expressing z directly as a function of t and

b. using the chain rule (1).

Solution

a. We have

$$z = x^2 + \frac{x}{y} = (t^2 - 3t)^2 + \frac{t^2 - 3t}{3t - 5}.$$

Thus

$$\frac{dz}{dt} = 2(t^2 - 3t)(2t - 3) + \frac{(3t - 5)(2t - 3) - (t^2 - 3t)3}{(3t - 5)^2}.$$

Setting $t = 2$, we obtain

$$\left.\frac{dz}{dt}\right|_{t=2} = 2(-2)(1) + \frac{(1)(1) - (-2)(3)}{1^2} = -4 + 7 = 3.$$

b. Using the chain rule (1), we have

$$\frac{dz}{dt} = \frac{\partial z}{\partial x}\frac{dx}{dt} + \frac{\partial z}{\partial y}\frac{dy}{dt} = \left(2x + \frac{1}{y}\right)(2t - 3) + \left(\frac{-x}{y^2}\right)3.$$

When $t = 2$, we see that $x = -2$ and $y = 1$. Substituting in these values, we obtain

$$\left.\frac{dz}{dt}\right|_{t=2} = (-4 + 1)(1) + \left(\frac{2}{1}\right)3 = -3 + 6 = 3. \quad \square$$

EXAMPLE 2 The volume V of a right circular cylinder of radius r and height h is given by $V = \pi r^2 h$. The volume is increasing at a rate of 72π in^3/min while the height is decreasing at a rate of 4 in./min. Find the rate of increase of the radius when the height is 3 in. and the radius is 6 in.

Solution We have

$$\frac{dV}{dt} = \frac{\partial V}{\partial r}\cdot\frac{dr}{dt} + \frac{\partial V}{\partial h}\cdot\frac{dh}{dt},$$

so

$$\frac{dV}{dt} = 2\pi r h\frac{dr}{dt} + \pi r^2\left(\frac{dh}{dt}\right).$$

We know that $dV/dt = 72\pi$ and $dh/dt = -4$. (The negative sign occurs because h is decreasing.) We want to find dr/dt when $r = 6$ and $h = 3$. Substituting, we obtain

$$72\pi = 2\pi(6)(3)\frac{dr}{dt} + \pi(6^2)(-4),$$

so

$$36\pi\frac{dr}{dt} = 216\pi.$$

Hence $dr/dt = 216\pi/36\pi = 6$ in./min. $\square$

Now let $z = f(x, y)$, $x = g_1(s, t)$, and $y = g_2(s, t)$. This time z appears as the composite function of *two* variables, s and t, and we are interested in the *partial* derivatives $\partial z/\partial s$ and $\partial z/\partial t$. But Eq. (2) is still valid, and we divide by the increment Δt and take the limit as s is held constant, to find $\partial z/\partial t$. That is, we obtain from Eq. (2)

$$\begin{aligned}\frac{\partial z}{\partial t} = \lim_{\Delta t \to 0}\frac{\Delta z}{\Delta t} &= f_x(x, y)\frac{\partial x}{\partial t} + f_y(x, y)\frac{\partial y}{\partial t} + 0\cdot\frac{\partial x}{\partial t} + 0\cdot\frac{\partial y}{\partial t}\\ &= \frac{\partial z}{\partial x}\frac{\partial x}{\partial t} + \frac{\partial z}{\partial y}\frac{\partial y}{\partial t}.\end{aligned}$$

Thus the derivatives dx/dt and dy/dt in formula (1) simply become partial derivatives in this case. There are so many different types of situations where a chain rule applies that it is awkward to state one all-inclusive theorem. We state one special case as a theorem.

THEOREM 16.3 Chain rule

Let $w = f(x, y, z)$ be a differentiable function. Let x, y, and z all be differentiable functions of s and t. Then w is a differentiable composite function of s and t with first partial derivatives

$$\frac{\partial w}{\partial s} = \frac{\partial w}{\partial x}\frac{\partial x}{\partial s} + \frac{\partial w}{\partial y}\frac{\partial y}{\partial s} + \frac{\partial w}{\partial z}\frac{\partial z}{\partial s}$$

and

$$\frac{\partial w}{\partial t} = \frac{\partial w}{\partial x}\frac{\partial x}{\partial t} + \frac{\partial w}{\partial y}\frac{\partial y}{\partial t} + \frac{\partial w}{\partial z}\frac{\partial z}{\partial t}.$$

EXAMPLE 3 Consider the situation where

$$w = f(x, y, z) = xy^2 + ze^{x^2},$$

while $x = u$, $y = v - 1$, and $z = uv$. Compute $\partial w/\partial u$ at $(0, 2)$ in two ways:

a. by expressing w directly as a function of u and v and

b. by using the chain rule.

Solution

a. Expressing w directly as a function of u and v, we have

$$\begin{aligned} w &= f(u, v - 1, uv) \\ &= u(v - 1)^2 + uve^{u^2}. \end{aligned}$$

Thus

$$\frac{\partial w}{\partial u} = (v - 1)^2 + 2u^2ve^{u^2} + ve^{u^2}.$$

Therefore

$$\begin{aligned} \left.\frac{\partial w}{\partial u}\right|_{(0,2)} &= 1 + 0 + 2 \cdot e^0 \\ &= 1 + 2 = 3. \end{aligned}$$

b. To use the chain rule, we note that when $(u, v) = (0, 2)$,

$$(x, y, z) = (0, 1, 0).$$

If $w = xy^2 + ze^{x^2}$, then

$$\begin{aligned} \frac{\partial w}{\partial u} &= \frac{\partial w}{\partial x} \cdot \frac{\partial x}{\partial u} + \frac{\partial w}{\partial y} \cdot \frac{\partial y}{\partial u} + \frac{\partial w}{\partial z} \cdot \frac{\partial z}{\partial u} \\ &= (y^2 + 2xze^{x^2})(1) + (2xy)(0) + (e^{x^2})(v). \end{aligned}$$

Thus

$$\left.\frac{\partial w}{\partial u}\right|_{(u,v)=(0,2)} = (1+0)(1) + 0 + (1)(2) = 1 + 2 = 3. \quad \square$$

EXAMPLE 4 Suppose $w = f(u, v)$ is a differentiable function. Let $u = ax + by$ and $v = ax - by$. Show that

$$\frac{\partial w}{\partial x}\frac{\partial w}{\partial y} = ab\left[\left(\frac{\partial w}{\partial u}\right)^2 - \left(\frac{\partial w}{\partial v}\right)^2\right].$$

Solution We have

$$\frac{\partial w}{\partial x} = \frac{\partial w}{\partial u}\frac{\partial u}{\partial x} + \frac{\partial w}{\partial v}\frac{\partial v}{\partial x} = a\frac{\partial w}{\partial u} + a\frac{\partial w}{\partial v}.$$

Also

$$\frac{\partial w}{\partial y} = \frac{\partial w}{\partial u}\frac{\partial u}{\partial y} + \frac{\partial w}{\partial v}\frac{\partial v}{\partial y} = b\frac{\partial w}{\partial u} - b\frac{\partial w}{\partial v}.$$

Then

$$\frac{\partial w}{\partial x}\frac{\partial w}{\partial y} = \left(a\frac{\partial w}{\partial u} + a\frac{\partial w}{\partial v}\right)\left(b\frac{\partial w}{\partial u} - b\frac{\partial w}{\partial v}\right) = ab\left[\left(\frac{\partial w}{\partial u}\right)^2 - \left(\frac{\partial w}{\partial v}\right)^2\right]. \quad \square$$

We will consider a case involving subscripted variables. Suppose that $y = f(\mathbf{x}) = f(x_1, x_2, x_3)$ and $x_1 = g_1(t_1, t_2)$, $x_2 = g_2(t_1, t_2)$, and $x_3 = g_3(t_1, t_2)$. Then

$$\boxed{\frac{\partial y}{\partial t_1} = \frac{\partial y}{\partial x_1}\frac{\partial x_1}{\partial t_1} + \frac{\partial y}{\partial x_2}\frac{\partial x_2}{\partial t_1} + \frac{\partial y}{\partial x_3}\frac{\partial x_3}{\partial t_1}.} \quad \textbf{(3)}$$

Suppose we introduce the Leibniz-type notation

$$\frac{\partial y}{\partial \mathbf{x}} = \frac{\partial y}{\partial x_1}\mathbf{i} + \frac{\partial y}{\partial x_2}\mathbf{j} + \frac{\partial y}{\partial x_3}\mathbf{k}$$

and

$$\frac{\partial \mathbf{x}}{\partial t_1} = \frac{\partial x_1}{\partial t_1}\mathbf{i} + \frac{\partial x_2}{\partial t_1}\mathbf{j} + \frac{\partial x_3}{\partial t_1}\mathbf{k}.$$

In both of these Leibniz-type notations, the boldface part of the symbol indicates the variable whose subscripts change to give the components of the vectors. Thus Eq. (3) appears as a dot product

$$\boxed{\frac{\partial y}{\partial t_1} = \frac{\partial y}{\partial \mathbf{x}} \cdot \frac{\partial \mathbf{x}}{\partial t_1}.} \quad \textbf{(4)}$$

We have recovered our old chain rule formula by using vector notations, subscripted variables, and a dot product. Using these subscripted variables illuminates the structure of these formulas for functions of more than one variable.

EXAMPLE 5 If $y = f(\mathbf{x}) = x_1^2 x_2^3$, where $x_1 = t_1^2 - 2t_2 + t_3$ and $x_2 = t_1 t_2 - t_3^2$, use Eq. (4) to find $\partial y/\partial t_1$ where $(t_1, t_2, t_3) = (-1, 1, 2)$.

Solution When $(t_1, t_2, t_3) = (-1, 1, 2)$, we see that $(x_1, x_2) = (1, -5)$. Now

$$\frac{\partial y}{\partial \mathbf{x}} = \frac{\partial y}{\partial x_1}\mathbf{i} + \frac{\partial y}{\partial x_2}\mathbf{j} = 2x_1 x_2^3 \mathbf{i} + 3x_1^2 x_2^2 \mathbf{j},$$

so

$$\left.\frac{\partial y}{\partial \mathbf{x}}\right|_{(1,-5)} = -250\mathbf{i} + 75\mathbf{j}.$$

We also see that

$$\frac{\partial \mathbf{x}}{\partial t_1} = \frac{\partial x_1}{\partial t_1}\mathbf{i} + \frac{\partial x_2}{\partial t_1}\mathbf{j} = 2t_1\mathbf{i} + t_2\mathbf{j}$$

so

$$\left.\frac{\partial \mathbf{x}}{\partial t_1}\right|_{(-1,1,2)} = -2\mathbf{i} + \mathbf{j}.$$

Thus when $(t_1, t_2, t_3) = (-1, 1, 2)$, we have

$$\frac{\partial y}{\partial t_1} = \frac{\partial y}{\partial \mathbf{x}} \cdot \frac{\partial \mathbf{x}}{\partial t_1} = (-250\mathbf{i} + 75\mathbf{j}) \cdot (-2\mathbf{i} + \mathbf{j})$$

$$= 500 + 75 = 575. \quad \square$$

SUMMARY

1. If $z = f(x, y)$ and $x = g_1(t)$, $y = g_2(t)$, then

$$\frac{dz}{dt} = \frac{\partial z}{\partial x}\frac{dx}{dt} + \frac{\partial z}{\partial y}\frac{dy}{dt}.$$

With subscripted variables, if $y = f(\mathbf{x}) = f(x_1, x_2, x_3)$ and $x_1 = g_1(t)$, $x_2 = g_2(t)$, $x_3 = g_3(t)$, then

$$\frac{dy}{dt} = \frac{\partial y}{\partial x_1}\frac{dx_1}{dt} + \frac{\partial y}{\partial x_2}\frac{dx_2}{dt} + \frac{\partial y}{\partial x_3}\frac{dx_3}{dt}.$$

2. If $z = f(x, y)$ and $x = g_1(s, t)$, $y = g_2(s, t)$, then

$$\frac{\partial z}{\partial t} = \frac{\partial z}{\partial x}\frac{\partial x}{\partial t} + \frac{\partial z}{\partial y}\frac{\partial y}{\partial t}.$$

3. In vector notation, if $y = f(\mathbf{x}) = f(x_1, x_2, x_3)$ while $x_1 = g_1(t_1, t_2)$, $x_2 = g_2(t_1, t_2)$, and $x_3 = g_3(t_1, t_2)$, then

$$\frac{\partial y}{\partial t_k} = \frac{\partial y}{\partial x_1}\frac{\partial x_1}{\partial t_k} + \frac{\partial y}{\partial x_2}\frac{\partial x_2}{\partial t_k} + \frac{\partial y}{\partial x_3}\frac{\partial x_3}{\partial t_k}.$$

The partial derivative $\partial y/\partial t_k$ is given by the dot product

$$\frac{\partial y}{\partial t_k} = \frac{\partial y}{\partial \mathbf{x}} \cdot \frac{\partial \mathbf{x}}{\partial t_k},$$

where

$$\frac{\partial y}{\partial \mathbf{x}} = \frac{\partial y}{\partial x_1}\mathbf{i} + \frac{\partial y}{\partial x_2}\mathbf{j} + \frac{\partial y}{\partial x_3}\mathbf{k}$$

and

$$\frac{\partial \mathbf{x}}{\partial t_k} = \frac{\partial x_1}{\partial t_k}\mathbf{i} + \frac{\partial x_2}{\partial t_k}\mathbf{j} + \frac{\partial x_3}{\partial t_k}\mathbf{k}.$$

EXERCISES

1. Let $z = x^2 + (1/y^2)$, $x = t^2$, and $y = t + 1$.
   a) Find x, y, and z when $t = 1$.
   b) Find $dz/dt|_{t=1}$, using a chain rule.
   c) Express z as a function of t by substitution.
   d) Find $dz/dt|_{t=1}$ by differentiating your answer to part (c).
2. Let $w = f(u, v) = u^2/v$, $u = s^2 - 3s + 1$, and $v = s^3 - s^2$.
   a) Find u, v, and w when $s = 2$.
   b) Find $dw/ds|_{s=2}$ using a chain rule.
   c) Express w as a function of s by substitution.
   d) Find $dw/ds|_{s=2}$ by differentiating your answer to part (c).
3. Let $z = f(x, y) = xy^2$, $x = r\cos\theta$, and $y = r\sin\theta$.
   a) Find x, y, and z when $(r, \theta) = (2, 3\pi/4)$.
   b) Find $\partial z/\partial r$ and $\partial z/\partial\theta$ when $(r, \theta) = (2, 3\pi/4)$.
   c) Express z as a function of r and θ by substitution.
   d) Find $\partial z/\partial r$ and $\partial z/\partial\theta$ when $(r, \theta) = (2, 3\pi/4)$ by differentiating your answer to part (c).
4. Let $w = f(x, y, z) = xz^2 + y/z$, $x = 2s + 3t$, $y = s^2 - t^2$, and $z = s^2t$.
   a) Find x, y, z, and w when $(s, t) = (-1, 1)$.
   b) Find $\partial w/\partial s$ and $\partial w/\partial t$ when $(s, t) = (-1, 1)$.
   c) Express w as a function of s and t by substitution.
   d) Find $\partial w/\partial s$ and $\partial w/\partial t$ when $(s, t) = (-1, 1)$ by differentiating your answer to part (c).

In Exercises 5 through 14, use a chain rule to find the indicated derivative at the given point.

5. $z = x^2 - 2xy + xy^3$, $x = t^3 + 1$, $y = 1/t$; find dz/dt when $t = 1$.
6. $w = xy^2 + z^3$, $x = 2s - 1$, $y = s^3$, $z = s - 4$; find dw/ds when $s = -1$.
7. $w = \sin(uv)$, $u = x^2 - 2x$, $v = x^3 - 5x$; find dw/dx when $x = 2$.
8. $w = x\sin(yz)$, $x = 2t + 1$, $y = 3t^2$, $z = \pi t/2$; find dw/dt when $t = -1$.
9. $z = x^2/y^3$, $x = 2t + 3s$, $y = 3t^3 + 2s$; find $\partial z/\partial t$ when $(t, s) = (-1, 1)$.
10. $w = \tan^{-1}(uv)$, $u = 2x + 3y$, $v = x^2 + y^2$; find $\partial w/\partial y$ when $(x, y) = (-1, 1)$.
11. $w = x^2 + yz$, $x = uv$, $y = u - v$, $z = 2u^2v$; find $\partial w/\partial u$ when $(u, v) = (-1, 2)$.
12. $y = x_1^2 - 3x_1x_2$, $x_1 = (t_1 + 1)/t_2$, $x_2 = t_2e^{t_1}$; find $\partial y/\partial t_1$ when $(t_1, t_2) = (0, -1)$.
13. $z = y_1e^{y_2}$, $y_1 = x_1 - x_2 - x_3$, $y_2 = 2x_1 + x_3$; find $\partial z/\partial x_2$ when $(x_1, x_2, x_3) = (1, 3, -2)$.
14. $w = y_1^2 + \sin(y_2y_3) - e^{y_1}$, $y_1 = x_1x_3$, $y_2 = \ln(x_3^2 + 1)$, $y_3 = x_2\cos x_3$; find $\partial w/\partial x_2$, where $(x_1, x_2, x_3) = (-1, 2, 0)$.

In Exercises 15 through 18, let $z = f(x, y)$ be a differentiable function, and let $x = t^2 - 2s$ and $y = 3t + s^2$.

15. Find $f_x(-1, 4)$ if $f_y(-1, 4) = 5$ and $\partial z/\partial t = 3$ when $(t, s) = (1, 1)$.
16. Find $f_y(1, 3)$ if $f_x(1, 3) = 2$ and $\partial z/\partial t = -4$ when $(t, s) = (1, 0)$.
17. Find $f_x(-1, 4)$ and $f_y(-1, 4)$ if $\partial z/\partial t = 2$ and $\partial z/\partial s = -1$ when $(t, s) = (1, 1)$.
18. Find $f_x(1, 3)$ and $f_y(1, 3)$ if $\partial z/\partial t = 0$ and $\partial z/\partial s = -1$ when $(t, s) = (1, 0)$.
19. If the radius of a circular cylinder is increasing at a rate of 4 in./min while the length is increasing at a rate of 8 in./min, find the rate of change of the volume of the cylinder when the radius is 10 in. and the length is 50 in.
20. The voltage drop V, measured in volts, across a certain conductor of variable resistance R ohms is IR, where I, measured in amperes, is the current flowing through the conductor. The current increases at a constant rate of 2 amp/sec while the voltage drop is kept constant by decreasing the resistance as the current increases. Find the rate of change of the resistance when $I = 5$ amp and $R = 1000$ ohms.
21. The moment of inertia I about an axis of a body of mass m and distance s from the axis is given by $I = ms^2$. Find the rate of change of the moment of inertia about the axis when $s = 50$ if the mass remains constant while the distance s is decreasing at a rate of 3 units length per unit time.
22. Answer Exercise 21 if the body is gaining mass at a rate of 2 units mass per unit time and has mass 20 when $s = 50$, while the other data remain the same.
23. The pressure P in lb/ft^2 of a certain gas at temperature T degrees in a container of variable volume V ft^3 is given by $P = 8T/V$. The temperature is increased at a rate of 5°/min while the volume of the container is increased at a rate of 2 ft^3/min. If, at time t_0, the temperature was 20° and the volume was 10 ft^3, find the rate of change of the pressure 5 min later.

24. By Newton's law of gravitation, the force of attraction between two bodies of masses m_1 and m_2 is Gm_1m_2/s^2, where s is the distance between the bodies and G is the universal gravitational constant. Find the rate of change of the force of attraction for two bodies of constant masses of 10^4 and 10^7 units that are 10^4 units distance apart and are separating at a rate of 10^2 units distance per unit time. (Assume the given units are compatible and don't worry about the value of G or the name of the units in the answer.)

25. Repeat Exercise 24 if the first body is gaining mass at the rate of 30 units mass per unit time and the second body is losing mass at the rate of 80 units mass per unit time, while the other data remain the same.

In Exercises 26 through 34, assume that all functions encountered satisfy enough differentiability conditions to enable you to use any chain rule you wish.

26. If $w = f(u)$ and $u = ax + by$, show that $b(\partial w/\partial x) = a(\partial w/\partial y)$.

27. Obtain a result similar to that in Exercise 26 for $w = f(u)$ and $u = ax + by + cz$.

28. If $w = f(u, v)$, $u = x + y$, and $v = 2x - 2y$, show that

$$\frac{\partial w}{\partial x} \cdot \frac{\partial w}{\partial y} = \left(\frac{\partial f}{\partial u}\right)^2 - 4\left(\frac{\partial f}{\partial v}\right)^2.$$

29. If $w = f(u)$ and $u = xy^2$, show that $2x(\partial w/\partial x) - y(\partial w/\partial y) = 0$.

30. If $w = f(u) + g(v)$, $u = ax + by$, and $v = ax - by$, show that $b^2(\partial^2 w/\partial x^2) = a^2(\partial^2 w/\partial y^2)$.

31. If $w = f(u)$ and $u = x/y$ show that $x(\partial w/\partial x) + y(\partial w/\partial y) = 0$.

32. Let $z = f(x, y)$ and let x and y be functions of t. Find a formula for d^2z/dt^2.

33. Let $z = f(x, y)$ and let x and y be functions of t and s. Find a formula for $\partial^2 z/\partial s^2$.

34. Let $z = f(x, y)$ and let x and y be functions of t and s. Find a formula for $\partial^2 z/\partial s\, \partial t$.

35. If f is a differentiable function of two variables and $f(tx, ty) = t^2 f(x, y)$ for all t, show that $x \cdot f_x(x, y) + y \cdot f_y(x, y) = 2 \cdot f(x, y)$. (Such a function is *homogeneous of degree* 2.) [*Hint:* Differentiate the equation $f(tx_0, ty_0) = t^2 f(x_0, y_0)$ with respect to t, and put $t = 1$.]

36. Generalize the conclusion in Exercise 35 in the case that $f(tx, ty) = t^k f(x, y)$ for all t. (Such a function is *homogeneous of degree k*.)

37. Show that the result in Exercise 31 is a special case of your generalization in Exercise 36.

38. Generalize the conclusion in Exercise 36 for a differentiable function of n variables.

16.5 THE DIRECTIONAL DERIVATIVE

DERIVATIVES IN ALL DIRECTIONS

Let (a_1, a_2) be a point in the x,y-plane in the domain of $f(x, y)$. We want to find the instantaneous rate at which $f(x, y)$ increases per unit change along a direction given by a unit vector $\mathbf{u} = u_1\mathbf{i} + u_2\mathbf{j}$ at (a_1, a_2), as indicated in Fig. 16.25. We think of taking an s-axis, with origin at (a_1, a_2) in the plane, in the direction given by $\mathbf{u}$, as shown in the figure. Figure 16.26 shows that the curve on the surface $z = f(x, y)$ lying directly over an s-axis is the graph of a function $z = h(s)$. The rate of change of z with respect to s at (a_1, a_2) is the derivative $h'(s) = dz/ds$ evaluated at $s = 0$. If we choose the scale on the s-axis to be the same as on the coordinate axes, then $h'(s)$ gives the rate of change of z in the direction given by the vector $\mathbf{u}$ at (a_1, a_2). This derivative $h'(s)$ is then called the **directional derivative** of f at (a_1, a_2) in the direction given by $\mathbf{u}$. We use the notation $f_{\mathbf{u}}(a_1, a_2)$.

Figure 16.25 **u** gives a direction at (a_1, a_2).

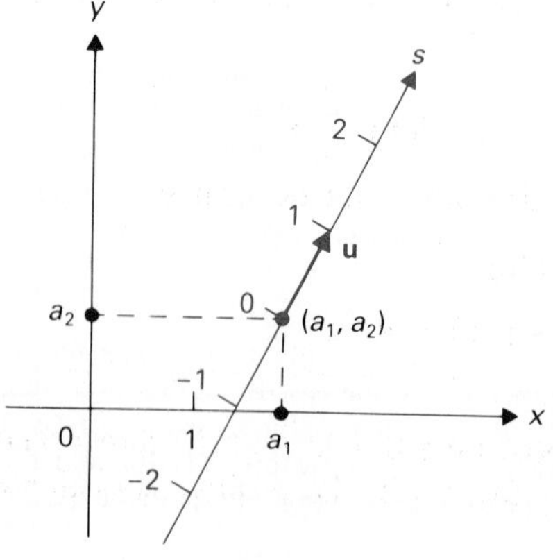

We can compute dz/ds using a chain rule, because z is a function of x and y, and it is easy to express x and y in terms of s. It is crucial to remember, however, that we want the *scale* on the s-axis to be the same as on our coordinate axes. We want dz/ds to be the rate of change of z with respect to *distance*. Let the origin on the s-axis be at the point (a_1, a_2), as shown in Fig. 16.27. *Since* **u** *is a unit vector*, we can measure off our desired scale on the s-axis in terms of multiples of **u**, where we think of **u** as starting from (a_1, a_2). As indicated in Fig. 16.27, for a point (x, y) on the s-axis corre-

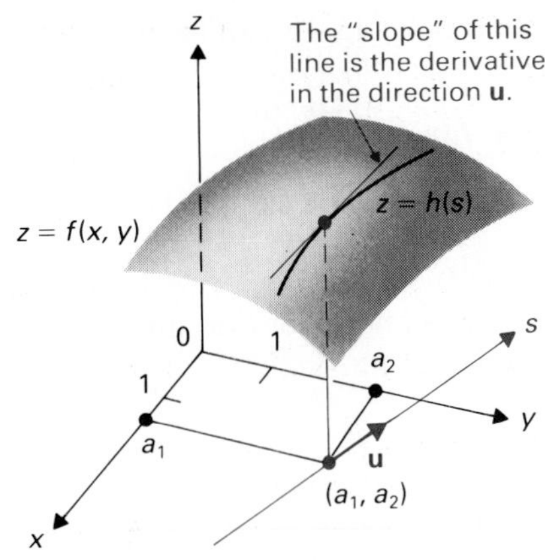

Figure 16.26 The directional derivative dz/ds is the "slope" of the surface in the direction **u**.

sponding to a value s, we have

$$\begin{aligned} x\mathbf{i} + y\mathbf{j} &= (a_1\mathbf{i} + a_2\mathbf{j}) + s\mathbf{u} \\ &= a_1\mathbf{i} + a_2\mathbf{j} + s(u_1\mathbf{i} + u_2\mathbf{j}) \\ &= (a_1 + u_1 s)\mathbf{i} + (a_2 + u_2 s)\mathbf{j}. \end{aligned}$$

Therefore,

$$x = a_1 + u_1 s, \qquad y = a_2 + u_2 s.$$

Then the chain rule shows that

$$\frac{dz}{ds} = \frac{\partial z}{\partial x}\frac{dx}{ds} + \frac{\partial z}{\partial y}\frac{dy}{ds} = \frac{\partial z}{\partial x} \cdot u_1 + \frac{\partial z}{\partial y} \cdot u_2. \tag{1}$$

When $s = 0$, we have

$$\left.\frac{dz}{ds}\right|_{s=0} = u_1 f_x(a_1, a_2) + u_2 f_y(a_1, a_2). \tag{2}$$

If $w = f(x, y, z)$, then the directional derivative of w at $\mathbf{a} = (a_1, a_2, a_3)$ in the direction given by the *unit* vector $\mathbf{u} = u_1\mathbf{i} + u_2\mathbf{j} + u_3\mathbf{k}$ is, of course,

$$\left.\frac{dw}{ds}\right|_{s=0} = u_1 f_x(\mathbf{a}) + u_2 f_y(\mathbf{a}) + u_3 f_z(\mathbf{a}). \tag{3}$$

We also denote the directional derivative given in Eq. (2) by $f_{\mathbf{u}}(\mathbf{a})$, the derivative of $f(x, y)$ at $\mathbf{a} = (a_1, a_2)$ in the direction given by the unit vector $\mathbf{u} = u_1\mathbf{i} + u_2\mathbf{j}$. Thus

$$f_{\mathbf{u}}(\mathbf{a}) = u_1 f_x(\mathbf{a}) + u_2 f_y(\mathbf{a}). \tag{4}$$

Using subscripted variables, if $y = f(\mathbf{x}) = f(x_1, x_2, x_3)$, while $\mathbf{a} = (a_1, a_2, a_3)$ and $\mathbf{u} = u_1\mathbf{i} + u_2\mathbf{j} + u_3\mathbf{k}$, then

$$f_{\mathbf{u}}(\mathbf{a}) = u_1 f_{x_1}(\mathbf{a}) + u_2 f_{x_2}(\mathbf{a}) + u_3 f_{x_3}(\mathbf{a}). \tag{5}$$

Recall that the gradient vector $\nabla f(\mathbf{a})$ for a function $f(x, y)$ at a point $\mathbf{a} = (a_1, a_2)$ is given by

$$\nabla f(\mathbf{a}) = f_x(\mathbf{a})\mathbf{i} + f_y(\mathbf{a})\mathbf{j}. \tag{6}$$

Since $\mathbf{u} = u_1\mathbf{i} + u_2\mathbf{j}$, we see that Eq. (4) can be written as a dot product,

$$f_{\mathbf{u}}(\mathbf{a}) = \nabla f(\mathbf{a}) \cdot \mathbf{u}. \tag{7}$$

We will always think in terms of Eq. (7) when actually computing a directional derivative. Since **u** is a unit vector, we see from Eq. (7) that $f_{\mathbf{u}}(\mathbf{a})$ is the scalar component of $\nabla f(\mathbf{a})$ along **u**.

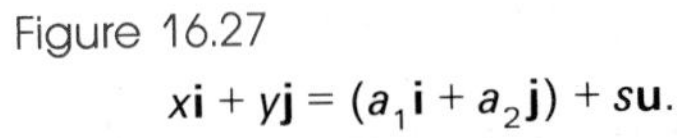

Figure 16.27
$x\mathbf{i} + y\mathbf{j} = (a_1\mathbf{i} + a_2\mathbf{j}) + s\mathbf{u}$.

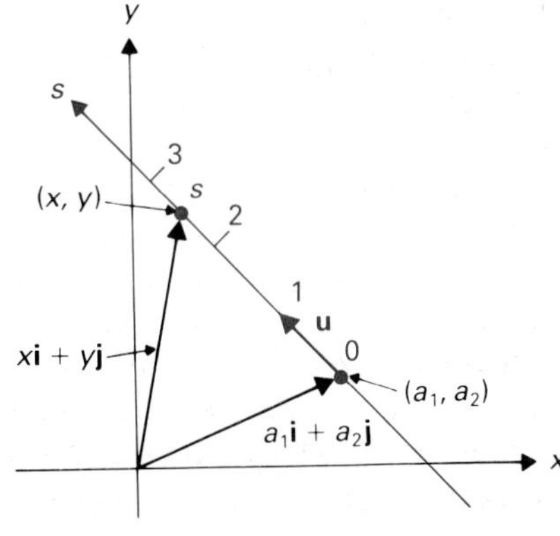

> To find a directional derivative of f at a point **a**:
>
> **Step 1** Find a *unit* vector **u** in the direction of the derivative.
>
> **Step 2** Find the gradient vector $\nabla f(\mathbf{a})$.
>
> **Step 3** The directional derivative is $f_{\mathbf{u}}(\mathbf{a}) = \nabla f(\mathbf{a}) \cdot \mathbf{u}$.

EXAMPLE 1 Use the steps in the preceding box to find the directional derivatives of a function $f(x, y)$ at $\mathbf{a} = (a_1, a_2)$ in the positive coordinate directions.

Solution A unit vector in the direction of increasing x is $\mathbf{i}$ and a unit vector in the direction of increasing y is $\mathbf{j}$. Since

$$\nabla f(\mathbf{a}) = f_x(\mathbf{a})\mathbf{i} + f_y(\mathbf{a})\mathbf{j},$$

we have

$$f_{\mathbf{i}}(\mathbf{a}) = \nabla f(\mathbf{a}) \cdot \mathbf{i} = f_x(\mathbf{a}) \qquad \text{and} \qquad f_{\mathbf{j}}(\mathbf{a}) = \nabla f(\mathbf{a}) \cdot \mathbf{j} = f_y(\mathbf{a}).$$

Of course this just verifies our intuitive understanding of $f_x(\mathbf{a})$ and $f_y(\mathbf{a})$ as derivatives in the positive coordinate directions. □

EXAMPLE 2 Let f be defined by

$$f(x, y) = x^2 + 3xy^2.$$

Find the directional derivative of f at $(1, 2)$ in the direction toward the origin.

Solution A vector in the direction from $(1, 2)$ to $(0, 0)$ is $-\mathbf{i} - 2\mathbf{j}$, so a *unit* vector in this direction is therefore

$$\mathbf{u} = -\frac{1}{\sqrt{5}}\mathbf{i} - \frac{2}{\sqrt{5}}\mathbf{j}.$$

We find that

$$f_x(1, 2) = (2x + 3y^2)\big|_{(1,2)} = 14$$

and

$$f_y(1, 2) = 6xy\big|_{(1,2)} = 12.$$

Thus

$$\nabla f(1, 2) = 14\mathbf{i} + 12\mathbf{j}.$$

Therefore the directional derivative is given by

$$f_{\mathbf{u}}(1, 2) = \nabla f(1, 2) \cdot \mathbf{u} = -\frac{1}{\sqrt{5}}(14) + \left(-\frac{2}{\sqrt{5}}\right)(12) = -\frac{38}{\sqrt{5}}. \quad \square$$

EXAMPLE 3 Suppose the temperature at a point in space is given by

$$T(x, y, z) = \frac{180}{x^2 + y^2 + z^2} \text{ degrees.}$$

Find the rate of change of temperature with respect to distance at $(1, -1, 4)$ in the direction toward the point $(2, -3, 2)$.

Solution The vector from $(1, -1, 4)$ to $(2, -3, 2)$ is $\mathbf{i} - 2\mathbf{j} - 2\mathbf{k}$. Since this vector has length $\sqrt{1^2 + (-2)^2 + (-2)^2} = 3$, a unit vector with this direction is

$$\mathbf{u} = \frac{1}{3}\mathbf{i} - \frac{2}{3}\mathbf{j} - \frac{2}{3}\mathbf{k}.$$

We find that

$$\left.\frac{\partial T}{\partial x}\right|_{(1,-1,4)} = \left.\frac{180(-2x)}{(x^2 + y^2 + z^2)^2}\right|_{(1,-1,4)} = \frac{-360}{18^2} = -\frac{10}{9}.$$

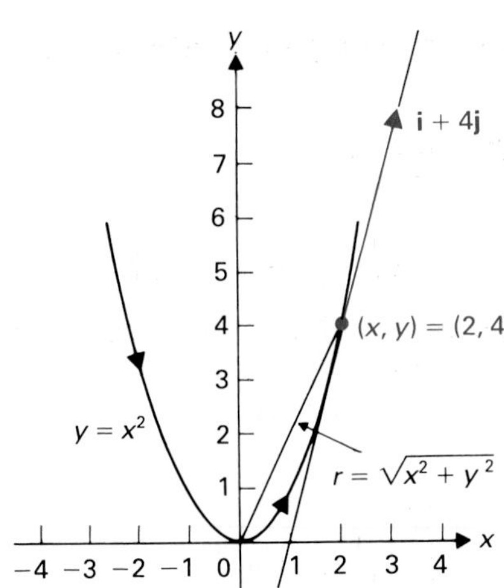

Figure 16.28 As a body moves on $y = x^2$ in the direction of increasing x, its direction at (2, 4) is given by $\mathbf{i} + 4\mathbf{j}$.

Similarly, we find that

$$\left.\frac{\partial T}{\partial y}\right|_{(1,-1,4)} = \frac{10}{9} \quad \text{and} \quad \left.\frac{\partial T}{\partial z}\right|_{(1,-1,4)} = -\frac{40}{9}.$$

Thus

$$\nabla T(1, -1, 4) = -\frac{10}{9}\mathbf{i} + \frac{10}{9}\mathbf{j} - \frac{40}{9}\mathbf{k}.$$

Consequently, the rate of change of temperature is

$$T_{\mathbf{u}}(1, -1, 4) = \nabla T(1, -1, 4) \cdot \mathbf{u} = -\frac{10}{27} - \frac{20}{27} + \frac{80}{27} = \frac{50}{27}. \quad \square$$

EXAMPLE 4 A body in the plane moves on the parabola $y = x^2$ in the direction of increasing x, as shown in Fig. 16.28. Find the rate of change with respect to arc length s of the distance r of the body from the origin when the body is at the point (2, 4).

Solution We need the directional derivative of $r = f(x, y) = \sqrt{x^2 + y^2}$ at (2, 4) in the direction tangent to the parabola and in which x is increasing. Since $dy/dx = 2x = 4$ when $x = 2$, we see that a vector tangent to the curve is $\mathbf{i} + 4\mathbf{j}$. Since this vector has positive coefficient of $\mathbf{i}$, it points in the direction tangent to the curve and in which x increases. A unit vector in this direction is

$$\mathbf{u} = \frac{1}{\sqrt{17}}\mathbf{i} + \frac{4}{\sqrt{17}}\mathbf{j}.$$

Now

$$\frac{\partial f}{\partial x} = \frac{x}{\sqrt{x^2 + y^2}} \quad \text{and} \quad \frac{\partial f}{\partial y} = \frac{y}{\sqrt{x^2 + y^2}}.$$

At $(x, y) = (2, 4)$, we see that

$$\nabla f(2, 4) = \frac{2}{\sqrt{20}}\mathbf{i} + \frac{4}{\sqrt{20}}\mathbf{j} = \frac{1}{\sqrt{5}}\mathbf{i} + \frac{2}{\sqrt{5}}\mathbf{j}.$$

Thus

$$f_{\mathbf{u}}(2, 4) = \nabla f(2, 4) \cdot \mathbf{u} = \frac{1}{\sqrt{85}} + \frac{8}{\sqrt{85}} = \frac{9}{\sqrt{85}}. \quad \square$$

Figure 16.29

$\nabla f(\mathbf{a}) \cdot \mathbf{u} = |\nabla f(\mathbf{a})||\mathbf{u}|(\cos\theta)$
$= |\nabla f(\mathbf{a})|(\cos\theta)$

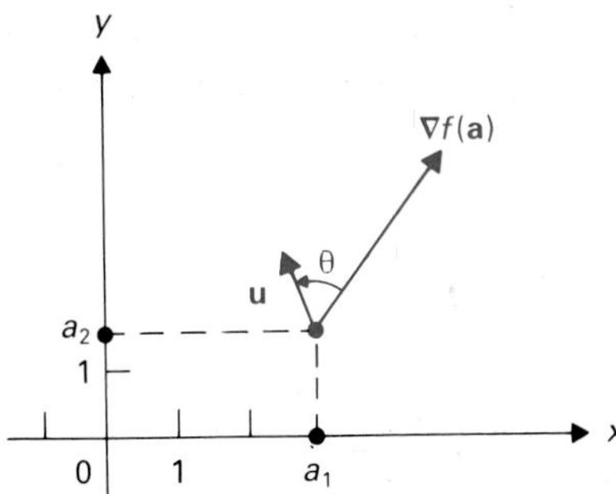

THE MAGNITUDE AND DIRECTION OF ∇f

We saw that

$$f_{\mathbf{u}}(\mathbf{a}) = \nabla f(\mathbf{a}) \cdot \mathbf{u}. \tag{8}$$

Recall that a dot product of two vectors is equal to the product of their lengths and the cosine of the angle θ between them. See Fig. 16.29. Thus

$$f_{\mathbf{u}}(\mathbf{a}) = |\nabla f(\mathbf{a})||\mathbf{u}|(\cos\theta) = |\nabla f(\mathbf{a})|(\cos\theta), \tag{9}$$

since $|\mathbf{u}| = 1$. Now $|\nabla f(\mathbf{a})|(\cos\theta)$ is a function of θ at the point $\mathbf{a}$. It has its maximum value when $\cos\theta = 1$, and that maximum is then $|\nabla f(\mathbf{a})|$. This shows that

> the magnitude of $\nabla f(\mathbf{a})$ is the maximum of the directional derivatives of f at $\mathbf{a}$ in all possible directions.

Since $f_{\mathbf{u}}(\mathbf{a})$ in Eq. (9) assumes this maximum value when $\cos\theta = 1$, so that the angle θ between $f(\mathbf{a})$ and $\mathbf{u}$ is zero, we see that

> $\nabla f(\mathbf{a})$ points in the direction of maximum rate of change of f at the point $\mathbf{a}$.

We now understand the use of the term "gradient vector" for $\nabla f(\mathbf{a})$. It points in the direction of maximum steepness or "grade" of the graph of the function f at the point and has length equal to the slope or "grade" of the climb in that direction.

EXAMPLE 5 Find the maximum rate of change with respect to distance, taken over all possible directions, of $f(x, y) = x^2e^y + y\tan^{-1}z$ at the point $(1, 0, 0)$.

Solution The gradient vector of f at $(1, 0, 0)$ is given by

$$\nabla f(1, 0, 0) = f_x(1, 0, 0)\mathbf{i} + f_y(1, 0, 0)\mathbf{j} + f_z(1, 0, 0)\mathbf{k} = 2\mathbf{i} + \mathbf{j}.$$

Thus the direction of maximum rate of increase of $f(x, y, z)$ at $(1, 0, 0)$ is given by the vector $2\mathbf{i} + \mathbf{j}$, and this maximum rate of increase is

$$|\nabla f(1, 0, 0)| = |2\mathbf{i} + \mathbf{j}| = \sqrt{5}. \quad \square$$

We observe from Eq. (9) that $f_{\mathbf{u}}(\mathbf{a})$ is minimum at $\mathbf{a}$ when $\cos\theta = -1$, when $\theta = \pi$. Of course the direction is then opposite to $\nabla f(\mathbf{a})$. Thus $-\nabla f(\mathbf{a})$ points in the direction of *minimum* rate of *increase* (or *maximum* rate of *decrease*) of f at $\mathbf{a}$, and the rate of change in that direction is $-|\nabla f(\mathbf{a})|$. This is intuitively clear. If a function has a maximum rate of change of 5 in some direction, then the rate of change in the opposite direction is -5, and that is the minimum rate of change. If $f(x, y)$ describes altitude above sea level, then $-\nabla f$ describes the direction for the expert ski trail from the top of a mountain to the bottom, assuming there are no hollows or cliffs!

EXAMPLE 6 Consider the points (x, y, z) in the first octant in space, where x, y, and z are all positive. The volume of a rectangular box having length x, width y, and height z is $V = xyz$. We wish to travel in space from $(6, 3, 2)$ in the direction corresponding to the greatest rate of *decrease* of volume of the box with respect to distance traveled in space. Describe the relative rates of decrease of the length, width, and height of the box as we leave the point $(6, 3, 2)$ in that direction.

Solution The volume of a box of length x, width y, and height z is

$$V = f(x, y, z) = xyz.$$

We see that

$$\nabla f(6, 3, 2) = (yz\mathbf{i} + xz\mathbf{j} + xy\mathbf{k})|_{(6,3,2)} = 6\mathbf{i} + 12\mathbf{j} + 18\mathbf{k}.$$

Then

$$-\nabla f(6, 3, 2) = -6\mathbf{i} - 12\mathbf{j} - 18\mathbf{k}.$$

This vector points in the direction for maximum rate of decrease of volume. To achieve this maximum rate of decrease, we should start at (6, 3, 2) to decrease the width twice as rapidly as the length and decrease the height three times as rapidly as the length. □

SUMMARY

1. The directional derivative of a function f at a point $\mathbf{a}$ in the direction of a unit vector $\mathbf{u}$ is the instantaneous rate of change at $\mathbf{a}$ of function values z with respect to distance s as we travel in the direction given by $\mathbf{u}$. Notations are $dz/ds|_{\mathbf{a}}$, $f_{\mathbf{u}}(\mathbf{a})$, and $D_{\mathbf{u}}f(\mathbf{a})$.
2. If f is differentiable at $\mathbf{a}$, the directional derivative just described is found by the formula

$$f_{\mathbf{u}}(\mathbf{a}) = \nabla f(\mathbf{a}) \cdot \mathbf{u}.$$

It is very important to remember that $\mathbf{u}$ must be a *unit* vector.
3. The gradient vector $\nabla f(\mathbf{a})$ points in the direction of maximum rate of increase of the function at $\mathbf{a}$ and has magnitude equal to that rate of increase.

EXERCISES

In Exercises 1 through 20, find the directional derivative of the indicated function f at the given point in the given direction.

1. $f(x, y) = x^2 - 3xy$ at $(1, -2)$ in the direction of $3\mathbf{i} - 4\mathbf{j}$
2. $f(x, y) = \sin(xy)$ at $(\pi, -1)$ in the direction of $\mathbf{i} - \mathbf{j}$
3. $f(x, y, z) = xy/z$ at $(2, -3, -1)$ in the direction of $-\mathbf{i} + 2\mathbf{j} + 2\mathbf{k}$
4. $f(x, y, z) = x^2 + 2xz^2$ at $(1, 5, -1)$ in the direction of $\mathbf{i} - \mathbf{j} + \mathbf{k}$
5. $f(x, y) = x^2 - 3xy^3$ at $(-2, 1)$ toward the origin
6. $f(x, y) = \sin^{-1}(x/y)$ at $(3, 5)$ toward $(4, 4)$
7. $f(x_1, x_2, x_3) = x_1x_2^2e^{x_3}$ at $(1, 3, 0)$ toward $(1, 3, -1)$
8. $f(x_1, x_2, x_3) = x_2\tan^{-1}(x_1x_3)$ at $(-1, 2, 1)$ in the direction toward $(3, 1, -1)$
9. $f(x, y) = \tan^{-1}(x^2 + y^2)$ at $(1, -2)$ in the direction of increasing y along the line $y = 3x - 5$
10. $f(x, y) = x^2e^{xy}$ at $(3, 0)$ in the direction of decreasing x along the normal to the line $3x - 2y = 9$
11. $f(x, y) = xe^{x+y}$ at $(-3, 3)$ in the direction of decreasing y normal to the curve $y = x^2 + 3x + 3$
12. $f(x, y) = x^2 + y\sinh(xy)$ at $(2, 0)$ in the direction of increasing x tangent to the curve $y = \sqrt{x + 7} - 3$
13. $f(x, y, z) = x^2z/y$ at $(2, -1, 3)$ in the direction of increasing z normal to the plane $x + 2y - 2z = -6$
14. $f(x, y, z) = x^2 + 3yz$ at $(1, -1, 2)$ in the direction of increasing y normal to the surface $z = x^2 + y^2$
15. $f(x, y, z) = x^2y - 3xz^2$ at $(1, -1, 1)$ in the direction of decreasing x normal to the surface $z = xy^2$
16. $f(x, y, z) = x^2 + \ln(y^2 + z)$ at $(3, 2, -2)$ in the direction of decreasing z normal to the surface $z = xy - 3x + 4y - 7$
17. $f(x, y, z) = z/(x^2 + y^2)$ at $(-1, 1, 3)$ along the line $x = -1 - 2t$, $y = 1 + t$, $z = 3 + 2t$ in the direction of decreasing t
18. $f(x, y, z) = y^2\sin(xz^2)$ at $(0, -2, 2)$ along the line $x = 6t$, $y = -2 + 2t$, $z = 2 - 3t$ in the direction of increasing t
19. $f(x, y, z) = xy^2e^z$ at $(2, -1, 0)$ in the direction of increasing y along the line of intersection of the planes $x - y + z = 3$ and $2x + y - z = 3$
20. $f(x, y, z) = (xy + y^2)/z$ at $(2, -1, -1)$ in the direction of decreasing x along the line of intersection of the planes $3x - y + z = 6$ and $2x - 3y - z = 8$
21. One side of the roof of a cathedral has the shape of the plane $z = 30 + 0.8y$ for $0 \le x \le 100$ and $0 \le y \le 40$. Find the rate at which the height to the roof is increasing over the head of a person at $(35, 25, 0)$ walking toward $(0, 40, 0)$.
22. If the temperature in space is $(y - z)/(1 + x^2 + y^2 + z^2)$ degrees at the point (x, y, z), find the rate of change of temperature at $(1, 0, -1)$ in the direction toward $(1, 1, 1)$.

In Exercises 23 through 26, find (a) a vector in the direction in which the function increases most rapidly at the given point, (b) that maximum rate of increase, (c) a vector in the direction for minimum rate of change at the point, and (d) the minimum rate of change.

23. $f(x, y) = x^2 + x \ln(y)$ at $(2, 1)$

24. $f(x, y) = xy^2 + e^{xy}$ at $(0, 2)$

25. $f(x, y, z) = x^2y + \tan^{-1}(xz)$ at $(1, -2, 1)$

26. $f(x, y, z) = (x + y)/z^2$ at $(2, -1, 2)$

27. Let f be differentiable at $\mathbf{a}$ and have a local maximum at $\mathbf{a}$. What can be said concerning the directional derivatives of f in all directions at $\mathbf{a}$?

28. Give an example to show that it is possible to have $f(x, y)$ such that $f_{\mathbf{u}}(0, 0) = 0$ for all unit vectors $\mathbf{u}$ and still have neither a local maximum nor a local minimum for f at $(0, 0)$.

29. Consider a surface in space with equation $f(x, y, z) = c$, where c is some constant.

a) Argue that since the derivative is a rate of change, the directional derivative of f at any point on the surface in any direction tangent to the surface should be zero.

b) Argue from part (a) that ∇f should be normal to the surface at each point on the surface.

You know that for a suitable function f, the gradient vector $\nabla f(\mathbf{a})$ points in the direction of maximum rate of change of $f(\mathbf{x})$ at $\mathbf{a}$ if $\nabla f(\mathbf{a}) \neq \mathbf{0}$. If $\nabla f(\mathbf{a}) = \mathbf{0}$, then the rate of increase of $f(\mathbf{x})$ in all directions at $\mathbf{a}$ is zero. In Exercises 30 through 36, find the direction or directions in which you should travel from the given point to attain maximum increase $f(\mathbf{x})$ over a short distance. The text gives no instruction in this; you just think about it in each case.

30. $f(x) = x^2$ at 0

31. $f(x) = x^3$ at 0

32. $f(x, y) = x^2 + y^2$ at $(0, 0)$

33. $f(x, y) = x^2 - y^2$ at $(0, 0)$

34. $f(x, y) = x^2 + 2xy + y^2$ at $(0, 0)$

35. $f(x, y) = x^3 - x^2 + y^2$ at $(0, 0)$

36. $f(x, y, z) = 5$ at $(1, -2, 7)$

16.6 DIFFERENTIATION OF IMPLICIT FUNCTIONS

Section 3.6 treated differentiation of implicitly defined functions of one variable, finding dy/dx given an equation of the form $G(x, y) = c$. We refresh our memory with an example.

EXAMPLE 1 Let $x^2y^3 + 2xy^2 - x^3 = 3$ and find dy/dx by differentiating implicitly.

Solution We have

$$x^2 \cdot 3y^2 \frac{dy}{dx} + 2xy^3 + 2x \cdot 2y \frac{dy}{dx} + 2y^2 - 3x^2 = 0,$$

so

$$\frac{dy}{dx} = \frac{3x^2 - 2xy^3 - 2y^2}{3x^2y^2 + 4xy}. \quad \square$$

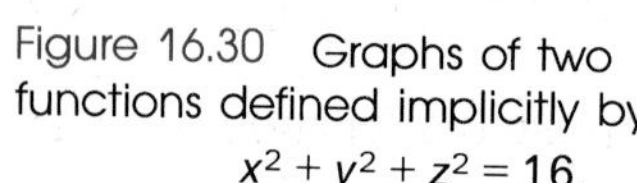
Figure 16.30 Graphs of two functions defined implicitly by $x^2 + y^2 + z^2 = 16$.

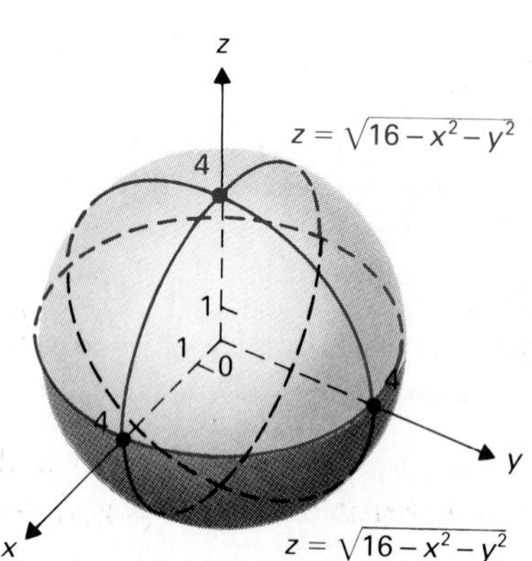

An equation of the form $G(x, y, z) = c$ may define z implicitly as one or more functions of both x and y. For example, $x^2 + y^2 + z^2 = 16$ gives rise to the functions

$$z = \sqrt{16 - x^2 - y^2} \quad \text{and} \quad z = -\sqrt{16 - x^2 - y^2},$$

whose graphs are shaded in different colors in Fig. 16.30. We will not describe exactly when $G(x, y, z) = c$ gives such implicitly defined functions. For functions we will use here, the solution set of $G(x, y, z) = c$ is a surface in space, and a piece of such a surface containing a point (x_0, y_0, z_0) can often be regarded as the graph of a function $z = f(x, y)$ such that $z_0 = f(x_0, y_0)$. In this case, we would like to find $\partial z/\partial x$ and $\partial z/\partial y$. This can again be accomplished by implicit differentiation.

EXAMPLE 2 Let $x^2z + yz^3 - 2xy^2 = -9$. Find $\partial z/\partial x$ at the point $(1, -2, 1)$ on the surface.

Solution We differentiate implicitly with respect to x, thinking of y as a constant, obtaining

$$x^2 \frac{\partial z}{\partial x} + z \cdot 2x + y \cdot 3z^2 \frac{\partial z}{\partial x} - 2y^2 = 0.$$

Then

$$\frac{\partial z}{\partial x} = \frac{2y^2 - 2xz}{x^2 + 3yz^2},$$

so

$$\left.\frac{\partial z}{\partial x}\right|_{(1,-2,1)} = \frac{8 - 2}{1 - 6} = -\frac{6}{5}. \quad \square$$

There is an easy formula that avoids the technique of implicit differentiation. Suppose $G(x, y, z) = c$ defines $z = f(x, y)$. Then the equations

$$w = G(x, y, z), \tag{1}$$

$$x = x, \qquad y = y, \qquad z = f(x, y) \tag{2}$$

define w as a composite function of the *two* variables x and y. Furthermore, as a function of the two independent variables x and y in Eqs. (2), w remains constant at c, because

$$w = G(x, y, z) = G(x, y, f(x, y)) = c$$

since $z = f(x, y)$ was chosen so that $G(x, y, z) = c$. Thus $\partial w/\partial x = 0$ and $\partial w/\partial y = 0$. By the chain rule for the composition of functions given by Eqs. (1) and (2),

$$\frac{\partial w}{\partial x} = \frac{\partial G}{\partial x}\frac{\partial x}{\partial x} + \frac{\partial G}{\partial y}\frac{\partial y}{\partial x} + \frac{\partial G}{\partial z}\frac{\partial z}{\partial x} = \frac{\partial G}{\partial x} \cdot 1 + \frac{\partial G}{\partial y} \cdot 0 + \frac{\partial G}{\partial z}\frac{\partial z}{\partial x}. \tag{3}$$

(Note that $\partial y/\partial x = 0$ from Eq. 2, where y is regarded as a function of x and y.) Since $\partial w/\partial x = 0$, we obtain from Eq. (3)

$$\frac{\partial G}{\partial x} + \frac{\partial G}{\partial z}\frac{\partial z}{\partial x} = 0,$$

so

$$\boxed{\frac{\partial z}{\partial x} = -\frac{\partial G/\partial x}{\partial G/\partial z}.} \tag{4}$$

Formula (4) for $\partial z/\partial x$ is valid wherever $\partial G/\partial z \neq 0$. It can be shown that, for a nice function G, the condition $\partial G/\partial x \neq 0$ guarantees that $G(x, y, z) = c$ does define z implicitly as a function of x and y.

EXAMPLE 3 Do Example 2 again using formula (4).

Solution From

$$G(x, y, z) = x^2z + yz^3 - 2xy^2 = -9,$$

we obtain

$$\frac{\partial G}{\partial x} = 2xz - 2y^2 \quad \text{and} \quad \frac{\partial G}{\partial z} = x^2 + 3yz^2,$$

so

$$\frac{\partial z}{\partial x} = -\frac{2xz - 2y^2}{x^2 + 3yz^2} = \frac{2y^2 - 2xz}{x^2 + 3yz^2}.$$

The derivative is then computed at $(1, -2, 1)$ as in Example 2, but the messy implicit differentiation is gone. □

The formula

$$\boxed{\frac{\partial z}{\partial x} = -\frac{\partial G/\partial x}{\partial G/\partial z}}$$

is fairly easy to remember in this Leibniz notation; the ∂G's "cancel" to give what we want, *but we must remember the minus sign also.* This is one place where the Leibniz notation lets us down just a bit.

Of course, a similar argument shows that

$$\boxed{\frac{\partial z}{\partial y} = -\frac{\partial G/\partial y}{\partial G/\partial z}.}$$

Indeed, if $G(x_1, x_2, x_3) = c$ and we solve for x_i in terms of the other x's, similar arguments show that

$$\boxed{\frac{\partial x_i}{\partial x_j} = -\frac{\partial G/\partial x_j}{\partial G/\partial x_i}, \qquad i \neq j.} \tag{5}$$

We can even use the formula to solve the implicit differentiation problems in Section 3.6, finding dy/dx if $G(x, y) = c$, for

$$\boxed{\frac{dy}{dx} = -\frac{\partial G/\partial x}{\partial G/\partial y}.} \tag{6}$$

EXAMPLE 4 Solve Example 1 again using Eq. (6), finding dy/dx if $G(x, y) = x^2y^3 + 2xy^2 - x^3 = 3$.

Solution We have

$$\frac{\partial G}{\partial x} = 2xy^3 + 2y^2 - 3x^2 \quad \text{and} \quad \frac{\partial G}{\partial y} = 3x^2y^2 + 4xy,$$

so

$$\frac{dy}{dx} = -\frac{2xy^3 + 2y^2 - 3x^2}{3x^2y^2 + 4xy} = \frac{3x^2 - 2xy^3 - 2y^2}{3x^2y^2 + 4xy},$$

as obtained in Example 1. □

Recall that a vector normal to a surface $z = f(x, y)$ given implicitly by $G(x, y, z) = c$ is

$$\frac{\partial z}{\partial x}\mathbf{i} + \frac{\partial z}{\partial y}\mathbf{j} - \mathbf{k} = -\frac{\partial G/\partial x}{\partial G/\partial z}\mathbf{i} - \frac{\partial G/\partial y}{\partial G/\partial z}\mathbf{j} - \mathbf{k}. \tag{7}$$

Multiplying the vector in Eq. (7) by $-\partial G/\partial z$, we obtain, as a vector normal to the surface $G(x, y, z) = c$,

$$\frac{\partial G}{\partial x}\mathbf{i} + \frac{\partial G}{\partial y}\mathbf{j} + \frac{\partial G}{\partial z}\mathbf{k}. \tag{8}$$

But Eq. (8) is the gradient vector ∇G. The surface $G(x, y, z) = c$ is called a **level surface** of the function $G(x, y, z)$, for it consists of all points (x, y, z) where the function attains the "level" c. In summary,

> the gradient vector of a function at a point is normal to the level surface of the function through that point. **(9)**

Of course, for a function $G(x, y)$ of just two variables, $G(x, y) = c$ is a curve in the plane, called a **level curve** of G. Again, the gradient vector ∇G is perpendicular to the level curve at each point. This follows at once from the statement (9) if we consider the solution set of $G(x, y) = c$ in space, which is a vertical cylinder intersecting the x, y-plane in the curve $G(x, y) = c$.

EXAMPLE 5 Use the statement (9) to find a vector normal to the curve $y = x^2 - 1$ at the point (2, 3).

Solution The curve $y = x^2 - 1$ is a level curve in the plane of $G(x, y) = y - x^2$. Thus a perpendicular vector to the curve at (2, 3) is given by $\nabla G(2, 3)$. We obtain

$$\frac{\partial G}{\partial x} = -2x \quad \text{and} \quad \frac{\partial G}{\partial y} = 1,$$

so $\nabla G = -2x\mathbf{i} + \mathbf{j}$, and $\nabla G(2, 3) = -4\mathbf{i} + \mathbf{j}$. The curve and this vector are shown in Fig. 16.31. □

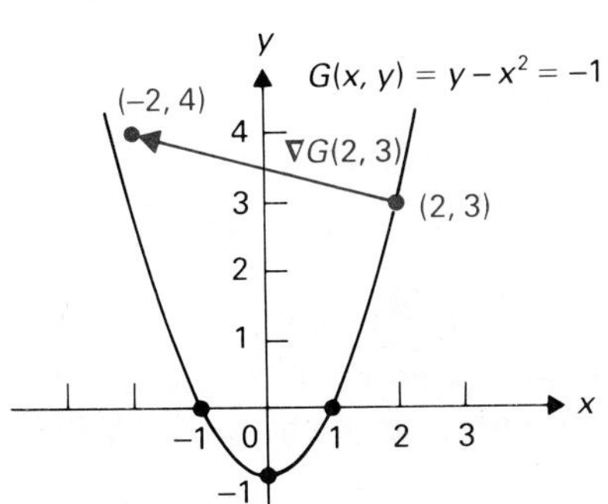

Figure 16.31 $\nabla G(2, 3) = -4\mathbf{i} + \mathbf{j}$ is normal to the level curve of

$$G(x, y) = y - x^2$$

through (2, 3).

EXAMPLE 6 Find the tangent plane and normal line to the surface $x^2yz + x^2y^3 + \sin(x^2z) = 8$ at the point $(-1, 2, 0)$.

Solution Setting $G(x, y, z) = x^2yz + x^2y^3 + \sin(x^2z)$, we have

$$G_x(-1, 2, 0) = [2xyz + 2xy^3 + 2xz\cos(x^2z)]|_{(-1,2,0)} = -16,$$
$$G_y(-1, 2, 0) = (x^2z + 3x^2y^2)|_{(-1,2,0)} = 12,$$
$$G_z(-1, 2, 0) = [x^2y + x^2\cos(x^2z)]|_{(-1,2,0)} = 3.$$

Therefore the normal vector (8) is $-16\mathbf{i} + 12\mathbf{j} + 3\mathbf{k}$. The tangent plane at $(-1, 2, 0)$ has equation

$$16x - 12y - 3z = -40,$$

and the normal line has parametric equations

$$x = -1 - 16t, \qquad y = 2 + 12t, \qquad z = 3t. \quad \square$$

Let $T(x, y, z)$ give the temperature at points (x, y, z) in space. A surface $T(x, y, z) = c$, where the temperature remains constant at c, is called an **isothermal surface.**

EXAMPLE 7 Let the temperature at (x, y, z) be given by

$$T(x, y, z) = \frac{100}{2 + x^2 + y^2 + z^2 + 2y} \text{ degrees.}$$

Find:

a. the temperature at $(2, 1, -1)$,

b. the equation of the isothermal surface through $(2, 1, -1)$, and sketch the surface,

c. the direction for maximum rate of change of temperature at $(2, 1, -1)$ and the rate of change in that direction.

Solution

a. The temperature at $(2, 1, -1)$ is

$$T(2, 1, -1) = \frac{100}{2 + 4 + 1 + 1 + 2} = \frac{100}{10} = 10 \text{ degrees.}$$

b. The equation of the isothermal surface through $(2, 1, -1)$ is $T(x, y, z) = 10$, or

$$\frac{100}{2 + x^2 + y^2 + z^2 + 2y} = 10,$$

$$2 + x^2 + y^2 + z^2 + 2y = 10,$$

$$x^2 + (y^2 + 2y) + z^2 = 8,$$

$$x^2 + (y + 1)^2 + z^2 = 1 + 8 = 9.$$

The surface is therefore a sphere with center $(0, -1, 0)$ and radius 3, sketched in Fig. 16.32.

c. The gradient $\nabla T(2, 1, -1)$ gives the direction for maximum rate of change of temperature at the point $(2, 1, -1)$. We have

$$\frac{\partial T}{\partial x} = \frac{-200x}{(2 + x^2 + y^2 + z^2 + 2y)^2},$$

$$\frac{\partial T}{\partial y} = \frac{-200y - 200}{(2 + x^2 + y^2 + z^2 + 2y)^2},$$

and

$$\frac{\partial T}{\partial z} = \frac{-200z}{(2 + x^2 + y^2 + z^2 + 2y)^2}.$$

We easily find that

$$\nabla T(2, 1, -1) = -4\mathbf{i} - 4\mathbf{j} + 2\mathbf{k},$$

so $-2\mathbf{i} - 2\mathbf{j} + \mathbf{k}$ is a vector at $(2, 1, -1)$ pointing in the direction of maximum rate of change. This maximum rate of change is $|\nabla T| = \sqrt{4^2 + 4^2 + 2^2} = \sqrt{36} = 6$. □

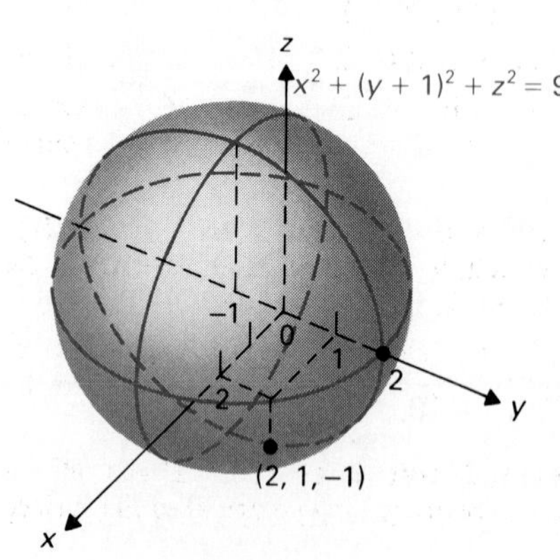

Figure 16.32 The isothermal surface of $T(x, y, z) = 100/(2 + x^2 + y^2 + z^2 + 2y)$ through $(2, 1, -1)$.

SUMMARY

1. If $G(x, y, z) = c$ defines $z = f(x, y)$, then $\partial z/\partial x$ and $\partial z/\partial y$ can be found using either implicit differentiation or the formulas
$$\frac{\partial z}{\partial x} = -\frac{\partial G/\partial x}{\partial G/\partial z} \quad \text{and} \quad \frac{\partial z}{\partial y} = -\frac{\partial G/\partial y}{\partial G/\partial z}.$$
2. If $G(x, y) = c$ defines $y = f(x)$, then
$$\frac{dy}{dx} = -\frac{\partial G/\partial x}{\partial G/\partial y}.$$
3. If $G(\mathbf{x}) = G(x_1, x_2, x_3) = c$ defines x_i as a function of the other x's, then $\partial x_i/\partial x_j$ for $i \neq j$ can be found by implicit differentiation or by using the formula
$$\frac{\partial x_i}{\partial x_j} = -\frac{\partial G/\partial x_j}{\partial G/\partial x_i}.$$
4. Given a level surface $G(x, y, z) = c$, the gradient vector
$$\nabla G = \frac{\partial G}{\partial x}\mathbf{i} + \frac{\partial G}{\partial y}\mathbf{j} + \frac{\partial G}{\partial z}\mathbf{k}$$
is normal to the surface at each point on the surface.

EXERCISES

In Exercises 1 through 16, find the desired derivative using formulas like those in Eqs. (4) and (5).

1. $dy/dx|_{(1,-1)}$ if $x^2y - x\sin(\pi y) - y^3 = 0$
2. $dx/dy|_{(0,2)}$ if $e^{xy} - (3xy + 2y)^3 = -63$
3. $dx/dy|_{(1,1)}$ if $3x^2y + \ln(xy) = 3$
4. $dy/dx|_{(2,-1)}$ if $(2x + y)^3 + (x^2/y) = 23$
5. $\partial z/\partial x|_{(1,-1,2)}$ if $(2xy/z) + z/(x^2 + y^2) = 0$
6. $\partial z/\partial x|_{(-1,2,0)}$ if $x\sin(yz) - 3x^2z + ye^z = 2$
7. $\partial u/\partial v$ if $u^2w^3 - 2u^3v^2 = 3v - 4$
8. $\partial v/\partial w$ if $uv^2 - \tan^{-1}(vw^2) = 8$
9. $\partial x_3/\partial x_2|_{(1,-1,1)}$ if $x_1^2 - 2x_2x_3^4 = 3x_1x_3^2$
10. $\partial x_1/\partial x_3|_{(1,2,\pi)}$ if $\sin(x_1x_3) - x_1^3x_2^2 = -4$
11. $\partial y/\partial z|_{(0,1,3)}$ if $\tan^{-1}(x + y) + \ln(xz + y) = \pi/4$
12. $\partial x/\partial z|_{(3,0,0)}$ if $\sin^{-1}z + x^2e^{y^2} = 12 - xe^y$
13. $\partial y/\partial x|_{(2,0,2)}$ if $\ln(xy + 1) + 3xz^2 = 24 + \tan(x - z)$
14. $\partial u/\partial y$ if $x^2y^3 - xu^2 + 2vu^3 - v^2y^4 = 1$
15. $\partial x/\partial w$ if $x^2\sin(vw) - e^{xu} = 0$
16. $\partial y/\partial v$ if $\ln(xy) + 3xv^2 - e^{yv^2} = 0$

In Exercises 17 through 22, use implicit differentiation to find the desired derivative.

17. $\partial x/\partial z|_{(-1,0,1)}$ if $\cos(x^2y) - yz^3 + xz^2 = 0$
18. $\partial x_2/\partial x_3|_{(0,1,-1)}$ if $x_1x_2^3 - 3x_2x_3 + x_1x_3^4 = 3$
19. $\partial z/\partial y|_{(-1,2,0)}$ if $e^{xyz} - \ln(xz + 1) = 1$
20. $\partial x_3/\partial x_1|_{(1,2,0)}$ if $x_2^2 - 4x_1x_2^2 = 7x_1x_3 - 12x_1^3$
21. $\partial u/\partial v$ if $x^2v + yu^2 - (xv/u) = 1$
22. $\partial v/\partial x$ if $u^2vx^3 - \tan^{-1}(2x + u) = 6$

In Exercises 23 through 28, find the equation of the tangent line or plane and parametric equations of the normal line to the given curve or surface at the given point.

23. $x^3y - 3y^2 = -2$ at $(1, 1)$
24. $xy^3 - (x^2/y) = -2$ at $(2, 1)$
25. $3xe^y + xy^3 = 2 + x$ at $(1, 0)$
26. $x\sin y + x^2e^z = 4$ at $(2, \pi, 0)$
27. $xz^2 + (2x - z)^2/y^3 = 19$ at $(2, 1, 3)$
28. $\sin(x^2y) + (3x + 2z)^5 = x$ at $(-1, 0, 1)$
29. Find the directional derivative of $f(x, y) = x^2 - 3xy^3$ at $(-1, 1)$ in the direction of increasing x normal to the curve $xy^3 - 3xy = 2$.
30. Find the directional derivative of $f(x, y) = (x/y^2) + (x + 2y)^3$ at $(3, -1)$ in the direction of increasing y normal to the curve $x^2 - 3xy + y^3 = 17$.
31. Find the directional derivative of $f(x, y, z) = (xy + y^2z)/x^3$ at $(1, 2, -1)$ in the direction of decreasing z normal to the surface $x^2z - yz^3 = 1$.
32. Find the directional derivative of $f(x, y, z) = x^2e^{yz}$ at $(1, 0, 2)$ in the direction of decreasing y normal to the surface $z + xyz^3 = 2$.

In Exercises 33 through 36, the function T gives the temperature in the plane or in space. Find:
a) the temperature at the given point,
b) the equation of the isothermal curve or surface through the given point,
c) the direction for maximum rate of change of temperature at the given point,
d) the maximum change of temperature at the given point.

33. $T(x, y) = 80/(2 + x^2 + y^2)$ at $(2, -2)$

34. $T(x, y) = (100 + x^2 + y^2)/(15 + x^4 + y^4)$ at $(1, 1)$

35. $T(x, y, z) = 27/(1 + x^2 + y^2) + 36/(1 + y^2 + z^2)$ at $(1, -1, 1)$

36. $T(x, y, z) = \ln(1 + x^2 + y^2 + z^2)$ at $(1, -1, 2)$

37. Show that if $G(x, y)$ has continuous second partial derivatives and if the curve $G(x, y) = c$ defines y as a twice-differentiable function of x, then at a point (x, y) where $G_y(x, y) \neq 0$, we have

$$\frac{d^2y}{dx^2} = -\frac{G_y^2 G_{xx} - 2G_x G_y G_{xy} + G_x^2 G_{yy}}{G_y^3}.$$

38. Use the result in Exercise 37 to compute $d^2y/dx^2|_{(0,1)}$ for y defined implicitly as a function of x by $x^3 - 3xy^2 + 4y^3 = 4$.

39. Let $G(x, y, z)$ have continuous second partial derivatives, and let $G(x, y, z) = c$ define z as a function of x and y. Find a formula like that in Exercise 37 for $\partial^2 z/\partial y^2$.

40. Repeat Exercise 39, but find a formula for $\partial^2 z/\partial x\,\partial y$.

41. Show that the curves

$$5x^4y - 10x^2y^3 + y^5 = 4$$

and

$$x^5 - 10x^3y^2 + 5xy^4 = -4$$

are orthogonal at all points of intersection.

42. Show that the surfaces

$$x^2 - 2y^2 + z^2 = 0 \quad \text{and} \quad xyz = 1$$

are orthogonal at all points of intersection. (Surfaces are *orthogonal* at a point of intersection if their normal lines at at that point are orthogonal.)

43. Show that the surfaces

$$x + y^2 + 2z^3 = 4$$

and

$$12x - 3(\ln y) + z^{-1} = 13$$

are orthogonal at all points of intersection.

EXERCISE SETS FOR CHAPTER 16

Review Exercise Set 16.1

1. If $w = f(x, y, z) = xz^2 + y^2e^{xz}$, find the following.
a) $\partial w/\partial x$ b) $\partial^2 w/\partial x^2$ c) $\partial^2 w/\partial x\,\partial y$

2. Find the equation of the plane tangent to the surface

$$z = \frac{x + y}{x - y}$$

at the point $(2, 1, 3)$.

3. Let $w = f(x, y, z) = x^2y - xy^2z^3$. Find the gradient vector $\nabla f(1, -1, 2)$.

4. Use a differential to estimate

$$\frac{2.05^4 - 3.97^2}{1.08^5}.$$

5. Let $z = f(x, y) = e^{x^2y}$, while $x = 2st$ and $y = s^2 + t^3$. Use a chain rule to find $\partial z/\partial t$ at the points $(s, t) = (1, -1)$.

6. Let $z = f(x, y)$ be differentiable, and let x and y be differentiable functions of t. Find a formula for d^2z/dt^2.

7. Find the directional derivative of $z = f(x, y) = x^2y^3 - 3y^2$ at $(-3, 4)$ in the direction toward the origin.

8. Find the direction for maximum rate of increase of $f(x, y, z) = y^2 \sin xz^2$ at the point $(0, 3, -1)$ and the magnitude of this maximum rate of increase.

9. Let y be defined implicitly as a function of x and z by $x^2y - 3xz^2 + y^3 = -13$. Find $\partial y/\partial x$ at the point $(1, -2, 1)$.

10. Find parametric equations of the line normal to the surface $x^2y - 3xz^2 + y^3 = -13$ at the point $(1, -2, 1)$.

Review Exercise Set 16.2

1. If $y = x_1 \sin x_2x_3 - x_2^2x_1^3$, find the following.
a) $\partial y/\partial x_2$ b) $\partial^2 y/\partial x_1\,\partial x_3$

2. Find the equation of the line normal to the surface

$$z = \frac{x^2 - y}{x + y}$$

at the point $(2, 1, 1)$.

3. Let $y = f(x_1, x_2) = x_2^3 + x_1 \cos x_2$. Find the gradient vector $\nabla f(1, 0)$.

4. Use a differential to estimate $2.03^2 \cos(-0.05)$.

5. Let $z = f(u, v) = u^2/v^3$, and let $u = t^2 - 3t - 7$ and $v = t^3 + 7$. Use a chain rule to find dz/dt when $t = -2$.

6. Let $z = f(x, y)$, while $x = g_1(t)$ and $y = g_2(t)$. If $g_1(1) = -3$ $g_2(1) = 4$, while

$$\left.\frac{dz}{dt}\right|_{t=1} = 2, \qquad \left.\frac{dx}{dt}\right|_{t=1} = 4,$$

$$\left.\frac{dy}{dt}\right|_{t=1} = -1, \qquad \left.\frac{\partial z}{\partial x}\right|_{(-3,4)} = 3,$$

find $f_y(-3, 4)$.

7. Find the directional derivative of $x^3y + (x/y)$ at $(-1, 1)$ in the direction toward $(2, -3)$.

8. Find the direction of the maximum rate of increase of $f(x, y) = x^2e^{xy}$ at the point $(3, 0)$, and then find the magnitude of this rate of change.

9. If x is defined implicitly as a function of y and z by $y^2 \cos x + xz^3 - 3y^2z = -5$, find $\partial x/\partial z$ at $(0, 1, 2)$.

10. Find the equation of the plane tangent to the surface $x^3y + y^3z + z^3x = -1$ at the point $(1, 1, -1)$.

Review Exercise Set 16.3

1. If the temperature at a point in space is $T(x, y, z) = 500/(1 + x^2 + y^2 + z^2)$, find the rate of change of temperature at $(-1, 2, -2)$ in the positive coordinate directions.

2. Find all points on the surface $z = x^2 - 2y^3$ where the normal line is parallel to the line through $(-1, 4, 2)$ and $(1, -2, 0)$.

3. Find the differential df at $(0, -1, 2)$ of $f(x, y, z) = z \sin(xz) + z^2e^{xy}$.

4. A right circular cone has base radius 3 ft and altitude 6 ft. Use a differential to estimate the change in r to produce an increase of 0.5 ft^3 in volume if the altitude is decreased by 4 in.

5. Let z be a differentiable function of u and v, let u and v be differentiable functions of x and y, and let x and y be differentiable functions of t. Give the chain rule formula for dz/dt.

6. Let $z = f(x, y)$ be differentiable, and let $x = t^2 + 2t$ and $y = (t + 1)/t$. If $dz/dt = 5$ when $t = -1$, find $f_y(-1, 0)$.

7. A circular pond covers the disk $x^2 + y^2 \leq 100{,}000$. The depth of water in the pond at a point (x, y) in the disk is $100 - (x^2 + y^2)^{2/5}$ in the same units as the coordinates. Find the rate of change of depth with respect to distance for a swimmer at the point $(4, -4)$ swimming toward the point $(-26, -44)$.

8. For the temperature function in Exercise 1,
   a) find a vector in the direction for maximum rate of change at the point $(-1, 2, -2)$,
   b) find the maximum rate of change over all possible directions at $(-1, 2, -2)$.

9. Find $\partial z/\partial y$ if $x^2y + xz^3 + y^2z^2 = 3$.

10. a) Find the equation of the level surface of $G(x, y, z) = x^3y - xyz^2 + z^3$ through the point $(-1, 1, 2)$.
    b) Find the equation of the tangent plane to the level surface in part (a) at $(-1, 1, 2)$.

Review Exercise Set 16.4

1. Find $\partial z/\partial x$ if $z = x^2e^{\sin(xy)}$.

2. Find all points on the surface $z = x^2y + y$ where the tangent plane is parallel to the plane $3x - 10y + 2z = 7$.

3. Use a differential to estimate $\sqrt{2.95^3 + 3.01^2}$.

4. If $z = u^2 - 3uv$ while $u = (x + y)/(y + 1)$ and $v = xy$, use a chain rule to compute $\partial z/\partial y$ at the point $(x, y) = (3, 1)$.

5. Let z be defined implicitly as a function of x and y near $(2, 1, -1)$ by $x^2z + xyz^3 = -6$. If $x = t^3 - 3t$ and $y = 3t^2 - 6t + 1$, use a chain rule to find dz/dt when $t = 2$.

6. Let $z = f(x, y)$ be a differentiable function, and let $x = ts - 2t$ and $y = t^2 - 3ts$. If $\partial z/\partial t = 3$ and $\partial z/\partial s = -4$ at $(t, s) = (2, 1)$ find $f_x(-2, -2)$ and $f_y(-2, -2)$.

7. Find the directional derivative of $w = f(x, y, z) = xy^2z^3$ at $(2, -1, 1)$ in the direction toward $(0, 1, 0)$.

8. Find all points (x, y) where $z = f(x, y) = x^2 - y^2 = 8$ and where the maximum value of the directional derivative of f over all possible directions is also 8.

9. Let z be defined implicitly as a differentiable function of x and y by $G(x, y, z) = 0$, and let x and y be defined implicitly as differentiable functions of u and v by $H(x, u, v) = 0$ and $K(y, u, v) = 0$. Find a formula for $\partial z/\partial u$.

10. a) Find the equation of the level curve of $G(x, y) = x^3 + 2xy + e^{xy}$ that passes through the point $(2, 0)$.
    b) Find parametric equations of the normal line to the level curve in part (a) at $(2, 0)$.

More Challenging Exercises 16

1. Find the appropriate "subtract-and-add trick" that could be used to prove the three-variable analogue of Theorem 16.2 for a function $w = f(x, y, z)$. That is, give the analogue of Eq. (11) of Section 16.3 for $w = f(x, y, z)$.

Exercises 2 through 9 introduce power series representation for a function of two variables.

2. Consider the polynomial function

$$\begin{aligned} P(x, y) = {} & a_{00} + a_{10}(x - x_0) + a_{01}(y - y_0) + \cdots \\ & + a_{ij}(x - x_0)^i(y - y_0)^j + \cdots \\ & + a_{0n}(y - y_0)^n \end{aligned}$$

for all nonnegative integers i and j, where $i + j \leq n$. Let $f(x, y)$ have continuous partial derivatives of all orders less than or equal to n at (x_0, y_0). Find a formula for a_{ij} if $P(x, y)$ has the same partial derivatives as $f(x, y)$ of all orders less than or equal to n at (x_0, y_0); that is, if

$$\left.\frac{\partial^{i+j}P}{\partial x^i\,\partial y^j}\right|_{(x_0, y_0)} = \left.\frac{\partial^{i+j}f}{\partial x^i\,\partial y^j}\right|_{(x_0, y_0)}.$$

We define the ***n*th Taylor polynomial** $T_n(x, y)$ for $f(x, y)$ at (x_0, y_0) to be

$$T_n(x, y) = \sum_{1 + j \leq n}\left[\frac{1}{i!j!}\cdot\left.\frac{\partial^{i+j}f}{\partial x^i\,\partial y^j}\right|_{(x_0, y_0)}\cdot(x - x_0)^i(y - y_0)^j\right]. \quad (1)$$

---

3. Find the Taylor polynomial $T_3(x, y)$ for $\sin(x + y)$ at $(0, 0)$.

4. Explain how the answer to Exercise 3 is predictable in terms of Taylor series for a function of one variable.

5. Following the idea of Exercise 4, predict the polynomial $T_4(x, y)$ for e^{xy} at $(0, 0)$. Then verify your answer by computing Eq. (1) with $n = 4$.

6. Find the Taylor polynomial $T_4(x, y)$ for e^{xy} at $(0, 1)$.

Taylor's theorem Let $f(x, y)$ have continuous partial derivatives of all orders less than or equal to $(n + 1)$ in some disk with center at (x_0, y_0). Then for each $(x, y) \neq (x_0, y_0)$ in the disk, there exists (c_1, c_2), depending on x and y and strictly between (x_0, y_0) and (x, y) on the line segment joining them, such that

$$E_n(x, y) = \sum_{\substack{i, j \\ i + j = n + 1}}\left[\frac{1}{i!j!}\cdot\left.\frac{\partial^{i+j}f}{\partial x^i\,\partial y^j}\right|_{(c_1, c_2)}\cdot(x - x_0)^i(y - y_0)^j\right],$$

where $E_n(x, y) = f(x, y) - T_n(x, y)$.

---

7. Estimate $1.02^2 \ln 0.97$ using a differential, and then use Taylor's theorem to find a bound for the error.

8. Use the Taylor polynomial $T_2(x, y)$ at $(0, 1)$ for $y^3 \cos x$ to estimate $1.03^3 \cos(-0.02)$, and then find a bound for the error.

9. Let f be a function of two variables with continuous coordinate derivatives of all orders less than or equal to n. Let us introduce the *operation notations*

$$\left(\frac{\partial}{\partial x}h + \frac{\partial}{\partial y}k\right)f = \frac{\partial f}{\partial x}h + \frac{\partial f}{\partial y}k,$$

$$\left(\frac{\partial}{\partial x}h + \frac{\partial}{\partial y}k\right)^2 f = \frac{\partial^2 f}{\partial x^2}h^2 + 2\frac{\partial^2 f}{\partial x\,\partial y}hk + \frac{\partial^2 f}{\partial y^2}k^2,$$

$$\vdots \qquad\qquad\qquad \vdots$$

$$\left(\frac{\partial}{\partial x}h + \frac{\partial}{\partial y}k\right)^n f = \sum_{i=0}^{n}\binom{n}{i}\frac{\partial^n f}{\partial x^i\,\partial y^{n-i}}h^i k^{n-i}.$$

Show that the nth Taylor polynomial for f at (x_0, y_0) is given by

$$T_n(x, y) = f(x_0, y_0) + \sum_{i=1}^{n}\frac{1}{i!}\left(\frac{\partial}{\partial x}\bigg|_{(x_0, y_0)}(x - x_0) + \frac{\partial}{\partial y}\bigg|_{(x_0, y_0)}(y - y_0)\right)^i f.$$

APPLICATIONS OF PARTIAL DERIVATIVES 17

17.1 MAXIMA AND MINIMA

In Section 2.5, we stated that a *continuous* function $f(x)$ assumes a maximum value and also a minimum value on each *closed interval* in its domain. The situation for functions of more than one variable is similar, although more complicated to explain carefully; we make a few intuitive introductory remarks. We consider a region in the plane to be **bounded** if it is contained inside some sufficiently large circle, so that it doesn't extend infinitely far in any direction. Such a region is **closed** if it contains all points that we would consider to be part of the *boundary* of the region. (We will not give a formal definition of the boundary.) It can be shown that a *continuous* function $f(x, y)$ assumes a maximum value and also a minimum value on each *closed, bounded* region in its domain. This is the analogue for $f(x, y)$ of the result stated above for $f(x)$. The boundary of $[a, b]$ is considered to consist of the two points a and b, so $[a, b]$ is *closed* since it includes those points. We restricted the theorem for $f(x)$ to an *interval*, which does not extend to infinity and is hence *bounded*. The theorem for functions of more than two variables is similar in every way.

In Section 5.3, we learned to find such extrema for $f(x)$ on $[a, b]$: We found *local extrema* inside the interval and compared function values there with the values $f(a)$ and $f(b)$ at points on the boundary. In the present section, we consider only the problem of finding the *local extrema* for a function of two or more variables, and we restrict ourselves to quite simple cases. Sometimes these local extrema are obviously maximum or minimum values for a function in its entire domain.

FINDING LOCAL EXTREMA

Let $z = f(x, y)$ and suppose $f(x, y) \leq f(x_0, y_0)$ for all (x, y) inside some sufficiently small circle with center at (x_0, y_0). Then $f(x, y)$ has a **local maximum** or **relative maximum** of $f(x_0, y_0)$ at the point (x_0, y_0). See Fig. 17.1. Of course, if the inequality were reversed so that $f(x, y) \geq f(x_0, y_0)$ for all such (x_0, y_0), then $f(x_0, y_0)$ would be a **local minimum** or **relative minimum.**

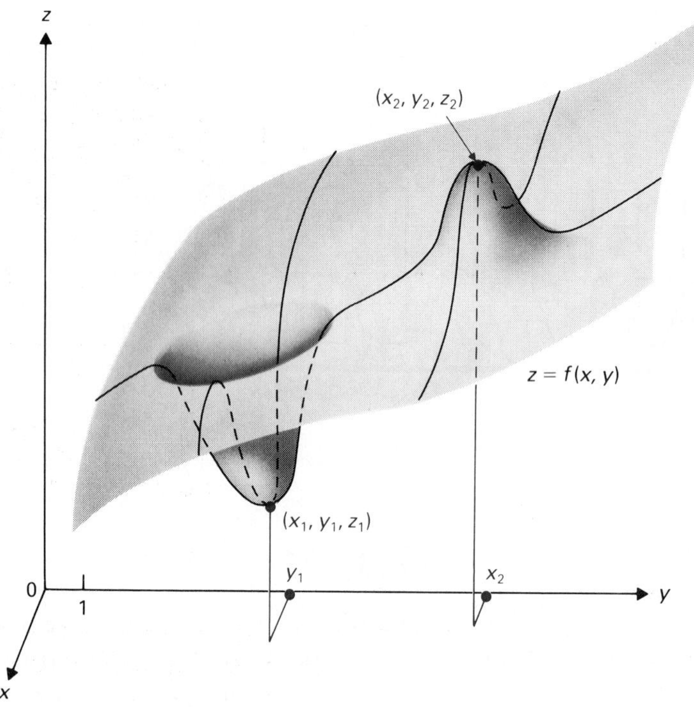

Figure 17.1 $f(x, y)$ has a local minimum of z_1 at (x_1, y_1) and a local maximum of z_2 at (x_2, y_2).

This is an obvious generalization of the same notions for a function of one variable; and still further generalizations to functions of three variables are clear.

We want to find such local maxima and minima of $z = f(x, y)$. Suppose that $f_x(x, y)$ and $f_y(x, y)$ exist. If $f(x_0, y_0)$ is a local maximum, then the function $g(x) = f(x, y_0)$ shown in Fig. 17.2 has a local maximum $g(x_0)$ at x_0. Consequently

$$g'(x_0) = f_x(x_0, y_0) = 0.$$

Also, we must have a local maximum at y_0 of $h(y) = f(x_0, y)$, so

$$h'(y_0) = f_y(x_0, y_0) = 0.$$

A similar argument shows that first partial derivatives must be zero if $f(x_0, y_0)$ is a local minimum, and, indeed, the same results hold for functions of more than two variables.

THEOREM 17.1 Partial derivatives at local extrema

If a function has first-order partial derivatives, then these derivatives will be zero at any point where the function has a local maximum or a local minimum.

Consequently, we can find all *candidates* for local extrema of differentiable functions by finding those points where all first-order partial derivatives are zero simultaneously. We regard such points as **critical points.**

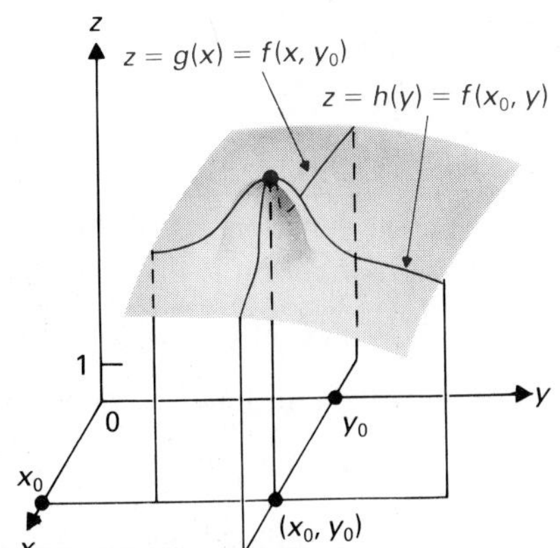

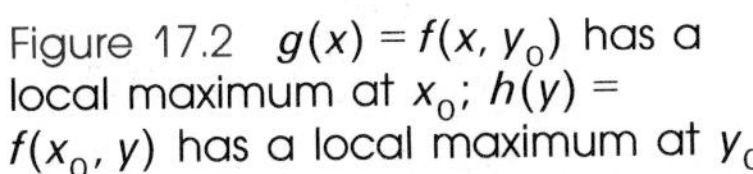
Figure 17.2 $g(x) = f(x, y_0)$ has a local maximum at x_0; $h(y) = f(x_0, y)$ has a local maximum at y_0.

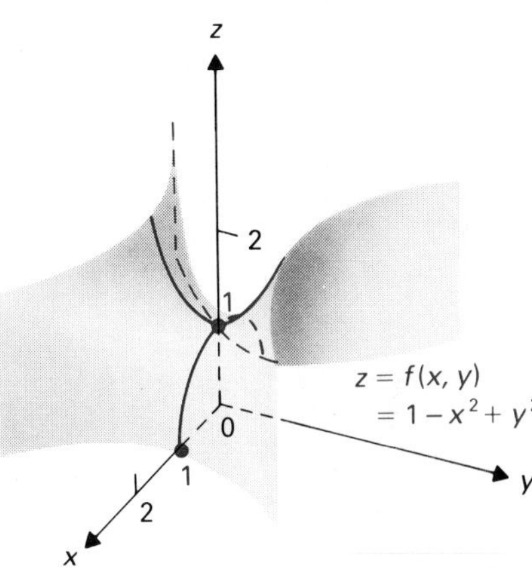

Figure 17.3 $f_x(0, 0) = f_y(0, 0)$, but f has neither a local maximum nor a local minimum at (0, 0).

So far the situation is just the same as for a function of one variable. From here on, more complicated things can happen for functions of two or more variables, as the following example shows.

EXAMPLE 1 Test $f(x, y) = 1 - x^2 + y^2$ for local maxima or minima.

Solution Using Theorem 17.1, we find that

$$f_x = -2x = 0 \quad \text{only if } x = 0 \qquad \text{and} \qquad f_y = 2y = 0 \quad \text{only if } y = 0.$$

Thus (0, 0) is our only candidate for a local extremum point. Clearly f has neither a local maximum nor a local minimum at (0, 0), for $f(0, 0) = 1$, but $f(x_1, 0) < 1$ and $f(0, y_1) > 1$ for nonzero x_1 and y_1 close to zero. The graph of f is shown in Fig. 17.3. □

Figure 17.4 The graph of $f(x, y) = y^3$ has a "saddle point" at every point on the x-axis, $f_x = f_y = 0$ there, but there is neither a local maximum nor a local minimum.

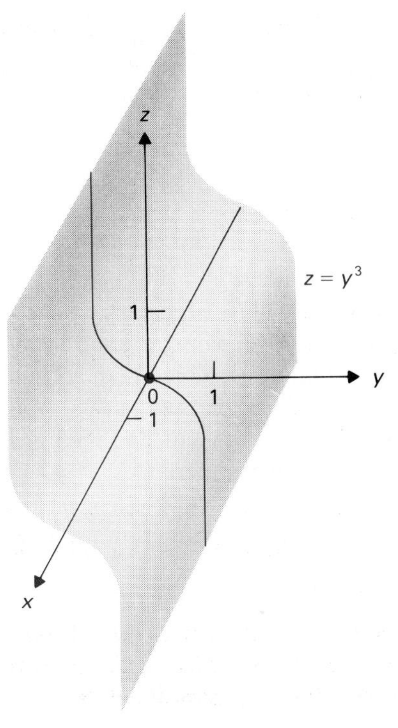

The surface in Fig. 17.3 is shaped like a saddle near the origin. It is customary to refer to a point of the graph of $z = f(x, y)$ where $f_x = 0$ and $f_y = 0$, but where there is no local extremum, as a **saddle point.** Figure 17.4 shows the graph of $z = f(x, y) = -y^3$. We easily see that $f_x = f_y = 0$ where $y = 0$, which corresponds to all points of the graph on the entire x-axis. The surface does not actually look like a saddle near any of these points, but it is nevertheless customary to call them all *saddle points.*

For a function of one variable with first derivative zero at x_0, the sign of a nonzero second derivative at x_0 determines whether the function has a local minimum or a local maximum at x_0. The situation is not so simple for functions of more than one variable, as we show in the next example.

EXAMPLE 2 Show that $f(x, y) = x^2 + 4xy + y^2$ has the property that

$$f_x(0, 0) = 0, \qquad f_y(0, 0) = 0,$$

$$f_{xx}(0, 0) > 0, \qquad f_{xy}(0, 0) > 0, \qquad f_{yy}(0, 0) > 0,$$

but that the function has no local maximum or minimum at (0, 0).

Solution We easily compute that

$$f_x(0, 0) = (2x + 4y)|_{(0,0)} = 0, \qquad f_y(0, 0) = (4x + 2y)|_{(0,0)} = 0,$$

$$f_{xx} = 2 > 0, \qquad f_{xy} = 4 > 0, \qquad f_{yy} = 2 > 0.$$

However, the function f still does not have a local extremum at (0, 0). To see this, note that on the line $y = x$, we have

$$f(x, y) = f(x, x) = x^2 + 4x^2 + x^2 = 6x^2 \geq 0,$$

so $f(x, y) > 0$ on the line $y = x$ except at (0, 0). However, on the line $y = -x$,

$$f(x, y) = f(x, -x) = x^2 - 4x^2 + x^2 = -2x^2 \leq 0,$$

so $f(x, y) < 0$ on the line $y = -x$ except at (0, 0). Since $f(0, 0) = 0$, we see that f has neither a local maximum nor a local minimum at (0, 0). □

We state without proof a second-order derivative test for a local maximum or minimum of a function of two variables. A proof can be found in any advanced calculus text.

THEOREM 17.2 Test for local extrema

Let $f(x, y)$ be a function of two variables with continuous partial derivatives of orders less than or equal to 2 in some disk with center (x_0, y_0), and suppose that

$$f_x(x_0, y_0) = f_y(x_0, y_0) = 0,$$

while not all second-order partial derivatives are zero at (x_0, y_0). Let

$$A = f_{xx}(x_0, y_0), \qquad B = f_{xy}(x_0, y_0), \qquad \text{and} \qquad C = f_{yy}(x_0, y_0).$$

1. If $AC - B^2 > 0$, the function has a local maximum at (x_0, y_0) if $A < 0$ and a local minimum there if $A > 0$.
2. If $AC - B^2 < 0$, the function has neither a local maximum nor a local minimum at (x_0, y_0); it has a saddle point there.
3. If $AC - B^2 = 0$, more work is needed to determine whether the function has a local maximum, a local minimum, or neither at (x_0, y_0).

Referring to Example 2, we see that there $A = 2$, $B = 4$, and $C = 2$, so $AC - B^2 = 4 - 16 = -12 < 0$. Consequently, $f(x, y) = x^2 + 4xy + y^2$ had neither a local maximum nor a local minimum at (0, 0); the graph has a saddle point there. In Exercise 27, we ask you to demonstrate part 3 of Theorem 17.2 by giving examples of three functions, each having the origin as a critical point with $AC - B^2 = 0$ there. One function is to have a local maximum at (0, 0), another is to have a local minimum there, and the third is to have neither a local maximum nor local minimum there.

EXAMPLE 3 Find all local minima and maxima of the function f where

$$f(x, y) = x^2 - 2xy + 2y^2 - 2x + 2y + 4.$$

Solution We have

$$f_x(x, y) = 2x - 2y - 2 \qquad \text{and} \qquad f_y(x, y) = -2x + 4y + 2.$$

In order for both partial derivatives to be zero, we must have

$$2x - 2y - 2 = 0,$$
$$-2x + 4y + 2 = 0.$$

Adding these equations, we find that

$$2y = 0, \qquad \text{or} \qquad y = 0.$$

Then we must have $x = 1$. Therefore $(1, 0)$ is the only candidate for a point where f can have a local maximum or minimum. Computing second partial derivatives, we have

$$A = f_{xx}(1, 0) = 2, \qquad B = f_{xy}(1, 0) = -2, \qquad C = f_{yy}(1, 0) = 4.$$

Therefore $AC - B^2 = 8 - 4 = 4 > 0$. Since $A > 0$, we know by Theorem 17.2 that f has a local minimum of $f(1, 0) = 3$ at $(1, 0)$. □

EXAMPLE 4 Find all local maxima and minima of the function f where

$$f(x, y) = x^3 - y^3 - 3xy + 4.$$

Solution We have

$$f_x(x, y) = 3x^2 - 3y \qquad \text{and} \qquad f_y(x, y) = -3y^2 - 3x.$$

Setting these derivatives equal to zero, we obtain from the first one $y = x^2$, and the substituting into the second one we have $x^4 + x = 0$. Now

$$\begin{aligned} x^4 + x &= x(x^3 + 1) \\ &= x(x + 1)(x^2 - x + 1) = 0 \end{aligned}$$

has only $x = 0$ and $x = -1$ as real solutions.

At $x = 0$, we obtain $y = 0$, and

$$A = f_{xx}(0, 0) = 0, \qquad B = f_{xy}(0, 0) = -3, \qquad C = f_{yy}(0, 0) = 0.$$

In this case, $AC - B^2 = -9 < 0$, so f has neither a local maximum nor a local minimum at $(0, 0)$.

At $x = -1$, we have $y = 1$, and

$$A = f_{xx}(-1, 1) = -6, \qquad B = f_{xy}(-1, 1) = -3,$$
$$C = f_{yy}(-1, 1) = -6,$$

so $AC - B^2 = 36 - 9 = 27 > 0$. Since $A < 0$, we see in this case that f has a local maximum of $f(-1, 1) = 5$ at $(-1, 1)$. □

The technique described in Theorems 17.1 and 17.2 and used in Examples 3 and 4 can be a lot of work in some cases. Sometimes the answer is obvious without the work.

EXAMPLE 5 Find all local maxima and minima of the function

$$f(x, y) = \sin(xy).$$

Solution We could find where $f_x = f_y = 0$ and then check $AC - B^2$ at those candidate points. We would discover it involves more work than in Examples 3 and 4, and we would also run into some cases where $AC - B^2 = 0$.

We can easily find the answer by inspection. We know that $\sin(xy)$ has maxima of 1 where $xy = (\pi/2) + 2n\pi$, which occurs on the hyperbolas drawn in color in Fig. 17.5. Minima of -1 occur where $xy = (3\pi/2) + 2n\pi$, which occur on the hyperbolas shown in black in Fig. 17.5. □

We will do little toward finding local extrema for functions of more than two variables. Theorem 17.1 is true for any number of variables; all first-order partial derivatives must be zero at a local extremum. However, Theorem 17.2 is a specialized test for functions of just two variables.

There is one situation we can handle for a function f of more than two variables. That is the case where the function can be expressed as a sum of functions of at most two variables, with no variable appearing in more than one of the summand functions. For example, perhaps

$$f(x, y, z) = g(x) + h(y) + k(z), \tag{1}$$

or

$$f(x, y, z) = g(x, y) + h(z), \tag{2}$$

or

$$f(w, x, y, z) = g(w, y) + h(x, z). \tag{3}$$

In cases like Eqs. (1), (2), and (3), it is easy to see that f can have a local maximum only at a point where each individual summand function also has a local maximum for corresponding coordinate values. Take, for illustration, case (2), where

$$f(x, y, z) = g(x, y) + h(z).$$

It is obvious that if $f(x_0, y_0, z_0)$ is a local maximum, then $g(x_0, y_0)$ and $h(z_0)$

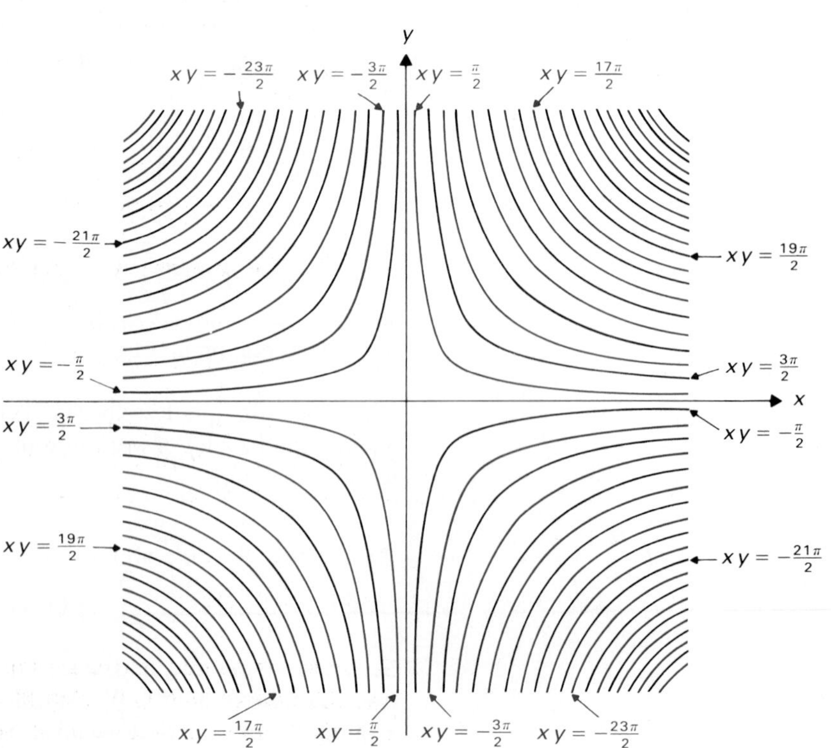

Figure 17.5 $\sin(xy)$ assumes a maximum of 1 on the red curves

$$xy = \pi/2 + 2n\pi$$

and a minimum of -1 on the black curves $xy = 3\pi/2 + 2n\pi$.

must also be local maxima of those functions, and conversely. If, however, $g(x_0, y_0)$ gives a local maximum and $h(z_0)$ a local minimum, then $f(x_0, y_0, z) \geq f(x_0, y_0, z_0)$ for z close to z_0, while $f(x, y, z_0) \leq f(x_0, y_0, z_0)$ for x close to x_0 and y close to y_0. By such arguments, we see that

> a function of the type in Eq. (1), (2), or (3) has a local extremum at a point if and only if each of the summand functions has the same type of local extremum at the points with corresponding coordinate values.

EXAMPLE 6 Find all local maxima and minima of the function

$$f(x, y, z) = x^3 + y^2 + z^2 - 12x - 2y + 6z.$$

Solution This function is of the form of Eq. (1), and we may write

$$f(x, y, z) = (x^3 - 12x) + (y^2 - 2y) + (z^2 + 6z).$$

We find that

$$f_x = 3x^2 - 12 = 0, \qquad \text{if } x = \pm 2,$$
$$f_y = 2y - 2 = 0, \qquad \text{if } y = 1,$$
$$f_z = 2z + 6 = 0, \qquad \text{if } z = -3.$$

Thus the only candidates for local extrema are

$$(2, 1, -3) \qquad \text{and} \qquad (-2, 1, -3).$$

Since $f_{xx} = 6x$, we easily see that $x^3 - 12x$ has a local maximum where $x = -2$ and a local minimum where $x = 2$. Since $f_{yy} = 2 > 0$ and $f_{zz} = 2 > 0$, we see that $y^2 - 2y$ and $z^2 + 6z$ have local minima where $y = 1$ and $z = -3$, respectively. The point where all three summand functions have the same type of extremum is therefore $(2, 1, -3)$, where they all have a local minimum. Thus $f(x, y, z)$ has a local minimum of -26 at $(2, 1, -3)$. There is neither a local minimum nor a local maximum at $(-2, 1, -3)$. □

EXAMPLE 7 Find all local maxima and minima of the function

$$f(x, y, z) = x^3 - y^3 - z^3 - 3xy + 3z + 4.$$

Solution We can write the function in the form of Eq. (2) as

$$f(x, y, z) = (x^3 - y^3 - 3xy + 4) + (3z - z^3).$$

We saw in Example 4 that $x^3 - y^3 - 3xy + 4$ has a local maximum of 5 at $(x, y) = (-1, 1)$ as its only local extremum. We easily see that $3z - z^3$ has a local minimum where $z = -1$ and a local maximum where $z = 1$. Thus the only local extremum of $f(x, y, z)$ occurs at $(-1, 1, 1)$, where the function has a local maximum of $5 + 2 = 7$. □

APPLICATIONS

As an aid to solving maximum/minimum word problems involving several variables, we give a step-by-step procedure in the box on page 742, modifying slightly the outline we gave in Section 5.6.

SOLVING A MAXIMUM/MINIMUM WORD PROBLEM

Step 1 Draw a figure where appropriate and assign letter variables. Decide what to maximize or minimize.

Step 2 Express this quantity to be maximized or minimized as a function f of *independent* variables. (You may need algebra to do this.)

Step 3 Find all points where the first partial derivatives of f are simultaneously zero.

Step 4 Decide whether the desired maximum or minimum occurs at one of the points you found in Step 3. Frequently, it is clear that a maximum or minimum exists from the nature of the problem, and if Step 3 gives only one candidate, that candidate provides the answer. If f is a function of two variables, try the $AC - B^2$ test in Theorem 17.2.

Step 5 Put the answer in the requested form.

EXAMPLE 8 Show that the rectangular box of given volume V_0 with minimum surface area is a cube.

Solution

Step 1 We take the box with one corner at the origin and opposite corner at (x, y, z), as shown in Fig. 17.6. The box then has dimensions x, y, and z and volume $V_0 = xyz$. We wish to minimize the surface area S of the box.

Step 2 We have $S = 2xy + 2xz + 2yz$. Since $xyz = V_0$, we have $z = V_0/(xy)$, and we can express S as a function of independent variables x and y by

$$S = f(x, y) = 2xy + \frac{2V_0}{y} + \frac{2V_0}{x} \qquad \text{for } x, y \text{ both} > 0.$$

Step 3 We have

$$\frac{\partial S}{\partial x} = 2y - \frac{2V_0}{x^2} \quad \text{and} \quad \frac{\partial S}{\partial y} = 2x - \frac{2V_0}{y^2}.$$

Setting these partial derivaties equal to zero, we obtain the system

$$\begin{cases} 2x^2y - 2V_0 = 0, \\ 2xy^2 - 2V_0 = 0, \end{cases} \quad \text{or} \quad \begin{cases} x^2y = V_0, \\ xy^2 = V_0. \end{cases}$$

Thus $x^2y = xy^2$, so $x = y$. Then $x^3 = y^3 = V_0$, so $x = y = \sqrt[3]{V_0}$. Finally, $z = V_0/(xy) = V_0/(V_0^{2/3}) = \sqrt[3]{V_0}$, so $x = y = z = \sqrt[3]{V_0}$.

Step 4 This is a problem where it is obvious that a minimum does exist, and we have only one candidate for it. That is, it is obvious that a box with huge base and very small altitude has a large surface area, and so on. Alternatively, with $x = y = \sqrt[3]{V_0}$, the test of Theorem 17.2 gives $AC - B^2 = 4 \cdot 4 - 2^2 = 12 > 0$, so we have a minimum.

Step 5 Since $x = y = z = \sqrt[3]{V_0}$, our box of minimum surface area is indeed a cube. □

Figure 17.6 Rectangular box with opposite vertices (0, 0, 0) and (x, y, z).

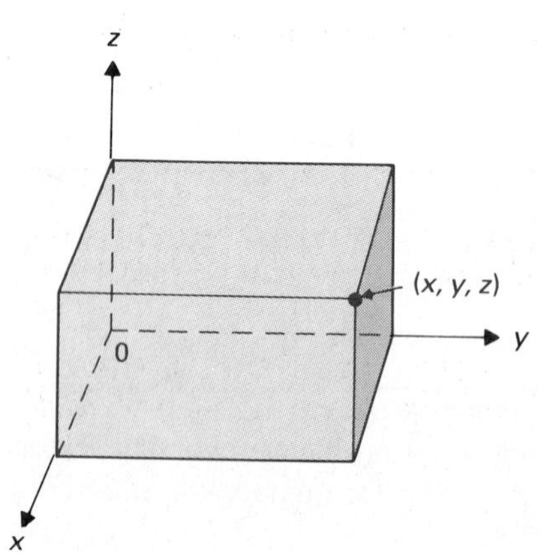

EXAMPLE 9 Find the point on the plane $x - 2y + z = 5$ that is closest to $(5, -3, 6)$.

Solution

Step 1 We want to minimize the distance s from $(5, -3, 6)$ to a point (x, y, z) on the plane $x - 2y + z = 5$.

Step 2 We have $s = \sqrt{(x-5)^2 + (y+3)^2 + (z-6)^2}$. Minimizing s is equivalent to minimizing

$$s^2 = (x-5)^2 + (y+3)^3 + (z-6)^2,$$

which eliminates the radical from the problem. Since $z = 5 - x + 2y$, we obtain

$$\begin{aligned} s^2 = f(x, y) &= (x-5)^2 + (y+3)^2 + (5 - x + 2y - 6)^2 \\ &= (x-5)^2 + (y+3)^2 + (-x + 2y - 1)^2 \end{aligned}$$

as a function of two independent variables x and y.

Step 3 We have

$$\frac{\partial(s^2)}{\partial x} = 2(x-5) - 2(-x + 2y - 1)$$

and

$$\frac{\partial(s^2)}{\partial y} = 2(y+3) + 4(-x + 2y - 1).$$

Setting these partial derivatives equal to zero, we obtain the system

$$\begin{cases} 4x - 4y - 8 = 0, \\ -4x + 10y + 2 = 0, \end{cases} \quad \text{or} \quad \begin{cases} 2x - 2y = 4, \\ -2x + 5y = -1. \end{cases}$$

Adding these equations, we obtain $3y = 3$, so $y = 1$. We then find that $x = 3$ and $z = 5 - x + 2y = 4$.

Step 4 It is geometrically obvious that there is a point on the plane $x - 2y + z = 5$ closest to $(5, -3, 6)$, and we have only the one candidate $(3, 1, 4)$. Alternatively, we can use Theorem 17.2, and we find that $AC - B^2 = 4 \cdot 10 - (-4)^2 = 24 > 0$. Since $A = 4 > 0$, we have a minimum.

Step 5 We are asked to find the point on the plane that is closest to $(5, -3, 6)$, rather than the minimum distance, so our answer is $(3, 1, 4)$. □

SUMMARY

1. If a function has a local maximum or minimum at a point where the first-order partial derivatives exist, then these first-order partial derivatives must be zero there.
2. Let $f(x, y)$ be a function of two variables with continuous partial derivatives of orders less than or equal to 2 in a neighborhood of (x_0, y_0), and suppose that

$$f_x(x_0, y_0) = f_y(x_0, y_0) = 0,$$

while not all second-order partial derivatives are zero at (x_0, y_0). Let

$$A = f_{xx}(x_0, y_0), \qquad B = f_{xy}(x_0, y_0), \qquad \text{and} \qquad C = f_{yy}(x_0, y_0).$$

Then the function $f(x, y)$ has either a local maximum or a local minimum at (x_0, y_0) if $AC - B^2 > 0$. In this case, the function has a local minimum if $A > 0$ and a local maximum if $A < 0$. If $AC - B^2 < 0$, then $f(x, y)$ has neither a local maximum nor a local minimum at (x_0, y_0) but rather a saddle point there. If $AC - B^2 = 0$, the function may or may not have a local extremum at (x_0, y_0).

3. A function that can be expressed as a sum of functions of distinct variables has a local extremum at a point if and only if each summand function has the same type of local extremum for those values of the variables.

4. An abbreviated outline to aid in solving extremum word problems is as follows:

Step 1 Decide what to maximize or minimize.

Step 2 Express it as a function f of *independent* variables.

Step 3 Find where all first partial derivatives of f are simultaneously zero.

Step 4 Test for extrema at the points found in Step 3.

Step 5 Put the answer in the requested form.

EXERCISES

In Exercises 1 through 26, find all local maxima and minima of the function.

1. $x^2 + y^2 - 4$
2. $xy + 3$
3. $x^2 + y^2 + 4x - 2y + 3$
4. $x^2 - y^2 + 2x + 8y - 7$
5. $x^2 + y^2 + 4xy - 2x + 6y$
6. $3x^2 + y^2 - 3xy + 6x - 4y$
7. $x^3 + 2y^3 - 3x^2 - 24y + 16$
8. $x^3 + y^3 + 3xy - 6$
9. $x^4 - 2x^2y + 2y^2 - 8y$
10. $x^3 - 3xy + y^2 - 2y + 2$
11. $4y^3 + 6xy - x^2 - 4x + 10$
12. $x^3 - 3x^2y - y^2 + 10y - 3$
13. $\ln(x^2 + 2xy + 2y^2 - 2x - 8y + 20)$ [*Hint:* Use the fact that $\ln u$ is an increasing function of u.]
14. $e^{4xy - x^2 - 5y^2 + 2y + 6}$ [*Hint:* Similar to that in Exercise 13.]
15. $\cos(xy - 1)$
16. $\sin(y + x^2)$
17. $6/(2x^2 + 2xy + y^2 - 2x + 3)$
18. $8/(4xy - 4x^2 - 2y^2 - 4y)$
19. $x^4 + 2y^2 + 3z^2 - 2x^2 + 4y - 12z + 3$
20. $x^4 + y^2 + 2z^2 - 2x^2 - 4y - 12z + 5$
21. $x^4 + y^4 - z^4 - 2x^2 - 8y^2 + 2z^2 + 10$
22. $2x^2 - 2y^2 + 4yz - 3z^2 - x^4 + 5$
23. $z^4 - 3x^2 - 4y^2 + 4xy - 2z^2 - 11$
24. $3x^2 + 5y^2 + z^2 - 8xy + 6x - 8y - 4z + 9$
25. $x^4 - 2x^2y - 6x^2 + 2y^2 + 2z^2 + w^2 + 2zw + 4y + 4z + 18$
26. $z^4 - 2xz^2 + x^2 + y^2 + z^2 + 3w^2 - 4yw - 2z + 2w - 5$
27. Consider a function f of two variables having continuous partial derivatives of orders less than or equal to 2 with $f_x(0, 0) = f_y(0, 0) = 0$, and let

$$A = f_{xx}(0, 0), \qquad B = f_{xy}(0, 0), \qquad \text{and} \qquad C = f_{yy}(0, 0).$$

Suppose that $AC - B^2 = 0$ with A, B, and C not all zero.
a) Give an example of such a function f with a local maximum $(0, 0)$.
b) Give an example of such a function f with a local minimum at $(0, 0)$.

c) Give an example of such a function f with neither a local maximum nor a local minimum at (0, 0).
d) What is the significance of this exercise?

In Exercises 28 through 32, use common sense to find a point at which the function assumes its maximum value on the square where $-1 \leq x \leq 1$ and $-1 \leq y \leq 1$. Then find a point where it assumes its minimum value on this square.

28. $x^2 + y^2$ **29.** xy **30.** $y - 2x$

31. $x^2 + y^2 - xy$ **32.** $x^2 - y^2 + y$

Work Exercises 33 through 43 by the methods of this section.

33. Find the point on the plane $4x - 3y + z = 10$ closest to the points $(17, -7, -1)$.

34. Find the point on the plane $3x - 2y + 4z = -7$ closest to the point $(-5, 7, -9)$.

35. A rectangular box with open top is to have volume 108 ft^3. Find the dimensions of the box for minimum surface area.

36. A rectangular box with open top is to have volume 192 ft^3. Find the dimensions for minimum cost if material for the ends costs \$6/ft^2, for the side \$3/ft^2, and for the bottom \$9/ft^2.

37. Find three positive real numbers whose sum is 9 and whose product is maximum.

38. Find the minimum distance from the origin to a point on the surface $xyz^2 = 32$.

39. Find the minimum distance from the origin to a point on the surface $x^2yz^3 = 162\sqrt{3}$.

40. Find the maximum possible volume of a rectangular box having one vertex at the origin and the diagonally opposite vertex at a point (x, y, z) in the first octant, as shown in Fig. 17.6, if (x, y, z) lies on the plane $x + 4y + 3z = 18$.

41. Find the volume of the rectangular box of maximum volume that can be inscribed in the ellipsoid $(x^2/a^2) + (y^2/b^2) + (z^2/c^2) = 1$ with sides parallel to the coordinate planes.

42. A long rectangular sheet of tin 12 in. wide is to be bent to form a gutter with cross section in the shape of an isosceles trapezoid, as shown in Fig. 17.7. Find the length x and the angle θ shown in the figure if the gutter has maximum carrying capacity.

Figure 17.7 Cross section of gutter.

43. *Method of least squares:* Find the line $y = mx + b$ that most nearly contains the points (2, 3), (4, 2), and (6, 4) in the sense that the sum of the squares of the *deviations* d_1, d_2, and d_3 shown in Fig. 17.8 is minimum. [*Hint:* The unknowns are m and b.] (This method is often used to find the best linear fit to measure data that it is believed should be linear but that are actually not linear due to measurement errors.)

44. Consider the line $y = mx + b$ that best fits data points (x_1, y_1), $(x_2, y_2), \ldots, (x_n, y_n)$ in the least-squares sense of Exercise 43. Show that m and b satisfy the system of equations

$$\begin{cases} Am + nb = B, \\ Cm + Ab = D, \end{cases}$$

where $A = \sum_{i=1}^n x_i$, $B = \sum_{i=1}^n y_i$, $C = \sum_{i=1}^n x_i^2$, and $D = \sum_{i=1}^n x_i y_i$.

45. Use the result of Exercise 44 to find the best linear fit to the data points $(-2, 4)$, $(-1, 3)$, $(1, 0)$, $(3, -1)$, $(4, -2)$ in the least-squares sense.

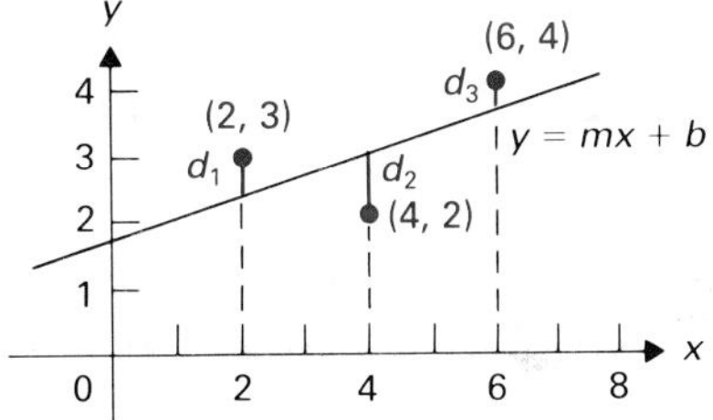

Figure 17.8 Deviations d_1, d_2, d_3 from the line $y = mx + b$ to three data points.

17.2 LAGRANGE MULTIPLIERS

The function $x^2 + y^2$ does not assume a maximum value in the entire plane, because x and y may become arbitrarily large there, so $x^2 + y^2$ may also become arbitrarily large. However, $x^2 + y^2$ does assume a maximum value for points (x, y) restricted to the ellipse $(x^2/4) + (y^2/9) = 1$; the ellipse is a closed and bounded subset of the plane. In this section, we present a useful method for finding the maximum or minimum value assumed by a function, subject to some restriction on the domain of the function. This problem is of practical importance. For example, one may want to find the maximum current that may flow in a circuit containing several resistors under the restriction that the sum of the resistances have some constant value.

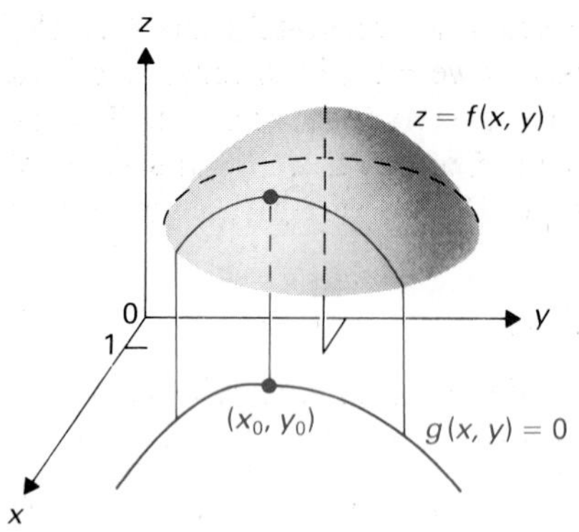

Figure 17.9 $f(x, y)$ restricted to the curve $g(x, y) = 0$ has a local maximum at (x_0, y_0).

Suppose we want to maximize or minimize a function $f(x, y)$ subject to some relation $g(x, y) = 0$. The relation $g(x, y) = 0$ is called a **side condition** or a **constraint.** Geometrically, $g(x, y) = 0$ describes a curve in the x,y-plane, shown in Fig. 17.9. We wish to find local extrema of $f(x, y)$ where we restrict (x, y) to points on this curve. If $f(x_0, y_0)$ is a local maximum subject to $g(x, y) = 0$ as in Fig. 17.9, then $f(x_0, y_0) \geq f(x, y)$ for points (x, y) sufficiently close to (x_0, y_0) *but on the curve* $g(x, y) = 0$ *through* (x_0, y_0). It need *not* be true that $f(x_0, y_0) \geq f(x, y)$ for (x, y) close to (x_0, y_0) but off the curve.

This is really not a new concept for us. We first met problems of this type in Section 5.6. There we solved $g(x, y) = 0$ for one of the variables in terms of the other and then substituted to express f as a function of just one of the variables, x or y, along the curve. We give an example as a reminder.

EXAMPLE 1 Describe a method to find the dimensions of the rectangle of maximum area that can be inscribed in a semicircle of radius a.

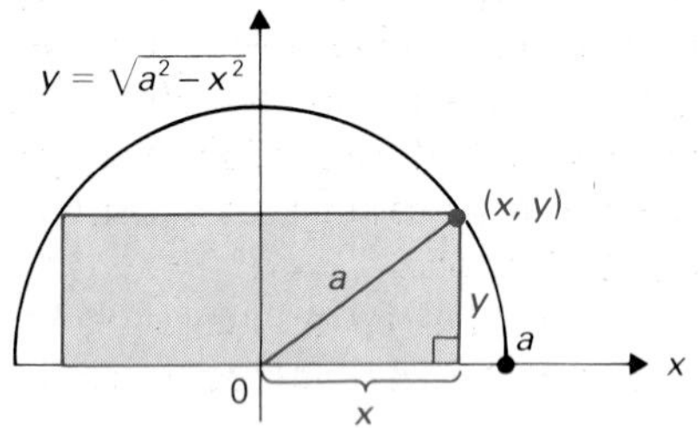

Figure 17.10 Rectangle inscribed in a semicircle of radius a.

Solution From Fig. 17.10, we see that the problem is to maximize $f(x, y) = 2xy$ subject to the constraint $x^2 + y^2 = a^2$. In Section 5.6, we would have solved this problem by using the constraint to express the area as a function of one *independent* variable as follows:

$$\text{Area} = 2xy,$$

$$x^2 + y^2 - a^2 = 0, \qquad \text{so} \quad y = \sqrt{a^2 - x^2},$$

$$\text{Area} = 2x\sqrt{a^2 - x^2}.$$

We then took the derivative of this area function, set it equal to zero, solved that equation, and so on. We continue this example in a moment. □

When discussing word problems in Section 17.1, we generalized the technique of Example 1 to cases involving more variables. We expressed the quantity we wish to maximize or minimize as a function of *independent* variables. As in Example 1, this involves solving constraint equations for one variable in terms of others. In itself, that algebraic problem can be tough to execute. The method of Lagrange multipliers that we now present avoids this algebraic problem with the constraint equations.

We wish to find local extrema of $f(x, y)$ subject to $g(x, y) = 0$. The constraint $g(x, y) = 0$ is a curve in the plane. Since this curve is a level curve of $g(x, y)$, we know that

$$\nabla g = \frac{\partial g}{\partial x}\mathbf{i} + \frac{\partial g}{\partial y}\mathbf{j} \tag{1}$$

is perpendicular to the curve at each point on the curve. See Fig. 17.11. Turning to the function $f(x, y)$ to be maximized or minimized, we know that

$$\nabla f = \frac{\partial f}{\partial x}\mathbf{i} + \frac{\partial f}{\partial y}\mathbf{j} \tag{2}$$

points in the direction of maximum increase of the function $f(x, y)$ at each point. Furthermore, if $\mathbf{u}$ is a unit vector,

$$\nabla f \cdot \mathbf{u} = \text{Directional derivative of } f \text{ in the direction } \mathbf{u}. \tag{3}$$

Now at a point (x_0, y_0) on the curve $g(x, y) = 0$, where $f(x, y)$ has a local

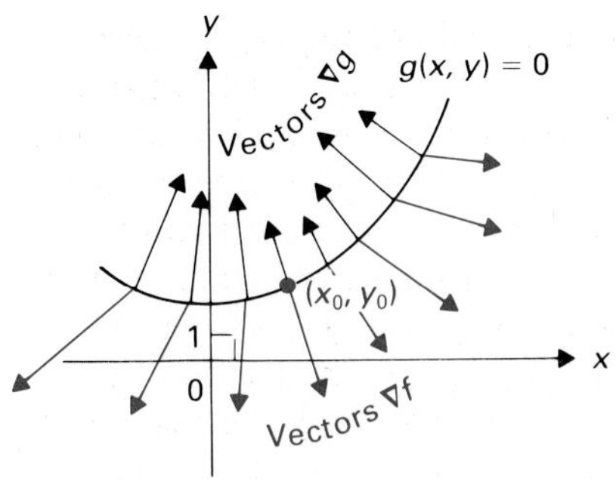

Figure 17.11 ∇f is parallel to ∇g at (x_0, y_0).

extremum *when considered only on the curve*, the directional derivative of f along the curve must be zero. (This is clear if we regard the curve as a bent s-axis and regard f as a function of the one variable s on the curve.) That is, ∇f must be normal to the curve at that point. Consequently, ∇f and ∇g must be *parallel* at (x_0, y_0), as illustrated in Fig. 17.11. Therefore there must exist λ such that

$$\nabla f = \lambda(\nabla g). \tag{4}$$

Equation (4), together with $g(x, y) = 0$, leads to the three conditions

$$\boxed{\frac{\partial f}{\partial x} = \lambda\frac{\partial g}{\partial x}, \qquad \frac{\partial f}{\partial y} = \lambda\frac{\partial g}{\partial y}, \qquad g(x, y) = 0.} \tag{5}$$

We then solve the three Eqs. (5) for the three unknowns x, y, and λ to find candidates (x, y) for points where local extrema occur.

You may quite properly observe that solving Eqs. (5) simultaneously looks as hard as using the substitution technique in Example 1. For the problems we can handle with pencil and paper, that is usually true. However, using a computer, techniques such as Newton's method are available to solve a system of n equations in n unknowns, even if the equations are not linear. See the starred subsection of Section 16.2. The substitution technique, requiring us to express the quantity we want to maximize or minimize as a function of independent variables, is not as feasible on a computer at this time.

Equations (5) are the conditions of the *method of Lagrange multipliers*. The method itself is nothing more than a handy device for obtaining the conditions (5). Let

$$\boxed{L(x, y, \lambda) = f(x, y) - \lambda g(x, y).} \tag{6}$$

The conditions (5) are then equivalent to the following conditions, in the same order:

$$\boxed{\frac{\partial L}{\partial x} = 0, \qquad \frac{\partial L}{\partial y} = 0, \qquad \frac{\partial L}{\partial \lambda} = 0.} \tag{7}$$

The variable λ is called the *Lagrange multiplier*.

A point satisfying Eqs. (7) is a *candidate* for a point where $f(x, y)$ has a local maximum or minimum value, subject to $g(x, y) = 0$. If we know that such a local maximum or minimum exists, and if we can find all solutions of Eqs. (7), then computation of $f(x, y)$ at those points may indicate which is the desired extremum.

EXAMPLE 2 Continue Example 1, and solve the problem of maximizing the function $f(x, y) = 2xy$, subject to $x^2 + y^2 - a^2 = 0$, using a Lagrange multiplier.

Solution First we form

$$L(x, y, \lambda) = 2xy - \lambda(x^2 + y^2 - a^2).$$

Then we set

$$\frac{\partial L}{\partial x} = 2y - 2x\lambda = 0, \tag{8}$$

$$\frac{\partial L}{\partial y} = 2x - 2y\lambda = 0, \tag{9}$$

$$\frac{\partial L}{\partial \lambda} = -x^2 - y^2 + a^2 = 0. \tag{10}$$

Substituting $y = x\lambda$ and $x = y\lambda$ in Eq. (10), we obtain

$$-y^2\lambda^2 - x^2\lambda^2 + a^2 = 0, \qquad \text{or} \qquad -(x^2 + y^2)\lambda^2 + a^2 = 0.$$

From Eq. (10), we then know that

$$-a^2\lambda^2 + a^2 = 0 \qquad \text{or} \qquad a^2(-\lambda^2 + 1) = 0,$$

so

$$\lambda^2 = 1 \qquad \text{and} \qquad \lambda = \pm 1.$$

The value $\lambda = -1$ would give $y = -x$, which is impossible for our geometric problem. Thus $\lambda = 1$, and $y = x$. From Eq. (10), we obtain

$$2x^2 = a^2, \qquad \text{so} \qquad x = \frac{a}{\sqrt{2}},$$

so $(x, y) = (a/\sqrt{2}, a/\sqrt{2})$ gives the maximum. That is, we know from geometry that a maximum exists, and we found only one candidate for it. □

EXAMPLE 3 Find local extrema of $f(x, y) = 3x^2 + y^3$ on the circle $x^2 + y^2 = 9$.

Solution We form

$$L(x, y, \lambda) = (3x^2 + y^3) - \lambda(x^2 + y^2 - 9).$$

The conditions (7) become

$$\begin{aligned} L_x &= 6x - 2\lambda x = 0, \\ L_y &= 3y^2 - 2\lambda y = 0, \\ L_\lambda &= -(x^2 + y^2 - 9) = 0, \end{aligned} \qquad \text{or} \qquad \begin{aligned} 2x(3 - \lambda) &= 0, \\ y(3y - 2\lambda) &= 0, \\ x^2 + y^2 &= 9. \end{aligned}$$

The first equation tells us that either $x = 0$ or $\lambda = 3$.

Case $x = 0$ From $x^2 + y^2 = 9$, we have $y = \pm 3$. If $\lambda = \pm\frac{9}{2}$, the system is satisfied. This gives the candidate points

$$(0, 3) \qquad \text{and} \qquad (0, -3).$$

Case $\lambda = 3$ From the second equation, we have $y = 0$ or $y = 2$, giving the candidate points

$$(3, 0), \qquad (-3, 0), \qquad (\sqrt{5}, 2), \qquad \text{and} \qquad (-\sqrt{5}, 2).$$

It is convenient to sketch the circle $x^2 + y^2 = 9$ and label the candidate points

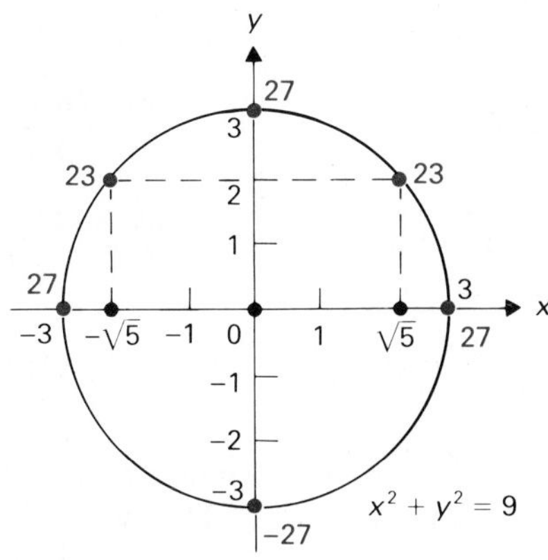

Figure 17.12 Values of $f(x, y) = 3x^2 + y^3$ at candidate points for local extrema on $x^2 + y^2 = 9$.

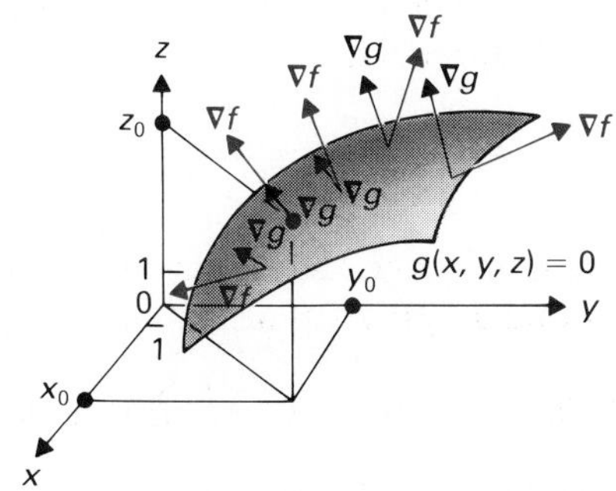

Figure 17.13 ∇f is parallel to ∇g at (x_0, y_0, z_0).

with the function values $f(x, y)$ there, as shown in color in Fig. 17.12. We see that $f(x, y)$ restricted to the circle has

local maxima of 27 at $(-3, 0)$, $(0, 3)$, and $(3, 0)$

and

local minima of 23 at $(\pm\sqrt{5}, 2)$ and of -27 at $(0, -3)$. □

We can make analogous use of Lagrange multipliers to handle situations with more variables or more constraints. For example, to maximize $f(x, y, z)$ subject to $g(x, y, z) = 0$, the gradient

$$\nabla f = \frac{\partial f}{\partial x}\mathbf{i} + \frac{\partial f}{\partial y}\mathbf{j} + \frac{\partial f}{\partial z}\mathbf{k}$$

must be perpendicular to the level surface $g(x, y, z) = 0$, and, consequently, ∇f must be parallel to ∇g, so again $\nabla f = \lambda(\nabla g)$. See Fig. 17.13. The four conditions

$$\frac{\partial f}{\partial x} = \lambda\frac{\partial g}{\partial x}, \qquad \frac{\partial f}{\partial y} = \lambda\frac{\partial g}{\partial y}, \qquad \frac{\partial f}{\partial z} = \lambda\frac{\partial g}{\partial z}, \qquad g(x, y, z) = 0$$

in x, y, z, and λ can again be concisely expressed as

$$\boxed{\frac{\partial L}{\partial x} = 0, \qquad \frac{\partial L}{\partial y} = 0, \qquad \frac{\partial L}{\partial z} = 0, \qquad \frac{\partial L}{\partial \lambda} = 0,} \tag{11}$$

where

$$\boxed{L(x, y, z, \lambda) = f(x, y, z) - \lambda g(x, y, z).}$$

EXAMPLE 4 Use Lagrange multipliers to find the point on the plane $2x - 2y + z = 4$ that is closest to the origin.

Solution We want to minimize $\sqrt{x^2 + y^2 + z^2}$ subject to $2x - 2y + z - 4 = 0$. To make things easier, we minimize the square of the distance $x^2 + y^2 + z^2$ subject to $2x - 2y + z - 4 = 0$.

Let $L(x, y, z, \lambda) = x^2 + y^2 + z^2 - \lambda(2x - 2y + z - 4)$. Then the conditions (11) are

$$\frac{\partial L}{\partial x} = 2x - 2\lambda = 0,$$

$$\frac{\partial L}{\partial y} = 2y + 2\lambda = 0,$$

$$\frac{\partial L}{\partial z} = 2z - \lambda = 0,$$

$$\frac{\partial L}{\partial \lambda} = -2x + 2y - z + 4 = 0.$$

Substituting the values for $2x$, $2y$, and z from the first three conditions into

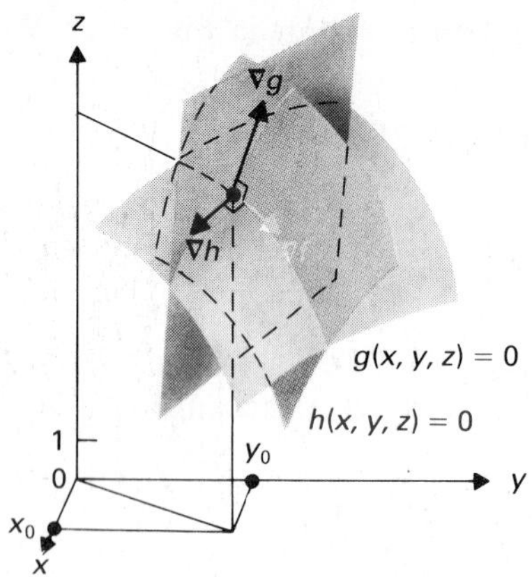

Figure 17.14 ∇f lies in the plane normal to the curve of intersection. The plane contains ∇g and ∇h.

the fourth condition yields

$$-2\lambda - 2\lambda - \frac{\lambda}{2} + 4 = 0, \qquad \text{or} \qquad -\frac{9}{2}\lambda + 4 = 0, \qquad \text{or} \qquad \lambda = \frac{8}{9}.$$

Therefore $x = \frac{8}{9}$, $y = -\frac{8}{9}$, and $z = \frac{4}{9}$, so $(\frac{8}{9}, -\frac{8}{9}, \frac{4}{9})$ is the desired point, and the distance from this point to the origin is

$$\sqrt{\frac{64 + 64 + 16}{81}} = 4 \cdot \sqrt{\frac{4 + 4 + 1}{81}} = 4 \cdot \sqrt{\frac{1}{9}} = \frac{4}{3}. \qquad \square$$

For another case, suppose we wish to maximize or minimize $f(x, y, z)$ subject to *two* constraints $g(x, y, z) = 0$ and $h(x, y, z) = 0$. The two constraints $g(x, y, z) = 0$ and $h(x, y, z) = 0$ define the curve of intersection of the two level surfaces. See Fig. 17.14. At a point where $f(x, y, z)$ is maximum or minimum, subject to these constraints, the directional derivative along this curve must be zero. Consequently, ∇f must be perpendicular to the curve. Since both ∇g and ∇h are perpendicular to the curve, it must be that ∇f lies in the plane determined by the vectors* ∇g and ∇h. This time there are *two* Lagrange multipliers. We must have

$$\nabla f = \lambda_1(\ g) + \lambda_2(\nabla h)$$

for some λ_1 and λ_2. This leads to

$$\frac{\partial f}{\partial x} = \lambda_1 \frac{\partial g}{\partial x} + \lambda_2 \frac{\partial h}{\partial x}, \qquad \frac{\partial f}{\partial y} = \lambda_1 \frac{\partial g}{\partial y} + \lambda_2 \frac{\partial h}{\partial y}, \qquad \frac{\partial f}{\partial z} = \lambda_1 \frac{\partial g}{\partial z} + \lambda_2 \frac{\partial h}{\partial z},$$

and

$$g(x, y, z) = 0, \qquad h(x, y, z) = 0.$$

These are five equations in five unknowns and can be written as

$$\boxed{\frac{\partial L}{\partial x} = 0, \quad \frac{\partial L}{\partial y} = 0, \quad \frac{\partial L}{\partial z} = 0, \quad \frac{\partial L}{\partial \lambda_1} = 0, \quad \frac{\partial L}{\partial \lambda_2} = 0,} \qquad \textbf{(12)}$$

where

$$\boxed{L(x, y, z, \lambda_1, \lambda_2) = f(x, y, z) - \lambda_1 g(x, y, z) - \lambda_2 h(x, y, z).}$$

Of course, solving these equations (12) can be very messy.

EXAMPLE 5 To illustrate Lagrange multipliers when there are two constraints, find the point on the line of intersection of the planes $x - y = 2$ and $x - 2z = 4$ that is closest to the origin.

Solution This time we want to minimize $x^2 + y^2 + z^2$ subject to the side conditions $x - y - 2 = 0$ and $x - 2z - 4 = 0$. We form

$$L(x, y, z, \lambda_1, \lambda_2) = x^2 + y^2 + z^2 - \lambda_1(x - y - 2) - \lambda_2(x - 2z - 4).$$

* One hopes these will not be parallel.

The conditions (12) are

$$\frac{\partial L}{\partial x} = 2x - \lambda_1 - \lambda_2 = 0,$$

$$\frac{\partial L}{\partial y} = 2y + \lambda_1 = 0,$$

$$\frac{\partial L}{\partial z} = 2z + 2\lambda_2 = 0,$$

$$\frac{\partial L}{\partial \lambda_1} = -x + y + 2 = 0,$$

$$\frac{\partial L}{\partial \lambda_2} = -x + 2z + 4 = 0.$$

The second and third conditions give

$$\lambda_1 = -2y \qquad \text{and} \qquad \lambda_2 = -z,$$

so the first condition becomes $2x + 2y + z = 0$. We then have

$$\begin{aligned} 2x + 2y + z &= 0, \\ -x + y &= -2, \\ -x + 2z &= -4. \end{aligned}$$

The last two equations may be written as $y = x - 2$ and $z = (x - 4)/2$. Substitution of these values into the first equation gives

$$2x + 2(x - 2) + \frac{x - 4}{2} = 0, \qquad \text{or} \qquad \frac{9}{2}x - 6 = 0, \qquad \text{or} \qquad x = \frac{4}{3}.$$

Consequently, $y = -\frac{2}{3}$ and $z = -\frac{4}{3}$. The desired point is therefore $(\frac{4}{3}, -\frac{2}{3}, -\frac{4}{3})$. □

SUMMARY

1. To maximize or minimize a function $f(x, y)$ subject to a constraint $g(x, y) = 0$, form $L(x, y, \lambda) = f(x, y) - \lambda g(x, y)$. Candidates for points (x, y) where extrema occur are such that x, y, and λ satisfy the three simultaneous conditions

$$\frac{\partial L}{\partial x} = 0, \qquad \frac{\partial L}{\partial y} = 0, \qquad \frac{\partial L}{\partial \lambda} = 0.$$

For a function $f(x, y, z)$ subject to a constraint $g(x, y, z) = 0$, form $L(x, y, z, \lambda) = f(x, y, z) - \lambda g(x, y, z)$ and solve simultaneously the four conditions

$$\frac{\partial L}{\partial x} = 0, \qquad \frac{\partial L}{\partial y} = 0, \qquad \frac{\partial L}{\partial z} = 0, \qquad \frac{\partial L}{\partial \lambda} = 0.$$

2. If more than one constraint is present, use Lagrange multipliers, λ_1, $\lambda_2, \ldots$, equal in number to the number of constraints. To illustrate with

$f(x, y, z)$ subject to constraints $g(x, y, z) = 0$ and $h(x, y, z) = 0$, form $L(x, y, z, \lambda_1, \lambda_2) = f(x, y, z) - \lambda_1 g(x, y, z) - \lambda_2 h(x, y, z)$. The conditions are then

$$\frac{\partial L}{\partial x} = 0, \quad \frac{\partial L}{\partial y} = 0, \quad \frac{\partial L}{\partial z} = 0, \quad \frac{\partial L}{\partial \lambda_1} = 0, \quad \frac{\partial L}{\partial \lambda_2} = 0.$$

EXERCISES

In Exercises 1 through 16, find local extrema of the function f subject to the given constraints, using Lagrange multipliers.

1. $f(x, y) = x + y$; constraint $x^2 + y^2 = 4$
2. $f(x, y) = xy$; constraint $x^2 + y^2 = 4$
3. $f(x, y) = x - y$; constraint $x^2 + 2y^2 = 24$
4. $f(x, y) = x + 2y$; constraint $2x^2 + y^2 = 36$
5. $f(x, y) = x^2 - 3y^2$; constraint $x^2 + 4y^2 = 36$
6. $f(x, y) = x^3 + 2y^2$; constraint $x^2 + y^2 = 4$
7. $f(x, y) = x^2 + 4y^3$; constraint $2x + 4y = 2$
8. $f(x, y, z) = x^2 + y^2 - z^2$; constraint $2x - y + 3z = 4$
9. $f(x, y, z) = x^2 - 2y^2 - 2z^2$; constraint $3x - y - z = 32$
10. $f(x, y, z) = x + y - z$; constraint $x^2 + y^2 + z^2 = 12$
11. $f(x, y, z) = 2x - 3y + z$; constraint $x^2 + y^2 + 3z^2 = 120$
12. $f(x, y, z) = x^2 + y^2 + z^2$; constraints $x^2 + y^2 = 1, y - z = 0$
13. $f(x, y, z) = x - y + z^2$; constraints $y^2 + z^2 = 1, x + y = 2$
14. $f(x, y, z) = xyz$; constraints $x - y = 0, z = y - 2$
15. $f(x, y, z) = x^2 + y^2 + z^2$; constraints $z^2 = x^2 + y^2$, $x - y + z = 1$
16. $f(x, y, z) = xyz$; constraints $x^2 + y^2 + z^2 = 8, x + y = 0$

Solve Exercises 17 through 25 using Lagrange multipliers.

17. Find the point on the plane $2x - 3y + 6z = 5$ that is closest to the origin.
18. Find the point on the plane $3x - 2y + z = 5$ that is closest to $(4, -8, 5)$.
19. A rectangular box has a square base and open top. Find the dimensions for minimum surface area if the volume is to be 108 in^3.
20. An open rectangular box has ends costing \$6/ft^2, sides costing \$4/ft^2, and a bottom costing \$10/ft^2. Find the dimensions for minimum cost if the volume of the box is 120 ft^3.
21. Find the maximum possible volume of a right circular cone inscribed in a sphere of radius a.
22. The sides of a closed cylindrical container cost twice as much per square foot as the ends. Find the ratio of the radius to the altitude of the cylinder for the cheapest such container having a fixed volume.
23. A pentagon consists of a rectangle surmounted by an isosceles triangle. Find the dimensions of the pentagon having the maximum area if the perimeter is to be P.
24. Find the point on the curve of intersection of $x^2 + z^2 = 4$ and $x - y = 8$ that is farthest from the origin.
25. Find the point on the line of intersection of the planes $x - y = 4$ and $y + 3z = 6$ that is closest to $(-1, 3, 2)$.

17.3 EXACT DIFFERENTIAL FORMS

Recall that if $H(x, y)$ has first partial derivatives, then the differential dH is given by

$$dH = H_x(x, y)\,dx + H_y(x, y)\,dy. \tag{1}$$

In this section, we consider expressions similar to Eq. (1), of the form

$$P(x, y)\,dx + Q(x, y)\,dy \tag{2}$$

for two variables, or

$$P(x, y, z)\,dx + Q(x, y, z)\,dy + R(x, y, z)\,dz \tag{3}$$

for three variables. We will see that the expressions (2) and (3) need not always be differentials of a function. For this reason, we consider them to be *differential forms* rather than *differentials*.

DEFINITION 17.1 Differential form

A **differential form** is an expression of type (2) or (3) for two or three variables and of type

$$f_1(\mathbf{x})\,dx_1 + f_2(\mathbf{x})\,dx_2 + \cdots + f_n(\mathbf{x})\,dx_n$$

for n variables, where $\mathbf{x} = (x_1, x_2, \ldots, x_n)$.

In Section 6.4, we saw how to solve some differential equations by separating variables. For example, we can solve the equation $dy/dx = x^2/y^2$ as follows:

$$\frac{dy}{dx} = \frac{x^2}{y^2},$$

$$y^2\,dy = x^2\,dx,$$

$$\frac{y^3}{3} = \frac{x^3}{3} + C.$$

However, if we try that technique to solve the differential equation $dy/dx = -3x^2y/(x^3 + 6y)$, we run into a problem. We have

$$\frac{dy}{dx} = \frac{-3x^2y}{x^3 + 6y}, \tag{4}$$

$$(x^3 + 6y)\,dy = -3x^2y\,dx.$$

We cannot "separate" the variables. However, let us write the equation in the form

$$3x^2y\,dx + (x^3 + 6y)\,dy = 0. \tag{5}$$

The left-hand side of Eq. (5) is a differential form. If it is the differential of a function $H(x, y)$, then Eq. (5) can be written as

$$dH = 0.$$

If we set $H(x, y) = C$, we then have the implicit general solution of the differential equation (4). That is, any differentiable function $y = f(x)$ defined implicitly by $H(x, y) = C$ is a solution of the differential equation (4).

EXAMPLE 1 Find $H(x, y)$ such that

$$dH = (3x^2y)\,dx + (x^3 + 6y)\,dy$$

and solve the differential equation (4).

Solution From Eq. (1), we see we must have

$$\frac{\partial H}{\partial x} = 3x^2y \qquad \text{and} \qquad \frac{\partial H}{\partial y} = x^3 + 6y.$$

We can attempt to find $H(x, y)$ by *partial integration*, reversing the operation of partial differentiation.

Since $\partial H/\partial x = 3x^2y$, to find $H(x, y)$ we integrate $3x^2y$ with respect to x only, *treating y as a constant*. An indefinite integral is defined only up to a *constant*. For partial integration with respect to x, such a "constant" may be any function of y only. We obtain

$$H(x, y) = \int 3x^2y\,dx = x^3y + h(y), \tag{6}$$

where $h(y)$ is a function of y only.

Similarly, since we must have $\partial H/\partial y = x^3 + 6y$, we obtain

$$H(x, y) = \int (x^3 + 6y)\,dy = x^3y + 3y^2 + k(x), \tag{7}$$

where $k(x)$ may be any function of x only.

Our task is now to find $h(y)$ and $k(x)$ so that the expressions (6) and (7) are identical. Clearly, we can take $h(y) = 3y^2$ and $k(x) = 0$. Thus $H(x, y) = x^3y + 3y^2$ is a function such that dH is the given differential form. The differential equation (4) becomes $dH = 0$, and the equation has the implicit general solution $x^3y + 3y^2 = C$. □

DEFINITION 17.2 **Exact differential**

A differential form

$$P(x, y)\,dx + Q(x, y)\,dy \tag{8}$$

is called an **exact differential** if there exists a function $H(x, y)$ such that the differential form is equal to dH.

In other words, an *exact differential* really is the *differential* of some function. Of course, the analogue of Definition 17.2 is made for functions of more than two variables.

Recall that the gradient ∇H of $H(x, y)$ is given by

$$\nabla H = H_x(x, y)\mathbf{i} + H_y(x, y)\mathbf{j}.$$

We see immediately that $P(x, y)\,dx + Q(x, y)\,dy$ is the differential dH of $H(x, y)$ if and only if $P(x, y)\mathbf{i} + Q(x, y)\mathbf{j}$ is the gradient ∇H of $H(x, y)$. That is, our work in this section with differential forms can be phrased in terms of gradient vectors. We will be using the gradient interpretation in Sections 17.5 and Chapter 19.

Since

$$dH = \frac{\partial H}{\partial x}\,dx + \frac{\partial H}{\partial y}\,dy, \tag{9}$$

we see that, in order for Eq. (9) to be the same differential form as (8), we must have

$$\boxed{\frac{\partial H}{\partial x} = P(x, y) \quad \text{and} \quad \frac{\partial H}{\partial y} = Q(x, y).} \tag{10}$$

Now if $P(x, y)$ and $Q(x, y)$ have continuous partial derivatives, so that

$H(x, y)$ would have to have continuous second partial derivatives, we would have

$$\frac{\partial^2 H}{\partial y\, \partial x} = \frac{\partial^2 H}{\partial x\, \partial y}. \tag{11}$$

From Eqs. (10) and (11), we see that, in order for form (8) to be an exact differential, we must have

$$\frac{\partial P}{\partial y} = \frac{\partial^2 H}{\partial y\, \partial x} = \frac{\partial^2 H}{\partial x\, \partial y} = \frac{\partial Q}{\partial x}.$$

THEOREM 17.3 Necessary condition for exactness

If $P(x, y)$ and $Q(x, y)$ have continuous partial derivatives, then the differential form $P(x, y)\,dx + Q(x, y)\,dy$ can be exact only if

$$\frac{\partial P}{\partial y} = \frac{\partial Q}{\partial x}. \tag{12}$$

EXAMPLE 2 Test the differential form $x^2 y\,dx + (x^2 - y^2)\,dy$ for exactness.

Solution Here $P(x, y) = x^2 y$ and $Q(x, y) = x^2 - y^2$. Now

$$\frac{\partial P}{\partial y} = \frac{\partial (x^2 y)}{\partial y} = x^2 \quad \text{and} \quad \frac{\partial Q}{\partial x} = \frac{\partial (x^2 - y^2)}{\partial x} = 2x.$$

Thus $\partial P/\partial y \neq \partial Q/\partial x$, so condition (12) is not satisfied. Consequently the differential form cannot be exact. □

EXAMPLE 3 Test the differential form $(2xy + y^2)\,dx + (x^2 + 2xy)\,dy$ for exactness.

Solution This time

$$\frac{\partial P}{\partial y} = \frac{\partial (2xy + y^2)}{\partial y} = 2x + 2y \quad \text{and} \quad \frac{\partial Q}{\partial x} = \frac{\partial (x^2 + 2xy)}{\partial x} = 2x + 2y.$$

Thus $\partial P/\partial y = \partial Q/\partial x$, so condition (12) is satisfied. In this case, a little computation as in Example 1 shows that our differential form is indeed exact; it is dH for $H(x, y) = x^2 y + xy^2$, because then

$$dH = \frac{\partial H}{\partial x}\,dx + \frac{\partial H}{\partial y}\,dy = (2xy + y^2)\,dx + (x^2 + 2xy)\,dy. \quad \square$$

The preceding example suggests that perhaps condition (12) not only is a *necessary* condition for $P(x, y)\,dx + Q(x, y)\,dy$ to be exact but also is a *sufficient* condition, at least if $P(x, y)$ and $Q(x, y)$ have continuous first partial derivatives. This is *not* always true, but it is true if the domain in the plane where $P(x, y)$ and $Q(x, y)$ are both defined has no "holes" in it. It can be demonstrated that

$$\frac{y}{x^2 + y^2}\,dx - \frac{x}{x^2 + y^2}\,dy,$$

which does satisfy condition (12), is not exact in its entire domain, which consists of the plane minus the origin. (See the final exercise in this chapter.) Here, the origin is a "hole" in the domain.

We will not concern ourselves with the niceties of finding as large regions as possible where a differential form that satisfies condition (12) is exact. We indicate informally that, if $P(x, y)\,dx + Q(x, y)\,dy$ is defined throughout some disk containing a point and $P(x, y)$ and $Q(x, y)$ are differentiable there with $\partial P/\partial y = \partial Q/\partial x$, then the differential form is exact in that disk. Our argument will indicate a four-step procedure to find all functions $H(x, y)$ such that $P(x, y)\,dx + Q(x, y)\,dy = dH$.

Since we must have $\partial H/\partial x = P(x, y)$, we let $G(x, y)$ be some antiderivative of $P(x, y)$ with respect to x only, treating y as a constant. Such a function $G(x, y)$ is found by *partial integration* with respect to x, holding y constant, as in Example 1. Now an indefinite integral is defined only up to a constant. Since y is treated as a constant in this integration with respect to x, we see that the most general partial antiderivative of $P(x, y)$ with respect to x is of the form $G(x, y) + h(y)$ for an arbitrary function $h(y)$ of y only.

Step 1 Compute

$$H(x, y) = \int P(x, y)\,dx = G(x, y) + h(y), \tag{13}$$

where $G(x, y)$ is any computed antiderivative of $P(x, y)$ with respect to x only, treating y as a constant.

Our problem is to determine $h(y)$ such that $\partial H/\partial y = Q(x, y)$. Now $\partial H/\partial y = \partial G/\partial y + h'(y)$, so $\partial G/\partial y + h'(y) = Q(x, y)$, or

$$h'(y) = Q(x, y) - \frac{\partial G}{\partial y}. \tag{14}$$

If $Q(x, y) - \partial G/\partial y$ is a continuous function of *y only*, we could set $h'(y)$ equal to it and integrate to find the desired $h(y) = \int h'(y)\,dy$. Now Eq. (14) is a function of y only provided that

$$\frac{\partial}{\partial x}\left(Q(x, y) - \frac{\partial G}{\partial y}\right) = 0, \tag{15}$$

that is, if

$$\frac{\partial Q}{\partial x} - \frac{\partial^2 G}{\partial x\,\partial y} = 0. \tag{16}$$

But if $G(x, y)$ has continuous second-order partial derivatives throughout the neighborhood where we are working, then

$$\frac{\partial^2 G}{\partial x\,\partial y} = \frac{\partial^2 G}{\partial y\,\partial x} = \frac{\partial}{\partial y}\left(\frac{\partial G}{\partial x}\right) = \frac{\partial}{\partial y}(P(x, y)) = \frac{\partial P}{\partial y}.$$

Under these conditions, condition (16) is satisfied, for the left-hand side becomes

$$\frac{\partial Q}{\partial x} - \frac{\partial P}{\partial y},$$

which is zero by assumption. Thus we can complete our steps to find $H(x, y)$, as follows.

Step 2	Compute $\partial H/\partial y = \partial G/\partial y + h'(y)$, set it equal to $Q(x, y)$, and solve for $h'(y)$.
Step 3	Integrate to find $h(y) = \int h'(y)\,dy$.
Step 4	The final answer is $H(x, y) = G(x, y) + h(y) + C$.

We state a theorem that seems reasonable in view of the work we just completed.

THEOREM 17.4 Test for exactness

If $P(x, y)$ and $Q(x, y)$ have continuous first partial derivatives in some disk, and if $\partial P/\partial x = \partial Q/\partial y$ in that disk, then there exists a function $H(x, y)$ defined in the disk such that $dH = P(x, y)\,dx + Q(x, y)\,dy$.

EXAMPLE 4 Use the steps of the argument preceding Theorem 17.4 to show that

$$(2xy^3 + 6x)\,dx + (3x^2y^2 + 4y^3)\,dy$$

is an exact differential dH and to find the function $H(x, y)$.

Solution Note that

$$\frac{\partial P}{\partial y} = \frac{\partial(2xy^3 + 6x)}{\partial y} = 6xy^2 \qquad \text{and} \qquad \frac{\partial Q}{\partial x} = \frac{\partial(3x^2y^2 + 4y^3)}{\partial x} = 6xy^2,$$

so the differential form is indeed exact. Then

Step 1 $H(x, y) = \int P(x, y)\,dx = \int (2xy^3 + 6x)\,dx = x^2y^3 + 3x^2 + h(y)$.

Step 2 $\partial H/\partial y = 3x^2y^2 + h'(y)$, so

$$3x^2y^2 + h'(y) = 3x^2y^2 + 4y^3, \qquad \text{and} \qquad h'(y) = 4y^3.$$

Step 3 $h(y) = \int 4y^3\,dy = y^4 + C$.

Step 4 $H(x, y) = x^2y^3 + 3x^2 + y^4 + C$. □

For a differential form in three variables

$$P(x, y, z)\,dx + Q(x, y, z)\,dy + R(x, y, z)\,dz$$

to be exact, the corresponding conditions are

$$\boxed{\frac{\partial P}{\partial y} = \frac{\partial Q}{\partial x}, \qquad \frac{\partial P}{\partial z} = \frac{\partial R}{\partial x}, \qquad \frac{\partial Q}{\partial z} = \frac{\partial R}{\partial y}.} \tag{17}$$

Using the notation of subscripted variables, if $\mathbf{x} = (x_1, x_2, x_3)$, then

$$f_1(\mathbf{x})\,dx_1 + f_2(\mathbf{x})\,dx_2 + f_3(\mathbf{x})\,dx_3$$

is exact in some neighborhood of each point if

$$\boxed{\frac{\partial f_i}{\partial x_j} = \frac{\partial f_j}{\partial x_i} \quad \text{for all } i \text{ and } j.} \tag{18}$$

Again, the necessity of these relations follows from equality of mixed second-order partial derivatives. See Exercise 21.

EXAMPLE 5 Consider the differential form

$$(yz^2 - 6x \sin z)\, dx + (xz^2 - 3y^2 \cos z)\, dy$$
$$+ (2xyz - 3x^2 \cos z + y^3 \sin z)\, dz,$$

which can easily be checked to satisfy the conditions (17). Find $H(x, y, z)$ so that the differential form is dH.

Solution We proceed in a fashion analogous to the steps preceding Theorem 17.4. From $\partial H/\partial x = yz^2 - 6x \sin z$, we have

$$H(x, y, z) = \int (yz^2 - 6x \sin z)\, dx$$
$$= xyz^2 - 3x^2 \sin z + h(y, z).$$

Then $\partial H/\partial y = xz^2 + \partial h/\partial y$ yields

$$xz^2 + \frac{\partial h}{\partial y} = xz^2 - 3y^2 \cos z,$$

so

$$\frac{\partial h}{\partial y} = -3y^2 \cos z,$$

and

$$h(y, z) = \int -3y^2 \cos z\, dy$$
$$= -y^3 \cos z + k(z).$$

We are now down to

$$H(x, y, z) = xyz^2 - 3x^2 \sin z - y^3 \cos z + k(z).$$

Finally, $\partial H/\partial z = 2xyz - 3x^2 \cos z + y^3 \sin z + k'(z)$ yields

$$2xyz - 3x^2 \cos z + y^3 \sin z = 2xyz - 3x^2 \cos z + y^3 \sin z + k'(z),$$

so $k'(z) = 0$ and $k(z) = C$. Thus

$$H(x, y, z) = xyz^2 - 3x^2 \sin z - y^3 \cos z + C. \quad \square$$

In Examples 4 and 5, we followed the steps of the argument preceding Theorem 17.4 to find the function H. Students often find the technique of Example 1 more natural. This technique is used in the remaining examples.

EXAMPLE 6 Find $H(x, y)$ such that

$$(2xy + ye^{xy} - 4x)\, dx + (x^2 + xe^{xy} - 3 \sin y)\, dy$$

is dH, if the differential form is exact.

Solution If the form is dH, we must have

$$H_x(x, y) = 2xy + ye^{xy} - 4x,$$

so

$$H(x, y) = \int (2xy + ye^{xy} - 4x)\, dx = x^2y + e^{xy} - 2x^2 + h(y) \qquad \textbf{(19)}$$

for some function $h(y)$ of y only. Similarly,

$$H(x, y) = \int (x^2 + xe^{xy} - 3 \sin y)\, dy = x^2y + e^{xy} + 3 \cos y + k(x) \qquad \textbf{(20)}$$

for some function $k(x)$ of x only. Comparison of forms (19) and (20) shows that we may take

$$H(x, y) = x^2y + e^{xy} - 2x^2 + 3 \cos y + C$$

for any constant C, for this function is of both forms (19) and (20). Thus our differential form is exact, and we found $H(x, y)$. □

EXAMPLE 7 Find $H(x, y, z)$ such that

$$\left(2xyz + \frac{1}{z}\right) dx + (x^2z)\, dy + \left(x^2y - \frac{x}{z^2}\right) dz$$

is dH, if the differential form is exact.

Solution Arguing as in Example 6, we must have

$$H(x, y, z) = \int \left(2xyz + \frac{1}{z}\right) dx = x^2yz + \frac{x}{z} + h(y, z),$$

$$H(x, y, z) = \int (x^2z)\, dy = x^2yz + k(x, z),$$

and

$$H(x, y, z) = \int \left(x^2y - \frac{x}{z^2}\right) dz = x^2yz + \frac{x}{z} + g(x, y).$$

Comparison of these three expressions shows that

$$H(x, y, z) = x^2yz + \frac{x}{z} + C$$

satisfies all three conditions for any constant C. □

EXAMPLE 8 Find $H(u, v)$ such that

$$(v^2 - 2)\, du + (2uv + 3u)\, dv$$

is dH, if the differential form is exact.

Solution We must have

$$H(u, v) = \int (v^2 - 2)\, du = uv^2 - 2u + h(v)$$

and

$$H(u, v) = \int (2uv + 3u)\, dv = uv^2 + 3uv + k(u).$$

We see that this differential form is not exact, since there is no function $h(v)$ of v only that could supply the summand $3uv$ needed in the second expression for $H(u, v)$. □

EXAMPLE 9 Find the solution of the differential equation

$$\frac{dy}{dx} = -\frac{y}{x} - 2$$

that goes through the point (1, 2).

Solution We rewrite the equation:

$$\frac{dy}{dx} = -\frac{y}{x} - 2 = -\frac{y + 2x}{x},$$

$$x\,dy = -(y + 2x)\,dx,$$

$$(y + 2x)\,dx + x\,dy = 0. \tag{21}$$

We can solve this equation if the differential form (21) is an exact differential dH. We must have

$$H(x, y) = \int (y + 2x)\,dx = xy + x^2 + h(y)$$

and

$$H(x, y) = \int x\,dy = xy + k(x).$$

We see that we may take

$$H(x, y) = xy + x^2,$$

and Eq. (21) then becomes $dH = 0$. Thus the general implicit solution is $H(x, y) = C$, or

$$xy + x^2 = C.$$

Evaluating C to find the solution containing the point (1, 2), we have

$$1 \cdot 2 + 1^2 = C, \qquad \text{so} \qquad C = 3.$$

Thus

$$xy + x^2 = 3, \qquad \text{or} \qquad y = \frac{3 - x^2}{x}$$

is the desired solution. □

SUMMARY

1. A differential form $P(x, y)\,dx + Q(x, y)\,dy$ is exact if it is the differential of some function $H(x, y)$. A differential form $f_1(\mathbf{x})\,dx_1 + f_2(\mathbf{x})\,dx_2 + f_3(\mathbf{x})\,dx_3$ is exact if it is the differential of some function $H(\mathbf{x})$.
2. If P and Q are continuously differentiable throughout a disk, then $P(x, y)\,dx + Q(x, y)\,dy$ is exact there if and only if $\partial P/\partial y = \partial Q/\partial x$.

Similarly, $P\,dx + Q\,dy + R\,dz$ is exact if and only if $\partial P/\partial y = \partial Q/\partial x$, $\partial P/\partial z = \partial R/\partial x$, and $\partial Q/\partial z = \partial R/\partial y$.

3. If $P(x, y)\,dx + Q(x, y)\,dy = dH$, then $H(x, y)$ may be found as follows:

Step 1 Compute $H(x, y) = \int P(x, y)\,dx = G(x, y) + h(y)$, where $G(x, y)$ is any partial antiderivative of $P(x, y)$ with respect to x, treating y as a constant.

Step 2 Compute $\partial H/\partial y = \partial G/\partial y + h'(y)$, set it equal to $Q(x, y)$, and solve for $h'(y)$.

Step 3 Integrate to find $h(y) = \int h'(y)\,dy$.

Step 4 Then $H(x, y) = G(x, y) + h(y) + C$.

See Examples 4 and 5. A similar technique is valid for differential forms involving more variables.

4. Another technique for finding $H(x, y)$ if $dH = P(x, y)\,dx + Q(x, y)\,dy$ is as follows:

Step 1 Compute $H(x, y) = \int P(x, y)\,dx = G(x, y) + h(y)$ and $H(x, y) = \int Q(x, y)\,dy = F(x, y) + k(x)$.

Step 2 Compare the results of Step 1 and find a function $H(x, y)$ of both types.

See Examples 6 through 8. A similar technique is valid for differential forms with more variables.

5. If a differential equation $P(x, y)\,dx + Q(x, y)\,dy = 0$ can be written in the form $dH = 0$, then the implicit general solution is $H(x, y) = C$.

EXERCISES

In Exercises 1 through 10, test whether the given differential form is exact inside disks using the criterion described in number 2 of the Summary. If it is exact, find the most general function H such that the differential form is dH, using the technique described in number 3 of the Summary.

1. $x^2\,dx - y\,dy$
2. $2xy\,dx + x^2\,dy$
3. $(3x - 2y)\,dx + (2x - 3y)\,dy$
4. $2xy\,dx + (x^2 - e^{-y})\,dy$
5. $(y \sec^2 xy)\,dy + (1 + x \sec^2 xy)\,dy$
6. $(2xy^3 - 3)\,dx + (3x^2y^2 + 4y)\,dy$
7. $\dfrac{-y}{x^2 + z^2}\,dy + \dfrac{x}{x^2 + z^2}\,dz$ for $x > 0$
8. $\dfrac{-z}{y^2 + z^2}\,dy + \dfrac{y}{y^2 + z^2}\,dz$ for $z > 0$
9. $(2xyz - 3y^2 + 2z^3)\,dx + (x^2z - 6xy)\,dy + (x^2y + 8z + 6xz^2)\,dz$
10. $(yz \cos xyz - 3z^2)\,dx + (xz \cos xyz + 3y^2)\,dy + (xy \cos xyz - 6xz)\,dz$

In Exercises 11 through 20, use the technique described in number 4 of the Summary to find the most general function H such that the given differential form is dH, if the form is exact.

11. $x^2z\,dx - yz\,dy$
12. $2xz\,dx + x^2\,dz$
13. $\cos y\,dx + (1 - x \sin y)\,dy$
14. $y^2\,dx + \left(\dfrac{1}{y} + 2xy\right)dy$
15. $(e^y - y \cos xy)\,dx + (xe^y - x \cos xy)\,dy$
16. $\left[\dfrac{2u}{v} - \dfrac{v}{(u + v)^2} + \dfrac{1}{u^2}\right]du + \left[\dfrac{u}{(u + v)^2} - \dfrac{u^2}{v^2} - \dfrac{3}{v^2}\right]dv$
17. $\left[-2z \sin(2wz) + \dfrac{1}{(1 + w)^2}\right]dw + \left(\dfrac{1}{z} - 4w \sin wz \cos wz\right)dz$
18. $\left(x_2^3 - \dfrac{1}{x_3^2}\right)dx_1 + (3x_1x_2^2 + 4x_2x_3)\,dx_2 + \left(2x_2^2 + \dfrac{x_1}{x_3^3}\right)dx_3$
19. $(ze^{yz})\,dx + (xz^2e^{yz})\,dy + (xyze^{yz})\,dz$
20. $(y \sin yz)\,dx + (xyz \cos yz + x \sin yz)\,dy + (xy^2 \cos yz)\,dz$

21. Let $\mathbf{x} = (x_1, x_2, \ldots, x_n)$ and let $f_1(\mathbf{x}), f_2(\mathbf{x}), \ldots, f_n(\mathbf{x})$ have continuous first partial derivatives. Show that if

$$f_1(\mathbf{x})\,dx_1 + f_2(\mathbf{x})\,dx_2 + \cdots + f_n(\mathbf{x})\,dx_n$$

is an exact differential form, then $\partial f_i/\partial x_j = \partial f_j/\partial x_i$ for each pair of subscripts i and j, where $1 \le i < j \le n$.

22. Consider a differential equation in x and y involving continuous functions that can be solved by separation of variables. Show that the equation can also be written in the form $dH = 0$ for some function $H(x, y)$.

In Exercises 23 through 30, find the solution of the given differential equation containing the given point.

23. $dy/dx = -y/(x + 2y)$; point $(-1, 3)$

24. $dy/dx = (3x^2 - y^2)/2xy$; point $(2, -3)$

25. $dy/dx = (2x - y \sin xy)/(x \sin xy)$; point $(1, \pi/2)$

26. $dy/dx = -(3x^2 + e^y)/(xe^y)$; point $(3, 0)$

27. $dy/dx = (2x + \sin y)/(3y^2 - x \cos y)$; point $(1, 0)$

28. $dy/dx = -(y^3 + y \sin xy)/(3xy^2 + x \sin xy + e^y)$; point $(0, 0)$

29. $dy/dx = (y + 2xy^2)/x$; point $(2, -1)$ [*Hint:* After writing the equation in differential form, divide by y^2.]

30. $dy/dx = -y/(x + x^2y)$; point $(1, e)$ [*Hint:* After writing the equation in differential form, divide by x^2y^2.]

17.4 LINE INTEGRALS

An integral $\int_a^b f(x)\,dx$ should be regarded as one-dimensional. The function f is integrated over a line segment, which is one-dimensional. In this section, we continue our study of one-dimensional integrals. We no longer restrict ourselves to a line segment but integrate over a portion of a curve γ in the plane or in space. Such integrals are called *line integrals*: the word "line" in this context should be viewed as indicating "dimension one" rather than meaning "straight."

Recall that a *smooth curve* γ in the plane can be described parametrically by

$$x = h(t), \qquad y = k(t), \qquad \text{for } a \le t \le b, \tag{1}$$

where the functions h and k have continuous derivatives and those derivatives are never simultaneously zero for $a < t < b$. The point A corresponding to $t = a$ is the **initial point** of γ, and the point B, where $t = b$, is the **terminal point.** We give a few illustrations of how integrals over such curves arise naturally.

Figure 17.15 Partition element of γ of length Δs.

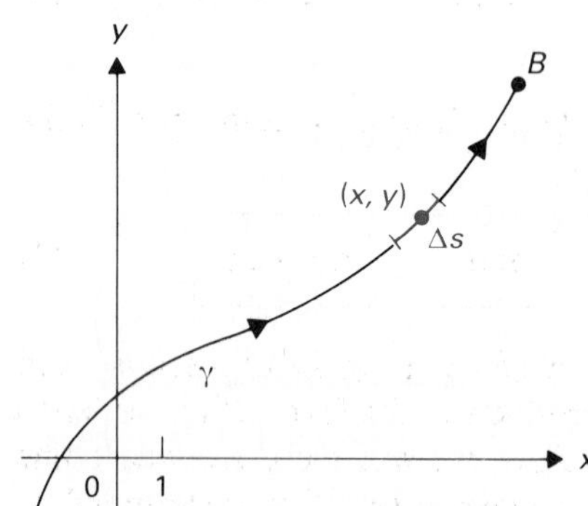

Application 1 Consider a wire covering the smooth curve γ in Fig. 17.15. Suppose the mass density of the wire at a point (x, y) on the curve is $\sigma(x, y)$. Let γ be partitioned into *short* pieces over which $\sigma(x, y)$ remains almost constant. Suppose the arc of length Δs shown in Fig. 17.15 is such a piece. For the point (x, y) shown in Fig. 17.15, the approximate mass of this piece of wire of length Δs is $\sigma(x, y) \cdot \Delta s$. If we add up such products over a partition of the curve, we obtain an approximation to the mass of the wire. It is clear from our work in Chapter 6 that the actual mass of the wire should be given by an integral of $\sigma(x, y)$ over γ with respect to arc length. We denote this integral by $\int_\gamma \sigma(x, y)\,ds$, so we have

$$\text{Mass} = \int_\gamma \sigma(x, y)\,ds.$$

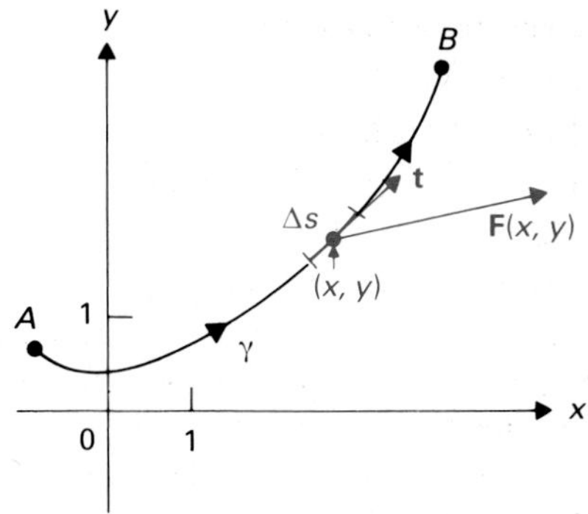

Figure 17.16 Force vector $\mathbf{F}(x, y)$ and unit tangent vector $\mathbf{t}$ at (x, y) in a partition element.

Application 2 Let $\kappa(x, y)$ be the curvature of γ in Fig. 17.15 at (x, y). If we sum terms of the form $\kappa(x, y) \cdot \Delta s$ over a partition of γ, we get an estimate for a *total measure* of the bend of the *entire curve*. Passing to an integral, we define

$$\textbf{Total curvature of } \gamma = \int_\gamma \kappa(x, y)\, ds.$$

Application 3 Let $\mathbf{F}(x, y) = P(x, y)\mathbf{i} + Q(x, y)\mathbf{j}$ be a force vector at each point (x, y) in a portion of the plane containing the curve γ in Fig. 17.16. Let $\mathbf{t}$ be the unit vector tangent to the curve at (x, y) in the direction corresponding to increasing time t. Then $\mathbf{F} \cdot \mathbf{t}$ is the scalar component of $\mathbf{F}$ along the curve. The product $(\mathbf{F} \cdot \mathbf{t})(\Delta s)$ for the curve in Fig. 17.16 is the approximate work done by $\mathbf{F}$ acting on a body moving on the arc of length Δs. Clearly the *total work* done on the body by $\mathbf{F}$ along the *entire curve* γ is

$$\text{Work} = \int_\gamma (\mathbf{F} \cdot \mathbf{t})\, ds. \qquad \textbf{(2)}$$

Recall from Section 15.1 that

$$\mathbf{t} = \frac{1}{|\mathbf{v}|}\mathbf{v} = \frac{1}{ds/dt}\left(\frac{dx}{dt}\mathbf{i} + \frac{dy}{dt}\mathbf{j}\right) = \frac{dx}{ds}\mathbf{i} + \frac{dy}{ds}\mathbf{j}.$$

Since $\mathbf{F}(x, y) = P(x, y)\mathbf{i} + Q(x, y)\mathbf{j}$, we have

$$\mathbf{F} \cdot \mathbf{t} = P(x, y)\frac{dx}{ds} + Q(x, y)\frac{dy}{ds}.$$

Substituting this expression in formula (2), we can also express the work as

$$\text{Work} = \int_\gamma (P(x, y)\, dx + Q(x, y)\, dy). \qquad \textbf{(3)}$$

This leads us to consider also integrals of functions with respect to x or y along a curve γ.

In the preceding applications, we have met integrals of the forms

$$\int_\gamma f(x, y)\, ds, \qquad \int_\gamma f(x, y)\, dx, \qquad \text{and} \qquad \int_\gamma f(x, y)\, dy. \qquad \textbf{(4)}$$

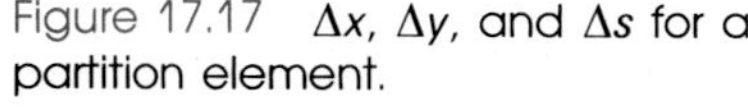
Figure 17.17 Δx, Δy, and Δs for a partition element.

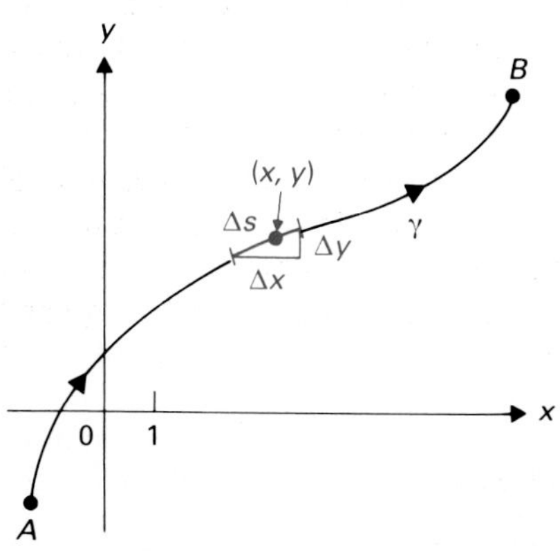

It is important to remember that the integrals of forms (4) come from limits of sums of terms of the form

$$f(x, y) \cdot \Delta s, \qquad f(x, y) \cdot \Delta x, \qquad \text{and} \qquad f(x, y) \cdot \Delta y, \qquad \textbf{(5)}$$

for the point (x, y) and Δx, Δy, and Δs shown for γ in Fig. 17.17. As illustrated in the above applications, we think of products of the forms (5) in order to recognize what integral is appropriate. Also, we will see cases where line integrals can be evaluated quickly by recognizing that they correspond to limits of sums of terms of the forms (5).

To evaluate integrals of the forms (4), simply use the parametric equations

$$x = h(t), \qquad y = k(t), \qquad \text{for } a \le t \le b$$

for γ, and substitute as follows:

$$\boxed{\begin{aligned} f(x, y) &= f(h(t), k(t)), \\ dx &= h'(t)\,dt, \\ dy &= k'(t)\,dt, \\ ds &= \sqrt{\left(\frac{dx}{dt}\right)^2 + \left(\frac{dy}{dt}\right)^2}\,dt. \end{aligned}}$$

Then integrate the resulting expression from $t = a$ to $t = b$.

EXAMPLE 1 Let $f(x, y) = x^2y$, and let γ be the semicircular curve

$$\gamma: x = a\cos t, \quad y = a\sin t \qquad \text{for } 0 \le t \le \pi.$$

Find $\int_\gamma f(x, y)\,ds$ and $\int_\gamma f(x, y)\,dy$.

Solution Now

$$ds = \sqrt{\left(\frac{dx}{dt}\right)^2 + \left(\frac{dy}{dt}\right)^2}\,dt = \sqrt{a^2\sin^2 t + a^2\cos^2 t}\,dt = \sqrt{a^2}\,dt = a\,dt.$$

Thus

$$\begin{aligned} \int_\gamma f(x, y)\,ds &= \int_0^\pi (a^2\cos^2 t)(a\sin t)(a\,dt) \\ &= a^4\int_0^\pi \cos^2 t\sin t\,dt = -a^4\frac{\cos^3 t}{3}\bigg]_0^\pi \\ &= (-a^4)\left(-\frac{1}{3}\right) - (-a^4)\left(\frac{1}{3}\right) = \frac{2a^4}{3}. \end{aligned}$$

For the other integral, we note that $dy = a\cos t\,dt$ and obtain

$$\begin{aligned} \int_\gamma f(x, y)\,dy &= \int_\gamma (a^2\cos^2 t)(a\sin t)(a\cos t\,dt) \\ &= a^4\int_0^\pi \cos^3 t\sin t\,dt = -a^4\frac{\cos^4 t}{4}\bigg]_0^\pi \\ &= (-a^4)\left(\frac{1}{4}\right) - (-a^4)\left(\frac{1}{4}\right) = 0. \quad \square \end{aligned}$$

Let γ be a smooth plane curve given parametrically by

$$x = h(t), \qquad y = k(t), \qquad \text{for } a \le t \le b.$$

We refer to the set of points $(h(t),\ k(t))$ in the plane for $a \le t \le b$ as the **trace** (or **path set**) **of** γ. The parametric equations $x = h(t)$, $y = k(t)$ are involved as part of γ. If we change the parametric equations, we consider that we have a curve *different* from γ, even if it has the same initial point A, terminal point B, and trace that γ has. However, the following theorem can be shown to be true.

THEOREM 17.5 Independence of parametrization

Let γ be a smooth curve and let $f(x, y)$ be a continuous function with domain containing the trace of γ. Then the value of an integral $\int_\gamma f(x, y)\, ds$, $\int_\gamma f(x, y)\, dx$, or $\int_\gamma f(x, y)\, dy$ depends only on the initial point A, terminal point B, and the trace of γ. That is, two different parametrizations having the same trace from A to B yield the same values for these integrals.

This independence of parametrization seems reasonable when we regard the integrals as limits of sums of terms of the forms (5). Each term in (5) is defined only in terms of the function $f(x, y)$ and the trace of γ. We illustrate Theorem 17.5 in the following example.

EXAMPLE 2 Consider the smooth curves

$$\gamma_1 : x = t, \quad y = t^2 \qquad \text{for } 0 \le t \le 1$$

and

$$\gamma_2 : x = \sin t, \quad y = \sin^2 t \qquad \text{for } 0 \le t \le \frac{\pi}{2}.$$

Both γ_1 and γ_2 are smooth curves from (0, 0) to (1, 1) and have as trace the portion of the parabola $y = x^2$ for $0 \le x \le 1$, shown in Fig. 17.18. Show that $\int_{\gamma_1} x\, ds = \int_{\gamma_2} x\, ds$.

Solution For γ_1, we have

$$x = t \qquad \text{and} \qquad ds = \sqrt{\left(\frac{dx}{dt}\right)^2 + \left(\frac{dy}{dt}\right)^2} = \sqrt{1 + 4t^2}\, dt.$$

Therefore

$$\int_{\gamma_1} x\, ds = \int_0^1 t\sqrt{1 + 4t^2}\, dt = \frac{1}{8}\cdot\frac{2}{3}(1 + 4t^2)^{3/2}\Big]_0^1 = \frac{1}{12}(17^{3/2} - 1).$$

For γ_2, we have

$$x = \sin t \qquad \text{and} \qquad ds = \sqrt{\cos^2 t + 4\sin^2 t \cos^2 t}\, dt.$$

Thus

$$\int_{\gamma_2} x\, ds = \int_0^{\pi/2} \sin t \cos t\sqrt{1 + 4\sin^2 t}\, dt$$

$$= \frac{1}{8}\cdot\frac{2}{3}(1 + 4\sin t)^{3/2}\Big]_0^{\pi/2} = \frac{1}{12}(17^{3/2} - 1). \quad \square$$

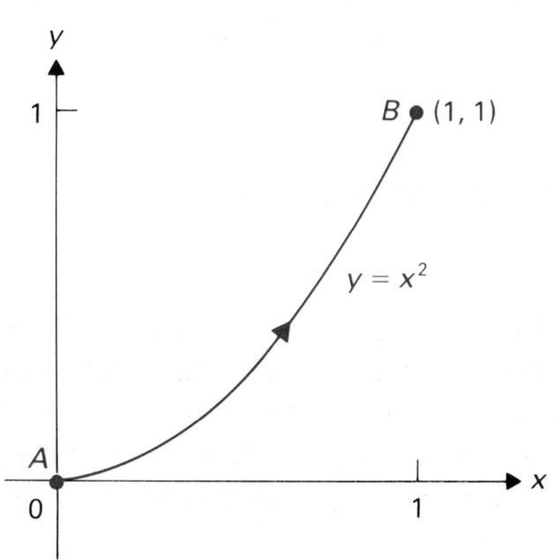

Figure 17.18 Trace of γ_1: $x = t$, $y = t^2$, $0 \le t \le 1$; γ_2: $x = \sin t$, $y = \sin 2t$, $0 \le t \le \pi/2$.

We have discussed line integrals of $f(x, y)$ over a smooth curve γ in the plane. Generalizations to integrals of a continuous function $f(x, y, z)$ over a smooth curve γ in space are obvious, and we will feel free to employ such integrals with no additional discussion.

The remainder of this section is devoted to examples involving line integrals over curves that are smooth or composed of a number of smooth pieces joined together. In view of Theorem 17.5, we may consider a line integral over a path set from a point A to a point B without specifying a parametrization. To evaluate such an integral, we have to be able to find a

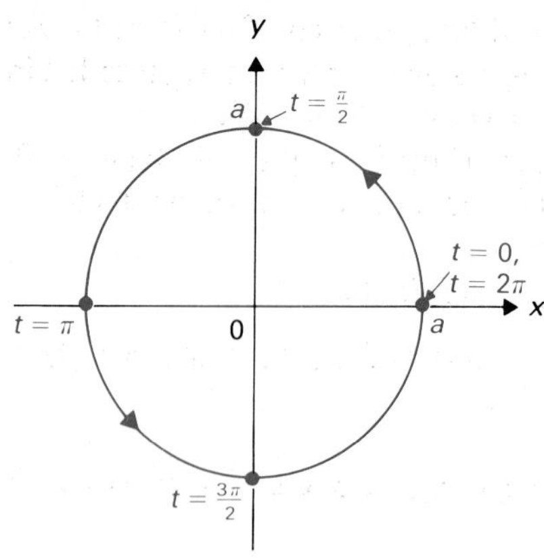

Figure 17.19 The circle $x = a \cos t$, $y = a \sin t$, $0 \le t \le 2\pi$.

smooth parametrization with the path set as trace. We have found that students need plenty of practice doing this. In particular, remember that

$$x = a \cos t, \qquad y = a \sin t \qquad \text{for } 0 \le t \le 2\pi$$

is one *parametrization of the circle* with center (0, 0) and radius a. In this parametrization, the circle is traced once counterclockwise, with both initial and terminal point $(a, 0)$, as shown in Fig. 17.19. A *parametrization of the line segment* in Fig. 17.20 with initial point (a_1, a_2) and terminal point (b_1, b_2) is found by taking the parametric equations for the line, using our familiar steps but restricting t to [0, 1]. We obtain

Point (a_1, a_2)

∥ *vector* $(b_1 - a_1)\mathbf{i} + (b_2 - a_2)\mathbf{j}$

Segment $x = a_1 + (b_1 - a_1)t$, $y = a_2 + (b_2 - a_2)t$ for $0 \le t \le 1$

This section is devoted to computational technique. The next section continues with applications and relates the notions of exact differentials and line integrals.

Our next example illustrates how we may, on occasion, easily evaluate a line integral by regarding it as a limit of sums of products of one of the forms (5).

EXAMPLE 3 Let γ be a smooth curve with initial point (3, 2), terminal point (0, 2), and having as trace the line segment from (3, 2) to (0, 2), shown in Fig. 17.21. Find

$$\int_\gamma (y^3 + 3y)\, ds, \qquad \int_\gamma (y^3 + 3y)\, dx, \qquad \text{and} \qquad \int_\gamma (y^3 + 3y)\, dy.$$

Solution Note that $y^3 + 3y$ has the constant value $2^3 + 3 \cdot 2 = 14$ on the trace of γ. When integrating with respect to s along γ, we always regard ds (and Δs) as *positive*. To find $\int_\gamma (y^3 + 3y)\, ds$, we think of adding terms of the form $14(\Delta s)$ along γ. Clearly we obtain $14 \cdot 3 = 42$, as the curve γ has length 3.

Figure 17.20 The line segment $x = a_1 + (b_1 - a_1)t$, $y = a_2 + (b_2 - a_2)t$, $0 \le t \le 1$.

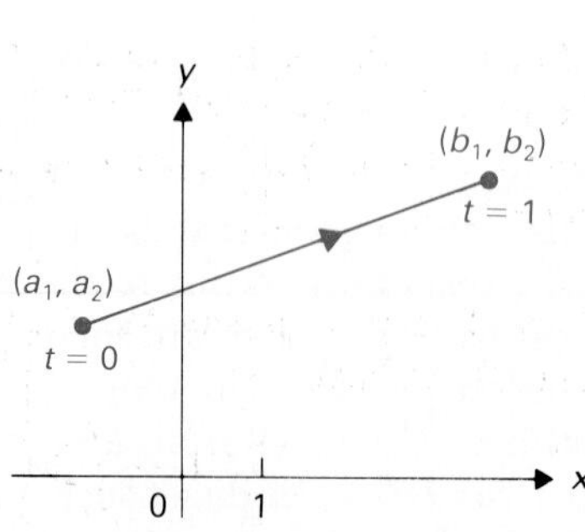

Figure 17.21 Trace of γ from (3, 2) to (0, 2).

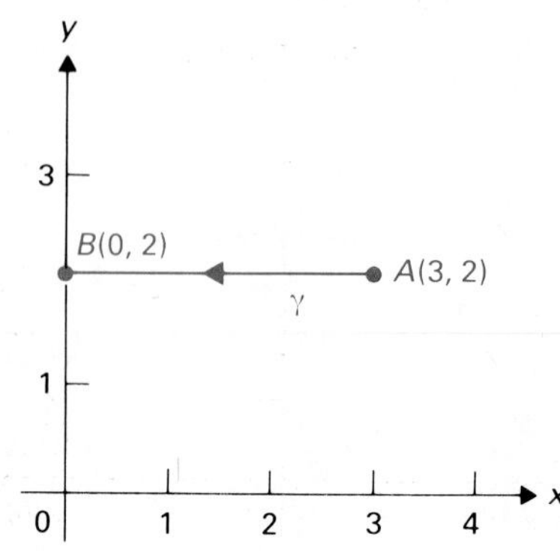

To find $\int_\gamma (y^3 + 3y)\,dx$, we think of adding products of the form $14 \cdot \Delta x$ along γ. Now Δx is *negative* as we go along γ, since γ goes from right to left in Fig. 17.21. Thus we obtain $\int_\gamma (y^3 + 3y)\,dx = 14(-3) = -42$.

Finally, $\Delta y = 0$ for an element of a partition of γ, so $\int_\gamma (y^3 + 3y)\,dy = 0$.

All these integrals could have been found using a smooth parametrization of γ, like

$$x = 3 - 3t, \qquad y = 2, \qquad \text{for } 0 \le t \le 1,$$

but they were easier to evaluate from our understanding of the integrals (4) as limits of sums of terms in the forms (5). □

EXAMPLE 4 Find the mass of a wire covering the circle $x^2 + y^2 = 1$ with mass density $\sigma(x, y) = ky^2$.

Solution The equations

$$x = h(t) = \cos t, \qquad y = k(t) = \sin t \qquad \text{for } 0 \le t \le 2\pi$$

parametrize the circle γ. Thus

$$m = \int_\gamma \sigma(x, y)\,ds = \int_\gamma ky^2\,ds = \int_0^{2\pi} k \sin^2 t\sqrt{(-\sin t)^2 + (\cos t)^2}\,dt$$

$$= k\int_0^{2\pi} \sin^2 t\,dt = k\left(\frac{t}{2} - \frac{\sin 2t}{4}\right)\Bigg]_0^{2\pi} = k\pi. \quad \square$$

EXAMPLE 5 Find the total curvature of a circle of radius a.

Solution We know the curvature of the circle at any point is $\kappa = 1/a$. If γ is a parametrization of the circle, we then know the total curvature is given by $\int_\gamma (1/a)\,ds = (1/a)\int_\gamma ds$. Clearly, $\int_\gamma ds$ gives the arc length $2\pi a$ around the circle. Therefore the total curvature is $(1/a)(2\pi a) = 2\pi$. □

EXAMPLE 6 Find $\int_\gamma f(x, y)\,dx$ and $\int_\gamma f(x, y)\,dy$ if $f(x, y) = xy^2$ and γ is the curve joining (0, 0) and (1, 1) defined by

$$x = t, \qquad y = t^2 \qquad \text{for } 0 \le t \le 1.$$

Solution Now $f(x, y) = xy^2$, so that $f(t, t^2) = t^5$. Therefore

$$\int_\gamma f(x, y)\,dx = \int_0^1 t^5 \cdot 1\,dt = \frac{t^6}{6}\Bigg]_0^1 = \frac{1}{6},$$

while

$$\int_\gamma f(x, y)\,dy = \int_0^1 t^5 \cdot 2t\,dt = \frac{2t^7}{7}\Bigg]_0^1 = \frac{2}{7}. \quad \square$$

Figure 17.22
$\gamma: x = a\cos t,$
$y = -a\sin t,\ \pi \le t \le 2\pi.$

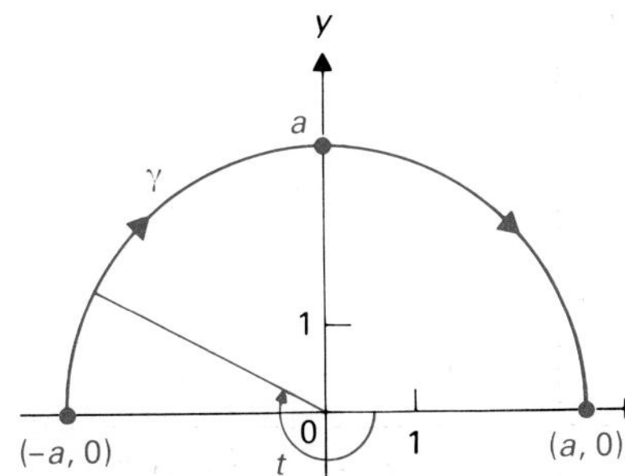

EXAMPLE 7 Let a force vector at each point (x, y) in the plane be given by $\mathbf{F}(x, y) = (x + y)\mathbf{i} + y^2\mathbf{j}$. Find the work done by this force on a body moving from $(-a, 0)$ to $(a, 0)$ along the top arc of the circle with center (0, 0), as shown in Fig. 17.22.

Solution Let γ be a smooth parametrization of this semicircle. From formula (3), we have

$$\text{Work} = \int_\gamma [(x + y)\,dx + y^2\,dy].$$

We need a smooth parametrization of this semicircle. We know that $x = a\cos t$, $y = a\sin t$ parametrizes the circle in a *counterclockwise* direction. Since we want to go *clockwise*, we replace t by $-t$. Using some trigonometric identities, we see that we can consider γ to be given by

$$x = a\cos t, \qquad y = -a\sin t \qquad \text{for } \pi \le t \le 2\pi.$$

Then

$$\begin{aligned}
&\int_\gamma [(x+y)\,dx + y^2\,dy] \\
&\quad = \int_\pi^{2\pi} [(a\cos t - a\sin t)(-a\sin t) + (a^2\sin^2 t)(-a\cos t)]\,dt \\
&\quad = a^2 \int_\pi^{2\pi} (-\sin t\cos t + \sin^2 t - a\sin^2 t\cos t)\,dt \\
&\quad = a^2 \int_\pi^{2\pi} \left[-\sin t\cos t + \frac{1}{2}(1-\cos 2t) - a\sin^2 t\cos t\right] dt \\
&\quad = a^2\left(-\frac{\sin^2 t}{2} + \frac{t}{2} - \frac{1}{4}\sin 2t - \frac{a}{3}\sin^3 t\right)\Bigg]_\pi^{2\pi} \\
&\quad = a^2\left(\frac{2\pi}{2}\right) - a^2\left(\frac{\pi}{2}\right) = \frac{\pi a^2}{2}. \quad \square
\end{aligned}$$

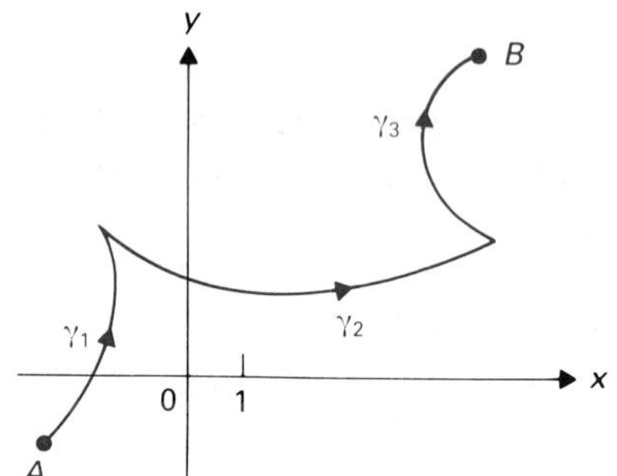

Figure 17.23 Piecewise-smooth curve $\gamma = \gamma_1 + \gamma_2 + \gamma_3$.

Finally, it is often useful to relax the condition that γ be smooth. (Remember that a curve is not smooth where it has a sharp point.) We will allow γ to be any curve that consists of a finite number of smooth arcs joined together, as illustrated in Fig. 17.23. Such a curve is called **piecewise smooth.** Let γ be a curve with smooth pieces γ_1, γ_2, and γ_3, as shown in the figure. It is natural to write $\gamma = \gamma_1 + \gamma_2 + \gamma_3$. We define

$$\int_\gamma f(x,y)\,ds = \int_{\gamma_1} f(x,y)\,ds + \int_{\gamma_2} f(x,y)\,ds + \int_{\gamma_3} f(x,y)\,ds,$$

and the integrals with respect to dx and dy are similarly defined.

EXAMPLE 8 Find $\int_\gamma xy^2\,dx$ and $\int_\gamma xy^2\,dy$ if $\gamma = \gamma_1 + \gamma_2$ is the piecewise-smooth curve joining $A(0, 0)$ and $B(1, 1)$ shown in Fig. 17.24.

Figure 17.24 $\gamma = \gamma_1 + \gamma_2$.

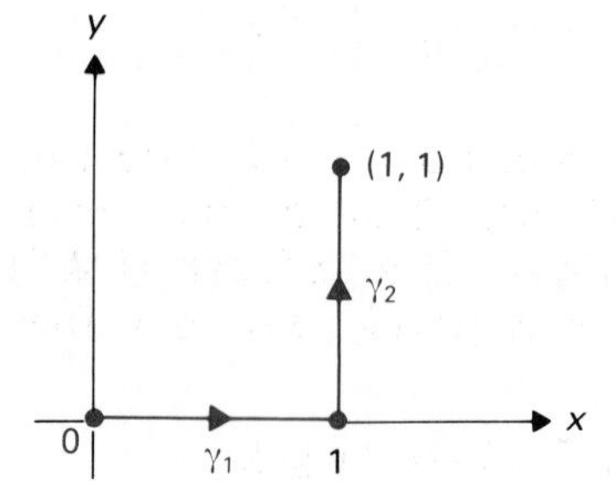

Solution Here, γ_1 is the straight-line segment from (0, 0) to (1, 0), and γ_2 is the vertical-line segment from (1, 0) to (1, 1).

Since the value of xy^2 on γ_1 is zero at each point, we see that

$$\int_{\gamma_1} xy^2\,dx = 0.$$

On γ_2, every short piece corresponds to $\Delta x = 0$, so also

$$\int_{\gamma_2} xy^2\,dx = 0.$$

Consequently,

$$\int_\gamma xy^2\,dx = \int_{\gamma_1} xy^2\,dx + \int_{\gamma_2} xy^2\,dx = 0 + 0 = 0.$$

Note that we discovered that this integral was zero by thinking in terms of the contributions to the typical sums. We did not even bother to parametrize γ_1 or γ_2.

Turning to $\int_\gamma xy^2\,dy$, we see that a little piece of γ_1 corresponds to a change $\Delta y = 0$, so

$$\int_{\gamma_1} xy^2\,dy = 0.$$

But xy^2 is nonzero on most of γ_2, and $\Delta y \neq 0$ there, so there will be nonzero terms contributing to $\int_{\gamma_2} xy^2\,dy$. We may take as parametrization of γ_2

$$x = 1, \qquad y = t \qquad \text{for } 0 \le t \le 1.$$

Then $dy = dt$, and

$$\int_{\gamma_2} xy^2\,dy = \int_{\gamma_2} 1 \cdot t^2\,dt = \frac{t^3}{3}\Big]_0^1 = \frac{1}{3},$$

so

$$\int_\gamma xy^2\,dy = \int_{\gamma_1} xy^2\,dy + \int_{\gamma_2} xy^2\,dy = 0 + \frac{1}{3} = \frac{1}{3}. \quad \square$$

We conclude with a space example.

EXAMPLE 9 Let γ_1 be the arc of the circle in the x,z-plane from (2, 0, 0) to (0, 0, 2), and let γ_2 be the straight-line segment from (0, 0, 2) to (1, 3, 1), as shown in Fig. 17.25. Find $\int_{\gamma_1+\gamma_2} (x + y + z)\,dz$.

Solution A parametrization of γ_1 is given by

$$x = 2\cos t, \qquad y = 0, \qquad z = 2\sin t, \qquad \text{for } 0 \le t \le \frac{\pi}{2}.$$

Then

$$\begin{aligned}\int_{\gamma_1} (x + y + z)\,dz &= \int_0^{\pi/2} (2\cos t + 2\sin t)(2\cos t)\,dt \\ &= 4\int_0^{\pi/2} (\cos^2 t + \sin t\cos t)\,dt \\ &= 4\int_0^{\pi/2} \left[\frac{1}{2}(1 + \cos 2t) + \sin t\cos t\right] dt \\ &= \left[2\left(t + \frac{1}{2}\sin 2t\right) + 2\sin^2 t\right]_0^{\pi/2} \\ &= \left[2\left(\frac{\pi}{2}\right) + 2\right] - 0 = \pi + 2.\end{aligned}$$

Figure 17.25 $\gamma = \gamma_1 + \gamma_2$.

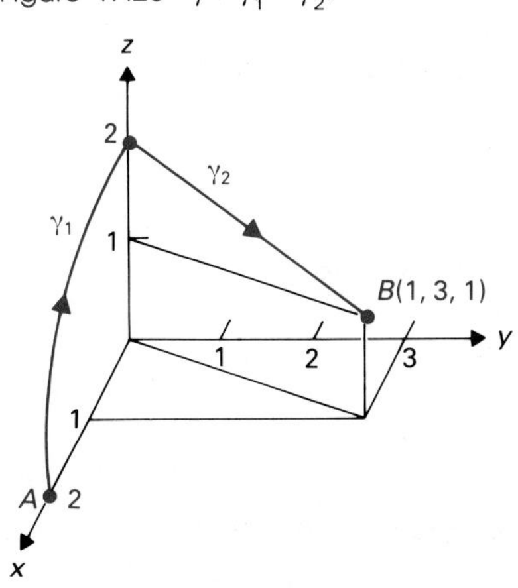

We now turn to γ_2. It is always easy to parametrize a line segment, where all we need is a *point* and a *parallel vector*. Take the initial point (0, 0, 2) of γ_2 as the *point* and the vector $\mathbf{i} + 3\mathbf{j} - \mathbf{k}$ from (0, 0, 2) to (1, 3, 1) as the *direction vector*. The line segment is then covered for $0 \le t \le 1$. Thus γ_2 is given by

$$x = t, \qquad y = 3t, \qquad z = 2 - t, \qquad \text{for } 0 \le t \le 1.$$

Then

$$\int_{\gamma_2} (x + y + z)\,dz = \int_0^1 (3t + 2)(-1)\,dt$$

$$= \left(-\frac{3t^2}{2} - 2t\right)\Big]_0^1$$

$$= -\frac{3}{2} - 2 = -\frac{7}{2}.$$

Therefore $\int_{\gamma_1+\gamma_2} (x + y + z)\,dz = (\pi + 2) - \frac{7}{2} = \pi - \frac{3}{2}$. □

SUMMARY

Throughout the following, γ is a smooth curve in the plane given by $x = h(t), y = k(t)$ for $a \le t \le b$, where $h(t)$ and $k(t)$ have continuous derivatives. Also, $f(x, y)$ is continuous.

1. $\int_\gamma f(x, y)\,ds$ is computed as $\int_a^b f(h(t), k(t))\sqrt{(dx/dt)^2 + (dy/dt)^2}\,dt$.
2. $\int_\gamma f(x, y)\,dx$ is computed as $\int_a^b f(h(t), k(t))(dx/dt)\,dt$.
3. $\int_\gamma f(x, y)\,dy$ is computed as $\int_a^b f(h(t), k(t))(dy/dt)\,dt$.
4. The value obtained in computing a line integral of a continuous function along the trace of a smooth curve from A to B is independent of the smooth parametrization used for the trace.
5. The mass of a wire covering γ with density function $\sigma(x, y)$ is $\int_\gamma \sigma(x, y)\,ds$.
6. The total curvature of a curve γ is $\int_\gamma \kappa\,ds$.
7. The work done by a force $\mathbf{F}(x, y) = P(x, y)\mathbf{i} + Q(x, y)\mathbf{j}$ acting on a body as it moves along γ is $\int_\gamma (\mathbf{F} \cdot \mathbf{t})\,ds = \int_\gamma (P(x, y)\,dx + Q(x, y)\,dy)$.
8. A line integral over a piecewise-smooth curve is the sum of the integrals over the smooth pieces.

EXERCISES

In Exercises 1 through 10, the curves are

γ_1 given by $x = t, y = t^2$ for $0 \le t \le 1$,

γ_2 given by $x = t + 1$, $y = 2t + 1$ for $0 \le t \le 1$,

γ_3 given by $x = \sin t$, $y = \cos t$, $z = -2t$ for $0 \le t \le \pi$.

Compute the indicated integrals.

1. $\int_{\gamma_1} x\,ds$
2. $\int_{\gamma_2} \sqrt{y}\,ds$
3. $\int_{\gamma_2} (x + y)\,dx$
4. $\int_{\gamma_1} (x - y)\,dy$
5. $\int_{\gamma_2} (x^2\,dx - y\,dy)$
6. $\int_{\gamma_1} (xe^y\,dx + 3x^2y\,dy)$
7. $\int_{\gamma_3} xy^2\,ds$
8. $\int_{\gamma_3} (x\,dy + y\,dx + x\,dz)$
9. $\int_{\gamma_3} (xz\,dx + yz\,dy - 2xy\,dz)$
10. $\int_{\gamma_1+\gamma_2} [(x^2 - 3y)\,dx - (y^2 + 3x)\,dy]$

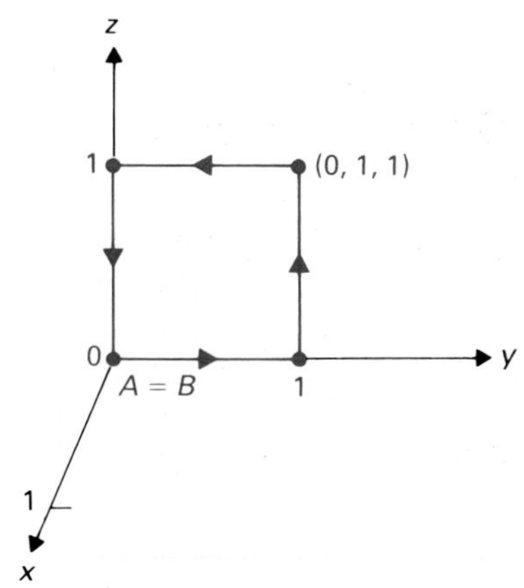

Figure 17.26 γ traces the square once counterclockwise.

11. Let γ be the space curve given by $x = 3t^2$, $y = 2t^3$, $z = 3t$ for $0 \leq t \leq 1$, and let $F(x, y, z) = 3x^2yz$. Compute $\int_\gamma F(x, y, z)\, dx$.

12. Repeat Exercise 11, but compute $\int_\gamma F(x, y, z)\, dy$.

13. Let γ be the space curve given by $x = \sin t$, $y = \cos t$, $z = t$ for $0 \leq t \leq \pi/4$, and let $F(x, y, z) = 3x^2yz$. Compute $\int_\gamma F(x, y, z)\, ds$.

14. Repeat Exercise 13, but compute $\int_\gamma F(x, y, z)\, dx$.

15. Let γ be a smooth curve having as trace the straight-line segment from (1, 0) to (3, 4). Find $\int_\gamma (x^2 + y^2)\, ds$.

16. Let $\gamma = \gamma_1 + \gamma_2$ be a piecewise-smooth curve where γ_1 has as trace the straight-line segment from (0, 0) to (2, 0) and γ_2 has as trace the straight-line segment from (2, 0) to (4, 2). Find $\int_\gamma (x^2 + xy)\, dx$.

17. Let $\gamma = \gamma_1 + \gamma_2$ be a piecewise-smooth curve where γ_1 has as trace the arc of the parabola $y = x^2$ from (0, 0) to (1, 1) and γ_2 has as trace the arc of the parabola $y = 2 - x^2$ from (1, 1) to (0, 2). Find $\int_\gamma (x + 2y)\, dy$.

18. Let $\gamma = \gamma_1 + \gamma_2$ be a piecewise-smooth curve where γ_1 has as trace the shorter arc of the circle $x^2 + y^2 = 4$ from (2, 0) to (0, 2) and γ_2 has as trace the straight-line segment from (0, 2) to (2, 0). Find $\int_\gamma (x\, dx - x^2\, dy)$.

19. Let γ be a smooth curve having as trace the straight-line segment from (2, −1, 3) to (0, 1, 4). Find $\int_\gamma (x + y + z^2)\, ds$.

20. Repeat Exercise 19, but find $\int_\gamma (x\, dx - y\, dy)$.

21. Let $\gamma = \gamma_1 + \gamma_2$ be a piecewise-smooth curve where γ_1 has as trace the straight-line segment from (0, 0, 0) to (1, 2, −1) and γ_2 has as trace the straight-line segment from (1, 2, −1) to (2, 1, 3). Find $\int_\gamma (x\, dy - y\, dx + z\, dz)$.

22. Let γ be a piecewise-smooth curve having as trace the unit square in the y,z-plane shown in Fig. 17.26, having the origin as both initial and terminal point. Find

$$\int_\gamma (yz\, dx - xz\, dy + xy\, dx).$$

23. Repeat Exercise 22, but find $\int_\gamma (xy\, dx - yz\, dy + y^2\, dz)$.

24. Repeat Exercise 22, but find $\int_\gamma (xy - yz + y^2)\, ds$.

25. Let $\gamma = \gamma_1 + \gamma_2 + \gamma_3$, where γ_1 has as trace the shorter arc of the circle with center at the origin from (0, −2, 0) to (2, 0, 0), γ_2 has as trace the vertical line segment from (2, 0, 0) to (2, 0, 1), and γ_3 has as trace the shorter arc of the circle with center (0, 0, 1) from (2, 0, 1) to (0, 2, 1). Find $\int_\gamma (x + 2y + z)\, ds$.

26. Let the mass density of a thin wire covering the portion of the parabola $y = x^2$ from (0, 0) to (2, 4) be $\sigma(x, y) = xy$. Find the total mass of the wire.

27. Let the mass density of a thin wire covering the semicircle $y = \sqrt{4 - x^2}$ be $\sigma(x, y) = y$. Find the mass of the wire.

28. Find the total curvature of the plane curve $y = x^2$ from (0, 0) to (2, 4).

29. Find the total curvature of the *entire* parabola $y = ax^2 + bx + c$.

30. Find the total curvature of the portion of the helix

$$x = a \cos t, \qquad y = a \sin t, \qquad z = t,$$

where $0 \leq t \leq 2\pi$.

In Exercises 31 through 34, find the work done by the given force $\mathbf{F}(x, y)$ or $\mathbf{F}(x, y, z)$ on a body as it moves along the given curve.

31. $\mathbf{F}(x, y) = xy\mathbf{i} - y\mathbf{j}$; γ having as trace the straight-line segment from (1, 2) to (3, −1)

32. $\mathbf{F}(x, y) = y\mathbf{i} - x\mathbf{j}$; γ having as trace the circle $x^2 + y^2 = a^2$ traced once clockwise

33. $\mathbf{F}(x, y, z) = x\mathbf{i} + y\mathbf{j} - z\mathbf{k}$; γ having as trace the straight-line segment from (1, −1, 2) to (2, 1, 4)

34. $\mathbf{F}(x, y, z) = x\mathbf{i} + z\mathbf{j} - y\mathbf{k}$; γ having as trace the rectangle shown in Fig. 17.27 with (0, 0, 0) as both initial and terminal point

Figure 17.27 γ traces the rectangle once in the direction of the arrows.

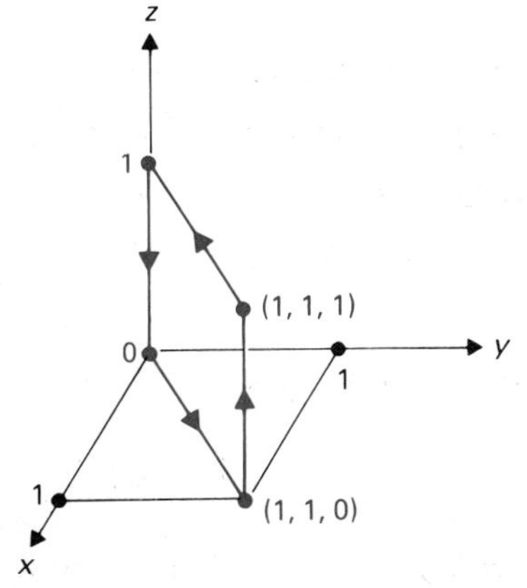

17.5 INTEGRATION OF VECTOR FIELDS ALONG CURVES

This section introduces the notion of a *vector field* and indicates a few places where vector fields occur naturally in physics and fluid dymanics. Using the illustration of work done by a force field in moving a body along a path, we will see that we are led to consider *integrals of vector fields along curves.* We will immediately obtain line integrals of the type studied in the preceding section. The section closes with a study of vector fields having the property that their line integrals along a curve depend only on the initial and terminal points of the curve and not on the trace of the curve between these points. We will show that the earth's gravitational field is such a force field, obtaining one form of the *law of conservation of energy.*

THE NOTION OF A VECTOR FIELD

A **vector field** on a region G in the plane assigns to each point (x, y) in G a vector in the plane, which we visualize as emanating from (x, y). A vector field

$$\mathbf{F}(x, y) = P(x, y)\mathbf{i} + Q(x, y)\mathbf{j}$$

is *continuous* if P and Q are continuous functions. We use a boldface letter **F** to indicate that we are considering a vector field; we would use $\vec{F}$ in written work.

Illustration 1 *Gradient vector field* We encountered the notion of a vector field in the plane in Chapter 16, where we discussed the gradient ∇f of a differentiable function f of two variables. To each point (x, y) in the domain of f, we assign the gradient vector

$$\nabla f = \frac{\partial f}{\partial x}\mathbf{i} + \frac{\partial f}{\partial y}\mathbf{j}.$$

Recall that the direction at (x_0, y_0) of the gradient vector there is normal to the level curve of f through (x_0, y_0). Also, the gradient vector points in the direction of maximum rate of increase of $f(x, y)$ at (x_0, y_0) and has length equal to this maximum rate of increase.

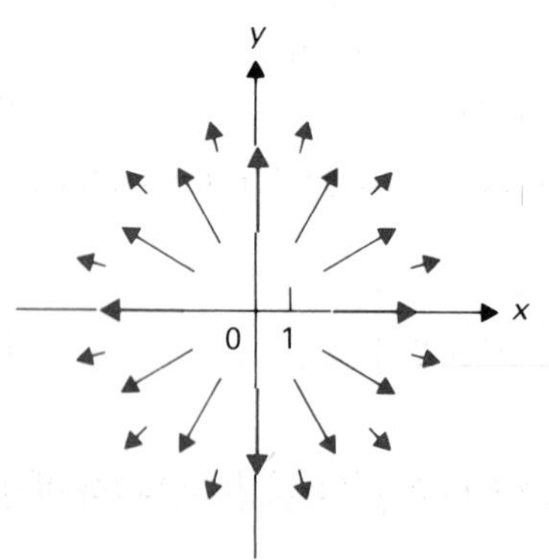

Figure 17.28 Force field of repulsion by a charge at the origin.

Illustration 2 *Force field* An electrical charge at the origin in the plane exerts a force of repulsion on a like charge at any other point (x, y). This force at (x, y) may be represented by a vector of length equal to the magnitude of the force and having the direction of the force, namely, away from the origin. The vector field of these vectors for $(x, y) \neq (0, 0)$ is the *force field* of the charge. This vector field is continuous and grows weaker (with shorter vectors) as the distance from the origin increases (see Fig. 17.28).

Illustration 3 *Flux field* If a region G in the plane is covered with a flowing liquid or gas, then we can associate with each (x, y) in G and time t the velocity vector $\mathbf{V}(x, y)$ of the fluid flow at (x, y) at time t. This vector has the *direction* of the flow and *length* equal to the speed of the flow. The vector field $\mathbf{V}(x, y)$ is the *velocity field of the flow at time t.*

Let $\rho(x, y)$ be the mass density of the fluid or gas at (x, y) at time t. The vector field $\mathbf{F}(x, y) = \rho(x, y)\mathbf{V}(x, y)$ is the *flux field of the flow at time t*. The

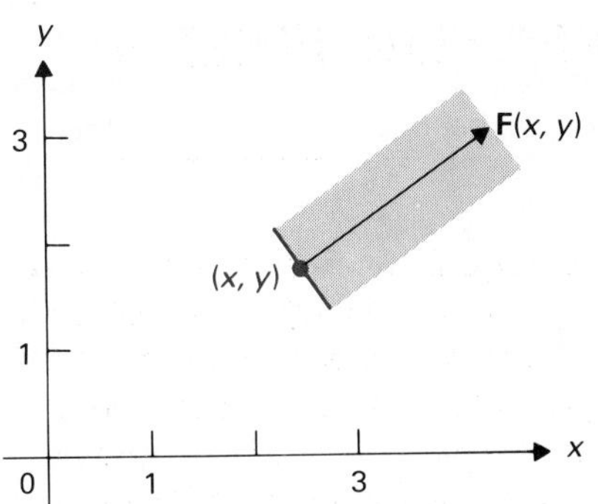

Figure 17.29 $|\mathbf{F}(x, y)|$ is the mass flowing across the unit line segment if the flow everywhere were the same as at (x, y).

physical interpretation of $\mathbf{F}(x, y)$ is as follows. At (x, y), take a unit line segment orthogonal to $\mathbf{F}(x, y)$, as shown in Fig. 17.29. If the flow at all points on this line segment were just as it is at (x, y) at this time t, and if it did not change with time, then $|\mathbf{F}(x, y)|$ is the mass of fluid or gas flowing across the line segment per unit time. Of course, $\mathbf{F}(x, y)$ points in the direction of the flow at (x, y).

If the flux field does not vary with time, the flow is called *steady state*.

INTEGRAL OF A VECTOR FIELD ALONG A CURVE

Let $\mathbf{F}(x, y) = P(x, y)\mathbf{i} + Q(x, y)\mathbf{j}$ be a continuous vector field on a region G of the plane, and let

$$x = h(t), \qquad y = k(t), \qquad \text{for } a \leq t \leq b$$

be a smooth curve γ lying in G. Recall that the position vector to a point on γ at time t is

$$\mathbf{r} = x\mathbf{i} + y\mathbf{j} = h(t)\mathbf{i} + k(t)\mathbf{j}. \tag{1}$$

In the preceding section, we recalled that the *unit tangent vector* $\mathbf{t}$ (in the direction corresponding to increasing time) at each point on γ is given by

$$\mathbf{t} = \frac{dx}{ds}\mathbf{i} + \frac{dy}{ds}\mathbf{j}. \tag{2}$$

The scalar component of $\mathbf{F}$ along $\mathbf{t}$ is $\mathbf{F} \cdot \mathbf{t}$. The integral of this tangential scalar component of $\mathbf{F}$ along γ is

$$\int_\gamma (\mathbf{F} \cdot \mathbf{t})\, ds, \tag{3}$$

which may be written in the form

$$\int_\gamma (P(x, y)\, dx + Q(x, y)\, dy), \tag{4}$$

as we saw in the preceding section.

Now the differential vector

$$\boxed{d\mathbf{r} = dx\,\mathbf{i} + dy\,\mathbf{j}} \tag{5}$$

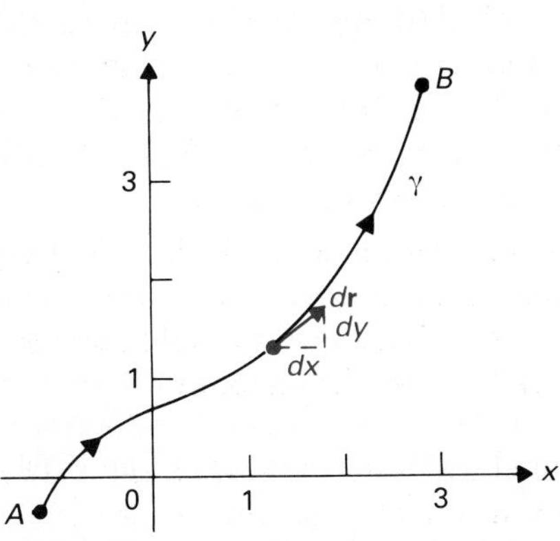

Figure 17.30 The tangent vector $d\mathbf{r} = dx\,\mathbf{i} + dy\,\mathbf{j}$.

shown in Fig. 17.30 has direction tangent to the curve and length $ds = \sqrt{(dx)^2 + (dy)^2}$. Note that

$$\begin{aligned} \mathbf{F} \cdot d\mathbf{r} &= (P(x, y)\mathbf{i} + Q(x, y)\mathbf{j}) \cdot (dx\,\mathbf{i} + dy\,\mathbf{j}) \\ &= P(x, y)\, dx + Q(x, y)\, dy. \end{aligned} \tag{6}$$

Comparing integral (4) and Eq. (6), we write the **integral of the vector field F along** γ as

$$\boxed{\int_\gamma (\mathbf{F} \cdot \mathbf{t})\, ds = \int_\gamma (\mathbf{F} \cdot d\mathbf{r}).} \tag{7}$$

In practice, Eq. (7) is calculated by expressing both $\mathbf{F}$ and $d\mathbf{r}$ in terms of the parameter t and dt.

If **F** is a force field, then the component of **F** tangent to a smooth curve γ is the portion of the force that acts to move a body along γ. The *work* done by this force field in moving a body along γ is then the integral with respect to arc length of this component, which we have seen is $\int_\gamma \mathbf{F} \cdot d\mathbf{r}$. Thus we may now write

$$\boxed{\text{Work} = \int_\gamma \mathbf{F} \cdot d\mathbf{r}.} \tag{8}$$

EXAMPLE 1 Find the work done by the force field

$$\mathbf{F}(x, y) = x^2\mathbf{i} + y^2\mathbf{j}$$

in moving a body from (0, 0) to (1, 1) if the position (x, y) of the body at time t on a curve γ is given by

$$x = t, \qquad y = t^2 \qquad \text{for } 0 \le t \le 1.$$

Solution In terms of t, we have

$$\mathbf{F}(x, y) = \mathbf{F}(t, t^2) = t^2\mathbf{i} + t^4\mathbf{j},$$

and

$$d\mathbf{r} = dx\,\mathbf{i} + dy\mathbf{j} = (\mathbf{i} + 2t\mathbf{j})\,dt.$$

Thus

$$\int_\gamma \mathbf{F} \cdot d\mathbf{r} = \int_0^1 (t^2\mathbf{i} + t^4\mathbf{j}) \cdot (\mathbf{i} + 2t\mathbf{j})\,dt$$

$$= \int_0^1 (t^2 + 2t^5)\,dt = \left(\frac{t^3}{3} + \frac{2t^6}{6}\right)\Bigg]_0^1 = \frac{1}{3} + \frac{1}{3} = \frac{2}{3}. \quad \square$$

Everything we have mentioned has an obvious analogue in space. We will feel free to make use of the space analogues without further discussion.

Figure 17.31 The curves γ and $-\gamma$: (a) γ: $x = h(t)$, $y = k(t)$, $0 \le t \le 1$; (b) $-\gamma$: $x = h(1 - t)$, $y = k(1 - t)$, $0 \le t \le 1$.

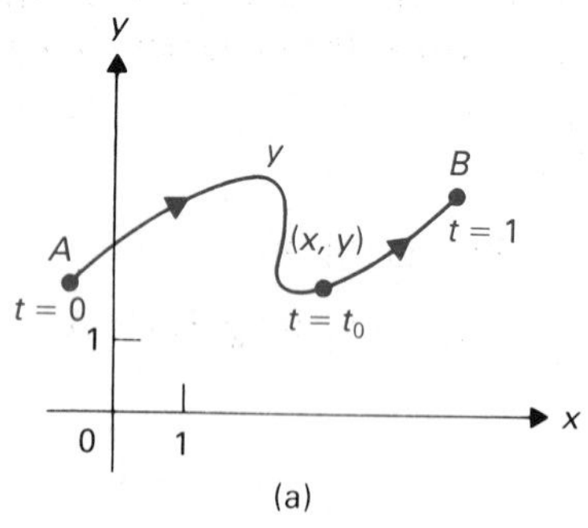

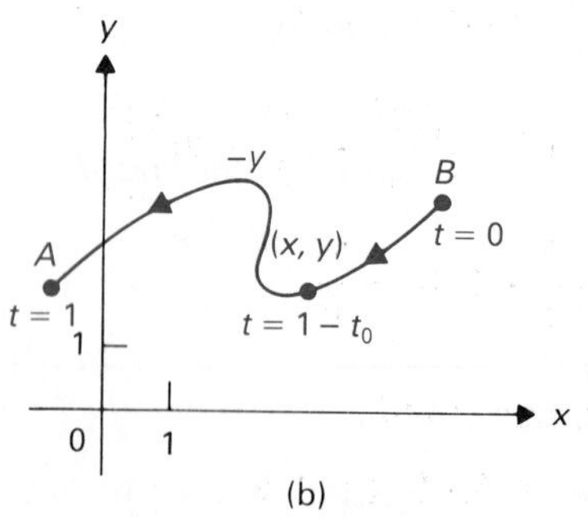

THE CURVE $-\gamma$

Let γ be a smooth curve. Now line integrals of vector fields along γ depend only on the trace of γ and not on the parametrization. Thus when working with γ, we can always assume that $0 \le t \le 1$ rather than $a \le t \le b$. We therefore suppose that γ is given by

$$\boxed{\gamma: x = h(t), \quad y = k(t) \qquad \text{for } 0 \le t \le 1,} \tag{9}$$

with initial point A and terminal point B, as shown in Fig. 17.31(a). We denote by $-\gamma$ the curve in Fig. 17.31(b), which has the same trace as γ but is traveled in the opposite direction from B to A for $0 \le t \le 1$. In terms of time, a body moving on $-\gamma$ appears as though we took a moving picture of the body moving on γ and then ran the film backward. Wherever the body on γ appears at time t, the body on $-\gamma$ appears at time $1 - t$, as illustrated in Fig. 17.31.

Thus parametric equations for $-\gamma$ are given by

$$-\gamma : x = h(1 - t), \quad y = k(1 - t) \qquad \text{for } 0 \le t \le 1. \tag{10}$$

Note that

$$\frac{d}{dt}(h(1 - t)) = h'(1 - t)(-1) = -h'(1 - t).$$

If we let $u = 1 - t$ so that $du = -dt$, then

$$\begin{aligned}\int_{-\gamma} P(x, y)\, dx &= \int_{t=0}^{1} P(h(1 - t), k(1 - t))\cdot(-h'(1 - t))\, dt\\ &= \int_{u=1}^{0} P(h(u), k(u))\cdot(-h'(u))(-du)\\ &= \int_{1}^{0} P(h(u), k(u))\cdot h'(u)\, du\\ &= -\int_{0}^{1} P(h(u), k(u))\cdot h'(u)\, du = -\int_{\gamma} P(x, y)\, dx.\end{aligned}$$

A similar argument shows that $\int_{-\gamma} Q(x, y)\, dy = -\int_{\gamma} Q(x, y)\, dy$. Thus we see that

$$\int_{-\gamma} (\mathbf{F}\cdot d\mathbf{r}) = -\int_{\gamma} (\mathbf{F}\cdot d\mathbf{r}). \tag{11}$$

Another way of seeing that Eq. (11) holds is to note that the unit tangent vectors at a point on γ and at the same point on $-\gamma$ have opposite direction. Since $\mathbf{F}\cdot d\mathbf{r} = (\mathbf{F}\cdot\mathbf{t})\, ds$, and ds is always positive, we see that the sign of $\mathbf{F}\cdot d\mathbf{r}$ at a point on $-\gamma$ is opposite to its sign at the same point on γ.

EXAMPLE 2 Let γ be given by

$$\gamma : x = t - 1, \quad y = 2t + 3 \qquad \text{for } 0 \le t \le 1.$$

Compute $\int_{\gamma} (xy\, dx - y\, dy)$ and $\int_{-\gamma} (xy\, dx - y\, dy)$, illustrating Eq. (11).

Solution We have

$$\begin{aligned}\int_{\gamma} (xy\, dx - y\, dy) &= \int_{0}^{1} [(2t^2 + t - 3) - (2t + 3)2]\, dt\\ &= \int_{0}^{1} (2t^2 - 3t - 9)\, dt = \left(\frac{2t^3}{3} - \frac{3t^2}{2} - 9t\right)\Big]_0^1\\ &= \frac{2}{3} - \frac{3}{2} - 9 = -\frac{5}{6} - 9 = -\frac{59}{6}.\end{aligned}$$

From Eq. (10), we see that $-\gamma$ has the parametrization

$$-\gamma : x = -t, \quad y = 5 - 2t \qquad \text{for } 0 \le t \le 1.$$

Then

$$\int_{-\gamma} (xy\,dx - y\,dy) = \int_0^1 [(2t^2 - 5t)(-1) - (5 - 2t)(-2)]\,dt$$

$$= \int_0^1 (-2t^2 + t + 10)\,dt = \left(-\frac{2t^3}{3} + \frac{t^2}{2} + 10t\right)\Big]_0^1$$

$$= -\frac{2}{3} + \frac{1}{2} + 10 = -\frac{1}{6} + 10 = \frac{59}{6},$$

which illustrates Eq. (11). □

INDEPENDENCE OF PATH

Suppose we pick up a box of books in the front hall, carry it up to the second floor, and set it on a desk. We have done some work against the gravitational force field of the earth. Now suppose we had picked up the box in the front hall, carried it up two flights to the third floor, changed our mind, and brought it back down to the second floor and set it on the desk. A physicist will tell us that we would have done the same amount of work on the box of books going via the third floor as when we stopped at the second floor the first time. The work done depends only on the weight of the box, its initial position, and its final position. It *does not depend on the path* used to transport the box. This is a very important property of the earth's gravitational field. *Not all vector fields have this property.* For what vector fields does this property hold?

Let $\mathbf{F}(x, y) = P(x, y)\mathbf{i} + Q(x, y)\mathbf{j}$ be a continuous vector field. For convenience, we shorten our notation to $\mathbf{F} = P\mathbf{i} + Q\mathbf{j}$ and $\mathbf{F} \cdot d\mathbf{r} = P\,dx + Q\,dy$. We are interested in the following question:

> Under what conditions does **F** have the property that for any piecewise-smooth curve γ in the domain of **F**, the integral $\int_\gamma (\mathbf{F} \cdot d\mathbf{r})$ depends only on the initial point A and terminal point B of γ and not on the trace of γ from A to B?

DEFINITION 17.3 Independence of path

If **F** has the property just described, then we say that $\int_\gamma (\mathbf{F} \cdot d\mathbf{r})$ is **independent of the path.**

Let γ_1 and γ_2 be two piecewise-smooth curves from A to B as in Fig. 17.32. If $\int_\gamma (\mathbf{F} \cdot d\mathbf{r})$ is independent of the path, then we have

$$\int_{\gamma_1} (\mathbf{F} \cdot d\mathbf{r}) = \int_{\gamma_2} (\mathbf{F} \cdot d\mathbf{r}). \tag{12}$$

It follows that

$$\int_{\gamma_1 - \gamma_2} (\mathbf{F} \cdot d\mathbf{r}) = \int_{\gamma_1} (\mathbf{F} \cdot d\mathbf{r}) + \int_{-\gamma_2} (\mathbf{F} \cdot d\mathbf{r})$$

$$= \int_{\gamma_1} (\mathbf{F} \cdot d\mathbf{r}) - \int_{\gamma_2} (\mathbf{F} \cdot d\mathbf{r}) = 0. \tag{13}$$

Now $\lambda = \gamma_1 - \gamma_2$ is a piecewise-smooth curve with *both initial and terminal point A*. We call such a curve a **closed curve** or a **loop.** An integral around a closed curve λ is often denoted by $\oint_\lambda$ to call attention to the fact that λ is actually a loop.

Any piecewise-smooth loop from A to A can be regarded as $\gamma_1 - \gamma_2$ for piecewise-smooth curves γ_1 and γ_2. Simply choose a point B on the loop, as shown in Fig. 17.32, to define γ_1 and γ_2. From Eq. (13), we obtain at once the following theorem.

THEOREM 17.6 Independence of path and integrals over loops

Let $\mathbf{F}(x, y)$ be a continuous vector field. The integral $\int_\gamma (\mathbf{F} \cdot d\mathbf{r})$ is independent of the path for every piecewise-smooth curve γ in the domain of F if and only if for every loop λ in the domain of $\mathbf{F}$, we have $\oint_\lambda (\mathbf{F} \cdot d\mathbf{r}) = 0$.

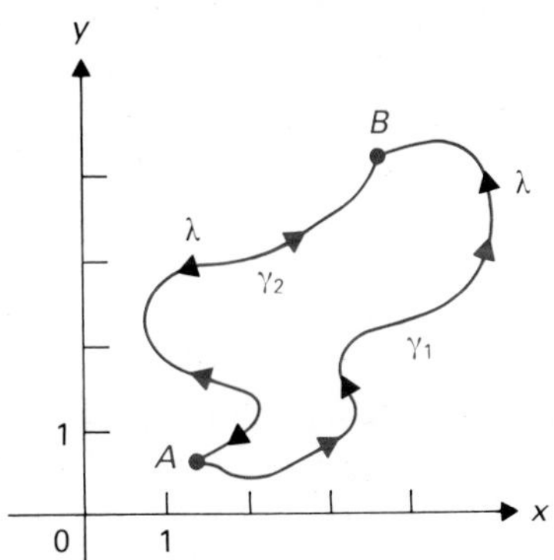

Figure 17.32 If γ_1 and γ_2 go from A to B, then $\lambda = \gamma_1 - \gamma_2$ is a loop from A to A.

EXAMPLE 3 Suppose that $\lambda = \gamma_1 - \gamma_2$ is a loop, and suppose that $\int_{\gamma_1} (\mathbf{F} \cdot d\mathbf{r}) = 18$ and $\int_{\gamma_2} (\mathbf{F} \cdot d\mathbf{r}) = 24$. Find $\oint_\lambda (\mathbf{F} \cdot d\mathbf{r})$ and $\oint_{-\lambda} (\mathbf{F} \cdot d\mathbf{r})$.

Solution We have

$$\begin{aligned}\oint_\lambda (\mathbf{F} \cdot d\mathbf{r}) &= \int_{\gamma_1 - \gamma_2} (\mathbf{F} \cdot d\mathbf{r}) = \int_{\gamma_1} (\mathbf{F} \cdot d\mathbf{r}) + \int_{-\gamma_2} (\mathbf{F} \cdot d\mathbf{r}) \\ &= \int_{\gamma_1} (F \cdot d\mathbf{r}) - \int_{\gamma_2} (\mathbf{F} \cdot d\mathbf{r}) \\ &= 18 - 24 = -6.\end{aligned}$$

By Eq. (11), we have $\oint_{-\lambda} (\mathbf{F} \cdot d\mathbf{r}) = -\oint_\lambda (\mathbf{F} \cdot d\mathbf{r}) = -(-6) = 6.$ □

If γ is a smooth curve joining A and B in the plane and if $P(x, y)\,dx + Q(x, y)\,dy$ is an *exact* differential, then

$$\int_\gamma [P(x, y)\,dx + Q(x, y)\,dy]$$

depends only on A and B and is independent of which smooth curve γ is chosen joining A and B. To see that this is so, let the curve γ be given by $x = h(t), y = k(t)$ for $a \le t \le b$. Also, let $H(x, y)$ be such that $dH = P(x, y)\,dx + Q(x, y)\,dy$. Then

$$\begin{aligned}\int_\gamma [P(x, y)\,dx + Q(x, y)\,dy] &= \int_a^b \left(P(x, y)\frac{dx}{dt} + Q(x, y)\frac{dy}{dt}\right) dt \\ &= \int_a^b \frac{d}{dt}(H(h(t), k(t)))\,dt \\ &= H(h(b), k(b)) - H(h(a), k(a)).\end{aligned}$$

Thus the integral depends only on the endpoints $A = (h(a), k(a))$ and $B = (h(b), k(b))$. We summarize in a theorem.

THEOREM 17.7 Integration of exact differentials

The line integral $\int_\gamma (P\,dx + Q\,dy)$ of an exact differential $dH = P\,dx + Q\,dy$ is independent of the path. Furthermore,

$$\int_\gamma (P\,dx + Q\,dy) = H(B) - H(A), \tag{14}$$

where A is the initial point and B the terminal point of γ.

EXAMPLE 4 Let γ be given by

$$\gamma\colon x = t^2 - 2t, \quad y = \frac{t^4 + 3}{4} \quad \text{for } 1 \le t \le 3.$$

Find $\int_\gamma [(2xy - 4x)\,dx + (x^2 + 2y)\,dy]$.

Solution Using the methods of Section 17.3, we easily see that $(2xy - 4x)\,dx + (x^2 + 2y)\,dy$ is the exact differential $d(x^2y - 2x^2 + y^2)$. The initial point of γ when $t = 1$ is $A = (-1, 1)$, and the terminal point when $t = 3$ is $B = (3, 21)$. We let $H(x, y) = x^2y - 2x^2 + y^2$, and by Theorem 17.7 we see that the desired integral is given by

$$\begin{aligned} H(3, 21) - H(-1, 1) &= (9 \cdot 21 - 18 + 21^2) - (1 - 2 + 1) \\ &= 189 - 18 + 441 = 612. \quad \square \end{aligned}$$

The analogue for space of all our work in this section holds also.

EXAMPLE 5 Let γ be a piecewise-smooth curve with initial point $(3, -1, 2)$ and terminal point $(1, -1, 0)$ in space. Find

$$\int_\gamma (yz\,dx + xz\,dy + xy\,dz).$$

Solution Using the methods of Section 17.3, we easily see that the differential form being integrated is dH for $H(x, y, z) = xyz$. By the space analogue of Theorem 17.7, the desired integral is

$$H(1, -1, 0) - H(3, -1, 2) = 0 - (-6) = 6. \quad \square$$

Sometimes Theorem 17.7 can be used to help find $\int_\gamma (P\,dx + Q\,dy)$, even when $P\,dx + Q\,dy$ is not an exact differential form. The next example illustrates what we mean.

EXAMPLE 6 Let γ be given by

$$\gamma\colon x = t^2 - 2t, \quad y = t^3 \qquad \text{for } 0 \le t \le 2.$$

Find $\int_\gamma [(3x^2y^2 - 3)\,dx + (2x^3y + x + 2y)\,dy]$.

Solution A check shows that the differential form to be integrated along γ is not exact. However, it may be written as

$$[(3x^2y^2 - 3)\,dx + (2x^3y + 2y)\,dy] + x\,dy,$$

where the differential form in the square bracket is exact and is dH for

$H(x, y) = x^3y^2 - 3x + y^2$. The initial point of γ is (0, 0), and the terminal point is (0, 8). Thus we see that the desired integral can be evaluated as

$$\begin{aligned} H(0, 8) - H(0, 0) + \int_\gamma x\,dy &= 64 - 0 + \int_0^2 (t^2 - 2t)3t^2\,dt \\ &= 64 + \int_0^2 (3t^4 - 6t^3)\,dt \\ &= 64 + \left(\frac{3t^5}{5} - \frac{6t^4}{4}\right)\bigg]_0^2 \\ &= 64 + \frac{96}{5} - 24 = \frac{296}{5}. \quad \square \end{aligned}$$

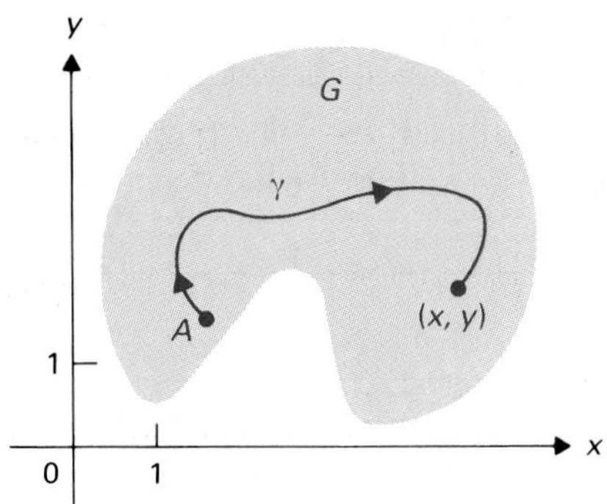

Figure 17.33
$H(x, y) = \int_\gamma (P\,dx + Q\,dy)$.

We now show that independence of the path for line integrals of $\mathbf{F} = P\mathbf{i} + Q\mathbf{j}$ implies that $P\,dx + Q\,dy$ is exact. Let a region G be such that any two points in it can be joined by a piecewise-smooth curve and also such that for every point in G there is some small disk centered at that point contained entirely in G. Such a region G is a **connected open region.**

Let A be any point in the region G. At any point (x, y) in G, we define $H(x, y)$ as follows. Let γ be a piecewise-smooth curve in G from A to (x, y), as shown in Fig. 17.33. We define

$$H(x, y) = \int_\gamma (P\,dx + Q\,dy). \tag{15}$$

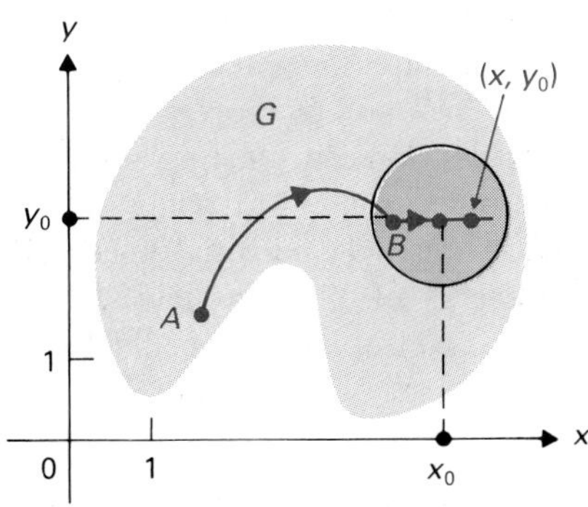

Figure 17.34
$H(x, y_0) = H(B) + \int_B^{(x, y_0)} P(x, y_0)\,dx$.

By assumption, the value $H(x, y)$ is independent of the path γ chosen from A to (x, y).

We wish to demonstrate that at any point (x_0, y_0) in G, we have $H_x = P$ and $H_y = Q$, which will show that $P\,dx + Q\,dy$ is exact. Now some disk centered at (x_0, y_0) is contained in the open region G. Referring to Fig. 17.34, let B be a point at the left end of a horizontal line segment with midpoint (x_0, y_0) and contained in the disk. Let us denote by $\int_B^{(x, y_0)} (P\,dx + Q\,dy)$ the integral of the differential form along this line segment from B to a point (x, y_0) on it. See Fig. 17.34. *Since the integral of $P\,dx + Q\,dy$ is independent of the path*, we could use the path in Fig. 17.34 to find $H(x, y_0)$, obtaining

$$H(x, y_0) = H(B) + \int_B^{(x, y_0)} (P\,dx + Q\,dy) = H(B) + \int_B^{(x, y_0)} P(x, y_0)\,dx,$$

since $dy = 0$ on the line segment. By the fundamental theorem of calculus, we obtain

$$H_x(x_0, y_0) = P(x_0, y_0).$$

A similar argument using Fig. 17.35 shows that $H_y(x_0, y_0) = Q(x_0, y_0)$. Thus $dH = P\,dx + Q\,dy$, so $P\,dx + Q\,dy$ is an exact differential form.

THEOREM 17.8 Independence of path and exact differentials

If $\int_\gamma (P\,dx + Q\,dy)$ is independent of the path in a connected open region G, then $P\,dx + Q\,dy$ is an exact differential form.

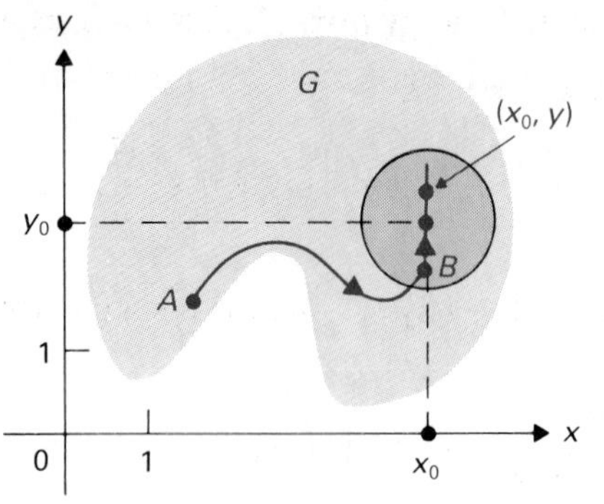

Figure 17.35
$H(x_0, y) = H(B) + \int_B^{(x_0, y)} Q(x_0, y)\, dy.$

Theorems 17.7 and 17.8 taken together show that

> in a connected open region G, the integral $\int_\gamma (P\,dx + Q\,dy)$ is independent of the path if and only if $P\,dx + Q\,dy$ is an exact differential form.

In view of Theorem 17.6, we then obtain

> $\oint_\lambda (P\,dx + Q\,dy) = 0$ for every loop λ in a connected open region G if and only if $P\,dx + Q\,dy$ is an exact differential form.

Recall that $dH = P\,dx + Q\,dy$ if and only if $\nabla H = P\mathbf{i} + Q\mathbf{j}$. Thus we also have

> line integrals of a vector field $\mathbf{F} = P\mathbf{i} + Q\mathbf{j}$ are independent of the path in a connected open region G if and only if $\mathbf{F} = \nabla H$ for some function H.

Of course, space analogues of all these statements are also true.

APPLICATIONS

Let $\mathbf{F}$ be a continuous vector field on a connected open region in the plane or in space. If $\mathbf{F} = \nabla H$, then $\oint_\lambda (\mathbf{F} \cdot d\mathbf{r}) = 0$ around any loop λ in G. That is, the work done by the force field $\mathbf{F}$ moving a body around any loop in G is zero. Such a force field $\mathbf{F}$ is **conservative.** No work is done by such a force field acting on a body unless the position of the body is changed. The function $-H(x, y)$ in the plane, or $-H(x, y, z)$ in space, is called a **potential function of F.** A potential function gives the **potential energy** of the force field at each point. By Theorem 17.7, the work done by the conservative force field $\mathbf{F}$ in moving a body from A to B is

$$H(B) - H(A) = -H(A) - (-H(B)),$$

which is the *loss of potential energy from A to B.*

As mentioned in Section 17.3, $P\mathbf{i} + Q\mathbf{j}$ is a gradient vector field in a region having no holes in it if P and Q have continuous first partial derivatives and $\partial P/\partial y = \partial Q/\partial x$.

EXAMPLE 7 Consider the force field

$$\mathbf{F}(x, y) = (2x + \cos y)\mathbf{i} + (y - x \sin y)\mathbf{j}.$$

Show that $\mathbf{F}$ is conservative, find a potential function $-H(x, y)$, and find the loss in potential energy as a body is moved by the field from $(3, 0)$ to $(-2, \pi)$.

Solution We have

$$\frac{\partial(2x + \cos y)}{\partial y} = -\sin y = \frac{\partial(y - x \sin y)}{\partial x},$$

so the field $\mathbf{F}$ is conservative. As in Section 17.3, we easily find that $\mathbf{F} = \nabla H$, where $H(x, y) = x^2 + x \cos y + (y^2/2)$. Therefore

$$-H(x, y) = -x^2 - x \cos y - \frac{y^2}{2}$$

is a potential function. The loss in potential energy from $(3, 0)$ to $(-2, \pi)$ is

$$\begin{aligned} -H(3, 0) - (-H(-2, \pi)) &= H(-2, \pi) - H(3, 0) \\ &= \left(4 + 2 + \frac{\pi^2}{2}\right) - (9 + 3) = -6 + \frac{\pi^2}{2}. \quad \square \end{aligned}$$

The force of gravitational attraction of a body in space toward a mass at the origin is given by the force field

$$\mathbf{F}(x, y, z) = \frac{-K}{(x^2 + y^2 + z^2)^{3/2}}(x\mathbf{i} + y\mathbf{j} + z\mathbf{k})$$

for some constant K. Force vectors are directed toward the origin and are of magnitude inversely proportional to the square of the distance from the origin. It is easily checked that $\mathbf{F} = \nabla H$ where $H(x, y, z) = K/\sqrt{x^2 + y^2 + z^2}$. This shows that the gravitational attraction field $\mathbf{F}$ is a conservative force field in all of space except the origin, which is a connected open region. The work done by gravitational attraction in moving a body around any loop in space is zero. This is one aspect of the *law of conservation of energy* of physics. The potential function $-K/\sqrt{x^2 + y^2 + z^2}$ is the *Newtonian potential.*

Let F be the flux field of a steady-state flow in the plane or in space. If λ is a loop, then $\oint_\lambda (\mathbf{F} \cdot d\mathbf{r})$ is the **circulation of the flow around** λ. If $\mathbf{F} = \nabla H$, then the circulation of the flow around any loop is zero. Such a flux field $\mathbf{F}$ is called **irrotational.**

EXAMPLE 8 Find all functions $P(x, y)$ such that the flux field

$$\mathbf{F}(x, y) = P(x, y)\mathbf{i} + (xy^2 + 2x - 3y)\mathbf{j}$$

in the plane is irrotational.

Solution Our work in Section 17.3 shows that $\mathbf{F}$ is irrotational if and only if

$$\frac{\partial P}{\partial y} = \frac{\partial(xy^2 + 2x - 3y)}{\partial x} = y^2 + 2.$$

Therefore we must have

$$P(x, y) = \int (y^2 + 2)\, dy = \frac{y^3}{3} + 2y + k(x)$$

for some function $k(x)$ of x only. $\square$

SUMMARY

Throughout the following, γ is a piecewise-smooth curve, and $\mathbf{F}(x, y) = P(x, y)\mathbf{i} + Q(x, y)\mathbf{j}$ is a continuous vector field.

1. The integral of the tangential component of the vector field $\mathbf{F}$ along γ is

$$\int_\gamma (\mathbf{F} \cdot \mathbf{t})\, ds = \int_\gamma \mathbf{F} \cdot d\mathbf{r} = \int_\gamma (P\, dx + Q\, dy).$$

2. If $\mathbf{F}$ is a force field, then $\int_\gamma \mathbf{F} \cdot d\mathbf{r}$ is the work done by the force on a body moving along γ.

3. For

$$\gamma : x = h(t), \quad y = k(t) \qquad \text{for } 0 \le t \le 1,$$

the curve $-\gamma$ has the same trace traveled in the opposite direction, and

$$-\gamma : x = h(1 - t), \quad y = k(1 - t) \qquad \text{for } 0 \le t \le 1.$$

4. $\int_{-\gamma} (\mathbf{F} \cdot d\mathbf{r}) = -\int_{\gamma} (\mathbf{F} \cdot d\mathbf{r})$
5. $\int_{\gamma} (\mathbf{F} \cdot d\mathbf{r})$ is independent of the path in a connected open region G, depending only on the initial point A and terminal point B for γ in G, if and only if $\oint_{\lambda} (\mathbf{F} \cdot d\mathbf{r}) = 0$ for every loop λ in G. (A loop or closed curve is one whose initial and terminal points coincide.)
6. $\int_{\gamma} (\mathbf{F} \cdot d\mathbf{r})$ is independent of the path in a connected open region G if and only if $P\,dx + Q\,dy$ is an exact differential form dH in G or, equivalently, if and only if $\mathbf{F}$ is a gradient vector field ∇H.
7. A force field $\mathbf{F}$ is called conservative if $\mathbf{F}$ is a gradient vector field ∇H. The function $-H(x, y)$ is then a potential function of $\mathbf{F}$.
8. If $\mathbf{F}$ is the flux field of a flow, then $\oint_{\lambda} \mathbf{F} \cdot d\mathbf{r}$ is the circulation of the flow around the loop λ. If $\mathbf{F}$ is a gradient field ∇H, then the field is irrotational.
9. Analogues of all the above hold for continuous vector fields and piecewise-smooth curves in space.

EXERCISES

In Exercises 1 through 6, the curves are

γ_1 given by $x = t, \quad y = t^2 \qquad$ for $0 \le t \le 1$,

γ_2 given by $x = t + 1, \quad y = 2t + 1 \qquad$ for $0 \le t \le 1$,

and the vector fields are

$$\mathbf{F}(x, y) = x^2\mathbf{i} + y^2\mathbf{j} \qquad \text{and} \qquad \mathbf{G}(x, y) = xy\mathbf{i} - x^2y\mathbf{j}.$$

Find the indicated integral.

1. $\int_{\gamma_1} \mathbf{F} \cdot d\mathbf{r}$
2. $\int_{\gamma_2} \mathbf{G} \cdot d\mathbf{r}$
3. $\int_{\gamma_1} (\mathbf{F} - \mathbf{G}) \cdot d\mathbf{r}$
4. $\int_{\gamma_2} (2\mathbf{F} + \mathbf{G}) \cdot d\mathbf{r}$
5. $\int_{\gamma_1 + \gamma_2} (\mathbf{F} + \mathbf{G}) \cdot d\mathbf{r}$
6. $\int_{-\gamma_1} \mathbf{G} \cdot d\mathbf{r}$
7. Let γ be the curve in the plane that consists of the line segment from $(-1, 2)$ to $(4, 3)$ traveled at constant speed for $0 \le t \le 1$. Find parametric equations for $-\gamma$.
8. Let γ be the curve in space that consists of the line segment from $(3, -1, 4)$ to $(2, 1, 0)$ traveled at constant speed for $0 \le t \le 1$. Find parametric equations for $-\gamma$.

Exercises 9 through 14 relate to the curves

$\gamma_1 : x = t, \quad y = t^2 \qquad$ for $0 \le t \le 1$,

$\gamma_2 : x = 1 + t, \quad y = 1 + 6t \qquad$ for $0 \le t \le 1$,

$\gamma_3 : x = t^2 - t, \quad y = 2t^2 + t - 3 \qquad$ for $1 \le t \le 2$.

Decide whether or not the given expression is a single piecewise-smooth curve. If it is, find the initial point and the terminal point, and state whether or not it is a loop.

9. $\gamma_1 + \gamma_2$
10. $\gamma_1 + \gamma_2 + \gamma_3$
11. $\gamma_1 + \gamma_2 - \gamma_3$
12. $-\gamma_2 - \gamma_1 + \gamma_3$
13. $-\gamma_2 - \gamma_1$
14. $-\gamma_1 + \gamma_3 - \gamma_2$
15. Let $\mathbf{F}(x, y) = xy\mathbf{i} + x^2\mathbf{j}$ and let γ be the shorter arc from $(3, 0)$ to $(0, 3)$ of a circle with center at the origin. Find $\int_{\gamma} \mathbf{F} \cdot d\mathbf{r}$.
16. Let $F(x, y, z) = (x^2 + z)\mathbf{i} + (x + y^2)\mathbf{j} + xz\mathbf{k}$ and let $\gamma = \gamma_1 + \gamma_2 + \gamma_3$ be the path consisting of three straight-line segments from $(0, 0, 0)$ to $(1, 0, 0)$ to $(1, 1, 1)$ to $(0, 1, 2)$. Find $\int_{\gamma} \mathbf{F} \cdot d\mathbf{r}$.
17. Let γ be the space curve given by $x = 3t^2$, $y = 2t^3$, $z = 3t$ for $0 \le t \le 1$ and let $F(x, y, z) = 3x^2yz$.
    a) Compute $\int_{\gamma} dF$ by integration.
    b) Compute the integral again, using the fact that dF is an exact differential and Theorem 17.7.
18. Repeat Exercise 17(b) for the space curve γ given by $x = \sin t$, $y = \cos t$, $z = t$ for $0 \le t \le \pi/4$.

In Exercises 19 through 30, find $\int_{\gamma} \mathbf{F} \cdot d\mathbf{r}$ for the given vector field $\mathbf{F}$ along the given curve γ. Use the fact that $\mathbf{F}$ is either a gradient vector field or "almost" a gradient vector field. See Example 6.

19. $\mathbf{F}(x, y) = xy^2\mathbf{i} + x^2y\mathbf{j}$;
$\gamma : x = 3t^2, y = 2t^3$ for $0 \le t \le 1$

20. $\mathbf{F}(x, y) = ((1/y) - x)\mathbf{i} + (2y - (x/y^2))\mathbf{j}$;
$\gamma : x = 2t + 3, y = t^2 + 1$ for $0 \le t \le 1$

21. $\mathbf{F}(x, y) = (y^2 + 2x)\mathbf{i} + (2xy + x)\mathbf{j}$;
$\gamma : x = \sin^2 t, y = \sin t$ for $0 \le t \le 3\pi/2$

22. $\mathbf{F}(x, y) = (ye^{xy} + 2xy^2 - y^2)\mathbf{i} + (xe^{xy} + 2x^2y)\mathbf{j}$;
$\gamma: x = \cos 2t, y = \sin t$ for $0 \le t \le \pi/2$

23. $\mathbf{F}(x, y) = (2y \sin xy \cos xy + x)\mathbf{i} + (2x \sin xy \cos xy - x)\mathbf{j}$;
$\gamma : x = 1 + \cos t, y = 1 + \sin t$ for $0 \le t \le 2\pi$

24. $\mathbf{F}(x, y) = (e^y + 2xy^3 - 3 \sin x)\mathbf{i} + (xe^y + 3x^2y^2 - \cos y^2)\mathbf{j}$;
$\gamma : x = 2 + 3 \sin t, y = -4 + 2 \sin t$ for $0 \le t \le 2\pi$

25. $\mathbf{F}(x, y, z) = 2xyz\mathbf{i} + x^2z\mathbf{j} + x^2y\mathbf{k}$;
$\gamma : x = 2t + 1, y = t^2, z = t - 3$ for $1 \le t \le 2$

26. $\mathbf{F}(x, y, z) = (y^2 - 6xz)\mathbf{i} + 2xy\mathbf{j} - 3x^2\mathbf{k}$;
$\gamma : x = t^3 - 2t, y = (t + 2)/(t - 2), z = 3t - 1$ for $0 \le t \le 1$

27. $\mathbf{F}(x, y, z) = (y/z)\mathbf{i} + ((x/z) + z^2)\mathbf{j} + (2yz - (xy/z^2))\mathbf{k}$;
$\gamma : x = 4 + \sin \pi t, y = -3 - \cos \pi t, z = t^2 - 2t + 2$ for $0 \le t \le 2$

28. $\mathbf{F}(x, y, z) = (yz + 2x)\mathbf{i} + (xz - y)\mathbf{j} + (xy + y - z)\mathbf{k}$;
$\gamma : x = \cos t, y = \sin t, z = t^2 - \pi^2$ for $-\pi \le t \le \pi$

29. $\mathbf{F}(x, y, z) = (x + y)\mathbf{i} + ye^y\mathbf{j} - (z^3 \sin z)\mathbf{k}$;
$\gamma : x = \cos 2t, y = -\sin 2t, z = \cos t$ for $0 \le t \le 2\pi$

30. $\mathbf{F}(x, y, z) = (z + \sin x^2)\mathbf{i} + \cos^2 y^3\mathbf{j} + y\mathbf{k}$;
$\gamma : x = t^2 + 1, y = \sin \pi t, z = \cos \pi t$ for $-1 \le t \le 1$

31. Let the position of a body moving in the plane be $(\cos t, \sin 2t)$ at time t, and let the body be subject to the force field $x\mathbf{i} + y\mathbf{j}$. Find the work done by the force field on the body from time $t = 0$ to time $t = \pi/2$.

32. Let the position of a moving body in space be $(3t^2, 2t^3, 3t)$ at time t, and let the body be subject to the force field $x\mathbf{i} + z\mathbf{j} + y\mathbf{k}$. Find the work done by the force field on the body from time $t = 1$ to time $t = 2$.

33. Let the flux field of a steady-state fluid flow in the plane be $x\mathbf{i} - y\mathbf{j}$. Find the circulation of the flow around the circle $x = \cos t, y = \sin t$ for $0 \le t \le 2\pi$.

34. Let the flux field of a steady-state fluid flow in the plane be $xy\mathbf{i} - x\mathbf{j}$. Find the circulation of the flow around the ellipse $x = 2 \cos t, y = 3 \sin t$ for $0 \le t \le 2\pi$.

35. Consider the conservative force field

$$\mathbf{F}(x, y, z) = (2x + z^2)\mathbf{i} - 4y\mathbf{j} + 2xz\mathbf{k}.$$

Find the loss of potential energy from $(2, 1, 0)$ to $(3, -1, 1)$.

36. Consider the conservative force field

$$\mathbf{F}(x, y, z) = (y^2z - 2x)\mathbf{i} + (2xyz - z)\mathbf{j} + (xy^2 - y + 8z)\mathbf{k}.$$

Find the loss of potential energy from $(1, -1, 1)$ to $(2, -3, 1)$.

EXERCISE SETS FOR CHAPTER 17

Review Exercise Set 17.1

1. Find all local maxima and minima of the function

$$f(x, y) = 2x^2 - xy + y - y^2 - 7x + 3.$$

2. Use Lagrange multipliers to maximize $x + y^2$ over the ellipse $x^2 + 4y^2 = 4$.

3. Use the method of Lagrange multipliers to find the point on the intersection of the spheres $x^2 + y^2 + z^2 = 4$ and $(x - 2)^2 + y^2 + z^2 = 12$ that is closest to $(1, 1, 0)$. Find also the point farthest from $(1, 1, 0)$.

4. Let $P(x, y)$ and $Q(x, y)$ have continuous partial derivatives in a region without any "holes" in it. State a criterion for $P(x, y)\,dx + Q(x, y)\,dy$ to be an exact differential.

5. Test whether

$$(y^2e^x + 3x^2y)\,dx + (2ye^x + x^3 + \sin y)\,dy$$

is an exact differential, and, if it is, find $H(x, y)$ such that the differential is dH.

6. Find the total curvature of the arc of the graph of $y = \sin x$ for $0 \le x \le \pi$.

7. Evaluate $\int_\gamma x\,ds$ if γ is the plane curve given by $x = 8t$, $y = 3t^2$ for $0 \le t \le 1$.

8. Consider the force field in space $\mathbf{F}(x, y, z) = xy\mathbf{i} + (z/x)\mathbf{j} + y^2z\mathbf{k}$. Find the work done by the field in moving a body along the curve $x = t, y = t, z = t - 1$, from $t = 1$ to $t = 2$.

9. Let γ be a smooth curve from $(-1, 2)$ to $(3, 4)$. When is $\int_\gamma (P(x, y)\,dx + Q(x, y)\,dy)$ independent of the choice of path joining these points?

10. Let $\mathbf{F}(x, y) = (3x^2 - 2xy)\mathbf{i} + (4y - x^2)\mathbf{j}$, and let

$$\gamma : x = t^2 - 2t, \quad y = 3t - 2 \quad \text{for } 0 \le t \le 1.$$

Find $\int_\gamma \mathbf{F} \cdot d\mathbf{r}$.

Review Exercise Set 17.2

1. Find all relative maxima and minima of the function

$$f(x, y) = xy - x^2 - 2y^2 + 3x - 5y - 6.$$

2. Use the method of Lagrange multipliers to maximize $x^2 + y + z^2$ over the ellipsoid $x^2 + y^2 + 4z^2 = 4$.

3. Use the method of Lagrange multipliers to find the point on the line of intersection of the planes

$$x - 2y + z = 4 \quad \text{and} \quad 2x + y - z = 8$$

that is closest to the origin.

4. Find c and k such that the differential

$$(y^2 + kxy)\,dx + (3x^2 + cxy)\,dy$$

is exact.

5. Find $H(x, y)$ such that dH is
$$(2x \sin y - 3x^2 y)\,dx + (x^2 \cos y - x^3 + 4y^2)\,dy.$$

6. A wire of variable density $\rho(x, y) = 2x$ covers the curve
$$x = t^2 - 1, \qquad y = 3t + 1 \qquad \text{for } 1 \le t \le 2.$$
Express as an integral the moment of the wire about the y-axis.

7. If γ is the curve $x = \sin t$, $y = 2 \cos t$ for $0 \le t \le \pi/4$, and $\mathbf{F}(x, y)$ is the vector field $xy\mathbf{i} - (x/y)\mathbf{j}$, find $\int_\gamma \mathbf{F} \cdot d\mathbf{r}$.

8. Note that
$$\mathbf{F}(x, y) = (x^2 - 4xy)\mathbf{i} + (y^2 - 2x^2)\mathbf{j}$$
is a conservative force field. Find $\int_\gamma \mathbf{F} \cdot d\mathbf{r}$ where γ is any smooth curve from $(-1, 1)$ to $(1, -2)$.

9. If the flux field of a fluid is $xy\mathbf{i} + 2y\mathbf{j}$, find the circulation of the fluid about the unit circle $x^2 + y^2 = 1$ in the counterclockwise direction. [*Hint*: Take the parametrization $x = \cos t$, $y = \sin t$ for $0 \le t \le 2\pi$.]

10. Let γ be given by $x = \cos t$, $y = \sin t$ for $0 \le t \le 2\pi$. Find $\int_{-\gamma} [(x^2 + y)\,dx - (y^3 + y)\,dy]$.

Review Exercise Set 17.3

1. Find all local maxima and local minima of the function $f(x, y, z) = z^3 + x^2 + 3y^2 - 3z^2 - 2xy + 3$.

2. Find the maximum possible volume of a rectangular box having one vertex at $(0, 0, 0)$ and the diagonally opposite vertex at a point in the first octant on the plane $x + y + 2z = 6$.

3. Use the method of Lagrange multipliers to maximize $2x + 3y - 4$ on the circle $x^2 + y^2 = 13$.

4. Use the method of Lagrange multipliers to find the area of the largest isosceles triangle that can be inscribed in a circle of radius a.

5. Find the general solution of the differential equation
$$\frac{dy}{dx} = \frac{3 - 2xy}{x^2 + 1}.$$

6. If possible, find $H(x, y, z)$ such that
$$dH = (2xyz - 3x^2 y)\,dx + (x^2 z - x^3 + 2y)\,dy + (x^2 y + 2y^2 z)\,dz.$$

7. Find the total curvature of the graph of $y = x^n$ for any integer $n \ge 2$.

8. Find $\int_\gamma y^2\,dx$ where the trace of γ is the line segment from $(-1, 2)$ to $(2, 3)$.

9. Let γ be given by $x = 2 \cos \pi t$, $y = 3 \sin \pi t$ for $0 \le t \le 1$. Find $-\gamma$.

10. Consider the conservative force field $\mathbf{F}(x, y) = (x^2 + y^2)\mathbf{i} + 2xy\mathbf{j}$. Find the loss in potential energy of the field from $(3, 1)$ to $(1, -1)$.

Review Exercise Set 17.4

1. Find the minimum distance from the origin to a point on the surface $xy^2z^2 = 4$.

2. Find all local maxima and local minima of
$$x^2 - y^2 + 2z^2 + 2xz + 6x + 4y - 2z + 5.$$

3. Use the method of Lagrange multipliers to find all local extrema of $x^2 + y^2 + z^2$ subject to the constraints $x^2 + y^2 = 4$ and $x + z = 1$.

4. The strength of a beam of rectangular cross section is directly proportional to the width and to the square of the depth. Use the method of Lagrange multipliers to find the dimensions of the strongest beam that can be cut from a circular log of radius a inches.

5. Find the implicit solution of the differential equation $dy/dx = (2 - 3x^2 y)/(x^3 + 2y)$ that contains the point $(-1, 4)$.

6. Let $P(x, y)$ and $Q(x, y)$ have continuous first partial derivatives in the entire plane. If both $P\,dx + Q\,dy$ and $P\,dx - Q\,dy$ are exact differential forms, what can be said concerning P and Q?

7. Let γ_1 have as path set the line segment from $(0, 0)$ to $(1, -2)$, and let γ_2 have as path set the line segment from $(1, -2)$ to $(1, 0)$. Find $\int_{\gamma_1 + \gamma_2} xy\,ds$.

8. Find the work done by the force field $\mathbf{F}(x, y) = xy\mathbf{i} - y\mathbf{j}$ acting on a body moving from $(0, 0)$ to $(1, 1)$ along the parabola $y = x^2$.

9. Let the flux field of a flow in space be
$$\mathbf{F}(x, y, z) = (2xyz + x - y)\mathbf{i} + (x^2 z + y)\mathbf{j} + (x^2 y - e^z)\mathbf{k}.$$
Find the circulation of the flow around the loop
$$\lambda: x = a \cos t, \quad y = b \sin t, \quad z = c \sin t \qquad 0 \le t \le 2\pi.$$

10. The potential energy of the conservative force field
$$\mathbf{F}(x, y) = (2xy^3 - e^y)\mathbf{i} + (3x^2y^2 - xe^y)\mathbf{j}$$
at $(3, 0)$ is 15. Find the potential function associated with $\mathbf{F}$.

More Challenging Exercises 17

Exercises 1 through 3 are designed to shed a little light on the criterion $AC - B^2 > 0$ given in Theorem 17.2 for a local maximum or local minimum of $f(x, y)$.

1. Taylor's theorem for functions $f(x, y)$ of two variables was stated in the last exercise set at the end of Chapter 16.
   a) Write the expression for $E_1(x, y)$ given in the theorem.
   b) Suppose $f_x(x_0, y_0) = f_y(x_0, y_0) = 0$. Show that
      i) $f(x_0, y_0)$ is a local minimum if $E_1(x, y) \ge 0$ for all (x, y) in some small disk with center at (x_0, y_0);
      ii) $f(x_0, y_0)$ is a local maximum if $E_1(x, y) \le 0$ for all (x, y) in some small disk with center at (x_0, y_0);
      iii) $f(x_0, y_0)$ is neither a local minimum nor a local maximum if $E_1(x, y)$ assumes both positive and negative values in every small disk with center at (x_0, y_0).

2. Let $f(x, y)$ have continuous second partial derivatives, and let $A = f_{xx}(x_0, y_0)$, $B = f_{xy}(x_0, y_0)$, and $C = f_{yy}(x_0, y_0)$. Convince yourself that if $A(\Delta x)^2 + 2B(\Delta x)(\Delta y) + C(\Delta y)^2 > 0$ for all choices of $(\Delta x, \Delta y) \ne (0, 0)$, then $E_1(x, y) \ge 0$ for all (x, y) in some small disk with center at (x_0, y_0). [*Hint*: Think of $\Delta x = x - x_0$, $\Delta y = y - y_0$, and set $\Delta y = t(\Delta x)$.] What would be true if $A(\Delta x)^2 + 2B(\Delta x)(\Delta y) + C(\Delta y)^2 > 0$ for $(\Delta x, \Delta y) \ne (0, 0)$? If it were sometimes positive and sometimes negative?

3. Show that $A(\Delta x)^2 + 2B(\Delta x)(\Delta y) + C(\Delta y)^2 > 0$ for all $(\Delta x, \Delta y) \ne (0, 0)$ if $AC - B^2 > 0$ and $A > 0$ or $C > 0$. [*Hint*:

Note that

$$A(A(\Delta x)^2 + 2B(\Delta x)(\Delta y) + C(\Delta y)^2) = (A(\Delta x) + B(\Delta y))^2 + (AC - B^2)(\Delta y)^2.]$$

4. Suppose $f(x, y)$ has continuous partial derivatives of all orders less than or equal to n. Suppose, further, that all partial derivatives of order less than n are zero at (x_0, y_0), but some partial derivative of order n is nonzero at (x_0, y_0).
a) Show that if n is odd, then $f(x, y)$ has neither a local minimum nor a local maximum at (x_0, y_0).
b) If n is even, show that Exercise 1(b) holds with $E_{n-1}(x, y)$ in place of $E_1(x, y)$.

5. Let P, Q, R, and S be four functions of the four variables w, x, y, and z, all having continuous partial derivatives. State the criteria for $P\,dw + Q\,dx + R\,dy + S\,dz$ to be an exact differential in some region of four-space having no holes in it.

6. Using subscript notation, let $\mathbf{x} = (x_1, \ldots, x_n)$ and let $F_1, F_2, \ldots, F_n$ be n functions of $\mathbf{x}$ having continuous partial derivatives. Repeat Exercise 5, giving criteria for $F_1\,dx_1 + F_2\,dx_2 + \cdots + F_n\,dx_n$ to be exact in a suitable region. (It is much easier in this notation.)

7. Consider the differential

$$\frac{y}{x^2 + y^2}\,dx - \frac{x}{x^2 + y^2}\,dy, \qquad (x, y) \neq (0, 0).$$

a) Show that this differential satisfies the criterion $\partial P/\partial y = \partial Q/\partial x$ for all $(x, y) \neq (0, 0)$.
b) Show that, if the given differential were dH for $H(x, y)$ and all $(x, y) \neq (0, 0)$, then we would have to have

$$H(x, y) = \tan^{-1}\left(\frac{x}{y}\right) + A \quad \text{for } y > 0 \text{ and some } A$$

and

$$H(x, y) = \tan^{-1}\left(\frac{x}{y}\right) + B \quad \text{for } y < 0 \text{ and some } B.$$

Show that there are no choices for A and B and no way of defining $H(x, y)$ on the positive and negative portions of the x-axis that would give a differentiable function $H(x, y)$ for all $(x, y) \neq (0, 0)$. (This illustrates that the region should have no holes in it in order for the partial derivative criterion for exactness to guarantee exactness in the entire region.) [*Hint:* Show that we would have to have $B = A + \pi$, and $H(x, 0) = A + \pi/2$ for $x > 0$. Then show that it becomes impossible to define $H(x, 0)$ for $x < 0$ to make H even continuous on the negative x-axis, to say nothing of differentiable.]

18 MULTIPLE INTEGRALS

18.1 INTEGRALS OVER RECTANGULAR REGIONS

INTEGRALS AS LIMITS OF SUMS

We have tried to emphasize how important it is to regard an integral as a limit of sums. In an application involving an integral, we always consider what quantities we wish to add in order to arrive at the appropriate integral.

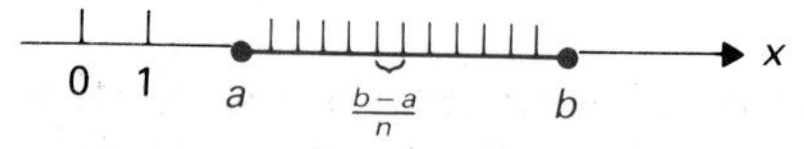

Figure 18.1 Partition of $[a, b]$ into $n = 12$ equal-length subintervals.

Table 18.1 The integral as a limit of Riemann sums

	Integral of f over G	$\int_a^b f(x)\,dx$
Step 1	Partition G into subregions of small size such that values of f do not vary much over any single subregion.	Partition $[a, b]$ into n subintervals of equal length $(b-a)/n$. See Fig. 18.1.
Step 2	For each subregion, form the product of the value of f at some point in the subregion and the size of the subregion.	Form the products $$f(x_i)\cdot\left(\frac{b-a}{n}\right)$$ for x_i in the ith subinterval. See Fig. 18.2.
Step 3	Add the products obtained in Step 2. The result is an approximation to the integral.	Find the Riemann sum $$\mathcal{S}_n = \sum_{i=1}^{n}\left(f(x_i)\cdot\frac{b-a}{n}\right) = \frac{b-a}{n}\sum_{i=1}^{n} f(x_i).$$ See Fig. 18.3.
Step 4	The integral is the limit of sums found in Step 3 as partitions become finer and finer in such a way that the variation of values of f on partition elements approaches zero.	$$\int_a^b f(x)\,dx = \lim_{n\to\infty}\mathcal{S}_n.$$

Figure 18.2 Strip of area $f(x_i)\cdot(b-a)/n$.

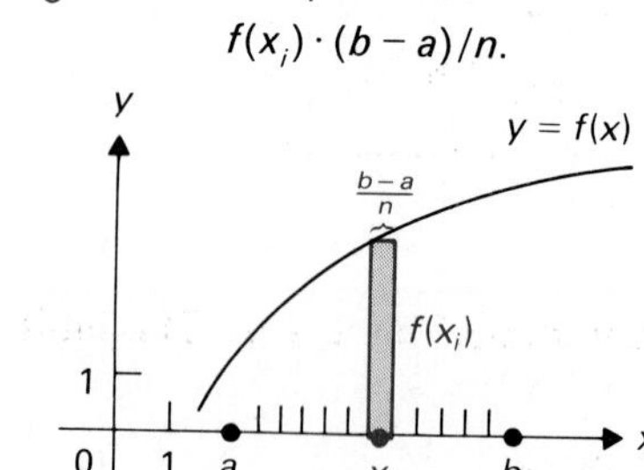

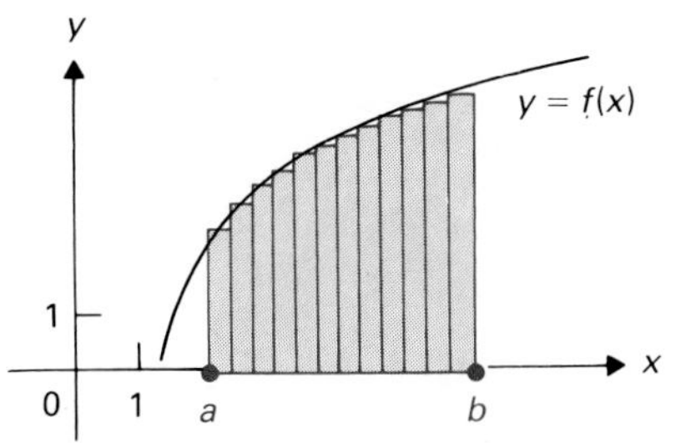

Figure 18.3 A Riemann sum $\mathscr{S}_n$ for $n = 12$.

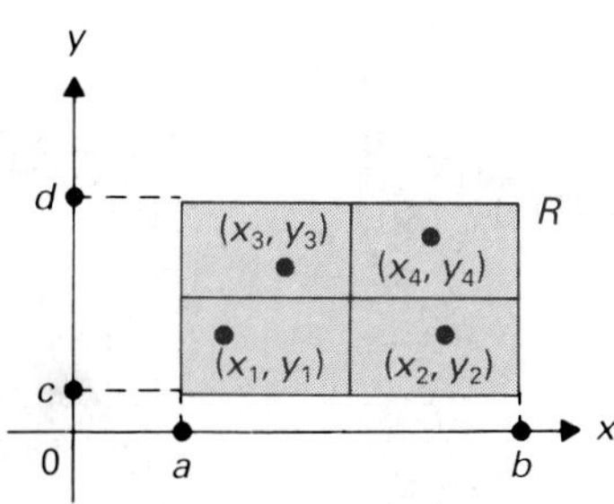

Figure 18.4 Partition of a rectangular region R for a Riemann sum with $n = 2$.

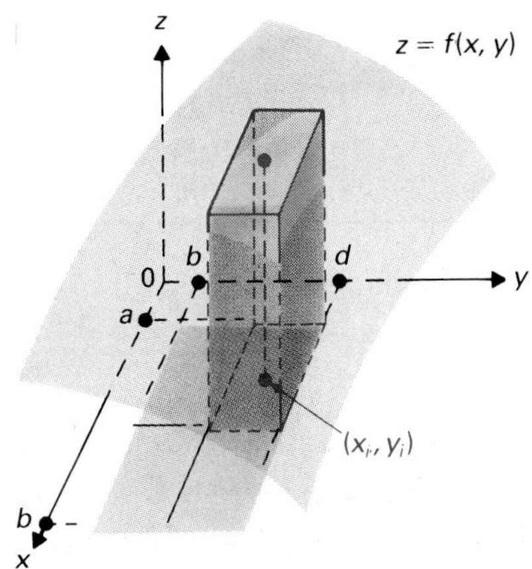

Figure 18.5 Column of volume $f(x_i, y_i) \cdot [(b-a)(d-c)]/n$ for $n = 2$.

Let f be a continuous real-valued function defined on a region G. In Table 18.1 we give an outline showing how we should regard the integral of f over G as a limit of sums. We illustrate each step using the one-dimensional case $\int_a^b f(x)\,dx$, with which we are familiar.

We apply the above steps to describe the integral of a continuous function $f(x, y)$ over a rectangular region R, where $a \leq x \leq b$ and $c \leq y \leq d$, as shown in Fig. 18.4.

Step 1 We partition $[a, b]$ into n subintervals of equal length and then do the same for $[c, d]$. This gives rise to a partition of the region R into n^2 subrectangles, as illustrated in Fig. 18.4 for $n = 2$. Each subrectangle has area

$$\frac{(b-a)(d-c)}{n^2}.$$

Step 2 Numbering the subrectangles from 1 to n^2 in some convenient fashion, we let (x_i, y_i) be a point in the ith subrectangle, as shown in Fig. 18.4. We then form the n^2 products

$$f(x_i, y_i) \cdot \frac{(b-a)(d-c)}{n^2}.$$

Each such product gives the volume of a column, as illustrated in Fig. 18.5.

Figure 18.6 A Riemann sum $\mathscr{S}_n$ for $n = 2$.

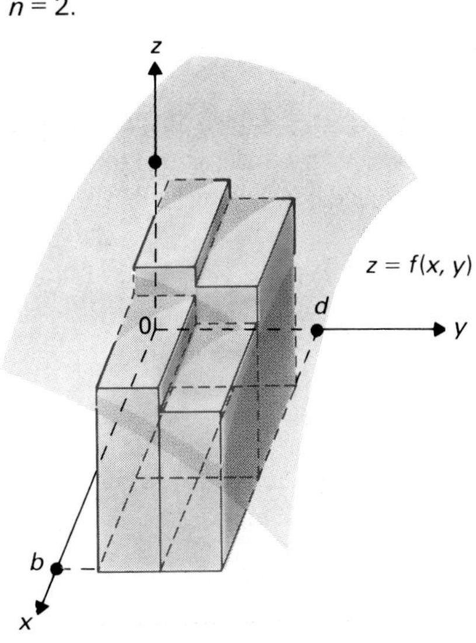

Step 3 We form the **Riemann sum**

$$\boxed{\mathscr{S}_n = \frac{(b-a)(d-c)}{n^2} \cdot \sum_{i=1}^{n^2} f(x_i, y_i).} \quad (1)$$

See Fig. 18.6.

Step 4 The integral is the limit of the Riemann sums $\mathscr{S}_n$ as $n \to \infty$. This limit can be shown to exist.

We have been led to the following definition.

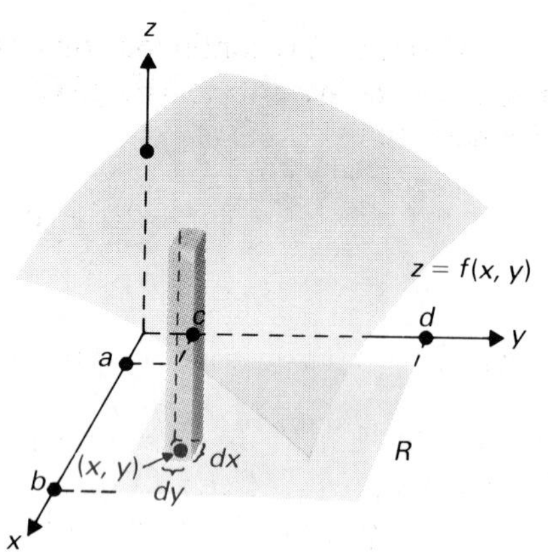

Figure 18.7 $f(x, y)\,dx\,dy$ is the volume of the column.

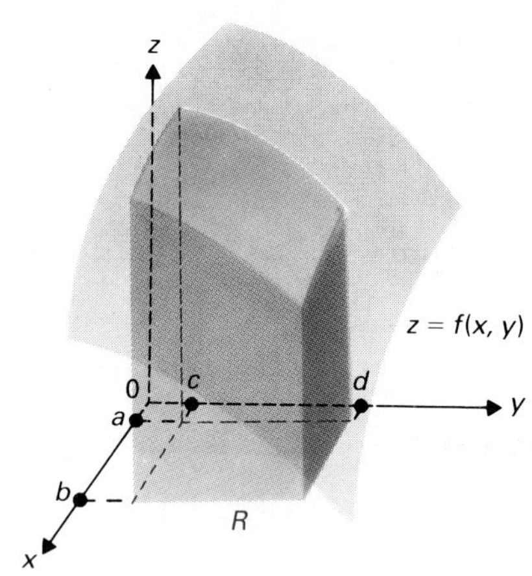

Figure 18.8 $\iint_R f(x, y)\,dx\,dy$ is the volume of the solid.

DEFINITION 18.1 Definite integral

Let $f(x, y)$ be continuous in the rectangular region R, where $a \le x \le b$ and $c \le y \le d$. The **definite integral** of f **over** R is

$$\iint_R f(x, y)\,dx\,dy = \lim_{n\to\infty} \mathcal{S}_n$$

for the Riemann sums $\mathcal{S}_n$ in Eq. (1).

Figure 18.9 $\iint_R f(x, y)\,dx\,dy$ = (Volume of black solid) − (Volume of red solid).

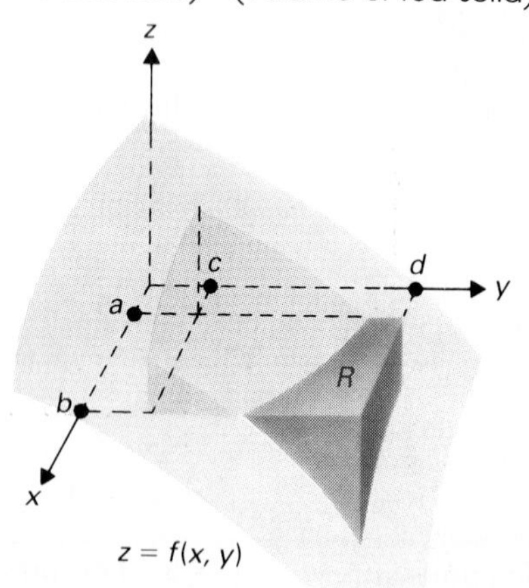

The double integral sign $\iint$ in Definition 18.1 is used to indicate that the domain of f is two-dimensional. The geometric meaning of the integrand $f(x, y)\,dx\,dy$ is shown in Fig. 18.7.

It is geometrically clear that if $f(x, y) \ge 0$, then $\iint_R f(x, y)\,dx\,dy$ is equal to the volume of the solid having R as base, vertical-plane sides, and a portion of the surface $z = f(x, y)$ as top, as illustrated in Fig. 18.8. If $f(x, y)$ is sometimes positive and sometimes negative, then $\iint_R f(x, y)\,dx\,dy$ gives the volume of the portion of the solid above the x,y-plane minus the volume of the portion below the x,y-plane, as shown in Fig. 18.9.

EXAMPLE 1 Let R be the rectangular region $1 \le x \le 3$, $0 \le y \le 1$ shown in Fig. 18.10. Estimate $\iint_R (x^2 + 3xy)\,dx\,dy$ by finding $\mathcal{S}_2$, taking as points (x_i, y_i) the midpoints of the subrectangles.

Solution The partition and points are shown in Fig. 18.10. The Riemann sum $\mathcal{S}_2$ in Eq. (1) becomes

$$\mathcal{S}_2 = \frac{(3-1)(1-0)}{2^2}\left[f\left(\frac{3}{2}, \frac{1}{4}\right) + f\left(\frac{5}{2}, \frac{1}{4}\right) + f\left(\frac{3}{2}, \frac{3}{4}\right) + f\left(\frac{5}{2}, \frac{3}{4}\right)\right]$$

$$= \frac{2}{4}\left[\left(\frac{9}{4} + \frac{9}{8}\right) + \left(\frac{25}{4} + \frac{15}{8}\right) + \left(\frac{9}{4} + \frac{27}{8}\right) + \left(\frac{25}{4} + \frac{45}{8}\right)\right]$$

$$= \frac{1}{2}\left(\frac{68}{4} + \frac{96}{8}\right) = \frac{1}{2}(17 + 12) = \frac{29}{2}. \quad \square$$

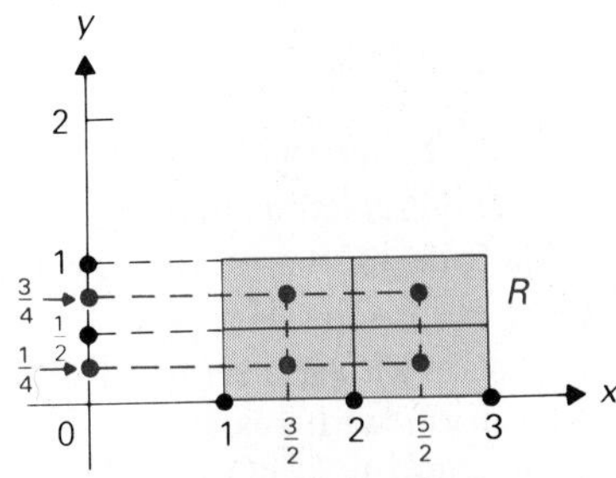

Figure 18.10 Partition for the midpoint sum $\mathscr{S}_2$.

If $f(x, y, z)$ is continuous on a rectangular box $a \leq x \leq b$, $c \leq y \leq d$, $r \leq z \leq s$ in space, then $\iiint_R f(x, y, z)\,dx\,dy\,dz$ is defined as the limit of Riemann sums

$$\mathscr{S}_n = \frac{(b-a)(d-c)(s-r)}{n^3} \cdot \sum_{i=1}^{n} f(x_i, y_i, z_i) \tag{2}$$

obtained by following the steps in our outline, just as we obtained Eq. (1).

EXAMPLE 2 Use a calculator and the midpoint Riemann sum $\mathscr{S}_2$ to estimate $\iiint_R x^{y+2z}\,dx\,dy\,dz$, where R is the rectangular region $1 \leq x \leq 2$, $0 \leq y \leq 2$, $0 \leq z \leq 1$ shown in Fig. 18.11(a).

Solution Figure 18.11(b) indicates how to find the midpoint coordinates. We have

$$\frac{(b-a)(d-c)(s-r)}{n^3} = \frac{(2-1)(2-0)(1-0)}{2^3} = \frac{1}{4}.$$

We should sum the values of $f(x, y, z) = x^{y+2z}$ at the eight points

$$\left(\frac{5}{4}, \frac{1}{2}, \frac{1}{4}\right), \quad \left(\frac{5}{4}, \frac{1}{2}, \frac{3}{4}\right), \quad \left(\frac{5}{4}, \frac{3}{2}, \frac{1}{4}\right), \quad \left(\frac{5}{4}, \frac{3}{2}, \frac{3}{4}\right),$$
$$\left(\frac{7}{4}, \frac{1}{2}, \frac{1}{4}\right), \quad \left(\frac{7}{4}, \frac{1}{2}, \frac{3}{4}\right), \quad \left(\frac{7}{4}, \frac{3}{2}, \frac{1}{4}\right), \quad \left(\frac{7}{4}, \frac{3}{2}, \frac{3}{4}\right)$$

and multiply this sum by $\frac{1}{4}$. Our calculator gives

$$\frac{1}{4}(19.5625) = 4.890625$$

for this Riemann sum. □

Of course, better accuracy in Example 2 can be obtained by using a value of n greater than 2, but the number n^3 of points quickly becomes large. To handle $\mathscr{S}_{10}$ with 1000 points, we would need to write a short computer program rather than rely on our calculator.

Figure 18.11 Partition for $\mathscr{S}_2$ using midpoints: (a) the box

$$1 \leq x \leq 2,\ 0 \leq y \leq 2,\ 0 \leq z \leq 1$$

partitioned for $\mathscr{S}_2$; (b) finding midpoint coordinates for $\mathscr{S}_2$.

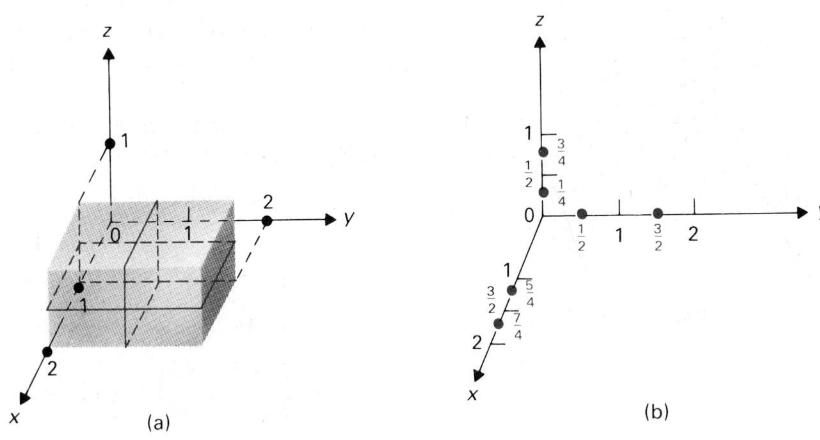

ITERATED INTEGRALS

How can we compute accurately an integral $\iint_R f(x, y)\,dx\,dy$ just defined? We need something like our fundamental theorem of calculus in Section 6.3, but for multiple integrals. Theorem 18.1, which follows in a moment, is our *fundamental theorem for computation*. The theorem states that $\iint_R f(x, y)\,dx\,dy$ can be computed by performing a *partial integration with respect to* x (just as we did in Section 17.3), followed by an integration with respect to y. An integral computed by successive integration steps in this fashion is called an *iterated integral.*

Let $f(x, y)$ be a continuous function on a rectangular region $a \le x \le b$, $c \le y \le d$ in the plane. The *iterated integral* $\int_c^d \int_a^b f(x, y)\,dx\,dy$ is evaluated from "inside out" as

$$\boxed{\int_c^d \int_a^b f(x, y)\,dx\,dy = \int_{y=c}^{d} \left(\int_{x=a}^{b} f(x, y)\,dx \right) dy.} \qquad (3)$$

That is, first integrate $f(x, y)$ with respect to x only and evaluate from $x = a$ to $x = b$. This gives a function of y only, which is then integrated from $y = c$ to d.

EXAMPLE 3 Find $\int_0^1 \int_1^3 xy^2\,dx\,dy$.

Solution We have

$$\int_0^1 \int_1^3 xy^2\,dx\,dy = \int_0^1 \left(\frac{x^2}{2} y^2 \Big]_{x=1}^{x=3} \right) dy = \int_0^1 \left(\frac{9}{2} y^2 - \frac{1}{2} y^2 \right) dy$$

$$= \int_0^1 4y^2\,dy = \frac{4}{3} y^3 \Big]_0^1 = \frac{4}{3}. \quad \square$$

An iterated integral $\int_a^b \int_c^d f(x, y)\,dy\,dx$ is again computed from "inside out" as

$$\boxed{\int_a^b \int_c^d f(x, y)\,dy\,dx = \int_{x=a}^{b} \left(\int_{y=c}^{d} f(x, y)\,dy \right) dx.} \qquad (4)$$

The first integration is with respect to y only, and when it is evaluated from $y = c$ to d, a function of x only is obtained. This function is then integrated from $x = a$ to b.

EXAMPLE 4 Compute the iterated integral

$$\int_1^3 \int_0^1 xy^2\,dy\,dx,$$

which is the integral in Example 3 in the "reverse order."

Solution We have

$$\int_1^3 \int_0^1 xy^2\,dy\,dx = \int_1^3 \left(\frac{xy^3}{3} \Big]_{y=0}^{y=1} \right) dx = \int_1^3 \left(\frac{x}{3} - 0 \right) dx = \frac{x^2}{6} \Big]_1^3$$

$$= \frac{9}{6} - \frac{1}{6} = \frac{4}{3}. \quad \square$$

Note that the iterated integrals in Examples 3 and 4 are equal; the *order* of integration does not seem to matter. This can be shown to be the case.

We will state the fundamental theorem for computation and then attempt to make it seem reasonable by geometric arguments.

THEOREM 18.1 Fundamental theorem for computation

Let $f(x, y)$ be continuous for (x, y) in the rectangular region R where $a \le x \le b$ and $c \le y \le d$. Then

$$\int_c^d \int_a^b f(x, y)\,dx\,dy = \iint_R f(x, y)\,dx\,dy.$$

Geometric explanation The Leibniz notation suggests the following interpretation of $\int_c^d \int_a^b f(x, y)\,dx\,dy$. If we form $\int_a^b f(x, y_j)\,dx$, thinking of y_j as a constant, we obtain the area of the vertical shaded plane region shown in Fig. 18.12, which lies in the plane $y = y_j$ and under the surface $z = f(x, y)$ between $x = a$ and $x = b$. Multiplying by dy, we obtain the volume of the slab shown in Fig. 18.13, and the iterated integral

$$\int_c^d \left(\int_a^b f(x, y)\,dx \right) dy$$

thus adds up the volumes of the slabs from $y = c$ to $y = d$ as $dy \to 0$. Clearly the result for $f(x, y) \ge 0$ is the volume of the three-dimensional region under the surface $z = f(x, y)$ and over the rectangle R, which is precisely the geometric interpretation of $\iint_R f(x, y)\,dx\,dy$. Similar geometric considerations indicate that $\int_a^b \int_c^d f(x, y)\,dy\,dx$ must also be equal to $\iint_R f(x, y)\,dx\,dy$. We leave the sketching of figures like Figs. 18.12 and 18.13 to the exercises (see Exercise 2).

Theorem 18.1 is a powerful tool for computing $\iint_R f(x, y)\,dx\,dy$ when the integration can be carried out.

EXAMPLE 5 Let R be the rectangular region $1 \le x \le 3$, $0 \le y \le 1$. In Example 1, we estimated $\iint_R (x^2 + 3xy)\,dx\,dy$ using the midpoint Riemann sum $\mathscr{S}_2$. Find the exact value of this integral.

Figure 18.12 $\int_a^b f(x, y_j)\,dx$ is the area of the vertical surface.

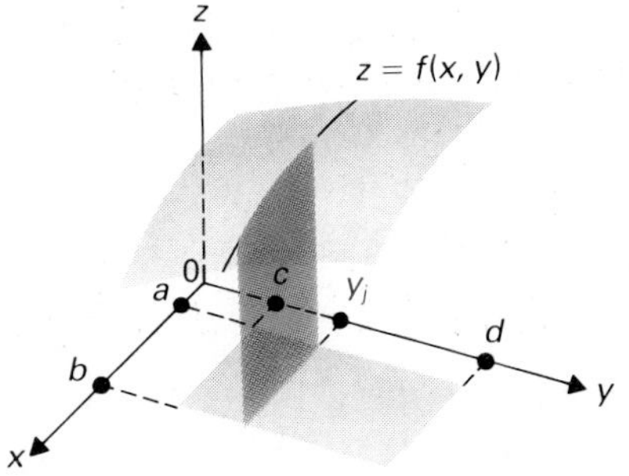

Figure 18.13 $(\int_a^b f(x, y)\,dx)\,dy$ is the volume of the vertical slab.

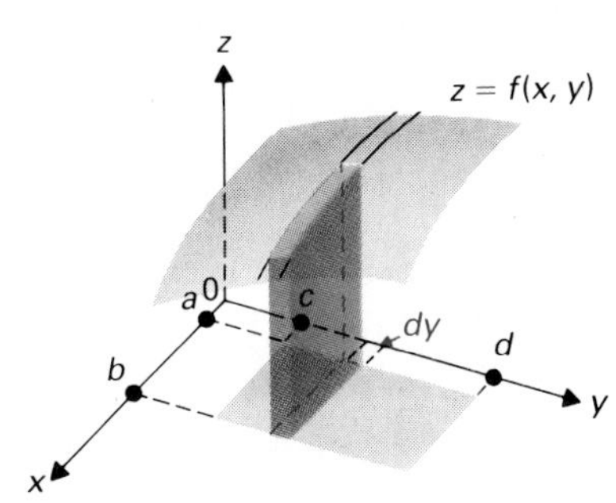

Solution By Theorem 18.1, the integral is

$$\begin{aligned}
\int_0^1 \int_1^3 (x^2 + 3xy)\,dx\,dy &= \int_0^1 \left(\int_{x=1}^3 (x^2 + 3xy)\,dx \right) dy \\
&= \int_0^1 \left(\frac{x^3}{3} + \frac{3}{2}x^2 y \right) \Bigg]_{x=1}^3 dy \\
&= \int_0^1 \left[\left(9 + \frac{27}{2} y \right) - \left(\frac{1}{3} + \frac{3}{2} y \right) \right] dy \\
&= \int_0^1 \left(\frac{26}{3} + 12y \right) dy \\
&= \left(\frac{26}{3} y + 6y^2 \right) \Bigg]_0^1 = \frac{26}{3} + 6 \\
&= \frac{26 + 18}{3} = \frac{44}{3} \approx 14.67.
\end{aligned}$$

In Example 1, we obtained as estimate $\mathcal{S}_2 = 14.5$. □

We would generally omit the large parentheses in the first computation line of Example 5 in computing an iterated integral like $\int_a^b \int_c^d f(x, y)\,dy\,dx$. Just remember to compute from "inside out" and be sure to evaluate using the correct variables. We give another illustration.

EXAMPLE 6 Find $\int_0^4 \int_0^\pi x^2 \sin y\,dy\,dx$.

Solution We have

$$\begin{aligned}
\int_0^4 \int_0^\pi x^2 \sin y\,dy\,dx &= \int_0^4 (-x^2 \cos y) \Big]_0^\pi dx = \int_0^4 [-x^2(-1) + x^2(1)]\,dx \\
&= \int_0^4 2x^2\,dx = \frac{2x^3}{3} \Big]_0^4 = \frac{128}{3}. \quad \square
\end{aligned}$$

To evaluate $\iiint_R f(x, y, z)\,dx\,dy\,dz$ over a rectangular box R, where $a \le x \le b$, $c \le y \le d$, and $r \le z \le s$, we may use a triple iterated integral

$$\boxed{\int_r^s \int_c^d \int_a^b f(x, y, z)\,dx\,dy\,dz = \int_r^s \left(\int_{y=c}^d \left(\int_{x=a}^b f(x, y, z)\,dx \right) dy \right) dz.} \quad \textbf{(5)}$$

Again, the integral is computed from "inside out." Also, it can be shown that the same value is obtained for all orders of integration; that is,

$$\int_r^s \int_c^d \int_a^b f(x, y, z)\,dx\,dy\,dz = \int_a^b \int_r^s \int_c^d f(x, y, z)\,dy\,dz\,dx,$$

and so on. We ask you to discuss possible orders of integration in Exercise 4.

EXAMPLE 7 Let R be the rectangular box $1 \le x \le 3$, $0 \le y \le 2$, $1 \le z \le 2$ in space. Find $\iiint_R [(xy - y^2)/z]\,dx\,dy\,dz$.

Solution We compute the iterated integral

$$\int_1^2 \int_0^2 \int_1^3 \left(\frac{xy - y^2}{z}\right) dx\, dy\, dz = \int_1^2 \int_0^2 \frac{1}{z}\left(\frac{x^2y}{2} - xy^2\right)\bigg]_{x=1}^3 dy\, dz$$

$$= \int_1^2 \int_0^2 \frac{1}{z}\left[\left(\frac{9}{2}y - 3y^2\right) - \left(\frac{y}{2} - y^2\right)\right] dy\, dz$$

$$= \int_1^2 \int_0^2 \frac{1}{z}(4y - 2y^2)\, dy\, dz$$

$$= \int_1^2 \frac{1}{z}\left(2y^2 - \frac{2y^3}{3}\right)\bigg]_{y=0}^2 dz$$

$$= \int_1^2 \frac{1}{z}\left(8 - \frac{16}{3}\right) dz = \frac{8}{3} \ln|z|\bigg]_1^2$$

$$= \frac{8}{3}(\ln 2). \quad \square$$

We should not think that numerical methods of integration such as midpoint Riemann sums are a waste of time. Let R be the rectangular box $1 \le x \le 2$, $0 \le y \le 2$, $0 \le z \le 1$. In Example 2, we estimated

$$\iiint_R x^{y+2z}\, dx\, dy\, dz$$

using $\mathscr{S}_2$. An iterated integral would be $\int_0^1 \int_0^2 \int_1^2 x^{y+2z}\, dx\, dy\, dz$. We cannot perform the integration to evaluate this integral. The first integration is not bad, but it leads to $\int [(2^{y+2z+1} - 1)/(y + 2z + 1)]\, dy$ for the second integral. We then turn to a numerical method.

Sometimes, it is best to carry integration as far as possible and to use a numerical estimate the rest of the way. An example follows.

EXAMPLE 8 Estimate

$$\int_1^2 \int_0^1 \int_1^3 xe^{yz}\, dx\, dy\, dz.$$

Solution We have

$$\int_1^2 \int_0^1 \int_1^3 xe^{yz}\, dx\, dy\, dz = \int_1^2 \int_0^1 \left(\frac{x^2}{2} e^{yz}\right)\bigg]_{x=1}^3 dy\, dz = \int_1^2 \int_0^1 4e^{yz}\, dy\, dz$$

$$= \int_1^2 \frac{4}{z} e^{yz}\bigg]_{y=0}^1 dz = \int_1^2 \frac{4}{z}(e^z - 1)\, dz.$$

At this stage, we are stuck. Using Simpson's rule for $n = 20$, we obtain from our calculator

$$\int_1^2 \frac{4}{z}(e^z - 1)\, dz \approx 9.463877538.$$

Simpson's rule with $n = 40$ yields 9.463877443, which we are confident is close to the correct answer. $\square$

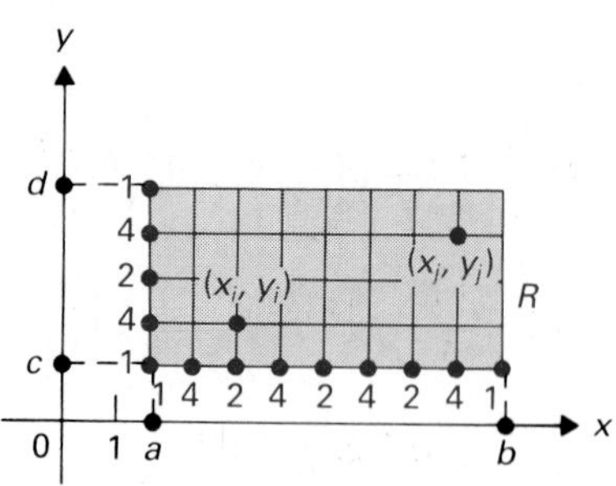

Figure 18.14 Partition of R for Simpson's rule with $m = 8$, $n = 4$; the coefficient of $f(x_j, y_j)$ is $2 \cdot 4 = 8$, and of $f(x_j, y_j)$ is $4 \cdot 4 = 16$

*SIMPSON'S RULE FOR MULTIPLE INTEGRALS

Numerical estimation of multiple integrals is a much tougher job than for an integral of a function of one variable. As the dimension of the domain of f increases, the number of points required for accurate estimation increases exponentially.

We describe Simpson's rule for estimating $\int_a^b \int_c^d f(x, y)\,dx\,dy$. An analogous method is valid for triple integrals. In Fig. 18.14, we show a rectangular region $a \le x \le b$, $c \le y \le d$. We divide $[a, b]$ into m equal subintervals and $[c, d]$ into n equal subintervals, where m and n are both even. In Fig. 18.14, $m = 8$ and $n = 4$. This gives us a natural grid on the rectangle. At the points along the bottom and left edges of the rectangle, we write the coefficients $1, 4, 2, 4, 2, 4, 2, \ldots, 4, 1$ of function values that appear in Simpson's rule for a function of one variable. For two variables, we follow these steps.

Step 1 Find $f(x_i, y_i)$ at each of the $(m + 1)(n + 1)$ grid points in the rectangle.

Step 2 Multiply each function value $f(x_i, y_i)$ in Step 1 by the *product* of the coefficient at the left of (x_i, y_i) and the one underneath (x_i, y_i), as illustrated in Fig. 18.14.

Step 3 Add the results in Step 2.

Step 4 Multiply the sum in Step 3 by

$$\frac{b - a}{3m} \cdot \frac{d - c}{3n}$$

to obtain the estimate for the integral.

Figure 18.15 Areas of vertical surfaces estimated using Simpson's rule on grid lines parallel to the x-axis.

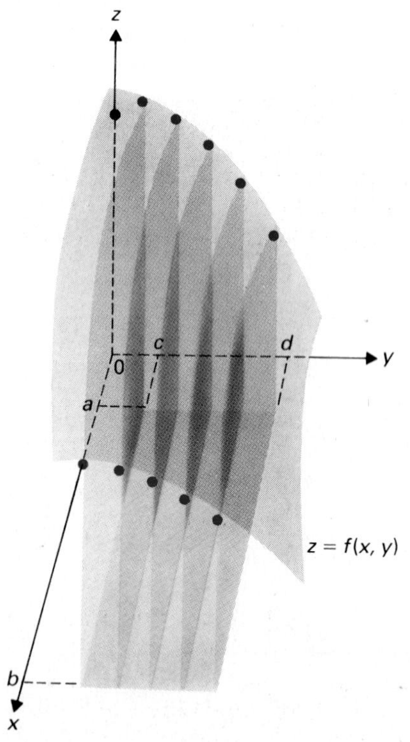

It is not difficult to see why the above steps give an estimate for the integral. If we add the products

$$\text{Bottom coefficient} \cdot f(x_i, y_i)$$

separately for each row of the grid and multiply each such sum by $(b - a)/(3m)$, we obtain our old Simpson's rule estimate for the areas of the vertical surfaces up to $z = f(x, y)$, as shown in Fig. 18.15. To estimate the volume, we then use our old Simpson's rule in the y-direction. That is, we add these surface areas, multiplied by the coefficients at the left of the rectangle, and then multiply the sum by $(d - c)/(3n)$. We can easily convince ourselves that this is equivalent to the steps outlined above.

EXAMPLE 9 Use a calculator and Simpson's rule with $m = n = 2$ to estimate

$$\int_0^1 \int_0^1 (x + 1)^y\,dx\,dy.$$

Solution Let $f(x, y) = (x + 1)^y$. Using Fig. 18.16 and the steps outlined

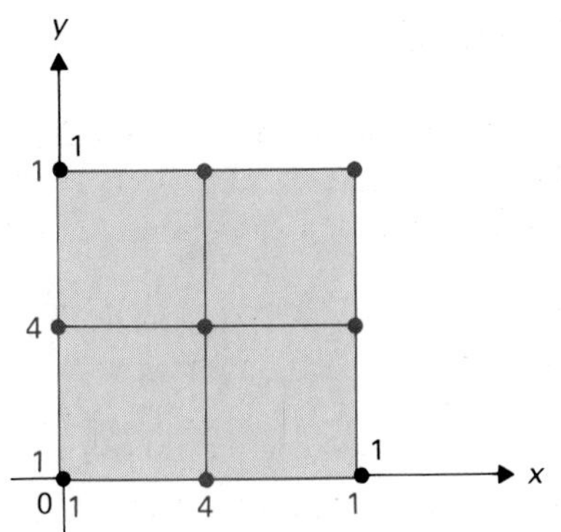

Figure 18.16 Grid points and coefficients for Simpson's rule with $m = n = 2$.

above, we obtain the estimate

$$\int_0^1 \int_0^1 (x + 1)^y \, dx \, dy \approx \frac{1}{6} \cdot \frac{1}{6} \left[f(0, 0) + 4f\left(\frac{1}{2}, 0\right) + f(1, 0) + 4f\left(0, \frac{1}{2}\right) + 16f\left(\frac{1}{2}, \frac{1}{2}\right) + 4f\left(1, \frac{1}{2}\right) + f(0, 1) + 4f\left(\frac{1}{2}, 1\right) + f(1, 1) \right] \approx 1.229243672.$$

We also estimated using Simpson's rule with $m = n = 4$ and obtained as estimate 1.229271648. Once again, it appears that Simpson's rule gives quite accurate estimates for low values of m and n. □

A better way to estimate the integral in Example 9 is to carry out the first integration, which is feasible, and then use Simpson's rule for a function of a single variable the rest of the way. We do this in the next example.

EXAMPLE 10 Estimate $\int_0^1 \int_0^1 (x + 1)^y \, dx \, dy$ by the method just described.

Solution We have

$$\int_0^1 \int_0^1 (x + 1)^y \, dx \, dy = \int_0^1 \left. \frac{(x + 1)^{y+1}}{y + 1} \right]_0^1 dy = \int_0^1 \frac{2^{y+1} - 1}{y + 1} \, dy.$$

Taking up our calculator, we use Simpson's rule with $n = 20$ for this integral and obtain as estimate 1.229274137. With $n = 40$, we obtain as estimate 1.229274135. This indicates that our estimate in Example 9 with $m = n = 2$ was accurate to four significant figures, and the estimate with $n = m = 4$ was accurate to six significant figures. □

SUMMARY

Let $f(x, y)$ be continuous in the rectangular region R, where $a \le x \le b$ and $c \le y \le d$. Partition R into n^2 subrectangles by subdividing both intervals $[a, b]$ and $[c, d]$ into n subintervals of equal length. Number the subrectangles from 1 to n^2, and let (x_i, y_i) be a point in the ith subrectangle.

1. A Riemann sum

$$\mathscr{S}_n = \frac{(b - a)(d - c)}{n^2} \cdot \sum_{i=1}^{n^2} f(x_i, y_i)$$

2. $\iint_R f(x, y) \, dx \, dy = \lim_{n \to \infty} \mathscr{S}_n$

3. Analogues of numbers 1 and 2 are valid for continuous $f(x, y, z)$ and a rectangular box R in space.

4. If $f(x, y) \ge 0$ in R, then $\iint_R f(x, y) \, dx \, dy$ is the volume of the solid with base R, vertical plane sides, and a portion of the graph of $z = f(x, y)$ as top.

5. $\int_c^d \int_a^b f(x, y) \, dx \, dy$ and $\int_r^s \int_c^d \int_a^b f(x, y, z) \, dx \, dy \, dz$ are computed from the "inside outward." For example,

$$\int_c^d \int_a^b f(x, y) \, dx \, dy = \int_c^d \left(\int_{x=a}^b f(x, y) \, dx \right) dy.$$

6. The value of these iterated integrals does not depend on the order of integration. For example,

$$\int_r^s \int_c^d \int_a^b f(x, y, z)\,dx\,dy\,dz = \int_a^b \int_r^s \int_c^d f(x, y, z)\,dy\,dz\,dx.$$

7. *Fundamental theorem:*

$$\iint_R f(x, y)\,dx\,dy = \int_c^d \int_a^b f(x, y)\,dx\,dy,$$

and

$$\iiint_R f(x, y, z)\,dx\,dy\,dz$$

can also be evaluated as an iterated integral.

8. If an iterated integral cannot be evaluated by integration, it may be estimated using a midpoint Riemann sum $\mathcal{S}_n$. If only the last step of integration cannot be done, Simpson's rule can be used at that stage to obtain an estimate.

*9. If none of the integration steps can be done for an iterated integral, Simpson's rule for multiple integrals, explained on pages 794–795, can be used to estimate the integral.

EXERCISES

1. We have treated only a very restrictive case of multiple integrals of continuous functions in this section. Can you guess the restriction to which we are referring?
2. Sketch figures similar to Figs. 18.12 and 18.13 to illustrate geometrically that $\int_a^b \int_c^d f(x, y)\,dy\,dx = \iint_R f(x, y)\,dx\,dy$, where R is the rectangle $a \le x \le b$, $c \le y \le d$ and f is continuous over R.
3. We have defined iterated integrals $\int_c^d \int_a^b f(x, y)\,dx\,dy$ and $\int_a^b \int_c^d f(x, y)\,dy\,dx$ over R where $a \le x \le b$, $c \le y \le d$ for f continuous on R. We could also consider

$$\int_d^c \int_a^b f(x, y)\,dx\,dy,$$

where, again, the integral is to be computed from the "inside outward." List all such iterated integrals of f arising from R, and compare the values of these integrals. [*Hint:* There are eight of them.]

4. Let f be a continuous function of three variables with domain containing a rectangular box $a_1 \le x \le b_1$, $a_2 \le y \le b_2$, $a_3 \le z \le b_3$.
   a) Show that there are six possible "orders of integration" for an iterated integral, where each integral has some a_i for lower limit and b_i for upper limit.
   b) How many iterated integrals can you form for f over the box if the a_i's are not restricted to lower limits and the b_i's are not restricted to upper limits? (See Exercise 3.)

In Exercises 5 through 12, estimate the given integral by computing the indicated midpoint Riemann sum $\mathcal{S}_n$.

5. $\int_1^5 \int_0^2 (x + y)\,dx\,dy$ using $\mathcal{S}_2$
6. $\int_0^4 \int_{-1}^1 xy\,dx\,dy$ using $\mathcal{S}_2$
7. $\int_0^4 \int_1^3 \frac{x}{y}\,dy\,dx$ using $\mathcal{S}_2$
8. $\int_0^2 \int_1^5 \frac{y}{x + y}\,dy\,dx$ using $\mathcal{S}_2$
9. $\int_0^4 \int_1^5 \frac{xy}{x + y}\,dx\,dy$ using $\mathcal{S}_4$
10. $\int_1^9 \int_0^8 (x + xy)\,dy\,dx$ using $\mathcal{S}_4$
11. $\int_1^3 \int_0^2 \int_0^2 xyz^2\,dx\,dy\,dz$ using $\mathcal{S}_2$
12. $\int_1^3 \int_1^3 \int_1^3 \frac{1}{xyz}\,dx\,dy\,dz$ using $\mathcal{S}_2$

In Exercises 13 through 32, compute the given iterated integral.

13. $\int_1^5 \int_0^2 (x + y)\,dx\,dy$
14. $\int_0^4 \int_{-1}^1 xy\,dx\,dy$

15. $\int_0^4 \int_1^3 \frac{x}{y}\, dy\, dx$

16. $\int_1^9 \int_0^8 (x + xy)\, dy\, dx$

17. $\int_1^4 \int_0^2 (x + y^2)\, dx\, dy$

18. $\int_1^3 \int_{-1}^2 x^2 y\, dy\, dx$

19. $\int_0^{\pi} \int_0^{\pi/2} x \sin y\, dx\, dy$

20. $\int_0^2 \int_0^{\pi/2} x \sin y\, dy\, dx$

21. $\int_0^2 \int_0^{\pi} x \sin^2 y\, dy\, dx$

22. $\int_1^{e^2} \int_1^e \ln(xy)\, dx\, dy$

23. $\int_1^3 \int_0^1 xye^x\, dx\, dy$

24. $\int_0^1 \int_1^5 \frac{y+1}{e^x}\, dy\, dx$

25. $\int_0^1 \int_0^1 \frac{y}{1 + x^2y^2}\, dx\, dy$

26. $\int_0^{\sqrt{2}} \int_0^1 \frac{x}{\sqrt{4 - x^2y^2}}\, dy\, dx$

27. $\int_0^1 \int_2^3 \int_{-1}^4 xy^2z\, dx\, dy\, dz$

28. $\int_0^1 \int_{-1}^1 \int_1^2 (x^2 + yz)\, dz\, dx\, dy$

29. $\int_{-1}^3 \int_0^{\ln 2} \int_0^4 xze^y\, dx\, dy\, dz$

30. $\int_0^2 \int_0^1 \int_0^1 xyz\sqrt{2 - x^2 - y^2}\, dx\, dy\, dz$

31. $\int_0^{\pi} \int_0^1 \int_0^1 yz^2 \sin(xyz)\, dx\, dy\, dz$

32. $\int_0^1 \int_0^1 \int_0^1 xy^3e^{xyz}\, dz\, dx\, dy$

In Exercises 33 through 40, estimate the iterated integral by integrating as far as you can and then using Simpson's rule with a calculator or computer to estimate the final integral.

33. $\int_0^2 \int_1^3 x^y\, dx\, dy$

34. $\int_0^{\sqrt{\pi}} \int_0^1 x^2 \cos(x^2y)\, dy\, dx$

35. $\int_0^{\pi} \int_0^1 \frac{x \sin x}{1 + x^2y^2}\, dy\, dx$

36. $\int_0^1 \int_0^2 y^2e^{x+y^2}\, dx\, dy$

37. $\int_1^3 \int_1^2 (xy)^x\, dy\, dx$

38. $\int_0^1 \int_0^1 \int_0^1 e^{x+y+z^2}\, dx\, dy\, dz$

39. $\int_1^3 \int_1^2 \int_0^1 (xy)^z\, dx\, dy\, dz$

40. $\int_1^3 \int_1^2 \int_0^1 (x + y)^z\, dx\, dy\, dz$

*In Exercises 41 through 44, use a calculator or computer and Simpson's rule for double integrals with the given values of m and n to estimate the given integral.

*41. $\int_0^1 \int_0^1 e^{x^2+y^2}\, dx\, dy,\ m = n = 2$

*42. Repeat Exercise 41 for $m = n = 4$.

*43. $\int_2^4 \int_1^2 \frac{x^4 + y^4}{1 + 2\sqrt{x} + 3\sqrt{y}}\, dx\, dy$, $m = 2$ intervals for $[1, 2]$, $n = 4$ intervals for $[2, 4]$

*44. Repeat Exercise 43 using $m = 4$ intervals for $[1, 2]$ and $n = 8$ intervals for $[2, 4]$.

18.2 INTEGRALS OVER MORE GENERAL REGIONS

INTEGRALS AS LIMITS OF SUMS

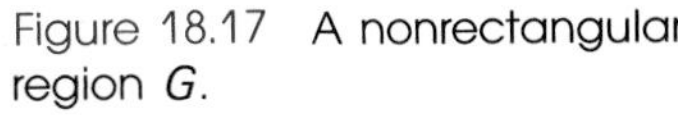

Figure 18.17 A nonrectangular region G.

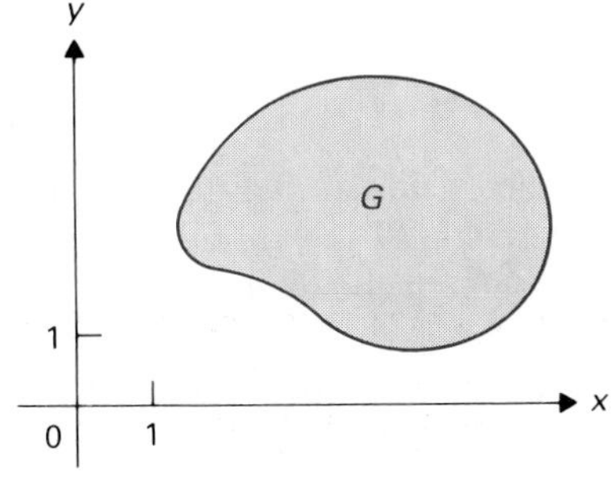

A region G in the plane is **bounded** if it is contained in some sufficiently large rectangle. Naively, bounded regions are those that do not extend to infinity in any direction. We will consider "nice" bounded regions, like the region G in Fig. 18.17, where we have a natural idea of the *boundary* (bounding curves) of the region and the *interior* (inside) of the region. A region is **closed** if the boundary is considered to be part of the region.

Let G be a closed, bounded region in the plane of the nice type just mentioned. We indicate how to regard

$$\iint_G f(x, y)\, dx\, dy$$

as a limit of sums. We need only show how to perform Step 1 of the four-step outline in Section 18.1.

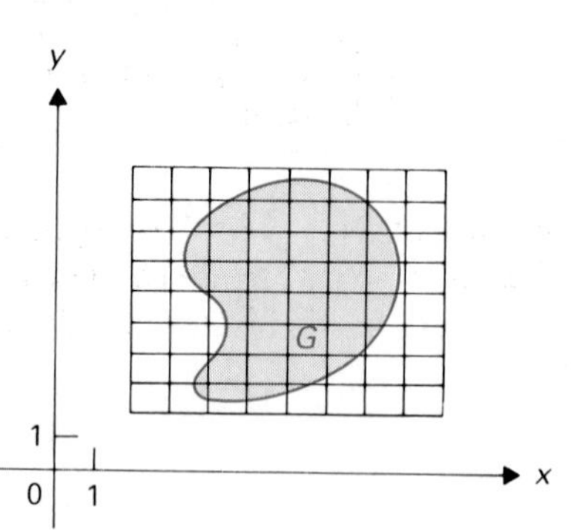

Figure 18.18 The grid on the rectangle ($n = 8$) gives a partition of G.

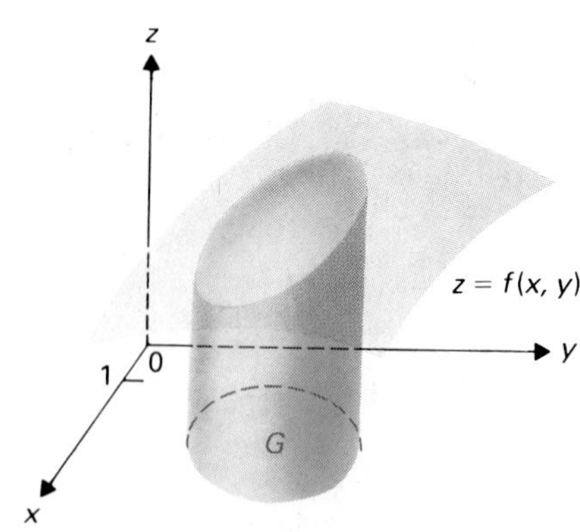

Figure 18.19 $\iint_G f(x, y)\,dx\,dy$ is the volume of the red-shaded solid.

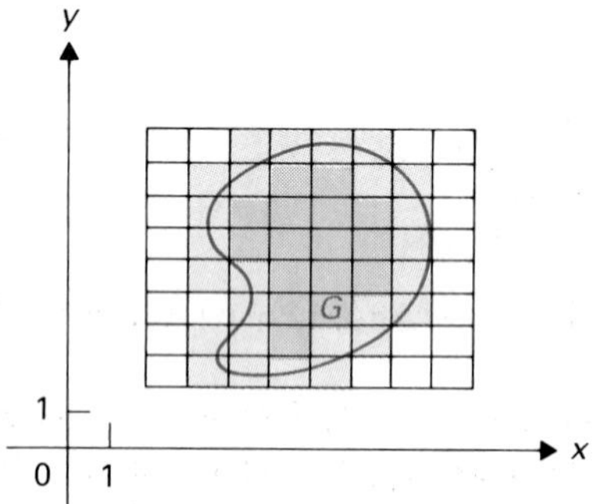

Figure 18.20 $\iint_G f(x, y)\,dx\,dy$ equals the limit as $n \to \infty$ of sums taken over just the red-shaded rectangles.

Step 1: Partition G into subregions of small size. Since G is bounded, we can enclose it in some rectangle with sides parallel to the coordinate axes, as shown in Fig. 18.18. We subdivide this rectangle into n^2 subrectangles precisely as in Section 18.1. The grid on the rectangle then chops the region G into small pieces, shaded red in Fig. 18.18. This completes Step 1.

For Step 2, multiply the area of each subregion of G by $f(x, y)$ for some point (x, y) in the subregion. In Step 3, we add these products, obtaining an approximation to $\iint_G f(x, y)\,dx\,dy$. The **definite integral** $\iint_G f(x, y)\,dx\,dy$ is defined as the limit of these approximations as $n \to \infty$. It is clear that for $f(x, y) \geq 0$, the integral $\iint_G f(x, y)\,dx\,dy$ is equal to the volume of the solid having base G and part of the surface $z = f(x, y)$ as top, as in Fig. 18.19.

Accurate estimation of an integral using such sums is generally not feasible, since some subregions in these partitions of G are not rectangular, and their areas may not be known. It can be shown that if the total length of the curves bounding G is finite, then the integral is actually equal to the limit of sums taken over only those subregions of G that are entire rectangles. Such subregions are shaded red in Fig. 18.20. Specifically, it can be shown that if the boundary has finite length, the sum of the areas of the rectangles containing portions of the boundary, shaded black in Fig. 18.20, approaches zero as $n \to \infty$. Thus omitting them does not affect the limit of the approximating sums. However, for small values of n that we like to use for approximations, as in Fig. 18.20, omission of these subregions can ruin the accuracy of our estimate. We will not bother to compute such sums as estimates. At the end of this section, we show how to use Simpson's rule if it is necessary to estimate integrals over regions that are not rectangular.

Figure 18.21 Upper and lower boundary curves of G.

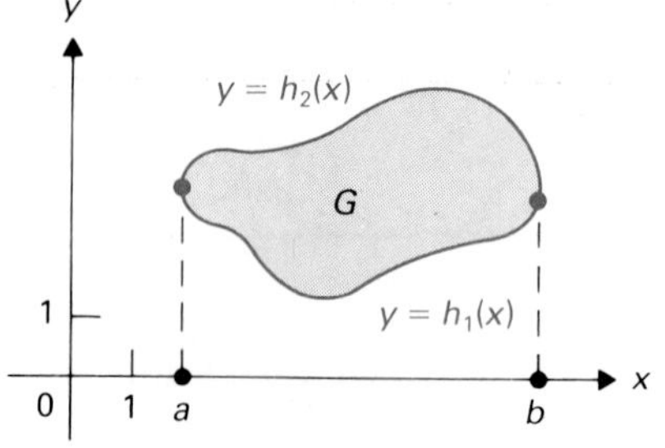

ITERATED INTEGRALS OVER REGIONS

The integrals just defined are usually computed using an iterated integral, much as in Section 18.1. Consider the plane region G shown in Fig. 18.21, where $a \leq x \leq b$ and where the "lower" portion of the boundary is the curve $y = h_1(x)$ and the "top" portion is the curve $y = h_2(x)$. We assume that h_1 and h_2 are continuous functions. If f is continuous on G, we form the iterated integral

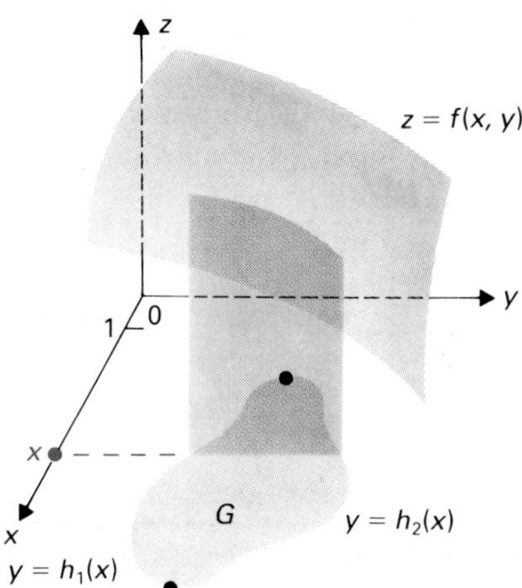

Figure 18.22 $\int_{h_1(x)}^{h_2(x)} f(x, y)\,dy$ is the area of the color-shaded surface.

$$\boxed{\int_a^b \int_{h_1(x)}^{h_2(x)} f(x, y)\,dy\,dx = \int_a^b \left(\int_{h_1(x)}^{h_2(x)} f(x, y)\,dy\right) dx.} \quad (1)$$

The computation of Eq. (1) is again to be made from the "inside outward," as indicated by the parentheses. The first integration is performed with respect to y only, and the first limits depend on x. For a particular value x, the integral

$$\int_{h_1(x)}^{h_2(x)} f(x_i, y)\,dy$$

gives us the area of the vertical color-shaded surface in Fig. 18.22. This area is then multiplied by dy, giving the volume of the slab shaded in color in Fig. 18.23. Integrating from a to b gives us the volume of the solid with base G and top a portion of the graph of $z = f(x, y)$, like the solid shown in Fig. 18.19. This same volume is given by $\iint_G f(x, y)\,dx\,dy$. We therefore have a "fundamental theorem for computation"

$$\boxed{\iint_G f(x, y)\,dx\,dy = \int_a^b \int_{h_1(x)}^{h_2(x)} f(x, y)\,dy\,dx,} \quad (2)$$

which is the analogue of Theorem 18.1 in the last section, but for more general regions than a rectangle.

We could also integrate in the reverse order, in which case we consider the "left" and "right" boundary curves of G to be given by $x = k_1(y)$ and $x = k_2(y)$ for $c \le y \le d$, as shown in Fig. 18.24. Our integral then takes the form

$$\boxed{\int_c^d \int_{k_1(y)}^{k_2(y)} f(x, y)\,dx\,dy = \int_c^d \left(\int_{k_1(y)}^{k_2(y)} f(x, y)\,dx\right) dy.}$$

Figure 18.23 $\int_{h_1(x)}^{h_2(x)} f(x, y)\,dy\,dx$ is the volume of the slab.

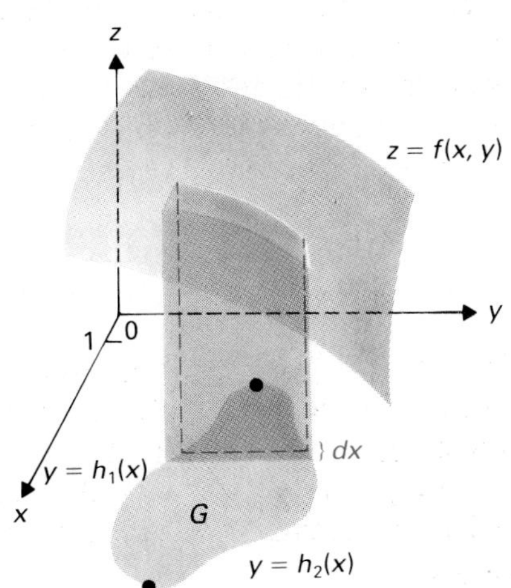

Figure 18.24 Left and right boundary curves of G.

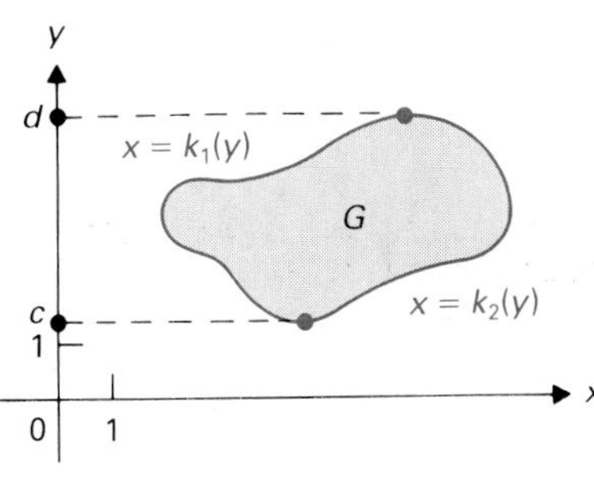

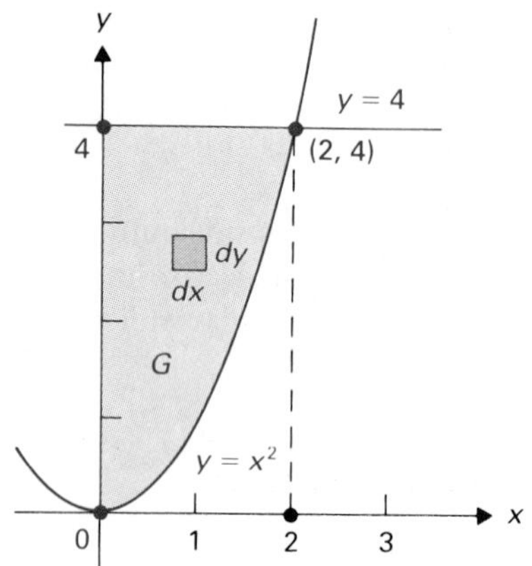

Figure 18.25 The region G in the first quadrant bounded by $y = x^2$, $y = 4$, and the y-axis.

Analogous iterated integrals can be formed for continuous functions of three variables over a suitable region in space.

In forming an iterated integral to compute $\iint_G f(x, y)\,dx\,dy$, *draw a sketch of the region* G. The appropriate limits for the iterated integral are found by studying the sketch.

EXAMPLE 1 Let G be the plane region in the first quadrant bounded by $y = x^2$, $y = 4$, and the y-axis. Find $\iint_G x^2y\,dx\,dy$.

Solution The region G is shown in Fig. 18.25, where we also draw a differential rectangle of area $dA = dx\,dy$. We show how to find and compute appropriate iterated integrals for both orders of integration. In both cases,

we think of summing contributions $f(x, y)\,dx\,dy$ over G.

Case 1 dx dy-order If we integrate first with respect to x, holding y constant, we are adding contributions $x^2y\,dx\,dy$ in the x-direction, along the strip shown in Fig. 18.26. In terms of its y-height, the left end of the strip has x-coordinate zero and the right end has x-coordinate $x = \sqrt{y}$. Thus we arrive at

$$\int_0^{\sqrt{y}} x^2y\,dx\,dy$$

as the total contribution along the strip in Fig. 18.26. We then sum such strip contributions in the y-direction to sweep out the entire region G. The smallest y-value is zero and the largest is 4. Thus we obtain the iterated integral

$$\int_0^4 \int_0^{\sqrt{y}} x^2y\,dx\,dy = \int_0^4 \frac{x^3}{3}y\bigg|_{x=0}^{\sqrt{y}}\,dy$$

$$= \int_0^4 \frac{y^{5/2}}{3}\,dy = \frac{1}{3}\cdot\frac{2}{7}y^{7/2}\bigg]_0^4 = \frac{2}{21}4^{7/2} = \frac{256}{21}.$$

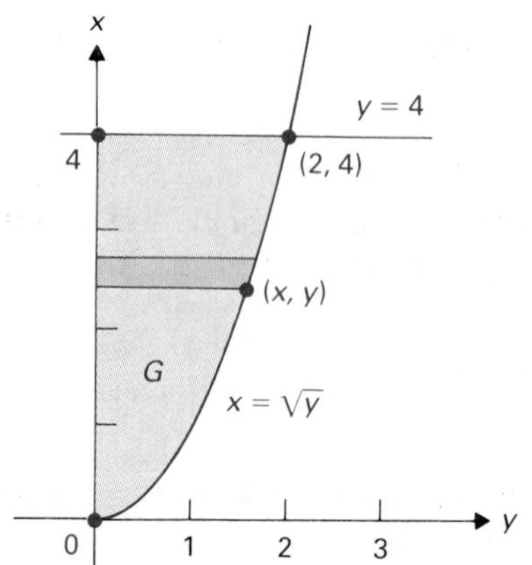

Figure 18.26 $\int_0^{\sqrt{y}} x^2y\,dx\,dy$ sums the contributions $x^2y\,dx\,dy$ on this strip.

Case 2 dy dx-order Integrating first with respect to y, we sum contributions $x^2y\,dy\,dx$ along the vertical strip shown in Fig. 18.27. In terms of the left–right position x of the strip, the y-value at the bottom is $y = x^2$ and the y-value at the top is $y = 4$. We are led to

$$\int_{x^2}^4 x^2y\,dy\,dx.$$

We then add the contributions of these strips in the x-direction. Since the smallest x-value is zero and the largest is 2, we obtain the integral

$$\int_0^2 \int_{x^2}^4 x^2y\,dy\,dx = \int_0^2 \left(\int_{x^2}^4 x^2y\,dy\right)dx = \int_0^2 x^2\frac{y^2}{2}\bigg]_{y=x^2}^{y=4} dx$$

$$= \int_0^2 \left(8x^2 - \frac{1}{2}x^6\right)dx$$

$$= \left(\frac{8}{3}x^3 - \frac{1}{14}x^7\right)\bigg]_0^2$$

$$= \frac{64}{3} - \frac{64}{7} = \frac{256}{21}. \quad \square$$

Figure 18.27 $\int_{x^2}^4 x^2y\,dy\,dx$ sums the contributions $x^2y\,dx\,dy$ over this strip.

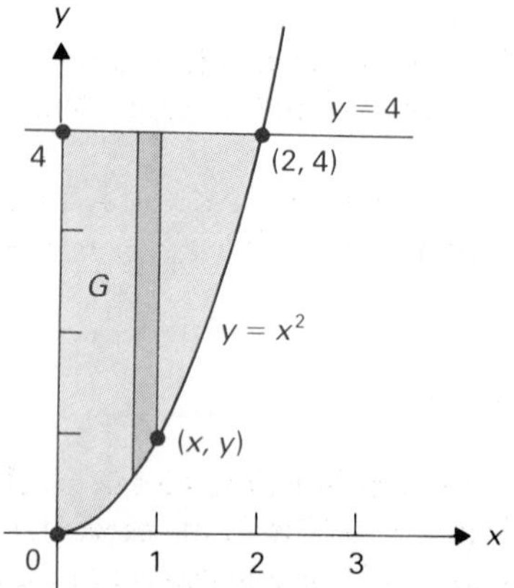

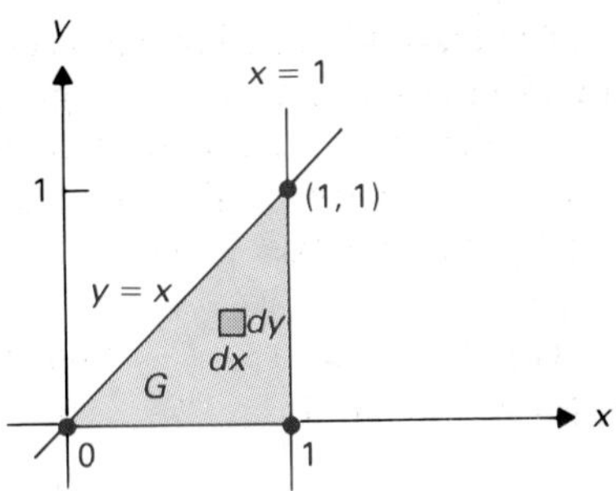

Figure 18.28 Differential area $dA = dx\,dy$ in G.

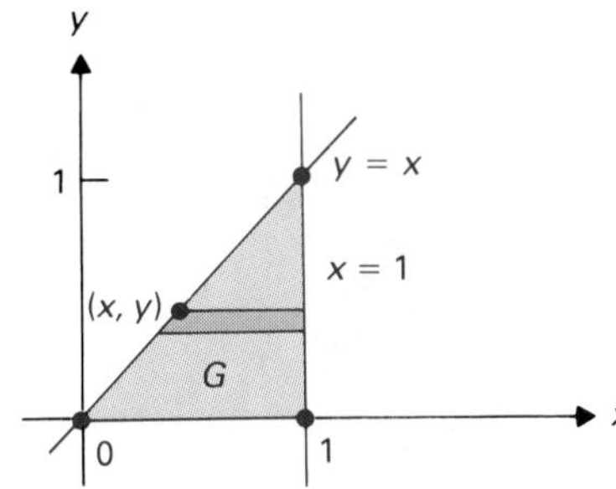

Figure 18.29 $\int_y^1 e^{x^2}\,dx\,dy$ sums $e^{x^2}\,dx\,dy$ over this strip.

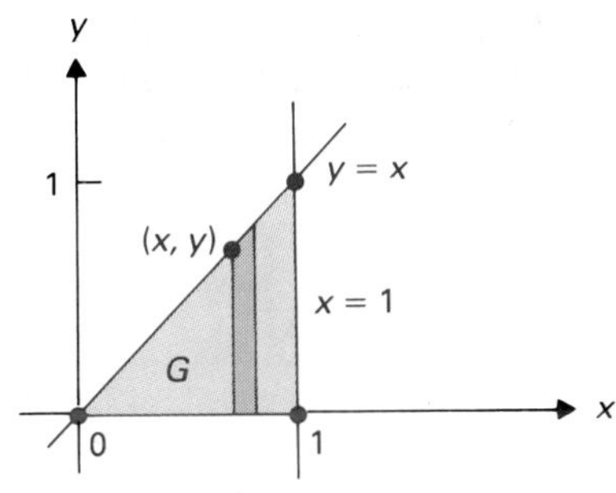

Figure 18.30 $\int_0^x e^{x^2}\,dy\,dx$ sums $e^{x^2}\,dx\,dy$ over this strip.

When we are evaluating $\iint_G f(x, y)\,dx\,dy$ using an iterated integral, one order of integration might be impossible, while the other might be easy. We give an example.

EXAMPLE 2 Let G be the region bounded by $y = x$, $x = 1$, and the x-axis. Find $\iint_G e^{x^2}\,dx\,dy$.

Solution The region G is shown in Fig. 18.28. If we integrate first in the x-direction, over the horizontal strip in Fig. 18.29, then an argument analogous to that in Example 1 shows that the appropriate iterated integral is

$$\int_0^1 \int_y^1 e^{x^2}\,dx\,dy.$$

However, we can't find $\int e^{x^2}\,dx$.

The other order of integration, with respect to y first, sums along the vertical strip in Fig. 18.30. This leads to the integral

$$\int_0^1 \int_0^x e^{x^2}\,dy\,dx = \int_0^1 ye^{x^2}\Big]_{y=0}^{x} dx = \int_0^1 xe^{x^2}\,dx = \frac{1}{2}e^{x^2}\Big]_0^1 = \frac{1}{2}(e - 1). \quad \square$$

We give an example involving a triple iterated integral.

Figure 18.31 Differential volume $dV = dx\,dy\,dz$ in the region G.

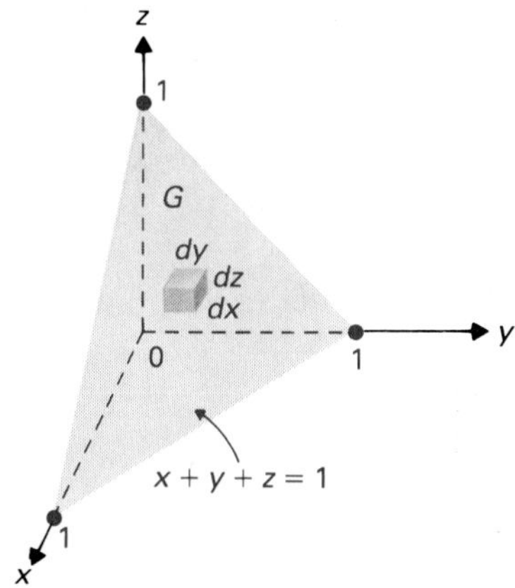

EXAMPLE 3 Let G be the region in the first octant bounded by the coordinate planes and the plane $x + y + z = 1$. Find $\iiint_G x\,dx\,dy\,dz$.

Solution As always, we draw a sketch of G, shown in Fig. 18.31. The figure also shows a differential box of volume $dV = dx\,dy\,dz$ in the region. We will form a $dz\,dx\,dy$-order iterated integral.

Integration with respect to z adds the contributions $x\,dz\,dx\,dy$ up the column in Fig. 18.32. At the bottom of the column, $z = 0$, while $z = 1 - x - y$ at the top. This leads to

$$\int_0^{1-x-y} x\,dz\,dx\,dy.$$

We now add these column contributions in the x-direction, along the slab shown in Fig. 18.33. The back of the slab is at $x = 0$. The front comes out to the line $x + y = 1$ in the x,y-plane. The equation of this line was obtained by

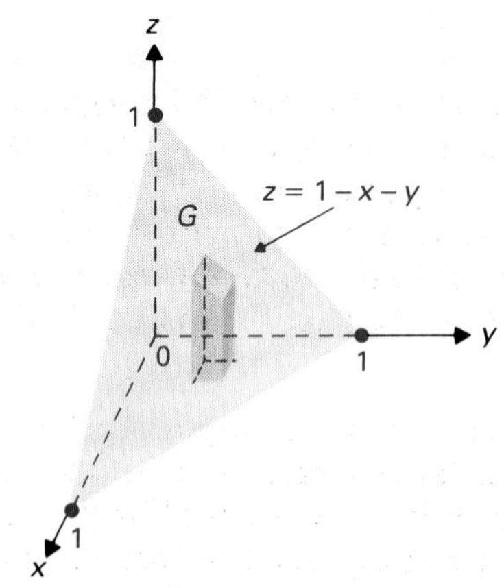

Figure 18.32
$\int_0^{1-x-y} x\,dz\,dx\,dy$
sums $x\,dx\,dy\,dz$ over this column.

setting $z = 0$ in the equation $x + y + z = 1$ of the front plane bounding the region. Thus $x = 1 - y$ at the front of the slab, and we are led to

$$\int_0^{1-y}\int_0^{1-x-y} x\,dz\,dx\,dy.$$

Finally, we add the contributions of these slabs in the y-direction, from the smallest y-value of zero to the largest y-value of 1. We obtain the iterated integral

$$\begin{aligned}
\int_0^1\int_0^{1-y}\int_0^{1-x-y} x\,dz\,dx\,dy &= \int_0^1\int_0^{1-y} xz\Big]_{z=0}^{1-x-y} dx\,dy \\
&= \int_0^1\int_0^{1-y} (x - x^2 - xy)\,dx\,dy \\
&= \int_0^1\int_0^{1-y} [x(1-y) - x^2]\,dx\,dy \\
&= \int_0^1 \left(\frac{x^2}{2}(1-y) - \frac{x^3}{3}\right)\Big]_{x=0}^{1-y} dy \\
&= \int_0^1 \left[\frac{(1-y)^3}{2} - \frac{(1-y)^3}{3}\right] dy \\
&= \int_0^1 \frac{1}{6}(1-y)^3\,dy \\
&= -\frac{1}{24}(1-y)^4\Big]_0^1 = 0 + \frac{1}{24} = \frac{1}{24}. \quad \square
\end{aligned}$$

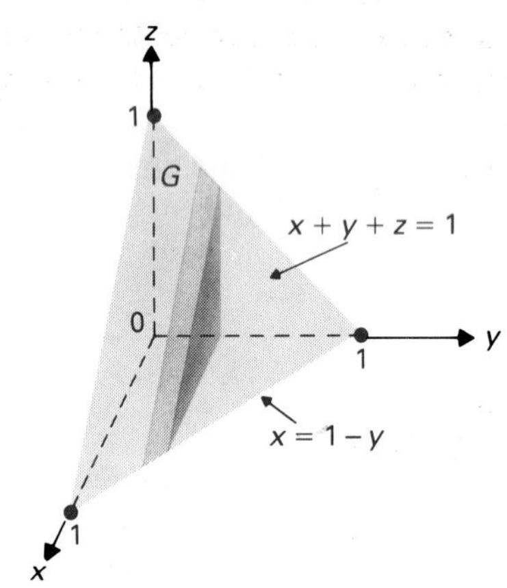

Figure 18.33
$\int_0^{1-y}\int_0^{1-x-y} x\,dz\,dx\,dy$
sums $x\,dx\,dy\,dz$ over this slab.

CHANGING THE ORDER OF INTEGRATION

It is important to be able to change the order of integration in a given iterated integral. An outline of a procedure by which this may be done for double integrals is as follows.

Step 1 *Draw a sketch* and shade the region of the plane over which integration takes place. Draw a little "differential rectangle" of dimensions dx by dy in the shaded region.

Step 2 Convert your given iterated integral into one with the order of integration reversed *by looking at your sketch* and writing down the appropriate limits. The limits on the inside integral sign are functions of the remaining variable of integration, while the limits on the outside integral sign are always constants.

Similar steps may be followed to change the order of integration in a triple iterated integral.

EXAMPLE 4 Reverse the order of integration for the integral

$$\int_0^2\int_{x^3}^{4x} x^2 y\,dy\,dx.$$

Solution

Step 1 Starting with the limits on the inside integral, the first integration with respect to y goes from $y = x^3$ to $y = 4x$, so we sketch these curves in Fig. 18.34. Since this first integration was with respect to y, we think of $y = x^3$ and $y = 4x$ as forming the bottom and top boundaries of our region. Now the final integration with respect to x goes only from $x = 0$ to $x = 2$. Thus the region is the one in the first quadrant only, shaded in Fig. 18.34.

Step 2 To reverse the order of integration, we wish to integrate first with respect to x. Pushing our differential rectangle as far to the left (negative x-direction) as it will go, it is always stopped by the line $y = 4x$. Since these inside limits must be expressed as functions of y, we write $y = x^3$ as $x = \sqrt[3]{y}$ and $y = 4x$ as $x = y/4$. This leads to

$$\int_{y/4}^{\sqrt[3]{y}} x^2 y \, dx \, dy.$$

This integral corresponds to adding the contributions $x^2 y \, dx \, dy$ over the horizontal strip shown in Fig. 18.35. Now we must add the contributions of these strips from a strip at the bottom of the region, where $y = 0$, to a strip at the top, where $y = 8$. Thus we arrive at

$$\int_0^8 \int_{y/4}^{\sqrt[3]{y}} x^2 y \, dx \, dy$$

as the desired integral. □

EXAMPLE 5 Convert

$$\int_0^2 \int_{(3x/2)-3}^{0} \int_0^{2-x+(2y/3)} xyz^2 \, dz \, dy \, dx$$

to the order $\iiint xyz^2 \, dx \, dz \, dy$.

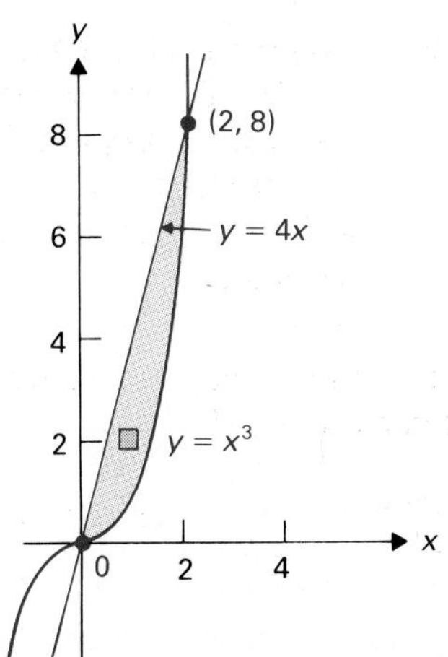

Figure 18.34 Differential area $dA = dx\, dy$.

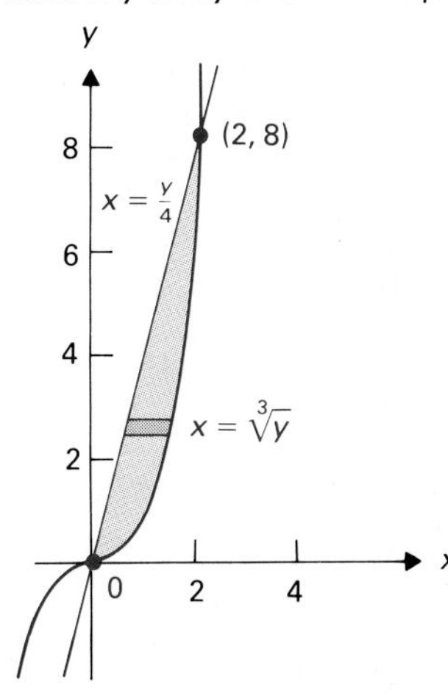

Figure 18.35 $\int_{y/4}^{\sqrt[3]{y}} x^2 y \, dx \, dy$ sums $x^2 y \, dx \, dy$ over this strip.

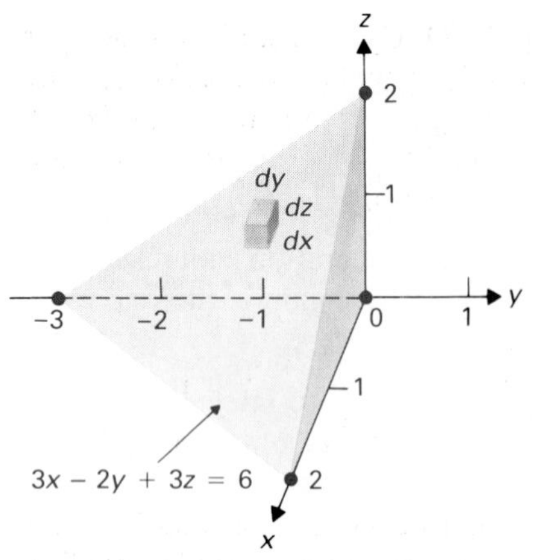

Figure 18.36 Differential volume $dV = dx\,dy\,dz$.

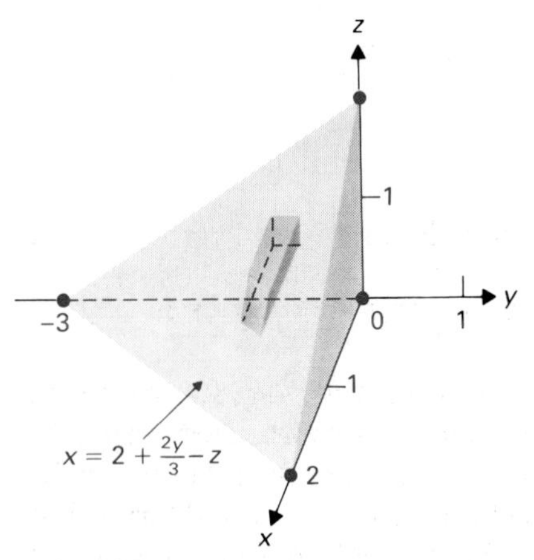

Figure 18.37 $\int_0^{2+(2y/3)-z} xyz^2\,dx\,dz\,dy$ sums $xyz^2\,dx\,dy\,dz$ over this column.

Solution

Step 1 The inside limits with respect to z show that the bottom of the region is the plane $z = 0$ and the top is the plane $z = 2 - x + (2y/3)$, or $3x - 2y + 3z = 6$. We sketch these planes in Fig. 18.36. The remaining limits of integration then show that the region of integration is the tetrahedron shaded in the figure.

Step 2 In the new order, we wish to integrate first in the x-direction, from $x = 0$ out to the plane $3x - 2y + 3z = 6$, where $x = 2 + (2y/3) - z$. Thus we start with

$$\int_0^{2+(2y/3)-z} xyz^2\,dx\,dz\,dy$$

corresponding to the horizontal box in Fig. 18.37. The next z-limits may be found from the back triangle of the region, in the y,z-plane, which has as base the line where $z = 0$ and as top the line where $-2y + 3z = 6$, or $z = 2 + (2y/3)$, obtained by setting $x = 0$ in $3x - 2y + 3z = 6$. We have arrived so far at

$$\int_0^{2+(2y/3)} \int_0^{2+(2y/3)-z} xyz^2\,dx\,dz\,dy,$$

corresponding to Fig. 18.38. Finally, the constant y-limits go from the minimum y-value of -3 to the maximum y-value of zero, so the desired integral is

$$\int_{-3}^{0} \int_0^{2+(2y/3)} \int_0^{2+(2y/3)-z} xyz^2\,dx\,dz\,dy. \quad \square$$

People often have trouble finding the appropriate limits of integration in setting up an iterated integral. Note in particular that

> the *final* limits (on the left-hand integral sign) are always constant, and other limits may be functions of only those variables with respect to which integration will be performed *later*.
>
> *Wrong:* $\displaystyle\int_y^{y^2} \int_1^3 dy\,dx, \qquad \int_1^3 \int_x^{x+5} \int_z^{z-y} yz^2\,dx\,dy\,dz.$

AREAS AND VOLUMES BY MULTIPLE INTEGRATION

Let G be a closed, bounded region in the plane. It is geometrically clear that for the constant function 1, the integral

$$\iint_G 1\,dx\,dy$$

gives the area of the region. The integral is usually computed using an iterated integral. We give two examples.

EXAMPLE 6 We solve using "double integrals," an area problem we could easily have solved in Chapter 7. Find the area of the plane region in the first quadrant of the plane bounded by the curves $y = x^3$ and $y = \sqrt{x}$.

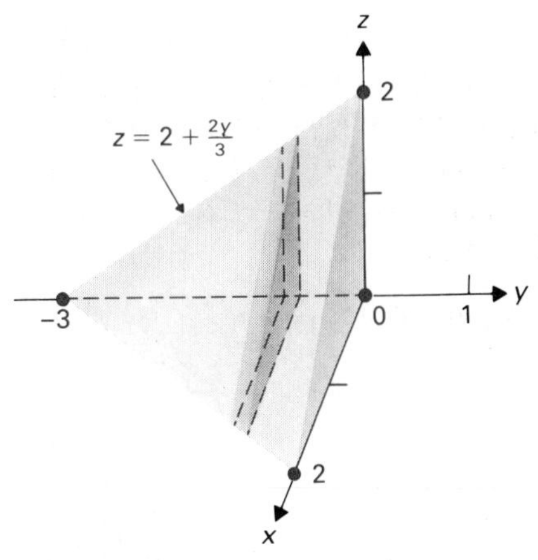

Figure 18.38

$\int_0^{2+(2y/3)} \int_0^{2+(2y/3)-z} xyz^2\, dx\, dz\, dy$ sums $xyz^2\, dx\, dy\, dz$ over this slab.

Solution The region is sketched in Fig. 18.39. Using Leibniz notation, we think of $dx\, dy = dy\, dx$ as the area of a small rectangle in the region and use an integral to add up the areas of such rectangles over the region as $dx \to 0$ and $dy \to 0$. The "lower" and "upper" bounding curves of the region are $y = x^3$ and $y = \sqrt{x}$, respectively, so our iterated integral is

$$\int_0^1 \int_{x^3}^{\sqrt{x}} dy\, dx = \int_0^1 \left(\int_{x^3}^{\sqrt{x}} dy \right) dx = \int_0^1 y \Big]_{x^3}^{\sqrt{x}} dx = \int_0^1 (\sqrt{x} - x^3)\, dx$$

$$= \left(\frac{2}{3} x^{3/2} - \frac{x^4}{4} \right) \Bigg]_0^1 = \frac{2}{3} - \frac{1}{4} = \frac{5}{12}.$$

Note that the integral $\int_0^1 (\sqrt{x} - x^3)\, dx$ that occurs in the middle of the computation above is the one we would have started with in Chapter 7 to compute the area.

If we wish to compute the iterated integral in the other order, we must find the "left" and "right" bounding curves of our region, and we obtain $x = y^2$ and $x = y^{1/3}$, respectively. The iterated integral in this order is therefore

$$\int_0^1 \int_{y^2}^{y^{1/3}} dx\, dy = \int_0^1 x \Big]_{y^2}^{y^{1/3}} dy = \int_0^1 (y^{1/3} - y^2)\, dy$$

$$= \left(\frac{3}{4} y^{4/3} - \frac{y^3}{3} \right) \Bigg]_0^1 = \frac{3}{4} - \frac{1}{3} = \frac{5}{12}.$$

Of course, we obtained the same answer. □

EXAMPLE 7 Find the volume of the region in space bounded above by the surface $z = 1 - x^2 - y^2$, on the sides by the planes $x = 0$, $y = 0$, $x + y = 1$, and below by the plane $z = 0$. The region is sketched in Fig. 18.40.

Solution We think of adding the volumes of small rectangular boxes with edges having lengths dx, dy, and dz. An attempt to find our iterated integral by integrating in the x-direction first leads to problems in the x-limits, for the "lower" boxes would have to be summed from the plane $x = 0$ forward to the plane $x = 1 - y$, while the "higher" boxes would have to be summed

Figure 18.39 Differential area $dA = dx\, dy$.

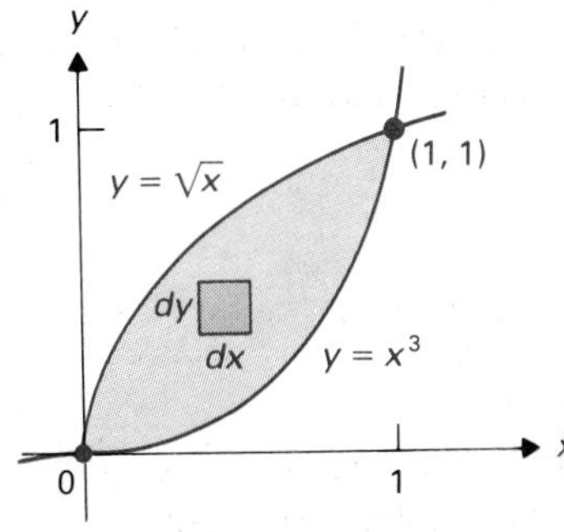

Figure 18.40 Differential volume $dV = dx\, dy\, dz$.

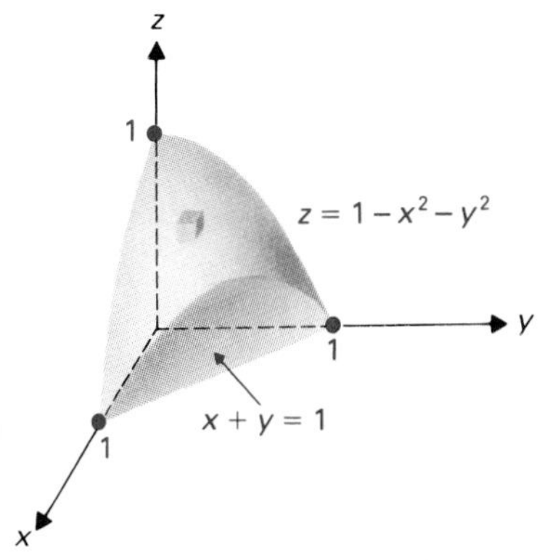

from $x = 0$ forward to the surface $x = \sqrt{1 - y^2 - z}$, as indicated in Fig. 18.41. The same problem occurs if we integrate first in the y-direction. Thus we integrate in the order z, x, y and use Fig. 18.42 to obtain the iterated integral

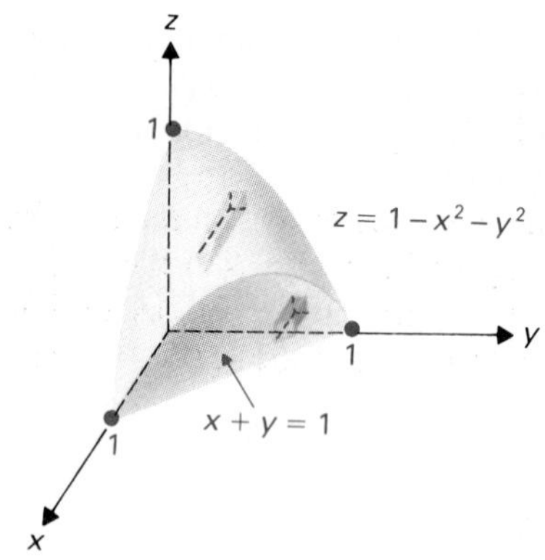

Figure 18.41 These columns in the x-direction have different front boundary surfaces.

$$\begin{aligned}
\int_0^1 \int_0^{1-y} \int_0^{1-x^2-y^2} dz\,dx\,dy &= \int_0^1 \int_0^{1-y} z\Big]_0^{1-x^2-y^2} dx\,dy \\
&= \int_0^1 \int_0^{1-y} (1 - x^2 - y^2)\,dx\,dy \\
&= \int_0^1 \left(x - \frac{x^3}{3} - y^2 x\right)\Big]_0^{1-y} dy \\
&= \int_0^1 \left((1 - y) - \frac{(1 - y)^3}{3} - y^2(1 - y)\right) dy \\
&= \int_0^1 \left(1 - y - y^2 + y^3 - \frac{(1 - y)^3}{3}\right) dy \\
&= \left(y - \frac{y^2}{2} - \frac{y^3}{3} + \frac{y^4}{4} + \frac{(1 - y)^4}{12}\right)\Big]_0^1 \\
&= 1 - \frac{1}{2} - \frac{1}{3} + \frac{1}{4} - \frac{1}{12} \\
&= 1 - \frac{6 + 4 - 3 + 1}{12} = 1 - \frac{8}{12} = \frac{1}{3}.
\end{aligned}$$

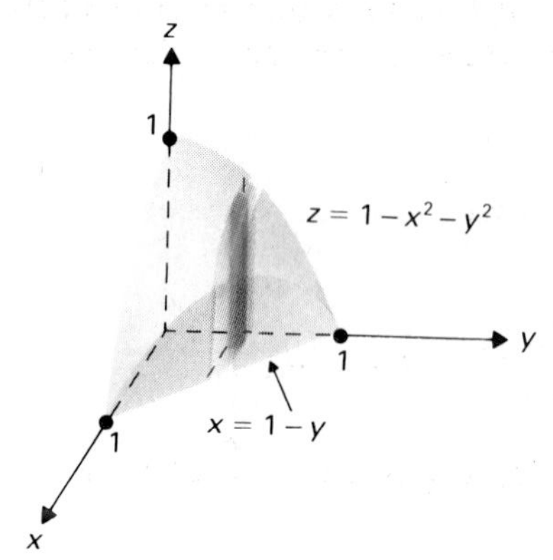

Figure 18.42
Column: $\int_0^{1-x^2-y^2} dz\,dx\,dy$;
slab: $\int_0^{1-y}\int_0^{1-x^2-y^2} dz\,dx\,dy$.

Note that we could have started with the integral

$$\int_0^1 \int_0^{1-y} (1 - x^2 - y^2)\,dx\,dy$$

that appears in the computation if we just thought of our region as lying under the surface $z = 1 - x^2 - y^2$ and over the triangular region in the x,y-plane bounded by the coordinate axes and the line $x + y = 1$. □

*SIMPSON'S RULE

We give an example that indicates how to estimate a double iterated integral if we can't perform even the first integration. An analogous method can be used for triple integrals.

Figure 18.43 Table 18.2 contains
$\int_0^{\sqrt{y_i}} \sin\sqrt{xy_i}\,dx,$
which is across each horizontal segment.

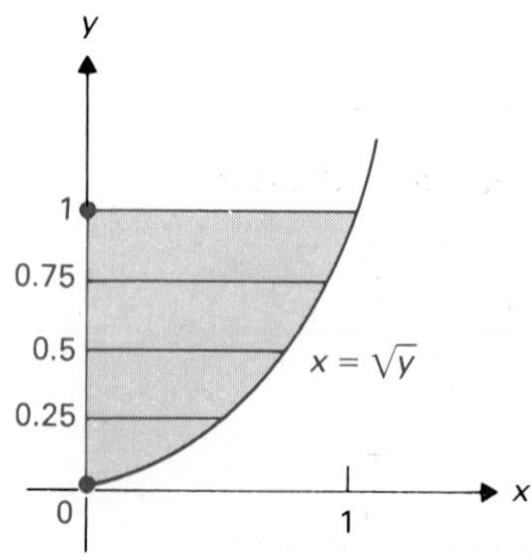

EXAMPLE 8 Use a calculator or computer to estimate

$$\int_0^1 \int_0^{\sqrt{y}} \sin(\sqrt{xy})\,dx\,dy.$$

Solution The region of integration is sketched in Fig. 18.43. We decide to use Simpson's rule first in the x-direction with $m = 4$ and then in the y-direction with $n = 4$. We have labeled the points on the y-axis corresponding to partitioning the interval [0, 1] into $n = 4$ equal subintervals. For each of these five values

$$y_i = 0, 0.25, 0.5, 0.75, 1,$$

we estimate the integral

$$\int_0^{\sqrt{y_i}} \sin(\sqrt{xy_i})\, dx$$

using Simpson's rule with $m = 4$. The results are shown in Table 18.2. We then use Simpson's rule with $n = 4$ to "integrate" the entries in the second column of the table. That is, we multiply each entry by the corresponding Simpson's coefficient in the third column, add these products, and multiply the sum by

$$\frac{1-0}{3n} = \frac{1}{12}.$$

The result is 0.2748010417, which is our estimate for the integral. We also worked it out on our calculator for $m = n = 8$ and obtained 0.2774440753. The reason this technique works was indicated in the final starred portion of Section 18.1. □

For iterated integrals where all but the final integration can be performed, we use Simpson's rule at that last stage to estimate the integral.

EXAMPLE 9 Estimate $\int_0^1 \int_{x^2}^x e^{x^2}\, dy\, dx$.

Solution We have

$$\int_0^1 \int_{x^2}^x e^{x^2}\, dy\, dx = \int_0^1 ye^{x^2}\Big]_{x^2}^x dx = \int_0^1 (xe^{x^2} - x^2e^{x^2})\, dx$$

$$= \frac{1}{2}e^{x^2}\Big]_0^1 - \int_0^1 x^2e^{x^2}\, dx = \frac{1}{2}(e-1) - \int_0^1 x^2e^{x^2}\, dx.$$

Simpson's rule with $n = 40$ shows that

$$\int_0^1 x^2e^{x^2}\, dx \approx 0.6278154418.$$

Thus we obtain the estimate

$$\frac{1}{2}(e-1) - 0.6278154418 = 0.2313254725$$

for the integral. □

Table 18.2

y_i	$\int_0^{\sqrt{y_i}} \sin(\sqrt{xy_i})\, dx$	Simpson's coefficients
0	0	1
0.25	0.1145903722	4
0.5	0.2662387945	2
0.75	0.4286623777	4
1.0	0.5921239119	1

SUMMARY

1. A bounded plane region G is one that can be enclosed in a rectangle. The region is closed if the region includes its boundary.
2. Let a rectangle containing G be partitioned into n^2 subrectangles of equal size, as in Section 18.1. This rectangular grid partitions G also. Sums approximating $\iint_G f(x, y)\,dx\,dy$ are defined as in Section 18.1. The integral is defined as the limit as $n \to \infty$ of these approximating sums. For a nonrectangular region G, these sums may be hard to actually compute.
3. *Fundamental theorem:* Iterated integrals are used for computation of an integral over a region.
4. Draw a sketch and work from that sketch when changing the order of integration.
5. The area of a region G is found by integrating the constant function 1 over G.

*6. Simpson's rule for estimating double integrals over nonrectangular regions is illustrated in Example 8.

7. Analogous work is valid in space.

EXERCISES

In Exercises 1 through 10, compute the given iterated integral.

1. $\int_0^1 \int_0^{\sqrt{1-y^2}} 4xy\,dx\,dy$
2. $\int_0^{\pi} \int_0^{\sin x} x\,dy\,dx$
3. $\int_1^e \int_0^{\ln x} \frac{y}{x}\,dy\,dx$
4. $\int_9^1 \int_{-\sqrt{y}}^{y} (x + y^2)\,dx\,dy$
5. $\int_0^2 \int_{1-x}^{1+x} x^2\,dy\,dx$
6. $\int_0^{\sqrt{2}} \int_y^{\sqrt{4-y^2}} y\,dx\,dy$
7. $\int_0^1 \int_0^{1-y} \int_0^{x^2+y^2} y\,dz\,dx\,dy$
8. $\int_0^a \int_0^{\sqrt{a^2-y^2}} \int_0^{\sqrt{a^2-x^2-y^2}} x\,dz\,dx\,dy$
9. $\int_0^2 \int_0^{\sqrt{4-z^2}} \int_{y^2+z^2-4}^{4-y^2-z^2} 1\,dx\,dy\,dz$
10. $\int_0^1 \int_0^z \int_0^{y+z} yz\,dx\,dy\,dz$

In Exercises 11 through 22, find the integral of the given function f over the indicated region G.

11. $f(x, y) = y$; G the plane region bounded by $y = 1 - x^2$ and the x-axis
12. $f(x, y) = xy$; G the quarter-disk $x^2 + y^2 \le a^2$ in the first quadrant
13. $f(x, y) = x$; G the plane region bounded by $y = x^2$ and $y = x + 2$
14. $f(x, y) = e^{x^2}$; G the plane region bounded by $y = x, y = -x$, and $x = 4$
15. $f(x, y) = \sin y^2$; G the plane region bounded by $y = x$, $y = \sqrt{\pi}$, and $x = 0$
16. $f(x, y) = y \sin^2 x^2$; G the plane region in the first quadrant bounded by $x = y^2$, $x = \sqrt{\pi}$, and $y = 0$
17. $f(x, y) = ye^{x^2}$; G the plane region bounded by $y = x^{3/2}$, $y = 0$, and $x = 1$
18. $f(x, y, z) = y$; G the region in space bounded by $z = 1 - y^2$, $x = 0$, $x = 4$, and $z = 0$ [*Hint:* This integral can be found immediately by symmetry.]
19. $f(x, y, z) = y^2z$; G the region in space bounded by $x = 4 - y^2 - z^2$ and $x = 0$ [*Hint:* This integral can be found immediately by symmetry.]
20. $f(x, y, z) = x$; G the region in space bounded by the coordinate planes and the plane $x + y + 2z = 2$
21. $f(x, y, z) = yz$; G the region in space bounded by $z = x^2$, $z = 1$, $y = 0$, and $y = 2$
22. $f(x, y, z) = x$; G the region in space bounded by the coordinate planes and the planes $x + y = 1$ and $z = 4$

In Exercises 23 through 32, sketch the region of integration for the given iterated integral and then write an equal iterated integral (or an equal sum of iterated integrals) with the integration in the "reverse order." (You are not asked to compute the integral.)

23. $\int_0^1 \int_{-\sqrt{1-y^2}}^{\sqrt{1-y^2}} 4xy\,dx\,dy$
24. $\int_0^2 \int_0^{x^2} (\sin xy)\,dy\,dx$

25. $\int_0^1 \int_1^{e^y} x^2 \, dx \, dy$

26. $\int_0^1 \int_{x^2}^{x} y \cos x \, dy \, dx$

27. $\int_{-3}^{1} \int_{-\sqrt{1-y}}^{\sqrt{1-y}} e^{xy} \, dx \, dy$

28. $\int_0^1 \int_{2y-2}^{1-y} x \cos^2 y \, dx \, dy$

29. $\int_{-1}^{1} \int_{1-y^2}^{y^2-1} x^2 y^3 \, dx \, dy$

30. $\int_0^4 \int_{-x}^{x} x^2 y^2 \, dy \, dx$

31. $\int_0^{\pi/2} \int_0^{(\pi/4)(\sin x)} (x + y) \, dy \, dx$

32. $\int_0^{\pi} \int_0^{\sin x} x^2 y \, dy \, dx$

In Exercises 33 through 38, express the given triple integral in the indicated order. Do not integrate.

33. $\int_{-1}^{1} \int_{-\sqrt{1-x^2}}^{\sqrt{1-x^2}} \int_0^{1-x^2-y^2} xyz^2 \, dz \, dy \, dx$ in $dx \, dy \, dz$-order

34. $\int_{-2}^{1} \int_0^{\pi/2} \int_0^{\cos z} x \sin yz \, dx \, dz \, dy$ in $dy \, dz \, dx$-order

35. $\int_{-1}^{0} \int_0^{1+x} \int_0^{1+x-y} \sin(xy) \, dz \, dy \, dx$ in $dx \, dy \, dz$-order

36. $\int_{-1}^{0} \int_{-1-y}^{0} \int_0^{1+x+y} 2xy \, dz \, dx \, dy$ in $dy \, dz \, dx$-order

37. $\int_0^2 \int_1^3 \int_0^{4-y^2} z^2 \, dz \, dx \, dy$ in $dx \, dy \, dz$-order

38. $\int_2^5 \int_{-4}^{0} \int_0^{16-x^2} (x + 1) \, dz \, dx \, dy$ in $dx \, dy \, dz$-order

In Exercises 39 through 46, use an iterated integral to find the area or volume of the indicated region.

39. The region of the plane bounded by the curves $y = 0$, $y = 1 + x$, and $y = \sqrt{1 - x}$

40. The region of the plane bounded by the curves $y = \sin x$, $x = \pi/2$, and $y = x$

41. The region of the plane bounded by $y = \ln x$, $y = 1 - x$, and $y = 1$.

42. The region of the plane bounded by $y = e^x$, $x + y = e + 1$, and $x = 0$.

43. The region of space bounded by $z = 2x^2$, $z = 8$, $y = 0$, and $y = 4$.

44. The region of space bounded by $z = \cos x$ for $-\pi/2 \leq x \leq \pi/2$, $z = 0$, $y = -1$, and $y = 2$.

45. The region of space bounded by $z = 4 - x^2 - y^2$ and $z = x^2 + y^2 - 4$.

46. The region of space bounded by $z = 1 + x^2 + y^2$, $y = 1 - x^2$, $y = 0$, and $z = 0$.

In Exercises 47 through 50, carry the integration as far as you can and then use a calculator or computer and Simpson's rule to estimate the integral of the given function f over the indicated region.

47. $f(x, y) = (x + 1)^y$ over the plane region bounded by $y = 0$, $y = x^2$, and $x = 1$

48. $f(x, y) = e^{-y-x^2}$ over the plane region bounded by $x = 0$, $y = 0$, and $x + y = 1$

49. $f(x, y) = \sin(xy^2)$ over the plane region bounded by $x = 1 - y^2$ and $x = 0$

50. $f(x, y) = (100x^3y^2)/(1 + x^4y^2)$ over the quarter-disk $x^2 + y^2 = 1$ in the first quadrant.

*In Exercises 51 through 54, use Simpson's rule, illustrated in Example 8, to estimate the integral, taking $m = n = 4$.

***51.** $\int_0^1 \int_0^{\sqrt{1-y^2}} e^{-x^2-y^2} \, dx \, dy$

***52.** $\int_0^{0.4} \int_0^{0.4-y} [\ln (3 + x)]^y \, dx \, dy$

***53.** $\int_0^1 \int_0^x \sin \left(\frac{1 + x^2}{1 + y^2}\right) dy \, dx$

***54.** $\int_0^1 \int_{x^3}^{x} \left(\frac{1 + x}{1 + y}\right)^{xy} dy \, dx$

18.3 INTEGRATION IN POLAR AND CYLINDRICAL COORDINATES

DOUBLE INTEGRALS IN POLAR COORDINATES

Consider polar r,θ-coordinates in the plane. Recall that the transformation from polar r,θ-coordinates to rectangular x,y-coordinates is given by

$$x = r \cos \theta, \qquad y = r \sin \theta, \tag{1}$$

as indicated in Fig. 18.44.

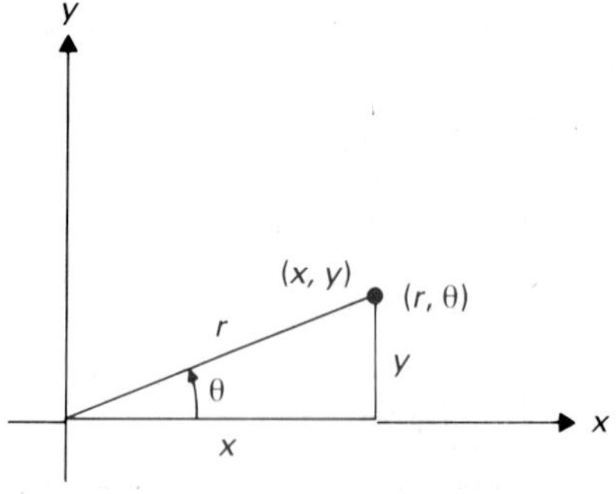

Figure 18.44 $x = r\cos\theta,\ y = r\sin\theta$.

Suppose we wish to integrate a continuous function over a region G in the plane. We may express our function either in terms of the independent variables x and y or in terms of r and θ. If the function is expressed by $f(x, y)$ in terms of x and y, then in terms of r,θ-coordinates it is

$$h(r, \theta) = f(r\cos\theta, r\sin\theta).$$

The differential area element, $dx\,dy$, should be replaced by one in polar coordinates. Now the rectangular area element, $dx\,dy$, is obtained by starting at a point in the region, increasing x by dx, increasing y by dy, then decreasing x by dx, and finally decreasing y by dy. We show this in Fig. 18.45, where the arrows indicate the way we generate this cartesian differential area element, starting at (x, y).

Figure 18.46 shows the generation of the polar differential area element in the same fashion. We start at a point (r, θ), increase r by dr, increase θ by $d\theta$, decrease r by dr, and finally decrease θ by $d\theta$.

If we pretend that the shaded differential area element in Fig. 18.46 is a rectangle, then its area becomes $(r\,d\theta)\,dr$, and we have

$$\boxed{dA = r\,dr\,d\theta} \qquad \textbf{(2)}$$

for polar coordinates. *It is important to note that dA is* not *just $dr\,d\theta$.*

We have attempted to make it seem reasonable that the integral of $f(x, y)$ over a region G is given by

$$\boxed{\iint_G f(x, y)\,dx\,dy = \iint_G f(r\cos\theta, r\sin\theta)\,r\,dr\,d\theta} \qquad \textbf{(3)}$$

in polar coordinates. A careful proof is left to a more advanced course.

> When finding the appropriate polar coordinate limits for an iterated integral over a region G, it is important to draw a sketch of G.

Suppose we integrate in the $dr\,d\theta$-order. Then the first integral

$$\int_{g_1(\theta)}^{g_2(\theta)} h(r, \theta) r\,dr\,d\theta$$

Figure 18.45 In cartesian coordinates, $dA = (dx)(dy)$.

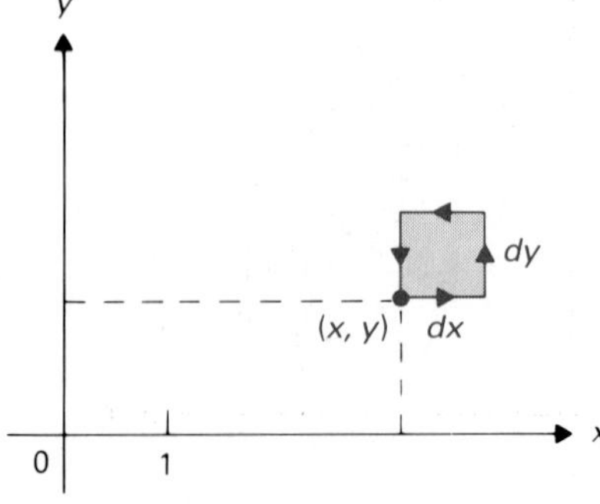

Figure 18.46 In polar coordinates, $dA = (r\,d\theta)(dr)$.

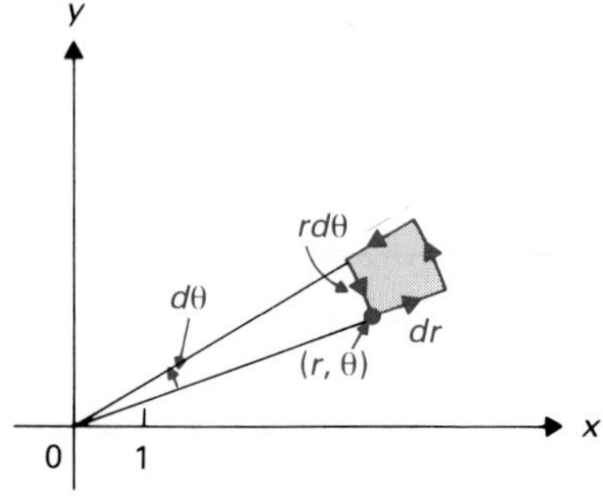

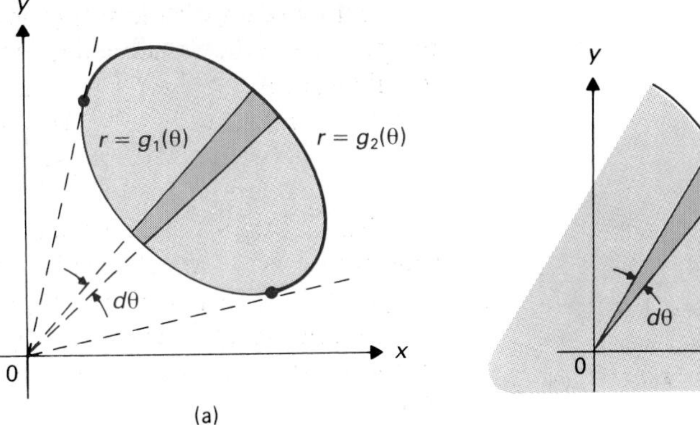

Figure 18.47 (a) $\int_{g_1(\theta)}^{g_2(\theta)} h(r, \theta)r\, dr\, d\theta$ sums contributions over this tapered strip; (b) $\int_0^{g(\theta)} h(r, \theta)r\, dr\, d\theta$ sums contributions over this wedge.

adds up contributions $h(r, \theta)r\, dr\, d\theta$ over the tapered strip shown in Fig. 18.47(a). Often, the lower limit is zero, in which case the strip becomes a wedge from the origin, shown in Fig. 18.47(b).

Suppose, on the other hand, we integrate in the $d\theta\, dr$-order. Then the first integral

$$\int_{k_1(r)}^{k_2(r)} h(r, \theta)r\, d\theta\, dr$$

gives the curved strip in Fig. 18.48. In practice, we almost always integrate in the $dr\, d\theta$-order.

Recall that integration of the constant function 1 over a region gives the area of the region.

EXAMPLE 1 Find the area of the region bounded by the first loop of the spiral $r = \theta$, for $0 \leq \theta \leq 2\pi$, and the positive x-axis.

Solution The region is sketched in Fig. 18.49. From the figure, we obtain the integral

$$\int_0^{2\pi} \int_0^{\theta} 1 \cdot r\, dr\, d\theta = \int_0^{2\pi} \frac{r^2}{2}\bigg|_0^{\theta} d\theta = \int_0^{2\pi} \frac{\theta^2}{2}\, d\theta = \frac{\theta^3}{6}\bigg]_0^{2\pi} = \frac{8\pi^3}{6} = \frac{4}{3}\pi^3.$$

Note that the integral $\int_0^{2\pi} (1/2)\theta^2\, d\theta$ in the computation is the one we would have started with in Chapter 13 to find this area. □

Figure 18.48 $\int_{k_1(r)}^{k_2(r)} h(r, \theta)r\, dr\, d\theta$ sums contributions over this curved strip.

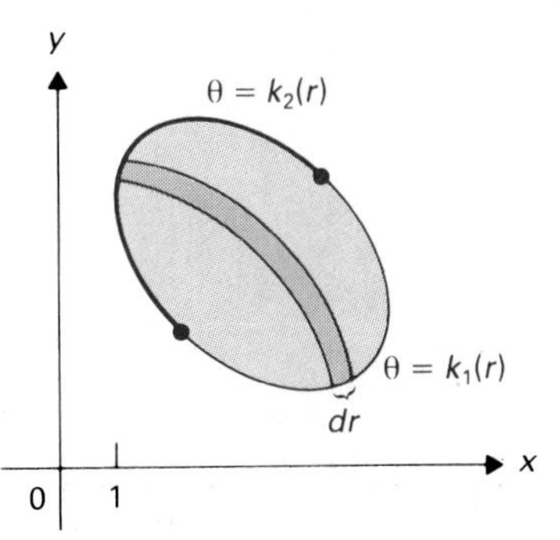

Figure 18.49

$dA = r\, dr\, d\theta$; $A = \int_0^{2\pi}\int_0^{\theta} r\, dr\, d\theta$.

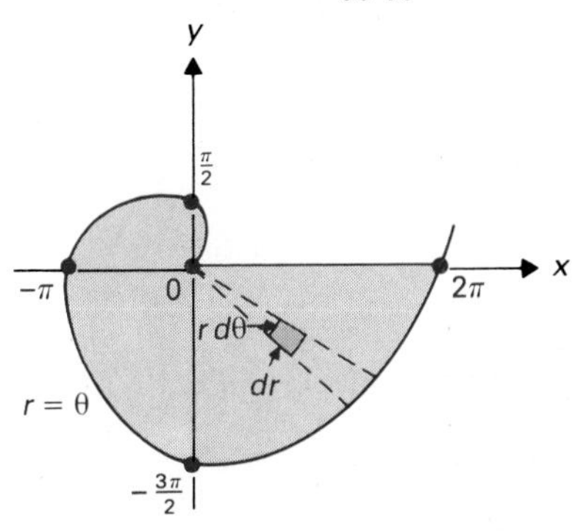

EXAMPLE 2 Find the area of the region G bounded by the cardioid $r = 1 + \cos\theta$, shown in Fig. 18.50, by integrating the constant function 1 over this region.

Solution By symmetry, we can find the area of the upper half of the region and double it for our answer. From Fig. 18.50, we see that the area is given by

$$2\int_0^{\pi}\int_0^{1+\cos\theta} 1\cdot r\,dr\,d\theta = 2\int_0^{\pi}\frac{1}{2}r^2\bigg]_0^{1+\cos\theta} d\theta = 2\int_0^{\pi}\frac{1}{2}(1+\cos\theta)^2\,d\theta$$

$$= \int_0^{\pi}(1 + 2\cos\theta + \cos^2\theta)\,d\theta$$

$$= \left(\theta + 2\sin\theta + \frac{\theta}{2} + \frac{\sin 2\theta}{4}\right)\bigg]_0^{\pi}$$

$$= \pi + \frac{\pi}{2} = \frac{3\pi}{2}.$$

Note that the integral $2\int_0^{\pi}(1/2)(1+\cos\theta)^2\,d\theta$ in our computation is the one we would have started with in Chapter 13 to find the area. □

EXAMPLE 3 Integrate the polar function $h(r, \theta) = 1/r$ over the region inside the cardioid $r = a(1 + \sin\theta)$ and outside the circle $r = a$.

Solution The region of integration is shown in Fig. 18.51. From the figure, we obtain the integral

$$\int_0^{\pi}\int_a^{a(1+\sin\theta)}\frac{1}{r}\cdot r\,dr\,d\theta = \int_0^{\pi}\int_a^{a(1+\sin\theta)} 1\,dr\,d\theta$$

$$= \int_0^{\pi} r\bigg]_a^{a(1+\sin\theta)} d\theta = \int_0^{\pi}[a(1+\sin\theta) - a]\,d\theta$$

$$= \int_0^{\pi} a\sin\theta\,d\theta = a(-\cos\theta)\bigg]_0^{\pi} = a(1) - a(-1) = 2a. \quad □$$

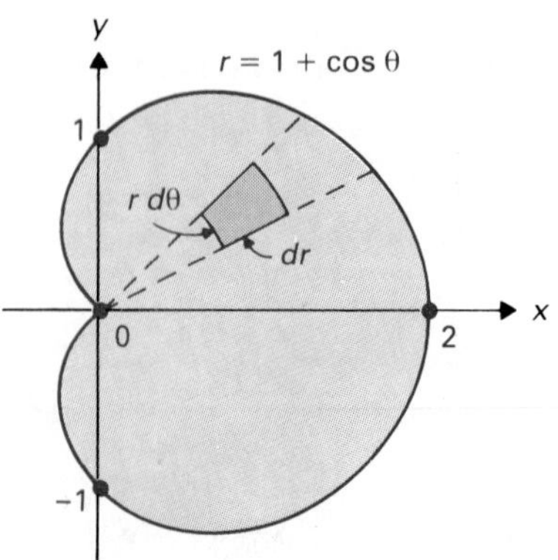

Figure 18.50
$dA = r\,dr\,d\theta$; $A = \int_0^{2\pi}\int_0^{1+\cos\theta} r\,dr\,d\theta$.

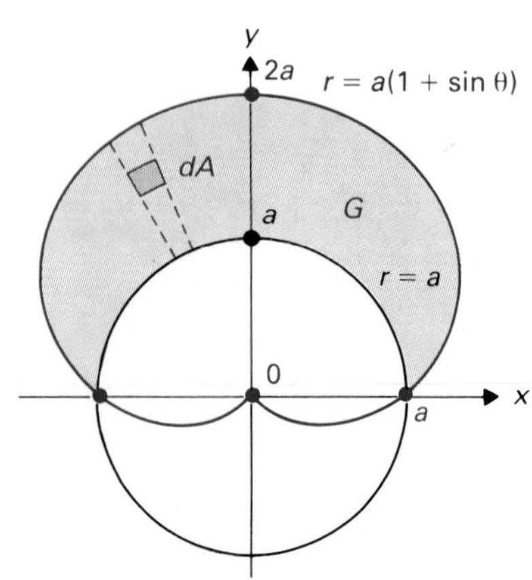

Figure 18.51
$\iint_G h(r,\theta)r\,dr\,d\theta = \int_0^{\pi}\int_a^{a(1+\sin\theta)} h(r,\theta)r\,dr\,d\theta$

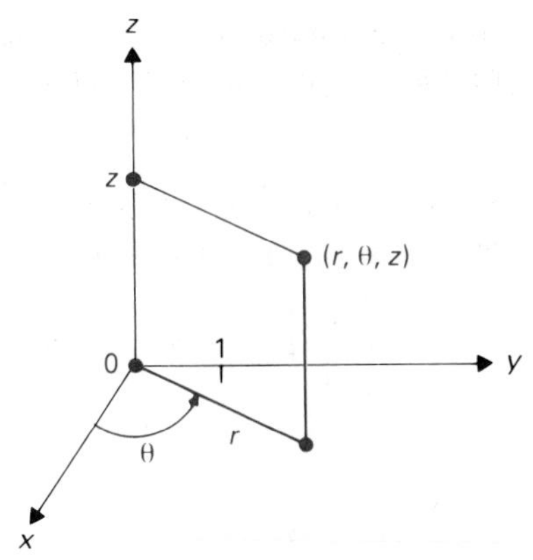

Figure 18.52 (r, θ, z) in cylindrical coordinates.

TRIPLE INTEGRALS IN CYLINDRICAL COORDINATES

Recall that cylindrical r,θ,z-coordinates for space are formed by taking polar r,θ-coordinates in the x,y-plane and the usual rectangular z-coordinate, as indicated in Fig. 18.52. The solution set of $r = a$ is the cylinder $x^2 + y^2 = a^2$, shown in Fig. 18.53. Transformation to cylindrical coordinates is especially useful in integrating a cartesian expression involving $x^2 + y^2$, or in integrating over regions bounded by surfaces with simple cylindrical coordinate equations. If an integral involves $x^2 + z^2$, we may take "cylindrical r,θ,y-coordinates," corresponding to polar r,θ-coordinates in the x,z-plane.

The differential volume element in cylindrical coordinates is shown in Fig. 18.54. The base of this differential element has the same size as the polar differential area element, as shown in the figure. Thus the cylindrical coordinate differential volume dV is

$$dV = r\,dr\,d\theta\,dz. \tag{4}$$

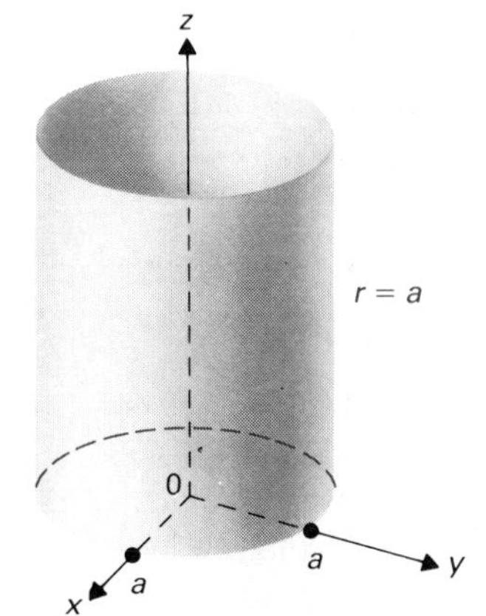

Figure 18.53 The cylinder $x^2 + y^2 = a^2$, or $r = a$.

If $h(r, \theta, z)$ is a continuous cylindrical coordinate function defined on a region G in space, then the integral of $h(r, \theta, z)$ over G is

$$\iiint_G h(r, \theta, z) \cdot r\,dz\,dr\,d\theta. \tag{5}$$

The $dz\,dr\,d\theta$-order in integral (5) is often the most convenient order of integration.

EXAMPLE 4 Integrate $f(x, y, z) = x^2z$ over the region in space bounded by $x^2 + y^2 = 4$, $z = 0$, and $z = 4$.

Solution The region is a solid cylinder, shown in Fig. 18.55. We change to cylindrical coordinates. Now $x^2z = (r\cos\theta)^2z$. Thus we integrate $h(r, \theta, z) = r^2(\cos^2\theta)z$ over the cylinder. In Fig. 18.55, we have shown a differential volume element in the cylinder. We will integrate in the $dz\,dr\,d\theta$-order.

Figure 18.54 $dV = (r\,d\theta)(dr)(dz)$ in cylindrical coordinates.

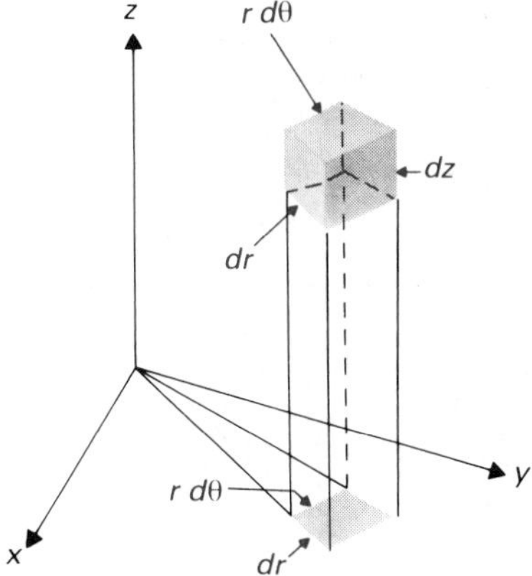

Figure 18.55 The solid cylinder $x^2 + y^2 \leq 4$, $0 \leq z \leq 4$.

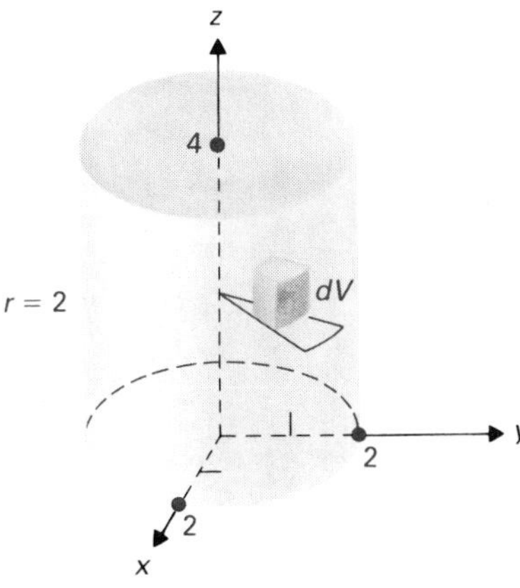

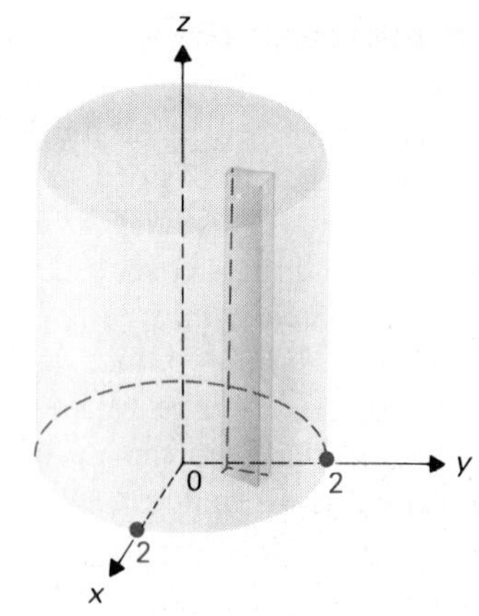

Figure 18.56
$\int_0^4 h(r, \theta, z) r\, dz\, dr\, d\theta$ goes over this column.

Integrating first with respect to z, we see that the smallest z-value is always zero, and the largest is always 4. We are led to the integral

$$\int_0^4 [r^2(\cos^2\theta)z] \cdot r\, dz\, dr\, d\theta,$$

which sums contributions up the column shown in Fig. 18.56. The cylinder has cylindrical coordinate equation $r = 2$, and we next arrive at

$$\int_0^2 \int_0^4 r^2(\cos^2\theta) \cdot r\, dz\, dr\, d\theta,$$

which sums contributions over the wedge shown in Fig. 18.57. Finally, we spin this wedge "all the way around" by letting θ increase from zero to 2π, obtaining

$$\int_0^{2\pi} \int_0^2 \int_0^4 [r^2(\cos^2\theta)z] \cdot r\, dz\, dr\, d\theta = \int_0^{2\pi} \int_0^2 r^3(\cos^2\theta) \frac{z^2}{2}\bigg]_{z=0}^4 dr\, d\theta$$

$$= \int_0^{2\pi} \int_0^2 8r^3 \cos^2\theta\, dr\, d\theta = \int_0^{2\pi} 2r^4\bigg]_0^2 \cos^2\theta\, d\theta = \int_0^{2\pi} 32 \cos^2\theta\, d\theta$$

$$= \int_0^{2\pi} 16(1 + \cos 2\theta)\, d\theta = (16\theta + 8 \sin 2\theta)\bigg]_0^{2\pi} = 32\pi. \quad \square$$

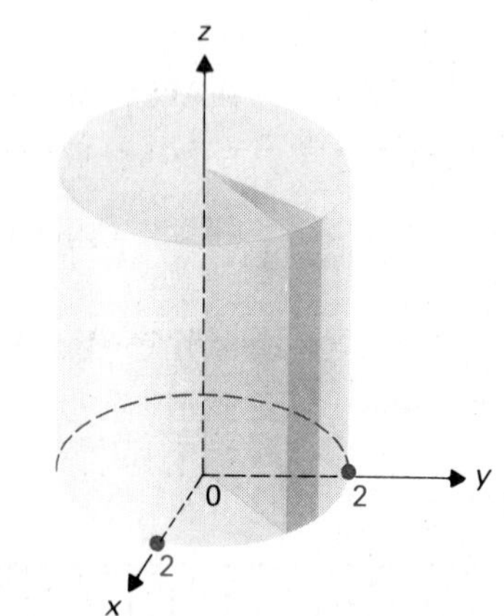

Figure 18.57
$\int_0^2\int_0^4 h(r, \theta, z) r\, dz\, dr\, d\theta$ goes over this wedge.

EXAMPLE 5 Find the volume of the solid in space bounded by $x^2 + z^2 = 9$, $y = 0$, and $y + z = 8$.

Solution We sketch the solid in Fig. 18.58. Since $x^2 + z^2 = 9$ is a circular cylinder perpendicular to the x,z-plane, we use cylindrical coordinates (r, θ, y), thinking of (r, θ) as polar coordinates in the x,z-plane, with $z = r \cos\theta$ and $x = r \sin\theta$. (We let z be the $r \cos\theta$ rather than x since the z,x,y-order gives a *right-hand* coordinate system, just like the familiar x,y,z-order.) Thus $y + z = 8$ becomes $y + r\cos\theta = 8$. From Fig. 18.58, we see that the volume is given by the integral

$$\int_0^{2\pi} \int_0^3 \int_0^{8 - r\cos\theta} 1 \cdot r\, dy\, dr\, d\theta = \int_0^{2\pi} \int_0^3 ry\bigg]_{y=0}^{8 - r\cos\theta} dr\, d\theta$$

$$= \int_0^{2\pi} \int_0^3 (8r - r^2 \cos\theta)\, dr\, d\theta = \int_0^{2\pi} \left(4r^2 - \frac{r^3}{3} \cos\theta\right)\bigg]_{r=0}^3 d\theta$$

$$= \int_0^{2\pi} (36 - 9 \cos\theta)\, d\theta = (36\theta - 9 \sin\theta)\bigg]_0^{2\pi} = 72\pi. \quad \square$$

EXAMPLE 6 Let G be the region bounded above by $z = 1 + x^2 + y^2$, below by $z = 0$, and on the sides by $x^2 + y^2 = 4$, as indicated in Fig. 18.59. Integrate $f(x, y, z) = x - y + z^2$ over G.

Solution Our x,y,z-integral would be

$$\int_{-2}^2 \int_{-\sqrt{4-y^2}}^{\sqrt{4-y^2}} \int_0^{1+x^2+y^2} (x - y + z^2)\, dz\, dx\, dy,$$

which is not too pleasant to evaluate. However, changing to cylindrical

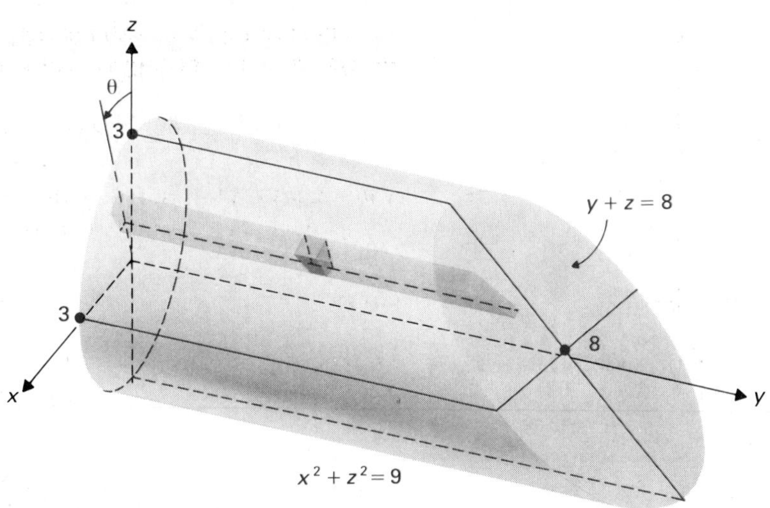

Figure 18.58 With cylindrical coordinates (r, θ, y), we have $dV = dy(r\,dr\,d\theta)$.

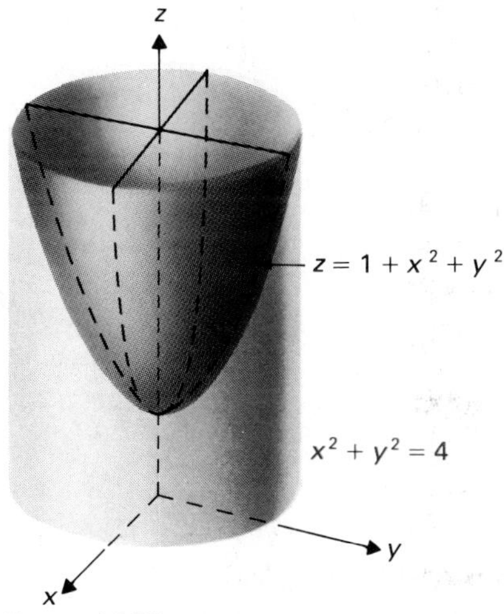

Figure 18.59 Solid bounded above by $z = 1 + x^2 + y^2$, on the sides by $x^2 + y^2 = 4$, and below by $z = 0$.

coordinates, we obtain the integral

$$\begin{aligned}
&\int_0^{2\pi}\int_0^2\int_0^{1+r^2} (r\cos\theta - r\sin\theta + z^2)r\,dz\,dr\,d\theta \\
&= \int_0^{2\pi}\int_0^2 \left((\cos\theta - \sin\theta)r^2 z + r\frac{z^3}{3}\right)\Bigg]_{z=0}^{z=1+r^2} dr\,d\theta \\
&= \int_0^{2\pi}\int_0^2 \left((\cos\theta - \sin\theta)(r^2)(1+r^2) + \frac{1}{3}r(1+r^2)^3\right) dr\,d\theta \\
&= \int_0^{2\pi} \left((\cos\theta - \sin\theta)\left(\frac{r^3}{3} + \frac{r^5}{5}\right) + \frac{1}{6}\cdot\frac{(1+r^2)^4}{4}\right)\Bigg]_{r=0}^{r=2} d\theta \\
&= \int_0^{2\pi} \left((\cos\theta - \sin\theta)\left(\frac{8}{3} + \frac{32}{5}\right) + \frac{5^4 - 1}{24}\right) d\theta \\
&= \int_0^{2\pi} \left((\cos\theta - \sin\theta)\frac{136}{15} + \frac{624}{24}\right) d\theta \\
&= \left(\frac{136}{15}(\sin\theta + \cos\theta) + \frac{624}{24}\theta\right)\Bigg]_0^{2\pi} \\
&= \frac{136}{15}(1-1) + \frac{624}{24}2\pi = \frac{624}{12}\pi = 52\pi.
\end{aligned}$$

Perhaps you would like to perform the integration in rectangular coordinates and compare it with our cylindrical coordinate computation! □

SUMMARY

1. The area of the differential plane element in polar coordinates is
$$dA = r\cdot dr\,d\theta.$$
2. An integral $\iint f(x, y)\,dx\,dy$ with appropriate x,y-limits becomes in polar coordinates $\iint f(r\cos\theta, r\sin\theta)\cdot r\,dr\,d\theta$ with appropriate polar limits for the region.

3. The volume of the differential solid element in cylindrical coordinates is $dV = r \cdot dz\, dr\, d\theta$.
4. An integral $\iiint f(x, y, z)\, dx\, dy\, dz$ with appropriate x,y,z-limits becomes, in cylindrical coordinates,

$$\iiint f(r\cos\theta, r\sin\theta, z) \cdot r\, dz\, dr\, d\theta$$

with appropriate cylindrical coordinate limits for the region.

EXERCISES

1. An integral in polar coordinates is of the form

$$\int_\alpha^\beta \int_a^b h(r, \theta) r\, dr\, d\theta$$

for $0 \le \alpha < \theta \le 2\pi$ and $0 \le a < b$. Sketch the region of integration.

In Exercises 2 through 10, find the area of the plane region, using double integration in polar coordinates.

2. The disk $x^2 + y^2 \le a^2$
3. The annular ring $a^2 \le x^2 + y^2 \le b^2$ for $0 \le a \le b$
4. The region bounded by the positive x-axis for $2\pi a \le x \le 4\pi a$ and the portion of the spiral $r = a\theta$ for $2\pi \le \theta \le 4\pi$
5. The region inside the cardioid $r = a(1 + \cos\theta)$ and outside the circle $r = a$
6. The region inside one loop of the four-leaved rose $r = a\sin 2\theta$
7. The region in the first quadrant bounded by $x^2 + y^2 = a^2$, $y = 0$, and $x = a/2$
8. The region inside the larger loop and outside the smaller loop of the limaçon $r = a(1 + 2\cos\theta)$
9. The region inside the circle $r = a\cos\theta$ and outside the cardioid $r = a(1 - \cos\theta)$
10. The total region inside the lemniscate $r^2 = 2a^2\cos 2\theta$ and outside the circle $r = a$
11. Find the integral of the polar function $h(r, \theta) = r\sin^2\theta$, $r \ge 0$, over the closed disk bounded by $r = a$.
12. Find the integral of the polar function $h(r, \theta) = \cos\theta$, $r \ge 0$, over the region bounded by the cardioid $r = a(1 + \sin\theta)$.

In Exercises 13 through 20, change the integral to polar coordinates and then evaluate the transformed version.

13. $\displaystyle\int_0^a \int_0^{\sqrt{a^2 - y^2}} xy\, dx\, dy$
14. $\displaystyle\int_{-2}^2 \int_{-\sqrt{4-y^2}}^{\sqrt{4-y^2}} (x^2 + y^2)\, dx\, dy$
15. $\displaystyle\int_{-1}^1 \int_{-\sqrt{1-y^2}}^{\sqrt{1-y^2}} e^{-x^2-y^2}\, dx\, dy$
16. $\displaystyle\int_0^2 \int_{-\sqrt{4-x^2}}^{\sqrt{4-x^2}} \frac{1}{1 + x^2 + y^2}\, dy\, dx$
17. $\displaystyle\int_0^{3/\sqrt{2}} \int_y^{\sqrt{9-y^2}} y\, dx\, dy$
18. $\displaystyle\int_{-2}^2 \int_2^{\sqrt{8-x^2}} \frac{1}{y}\, dy\, dx$
19. $\displaystyle\int_0^4 \int_0^x (x^2 + y^2)\, dy\, dx$
20. $\displaystyle\int_0^2 \int_{-y}^y (x^2 + y^2)\, dx\, dy$
21. Express the volume of the ball $x^2 + y^2 + z^2 \le a^2$ as an integral (do not integrate) in the integration order
    a) $dz\, dr\, d\theta$ b) $dz\, d\theta\, dr$
    c) $dr\, d\theta\, dz$ d) $dr\, dz\, d\theta$
    e) $d\theta\, dr\, dz$ f) $d\theta\, dz\, dr$

In Exercises 22 through 28, find the volume of the given region in space, using triple integration in cylindrical coordinates.

22. The region bounded by the paraboloid $z = 4 - x^2 - y^2$ and the plane $z = 0$
23. The region bounded by the paraboloids $z = 9 - x^2 - y^2$ and $z = x^2 + y^2 - 9$
24. The region bounded by $x^2 + y^2 = 1$, $z = 0$ and $x + z = 4$
25. The region bounded by the paraboloid $x = 8 - y^2 - z^2$, the cylinder $y^2 + z^2 = 4$, and the plane $x = -1$ [*Hint:* Use x,r,θ-coordinates.]
26. The region bounded by the hemisphere $z = \sqrt{a^2 - x^2 - y^2}$ and the plane $z = b$ for $0 \le b < a$
27. The region inside the semicircular cylinder bounded by $x = \sqrt{4 - z^2}$ and $x = 0$ and bounded on the ends by $y = 0$ and the hemisphere $y = \sqrt{16 - x^2 - z^2}$. [*Hint:* Use "cylindrical coordinates" (r, y, θ).]
28. The region bounded by the paraboloid $z = x^2 + y^2$, the plane $z = 0$, and the cylinder $x^2 + y^2 = 2x$
29. Find the integral of the cylindrical coordinate function $h(r, \theta, z) = rz\cos^2\theta$ for $r \ge 0$ over the region in space bounded by $r = a$, $z = 0$, and $z = 4$.

30. Find the integral of the cylindrical coordinate function $h(r, \theta, z) = rz^2$ for $r \geq 0$ over the region in space bounded by the cone $z^2 = x^2 + y^2$ and the plane $z = 4$.

In Exercises 31 through 36, transform the integral to cylindrical coordinates and then evaluate the transformed version.

31. $\displaystyle\int_0^1 \int_{-\sqrt{1-x^2}}^{\sqrt{1-x^2}} \int_0^3 x\,dz\,dy\,dx$

32. $\displaystyle\int_0^2 \int_{-\sqrt{4-y^2}}^{\sqrt{4-y^2}} \int_{1-y}^4 y^2\,dz\,dx\,dy$

33. $\displaystyle\int_{-2}^2 \int_{-\sqrt{4-x^2}}^{\sqrt{4-x^2}} \int_0^{\sqrt{x^2+y^2}} z\,dz\,dy\,dx$

34. $\displaystyle\int_{-3}^3 \int_{-\sqrt{9-y^2}}^{\sqrt{9-y^2}} \int_{\sqrt{x^2+y^2}}^{12-x^2-y^2} z\,dz\,dx\,dy$

35. $\displaystyle\int_{-\sqrt{2}}^{\sqrt{2}} \int_{-\sqrt{2-z^2}}^{\sqrt{2-z^2}} \int_0^{1+y^2+z^2} z^2\,dx\,dy\,dz$

36. $\displaystyle\int_0^2 \int_0^{\sqrt{4-x^2}} \int_0^{\sqrt{x^2+z^2}} y\,dy\,dz\,dx$

18.4 INTEGRATION IN SPHERICAL COORDINATES

Recall the spherical coordinate system, where a point has coordinates (ρ, ϕ, θ) as indicated in Fig. 18.60. The coordinate ρ is the length of the line segment joining the point and the origin, ϕ is the angle from the z-axis to this line segment, and θ is the same angle as in cylindrical coordinates. Note that $\rho = a$ describes a sphere with center at the origin and radius a, as indicated in Fig. 18.61.

Since ρ is the distance from the point to the origin, it is obvious that

$$\rho^2 = x^2 + y^2 + z^2, \tag{1}$$

and, consequently, transformation to spherical coordinates is useful in the triple integration of cartesian expressions involving $x^2 + y^2 + z^2$, or integrals over regions bounded in part by spherical surfaces. Since $\phi = \alpha$ is a cone, we expect ρ,ϕ,θ-coordinates to be useful when cones are involved.

We need to express x, y, and z in terms of spherical ρ,ϕ,θ-coordinates so that we can express an integral $\iiint_G f(x, y, z)\,dx\,dy\,dz$ in terms of spherical

Figure 18.60 Spherical coordinates.

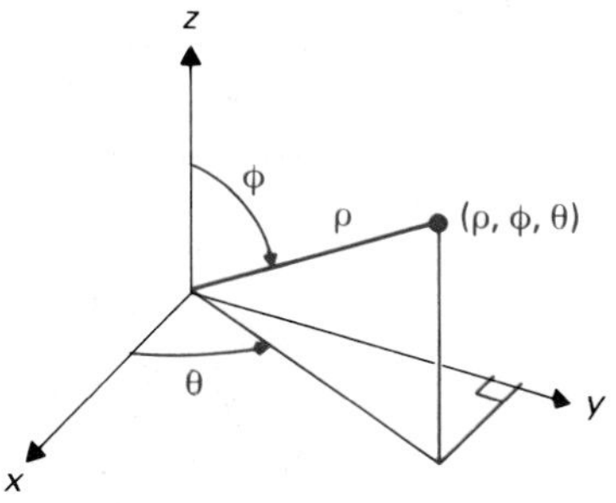

Figure 18.61 The sphere $\rho = a$.

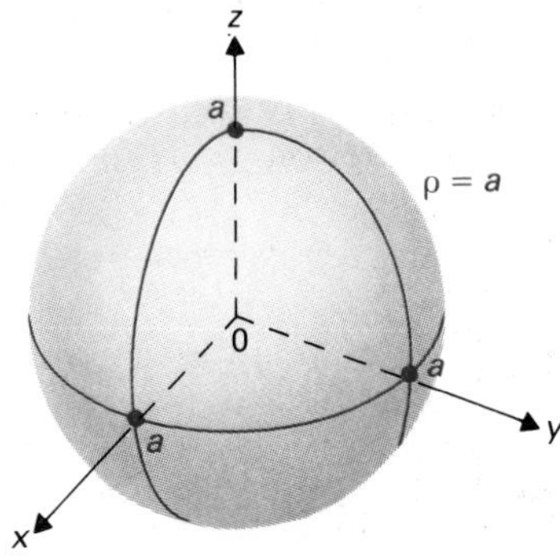

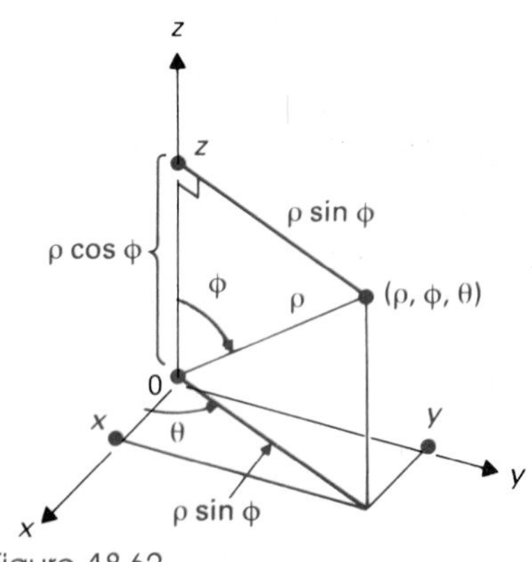

Figure 18.62
$x = \rho \sin\phi \cos\theta,$
$y = \rho \sin\phi \sin\theta,\ z = \rho\cos\phi.$

coordinates. From Fig. 18.62, we easily recall that

$$\begin{aligned} x &= \rho \sin\phi \cos\theta, \\ y &= \rho \sin\phi \sin\theta, \\ z &= \rho \cos\phi. \end{aligned} \tag{2}$$

Increasing and decreasing ρ, ϕ, and θ by amounts $d\rho$, $d\phi$, and $d\theta$, we generate the differential element shown in Fig. 18.63. The volume of this element is approximately $(\rho \sin\phi\, d\theta)(d\rho)(\rho\, d\phi)$, as shown in the figure. Therefore, the differential solid element in spherical coordinates has volume

$$\boxed{dV = \rho^2 \sin\phi\, d\rho\, d\phi\, d\theta.} \tag{3}$$

A real proof that Eq. (3) is appropriate is left to an advanced calculus course.

Remember that we restrict ϕ to the range $0 \le \phi \le \pi$; thus $\sin\phi \ge 0$.

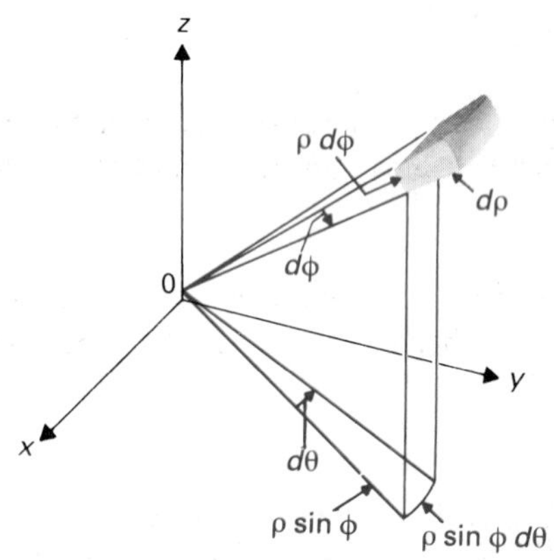

Figure 18.63 Solid differential element of volume
$dV = (\rho\, d\phi)(d\rho)(\rho \sin\phi\, d\theta)$
$= \rho^2 \sin\phi\, d\rho\, d\phi\, d\theta.$

EXAMPLE 1 Find the volume of the ball bounded by the sphere $x^2 + y^2 + z^2 = a^2$, which has spherical coordinate equation $\rho = a$.

Solution We integrate the constant function 1 over this region, using the volume element (3) and limits in spherical coordinates. We integrate in the $d\rho\, d\phi\, d\theta$-order, which frequently is the most convenient order. We think of the first integral, with respect to ρ, as adding up our volume elements to give the spike shown in Fig. 18.64. The next integration with respect to ϕ adds up the volumes of these spikes to give the volume of the wedge in Fig. 18.65, and the final integration with respect to θ from zero to 2π adds up the volumes of these wedges to give the volume of the entire ball. Computing, we have

$$\begin{aligned} \int_0^{2\pi}\int_0^{\pi}\int_0^{a} (1)\rho^2 \sin\phi\, d\rho\, d\phi\, d\theta &= \int_0^{2\pi}\int_0^{\pi} \frac{\rho^3}{3}\sin\phi\,\Big]_{\rho=0}^{\rho=a} d\phi\, d\theta \\ &= \int_0^{2\pi}\int_0^{\pi} \frac{a^3}{3}\sin\phi\, d\phi\, d\theta \\ &= \frac{a^3}{3}\int_0^{2\pi} -\cos\phi\,\Big]_0^{\pi} d\theta \\ &= \frac{a^3}{3}\int_0^{2\pi} [-(-1)+1]\, d\theta \\ &= \frac{a^3}{3}\, 2\theta\,\Big]_0^{2\pi} = \frac{a^3}{3}4\pi = \frac{4}{3}\pi a^3. \quad \square \end{aligned}$$

Figure 18.64 Integration with respect to ρ.

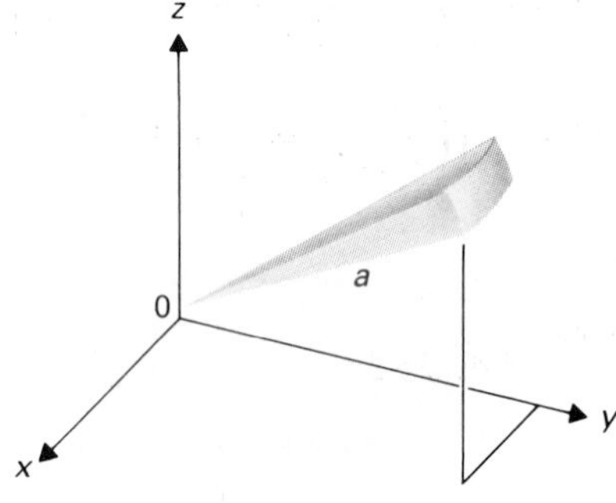

EXAMPLE 2 Integrate the function $f(x, y, z) = z$ over the half-ball bounded above by $z = \sqrt{1 - x^2 - y^2}$ and below by $z = 0$, as indicated in Fig. 18.66.

Solution In spherical coordinates, $z = \rho\cos\phi$ by Eqs. (2), so our integral in

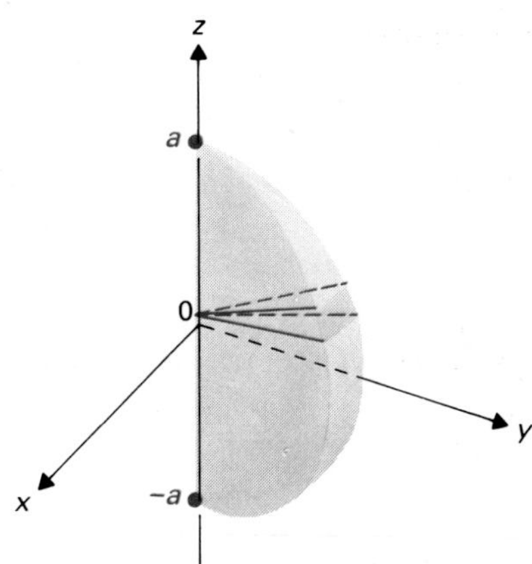

Figure 18.65 Subsequent integration with respect to ϕ.

spherical coordinate form is

$$\int_0^{2\pi}\int_0^{\pi/2}\int_0^1 (\rho\cos\phi)(\rho^2\sin\phi)\,d\rho\,d\phi\,d\theta$$

$$= \int_0^{2\pi}\int_0^{\pi/2} \frac{\rho^4}{4}\cos\phi\sin\phi\Big]_{\rho=0}^{\rho=1} d\phi\,d\theta$$

$$= \int_0^{2\pi}\int_0^{\pi/2} \frac{1}{4}\cos\phi\sin\,d\phi\,d\theta$$

$$= \frac{1}{4}\int_0^{2\pi} \frac{\sin^2\phi}{2}\Big]_0^{\pi/2} d\theta$$

$$= \frac{1}{4}\int_0^{2\pi}\frac{1}{2}\,d\theta = \frac{1}{8}\int_0^{2\pi} d\theta = \frac{1}{8}\theta\Big]_0^{2\pi} = \frac{1}{8}2\pi = \frac{\pi}{4}.$$

Note that the rectangular coordinate integral of z over this half-ball is

$$\int_{-1}^{1}\int_{-\sqrt{1-y^2}}^{\sqrt{1-y^2}}\int_0^{\sqrt{1-x^2-y^2}} z\,dz\,dx\,dy.$$

This integral is less pleasant to compute! □

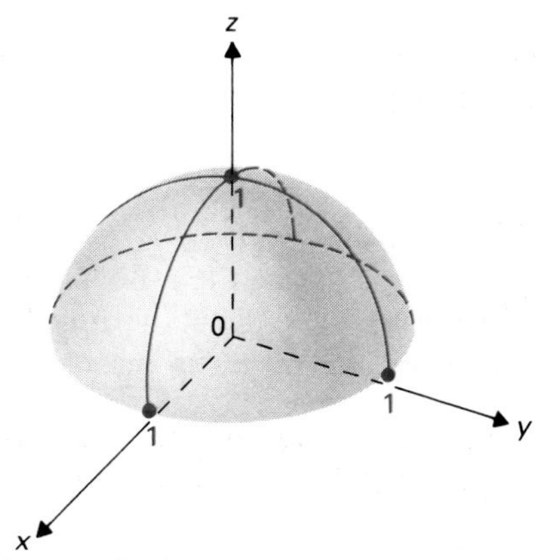

Figure 18.66 Half-ball bounded by $z = \sqrt{1 - x^2 - y^2}$ and $z = 0$.

EXAMPLE 3 Integrate in spherical coordinates to derive the formula $V = \frac{1}{3}\pi a^2 h$ for the volume of a solid right circular cone of altitude h and radius of base a.

Solution The solid bounded by $a^2z^2 = h^2(x^2 + y^2)$ and $z = h$ is such a cone, as shown in Fig. 18.67. The plane $z = h$ becomes $\rho\cos\phi = h$, and the integral is

$$\int_0^{2\pi}\int_0^{\tan^{-1}(a/h)}\int_0^{h/\cos\phi} p^2\sin\phi\,d\rho\,d\phi\,d\theta$$

$$= \int_0^{2\pi}\int_0^{\tan^{-1}(a/h)} \frac{\rho^3}{3}\sin\phi\Big]_{\rho=0}^{\rho=h/\cos\phi} d\phi\,d\theta$$

$$= \int_0^{2\pi}\int_0^{\tan^{-1}(a/h)} \frac{h^3}{3}(\cos\phi)^{-3}\sin\phi\,d\phi\,d\theta$$

$$= \int_0^{2\pi} -\frac{h^3}{3}\cdot\frac{\sec^2\phi}{-2}\Big]_0^{\tan^{-1}(a/h)} d\theta$$

$$= \int_0^{2\pi} -\frac{h^3}{3}\left(\frac{a^2+h^2}{-2h^2} + \frac{1}{2}\right) d\theta$$

$$= 2\pi\left(-\frac{h^3}{3}\right)\left(\frac{a^2+h^2-h^2}{-2h^2}\right)$$

$$= 2\pi\left(-\frac{h^3}{3}\right)\left(\frac{a^2}{-2h^2}\right) = \frac{1}{3}\pi a^2 h. \quad □$$

Figure 18.67 Cone with base radius a and altitude h; the cone has equation $\phi = \tan^{-1}(a/h)$.

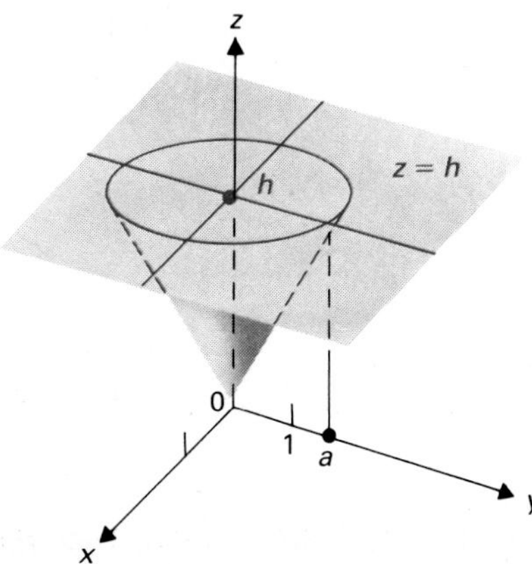

EXAMPLE 4 Transform the integral

$$\int_{-2}^{2}\int_{-\sqrt{4-y^2}}^{\sqrt{4-y^2}}\int_0^{\sqrt{x^2+y^2}} x^2\,dz\,dx\,dy$$

to spherical coordinates.

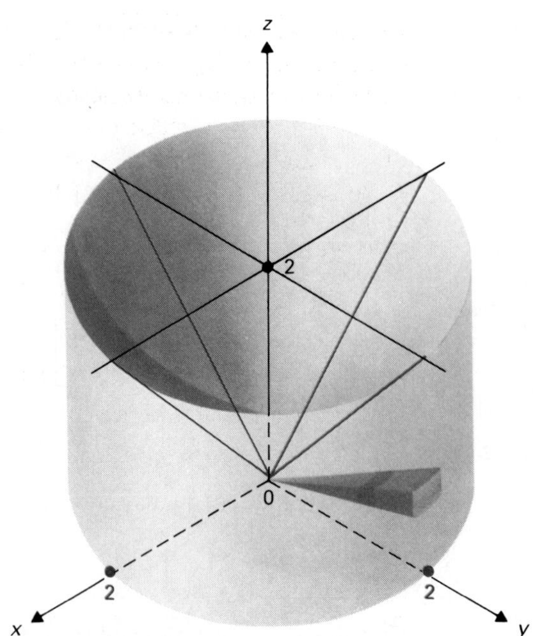

Figure 18.68 Region bounded by the cone $z = x^2 + y^2$ (or $\phi = \pi/4$), the cylinder $x^2 + y^2 = 4$ (or $\rho = 2 \csc \theta$), and the plane $z = 0$ (or $\phi = \pi/2$).

Solution The lower z-limit, $z = 0$, gives the x,y-plane, while $z = \sqrt{x^2 + y^2}$ is the upper half of the cone $z^2 = x^2 + y^2$. We recognize the x-limits and y-limits as corresponding to integrating over a disk with center at the origin in the x,y-plane and radius 2. Thus the region of integration is the one in Fig. 18.68.

In Fig. 18.68, we have drawn the spherical coordinate differential volume element. If we integrate first with respect to ρ, we see that ρ goes from zero to the cylinder $x^2 + y^2 = 4$. In spherical coordinates, this cylinder has equation

$$\rho^2 \sin^2\phi \cos^2\theta + \rho^2 \sin^2\phi \sin^2\theta = 4,$$

or $\rho^2 \sin^2\phi = 4$. Thus $\rho = 2/(\sin \theta) = 2 \csc \theta$. Consequently we start our integral with

$$\int_0^{2 \csc \phi} (\rho^2 \sin^2\phi \cos^2\theta)\rho^2 \sin \phi \, d\rho \, d\phi \, d\theta.$$

Since the cone makes a 45° angle with the z-axis, ϕ goes from $\pi/4$ to $\pi/2$. Finally, we let θ go from zero to 2π and obtain

$$\int_0^{2\pi} \int_{\pi/4}^{\pi/2} \int_0^{2 \csc \phi} \rho^4 \sin^3\phi \cos^2\theta \, d\rho \, d\phi \, d\theta. \quad \square$$

EXAMPLE 5 Transform the cylindrical coordinate integral

$$\int_0^{2\pi} \int_0^{2} \int_r^{\sqrt{8 - r^2}} r^2 z \, dz \, dr \, d\theta$$

to spherical coordinates.

Solution The lower z-limit, $z = r = \sqrt{x^2 + y^2}$, and the upper z-limit,

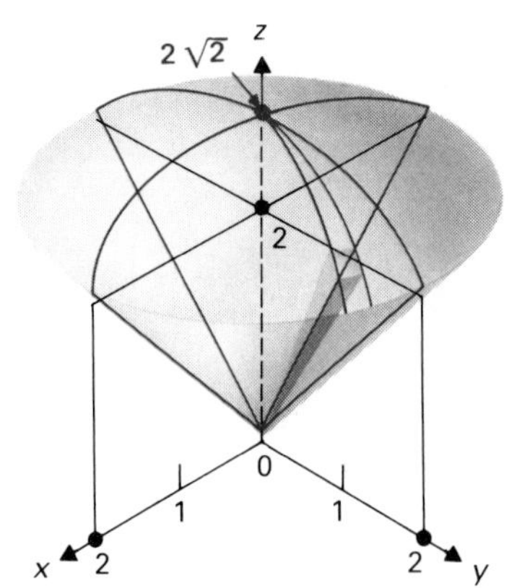

Figure 18.69 Region bounded by the hemisphere $z = \sqrt{8 - r^2}$ (or $\rho = 2\sqrt{2}$) and the cone $z = r$ (or $\phi = \pi/4$).

$z = \sqrt{8 - r^2} = \sqrt{8 - x^2 - y^2}$, show that the region of integration is bounded below by a cone and above by the sphere shown in Fig. 18.69. Squaring the equations, we have $z^2 = r^2$ and $z^2 = 8 - r^2$. These yield $2z^2 = 8$, so $z = 2$. The intersection of these surfaces is thus in the plane $z = 2$ and has r,θ-equation $r = 2$, which is a circle of radius 2. The r-limit and θ-limit in the given integral then show that the region of integration is the entire solid bounded by the cone below and the hemisphere above, as shown in Fig. 18.69.

As indicated by Fig. 18.69, the ρ-limits for an integral in $d\rho\, d\phi\, d\theta$-order are from 0 to $\sqrt{8} = 2\sqrt{2}$. The cone makes an angle of 45° with the z-axis. Now

$$r^2 z\, dz\, dr\, d\theta = (rz) \cdot r\, dz\, dr\, d\theta = (rz)\, dV.$$

Since $rz = (\rho \sin\phi)(\rho \cos\phi) = \rho^2 \sin\phi \cos\phi$, we see that the integral in spherical coordinates becomes

$$\int_0^{2\pi} \int_0^{\pi/4} \int_0^{2\sqrt{2}} \rho^4 \sin^2\phi \cos\phi\, d\rho\, d\phi\, d\theta. \quad \square$$

SUMMARY

1. The meaning of spherical ρ,ϕ,θ-coordinates is given in Fig. 18.60.
2. Transformation from x,y,z-coordinates to ρ,ϕ,θ-coordinates is accomplished by
$$x = \rho \sin\phi \cos\theta,$$
$$y = \rho \sin\phi \sin\theta,$$
$$z = \rho \cos\phi.$$
3. The differential solid element in spherical coordinates has volume $dV = \rho^2 \sin\phi\, d\rho\, d\phi\, d\theta$.
4. An integral $\iiint f(x, y, z)\, dx\, dy\, dz$ with appropriate x,y,z-limits becomes, in spherical coordinates,
$$\iiint f(\rho \sin\phi \cos\theta, \rho \sin\phi \sin\theta, \rho \cos\phi) \cdot \rho^2 \sin\phi\, d\rho\, d\phi\, d\theta,$$
with appropriate spherical coordinate limits for the region chosen so that $0 \le \phi \le \pi$.

EXERCISES

In Exercises 1 through 8, find the volume of the given region by triple integration in spherical coordinates.

1. The region bounded by the cone $z^2 = x^2 + y^2$ and the hemisphere $z = \sqrt{16 - x^2 - y^2}$
2. The larger region bounded by the sphere $x^2 + y^2 + z^2 = 9$ and the cone $z = -\sqrt{x^2 + y^2}$
3. The region bounded above by the cone $z = \sqrt{x^2 + y^2}$, below by the cone $z = -2\sqrt{x^2 + y^2}$, and on the sides by the sphere $x^2 + y^2 + z^2 = 9$
4. The region between the cones $z^2 = x^2 + y^2$ and $3z^2 = x^2 + y^2$ and below the hemisphere $z = \sqrt{4 - x^2 - y^2}$
5. The region bounded by the hemisphere $y = \sqrt{4 - x^2 - z^2}$ and the planes $y = x$ and $y = \sqrt{3}x$
6. The region bounded on the sides by the cylinder $x^2 + y^2 = 4$, above by the cone $z = \sqrt{x^2 + y^2}$, and below by the plane $z = 0$
7. The region bounded by the hemisphere $z = \sqrt{a^2 - x^2 - y^2}$ and the plane $z = b$ for $0 \le b < a$

8. The region bounded below by the plane $z = b$, above by the plane $z = c$, and on the sides by the sphere $x^2 + y^2 + z^2 = a^2$ for $-a < b < c < a$ [*Hint:* Express as a difference of integrals.]
9. Find the integral of the spherical coordinate function $h(\rho, \phi, \theta) = \rho^2$ over the ball bounded by the sphere $x^2 + y^2 + z^2 = a^2$.
10. Find the integral of the spherical coordinate function $h(\rho, \phi, \theta) = \rho^2 \cos \phi$ over the region bounded by the cone $z^2 = x^2 + y^2$ and the hemisphere $z = \sqrt{4 - x^2 - y^2}$.

In Exercises 11 through 20, transform the integrals into ones that use spherical coordinates. Do not evaluate the integrals.

11. $\displaystyle\int_{-2}^{2}\int_{-\sqrt{4-y^2}}^{\sqrt{4-y^2}}\int_{\sqrt{x^2+y^2}}^{\sqrt{8-x^2-y^2}} z\,dz\,dx\,dy$
12. $\displaystyle\int_{0}^{1}\int_{-\sqrt{1-x^2}}^{\sqrt{1-x^2}}\int_{3\sqrt{x^2+y^2}}^{\sqrt{10-x^2-y^2}} x\,dz\,dy\,dx$
13. $\displaystyle\int_{0}^{3}\int_{0}^{\sqrt{9-x^2}}\int_{\sqrt{(x^2+y^2)/3}}^{\sqrt{3(x^2+y^2)}} xy\,dz\,dx\,dy$
14. $\displaystyle\int_{-2}^{0}\int_{-\sqrt{4-y^2}}^{0}\int_{-\sqrt{3(x^2+y^2)}}^{\sqrt{x^2+y^2}} x^2\,dz\,dx\,dy$
15. $\displaystyle\int_{-1}^{0}\int_{0}^{\sqrt{1-x^2}}\int_{3}^{\sqrt{10-x^2-y^2}} (y-x)\,dz\,dy\,dx$
16. $\displaystyle\int_{0}^{2\pi}\int_{0}^{4}\int_{r}^{4} r^2\,dz\,dr\,d\theta$
17. $\displaystyle\int_{0}^{\pi}\int_{0}^{3}\int_{0}^{\sqrt{3}r} r^2z\,dz\,dr\,d\theta$
18. $\displaystyle\int_{0}^{\pi/2}\int_{0}^{4}\int_{-r}^{r/\sqrt{3}} r^3\,dz\,dr\,d\theta$
19. $\displaystyle\int_{0}^{2\pi}\int_{2}^{4}\int_{0}^{r} dz\,dr\,d\theta$
20. $\displaystyle\int_{0}^{2\pi}\int_{0}^{2}\int_{0}^{1} rz\,dz\,dr\,d\theta$

18.5 MOMENTS AND CENTERS OF MASS

We introduced moments and centroids in Sections 7.8 and 7.9, treating only the special cases that can be handled with an integral of a function of one variable. We can give a better presentation now, using multiple integrals.

MASS

Imagine a physical body to occupy some region G in space. The *mass m* of the body is a numerical measure of the "amount of material" it contains. Near the surface of the earth, the *weight* of a body is mg, where g is the gravitational acceleration; one slug of mass weighs about 32 pounds.

The *mass density* of the body is the *mass per unit volume*. If the body is not homogeneous, the mass density may vary and be a function of the position within the body. To say that the mass density at a point (x_0, y_0, z_0) is $\sigma(x_0, y_0, z_0)$ is to say that if the body had the same composition everywhere that it has at (x_0, y_0, z_0), then its mass would be

$$\sigma(x_0, y_0, z_0)\cdot \text{Volume of } G.$$

Figure 18.70 For mass density $\sigma(x, y, z)$, the differential element has approximate mass $m = \sigma(x, y, z)\,dx\,dy\,dz$.

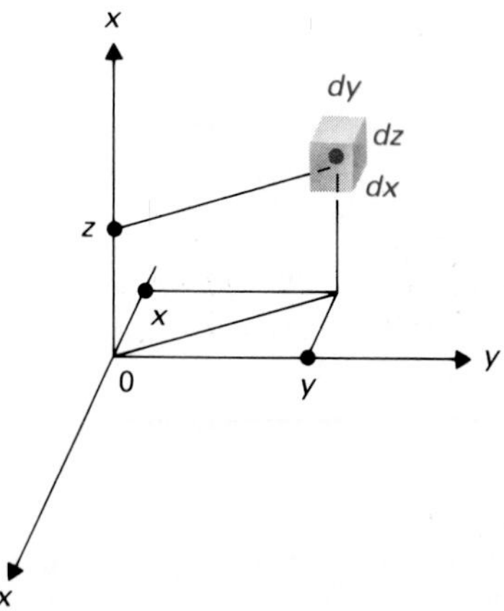

Suppose the mass density $\sigma(x, y, z)$ is a *continuous* function for (x, y, z) in G. If (x, y, z) is a point in the small box with edges of lengths dx, dy, and dz, shown in Fig. 18.70, then the approximate mass of the material in this box is

$$dm = \sigma(x, y, z)\,dx\,dy\,dz.$$

If we add all these small amounts of mass with an integral as dx, dy, and dz

approach zero, we obtain

$$m = \iiint_G \sigma(x, y, z)\, dx\, dy\, dz. \tag{1}$$

Mass of the body

Of course, in cylindrical and spherical coordinates, our differential volume elements are $r\, dz\, dr\, d\theta$ and $\rho^2 \sin\phi\, d\rho\, d\phi\, d\theta$, respectively.

EXAMPLE 1 Let a rectangular plate covering the rectangle $0 \le x \le 2$, $0 \le y \le 1$ in the x,y-plane have mass density $\sigma(x, y) = x + y^2$ at (x, y). Find the mass of the plate.

Solution Consider a differential element of the plate containing a point (x, y) and having area $dA = dx\, dy$. The mass of this differential element is

$$dm = \sigma(x, y)\, dA = (x + y^2)\, dx\, dy.$$

Thus the mass of the plate is given by the integral

$$\begin{aligned} m &= \int_0^1 \int_0^2 (x + y^2)\, dx\, dy = \int_0^1 \left(\frac{x^2}{2} + xy^2\right)\Bigg]_{x=0}^{2} dy \\ &= \int_0^1 (2 + 2y^2)\, dy = \left(2y + \frac{2}{3}y^3\right)\Bigg]_0^1 = \left(2 + \frac{2}{3}\right) - 0 = \frac{8}{3}. \quad \square \end{aligned}$$

EXAMPLE 2 Let the mass density of a ball of radius a be proportional to the distance from the center of the ball. Find the mass of the ball if the mass density at a distance 1 unit from the center is k.

Solution If we take the center of the ball as origin, then the mass density is given by $k\sqrt{x^2 + y^2 + z^2}$. It is natural to use spherical coordinates in integrating over a ball; in terms of spherical coordinates, the mass density is given by

$$k\sqrt{x^2 + y^2 + z^2} = k\rho.$$

The mass is then

$$\begin{aligned} m &= \int_0^{2\pi} \int_0^{\pi} \int_0^{a} k\rho \cdot \rho^2 \sin\phi\, d\rho\, d\phi\, d\theta \\ &= \int_0^{2\pi} \int_0^{\pi} k\frac{\rho^4}{4}\Bigg]_0^a \sin\phi\, d\phi\, d\theta \\ &= \int_0^{2\pi} \int_0^{\pi} k\frac{a^4}{4} \sin\phi\, d\phi\, d\theta = \frac{ka^4}{4} \int_0^{2\pi} -\cos\phi\Bigg]_0^{\pi} d\theta \\ &= \frac{ka^4}{4} \int_0^{2\pi} [-(-1) + 1]\, d\theta \\ &= \frac{2ka^4}{4} \int_0^{2\pi} d\theta = \frac{ka^4}{2}\theta\Bigg]_0^{2\pi} = k\pi a^4. \quad \square \end{aligned}$$

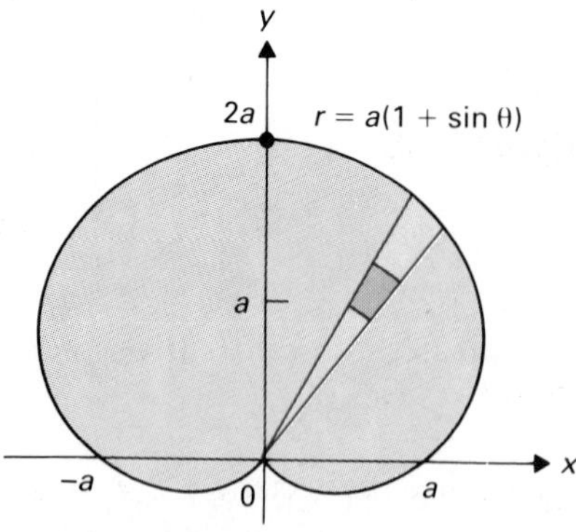

Figure 18.71 For $\sigma(x, y) = k|x|$, the differential element has approximate mass $k|r\cos\theta|r\,dr\,d\theta$.

If we are dealing with a flat sheet of material of constant thickness that is homogeneous in the direction perpendicular to the sheet, we often use mass per unit *area* as mass density.

EXAMPLE 3 Let a flat sheet of material of constant thickness cover the region bounded by the cardioid $r = a(1 + \sin\theta)$ shown in Fig. 18.71, and let the area mass density of the sheet be proportional to the distance from the y-axis. Find the mass of the body.

Solution The mass density is

$$\begin{aligned}\sigma(x, y) &= k|x| \\ &= k|r\cos\theta|,\end{aligned}$$

where k is a constant of proportionality. We make use of the symmetry of both $\sigma(x, y)$ and the cardioid about the y-axis. The mass is given by the integral

$$\begin{aligned}m &= 2\int_{-\pi/2}^{\pi/2}\int_0^{a(1+\sin\theta)} k(r\cos\theta)r\,dr\,d\theta \\ &= 2k\int_{-\pi/2}^{\pi/2} \frac{r^3}{3}\bigg]_0^{a(1+\sin\theta)} \cos\theta\,d\theta \\ &= 2k\int_{-\pi/2}^{\pi/2} \frac{a^3}{3}(1+\sin\theta)^3\cos\theta\,d\theta \\ &= \frac{2}{3}ka^3\frac{(1+\sin\theta)^4}{4}\bigg]_{-\pi/2}^{\pi/2} \\ &= \frac{2}{3}ka^3\frac{2^4}{4} - \frac{2}{3}ka^3(0) \\ &= \frac{8ka^3}{3}. \quad \square\end{aligned}$$

FIRST MOMENTS

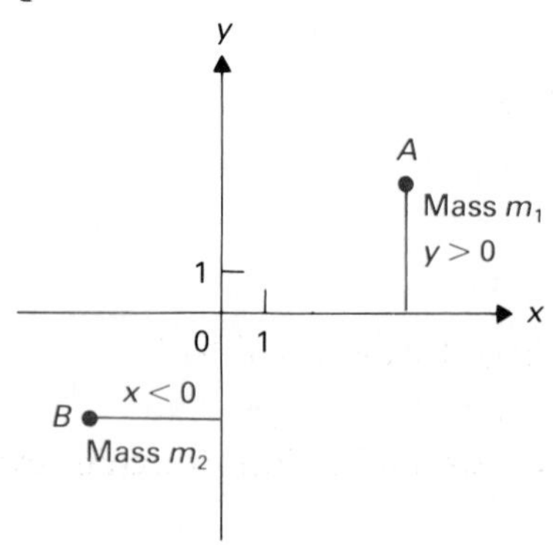

Figure 18.72 Mass of m_1 at A has moment $m_1 y > 0$ about the x-axis; mass of m_2 at B has moment $m_2 x < 0$ about the y-axis.

The **first moment** (or simply the **moment**) about an axis in a plane of a "point mass" in the plane is the product of the mass and the *signed* distance from the axis. See Fig. 18.72. If the point mass is in space, we consider the first moment about a plane, which is the product of the mass and the signed distance from the plane. See Fig. 18.73.

Now consider a body whose mass is not concentrated at one point (which is usually the case). We compute the first moment by adding up products of the masses of small pieces and the signed distances of the pieces from the axis (or plane) and taking the limit as the pieces become smaller and smaller. Of course, this leads to an integral. For a flat sheet of material in the plane, we let M_x and M_y be the first moments about the x-axis and y-axis,

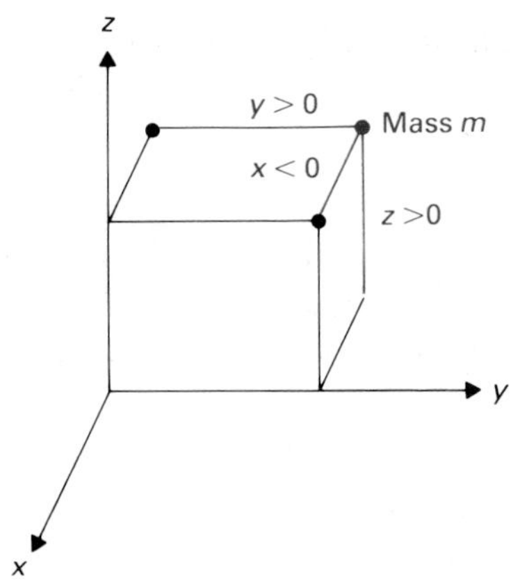

Figure 18.73 The point of mass m has moments $mz > 0$ about the x,y-plane; $my > 0$ about the x,z-plane; $mx < 0$ about the y,z-plane.

respectively. We then have

$$M_x = \iint_G y \cdot \sigma(x, y)\, dx\, dy,$$
$$M_y = \iint_G x \cdot \sigma(x, y)\, dx\, dy. \tag{2}$$

First moments

For a body occupying a region G in space, the first moments M_{xy}, M_{yx}, and M_{xz} about the x,y-plane, y,z-plane, and x,z-plane, respectively, are given by

$$M_{xy} = \iiint_G z \cdot \sigma(x, y, z)\, dx\, dy\, dz,$$
$$M_{yz} = \iiint_G x \cdot \sigma(x, y, z)\, dx\, dy\, dz, \tag{3}$$
$$M_{xz} = \iiint_G y \cdot \sigma(x, y, z)\, dx\, dy\, dz.$$

First moments

We illustrate with two examples.

EXAMPLE 4 Find the first moments about the x-axis and y-axis of a flat sheet of material covering the region bounded by the cardioid $r = a(1 + \sin\theta)$ shown in Fig. 18.74, if the area mass density is the constant k.

Solution By symmetry, the moment about the y-axis is zero, since the mass of a small piece is multiplied by the *signed* distance from the axis; a positive contribution of a piece on the right-hand side of the y-axis in Fig. 18.74 is counterbalanced by the negative contribution of the symmetric piece on the left-hand side. Since the signed distance from a point (x, y) to the x-axis is $y = r\sin\theta$, we obtain

Figure 18.74 For $\sigma(x, y)$ symmetric about the y-axis the sum of contributions of these two differential elements to M_y is zero.

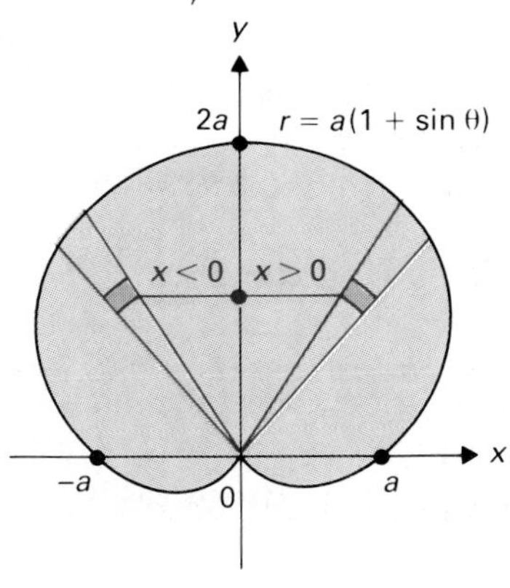

$$\begin{aligned} M_x &= 2\int_{-\pi/2}^{\pi/2}\int_0^{a(1+\sin\theta)} (r\sin\theta)kr\, dr\, d\theta \\ &= 2k\int_{-\pi/2}^{\pi/2} \frac{r^3}{3}\bigg]_0^{a(1+\sin\theta)} \sin\theta\, d\theta \\ &= 2k\int_{-\pi/2}^{\pi/2} \frac{a^3(1+\sin\theta)^3}{3}\sin\theta\, d\theta \\ &= \frac{2ka^3}{3}\int_{-\pi/2}^{\pi/2} (\sin\theta + 3\sin^2\theta + 3\sin^3\theta + \sin^4\theta)\, d\theta. \end{aligned}$$

Since $\sin\theta = -\sin(-\theta)$ and $\sin^3\theta = -\sin^3(-\theta)$, their integrals over

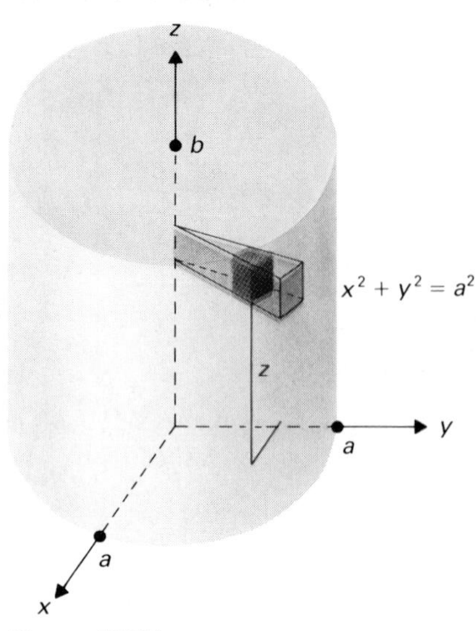

Figure 18.75
$\sigma(x, y, z) = kz$;
$M_{xy} = \int_0^{2\pi}\int_0^a\int_0^b z(kz)r\,dz\,dr\,d\theta$.

$[-\pi/2, \pi/2]$ are zero. Our integral reduces to

$$\frac{2ka^3}{3}\int_{-\pi/2}^{\pi/2}(3\sin^2\theta + \sin^4\theta)\,d\theta$$

$$= \frac{2ka^3}{3}\left(\frac{3\theta}{2} - \frac{3\sin 2\theta}{4} + \frac{3\theta}{8} - \frac{\sin 2\theta}{4} + \frac{\sin 4\theta}{32}\right)\Bigg]_{-\pi/2}^{\pi/2}$$

$$= \frac{2ka^3}{3}\left[\frac{3\pi}{4} + \frac{3\pi}{16} - \left(-\frac{3\pi}{4} - \frac{3\pi}{16}\right)\right]$$

$$= \frac{2ka^3}{3}\left(\frac{3\pi}{2} + \frac{3\pi}{8}\right) = \frac{2ka^3}{3}\cdot\frac{15\pi}{8} = \frac{5k\pi a^3}{4}. \quad \square$$

EXAMPLE 5 Let a solid in space be bounded by the cylinder $x^2 + y^2 = a^2$ and the planes $z = 0$ and $z = b$. If the mass density at a height z above the x,y-plane is kz, find the first moments of the solid about the coordinate planes.

Solution The solid is shown in Fig. 18.75. Symmetry shows at once that

$$M_{xz} = M_{yz} = 0.$$

We use cylindrical coordinates and obtain

$$M_{xy} = \int_0^{2\pi}\int_0^a\int_0^b (z)(kz)r\,dz\,dr\,d\theta = k\int_0^{2\pi}\int_0^a \frac{z^3}{3}\Bigg]_0^b r\,dr\,d\theta$$

$$= k\int_0^{2\pi}\int_0^a \frac{b^3}{3}r\,dr\,d\theta = \frac{kb^3}{3}\int_0^{2\pi}\frac{r^2}{2}\Bigg]_0^a d\theta$$

$$= \frac{kb^3}{3}\int_0^{2\pi}\frac{a^2}{2}\,d\theta = \frac{ka^2b^3}{6}\theta\Bigg]_0^{2\pi} = \frac{k\pi a^2b^3}{3}. \quad \square$$

SECOND MOMENTS

The **second moment** I (or **moment of inertia**) about an axis of a "point mass" is the product of the mass and the *square* of the distance from the axis. A moment of inertia is used in computing kinetic energy of rotation, which is given by the formula

$$\text{K.E.} = \tfrac{1}{2}I\omega^2,$$

where ω is the angular speed of rotation. Computation of a moment of inertia often is accomplished by integration. We illustrate with an example.

EXAMPLE 6 Find the moment of inertia of a homogeneous ball of radius a and constant mass density k about a diameter.

Solution We take the center of the ball at the origin and let the z-axis be the diameter about which the moment of inertia is to be computed. The distance from a point (ρ, ϕ, θ) in spherical coordinates to the z-axis is easily seen to

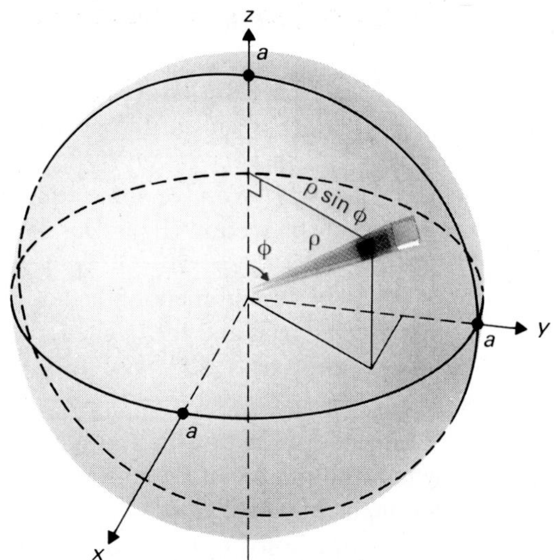

Figure 18.76
$\sigma(x, y, z) = k$, $I_z = \int_0^{2\pi}\int_0^{\pi}\int_0^a (\rho \sin \phi)^2 \cdot k \cdot \rho^2 \sin \phi \, d\rho \, d\phi \, d\theta$.

be $\rho \sin \phi$ (see Fig. 18.76). Thus

$$I = \int_0^{2\pi}\int_0^{\pi}\int_0^a (\rho \sin \phi)^2 k \rho^2 \sin \phi \, d\rho \, d\phi \, d\theta$$

$$= k \int_0^{2\pi}\int_0^{\pi} \frac{\rho^5}{5}\bigg]_0^a \sin^3\phi \, d\phi \, d\theta = k \int_0^{2\pi}\int_0^{\pi} \frac{a^5}{5} \sin^3\phi \, d\phi \, d\theta$$

$$= \frac{ka^5}{5}\int_0^{2\pi}\int_0^{\pi}(1 - \cos^2\phi)\sin\phi \, d\phi \, d\theta$$

$$= \frac{ka^5}{5}\int_0^{2\pi}\left(-\cos\phi + \frac{\cos^3\phi}{3}\right)\bigg]_0^{\pi} d\theta$$

$$= \frac{ka^5}{5}\int_0^{2\pi}\left[-(-1) - \frac{1}{3} - \left(-1 + \frac{1}{3}\right)\right] d\theta = \frac{ka^5}{5}\int_0^{2\pi}\frac{4}{3}\,d\theta = \frac{4ka^5}{15}\,\theta\bigg]_0^{2\pi}$$

$$= \frac{8\pi k a^5}{15}. \quad \square$$

Figure 18.77 (a) Center of mass of a flat body; (b) center of mass of the same body in a different position.

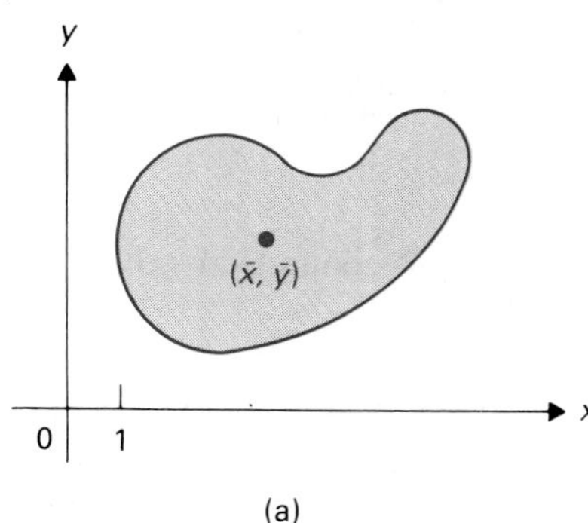

(a)

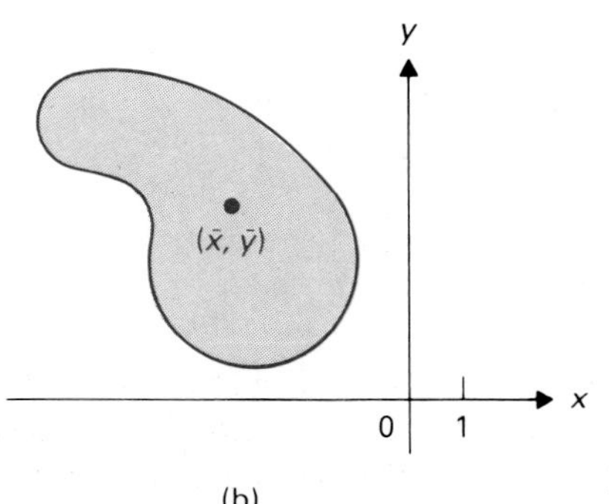

(b)

CENTERS OF MASS AND CENTROIDS

Consider a flat sheet of material in the plane. The **center of mass** of the sheet is the point at which we can consider all the mass to be concentrated for computation of first moments about the coordinate axes (see Fig. 18.77). Thus if the center of mass is $(\bar{x}, \bar{y})$ and the body has mass m, we must have

$$M_x = m\bar{y} \qquad \text{and} \qquad M_y = m\bar{x}.$$

Hence

$$\bar{x} = \frac{M_y}{m} \qquad \text{and} \qquad \bar{y} = \frac{M_x}{m}. \qquad \textbf{(4)}$$

Center $(\bar{x}, \bar{y})$ of mass

It is a fact that

> the location (4) of the center of mass in relation to the body is independent of the position of the body in the plane. **(5)**

See Fig. 18.77. We give some exercises that indicate the reason for this at the end of the section (Exercises 28 and 29). Also

> the first moment of the body about *any* axis is the product of its mass and the (signed) distance from its center of mass to the axis. **(6)**

We should warn you that, in general, there is *no* single point in a body at which the mass can be considered to be concentrated for computation of moments of inertia about *every* axis (see Exercise 30).

For a body in space, coordinates of the center of mass $(\bar{x}, \bar{y}, \bar{z})$ are given by

$$\bar{x} = \frac{M_{yz}}{m}, \qquad \bar{y} = \frac{M_{xz}}{m}, \qquad \text{and} \qquad \bar{z} = \frac{M_{xy}}{m}. \tag{7}$$

Center* $(\bar{x}, \bar{y}, \bar{z})$ *of mass

in analogy with Eqs. (4). Analogous statements to Eqs. (5) and (6) hold for a body in space.

If a body is homogeneous with constant mass density, the center of mass is also called the **centroid of the body,** or the **centroid of the region** that the body occupies.

To compute the center of mass, we simply form the quotients of the first moments by the mass, and we have illustrated how to compute mass and first moments.

EXAMPLE 7 Find the centroid of the plane region bounded by the curves $y = x^2$ and $y = \sqrt{x}$.

Solution The region is shown in Fig. 18.78. By symmetry, we see that $\bar{x} = \bar{y}$. We compute $\bar{x}$.

For a centoid of a region, we assume constant mass density, which we may take to be 1. Then we have

Figure 18.78 By symmetry of the region about $y = x$, we see $\bar{x} = \bar{y}$.

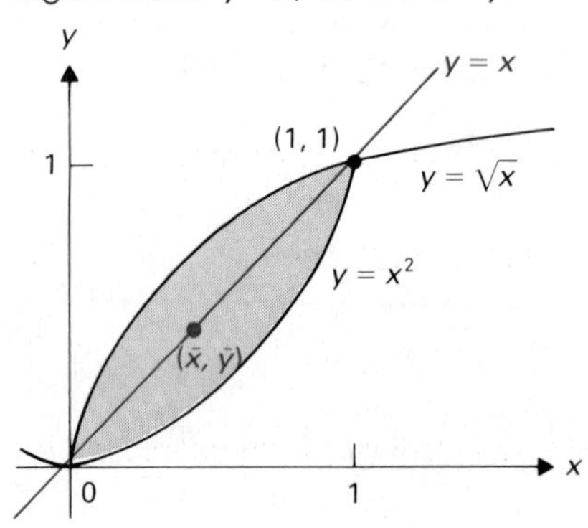

$$m = \int_0^1 \int_{x^2}^{\sqrt{x}} 1 \cdot dy\, dx = \int_0^1 y\Big]_{x^2}^{\sqrt{x}} dx = \int_0^1 (\sqrt{x} - x^2)\, dx$$

$$= \left(\frac{2}{3}x^{3/2} - \frac{x^3}{3}\right)\Big]_0^1 = \frac{2}{3} - \frac{1}{3} = \frac{1}{3}.$$

Also

$$M_y = \int_0^1 \int_{x^2}^{\sqrt{x}} x \cdot 1\, dy\, dx = \int_0^1 xy\Big]_{x^2}^{\sqrt{x}} dx = \int_0^1 (x^{3/2} - x^3)\, dx$$

$$= \left(\frac{2}{5}x^{5/2} - \frac{x^4}{4}\right)\Big]_0^1 = \frac{2}{5} - \frac{1}{4} = \frac{3}{20}.$$

Thus

$$\bar{x} = \frac{M_y}{m} = \frac{3/20}{1/3} = \frac{9}{20},$$

and $(\bar{x}, \bar{y}) = (\frac{9}{20}, \frac{9}{20})$. □

EXAMPLE 8 Consider a solid bounded by the cylinder $x^2 + y^2 = a^2$ and the planes $z = 0$ and $z = b$. Suppose the mass density of the solid at the point (x, y, z) is kz. Find the center of mass of the solid.

Solution In Example 5, we found that $M_{xz} = M_{yz} = 0$ and $M_{xy} = k\pi a^2 b^3/3$. It only remains to compute the mass, which is given by the integral

$$\begin{aligned} m &= \int_0^{2\pi}\int_0^a\int_0^b kzr\,dz\,dr\,d\theta = k\int_0^{2\pi}\int_0^a \frac{z^2}{2}\bigg]_0^b r\,dr\,d\theta = k\int_0^{2\pi}\int_0^a \frac{b^2}{2}r\,dr\,d\theta \\ &= \frac{kb^2}{2}\int_0^{2\pi}\frac{r^2}{2}\bigg]_0^a d\theta = \frac{kb^2}{2}\int_0^{2\pi}\frac{a^2}{2}\,d\theta = \frac{ka^2b^2}{4}\theta\bigg]_0^{2\pi} = \frac{k\pi a^2 b^2}{2}. \end{aligned}$$

Hence

$$\bar{z} = \frac{M_{xy}}{m} = \frac{(k\pi a^2 b^3/3)}{(k\pi a^2 b^2/2)} = \frac{2}{3}b,$$

so the center of mass is at the point $(0, 0, 2b/3)$. □

EXAMPLE 9 Find the first moment of the body in Example 8 about the plane $z = -b$.

Solution From Example 8, we know that the body has mass $m = k\pi a^2 b^2/2$ and center of mass $(0, 0, 2b/3)$. The distance from this center of mass to the plane $z = -b$ is $(2b/3) - (-b) = 5b/3$. Using property (6) for a body in space, we see that the desired first moment is given by

$$\frac{k\pi a^2 b^2}{2}\cdot\frac{5b}{3} = \frac{5k\pi a^2 b^3}{6}. \quad □$$

SUMMARY

Let a body in space occupy a region G.

1. The mass of the body is $m = \iiint_G \sigma(x, y, z)\,dx\,dy\,dz$, where $\sigma(x, y, z)$ is the mass density of the body at (x, y, z).
2. The first moment of the body about a plane is

$$\iiint_G (\text{Signed distance to plane})\cdot\sigma(x, y, z)\,dx\,dy\,dz,$$

where the distance is from the differential element of volume $dx\,dy\,dz$ to the plane.

3. The second moment or moment of inertia of the body about an axis is

$$\iiint_G (\text{Distance from axis})^2\cdot\sigma(x, y, z)\,dx\,dy\,dz,$$

where the distance is from the differential element of volume $dx\,dy\,dz$ to the axis.

4. The center of mass of the body is $(\bar{x}, \bar{y}, \bar{z})$, where

$$\bar{x} = \frac{M_{yz}}{m}, \qquad \bar{y} = \frac{M_{xz}}{m}, \qquad \bar{z} = \frac{M_{xy}}{m}.$$

Here M_{yz} is the first moment about the y,z-plane, and so on.

Analogous formulas give these quantities for a body occupying a region in the plane.

EXERCISES

1. Let the area mass density at a point (x, y) of a flat body covering the square $0 \le x \le 1, 0 \le y \le 1$ be xy^2.
a) Find the mass of the body.
b) Find the center of mass of the body.

2. Consider a flat body covering the plane region bounded by $y = x^2$ and $x = y^2$, and let the area mass density of the body at a point (x, y) be xy. Find the center of mass of the body. [*Hint:* Use symmetry.]

Exercises 3 through 6 concern a thin plate covering the plane disk $x^2 + y^2 \le a^2$. Let the mass density be proportional to the distance from the center of the disk, with mass density k at a distance 1 unit from the center.

3. Find the mass of the plate.

4. Find the first moment of the plate about the line $x = -a$.

5. Find the absolute value of the first moment of the plate about the line $x + y = 2a$.

6. Find the moment of inertia of the plate about an axis perpendicular to the plane through the origin.

7. a) Find the x,y-coordinates of the centroid of the sector of the circle $0 \le r \le a$, where $0 \le \theta \le \alpha$ for $0 < \alpha \le 2\pi$.
b) Find the limiting position of the centroid in part (a) as $\alpha \to 0$.

8. Find the centroid of the plane region inside the cardioid $r = a(1 + \cos\theta)$ and outside the circle $r = a$.

Exercises 9 through 12 concern a thin plate of mass density $\sigma(x, y)$ covering the upper half-disk $x^2 + y^2 \le a^2, y \ge 0$.

9. Find the center of mass if $\sigma(x, y) = k$.

10. Find the center of mass if $\sigma(x, y) = kx^2y$.

11. Find the center of mass if $\sigma(x, y) = k\sqrt{x^2 + y^2}$.

12. a) Find the center of mass if $\sigma(x, y) = k(x^2 + y^2)^{n/2}$ for $n \ge 0$.
b) Find the limiting position of the center of mass in part (a) as $n \to \infty$.

Exercises 13 through 17 concern a solid cone in space bounded by $z = a$ and $z = \sqrt{x^2 + y^2}$. Let the mass density be proportional to the distance from the z-axis, with mass density k 1 unit away from the z-axis.

13. Find the mass of the solid.

14. Find the center of mass of the solid.

15. Find the absolute value of the first moment of the solid about the plane $x = a$.

16. Find the absolute value of the first moment of the solid about the plane $z = -a$.

17. Find the moment of inertia of the solid about the z-axis.

Exercises 18 through 21 concern a solid in space bounded by the cylinder $y = a^2 - z^2$ and the planes $y = 0$, $x = 0$, and $x = b$. Let the mass density of the solid be $\sigma(x, y, z) = ky$.

18. Find the mass of the solid.

19. Find the centroid of the solid.

20. Find the absolute value of the first moment of the solid about the plane $x + y - 2z = 4$.

21. Express as an integral the moment of inertia of the solid about the y-axis.

22. Find the moment of inertia of a solid ball in space of radius a and constant mass density k about a line tangent to the ball.

23. Find the centroid of the hemispherical region in space bounded by $z = \sqrt{a^2 - x^2 - y^2}$ and $z = 0$.

24. Find the centroid of the region in space bounded by $z = 0$, $x^2 + y^2 = 4$, and $z = 1 + x^2 + y^2$.

25. Find the centroid of the region in space bounded by $x = 0$ and $x = 4 - y^2 - z^2$.

Exercises 26 and 27 concern a solid of mass density $\sigma(\rho, \phi, \theta)$ that occupies the region in space bounded above by the sphere $\rho = a$ and below by the cone $\phi = \alpha$, where $0 < \alpha \le \pi$.

26. a) Find the center of mass if $\sigma(\rho, \phi, \theta) = k$.
b) Find the limiting position of the center of mass in part (a) as $\alpha \to 0$.
c) Use the answer to part (a) to find the centroid of the half-ball $0 \le z \le \sqrt{a^2 - x^2 - y^2}$.
d) Use the answer to part (a) to find the centroid of the solid bounded below by $z = \sqrt{x^2 + y^2}$ and above by $z = \sqrt{a^2 - x^2 - y^2}$.

27. a) Find the center of mass if $\sigma(\rho, \phi, \theta) = k\rho^n$ for $n \ge 0$.
b) Find the limiting position of the center of mass in part (a) as $n \to \infty$.

28. a) Show that the first moment of a body in the plane about the line $x = -a$ is $M_y + ma$. (This is known as the *parallel axis theorem.*)
    b) Let a new origin (h, k) be chosen in the plane and let the x'-axis be the line $y = k$ and the y'-axis the line $x = h$. Argue from part (a) that the same location for the center of mass of a body in the plane, relative to the body, is obtained whether one computes coordinates of the center using the x-axis and y-axis or using the x'-axis and y'-axis.
29. State the analogue for space of Exercise 28(a).
30. Let a flat body of constant mass density k cover the unit square $0 \le x \le 1, 0 \le y \le 1$, in the plane.
    a) Find the moment of inertia of the body about the y-axis.
    b) Find the moment of inertia of the body about the line $x = -a$.
    c) Find a point (x_1, y_1) in the body such that the moment of inertia of the body about either the x-axis or the y-axis is the product of the mass and the square of the distance from (x_1, y_1) to the axis.
    d) Find a point (x_2, y_2) in the body such that the moment of inertia of the body about either the line $x = -a$ or the line $y = -a$ is the product of the mass and the square of the distance from (x_2, y_2) to the line.
    e) Compare the answers to parts (c) and (d), and comment on the result.
31. The **radius of gyration** R of a body about an axis is defined by

$$R = \sqrt{\frac{I}{m}},$$

so that $I = mR^2$.
    a) From the answer $k/3$ to Exercise 30(a), what is the radius of gyration about the y-axis of a homogeneous flat body covering the square $0 \le x \le 1, 0 \le y \le 1$?
    b) From the answer

$$\frac{k}{3}[(a + 1)^3 - a^3]$$

to Exercise 30(b), what is the radius of gyration about the line $x = -a$ of a homogeneous flat body covering this square?

18.6 SURFACE AREA

Let $z = f(x, y)$ be a function of two variables with continuous partial derivatives. Let G be a closed bounded region in the domain of f. The graph of f over G is then a surface in space, as indicated in Fig. 18.79. We attempt to find the area of this surface.

The situation is analogous to finding the length of a curve lying over an interval $[a, b]$ on the x-axis, as shown in Fig. 18.80. In that case, we use a short tangent-line segment of length ds to approximate a short length of curve. The tangent segment of length ds lies over a segment of length dx on the x-axis, as shown in Fig. 18.80. We have a formula for ds in terms of dx, namely

$$ds = \sqrt{1 + (f'(x))^2}\, dx.$$

Figure 18.79 Portion of the surface $z = f(x, y)$ lying over G.

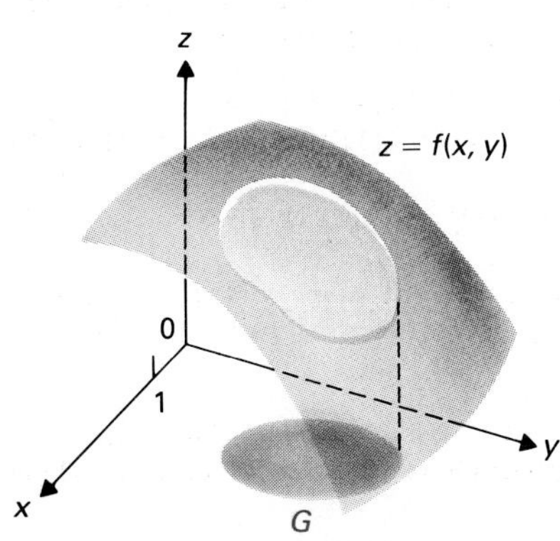

For surfaces, we approximate the area of a small piece of surface by the area of a piece of a tangent plane to the surface. We would like to find a formula for the area dS of a piece of tangent plane lying over a differential rectangle of area $dx\, dy$ in the x,y-plane.

Figure 18.81 shows the tangent plane to be $z = f(x, y)$ at a point (x, y, z) and also shows a parallelogram in the tangent plane that lies over a differential rectangle of area $dx\, dy$ in the x,y-plane. We can find the vectors along the two edges of the parallelogram shown in Fig. 18.81. The vector lying over the line segment of length dx in the x,y-plane must have dx as coefficient of $\mathbf{i}$. Since $\partial z/\partial x$ gives the "slope" in the x-direction of the tangent plane, we see that this vector is

$$(dx)\mathbf{i} + \left(\frac{\partial z}{\partial x}\, dx\right)\mathbf{k}.$$

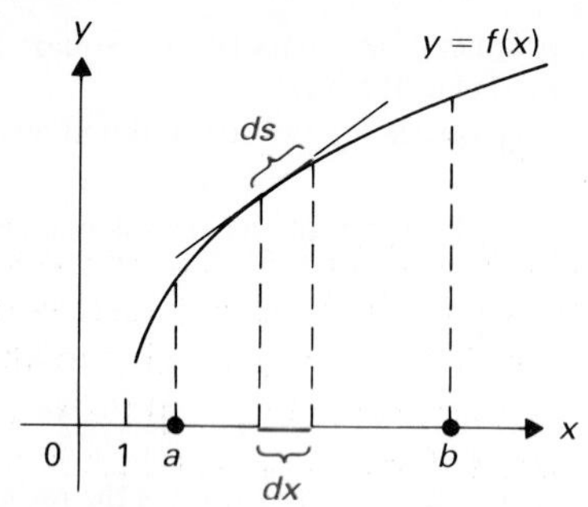

Figure 18.80 Approximation *ds* to arc length lying over the segment of length *dx*.

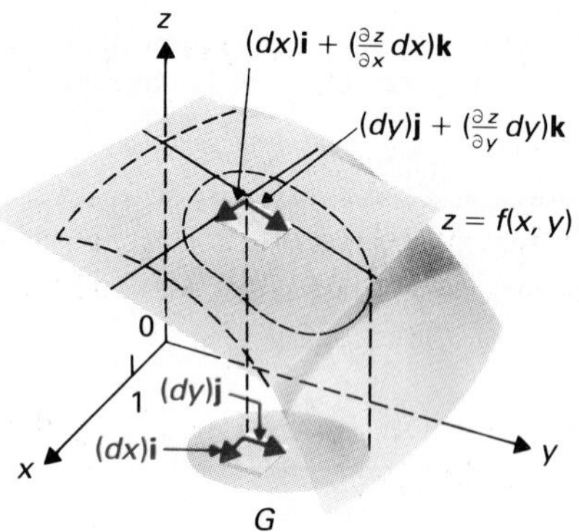

Figure 18.81 The area *dS* of the parallelogram in the tangent plane is the magnitude of the cross product of the vectors shown.

A similar argument shows that the vector lying over the line segment of length dy is

$$(dy)\mathbf{j} + \left(\frac{\partial z}{\partial y}\,dy\right)\mathbf{k}.$$

The area of this parallelogram is the magnitude of the cross product of the vectors. Computing, we obtain the cross product

$$\begin{vmatrix} \mathbf{i} & \mathbf{j} & \mathbf{k} \\ dx & 0 & \dfrac{\partial z}{dx}\,dx \\ 0 & dy & \dfrac{\partial z}{\partial y}\,dy \end{vmatrix} = -\left(\frac{\partial z}{dx}\,dx\,dy\right)\mathbf{i} - \left(\frac{\partial z}{\partial y}\,dx\,dy\right)\mathbf{j} + (dx\,dy)\mathbf{k}.$$

The length of this cross product is the *differential element dS of surface area*, so

$$\boxed{dS = \sqrt{\left(\frac{\partial z}{\partial x}\right)^2 + \left(\frac{\partial z}{\partial y}\right)^2 + 1}\,dx\,dy.} \qquad \textbf{(1)}$$

Consequently, we have

$$\boxed{\text{Surface area} = \iint_G \sqrt{\left(\frac{\partial z}{\partial x}\right)^2 + \left(\frac{\partial z}{\partial y}\right)^2 + 1}\,dx\,dy.} \qquad \textbf{(2)}$$

EXAMPLE 1 Find the area of the sphere $x^2 + y^2 + z^2 = a^2$.

Solution We find the area of the top hemisphere $z = \sqrt{a^2 - x^2 - y^2}$ and then double the result for our final answer. Computing, we find that

$$\frac{\partial z}{\partial x} = \frac{-x}{\sqrt{a^2 - x^2 - y^2}} \qquad \text{and} \qquad \frac{\partial z}{\partial y} = \frac{-y}{\sqrt{a^2 - x^2 - y^2}}.$$

Hence

$$\begin{aligned} dS = \sqrt{\left(\frac{\partial z}{\partial x}\right)^2 + \left(\frac{\partial z}{\partial y}\right)^2 + 1}\,dx\,dy &= \sqrt{\frac{x^2 + y^2}{a^2 - x^2 - y^2} + 1}\,dx\,dy \\ &= \sqrt{\frac{x^2 + y^2 + a^2 - x^2 - y^2}{a^2 - x^2 - y^2}}\,dx\,dy \\ &= \sqrt{\frac{a^2}{a^2 - x^2 - y^2}}\,dx\,dy \\ &= \frac{a}{\sqrt{a^2 - x^2 - y^2}}\,dx\,dy. \end{aligned}$$

Since we will be integrating over the disk $x^2 + y^2 \le a^2$ (or $r \le a$), and $x^2 + y^2 = r^2$ appears in our computation of dS, we change to polar coordinates, where

$$\frac{a}{\sqrt{a^2 - x^2 - y^2}} = \frac{a}{\sqrt{a^2 - r^2}}.$$

We thus form the integral

$$\int_0^{2\pi}\int_0^a \frac{a}{\sqrt{a^2 - r^2}}\, r\, dr\, d\theta.$$

We should point out that our integrand is undefined for $r = a$, so this is an improper integral of two variables, which we really have not discussed. (Geometrically, this happens because our surface is perpendicular to the x,y-plane at $r = a$, so $\partial z/\partial x$ and $\partial z/\partial y$ are undefined there.) In straightforward analogy with our improper integrals of a function of one variable, we compute

$$\begin{aligned}\int_0^{2\pi}\left(\lim_{h\to a^-}\int_0^h \frac{a}{\sqrt{a^2 - r^2}}\, r\, dr\right) d\theta &= \int_0^{2\pi} \lim_{h\to a^-} \left(-a\sqrt{a^2 - r^2}\right)\Big]_0^h\, d\theta \\ &= \int_0^{2\pi}\left[\lim_{h\to a^-}\left(-a\sqrt{a^2 - h^2}\right) + a^2\right] d\theta \\ &= \int_0^{2\pi} a^2\, d\theta = a^2\theta\Big]_0^{2\pi} = 2\pi a^2.\end{aligned}$$

Doubling, we obtain $4\pi a^2$ as the area of the sphere. □

The radical appearing in formula (1) for dS often makes integration to find surface area difficult. Sometimes it is necessary to use a table or to find a numerical estimate for the surface area.

EXAMPLE 2 Find the area of the portion of the surface $z = x^2 + y$ that lies over the region $0 \le x \le 1, 0 \le y \le 1$ in the x,y-plane.

Solution We see from formula (1) that

$$dS = \sqrt{4x^2 + 1 + 1}\, dx\, dy = \sqrt{2 + 4x^2}\, dx\, dy.$$

Integrating first with respect to y and then using a table, we find that the surface area is

$$\begin{aligned}\int_0^1\int_0^1 \sqrt{2 + 4x^2}\, dy\, dx &= \int_0^1 \sqrt{2 + 4x^2}\, y\Big]_0^1 dx = \int_0^1 \sqrt{2 + 4x^2}\, dx \\ &= \int_0^1 \sqrt{1 + (\sqrt{2}x)^2}\,\sqrt{2}\, dx \\ &= \left(\frac{\sqrt{2}x}{2}\sqrt{1 + (\sqrt{2}x)^2}\right. \\ &\qquad \left. + \frac{1}{2}\ln\left(\sqrt{2}x + \sqrt{1 + (\sqrt{2}x)^2}\right)\right]_0^1 \\ &= \frac{\sqrt{2}}{2}\sqrt{3} + \frac{1}{2}\ln(\sqrt{2} + \sqrt{3}) - 0 \\ &= \frac{1}{2}\left(\sqrt{6} + \ln(\sqrt{2} + \sqrt{3})\right).\end{aligned}$$

□

Sometimes it is more convenient to integrate over the projection of a surface on the x,z-plane or y,z-plane rather than the x,y-plane. We then make the obvious modification in formula (1) for dS.

EXAMPLE 3 Use a calculator or computer to estimate the area of the portion of the surface $x + y + z^3 = 5$, where $0 \le z \le 1$ and $-1 \le y \le 0$.

Solution We write the surface equation in the form $x = f(y, z) = 5 - y - z^3$. Interchanging the roles of x and z in formula (1), we have

$$dS = \sqrt{\left(\frac{\partial x}{\partial z}\right)^2 + \left(\frac{\partial x}{\partial y}\right)^2 + 1}\, dy\, dz = \sqrt{9z^4 + 1 + 1}\, dy\, dz.$$

Thus the surface area is given by

$$\int_0^1 \int_{-1}^0 \sqrt{9z^4 + 2}\, dy\, dz = \int_0^1 \sqrt{9z^4 + 2}\, y\Big]_{-1}^0 dz = \int_0^1 \sqrt{9z^4 + 2}\, dz.$$

Using Simpson's rule with $n = 40$, we obtain the estimate 1.870345415 for this integral and the surface area. □

Sometimes a surface is given in the form $F(x, y, z) = 0$ rather than in the form $z = f(x, y)$. Recall that if F has continuous partial derivatives and $\partial F/\partial z$ does not assume the value zero in a neighborhood of a point, then the surface does define an implicit function $z = f(x, y)$ in a neighborhood of the point, and, furthermore,

$$\frac{\partial z}{\partial x} = -\frac{\partial F/\partial x}{\partial F/\partial z} \quad \text{and} \quad \frac{\partial z}{\partial y} = -\frac{\partial F/\partial y}{\partial F/\partial z}.$$

We then obtain

$$\sqrt{\left(\frac{\partial z}{\partial x}\right)^2 + \left(\frac{\partial z}{\partial y}\right)^2 + 1} = \sqrt{\left(\frac{\partial F/\partial x}{\partial F/\partial z}\right)^2 + \left(\frac{\partial F/\partial y}{\partial F/\partial z}\right)^2 + 1}$$
$$= \frac{\sqrt{(\partial F/\partial x)^2 + (\partial F/\partial y)^2 + (\partial F/\partial z)^2}}{|\partial F/\partial z|}.$$

This gives us the formula

$$\boxed{dS = \frac{\sqrt{F_x^2 + F_y^2 + F_z^2}}{|F_z|}\, dx\, dy} \qquad (3)$$

in place of formula (1). Of course, in using formula (3) to find surface area, we have to express all the partial derivatives as functions of x and y only, which may be difficult. We repeat Example 1, finding the area of a sphere, but using formula (3) to find dS.

EXAMPLE 4 Find the area of the sphere $x^2 + y^2 + z^2 = a^2$, using formula (3) to compute dS.

Solution We let $F(x, y, z) = x^2 + y^2 + z^2 - a^2$. From formula (3), we see that

$$dS = \frac{\sqrt{(2x)^2 + (2y)^2 + (2z)^2}}{|2z|}\, dx\, dy = \frac{\sqrt{4(x^2 + y^2 + z^2)}}{|2z|}\, dx\, dy$$
$$= \frac{\sqrt{4a^2}}{2|z|}\, dx\, dy = \frac{a}{|z|}\, dx\, dy = \frac{a}{\sqrt{a^2 - x^2 - y^2}}\, dx\, dy.$$

We have obtained the same formula for dS as in Example 1, and from this point the computations are identical. □

SUMMARY

1. The area of a surface consisting of part of a graph $z = f(x, y)$ is equal to the integral

$$\iint_G \sqrt{\left(\frac{\partial z}{\partial x}\right)^2 + \left(\frac{\partial z}{\partial y}\right)^2 + 1}\, dx\, dy$$

evaluated over the region G in the x,y-plane under the surface.

2. The area of a surface consisting of part of a surface $F(x, y, z) = c$ is equal to the integral

$$\iint_G \frac{\sqrt{(\partial F/\partial x)^2 + (\partial F/\partial y)^2 + (\partial F/\partial z)^2}}{|\partial F/\partial z|}\, dx\, dy$$

evaluated over the region G in the x,y-plane under the surface.

3. The differential element of surface area is

$$dS = \sqrt{\left(\frac{dz}{\partial x}\right)^2 + \left(\frac{dz}{dy}\right)^2 + 1}\, dx\, dy \qquad \text{for the graph of } z = f(x, y)$$

and

$$dS = \frac{\sqrt{(\partial F/\partial x)^2 + (\partial F/\partial y)^2 + (\partial F/\partial z)^2}}{|\partial F/\partial z|}\, dx\, dy$$

$$= \frac{|\nabla F|}{|\partial F/\partial z|} \qquad \text{for } F(x, y, z) = c.$$

EXERCISES

In Exercises 1 through 16, find the area of the indicated surface. Use the integral table where necessary.

1. The portion of the plane $x + 2y + 2z = 10$ inside the cylinder $x^2 + y^2 = 4$
2. The portion of the plane $6x + 3y + 2z = 100$ inside the cylinder $y^2 + z^2 = 9$
3. The portion of the plane $4x + 3y + 5z = 50$ lying over the region in the x,y-plane bounded by $y = x$ and $y = x^2$
4. The portion of the surface $x + 2y + 2z^{3/2} = 100$ where $0 \le x \le 4$ and $0 \le z \le 3$
5. The portion of the surface $4x^{3/2} + 3y + 4z = 8$, where $0 \le x \le 4$ and $0 \le y \le 2$
6. The portion of the surface $z = 4x^2 + 3y$, where $0 \le x \le 1$ and $0 \le y \le 1$
7. The portion of the surface $4x + 3y^2 + 2z = 10$, where $0 \le y \le 1$ and $0 \le z \le 2$
8. The portion of the surface $z = \frac{2}{3}(x^{3/2} + y^{3/2})$, where $0 \le x \le 1$ and $0 \le y \le 2$
9. The portion of the surface $z = x^2 + y^2$ inside the cylinder $x^2 + y^2 = a^2$
10. The portion of the surface $4x - y^2 + z^2 = 8$ inside the cylinder $y^2 + z^2 = 25$
11. The portion of the hemisphere $z = -\sqrt{25 - x^2 - y^2}$ inside the cylinder $x^2 + y^2 = 9$
12. The portion of the sphere $x^2 + y^2 + z^2 = a^2$ inside the cone $z = \sqrt{x^2 + y^2}$
13. The portion of the surface $ax = z^2 - y^2$ inside the cylinder $y^2 + z^2 = a^2$
14. The portion of the sphere $x^2 + y^2 + z^2 = a^2$ inside the cylinder $x^2 + z^2 = az$
15. The surface of the solid bounded by $z = 4 - x^2 - y^2$ and by $z = -4 + x^2 + y^2$

16. The surface of the solid bounded above by $z = 4 - x^2 - y^2$, below by $z = -\sqrt{8 - x^2 - y^2}$, and on the sides by $x^2 + y^2 = 4$

In Exercises 17 through 20, use a calculator or computer and Simpson's rule to estimate the area of the indicated surface.

17. The portion of the surface $4x + 3y^3 + 5z = 100$, where $0 \le x \le 5$ and $0 \le y \le 1$

18. The portion of the surface $3x + 4y^{3/2} + z^3 = 16$, where $0 \le y \le 4$ and $0 \le z \le 1$

19. The portion of the hemisphere $z = \sqrt{25 - x^2 - y^2}$, where $-1 \le x \le 1$ and $-1 \le y \le 1$

20. The portion of the paraboloid $z = x^2 + y^2$, where $-2 \le x \le 2$ and $-2 \le y \le 2$

EXERCISE SETS FOR CHAPTER 18

Review Exercise Set 18.1

1. Find the Riemann sum $\mathscr{S}_2$, using midpoints of subrectangles, approximating the integral of the function $f(x, y) = xy^2$ over the rectangle $1 \le x \le 3$, $-1 \le y \le 3$.
2. Compute $\int_{-1}^{2} \int_{1}^{4} (3xy^2 - 2y)\, dx\, dy$.
3. Express $\int_{-2}^{2} \int_{x^2}^{4} (x^2 - 3xy)\, dy\, dx$ in the form

$$\iint (x^2 - 3xy)\, dx\, dy,$$

reversing the order of integration. Do not compute either integral.
4. Compute, using an iterated integral, the area of the region in the plane bounded by $y = x^2$ and $y = x$.
5. Express as an iterated integral the volume of the region in space bounded below by $z = x^2 + y^2$ and above by the hemisphere $z - 4 = \sqrt{4 - x^2 - y^2}$. Do not compute the integral.
6. Find the integral of the polar function $h(r, \theta) = r \cos \theta$, $r \ge 0$, over the circle $r = 2a \cos \theta$.
7. Use triple integration in cylindrical coordinates to find the volume of the solid bounded by the paraboloids $z = x^2 + y^2$ and $z = 8 - x^2 - y^2$.
8. Find the integral of the spherical coordinate function $h(\rho, \phi, \theta) = \rho \cos^2\theta$ over the ball $0 \le \rho \le a$.
9. Find the moment of inertia of a solid ball $x^2 + y^2 + z^2 \le a^2$ about the z-axis if the mass density of the ball at (x, y, z) is given by $|z|$.
10. Find the area of the portion of the surface of the sphere $x^2 + y^2 + z^2 = a^2$ lying above the plane $z = b$ for $0 \le b \le a$.

Review Exercise Set 18.2

1. Find the Riemann sum $\mathscr{S}_2$, using midpoints of the subrectangles, approximating the integral of the function $f(x, y) = 3x - 2y$ over the rectangle $-1 \le x \le 3$, $1 \le y \le 3$.
2. Compute $\int_{4}^{-2} \int_{-1}^{3} (2xy - 3y^2)\, dy\, dx$.
3. Express $\int_{0}^{2} \int_{0}^{\sqrt{4-y^2}} \int_{x^2+y^2}^{4} x^2 z\, dz\, dx\, dy$ in the form

$$\iiint x^2 z\, dx\, dy\, dz,$$

changing the order of integration. Do not evaluate either integral.
4. Compute, using an iterated integral, the area of the region in the plane bounded by $y = 1/x$ and $x + y = \frac{5}{2}$.
5. Compute, using an iterated integral, the volume of the region in space bounded below by $z = 0$, on the sides by $x = 0$, $y = 0$, and $x + y = 2$, and above by $z = x^2 + y^2$.
6. Using double integration in polar coordinates, find the area of one loop of the rose $r = \cos 3\theta$.
7. Find the integral of the cylindrical coordinate function $h(r, \theta, z) = rz \sin^2\theta$, $r \ge 0$, over the region bounded above by $z = 4 + x^2 + y^2$, below by $z = 0$, and on the sides by $x^2 + y^2 = 4$.
8. Use triple integration in spherical coordinates to find the volume of the solid bounded by the cone $z^2 = 3x^2 + 3y^2$ and the hemisphere $z = \sqrt{16 - x^2 - y^2}$.
9. Find the centroid of the region in space bounded by $z = -1$ and $z = 3 - x^2 - y^2$.
10. Find the area of the surface $z = 16 - x^2 - y^2$ that lies above the plane $z = 12$.

Review Exercise Set 18.3

1. Use a calculator or computer and Simpson's rule to estimate

$$\int_0^2 \int_0^1 \frac{y}{1 + \sin x^2}\, dx\, dy.$$

2. Compute $\int_0^1 \int_1^2 (x/y^2)\, dy\, dx$.
3. Express $\int_{-2}^{2} \int_0^{\sqrt{4-y^2}} \int_0^{4-x^2-y^2} yz\, dz\, dx\, dy$ in the $dx\, dy\, dz$-order.
4. Find the volume of the region in space bounded above by $z = 2x + y$, on the sides by $x = 0$, $y = 0$, and $x + y = 1$, and below by $z = -1$.
5. Transform $\int_0^1 \int_x^1 (x^2 + y^2)^{3/2}\, dy\, dx$ into polar coordinates. Do not evaluate the integral.
6. Transform the integral $\int_{\pi/2}^{\pi} \int_0^1 \int_0^{r^2} r^2\, dz\, dr\, d\theta$ into rectangular coordinates. Do not evaluate the integral.
7. Integrate the spherical coordinate function $f(\rho, \phi, \theta) = \rho^2 \cos^2\theta$ over the ball $\rho \le a$.

8. Let a thin plate cover the disk $r \le 4$ and have mass density $\sigma(r, \theta) = kr^2$. Find the absolute first moment of the plate about the line $3x + 4y = 100$.
9. Find the centroid of the region bounded below by $z = 0$, above by $z = 1 - x^2$, and on the ends by $y = 0$ and $y = 4$.
10. Find the area of the portion of the plane $x - 3y + 2z = 4$ inside the elliptic cylinder $(x^2/4) + (y^2/9) = 1$.

Review Exercise Set 18.4

1. Use Simpson's rule with a calculator or computer to estimate

$$\int_0^3 \int_0^1 \frac{x^2y}{y + \cos^2 y}\, dy\, dx.$$

2. Compute $\int_1^3 \int_0^2 (x + y)/(y + 1)\, dx\, dy$.
3. Compute $\int_0^1 \int_y^1 e^{x^2+1}\, dx\, dy$.
4. Express $\int_0^1 \int_0^3 \int_{\sqrt{1-z}}^{1-z} yz^2\, dx\, dy\, dz$ in the $dz\, dx\, dy$-order.
5. Find the volume of the region in space bounded by $z = x^2 + y^2$ and $z = \sqrt{x^2 + y^2}$.
6. Transform $\int_0^{2\pi} \int_0^1 \int_{r^2}^{r} r\, dz\, dr\, d\theta$ into spherical coordinates. Do not evaluate the integral.
7. Find the volume of the region bounded above by $z = \sqrt{16 - x^2 - y^2}$ and below by $z = \sqrt{3x^2 + 3y^2}$.
8. A thin plate covers the cardioid $r = a(1 + \cos\theta)$. If the mass density of the plate is $\sigma(r, \theta) = kr$, find the mass of the plate.
9. The mass density of a solid bounded above by $z = \sqrt{4 - x^2 - y^2}$ and below by $z = \sqrt{x^2 + y^2}$ is $\sigma(\rho, \phi, \theta) = k\rho$. Find the moment of inertia of the solid about the z-axis.
10. Find the area of the portion of the sphere $x^2 + y^2 + z^2 = 16$ inside the cone $\sqrt{3}z = \sqrt{x^2 + y^2}$.

More Challenging Exercises 18

1. Use integration to find the four-dimensional "volume" of the "4-ball of radius a" consisting of all (x, y, z, w) such that $x^2 + y^2 + z^2 + w^2 \le a^2$. [*Hint:* Use (ρ, ϕ, θ, w) co-ordinates where ρ, ϕ, and θ are the usual spherical coordinates replacing x, y, and z. This is analogous to cylindrical co-ordinates (r, θ, z) in space compared with the polar co-ordinates (r, θ) in the plane.]
2. Work Exercise 1 again, but this time use Pappus' theorem and the fact that the centroid of the half-ball $x^2 + y^2 + z^2 \le a^2$, where $z \ge 0$, is $(0, 0, 3a/8)$. (See Exercise 23 of Section 18.5.)
3. Find the three-dimensional volume of the "3-sphere of radius a" consisting of all (x, y, z, w) such that $x^2 + y^2 + z^2 + w^2 = a^2$. [*Hint:* Recall that for a 3-ball $x^2 + y^2 + z^2 \le a^2$ of volume $(4/3)\pi a^3$, the area of the surface of the 2-sphere $x^2 + y^2 + z^2 = a^2$ is $4\pi a^2$. Consider the relation between $V = (4/3)\pi a^3$ and $A = 4\pi a^2$ in terms of approximation of the volume V by a differential for a small change in radius. Then jump up one dimension and answer the new problem with essentially no work, using the answer to Exercise 1.]

If we partition the rectangle $0 \le x \le 4$, $-2 \le y \le 6$ into n^2 subrectangles of equal areas, then each subrectangle has area $32/n^2$. Using Riemann sums with the upper right corner of each subrectangle as the place to evaluate the function, we see that

$$\int_{-2}^{6} \int_0^4 x^2y\, dx\, dy = \lim_{n\to\infty} \sum_{i,j=1}^{n} \left[\left(\frac{4i}{n}\right)^2\left(-2 + \frac{8j}{n}\right)\left(\frac{32}{n^2}\right)\right].$$

Exercises 4 through 7 give you some practice in writing double integrals as limits of double sums and in estimating some double sums using integrals.

4. Write a sum whose limit as $n \to \infty$ is equal to

$$\int_1^7 \int_{-3}^0 (x + 4y)^3\, dx\, dy.$$

5. Repeat Exercise 4 for $\int_5^{10} \int_{-2}^2 (x^2 + 3xy)\, dy\, dx$.
6. Use an integral to estimate

$$\sum_{i,j=1}^{100} \left[\frac{3i}{100}\left(-1 + \left(\frac{2j}{100}\right)^2\right)\frac{6}{10{,}000}\right].$$

7. Use an integral to estimate $(8/10^7)\sum_{i,j=1}^{10} i^2j^3$.

Exercises 8 through 11 deal with a simple algebraic device that is useful in multivariable calculus. We define a type of product of differential expressions, which is written using the symbol $\wedge$. All the properties of regular multiplication hold for this $\wedge$-product except that

$$dx \wedge dy = -dy \wedge dx$$

while

$$dx \wedge dx = dy \wedge dy = 0.$$

With three variables, we have the logical extension of such relations. For example,

$$dx \wedge dy \wedge dz = -dx \wedge dz \wedge dy$$

and

$$dy \wedge dz \wedge dy = 0.$$

Roughly speaking, a term containing the $\wedge$-product differentials like dx, dy and so on, is zero if two differentials are the same and changes sign if two differentials are interchanged. For a further illustration,

$$\begin{aligned}(x^2\, dx + y\, dy) \wedge (2y\, dx - x^2y\, dy) &= 2x^2y\, dx \wedge dx - x^4y\, dx \wedge dy + 2y^2\, dy \wedge dx \\ &\quad - x^2y^2\, dy \wedge dy \\ &= -x^4y\, dx \wedge dy + 2y^2\, dy \wedge dx \\ &= -x^4y\, dx \wedge dy - 2y^2\, dx \wedge dy \\ &= -(x^4y + 2y^2)\, dx \wedge dy.\end{aligned}$$

8. Compute

$$(x\, dx + x^2y\, dy - xz\, dz) \wedge (yz\, dx + xz\, dy - z^2\, dz),$$

as in the preceding illustration, simplifying as much as possible.

Exercises 9 through 11 indicate that this new multiplication is useful in changing variables in multiple integrals.

9. Recall that for polar coordinates, $x = r \cos \theta$ and $y = r \sin \theta$.
a) Compute dx and dy in terms of the polar coordinate variables.
b) Using part (a), compute $dx \wedge dy$ in terms of polar coordinates, simplifying as much as possible.

10. Recall that, for spherical coordinates,

$$x = \rho \sin \phi \cos \theta, \qquad y = \rho \sin \phi \sin \theta,$$
$$z = \rho \cos \phi.$$

a) Compute dx, dy, and dz in terms of spherical coordinates.
b) Using part (a), compute $dx \wedge dy \wedge dz$ in terms of spherical coordinates, simplifying as much as possible.

11. Consider $\iint_G (x - y)^4(3x + 2y)^5 \, dx \, dy$, where G is the parallelogram bounded by $x - y = 1$, $x - y = 3$, $3x + 2y = -1$, and $3x + 2y = 2$. This integral is tough to evaluate in x,y-coordinates. Exercises 9 and 10 should give you the clue as to how to proceed. Make the variable substitution $u = x - y$, $v = 3x + 2y$. Solve for x and y in terms of u and v, and compute dx and dy. Then compute $dx \, dy = dx \wedge dy$ in terms of du and dv. Form the new integral in terms of u,v-coordinates; be sure you change to u,v-limits. Evaluate the integral.

GREEN'S, DIVERGENCE, AND STOKES' THEOREMS

19

This chapter gives an intuitive introduction to some fundamental integral theorems of vector calculus. Precise statements and proofs of the most general cases of these theorems are not appropriate for a first calculus course. The theorems extend the applications to *work* and *fluid dynamics* that we introduced in Section 17.5.

19.1 PHYSICAL MODELS FOR GREEN'S THEOREM AND THE DIVERGENCE THEOREM

The theorems to be studied in this chapter are all of the same nature. They all assert:

The integral of some quantity over the boundary of a region is equal to the integral of a related quantity over the region itself.	**(1)**

Weeks could be spent discussing what is meant by a region and by its boundary. Our calculus course is almost over, and we have only a few days to spend on this material. We will be very intuitive. Throughout this section, we assume that all the functions considered have continuous partial derivatives so that we can consider integrals of these partial derivatives.

A NEW LOOK AT THE FUNDAMENTAL THEOREM OF CALCULUS

We will start off with a new look at the fundamental theorem of calculus in Theorem 19.1. We then describe a physical demonstration of Theorem 19.1. This demonstration will generalize to give us Green's theorem and the divergence theorem.

THEOREM 19.1 Fundamental theorem of calculus

If $f(x)$ has a continuous derivative $f'(x)$ for all x in the one-dimensional region $[a, b]$, then

$$f(b) - f(a) = \int_a^b f'(x)\,dx.$$

When stated this way, the fundamental theorem is of the type described in statement (1). Surely $f(x)$ and $f'(x)$ are "related quantities," and the endpoints a and b can be viewed as the boundary of the one-dimensional region $[a, b]$. By suitable definition, $f(b) - f(a)$ can be considered to be the "integral" of $f(x)$ over this two-point boundary.

For a physical demonstration of Theorem 19.1, imagine a gas flowing through a long cylinder *with cross-section area* 1 and reaching from a to b on the x-axis, as shown in Fig. 19.1. In this idealistic situation, we will suppose that the velocity and mass density of the gas depend only on the location x along the cylinder and are independent of the up-down and front-back locations within a cross section of the cylinder. This is a model for *one-dimensional flow*. Both the velocity and mass density may vary with time, but we will concentrate on one particular instant and consider how the total mass of gas in the cylinder would change if it were always to flow just as it did at that instant.

The velocity could be represented by a vector function

$$\mathbf{V} = v(x)\mathbf{i}$$

and the mass density by a scalar function $\rho(x)$. Recall from Section 17.5 that the vector function

$$\mathbf{F} = \rho(x)v(x)\mathbf{i} = f(x)\mathbf{i}$$

is called the **flux vector** of the flow. If the flow does not vary with time, then since the cylinder has 1 as cross-sectional area, $f(x)$ is a (signed) measure of the mass of gas that goes past x in 1 unit of time. Since $\mathbf{i}$ is directed toward the right and since $\mathbf{F} = f(x)\mathbf{i}$, we see that $f(b)$ is the mass of gas *leaving* the right end of the cylinder in 1 unit time. That is, the gas is actually leaving at b if $f(b) > 0$ and is entering at b if $f(b) < 0$. Similarly, $f(a)$ is the mass of gas *entering* at the left end per unit time, actually entering if $f(a) > 0$ and leaving if $f(a) < 0$. No gas is entering or leaving through the walls of the cylinder, only at the ends. Thus we see that

$$f(b) - f(a) = \begin{cases}\text{Decrease of mass of gas in the}\\ \text{cylinder per unit time.}\end{cases} \quad \textbf{(2)}$$

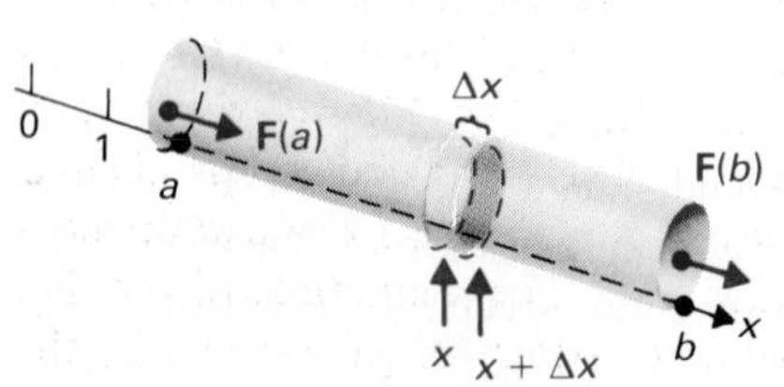

Figure 19.1 Flux vector $\mathbf{F}(x)$ for a one-dimensional flow through a cylinder of cross-section area 1.

The same reasoning applied to the short element of cylinder from x to $x + \Delta x$ in Fig. 19.1 shows that

$$f(x + \Delta x) - f(x) = \begin{cases}\text{Decrease of mass of gas in this}\\ \text{cylindrical element per unit time.}\end{cases}$$

Therefore

$$\frac{f(x + \Delta x) - f(x)}{\Delta x} = \begin{cases}\text{Average decrease in mass of gas per}\\ \text{unit length of cylinder per unit time}\end{cases}$$

and

$$f'(x) = \lim_{\Delta x \to 0} \frac{f(x + \Delta x) - f(x)}{\Delta x}$$

$$= \begin{cases} \text{Decrease of mass of gas measured at} \\ x \text{ per unit length per unit time.} \end{cases}$$

Consequently $\int_a^b f'(x)\,dx$ has the following interpretation: The product $f'(x) \cdot dx$ may be viewed as the decrease in 1 unit time of mass of the gas over a short length dx of cylinder at x. Thus

$$\int_a^b f'(x)\,dx = \begin{cases} \text{Decrease of mass of the gas over the} \\ \text{entire cylinder in one unit time.} \end{cases} \qquad \textbf{(3)}$$

Comparison of Eqs. (2) and (3) shows that

$$f(b) - f(a) = \int_a^b f'(x)\,dx,$$

which is, of course, the fundamental theorem (Theorem 19.1).

EXAMPLE 1 Suppose gas is flowing in the cylinder $y^2 + z^2 = 4$ for $-2 \le x \le 4$, with flux vector $\mathbf{F} = (3x - x^2)\mathbf{i}$ at a certain instant. Find the rate of decrease of mass of gas in the cylinder at that instant.

Solution By Eq. (2), the decrease of mass of gas per unit time in a cylinder *of cross-section area* 1 would be $f(b) - f(a) = f(4) - f(-2)$ for $f(x) = 3x - x^2$. Now $f(4) - f(-2) = -4 - (-10) = 6$. The given cylinder has cross-section area $\pi r^2 = \pi 2^2 = 4\pi$. Thus the mass of gas leaving per unit time is $6 \cdot 4\pi = 24\pi$. □

A PHYSICAL DEMONSTRATION OF GREEN'S THEOREM

The same ideas as those just presented, but with a two-dimensional flow, lead us to Green's theorem. Imagine gas to be flowing between two identical parallel plates, placed 1 *unit distance apart* with the lower plate in the x,y-plane, as in Fig. 19.2. This time we suppose the *gas flow* is *two-dimensional*, so its velocity vector

$$\mathbf{V} = v_1(x, y)\mathbf{i} + v_2(x, y)\mathbf{j}$$

has no **k**-component and also does not depend on the position $0 \le z \le 1$ between the plates. Also, we assume that the mass density $\rho(x, y)$ does not depend on z. The flux vector

$$\mathbf{F} = \rho(x, y)\mathbf{V} = P(x, y)\mathbf{i} + Q(x, y)\mathbf{j},$$

where $P(x, y) = \rho(x, y)v_1(x, y)$ and $Q(x, y) = \rho(x, y)v_2(x, y)$, again measures the mass of flow of gas at each point per unit time. This flux vector has as direction the direction of the flow. Recall from Section 17.5 the meaning of $|\mathbf{F}|$, which we give again here. Imagine a small square placed perpendicular to $\mathbf{F}$ at a point, with the flux vector at the center of the square. A certain mass of gas flows through this square per unit time. The magnitude of the flux vector $\mathbf{F}$ is the limit of the quotients of these masses of gas divided by the

Figure 19.2 Parallel plates 1 unit apart.

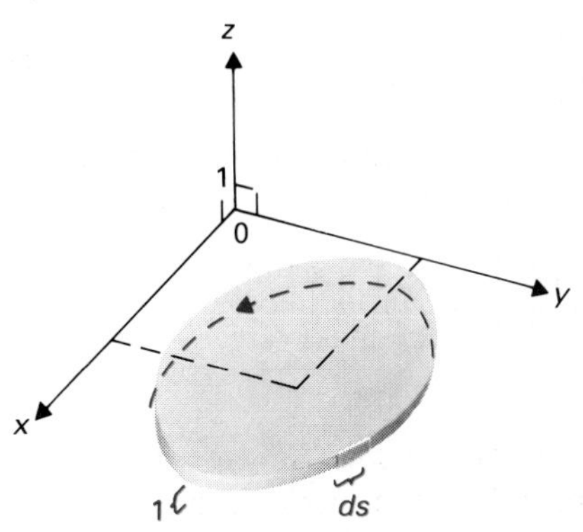

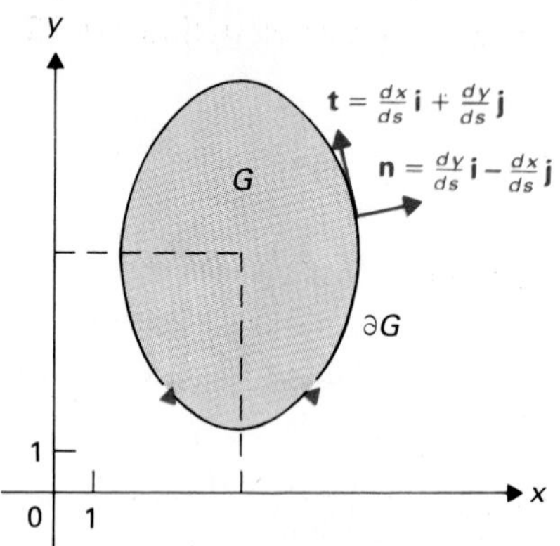

Figure 19.3 Unit tangent and normal vectors to ∂G.

areas of the squares, as the areas approach zero. That is, if the flow everywhere were the same as at (x_0, y_0), then $|\mathbf{F}(x_0, y_0)|$ would be the mass of gas flowing through a square of unit area, placed perpendicular to the flow, in 1 unit time.

Figure 19.3 shows the region G in the x, y-plane occupied by the lower of the two plates. The boundary of the region is denoted by ∂G, read "the boundary of G." Imagine that we travel along this curve, ∂G, so that the region G lies on our left-hand side, as indicated by the arrows on the curve. From previous work, we know that

$$\boxed{\mathbf{t} = \frac{dx}{ds}\mathbf{i} + \frac{dy}{ds}\mathbf{j}}$$

is a unit vector tangent to this curve at each point. It is then easy to see that

$$\boxed{\mathbf{n} = \frac{dy}{ds}\mathbf{i} - \frac{dx}{ds}\mathbf{j}}$$

is a unit vector normal to the curve and directed *outward* from the region.

Now look back at Fig. 19.2. How much gas is flowing out of the region between the plates through the little strip of height 1 and width ds along ∂G, shown in the figure? This outward flow is measured by the *normal scalar* component $\mathbf{F} \cdot \mathbf{n}$ of the flux vector. Thus the mass of gas per unit time coming through this strip of area ds (recall that the plates are 1 unit apart) is approximately $(\mathbf{F} \cdot \mathbf{n})\, ds$. Consequently, the total mass of gas leaving the region between the plates per unit time is

$$\oint_{\partial G} (\mathbf{F} \cdot \mathbf{n})\, ds, \tag{4}$$

where $\oint_{\partial G}$ denotes the line integral once around the boundary of G in the direction given by the arrows in Fig. 19.3. But

$$\begin{aligned}\mathbf{F} \cdot \mathbf{n} &= [P(x, y)\mathbf{i} + Q(x, y)\mathbf{j}] \cdot \left(\frac{dy}{ds}\mathbf{i} - \frac{dx}{ds}\mathbf{j}\right)\\ &= P(x, y)\frac{dy}{ds} - Q(x, y)\frac{dx}{ds}.\end{aligned}$$

Therefore integral (4) becomes

$$\oint_{\partial G} [P(x, y)\, dy - Q(x, y)\, dx] = \begin{cases}\text{Mass of gas leaving the region}\\ \text{between the plates per unit time.}\end{cases} \tag{5}$$

Figure 19.4 Parallel circular plates $x^2 + y^2 \leq 9$, where $z = 0$, $z = 4$.

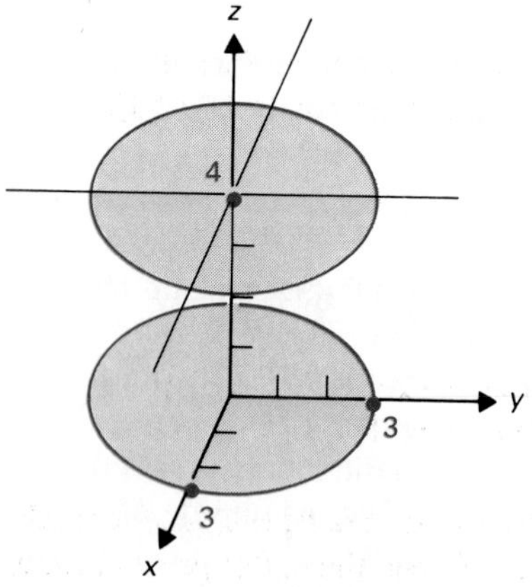

EXAMPLE 2 Consider two parallel circular plates lying over $x^2 + y^2 \leq 9$ in the x, y-plane, one in the x, y-plane and one in the plane $z = 4$, as shown in Fig. 19.4. Suppose the flux vector of a two-dimensional flow is $\mathbf{F} = x\mathbf{i} + y\mathbf{j}$. Find the mass of gas leaving the region between the plates per unit time.

Solution Let G be the disk $x^2 + y^2 \leq 9$ in the x, y-plane. Since our plates are 4 units apart, we see from formula (5) that the mass of gas leaving the region between the plates per unit time is

$$4\oint_{\partial G} (x\, dy - y\, dx).$$

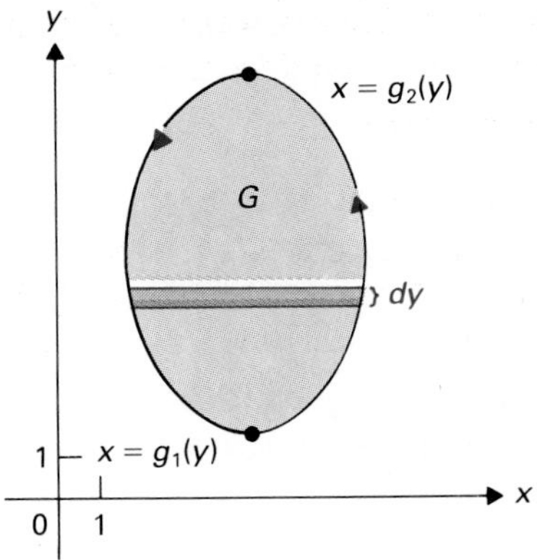

Figure 19.5 Strip of width dy.

Taking $x = 3\cos t$, $y = 3\sin t$ for $0 \le t \le 2\pi$ as parametrization for ∂G, we have

$$4\oint_{\partial G}(x\,dy - y\,dx) = 4\int_0^{2\pi}[(3\cos t)(3\cos t) - (3\sin t)(-3\sin t)]\,dt$$
$$= 4\int_0^{2\pi} 9(\cos^2 t + \sin^2 t)\,dt$$
$$= 36\int_0^{2\pi} dt = 72\pi. \quad \square$$

We now compute in another way the mass of gas leaving, namely, by considering separately the contributions of $P(x, y)\mathbf{i}$ and $Q(x, y)\mathbf{j}$ to the flux vector $\mathbf{F}$. The vector $P(x, y)\mathbf{i}$ is the horizontal component of the flux. The region between the two plates over the strip of width dy shown in Fig. 19.5 is a cylinder having area of cross section $(dy)(1) = dy$. Referring back to formula (3), where we discussed flow along a cylinder, and considering just the vector component $P(x, y)\mathbf{i}$,

$$\int_{g_1(y)}^{g_2(y)} \frac{\partial P}{\partial x}(x, y)\,dx = \begin{cases}\text{Mass of gas leaving the ends of the cylinder}\\ \text{per unit area cross section per unit time.}\end{cases}$$

Since the cylinder has cross section area dy (the plates are still 1 unit apart), we have

$$\left(\int_{g_1(y)}^{g_2(y)} \frac{\partial P}{\partial x}(x, y)\,dx\right) dy = \begin{cases}\text{Mass of gas leaving the cylinder}\\ \text{at the ends per unit time.}\end{cases}$$

Therefore

$$\int_c^d \left(\int_{g_1(y)}^{g_2(y)} \frac{\partial P}{\partial x}(x, y)\,dx\right) dy$$
$$= \iint_G \frac{\partial P}{\partial x}(x, y)\,dx\,dy$$
$$= \begin{cases}\text{Mass of gas leaving the region between}\\ \text{the plates per unit time due to } P(x, y)\mathbf{i}.\end{cases}$$

A similar computation with a vertical cylindrical strip of width dx and the component $Q(x, y)\mathbf{j}$ of $\mathbf{F}$ shows that

$$\iint_G \frac{\partial Q}{\partial y}(x, y)\,dx\,dy = \begin{cases}\text{Mass of gas leaving the region between}\\ \text{the plates per unit time due to } Q(x, y)\mathbf{j}.\end{cases}$$

Thus

$$\iint_G \left[\frac{\partial P}{\partial x}(x, y) + \frac{\partial Q}{\partial y}(x, y)\right] dx\,dy$$
$$= \begin{cases}\text{Mass of gas leaving the region}\\ \text{between the plates per unit time.}\end{cases} \tag{6}$$

EXAMPLE 3 For the plates and flux vector in Example 2, use formula (6) to find the mass of gas leaving the region between the plates per unit time.

Solution Since the plates are 4 units apart, we see from formula (6) that the answer is given by

$$4\iint_G (1+1)\,dx\,dy = 8\iint_G 1\cdot dx\,dy.$$

Since G is a disk of radius 3 with area 9π, our answer becomes $8\cdot 9\pi = 72\pi$. □

Of course, we obtained the same answer in Example 2 using formula (5) as we obtained in Example 3 using formula (6). Comparing formulas (5) and (6), we have

$$\iint_G \left[\frac{\partial P}{\partial x}(x,y) + \frac{\partial Q}{\partial y}(x,y)\right] dx\,dy = \oint_{\partial G} [P(x,y)\,dy - Q(x,y)\,dx], \qquad \textbf{(7)}$$

which is again a relation of the form (1). Equation (7) is known as Green's theorem.

THEOREM 19.2 Green's theorem, divergence form

For a suitable plane region G, and for functions $P(x, y)$ and $Q(x, y)$ with continuous partial derivatives,

$$\oint_{\partial G} (P\,dy - Q\,dx) = \iint_G \left(\frac{\partial P}{\partial x} + \frac{\partial Q}{\partial y}\right) dx\,dy. \qquad \textbf{(8)}$$

In the line integral, ∂G is traced in the direction that keeps G on the left.

Equation (8) is often stated in vector form, reflecting the physical demonstration we gave for it. Let $\mathbf{F} = P(x, y)\mathbf{i} + Q(x, y)\mathbf{j}$ be a flux vector. We make use of the symbolic operator

$$\nabla = \frac{\partial}{\partial x}\mathbf{i} + \frac{\partial}{\partial y}\mathbf{j},$$

where ∇ is read "del." Symbolically, we have

$$\begin{aligned}\nabla\cdot\mathbf{F} &= \left(\frac{\partial}{\partial x}\mathbf{i} + \frac{\partial}{\partial y}\mathbf{j}\right)\cdot[P(x,y)\mathbf{i} + Q(x,y)\mathbf{j}]\\ &= \frac{\partial}{\partial x}P(x,y) + \frac{\partial}{\partial y}Q(x,y)\\ &= \frac{\partial P}{\partial x} + \frac{\partial Q}{\partial y}.\end{aligned}$$

Also, we have seen that if $\mathbf{n}$ is a unit normal vector to the boundary, then

$$\oint_{\partial G} (P\,dy - Q\,dx) = \oint_{\partial G} (\mathbf{F}\cdot\mathbf{n})\,ds.$$

Thus Eq. (8) becomes

$$\boxed{\oint_{\partial G} (\mathbf{F}\cdot\mathbf{n})\,ds = \iint_G (\nabla\cdot\mathbf{F})\,dx\,dy.} \qquad \textbf{(9)}$$

The scalar $\nabla \cdot \mathbf{F}$ is the **divergence** of **F**. It measures the rate at which the gas diverges (leaves) at each point.

We labeled Theorem 19.2 the *divergence form* of Green's theorem. The reason for the adjective "divergence" has just been explained. In the next section, we will present precisely the same theorem but in another notation that arises by considering the way the gas "rotates" around the boundary of G. We will call this the *rotation form* of the theorem. In this section, we gave only a physical argument for the validity of the theorem. When we present the rotation form, we will supply a mathematical proof.

In the exercises of this section, you are asked to illustrate this divergence form, Theorem 19.2, of Green's theorem. Here is an illustration to use as a model.

EXAMPLE 4 Illustrate the divergence form of Green's theorem for the functions

$$P(x, y) = x^3 \qquad \text{and} \qquad Q(x, y) = 2x + y^3$$

over the disk G given by $x^2 + y^2 \leq a^2$.

Solution Now

$$\frac{\partial P}{\partial x} = 3x^2 \qquad \text{and} \qquad \frac{\partial Q}{\partial y} = 3y^2,$$

so

$$\iint_G \left(\frac{\partial P}{\partial x} + \frac{\partial Q}{\partial y}\right) dx\, dy = \iint_G 3(x^2 + y^2)\, dx\, dy.$$

Changing to polar coordinates, we have

$$\iint_G 3(x^2 + y^2)\, dx\, dy = \int_0^{2\pi} \int_0^a 3r^2 \cdot r\, dr\, d\theta$$

$$= \int_0^{2\pi} 3\frac{r^4}{4}\bigg]_0^a d\theta = 3\frac{a^4}{4}\theta\bigg]_0^{2\pi} = \frac{3a^4\pi}{2}.$$

The boundary ∂G may be parametrized by $x = a\cos\theta$, $y = a\sin\theta$ for $0 \leq \theta \leq 2\pi$. The integral around the boundary becomes

$$\oint_{\partial G} (P\, dy - Q\, dx) = \oint_{\partial G} [x^3\, dy - (2x + y^3)\, dx]$$

$$= \int_0^{2\pi} [a^3 \cos^3\theta \cdot a\cos\theta\, d\theta$$

$$- (2a\cos\theta + a^3\sin^3\theta)(-a\sin\theta\, d\theta)]$$

$$= \int_0^{2\pi} [a^4(\cos^4\theta + \sin^4\theta) + 2a^2(\sin\theta\cos\theta)]\, d\theta$$

$$= \int_0^{2\pi} \left[a^4\left(\left(\frac{1 + \cos 2\theta}{2}\right)^2 + \left(\frac{1 - \cos 2\theta}{2}\right)^2\right) + 2a^2 \sin\theta\cos\theta\right] d\theta$$

$$= \int_0^{2\pi} \left[a^4\left(\frac{\cos^2 2\theta}{2} + \frac{1}{2}\right) + 2a^2\sin\theta\cos\theta\right] d\theta$$

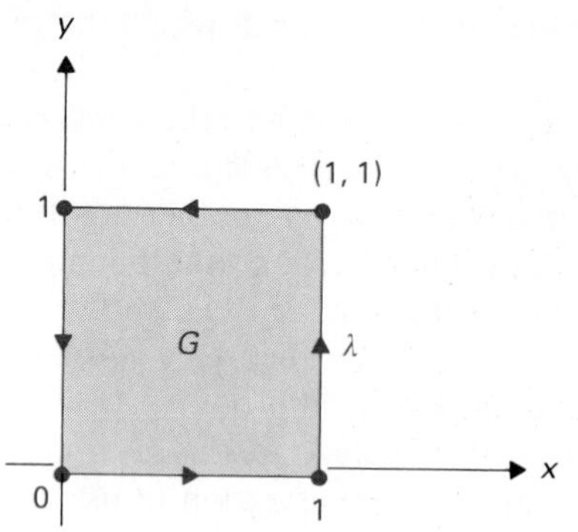

Figure 19.6 The loop $\lambda = \partial G$.

$$= \int_0^{2\pi} \left[a^4\left(\frac{1 + \cos 4\theta}{4} + \frac{1}{2}\right) + 2a^2 \sin\theta \cos\theta \right] d\theta$$

$$= \left[a^4\left(\frac{1}{4}\theta + \frac{\sin 4\theta}{16} + \frac{1}{2}\theta\right) + a^2 \sin^2\theta \right]_0^{2\pi}$$

$$= a^4\left(\frac{2\pi}{4} + \frac{2\pi}{2}\right) - 0 = a^4 \cdot \frac{3\pi}{2} = \frac{3\pi a^4}{2}.$$

The same answer was obtained, illustrating Green's theorem. Obviously, the area integral was much easier to evaluate than the line integral in this case. □

Often Green's theorem is used to transform a line integral around a loop into a more easily computed area integral, or occasionally vice versa.

EXAMPLE 5 Let $P(x, y) = xy^2 - 3xy$ and $Q(x, y) = y/(1 + x^2)$. Compute $\oint_\lambda (P\,dy - Q\,dx)$, where λ is the boundary of the unit square in Fig. 19.6, having opposite vertices at (0, 0) and (1, 1), traversed in the counterclockwise direction.

Solution Let G be the two-dimensional region enclosed by this square, so that $\lambda = \partial G$. By Green's theorem, we see that

$$\oint_\lambda (P\,dy - Q\,dx) = \iint_G \left(\frac{\partial P}{\partial x} + \frac{\partial Q}{\partial y}\right) dx\,dy$$

$$= \int_0^1 \int_0^1 \left(y^2 - 3y + \frac{1}{x^2 + 1}\right) dx\,dy$$

$$= \int_0^1 (y^2 x - 3xy + \tan^{-1} x)\Big]_0^1 dy$$

$$= \int_0^1 \left(y^2 - 3y + \frac{\pi}{4}\right) dy$$

$$= \left(\frac{y^3}{3} - \frac{3y^2}{2} + \frac{\pi}{4} y\right)\Big]_0^1$$

$$= \frac{1}{3} - \frac{3}{2} + \frac{\pi}{4} = -\frac{7}{6} + \frac{\pi}{4}. \quad \square$$

THE DIVERGENCE THEOREM

Figure 19.7 dS and $\mathbf{n}$ for $\iint_G (\mathbf{F}\cdot\mathbf{n})\,dS$; column for $\iiint_G (\partial R/\partial z)\,dx\,dy\,dz$.

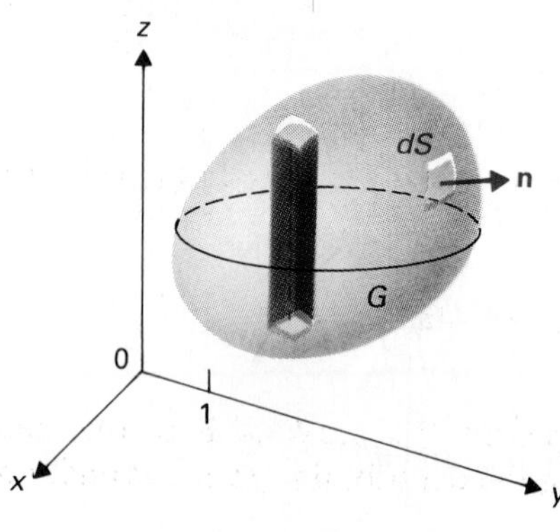

If we repeat the gas flow arguments we made earlier, but in a three-dimensional setting, we obtain the divergence theorem. Let G now be a region in space with boundary ∂G, as illustrated in Fig. 19.7. This time ∂G is a *surface*. Imagine gas flowing in space and able to flow in and out of G without impediment. (You can think of the boundary of G as not being physically present, or, if you prefer, consider it to be a netting through which gas can flow unhindered.) Let

$$\mathbf{F} = P(x, y, z)\mathbf{i} + Q(x, y, z)\mathbf{j} + R(x, y, z)\mathbf{k}$$

be the flux vector of the flow at (x, y, z) in the region. Let $\mathbf{n}$ be a unit normal vector outward from the boundary surface at a point and let a little differential

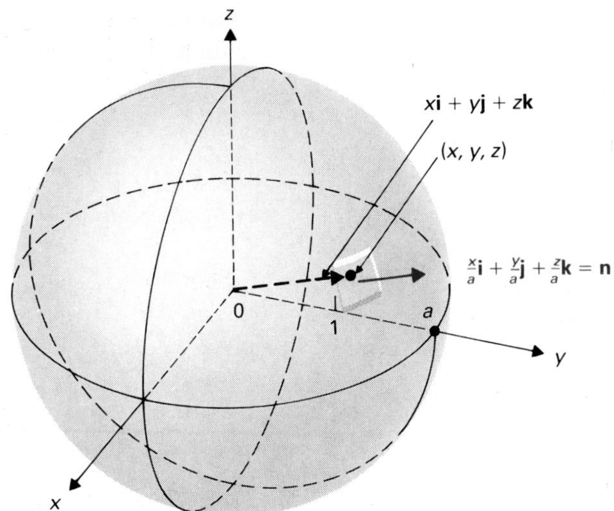

Figure 19.8 $x\mathbf{i} + y\mathbf{j} + z\mathbf{k}$ goes from $(0, 0, 0)$ to (x, y, z) and has length a; $\mathbf{n} = (1/a)(x\mathbf{i} + y\mathbf{j} + z\mathbf{k})$.

piece of surface at that point have area dS. Then the mass of gas flowing out of the region G per unit time is given by

$$\iint_{\partial G} (\mathbf{F} \cdot \mathbf{n})\, dS, \tag{10}$$

by reasoning like that in the previous subsection.

EXAMPLE 6 Let the flux vector for a given flow in space be

$$\mathbf{F}(x, y, z) = x\mathbf{i} + y\mathbf{j} + z\mathbf{k}.$$

Find the mass of gas leaving the ball $x^2 + y^2 + z^2 \le a^2$ per unit time.

Solution If G is the given ball, we need to find

$$\iint_{\partial G} (\mathbf{F} \cdot \mathbf{n})\, dS.$$

Now ∂G is the sphere $x^2 + y^2 + z^2 = a^2$, and at a point on the sphere, a unit normal vector outward is clearly $\mathbf{n} = (x/a)\mathbf{i} + (y/a)\mathbf{j} + (z/a)\mathbf{k}$, as shown in Fig. 19.8. (We could also find $\mathbf{n}$ using the fact that $\nabla(x^2 + y^2 + z^2)$ is orthogonal to $x^2 + y^2 + z^2 = a^2$.) Then

$$\begin{aligned}\mathbf{F} \cdot \mathbf{n} &= (x\mathbf{i} + y\mathbf{j} + z\mathbf{k}) \cdot \left(\frac{x}{a}\mathbf{i} + \frac{y}{a}\mathbf{j} + \frac{z}{a}\mathbf{k}\right)\\ &= \frac{1}{a}(x^2 + y^2 + z^2) = \frac{1}{a}(a^2) = a\end{aligned}$$

at each point on the sphere. Since the area of the surface of the sphere is $4\pi a^2$, we have

$$\iint_{\partial G} (\mathbf{F} \cdot \mathbf{n})\, dS = \iint_{\partial G} a\, dS = a \cdot 4\pi a^2 = 4\pi a^3. \quad \square$$

The contribution of the z-component $R(x, y, z)\mathbf{k}$ of $\mathbf{F}$ to the mass leaving G can be computed by finding the contribution in the cylinder of cross-section area $dx\,dy$, shown in Fig. 19.7, and then adding these contributions over the entire region G. Obviously we obtain

$$\iiint_G \frac{\partial R}{\partial z}\,dz\,dx\,dy$$

by reasoning just like that in the previous subsection. The total mass of gas leaving the region per unit time is thus

$$\iiint_G \left(\frac{\partial P}{\partial x} + \frac{\partial Q}{\partial y} + \frac{\partial R}{\partial z}\right) dx\,dy\,dz, \tag{11}$$

which may also be written

$$\iiint_G \nabla \cdot \mathbf{F}\,dx\,dy\,dz, \tag{12}$$

where this time

$$\nabla = \frac{\partial}{\partial x}\mathbf{i} + \frac{\partial}{\partial y}\mathbf{j} + \frac{\partial}{\partial z}\mathbf{k}.$$

EXAMPLE 7 Repeat Example 6, but use integral (11) to find the mass of gas leaving the ball per unit time.

Solution Since $\mathbf{F} = x\mathbf{i} + y\mathbf{j} + z\mathbf{k}$, we have

$$\frac{\partial P}{\partial x} + \frac{\partial Q}{\partial y} + \frac{\partial R}{\partial z} = 1 + 1 + 1 = 3.$$

Thus integral (11) becomes

$$\iiint_G 3\,dx\,dy\,dz.$$

This integral obviously equals three times the volume of the ball G with radius a. The answer is therefore $3 \cdot \frac{4}{3}\pi a^3 = 4\pi a^3$. □

Of course, the answer found in Example 6 using integral (10) is the same as that found in Example 7 using integral (11). Comparison of integrals (10) and (12) gives us the *divergence theorem*; $\nabla \cdot \mathbf{F}$ is called the **divergence** of $\mathbf{F}$.

THEOREM 19.3 Divergence theorem

If $\mathbf{F} = P(x, y, z)\mathbf{i} + Q(x, y, z)\mathbf{j} + R(x, y, z)\mathbf{k}$ is a vector field with continuously differentiable components over a suitable region G in space, then

$$\iint_{\partial G} (\mathbf{F} \cdot \mathbf{n})\,dS = \iiint_G (\nabla \cdot \mathbf{F})\,dx\,dy\,dz. \tag{13}$$

The divergence theorem is illustrated by Examples 6 and 7. The exercises for this section involve the divergence theorem only for very simple cases; they concentrate primarily on Green's theorem. More work is done with the divergence theorem in Section 19.3; in particular, we discuss the surface integral $\iint_{\partial G} (\mathbf{F} \cdot \mathbf{n})\,dS$ in more detail there. However, we wanted to present the theorem here, while we have the gas flow argument well in mind. We see that the fundamental theorem, the divergence form of Green's theorem, and the divergence theorem are essentially the same theorem, but in different dimensions.

SUMMARY

1. The theorems to be studied in this chapter all relate the integral of a quantity over the boundary of a region to the integral of a related quantity over the region itself.
2. *Green's theorem, divergence form:* If $P(x, y)$ and $Q(x, y)$ are continuously differentiable throughout a suitable plane region G with boundary ∂G, then
$$\oint_{\partial G} (P\,dy - Q\,dx) = \iint_G \left(\frac{\partial P}{\partial x} + \frac{\partial Q}{\partial y}\right) dx\,dy.$$
In the line integral, ∂G is traced in the direction that keeps G on the left.
3. *Vector statement of Green's theorem:* If $\mathbf{F} = P(x, y)\mathbf{i} + Q(x, y)\mathbf{j}$ is a continuously differentiable vector field throughout a suitable plane region G with boundary ∂G, then
$$\oint_{\partial G} (\mathbf{F} \cdot \mathbf{n})\,ds = \iint_G (\nabla \cdot \mathbf{F})\,dx\,dy,$$
where $\mathbf{n}$ is the outward unit normal vector at each point of ∂G and ∇ is the symbolic operator
$$\nabla = \frac{\partial}{\partial x}\mathbf{i} + \frac{\partial}{\partial y}\mathbf{j}.$$
4. *Divergence theorem:* Let $\mathbf{F} = P(x, y, z)\mathbf{i} + Q(x, y, z)\mathbf{j} + R(x, y, z)\mathbf{k}$ be a continuously differentiable vector field over a suitable region G in space with boundary surface ∂G. Then
$$\iint_{\partial G} (\mathbf{F} \cdot \mathbf{n})\,dS = \iiint_G (\nabla \cdot \mathbf{F})\,dx\,dy\,dz,$$
where $\mathbf{n}$ is the unit normal vector outward at each point of ∂G, and dS is the differential element of surface area, and ∇ is the symbolic operator
$$\nabla = \frac{\partial}{\partial x}\mathbf{i} + \frac{\partial}{\partial y}\mathbf{j} + \frac{\partial}{\partial z}\mathbf{k}.$$

EXERCISES

1. Gas is flowing in the cylinder $y^2 + z^2 = 1$ for $0 \le x \le 8$, with flux vector at a certain instant equal to $\mathbf{F} = (4 + x^{2/3})\mathbf{i}$. Find the rate of decrease of mass of the gas in the cylinder at that instant.
2. Repeat Exercise 1 if the flux vector is
$$\mathbf{F} = \frac{16}{2 + x^{1/3}}\,\mathbf{i}.$$
3. Gas is flowing in the cylinder $-1 \le x \le 1$, $0 \le z \le 3$ for $-1 \le y \le 10$, with flux vector at a certain instant equal to $\mathbf{F} = (3y - 2)\mathbf{j}$. Find the rate of *increase* of mass of gas in the cylinder at that instant.
4. Gas is flowing in the cylinder bounded by $z = 1 - y^2$ and $z = 0$ for $0 \le x \le 4$. If the flux vector at a certain instant is $\mathbf{F} = (x^2 + 3x + 1)\mathbf{i}$, find the rate of *increase* of gas in the cylinder at that instant.

In Exercises 5 through 12, illustrate the divergence form of Green's theorem by computing

$$\oint_{\partial G} (P\,dy - Q\,dx) \qquad \text{and} \qquad \iint_G \left(\frac{\partial P}{\partial x} + \frac{\partial Q}{\partial y}\right) dx\,dy$$

for the given $P(x, y)$, $Q(x, y)$ and the region G.

5. $P(x, y) = x$, $Q(x, y) = 2y$, where G is the square region with vertices $(1, 1)$, $(-1, 1)$, $(-1, -1)$, and $(1, -1)$
6. $P(x, y) = x^2y^2$, $Q(x, y) = 2x - 3y$, where G is the square region bounded by $x = 0$, $x = 1$, $y = 0$, and $y = 1$
7. $P(x, y) = y^2x$, $Q(x, y) = x^2y$, where G is the disk $x^2 + y^2 \le 16$
8. $P(x, y) = x^2y$, $Q(x, y) = xy^2$, where G is the disk $x^2 + y^2 \le 4$
9. $P(x, y) = y$, $Q(x, y) = xy$, where G is the region bounded by $y = x^2$ and $y = x$
10. $P(x, y) = x^2y + x$, $Q(x, y) = y^2 + 4x$, where G is the region bounded by $y = x^2$ and $y = 1$

As indicated by Examples 4 and 5, it may be that one of the integrals appearing in the divergence form of Green's theorem is much easier to compute than the other. In Exercises 11 through 24, find the indicated quantity by computing whichever integral is easier.

11. $P(x, y) = xe^y$, $Q(x, y) = xy^2$, where G is the triangular region bounded by $x = 0$, $y = 0$, and $x + y = 2$
12. $P(x, y) = x/(1 + y^2)$, $Q(x, y) = y/(1 + x^2)$, where G is the triangular region bounded by $y = x$, $x = 1$, and $y = 0$
13. Two parallel identical plates are 3 units apart with the lower one having as boundary the circle $(x - 3)^2 + y^2 = 16$ in the x,y-plane. Gas flowing between the plates has as flux vector

$$(2x + y^3)\mathbf{i} + (e^x - 4y)\mathbf{j}$$

at a certain instant. Find the rate of decrease of mass of gas between the plates at that instant.

14. Repeat Exercise 13 if the plates are 2 units apart covering $x^2 + y^2 \le 4$ and the flux vector is $\mathbf{F} = (xy^2 + 3x)\mathbf{i} + (x^2y - y)\mathbf{j}$.
15. Find the integral of the outward normal component of the vector field $\mathbf{F} = (3x - 2y)\mathbf{i} + (5x + 7y)\mathbf{j}$ *clockwise* around the rectangular region bounded by $x = 0$, $x = 3$, $y = 0$, and $y = 4$.
16. Repeat Exercise 15 if $\mathbf{F} = (2x + 4y^2)\mathbf{i} + (8x^2 - 5y)\mathbf{j}$.
17. Find the integral of the normal component of the vector field

$$\mathbf{F} = (x^2y + xy^2)\mathbf{i} + (xy)\mathbf{j}$$

counterclockwise around the triangle bounded by

$$x = 0, \quad y = 0, \quad \text{and} \quad y = 1 - x.$$

18. Find $\oint_\lambda (x^2y\,dy - y^2\,dx)$, where λ is the boundary of the triangle bounded by $y = x$, $y = -x$, and $x = 1$, traced counterclockwise.
19. Find $\oint_\lambda (2xy\,dx + 3x^2\,dy)$, where λ is the boundary of the region bounded by $y = \sqrt{x}$ and $y = x^2$, traced counterclockwise.
20. Find

$$\oint_\lambda \left(\frac{3^y}{y - 2}\,dx + \frac{2^x}{x + 1}\,dy\right),$$

where λ is the boundary of the square with vertices $(0, 0)$, $(0, 1)$, $(1, 0)$, and $(1, 1)$, traced counterclockwise.

21. Find

$$\oint_\lambda \left(\frac{y}{y^3 + 1}\,dx - \frac{x^2 + 4}{x^4 + 1}\,dy\right)$$

for λ as in Exercise 20.

22. Find

$$\oint_\lambda \left(\frac{y + 1}{y^2 + 3}\,dx - \frac{x + 1}{x^2 + 3}\,dy\right)$$

for the boundary λ of the triangle with vertices $(0, 0)$, $(1, 0)$, and $(1, 1)$ traced counterclockwise.

23. Find the integral of the divergence of the vector field $x\mathbf{i} + y\mathbf{j}$ over the ellipse $(x^2/a^2) + (y^2/b^2) = 1$. [*Hint:* A parametrization of the ellipse is

$$x = a\cos t, \qquad y = b\sin t$$

for $0 \le t \le 2\pi$.]

24. Find the integral of the normal component of the gradient vector field of the function $f(x, y) = x^2/y^2$ counterclockwise around the boundary of the rectangular region bounded by

$$x = 0, \quad x = 4, \quad y = 1, \quad \text{and} \quad y = 3.$$

In Exercises 25 through 29, let G be a region where the divergence form of Green's theorem applies.

25. Show that the area of G is $\frac{1}{2}\oint_{\partial G}(x\,dy - y\,dx)$.
26. If $\int_{\partial G}[(3x + 2y)\,dy - (5y - 4x)\,dx] = 24$, find the area of G.
27. Repeat Exercise 26 if

$$\oint_{\partial G}[(y^3 - 4xy - 2x)\,dy - (x^4 + 2y^2 + 7y)\,dx] = 50.$$

28. If $\oint_{\partial G}(x\,dy - y\,dx) = 3$, $\oint_{\partial G}[y\,dx + (x^2 + x)\,dy] = 10$, and $\oint_{\partial G}[(4y^2 + 2y)\,dx + 2x\,dy] = 16$, find the centroid of G.
29. If

$$\oint_{\partial G}[(3x - y)\,dy - (x + 2y)\,dx] = 10,$$

$$\oint_{\partial G}[(4x - y)\,dx + (3x^2 - 2x)\,dy] = -18,$$

and

$$\oint_{\partial G}[(x^3 + 4y)\,dx + (4xy - y^3 + 2x)\,dy] = 25,$$

find the centroid of G.

30. Suppose that $\mathbf{F} = P(x, y)\mathbf{i} + Q(x, y)\mathbf{j}$ is a vector field such that $\partial P/\partial x = -\partial Q/\partial y$. Show that the integral of the normal component of $\mathbf{F}$ around ∂G is zero for any region G in the domain of $\mathbf{F}$ for which Green's theorem applies.

31. Let G be a region where Green's theorem applies. If $f(x, y)$ has G in its domain, find a line integral around ∂G equal to $\iint_G (\nabla \cdot \nabla f)\, dx\, dy$.

32. Follow the steps indicated to illustrate the divergence theorem for the vector field $\mathbf{F} = x^2z^2\mathbf{i} + xz\mathbf{j} + xy^2(z + 1)\mathbf{k}$ and the unit cube G with one vertex at the origin and the diagonally opposite vertex at (1, 1, 1).

a) i) Compute $\iint (\mathbf{F} \cdot \mathbf{n})\, dS$ over the square face of G in the x,y-coordinate plane, where $z = 0$. Here

$$\mathbf{n} = -\mathbf{k} \quad \text{and} \quad dS = dx\, dy.$$

ii) Compute $\iint (\mathbf{F} \cdot \mathbf{n})\, dS$ over the square face of G in the plane $z = 1$. Here

$$\mathbf{n} = \mathbf{k} \quad \text{and} \quad dS = dx\, dy.$$

iii) Proceed as in parts (i) and (ii) to find the integrals $\iint (\mathbf{F} \cdot \mathbf{n})\, dS$ over the square faces of G in the planes $y = 0$ and $y = 1$.

iv) Proceed as in parts (i) and (ii) to find the integrals $\iint (\mathbf{F} \cdot \mathbf{n})\, dS$ over the square faces of G in the planes $x = 0$ and $x = 1$.

v) Add up the six answers found in parts (i) through (iv) to give $\iint_{\partial G} (\mathbf{F} \cdot \mathbf{n})\, dS$.

b) Compute $\iiint_G (\nabla \cdot \mathbf{F})\, dx\, dy\, dz$; the answer should be the same as in part (a)(v).

In Exercises 33 through 36, use the divergence theorem to compute $\iint_{\partial G} (\mathbf{F} \cdot \mathbf{n})\, dS$.

33. $\mathbf{F} = (x + y)\mathbf{i} + (z^2 - 2y)\mathbf{j} + (x^2 - y^2 + 4z)\mathbf{k}$ and G is the ball $x^2 + y^2 + z^2 \le a$.

34. $\mathbf{F} = xz\mathbf{i} + x^2\mathbf{j} + y^2\mathbf{k}$ and G is the region bounded by $z = x^2 + y^2$ and $z = 1$.

35. $\mathbf{F} = xz\mathbf{i} + xy\mathbf{j} + 2z\mathbf{k}$ and G is the region bounded by $z = 4 - x^2 - y^2$ and $z = x^2 + y^2 - 4$.

36. $\mathbf{F} = xy^2\mathbf{i} + yz^2\mathbf{j} + x^2z\mathbf{k}$ and G is the region bounded by $z = \sqrt{a^2 - x^2 - y^2}$ and $z = \sqrt{x^2 + y^2}$.

19.2 GREEN'S THEOREM AND APPLICATIONS

In this section, we give the *rotation interpretation* of Green's theorem and give a mathematical proof of the theorem. We will first prove the theorem for a *simple region* and then generalize the proof to integrals over more general regions. We will see that the material presented here is closely related to the work we did in Section 17.3 on exact differential forms.

Once again, we emphasize that the fundamental theorem of calculus, Green's theorem, the divergence theorem, and Stokes' theorem (to be presented in Section 19.3) are all theorems of the same type. Indeed, the more challenging exercises at the end of this chapter indicate that they are all special cases of a very general theorem, the generalized Stokes' theorem.

Figure 19.9 Two simple regions.

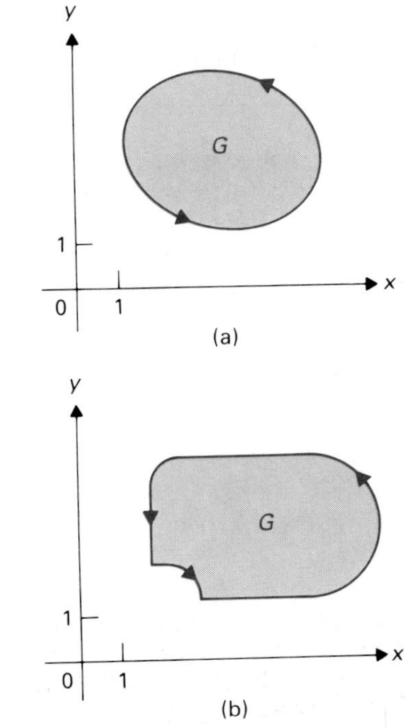

PROOF OF GREEN'S THEOREM

Again, we just assume we have the correct notion of a plane region G and its boundary ∂G, traced so that G is on the left. A region is **bounded** if it all lies inside some sufficiently large square and is **closed** if the boundary is considered to be part of the region.

It is usual to prove Green's theorem first for a special type of region G and then extend to more general regions by decomposing them into regions of the special type. We will call a region G **simple** if any line parallel to one of the coordinate axes *crosses* the boundary of G in at most two points. (We allow such a line to *coincide* with the boundary for a whole interval.) The regions shown in Fig. 19.9 are simple.

Note that the *rotation form* of Green's theorem in Theorem 19.4 below differs slightly in notation from the *divergence form* (Theorem 19.2) of the last section. In Section 19.1, we saw that the integral of the *outward normal component* of a flux vector field $\mathbf{F} = P(x, y)\mathbf{i} + Q(x, y)\mathbf{j}$ around ∂G is given by

$$\oint_{\partial G} (\mathbf{F} \cdot \mathbf{n})\, ds = \oint_{\partial G} (P\, dy - Q\, dx). \tag{1}$$

The integral measures the total *divergence* of the flow through ∂G. In Section 17.5, we saw that the integral of the *tangential component* of this flux vector field around ∂G is given by

$$\oint_{\partial G} (\mathbf{F} \cdot \mathbf{t})\, ds = \oint_{\partial G} (\mathbf{F} \cdot d\mathbf{r}) = \oint_{\partial G} (P\, dx + Q\, dy). \tag{2}$$

This integral measures the *rotation* of the flow around ∂G.

Both Eqs. (1) and (2) involve the integral of a differential form in two variables over ∂G. Green's theorem simply makes an assertion about the integral of such a differential form around a loop, and the theorem should be regarded in that way, independent of the physical interpretation and notation used. In this section, we will be primarily concerned with the *rotational* application involving $\oint_{\partial G}(\mathbf{F} \cdot \mathbf{t})\, ds$, and thus we change to the notation in Eq. (2), involving $\oint_{\partial G}(P\, dx + Q\, dy)$. To obtain the statement of Theorem 19.4 from Theorem 19.2, simply replace $Q(x, y)$ in Theorem 19.2 by $-P(x, y)$ and replace $P(x, y)$ there by $Q(x, y)$.

THEOREM 19.4 Green's theorem for simple regions, rotation form

Let G be a bounded, simple, closed region in the plane with boundary ∂G. If $P(x, y)$ and $Q(x, y)$ are continuously differentiable functions defined on G, then

$$\oint_{\partial G} [P(x, y)\, dx + Q(x, y)\, dy] = \iint_G \left(\frac{\partial Q}{\partial x} - \frac{\partial P}{\partial y}\right) dx\, dy. \tag{3}$$

In the line integral, ∂G is traced in the direction that keeps G on the left.

Proof. Since G is a simple region, ∂G can be split into a "top curve" and a "bottom curve," possibly separated by straight-line "sides," as shown in Fig. 19.10. The top and bottom curves may be parametrized by the parameter x; we let the equations of these curves be

$$y = u(x) \quad \text{for the bottom curve}$$

and

$$y = v(x) \quad \text{for the top curve,}$$

as shown in Fig. 19.10. The integral $\int P(x, y)\, dx$ over any portion of ∂G consisting of vertical line segments is zero, since x remains constant, so $dx = 0$ on any such segment. Consequently, from Fig. 19.10, we obtain

$$\oint_{\partial G} P(x, y)\, dx = \int_a^b P(x, u(x))\, dx + \int_b^a P(x, v(x))\, dx, \tag{4}$$

where the right-to-left integral sign $\int_b^a$ occurs since the top curve is traced

Figure 19.10 Top-bounding curve is $y = v(x)$; bottom-bounding curve is $y = u(x)$.

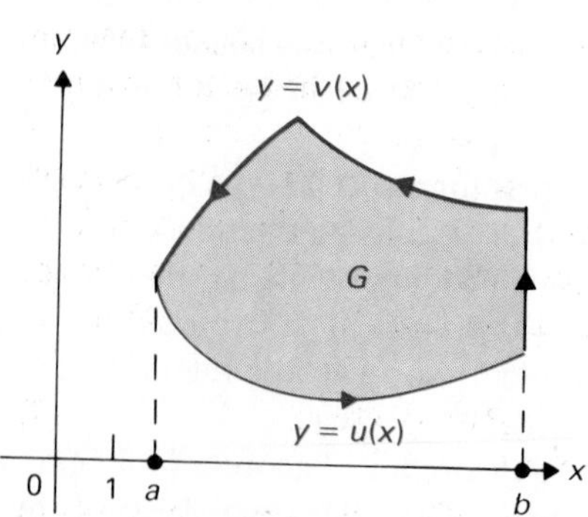

from right to left. From Eq. 4, we have

$$\oint_{\partial G} P(x, y)\, dx = \int_a^b [P(x, u(x)) - P(x, v(x))]\, dx$$
$$= \int_a^b -P(x, y)\Big]_{y=u(x)}^{y=v(x)} dx \tag{5}$$
$$= \int_a^b \int_{u(x)}^{v(x)} -\frac{\partial P}{\partial y}\, dy\, dx = \iint_G -\frac{\partial P}{\partial y}\, dx\, dy.$$

A similar argument, which we ask you to give in Exercise 6, shows that

$$\oint_{\partial G} Q(x, y)\, dy = \iint_G \frac{\partial Q}{\partial x}\, dx\, dy. \tag{6}$$

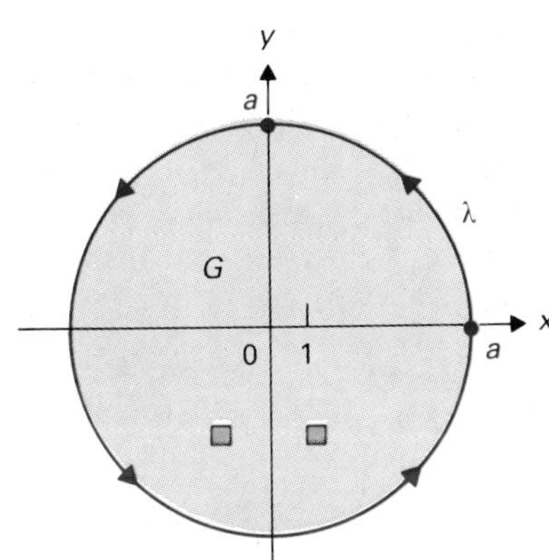

Figure 19.11 $\iint_G x\, dx\, dy = 0$ since contributions on these differential elements cancel each other.

From Eqs. (5) and (6), we have

$$\oint_{\partial G} [P(x, y)\, dx + Q(x, y)\, dy] = \iint_G \left(\frac{\partial Q}{\partial x} - \frac{\partial P}{\partial y}\right) dx\, dy,$$

which is the assertion of the theorem. •

EXAMPLE 1 Let λ be the loop $x = a \cos t$, $y = a \sin t$ for $0 \le t \le 2\pi$. Use the rotation form of Green's theorem to find

$$\oint_\lambda [(2y + xy)\, dx + (3x - x^2)\, dy].$$

Solution The loop λ is precisely ∂G for the simple region G consisting of the disk $x^2 + y^2 \le a^2$. By Theorem 19.4, the desired integral is equal to

$$\iint_G \left[\frac{\partial(3x - x^2)}{\partial x} - \frac{\partial(2y + xy)}{\partial y}\right] dx\, dy = \iint_G (3 - 2x - 2 - x)\, dx\, dy$$
$$= \iint_G (1 - 3x)\, dx\, dy.$$

Now $\iint_G 1 \cdot dx\, dy = \pi a^2$, the area of G. Also, $\iint_G x\, dx\, dy = 0$, since contributions from differential elements located symmetrically with respect to the y-axis cancel each other, as indicated in Fig. 19.11. Thus the answer is πa^2. □

Figure 19.12 Breaking a region into two simple regions.

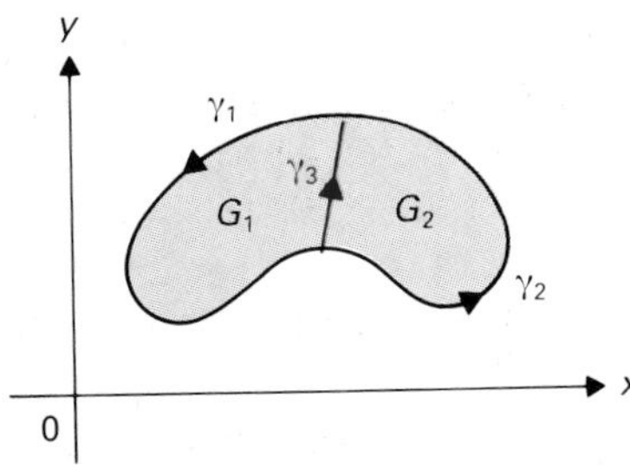

The result in Theorem 19.4 can easily be extended to a region that can be decomposed into a finite number of simple regions. Consider, for example, the region shown in Fig. 19.12. This region G is not simple, but it may be decomposed into the two simple regions G_1 and G_2 separated by the curve γ_3 shown in the figure. We recall some convenient algebraic notation. If γ_3 is the curve traced in the direction of the arrow in Fig. 19.12, let $-\gamma_3$ be the curve traced in the opposite direction. We know from Section 17.5 that

$$\int_{-\gamma_3} (P\, dx + Q\, dy) = -\int_{\gamma_3} (P\, dx + Q\, dy). \tag{7}$$

Symbolically, we write $\partial G_1 = \gamma_1 + \gamma_3$ and $\partial G_2 = \gamma_2 - \gamma_3$. In view of

Eq. (7), we have symbolically

$$\oint_{\partial G_1} + \oint_{\partial G_2} = \oint_{\gamma_1+\gamma_3} + \oint_{\gamma_2-\gamma_3} = \int_{\gamma_1} + \int_{\gamma_3} + \int_{\gamma_2} - \int_{\gamma_3} = \oint_{\gamma_1+\gamma_2} = \oint_{\partial G}, \tag{8}$$

where we have omitted the integrand $P\,dx + Q\,dy$ for brevity. Theorem 19.4 applied to the simple regions G_1 and G_2 tells us that

$$\oint_{\partial G_1} (P\,dx + Q\,dy) = \iint_{G_1} \left(\frac{\partial Q}{\partial x} - \frac{\partial P}{\partial y}\right) dx\,dy \tag{9}$$

and

$$\oint_{\partial G_2} (P\,dx + Q\,dy) = \iint_{G_2} \left(\frac{\partial Q}{\partial x} - \frac{\partial P}{\partial y}\right) dx\,dy. \tag{10}$$

Adding Eqs. (9) and (10), we see from Eq. (8) that

$$\oint_{\partial G} (P\,dx + Q\,dy) = \iint_{G} \left(\frac{\partial Q}{\partial x} - \frac{\partial P}{\partial y}\right) dx\,dy,$$

which is the rotation form of Green's theorem for the region G. Similar arguments can be used for any decomposition of a region into a finite number of simple regions, and we state the result as a corollary.

COROLLARY Green's theorem for general regions, rotation form

Let G be a bounded, closed region in the plane that can be decomposed into a finite number of simple regions. If $P(x, y)$ and $Q(x, y)$ are continuously differentiable on G, then

$$\oint_{\partial G} (P\,dx + Q\,dy) = \iint_{G} \left(\frac{\partial Q}{\partial x} - \frac{\partial P}{\partial y}\right) dx\,dy. \tag{11}$$

In the line integral, ∂G is traced in the direction that keeps G on the left.

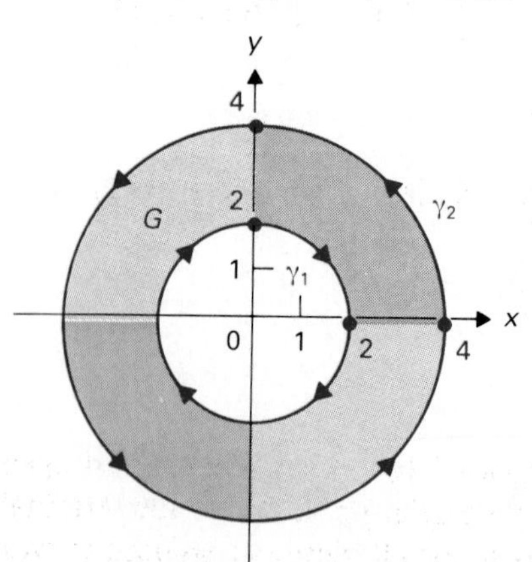

Figure 19.13 Annular region broken into four simple regions by the axes.

EXAMPLE 2 Let G be the annular region between the circles $x^2 + y^2 = 4$ and $x^2 + y^2 = 16$ shown in Fig. 19.13. The direction in which the boundary circles should be traced to keep G on the left is indicated in the figure. Now G is decomposed into four simple regions by the coordinate axes. Illustrate Green's theorem 19.4 for G using $P(x, y) = xy$ and $Q(x, y) = -x$.

Solution Let γ_1 be the clockwise-traced inner circle parametrized by

$$x = h_1(t) = 2\cos t, \qquad y = k_1(t) = -2\sin t \qquad \text{for } 0 \le t \le 2\pi,$$

and let γ_2 be the counterclockwise-traced outer circle parmetrized by

$$x = h_2(t) = 4\cos t, \qquad y = k_2(t) = 4\sin t \qquad \text{for } 0 \le t \le 2\pi.$$

Then

$$\int_{\gamma_1+\gamma_2} (xy\,dx - x\,dy) = \int_{\gamma_1} (xy\,dx - x\,dy) + \int_{\gamma_2} (xy\,dx - x\,dy)$$

$$= \int_0^{2\pi} (8 \cos t \sin^2 t + 4 \cos^2 t)\, dt$$

$$+ \int_0^{2\pi} (-64 \cos t \sin^2 t - 16 \cos^2 t)\, dt$$

$$= \int_0^{2\pi} (-56 \cos t \sin^2 t - 12 \cos^2 t)\, dt$$

$$= \left(-56 \frac{\sin^3 t}{3} - 12\left(\frac{t}{2} + \frac{\sin 2t}{4}\right)\right)\Bigg]_0^{2\pi}$$

$$= -12 \frac{2\pi}{2} = -12\pi.$$

On the other hand, we obtain, using polar coordinates,

$$\iint_G \left[\frac{\partial(-x)}{\partial x} - \frac{\partial(xy)}{\partial y}\right] dx\, dy = \iint_G (-1 - x)\, dx\, dy$$

$$= \int_0^{2\pi} \int_2^4 (-1 - r \cos \theta) r\, dr\, d\theta$$

$$= \int_0^{2\pi} \left(-\frac{r^2}{2} - \frac{r^3}{3} \cos \theta\right)\Bigg]_{r=2}^{r=4} d\theta$$

$$= \int_0^{2\pi} \left(-6 - \frac{56}{3} \cos \theta\right) d\theta$$

$$= \left(-6\theta - \frac{56}{3} \sin \theta\right)\Bigg]_0^{2\pi} = -12\pi.$$

The same answer was obtained for each integral, illustrating Green's theorem. □

EXAMPLE 3 Let $P(x, y)$ and $Q(x, y)$ have continuous partial derivatives in the annular region G, where $4 \le x^2 + y^2 \le 16$, shown in Fig. 19.13. Suppose the counterclockwise integral of $P\, dx + Q\, dy$ around $x^2 + y^2 = 4$ is 5 and around $x^2 + y^2 = 16$ is -13. Find $\iint_G (\partial Q/\partial x - \partial P/\partial y)$.

Solution For γ_1 and γ_2 shown in Fig. 19.13, we are given that

$$\oint_{-\gamma_1} (P\, dx + Q\, dy) = 5 \quad \text{and} \quad \oint_{\gamma_2} (P\, dx + Q\, dy) = -13.$$

By Green's theorem,

$$\iint_G \left(\frac{\partial Q}{\partial x} - \frac{\partial P}{\partial y}\right) dx\, dy = \oint_{\gamma_1} (P\, dx + Q\, dy) + \oint_{\gamma_2} (P\, dx + Q\, dy)$$

$$= -5 - 13 = -18. \quad \square$$

INDEPENDENCE OF PATH

Recall that $\int_\gamma (P\, dx + Q\, dy)$ is *independent of the path* in a connected open region if the integral depends only on the initial point A and the terminal point B of γ, not on its trace from A to B, for all piecewise-smooth curves

γ. We recall as a theorem some things we proved or stated in Sections 17.3 and 17.5. We should keep them well in mind here, for they are closely related to the rotation form of Green's theorem. All curves mentioned in the theorem are to be piecewise smooth.

THEOREM 19.5 Equivalents to independence of path

If $P(x, y)$ and $Q(x, y)$ have continuous first partial derivatives in a connected open region, the following are equivalent.

A. $\int_\gamma (P\,dx + Q\,dy)$ is independent of the path.
B. $\oint_\lambda (P\,dx + Q\,dy) = 0$ around every loop λ.
C. $P\,dx + Q\,dy$ is an exact differential form.
D. The vector field $\mathbf{F} = P\mathbf{i} + Q\mathbf{j}$ is ∇H for some function $H(x, y)$.

If the region also has no holes in it, the above are also equivalent to

E. $\partial P/\partial y = \partial Q/\partial x$.

The only part of Theorem 19.5 that we did not demonstrate in Chapter 17 is that (E) implies (A) through (D) if the region has no holes in it. We can partially close this gap now. Note that the rotation form of Green's theorem shows that (E) implies (B) for any loop λ that is ∂G for a region G where Green's theorem applies. For if $\partial P/\partial y = \partial Q/\partial x$ in such a region, then $\partial Q/\partial x - \partial P/\partial y = 0$, so

$$0 = \iint_G \left(\frac{\partial Q}{\partial x} - \frac{\partial P}{\partial y}\right) dx\,dy = \oint_{\partial G} (P\,dx + Q\,dy) = \oint_\lambda (P\,dx + Q\,dy).$$

The rotation form of Green's theorem can be an important tool for computing $\oint_\lambda (P\,dx + Q\,dy)$ in a case where $\partial P/\partial y = \partial Q/\partial x$ but where the loop λ goes around a "hole" in the domain of definition of P and Q. Suppose, for example, that $\partial P/\partial y = \partial Q/\partial x$ except at (x_0, y_0), and let λ_1 and λ_2 be two loops going once around (x_0, y_0) counterclockwise, as shown in Fig. 19.14. If the shaded region G between λ_1 and λ_2 in the figure is a region where Green's theorem applies, then

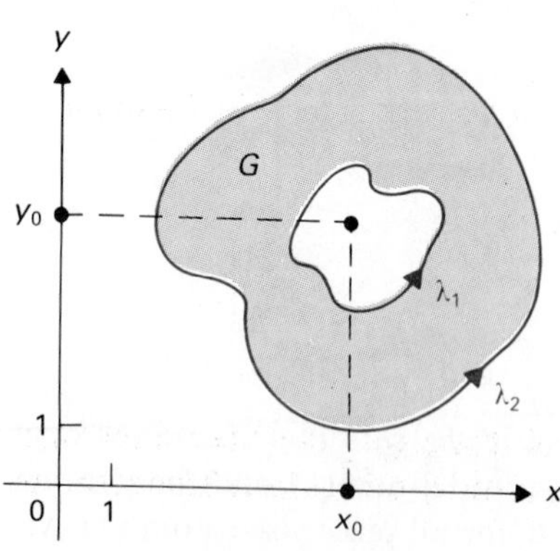

Figure 19.14 Two loops going once around (x_0, y_0) counterclockwise.

$$\begin{aligned}\oint_{\lambda_2} (P\,dx + Q\,dy) - \oint_{\lambda_1} (P\,dx + Q\,dy) &= \oint_{\partial G} (P\,dx + Q\,dy)\\ &= \iint_G \left(\frac{\partial Q}{\partial x} - \frac{\partial P}{\partial y}\right) dx\,dy\\ &= \iint_G 0\,dx\,dy = 0.\end{aligned}$$

Thus

$$\oint_{\lambda_1} (P\,dx + Q\,dy) = \oint_{\lambda_2} (P\,dx + Q\,dy).$$

That is, the value of this line integral around a loop encircling such a "hole" is independent of the loop. It can therefore be computed by choosing a special such loop to make the computation as easy as possible, perhaps a circle or a square.

EXAMPLE 4 Let λ be a loop going once around the origin counterclockwise. Find

$$\oint_\lambda \left(\frac{y}{x^2+y^2}\,dx - \frac{x}{x^2+y^2}\,dy \right).$$

Solution We take $P(x, y) = y/(x^2 + y^2)$ and $Q(x, y) = -x/(x^2 + y^2)$. We easily find that

$$\frac{\partial P}{\partial y} = \frac{x^2 - y^2}{(x^2+y^2)^2} = \frac{\partial Q}{\partial x},$$

but P and Q are not defined at $(0, 0)$. We may compute the desired integral by choosing any loop λ going once around $(0, 0)$ counterclockwise. We illustrate with both a circle and a square.

Circle We use the circle $\lambda_1: x = \cos t,\ y = \sin t,\ 0 \le t \le 2\pi$. Then

$$\oint_{\lambda_1} (P\,dx + Q\,dy) = \int_0^{2\pi} \left[\frac{\sin t}{1}(-\sin t) - \frac{\cos t}{1}(\cos t) \right] dt$$

$$= \int_0^{2\pi} (-\sin^2 t - \cos^2 t)\,dt = \int_0^{2\pi} (-1)\,dt = -2\pi.$$

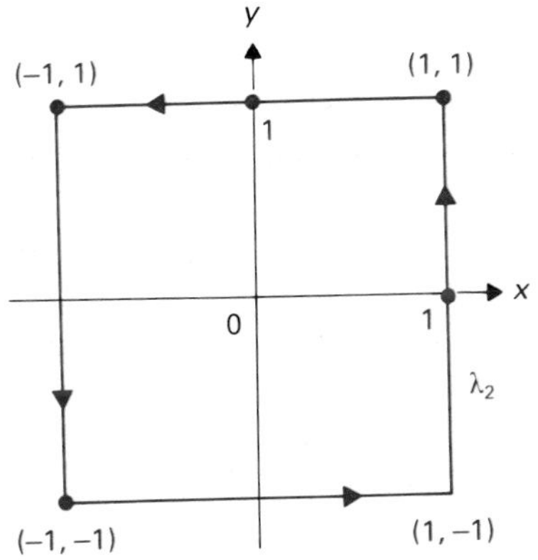

Figure 19.15 Square loop λ_2.

Square We choose the square λ_2 with vertices $(1, 1), (-1, 1), (-1, -1)$, and $(1, -1)$, shown in Fig. 19.15. On the right-hand side of the square, $x = 1$ and $dx = 0$, so our integral up this side is

$$\int_{-1}^{1} \frac{-1}{1+y^2}\,dx = -\tan^{-1} y \Big]_{-1}^{1} = -\frac{\pi}{4} + \left(-\frac{\pi}{4}\right) = -\frac{\pi}{2}.$$

Along the top of the square, $y = 1$ and $dy = 0$. To integrate across the top from right to left (counterclockwise), we let x go from 1 to -1, obtaining

$$\int_1^{-1} \frac{1}{1+x^2}\,dx = \tan^{-1} x \Big]_1^{-1} = -\frac{\pi}{4} - \frac{\pi}{4} = -\frac{\pi}{2}.$$

In a similar fashion, we find that the integrals down the left-hand side and across the bottom of the square are also each $-\pi/2$. Thus

$$\oint_{\lambda_2} (P\,dx + Q\,dy) = 4\left(-\frac{\pi}{2}\right) = -2\pi. \quad \square$$

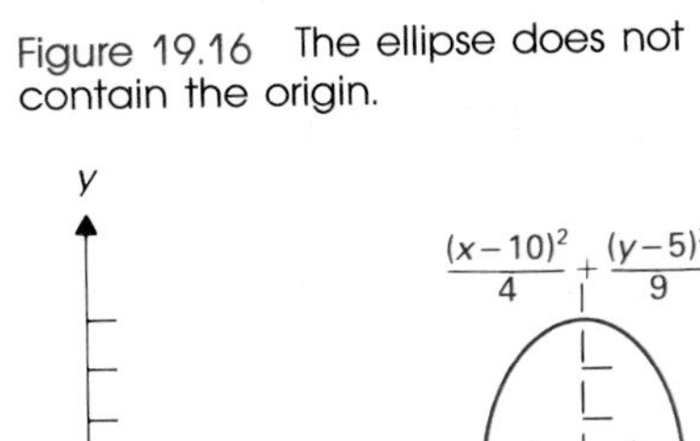

Figure 19.16 The ellipse does not contain the origin.

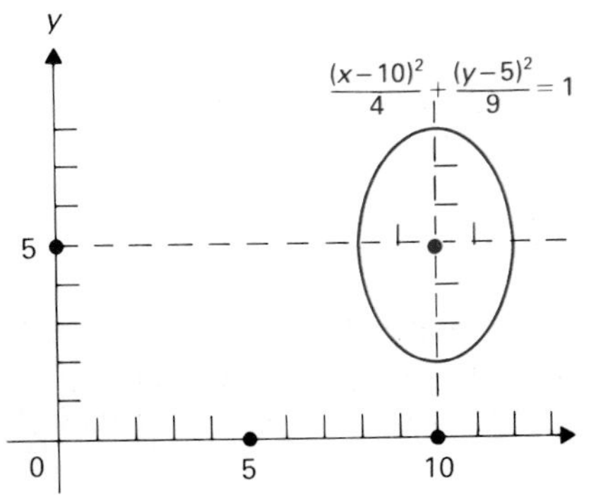

EXAMPLE 5 Find the integral of the differential form in Example 4 around the ellipse

$$\frac{(x-10)^2}{4} + \frac{(y-5)^2}{9} = 1$$

shown in Fig. 19.16.

Solution Since the ellipse does not encircle the origin, the differential form is exact throughout the entire region bounded by this ellipse. Therefore its integral around the ellipse is zero. $\square$

APPLICATION TO CIRCULATION OF A FLOW

Let $\mathbf{F} = P(x, y)\mathbf{i} + Q(x, y)\mathbf{j}$ be the flux vector of a flow over a region G. We have seen a physical interpretation of Green's theorem in terms of the divergence of the flow, which measures the rate at which mass is leaving G. This interpretation arises from integrating the outward *normal* component of the flux over the boundary of G.

Recall from Section 17.5 that the integral of the *tangential* component of the flux over the boundary measures the rotation or **circulation** of the flow around the boundary of G. A unit tangent vector is

$$\mathbf{t} = \frac{dx}{ds}\mathbf{i} + \frac{dy}{ds}\mathbf{j}.$$

The circulation around ∂G is

$$\oint_{\partial G} (\mathbf{F} \cdot \mathbf{t})\, ds = \oint_{\partial G} \left(P\frac{dx}{ds} + Q\frac{dy}{ds}\right) ds = \oint_{\partial G} (P\,dx + Q\,dy)$$
$$= \oint_{\partial G} \mathbf{F} \cdot d\mathbf{r}.$$

By Theorem 19.4, we have

$$\text{Circulation of flow around } \partial G = \oint_{\partial G} (\mathbf{F} \cdot \mathbf{t})\, ds = \oint_{\partial G} (P\,dx + Q\,dy)$$
$$= \iint_G \left(\frac{\partial Q}{\partial x} - \frac{\partial P}{\partial y}\right) dx\,dy.$$

Let us take a very small disk D_ε of radius ε centered at a point (x_0, y_0) in G, as shown in Fig. 19.17. For ε very small, we have

$$\int_{\partial D_\varepsilon} (P\,dx + Q\,dy) = \iint_{D_\varepsilon} \left(\frac{\partial Q}{\partial x} - \frac{\partial P}{\partial y}\right) dx\,dy$$
$$\approx \pi\varepsilon^2 \left(\frac{\partial Q}{\partial x} - \frac{\partial P}{\partial y}\right)\Bigg]_{(x_0, y_0)}.$$

Thus

$$\left(\frac{\partial Q}{\partial x} - \frac{\partial P}{\partial y}\right)\Bigg]_{(x_0, y_0)} \approx \frac{\text{Circulation around } \partial D_\varepsilon}{\text{Area of } D_\varepsilon},$$

Figure 19.17 Disk D_ε of radius ε with center (x_0, y_0).

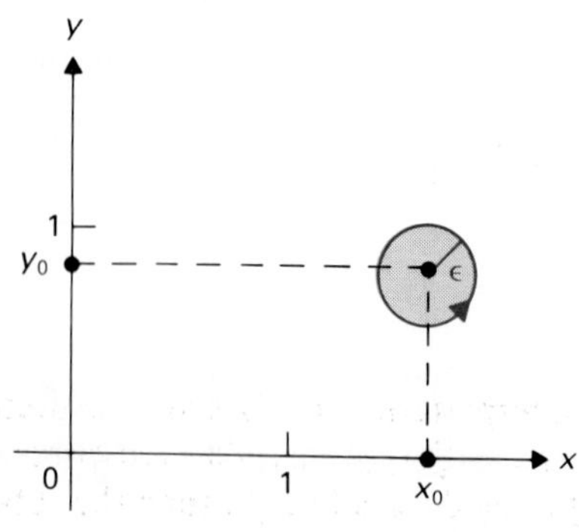

and we expect the approximation to become very accurate as $\varepsilon \to 0$. Thus $\partial Q/\partial x - \partial P/\partial y$ can be viewed as measuring the *circulation per unit area* at each point (x, y). It thus measures the tendency of the flow to rotate or *curl* at each point. For this reason, we write

$$\boxed{\text{Curl } \mathbf{F} = \frac{\partial Q}{\partial x} - \frac{\partial P}{\partial y}.^*}$$

* In Section 19.3, we will define the *vector* **curl F** for a vector field in space. Our present *scalar* curl **F** is really the **k**-component of the vector **curl F**, if we regard **F** as being in space. Our scalar curl **F** is a temporary, expedient device.

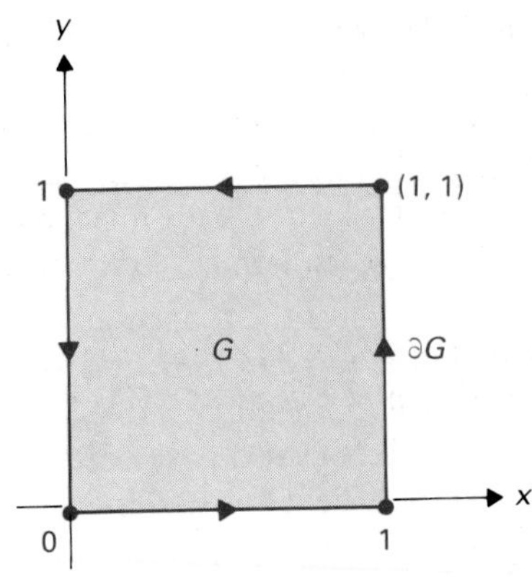

Figure 19.18 Unit square region G.

We then obtain

$$\oint_{\partial G} (\mathbf{F} \cdot \mathbf{t})\, ds = \iint_G (\text{Curl } \mathbf{F})\, dx\, dy. \tag{12}$$

If curl $\mathbf{F} = 0$ at all (x, y) in G, then the integral of the tangential component of $\mathbf{F}$ is zero about any closed path bounding a suitable region lying within G. That is, the total rotation or circulation around any such closed path is zero. Such a flow is called **irrotational.**

EXAMPLE 6 Let

$$\mathbf{F} = \frac{y}{1 + y^3}\mathbf{i} + \frac{3x}{2 + x^3}\mathbf{j},$$

and let G be the region bounded by the square with vertices $(0, 0)$, $(1, 0)$, $(1, 1)$, and $(0, 1)$, shown in Fig. 19.18. Find $\iint_G (\text{Curl } \mathbf{F})\, dx\, dy$.

Solution Computation of

$$\text{Curl } \mathbf{F} = \partial\left(\frac{3x}{2 + x^3}\right)\Big/\partial x - \partial\left(\frac{y}{1 + y^3}\right)\Big/\partial y$$

is not particularly pleasant, so we try using Green's theorem. Now

$$\iint_G (\text{Curl } \mathbf{F})\, dx\, dy = \oint_{\partial G} (\mathbf{F} \cdot \mathbf{t})\, ds = \oint_{\partial G} (\mathbf{F} \cdot d\mathbf{r})$$

$$= \oint_{\partial G} \left(\frac{y}{1 + y^3}\, dx + \frac{3x}{2 + x^3}\, dy\right).$$

On the right-hand side of the square, $x = 1$ and $dx = 0$, so the line integral up this side becomes

$$\int_0^1 \frac{3}{2 + 1}\, dy = \int_0^1 1\, dy = 1.$$

Similarly, we obtain

$$\int_1^0 \frac{1}{1 + 1}\, dx = \int_1^0 \frac{1}{2}\, dx = -\frac{1}{2}, \qquad \text{Across the top}$$

$$\int_1^0 \frac{0}{2 + 0}\, dy = 0, \qquad \text{Down the left}$$

$$\int_0^1 \frac{0}{1 + 0}\, dx = 0. \qquad \text{Across the bottom}$$

Thus the desired integral has the value

$$1 - \frac{1}{2} + 0 + 0 = \frac{1}{2}. \quad \square$$

INCOMPRESSIBLE FLOW

Think back again to the divergence interpretation of Green's theorem (Theorem 19.2). We see that if $\mathbf{F} = P(x, y)\mathbf{i} + Q(x, y)\mathbf{j}$ is a flux vector such that the divergence $\nabla \cdot \mathbf{F} = \partial P/\partial x + \partial Q/\partial y = 0$ for all (x, y), then the total flux of mass outward (or inward) through the boundary of any suitable region is zero. Such a flow is said to be **incompressible.**

APPLICATION TO WORK

We review the notions of work and potential energy. If $\mathbf{F} = P(x, y)\mathbf{i} + Q(x, y)\mathbf{j}$ is regarded as a force field, then the integral of the tangential component of $\mathbf{F}$ along a curve is the work done by the force in moving a body along the curve. Once again, if $\partial Q/\partial x - \partial P/\partial y = 0$ for all (x, y), then the work done by the force in moving a body around any closed path bounding a suitable region is zero. In this context, such a force field $\mathbf{F}$ is said to be **conservative.** The work done by such a force field along a curve depends only on the endpoints of the curve; that is, we have independence of path. Note that $P\,dx + Q\,dy$ is then the exact differential of some function $H(x, y)$. Then $u(x, y) = -H(x, y)$ is called a **potential function** of the force field. The potential function gives the **potential energy** of the field at each point (x, y). By our work in Section 17.5, the work done by the force in moving a body from A to B is then $H(B) - H(A) = -H(A) - (-H(B)) = u(A) - u(B)$.

SUMMARY

1. *Green's theorem, rotation form:* If $P(x, y)$ and $Q(x, y)$ are continuously differentiable, and if G can be decomposed into a finite number of simple regions, then

$$\oint_{\partial G} (P\,dx + Q\,dy) = \iint_G \left(\frac{\partial Q}{\partial x} - \frac{\partial P}{\partial y}\right) dx\,dy.$$

In the line integral, ∂G is traced in the direction that keeps G on the left.

2. If $P(x, y)$ and $Q(x, y)$ have continuous first partial derivatives in a connected, open region, the following are equivalent.

A. $\int_\gamma (P\,dx + Q\,dy)$ is independent of the path.

B. $\oint_\lambda (P\,dx + Q\,dy) = 0$ around every loop λ.

C. $P\,dx + Q\,dy$ is an exact differential form.

D. The vector field $\mathbf{F} = P\mathbf{i} + Q\mathbf{j}$ is a gradient field ∇H for some function $H(x, y)$.

If the region also has no holes in it, the above are also equivalent to

E. $\partial P/\partial y = \partial Q/\partial x$.

3. If $\partial P/\partial y = \partial Q/\partial x$ except at a "hole" in a region, then $\oint_\lambda (P\,dx + Q\,dy)$ is independent of the loop λ going once counterclockwise around the hole.

4. If $\mathbf{F} = P(x, y)\mathbf{i} + Q(x, y)\mathbf{j}$ is the flux vector of a flow, then

$$\oint_{\partial G} (\mathbf{F} \cdot \mathbf{t})\,ds = \iint_G (\text{Curl } \mathbf{F})\,dx\,dy,$$

where curl $\mathbf{F} = \partial Q/\partial x - \partial P/\partial y$. If curl $\mathbf{F} = 0$, then the flow is called irrotational.

5. A flow with flux vector $\mathbf{F} = P(x, y)\mathbf{i} + Q(x, y)\mathbf{j}$ is incompressible if $\nabla \cdot \mathbf{F} = \partial P/\partial x + \partial Q/\partial y = 0$.
6. The work done by a force field $\mathbf{F}$ along γ is $\int_\gamma (\mathbf{F} \cdot \mathbf{t})\, ds = \int_\gamma (\mathbf{F} \cdot d\mathbf{r})$.
7. A force field $\mathbf{F} = P(x, y)\mathbf{i} + Q(x, y)\mathbf{j}$ is conservative if $\partial Q/\partial x - \partial P/\partial y = 0$. For a conservative force field in a region with no holes in it, the work done is independent of the path, and $u(x, y) = -H(x, y)$ such that $d(H) = P\,dx + Q\,dy$ is a potential function of the force field. The work done by the force field in moving a body from A to B is then $H(B) - H(A) = u(A) - u(B)$.

EXERCISES

1. Let $P(x, y)$ and $Q(x, y)$ have continuous first partial derivatives. Let $\mathbf{F} = P\mathbf{i} + Q\mathbf{j}$, and let G be a region in the domain of $\mathbf{F}$ to which the rotation form of Green's theorem applies. Mark each of the following true or false.
   ______ a) The theorem is concerned only with the integral of the tangential component of a vector field over ∂G.
   ______ b) The theorem is concerned only with the integral of the normal component of a vector field over ∂G.
   ______ c) The theorem is concerned with the integral of a differential form in two variables over ∂G.
   ______ d) The theorem asserts that $\oint_{\partial G} (P\,dx + Q\,dy) = \iint_G [(\partial P/\partial x) + (\partial Q/\partial y)]\, dx\, dy$.
   ______ e) The theorem asserts that $\oint_{\partial G} (P\,dx + Q\,dy) = \iint_G [(\partial Q/\partial x) - (\partial P/\partial y)]\, dx\, dy$.
   ______ f) The rotation form of Green's theorem is mathematically equivalent to the divergence form.
   ______ g) The integral of the tangential component of a force field $\mathbf{F}$ around any loop is always zero by the law of conservation of energy.
   ______ h) The integral of the tangential component of a conservative force field $\mathbf{F}$ around any loop is zero.
   ______ i) If $\partial P/\partial y = \partial Q/\partial x$, then $\oint_\lambda (\mathbf{F} \cdot d\mathbf{r}) = 0$ for any loop λ.
   ______ j) If $\partial P/\partial y = \partial Q/\partial x$, then $\oint_\lambda (\mathbf{F} \cdot d\mathbf{r}) = 0$ for any loop λ that does not encircle any holes in the domain of $\mathbf{F}$.
2. Illustrate the rotation form of Green's theorem, Theorem 19.4, for $P(x, y) = x - 2y$, $Q(x, y) = y + x$, and G the annular region $1 \le x^2 + y^2 \le 4$.
3. Let G be the region between the two squares with center at the origin and sides of lengths 2 and 4. Illustrate Green's theorem 19.4 for $P(x, y) = y^3$ and $Q(x, y) = -x^3$; that is, for the vector field $\mathbf{F} = y^3\mathbf{i} - x^3\mathbf{j}$.
4. Let $\mathbf{F}$ be a vector field in the plane. Suppose the integrals $\oint (\mathbf{F} \cdot \mathbf{t})\, ds$ taken counterclockwise about the circles $x^2 + y^2 = 100$ and $x^2 + y^2 = 25$ are 35 and -24, respectively. Find $\iint_G (\text{Curl } \mathbf{F})\, dx\, dy$, where G is the region between the two circles.
5. Let $\mathbf{F}$ be a vector field in the plane, and let G be the annular region $1 \le x^2 + y^2 \le 4$. Let $\iint_G (\text{Curl } \mathbf{F})\, dx\, dy = 10$ and $\oint_{\lambda_1} (\mathbf{F} \cdot d\mathbf{r}) = 7$, where λ_1 is the circle $x^2 + y^2 = 4$ traced once clockwise. Find $\oint_{\lambda_2} (\mathbf{F} \cdot d\mathbf{r})$, where λ_2 is the circle $x^2 + y^2 = 1$ traced once counterclockwise.
6. Prove Eq. (6) of the proof of the rotation form of Green's theorem in the text.
7. Let $\mathbf{E}$ and $\mathbf{F}$ be two vector fields on a simple, closed, bounded region G. Suppose that $\mathbf{E} = \mathbf{F}$ at each point on ∂G.
   a) Show that
   $$\iint_G (\nabla \cdot \mathbf{E})\, dx\, dy = \iint_G (\nabla \cdot \mathbf{F})\, dx\, dy.$$
   b) Show that
   $$\iint_G (\text{Curl } \mathbf{E})\, dx\, dy = \iint_G (\text{Curl } \mathbf{F})\, dx\, dy.$$
8. Let G be a simple, closed, bounded region, and let $\mathbf{F}$ be a vector field on G.
   a) Suppose $\oint_\lambda (\mathbf{F} \cdot \mathbf{t})\, ds = 0$ for every closed curve λ lying in G. Argue as best you can using just Green's theorem that curl $\mathbf{F} = 0$ at each (x, y) in G.
   b) State a result similar to that in part (a) under the hypothesis $\oint_\lambda (\mathbf{F} \cdot \mathbf{n})\, ds = 0$.
9. Let a plane region G, which can be decomposed into a finite number of simple regions, be bounded by a piecewise-smooth curve λ. Using Green's theorem 19.4, show that the line integral $\oint_\lambda (-y\,dx + x\,dy)$ is equal to twice the area of G.

In Exercises 10 through 20, let λ be the ellipse $x = 2\cos t$, $y = 3\sin t$ for $0 \le t \le 2\pi$. For the given vector field $\mathbf{F}$, find $\oint_\lambda (\mathbf{F} \cdot \mathbf{t})\, ds$ in the easiest way you can. You may use the fact that the ellipse $(x^2/a^2) + (y^2/b^2) = 1$ has area πab. Be sure to use the symmetry of the ellipse about the axes and origin, where such symmetry is helpful.

10. $\mathbf{F} = x^2\mathbf{i} + (y^3 - 2y)\mathbf{j}$
11. $\mathbf{F} = (x^2 + 2y)\mathbf{i} + (y^2 - 3x)\mathbf{j}$

12. $\mathbf{F} = (2xy - y)\mathbf{i} - (3x + x^2)\mathbf{j}$

13. $\mathbf{F} = (x^2y^2 - y)\mathbf{i} + (3x^2y^2 + 2x)\mathbf{j}$

14. $\mathbf{F} = (3xy^3 - 2x^3y - xy^2)\mathbf{i} + (x^4y^2 + x^3y^3 - 2x^2y^3)\mathbf{j}$

15. $\mathbf{F} = \dfrac{-2y}{x^2 + y^2}\mathbf{i} + \dfrac{2x}{x^2 + y^2}\mathbf{j}$

16. $\mathbf{F} = \dfrac{x}{x^2 + y^2}\mathbf{i} + \dfrac{y}{x^2 + y^2}\mathbf{j}$

17. $\mathbf{F} = \dfrac{x + y}{x^2 + y^2}\mathbf{i} + \dfrac{y - x}{x^2 + y^2}\mathbf{j}$

18. $\mathbf{F} = \dfrac{1 - y}{x^2 + (y - 1)^2}\mathbf{i} + \dfrac{x}{x^2 + (y - 1)^2}\mathbf{j}$

19. $\mathbf{F} = \dfrac{x^2y^2 - y}{x^2 + y^2}\mathbf{i} + \dfrac{x - xy}{x^2 + y^2}\mathbf{j}$

20. $\mathbf{F} = \dfrac{4 - y}{x^2 + (y - 4)^2}\mathbf{i} + \dfrac{x}{x^2 + (y - 4)^2}\mathbf{j}$

In Exercises 21 through 27, let $P(x, y)$ and $Q(x, y)$ have continuous first partial derivatives except at (0, 0), (0, 4), and (6, 0), and let $\partial Q/\partial x = \partial P/\partial y$ except at those points. Let $\oint (P\,dx + Q\,dy)$ have these values taken counterclockwise once around these circles:

$$-5 \text{ taken around } x^2 + y^2 = 1,$$

$$3 \text{ taken around } x^2 + (y - 4)^2 = 1,$$

$$9 \text{ taken around } (x - 6)^2 + y^2 = 1.$$

Find $\oint_\lambda (P\,dx + Q\,dy)$ for the given loop λ.

21. λ given by $x = 5\cos t$, $y = 5\sin t$ for $0 \le t \le 2\pi$

22. λ given by $x = 3 + 4\cos t$, $y = 4\sin t$ for $0 \le t \le 2\pi$

23. λ given by $x = \cos t$, $y = \sin t$ for $0 \le t \le 4\pi$

24. λ given by $x = 8\cos t$, $y = 8\sin t$ for $0 \le t \le 2\pi$

25. λ given by $x = 3\cos t$, $y = 2 - 3\sin t$ for $0 \le t \le 2\pi$

26. λ given by $x = 2 + 3\cos t$, $y = 3 - 3\sin t$ for $0 \le t \le 4\pi$

27. λ given by $x = 2 + 2\cos t$, $y = 2 + 2\sin t$ for $0 \le t \le 2\pi$

28. Show that if $f(x, y)$ has continuous second partial derivatives, then curl $\nabla f = 0$.

29. Show that if ∇f is the flux vector of an incompressible flow, then f satisfies Laplace's equation $f_{xx} + f_{yy} = 0$.

30. Use the appropriate form of Green's theorem to find the work done by the force field $\mathbf{F}(x, y) = (x^2 + y^2)\mathbf{i} - 2xy\mathbf{j}$ in moving a body counterclockwise around the square with vertices at (0, 0), (1, 0), (1, 1), and (0, 1).

31. Find all functions $g(x, y)$ such that the force field $\mathbf{F}(x, y) = 3xy^2\mathbf{i} + g(x, y)\mathbf{j}$ is conservative.

32. Let the flux vector of a plane flow be $\mathbf{F}(x, y) = (x^2 + y^2)\mathbf{i} + 2xy\mathbf{j}$.
a) Find the divergence of $\mathbf{F}$.
b) Use the divergence form of Green's theorem to find the flux of the flow across the border of the square with vertices (0, 0), (1, 0), (1, 1), and (0, 1).

33. For the flow in Exercise 32, use the rotation form of Green's theorem to find the circulation of the flow counterclockwise around the circle $x^2 + y^2 = 4$.

34. Show that every constant force field is conservative. What can be said concerning the work done by a constant force field in moving a body from a point A to a point B?

35. Show that every constant flux field in the plane gives a flow that is both irrotational and incompressible. Give physical interpretations of this result.

36. Let $\mathbf{E}$ and $\mathbf{F}$ be conservative force fields in a plane region G.
a) Show that for any numbers a and b, the force field $a\mathbf{E} + b\mathbf{F}$ is conservative.
b) If u is a potential function of $\mathbf{E}$ and v a potential function of $\mathbf{F}$ in G, describe all potential functions of $a\mathbf{E} + b\mathbf{F}$.

37. Consider the force field $\mathbf{F} = 2xy\mathbf{i} + x^2\mathbf{j}$ in the plane.
a) Show that the field is conservative.
b) Find a potential function for the field.
c) Find the potential energy $u(x, y)$ of the field such that $u(0, 0) = 5$.
d) Find the work done by the field in moving a body from the point (1, −1) to the point (2, 1).

38. Let $\mathbf{F}$ be a force field in the plane, and let the position of a body in the plane at time t be given by $x = h_1(t)$, $y = h_2(t)$ for $a \le t \le b$. Let the position be A when $t = a$ and B when $t = b$. Let $W(A, B)$ be the work done by the field in moving the body from A to B for $a \le t \le b$. Show that

$$W(A, B) = \tfrac{1}{2}m|\mathbf{v}(b)|^2 - \tfrac{1}{2}m|\mathbf{v}(a)|^2 = k(b) - k(a),$$

where $k(t) = \frac{1}{2}m|\mathbf{v}(t)|^2$ is the *kinetic energy* of the body at time t. [*Hint:* Use $\mathbf{F}(t) = m\mathbf{a}(t) = m\mathbf{v}'(t)$ and $d\mathbf{r}/dt = \mathbf{v}(t)$ to express $W(A, B) = \int_a^b (\mathbf{F} \cdot d\mathbf{r})$ in terms of $\mathbf{v}$.]

39. Continuing Exercise 38, show that if $\mathbf{F}$ is the force field for a potential function u, then $u(A) + k(A) = u(B) + k(B)$, so the sum of the potential and kinetic energies remains constant. (Note that it is permissible to write u and k as functions of positions in the plane since $\mathbf{F}$ is a conservative force field, so the work in Exercise 38 is independent of the path.) This is the reason why such a force field is called *conservative*.

40. A positive electric charge in the plane at the origin exerts a force of attraction on a negative unit charge in the plane that is inversely proportional to the square of the distance between the charges. Let the force of attraction on a negative unit charge at (1, 0) be k units.
a) Find the force field created by the charge at the origin.
b) Show that the force field in part (a) is conservative by finding a potential function; the function is the *Newtonian potential*.
c) Using the answer to part (b), find the work done by the field in moving a unit negative charge from (0, 2) to (2, 3).

41. Let u be a potential function of a conservative force field $\mathbf{F}$ in the plane. Level curves of u are known as *equipotential curves*.
a) What geometric relation exists between the field $\mathbf{F}$ and its equipotential curves?
b) If γ_1 and γ_2 are equipotential curves, show that the work done by the field $\mathbf{F}$ in moving a body from a point P_1 on γ_1 to a point P_2 on γ_2 is independent of the choices of P_1 on γ_1 and P_2 on γ_2.

19.3 STOKES' THEOREM

Green's theorem is concerned with integrals over a plane region. Stokes' theorem is a generalization of the rotation form of Green's theorem to more general two-dimensional regions, which do not necessarily lie in a plane. That is, Stokes' theorem is concerned with integrals over a *surface* in space. We will not prove Stokes' theorem; the proof is best left to a more advanced course. Of course, we have proved the special case of it, the rotation form of Green's theorem, in the case where the "surface" lies in the x, y-plane.

INTEGRATION OVER A SURFACE

To prepare for Stokes' theorem, we discuss integration of the normal component of a vector field over a surface in space. We phrase this discussion in the context of the divergence theorem, presented in Section 19.1. Recall the statement of the theorem.

> If $\mathbf{F} = P\mathbf{i} + Q\mathbf{j} + R\mathbf{k}$ is a continuously differentiable vector field over a suitable region G in space, then
>
> $$\iint_{\partial G} (\mathbf{F} \cdot \mathbf{n})\, dS = \iiint_G (\nabla \cdot \mathbf{F})\, dx\, dy\, dz. \tag{1}$$
>
> In the surface integral, $\mathbf{n}$ is the outward normal unit vector to the surface ∂G at each point.
>
> ***Divergence theorem***

We study the integral of the normal component of a vector field $\mathbf{F} = P\mathbf{i} + Q\mathbf{j} + R\mathbf{k}$ over a surface. We can compute $\mathbf{F} \cdot \mathbf{n}$ if we know $\mathbf{n}$; presumably $\mathbf{F}$ is given. Now if the surface is given in the form $z = f(x, y)$, a normal vector is

$$\frac{\partial f}{\partial x}\mathbf{i} + \frac{\partial f}{\partial y}\mathbf{j} - \mathbf{k},$$

so

$$\mathbf{n} = \pm \frac{1}{\sqrt{(f_x)^2 + (f_y)^2 + 1}} (f_x\mathbf{i} + f_y\mathbf{j} - \mathbf{k}). \tag{2}$$

It is usually best to see which sign is appropriate for the *outward* normal by sketching. From Section 18.6, we know that

$$dS = \sqrt{(f_x)^2 + (f_y)^2 + 1}\, dx\, dy \tag{3}$$

is the differential surface area.

Suppose, on the other hand, that the surface ∂G is given in terms of $f(x, y, z) = c$. Then a normal vector is $f_x\mathbf{i} + f_y\mathbf{j} + f_z\mathbf{k}$, so

$$\mathbf{n} = \pm \frac{1}{\sqrt{(f_x)^2 + (f_y)^2 + (f_z)^2}} (f_x\mathbf{i} + f_y\mathbf{j} + f_z\mathbf{k}). \tag{4}$$

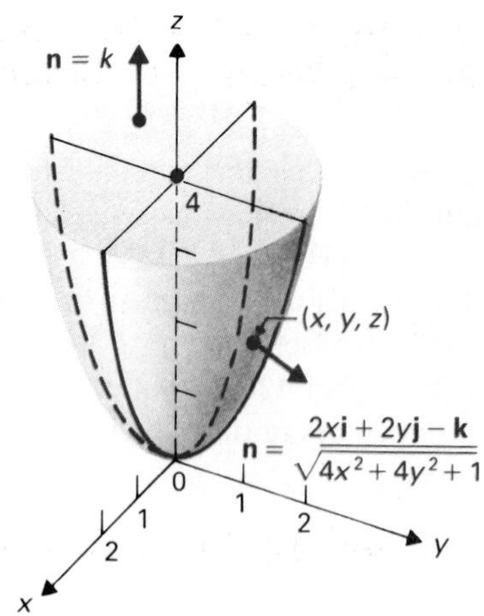

Figure 19.19 Outward unit normal vectors to the region bounded by $z = x^2 + y^2$ and $z = 4$.

This time

$$dS = \frac{\sqrt{(f_x)^2 + (f_y)^2 + (f_z)^2}}{|f_z|}\, dx\, dy. \tag{5}$$

The radicals appearing in dS in Eqs. (3) and (5) might be expected to make integration over a surface difficult. However, we note that the same radicals appear in the denominators for $\mathbf{n}$ in Eqs. (2) and (4) and consequently appear in the denominator of $\mathbf{F} \cdot \mathbf{n}$. Thus these radicals cancel each other in the computation of $(\mathbf{F} \cdot \mathbf{n})\, dS$, so integration of a normal component of a vector field over a surface may not be especially difficult.

EXAMPLE 1 Illustrate the divergence theorem for the vector field $\mathbf{F} = x^2\mathbf{i} + y^2\mathbf{j} + z^2\mathbf{k}$ over the region G bounded by $z = x^2 + y^2$ and $z = 4$, shown in Fig. 19.19.

Solution We first compute $\iint_{\partial G} (\mathbf{F} \cdot \mathbf{n})\, dS$. Starting with the portion of the paraboloid $z = x^2 + y^2$, we find from Eq. (2) that

$$\mathbf{n} = \frac{1}{\sqrt{4x^2 + 4y^2 + 1}}(2x\mathbf{i} + 2y\mathbf{j} - \mathbf{k}).$$

Our sketch in Fig. 19.19 shows that the negative coefficient of $\mathbf{k}$ is appropriate for the outward normal. From Eq. (3), we have

$$dS = \sqrt{4x^2 + 4y^2 + 1}\, dx\, dy.$$

Therefore

$$(\mathbf{F} \cdot \mathbf{n})\, dS = (2x^3 + 2y^3 - z^2)\, dx\, dy.$$

Since $z = x^2 + y^2$, we obtain

$$(\mathbf{F} \cdot \mathbf{n})\, dS = [2x^3 + 2y^3 - (x^2 + y^2)^2]\, dx\, dy.$$

The projection of this portion of the paraboloid $z = x^2 + y^2$ for $0 \le z \le 4$ onto the x,y-plane is the disk $x^2 + y^2 \le 4$. The integral of $2x^3 + 2y^3$ over this disk is zero by symmetry, so we are left with the integral of $-(x^2 + y^2)^2$ over the disk. Changing to polar coordinates, we obtain the integral

$$\int_0^{2\pi}\int_0^2 (-r^4) \cdot r\, dr\, d\theta = \int_0^{2\pi} -\frac{r^6}{6}\Big]_0^2 d\theta = \int_0^{2\pi} -\frac{64}{6}\, d\theta$$
$$= 2\pi\left(-\frac{64}{6}\right) = -\frac{64}{3}\pi.$$

We now turn to the integral of $\mathbf{F} \cdot \mathbf{n}$ over the disk $x^2 + y^2 \le 4$ in the plane $z = 4$. Clearly, $\mathbf{n} = \mathbf{k}$ is the outward unit normal vector, as shown in Fig. 19.19. Then $\mathbf{F} \cdot \mathbf{n} = z^2 = 4^2 = 16$. The integral of $\mathbf{F} \cdot \mathbf{n}$ over the disk is then

$$16 \cdot \text{Area of the disk} = 16 \cdot 4\pi = 64\pi.$$

Adding the integrals of $\mathbf{F} \cdot \mathbf{n}$ over these two portions of ∂G, we have

$$\iint_{\partial G} (\mathbf{F} \cdot \mathbf{n})\, dS = -\frac{64}{3}\pi + 64\pi = \frac{128}{3}\pi.$$

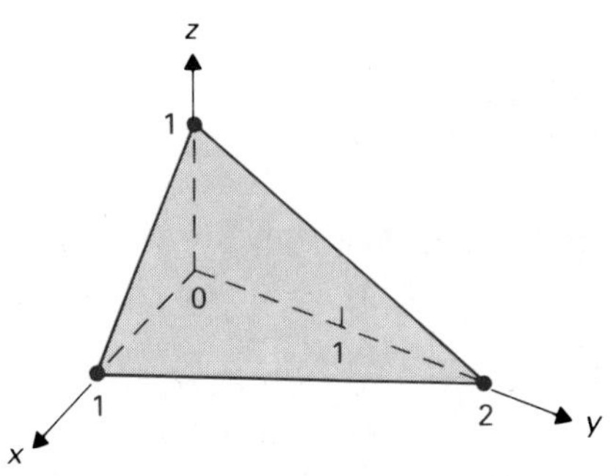

Figure 19.20 Tetrahedron with vertices (0, 0, 0), (1, 0, 0), (0, 2, 0), and (0, 0, 1).

We now compute $\iiint_G(\nabla \cdot \mathbf{F})\,dx\,dy\,dz = \iiint_G(2x + 2y + 2z)\,dx\,dy\,dz$. By symmetry, the integral over G of $2x + 2y$ is zero. We change to cylindrical coordinates and find that

$$\iiint_G 2z\,dz\,dx\,dy = \int_0^{2\pi}\int_0^2\int_{r^2}^4 2z\,r\,dz\,dr\,d\theta = \int_0^{2\pi}\int_0^2 z^2\Big]_{r^2}^4 r\,dr\,d\theta$$

$$= \int_0^{2\pi}\int_0^2 (16 - r^4)r\,dr\,d\theta = \int_0^{2\pi}\left(8r^2 - \frac{r^6}{6}\right)\Big]_0^2 d\theta$$

$$= 2\pi\left(32 - \frac{64}{6}\right) = \pi\left(64 - \frac{64}{3}\right) = \frac{128\pi}{3}. \quad \square$$

EXAMPLE 2 Illustrate the divergence theorem, Eq. (1), for the vector field

$$\mathbf{F} = x\mathbf{i} + y\mathbf{j} + 2z\mathbf{k}$$

over the region G, which is the tetrahedron with vertices (0, 0, 0), (1, 0, 0), (0, 2, 0), and (0, 0, 1) shown in Fig. 19.20.

Solution The volume integral is

$$\iiint_G\left(\frac{\partial P}{\partial x} + \frac{\partial Q}{\partial y} + \frac{\partial R}{\partial z}\right)dx\,dy\,dz = \iiint_G (1 + 1 + 2)\,dx\,dy\,dz$$

$$= 4(\text{Volume of tetrahedron})$$

$$= 4 \cdot \frac{1}{3} \cdot 1 \cdot 1 = \frac{4}{3}.$$

Now we turn to the integral $\iint_{\partial G}(\mathbf{F} \cdot \mathbf{n})\,dS$ over the four triangles that form the surface of the tetrahedron. For the triangle in the x,y-plane, $\mathbf{n} = -\mathbf{k}$, $z = 0$, and $dS = dx\,dy$, so the integral becomes

$$\int_0^1\int_0^{2-2x}[(x\mathbf{i} + y\mathbf{j}) \cdot (-\mathbf{k})]\,dy\,dx = \int_0^1\int_0^{2-2x} 0\,dy\,dx = 0.$$

The integrals over the triangles in the x,z-plane and in the y,z-plane are similarly zero.

We must now find the equation of the front plane of the region, determined by the points (1, 0, 0), (0, 2, 0), and (0, 0, 1). A vector perpendicular to the plane is

$$(-\mathbf{i} + 2\mathbf{j}) \times (-\mathbf{i} + \mathbf{k}) = \begin{vmatrix} \mathbf{i} & \mathbf{j} & \mathbf{k} \\ -1 & 2 & 0 \\ -1 & 0 & 1 \end{vmatrix} = 2\mathbf{i} + \mathbf{j} + 2\mathbf{k},$$

so the equation of the plane is

$$2x + y + 2z = 2.$$

For this front triangle, we see from the equation $2x + y + 2z = 2$ of its plane that

$$\mathbf{n} = \frac{2}{3}\mathbf{i} + \frac{1}{3}\mathbf{j} + \frac{2}{3}\mathbf{k}$$

and, from Eq. (5),

$$dS = \left(\frac{\sqrt{4+1+4}}{2}\right) dx\,dy = \left(\frac{3}{2}\right) dx\,dy.$$

Therefore the surface integral over the front face of the tetrahedron becomes

$$\int_0^1 \int_0^{2-2x} \underbrace{\left[\frac{2}{3}x + \frac{1}{3}y + \frac{4}{3}\left(1 - x - \frac{y}{2}\right)\right]}_{\mathbf{F}\cdot\mathbf{n}} \underbrace{\frac{3}{2}\,dy\,dx}_{dS}$$

$$= \int_0^1 \int_0^{2-2x} \left(2 - x - \frac{y}{2}\right) dy\,dx$$

$$= \int_0^1 \left[2(2-2x) - x(2-2x) - \frac{1}{4}(2-2x)^2\right] dx$$

$$= \int_0^1 (3 - 4x + x^2)\,dx = \left(3x - 2x^2 + \frac{x^3}{3}\right)\Big]_0^1$$

$$= 3 - 2 + \frac{1}{3} = \frac{4}{3}.$$

Thus both sides of Eq. (1) are equal to $\frac{4}{3}$, illustrating the divergence theorem. □

THE CURL OF A VECTOR FIELD

Let $\mathbf{F} = P\mathbf{i} + Q\mathbf{j} + R\mathbf{k}$ be a differentiable vector field in space. The **curl of F** is the *vector* defined by the symbolic determinant

$$\textbf{Curl F} = \nabla \times \mathbf{F} = \begin{vmatrix} \mathbf{i} & \mathbf{j} & \mathbf{k} \\ \frac{\partial}{\partial x} & \frac{\partial}{\partial y} & \frac{\partial}{\partial z} \\ P & Q & R \end{vmatrix}.$$

Note that **curl F** is a *vector*. Our definition of curl **F** as a scalar quantity in the preceding section was a temporary expedient in discussing rotation of a flow in the plane and is explained in Example 4 below.

EXAMPLE 3 Find **curl F** if $\mathbf{F} = xy\mathbf{i} + y^2\mathbf{j} + yz^2\mathbf{k}$.

Solution We have

$$\textbf{Curl F} = \nabla \times \mathbf{F} = \begin{vmatrix} \mathbf{i} & \mathbf{j} & \mathbf{k} \\ \frac{\partial}{\partial x} & \frac{\partial}{\partial y} & \frac{\partial}{\partial z} \\ xy & y^2 & yz^2 \end{vmatrix} = z^2\mathbf{i} + 0\mathbf{j} - x\mathbf{k}$$

$$= z^2\mathbf{i} - x\mathbf{k}. \quad \square$$

EXAMPLE 4 Find **curl F** if $R(x, y, z) = 0$, so $\mathbf{F} = P\mathbf{i} + Q\mathbf{j}$.

Solution We have

$$\textbf{Curl F} = \nabla \times \mathbf{F} = \begin{vmatrix} \mathbf{i} & \mathbf{j} & \mathbf{k} \\ \dfrac{\partial}{\partial x} & \dfrac{\partial}{\partial y} & \dfrac{\partial}{\partial z} \\ P & Q & 0 \end{vmatrix} = -\frac{\partial Q}{\partial z}\mathbf{i} + \frac{\partial P}{\partial z}\mathbf{j} + \left(\frac{\partial Q}{\partial x} - \frac{\partial P}{\partial y}\right)\mathbf{k}.$$

We see that what we called curl **F** in Section 19.2 for $\mathbf{F} = P\mathbf{i} + Q\mathbf{j}$ was really the **k**-component of **curl F**, viewed as a vector in space. □

STOKES' THEOREM

We make no attempt to prove Stokes' theorem but will, roughly, state it. It is the generalization of Green's theorem to a two-dimensional surface G in space. For example, G might be the surface shown in Fig. 19.21. (The theorem is not true for some "one-sided" surfaces, which we will not describe.) Figure 19.21 also shows a unit normal vector **n** at a point on G and indicates by arrows a direction around ∂G.

> For Stokes' theorem, the direction for **n** and the direction around ∂G are always related so that as the fingers of the right-hand curve in the direction around ∂G, the thumb points in the direction of **n**.
>
> ***Right-hand rule***

This right-hand rule is illustrated in Fig. 19.22. To tie this in with G and ∂G for the rotation form of Green's theorem in the plane, note that this means that when walking along ∂G with our head in the direction of **n**, we have the region G on our left.

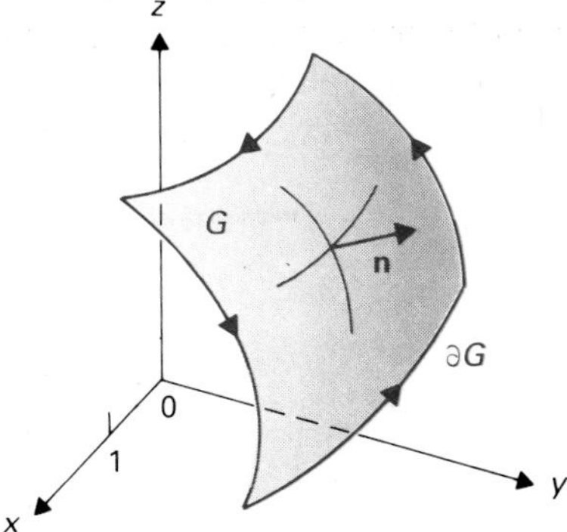

Figure 19.21 Surface G with directed boundary curve ∂G and unit normal vector **n**.

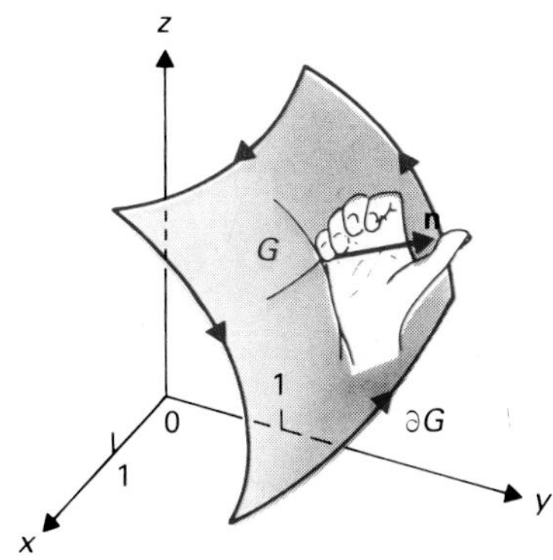

Figure 19.22 Right-hand rule relating the direction of **n** and the direction around ∂G.

THEOREM Stokes' theorem

19.6

For a suitable surface G in space and a continuously differentiable vector field $\mathbf{F} = P\mathbf{i} + Q\mathbf{j} + R\mathbf{k}$, we have

$$\oint_{\partial G} (\mathbf{F} \cdot \mathbf{t})\, ds = \iint_G [(\nabla \times \mathbf{F}) \cdot \mathbf{n}]\, dS. \qquad \textbf{(6)}$$

In the line integral, ∂G is traced in the direction with respect to $\mathbf{n}$ given by the right-hand rule.

In words, the integral of the tangential component of the vector field around ∂G is equal to the integral of the normal component of **curl F** over the surface G. If $\mathbf{F}$ is the flux field of a flow, this has the interpretation that the circulation around ∂G equals the integral of the normal component of the **curl** over G. The **curl** measures the rotation of the flow at each point, and the normal component measures the portion of this rotation that acts *tangent* to the surface. Remember that the direction of a *tangent* plane to a surface is specified by a *normal* direction to the surface.

EXAMPLE 5 Write Eq. (6) in Stokes' theorem in the case that the surface G lies in the x, y-plane and $\mathbf{F} = P\mathbf{i} + Q\mathbf{j}$, taking $\mathbf{n} = \mathbf{k}$.

Solution If G lies in the x, y-plane and $\mathbf{n} = \mathbf{k}$, then $dS = dx\, dy$. If $\mathbf{F} = P\mathbf{i} + Q\mathbf{j}$, then, as shown in Example 4,

$$\nabla \times \mathbf{F} = -\frac{\partial Q}{\partial z}\mathbf{i} + \frac{\partial P}{\partial z}\mathbf{j} + \left(\frac{\partial Q}{\partial x} - \frac{\partial P}{\partial y}\right)\mathbf{k}.$$

Equation (6) then becomes

$$\oint_{\partial G} (\mathbf{F} \cdot \mathbf{t})\, ds = \iint_G \left(\frac{\partial Q}{\partial x} - \frac{\partial P}{\partial y}\right) dx\, dy.$$

This is the rotation form of Green's theorem, which is thus a special case of Stokes' theorem. □

EXAMPLE 6 Illustrate Stokes' theorem if the surface G is the hemisphere $z = \sqrt{a^2 - x^2 - y^2}$, shown in Fig. 19.23, and

$$\mathbf{F} = yz\mathbf{i} - xz\mathbf{j} + 3\mathbf{k}.$$

Solution Now

$$\mathbf{n} = \frac{1}{a}(x\mathbf{i} + y\mathbf{j} + z\mathbf{k}),$$

and Fig. 19.24 shows that we could use spherical coordinates and take

$$dS = (a \sin\phi\, d\theta)(a\, d\phi) = a^2 \sin\phi\, d\phi\, d\theta.$$

Also

$$\nabla \times \mathbf{F} = \begin{vmatrix} \mathbf{i} & \mathbf{j} & \mathbf{k} \\ \dfrac{\partial}{\partial x} & \dfrac{\partial}{\partial y} & \dfrac{\partial}{\partial z} \\ yz & -xz & 3 \end{vmatrix} = x\mathbf{i} + y\mathbf{j} - 2z\mathbf{k}.$$

Figure 19.23 Unit normal vector $\mathbf{n} = (1/a)(x\mathbf{i} + y\mathbf{j} + z\mathbf{k})$ to the hemisphere G; direction around ∂G.

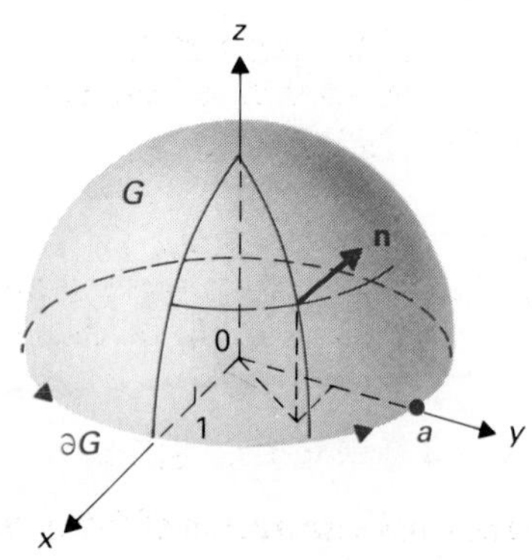

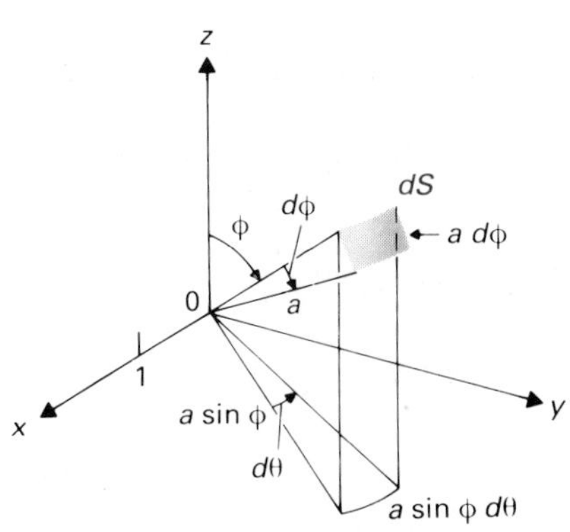

Figure 19.24 In spherical coordinates, we may take $dS = (a \sin \phi \, d\theta)(a \, d\phi)$ on the surface of a sphere of radius a.

Consequently,

$$(\nabla \times \mathbf{F}) \cdot \mathbf{n} = \frac{1}{a}(x^2 + y^2 - 2z^2) = \frac{1}{a}(x^2 + y^2 + z^2 - 3z^2)$$

$$= \frac{1}{a}(a^2 - 3z^2) = \frac{1}{a}(a^2 - 3a^2 \cos^2\phi)$$

$$= a(1 - 3\cos^2\phi).$$

Then

$$\iint_G [(\nabla \times \mathbf{F}) \cdot \mathbf{n}] \, dS = \int_0^{2\pi} \int_0^{\pi/2} a(1 - 3\cos^2\phi)a^2 \sin\phi \, d\phi \, d\theta$$

$$= a^3 \int_0^{2\pi} (-\cos\phi + \cos^3\phi)\Big]_0^{\pi/2} d\theta$$

$$= a^3 \int_0^{2\pi} 0 \, d\theta = 0.$$

Turning to $\oint_{\partial G} (\mathbf{F} \cdot \mathbf{t}) \, ds$, we parametrize ∂G by $x = a \cos t$ and $y = a \sin t$ for $0 \le t \le 2\pi$. Then

$$\mathbf{t} = \frac{dx}{ds}\mathbf{i} + \frac{dy}{ds}\mathbf{j} + 0\mathbf{k}.$$

Since ∂G lies in the plane $z = 0$, we see that there $\mathbf{F} = 0\mathbf{i} + 0\mathbf{j} + 3\mathbf{k}$. Therefore $\mathbf{F} \cdot \mathbf{t} = 0$, so $\oint_{\partial G} (\mathbf{F} \cdot \mathbf{t}) \, ds = 0$ also, illustrating Stokes' theorem. □

We have seen that it is sometimes convenient to use a form of Green's theorem or the divergence theorem to change from an integral over the boundary of a region to an integral over the region itself, or vice versa. Computing the integral is sometimes simplified in this way. Of course, Stokes' theorem can be used in the same way. In fact, Stokes' theorem presents even more possibilities. The problem of computing $\iint_G [(\nabla \times \mathbf{F}) \cdot \mathbf{n}] \, dS$ over a surface G can of course be transformed into computing $\oint_{\partial G} (\mathbf{F} \cdot \mathbf{t}) \, ds$. But it can also be transformed into computing $\iint_{G'} [(\nabla \times \mathbf{F}) \cdot \mathbf{n}] \, dS$ for *any suitable surface G' for which $\partial G' = \partial G$*, since by Stokes' theorem both integrals equal $\oint_{\partial G} (\mathbf{F} \cdot \mathbf{t}) \, ds$. The second solutions in our final examples illustrate this possibility.

Figure 19.25
$G: z = x^2 + y^2, 0 \le z \le 4$;
$G': x^2 + y^2 = 4, z = 4; \partial G = \partial G'$.

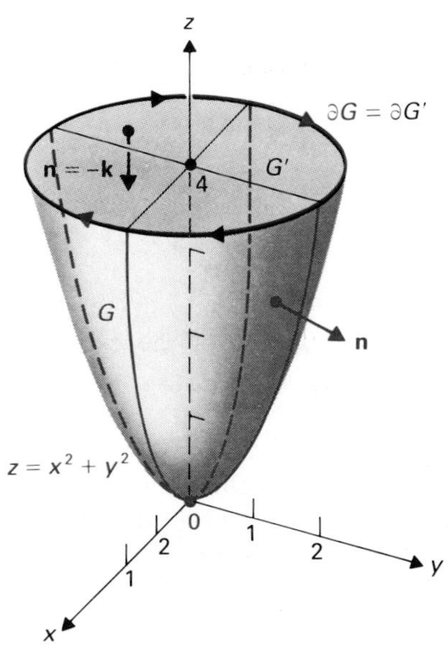

EXAMPLE 7 Let G be the surface $z = x^2 + y^2$ for $0 \le z \le 4$, and let $\mathbf{n}$ be the unit normal vector pointing out the convex side of the surface. Let $\mathbf{F} = yz\mathbf{i} + xz^2\mathbf{j} - xyz\mathbf{k}$. Find $\iint_G [(\nabla \times \mathbf{F}) \cdot \mathbf{n}] \, dS$.

Solution 1 By Stokes' theorem, we may instead compute $\oint_{\partial G} (\mathbf{F} \cdot \mathbf{t}) \, ds = \oint_{\partial G} (\mathbf{F} \cdot d\mathbf{r})$. The surface is sketched in Fig. 19.25. We see that ∂G is the circle $x^2 + y^2 = 4$ in the plane $z = 4$, traced *clockwise* as viewed from above. A parametrization of ∂G is therefore

$$x = 2\cos t, \qquad y = -2\sin t, \qquad z = 4 \qquad \text{for } 0 \le t \le 2\pi.$$

(The minus sign in the y-equation is due to the *clockwise* tracing of the circle.)

In the plane $z = 4$, we have $\mathbf{F} = 4y\mathbf{i} + 16x\mathbf{j} - 4xy\mathbf{k}$. Now $dz = 0$, and we see that

$$\begin{aligned}\oint_{\partial G} (\mathbf{F} \cdot d\mathbf{r}) &= \oint_{\partial G} (4y\,dx + 16x\,dy)\\ &= \int_0^{2\pi} [(-8 \sin t)(-2 \sin t) + (32 \cos t)(-2 \cos t)]\,dt\\ &= \int_0^{2\pi} (16 \sin^2 t - 64 \cos^2 t)\,dt\\ &= \int_0^{2\pi} [8(1 - \cos 2t) - 32(1 + \cos 2t)]\,dt\\ &= 8\int_0^{2\pi} (-3 - 5 \cos 2t)\,dt = 8\left(-3t - \frac{5}{2} \sin 2t\right)\Bigg]_0^{2\pi}\\ &= 8(-6\pi) = -48\pi.\end{aligned}$$

Solution 2 The disk G', where $0 \le x^2 + y^2 \le 4$, in the plane $z = 4$ with unit normal vector $\mathbf{n} = -\mathbf{k}$ has the same boundary as G. See Fig. 19.25. Thus we may also compute $\iint_{G'} [(\nabla \times \mathbf{F}) \cdot \mathbf{n}]\,dS$ to find the desired integral. We find that

$$\nabla \times \mathbf{F} = \begin{vmatrix} \mathbf{i} & \mathbf{j} & \mathbf{k} \\ \dfrac{\partial}{\partial x} & \dfrac{\partial}{\partial y} & \dfrac{\partial}{\partial z} \\ yz & xz^2 & -xyz \end{vmatrix}$$

$$= (-xz - 2xz)\mathbf{i} + (y + yz)\mathbf{j} + (z^2 - z)\mathbf{k}.$$

In the plane $z = 4$, we have

$$\nabla \times \mathbf{F} = -12x\mathbf{i} + 5y\mathbf{j} + 12\mathbf{k}.$$

Since $\mathbf{n} = -\mathbf{k}$, we find that

$$(\nabla \times \mathbf{F}) \cdot \mathbf{n} = -12.$$

On this disk G' parallel to the x,y-plane, we have $dS = dx\,dy$. Thus

$$\iint_G [(\nabla \times \mathbf{F}) \cdot \mathbf{n}]\,dS = \iint_{G'} [(\nabla \times \mathbf{F}) \cdot \mathbf{n}]\,dx\,dy = \iint_{G'} (-12)\,dx\,dy$$

$$= -12 \cdot \text{Area of } G' = -12 \cdot 4\pi = -48\pi. \quad \square$$

EXAMPLE 8 Let G be the cylinder $x^2 + y^2 = 4$ for $1 \le x \le 5$, and let $\mathbf{F} = yz\mathbf{i} - xz\mathbf{j} + xyz\mathbf{k}$. If $\mathbf{n}$ points out the convex side of the cylinder, compute $\iint_G [(\nabla \times \mathbf{F}) \cdot \mathbf{n}]\,dS$.

Figure 19.26 The boundary of this cylinder consists of two circles traced in opposite directions.

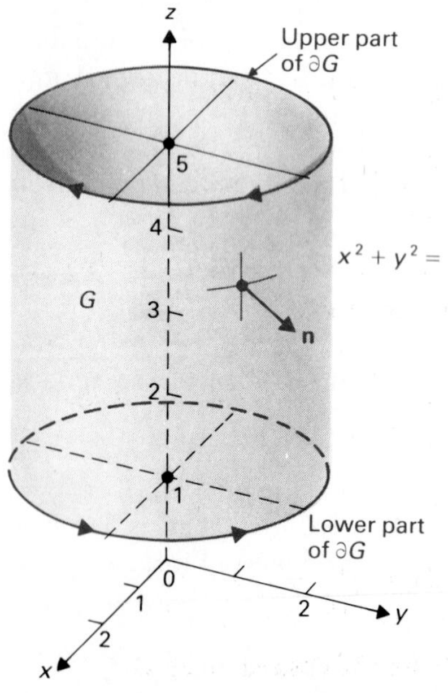

Solution 1 The cylinder is shown in Fig. 19.26. Note that ∂G consists of *two* circles. The upper boundary circle $x^2 + y^2 = 4$ in the plane $z = 5$ is traced *clockwise* when viewed from above, according to the right-hand rule. The lower boundary circle $x^2 + y^2 = 4$ in the plane $z = 1$ is traced counterclockwise, viewed from above.

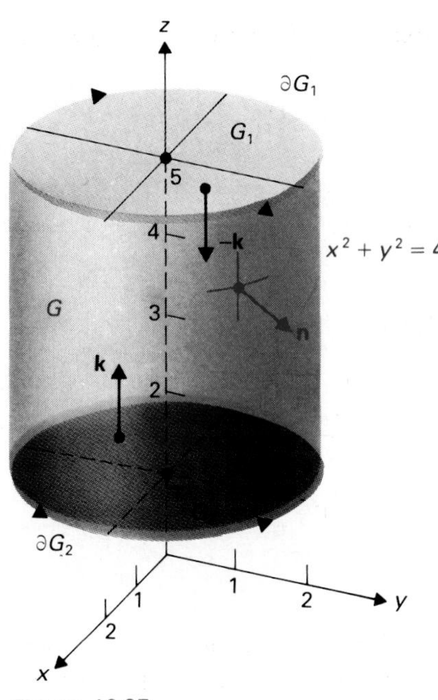

Figure 19.27 $\partial G = \partial G_1 + \partial G_2$.

By Stokes' theorem,

$$\iint_G [(\nabla \times \mathbf{F}) \cdot \mathbf{n}]\, dS = \oint_{\partial G} (\mathbf{F} \cdot d\mathbf{r}).$$

Now in the plane $z = 5$, we have $\mathbf{F} = 5y\mathbf{i} - 5x\mathbf{j} + 5xy\mathbf{k}$. Since $dz = 0$ on both boundary circles, we have for the upper circle

$$\mathbf{F} \cdot d\mathbf{r} = 5(y\,dx - x\,dy).$$

We wish to integrate this expression *clockwise* around the circle $x^2 + y^2 = 4$. By the rotation form of Green's theorem, this integral is

$$\begin{aligned} -5 \oint_{x^2+y^2=4} (y\,dx - x\,dy) &= -5 \iint_{x^2+y^2 \le 4} (-1 - 1)\, dx\, dy \\ &= 10 \iint_{x^2+y^2 \le 4} 1 \cdot dx\, dy \\ &= 10 \cdot 4\pi = 40\pi. \end{aligned}$$

Similarly, in the plane $z = 1$, we have $\mathbf{F} = y\mathbf{i} - x\mathbf{j} + xy\mathbf{k}$. Since the part of ∂G in this plane is the circle $x^2 + y^2 = 4$ traced counterclockwise, the rotation form of Green's theorem gives us the integral value

$$\oint_{x^2+y^2=4} (y\,dx - x\,dy) = \iint_{x^2+y^2 \le 4} -2\, dx\, dy = -2(4\pi) = -8\pi.$$

Thus our answer is $40\pi + (-8\pi) = 32\pi$.

Solution 2 Alternatively, we could let G_1 be the disk at the top of the cylinder with unit normal vector $-\mathbf{k}$, and G_2 the disk at the bottom with unit normal vector $\mathbf{k}$, as shown in Fig. 19.27. As explained before Example 7, we have

$$\begin{aligned} \iint_G [(\nabla \times \mathbf{F}) \cdot \mathbf{n}]\, dS &= \iint_{G_1} [(\nabla \times \mathbf{F}) \cdot (-\mathbf{k})]\, dS \\ &\quad + \iint_{G_2} [(\nabla \times \mathbf{F}) \cdot \mathbf{k}]\, dS. \end{aligned}$$

Since we are going to compute a dot product with $\pm\mathbf{k}$, we need only compute the $\mathbf{k}$-component of $\nabla \times \mathbf{F}$, which we easily find to be $-2z\mathbf{k}$. Then

$$\iint_{G_1} [(\nabla \times \mathbf{F}) \cdot (-\mathbf{k})]\, dS = \iint_{G_1} 2z\, dS = \iint_{G_1} 10\, dS = 10 \cdot 4\pi = 40\pi.$$

Similarly,

$$\iint_{G_1} [(\nabla \times \mathbf{F}) \cdot \mathbf{k}]\, dS = \iint_{G_2} -2z\, dS = \iint_{G_1} -2\, dS = -8\pi.$$

This again gives the value $40\pi - 8\pi = 32\pi$ for the desired integral. □

SUMMARY

Let $\mathbf{F} = P\mathbf{i} + Q\mathbf{j} + R\mathbf{k}$ be a continuously differentiable vector field.

1. *Divergence theorem:* If G is a suitable three-dimensional region with boundary surface ∂G and $\mathbf{n}$ is a unit outward normal vector, then

$$\iint_{\partial G} (\mathbf{F} \cdot \mathbf{n})\, dS = \iiint_G (\nabla \cdot \mathbf{F})\, dx\, dy\, dz.$$

2. Unit normal vectors $\mathbf{n}$ to a surface are found using the formulas

$$\mathbf{n} = \frac{1}{\sqrt{f_x^2 + f_y^2 + 1}}(f_x\mathbf{i} + f_y\mathbf{j} - \mathbf{k}) \qquad \text{for } z = f(x, y)$$

and

$$\mathbf{n} = \frac{1}{\sqrt{f_x^2 + f_y^2 + f_z^2}}(f_x\mathbf{i} + f_y\mathbf{j} + f_z\mathbf{k}) \qquad \text{for } f(x, y, z) = c.$$

3. **Curl F** $= \nabla \times \mathbf{F}$

4. *Stokes' theorem:* If G is a suitable surface in space, and if the direction of the unit normal vector $\mathbf{n}$ and the direction around the boundary ∂G are related by the right-hand rule, then

$$\oint_{\partial G} (\mathbf{F} \cdot \mathbf{t})\, ds = \iint_G [(\nabla \times \mathbf{F}) \cdot \mathbf{n}]\, dS.$$

5. When the value of the integrals in Stokes' theorem is desired, remember the integrals are also equal to $\iint_{G'} [(\nabla \times \mathbf{F}) \cdot \mathbf{n}]\, dS$ for any suitable surface G' for which $\partial G' = \partial G$.

EXERCISES

In Exercises 1 through 7, find **curl F**.

1. $\mathbf{F} = x\mathbf{i} - 2x\mathbf{j} + 5z\mathbf{k}$
2. $\mathbf{F} = x^2\mathbf{i} + y^2\mathbf{j} + z^2\mathbf{k}$
3. $\mathbf{F} = xyz\mathbf{i} - 2y^2z\mathbf{j} + 3z^4\mathbf{k}$
4. $\mathbf{F} = \frac{y}{x}\mathbf{i} - \frac{z}{y}\mathbf{j} + \frac{x}{z}\mathbf{k}$
5. $\mathbf{F} = \frac{y}{z}\mathbf{i} + \frac{z}{x}\mathbf{j} + \frac{x}{y}\mathbf{k}$
6. $\mathbf{F} = e^{xy}\mathbf{i} + e^{-yz}\mathbf{j} + e^{x^2}\mathbf{k}$
7. $\mathbf{F} = (\sin xy)\mathbf{i} + xe^y\mathbf{j} + (\ln yz)\mathbf{k}$

In Exercises 8 through 11, illustrate the divergence theorem for the given region G and vector field $\mathbf{F}$.

8. G the ball $x^2 + y^2 + z^2 \le a^2$, $\mathbf{F} = x\mathbf{i} + y\mathbf{j} + z\mathbf{k}$
9. G the region bounded by $z = 1 - x^2 - y^2$ and $z = 0$, $\mathbf{F} = y\mathbf{j}$
10. G the region bounded by $z^2 = x^2 + y^2$ and $z = 4$, $\mathbf{F} = y\mathbf{i} - x\mathbf{j} - z\mathbf{k}$
11. G the half-ball $x^2 + y^2 + z^2 \le a^2$ for $z \ge 0$, $\mathbf{F} = -y\mathbf{i} + x\mathbf{j} + \mathbf{k}$

In Exercises 12 through 15, illustrate Stokes' theorem for the given surface G, direction of $\mathbf{n}$, and vector field $\mathbf{F}$.

12. G the paraboloid $z = x^2 + y^2$ for $0 \le z \le 1$, $\mathbf{n}$ pointing out the concave side, $\mathbf{F} = -y\mathbf{i} + x\mathbf{j} + z\mathbf{k}$
13. G the cone $z^2 = x^2 + y^2$ for $0 \le z \le 4$, $\mathbf{n}$ pointing out the convex side, $\mathbf{F} = -yz^2\mathbf{i} + xz^2\mathbf{j} + 2xyz\mathbf{k}$
14. G the portion of the sphere $x^2 + y^2 + z^2 = 5$ on or above the plane $z = 1$, $\mathbf{n}$ pointing out the convex side, $\mathbf{F} = 2y\mathbf{i} + 3x\mathbf{j} + xz\mathbf{k}$
15. G the triangular region consisting of the portion of the plane $2x + y + 2z = 2$ in the first octant, $\mathbf{n}$ with positive $\mathbf{k}$-coefficient, $\mathbf{F} = yz\mathbf{i} + y^2\mathbf{j} + xy\mathbf{k}$

In Exercises 16 through 27, compute the indicated integral in the easiest way you can. Make use of the divergence theorem or Stokes' theorem where appropriate, including the use of Stokes' theorem in number 5 of the Summary. You may use symmetry and the fact that the area of the region bounded by the ellipse $(x^2/a^2) + (y^2/b^2) = 1$ is πab.

16. $\iint_G [(\nabla \times \mathbf{F}) \cdot \mathbf{n}]\, dS$, where G is the surface $z = 4 - x^2 - y^2$ for $z \ge 0$, $\mathbf{n}$ has positive $\mathbf{k}$-coefficient, and $\mathbf{F} = x^2y\mathbf{i} + xyz\mathbf{j} + xz^2\mathbf{k}$
17. $\iint_G [(\nabla \times \mathbf{F}) \cdot \mathbf{n}]\, dS$, where G is the surface $36z = 4x^2 + 9y^2$

for $0 \le z \le 1$, **n** has positive **k**-coefficient, and $\mathbf{F} = y\mathbf{i} + 3x\mathbf{j} - xyz\mathbf{k}$

18. $\iint_G [(\nabla \times \mathbf{F}) \cdot \mathbf{n}]\, dS$, where G is the cone $z^2 = x^2 + y^2$ for $0 \le z \le 3$, **n** has negative k-coefficient, and $\mathbf{F} = -yz\mathbf{i} + xz\mathbf{j} - [z/(1 + x^2 + y^2)]\mathbf{k}$

19. $\iint_G [(\nabla \times \mathbf{F}) \cdot \mathbf{n}]\, dS$, where G is the cylinder $x^2 + y^2 = 4$ for $0 \le z \le 3$, **n** points out the convex side, and $\mathbf{F} = -yz^2\mathbf{i} + xz^2\mathbf{j} + 2xyz\mathbf{k}$

20. $\iint_G [(\nabla \times \mathbf{F}) \cdot \mathbf{n}]\, dS$, where G is the cylinder $x^2 + y^2 = a^2$ for $0 \le z \le b$, **n** points out the convex side, and $\mathbf{F} = xz\mathbf{i} + (4 + z^2)\mathbf{j} + xe^{yz}\mathbf{k}$

21. $\iint_G (F \cdot \mathbf{n})\, dS$, where G is the sphere $x^2 + y^2 + z^2 = a^2$, **n** is the outward normal, and $\mathbf{F} = 4x\mathbf{i} + 5y\mathbf{j} - 3z\mathbf{k}$

22. $\iint_G (F \cdot \mathbf{n})\, dS$, where G is the surface of the region bounded by the cone $z = \sqrt{x^2 + y^2}$ and the plane $z = 2$, **n** points outward, and $\mathbf{F} = xy\mathbf{i} + yz\mathbf{j} + xyz^2\mathbf{k}$

23. $\iint_G (\mathbf{F} \cdot \mathbf{n})\, dS$, where G is the surface of the region bounded by $z = 9 - x^2 - y^2$ and $z = x^2 + y^2 - 9$, **n** points inward, and $\mathbf{F} = (xy + 3x)\mathbf{i} + (y - 2xyz^2)\mathbf{j} + (z + 4xyz^3)\mathbf{k}$

24. $\iint_G (\mathbf{F} \cdot \mathbf{n})\, dS$, where G is the surface of the region bounded by $y = x^2 + z^2 - 2$ and $y = -1$, **n** points inward, and $\mathbf{F} = (x^2y - 3xz)\mathbf{i} + (xyz + 3x^2y)\mathbf{j} + (x^3y^2z + z^3)\mathbf{k}$

25. $\iint_G [(\nabla \times \mathbf{F}) \cdot \mathbf{n}]\, dS$, where G is the cone $4x^2 = y^2 + 4z^2$ for $-1 \le x \le 2$, **n** points out the convex side, and $\mathbf{F} = x^3y^2\mathbf{i} + xyz^2\mathbf{j} + x^3y\mathbf{k}$

26. $\iint_G [(\nabla \times \mathbf{F}) \cdot \mathbf{n}]\, dS$, where G is the portion of the ellipsoid $x^2/4 + y^2 + z^2/9 = 5$ for $y \le 2$, **n** points out the convex side, and $\mathbf{F} = xy^2z^2\mathbf{i} + e^{xyz}\mathbf{j} + 4xy^2\mathbf{k}$

27. $\iint_G [(\nabla \times \mathbf{F}) \cdot \mathbf{n}]\, dS$, where G is the portion of the cylinder $x^2 + z^2 - 2x + 4z = 4$ for $-2 \le y \le 3$, **n** points out the concave side, and $\mathbf{F} = y^3z\mathbf{i} + 2xy\mathbf{j} - xy^2\mathbf{k}$

28. Let G be a three-dimensional region where the divergence theorem applies. Use the theorem to show that

$$\text{Volume of } G = \frac{1}{3}\iint_{\partial G} [(x\mathbf{i} + y\mathbf{j} + z\mathbf{k}) \cdot \mathbf{n}]\, dS.$$

29. Let **F** be a continuously differentiable vector field on a surface G in space.
a) Show that if **F** is normal to the unit tangent vector **t** at each point on ∂G, then $\iint_G [(\textbf{Curl F}) \cdot \mathbf{n}]\, dS = 0$.
b) Show that if the vector **curl F** is tangent to G at each point of G, then $\oint_{\partial G} (\mathbf{F} \cdot \mathbf{t})\, ds = 0$.

30. Use the divergence theorem for the region $a^2 \le x^2 + y^2 + z^2 \le b^2$ to show that the flux of the gradient field **F** of the Newtonian potential function $(x^2 + y^2 + z^2)^{-1/2}$ across a sphere with center at the origin is independent of the radius of the sphere.

31. A vector field **F** in a region of space is *irrotational* if $\textbf{curl F} = \mathbf{0}$ throughout the region. Show that the circulation of an irrotational vector field about every piecewise-smooth closed path bounding a suitable surface G lying within the region is zero.

32. A vector field **F** on a region of space is *incompressible* if $\nabla \cdot \mathbf{F} = 0$ throughout the region. Show that the flux of an incompressible field **F** across the boundary of each ball inside the region is zero.

33. a) Show that $\nabla \cdot (\textbf{Curl F}) = 0$ for a twice continuously differentiable vector field in space.
b) Show that $\nabla \times (\nabla f) = \mathbf{0}$ for a twice continuously differentiable function of three variables.

34. Often $\nabla \cdot (\nabla f)$ is written as $\nabla^2 f$. Show that for a suitable region G in space, we have

$$\iiint_G (\nabla^2 f)\, dx\, dy\, dz = \iint_{\partial G} (\nabla f \cdot \mathbf{n})\, dS$$

for a twice continuously differentiable function f of three variables.

35. Argue as best you can from Stokes' theorem that if G is a surface that is the entire boundary of a suitable three-dimensional region, then

$$\iint_G [(\nabla \times \mathbf{F}) \cdot \mathbf{n}]\, dS = 0$$

for every continuously differentiable vector field **F**.

36. Let **F** be a twice continuously differentiable vector field and let G be a three-dimensional region in space where the divergence theorem can be applied. Show that

$$\iint_{\partial G} [(\textbf{Curl F}) \cdot \mathbf{n}]\, dS = \iiint_G \nabla \cdot (\textbf{Curl F})\, dx\, dy\, dz.$$

37. Use the results of Exercises 35 and 36 to show that for **F** and G as described in Exercise 36,

$$\iiint_G \nabla \cdot (\textbf{Curl F})\, dx\, dy\, dz = 0.$$

38. Argue as best you can from Exercise 37, with no computation, that $\nabla \cdot (\nabla \times \mathbf{F}) = 0$ for every twice continuously differentiable vector field **F**. You were asked to show this by computation in Exercise 33(a).

EXERCISE SETS FOR CHAPTER 19

Review Exercise Set 19.1

1. State the rotation form of Green's theorem.
2. Use the rotation form of Green's theorem to compute

$$\oint_\lambda [xy\, dx + (x^3 + y^3)\, dy],$$

where λ is the boundary of the region G bounded by $y = x^2$ and $y = 4$.

3. State the vector form of the divergence theorem and explain its interpretation for the flux vector **F** of a flow.
4. Let **F** be the vector field

$$\mathbf{F} = x^3\mathbf{i} + y^3\mathbf{j} + z^2\mathbf{k}.$$

Use the divergence theorem to compute $\iint_{\partial G} (\mathbf{F} \cdot \mathbf{n})\, dS$, where G is the region bounded by $z = x^2 + y^2$ and $z = 4$.

5. Let $\mathbf{F} = P\mathbf{i} + Q\mathbf{j}$ be the continuously differentiable flux vector for a flow in the plane. Give the conditions for $\mathbf{F}$ to be (a) irrotational, (b) incompressible.

6. Let $\mathbf{F} = xyz^2\mathbf{i} + 2yz\mathbf{j} - x^3\mathbf{k}$.
a) Find **curl F**. b) Find the divergence of **F**.

7. Let $\mathbf{F} = 4\mathbf{i} + xz\mathbf{j} - xy\mathbf{k}$. Use Stokes' theorem to find $\iint_G [(\mathbf{Curl\ F}) \cdot \mathbf{n}]\, dS$, where G is the surface $x = y^2 + z^2$ for $0 \le x \le 9$, while vectors $\mathbf{n}$ are chosen so that their $\mathbf{i}$-components are positive.

Review Exercise Set 19.2

1. Give a vector statement of the rotation form of Green's theorem.

2. Use the rotation form of Green's theorem to find the circulation of the flow with flux vector

$$\mathbf{F} = x\mathbf{i} + 3xy\mathbf{j}$$

about the boundary of the plane region bounded by

$$x = y^2 \quad \text{and} \quad y = x - 2.$$

3. Let a ball of mass density 1 be bounded by the spherical surface with equation

$$x^2 + y^2 + z^2 = a^2.$$

Show that the moment of inertia of the ball about the z-axis is equal to

$$\iint_G \left[\left(\frac{x^3}{3}\mathbf{i} + \frac{y^3}{3}\mathbf{j} + 0\mathbf{k}\right) \cdot \mathbf{n}\right] dS,$$

where $\mathbf{n}$ is the outward normal to the sphere.

4. a) Give the condition for a force field $\mathbf{F} = P\mathbf{i} + Q\mathbf{j}$ in the plane to be conservative.
b) Show the force field $\mathbf{F} = y^2\mathbf{i} + 2xy\mathbf{j}$ is conservative, and find a potential function u.
c) Find the work done by the force field in part (b) in moving a body from $(-1, 2)$ to $(3, -1)$.

5. Find **curl F** if $F = x\mathbf{i} + yz\mathbf{j} + xyz\mathbf{k}$.

6. Let G be the surface $z = 4 + \sqrt{9 - x^2 - y^2}$ with normal vectors $\mathbf{n}$ having positive $\mathbf{k}$-component. Let G' be the portion of the sphere $x^2 + y^2 + z^2 = 25$, where $z \le 4$, with outward normal vectors $\mathbf{n}$. Let $\mathbf{F}$ be a vector field. What relation holds between

$$\iint_G [(\mathbf{Curl\ F}) \cdot \mathbf{n}]\, dS$$

and

$$\iint_{G'} [(\mathbf{Curl\ F}) \cdot \mathbf{n}]\, dS?$$

Why?

7. Use Stokes' theorem to find $\iint_G [(\mathbf{Curl\ F}) \cdot \mathbf{n}]\, dS$ where $\mathbf{F} = xy\mathbf{i} + y\mathbf{j} + yz\mathbf{k}$ and G is the surface of the cubical box with no top and having diagonally opposite vertices at $(0, 0, 0)$ and $(1, 1, 1)$. Let $\mathbf{n}$ be the outward normal.

Review Exercise Set 19.3

1. Let G be a region in the x,y-plane where Green's theorem applies. If the counterclockwise integral

$$\oint_{\partial G} (3y\, dx - 5x\, dy) = -28,$$

find the area of G.

2. Use the rotation form of Green's theorem to find the counterclockwise integral $\oint_{\partial G}(\mathbf{F} \cdot \mathbf{t})\, ds$ if G is the region $0 \le x \le 2$, $0 \le y \le 1$, and $\mathbf{F} = -3xy\mathbf{i} + 4x^2y^3\mathbf{j}$.

3. Let $P(x, y) = (y^2 + 4x)/(y + 1)$ and $Q(x, y) = (3^x + y^2)/(x + 2)$. Use the rotation form of Green's theorem to find

$$\int_0^1 \int_0^1 \left[\frac{\partial Q}{\partial x} - \frac{\partial P}{\partial y}\right] dx\, dy.$$

4. Find all functions $Q(x, y)$ such that $\mathbf{F} = x^2e^y\mathbf{i} + Q(x, y)\mathbf{j}$ is the flux field of an incompressible flow in the plane.

5. Find all functions $P(x, y)$ such that $\mathbf{F} = P(x, y)\mathbf{i} + (x^2y^3 - 2x)\mathbf{j}$ is the flux field of an irrotational flow in the plane.

6. Let G be a region in space where the divergence theorem applies. Let $\mathbf{F} = 3x\mathbf{i} - 2y\mathbf{j} + 4z\mathbf{k}$. If $\iint_{\partial G} (\mathbf{F} \cdot \mathbf{n})\, dS = 20$, find the volume of G.

7. Let G be the cylindrical surface $x^2 + y^2 = 9$ for $0 \le z \le 2$, and let $\mathbf{F} = 4yz\mathbf{i} + 2xz\mathbf{j} - xy^2\mathbf{k}$. Use Stokes' theorem to find $\oint_{\partial G} [(\nabla \times \mathbf{F}) \cdot \mathbf{n}]\, dS$, where $\mathbf{n}$ points out the convex side of the cylinder.

Review Exercise Set 19.4

1. Let the flux field for a flow in the x,y-plane be $\mathbf{F} = x^3\mathbf{i} + y^3\mathbf{j}$. Let λ be the circle $x^2 + y^2 = 9$ traced once counterclockwise. Find $\oint_\lambda (\mathbf{F} \cdot \mathbf{n})\, ds$.

2. Let $\mathbf{F} = P(x)\mathbf{i} + Q(y)\mathbf{j}$ be a flux field for a flow in the plane. Under what conditions on P and Q is the flow (a) incompressible, (b) irrotational?

3. Let G be a region in the plane where the divergence form of Green's theorem applies. Let $\mathbf{F} = 2xy\mathbf{i} + x^3\mathbf{j}$. If

$$\iint_G (\nabla \cdot \mathbf{F})\, dx\, dy = 12,$$

find the moment of G about the x-axis.

4. Let G be a region in the plane and let $\mathbf{F}(x, y) = \nabla H(x, y)$ be a continuously differentiable vector field. Find $\iint_G (\mathbf{Curl\ F})\, dx\, dy$.

5. Let $\mathbf{F} = (xy/z)\mathbf{i} - (x^2/y)\mathbf{j} + (yz/x^2)\mathbf{k}$. Find **curl F**.

6. Let G be the ball $x^2 + y^2 + z^2 \le a^2$. Let $\mathbf{F} = 2xyz^2\mathbf{i} - x^3y^2\mathbf{j} + z^3\mathbf{k}$. Find $\iint_{\partial G} (\mathbf{F} \cdot \mathbf{n})\, dS$, where $\mathbf{n}$ is the outward unit normal vector.

7. Let G be the portion of the cone $z = \sqrt{x^2 + y^2}$ for $1 \le z \le 4$, and let $\mathbf{n}$ point out the concave side of the cone. Let $\mathbf{F} = -y^3z\mathbf{i} + x^3z\mathbf{j} + e^{xyz}\mathbf{k}$. Find $\iint_G [(\mathbf{Curl\ F}) \cdot \mathbf{n}]\, dS$.

More Challenging Exercises 19

1. Let G be a suitable surface in space and let $P(x, y, z)$, $Q(x, y, z)$, and $R(x, y, z)$ be continuously differentiable functions. Stokes'

theorem is often stated in the form

$$\int_{\partial G} (P\,dx + Q\,dy + R\,dz)$$
$$= \iint_G \left[\left(\frac{\partial R}{\partial y} - \frac{\partial Q}{\partial z} \right) dy\,dz + \left(\frac{\partial P}{\partial z} - \frac{\partial R}{\partial x} \right) dz\,dx + \left(\frac{\partial Q}{\partial x} - \frac{\partial P}{\partial y} \right) dx\,dy \right].$$

Convince yourself that this is equivalent to the statement

$$\int_{\partial G} (\mathbf{F} \cdot \mathbf{t})\,ds = \iint_G [(\nabla \times \mathbf{F}) \cdot \mathbf{n}]\,dS$$

for $\mathbf{F} = P\mathbf{i} + Q\mathbf{j} + R\mathbf{k}$.

Exercises 2 through 8 use the notation of the $\wedge$-product introduced in the last exercise set of Chapter 18. Review that material before proceeding.

The table below indicates what is meant by a *differential form* ω, the *order* of the form, and its *exterior derivative* $d\omega$. Here P, Q, and R are differentiable functions of two or three variables (we give the table for three variables). For $P(x, y, z)$,

$$dP = \frac{\partial P}{\partial x} dx + \frac{\partial P}{\partial y} dy + \frac{\partial P}{\partial z} dz$$

as usual.

2. Show that if $\omega = P\,dx \wedge dy \wedge dz$, then $d\omega = 0$.

3. Note that if ω has order r, then $d\omega$ has order $r + 1$. Compute $d\omega$, simplifying as much as possible.
 a) $\omega = xy\,dx - x^2\,dy$
 b) $\omega = xz\,dx \wedge dy - yz^3\,dx \wedge dz$

All the main theorems in this chapter can be expressed in the following form, as Exercises 4 through 8 ask you to show.

Generalized Stokes' theorem If ω is a differential form of order r with continuously differentiable coefficients P, Q, and so on, and if G is a suitable $(r + 1)$-dimensional region, then

$$\int_{\partial G} \omega = \int_G d\omega. \tag{1}$$

In Eq. (1), a single integral sign is used for each integral, rather than double or triple integral signs as we have used in the past. The integrals of differential forms are defined so that

$$\int_G P(x, y)\,dx \wedge dy = \iint_G P(x, y)\,dx\,dy = -\int_G P(x, y)\,dy \wedge dx,$$

$$\int_G P(x, y, z)\,dx \wedge dy \wedge dz = \iiint_G P(x, y, z)\,dx\,dy\,dz$$
$$= -\int_G P(x, y, z)\,dz \wedge dy \wedge dx,$$

and so on.

4. If G is a one-dimensional region consisting of a curve joining point A to point B, then ∂G is the symbolic expression $B - A$. The integral of a function f (a form of order 0) over ∂G is defined to be $f(B) - f(A)$. Show that if $\omega = f(x)$ and G is the line segment $[a, b]$ from a to b, then Eq. (1) reduces to the fundamental theorem of calculus.

5. State Eq. (1) for the special case of $\omega = f(x, y, z)$ and G a curve γ joining $A = (a_1, a_2, a_3)$ to $B = (b_1, b_2, b_3)$ in space. (See Exercise 4.)

6. State Eq. (1) for $\omega = P(x, y)\,dx + Q(x, y)\,dy$ in the plane and G a plane region. Simplify the statement as much as possible. What familiar theorem do you obtain?

7. State Eq. (1) for $\omega = P\,dx + Q\,dy + R\,dz$ and G a surface in space. Simplify the statement as much as possible. What theorem do you obtain?

8. a) State Eq. (1) for $\omega = P\,dy \wedge dz + Q\,dz \wedge dx + R\,dx \wedge dy$, simplifying as much as possible.
 b) Convince yourself that the statement obtained in part (a) is equivalent to the divergence theorem.

Differential forms ω

Order of ω	ω	$d\omega$
0	$P(x, y, z)$	$dP = \frac{\partial P}{\partial x} dx + \frac{\partial P}{\partial y} dy + \frac{\partial P}{\partial z} dz$
1	$P\,dx + Q\,dy + R\,dz$	$dP \wedge dx + dQ \wedge dy + dR \wedge dz$
2	$P\,dy \wedge dz + Q\,dz \wedge dx + R\,dx \wedge dy$	$dP \wedge dy \wedge dz + dQ \wedge dz \wedge dx + dR \wedge dx \wedge dy$
3	$P\,dx \wedge dy \wedge dz$	$dP \wedge dx \wedge dy \wedge dz$

20 DIFFERENTIAL EQUATIONS

A problem in calculus encountered by a scientist is probably most apt to appear in the form of a differential equation. Applications of differential equations abound in the physical, biological, and social sciences. Consequently, from the practical point of view, the study of differential equations is one of the most important branches of mathematics. Many schools offer several entire courses devoted to this special topic.

This final chapter of our text introduces a few techniques for solving certain simple differential equations that are frequently encountered. While we present a number of elementary applications, the emphasis of the chapter is on mathematical techniques for solving equations. We cannot do justice to the great wealth of applications in such an abbreviated introduction to differential equations within the space limitations of a calculus text. We suggest that the interested student consult a text on differential equations and applications. For example, in the text by Finizio and Ladas, *Ordinary Differential Equations with Modern Applications*, Wadsworth, 2nd edition, 1982, a list of applications is given at the start of the book, before the preface. The list is coded according to whether the text takes up an application in detail, in little detail, or as an example or exercise. A few minutes of looking through that list and browsing through a few of the applications will convince any calculus student of the importance of differential equations in all branches of science.

20.1 INTRODUCTION

THE NOTION OF A DIFFERENTIAL EQUATION

A **differential equation** is an equation that involves derivatives (or differentials) of an "unknown function" f. To solve the differential equation is to find all possible functions f for which the equation is true. We first encountered differential equations in Section 5.9, under the guise of an antidifferentiation problem. For example, the **general solution** of the equation

$$\frac{dy}{dx} = x^2$$

is

$$y = f(x) = \frac{1}{3}x^3 + C,$$

where C is an arbitrary constant, for this expression embodies *all* solutions of $dy/dx = x^2$. This equation is of *order* 1, since only derivatives of order 1 (that is, first derivatives) appear in the equation. The **order** of an equation is r if the equation involves an rth derivative, such as d^ry/dx^r but no derivatives of higher order. Thus the equation

$$\frac{d^2y}{dx^2} = x^2$$

is of second order. Solving the equation, we have

$$\frac{dy}{dx} = \frac{1}{3}x^3 + C_1,$$

so

$$y = f(x) = \frac{1}{12}x^4 + C_1x + C_2$$

is the general solution. Without going into detail, we state that one expects the general solution of a differential equation of order n to require n integrations and hence to contain n arbitrary constants.

A differential equation of order 1 need not be of the form $dy/dx = g(x)$. Recall that we considered equations such as

$$\frac{dy}{dx} = y^2x$$

in Section 6.4 and solved them by separating the variables:

$$\frac{dy}{y^2} = x\,dx, \qquad \int y^{-2}\,dy = \int x\,dx, \qquad \frac{y^{-1}}{-1} = \frac{x^2}{2} + C,$$

$$y = f(x) = \frac{-1}{(x^2/2) + C} = \frac{-2}{x^2 + 2C}.$$

We will review this technique in Section 20.2.

GEOMETRIC INTERPRETATION OF THE EQUATION $y' = F(x, y)$

Let a function F of two variables be given and consider the differential equation $y' = F(x, y)$. If $y = h(x)$ is a solution of this equation, then for each x in the domain of h, we must have

$$h'(x) = F(x, y) = F(x, h(x)).$$

Now $h'(x)$ can be interpreted geometrically as the slope of the tangent line to the graph of h at the point $(x, y) = (x, h(x))$. Thus if F is a known func-

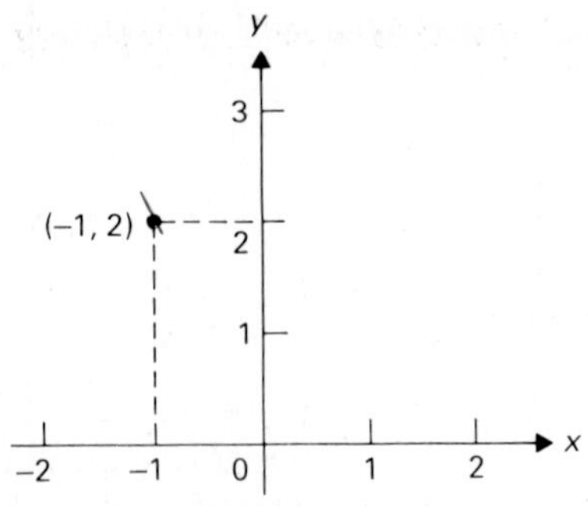

Figure 20.1 Short line segment of slope −2 at (−1, 2).

tion, the equation $y' = F(x, y)$ allows us to compute slopes of tangents to solutions at points (x, y) in the domain of F.

EXAMPLE 1 Find the slope at $(-1, 2)$ of the tangent line to the solution of $y' = xy^2/(x^2 + 1)$ through that point. Sketch a short tangent-line segment to the solution at that point.

Solution The slope is given by

$$\left.\frac{xy^2}{x^2 + 1}\right|_{(-1,2)} = \frac{-4}{2} = -2.$$

A short line segment of that slope at $(-1, 2)$ is shown in Fig. 20.1. □

If we place a short line segment of slope $m = F(x, y)$ at each point (x, y), we obtain the *direction field* of the equation $y' = F(x, y)$. By actually drawing a few of these line segments in a direction field, we may be able to obtain graphically some information about the solutions of the differential equation. We illustrate with two examples.

EXAMPLE 2 Sketch the direction field of the differential equation $y' = -x/y$.

Solution A useful device in sketching direction fields is to find all points (x, y) where the segments in the field have a particular slope c. For an equation $y' = F(x, y)$, the set of such points is the curve

$$F(x, y) = c,$$

and for simple functions F this curve may be easy to sketch. In our case,

$$F(x, y) = \frac{-x}{y},$$

and the equation $F(x, y) = c$ takes the form

$$\frac{-x}{y} = c, \qquad \text{or} \qquad y = \frac{-1}{c}x, \qquad \text{where } y \neq 0. \tag{1}$$

This is an equation of a line through the origin with the origin omitted. Thus at all points but $(0, 0)$ on the line $y = -x/c$, the segments in the direction field have slope $m = c$. For example, all segments have slope -1 on the line $y = x$; all segments have slope 1 on the line $y = -x$; all segments have slope 2 on the line $y = (-\frac{1}{2})x$, and so on. Putting $c = 0$ in the first equation in (1), we see that the segments at all points but $(0, 0)$ on the line $x = 0$ (that is, on the y-axis), have slope zero. We have sketched the direction field of $y' = -x/y$ in Fig. 20.2, labeling some lines through the origin and indicating the slopes m of the segments of the direction field on these lines.

From Fig. 20.2, we would guess that solutions of the equation $y' = -x/y$ have as graphs portions of certain curves about the origin; we might even guess that they are the circles

$$x^2 + y^2 = r^2,$$

which are level curves of the function

$$G(x, y) = x^2 + y^2.$$

Figure 20.2 Direction field of $y' - (-x/y)$; solutions are circles.

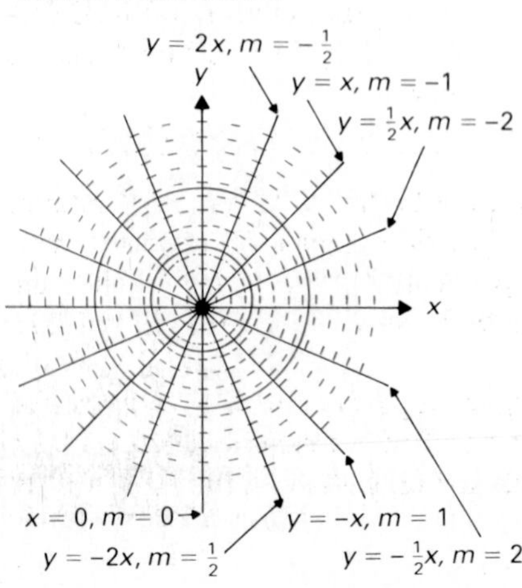

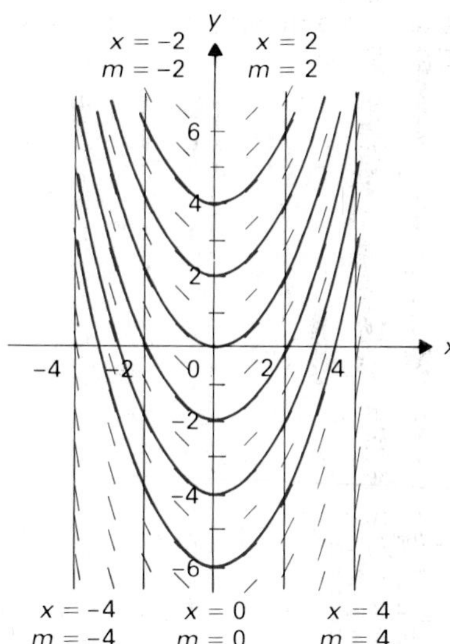

Figure 20.3 Direction field of $y' = x$; solutions are parabolas.

We can easily check our conjecture. We find that for y defined implicitly as a function of x by

$$G(x, y) = x^2 + y^2 = a^2,$$

we have

$$y' = \frac{dy}{dx} = -\frac{G_x(x, y)}{G_y(x, y)} = -\frac{2x}{2y} = -\frac{x}{y},$$

so our guess was correct. □

A computer with graphics capability can easily plot a direction field for an equation $y' = F(x, y)$.

EXAMPLE 3 Sketch the direction field of the differential equation $y' = x$.

Solution The direction field of the differential equation $y' = x$ is sketched in Fig. 20.3. This time, segments of equal slope $m = c$ lie on the vertical line $x = c$. Again, we have estimated a couple of graphs of solutions from the direction field; this time, the solution curves look like parabolas. Of course, we can easily solve $y' = x$ and obtain as general solution

$$y = \frac{x^2}{2} + C,$$

which is indeed a collection of parabolas. □

AN EXISTENCE THEOREM

We state without proof an existence theorem for solutions of a differential equation $y' = F(x, y)$. The hypotheses of the theorem could be weakened somewhat and the conclusion would still hold; however, the statement suffices for our needs.

THEOREM 20.1 Existence theorem for $y' = F(x, y)$

Let F be a continuous function of two variables with domain containing a neighborhood $(x - x_0)^2 + (y - y_0)^2 < r^2$ of (x_0, y_0) in which F_y exists and is continuous. Then there exists a number $c > 0$ and a differentiable function $h(x)$ for $x_0 - c < x < x_0 + c$ such that $y = h(x)$ is a solution of the differential equation $y' = F(x, y)$ and such that $y_0 = h(x_0)$. Furthermore, h is the unique such function with domain $x_0 - c < x < x_0 + c$.

Figure 20.4 Solution $y = h(x)$ of $y' = F(x, y)$ for $x_0 - c < x < x_0 + c$.

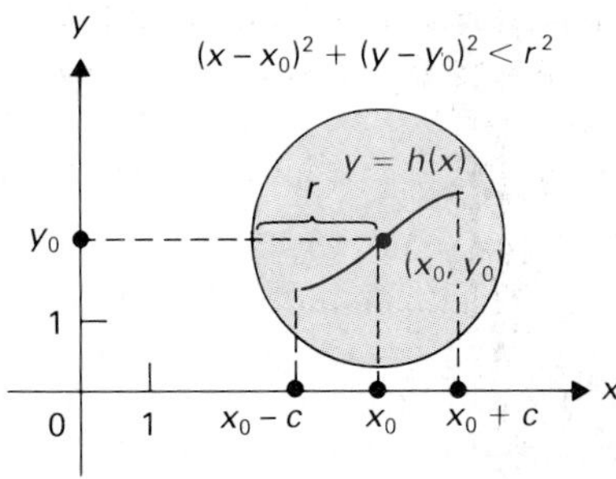

Theorem 20.1 essentially asserts the existence of a (unique) solution of $y' = F(x, y)$ through any point (x_0, y_0), provided that F is sufficiently well behaved in some neighborhood of (x_0, y_0). A sketch illustrating the theorem is given in Fig. 20.4, and the direction fields shown in Figs. 20.2 and 20.3 also illustrate the theorem.

In finding the solution $y = h(x)$ through (x_0, y_0) of a differential equation $y' = F(x, y)$, one often attempts to find the *general solution* of the differential equation. The general solution frequently can be expressed in the form

$$G(x, y) = C,$$

where C is an arbitrary constant. (Recall that we would expect the general solution of a first-order differential equation to contain an arbitrary constant.) Each individual solution then appears as a level curve of the function G, and the solution

$$y = h(x) \qquad \text{such that } h(x_0) = y_0$$

is given by the level curve of G through (x_0, y_0), namely,

$$G(x, y) = G(x_0, y_0).$$

Thus the *initial condition* $y_0 = h(x_0)$ can be used to determine the value of the arbitrary constant in the general solution to give the desired *particular solution* through (x_0, y_0).

EXAMPLE 4 Find the solution $y = h(x)$ of the differential equation $y' = x^2$ that satisfies the initial condition $h(1) = 3$.

Solution Of course the general solution of $y' = x^2$ is

$$y = \frac{x^3}{3} + C.$$

Setting $x = 1$ and $y = 3$, we find that we have

$$3 = \frac{1}{3} + C, \qquad \text{so} \quad C = 3 - \frac{1}{3} = \frac{8}{3}.$$

Thus the desired particular solution is

$$y = h(x) = \frac{x^3}{3} + \frac{8}{3}. \quad \square$$

NUMERICAL SOLUTION OF $y' = F(x, y)$

Consider the differential equation

$$\frac{dy}{dx} = \frac{\sin x^2}{1 + y^2}. \tag{2}$$

The variables can be separated in this equation to yield

$$(1 + y^2)\,dy = (\sin x^2)\,dx,$$

as in Section 6.4. However, we cannot integrate $\sin x^2$ to solve the equation in terms of elementary functions. We would like to be able to approximate the solution through a point (x_0, y_0) for $x_0 \le x \le b$.

Figure 20.5 Points x_i partitioning $[x_0, b]$ into n subintervals of equal length Δx.

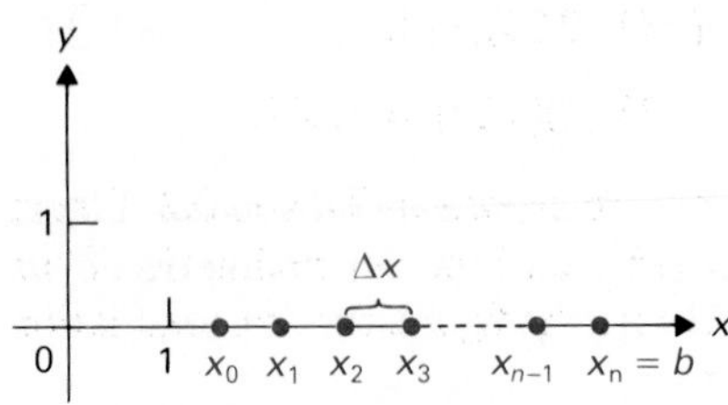

Divide the interval $[x_0, b]$ into n subintervals of equal length by points $x_0 < x_1 < x_2 < \cdots < x_n = b$, as shown in Fig. 20.5. We let Δx be the length of each subinterval, so $\Delta x = x_i - x_{i-1}$ for $i = 1, 2, \ldots, n$.

Let $y = h(x)$ be the solution through (x_0, y_0) of $y' = F(x, y)$, as described in Theorem 20.1. We try to approximate the values $h(x_1), h(x_2), \ldots, h(x_n)$. Let y_i be an approximation of $h(x_i)$. Then the polygonal curve joining (x_0, y_0) to (x_1, y_1), (x_1, y_1) to (x_2, y_2), and so on, shown in Fig. 20.6, becomes an approximation to the desired solution $y = h(x)$ over $[x_0, b]$, as illustrated in the figure. The better the approximations y_i to $h(x_i)$ and the larger the value of n, the more accurate we expect this approximation to be.

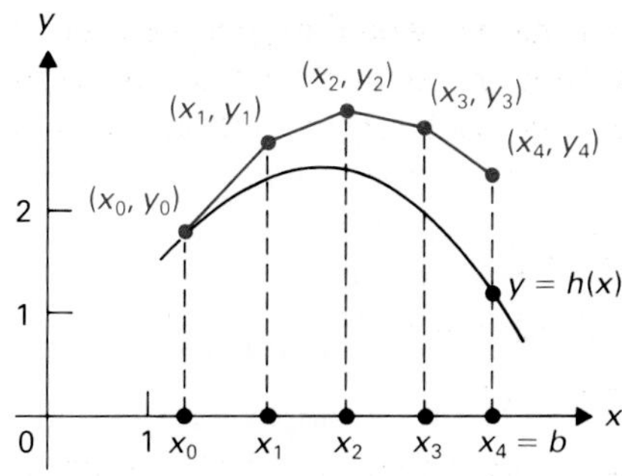

Figure 20.6 Polygonal approximation of the solution $y = h(x)$ through (x_0, y_0) over $[x_0, b]$ for $n = 4$.

There are many methods for finding such numerical approximations of a solution of $y' = F(x, y)$. We present three examples, each finding an approximation of the solution through $(x_0, y_0) = (0, 1)$ of the equation

$$y' = xy$$

on the interval [0, 2]. Each example explains and illustrates a different method of numerical approximation. In each example, we take $n = 8$. A computer or calculator is of course very handy.

The equation $y' = xy$ is one that we can solve using the methods of Section 6.4:

$$\frac{dy}{dx} = xy, \qquad \frac{dy}{y} = x\,dx,$$

$$\ln|y| = \frac{x^2}{2} + C, \qquad |y| = e^C \cdot e^{x^2/2}, \qquad y = Ke^{x^2/2}$$

for some constant K. The particular solution through $(x_0, y_0) = (0, 1)$ occurs when $1 = Ke^0 = K$, or when $K = 1$. Thus the true solution is

$$y = e^{x^2/2}.$$

Table 20.1 shows computed values of $e^{x_i^2/2}$ for the points x_i that divide the interval [0, 2] into $n = 8$ subintervals of equal length. We deliberately chose an equation that we could actually solve for Examples 5 through 7 below so we could see the accuracy of the approximations.

Table 20.1

i	x_i	$h(x_i) = e^{x_i^2/2}$
0	0	1.0
1	0.25	1.031743407
2	0.5	1.133148453
3	0.75	1.324784759
4	1.0	1.648721271
5	1.25	2.184200811
6	1.5	3.080216849
7	1.75	4.623953153
8	2.0	7.389056099

EXAMPLE 5 *Euler's method.* Approximate the solution of $y' = xy$ through $(x_0, y_0) = (0, 1)$ on the interval [0, 2], using Euler's method with $n = 8$.

Solution Euler's method approximates y_1 from y_0 using a differential, that is, the solution curve is approximated by its tangent line at (x_0, y_0). The method then approximates y_2 from y_1 the same way, and so on. For a differential equation $y' = F(x, y)$, the *slope* of the solution through (x_i, y_i) is $F(x_i, y_i)$. Thus, we have the Euler's method recursion formula

$$\boxed{y_{i+1} = y_i + F(x_i, y_i)\,\Delta x,} \tag{3}$$

arising from the differential approximation formula $h(x_i + dx) \approx h(x_i) + h'(x_i)\,dx$.

For our particular problem in this example, we have $F(x, y) = xy$ and $\Delta x = (2 - 0)/8 = 0.25$. The recursion formula (3) then becomes

$$y_{i+1} = y_i + x_i y_i(0.25). \tag{4}$$

Of course, $x_{i+1} = x_i + \Delta x = x_i + 0.25$. Starting with $(x_0, y_0) = (0, 1)$, we obtain $x_1 = 0.25$, $x_2 = 0.5$, and

$$y_1 = y_0 + x_0 y_0(0.25) = 1 + (0)(1)(0.25) = 1,$$

$$y_2 = y_1 + x_1 y_1(0.25) = 1 + (0.25)(1)(0.25) = 1.0625,$$

and so on. The results of the computation are shown in Table 20.2. Figure 20.7 shows the actual solution $y = h(x) = e^{x^2/2}$ and the approximation from Table 20.2. As you can see from Table 20.1 and the figure, the approximation is not very good for this small value of n. □

Table 20.2 Euler's method

i	x_i	y_i
0	0	1.0
1	0.25	1.0
2	0.5	1.0625
3	0.75	1.1953125
4	1.0	1.419433594
5	1.25	1.774291992
6	1.5	2.32875824
7	1.75	3.20204258
8	2.0	4.602936208

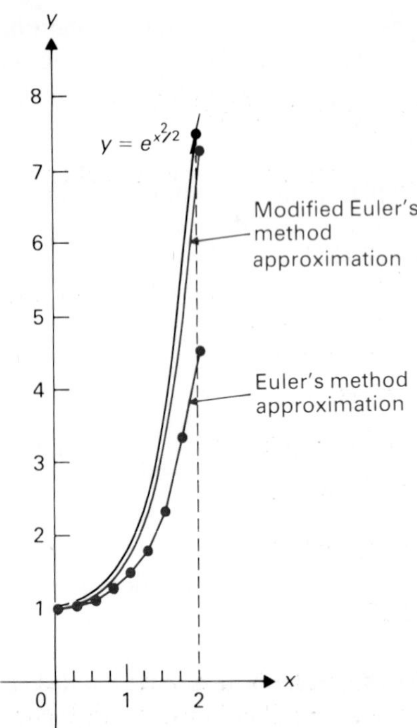

Figure 20.7 Approximations of the solution $y = e^{x^2/2}$ through (0, 1) of $y' = xy$ over [0, 2] for $n = 8$.

EXAMPLE 6 *Modified Euler's method.* Repeat Example 5, using a modified Euler's method.

Solution In Euler's method, we approximated the slope of $h(x)$ over $[x_0, x_1]$ by its value $F(x_0, y_0)$ at the left-hand endpoint. To get a better approximation for this slope over the entire interval, we proceed as follows.

Step 1 Find $\bar{y}_1 = y_0 + F(x_0, y_0) \cdot \Delta x$. (Thus $\bar{y}_1$ is the y_1 of Euler's method.)

Step 2 Find the average $\frac{1}{2}(F(x_0, y_0) + F(x_1, \bar{y}_1))$ of slopes of solutions through (x_0, y_0) and $(x_1, \bar{y}_1)$.

Step 3 Use the average slope from Step 2 to compute y_1, as in Euler's method.

Thus the recursion formulas for computation of y_{i+1} in this modification of Euler's method become

$$\boxed{\begin{aligned} \bar{y}_{i+1} &= y_i + F(x_i, y_i) \cdot \Delta x, \\ y_{i+1} &= y_i + \tfrac{1}{2}(F(x_i, y_i) + F(x_{i+1}, \bar{y}_{i+1}))\,\Delta x. \end{aligned}} \tag{5}$$

For the equation $y' = xy$ in this example, the relations (5) become

$$\begin{aligned} \bar{y}_{i+1} &= y_i + x_i y_i \cdot \Delta x, \\ y_{i+1} &= y_i + \tfrac{1}{2}(x_i y_i + x_{i+1}\bar{y}_{i+1}) \cdot \Delta x. \end{aligned} \tag{6}$$

Starting with $(x_0, y_0) = (1, 0)$ and $\Delta x = 0.25$, we have

$$\bar{y}_1 = 1 + 0 \cdot 1 \cdot 0.25 = 1,$$

$$y_1 = 1 + \tfrac{1}{2}(0 \cdot 1 + 0.25 \cdot 1) \cdot 0.25 = 1.03125,$$

and so on. Table 20.3 shows the values y_i obtained using this modified Euler's method, and the approximation of the solution is sketched in Fig. 20.7. From the exact values shown in Table 20.1, we see that this modification produced a much better approximation to the solution than Euler's method in Example 5. □

EXAMPLE 7 *Taylor's method of order 2.* Repeat Example 5, using Taylor's method of order 2.

Solution Recall that the *first* Taylor approximation

$$h(x_1) \approx T_1(x_1) = h(x_0) + h'(x_0)(x_1 - x_0)$$

is equivalent to approximating $h(x_1)$ using differentials, with $dx = \Delta x = x_1 - x_0$. Thus Euler's method is the same as Taylor's method of order 1. For Taylor's method of order 2, we use the *second* Taylor polynomial $T_2(x)$. That is, we use the approximation

$$\begin{aligned} y_1 = T_2(x_1) &= h(x_0) + h'(x_0)(x_1 - x_0) + \frac{h''(x_0)}{2!}(x_1 - x_0)^2 \\ &= h(x_0) + h'(x_0)(\Delta x) + \tfrac{1}{2}h''(x_0)(\Delta x)^2. \end{aligned}$$

We know that $h'(x) = F(x, y)$, and we need a formula for the second de-

Table 20.3 Modified Euler's method

i	x_i	y_i
0	0	1.0
1	0.25	1.03125
2	0.5	1.131958008
3	0.75	1.322091579
4	1.0	1.642285634
5	1.25	2.168330251
6	1.5	3.040744375
7	1.75	4.52548284
8	2.0	7.141777607

rivative $h''(x)$. By the chain rule, we have

$$h''(x) = F_x(x, y) + F_y(x, y)\frac{dy}{dx} = F_x(x, y) + F_y(x, y) \cdot F(x, y).$$

Thus the recursion formula to compute y_{i+1} using Taylor's method of order 2 becomes

$$\boxed{\begin{aligned} y_{i+1} &= y_i + F(x_i, y_i)(\Delta x) \\ &\quad + \tfrac{1}{2}(F_x + F_y \cdot F)|_{(x_i, y_i)} \cdot (\Delta x)^2. \end{aligned}} \qquad \textbf{(7)}$$

For our equation $y' = xy$, formula (7) becomes

$$y_{i+1} = y_i + x_i y_i(\Delta x) + \tfrac{1}{2}(y_i + x_i^2 y_i)(\Delta x)^2. \qquad \textbf{(8)}$$

Starting with $(x_0, y_0) = (0, 1)$ and $\Delta x = 0.25$, we have

$$\begin{aligned} y_1 &= 1 + (0)(1)(0.25) + \tfrac{1}{2}(1 + 0)0.25^2 \\ &= 1 + \frac{0.25^2}{2} = 1.03125, \end{aligned}$$

Table 20.4 Taylor's order 2 method

i	x_i	y_i
0	0	1.0
1	0.25	1.03125
2	0.5	1.129943848
3	0.75	1.31532526
4	1.0	1.626173613
5	1.25	2.134352867
6	1.5	2.972253113
7	1.75	4.388717487
8	2.0	6.865942788

and so on. The results of continuing to use formula (8) are shown in Table 20.4. For this solution of our equation, the approximation is much better than Euler's method in Example 5 but not quite as good as the modified Euler method in Example 6. □

Numerical solution of differential equations has great practical importance. Many other methods have been devised. A computer is essential as an aid for even the methods given here, as soon as n has any size at all.

We will not discuss accuracy of approximations, except to say that we expect the approximations to become better as n increases. We used our programmable calculator to estimate $h(2)$ for the solution through $(0, 1)$ of the equation $y' = xy$, using $n = 50$, 100, and 200, and the modified Euler's method of Example 6. We obtained the values

$$y_{50} = 7.381434453, \qquad y_{100} = 7.387118206, \qquad y_{200} = 7.38856756.$$

Comparison with Table 20.1 indicates that with $n = 200$, we have about four-significant-figure accuracy. However, if n is excessively large, roundoff error can accumulate to the point where the approximation becomes poor compared with that for a smaller value of n.

SUMMARY

1. A differential equation is one involving derivatives (or differentials) of an unknown function f. The order of the equation is that of the highest-order derivative that appears. The general solution is the expression, containing as many arbitrary constants as the order of the equation, that yields functions $f(x)$ satisfying the equation as the constants assume all values.
2. The equation $y' = F(x, y)$ can be viewed geometrically as specifying the slope at each point (x, y) of any solution of the equation that passes

through that point. If a short line segment of slope $F(x, y)$ is placed at (x, y), we obtain the direction field of the equation.

3. Let F be a continuous function of two variables with domain containing a neighborhood of (x_0, y_0) in which F_y exists and is continuous. Then there exists a number $c > 0$ and a differentiable function $h(x)$ for $x_0 - c < x < x_0 + c$ such that $y = h(x)$ is a solution of the differential equation $y' = F(x, y)$ and such that $y_0 = h(x_0)$. Furthermore, h is the unique such function with domain $x_0 - c < x < x_0 + c$.

4. Numerical approximation points (x_i, y_i) for the solution of $y' = F(x, y)$ through (x_0, y_0) over $[x_0, b]$ can be found as follows. Let n be a positive integer, and let $\Delta x = (b - x_0)/n$. The point (x_0, y_0) is given. For $i = 0, 1, \ldots, n - 1$,

$$x_{i+1} = x_0 + (i + 1)(\Delta x).$$

Euler's method

$$y_{i+1} = y_i + F(x_i, y_i)(\Delta x)$$

Modified Euler's method

$$\bar{y}_{i+1} = y_i + F(x_i, y_i)(\Delta x),$$

$$y_{i+1} = y_i + \tfrac{1}{2}(F(x_i, y_i) + F(x_{i+1}, \bar{y}_{i+1}))(\Delta x)$$

Taylor's order 2 method

$$y_{i+1} = y_i + F(x_i, y_i)(\Delta x) + \tfrac{1}{2}(F_x + F_y \cdot F)|_{(x_i, y_i)} \cdot (\Delta x)^2$$

EXERCISES

In Exercises 1 through 8, sketch the direction field of the differential equation and estimate a few solution curves, as in Figs. 20.2 and 20.3.

1. $y' = y$

2. $y' = -y$

3. $y' = x^2 + y^2$

4. $y' = \dfrac{1}{x^2 + y^2}$

5. $y' = xy$

6. $y' = x + y$

7. $y' = \dfrac{x + y}{x - y}$

8. $y' = y^2$

9. This exercise shows that the number c in our existence theorem (Theorem 20.1) may be very small compared with the radius r of the neighborhood of (x_0, y_0) in which F_y exists and is continuous. Consider the differential equation

$$y' = 1 + y^2 = F(x, y).$$

a) For what values of r is it true that F and F_y are continuous within the neighborhood $x^2 + y^2 < r^2$ of $(0, 0)$?

b) What is the largest value of c such that there exists a differentiable function $h(x)$ for $-c < x < c$ that is a solution of $y' = 1 + y^2$ with $h(0) = 0$? [*Hint:* Solve the equation $y' = 1 + y^2$, and examine the solution through $(0, 0)$.]

10. a) Check that

$$y = h_1(x) = -x^2/4$$

and

$$y = h_2(x) = 1 - x$$

are both solutions of the differential equation

$$y' = \frac{-x + (x^2 + 4y)^{1/2}}{2}$$

for $x \geq 2$, and that $h_1(2) = h_2(2) = -1$.

b) Why doesn't the result in part (a) contradict the uniqueness statement in Theorem 20.1?

Exercises 11 through 22 are designed to be feasible without a calculator or computer. Using the indicated method and value of n, find approximation points for the desired solution of the differential equation, partitioning the given interval into n subintervals of equal length.

11. Euler's method, $n = 2$, solution of $y' = x + y$ through $(0, 1)$ over the interval $[0, 1]$

12. Euler's method, $n = 4$, solution of $y' = x - y$ through $(1, 1)$ over the interval $[1, 2]$

13. Euler's method, $n = 4$, solution of $y' = 2x + y$ through $(1, 4)$ over the integral $[1, 3]$
14. Euler's method, $n = 4$, solution of $y' = -xy$ through $(0, 3)$ over the interval $[0, 4]$
15. Modified Euler's method, data as in Exercise 11
16. Modified Euler's method, $n = 2$, other data as in Exercise 12
17. Modified Euler's method, $n = 2$, other data as in Exercise 13
18. Modified Euler's method, $n = 2$, solution in Exercise 14 over the interval $[0, 2]$
19. Taylor's method of order 2, data as in Exercise 11
20. Taylor's method of order 2, $n = 2$, other data as in Exercise 12
21. Taylor's method of order 2, $n = 2$, solution in Exercise 14 over the interval $[0, 2]$
22. Taylor's method of order 2, $n = 2$, solution of $y' = \sin(xy)$ through $(0, 4\pi)$ over the interval $[0, 1]$
23. A modified Euler's method was described in Example 6. A slightly different modification is to compute $\bar{y}_{i+1}$ as in Example 6 and then use the slope at the midpoint of the line segment joining (x_i, y_i) and $(x_{i+1}, \bar{y}_{i+1})$ to find y_{i+1}, rather than using the average of the slopes at those two points. Give the analogue of formula (5) for this modification.
24. Give the analogue of formula (7) for Taylor's method of order 3.

In Exercises 25 through 30, use a programmable calculator or a computer. Use the indicated method and value of n to estimate the value y_n of the solution at the right-hand endpoint of the given interval.

25. Euler's method, $n = 20$, solution of $y' = -x/y$ through $(0, 3)$ on the interval $[0, 2]$ (The method of Section 6.4 shows that the exact answer is $\sqrt{5}$.)
26. Euler's method, $n = 40$, solution of $y' = (\sin x^2)/(y^2 + 1)$ through $(0, 1)$ on the interval $[0, 4]$
27. Exercise 25 using the modified Euler's method
28. Exercise 26 using the modified Euler's method
29. Exercise 25 using Taylor's method of order 2
30. Exercise 26 using Taylor's method of order 2

20.2 VARIABLES-SEPARABLE AND HOMOGENEOUS EQUATIONS

VARIABLES-SEPARABLE EQUATIONS

Recall from Section 6.4 that if a differential equation $y' = F(x, y)$ can be expressed in the form

$$y' = \frac{dy}{dx} = \frac{f(x)}{g(y)},$$

then the variables can be separated and the equation solved as follows:

$$g(y)\,dy = f(x)\,dx,$$

$$\int g(y)\,dy = \int f(x)\,dx.$$

That is, one puts all terms involving y (including dy) on the left-hand side of the equation and all terms involving x (including dx) on the right-hand side. The variables are then "separated," and the equation is solved by finding two indefinite integrals of functions of a single variable.

EXAMPLE 1 Solve

$$\frac{dy}{dx} = \frac{\ln x}{xy^2}.$$

Solution The equation is of variables-separable type, and

$$y^2\,dy = \frac{\ln x}{x}\,dx,$$

so

$$\int y^2\, dy = \int \frac{\ln x}{x}\, dx,$$

and

$$\frac{1}{3}y^3 = \frac{(\ln x)^2}{2} + C$$

is the general solution. We may also express this as

$$2y^3 = 3(\ln x)^2 + 6C,$$

or, since $6C$ may again be any arbitrary constant, it is acceptable to simply write

$$2y^3 = 3(\ln x)^2 + C.$$

Such informality with arbitrary constants is conventional. □

The particular solution $y = h(x)$ of a differential equation $y' = F(x, y)$ such that $y_0 = h(x_0)$ is often called the solution of the initial-value problem:

$$y' = F(x, y), \quad y(x_0) = y_0.$$

Initial-value problem

EXAMPLE 2 Solve the initial-value problem

$$\frac{dy}{dx} = \frac{(x+1)(\sin x)}{\tan y}, \qquad y(0) = \pi.$$

Solution The equation is of variables-separable type, and we have

$$\tan y\, dy = (x + 1)(\sin x)\, dx,$$

$$\int \frac{\sin y}{\cos y}\, dy = \int (x \sin x + \sin x)\, dx.$$

Integration by parts shows that $\int x \sin x\, dx = -x \cos x + \sin x + C$. Thus we obtain

$$-\ln|\cos y| = -x \cos x + \sin x - \cos x + C.$$

Setting $x = 0$ and $y = \pi$ so that $y(0) = \pi$, we have

$$-\ln 1 = 0 + 0 - 1 + C,$$

so

$$C = 1 - \ln 1 = 1 - 0 = 1.$$

Thus our solution of the initial-value problem is defined implicitly by

$$-\ln|\cos y| = -x \cos x + \sin x - \cos x + 1,$$

or

$$\ln|\cos y| = (x + 1)(\cos x) - \sin x - 1. \quad \square$$

Section 8.5 gave an important application of one type of variables-separable equation, $dQ/dt = kQ$. Recall that the solution $Q = Q_0 e^{kt}$ gives the growth (or decay) of the quantity Q.

For another application, consider a body of mass m falling freely near the surface of the earth. The body is subject to a downward force mg due to gravity and an upward force due to air resistance. The body attains *terminal velocity* when these two forces are of equal magnitude, so the velocity no longer changes.

EXAMPLE 3 Suppose a body of mass m falling freely near the surface of the earth is subject to a force of air resistance of magnitude $k|v|$, where $k > 0$ and v is the (signed) scalar velocity. Find v at time t if $v = 0$ when $t = 0$, and find the terminal velocity.

Solution Let us consider positive velocity v as directed downward, in the direction of motion of the falling body. By Newton's second law, the force acting on the body is

$$ma = m\frac{dv}{dt} = mg - kv,$$

where $a = dv/dt$ is the acceleration of the body. This is a differential equation in v and t of variables-separable type. We obtain

$$\frac{m\,dv}{mg - kv} = dt,$$

$$-\frac{m}{k}\ln|mg - kv| = t + C.$$

Setting $t = v = 0$, we see that $C = -(m/k)\cdot\ln(mg)$, so

$$-\frac{m}{k}\ln|mg - kv| = t - \frac{m}{k}\ln(mg),$$

$$\ln|mg - kv| = -\frac{k}{m}t + \ln(mg),$$

$$|mg - kv| = mge^{-kt/m}.$$

From the physical situation, we know that $mg - kv \geq 0$. Thus we can drop the absolute-value symbol, and we find that

$$kv = mg - mge^{-kt/m},$$

so

$$v = \frac{mg}{k}(1 - e^{-kt/m}). \qquad \textbf{(1)}$$

Equation (1) is the desired expression for the velocity in terms of m, g, k, and t.

To find the terminal velocity, we could proceed in two ways. Physically, terminal velocity is achieved when the forces mg of gravity and kv of air resistance are of equal magnitude, or when $v = mg/k$. Alternatively, we can let $t \to \infty$ in Eq. (1), so that $e^{-kt/m} \to 0$, and see that v approaches mg/k. Equation (1) shows that the body never actually attains terminal velocity. However, it approaches it quite rapidly, since the exponential $e^{-kt/m}$ decreases rapidly as t increases. □

HOMOGENEOUS EQUATIONS

The differential equation $y' = F(x, y)$ is **homogeneous** if $F(x, y)$ can be expressed in the form

$$F(x, y) = g\left(\frac{y}{x}\right).$$

A criterion for homogeneity is

$$F(kx, ky) = F(x, y) \qquad \text{for all } k \neq 0. \tag{2}$$

If Eq. (2) is true, then taking $k = 1/x$, we obtain

$$F(x, y) = F\left(\frac{1}{x}x, \frac{1}{x}y\right) = F\left(1, \frac{y}{x}\right) = g\left(\frac{y}{x}\right).$$

Conversely, if $F(x, y) = g(y/x)$, then

$$F(kx, ky) = g\left(\frac{ky}{kx}\right) = g\left(\frac{y}{x}\right) = F(x, y).$$

Such a function $F(x, y)$ is called *homogeneous of degree* 0. (If $F(kx, ky) = k^r F(x, y)$, then $F(x, y)$ is *homogeneous of degree r*.)

EXAMPLE 4 Show that the differential equation

$$y' = F(x, y) = \frac{x^2 + y^2}{2x^2}$$

is homogeneous, and express the equation in the form $y' = g(y/x)$.

Solution We see that

$$F(kx, ky) = \frac{(kx)^2 + (ky)^2}{2(kx)^2} = \frac{k^2x^2 + k^2y^2}{2k^2x^2}$$
$$= \frac{x^2 + y^2}{2x^2} = F(x, y).$$

Thus condition (2) is satisfied, so the equation is homogeneous. We find that

$$y' = \frac{x^2 + y^2}{2x^2} = \frac{x^2}{2x^2} + \frac{y^2}{2x^2} = \frac{1}{2} + \frac{1}{2}\left(\frac{y}{x}\right)^2. \quad \square$$

Suppose we write a differential equation in the form

$$M(x, y)\,dx + N(x, y)\,dy = 0.$$

This equation is homogeneous whenever $M(x, y)$ and $N(x, y)$ are both homogeneous *of the same degree*. For suppose $M(kx, ky) = k^r M(x, y)$ and $N(kx, ky) = k^r N(x, y)$. We can write the differential equation as

$$y' = \frac{dy}{dx} = \frac{-M(x, y)}{N(x, y)} = F(x, y),$$

and

$$F(kx, ky) = \frac{-M(kx, ky)}{N(kx, ky)} = \frac{-k^r M(x, y)}{k^r N(x, y)} = \frac{-M(x, y)}{N(x, y)} = F(x, y).$$

This shows that the differential equation is indeed homogeneous. For example, the differential equation in Example 4 above can be written in the form

$$(x^2 + y^2)\,dx + (-2x^2)\,dy = 0,$$

and both $-2x^2$ and $x^2 + y^2$ are homogeneous functions of the same degree, 2.

If

$$\frac{dy}{dx} = g\left(\frac{y}{x}\right), \tag{3}$$

then the substitution $y = vx$ yields an equation in v and x of variables-separable type. From $y = vx$, we have

$$\frac{dy}{dx} = v \cdot 1 + x\frac{dv}{dx} \qquad \text{and} \qquad \frac{y}{x} = v.$$

The Eq. (3) becomes

$$v + x\frac{dv}{dx} = g(v) \qquad \text{or} \qquad x\frac{dv}{dx} = g(v) - v,$$

or

$$\frac{dv}{g(v) - v} = \frac{dx}{x}, \tag{4}$$

and the variables are separated. Equation (4) can be integrated to find the solution in terms of x and v and then v replaced by y/x to find the solution in terms of x and y.

EXAMPLE 5 Solve the equation of Example 4,

$$\frac{dy}{dx} = \frac{x^2 + y^2}{2x^2} = \frac{1}{2} + \frac{1}{2}\left(\frac{y}{x}\right)^2.$$

Solution The substitution $y = vx$ yields

$$v + x\frac{dv}{dx} = \frac{1}{2} + \frac{1}{2}v^2,$$

so

$$x\frac{dv}{dx} = \frac{1 + v^2}{2} - v = \frac{v^2 - 2v + 1}{2}.$$

Then

$$\frac{dv}{v^2 - 2v + 1} = \frac{dx}{2x}, \qquad \text{so} \qquad (v - 1)^{-2}\,dv = \frac{1}{2}\cdot\frac{dx}{x}.$$

Hence

$$\int (v - 1)^{-2}\,dv = \frac{1}{2}\int\frac{dx}{x}, \qquad \text{so} \qquad \frac{(v - 1)^{-1}}{-1} = \frac{1}{2}\ln|x| + C.$$

Substituting $v = y/x$, we obtain

$$\frac{-1}{(y/x) - 1} = \frac{1}{2}\ln|x| + C, \qquad \text{or} \qquad \frac{-2x}{y - x} = \ln|x| + C,$$

as our general solution relation. □

EXAMPLE 6 Solve the initial-value problem

$$y' = \frac{x+y}{y-x}, \qquad y(1) = 3.$$

Solution Our equation is homogeneous, and the substitution $y = vx$ yields

$$v + x\frac{dv}{dx} = \frac{x+vx}{vx-x} = \frac{1+v}{v-1},$$

$$x\frac{dv}{dx} = \frac{1+v}{v-1} - v = -\frac{v^2-2v-1}{v-1},$$

$$\frac{v-1}{v^2-2v-1}\,dv = -\frac{dx}{x},$$

$$\frac{1}{2}\frac{2(v-1)}{v^2-2v-1}\,dv = -\frac{dx}{x},$$

$$\frac{1}{2}\ln|v^2-2v-1| = -\ln|x| + C,$$

$$\frac{1}{2}\ln\left|\frac{y^2}{x^2} - 2\frac{y}{x} - 1\right| = -\ln|x| + C.$$

Setting $x = 1$ and $y = 3$ to satisfy $y(1) = 3$, we have

$$\frac{1}{2}\ln|9-6-1| = 0 + C, \qquad \text{so} \qquad C = \frac{1}{2}(\ln 2).$$

Thus the desired solution is

$$\frac{1}{2}\ln\left|\frac{y^2}{x^2} - 2\frac{y}{x} - 1\right| = -\ln|x| + \frac{1}{2}(\ln 2).$$

This solution is perfectly acceptable. Using properties of logarithms, we could also see that the solution can be written as

$$|y^2 - 2xy - x^2| = 2. \quad \square$$

APPLICATION TO GEOMETRY

Let G be a function of two variables with continuous partial derivatives. The relation $G(x, y) = C$ for an arbitrary constant C gives a *family of curves*, one curve for each value of C. A second family of curves $H(x, y) = K$ is **orthogonal** to the first family (or consists of **orthogonal trajectories** of the first family) if every curve of the first family is perpendicular to every curve of the second family at all points of intersection.

The problem of finding a family of curves orthogonal to a given family $G(x, y) = C$ is a problem in differential equations and can sometimes be solved by the techniques described in this section. We illustrate with an example.

EXAMPLE 7 Find the family of orthogonal trajectories to the hyperbolas $x^2 - y^2 = C$.

Solution Let $G(x, y) = x^2 - y^2$. The slope of a curve $x^2 - y^2 = C$ at a point (x, y) is given by

$$\frac{dy}{dx} = -\frac{\partial G/\partial x}{\partial G/\partial y} = -\frac{2x}{-2y} = \frac{x}{y}.$$

The slope of a curve in the orthogonal family at (x, y) is the negative reciprocal, $-y/x$. Thus our orthogonal family of curves consists of the solutions of the differential equation

$$\frac{dy}{dx} = -\frac{y}{x}.$$

Then

$$\frac{dy}{y} + \frac{dx}{x} = 0,$$

so

$$\ln|x| + \ln|y| = K. \tag{5}$$

We may write Eq. (5) in the form

$$\ln|xy| = K.$$

Then $e^{\ln|xy|} = |xy| = e^K$, so $xy = \pm e^K$. Now as K runs through all constants, $\pm e^K$ runs through all constants except zero. A special examination shows that the curves $x = 0$ and $y = 0$ are also orthogonal to the hyperbolas $x^2 - y^2 = C$; changing notation, we obtain, as orthogonal family,

$$xy = K.$$

This is also a family of hyperbolas, with their axes tipped at 45° from the x,y-axes. Figure 20.8 shows some of the curves of these two orthogonal families. □

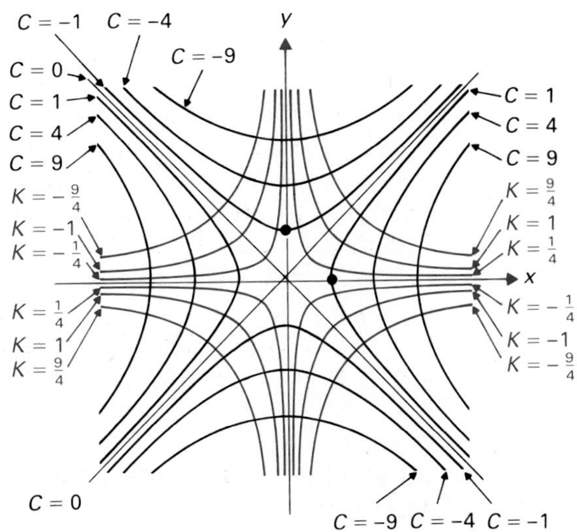

Figure 20.8 Orthogonal trajectories of hyperbolas $x^2 - y^2 = C$ are hyperbolas $xy = K$.

SUMMARY

1. A differential equation $dy/dx = F(x, y)$ is of variables-separable type if all the terms involving y, including dy, can be placed on the left-hand side and all the terms involving x, including dx, on the right-hand side. The equation is then solved by integrating each side of this separated equation.
2. The equation $dy/dx = F(x, y)$ is homogeneous if $F(x, y)$ can be written in the form $g(y/x)$. This is the case if and only if $F(kx, ky) = F(x, y)$ for all $k \neq 0$.
3. If the substitution $y = vx$ is made in a homogeneous equation $dy/dx = g(y/x)$, so that
$$\frac{dy}{dx} = v + x\frac{dv}{dx} \quad \text{and} \quad g\left(\frac{y}{x}\right) = g(v),$$
then the new equation in x and v is of variables-separable type. Solving the new equation and then replacing v by y/x yields the solution of the original equation.
4. For a family of curves $G(x, y) = C$, the family of orthogonal trajectories is found as follows:

Step 1 Compute
$$\frac{dy}{dx} = -\frac{\partial G/\partial x}{\partial G/\partial y}$$
to find the slope of a curve of the first family at (x, y).

Step 2 Solve the differential equation
$$\frac{dy}{dx} = \frac{\partial G/\partial y}{\partial G/\partial x}.$$
The solution is the family of orthogonal trajectories.

EXERCISES

In Exercises 1 through 16, find the general solution of the differential equation.

1. $\dfrac{dy}{dx} = xy^2$
2. $\dfrac{dy}{dx} = e^x \tan y$
3. $\dfrac{dy}{dx} = x^2 + x^2y^2$
4. $\dfrac{dy}{dx} = \sin x \cos^2 y$
5. $\dfrac{dy}{dx} = x(1 - y^2)^{1/2}$
6. $\dfrac{dy}{dx} = y (\ln y) \cos^2 x$
7. $\dfrac{dy}{dx} = \dfrac{x^2 + y^2}{2xy}$
8. $(x + y)\, dy = (y - x)\, dx$
9. $x\, dy = (x + y)\, dx$
10. $x\, dy = [y + x \sin (y/x)]\, dx$
11. $\dfrac{dy}{dx} = \dfrac{x^2 + y^2}{xy}$
12. $\dfrac{dy}{dx} = \dfrac{2x^3 + y^3}{xy^2}$
13. $\dfrac{dy}{dx} = e^{y/x} + \dfrac{y}{x}$
14. $\dfrac{dy}{dx} = \dfrac{x}{y} - 3\dfrac{y}{x}$
15. $\dfrac{dy}{dx} = \dfrac{4x + y}{x + y}$
16. $\dfrac{dy}{dx} = \dfrac{x - 2y}{y - 2x}$

In Exercises 17 through 26, find the solution of the given initial-value problem.

17. $y' = 1 + x,\ y(1) = -1$
18. $y' = x \sin y,\ y(2) = \dfrac{\pi}{2}$
19. $y' = \dfrac{x}{y},\ y(2) = -3$
20. $y' = x^2y + 2xy,\ y(2) = 1$
21. $y' = \dfrac{1 + y^2}{xy},\ y(1) = 5$
22. $y' = y^3e^{2x},\ y(0) = 2$
23. $y' = \dfrac{x + y}{x},\ y(1) = 2$
24. $y' = \dfrac{x + y}{x - y},\ y(1) = 3$

25. $y' = \cos\left(\frac{y}{x}\right) + \frac{y}{x}$, $y(\pi) = 0$

26. $y' = \tan\left(\frac{y}{x}\right) + \frac{y}{x}$, $y(4) = \pi$

In Exercises 27 through 32, find the family of orthogonal trajectories of the given family of curves. Sketch both families as in Fig. 20.8.

27. $y - x = C$

28. $y - x^2 = C$

29. $y^2 + x = C$

30. $x^2 + y^2 = C$

31. $y = C(x + 1)$

32. $x^2 + y^2 = Cx$

20.3 EXACT EQUATIONS

The notion of an exact differential equation is closely related to the idea of an exact differential form, which we studied in Section 17.3. You may wish to review that section before proceeding.

SOLUTION OF AN EXACT DIFFERENTIAL EQUATION

The differential equation

$$\frac{dy}{dx} = -\frac{P(x, y)}{Q(x, y)}$$

can be written in the form

$$P(x, y)\, dx + Q(x, y)\, dy = 0. \tag{1}$$

Suppose $P(x, y)\, dx + Q(x, y)\, dy$ is an exact differential form, discussed in Section 17.3. If $dG = P\, dx + Q\, dy$, Eq. (1) becomes

$$dG(x, y) = 0,$$

so the solution of Eq. (1) is

$$G(x, y) = C.$$

In the study of differential equations, it is customary to use $M(x, y)$ in place of $P(x, y)$ and $N(x, y)$ in place of $Q(x, y)$ in Eq. (1). We make this change now to conform to notation you are apt to find elsewhere.

DEFINITION 20.1 Exact equation

A differential equation

$$M(x, y)\, dx + N(x, y)\, dy = 0 \tag{2}$$

is **exact** if $M(x, y)\, dx + N(x, y)\, dy$ is an exact differential form.

Recall that if M and N have continuous partial derivatives in a connected region of the plane with no holes in it, then Eq. (2) is an exact form if and only if

$$\boxed{\frac{\partial M}{\partial y} = \frac{\partial N}{\partial x}.} \tag{3}$$

If Eq. (3) is satisfied and if G is a function of two variables such that

$$dG = M(x, y)\,dx + N(x, y)\,dy,$$

then the differential equation (2) has as general solution

$$\boxed{G(x, y) = C.} \tag{4}$$

The technique used to solve an exact differential equation $M\,dx + N\,dy = 0$ is precisely the partial integration technique employed to find a function $G(x, y)$ such that $dG = M(x, y)\,dx + N(x, y)\,dy$, described in Section 17.3. We illustrate with an example.

EXAMPLE 1 Solve $3x^2y\,dx + (x^3 - y^2)\,dy = 0$.

Solution The differential equation

$$M\,dx + N\,dy = 3x^2y\,dx + (x^3 - y^2)\,dy = 0$$

is exact since

$$\frac{\partial M}{\partial y} = \frac{\partial(3x^2y)}{\partial y} = 3x^2 \quad \text{and} \quad \frac{\partial N}{\partial x} = \frac{\partial(x^3 - y^2)}{\partial x} = 3x^2.$$

Setting $\partial G/\partial x = 3x^2y$, we find that

$$G(x, y) = x^3y + h(y).$$

Then

$$\frac{\partial G}{\partial y} = x^3 + h'(y) = x^3 - y^2.$$

Consequently,

$$h'(y) = -y^2 \quad \text{and} \quad h(y) = \frac{-y^3}{3}.$$

This shows that

$$G(x, y) = x^3y - \frac{y^3}{3}$$

has differential $3x^2y\,dx + (x^3 - y^2)\,dy$, so the general solution of our equation is

$$x^3y - \frac{y^3}{3} = C. \quad \square$$

EXAMPLE 2 Solve the initial-value problem

$$(ye^{xy} + \sin x)\,dx + (xe^{xy} + \cos y)\,dy = 0, \qquad y(0) = \frac{\pi}{2}.$$

Solution Using condition (3), we can easily see that this equation is exact. We use another technique described in Section 17.3 to find a function $G(x, y)$ such that the equation is $dG = 0$. We must have

$$\frac{\partial G}{\partial x} = ye^{xy} + \sin x \quad \text{and} \quad \frac{\partial G}{\partial y} = xe^{xy} + \cos y.$$

Thus $G(x, y)$ must be simultaneously of the form

$$G(x, y) = \int (ye^{xy} + \sin x)\, dx = e^{xy} - \cos x + h(y) \tag{5}$$

and

$$G(x, y) = \int (xe^{xy} + \cos y)\, dy = e^{xy} + \sin y + k(x). \tag{6}$$

Comparison of Eqs. (5) and (6) show that we may take

$$G(x, y) = e^{xy} - \cos x + \sin y.$$

Therefore the general solution $G(x, y) = C$ is

$$e^{xy} - \cos x + \sin y = C.$$

We substitute $x = 0$ and $y = \pi/2$ to find C so that $y(0) = \pi/2$. We have

$$e^0 - 1 + 1 = C, \qquad \text{so} \qquad C = e^0 = 1.$$

The desired solution is therefore

$$e^{xy} - \cos x + \sin y = 1. \quad \square$$

INTEGRATING FACTORS

Let a differential equation

$$y' = F(x, y) = \frac{-M(x, y)}{N(x, y)}$$

be written in the form

$$M(x, y)\, dx + N(x, y)\, dy = 0, \tag{7}$$

and let Eq. (7) have general solution

$$G(x, y) = C. \tag{8}$$

From Eq. (8) we see that, at any point on a solution curve $y = h(x)$, we have

$$y' = -\frac{\partial G/\partial x}{\partial G/\partial y}.$$

Thus we must have

$$\frac{\partial G/\partial x}{\partial G/\partial y} = \frac{M(x, y)}{N(x, y)}. \tag{9}$$

From Eq. (9), we obtain

$$\frac{\partial G/\partial x}{M(x, y)} = \frac{\partial G/\partial y}{N(x, y)} = \mu(x, y), \tag{10}$$

where we let $\mu(x, y)$ be the common ratio in Eq. (10). From Eq. (10), we obtain

$$\frac{\partial G}{\partial x} = \mu(x, y)M(x, y) \qquad \text{and} \qquad \frac{\partial G}{\partial y} = \mu(x, y)N(x, y). \tag{11}$$

The Eqs. (11) show that the equation

$$\mu(x, y)M(x, y)\, dx + \mu(x, y)N(x, y)\, dy = 0, \tag{12}$$

obtained by multiplying Eq. (7) by $\mu(x, y)$, is exact.

DEFINITION Integrating factor

20.2

A function $\mu(x, y)$ is an **integrating factor** for a differential equation $M(x, y)\,dx + N(x, y)\,dy = 0$ if

$$\mu(x, y)M(x, y)\,dx + \mu(x, y)N(x, y)\,dy = 0$$

is an exact differential equation.

Our work prior to this definition shows that if the differential equation (7) has a general solution $G(x, y) = C$, then the equation has an integrating factor $\mu(x, y)$. Multiplication of an equation $M(x, y)\,dx + N(x, y)\,dy = 0$ by such a factor $\mu(x, y)$ does not change the solutions of the equation. If we can find an integrating factor, then we can multiply the equation by the factor. The equation is then exact and can be solved by the techniques used in Examples 1 and 2.

Integrating factors are by no means unique as the following example shows.

EXAMPLE 3 Show that both x and $1/(x^2y)$ are integrating factors for the equation

$$3xy\,dx + x^2\,dy = 0.$$

Solution The differential equation has x as an integrating factor, for

$$\begin{aligned} x(3xy\,dx + x^2\,dy) &= 3x^2y\,dx + x^3\,dy \\ &= d(x^3y). \end{aligned}$$

Another integrating factor is $1/x^2y$, for

$$\frac{1}{x^2y}(3xy\,dx + x^2\,dy) = \frac{3}{x}\,dx + \frac{1}{y}\,dy = d(3\ln|x| + \ln|y|). \quad \square$$

Exercise 25 indicates that, in general, Eq. (7) can be expected to have an infinite number of integrating factors.

In Exercises 26 and 27, we give a partial differential equation involving $M(x, y)$ and $N(x, y)$ that a function μ must satisfy to be an integrating factor of Eq. (7). From this differential equation, one can characterize the differential equations (7) that have certain types of integrating factors, such as factors that are functions of just x or of just y. Some results and illustrations along these lines are given in Exercises 28 through 30. In the following subsection, we indicate how one may sometimes find an integrating factor by "inspection."

FINDING INTEGRATING FACTORS BY INSPECTION

Integrating factors for certain differential equations of the form of Eq. (7) can be found by inspection; facility requires a certain amount of practice. One useful technique is to watch for expressions such as $y\,dx + x\,dy$ that are themselves exact or have obvious integrating factors; note that

$$d(xy) = y\,dx + x\,dy.$$

Thus the presence of $y\,dx + x\,dy$ suggests that a function of xy might be an integrating factor. Similarly, the differentials

$$d\left(\frac{x}{y}\right) = \frac{1}{y^2}(y\,dx - x\,dy) \quad \text{and} \quad d\left(\frac{y}{x}\right) = \frac{1}{x^2}(x\,dy - y\,dx)$$

suggest watching for the expressions $y\,dx - x\,dy$ and $x\,dy - y\,dx$, whose presence would lead us to try integrating factors of the form

$$\frac{1}{y^2} f\left(\frac{x}{y}\right) \quad \text{and} \quad \frac{1}{x^2} g\left(\frac{y}{x}\right),$$

respectively. We illustrate with some examples.

EXAMPLE 4 Solve the differential equation

$$y\,dx + (x + x^2y)\,dy = 0.$$

Solution The expression $y\,dx + x\,dy$ contained in this equation suggests an integrating factor that is a function of xy. If we divide through the equation by $(xy)^2$, then the term $x^2y\,dy$ becomes $(1/y)\,dy$, which can be integrated. Thus we take as integrating factor $1/(xy)^2$ and obtain the equation

$$\frac{y\,dx + x\,dy}{(xy)^2} + \frac{x^2y}{(xy)^2}\,dy = 0,$$

or

$$(xy)^{-2}\,d(xy) + \frac{1}{y}\,dy = 0.$$

Integration of the last equation yields

$$-\frac{1}{xy} + \ln|y| = C$$

as general solution. □

EXAMPLE 5 Solve the differential equation

$$(xy^4 + y)\,dx - x\,dy = 0.$$

Solution The presence of $y\,dx - x\,dy$ suggests an integrating factor of the form $(1/y^2)f(x/y)$. To "eliminate" the y^4 from $xy^4\,dx$ for integration, we use $(1/y^2)(x/y)^2$ as integrating factor and obtain

$$\frac{1}{y^2}\left(\frac{x}{y}\right)^2 xy^4\,dx + \frac{1}{y^2}\left(\frac{x}{y}\right)^2(y\,dx - x\,dy) = 0,$$

or

$$x^3\,dx + \left(\frac{x}{y}\right)^2 d\left(\frac{x}{y}\right) = 0.$$

Integrating this last equation, we obtain

$$\frac{x^4}{4} + \frac{1}{3}\left(\frac{x}{y}\right)^3 = C$$

as general solution. □

EXAMPLE 6 Solve the initial-value problem

$$y' = \frac{2y}{xy^4 - x}, \qquad y(1) = -2.$$

Solution We may write the equation

$$\frac{dy}{dx} = \frac{2y}{xy^4 - x}$$

as

$$2y\,dx + x\,dy - xy^4\,dy = 0. \tag{13}$$

The terms $2y\,dx + x\,dy$ are not quite $d(xy)$ because of the factor 2. To obtain a factor 2 in the dx-term, we form instead $d(x^2y) = 2xy\,dx + x^2\,dy$. This suggests that we multiply Eq. (13) by x and obtain

$$2xy\,dx + x^2\,dy - x^2y^4\,dy = 0.$$

We now try an integrating factor of the form $g(x^2y)$. To eliminate the x^2 factor from $x^2y^4\,dy$, we use the factor $1/(x^2y)$ and have

$$\frac{d(x^2y)}{x^2y} - y^3\,dy = 0.$$

Integrating, we have

$$\ln|x^2y| - \frac{y^4}{4} = C.$$

Setting $x = 1$ and $y = -2$, we have

$$\ln 2 = 4 + C, \qquad \text{so} \qquad C = -4 + \ln 2.$$

Thus our desired solution is

$$\ln|x^2y| - \frac{y^4}{4} = -4 + \ln 2. \qquad \square$$

SUMMARY

1. The differential equation $M(x, y)\,dx + N(x, y)\,dy = 0$ is exact if and only if $\partial M/\partial y = \partial N/\partial x$, and $M(x, y)$ and $N(x, y)$ satisfy suitable conditions. In this case, find $G(x, y)$ such that

$$dG = M(x, y)\,dx + N(x, y)\,dy$$

as described in Section 17.3, and the solution of the given equation is $G(x, y) = C$.

2. If $M(x, y)\,dx + N(x, y)\,dy = 0$ has a solution $G(x, y) = C$, then there exists a function (integrating factor) $\mu(x, y)$ such that

$$\mu(x, y)M(x, y)\,dx + \mu(x, y)N(x, y)\,dy = 0$$

is an exact equation and can be solved as in number 1 above. Integrating factors can sometimes be found by inspection; attempt to create obvious exact differentials such as

$$d(xy) = x\,dy + y\,dx,$$

$$d\left(\frac{x}{y}\right) = \frac{1}{y^2}(y\,dx - x\,dx),$$

$$d\left(\frac{y}{x}\right) = \frac{1}{x^2}(x\,dy - y\,dx).$$

See Examples 4, 5, and 6 in the text for illustrations.

EXERCISES

In Exercises 1 through 6, verify that the equation is exact and find the general solution.

1. $\cos y\,dx + (1 - x\sin y)\,dy = 0$
2. $2xy\,dx + (x^2 - e^{-y})\,dy = 0$
3. $y^2\,dx + \left(\frac{1}{y} + 2xy\right)dy = 0$
4. $(y\sec^2 xy)\,dx + (1 + x\sec^2 xy)\,dy = 0$
5. $(e^y - y\cos xy)\,dx + (xe^y - x\cos xy)\,dy = 0$
6. $(2xy^3 - 3)\,dx + (3x^2y^2 + 4y)\,dy = 0$

In Exercises 7 through 14, find an integrating factor by inspection and solve the differential equation.

7. $(xy^2 + y)\,dx + (x - 3x^2)\,dy = 0$
8. $(x^2y^2 + y^2)\,dx + (2xy + x)\,dy = 0$
9. $(xy^2 + y)\,dx - (x + y)\,dy = 0$
10. $(4 - y)\,dx + (x + 3x^2)\,dy = 0$
11. $(4 + y)\,dx - (x + 3x^2)\,dy = 0$
12. $\left(\frac{x^2}{y} + y\right)dx + (2x - e^y)\,dy = 0$
13. $(1 + 2x^2y^3)\,dx + (xy^2 + 3x^3y^2)\,dy = 0$
14. $(x^2y^2 + 2xy)\,dx - (x^2 + 3)\,dy = 0$

In Exercises 15 through 24, solve the given initial-value problem.

15. $[y^2\cos(xy^2) - x + 2] + [2xy\cos(xy^2) + 3e^y]\,dy = 0,$ $y(0) = 0$
16. $\left(\frac{1}{y} - ye^x + x\right)dx - \left(\frac{x}{y^2} + e^x - 3\right)dy = 0,\ y(0) = 4$
17. $\left(-\frac{y}{x^2}e^{y/x} - \sin x\right)dx + \left(\frac{1}{x}e^{y/x} + \cos y\right)dy = 0,$ $y(\pi) = -\pi$
18. $\left(\frac{1}{x+y} - \frac{1}{x}\right)dx + \left(\frac{1}{x+y} + \frac{1}{y}\right)dy = 0,\ y(1) = 1$
19. $(x^2y^2 + y)\,dx + (x - 3x^2y)\,dy = 0,\ y(1) = 1$
20. $(y + x^2y^2)\,dx - (x + 1)\,dy = 0,\ y(0) = 2$
21. $(y + 1)\,dx - (x + x^2y^3)\,dy = 0,\ y(1) = -1$
22. $(2xy + 2xy^2)\,dx - (x^2 + 4y^3)\,dy = 0,\ y(-1) = 4$
23. $(2xy - 1)\,dx - (x^2 + x^4)\,dy = 0,\ y(1) = 0$
24. $(y + x^3y^3)\,dx + 2x\,dy = 0,\ y(2) = 1$
25. Let $M(x, y)\,dx + N(x, y)\,dy = 0$ have general solution $G(x, y) = C$ and let μ be an integrating factor such that $dG = \mu M\,dx + \mu N\,dy$. Show that if f is any continuous function of one variable, then $\mu(x, y)\cdot f(G(x, y))$ is an integrating factor. (This indicates that the differential equation (7) can be expected to have an infinite number of integrating factors.)
26. Show that μ is an integrating factor of a differential equation $M\,dx + N\,dy = 0$, where M and N are continuously differentiable functions in a suitable region, if and only if μ is a solution of the partial differential equation
$$N\frac{\partial u}{\partial x} - M\frac{\partial \mu}{\partial y} = \mu\left(\frac{\partial M}{\partial y} - \frac{\partial N}{\partial x}\right).$$
[*Hint:* Apply the condition for $\mu M\,dx + \mu N\,dy$ to be exact.]
27. Show that the partial differential equation in Exercise 26 can be written in the form
$$N\frac{\partial(\ln|\mu|)}{\partial x} - M\frac{\partial(\ln|\mu|)}{\partial y} = \frac{\partial M}{\partial y} - \frac{\partial N}{\partial x}.$$
28. Using Exercises 26 and 27, show that if M and N are continuously differentiable in a suitable region, then $M\,dx + N\,dy = 0$ has an integrating factor μ that is a function of x only (that is, $\partial\mu/\partial y = 0$) if and only if $(1/N)\cdot(\partial M/\partial y - \partial N/\partial x)$ is a function of x only, say $f(x)$, and that the integrating factor μ is then of the form $\mu(x) = Ke^{\int f(x)\,dx}$.
29. Use the result in Exercise 28 to solve the differential equations
a) $(xy^2 + x + y^2 + 1)\,dx + (2x^2y - 3xy^2 + 2xy - 3y^2)\,dy = 0$
b) $(xy - y)\,dx + (x^2 + x\cos y - x - \cos y)\,dy = 0.$
30. State a result analogous to that in Exercise 28 in the case that $M\,dx + N\,dy = 0$ has an integrating factor that is a function of y only.

20.4 FIRST-ORDER LINEAR EQUATIONS

In this section, we are concerned with finding all solutions in a neighborhood of some point x_0 of the first-order *linear* differential equation

$$p_0(x)y' + p_1(x)y = q(x), \tag{1}$$

where $p_0(x)$, $p_1(x)$, and $q(x)$ are known functions defined in some neighborhood of x_0. The equation is called *linear* since each summand contains at most one factor y or y', raised to just the first power.

We will restrict our discussion to the case where the coefficient $p_0(x)$ of y' in Eq. (1) takes on only nonzero values throughout some neighborhood of x_0. We may then divide Eq. (1) by $p_0(x)$ and set $p(x) = p_1(x)/p_0(x)$ and $g(x) = q(x)/p_0(x)$ to write Eq. (1) in the simpler form

$$y' + p(x)y = g(x). \tag{2}$$

We assume that the functions $p(x)$ and $g(x)$ in Eq. (2) are continuous near x_0 so that we can integrate them. This differential equation (2) is one whose solutions can be found easily; we can actually obtain a formula for the general solution.

The trick is to find a continuous function $\mu(x)$ that is nonzero in the neighborhood and that has the property

$$\mu(x)y' + \mu(x)p(x)y = v' \tag{3}$$

for some function v. (Such a function μ is an *integrating factor*.) Upon multiplication by $\mu(x)$, Eq. (2) would then reduce to the equivalent equation

$$v' = \mu(x)g(x), \tag{4}$$

which can be solved by a single integration. It has been found that

$$\mu(x) = e^{\int p(x)\,dx} \tag{5}$$

for any choice of antiderivative of $p(x)$ is such an integrating factor. (Observe that $\mu(x)$ is continuous and $\mu(x) \neq 0$.) To see that μ is indeed an integrating factor, note that if

$$v = \mu(x) \cdot y, \tag{6}$$

then

$$\begin{aligned} v' = \frac{d[\mu(x)]}{dx} \cdot y + \mu(x) \cdot y' &= \frac{d[e^{\int p(x)\,dx}]}{dx} \cdot y + \mu(x) \cdot y' \\ &= e^{\int p(x)\,dx} \cdot p(x) \cdot y + \mu(x) \cdot y' \\ &= \mu(x) \cdot p(x) \cdot y + \mu(x) \cdot y', \end{aligned}$$

which is Eq. (3). Thus multiplying Eq. (2) by $\mu(x)$, we obtain

$$v' = (\mu(x)y)' = \mu(x)g(x), \tag{7}$$

which yields

$$\mu(x)y = \int \mu(x)g(x)\,dx. \tag{8}$$

From Eq. (8), we obtain

$$y = \frac{1}{\mu(x)} \int \mu(x)g(x)\,dx. \tag{9}$$

If we take $\int_{x_0}^{x} \mu(t)g(t)\,dt$ as a particular antiderivative of $\mu(x)g(x)$, then Eq. (9) becomes

$$y = \frac{1}{\mu(x)} \left(\int_{x_0}^{x} \mu(t)g(t)\,dt + C \right). \tag{10}$$

We have almost proved the following theorem.

THEOREM 20.2 Existence theorem for $y' + p(x)y = g(x)$

Let $p(x)$ and $g(x)$ be continuous in a neighborhood of x_0, and let y_0 be any real number. Then there exists a *unique* solution $y = f(x)$ of the differential equation $y' + p(x)y = g(x)$ such that $y(x_0) = y_0$, and this solution satisfies the differential equation throughout the neighborhood. Furthermore, the general solution of the differential equation is

$$y = \frac{1}{\mu(x)} \int \mu(x)g(x)\,dx,$$

where

$$\mu(x) = e^{\int p(x)\,dx}.$$

Proof. We have already seen in Eq. (9) that the general solution is as stated in the theorem and that these solutions are valid for all x in the neighborhood. It remains only to demonstrate the existence and uniqueness of a particular solution through the point (x_0, y_0). But putting $x = x_0$ and $y = y_0$ in the form of the general solution given in Eq. (10), we obtain the relation

$$y_0 = \frac{1}{\mu(x_0)} C, \tag{11}$$

and hence $C = y_0 \mu(x_0)$ yields the only solution through (x_0, y_0). •

EXAMPLE 1 We showed in Section 8.5 that the general solution of the differential equation $y' = ky$ is $y = Ae^{kx}$, where A is an arbitrary constant that controls $y(0)$. Obtain it again by using the theorem.

Solution Our equation is $y' - ky = 0$, so we have $p(x) = -k$ and $g(x) = 0$. Our integrating factor is thus

$$\mu(x) = e^{\int -k\,dx} = e^{-kx}.$$

The general solution is therefore

$$y = \frac{1}{e^{-kx}} \int e^{-kx} \cdot 0\,dx = \frac{1}{e^{-kx}} C = Ce^{kx}.$$

This coincides with our previous result. □

EXAMPLE 2 Find the particular solution of the differential equation

$$y' + 3xy = x,$$

which passes through the point (0, 4).

Solution Here $p(x) = 3x$ and our integrating factor is

$$\mu(x) = e^{\int 3x\,dx} = e^{3x^2/2}.$$

By Theorem 20.2, the general solution is

$$y = \frac{1}{e^{3x^2/2}} \int xe^{3x^2/2}\,dx = \frac{1}{e^{3x^2/2}}\left(\frac{1}{3}e^{3x^2/2} + C\right) = \frac{1}{3} + Ce^{-3x^2/2}.$$

Putting $y = 4$ and $x = 0$, we find that

$$4 = \frac{1}{3} + C \cdot 1,$$

so $C = \frac{11}{3}$, and the desired particular solution is

$$y = \frac{1}{3} + \frac{11}{3}e^{-3x^2/2}. \quad \square$$

The only problem in using Theorem 20.2 to solve $y' + p(x)y = g(x)$ is that the solution contains two integrals, namely

$$\mu(x) = e^{\int p(x)\,dx} \qquad \text{and} \qquad \int \mu(x)g(x)\,dx.$$

Sometimes it is impossible to evaluate one of these integrals in terms of elementary functions, even if $p(x)$ and $g(x)$ are themselves elementary functions. For example, if $p(x) = -x$ and $g(x) = 1$, then

$$\mu(x) = e^{-x^2/2} \qquad \text{and} \qquad \int \mu(x)g(x)\,dx = \int e^{-x^2/2}\,dx.$$

This last integral cannot be evaluated in terms of elementary functions. However, we have seen how this integral can be expressed as an infinite series. (This particular integral is so important in the theory of probability that $\int_0^x e^{-t^2/2}\,dt$ has actually been tabulated for many values of x.) There are many numerical methods for estimating an integral, so the presence of integrals in the general solution of the first-order linear differential equation is not really a serious problem in practical applications. Equations (10) and (11) show that the solution of $y' + p(x) \cdot y = g(x)$ through (x_0, y_0) is

$$\boxed{y = \frac{1}{\mu(x)}\left(\int_{x_0}^{x} \mu(t)g(t)\,dt + \mu(x_0) \cdot y_0\right).} \qquad \textbf{(12)}$$

EXAMPLE 3 Let $y = h(x)$ be the solution of the initial-value problem

$$y' - xy = 1, \qquad y(1) = 4.$$

Estimate $h(2)$ using a programmable calculator or a computer and Simpson's rule with $n = 40$.

Solution For the equation $y' - xy = 1$, we have $p(x) = -x$, so

$$\mu(x) = e^{\int -x\,dx} = e^{-x^2/2}.$$

Using Eq. (12), the solution $y = h(x)$ through $(x_0, y_0) = (1, 4)$ is

$$y = h(x) = \frac{1}{e^{-x^2/2}}\left(\int_1^x e^{-t^2/2} \cdot 1\,dt + e^{-1/2} \cdot 4\right).$$

Therefore

$$h(2) = \frac{1}{e^{-2}}\left(\int_1^2 e^{-t^2/2}\,dt + 4\sqrt{e}\right) = e^2 \int_1^2 e^{-t^2/2}\,dt + 4e^{5/2}.$$

Our calculator and Simpson's rule with $n = 40$ yield

$$h(2) = e^2 \int_1^2 e^{-t^2/2}\,dt + 4e^{5/2} \approx e^2(0.3406636182) + 4e^{5/2}$$

$$\approx 51.24715843. \quad \square$$

The exercises give some applications of these first-order linear differential equations to electric circuits, Newton's law of cooling, and mixing problems. We close with an example of a mixing problem to serve as a model for those exercises.

EXAMPLE 4 A tank initially contains 50 gal of brine in which 5 lb of salt is dissolved. Brine containing 1 lb of salt per gallon flows into the tank at a constant rate of 4 gal/min. The concentration of the brine in the tank is kept uniform throughout the tank by stirring. While the brine is entering the tank, brine from the tank is also being drawn off at a rate of 2 gal/min. Find the amount of salt in the tank after 25 min.

Solution We let $y = f(t)$ be the number of pounds of salt in the tank at time t. Clearly,

$$\frac{dy}{dt} = (\text{Rate salt enters the tank}) - (\text{Rate salt leaves the tank}). \quad \textbf{(13)}$$

Since 4 gal/min of brine containing 1 lb of salt per gallon is entering the tank, we see that

$$\text{Rate salt enters the tank} = 4 \text{ lb/min}.$$

Now the brine is drawn off at a rate of 2 gal/min and enters at a rate of 4 gal/min, so after t min the number of gallons of brine in the tank must be $50 + 2t$. The concentration of salt per gallon in the tank is therefore

$$\frac{y}{50 + 2t} \text{ lb/gal} \quad \text{at time } t.$$

Since 2 gal/min are being drawn off, we see that

$$\text{Rate salt leaves the tank} = 2 \cdot \frac{y}{50 + 2t} = \frac{y}{25 + t} \text{ lb/min}$$

at time t. The differential equation (13) thus becomes

$$\frac{dy}{dt} = 4 - \frac{y}{25 + t}, \quad \text{or} \quad \frac{dy}{dt} + \frac{1}{25 + t}y = 4.$$

We take as integrating factor

$$\mu(t) = e^{\int [1/(25+t)]\,dt} = e^{\ln(25+t)} = 25 + t.$$

From Eq. (9), we obtain the general solution

$$y = \frac{1}{25+t}\int (25+t)4\,dt$$

$$= \frac{1}{25+t}(100t + 2t^2 + C).$$

Since $y = 5$ lb when $t = 0$ min, we have

$$5 = \frac{1}{25}C, \qquad \text{so} \qquad C = 125.$$

Consequently,

$$y = \frac{1}{25+t}(2t^2 + 100t + 125),$$

so when $t = 25$ min we have

$$y = \frac{1}{50}[2(25)^2 + 100(25) + 125] = 25 + 50 + 2.5 = 77.5 \text{ lb}.$$

This gives us the answer to our problem. □

SUMMARY

1. The general solution of the differential equation

$$y' + p(x)y = g(x),$$

where $p(x)$ and $g(x)$ satisfy suitable conditions described in the text, can be found as follows.

Step 1 Compute the integrating factor

$$\mu(x) = e^{\int p(x)\,dx},$$

where $\int p(x)\,dx$ is any particular antiderivative of $p(x)$.

Step 2 Upon multiplication by $\mu(x)\,dx$, the equation $y' + p(x)y = g(x)$ becomes

$$d(\mu(x)\cdot y) = \mu(x)g(x)\,dx.$$

Step 3 The solution is then

$$\mu(x)\cdot y = \int \mu(x)g(x)\,dx,$$

or

$$y = \frac{1}{\mu(x)}\int \mu(x)g(x)\,dx.$$

2. The solution of $y' + p(x) \cdot y = g(x)$ through (x_0, y_0) is given by

$$y = h(x) = \frac{1}{\mu(x)} \left(\int_{x_0}^{x} \mu(t)g(t)\, dt + \mu(x_0) \cdot y_0 \right),$$

where $\mu(x) = e^{\int p(x)\, dx}$. If $\mu(x)$ can be computed, then a calculator or computer and Simpson's rule can be used to estimate $h(x_1)$.

EXERCISES

In Exercises 1 through 10, find the general solution of the given differential equation.

1. $y' - xy = 0$

2. $y' - 3y = 2$

3. $y' + y = 3e^x$

4. $y' + 2y = x + e^{-3x}$

5. $y' + 2y = xe^{-2x} + 3$

6. $xy' + y = 2x \sin x$; $x > 0$ [*Hint:* Reduce to the form of Eq. (2) by dividing by x.]

7. $y' + (\cot x)y = 3x + 1, 0 < x < \pi$

8. $y' + (\sin x)y = 3 \sin x$

9. $y' + 2xy = x$

10. $y' - 2xy = x^3$

In Exercises 11 through 18, find the solution of the given initial-value problem.

11. $y' - 3y = x + 2, y(0) = -1$

12. $xy' - y = x^3, y(1) = 5$

13. $(1 + x^2)y' + y = 3, y(0) = 2$

14. $y' - (\cos 2x)y = \cos 2x, y(\pi/2) = 3$

15. $y' - xy = x, y(0) = -3$

16. $y' + 2xy = e^{-x^2}, y(0) = 4$

17. $(1 + \sin x)y' + (\cos x)y = \sin^2 x, y(\pi/2) = 3$

18. $x^2y' + y = 1, y(1) = 0$

19. Find the general solution of the differential equation $y' - xy = x^2$ by using Theorem 20.2 and expressing the integral in series form.

20. Find the general solution of the differential equation

$$y' + 2x(\cot x^2)y = 3, 0 < x < \sqrt{\pi}$$

by using Theorem 20.2 and expressing the integral in series form.

21. Find the plane curve through (2, 1) whose slope at (x, y) is $2 + (y/x)$.

22. Find the plane curve through $(1, -1)$ whose slope at (x, y) is $3 - (y/x)$.

23. A tank initially contains 60 gal of brine in which 15 lb of salt is dissolved. Brine containing 0.5 lb of salt per gallon flows into the tank at a constant rate of 6 gal/min. The concentration of brine in the tank is kept uniform by stirring, and the brine is drawn off at a rate of 3 gal/min. Find the amount of salt in the tank after 10 min.

24. A tank initially contains 40 gal of brine containing 0.5 lb of salt per gallon. Brine containing 1 lb of salt per gallon flows into the tank at a rate of 2 gal/min, while brine is drawn off at a rate of 6 gal/min. If the concentration of the solution in the tank is kept uniform by stirring, find
a) the amount of salt in the tank after 5 min,
b) the amount of salt in the tank after 10 min.

25. If $i(t)$ is the current at time t in an electrical circuit with constant resistance R, constant inductance L, and variable electromotive force $E(t)$, then it can be shown that

$$L\frac{di}{dt} + Ri = E(t).$$

a) Let the current at time $t = 0$ be i_0, and let $E(t)$ be a continuous function of the time t. Find an expression for i at time t.
b) Show that if E is constant, then for large values of t, we have $i \approx E/R$ (so that Ohm's law is approximately true after a long time).
c) Describe the current after a long period of time if E diminishes exponentially, that is, if $E(t) = E_0 e^{-kt}$, where

$$E_0 = E(0) \quad \text{and} \quad k > 0.$$

d) Describe the behavior of the current as $t \to \infty$ if E is abruptly cut off at time t_0, so that $E(t) = 0$ for $t > t_0$.

26. According to Newton's law of cooling, the rate at which a body in a medium changes temperature is proportional to the difference between its temperature and the temperature of the medium.
a) Assuming that the temperature of the medium remains constant at a degrees and that the temperature T_0 of the body at time $t = 0$ is higher than that of the medium, express the temperature T of the body as a function of the time t for $t > 0$. (Let the constant of proportionality in Newton's law be $-k$.)
b) For $k > 0$, what is the approximate temperature of the body as $t \to \infty$.

27. If n is a constant different from 0 or 1, then the equation

$$y' + p(x)y = g(x)y^n$$

is known as *Bernoulli's equation.* (For both $n = 0$ and $n = 1$, the equation is linear and can be solved as described in this section.)

a) Show that the substitution $v = y^{1-n}$ enables us to reduce the solution of Bernoulli's equation to a differential equation for v that is linear.

b) Use the result of part (a) and Theorem 20.2 to solve the differential equation

$$y' - 2xy = 5xy^3.$$

28. Use Exercise 27 to solve the initial-value problem

$$y' + \frac{2}{x}y = \frac{1}{y}, \quad y(1) = -2.$$

In Exercises 29 through 34, use a programmable calculator or computer and Simpson's rule to estimate $h(x_1)$ if $y = h(x)$ is the solution of the given initial-value problem.

29. $y' + 2xy = \sin x,\ y(0) = 1,\ x_1 = 0.5$

30. $y' - x^2y = 1,\ y(1) = 4,\ x_1 = 2$

31. $y' - (\sin x)y = x^2,\ y(0) = 3,\ x_1 = -\dfrac{\pi}{4}$

32. $y' + \sqrt{x}\,y = x,\ y(4) = 2,\ x_1 = 3$

33. $y' - \dfrac{1}{2\sqrt{x}}y = x^3,\ y(4) = 0,\ x_1 = 4.5$

34. $y' + \dfrac{1}{1+x^2}y = x^2,\ y(1) = -1,\ x_1 = 2$

20.5 HOMOGENEOUS LINEAR EQUATIONS WITH CONSTANT COEFFICIENTS

We now turn to differential equations of order greater than 1. The **order** of a differential equation is the order of the highest-order derivative it contains. In this section and the next, we study linear differential equations with constant coefficients. Such equations occur frequently in applications.

THE EXISTENCE THEOREM

The *general linear differential equation of order n* is

$$p_0(x)y^{(n)} + p_1(x)y^{(n-1)} + \cdots + p_{n-1}(x)y' + p_n(x)y = g(x).$$

The equation is called *linear* since each term contains at most a single factor that is y or a derivative of y to only the first power. Theorem 20.3 below describes the existence of solutions in the neighborhood of a point x_0 throughout which $p_0(x) \neq 0$. Where $p_0(x) \neq 0$, we may divide the equation by $p_0(x)$ and obtain 1 as coefficient of $y^{(n)}$. Thus it is no loss of generality at such a point x_0 to suppose that the equation has the form

$$y^{(n)} + p_1(x)y^{(n-1)} + \cdots + p_{n-1}(x)y' + p_n(x)y = g(x). \quad \textbf{(1)}$$

We state the main existence theorem without proof.

THEOREM 20.3 Existence theorem for the general linear equation

If $p_1(x), \ldots, p_n(x), g(x)$ are continuous in a neighborhood of x_0 and if $a_0, a_1, \ldots, a_{n-1}$ are any constants, then there is a *unique* solution $y = f(x)$ of Eq. (1) that is valid throughout the neighborhood and has the property

$$y(x_0) = a_0,\ y'(x_0) = a_1, \ldots, y^{(n-1)}(x_0) = a_{n-1}.$$

The conditions

$$y(x_0) = a_0,\ y'(x_0) = a_1, \ldots,\ y^{(n-1)}(x_0) = a_{n-1}$$

for the solution whose existence is asserted in Theorem 20.3 are called **initial conditions.** They all involve the values of y and its derivatives at the same *initial point* x_0. Equation (1) and these initial conditions constitute an **initial-value problem.** Theorem 20.3 tells us that this initial-value problem has a unique solution, provided the coefficients $p_i(x)$ and $g(x)$ in Eq. (1) are continuous.

In the case to be considered in the remainder of this section, the coefficient functions $p_1(x), \ldots, p_n(x)$ are constant functions and $g(x) = 0$. Then Eq. (1) takes the form

$$y^{(n)} + b_1 y^{(n-1)} + \cdots + b_{n-1} y' + b_n y = 0. \tag{2}$$

Equation (2) is a linear *homogeneous* differential equation with *constant coefficients.* In this section, we will see how the general solution of Eq. (2) can be obtained in terms of elementary functions by very simple algebraic means.

POLYNOMIALS IN THE OPERATOR D

We may write Eq. (2) in the form

$$D^n y + b_1 D^{n-1} y + \cdots + b_{n-1} D y + b_n y = 0, \tag{3}$$

where of course $Dy = y'$, $D^2 y = y''$, and so on. Proceeding purely formally, it is natural to factor out y in Eq. (3) and write the equation in the form

$$(D^n + b_1 D^{n-1} + \cdots + b_{n-1} D + b_n) y = 0, \tag{4}$$

or, more briefly,

$$P(D)y = 0, \tag{5}$$

where $P(D)$ is the polynomial $D^n + b_1 D^{n-1} + \cdots + b_{n-1} D + b_n$ in D.

Suppose that the polynomial $P(D)$ factors (in the sense of polynomial factorization) so that $P(D) = Q_1(D)Q_2(D)$ for polynomials $Q_1(D)$ and $Q_2(D)$ in D. It is not difficult to show that

$$P(D)y = Q_1(D)(Q_2(D)y), \tag{6}$$

where $y = f(x)$ has derivatives of all orders less than or equal to n. A careful proof of Eq. (6) can be given using mathematical induction; you will probably find it sufficiently convincing if we compute a special case to illustrate what we mean.

EXAMPLE 1 Illustrate that $D^2 - 3D + 2 = (D - 1)(D - 2)$, that is, that these polynomial operators have the same effect on any twice-differentiable function $y = f(x)$.

Solution Computing, we find that for a twice-differentiable function $y = f(x)$, we have

$$\begin{aligned}(D - 1)((D - 2)y) &= (D - 1)(Dy - 2y)\\ &= D(Dy - 2y) - 1(Dy - 2y)\\ &= D^2y - D(2y) - Dy + 2y\\ &= D^2y - 2Dy - Dy + 2y\\ &= D^2y - 3Dy + 2y\\ &= (D^2 - 3D + 2)y. \quad \square\end{aligned}$$

The result in Eq. (6) can easily be extended to more than two factors. Since polynomial multiplication is commutative (that is, does not depend on the order of multiplication), we see at once from Eq. (6) that

$$Q_1(D)(Q_2(D)y) = Q_2(D)(Q_1(D)y) \tag{7}$$

for polynomials $Q_1(D)$ and $Q_2(D)$. Equation (7) can also be extended to any number of factors in any order.

CASE 1: $P(D)$ FACTORS INTO DISTINCT LINEAR FACTORS

We first consider a differential equation of the form

$$P(D)y = 0, \tag{8}$$

where $P(D)$ is a product of *distinct* linear factors. In this case, we have

$$P(D) = (D - r_1)(D - r_2) \cdots (D - r_n), \tag{9}$$

where $r_i \neq r_j$ for $i \neq j$. Such an equation can be solved by repeated application of Theorem 20.2. The technique is best illustrated by an example.

EXAMPLE 2 Solve the homogeneous equation $y'' - 3y' + 2 = 0$.

Solution Our equation can be written in the form

$$(D^2 - 3D + 2)y = (D - 1)(D - 2)y = 0. \tag{10}$$

If we let $u = (D - 2)y$, then we must have

$$(D - 1)u = 0. \tag{11}$$

Equation (11) is a linear first-order equation for u, and by Section 20.4 we have

$$u = \frac{1}{e^{-x}} \int e^{-x} \cdot 0\, dx = C_1 e^x.$$

We then solve the equation

$$(D - 2)y = u = C_1 e^x$$

and obtain

$$\begin{aligned}y &= \frac{1}{e^{-2x}} \int e^{-2x}(C_1 e^x)\, dx\\ &= \frac{1}{e^{-2x}}(-C_1 e^{-x} + C_2)\\ &= -C_1 e^x + C_2 e^{2x}.\end{aligned}$$

If we write our linear factors in the reverse order so that Eq. (10) becomes

$$(D - 2)(D - 1)y = 0$$

and make the substitution $(D - 1)y = v$, we obtain first the e^{2x} part of the solution, namely $v = C_1e^{2x}$. The equation $(D - 1)y = v$ then yields the same solution $y = C_1e^{2x} + C_2e^x$. □

It is clear that if $D - r$ is a factor of $P(D)$, then some solutions of $P(D)y = 0$ are given by $y = Ce^{rx}$, since

$$(D - r)(Ce^{rx}) = rCe^{rx} - rCe^{rx} = 0.$$

The argument in Example 2 can obviously be extended to give the following theorem.

THEOREM 20.4 **Characteristic equation with distinct roots**

Let $P(D) = (D - r_1)(D - r_2) \cdots (D - r_n)$, where $r_i \neq r_j$ for $i \neq j$. Then the general solution of the differential equation $P(D)y = 0$ is

$$y = C_1e^{r_1x} + C_2e^{r_2x} + \cdots + C_ne^{r_nx}.$$

EXAMPLE 3 Solve the differential equation

$$3y''' - 2y'' - y' = 0.$$

Solution We write our equation in the form

$$(3D^3 - 2D^2 - D)y = D(3D + 1)(D - 1)y = 0.$$

While Theorem 20.4 treats the case where the coefficient of D^n is 1, the solutions of our differential equation remain the same if we divide through by 3. We see that the numbers r_i in Theorem 20.4 can be characterized as solutions of the polynomial equation

$$P(r) = 0, \qquad \textbf{(12)}$$

whether the coefficient of D^n is 1 or not. This polynomial equation (12) is the **auxiliary** or **characteristic equation** of the differential equation $P(D)y = 0$. Our characteristic equation in this example is

$$3r^3 - 2r^2 - r = r(3r + 1)(r - 1) = 0$$

and has as solutions $r_1 = 0, r_2 = -\frac{1}{3}$, and $r_3 = 1$. Thus our general solution is

$$y = C_1e^{0x} + C_2e^{-x/3} + C_3e^x = C_1 + C_2e^{-x/3} + C_3e^x. \quad \square$$

EXAMPLE 4 Solve the initial-value problem

$$y''' - y'' - 6y' = 0, \qquad y(0) = 1, \quad y'(0) = -2, \quad y''(0) = 0.$$

Solution The characteristic equation of the differential equation is

$$r^3 - r^2 - 6r = 0, \qquad \text{or} \qquad r(r - 3)(r + 2) = 0.$$

This characteristic equation has solutions $r = -2, 0, 3$. The general solution of the differential equation is therefore

$$y = C_1 e^{-2x} + C_2 + C_3 e^{3x}.$$

To make use of the initial conditions involving y' and y'', we differentiate this general solution and obtain

$$y' = -2C_1 e^{-2x} + 3C_3 e^{3x} \qquad \text{and} \qquad y'' = 4C_1 e^{-2x} + 9C_3 e^{3x}.$$

The initial conditions then give these equations in C_1, C_2, and C_3:

$$\begin{aligned} C_1 + C_2 + C_3 &= 1, & y(0) &= 1 \\ -2C_1 \qquad + 3C_3 &= -2, & y'(0) &= -2 \\ 4C_1 \qquad + 9C_3 &= 0. & y''(0) &= 0 \end{aligned}$$

Solving the last two equations for C_1 and C_3, we have

$$\begin{aligned} -4C_1 + 6C_3 &= -4 \\ 4C_1 + 9C_3 &= 0 \\ \hline 15C_3 &= -4, \end{aligned} \qquad C_3 = \frac{-4}{15}.$$

Then

$$C_1 = \frac{1}{2}(3C_3 + 2) = \frac{1}{2}\left(-\frac{12}{15} + \frac{30}{15}\right) = \frac{9}{15}.$$

The first equation, corresponding to $y(0) = 1$, then gives

$$\frac{9}{15} + C_2 - \frac{4}{15} = 1, \qquad C_2 = 1 - \frac{5}{15} = \frac{2}{3}.$$

Our desired particular solution is therefore

$$y = \frac{9}{15} e^{-2x} + \frac{2}{3} - \frac{4}{15} e^{3x}. \quad \square$$

CASE 2: REPEATED LINEAR FACTORS

We now take up the case where the polynomial $P(D)$ factors into a product of the form

$$P(D) = (D - r_1)^{n_1}(D - r_2)^{n_2} \cdots (D - r_m)^{n_m}, \tag{13}$$

where, of course, $n_1 + n_2 + \cdots + n_m = n$.

The general solution of $P(D)y = 0$ can again be obtained by repeated application of Theorem 20.2. To see what form the solution now takes, we consider a simple case where $P(D) = (D - r)^2$. Let $u = (D - r)y$, so that the equation $(D - r)^2 y = 0$ becomes $(D - r)u = 0$. We then obtain $u = C_1 e^{rx}$. Therefore

$$(D - r)y = C_1 e^{rx}$$

and

$$\begin{aligned} y &= \frac{1}{e^{-rx}} \int e^{-rx} C_1 e^{rx}\, dx = \frac{1}{e^{-rx}} \int C_1\, dx = \frac{1}{e^{-rx}}(C_1 x + C_2) \\ &= e^{rx}(C_1 x + C_2). \end{aligned}$$

It is easy to check that the general solution of $(D - r)^3 y = 0$ is

$$y = e^{rx}(C_1x^2 + C_2x + C_3).$$

The factors in Eq. (13) can be written in any order. For the given order, the first n_1 iterations to solve $P(D)y = 0$ show that

$$e^{r_1x}(C_1x^{n_1-1} + C_2x^{n_1-2} + \cdots + C_{n_1})$$

forms a portion of the general solution. We consider that we have proved the following theorem.

THEOREM 20.5 Characteristic equation with multiple roots

Let $P(D) = (D - r_1)^{n_1} \cdots (D - r_m)^{n_m}$, where $n_1 + \cdots + n_m = n$. Then the general solution of $P(D)y = 0$ is

$$\begin{aligned} y = e^{r_1x}&(C_1x^{n_1-1} + C_2x^{n_1-2} + \cdots + C_{n_1}) + \cdots \\ &+ e^{r_mx}(C_{n-n_m+1}\,x^{n_m-1} + C_{n-n_m+2}\,x^{n_m-2} + \cdots + C_n). \end{aligned}$$

EXAMPLE 5 Solve $y''' + 2y'' + y' = 0$.

Solution The characteristic equation of $y''' + 2y'' + y' = 0$ is

$$r^3 + 2r^2 + r = r(r + 1)^2 = 0,$$

and we see that the general solution of the equation is

$$y = C_1e^{0x} + e^{-x}(C_2x + C_3) = C_1 + e^{-x}(C_2x + C_3). \quad \square$$

CASE 3: $P(D)$ CONTAINS QUADRATIC FACTORS

It is a theorem of algebra that every polynomial with real coefficients can be factored into a product of linear and quadratic factors with real coefficients. Namely, it can be shown that a polynomial can be factored into a product of linear factors if one allows complex coefficients. (This is the *fundamental theorem of algebra.*) If $D - (a + bi)$ is a factor, then $D - (a - bi)$ can be shown to be a factor also, and the product $D^2 - 2aD + (a^2 + b^2)$ is therefore a quadratic factor of $P(D)$ with real coefficients. We are interested in discovering what such a quadratic factor contributes to the general solution of $P(D)y = 0$.

Proceeding purely formally with complex numbers, we might expect the general solution of

$$[D^2 - 2aD + (a^2 + b^2)]y = [D - (a + bi)][D - (a - bi)]y = 0 \qquad \textbf{(14)}$$

to be

$$\begin{aligned} y &= C_1e^{(a+bi)x} + C_2e^{(a-bi)x} \\ &= C_1e^{ax}e^{bix} + C_2e^{ax}e^{-bix} \\ &= e^{ax}(C_1e^{i(bx)} + C_2e^{-i(bx)}). \end{aligned} \qquad \textbf{(15)}$$

In Exercise 43 of Section 11.3, we asked you to verify formally Euler's formula

$$e^{ix} = \cos x + i \sin x.$$

Using this formula and proceeding formally from Eq. (15), we obtain

$$\begin{aligned} y &= e^{ax}[C_1(\cos bx + i \sin bx) + C_2(\cos(-bx) + i \sin(-bx))] \\ &= e^{ax}[(C_1 + C_2)\cos bx + (C_i i - C_2 i)\sin bx]. \end{aligned} \qquad \textbf{(16)}$$

Replacing the arbitrary constant $C_1 + C_2$ by C_1 and replacing $C_1 i - C_2 i$ by C_2, we obtain from Eq. (16)

$$y = e^{ax}(C_1 \cos bx + C_2 \sin bx). \qquad \textbf{(17)}$$

We conjecture that Eq. (17) is the general solution of Eq. (14). The preceding use of complex numbers can be justified, and our conjecture is indeed correct.

We could compute directly that

$$[D^2 - 2aD + (a^2 + b^2)][e^{ax}(C_1 \cos bx + C_2 \sin bx)] = 0$$

(see Exercise 1). From our work with repeated linear factors, we would guess that a repeated quadratic factor $[D_2 - 2aD + (a^2 + b^2)]^2$ would give rise to a repetition of Eq. (17) with an additional factor x in the general solution. This can also be verified.

We can now solve any homogeneous linear differential $P(D)y = 0$ with constant coefficients, provided we are able to factor the polynomial $P(D)$ into quadratic and linear factors.

EXAMPLE 6 Solve the differential equation

$$D(D - 1)^2(D^2 + 2D + 4)y = 0.$$

Solution The characteristic equation is $r(r - 1)^2(r^2 + 2r + 4) = 0$, which has a root $r_1 = 0$, a double root $r_2 = 1$, and complex roots $-1 \pm i\sqrt{3}$ that are obtained by solving $r^2 + 2r + 4 = 0$ by the quadratic formula. The general solution of our equation is therefore

$$y = C_1 + e^x(C_2 x + C_3) + e^{-x}(C_4 \cos\sqrt{3}x + C_5 \sin\sqrt{3}x). \quad \square$$

EXAMPLE 7 Solve the differential equation $(D^6 + 4D^4 + 4D^2)y = 0$.

Solution We turn to the characteristic equation

$$r^6 + 4r^4 + 4r^2 = r^2(r^2 + 2)^2 = 0.$$

Here $r_1 = 0$, $r_2 = \sqrt{2}i$, and $r_3 = -\sqrt{2}i$ are all double roots. The general solution of our differential equation is

$$y = C_1 x + C_2 + (\cos\sqrt{2}x)(C_3 x + C_4) + (\sin\sqrt{2}x)(C_5 x + C_6). \quad \square$$

EXAMPLE 8 Solve the initial-value problem $y'' + y = 0$, $y(\pi/2) = 0$, $y'(\pi/2) = 1$.

Solution The characteristic equation is

$$r^2 + 1 = 0,$$

which has complex roots $r = 0 \pm i$. The general solution of the differential equation is thus

$$y = e^{0x}(C_1 \cos x + C_2 \sin x), \qquad \text{or} \qquad y = C_1 \cos x + C_2 \sin x.$$

To make use of the initial condition $y'(\pi/2) = 1$, we differentiate and obtain

$$y' = -C_1 \sin x + C_2 \cos x.$$

The initial conditions then give rise to these equations:

$$C_1 \cdot 0 + C_2 \cdot 1 = 0, \qquad y(\pi/2) = 0$$
$$-C_1 \cdot 1 + C_2 \cdot 0 = 1. \qquad y'(\pi/2) = 1$$

Thus $C_2 = 0$ and $C_1 = -1$, so our desired particular solution is

$$y = -\cos x. \quad \square$$

SUMMARY

1. The differential equation

$$y^{(n)} + b_1 y^{(n-1)} + b_2 y^{(n-2)} + \cdots + b_{n-1}y' + b_n y = 0$$

is a linear homogeneous differential equation with constant coefficients b_i. We let

$$P(D) = D^n + b_1 D^{n-1} + b_2 D^{n-2} + \cdots + b_{n-1}D + b_n$$

be the polynomial in the differential operator D (standing for differentiation). The equation then is symbolically written

$$P(D)y = 0.$$

2. If $P(D) = (D - r_1)(D - r_2) \cdots (D - r_n)$, where $r_i \neq r_j$ for $i \neq j$, then the general solution of $P(D)y = 0$ is

$$y = C_1 e^{r_1 x} + C_2 e^{r_2 x} + \cdots + C_n e^{r_n x}.$$

3. Let $P(D) = (D - r_1)^{n_1} \cdots (D - r_m)^{n_m}$, where $n_1 + \cdots + n_m = n$. Then the general solution of $P(D)y = 0$ is

$$y = e^{r_1 x}(C_1 x^{n_1 - 1} + C_2 x^{n_1 - 2} + \cdots + C_{n_1}) + \cdots + e^{r_m x}(C_{n - n_m + 1} x^{n_m - 1} + C_{n - n_m + 2} x^{n_m - 2} + \cdots + C_n).$$

4. An irreducible quadratic factor in $P(D)$ having as complex roots $a + bi$ and $a - bi$ gives rise to a summand of the general solution of $P(D)y = 0$ of the form

$$e^{ax}(C_1 \cos bx + C_2 \sin bx).$$

EXERCISES

1. Convince yourself by direct computation that

$$[D^2 - 2aD + (a^2 + b^2)][e^{ax}(C_1 \cos bx + C_2 \sin bx)] = 0.$$

In Exercises 2 through 20, find the general solution of the given differential equation.

2. $y' + 3y = 0$
3. $2y' + 4y = 0$
4. $y'' + 4y' + 3y = 0$
5. $4y'' + 12y' + 5y = 0$
6. $y'' - 6y' + 9y = 0$
7. $4y''' + 4y'' + y' = 0$
8. $y''' - 3y'' = 0$
9. $y'' + 3y = 0$
10. $y'' + 2y' + 6y = 0$
11. $y''' - y = 0$
12. $(D^2 + 4)y = 0$
13. $(D^3 + 9D)y = 0$
14. $(D^4 - 16)y = 0$
15. $(D^4 + 2D^2 + 1)y = 0$

16. $D(D-3)^2(D^2+1)y=0$

17. $D^2(D+2)(D^2+2)y=0$

18. $D^3(D^2+1)^2y=0$

19. $(D+1)^2(D^2+D+2)y=0$

20. $D^2(D+5)(D^2+3D+5)^2y=0$

In Exercises 21 through 30, find the solution of the initial-value problem.

21. $y''-5y'+6y=0,\ y(0)=1,\ y'(0)=-1$

22. $y''-y=0,\ y(0)=0,\ y'(0)=1$

23. $y'''-y''=0,\ y(0)=1,\ y'(0)=0,\ y''(0)=3$

24. $y'''-4y'=0,\ y(0)=-1,\ y'(0)=2,\ y''(0)=0$

25. $y^{(4)}-9y''=0,\ y(0)=0,\ y'(0)=2,\ y''(0)=0,\ y'''(0)=-1$

26. $y''+y=0,\ y(\pi/2)=3,\ y'(\pi/2)=-2$

27. $y''+4y=0,\ y(\pi)=3,\ y'(\pi)=-4$

28. $y'''+y'=0,\ y(0)=-1,\ y'(0)=2,\ y''(0)=-3$

29. $y'''-8y=0,\ y(0)=2,\ y'(0)=0,\ y''(0)=4$

30. $y'''-2y''+2y'=0,\ y(0)=1,\ y'(0)=-1,\ y''(0)=2$

20.6 THE NONHOMOGENEOUS CASE; APPLICATIONS

In the first two parts of this section, we consider the problem of finding the solutions of a linear differential equation with constant coefficients of the form

$$y^{(n)}+b_1y^{(n-1)}+\cdots+b_{n-1}y'+b_ny=g(x)$$

for a continuous function $g(x)$. The existence theorem was stated in the preceding section. If $g(x)=0$, the equation is called **homogeneous.**

THE FORM OF THE SOLUTION

Writing the equation in the form

$$P(D)y=g(x), \tag{1}$$

suppose that $C_1y_1(x)+\cdots+C_ny_n(x)$ is the general solution of the *homogeneous* equation obtained from Eq. (1) by replacing $g(x)$ by zero. Suppose also that $f(x)$ is any particular solution of $P(D)y=g(x)$. Then

$$\boxed{y=C_1y_1(x)+\cdots+C_ny_n(x)+f(x)} \tag{2}$$

gives solutions of Eq. (1), because

$$\begin{aligned}P(D)y&=P(D)(C_1y_1(x)+\cdots+C_ny_n(x))+P(D)f(x)\\&=0+g(x)=g(x).\end{aligned}$$

Also, if $h(x)$ is any solution of Eq. (1), then

$$P(D)(h(x)-f(x))=P(D)h(x)-P(D)f(x)=g(x)-g(x)=0,$$

so $h(x)-f(x)$ is a solution of the homogeneous equation obtained by setting $g(x)=0$. Thus

$$h(x)=(\text{A solution of } P(D)y=0)+f(x),$$

so any solution $h(x)$ is of the form of Eq. (2). Thus Eq. (2) is the general solution of Eq. (1). We have proved the following theorem.

THEOREM 20.6 Linearity property

Let $P(D)y = g(x)$ be a linear differential equation with constant coefficients. Let $y = H(x)$ be the general solution of the homogeneous equation $P(D)y = 0$, and let $y = f(x)$ be any solution of $P(D)y = g(x)$. Then the general solution of $P(D)y = g(x)$ is $y = H(x) + f(x)$.

In Section 20.5, we showed how to find solutions of the homogeneous equation. We therefore focus our attention on finding a particular solution $y = f(x)$ of Eq. (1). Two methods are presented.

SUCCESSIVE REDUCTION TO FIRST-ORDER EQUATIONS

Let $P(D)$ be a polynomial in D that factors into (not necessarily distinct) linear factors, so that

$$P(D) = (D - r_1)(D - r_2) \cdots (D - r_n).$$

Consider an equation of the form

$$P(D)y = (D - r_1)(D - r_2) \cdots (D - r_n)y = g(x). \tag{3}$$

We can solve Eq. (3) by the method illustrated in Example 2 of Section 20.5. This technique consists of setting

$$(D - r_2) \cdots (D - r_n)y = u, \tag{4}$$

so that Eq. (3) becomes

$$(D - r_1)u = g(x). \tag{5}$$

Equation (5) is linear in u and can be solved using the method of Section 20.4. We have thus reduced our problem to the solution of Eq. (4), which is a differential equation of order $n - 1$. Repetition of this process n times enables us to find y.

Actually, the procedure just outlined yields the general solution of Eq. (3). Since it is easy for us to write down "most" of this general solution, namely the part in Eq. (2) involving the arbitrary constants, it is more efficient when using this technique to take all arbitrary constants equal to zero and obtain just a particular solution of Eq. (3).

EXAMPLE 1 Solve the differential equation

$$y'' - 3y' + 2y = x + 1.$$

Solution Our equation can be written in the form

$$(D - 1)(D - 2)y = x + 1.$$

We let $(D - 2)y = u$ and find a particular solution of $(D - 1)u = x + 1$. Integrating by parts or using tables, we obtain as particular solution, taking zero for all constants of integration,

$$u = \frac{1}{e^{-x}} \int e^{-x}(x + 1)\,dx = \frac{1}{e^{-x}}(-xe^{-x} - 2e^{-x})$$
$$= -x - 2.$$

Solving $(D - 2)y = u = -x - 2$ by the same method, we obtain as a particular solution of our original equation

$$y = \frac{1}{e^{-2x}} \int -e^{-2x}(x + 2)\, dx = \frac{1}{e^{-2x}} \left(\frac{1}{2} x e^{-2x} + \frac{5}{4} e^{-2x} \right)$$

$$= \frac{1}{2} x + \frac{5}{4}.$$

Our general solution is therefore

$$y = C_1 e^x + C_2 e^{2x} + \frac{1}{2} x + \frac{5}{4}. \quad \square$$

THE METHOD OF UNDETERMINED COEFFICIENTS

Sometimes a particular solution of $P(D)y = g(x)$ can be determined by inspection.

EXAMPLE 2 Solve the equation

$$y'' + 4y' + 4y = 12.$$

Solution Obviously a particular solution of $y'' + 4y' + 4y = 12$ is $y = 3$; just check that $y = 3$ satisfies the equation. Since the characteristic equation is $r^2 + 4r + 4 = (r + 2)^2 = 0$, we see that the general solution is

$$y = e^{-2x}(C_1 x + C_2) + 3. \quad \square$$

EXAMPLE 3 Find by inspection a particular solution of the equation $y'' - 3y' + 2y = x + 1$, which we solved in Example 1.

Solution We try to find $y = f(x)$ such that we obtain $x + 1$ upon computing $y'' - 3y' + 2y$. Clearly $y = x/2$ will give the desired amount of x; namely for $y = x/2$, we have

$$y'' - 3y' + 2y = D^2\left(\frac{x}{2}\right) - 3D\left(\frac{x}{2}\right) + 2\left(\frac{x}{2}\right) = 0 - \frac{3}{2} + x.$$

To get the desired constant 1 rather than $-\frac{3}{2}$, we need to add to our $x/2$ a constant that when *doubled* yields $1 + \frac{3}{2} = \frac{5}{2}$, since our equation contains $2y$. Thus the needed constant is $\frac{5}{4}$, and a particular solution is $y = (x/2) + \frac{5}{4}$, as obtained in Example 1. $\square$

A more systematic attack suggested by Example 3 leads to the method of undetermined coefficients.

EXAMPLE 4 Solve the differential equation

$$y'' + 2y' - 3y = 2x - 17.$$

Solution From Example 3, we would guess that $y = ax + b$ should be a solution for some a and b. For $y = ax + b$, we have

$$y = ax + b, \qquad y' = a, \qquad y'' = 0.$$

Substituting these expressions into the given differential equation, we see that we must have

$$2a - 3ax - 3b = 2x - 17.$$

Equating the coefficients of x and the constant terms, we obtain

$$-3a = 2 \quad \text{and} \quad 2a - 3b = -17.$$

Thus $a = -\frac{2}{3}$ and $b = \frac{47}{9}$. A particular solution is therefore

$$y = -\frac{2}{3}x + \frac{47}{9}.$$

The general solution is then easily found to be

$$y = C_1e^{-3x} + C_2e^x - \frac{2}{3}x + \frac{47}{9}. \quad \square$$

The *method of undetermined coefficients* illustrated in Example 4 works well when the function $g(x)$ on the right-hand side of Eq. (1) and derivatives of all orders of $g(x)$ involve sums of only a *finite number* of different functions, except for constant factors. For example, a polynomial of degree n and its derivatives of all orders involve only sums of constant multiples of the finite number of functions $1, x, x^2, \ldots, x^n$. Also, derivatives of $\sin ax$ or $\cos ax$ are just constant multiples of these same trigonometric functions. Derivatives of the exponential function e^{ax} involve only constant multiples of e^{ax}. If $g(x)$ contains only sums and products of these functions, then the method of undetermined coefficients is often useful. Suppose now that $g(x)$ is of this type.

1. Suppose none of the summands in $g(x)$ or any of their derivatives is a solution of the homogeneous equation where $g(x)$ is replaced by zero. Then one takes as a trial particular solution a sum of all the summands in $g(x)$ and derivatives of all orders of these summands, with coefficients a, b, c, and so on, to be determined.
2. Suppose some summand of $g(x)$ or a derivative of some order of a summand is a solution of the homogeneous equation corresponding to an s-fold root of the characteristic equation. Then one should start with a trial particular solution in which that summand of $g(x)$ is multiplied by x^s before taking derivatives of it.

This seemingly complicated procedure is quite simple (although frequently tedious) in practice and is best understood by further examples.

EXAMPLE 5 Solve the differential equation

$$y'' - 3y' + 2y = 2\cos 4x.$$

Solution Now $\cos 4x$ is not part of the solution of the homogeneous equation $y'' - 3y' + 2y = 0$, so we try as particular solution

$$y = a\cos 4x + b\sin 4x,$$

which is a sum with coefficients a and b of $\cos 4x$ and all different types of derivatives of cos $4x$. We then obtain

$$y = a\cos 4x + b\sin 4x,$$

$$y' = 4b\cos 4x - 4a\sin 4x,$$

$$y'' = -16a\cos 4x - 16b\sin 4x.$$

Multiplying the first equation by 2, the second by -3, the last by 1, and adding, we see from our original differential equation that we must have

$$\begin{aligned} 2\cos 4x &= (2a - 12b - 16a)\cos 4x + (2b + 12a - 16b)\sin 4x \\ &= (-14a - 12b)\cos 4x + (12a - 14b)\sin 4x. \end{aligned}$$

We therefore must have

$$-14a - 12b = 2,$$

$$12a - 14b = 0,$$

which yields upon solution $a = -\frac{7}{85}$ and $b = -\frac{6}{85}$. Our general solution is therefore

$$y = C_1e^x + C_2e^{2x} - \frac{7}{85}\cos 4x - \frac{6}{85}\sin 4x. \quad \square$$

EXAMPLE 6 Solve the differential equation

$$y'' - 2y' + y = e^x.$$

Solution In this case, e^x is part of the solution of the homogeneous equation corresponding to the double root $r = 1$ of the characteristic equation $r^2 - 2r + 1 = (r - 1)^2 = 0$. We therefore multiply by x^2 and start with x^2e^x and functions obtained by differentiating it; that is, we first take as a trial particular solution

$$y = ax^2e^x + bxe^x + ce^x.$$

But since e^x and xe^x are solutions of the homogeneous equation, the bxe^x and ce^x portions will contribute zero when we compute $y'' - 2y' + y$. We thus simply take $y = ax^2e^x$ as trial solution, and we find that

$$y = ax^2e^x,$$

$$y' = ax^2e^x + 2axe^x,$$

$$y'' = ax^2e^x + 4axe^x + 2ae^x.$$

Multiplying the first equation by 1, the second by -2, the last by 1, and adding, we obtain

$$e^x = 0(x^2e^x) + 0(xe^x) + 2ae^x = 2ae^x,$$

so we must have $2a = 1$ or $a = \frac{1}{2}$. Our general solution is therefore

$$y = e^x(C_1x + C_2) + \frac{1}{2}x^2e^x. \quad \square$$

APPLICATIONS

Let us consider motion of a body of mass m along a straight line, which we consider to be an s-axis. By Newton's second law of motion, we have

$$F(t) = m\frac{d^2s}{dt^2},$$

where $F(t)$ is the force at time t acting on the body and directed along the line. Frequently, one denotes a derivative with respect to time t by a dot over a variable rather than by a prime. Thus we let $\dot{s} = ds/dt$, $\ddot{s} = d^2s/dt^2$, and so on.

There are many physical situations in which the force on the body has a certain basic component $g(t)$ at time t together with components due to the velocity and position of the body. If these additional components due to velocity and position are constant multiples of the velocity and position, then by Newton's law, we have

$$\ddot{s} = k_1\dot{s} + k_2 s + \frac{1}{m}g(t) \tag{6}$$

for constants k_1 and k_2. The differential equation (6) is linear with constant coefficients and might be solved by the methods we have presented. We illustrate several particular cases.

EXAMPLE 7 *Free fall in a vacuum.* We consider a body falling freely near the surface of a planet without atmosphere and with gravitational acceleration g, which is essentially constant near the surface. The motion of the body is then governed by the differential equation

$$\ddot{s} = -g, \tag{7}$$

where s is measured upward from the surface of the planet. The general solution of Eq. (7) is easily found to be

$$s = C_1 t + C_2 - \frac{1}{2}gt^2,$$

and the constants C_1 and C_2 can be determined if the position s and velocity $\dot{s}$ of the body are known at a particular time t_0. □

EXAMPLE 8 *Free fall in a medium.* Let a body be falling freely through a medium (atmosphere) near the surface of a planet with gravitational acceleration g near its surface. Suppose that, due to the medium, the motion of the body is retarded by a force proportional to its velocity. The motion of the body is then governed by the differential equation

$$\ddot{s} = -k\dot{s} - g, \tag{8}$$

where s is measured upward from the planet and $k > 0$. The characteristic equation of (8) is $r^2 + kr = 0$, and a particular solution is easily found to be $s = -(g/k)t$. The general solution is therefore

$$s = C_1 + C_2 e^{-kt} - \frac{g}{k}t. \tag{9}$$

Again, C_1 and C_2 can be determined by the position s and velocity $\dot{s}$ at a particular time t_0.

Differentiating Eq. (9), we obtain for the velocity

$$\dot{s} = -kC_2 e^{-kt} - \frac{g}{k}.$$

As t becomes large, the term $-kC_2 e^{-kt}$ becomes very small; the *terminal velocity* of the body is thus $-g/k$. Note that the body approaches this terminal velocity exponentially (quite rapidly). In Example 3 of Section 20.2, we

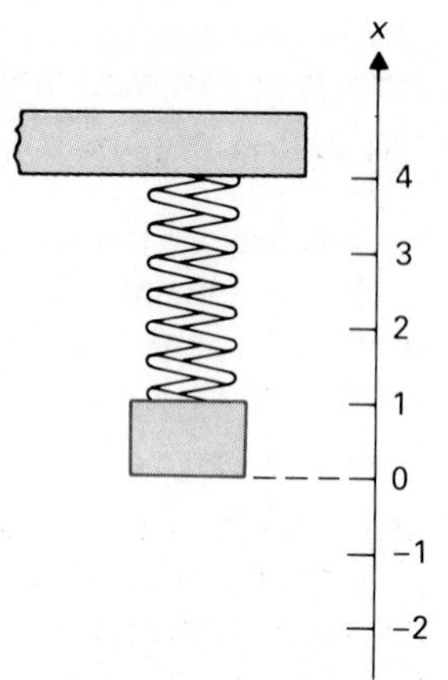

Figure 20.9 Body suspended by a spring in rest position at $s = 0$.

obtained mg/k as terminal velocity. In that example, we took positive direction downward, which accounts for the difference in sign. Also, k/m in that example played the role of k in this example, which accounts for the presence of m in the formula mg/k obtained in Section 20.2. □

EXAMPLE 9 *Undamped vibrating spring.* Consider a body of mass m hanging on a spring, shown in natural position of rest at $s = 0$ on the vertical s-axis in Fig. 20.9. If the spring is stretched (or compressed) from this natural position, then the spring exerts a restoring force proportional to the displacement of the body, assuming that the elastic limit of the spring is not exceeded. If the body is set in vertical motion and this restoring force is the only force on the system, then the motion of the body must satisfy the differential equation

$$m\ddot{s} = -ks \tag{10}$$

for a spring constant $k > 0$. The characteristic equation for (10) is $mr^2 + k = 0$, and we see that the general solution is

$$s = C_1 \cos\left(\sqrt{\frac{k}{m}}\,t\right) + C_2 \sin\left(\sqrt{\frac{k}{m}}\,t\right), \tag{11}$$

where C_1 and C_2 can be determined by the position and velocity of the body at a particular time t_0.

If we set $A = \sqrt{C_1^2 + C_2^2}$, then Eq. (11) can be written

$$s = A\left[\frac{C_1}{A}\cos\left(\sqrt{\frac{k}{m}}\,t\right) + \frac{C_2}{A}\sin\left(\sqrt{\frac{k}{m}}\,t\right)\right].$$

Since $(C_1/A)^2 + (C_2/A)^2 = 1$, there exists θ where $0 \le \theta < 2\pi$ and where $\sin\theta = C_1/A$ and $\cos\theta = C_2/A$. Thus

$$\begin{aligned} s &= A\left(\sin\theta\cos\sqrt{\frac{k}{m}}\,t + \cos\theta\sin\sqrt{\frac{k}{m}}\,t\right) \\ &= A\sin\left(\sqrt{\frac{k}{m}}\,t + \theta\right). \end{aligned} \tag{12}$$

Equation (12) shows that our vibratory motion is sinusoidal, with amplitude A. Such sinusoidal undamped vibratory motion is frequently referred to as "simple harmonic motion." □

EXAMPLE 10 *Damped vibrating spring.* Suppose the spring in Fig. 20.9 has a damping mechanism attached that exerts a force against the direction of motion and proportional to the velocity of the body. The differential equation then becomes

$$m\ddot{s} = -ks - c\dot{s} \tag{13}$$

for $k > 0$ and $c > 0$. The character of the general solution of this equation $m\ddot{s} + c\dot{s} + ks = 0$ depends on the relative sizes of m, c, and k and on the initial conditions. In the exercises, we ask you to describe the nature of particular motions in the *overdamped case* where $c^2 > 4km$, the *critically damped case* where $c^2 = 4km$, and the *underdamped case* where $c^2 < 4km$ (see Exercises 27, 28, and 29). □

SUMMARY

1. The general solution of a linear equation with constant coefficients
$$P(D)y = g(x)$$
is of the form
$$y = (\text{General solution of } P(D)y = 0) + f(x),$$
where $f(x)$ is any particular solution of $P(D)y = g(x)$.
2. If $P(D) = (D - r_1)(D - r_2) \cdots (D - r_n)$, then $P(D)y = g(x)$ can be solved by setting $u = (D - r_2) \cdots (D - r_n)y$, so $P(D)y = g(x)$ becomes $(D - r_1)u = g(x)$. Solve for u, and apply the same technique again to $(D - r_2) \cdots (D - r_n)y = u$, and so on, to find the general solution.
3. The method of undetermined coefficients to find a particular solution $f(x)$ of $P(D)y = g(x)$ is too complicated to describe in a summary. See Examples 5 and 6 and the boxed explanation preceding them.

EXERCISES

In Exercises 1 through 8, find the general solution of the differential equation by finding a particular solution by the method of successive reduction to first-order equations, as in Example 1.

1. $y'' - 2y' - 3y = 4x$
2. $y'' - 4y = e^x$
3. $y'' - y = e^x$
4. $y'' - y = \cos x$
5. $y'' - y' = 1$
6. $y'' + y' = x$
7. $y'' + 2y' + y = e^{-x}$
8. $y''' - y'' = x^2$

In Exercises 9 through 24, find the general solution of the differential equation by the method of undetermined coefficients.

9. $y'' - 3y' + 2y = x - 3$
10. $y'' - y = x + e^{2x}$
11. $y'' + 4y = \sin x$
12. $y'' + y = \sin 2x$
13. $y'' - 2y' = 2 \sin x + 3e^x$
14. $y'' + 2y' + y = x^2$
15. $y'' + y = xe^x$
16. $y'' - 4y = x \sin 2x$
17. $y'' - y = x \sin x + \sin x$
18. $y'' + 4y' + 4y = e^{2x} \cos x$
19. $y'' + 4y' = x^2$
20. $y'' - 4y' - 5y = x^2 + 2e^{-x}$
21. $y''' + 3y'' = x + e^{3x}$
22. $y''' - y'' = x^2 + e^x$
23. $(D^4 - 2D^3)y = x^3 + 3x^2$
24. $(D - 1)^3y = 4 - 3e^x$

25. A body of mass 2 slugs is dropped from an altitude of 3000 ft above the Atlantic Ocean. Suppose the motion of the body is retarded by a force due to air resistance and of magnitude $v/2$ lb, where v is the velocity of the body measured in ft/sec. Find the height s of the body above the ocean as a function of time t, and find the terminal velocity of the body. (Take $g = 32$ ft/sec^2.)
26. A body of mass 1 slug (weighing 32 lb) is attached to a vertical spring and stretches the spring 2 ft. The body is then raised 1 ft and released.
    a) Find the displacement s of the body from its natural position at rest at the end of the spring as a function of the time after it was released. (For example, $s = 1$ when $t = 0$.)
    b) What is the amplitude of this oscillatory motion?
    c) What is the period of this oscillatory motion? (The period is the time required for one complete oscillation.)
27. For a damped vibrating spring with spring constant k and with a weight of mass m attached, suppose the weight is raised a height h from its position at rest and then released. Show that if $c^2 > 4km$, the weight eases back to its original position of rest without crossing this position. See Eq. (13).
28. With reference to Exercise 27, suppose now that $c^2 = 4km$ and that the weight is given an initial velocity $v_0 < -ch/2m$ toward the position of rest. Show that the weight crosses its original position of rest and then eases back to this position.
29. With reference to Exercise 27, show that if $c^2 < 4km$, then the motion of the weight about its position of rest is oscillatory but has amplitude decreasing exponentially to zero as time increases.
30. It can be shown that if an electric circuit has a (constant) resistance R, a (constant) inductance L, (constant) capacitance C, and variable impressed electromotive force $E(t)$, then the charge $Q(t)$ on the capacitor at time t is governed by the differential equation
$$L\ddot{Q} + R\dot{Q} + \frac{1}{C}Q = E(t).$$
Suppose the electromotive force $E(t)$ is abruptly cut off, so that $E(t) = 0$ for $t > t_0$. Use Exercises 27, 28, and 29 to discuss the possibilities for the behavior of the charge Q after time t_0.
31. *Resonance.* Some possibilities for the behavior of the solutions of a differential equation $a\ddot{s} + b\dot{s} + cs = 0$, where a, b, $c > 0$, are indicated in Exercises 27, 28, and 29. This exercise

exhibits a phenomenon that may appear in the solutions of the differential equation $a\ddot{s} + b\dot{s} + cs = \sin kt$. A physical model for such an equation is given by the electric circuit described in Exercise 30, where the impressed electromotive force $E(t)$ is sinusoidal.

Suppose the "damping factor" b in our equation is zero. Show that if $k = 0$, the motion described by the resulting (homogeneous) equation is oscillatory with period $2\pi\sqrt{a}/\sqrt{c}$. Then show that if the impressed sinusoidal force $\sin kt$ has the same period so that $k = \sqrt{c/a}$, the amplitude of the motion increases without bound as time increases. (If the damping factor b is nonzero but quite small, the amplitude can still get quite large if $k = \sqrt{4ac - b^2}/2a$. This phenomenon is known as "resonance" and can be very destructive. A group of men marching in step should be instructed to break step when crossing a bridge, on the outside chance that a frequency in their step might be the same as the natural frequency of vibration of the bridge.)

20.7 SERIES SOLUTIONS BY THE METHOD OF SUCCESSIVE DIFFERENTIATION

So far we have discussed techniques for finding solutions of differential equations of very special types. While those types do occur in many applications, more complicated differential equations also occur naturally. Sometimes, a *series solution* seems to be the only hope for finding a "formula" for a desired solution to a differential equation. In the final two sections of this chapter, we discuss series solutions.

THE NATURE OF A SERIES SOLUTION

Consider the differential equation

$$y' = g(x). \tag{1}$$

If the function g in Eq. (1) is analytic at x_0 and if we can find a power series in $x - x_0$ that represents g in a neighborhood of x_0, then we can find the general solution of Eq. (1) near x_0 by integrating the series term by term. Since some elementary functions g do not have elementary functions as antiderivatives, series solutions of Eq. (1) can be very useful. We should think of a series solution in powers of $x - x_0$ as describing the behavior of y for x near x_0.

EXAMPLE 1 We have stated that the function e^{x^2} does not have an elementary function as antiderivative. Find a series solution of the differential equation $y' = e^{x^2}$.

Solution If we replace x by x^2 in the well-known series for e^x, our differential equation becomes

$$y' = e^{x^2} = 1 + x^2 + \frac{x^4}{2!} + \frac{x^6}{3!} + \cdots + \frac{x^{2n}}{n!} + \cdots.$$

The series converges to e^{x^2} for all x. Integrating, we obtain as general solution,

$$y = C + x + \frac{x^3}{3} + \frac{x^5}{5\cdot 2!} + \frac{x^7}{7\cdot 3!} + \cdots + \frac{x^{2n+1}}{(2n+1)n!} + \cdots$$

for all x. By adjusting the arbitrary constant C, we can find a particular solution having any desired value at the origin. For example, $C = 0$ gives a solution f_1, where $f_1(0) = 0$, while $C = -5$ gives a solution f_2, where $f_2(0) = -5$. □

EXAMPLE 2 Find the general series solution of $y'' = e^{x^2}$.

Solution From Example 1, we see that, starting with the equation $y'' = e^{x^2}$, we obtain, by integrating once,

$$y' = C_1 + x + \frac{x^3}{3} + \frac{x^5}{5 \cdot 2!} + \frac{x^7}{7 \cdot 3!} + \cdots + \frac{x^{2n+1}}{(2n+1)n!} + \cdots,$$

for all x. Another integration yields the general solution

$$y = C_2 + C_1 x + \frac{x^2}{2} + \frac{x^4}{4 \cdot 3} + \frac{x^6}{6 \cdot 5 \cdot 2!} + \frac{x^8}{8 \cdot 7 \cdot 3!} + \cdots$$

$$+ \frac{x^{2n+2}}{(2n+2)(2n+1)n!} + \cdots$$

for all x. If $y = f(x)$, then $C_1 = f'(0)$ and $C_2 = f(0)$; both the value of a particular solution at zero and the value of its derivative at zero are controlled by the two arbitrary constants C_2 and C_1. □

We should mention that not every differential equation has a power series solution in $x - x_0$ for general solution, even if the equation has a general solution in a neighborhood of x_0. In Section 11.3, we mentioned that the function

$$g(x) = \begin{cases} e^{-1/x^2} & \text{for } x \neq 0, \\ 0 & \text{for } x = 0 \end{cases} \tag{2}$$

has derivatives of all orders equal to zero at $x_0 = 0$, so its associated Taylor series is the zero series. Since $g(x) \neq 0$ for $x \neq 0$, the function is not represented by a series in any neighborhood of $x_0 = 0$. Therefore, no antiderivative could be represented by a series in a neighborhood of zero either. Hence there is no power series solution at $x_0 = 0$ for the differential equation $y' = g(x)$. However, since $g(x)$ is continuous, the equation $y' = g(x)$ does have a general solution, namely,

$$y = \int_0^x g(t)\,dt + C.$$

HOMOGENEOUS LINEAR EQUATIONS

The equation

$$y^{(n)} + p_1(x)y^{(n-1)} + \cdots + p_{n-1}(x)y' + p_n(x)y = 0, \tag{3}$$

where $p_1, \ldots, p_n$ are known functions, is a homogeneous linear differential equation of order n with coefficients $p_1(x), \ldots, p_n(x)$. We state without proof the main theorem concerning the existence of analytic solutions $y = f(x)$ of Eq. (3) in a neighborhood of a point x_0. (See also Theorem 20.3 in Section 20.5.) Since Eq. (3) is of order n, we would expect the general solution to have n arbitrary constants, which could be adjusted to control the values

$$f(x_0),\quad f'(x_0), \ldots, \quad f^{(n-1)}(x_0).$$

As before, we will let

$$y(x_0) = f(x_0), \qquad y'(x_0) = f'(x_0), \quad \ldots, \quad y^{(n-1)}(x_0) = f^{(n-1)}(x_0).$$

THEOREM 20.7 Existence of analytic solutions

Let the coefficients $p_1(x), \ldots, p_n(x)$ of Eq. (3) be analytic at x_0 with power series expansions in $x - x_0$ that represent them in a neighborhood $x_0 - r < x < x_0 + r$. Then every solution of Eq. (3) in this neighborhood is analytic at x_0, and for any constants $a_0, \ldots, a_{n-1}$, there exist *unique* constants a_n, $a_{n+1}, \ldots$ such that the series $\sum_{k=0}^{\infty} a_k(x - x_0)^k$ represents a solution of Eq. (3) in this neighborhood. Furthermore, if we regard $a_0, \ldots, a_{n-1}$ as arbitrary constants, there exist n functions $y_1(x), \ldots, y_n(x)$ such that the general solution of Eq. (3) in this neighborhood is

$$a_0 y_1(x) + \cdots + a_{n-1} y_n(x). \tag{4}$$

In the following subsection, we present a technique for finding series solutions of Eq. (3). Another technique and a discussion of the nonhomogeneous case (where the right-hand side of Eq. 3 is nonzero) appear in Section 20.8.

SOLVING EQ. (3) BY DIFFERENTIATING

Let $y = f(x)$ give a solution of Eq. (3) in $x_0 - r < x < x_0 + r$ such that

$$y(x_0) = a_0, \qquad y'(x_0) = a_1, \qquad \ldots, \qquad y^{(n-1)}(x_0) = (n-1)!\, a_{n-1},$$

for constants $a_0, \ldots, a_{n-1}$. By Theorem 20.7, if the coefficients $p_i(x)$ are analytic in $x_0 - r < x < x_0 + r$, then there exist constants $a_n, a_{n+1}, \ldots$ such that

$$y = \sum_{k=0}^{\infty} a_k(x - x_0)^k$$

for $x_0 - r < x < x_0 + r$. By our uniqueness theorem for series expansions, the coefficient a_k in the series must be the kth Taylor coefficient, so

$$a_k = \frac{y^{(k)}(x_0)}{k!} \tag{5}$$

for $k = 0, 1, \ldots$. We are assuming that the coefficients $p_1(x), \ldots, p_n(x)$ in Eq. (3) are known functions that are analytic at x_0 and that $a_0, \ldots, a_{n-1}$ are known constants. Our job is to find

$$a_n = \frac{y^{(n)}(x_0)}{n!}, \qquad a_{n+1} = \frac{y^{(n+1)}(x_0)}{(n+1)!}, \qquad \ldots.$$

Obviously it suffices to find the derivatives $y^{(i)}(x_0)$ for $i \geq n$. From Eq. (3), we obtain

$$y^{(n)} = -p_1(x)y^{(n-1)} - \cdots - p_n(x)y, \tag{6}$$

and we can use Eq. (6) to compute $y^{(n)}(x_0)$, since we know the values at x_0 of all the functions that appear on the right-hand side of Eq. (6). We then differentiate Eq. (6) to obtain

$$y^{(n+1)} = -(p_1(x)y^{(n)} + p_1'(x)y^{(n-1)}) - \cdots - (p_n(x)y' + p_n'(x)y). \tag{7}$$

Since we have computed $y^{(n)}(x_0)$, we now know the values at x_0 of all the functions that appear on the right-hand side of Eq. (7), so we can compute $y^{(n+1)}(x_0)$. By differentiating Eq. (7), we can compute $y^{(n+2)}(x_0)$, and so on.

EXAMPLE 3 Let $y = h(x)$ be the solution of the initial-value problem

$$y'' + xy' - x^2y = 0, \qquad y(1) = 3, \qquad y'(1) = -2.$$

Find $y^{(4)}(1)$.

Solution We have $y'' = -xy' + x^2y$, so

$$y''(1) = (-1)(-2) + 1(3) = 5.$$

Differentiating, we obtain

$$y''' = -xy'' - y' + x^2y' + 2xy = -xy'' + (x^2 - 1)y' + 2xy,$$

so

$$y'''(1) = -1(5) + 0(-2) + 2(3) = 1.$$

Differentiating again, we obtain

$$\begin{aligned} y^{(4)} &= -xy''' + (x^2 - 2)y'' + 2xy' + 2xy' + 2y \\ &= -xy''' + (x^2 - 2)y'' + 4xy' + 2y. \end{aligned}$$

Therefore

$$y^{(4)}(1) = (-1)(1) + (-1)(5) + 4(-2) + 6 = -8. \quad \square$$

By differentiating as in Example 3, we can find the Taylor coefficients (5) for a series solution of the differential equation (6). If all the $p_i(x)$ are constant functions so that $p_i'(x) = 0$ for $i = 1, \ldots, n$, then the differentiation as in Eq. (7) is quite easy. We know that if Eq. (3) has constant coefficients, the general solution can be expressed in terms of elementary functions, and we know how to compute those functions quite easily. Thus the series technique is important chiefly in case the $p_i(x)$ are not all constant functions. However, constant-coefficient cases give nice, easy illustrations of the differentiation technique and of Theorem 20.7.

EXAMPLE 4 We showed in Section 20.4 that the general solution of the differential equation

$$y' = ky \tag{8}$$

is $y = Ae^{kx}$, where A is an arbitrary constant that controls $y(0)$. Derive this general solution by series methods. Note that Eq. (8) is of the form (3), where $n = 1$ and $p_1(x) = -k$.

Solution We try to find a series $y = \sum_{n=0}^{\infty} a_n x^n$ at $x_0 = 0$ that is a solution of Eq. (8) such that $y(0) = a_0$. We give the equations obtained from Eq. (8) and repeated differentiation of Eq. (8) in a column at the left; the values

$y(0)$, $y'(0)$, $y''(0)$, and so on are given in a column at the right.

$$
\begin{array}{ll}
 & y(0) = a_0 \\
y' = ky & y'(0) = ka_0 \\
y'' = ky' & y''(0) = k(ka_0) = k^2 a_0 \\
y''' = ky'' & y'''(0) = k(k^2 a_0) = k^3 a_0 \\
\vdots & \vdots \\
y^{(n)} = ky^{(n-1)} & y^{(n)}(0) = k^n a_0 \\
\vdots & \vdots
\end{array}
$$

From Eq. (5), we have $a_n = y^{(n)}(0)/n!$, so we obtain for our solution

$$
\begin{aligned}
y &= a_0 + ka_0 x + \frac{k^2 a_0}{2!} x^2 + \frac{k^3 a_0}{3!} x^3 + \cdots + \frac{k^n a_0}{n!} x^n + \cdots \\
&= a_0 \left[1 + kx + \frac{(kx)^2}{2!} + \frac{(kx)^3}{3!} + \cdots + \frac{(kx)^n}{n!} + \cdots \right] \\
&= a_0 e^{kx}.
\end{aligned}
$$

Thus we have, as general solution, $y = a_0 e^{kx}$ for all x, where a_0 may be any constant. Note that we have given the general solution in the form described in the last sentence of Theorem 20.7, where $y_1(x) = e^{kx}$. □

EXAMPLE 5 Find the general solution of the equation $y'' + y = 0$ in series form at $x_0 = 0$.

Solution For this second-degree equation, we let $a_0 = y(0)$ and $a_1 = y'(0)$ be "arbitrary," and compute further derivatives of y at zero using two columns as in the preceding example.

$$
\begin{array}{ll}
 & y(0) = a_0 \\
 & y'(0) = a_1 \\
y'' = -y & y''(0) = -a_0 \\
y''' = -y' & y'''(0) = -a_1 \\
y^{(4)} = -y'' & y^{(4)}(0) = a_0 \\
y^{(5)} = -y''' & y^{(5)}(0) = a_1 \\
y^{(6)} = -y^{(4)} & y^{(6)}(0) = -a_0 \\
y^{(7)} = -y^{(5)} & y^{(7)}(0) = -a_1 \\
\vdots & \vdots
\end{array}
$$

The general solution is therefore

$$
\begin{aligned}
y &= a_0 + a_1 x - \frac{a_0}{2!} x^2 - \frac{a_1}{3!} x^3 + \frac{a_0}{4!} x^4 + \frac{a_1}{5!} x^5 - \frac{a_0}{6!} x^6 - \frac{a_1}{7!} x^7 + \cdots \\
&= a_0 \left(1 - \frac{x^2}{2!} + \frac{x^4}{4!} - \frac{x^6}{6!} \cdots \right) + a_1 \left(x - \frac{x^3}{3!} + \frac{x^5}{5!} - \frac{x^7}{7!} + \cdots \right) \\
&= a_0 \cos x + a_1 \sin x
\end{aligned}
$$

for all x. Note that we have expressed our solution in the form $a_0 y_1(x) +$

$a_1 y_2(x)$ described in Theorem 20.7, where $y_1(x) = \cos x$ and $y_2(x) = \sin x$. □

We give examples that illustrate the differentiation method when the functions $p_i(x)$ are not all constant. None of our previous methods would solve such equations of order greater than 1.

EXAMPLE 6 Find the general solution at $x_0 = 0$ of the differential equation

$$y'' - xy = 0$$

Solution We let $y(0) = a_0$ and $y'(0) = a_1$, and compute in two columns as before.

$$\begin{aligned}
& & y(0) &= a_0 \\
& & y'(0) &= a_1 \\
y'' &= xy & y''(0) &= 0 \cdot a_0 = 0 \\
y''' &= xy' + y & y'''(0) &= a_0 \\
y^{(4)} &= xy'' + y' + y' = xy'' + 2y' & y^{(4)}(0) &= 2a_1 \\
y^{(5)} &= xy''' + 3y'' & y^{(5)}(0) &= 0 \\
y^{(6)} &= xy^{(4)} + 4y''' & y^{(6)}(0) &= 4a_0 \\
y^{(7)} &= xy^{(5)} + 5y^{(4)} & y^{(7)}(0) &= 5(2a_1) = 10a_1 \\
&\vdots & &\vdots \\
y^{(n)} &= xy^{(n-2)} + (n-2)y^{(n-3)}
\end{aligned}$$

Here we find that

$$y^{(n)}(0) = \begin{cases} (4)(7)(10)\cdots(n-2)a_0, & \text{if } n = 3m, \\ (2)(5)(8)\cdots(n-2)a_1 & \text{if } n = 3m+1, \\ 0 & \text{if } n = 3m+2. \end{cases}$$

The general solution is therefore

$$\begin{aligned}
y &= a_0 + a_1 x + 0x^2 + \frac{a_0}{3!}x^3 + \frac{2a_1}{4!}x^4 + 0x^5 + \frac{4a_0}{6!}x^6 + \frac{2 \cdot 5a_1}{7!}x^7 + \cdots \\
&= a_0\left(1 + \frac{x^3}{3!} + \frac{4x^6}{6!} + \frac{4 \cdot 7x^9}{9!} + \cdots \right. \\
&\qquad \left. + \frac{(4)(7)(10)\cdots(3n-2)}{(3n)!}x^{3n} + \cdots\right) \\
&\quad + a_1\left(x + \frac{2x^4}{4!} + \frac{2 \cdot 5x^7}{7!} + \frac{2 \cdot 5 \cdot 8x^{10}}{10!} + \cdots \right. \\
&\qquad \left. + \frac{(2)(5)(8)\cdots(3n-1)}{(3n+1)!}x^{3n+1} + \cdots\right)
\end{aligned}$$

for all x. Here we have again expressed the solution in the form $a_0 y_1(x) + a_1 y_2(x)$ described in Theorem 20.7. □

EXAMPLE 7 Find as many terms as convenient of the series solution in a neighborhood of $x_0 = 1$ of the initial-value problem

$$y'' + xy' - x^2y = 0, \qquad y(1) = 3, \qquad y'(1) = -2.$$

Solution Our series solution will have the form

$$\sum_{n=0}^{\infty} \frac{y^{(n)}(1)}{n!}(x - 1)^n. \tag{9}$$

In Example 3, we found that

$$y(1) = 3, \quad y'(1) = -2, \quad y''(1) = 5,$$
$$y'''(1) = 1, \quad \text{and} \quad y^{(4)}(1) = -8.$$

We will find two more derivatives, taking as our starting point the equation

$$y^{(4)} = -xy''' + (x^2 - 2)y'' + 4xy' + 2y$$

obtained in Example 3. Differentiating, we have

$$\begin{aligned} y^{(5)} &= -xy^{(4)} - y''' + (x^2 - 2)y''' + 2xy'' + 4xy'' + 4y' + 2y' \\ &= -xy^{(4)} + (x^2 - 3)y''' + 6xy'' + 6y', \end{aligned}$$

so

$$y^{(5)}(1) = (-1)(-8) + (-2)(1) + 6(5) + 6(-2) = 24.$$

Differentiating once more, we obtain

$$\begin{aligned} y^{(6)} &= -xy^{(5)} - y^{(4)} + (x^2 - 3)y^{(4)} + 2xy''' + 6xy''' + 6y'' + 6y'' \\ &= -xy^{(5)} + (x^2 - 4)y^{(4)} + 8xy''' + 12y''. \end{aligned}$$

Therefore

$$y^{(6)}(1) = (-1)(24) + (-3)(-8) + 8(1) + 12(5) = 68.$$

When substituted in Eq. (9), the derivatives we have computed yield the beginning

$$3 - 2(x - 1) + \frac{5}{2}(x - 1)^2 + \frac{1}{6}(x - 1)^3 - \frac{1}{3}(x - 1)^4$$
$$+ \frac{1}{5}(x - 1)^5 + \frac{17}{180}(x - 1)^6 + \cdots$$

of the series solution for our initial-value problem. □

SUMMARY

A series solution of

$$y^{(n)} + p_1(x)y^{(n-1)} + \cdots + p_{n-1}(x)y' + p_n(x)y = 0$$

in a neighborhood of x_0 is of the form

$$\sum_{k=0}^{\infty} a_k(x - x_0)^k,$$

where $a_0, \ldots, a_{n-1}$ may be arbitrary constants, and

$$a_k = \frac{y^{(k)}(x_0)}{k!}.$$

Find a_n by solving the equation for $y^{(n)}$,

$$y^{(n)} = -p_1(x)y^{(n-1)} - \cdots - p_{n-1}(x)y' - p_n(x)y,$$

and evaluate at x_0 and then divide by $n!$ Then differentiate this equation for $y^{(n)}$ to obtain an expression for $y^{(n+1)}$ and evaluate it at x_0 and divide by $(n+1)!$ to find a_{n+1}. Continue this differentiation and evaluation as far as desired to obtain successive terms of the series solution.

EXERCISES

In Exercises 1 through 8, find the indicated derivative of the solution $y = h(x)$ of the given initial-value problem.

1. $y' - xy = 0$, $y(1) = 4$; find $y''(1)$

2. $y'' - y' + x^2y = 0$, $y(2) = 1$, $y'(2) = -3$; find $y'''(2)$

3. $y'' + e^xy' - (\cos x)y = 0$, $y(0) = -1$, $y'(0) = 2$; find $y'''(0)$

4. $y''' + (\sin x)y' - 3(\cos x)y = 0$, $y(0) = -1$, $y'(0) = 2$, $y''(0) = 0$; find $y^{(4)}(0)$

5. $y'' - x^2y' + 2y = 0$, $y(0) = 1$, $y'(0) = -2$; find $y^{(4)}(0)$

6. $y'' - xy' = 0$, $y(1) = 2$, $y'(1) = -4$; find $y^{(4)}(1)$

7. $y''' + xy' = 0$, $y(2) = 0$, $y'(2) = -1$, $y''(2) = 3$; find $y^{(5)}(2)$

8. $y''' + x^3y'' - xy = 0$, $y(-1) = 3$, $y'(-1) = -2$, $y''(-1) = 5$; find $y^{(5)}(-1)$

In Exercises 9 through 18, find, by the method of repeated differentiation, as many terms as you conveniently can of the general series solution of the differential equation in a neighborhood of the indicated point x_0. Express your answer in the form $a_0y_1(x) + \cdots + a_{n-1}y_n(x)$ described in Theorem 20.7, and express the functions $y_i(x)$ in terms of elementary functions wherever you can.

9. $y' + xy = 0$; $x_0 = 0$

10. $y'' - y' = 0$; $x_0 = 1$

11. $y'' - 3y' + 2y = 0$; $x_0 = -1$

12. $y'' - xy' = 0$; $x_0 = 0$

13. $y'' + xy' - y = 0$; $x_0 = 0$

14. $y'' + (\sin x)y' + (\cos x)y = 0$; $x_0 = 0$

15. $(x + 1)y' - y = 0$; $x_0 = 0$

16. $y''' - xy' = 0$; $x_0 = -1$

17. $y^{(4)} + x^2y'' = 0$; $x_0 = 2$

18. $y''' - xy'' + y' + x^2y = 0$; $x_0 = 1$

In Exercises 19 through 24, find as many terms of the series solution of the given initial-value problem as you conveniently can.

19. $y'' + y = 0$, $y(1) = -1$, $y'(1) = 2$

20. $y'' + y' - y = 0$, $y(0) = 1$, $y'(0) = 3$

21. $y^{(4)} - 2y'' + y = 0$, $y(0) = -2$, $y'(0) = 1$, $y''(0) = -1$, $y'''(0) = 3$

22. $y'' - (1/x)y' + x^2y = 0$, $y(1) = 2$, $y'(1) = 0$

23. $y''' - y'' + xy' = 0$, $y(0) = 1$, $y'(0) = 2$, $y''(0) = 0$

24. $y^{(4)} - xy''' + y' = 0$, $y(-1) = 0$, $y'(-1) = 0$, $y''(-1) = 2$, $y'''(-1) = 1$

20.8 SERIES SOLUTIONS BY THE METHOD OF UNDETERMINED COEFFICIENTS

In the preceding section, we gave a method for finding a series solution of a homogeneous linear differential equation. We simply use the equation to compute higher and higher derivatives of a solution, which provide us with the coefficients in the Taylor series for the solution. That technique also works perfectly well for nonhomogeneous differential equations. We can still compute the higher-order derivatives, using the given differential equation.

This section presents an alternative method of computing a series solution, the *method of undetermined coefficients.* After presenting the method, we give examples showing the use of both methods.

THE METHOD OF UNDETERMINED COEFFICIENTS

We consider again the homogeneous linear differential equation

$$y^{(n)} + p_1(x)y^{(n-1)} + \cdots + p_{n-1}(x)y' + p_n(x)y = 0, \tag{1}$$

with coefficients that are analytic in $x_0 - r < x < x_0 + r$. The method of undetermined coefficients for solving Eq. (1) consists of substituting $y = \sum_{k=0}^{\infty} a_k(x - x_0)^k$ in the left-hand side and determining the constants a_n, $a_{n+1}, \ldots$ in terms of $a_0, \ldots, a_{n-1}$, so that the resulting expression is indeed zero. The technique is best illustrated by an example. We take $x_0 = 0$. The same procedure can be used at any point x_0 to obtain a series in $x - x_0$; however, all coefficients $p_i(x)$ must be expanded in power series centered at x_0.

EXAMPLE 1 Find the general power series solution at $x_0 = 0$ of the equation

$$y'' - 2xy' + y = 0. \tag{2}$$

Solution We let

$$y = \sum_{n=0}^{\infty} a_n x^n = a_0 + a_1 x + a_2 x^2 + \cdots + a_n x^n + \cdots.$$

We wish to compute $y'' - 2xy' + y$. To compute $-2xy'$, we differentiate to find

$$y' = a_1 + 2a_2 x + 3a_3 x^2 + \cdots + na_n x^{n-1} + \cdots$$

and multiply by $-2x$. We then easily compute the sum $y'' + (-2xy) + y$ by writing out the summands, *keeping like powers of x aligned* as when adding polynomials:

$$\begin{aligned}
y &= a_0 && + a_1 x + \cdots && + a_n x^n + \cdots \\
-2xy' &= && -2a_1 x - \cdots && -2na_n x^n + \cdots \\
y'' &= 2a_2 && + 3\cdot 2a_3 x + \cdots && + (n+2)(n+1)a_{n+2}x^n + \cdots \\
\text{(add)} & && && \\
\hline
0 &= (a_0 + 2a_2) + (6a_3 - a_1)x + \cdots && && \\
& && + [(n+2)(n+1)a_{n+2} - (2n-1)a_n]x^n + \cdots
\end{aligned}$$

The uniqueness of power series shows that, in order for the series we obtained by adding to represent the constant function zero, we must have

$$a_0 + 2a_2 = 0, \quad 6a_3 - a_1 = 0, \quad \ldots, \quad (n+2)(n+1)a_{n+2} - (2n-1)a_n = 0, \quad \ldots$$

or

$$a_2 = -\frac{a_0}{2}, \quad a_3 = \frac{a_1}{6}, \quad \ldots, \quad a_{n+2} = \frac{2n-1}{(n+2)(n+1)}a_n, \quad \ldots.$$

The relation

$$a_{n+2} = \frac{2n-1}{(n+2)(n+1)} a_n \tag{3}$$

is a *recursion relation* and can be used to compute all the coefficients in terms of a_0 and a_1. We obtain as general solution

$$\begin{aligned} y &= a_0 + a_1 x - \frac{a_0}{2}x^2 + \frac{a_1}{3\cdot 2}x^3 + \frac{3}{4\cdot 3}\left(-\frac{a_0}{2}\right)x^4 \\ &\quad + \frac{5}{5\cdot 4}\left(\frac{a_1}{3\cdot 2}\right)x^5 + \cdots \\ &= a_0 + a_1 x - \frac{a_0}{2!}x^2 + \frac{a_1}{3!}x^3 - \frac{3a_0}{4!}x^4 + \frac{5a_1}{5!}x^5 \\ &\quad - \frac{7\cdot 3a_0}{6!}x^6 + \frac{9\cdot 5a_1}{7!}x^7 - \cdots \\ &= a_0\left(1 - \frac{1}{2!}x^2 - \frac{3}{4!}x^4 - \frac{7\cdot 3}{6!}x^6 - \frac{11\cdot 7\cdot 3}{8!}x^8 - \cdots\right) \\ &\quad + a_1\left(x + \frac{1}{3!}x^3 + \frac{5}{5!}x^5 + \frac{9\cdot 5}{7!}x^7 + \frac{13\cdot 9\cdot 5}{9!}x^9 + \cdots\right) \end{aligned}$$

for all x. Again, we have expressed our solution in the form $a_0 y_1(x) + a_1 y_2(x)$ given in Theorem 20.7 (Section 20.7). □

THE NONHOMOGENEOUS CASE

The equation

$$y^{(n)} + p_1(x)y^{(n-1)} + \cdots + p_{n-1}(x)y' + p_n(x)y = g(x) \tag{4}$$

for a nonzero function g is a nonhomogeneous linear differential equation of order n with coefficients $p_1(x), \ldots, p_n(x)$. Equation (4) differs from the homogeneous equation (1) only in that the right-hand side of Eq. (4) is not zero. As in Theorem 20.7, it is true that if $p_1(x), \ldots, p_n(x)$, and $g(x)$ are all analytic at x_0 with power series expansions in $x - x_0$ representing them in some neighborhood of x_0, then every solution of Eq. (4) in that neighborhood is analytic at x_0, and for any constants $a_1, \ldots, a_{n-1}$ there exist *unique* constants $a_n, a_{n+1}, \ldots$ such that the series $y = \sum_{k=0}^{\infty} a_k(x - x_0)^k$ represents a solution of Eq. (4) in that neighborhood. As in Section 20.6, it is easy to see that the solution of Eq. (4) is of the form

$$\boxed{y = (\text{Solution of homogeneous equation where } g(x) = 0) + f(x),}$$

where $f(x)$ is any particular solution of Eq. (4).

Both the technique of differentiation described in Section 20.7 and the technique of undetermined coefficients can be used to find series solutions of Eq. (4). We give an illustration using the method of undetermined coefficients.

EXAMPLE 2 Use the method of undetermined coefficients to find the general series solution at $x_0 = 0$ of

$$y'' - 2xy' + y = x, \tag{5}$$

which is similar to Eq. (2) of Example 1, but with x in place of zero on the right-hand side.

Solution We let $y = \sum_{n=0}^{\infty} a_n x^n$ as in Example 1, and we have exactly the same computations, except that this time when we add to compute $y'' - 2xy' + y$, we must have

$$\begin{aligned} x = (a_0 + 2a_2) + (6a_3 - a_1)x + \cdots \\ + [(n + 2)(n + 1)a_{n+2} - (2n - 1)a_n]x^n + \cdots. \end{aligned}$$

This time, we obtain

$$0 = a_0 + 2a_2, \qquad 1 = 6a_3 - a_1, \qquad \ldots, \qquad 0 = (n + 2)(n + 1)a_{n+2} - (2n - 1)a_n, \qquad \ldots,$$

so $a_2 = -a_0/2$ and $a_3 = (a_1 + 1)/6$, while for $n > 1$, we have again the recursion relation

$$a_{n+2} = \frac{2n - 1}{(n + 2)(n + 1)} a_n.$$

Our general solution is therefore

$$\begin{aligned} y &= a_0 + a_1 x - \frac{a_0}{2}x^2 + \frac{a_1 + 1}{3 \cdot 2}x^3 + \frac{3}{4 \cdot 3}\left(-\frac{a_0}{2}\right)x^4 \\ &\quad + \frac{5}{5 \cdot 4}\left(\frac{a_1 + 1}{3 \cdot 2}\right)x^5 + \cdots \\ &= a_0 + a_1 x - \frac{a_0}{2!}x^2 + \frac{a_1 + 1}{3!}x^3 - \frac{3a_0}{4!}x^4 + \frac{5(a_1 + 1)}{5!}x^5 - \cdots \\ &= a_0\left(1 - \frac{1}{2!}x^2 - \frac{3}{4!}x^4 - \frac{7 \cdot 3}{6!}x^6 - \frac{11 \cdot 7 \cdot 3}{8!}x^8 - \cdots\right) \\ &\quad + a_1\left(x + \frac{1}{3!}x^3 + \frac{5}{5!}x^5 + \frac{9 \cdot 5}{7!}x^7 + \frac{13 \cdot 9 \cdot 5}{9!}x^9 + \cdots\right) \\ &\quad + \left(\frac{1}{3!}x^3 + \frac{5}{5!}x^5 + \frac{9 \cdot 5}{7!}x^7 + \frac{13 \cdot 9 \cdot 5}{9!}x^9 + \cdots\right) \end{aligned}$$

for all x. Note that this is the solution found in Example 1 to the homogeneous equation $y'' - 2xy' + y = 0$ plus the particular solution

$$y = \frac{1}{3!}x^3 + \frac{5}{5!}x^5 + \frac{9 \cdot 5}{7!}x^7 + \frac{13 \cdot 9 \cdot 5}{9!}x^9 + \cdots. \quad \square$$

We now give an illustration using the differentiation technique of Section 20.7 with a nonhomogeneous linear equation.

EXAMPLE 3 Find the general series solution at $x_0 = 1$ of

$$y'' - x^2 y = \frac{1}{x},$$

using the differentiation technique of Section 20.7.

Solution We let $y(1) = a_0$ and $y'(1) = a_1$ and compute in two columns as in Section 20.7. We easily find

$$\begin{aligned}
& & y(1) &= a_0 \\
& & y'(1) &= a_1 \\
y'' &= x^2 y + 1/x & y''(1) &= a_0 + 1 \\
y''' &= x^2 y' + 2xy - 1/x^2 & y'''(1) &= 2a_0 + a_1 - 1 \\
y^{(4)} &= x^2 y'' + 4xy' + 2y + 2/x^3 & y^{(4)}(1) &= 3a_0 + 4a_1 + 3 \\
y^{(5)} &= x^2 y''' + 6xy'' + 6y' - 6/x^4 & y^{(5)}(1) &= 8a_0 + 7a_1 - 1 \\
y^{(6)} &= x^2 y^{(4)} + 8xy''' + 12y'' + 24/x^5 & y^{(6)}(1) &= 31a_0 + 12a_1 + 31 \\
y^{(7)} &= x^2 y^{(5)} + 10xy^{(4)} & y^{(7)}(1) &= 78a_0 + 67a_1 - 111 \\
&\quad + 20y''' - 120/x^6 & &
\end{aligned}$$

From the right-hand column, we find that the general solution is

$$\begin{aligned}
y = a_0 \bigg[& 1 + \frac{1}{2!}(x-1)^2 + \frac{2}{3!}(x-1)^3 + \frac{3}{4!}(x-1)^4 \\
& + \frac{8}{5!}(x-1)^5 + \frac{31}{6!}(x-1)^6 + \frac{78}{7!}(x-1)^7 + \cdots \bigg] \\
+ a_1 \bigg[& (x-1) + \frac{1}{3!}(x-1)^3 + \frac{4}{4!}(x-1)^4 + \frac{7}{5!}(x-1)^5 \\
& + \frac{12}{6!}(x-1)^6 + \frac{67}{7!}(x-1)^7 + \cdots \bigg] \\
+ \bigg[& \frac{1}{2!}(x-1)^2 - \frac{1}{3!}(x-1)^3 + \frac{3}{4!}(x-1)^4 - \frac{1}{5!}(x-1)^5 \\
& + \frac{31}{6!}(x-1)^6 - \frac{111}{7!}(x-1)^7 + \cdots \bigg]. \quad \square
\end{aligned}$$

EXAMPLE 4 Use the method of undetermined coefficients to find as many terms as one conveniently can of the series solution of the initial-value problem

$$y'' - x^2 y' + y = x^2 - 2x, \qquad y(0) = 0, \qquad y'(0) = 1.$$

Solution We let

$$y = a_0 + a_1 x + a_2 x^2 + a_3 x^3 + \cdots + a_n x^n + \cdots.$$

Keeping like powers aligned, we find that

$$\begin{array}{rcccccccc}
y = & a_0 & + & a_1 x & + & a_2 x^2 & + \cdots & & + a_n x^n + \cdots \\
-x^2 y' = & & & & & -a_1 x^2 & + \cdots & & -(n-1)a_{n-1}x^n + \cdots \\
y'' = & 2a_2 & + & 6a_3 x & + & 12a_4 x^2 & + \cdots & + & (n+2)(n+1)a_{n+2}x^n + \cdots \\
\text{(add)} & & & & & & & &
\end{array}$$

$$x^2 - 2x = (a_0 + 2a_2) + (a_1 + 6a_3)x + (-a_1 + a_2 + 12a_4)x^2 + \cdots + [-(n-1)a_{n-1} + a_n + (n+2)(n+1)a_{n+2}]x^n + \cdots.$$

Equating coefficients of x, we see that

$$a_0 + 2a_2 = 0, \quad a_1 + 6a_3 = -2, \quad -a_1 + a_2 + 12a_4 = 1, \tag{6}$$

while the coefficient of x^n is zero for $n \geq 3$. Setting the coefficient of x^n equal to zero gives the recursion relation

$$a_{n+2} = \frac{(n-1)a_{n-1} - a_n}{(n+2)(n+1)} \quad \text{for } n \geq 3. \tag{7}$$

From the given initial values, we have $a_0 = 0$ and $a_1 = 1$. Using Eqs. (6), we then find that

$$a_0 = 0, \quad a_1 = 1, \quad a_2 = 0, \quad a_3 = -\frac{1}{2}, \quad a_4 = \frac{1}{6}.$$

The recursion relation (7) now takes over, starting with $n = 3$, to yield

$$a_5 = \frac{2 \cdot 0 - (-1/2)}{5 \cdot 4} = \frac{1}{40}, \qquad n = 3$$

$$a_6 = \frac{3(-1/2) - (1/6)}{6 \cdot 5} = \frac{-10/6}{30} = -\frac{1}{18}, \qquad n = 4$$

$$a_7 = \frac{4(1/6) - (1/40)}{7 \cdot 6} = \frac{77/120}{42} = \frac{11}{720}. \qquad n = 5$$

The desired series solution therefore begins

$$y = x - \frac{1}{2}x^3 + \frac{1}{6}x^4 + \frac{1}{40}x^5 - \frac{1}{18}x^6 + \frac{11}{720}x^7 + \cdots. \quad \square$$

SUMMARY

1. One may be able to find a series solution in a neighborhood of zero for a linear differential equation by substituting

$$y = \sum_{k=0}^{\infty} a_k x^k$$

in the equation and equating coefficients of like powers of x on the two sides of the equation. (All coefficients and functions in the equation must first be expressed in power series at $x = 0$.) Hopefully, one may obtain a recursion relation to determine a_n in terms of earlier coefficients in the series.

2. It may be possible to obtain a series solution using the differentiation technique described in the last section for this nonhomogeneous case also.

EXERCISES

In Exercises 1 through 8, find, by the method of undetermined coefficients, as many terms as you conveniently can of the general series solution of the differential equation in a neighborhood of the indicated point x_0. Express your answer in the form $a_0 y_1(x) + \cdots + a_{n-1} y_n(x) + f(x)$.

1. $y'' + xy' + 2y = 0; x_0 = 0$
2. $y^{(4)} - x^2 y'' = 0; x_0 = 0$
3. $y'' + 2xy' - xy = 0; x_0 = 0$
4. $y'' + 3(x - 1)y' - 2y = 0; x_0 = 1$
5. $y'' - 2xy' = x^2; x_0 = 0$
6. $y'' + x^2 y = x + 3; x_0 = 0$
7. $y''' - xy' = x + 1; x_0 = 0$
8. $y''' - x^2 y'' + 2y = x^2 - 3x; x_0 = 0$

In Exercises 9 through 16, use any method you please to find as many terms as you conveniently can of the general series solution of the differential equation in a neighborhood of x_0.

9. $y'' = \sin x^2; x_0 = 0$
10. $y'' = x \cos x^2; x_0 = 0$
11. $y'' = x^2 \tan^{-1} x; x_0 = 0$
12. $(x^2 + 1)y'' + 2xy' - y = 0; x_0 = 0$
13. $xy''' - y'' = 0; x_0 = 1$
14. $y'' - y = x^2; x_0 = 0$
15. $y'' + y = \sin x; x_0 = 0$
16. $y'' - xy' + 2y = x^2 + 2; x_0 = 0$

In Exercises 17 through 20, use any method you please to find as many terms as you conveniently can of the series solution at $x_0 = 0$ of the given initial-value problem.

17. $y'' - xy' = x^2, y(0) = 1, y'(0) = 0$
18. $y'' + 2xy = x + 1, y(0) = -1, y'(0) = 1$
19. $y''' - x^2 y'' = x^3, y(0) = 1, y'(0) = 0, y''(0) = 0$
20. $y''' - xy' + x^2 y = x^2 - x, y(0) = 0, y'(0) = 1, y''(0) = -1$

EXERCISE SETS FOR CHAPTER 20

Review Exercise Set 20.1

1. Sketch the direction field of the differential equation
$$y' = y - x$$
and sketch a few solution curves.
2. Find the family of orthogonal trajectories of $x^2 - y^2 = C$.
3. Find the particular solution of $dy/dx = x \tan y$ such that $y = \pi/4$ when $x = 0$.
4. Find the general solution of $dy/dx = (x + y)/x$.
5. Find the general solution of $(x^2 + y^2)\,dx + 2xy\,dy = 0$.
6. Find the general solution of $(x + y^2 + y)\,dy - y\,dx = 0$.
7. Find the general solution of $y' + (\sin x)y = \sin x$.
8. Find the general solution of $y''' - 6y'' + 9y' = 0$.
9. Find the general solution of $y'' - 2y' + 5y = e^{3x}$.
10. Find the terms of degree less than or equal to 7 of the series solution about $x_0 = 0$ of $y'' - 2xy' + y = 0$.

Review Exercise Set 20.2

1. State the existence theorem for solutions of a differential equation $y' = F(x, y)$ through a point (x_0, y_0).
2. Find the family of orthogonal trajectories of $y = 4x^2 + C$.
3. Find the general solution of $dy/dx = (y - x)/(y + x)$.
4. Find the solution of
$$(2x \sin y)\,dx + (x^2 \cos y)\,dy = 0$$
such that $y = \pi/2$ when $x = 3$.
5. Find the general solution of
$$(x + \cos y)\,dy + (y + \sin x)\,dx = 0.$$
6. Find the general solution of $y' + (\cot x)y = x$.
7. Find the general solution of $y''' - 2y'' = 0$.
8. Find the general solution of
$$(D^2 - 4)(D^2 + 3)y = 0.$$
9. Find the general solution of
$$y'' - 3y' + 2y = \sin x.$$
10. Find the terms of degree less than or equal to 4 of the series solution about $x_0 = 0$ of
$$y'' - 2xy' + xy = 1 - \sin x.$$

Review Exercise Set 20.3

1. Use the modified Euler's method with $n = 2$ to find approximation points for the solution of $y' = x^2 + y$ through $(0, 1)$ over the interval $[0, 2]$.
2. Solve the initial-value problem $y' = x \cos y, y(4) = 0$.
3. Find the general solution of
$$\frac{dy}{dx} = \frac{2x + y}{x - y}.$$
4. Find the general solution of
$$(2xy - \cos x)\,dx + (x^2 - \sqrt{y})\,dy = 0.$$
5. Find the general solution of
$$(y - xy^2)\,dx + (x + 4x^2 y^2)\,dy = 0.$$

6. Solve the initial-value problem $y' - e^x y = e^x$, $y(0) = 2$.

7. Find the general solution of the differential equation $y''' - 2y'' + y' = 0$.

8. Solve the initial-value problem $y'' - 3y' + 2y = x$, $y(0) = 1$, $y'(0) = -1$.

9. Find the terms of degree less than or equal to 6 of the general series solution at $x_0 = 0$ of $y'' - xy' + 2y = 0$.

10. Find the terms of degree less than or equal to 6 of the series solution at $x_0 = 0$ of the initial-value problem $y'' + xy' - 2y = x^2$, $y(0) = 1$, $y'(0) = -3$.

Review Exercise Set 20.4

1. Use a programmable calculator or computer and the modified Euler's method with $n = 20$ to estimate $h(1)$ if $y = h(x)$ is the solution of the initial-value problem $y' = x + \sqrt{y}$, $y(0) = 2$.

2. Repeat the preceding exercise using Taylor's method of order 2 and $n = 40$.

3. Find the general solution of

$$\frac{dy}{dx} = \frac{x + 2y}{2x + y}.$$

4. Find the solution of the initial-value problem

$$(ye^x + \cos x)\,dx + (e^x - \sin y)\,dy = 0, \qquad y(0) = 0.$$

5. Find the general solution of $y' - 2xy = 3x$.

6. Find the general solution of $(D^3 + 4D)y = 0$.

7. Solve the initial-value problem $(D^2 - 4D + 5)y = 0$, $y(0) = 1$, $y'(0) = -1$.

8. Find the general solution of the differential equation $(D^3 - D^2)y = x^2$.

9. Find the terms of degree less than or equal to 6 of the general series solution at $x_0 = 0$ of the differential equation $y'' - x^2 y' + xy = 0$.

10. Find the terms of degree less than or equal to 6 of the series solution at $x_0 = 1$ of the initial-value problem

$$y'' + (x^2 - 1)y' + xy = x^3, \quad y(1) = 1, \quad y'(1) = -1.$$

APPENDIXES
ANSWERS
INDEX

APPENDIX

SUMMARY OF ALGEBRA AND GEOMETRY

1

A. ALGEBRA

1. INEQUALITIES

Notation	*Read*
$a < b$	a is less than b
$a > b$	a is greater than b
$a \le b$	a is less than or equal to b
$a \ge b$	a is greater than or equal to b

Laws

$a \le a$

If $a \le b$ and $b \le a$, then $a = b$.

If $a \le b$ and $b \le c$, then $a \le c$.

If $a \le b$, then $a + c \le b + c$.

If $a \le b$ and $c \le d$, then $a + c \le b + d$.

If $a \le b$ and $c > 0$, then $ac \le bc$.

If $a \le b$ and $c < 0$, then $bc \le ac$.

2. ABSOLUTE VALUE

Notation	*Read*
$\|a\|$	The absolute value of a

Properties

$$|a| = \begin{cases} a, & \text{if } a \ge 0 \\ -a, & \text{if } a < 0 \end{cases}$$

$|a| \ge 0$, and $|a| = 0$ if and only if $a = 0$

$|ab| = |a| \cdot |b|$

$|a + b| \le |a| + |b|$

$|a - b| \ge |a| - |b|$

$|a - b| =$ distance from a to b on the number line

3. ARITHMETIC OF RATIONAL NUMBERS

$$\frac{a}{b} + \frac{c}{d} = \frac{ad + bc}{bd}, \qquad \frac{a}{b} \cdot \frac{c}{d} = \frac{ac}{bd}, \qquad \frac{a/b}{c/d} = \frac{ad}{bc}$$

4. LAWS OF SIGNS

$$(-a)(b) = a(-b) = -(ab), \qquad (-a)(-b) = ab$$

5. DISTRIBUTIVE LAWS

$$a(b + c) = ab + ac, \qquad (a + b)c = ac + bc$$

6. LAWS OF EXPONENTS

$$a^m a^n = a^{m+n}, \qquad (ab)^m = a^m b^m, \qquad (a^m)^n = a^{mn}$$

$$a^{m/n} = \sqrt[n]{a^m} = (\sqrt[n]{a})^m, \qquad a^{-n} = \frac{1}{a^n}, \qquad \frac{a^m}{a^n} = a^{m-n}$$

7. ARITHMETIC INVOLVING ZERO

$a \cdot 0 = 0 \cdot a = 0$ for any number a

$a + 0 = 0 + a = a$ for any number a

$\frac{0}{a} = 0$

$a^0 = 1$ and $0^a = 0$ if $a > 0$

8. ARITHMETIC INVOLVING ONE

$1 \cdot a = a \cdot 1 = a$ for any number a

$\frac{a}{1} = a^1 = a$ for any number a

$1^n = 1$ for any integer n

9. BINOMIAL THEOREM

$$(a + b)^n = a^n + na^{n-1}b + \frac{n(n-1)}{1 \cdot 2} a^{n-2}b^2 + \frac{n(n-1)(n-2)}{1 \cdot 2 \cdot 3} a^{n-3}b^3 + \cdots + nab^{n-1} + b^n,$$

where n is a positive integer

10. QUADRATIC FORMULA

If $a \neq 0$, then the solutions of the quadratic equation $ax^2 + bx + c = 0$ are given by the formula

$$x = \frac{-b \pm \sqrt{b^2 - 4ac}}{2a}.$$

B. GEOMETRY

In the formulas that follow,

A = Area | b = Length of base | h = Altitude

s = Slant height | r = Radius | V = Volume

C = Circumference | S = Surface area or lateral area

1. Triangle

$$A = \tfrac{1}{2}bh$$

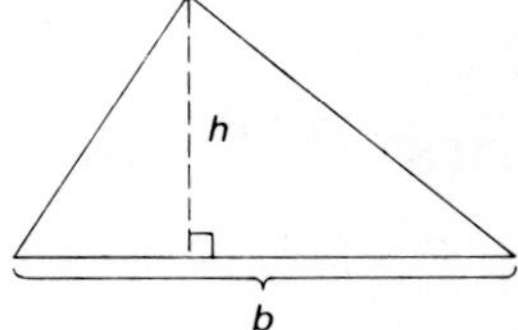

2. Similar triangles

$$\frac{a'}{a} = \frac{b'}{b} = \frac{c'}{c}$$

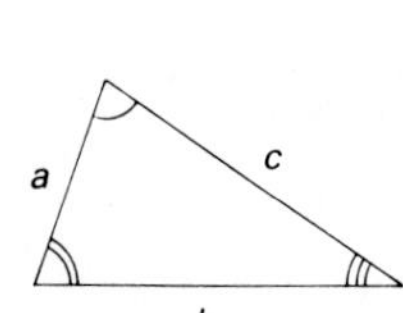

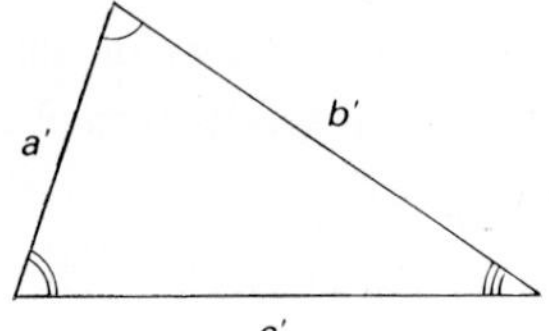

3. Pythagorean theorem

$$c^2 = a^2 + b^2$$

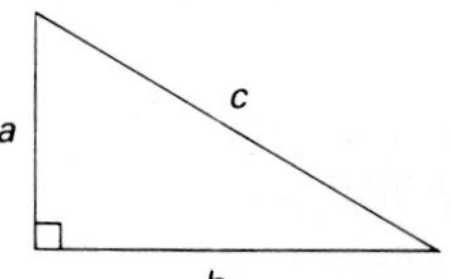

4. Parallelogram

$$A = bh$$

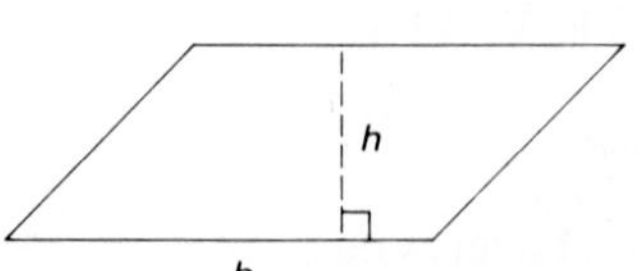

5. Trapezoid

$$A = \tfrac{1}{2}(b_1 + b_2)h$$

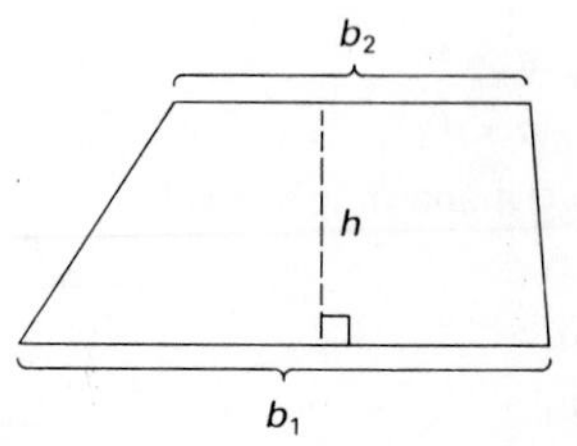

6. Circle

$$A = \pi r^2, \qquad C = 2\pi r$$

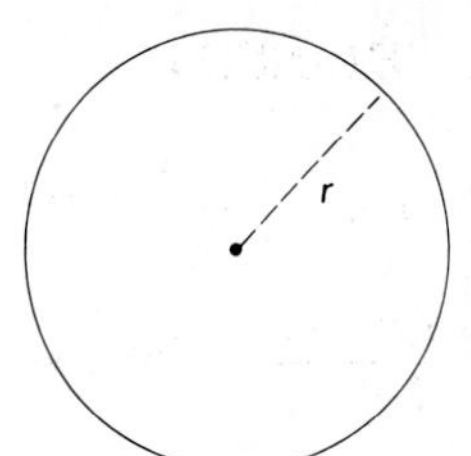

7. Right circular cylinder

$$V = \pi r^2 h, \qquad S = 2\pi rh$$

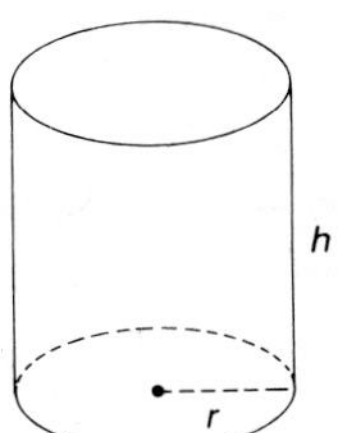

8. Right circular cone

$$V = \tfrac{1}{3}\pi r^2 h, \qquad S = \pi rs$$

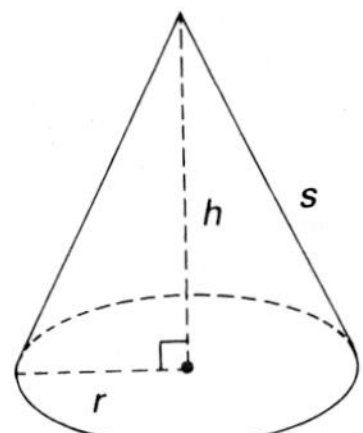

9. Sphere

$$V = \tfrac{4}{3}\pi r^3, \qquad S = 4\pi r^2$$

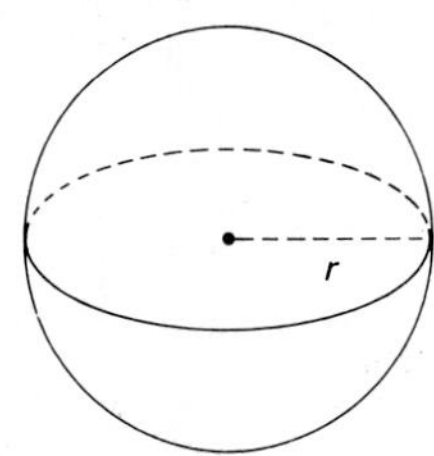

APPENDIX 2

TABLES OF FUNCTIONS

Table 1 Natural trigonometric functions

Angle					Angle				
Degree	Radian	Sine	Cosine	Tangent	Degree	Radian	Sine	Cosine	Tangent
0°	0.000	0.000	1.000	0.000					
1°	0.017	0.017	1.000	0.017	46°	0.803	0.719	0.695	1.036
2°	0.035	0.035	0.999	0.035	47°	0.820	0.731	0.682	1.072
3°	0.052	0.052	0.999	0.052	48°	0.838	0.743	0.669	1.111
4°	0.070	0.070	0.998	0.070	49°	0.855	0.755	0.656	1.150
5°	0.087	0.087	0.996	0.087	50°	0.873	0.766	0.643	1.192
6°	0.105	0.105	0.995	0.105	51°	0.890	0.777	0.629	1.235
7°	0.122	0.122	0.993	0.123	52°	0.908	0.788	0.616	1.280
8°	0.140	0.139	0.990	0.141	53°	0.925	0.799	0.602	1.327
9°	0.157	0.156	0.988	0.158	54°	0.942	0.809	0.588	1.376
10°	0.175	0.174	0.985	0.176	55°	0.960	0.819	0.574	1.428
11°	0.192	0.191	0.982	0.194	56°	0.977	0.829	0.559	1.483
12°	0.209	0.208	0.978	0.213	57°	0.995	0.839	0.545	1.540
13°	0.227	0.225	0.974	0.231	58°	1.012	0.848	0.530	1.600
14°	0.244	0.242	0.970	0.249	59°	1.030	0.857	0.515	1.664
15°	0.262	0.259	0.966	0.268	60°	1.047	0.866	0.500	1.732
16°	0.279	0.276	0.961	0.287	61°	1.065	0.875	0.485	1.804
17°	0.297	0.292	0.956	0.306	62°	1.082	0.883	0.469	1.881
18°	0.314	0.309	0.951	0.325	63°	1.100	0.891	0.454	1.963
19°	0.332	0.326	0.946	0.344	64°	1.117	0.899	0.438	2.050
20°	0.349	0.342	0.940	0.364	65°	1.134	0.906	0.423	2.145
21°	0.367	0.358	0.934	0.384	66°	1.152	0.914	0.407	2.246
22°	0.384	0.375	0.927	0.404	67°	1.169	0.921	0.391	2.356
23°	0.401	0.391	0.921	0.424	68°	1.187	0.927	0.375	2.475
24°	0.419	0.407	0.914	0.445	69°	1.204	0.934	0.358	2.605
25°	0.436	0.423	0.906	0.466	70°	1.222	0.940	0.342	2.748
26°	0.454	0.438	0.899	0.488	71°	1.239	0.946	0.326	2.904
27°	0.471	0.454	0.891	0.510	72°	1.257	0.951	0.309	3.078
28°	0.489	0.469	0.883	0.532	73°	1.274	0.956	0.292	3.271
29°	0.506	0.485	0.875	0.554	74°	1.292	0.961	0.276	3.487
30°	0.524	0.500	0.866	0.577	75°	1.309	0.966	0.259	3.732
31°	0.541	0.515	0.857	0.601	76°	1.326	0.970	0.242	4.011
32°	0.559	0.530	0.848	0.625	77°	1.344	0.974	0.225	4.332
33°	0.576	0.545	0.839	0.649	78°	1.361	0.978	0.208	4.705

Table 1 Natural trigonometric functions (*continued*)

Angle					Angle				
Degree	Radian	Sine	Cosine	Tangent	Degree	Radian	Sine	Cosine	Tangent
34°	0.593	0.559	0.829	0.675	79°	1.379	0.982	0.191	5.145
35°	0.611	0.574	0.819	0.700	80°	1.396	0.985	0.174	5.671
36°	0.628	0.588	0.809	0.727	81°	1.414	0.988	0.156	6.314
37°	0.646	0.602	0.799	0.754	82°	1.431	0.990	0.139	7.115
38°	0.663	0.616	0.788	0.781	83°	1.449	0.993	0.122	8.144
39°	0.681	0.629	0.777	0.810	84°	1.466	0.995	0.105	9.514
40°	0.698	0.643	0.766	0.839	85°	1.484	0.996	0.087	11.43
41°	0.716	0.656	0.755	0.869	86°	1.501	0.998	0.070	14.30
42°	0.733	0.669	0.743	0.900	87°	1.518	0.999	0.052	19.08
43°	0.750	0.682	0.731	0.933	88°	1.536	0.999	0.035	28.64
44°	0.768	0.695	0.719	0.966	89°	1.553	1.000	0.017	57.29
45°	0.785	0.707	0.707	1.000	90°	1.571	1.000	0.000	

Table 2 Exponential functions

x	e^x	e^{-x}	x	e^x	e^{-x}
0.00	1.0000	1.0000	2.5	12.182	0.0821
0.05	1.0513	0.9512	2.6	13.464	0.0743
0.10	1.1052	0.9048	2.7	14.880	0.0672
0.15	1.1618	0.8607	2.8	16.445	0.0608
0.20	1.2214	0.8187	2.9	18.174	0.0550
0.25	1.2840	0.7788	3.0	20.086	0.0498
0.30	1.3499	0.7408	3.1	22.198	0.0450
0.35	1.4191	0.7047	3.2	24.533	0.0408
0.40	1.4918	0.6703	3.3	27.113	0.0369
0.45	1.5683	0.6376	3.4	29.964	0.0334
0.50	1.6487	0.6065	3.5	33.115	0.0302
0.55	1.7333	0.5769	3.6	36.598	0.0273
0.60	1.8221	0.5488	3.7	40.447	0.0247
0.65	1.9155	0.5220	3.8	44.701	0.0224
0.70	2.0138	0.4966	3.9	49.402	0.0202
0.75	2.1170	0.4724	4.0	54.598	0.0183
0.80	2.2255	0.4493	4.1	60.340	0.0166
0.85	2.3396	0.4274	4.2	66.686	0.0150
0.90	2.4596	0.4066	4.3	73.700	0.0136
0.95	2.5857	0.3867	4.4	81.451	0.0123
1.0	2.7183	0.3679	4.5	90.017	0.0111
1.1	3.0042	0.3329	4.6	99.484	0.0101
1.2	3.3201	0.3012	4.7	109.95	0.0091
1.3	3.6693	0.2725	4.8	121.51	0.0082
1.4	4.0552	0.2466	4.9	134.29	0.0074
1.5	4.4817	0.2231	5	148.41	0.0067
1.6	4.9530	0.2019	6	403.43	0.0025
1.7	5.4739	0.1827	7	1096.6	0.0009
1.8	6.0496	0.1653	8	2981.0	0.0003
1.9	6.6859	0.1496	9	8103.1	0.0001
2.0	7.3891	0.1353	10	22026	0.00005
2.1	8.1662	0.1225			
2.2	9.0250	0.1108			
2.3	9.9742	0.1003			
2.4	11.023	0.0907			

Table 3 Natural logarithms

n	$\log_e n$	n	$\log_e n$	n	$\log_e n$
0.0	*	4.5	1.5041	9.0	2.1972
0.1	7.6974	4.6	1.5261	9.1	2.2083
0.2	8.3906	4.7	1.5476	9.2	2.2192
0.3	8.7960	4.8	1.5686	9.3	2.2300
0.4	9.0837	4.9	1.5892	9.4	2.2407
0.5	9.3069	5.0	1.6094	9.5	2.2513
0.6	9.4892	5.1	1.6292	9.6	2.2618
0.7	9.6433	5.2	1.6487	9.7	2.2721
0.8	9.7769	5.3	1.6677	9.8	2.2824
0.9	9.8946	5.4	1.6864	9.9	2.2925
1.0	0.0000	5.5	1.7047	10	2.3026
1.1	0.0953	5.6	1.7228	11	2.3979
1.2	0.1823	5.7	1.7405	12	2.4849
1.3	0.2624	5.8	1.7579	13	2.5649
1.4	0.3365	5.9	1.7750	14	2.6391
1.5	0.4055	6.0	1.7918	15	2.7081
1.6	0.4700	6.1	1.8083	16	2.7726
1.7	0.5306	6.2	1.8245	17	2.8332
1.8	0.5878	6.3	1.8405	18	2.8904
1.9	0.6419	6.4	1.8563	19	2.9444
2.0	0.6931	6.5	1.8718	20	2.9957
2.1	0.7419	6.6	1.8871	25	3.2189
2.2	0.7885	6.7	1.9021	30	3.4012
2.3	0.8329	6.8	1.9169	35	3.5553
2.4	0.8755	6.9	1.9315	40	3.6889
2.5	0.9163	7.0	1.9459	45	3.8067
2.6	0.9555	7.1	1.9601	50	3.9120
2.7	0.9933	7.2	1.9741	55	4.0073
2.8	1.0296	7.3	1.9879	60	4.0943
2.9	1.0647	7.4	2.0015	65	4.1744
3.0	1.0986	7.5	2.0149	70	4.2485
3.1	1.1314	7.6	2.0281	75	4.3175
3.2	1.1632	7.7	2.0412	80	4.3820
3.3	1.1939	7.8	2.0541	85	4.4427
3.4	1.2238	7.9	2.0669	90	4.4998
3.5	1.2528	8.0	2.0794	95	4.5539
3.6	1.2809	8.1	2.0919	100	4.6052
3.7	1.3083	8.2	2.1041		
3.8	1.3350	8.3	2.1163		
3.9	1.3610	8.4	2.1282		
4.0	1.3863	8.5	2.1401		
4.1	1.4110	8.6	2.1518		
4.2	1.4351	8.7	2.1633		
4.3	1.4586	8.8	2.1748		
4.4	1.4816	8.9	2.1861		

* Subtract 10 from these entries.

ANSWERS TO ODD-NUMBERED EXERCISES

CHAPTER 1

Section 1.1

1. a) 0 1 2 3 4 x b) −3 −2 −1 0 1 2 3 x

c) −3 −2 −1 0 1 2 3 x (No solution set) **3.** a) 3 b) 5 c) 3 **5.** a) 2 b) -2 **7.** $x > -2$

9. The distance from $(a+b)/2$ to a is

$$\left|a - \frac{a+b}{2}\right| = \left|\frac{a}{2} - \frac{b}{2}\right| = \left|\frac{a-b}{2}\right|.$$

Similarly, the distance from $(a+b)/2$ to b is

$$\left|b - \frac{a+b}{2}\right| = \left|\frac{b}{2} - \frac{a}{2}\right| = \left|\frac{b-a}{2}\right| = \left|\frac{a-b}{2}\right|.$$

Thus $(a+b)/2$ is equidistant from a and b.

11. a) -4.5 b) $-\sqrt{2}/2$ c) $(\sqrt{2}+\pi)/2$ **13.** a) -10 b) -8 **15.** $[2, 4]$ **17.** $[-5, 2]$ **19.** $[\frac{7}{2}, 9]$ **21.** $[-4, 4]$ **23.** $[-5, 3]$ **25.** $[1, \frac{13}{3}]$ **27.** a)

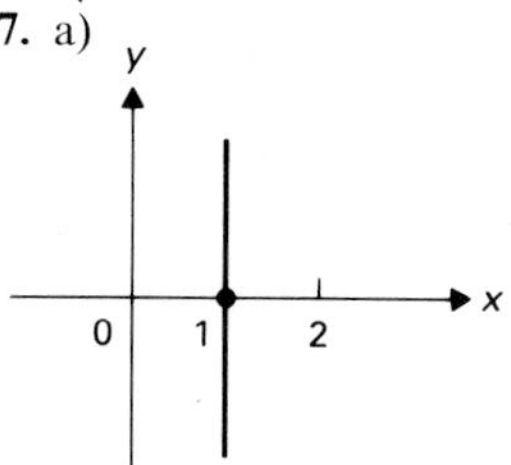

b)

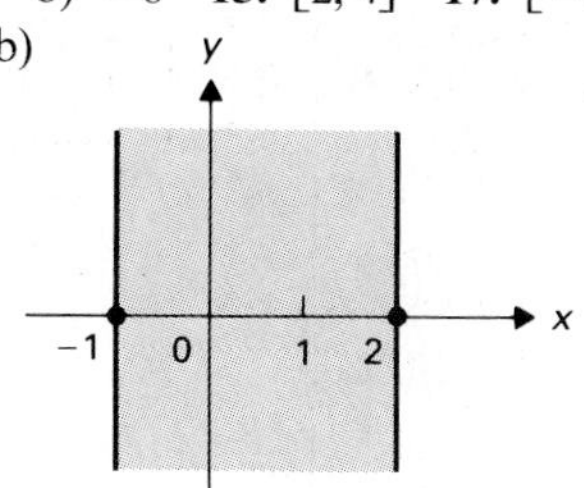

c)

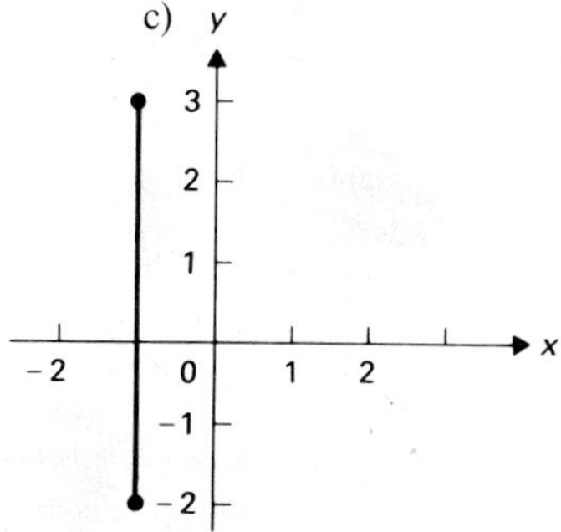

d)

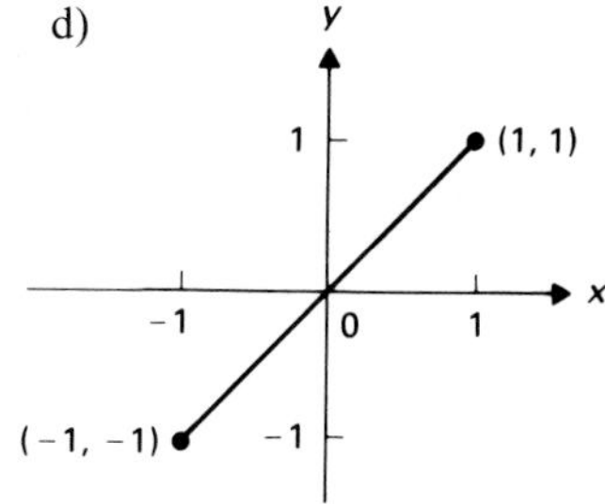

29. a) $(1, -3)$ b) $(2, 6)$ **31.** a) $\sqrt{89}$ b) $\sqrt{171}$

33. The distance from $(-1, 3)$ to $(4, 1)$ is $\sqrt{25+4} = \sqrt{29}$. The distance from $(-1, 3)$ to $(1, 8)$ is $\sqrt{4+25} = \sqrt{29}$. The distance from $(4, 1)$ to $(1, 8)$ is $\sqrt{9+49} = \sqrt{58}$. Since $(\sqrt{58})^2 = (\sqrt{29})^2 + (\sqrt{29})^2$, the Pythagorean relation holds and we have a right triangle. We have Area $= \frac{1}{2}$(Product of legs) $= \frac{1}{2}\sqrt{29}\sqrt{29} = \frac{29}{2}$.

35. $\sqrt{145}$ miles **37.** $\sqrt{1553}/4 \approx 9.85$ miles **39.** -0.39349 **41.** -4.4430 **43.** 12.82252

Section 1.2

1. $x^2 + y^2 = 25$ **3.** $(x-3)^2 + (y+4)^2 = 30$ **5.** Center $(2, 3)$, radius 6 **7.** Center $(-1, -4)$, radius $5\sqrt{2}$ **9.** Center $(-4, 0)$, radius 5 **11.** $(x+4)^2 + (y-4)^2 = 16$ **13.** $(x-2)^2 + (y+3)^2 = 58$ **15.** $(3, -3)$

17. a) $(1, -1)$ b) $(9, -2)$ c) $(12, 0)$ **19.** $\frac{1}{11}$ **21.** $\frac{4}{5}$ **23.** -9 **25.** $\frac{7}{3}$ **27.** $-\frac{2}{3}$
29. a) A parallelogram b) A rectangle (actually a square) **31.** a) A parallelogram b) Not a rectangle **33.** a) 19 b) -6
35. Collinear **37.** Not collinear **39.** $(\sqrt{10}, 3\sqrt{10})$ **41.** $\frac{1}{2}$ **43.** $(7, -1)$
45. Slope $= \frac{9}{5} =$ Increase in Fahrenheit temperature per degree increase in Celsius temperature.
47. a) Type 3 is best with a change (slope) of 7.5°/inch. b) Type 1 is worst with a change of 3.5°/inch.
49. Center $(-1.5788, 0.6177)$, radius 2.49256 **51.** -9.06908 **53.** 30.58081

Section 1.3

1. $y - 4 = 5(x + 1)$, or $y = 5x + 9$ **3.** $y - 2 = 0(x - 4)$, or $y = 2$ **5.** $y + 5 = -\frac{6}{5}(x - 4)$, or $5y + 6x = -1$ **7.** $x = -3$
9. $2x + 3y = -1$ **11.** $4x - y = -11$ **13.** $y = 5x + 15$ **15.** $3y + 2x = 26$ **17.** $4y - 3x = 25$ **19.** $2x + y = 1$
21. Perpendicular **23.** Neither **25.** Slope 1, x-intercept 7, y-intercept -7 **27.** Slope undefined, x-intercept 4, no y-intercept

29.

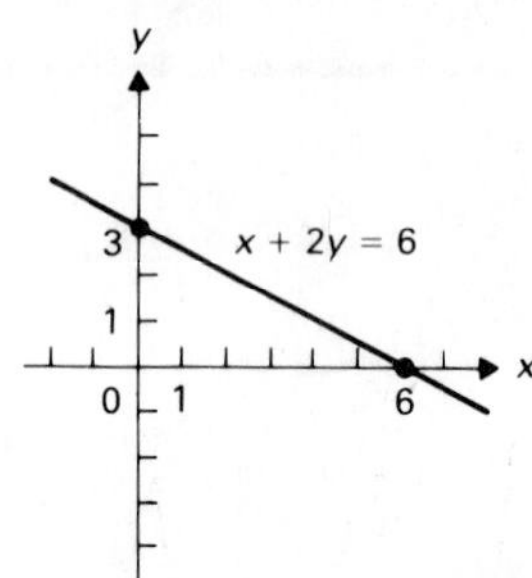

31.

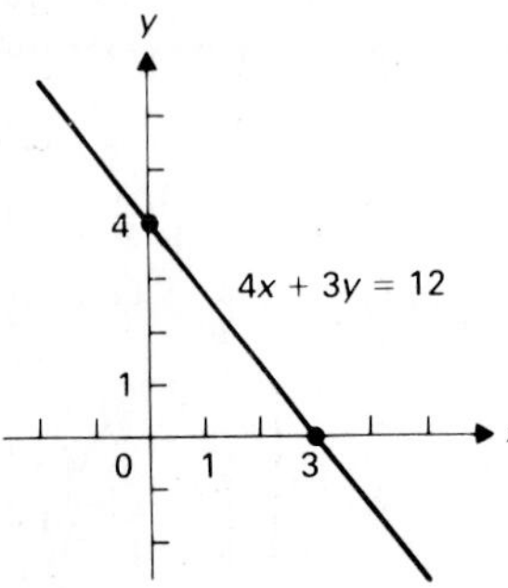

33.

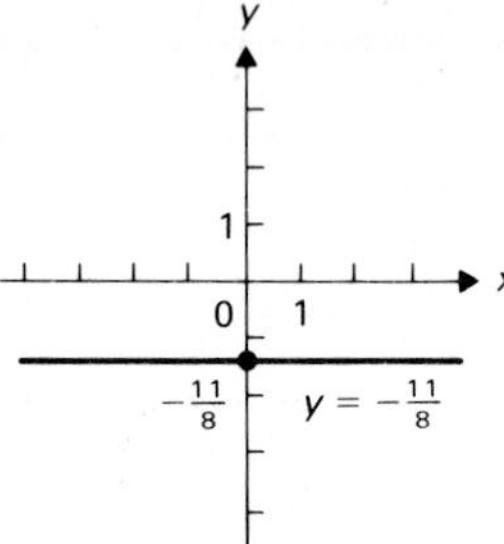

35. $(-52, 37)$ **37.** 2

39.

Side	*Equation of the perpendicular bisector*
$(-a, 0)$ to $(a, 0)$	$x = 0$ (the y-axis)
$(-a, 0)$ to (b, c)	$y - \frac{c}{2} = -\frac{a+b}{c}\left(x - \frac{b-a}{2}\right)$
$(a, 0)$ to (b, c)	$y - \frac{c}{2} = \frac{a-b}{c}\left(x - \frac{a+b}{2}\right)$

The first two meet at $\left(0, \frac{b^2 - a^2}{2c} + \frac{c}{2}\right)$

The equation of the third side is also satisfied by the point $\left(0, \frac{b^2 - a^2}{2c} + \frac{c}{2}\right)$, so all three perpendicular bisectors meet at this point.

41. $x^2 + y^2 + 6x - 4y = -3$ **43.** $d = 13 + \frac{3}{2}(t - 3)$ in. **45.** $W = \frac{1}{6}d$ **47.** a) $d = \frac{1}{20}t$ b) 4:20 P.M.

Section 1.4

1. $V = x^3, x > 0$ **3.** $A = s^2/(4\pi), s > 0$ **5.** $V = d^3/(3\sqrt{3}), d > 0$ **7.** $s = (\sqrt{34})t, t \geq 0$ **9.** a) 1 b) 6 c) 6
11. a) $-\frac{5}{3}$ b) -1 c) 5 **13.** a) 1 b) 11 c) $\frac{1}{2}$ d) $\frac{63}{4}$ **15.** a) $4 + 4(\Delta x) + (\Delta x)^2$ b) $4(\Delta x) + (\Delta x)^2$ c) $4 + \Delta x, \Delta x \neq 0$
17. a) $1/(-3 + \Delta t)$ b) $\Delta t/[3(-3 + \Delta t)]$ c) $1/[3(-3 + \Delta t)], \Delta t \neq 0$ **19.** $x \neq 0$ **21.** $x \neq 1, 2$ **23.** $u \leq -1$ or $u \geq 1$
25. $x \geq 4$ **27.** $x \neq \pm 1$ **29.** $2, x \neq 0$ **31.** $t - 2, t \neq -2$ **33.** $\begin{cases} -1, & x < 0 \\ 1, & x > 0 \end{cases}$ **35.** $4 + \Delta x, \Delta x \neq 0$ **37.** a) 1 b) $\frac{1}{2}$ c) $-\frac{3}{2}$

39.

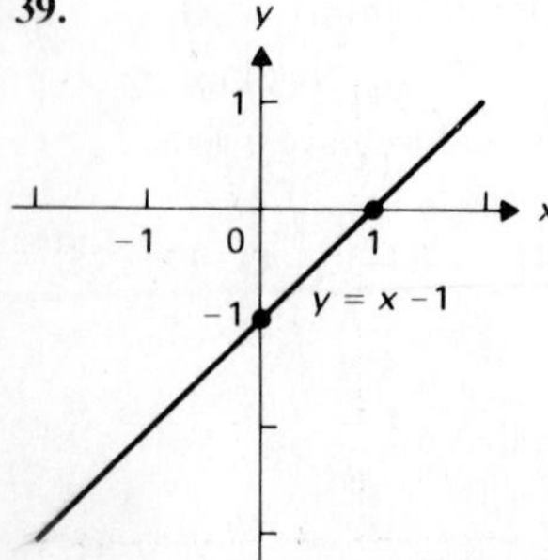

41.

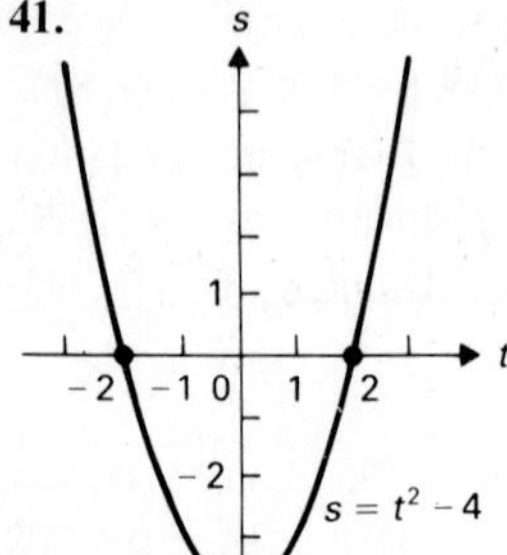

43.

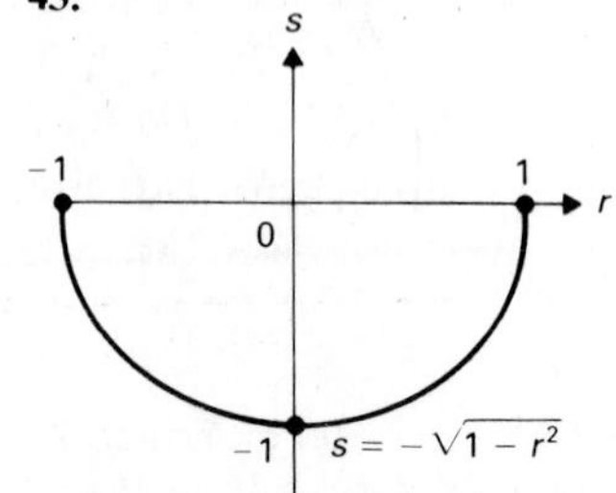

45.

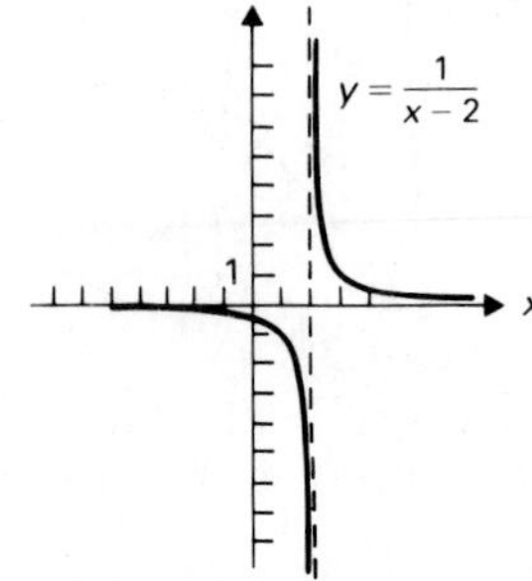

47.

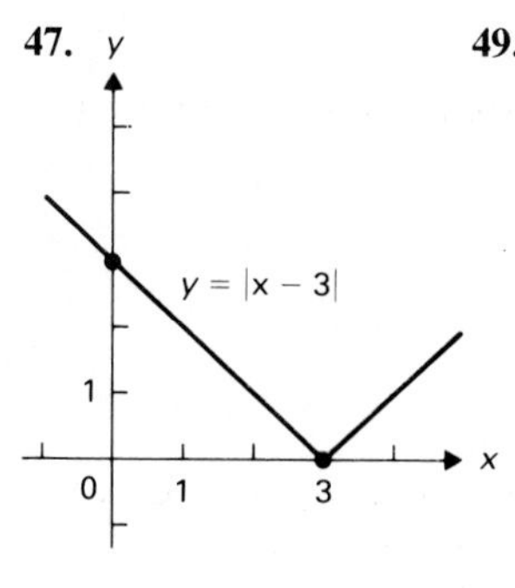

49.

x	y
-1	0
$-\frac{1}{2}$	$-\frac{1}{3}$
0	-1
$\frac{1}{2}$	-3
$\frac{3}{4}$	-7
$\frac{7}{8}$	-15
$\frac{9}{8}$	17
$\frac{5}{4}$	9
$\frac{3}{2}$	5
2	3
$\frac{5}{2}$	$\frac{7}{3}$
3	2

$y = \frac{x+1}{x-1}$

51.

x	$(x + 1)/\sqrt{x^3 + 1}$
0	1.0
1	1.4142
2	1.0
3	0.7559
4	0.6202
5	0.5345
6	0.4752
7	0.4313
8	0.3974
9	0.3701
10	0.3477

$y = \frac{(x+1)}{\sqrt{x^3+1}}$

Section 1.5

1. $y = 3x$

3.

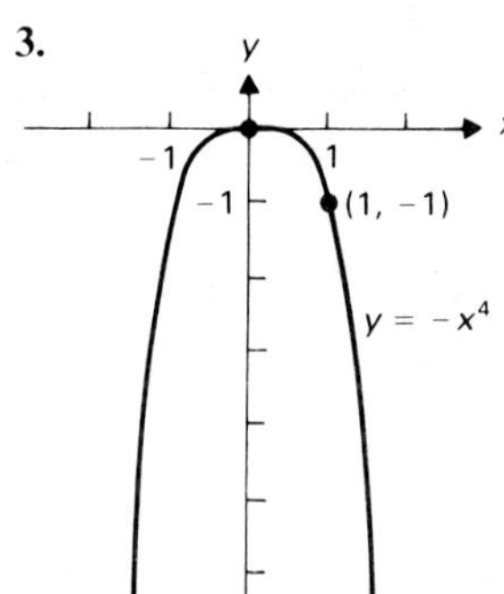

5.

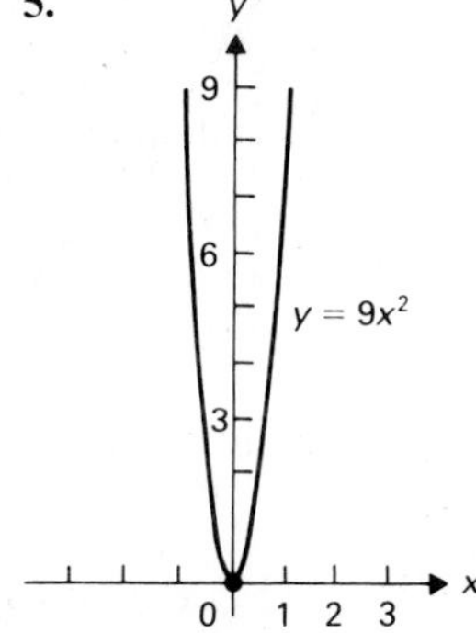

7.

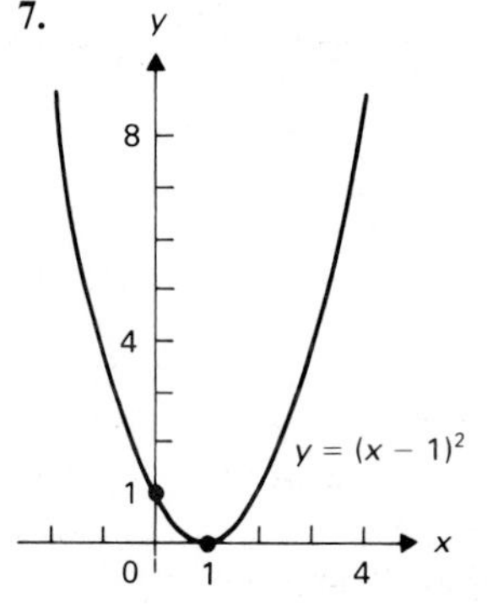

9.

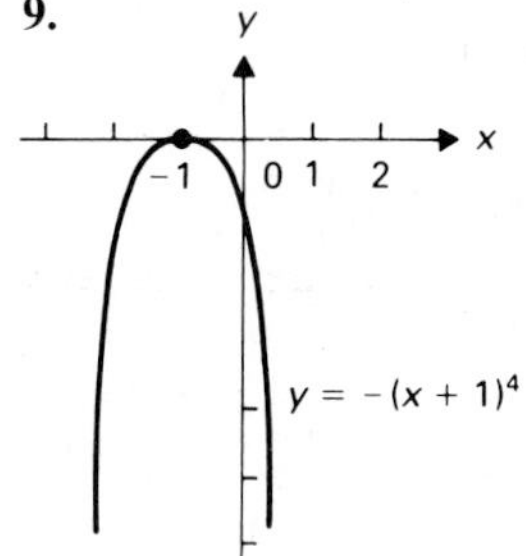

11.

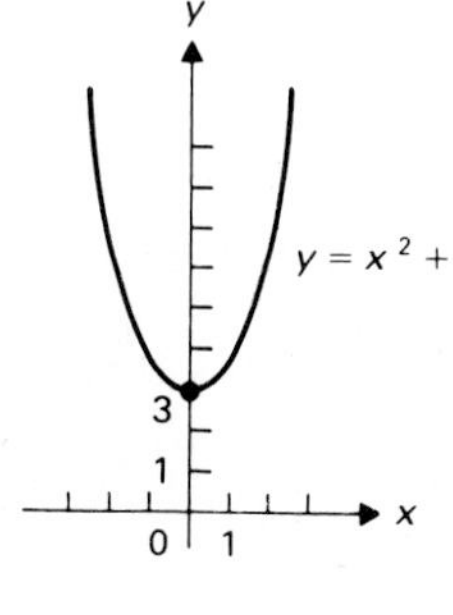

13.

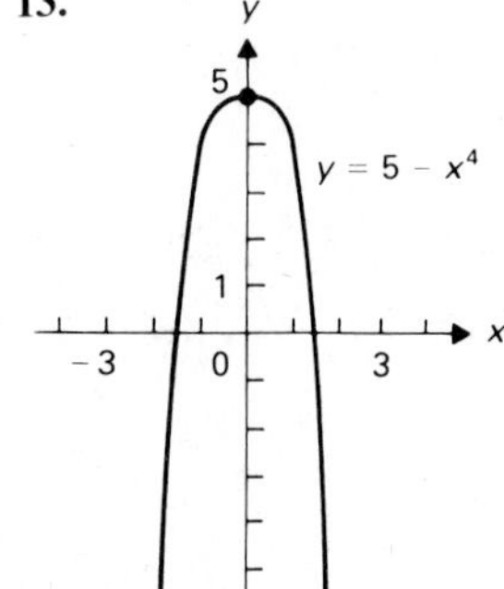

15.

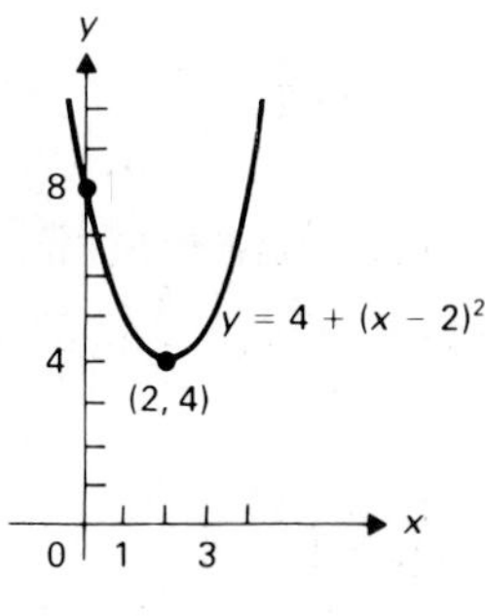

17.

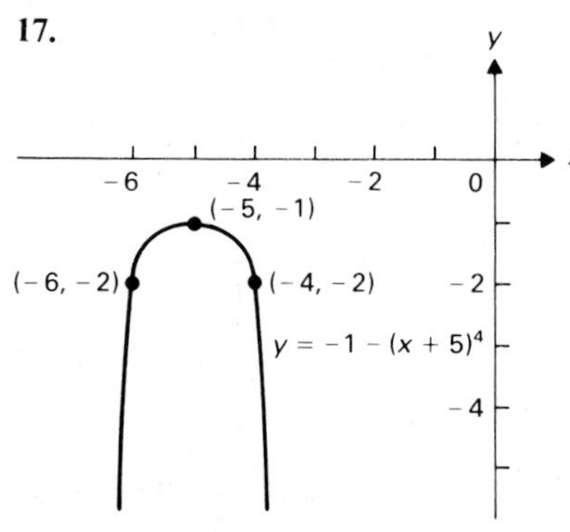

19.

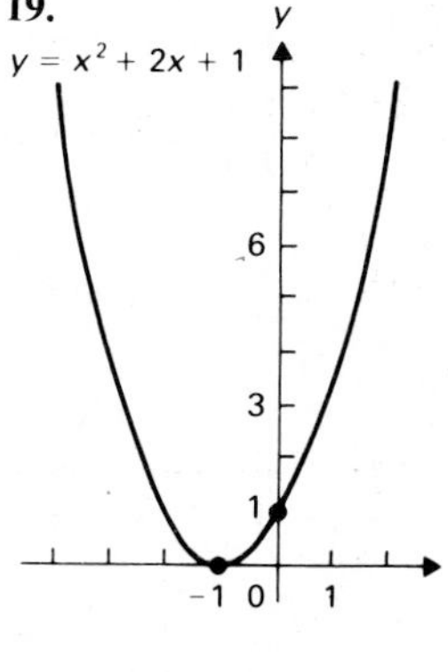

21.

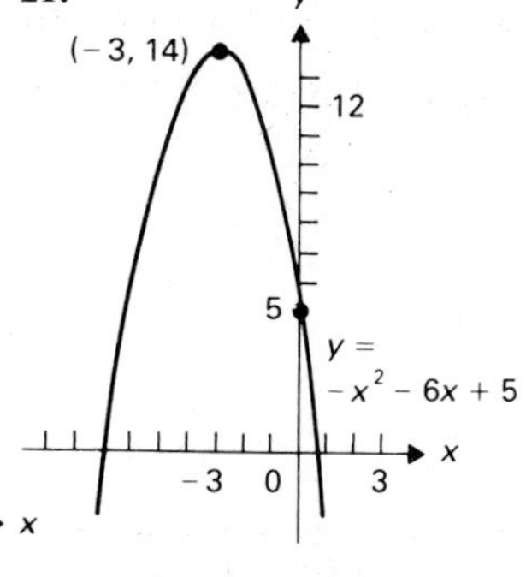

23.

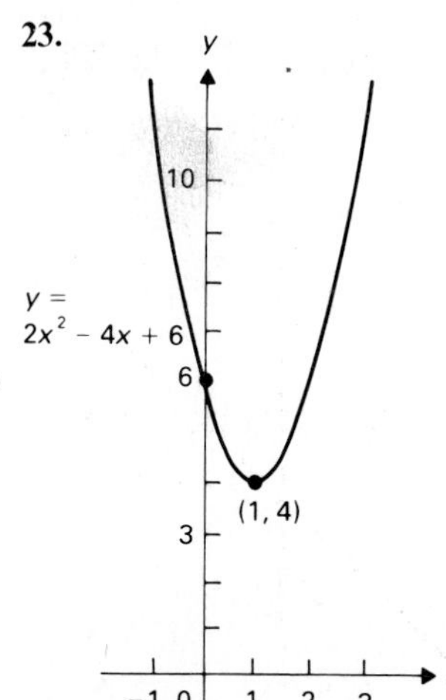

25.

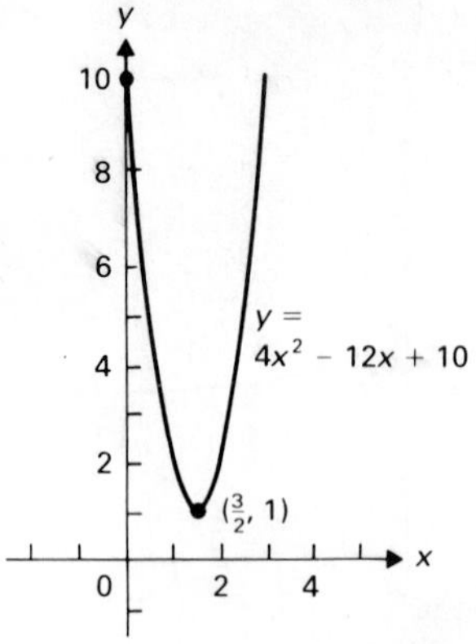

27.

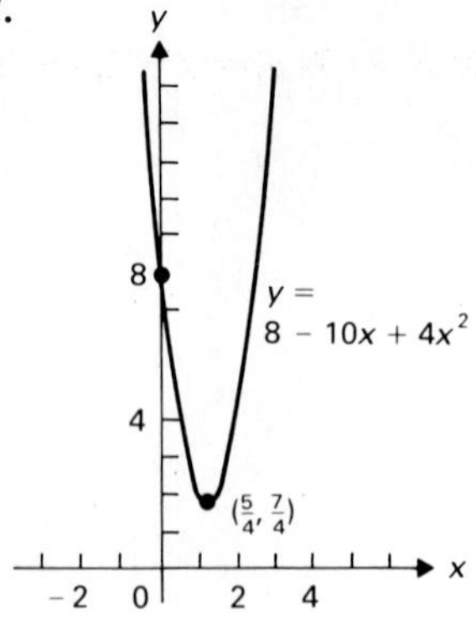

29. No.

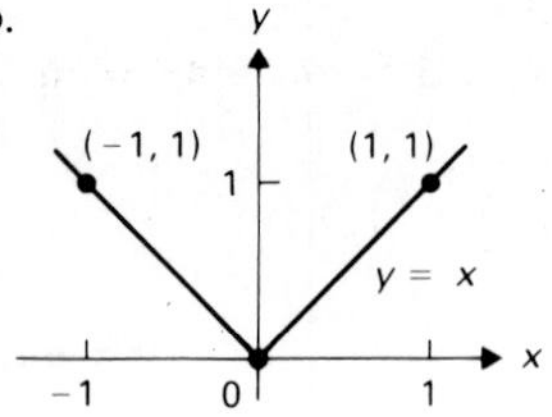

Review Exercise Set 1.1

1. a) 7 b)

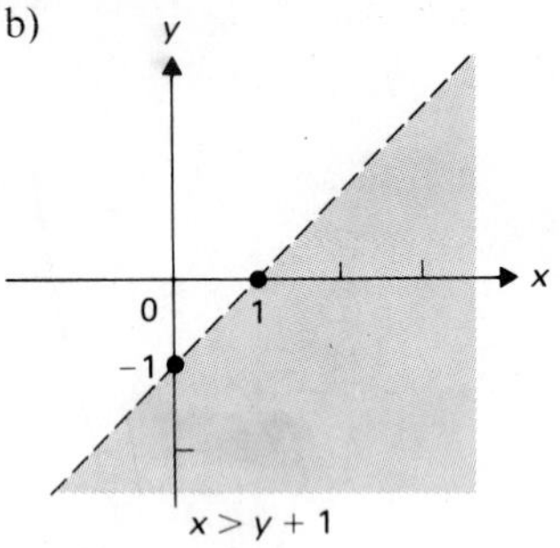

3. a) $(x-2)^2 + (y+1)^2 = 53$ b)

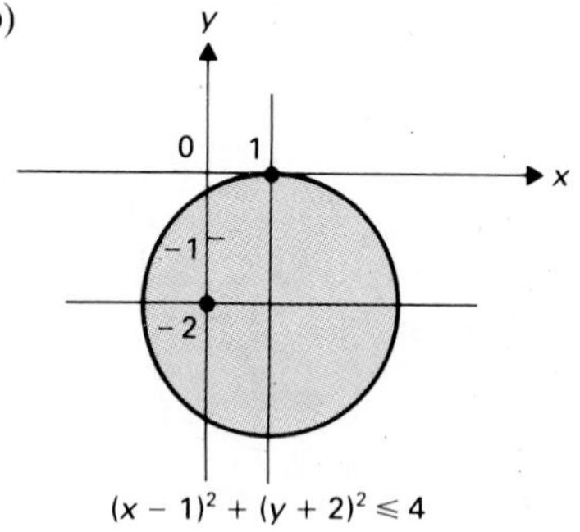

5. a) $x = -4$ b) $x - 3y = -7$ **7.** a) $x \neq 0, x \neq 5$ b) $\frac{6}{7}$ **9.**

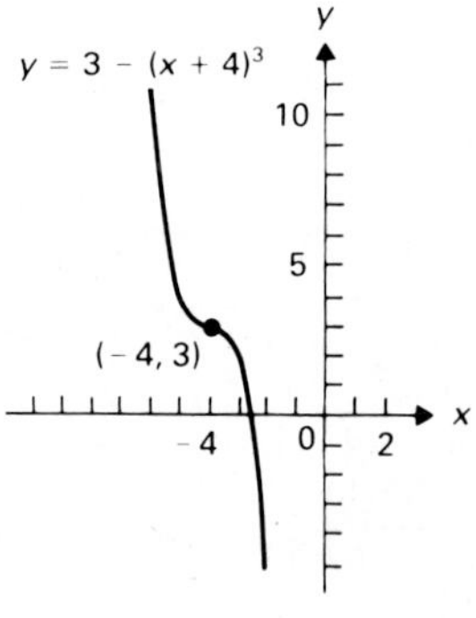

Review Exercise Set 1.2

1. a) 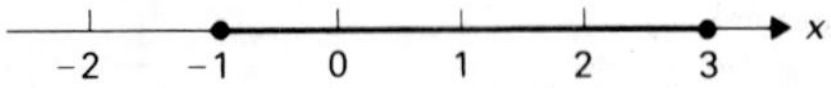

b) −1.6 **3.** a) $(x-1)^2 + (y-5)^2 = 10$ b) Center $(3, -4)$, radius 6

5. a) $4y = x + 17$ b) $x = 3$ **7.** a) $-5 \le x \le 5$ b) 4 c)

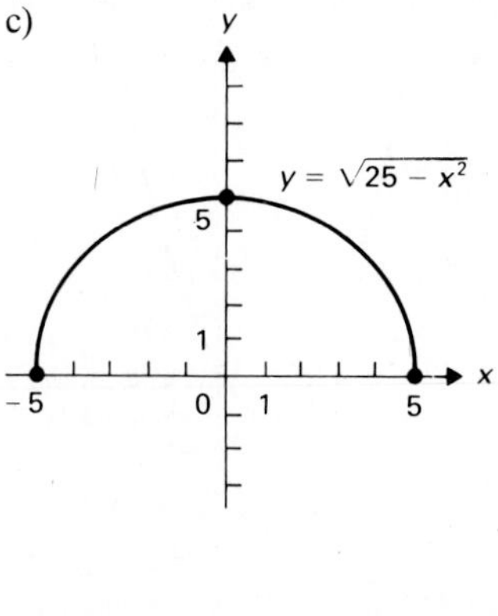

9.

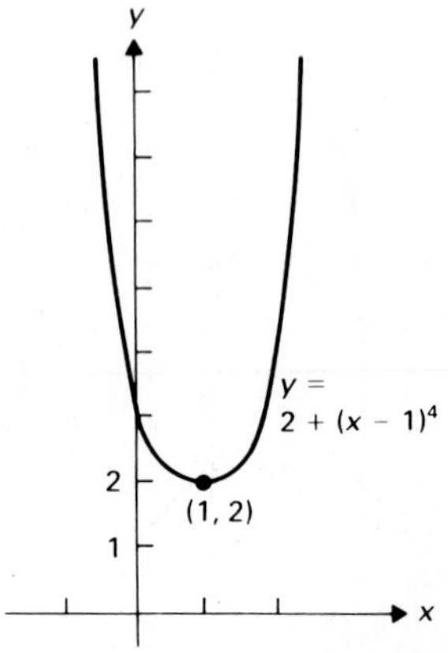

Review Exercise Set 1.3

1. a) 1 b) $[-\frac{1}{2}, \frac{9}{2}]$ **3.** $(x + 1)^2 + (y - 4)^2 = 36$ **5.** $4x + 3y = 4$ **7.** a) $x \neq -2, x \neq 1$ b) $f(x) = x/(x + 2), x \neq 1$

9.

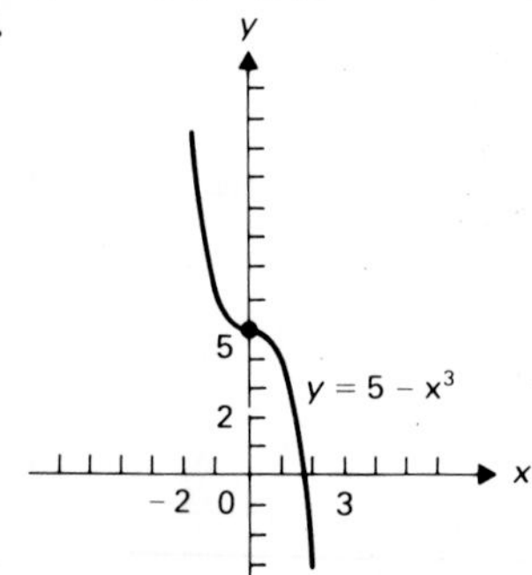

Review Exercise Set 1.4

1. a) −1 0 1 2 3 4 5 x b) $[-2, 6]$ **3.** $(x - 7)^2 + (y - 6)^2 = 36$ **5.** Perpendicular **7.** a) 1 b) -7 c) 7

9.

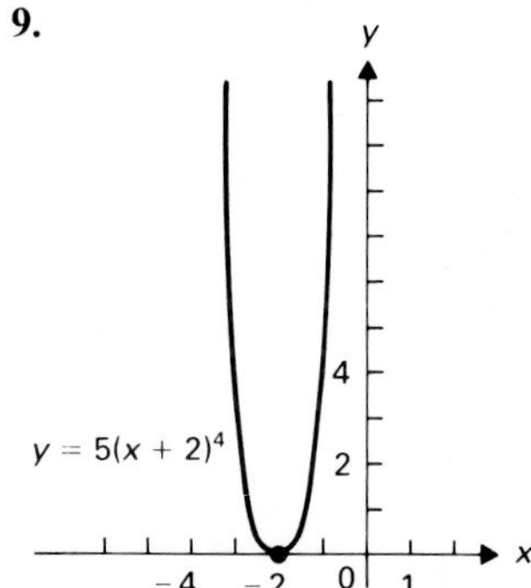

More Challenging Exercises 1

1. a) Adding:

$$\begin{array}{c} -|a| \leq a \leq |a| \\ -|b| \leq b \leq |b| \\ \hline -(|a| + |b|) \leq a + b \leq |a| + |b| \end{array}$$

so $|a + b| \leq |a| + |b|$.

b) From part (a), we have $|a| = |(a - b) + b| \leq |a - b| + |b|$, so $|a - b| \geq |a| - |b|$.

3. From Exercise 2,

$$2(a_1a_2 + b_1b_2) \leq 2\sqrt{a_1^2 + b_1^2} \cdot \sqrt{a_2^2 + b_2^2}.$$

Adding $a_1^2 + a_2^2 + b_1^2 + b_2^2$ to both sides, we obtain

$$(a_2 + a_1)^2 + (b_2 + b_1)^2 \leq (\sqrt{a_1^2 + b_1^2} + \sqrt{a_2^2 + b_2^2})^2.$$

5. a) Do not intersect b) $10x + 2y = 9$ **7.** a) 4 b) 5 **9.** $2 - \sqrt{5} < x < 2 + \sqrt{5}$ **11.** $18\sqrt{5}/5$

13.

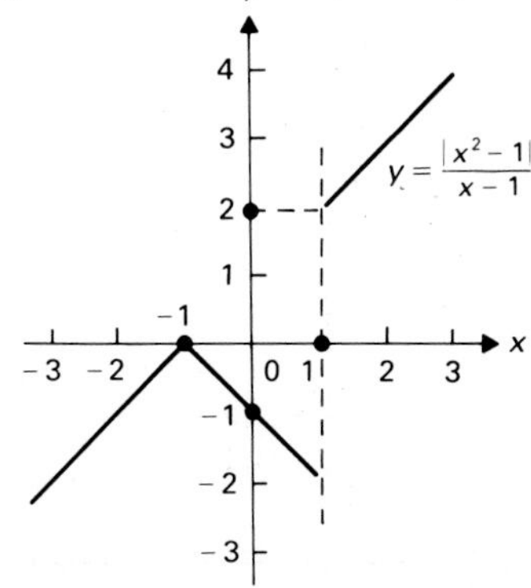

15. $g(y) = (3y + 7)/(2 - y)$

CHAPTER 2

Section 2.1

1. 8.01 **3.** -4.2 **5.** 0.31 **7.** 0.001 **9.** 0.24922 **11.** 8 **13.** -4 **15.** 0 **17.** 0 **19.** $\frac{1}{4}$ **21.** $y = 2x - 1$ **23.** $y = -\frac{1}{4}x + 1$
25. a) 26 mph b) 23 mph c) 21.5 mph d) About 20 mph
27. a) $0 \le t \le 4$ b) 64 ft/sec c) Speed $= 32t$ ft/sec d) $t = 2$ sec **29.** a) -0.06 dynes/sec b) -0.02 dynes/sec **31.** 3.42552
33. 1.34164 **35.** 9.88751 **37.** 2.582

Section 2.2

1. 3 **3.** 0 **5.** 2 **7.** $\frac{1}{2}$ **9.** Does not exist **11.** 0 **13.** 2 **15.** $-\frac{2}{3}$ **17.** $\frac{2}{5}$ **19.** $\lim_{s\to 1}\left|\frac{2(s-1)}{(s-1)^2}\right| = \infty$ **21.** -1 **23.** 2 **25.** 0
27. $-\frac{1}{9}$ **29.** a) -2 b) 0 c) 9 **31.** a) 2 b) Does not exist c) 2 **33.** a) $\frac{1}{8}$ b) ∞ c) $\frac{1}{3}$ **35.** a) $\frac{1}{2}$ b) ∞ c) ∞
37. No points near -3 except -3 itself are in the domain of $\sqrt{-(x+3)^2}$.
39. No points near zero are in the domain of $1/\sqrt{x^2 - 9}$. **41.** Any $\delta' > 0$ **43.** $0 < \delta \le \varepsilon/3$ **45.** $0 < \delta \le 3\varepsilon$ **47.** -0.0185
49. 2.718 **51.** Does not exist **53.** -1

Section 2.3

1. $\lim_{x\to 2}\left|\frac{1}{2-x}\right| = \infty$ **3.** ∞ **5.** ∞ **7.** $-\infty$ **9.** $-\infty$ **11.** ∞ **13.** ∞ **15.** ∞ **17.** $-\infty$ **19.** 1 **21.** -1 **23.** 0 **25.** $-\infty$

27. -2 **29.** $-\infty$ **31.** 0 **33.** $-\infty$ **35.** ∞ **37.** 48 ft

39. Vertical asymptote $x = 2$, horizontal asymptote $y = 0$, y-intercept $-\frac{1}{2}$

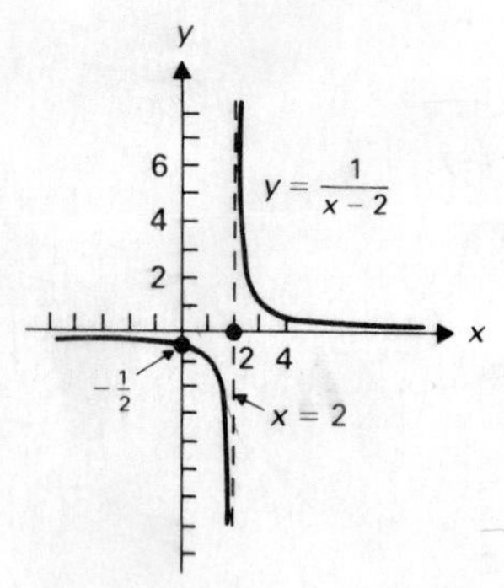

41. Vertical asymptote $x = 0$, horizontal asymptote $y = 1$, x-intercept 1

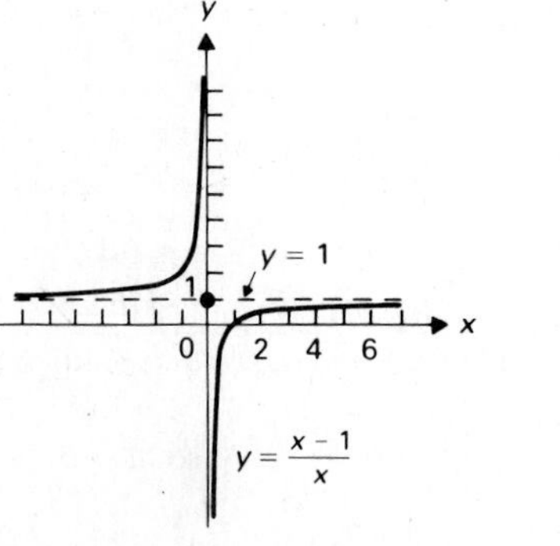

43. Vertical asymptote $x = 1$, horizontal asymptote $y = 2$, x-intercept -2, y-intercept -4

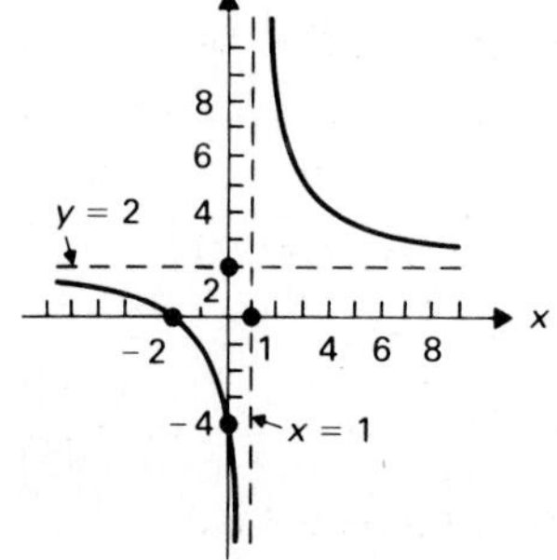

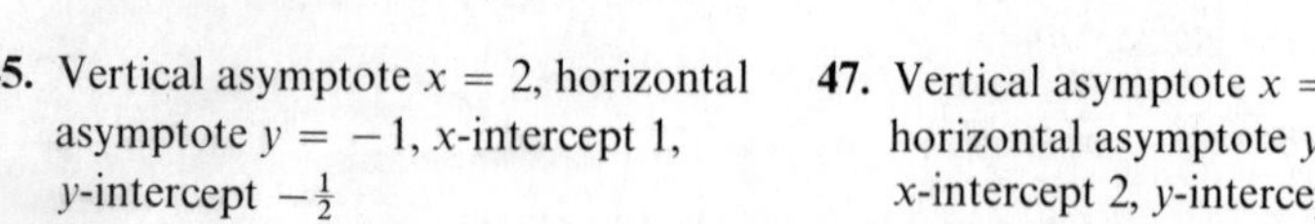

45. Vertical asymptote $x = 2$, horizontal asymptote $y = -1$, x-intercept 1, y-intercept $-\frac{1}{2}$

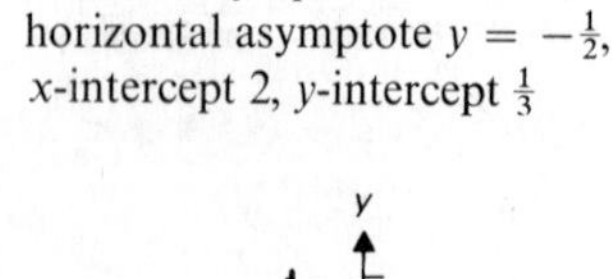

47. Vertical asymptote $x = -3$, horizontal asymptote $y = -\frac{1}{2}$, x-intercept 2, y-intercept $\frac{1}{3}$

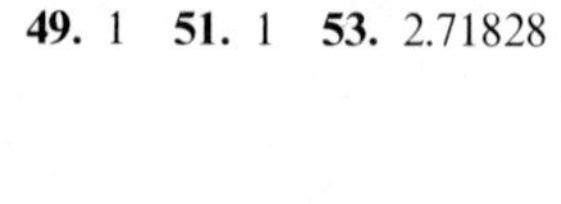

49. 1 **51.** 1 **53.** 2.71828

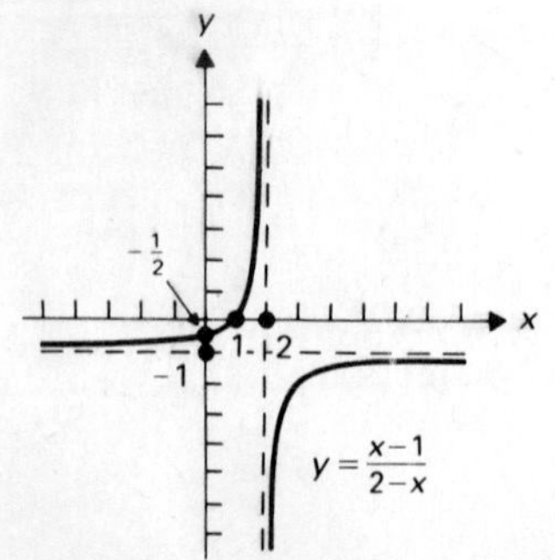

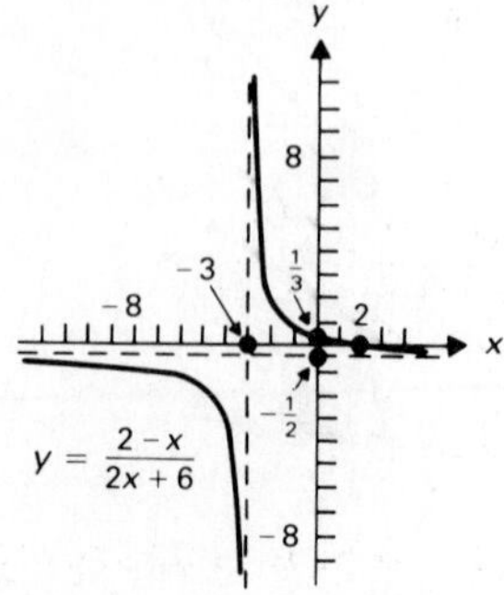

Section 2.4

1.

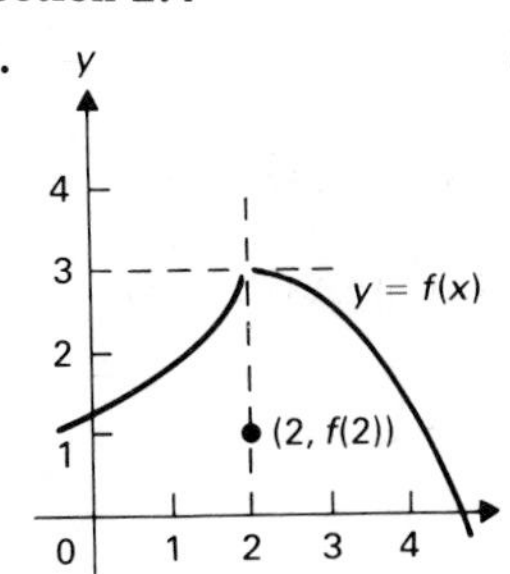

3.

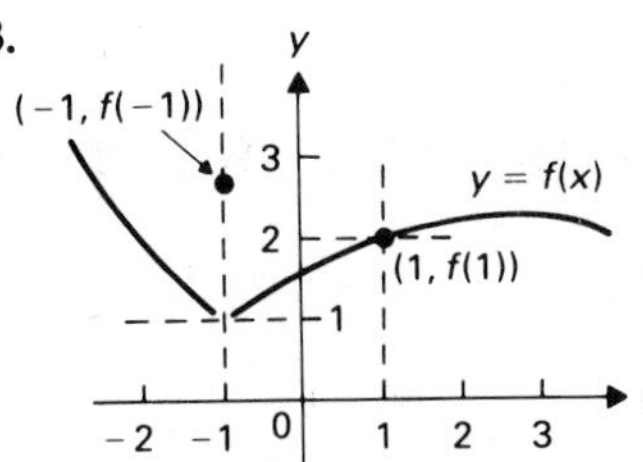

5.

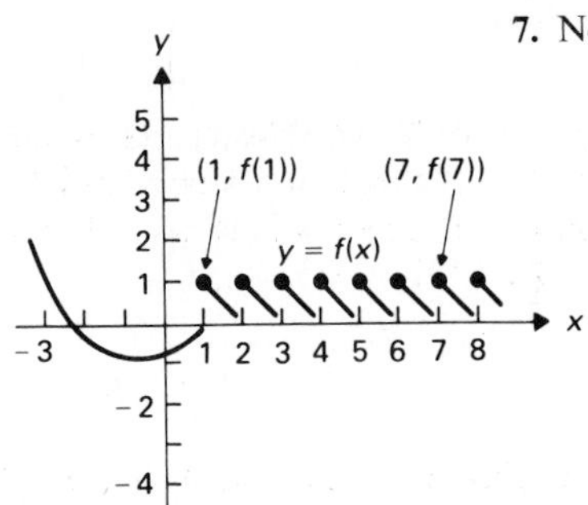

7. No; $\lim_{x\to 2} f(x) = -8 \neq 8$.

9. a) Yes; $\lim_{x\to 3^+} f(x) = 0 = \lim_{x\to 3^-} f(x) = f(3)$. b) Yes; $\lim_{x\to 1^+} f(x) = -2 = \lim_{x\to 1^-} f(x) = f(1)$. c) Yes; continuous where $x \neq 1, 3$ as part of a polynomial function and continuous at 1 and 3 by parts (a) and (b).

11. i) If your birth weight was 7 lb and you now weigh 138 lb, at some time you weighed exactly 71.583 lb. ii) If you are driving on a trip and going 47 mph at 1:00 P.M. and 63 mph at 2:00 P.M., then at some instant between 1:00 and 2:00 P.M. you were going exactly 53.4 mph. iii) If a tire with pressure 15 lb/in^2 is inflated to a pressure of 30 lb/in^2, then at some instant during inflation the pressure was 23 lb/in^2.

13. Since $f(a)$ and $f(b)$ have opposite sign, zero is between $f(a)$ and $f(b)$. By Theorem 2.3, we have $f(x_0) = 0$ for some x_0 in $[a, b]$.

15. Let $f(x) = x^4 + 3x^3 + x + 4$. a) $f(-2) = -6$, $f(0) = 4$; by the corollary, $f(x) = 0$ has a solution in $[-2, 0]$. b) No; $f(x) > 0$ for $x \geq 0$ as a sum of positive quantities.

17. No; $x^2 - 4x + 5 = (x - 2)^2 + 1 > 0$ for all x. **19.** F F T F T

21. If $f(S/2) = 96$, we are done. Suppose $f(S/2) \neq 96$. Then $f(0) = 96 -$ (Speed at $S/2$) and $f(S/2) =$ (Speed at $S/2$) $- 96 = -f(x)$. Then $f(0)$ and $f(S/2)$ have opposite sign, so for some c, where $0 < c < S/2$, we have $f(c) =$ (Speed at c) $-$ (Speed at $c + S/2$) $= 0$. Thus (Speed at c) $=$ (Speed at $c + S/2$).

23. a) $\lim_{x\to\infty} f(x) = \lim_{x\to-\infty} f(x) = \infty$ b) Find $K_1 > 0$ such that $f(x) > f(0)$ for $x > K_1$, and find $K_2 > 0$ such that $f(x) > f(0)$ for $x < -K_2$. Let C be the maximum of K_1 and K_2. c) By Theorem 2.4, f assumes a minimum over $[-C, C]$. Since $f(x) > f(0)$ for $|x| > C$, this minimum value must also be minimum assumed for all x.

Review Exercise Set 2.1

1. $-\frac{2}{3}$ **3.** $-\infty$ **5.** 0 **7.** ∞ **9.** No, for $\lim_{x\to -3} f(x) = \lim_{x\to -3} \dfrac{(x+3)(x-3)}{x+3} = -6 \neq f(3)$.

Review Exercise Set 2.2

1. $-2/(2x_1 + 1)^2$ **3.** $\frac{1}{4}$ **5.** 0 **7.** $\frac{7}{4}$

9. a) 6 b) 6 c) Yes; $\lim_{x\to 5} f(x) = 6 = f(5)$. d) Yes; it is continuous at 5 by parts (*a*) – (*c*), and is continuous at every other point in its domain as a rational function near that point.

Review Exercise Set 2.3

1. $1/\sqrt{2x_1 - 3}$ **3.** a) Does not exist b) 0 **5.** $-\infty$ **7.** $-\infty$

9. Yes; it is continuous as a rational function at every point in its domain.

Review Exercise Set 2.4

1. $[1/2(\sqrt{x_1})] + 1$ **3.** $-\frac{4}{5}$ **5.** $-\infty$ **7.**

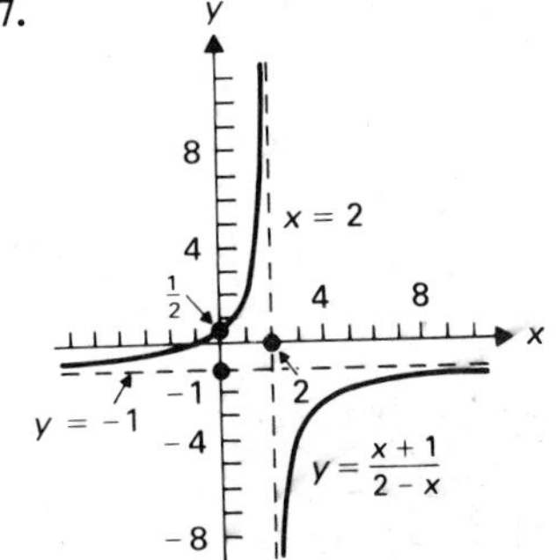

9. $\text{Lim}_{x\to -1^+} f(x) = 0 = \lim_{x\to -1^-} f(x) \neq 1 = f(-1)$. Therefore $f(x)$ is not continuous at -1.

More Challenging Exercises 2

1. 1 **3.** 2 **5.** 0 **7.** ∞ **9.** 1

11. a) For some $\varepsilon > 0$, there does not exist $\delta > 0$. b) For some apple blossom, there exists no apple. d) Find one $\varepsilon > 0$ such that for every $\delta > 0$, there is an x_δ such that $0 < |x_\delta - a| < \delta$, but $|f(x_\delta) - c| \geq \varepsilon$.

13. Let $\varepsilon > 0$ be given. Find $\delta_1 > 0$ such that

$$L - \frac{\varepsilon}{2} < f(x) < L + \frac{\varepsilon}{2} \quad \text{if} \quad 0 < |x - a| < \delta_1$$

and $\delta_2 > 0$ such that

$$M - \frac{\varepsilon}{2} < g(x) < M + \frac{\varepsilon}{2} \quad \text{if} \quad 0 < |x - a| < \delta_2.$$

Let δ be the minimum of δ_1 and δ_2, so both inequalities hold if $0 < |x - a| < \delta$. Adding the inequalities, we have

$$L + M - \varepsilon < f(x) + g(x) < L + M + \varepsilon \quad \text{if} \quad 0 < |x - a| < \delta.$$

15. a) Incorrect; replace "$|f(x) - a|$" by "$|f(x) - f(a)|$." b) Incorrect; delete "$0 <$." c) Correct d) Incorrect; replace "some" by "each." e) Incorrect; replace "$\leq$" by "$<$." f) Correct

17. Let

$$f(x) = \begin{cases} 1 & \text{for } x \geq 2, \\ 0 & \text{for } x < 2, \end{cases} \quad \text{and} \quad g(x) = \begin{cases} 0 & \text{for } x \geq 2, \\ 10 & \text{for } x < 2. \end{cases}$$

Then $f(x)g(x) = 0$ for all x.

CHAPTER 3

Section 3.1

1. $2x - 3$ **3.** $-2(2x + 3)^2$ **5.** $1/(x + 1)^2$ **7.** 3 **9.** $14x^6 + 8x$ **11.** $x - \frac{3}{2}$ **13.** $324x^3 - 160x^4$ **15.** $4x^3 + 12x^2 + 8x$ **17.** $36x^2 + 40x$ **19.** $x^2 - \frac{1}{3}$ **21.** a) 2 b) 0 **23.** a) 28 b) -8 **25.** a) -1 b) $-\frac{3}{2}$ **27.** a) $y = -14x - 11$ b) $14y = x + 240$ **29.** a) $y = 21x - 32$ b) $x + 21y = 212$ **31.** a) $y = -4x - 1$ b) $4y = x + 13$

33. a) $1/(2\sqrt{x})$

b)
$$\lim_{\Delta x \to 0} \frac{\sqrt{x + \Delta x} - \sqrt{x}}{\Delta x} = \lim_{\Delta x \to 0} \frac{(\sqrt{x + \Delta x} - \sqrt{x})(\sqrt{x + \Delta x} + \sqrt{x})}{\Delta x(\sqrt{x + \Delta x} + \sqrt{x})}$$
$$= \lim_{\Delta x \to 0} \frac{x + \Delta x - x}{\Delta x(\sqrt{x + \Delta x} + \sqrt{x})}$$
$$= \lim_{\Delta x \to 0} \frac{1}{\sqrt{x + \Delta x} + \sqrt{x}} = \frac{1}{2\sqrt{x}}$$

c) $\dfrac{3}{2\sqrt{x}} - 4x$ d) $\dfrac{\sqrt{5}}{2\sqrt{x}} - \dfrac{\sqrt{7}}{2\sqrt{x}}$

35. a) 12 in^3/sec b) 75 in^3/sec **37.** a) 256π in^2 b) 256π in^2/sec **39.** $f(x) = |x - 3| + |x + 3|$ **41.** $f(x) = |x|, x_1 = 0$ **43.** -0.108577 **45.** -12.51929 **47.** 45.545

Section 3.2

1. $6x + 17$ **3.** $2x/3$ **5.** $-3/x^2$ **7.** $12x^2 + (4/x^3)$ **9.** $(x^2 - 1)(2x + 1) + (x^2 + x + 2)(2x)$

11. $(x^2 + 1)[(x - 1)3x^2 + (x^3 + 3)] + [(x - 1)(x^3 + 3)](2x)$ **13.** $\left(\dfrac{1}{x^2} - \dfrac{4}{x^3}\right)2 + (2x + 3)\left(\dfrac{-2}{x^3} + \dfrac{12}{x^4}\right)$

15. $4 + (3/x^2)$ **17.** $[(x + 3)2x - (x^2 - 2)]/(x + 3)^2$

19. $[(x^2 + 2)((x^2 + 9) + (x - 3)2x) - (x^2 + 9)(x - 3)2x]/(x^2 + 2)^2$

21. $\dfrac{(x - 1)(4x^2 + 5)[(2x + 3)2x + (x^2 - 4)2] - (2x + 3)(x^2 - 4)[(x - 1)8x + (4x^2 + 5)]}{(x - 1)^2(4x^2 + 5)^2}$

23. $\dfrac{x + 1}{2x + 3}\left(\dfrac{-1}{x^2} + \dfrac{2}{x^3}\right) + \left(\dfrac{1}{x} - \dfrac{1}{x^2}\right)\dfrac{1}{(2x + 3)^2}$ **25.** $x \cos x + \sin x$ **27.** $2 \sin x \cos x$ **29.** $\dfrac{(\cos x)^2 + (\sin x)^2}{(\cos x)^2}$

31. $\dfrac{3x^2 \sin x - x^3 \cos x}{(\sin x)^2}$ **33.** $\dfrac{(x^2 - 4x) \cos x - (2x - 4) \sin x}{(x^2 - 4x)^2}$ **35.** -3 **37.** 2 **39.** $-1380/169$ **41.** $\frac{41}{6}$

43. $f(4) = \frac{28}{17}, f'(4) = -\frac{1}{17}$ **45.** $f'(3) = -5, g'(3) = 2$ **47.** Tangent line: $x + y = 2$; normal line: $y - x = 0$

49. Tangent line: $y + 5x = -3$; normal line: $5y - x = -15$

Section 3.3

1. $dy = \frac{1}{(x+1)^2}\,dx$ **3.** $dA = 2\pi r\,dr$ **5.** $dx = \frac{-4t}{(t^2-1)^2}\,dt$ **7.** $dV = 3\pi r^2\,dr$ **9.** 0.99 **11.** $\frac{47.68}{9} \approx 5.298$ **13.** -0.4
15. 10.05 **17.** 3.975 **19.** $\frac{23}{12}$ **21.** $\frac{93}{46}$
23. a) $3/\pi$ ft b) The estimate is exact, because the circumference of the earth is a *linear* function of the radius. **25.** $\frac{7}{24}$ ft^2
27. $25/(68\pi)$ ft **29.** 6% **31.** 0.5% **33.** $\varepsilon = \Delta x$; $\lim_{\Delta x\to 0} \varepsilon = \lim_{\Delta x\to 0} \Delta x = 0$ **35.** 3.0541 **37.** 4.1803 **39.** 1.9991

Section 3.4

1. a) 3 b) 3 **3.** a) 1404 b) 1404 **5.** $12(3x+2)^3$ **7.** $9x^2(x^2+3x)^2(x^3-1)^2 + 2(x^3-1)^3(x^2+3x)(2x+3)$
9. $\frac{16x(4x^2+1)^2 - 16x(8x^2-2)(4x^2+1)}{(4x^2+1)^4}$ **11.** $8x - 2 + 5x^{2/3}$ **13.** $-\frac{1}{2}x^{-3/2}$ **15.** $\frac{2}{3}x^{-1/3} + \frac{1}{5}x^{-4/5}$ **17.** $(2x+1)^{-1/2}$
19. $-(5x^2+10x)^{-3/2}(5x+5)$ **21.** $x(x^2+1)^{-1/2}$ **23.** $\frac{\frac{1}{2}(x+1)x^{-1/2} - \sqrt{x}}{(x+1)^2}$
25. $\frac{8}{3}\sqrt{3x+4}(4x+2)^{-1/3} + \frac{3}{2}(4x+2)^{2/3}(3x+4)^{-1/2}$
27. $\sqrt{2x+1}\left[\frac{2(2x+5)(4x^2-3x)(8x-3) - 2(4x^2-3x)^2}{(2x+5)^2}\right] + \frac{(4x^2-3x)^2}{2x+5}(2x+1)^{-1/2}$ **29.** $2\cos 2x$ **31.** $3\sin^2 x\cos x$
33. $\frac{1}{2}(x+\sin x)^{-1/2}(1+\cos x)$ **35.** -2 **37.** 40 **39.** $4y + 3x = 25$ **41.** $3y - x = 5$ **43.** 70.4875 **45.** 8.6875

Section 3.5

1. $y' = 5x^4 - 12x^3$, $y'' = 20x^3 - 36x^2$, $y''' = 60x^2 - 72x$
3. $y' = (1/\sqrt{5})(-\frac{1}{2})x^{-3/2}$, $y'' = (1/\sqrt{5})(\frac{3}{4})x^{-5/2}$, $y''' = (1/\sqrt{5})(-\frac{15}{8})x^{-7/2}$
5. $y' = x(x^2+1)^{-1/2}$, $y'' = -x^2(x^2+1)^{-3/2} + (x^2+1)^{-1/2}$, $y''' = 3x^3(x^2+1)^{-5/2} - 3x(x^2+1)^{-3/2}$
7. $y' = (x+1)^{-2}$, $y'' = -2(x+1)^{-3}$, $y''' = 6(x+1)^{-4}$ **9.** $-1/(10\sqrt{10})$ **11.** $\frac{3}{4}$ **13.** Velocity 2; acceleration 18
15. a) $x = -10$ when $t = 0$; clearly x increases as t increases, and $\lim_{t\to\infty} x = 10$, so the distance traveled is always less than $10 - (-10) = 20$. b) $v = 40t/(t^2+1)^2$, $t \geq 0$ c) The body is always moving in the direction of increasing x for $t > 0$.
17. a) $v = -32t + 48$ ft/sec b) $a = -32$ ft/sec^2 c) 48 ft/sec d) $t = \frac{3}{2}$ sec e) 36 ft f) $0 \leq t \leq 3$
19. Decreasing **21.** Increasing, downward **23.** Increasing, upward **25.** Speed $= \sqrt{65/4}$; slope $= 8$
27. Speed $= \frac{1}{4}$; slope undefined **29.** Speed $= 2\sqrt{34/25}$; slope $= -\frac{3}{5}$ **31.** $3x + y = 3$ **33.** $x + 3y = 12$
35. $dy/dx = -\frac{1}{2}$; $d^2y/dx^2 = \frac{3}{4}$ **37.** a) 4.2 m/sec b) 4.2 m/sec c) 25 m d) $\frac{10 + 5\sqrt{10}}{7} \approx 3.687$ sec
39. a) 122.5 m b) 300 m c) $\sqrt{3301} \approx 57.454$ m/sec **41.** $f''(1) \approx 16.25$; $f'''(1) \approx 16.875$ **43.** $f''(3) \approx 3.84362$; $f'''(3) \approx 2.6642$
45. $f''(1) \approx 14.46699$; $f'''(1) \approx 25.72$

Section 3.6

1. $\frac{3}{4}$ **3.** 1 **5.** $-\frac{5}{3}$ **7.** -3 **9.** $\frac{7}{5}$ **11.** 2 **13.** $-\frac{20}{33}$ **15.** $-\frac{3}{4}$ **17.** $-\frac{160}{97}$
19. Tangent line: $5y - 2x = 1$; normal line: $2y + 5x = 12$ **21.** $x + 16y = 40$ **23.** $(2, 4), (-2, -4)$ **25.** $(4, 2)$ **27.** $\frac{29}{30}$
29. $-\frac{7}{27}$ **31.** 4
33. Computation shows that the slope of the first curve is -1 at $(2, 4)$, while the second curve has slope 1 there. The curves are orthogonal.
35. Let (x_0, y_0) be a point of intersection. Since both c and k are nonzero, neither x_0 nor y_0 is zero at a point of intersection. By implicit differentiation, the slope of $y^2 - x^2 = c$ at (x_0, y_0) is x_0/y_0, while the slope of $xy = k$ is $-y_0/x_0$. The curves are orthogonal.

Review Exercise Set 3.1

1. a) $f'(x_1) = \lim_{\Delta x\to 0} \frac{f(x_1+\Delta x) - f(x_1)}{\Delta x}$ b) $f'(x) = \lim_{\Delta x\to 0} \frac{(x+\Delta x)^2 - 3(x+\Delta x) - (x^2-3x)}{\Delta x}$

$$= \lim_{\Delta x\to 0} \frac{x^2 + 2x\cdot\Delta x + (\Delta x)^2 - 3x - 3\cdot\Delta x - x^2 + 3x}{\Delta x}$$

$$= \lim_{\Delta x\to 0} \frac{\Delta x(2x - 3 + \Delta x)}{\Delta x} = \lim_{\Delta x\to 0} (2x - 3 + \Delta x) = 2x - 3$$

3. $(x^2-3x)(12x^2-2) + (4x^3-2x+17)(2x-3)$ **5.** 150.4 **7.** $\frac{1}{2}(x^2-17x)^{-1/2}(2x-17)$ **9.** $32y - x = 67$

Review Exercise Set 3.2

1. $$f'(x) = \lim_{\Delta x \to 0} \frac{[1/(2(x + \Delta x) + 1)] - [1/(2x + 1)]}{\Delta x}$$
$$= \lim_{\Delta x \to 0} \frac{(2x + 1 - 2x - 2 \cdot \Delta x - 1)/[(2(x + \Delta x) + 1)(2x + 1)]}{\Delta x}$$
$$= \lim_{\Delta x \to 0} \frac{-2}{[2(x + \Delta x) + 1](2x + 1)} = \frac{-2}{(2x + 1)^2}$$

3. $\dfrac{x^2 + 3}{x} + 2x(\ln x)$ **5.** $dy = (6x - 6)\,dx$ **7.** $-\frac{15}{4}$ **9.** a) $t = 1, 2$ b) $t = 0, 6$ c) $6\sqrt{17}$

Review Exercise Set 3.3

1. $y = 2x$ **3.** $\frac{14}{5}$ **5.** $-\frac{512}{625}$ **7.** a) 4 b) 3 c) 5 **9.** $\frac{13}{152}$

Review Exercise Set 3.4

1. $x + 8y = -22$ **3.** $[(x^2 + 3)/\sqrt{2x + 4}] + 2x\sqrt{2x + 4}$ **5.** **7.** -57 **9.** $(-7, 1)$ and $(3, 1)$

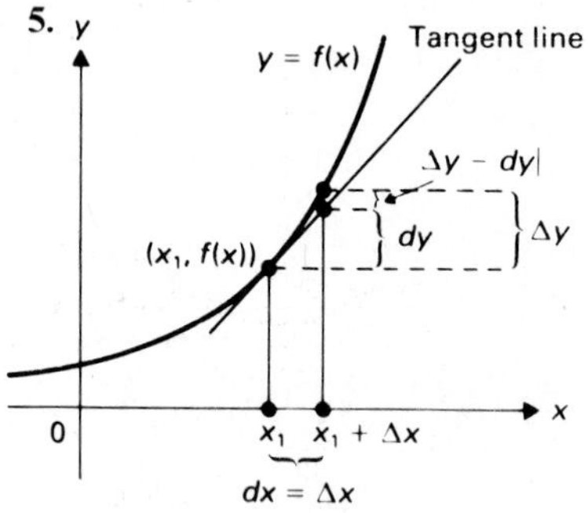

More Challenging Exercises 3

1. 20 **3.** 24 **5.** 8 **7.** $(fg)'(a) = f(a)g'(a) + f'(a)g(a) = 0 \cdot g'(a) + f'(a) \cdot 0 = 0$

9. Let $p(x) = (x - a)^2 q(x)$. Then $p(a) = 0 \cdot q(a) = 0$. Also

$$p'(x) = (x - a)^2 q'(x) + 2(x - a)q(x).$$

so

$$p'(a) = 0 \cdot q'(a) + 0 \cdot q(a) = 0.$$

11. $y = 2x + 2$, $y = 6x - 14$

13. From $y = f(x)/g(x)$, we have $y \cdot g(x) = f(x)$. Then

$$y \cdot g'(x) + \frac{dy}{dx} g(x) = f'(x),$$

so

$$\frac{dy}{dx} = \frac{f'(x) - y \cdot g'(x)}{g(x)} = \frac{f'(x) - (f(x)/g(x))g'(x)}{g(x)}$$
$$= \frac{f'(x)g(x) - f(x)g'(x)}{g(x)^2}.$$

15. $$\left.\frac{d(f(g(h(t))))}{dt}\right|_{t=t_1} = f'(g(h(t_1))) \cdot g'(h(t_1)) \cdot h'(t_1)$$

CHAPTER 4

Section 4.1

1. $\sqrt{3}/2$ **3.** $-1/\sqrt{3}$ **5.** $-\sqrt{2}$ **7.** $2/\sqrt{3}$ **9.** Undefined **11.** Undefined **13.** $-\sqrt{2}$ **15.** $-\sqrt{2}$ **17.** -2 **19.** 0 **21.** 1 **23.** 0 **25.** -1 **27.** Undefined **29.** $\sqrt{2}$ **31.** $2\sqrt{2}/3$ **33.** $-1/\sqrt{24}$ **35.** $-\sqrt{15}$ **37.** $-4\sqrt{5}/9$ **39.** $-\frac{7}{8}$

41. a) $(u, -v)$ b) $\sin(-x) = -v = -\sin x$; $\cos(-x) = u = \cos x$

Solution

a. We are given that two points (t, P) satisfying the linear relation are (10, 500) and (12, 640).

Point (10, 500)

Slope $\dfrac{\Delta P}{\Delta t} = \dfrac{640 - 500}{12 - 10} = \dfrac{140}{2} = 70$

Linear Relation $(P - 500) = 70(t - 10)$, or $P = 70t - 200$

b. From part (a), we find that when $t = 0$, the accumulated profit is -200. This means that the cost to start the company was \$200,000. □

SUMMARY

1. A vertical line has equation $x = a$.
2. A horizontal line has equation $y = b$.
3. To find the equation of a line, find one point (x_1, y_1) on the line and the slope m of the line. The equation is then

$$y - y_1 = m(x - x_1).$$

4. The line $y = mx + b$ has slope m and y-intercept b.

EXERCISES

In Exercises 1 through 20, find the equation of the indicated line.

1. Through $(-1, 4)$ with slope 5
2. Through $(2, -3)$ with slope -3
3. Through $(4, 2)$ with slope 0
4. Through $(-2, 1)$ with slope undefined
5. Through $(4, -5)$ and $(-1, 1)$
6. Through $(2, 5)$ and $(-3, 5)$
7. Through $(-3, 4)$ and $(-3, -1)$
8. Through $(-1, -2)$ and $(4, -3)$
9. Through $(-2, 1)$ parallel to $2x + 3y = 7$
10. Through $(4, -2)$ parallel to $4x - 2y = 5$
11. Through $(-3, -1)$ perpendicular to $x + 4y = 8$
12. Through $(4, -5)$ perpendicular to $3x + 2y = -6$
13. Through $(-2, 5)$ with x-intercept -3
14. Through $(1, -4)$ with y-intercept 6
15. The perpendicular bisector of the line segment from $(-1, 5)$ to $(3, 11)$
16. The perpendicular bisector of the line segment from $(2, -3)$ to $(8, 5)$.
17. Tangent to the circle $x^2 + y^2 = 25$ at $(-3, 4)$
18. Tangent to the circle $x^2 + y^2 + 2x - 4y = -4$ at $(-1, 3)$
19. Through the centers of the circles $x^2 + y^2 - 2x + 2y = 7$ and $x^2 + y^2 - 4x + 6y = 0$
20. Through the centers of the circles $x^2 + y^2 - x + 2y = 4$ and $x^2 + y^2 + 3x - 4y = 3$

In Exercises 21 through 24, determine whether the two lines are parallel, perpendicular, or neither.

21. $x + y = 6$, $2x - 2y = 7$
22. $3x + y = 8$, $12x + 4y = -5$
23. $4x - 3y = 6$, $3x - 4y = 8$
24. $6x - 2y = 7$, $x + 3y = 4$

In Exercises 25 through 28, find the slope, x-intercept, and y-intercept of the given line.

25. $x - y = 7$
26. $y = 11$
27. $x = 4$
28. $7x - 13y = 8$

In Exercises 29 through 34, sketch the line with the given equation.

29. $x + 2y = 6$
30. $2x - y = 4$
31. $4x + 3y = 12$
32. $4x = 9$
33. $8y = -11$
34. $3x - 2y = 0$

35. Find the point of intersection of the line $2x + 3y = 7$ and the line $3x + 4y = -8$.

36. Find the point of intersection of the line $2x - 5y = 4$ and the line $3x - 9y = 1$.

37. Find the distance from the point $(-2, 1)$ to the line $3x + 4y = 8$. [*Hint:* Find the point where the line through $(-2, 1)$ and perpendicular to $3x + 4y = 8$ meets $3x + 4y = 8$.]

38. Find the distance from the point $(-4, 3)$ to the line $x - y = 6$. (See the hint in Exercise 37.)

39. Show that the perpendicular bisectors of the sides of a triangle meet at a point. [*Hint:* Let the vertices of the triangle be $(-a, 0)$, $(a, 0)$, and (b, c).]

40. Find the equation of the circle through the points $(1, 5)$, $(2, 4)$, and $(-2, 6)$. [*Hint:* The center of the circle lies on the perpendicular bisector of each chord.]

41. Find the equation of the circle through the points $(0, 3)$, $(-2, 5)$, and $(-4, 5)$. (See the hint in Exercise 40.)

42. Referring to Exercise 45 of the preceding section, find the linear relation giving the temperature F in degrees Fahrenheit corresponding to a temperature of C degrees Celsius.

43. A snowstorm starts at 3:00 A.M. and continues until 11:00 A.M. If there was 13 in. of old snow on the ground at the start of the storm and the new snow accumulates at a constant rate of $\frac{3}{2}$ in. per hour, find the depth d in inches at time of day t for $3 \leq t \leq 11$.

44. For a gas confined in a container of fixed volume, the pressure P (in g/cm^2) exerted by the gas on the container and the temperature T (in degrees Celsius) are linearly related. If the pressure is 50 g/cm^2 at a temperature of 16° C and 65 g/cm^2 at a temperature of 21° C, find the linear relation between P and T.

45. A spring is suspended vertically from a beam. The weight W (in kg) required to stretch a spring is linearly related to the distance d (in cm) the spring is stretched from its natural length, as long as the spring is not stretched too far. The spring is stretched 10 cm from its natural length when a weight is attached. An additional weight of 2 kg is required to stretch the spring an additional 12 cm. Find the linear relation between W and d.

46. The electric resistance R (in ohms) of a certain resistor is linearly related to the temperature T (in degrees Celsius). If $R = 3$ ohms when $T = 10°$ C and $R = 3.003$ ohms when $T = 16°$ C, find
a) the linear relation between R and T,
b) the resistance R at 0° C.

47. Water drips from a leaking faucet at a constant rate. A cylindrical pail is set under the faucet. The depth d (in cm) of water in the pail is linearly related to the elapsed time t (in min). The depth at 5:00 P.M. is 2 cm and the depth at 6:00 P.M. is 5 cm. Find
a) the linear relation between d and t,
b) the time when the pail was set under the faucet.

1.4 FUNCTIONS AND THEIR GRAPHS

FUNCTIONS

The area enclosed by a circle is a *function* of the radius of the circle, meaning that the area depends on and varies with the radius. If a numerical value for the radius is given, the area enclosed by the circle is determined. For example, if the radius is 3 units, then the area is 9π square units. Similarly, the area of a rectangular region is a *function* of both the length and the width of the rectangle; that is, the area depends on and varies with these quantities. If the length of a rectangle is 5 units and the width is 3 units, the rectangle encloses a region that has an area of 15 square units.

The study of how one quantity Q depends on and varies with other quantities is one of the major concerns of science. A rule that specifies Q for each possibility for the other quantities is exceedingly useful. Such rules are called *functions*.

DEFINITION 1.1 Function

A **function** f is a rule that assigns to each element x of some set X exactly one element y of a set Y. We use the notation $y = f(x)$, read "y equals f of x."

43. a) $(v, -u)$ b) $\sin\left(x - \dfrac{\pi}{2}\right) = -u = -\cos x$; $\cos\left(x - \dfrac{\pi}{2}\right) = v = \sin x$ **45.** $\sec(-x) = \dfrac{1}{\cos(-x)} = \dfrac{1}{\cos x} = \sec x$

47. $\sin\left(x - \dfrac{\pi}{2}\right) = \sin x \cos\left(-\dfrac{\pi}{2}\right) + \cos x \sin\left(-\dfrac{\pi}{2}\right) = (\sin x)(0) + (\cos x)(-1) = -\cos x$

49. $\sec\left(x - \dfrac{\pi}{2}\right) = \dfrac{1}{\cos[x - (\pi/2)]} = \dfrac{1}{(\cos x)\cos(-\pi/2) - (\sin x)\sin(-\pi/2)}$

$$= \frac{1}{(\cos x)(0) - (\sin x)(-1)} = \frac{1}{\sin x} = \csc x$$

51. $\cos 2x = \cos^2 x - \sin^2 x = \cos^2 x - (1 - \cos^2 x) = 2\cos^2 x - 1$,
$\cos 2x = \cos^2 x - \sin^2 x = (1 - \sin^2 x) - \sin^2 x = 1 - 2\sin^2 x$

53. $\frac{3}{10}$ **55.** $\sqrt{74 - 35\sqrt{2}} \approx 4.95$ **57.** $\sqrt{1 - (\frac{37}{40})^2} \approx 0.38$

Section 4.2

1. Amplitude 3, period 2π

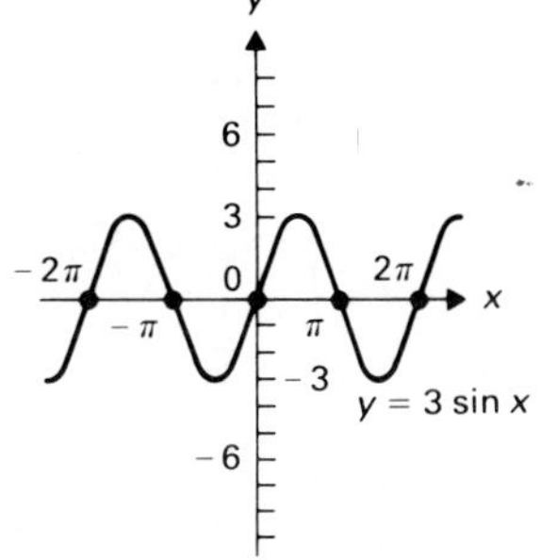

3. Amplitude $\frac{1}{2}$, period 2π

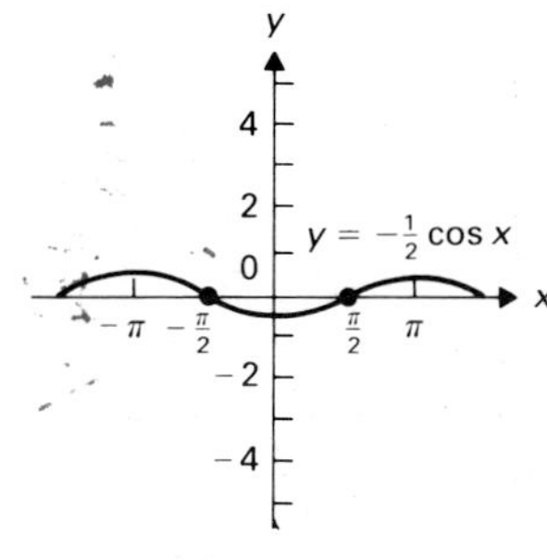

5. Amplitude 1, period 2π

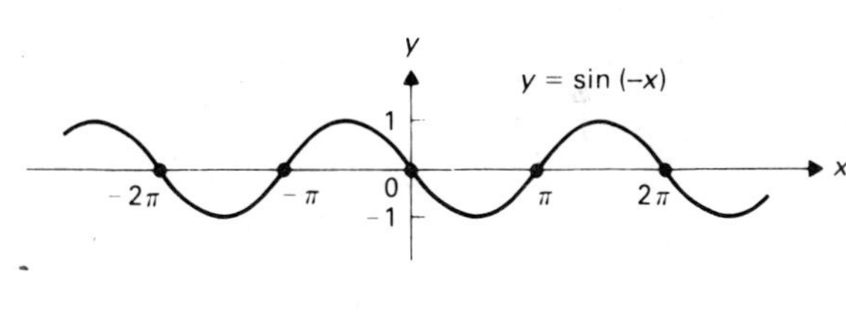

7. Amplitude 3, period $2\pi/3$

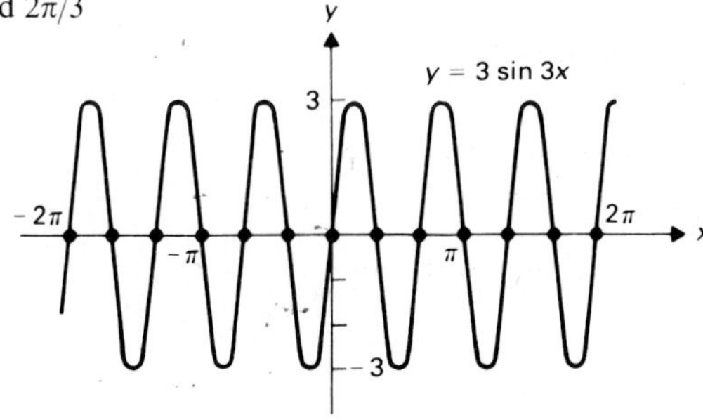

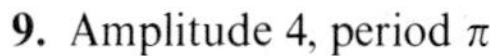

9. Amplitude 4, period π

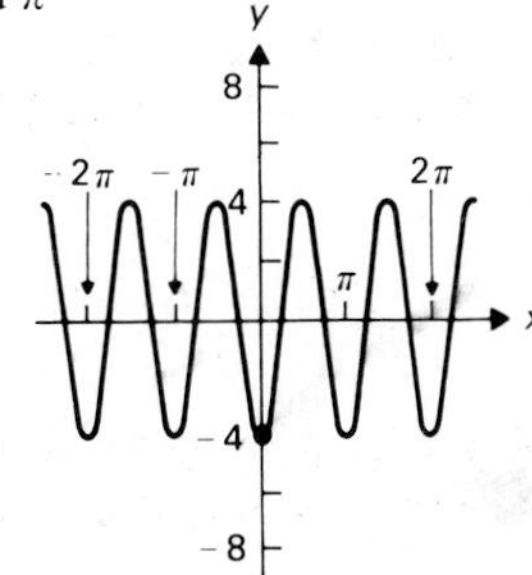

11. Amplitude 2, period 2π

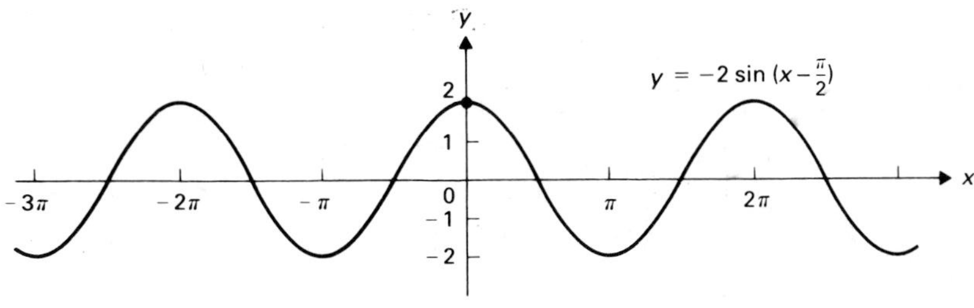

13. Amplitude 5, period 4π

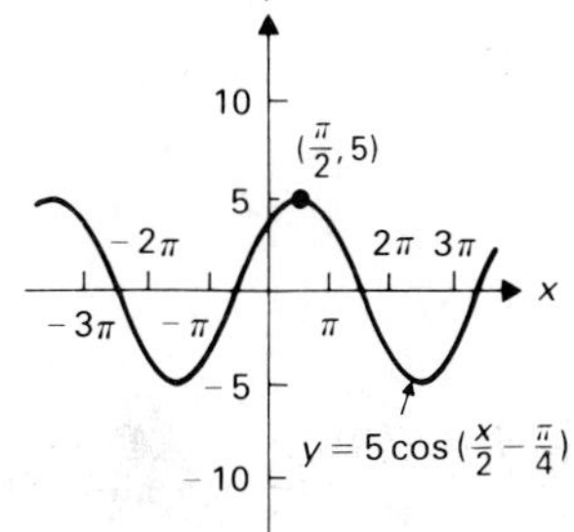

15. Amplitude 5, period 8π

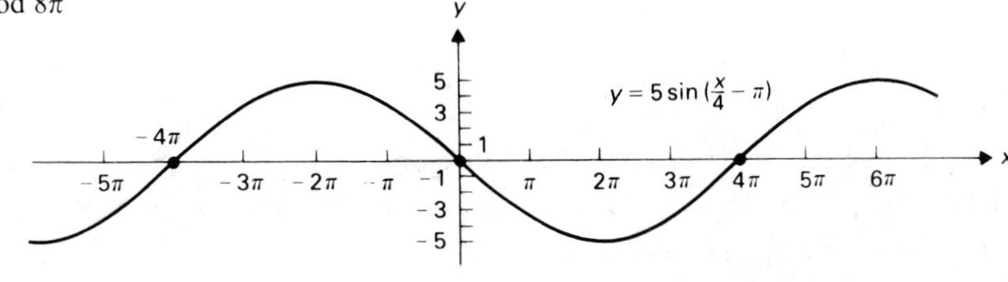

17. Period π

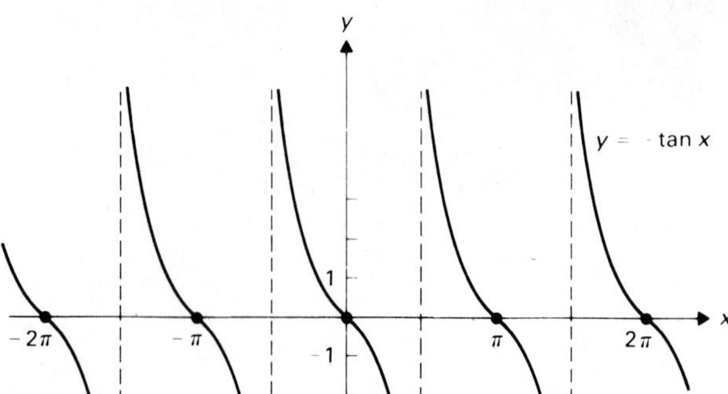

19. Period 2π

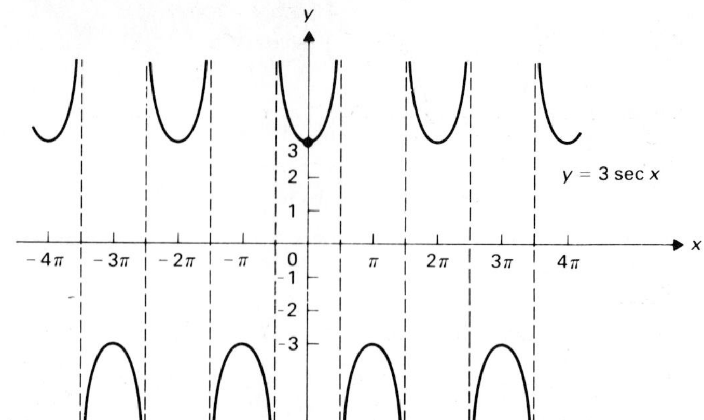

21. Period π

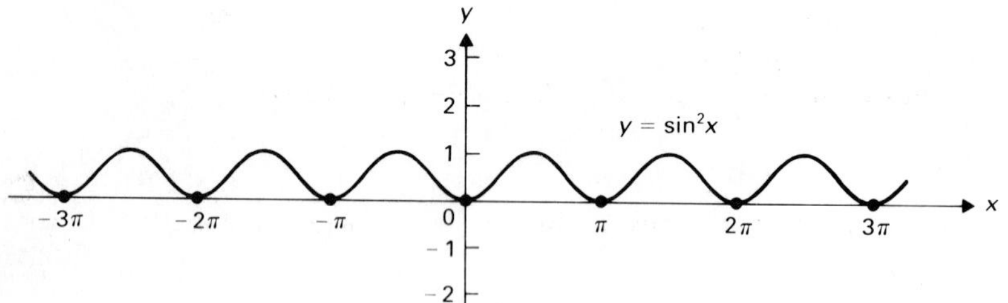

23. Period π

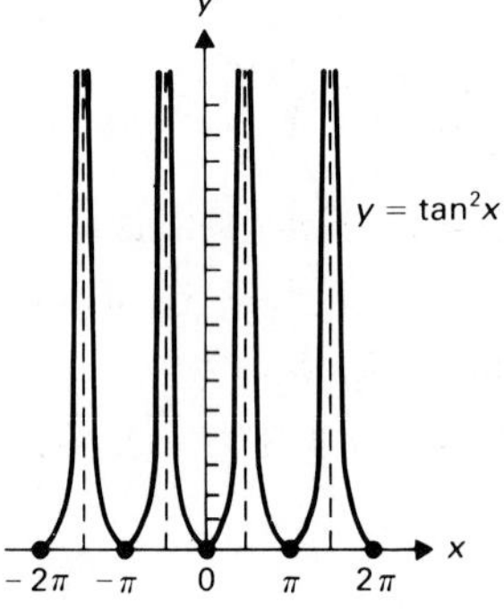

25.

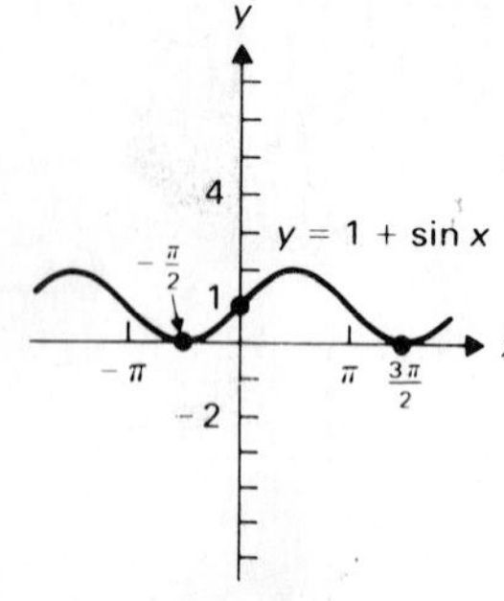

27.

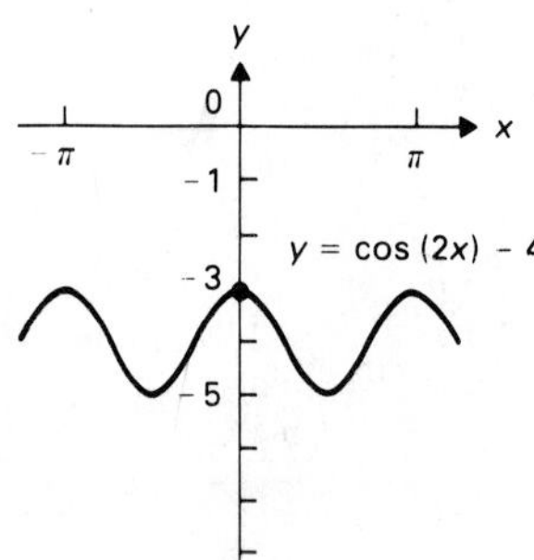

29.

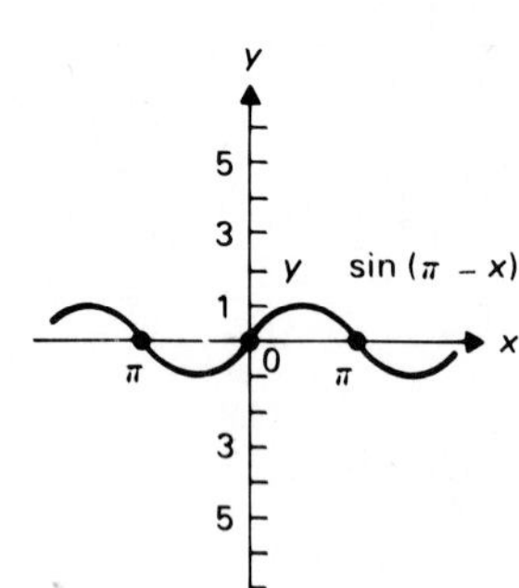

31.

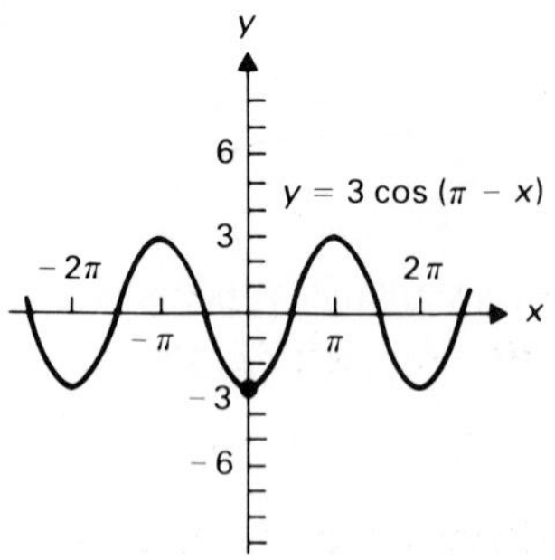

33.

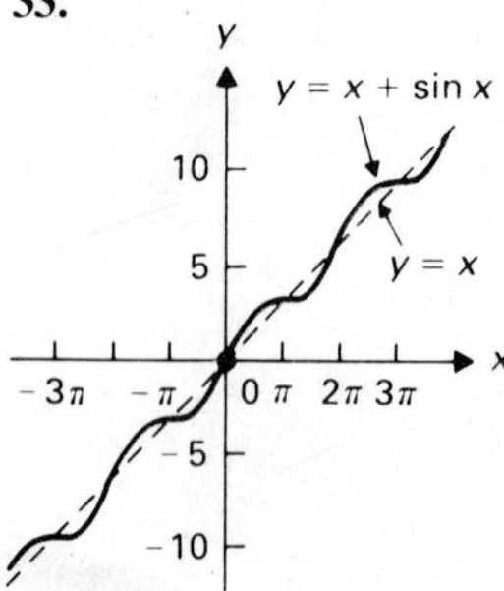

35.

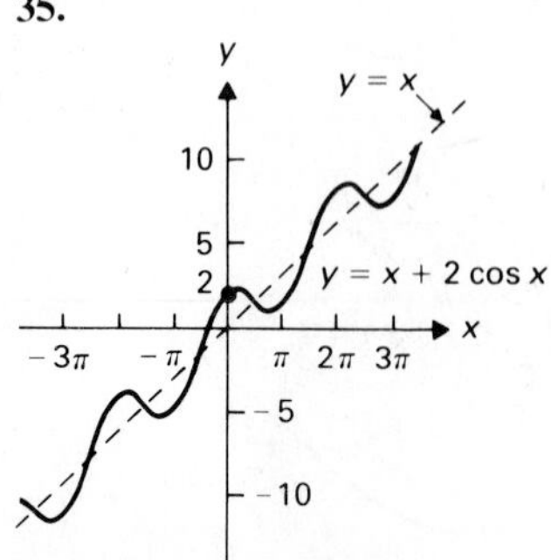

37.

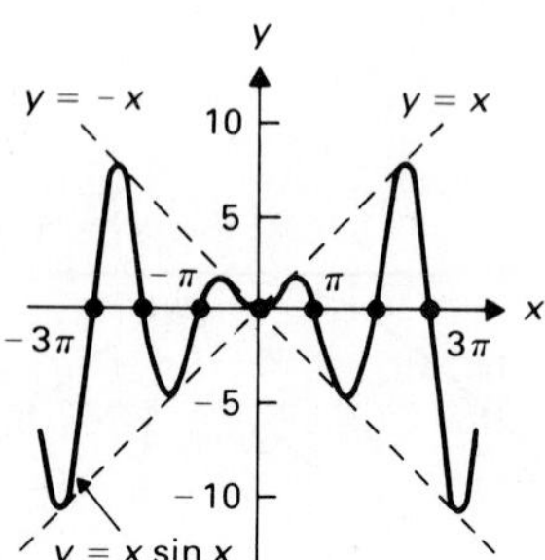

39.

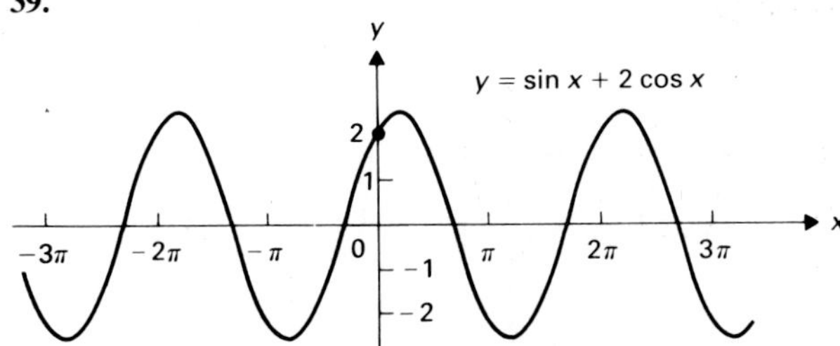

41.

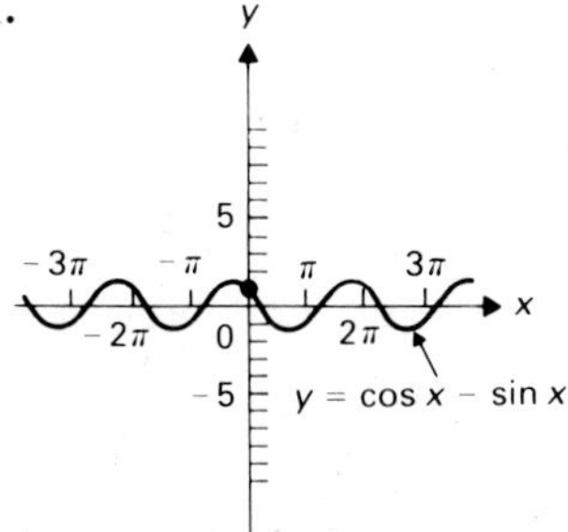

Section 4.3

1. Does not exist **3.** 2 **5.** 1 **7.** 1 **9.** 3 **11.** $\frac{1}{4}$ **13.** Does not exist **15.** 1 **17.** $-\infty$

19. $y = \sec x = (\cos x)^{-1}; \dfrac{dy}{dx} = -1(\cos x)^{-2}(-\sin x) = \dfrac{1}{\cos x}\dfrac{\sin x}{\cos x} = \sec x \tan x.$ **21.** $-x \sin x + \cos x$

23. $(x^2 + 3x) \sec x \tan x + (2x + 3) \sec x$ **25.** $2 \sin x \cos x$ **27.** $2 \sec^2 x \tan x$ **29.** $(\cot x + x \csc^2 x)/\cot^2 x$

31. $2 \cos 2x$ **33.** $6 \cos (2x - 3) \sin (2x - 3)$ **35.** $-2 \sin^3 x \cos x + 2 \sin x \cos^3 x$ **37.** $\frac{1}{2}(1 + 2 \cot^2 x)^{-1/2}(-4 \cot x \csc^2 x)$

39. $3[\cos (\tan 3x)] \sec^2 3x$ **41.** 1 **43.** 0 **45.** 0 **47.** $y - \dfrac{1}{\sqrt{2}} = \dfrac{1}{\sqrt{2}}\left(x - \dfrac{\pi}{4}\right)$ **49.** $2y + x = -1$ **51.** $\dfrac{180 + \sqrt{3}\pi}{360}$

53. $\dfrac{1}{2} + \dfrac{\pi}{180}$ **55.** a) 5 cm b) $-\dfrac{5\pi}{2}$ cm/sec c) $\dfrac{5\pi}{2}$ cm/sec d) $\dfrac{5\sqrt{3}\pi^2}{2}$ cm/sec^2 **57.** $a = 0$ cm, $b = 3/\pi$ cm

Review Exercise Set 4.1

1. a) $-1/\sqrt{3}$ b) $-1/\sqrt{2}$ **3.** $(\sqrt{3} - 1)/(2\sqrt{2})$ **5.** $\pi/3$ **7.** a) $6 \sin^2 2x \cos 2x$ b) $-3x^4 \csc x^3 \cot x^3 + 2x \csc x^3$

Review Exercise Set 4.2

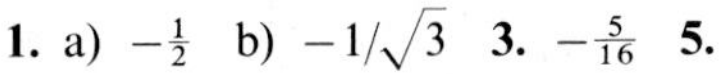

1. a) $-\frac{1}{2}$ b) $-1/\sqrt{3}$ **3.** $-\frac{5}{16}$ **5.**

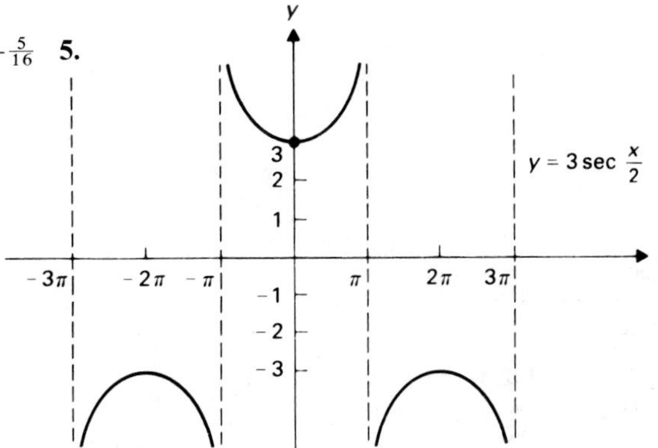

7. a) $2x \sec^2 (x^2 + 1)$ b) $\dfrac{(x - 4)(2 \sin x \cos x) - \sin^2 x}{(x - 4)^2}$

Review Exercise Set 4.3

1. a) -1 b) $\sqrt{2}$ **3.** $2\sqrt{5 - 2\sqrt{3}} \approx 2.479$ **5.** $2\pi/3$ **7.** a) $\dfrac{(\cos x) + (x + 1)(\sin x)}{\cos^2 x}$ b) $\dfrac{y \sin (xy)}{2y - x \sin (xy)}$

Review Exercise Set 4.4

1. a) -1 b) $-\sqrt{3}/2$ **3.** $5\sqrt{2}$ **5.** Period 4π, amplitude 2

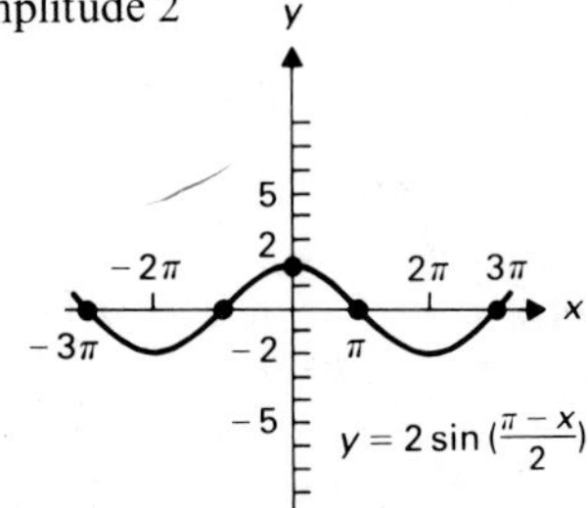

7. a) $3 \sin 2x \tan^3 x \sec^2 x + 2 \cos 2x \tan^3 x$ b) $\dfrac{y \cos x - \sin y}{x \cos y - \sin x}$

More Challenging Exercises 4

1. 1 **3.** 0 **5.** 2 **7.** 1 **9.** 2

CHAPTER 5

Section 5.1

1. $1/(4\pi)$ ft/sec **3.** $10\sqrt{3}$ in^2/min **5.** a) 0.1 mi/sec b) 0.5 mi/sec **7.** $125\pi/12$ cm^2/min **9.** 4 volts/min **11.** $12/\sqrt{65}$ ft/sec **13.** $\frac{100}{13}$ ft/sec **15.** $\frac{5}{6}$ units/sec **17.** $\frac{15}{7}$ ft/sec **19.** $-\sqrt{2}/14$ rad/min **21.** $\frac{8}{63}$ rad/sec
23. Show: dr/dt = constant. Given: $dV/dt = kS = k(4\pi r^2)$. Then

$$V = \frac{4}{3}\pi r^3, \qquad \frac{dV}{dt} = 4\pi r^2\frac{dr}{dt}, \qquad k \cdot 4\pi r^2 = 4\pi r^2\frac{dr}{dt}, \qquad \frac{dr}{dt} = k.$$

25. $3/(4\pi)$ in/sec **27.** $-\pi l^3/(60\sqrt{2})$ units^3/sec **29.** $365/\sqrt{89}$ ft/sec

Section 5.2

1. $f(-3) = f(2) = 10$, $c = -\frac{1}{2}$ **3.** $f(0) = f(\pi) = 0$, $c = \pi/2$ **5.** $f(1) = f(5) = 3$, $c = 2 + \sqrt{7/3}$
7. Let $f(x) = x^2 - 3x + 1$. Then $f(2) = -1$ and $f(6) = 19$ have opposite sign, so there is at least one solution. But $f'(x) = 2x - 3 = 0$ when $x = \frac{3}{2}$, which is not in $[2, 6]$, so there is at most one solution.
9. Let $f(x) = \sin x - x/4$. Then $f(\pi/2) = 1 - \pi/8 > 0$ and $f(\pi) = 0 - \pi/4 < 0$, so there is at least one solution. But $f'(x) = (\cos x) - \frac{1}{4} < 0$ in $[\pi/2, \pi]$, so there is at most one solution.
11. Let $f(x) = 0$ at points $x_1, x_2, \ldots, x_r$, listed in increasing order, in $[a, b]$. Then $f(x_i) = f(x_{i+1}) = 0$ for $i = 1, 2, \ldots, r - 1$. By Rolles' theorem, there exists c_i where $x_i < c_i < x_{i+1}$ such that $f'(c_i) = 0$. This holds for $i = 1, 2, \ldots, r - 1$.
13. All c, where $1 < c < 4$ **15.** $\sqrt{3}$ **17.** $-\frac{5}{4}$
19. a) It is the average rate of change of $f(x)$ over $[a, b]$. b) It is the instantaneous rate of change of $f(x)$ at c. c) If f is continuous on $[a, b]$ and differentiable for $a < x < b$, then there exists c, where $a < c < b$, such that the instantaneous rate of change of $f(x)$ at c is the same as the average rate of change of $f(x)$ over $[a, b]$.
21. We easily compute that

$$\frac{f(x_2) - f(x_1)}{x_2 - x_1} = \frac{a(x_2^2 - x_1^2) + b(x_2 - x_1)}{x_2 - x_1} = a(x_2 + x_1) + b$$

and

$$f'\left(\frac{x_2 + x_1}{2}\right) = 2a\left(\frac{x_2 + x_1}{2}\right) + b = a(x_2 + x_1) + b$$

also. Example 3 illustrates this.
23. By the mean-value theorem applied to $[a, x]$ for $a < x \le b$, $(f(x) - f(a))/(x - a) = f'(c)$ for some c, where $a < c < x$. Then $m \le (f(x) - f(a))/(x - a) \le M$, so $m(x - a) \le f(x) - f(a) \le M(x - a)$ for $a < x \le b$. Clearly this also holds at $x = a$.
25. $-9 - 5x \le f(x) \le 11 + 5x$ for x in $[-2, 4]$ **27.** $6 + 2x \le f(x) \le 6 + 5x$ for x in $[0, 6]$
29. $-6 + 5x \le f(x)$ for x in $[0, 2]$ **31.** $-4 - x \le f(x) \le -4 + 4x$ for $x \ge 0$
33. $1 + 3x \le f(x) \le 1 + 7x$ for $x \ge 0$, $1 + 7x \le f(x) \le 1 + 3x$ for $x < 0$

Section 5.3

1. F T T T F **3.** $x_0 = 0$ is not in the domain of f. **5.** Local maximum at critical point 1
7. Local minimum at critical point 1 **9.** Critical point 0, no local extremum **11.** Local minimum at critical point 0
13. Local maximum at critical point 0 **15.** Local maximum at critical point -1; local minimum at critical point 1
17. Local minima at critical points $n\pi$ for all integers n; local maxima at critical points $n\pi + (\pi/2)$ for all integers n
19. Local maxima at critical points $n\pi$ for all integers n; local minima at critical points $(\pi/2) + n\pi$ for all integers n
21. Maximum 16; minimum 0 **23.** Maximum 1; minimum $\frac{1}{5}$
25. a) Maximum 2; minimum -6 b) Maximum -3; minimum -7 c) Maximum -3; minimum -7
27. a) Maximum $\frac{7}{5}$; minimum $\frac{13}{10}$ b) Maximum $\frac{3}{2}$; minimum 1 c) Maximum 1; minimum $\frac{1}{2}$ d) Maximum $\frac{3}{2}$; minimum $\frac{1}{2}$
29. a) Maximum $\sqrt{2}$; minimum 1 b) Maximum $\sqrt{2}$; minimum -1 c) Maximum $\sqrt{2}$; minimum -1 d) Maximum 1; minimum $-\sqrt{2}$
31. Maximum 1; minimum 0
33. a) Maximum 1600 ft; minimum 1200 ft b) Maximum 0 ft/sec; minimum -160 ft/sec c) Maximum 160 ft/sec; minimum 0 ft/sec.
35. a) Maximum 1400 ft; minimum 1336 ft b) Maximum 64 ft/sec; minimum -32 ft/sec c) Maximum 64 ft/sec; minimum 0 ft/sec

37. We may write

$$f(x) = x^n\left(a_n + \frac{a_{n-1}}{x} + \cdots + \frac{a_1}{x^{n-1}} + \frac{a_0}{x^n}\right), \qquad x \neq 0.$$

Then $\lim_{x\to\infty} f(x) = \lim_{x\to-\infty} f(x) = \infty$. Consequently, there exists $x > 0$ such that $f(x) > f(1)$ if $|x| > c$. Then the minimum value assumed on $[-c, c]$, which exists by Theorem 5.2, is also the minimum value assumed on the whole x-axis.

39. -14 **41.** -27

Section 5.4

1. Increasing $x > 3$; decreasing $x < 3$ **3.** Increasing $x < -3$, $x > 1$; decreasing $-3 < x < 1$
5. Increasing $-2 < x < 2$; decreasing $x < -2$, $x > 2$ **7.** Increasing $-\infty < x < \infty$ **9.** Decreasing $x < -2$, $x > 2$
11. Decreasing $x < -1$, $-1 < x < 1$, $x > 1$ **13.** Increasing $-2 < x < 0$, $x > 2$; decreasing $x < -2$, $0 < x < 2$ **15.** -8
17. a) $b/a = -4$ b) $a = 3$, $b = -12$ c) No **19.** No **21.** a) $-1 < x < 3$, $x > 3$ b) $x < -1$ c) None d) -1
23. a) $x < -2$, $x > 1$ b) $-2 < x < 0$, $0 < x < 1$ c) None d) None
25. a) $-2 < x < 0$, $0 < x < 1$, $x > 4$ b) $x < -2$, $1 < x < 4$ c) None d) 4 **27.** $n - 1$ **29.** $n/2$, $n/2 - 1$ **31.** F F T T F
33. T F T F F **35.**

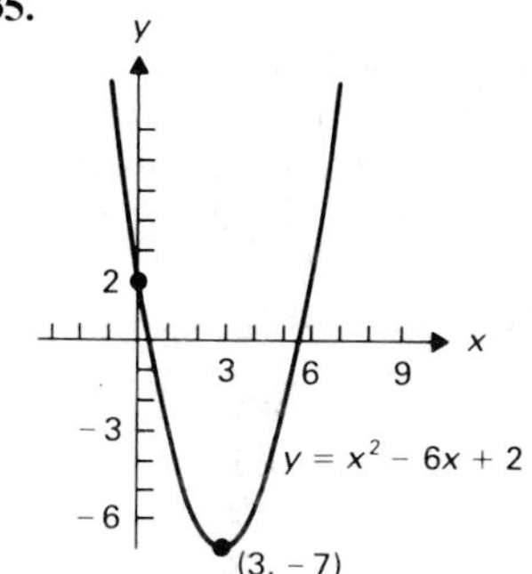

37.

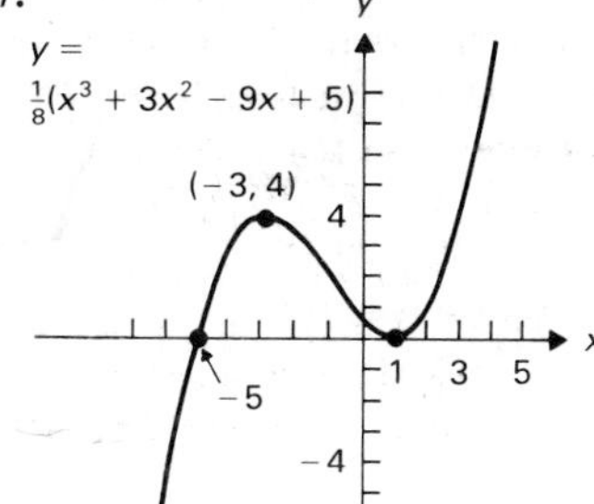

39.

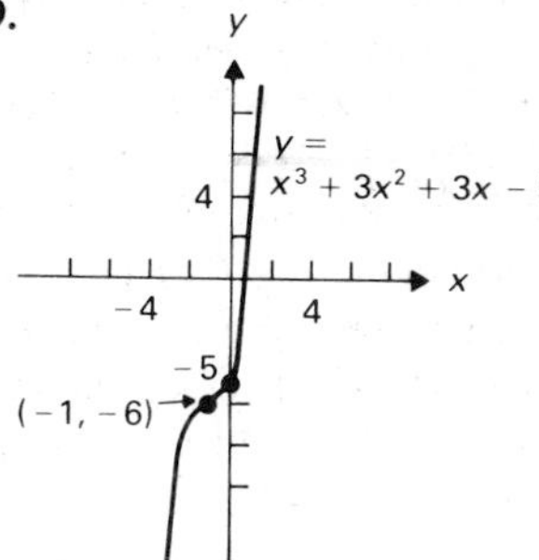

41.

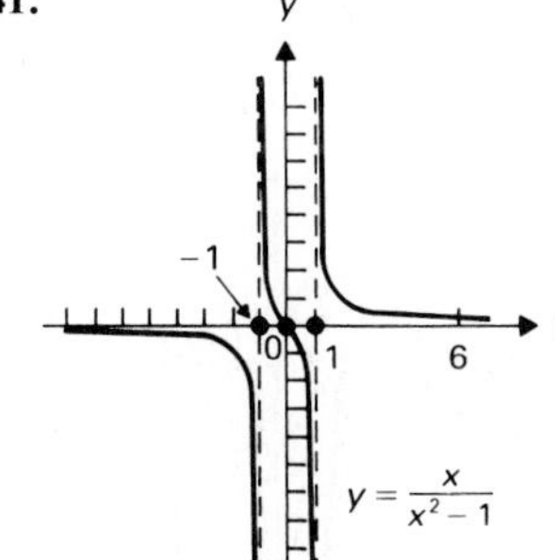

43.

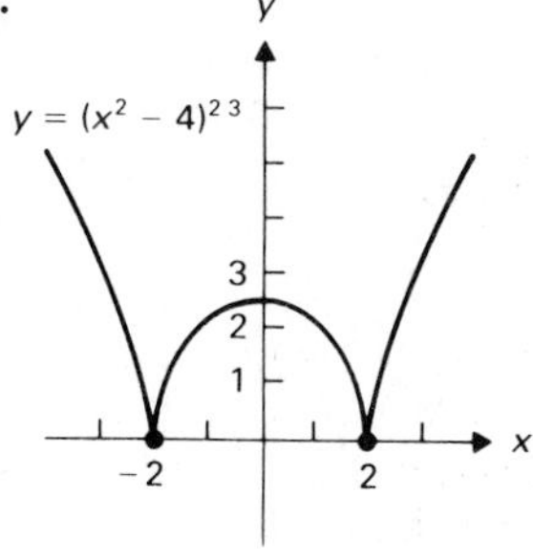

Section 5.5

1. T F F T T F T F T T **3.**

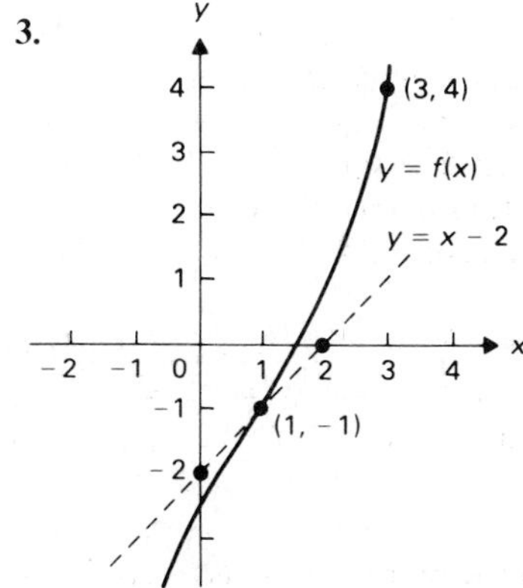

5. No. The tangent line to the graph of f at $(0, 0)$ passes through $(1, 1)$ and is below the graph of f since $f''(x) > 0$. Thus the graph of f must pass above $(1, 1)$, so $f(1) > 1$. Alternatively, $f''(x) > 0$ means $f'(x)$ is increasing, and thus $f'(x) > 1$ if $x > 0$. If $f(1) = 1$, then by the mean-value theorem applied to $[0, 1]$, we have $f'(c) = 1$ for some c, where $0 < c < 1$, which is a contradiction.

7. a) $-\infty < x < \infty$ b) None **9.** a) $x > -1$ b) $x < -1$ **11.** a) $x > 1$ b) $x < 1$ **13.** a) $x > 0$ b) $x < 0$
15. a) $x < -1/\sqrt{3}, x > 1/\sqrt{3}$ b) $-1/\sqrt{3} < x < 1/\sqrt{3}$ **17.** a) $(2n - \frac{1}{2})\pi < x < (2n + \frac{1}{2})\pi$ b) $(2n + \frac{1}{2})\pi < x < (2n + \frac{3}{2})\pi$
19. a) $x < -1, -1 < x < 0, x > 1$ b) $0 < x < 1$ **21.** a) $-3 < x < 0, x > 1$ b) $x < -3, 0 < x < 1$
23. a) $-3 < x < 0,\ x > 1$ b) $x < -3,\ 0 < x < 1$ **25.** a) $x > 2$ b) $x < -1,\ -1 < x < 0,\ 0 < x < 2$ **27.** F T F T F
29. T F T T T **31.** a) 4 at $x = 0$ b) None c) None d)

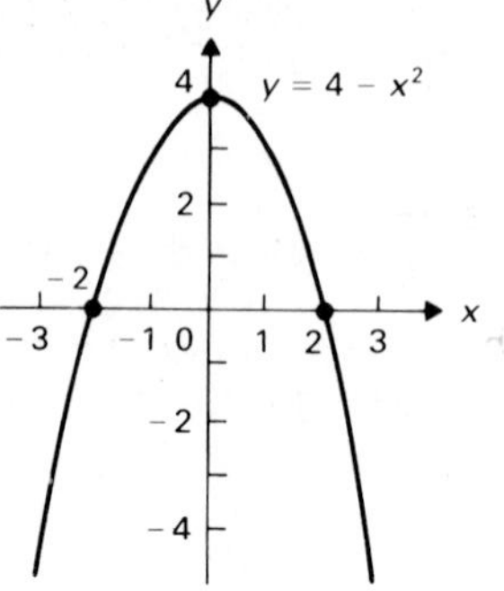

33. a) None b) None c) None d)

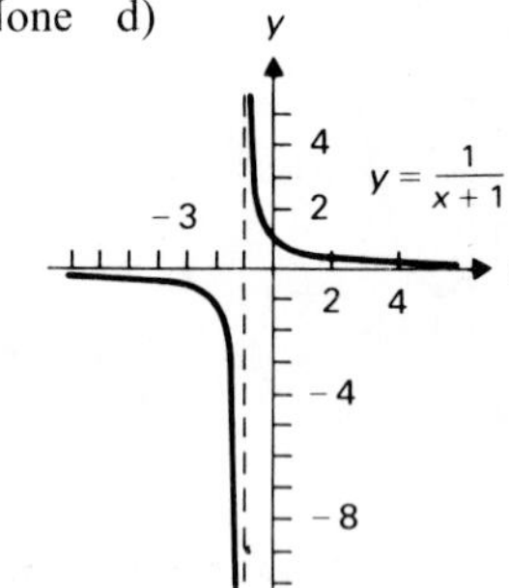

35. a) None b) None c) $(-1, -\frac{19}{3})$ d)

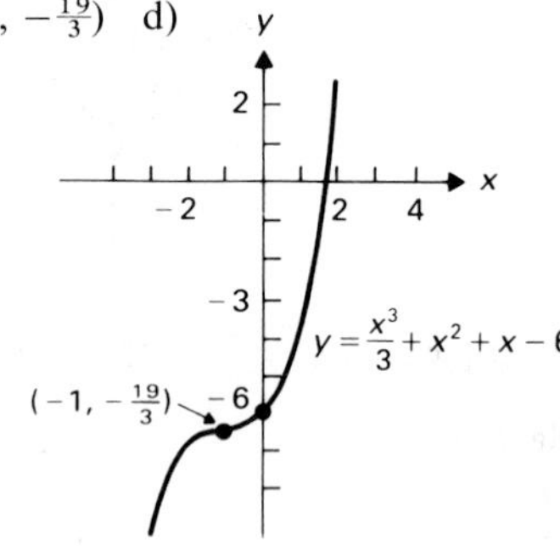

37. a) None b) -2 at $x = 1$ c) None d)

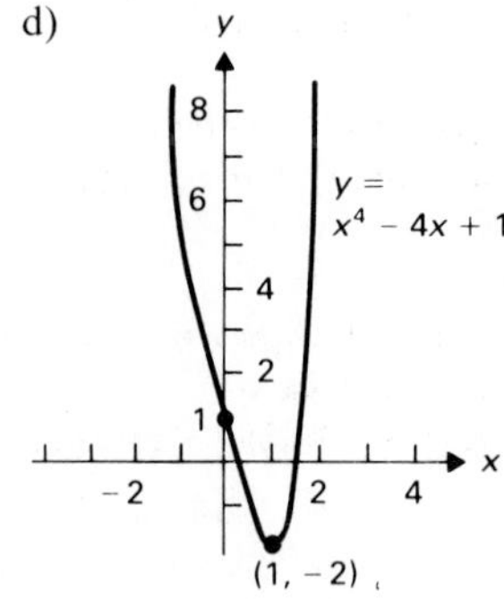

39. a) None b) -1 at $x = 1$
c) $(0, 0), \left(\frac{2}{3}, \frac{-16}{27}\right)$ d)

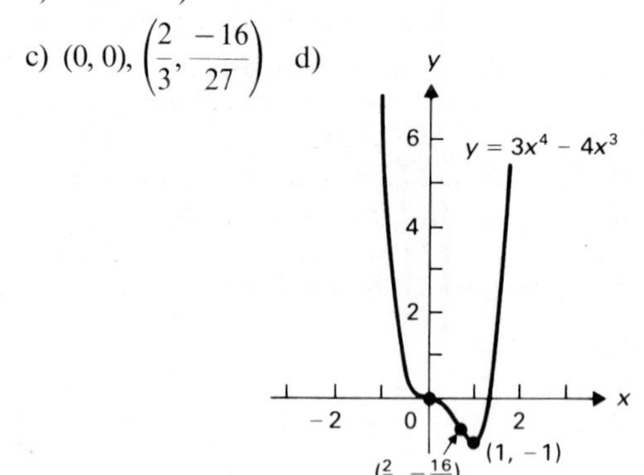

41. a) $2n\pi + \frac{2}{3}\pi + \sqrt{3}$ at $x = 2n\pi + \frac{2}{3}\pi$
b) $2n\pi + \frac{4}{3}\pi - \sqrt{3}$ at $x = 2n\pi + \frac{4}{3}\pi$
c) $(n\pi, n\pi)$ d)

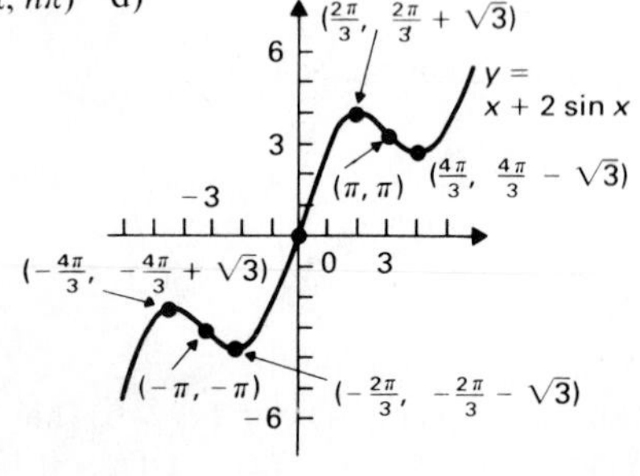

43.

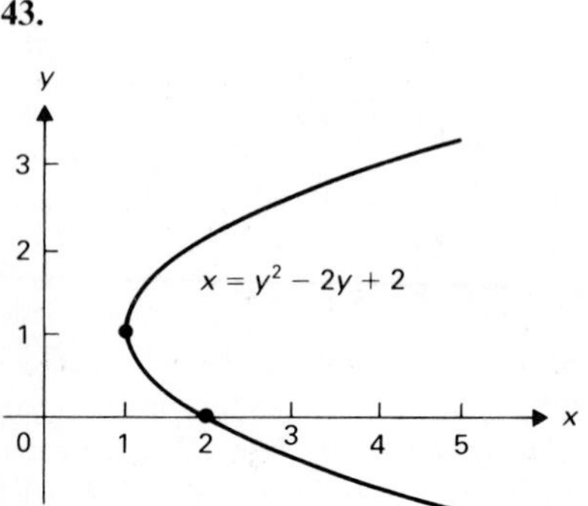

45.

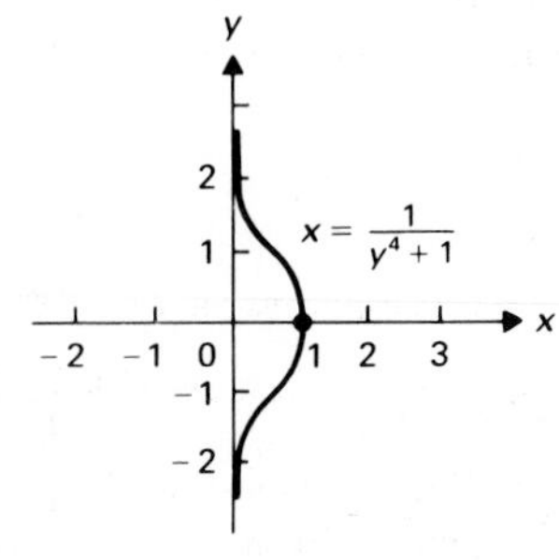

Section 5.6

1. 25 ft **3.** $x = 2, y = 4$ **5.** 6, 6 **7.** $8\sqrt{2}$ units^2 **9.** 42 units^2 **11.** 108 in^2 **13.** a) 8192 in^3 b) $32,768/\pi$ in^3 **15.** 60,000 ft^2 **17.** 4 in. wide by 2 in. high **19.** $\sqrt{2}a$ by $a/\sqrt{2}$ **21.** $4a/3$ **23.** $a^2\sqrt{3}/8$ units^2 **25.** 16 in. wide by 24 in. high **27.** $2/\sqrt{5}$ miles **29.** $a/(\sqrt[3]{c} + 1)$ ft **31.** a) Cut into two 50-in. pieces. b) Bend it all into one triangle. **33.** a) $(2 + \pi)/4$ b) $(4 + \pi)/8$ c) 4 ft **35.** 10:36 A.M. **37.** Squares of sides $(a + b - \sqrt{a^2 + b^2 - ab})/6$ **39.** a) $x = (A + B - b - t)/(2B + 2c)$ b) $t = (A + B - b)/2$ **41.** $(a^{2/3} + b^{2/3})^{3/2}$ ft **43.** $24\sqrt{3}$ ft

Section 5.7

1. $\frac{97}{56}$ **3.** $\frac{49}{20}$ **5.** $\frac{32{,}257}{12{,}192} \approx 2.6458$ **7.** $\frac{3{,}958}{2{,}178} \approx 1.8173$ **9.** $\frac{59}{86}$ **11.** $\frac{333}{440}$

13. a)

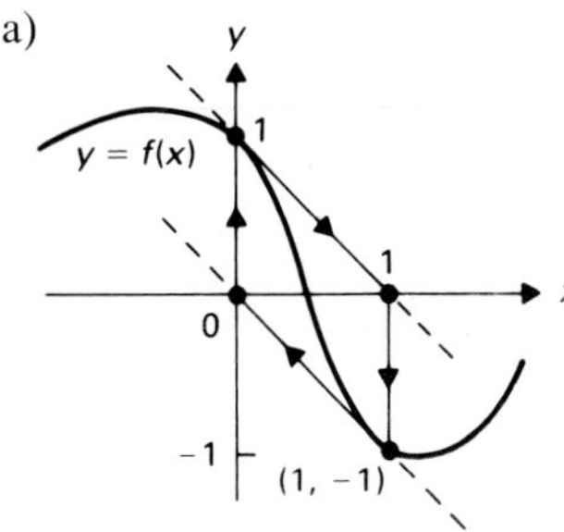

b) i) $d = 1$ ii) $a + b + c + d = -1$ iii) $c = -1$ iv) $3a + 2b + c = -1$ v) $f(x) = 2x^3 - 3x^2 - x + 1$ **15.** -2.387687 **17.** 2.924017738 **19.** 0.7390851332

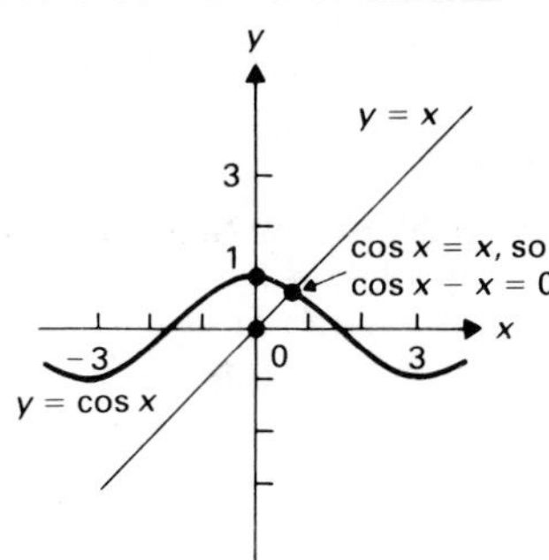

21. 3.169925001 **23.** 2.129372483 **25.** 3.352823623

Section 5.8

1. a) \$540/unit b) \$800/unit c) \$260/unit d) \$335/unit **3.** a) \$90/stove b) 1000 stoves for \$40,000 profit
5. a) \$78/cord b) \$79/cord c) 500 cords
7. a) 0 cords b) 1 cord. The average revenue approaches a maximum of \$80/cord as $x \to 0$, but 0 cords produce 0 revenue. The average revenue function is not defined when $x = 0$.
9. (Average cost) > (Marginal cost) would mean that producing another unit would reduce the average cost. Similarly, (Average cost) < (Marginal cost) would mean that producing 1 less unit would reduce the average cost. Thus average cost can only be minimum when it equals marginal cost.
11. a) \$$\frac{2}{9}$ per dollar income b) \$$\frac{7}{9}$ per dollar income
13. a) 0.373 b) There is no minimum or maximum average tax; it always increases. However, it never exceeds 0.6.
15. 10 orders of 20 refrigerators each **17.** Order 10 lots of 200 trees each.
19. a) Populations of 0 and of 60 b) A population of 30 rabbits will yield a maximum sustainable harvest of 75 rabbits per year.

Section 5.9

1. $2x + C$ **3.** $\frac{x^3}{3} - \frac{3x^2}{2} + 2x + C$ **5.** $\frac{2}{3}x^{3/2} + C$ **7.** $2x^2 + \frac{2}{3}x^{3/2} + C$ **9.** $-\frac{1}{x} + \frac{3}{2}x^2 + x + C$ **11.** $-\frac{1}{x} - \frac{1}{2x^2} + C$ **13.** $\frac{6}{5}x^{5/2} - \frac{4}{3}x^{3/2} + 2\sqrt{x} + C$ **15.** $\frac{x^4}{4} + \frac{x^2}{2} + C$ **17.** $\frac{2}{3}(x - 1)^{3/2} + C$ **19.** $2\sqrt{x + 1} + C$ **21.** $-\cos x + C$ **23.** $-\frac{5}{8}\cos 8x + C$ **25.** $-\frac{1}{4}\sin(2 - 4x) + C$ **27.** $-2\cot 2x + C$ **29.** $-\frac{1}{4}\csc 4x + C$ **31.** $8x - 19$ **33.** $\frac{x^2}{2} - \cos x + 4$ **35.** $\frac{1}{2}x^2 - \frac{2}{3}x^{3/2} + \frac{1}{6} + \pi$ **37.** $y = F(x) = x^2 - 2x + 1$ **39.** $y = F(x) = \frac{1}{4}\sin 2x + \frac{3}{2}x - 1$
41. The general solution is $y = F(x) + C_1x + C_2$. The boundary conditions give the simultaneous equations

$$C_1a + C_2 = \alpha - F(a),$$

$$C_1b + C_2 = \beta - F(b),$$

which always have unique solutions C_1 and C_2 if $a \neq b$.

43. $y = G(x) = -\sin x + \frac{2}{\pi}x$ **45.** a) $v = -32t + v_0$ b) $y = -16t^2 + v_0t + y_0$

47. $\frac{d^2s}{dt^2} = -\frac{k}{m}\frac{ds}{dt}$; when $t = 0$, $s = 0$ and $ds/dt = 80$ **49.** $\frac{dP}{dt} = kP$

Review Exercise Set 5.1

1. $3/(10\pi)$ ft/min **3.** a) $x > 2$ b) $x < -1, -1 < x < 2$ **5.** a) $x < -1, \frac{1}{5} < x < 2, x > 2$ b) $-1 < x < \frac{1}{5}$
7. $r = 4\sqrt{6}$ units **9.** a) 25 b) 249.5

Review Exercise Set 5.2

1. $3\sqrt{3}/(2\sqrt{7})$ ft/min **3.** a) $x < -2, -2 < x < 0$ b) $0 < x < 2, x > 2$ **5.** a) $x > 2$ b) $x < 2$ **7.** $256/(3\sqrt{3})$ units^2
9. \$29,300

Review Exercise Set 5.3

1. $\frac{1}{16}$ amp/sec **3.** a) None b) -1 **5.** $\frac{11}{7}, 5$ **7.** $\sqrt{3}a^2/8$ ft^2 **9.** 80 pairs

Review Exercise Set 5.4

1. $5\sqrt{3}/4$ ft^2/min **3.** a) None b) $-1, 4$ **5.** a) $-3 < x < 0, x > 2$ b) $x < -3, 0 < x < 2$ **7.** 2, 4 **9.** 6250

More Challenging Exercises 5

1. Use 400 ft of fence for a square enclosure of 10,000 ft^2 and the rest for the circular enclosure. Total enclosed: 38,648 ft^2.
3. If f is n times differentiable for $a \le x \le b$ and $f(x)$ assumes the same value at $n + 1$ distinct points in $[a, b]$, then $f^{(n)}(c) = 0$ for some c, where $a < c < b$.
5. Jog all the way.
7. This sequence will occur if $f(0) = -1, f'(0) = 1, f(1) = -1, f'(1) = 1, f(2) = -2, f'(2) = -1$. These six conditions should be able to be satisfied by a polynomial function $f(x) = ax^5 + bx^4 + cx^3 + dx^2 + ex + f$ with six coefficients.

CHAPTER 6

Section 6.1

1. $a_0 + a_1 + a_2 + a_3$ **3.** $a_2 + a_4 + a_6 + a_8$ **5.** $c + c^2 + c^3 + c^4 + c^5$ **7.** 30 **9.** 35 **11.** 44 **13.** $\sum_{i=1}^{3} a_i b_{i+1}$ **15.** $\sum_{i=1}^{3} a_i^{i+1}$

17. $\sum_{i=1}^{3} a_i^{b_{3i}}$

19.
$$\begin{aligned}\sum_{i=1}^{n}(a_i + b_i)^2 &= (a_1 + b_1)^2 + \cdots + (a_n + b_n)^2\\ &= a_1^2 + 2a_1b_1 + b_1^2 + \cdots + a_n^2 + 2a_nb_n + b_n^2\\ &= a_1^2 + \cdots + a_n^2 + 2(a_1b_1 + \cdots + a_nb_n) + b_1^2 + \cdots + b_n^2\\ &= \sum_{i=1}^{n} a_i^2 + 2\sum_{i=1}^{n} a_ib_i + \sum_{i=1}^{n} b_i^2\end{aligned}$$

21. 1.575 **23.** $S_2 = 5, s_2 = 1$ **25.** $S_4 \approx 0.76, s_4 \approx 0.63$ **27.** 153 **29.** π **31.** 40 ft · lb **33.** 46 ft **35.** 1.89 **37.** 2.006
39. 3.142 **41.** 19.703 **43.** 7.45588

Section 6.2

1. a)

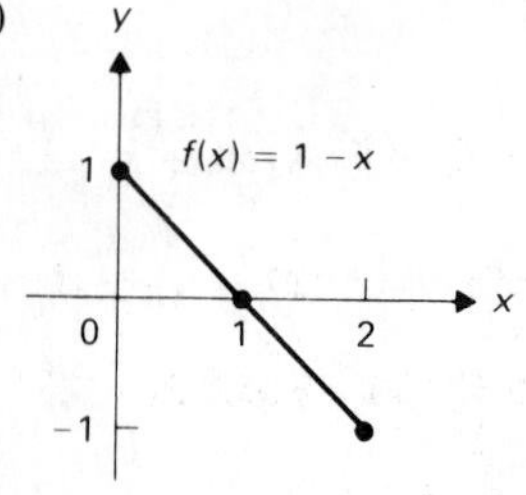

b) 1 c) -1 d) 0 **3.** 2

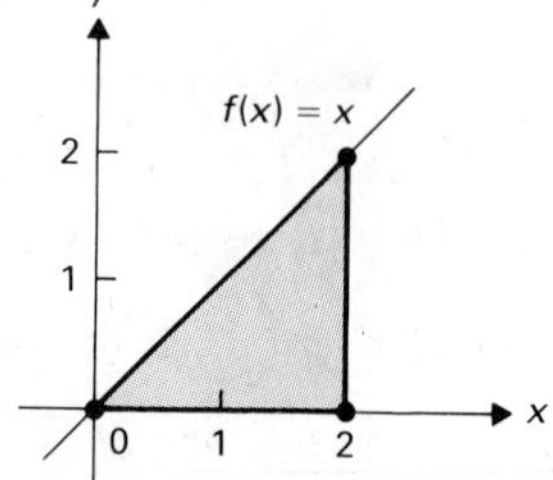

5. 14

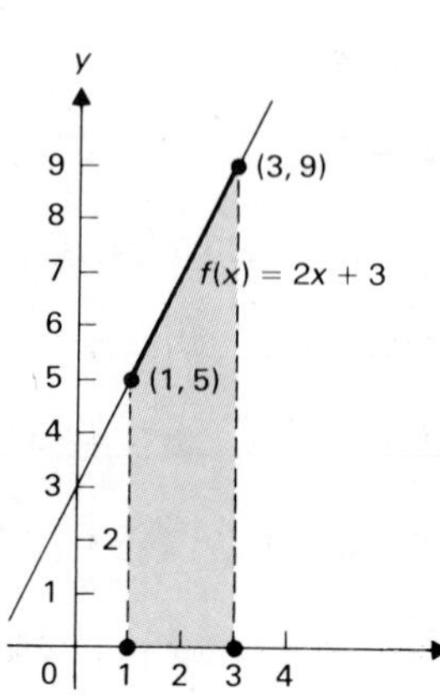

7. 2

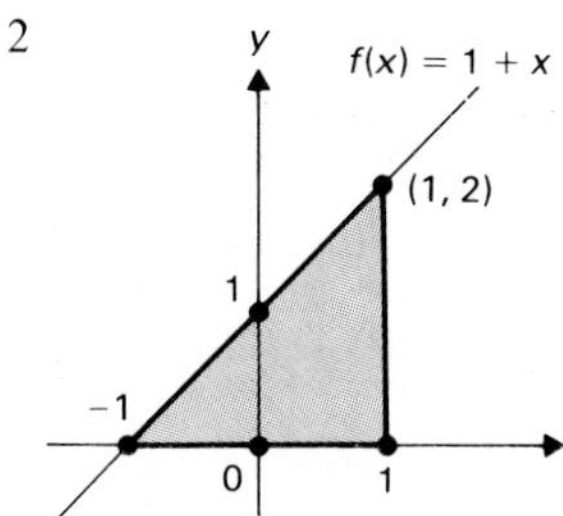

9. $9\pi/2$

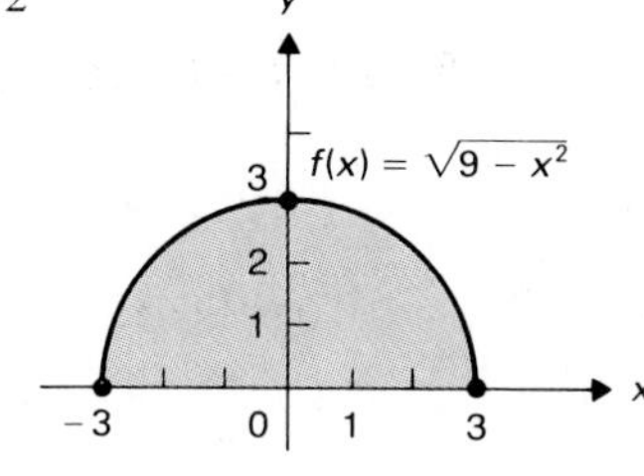

11. $12 + 4\pi$

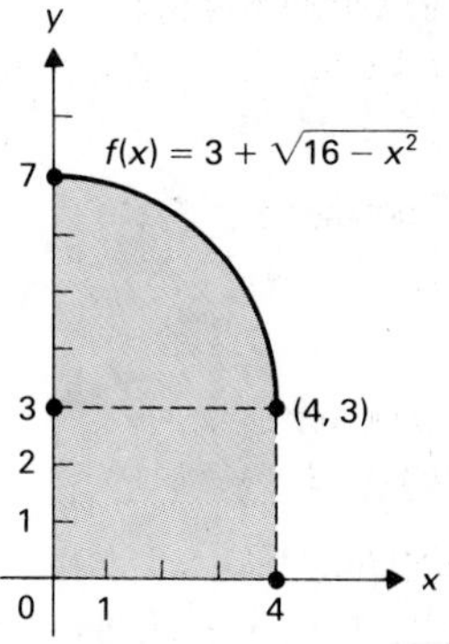

13.

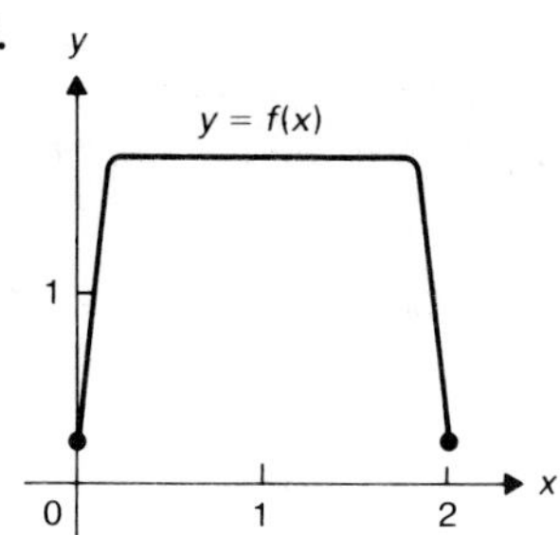

15. 1 **17.** 1 **19.** 0 **21.** -6 **23.** 2 **25.** 4 **27.** 4 **29.** $(\pi^2/4) - 3$ **31.** 6 **33.** 3 **35.** -1 **37.** -10 **39.** 5 **41.** $-\frac{3}{2}$ **43.** -2 **45.** $\frac{1}{4}$ **47.** $-\frac{9}{4}$ **49.** 2.046696 **51.** 0.459481

Section 6.3

1. Refer to Theorem 6.3 to check your answer. **3.** $\frac{1}{4}$ **5.** $\frac{20}{3}$ **7.** $\frac{45}{4}$ **9.** $-\frac{3}{8}$ **11.** $\frac{14}{9}$ **13.** 2 **15.** 3 **17.** $3\sqrt{2}$ **19.** $2/\sqrt{3}$ **21.** 1 **23.** -20 **25.** $\frac{3}{2}$ **27.** $-\pi/2$ **29.** $2 + (\pi/2)$ **31.** π **33.** $\pi + (3\pi^3/4)$ **35.** $(3\pi^2/2) - 2\pi$ **37.** 8 **39.** $1/\sqrt{2}$ **41.** $\frac{88}{15}$

43. 36

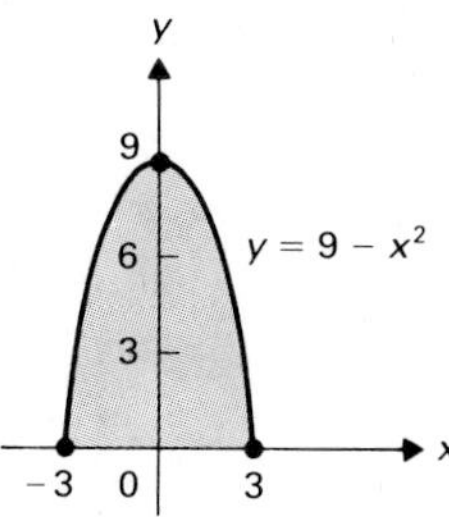

45. $\frac{256}{5}$

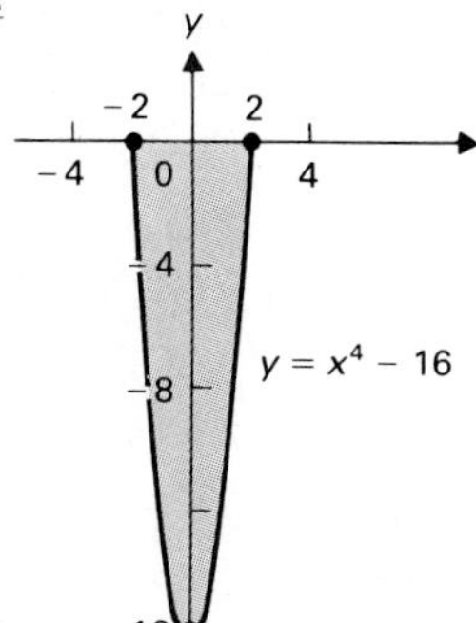

47. $\sqrt{t^2 + 1}$ **49.** $-1/(1 + t^2)$ **51.** $2\sqrt{3 + 4t^2}$ **53.** $-\sqrt{t^2 + 1}$ **55.** $2t\sqrt{t^4 - 6t^2 + 10}$

Section 6.4

1. $\frac{1}{4}x^4 + \frac{4}{3}x^3 + C$ **3.** $\frac{1}{6}(x + 1)^6 + C$ **5.** $\dfrac{-1}{4(4x + 1)} + C$ **7.** $\dfrac{-1}{18x^2 + 6} + C$ **9.** $\dfrac{1}{12(4 - 3x^2)^2} + C$ **11.** $4\sqrt{x^2 + x} + C$

13. $\dfrac{-2}{3(\sqrt{x} + 1)^3} + C$ **15.** $\frac{1}{12}(x^3 + 4)^4 + C$ **17.** $\frac{1}{3}\sin 3x + C$ **19.** $\frac{1}{2}\sin(x^2 + 1) + C$ **21.** $\frac{1}{2}\sin^2 x + C$ **23.** $-\frac{1}{16}\cos^4 4x + C$

25. $\frac{1}{2}\tan x^2 + C$ **27.** $\frac{1}{2}\tan^2 x + C$ **29.** $\frac{1}{8}\tan^4 x^2 + C$ **31.** $(1 + \cos x)^{-1} + C$ **33.** $\frac{1}{4}\sin 4x + C$ **35.** $\frac{1}{12}\sec^4 3x + C$

37. $\frac{2}{9}(1 + \sec 3x)^{3/2} + C$, $\sec 3x > 0$ **39.** $-\frac{1}{10}\csc^5 2x + C$ **41.** $\frac{1}{2}x + \frac{1}{4}\sin 2x + C$ **43.** $\dfrac{1}{y} = -\dfrac{x^3}{3} + C$

45. $\dfrac{1}{s} = -\dfrac{1}{3}(1 + t^2)^{3/2} + C$ **47.** $\frac{2}{5}\sqrt{5v - 7} = \frac{1}{2}\sin t^2 + C$ **49.** $-\cot y = \tan x + C$ **51.** $\dfrac{-1}{y^2} = x^2 - 1$

53. $-\cos 2y = 2\sin x + 1$ **55.** $3\sqrt{r^2 + 1} = (2s + 3)^{3/2} - 119$

Section 6.5

1. $\frac{56}{3}$ **3.** $[-3/\sqrt{4+x^2}] + C$ **5.** $-\sqrt{(10-x)/x} + C$ **7.** $\dfrac{4(7x+6)}{147}\sqrt{3-7x} + C$ **9.** $\dfrac{6x^2-1}{120}\sqrt{(4x^2+1)^3} + C$

11. $\dfrac{(4x^2-1)^8}{32}\left(\dfrac{4x^2-1}{9}+\dfrac{1}{8}\right) + C$ **13.** $-\dfrac{\sqrt{16-x^4}}{32x^2} + C$ **15.** $-\dfrac{1}{4}\sqrt{\dfrac{4-x^2}{x^2}} + C$ **17.** $\dfrac{\pi}{2}$

19. $\dfrac{-\sin^3 x\cos x}{4} + \dfrac{3}{4}\left(\dfrac{x}{2} - \dfrac{\sin 2x}{3}\right) + C$ **21.** $-\dfrac{\cos 5x}{10} + \dfrac{\cos x}{2} + C$ **23.** $\dfrac{1}{4}\cdot\dfrac{\sin^2 4x}{2} + C$

25. $\dfrac{\sin^5 x\cos x}{6} - \dfrac{\sin^3 x\cos x}{24} + \dfrac{1}{8}\left(\dfrac{x}{2} - \dfrac{\sin 2x}{4}\right) + C$ **27.** $-\dfrac{1}{2}\cdot\cot\dfrac{x^2}{2} + C$ **29.** $\dfrac{1}{16}\cdot\cos 4x + \dfrac{x}{4}\cdot\sin 4x + C$

31. $x^3\sin x + 3x^2\cos x - 6(\cos x + x\sin x) + C$ **33.** $\dfrac{1}{3}\cdot\tan 3x - x + C$ **35.** $\dfrac{\tan^3 2x}{6} - \left(\dfrac{1}{2}\cdot\tan 2x - x\right) + C$

37. $\dfrac{1}{3}\left(\dfrac{x^3}{2} + \dfrac{\sin 2x^3}{4}\right) + C$ **39.** $\dfrac{(2\sin x+4)^9}{4}\left(\dfrac{\sin x+2}{5} - \dfrac{4}{9}\right) + C$ **41.** $-\dfrac{\sqrt{4-\sin^2 x}}{4\sin x} + C$

43. $-\dfrac{\sqrt{\sin^2 2x+9}}{18\sin 2x} + C$ **45.** $-\dfrac{6\sec 2x+1}{120}\sqrt{(1-4\sec 2x)^3} + C$ **47.** $-\dfrac{\sqrt{25+4\tan^2 x}}{25\tan x} + C$

49. $-\dfrac{1}{6}\sqrt{\dfrac{4-\sec 3x}{\sec 3x}} + C$

Section 6.6

1. 1.3524 **3.** 1.425 **5.** a) 1.737 b) 1.7321 **7.** 3.1416 **9.** $\dfrac{73\pi}{180} \approx 1.2741$ **11.** 1.1167 **13.** 0.5917 **15.** 3.6219

17. $\int_{-1}^{1}|x|\,dx$ for $n = 2$ is found exactly by the rectangular rule and the trapezoidal rule, but not by Simpson's rule.
19. 3.057355539 **21.** 3.1412554158 **23.** 33.02739148 **25.** 0.89508189 **27.** 0.8820997338 **29.** 6.043173883

Review Exercise Set 6.1

1. $\dfrac{19}{20}$ **3.** $\dfrac{\sqrt{2t+t^2}}{2\sqrt{t}}$ **5.** -2 **7.** a) $\frac{1}{4}\sin^2 2x + C$ b) $\frac{1}{3}\sec^3 x + C$ **9.** $\dfrac{12\sqrt{3}-20}{3}$

Review Exercise Set 6.2

1. $\dfrac{\pi}{2}\left[\sin\dfrac{\pi}{8} + \sin\dfrac{3\pi}{8} + \sin\dfrac{5\pi}{8} + \sin\dfrac{7\pi}{8}\right]$ **3.** $2\sin^2 2t + \sin^2 t$ **5.** -5 **7.** a) $-\frac{1}{9}\csc^3 3x + C$ b) $-2\sqrt{\csc x} + C$
9. $(8 + 3\pi)/64$

Review Exercise Set 6.3

1. $\frac{15}{2}$ **3.** $12 + \dfrac{9\pi}{2}$ **5.** 1 **7.** a) $\dfrac{-1}{9(x^3+1)^3} + C$ b) $-2\sqrt{\cos x} + C$ **9.** $-\dfrac{2\sqrt{9+x^3}}{27x^{3/2}} + C$

Review Exercise Set 6.4

1. $\frac{53}{64}$ **3.** $9\sqrt{6\pi}/4$ **5.** $\frac{32}{3}$ **7.** a) $\dfrac{(4x^3+2)^6}{72} + C$ b) $\dfrac{\sec 3x}{3} + C$ **9.** $-\dfrac{\sqrt{9-4\sin^2 x}}{9\sin x} + C$

More Challenging Exercises 6

1.

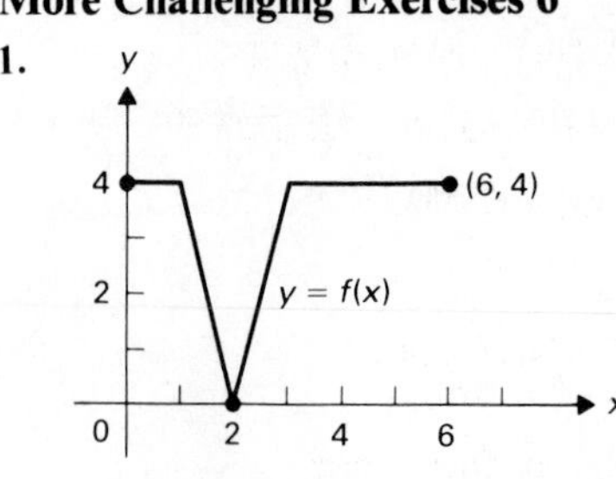

3. $$S_n = \frac{1}{n}\left[f\left(\frac{1}{n}\right) + f\left(\frac{2}{n}\right) + f\left(\frac{3}{n}\right) + \cdots + f\left(\frac{n-1}{n}\right) + f(1)\right],$$
$$s_n = \frac{1}{n}\left[f(0) + f\left(\frac{1}{n}\right) + f\left(\frac{2}{n}\right) + \cdots + f\left(\frac{n-1}{n}\right)\right]$$
(Subtract)
$$S_n - s_n = \frac{1}{n}[f(1) - f(0)] = f(1)/n$$

5. $\dfrac{b-a}{n}[f(a) - f(b)]$ **7.** $\frac{16}{3}$ **9.** $\frac{2}{5}$ **11.** $\frac{16}{5}$

CHAPTER 7

Section 7.1

1. $\frac{32}{3}$ **3.** $\frac{1}{6}$ **5.** $\frac{4}{15}$ **7.** $\frac{44}{15}$ **9.** $\frac{9}{2}$ **11.** $\frac{9}{2}$ **13.** $\frac{7}{15}$ **15.** $\dfrac{4}{3}+\dfrac{\pi}{2}$ **17.** $\frac{8}{3}$ **19.** 4 **21.** $\dfrac{8\pi}{3}-2\sqrt{3}$ **23.** $\dfrac{\pi-2}{4}$ **25.** $\frac{2}{3}\pi^{3/2}-2$

27. $2\displaystyle\int_0^2(\sqrt{68-y^2}-2y^2)\,dy$ **29.** $25\pi-\displaystyle\int_{-4}^{3}\left(\sqrt{25-x^2}-\frac{x+25}{7}\right)dx$ **31.** $8-4\sqrt{2}$ **33.** $-\frac{1}{3}$ **35.** $5^{1/4}$

37. $F(t)$ is continuous on $[a, b]$ and differentiable for $a < t < b$. Thus

$$\frac{F(b)-F(a)}{b-a}=F'(c)$$

for some c, where $a < c < b$. Now $F'(t) = f(t)$ by Theorem 6.3. Since $F(b) = \int_a^b f(x)\,dx$ and $F(a) = \int_a^a f(x)\,dx = 0$, we have

$$\frac{1}{b-a}\int_a^b f(x)\,dx = f(c)$$

for some c, where $a < c < b$.

39. 21.9919 **41.** 71.40633459 ($n = 10$) **43.** 0.1356975072

Section 7.2

1. $16\pi/15$ **3.** $4\sqrt{3}\pi$ **5.** $128\pi/5$ **7.** $\pi/6$ **9.** 2π **11.** $\pi^2/2$ **13.** $\dfrac{\pi}{2}(8+3\pi)$ **15.** $\pi/2$ **17.** $8\pi/3$ **19.** $\displaystyle\int_0^h\pi\left(\frac{r}{h}x\right)^2dx=\frac{\pi}{3}r^2h$

21. $4a^3/\sqrt{3}$ **23.** $\dfrac{\pi}{3}(2a^3+b^3-3a^2b)$

25. The area of the base, where $x = 0$, is ch^2. The volume is $\displaystyle\int_0^h c(h-x)^2\,dx=\frac{ch^3}{3}=\frac{h}{3}\cdot ch^2$.

27. 6.953900748 ($n = 20$) **29.** 226.2117401 ($n = 20$) **31.** 10.55596978

Section 7.3

1. $16\pi/15$ **3.** $\pi/6$ **5.** $128\pi/5$ **7.** $3\pi/10$ **9.** $27\pi/2$ **11.** $4\pi^2$ **13.** $6\pi^2$ **15.** $\displaystyle\int_0^r 2\pi y\left(h-\frac{h}{r}y\right)dy=\frac{\pi}{3}r^2h$ **17.** $2\pi^2a^2b$

19. 6.781044926 **21.** 75.96187138 **23.** 266.5887378

Section 7.4

1. $\frac{8}{27}(10^{3/2}-1)$ **3.** $\frac{1}{6}(125-13^{3/2})$ **5.** $\frac{1}{27}(40^{3/2}-13^{3/2})$ **7.** $\frac{53}{6}$ **9.** $\frac{221}{120}$ **11.** 2π **13.** $3a/2$ **15.** 5

17. $\dfrac{816-432\sqrt{3}}{105}\approx 0.6453$ **19.** $\sqrt{5}\pi/180$ **21.** $\sqrt{17}/20$ **23.** 0.25 **25.** 124.1324579 ($n = 20$) **27.** 3.82019779 ($n = 20$)

29. 11.54129197 ($n = 20$)

Section 7.5

1. $\sqrt{17}\pi$ **3.** $5\sqrt{2}\pi$ **5.** $\dfrac{\pi}{27}(145\sqrt{145}+10\sqrt{10}-2)$ **7.** $2\pi\dfrac{24\sqrt{3}+4}{15}$ **9.** 112π **11.** $\dfrac{12289\pi}{192}$ **13.** $4\pi a^2$ **15.** $41\pi/12$

17. $\dfrac{\pi}{6}(37^{3/2}-1)$ **19.** $\dfrac{\pi}{3}(29^{3/2}-13^{3/2})$ **21.** $2\pi\displaystyle\int_0^h\frac{r}{h}x\sqrt{1+\left(\frac{r^2}{h^2}\right)}\,dx=\pi r\sqrt{r^2+h^2}=\pi rs$ **23.** 53.226 **25.** 39419 **27.** 123.77

29. 33.582

Section 7.6

1. a) -6 b) 6 **3.** a) $-\frac{16}{3}$ b) 8 **5.** a) $\frac{5}{6}$ b) $\frac{5}{6}$ **7.** a) $\frac{3}{2}$ b) $\frac{11}{6}$ **9.** a) 0 b) 10 **11.** a) $4/\pi$ b) $4\sqrt{2}/\pi$

13. a) $\frac{25}{2}$ b) $\frac{25}{2}$ **15.** a) $3t$ b) 6 **17.** a) $1-\cos t$ b) $(3\pi+2)/2$ **19.** a) $3t^2+\dfrac{1}{2(t+1)^2}+\dfrac{3}{2}$ b) $\frac{34}{3}$ **21.** $9\pi/2$

23. $\frac{4}{3}(2\sqrt{2}-1)$ **25.** 27.328 **27.** 49.573

Section 7.7

1. 64 ft · lb **3.** 8 **5.** a) 9 ft · lb b) 36 ft · lb **7.** 12480π ft · lb **9.** 3 ft **11.** 2325 ft · lb **13.** $17k/15$ units

15. $W=\displaystyle\int_{x_{t_1}}^{x_{t_2}}F(x)\,dx=\int_{x_{t_1}}^{x_{t_2}}\left(m\frac{dv}{dx}v\right)dx=\int_{v_{t_1}}^{v_{t_2}}mv\,dv=\frac{1}{2}mv^2\Big]_{v_{t_1}}^{v_{t_2}}=\frac{1}{2}mv_{t_2}^2-\frac{1}{2}mv_{t_1}^2=$ Change in kinetic energy

17. $(1497.6)\pi$ lb **19.** 55,328 lb **21.** 550,368 lb **23.** $(998.4)\pi$ lb **25.** 5148π lb

Section 7.8

1. 17 **3.** 4 **5.** 29 **7.** 16 **9.** $\frac{8}{3}$ **11.** 64 **13.** 0

15. i) $k\,dx$ ii) $\left(\frac{a+b}{2}+s\right)^2(k\,dx)$ and $\left(\frac{a+b}{2}-s\right)^2(k\,dx)$ iii) $\left[\left(\frac{a+b}{2}\right)^2+2s^2\right]k\,dx \neq \left(\frac{a+b}{2}\right)^2(2k\,dx)$, so the mass $2k\,dx$ cannot be considered to be concentrated at the midpoint to compute I.

17. $(4+3\pi)/3$ **19.** $\frac{13}{4}$ **21.** a) $ka^2/2$ b) $ka^3/3$ **23.** $\frac{5}{54}(37^{3/2}-1)$ **25.** $k\pi a^4/2$

27. $M = \int_{l_1}^{l_2} \rho\cdot(x+a)\,dA = \int_{l_1}^{l_2} \rho x\,dA + a\int_{l_1}^{l_2} \rho\,dA = M_y + a\cdot m$ **29.** $1709.8k$

Section 7.9

1. A plane annular region (a disk with a hole in it) **3.** $(\frac{5}{9}, \frac{1}{2})$ **5.** $(0, \frac{3}{5})$ **7.** $\left(\frac{a}{3}, \frac{b}{3}\right)$ **9.** $(0, \frac{244}{91})$ **11.** $(\frac{12}{5}, 0)$ **13.** $(0, 4.7943)$

15. $\left(\frac{\pi}{2}, 0.6009\right)$ **17.** $\frac{3\pi-1}{3}\sqrt{2\rho}$

19. a) $k/3$ b) $\frac{k}{3}[(1+a)^3-a^3]$ c) $(1/\sqrt{3}, 1/\sqrt{3})$ d) (x_2, y_2), where $x_2 = y_2 = (1/\sqrt{3})\sqrt{(1+a)^3-a^3}-a$ e) The point obtained in part (d) depends on the value of a and is different from the point in part (c) if $a \neq 0$. This shows that there is no single point where mass can be considered to be concentrated for computations of moments of inertia about *all* axes.

21. a) $\pi^2-\frac{4\pi}{3}$ b) $\frac{4\pi}{3}+\pi^2$ c) $2\pi^2-\frac{4\pi}{3}$ d) $2\pi^2$ e) $\frac{6\pi-2}{3\pi}\sqrt{2\pi}$ **23.** $(r/3, h/3)$

Review Exercise Set 7.1

1. $\frac{4}{3}$ **3.** 144 ft·lb **5.** $\frac{7\pi}{9\sqrt{3}}$ **7.** $2\int_0^1 (y^3+3y^2)\sqrt{1-y}\,dy$

Review Exercise Set 7.2

1. $\frac{125}{6}$ **3.** 208,000 lb **5.** $9\sqrt{2\pi}$ **7.** $\frac{20}{3\pi}m$

Review Exercise Set 7.3

1. $4(\sqrt{2}-1)$ **3.** $\frac{k}{20}$ ft·lb **5.** $\frac{\pi}{6}(17\sqrt{17}-1)$ **7.** 0.1413

Review Exercise Set 7.4

1. $\frac{256}{5}$ **3.** $(249.6)\sqrt{2}$ lb **5.** a) $2\pi\int_0^{\pi}(2\sin t+1)\sqrt{4+5\sin^2 t}\,dt$ b) 117.5151208 **7.** $k(\pi^2-4)$

More Challenging Exercises 7

1. $(\pi/\sqrt{2})\int_{-2}^{2}(x^2-2x+8)(4-x^2)\,dx$ or $2\sqrt{2}\pi\int_0^4 \sqrt{y}(y+2)\,dy$; $\frac{704\sqrt{2}\pi}{15}$

3. $\frac{2\pi}{5}\int_4^5 (2x^2+3x+62)(25-x^2)\,dx + \frac{18\pi}{5}\int_0^4 (4x^2+3x+30)\,dx$ **5.** $\frac{256\sqrt{2}}{5}$ **7.** $\frac{200}{3}$ hr

CHAPTER 8

Section 8.1

1. Invertible; $f^{-1}(x) = x+1$ **3.** Invertible; $f^{-1}(x) = 3-x$ **5.** Not invertible **7.** Not invertible

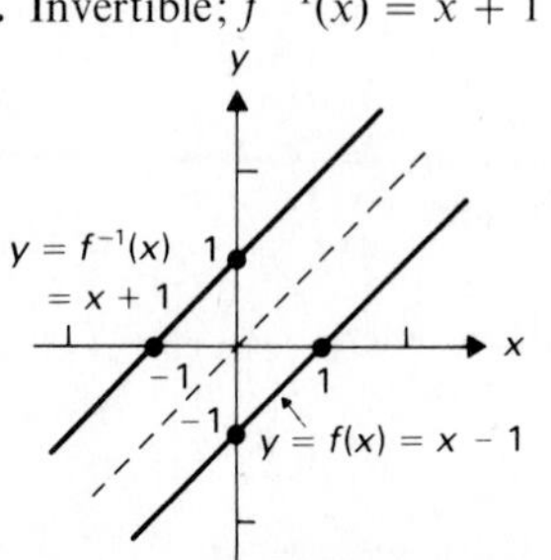

9. Invertible; $f^{-1}(x) = \sqrt[3]{x-1}$

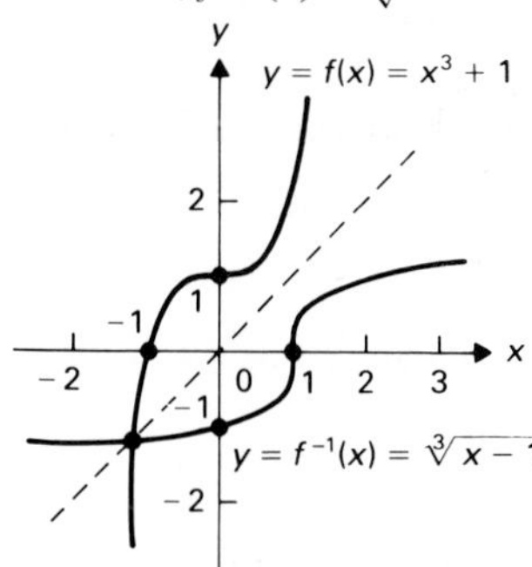

11. Invertible; $f^{-1}(x) = x^2, x \geq 0$

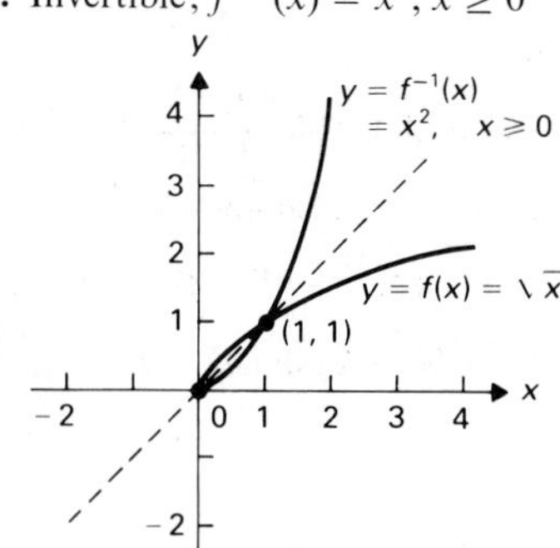

13. Not invertible

15. Invertible; $f^{-1}(x) = \dfrac{x+1}{1-x}$

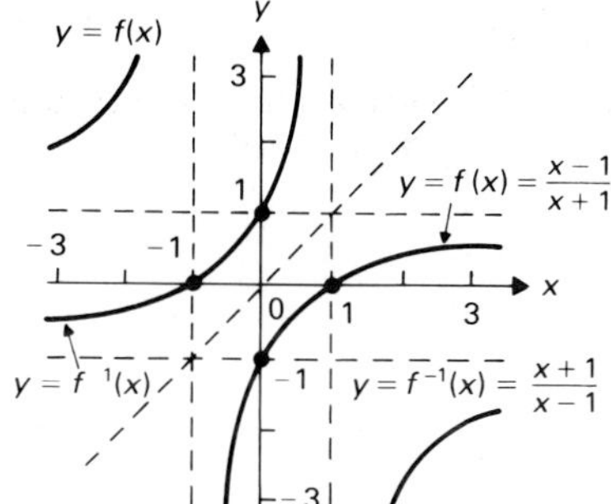

17. Not invertible **19.** Not invertible **21.** Invertible **23.** Not invertible

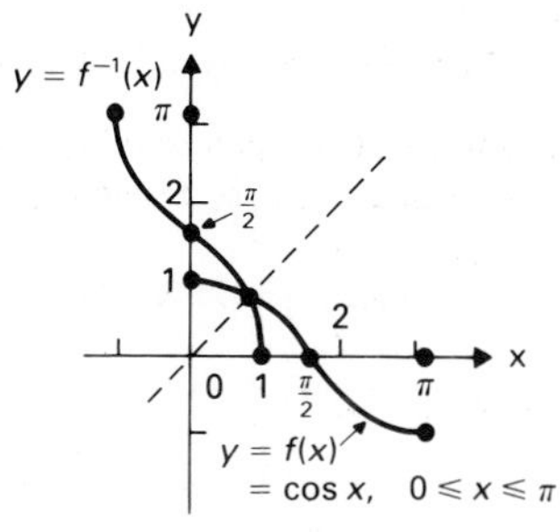

25. Invertible

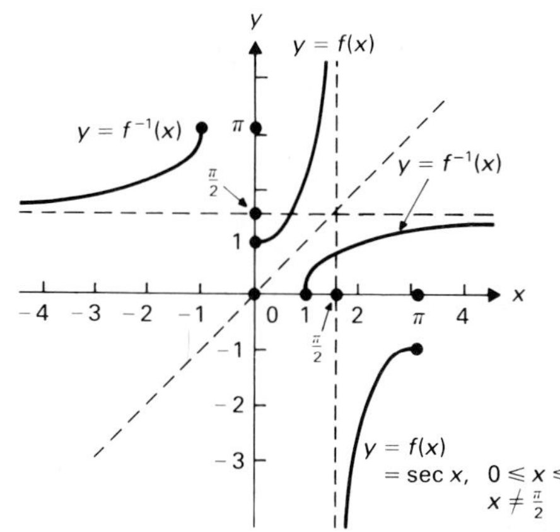

27. T T T F F T T F T T **29.** -4 **31.** $\frac{1}{2}$ **33.** 1 **35.** $\frac{1}{2}$ **37.** 3

39. a) $f(x)$ is decreasing b) $[-1, 1]$ c) $-1\sqrt{1-x^2}$ d) $-1/\sqrt{-x^2-x}$

41. a) $g(x)$ is decreasing b) All x c) $-1/(1+x^2)$ d) $-2x/(1+x^4)$

43. a) $f(x)$ is one to one b) $x \geq 1, x \geq -1$ c) $-1/(x\sqrt{x^2-1})$ d) $-1/(2x\sqrt{x-1})$

Section 8.2

1. Yes; it is differentiable. **3.** $y = x - 1$ **5.**

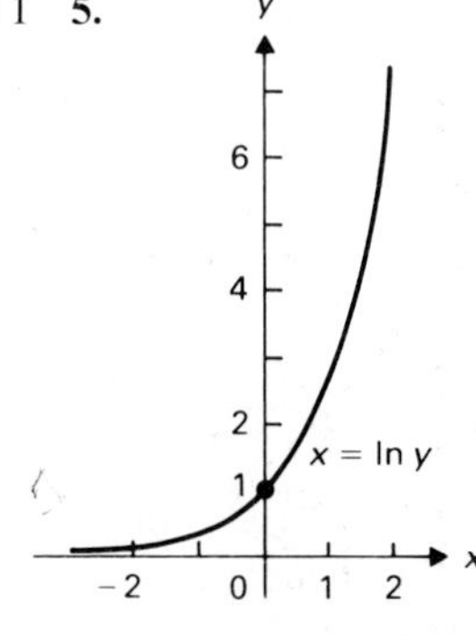

7.

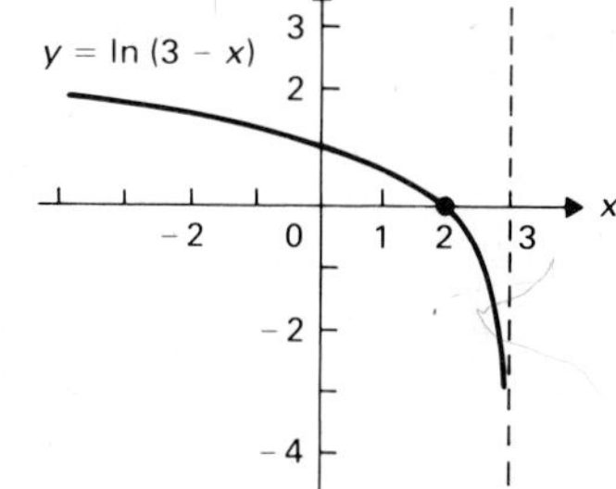

9. -1.1 **11.** 2.1 **13.** -0.3

15. 0.55 **17.** $3/(3x+2),\ x > -\frac{2}{3}$ **19.** $3/x,\ x > 0$ **21.** $-2\tan x$ **23.** $2(\ln x)\cdot\frac{1}{x}$ **25.** $(9x^2-4)/(6x^3-8x),\ 3x^3-4x>0$

27. $\tan x + \frac{\sec^2 x}{\tan x},\ \sec x \tan x > 0$ **29.** $-2\tan x + 6\cot 2x,\ \sin 2x > 0$ **31.** $\frac{1}{\ln x}\cdot\frac{1}{x}$ **33.** $-\sin(\ln x)\cdot\frac{1}{x}$ **35.** $\ln|x+1| + C$

37. $-\frac{1}{2}\ln|\cos 2x| + C$ **39.** $\frac{1}{2}\ln(x^2+1) + C$ **41.** $\frac{1}{2}(\ln x)^2 + C$ **43.** $\frac{1}{2}\ln\left|\ln|x|\right| + C$ **45.** $-\frac{1}{8}\ln|1-4x^2| + C$

47. $-\ln|1+\cos x| + C$ **49.** $1-\ln 2$ **51.** $1+\ln 2$ **53.** $\frac{1}{3}\ln\left(\frac{3+\sqrt{10}}{2}\right)$ **55.** $\frac{1}{2}\ln(\sqrt{2}+1)$

57. a) Since $(1/\sqrt{x}) > 1/x$ for $x > 1$, the definition of the integral shows at once that $\int_1^t (1/\sqrt{x})\,dx > \int_1^t (1/x)\,dx$ for $t > 1$. We obtain

$$2\sqrt{x}\Big]_1^t > \ln|x|\Big]_1^t,$$

or $2\sqrt{t}-2 > \ln t$ for $t > 1$. b) Replacing t by x and dividing by x, we have $(2\sqrt{x}-2)/x > (\ln x)/x$ for $x > 1$. Then

$$\lim_{x\to\infty}\frac{\ln x}{x} \le \lim_{x\to\infty}\frac{2\sqrt{x}-2}{x} = 0, \qquad \text{so} \qquad \lim_{x\to\infty}\frac{\ln x}{x} = 0.$$

c) From part (b), we have

$$\lim_{x\to 0+}[x(\ln x)] = \lim_{u\to\infty}\left(\frac{1}{u}\cdot\ln\frac{1}{u}\right) = \lim_{u\to\infty}\frac{-\ln u}{u} = -\lim_{u\to\infty}\frac{\ln u}{u} = 0.$$

59. 3.1461932 **61.** 2.219107149 **63.** -0.1875

Section 8.3

1. 2 **3.** 0 **5.** -2 **7.** 6 **9.** 16 **11.** 9 **13.** $2e^{2x}$ **15.** $e^{2x}\cos x + 2e^{2x}\sin x$ **17.** $\frac{e^x}{x} + e^x(\ln 2x)$ **19.** $e^{\sec x}\sec x\tan x$

21. $\frac{-e^{1/x}}{x^2}$ **23.** $-e^{-x}$ **25.** $\frac{e^{\sin x}(\cos^2 x + \sin x)}{\cos^2 x}$ **27.** $\frac{-e^{-x^2}(2x^2+2x+1)}{(x+1)^2}$ **29.** 4 **31.** $-e^{\cos x} + C$ **33.** $\ln(1+e^x) + C$

35. $\ln|e^x - e^{-x}| + C$ **37.** $\frac{2}{3}(e^x+1)^{3/2} + C$ **39.** $\frac{ex^2}{2} + C$ **41.**

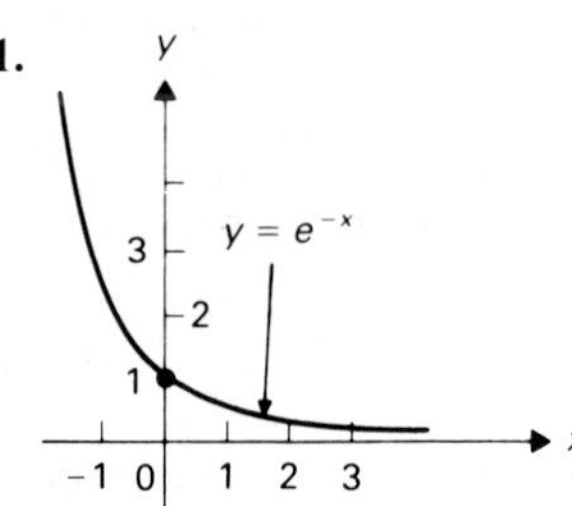

43.

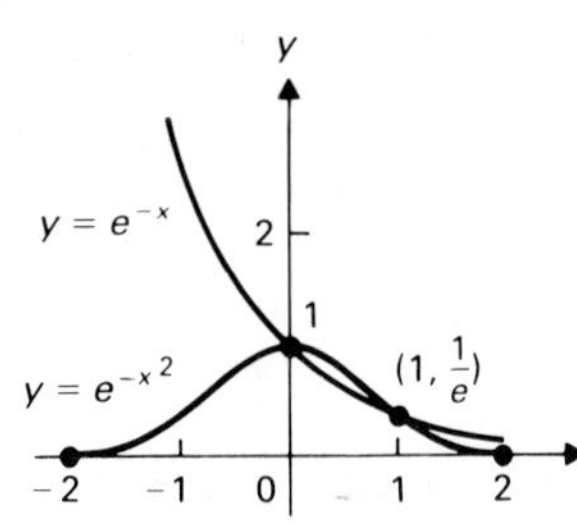

45. $m = 2, -2$ **47.** 2.070579905 **49.** 1.309799586 **51.** -0.90356 **53.** 0.4890435

Section 8.4

1. 5 **3.** 1 **5.** $\frac{1}{2}$ **7.** 9 **9.** 6 **11.** 64 **13.** 5 **15.** $3(\ln 2)2^{3x}$ **17.** $3^{2x-1}[2(\ln 3)x^2 + 2x]$ **19.** $(\ln 18)2^x\cdot 3^{2x}$ **21.** $1/[(\ln 10)x]$

23. $(\cos x)/(\ln 2)$ **25.** $\frac{1}{\ln 5}\left(\frac{2}{2x+1} - \frac{1}{x}\right)$ **27.** $\frac{1}{\ln 4} + \log_4 x$ **29.** $(\sin x)^x[x\cot x + \ln(\sin x)]$

31. $-\sin(x^x)[x^x\cdot(\ln x + 1)]$ **33.** $2^x\cdot 3^x\cdot[(\ln 2) + 2(\ln 3)]$ **35.** $\frac{5^{x^2}}{7x}[2x(\ln 5) - (\ln 7)]$ **37.** $x^2e^x(\cos x)\left(\frac{2}{x} + 1 - \tan x\right)$

39. $(x^2+1)\sqrt{2x+3}(x^3-2x)\left(\frac{2x}{x^2+1} + \frac{1}{2x+3} + \frac{3x^2-2}{x^3-2x}\right)$ **41.** $2e^{\sqrt{x}} + C$ **43.** $\frac{1}{6(\ln 5)}5^{3x^2} + C$

45. $\frac{1}{\ln 2}\cdot\ln(1+2^x) + C$ **47.** $\frac{1}{(\ln 2)^2}2^{(2^x)} + C$

49. Let $y = \log_a x$. Then $x = a^y$. Differentiating implicitly, we have

$$1 = a^y(\ln a)\cdot\frac{dy}{dx}, \quad \text{so} \quad \frac{dy}{dx} = \frac{1}{(\ln a)a^y} = \frac{1}{(\ln a)x}.$$

51.

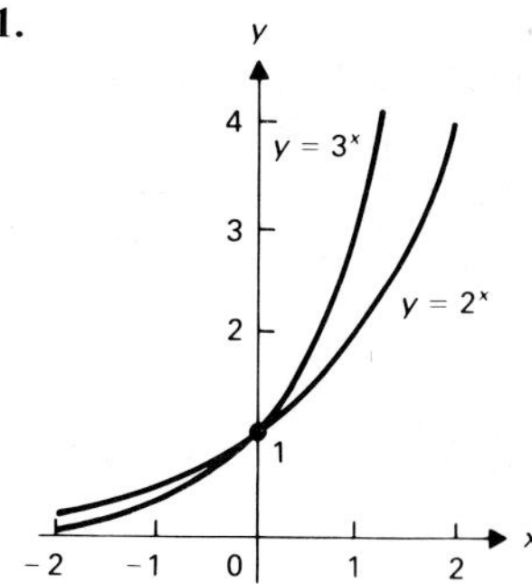

53. All graphs go through the point (0, 1). If $0 < a < 1$, the function is decreasing, with $\lim_{x\to-\infty} a^x = \infty$ and $\lim_{x\to\infty} a^x = 0$. If $a = 1$, the graph is the line $y = 1$. If $a > 1$, the function is increasing and the graph has shape similar to the graph of e^x, with $\lim_{x\to-\infty} a^x = 0$ and $\lim_{x\to\infty} a^x = \infty$.
55. 2.231829 **57.** 2.484709796 **59.** 3.0457

Section 8.5

1. $\dfrac{1600 \ln(\frac{3}{2})}{\ln 2}$ yr **3.** $\dfrac{\ln 2000}{\ln 2.4} \approx 8.68$ days **5.** a) 80 yr b) 80 yr **7.** a) $\sqrt[4]{3}$ b) $\sqrt[4]{3}$ **9.** 12.6 yr **11.** \$29,313.65
13. $\frac{25}{2}(\ln 2)\% \approx 8.66\%$ **15.** a) $1985 + 20(\ln 50) \approx 2063$ b) $1985 + 20(\ln 50{,}000) \approx 2201$ c) Have a "decimal point day" when all wages and prices are divided by 10 every time the average yearly wage reaches \$100,000
17. a) $(2000)e^{2.25} \approx \$18{,}975$ b) 3.9 **19.** $\dfrac{200}{e}$ lb **21.** $200 \cdot e^{-1.2} \approx 60.24$ lb **23.** $\dfrac{2560}{9 \ln(\frac{4}{3})}$ ft **25.** $40 + (\frac{3}{5})^{8/3}$ degrees
27. Year 4,029

Section 8.6

1. $\pi/2$ **3.** $-\pi/3$ **5.** $7\pi/6$ **7.** $\pi/4$ **9.** $3\pi/2$ **11.** π **13.** 0 **15.** $5\pi/4$ **17.** $-\pi/4$ **19.** $\pi/2$ **21.** $\pi/2$ **23.** 0 **25.** $2/\sqrt{1-4x^2}$
27. $1/[2\sqrt{x}(1+x)]$ **29.** $1/[x\sqrt{(1/x)^2-1}]$ **31.** $6(\tan^{-1} 2x)^2/(1+4x^2)$ **33.** $-1/[(1+x^2)(\tan^{-1} x)^2]$
35. $2(x + \sin^{-1} 3x)[1 + (3/\sqrt{1-9x^2})]$ **37.** $\pi/2$ **39.** π **41.** $\frac{1}{3}\sec^{-1} 2x + C$ **43.** 0.4889 **45.** 0.2145 **47.** 0.9463 **49.** 0.472
51. $\pi/2$ **53.** $\dfrac{2\sqrt{3}}{3} - \dfrac{\pi}{6}$ **55.** $\pi/6$ **57.** $\dfrac{2\pi}{3} + \dfrac{8}{3}$
59. Let $y = \tan^{-1} x$. Then $x = \tan y$ and $1 = (\sec^2 y)(dy/dx)$, so

$$\frac{dy}{dx} = \frac{1}{\sec^2 y} = \frac{1}{1+\tan^2 y} = \frac{1}{1+x^2}.$$

61. Let $y = \sec^{-1} x$. Then $x = \sec y$ and $1 = (\sec y \tan y)(dy/dx)$, so

$$\frac{dy}{dx} = \frac{1}{\sec y \tan y} = \frac{1}{\sec y\sqrt{\sec^2 y - 1}} = \frac{1}{x\sqrt{x^2-1}}.$$

(It is appropriate to substitute the *positive* quantity $\sqrt{\sec^2 y - 1}$ for $\tan y$ since $0 \le y < \pi/2$, or $\pi \le y < 3\pi/2$.)

63. a) $\sec^{-1}(-\sqrt{2}) = \dfrac{5\pi}{4}$, while $\cos^{-1}(-1/\sqrt{2}) = 3\pi/4$. b) For the "new" *inverse secant*, let $y = \sec^{-1} x$, so $x = \sec y$ where $0 \le y < \pi/2$, or $\pi/2 < y \le \pi$. Now

$$\frac{dy}{dx} = \frac{1}{dx/dy} = \frac{1}{\sec y \tan y}.$$

For $x > 1$ so that $0 \le y < \pi/2$, we have $\tan y > 0$, so $\tan y = \sqrt{\sec^2 y - 1} = \sqrt{x^2-1}$. For $x < -1$ so that $\pi/2 < y \le \pi$, we have $\tan y < 0$, so $\tan y = -\sqrt{\sec^2 y - 1} = -\sqrt{x^2-1}$. Thus

$$\frac{dy}{dx} = \begin{cases} \dfrac{1}{x\sqrt{x^2-1}} & \text{for } x > 1, \\[2ex] \dfrac{1}{-x\sqrt{x^2-1}} & \text{for } x < -1, \end{cases} = \frac{1}{|x|\sqrt{x^2-1}}.$$

Section 8.7

1. Divide the relation $\cosh^2 x - \sinh^2 x = 1$ by $\cosh^2 x$. **3.** $\sinh(-x) = \dfrac{e^{-x} - e^{-(-x)}}{2} = \dfrac{-e^x + e^{-x}}{2} = -\sinh x$

5.
$$\sinh x \cosh y + \cosh x \sinh y = \frac{e^x - e^{-x}}{2}\cdot\frac{e^y + e^{-y}}{2} + \frac{e^x + e^{-x}}{2}\cdot\frac{e^y - e^{-y}}{2}$$
$$= \frac{2e^{x+y} - 2e^{-x-y}}{4} = \frac{e^{x+y} - e^{-x-y}}{2}$$
$$= \sinh(x + y)$$

7. $\sinh 2x = 2\sinh x\cosh x$; $\cosh 2x = \cosh^2 x + \sinh^2 x$

9. $D(\cosh x) = D\left(\dfrac{e^x + e^{-x}}{2}\right) = \dfrac{1}{2}D(e^x + e^{-x}) = \dfrac{1}{2}(e^x - e^{-x}) = \sinh x$

11. $D(\operatorname{sech} x) = D\left(\dfrac{1}{\cosh x}\right) = \dfrac{-\sinh x}{\cosh^2 x} = -\tanh x \operatorname{sech} x$

13. Let $y = \cosh^{-1} x$, so $x = \cosh y$. Then
$$\frac{dy}{dx} = \frac{1}{dx/dy} = \frac{1}{\sinh y} = \frac{1}{\sqrt{\cosh^2 y - 1}} = \frac{1}{\sqrt{x^2 - 1}}, \qquad x > 1.$$
(Since $y = \cosh^{-1} x \geq 0$, $\sinh y \geq 0$, so the *positive* square root was appropriate.)

15. Let $y = \coth^{-1} x$, so $x = \coth y$. Then
$$\frac{dy}{dx} = \frac{1}{dx/dy} = \frac{-1}{\operatorname{csch}^2 y} = \frac{-1}{\coth^2 y - 1} = \frac{-1}{x^2 - 1} = \frac{1}{1 - x^2}, \qquad |x| > 1.$$

17. Let $y = \operatorname{csch}^{-1} x$, so $x = \operatorname{csch} y$. Then
$$\frac{dy}{dx} = \frac{1}{dx/dy} = \frac{-1}{\operatorname{csch} y \coth y} = \frac{-1}{(\operatorname{csch} y)(\pm\sqrt{1 + \operatorname{csch}^2 y})} = \frac{-1}{(x)(\pm\sqrt{1 + x^2})}.$$
If $x > 0$, then $y = \operatorname{csch}^{-1} x > 0$, so $\coth y > 0$ and the *plus sign* is appropriate. If $x < 0$, then $y = \operatorname{csch}^{-1} x < 0$ so $\coth y < 0$ and the *minus sign* is appropriate. These two cases are both covered by the formula
$$D(\operatorname{csch}^{-1} x) = \frac{-1}{(|x|\sqrt{1 + x^2})}.$$

19. $2x\sinh(x^2)$ **21.** $-\dfrac{\operatorname{sech}\sqrt{x}\tanh\sqrt{x}}{2\sqrt{x}}$ **23.** $-\dfrac{\operatorname{csch}(\ln x)\coth(\ln x)}{x}$ **25.** $2\sinh^3 x\cosh x + 2\cosh^3 x\sinh x$

27. $\dfrac{2}{\sqrt{1 + 4x^2}}$ **29.** $2\sec 2x$ **31.** $\dfrac{-1}{x\sqrt{1 - x^4}}$ **33.** $\dfrac{e^{2x}}{\sqrt{1 + e^{2x}}} + e^x\sinh^{-1}(e^x)$ **35.** $\dfrac{-3\operatorname{csch}^2 3x}{1 - \coth^2 3x}$, $|\cot 3x| < 1$

37. $\ln|\sinh x| + C$ **39.** $\frac{1}{3}\cosh(3x + 2) + C$ **41.** $\frac{1}{3}\tanh 3x + C$ **43.** $\ln|\sinh x| + C$ **45.** $\frac{1}{2}(\ln\frac{4}{3})$ **47.** $\frac{1}{4}\sqrt{1 + 4x^2} + C$

49. $-\operatorname{sech}^{-1}(e^x) + C$ **51.** $\dfrac{\sqrt{5}}{2} + 2\sinh^{-1}\left(\dfrac{1}{2}\right) = \dfrac{\sqrt{5}}{2} + 2\ln\left(\dfrac{1 + \sqrt{5}}{2}\right)$

53. $\sqrt{16 + x^2} - 4\sinh^{-1}\left|\dfrac{4}{x}\right| + C = \sqrt{16 + x^2} - 4\ln\left(\dfrac{4 + \sqrt{16 + x^2}}{x}\right) + C$

55. $-\dfrac{9}{2}\sinh^{-1}\left(\dfrac{\sin x}{3}\right) + \dfrac{\sin x\sqrt{9 + \sin^2 x}}{2} + C = \dfrac{(\sin x)}{2}\sqrt{9 + \sin^2 x} - \dfrac{9}{2}\ln(\sin x + \sqrt{9 + \sin^2 x}) + C$

57. $8\cosh^{-1}\left(\dfrac{e^x}{4}\right) + \dfrac{e^x}{2}\sqrt{e^{2x} - 16} + C = \dfrac{e^x}{2}\sqrt{e^{2x} - 16} + 8\ln|e^x + \sqrt{e^{2x} - 16}| + C$ **59.** $\dfrac{\cosh^2 4x\sinh 4x}{12} + \dfrac{1}{6}\sinh 4x + C$

61. Let $x = \tanh y = \dfrac{e^y - e^{-y}}{e^y + e^{-y}}$. Then $xe^y + xe^{-y} = e^y - e^{-y}$ and $xe^{2y} + x = e^{2y} - 1$, so $(x - 1)e^{2y} = -x - 1$ and $e^{2y} = \dfrac{1 + x}{1 - x}$.

Then $2y = 2\tanh^{-1} x = \ln\left(\dfrac{1 + x}{1 - x}\right)$, and $\tanh^{-1} x = \dfrac{1}{2}\ln\left(\dfrac{1 + x}{1 - x}\right)$.

63. Let $x = \text{sech}\, y = \dfrac{2}{e^y + e^{-y}}$. Then $xe^y + xe^{-y} = 2$, so $xe^{2y} - 2e^y + x = 0$ and $e^y = \dfrac{2 \pm \sqrt{4 - 4x^2}}{2x} = \dfrac{1 + \sqrt{1 - x^2}}{x}$. (The positive sign is appropriate since $y \geq 0$ if $y = \text{sech}^{-1} x$.) Then $y = \text{sech}^{-1} x = \ln\left(\dfrac{1 + \sqrt{1 - x^2}}{x}\right)$.

Review Exercise Set 8.1

1. a) $\ln x = \displaystyle\int_1^x \frac{1}{t}\, dt, x > 0$ b)

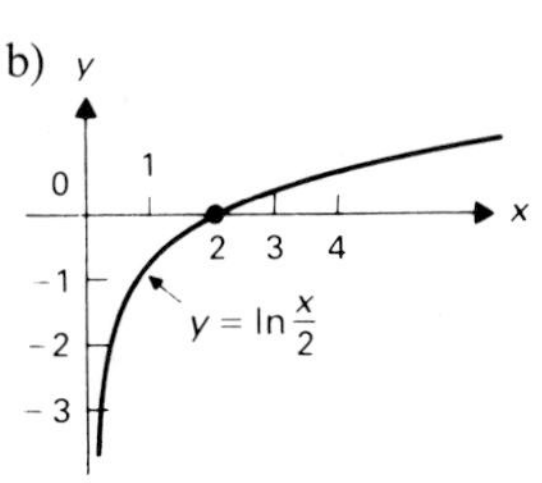

3. a) $e^{\tan x} \sec^2 x$ b) $\frac{7}{3}$ **5.** a) $-\dfrac{\ln 3}{\ln 24}$ b) 3 **7.** a) $-\pi/6$ b) $-\pi/3$ **9.** a) $\sinh x = \dfrac{e^x - e^{-x}}{2}$ b) $-6 \,\text{sech}^3\, 2x \tanh 2x$

Review Exercise Set 8.2

1. a) It is the unique number e satisfying $\ln e = 1$. b)

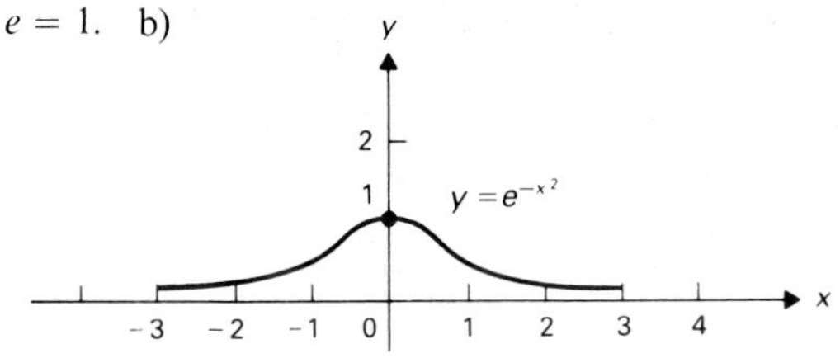

3. a) $\dfrac{e^{\sin^{-1} x}}{\sqrt{1 - x^2}}$ b) $\frac{1}{3}e^{\tan 3x} + C$

5. a) $10^{-1/5}$ b) 49 **7.** a) $\pi/3$ b) $4\pi/3$ **9.** a)

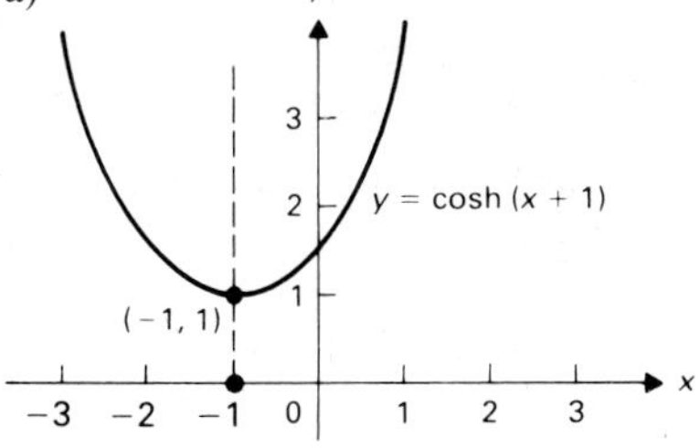

b) $-6 \coth^2(2x + 1) \,\text{csch}^2(2x + 1)$

Review Exercise Set 8.3

1. a) e^{500} b) $1/e^4$ **3.** a) $-e^{2x} \sin x + 2e^{2x} \cos x$ b) $x + e^{-x} + C$

5. a) $(\ln 3)3^{\sin x}(\cos x)$ b) $\dfrac{3^x \cdot 4x^3}{5x}[(\ln 3) + (\ln 4)3x^2 - (\ln 5)]$ **7.** a) $3\pi/4$ b) $-\pi/2$ **9.** a) $\dfrac{e^x + e^{-x}}{2}$ b) $\dfrac{e^2 - 1}{2e}$

Review Exercise Set 8.4

1. a) $1/\sqrt{5}$ b) $\dfrac{\ln 4}{\ln 0.6}$ **3.** a) $\dfrac{e^{\sqrt{x}}}{2\sqrt{x}} \sec^2(e^{\sqrt{x}})$ b) $16e$ **5.** a) Use Simpson's rule. b) 1241.06646 ($n = 40$)

7. a) $2\pi/3$ b) $-\pi/2$ **9.** $\ln(2x + \sqrt{4x^2 + 1})$

More Challenging Exercises 8

1. $x = 0, \ln 2$

3. $\ln n = \displaystyle\int_1^n \frac{1}{x}\, dx$. For this integral,

$$s_{n-1} = \frac{1}{2} + \frac{1}{3} + \cdots + \frac{1}{n} \quad \text{and} \quad S_{n-1} = 1 + \frac{1}{2} + \frac{1}{3} + \cdots + \frac{1}{n-1}.$$

5. $2 \ln a$ **7.** If x is large enough, then e^x is greater than $f(x)$ for any particular polynomial function f.

CHAPTER 9

Section 9.1

1. $\cos x + x \sin x + C$ **3.** $\frac{1}{3}xe^{3x} - \frac{1}{9}e^{3x} + C$ **5.** $-\frac{1}{4}x^2 \cos 2x^2 + \frac{1}{8}\sin 2x^2 + C$ **7.** $-x^2 \cos x + 2x \sin x + 2\cos x + C$

9. $\frac{1}{2}x^2 e^{x^2} - \frac{x}{2}e^{x^2} + C$ **11.** $x \tan x - \frac{x^2}{2} + \ln|\cos x| + C$ **13.** $\frac{x}{2}\tan 2x + \frac{1}{4}\ln|\cos 2x| + C$

15. $\frac{x^2}{2}\sqrt{1+2x^2} - \frac{1}{6}(1+2x^2)^{3/2} + C$ **17.** $\frac{x^4}{3}(1+x^2)^{3/2} - \frac{4x^2}{15}(1+x^2)^{5/2} + \frac{8}{105}(1+x^2)^{7/2} + C$

19. $\frac{-x^2}{2+2x^2} + \frac{1}{2}\ln(1+x^2) + C$ **21.** $\frac{x^2}{2}(\ln x) - \frac{x^2}{4} + C$ **23.** $x\ln(1+x) - x + \ln(1+x) + C$

25. $x\ln(1+x^2) - 2x + 2\tan^{-1}x + C$ **27.** $x\cos^{-1}x - \sqrt{1-x^2} + C$ **29.** $x\tan^{-1}3x - \frac{1}{6}\ln(1+9x^2) + C$

31. $\frac{x^2}{2}\tan^{-1}x - \frac{1}{2}x + \frac{1}{2}\tan^{-1}x + C$ **33.** $\frac{x^3}{3}\sin^{-1}x + \frac{x^3}{3}\sqrt{1-x^2} + \frac{2}{9}(1-x^2)^{3/2} + C$

35. $\frac{1}{2}[x\cos(\ln x) - x\sin(\ln x)] + C$ **37.** $\frac{e^{ax}}{a^2+b^2}(a\sin bx - b\cos bx) + C$ **39.** See formula 16 in the integral table.

41. See formula 86 in the integral table. **43.** See formula 67 in the integral table. **45.** See formula 92 in the integral table.

Section 9.2

1. $\ln|x| + 1/x + C$ **3.** $\frac{x^3}{3} - \frac{x^2}{2} - x + \ln|x+1| + C$ **5.** $x^2/2 + \frac{1}{2}\ln|x^2+1| + \tan^{-1}x + C$ **7.** $\frac{1}{2}\ln\left|\frac{x-1}{x+1}\right| + C$

9. $\frac{1}{2}x^2 + \ln|(x+2)(x-2)^3| + C$ **11.** $\frac{x^3}{3} + \frac{3x^2}{2} + 5x + 14\ln|x-2| - 5\ln|x-1| + C$

13. $-\frac{4}{3}\ln|x| + \frac{13}{12}\ln|x-3| + \frac{1}{4}\ln|x+1| + C$ **15.** $x + \frac{5}{2}\ln|x+3| - \frac{1}{2}\ln|x-1| - \ln|x+1| + C$

17. $\ln\left|\frac{x+1}{x}\right| - \frac{2}{x+1} + C$ **19.** $\frac{-1}{3(x+1)^2} - \frac{8}{9(x+1)} + \frac{1}{27}\ln\left|\frac{x-2}{x+1}\right| + C$

21. $\frac{-3}{x} - \ln|x| + 2\ln|x-2| + 4\ln|x+2| + C$ **23.** $4\tan^{-1}x + \frac{1}{2}\ln(x^2+1) + 5\ln|x-1| + C$

25. $\ln\left|\frac{(x-2)^2}{x+2}\right| + \frac{1}{2}\ln(x^2+2) - \frac{1}{\sqrt{2}}\tan^{-1}\left(\frac{x}{\sqrt{2}}\right) + C$ **27.** $\frac{5}{6}\ln|3x^2-4x+2| - \frac{\sqrt{2}}{3}\tan^{-1}\left(\frac{3x-2}{\sqrt{2}}\right) + C$

29. $\frac{2x+5}{2(x^2-2x+2)} + \tan^{-1}(x-1) + C$ **31.** $\frac{x-1}{x^2-x+1} + \frac{1}{2}\ln|x^2-x+1| + \sqrt{3}\tan^{-1}\left(\frac{2x-1}{\sqrt{3}}\right) + C$

33. $-\frac{3}{x} + \frac{2-x}{2(x^2+1)} - \frac{3x}{4x^2+4} - \frac{3}{4}\tan^{-1}x + C$

Section 9.3

1. $u = x \qquad dv = (x+1)^{1/2}\,dx$

$du = dx \qquad v = \frac{2}{3}(x+1)^{3/2}$

$$\int x\sqrt{x+1}\,dx = \frac{2}{3}x(x+1)^{3/2} - \int \frac{2}{3}(x+1)^{3/2}\,dx = \frac{2}{3}x(x+1)^{3/2} - \frac{4}{15}(x+1)^{5/2} + C$$

3. $\frac{2}{3}(1+x)^{3/2} - 2\sqrt{1+x} + C$ **5.** $\frac{2}{7}(4+x)^{7/2} - \frac{16}{5}(4+x)^{5/2} + \frac{32}{3}(4+x)^{3/2} + C$ **7.** $x - 2\sqrt{x} + 2\ln|\sqrt{x}+1| + C$

9. $x + 2 - 4\sqrt{x+2} + 4\ln|\sqrt{x+2}+1| + C$

11. $2(x+1)^{3/2} - 2\sqrt{x+1} + C$ **13.** $-\frac{116}{15}$ **15.** $\sqrt{1+x^2} + C$ **17.** $\frac{1}{3}(1+x^2)^{3/2} - \sqrt{1+x^2} + C$

19. $\frac{1}{7}(x^2+1)^{7/2} - \frac{1}{5}(x^2+1)^{5/2} + C$ **21.** $\frac{1}{3}(x^2+8)^{3/4} - 9\sqrt{x^2+8} + C$ **23.** $\frac{-1}{\sqrt{x^2-1}} - \frac{1}{3(x^2-1)^{3/2}} + C$ **25.** $\frac{64}{3}$

27. $\frac{13}{3}$ **29.** $\frac{4}{7}x^{7/4} + \frac{2}{3}x^{3/2} + \frac{8}{5}x^{5/4} + 2x + \frac{8}{3}x^{3/4} + 4x^{1/2} + 8x^{1/4} + 8\ln|x^{1/4}-1| + C$

31. $\frac{6}{7}x^{7/6} - \frac{3}{2}x^{2/3} + 6x^{1/6} - 2\ln|1+x^{1/6}| + \ln|x^{1/3} - x^{1/6} + 1| - 2\sqrt{3}\tan^{-1}\left(\frac{2x^{1/6}-1}{\sqrt{3}}\right) + C$

33. $\frac{e^{\cos^{-1}x}}{2}(x - \sqrt{1-x^2}) + C$ **35.** $\frac{e^{\sin^{-1}x}}{10}[2x\sqrt{1-x^2} - 2(1-2x^2)] + C$ **37.** $\frac{e^{\pi/2}+1}{2}$

Section 9.4

1. $\ln\left|\dfrac{1+\tan(x/2)}{1-\tan(x/2)}\right| + C$ **3.** $-\ln|1+\cos x| + C$ **5.** $-\dfrac{2}{\sqrt{3}}\tan^{-1}\left[\sqrt{3}\tan\left(\dfrac{\pi}{4}-\dfrac{x}{2}\right)\right] + C$

7. $\dfrac{1}{2}\ln\left|\tan\dfrac{x}{2}\right| - \dfrac{1}{4}\tan^2\left(\dfrac{x}{2}\right) + C$ **9.** $-\cot x + \csc x + \ln|\sin x| + \ln|\csc x + \cot x| + C$

11. $\dfrac{x}{2} - \dfrac{1}{4}\ln|\cos 2x| - \dfrac{1}{4}\ln|\sec 2x + \tan 2x| + C$ **13.** $\dfrac{1}{4}\ln|\sec 2x + \tan 2x| + \dfrac{x}{2} + \dfrac{1}{4}\ln|\cos 2x| + C$

15. $\dfrac{1}{5}\ln\left|\dfrac{3\tan(x/2)-1}{\tan(x/2)+3}\right| + C$

Section 9.5

1. Using $\sin^2 x + \cos^2 x = 1$, we have $\cos 2x = \cos^2 x - \sin^2 x = 2\cos^2 x - 1 = 1 - 2\sin^2 x$. Then

$$\cos^2 x = \frac{1+\cos 2x}{2} \quad \text{and} \quad \sin^2 x = \frac{1-\cos 2x}{2}$$

from which formulas (1) follow at once.

3. $\dfrac{\sin^3 x}{3} - \dfrac{\sin^5 x}{5} + C$ **5.** $\frac{1}{5}\sin^5 x - \frac{2}{7}\sin^7 x + \frac{1}{9}\sin^9 x + C$ **7.** $-\frac{1}{3}(\cos 2x)^{3/2} + \frac{1}{7}(\cos 2x)^{7/2} + C$

9. $\dfrac{-2}{\sqrt{\sin x}} - \dfrac{2}{3}(\sin x)^{3/2} + C$ **11.** $\dfrac{x}{2} + \dfrac{1}{20}\sin 10x + C$ **13.** $\frac{3}{8}x + \frac{1}{8}\sin 4x + \frac{1}{64}\sin 8x + C$

15. $\frac{5}{16}x - \frac{1}{4}\sin 2x + \frac{3}{64}\sin 4x + \frac{1}{48}\sin^3 2x + C$ **17.** $-\cot x + C$ **19.** $\tan x + \frac{1}{3}\tan^3 x + C$ **21.** $\dfrac{\tan^3 x}{3} + C$ **23.** $\dfrac{\tan^5 x}{5} + C$

25. $\frac{1}{6}\tan^2 3x + C$ **27.** $-\tan x + \dfrac{\tan^3 x}{3} + x + C$ **29.** $\frac{1}{12}\tan^6 2x - \frac{1}{8}\tan^4 2x + \frac{1}{4}\tan^2 2x + \ln|\cos 2x| + C$

31. $-\frac{1}{4}\csc^4 x + C$ **33.** $-\frac{1}{2}(\csc x\cot x + \ln|\csc x + \cot x|) + C$

35. $\frac{1}{4}\sec^3 x\tan x + \frac{3}{8}(\sec x\tan x + \ln|\sec x + \tan x|) + C$

37.
$$\begin{aligned}\int \cot^n ax\,dx &= \int \cot^{n-2} ax(\csc^2 ax - 1)\,dx \\ &= \int \cot^{n-2} ax\csc^2 ax\,dx - \int \cot^{n-2} ax\,dx \\ &= -\frac{\cot^{n-1} ax}{a(n-1)} - \int \cot^{n-2} ax\,dx\end{aligned}$$

39.
$$\begin{aligned}\int \csc^n ax\,dx &= \int \csc^{n-2} ax(1 + \cot^2 ax)\,dx \\ &= \int \csc^{n-2} ax\,dx + \int \csc^{n-2} ax\cot^2 ax\,dx\end{aligned}$$

Let

$$u = \cot ax \qquad dv = \csc^{n-2} ax\cot ax\,dx$$

$$du = -a\csc^2 ax\,dx \qquad v = -\frac{\csc^{n-2} ax}{a(n-2)}$$

for integration by parts. We obtain

$$\int \csc^{n-2} ax\cot^2 ax\,dx = -\frac{\csc^{n-2} ax\cot ax}{a(n-2)} - \frac{1}{n-2}\int \csc^n ax\,dx.$$

Thus

$$\int \csc^n ax\,dx = \int \csc^{n-2} ax\,dx - \frac{\csc^{n-2} ax\cot ax}{a(n-2)} - \frac{1}{n-2}\int \csc^n ax\,dx.$$

Solving this equation for $\int \csc^n ax\,dx$ gives the formula.

41. For integration by parts, we let

$$u = \cos^{n-1} ax \qquad dv = \cos ax\, dx$$

$$du = -(n-1)a \cos^{n-2} ax \sin ax\, dx \qquad v = \frac{1}{a} \sin ax.$$

Then

$$\int \cos^n ax\, dx = \frac{1}{a} \cos^{n-1} ax \sin ax + (n-1) \int \cos^{n-2} ax \sin^2 ax\, dx$$

$$= \frac{1}{a} \cos^{n-1} ax \sin ax - (n-1) \int \cos^n ax\, dx + (n-1) \int \cos^{n-2} ax\, dx.$$

Solving this equation for $\int \cos^n ax\, dx$ gives the formula.

Section 9.6

1. $x = a \sin t$, $\sqrt{a^2 - x^2} = \sqrt{a^2 - a^2 \sin^2 t} = \sqrt{a^2 \cos^2 t}$. Now $-\pi/2 \le t \le \pi/2$, so $\cos t \ge 0$. Thus $\sqrt{a^2 \cos^2 t} = a \cos t$. **3.** $x = 3 \sin t$, $\int 9 \cos^2 t\, dt$ **5.** $x = 4 \sec t$, $\int \sec t\, dt$ **7.** $x = \sec t$, $\int (\sec^2 t + 2 \sec t)\, dt$ **9.** $x = \frac{1}{2} \tan t$, $\int \frac{1}{2} \sec^3 t\, dt$ **11.** $x = \frac{1}{4} \sin t$, $\int \frac{1}{1024} \sin^4 t\, dt$ **13.** $x - \frac{3}{\sqrt{5}} \tan t$, $\int \frac{\sqrt{5}}{45} \cos t\, dt$ **15.** $x = \sqrt{\frac{2}{3}} \tan t$, $\int \frac{2}{\sqrt{3}} \sec^3 t\, dt$ **17.** $x - 2 = \sec t$, $\int \sec t \tan^2 t\, dt$ **19.** $x + 1 = \sin t$, $\int (-1 + \sin t)^2\, dt$ **21.** $\sqrt{3}x + \frac{2}{\sqrt{3}} = \frac{4\sqrt{7}}{3} \sin t$, $\int \left(\frac{1}{\sqrt{3}}\right) dt$ **23.** $u = e^x$, $\int \frac{du}{\sqrt{15 + 2u - u^2}}$, $u - 1 = 4 \sin t$, $\int 1\, dt$ **25.** $\sqrt{1 + x^2} + C$ **27.** $\frac{4}{3}$ **29.** $\frac{136}{15}$ **31.** $\sqrt{5 + x^2} + C$ **33.** $\sqrt{x^2 - 16} - \ln |x + \sqrt{x^2 - 16}| + C$ **35.** $\frac{5\sqrt{6}}{2} + \frac{1}{4} \ln (5 + 2\sqrt{6})$ **37.** $\frac{-1}{\sqrt{9 + x^2}} + \frac{3}{(9 + x^2)^{3/2}} + C$ **39.** $-\frac{1}{3}(1 - e^{2x})^{3/2} + \frac{1}{5}(1 - e^{2x})^{5/2} + C$ **41.** $\frac{1}{2}[\sqrt{2} + \ln (1 + \sqrt{2})]$ **43.** $\pi/6$ **45.** $2 \ln |\sqrt{x^2 - 4x + 5} + x - 2| + \sqrt{x^2 - 4x + 5} + C$

Section 9.7

1. $\lim_{h \to 1-} \int_0^h \frac{dx}{x^2 - 1}$ **3.** $\lim_{h \to 1+} \int_h^2 \frac{dx}{x - 1} + \lim_{h \to \infty} \int_2^h \frac{dx}{x - 1}$ **5.** $\lim_{h \to -\infty} \int_h^0 e^{-x^2}\, dx + \lim_{h \to \infty} \int_0^h e^{-x^2}\, dx$ **7.** Not improper **9.** $\lim_{h \to 0+} \int_h^1 \frac{dx}{\sqrt{x + x^2}} + \lim_{h \to \infty} \int_1^h \frac{dx}{\sqrt{x + x^2}}$ **11.** $\lim_{h \to (\pi/2)-} \int_0^h \sec x\, dx + \lim_{h \to (\pi/2)+} \int_h^\pi \sec x\, dx + \lim_{h \to (3\pi/2)-} \int_\pi^h \sec x\, dx + \lim_{h \to (3\pi/2)+} \int_h^{2\pi} \sec x\, dx$ **13.** $\frac{1}{2}$ **15.** $\pi/2$ **17.** 1 **19.** $2\sqrt{2}$ **21.** $\frac{10}{3}$ **23.** 1 **25.** Diverges **27.** Diverges **29.** For $p \ne 1$, $\int_1^\infty x^{-p}\, dx = \lim_{h \to \infty} \frac{x^{-p+1}}{1 - p}\Big]_1^h = \lim_{h \to \infty} \left(\frac{h^{1-p}}{1 - p} - \frac{1}{1 - p}\right)$. Now $\lim_{h \to \infty} \frac{h^{1-p}}{1 - p} = \begin{cases} 0 & \text{if } p > 1, \\ \infty & \text{if } p < 1. \end{cases}$

Thus the given integral converges for $p > 1$ and diverges for $p < 1$. For $p = 1$,

$$\int_1^\infty \frac{1}{x}\, dx = \lim_{h \to \infty} (\ln x)\Big]_1^h = \lim_{h \to \infty} (\ln h) = \infty,$$

so the integral diverges for $p \le 1$.

31. a) No b) $V = \int_1^\infty \frac{\pi}{x^2}\, dx = \lim_{h \to \infty} -\frac{\pi}{x}\Big]_1^h = \pi \lim_{h \to \infty} \left(-\frac{1}{h} + 1\right) = \pi$ c) No **33.** $\sqrt{2}$ **35.** $\frac{3k}{2}$ units **37.** a) $\int_0^\infty \frac{ds}{v} = \int_0^\infty \frac{ds}{Ae^{ks}} = \int_0^\infty \frac{1}{A} e^{-ks}\, ds = \lim_{h \to \infty} \frac{-1}{kA} e^{-ks}\Big]_0^h = \frac{1}{kA}$, which is finite. b) $\frac{dv}{dt} = \frac{dv}{ds} \cdot \frac{ds}{dt} = Ake^{ks} \cdot v = kv^2$ c) If $v = a(1 + s^2)$, which increases more slowly than Ae^{ks} for $k > 0$, the integral $\int_0^\infty (ds/v)$ still converges.

Section 9.8

1. Diverges **3.** Converges **5.** Diverges **7.** Diverges **9.** Converges **11.** Diverges **13.** Converges **15.** Diverges **17.** Diverges **19.** Diverges **21.** Converges **23.** Diverges **25.** Diverges **27.** Converges **29.** Converges **31.** Diverges **33.** Converges **35.** Diverges **37.** Converges **39.** Diverges **41.** Absolutely convergent **43.** Not absolutely convergent **45.** Absolutely convergent **47.** Not absolutely convergent **49.** Absolutely convergent

Review Exercise Set 9.1

1. $\dfrac{x^2}{3}\sin 3x - \dfrac{2}{27}\sin 3x + \dfrac{2x}{9}\cos 3x + C$ **3.** $-\frac{5}{2}\ln|2x+3| + 5\ln|x-5| + C$

5. $6(\frac{1}{7}x^{7/6} - \frac{1}{6}x + \frac{1}{5}x^{5/6} - \frac{1}{4}x^{2/3} + \frac{1}{3}x^{1/2} - \frac{1}{2}x^{1/3} + x^{1/6} - \ln|x^{1/6}+1|) + C$

7. $-\frac{1}{2}\cos 2x + \frac{1}{3}\cos^3 2x - \frac{1}{10}\cos^5 2x + C$ **9.** $\frac{1}{9}\tan^3 3x + \frac{1}{15}\tan^5 3x + C$ **11.** $\dfrac{x}{2}\sqrt{16-9x^2} + \dfrac{8}{3}\sin^{-1}\left(\dfrac{3x}{4}\right) + C$

13. a) Diverges b) Diverges c) Converges

Review Exercise Set 9.2

1. $\dfrac{x^2}{2}\ln x - \dfrac{x^2}{4} + C$ **3.** $2\ln|x-3| + \dfrac{3}{2}\tan^{-1} 2x + C$

5. $-\frac{4}{5}(x-1)^{5/4} - 3(x-1) - \frac{8}{3}(x-1)^{3/4} - 12(x-1)^{1/2} - 72(x-1)^{1/4} - 216\ln|(x-1)^{1/4} - 3| + C$

7. $\frac{1}{3}\sin^3 x - \frac{1}{5}\sin^5 x + C$ **9.** $-\frac{1}{3}\cot^3 x - \frac{1}{5}\cot^5 x + C$ **11.** $2\sin^{-1}\left(\dfrac{x}{2}\right) - \dfrac{1}{2}x\sqrt{4-x^2} + C$

13. a) Diverges b) Converges c) Diverges

Review Exercise Set 9.3

1. $\dfrac{x^2}{4} - \dfrac{1}{4}x\sin 2x - \dfrac{1}{8}\cos 2x + C$ **3.** $\dfrac{x^2}{2} - \ln|x^2-1| + \ln\left|\dfrac{x-1}{x+1}\right| + C$ **5.** $\dfrac{1}{\sqrt{3}}\ln\left|\dfrac{2-\sqrt{3}+\tan(x/2)}{2+\sqrt{3}+\tan(x/2)}\right| + C$

7. $\dfrac{x}{2} - \dfrac{1}{16}\sin 8x + C$ **9.** $\frac{1}{4}(\sec 2x\tan 2x + \ln|\sec 2x + \tan 2x|) + C$ **11.** $1/(\ln 2)$

13. a) Converges b) Diverges c) Converges

Review Exercise Set 9.4

1. $x\sin x - \dfrac{x}{3}\sin^3 x + \dfrac{2}{3}\cos x + \dfrac{1}{9}\cos^3 x + C$ **3.** $\dfrac{x^2}{2} + \ln\left|\dfrac{x}{x^2+1}\right| + C$ **5.** $\dfrac{1}{6}\ln\left|\dfrac{\sqrt{9-x^2}-3}{\sqrt{9-x^2}+3}\right| + C$

7. $-\frac{1}{4}\cot^4 x + C$ **9.** $\frac{1}{4}\sec^4 x$ **11.** $\sin^{-1}\left(\dfrac{x+1}{3}\right) + C$ **13.** a) Converges b) Converges c) Diverges

More Challenging Exercises 9

1. $\displaystyle\int_0^x f'(t)\,dt = f(t)\Big]_0^x = f(x) - f(0)$. On the other hand, taking

$$u = f'(t) \qquad dv = 1\cdot dt$$
$$du = f''(t)\,dt \qquad v = t - x,$$

we obtain

$$\int_0^x f'(t)\,dt = (t-x)f'(t)\Big]_0^x - \int_0^x f''(t)(t-x)\,dt$$
$$= 0 - (-x)f'(0) + \int_0^x f''(t)(x-t)\,dt.$$

Thus $f(x) - f(0) = f'(0)x + \int_0^x f''(t)(x-t)\,dt$, which gives the desired formula.

3. Suppose $\lim_{x\to\infty} f(x) = a > 0$. Then $f(x) > a/2$ for $x > b$ for some b. Thus $\lim_{h\to\infty}\int_b^h f(x)\,dx > \lim_{h\to\infty}\int_b^h (a/2)\,dx = \lim_{h\to\infty}(a/2)(h-b) = \infty$, so $\int_0^\infty f(x)\,dx$ diverges. A similar argument shows that $\int_0^\infty f(x)\,dx$ diverges if $\lim_{x\to\infty} f(x) = a < 0$. We have proved the equivalent contrapositive of the statement in the exercise.

5. a) If $f(x) = x$, then $\int_{-h}^h f(x)\,dx = \int_{-h}^h x\,dx = 0$, so the Cauchy principal value of $\int_{-\infty}^\infty f(x)\,dx$ is zero, although $\int_{-\infty}^\infty f(x)\,dx$ diverges. b) If $\int_{-\infty}^\infty f(x)\,dx$ converges, then $\int_0^\infty f(x)\,dx$ and $\int_{-\infty}^0 f(x)\,dx$ converge, so $\lim_{h\to\infty}\int_0^h f(x)\,dx$ and $\lim_{h\to\infty}\int_{-h}^0 f(x)\,dx$ exist. Then $\int_{-\infty}^\infty f(x)\,dx = \lim_{h\to\infty}\int_0^h f(x)\,dx + \lim_{h\to\infty}\int_{-h}^0 f(x)\,dx = \lim_{h\to\infty}[\int_0^h f(x)\,dx + \int_{-h}^0 f(x)\,dx] = \lim_{h\to\infty}\int_{-h}^h f(x)\,dx$, which is the Cauchy principal value of $\int_{-\infty}^\infty f(x)\,dx$.

7. Let $f(x) = 0$ for x not in $[n - 1/10^{2n}, n + 1/10^{2n}]$ for each integer $n > 0$, and let the graph of f over $[n - 1/10^{2n}, n + 1/10^{2n}]$ consist of the line segment joining $(n - 1/10^{2n}, 0)$ and $(n, 10^n)$ and the line segment joining $(n, 10^n)$ and $(n + 1/10^{2n}, 0)$. The area inside this nth "spike" is $10^n(1/10^{2n}) = 1/10^n$. Thus

$$\int_0^{\infty} f(x)\,dx = \frac{1}{10} + \frac{1}{100} + \frac{1}{1000} + \cdots = 0.1 + 0.01 + 0.001 + \cdots = 0.1111\cdots.$$

CHAPTER 10

Section 10.1

1. We have $\lim_{n\to\infty} a_n = \infty$ if for each real number γ there exists an integer N such that $a_n > \gamma$ provided that $n > N$.
3. Converges to 0 **5.** Converges to 1 **7.** Converges to 2 **9.** Diverges to ∞ **11.** Converges to 0 **13.** Diverges **15.** 0
17. 1 **19.** ∞ **21.** 0 **23.** 0 **25.** Diverges **27.** 0 **29.** 0 **31.** ∞ **33.** $\frac{6}{5}$ **35.** 4 **37.** 120 ft
39. Let $\varepsilon > 0$ be given. Let $N = 5$. Then $|a_n - 1| = |1 - 1| = 0 < \varepsilon$ if $n > 5$, so the sequence converges to 1. (Any positive integer will do for N.)
41. Let $\varepsilon > 0$ be given, and find a positive integer N such that $N > 1/\varepsilon^2$. Then $\varepsilon^2 > 1/N$, so $1/\sqrt{N} < \varepsilon$. If $n > N$, then

$$\left|\frac{1}{\sqrt{n+1}} - 0\right| = \frac{1}{\sqrt{n+1}} < \frac{1}{\sqrt{n}} < \frac{1}{\sqrt{N}} < \varepsilon,$$

so the sequence converges to zero.
43. Suppose the sequence has limit c. Then, with $\varepsilon = \frac{1}{4}$, there exists a positive integer N such that $|a_n - c| < \frac{1}{4}$ for $n > N$. In particular, $|a_{N+1} - c| < \frac{1}{4}$ and $|a_{N+2} - c| < \frac{1}{4}$, which implies $|a_{N+1} - a_{N+2}| < \frac{1}{2}$. However, $|a_{N+1} - a_{N+2}| = 1$, so our assumption that the sequence has limit c must be false.
45. Let $\varepsilon > 0$ be given. We may assume $\varepsilon < 1$. Then $|a|^n < \varepsilon$ if $n \cdot \ln|a| < \ln \varepsilon$; that is, if $n > (\ln \varepsilon)/(\ln|a|)$. Thus we can let N be the largest integer less than $(\ln \varepsilon)/(\ln|a|)$.
47. 0.36787944 (which equals $1/e$) **49.** 1 **51.** $\frac{1}{2}$

Section 10.2

1. a) 1 b) 1 c) 0 d) 0 e) 0 f) 0 g) 1 **3.** $a_1 = \frac{1}{2}, a_2 = -\frac{1}{6}, a_3 = -\frac{1}{12}, a_4 = -\frac{1}{20}, a_5 = -\frac{1}{30}$ **5.** Diverges to ∞
7. Converges to 1 **9.** Diverges to $-\infty$ **11.** Diverges to ∞ **13.** Converges to $-\frac{4}{3}$ **15.** Converges to $e^2/(e^2 - 1)$
17. Diverges **19.** Converges to $\frac{2}{21}$ **21.** Converges to $4/(1 + \ln 2)$ **23.** Converges to $\frac{7}{2}$ **25.** Converges to 6
27. Converges to 13 **29.** 60 ft
31. If a is the time required to travel the first mile, then the time required to travel the first mile plus the time required to travel the second plus ... is $a + ka + k^2a + k^3a + \cdots = a/(1 - k)$ since $|k| < 1$.
33. a) Since $\ln[(n + 1)/n] = \ln(n + 1) - \ln(n)$, we see that $s_n = \ln(n + 1) - \ln(1) = \ln(n + 1)$.
b) $\lim_{n\to\infty} s_n = \lim_{n\to\infty} \ln(n + 1) = \infty$, so the series diverges.
35. $\frac{11}{6}$ **37.** a) $\dfrac{1}{n^2 - 1} = \dfrac{\frac{1}{2}}{n - 1} - \dfrac{\frac{1}{2}}{n + 1}$ b) $\frac{7}{24}$ **39.** 1, 2, 3, 4, ..., n, ... and $-1, -2, -3, -4, \ldots, -n, \ldots$ **41.** It diverges.
43. $\frac{247}{30}$ **45.** $\frac{4}{33}$ **47.** $\frac{27287339}{99900}$
49. a) Impossible for us to answer. b) Any number with a nonzero digit in some decimal place when added to 4.12999999999... will yield a number greater than 4.13000000.... c) $\dfrac{412}{100} + \dfrac{\frac{9}{1000}}{\frac{9}{10}} = \dfrac{412}{100} + \dfrac{1}{100} = \dfrac{413}{100}$

Section 10.3

1. Converges to $\sqrt{5}$ **3.** Converges to $\frac{3}{4} + \sqrt{2} - \sqrt{3}$ **5.** T F F T F T T F T T **7.** Converges; SCT with $\sum_{n=1}^{\infty} (\frac{1}{2})^n$
9. Converges; SCT with $\sum_{n=2}^{\infty} (\frac{1}{2})^n$ **11.** Converges; SCT with $\sum_{n=1}^{\infty} (\frac{1}{2})^n$ **13.** Diverges; $\lim_{n\to\infty} a_n = \frac{1}{3} \neq 0$
15. Converges; SCT with $\sum_{n=1}^{\infty} (\frac{1}{5})^n$ **17.** Converges; sum of two convergent geometric series
19. Diverges; LCT with the harmonic series **21.** Converges; LCT with a geometric series with ratio $1/e^2$
23. Diverges; $a_n \geq 1$ for all n **25.** Diverges; $a_n \geq 1$ for all n **27.** Diverges; LCT with geometric series with ratio $1/(\ln 2) > 1$
29. Diverges; LCT with $\sum_{n=1}^{\infty} \cos^2 n$; $\lim_{n\to\infty} \cos^2 n \neq 0$ **31.** Diverges; LCT with $\sum_{n=1}^{\infty} (1/n)$
33. Converges; SCT with $\sum_{n=1}^{\infty} (\frac{1}{2})^n$ **35.** Converges; SCT with $\sum_{n=1}^{\infty} (\frac{2}{3})^n$ **37.** Diverges; $\lim_{n\to\infty} a_n \neq 0$
39. Diverges; $\lim_{n\to\infty} a_n = 1 \neq 0$ **41.** Converges; SCT with $\sum_{n=1}^{\infty} (\frac{1}{2})^n$

Section 10.4

1. Diverges **3.** Diverges **5.** Converges **7.** $\displaystyle\int_2^{\infty} \frac{dx}{x(\ln x)} = \lim_{t\to\infty} \ln(\ln x)\Big]_2^t = \lim_{t\to\infty} \ln(\ln t) - \ln(\ln 2) = \infty$ **9.** Converges

11. Diverges **13.** Converges **15.** Converges **17.** Converges **19.** Diverges **21.** Diverges **23.** Diverges **25.** Converges **27.** Converges **29.** Diverges **31.** Diverges **33.** Converges **35.** Converges **37.** Converges **39.** Converges **41.** Converges **43.** Diverges **45.** Diverges **47.** Converges **49.** e^{1000} **51.** 2.0762 **53.** 2.607 **55.** 10.544

Section 10.5

1. $1 - 1 + 1 - 1 + 1 - 1 + \cdots$ **3.** $1 + \frac{1}{2} + \frac{1}{3} + \frac{1}{4} + \frac{1}{5} + \cdots$ **5.** F T T F F F **7.** Converges absolutely **9.** Divergent **11.** Conditionally convergent **13.** Divergent **15.** Absolutely convergent **17.** Absolutely convergent **19.** Divergent **21.** Conditionally convergent **23.** Conditionally convergent **25.** Conditionally convergent **27.** Divergent **29.** Absolutely convergent **31.** Conditionally convergent **33.** Absolutely convergent **35.** Divergent **37.** Conditionally convergent **39.** Conditionally convergent

41. a) Since $a_n = u_n + v_n$, this follows from Corollary 1 of Theorem 10.2 (Section 10.2). b) If say $\sum_{n=1}^{\infty} u_n$ converges, then convergence of $\sum_{n=1}^{\infty} a_n$ would imply convergence of $\sum_{n=1}^{\infty} (a_n - u_n) = \sum_{n=1}^{\infty} v_n$ by Corollary 1 of Theorem 10.2 (Section 10.2). c) Let $\sum_{n=1}^{\infty} a_n = 1 - \frac{1}{2} + \frac{1}{3} - \frac{1}{4} + \frac{1}{5} - \frac{1}{6} + \cdots$.

43. 99 **45.** 4 **47.** 6 **49.** -0.823 **51.** -3.645 **53.** -0.63214

Review Exercise Set 10.1

1. See Definition 10.2 of Section 10.1. **3.** 48

5. a) Diverges; $\lim_{n\to\infty} a_n = 1 \neq 0$ b) Diverges; behaves like $\sum_{n=2}^{\infty} (1/n)$ by the LCT

7. The hypotheses of the theorem are satisfied.

$$\int_2^{\infty} \frac{1}{x^2 - x}\,dx = \int_2^{\infty} \left(-\frac{1}{x} + \frac{1}{x-1}\right) dx = \lim_{h\to\infty} \Big(-\ln|x| + \ln|x-1|\Big]_2^h$$

$$= \lim_{h\to\infty} \left(\ln\left|\frac{h-1}{h}\right| - \ln\frac{1}{2}\right) = \ln 1 - \ln\frac{1}{2} = \ln 2$$

Converges

9. a) Conditionally convergent: satisfies alternating series test, but diverges absolutely by the LCT with $\sum_{n=1}^{\infty} (1/\sqrt{n})$ b) Converges absolutely by the integral test.

Review Exercise Set 10.2

1. Note that $(n-1)/n = 1 - (1/n)$. Let $\varepsilon > 0$ be given. Find N such that $1/N < \varepsilon$. Then if $n > N$, we have $(1/n) < \varepsilon$, so $|(1/n)| = |[1 - (1/n)] - 1| < \varepsilon$. Thus $\{(n-1)/n\}$ converges to 1.

3. 30 ft

5. a) Converges; ratio test gives a ratio of $\frac{1}{3} < 1$. b) Converges; sum of convergent geometric series with ratios of $\frac{3}{5}$ and $\frac{4}{5}$.

7. The hypotheses are satisfied.

$$\int_2^{\infty} \frac{x+1}{x^2+1}\,dx = \lim_{h\to\infty} \left(\frac{1}{2}\ln|x^2+1| + \tan^{-1} x\right)\Big]_2^h$$

$$= \lim_{h\to\infty} \left[\frac{1}{2}\ln|h^2+1| + \tan^{-1} h - \frac{1}{2}\ln(5) - \tan^{-1} 2\right] = \infty.$$

Diverges

9. a) Conditionally convergent; satisfies the alternating series test, but diverges absolutely by the integral test b) Converges absolutely by the SCT for the terms are at most $1/n^{3/2}$

Review Exercise Set 10.3

1. Given $K > 0$, there exists a positive integer N such that $a_n > K$ for $n > N$. **3.** $-\frac{25}{6}$

5. a) Converges; LCT with $\sum_{n=1}^{\infty} (1/n^{3/2})$ b) Converges; LCT with $\sum_{n=1}^{\infty} (1/n^2)$ **7.** See Theorem 10.7.

9. a) Absolutely convergent; LCT with $\sum_{n=1}^{\infty} (1/n^{3/2})$ b) Conditionally convergent; $\{\sin(3\pi n/2)\} = -1, 0, 1, 0, -1, 0, \ldots$. The series satisfies the alternating series test but diverges absolutely since it then contains every other term of the harmonic series.

Review Exercise Set 10.4

1. $s_n = 2, 3, 4, 5, 6, 7, \ldots$

$t_n = 1, 0, -1, -2, -3, -4, \ldots$

3. $\frac{3}{5}$ **5.** a) Diverges; $\lim_{n\to\infty} a_n = \infty$ b) Diverges; $= \sum_{n=1}^{\infty} (1/n^{\ln 2.7})$, p-series, $p = \ln(2.7) < 1$

7. $\displaystyle\int_1^{\infty} \frac{e^x}{1 + e^{2x}}\,dx = \frac{\pi}{2} - \tan^{-1} e$; converges

9. a) Diverges; $\lim_{n\to\infty} |a_n| = \infty$ b) Conditionally convergent; satisfies the alternating series test, but diverges absolutely as a p-series with $p = \frac{1}{2}$.

More Challenging Exercises 10

1. If $s_n = -t_n$, then $\{s_n\}$ is monotone increasing. By the fundamental property, either $\lim_{n\to\infty} s_n = \infty$, so that $\lim_{n\to\infty} - s_n = \lim_{n\to\infty} t_n = -\infty$, or $\lim_{n\to\infty} s_n = c$, so that $\lim_{n\to\infty} t_n = -c = d$.
3. From $m < b_n/a_n < M$ for all sufficiently large n, we have $ma_n < b_n < Ma_n$ for all $n > N$, for some integer N. Now $\sum_{n=1}^{\infty} ma_n$ and $\sum_{n=1}^{\infty} Ma_n$ converge if and only if $\sum_{n=1}^{\infty} a_n$ converges. It follows at once by the SCT that $\sum_{n=1}^{\infty} a_n$ and $\sum_{n=1}^{\infty} b_n$ either both converge or both diverge.
5. Pick up *positive* ($\neq 0$) terms u_n of $\sum_{n=1}^{\infty} u_n$ in order, until a partial sum >17 is obtained. Then pick up *negative* ($\neq 0$) terms v_n in order until the partial sum becomes <17. Then pick up subsequent *positive* terms u_n in order until the partial sum becomes >17 again, then subsequent *negative* terms v_n until the partial sum becomes <17, and so on.
7. Pick up *positive* ($\neq 0$) terms u_n in order until the partial sum becomes >1. Then pick up the first single *negative* ($\neq 0$) term v_n. Then pick up *positive* terms until the partial sum becomes >2, and then the next single *negative* term v_n. Then pick up *positive* terms u_n until the partial sum becomes >3, and then pick up the next single *negative* term, and so on.
9. Pick up groups of successive *positive* ($\neq 0$) terms u_n and groups of successive *negative* ($\neq 0$) terms v_n alternately so that the partial sums, after picking up the groups, become successively ≥ 15, ≤ -6, ≥ 15, ≤ -6, and so on.

CHAPTER 11

Section 11.1

1. $R = 1; -1 \leq x < 1$ **3.** $R = 1; -1 < x \leq 1$ **5.** $R = 1; 0 \leq x \leq 2$ **7.** $R = 5; -6 \leq x < 4$ **9.** $R = 3; -3 < x < 3$
11. $R = 4; -4 \leq x \leq 4$ **13.** $R = 1; 0 \leq x < 2$ **15.** $R = \infty; -\infty < x < \infty$ **17.** $R = 0; x = -5$ **19.** $R = \frac{1}{2}; -\frac{7}{2} < x \leq -\frac{5}{2}$
21. $R = \frac{3}{2}; \frac{1}{2} \leq x \leq \frac{7}{2}$ **23.** $R = \frac{3}{2}; -\frac{9}{2} \leq x < -\frac{3}{2}$ **25.** $R = \frac{1}{3}; -\frac{7}{3} \leq x \leq -\frac{5}{3}$ **27.** $R = \frac{1}{2}; -\frac{5}{2} \leq x \leq -\frac{3}{2}$
29. $\sum_{n=0}^{\infty} \frac{(x - \frac{5}{2})^n}{(\frac{3}{2})^n} = \sum_{n=0}^{\infty} \frac{(2x - 5)^n}{3^n}$ **31.** $\sum_{n=1}^{\infty} \frac{(x + 3)^n}{n \cdot 2^n}$ **33.** $1 + x^2 + x^4 + \cdots + x^{2n} + \cdots$
35. $-5 - 5 \cdot 3x - 5 \cdot 3^2x^2 - \cdots - 5 \cdot 3^n x^n - \cdots$ **37.** $\frac{1}{2} + \frac{1}{2 \cdot 4}x + \frac{1}{2 \cdot 4^2}x^2 + \cdots + \frac{1}{2 \cdot 4^n}x^n + \cdots$
39. $2x^2 - 3x^3 + \frac{3^2}{2}x^4 - \frac{3^3}{2^2}x^5 + \cdots + \frac{(-1)^n 3^n}{2^{n-1}}x^{n+2} + \cdots$ **41.** $\frac{1}{2} - \frac{(x-1)}{2^2} + \frac{(x-1)^2}{2^3} - \cdots + (-1)^n \frac{(x-1)^n}{2^{n+1}} + \cdots$
43. $\frac{1}{4} - \frac{1}{4 \cdot 8}(x - 2) + \frac{1}{4 \cdot 8^2}(x - 2)^2 + \cdots + \frac{(-1)^n}{4 \cdot 8^n}(x - 2)^n + \cdots$
45. $-\frac{(x-3)^2}{2} + \frac{(x-3)^3}{2} - \frac{(x-3)^4}{2} + \cdots + (-1)^{n+1}\frac{(x-3)^{n+2}}{2} + \cdots$
47. Letting $u = 1/x$, we see that $\sum_{n=0}^{\infty} a_n u^n$ converges for $u = 1/c$. By Theorem 11.1, it must converge for all u such that $|u| < |1/c|$. The given series thus converges for all x such that $|1/x| < |1/c|$, that is, for $|x| > |c|$.
49. $x < -2, x > 2$ **51.** No x **53.** $x \leq -2, x \geq 6$ **55.** $x < \frac{8}{3}, x > \frac{16}{3}$

Section 11.2

1. If $f'(x)$ exists for $x_0 - h < x < x_0 + h$, then for each such x there exists c between x and x_0 such that $f(x) = f(x_0) + f'(c)(x - x_0)$. This is the mean-value theorem.
3. $2 + 2(x + 1)$ **5.** $12 - 5(x + 2)$ **7.** $9 - 5(x + 2) + \frac{1}{2}(x + 2)^2$ **9.** $-47 + 27(x + 3) - 4(x + 3)^2$
11. $4 - 7(x + 1) + 6(x + 1)^2 - 4(x + 1)^3 + (x + 1)^4$ **13.** $-113 + 198(x + 2) - 135(x + 2)^2 + 42(x + 2)^3 - 5(x + 2)^4$
15. $1 - \frac{x^2}{2!} + \frac{x^4}{4!} - \frac{x^6}{6!} + \frac{x^8}{8!} - \frac{x^{10}}{10!}$ **17.** $1 - \frac{(x - \pi/2)^2}{2!} + \frac{(x - \pi/2)^4}{4!} - \frac{(x - \pi/2)^6}{6!}$
19. $1 - (x - 1) + (x - 1)^2 - (x - 1)^3 + (x - 1)^4 - (x - 1)^5 + (x - 1)^6$ **21.** $-1 + x + \frac{3}{2}x^2 - \frac{1}{24}x^4$
23. $1 + x + \frac{x^2}{2!} + \frac{7}{6}x^3 + \frac{x^4}{4!} + \frac{x^5}{5!} + \frac{x^6}{6!} + \frac{x^7}{7!} + \frac{x^8}{8!}$ **25.** $1 + 2x + \frac{1}{2}x^3 - \frac{5}{24}x^4$
27. a) $2x - \frac{8x^3}{3!} + \frac{32x^5}{5!} - \frac{128x^7}{7!}$ b) The polynomial may be obtained by replacing x by $2x$ in the answer to Example 3.
29. a) $1 + x^2$ b) The polynomial is the portion of degree less than or equal to 2 of the polynomial obtained by replacing x by x^2 in the answer to Example 5. c) $1 + x^2 + x^4 + x^6 + x^8$
31. a) $1 - \frac{1}{2}(x - 1) + \frac{1}{4}(x - 1)^2 - \frac{1}{8}(x - 1)^3 + \frac{1}{16}(x - 1)^4$ b) $1 - \frac{x - 1}{2} + \frac{(x-1)^2}{2^2} - \frac{(x-1)^3}{2^3} + \frac{(x-1)^4}{2^4} - \cdots$
c) They agree through the term involving $(x - 1)^4$.
33. By the chain rule, $g^{(m)}(x_0) = c^m f^{(m)}(x_0)$ for $m \leq n$, so the coefficient of x^m in the Taylor polynomial for g is c^m times the coefficient of x^m in the Taylor polynomial $T_n(x)$ for f. But this multiplication by c^m can also be achieved by forming $T_n(cx)$.

35. a) 26.46 b) 0.0036 **37.** a) 10π ft^3 b) $\frac{5\pi}{36}$ ft^3 **39.** a) $\frac{25\pi}{18}$ ft^3 b) $\frac{4\pi}{3}\left(\frac{1}{12}\right)^3$ ft^3 c) 0 ft^3 **41.** a) $\frac{\pi}{90}$ b) $\frac{8\pi^3}{(9)(90)^3}$
43. a) 2.7182788 b) 0.0000083 **45.** a) 1.543055556 b) 0.0000434028 (Using $e^c < 3, e^{-c} < \frac{1}{2}$)
47. $n = 9$; the same value 9 of n is obtained.

49. $|E_n(x)| = \left|\frac{e^{-c}x^{n+1}}{(n+1)!}\right|$, and $e^{-c} < e^{|x|}$ while $\lim_{n\to\infty} \frac{x^{n+1}}{(n+1)!} = 0$; thus $\lim_{n\to\infty} |E_n(x)| = 0$.

51. $|E_n(x)| \leq \left|\frac{x^{n+1}}{(n+1)!}\right|$, so $\lim_{n\to\infty} |E_n(x)| = 0$ **53.** $|E_n(x)| \leq \left|\frac{2e^{|x|}}{2}\cdot\frac{x^{n+1}}{(n+1)!}\right|$ and $\lim_{n\to\infty} \frac{x^{n+1}}{(x+1)!} = 0$; thus $\lim_{n\to\infty} |E_n(x)| = 0$.

55. $|E_n(x)| = \left|\frac{n!}{c^{n+1}}\cdot\frac{(x-1)^{n+1}}{(n+1)!}\right| \leq \left|\frac{1^{n+1}}{1^{n+1}(n+1)}\right| = \frac{1}{n+1}$, so $\lim_{n\to\infty} |E_n(x)| = 0$

57. $|E_n(x)| = \left|\frac{1\cdot 1\cdot 3\cdot 5\cdot 7\cdots(2n-1)}{2^{n+1}(c+2)^{(2n+1)/2}}\cdot\frac{x^{n+1}}{(n+1)!}\right| \leq \left|\frac{1}{n+1}\cdot\frac{2^{n+1}}{2^{(2n+1)/2}}\right| = \frac{\sqrt{2}}{n+1}$, so $\lim_{n\to\infty} |E_n(x)| = 0$.

Section 11.3

1. F T T T T F T **3.** $1 + x + \frac{3}{2}x^2 + \frac{x^3}{3!} + \cdots + \frac{x^n}{n!} + \cdots$ for $n \geq 3$ **5.** $x^2 - \frac{x^4}{3!} + \cdots + (-1)^n \frac{x^{2n+2}}{(2n+1)!} + \cdots$ for $n \geq 1$

7. $x + x^2 + x^3 + \cdots + x^{n+1} + \cdots$ **9.** $1 - \frac{x^6}{2!} + \cdots + (-1)^n \frac{x^{6n}}{(2n)!} + \cdots$ **11.** $1 + x - \frac{x^3}{3} - \frac{x^4}{6} - \frac{x^5}{30} - \cdots$

13. $1 - 2x + 3x^2 - \cdots + (-1)^n(n+1)x^n + \cdots$ **15.** $1 + \frac{x^2}{2!} + \frac{5x^4}{4!} + \cdots$ **17.** $1 + 2x + \frac{5}{2}x^2 + \frac{8}{3}x^3 + \cdots$

19. $x + \frac{x^3}{3!} + \frac{x^5}{5!} + \cdots + \frac{x^{2n+1}}{(2n+1)!} + \cdots$ **21.** $1 - x + \frac{x^2}{2!} - \frac{x^3}{3!} + \cdots + (-1)^n \frac{x^n}{n!} + \cdots$ **23.** $x + \frac{5x^3}{6} + \cdots$ **25.** -24 **27.** $\frac{3}{4}$

29. 0 **31.** 240 **33.** a) $x - \frac{2}{3}x^3 + \frac{2}{15}x^5 + \cdots$ b) $x - \frac{2}{3}x^3 + \frac{2}{15}x^5 - \cdots + (-1)^n(2^{2n})\frac{x^{2n+1}}{(2n+1)!} + \cdots$

35. $1/x = -1/[1 - (x+1)]$; $-1 - (x+1) - (x+1)^2 - \cdots - (x+1)^n - \cdots$
37. a) Replace x by $-x$ in Eq. (4). b) The alternating series test is satisfied.

c) $|E_n(1)| = \left|\frac{n!}{(1+c)^{n+1}}\cdot\frac{1^{n+1}}{(n+1)!}\right| = \left|\frac{1}{(1+c)^{n+1}(n+1)}\right| \leq \left|\frac{1}{n+1}\right|$, so $\lim_{n\to\infty} E_n(1) = 0$. It now follows from part (a) that the alternating harmonic series converges to ln 2.

39. $C + x + \frac{x^3}{3} + \frac{x^5}{5\cdot 2!} + \frac{x^7}{7\cdot 3!} + \cdots + \frac{x^{2n+1}}{(2n+1)n!} + \cdots$ **41.** $\pi + x - \frac{x^5}{5\cdot 2!} + \frac{x^9}{9\cdot 4!} - \cdots + (-1)^n \frac{x^{4n+1}}{(4n+1)(2n)!} + \cdots$

43. a) $e^{ix}: 1 + ix - \frac{x^2}{2!} - i\frac{x^3}{3!} + \frac{x^4}{4!} + i\frac{x^5}{5!} - \frac{x^6}{6!} - \cdots + (i)^n \frac{x^n}{n!} + \cdots$

$e^{-ix}: 1 - ix - \frac{x^2}{2!} + i\frac{x^3}{3!} + \frac{x^4}{4!} - i\frac{x^5}{5!} - \frac{x^6}{6!} + \cdots + (-i)^n \frac{x^n}{n!} + \cdots$

b) and c) This is obvious from part (a) and the series (2) and (3) for $\sin x$ and $\cos x$. d) $\cos x = (e^{ix} + e^{-ix})/2$; $\sin x = (e^{ix} - e^{-ix})/2i$ e) They are the same except for the presence of i in certain places.
45. $2x + [1/(1+x)]$ **47.** $3x + \cos x$ **49.** $\sin x + \cos x$ **51.** $(e^x - 1) + [1/(1-x)]$ **53.** $2/(1-x)^3$
55. $[1/(1+2x^2)] + 2x^2 - 1 = 4x^4/(1+2x^2)$

Section 11.4

1. 1 **3.** 2 **5.** $-\infty$ **7.** 0 **9.** 1 **11.** 0 **13.** 0 **15.** 0 **17.** 0 **19.** -1 **21.** ∞ **23.** $\frac{1}{2}$ **25.** -1 **27.** ∞ **29.** 0 **31.** $\frac{1}{2}e^{\pi/2}$
33. 1 **35.** 1 **37.** e^2 **39.** 1 **41.** 0 **43.** ∞ **45.** ∞ **47.** e^3 **49.** ∞
51. a) Now $h(x)$ is continuous (differentiable) in $[a, b]$ (for $a < x < b$) as a difference of products of constants and functions that are continuous (differentiable) there. Also $h(a) = h(b) = f(b)g(a) - g(b)f(a)$. b) Since $h'(x) = (f(b) - f(a))g'(x) - (g(b) - g(a))f'(x)$, this conclusion is immediate from Rolle's theorem. c) If $g(a) = g(b)$, then by Rolle's theorem, $g'(c) = 0$ for some c between a and b, contradicting the hypothesis that $g'(x) \neq 0$ for $a < x < b$. Thus $g(b) - g(a) \neq 0$. d) This is immediate from parts (b) and (c) and the hypothesis that $g'(x) \neq 0$ for $a < x < b$.

53. a) $\lim_{u\to 0^+}(1/u) = \infty$, so $x \to \infty$ as $u \to 0^+$. Thus

$$\lim_{u\to 0+}\frac{F(u)}{G(u)} = \lim_{x\to\infty}\frac{f(x)}{g(x)},$$

which we know to be of 0/0-type.

b) $$\lim_{x\to\infty}\frac{f(x)}{g(x)} = \lim_{u\to 0+}\frac{F(u)}{G(u)} = \lim_{u\to 0+}\frac{F'(u)}{G'(u)} = \lim_{u\to 0+}\frac{f'(1/u)(-1/u^2)}{g'(1/u)(-1/u^2)}$$

$$= \lim_{u\to 0+}\frac{f'(1/u)}{g'(1/u)} = \lim_{x\to\infty}\frac{f'(x)}{g'(x)}$$

55. 0
57. 0 (The calculator leads us to believe that the limit is ∞. We can't compute $(\ln x)^{100}/x$ for large enough x to see what happens. If we were suspicious of this, we might turn to its logarithm, $100\cdot\ln(\ln x) - \ln x$, and try that. We still can't tell what happens substituting values of x. But if we substitute $t = \ln x$ and compute $100\cdot\ln(t) - t$ for large t, we discover this approaches $-\infty$ as $t\to\infty$, so the original quotient approaches zero. However, it is unlikely we would be suspicious of the original calculation, and the calculator leads us to an answer just about as wrong as we could get!) **59.** 20.08554

Section 11.5

1. 6 **3.** 8 **5.** 1 **7.** $\frac{35}{16}$ **9.** 1 **11.** $\frac{5}{81}$ **13.** $1 + 0x + 0x^2 + \cdots + 0x^n + \cdots;\ \binom{0}{k} = 1$ for $k = 0$ and 0 for $k > 0$
15. $1 - 2x + 3x^2 - 4x^3 + 5x^4 - \cdots;\ R = 1$ **17.** $1 - \frac{1}{3}x^2 - \frac{1}{9}x^4 - \frac{5}{81}x^6 - \frac{10}{243}x^8 - \cdots;\ R = 1$
19. $1 + 5x + 5x^2 - \frac{5}{3}x^3 + \frac{5}{3}x^4 - \cdots;\ R = \frac{1}{3}$ **21.** $\frac{1}{5} + \frac{1}{10}\left(\frac{x}{5}\right)^2 + \frac{3}{40}\left(\frac{x}{5}\right)^4 + \frac{1}{16}\left(\frac{x}{5}\right)^6 + \frac{7}{128}\left(\frac{x}{5}\right)^8 + \cdots;\ R = 5$
23. $16 + \frac{8}{3}x^3 + \frac{1}{18}x^6 - \frac{1}{648}x^9 + \frac{80}{243}\left(\frac{x}{2}\right)^{12} - \cdots;\ R = 2$ **25.** Find a partial sum of $(1 - x)^{1/2}$ with $x = \frac{1}{3}$.
27. Find a partial sum of $(1 - x)^{-1/3}$ with $x = \frac{1}{2}$. **29.** $x - \frac{1}{6}x^3 + \frac{3}{40}x^5 - \frac{5}{112}x^7 + \cdots$ **31.** $x + \frac{1}{2}x^3 + \frac{3}{8}x^5 + \frac{5}{16}x^7 + \cdots$
33. $1 + \frac{3}{2}x + \frac{11}{8}x^2 + \frac{23}{16}x^3 + \frac{179}{128}x^4 + \cdots$ **35.** $1 + \frac{3}{2}x - \frac{1}{8}x^2 + \frac{3}{16}x^3 - \frac{37}{128}x^4 + \cdots$
37. $\frac{1}{2}x^3 + \frac{1}{16}x^5 + \frac{3}{2^8}x^7 + \frac{5}{2^{11}}x^9 + \cdots$ **39.** $\frac{1}{3} - \frac{1}{7\cdot 3!} + \frac{1}{11\cdot 5!} - \frac{1}{15\cdot 7!}$ with error less than $\frac{1}{19\cdot 9!}$
41. $\frac{1}{4} - \frac{1}{8\cdot 2!} + \frac{1}{12\cdot 4!} - \frac{1}{16\cdot 6!} + \frac{1}{20\cdot 8!}$ with error less than $\frac{1}{24\cdot 10!}$
43. $\frac{1}{10} - \frac{1}{3\cdot 10^3} + \frac{1}{5\cdot 2!\cdot 10^5} - \frac{1}{7\cdot 3!\cdot 10^7}$ with error less than $\frac{1}{9\cdot 4!\cdot 10^9}$ **45.** $\frac{1}{2} - \frac{1}{7\cdot 2^8} \approx 0.49944$ with error less than $\frac{1}{39\cdot 2^{16}}$
47. $2\left[\frac{1}{2} - \frac{1}{10}\left(\frac{1}{2}\right)^5 + \frac{1}{24\cdot 9}\left(\frac{1}{2}\right)^9\right] \approx 0.99376808$ with error less than $\frac{2}{720\cdot 13}\left(\frac{1}{2}\right)^{13} \approx 0.000000026$
49. Checking coefficients of x^k, we need only show that

$$p\binom{p}{k} = \left[(k+1)\binom{p}{k+1}\right] + \left[k\binom{p}{k}\right].$$

We have

$$(k+1)\binom{p}{k+1} + k\binom{p}{k} = (k+1)\frac{p(p-1)\cdots(p-k)}{(k+1)(k)\cdots(2)(1)} + k\frac{p(p-1)\cdots(p-k+1)}{k!}$$

$$= \frac{[p(p-1)\cdots(p-k)] + k[p(p-1)\cdots(p-k+1)]}{k!}$$

$$= \frac{p(p-1)\cdots(p-k+1)}{k!}(p-k+k) = \binom{p}{k}p.$$

51. Series 1.462650; Simpson's rule 1.462654; difference 0.000004

Review Exercise Set 11.1

1. $-8 \le x < -2$ **3.** $\frac{\sqrt{3}}{2} + \frac{1}{2}\left(x - \frac{\pi}{3}\right) - \frac{\sqrt{3}}{4}\left(x - \frac{\pi}{3}\right)^2 - \frac{1}{12}\left(x - \frac{\pi}{3}\right)^3 + \frac{\sqrt{3}}{48}\left(x - \frac{\pi}{3}\right)^4$

5. a) A function is analytic if it can be represented by a power series in a neighborhood of each point in its domain.
b) $x + x^3 + x^5 + x^7 + \cdots + x^{2n+1} + \cdots$
7. 6 **9.** $1 + \frac{1}{2}x^2 - \frac{1}{8}x^4 + \frac{1}{16}x^6 - \frac{5}{128}x^8$

Review Exercise Set 11.2

1. $\dfrac{-\sqrt{3}-1}{2} \le x \le \dfrac{\sqrt{3}-1}{2}$ **3.** $-2\left(x - \dfrac{\pi}{4}\right) + \dfrac{4}{3}\left(x - \dfrac{\pi}{4}\right)^3$ **5.** $x^5 - \dfrac{x^7}{2} + \dfrac{x^9}{3} - \dfrac{x^{11}}{4} + \cdots + (-1)^n \dfrac{x^{2n+5}}{n+1} + \cdots$ **7.** $-\frac{1}{2}$
9. $1 + \dfrac{x}{4} + \dfrac{3}{32}x^2 + \dfrac{5}{128}x^3$

Review Exercise Set 11.3

1. $-\infty < x < \infty$ **3.** $e^2 + 2e^2(x-1) + 2e^2(x-1)^2 + \frac{4}{3}e^2(x-1)^3 + \frac{2}{3}e^2(x-1)^4$ **5.** $8 \cdot 10!$ **7.** 6
9. $4 + \frac{1}{3}x - \frac{1}{144}x^2 + \frac{1}{2592}x^3 + \cdots$; $R = 8$

Review Exercise Set 11.4

1. $x = 3$ **3.** $-1 - 16(x+1) + 17(x+1)^2 - 7(x+1)^3 + (x+1)^4$
5. $-(x-1) + \frac{1}{2}(x-1)^2 - \frac{1}{4}(x-1)^4 + \frac{1}{4}(x-1)^5 - \frac{1}{8}(x-1)^6 + \cdots$ **7.** $\frac{1}{3}$ **9.** $\frac{5}{243}$

More Challenging Exercises 11

1. $1/e$ **3.** ∞ **5.** $\frac{1}{2}$ **7.** 0 **9.** $\frac{4}{3}$

CHAPTER 12

Section 12.1

1.

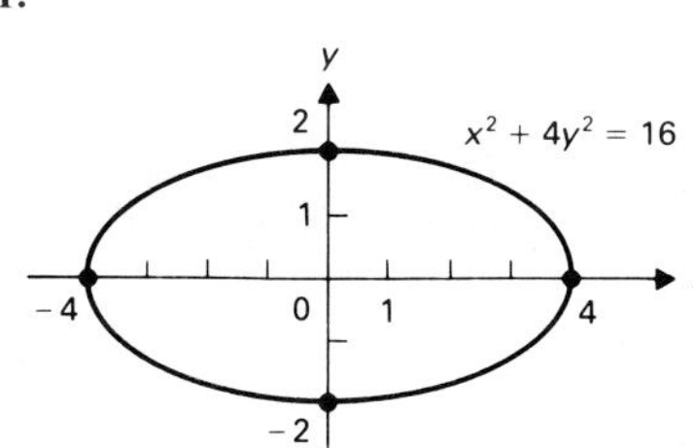

3.

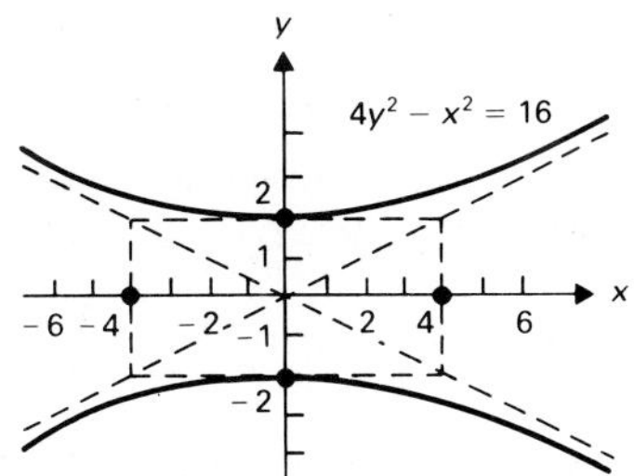

5.

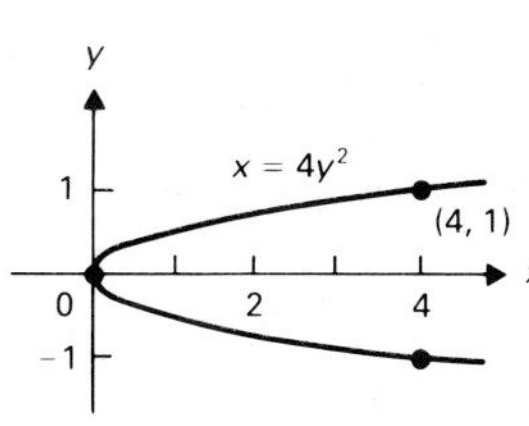

7.

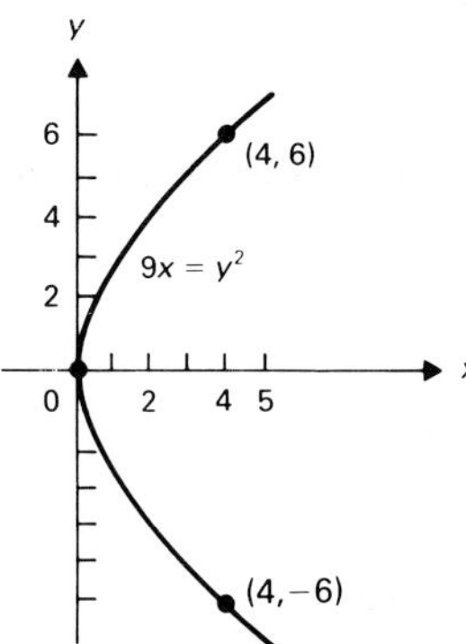

9.

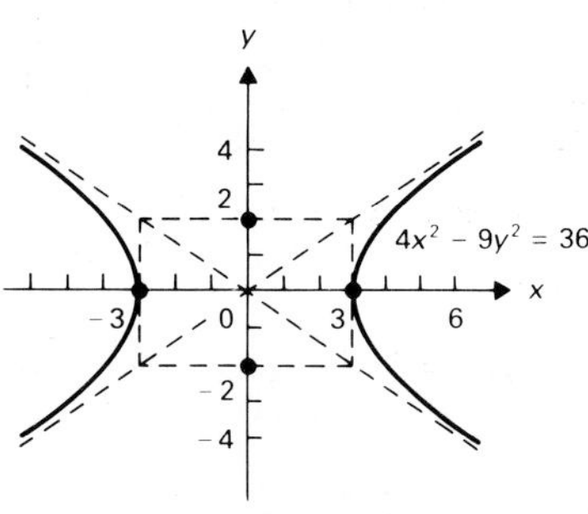

11.

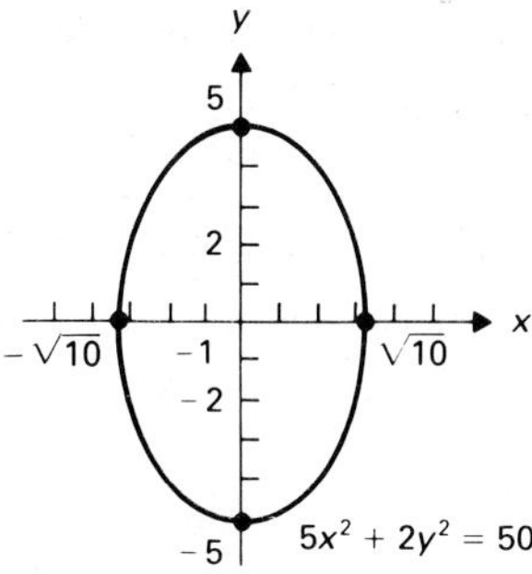

13.

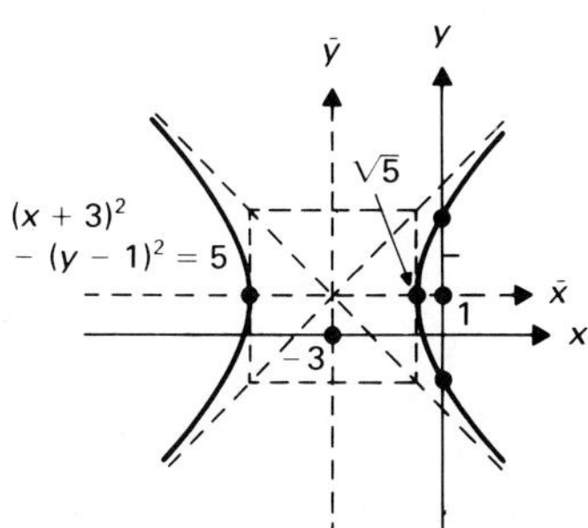

15.

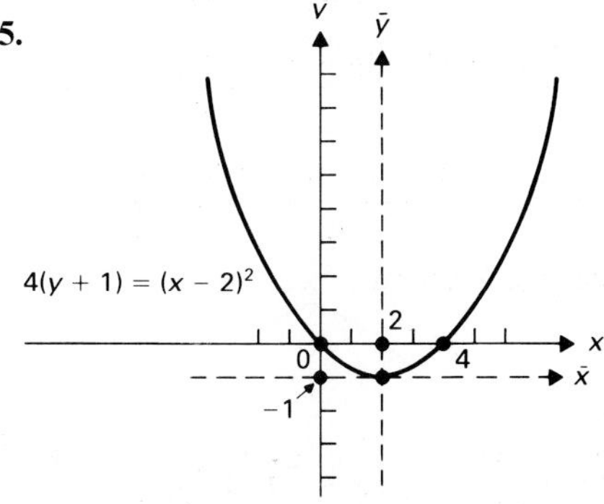

17.

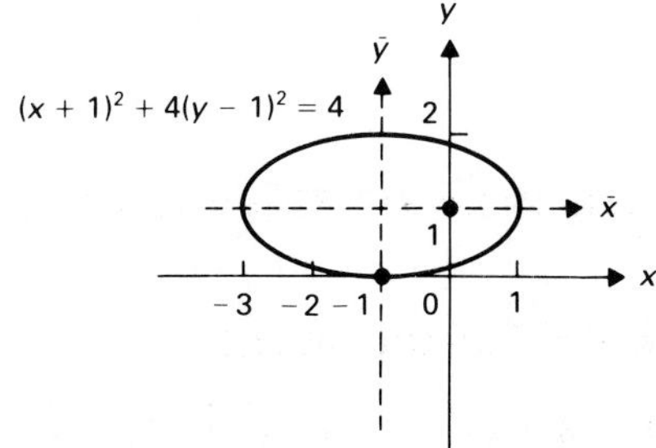

19.

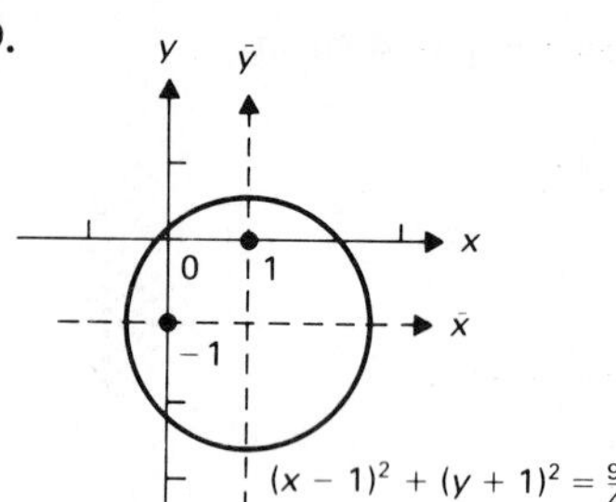

21. No points satisfy this equation.

23.

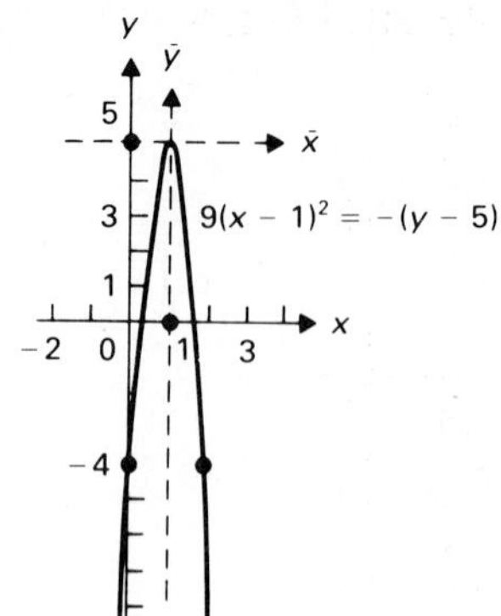

25.

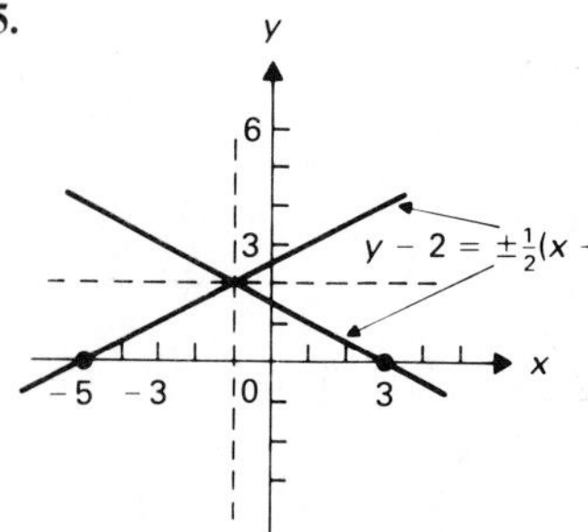

27.

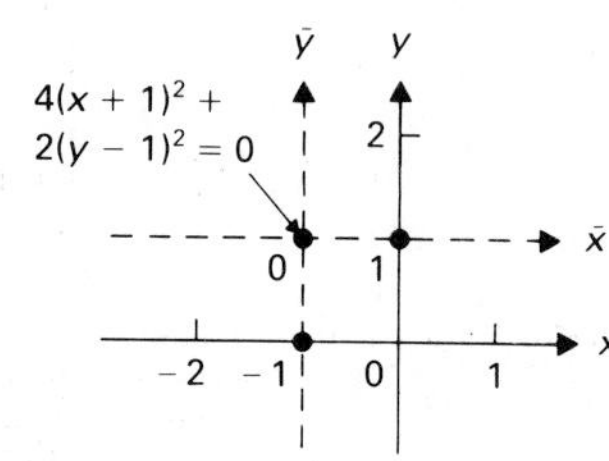

29.

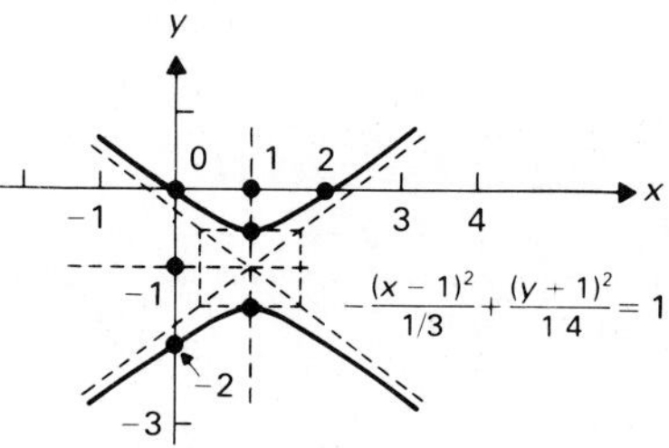

31. The point $(\cosh t, \sinh t)$ lies on the hyperbola $x^2 - y^2 = 1$ for each real number t.

Section 12.2

1. Let P be a point on the hyperbola a distance d_1 from F_1 and a distance d_2 from F_2. Suppose $d_1 > d_2$. Then $d_1 - d_2 = 2a$. The triangle with vertices P, F_1, and F_2 has sides of lengths d_1, d_2, and $2c$. As suggested in the exercise, we then know that $d_1 < d_2 + 2c$. Therefore $d_1 - d_2 < 2c$, so $2a < 2c$ and $a < c$.
3. Let (x, y) be on the ellipse. Then

$$\left(\frac{\text{Distance to focus}}{\text{Distance to directrix}}\right)^2 = \frac{(x-c)^2 + y^2}{[x - (a^2/c)]^2} = \frac{(x-c)^2 + b^2[1 - (x^2/a^2)]}{[x - (a^2/c)]^2}$$

$$= \frac{(x-c)^2 + (a^2 - c^2)[1 - (x^2/a^2)]}{[x - (a^2/c)]^2}$$

$$= \frac{x^2 - 2cx + c^2 + a^2 - x^2 - c^2 + (c^2/a^2)x^2}{[x - (a^2/c)]^2}$$

$$= \frac{(c^2/a^2)x^2 - 2cx + a^2}{[x - (a^2/c)]^2} = \frac{(c^2/a^2)[x^2 - 2(a^2/c)x + (a^4/c^2)]}{[x - (a^2/c)]^2} = \frac{c^2}{a^2}.$$

5. Foci $(\pm\sqrt{21}, 0)$, directrices $x = \pm(25/\sqrt{21})$

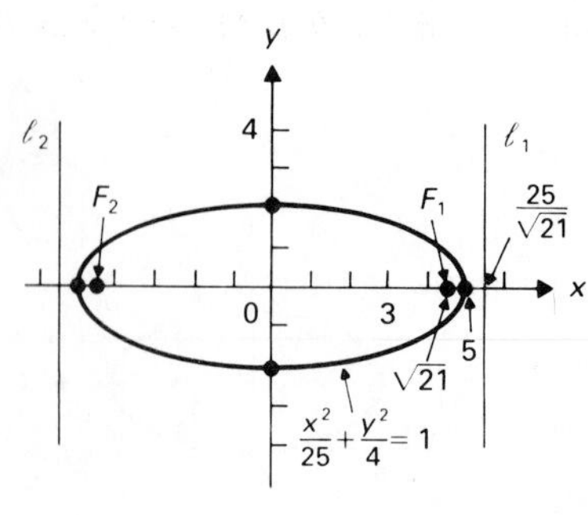

7. Foci $(0, \pm\sqrt{13})$, directrices $y = \pm(9/\sqrt{13})$

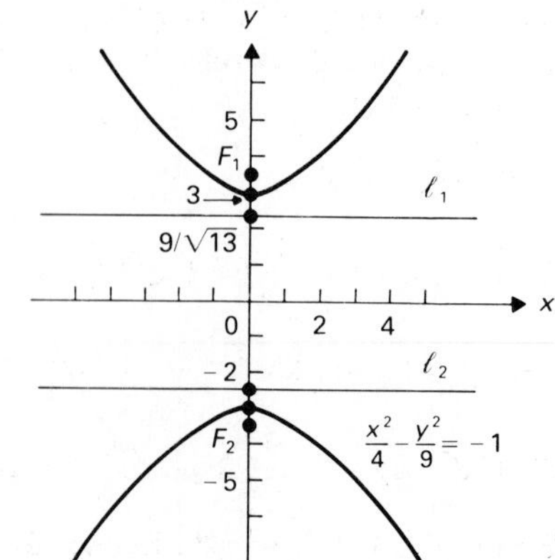

9. Focus $(-2, 0)$, directrix $x = 2$

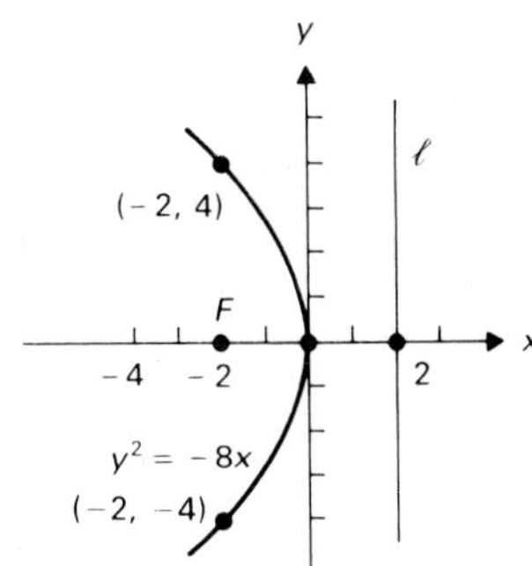

11. Foci $(2, -3 \pm 3\sqrt{3})$, directrices $y = -3 \pm 4\sqrt{3}$

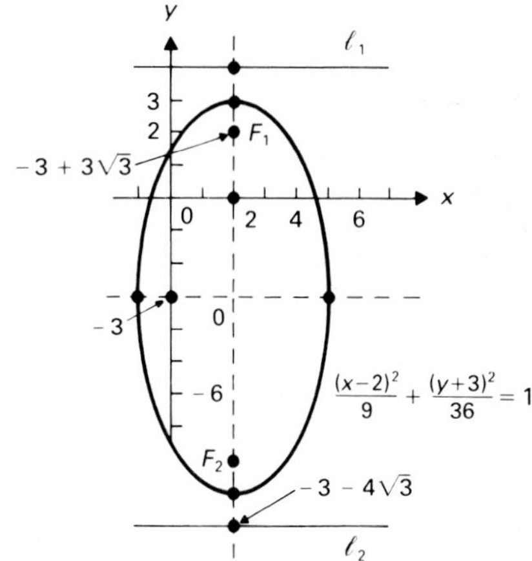

13. Foci $(1 \pm 2\sqrt{10}, -1)$, directrices $x = 1 \pm 18/\sqrt{10}$

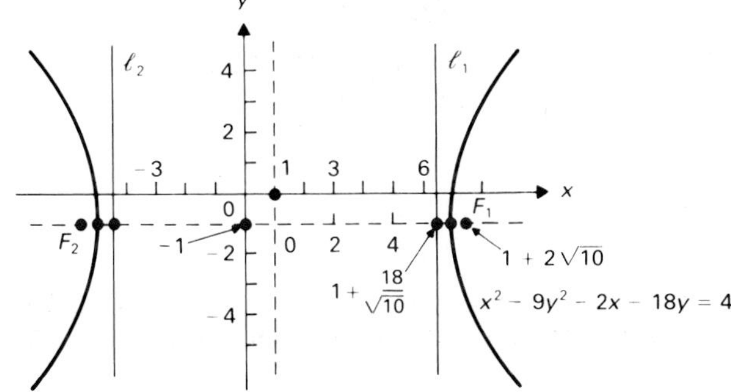

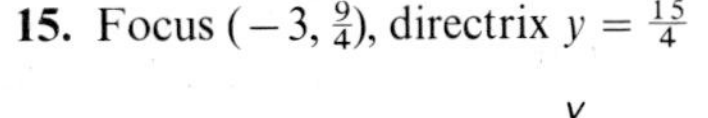

15. Focus $(-3, \frac{9}{4})$, directrix $y = \frac{15}{4}$

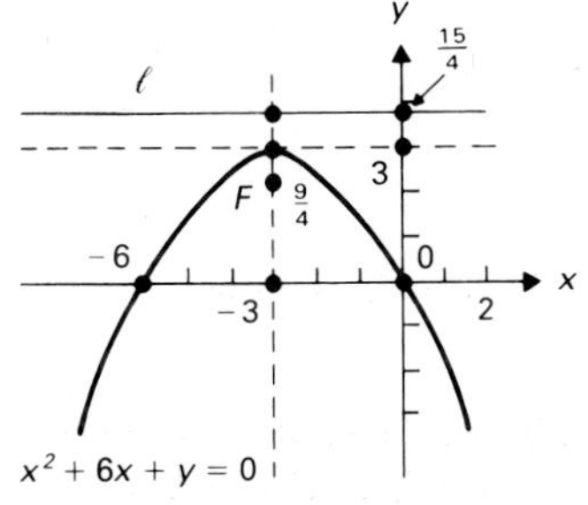

17. πab **19.** $2\pi^2 abs$ **21.** $3x^2 - y^2 - 10x + 4y = 1$ **23.** $5x^2 + 9y^2 + 26x - 36y + 41 = 0$ **25.** $x^2 + 4x + 2y = 1$

27. $16x^2 - 24xy + 9y^2 + 56x - 42y + 49 = 0$ **29.** $\frac{x^2}{36} + \frac{y^2}{27} = 1$ **31.** $\frac{x^2}{8} - \frac{y^2}{8} = 1$ **33.** $y^2 = 12x$

35. $\frac{(x+1)^2}{12} + \frac{(y-2)^2}{16} = 1$ **37.** $(x+5)^2 = -8(y-2)$

39. The line $x = p$ intersects $y^2 = 4px$ at $(0, 2p)$ and $(0, -2p)$. The distance between these points is $4p$.

41. a) The two horizontal lines $y = \pm b$ b) The foci recede to ∞ on the x-axis. c) The directrices recede to ∞. d) $e \to 1$ as $a \to \infty$

43. a) The empty set b) The foci recede to infinity on the x-axis. c) The directrices recede to infinity. d) $e \to 1$ as $a \to \infty$

45. a) The hyperbolas approach the y-axis, $x = 0$. b) The foci approach $(\pm b, 0)$. c) The directrices approach $x = 0$. d) $e \to \infty$ as $a \to 0$

Section 12.3

1. $x' = x\cos\theta + y\sin\theta$, $y' = -x\sin\theta + y\cos\theta$

3. Using Eq. (5) and trigonometric identities, we obtain $B'^2 - 4A'C' = (C - A)^2\sin^2 2\theta + 2(C - A)B\sin 2\theta\cos 2\theta + B^2\cos^2 2\theta - 4A^2\cos^2\theta\sin^2\theta + 4AB\sin\theta\cos^3\theta - 4AC\cos^4\theta - 4AB\sin^3\theta\cos\theta + 4B^2\sin^2\theta\cos^2\theta - 4BC\sin\theta\cos^3\theta - 4AC\sin^4\theta + 4CB\sin^3\theta\cos\theta - 4C^2\sin^2\theta\cos^2\theta = B^2 - 4AC$.

5. a) $\theta = \pi/6$ b) $x = \frac{1}{2}(\sqrt{3}x' - y')$, $y = \frac{1}{2}(x' + \sqrt{3}y')$

7. a) $\theta = \frac{1}{2}\cot^{-1}\frac{1}{2\sqrt{2}}$ b) $x = \frac{1}{\sqrt{3}}(\sqrt{2}x' - y')$, $y = \frac{1}{\sqrt{3}}(x' + \sqrt{2}y')$

9. a) $\theta = \frac{1}{2}\cot^{-1}\frac{-7}{4\sqrt{2}}$ b) $x = \frac{1}{3}(x' - 2\sqrt{2}y')$, $y = \frac{1}{3}(2\sqrt{2}x' + y')$ **11.** $x'^2 - 5y'^2 = -10$

13. $5x'^2 + y'^2 - 4\sqrt{3}x' - 4y' = 36$ **15.** $11x'^2 + y'^2 + \frac{6}{\sqrt{10}}x' - \frac{2}{\sqrt{10}}y' = 24$

17. Hyperbola **19.** Hyperbola **21.** Parabola **23.** Ellipse **25.** Ellipse **27.** $A' = 3, C' = -1$ or $A' = -1, C' = 3$

29. Rotate through $\theta = 45^\circ$. Then $(x' + 1)^2 = -4(y' - 1)$, so $\bar{x}^2 = -4\bar{y}$.

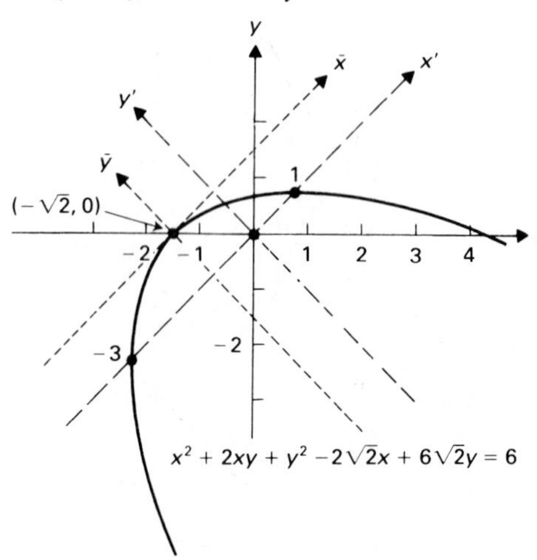

31. Rotate through $\theta = \cos^{-1}(2/\sqrt{5})$. Then $\dfrac{(x' - 4)^2}{8} - \dfrac{(y' - 3)^2}{12} = 1$, so $\dfrac{\bar{x}^2}{8} - \dfrac{\bar{y}^2}{12} = 1$.

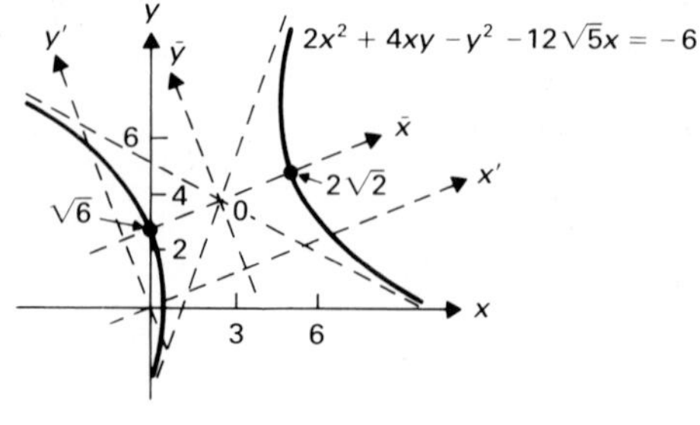

33. This is immediate from Eq. (3) of the text. **35.** $-7.385x'^2 + 3.385y'^2 + 2.495x' - 7.667y' - 10 = 0$
37. $0.678x'^2 - 3.507y'^2 - 0.097x' + 3.161y' - 5 = 0$

Section 12.4

1. For an ellipse, we have $c = \sqrt{a^2 - b^2} = a\sqrt{1 - (b/a)^2}$. It follows that if b/a is very near zero, we have $c/a \approx 1$.

3. As in Exercise 2, we could show that the point on an ellipse farthest from a given focus is the point farthest away on the major axis. The exercise now follows immediately from the fact that the center of the earth is at the focus of the ellipse.

5. Let the parabola be $y^2 = 4px$. The slope of the normal to the parabola at (x, y) is easily found to be $-y/(2p)$. The slope of the line from $(p, 0)$ to (x, y) is $y/(x - p)$. The tangent of an angle between lines whose slopes are m_1 and m_2 is $(m_1 - m_2)/(1 + m_1m_2)$ by the formula for $\tan(\theta_1 - \theta_2)$. Referring to the figure, we then see that

$$\tan\alpha = \frac{y/(x-p) + y/(2p)}{1 + [-y/(2p)][y/(x-p)]} = \frac{2py + xy - py}{2px - 2p^2 - y^2} = \frac{py + xy}{2px - 2p^2 - 4px} = \frac{y(p + x)}{-2p(p + x)} = \frac{-y}{2p}$$

and

$$\tan\beta = \frac{-[y/(2p)] - 0}{1 + [-y/(2p)](0)} = \frac{-y}{2p}.$$

Thus $\alpha = \beta$.

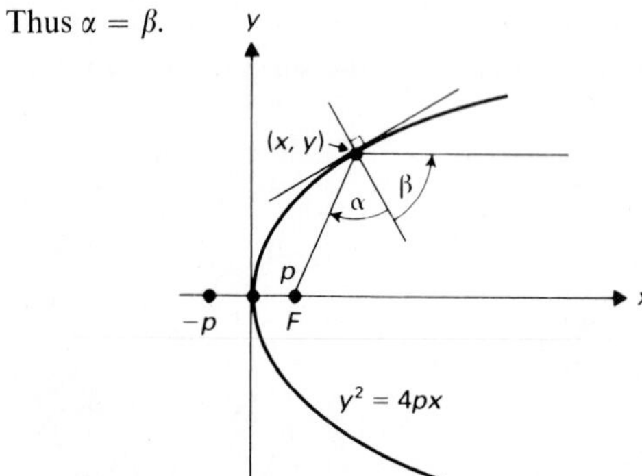

Section 12.5

1.

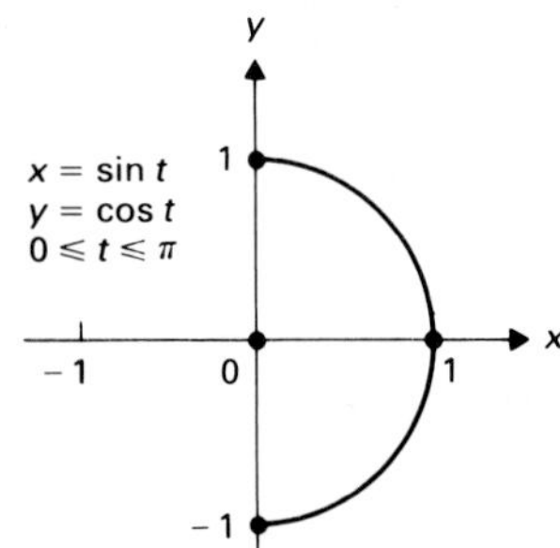

3.

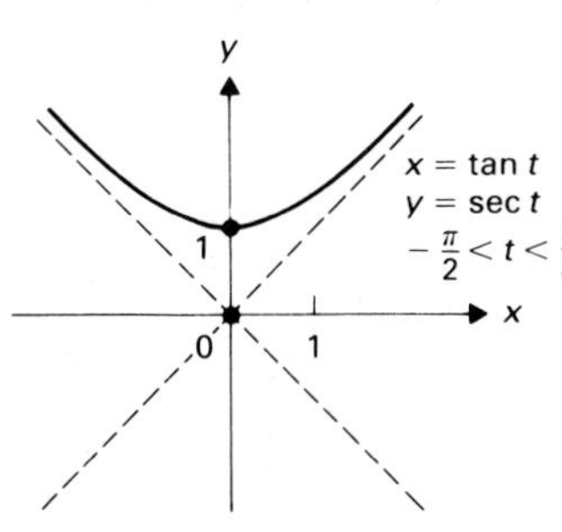

5.

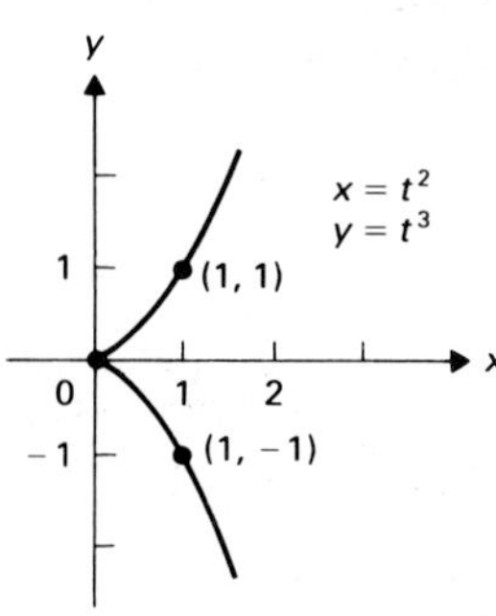

7.

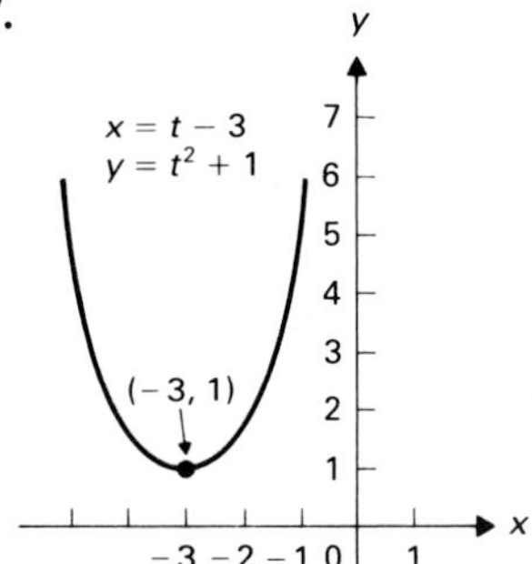

9.

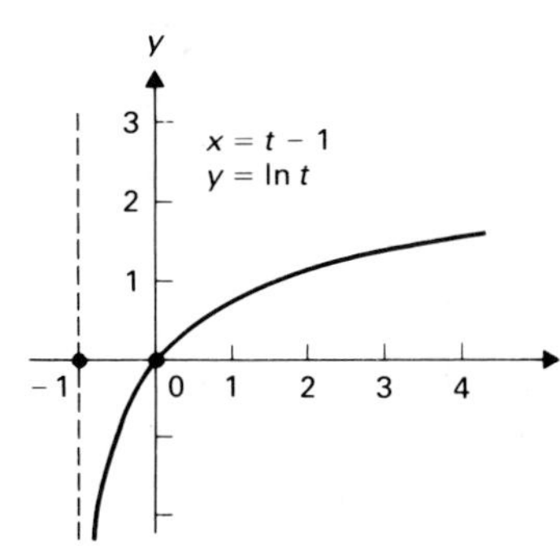

11.

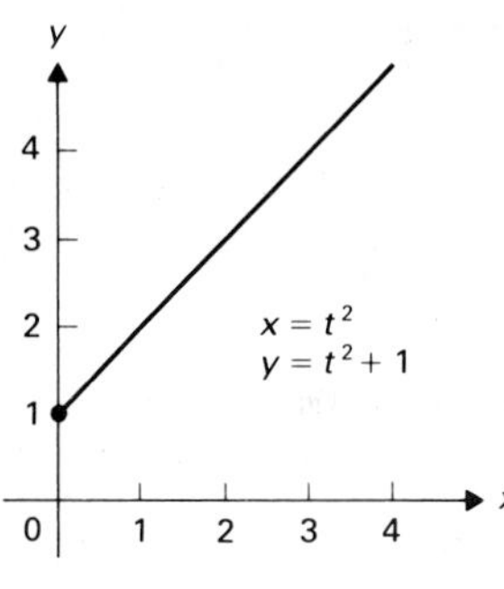

13. $x = \ln m$, $y = m$ for $0 < m < \infty$ **15.** $x = s/6$, $y = s^3/216$ for all s **17.** $x = 3 \cot \theta$, $y = 3$ for $0 < \theta \le \pi/2$

19. a) Yes b) No c) Yes d) No
In parts (a) and (c), different points correspond to different values of the derivative, while in parts (b) and (d), (x, y) and $(-x, y)$ correspond to the same values of the derivative.

21. $x = ma/\sqrt{1 + m^2}$, $y = a/\sqrt{1 + m^2}$ for $-1 \le m \le 1$ **23.** $x = a\theta - b \sin \theta$, $y = a - b \cos \theta$ **25.** $-\frac{2}{3}$ **27.** 1

29. $(8, -26), (-4, 6)$ **31.** $(-1, e - 1)$ **33.** $2\sqrt{2}$ **35.** $\frac{3}{4}$ **37.** -36 **39.** $\dfrac{16 \sin 2t}{81}$ **41.** 24 **43.** $3a/2$ **45.** 4.148 **47.** 13.1

Section 12.6

1. 1 **3.** $1/(2\sqrt{2})$ **5.** $1/\sqrt{2}$ **7.** $12/(5\sqrt{5})$ **9.** $\frac{1}{5}$ **11.** $\frac{4}{25}$ **13.** $\frac{5}{16}$ **15.** 2

17. Where $d\phi/ds$ is positive, ϕ increases as s increases, and increasing ϕ corresponds to counterclockwise rotation of the tangent line, so that the curve must bend to the left. Where $d\phi/ds$ is negative, ϕ decreases as s increases, and the curve bends to the right.

19. Zero

21. If the curvature is zero at each point, then by formula (7) of the text, we have $d^2y/dx^2 = 0$ at each point. It follows from this differential equation that $y = Ax + B$ for some constants A and B.

23. $x^2 + y^2 - 4x + 2y = 0$ **25.** 0 **27.** a) $(\pm 1/(45)^{1/4}, \pm 1/(45)^{3/4})$ b) $(0, 0)$ **29.** $(1, 0)$ **31.** $(x - 2)^2 + (y - 2)^2 = 2$

33. $(x + \frac{11}{2})^2 + (y + \frac{16}{3})^2 = \frac{2197}{36}$ **35.** $x = -4t^3$, $y = \frac{1}{2} + 3t^2$

Review Exercise Set 12.1

1.1

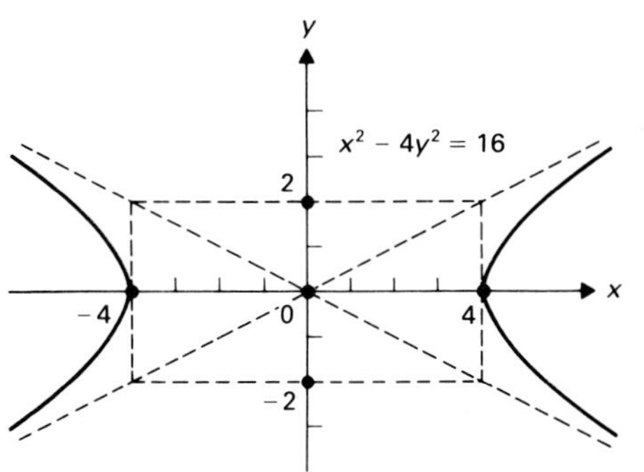

1.3

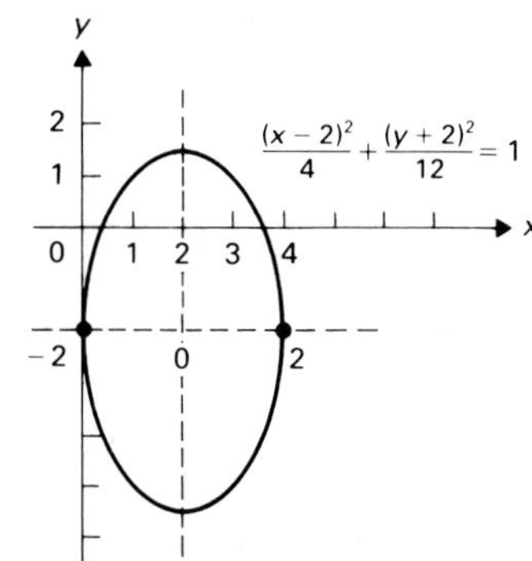

1. $3x^2 - y^2 + 14x + 4y = 9$ **3.** a) Hyperbola b) Parabola c) Ellipse **5.** $\left.\dfrac{dy}{dx}\right|_{t=1} = \dfrac{1}{2}, \left.\dfrac{d^2y}{dx^2}\right|_{t=1} = \dfrac{5}{4}$ **7.** $\int_0^3 \sqrt{4t^2 + \cos^2 t}\, dt$
9. $6\sqrt{3}/10^{3/2}$

Review Exercise Set 12.2

1.1

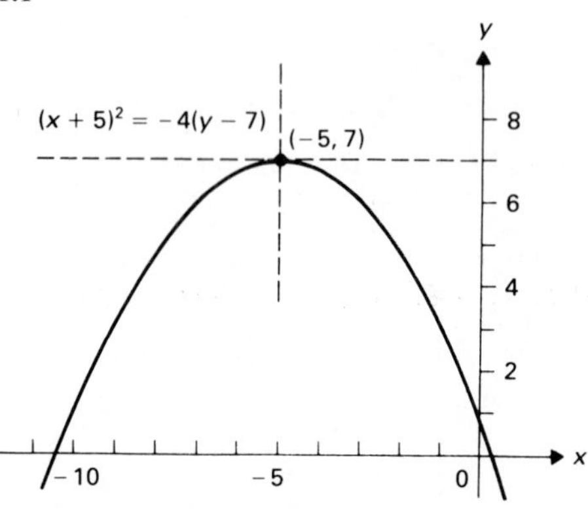

1.3

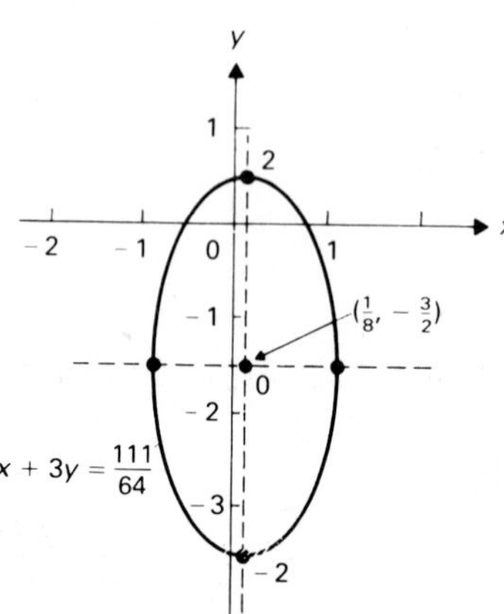

1. $9x^2 + 8y^2 + 18x - 14y = -14$ **3.** $\theta = \pi/4$; $(y' - 1)^2 = x' + 3$

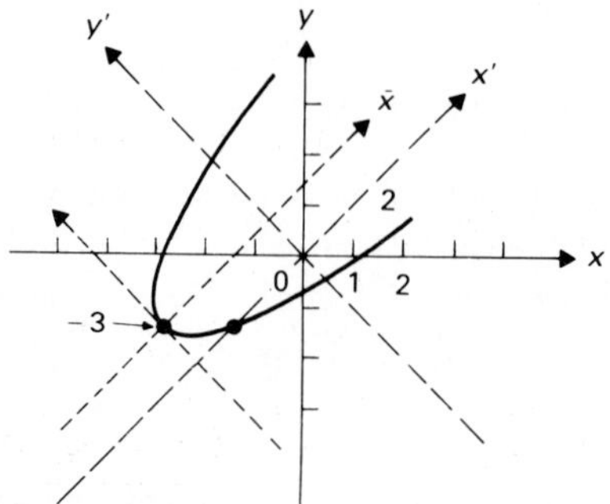

5. $\left.\dfrac{dy}{dx}\right|_{t=\pi/3} = -2\sqrt{3}, \left.\dfrac{d^2y}{dx^2}\right|_{t=\pi/3} = -4$ **7.** $(15, 10\sqrt{15})$ **9.** $2/29^{3/2}$

Review Exercise Set 12.3

1.1

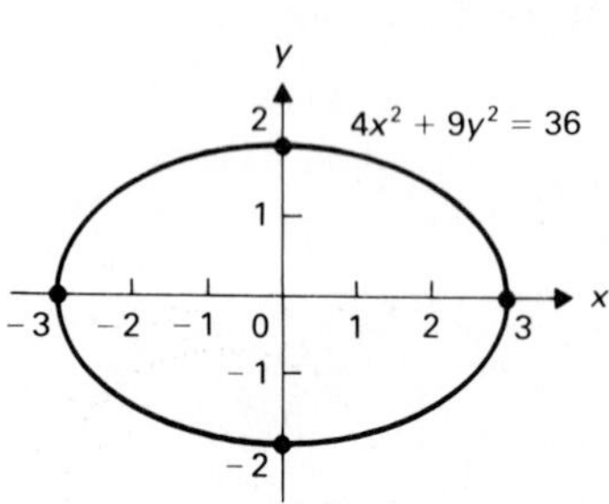

1.3

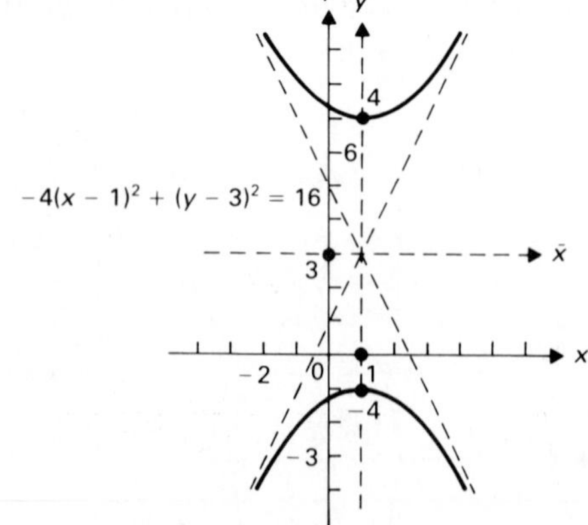

1. See Definitions 12.2 and 12.5.
3. Let $a = \sqrt{(\sqrt{2} - 1)/(2\sqrt{2})}$ and $b = \sqrt{(\sqrt{2} + 1)/(2\sqrt{2})}$. Then $x = bx' - ay'$ and $y = ax' + by'$. **5.** $27/(16t^4)$ **7.** 8.6670
9. $(-\frac{1}{2} \ln 2, 1/\sqrt{2})$

Review Exercise Set 12.4

1.1

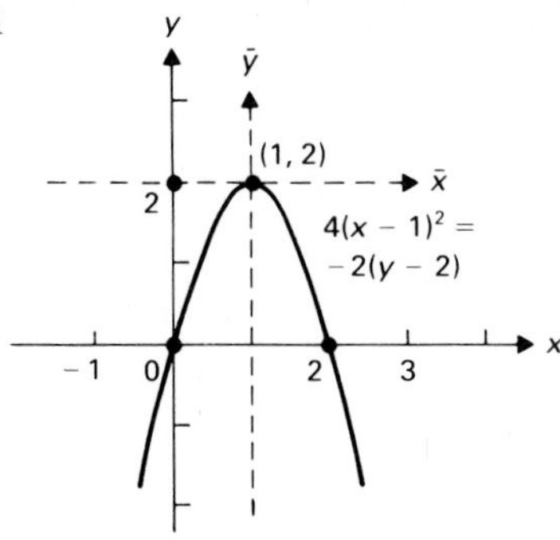

1.3 Empty curve

1. $x^2 - 2x - 8y = 7$ **3.** a) Parabola b) Hyperbola c) Ellipse **5.** (4, 8) **7.** $\frac{1}{27}(85\sqrt{85} - 8)$ **9.** $(x + 2)^2 + (y - 4)^2 = 1$

More Challenging Exercises 12

1. Since $1 + f'(x)^2 \geq 1$, we have

$$|\kappa(x)| = \left| -\frac{f''(x)}{(1 + f'(x)^2)^{3/2}} \right| \leq |f''(x)|.$$

3. $15a/8$

CHAPTER 13

Section 13.1

1. $(2\sqrt{2}, 2\sqrt{2})$ **3.** $(0, -6)$ **5.** $(-\sqrt{2}, -\sqrt{2})$ **7.** $(4, 0)$ **9.** $(0, 0)$ **11.** $\left(2\sqrt{2}, \frac{\pi}{4} + 2n\pi\right)$ and $\left(-2\sqrt{2}, \frac{5\pi}{4} + 2n\pi\right)$

13. $(0, \theta)$ for all θ **15.** $\left(5, \tan^{-1}\left(\frac{-4}{3}\right) + 2n\pi\right), \left(-5, \tan^{-1}\left(\frac{-4}{3}\right) + \pi + 2n\pi\right)$

17. $(\sqrt{13}, \tan^{-1}\frac{3}{2} + \pi + 2n\pi)$ and $(-\sqrt{13}, \tan^{-1}\frac{3}{2} + 2n\pi)$ **19.** $\left(4, -\frac{\pi}{6} + 2n\pi\right), \left(-4, \frac{5\pi}{6} + 2n\pi\right)$

21. $2r\cos\theta + 3r\sin\theta = 5$ **23.** $r^2 = \sec 2\theta$ **25.** $r^2 = 4/(1 + \cos^2\theta)$ **27.** $r^3 = -4\csc\theta$ **29.** $r = -4\sin\theta$ **31.** $x^2 + y^2 = 64$
33. $x^2 + y^2 = -3ay$ **35.** $x = -3$ **37.** $(x^2 + y^2)^{3/2} = -2ax$ **39.** $x^2y + y^3 = 4$
41. $d = \sqrt{(x_2 - x_1)^2 + (y_2 - y_1)^2} = \sqrt{(r_2\cos\theta_2 - r_1\cos\theta_1)^2 + (r_2\sin\theta_2 - r_1\sin\theta_1)^2}$
$= \sqrt{r_1^2 + r_2^2 - 2r_1r_2(\cos\theta_1\cos\theta_2 + \sin\theta_1\sin\theta_2)}$
$= \sqrt{r_1^2 + r_2^2 - 2r_1r_2\cos(\theta_1 - \theta_2)}$
43. $r = 18/(r - 3\cos\theta)$ **45.** $e = \frac{3}{2}$; $x = -\frac{5}{3}$ **47.** $\frac{64}{3}$
49. If we take polar r,ϕ-coordinates with the positive y-axis as polar axis, the equation is $r = qe/(1 - \cos\phi)$. Now $\phi = \theta + \pi/2$, so $\cos\phi = \cos[\theta + (\pi/2)] = -\sin\theta$ and $r = qe/(1 + e\sin\theta)$.
51. $r = qe/(1 - e\sin\theta)$

Section 13.2

1.

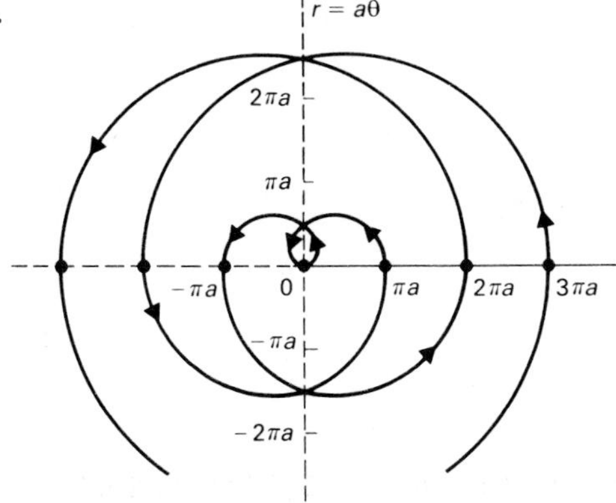

3.

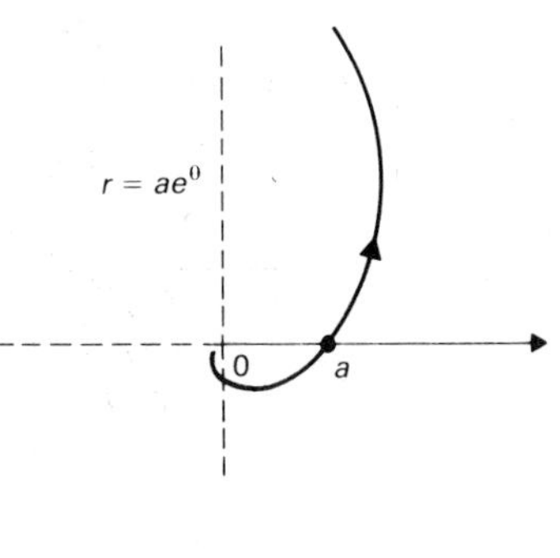

5.

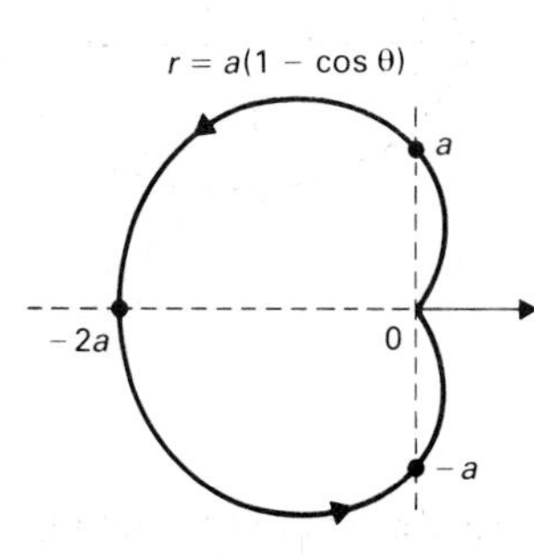

7.

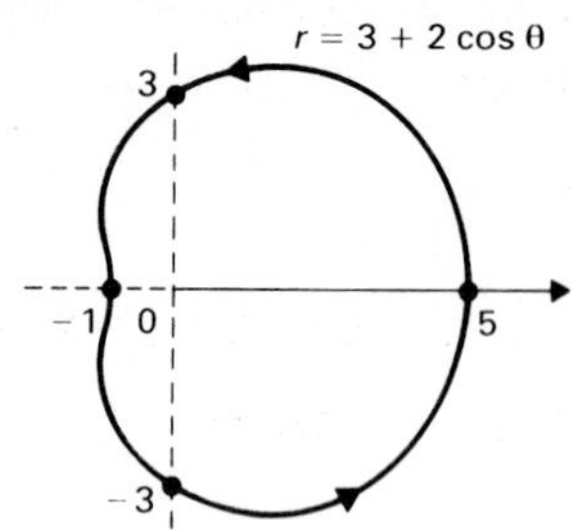

9.

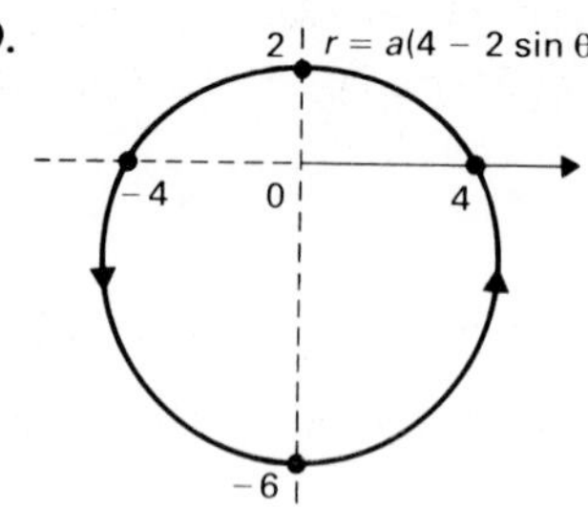

11.

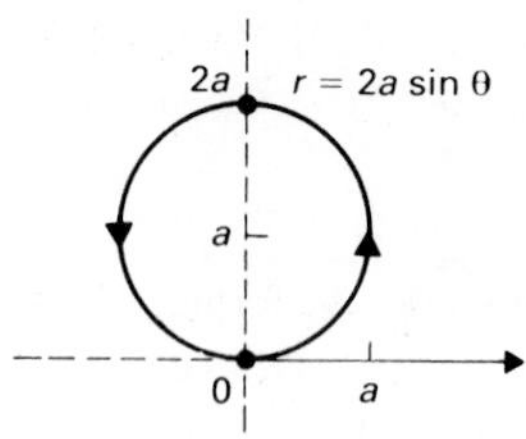

13.

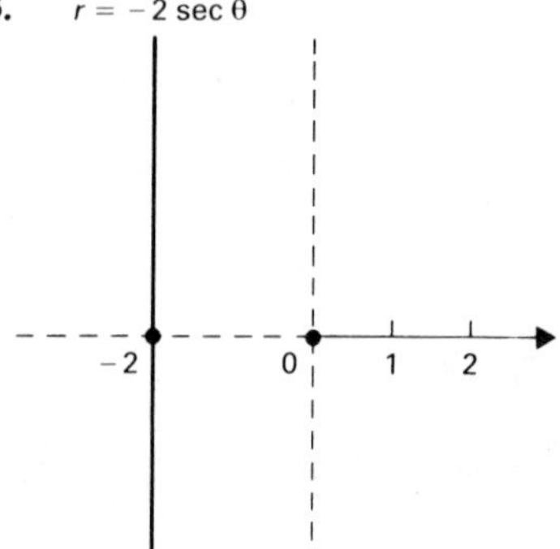

15.

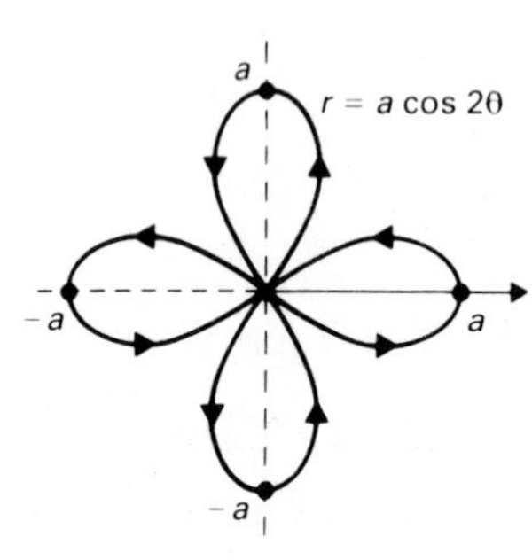

17.

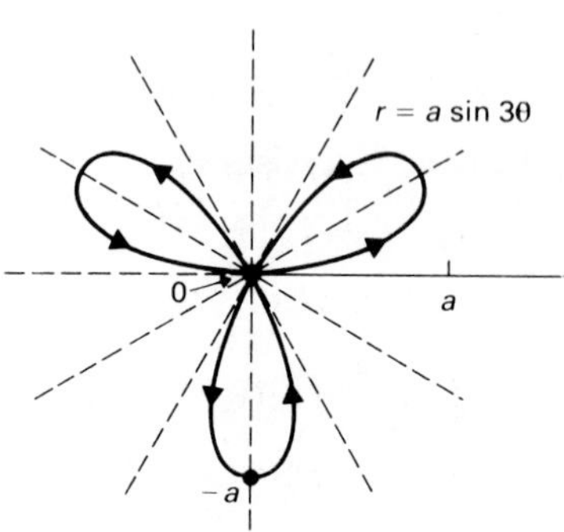

19.

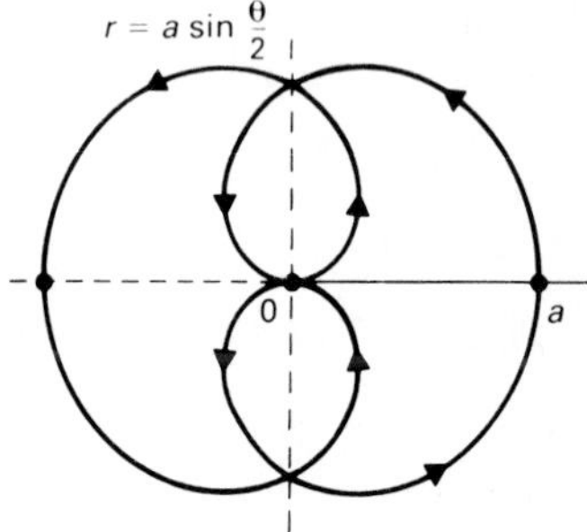

21.

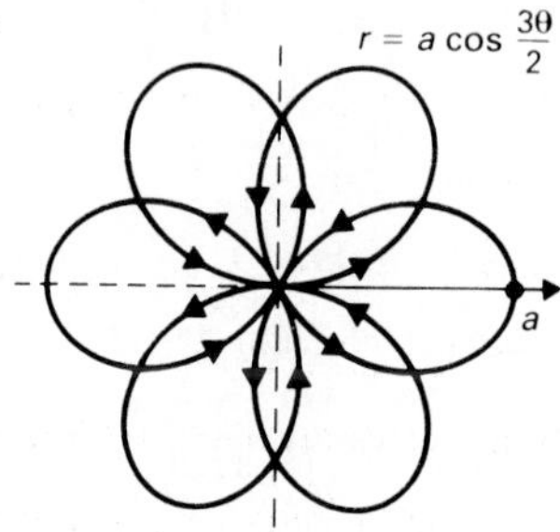

23.

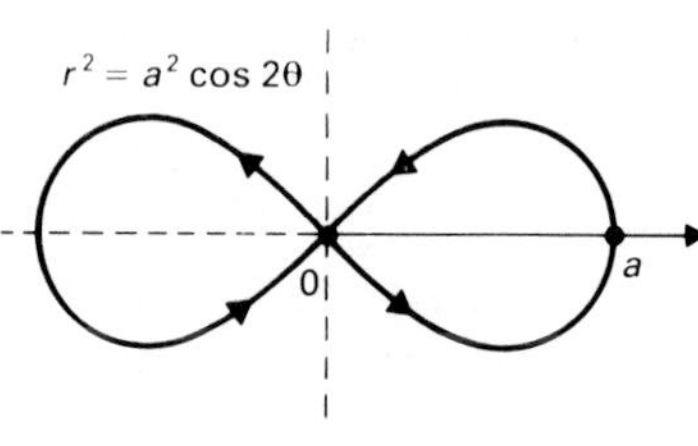

25.

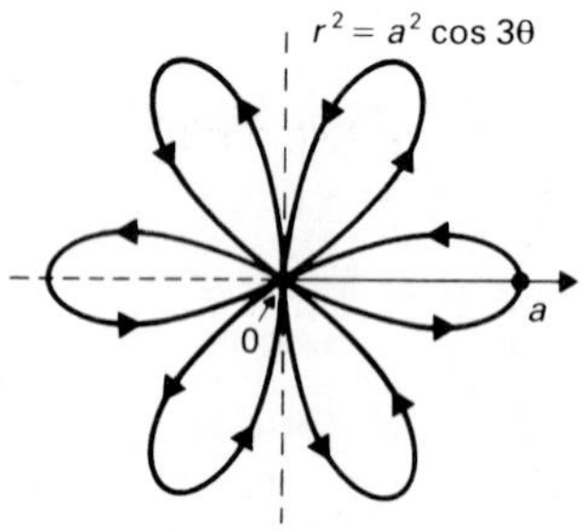

27.

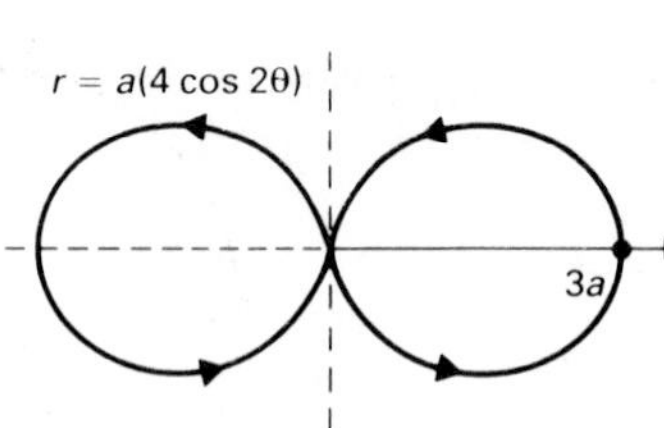

29.

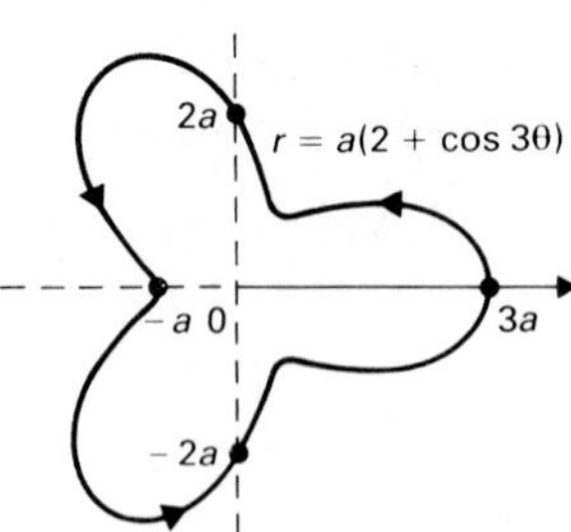

31. a) $r - 2 - \csc\theta,\ r = 2 - \dfrac{1}{\sin\theta},\ r\sin\theta = 2\sin\theta - 1,\ r\sin\theta = \dfrac{2}{\csc\theta} - 1,\ y = \dfrac{2}{2-r} - 1$

b) $\lim_{r\to\infty} y = \lim_{r\to\infty}\left(\dfrac{2}{2-r} - 1\right) = -1$ c)

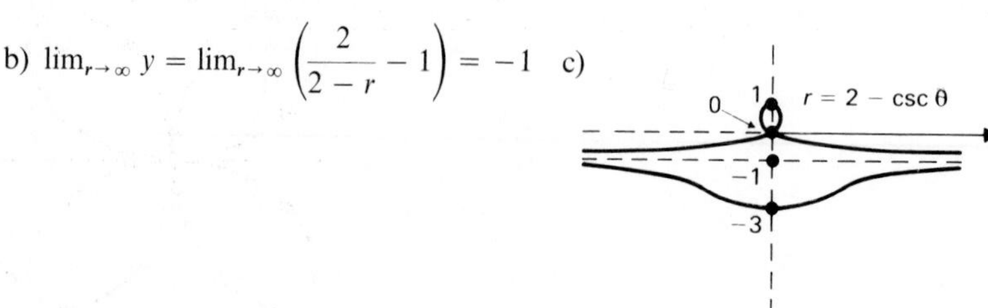

33. a) $x = a(\cos\theta + \cos^2\theta)$, $y = a(\sin\theta + \sin\theta\cos\theta)$ b) -1 **35.** $(a/2, 2\pi/3)$, $(a/2, 4\pi/3)$, $(2a, 0)$ **37.** $(\frac{1}{4}, \pi/4)$, $(-\frac{1}{4}, -\pi/4)$ **39.** $(a/\sqrt{2}, \pi/4)$, $(0, 0)$ **41.** $(a, \pi/12)$, $(a, 5\pi/12)$, $(a, 13\pi/12)$, $(a, 17\pi/12)$ **43.** $(\pm a, \pi/6)$, $(\pm a, 5\pi/6)$, $(\pm a, \pi/3)$, $(\pm a, 2\pi/3)$ **45.** $(\sqrt{3}/2, \pm\pi/3)$, $(\sqrt{3}/2, \pm 4\pi/3)$

Section 13.3

1. πa^2 (No. We used the formula for the area of a circle in deriving $dA = \frac{1}{2}r^2\,d\theta$.) **3.** $\dfrac{a^2}{4}(e^{2\pi} - 1)$ **5.** $\frac{3}{2}\pi a^2$ **7.** $\frac{1}{8}\pi a^2$ **9.** $\pi a^2/6$ **11.** $2a^2$ **13.** $a^2(2\pi/3 + \sqrt{3})$ **15.** $a^2(\pi + 3\sqrt{3})$ **17.** a^2 **19.** $99\pi a^2/8$ **21.** $a^2(19\pi/2 + 12)$ **23.** 4.80956 **25.** 53.4396

Section 13.4

1. $-\sqrt{3}$ **3.** $-2/\pi$ **5.** $\left(a\cos\dfrac{3\pi}{8}, \dfrac{3\pi}{8}\right)$, $\left(a\cos\dfrac{7\pi}{8}, \dfrac{7\pi}{8}\right)$

7. $\left(\dfrac{3+\sqrt{33}}{4}, \sin^{-1}\left(\dfrac{\sqrt{33}-1}{8}\right)\right)$, $\left(\dfrac{3+\sqrt{33}}{4}, \pi - \sin^{-1}\left(\dfrac{\sqrt{33}-1}{8}\right)\right)$,

$\left(\dfrac{3-\sqrt{33}}{4}, \sin^{-1}\left(\dfrac{\sqrt{33}+1}{-8}\right)\right)$, $\left(\dfrac{3-\sqrt{33}}{4}, \pi - \sin^{-1}\left(\dfrac{\sqrt{33}+1}{-8}\right)\right)$

9. $-\tan^{-1}(-\pi/2) \approx 57.5°$ **11.** $\tan^{-1} 2 \approx 63.4°$ **13.** $\tan\psi = \dfrac{a\cdot e^\theta}{a\cdot e^\theta} = 1$ for all θ; $\psi = 45°$ for all θ

15. $\dfrac{\pi}{2} - \tan^{-1}\left(\dfrac{-\sqrt{3}}{6}\right) \approx 106.1°$ **17.** $\tan^{-1}(\frac{4}{3}) \approx 53.1°$ **19.** $8a$ **21.** $6a\displaystyle\int_0^{\pi/6}\sqrt{1 + 8\cos^2 3\theta}\,d\theta$ **23.** $128\pi/5$

25. If a dot over a variable denotes differentiation with respect to a parameter, then

$$\kappa = \left|\frac{\dot{x}\ddot{y} - \dot{y}\ddot{x}}{(\dot{x}^2 + \dot{y}^2)^{3/2}}\right|.$$

Now $x = f(\theta)\cos\theta$ and $y = f(\theta)\sin\theta$. Thus, taking θ as parameter, we have

$$\dot{x} = -f(\theta)\sin\theta + f'(\theta)\cos\theta$$

$$\ddot{x} = -f(\theta)\cos\theta - 2f'(\theta)\sin\theta + f''(\theta)\cos\theta$$

$$\dot{y} = f(\theta)\cos\theta + f'(\theta)\sin\theta$$

$$\ddot{y} = -f(\theta)\sin\theta + 2f'(\theta)\cos\theta + f''(\theta)\sin\theta.$$

Substituting these expressions in the formula for κ and simplifying yields the desired result.

27. $a/3$ **29.** $6.68245a$ **31.** 501.668

Review Exercise Set 13.1

1. $(2, 5\pi/6 + 2n\pi)$, $(-2, 11\pi/6 + 2n\pi)$ **3.** $r^2 = 1/(4 + 5\sin^2\theta)$ **5.**

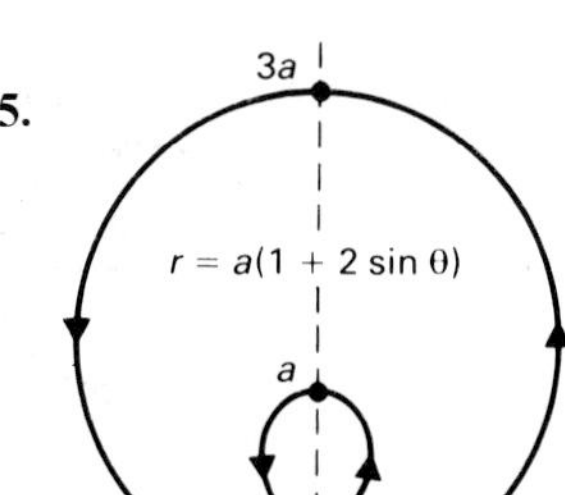

7. $a^2(1 + \pi/16)$ **9.** $\pi/2 - \tan^{-1}(2/\sqrt{3})$

Review Exercise Set 13.2

1. $(-\sqrt{2}, 3\pi/4 + 2n\pi)$, $(\sqrt{2}, 7\pi/4 + 2n\pi)$ **3.** $(x^2 + y^2)^2 = 2(x^2 + y^2) + 2xy$, $(x, y) \neq (0, 0)$ **5.**

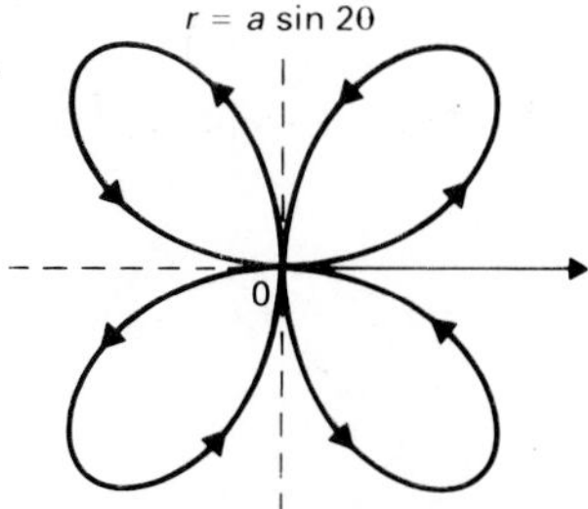

7. $3\pi/2$ **9.** $\pi/2$

Review Exercise Set 13.3

1. $(8, 4\pi/3 + 2n\pi), (-8, \pi/3 + 2n\pi)$ **3.** $(x^2 + y^2)^2 = x^2 - y^2$ **5.**

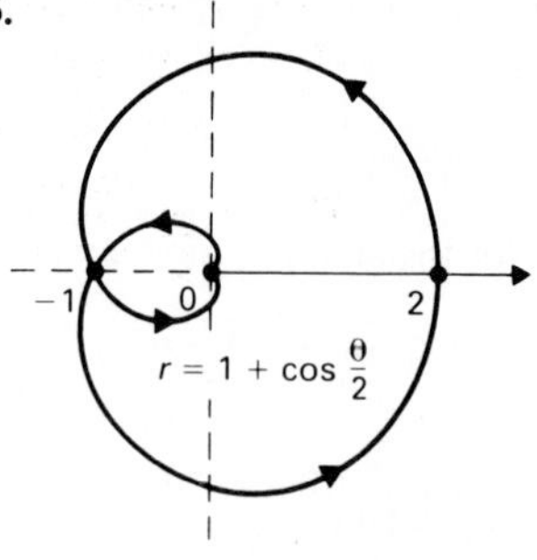

7. π **9.** $\pi/3$

Review Exercise Set 13.4

1. $(0, \theta)$ for all θ **3.** $r = 5/(\cos\theta - 3\sin\theta)$ **5.**

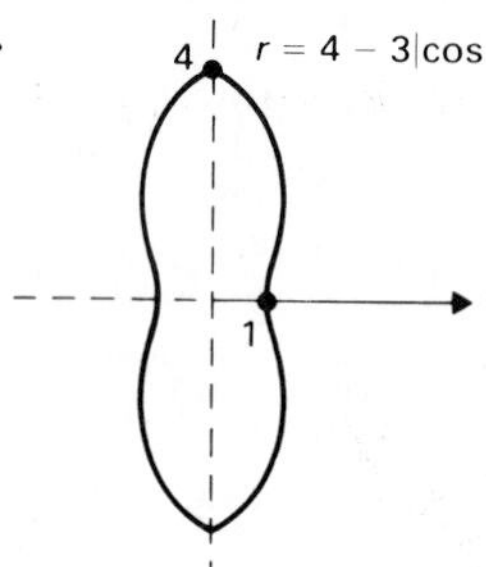

7. $\pi a^2/2$ **9.** 15.4038

More Challenging Exercise 13

1. a) (0, 0) b) 2 units time c) $r = \sqrt{2}e^{\pi/4}e^{-\theta}, \theta \geq \pi/4$ d) 2 units e) It will get dizzy.

CHAPTER 14

Section 14.1

1.

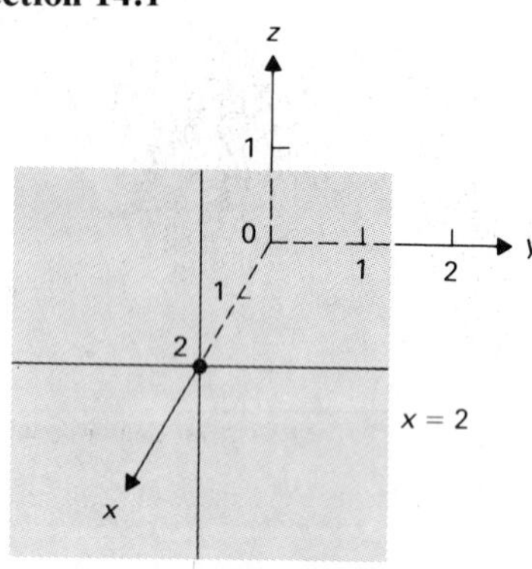

3.

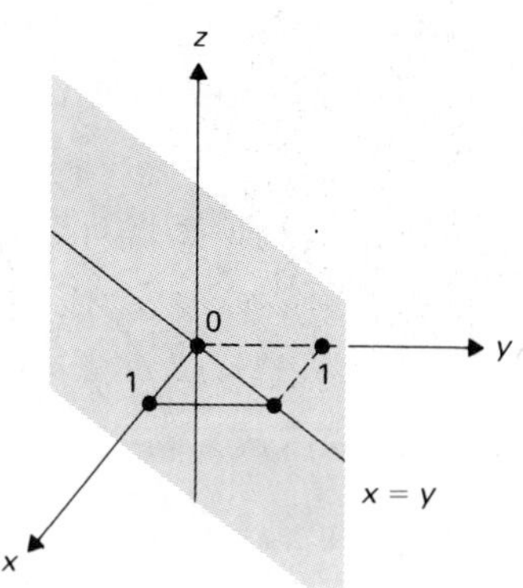

5.

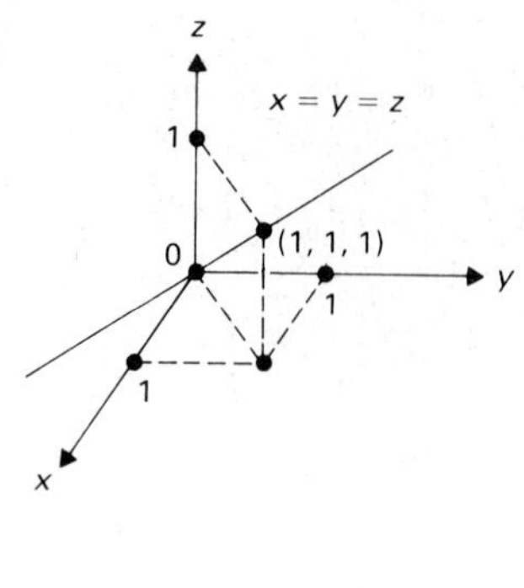

7.

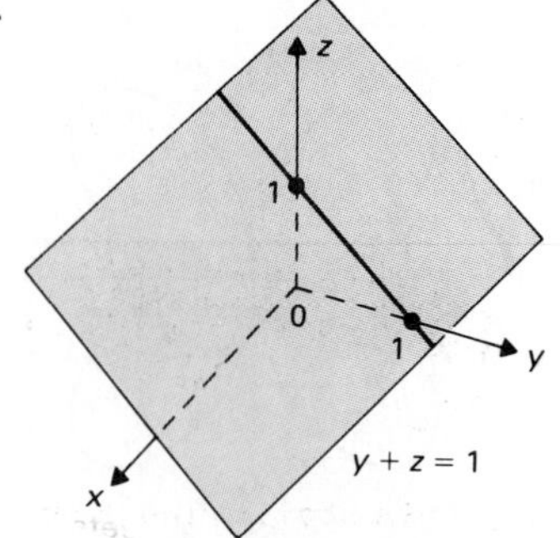

9. $(2, 1, -4)$ **11.** $(-1, 0, -3)$ **13.** $(0, 0, 2)$
15. a) $(2, -4, 0)$ b) $(-2, 2, -1)$ c) $(6, -3, 1)$
17. $2\sqrt{6}$ **19.** $3\sqrt{2}$ **21.** $(x + 1)^2 + (y - 2)^2 + (z - 4)^2 = 19$
23. a) Center $(1, -1, 0)$, radius $\sqrt{2}$ b) Center $(3, 2, -4)$, radius 5
25. $(2, -\pi/3 + 2n\pi, 5), (-2, 2\pi/3 + 2n\pi, 5)$

27.

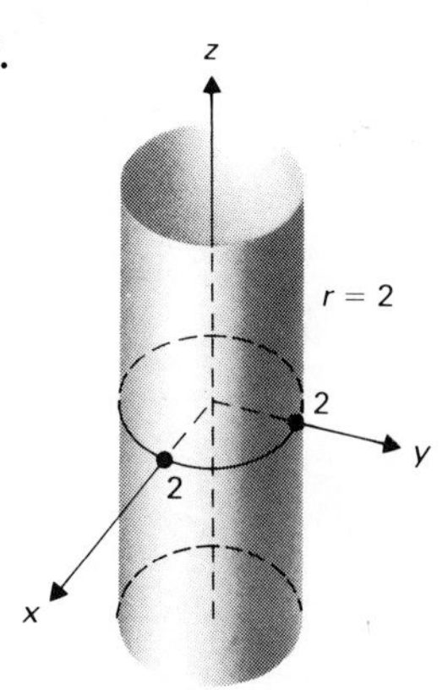

29. (1, 1, 3) **31.** (0, 0, −1) **33.** $(\sqrt{2}, 3\pi/4, -5)$ **35.** $(1, \pi/2, 0)$
37. $(\sqrt{3}, \cos^{-1}(1/\sqrt{3}), \pi/4)$ **39.** $(-\sqrt{2}, 0, \sqrt{2})$ **41.** $(2, 2\sqrt{3}, 0)$
43. $r = \rho \sin \phi, \theta = \theta, z = \rho \cos \phi$ **45.** $(3, \cos^{-1}(-2/\sqrt{3}), \tan^{-1} 2)$
47. $(2, 5\pi/6, \pi/2)$ **49.** $(2, 5\pi/6, 4)$ **51.** $(-\frac{1}{4}, -\sqrt{3}/4, -\sqrt{3}/2)$ **53.** $\rho = 5$ **55.** 78π
57. $\rho = \dfrac{\csc \phi}{\cos \theta + \sin \theta}$ **59.** 70 ft

Section 14.2

1. A plane $z - z_0$ intersects the surface in an ellipse if $z_0 > c$. Planes $x = x_0$ and $y = y_0$ intersect the surface in hyperbolas.

3.

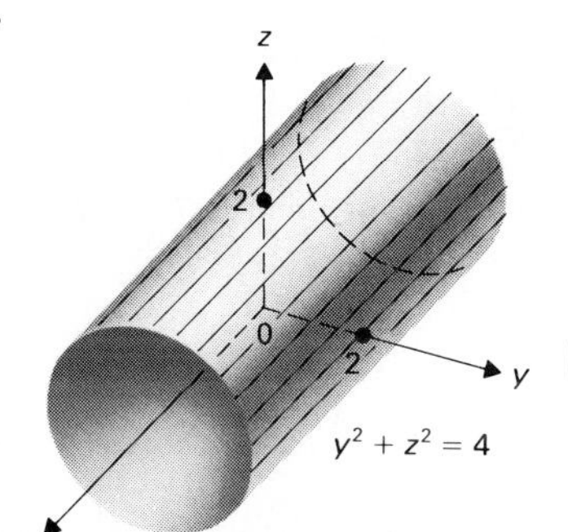

Right circular cylinder

5.

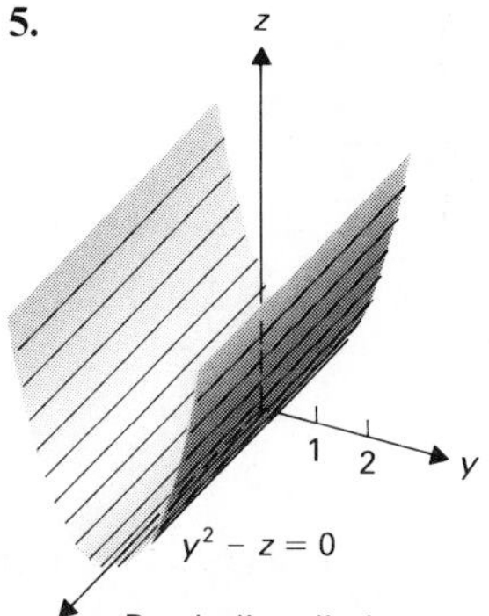

Parabolic cylinder

7.

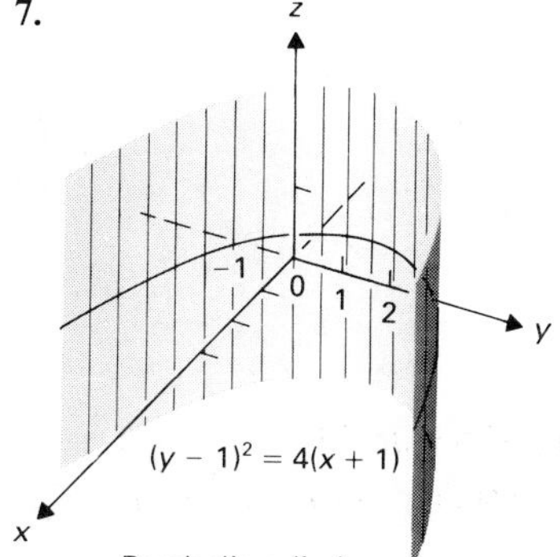

Parabolic cylinder

9.

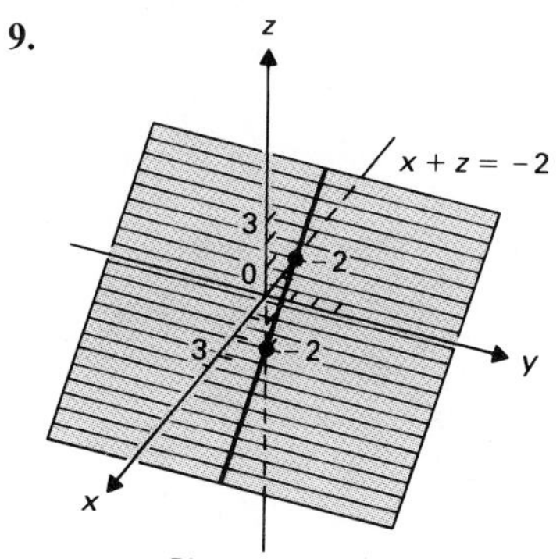

Planar cylinder

11.

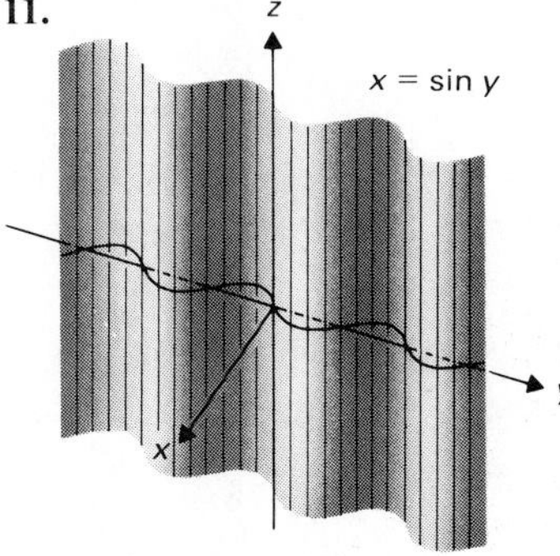

Sinusoidal cylinder

13.

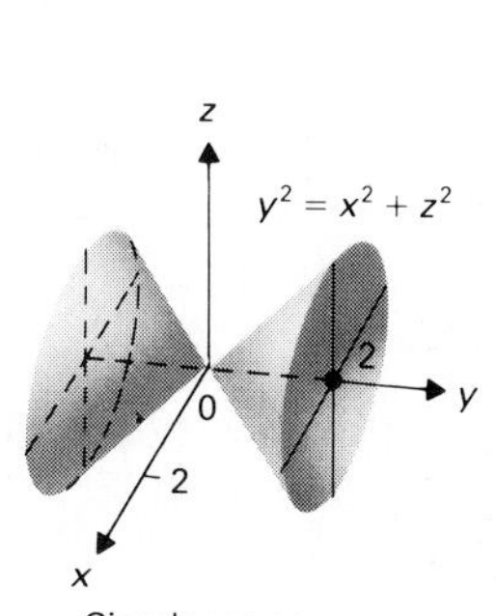

Circular cone

15.

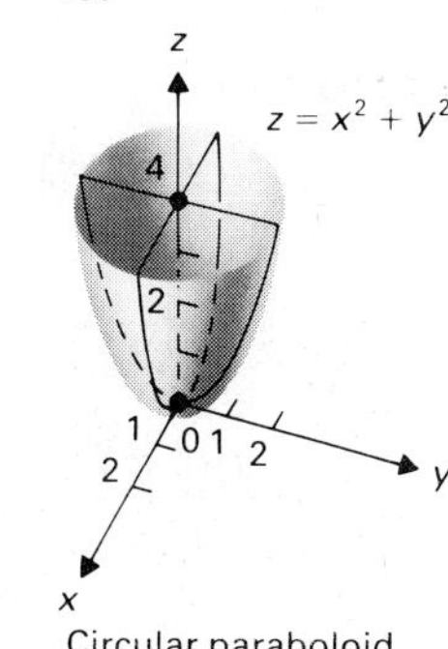

Circular paraboloid

17.

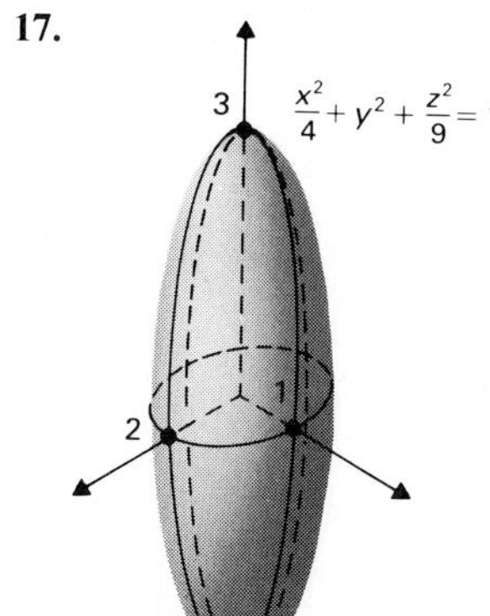

Ellipsoid

19.

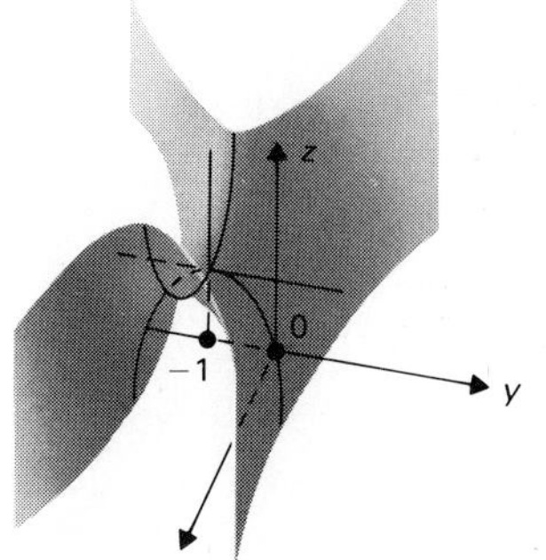

Hyperbolic paraboloid

21.

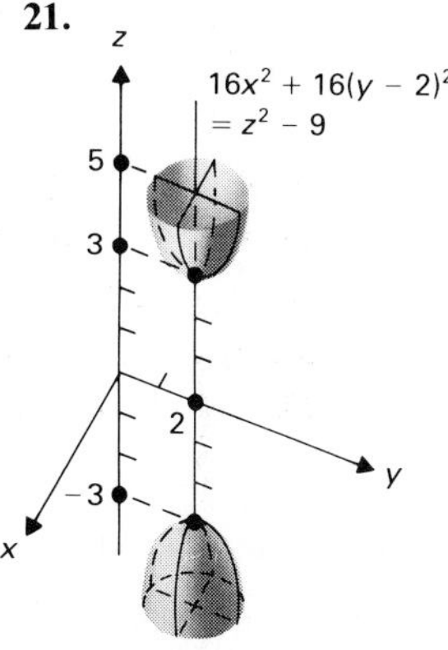

Hyperboloid of two sheets

23.

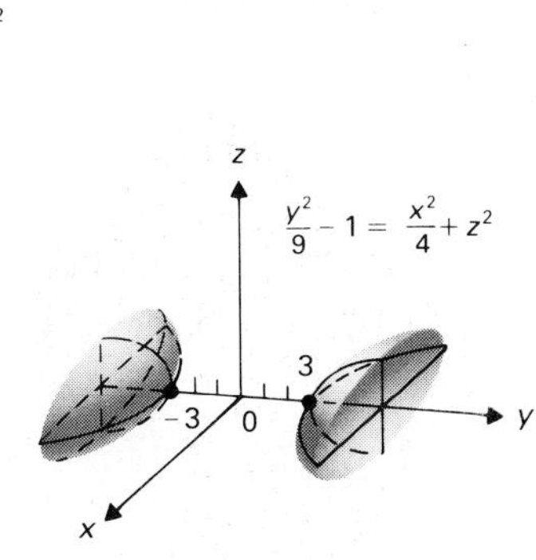

Hyperboloid of two sheets

25.

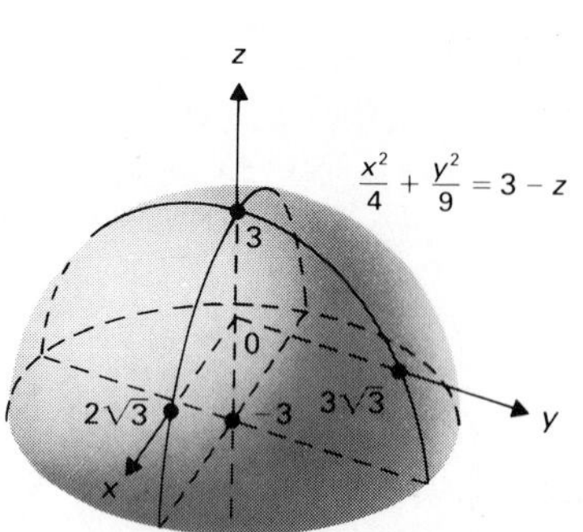

Elliptic paraboloid

27.

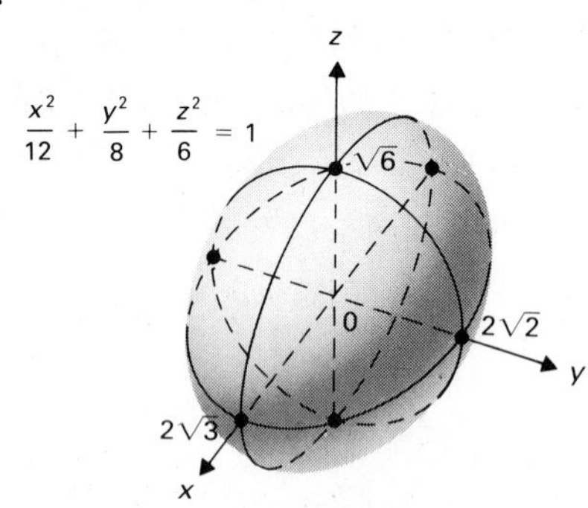

Ellipsoid

29.

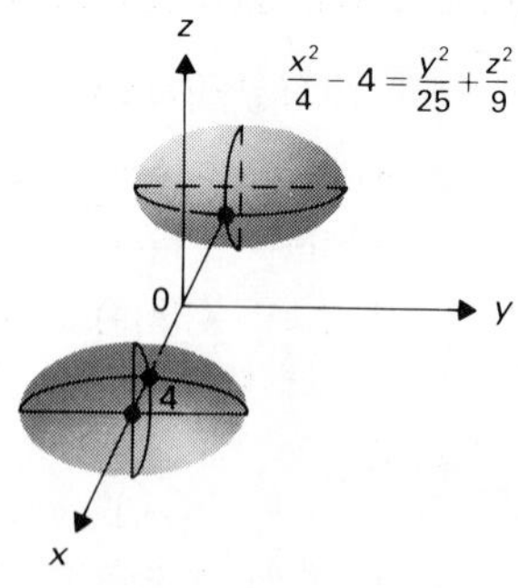

Hyperboloid of two sheets

31. Elliptic paraboloid **33.** Hyperboloid of two sheets **35.** Hyperbolic paraboloid **37.** Point ellipsoid **39.** Hyperbolic cylinder **41.** Elliptic cylinder **43.** Parabolic cylinder **45.** Hyperboloid of two sheets

Section 14.3

1.

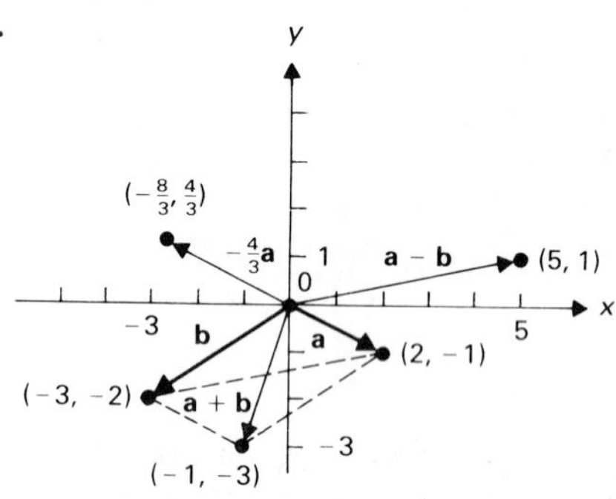

3. $-3\mathbf{i} + 9\mathbf{j} - 6\mathbf{k}$ **5.** $3\mathbf{i} + 3\mathbf{j} - 3\mathbf{k}$ **7.** $25\mathbf{i} + 9\mathbf{j} - 16\mathbf{k}$ **9.** $68\mathbf{i} + 36\mathbf{j} - 44\mathbf{k}$ **11.** $\sqrt{17}$ **13.** $\sqrt{17} + 3\sqrt{2}$ **15.** $3\sqrt{2} - \sqrt{17}$ **17.** $5\mathbf{i} - \mathbf{j}$ **19.** $7\mathbf{i} - 4\mathbf{j} - 11\mathbf{k}$ **21.** Perpendicular **23.** Neither **25.** Parallel, same direction **27.** 3 **29.** 0

31. $-\frac{3}{2}$ **33.** None **35.** -3 **37.** $\frac{1}{\sqrt{11}}(\mathbf{i} - \mathbf{j} + 3\mathbf{k})$

39. $\frac{1}{\sqrt{5}}(\mathbf{i} + 2\mathbf{j})$ and $\frac{1}{\sqrt{2}}(\mathbf{i} + \mathbf{k})$ (Many other answers are possible.)

41. $\frac{4}{\sqrt{17}}\mathbf{i} + \frac{1}{\sqrt{17}}\mathbf{j}$ **43.** $x + y = -4$ **45.** It stays at $(-1, 4)$.

47. $(3, -2, 4) - (1, -1, 4) = (2, -1, 0) = (-2, 1, 6) - (-4, 2, 6)$; also, $(3, -2, 4) - (-2, 1, 6) = (5, -3, -2) = (1, -1, 4) - (-4, 2, 6)$

49. $\mathbf{a} + \mathbf{b} = (a_1 + b_1)\mathbf{i} + (a_2 + b_2)\mathbf{j} + (a_3 + b_3)\mathbf{k}$
$= (b_1 + a_1)\mathbf{i} + (b_2 + a_2)\mathbf{j} + (b_3 + a_3)\mathbf{k} = \mathbf{b} + \mathbf{a}$

51. $(r + s)\mathbf{a} = [(r + s)a_1]\mathbf{i} + [(r + s)a_2]\mathbf{j} + [(r + s)a_3]\mathbf{k}$
$= (ra_1 + sa_1)\mathbf{i} + (ra_2 + sa_2)\mathbf{j} + (ra_3 + sa_3)\mathbf{k} = r\mathbf{a} + s\mathbf{a}$

Section 14.4

1. $\pi/2 = 90°$ **3.** $3\pi/4 = 135°$ **5.** $7\pi/12 = 105°$ **7.** $11\pi/30 = 66°$ **9.** $\pi = 180°$ **11.** $\cos^{-1}(-\frac{1}{3}) \approx 109.47°$
13. $\cos^{-1}(4/\sqrt{65}) \approx 60.26°$ **15.** $\cos^{-1}(\frac{3}{5}) \approx 53.13°$ **17.** $4\mathbf{i} - 8\mathbf{j} + 8\mathbf{k}$; 12 **19.** $\frac{3}{26}(\mathbf{i} + 3\mathbf{j} + 4\mathbf{k})$, $3/\sqrt{26}$ **21.** $\mathbf{0}$, 0

23. $\mathbf{i} + \mathbf{j}$; $\sqrt{2}$ **25.** $(\mathbf{a} - \mathbf{c}) \cdot \mathbf{b} = \left(\mathbf{a} - \frac{\mathbf{a} \cdot \mathbf{b}}{|\mathbf{b}|^2}\mathbf{b}\right) \cdot \mathbf{b} = \mathbf{a} \cdot \mathbf{b} - \frac{\mathbf{a} \cdot \mathbf{b}}{|\mathbf{b}|^2}(\mathbf{b} \cdot \mathbf{b}) = \mathbf{a} \cdot \mathbf{b} - \frac{\mathbf{a} \cdot \mathbf{b}}{|\mathbf{b}|^2}|\mathbf{b}|^2 = \mathbf{a} \cdot \mathbf{b} - \mathbf{a} \cdot \mathbf{b} = 0$

27. $300/\sqrt{37}$ **29.** $100/\sqrt{3}$ lb

31. a) $(\sin \alpha)\mathbf{i} + (\cos \alpha)\mathbf{j}$ is a unit vector along the right-hand rope. Thus $\mathbf{F}_1 = T_1[(\sin \alpha)\mathbf{i} + (\cos \alpha)\mathbf{j}]$
b) $-T_2(\sin \beta)\mathbf{i} + T_2(\cos \beta)\mathbf{j}$
c) $T_1(\sin \alpha) - T_2(\sin \beta) = 0$, $T_1(\cos \alpha) + T_2(\cos \beta) = 100$
d) $T_1 = \frac{100\sqrt{2}}{\sqrt{3} + 1}$, $T_2 = \frac{200}{\sqrt{3} + 1}$

33. Using the hint, we show $\left|\frac{\mathbf{a} + \mathbf{b}}{2}\right|^2 = \left|\frac{\mathbf{a} - \mathbf{b}}{2}\right|^2$.

$$\left|\frac{\mathbf{a} + \mathbf{b}}{2}\right|^2 = \left(\frac{\mathbf{a} + \mathbf{b}}{2}\right) \cdot \left(\frac{\mathbf{a} + \mathbf{b}}{2}\right) = \frac{1}{4}(\mathbf{a} \cdot \mathbf{a} + 2\mathbf{a} \cdot \mathbf{b} + \mathbf{b} \cdot \mathbf{b});$$

$$\left|\frac{\mathbf{a} - \mathbf{b}}{2}\right|^2 = \left(\frac{\mathbf{a} - \mathbf{b}}{2}\right) \cdot \left(\frac{\mathbf{a} - \mathbf{b}}{2}\right) = \frac{1}{4}(\mathbf{a} \cdot \mathbf{a} - 2\mathbf{a} \cdot \mathbf{b} + \mathbf{b} \cdot \mathbf{b}).$$

Now $\mathbf{a}\cdot\mathbf{b}=0$ since the triangle is a right triangle. Thus $\left|\dfrac{\mathbf{a}+\mathbf{b}}{2}\right|^2=\left|\dfrac{\mathbf{a}-\mathbf{b}}{2}\right|^2=\dfrac{1}{4}(\mathbf{a}\cdot\mathbf{a}+\mathbf{b}\cdot\mathbf{b})$.

35. Let θ_1 be the angle from $\mathbf{a}$ to $\mathbf{v}=|\mathbf{a}|\mathbf{b}+|\mathbf{b}|\mathbf{a}$ and θ_2 the angle from $\mathbf{b}$ to $\mathbf{v}$. Then

$$\cos\theta_1=\frac{\mathbf{a}\cdot(|\mathbf{a}|\,\mathbf{b}+|\mathbf{b}|\mathbf{a})}{|\mathbf{a}|\,||\mathbf{a}|\mathbf{b}+|\mathbf{b}|\mathbf{a}|}=\frac{|\mathbf{a}|(\mathbf{a}\cdot\mathbf{b})+|\mathbf{b}||\mathbf{a}|^2}{|\mathbf{a}|\,||\mathbf{a}|\mathbf{b}+|\mathbf{b}|\mathbf{a}|}=\frac{\mathbf{a}\cdot\mathbf{b}+|\mathbf{a}||\mathbf{b}|}{||\mathbf{a}|\mathbf{b}+|\mathbf{b}|\mathbf{a}|},\ \text{and}$$

$$\cos\theta_2=\frac{\mathbf{b}\cdot(|\mathbf{a}|\mathbf{b}+|\mathbf{b}|\mathbf{a})}{|\mathbf{b}|\,||\mathbf{a}|\mathbf{b}+|\mathbf{b}|\mathbf{a}|}=\frac{|\mathbf{a}||\mathbf{b}|^2+|\mathbf{b}|(\mathbf{a}\cdot\mathbf{b})}{|\mathbf{b}|\,||\mathbf{a}|\mathbf{b}+|\mathbf{b}|\mathbf{a}|}=\frac{|\mathbf{a}||\mathbf{b}|+\mathbf{a}\cdot\mathbf{b}}{||\mathbf{a}|\mathbf{b}+|\mathbf{b}|\mathbf{a}|}=\cos\theta_1.$$

Thus $\theta_1=\theta_2$.

37. Let $\mathbf{a}=a_1\mathbf{i}+a_2\mathbf{j}+a_3\mathbf{k}$ and $\mathbf{b}=b_1\mathbf{i}+b_2\mathbf{j}+b_3\mathbf{k}$. Then $\mathbf{a}\cdot\mathbf{b}=a_1b_1+a_2b_2+a_3b_3=b_1a_1+b_2a_2+b_3a_3=\mathbf{b}\cdot\mathbf{a}$.

39. Let $\mathbf{a}=a_1\mathbf{i}+a_2\mathbf{j}+a_3\mathbf{k}$ and $\mathbf{b}=b_1\mathbf{i}+b_2\mathbf{j}+b_3\mathbf{k}$. Then $(r\mathbf{a})\cdot\mathbf{b}=(ra_1)b_1+(ra_2)b_2+(ra_3)b_3=a_1(rb_1)+a_2(rb_2)+a_3(rb_3)$ $=\mathbf{a}\cdot(r\mathbf{b})=r(a_1b_1+a_2b_2+a_3b_3)=r(\mathbf{a}\cdot\mathbf{b})$.

Section 14.5

1. -15 **3.** 15 **5.** 61 **7.** -12

9. a)
$$\begin{vmatrix}a_1&a_2&a_3\\a_1&a_2&a_3\\c_1&c_2&c_3\end{vmatrix}=a_1\begin{vmatrix}a_2&a_3\\c_2&c_3\end{vmatrix}-a_2\begin{vmatrix}a_1&a_3\\c_1&c_3\end{vmatrix}+a_3\begin{vmatrix}a_1&a_2\\c_1&c_2\end{vmatrix}$$
$$=a_1a_2c_3-a_1a_3c_2-a_1a_2c_3+a_2a_3c_1+a_1a_3c_2-a_2a_3c_1=0$$

b)
$$\begin{vmatrix}a_1&a_2&a_3\\b_1&b_2&b_3\\a_1&a_2&a_3\end{vmatrix}=a_1\begin{vmatrix}b_2&b_3\\a_2&a_3\end{vmatrix}-a_2\begin{vmatrix}b_1&b_3\\a_1&a_3\end{vmatrix}+a_3\begin{vmatrix}b_1&b_2\\a_1&a_2\end{vmatrix}$$
$$=a_1a_3b_2-a_1a_2b_3-a_2a_3b_1+a_1a_2b_3+a_2a_3b_1-a_1a_3b_2=0$$

11. $-6\mathbf{i}+3\mathbf{j}+5\mathbf{k}$ **13.** $\mathbf{0}$ **15.** $22\mathbf{i}+18\mathbf{j}+2\mathbf{k}$ **17.** 11 **19.** $\sqrt{374}$ **21.** $9\sqrt{2}$ **23.** 16 **25.** $\sqrt{166}/2$ **27.** 24 **29.** $\frac{20}{7}$
31. $\mathbf{a}\cdot(\mathbf{b}\times\mathbf{c})=-8$, $\mathbf{a}\times(\mathbf{b}\times\mathbf{c})=2\mathbf{i}+2\mathbf{j}$ **33.** $\mathbf{a}\cdot(\mathbf{b}\times\mathbf{c})=-27$, $\mathbf{a}\times(\mathbf{b}\times\mathbf{c})=24\mathbf{i}+54\mathbf{j}-21\mathbf{k}$ **35.** 0 **37.** 6 **39.** $\frac{175}{6}$
41. $\frac{71}{3}$ **43.** $(\mathbf{a}\times\mathbf{b})\times\mathbf{c}=(\mathbf{a}\cdot\mathbf{c})\mathbf{b}-(\mathbf{b}\cdot\mathbf{c})\mathbf{a}$
45. Computation gives the determinant of a matrix with the first two rows the same, which is thus zero.
47. a) The vector $\mathbf{a}\times(\mathbf{b}\times\mathbf{c})$ is perpendicular to $\mathbf{b}\times\mathbf{c}$, which is in turn a vector perpendicular to the plane containing $\mathbf{b}$ and $\mathbf{c}$. Thus $\mathbf{a}\times(\mathbf{b}\times\mathbf{c})$ lies in this plane. b) The argument is just like that in part (a). c) From parts (a) and (b), equal products $\mathbf{a}\times(\mathbf{b}\times\mathbf{c})$ and $(\mathbf{a}\times\mathbf{b})\times\mathbf{c}$ would have to be parallel to $\mathbf{b}$. A quick sketch shows that $\mathbf{a}\times(\mathbf{b}\times\mathbf{c})$ is not, in general, parallel to $\mathbf{b}$.

Section 14.6

1. $x=3-8t,\ y=-2+4t$ **3.** $x=4+t,\ y=-1-3t,\ z=t$ **5.** $x=1+t,\ y=-1,\ z=1+4t$
7. $x=1-2t,\ y=-4+5t$ **9.** $x=3-2t,\ y=-1+3t,\ z=4-8t$ **11.** $x=3+2t,\ y=-1+t$
13. $x=-1+3t,\ y=5+t$ **15.** $x=-1+3t,\ y=4+2t$ **17.** $x=-1+9t,\ y=4-3t,\ z=-4t$
19. $x=4-5t,\ y=-1+5t,\ z=3+3t$ **21.** $x=2,\ y=1+t,\ z=4$ **23.** $x=-1+3t,\ y=2t,\ z=3+3+9t$ **25.** $(7,3)$
27. $(5,-2,0)$ **29.** $(-1,4)$ **31.** $(0,-2)$ **33.** No intersection **35.** $\cos^{-1}(2/\sqrt{5})$ **37.** $\cos^{-1}(2/\sqrt{210})$ **39.** $\cos^{-1}(1/\sqrt{3})$
41. $\cos\alpha=d_1/\sqrt{d_1^2+d_2^2+d_3^2}$, $\cos\beta=d_2/\sqrt{d_1^2+d_2^2+d_3^2}$, $\cos\gamma=d_3/\sqrt{d_1^2+d_2^2+d_3^2}$ **43.** $(\frac{95}{17},\frac{91}{17})$
45. $\left(\dfrac{43}{11},\dfrac{58}{11},\dfrac{-1}{11}\right)$ **47.** $(\frac{5}{2},3,1)$ **49.** $(\frac{7}{3},3,6)$
51. By the work in the text, the line segment is given by $x_1=a_1+(b_1-a_1)t$, $x_2=a_2+(b_2-a_2)t$, $x_3=a_3+(b_3-a_3)t$ for $0\le t\le 1$. These equations can be rewritten $x_1=(1-t)a_1+tb_1$, $x_2=(1-t)a_2+tb_2$, $x_3=(1-t)a_3+tb_3$ for $0\le t\le 1$.
53. $(\frac{2}{2},\frac{5}{3},\frac{4}{3})$ **55.** $\left(-1-\dfrac{12}{\sqrt{10}},\,5+\dfrac{4}{\sqrt{10}},\,6\right)$

Section 14.7

1. $x-2y+z=-7$ **3.** $x=2$ **5.** $4x-2y+z=22$ **7.** $4x-3y+4z=-3$ **9.** $7x-y-2z=-5$
11. $2x-3y+z=-2$ **13.** $x-2y+7z=10$ **15.** $x+y+z=1$ **17.** $5x+7y-9z=-28$ **19.** $5x+3y-3z=-2$

21. $7x - 23y - z = -101$ **23.** $8x + 4y + z = 6$ **25.** $35x + 14y + 5z = 141$ **27.** $5x - 12y + 22z = 13$ **29.** $x + y + z = 6$
31. The lines don't intersect. **33.** $\pi/2$ **35.** $\pi/2 - \cos^{-1}(\frac{16}{21}) \approx 49.63°$ **37.** $(3, 6, -10)$ **39.** Empty intersection
41. $x = -2 + t, y = 1 - 2t, z = 4t$ **43.** $x = 1 + 4t, y = -1 - 3t, z = 3 + t$ **45.** $x = 3 + 4t, y = \frac{11}{2} + 7t, z = 2 - 2t$ **47.** 4
49. $7/\sqrt{17}$ **51.** 2 **53.** $1728\pi/\sqrt{5}$ **55.** $-7x + y + 11 = 0$

Review Exercise Set 14.1

1. a) $2\sqrt{22}$ b) $(x + 1)^2 + (y - 1)^2 + (z - 6)^2 = 21$ **3.**

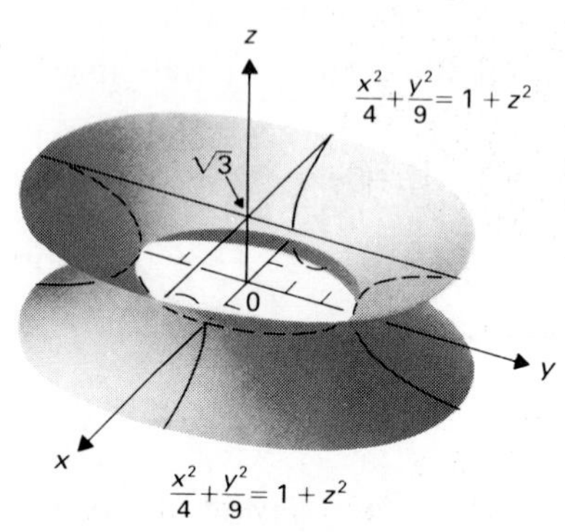

Hyperboloid of one sheet

5. $\cos^{-1}(-\frac{5}{14})$ **7.** $-\mathbf{i} - 13\mathbf{j} - 9\mathbf{k}$ **9.** $-4\mathbf{i} - 2\mathbf{j} + 6\mathbf{k}$ **11.** $x = t, y = -2t, z = 3t$ **13.** $\frac{11}{3}$ **15.** $(\frac{37}{16}, \frac{59}{16}, \frac{31}{16})$

Review Exercise Set 14.2

1. a) $1 \pm 5\sqrt{5}$ b) Center $(1, 0, -2)$, radius 3 **3.** a)

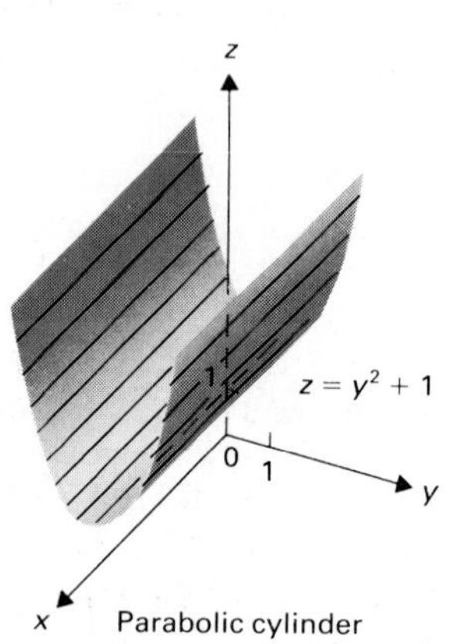

Parabolic cylinder

b)

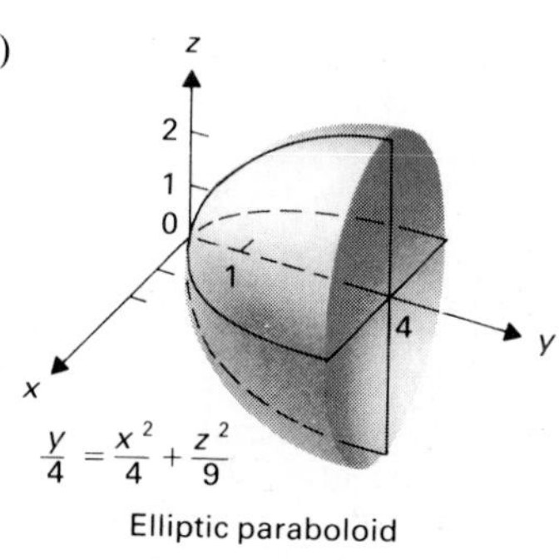

Elliptic paraboloid

5. $\cos^{-1}(7/(3\sqrt{11}))$ **7.** a) 0 b) **b** c) **0** **9.** 52 **11.** $x = -1 + 4t, y = 5 - 6t, z = 2 + 2t$ **13.** $19/\sqrt{10}$
15. $14x + 13y + 10z = 45$

Review Exercise Set 14.3

1. $(x + 2)^2 + (y - 1)^2 + (z - 4)^2 = 14$ **3.**

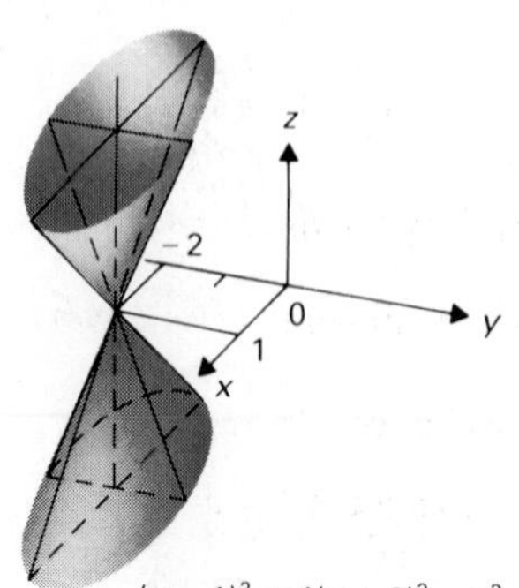

Elliptic cone

5. $\sqrt{6}$ **7.** $7\mathbf{i} - 34\mathbf{j} + 14\mathbf{k}$ **9.** $\sqrt{109}$
11. $(0, -11, -5)$
13. $y + z = 2$ **15.** $6, -18$

Review Exercise Set 14.4

1. $-1 \pm \sqrt{3}$ **3.**

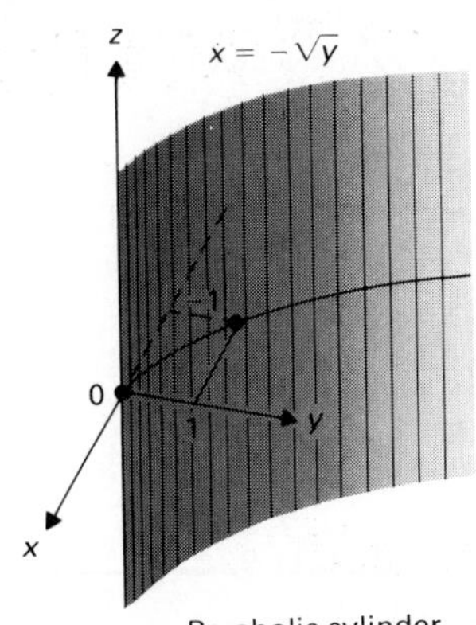

Parabolic cylinder

5. $\pm 2\sqrt{5}$ **7.** a) -1 b) $\mathbf{0}$ c) $\mathbf{b}$ d) 2 **9.** $\frac{67}{8}, \frac{27}{8}$
11. $x = -1 + 2t$, $y = 3 + t$ **13.** $2x - y - z = 0$ **15.** $1664\pi/\sqrt{17}$

More Challenging Exercises 14

1. $47/(4\sqrt{14})$ **3.** $2x - 3y - z = 3$ **5.** 6
7. Theorem 14.2 applied to $(\mathbf{a} - x\mathbf{b}) \cdot (\mathbf{a} - x\mathbf{b}) \geq 0$ yields $(\mathbf{b} \cdot \mathbf{b})x^2 - 2(\mathbf{a} \cdot \mathbf{b})x + \mathbf{a} \cdot \mathbf{a} \geq 0$ for all x. Thus the equation $(\mathbf{b} \cdot \mathbf{b})x^2 - 2(\mathbf{a} \cdot \mathbf{b})x + (\mathbf{a} \cdot \mathbf{a}) = 0$ does *not* have two distinct real roots, so by the quadratic formula,

$$(-2(\mathbf{a} \cdot \mathbf{b}))^2 - 4(\mathbf{b} \cdot \mathbf{b})(\mathbf{a} \cdot \mathbf{a}) \leq 0,$$
$$4(\mathbf{a} \cdot \mathbf{b})^2 \leq 4(\mathbf{a} \cdot \mathbf{a})(\mathbf{b} \cdot \mathbf{b}),$$
$$(\mathbf{a} \cdot \mathbf{b})^2 \leq (\mathbf{a} \cdot \mathbf{a})(\mathbf{b} \cdot \mathbf{b}).$$

Since $\|\mathbf{a}\|^2 = \mathbf{a} \cdot \mathbf{a}$ and $\|\mathbf{b}\|^2 = \mathbf{b} \cdot \mathbf{b}$, this yields $|\mathbf{a} \cdot \mathbf{b}| \leq \|\mathbf{a}\| \, \|\mathbf{b}\|$.

CHAPTER 15

Section 15.1

1. By the chain rule and Eq. (8), we have

$$\frac{d\mathbf{r}}{ds} = \frac{dt}{ds}\frac{d\mathbf{r}}{dt} = \frac{dt}{ds}\mathbf{v} = \frac{dt}{ds}\left(\frac{ds}{dt}\mathbf{t}\right) = \mathbf{t}.$$

3. a) $2\mathbf{i} + 3\mathbf{j}$ b) $\sqrt{13}$ c) $0\mathbf{i} + 0\mathbf{j} = \mathbf{0}$ **5.** a) $-\mathbf{i}$ b) 1 c) $-4\mathbf{j}$ **7.** a) $\mathbf{i}$ b) 1 c) $-\mathbf{i} + \mathbf{j}$
9. a) $-\mathbf{i} - 2\mathbf{j}$ b) $\sqrt{5}$ c) $2\mathbf{i} + 6\mathbf{j}$ **11.** a) $4\mathbf{i} + 12\mathbf{j} - \mathbf{k}$ b) $\sqrt{161}$ c) $2\mathbf{i} + 12\mathbf{j}$
13. a) $-\mathbf{i} + \mathbf{j} + 2\mathbf{k}$ b) $\sqrt{6}$ c) $2\mathbf{i} - 2\mathbf{j} + 2\mathbf{k}$ **15.** $\mathbf{a}_{\text{tan}} = \frac{70}{17}\mathbf{i} - \frac{42}{17}\mathbf{j}$, $\mathbf{a}_{\text{nor}} = -\frac{36}{17}\mathbf{i} - \frac{60}{17}\mathbf{j}$; speed is decreasing.
17. $\mathbf{a}_{\text{tan}} = \frac{3}{2}\mathbf{i} - \frac{3}{2}\mathbf{j}$, $\mathbf{a}_{\text{nor}} = \frac{1}{2}\mathbf{i} + \frac{1}{2}\mathbf{j}$; speed is decreasing.
19. $\mathbf{a}_{\text{tan}} = \dfrac{-4\pi}{1 + 4\pi^2}\mathbf{i} - \dfrac{8\pi^2}{1 + 4\pi^2}\mathbf{k}$, $\mathbf{a}_{\text{nor}} = \dfrac{4\pi}{1 + 4\pi^2}\mathbf{i} - 4\mathbf{j} - \dfrac{2}{1 + 4\pi^2}\mathbf{k}$; speed is increasing. **21.** $x = 1 + t$, $y = t$, $z = 0$
23. $x = \frac{1}{2} - \frac{1}{2}t$, $y = 8 - 8t$, $z = -1 + 3t$ **25.** $x + y = 1$ **27.** $2x + 32y - 6z = 135$ **29.** $2\pi\sqrt{4a^2 + 1}$ **31.** 140 **33.** 8.23730
35. $\mathbf{a} \cdot \mathbf{b} = 2t^3 + 3t^4 + t^3 = 3t^3 + 3t^4$; $\dfrac{d(\mathbf{a} \cdot \mathbf{b})}{dt} = 9t^2 + 12t^3$. Also

$$\frac{d(\mathbf{a} \cdot \mathbf{b})}{dt} = \mathbf{a} \cdot \frac{d\mathbf{b}}{dt} + \frac{d\mathbf{a}}{dt} \cdot \mathbf{b} = [t^2\mathbf{i} - (3t + 1)\mathbf{j}] \cdot (2\mathbf{i} - 3t^2\mathbf{j})$$
$$+ (2t\mathbf{i} - 3\mathbf{j}) \cdot (2t\mathbf{i} - t^3\mathbf{j}) = (2t^2 + 9t^3 + 3t^2) + (4t^2 + 3t^3) = 9t^2 + 12t^3.$$

37. By Example 8, $x = (v_0 \cos \theta)t$, $y = (v_0 \sin \theta)t - 16t^2$. Eliminating t, we have $y = (\tan \theta)x - (16/v_0^2)(\sec^2 \theta)x^2$, which is the equation of a parabola.
39. Substituting $t = (v_0 \sin \theta)/16$ from Exercise 38 into $x = (v_0 \cos \theta)t$, we have

$$x = \frac{v_0^2 \sin \theta \cos \theta}{16} = \frac{v_0^2 \sin 2\theta}{32}$$

as range.

41. 128 ft/sec **43.** $v_0 = 80\sqrt{10}$ ft/sec; $\theta = \tan^{-1}\left(\frac{1}{2}\right) \approx 26.57°$

45. $\mathbf{r} = a(\cos 2\pi\beta t)\mathbf{i} + a(\sin 2\pi\beta t)\mathbf{j}$; $\mathbf{F} = m\mathbf{a} = m\dfrac{d^2\mathbf{r}}{dt^2} = 4m\pi^2\beta^2(-\mathbf{r})$ is directed toward the center of the circle; $|\mathbf{F}| = 4m\pi^2 a\beta^2$.

Section 15.2

1. $\mathbf{i}$ **3.** 0 **5.** $9b/(4a^2)$ **7.** $-\frac{3}{5}\mathbf{i} + \frac{4}{5}\mathbf{j}$ **9.** $\frac{2}{5}$ **11.** $(2/\sqrt{5})\mathbf{i} + (1/\sqrt{5})\mathbf{j}$ **13.** $7/\sqrt{5}$ **15.** $6/(5\sqrt{5})$ **17.** 3 **19.** $2x - z = 0$ **21.** $2\sqrt{5}/3$ **23.** $(1/\sqrt{6})\mathbf{i} + (1/\sqrt{6})\mathbf{j} + (2/\sqrt{6})\mathbf{k}$ **25.** $(1/\sqrt{30})\mathbf{i} - (5/\sqrt{30})\mathbf{j} + (2/\sqrt{30})\mathbf{k}$ **27.** $5\sqrt{6}/3$ **29.** $\sqrt{30}/18$ **31.** $\sqrt{2}$

33. $x + y - z = -1 - \ln 2$ **35.** $2\sqrt{6}$ **37.** If $\kappa = |d\mathbf{t}/ds| = 0$, then $\mathbf{n} = \dfrac{d\mathbf{t}/ds}{|d\mathbf{t}/ds|}$ is not defined. **39.** a) $-\sqrt{26}/13$ b) $5/(13\sqrt{26})$

Section 15.3

1. $\mathbf{u}_r = -\mathbf{i}$, $\mathbf{u}_\theta = -\mathbf{j}$ **3.** $\sqrt{2}$ **5.** $9/(2\sqrt{2})$

7. a) $\mathbf{v} = -\sqrt{3}a\mathbf{u}_r + a\mathbf{u}_\theta$, $\mathbf{a} = -2a\mathbf{u}_r - 2\sqrt{3}a\mathbf{u}_\theta$ b) $\mathbf{v} = -\sqrt{3}a\mathbf{i} - a\mathbf{j}$, $\mathbf{a} = 2a\mathbf{i} - 2\sqrt{3}a\mathbf{j}$ **9.** 0

11. $\mathbf{u}_r = -\dfrac{\sqrt{3}}{2}\mathbf{i} + \dfrac{1}{2}\mathbf{j}$, $\mathbf{u}_\theta = -\dfrac{1}{2}\mathbf{i} - \dfrac{\sqrt{3}}{2}\mathbf{j}$ **13.** $\sqrt{3}a$ **15.** $\dfrac{\sqrt{3}}{2a}$

17. From Eq. (32), we have $T = 2\pi a^{3/2}/\sqrt{GM} = (\pi/\sqrt{2GM})(2a)^{3/2}$, so T can be computed if the length $2a$ of the major axis is known.

19. No. By Eq. (32), $a^3/T^2 = GM/(4\pi^2)$, and the mass "M" of the sun is different from the mass "M" of Jupiter.

21. As in Eq. (11) of the text, we have $dA/dt = \frac{1}{2}r^2\dot{\theta}$, so by Exercise 20, $\frac{1}{2}r^2\dot{\theta} = \beta$ and $r^2\dot{\theta} = 2\beta$.

23. Since $\mathbf{a} = (\ddot{r} - r\dot{\theta}^2)\mathbf{u}_r + (r\ddot{\theta} + 2\dot{r}\dot{\theta})\mathbf{u}_\theta$, we see from Exercise 22 that we must have $\mathbf{a} = (\ddot{r} - r\dot{\theta}^2)\mathbf{u}_r$, which is directed along the ray joining the sun and the planet.

25. From Exercise 24 we obtain, upon multiplication by r, $\dot{r}[r(1 - e\cos\theta)] + r^2\dot{\theta}e\sin\theta = 0$, or $\dot{r}B + 2\beta e\sin\theta = 0$.

27. By Exercise 21, we have

$$r\dot{\theta}^2 = \frac{(r^2\dot{\theta})^2}{r^3} = \frac{(2\beta)^2}{r^3} = \frac{4\beta^2}{r^3}.$$

From Exercise 26, we then have

$$\ddot{r} - r\dot{\theta}^2 = \frac{4\beta^2}{r^3} - \frac{4\beta^2}{r^2B} - \frac{4\beta^2}{r^3} = -\frac{4\beta^2}{r^2B} = -\frac{4\beta^2}{B}\cdot\frac{1}{r^2}.$$

29. Referring to Exercise 24 and Fig. 15.21, we have

$$2a = r|_{\theta=0} + r|_{\theta=\pi} = \frac{B}{1-e} + \frac{B}{1+e} = \frac{2B}{1-e^2}.$$

Thus $B = a(1 - e^2)$. Also,

$$c = a - r|_{\theta=\pi} = \frac{B}{1-e^2} - \frac{B}{1+e} = \frac{B}{1-e^2} - \frac{B(1-e)}{1-e^2} = \frac{Be}{1-e^2}.$$

Then

$$b^2 = a^2 - c^2 = \frac{B^2}{(1-e^2)^2} - \frac{B^2e^2}{(1-e^2)^2} = \frac{B^2}{1-e^2} = \frac{a^2(1-e^2)^2}{1-e^2} = a^2(1-e^2).$$

Thus $b = a\sqrt{1 - e^2}$.

31. Since the area of an ellipse of major axis $2a$ and minor axis $2b$ is πab, we have $\pi ab = \beta T$, so $\beta = \pi ab/T$.

33. Since by Kepler's third law we have $a^3/T^2 = k$, a constant, we have, by Exercise 32, $|\mathbf{F}| = \dfrac{4\pi^2k}{r^2}m$, so the force per unit mass is $4\pi^2k/r^2$.

Section 15.4

1. 0 **3.** $\sqrt{2}/[2(1 + 2t^2)^{3/2}]$ **5.** 0 **7.** $\frac{1}{3}(2\mathbf{i} - 2\mathbf{j} + \mathbf{k})$ **9.** $\frac{2}{27}$ **11.** $(1/\sqrt{5})(2\mathbf{i} - \mathbf{k})$ **13.** $2\sqrt{5}/27$ **15.** $(1/\sqrt{5})(2\mathbf{i} - \mathbf{k})$ **17.** $\sqrt{5}(3\sqrt{6})$ **19.** $(1/\sqrt{3})(\mathbf{i} + \mathbf{j} - \mathbf{k})$ **21.** $\sqrt{6}$ **23.** $(1/\sqrt{13})(-3\mathbf{j} + 2\mathbf{k})$ **25.** $\frac{1}{13}$

27. From $\dot{\mathbf{r}} = \dot{s}\mathbf{t}$, we obtain $\ddot{\mathbf{r}} = \ddot{s}\mathbf{t} + \dot{s}\dot{\mathbf{t}} = \ddot{s}\mathbf{t} + \dot{s}\left(\dfrac{ds}{dt}\dfrac{d\mathbf{t}}{ds}\right) = \ddot{s}\mathbf{t} - \dot{s}^2\kappa\mathbf{n}$.

29. a) Since **t**, **n**, **b** is an orthogonal unit right-hand triple, we have from Exercise 27

$$\dot{\mathbf{r}} \times \ddot{\mathbf{r}} = \begin{vmatrix} \mathbf{t} & \mathbf{n} & \mathbf{b} \\ \dot{s} & 0 & 0 \\ \ddot{s} & \dot{s}^2\kappa & 0 \end{vmatrix} = 0\mathbf{t} + 0\mathbf{n} + \dot{s}^3\kappa\mathbf{b} = \dot{s}^3\kappa\mathbf{b}.$$

b) From the remarks and result in part (a) and the result in Exercise 28, we have $(\dot{\mathbf{r}} \times \ddot{\mathbf{r}}) \cdot \dddot{\mathbf{r}} = 0(\dddot{s} - \dot{s}^3\kappa^2) + 0(3\dot{s}\ddot{s} + \dot{s}^2\dot{\kappa}) + (\dot{s}^3\kappa)(\dot{s}^3\kappa\tau) = \dot{s}^6\kappa^2\tau$.

31. $\boldsymbol{\delta} \times \mathbf{t} = (\tau\mathbf{t} + \kappa\mathbf{b}) \times \mathbf{t} = \tau(\mathbf{t} \times \mathbf{t}) + \kappa(\mathbf{b} \times \mathbf{t}) = \tau\mathbf{0} + \kappa\mathbf{n} = \kappa\mathbf{n}$

$\boldsymbol{\delta} \times \mathbf{n} = (\tau\mathbf{t} + \kappa\mathbf{b}) \times \mathbf{n} = \tau(\mathbf{t} \times \mathbf{n}) + \kappa(\mathbf{b} \times \mathbf{n}) = \tau\mathbf{b} + \kappa(-\mathbf{t}) = -\kappa\mathbf{t} + \tau\mathbf{b}$

$\boldsymbol{\delta} \times \mathbf{b} = (\tau\mathbf{t} + \kappa\mathbf{b}) \times \mathbf{b} = \tau(\mathbf{t} \times \mathbf{b}) + \kappa(\mathbf{b} \times \mathbf{b}) = \tau(-\mathbf{n}) + \kappa\mathbf{0} = -\tau\mathbf{n}$

33. The Frenet formula $d\mathbf{t}/ds = \kappa\mathbf{n}$ yields $d\mathbf{t}/ds = \mathbf{0}$ if $\kappa = 0$. But then **t** is constant, so $\mathbf{t} = c_1\mathbf{i} + c_2\mathbf{j} + c_3\mathbf{k}$. Thus $d\mathbf{r}/ds = c_1\mathbf{i} + c_2\mathbf{j} + c_3\mathbf{k}$, so in terms of the parameter s, we obtain $\mathbf{r} = (c_1 s + d_1)\mathbf{i} + (c_2 s + d_2)\mathbf{j} + (c_3 s + d_3)\mathbf{k}$, and the curve has parametric equations $x = c_1 s + d_1$, $y = c_2 s + d_2$, $z = c_3 s + d_3$ and is a line.

35. Following the hint, we have

$$\frac{dw}{ds} = \mathbf{t} \cdot \frac{d\bar{\mathbf{t}}}{ds} + \frac{d\mathbf{t}}{ds} \cdot \bar{\mathbf{t}} + \mathbf{n} \cdot \frac{d\bar{\mathbf{n}}}{ds} + \frac{d\mathbf{n}}{ds} \cdot \bar{\mathbf{n}} + \mathbf{b} \cdot \frac{d\bar{\mathbf{b}}}{ds} + \frac{d\mathbf{b}}{ds} \cdot \bar{\mathbf{b}}$$

$$= \mathbf{t} \cdot (\kappa\bar{\mathbf{n}}) + \kappa\mathbf{n} \cdot \bar{\mathbf{t}} + \mathbf{n} \cdot (-\kappa\bar{\mathbf{t}} + \tau\bar{\mathbf{b}}) + (-\kappa\mathbf{t} + \tau\mathbf{b}) \cdot \bar{\mathbf{n}} + \mathbf{b} \cdot (-\tau\bar{\mathbf{n}}) + (-\tau\mathbf{n}) \cdot \bar{\mathbf{b}}$$

$$= 0.$$

Thus w is a constant, and since **t**, **n**, and **b** are equal to their respective barred counterparts at zero, we have $w(0) = 1 + 1 + 1 = 3$. Since $\mathbf{t}, \bar{\mathbf{t}}, \mathbf{n}, \bar{\mathbf{n}}, \mathbf{b}, \bar{\mathbf{b}}$ are unit vectors, we have $\mathbf{t} \cdot \bar{\mathbf{t}} \le 1$ and $\mathbf{t} \cdot \bar{\mathbf{t}} = 1$ if and only if $\mathbf{t} = \bar{\mathbf{t}}$, with similar results holding for **n** and $\bar{\mathbf{n}}$ and for **b** and $\bar{\mathbf{b}}$. Thus $w = 3$ implies $\mathbf{t} = \bar{\mathbf{t}}$, $\mathbf{n} = \bar{\mathbf{n}}$, and $\mathbf{b} = \bar{\mathbf{b}}$ for all values of the parameter s. From $\mathbf{t} = \bar{\mathbf{t}}$, we have $d\mathbf{r}/ds = d\bar{\mathbf{r}}/ds$, so $\mathbf{r}(s) = \bar{\mathbf{r}}(s) + \mathbf{c}$, and from $\mathbf{r}(0) = \bar{\mathbf{r}}(0)$, we have $\mathbf{c} = \mathbf{0}$. Thus $\mathbf{r}(s) = \bar{\mathbf{r}}(s)$, so our curves are the same in terms of the arc length parameter.

Review Exercise Set 15.1

1. $\mathbf{i} + (1/\sqrt{2})\mathbf{j}$ **3.** $1/\sqrt{2}$ **5.** $-(1/\sqrt{2})\mathbf{j}$ **7.** $-4\mathbf{i}$ **9.** $\mathbf{v} = \frac{1}{4}\mathbf{u}_r + \frac{3}{4}\mathbf{u}_\theta$, $\mathbf{a} = -\frac{10}{32}\mathbf{u}_r + \frac{15}{128}\mathbf{u}_\theta$

Review Exercise Set 15.2

1. $\mathbf{i} + (3/\sqrt{2})\mathbf{j} + (3/\sqrt{2})\mathbf{k}$ **3.** $3\sqrt{2}x - y - z = \dfrac{3\sqrt{2}}{4}\pi$ **5.** $-(1/\sqrt{2})\mathbf{j} + (1/\sqrt{2})\mathbf{k}$ **7.** $\left(\dfrac{1}{\sqrt{10}}\right)\left(3\mathbf{i} - \dfrac{1}{\sqrt{2}}\mathbf{j} - \dfrac{1}{\sqrt{2}}\mathbf{k}\right)$

9. A (possibly degenerate) ellipse, parabola, or hyperbola with focus at the center of the force field; that is, a second-degree plane curve.

Review Exercise Set 15.3

1. $2\sqrt{2}$ **3.** $(1/\sqrt{2})(-\mathbf{i} - \mathbf{j})$ **5.** $8\sqrt{2}$

7. $\mathbf{a} = \ddot{s}\mathbf{t} + \kappa\dot{s}^2\mathbf{n} = \kappa\dot{s}^2\mathbf{t} + \kappa\dot{s}^2\mathbf{n} = \kappa\dot{s}^2(\mathbf{t} + \mathbf{n})$. Then **a**, and hence $\mathbf{F} = m\mathbf{a}$, has the direction of $\mathbf{t} + \mathbf{n}$, which makes an angle of 45° with **t** since **t** and **n** are orthogonal unit vectors.

9. $\mathbf{a} = (\ddot{r} - r\dot{\theta}^2)\mathbf{u}_r + (r\ddot{\theta} + 2\dot{r}\dot{\theta})\mathbf{u}_\theta$

Review Exercise Set 15.4

1. $\sqrt{17}$ **3.** $\dfrac{-22}{17}(3\mathbf{i} + 2\mathbf{j} - 2\mathbf{k})$ **5.** $2x + 3y + 6z = -2$ **7.** $14/(17\sqrt{17})$ **9.** 6 hr

CHAPTER 16

Section 16.1

1. $f_x = 3, f_y = 4$ **3.** $f_x = 2x, f_y = 2y$ **5.** $f_x = \dfrac{e^{x/y}}{y}, f_y = \dfrac{-xe^{x/y}}{y^2}$ **7.** $f_x = y^2 + \dfrac{6x}{y^3}, f_y = 2xy - \dfrac{9x^2}{y^4}$

9. $f_x = (x^2 + 2xy)(2x) + (y^3 + x^2)(2x + 2y)$, $f_y = (x^2 + 2xy)(3y^2) + (y^3 + x^2)(2x)$

11. $f_x = \dfrac{2xy - 2xy^2}{(x^2 + y)^2}, f_y = \dfrac{2y(x^2 + y) - (x^2 + y^2)}{(x^2 + y)^2}$ **13.** $f_x = 2x \sec^2(x^2 + y^2), f_y = 2y \sec^2(x^2 + y^2)$

15. $f_x = 2xye^{xy^2}\sec(x^2y)\tan(x^2y) + y^2e^{xy^2}\sec(x^2y)$, $f_y = x^2e^{xy^2}\sec(x^2y)\tan(x^2y) + 2xye^{xy^2}\sec(x^2y)$
17. $f_x = \dfrac{2\cot y^2}{2x+y}$, $f_y = -2y\ln(2x+y)\csc^2 y^2 + \dfrac{\cot y^2}{2x+y}$ **19.** $f_x = 3y\sec^3 x\tan x + y^2$, $f_y = \sec^3 x + 2xy$
21. $f_x = \dfrac{y^2}{1+x^2y^4}$, $f_y = \dfrac{2xy}{1+x^2y^4}$ **23.** 1 **25.** 18 **27.** -48 **29.** $-\frac{1}{2}$ **31.** 0 **33.** $-1/z^2$ **35.** 0 **37.** 0 **39.** $f_{xy} = f_{yx} = 2x$
41. $f_{xy} = f_{yx} = 0$ **43.** $f_x = y\cos xy$, $f_{xx} = -y^2\sin xy$, $f_y = x\cos xy$, $f_{yy} = -x^2\sin xy$, $x^2f_{xx} = -x^2y^2\sin xy = y^2f_{yy}$
45. 3π ft^3/(unit increase in altitude) **47.** $2k/27$ **49.** East

Section 16.2

1. $3x + 16y - z = 21$ **3.** $3x - y - 4z = 0$ **5.** $8x_1 - 28x_2 + x_3 = -32$ **7.** $x = 1 + 2t$, $y = -1 + t$, $z = t$
9. $x = 2 + 9t$, $y = -3 + 4t$, $z = 6 - t$ **11.** $x = -3 + 3t$, $y = 2 - 8t$, $z = 5 + 5t$ **13.** $(-\frac{3}{2}, 1, \frac{13}{4})$ **15.** $(-\frac{5}{3}, \frac{2}{3}, -\frac{10}{9})$
17. $(2, \frac{1}{2}, 2)$, $(-2, -\frac{1}{2}, -2)$ **19.** 1.96 **21.** $1 - (\pi/90)$ **23.** 103 in^3 **25.** 85/12 ft/sec^2 **27.** 13/32 ft^3 **29.** 2.67% **31.** $3a$%
33. $\left(\dfrac{108}{17}, \dfrac{-28}{17}\right)$ **35.** $(\frac{4}{3}, \frac{5}{6})$ **37.** $(-\frac{1}{4}, \frac{3}{4})$
39. Multiply the first equation in (10) in the text by $g_y(a_1, b_1)$ and the second by $-f_y(a_1, b_1)$ and add. We obtain

$$(f_xg_y - f_yg_x)|_{(a_1,b_1)} \cdot \Delta x = (-f \cdot g_y + g \cdot f_y)|_{(a_1,b_1)}$$

so

$$a_2 = a_1 + \Delta x = a_1 - \left.\left(\frac{f \cdot g_y - g \cdot f_y}{f_xg_y - f_yg_x}\right)\right|_{(a_1,b_1)}.$$

The second formula follows similarly.
41. (10.04950857, -0.0495085669) **43.** (-0.5321332411, -1.62529471) **45.** ($-0.$ 0781627166, 1.817072385)

Section 16.3

1. 0 **3.** $z = 0$
5. For $\Delta y = \Delta x$, Eq. (17) at $(x_0, y_0) = (0, 0)$ becomes

$$\sqrt{2|\Delta x||\Delta x|} = \varepsilon_1\,\Delta x + \varepsilon_2\,\Delta x \quad \text{or} \quad \sqrt{2}|\Delta x| = \Delta x(\varepsilon_1 + \varepsilon_2).$$

Thus $|\varepsilon_1 + \varepsilon_2| = \sqrt{2}$, so we do not have $\varepsilon_1 \to 0$ and $\varepsilon_2 \to 0$ as $\Delta x \to 0$ and $\Delta y \to 0$. Therefore f is not differentiable at (0, 0).
7. $\varepsilon_1 = \Delta x$, $\varepsilon_2 = \Delta y$, which both approach zero as $\Delta x \to 0$ and $\Delta y \to 0$.
9. $\varepsilon_1 = \dfrac{\Delta x - \Delta y + (\Delta x)(\Delta y)}{(-1 + \Delta x)(1 + \Delta y)}$, $\varepsilon_2 = -\dfrac{\Delta y}{1 + \Delta y}$, which both approach zero as $\Delta x \to 0$ and $\Delta y \to 0$. **11.** $132\,dx + 56\,dy$
13. $dx - 3\,dy$ **15.** $(\frac{1}{2} + \pi)\,dx + (\frac{1}{4} + \pi e^{4\pi})\,dy + (4e^{4\pi} + 2)\,dz$ **17.** 2.04 **19.** -0.05 **21.** $-6\mathbf{j}$ **23.** $\frac{1}{3}\mathbf{i} - \frac{1}{3}\mathbf{j} - \frac{2}{9}\mathbf{k}$ **25.** $-\frac{5}{2}\mathbf{i}$
27. Points on a circle with center $(-1, 0)$ and radius $\frac{1}{2}$ **29.** Points on the line $3x - 4y = -3$
31. Points on the ellipsoid $\dfrac{x^2}{4} + \dfrac{y^2}{4} + z^2 = 1$ **33.** Points in the plane $3x - y - 4z = 0$

Section 16.4

1. a) $x = 1$, $y = 2$, $z = \frac{5}{4}$ b) $\frac{15}{4}$ c) $z = t^4 + (t+1)^{-2}$ d) $\frac{15}{4}$
3. a) $x = -\sqrt{2}$, $y = \sqrt{2}$, $z = -2\sqrt{2}$ b) $\partial z/\partial r = -3\sqrt{2}$, $\partial z/\partial\theta = 2\sqrt{2}$ c) $z = r^3\sin^2\theta\cos\theta$ d) $\partial z/\partial r = -3\sqrt{2}$, $\partial z/\partial\theta = 2\sqrt{2}$
5. 7 **7.** -4 **9.** -31 **11.** 20 **13.** -1 **15.** -6 **17.** $f_x = \frac{7}{10}$, $f_y = \frac{1}{5}$ **19.** 4800π in^3/min **21.** $-300m$ units/unit time
23. $\frac{1}{5}$ lb/ft^2/min **25.** $-17.008G$ **27.** $bc(\partial w/\partial x) = ac(\partial w/\partial y) = ab(\partial w/\partial z)$
29. We have

$$\frac{\partial w}{\partial x} = f'(u) \cdot y^2 \quad \text{and} \quad \frac{\partial w}{\partial y} = f'(u) \cdot 2xy,$$

from which the desired result follows at once.
31. We have

$$\frac{\partial w}{\partial x} = f'(u)\left(\frac{1}{y}\right) \quad \text{and} \quad \frac{\partial w}{\partial y} = f'(u)\left(\frac{-x}{y^2}\right),$$

from which we at once obtain the desired result.

33. $\dfrac{\partial^2 z}{\partial s^2} = \dfrac{\partial z}{\partial x}\dfrac{\partial^2 x}{\partial s^2} + \dfrac{\partial z}{\partial y}\dfrac{\partial^2 y}{\partial s^2} + \dfrac{\partial^2 z}{\partial x^2}\left(\dfrac{\partial x}{\partial s}\right)^2 + 2\dfrac{\partial^2 z}{\partial x\,\partial y}\dfrac{\partial x}{\partial s}\dfrac{\partial y}{\partial s} + \dfrac{\partial^2 z}{\partial y^2}\left(\dfrac{\partial y}{\partial s}\right)^2$

35. Following the hint, we obtain $f_x(tx_0, ty_0)\cdot x_0 + f_y(tx_0, ty_0)\cdot y_0 = 2t\cdot f(x_0, y_0)$, so, letting $t = 1$, we obtain

$$x_0 f_x(x_0, y_0) + y_0 f_y(x_0, y_0) = 2\cdot f(x_0, y_0).$$

Since x_0 and y_0 could be arbitrary, we are done.

37. Let $w = g(x, y) = f(x/y)$. Then $g(tx, ty) = t^0 g(x, y)$, so by Exercise 36, $x\cdot\dfrac{\partial w}{\partial x} + y\cdot\dfrac{\partial w}{\partial y} = 0\cdot g(x, y) = 0$.

Section 16.5

1. $\frac{36}{5}$ **3.** $\frac{5}{3}$ **5.** $-32/\sqrt{5}$ **7.** -9 **9.** $-5/(13\sqrt{10})$ **11.** $9/\sqrt{10}$ **13.** $\frac{28}{3}$ **15.** $1/\sqrt{6}$ **17.** $\frac{7}{6}$ **19.** $-\sqrt{2}$ **21.** $2.4/\sqrt{58}$
23. a) $2\mathbf{i} + \mathbf{j}$ b) $2\sqrt{5}$ c) $-2\mathbf{i} - \mathbf{j}$ d) $-2\sqrt{5}$ **25.** a) $-7\mathbf{i} + 2\mathbf{j} + \mathbf{k}$ b) $\sqrt{54}/2$ c) $7\mathbf{i} - 2\mathbf{j} - \mathbf{k}$ d) $-\sqrt{54}/2$
27. Directional derivatives in all directions will be zero.
29. a) Along the surface, $f(x, y, z)$ has constant value c, so it has zero rate of change in a direction tangent to the surface. b) For a point $\mathbf{a}$ on the surface, $\nabla f(\mathbf{a})\cdot\mathbf{u} = 0$ for any vector $\mathbf{u}$ tangent to the surface at $\mathbf{a}$, so $\nabla f(\mathbf{a})$ is normal to the tangent plane to the surface at $\mathbf{a}$.
31. $\mathbf{i}$ **33.** $\mathbf{i}$ and $-\mathbf{i}$ **35.** $\mathbf{j}$ and $-\mathbf{j}$

Section 16.6

1. $2/(\pi - 2)$ **3.** $-\frac{4}{7}$ **5.** 2 **7.** $(4u^3v + 3)/(2uw^3 - 6u^2v^2)$ **9.** 1 **11.** 0 **13.** $-\frac{11}{2}$ **15.** $-\dfrac{x^2v\cos(vw)}{2x\sin(vw) - ue^{xu}}$ **17.** 2 **19.** 0
21. $(xu - x^2u^2)/(2yu^3 + xv)$ **23.** Tangent line: $3x - 5y = -2$; normal line: $x = 1 + 3t$, $y = 1 - 5t$
25. Tangent line: $2x + 3y = 2$; normal line: $x = 1 + 2t$, $y = 3t$
27. Tangent plane: $13x - 3y + 10z = 53$; normal line: $x = 2 + 13t$, $y = 1 - 3t$, $z = 3 + 10t$ **29.** -5 **31.** $-39/\sqrt{30}$

33. a) 8 b) $x^2 + y^2 = 8$ c) $-\mathbf{i} + \mathbf{j}$ d) $16\sqrt{2}/5$

35. a) 21 b) $\dfrac{27}{1 + x^2 + y^2} + \dfrac{36}{1 + y^2 + z^2} = 21$ c) $-3\mathbf{i} + 7\mathbf{j} - 4\mathbf{k}$ d) $2\sqrt{74}$
37. Differentiating the relation $dy/dx = -G_x/G_y$ using the quotient rule and the chain rule, we have

$$\frac{d^2y}{dx^2} = -\frac{G_y\left(G_{xx} + G_{xy}\cdot\dfrac{dy}{dx}\right) - G_x\left(G_{yx} + G_{yy}\cdot\dfrac{dy}{dx}\right)}{G_y^2}$$

$$= -\frac{G_y\left(G_{xx} + G_{xy}\dfrac{-G_x}{G_y}\right) - G_x\left(G_{yx} + G_{yy}\dfrac{-G_x}{G_y}\right)}{G_y^2} = -\frac{G_y^2G_{xx} - 2G_xG_yG_{xy} + G_x^2G_{yy}}{G_y^3}.$$

39. $-\dfrac{G_z^2G_{yy} - 2G_yG_zG_{yz} + G_y^2G_{zz}}{G_z^3}$

41. At a point (x, y) of intersection, a vector normal to the first curve is $(20x^3y - 20xy^3)\mathbf{i} + (5x^4 - 30x^2y^2 + 5y^4)\mathbf{j}$ while a vector normal to the second curve is $(5x^4 - 30x^2y^2 + 5y^4)\mathbf{i} + (-20x^3y + 20xy^3)\mathbf{j}$. These vectors are perpendicular, for their dot product is zero.
43. At a point (x, y, z) of intersection, a vector normal to the first surface is $\mathbf{i} + 2y\mathbf{j} + 6z^2\mathbf{k}$, while a vector normal to the second surface is $12\mathbf{i} - (3/y)\mathbf{j} - (1/z^2)\mathbf{k}$. These vectors are perpendicular, for their dot product is zero.

Review Exercise Set 16.1

1. a) $z^2 + y^2ze^{xz}$ b) $y^2z^2e^{xz}$ c) $2yze^{xz}$ **3.** $-10\mathbf{i} + 17\mathbf{j} - 12\mathbf{k}$ **5.** 12 **7.** $-\frac{2784}{5}$ **9.** $\frac{7}{13}$

Review Exercise Set 16.2

1. a) $x_1x_3\cos x_2x_3 - 2x_2x_1^3$ b) $x_2\cos x_2x_3$ **3.** $\mathbf{i}$ **5.** -282 **7.** $\frac{12}{5}$ **9.** $\frac{3}{8}$

Review Exercise Set 16.3

1. $T_x = 10$, $T_y = -20$, $T_z = 20$ **3.** $4\,dz$ **5.** $\dfrac{dz}{dt} = \dfrac{\partial z}{\partial u}\left(\dfrac{\partial u}{\partial x}\dfrac{dx}{dt} + \dfrac{\partial u}{\partial y}\dfrac{dy}{dt}\right) + \dfrac{\partial z}{\partial v}\left(\dfrac{\partial v}{\partial x}\dfrac{dx}{dt} + \dfrac{\partial v}{\partial y}\dfrac{dy}{dt}\right)$ **7.** $-\frac{2}{25}$
9. $-(x^2 + 2yz^2)/(3xz^2 + 2y^2z)$

Review Exercise Set 16.4

1. $e^{\sin(xy)}[x^2y\cos(xy) + 2x]$ **3.** 5.8925 **5.** $\frac{57}{10}$ **7.** $-\frac{16}{3}$ **9.** $\dfrac{\partial z}{\partial u} = \dfrac{G_x H_u}{G_z H_x} + \dfrac{G_y K_u}{G_z K_y}$

More Challenging Exercises 16

1. $f(x + \Delta x, y + \Delta y, z + \Delta z) - f(x, y, z) = [f(x + \Delta x, y + \Delta y, z + \Delta z) - f(x, y + \Delta y, z + \Delta z)]$
$+ [f(x, y + \Delta y, z + \Delta z) - f(x, y, z + \Delta z)]$
$+ [f(x, y, z + \Delta z) - f(x, y, z)]$

3. $(x + y) - [(x + y)^3/3!]$ **5.** $1 + xy + \frac{1}{2}x^2y^2$ **7.** -0.03 with error ≤ 0.0037

9. The binomial theorem of algebra states that

$$(a + b)^n = \sum_{i=1}^{n} \binom{n}{i} a^i b^{n-i},$$

where

$$\binom{n}{i} = \frac{n!}{i!(n-i)!}.$$

Thus the coefficient of $(x - x_0)^h(y - y_0)^k$ in the given expression for $T_n(x, y)$ is

$$\frac{1}{(h+k)!}\binom{h+k}{h} \left.\frac{\partial^{h+k} f}{\partial x^h\, \partial_y^k}\right|_{(x_0, y_0)}.$$

Since

$$\frac{1}{(h+k)!}\binom{h+k}{h} = \frac{1}{(h+k)!}\cdot\frac{(h+k)!}{h!k!} = \frac{1}{h!k!},$$

the result follows immediately.

CHAPTER 17

Section 17.1

1. Local minimum of -4 at $(0, 0)$; no local maximum **3.** Local minimum of -2 at $(-2, 1)$; no local maximum
5. No local minimum or maximum **7.** Local maximum of 48 at $(0, -2)$; local minimum of -20 at $(2, 2)$
9. Local minimum of -16 at $(\pm 2, 4)$; no local maximum **11.** Local maximum of 42 at $(-8, -2)$; no local minimum
13. Local minimum of ln 10 at $(-2, 3)$; no local maximum
15. Local maximum of 1 where $xy = 1 + 2n\pi$; local minimum of -1 where $xy = 1 + (2n + 1)\pi$
17. Local maximum of 3 at $(1, -1)$ **19.** No local maximum; local minimum of -12 at $(\pm 1, -1, 2)$
21. Local maximum of 11 at $(0, 0, \pm 1)$; local minimum of -7 at $(\pm 1, -2, 0)$ and $(\pm 1, 2, 0)$ **23.** Local maximum of -11 at $(0, 0, 0)$
25. Local minimum of 4 at $(x, y, z, w) = (\pm 2, 1, -2, 2)$
27. a) $-x^2 - y^4$ b) $x^2 + y^4$ c) $x^2 + y^3$ d) If $AC - B^2 = 0$, further examination of f is needed to determine its behavior at (x_0, y_0).
29. Maximum assumed at $(1, 1)$ and $(-1, -1)$; minimum assumed at $(1, -1)$ and $(-1, 1)$
31. Maximum assumed at $(1, -1)$ and $(-1, 1)$; minimum assumed at $(0, 0)$. [Seen from $x^2 + y^2 - xy = (\frac{1}{2}x - y)^2 + \frac{3}{4}x^2$.]
33. $(5, 2, -4)$ **35.** Base 6 ft by 6 ft, height 3 ft **37.** 3, 3, and 3 **39.** $3\sqrt{2}$ **41.** $8abc/(3\sqrt{3})$ **43.** $y = \frac{1}{4}x + 2$ **45.** $y = -x + \frac{9}{5}$

Section 17.2

1. Local maximum $2\sqrt{2}$ at $(\sqrt{2}, \sqrt{2})$; local minimum $-2\sqrt{2}$ at $(-\sqrt{2}, -\sqrt{2})$ **3.** Local maximum 6 at $(4, -2)$; local minimum -6 at $(-4, 2)$ **5.** Local maximum 36 at $(\pm 6, 0)$; local minimum -27 at $(0, \pm 3)$
7. Local maximum 5 at $(3, -1)$; local minimum 7/27 at $(1/3, 1/3)$ **9.** Local maximum 128 at $(12, 2, 2)$
11. Local maximum 40 at $(6, -9, 1)$; local minimum -40 at $(-6, 9, -1)$
13. Local maximum 4 at $(3, -1, 0)$; local minimum 0 at $(1, 1, 0)$
15. Local maximum $\dfrac{2}{3 - \sqrt{2}}$ at $\left(\dfrac{1}{2 - \sqrt{2}}, \dfrac{-1}{2 - \sqrt{2}}, \dfrac{-\sqrt{2}}{2 - \sqrt{2}}\right)$; local minimum $\dfrac{2}{3 + \sqrt{2}}$ at $\left(\dfrac{1}{2 + \sqrt{2}}, \dfrac{-1}{2 + \sqrt{2}}, \dfrac{-\sqrt{2}}{2 + \sqrt{2}}\right)$
17. $\left(\dfrac{10}{49}, \dfrac{-15}{49}, \dfrac{30}{49}\right)$ **19.** Base 6 ft by 6 ft; height 3 ft **21.** $\frac{32}{81}\pi a^3$
23. Base of rectangle $(2 - \sqrt{3})P$; sides of rectangle $\dfrac{3 - \sqrt{3}}{6}P$; sides of the triangle $\dfrac{2\sqrt{3} - 3}{3}P$ **25.** $\left(\dfrac{58}{19}, \dfrac{-18}{19}, \dfrac{44}{19}\right)$

Section 17.3

1. Exact; $\frac{x^3}{3} - \frac{y^2}{2} + C$ **3.** Not exact **5.** Exact; $\tan xy + y + C$ **7.** Not exact
9. Exact; $x^2yz - 3xy^2 + 2xz^3 + 4z^2 + C$ **11.** Not exact **13.** $x \cos y + y + C$ **15.** $xe^y - \sin xy + C$
17. $\cos 2wz - \frac{1}{1+w} + \ln|z| + C$ **19.** Not exact
21. Let the given form be $dF(\mathbf{x})$. Then

$$\frac{\partial f_i}{\partial x_j} = \frac{\partial^2 F}{\partial x_j\,\partial x_i} = \frac{\partial^2 F}{\partial x_i\,\partial x_j} = \frac{\partial f_j}{\partial x_i}.$$

23. $xy + y^2 = 6$ **25.** $x^2 + \cos xy = 1$ **27.** $x \sin y + x^2 - y^3 = 1$ **29.** $\frac{x}{y} + x^2 = 2$

Section 17.4

1. $\frac{1}{12}(5^{3/2} - 1)$ **3.** $\frac{7}{2}$ **5.** $-\frac{5}{3}$ **7.** $2\sqrt{5}/3$ **9.** 0 **11.** $\frac{486}{5}$ **13.** $\frac{\pi}{8} + \frac{5}{6} - \frac{2\sqrt{2}}{3}$ **15.** $58\sqrt{5}/3$ **17.** $\frac{16}{3}$ **19.** 40 **21.** $\frac{3}{2}$ **23.** $\frac{3}{2}$
25. $\frac{2\pi + 21}{2}$ **27.** 8 **29.** π **31.** $\frac{5}{2}$ **33.** $-9/2$

Section 17.5

1. $\frac{2}{3}$ **3.** $\frac{3}{4}$ **6.** $\frac{53}{12}$ **7.** $x = 4 - 5t$, $y = 3 - t$ for $0 \le t \le 1$ **9.** Initial point (0, 0), terminal point (2, 7); not a loop
11. Initial and terminal point (0, 0); a loop **13.** Initial point (2, 7), terminal point (0, 0); not a loop **15.** 9 **17.** 162 **19.** 18
21. $\frac{5}{3}$ **23.** $-\pi$ **25.** -82 **27.** 0 **29.** 2π **31.** $-\frac{1}{2}$ **33.** 0 **35.** 8

Review Exercise Set 17.1

1. There are none. **3.** Closest $(-1, \sqrt{3}, 0)$; farthest $(-1, -\sqrt{3}, 0)$ **5.** $y^2e^x + x^3y - \cos y + C$ **7.** $\frac{976}{27}$
9. When $P(x, y)\,dx + Q(x, y)\,dy$ is an exact differential. If P and Q have continuous partial derivatives and their common domain has no holes in it, then it is true if and only if $\partial P/\partial y = \partial Q/\partial x$.

Review Exercise Set 17.2

1. Relative maximum of -2 at $(1, -1)$ **3.** $\left(\frac{136}{35}, \frac{-12}{35}, \frac{-20}{35}\right)$ **5.** $x^2 \sin y - x^3y + \frac{4}{3}y^3 + C$ **7.** $\ln(\sqrt{2} + 1) + \frac{2 - 2\sqrt{2}}{3}$
9. 0

Review Exercise Set 17.3

1. Local minimum -1 at (0, 0, 2) **3.** 9 **5.** $y = (C + 3x)/(x^2 + 1)$ **7.** π **9.** $x = -2\cos \pi t$, $y = 3 \sin \pi t$ for $0 \le t \le 1$

Review Exercise Set 17.4

1. $\sqrt{5}$ **3.** Local maxima of 13 at $(-2, 0, 3)$ and of 5 at $(2, 0, -1)$; local minimum of 4 at $(1, \pm\sqrt{3}, 0)$ **5.** $x^3y + y^2 - 2x = 14$
7. $\frac{-2\sqrt{5}}{3} - 2$ **9.** πab

More Challenging Exercises 17

1. a) $E_1(x, y) = \frac{1}{2}[f_{xx}(c_1, c_2)\cdot(x - x_0)^2 + 2f_{xy}(c_1, c_2)\cdot(x - x_0)(y - y_0) + f_{yy}(c_1, c_2)\cdot(y - y_0)^2]$
b) We have then $f(x, y) = f(x_0, y_0) + E_1(x, y)$, from which the result is immediate.
3. Consider

$$[A(\Delta x) + B(\Delta y)]^2 + (AC - B^2)(\Delta y)^2. \quad (1)$$

Suppose $A \ne 0$. If $AC - B^2 > 0$, then formula (1) is greater than zero for all $(\Delta x, \Delta y) \ne (0, 0)$. If $AC - B^2 < 0$, then for $\Delta y \ne 0$ and $\Delta x = (-B/A)\cdot\Delta y$, we see that formula (1) is less than zero, while for $\Delta y = 0$ and $\Delta x \ne 0$, formula (1) is greater than zero. If $AC - B^2 = 0$, then formula (1) assumes the value zero at points where $A(\Delta x) + B(\Delta y) = 0$. With the hint, this shows that

$$A(\Delta x)^2 + 2B(\Delta x)(\Delta y) + C(\Delta y)^2 \quad (2)$$

is greater than zero for all $(\Delta x, \Delta y) \ne (0, 0)$ if $A > 0$ and $AC - B^2 > 0$. Similarly, if $A < 0$ and $AC - B^2 > 0$, then formula (2) is less than zero for $(\Delta x, \Delta y) \ne (0, 0)$. Clearly if $A \ne 0$ and $AC - B^2 < 0$, then formula (1) can assume both positive and negative values, and thus formula (2) can also. If $A = 0$ and $C \neq 0$, then a similar argument can be made using

$$C[A(\Delta x)^2 + 2B(\Delta x)(\Delta y) + C(\Delta y)^2] = (AC - B^2)(\Delta x)^2 + [B(\Delta x) + C(\Delta y)]^2.$$

If both A and C are zero, then formula (2) reduces to $2B(\Delta x)(\Delta y)$, which can assume both positive and negative values if $B \neq 0$.

5. $\partial P/\partial x = \partial Q/\partial w$, $\partial P/\partial y = \partial R/\partial w$, $\partial P/\partial z = \partial S/\partial w$, $\partial Q/\partial y = \partial R/\partial x$, $\partial Q/\partial z = \partial S/\partial x$, $\partial R/\partial z = \partial S/\partial y$

7. a) $\dfrac{\partial P}{\partial y} = \dfrac{\partial Q}{\partial x} = \dfrac{x^2 - y^2}{(x^2 + y^2)^2}$ b) $\displaystyle\int \frac{y}{x^2 + y^2}\,dx = \tan^{-1}\left(\frac{x}{y}\right) + A$ and $\displaystyle\int \frac{-x}{x^2 + y^2}\,dy = \tan^{-1}\left(\frac{x}{y}\right) + B$ show that $H(x, y)$ has the desired form for $y \neq 0$. Since $\lim_{y\to 0+} \tan^{-1}(1/y) + A = (\pi/2) + A$ and $\lim_{y\to 0-} \tan^{-1}(1/y) + B = -(\pi/2) + B$, we see we must have $(\pi/2) + A = -(\pi/2) + B$, so $B = A + \pi$. We then see that we must define $H(x, 0) = A + (\pi/2)$ for $x > 0$, in order to have $H(x, y)$ continuous at $(x, 0)$ for $x > 0$. But then

$$\lim_{y\to 0+} H(-1, y) = A - \frac{\pi}{2} \quad \text{while} \quad \lim_{y\to 0-} H(-1, y) = A + \frac{3\pi}{2},$$

so it is impossible to define $H(x, 0)$ for $x < 0$ to make $H(x, y)$ continuous there.

CHAPTER 18

Section 18.1

1. We have integrated only over a *rectangular* region.

3. $$\int_c^d \int_a^b f(x, y)\,dx\,dy = \int_d^c \int_b^a f(x, y)\,dx\,dy = \int_a^b \int_c^d f(x, y)\,dy\,dx = \int_b^a \int_d^c f(x, y)\,dy\,dx$$

$$= -\int_c^d \int_b^a f(x, y)\,dx\,dy = -\int_d^c \int_a^b f(x, y)\,dx\,dy = -\int_a^b \int_d^c f(x, y)\,dy\,dx = -\int_b^a \int_c^d f(x, y)\,dy\,dx$$

5. 32 **7.** $\frac{128}{15}$ **9.** 17.022 **11.** 34 **13.** 32 **15.** 8(ln 3) **17.** 48 **19.** $\pi^2/4$ **21.** π **23.** 4 **25.** $\dfrac{\pi}{4} - \dfrac{1}{2}(\ln 2)$ **27.** $\frac{95}{4}$ **29.** 32

31. $\dfrac{\pi^2}{2} - 2$ **33.** 8.864686 **35.** 1.873496 **37.** 41.3772 **39.** 1.598444 **41.** 2.17780752 **43.** 23.66992474

Section 18.2

1. $\frac{1}{2}$ **3.** $\frac{1}{6}$ **5.** 8 **7.** $\frac{1}{15}$ **9.** 4π **11.** $\frac{8}{15}$ **13.** $\frac{9}{4}$ **15.** 1 **17.** $\frac{1}{4}$ **19.** 0 **21.** $\frac{8}{5}$ **23.** $\displaystyle\int_{-1}^{1}\int_{0}^{\sqrt{1-x^2}} 4xy\,dy\,dx$ **25.** $\displaystyle\int_{1}^{e}\int_{\ln x}^{1} x^2\,dy\,dx$

27. $\displaystyle\int_{-2}^{2}\int_{-3}^{1-x^2} e^{xy}\,dy\,dx$ **29.** $\displaystyle\int_{-1}^{0}\int_{-\sqrt{1+x}}^{\sqrt{1+x}} x^2y^3\,dy\,dx + \int_{0}^{1}\int_{-\sqrt{1-x}}^{\sqrt{1-x}} x^2y^3\,dy\,dx$ (= 0 by symmetry)

31. $\displaystyle\int_{0}^{\pi/4}\int_{\sin^{-1}(4y/\pi)}^{\pi/2} (x + y)\,dx\,dy$ **33.** $\displaystyle\int_{0}^{1}\int_{-\sqrt{1-z}}^{\sqrt{1-z}}\int_{-\sqrt{1-y^2-z}}^{\sqrt{1-y^2-z}} xyz^2\,dx\,dy\,dz$ **35.** $\displaystyle\int_{0}^{1}\int_{0}^{1-z}\int_{y+z-1}^{0} \sin(xy)\,dx\,dy\,dz$

37. $\displaystyle\int_{0}^{4}\int_{0}^{\sqrt{4-z}}\int_{1}^{3} z^2\,dx\,dy\,dz$ **39.** $\frac{7}{6}$ **41.** $e - \frac{3}{2}$ **43.** $\frac{256}{3}$ **45.** 16π **47.** 0.4045 **49.** 0.025300 **51.** 0.4914652902

53. 0.4667813944

Section 18.3

1.

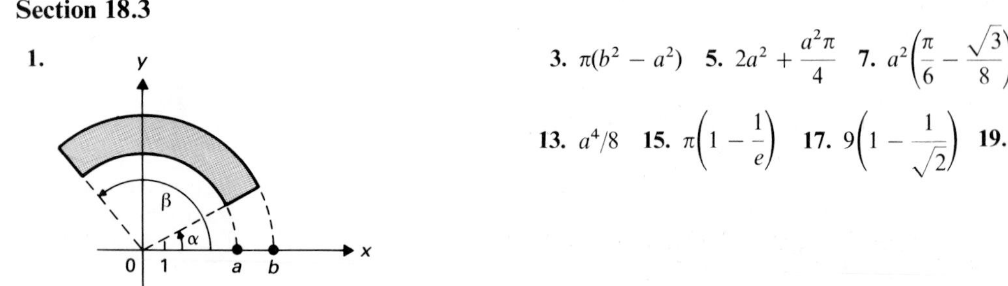

3. $\pi(b^2 - a^2)$ **5.** $2a^2 + \dfrac{a^2\pi}{4}$ **7.** $a^2\left(\dfrac{\pi}{6} - \dfrac{\sqrt{3}}{8}\right)$ **9.** $a^2\left(\sqrt{3} - \dfrac{\pi}{3}\right)$ **11.** $\dfrac{\pi a^3}{3}$

13. $a^4/8$ **15.** $\pi\left(1 - \dfrac{1}{e}\right)$ **17.** $9\left(1 - \dfrac{1}{\sqrt{2}}\right)$ **19.** 256/3

21. a) $\displaystyle\int_0^{2\pi}\int_0^{a}\int_{-\sqrt{a^2-r^2}}^{\sqrt{a^2-r^2}} r\,dz\,dr\,d\theta$ b) $\displaystyle\int_0^{a}\int_0^{2\pi}\int_{-\sqrt{a^2-r^2}}^{\sqrt{a^2-r^2}} r\,dz\,d\theta\,dr$ c) $\displaystyle\int_{-a}^{a}\int_0^{2\pi}\int_0^{\sqrt{a^2-z^2}} r\,dr\,d\theta\,dz$ d) $\displaystyle\int_0^{2\pi}\int_{-a}^{a}\int_0^{\sqrt{a^2-z^2}} r\,dr\,dz\,d\theta$

e) $\displaystyle\int_{-a}^{a}\int_0^{\sqrt{a^2-z^2}}\int_0^{2\pi} r\,d\theta\,dr\,dz$ f) $\displaystyle\int_0^{a}\int_{-\sqrt{a^2-r^2}}^{\sqrt{a^2-r^2}}\int_0^{2\pi} r\,d\theta\,dz\,dr$

23. 81π **25.** 28π **27.** $\frac{\pi}{3}(64 - 24\sqrt{3})$ **29.** $8\pi a^3/3$ **31.** 2 **33.** 4π **35.** $7\pi/3$

Section 18.4

1. $\frac{128\pi}{3}\left(1 - \frac{1}{\sqrt{2}}\right)$ **3.** $18\pi\left(\frac{2}{\sqrt{5}} + \frac{1}{\sqrt{2}}\right)$ **5.** $4\pi/9$ **7.** $\frac{\pi}{3}(2a^3 - 3a^2b + b^3)$ **9.** $4\pi a^5/5$

11. $\int_0^{2\pi}\int_0^{\pi/4}\int_0^{2\sqrt{2}} \rho^3 \sin\phi \cos\phi \, d\rho \, d\phi \, d\theta$ **13.** $\int_0^{\pi/2}\int_{\pi/6}^{\pi/3}\int_0^{3\csc\phi} \rho^4 \sin^3\phi \sin\theta \cos\theta \, d\rho \, d\phi \, d\theta$

15. $\int_{\pi/2}^{\pi}\int_0^{\tan^{-1}(1/3)}\int_0^{\sqrt{10}} \rho^3(\sin^2\phi)(\sin\theta - \cos\theta)\, d\rho \, d\phi \, d\theta$ **17.** $\int_0^{\pi}\int_{\pi/6}^{\pi/2}\int_0^{3\csc\phi} \rho^4 \sin^2\phi \cos\phi \, d\rho \, d\phi \, d\theta$

19. $\int_0^{2\pi}\int_{\pi/4}^{\pi/2}\int_{2\csc\phi}^{4\csc\phi} \rho \, d\rho \, d\phi \, d\theta$

Section 18.5

1. a) $\frac{1}{6}$ b) $(\frac{2}{3}, \frac{3}{4})$ **3.** $2k\pi a^3/3$ **5.** $2\sqrt{2}k\pi a^4/3$ **7.** a) $\left(\frac{2a\sin\alpha}{3\alpha}, \frac{2a(1-\cos\alpha)}{3\alpha}\right)$ b) $\left(\frac{2a}{3}, 0\right)$ **9.** $\left(0, \frac{4a}{3\pi}\right)$ **11.** $\left(0, \frac{3a}{2\pi}\right)$

13. $k\pi a^4/6$ **15.** $k\pi a^5/6$ **17.** $k\pi a^6/15$ **19.** $\left(\frac{b}{2}, \frac{4a^2}{7}, 0\right)$ **21.** $2\int_0^b\int_0^a\int_0^{a^2-z^2} (x^2 + z^2)ky \, dy \, dz \, dx$ **23.** $\left(0, 0, \frac{3a}{8}\right)$ **25.** $\left(\frac{4}{3}, 0, 0\right)$

27. a) $\left(0, 0, \frac{a(n+3)\sin^2\alpha}{2(n+4)(1-\cos\alpha)}\right)$ b) $\left(0, 0, \frac{a\sin^2\alpha}{2(1-\cos\alpha)}\right)$

29. The first moment of a body in space about the plane $z = -a$ is $M_{xy} + ma$. **31.** a) $\frac{1}{\sqrt{3}}$ b) $\frac{1}{\sqrt{3}}\sqrt{(a+1)^2 - a^2}$

Section 18.6

1. 6π **3.** $\sqrt{2}/6$ **5.** $\frac{518}{27}$ **7.** $\frac{1}{2}\sqrt{14} + \frac{5}{6}\ln\left(\frac{3+\sqrt{14}}{\sqrt{5}}\right)$ **9.** $\frac{\pi}{6}[(1 + 4a^2)^{3/2} - 1]$ **11.** 10π **13.** $\frac{\pi a^2}{6}(5\sqrt{5} - 1)$

15. $\frac{\pi}{3}(17^{3/2} - 1)$ **17.** 37.2448 **19.** 4.05488

Review Exercise Set 18.1

1. 32 **3.** $\int_0^4\int_{-\sqrt{y}}^{\sqrt{y}} (x^2 - 3xy)\, dx \, dy$

5. $\int_{-2}^{2}\int_{-\sqrt{4-x^2}}^{\sqrt{4-x^2}}\int_{x^2+y^2}^{4+\sqrt{4-x^2-y^2}} 1 \cdot dz \, dy \, dx = 4\int_0^2\int_0^{\sqrt{4-x^2}}\int_{x^2+y^2}^{4+\sqrt{4-x^2-y^2}} 1 \cdot dz \, dy \, dx = \int_0^{2\pi}\int_0^2\int_{r^2}^{4+\sqrt{4-r^2}} r \, dz \, dr \, d\theta$ **7.** 16π **9.** $\pi a^6/6$

Review Exercise Set 18.2

1. -8 **3.** $\int_0^4\int_0^{\sqrt{z}}\int_0^{\sqrt{z-y^2}} x^2 z \, dx \, dy \, dz$ **5.** $\frac{8}{3}$ **7.** $2^8 \cdot 23\pi/105$ **9.** $(0, 0, \frac{1}{3})$

Review Exercise Set 18.3

1. 1.58663 **3.** $\int_0^4\int_{-\sqrt{4-z}}^{\sqrt{4-z}}\int_0^{\sqrt{4-z-y^2}} yz \, dx \, dy \, dz$ **5.** $\int_{\pi/4}^{\pi/2}\int_0^{\csc\theta} r^4 \, dr \, d\theta$ **7.** $2\pi a^5/5$ **9.** $(0, 2, \frac{1}{5})$

Review Exercise Set 18.4

1. 3.55495 **3.** $\frac{1}{2}(e^2 - e)$ **5.** $\pi/6$ **7.** $\frac{128\pi}{3}\left(1 - \frac{\sqrt{3}}{2}\right)$ **9.** $\frac{64k\pi}{3}\left(\frac{2}{3} - \frac{5}{6\sqrt{2}}\right)$

More Challenging Exercises 18

1. $\pi^2 a^4/2$ **3.** $2\pi^2 a^3$ **5.** $\sum_{i,j=1}^{n}\left[\left(5 + \frac{5i}{n}\right)^2 + 3\left(5 + \frac{5i}{n}\right)\left(-2 + \frac{4j}{n}\right)\right] \cdot \frac{20}{n^2}$ **7.** $\frac{2}{3}$

9. a) $dx = \cos\theta \, dr - r\sin\theta \, d\theta$, $dy = \sin\theta \, dr + r\cos\theta \, d\theta$ b) $r dr \wedge d\theta$ **11.** 2541/25

CHAPTER 19

Section 19.1

1. 4π units mass/unit time **3.** -198 **5.** Both integrals equal 12. **7.** Both integrals equal 128π. **9.** Both integrals equal $\frac{1}{12}$. **11.** $e^2 - \frac{5}{3}$. **13.** -96 units mass/unit time (The plates are *three* units apart.) **15.** -120 **17.** $\frac{1}{3}$ **19.** $\frac{3}{5}$ **21.** 1 **23.** $2\pi ab$
25. By Green's theorem,

$$\frac{1}{2}\oint_{\partial G}(x\,dy - y\,dx) = \frac{1}{2}\iint_G (1+1)\,dx\,dy = \iint_G 1\cdot dx\,dy = \text{Area of } G.$$

27. 10 **29.** $\left(\frac{-4}{3}, \frac{29}{8}\right)$ **31.** $\oint_{\partial G}\left(\frac{\partial f}{\partial x}dy - \frac{\partial f}{\partial y}dx\right)$ **33.** $4\pi a^3$ **35.** 32π

Section 19.2

1. F F T F T T F T F T **3.** Both integrals equal -120. **5.** -17
7. a) By Green's theorem (as in Eq. 9 of Section 19.1),

$$\iint_G (\nabla\cdot\mathbf{E})\,dx\,dy = \oint_{\partial G}(\mathbf{E}\cdot\mathbf{n})\,ds = \oint_{\partial G}(\mathbf{F}\cdot\mathbf{n})\,ds = \iint_G(\nabla\cdot\mathbf{F})\,dx\,dy.$$

b) By Green's theorem (as in Eq. 12 of Section 19.2),

$$\iint_G (\text{curl }\mathbf{E})\,dx\,dy = \oint_{\partial G}(\mathbf{E}\cdot\mathbf{t})\,ds = \oint_{\partial G}(\mathbf{F}\cdot\mathbf{t})\,ds = \iint_G(\text{curl }\mathbf{F})\,dx\,dy.$$

9. Let $P(x, y) = -y$ and $Q(x, y) = x$. By Green's theorem, $\oint_\lambda(-y\,dx + x\,dy) = \iint_G(\partial x/\partial x - \partial(-y)/\partial y)\,dx\,dy = \iint_G 2\,dx\,dy = 2\iint_G 1\,dx\,dy = 2(\text{Area of } G)$.
11. -30π **13.** 18π **15.** 4π **17.** -2π **19.** 2π **21.** -2 **23.** -10 **25.** 2 **27.** 0
29. $\nabla f = f_x\mathbf{i} + f_y\mathbf{j}$. If ∇f is an incompressible flow, then

$$\nabla\cdot(\nabla f) = \frac{\partial(f_x)}{\partial x} + \frac{\partial(f_y)}{\partial y} = 0, \quad \text{so} \quad f_{xx} + f_{yy} = 0.$$

31. $g(x, y) = 3x^2y + h(y)$ **33.** 0
35. If the force field $\mathbf{F}(x, y) = P(x, y)\mathbf{i} + Q(x, y)\mathbf{j} = a\mathbf{i} + b\mathbf{j}$, then $\frac{\partial P}{\partial x} = \frac{\partial P}{\partial y} = 0$ and $\frac{\partial Q}{\partial x} = \frac{\partial Q}{\partial y} = 0$. Thus $\frac{\partial Q}{\partial x} - \frac{\partial P}{\partial y} = 0$, so the flow is irrotational, and $\frac{\partial P}{\partial x} + \frac{\partial Q}{\partial y} = 0$, so the flow is incompressible. With such a constant flow, mass is neither rotating nor collecting within any region.
37. a) Curl $\mathbf{F} = \frac{\partial(x^2)}{\partial x} - \frac{\partial(2xy)}{\partial y} = 2x - 2x = 0$, so $\mathbf{F}$ is conservative. b) $u(x, y) = -x^2y$ c) $u(x, y) = -x^2y + 5$ d) 5
39. If $W(A, B)$ is the work done by the force in moving a body from A to B, then using Exercise 38, we have $u(A) - u(B) = W(A, B) = k(B) - k(A)$. Then $u(A) + k(A) = u(B) + k(B)$.
41. a) Since $\mathbf{F} = \nabla(u) = \nabla(-H) = -\nabla H$, we see that $\mathbf{F}$ is perpendicular to level curves of $-H$, which are the same as the level curves of H, that is, the equipotential curves. b) The work done is $W(P_1, P_2) = H(P_2) - H(P_1)$. If P_1' is on γ_1 and P_2' is on γ_2, then $H(P_1') = H(P_1)$ and $H(P_2') = H(P_2)$ by the meaning of equipotential (level) curves. Thus $W(P_1', P_2') = H(P_2') - H(P_1') = W(P_1, P_2)$.

Section 19.3

1. $-2\mathbf{k}$ **3.** $2y^2\mathbf{i} + xy\mathbf{j} - xz\mathbf{k}$ **5.** $-\left(\frac{x}{y^2} + \frac{1}{x}\right)\mathbf{i} - \left(\frac{y}{z^2} + \frac{1}{y}\right)\mathbf{j} - \left(\frac{z}{x^2} + \frac{1}{z}\right)\mathbf{k}$ **7.** $\frac{1}{y}\mathbf{i} + (e^y - x\cos xy)\mathbf{k}$
9. Both integrals equal $\pi/2$. **11.** Both integrals equal zero. **13.** Both integrals equal -512π. **15.** Both integrals equal zero.
17. 12π **19.** -72π **21.** $8\pi a^3$ **23.** -405π **25.** -66π **27.** 216π
29. a) $\mathbf{F}\cdot\mathbf{t} = 0$, so $\oint_{\partial G}(\mathbf{F}\cdot\mathbf{t})\,ds = 0$, so $\iint_G[(\text{curl }\mathbf{F})\cdot\mathbf{n}]\,dS = 0$ by Stokes' theorem. b) $(\text{Curl }\mathbf{F})\cdot\mathbf{n} = 0$, so $\iint_G[(\text{curl }\mathbf{F})\cdot\mathbf{n}]\,dS = 0$, so $\oint_{\partial G}(\mathbf{F}\cdot\mathbf{t})\,ds = 0$.
31. Curl $\mathbf{F} = \mathbf{0}$, so $\iint_G[(\text{curl }\mathbf{F})\cdot\mathbf{n}]\,dS = 0$, so $\oint_{\partial G}(\mathbf{F}\cdot\mathbf{t})\,ds = 0$.
33. a) If $\mathbf{F} = F_1\mathbf{i} + F_2\mathbf{j} + F_3\mathbf{k}$, then

$$\text{Curl }\mathbf{F} = \begin{vmatrix} \mathbf{i} & \mathbf{j} & \mathbf{k} \\ \frac{\partial}{\partial x} & \frac{\partial}{\partial y} & \frac{\partial}{\partial z} \\ F_1 & F_2 & F_3 \end{vmatrix} = \left(\frac{\partial F_3}{\partial y} - \frac{\partial F_2}{\partial z}\right)\mathbf{i} + \left(\frac{\partial F_1}{\partial z} - \frac{\partial F_3}{\partial x}\right)\mathbf{j} + \left(\frac{\partial F_2}{\partial x} - \frac{\partial F_1}{\partial y}\right)\mathbf{k}.$$

Then

$$\nabla \cdot (\textbf{curl F}) = \left(\frac{\partial^2 F_3}{\partial x\, \partial y} - \frac{\partial^2 F_2}{\partial x\, \partial z}\right) + \left(\frac{\partial^2 F_1}{\partial y\, \partial z} - \frac{\partial^2 F_3}{\partial y\, \partial x}\right) + \left(\frac{\partial^2 F_2}{\partial z\, \partial x} - \frac{\partial^2 F_1}{\partial z\, \partial y}\right)$$

$$= \left(\frac{\partial^2 F_3}{\partial x\, \partial y} - \frac{\partial^2 F_3}{\partial y\, \partial x}\right) + \left(\frac{\partial^2 F_2}{\partial z\, \partial x} - \frac{\partial^2 F_2}{\partial x\, \partial z}\right) + \left(\frac{\partial^2 F_1}{\partial y\, \partial z} - \frac{\partial^2 F_1}{\partial z\, \partial y}\right)$$

$$= 0$$

if second partial derivatives are continuous.

b) $$\begin{vmatrix} \mathbf{i} & \mathbf{j} & \mathbf{k} \\ \frac{\partial}{\partial x} & \frac{\partial}{\partial y} & \frac{\partial}{\partial z} \\ \frac{\partial f}{\partial x} & \frac{\partial f}{\partial y} & \frac{\partial f}{\partial z} \end{vmatrix} = \left(\frac{\partial^2 f}{\partial y\, \partial z} - \frac{\partial^2 f}{\partial z\, \partial y}\right)\mathbf{i} + \left(\frac{\partial^2 f}{\partial z\, \partial x} - \frac{\partial^2 f}{\partial x\, \partial z}\right)\mathbf{j} + \left(\frac{\partial^2 f}{\partial x\, \partial y} - \frac{\partial^2 f}{\partial y\, \partial x}\right)\mathbf{k}$$

$$= 0\mathbf{i} + 0\mathbf{j} + 0\mathbf{k} = \mathbf{0}$$

if second partial derivatives are continuous.

35. Remove a very small "disk" from G and let the remaining surface be H. Then

$$\iint_G [(\nabla \times \mathbf{F}) \cdot \mathbf{n}]\, dS \approx \iint_H [(\nabla \times \mathbf{F}) \cdot \mathbf{n}]\, dS,$$

since only a very small part of G was removed. In fact, the integrals can be made as nearly equal as we please by removing a sufficiently small part of G. But

$$\iint_H [(\nabla \times \mathbf{F}) \cdot \mathbf{n}]\, dS = \oint_{\partial H} (\mathbf{F} \cdot \mathbf{t})\, ds \approx 0,$$

since ∂H is a very short curve. By removing a sufficiently small part of G, we can make ∂H as short as we please, and $\oint_{\partial H} (\mathbf{F} \cdot \mathbf{t})\, ds$ as close to zero as we please. Thus, taking the limit as smaller and smaller parts of G are removed, we see that $\iint_G [(\nabla \times \mathbf{F}) \cdot \mathbf{n}]\, dS$ must be zero.

37. $\iiint_G [\nabla \cdot (\textbf{curl F})]\, dx\, dy\, dz = \iint_{\partial G} [(\textbf{curl F}) \cdot \mathbf{n}]\, dS$ by Exercise 36. Now ∂G is a surface that is the entire boundary of the three-dimensional region G. By Exercise 35, $\iint_{\partial G} [(\textbf{curl F}) \cdot \mathbf{n}]\, dS = \iint_{\partial G} [(\nabla \times \mathbf{F}) \cdot \mathbf{n}]\, dS = 0$.

Review Exercise Set 19.1

1. See Theorem 19.4 of Section 19.2.

3. Let $\mathbf{F}$ be a continuously differentiable vector field, and let G be a suitable region in space. Then $\iiint_G (\nabla \cdot \mathbf{F})\, dx\, dy\, dz = \iint_{\partial G} (\mathbf{F} \cdot \mathbf{n})\, dS$, where $\mathbf{n}$ is the unit outward normal vector. For a flux vector $\mathbf{F}$, this states that the integral of the divergence of the flux vector is equal to the integral of the normal component of the flux vector over the surface. Roughly speaking, integrating the rate at which mass is leaving little volume elements throughout G gives the same result as finding the rate that mass leaves G through the surface ∂G.

5. a) $\partial Q/\partial x = \partial P/\partial y$ b) $\partial P/\partial x + \partial Q/\partial y = 0$ **7.** -162π

Review Exercise Set 19.2

1. Let $\mathbf{F}$ be a continuously differentiable vector field in the plane and let G be a plane region that can be decomposed into a finite number of simple subregions. Then $\iint_G (\text{curl } \mathbf{F})\, dx\, dy = \oint_{\partial G} (\mathbf{F} \cdot \mathbf{t})\, ds$.

3. By the divergence theorem,

$$\iint_{\partial G} \left[\left(\frac{x^3}{3}\mathbf{i} + \frac{y^3}{3}\mathbf{j} + 0\mathbf{k}\right) \cdot \mathbf{n}\right] dS = \iiint_G (x^2 + y^2)\, dx\, dy\, dz,$$

and the second integral gives the moment of inertia about the z-axis if mass density is 1. (The fact that G is a ball with center at the origin is irrelevant.)

5. $(xz - y)\mathbf{i} - yz\mathbf{j}$ **7.** $\frac{1}{2}$

Review Exercise Set 19.3

1. $\frac{7}{2}$ **3.** $\frac{17}{18}$ **5.** $\frac{1}{2}xy^4 - 2y + k(x)$ **7.** 36π

Review Exercise Set 19.4

1. $243\pi/2$ **3.** 6 **5.** $\dfrac{z}{x^2}\mathbf{i} + \left(\dfrac{2yz}{x^3} - \dfrac{xy}{z^2}\right)\mathbf{j} - \left(\dfrac{2x}{y} + \dfrac{x}{z}\right)\mathbf{k}$ **7.** $3069\pi/2$

More Challenging Exercises 19

1. A computation shows that

$$\nabla \times \mathbf{F} = \left(\frac{\partial R}{\partial y} - \frac{\partial Q}{\partial z}\right)\mathbf{i} + \left(\frac{\partial P}{\partial z} - \frac{\partial R}{\partial x}\right)\mathbf{j} + \left(\frac{\partial Q}{\partial x} - \frac{\partial P}{\partial y}\right)\mathbf{k}.$$

The normal unit vector $\mathbf{n}$ at (x, y, z) on G can be written as $\mathbf{n} = (\cos\alpha)\mathbf{i} + (\cos\beta)\mathbf{j} + (\cos\gamma)\mathbf{k}$, where α, β, and γ are the angles that $\mathbf{n}$ makes with the x-, y-, and z-axes. Consider the contribution $\iint_G [(\partial R/\partial y) - (\partial Q/\partial z)]\,(\cos\alpha)\,dS$ to $\iint_G [(\nabla \times \mathbf{F})\cdot\mathbf{n}]\,dS$. Since α is the angle between $\mathbf{n}$ and $\mathbf{i}$, it is also the angle between the tangent plane at (x, y, z) and the y,z-plane. Thus $(\cos\alpha)\,dS$ is the area of the projection of a surface element of area dS onto the y,z-plane. But this is the meaning of $dy\,dz$ in $\iint_G [(\partial R/\partial y) - (\partial Q/\partial z)]\,dy\,dz$, which thus equals $\iint_G [(\partial R/\partial y) - (\partial Q/\partial z)](\cos\alpha)\,dS$. A similar analysis with other components of $\nabla \times \mathbf{F}$ shows that

$$\iint_G [(\nabla \times \mathbf{F})\cdot\mathbf{n}]\,dS = \iint_G \left[\left(\frac{\partial R}{\partial y} - \frac{\partial Q}{\partial z}\right)dy\,dz + \left(\frac{\partial P}{\partial z} - \frac{\partial R}{\partial x}\right)dz\,dx + \left(\frac{\partial Q}{\partial x} - \frac{\partial P}{\partial y}\right)dx\,dy\right].$$

Since $\mathbf{t} = (dx/ds)\mathbf{i} + (dy/ds)\mathbf{j} + (dz/ds)\mathbf{k}$, it is clear that

$$\oint_{\partial G} (\mathbf{F}\cdot\mathbf{t})\,ds = \oint_{\partial G} (P\,dx + Q\,dy + R\,dz).$$

3. a) $-3x\,dx \wedge dy$ b) $(x + z^3)\,dx \wedge dy \wedge dz$ **5.** $f(B) - f(A) = \displaystyle\int_\gamma \left(\frac{\partial f}{\partial x}dx + \frac{\partial f}{\partial y}dy + \frac{\partial f}{\partial z}dz\right)$

7. $\displaystyle\oint_{\partial G} (P\,dx + Q\,dy + R\,dz) = \iint_G \left[\left(\frac{\partial R}{\partial y} - \frac{\partial Q}{\partial z}\right)dy\,dz + \left(\frac{\partial P}{\partial y} - \frac{\partial R}{\partial x}\right)dz\,dx + \left(\frac{\partial Q}{\partial x} - \frac{\partial P}{\partial y}\right)dx\,dy\right]$,

which is the form of Stokes' theorem given in Exercise 1.

CHAPTER 20

Section 20.1

1.

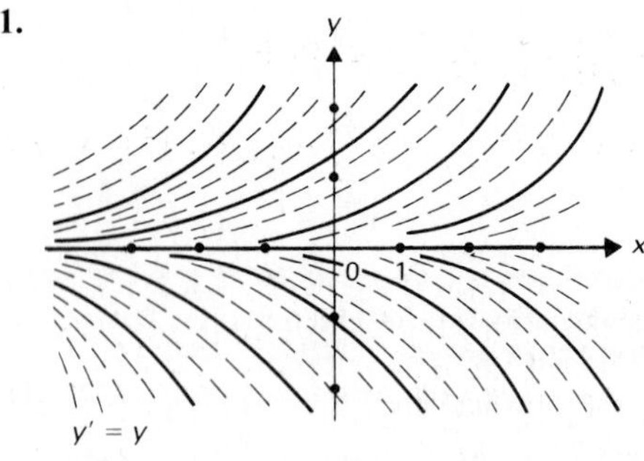

3.

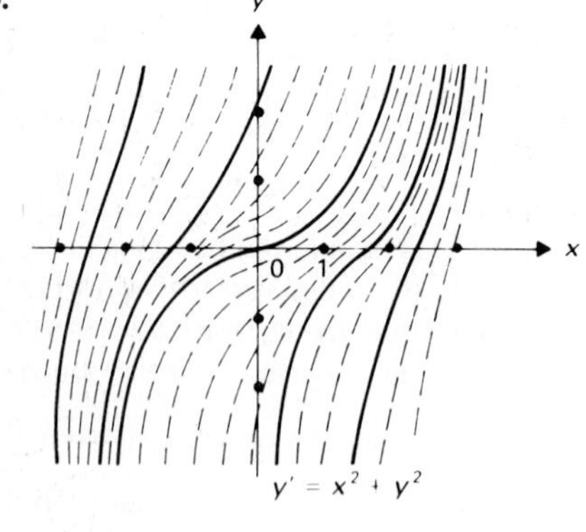

5.

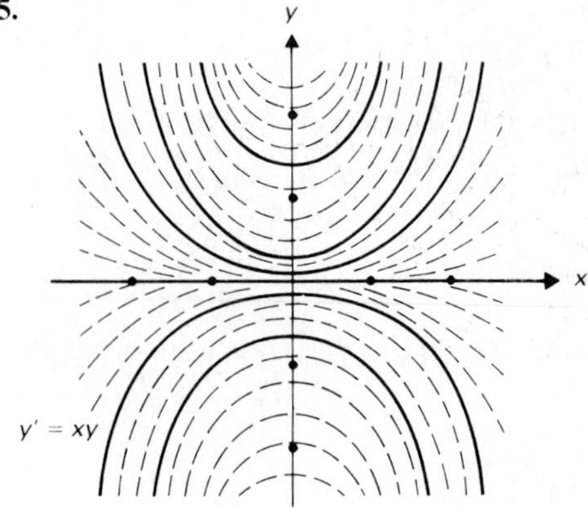

7.

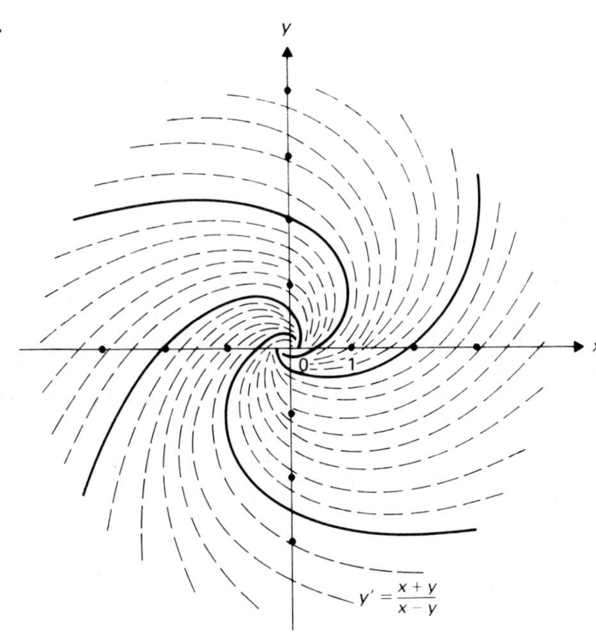

9. a) All $r > 0$ b) $\pi/2$ **11.** $(0, 1), (\frac{1}{2}, \frac{3}{2}), (1, \frac{5}{2})$ **13.** $(1, 4), (\frac{3}{2}, 7), (2, 12), (\frac{5}{2}, 20), (3, \frac{65}{2})$ **15.** $(0, 1), (\frac{1}{2}, \frac{7}{4}), (1, \frac{105}{32})$
17. $(1, 4), (2, 14), (3, 42)$ **19.** $(0, 1), (\frac{1}{2}, \frac{7}{4}), (1, \frac{105}{32})$ **21.** $(0, 3), (1, \frac{3}{2}), (2, 0)$
23. $\bar{y}_{i+1} = y_i + F(x_i, y_i)\,\Delta x;\ y_{i+1} = y_i + F\left(\dfrac{x_i + x_{i+1}}{2}, \dfrac{y_i + \bar{y}_{i+1}}{2}\right)\Delta x$ **25.** 2.288351556 **27.** 2.236091518 **29.** 2.236906681

Section 20.2

1. $-\dfrac{1}{y} = \dfrac{1}{2}x^2 + C$ **3.** $\tan^{-1} y = \dfrac{1}{3}x^3 + C$ **5.** $\sin^{-1} y = \dfrac{1}{2}x^2 + C$ **7.** $Cx\left(1 - \dfrac{y^2}{x^2}\right) = 1$ **9.** $\dfrac{y}{x} = \ln|x| + C$
11. $y^2 = x^2(2 \ln|x| + C)$ **13.** $e^{y/x} = \dfrac{-1}{\ln|x| + C}$ **15.** $3 \ln|y - 2x| + \ln|y + 2x| = C$ **17.** $y = x + \dfrac{x^2}{2} - \dfrac{5}{2}$ **19.** $y^2 = x^2 + 5$
21. $26x^2 - y^2 = 1$ **23.** $y = x(\ln|x| + 2x)$ **25.** $\sec\left(\dfrac{y}{x}\right) + \tan\left(\dfrac{y}{x}\right) = \dfrac{x}{\pi}$

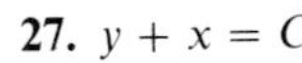
27. $y + x = C$

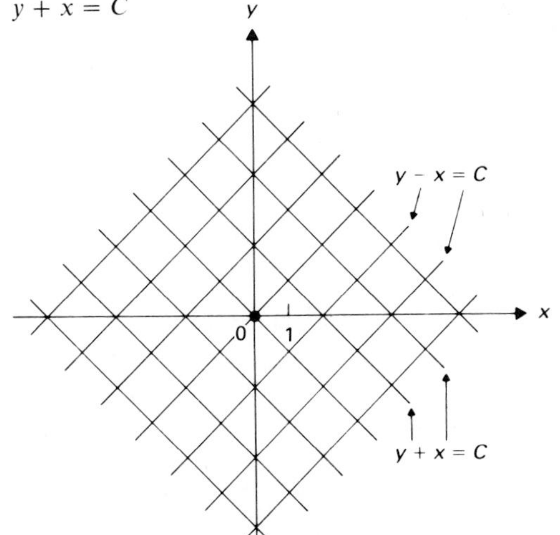

29. $y = Ce^{2x}$

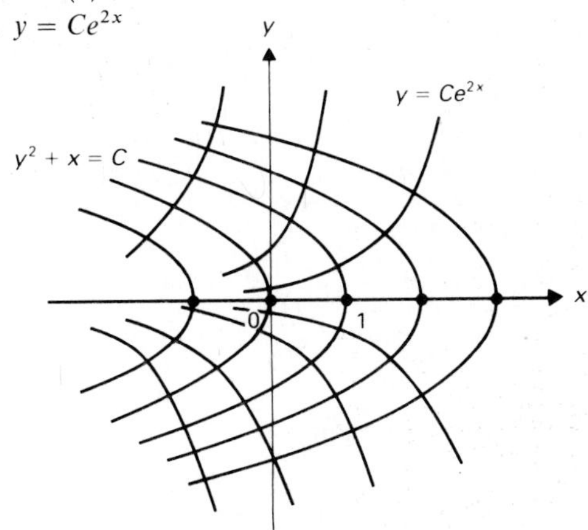

31. $(x+1)^2+y^2=C$

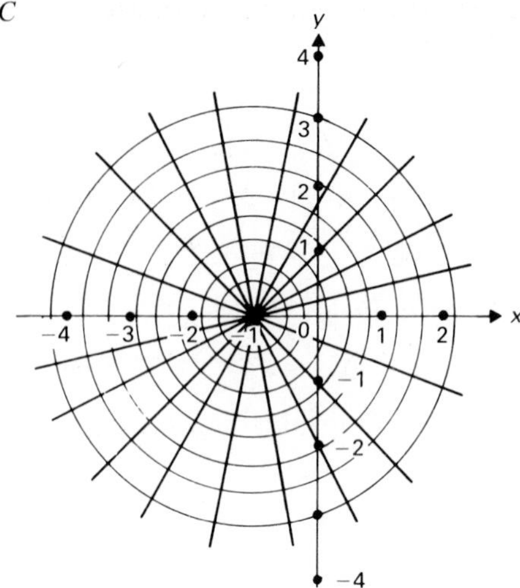

Section 20.3

1. $x\cos y+y=C$ **3.** $xy^2+\ln|y|=C$ **5.** $xe^y-\sin xy=C$ **7.** $\ln|x|-\frac{1}{xy}+\frac{3}{y}=C$ **9.** $\frac{x^2}{2}+\frac{x}{y}-\ln|y|=C$

11. $\frac{4}{x}+\frac{y}{x}+3y=C$ **13.** $\ln|x|+x^2y^3+\frac{y^3}{3}=C$ **15.** $\sin(xy^2)-\frac{x^2}{2}+2x+3e^y=3$ **17.** $e^{y/x}+\cos x+\sin y=\frac{1}{e}-1$

19. $x-\frac{1}{xy}-3\ln|y|=0$ **21.** $4\frac{y+1}{x}+y^4=1$ **23.** $\frac{-y}{x^2}+\frac{1}{3x^3}-y=\frac{1}{3}$

25. Let $F'=f$. Now

$$d(F(G(x,y)))=\frac{\partial(F(G(x,y)))}{\partial x}dx+\frac{\partial(F(G(x,y)))}{\partial y}dy$$

$$=F'(G(x,y))\frac{\partial G}{\partial x}dx+F'(G(x,y))\frac{\partial G}{\partial y}dy=f(G(x,y))\mu M\,dx+f(G(x,y))\mu N\,dy.$$

This shows that $f(G(x,y))\mu$ is an integrating factor.

27. From

$$N\frac{\partial\mu}{\partial x}-M\frac{\partial\mu}{\partial y}=\mu\left(\frac{\partial M}{\partial y}-\frac{\partial N}{\partial x}\right),$$

we obtain

$$N\left(\frac{1}{\mu}\frac{\partial\mu}{\partial x}\right)-M\left(\frac{1}{\mu}\frac{\partial\mu}{\partial y}\right)=\frac{\partial M}{\partial y}-\frac{\partial N}{\partial x}\quad\text{or}\quad N\frac{\partial(\ln|\mu|)}{\partial x}-M\frac{\partial(\ln|\mu|)}{\partial y}=\frac{\partial M}{\partial y}-\frac{\partial N}{\partial x}.$$

29. a) $xy^2+x-y^3=C$ b) $xy+\sin y=C$

Section 20.4

1. $y=Ce^{x^2/2}$ **3.** $y=Ce^{-x}+\frac{3}{2}e^x$ **5.** $y=Ce^{-2x}+\frac{x^2}{2}e^{-2x}+\frac{3}{2}$ **7.** $y=C\csc x-3x\cot x-\cot x+3,\ 0<x<\pi$

9. $y=Ce^{-x^2}+\frac{1}{2}$ **11.** $y=-\frac{2}{9}e^{3x}-\frac{x}{3}-\frac{7}{9}$ **13.** $y=-e^{-\tan^{-1}x}+3$ **15.** $y=-2e^{x^2/2}-1$

17. $y=\frac{1}{1+\sin x}\left(\frac{x}{2}-\frac{\sin 2x}{4}+6-\frac{\pi}{4}\right)$ **19.** $y=e^{x^2/2}\left(C+\frac{x^3}{3}-\frac{x^5}{5\cdot 2!}+\frac{x^7}{7\cdot 2^2\cdot 2!}-\frac{x^9}{9\cdot 2^3\cdot 3!}+\frac{x^{11}}{11\cdot 2^4\cdot 4!}-\cdots\right)$

21. $y=2x(\ln x)+\frac{1-4(\ln 2)}{2}x$ **23.** 35 lb

25. a) $i=e^{-Rt/L}\left(i_0+\frac{1}{L}\int_0^t e^{Ru/L}E(u)\,du\right)$ b) From part (a), we obtain

$$i=e^{-Rt/L}\left(i_0+\frac{E}{R}e^{Rt/L}-\frac{E}{R}\right)=\left(i_0-\frac{E}{R}\right)e^{-Rt/L}+\frac{E}{R}.$$

Thus $\lim_{t\to\infty} i=E/R$. c) i approaches zero. d) It decays exponentially from its value at $t=t_0$.

27. a) Differentiation of $v = y^{1-n}$ yields $v' = (1-n)y'/y^n$. Multiplying the given differential equation by $(1-n)/y^n$, we obtain $v' + (1-n)p(x)v = (1-n)g(x)$, which is linear. b) $y^2(Ce^{-2x^2} - \frac{5}{2}) = 1$

29. 0.8870206 **31.** 3.8392805 **33.** 40.77954219

Section 20.5

1. (This is a slightly tedious but routine exercise in differentiation.) **3.** $y = Ce^{-2x}$ **5.** $y = C_1e^{-5x/2} + C_2e^{-x/2}$

7. $y = C_1 + e^{-x/2}(C_2 + C_3x)$ **9.** $y = C_1 \cos\sqrt{3}x + C_2 \sin\sqrt{3}x$ **11.** $y = C_1e^x + e^{-x/2}\left(C_2 \cos\frac{\sqrt{3}}{2}x + C_3 \sin\frac{\sqrt{3}}{2}x\right)$

13. $y = C_1 + C_2 \cos 3x + C_3 \sin 3x$ **15.** $y = (C_1x + C_2)(\cos x) + (C_3x + C_4)(\sin x)$

17. $y = C_1 + C_2x + C_3e^{-2x} + C_4 \cos\sqrt{2}x + C_5 \sin\sqrt{2}x$ **19.** $y = e^{-x}(C_1 + C_2x) + e^{-x/2}\left(C_3 \cos\frac{\sqrt{7}}{2}x + C_4 \sin\frac{\sqrt{7}}{2}x\right)$

21. $y = -3e^{3x} + 4e^{2x}$ **23.** $y = -2 - 3x + 3e^x$ **25.** $y = \frac{19}{4}x - \frac{1}{54}e^{3x} + \frac{1}{54}e^{-3x}$ **27.** $y = 3\cos 2x - 2\sin 2x$

29. $y = e^{2x} + e^{-x}\left(\cos\sqrt{3}x - \frac{1}{\sqrt{3}}\sin\sqrt{3}x\right)$

Section 20.6

1. $y = C_1e^{-x} + C_2e^{3x} - \frac{4}{3}x + \frac{8}{9}$ **3.** $y = C_1e^x + C_2e^{-x} + \frac{1}{2}xe^x$ **5.** $y = C_1 + C_2e^x - x$ **7.** $y = e^{-x}(C_1 + C_2x) + \frac{1}{2}x^2e^{-x}$

9. $y = C_1e^x + C_2e^{2x} + \frac{x}{2} - \frac{3}{4}$ **11.** $y = C_1 \sin 2x + C_2 \cos 2x + \frac{1}{3}\sin x$

13. $y = C_1 + C_2e^{2x} - 3e^x - \frac{2}{5}\sin x + \frac{4}{5}\cos x$ **15.** $y = C_1 \sin x + C_2 \cos x + \frac{1}{2}xe^x - \frac{1}{2}e^x$

17. $y = C_1e^x + C_2e^{-x} - \frac{1}{2}(x \sin x + \sin x + \cos x)$ **19.** $y = C_1 + C_2e^{-4x} + \frac{1}{12}x^3 - \frac{1}{16}x^2 + \frac{1}{32}x$

21. $y = C_1 + C_2x + C_3e^{-3x} + \frac{1}{18}x^3 - \frac{1}{18}x^2 + \frac{1}{54}e^{3x}$ **23.** $y = C_1 + C_2x + C_3x^2 + C_4e^{2x} - \frac{1}{240}x^6 - \frac{3}{80}x^5 - \frac{3}{32}x^4 - \frac{3}{16}x^3$

25. Height: $s = 3512 - 512e^{-t/4} - 128t$; terminal velocity: -128 ft/sec

27. The solutions of the characteristic equation $mr^2 + cr + k = 0$ are

$$r_1 = \frac{-c + \sqrt{c^2 - 4km}}{2m}$$

and

$$r_2 = \frac{-c - \sqrt{c^2 - 4km}}{2m}$$

Since $c^2 > 4km$, we have $r_2 < r_1 < 0$. The general solution of $m\ddot{s} + c\dot{s} + ks = 0$ is then $s = C_1e^{r_1t} + C_2e^{r_2t}$. From the initial conditions $s(0) = h$ and $\dot{s}(0) = 0$, we obtain as solution for the given problem

$$s = \frac{-r_2h}{r_1 - r_2}e^{r_1t} + \frac{r_1h}{r_1 - r_2}e^{r_2t}.$$

Since $r_1 < 0$ and $r_2 < 0$, we have $\lim_{t\to\infty} s(t) = 0$, so the body approaches its position of rest as $t \to \infty$. For s to be zero, our solution shows we would need to have $e^{(r_1 - r_2)t} = r_1/r_2$. Since $r_1 - r_2 > 0$ and $0 < r_1/r_2 < 1$, we see that there is no $t > 0$ satisfying this condition, so the body never crosses its position of rest.

29. The solutions of the characteristic equation $mr^2 + cr + k = 0$ are complex, and if $a = \sqrt{4km - c^2}/2m$, the solutions of the equation have the form

$$s = e^{(-c/2m)t}[C_1 \cos at + C_2 \sin at]$$

for constants C_1 and C_2. The desired conclusion now follows at once.

31. If $k = 0$, our equation may be written as

$$\ddot{s} + \frac{c}{a}s = \frac{1}{a}\sin kt = 0.$$

We see that any solution of this equation is of the form

$$s = C_1 \cos\left(\sqrt{\frac{c}{a}}\,t\right) + C_2 \sin\left(\sqrt{\frac{c}{a}}\,t\right)$$

for suitable constants C_1 and C_2. The period of this oscillatory motion is

$$2\pi\sqrt{a}/\sqrt{c}.$$

If $k = \sqrt{c/a}$, then any solution of our equation is of the form

$$s = (C_1 + At)\cos\left(\sqrt{\frac{c}{a}}\,t\right) + C_2 \sin\left(\sqrt{\frac{c}{a}}\,t\right)$$

for suitable constants C_1, C_2, and $A \neq 0$. The amplitude $\sqrt{(C_1 + At)^2 + C_2^2}$ then increases without bound as $t \to \infty$.

Section 20.7

1. 8 **3.** 3 **5.** 0 **7.** -10

9. $y = a_0\left[1 - \frac{x^2}{2!} + \frac{3x^4}{4!} - \frac{3 \cdot 5x^6}{6!} + \cdots + (-1)^n \frac{(3)(5)\cdots(2n-1)}{(2n)!}x^{2n} + \cdots\right] = a_0 e^{-x^2/2}$

11. $y = a_0\left[1 - \frac{2}{2!}(x+1)^2 - \frac{6}{3!}(x+1)^3 - \cdots - \frac{2^n - 2}{n!}(x+1)^n + \cdots\right]$

$+ a_1\left[(x+1) + \frac{3}{2!}(x+1)^2 + \frac{7}{3!}(x+1)^3 + \cdots + \frac{2^n - 1}{n!}(x+1)^n + \cdots\right]$

$= a_0(2e^{x+1} - e^{2x+2}) + a_1(e^{2x+2} - e^{x+1})$

13. $y = a_0\left[1 + \frac{x^3}{3!} - \frac{2x^6}{6!} + \cdots + (-1)^{n+1}\frac{(2)(5)\cdots(3n-4)}{(3n)!}x^{3n} + \cdots\right]$

$+ a_1 x + a_2\left[\frac{x^2}{2!} - \frac{x^5}{5!} + \frac{4x^8}{8!} - \cdots + (-1)^{n+1}\frac{(4)(7)\cdots(3n-5)}{(3n-1)!}x^{3n-1} + \cdots\right]$ for $n \geq 2$

15. $y = a_0(x+1)$ **17.** $y = a_0 + a_1(x-2) + a_2\left[\frac{1}{2!}(x-2)^2 - \frac{4}{4!}(x-2)^4 - \frac{4}{5!}(x-2)^5 + \frac{14}{6!}(x-2)^6 + \cdots\right]$

$+ a_3\left[\frac{1}{3!}(x-2)^3 - \frac{4}{5!}(x-2)^5 - \frac{8}{6!}(x-2)^6 + \cdots\right]$

19. $y = -1 + 2(x-1) + \frac{1}{2!}(x-1)^2 - \frac{2}{3!}(x-1)^3 - \frac{1}{4!}(x-1)^4 + \frac{2}{5!}(x-1)^5 + \frac{1}{6!}(x-1)^6 - \frac{2}{7!}(x-1)^7 + \cdots$

21. $y = -2 + x - \frac{1}{2!}x^2 + \frac{1}{3!}x^3 + \frac{5}{5!}x^5 + \frac{1}{6!}x^6 + \frac{7}{7!}x^7 + \cdots$ **23.** $y = 1 + 2x - \frac{2}{4!}x^4 - \frac{2}{5!}x^5 - \frac{2}{6!}x^6 + \frac{6}{7!}x^7 + \frac{16}{8!}x^8 + \cdots$

Section 20.8

1. $y = a_0\left[1 - x^2 + \frac{x^4}{3} - \frac{x^6}{3 \cdot 5} + \cdots + (-1)^n \frac{x^{2n}}{(3)(5)\cdots(2n-1)} + \cdots\right]$

$+ a_1\left(x - \frac{x^3}{2} + \frac{x^5}{2 \cdot 4} - \frac{x^7}{2 \cdot 4 \cdot 6} + \cdots + (-)^n \frac{x^{2n+1}}{(2)(4)\cdots(2n)} + \cdots\right)$

3. $y = a_0(1 + \frac{1}{6}x^3 - \frac{1}{20}x^5 + \frac{1}{180}x^6 + \cdots) + a_1(x - \frac{1}{3}x^3 + \frac{1}{12}x^4 + \frac{1}{10}x^5 - \frac{1}{30}x^6 + \cdots)$

5. $y = a_0 + a_1(x + \frac{1}{3}x^3 + \frac{1}{10}x^5 + \frac{1}{42}x^7 + \frac{1}{216}x^9 + \cdots) + (\frac{1}{12}x^4 + \frac{1}{45}x^6 + \frac{1}{210}x^8 + \cdots)$

7. $y = a_0 + a_1\left(x + \frac{1}{24}x^4 + \frac{1}{7 \cdot 6 \cdot 6 \cdot 5}x^7 + \frac{1}{10 \cdot 9 \cdot 8 \cdot 6 \cdot 6 \cdot 5}x^{10} + \cdots\right) + a_2\left(x^2 + \frac{1}{30}x^5 + \frac{1}{8 \cdot 7 \cdot 6 \cdot 6}x^8 + \cdots\right)$

$+ \left(\frac{1}{6}x^3 + \frac{1}{24}x^4 + \frac{1}{240}x^6 + \frac{1}{7 \cdot 6 \cdot 6 \cdot 5}x^7 + \frac{3}{9 \cdot 8 \cdot 7 \cdot 6 \cdot 5 \cdot 4}x^9 + \frac{1}{10 \cdot 9 \cdot 8 \cdot 6 \cdot 6 \cdot 5}x^{10} + \cdots\right)$

9. $y = a_0 + a_1 x + \left(\frac{x^4}{12} - \frac{x^8}{56 \cdot 3!} + \frac{x^{12}}{132 \cdot 5!} - \cdots + (-1)^n \frac{x^{4n+4}}{(4n+4)(4n+3)(2n+1)!} + \cdots\right)$

11. $y = a_0 + a_1 x + \left(\frac{x^5}{5 \cdot 4} - \frac{x^7}{7 \cdot 6 \cdot 3} + \frac{x^9}{9 \cdot 8 \cdot 5} - \cdots + (-1)^n \frac{x^{2n+5}}{(2n+5)(2n+4)(2n+1)} + \cdots\right)$

13. $y = a_0 + a_1(x-1) + a_2[(x-1)^2 + \frac{1}{3}(x-1)^3]$

15. $y = a_0\left(1 - \frac{x^2}{2!} + \frac{x^4}{4!} - \frac{x^6}{6!} + \cdots + (-1)^n \frac{x^{2n}}{(2n)!} + \cdots\right)$

$+ a_1\left(x - \frac{x^3}{3!} + \frac{x^5}{5!} - \frac{x^7}{7!} + \cdots + (-1)^n \frac{x^{2n+1}}{(2n+1)!} + \cdots\right) + \left(\frac{x^3}{3!} - \frac{2x^5}{5!} + \frac{3x^7}{7!} - \frac{4x^9}{9!} + \cdots\right)$

$= a_0 \cos x + a_1 \sin x + (\frac{1}{2}\sin x - \frac{1}{2}x\cos x)$

17. $y = 1 + \frac{1}{4 \cdot 3}x^4 + \frac{1}{6 \cdot 5 \cdot 3}x^6 + \frac{1}{8 \cdot 7 \cdot 5 \cdot 3}x^8 + \cdots$ **19.** $y = 1 + \frac{6}{6!}x^6 + \frac{30 \cdot 6}{9!}x^9 + \cdots$

Review Exercise Set 20.1

1.

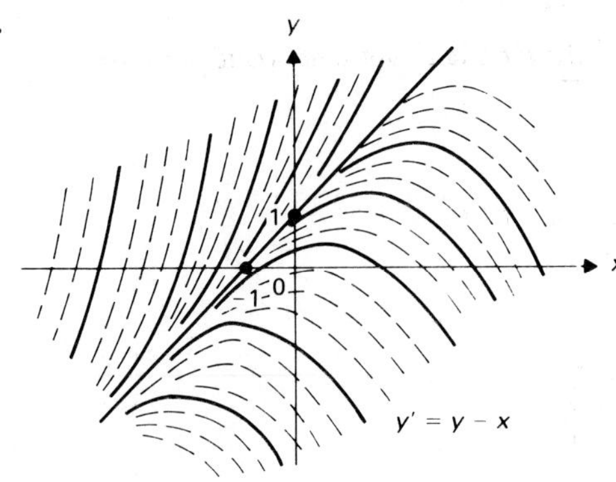

3. $\ln(2\sin^2 y) = x^2$ **5.** $x^3 + 3xy^2 = C$ **7.** $y = 1 + Ce^{\cos x}$
9. $y = e^x(C_1 \cos 2x + C_2 \sin 2x) + \frac{1}{8}e^{3x}$

Review Exercise Set 20.2

1. (See Theorem 20.1, Section 20.1.) **3.** $\frac{1}{2}\ln\left(\left(\frac{y}{x}\right)^2 + 1\right) + \tan^{-1}\left(\frac{y}{x}\right) = -\ln|x| + C$ **5.** $xy + \sin y - \cos x = C$
7. $y = C_1 + C_2x + C_3e^{2x}$ **9.** $y = C_1e^x + C_2e^{2x} + \frac{1}{10}\sin x + \frac{3}{10}\cos x$

Review Exercise Set 20.3

1. (0, 1), (1, 3), (2, $\frac{21}{2}$) **3.** $\frac{1}{\sqrt{2}}\tan^{-1}\left(\frac{y}{\sqrt{2}x}\right) - \ln|2x^2 + y^2| = C$ **5.** $\frac{-1}{xy} - \ln|x| + 4y = C$ **7.** $y = C_1 + e^x(C_2x + C_3)$
9. $y = a_0(1 - x^2 + \cdots) + a_1\left(x - \frac{x^3}{3!} - \frac{x^5}{5!} - \cdots\right)$

Review Exercise Set 20.4

1. 4.213029761 **3.** $-\frac{3}{2}\ln|x - y| + \frac{1}{2}\ln|x + y| = C$ **5.** $y = Ce^{x^2} - \frac{3}{2}$ **7.** $y = e^{2x}(\cos x - 3\sin x)$
9. $y = a_0\left(1 - \frac{x^3}{6} - \frac{x^6}{90} - \cdots\right) + a_1x$

INDEX

A

B

C

D

E

F

G

H

I

K

L

Q

R

U

V

W

Z

64. $\int \sin^2 ax\, dx = \frac{x}{2} - \frac{\sin 2ax}{4a} + C$

65. $\int \cos^2 ax\, dx = \frac{x}{2} + \frac{\sin 2ax}{4a} + C$

66. $\int \sin^n ax\, dx = \frac{-\sin^{n-1} ax \cos ax}{na} + \frac{n-1}{n} \int \sin^{n-2} ax\, dx$

67. $\int \cos^n ax\, dx = \frac{\cos^{n-1} ax \sin ax}{na} + \frac{n-1}{n} \int \cos^{n-2} ax\, dx$

68. (a) $\int \sin ax \cos bx\, dx = -\frac{\cos(a+b)x}{2(a+b)} - \frac{\cos(a-b)x}{2(a-b)} + C, \quad a^2 \neq b^2$

(b) $\int \sin ax \sin bx\, dx = \frac{\sin(a-b)x}{2(a-b)} - \frac{\sin(a+b)x}{2(a+b)} + C, \quad a^2 \neq b^2$

(c) $\int \cos ax \cos bx\, dx = \frac{\sin(a-b)x}{2(a-b)} + \frac{\sin(a+b)x}{2(a+b)} + C, \quad a^2 \neq b^2$

69. $\int \sin ax \cos ax\, dx = -\frac{\cos 2ax}{4a} + C$

70. $\int \sin^n ax \cos ax\, dx = \frac{\sin^{n+1} ax}{(n+1)a} + C, \quad n \neq -1$

71. $\int \frac{\cos ax}{\sin ax}\, dx = \frac{1}{a} \ln |\sin ax| + C$

72. $\int \cos^n ax \sin ax\, dx = -\frac{\cos^{n+1} ax}{(n+1)a} + C, \quad n \neq -1$

73. $\int \frac{\sin ax}{\cos ax}\, dx = -\frac{1}{a} \ln |\cos ax| + C$

74. $\int \sin^n ax \cos^m ax\, dx = -\frac{\sin^{n-1} ax \cos^{m+1} ax}{a(m+n)} + \frac{n-1}{m+n} \int \sin^{n-2} ax \cos^m ax\, dx, \quad n \neq -m$

(If $n = -m$, use No. 92.)

75. $\int \sin^n ax \cos^m ax\, dx = \frac{\sin^{n+1} ax \cos^{m-1} ax}{a(m+n)} + \frac{m-1}{m+n} \int \sin^n ax \cos^{m-2} ax\, dx, \quad m \neq -n$

(If $m = -n$, use No. 93.)

76. $\int \frac{dx}{b + c \sin ax} = \frac{-2}{a\sqrt{b^2 - c^2}} \tan^{-1}\left[\sqrt{\frac{b-c}{b+c}} \tan\left(\frac{\pi}{4} - \frac{ax}{2}\right)\right] + C, \quad b^2 > c^2$

77. $\int \frac{dx}{b + c \sin ax} = \frac{-1}{a\sqrt{c^2 - b^2}} \ln \left| \frac{c + b \sin ax + \sqrt{c^2 - b^2} \cos ax}{b + c \sin ax} \right| + C, \quad b^2 < c^2$

78. $\int \frac{dx}{1 + \sin ax} = -\frac{1}{a} \tan\left(\frac{\pi}{4} - \frac{ax}{2}\right) + C$

79. $\int \frac{dx}{1 - \sin ax} = \frac{1}{a} \tan\left(\frac{\pi}{4} + \frac{ax}{2}\right) + C$

80. $\int \frac{dx}{b + c \cos ax} = \frac{2}{a\sqrt{b^2 - c^2}} \tan^{-1}\left[\sqrt{\frac{b-c}{b+c}} \tan \frac{ax}{2}\right] + C, \quad b^2 > c^2$

81. $\int \frac{dx}{b + c \cos ax} = \frac{1}{a\sqrt{c^2 - b^2}} \ln \left| \frac{c + b \cos ax + \sqrt{c^2 - b^2} \sin ax}{b + c \cos ax} \right| + C, \quad b^2 < c^2$

82. $\int \frac{dx}{1 + \cos ax} = \frac{1}{a} \tan \frac{ax}{2} + C$

83. $\int \frac{dx}{1 - \cos ax} = -\frac{1}{a} \cot \frac{ax}{2} + C$

84. $\int x \sin ax\, dx = \frac{1}{a^2} \sin ax - \frac{x}{a} \cos ax + C$

85. $\int x \cos ax\, dx = \frac{1}{a^2} \cos ax + \frac{x}{a} \sin ax + C$

86. $\int x^n \sin ax\, dx = -\frac{x^n}{a} \cos ax + \frac{n}{a} \int x^{n-1} \cos ax\, dx$

87. $\int x^n \cos ax\, dx = \frac{x^n}{a} \sin ax - \frac{n}{a} \int x^{n-1} \sin ax\, dx$

88. $\int \tan ax\,dx = -\frac{1}{a}\ln|\cos ax| + C$

89. $\int \cot ax\,dx = \frac{1}{a}\ln|\sin ax| + C$

90. $\int \tan^2 ax\,dx = \frac{1}{a}\tan ax - x + C$

91. $\int \cot^2 ax\,dx = -\frac{1}{a}\cot ax - x + C$

92. $\int \tan^n ax\,dx = \frac{\tan^{n-1} ax}{a(n-1)} - \int \tan^{n-2} ax\,dx, \quad n \neq 1$

93. $\int \cot^n ax\,dx = -\frac{\cot^{n-1} ax}{a(n-1)} - \int \cot^{n-2} ax\,dx, \quad n \neq 1$

94. $\int \sec ax\,dx = \frac{1}{a}\ln|\sec ax + \tan ax| + C$

95. $\int \csc ax\,dx = -\frac{1}{a}\ln|\csc ax + \cot ax| + C$

96. $\int \sec^2 ax\,dx = \frac{1}{a}\tan ax + C$

97. $\int \csc^2 ax\,dx = -\frac{1}{a}\cot ax + C$

98. $\int \sec^n ax\,dx = \frac{\sec^{n-2} ax \tan ax}{a(n-1)} + \frac{n-2}{n-1}\int \sec^{n-2} ax\,dx, \quad n \neq 1$

99. $\int \csc^n ax\,dx = -\frac{\csc^{n-2} ax \cot ax}{a(n-1)} + \frac{n-2}{n-1}\int \csc^{n-2} ax\,dx, \quad n \neq 1$

100. $\int \sec^n ax \tan ax\,dx = \frac{\sec^n ax}{na} + C, \quad n \neq 0$

101. $\int \csc^n ax \cot ax\,dx = -\frac{\csc^n ax}{na} + C, \quad n \neq 0$

102. $\int \sin^{-1} ax\,dx = x\sin^{-1} ax + \frac{1}{a}\sqrt{1 - a^2x^2} + C$

103. $\int \cos^{-1} ax\,dx = x\cos^{-1} ax - \frac{1}{a}\sqrt{1 - a^2x^2} + C$

104. $\int \tan^{-1} ax\,dx = x\tan^{-1} ax - \frac{1}{2a}\ln(1 + a^2x^2) + C$

105. $\int x^n \sin^{-1} ax\,dx = \frac{x^{n+1}}{n+1}\sin^{-1} ax - \frac{a}{n+1}\int \frac{x^{n+1}dx}{\sqrt{1 - a^2x^2}}, \quad n \neq -1$

106. $\int x^n \cos^{-1} ax\,dx = \frac{x^{n+1}}{n+1}\cos^{-1} ax + \frac{a}{n+1}\int \frac{x^{n+1}dx}{\sqrt{1 - a^2x^2}}, \quad n \neq -1$

107. $\int x^n \tan^{-1} ax\,dx = \frac{x^{n+1}}{n+1}\tan^{-1} ax - \frac{a}{n+1}\int \frac{x^{n+1}dx}{1 + a^2x^2}, \quad n \neq -1$

INTEGRALS INVOLVING EXPONENTIAL AND LOGARITHMIC FUNCTIONS

108. $\int e^{ax}dx = \frac{1}{a}e^{ax} + C$

109. $\int b^{ax}dx = \frac{1}{a}\frac{b^{ax}}{\ln b} + C, \quad b > 0, \quad b \neq 1$

110. $\int xe^{ax}dx = \frac{e^{ax}}{a^2}(ax - 1) + C$

111. $\int x^n e^{ax}dx = \frac{1}{a}x^n e^{ax} - \frac{n}{a}\int x^{n-1}e^{ax}\,dx$

112. $\int x^n b^{ax}\,dx = \frac{x^n b^{ax}}{a\ln b} - \frac{n}{a\ln b}\int x^{n-1}b^{ax}\,dx, \quad b > 0, \quad b \neq 1$

113. $\int e^{ax}\sin bx\,dx = \frac{e^{ax}}{a^2 + b^2}(a\sin bx - b\cos bx) + C$

114. $\int e^{ax}\cos bx\,dx = \frac{e^{ax}}{a^2 + b^2}(a\cos bx + b\sin bx) + C$